Optimum Array Processing

Optimum Array Processing

Part IV of Detection, Estimation, and Modulation Theory

Harry L. Van Trees

A JOHN WILEY & SONS, INC., PUBLICATION

This text is printed on acid-free paper. ♾

For ordering and customer service, call 1-800-CALL WILEY.

Library of Congress Cataloging-in-Publication Data:

Van Trees, Harry L.

Optimum Array Processing. Part IV of Detection, Estimation, and Modulation Theory/
Harry L. Van Trees.

p. cm.

Includes bibliographical references and index.

ISBN 0-471-09390-4 (cloth: alk paper)

Printed in the United States of America

To Diane

For her continuing support and encouragement during the many years that this book was discussed, researched, and finally written. More importantly, for her loyalty, love, and understanding during a sequence of challenging periods,

and to

Professor Wilbur Davenport, whose book introduced me to random processes and who was a mentor, friend, and supporter during my career at Massachusetts Institute of Technology.

Contents

Preface

Array processing has played an important role in many diverse application areas. Most modern radar and sonar systems rely on antenna arrays or hydrophone arrays as an essential component of the system. Many communication systems utilize phased arrays or multiple beam antennas to achieve their performance objectives. Seismic arrays are widely used for oil exploration and detection of underground nuclear tests. Various medical diagnosis and treatment techniques exploit arrays. Radio astronomy utilizes very large antenna arrays to achieve resolution goals. It appears that the third generation of wireless systems will utilize adaptive array processing to achieve the desired system capacity. We discuss various applications in Chapter 1.

My interest in optimum array processing started in 1963 when I was an Assistant Professor at M.I.T. and consulting with Arthur D. Little on a sonar project for the U.S. Navy. I derived the optimum processor for detecting Gaussian plane-wave signals in Gaussian noise [VT66a], [VT66b]. It turned out that Bryn [Bry62] had published this result previously (see also Vanderkulk [Van63]). My work in array processing decreased as I spent more time in the general area of detection, estimation, and modulation theory.

In 1968, Part I of *Detection, Estimation, and Modulation Theory* [VT68] was published. It turned out to be a reasonably successful book that has been widely used by several generations of engineers. Parts II and III ([VT71a], [VT71b]) were published in 1971 and focused on specific application areas such as analog modulation, Gaussian signals and noise, and the radar-sonar problem. Part II had a short life span due to the shift from analog modulation to digital modulation. Part III is still widely used as a reference and as a supplementary text. In a moment of youthful optimism, I indicated in the Preface to Part III and in Chapter III-14 that a short monograph on optimum array processing would be published in 1971. The bibliography lists it as a reference, (*Optimum Array Processing*, Wiley, 1971), which has been subsequently cited by several authors. Unpublished class notes [VT69] contained much of the planned material. In a very loose sense, this text is the extrapolation of that monograph.

Throughout the text, there are references to Parts I and III of *Detection, Estimation, and Modulation Theory*. The referenced material is available in several other books, but I am most familiar with my own work. Wiley has republished Parts I and III [VT01a], [VT01b] in paperback in conjunction with the publication of this book so the material will be readily available.

A few comments on my career may help explain the thirty-year delay. In 1972, M.I.T. loaned me to the Defense Communications Agency in Washington, D.C., where I spent three years as the Chief Scientist and the Associate Director for Technology. At the end of this tour, I decided for personal reasons to stay in the Washington, D.C., area. I spent three years as an Assistant Vice-President at COMSAT where my group did the advanced planning for the INTELSAT satellites. In 1978, I became the Chief Scientist of the United States Air Force. In 1979, Dr.Gerald Dinneen, the former director of Lincoln Laboratories, was serving as Assistant Secretary of Defense for C^3I. He asked me to become his Principal Deputy and I spent two years in that position. In 1981, I joined M/A-COM Linkabit. Linkabit is the company that Irwin Jacobs and Andrew Viterbi started in 1969 and sold to M/A-COM in 1979. I started an Eastern operations, which grew to about 200 people in three years. After Irwin and Andy left M/A-COM and started Qualcomm, I was responsible for the government operations in San Diego as well as Washington, D.C. In 1988, M/A-COM sold the division. At that point I decided to return to the academic world.

I joined George Mason University in September of 1988. One of my priorities was to finish the book on optimum array processing. However, I found that I needed to build up a research center in order to attract young research-oriented faculty and doctoral students. This process took about six years. The C^3I Center of Excellence in Command, Control, Communications, and Intelligence has been very successful and has generated over $30 million in research funding during its existence. During this growth period, I spent some time on array processing, but a concentrated effort was not possible.

The basic problem in writing a text on optimum array processing is that, in the past three decades, enormous progress had been made in the array processing area by a number of outstanding researchers. In addition, increased computational power had resulted in many practical applications of optimum algorithms. Professor Arthur Baggeroer of M.I.T. is one of the leading contributors to array processing in the sonar area. I convinced Arthur, who had done his doctoral thesis with me in 1969, to co-author the optimum array processing book with me. We jointly developed a comprehensive outline. After several years it became apparent that the geographical distance and Arthur's significant other commitments would make a joint authorship difficult and we agreed that I would proceed by myself. Although the final outline has about a 0.25 correlation with the original outline, Arthur's collaboration in structuring the original outline and commenting on the results have played an important role in the process.

In 1995, I took a sabbatical leave and spent the year writing the first draft. I taught a one-year graduate course using the first draft in the 1996–1997 academic year. A second draft was used in the 1997–1998 academic year. A third draft was used by Professor Kristine Bell in the 1998–1999 academic year. Unlike the M.I.T. environment where I typically had 40–50 graduate students in my detection and estimation classes, our typical enrollment has been 8–10 students per class. However, many of these students were actively working in the array processing area and have offered constructive suggestions.

The book is designed to provide a comprehensive introduction to optimum array processing for students and practicing engineers. It will prepare the students to do research in the array processing area or to implement actual array processing systems. The book should also be useful to people doing current research in the field. We assume a background in probability theory and random processes. We assume that the reader is familiar with Part I of *Detection, Estimation, and Modulation Theory* [VT68], [VT01a] and parts of Part III [VT71b], [VT01b]. The first use of [VT68], [VT01a] is in Chapter 5, so that a detection theory course could be taken at the same time. We also assume some background in matrix theory and linear algebra. The book emphasizes the ability to work problems, and competency in MATLAB® is essential.

The final product has grown from a short monograph to a lengthy text. Our experience is that, if the students have the correct background and motivation, we can cover the book in two fifteen-week semesters.

In order to make the book more useful, Professor Kristine Bell has developed a Web site:

http://ite.gmu.edu/DetectionandEstimationTheory/

that contains material related to all four parts of the *Detection, Estimation, and Modulation Theory* series.

The *Optimum Array Processing* portion of the site contains:

(i) MATLAB® scripts for most of the figures in the book. These scripts enable the reader to explore different signal and interference environments and are helpful in solving the problems. The disadvantage is that a student can use them without trying to solve the problem independently. We hope that serious students will resist this temptation.

(ii) Several demos that allow the reader to see the effect of parameter changes on beam patterns and other algorithm outputs. Some of the demos for later chapters allow the reader to view the adaptive behavior of the system dynamically. The development of demos is an ongoing process.

(iii) An erratum and supplementary comments regarding the text will be updated periodically on the Web site. Errors and comments can be sent to either hlv@gmu.edu or kbell@gmu.edu.

(iv) Solutions, including MATLAB® scripts where appropriate, to many of the problems and some of the exams we have used. This part is password protected and is only available to instructors. To obtain a password, send an e-mail request to either hlv@gmu.edu or kbell@gmu.edu.

In order to teach the course, we created a separate LATEX file containing only the equations. By using Ghostview, viewgraphs containing the equations can be generated. A CD-rom with the file is available to instructors who have adopted the text for a course by sending me an e-mail at hlv@gmu.edu.

The book has relied heavily on the results of a number of researchers. We have tried to acknowledge their contributions. The end-of-chapter bibliographies contain over 2,000 references. Certainly the book would not have been possible without this sequence of excellent research results.

A number of people have contributed in many ways and it is a pleasure to acknowledge them. Andrew Sage, founding dean of the School of Information Technology and Engineering at George Mason University, provided continual encouragement in my writing efforts and extensive support in developing the C^{3}I Center. The current dean, Lloyd Griffiths, has also been supportive of my work.

A number of the students taking my course have offered constructive criticism and corrected errors in the various drafts. The following deserve explicit recognition: Amin Jazaeri, Hung Lai, Brian Flanagan, Joseph Herman, John Uber, Richard Bliss, Mike Butler, Nirmal Warke, Robert Zarnich, Xiaolan Xu, and Zhi Tian suffered through the first draft that contained what were euphemistically referred to as typos. Geoff Street, Stan Pawlukiewicz, Newell Stacey, Norman Evans, Terry Antler, and Xiaomin Lu encountered the second draft, which was significantly expanded. Roy Bethel, Paul Techau, Jamie Bergin, Hao Cheng, and Xin Zhang critiqued the third draft. The final draft was used in my Optimum Array Processing course during the 2000–2001 academic year. John Hiemstra, Russ Jeffers, Simon Wood, Daniel Bray, Ben Shapo, and Michael Hunter offered useful comments and corrections. In spite of this evolution and revision, there are probably still errors. Please send corrections to me at hlv@gmu.edu and they will be posted on the Web site.

Two Visiting Research Professors, Shulin Yang and Chen-yang Yang also listened to the course and offered comments. Drs. Shulin Yang, Chen-yang Yang, and Ms. Xin Zhang composed the book in LATEX and provided important editorial advice. Aynur Abdurazik and Muhammad Abdulla did the final LATEX version. Their competence and patience have been extraordinary. Joshua Kennedy and Xiaomin Lu drew many of the figures. Four of my graduate research assistants, Miss Zhi Tian, Miss Xiaolan Xu, Mr. Xiaomin Lu, and Miss Xin Zhang worked most of the examples in various chapters. Their help has been invaluable in improving the book.

A separate acknowledgment is needed for Professor Kristine Bell. She did her doctoral dissertation in the array processing area for Professor Yariv Ephraim and me, and she has continued to work with me on the text for several years. She has offered numerous insights into the material and into new developments in many areas. She also taught the two-semester course in 1998–1999 and developed many aspects of the material. Her development of the Web site adds to the pedagogical value of the book.

Several colleagues agreed to review the manuscript and offer criticisms. The group included many of the outstanding researchers in the array processing area. Dan Fuhrmann, Norman Owsley, Mats Viberg, and Mos Kaveh reviewed the entire book and offered numerous corrections and suggestions. In addition, they pointed out a number of useful references that I had missed. Petre Stoica provided excellent comments on Chapters 7–10, and two of his students, Erik Larsson and Richard Abrhamsson, provided additional comments. Louis Scharf, Ben Friedlander, Mati Wax, and John Buck provided constructive comments on various sections of the book. Don Tufts provided a large amount of historical material that was very useful. I appreciate the time that all of these colleagues took from their busy schedules. Their comments have improved the book.

Harry L. Van Trees

January 2002

Bibliography

[Bry62] F. Bryn. Optimum signal processing of three-dimensional array operating on Gaussian signals and noise. *J. Acoust. Soc. Amer.*, 34(3):289–297, March 1962.

[Van63] V. Vanderkulk. Optimum processing for acoustic arrays. *J. Brit. IRE*, 26(4):286–292, October 1963.

[VT66a] H. L Van Trees. Optimum processing for passive sonar arrays. *Proc. IEEE Ocean Electronics Symp.*, pages 41–65, Honolulu, Hawaii, 1966.

[VT66b] H. L. Van Trees. A unified theory for optimum array processing. Technical Report 4160866, Dept. of the Navy Naval Ship Systems Command, Arthur D. Little, Inc., Cambridge, MA, Aug. 1966.

[VT68] H. L. Van Trees. *Detection, Estimation, and Modulation Theory, Part I.* Wiley, New York, 1968.

[VT01a] H. L. Van Trees. *Detection, Estimation, and Modulation Theory, Part I.* Wiley Interscience, New York, 2001.

[VT69] H. L. Van Trees. *Multi-Dimensional and Multi-Variable Processes.* unpublished class notes, M.I.T, 1969.

[VT71a] H. L. Van Trees. *Detection, Estimation, and Modulation Theory, Part II.* Wiley, New York, 1971.

[VT71b] H. L. Van Trees. *Detection, Estimation, and Modulation Theory, Part III.* Wiley, New York, 1971.

[VT01b] H. L. Van Trees. *Detection, Estimation, and Modulation Theory, Part III.* Wiley Interscience, New York, 2001.

Optimum Array Processing

Chapter 1

Introduction

In Parts I, II, and III of *Detection, Estimation, and Modulation Theory* (DEMT) [VT68], [VT01a], [VT71a], [VT71b], [VT01b], we provide a reasonably complete discussion of several areas:

(i) Detection theory

In this case, we were concerned with detecting signals in the presence of Gaussian noise. The class of signals included known signals, signals with unknown parameters, and signals that are sample functions from Gaussian random processes. This problem was covered in Chapter I-4 and Chapters III-1 through III-5.

(ii) Estimation theory

In this case, we were concerned with estimating the parameters of signals in the presence of Gaussian noise. This problem was covered in Chapter I-4 and Chapters III-6 and III-7.

(iii) Modulation theory

In this case, we were concerned with estimating a continuous waveform (or the sampled version of it). If the signal has the waveform in it in a linear manner, then we have a linear estimation problem and obtain the Wiener filter or the Kalman-Bucy filter as the optimum estimator. This problem was covered in Chapter I-6. The case of nonlinear modulation is covered in Chapter I-5 and Volume II.

All of the results in the first three volumes consider signals and noises that could be characterized in the time domain (or equivalently, the frequency domain). In this book, we consider the case in which the signals and noises also have a spatial dependence. Therefore, we must characterize the signals and noises as space-time processes and solve the detection and estimation problems in the multidimensional space-time domain. This leads us to space-time processors. The spatial part of the processor is an **aperture** (or antenna) for the continuous space domain and an **array** for the discrete space domain. The focus of this book is on optimum array processing (and optimum aperture processing). The formal extension of the temporal results to the multidimensional problem is reasonably straightforward, but the implications of the results lead to a number of challenging questions.

In Section 1.1, we give a simple description of the array processing problem. In Section 1.2, we give a brief description of some representative applications in which arrays play a key role. In Section 1.3, we outline the structure of the array processing literature. In Section 1.4, we outline the organization of the book.

1.1 Array Processing

In this section, we introduce the array processing problem and discuss some of the issues that we will encounter in the text.

A representative array consisting of six sensors is shown in Figure 1.1. We can use this array to illustrate some of the issues. The four issues of interest are:

(A) Array configuration

(B) Spatial and temporal characteristics of the signal

(C) Spatial and temporal characteristics of the interference

(D) Objective of the array processing

The first issue is the array configuration. The array configuration consists of two parts. The first part is the antenna pattern of the individual elements. For example, in a transmitting RF array this will be a function of the physical configuration of the sensor and the current distribution on the elements. In many cases, we first assume that the elements have an **isotropic** pattern (i.e., uniform in all directions) and then incorporate the actual pattern later in the analysis.

The second part of the array configuration is the array geometry (i.e., the physical location of the elements). The array geometry is one part of the problem where we will focus our attention.

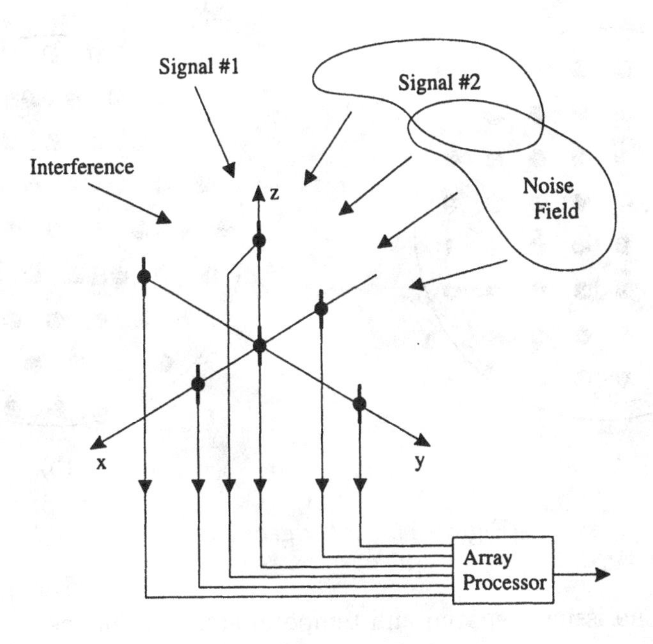

Figure 1.1 Array processing problem.

We can divide array geometries into three categories:[1]

A1 Linear

A2 Planar

A3 Volumetric (3-D)

Within each category we can develop a taxonomy. In linear arrays, the following cases are of interest:

A1.1 Uniform spacing

A1.2 Non-uniform spacing

A1.3 Random spacing

We will find that the total length of the array and the array spacing determine how the array geometry affects the problem.

In a planar array, the boundary of the array and the element geometry are important. For example, we can have a planar array whose boundary is circular and the element geometry could be either square or triangular as shown in Figure 1.2.

---

[1]The indexing notation is for convenience. We will not use it in our later discussions.

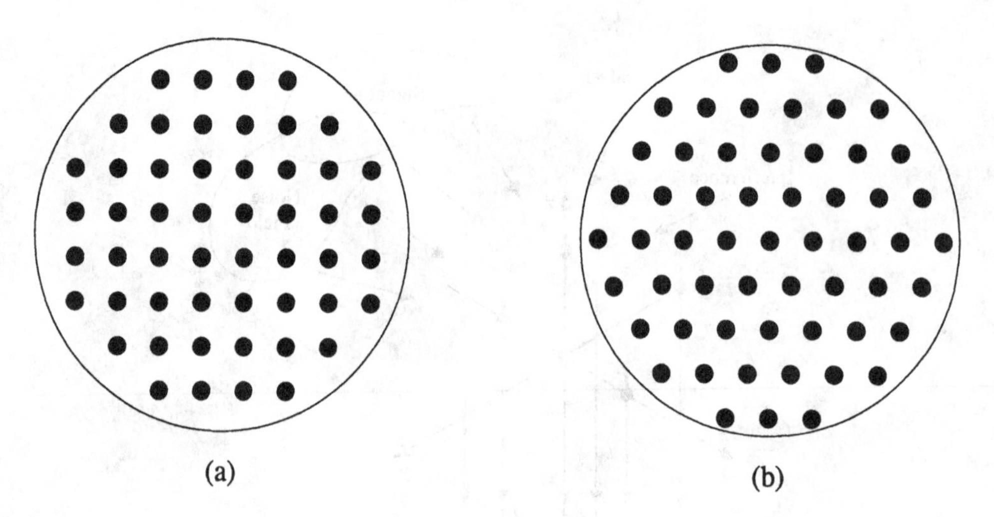

Figure 1.2 Array geometries.

The second issue is spatial and temporal structure of the signal. In the temporal domain we will encounter the same signal structures as in DEMT I and III:

BT1 Known signals

BT2 Signals with unknown parameters

BT3 Signals with known structure (e.g., QPSK)

BT4 Random signals

It is the spatial domain that is of most interest. The cases of interest here include:

BS1 Plane-wave signals from known directions

BS2 Plane-wave signals from unknown directions

BS3 Spatially spread signals

The third issue is the spatial and temporal structure of the noise (or interference). We will always include a "sensor noise" component that consists of a white Gaussian noise component that is statistically independent from sensor to sensor. The physical motivation for this inclusion is that noise of this type always exists. System design may reduce the sensor noise to a level that makes other noises dominant, but it will never be zero. From our results in DEMT I and III, we would anticipate that this white noise component will ensure that we do not obtain a singular solution.

The noise coming from external sources must be characterized both temporally and spatially. The temporal cases of most interest are:

CT1 Signals with unknown parameters (this includes the case where we know the modulation of an interfering signal, but not the information sequence)

CT2 Random signals

The spatial cases of most interest are:

CS1 One or more plane waves from known directions

CS2 One or more plane waves from unknown directions

CS3 Spatially spread interference

The fourth issue of interest is the objective of the array processing problem. Some representative objectives are:

D1 Detect the presence of a signal in the presence of noise and interference

D2 Demodulate the signal and estimate the information waveform (i.e., listen to the signal) in the presence of noise and interference

D3 A binary communication signal arrives over multiple paths, detect the information sequence

D4 Estimate the direction-of-arrival of multiple plane-wave signals in the presence of noise

D5 Construct temporal and spatial spectral estimate of the incoming signal and noise field

D6 Direct the transmitted signal to a specific spatial location

We will encounter other objectives in our discussion of applications in Section 1.2 and at various places in the text and problems. The above list serves to illustrate the wide variety of problems we will encounter.

We now consider various application areas in which arrays play an important role.

1.2 Applications

In this section we provide a brief discussion of seven areas in which arrays play an important role. These areas are:

(i) Radar

(ii) Radio astronomy

(iii) Sonar

(iv) Communications

(v) Direction-finding

(vi) Seismology

(vii) Medical diagnosis and treatment

Our discussion in each area is short. However, we provide a list of references that provide more detailed discussions. We also revisit these application areas in subsequent chapters and in the problems.

1.2.1 Radar

The radar area is the area in which antenna arrays were first used.[2] Skolnik has a good discussion of the use of phased arrays in radar in Chapter 8 of his book, *Introduction to Radar Systems* [Sko80], and some of our specific examples are taken from it.

Most radar systems are active systems and the antenna array is used for both transmission and reception of signals.

Although the concept of phased array antennas was known during World War I [Sou62], the first usage was in World War II. Examples of United States systems included fire control radars for Navy ships [OK72] and high resolution navigation and bombing radars [Rid47]. Examples of British usage included height-finding radars [Smi49]. The work at the M.I.T. Radiation Lab is described in the Radlab Series.

Current military systems include the PAVE PAWS radar, which is used for ballistic missile detection [Bro85], the AEGIS phased array antenna [Bro91], and numerous airborne systems. Non-military systems include air traffic control radars [EJS82].

[2]The usage of phased arrays in communication systems evolved in the same time period. We discuss communication applications in Section 1.2.4.

Ground- and ship-based radars can generally use a model in which both the signal and interference can be modeled as plane waves impinging on the array. In some environments we encounter multipath, but the plane-wave model is still valid.

Airborne radars looking at the ground have the additional complexity of reflections from the ground that is referred to as clutter (clutter models were discussed in Chapter 13 of DEMT III [VT71b], [VT01b]). Models must now include spatially spread interference. A discussion of this type of system is given in Ward [War94].

Other references that discuss various aspects of radar systems include Allen [All63], Reintjes and Coate [RC52], Skolnik [Sko70], Barton [Bar65], Berkowitz [Ber65], Di Franco and Rubin [FR68], Cook and Bernfeld [CB67], Barton and Ward [BW69], Reed [Ree69], and Haykin ([Hay80], Chapter 4 of [Hay85] and [SH92]).

1.2.2 Radio Astronomy

Antenna arrays are widely used in the radio (or radar) astronomy area. Books that provide good introductions to the area include Kraus [Kra66], Evans and Hagfors [EH68], Christiansen and Högbom [CH69], and Kellerman and Verschuur [KV73]. Yen in Chapter 5 of [Hay85] provides a good introduction.

A radio astronomy system is a passive system that is used to detect celestial objects and estimate their characteristics. These systems usually employ arrays with very long baselines. These baselines range from tens of kilometers to thousands of kilometers. Representative systems include the very large array (VLA) of the National Radio Astronomy Observatory [TCWN80] and Cambridge telescope [Ryl73].

Typical array configurations include:

- Linear arrays with unequal spacing
- Parallel linear arrays
- Circular arrays
- Arrays with three linear arms spaced at 120° with a common center

Some of the issues that must be considered in radio astronomy include the rotation of the earth during the signal processing period, different propagation characteristics through the ionosphere and troposphere at different array elements, and synchronization over long distances. We encounter various radio astronomy examples at different points in the text.

1.2.3 Sonar

Arrays are widely used in sonar systems. A good discussion of array processing in sonar systems is given by Baggeroer in Chapter 6 of [Opp78]. Several of our examples are taken from this reference. The journal article by Knight et al. [KPK81] (this article contains 253 references) and Owsley's chapter in [Hay85] also provide good introductions. It is useful to consider active and passive sonars separately.

An active sonar transmits acoustic energy into the water and processes received echos. The theory of active sonars has much in common with radars. However, a fundamental important difference between sonar and radar is that the propagation of acoustic energy in the ocean is significantly more complicated than the propagation of electromagnetic energy in the atmosphere. These propagation characteristics have a major influence on the design of sonar systems.

Propagation factors include spreading loss, absorption, and ducting. These factors will vary depending on the ranges of interest, the depth of the water, and the nature of the boundaries. Discussion of sound propagation is given in Urick [Uri67], Albers [Alb60], Burdic [Bur91], Horton [Hor57], and Hassab [Has89], as well as Section 6.2 of Baggeroer's chapter of [Opp78].

The noise background includes ambient noise, self noise, and reverberation noise. Ambient noise is acoustic noise generated by various sources in the ocean such as ships, industrial activity, sea life, precipitation, ice, and explosions. Typically, it is spread in both frequency and space. Self noise is generated by the platform. Examples include cavitation noise, flow noise, and machinery noise. Self noise may be either frequency spread or tonal and is normally concentrated spatially. Reverberation noise is due to reflections of the transmitted signal and is analogous to clutter in a radar system.

Other factors such as interaction with the boundaries, spatial coherence of the acoustic waves, and the severity of the ocean environment must also be taken into account.

Passive sonar systems listen to incoming acoustic energy and use it to estimate the temporal and spatial characteristics of the observed signal field. The most important application is the detection and tracking of submarines. Some representative configurations are shown in Figure 1.3. All of the comments about propagation and noise also apply to the passive sonar case. We analyze several passive sonar examples in the text.

Other sonar references include Wagstaff and Baggeroer [WB83], Officer [Off58], Tolstoy and Clay [TC66], Dyer [Dye70], Griffiths et al. [GSS73], Baggeroer [Bag76], Cron and Sherman [CS62], Cox [Cox73], and Blahut et

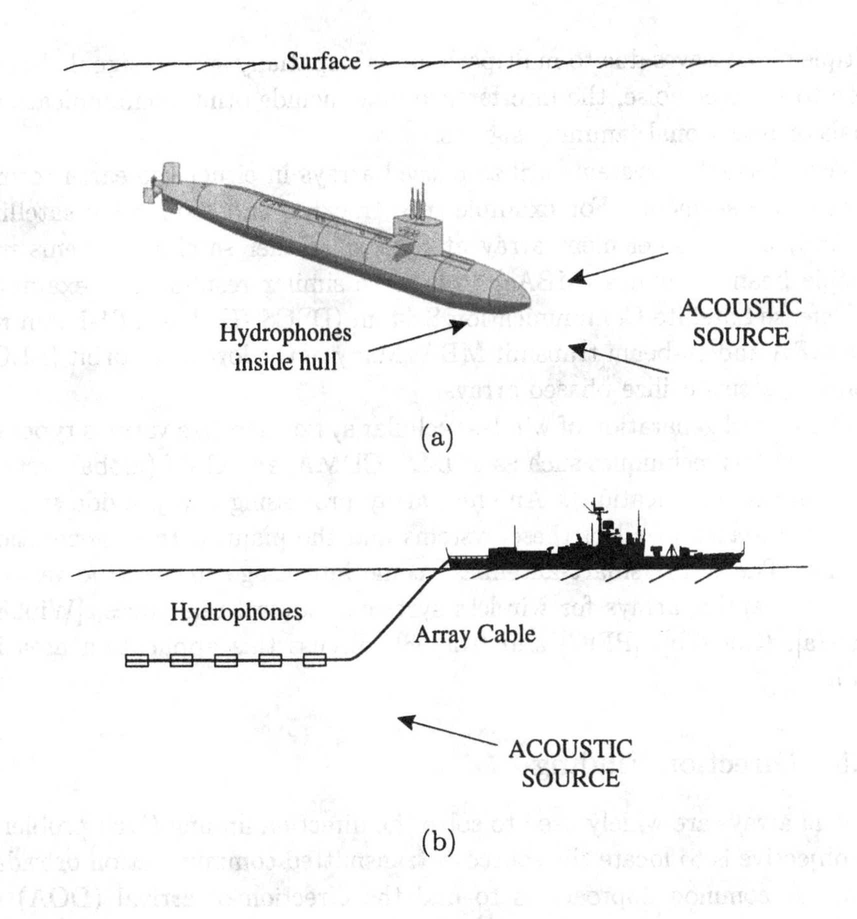

Figure 1.3 Examples of several passive sonar systems: (a) submarine mounted hydrophone arrays; (b) towed array.

al. [BMW91]. Current research is usually reported in the *Journal of the Acoustic Society of America* (JASA) and the *IEEE Transactions on Signal Processing.*

1.2.4 Communications

Antenna arrays are used in many communication systems. One of the first usages was for transatlantic shortwave communication in the 1930s [FF37]. Current usage spans the frequency bands from ELF to EHF and includes both terrestially based and satellite communications.

The communication signals are point sources but, due to the channel characteristics, they may arrive at the receiving array as a single plane wave, multiple plane waves due to multipath, or as a spatially spread signal. In addition to receiver noise, the interference may include other communications signals or intentional jamming signals.

Several satellite systems utilize phased arrays in either the earth terminal or space segment. For example, the tracking and data relay satellite (TDRSS) uses a 30-element array at S-band. Other satellite systems use multiple beam antennas (MBAs) to achieve similar results. For example, the Defense Satellite Communication System (DSCS III) has a 61-beam receive MBA and 19-beam transmit MBA. Many of the low earth orbit (LEO) satellite systems utilize phased arrays.

The second generation of wireless cellular systems utilize various types of multiple access techniques such as TDMA, CDMA, and GSM (global system for mobile communication). Antenna array processing can provide significant performance in all of these systems and the planned third-generation systems. The term "smart antennas" come into usage to describe various types of adaptive arrays for wireless systems. Several references, [Win98], [God97a], [God97b], [PP97] and [Rap98], discuss this application area in detail.

1.2.5 Direction Finding

Antenna arrays are widely used to solve the direction finding (DF) problem. The objective is to locate the source of transmitted communication or radar signal. A common approach is to find the direction of arrival (DOA) of the signal at two separated antenna arrays and locate the source at the intersection of the two lines of bearing. The estimation of DOAs is the central focus of Chapters 8 and 9, and we discuss the problem further at that point.

1.2.6 Seismology

There are two areas of seismology in which array processing plays an important role. The first area is the detection and location of underground nuclear explosions. The area received significant attention in the 1960s and 1970s and a number of results that were obtained for that area, such as Capon's minimum variance distortionless response (MVDR) beamformer, are used in many other areas.

The second area is exploration seismology and is the most important at the present time. Justice in Chapter 2 of [Hay85] has a detailed discussion of array processing in exploration seismology, and his chapter has

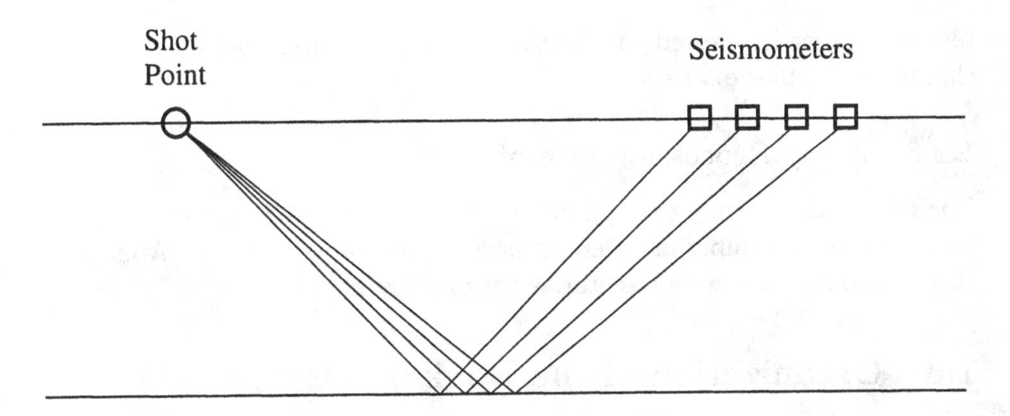

Figure 1.4 Seismic experiment.

322 references. Robinson and Treitel in Chapter 7 of [Opp78] also have a good discussion. The objective of the exploration process is to construct an image of the subsurface in which the structure and physical properties are described. As in the sonar case, the propagation characteristics of the acoustic signal in an inhomogeneous elastic medium has a dominant influence on the designs of the system.

A typical seismic experiment is shown in Figure 1.4. Acoustic energy is transmitted into the earth by a shot and reflected energy is received by a set of geophones arranged in linear array. Normally the earth is modeled as a stack of homogeneous layers and the array measures reflections from various layers. In the text we encounter various examples of seismic signal processing.

The references in Justice [Hay85] provide further discussions of seismic signal processing. Current research is reported in *Geophysics* and *Geophysical Prospecting.*

1.2.7 Tomography

Tomography is the cross-sectional imaging of objects from transmitted or reflected data. The object is illuminated from a number of different directions and data are collected at a receiving array. We then try to reconstruct the cross-sectional image from the data. Kak in Chapter 6 of [Hay85] has a good discussion of tomography. Tomography has had great success in the medical diagnosis area.

The processing algorithms used in tomography are different from those that we develop in the text, so we will not discuss them further. The interested reader is referred to [Hay85] and the references contained in that chapter for further discussion.

1.2.8 Array Processing Literature

Due to the wide variety of applications of array processing, the literature is spread across a number of different journals and conferences. In Appendix B, we have included a representative list of sources.

1.3 Organization of the Book

The book is organized into five parts. The first part consists of Chapters 2, 3, and 4. These three chapters discuss classical array analysis and synthesis techniques. We use the adjective "classical" because the techniques rely primarily on deterministic models and the theory was reasonably mature by the early 1970s. The techniques are important for two reasons:

(i) These techniques are still widely used in practical array applications.

(ii) They form the foundation for the statistical approaches that we use later in the book.

Chapter 2 introduces the basic definitions and relationships that are used to analyze and synthesize arrays. Our approach is to introduce the concept for an arbitrary array geometry. We then specialize the result to a uniform linear array and then further specialize the result to a uniform linear array with uniform weighting.

In Chapter 3, we return to linear arrays and provide a detailed discussion of the analysis and synthesis of linear arrays. In Chapter 4, we study the analysis and synthesis of planar and volumetric arrays.

The second part of the book consists of Chapter 5 and studies the characterization of space-time random processes. We develop second-moment theory for arbitrary processes and a complete characterization for Gaussian random space-time processes. We develop orthogonal expansions and review the concept of signal subspaces and noise subspaces that we utilized for the temporal problem in Chapter I-4 of DEMT I [VT68], [VT01a]. We introduce parametric spatial models and discuss their usage in array processing problems. The chapter provides the statistical models that will be used in the remainder of the text.

The third part of the book consists of Chapters 6 and 7 and studies waveform estimation. In Chapter 6, we derive optimum beamformers under the assumption of known statistics and analyze their performance. We investigate the sensitivity of the optimum beamformers to perturbations in the signal and noise model and the array description. These sensitivity results motivate the development of constrained processors that are more robust to model perturbations.

In Chapter 7, we consider the case in which the statistics must be determined from the data. This problem leads us into adaptive beamforming. We develop various adaptive algorithms and analyze their behavior.

The fourth part of the book consists of Chapters 8 and 9. These chapters consider the parameter estimation problem with emphasis on estimating the direction of arrival of incoming plane-wave signals. We first develop maximum likelihood estimators and compute bounds on the performance of any estimator. We then study a large variety of estimation algorithms and compare their performance to the maximum likelihood estimators and the bounds.

The fifth part of the book consists of Chapter 10 and contains a brief discussion of the optimum detection problem. Chapter 10 also contains a discussion of some of the areas that the book has not covered. There is an appendix that summarizes some of the matrix algebra results that we use in the text.

There are problems at the end of each chapter. As in DEMT I – III [VT68], [VT01a], [VT71a], [VT71b], [VT01b], it is necessary for the reader to solve problems in order to understand the material fully. Throughout the course and the book we emphasize the development of an ability to work problems. The problems range from routine manipulations to significant extensions of the material in the text. In many cases they are equivalent to journal articles currently being published. Only by working a fair number of them is it possible to appreciate the significance and generality of the results. Many of the problems require the use of mathematical computation package such as MATLAB®, Mathematica, or MAPLE. We assume that the student is experienced with one of these packages. The solutions on the Web site use MATLAB®.

The book can be covered in a two-semester graduate course for students with the appropriate background. Chapters 1 through 6 can be covered in the first semester. The common theme in these chapters is that the design is either deterministic or assumes that the necessary statistics are known. Chapters 7 through 10 can be covered in the second semester. The common theme in these chapters is that the algorithms obtain the necessary statistics from the data. This results in adaptive beamformers, adaptive detectors, and DOA estimators.

1.4 Interactive Study

The book is designed to be used in two different modes. The first mode is as a stand-alone text. By reading the text and doing a representative set of homework problems, one can understand the material. The difficulty with this mode is that we rely on a large number of examples to develop the material. By necessity, many of the examples will choose specific parameter values to demonstrate the point. It would be more desirable to be able to explore a family of parameters.

The second mode uses the Web site that was discussed in the Preface:

http://ite.gmu.edu/DetectionandEstimationTheory/

The contents of the Web site is described in the Preface. We anticipate that this second mode will be used by most serious readers and instructors.

Bibliography

[Alb60] V. O. Albers. *Underwater Acoustics Handbook.* Pennsylvania State Univesity Press, University Park, Pennsylvania, 1960.

[All63] J. Allen. The theory of arrays antennas with emphasis on radar applications. Technical Report 323, M.I.T. Lincoln Laboratory, Lexington, Massachusetts, July 1963.

[Bag76] A. B. Baggeroer. Space-time processes and optimal array processing. Technical Report 506, Navy Undersea Center, San Diego, California, December 1976.

[Bar65] D. K. Barton. *Radar System Analysis.* Prentice-Hall, Englewood Cliffs, New Jersey, 1965.

[Ber65] R. S. Berkowitz. *Modern Radar: Analysis, Evaluation, and System Design.* Wiley, New York, 1965.

[BMW91] R. E. Blahut, W. Miller, Jr., and C. H. Wilcox. *Radar and Sonar, Part I.* Springer-Verlag, New York, 1991.

[Bro85] E. Brookner. Phased array radars. *Sci. Amer.*, pp. 94–102, February 1985.

[Bro91] E. Brookner, editor. *Practical Phased-Array Antenna Systems.* Artech House, Boston, 1991.

[Bur91] W. S. Burdic. *Underwater Acoustic System Analysis.* Prentice-Hall, Englewood Cliffs, New Jersey, 2nd edition, 1991.

[BW69] D. K. Barton and H. R. Ward. *Handbook of Radar Measurement.* Prentice-Hall, New Jersey, 1969.

[CB67] C. E. Cook and M. Bernfeld. *Radar Signals, An Introduction to Theory and Application.* Academic Press, New York, 1967.

[CH69] W. N. Christiansen and J. A. Högbom. *Radiotelescopes.* Cambridge University Press, Cambridge, 1969.

[Cox73] H. Cox. Resolving power and sensitivity to mismatch of optimum array processors. *J. Acoust. Soc. of Amer.*, vol.54(3), pp. 771–785, September 1973.

[CS62] B. F. Cron and C. H. Sherman. Spatial-correlation functions for various noise models. *J. Acoust. Soc. Amer.*, vol.34, pp. 1732–1737, November 1962.

[Dye70] I. Dyer. Statistics of sound propagation in the ocean. *J. Acoust. Soc. Amer.*, vol.48, pp. 337–345, May 1970.

[EH68] J. V. Evans and T. Hagfors, editors. *Radar Astronomy*. McGraw-Hill, New York, 1968.

[EJS82] J. E. Evans, J. R. Johnson, and D.F. Sun. Application of advanced signal processing techniques to angle of arrival estimation in ATC navigation and surveillance systems. Technical Report, M.I.T. Lincoln Laboratory, Lexington, Massachusetts, June 1982.

[FF37] H. T. Friis and C. B. Feldman. A multiple unit steerable antenna for short-wave reception. *Bell Syst. Tech. J.*, vol.16, pp. 337–419, July 1937.

[FR68] J. V. Di Franco and W. L. Rubin. *Radar Detection*. Prentice-Hall, Englewood Cliffs, New Jersey, 1968.

[God97a] L. C. Godara. Application of antenna arrays to mobile communications, Part I: Performance improvement, feasibility, and system considerations. *Proc. IEEE*, vol.85(7), July 1997.

[God97b] L. C. Godara. Application of antenna arrays to mobile communications, Part II: Beamforming and direction-of-arrival considerations. *Proc. IEEE*, vol.85(8), August 1997.

[GSS73] J. W. R. Griffiths, P. Stocklin, and C. Van Schoonveld, editors. *Signal Processing*. Academic Press, New York, 1973.

[Has89] J. C. Hassab. *Underwater Signal and Data Processing*. CRC Press, Inc., Boca Raton, Florida, 1989.

[Hay80] S. Haykin, editor. *Array Processing: Applications to Radar*. Dowden, Hutchinson, & Ross, Stroudsburg, Pennsylvania, 1980.

[Hay85] S. Haykin. *Array Signal Processing*. Prentice-Hall, Englewood Cliffs, New Jersey, 1985.

[Hor57] J. W. Horton. *Fundamentals of Sonar*. U. S. Naval Institute, Washington, D. C., 1957.

[KPK81] W. C. Knight, R. G. Pridham, and S. M. Kay. Digital signal processing for sonar. *Proc. IEEE*, vol.69(11), pp. 1451–1507, November 1981.

[Kra66] J. D. Kraus. *Radio Astronomy*. McGraw-Hill, New York, 1966.

[KV73] K. I. Kellermann and G. L. Verschuur, editors. *Galactic and Extra-Galactic Radio Astronomy*. Springer-Verlag, New York, 1973.

[Off58] C. B. Officer. *Introduction to the Theory of Sound Transmission*. McGraw-Hill, New York, 1958.

[OK72] A. A. Oliner and G. H. Knittel, editors. *Phased Array Antennas*. Artech House, Boston, Massachusetts, 1972.

[Opp78] A. V. Oppenheim, editor. *Applications of Digital Signal Processing*. Prentice-Hall, Englewood Cliffs, New Jersey, 1978.

[PP97] A. J. Paulraj and C. B. Papadias. Space-time processing for wireless communications. *IEEE Signal Processing Mag.*, vol.14(6), November 1997.

[Rap98] T. S. Rappaport, editor. *Smart Antennas.* IEEE Press, New York, 1998.

[RC52] J. F. Reintjes and G. T. Coate. *Principles of Radar.* McGraw-Hill, New York, 1952.

[Ree69] J. E. Reed. The AN/FPS-85 radar system. *Proc. IEEE*, vol.57(3), pp. 324–335, March 1969.

[Rid47] L. N. Ridenour. *Radar System Engineering.* McGraw-Hill, New York, 1947.

[Ryl73] M. Ryle. The 5-km radio telescope at Cambridge. *Nature*, vol.239, pp. 435–438, October 1973.

[SH92] T. J. Shepherd, S. Haykin, and J. Litva, editors. *Radar Array Processing.* Springer-Verlag, New York, 1992.

[Sko70] M. I. Skolnik. *Radar Handbook.* McGraw-Hill, New York, 1970.

[Sko80] M. I. Skolnik. *Introduction to Radar Systems.* McGraw-Hill, New York, 1980.

[Smi49] R. A. Smith. *Aerials for Metre and Decimetre Wave-lengths.* Cambridge University Press, London, 1949.

[Sou62] G. C. Southworth. *Forty Years of Radio Research.* Gordon and Breach, New York, 1962.

[TC66] I. Tolstoy and C. S. Clay. *Ocean Acoustics: Theory and Experiment in Underwater Sound.* McGraw-Hill, New York, 1966.

[TCWN80] A. R. Thompson, B. G. Clark, C. M. Wade, and P. J. Napier. The very large array. *Astrophys. J. Suppl.*, vol.44, pp. 151–167, April 1980.

[Uri67] R. J. Urick. *Principles of Underwater Sound for Engineers.* McGraw-Hill, New York, 1967.

[VT68] H. L. Van Trees. *Detection, Estimation, and Modulation Theory, Part I.* Wiley, New York, 1968.

[VT01a] H. L. Van Trees. *Detection, Estimation, and Modulation Theory, Part I.* Wiley Interscience, New York, 2001.

[VT01b] H. L. Van Trees. *Detection, Estimation, and Modulation Theory, Part III.* Wiley Interscience, New York, 2001.

[VT71a] H. L. Van Trees. *Detection, Estimation, and Modulation Theory, Part II.* Wiley, New York, 1971.

[VT71b] H. L. Van Trees. *Detection, Estimation, and Modulation Theory, Part III.* Wiley, New York, 1971.

[War94] J. Ward. Space-time adaptive processing for airborne radar. Technical Report 1015, M.I.T Lincoln Laboratory, Lexington, Massachusetts, December 1994.

[WB83] R. A. Wagstaff and A. B. Baggeroer, editors. *High-Resolution Spatial Processing in Underwater Acoustics.* Naval Ocean Research and Development Activity, NSTL, Gulf Park, Mississippi, 1983.

[Win98] J. H. Winters. Smart Antennas for Wireless Systems. *IEEE Personal Commun.*, vol.1, pp. 23–27, February 1998.

Chapter 2

Arrays and Spatial Filters

2.1 Introduction

We assume that we have a signal or multiple signals that are located in some region of a space-time field. We also have noise and/or interference that is located in some region of a space-time field. In the applications of interest these regions have some overlap.

An array is used to filter signals in a space-time field by exploiting their spatial characteristics. This filtering may be expressed in terms of a dependence upon angle or wavenumber. Viewed in the frequency domain this filtering is done by combining the outputs of the array sensors with complex gains that enhance or reject signals according to their spatial dependence. Usually, we want to spatially filter the field such that a signal from a particular angle, or set of angles, is enhanced by a constructive combination and noise from other angles is rejected by destructive interference.

The design of arrays to achieve certain performance criteria involves trade-offs among the array geometry, the number of sensors, signal-to-noise, and signal-to-interference ratios, as well as a number of other factors.

There are two aspects of array design that determine their performance as spatial filters. First, their geometry establishes basic constraints upon their operation. Line arrays can resolve only one angular component. This leads to a cone of uncertainty and right/left ambiguities. Circular arrays have different patterns than crossed or planar arrays. Frequently the geometry is established by physical constraints and the designer may have limited freedom in specifying the array geometry.

The second aspect is the design of the complex weightings of the data at each sensor output. The choice of these weightings determines the spatial filtering characteristics of the array for a given geometry.

In this chapter we introduce the basic definitions and relationships that are used to analyze and synthesize arrays. Our approach is to introduce the concepts for an arbitrary array geometry. We then specialize the result to a uniform linear array and then further specialize the result to a uniform weighting. In Chapter 3, we return to linear arrays and provide a detailed discussion of the analysis and synthesis of linear arrays. In Chapter 4, we study the analysis and synthesis of planar and volume arrays.

This chapter is organized in the following manner. In Section 2.2, we introduce the frequency-wavenumber response function and beam pattern of an array. We employ wavenumber variables with dimensions of inverse length for a number of reasons. First, array coordinates and wavenumbers are conjugate Fourier variables, so Fourier transform operations are much simpler. Second, all the powerful properties of harmonic analysis as extended to homogenous processes can be used directly and the concept of an array as a spatial filter is most applicable. Third, angle variables specify array filter responses over a very restricted region of wavenumber space. While it does describe the response over the region for all real, propagating signals, that is, those space-time processes that implicitly satisfy a wave equation when one assigns a propagation speed and direction, there are a lot of advantages to considering the entire wavenumber space. The so-called virtual space, or wavenumber realm where real signals cannot propagate is very useful in the analysis of array performance.

In Section 2.3, we specialize these results to a uniform linear array and study the characteristics of the beam pattern. In Section 2.4, we further specialize these results to the case of a uniformly weighted linear array. This leads to a beam pattern that we refer to as the conventional beam pattern. It will play a fundamental role in many of our subsequent studies. In Section 2.5, we discuss array steering and show how it affects the beam pattern in wavenumber space and in angle space. In Section 2.6, we define three important performance measures:

(i) Directivity
(ii) Array gain
(iii) Tolerance function

These performance measures are utilized throughout our discussion.

The discussion in the first six sections assumes that the sensors are isotropic (i.e., their response is independent of the direction of arrival of the signal). In Section 2.7, we introduce the concept of pattern multiplication to accommodate non-isotropic sensors. In Section 2.8, we consider the case of a linear aperture and show how the performance of apertures and arrays are related. In Section 2.9, we give a brief summary of our development.

In Table 2.1, we have summarized the structure of the chapter. The various terms are defined at appropriate points in the chapter.[1]

The material in this chapter can be termed **classical array theory**, and it has been discussed in a number of books and articles. References that we have utilized include Kraus [Kra88], Balanis [Bal82], Elliott [Ell81], Johnson [Joh93], Milligan [Mil85], Ziomek [Zio95], Skolnik [Sko80], Stutzman and Thiele [ST81], and Weeks [Wee68].

The coordinate system of interest is shown in Figure 2.1. The relationships between rectangular and spherical coordinates is shown in Figure 2.1.

$$\begin{aligned} x &= r \sin\theta \cos\phi, \\ y &= r \sin\theta \sin\phi, \\ z &= r \cos\theta. \end{aligned} \tag{2.1}$$

The next set of figures shows various arrays and apertures placed in this coordinate system.

Figure 2.2 shows a linear array with equally spaced elements. The polar angle θ is the grazing angle with respect to the positive z-axis. In some cases the broadside angle $\bar{\theta}$ is a useful parameter

$$\bar{\theta} = \frac{\pi}{2} - \theta. \tag{2.2}$$

The position of the elements is denoted by p_{z_n},

$$p_{z_n} = \left(n - \frac{N-1}{2}\right) d, \quad n = 0, 1, \cdots, N-1, \tag{2.3}$$

where d is the interelement spacing.

Figure 2.3 shows a linear array with unequally spaced elements. In this case,

$$p_{z_n} = z_n, \tag{2.4}$$

where z_n is the z-coordinate of the nth element.

---

[1] We have included a structure chart at the beginning of Chapters 2–9. Its primary purpose is to serve as a graphical *a posteriori* reference for the reader so that, after reading the chapter, one can easily find a particular topic. A secondary purpose is to aid an instructor in planning the coverage of the material.

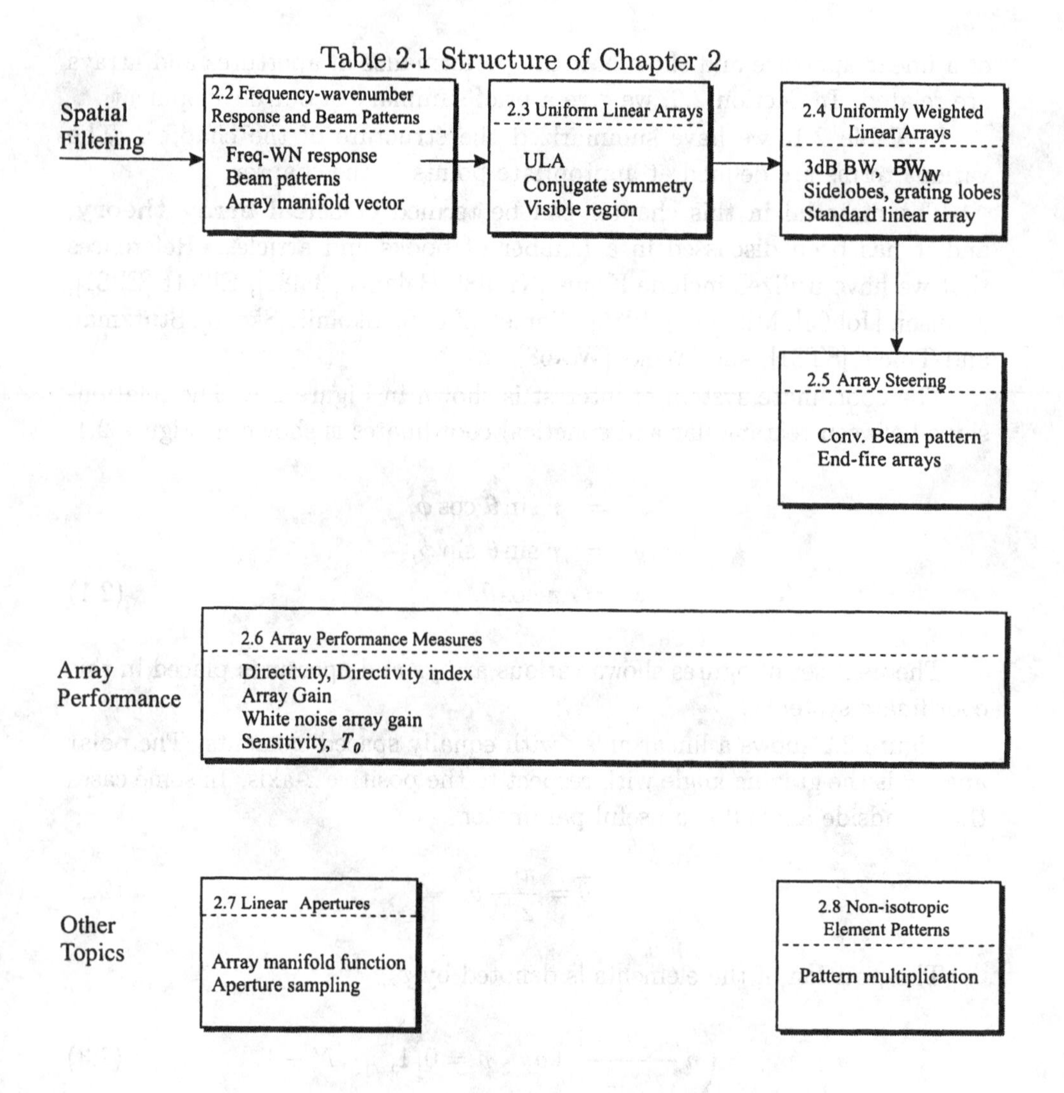

Table 2.1 Structure of Chapter 2

Figure 2.4 shows a continuous linear aperture along the z-axis. We would anticipate that if d is small and

$$L = Nd, \tag{2.5}$$

the array and aperture would have very similar performance. We demonstrate this relationship later. We also discuss how we sample a continuous aperture to obtain an array.

Figure 2.5 shows several examples of planar arrays that are of interest. Figure 2.6 shows the corresponding planar aperture. We define the coor-

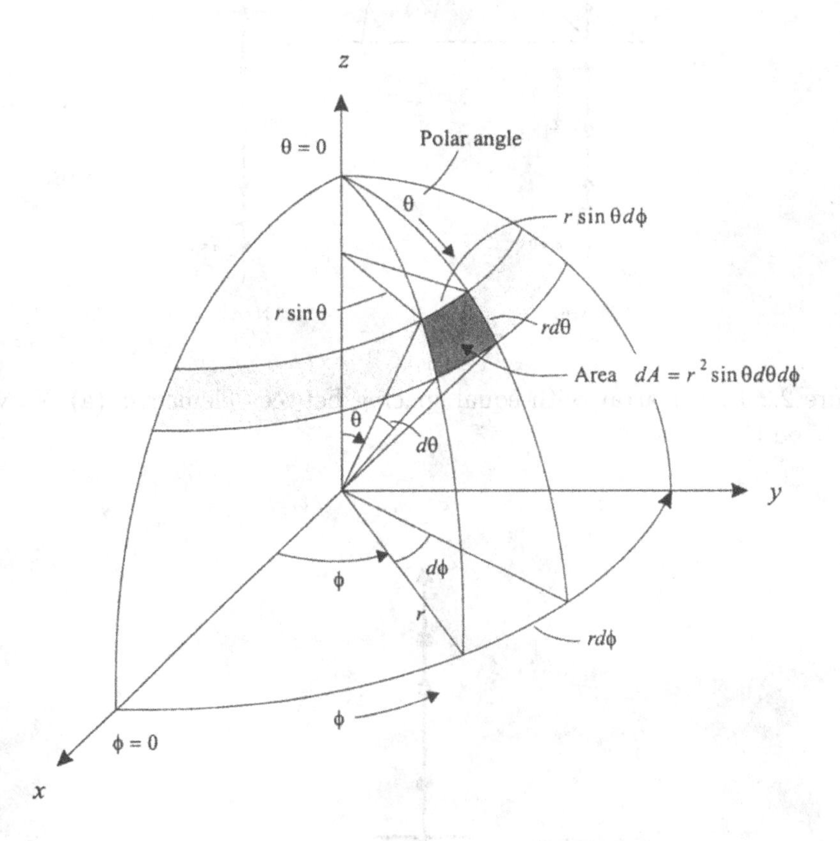

Figure 2.1 Spherical coordinate system.

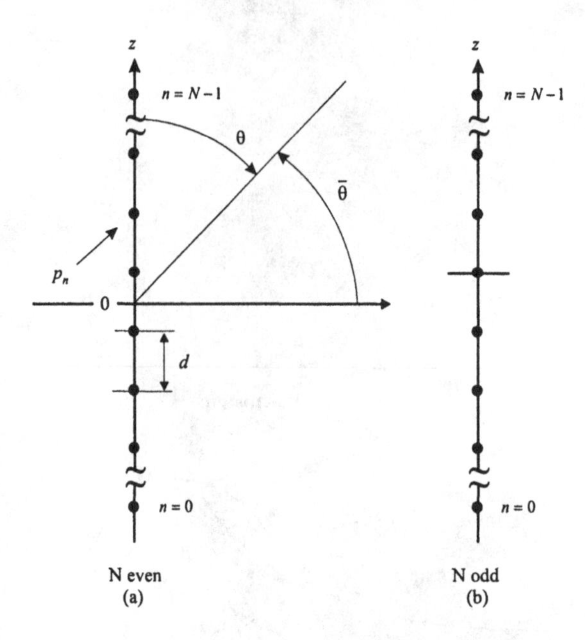

Figure 2.2 Linear array with equal spacing between elements: (a) N even; (b) N odd.

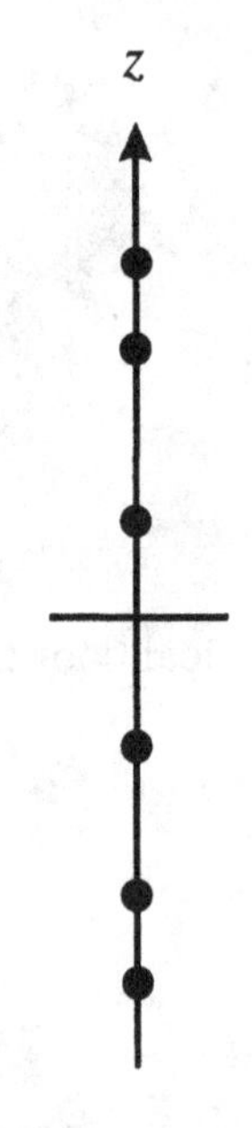

Figure 2.3 Linear array with unequal spacing between elements.

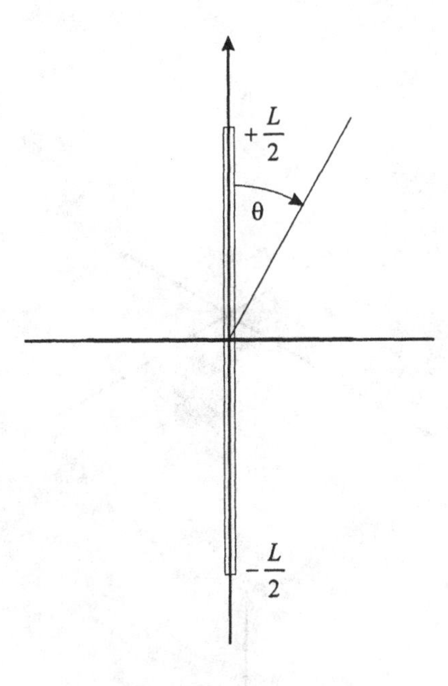

Figure 2.4 Linear aperture.

dinate system in detail in Chapter 4 where we analyze planar arrays and apertures.

Figure 2.7 shows some volume arrays of interest. We will discuss volume arrays and apertures in Chapter 4.

2.2 Frequency-wavenumber Response and Beam Patterns

In this section, we analyze the response of an array to an external signal field. The array consists of a set of isotropic sensors located at positions $\mathbf{p}_n$, as shown in Figure 2.8. The sensors spatially sample the signal field at the locations $\mathbf{p}_n : n = 0, 1, \cdots, N-1$. This yields a set of signals that we denote by the vector $\mathbf{f}(t, \mathbf{p})$

$$\mathbf{f}(t, \mathbf{p}) = \begin{bmatrix} f(t, \mathbf{p}_0) \\ f(t, \mathbf{p}_1) \\ \vdots \\ f(t, \mathbf{p}_{N-1}) \end{bmatrix}. \tag{2.6}$$

We process each sensor output by a linear, time-invariant filter with

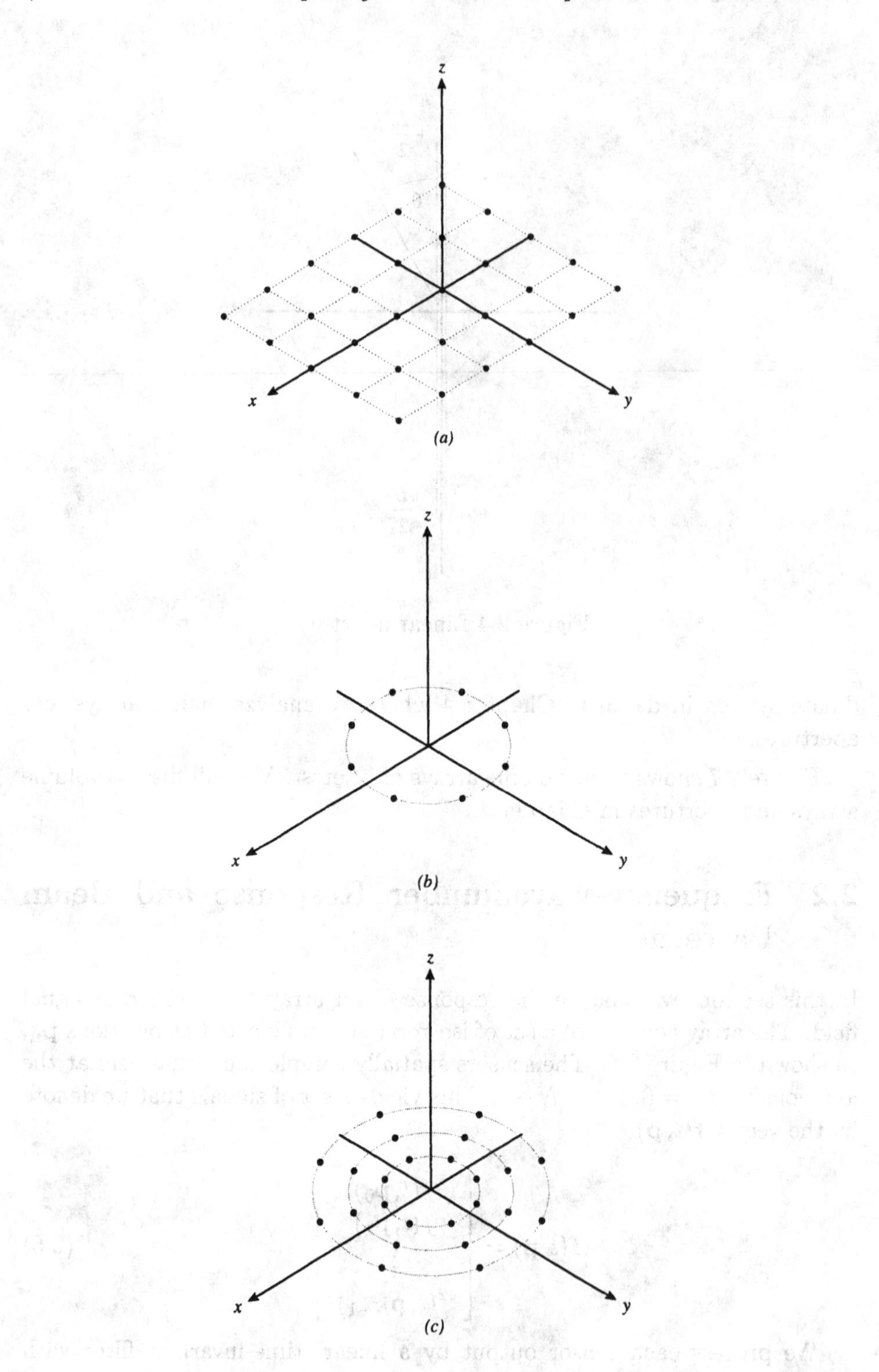
z
x
y
(a)
z
x
y
(b)
z
x
y
(c)

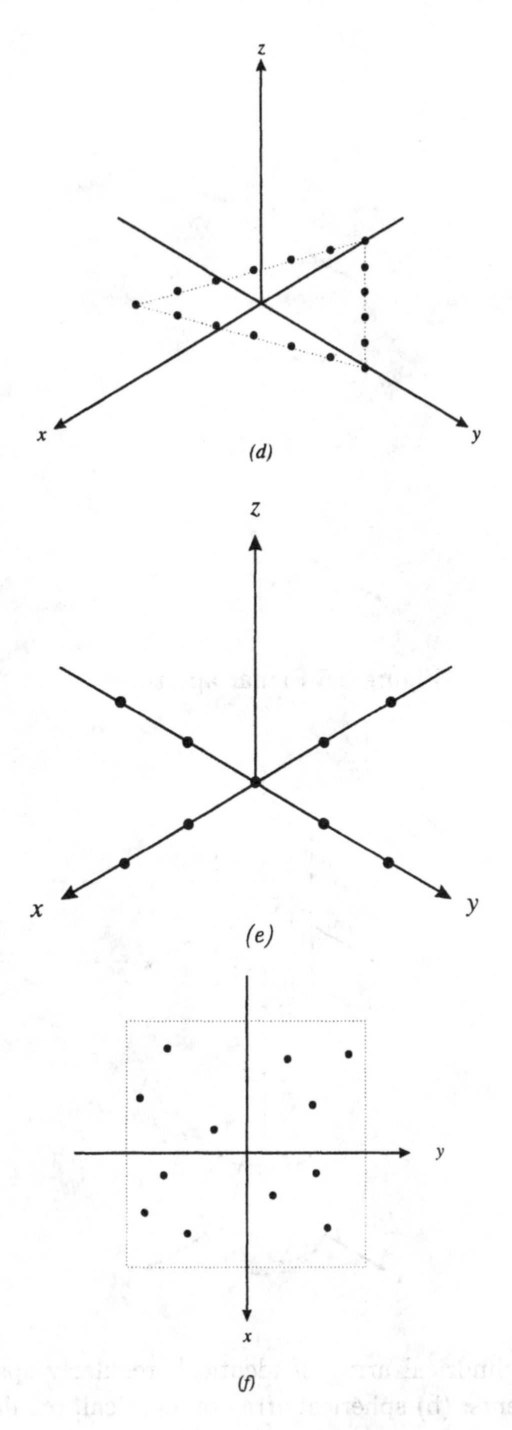

Figure 2.5 Planar arrays.

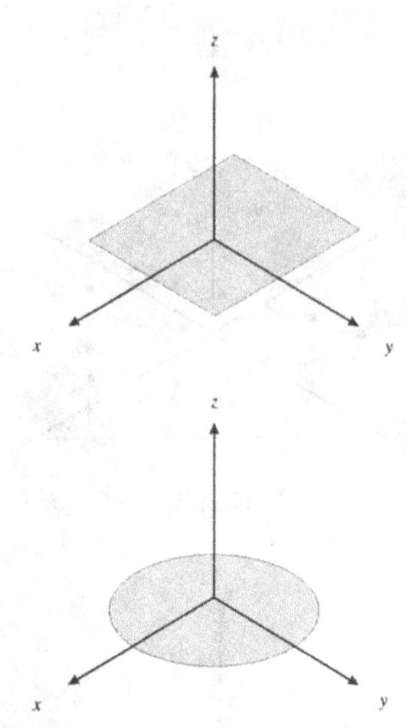

Figure 2.6 Planar apertures.

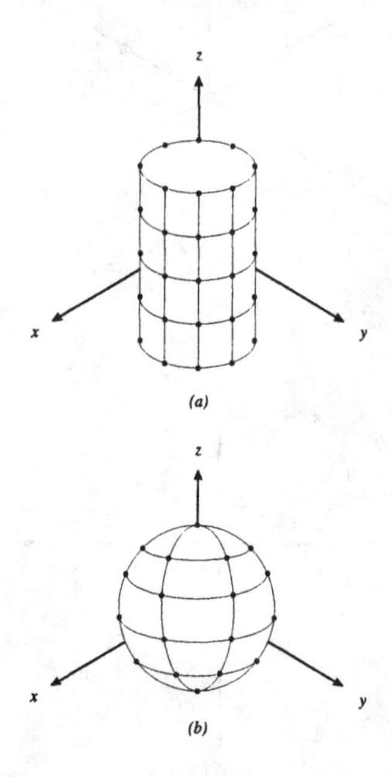

Figure 2.7 (a) Cylindrical array of identical, regularly spaced, omnidirectional point elements; (b) spherical array of identical, regularly spaced, omnidirectional point elements.

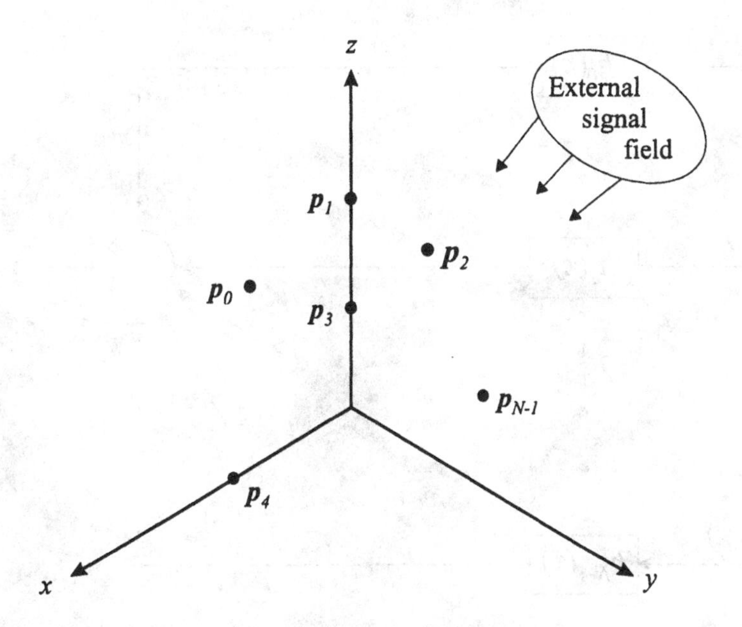

Figure 2.8 N-Element array.

impulse response $h_n(\tau)$ and sum the outputs to obtain the array output $y(t)$. The procedure is shown in Figure 2.9. We assume that the observation interval is long enough that it may be considered infinite. The output $y(t)$ can be written as a convolution integral,

$$y(t) = \sum_{n=0}^{N-1} \int_{-\infty}^{\infty} h_n(t-\tau)\, f_n(\tau, \mathbf{p}_n)\, d\tau. \tag{2.7}$$

This result can be written in vector notation as

$$y(t) = \int_{-\infty}^{\infty} \mathbf{h}^T(t-\tau)\, \mathbf{f}(\tau, \mathbf{p})\, d\tau, \tag{2.8}$$

where

$$\mathbf{h}(\tau) = \begin{bmatrix} h_0(\tau) \\ h_1(\tau) \\ \vdots \\ h_{N-1}(\tau) \end{bmatrix}. \tag{2.9}$$

The result in (2.8) is a straightforward extension of familiar scalar results to the vector model.

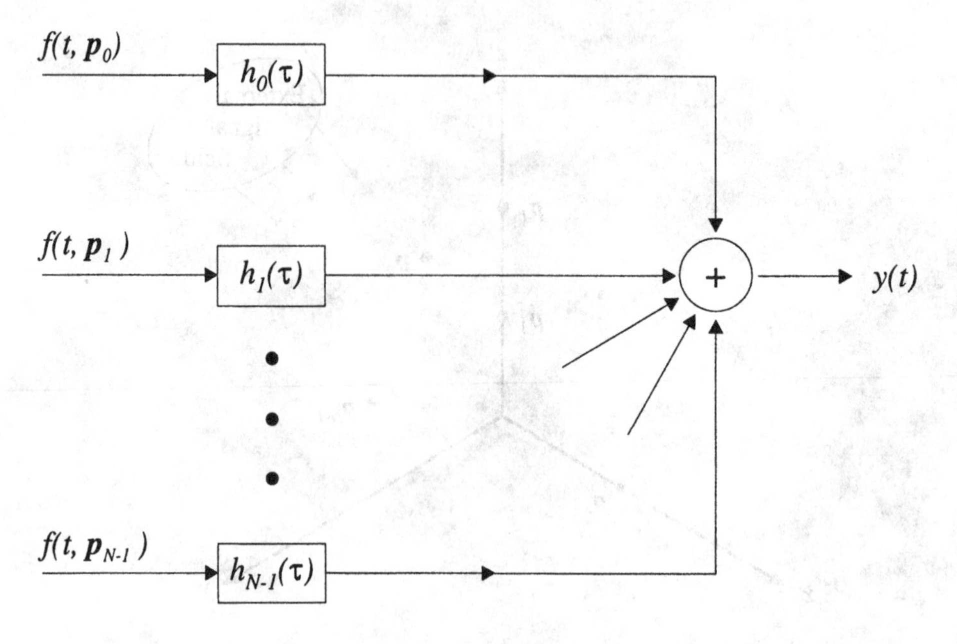

Figure 2.9 Array with linear processing.

Alternatively, we can write (2.8) in the transform domain as

$$\begin{aligned} Y(\omega) &= \int_{-\infty}^{\infty} y(t)e^{-j\omega t}dt \\ &= \mathbf{H}^T(\omega)\mathbf{F}(\omega), \end{aligned} \tag{2.10}$$

where

$$\mathbf{H}(\omega) = \int_{-\infty}^{\infty} \mathbf{h}(t)e^{-j\omega t}dt, \tag{2.11}$$

and

$$\mathbf{F}(\omega, \mathbf{p}) = \int_{-\infty}^{\infty} \mathbf{f}(t, \mathbf{p})e^{-j\omega t}dt. \tag{2.12}$$

In most cases, we suppress the $\mathbf{p}$ dependence on the left side of (2.12) and use $\mathbf{F}(\omega)$.

To illustrate a simple beamforming operation, consider the case shown in Figure 2.10. The input is a plane wave propagating in the direction $\mathbf{a}$ with temporal (radian) frequency ω. The time functions at the sensors due to this input can be written in two equivalent ways. The first way emphasizes the time delays corresponding to the time of arrival at the various sensors. If $f(t)$ is the signal that would be received at the origin of the coordinate

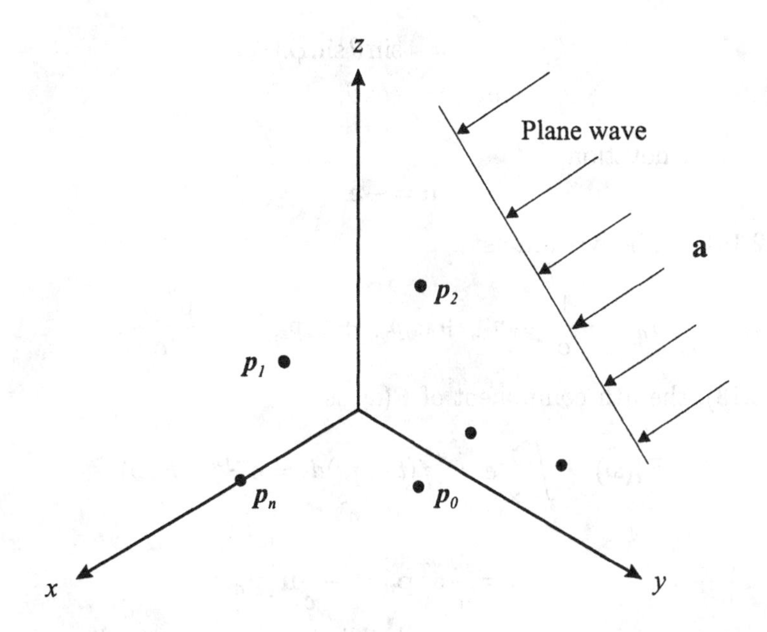

Figure 2.10 Array with plane-wave input.

system, then

$$\mathbf{f}(t, \mathbf{p}) = \begin{bmatrix} f(t-\tau_0) \\ f(t-\tau_1) \\ \vdots \\ f(t-\tau_{N-1}) \end{bmatrix}, \tag{2.13}$$

where

$$\tau_n = \frac{\mathbf{a}^T \mathbf{p}_n}{c}, \tag{2.14}$$

and c is the velocity of propagation in the medium and $\mathbf{a}$ is a unit vector that can be expressed as

$$\mathbf{a} = \begin{bmatrix} -\sin\theta\,\cos\phi \\ -\sin\theta\,\sin\phi \\ -\cos\theta \end{bmatrix}. \tag{2.15}$$

The minus sign arises because of the direction of $\mathbf{a}$. Then, τ_n is given by

$$\tau_n = -\frac{1}{c}\left[\sin\theta\,\cos\phi \cdot p_{x_n} + \sin\theta\,\sin\phi \cdot p_{y_n} + \cos\theta \cdot p_{z_n}\right]. \tag{2.16}$$

If we define direction cosines with respect to each axis as

$$u_x = \sin\theta\cos\phi, \tag{2.17}$$

$$u_y = \sin\theta\sin\phi, \tag{2.18}$$
$$u_z = \cos\theta, \tag{2.19}$$

then in vector notation,

$$\mathbf{u} = -\mathbf{a}. \tag{2.20}$$

Then (2.16) can be written as

$$\tau_n = -\frac{1}{c}\left[u_x p_{x_n} + u_y p_{y_n} + u_z p_{z_n}\right] = -\frac{\mathbf{u}^T \mathbf{p}_n}{c}. \tag{2.21}$$

From (2.13), the nth component of $\mathbf{F}(\omega)$ is

$$F_n(\omega) = \int_{-\infty}^{\infty} e^{-j\omega t} f(t-\tau_n)dt = e^{-j\omega\tau_n} F(\omega), \tag{2.22}$$

where

$$\omega\tau_n = \frac{\omega}{c}\mathbf{a}^T \mathbf{p}_n = -\frac{\omega}{c}\mathbf{u}^T \mathbf{p}_n. \tag{2.23}$$

For plane waves propagating in a locally homogeneous medium, we define the wavenumber $\mathbf{k}$ as

$$\mathbf{k} = \frac{\omega}{c}\mathbf{a} = \frac{2\pi}{\lambda}\mathbf{a}, \tag{2.24}$$

where λ is the wavelength corresponding to the frequency ω. Equivalently,

$$\mathbf{k} = -\frac{2\pi}{\lambda}\begin{bmatrix} \sin\theta\,\cos\phi \\ \sin\theta\,\sin\phi \\ \cos\theta \end{bmatrix} = -\frac{2\pi}{\lambda}\mathbf{u}. \tag{2.25}$$

The wave equation constrains the magnitude of the wavenumber,[2]

$$|\mathbf{k}| = \frac{\omega}{c} = \frac{2\pi}{\lambda}. \tag{2.26}$$

Therefore, only the direction of $\mathbf{k}$ varies. Comparing (2.14) and (2.24), we observe that

$$\omega\tau_n = \mathbf{k}^T \mathbf{p}_n. \tag{2.27}$$

Defining

$$\mathbf{v_k}(\mathbf{k}) = \begin{bmatrix} e^{-j\mathbf{k}^T\mathbf{p}_0} \\ e^{-j\mathbf{k}^T\mathbf{p}_1} \\ \vdots \\ e^{-j\mathbf{k}^T\mathbf{p}_{N-1}} \end{bmatrix}, \tag{2.28}$$

[2]The wave equation is developed in a number of references (e.g., [Bal82]).

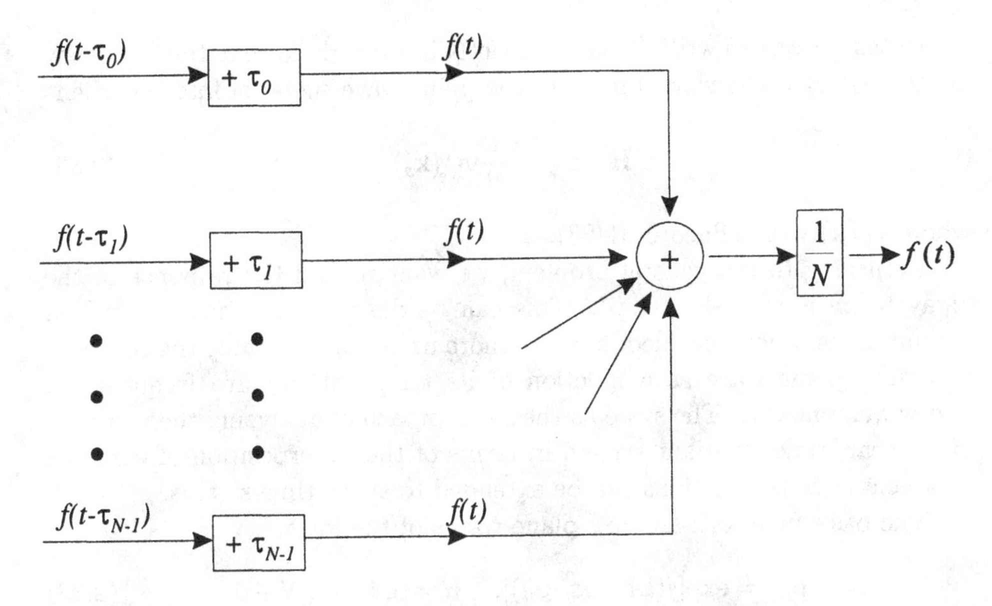

Figure 2.11 Delay-and-sum beamformer.

we can write $\mathbf{F}(\omega)$ as

$$\mathbf{F}(\omega) = F(\omega)\mathbf{v_k}(\mathbf{k}). \tag{2.29}$$

The vector $\mathbf{v_k}(\mathbf{k})$ incorporates all of the spatial characteristics of the array and is referred to the **array manifold vector**. It plays a central role in our discussion. The subscript $\mathbf{k}$ denotes that the argument is in $\mathbf{k}$-space. The subscript is to distinguish it from other variables we will use later as arguments to the array manifold vector.

In this case, we shift the inputs from each sensor so that the signals are aligned in time and add them. This operation is shown in Figure 2.11, where we have included a normalization factor $1/N$ so the output is $f(t)$. In this case,

$$h_n(\tau) = \frac{1}{N}\delta(\tau + \tau_n) \tag{2.30}$$

and

$$y(t) = f(t). \tag{2.31}$$

This processor is referred to as a **delay-and-sum beamformer** or the **conventional beamformer**. In practice we add a common delay in each channel so that the operations in Figure 2.11 are physically realizable.

Note that we can write (2.30) compactly in a matrix form in the frequency domain. If $\mathbf{k}_s$ is the wavenumber of the plane-wave signal of interest, then

$$\mathbf{H}^T(\omega) = \frac{1}{N}\mathbf{v}_{\mathbf{k}}^H(\mathbf{k}_s), \tag{2.32}$$

where $\mathbf{v}_{\mathbf{k}}(\mathbf{k})$ was defined in (2.28).

Returning to the general problem, we want to find the response of the array to an input field $f(t, \mathbf{p})$. This can be done by the convolution and summing operation specified, but it is more useful to determine the response to a unit plane wave as a function of its temporal (radian) frequency ω and wavenumber $\mathbf{k}$. The systems theory approach of analyzing the response of a linear, time-invariant system in terms of the superposition of complex exponential basis functions can be extended to space-time signals.

The basis functions are now plane waves of the form,

$$f_n(t, \mathbf{p}_n) = \exp[j(\omega t - \mathbf{k}^T \mathbf{p}_n)], \quad n = 0, 1, \cdots, N-1, \tag{2.33}$$

or

$$\mathbf{f}(t, \mathbf{p}) = e^{j\omega t}\, \mathbf{v}_{\mathbf{k}}(\mathbf{k}), \tag{2.34}$$

where $\mathbf{v}_{\mathbf{k}}(\mathbf{k})$ was defined in (2.28).

The response of the array processor of (2.8) to a plane wave is

$$y(t, \mathbf{k}) = \mathbf{H}^T(\omega)\, \mathbf{v}_{\mathbf{k}}(\mathbf{k})\, e^{j\omega t}, \tag{2.35}$$

where $\mathbf{H}(\omega)$ is the Fourier transform of $\mathbf{h}(\tau)$ in (2.9).

We emphasize the dependence of the output upon the input wavenumber $\mathbf{k}$ with the notation $y(t, \mathbf{k})$. The temporal dependence is a complex exponential at the same frequency as the input plane wave. Equation (2.35) can be written in the frequency domain as

$$Y(\omega, \mathbf{k}) = \mathbf{H}^T(\omega)\, \mathbf{v}_{\mathbf{k}}(\mathbf{k}). \tag{2.36}$$

Note that ω is a single frequency corresponding to the input frequency. The temporal spatial processing by the array is completely described by the term on the right side of (2.36). We define this term as

$$\boxed{\Upsilon(\omega, \mathbf{k}) \triangleq \mathbf{H}^T(\omega)\, \mathbf{v}_{\mathbf{k}}(\mathbf{k}),} \tag{2.37}$$

which we term the **frequency-wavenumber response function** of the array. It describes the complex gain of an array to an input plane wave with wavenumber $\mathbf{k}$ and temporal frequency ω, and has the same interpretation as a transfer function for a linear time-invariant system (we introduce $\Upsilon(\omega, \mathbf{k})$ because $Y(\omega, \mathbf{k})$ will be used later to describe the output due to arbitrary inputs). $\Upsilon(\omega, \mathbf{k})$ is defined over the entire $\mathbf{k}$ space. The second term in (2.37), $\mathbf{v}_{\mathbf{k}}(\mathbf{k})$, is the array manifold vector defined in (2.28).

The frequency-wavenumber response function describes the response to an arbitrary plane wave. In most physical applications there is a coupling between the temporal frequency ω and the spatial wavenumber $\mathbf{k}$ through the wave equation governing the propagation of the plane wave. Sometimes this can be a very simple relationship such as a plane wave in a homogeneous (and infinite) space; in other instances it can be quite complicated, such as the modal behavior in layered media that often occurs in underwater acoustics and seismology.

The **beam pattern** for an array is defined in the context of plane waves propagating in a locally homogeneous medium where one has the wave equation constraint given in (2.26). This constrains the magnitude of the wavenumber $\mathbf{k}$ as given in (2.26). The beam pattern is the frequency-wavenumber response function evaluated versus the direction, or

$$B(\omega : \theta, \phi) = \Upsilon(\omega, \mathbf{k})|_{\mathbf{k}=\frac{2\pi}{\lambda}\mathbf{a}(\theta,\phi)}, \tag{2.38}$$

where $\mathbf{a}(\theta, \phi)$ is a unit vector with spherical coordinate angles θ, ϕ. We see that the beam pattern is the frequency-wavenumber function evaluated on a sphere of radius $2\pi/\lambda$.

The beam pattern of an array is a key element in determining the array performance. In the next section, we develop the beam patterns for a uniformly weighted linear array.

In the text, we emphasize the case in which the $f(t, \mathbf{p}_n)$ are bandpass signals,

$$f(t, \mathbf{p}_n) = \sqrt{2}\Re e\left\{\tilde{f}(t, \mathbf{p}_n)e^{j\omega_c t}\right\}, \quad n = 0, \cdots, N-1, \tag{2.39}$$

where ω_c is the carrier frequency and $\tilde{f}(t, \mathbf{p}_n)$ is the complex envelope. We assume that the complex envelope is bandlimited to the region,

$$|\omega_L| \leq 2\pi B_s/2, \tag{2.40}$$

where

$$\omega_L \triangleq \omega - \omega_c, \tag{2.41}$$

and πB_s is a constant specifying the maximum bandwidth of the complex envelope.

For the plane wave in (2.13), (2.39) becomes

$$f(t, \mathbf{p}_n) = \sqrt{2} Re\left\{\tilde{f}(t-\tau_n)e^{j\omega_c(t-\tau_n)}\right\}, \quad n = 0, \cdots, N-1, \tag{2.42}$$

where τ_n is given by (2.21).

We now consider the travel time across the array. We define $\Delta T_{nm}(\mathbf{u})$ as the travel time between the n and m elements for a plane wave whose directional cosine is $\mathbf{u}$. Then,

$$\Delta T_{max} \triangleq \max_{n,m=0,\cdots,N-1;\mathbf{u}} \left\{\Delta T_{nm}(\mathbf{u})\right\}, \tag{2.43}$$

is the maximum travel time between any two elements in the array. For a linear array it would be the travel time between the two elements at the ends of the array for a signal arriving along the array axis (endfire).

We assume that the origin is located at the center of gravity of the array,

$$\sum_{n=0}^{N-1} \mathbf{p}_n = \mathbf{0}; \tag{2.44}$$

then all of the τ_n in (2.13) satisfy

$$\tau_n \leq \Delta T_{max}, \quad n = 0, \cdots, N-1. \tag{2.45}$$

In many cases of interest, the bandwidth of the complex envelope is small enough that

$$\tilde{f}(t-\tau_n) \simeq \tilde{f}(t), \quad n = 0, 1, \cdots, N-1. \tag{2.46}$$

In order for this approximation to be valid, we require

$$B_s \cdot \Delta T_{max} \ll 1. \tag{2.47}$$

We define bandpass signals whose complex envelopes satisfy (2.47) as **narrowband** signals. Later we will revisit this definition in the context of optimum processors and provide a better quantitative discussion. For the present, we use (2.47). Then (2.42) reduces to

$$f(t, \mathbf{p}_n) = \sqrt{2} Re\left\{\tilde{f}(t)e^{-j\omega_c\tau_n}e^{j\omega_c t}\right\}. \tag{2.48}$$

We see that, in the narrowband case, the delay is approximated by a phase shift. Therefore the delay-and-sum beamformer can be implemented by a set of phase shifts instead of delay lines. The resulting beamformer is shown in Figure 2.12. This implementation is commonly referred to as a **phased array** and is widely used in practice.

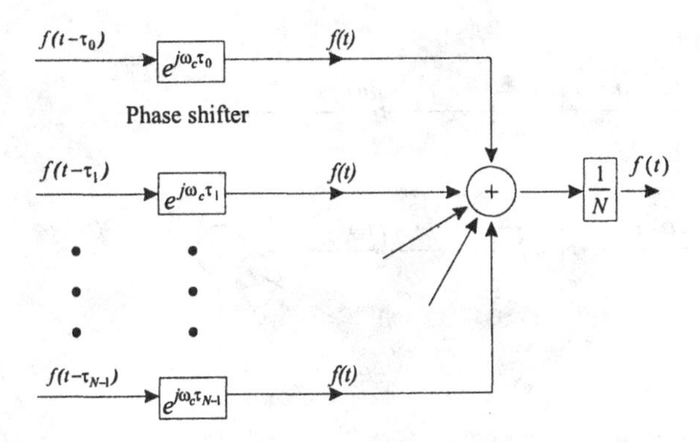

Figure 2.12 Narrowband beamformer implemented using phase shifters.

We will find that, in many applications, we want to adjust the gain and phase at the output of each sensor to achieve a desirable beam pattern. This leads to the narrowband model shown in Figure 2.13(a). The w_n^* are complex weights that are implemented as a cascade of a gain and phase shifter, as shown in Figure 2.13(b).

An alternative implementation is shown in Figure 2.13(c). In some cases, we implement the beamformer by performing a quadrature demodulation and applying the complex weights at baseband. We actually apply $Re[w_n^*]$ to one quadrature component and $Im[w_n^*]$ to the other quadrature component. The results are identical. We discuss these cases later in the text.

Defining the complex weight vector as

$$\mathbf{w}^H = \left[\begin{array}{cccc} w_0^* & w_1^* & \cdots & w_{N-1}^* \end{array}\right], \tag{2.49}$$

(2.35) becomes

$$y(t, \mathbf{k}) = \mathbf{w}^H \mathbf{v}_{\mathbf{k}}(\mathbf{k}) e^{j\omega t}, \tag{2.50}$$

and

$$\Upsilon(\omega, \mathbf{k}) = \mathbf{w}^H \mathbf{v}_{\mathbf{k}}(\mathbf{k}). \tag{2.51}$$

The definition in (2.49) is equivalent to

$$\mathbf{w}^H = \mathbf{H}^T(\omega_c). \tag{2.52}$$

The majority of the text focuses on the narrowband model. In Chapter 5, we show that one approach to processing broadband signals is to decompose them into narrower frequency bins by a discrete Fourier transform (DFT). Within each bin, the narrowband condition is satisfied and all of our narrowband results can be used directly.

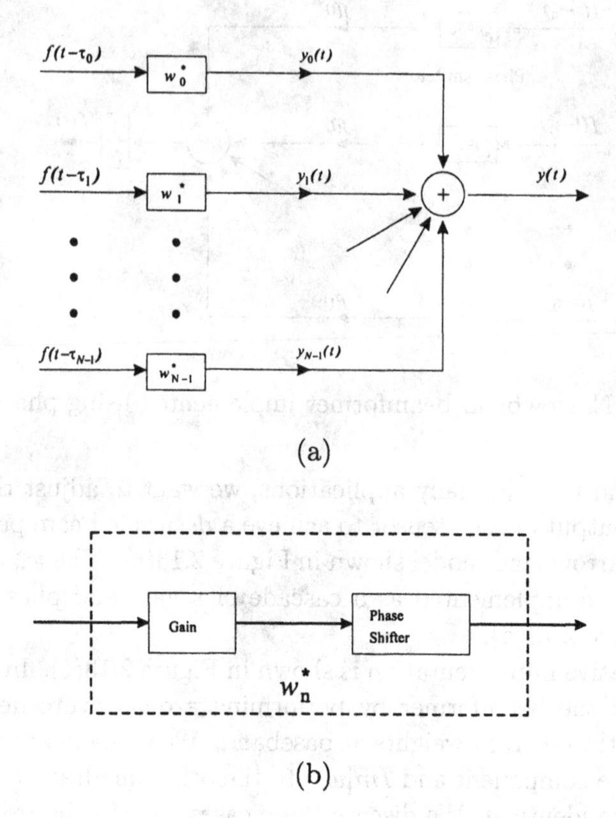

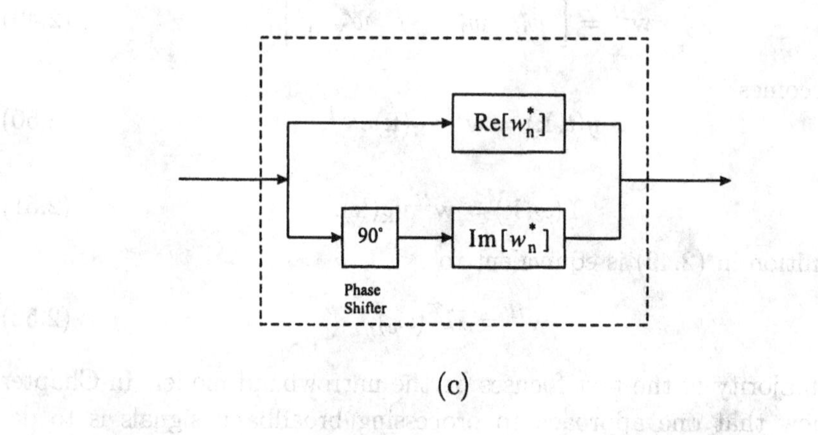

Figure 2.13 General narrowband beamformer.

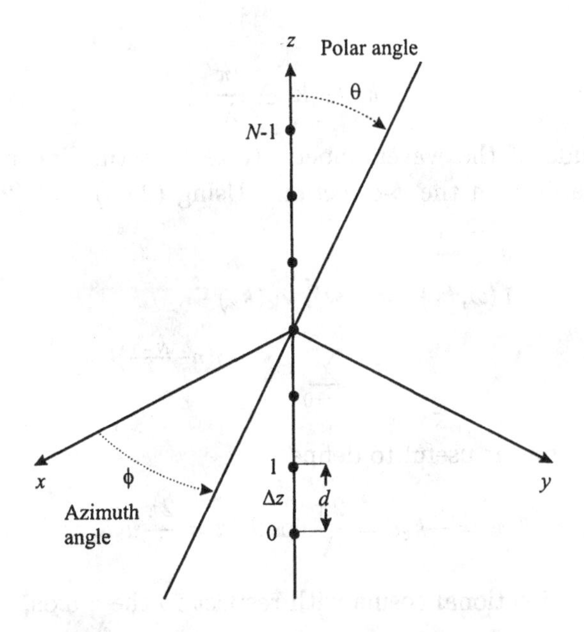

Figure 2.14 Linear array along z-axis.

2.3 Uniform Linear Arrays

The linear array of interest is shown in Figure 2.14. There are N elements located on the z-axis with uniform spacing equal to d. We refer to this type of array as a **uniform linear array (ULA)**. We have placed the center of the array at the origin of the coordinate system. This centering will lead to computational advantages and will be used throughout the text.

The locations of the elements are

$$p_{z_n} = \left(n - \frac{N-1}{2}\right) d, \quad n = 0, 1, \cdots, N-1, \tag{2.53}$$

and

$$p_{x_n} = p_{y_n} = 0. \tag{2.54}$$

To find the array manifold vector $\mathbf{v_k}(\mathbf{k})$, we substitute (2.53) and (2.54) into (2.28) to obtain

$$\mathbf{v_k}(k_z) = \left[\begin{array}{c:c:c:c} e^{j\left(\frac{N-1}{2}\right)k_z d} & e^{j\left(\frac{N-1}{2}-1\right)k_z d} & \cdots & e^{-j\left(\frac{N-1}{2}\right)k_z d} \end{array} \right]^T, \tag{2.55}$$

with

$$k_z = -\frac{2\pi}{\lambda}\cos\theta = -k_0\cos\theta, \tag{2.56}$$

where

$$k_0 \triangleq |\mathbf{k}| \triangleq \frac{2\pi}{\lambda} \tag{2.57}$$

is the magnitude of the wavenumber. Note that the linear array has no resolution capability in the ϕ-direction. Using (2.49) and (2.55) in (2.51) gives

$$\begin{aligned} \Upsilon(\omega, k_z) &= \mathbf{w}^H \mathbf{v_k}(k_z) \\ &= \sum_{n=0}^{N-1} w_n^* \, e^{-j\left(n-\frac{N-1}{2}\right)k_z d}. \end{aligned} \tag{2.58}$$

We will also find it useful to define

$$\psi = -k_z d = \frac{2\pi}{\lambda}\cos\theta \cdot d = \frac{2\pi}{\lambda} u_z d, \tag{2.59}$$

where u_z is the directional cosine with respect to the z-axis,

$$u_z = \cos\theta. \tag{2.60}$$

Using (2.59) in (2.58) gives

$$\Upsilon_\psi(\psi) = e^{-j\frac{N-1}{2}\psi} \sum_{n=0}^{N-1} w_n^* \, e^{jn\psi}. \tag{2.61}$$

We refer to $\Upsilon_\psi(\psi)$ as the frequency-wavenumber function in ψ-space. Both $\Upsilon(\omega, k_z)$ and $\Upsilon_\psi(\psi)$ are defined from $-\infty$ to ∞, but they only represent propagating signals in the region where $0 \le \theta \le \pi$ (or $-1 \le u_z \le 1$). This restriction implies $-\frac{2\pi d}{\lambda} \le \psi \le \frac{2\pi d}{\lambda}$ or $-\frac{2\pi}{\lambda} \le k_z \le \frac{2\pi}{\lambda}$. We refer to this as the **visible region**.

We observe that, if we define

$$z = e^{j\psi}, \tag{2.62}$$

$$\Upsilon_z(z) = z^{-\frac{N-1}{2}} \sum_{n=0}^{N-1} w_n^* \, z^n, \tag{2.63}$$

then (2.63) can be written as

$$\Upsilon_z(z) = z^{-\frac{N-1}{2}} \left(\sum_{n=0}^{N-1} w_n \, z^{-n} \right)^*. \tag{2.64}$$

The term

$$W(z) = \sum_{n=0}^{N-1} w_n z^{-n} \tag{2.65}$$

is familiar as the z-transform[3] and

$$\Upsilon_\psi(\psi) = \Upsilon_z(z)\big|_{z=e^{j\psi}} = \left(z^{-\frac{N-1}{2}} W^*(z)\right)\Big|_{z=e^{j\psi}} \tag{2.66}$$

is the frequency-wavenumber function in ψ-space. We exploit this relationship later in the text.

Although it may appear that we have introduced extra notation by writing the frequency-wavenumber function in three different ways ((2.58), (2.61) and (2.66)), we will find the different forms useful in different cases.

It is also useful to define the array manifold vector in θ and u space,

$$[\mathbf{v}_\theta(\theta)]_n = e^{j(n-\frac{N-1}{2})\frac{2\pi d}{\lambda}\cos\theta}, \quad n = 0, \cdots, N-1, \tag{2.67}$$

and

$$[\mathbf{v}_u(u)]_n = e^{j(n-\frac{N-1}{2})\frac{2\pi d}{\lambda}u}, \quad n = 0, \cdots, N-1. \tag{2.68}$$

We can also write the beam pattern in three forms. The key difference between the frequency-wavenumber function and the beam pattern is that the argument in the beam pattern is restricted to correspond to a physical angle θ. Thus,

$$B_\theta(\theta) = \mathbf{w}^H \mathbf{v}_\theta(\theta) = e^{-j\left(\frac{N-1}{2}\right)\frac{2\pi d}{\lambda}\cos\theta} \sum_{n=0}^{N-1} w_n^* \, e^{jn\frac{2\pi d}{\lambda}\cos\theta}, \quad 0 \le \theta \le \pi, \tag{2.69}$$

$$B_u(u) = \mathbf{w}^H \mathbf{v}_u(u) = e^{-j\left(\frac{N-1}{2}\right)\frac{2\pi d}{\lambda}u} \sum_{n=0}^{N-1} w_n^* \, e^{jn\frac{2\pi d}{\lambda}u}, \quad -1 \le u \le 1, \tag{2.70}$$

$$B_\psi(\psi) = \mathbf{w}^H \mathbf{v}_\psi(\psi) = e^{-j\left(\frac{N-1}{2}\right)\psi} \sum_{n=0}^{N-1} w_n^* \, e^{jn\psi}, \quad -\frac{2\pi d}{\lambda} \le \psi \le \frac{2\pi d}{\lambda}. \tag{2.71}$$

We suppress the subscript on $B(\cdot)$ when the variable is clear.

For uniform linear arrays, we normally write the array manifold vector in terms of ψ,

$$[\mathbf{v}_\psi(\psi)]_n = e^{j(n-\frac{N-1}{2})\psi}, \quad n = 0, 1, \cdots, N-1, \tag{2.72}$$

[3]The z-transform is discussed in a number of texts (e.g., Chapter 4 of [OS89]).

and

$$\boxed{\mathbf{v}_\psi(\psi) = \left[\begin{array}{ccccc} e^{-j\left(\frac{N-1}{2}\right)\psi} & e^{-j\left(\frac{N-3}{2}\right)\psi} & \cdots & e^{j\left(\frac{N-3}{2}\right)\psi} & e^{j\left(\frac{N-1}{2}\right)\psi} \end{array}\right]^T}. \tag{2.73}$$

We see that the array manifold vector for a uniform linear array exhibits a **conjugate symmetry**. For N even, if we define a $N/2$-dimensional vector $\mathbf{v}_{\psi_1}(\psi)$ corresponding to the first $N/2$ elements of $\mathbf{v}_\psi(\psi)$, then we can write

$$\mathbf{v}_\psi(\psi) = \left[\begin{array}{c} \mathbf{v}_{\psi_1}(\psi) \\ \hdashline \mathbf{J}\,\mathbf{v}^*_{\psi_1}(\psi) \end{array}\right], \tag{2.74}$$

where $\mathbf{J}$ is the exchange matrix defined in (A.125). For N odd, $\mathbf{v}_{\psi_1}(\psi)$ consists of the first $(N-1)/2$ elements and

$$\mathbf{v}_\psi(\psi) = \left[\begin{array}{c} \mathbf{v}_{\psi_1}(\psi) \\ \hdashline 1 \\ \hdashline \mathbf{J}\,\mathbf{v}^*_{\psi_1}(\psi) \end{array}\right]. \tag{2.75}$$

This conjugate symmetry will lead to computational savings and performance improvements in many applications. For example, if $\mathbf{w}$ is also conjugate symmetric, we can write, for N even,

$$\mathbf{w} = \left[\begin{array}{c} \mathbf{w}_1 \\ \hdashline \mathbf{J}\,\mathbf{w}_1^* \end{array}\right]. \tag{2.76}$$

The beam pattern in ψ-space is,

$$\begin{aligned} B_\psi(\psi) &= \mathbf{w}^H\,\mathbf{v}_\psi(\psi) \\ &= \left[\begin{array}{c:c} \mathbf{w}_1^H & \mathbf{w}_1^T\,\mathbf{J} \end{array}\right] \left[\begin{array}{c} \mathbf{v}_{\psi_1}(\psi) \\ \hdashline \mathbf{J}\,\mathbf{v}^*_{\psi_1}(\psi) \end{array}\right] \\ &= \mathbf{w}_1^H\,\mathbf{v}_{\psi_1}(\psi) + \mathbf{w}_1^T\,\mathbf{v}^*_{\psi_1}(\psi) \\ &= 2Re\left[\mathbf{w}_1^H\,\mathbf{v}_{\psi_1}(\psi)\right], \end{aligned} \tag{2.77}$$

so the beam pattern is a real function. Note that if we have real symmetric weightings, then (2.76) is also satisfied. A similar result follows for N

odd. Later, we will find that many other array geometries exhibit conjugate symmetry.

The form in (2.73) emphasizes the conjugate symmetry in $\mathbf{v}_\psi(\psi)$. We can also write $\mathbf{v}_\psi(\psi)$ as

$$\mathbf{v}_\psi(\psi) = e^{-j\frac{N-1}{2}\psi}\begin{bmatrix} 1 & e^{j\psi} & \cdots & e^{j(N-1)\psi} \end{bmatrix}^T. \tag{2.78}$$

This form emphasizes the Vandermonde structure (A.163) of $\mathbf{v}_\psi(\psi)$.

In Chapter 3, we develop techniques for choosing $\mathbf{w}$ in order to achieve a beam pattern with desirable properties. This is the most common approach. However, we will also develop techniques for synthesizing a desirable $B_\psi(\psi)$ without finding $\mathbf{w}$ in the process. Thus, our last step is to find the $\mathbf{w}$ that corresponds to a particular $B_\psi(\psi)$.

We start with the relation in (2.71),

$$B_\psi(\psi) = \mathbf{w}^H \mathbf{v}_\psi(\psi). \tag{2.79}$$

We assume that $B_\psi(\psi)$ is known and we want to find the $\mathbf{w}$ that generated it. Since $\mathbf{w}^H$ is a $1 \times N$ vector, we would anticipate that, if we know the value of $B_\psi(\psi)$ at N values of ψ, we can find $\mathbf{w}$.

We sample the beam pattern at N values of $\psi_i, i = 1, \cdots, N$. The ψ_i must be distinct but do not have to be spaced equally. We denote the beam pattern at the sample points as $B(\psi_i)$. From (2.79),

$$\mathbf{w}^H \mathbf{v}(\psi_i) = B(\psi_i), \quad i = 1, \cdots, N. \tag{2.80}$$

We define an $N \times N$ array manifold matrix,

$$\mathbf{V}(\psi) \triangleq \begin{bmatrix} \mathbf{v}(\psi_1) & \cdots & \mathbf{v}(\psi_N) \end{bmatrix}, \tag{2.81}$$

and a $1 \times N$ beam pattern matrix,

$$\mathbf{B} \triangleq \begin{bmatrix} B(\psi_1) & \cdots & B(\psi_N) \end{bmatrix}. \tag{2.82}$$

Then, (2.80) can be written as

$$\mathbf{w}^H \mathbf{V}(\psi) = \mathbf{B}, \tag{2.83}$$

or

$$\mathbf{V}^H(\psi)\mathbf{w} = \mathbf{B}^H. \tag{2.84}$$

Since $\mathbf{V}^H(\psi)$ is full rank,

$$\boxed{\mathbf{w} = \left[\mathbf{V}^H(\psi)\right]^{-1} \mathbf{B}^H}, \tag{2.85}$$

which is the desired result. Although the ψ_i are arbitrary, if they are too close, the array manifold vectors will approach linear dependence and there may be numerical problems with the inverse in (2.85). If we use uniform spacing of $2\pi/N$, we will derive an algorithm in Section 3.3 that is computationally more efficient.

A particular of interest is the case in which we have specified the beam pattern with $N-1$ zeros. If we assume the array is steered to broadside, we let

$$\psi_1 = 0, \tag{2.86}$$

and $\psi_2, \psi_3, \cdots, \psi_N$ correspond to the zero locations. Assuming a normalized beam pattern,

$$\mathbf{B} = \begin{bmatrix} 1 & 0 & \cdots & 0 \end{bmatrix} = \mathbf{e}_1^T. \tag{2.87}$$

Then, (2.85) reduces to

$$\boxed{\mathbf{w} = \left[\mathbf{V}^H(\psi)\right]^{-1} \mathbf{e}_1}. \tag{2.88}$$

We will find these algorithms to be very useful in subsequent sections.

There are two points with the results in (2.85) and (2.88) that should be emphasized:

(i) We have assumed that $B_\psi(\psi)$ was defined by (2.79). In other words, it was generated by a complex $N \times 1$ vector $\mathbf{w}$. If we start with an arbitrary function $B_\psi(\psi)$ and use (2.85), we will generate a pattern that matches $B(\psi_i), i = 1, \cdots, N$ but will not necessarily match the function $B_\psi(\psi)$. We discuss this issue further in Section 3.3.

(ii) We introduced this result in the context of a uniform linear array. However, the derivation is valid for an N-element array with an arbitrary geometry.

We have developed the basic relationships between the array manifold vector, the weight vector, and the beam pattern. In the next section, we consider the special case of uniform weighting.

2.4 Uniformly Weighted Linear Arrays

We now restrict our attention to the uniform weighting case,

$$w_n = \frac{1}{N}, \quad n = 0, 1, \cdots, N-1. \tag{2.89}$$

We can also write (2.89) as

$$\mathbf{w} = \frac{1}{N}\mathbf{1}, \tag{2.90}$$

where $\mathbf{1}$ is the $N \times 1$ unity vector defined in Section A.3.1.

Thus, the frequency-wavenumber function can be written in ψ-space as[4]

$$\begin{aligned} \Upsilon_\psi(\psi) &= \frac{1}{N}\sum_{n=0}^{N-1} e^{j\left(n-\frac{N-1}{2}\right)\psi} \\ &= \frac{1}{N}e^{-j\left(\frac{N-1}{2}\right)\psi}\sum_{n=0}^{N-1} e^{jn\psi} \\ &= \frac{1}{N}e^{-j\left(\frac{N-1}{2}\right)\psi}\left[\frac{1-e^{jN\psi}}{1-e^{j\psi}}\right], \end{aligned} \tag{2.91}$$

or

$$\Upsilon_\psi(\psi) = \frac{1}{N}\frac{\sin\left(N\frac{\psi}{2}\right)}{\sin\frac{\psi}{2}}, \qquad -\infty < \psi < \infty. \tag{2.92}$$

We observe that $\Upsilon_\psi(\psi)$ is periodic with period 2π for N odd. If N is even, the lobes at $\pm 2\pi$, $\pm 6\pi$ are negative and period is 4π. The period of $|\Upsilon_\psi(\psi)|$ is 2π for any value of N. $\Upsilon_\psi(\psi)$ is plotted versus ψ in Figure 2.15 for $N = 11$. In Figure 2.16, we plot $|\Upsilon_\psi(\psi)|$ in dB, where

$$\Upsilon_{dB}(\psi) = 10\log_{10}|\Upsilon(\psi)|^2. \tag{2.93}$$

For arbitrary $\mathbf{w}$, $\Upsilon_\psi(\psi)$ is complex, so the phase should also be plotted; however, the symmetry of this particular array leads to a purely real quantity.

We can also write the frequency-wavenumber response in terms of k_z,

$$\Upsilon(\omega : k_z) = \frac{1}{N}\frac{\sin\left(Nk_z\frac{d}{2}\right)}{\sin\left(k_z\frac{d}{2}\right)}. \tag{2.94}$$

$\Upsilon(\omega : k_z)$ is periodic with period $2\pi/d$.

Note that the response function depends only upon the wavenumber component k_z and is periodic with respect to k_z at intervals of $2\pi/d$. The dependence solely upon k_z is a consequence of the linear array being one-dimensional so it can only resolve wavenumber components which have a projection in this direction.

---

[4] $\sum_{n=0}^{N-1} x^n = \frac{1-x^N}{1-x}$.

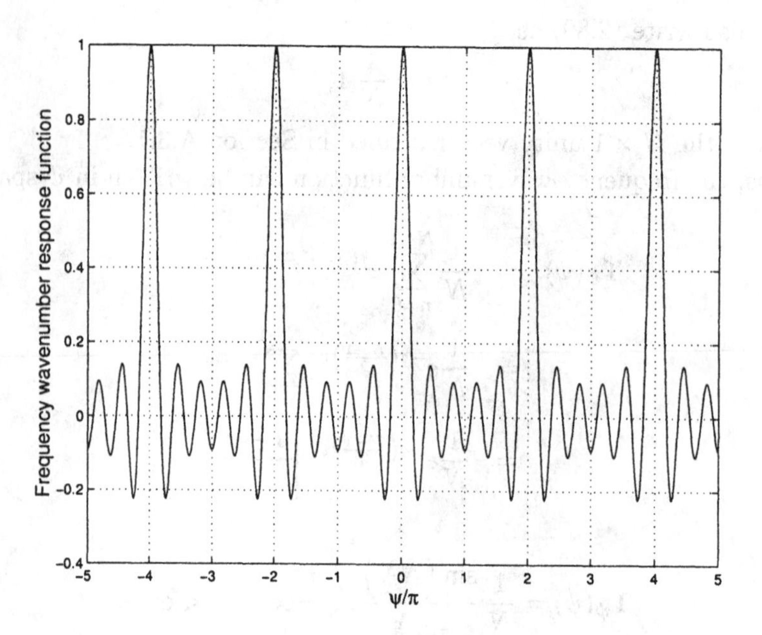

Figure 2.15 $\Upsilon(\psi) : \psi = \frac{2\pi}{\lambda}\, d \cos\theta,\, N = 11.$

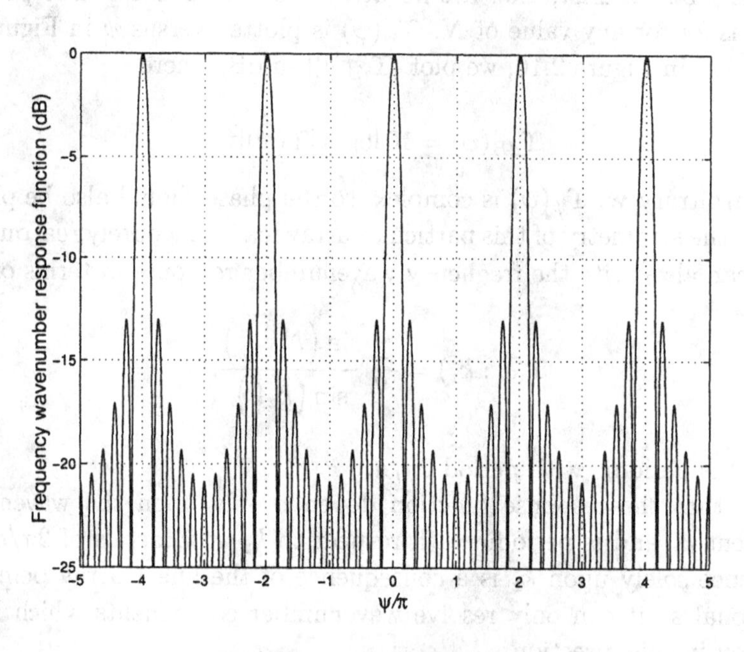

Figure 2.16 $|\Upsilon(\psi)|$ in dB.

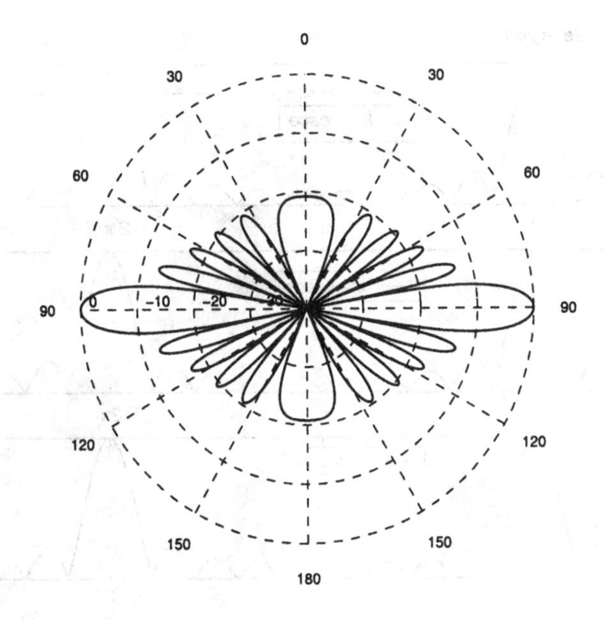

Figure 2.17 Polar plot of $B_\theta(\theta)$.

The beam pattern is given by

$$B_\theta(\theta) = \frac{1}{N} \frac{\sin(\frac{N}{2} \cdot \frac{2\pi}{\lambda} \cos\theta \cdot d)}{\sin(\frac{1}{2} \cdot \frac{2\pi}{\lambda} \cos\theta \cdot d)}, \quad 0 \leq \theta \leq \pi. \tag{2.95}$$

The beam pattern in u-space can be written as

$$B_u(u) = \frac{1}{N} \frac{\sin(\frac{\pi N d}{\lambda} u)}{\sin(\frac{\pi d}{\lambda} u)}, \quad -1 \leq u \leq 1. \tag{2.96}$$

The beam pattern is only defined over the region $(-1 \leq u \leq 1)$, the **visible region**. The beam pattern in ψ-space is

$$B_\psi(\psi) = \frac{1}{N} \frac{\sin(N\frac{\psi}{2})}{\sin(\frac{\psi}{2})}, \quad -\frac{2\pi d}{\lambda} \leq \psi \leq \frac{2\pi d}{\lambda}. \tag{2.97}$$

The functions $B_u(u)$ and $B_\psi(\psi)$ are sometimes referred to as the **array factor**. We will see the significance of this term when we look at non-isotropic sensors.

In Figure 2.17, we show a polar plot in dB of $B_\theta(\theta)$. If we plotted the beam pattern in three dimensions, the plot in Figure 2.17 would correspond

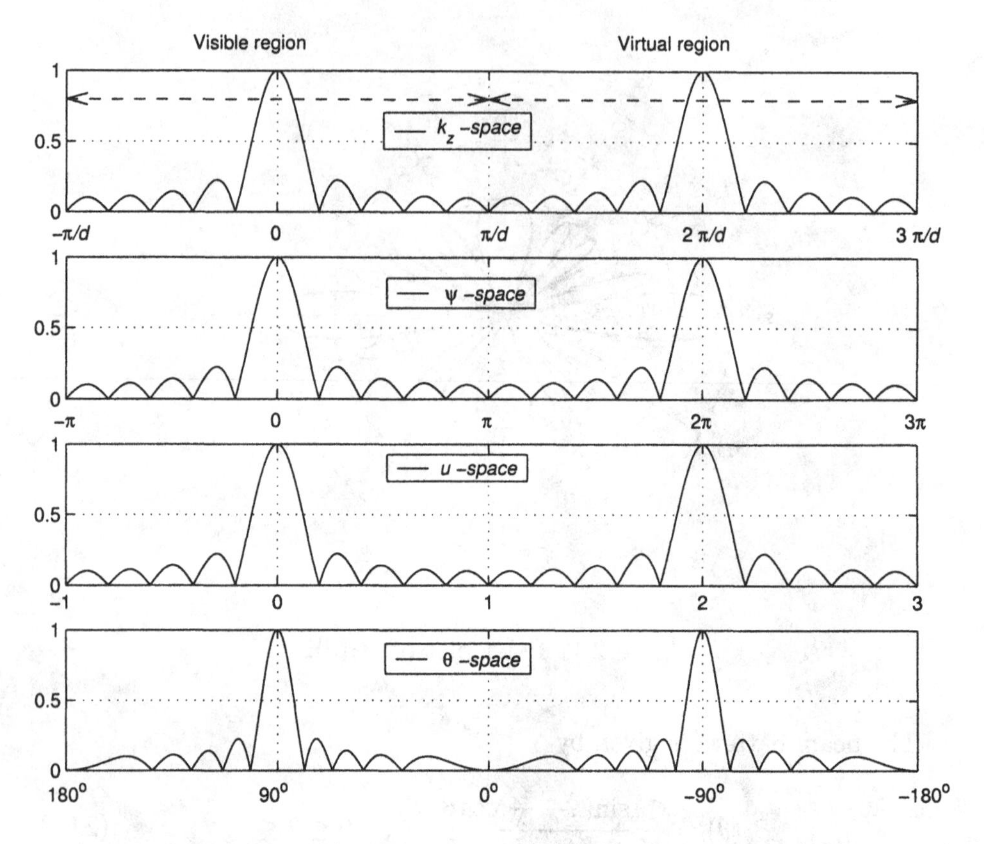

Figure 2.18 $|\Upsilon_\psi(\psi)|$ for a linear array with $d = \lambda/2$ and $N = 10$.

to a pattern cut along any value of ϕ, the azimuth angle. In Figure 2.18, we show the magnitude of the beam pattern versus different variables.

Although this is a simple example, we can use it to illustrate several important characteristics of linear arrays. The first set of characteristics describe the parameters of the beam pattern.

2.4.1 Beam Pattern Parameters

(i) 3-dB beamwidth (the half-power beamwidth, HPBW)
(ii) Distance to first null (twice this distance is BW_{NN})
(iii) Distance to first sidelobe
(iv) Height of first sidelobe
(v) Location of remaining nulls
(vi) Rate of decrease of sidelobes
(vii) Grating lobes

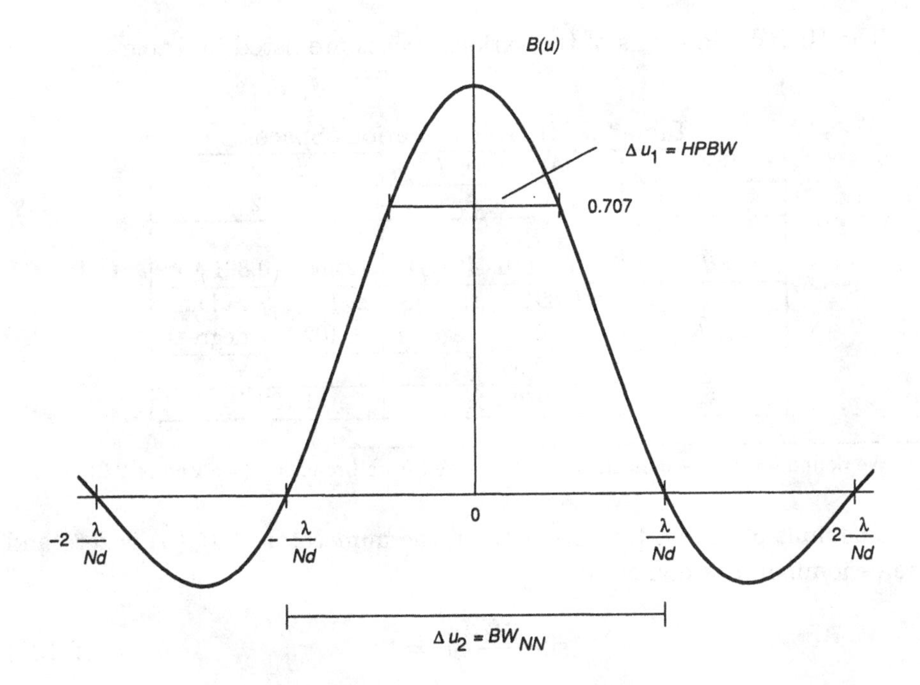

Figure 2.19 Main lobe of beam pattern.

To illustrate the first two points, consider the beam pattern near the origin as shown in Figure 2.19.

The 3-dB beamwidth is a measure of the width of the beam. It is defined to be the point where $|B_u(u)|^2 = 0.5$ or $|B_u(u)| = 1/\sqrt{2}$. We can find the half-power point in u-space by setting $B_u(u)$ in (2.96) equal to $1/\sqrt{2}$. Calculating this value as N increases, we find that, for $N \geq 10$, a good approximation is obtained by solving the equation,

$$\frac{\pi N d}{\lambda} u = 1.4 \tag{2.98}$$

(see Problem 2.4.7). Then,

$$\frac{\Delta u_1}{2} = 1.4 \frac{\lambda}{\pi N d} \tag{2.99}$$

or

$$\Delta u_1 = 0.891 \frac{\lambda}{Nd}. \tag{2.100}$$

We refer to this interval as the **half-power beamwidth (HPBW)**. As N increases, the coefficient in (2.100) reduces slightly. For $N > 30$, $0.886\lambda/Nd$ is a better approximation.

The HPBWs in terms of the various spaces are listed in Table 2.2.

Table 2.2 HPBWs in Various Spaces

Space	Arbitrary d	$d=\lambda/2$
u	$0.891\frac{\lambda}{Nd}$	$1.782\frac{1}{N}$
$\bar{\theta}$	$2\sin^{-1}\left(0.446\frac{\lambda}{Nd}\right)$	$2\sin^{-1}\left(0.891\frac{1}{N}\right)$
small $\bar{\theta}$	$\simeq 0.891\frac{\lambda}{Nd}$ radians $\simeq 51.05\frac{\lambda}{Nd}$ degrees	$\simeq 1.782\frac{1}{N}$ radians $\simeq 102.1\frac{1}{N}$ degrees
ψ	$0.891\frac{2\pi}{N}$	$0.891\frac{2\pi}{N}$
k_z	$0.891\frac{2\pi}{dN}$	$1.782\frac{2\pi}{\lambda N}$

We define $\bar{\theta} = \pi/2 - \theta$ as the angle measured from broadside (see Figure 2.2).

The nulls of the pattern occur when the numerator of $B_u(u)$ is zero and the denominator is non-zero:

$$\sin\left(\frac{\pi N d}{\lambda}u\right) = 0, \tag{2.101}$$

when

$$\frac{\pi N d}{\lambda}u = m\pi, \quad m = 1, 2, \cdots. \tag{2.102}$$

Thus the nulls occur when both

$$u = m\frac{\lambda}{Nd}, \quad m = 1, 2, \cdots, \tag{2.103}$$

and

$$u \neq m\frac{\lambda}{d}, \quad m = 1, 2, \cdots. \tag{2.104}$$

Thus, the first null occurs at λ/Nd and

$$\Delta u_2 = 2\frac{\lambda}{Nd}. \tag{2.105}$$

We refer to Δu_2 as the null-to-null beamwidth and denote it by BW_{NN}. One-half of the BW_{NN} is the distance to the first null ($0.5BW_{NN}$). This quantity provides a measure of the ability of the array to resolve two different plane waves. It is referred to as the **Rayleigh resolution limit**. Two plane waves are considered resolvable if the peak of the second beam pattern lies at or outside of the null of the first beam pattern (separation $\geq \Delta u_2/2$). Later, we look at statistical measures of an array's resolution capability.

Note that the linear array has no resolution capability in the azimuthal direction (ϕ) because it has no extent in either the x or y directions. We will discuss the resolution problem in detail in later sections.

The BW_{NN} in terms of the various spaces is listed in Table 2.3.

Table 2.3 BW_{NN} in Various Spaces

Space	Arbitrary d	$d = \lambda/2$
u	$2\frac{\lambda}{Nd}$	$\frac{4}{N}$
$\bar{\theta}$	$2\sin^{-1}\left(\frac{\lambda}{Nd}\right)$	$2\sin^{-1}\left(\frac{2}{N}\right)$
small $\bar{\theta}$	$\simeq 2\frac{\lambda}{Nd}$ radians	$\simeq \frac{4}{N}$ radians
ψ	$\frac{4\pi}{N}$	$\frac{4\pi}{N}$
k_z	$\frac{4\pi}{dN}$	$\frac{8\pi}{\lambda N}$

2.4.1.1 Location of sidelobes and the rate of decrease

The location of the maxima of the sidelobes occurs approximately when the numerator of (2.96) is a maximum:

$$\sin\left(\frac{N\psi}{2}\right) = 1. \tag{2.106}$$

Thus,

$$\frac{N\psi}{2} = \pm(2m+1)\frac{\pi}{2}, \quad m = 1, 2, \cdots \tag{2.107}$$

or

$$\psi = \pm\frac{2m+1}{N}\pi \tag{2.108}$$

and

$$u = \pm\frac{2m+1}{N}\frac{\lambda}{2d}. \tag{2.109}$$

The peak of the first sidelobe occurs at

$$\psi = \pm\frac{3\pi}{N}. \tag{2.110}$$

Since the numerator in (2.97) is approximately one at the maximum, the value at the maximum is given by

$$B_\psi\left(\pm\frac{3\pi}{N}\right) \cong \frac{1}{N\sin\left(\frac{3\pi}{2N}\right)}. \tag{2.111}$$

For large N, this can be further approximated by

$$B_\psi\left(\pm\frac{3\pi}{N}\right) \cong \frac{2}{3\pi} \tag{2.112}$$

or -13.5 dB. This implies that a signal 13.5 dB higher at these sidelobe locations would produce the same response as one at $k_z = 0$. The major sidelobes appear at $k_z = \pm(2m+1)\pi/Nd$ and the levels diminish as $1/(2m+1)$. For example, the next highest sidelobes are -17.9 dB. In practice this level of discrimination is generally not acceptable, so uniformly weighted arrays are seldom used. The issue of sidelobe control is especially important in both deterministic and adaptive array designs. We discuss the deterministic case in Chapter 3 and the adaptive case in Chapter 7.

2.4.1.2 Grating lobes

In Figure 2.20, we plot $|\Upsilon_u(u)|$ for various values of d/λ. It illustrates the important concept of "grating lobe," which is a lobe of the same height as the main lobe. Grating lobes occur when both the numerator and denominator of (2.97) equal zero. These appear at intervals,

$$\frac{\psi}{2} = m \cdot \pi, \tag{2.113}$$

or

$$\psi = m \cdot 2\pi, \tag{2.114}$$

or

$$u = m \cdot \frac{\lambda}{d}. \tag{2.115}$$

If the array spacing is greater than λ, then the peak of the grating lobe occurs within the region of propagating signals, that is, when $|u| \leq 1$. Here one has an ambiguity in terms of the peak response and only *a priori* information about the direction of the signal can resolve it.

In the next section, we discuss array steering. We find that steering causes the frequency-wavenumber function in u-space, $\Upsilon_u(u)$, to shift in u-space. This shift causes grating lobes to move into the visible region. We find that, if the array is required to steer $0° \leq \theta \leq 180°$, then we require,

$$\frac{d}{\lambda} \leq \frac{1}{2}, \tag{2.116}$$

or

$$d \leq \frac{\lambda}{2}. \tag{2.117}$$

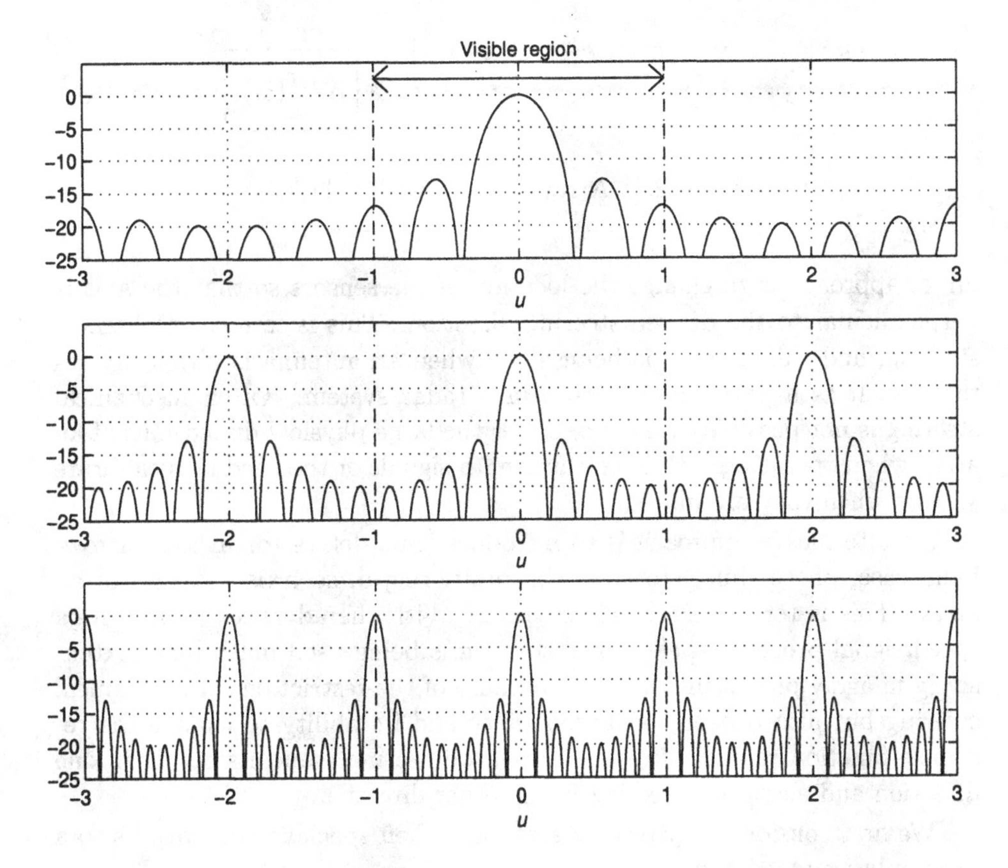

Figure 2.20 Effect of element spacing on beam pattern: (a) $d = \lambda/4$; (b) $d = \lambda/2$; (c) $d = \lambda$.

Normally, we consider arrays where $d \leq \lambda/2$ and assume that steering over the entire sphere is required. We refer to a uniform linear array with $d = \lambda/2$ as a **standard linear array**.

The problem of grating lobes is identical to the problem of aliasing in time series analysis, which occurs when we undersample the time domain waveform.

2.5 Array Steering

The discussions in Sections 2.2, 2.3, and 2.4 have considered arrays whose maximum response axis was at broadside, or $k_z = 0$. In most applications we want to be able to position, or steer, the response to an arbitrary wavenumber, or direction. There are two ways to accomplish this. The

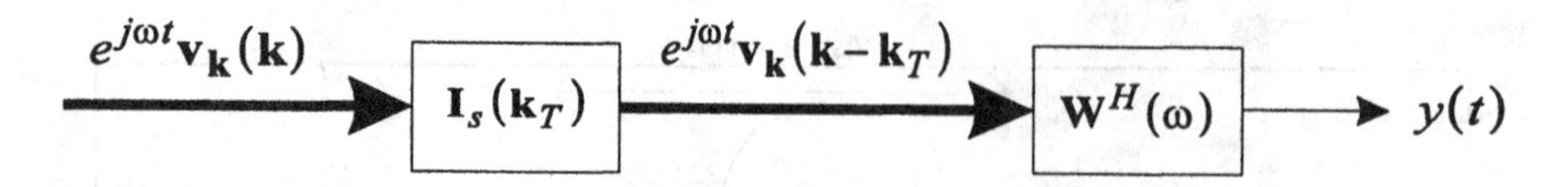

Figure 2.21 Array steering with delays.

direct approach is to change the location of the sensors so that the axis is perpendicular to the desired steering direction. This is termed mechanical steering, and this is what is being done when an antenna is physically rotated such as a parabolic aperture in a radar system. Often mechanical steering is not possible because of either the large physical dimensions of an array when operating with long wavelength signals or the need to recalibrate sensors when they are moved.

An alternative approach is to introduce time delays (or in the narrow-band case, phase shifts) to steer the **main response axis** (**MRA**) of an array. This is termed electronic steering. With the advances in very high speed signal processors, electronic steering is being used much more extensively in array processing, not only because of the restrictions of mechanical steering but also because of its flexibility and its ability to change the response function rapidly. In some arrays we use mechanical steering in one direction and electronic steering in the other direction.

We first consider an arbitrary array and then specialize our results to a uniformly weighted array.

Our simple example at the beginning of Section 2.2 illustrated the idea of steering the array in a specific direction. The effect of steering in wavenumber space is straightforward. Consider the processor in Figure 2.21.[5] The basis function input to the steering section is

$$f(t,\mathbf{p}) = e^{j\omega t}\,\mathbf{v_k}(\mathbf{k}). \tag{2.118}$$

We would like the output to be aligned when

$$\mathbf{k} = \mathbf{k}_T, \tag{2.119}$$

the "target" wavenumber. We refer to $\mathbf{k}_T$ as the **steering direction** or **main response axis** in **k**-space. We accomplish this with an $N \times N$ diagonal

[5]The convention in our figures is that the vector or matrix in a box pre-multiplies the input.

steering matrix,[6]

$$\mathbf{I}_s(\mathbf{k}_T) \triangleq \begin{bmatrix} e^{j\mathbf{k}_T^T\mathbf{p}_1} & 0 & \cdots & 0 \\ 0 & e^{j\mathbf{k}_T^T\mathbf{p}_2} & \cdots & 0 \\ 0 & \cdots & \ddots & 0 \\ 0 & \cdots & 0 & e^{j\mathbf{k}_T^T\mathbf{p}_N} \end{bmatrix}. \tag{2.120}$$

The resulting output is

$$f_s(t, \mathbf{p}) = e^{j\omega t}\, \mathbf{v}_{\mathbf{k}}(\mathbf{k} - \mathbf{k}_T), \tag{2.121}$$

and the overall frequency wavenumber response is

$$\Upsilon(\omega, \mathbf{k}|\mathbf{k}_T) = \Upsilon(\omega, \mathbf{k} - \mathbf{k}_T). \tag{2.122}$$

The array response function is simply displaced to be positioned about $\mathbf{k}_T$. This is one of the useful aspects of using wavenumber space in interpreting array response functions. If we consider the beam patterns in wavenumber space, we also get a simple displacement.

When we use uniform amplitude weighting, the two-step process in Figure 2.21 is unnecessary. We let

$$\mathbf{w} = \frac{1}{N}\mathbf{v}_{\mathbf{k}}(\mathbf{k}_T) \tag{2.123}$$

and

$$B_c(\mathbf{k} : \mathbf{k}_T) = \frac{1}{N}\mathbf{v}_{\mathbf{k}}^H(\mathbf{k}_T)\, \mathbf{v}_{\mathbf{k}}(\mathbf{k}). \tag{2.124}$$

We refer to $B_c(\mathbf{k} : \mathbf{k}_T)$ as the **conventional beam pattern**. We will find that this conventional beam pattern plays a fundamental role in many of the optimum processing schemes that we will develop in the sequel.

For a linear array, the conventional beam pattern can be written as

$$B_{\psi c}(\psi : \psi_T) = \frac{1}{N}\mathbf{v}_{\psi}^H(\psi_T)\, \mathbf{v}_{\psi}(\psi) \tag{2.125}$$

in ψ-space and

$$B_{uc}(u : u_T) = \frac{1}{N}\mathbf{v}_u^H(u_T)\, \mathbf{v}_u(u) \tag{2.126}$$

[6]Note that for the wideband case, we accomplish (2.120) with delays. Only in the narrowband case can the delays be replaced with phase shifts.

in u-space. For a uniform linear array,

$$B_{\psi}(\psi:\psi_T)=\frac{1}{N}\frac{\sin(N\frac{\psi-\psi_T}{2})}{\sin(\frac{\psi-\psi_T}{2})}. \tag{2.127}$$

The steering direction in ψ-space corresponds to an interelement phase shift of ψ_T. In u-space,

$$B_u(u:u_T)=\frac{1}{N}\frac{\sin[\frac{\pi Nd}{\lambda}(u-u_T)]}{\sin[\frac{\pi d}{\lambda}(u-u_T)]}. \tag{2.128}$$

Thus if we look at the expressions in (2.127), or (2.128), they all correspond to shifts in the pattern, but its shape is not changed. This property of shifting without distortion is one of many advantages of working in ψ-space or u-space.

As we steer the array so that the main response axis is aimed at $\bar{\theta}_0$, where $\bar{\theta}_0$ is the angle measured from broadside, the beam pattern shifts so that the center peak is at $u_0=\sin\bar{\theta}_0$. This shift causes the grating lobes to move.

In Figure 2.22, we show the effect of steering on the beam pattern. In Figure 2.22(a), we show the beam pattern for $d=2\lambda/3$ and $\bar{\theta}=30°$. We see that, at this steering angle, the grating lobe is at the edge of the visible region.

In Figure 2.22(b), we show the beam pattern for $d=\lambda/2$ and $\bar{\theta}=90°$. The grating lobe is at the edge of the visible region.

In general, we require

$$\frac{d}{\lambda}\leq\frac{1}{1+|\sin\bar{\theta}_{max}|}, \tag{2.129}$$

where $\bar{\theta}_{max}$ is the maximum angle to which the array will be required to steer, in order to avoid a grating lobe from moving into the visible region. This result follows from calculating the location of the first grating lobe as a function of d/λ with $\bar{\theta}_T=\bar{\theta}_{max}$. Thus, if the array is required to steer $-90°\leq\bar{\theta}\leq 90°$, we require

$$d\leq\frac{\lambda}{2}. \tag{2.130}$$

The behavior in ψ-space and u-space is useful. However, it is important to remember that the signals originate in a (θ,ϕ) space, and we need to understand the behavior in that space.

In θ-space (i.e., angle space),

$$B_{\theta c}(\theta:\theta_T)=\frac{1}{N}\frac{\sin[\frac{\pi Nd}{\lambda}(\cos\theta-\cos\theta_T)]}{\sin[\frac{\pi d}{\lambda}(\cos\theta-\cos\theta_T)]}. \tag{2.131}$$

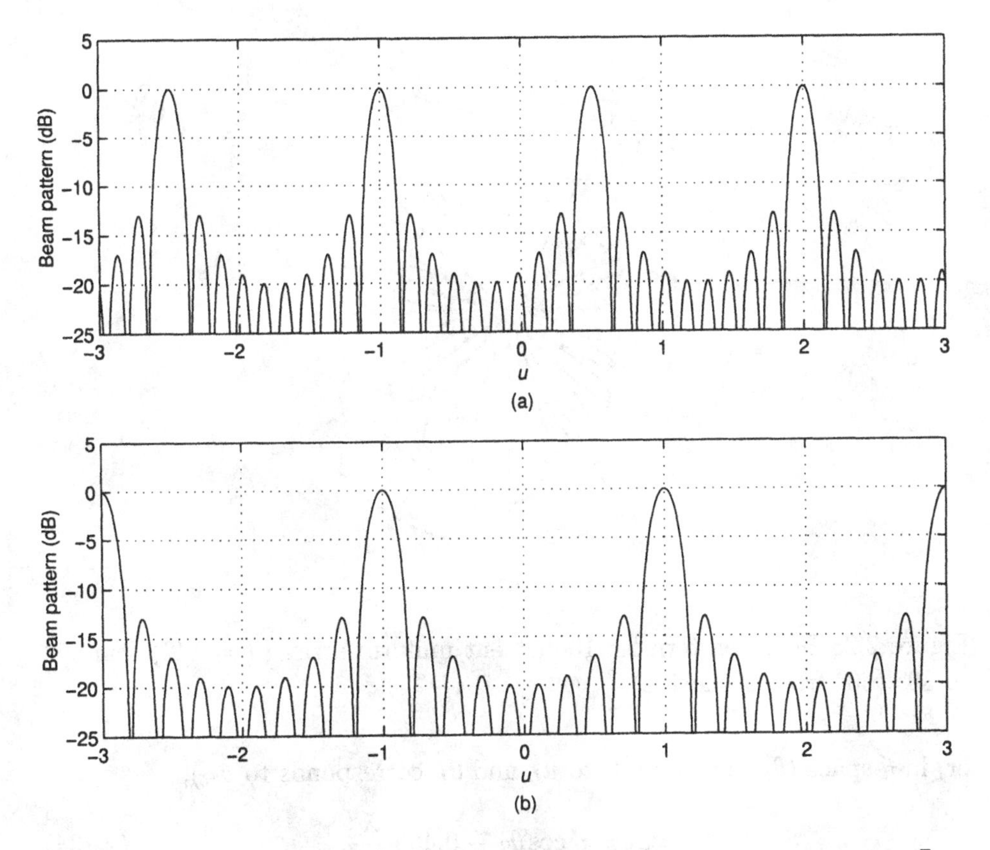

Figure 2.22 Effect of steering on the grating lobes: $N = 10$ (a) $d = 2\lambda/3, \bar{\theta} = 30°$; (b) $d = \lambda/2, \bar{\theta} = 90°$.

When we plot $B_{\theta c}(\theta : \theta_T)$ in θ-space, the shape of the pattern changes due to the $\cos\theta$ dependence. In Figure 2.23, we show the beam pattern for the case $\theta_T = 30°$ and $d = \lambda/2$. Comparing this pattern with the pattern in Figure 2.17, we see that the beamwidth of the main lobe has increased.

To investigate the behavior of the HPBW in θ-space, we use (2.131) and (2.100). The right half-power point in u-space is

$$u_R = u_T + 0.450\,\frac{\lambda}{Nd}, \tag{2.132}$$

and the left half-power point in u-space is

$$u_L = u_T - 0.450\,\frac{\lambda}{Nd}, \tag{2.133}$$

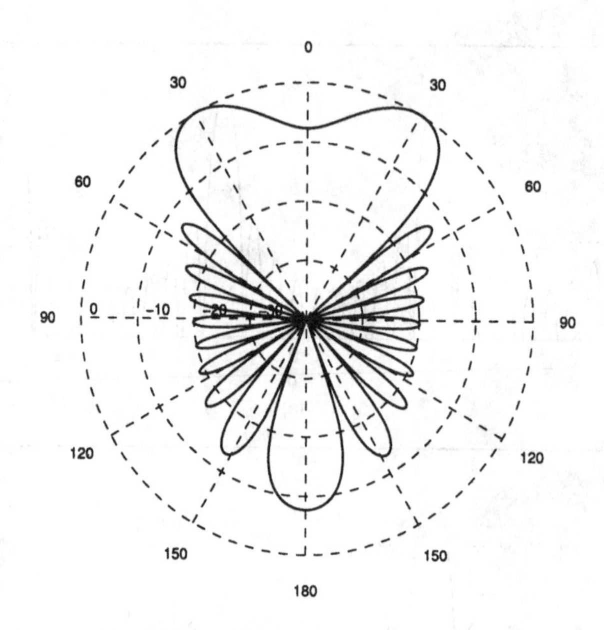

Figure 2.23 Beam pattern for 10-element uniform array ($d = \lambda/2$) scanned to 30° (60° from broadside).

or, in θ-space (θ_R corresponds to u_L and θ_L corresponds to u_R),

$$\cos\theta_R = \cos\theta_T - 0.450\frac{\lambda}{Nd}, \tag{2.134}$$

$$\cos\theta_L = \cos\theta_T + 0.450\frac{\lambda}{Nd}. \tag{2.135}$$

Thus, the half-power beamwidth in θ-space is

$$\theta_H = \theta_R - \theta_L = \cos^{-1}\left[\cos\theta_T - 0.450\frac{\lambda}{Nd}\right] - \cos^{-1}\left[\cos\theta_T + 0.450\frac{\lambda}{Nd}\right], \tag{2.136}$$

for $0 \leq \theta \leq \pi$, $\theta_L, \theta_R \geq 0$. Except for the case when $\theta_T = 0$ or π (endfire), θ_L is defined to be the half-power point closest to $\theta = 0$. As the beam is steered from broadside ($\theta_T = \pi/2$) toward the positive z-axis (endfire, $\theta_T = 0$), the beam broadens. At some point, θ_L as given by (2.135) equals 0. Beyond that point there is no half-power point on that side of the beam. Elliott [Ell81] refers to this point as the scan limit.

The beamwidths given by (2.136) and (2.138) were plotted by Elliott [Ell81] and are shown in Figure 2.24.

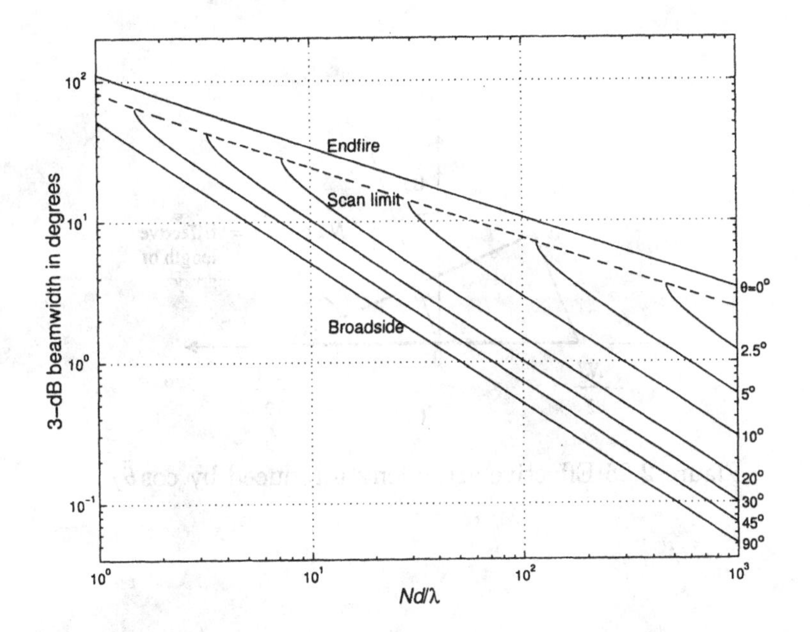

Figure 2.24 HPBW versus steering angle: standard linear array with uniform weighting.

When the beam is steered in the vicinity of broadside ($\bar{\theta}_T$ is small) and $Nd \gg \lambda$, θ_H will be small and we use a small angle expansion to obtain

$$\theta_H \simeq 0.891 \frac{\lambda}{Nd} \csc \bar{\theta}_T. \tag{2.137}$$

The behavior in (2.137) is apparent from Figure 2.25. The effective array length is reduced by $\cos \bar{\theta}_T$.

For $Nd \geq 5\lambda$, the result in (2.137) is in error by less than 0.2% near broadside and by less than 4% at the scan limit.

When $\theta_T = 0$ or π, the maximum response axis is pointed along the array axis and is referred to as an **endfire array**. The beam pattern for a standard 10-element endfire array is shown in Figure 2.26. In this case,

$$\theta_H = 2\cos^{-1}\left[1 - 0.450\frac{\lambda}{Nd}\right], \quad \theta_T = 0 \,\text{or}\, \pi. \tag{2.138}$$

We can rewrite (2.138) as

$$1 - \cos(\frac{\theta_H}{2}) = 0.450\frac{\lambda}{Nd} \tag{2.139}$$

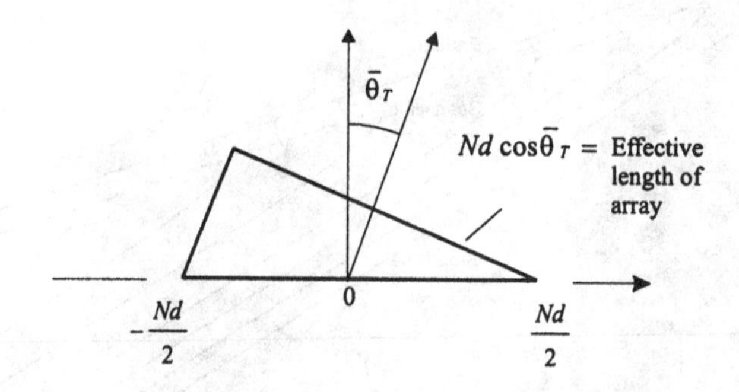

Figure 2.25 Effective array length reduced by $\cos\bar{\theta}_T$.

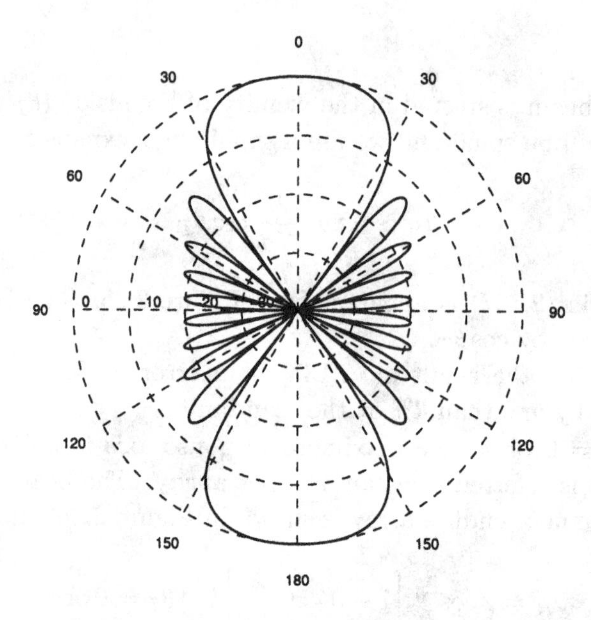

Figure 2.26 Beam pattern of a standard 10-element linear array with uniform amplitude weighting at endfire.

or

$$\sin(\frac{\theta_H}{4}) = \sqrt{\frac{0.450}{2} \cdot \frac{\lambda}{Nd}}. \tag{2.140}$$

For $Nd \gg \lambda$, θ_H is small and (2.140) becomes

$$\theta_H = 2\sqrt{0.890 \frac{\lambda}{Nd}}. \tag{2.141}$$

For $Nd \geq 5\lambda$, (2.141) is in error by less than 1%.

Similarly, the first null in the pattern is at

$$\theta_{null} = \sqrt{2 \frac{\lambda}{Nd}}. \tag{2.142}$$

Thus, the resolution of a linear array at endfire varies as the reciprocal of the square root of Nd/λ as contrasted to the linear dependence in the broadside case.

One can decrease the beamwidth of an endfire array by using a procedure proposed by Hansen and Woodyard [HW38]. This technique is described in Section 6.3.4 of [Bal82] and Section 4.7 of [Kra88]. We summarize the procedure in Problem 2.5.4.

2.6 Array Performance Measures

There are a number of performance measures by which one assesses the capabilities of an array. Each of the various measures attempts to quantify an important aspect of either the response of an array to the signal environment or of the sensitivity to an array design. We have already noted both the distance to the first nulls as a measure of the beamwidth or resolution of the frequency wavenumber response and the cardinal sidelobe levels of a uniformly weighted linear array. In this section we discuss three of the commonly used array performance measures:

(i) Directivity

(ii) Array gain versus spatially white noise (A_w)

(iii) Sensitivity and the tolerance factor

2.6.1 Directivity

A common measure of performance of an array or aperture is the **directivity**. We will define it for the general case and then specialize it to a linear array.

We define the **power pattern**, $P(\theta, \phi)$, to be the squared magnitude of the beam pattern $B(\omega : \theta, \phi)$

$$P(\theta, \phi) = |B(\omega : \theta, \phi)|^2, \tag{2.143}$$

where the frequency dependence of $P(\theta, \phi)$ is suppressed in the notation. Then the directivity D is defined as

$$D = \frac{P(\theta_T, \phi_T)}{\frac{1}{4\pi}\int_0^{\pi} d\theta \int_0^{2\pi} d\phi \sin\theta \cdot P(\theta, \phi)}, \tag{2.144}$$

where (θ_T, ϕ_T) is the steering direction or main response axis (MRA).

In a transmitting array or aperture, D represents the maximum radiation intensity (power per unit solid angle) divided by the average radiation intensity (averaged over the sphere).

In a receiving antenna, we find in Chapter 5 that the denominator represents the noise power at the array (or aperture) output due to isotropic noise (noise distributed uniformly over a sphere). The numerator will represent the power due to a signal arriving from (θ_T, ϕ_T). Thus, D can be interpreted as the array gain against isotropic noise.

If we assume that the weights are normalized so that $P(\theta_T, \phi_T) = 1$, then (2.144) can be written as

$$D = \left\{ \frac{1}{4\pi} \int_0^{\pi} d\theta \int_0^{2\pi} d\phi \sin\theta \cdot P(\theta, \phi) \right\}^{-1}. \tag{2.145}$$

For a linear array

$$B(\theta, \phi) = B(\theta), \tag{2.146}$$

so (2.145) becomes

$$D = \left\{ \frac{1}{2} \int_0^{\pi} |B(\theta)|^2 \sin\theta \, d\theta \right\}^{-1}. \tag{2.147}$$

This can be expressed in u-space as

$$D = \left\{ \frac{1}{2} \int_{-1}^{1} |B_u(u)|^2 \, du \right\}^{-1}. \tag{2.148}$$

In general, this expression must be evaluated numerically.

Using (2.70), the expression in (2.148) can be written in terms of the array weights as

$$D = \left\{ \frac{1}{2} \int_{-1}^{1} \sum_{n=0}^{N-1} w_n^* e^{jn\left(\frac{2\pi d}{\lambda}\right)(u-u_T)} \sum_{m=0}^{N-1} w_m e^{-jm\left(\frac{2\pi d}{\lambda}\right)(u-u_T)} du \right\}^{-1}. \quad (2.149)$$

where u_T is the steering direction in u-space. Rearranging terms and performing the integration gives

$$D = \left\{ \sum_{n=0}^{N-1} \sum_{m=0}^{N-1} w_m w_n^* e^{j\left(\frac{2\pi d}{\lambda}\right)(m-n)u_T} \operatorname{sinc}\left(\frac{2\pi d}{\lambda}(n-m)\right) \right\}^{-1}. \quad (2.150)$$

We define several matrices to obtain a compact expression. The n, m element of the **sinc** matrix is

$$[\mathbf{sinc}]_{nm} \triangleq \operatorname{sinc}\left(\frac{2\pi d}{\lambda}(n-m)\right). \quad (2.151)$$

The steering matrix in u-space is (2.120)

$$\mathbf{I}_{su} = \operatorname{diag}\left(1, e^{j\frac{2\pi d}{\lambda}u_T}, e^{j\frac{2\pi d}{\lambda}2u_T}, \cdots, e^{j\frac{2\pi d}{\lambda}(N-1)u_T}\right). \quad (2.152)$$

Then,

$$D = \mathbf{I}_{su}\mathbf{w}^H[\mathbf{sinc}]\mathbf{w}\mathbf{I}_{su}^H. \quad (2.153)$$

Normally, we include the steering in the weight vector,[7]

$$\mathbf{w}_s = \mathbf{w}\mathbf{I}_{su}^H. \quad (2.154)$$

Then

$$D = \mathbf{w}_s^H[\mathbf{sinc}]\mathbf{w}_s. \quad (2.155)$$

The standard linear array is a special case of interest. If $d = \lambda/2$, (2.150) reduces to

$$D = \left\{ \sum_{n=0}^{N-1} \sum_{m=0}^{N-1} w_m w_n^* e^{j\pi(m-n)u_T} \operatorname{sinc}\left(\pi(n-m)\right) \right\}^{-1}. \quad (2.156)$$

The sinc function equals one when $m = n$ and zero when $m \neq n$, so (2.156) reduces to

$$D = \left\{ \sum_{n=0}^{N-1} |w_n|^2 \right\}^{-1} = \left(\mathbf{w}^H\mathbf{w}\right)^{-2} = \left\{ \| \mathbf{w} \|^2 \right\}^{-1}, \quad (2.157)$$

[7]In most cases, the "s" subscript is dropped because we assume that **w** includes steering.

where

$$\| \mathbf{w} \| = (\mathbf{w}^H \mathbf{w})^{\frac{1}{2}}, \tag{2.158}$$

is the 2-norm of the vector $\mathbf{w}$ (A.36).

Thus, the directivity of a standard linear array is the reciprocal of the magnitude squared of the weight vector. The directivity does not depend on the steering direction. As the steering direction moves away from array broadside, the beam broadens but the circumference in the ϕ integration decreases.

For $d \neq \lambda/2$, we use the expression in (2.150) and the directivity will depend on the steering direction.

For a uniformly weighted standard linear array, $w_n = 1/N$, so

$$\sum_{n=0}^{N-1} |w_n|^2 = \frac{1}{N}, \tag{2.159}$$

so that

$$D = N. \tag{2.160}$$

Uniform weighting maximizes the directivity of the standard linear array. To show this, we constrain

$$\sum_{n=0}^{N-1} w_n = 1, \tag{2.161}$$

which guarantees that the beam pattern equals one for $u_T = 0$, and maximize $\sum_{n=0}^{N-1} |w_n|^2$. To perform the maximization, we write

$$F = \sum_{n=0}^{N-1} |w_n|^2 + \lambda \left(\sum_{n=0}^{N-1} w_n - 1 \right), \tag{2.162}$$

where λ is a Lagrange multiplier.[8] Differentiating with respect to w_n and setting the result to zero gives

$$w_n^* = -\lambda \tag{2.163}$$

or

$$w_n = -\lambda^*. \tag{2.164}$$

Substituting (2.164) into (2.161) gives $\lambda^* = -1/N$, so

$$w_n = \frac{1}{N}, \tag{2.165}$$

[8] See Appendix A (Section A.7.4) for a discussion of complex gradients and the differentiation of a non-analytic function.

which is the desired result.

In Chapter 3, we discuss non-uniform weightings. The directivity always decreases. It is a function of the length of array measured in half wavelengths plus a term due to the non-uniform weighting.

The directivity of a standard uniformly weighted linear array can be related to the HPBW or the beamwidth between the first nulls. From Table 2.3, we have

$$D = \frac{4}{BW_{NN}}, \tag{2.166}$$

where the BW_{NN} is expressed in u-space. Note that (2.166) applies to a uniformly weighted array. For other weightings (2.166) may not hold.

Frequently, we express the directivity in dB and refer to it as the **directivity index**,

$$DI = 10 \log_{10} D. \tag{2.167}$$

We can write the DI as

$$DI = 10 \log_{10} N + 10 \log_{10}(g(\mathbf{w})). \tag{2.168}$$

The second term is a function of the weights. For any non-uniform weighting ($w_n \neq N^{-1}$), the DI will be reduced.

2.6.2 Array Gain vs. Spatially White Noise (A_w)

One of the purposes of an array is to improve the **signal-to-noise ratio (SNR)** by adding signals coherently and noise incoherently. The improvement is measured by the **array gain**. It is an important measure of array performance that we discuss extensively in later chapters. The general definition must be deferred until we introduce the spectral covariance matrix, which describes the statistical concepts for describing the spatial properties of the noise processes; however, we can formulate a restricted definition here.

We assume that the input at each sensor consists of a plane wave arriving along the main response axis plus a noise process that is uncorrelated among the sensors (spatially white noise). Thus,

$$x_n(t) = f(t - \tau_n) + n_n(t), \quad n = 0, \cdots, N-1. \tag{2.169}$$

At each sensor, the signal spectrum-to-noise spectrum ratio at frequency ω is

$$SNR_{in}(\omega) \triangleq \frac{S_f(\omega)}{S_n(\omega)}, \tag{2.170}$$

where the subscript "*in*" denotes input and the noise spectrum at each sensor is assumed to be identical.

In order to find the output due to the signal and noise, we need an expression for the output spectrum in terms of the beamformer weights and the input spectral matrix. From (2.8),

$$y(t) = \int_{-\infty}^{\infty} \mathbf{h}^T(\tau)\,\mathbf{x}(t-\tau)\,d\tau. \tag{2.171}$$

We assume that $\mathbf{x}(t)$ is zero-mean and wide-sense stationary. The correlation function of the output $y(t)$ is

$$R_y(\tau) = E[y(t)y^*(t-\tau)]. \tag{2.172}$$

The spectrum of $y(t)$ is

$$S_y(\omega) = \int_{-\infty}^{\infty} e^{-j\omega\tau}\, R_y(\tau)\, d\tau. \tag{2.173}$$

Using (2.171) in (2.172) and the result in (2.173) gives

$$S_y(\omega) = \int_{-\infty}^{\infty} e^{-j\omega\tau} d\tau \int_{-\infty}^{\infty} \mathbf{h}^T(\alpha) d\alpha \int_{-\infty}^{\infty} E[\mathbf{x}(t-\alpha)\mathbf{x}^H(t-\tau-\beta)]\mathbf{h}^*(\beta) d\beta. \tag{2.174}$$

This can be rewritten as

$$S_y(\omega) = \int_{-\infty}^{\infty} d\alpha\, \mathbf{h}^T(\alpha) e^{-j\omega\alpha} \int_{-\infty}^{\infty} dz\, e^{-j\omega z}\mathbf{R}_{\mathbf{x}}(z) \int_{-\infty}^{\infty} d\beta\, e^{-j\omega\beta}\mathbf{h}^*(\beta), \tag{2.175}$$

which reduces to

$$S_y(\omega) = \mathbf{H}^T(\omega)\, \mathbf{S}_{\mathbf{x}}(\omega)\mathbf{H}^*(\omega). \tag{2.176}$$

Using (2.52) in (2.176) gives the desired results,

$$S_y(\omega) = \mathbf{w}^H\, \mathbf{S}_{\mathbf{x}}(\omega)\mathbf{w}, \tag{2.177}$$

for the narrowband beamformer.

To calculate the output due to the signal, we impose the constraint on $\mathbf{w}$ that,

$$\mathbf{w}^H\, \mathbf{v_k}(\mathbf{k}_s) = 1. \tag{2.178}$$

The constraint in (2.178) implies that any signal arriving along $\mathbf{k}_s$ will pass through the beamformer undistorted. We refer to the constraint in (2.178) as a **distortionless constraint**. It is used frequently in subsequent discussions.

We can write the input signal spectrum as

$$\mathbf{S_f}(\omega) = \mathbf{v_k}(\mathbf{k}_s)\, S_f(\omega)\, \mathbf{v_k}^H(\mathbf{k}_s). \tag{2.179}$$

Using (2.179) in (2.176), the output signal spectrum is

$$\begin{aligned} S_{y_s}(\omega) &= \mathbf{w}^H \mathbf{v_k}(\mathbf{k}_s)\, S_f(\omega)\, \mathbf{v_k}^H(\mathbf{k}_s)\, \mathbf{w} \\ &= S_f(\omega). \end{aligned} \tag{2.180}$$

The spectral output due to noise is

$$S_{y_n}(\omega) = \mathbf{w}^H \mathbf{S_n}(\omega)\mathbf{w}, \tag{2.181}$$

where $\mathbf{S_n}(\omega)$ is the spectral matrix of the input noise process. For the special case of spatial white noise and identical noise spectra at each sensor,

$$\mathbf{S_n}(\omega) = S_n(\omega)\, \mathbf{I}, \tag{2.182}$$

and

$$S_{y_n}(\omega) = \parallel \mathbf{w} \parallel^2\, S_n(\omega) = \sum_{n=0}^{N-1} |w_n|^2\, S_n(\omega). \tag{2.183}$$

Thus,

$$SNR_o(\omega) = \frac{1}{\sum_{n=0}^{N-1} |w_n|^2} \frac{S_f(\omega)}{S_n(\omega)}, \tag{2.184}$$

where the subscript "o" denotes output.

The array gain A_w reflects the improvement in *SNR* obtained by using the array. It is defined to be the ratio of the *SNR* at the output of the array to the *SNR* at an input sensor. The subscript "w" denotes the spatially uncorrelated noise input. The noise temporal frequency spectrum is not necessarily flat. Using (2.170) and (2.184),

$$A_w = \frac{SNR_o(\omega)}{SNR_{in}(\omega)} = \frac{1}{\sum_{n=0}^{N-1} |w_n|^2}, \tag{2.185}$$

or

$$A_w = \left(\sum_{n=0}^{N-1} |w_n|^2 \right)^{-1} = \parallel \mathbf{w} \parallel^{-2}. \tag{2.186}$$

Three observations with respect to (2.186) are useful:

(i) The result is valid for an arbitrary array geometry, as long as

$$|\mathbf{w}^H \mathbf{v_k}(\mathbf{k}_s)|^2 = 1. \tag{2.187}$$

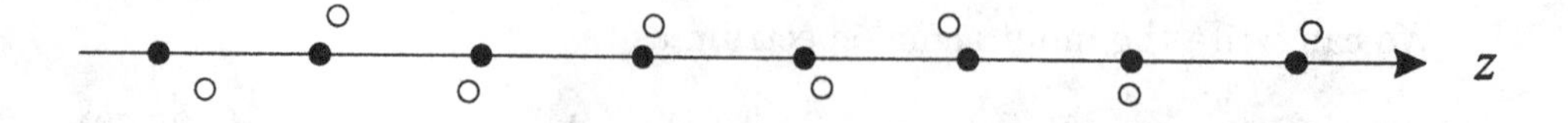

Figure 2.27 Array with position errors.

(ii) For a standard linear array (spacing $d = \lambda/2$), the white noise array gain is identical to the array directivity (2.157). After we discuss isotropic noise fields in Chapter 5, the reason for this will be clear.

(iii) For a uniform linear array with $d \neq \lambda/2$, D will not equal A_w. We will find the noise spectral matrix for an isotropic noise input is given by (2.151) so that D is the array gain A_{iso} for an isotropic noise input.

For a uniformly weighted array,

$$w_n = \frac{1}{N}, \quad n = 0, \cdots, N-1 \tag{2.188}$$

and $A_w = N$ (or $10 \log N$ in dB). A direct application of the Schwarz inequality shows that $A_w \leq N$. Therefore, if we are trying to maximize the array gain in the presence of spatially uncorrelated noise, a uniformly weighted array is optimum.

In Chapter 3, we will develop non-uniform weighting in order to improve sidelobe behavior. In these cases, we will be doing a trade-off between the loss in directivity and white noise array gain against improved sidelobe behavior.

2.6.3 Sensitivity and the Tolerance Factor

In later chapters, we analyze the sensitivity of optimum processors to gain and phase errors and imprecise positioning of the sensors. Here we summarize a typical result for the array shown in Figure 2.27. We design the processor so that, in the absence of array perturbations,

$$\mathbf{w}^H \cdot \mathbf{1} F(\omega) = F(\omega), \tag{2.189}$$

so that a signal from broadside passes through the array processing without distortion.

We consider the effect of filter perturbations and array location perturbations.

2.6.3.1 Filter perturbations

In this case we assume that the nominal matrix filter is $\mathbf{w}^n$. Denoting the ith component of $\mathbf{w}^n$ as w_i^n, we can write the nominal weight as

$$(w_i^n)^* = g_i^n e^{-j\phi_i^n} \tag{2.190}$$

and the actual weight as

$$\begin{aligned} w_i^* &= g_i e^{-j\phi_i} \\ &= g_i^n (1 + \Delta g_i) e^{-j(\phi_i^n + \Delta\phi_i)}, \end{aligned} \tag{2.191}$$

where the Δg_i and $\Delta\phi_i$ are random variables.

2.6.3.2 Array location perturbations

In this case we assume that the nominal array locations are $\mathbf{p}_i^n$ and that

$$\mathbf{p}_i = \mathbf{p}_i^n + \Delta\mathbf{p}_i. \tag{2.192}$$

Thus, we have the three variations from the nominal model:[9]

$$g_i = g_i^n (1 + \Delta g_i), \tag{2.193}$$

$$\phi_i = \phi_i^n + \Delta\phi_i, \tag{2.194}$$

$$\mathbf{p}_i = \mathbf{p}_i^n + \Delta\mathbf{p}_i. \tag{2.195}$$

The first two variations could result from changes in the gain and phase of the array sensors or from imperfect gain and phase in the processor filters. The last variation is caused by imperfect knowledge of the location of the array elements. We assume that the $\Delta g_i \; (i = 0, 1, \cdots, N-1)$, the $\Delta\phi_i \; (i = 0, 1, \cdots, N-1)$, and the $\Delta p_{xi}, \Delta p_{yi}, \Delta p_{zi} \; (i = 0, 1, \cdots, N-1)$ are statistically independent, zero-mean, Gaussian random variables. We analyze the behavior of the beam pattern in the presence of these variations.

The nominal beam pattern is

$$\begin{aligned} B^{(n)}(\mathbf{k}) &= (\mathbf{w}^n)^H \mathbf{v}(\mathbf{k}) \\ &= \sum_{i=0}^{N-1} g_i^n \exp\left(j\phi_i^n - j\mathbf{k}^T \mathbf{p}_i^n\right). \end{aligned} \tag{2.196}$$

[9] This case is from Gilbert and Morgan [GM55]

The actual beam pattern is a random function. The expectation of its magnitude squared can be written as

$$\begin{aligned}\overline{|B(\mathbf{k})|^2} &\triangleq E\left\{|B(\mathbf{k})|^2\right\} \\ &= E\left\{\sum_{i=0}^{N-1}\sum_{l=0}^{N-1} g_i \exp\left(j\phi_i - j\mathbf{k}^T\mathbf{p}_i\right)\right. \\ &\qquad \left.\cdot g_l \exp\left(-j\phi_l + j\mathbf{k}^T\mathbf{p}_l\right)\right\}\end{aligned} \tag{2.197}$$

or

$$\begin{aligned}\overline{|B(\mathbf{k})|^2} &= \sum_{i=0}^{N-1}\sum_{l=0}^{N-1} E\left\{g_i^n(1+\Delta g_i)g_l^n(1+\Delta g_l)\right. \\ &\quad \cdot \exp\left[j\left(\phi_i^n + \Delta\phi_i - \phi_l^n - \Delta\phi_l\right)\right] \\ &\quad \left.\cdot \exp\left[-j\mathbf{k}^T\left(\mathbf{p}_i^n + \Delta\mathbf{p}_i - \mathbf{p}_l^n - \Delta\mathbf{p}_l\right)\right]\right\}.\end{aligned} \tag{2.198}$$

Now define

$$\alpha_{il} = E\left\{(1+\Delta g_i)(1+\Delta g_l)\exp[j(\Delta\phi_i - \Delta\phi_l)]\right\} \tag{2.199}$$

and

$$\beta_{il}(\mathbf{k}) = E\left\{\exp[-j\mathbf{k}^T(\Delta\mathbf{p}_i - \Delta\mathbf{p}_l)]\right\}. \tag{2.200}$$

Using the independent Gaussian random variable assumption,

$$\alpha_{il} = \begin{cases}\exp\left(-\sigma_\phi^2\right), & i \neq l \\ 1+\sigma_g^2, & i = l\end{cases} \tag{2.201}$$

and

$$\beta_{il}(\mathbf{k}) = \begin{cases}\exp\left(-\sigma_p^2|\mathbf{k}|^2\right) = \exp\left(-\left[\frac{2\pi\sigma_p}{\lambda}\right]^2\right) \triangleq \exp\left(-\sigma_\lambda^2\right), & i \neq l, \\ 1, & i = l,\end{cases} \tag{2.202}$$

where

$$\sigma_\lambda \triangleq \frac{2\pi\sigma_p}{\lambda}. \tag{2.203}$$

In the expression for $\beta_{il}(\mathbf{k})$, we have assumed that the variance of each component of $\Delta\mathbf{p}_i$ is equal to σ_p^2. The term σ_λ^2 is the scaled variance measured in wavelengths. Then,

$$\overline{|B(\mathbf{k})|^2} = \sum_{\substack{i=0 \\ i\neq l}}^{N-1}\sum_{l=0}^{N-1} g_i^n g_l^n \exp\left(j\phi_i^n - j\phi_l^n\right)\exp\left[-j\mathbf{k}^T\left(\mathbf{p}_i^n - \mathbf{p}_l^n\right)\right]$$

$$\cdot \exp\left[-\left(\sigma_\phi^2+\sigma_\lambda^2\right)\right]+\sum_{i=0}^{N-1}\left(1+\sigma_g^2\right)\left(g_i^n\right)^2 . \tag{2.204}$$

Adding $(g_i^n)^2 \exp[-(\sigma_\phi^2+\sigma_\lambda^2)]$, $i=0,1,\cdots,N-1$ in the first term and subtracting the appropriate quantity from the second term gives

$$\begin{aligned}\overline{|B(\mathbf{k})|^2} &= |B^{(n)}(\mathbf{k})|^2 \exp[-(\sigma_\phi^2+\sigma_\lambda^2)] \\ &+ \sum_{i=0}^{N-1} (g_i^n)^2 \left\{(1+\sigma_g^2)-\exp\left[-(\sigma_\phi^2+\sigma_\lambda^2)\right]\right\},\end{aligned} \tag{2.205}$$

where σ_ϕ^2, σ_λ^2, and σ_g^2 denote the variance of the corresponding random variables.

The random variation has two effects. The first term attenuates the beam pattern uniformly in **k**. This uniformity is due to our assumption that the variations are not dependent on **k**. It means that the beam pattern has a statistical bias. The expected value of the pattern along the MRA is less than unity. The second term is more critical. Define

$$T_{se}=\sum_{i=0}^{N-1}|w_i^n|^2=\sum_{i=0}^{N-1}(g_i^n)^2 \tag{2.206}$$

as the **sensitivity function**. Then the second term becomes

$$\overline{|B_2(\mathbf{k})|^2}=T_{se}\{1+\sigma_g^2-\exp[-(\sigma_\phi^2+\sigma_\lambda^2)]\}, \tag{2.207}$$

which for small variances reduces to,

$$\overline{|B_2(\mathbf{k})|^2}=T_{se}\{\sigma_g^2+\sigma_\phi^2+\sigma_\lambda^2\}. \tag{2.208}$$

Note that T_{se} is the inverse of the array gain for white noise, (see (2.185))

$$T_{se}=[A_w]^{-1}=\| \mathbf{w} \|^2 . \tag{2.209}$$

Thus as the white noise array gain increases, the sensitivity decreases. For an N-element array, the maximum white noise gain is N and corresponds to uniform weighting. Thus, any array with non-uniform weighting will be more sensitive to parameter variations than the uniformly weighted array. The effect of the second term is to raise the expected value in the sidelobe region uniformly. This constant value across **k**-space can have a major impact. In many array designs, we would like to put a perfect null ($|B(\mathbf{k})|^2=0$) in the direction of an interfering signal. We look at techniques for doing this in

Section 3.7. The implication of the term in (2.208) or (2.207) is that, if any of the variances $(\sigma_g^2, \sigma_\phi^2, \sigma_\lambda^2)$ are non-zero, then we can not obtain a perfect null.

The level of the floor in the expected value of the power pattern will depend on the sensitivity function, T_{se}, and the variance of the perturbations. The effect is to limit the depth of the nulls in the pattern.[10]

As an example, suppose

$$\sigma_T^2 \triangleq (\sigma_g^2 + \sigma_\phi^2 + \sigma_\lambda^2) = 0.01. \tag{2.210}$$

Then, A_w must be greater than or equal to 100 in order to get -40-dB nulls in the pattern. This requires that a uniformly weighted array must contain at least 100 elements. An array with non-uniform weighting would require even more elements.

Later in the text, when we design optimum arrays, we often impose a **sensitivity constraint**,

$$T_{se} = \| \mathbf{w} \|^2 \leq T_o, \tag{2.211}$$

where T_o is a design constant to make the performance more robust to perturbations. The constraint in (2.211) is often referred to as a **white noise gain constraint**,

$$A_w = (\| \mathbf{w} \|^2)^{-1} \geq T_o^{-1}. \tag{2.212}$$

In many of the design problems that we consider later, we find that the constraint in (2.211) plays an important role.

2.6.4 Summary

In this section, we have developed three important array performance measures. We observe that the norm of the weight vector $\mathbf{w}$ has appeared in all three measures:

(i) For a standard linear array, the directivity is

$$D = \| \mathbf{w} \|^{-2} \leq N.$$

(ii) For any array geometry, the white noise array gain is

$$A_w = \| \mathbf{w} \|^{-2}.$$

[10] In mathematical terms, a null in the beam pattern means $B(\mathbf{k}) = 0$. However, in practice, the actual pattern has some non-zero value. The ratio of this value to the value at $B(\mathbf{k}_T)$ is referred to as the **null depth**.

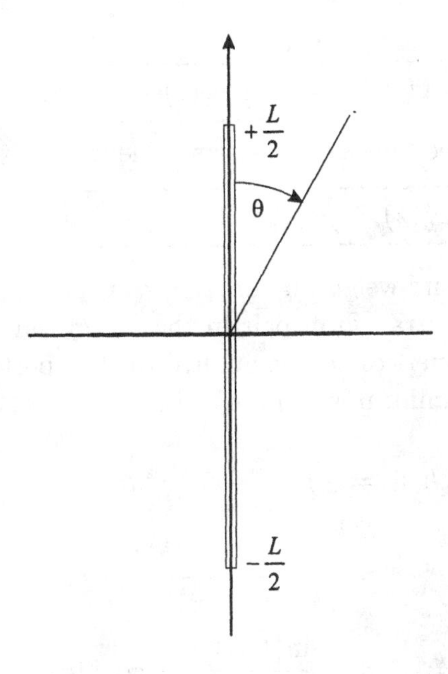

Figure 2.28 Linear aperture.

(iii) For any array geometry, the sensitivity function is

$$T_{se} = A_w^{-1} = \| \mathbf{w} \|^2 .$$

Later we will see that $\| \mathbf{w} \|^2$ will play a central role in many of our discussions.

2.7 Linear Apertures

2.7.1 Frequency-wavenumber Response

Consider the linear aperture shown in Figure 2.28. We assume that it is steered to broadside and has an aperture weighting function $w_a^*(z)$, where "a" denotes "aperture."

The frequency-wavenumber response function is given by

$$\Upsilon(\omega, k_z) = \int_{-L/2}^{L/2} dz\, w_a^*(z) e^{-jk_z z}. \tag{2.213}$$

The exponential term, $\exp(-jk_z z)$, is the **array manifold function** and is analogous to the array manifold vector in an array. Observing that

$$w_a^*(z) = 0, \quad |z| > \frac{L}{2}, \tag{2.214}$$

we have

$$\boxed{\Upsilon(\omega, k_z) = \int_{-\infty}^{\infty} w_a^*(z) e^{-jk_z z}\, dz,} \tag{2.215}$$

which is familiar as the Fourier transform. The inverse transform is

$$\boxed{w_a^*(z) = \frac{1}{2\pi}\int_{-\infty}^{\infty} \Upsilon(\omega, k_z) e^{jzk_z}\, dk_z.} \tag{2.216}$$

We see that the aperture weighting function and the frequency-wavenumber response are a Fourier transform pair in the $z - k_z$ variables. Thus all the Fourier transform properties can be applied to the aperture problem. For a linear aperture with uniform weighting,

$$\begin{aligned}
\Upsilon(\omega, k_z) &= \int_{-L/2}^{L/2} \frac{1}{L} e^{-jk_z z} dz \\
&= \frac{e^{j\frac{L}{2}k_z} - e^{-j\frac{L}{2}k_z}}{2j(k_z\frac{L}{2})} \\
&= \frac{\sin(\frac{L}{2}k_z)}{\frac{L}{2}k_z}, \qquad -\infty < k_z < \infty
\end{aligned} \tag{2.217}$$

or[11]

$$\Upsilon(\omega, k_z) = \text{sinc}(\frac{L}{2}k_z), \quad -\infty < k_z < \infty, \tag{2.218}$$

and, since $k_z = -(2\pi/\lambda)u$,

$$B_u(u) = \text{sinc}(\frac{\pi L}{\lambda} u), \quad -1 \le u \le 1. \tag{2.219}$$

The function is plotted in Figure 2.29.

It is useful to compare the result in (2.219) with the array result in (2.96). To find the equivalent length of aperture corresponding to an array of N elements, we equate the arguments of the sine functions in the numerators of (2.219) and (2.96),

$$\frac{\pi L}{\lambda} u = \frac{\pi N d}{\lambda} u. \tag{2.220}$$

This equality provides the same main lobe width and null spacing. Thus,

$$L = Nd. \tag{2.221}$$

This relationship is shown in Figure 2.30. The equivalent length aperture extends $d/2$ beyond the actual array length in each direction.

[11]We define $\text{sinc}\, x$ as $(\sin x)/x$. Some sources (e.g., MATLAB®) define $\text{sinc}\, x$ as $(\sin(\pi x))/(\pi x)$.

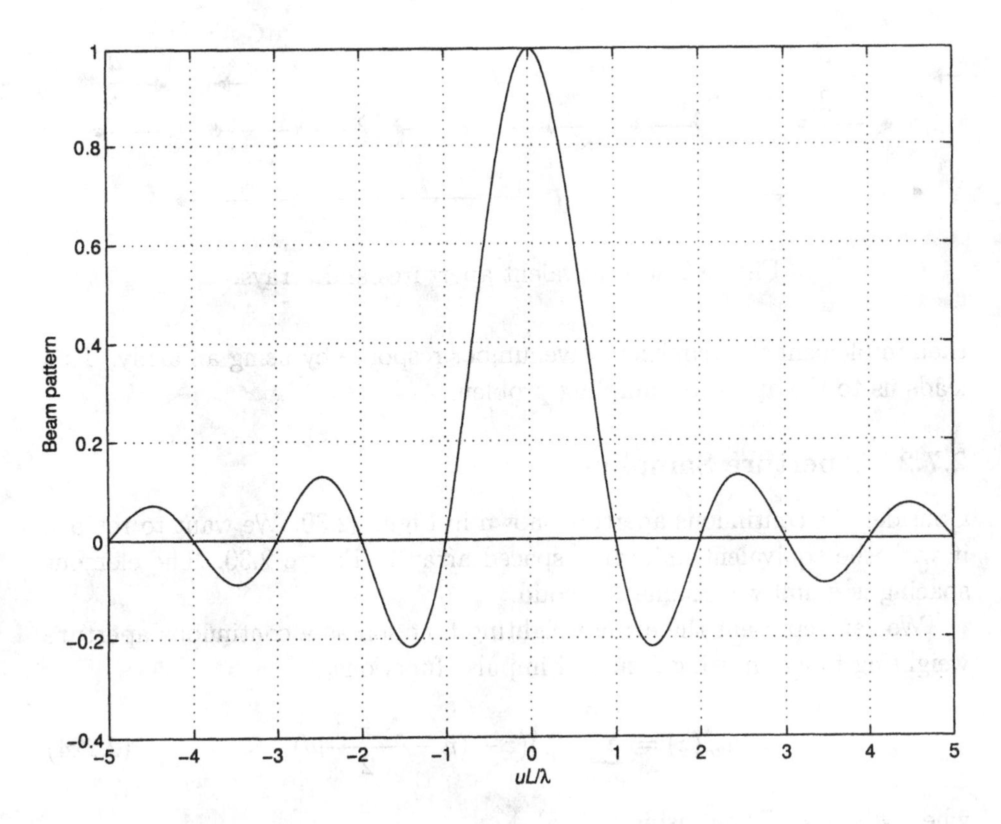

Figure 2.29 Beam pattern of uniformly weighted linear aperture.

One Fourier transform property of immediate use is the shifting property. The array in (2.213) is steered to broadside. To steer to k_{zT}, we have

$$\Upsilon(\omega, k_z : k_{zT}) = \Upsilon(\omega, k_z - k_{zT}). \tag{2.222}$$

Substituting into (2.216) gives

$$\begin{aligned} w_a^*(z : k_{zT}) &= \frac{1}{2\pi}\int_{-\infty}^{\infty} \Upsilon(\omega, k_z - k_{zT}) e^{jzk_z}\, dk_z \\ &= \frac{1}{2\pi}\int_{-\infty}^{\infty} \Upsilon(\omega, \Delta k) e^{jz(\Delta k + k_{zT})}\, d(\Delta k) \\ &= e^{jzk_{zT}}\, w_a^*(z). \end{aligned} \tag{2.223}$$

Thus, as we have seen with arrays, steering the beam in wavenumber space corresponds to a progressive phase shift in the weighting function.

In many cases, we want to start the analysis with the linear aperture and

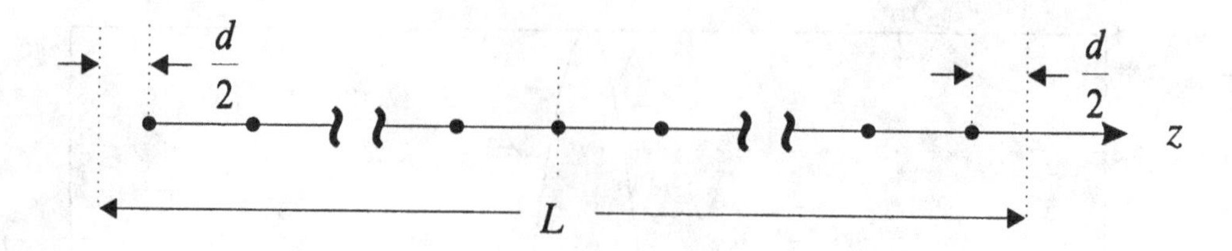

Figure 2.30 Equivalent apertures and arrays.

then implement its frequency wavenumber response by using an array. This leads us to the aperture sampling problem.

2.7.2 Aperture Sampling

Consider the continuous aperture shown in Figure 2.30. We want to replace it with the equivalent uniformly spaced array in Figure 2.30. The element spacing is d and we assume N is odd.

We can represent the array weighting function as a continuous aperture weighting function using a sum of impulse functions,

$$w_a^*(z) = \sum_{n=0}^{N-1} w_n^* \delta(z - (n - \frac{N-1}{2})d), \tag{2.224}$$

where $Nd = L$. Then, using (2.215),

$$\Upsilon(\omega, k_z) = \int_{-\infty}^{\infty} \sum_{n=0}^{N-1} w_n^* \delta(z - (n - \frac{N-1}{2})d) e^{-jk_z z} dz. \tag{2.225}$$

Integrating, we obtain

$$\Upsilon(\omega, k_z) = \sum_{n=0}^{N-1} w_n^* e^{-j(n-\frac{N-1}{2})k_z d}, \tag{2.226}$$

which is identical to (2.58).

If the continuous aperture weighting function is $w_a^*(z)$, then using a standard linear array ($d = \lambda/2$) with element weightings

$$w_n^* = w_a^*(z_n), \quad n = 0, 1, \cdots, N-1, \tag{2.227}$$

will not, in general, produce an identical beam pattern. Usually, the sampled pattern will have very similar main-lobe behavior, but its sidelobe behavior will be different. We see examples of this behavior in Chapter 3.

If our goal is to duplicate the beam pattern of continuous aperture, we can use smaller interelement spacing.

2.8 Non-isotropic Element Patterns

Our discussion up to this point has assumed that each element had an isotropic response. In many cases, each element will have a beam pattern. In other cases, groups of sensors may be combined into a subarray with a beam pattern. The subarrays are treated as elements in the overall array. It is straightforward to incorporate these element beam patterns through pattern multiplication.

We assume each element is a linear aperture and has an identical weighting, denoted by $w_{ae}^*(z)$. These elements are configured into a linear array at location z_n, $n = 0, 1, \cdots, N-1$. Note that the spacing does not have to be uniform. The array weightings are w_n^*. The total weighting function is,

$$w_a^*(z) = \sum_{n=0}^{N-1} w_n^* \cdot w_{ae}^*(z - z_n). \tag{2.228}$$

The resulting frequency-wavenumber function is

$$\begin{aligned} \Upsilon(\omega, k_z) &= \int_{-\infty}^{\infty} \sum_{n=0}^{N-1} w_n^* \cdot w_{ae}^*(z - z_n)\, e^{-jk_z z}\, dz \\ &= \sum_{n=0}^{N-1} w_n^*\, e^{-jk_z \cdot z_n} \int_{-\infty}^{\infty} w_{ae}^*(z_1) e^{-jk_z z_1} dz_1. \end{aligned} \tag{2.229}$$

The first term is familiar as the beam pattern of the array with isotropic elements. It is now convenient to denote it as the **array factor**

$$AF(k_z) \triangleq \sum_{n=0}^{N-1} w_n^* \cdot e^{-jk_z z_n}, \tag{2.230}$$

which is analogous to the expression in (2.58).

The second term is the element frequency-wavenumber function. Thus,

$$\Upsilon(\omega, k_z) = AF(k_z) \Upsilon_e(\omega, k_z). \tag{2.231}$$

In terms of beam patterns

$$B_u(u) = AF(u) B_{ue}(u), \tag{2.232}$$

where

$$k_z = -\frac{2\pi}{\lambda} \cos\theta = -\frac{2\pi}{\lambda} u. \tag{2.233}$$

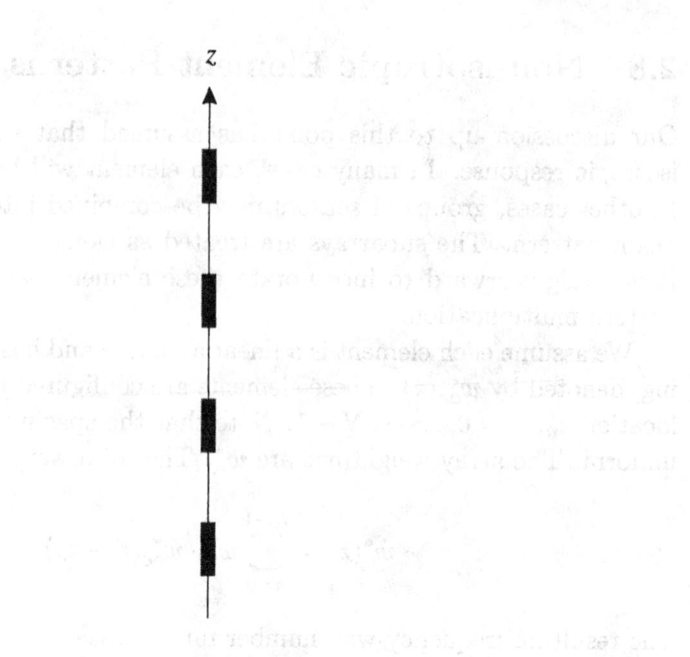

Figure 2.31 Array of colinear apertures.

Thus, the total beam pattern is the product of the array factor and the element beam pattern. This behavior is referred to as pattern multiplication.

The derivation for three dimensions follows easily:

$$\Upsilon(\omega, \mathbf{k}) = AF(\mathbf{k})\Upsilon_e(\omega, \mathbf{k}). \tag{2.234}$$

Three representative cases are shown in figures 2.31, 2.32, and 2.33.

In the first case, the element is colinear with the linear array axis, and the resulting pattern can be represented in k_z-space.

In the second case, the element is perpendicular, so that the resulting pattern must be represented in $\mathbf{k}$-space (see Problem 2.8.3).

In the third case, the array is a rectangular planar array. If we consider the sensors in the x-direction to be elements of a linear array along the z-axis and each column has identical weightings, then the total array factor is the product of the two array factors,

$$AF(\mathbf{k}) = AF_z(\mathbf{k}) \cdot AF_x(\mathbf{k}). \tag{2.235}$$

In addition to non-isotropic beam patterns, there are other sensor characteristics that should be considered in a particular physical problem. In some situations, there may be mutual coupling between the sensors. Several references (e.g., Balanis [Bal82], Yeh et al. [YLU89], Friedlander and Weiss

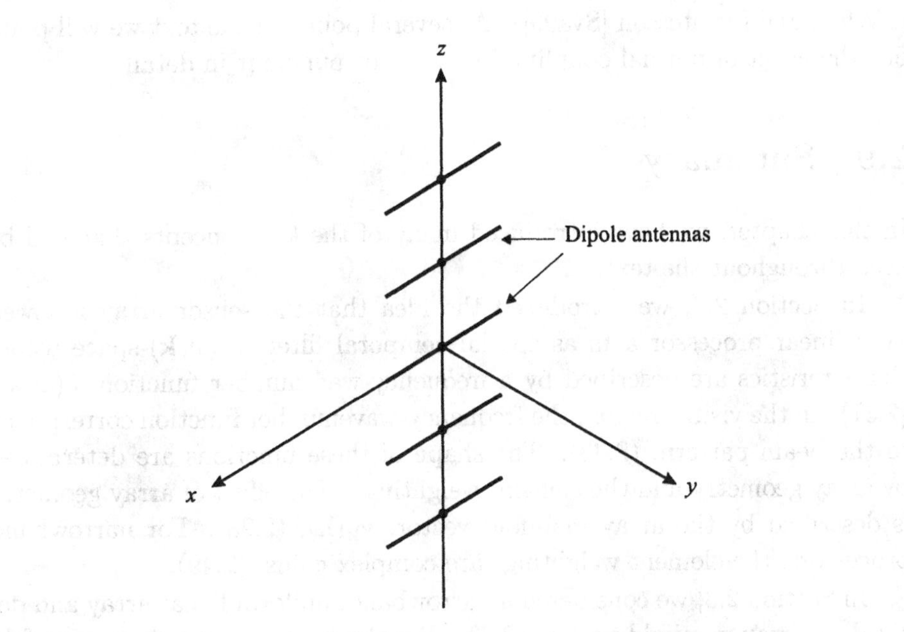

Figure 2.32 Array of dipoles.

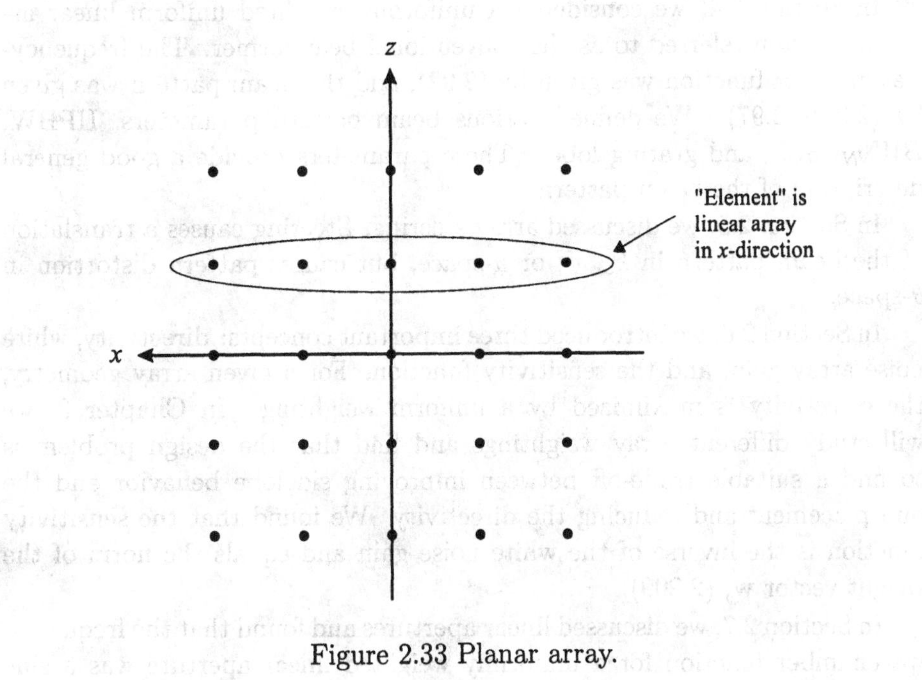

Figure 2.33 Planar array.

[FW91], and Svantesson [Sva99]). At several points in the text we will point out the effect of mutual coupling but will not pursue it in detail.

2.9 Summary

In this chapter, we have introduced many of the key concepts that will be used throughout the text.

In Section 2.2, we introduced the idea that the sensor array followed by a linear processor acts as spatial–temporal filter in $(\omega, \mathbf{k})$-space whose characteristics are described by a frequency-wavenumber function $\Upsilon(\omega, \mathbf{k})$, (2.37). In the visible region, the frequency-wavenumber function corresponds to the beam pattern, (2.38). The shape of these functions are determined by array geometry and the element weightings. The effect of array geometry is described by the array manifold vector, $\mathbf{v_k}(\mathbf{k})$, (2.28). For narrowband processors, the element weightings are complex gains, (2.49).

In Section 2.3, we considered a narrowband, uniform linear array and derived its array manifold vector, (2.73). We observed that the array manifold vector was conjugate symmetric, (2.74), (2.75). This symmetry will lead to computational simplifactions in several situations.

In Section 2.4, we considered a uniformly weighted uniform linear array that often referred to as the conventional beamformer. The frequency-wavenumber function was given by (2.92), and the beam pattern was given by (2.95)–(2.97). We defined various beam pattern parameters; HPBW, BW_{NN}, SLL, and grating lobes. These parameters provide a good general description of the beam pattern.

In Section 2.5, we discussed array steering. Steering causes a translation of the beam pattern in k_z, ψ, or u-space, but causes pattern distortion in θ-space.

In Section 2.6, we introduced three important concepts: directivity, white noise array gain, and the sensitivity function. For a given array geometry, the directivity is maximized by a uniform weighting. In Chapter 3, we will study different array weightings and find that the design problem is to find a suitable trade-off between improving sidelobe behavior and the null placement and reducing the directivity. We found that the sensitivity function is the inverse of the white noise gain and equals the norm of the weight vector $\mathbf{w}$, (2.209).

In Section 2.7, we discussed linear apertures and found that the frequency-wavenumber function for a uniformly weighted linear aperture was a sinc function, (2.218), (2.219). We introduced aperture sampling in order to approximate the aperture behavior with a uniform array.

In Section 2.8, we considered the case in which the sensor beam patterns are non-isotropic. We derived the pattern multiplication result.

This completes our discussion of uniformly weighted linear arrays. In Chapter 3 we discuss different weightings to improve the sidelobe performance of linear arrays.

2.10 Problems

The problems are divided corresponding to major sections in the chapter.

P2.3 Uniform Linear Arrays

Problem 2.3.1

Assume that N is even. Use the relationships

$$B_{\psi}(\psi) = 2Re\left[\mathbf{w}_1^H \, \mathbf{v}_{\psi_1}(\psi)\right] \tag{2.236}$$

from (2.77) to derive the beam pattern of a uniformly weighted uniform linear array.

Problem 2.3.2

When N is odd, we can partition the array manifold vector of a uniform linear array into three parts,

$$\mathbf{v}_{\psi}(\psi) = \begin{bmatrix} \mathbf{v}_{\psi_1}(\psi) \\ 1 \\ \mathbf{J}\,\mathbf{v}_{\psi_1}^*(\psi) \end{bmatrix}. \tag{2.237}$$

Repeat Problem 2.3.1 for this case.

Problem 2.3.3

Show that, if $\mathbf{w}$ is real and symmetric, then $B_{\psi}(\psi)$ for a uniform linear array is a real symmetric function. Consider both N even and N odd.

Problem 2.3.4

Assume that N is even. In some applications, we want to obtain an asymmetric beam pattern. Show that, if $\mathbf{v}_{\psi}(\psi)$ is conjugate symmetric and $\mathbf{w}$ is real asymmetric, that is,

$$\mathbf{w} = \begin{bmatrix} \mathbf{w}_1 \\ -\mathbf{J}\,\mathbf{w}_1^* \end{bmatrix}, \tag{2.238}$$

where $\mathbf{w}_1$ is real, then $B_{\psi}(\psi)$ will be an imaginary function.

Problem 2.3.5

Repeat Problem 2.3.4 for N odd.

P2.4 Uniformly Weighted Linear Arrays

Problem 2.4.1

(a) Construct a polar plot in dB of $B_\theta(\theta)$ for a standard 21-element linear array with uniform weighting for θ_T (the steering direction) equal to 0°, 15°, 30°, 60°, and 90°.

(b) Find the HPBW for each of these θ_T.

Problem 2.4.2

Assume N is even and

$$w_n = \begin{cases} -\frac{1}{N} & n = 0, 1, \cdots, \frac{N}{2} - 1, \\ \frac{1}{N} & n = \frac{N}{2}, \cdots, N - 1. \end{cases} \tag{2.239}$$

(a) Find the resulting beam pattern and plot the magnitude in dB and phase versus ψ for the case when $d = \lambda/2$.

(b) Find the slope of the beam pattern at $\psi = 0$.

Problem 2.4.3

Consider a standard 10-element linear array with uniform weighting. Assume that the nth sensor fails. Plot the resulting beam pattern for several values of n.

Problem 2.4.4

Assume that each of the 10 sensors in Problem 2.4.3 are equally likely to fail with probability, $P_n(F) = \frac{1}{10}(1 - \alpha)$, where α is a reliability parameter whose value is in the range [0, 1]. Assume that, at most, one failure occurs.

(a) Calculate the expected value of the beam pattern.

(b) Plot the result for $\alpha = 0$ and $\alpha = 0.9$.

(c) Does the expected value of the beam pattern provide a useful indication of the array behavior or do we have to consider the behavior on separate trials (as in Problem 2.4.3)?

Problem 2.4.5

Consider a standard 10-element linear array with uniform weighting. Assume that two sensors fail. Plot the resulting beam pattern for several values of n_1 and n_2.

Problem 2.4.6

Consider the non-uniform 4-element linear array whose sensor separations are d, $3d$, $2d$ where $d = \lambda/2$. The sensor outputs are weighted uniformly.

(a) Compute the beam pattern and BW_{NN}.

(b) Compare the results in (a) with a uniform 7-element array with spacing where $d = \lambda/2$. Discuss the behavior of the main lobe and the sidelobes.

Problem 2.4.7

In order to find the exact HPBW, we must solve

$$|B_u(u)|^2 = 0.5,$$

where $B_u(u)$ is given by (2.96).

(a) One approach to finding an approximate expression for the HPBW is to expand $|B_u(u)|^2$ in a second-order Taylor series around $u = 0$. Use the expression for $B_u(u)$ given in (2.126) to simplify the derivatives. Compare your result to the result in (2.98).

(b) Use (2.96) to find the exact result as a function of N and compare the result with (2.98).

Problem 2.4.8

The second central moment of an array weighting function is defined as

$$\sigma_A^2 = \frac{\sum_{m=-\frac{N-1}{2}}^{\frac{N-1}{2}} m^2 w_m}{\sum_{m=-\frac{N-1}{2}}^{\frac{N-1}{2}} w_m}, \quad N \text{ odd}, \tag{2.240}$$

where the origin is at the center of gravity,

$$\sum_{m=-\frac{N-1}{2}}^{\frac{N-1}{2}} m w_m = 0. \tag{2.241}$$

The second-moment beamwidth is defined as

$$\Delta u_4 = \frac{\lambda}{4\sigma_A d}. \tag{2.242}$$

(a) Find Δu_4 for a uniformly weighted linear array.

(b) Compare to the HPBW.

Problem 2.4.9

Consider an 8-element linear array with $d = 5\lambda/8$ and uniform weighting. Plot $B_u(u)$. Compare the resulting beam pattern to the beam pattern of a standard 10-element linear array.

P2.5 Array Steering

Problem 2.5.1

The conventional beam pattern for an array is given by

$$B_\psi(\psi : \psi_T) = \mathbf{v}^H(\psi_T)\,\mathbf{v}(\psi). \tag{2.243}$$

(a) Show that $B_\psi(\psi : \psi_T)$ is real if $\mathbf{v}(\psi)$ is conjugate symmetric.

(b) Give two examples of conjugate symmetric array manifolds in addition to the uniform linear array.

Problem 2.5.2

Verify the result in (2.141) and plot the percentage error versus N for various λ.

Problem 2.5.3

In an ordinary endfire array, the beam pattern is obtained by letting

$$\psi = \frac{2\pi d}{\lambda}\cos\theta - \psi_T, \tag{2.244}$$

where

$$\psi_T = \frac{2\pi d}{\lambda}. \tag{2.245}$$

If $d = \lambda/2$, there are two identical endfire lobes, as shown in Figure 2.26. One way to reduce the back lobe is to reduce the wavelength. The visible region is $2(2\pi d/\lambda)$ wide in ψ-space. The first null in the back lobe is at $2\pi/N$ in ψ-space. Therefore if we reduce the visible region from the $d = \lambda/2$ value of 2π by $2\pi/N$, the back lobe should be significantly decreased. Thus,

$$2(2\pi d/\lambda) \leq 2\pi - \frac{2\pi}{N} \tag{2.246}$$

or

$$d \leq \frac{\lambda}{2}(1 - \frac{1}{N}). \tag{2.247}$$

(a) Plot the beam pattern in u-space and θ-space for $N = 10$. In this case, $d = 0.45\lambda$ and $\psi_T = 0.9\pi$. Find the HPBW.

(b) Consider the case where only part of the back lobe is moved out of the visible region. Let

$$d = \frac{\lambda}{2}(1 - \frac{1}{2N}). \tag{2.248}$$

Plot the beam pattern in u-space and θ-space for $N = 10$. Find the HPBW.

Problem 2.5.4 (Hansen-Woodyard)

In order to make the main beam narrower, Hansen and Woodyard [HW38] proposed moving part of the main lobe out of the visible region by increasing the interelement phase shift ψ_T,

$$\psi_T = \left(\frac{2\pi d}{\lambda} + \frac{\pi}{N}\right). \tag{2.249}$$

However, the back lobe may move into the visible region unless d is decreased. To prevent the back lobe from becoming larger than the main lobe, we require $\psi_T < \pi$. Thus,

$$\frac{2\pi d}{\lambda} + \frac{\pi}{N} < \pi \tag{2.250}$$

or

$$d < \frac{\lambda}{2}(1 - \frac{1}{N}). \tag{2.251}$$

(a) Consider a 10-element linear array. Let $d = 0.45\lambda$. Plot the beam pattern in u-space and θ-space. Find the HPBW.

(b) Repeat for $d = 0.3\lambda$, $d = 0.35\lambda$, and $d = 0.4\lambda$.

(c) Show that the nulls occur at

$$\theta_0 = 2\sin^{-1}\left[\pm\sqrt{\frac{\lambda}{4Nd}(2m-1)}\right]. \tag{2.252}$$

(d) Show that, for long arrays,

$$\theta_0 \simeq \pm\sqrt{\frac{\lambda}{Nd}(2m-1)}. \tag{2.253}$$

and the first zero occurs at

$$\theta_{01} \simeq \pm\sqrt{\frac{\lambda}{Nd}}. \tag{2.254}$$

Thus the null-null beamwidth is 0.707 times the width of an ordinary endfire array.

Problem 2.5.5

(a) Consider a 10-element linear array pointed at endfire with $d = 3\lambda/8$. The progressive phase shift ψ_T is given by $\psi_T = 3\pi/4$. Plot $B_\theta(\theta)$.

(b) Repeat part (a) with $\psi_T = \pi\left[\frac{3}{4} + \frac{1}{N}\right]$.

P2.6 Array Performance Measures

Problem 2.6.1

Consider a uniform N-element linear array with isotropic elements pointed at broadside with uniform weighting.

(a) Plot the directivity versus d/λ over the range $0 \leq d/\lambda \leq 2.0$ for various N.

(b) An approximate expression for D at broadside is

$$D = 2\frac{Nd}{\lambda}, \quad 0 \leq d/\lambda \leq 1. \tag{2.255}$$

Superimpose this expression on the plot in part (a). Plot the error between the approximate expression and the exact value over the interval $0 \leq d/\lambda \leq 1$. Note that the expression is exact for $d = \lambda/2$ and that $D = N$.

Problem 2.6.2 (continuation)[12]

(a) Calculate the directivity of the array in Problem 2.4.6.

(b) Consider the general case pointed at broadside of a non-uniform linear array whose element locations are located on a grid whose points are separated by $d = \lambda/2$. Show that

$$D = \left(\sum_{n=0}^{N-1} |w_n|^2\right)^{-1}, \tag{2.256}$$

where

$$\sum_{n=0}^{N-1} w_n = 1. \tag{2.257}$$

Therefore $D = N$ when the weighting is uniform.

Problem 2.6.3

Consider a standard 11-element linear array with triangular weighting,

$$w_n = c\left(1 - \frac{2\left|n - \frac{N-1}{2}\right|}{N}\right), \quad n = 0, 1, \cdots, N-1, \tag{2.258}$$

[12] Continuation means that the problem assumes that either the previous problem or a specifically referenced problem has been read (or, in some cases, solved).

where

$$c = \left(\sum_{n=0}^{N-1} w_n \right)^{-1}. \tag{2.259}$$

(a) Compute the directivity when the array is steered to broadside.

(b) Generalize to an N-element array (N odd).

Problem 2.6.4

Consider a standard 5-element linear array pointed at broadside. We want to compare the following unnormalized weightings:

(a) 1, 1, 1, 1, 1

(b) 1, 2, 3, 2, 1

(c) 1, 4, 6, 4, 1

(d) 1, 1.61, 1.94, 1.61, 1

(e) 1, 2.41, 3.14, 2.41, 1

Normalize the weightings. Plot the beam patterns in u-space and θ-space on separate plots. Compute D; the HPBW in u-space and θ-space; the BW_{NN} in u-space and θ-space; and the height of the first sidelobe in dB. Discuss your results.

Problem 2.6.5

Repeat Problem 2.6.4 for an inverse triangular weighting,

3, 2, 1, 2, 3.

Discuss your result.

Problem 2.6.6

In order to find the directivity of a uniformly weighted linear array that is pointed at ψ_T, it is convenient to rewrite the beam pattern.

(a) Show that

$$B_\psi(\psi) = \frac{1}{N} \left\{ 1 + 2 \sum_{m=1}^{\frac{N-1}{2}} \cos m\psi \right\}, \quad N \text{ odd}, \tag{2.260}$$

and

$$B_\psi(\psi) = \frac{1}{N} \left\{ 2 \sum_{m=0}^{\frac{N}{2}-1} \cos(m - \frac{N-1}{2})\psi \right\}, \quad N \text{ even}, \tag{2.261}$$

where

$$\psi = \frac{2\pi d}{\lambda} \cos\theta - \psi_T = \frac{2\pi d}{\lambda} \cos\theta - \frac{2\pi d}{\lambda} \cos\theta_T. \tag{2.262}$$

(b) We then write

$$\left| \frac{\sin \frac{N\psi}{2}}{N \sin \frac{\psi}{2}} \right|^2 = \frac{1}{N} + \frac{2}{N^2} \sum_{m=1}^{N-1} (N-m) \cos m\psi. \tag{2.263}$$

Verify this expression for N=2, 3, 4, and 5.

(c) Show that

$$D=\left\{\frac{1}{N}+\frac{2}{N^2}\sum_{m=1}^{N-1}\frac{(N-m)}{mk_0d}\sin mk_0d\cos m\psi_T\right\}^{-1}, \tag{2.264}$$

where ψ_T is the progressive phase factor. Thus,

$$\psi = k_0 d\cos\theta-\psi_T=\frac{2\pi d}{\lambda}\cos\theta-\psi_T. \tag{2.265}$$

The above result is for a set of N isotropic sources with an element spacing of d and an interelement phase shift ψ_T (e.g., p.142 of [ST81]).

(d) Plot the directivity versus θ_T (in degrees) for a 5-element linear array. Consider $d = 0.3\lambda$, 0.4λ, 0.5λ, and 0.6λ.

(e) Repeat part d for $N = 10$.

Problem 2.6.7

Show that the directivity of an ordinary endfire array ($\psi_T = 2\pi d/\lambda$) is approximately

$$D \simeq 4Nd/\lambda. \tag{2.266}$$

Problem 2.6.8

Show that the directivity of an Hansen-Woodyard endfire array is approximately

$$D \simeq 7.28\, Nd/\lambda. \tag{2.267}$$

Problem 2.6.9

Consider a uniform 10-element linear array with uniform weighting pointed at endfire. Plot the directivity for the ordinary endfire array and the Hansen-Woodyard endfire array versus d/λ for the range $0.1 \leq d/\lambda \leq 0.6$.

When $D > N$, we refer to the array as superdirective. This problem shows a case of a practical superdirective array.

Problem 2.6.10

Consider the case of linear array whose element positions along the z-axis are z_n. The element phasings are linear with distance. Denote the phase of w_n as α_n. Then

$$\alpha_n = -\frac{2\pi}{\lambda} z_n\cos\theta_T. \tag{2.268}$$

(a) Show that the beam pattern can be written as

$$B_\theta(\theta)=\frac{\sum_{n=0}^{N-1}|w_n|\exp\left(j(\frac{2\pi}{\lambda}z_n\cos\theta+\alpha_n)\right)}{\sum_{n=0}^{N-1}|w_n|}. \tag{2.269}$$

(b) Assume the weights are normalized so that

$$\sum_{n=0}^{N-1}|w_n| = 1, \tag{2.270}$$

then

$$\Omega_A \triangleq 2\pi \int_0^{\pi} |B_\theta(\theta)|^2 \sin\theta \, d\theta. \tag{2.271}$$

Use the result in part (a) to show that

$$\Omega_A = 2\pi \sum_{n=0}^{N-1} \sum_{m=0}^{N-1} |w_m||w_n| e^{j(\alpha_n - \alpha_m)} \frac{\sin\left[\frac{2\pi}{\lambda}(z_n - z_m)\right]}{\frac{2\pi}{\lambda}(z_n - z_m)} \tag{2.272}$$

and

$$D = \frac{4\pi}{\Omega_A} = \left\{ \sum_{n=0}^{N-1} \sum_{m=0}^{N-1} |w_m||w_n| e^{j(\alpha_n - \alpha_m)} \frac{\sin\left[\frac{2\pi}{\lambda}(z_n - z_m)\right]}{\frac{2\pi}{\lambda}(z_n - z_m)} \right\}^{-1}. \tag{2.273}$$

Problem 2.6.11

Consider the special case of the perturbation model in Section 2.6.3 in which only the locations of the array elements are perturbed. Thus, (2.192) and the subsequent model applies. Assume that we have a standard N-element linear array along the z-axis and the only perturbations are in the z-direction.

(a) Find the expected value of the beam pattern as a function of w_i^n and σ_λ^2.

(b) Plot the result for a 10-element array and uniform weighting for various σ_λ^2.

Problem 2.6.12

Repeat Problem 2.6.11 for the case in which the only position perturbations are in the y-direction.

In part (b), plot the expected value of the beam pattern versus $u_z = \cos\theta$ for several values of ϕ.

Problem 2.6.13 (continuation Problem 2.6.4)

Calculate the sensitivity function for the five weightings in Problem 2.6.4.

Problem 2.6.14 (continuation)

(a) Repeat Problem 2.6.11 for the case of phase-only errors. Therefore, (2.194) applies.

(b) Compare your results with the results in Problem 2.6.11. Give an intuitive explanation of the comparison.

Problem 2.6.15

Consider a standard linear array designed for frequency f_c. We want to analyze the behavior for mismatched frequency. Assume the frequency of the incoming plane wave is f, where

$$f = \alpha f_c. \tag{2.274}$$

(a) Plot the broadside beam pattern for α =0.80, 0.90, 1.10, 1.20.

(b) Plot the directivity versus α over the range $0.5 < \alpha < 2.0$ for various scan directions: $\bar{\theta}_T = 0°, 15°, 30°, 45°$.

P2.7 Linear Apertures

Problem 2.7.1

Consider a linear aperture with $L = 5\lambda$. Assume that the weighting function is triangular.

(a) Find an expression for the beam pattern and plot it.

(b) How is the beam pattern in part (a) related to the beam pattern for uniform weighting?

(c) Compare the result in part (a) to an equivalent linear array with $d = \lambda/4$ and $d = \lambda/2$.

Problem 2.7.2

The second central moment of the aperture weighting is defined as

$$\sigma_w^2 = \frac{\int_L z^2 w(z)dz}{\int_L w(z)dz}, \tag{2.275}$$

where the origin is at the center of gravity. The second-moment beamwidth is defined as

$$\Delta u_4 = \frac{\lambda}{4\sigma_w}. \tag{2.276}$$

Find Δu_4 for a rectangular weighting.

P2.8 Non-isotropic Element Patterns

Problem 2.8.1

The beam pattern for a short dipole ($L < \lambda$) aligned with the z-axis is

$$B_{DP}(\theta) = \sin\theta. \tag{2.277}$$

(a) Find the beam pattern for the array in Figure 2.31.

(b) Plot your result for $N = 10$.

Problem 2.8.2

The beam pattern for a short dipole ($L < \lambda$) aligned with the x-axis is

$$B_{DP}(\theta, \phi) = \frac{\cos\left[\left(\frac{\pi}{2}\right)\sin\theta\cos\phi\right]}{\sqrt{1-\sin^2\theta\cos^2\phi}} \tag{2.278}$$

(e.g., pp.138–139 of [ST81]).

(a) Find the beam pattern for standard linear array along the z-axis with uniform weighting.

(b) Plot the beam pattern in the xz-plane and the yz-plane.

Problem 2.8.3 [ST81]

The directivity expression for a linear array with uniform weighting and non-isotropic elements is

$$D = \frac{1}{\frac{a_0}{N} + \frac{2}{N^2}\sum_{m=1}^{N-1}\frac{N-m}{mk_0d}(a_1 \sin mk_0d + a_2 \cos mk_0d)\cos m\psi_T} \tag{2.279}$$

where a_0, a_1, and a_2 are given in the table for various element patterns:

Element	$\|B_e(\theta,\phi)\|^2$	a_0	a_1	a_2
Isotropic	1	1	1	0
Collinear short dipoles	$\sin^2\theta$	$\frac{2}{3}$	$\frac{2}{(mk_0d)^2}$	$\frac{-2}{mk_0d}$
Parallel to x-axis short dipoles	$1-\sin^2\theta\cos^2\phi$	$\frac{2}{3}$	$1-\frac{1}{(mk_0d)^2}$	$\frac{1}{mk_0d}$

(a) Calculate the directivity for a 10-element linear array pointed at broadside with collinear short dipoles.

(b) Repeat for parallel short dipoles.

Problem 2.8.4

Consider the planar array in Figure 2.32 and assume the elements are isotropic. Assume that $N_x = 10$, $N_z = 10$, $d_x = \lambda/2$. Find the beam pattern $B_{\theta,\phi}(\theta,\phi)$ when the array is pointed at broadside.

Plot the beam pattern versus $\cos\theta$ for several values of ϕ.

Problem 2.8.5

Consider a uniformly spaced planar array in the xy-plane with isotropic elements.

(a) Find an expression for the beam pattern $B_{\theta,\phi}(\theta,\phi)$ when the array in pointed at broadside.

(b) Plot the beam pattern for $N_x = N_y = 10$ and $d_x = d_y = \lambda/2$. Plot $B_{\theta,\phi}(\theta,\phi)$ versus $\cos\theta$ for various values of ϕ.

(c) Repeat part (b) for the case in which the elements are short dipoles parallel to the z-axis.

Bibliography

[Bal82] C. A. Balanis. *Antenna Theory: Analysis and Design.* Wiley, New York, 1982.

[Ell81] R. S. Elliott. *Antenna Theory and Design.* Prentice-Hall, Englewood Cliffs, New Jersey, 1981.

[FW91] B. Friedlander and A. J. Weiss. Direction finding in the presence of mutual coupling. *IEEE Trans. Antennas Propag.*, vol.AP-39, pp. 273–284, March 1991.

[GM55] E.N. Gilbert and S.P. Morgan. Optimum design of directive antenna arrays subject to random variations. *Bell Syst. Tech. J.*, vol.34, pp. 637–663, May 1955.

[HW38] W.W. Hansen and J.R. Woodyard. A new principle in directional antenna design. *Proc. IRE*, vol.26, pp. 333–345, March 1938.

[Joh93] R. C. Johnson. *Antenna Engineering Handbook.* McGraw-Hill, New York, 3rd edition, 1993.

[Kra88] J. D. Kraus. *Antennas.* McGraw-Hill, New York, 2nd edition, 1988.

[Mil85] T. A. Milligan. *Modern Antenna Design.* McGraw-Hill, New York, 1985.

[OS89] A. V. Oppenheim and R. W. Schafer. *Discrete-Time Signal Processing.* Prentice-Hall, Englewood Cliffs, New Jersey, 1989.

[Sko80] M. I. Skolnik. *Introduction to Radar Systems.* McGraw-Hill, New York, 1980.

[ST81] W. L. Stutzman and G. A. Thiele. *Antenna Theory and Design.* Wiley, New York, 1981.

[Sva99] T. Svantesson. Modeling and estimation of mutual coupling in a uniform linear array of dipoles. *Proc. ICASSP*, vol.5, pp. 2961–2964, March 1999.

[VT71] H. L. Van Trees. *Detection, Estimation, and Modulation Theory, Part III.* Wiley, New York, 1971.

[VT01b] H. L. Van Trees. *Detection, Estimation, and Modulation Theory, Part III.* Wiley Interscience, New York, 2001.

[Wee68] W. L. Weeks. *Antenna Engineering.* McGraw-Hill, New York, 1968.

[YLU89] C. Yeh, M. Leou, and D. R. Ucci. Bearing estimations with mutual coupling present. *IEEE Trans. Antennas Propag.*, vol.AP-37, pp. 1332–1335, October 1989.

[Zio95] L. J. Ziomek. *Fundamentals of Acoustic Field Theory and Space-Time Signal Processing.* CRC Press, Boca Raton, Florida, 1995.

Chapter 3

Synthesis of Linear Arrays and Apertures

In this chapter we develop techniques for choosing the weighting of each sensor output in order to obtain a frequency-wavenumber response and beam pattern with desirable properties. The weighting process is also referred to as shading or tapering in the literature. In this chapter we restrict our attention to linear arrays. In Chapter 4, we consider planar array configurations.

The frequency-wavenumber response $\Upsilon(\omega, \mathbf{k})$ is the Fourier transform of the weights w_n, so there is a large body of mathematical results available. For linear arrays with equal spacing, $\Upsilon(\omega : \mathbf{k})$ has exactly the form of the discrete Fourier transform (DFT) so the techniques from equivalent temporal problems such as finite impulse response (FIR) filters and spectral estimation windows can be used directly. For planar arrays with equal spacing, the corresponding 2-D techniques can be used. For linear arrays whose sensors are located at arbitrary points on a line, the design problem is more difficult because we are no longer sampling on a uniform lattice.

Although there is a mathematical duality with the time domain problem, there are important differences between array processing and time domain processing that shape our conclusions:

(i) The spatial dimension of the array normally has an absolute constraint due to the structure supporting it (e.g., a radio tower, a mast on a ship, the fuselage of an airplane, a satellite bus, a towed array). Even if it is possible to extend the array, it will be much more expensive than obtaining more time samples.

(ii) The cost per sensor is significant in many cases because of the sensor itself and associated electronics. Therefore, even if space is available, we choose to increase the processing complexity in order to reduce the number of sensors. There is a large incentive to optimize processing performance.

(iii) In some cases, it is difficult to maintain amplitude and phase calibration of the sensors for a number of reasons (e.g., mutual coupling, environmental changes) and overall array calibration due to changes in sensor location. In Section 2.6, we saw how these changes placed a limit on the null depth in the beam pattern. We find many other cases in which calibration errors are the limiting factor in array performance. In contrast, time samples are very uniform in most applications.

(iv) In the time domain, for a given signal and noise model, one can vary the number of samples to change performance. However, in the array problem, we have two dimensions: N, the number of sensors, and K, the number of samples that can be varied to changed performance.

Therefore, although duality is an important factor that we will exploit, the array processing problem must be studied in its own context.

In this chapter, we examine a number of different techniques for design of linear, equally spaced arrays. Linear arrays are the traditional focus of texts on classical array processing. The design of weighting for linear arrays reveal many of the important concepts in array processing. Many of the ideas extend to more general geometries, although the mathematics becomes more involved (in some cases, the mathematics does not extend to higher dimensions).

Recall from our discussion in Chapter 1 that our ultimate goal is to design array processors that adapt their configuration to match the incoming data and are optimized in a statistical sense. However, it is important to have a thorough understanding of classical (or deterministic) beamformer design for several reasons:

(i) The classical array design provides a basis for comparison for any proposed adaptive design. If we derive some "optimum array processor," we should show its improvement over the classical array processor.

(ii) In many cases, we will find that the "optimum array processor" has one of the beamformers that we design using deterministic techniques as a basic building block in its implementation.

(iii) In some cases, understanding the deterministic design points out areas where statistical techniques may be useful.

We consider several approaches to the selection of weighting functions for linear apertures and linear arrays:

(i) Spectral Weightings (3.1)

This approach exploits the Fourier transform relationships between the frequency-wavenumber response function and the weighting function and the parallelism with windows and tapers used in the spectral analysis of time series.

(ii) Array Polynomials and the z-Transform (3.2)

This approach, which originated with Schelkunoff [Sch43], develops a polynomial representation of a linear array and leads to a z-transform relationship. We can then analyze and synthesize patterns by positioning the zeros of the array polynomial.

(iii) Pattern Sampling in Wavenumber Space (3.3)

This approach specifies the desired values of the pattern on a grid in wavenumber space. For apertures, it utilizes the sampling theorem. For arrays, it leads to a DFT relationship.

(iv) Minimum Beamwidth for Specified Sidelobe Level (3.4)

This approach attempts to find an array weighting function that minimizes the beamwidth for a given maximum sidelobe level. It leads us to the Dolph-Chebychev and Taylor weightings, which are widely used in practice.

(v) Least Squares Error Pattern Synthesis (3.5)

This approach specifies a desired pattern in frequency-wavenumber space and attempts to find a weighting function to achieve it. The approach uses the Fourier transform for apertures or the Fourier series for arrays to obtain a minimum mean-square error approximation to the desired pattern.

(vi) Minimax Design (3.6)

This approach utilizes a technique that was developed to design finite impulse response (FIR) filters. It specifies a maximum allowable variation in the height of the main lobe (e.g., $1 - \delta_p \leq B_\psi(\psi) \leq 1 + \delta_p$ and a maximum allowable height in the sidelobe region, δ_s, and finds a solution to meet these criteria.

(vii) Null Steering (3.7)

This approach assumes that there are certain points in wavenumber space where there are interfering signals (e.g., jammers). We design weightings so that the frequency-wavenumber response is zero in these directions.

(viii) Asymmetric Beams (3.8)

All of the beams discussed up to this point assume that the desired target direction is known. If we are required to estimate the target direction, then beam patterns with different characteristics, specifically a significant non-zero slope in the pointing direction, are useful. We develop this type of beam in Section 3.8 and discuss its properties.

(ix) Spatially Non-uniform Linear Arrays (3.9)

In this section, we discuss linear arrays with non-uniform element spacing. We develop several synthesis techniques. We also introduce the idea of minimally redundant linear arrays.

(x) Beamspace Processing (3.10)

In later chapters of the text, we find that, in many applications, it is useful to preprocess the array data to form a set of beams that span the space of interest and then do further processing in these output beams. We introduce this idea in Section 3.10 and develop it in detail in later chapters.

(xi) Broadband Arrays(3.11)

In this section, we develop linear array spacings that are useful when the signals of interest are broadband.

The structure of the chapter is shown in Table 3.1.

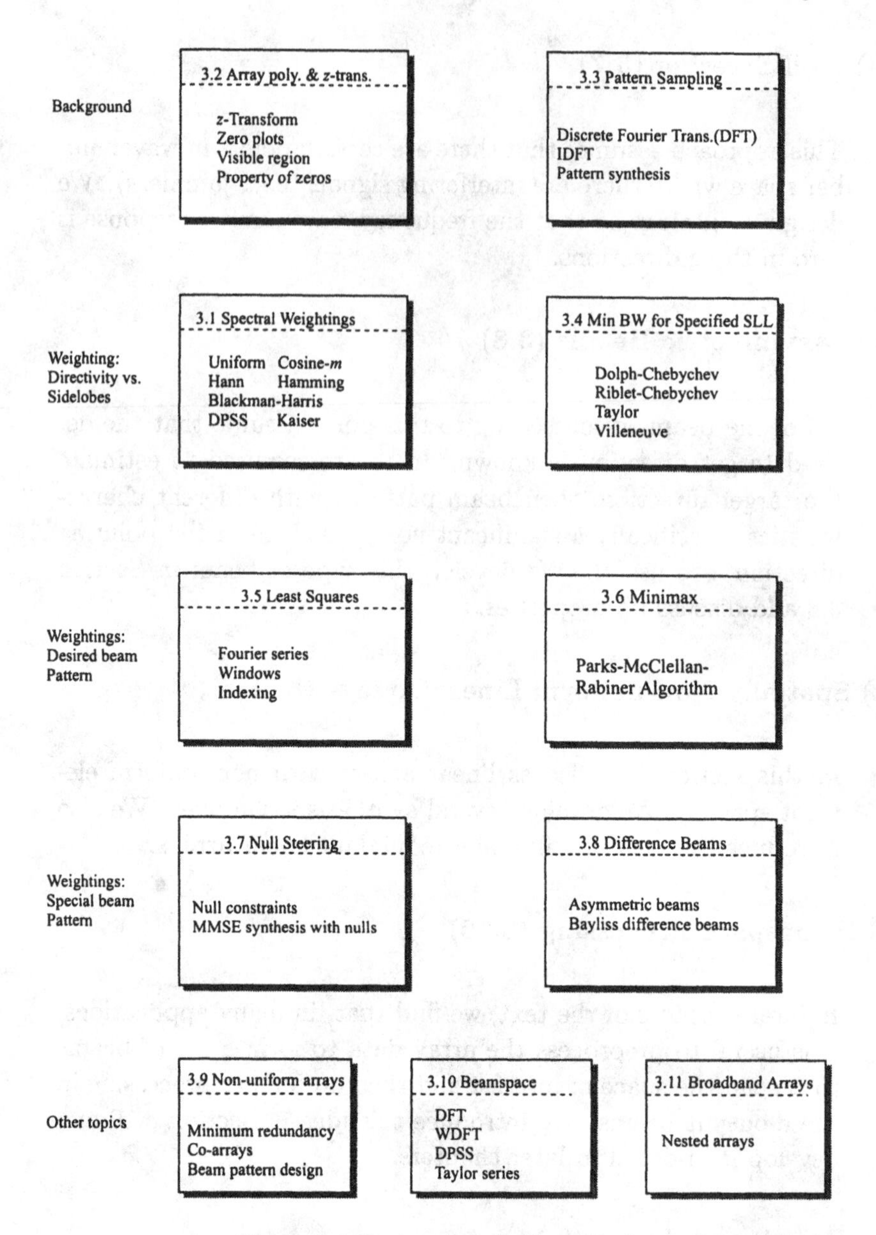

Table 3.1 Structure of Chapter 3.

The techniques are drawn from the classical antenna literature and the digital filter design literature. Representative texts in the antenna area are Elliot [Ell81], Kraus [Kra88], Balanis [Bal82], Milligan [Mil85], Johnson [Joh93], Steinberg [Ste76], Ziomek [Zio95], Ma [Ma74], Mailloux [Mai94], Weeks [Wee68], Stutzman and Thiele [ST81], and the four volume handbook edited by Lo and Lee [LL93a], [LL93b], [LL93c], [LL93d].

Representative texts in the digital filter area are Oppenheim and Schaefer [OS89], Proakis et al. [PRLN92], Rabiner and Gold [RG75], and Mitra and Kaiser [MK93]. We use examples from these references and other sources.

3.1 Spectral Weighting

There are several classes of array weightings that are equivalent to various windows or tapers used in the spectral analysis of time series. Their design depends explicitly upon the Fourier transform relationship for the weighting and the frequency-wavenumber response for a linear array with a sensor spacing less than or equal to $\lambda/2$; consequently, grating lobes and spatial aliasing become unimportant and Fourier transform theory for continuous functions can be used directly.

The starting point is the Fourier transform pair in (2.215) and (2.216),

$$\Upsilon(\omega, k) = \int_{-\infty}^{\infty} w_a^*(z)\, e^{-jkz}\, dz \tag{3.1}$$

and

$$w_a^*(z) = \frac{1}{2\pi}\int_{-\infty}^{\infty} \Upsilon(\omega, k) e^{jkz}\, dk, \tag{3.2}$$

where we have suppressed the z subscript on k_z.

As a result of this transform relationship there is a very large body of literature for weight design, or pattern synthesis, for arrays of this construction. Some of the more important considerations are the following:

- The Fourier uncertainty principle specifies a lower limit on the product of the mean-square aperture extent and the mean-square response width.[1] Specifically, we have

$$\sqrt{\overline{\Delta l^2}}\,\sqrt{\overline{\Delta k^2}} > 1/2, \tag{3.3}$$

where $\overline{\Delta l^2}$ and $\overline{\Delta k^2}$ are respectively the normalized mean-square widths of the weighting and the response and are given by

$$\overline{\Delta l^2} = \frac{\int_{-L/2}^{L/2} z^2 |w_a(z)|^2\, dz}{\int_{-L/2}^{L/2} |w_a(z)|^2\, dz}, \tag{3.4}$$

[1] The proof of the Fourier uncertainty principle requires zero-mean position of the weighting and of the response. The discussion refers to distributions about these means.

$$\overline{\Delta k^2} = \frac{\int_{-\infty}^{\infty} k^2 |\Upsilon(\omega, k)|^2 \, dk/2\pi}{\int_{-\infty}^{\infty} |\Upsilon(\omega, k)|^2 \, dk/2\pi}. \tag{3.5}$$

This uncertainty principle implies that a narrow, or a high-resolution, frequency-wavenumber response (e.g., small $\overline{\Delta k^2}$) requires a large mean-square weighting extent $\overline{\Delta l^2}$ that is bounded by $L^2/4$ for an array of finite extent. Consequently, there is a fundamental trade-off between the resolution of the response function and the extent of the aperture.

- Parseval's (or Plancherel's) theorem states

$$\int_{-\infty}^{\infty} |\Upsilon(\omega, k)|^2 \frac{dk}{2\pi} = \int_{-\infty}^{\infty} |w_a(z)|^2 dz. \tag{3.6}$$

This implies that high amplitude response functions such as those often generated for null placements or for sidelobe reduction lead to large values for the sum of the magnitude squared of the weighting. We have seen earlier that this decreases the white noise gain, A_w, and increases the sensitivity function, $T(\omega)$.

- The sidelobes decay asymptotically according to the order of the discontinuity in the aperture weighting. Uniform weighting leads to sidelobes that decay as $O(\frac{1}{n})$. Smooth weighting patterns lead to fast sidelobe decay. It is important to note that this is an asymptotic result and is not a statement about the maximum sidelobe level; there are several very useful weightings, such as Hamming and Taylor, that have step or derivative discontinuities. These weightings achieve low maximum sidelobes by using these discontinuities to cancel the high sidelobes near the main beam.

The value of this work extends beyond linear arrays. First, there are a number of important arrays that consist of sets of linear arrays; second, weightings for planar arrays are often pursued in terms of a product of linear array weightings in each dimension; finally, the design principles are often applicable to higher dimensions.

Our development in this section will focus on linear arrays. The analogous results for linear apertures follow directly. In Section 3.1, the problem of interest is to reduce the sidelobes while minimizing the increase in the main-lobe width (and the loss in directivity).

Our approach in this section is heuristic. We will try different weight vectors and analyze their performance. In Section 3.4, we will develop an analytic technique. We use the uniform weighting as a reference. We derived its beam pattern in Section 2.4 and repeat it for reference purposes.

3.1.1.1 Uniform weighting

From (2.77), the uniform weights are

$$w_n = \frac{1}{N}, \quad n = 0, \cdots, N-1. \tag{3.7}$$

The resulting beam pattern in u-space is

$$B_u(u) = \frac{1}{N}\frac{\sin(\frac{\pi N d}{\lambda}u)}{\sin(\frac{\pi d}{\lambda}u)}. \tag{3.8}$$

We focus our attention on a standard linear array, so (3.8) reduces to,

$$B_u(u) = \frac{1}{N}\frac{\sin(\frac{N\pi u}{2})}{\sin(\frac{\pi u}{2})}. \tag{3.9}$$

From (2.140), the directivity is

$$D = N. \tag{3.10}$$

All of the weightings that we are going to consider in this section are real and symmetric, so it is convenient to use the position of the nth element as the index,

$$\tilde{n} = n - \frac{N-1}{2}, \qquad n = 0, 1, \cdots, N-1,$$
$$\tilde{n} = -\frac{N-1}{2}, \cdots, \frac{N-1}{2}. \tag{3.11}$$

We first consider various weightings constructed from cosine functions.

3.1.1.2 Cosine weightings

We consider the case when N is odd. The cosine weighting is

$$w(\tilde{n}) = \sin(\frac{\pi}{2N})\cos(\pi\frac{\tilde{n}}{N}), \quad -\frac{N-1}{2} \le \tilde{n} \le \frac{N-1}{2}, \tag{3.12}$$

where the $\sin(\frac{\pi}{2N})$ term is a constant such that $B_u(0) = 1$. Writing the cosine in exponential form gives

$$w(\tilde{n}) = \sin(\frac{\pi}{2N})\left[\frac{e^{j\frac{\pi\tilde{n}}{N}} + e^{-j\frac{\pi\tilde{n}}{N}}}{2}\right], -\frac{N-1}{2} \le \tilde{n} \le \frac{N-1}{2}. \tag{3.13}$$

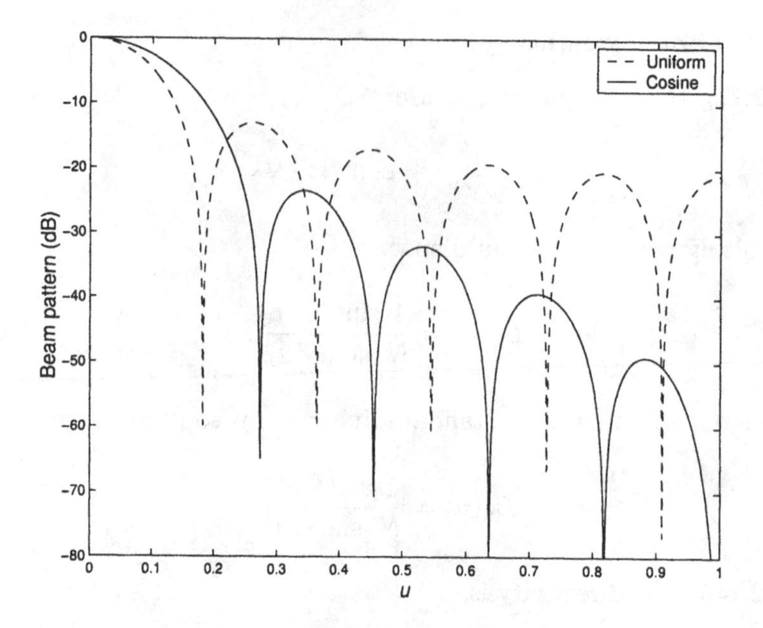

Figure 3.1 Beam pattern for uniform and cosine weighting: N=11.

Using the form of the array manifold vector in (2.70), the beam pattern is

$$B_u(u) = \frac{1}{2}\sin(\frac{\pi}{2N})\left\{\sum_{\tilde{n}=-\frac{N-1}{2}}^{\frac{N-1}{2}} e^{j\frac{\pi\tilde{n}}{N}}e^{-j\tilde{n}\pi u} + \sum_{\tilde{n}=-\frac{N-1}{2}}^{\frac{N-1}{2}} e^{-j\frac{\pi\tilde{n}}{N}}e^{-j\tilde{n}\pi u}\right\}. \quad (3.14)$$

The first term corresponds to a conventional beam pattern steered to $u_s = 1/N$, and the second term corresponds to a conventional beam pattern steered to $u_s = -1/N$. Therefore,

$$B_u(u) = \frac{1}{2}\sin(\frac{\pi}{2N})\left\{\frac{\sin\left(\frac{N\pi}{2}(u-\frac{1}{N})\right)}{\sin\left(\frac{\pi}{2}(u-\frac{1}{N})\right)} + \frac{\sin\left(\frac{N\pi}{2}(u+\frac{1}{N})\right)}{\sin\left(\frac{\pi}{2}(u+\frac{1}{N})\right)}\right\}. \quad (3.15)$$

We will find this superposition of shifted conventional beams to be a common characteristic of many of the patterns that we develop. We show the beam pattern for an 11-element array using the cosine weighting in Figure 3.1. We show the conventional beam pattern as a reference. The sidelobes have been reduced but the main lobe is wider. The parameters for the two beam patterns are:[2]

[2]The parameter D_N is the normalized directivity of the array. It is normalized with respect to the directivity of a uniformly weighted array. For the standard linear array it is also the normalized white noise gain.

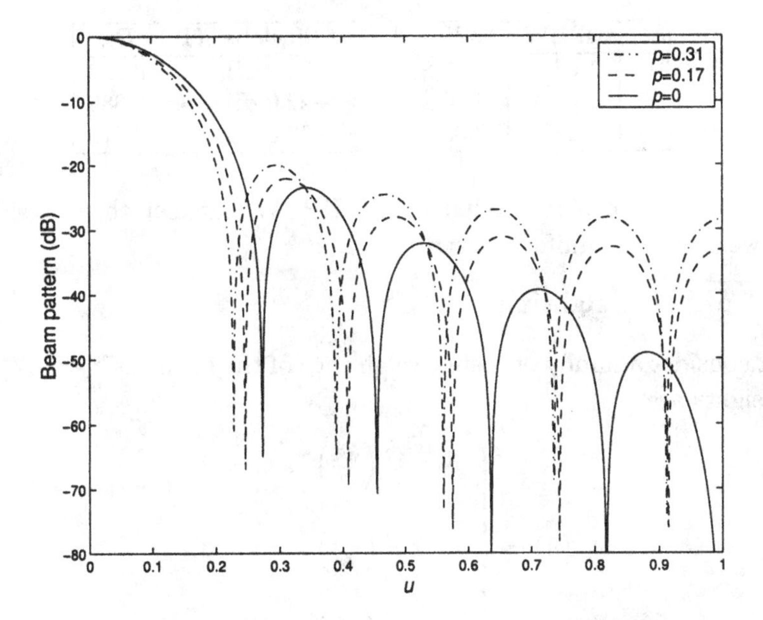

Figure 3.2 Beam pattern for raised cosine weighting: N=11.

Weighting	HPBW	BW_{NN}	First Sidelobe HT	D_N
Uniform	$0.89\frac{2}{N}$	$2.0\frac{2}{N}$	−13.0 dB	1
Cosine	$1.18\frac{2}{N}$	$3.0\frac{2}{N}$	−23.5 dB	0.816

3.1.1.3 Raised cosine

We can combine the rectangular uniform and the cosine weighting to obtain some of desireable features of each weighting. The corresponding array weighting is

$$w(\tilde{n}) = c(p)\left(p + (1-p)\cos\left(\pi\frac{\tilde{n}}{N}\right)\right), \quad \tilde{n} = -\frac{N-1}{2}, \cdots, \frac{N-1}{2}, \tag{3.16}$$

where

$$c(p) = \frac{p}{N} + \frac{(1-p)}{2}\sin\left(\frac{\pi}{2N}\right) \tag{3.17}$$

is a constant so that $B_u(0) = 1$.

The beam patterns for the raised cosine weighting for $p = 0.31, 0.17$, and 0 are shown in Figure 3.2. As p decreases, the height of the first sidelobe decreases and the width of the mainlobe increases. The beam patterns have the following parameters:

p	HPBW	BW_{NN}	First Sidelobe HT	D_N
0.31	$1.03\frac{2}{N}$	$2.50\frac{2}{N}$	−20.0 dB	0.928
0.17	$1.09\frac{2}{N}$	$2.70\frac{2}{N}$	−22.0 dB	0.886
0	$1.18\frac{2}{N}$	$3.00\frac{2}{N}$	−23.5 dB	0.816

Thus, we have been able to narrow the HPBW and keep the first sidelobe much lower than the uniform distribution.

3.1.1.4 Cosine^m weighting

We next consider a family of cosine weightings of the form $\cos^m(\pi\tilde{n}/N)$. The array weights are

$$w_m(\tilde{n}) = \begin{cases} c_2 \cos^2\left(\frac{\pi\tilde{n}}{N}\right), & m=2, \\ c_3 \cos^3\left(\frac{\pi\tilde{n}}{N}\right), & m=3, \\ c_4 \cos^4\left(\frac{\pi\tilde{n}}{N}\right), & m=4, \end{cases} \tag{3.18}$$

where c_2, c_3, and c_4 are normalization constants.

The weighting for $m = 2$ is sometimes called the Hann weighting.[3] Once again, the beam pattern is computed using the exponential form of cosine function.

The beam patterns are shown in Figure 3.3. As m increases, the sidelobes decrease but the main lobe widens. The parameters for the beam patterns are:

m	HPBW	BW_{NN}	First Sidelobe HT.	D_N
2	$1.44\frac{2}{N}$	$4\frac{2}{N}$	−31.4 dB	0.667
3	$1.66\frac{2}{N}$	$5\frac{2}{N}$	−39.4 dB	0.576
4	$1.85\frac{2}{N}$	$6\frac{2}{N}$	−46.7 dB	0.514

3.1.1.5 Raised cosine-squared weighting

The raised cosine-squared family of weightings is given by

$$\begin{aligned} w(\tilde{n}) &= c_2(p)\left\{p + (1-p)\cos^2\left(\frac{\pi\tilde{n}}{N}\right)\right\} \\ &= \frac{c_2(p)}{2}\left\{(1+p) + (1-p)\cos\left(\frac{2\pi\tilde{n}}{N}\right)\right\}, \end{aligned}$$

[3]The weighting is due to an Austrian meteorologist, von Hann. It is sometimes referred to as the Hanning weighting. We call it the Hann weighting.

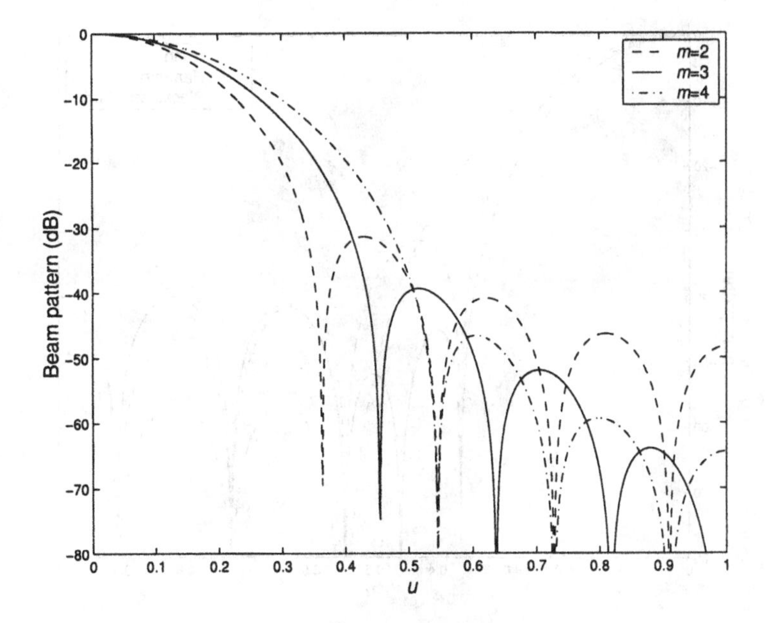

Figure 3.3 Beam patterns: cosine^m weighting, N=11.

where $c_2(p)$ is the normalization constant. We consider the general case in the problems. In the text, we consider a specific weighting known as the Hamming weighting.

3.1.1.6 Hamming weighting

The Hamming weighting exploits the characteristics of the rectangular pattern and the cosine-squared pattern to place a null at the peak of the first sidelobe. The weighting function is

$$w(\tilde{n}) = g_0 + g_1 \cos\left(\frac{2\pi\tilde{n}}{N}\right), \quad \tilde{n} = -\frac{N-1}{2}, \cdots, \frac{N-1}{2}. \tag{3.20}$$

The coefficients g_0 and g_1 are chosen to place a null at $u = 3/N$ and normalize the response at broadside to unity. The result is

$$w(\tilde{n}) = 0.54 + 0.46 \cos\left(\frac{2\pi\tilde{n}}{N}\right), \quad \tilde{n} = -\frac{N-1}{2} \leq n \leq \frac{N-1}{2}. \tag{3.21}$$

This corresponds to $p = 0.08$ in (3.19). The beam pattern is the sum of

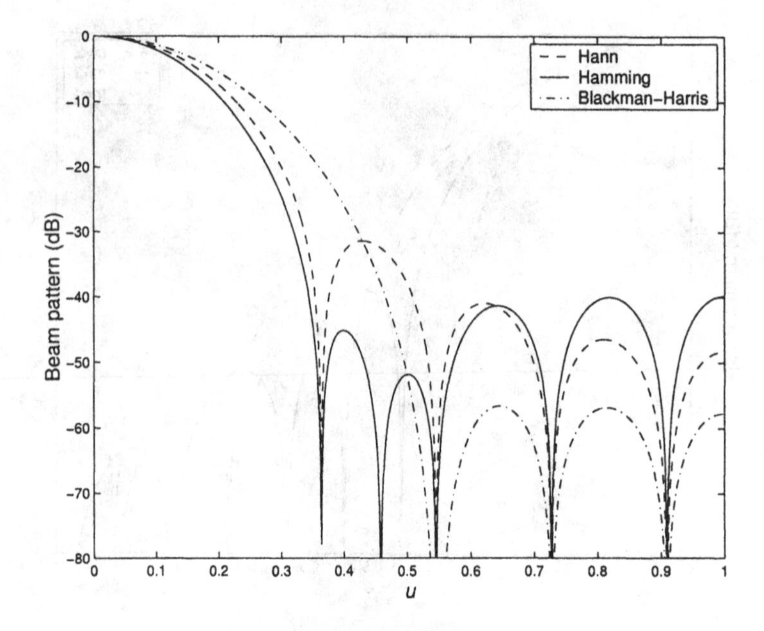

Figure 3.4 Beam patterns for Hann, Hamming, and Blackman-Harris weightings: N=11.

three conventional beam patterns:

$$B_u(u) = 0.54\,\frac{\sin\left(\frac{N\pi u}{2}\right)}{\sin\left(\frac{\pi u}{2}\right)} + 0.23\left[\frac{\sin\left(\frac{N\pi}{2}\left(u-\frac{2}{N}\right)\right)}{\sin\left(\frac{\pi}{2}\left(u-\frac{2}{N}\right)\right)} + \frac{\sin\left(\frac{N\pi}{2}\left(u+\frac{2}{N}\right)\right)}{\sin\left(\frac{\pi}{2}\left(u+\frac{2}{N}\right)\right)}\right]. \tag{3.22}$$

The Hamming weighting is shown in Figure 3.4. We also show the beam patterns for the Hann weighting (3.18) and the Blackman-Harris weighting, which we will derive next.

The first zero occurs at $u = 4/N$ and the height of the first non-zero sidelobe is -39.5 dB. There is a step discontinuity in the weighting that leads to an asymptotic falloff of the sidelobes of $O(\frac{1}{n})$; however, the first sidelobe is cancelled exactly and the remaining ones are low absolutely in spite of their relatively slow falloff. The directivity of the Hamming weighting is relatively high. In addition, we note that the beamwidth of the Hamming window is less than the beamwidth of the Hann weighting and its first sidelobe is lower. This is an exception to the general trend.

3.1.1.7 Blackman-Harris weighting

The Blackman-Harris weighting simply extends the procedure to higher order harmonics to provide nulls at the peaks of the first two sidelobes. The weighting function is

$$w(\tilde{n}) = 0.42 + 0.5\cos\left(\frac{2\pi\tilde{n}}{N}\right) + 0.08\cos\left(\frac{4\pi\tilde{n}}{N}\right), \tilde{n} = -\frac{N-1}{2}, \cdots, \frac{N-1}{2}. \tag{3.23}$$

The beam pattern is a sum of conventional beam patterns,

$$\begin{aligned} B_u(u) = \quad & 0.42\frac{\sin\left(\frac{N\pi u}{2}\right)}{\sin\left(\frac{\pi u}{2}\right)} + 0.25\left[\frac{\sin\left(\frac{N\pi}{2}\left(u-\frac{2}{N}\right)\right)}{\sin\left(\frac{\pi}{2}\left(u-\frac{2}{N}\right)\right)} + \frac{\sin\left(\frac{N\pi}{2}\left(u+\frac{2}{N}\right)\right)}{\sin\left(\frac{\pi}{2}\left(u+\frac{2}{N}\right)\right)}\right] \\ & +0.04\left[\frac{\sin\left(\frac{N\pi}{2}\left(u-\frac{4}{N}\right)\right)}{\sin\left(\frac{\pi}{2}\left(u-\frac{4}{N}\right)\right)} + \frac{\sin\left(\frac{N\pi}{2}\left(u+\frac{4}{N}\right)\right)}{\sin\left(\frac{\pi}{2}\left(u+\frac{4}{N}\right)\right)}\right], \end{aligned} \tag{3.24}$$

and is shown in Figure 3.4. The parameters for the beam patterns are:

Weighting	HPBW	BW_{NN}	First Sidelobe HT.	D_N
Hann	$1.44\frac{2}{N}$	$4.0\frac{2}{N}$	−31.4 dB	0.664
Hamming	$1.31\frac{2}{N}$	$4.0\frac{2}{N}$	−39.5 dB	0.730
Blackman-Harris	$1.65\frac{2}{N}$	$6.0\frac{2}{N}$	−56.6 dB	0.577

The weightings up to this point have been based on various sinusoidal functions. We now look at other types of array weightings in an attempt to improve the beamwidth–sidelobe trade-off.

3.1.1.8 Prolate spheroidal functions [4]

The problem of interest is to develop a weighting that will maximize the percentage of the total power that is concentrated in a given angular region. Thus, we want to maximize the ratio,

$$\alpha = \frac{\iint_{\Omega_1} |B(\theta,\phi)|^2 \sin\theta\, d\theta\, d\phi}{\int_0^{\pi} \sin\theta\, d\theta \int_0^{2\pi} d\phi\, |B(\theta,\phi)|^2}, \tag{3.25}$$

[4]Discrete prolate sequences have been applied to the FIR design problem by Tufts and Francis [TF70], Papoulis and Bertran [PB72], and Tufts [Tuf75]. They were applied to the aperture problem by Rhodes [Rho63]. Our discussion follows Prasad [Pra82].

where Ω_1 is a region around the mainbeam. For a linear array, this can be written as

$$\alpha = \frac{\int_{-\psi_0}^{\psi_0} |B_\psi(\psi)|^2 \, d\psi}{\int_{-\pi}^{\pi} |B_\psi(\psi)|^2 \, d\psi}, \tag{3.26}$$

where $\psi = \frac{2\pi}{\lambda} d \cos\theta$. From (2.51), we have

$$B_\psi(\psi) = \mathbf{w}^H \, \mathbf{v}_\psi(\psi). \tag{3.27}$$

The numerator can be written as

$$\begin{aligned} \alpha_N &= \int_{-\psi_0}^{\psi_0} \mathbf{w}^H \, \mathbf{v}_\psi(\psi) \, \mathbf{v}_\psi^H(\psi) \, \mathbf{w} \, d\psi \\ &= \mathbf{w}^H \left[\int_{-\psi_0}^{\psi_0} \mathbf{v}_\psi(\psi) \, \mathbf{v}_\psi^H(\psi) \, d\psi \right] \mathbf{w} \\ &= \mathbf{w}^H \, \mathbf{A} \, \mathbf{w}, \end{aligned} \tag{3.28}$$

where

$$\mathbf{A} \triangleq \int_{-\psi_0}^{\psi_0} \mathbf{v}_\psi(\psi) \, \mathbf{v}_\psi^H(\psi) \, d\psi. \tag{3.29}$$

The (m, n) element of $\mathbf{A}$ is

$$\int_{-\psi_0}^{\psi_0} e^{jm\psi} \, e^{-jn\psi} \, d\psi = \frac{2 \sin((m-n)\psi_0)}{(m-n)} = 2\psi_0 \text{sinc}((m-n)\psi_0). \tag{3.30}$$

Similarly, the denominator is

$$\begin{aligned} \alpha_D &= \int_{-\pi}^{\pi} \mathbf{w}^H \, \mathbf{v}_\psi(\psi) \, \mathbf{v}_\psi^H(\psi) \, \mathbf{w} \, d\psi \\ &= \mathbf{w}^H \, \mathbf{B} \, \mathbf{w}, \end{aligned} \tag{3.31}$$

where

$$\mathbf{B} = \int_{-\pi}^{\pi} \mathbf{v}_\psi(\psi) \, \mathbf{v}_\psi^H(\psi) \, d\psi = 2\pi \mathbf{I}. \tag{3.32}$$

Thus,

$$\alpha = \frac{\mathbf{w}^H \, \mathbf{A} \, \mathbf{w}}{2\pi \mathbf{w}^H \, \mathbf{w}}. \tag{3.33}$$

To maximize α, we find the eigenvalues and eigenvectors of the matrix

$$2\pi\lambda \mathbf{w} = \mathbf{A} \, \mathbf{w}, \tag{3.34}$$

and choose the eigenvector corresponding to the largest eigenvalue λ_{max}.[5] Using (3.30) in (3.34) and dropping the factor of 2, we have

$$\sum_{n=1}^{N} \frac{\sin((m-n)\psi_0)}{(m-n)} w_n = \pi\lambda w_m, \quad m = 1, 2, \cdots, N. \tag{3.35}$$

The weightings (or sequences) obtained by solving (3.35) are called **discrete prolate spheroidal sequences (DPSS)** and the corresponding beam patterns are called **discrete prolate spheroidal functions**. They are discussed in detail by Slepian [Sle78]. The discrete prolate spheroidal sequences are also referred to as **Slepian sequences**. The sequence corresponding to the largest eigenvalue is referred to as the first Slepian sequence. They are the discrete analog to the continuous case discussed by Slepian, Landau, and Pollack in a series of *Bell System Technical Journal* articles [SP61], [Sle64], [Sle65], [SS65], [LP61], and [LP62]. For our present application, we do not need most of their properties. We discuss the DPSS functions in more detail in Chapter 5.

We now consider a simple example to illustrate the results.

Example 3.1.1[6]

We consider an 11-element array and solve (3.35) for various values of ψ_0. In each case, the optimum weight vector $\mathbf{w}_0$ corresponds to the eigenvector corresponding to largest eigenvalue. The results are shown in Table 3.2. We show the first six normalized weights; the other five weights follow from symmetry. To simplify the plot we have normalized the weights so that $w_6 = 1$. The actual weights are normalized so that $B_u(0) = 1$.

Table 3.2 Normalized Weights Corresponding to Maximum Eigenvalues

ψ_0/π	w_1	w_2	w_3	w_4	w_5	w_6
0.025	0.975	0.984	0.991	0.996	0.999	1.000
0.06	0.865	0.912	0.950	0.978	0.994	1.000
0.10	0.678	0.785	0.875	0.943	0.986	1.000
0.20	0.274	0.466	0.665	0.839	0.958	1.000
0.40	0.043	0.168	0.391	0.670	0.907	1.000

In Figure 3.5, we show some representative discrete prolate spheroidal sequences for $\psi_0 = 0.1\pi$, 0.2π, and 0.4π.

The corresponding beam patterns are shown in Figure 3.6. As ψ_0 approaches zero, $\mathbf{w}_0$ approaches uniform weighting because we are maximizing the directivity. For $\psi_0 = 0.2\pi$, most of the energy is concentrated in the main beam with a slightly larger beamwidth and sidelobes of -20 dB and lower. For $\psi_0 = 0.4\pi$, the sidelobes are -53 dB, but the beam is much broader.

---

[5]The symbol λ denotes the eigenvalue. We also use λ for the wavelength but the meaning should be clear from the context.

[6]This example is similar to the result in Prasad [Pra82].

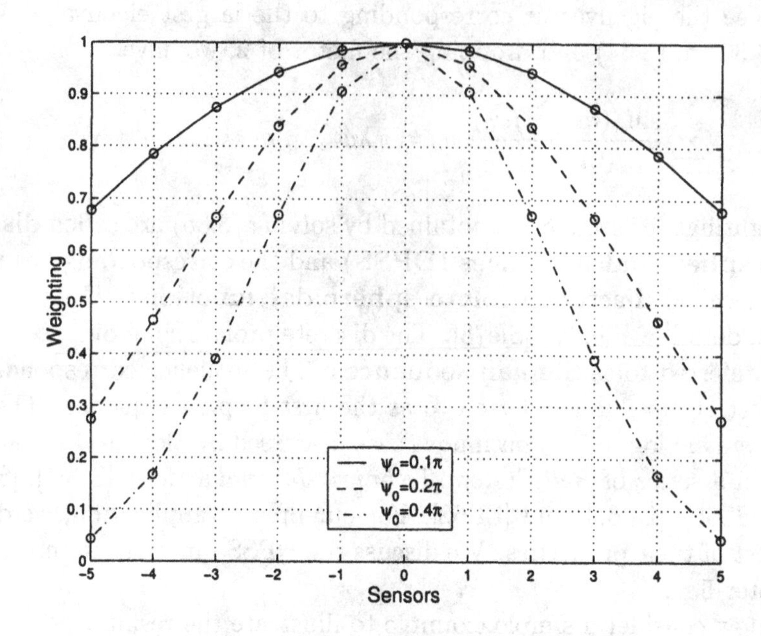

Figure 3.5 Discrete prolate spheroidal sequences: $\psi_0 = 0.1\pi$, 0.2π, and 0.4π.

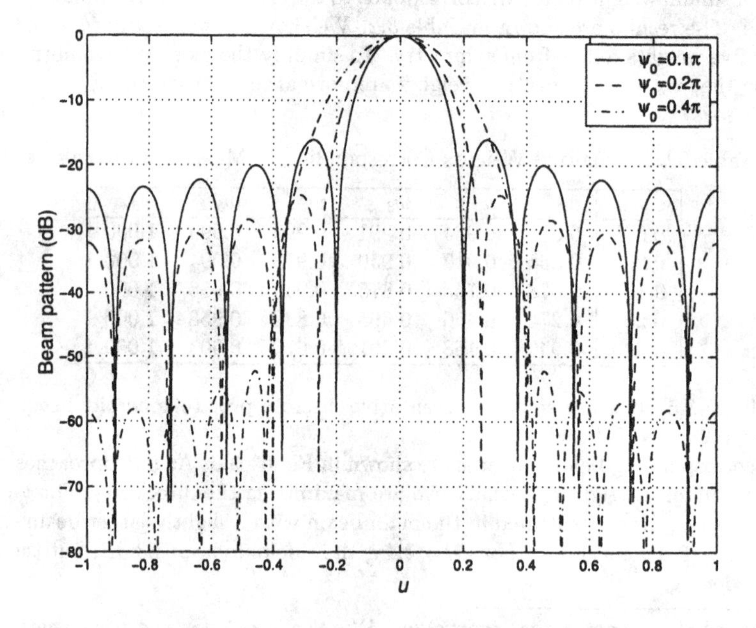

Figure 3.6 Beam patterns as a function of ψ_0: DPSS weighting, N=11.

The beam pattern parameters for the DPSS weightings are:

		HPBW	BW_{NN}	First Sidelobe HT.	D_N
DPSS	$\psi_0 = 0.1\pi$	$0.02/N$	$1.40\pi/N$	-15.6 dB	0.981
	$\psi_0 = 0.2\pi$	$2.20/N$	$1.79\pi/N$	-24.7 dB	0.869
	$\psi_0 = 0.4\pi$	$2.86/N$	$2.97\pi/N$	-52.2 dB	0.665

3.1.1.9 Kaiser weightings

Kaiser [Kai74] proposed a relatively simple approximation to the prolate spheroidal sequences using Bessel functions. The weighting has found widespread usage in spectral analysis, FIR filter design, and other fields.

The Kaiser weights are

$$w(\tilde{n}) = I_0\left(\beta\sqrt{1-\left[\frac{2\tilde{n}}{N}\right]^2}\right), \quad -\frac{N-1}{2} \leq \tilde{n} \leq \frac{N-1}{2}, \tag{3.36}$$

where $I_0(x)$ is the modified Bessel function of zero-order [AS65]. The parameter β specifies a beam pattern trade-off between the peak height of the sidelobes and the beamwidth of the main lobe.

We now consider two examples to illustrate the behavior.

Example 3.1.2

Consider a standard 11-element linear array. The normalized ($w_6 = 1$) weights for $\beta = 3$ and 6 are shown in Figure 3.7(a). The resulting beam patterns are shown in Figure 3.7(b). For $\beta = 3$, the HPBW is $2.52/N$, the BW_{NN} is $1.12\pi/N$, and the highest sidelobe is -26 dB. For $\beta = 6$, the HPBW is $2.86/N$, the BW_{NN} is $1.68\pi/N$, and the highest sidelobe is -47 dB. Note that as β decreases, the weighting function approaches the uniform weighting.

Example 3.1.3

In this case, we fix β at 3 and investigate the behavior for $N = 11$, 21, and 41. The beam patterns are shown in Figure 3.8. All three cases have the same maximum sidelobe. Changing N changes the value of u where this maximum occurs.

The beam pattern parameters for the Kaiser weightings for $N = 11$ are:

		HPBW	BW_{NN}	First Sidelobe HT.	D_N
Kaiser	$\beta = 3$	$2.18/N$	$1.75\pi/N$	-23.7 dB	0.882
	$\beta = 6$	$2.80/N$	$2.76\pi/N$	-44.4 dB	0.683

This completes our initial discussion of array weight vectors that are designed to allow trade-off between the beamwidth of the main lobe and the height of sidelobes.

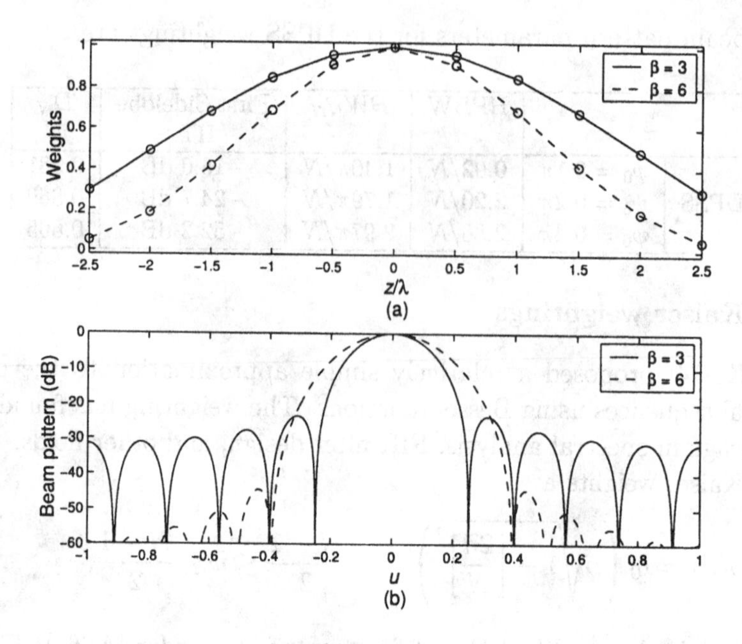

Figure 3.7 Kaiser weighting: (a) weighting for $\beta = 3$ and 6; (b) beam patterns for $\beta = 3$ and 6: N=11.

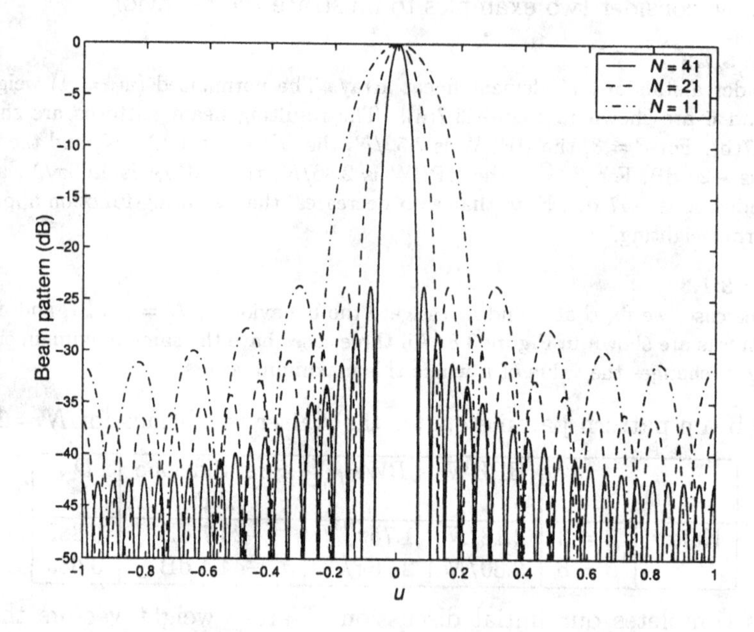

Figure 3.8 Beam patterns for Kaiser weighting; $\beta = 3$, $N = 11$, 21, and 41.

Other weight vectors are described in the comprehensive paper by Harris [Har78] on time-domain windows.

3.2 Array Polynomials and the z-Transform

For linear equally spaced arrays the beam pattern can be represented in terms of an **array polynomial.** In 1943, Schelkunoff [Sch43] utilized this representation to develop a theory of linear arrays. We utilize his work and introduce the z-transform representation.

3.2.1 z-Transform

From (2.71), the beam pattern can be written in ψ-space as

$$B_{\psi}(\psi) = e^{-j\left(\frac{N-1}{2}\right)\psi} \left(\sum_{n=0}^{N-1} w_n \, e^{-jn\psi}\right)^* . \tag{3.37}$$

Defining

$$z = e^{j\psi}, \tag{3.38}$$

we can write

$$B_z(z) = \sum_{n=0}^{N-1} w_n z^{-n}, \tag{3.39}$$

which is familiar as the z-transform. It maps the real variable ψ into a complex variable z with unit magnitude.

The transformation is shown in Figure 3.9. The variable ψ is the phase of the complex variable z. The beam pattern can be written as

$$\boxed{B_{\psi}(\psi) = \left[z^{-\frac{N-1}{2}} B_z^*(z)\right]_{z=e^{j\psi}} .} \tag{3.40}$$

Most discussions in the classical antenna literature focus on the case of real w_n. When we discuss optimum array design from a statistical standpoint in Chapter 6, we will usually have complex w_n. In Section 3.2.2, we restrict our attention to real weightings.[7] In Section 3.2.3, we discuss some properties of the beam pattern in the vicinity of the zeros of $B_z(z)$.

[7] In Section 3.7, we consider a design problem that leads to complex weights.

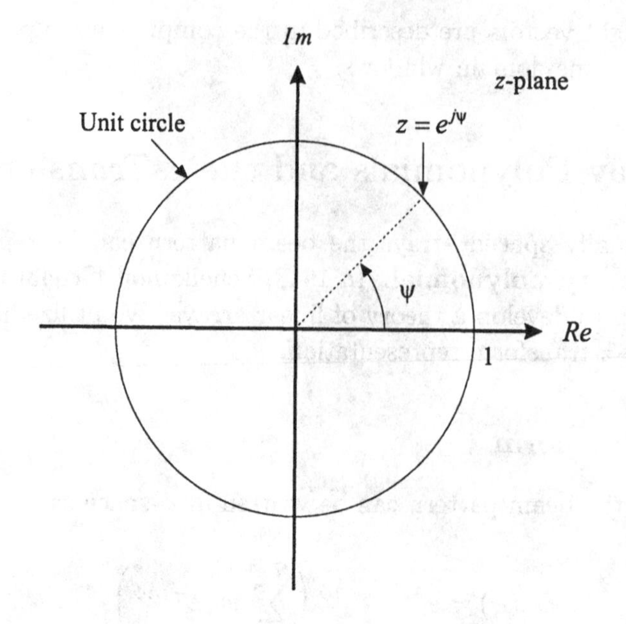

Figure 3.9 z-Transform.

3.2.2 Real Array Weights

In this section, we assume the w_n are real and symmetrical.[8] In the text, we analyze the case of symmetric weightings and N odd. We analyze the case of symmetric weightings and N even in the problems.

For symmetric weightings,

$$w(n) = w(N-1-n), \quad 0 \le n \le N-1. \tag{3.41}$$

The z-transform is

$$B_z(z) = \sum_{n=0}^{N-1} w_n z^{-n}. \tag{3.42}$$

Because of the symmetry we can define

$$M = \frac{N-1}{2}, \tag{3.43}$$

and write

$$\begin{aligned} B_z(z) &= z^{-M}\left\{w(M) + w(M-1)\left[z + z^{-1}\right]\right. \\ &\left. + w(M-2)\left[z^2 + z^{-2}\right] + \cdots + w(0)\left[z^M + z^{-M}\right]\right\}. \end{aligned} \tag{3.44}$$

[8] Many of the classical array weight vectors satisfy this assumption.

Letting $z = e^{j\psi}$,

$$\begin{aligned} B_z\left(e^{j\psi}\right) &= e^{-jM\psi}\left\{w(M) + 2w(M-1)\cos\psi \right. \\ &\quad \left. +2w(M-2)\cos 2\psi + \cdots + w(0)\cos(M\psi)\right\}. \end{aligned} \tag{3.45}$$

Using (3.40), the beam pattern is

$$B_\psi(\psi) = w(M) + 2\sum_{m=1}^{M-1} w(m)\cos(M-m)\psi. \tag{3.46}$$

We can also write (3.46) as

$$B_\psi(\psi) = \sum_{n=0}^{M} \alpha_n \cos n\psi, \tag{3.47}$$

where

$$\alpha_n = \begin{cases} w(M), & n = 0, \\ 2w(M-n), & n \neq 0, \text{for } N \text{ odd}. \end{cases} \tag{3.48}$$

We now explore the behavior of the zeros of $B_z(z)$. To find the zeros of $B_z(z)$, we observe that

$$B_z(z^{-1}) = z^{2M}\, B_z(z), \tag{3.49}$$

because of the symmetry of the coefficients. Therefore $B_z(z)$ and $B_z(z^{-1})$ have identical zeros. Since $B_z(z)$ has real coefficients, the zeros occur in complex conjugate pairs.

Thus, we can write $B_z(z)$ in factored form as[9]

$$B_z(z) = w(0)\, B_1(z)\, B_2(z)\, B_3(z), \tag{3.50}$$

where

$$B_1(z) = \prod_{i=1}^{N_1}\left[1 - \left(r_i + \frac{1}{r_i}\right) z + z^2\right], \tag{3.51}$$

contains zeros in reciprocal pairs on the real axis. If $B_1(z)$ contains a zero at $z = +1$ or $z = -1$, then it will appear in pairs since $B_1(z)$ is of even order.

$$B_2(z) = \prod_{i=1}^{N_2}\left[1 - (2\cos\theta_i)\, z + z^2\right] \tag{3.52}$$

[9] We drop the z subscript on the right side for simplicity.

contains zeros occurring in complex conjugate pairs on the unit circle at $z = e^{\pm j\theta_i}$.

$$\begin{aligned} B_3(z) &= \prod_{i=1}^{N_3}\left\{1 - \left[2\left(r_i + \frac{1}{r_i}\right)\cos\theta_i\right] z \right. \\ &\quad + \left[r_i^2 + \frac{1}{r_i^2} + 4\cos^2\theta_i\right] z^2 \\ &\quad \left. - \left[2\left(r_i + \frac{1}{r_i}\right)\cos\theta_i\right] z^3 + z^4\right\} \end{aligned} \tag{3.53}$$

contains zeros that occur in fours: complex conjugate pairs and reciprocal (with respect to the unit circle) pairs.

There are a total of $2M = N - 1$ zeros:

$$2M = 2\,N_1 + 2\,N_2 + 4\,N_3. \tag{3.54}$$

The **zeros on the unit circle** correspond to nulls in the beam pattern if they are in the visible region.

For N even,

$$M = \frac{N}{2} - 1, \tag{3.55}$$

and (3.50) becomes

$$B_z(z) = w(0)(1+z)B_1(z)B_2(z)B_3(z). \tag{3.56}$$

The $(1+z)$ corresponds to a zero at $z = -1$ and there are $2M$ additional zeros from $B_1(z)$, $B_2(z)$, and $B_3(z)$. There are always a total of $N-1$ zeros.

The magnitude of $B(z)$ is

$$|B(z)| = |w_{N-1}| \prod_{n=1}^{2M} |z - z_n|\,. \tag{3.57}$$

This magnitude can be expressed as the product of the distances from a point z on the unit circle to the roots. Since the point (1, 0) represents the MRA, the pattern is normalized by dividing by the product of the distances from (1, 0) to the roots. Thus,

$$|B(z)| = \frac{|z - z_1|\,|z - z_2| \cdots |z - z_{N-1}|}{|1 - z_1|\,|1 - z_2| \cdots |1 - z_{N-1}|}, \tag{3.58}$$

and

$$\arg B(z) = \sum \arg(z - z_n) - \sum \arg(1 - z_n). \tag{3.59}$$

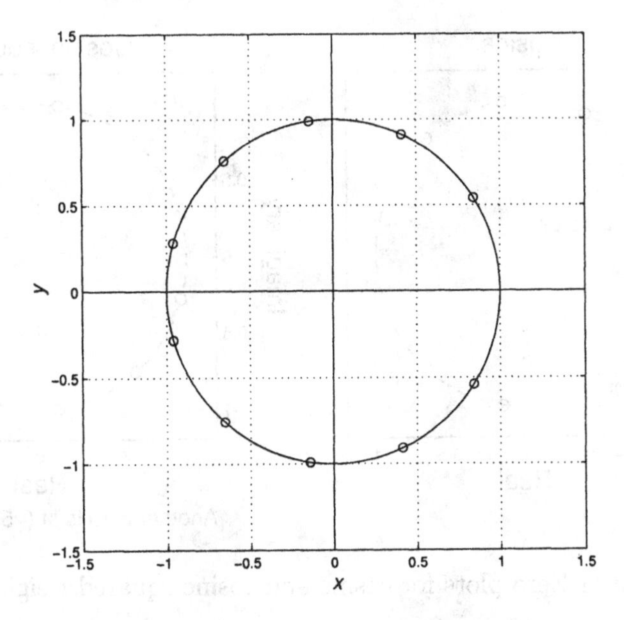

Figure 3.10 Zero plot for an 11-element array.

As z moves around the unit circle from (1, 0) in a counterclockwise direction, ψ is moving 0 to $+kd$. When $d = \lambda/2$, ψ is moving from 0 to π, u is moving from 0 to 1, and θ is moving from 90° to 0°.

Example 3.2.1

Consider a standard 11-element array with uniform weighting. From (2.101), the zeros are located at

$$u_n = \pm\frac{n}{N}\cdot\frac{\lambda}{d}, \quad n = 1, 2, \cdots, \frac{N-1}{2}, \tag{3.60}$$

or

$$u_n = \pm\frac{2n}{11}, \; n = 1, \cdots, 5. \tag{3.61}$$

The resulting z-plane plot is shown in Figure 3.10. Remember that the location of the first zero determines the beamwidth of the main lobe. Thus, we can develop techniques that constrain the first zero to be at a specified point and adjust the other zeros to obtain a desired pattern shape. Many of the commonly used patterns are developed in this manner.

In the next series of figures (Figures 3.11–3.13), we show the z-plane plots for some of the array weightings we derived in Section 3.1. They are grouped in the following sets:

- Figure 3.11: Cosine, cosine-squared
- Figure 3.12: Hamming, Blackman-Harris
- Figure 3.13: DPSS $(0.1\pi, 0.4\pi)$

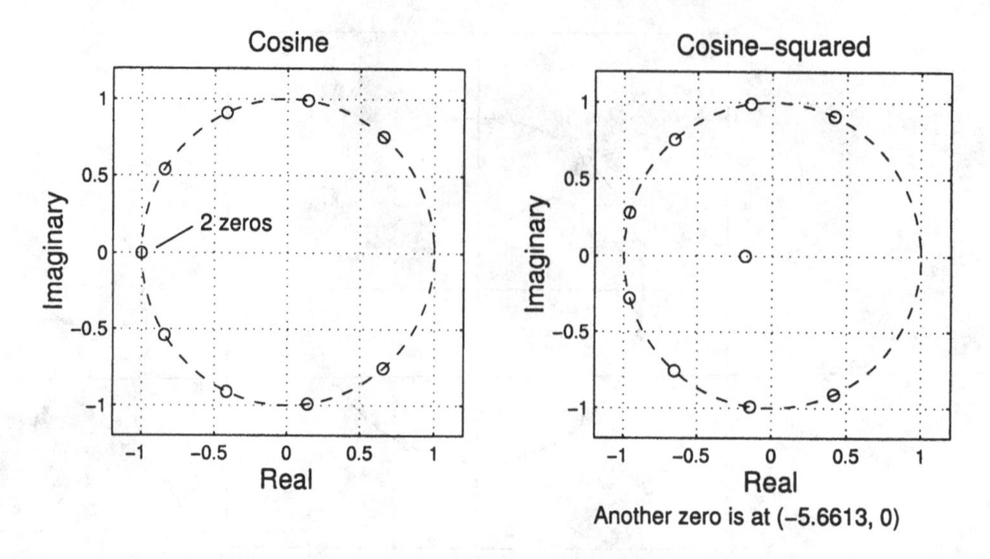

Figure 3.11 Zero plots for cosine and cosine-squared weightings.

In Figure 3.14, we show the effect of d on the z-plane plot for various values of d. The visible region represents the part of the circle corresponding to physically observable values of θ. The remainder of the circle corresponds to the virtual (or invisible) region that we have discussed previously.

For $d < \lambda/2$, the visible region does not cover the entire circle. For $d = \lambda/2$, the visible region corresponds exactly to the circle. For $d > \lambda/2$, the pattern overlaps and has grating lobes.

3.2.3 Properties of the Beam Pattern Near a Zero

In this section we discuss the behavior of the beam pattern in the vicinity of a zero.[10] Consider a linear aperture with uniform weighting. The beam pattern is

$$B_u(u) = \frac{\sin \alpha u}{\alpha u}, \tag{3.62}$$

where $\alpha = \pi L/\lambda$. The pattern in the vicinity the nth zero, u_n, is shown in Figure 3.15. It appears to be linear at the zero crossing. We want to recenter the pattern at u_n. Define

$$u_0 = u - u_n, \tag{3.63}$$

[10]This discussion follows pages 105–110 of Steinberg [Ste76].

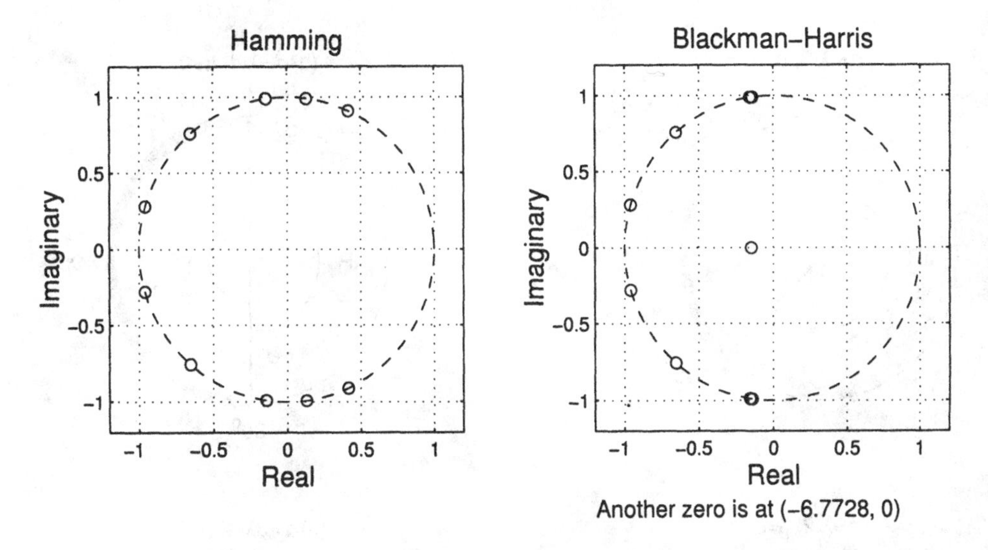

Figure 3.12 Zero plots for Hamming, Blackman-Harris weightings.[11]

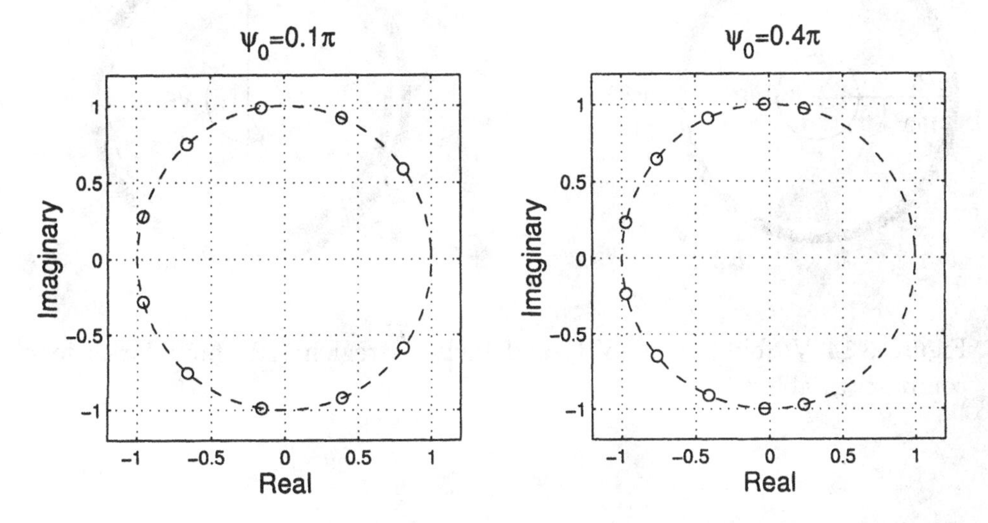

Figure 3.13 Zero plots for DPSS $(0.1\pi, 0.4\pi)$ weightings.

[11]Note that the two inner zeros are both double zeros.

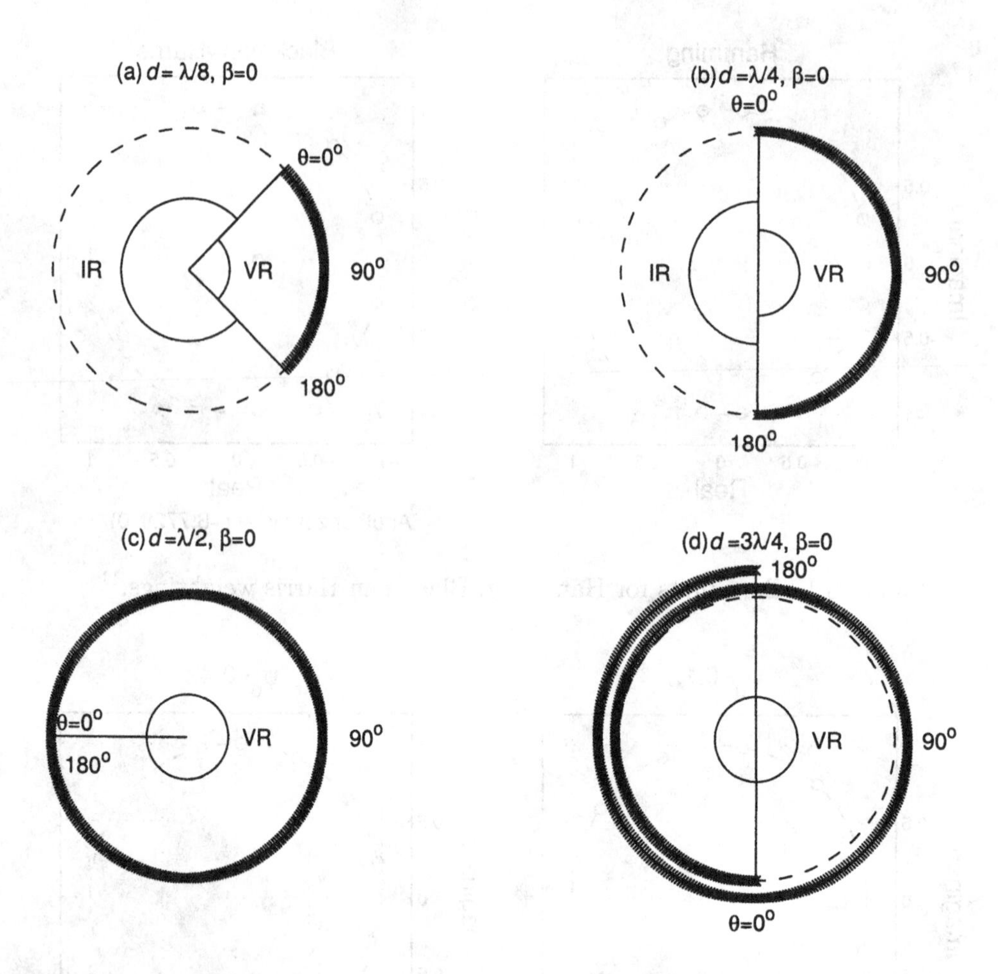

Figure 3.14 Visible region (VR) and invisible region (IR) boundaries for complex variable z.

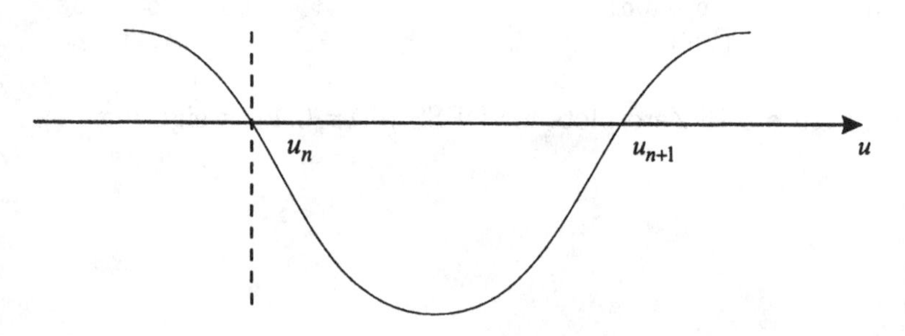

Figure 3.15 The beam pattern near a non-multiple zero.

and assume

$$u_n \gg \frac{\lambda}{L}. \tag{3.64}$$

Then,

$$\begin{aligned} B_{u_0}(u_0) &= B_u(u_0+u_n) = \frac{\sin\alpha(u_0+u_n)}{\alpha(u_0+u_n)} \\ &= \frac{\sin\alpha u_0 \, \cos\alpha u_n + \cos\alpha u_0 \, \sin\alpha u_n}{\alpha(u_0+u_n)}. \end{aligned} \tag{3.65}$$

Since $\sin\alpha u_n = 0$ and $\cos\alpha u_n = \pm 1$, we have

$$B_{u_0}(u_0) = \pm\frac{\sin\alpha u_0}{\alpha(u_0+u_n)}. \tag{3.66}$$

In the region near the zero,

$$\sin\alpha u_0 \cong \alpha u_0, \tag{3.67}$$

and

$$u_0 \ll u_n, \tag{3.68}$$

so that (3.66) becomes

$$B_{u_0}(u_0) \simeq \pm\frac{u_0}{u_n}. \tag{3.69}$$

Thus the pattern is linear near the zero and the slope is inversely proportional to the coordinate of the zero in u-space.

To investigate the behavior near multiple zeros, we use the approximate radiation pattern of a triangularly weighted aperture

$$B_{u_0}(u_0) = \left(\frac{\sin\alpha u_0}{\alpha u}\right)^2. \tag{3.70}$$

The behavior in the neighborhood of u_n is quadratic

$$B_{u_0}(u_0) \simeq \left(\frac{u_0}{u_n}\right)^2. \tag{3.71}$$

Similar results follow for higher order zeros. The magnitude of $|B_{u_0}(u_0)|$ is sketched in Figure 3.16.

Several comments are in order:

(i) A single zero creates a sharp null. If we are trying to null out an interfering source (perhaps a jammer), the performance will be sensitive to the exact location of the interfering source **and** its frequency.

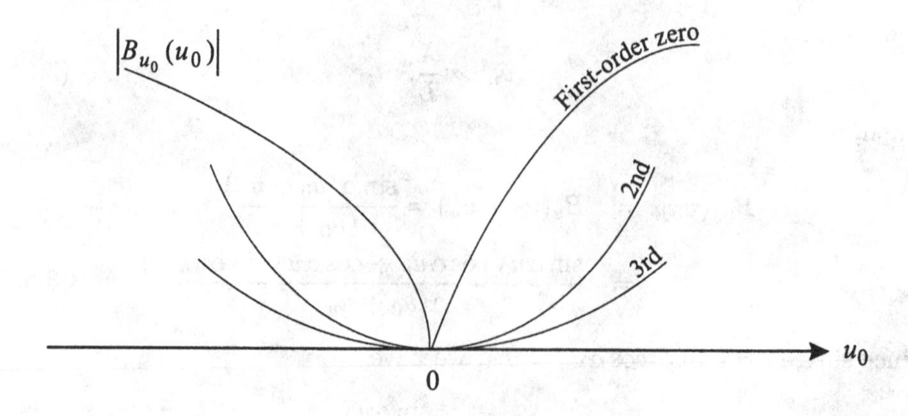

Figure 3.16 Beam pattern near first and higher order zeros.

(ii) However, if we are trying to estimate the location of a source, it will be useful to work in the vicinity of a null. We see later (in Section 3.8) that a difference beam does exactly that.

(iii) A double zero creates a wider null and is more robust to model variations or changes in the interferer's location. However, it utilizes 2 degrees of freedom per zero and limits the total number of nulls possible.

3.3 Pattern Sampling in Wavenumber Space

Woodward [Woo46] developed an antenna pattern synthesis technique based on the sampling theorem. We discuss it first for the case of continuous aperture and then for discrete arrays. For the discrete array case, it corresponds to a DFT relationship.

3.3.1 Continuous Aperture

Woodward's approach is based on the Fourier transform relationship between the wavenumber response and the aperture weighting. Rewriting (3.1) and (3.2) in u-space gives

$$\frac{1}{\lambda}\Upsilon_u(u) = \int_{-L/2}^{L/2} w^*(z_\lambda)\, e^{j2\pi u z_\lambda}\, dz_\lambda \tag{3.72}$$

and

$$w^*(z_\lambda) = \int_{-\infty}^{\infty} \frac{1}{\lambda}\Upsilon_u(u)\, e^{-j2\pi u z_\lambda}\, du, \tag{3.73}$$

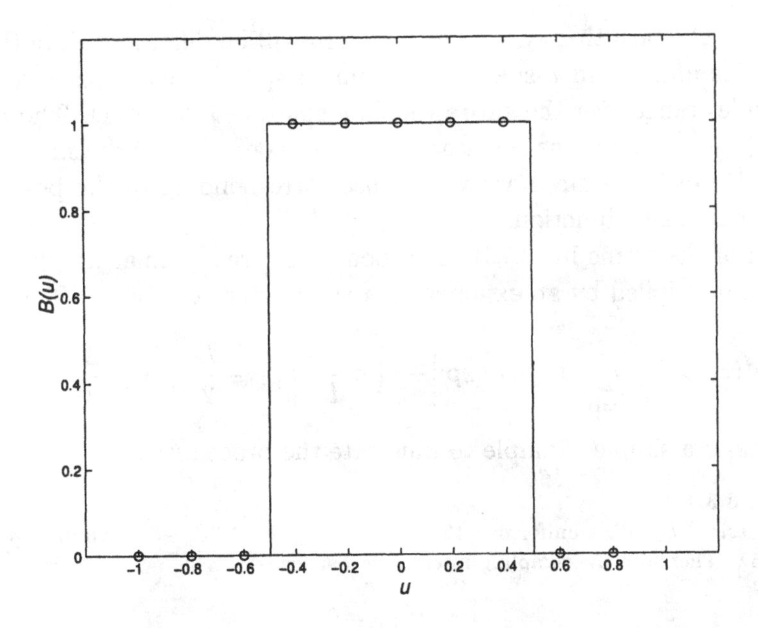

Figure 3.17 Pattern sampling.

where we have suppressed ω and $z_\lambda = z/\lambda$ is the normalized z-distance.

Since z_λ is limited to the range $\pm\frac{L}{2\lambda}$, we can specify the aperture illumination function by sampling the antenna pattern in wavenumber space at intervals of[12]

$$\Delta u_s = \frac{\lambda}{L}. \tag{3.74}$$

For example, if the desired pattern is the rectangular pattern shown in Figure 3.17, we would sample at λ/L intervals and obtain a set of samples,

$$B_u(u_m), \quad m = 0, 1, \cdots, N_s - 1, \tag{3.75}$$

where $u_m = (m\Delta u_s - 1), m = 0, 1, \cdots, N_s - 1$, and $N_s = \text{int}\lfloor 2L/\lambda \rfloor$ ($\text{int}\lfloor x \rfloor$ denotes the largest integer less than x). In Figure 3.17, $L = 5\lambda$, so $\Delta u_s = 0.2$ (see (2.214)). To reconstruct $B_u(u)$, we use a sum of shifted sinc functions,

$$B_u(u) = \sum_{m=0}^{N_s-1} B_u(u_m)\text{sinc}\left(\frac{\pi L}{\lambda}(u - u_m)\right). \tag{3.76}$$

In this case, we have chosen the initial sample at $u = -1.0$. In this example, we could choose the initial sample to be anywhere in the interval $-1.0 \leq$

[12]This is identical to the Nyquist sampling criterion in the time-frequency context. If the signal is bandlimited $[-W \leq f \leq W]$, then the sampling interval $T = 1/2W$.

$u \leq -0.8$. The resulting synthesized pattern will be different. Note that the sampling is uniform in u-space (wavenumber space), not θ-space, and that the samples range over the entire visible u-space ($-1 \leq u \leq 1$). The nulls of the $\text{sinc}(u - u_m)$ functions occur at Δu_s so the coefficient of the sinc function is the value of the desired array response corresponding to the peak of the corresponding sinc function.

Each of the terms in (3.76) corresponds to a rectangular amplitude distribution multiplied by an exponential term to steer the beam. Thus,

$$w^*(z) = \frac{1}{L} \sum_{m=0}^{N_s-1} B_u(m) \exp\left[-j\left(m\frac{z}{L}\right)\right], \quad -\frac{L}{2} \leq z \leq \frac{L}{2}. \tag{3.77}$$

We consider a simple example to illustrate the procedure.

Example 3.3.1

The desired $B_{d\theta}(\theta)$ is uniform in the range: $60° \leq \theta \leq 120°$, as shown in Figure 3.17. Let $L = 5\lambda$. Therefore the sampling interval in u-space is

$$\Delta u_s = \frac{\lambda}{5\lambda} = 0.2, \tag{3.78}$$

and there will be ten samples. The sample values are

$$B_u(u_m) = \begin{cases} 0, & m = 0, 1, 2, \\ 1, & m = 3, \cdots, 7, \\ 0, & m = 8, 9. \end{cases} \tag{3.79}$$

Note that the value at $m = 10$ is determined by the value at $m = 0$ by the periodicity. The resulting synthesized pattern is shown in Figure 3.18.

3.3.2 Linear Arrays

The Woodward approach can also be used for linear arrays. We discuss the approach from two viewpoints. In this section, we use a straightforward modification of the continuous aperture case. In Section 3.3.3, we introduce the DFT as an intermediate step.

We first consider the case of a standard N-element linear array ($d = \lambda/2$). To relate to the continuous aperture,

$$Nd = L. \tag{3.80}$$

The sampling interval in u-space is $2/N$. As in the aperture case, we can choose the initial sample point. For purposes of discussion in the text, we will assume the samples are taken symmetrically about $u = 0$. Other initial sampling points can also be used. Thus,

$$u_m = \frac{2}{N}\left(m - \frac{N-1}{2}\right), \quad m = 0, 1, \cdots, N-1. \tag{3.81}$$

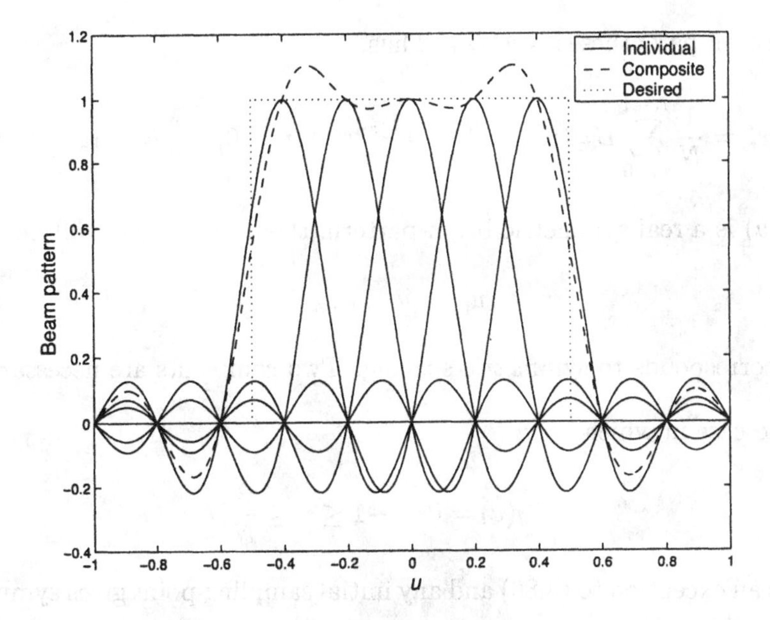

Figure 3.18 Desired and synthesized patterns using the Woodward synthesis procedure.

Because $d = \lambda/2$, the number of pattern samples in the visible region ($|u| \leq 1$) is equal to the number of elements. We consider the case when $d < \lambda/2$ in the next section.

Using (3.76), the component beam pattern can be written as

$$B_{u_m}(u) = B_{du}(u_m) \frac{\sin\left[\frac{N}{2}\pi(u - u_m)\right]}{N \sin\left[\frac{\pi}{2}(u - u_m)\right]}, \quad m = 0, 1, \cdots, N-1, \tag{3.82}$$

and the total beam pattern is

$$B_u(u) = \sum_{m=0}^{N-1} B_{du}(u_m) \frac{\sin\left[\frac{N}{2}\pi(u - u_m)\right]}{N \sin\left[\frac{\pi}{2}(u - u_m)\right]}. \tag{3.83}$$

In order to find the weight vector, we use (2.91) and (2.92) to rewrite (3.83) as

$$\begin{aligned} B_u(u) &= \sum_{m=0}^{N-1} B_{du}(u_m) \cdot \frac{1}{N} \sum_{k=0}^{N-1} e^{j(k-\frac{N-1}{2})\pi(u-u_m)} \\ &= \sum_{k=0}^{N-1} e^{j(k-\frac{N-1}{2})\pi u} \left(\sum_{m=0}^{N-1} B_{du}(u_m) \cdot \frac{1}{N} e^{-j(k-\frac{N-1}{2})\pi u_m} \right). \end{aligned} \tag{3.84}$$

The term in parentheses is $w^*(k)$. Thus,

$$w_n^* = \frac{1}{N} \sum_{m=0}^{N-1} B_{du}(u_m) e^{-j(n-\frac{N-1}{2})\pi u_m}, \quad n = 0, \cdots, N-1. \tag{3.85}$$

If $B_{du}(u)$ is a real symmetric beam pattern, then the weights will be real if

$$u_m = u_{N-1-m}, \tag{3.86}$$

which corresponds to symmetric spacing. Two comments are necessary:

(i) The case in which

$$B_d(u) = 0, \quad -1 \le u \le \frac{2}{N},$$

is an exception to (3.86) and any initial sampling point gives symmetric weights.

(ii) In many applications, the desired beam pattern is not symmetric, so we will have complex weights.

We also want to interpret these results in the context of the discrete Fourier transform.

3.3.3 Discrete Fourier Transform

In Section 3.2, we represented the beam pattern as the z-transform of the array weighting function. From (3.40),

$$B_\psi(\psi) = \left[z^{-\frac{N-1}{2}} B_z^*(z) \right]_{z=e^{j\psi}}, \tag{3.87}$$

and

$$B_z(z) = \sum_{n=0}^{N-1} w_n z^{-n}, \tag{3.88}$$

or

$$B_z(z)\big|_{z=e^{j\psi}} = B_\psi^*(\psi) e^{-j\frac{N-1}{2}\psi}. \tag{3.89}$$

We sample at

$$z_k = e^{j\left(k-\frac{N-1}{2}\right)\frac{2\pi}{N}}, \quad k = 0, 1, \cdots, N-1. \tag{3.90}$$

This corresponds to N samples symmetric about the origin and ψ ranges from $-\pi \leq \psi \leq \pi$.

$$\begin{aligned} B_{\psi}^{*}(\psi_k)e^{-j\frac{N-1}{2}\psi_k} &= \sum_{n=0}^{N-1} w_n z_k^{-n} \\ &= \sum_{n=0}^{N-1} w_n e^{-j\left(k-\frac{N-1}{2}\right)\frac{2\pi}{N}n} \\ &= \sum_{n=0}^{N-1} \left(w_n e^{jn\pi\left(\frac{N-1}{N}\right)}\right) e^{-jkn\frac{2\pi}{N}}, \end{aligned} \tag{3.91}$$

where

$$\psi_k = \left(k - \frac{N-1}{2}\right)\frac{2\pi}{N}, \quad k = 0, 1, \cdots, N-1. \tag{3.92}$$

Now define,

$$b_n \triangleq w_n e^{jn\pi\left(\frac{N-1}{N}\right)}, \tag{3.93}$$

and

$$B(k) = B_{\psi}^{*}(\psi_k)e^{-j\psi_k\left(\frac{N-1}{2}\right)}. \tag{3.94}$$

Then,

$$\boxed{B(k) = \sum_{n=0}^{N-1} b_n e^{-jkn\frac{2\pi}{N}}, \quad k = 0, 1, \cdots, N-1,} \tag{3.95}$$

which is called the **discrete Fourier transform**.

The relation in (3.91) is a linear transformation from the N-dimensional vector $\mathbf{b}$ to the N-dimensional vector $\mathbf{B}$,

$$\boxed{\mathbf{B} = \mathbf{F}\mathbf{b},} \tag{3.96}$$

where $\mathbf{F}$ is an $N \times N$ matrix, whose kl element is

$$[\mathbf{F}]_{kl} = (F_N)^{kl} = \left(e^{-j\frac{2\pi}{N}}\right)^{kl}, \tag{3.97}$$

where

$$F_N \triangleq e^{-j\frac{2\pi}{N}}. \tag{3.98}$$

To obtain the inverse, we multiply $B(k)$ by $e^{jkm\frac{2\pi}{N}}$, sum on k, and divide by N,

$$\frac{1}{N}\sum_{k=0}^{N-1} B(k)\, e^{jkm\frac{2\pi}{N}} = \frac{1}{N}\sum_{n=0}^{N-1} b_n \left(\sum_{k=0}^{N-1} e^{jk(m-n)\frac{2\pi}{N}}\right). \tag{3.99}$$

The sum in parentheses is N, if $m = n$, and 0 if $m \neq n$. Thus,

$$\boxed{b_n = \frac{1}{N}\sum_{k=0}^{N-1} B(k)\, e^{jkn\frac{2\pi}{N}}}, \tag{3.100}$$

which is the **inverse discrete Fourier transform (IDFT)**, and

$$w_n = b_n e^{-jn\pi\left(\frac{N-1}{N}\right)}. \tag{3.101}$$

In order to write (3.100) in vector notation, we first observe that

$$\mathbf{F}^H = \frac{1}{N}\mathbf{F}^{-1}. \tag{3.102}$$

Thus,

$$\boxed{\mathbf{b} = \frac{1}{N}\,\mathbf{F}^{-1}\,\mathbf{B}.} \tag{3.103}$$

We observe from (3.103) that, if we know $\mathbf{B}$, which is the vector of samples $B(k), k = 0, 1, \cdots, N-1$, then we can find $\mathbf{b}$ and $\mathbf{w}$. However, knowing $\mathbf{w}$ enables us to find the complete beam pattern $B_\psi(\psi)$. Therefore the pattern samples at $2\pi/N$ intervals completely determine the beam pattern.

We use this result in two different ways. In the first case, we have a beam pattern that we know was generated by a standard linear array. We encounter this case frequently in subsequent sections so we denote it as the **IDFT Weight Vector Determination** algorithm. The steps in the algorithm are:

(i) Sample the beam pattern at

$$\psi_k = \left(k - \frac{N-1}{2}\right)\frac{2\pi}{N}, \quad k = 0, 1, \cdots, N-1$$

to obtain $B_\psi(\psi_k)$.

(ii) Use (3.94) to find $B(k)$.

(iii) Find $\mathbf{b}$ as the IDFT of $B(k)$ using (3.100).

(iv) Use (3.101) to find $\mathbf{w}$.

In this case, the result will not depend on the initial sample point.

In the second case, we have a desired beam pattern that is not necessarily realizable by a standard linear array. We repeat steps (i)–(iv). We then use (2.65) to find the resulting beam pattern.

We consider two simple examples to illustrate the procedure.

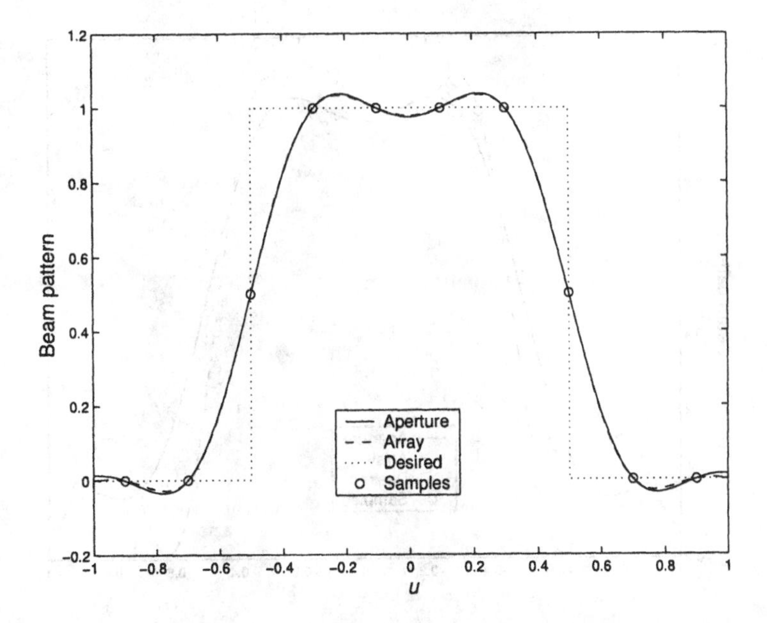

Figure 3.19 Array and aperture beam patterns using wavenumber sampling technique, symmetric sampling: $L = 5.0\lambda, d = \lambda/2, N = 10$.

Example 3.3.2 (continuation)

The array corresponding to the aperture in the previous example contains ten elements with spacing $d = \lambda/2$. The samples are taken symmetrically about $u = 0$. At the discontinuity (u=0.5) we use the value at the mid-point. The resulting pattern is shown in Figure 3.19. We see that the aperture and the array give essentially the same beam pattern.

Example 3.3.3 (continuation)

Consider the case when $N = 11$ and $d = \lambda/2$. The samples are taken symmetrically. The result is shown in Figure 3.20. The different sample separation causes a difference in the two beam patterns.

The discussion up to this point assumed a standard linear array so that $d = \lambda/2$. For the case when $d < \lambda/2$ a similar procedure can be utilized but there are some important changes. To illustrate the technique, we consider the case where

$$d = \frac{\lambda}{4}. \tag{3.104}$$

The DFT relations in (3.91) and (3.100) are still valid and we still sample at $2\pi/N$ intervals in ψ-space. However, we need to sample over a complete period of $\Upsilon_u(u)$, which in this case corresponds to $-2 \leq u \leq 2$ (recall Figure 2.20(a)). Thus, if we use a $2/N$ sampling interval, we use pattern samples

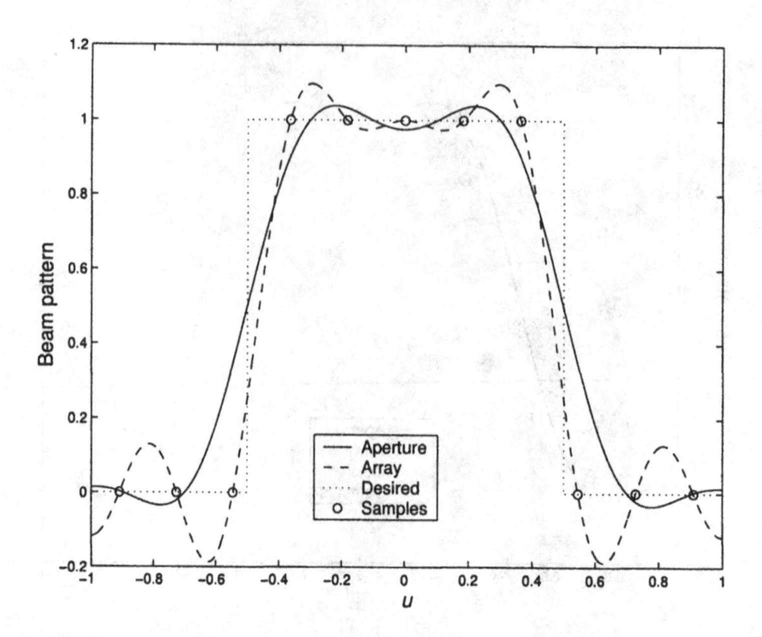

Figure 3.20 Array and aperture beam patterns using wavenumber sampling technique, symmetric sampling: $L = 5.0\lambda, d = \lambda/2, N = 11$.

outside the visible region to determine the weight vector and to find $B_\psi(\psi)$.

Example 3.3.4 (continuation)

Consider the same desired pattern with an array where $d = \lambda/4$ and $N = 20$. The result using symmetric sampling is shown in Figure 3.21.

3.3.4 Norms

In Section 2.6.4, we indicated the importance of the norm of the weight vector, $\|\mathbf{w}\|^2$. We want to relate the weight vector norm to the norm of the vector of beam pattern samples. From (3.96),

$$\mathbf{B}^H \mathbf{B} = \mathbf{b}^H \mathbf{F}^H \mathbf{F} \mathbf{b}. \tag{3.105}$$

Using (3.102), we obtain

$$\frac{1}{N}\mathbf{B}^H\mathbf{B} = \mathbf{b}^H\mathbf{b}. \tag{3.106}$$

Both $\mathbf{B}$ and $\mathbf{b}$ contain exponential factors that are not necessary in the norm relationship. From (3.93),

$$\mathbf{b} = \text{diag}\left[e^{jn\pi\left(\frac{N-1}{N}\right)}\right] \mathbf{w}. \tag{3.107}$$

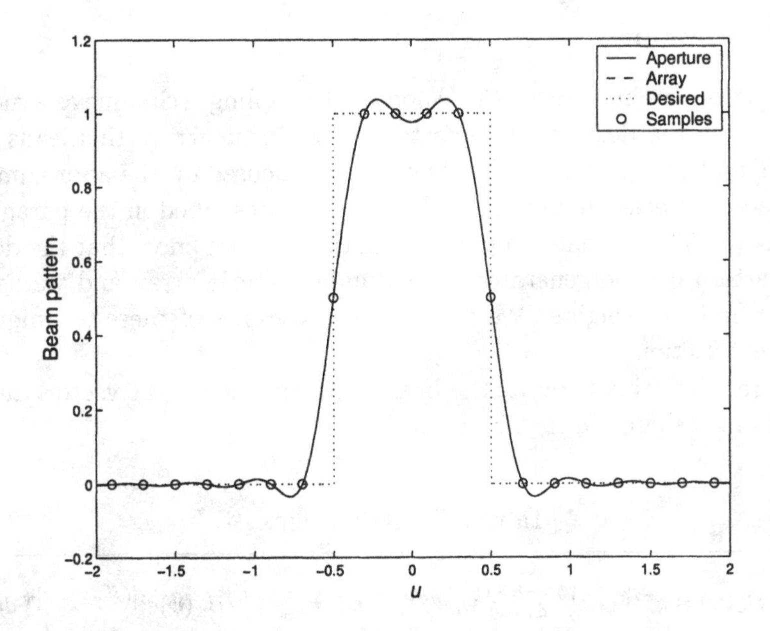

Figure 3.21 Array and aperture beam patterns using wavenumber sampling technique, symmetric samples from $-2.0 \leq u \leq 2.0$: $L = 5.0\lambda, d = \lambda/4, N = 20$.

Thus,

$$\mathbf{b}^H \mathbf{b} = \mathbf{w}^H \mathbf{w}. \tag{3.108}$$

Now define

$$[\mathbf{B}_\psi]_k = \mathbf{B}_\psi(\psi_k), \quad k = 0, 1, \cdots, N-1, \tag{3.109}$$

where ψ_k is given by (3.92). Then, from (3.94),

$$\mathbf{B} = \text{diag}\left\{e^{-j\psi_k\left(\frac{N-1}{2}\right)}\right\} \mathbf{B}_\psi^*. \tag{3.110}$$

Thus,

$$\mathbf{B}^H \mathbf{B} = \mathbf{B}_\psi^T \mathbf{B}_\psi^* = \mathbf{B}_\psi^H \mathbf{B}_\psi. \tag{3.111}$$

Using (3.108) and (3.111) in (3.106) gives

$$\boxed{\mathbf{w}^H \mathbf{w} = \frac{1}{N} \mathbf{B}_\psi^H \mathbf{B}_\psi,} \tag{3.112}$$

which is the desired result. Therefore, to find the directivity D, the white noise array gain A_w, or the sensitivity function T_{se}, we sample the beam pattern at the ψ_k given by (3.92), sum the magnitudes squared, and divide by N.

3.3.5 Summary

In this section, we introduced the Woodward sampling technique as a method to approximate a desired beam pattern. For linear arrays this leads us to the DFT and the inverse DFT. The examples focused on the approximation technique. The other usage of the IDFT that is described in the paragraphs following (3.103) is equally important. In this case we know that the desired beam pattern can be generated by a standard linear array and we use the IDFT to find the weights. We give further examples of these techniques in subsequent sections.

The relationships between the beam pattern, the weight vector, and the z-transform are summarized in Table 3.3.

Table 3.3 Relationships

1	$B_\psi(\psi) = e^{-j\left(\frac{N-1}{2}\right)\psi} \sum_{n=0}^{N-1} w_n^* e^{jn\psi}$	$w_n^* = \frac{1}{2\pi} \int_{-\pi}^{\pi} B_\psi(\psi) e^{j\left[\left(\frac{N-1}{2}\right)-n\right]\psi} \, d\psi$
2	z-Transform $B_\psi(\psi) = \left[z^{-\frac{N-1}{2}} B_z^*(z)\right]_{z=e^{j\psi}}$ (3.87)	
3	DFT $b_n = w_n e^{jn\pi\left(\frac{N-1}{N}\right)}$ (3.93) $B(k) = \sum_{n=0}^{N-1} b_n e^{-jkn\frac{2\pi}{N}}$ (3.95)	IDFT $b_n = \frac{1}{N} \sum_{k=0}^{N-1} B(k) e^{jkn\frac{2\pi}{N}}$ (3.100) $w_n = b_n e^{-jn\pi\left(\frac{N-1}{N}\right)}$ (3.101)

3.4 Minimum Beamwidth for Specified Sidelobe Level

3.4.1 Introduction

In this section, we consider the problem of minimizing the beamwidth for a given maximum sidelobe level.

The major focus of the section is on the Dolph-Chebychev weighting. It results in constant sidelobes and a beamwidth that is the minimum possible for the given sidelobe level. We first describe the classical Dolph-Chebychev synthesis technique and then show how it can be simplified by using the IDFT.

In many cases, a constant sidelobe behavior is not desirable. We develop a weighting due to Taylor that gives decreasing sidelobes.

We consider the case of linear uniformly spaced array pointed at broadside. The desired beam pattern is a real symmetric function, so the weights will be real and symmetric.

In this case it is useful to index the weights in a symmetric manner. For N odd, we define the weights as

$$a_m = w_n|_{n=m+\frac{N-1}{2}}, \quad m = -(N-1)/2, \cdots, (N-1)/2. \tag{3.113}$$

The assumption that $B(\psi)$ is a real symmetric function results in real symmetric weights,

$$a_m = a_{-m}, \tag{3.114}$$

and we can write $B(\psi)$ in a trigonometric form,

$$\boxed{B(\psi) = a_0 + 2\sum_{m=1}^{\frac{N-1}{2}} a_m \cos(m\psi), \quad N \text{ odd.}} \tag{3.115}$$

Similarly, for N even, we define

$$a_m = w_n|_{n=m-1+\frac{N}{2}}, \quad m = 1, 2, \cdots, \frac{N}{2}, \tag{3.116}$$

and

$$a_{-m} = a_m, \quad m = 1, 2, \cdots, \frac{N}{2}. \tag{3.117}$$

Then,

$$B(\psi) = \sum_{m=-\frac{N}{2}}^{-1} a_m\, e^{j[(m+\frac{1}{2})\psi]} + \sum_{m=1}^{\frac{N}{2}} a_m\, e^{j[(m-\frac{1}{2})\psi]}, \quad N \text{ even,} \tag{3.118}$$

which can be written as

$$\boxed{B(\psi) = 2\sum_{m=1}^{\frac{N}{2}} a_m \cos\left(\left(m - \tfrac{1}{2}\right)\psi\right), \quad N \text{ even.}} \tag{3.119}$$

The indexing for the cases of N odd and N even are shown in Figure 3.22.

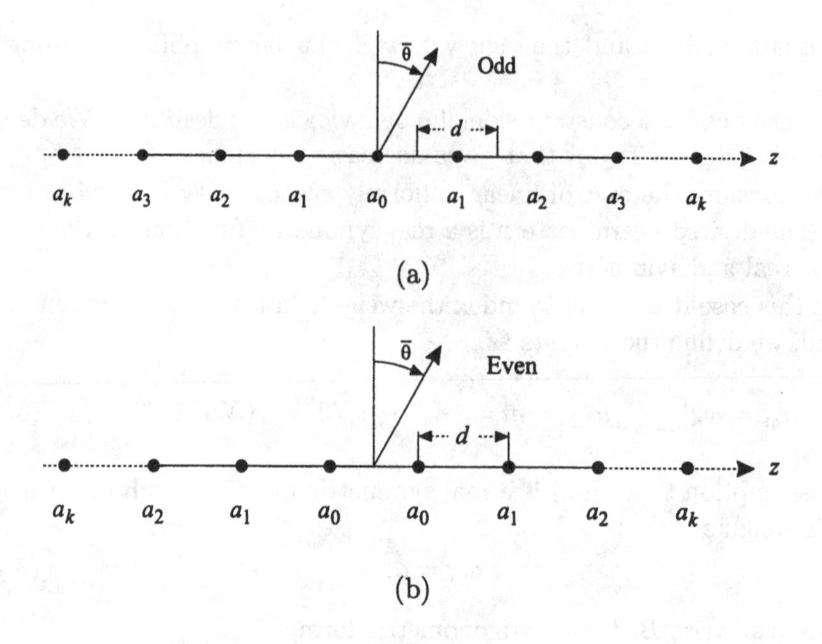

Figure 3.22 Linear broadside arrays of n isotropic sources with uniform spacing: (a) N odd; (b) N even.

3.4.2 Dolph-Chebychev Arrays

The spectral weightings control sidelobes by exploiting Fourier transform properties. However, they do not explicitly control the height of the sidelobes. For linear, equally spaced arrays there are methods to design weights that control sidelobes and beamwidths explicitly. The method that is discussed in this section was introduced by Dolph [Dol46], [Rib47] and is based upon the properties of Chebychev[13] polynomials. Our development follows that in Section 4.11 of Kraus [Kra88].

Consider the linear arrays shown in Figure 3.22. First, assume that there are an odd number of elements as shown in Figure 3.22(a), and that the main response axis is broadside. The weights are symmetric about the origin. In addition, we can use real weights because we want to design a real symmetric pattern.

Because the weights are real, the beam pattern can be written as (from (3.115))

$$B(\psi) = \alpha_0 + \alpha_1 \cos\psi + \alpha_2 \cos 2\psi$$

[13]There are various spellings of Chebychev in the literature (e.g., Tchebyscheff).

$$+\cdots+\alpha_{\frac{N-1}{2}}\cos\left[\left(\frac{N-1}{2}\right)\psi\right],\quad N\,\text{odd},\tag{3.120}$$

where

$$\psi=\frac{2\pi d}{\lambda}\cos\theta,\tag{3.121}$$

and α_n is defined as,

$$\alpha_n=\begin{cases}a_0, & n=0,\\ 2a_n, & n=1,\cdots,\frac{N-1}{2},\quad \text{for } N \text{ odd}.\end{cases}\tag{3.122}$$

We have deleted the ψ subscript on $B_\psi(\psi)$. This is the Fourier series expansion of the beam pattern. We can rewrite (3.120) as

$$B(\psi)=\sum_{k=0}^{\frac{N-1}{2}}\alpha_k\cos\left(2k\frac{\psi}{2}\right),\quad N\text{ odd}.\tag{3.123}$$

If N is even, the array is shown in Figure 3.22(b). Now, from (3.119),

$$B(\psi)=\alpha_1\cos\left(\frac{\psi}{2}\right)+\cdots+\alpha_{\frac{N}{2}}\cos\left[\left(\frac{N-1}{2}\right)\psi\right],\tag{3.124}$$

where

$$\alpha_n=2a_n,\quad n=1,\cdots,\frac{N}{2},\quad N\text{ even}.\tag{3.125}$$

Then, (3.124) can be written as

$$B(\psi)=\sum_{k=1}^{\frac{N}{2}}\alpha_k\cos\left[\left(\frac{2k-1}{2}\right)\psi\right],\quad N\text{ even}.\tag{3.126}$$

In this section, we want to derive an amplitude distribution that will produce a beam pattern with the minimum null-to-null beamwidth for a specified sidelobe level. The amplitude distribution is called the Dolph-Chebychev distribution.[14]

The first step is to show that $B(\psi)$ can be represented as a polynomial of order $N-1$. We represent the $\cos\left(m\frac{\psi}{2}\right)$ terms as a sum of $\cos^m\left(\frac{\psi}{2}\right)$ terms. We write

$$\begin{aligned}\exp\left[jm\frac{\psi}{2}\right] &= \cos\left(m\frac{\psi}{2}\right)+j\sin\left(m\frac{\psi}{2}\right)\\ &= \left[\cos\left(\frac{\psi}{2}\right)+j\sin\left(\frac{\psi}{2}\right)\right]^m.\end{aligned}\tag{3.127}$$

[14]The relationships in (3.123) and (3.126) were first given by Wolf [Wol37].

Expanding in a binomial series and taking the real part gives

$$\begin{aligned}\cos m\frac{\psi}{2} &= \cos^m\frac{\psi}{2} - \frac{m(m-1)}{2!}\cos^{m-2}\frac{\psi}{2}\sin^2\frac{\psi}{2} \\ &+\frac{m(m-1)(m-2)(m-3)}{4!}\cos^{m-4}\frac{\psi}{2}\sin^4\frac{\psi}{2} - \cdots.\end{aligned} \tag{3.128}$$

Putting $\sin^2(\psi/2) = 1 - \cos^2(\psi/2)$, and substituting particular values of m,[15] (3.128) reduces to the following:

$$\left.\begin{array}{ll} m=0, & \cos m\frac{\psi}{2} = 1 \\ m=1, & \cos m\frac{\psi}{2} = \cos\frac{\psi}{2} \\ m=2, & \cos m\frac{\psi}{2} = 2\cos^2\frac{\psi}{2} - 1 \\ m=3, & \cos m\frac{\psi}{2} = 4\cos^3\frac{\psi}{2} - 3\cos\frac{\psi}{2} \\ m=4, & \cos m\frac{\psi}{2} = 8\cos^4\frac{\psi}{2} - 8\cos^2\frac{\psi}{2} + 1 \end{array}\right\}. \tag{3.129}$$

Define

$$x = \cos\frac{\psi}{2}, \tag{3.130}$$

then (3.128) becomes

$$\left.\begin{array}{ll} \cos m\frac{\psi}{2} = 1, & \text{when} \quad m=0 \\ \cos m\frac{\psi}{2} = x, & \text{when} \quad m=1 \\ \cos m\frac{\psi}{2} = 2x^2-1, & \text{when} \quad m=2 \end{array}\right\}. \tag{3.131}$$

The polynomials of (3.131) are Chebychev polynomials, which may be designated in general by

$$T_m(x) = \cos\left(m\frac{\psi}{2}\right)\Big|_{\cos\left(\frac{\psi}{2}\right)=x}. \tag{3.132}$$

For particular values of m, the first eight Chebychev polynomials are

$$\left.\begin{array}{l} T_0(x) = 1 \\ T_1(x) = x \\ T_2(x) = 2x^2 - 1 \\ T_3(x) = 4x^3 - 3x \\ T_4(x) = 8x^4 - 8x^2 + 1 \\ T_5(x) = 16x^5 - 20x^3 + 5x \\ T_6(x) = 32x^6 - 48x^4 + 18x^2 - 1 \\ T_7(x) = 64x^7 - 112x^5 + 56x^3 - 7x \end{array}\right\}. \tag{3.133}$$

[15] m will equal $2k$ or $2k+1$ depending on whether N is odd or even.

We note that the degree of the polynomial in (3.133) is the same as the value of m.

The mth-degree Chebychev polynomial is defined as

$$T_m(x) = \begin{cases} \cos\left(m\cos^{-1}x\right), & |x| \leq 1, \\ \cosh\left(m\cosh^{-1}x\right), & x > 1, \\ (-1)^m \cosh\left(m\cosh^{-1}|x|\right), & x < -1. \end{cases} \tag{3.134}$$

The Chebychev polynomials are a set of functions that are orthogonal over the interval $-1 \leq x \leq 1$ with respect to the weighting function $w(x) = 1/\sqrt{1-x^2}$ (e.g., [AS65]),

$$\begin{aligned} \int_{-1}^{1} \frac{1}{\sqrt{1-x^2}} T_m(x)\, T_n(x)\, dx &= \\ \int_0^{\pi} T_m(\cos\theta)\, T_n(\cos\theta)\, d\theta &= c_m \delta_{mn}. \end{aligned} \tag{3.135}$$

The orthogonality constant c_m is

$$c_m = \begin{cases} \pi, & m = 0, \\ \frac{\pi}{2}, & m \neq 0. \end{cases} \tag{3.136}$$

The polynomials can be extended beyond the region $|x| \leq 1$ as defined in (3.134). The Chebychev polynomials of order $m = 2$ through $m = 4$ are shown in Figure 3.23. The following properties of Chebychev polynomials are useful for our development.

1. For $m \geq 2$,

$$T_m(x) = 2xT_{m-1}(x) - T_{m-2}(x), \tag{3.137}$$

where $T_0(x)$ and $T_1(x)$ are given in (3.133).

2. $T_m(x)$ has m real roots in the interval $|x| < 1$. The roots of the polynomials occur when $\cos\left(m(\psi/2)\right) = 0$ or when

$$m\frac{\psi}{2} = (2p-1)\frac{\pi}{2}, \quad p = 1, \cdots, m. \tag{3.138}$$

Thus, they are evenly spaced in ψ-space. The roots of x, designated x_p, are

$$x_p = \cos\left[(2p-1)\frac{\pi}{2m}\right]. \tag{3.139}$$

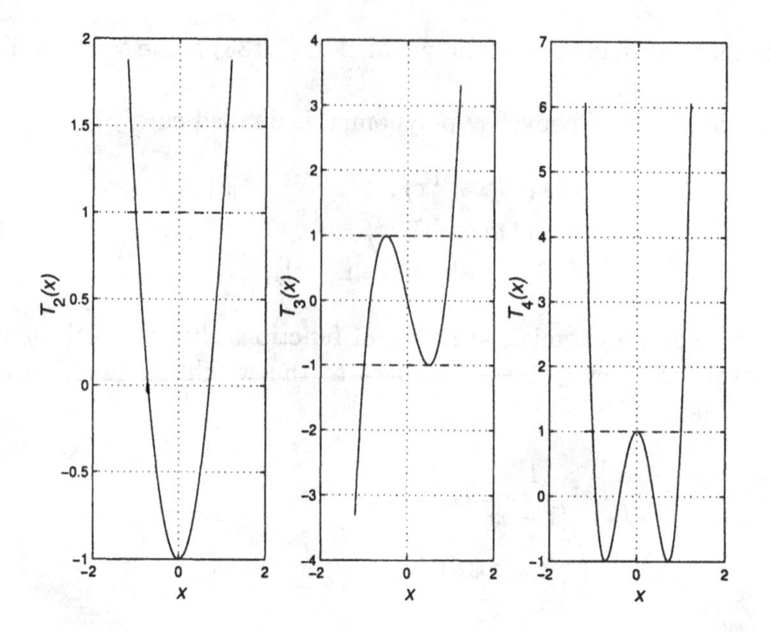

Figure 3.23 Chebychev polynomials: (a) $m = 2$; (b) $m = 3$; (c) $m = 4$.

3. $T_m(x)$ has alternating maxima and minima in the interval $-1 < x < 1$ that occur at

$$x_k = \cos\left(\frac{k\pi}{m}\right), \quad k = 1, 2, \cdots, m-1. \tag{3.140}$$

The magnitude of every maxima and minima is unity,

$$|T_m(x_k)| = 1. \tag{3.141}$$

Thus, the polynomials have an equal ripple characteristic in the interval $-1 < x < 1$.

4. All of the polynomials pass through the point (1, 1) and at $x = \pm 1, |T_m(\pm 1)| = 1$. For $x > 1$,

$$|T_m(x)| > 1. \tag{3.142}$$

From (3.129), the beam pattern, $B(\psi)$, for a symmetric, equally spaced array aimed at broadside (isotropic sources) is a polynomial of degree $N-1$ in the variable $\cos(\psi/2)$.

If we set this array polynomial equal to the Chebychev polynomial of the same degree $N-1$ and equate the array coefficients to the Chebychev

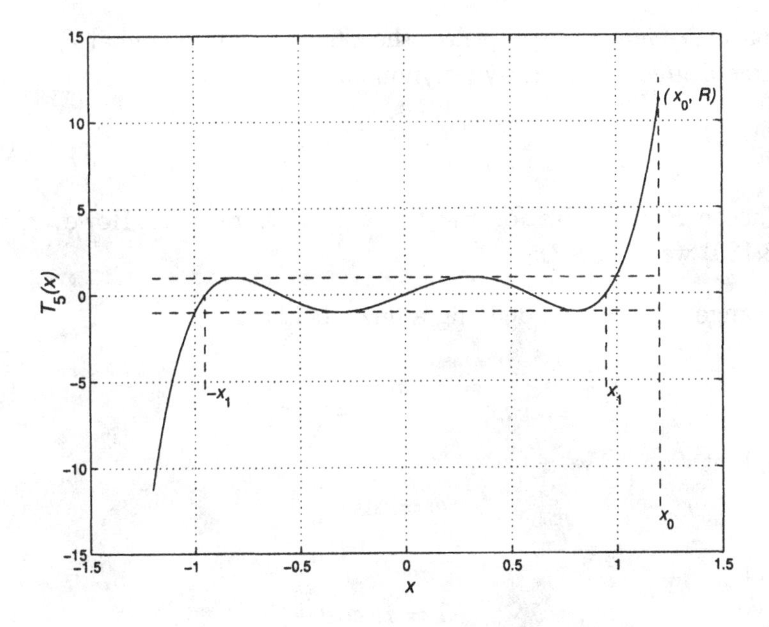

Figure 3.24 $T_5(x)$ with coordinate axes.

polynomial coefficients, then the beam pattern of the array will correspond to a Chebychev polynomial of degree $N-1$.

Dolph [Dol46] developed a procedure to utilize the properties of the Chebychev polynomials to develop an optimum pattern.

In the Dolph-Chebychev method the magnitude of the main lobe corresponds to the value of $T_m(x_0)$ where $x_0 > 1$ and the magnitude of the sidelobes is unity. We define the ratio of the main-lobe maximum to the sidelobe level as R,

$$R = \frac{\text{main-lobe maximum}}{\text{sidelobe level}}. \tag{3.143}$$

To illustrate this, consider the $T_5(x)$ curve in Figure 3.24. The point (x_0, R) on the $T_5(x)$ polynomial curve corresponds to the main-lobe maximum. From (3.134),

$$T_{N-1}(x_0) = \cosh\left((N-1)\cosh^{-1}x_0\right) = R, \quad |x_0| > 1, \tag{3.144}$$

or

$$x_0 = \cosh\left(\frac{1}{N-1}\cosh^{-1}R\right), \quad |x_0| > 1. \tag{3.145}$$

For example, if $R = 20$, the sidelobes will be 26 dB below the main-lobe maximum. The value $R = 31.62$ would lead to -30-dB sidelobes.

The synthesis procedure consists of five steps:

1. For an N-element array, select the Chebychev polynomial $T_m(x)$ of the same degree as the array polynomial. Thus,

$$m = N - 1. \tag{3.146}$$

2. Choose R and solve for x_0. Since $R > 1$, $x_0 > 1$. However, to use (3.130) we require $|x| < 1$.

3. Change the scale by defining a new abcissa w,

$$w = \frac{x}{x_0}, \tag{3.147}$$

and let

$$w = \cos\left(\frac{\psi}{2}\right), \tag{3.148}$$

and

$$x = x_0 \cos(\frac{\psi}{2}). \tag{3.149}$$

4. The beam pattern is

$$B(\psi) = \frac{1}{R} T_{N-1}(x_0 \cos(\frac{\psi}{2})). \tag{3.150}$$

The $1/R$ factor normalizes the beam pattern so that $B(0) = 1$.

5. The last step is to find the array weights to produce the beam pattern in (3.150).

The easiest way to find the weight vector is to find the zeros of the beam pattern and then use (2.88). The original zeros are given by (3.138),

$$\frac{\psi_{po}}{2} = \frac{(2p-1)\pi}{(N-1)2}, \quad p = 1, \cdots, N-1, \tag{3.151}$$

or in x-space,

$$x_p = \cos\left(\frac{(2p-1)\pi}{(N-1)2}\right), \quad p = 1, \cdots, N-1. \tag{3.152}$$

Changing the scale into w-space gives

$$w_p = \frac{1}{x_0} \cos\left(\frac{(2p-1)\pi}{(N-1)2}\right), \quad p = 1, \cdots, N-1. \tag{3.153}$$

Using (3.148) gives the zeros in ψ-space,

$$\psi_p = 2\cos^{-1}\left(\frac{1}{x_0}\cos\left(\frac{(2p-1)\pi}{(N-1)2}\right)\right), \quad p = 1, \cdots, N-1. \tag{3.154}$$

We then construct a $N \times N$ array manifold matrix, $\mathbf{V}(\psi)$, (2.81),

$$\mathbf{V}(\psi) = \left[\begin{array}{ccccccc} \mathbf{v}(0) & \mathbf{v}(\psi_1) & \cdots & \mathbf{v}(\psi_p) & \cdots & \mathbf{v}(\psi_{N-1}) \end{array}\right], \tag{3.155}$$

and use (2.88)

$$\mathbf{w} = \left[\mathbf{V}^H(\psi)\right]^{-1}\mathbf{e}_1, \tag{3.156}$$

which specifies the weight vector. We can translate from w_n to a_n, if desired, but it is not necessary.[16]

The results in (3.151)–(3.153) complete the Dolph-Chebychev synthesis process.

We now consider a typical example and then discuss the optimality of the result.

Example 3.4.1[17]

Consider an array of eight isotropic sources spaced at $d = \lambda/2$. Design a Dolph-Chebychev weighting with the sidelobes at -26 dB.

In this case, $R = 20$, so we set

$$T_7(x_0) = 20. \tag{3.157}$$

From (3.145),

$$x_0 = \cosh\left(\frac{1}{7}\cosh^{-1}(20)\right) = 1.142. \tag{3.158}$$

Then,

$$B(\psi) = \frac{1}{20}T_7\left(x_0 \cos\frac{\psi}{2}\right). \tag{3.159}$$

We use (3.156) to find the weights

$$\left.\begin{array}{l} w_0 = a_4 = 0.0633 \\ w_1 = a_3 = 0.1035 \\ w_2 = a_2 = 0.1517 \\ w_3 = a_1 = 0.1815 \end{array}\right\}. \tag{3.160}$$

A plot of $T_7(x)$ and the mapping in (3.159) is shown in Figure 3.25. As ψ moves from 0 to π, $B(\psi)$ moves from R through three zero crossings to 0 at $\psi = \pi$. Since the beam pattern is symmetric, this gives us a complete description.

The movement described in Figure 3.25 is summarized in Table 3.4.

---

[16]Stegen [Ste53] developed a procedure for finding an explicit formula for the coefficients. It corresponds to the IDFT approach in Section 3.3.3. We discuss this approach in Problem 3.4.18.

[17]This example is contained in [Kra88], but we use a simpler technique.

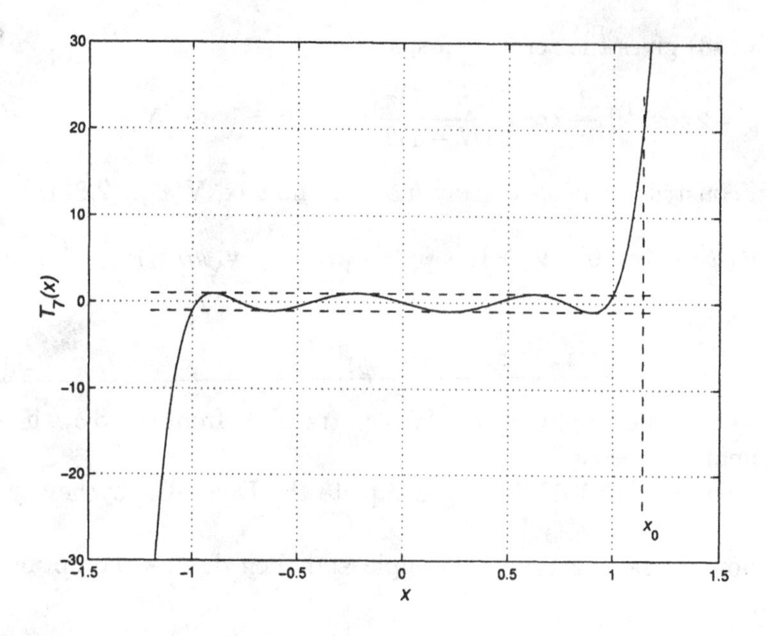

Figure 3.25 Chebychev polynomial of the seventh degree.

Table 3.4 Beam Pattern Movement

θ	0	$\frac{\pi}{2}$	π
$\psi = \pi \cos\theta$	π	0	$-\pi$
$x = x_0 \cos\left(\frac{\psi}{2}\right)$	0	x_0	0

We see that we have mapped the visible region in θ-space $(0, \pi)$ into a region in x-space $(0, x_0)$. We have not utilized the Chebychev polynomial in the region $(-1, 0)$. This is because we used an expression in $\psi/2$ and that the array spacing is $d = \lambda/2$.

For other values of d, the mapping is shown in Table 3.5.

Table 3.5 Mapping as a Function of d

θ	0	$\frac{\pi}{2}$	π
$\psi = \frac{2\pi}{\lambda} d \cos\theta$	$\frac{2\pi}{\lambda} d$	0	$-\frac{2\pi}{\lambda} d$
$x = x_0 \cos\left(\frac{\psi}{2}\right)$	$x_0 \cos\left(\frac{\pi d}{\lambda}\right)$	x_0	$x_0 \cos\left(\frac{\pi d}{\lambda}\right)$

For some representative values of d, the ranges on the x-axis are:

(i) $d = \frac{\lambda}{4}, \quad 0.707 x_0 \leq x \leq x_0,$

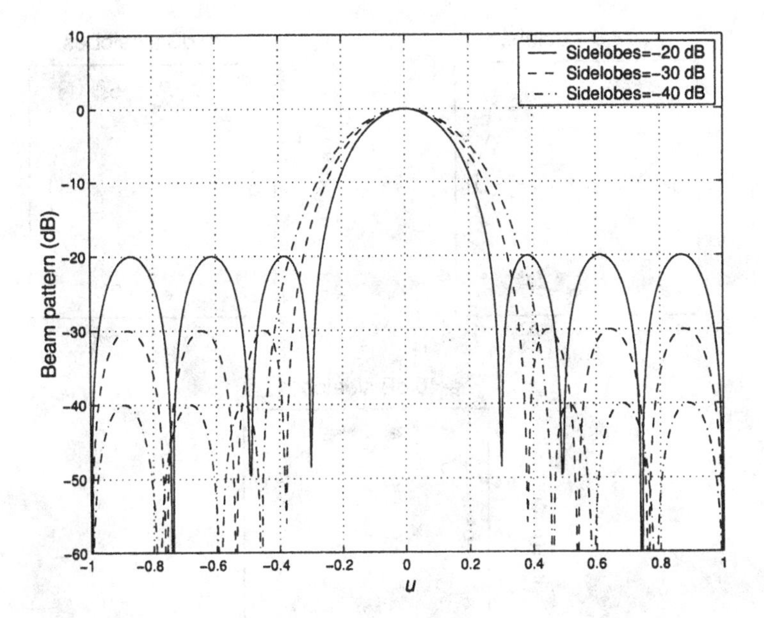

Figure 3.26 Beam patterns, Dolph-Chebychev weights: $N = 8$ (−20-, −30-, −40-dB sidelobes).

(ii) $d = \frac{\lambda}{2}, \quad 0 \leq x \leq x_0,$

(iii) $d = \frac{3\lambda}{4}, \quad -0.707x_0 \leq x \leq x_0.$

The maximum d that we can use corresponds to the point where the left end of the range is -1:

$$d = \frac{\lambda}{\pi} \cos^{-1}\left(-\frac{1}{x_0}\right). \tag{3.161}$$

In Figure 3.26, we show the beam patterns for an 8-element standard linear array for various sidelobe levels. As expected, the beamwidth increases as the sidelobe level decreases. In Figure 3.27, we show a z-plane plot of the zeros of the Dolph-Chebychev weighting for −20-dB, −30-dB, and −40-dB sidelobes.

In his original work, Dolph [Dol46] showed the optimality of the Dolph-Chebychev polynomials. In this case, the optimum beam pattern is defined to be the beam pattern with the smallest null-null beamwidth for a given sidelobe level. We discuss it in the context of $T_5(x)$ in Figure 3.24.[18] Consider another polynomial $P(x)$ of degree 5 that passes through (x_0, R) and

[18]This discussion follows p.170 of [Kra88]

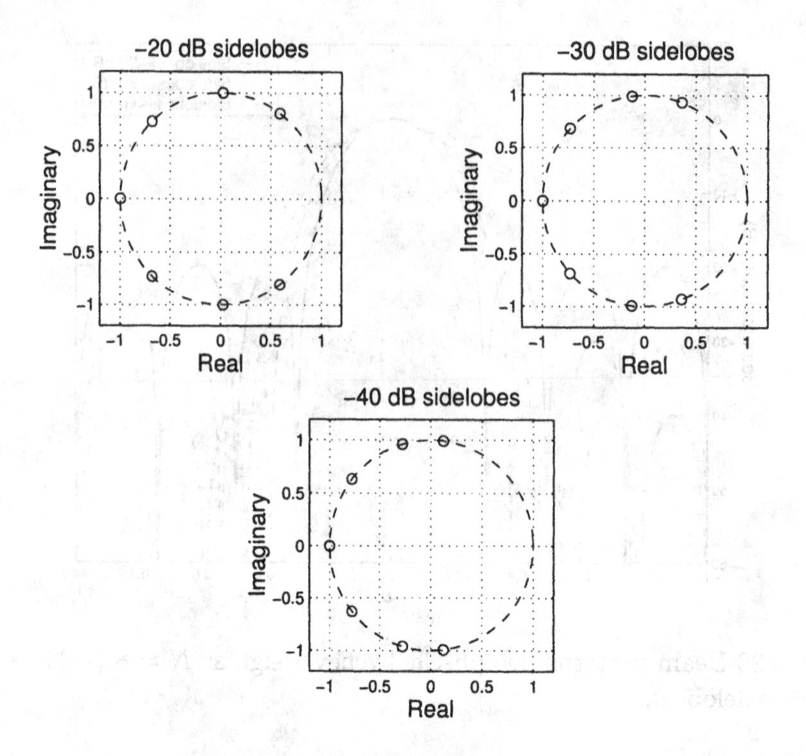

Figure 3.27 Zero plots of the Dolph-Chebychev weighting: (a) −20-dB sidelobes, (b) −30-dB sidelobes, (c) −40-dB sidelobes.

$x_1^{'}$ (the largest root) and, for $x < x_1^{'}$, it lies between +1 and −1. The null-null beamwidth is determined by $x_1^{'}$. We want to try and find a $P(x)$ whose values lies between $(1-\epsilon)$ and $-(1-\epsilon)$ for some positive ϵ. However, since $P(x)$ lies between ± 1 in the range $[-x_1^{'}, x_1^{'}]$, it must intersect $T_5(x)$ in at least $m+1 = 6$ points, including (x_0, R). But two polynomials of the same degree that intersect at $m+1$ points must be the same polynomial, so $P(x) = T_5(x)$. Thus, the Chebychev polynomial is optimum. Riblet [Rib47] subsequently showed that the procedure was only optimum for $d \geq \lambda/2$ and proposed an improved procedure for $d < \lambda/2$. The non-optimality for $d < \lambda/2$ is because of the mapping in Table 3.5. For $d < \lambda/2$, the range on the x-axis is not adequate to constrain the beam pattern everywhere in the visible region.

We describe the Riblet-Chebychev technique briefly. For $d = \lambda/2$, the two techniques give the same result.

Riblet-Chebychev weighting

In this case we use an expansion of $\cos\psi$ rather than $\cos(\psi/2)$. Assuming N is odd, we then match the array polynomial to a Chebychev polynomial of order $(N-1)/2$. In addition, we choose a mapping such that θ moves from 0 to π, so we use the entire range $(-1, x_0)$ of the Chebychev polynomial.

Thus, the mapping is

$$x = c_1 \cos\psi + c_2, \tag{3.162}$$

where

$$\psi = \frac{2\pi}{\lambda} d \cos\theta, \tag{3.163}$$

and, from (3.134),

$$x_0 = \cosh\left(\frac{2}{N-1}\cosh^{-1} R\right). \tag{3.164}$$

The required mapping is shown in Table 3.6.

Table 3.6 Mapping Requirement

θ	0	$\frac{\pi}{2}$	π
$\psi = \frac{2\pi}{\lambda} d \cos\theta$	$\frac{2\pi}{\lambda} d$	0	$-\frac{2\pi}{\lambda} d$
$x = c_1 \cos\psi + c_2$	-1	x_0	-1

To find the constants, we require

$$c_1 + c_2 = x_0, \tag{3.165}$$

$$c_1 \cos\left(\frac{2\pi}{\lambda} d\right) + c_2 = -1. \tag{3.166}$$

Solving for c_1 and c_2 and substituting into (3.162), we have

$$x = \frac{1}{1 - \cos\left(\frac{2\pi}{\lambda} d\right)} \left\{ (x_0 + 1)\cos\psi - \left[1 + x_0 \cos\left(\frac{2\pi}{\lambda} d\right)\right]\right\}. \tag{3.167}$$

For $d = \lambda/2$, (3.167) reduces to

$$x = \frac{1}{2}\left\{(x_0 + 1)\cos\psi + (x_0 - 1)\right\}. \tag{3.168}$$

One can show that the two approaches lead to identical arrays for $d = \lambda/2$. For $d < \lambda/2$, Riblet's approach leads to arrays with smaller beamwidth for a given sidelobe level. We consider an example to illustrate the behavior.

Example 3.4.2

Consider a 21-element linear array. We require −30-dB sidelobes and design the array using both the Dolph-Chebychev and Riblet-Chebychev procedures. We consider element

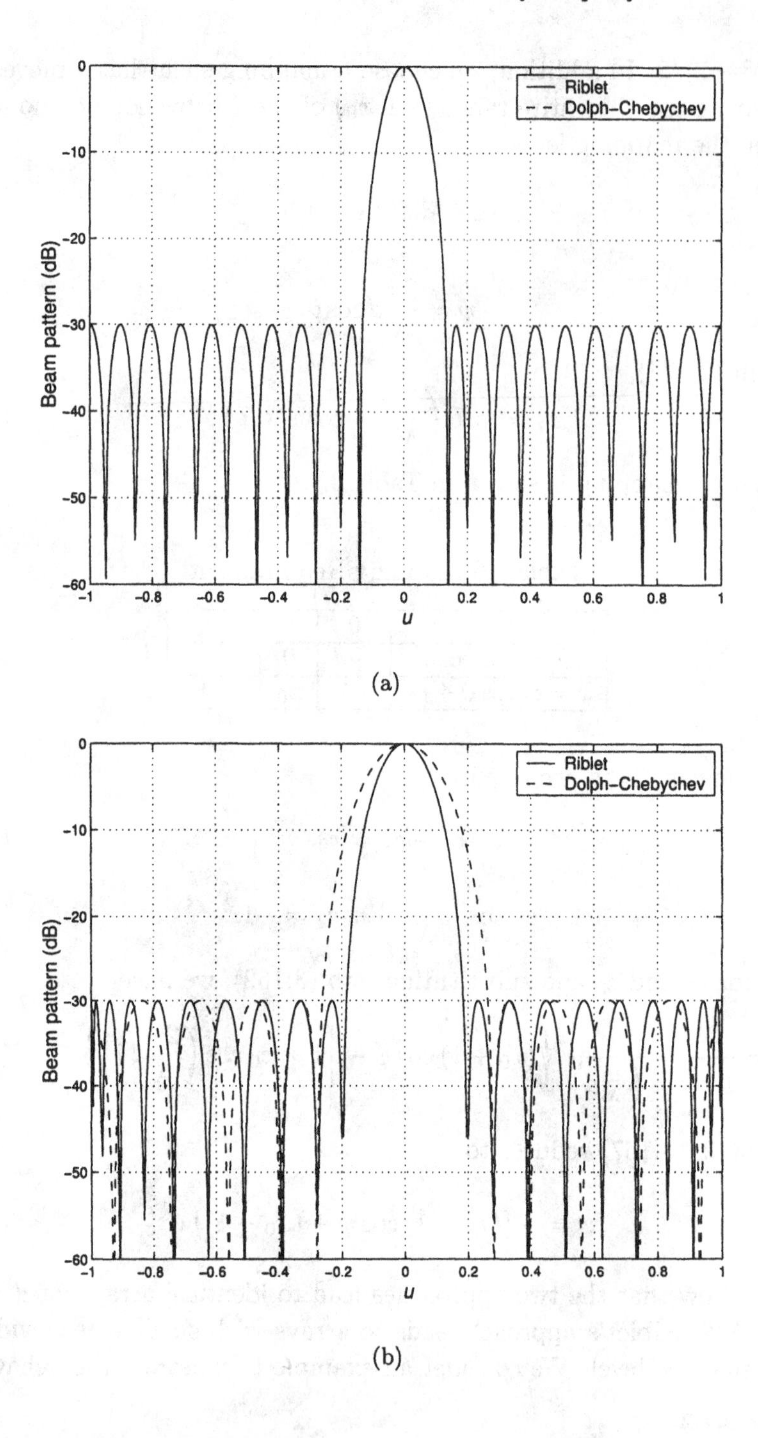

Figure 3.28 Beam patterns for Dolph-Chebychev and Riblet weighting: $N = 21$, -30-dB sidelobes: (a) $d = \lambda/2$, (b) $d = \lambda/4$.

spacings of $d = \lambda/2$ and $\lambda/4$. The results are shown in Figure 3.28. We see that there is significant difference in the main-lobe width.

Riblets' procedure is only applicable to the case when N is odd and $N \geq 7$.

Subsequent to Dolph's original work in 1946 [Dol46] and Riblet's discussion in 1947 [Rib47], work has been done to simplify the synthesis procedure and derive the beamwidth and directivity characteristics of Dolph -Chebychev arrays.

The Dolph-Chebychev weighting is used in a number of array applications because of its constant sidelobe behavior. In many applications, we would prefer that the sidelobes decay rather than remain constant. The motivation for decaying sidelobes is to reduce the effect of interferers that are located at angles that are significantly different from the MRA. In the next section, we discuss a weighting developed by Taylor [Tay53] to achieve this behavior.

3.4.3 Taylor Distribution

In [Tay53] and [Tay55], Taylor developed a technique that constrains the maximum sidelobe height and gives decaying outer sidelobes. The development is for a linear aperture. It can be applied to linear arrays by sampling the aperture weighting or by root-matching.[19] We can also solve for the array weighting directly (see Section 3.4.4).

Taylor starts with the uniform aperture weighting that we used in Section 2.7. The resulting beam pattern was derived and plotted in Figure 2.29. We define

$$v = \frac{L}{\lambda} u, \tag{3.169}$$

so that

$$B_v(v) = \frac{\sin(\pi v)}{\pi v} = \mathrm{sinc}(\pi v). \tag{3.170}$$

With this change of variables, all of the zeros of the pattern are located on the integers,

$$v = \pm 1, \pm 2, \cdots. \tag{3.171}$$

Taylor approaches the synthesis problem by moving the "inner" zeros to new locations on the unit circle in order to lower the inner sidelobes and leaving the outer zeros in the same location as the uniform distribution in order to maintain the $|v|^{-1}$ decay.

[19]This discussion follows Elliott [Ell81].

We first write (3.170) as an infinite product,

$$B_u(v) = \prod_{n=1}^{\infty}\left(1 - \frac{v^2}{n^2}\right). \tag{3.172}$$

We then define a new pattern,

$$B_T(v) = \frac{\sin(\pi v)}{\pi v}\frac{\prod_{n=1}^{\bar{n}-1}\left(1 - \frac{v^2}{v_n^2}\right)}{\prod_{n=1}^{\bar{n}-1}\left(1 - \frac{v^2}{n^2}\right)}. \tag{3.173}$$

The denominator removes the first $\bar{n}-1$ pairs of zeros and the numerator replaces them with $\bar{n}-1$ new pairs of zeros located at v_n, $n = 1, 2, \cdots, \bar{n}-1$.

Taylor showed that the new zeros should be located at

$$v_n = \bar{n}\left[\frac{A^2 + (n - \frac{1}{2})^2}{A^2 + (\bar{n} - \frac{1}{2})^2}\right]^{\frac{1}{2}}, \tag{3.174}$$

where

$$\cosh(\pi A) = R, \tag{3.175}$$

and R is the same as in the Dolph-Chebychev derivation. Thus, the maximum sidelobe height is $-20\log_{10}R$.

To find the corresponding aperture weighting, we recall from (2.215) that

$$B_T(v) = \int_{-\frac{L}{2}}^{\frac{L}{2}} w^*(z)\, e^{j\frac{2\pi}{L}vz}\, dz. \tag{3.176}$$

Since the weighting will be symmetric we can expand $w^*(z)$ in a Fourier cosine series:

$$w^*(z) = \sum_{m=0}^{\infty} c_m \cos\left(\frac{2m\pi z}{L}\right), \tag{3.177}$$

where c_m is a real constant. Using (3.177) in (3.176) gives,

$$B_T(v) = \int_{-\frac{L}{2}}^{\frac{L}{2}} \sum_{m=0}^{\infty} c_m \cos\left(\frac{2\pi mz}{L}\right) \cos\left(\frac{2\pi vz}{L}\right) dz. \tag{3.178}$$

We now choose integer values of the pattern $v = 0, 1, 2, \cdots$. Then, the integral will be zero unless $v = m$. Therefore,

$$L\, B_T(0) = c_0, \tag{3.179}$$

$$\frac{L}{2}B_T(m) = c_m, \quad m \leq \bar{n}-1, \tag{3.180}$$

and, from (3.173), $B_T(m) = 0$ for $m \geq \bar{n}$. Thus,

$$w^*(z) = \frac{1}{L}\left\{B_T(0) + 2\sum_{m=1}^{\bar{n}-1} B_T(m)\cos\left(\frac{2m\pi z}{L}\right)\right\}, \tag{3.181}$$

which is the desired result.

To find the corresponding weighting for an array, we can either sample $w^*(z)$ in (3.181) or we can match the zeros in (3.174).

To carry out the latter procedure, we define the zeros in u-space as,

$$u_n = \frac{\lambda}{Nd}\left\{\bar{n}\left[\frac{A^2 + (n-\frac{1}{2})^2}{A^2 + (\bar{n}-\frac{1}{2})^2}\right]^{\frac{1}{2}}\right\}. \tag{3.182}$$

The weight vector for a standard linear array is obtained by using (2.88) with

$$\psi_n = \frac{2\pi}{N}\left\{\bar{n}\left[\frac{A^2 + (n-\frac{1}{2})^2}{A^2 + (\bar{n}-\frac{1}{2})^2}\right]^{\frac{1}{2}}\right\}. \tag{3.183}$$

We now consider a simple example to illustrate the result.

Example 3.4.3

Consider a linear aperture of length $L = 10.5\lambda$ and a corresponding linear array with $d = \lambda/2$ so that $N = 21$. We require -30-dB sidelobes and utilize $\bar{n} = 6$. In Figure 3.29(a), we show the beam pattern for the aperture obtained from (3.173) and the beam pattern for the array obtained by sampling $w^*(z)$ given in (3.181). In Figure 3.29(b), we show the Dolph-Chebychev pattern from Figure 3.28(a) and the Taylor pattern.

The Taylor distribution is widely used in practice. After the desired sidelobe level is chosen, $\bar{n}$ has to be selected. If $\bar{n}$ is too large, the aperture weighting function will increase as z approaches L. Usually, $\bar{n}$ is chosen to have the largest value that does not cause $w^*(z)$ to increase as z increases (e.g., [Han98]).

There are several modifications to the Taylor pattern that are discussed in the antenna literature. For example, Elliott ([Ell75], [Ell81]) has developed techniques for designing modified Taylor patterns in which sidelobe heights can be individually specified. In Section 3.9, we develop techniques for controlling the sidelobe heights in specified regions.

The technique of assuming a continuous aperture and then finding a discretized weighting works well for large N. We can also solve the array problem directly using a technique invented by Villeneuve that is described in the next section.

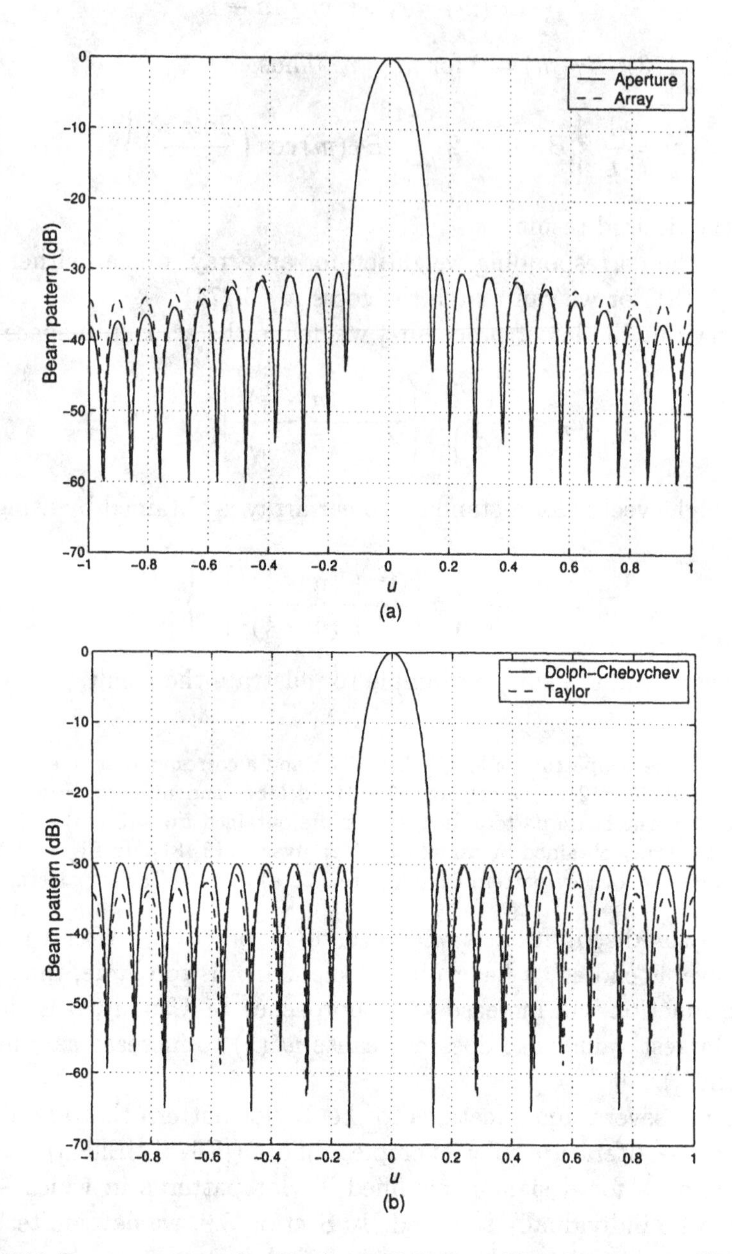

Figure 3.29 Beam pattern for Taylor weighting: $N = 21, \overline{n} = 6$, SLL $= -30$ dB (a) aperture and array; (b) array with Taylor weighting and Dolph-Chebychev weighting.

3.4.4 Villeneuve $\bar{n}$ Distribution

The Chebychev distribution for an array with $N = 2M + 1$ elements can be written as,

$$T_{2M}\left(x_0 \cos\frac{\psi}{2}\right) = cB(e^{j\psi}) = \prod_{p=1}^{2M}(z - z_p), \tag{3.184}$$

where

$$z = e^{j\psi}, \tag{3.185}$$

$$z_p = e^{j\psi_p}, \tag{3.186}$$

and c is a normalizing constant. We consider N odd in the text; the derivation for N even is similar.

The roots are given by (3.154),

$$\psi_p = 2\cos^{-1}\left(\frac{1}{x_0}\cos\left[(2p-1)\frac{\pi}{4M}\right]\right), \quad p = 1, 2, \cdots, 2M. \tag{3.187}$$

The beam pattern can be written as

$$B_\psi(\psi) = e^{jM\psi}\, 4^M \prod_{p=1}^{M} \sin\left(\frac{\psi - \psi_p}{2}\right)\sin\left(\frac{\psi + \psi_p}{2}\right), \tag{3.188}$$

which is the Chebychev pattern for $N = 2M + 1$ elements.

Villeneuve [Vil84] (e.g., [Kum92]) developed a technique for discrete arrays. He combines the better features of the uniform and Chebychev weightings. We start with the beam pattern for uniform weighting and replace the first $\bar{n} - 1$ roots with modified Dolph-Chebychev roots. The resulting beam pattern is

$$\begin{aligned} B_\psi(\psi) &= \frac{\sin\left(\frac{N}{2}\psi\right)}{\sin\left(\frac{\psi}{2}\right)} \\ &\quad \cdot \left[\frac{\prod_{n=1}^{\bar{n}-1}\sin\left(\frac{\psi-\psi_n'}{2}\right)\sin\left(\frac{\psi+\psi_n'}{2}\right)}{\prod_{n=1}^{\bar{n}-1}\sin\left(\frac{\psi-\psi_{un}}{2}\right)\sin\left(\frac{\psi+\psi_{un}}{2}\right)}\right]. \end{aligned} \tag{3.189}$$

The term in brackets modifies the beam pattern of the uniform weighting. In the numerator are the modified Chebychev roots corresponding to the first $\bar{n}-1$ interior sidelobes. In the denominator are the corresponding roots of the uniform weighting,

$$\psi_{un} = \frac{2\pi n}{N}, \quad n = 1, 2, \cdots, \bar{n} - 1. \tag{3.190}$$

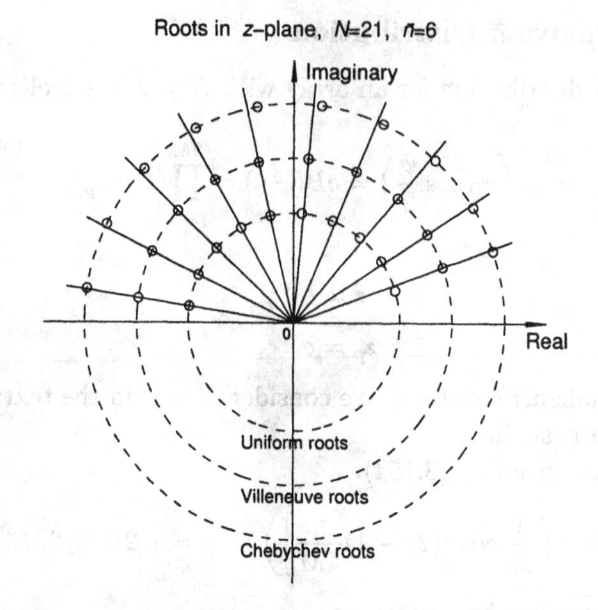

Figure 3.30 Roots of the three beam patterns. The three unit circles are shown with different radii. Only upper half-plane roots are shown.

The Chebychev roots are modified so there is not a jump at $n = \bar{n}$. This modification is accomplished by a progressive shift such that the $\bar{n}$th root of uniform weighting is unchanged.

The Chebychev zeros are given in (3.187). Each Chebychev zero is multiplied by

$$\sigma = \frac{2\pi\bar{n}}{N\psi_{\bar{n}}}, \tag{3.191}$$

where $\psi_{\bar{n}}$ is the $\bar{n}$th Chebychev root. Therefore,

$$\sigma = \frac{\pi\bar{n}}{N\cos^{-1}\left(\frac{1}{x_0}\cos\left[(2\bar{n}-1)\frac{\pi}{4M}\right]\right)}. \tag{3.192}$$

Then,

$$\psi'_n = \sigma\psi_n, \quad n = 1, \cdots, \bar{n}-1. \tag{3.193}$$

The remaining roots are uniform,

$$\psi_{un} = \frac{2\pi}{N}n, \quad n = \bar{n}, \cdots, \frac{N-1}{2}. \tag{3.194}$$

The resulting root patterns are shown in Figure 3.30 for $N = 21$. For clarity, we have shown three unit circles. We only show the roots in the upper half. The remaining roots are complex conjugates. The inner six (plus their six complex conjugates) Chebychev roots are moved outward. The amount of movement increases until, at $\bar{n} = 6$, the Villenueve root corresponds to the uniform root. The remaining roots are the uniform roots. The shift in roots is reasonably small, but it results in a significant pattern change. A beam pattern for a 21-element and 41-element array with −20-dB sidelobes are shown in Figure 3.31. It is essentially the same as the discretized Taylor beam pattern. For $N \leq 11$, there is some difference in the beam patterns and the Villeneuve technique is preferable.

The weight vector is obtained by using (3.193), (3.194) and their symmetrical values in either(2.88) or (3.188).

Villeneuve [Vil84] also derives the results for N even and provides further comparisons with the Taylor distribution. Hansen ([Han85], [Han92]) has studied the properties of the distribution. [Han98] has tables with detailed comparisons of Taylor and Villeneuve $\bar{n}$ beam patterns.

3.5 Least Squares Error Pattern Synthesis

In this section, we develop techniques for finding the least squares error approximation to a desired beam pattern. The technique is valid for an arbitrary array geometry. For notational simplicity, the 1-D case is discussed in the text.

The desired beam pattern is $B_d(\psi)$. The square error is defined as

$$\xi = \int_{-\pi}^{\pi} | B_d(\psi) - \mathbf{w}^H \mathbf{v}(\psi) |^2 \, d\psi. \tag{3.195}$$

A more general error expression includes a weighting function inside the integral. The expression in (3.195) is adequate for the present discussion. Taking the complex gradient with respect to $\mathbf{w}^H$ and setting the result equal to zero gives[20]

$$-\int_{-\pi}^{\pi} \mathbf{v}(\psi) B_d^*(\psi) + \left\{ \int_{-\pi}^{\pi} \mathbf{v}(\psi) \mathbf{v}^H(\psi) d\psi \right\} \mathbf{w}_o = \mathbf{0}. \tag{3.196}$$

Defining

$$\mathbf{A} = \int_{-\pi}^{\pi} \mathbf{v}(\psi) \mathbf{v}^H(\psi) d\psi, \tag{3.197}$$

[20]See Appendix A (Section A.7.4) for a discussion of complex gradients.

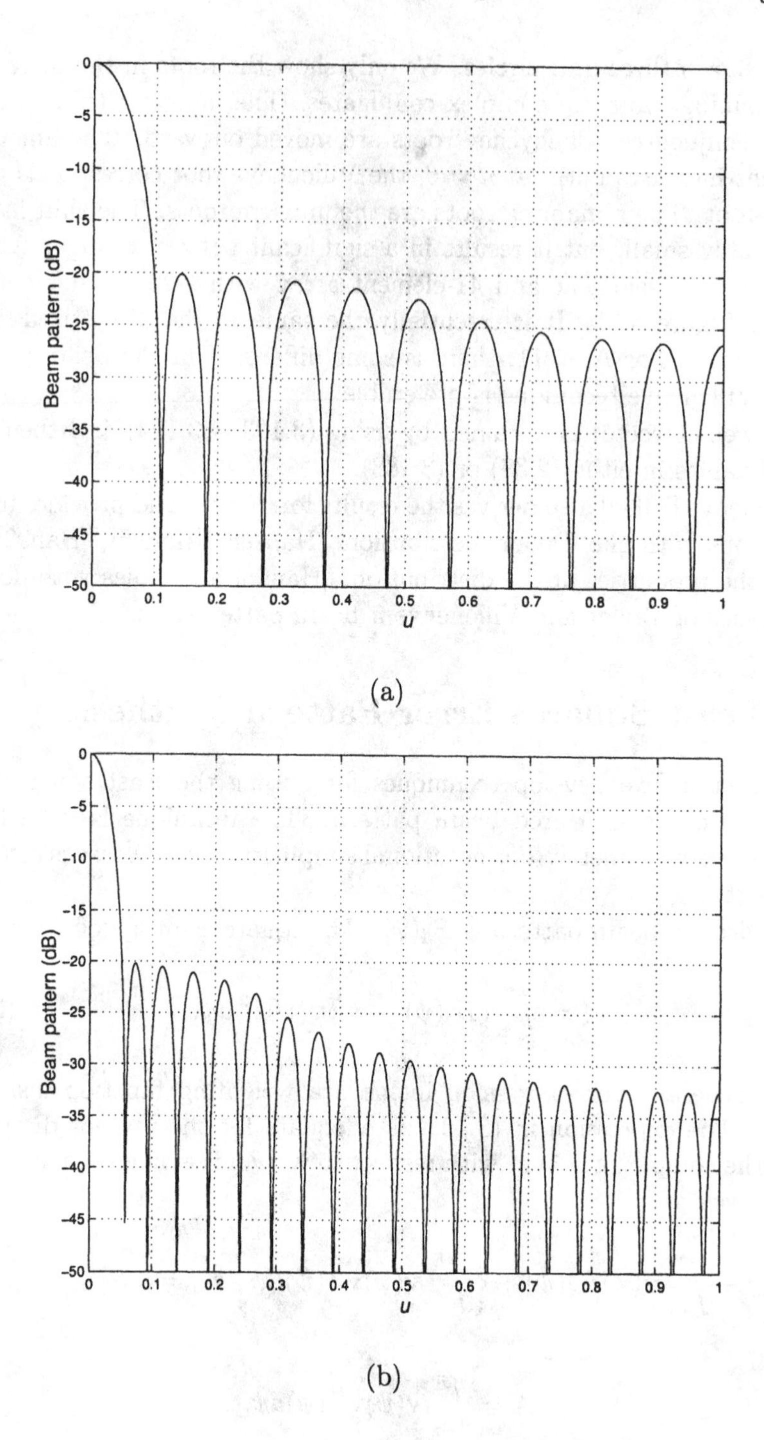

Figure 3.31 Beam pattern with Villeneuve weighting: $\bar{n} = 6$, SLL$=-20$ dB; (a) $N = 21$; (b) $N = 41$.

(3.196) can be written as

$$\mathbf{w}_o = \mathbf{A}^{-1} \int_{-\pi}^{\pi} \mathbf{v}(\psi) B_d^*(\psi)\, d\psi. \tag{3.198}$$

The result in (3.198) is an explicit expression for the least squares approximation to $B_d(\psi)$ and can be evaluated for a specific array manifold vector.

For the special case of a standard linear array, the array manifold vector is given by (2.70) and the nk element of $\mathbf{A}$ is

$$\begin{aligned} [\mathbf{A}]_{nk} &= \int_{-\pi}^{\pi} \exp\left\{ \left[j\,(n - \frac{N-1}{2}) - j\,(k - \frac{N-1}{2}) \right] \psi \right\} d\psi \\ &= \int_{-\pi}^{\pi} \exp\left[j(n-k)\psi \right] d\psi = 2\pi \delta_{nk}. \end{aligned} \tag{3.199}$$

Using (2.70) and (3.199) in (3.198) gives

$$\mathbf{w}_{no} = \frac{1}{2\pi} \int_{-\pi}^{\pi} \exp\left(j\,(n - \frac{N-1}{2})\,\psi \right) B_d^*(\psi)\, d\psi. \tag{3.200}$$

The weight vector is $\mathbf{w}^H$, whose components are

$$\mathbf{w}_{no}^* = \frac{1}{2\pi} \int_{-\pi}^{\pi} \exp\left\{ -j\,(n - \frac{N-1}{2})\,\psi \right\} B_d(\psi)\, d\psi. \tag{3.201}$$

For the special case of a real symmetric $B_d(\psi)$, it is convenient to use the symmetric indexing. For N odd,

$$a_{mo} = \frac{1}{2\pi} \int_{-\pi}^{\pi} B_d(\psi) e^{-jm\psi} d\psi. \tag{3.202}$$

For N even,

$$a_{mo} = \frac{1}{2\pi} \int_{-\pi}^{\pi} B_d(\psi) e^{-j(m-\frac{1}{2})\psi} d\psi. \tag{3.203}$$

In both cases,

$$a_{-mo} = a_{mo}. \tag{3.204}$$

The results in (3.202) and (3.203) are familiar as the Fourier series expansion of the desired beam pattern. The resulting beam pattern is

$$\hat{B}_d(\psi) = \sum_{m=-\frac{N-1}{2}}^{\frac{N-1}{2}} a_{mo} e^{jm\psi} \tag{3.205}$$

for N odd and a similar expression for N even.

In some cases the desired pattern $B_d(\psi)$ is constant from $-\psi_0 \leq \psi \leq \psi_0$ and zero elsewhere in ψ-space. In this case,

$$a_m = \frac{1}{2\pi}\int_{-\psi_0}^{\psi_0} e^{-jm\psi}\, d\psi = \frac{\sin m\psi_0}{m\pi}, \quad N \text{ odd}, \tag{3.206}$$

and

$$a_m = \frac{1}{2\pi}\int_{-\psi_0}^{\psi_0} e^{-j(m-\frac{1}{2})\psi}\, d\psi = \frac{\sin((m-\frac{1}{2})\psi_0)}{((m-\frac{1}{2})\psi_0)}, \quad N \text{ even}. \tag{3.207}$$

We consider a simple example to illustrate the technique.

Example 3.5.1

The desired $B_d(\psi)$ is shown in Figure 3.32. It is uniform from $60° \leq \theta \leq 120°$ or, since $\psi = \pi\cos\theta$,

$$-0.5\pi \leq \psi \leq 0.5\pi. \tag{3.208}$$

The results for $N = 10$ and 20 are shown in Figure 3.32. We also show the beam pattern using Woodward sampling.

We observe that the pattern has oscillatory overshoots at the discontinuities. This behavior is known as Gibbs phenomenon and is familiar from Fourier theory. We can use windows to decrease the overshoots.

To introduce the concept of windows, we write (3.205) as

$$\hat{B}_d(\psi) = \sum_{m=-\infty}^{\infty} a_{mo}\, R[m]\, e^{jm\psi}, \tag{3.209}$$

where $R[m]$ is a discrete rectangular window. We denote its Fourier transform by $B_R(\psi)$ and normalize it so $B_R(0)$ equals one. For N odd,

$$R[m] = \begin{cases} \frac{1}{N}, & m = -\frac{N-1}{2}, \cdots, 0, \cdots, \frac{N-1}{2} \\ 0, & \text{elsewhere.} \end{cases} \tag{3.210}$$

From Fourier transform properties, (3.209) corresponds to convolving $B_d(\psi)$ and $B_R(\psi)$ in ψ-space,

$$\hat{B}_d(\psi) = B_d(\psi) * B_R(\psi), \tag{3.211}$$

or

$$\hat{B}_d(\psi) = \frac{1}{2\pi}\int_{-\pi}^{\pi} B_d(\psi_x) B_R(\psi - \psi_x)\, d\psi_x, \tag{3.212}$$

where

$$B_R(\psi) = \frac{1}{N}\sum_{m=-\frac{N-1}{2}}^{\frac{N-1}{2}} e^{jm\psi} = \frac{1}{N}\frac{\sin\left(\frac{N}{2}\psi\right)}{\sin\left(\frac{\psi}{2}\right)}. \tag{3.213}$$

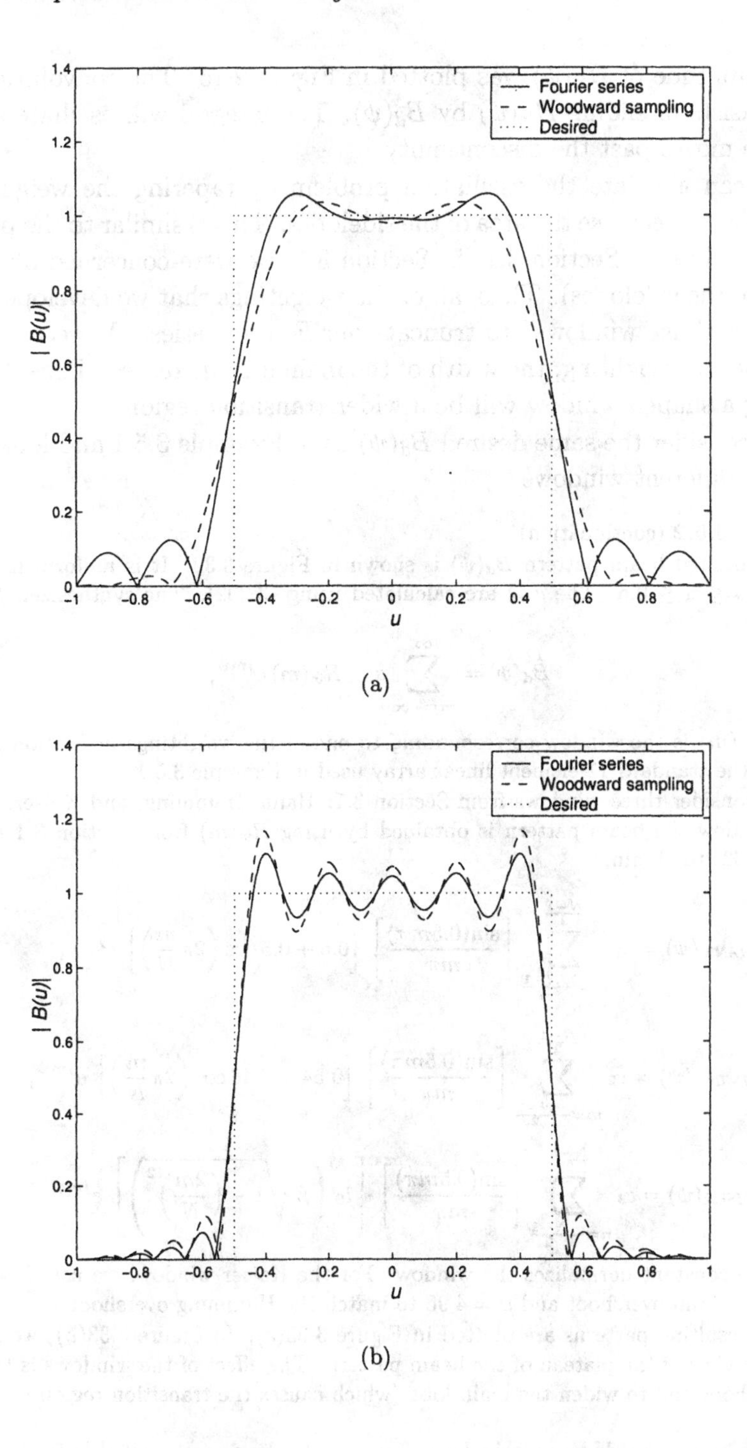

Figure 3.32 Desired beam pattern and synthesized patterns for linear array of 10 and 20 elements using the Fourier series method and the Woodward sampling technique: (a) N=10; (b) N=20.

This amplitude of $B_R(\psi)$ was plotted in Figure 2.15. The convolution process consists of sliding $B_R(\psi)$ by $B_d(\psi)$. The integral will oscillate as each sidelobe moves past the discontinuity.

We can alleviate the oscillation problem by tapering the weighting of $R[m]$ so as to decrease the area of the sidelobes. This is similar to the problem that we solved in Section 3.1 (in Section 3.1, we were concerned about the height of the sidelobes). Thus, all of the weightings that we developed there can be used as "windows" to truncate our Fourier series. We recall that as we shape the weighting, the width of the main lobe increases. Thus, the cost of using a shaped window will be a wider transition region.

We consider the same desired $B_d(\psi)$ as in Example 3.5.1 and look at the effect of different windows.

Example 3.5.2 (continuation)

The desired beam pattern $B_d(\psi)$ is shown in Figure 3.32. It is uniform in u-space from $-0.5 \leq u \leq 0.5$. The a_{mo} are calculated using (3.202). The synthesized $\hat{B}_d(\psi)$ is given by

$$\hat{B}_d(\psi) = \sum_{m=-\infty}^{\infty} a_{mo}\, R_w(m)\, e^{jm\psi}, \tag{3.214}$$

where $R_w(m)$ is the window corresponding to one of the weightings in Section 3.1. We consider the standard 11-element linear array used in Example 3.5.1.

We consider three windows from Section 3.1: Hann, Hamming, and Kaiser. For the Hann window the beam pattern is obtained by using $R_w(m)$ from Section 3.1 and a_{mo} from (3.202) to obtain,

$$B_{HANN}(\psi) = c_1 \sum_{m=-\frac{N-1}{2}}^{\frac{N-1}{2}} \left[\frac{\sin(0.5m\pi)}{m\pi}\right] \left[0.5 + 0.5\cos\left(2\pi\frac{m}{N}\right)\right] e^{jm\psi}, \tag{3.215}$$

$$B_{HAMMING}(\psi) = c_2 \sum_{m=-\frac{N-1}{2}}^{\frac{N-1}{2}} \left[\frac{\sin(0.5m\pi)}{m\pi}\right] \left[0.54 + 0.46\cos\left(2\pi\frac{m}{N}\right)\right] e^{jm\psi}, \tag{3.216}$$

$$B_{KAISER}(\psi) = c_3 \sum_{n=-\frac{N-1}{2}}^{\frac{N-1}{2}} \left[\frac{\sin(0.5m\pi)}{m\pi}\right] \left[I_0\left(\beta\sqrt{1-\left(\frac{2m}{N}\right)^2}\right)\right] e^{jm\psi}, \tag{3.217}$$

where the constant normalizes the window. For the Kaiser window, we use $\beta = 3.88$ to match the Hann overshoot and $\beta = 4.96$ to match the Hamming overshoot.

The resulting patterns are plotted in Figure 3.33(a). In Figure 3.33(b), we show an expanded view of the plateau of the beam pattern. The effect of the windows is to reduce the overshoot and to widen the main lobe (which causes the transition region to widen).

If we use the Kaiser window, we must choose a suitable value of β. Kaiser [KK66], [Kai74] developed a procedure to design FIR filters that can

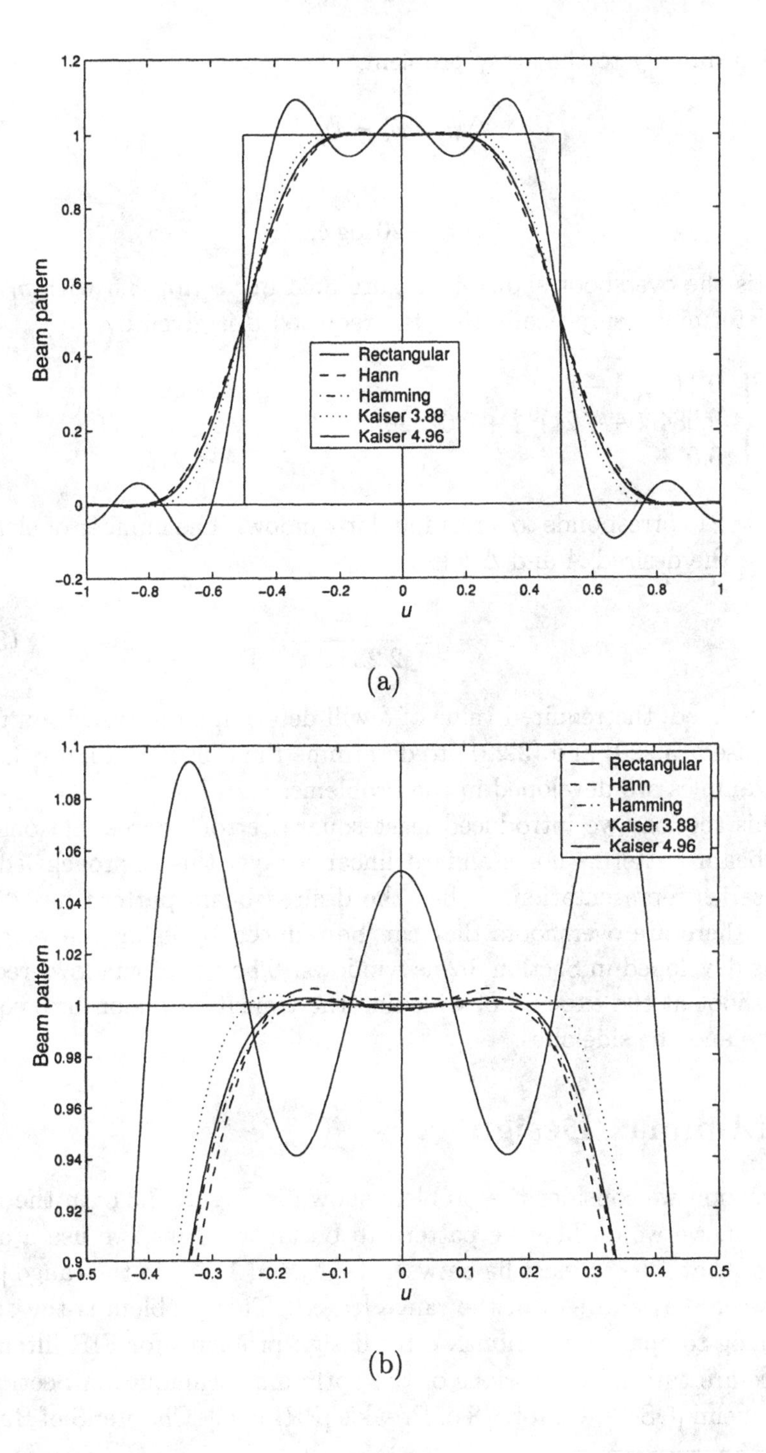

Figure 3.33 Beam patterns for various windows: $N = 11$.

be applied directly to the array problem.[21]

$$\Delta\psi = \psi_s - \psi_p, \tag{3.218}$$

and

$$A = -20 \log \delta, \tag{3.219}$$

where δ is the overshoot shown in Figure 3.34 in Section 3.6 with $\delta_p = \delta_s$. Kaiser determined empirically that the required β is given by

$$\beta = \begin{cases} 0.1102(A - 8.7), & A > 50, \\ 0.5842(A - 21)^{0.4} + 0.07886(A - 21), & 21 \le A \le 50, \\ 0.0, & A < 21, \end{cases} \tag{3.220}$$

where $A = 21$ corresponds to a rectangular window. The number of elements to achieve the desired A and $\Delta\psi$ is

$$N - 1 = \frac{A - 8}{2.285\,\Delta\psi}. \tag{3.221}$$

If N is fixed, the required value of δ will determine the transition region $\Delta\psi$. We used (3.220) and (3.221) to determine the β used in Figure 3.33(a). Other examples are developed in the problems.

In this section, we introduced least squares error approximations to a desired beam pattern. For standard linear arrays, this approach led to a Fourier series representation. When the desired beam pattern has discontinuities, there are overshoots that can be reduced by using the weighting functions developed in Section 3.2 as windows. The use of windows reduced the overshoot at the expense of widening the transition region between the main lobe and the sidelobes.

3.6 Minimax Design

In this section we consider the problem shown in Figure 3.34. In the main-lobe region, we would like the pattern to be unity. Thus, we use a design constraint that $B(e^{j\psi})$ must lie between $1 - \delta_p$ and $1 + \delta_p$ in the range $[0, \psi_p]$ and between $+\delta_s$ and $-\delta_s$ in the range $[\psi_s, \pi]$. This problem is the spatial filter analog to optimum minimax error design problems for FIR filters.

There are tutorial discussions of the optimum techniques in Section 7.6 of Oppenheim [OS89], Chapter 8 of Proakis [PRLN92], Chapter 3 of Rabiner

[21]Our discussion follows pp. 452–455 of [OS89]

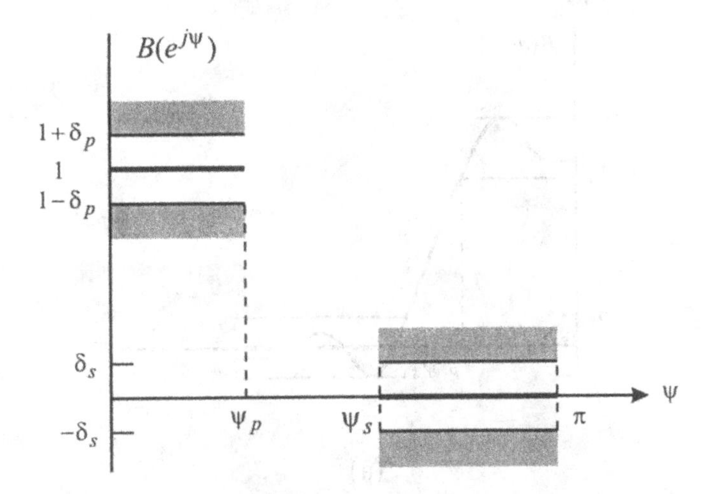

Figure 3.34 Tolerance scheme. [Adapted from [OS89]]

and Gold [RG75], and Chapter 4 of [MK93]. These discussions are based on the work of Parks and McClellan [PM72a], [PM72b] with subsequent contributions (e.g., McClellan et al. [MPR73], [MP73], [MR79], [RMP75]). Our discussion is similar to the above presentations.

At the same time that Parks and McClellan published [PM72a] and [PM72b], Hersey et al. [HTL72] published a paper on minimax design of digital filters that was more general because arbitrary upper and lower constraining functions can be used. The results were applied to linear arrays by Mucci et al. [MTL75]. Discussion of the history of these techniques is contained in Rabiner et al. [RMP75] and Tufts [Tuf75] (see also Farden and Scharf [FS74]).

We define a weighted error as

$$e_{pm}(\psi) = W_{pm}(\psi)\left[B_d(e^{j\psi}) - B(e^{j\psi})\right]. \tag{3.222}$$

We assume that $B_d(e^{j\psi})$ is a real symmetric function. The functions $e_{pm}(\psi)$, $W_{pm}(\psi)$, and $B_d(e^{j\psi})$ are defined only over closed sub-intervals of $0 \le \psi \le \pi$. For the N-odd case, they are defined over $[0, \psi_p]$ and $[\psi_s, \pi]$. We assume N, ψ_p, and ψ_s are fixed design parameters. Then, for the model in Figure 3.34,

$$B_d(e^{j\psi}) = \begin{cases} 1, & 0 \le \psi \le \psi_p, \\ 0, & \psi_s \le \psi \le \pi, \end{cases} \tag{3.223}$$

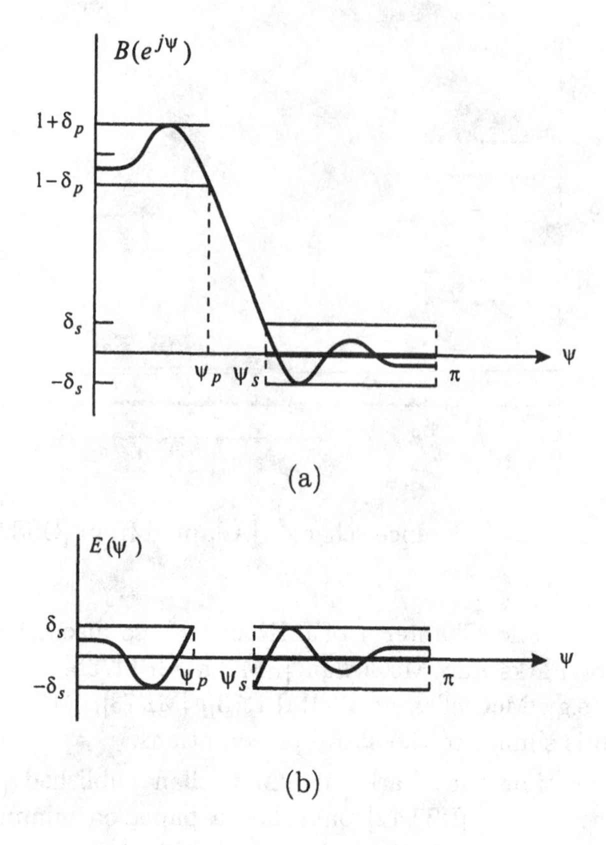

Figure 3.35 (a) Possible beam pattern meeting the specifications of Figure 3.34; (b) weighted error for the approximation in (a). [Adapted from [OS89]]

and the weighting function is

$$W_{pm}(\psi) = \begin{cases} \frac{1}{K}, & 0 \le \psi \le \psi_p, \\ 1, & \psi_s \le \psi \le \pi, \end{cases} \tag{3.224}$$

where $K = \delta_p/\delta_s$ and is a fixed parameter. The value of δ_p (or δ_s) is variable. A possible $B(e^{j\psi})$ that meets the desired criteria is shown in Figure 3.35(a). The corresponding error function is shown in Figure 3.35(b). We see that the maximum weighted approximation error is δ_s in both bands.

The criterion of interest is the minimax (or Chebychev) criterion. We seek a beam pattern that **minimizes** the maximum weighted approximation error.

$$\min_{\{w_n, n=0,1,\cdots,N-1\}} \left(\max_{\psi \in F} |e_{pm}(\psi)| \right), \tag{3.225}$$

where F is closed subset of $0 \leq \psi \leq \pi$:

$$[0 \leq \psi \leq \psi_p] \cup [\psi_s \leq \psi \leq \pi]. \tag{3.226}$$

Parks and McClellan [PM72a] formulate the problem as a polynomial approximation problem. Recall from Section 3.4.2, that we can write[22]

$$\cos(n\psi) = T_n(\cos\psi), \tag{3.227}$$

where $T_n(x)$ is the nth order Chebychev polynomial, and therefore

$$B(e^{j\psi}) = \sum_{k=0}^{L} c_k (\cos\psi)^k, \tag{3.228}$$

where the c_k are constants related to the original weights w_n and $L = (N-1)/2$. Letting

$$x = \cos\psi, \tag{3.229}$$

we have

$$B(e^{j\psi}) = P(x)\,|_{x=\cos\psi}, \tag{3.230}$$

where

$$P(x) = \sum_{k=0}^{L} c_k\, x^k. \tag{3.231}$$

There are several useful theorems [Che66] available to solve the approximation problem.[23] Parks and McClellan utilized the following theorem.

3.6.1 Alternation Theorem

Alternation Theorem[24] Let F_p denote the closed subset consisting of the disjoint union of closed subsets of the real axis x. $P(x)$ denotes an Lth-order polynomial

$$P(x) = \sum_{k=0}^{L} c_k\, x^k. \tag{3.232}$$

Also, $D_p(x)$ denotes a given desired function of x that is continuous on F_p; $W_{pm}(x)$ is a positive function, continuous on F_p, and $e_{pm}(x)$ denotes the weighted error

$$e_{pm}(x) = W_{pm}(x)\left[D_p(x) - P(x)\right]. \tag{3.233}$$

---

[22] Note that we are using a polynomial in $\cos\psi$ instead of $\cos\psi/2$.

[23] Other discussions of the approximation problem are contained in Rice [Ric64], Taylor [Tay69], and Taylor and Winter [TW70].

[24] This version of the alternation theorem follows [OS89] but is due to [PM72a].

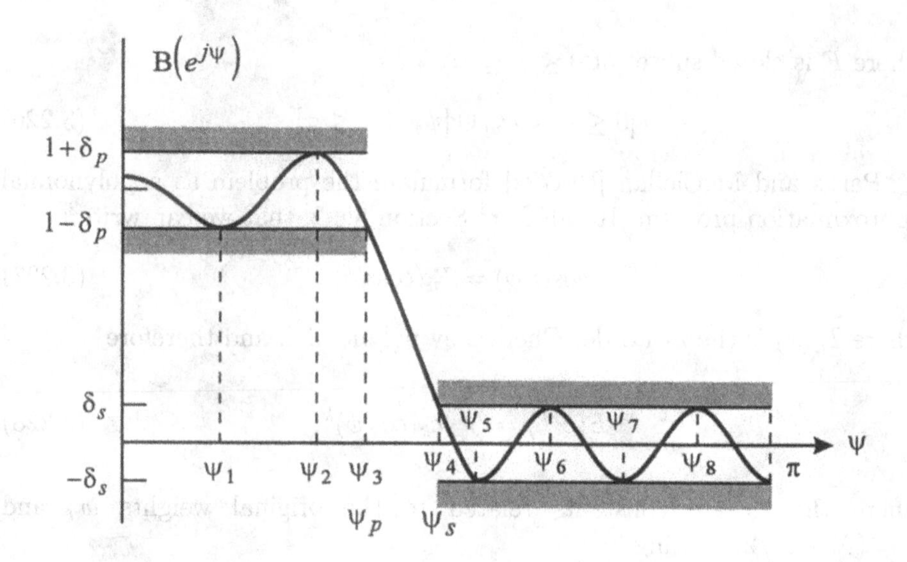

Figure 3.36 Typical example of a beam pattern approximation that is optimal according to the alternation theorem for $L = 7$. [Adapted from [OS89]]

The maximum error $\|E\|$ is defined as

$$\|E\| = \max_{x \in F_p} |e_{pm}(x)|. \tag{3.234}$$

A necessary and sufficient condition that $P(x)$ is the unique Lth-order polynomial that minimizes $\|E\|$ is that $e_{pm}(x)$ exhibit *at least* $(L+2)$ alternations, that is, there must exist at least $(L+2)$ values x_i in F_p such that $x_1 < x_2 < \cdots < x_{L+2}$ and such that $e_{pm}(x_i) = -e_{pm}(x_{i+1}) = \pm\|E\|$ for $i = 1, 2, \cdots, (L+2)$.

We now apply the alternation theorem to the array design. We let $x = \cos\psi$ and plot the polynomial versus ψ. A function that satisfies the Alternation Theorem for $L = 7$ is shown in Figure 3.36. We see that there are nine alternations of the error, occurring at $\psi_1, \psi_2, \cdots, \psi_8$ and π.

There are other possible functions that satisfy the Alternation Theorem (e.g., p. 472 of [OS89]).

3.6.2 Parks-McClellan-Rabiner Algorithm

From the alternation theorem, the optimum pattern $B_o(e^{j\psi})$ will satisfy the following set of equations,

$$W_{pm}(\psi_i)\left[B_d(e^{j\psi_i}) - B_o(e^{j\psi_i})\right] = (-1)^{i+1}\,\delta, \quad i = 1, 2, \cdots, L+2, \tag{3.235}$$

where δ is the optimum error and $B_o(e^{j\psi})$ is given by (3.228).

These equations lead to an iterative algorithm called the *Remez Multiple Exchange Algorithm* for finding the $B_o(e^{j\psi})$.

Step 1: Select an initial set of $\psi_i, i = 1, 2, \cdots, L+2$. ψ_p and ψ_s are included in the set.

Step 2: The set of equations (3.235) could be solved for c_k and δ. However, Parks and McClellan found a more efficient approach using a polynomial approximation.

(a) For a given set of ψ_i, δ is given by

$$\delta = \frac{\sum_{k=1}^{L+2} b_k B_d\left(e^{j\psi_k}\right)}{\sum_{k=1}^{L+2} \frac{b_k(-1)^{k+1}}{W_{pm}(\psi_k)}}, \tag{3.236}$$

where

$$b_k = \prod_{\substack{i=1 \\ i \neq k}}^{L+2} \frac{1}{x_k - x_i}, \tag{3.237}$$

and

$$x_i = \cos\psi_i. \tag{3.238}$$

(b) Since $B_o(e^{j\psi})$ is an Lth-order trigonmetric polynomial, we can interpolate it through $L+1$ of the known $L+2$ values. Parks and McClellan used a Lagrange interpolation formula to obtain

$$B_o(e^{j\psi}) = P(\cos\psi) = \frac{\sum_{k=1}^{L+1} \left[\frac{d_k}{x - x_k}\right] f_k}{\sum_{k=1}^{L+1} \left[\frac{d_k}{x - x_k}\right]}, \tag{3.239}$$

where $x = \cos\psi$ and $x_k = \cos\psi_k$,

$$f_k = B_d\left(e^{j\psi_k}\right) - \frac{(-1)^{k+1}\delta}{W_{pm}(\psi_k)}, \tag{3.240}$$

and

$$d_k = \prod_{\substack{i=1 \\ i \neq k}}^{L+1} \frac{1}{x_k - x_i} = \frac{b_k}{x_k - x_{L+2}}, \tag{3.241}$$

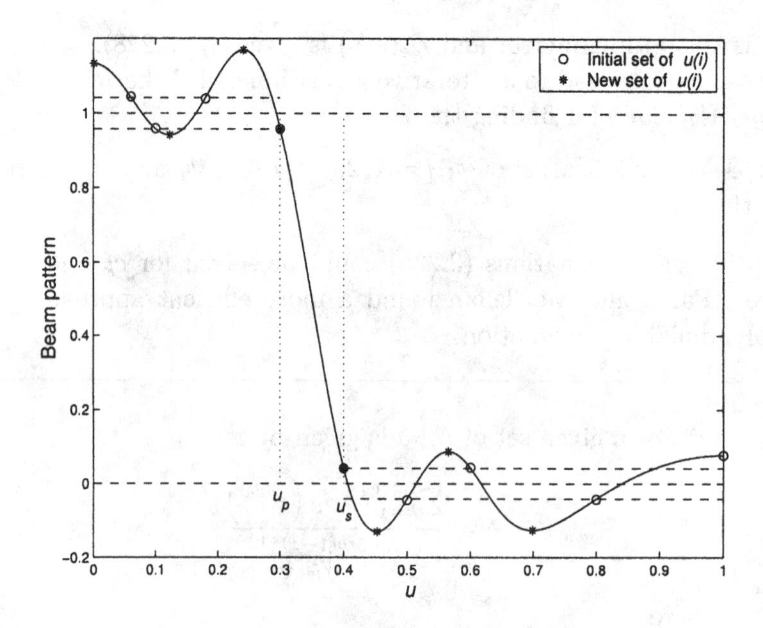

Figure 3.37 Illustration of the Parks-McClellan algorithm for equiripple approximation. [Adapted from [OS89]]

where we use $\psi_1, \psi_2, \cdots, \psi_{L+1}$ to find the polynomial, the value at ψ_{L+2} will be correct because $B_o(e^{j\psi})$ in (3.239) satisfies (3.235). A typical result is shown in Figure 3.37 after the first iteration.

Step 3: The original set $\psi_1, \psi_2, \cdots, \psi_{L+1}$ is exchanged for a completely new set $\psi_1', \psi_2', \cdots, \psi_{L+1}'$ (ψ_p and ψ_s are still included). The new ψ_i are defined by the $(L+2)$ largest peaks of the plotted curve. There are at most $L-1$ local minima and maxima in the open intervals $(0 < \psi < \psi_p)$ and $(\psi_s < \psi < \pi)$. If there are $L-1$, then the remaining point can be either 0 or π; one chooses the largest error point for the next iteration.

Step 4: The iteration is continued in this manner until the change in δ between iterations falls below some small pre-selected amount. The result is $B_o(e^{j\psi})$.

Notice that the array weightings were not computed as part of the design process. The calculation of the weightings is straightforward using the IDFT developed in Section 3.3.

We consider an example to illustrate the technique.

Example 3.6.1

Consider a standard 11-element linear array. The desired beam pattern is shown in Figure 3.17. In Figure 3.38, we show the resulting beam patterns using the Parks-McClellan, Woodward, and Fourier algorithms. We see that the Parks-McClellan algorithm gives uniform ripples with a slight increase in the transition region.

3.6.3 Summary

In this section, we have given a brief discussion of the Parks-McClellan (PM) algorithm for minimax optimization. The references at the beginning of the section give more comprehensive discussions and should be utilized if the algorithm is going to be used in practice. Lewis et al. [LMT76] developed a technique that mixes least squares and minimax techniques that may be useful in some applications.

In Sections 3.1 through 3.6, we developed three major topics:

(i) Synthesis of single MRA arrays

(ii) Synthesis of desired beam patterns

(iii) Relationships between beam patterns, weight vectors, and zeros of the array polynomial

In the single MRA problem, we assume that the signal of interest is a plane wave arriving from a specific direction (expressed in either angle space or wavenumber space). We want to design a beam pattern with good directivity (corresponding to a narrow main lobe) and low sidelobes. Our design procedures trades-off main-lobe width and sidelobe level. In Section 3.1, we used a cosine building block and developed several useful weight vectors. The Hamming weighting provided a good compromise between main-lobe expansion and sidelobe height. We then considered the case in which there was a region of interest around the main lobe. The DPSS and Kaiser weightings were effective for this application. In Section 3.4, we developed Dolph-Chebychev and Riblet-Chebychev weightings to provide uniform-height sidelobes and a main lobe whose width was the minimum possible for the given sidelobe height. We also developed Taylor and Villenueve weightings that had decreasing sidelobes. This collection of weightings is the most widely used for the deterministic synthesis of uniform linear arrays.

In Sections 3.3, 3.5, and 3.6, we studied the problem of synthesizing a desired beam pattern. We developed three approaches; the Woodward sampling approach, the least squares error approach, and the minimax algorithm. Each approach has advantages and disadvantages, and the appropriate technique will depend on the specific desired beam pattern. We revisit

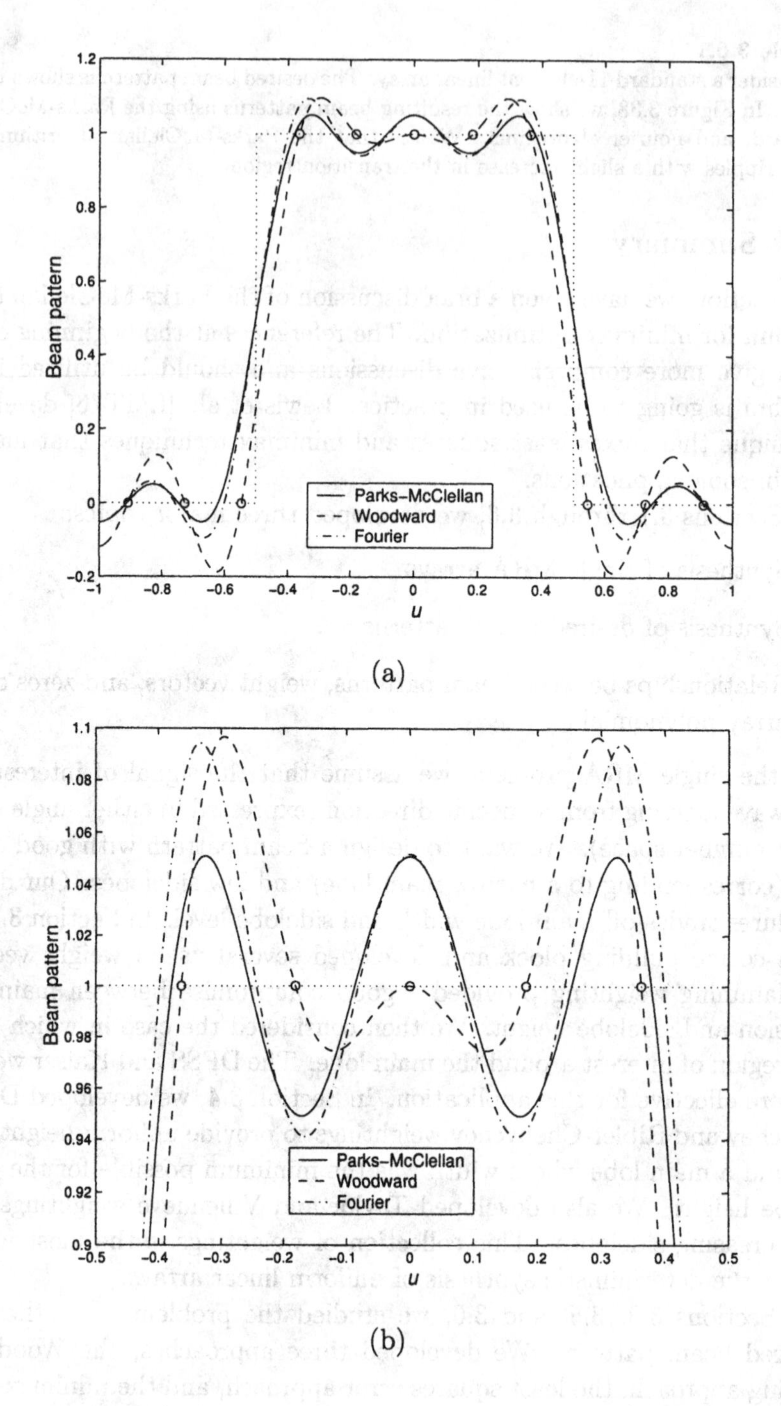

Figure 3.38 Comparison of various beam pattern synthesis techniques.

the synthesis problem in Sections 3.7 and 3.9.3. In Section 3.7, we generalize the least squares error solution to include linear constraints. In Section 3.9.3, we solve the minimax problem for arbitrary arrays.

In the course of the discussion we introduced various tools, such as the array polynomial and its zeros, the DFT, and the IDFT. All of these tools are used in many subsequent discussions.

All of the discussion up to this point did not assume any detailed knowledge of the interference environment. Using a beam pattern with constant sidelobe levels tacitly assumes that the interference is equally likely to arrive anywhere in the sidelobe region. In the next section we consider a model in which we have some knowledge of the location of the interference.

3.7 Null Steering

In our discussion of array design, we have seen the importance of pattern nulls. In many applications, we want to guarantee that the pattern will have a null in a given direction. In radar or communications applications, a jamming signal may be located at a specific wavenumber and we want to eliminate its effect.

3.7.1 Null Constraints

For an arbitrary array, to put a null at a given wavenumber $\mathbf{k}_J$, we require

$$B(\mathbf{k}_J) = \mathbf{w}^H \, \mathbf{v}_{\mathbf{k}}(\mathbf{k}_J) = 0, \tag{3.242}$$

where

$$\mathbf{v}_{\mathbf{k}}(\mathbf{k}_J) = \begin{bmatrix} e^{-j\mathbf{k}_J^T \mathbf{p}_1} \\ e^{-j\mathbf{k}_J^T \mathbf{p}_2} \\ \vdots \\ e^{-j\mathbf{k}_J^T \mathbf{p}_N} \end{bmatrix}. \tag{3.243}$$

For a uniformly spaced linear array with N odd, this reduces to

$$0 = \sum_{n=-\frac{N-1}{2}}^{\frac{N-1}{2}} w_n^* \, e^{jn\psi_J}. \tag{3.244}$$

Symmetric indexing is more convenient for this problem. We denote these weights as $w_n, n = -\frac{N-1}{2}, \cdots, \frac{N-1}{2}$.

We can then choose the w_i to synthesize a desired pattern subject to the constraint in (3.244). We now consider several ways to implement null constraints.

3.7.2 Least Squares Error Pattern Synthesis with Nulls

In this section we consider the problem of finding the best least squares error approximation to a desired pattern subject to a set of null constraints.[25] We develop the solution for an arbitrary array geometry and then consider some examples of linear arrays.

We assume that there is some desired beam pattern that can be synthesized by a discrete array.

$$B_d(\mathbf{k}) = \mathbf{w}_d^H \, \mathbf{v}_{\mathbf{k}}(\mathbf{k}). \tag{3.245}$$

We approximate it by a second pattern that has a set of constraints imposed on it. This pattern is denoted by

$$B(\mathbf{k}) = \mathbf{w}^H \, \mathbf{v}_{\mathbf{k}}(\mathbf{k}). \tag{3.246}$$

We minimize the squared error between the desired pattern $B_d(\mathbf{k})$ and the constrained pattern,

$$\varepsilon = \int\!\!\int \left| B_d(\mathbf{k}) - B(\mathbf{k}) \right|^2 d\mathbf{k}. \tag{3.247}$$

Substituting (3.245) and (3.246) into (3.247) and performing the integration gives,

$$\varepsilon = \| \, \mathbf{w}_d - \mathbf{w} \, \|^2 . \tag{3.248}$$

The restriction that $B_d(\mathbf{k})$ be synthesizable by a discrete array is for convenience. If the actual desired pattern is not synthesizable in this form, we let $B_d(\mathbf{k})$ be the least squares error approximation to it. For a linear equally spaced array, $B_d(\mathbf{k})$ would be obtained by the Fourier series approach of Section 3.4.

We consider constraints on the beam pattern and its derivatives at various values of $\mathbf{k}$. Normally we would include a constraint on the array response along the MRA. If the array is pointed at $\mathbf{k}_T$, then the constraint

$$\mathbf{w}^H \mathbf{v}_{\mathbf{k}}(\mathbf{k}_T) = 1 \tag{3.249}$$

is referred to as a distortionless constraint. Any plane-wave signal arriving along $\mathbf{k}_T$ will pass through the array processing undistorted. We use this

---

[25]This general formulation appeared in the adaptive array context in Applebaum [AC76]. The specific approach used here was done by Steyskal [Ste82] for linear arrays. The generalization to arbitrary arrays is straightforward. This problem is a special case of the problem of finding a weighted least squares error approximation to a desired beam pattern subject to a set of constraints (not necessarily zero). A solution to this problem is given in Mucci et al. [MTL76].

constraint extensively starting in Chapter 6. In this section, we focus on null constraints. If we only consider null constraints in the sidelobe region, we can omit (3.249) and get a slightly simpler derivation.

The first type of constraint is a null constraint. From (3.246)

$$\mathbf{w}^H \mathbf{v_k}(\mathbf{k}_i) = 0, \quad i = 1, 2, \cdots, M_0. \tag{3.250}$$

We refer to this as a zero-order constraint (or zero-order null) and define an $N \times M_0$ constraint matrix, $\mathbf{C}_0$

$$\mathbf{C}_0 = \left[\mathbf{v_k}(\mathbf{k}_1) \,\vdots\, \mathbf{v_k}(\mathbf{k}_2) \,\vdots\, \cdots \,\vdots\, \mathbf{v_k}(\mathbf{k}_{M_0})\right]. \tag{3.251}$$

The second type of constraint is the first derivative of the beam pattern with respect to $\mathbf{k}$. For a linear array, this corresponds to

$$\frac{d}{dk}B(k)\bigg|_{k=k_i} = \mathbf{w}^H \left[\frac{d}{dk}\mathbf{v_k}(k)\right]_{k=k_i} \triangleq \mathbf{w}^H \mathbf{d}_1(k)\bigg|_{k=k_i}, \quad i \in \Omega_1, \tag{3.252}$$

where Ω_1 is a subset of the M_0 locations where we want the derivative to equal zero and contains M_1 points.[26]

We assume that some of the nulls have derivatives set to zero while others do not. Thus, for a linear array, we define an $N \times M_1$ constraint matrix, $\mathbf{C}_1$:

$$\mathbf{C}_1 = \left[\mathbf{d}_1(\mathbf{k}_1) \,\vdots\, \mathbf{d}_1(\mathbf{k}_2) \,\vdots\, \cdots \,\vdots\, \mathbf{d}_1(\mathbf{k}_{M_1})\right]. \tag{3.253}$$

If we have a 2-D array, there will be a derivative with respect to two components of the wavenumber so $\mathbf{C}_1$ will be $2M_1$-dimensional.

The nth type of constraint is the nth derivative of the beam pattern with respect to $\mathbf{k}$. For a linear array, this corresponds to

$$\frac{d^n}{dk^n}B(k)\bigg|_{k=k_i} = \mathbf{w}^H \left[\frac{d^n}{dk^n}\mathbf{v_k}(k)\right]_{k=k_i} \triangleq \mathbf{w}^H \mathbf{d}_n(k)\bigg|_{k=k_i}, \quad i \in \Omega_n \tag{3.254}$$

where the set Ω_n is a subset of Ω_{n-1} and contains M_n points. For $n = 2$,

$$\mathbf{C}_2 = \left[\mathbf{d}_2(\mathbf{k}_1) \,\vdots\, \mathbf{d}_2(\mathbf{k}_2) \,\vdots\, \cdots \,\vdots\, \mathbf{d}_2(\mathbf{k}_{M_2})\right]. \tag{3.255}$$

For a 2-D array, $\mathbf{C}_2$ will be $3M_2$-dimensional. In practice, constraints beyond $\mathbf{C}_2$ are seldom used.

---

[26] A derivative with respect to a scalar wavenumber k is indicated. In many cases, a derivative with respect to ψ or u will be used.

Then, the total $\mathbf{C}$ matrix is an $N \times M_c$ matrix,

$$\mathbf{C} = \left[\ \mathbf{C}_0 \ \vdots \ \mathbf{C}_1 \ \vdots \ \mathbf{C}_2\ \right], \tag{3.256}$$

where M_c is the total number of constraints and $M_c < N$. We assume that the columns of $\mathbf{C}$ are linearly independent. The solution to the optimization problem requires the inversion of $\mathbf{C}^H\mathbf{C}$. If the columns of $\mathbf{C}$ are linearly independent, then $\mathbf{C}^H\mathbf{C}$ will not be singular. If the M_n are not selected carefully, there is no guarantee that the columns will be independent. In some cases, the columns may be independent, but several columns may be highly correlated, which will cause $\mathbf{C}^H\mathbf{C}$ to have a poor condition number. One approach to alleviate this problem is to use a singular value decomposition on the right-hand side of (3.256) and retain only the dominant singular value terms to form $\mathbf{C}$. This guarantees that $\mathbf{C}^H\mathbf{C}$ is not singular.[27]

Before solving the optimization problem, we develop the explicit form of the constraint matrices for a standard linear array with N elements (N odd).

The array manifold vector is

$$\mathbf{v}_u(u) = \mathbf{v}_{\mathbf{k}}(k)\big|_{k=\pi u} = \begin{bmatrix} e^{-j\left(\frac{N-1}{2}\right)\pi u} \\ \vdots \\ e^{j\left(\frac{N-1}{2}\right)\pi u} \end{bmatrix}. \tag{3.257}$$

Thus,

$$B_u(u) = \sum_{n=-\frac{N-1}{2}}^{\frac{N-1}{2}} w_n^*\, e^{jn\pi u} = \mathbf{w}^H\, \mathbf{v}_u(u), \tag{3.258}$$

and

$$\begin{aligned} \frac{d^n}{du^n}\, B_u(u) &= \sum_{m=-\frac{N-1}{2}}^{\frac{N-1}{2}} w_m^*\, (jm\pi)^n\, e^{jm\pi u} \\ &= \mathbf{w}^H\, \mathbf{d}_n(u), \end{aligned} \tag{3.259}$$

where $\mathbf{d}_n(u)$ is a $N \times 1$ matrix whose mth element is

$$\mathbf{d}_n(u)]_m = (jm\pi)^n\, e^{jm\pi u}. \tag{3.260}$$

Note that $\mathbf{v}_u(u)$ and the even-numbered derivatives are conjugate symmetric and the odd-numbered derivatives are conjugate asymmetric.

We now solve the optimization problem.

[27] N. Owsley (private communication).

Optimization

We want to optimize the squared weighting error subject to the constraint that

$$\mathbf{w}^H \mathbf{C} = \mathbf{0}, \tag{3.261}$$

where $\mathbf{C}$ is the constraint matrix in (3.256). We require the columns of $\mathbf{C}$ to be linearly independent. Using Lagrange multipliers we want to minimize G, where[28]

$$G = \left[\mathbf{w}_d^H - \mathbf{w}^H\right]\left[\mathbf{w}_d - \mathbf{w}\right] + \mathbf{w}^H \mathbf{C}\boldsymbol{\lambda} + \boldsymbol{\lambda}^H \mathbf{C}^H \mathbf{w}, \tag{3.262}$$

where $\boldsymbol{\lambda}$ is a $M_c \times 1$ Lagrange multiplier vector. Taking the gradient with respect to $\mathbf{w}$ gives

$$-\mathbf{w}_d^H + \mathbf{w}_o^H + \boldsymbol{\lambda}^H \mathbf{C}^H = \mathbf{0}, \tag{3.263}$$

or

$$\mathbf{w}_o^H = \mathbf{w}_d^H - \boldsymbol{\lambda}^H \mathbf{C}^H. \tag{3.264}$$

We solve for the Lagrange multiplier vector by using the constraint,

$$\mathbf{w}_o^H \mathbf{C} = \mathbf{0}. \tag{3.265}$$

Thus,

$$\mathbf{w}_d^H \mathbf{C} - \boldsymbol{\lambda}^H \mathbf{C}^H \mathbf{C} = \mathbf{0}. \tag{3.266}$$

The matrix $\mathbf{C}^H\mathbf{C}$ is not singular due to the assumption of linearly independent columns. Then, we can write

$$\boldsymbol{\lambda}^H = \mathbf{w}_d^H \mathbf{C}\left[\mathbf{C}^H \mathbf{C}\right]^{-1}. \tag{3.267}$$

Although the inverse of $[\mathbf{C}^H\mathbf{C}]$ exists, the condition number of the matrix may be poor if the columns of $\mathbf{C}$ approach dependency (e.g., null directions too close to each other).

The optimum weighting is

$$\mathbf{w}_o^H = \mathbf{w}_d^H \left(\mathbf{I}_N - \mathbf{C}\left[\mathbf{C}^H \mathbf{C}\right]^{-1} \mathbf{C}^H\right). \tag{3.268}$$

The matrix

$$\mathbf{P_C} = \mathbf{C}\left[\mathbf{C}^H \mathbf{C}\right]^{-1} \mathbf{C}^H \tag{3.269}$$

[28]See Appendix A (Section A.7.4) for a discussion of complex gradients.

is the projection matrix onto the constraint subspace. Thus $\mathbf{w}_o^H$ is the component of $\mathbf{w}_d^H$ in the subspace orthogonal to the constraint subspace. Thus (3.268) can be written as

$$\mathbf{w}_o^H = \mathbf{w}_d^H \, \mathbf{P}_{\mathbf{C}}^{\perp}, \tag{3.270}$$

and

$$\mathbf{w}_e^H = \mathbf{w}_d^H - \mathbf{w}_o^H = \mathbf{w}_d^H \, \mathbf{P}_{\mathbf{C}}. \tag{3.271}$$

The orthogonal behavior of $\mathbf{w}_e^H$ is familiar from other optimization problems.

A second interpretation of (3.270) is also of interest. We can write (3.268) as

$$\begin{aligned} \mathbf{w}_o^H &= \mathbf{w}_d^H - \left(\mathbf{w}_d^H \mathbf{C} \left[\mathbf{C}^H \mathbf{C} \right]^{-1} \right) \mathbf{C}^H \\ &= \mathbf{w}_d^H - \mathbf{a} \, \mathbf{C}^H, \end{aligned} \tag{3.272}$$

where $\mathbf{a}$ is a $1 \times M_c$ weighting vector. Thus, the optimum weight vector consists of the desired weight vector minus a weighted sum of the constraint vectors. The resulting beam pattern is

$$\begin{aligned} B_o(u) &= \left[\mathbf{w}_d^H - \mathbf{a} \, \mathbf{C}^H \right] \mathbf{v}(u) \\ &= B_d(u) - \mathbf{a} \, \mathbf{C}^H \, \mathbf{v}(u). \end{aligned} \tag{3.273}$$

For the zero-order constraints (i.e., null-only), the second term in (3.273) is a sum of conventional beam patterns steered at the wavenumber of the interferer. Thus,

$$B_o(u) = B_d(u) - \sum_{m=1}^{M_0} a_m \, B_c \left(u - u_m \right). \tag{3.274}$$

Note that, since $B_o(u_k) = 0$

$$B_d(u_k) = \sum_{m=1}^{M_0} a_m \, B_c(u_k - u_m), \quad k = 1, \cdots, M_0. \tag{3.275}$$

For the linear array,

$$B_c(u - u_m) = \frac{\sin \left[N\pi \frac{(u - u_m)}{2} \right]}{\sin \left[\pi \frac{(u - u_m)}{2} \right]}. \tag{3.276}$$

Similarly, for an nth-order derivative constraint, the cancellation beam patterns are derivatives of the conventional beam pattern,

$$B_c^{(n)}(u-u_m) = \frac{d^n}{du^n}\frac{\sin\left[N\pi\frac{(u-u_m)}{2}\right]}{\sin\left[\pi\frac{(u-u_m)}{2}\right]}. \tag{3.277}$$

Thus the beam pattern of the optimum processor is just the desired beam pattern $B_d(u)$ minus a weighted sum of conventional beam patterns and derivatives of conventional beam patterns centered at the null locations.

We observe that the same result holds for arbitrary arrays. For example, for zero-order nulls, from (3.251),

$$\mathbf{C}_0 = \left[\ \mathbf{v}_{\mathbf{k}}(\mathbf{k}_1) \ \vdots\ \mathbf{v}_{\mathbf{k}}(\mathbf{k}_2) \ \vdots \ \cdots \ \vdots\ \mathbf{v}_{\mathbf{k}}(\mathbf{k}_{M_0})\ \right]. \tag{3.278}$$

and (3.246) and (3.272) becomes

$$\begin{aligned} B_o(\mathbf{k}) &= \mathbf{w}_o^H \mathbf{v}_{\mathbf{k}}(\mathbf{k}) \\ &= B_d(\mathbf{k}) - \mathbf{a}\,\mathbf{C}_0^H\,\mathbf{v}_{\mathbf{k}}(\mathbf{k}) \\ &= B_d(\mathbf{k}) - \sum_{m=1}^{M_0} a_m\,B_c(\mathbf{k}-\mathbf{k}_m), \end{aligned} \tag{3.279}$$

where a_m is the mth element of the $1 \times M_0$ matrix,

$$\mathbf{a} = \mathbf{w}_d^H\,\mathbf{C}_0\left[\mathbf{C}_0^H\,\mathbf{C}_0\right]^{-1}. \tag{3.280}$$

Note that the discussion in (3.270)–(3.280) is useful in interpreting the result. We use (3.268) to find $\mathbf{w}_o^H$.

The resulting pattern error is

$$\varepsilon_0 = \mathbf{w}_e^H\,\mathbf{w}_e, \tag{3.281}$$

where $\mathbf{w}_e$ was defined in (3.271). Using (3.271) in (3.281) and recalling that $\mathbf{P_C P_C} = \mathbf{P_C}$, we obtain

$$\varepsilon_0 = \mathbf{w}_d^H\,\mathbf{C}\left[\mathbf{C}^H\,\mathbf{C}\right]^{-1}\mathbf{C}^H\,\mathbf{w}_d = \mathbf{w}_d^H\mathbf{P_C}\mathbf{w}_d. \tag{3.282}$$

We now consider several examples to illustrate the application of these results.

Example 3.7.1[29]

We consider a 21-element linear array spaced at $\lambda/2$. The desired pattern corresponds to uniform weighting ($w_n = 1/N$). We put a zero-order, first-order, and second-order null

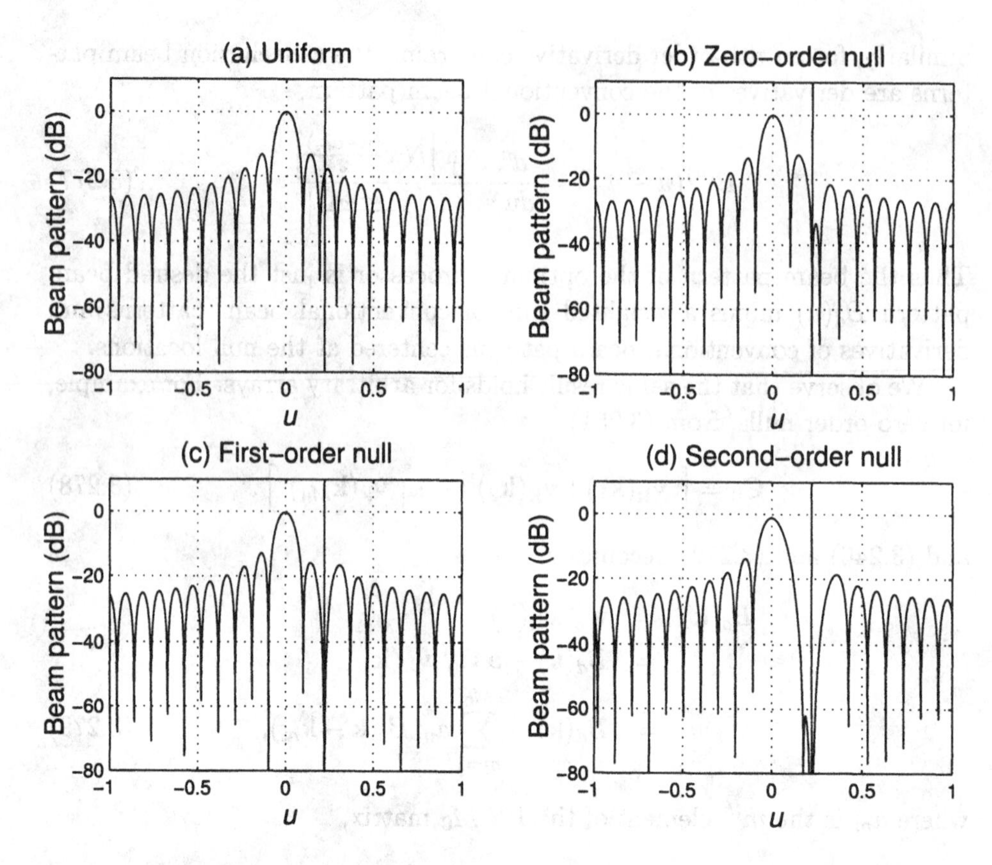

Figure 3.39 (a) Initial sinc-pattern; (b) pattern with a null of zero order imposed at $u = 0.22$; (c) with null of first order; (d) with null of second order.

at $u = 0.22$. The only important aspect of this location is that it is outside the main lobe. The results are shown in Figure 3.39.

Example 3.7.2 (continuation)

We consider the same array as in the preceding example. We now place three zero-order nulls at $u_1 = 0.21, u_2 = 0.22$, and $u_3 = 0.23$. Note that the nulls in the conventional pattern are spaced at $2/21 = 0.095$ so that the constraint vectors are not orthogonal.

The resulting pattern is shown in Figure 3.40. We have reduced the highest sidelobe in the sector $0.18 \leq u \leq 0.26$ to -63 dB.

Example 3.7.3

In this example, we consider a 41-element linear array spaced at $\lambda/2$. The desired pattern is a Chebychev pattern with -40-dB sidelobes.

---

[29]This sequence of examples (3.7.1–3.7.3) is due to Steyskal [Ste82].

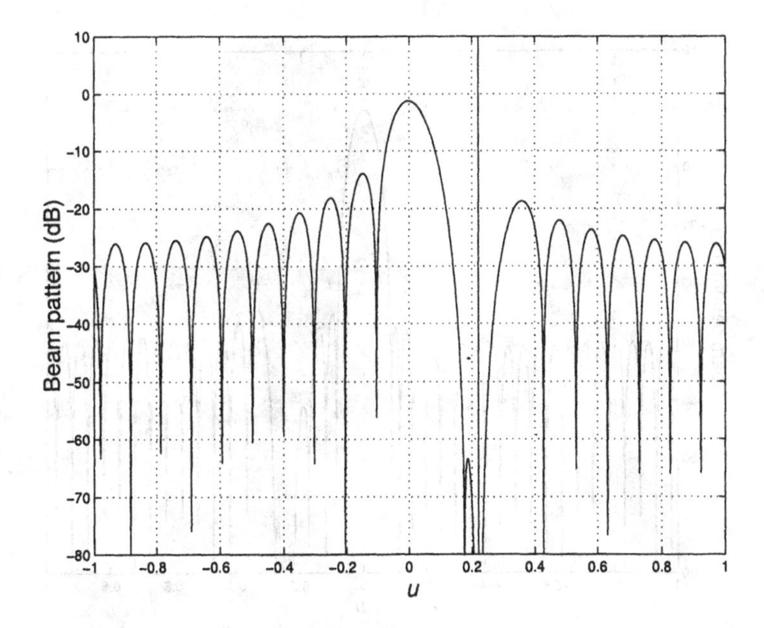

Figure 3.40 Sinc-pattern with three nulls equispaced over the sector (0.18, 0.26).

In case 1, we place four zero-order nulls at 0.22, 0.24, 0.26, and 0.28. The resulting pattern is shown in Figure 3.41.

In case 2, we place eight zero-order nulls spaced at $\Delta u = 0.02$ in the sector (0.22, 0.36). The result is shown in Figure 3.42.

We revisit the problem of constrained optimization several times in the text. In Chapter 6, we derive similar results for different constraints. In Chapter 7, we study adaptive arrays to achieve constrained optimization.

3.8 Asymmetric Beams

In many array applications, we want to measure the direction of arrival of an incoming plane wave. One method for doing this is to utilize an asymmetric beam that has a slope at the steering angle that can be used in a closed loop system to point the steering angle in the direction of the plane wave. This technique is the basis of many monopulse radars (e.g., [Sko80]). The beam pattern design problem consists of finding the maximum (or acceptable) slope of the beam pattern at the origin subject to a suitable constraint on

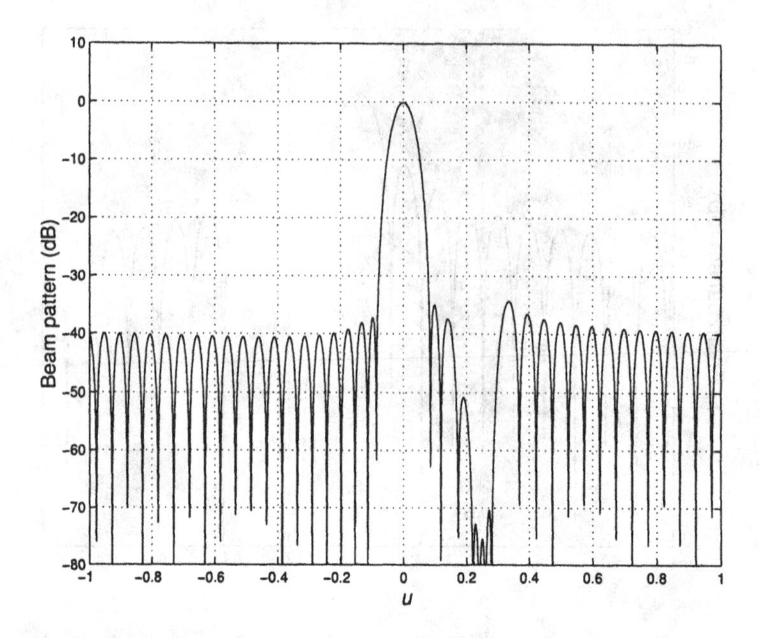

Figure 3.41 Initial 40-dB Chebychev pattern with four nulls equispaced over the sector $(0.22, 0.28)$.

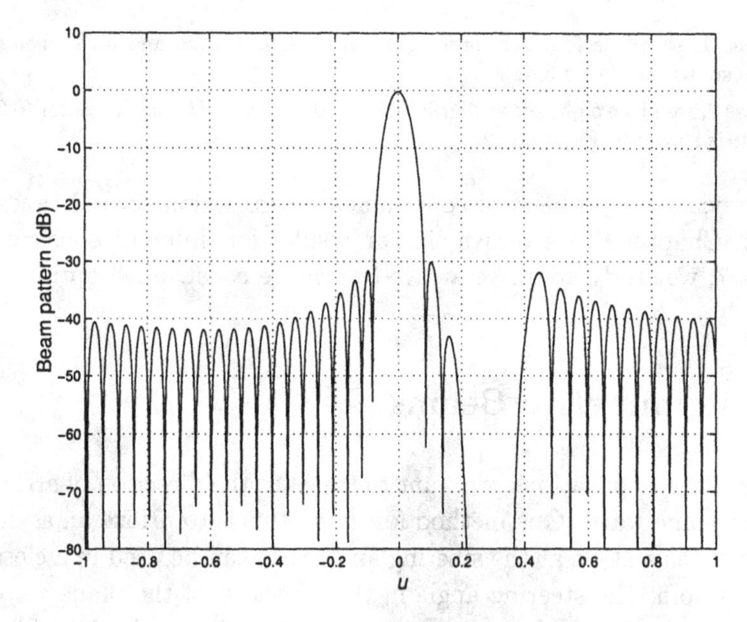

Figure 3.42 Initial 40-dB Chebychev pattern with eight nulls equispaced over the sector $(0.22, 0.36)$.

the sidelobes. We define an asymmetric[30] array weighting as one in which

$$w(-n) = \begin{cases} -w(n), & n = -\frac{N-1}{2}, \cdots, -1, 1, \cdots, \frac{N-1}{2}: \quad N \text{ odd}, \\ 0, & n = 0: \quad N \text{ odd}, \end{cases} \tag{3.283}$$

and

$$w(-n) = -w(n), \quad n = -\frac{N}{2}, \cdots, -1, 1, \cdots, \frac{N}{2}: \ N \text{ even}, \tag{3.284}$$

where we have used a symmetric indexing and assume that the weights are real.

In this section, we look at several examples of difference beams. The beam pattern is

$$B_a(\psi) = \sum_{n=-\frac{N}{2}}^{\frac{N}{2}-1} w_n \, e^{j(n+\frac{1}{2})\psi}, \tag{3.285}$$

for N even. Using (3.284) in (3.285) gives

$$\begin{aligned} B_a(\psi) &= \sum_{m=1}^{\frac{N}{2}} w_m e^{j(m-\frac{1}{2})\psi} + \sum_{m=1}^{\frac{N}{2}} (-w_m) e^{-j(m-\frac{1}{2})\psi} \\ &= 2j \sum_{m=1}^{\frac{N}{2}} w_m \sin(m - \frac{1}{2})\psi. \end{aligned} \tag{3.286}$$

The simplest case is uniform asymmetric weighting

$$w_n = \frac{1}{N}, \quad n \geq 1. \tag{3.287}$$

Then (3.286) becomes

$$B_a(\psi) = \frac{e^{-j\frac{\psi}{2}}}{N} \sum_{m=1}^{\frac{N}{2}} e^{jm\psi} - \frac{e^{j\frac{\psi}{2}}}{N} \sum_{m=1}^{\frac{N}{2}} e^{-jm\psi}. \tag{3.288}$$

The beam pattern in (3.288) is the difference between two shifted conventional beams so we refer to it as a difference beam. The corresponding beam with a plus sign is referred to a sum beam. For uniform weighting, the sum

---

[30]The dictionary definition of asymmetric is "non-symmetric." Our definition is in (3.283) and (3.284).

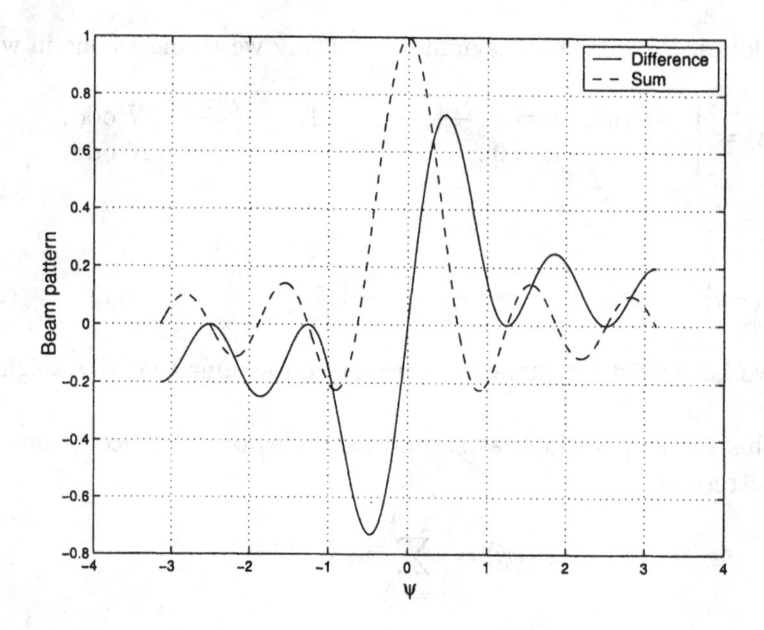

Figure 3.43 Beam pattern versus ψ: Difference beam with uniform weighting, $N = 10$.

beam is the familiar conventional beam pattern. The expression in (3.288) can be written as

$$B_a(\psi) = \frac{2j}{N}\frac{\sin^2\left(\frac{N}{4}\psi\right)}{\sin\left(\frac{\psi}{2}\right)}, \tag{3.289}$$

for N even. Proceeding in a similar manner, we obtain

$$B_a(\psi) = \frac{2j}{N}\frac{\sin\left(\frac{N+1}{4}\psi\right)\sin\left(\frac{N-1}{4}\psi\right)}{\sin\left(\frac{\psi}{2}\right)}, \tag{3.290}$$

for N odd.

The beam pattern of the difference beam is a purely imaginary function. We write

$$B_a(\psi) = j\, B_{aI}(\psi), \tag{3.291}$$

and plot $B_{aI}(\psi)$ in Figure 3.43 for a standard linear array with $N = 10$. The corresponding sum beam pattern is also plotted.

Several observations are useful:

(a) The sum and difference beams are orthogonal (this follows from the Fourier transform relationship).

(b) The difference beam is a weighted sum of two shifted conventional beams (see (3.288)).

(c) The first zero of the difference beam is at

$$\psi = \begin{cases} \frac{4\pi}{N+1}, & N \text{ odd}, \\ \\ \frac{4\pi}{N}, & N \text{ even}. \end{cases} \tag{3.292}$$

(d) The first sidelobe of the difference beam is 10 dB below the main lobe.

(e) The slope at the origin is

$$\left.\frac{dB_{aI}(\psi)}{d\psi}\right|_{\psi=0} = \frac{N}{4}. \tag{3.293}$$

The usage of difference beams for angle estimation (and tracking) is common. They are implemented in monopulse radars and phased array radars (e.g., [Sko80], [Sko90], [Bro88], [Bar89]).

Many useful difference beam patterns can be constructed as linear combinations of shifted conventional beam patterns. We can also use linear combinations of shifted beams designed using the techniques developed in Sections 3.1 and 3.5.

In the sum beam case we discussed design procedures due to Dolph and Taylor that allowed us to control the sidelobe behavior while maintaining the beamwidth of the main lobe. Zolotarev, a student of Chebychev, developed a class of odd polynomials that give an equal-ripple approximation over a given interval.[31] McNamara [McN93] used these polynomials to produce difference patterns with constant sidelobes. A discrete $\bar{n}$ difference pattern analogous to the discrete $\bar{n}$ Villenueve sum pattern was also developed by McNamara [McN94]. The Taylor distribution for sum beams was developed for a continuous aperture. Analogous procedures have been developed for difference beams by Bayliss [Bay68], in which he controls the sidelobe height. Discussions of his technique are available in Elliott [Ell81] and Mailloux [Mai94]. The reader is referred to these references.

Other properties of difference beams and their applications are discussed in the problems and at various points in the text.

[31]This discussion follows Hansen [Han98].

3.9 Spatially Non-uniform Linear Arrays

3.9.1 Introduction

The discussion of arrays up to this point has focused on linear arrays with uniform spacing. In many applications, the arrays are linear but have non-uniform spacing.

One application in which we encounter non-uniform arrays is the thinned or sparse array problem. In this case, we start with an N-element uniform linear array or linear aperture of length L that has a desired weighting and associated beam pattern. We then construct a linear array with fewer elements that retains the desirable features of the beam pattern. In some cases, we allow the elements to be in any location on the line. In other cases, we restrict their positions to a uniform grid. The motivation is to reduce the cost and complexity of the array by having fewer sensors. There are a large number of references in this area (e.g., [LL59], [KPT60], [San60], [Unz60], [Maf62], [Ish62], [Wil62], [Sko69], [IC65], [Har61], [Lo63], and [Ste76]).

A second application is the case in which the array locations are random along a segment of the axis (or, more generally, in an area in a plane or a volume in three dimensions). In this case, there are two categories. In the first category, the nominal locations are deterministic but the actual locations vary in a random manner. This is a generalization of our discussion of sensitivity and tolerance factors in Section 2.3.3. An example of this model is the dropping of sonobouys in the ocean to locate submarines. It also includes such problems as random removal or failure of elements. There is a large amount of literature in this area (e.g., [All61], [MC63], [Ruz52] and [GM55]).

In the second category there are arrays in which the elements are placed at random over some segment of the axis according to some probability density as part of the design procedure. This is known as statistical density tapering. There is a large amount of literature in this area (e.g., [Lo64a], [Lo64b], [Lo68], [PL69], [AL69], [Ste72] and [Ste76]).

There are a number of other references that discuss various issues concerning non-uniform arrays. We will limit our discussion to two topics that we will use later in the text.

In Section 3.9.2, we discuss a class of non-uniform linear arrays called minimum redundancy arrays. The reason for the name will be clear when we discuss these arrays.

In Section 3.9.3, we assume that the element locations are given and derive an algorithm for designing a desired beam pattern. In essence, this

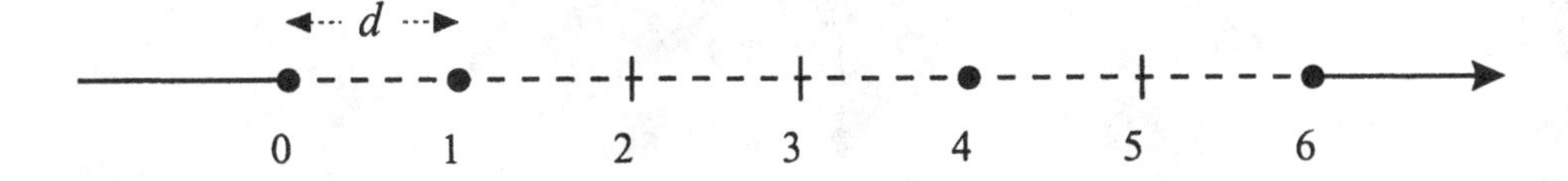

Figure 3.44 "Ideal" MRLA.

algorithm generalizes the minimax algorithms of Section 3.6 to the arbitrary array case.

3.9.2 Minimum Redundancy Arrays

In this section, we consider a class of non-uniformly spaced linear arrays referred to as minimum redundancy linear arrays (MRLA). We restrict our attention to the case in which the arrays are constructed on an underlying grid structure with grid spacing d.

MRLAs are designed so that the number of sensor pairs that have the same spatial correlation lag is made as small as possible. An example of an "ideal" MRLA is shown in Figure 3.44. This is a 4-element array whose aperture length is equivalent to a 7-element standard linear array.

We see that this configuration allows us to estimate

$$E\left[x(t,id)x^*(t,jd)\right] \triangleq R_x\left((i-j)d\right) \tag{3.294}$$

for at least one $(i-j)$ combination from 0 to 6. For example,

SENSOR LOCATIONS	LAG
0–1	d
4–6	$2d$
1–4	$3d$
0–4	$4d$
1–6	$5d$
0–6	$6d$

For the moment, we will assume that our estimate of $R_x((i-j)d)$ is correct.[32] If we denote the sensor outputs of the 7-element standard linear array by the 7×1 vector $\mathbf{x}(t)$, then the correlation matrix is a 7×7 matrix,

$$\mathbf{R_x} = E\left[\mathbf{x}(t)\mathbf{x}^H(t)\right], \tag{3.295}$$

[32]Our discussion at this point is heuristic because we have not developed the appropriate statistical model. We revisit the issue in Chapter 5.

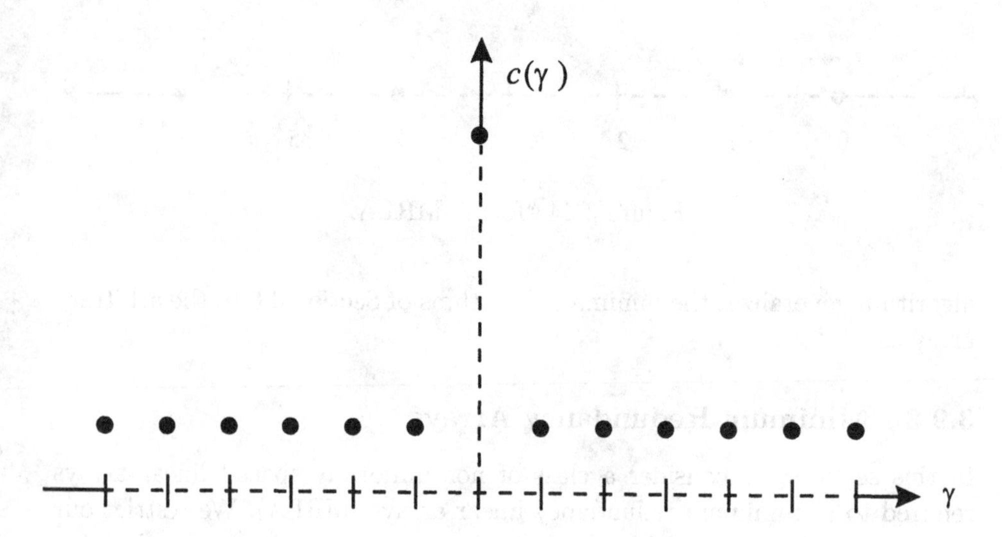

Figure 3.45 Co-array for the array in Figure 3.44.

whose elements are all of the form

$$R_x\left((i-j)d\right), \quad i=0,\cdots,6; j=0,\cdots,6. \tag{3.296}$$

Thus, the 4-element array allows us to measure all of the elements in the correlation matrix of a 7-element standard array. We find that many of our optimum processing algorithms are based on $\mathbf{R_x}$. Thus, there is a possibility that the 4-element MRLA might have similar performance to a 7-element standard linear array.

We explore the statistical significance of this conjecture in Chapters 5 and 7. In our present discussion, we consider the beam pattern behavior.

We denote the aperture length as N_a and it is measured in terms of number of grid intervals. For example, in Figure 3.44, $N_a = 6$ and $N = 4$.

In order to compute the number of times each spatial correlation lag is contained in an array, we assume the elements are uniformly weighted and compute the correlation of $\mathbf{w}$ with itself.

$$c(\gamma) \triangleq \sum_{|m-n|=\gamma} w_m w_n^*. \tag{3.297}$$

The resulting function is called the **co-array** (e.g., [Hau68], [Kre71], or [Bag76]) and is a symmetric function. The co-array for the array in Figure 3.44 is shown in Figure 3.45.

In Chapter 5, we argue that, from the standpoint of efficient spatial sampling, we would like the co-array to equal one except at the origin. If we could find an array with this property then

$$N_a = \frac{N(N-1)}{2}. \tag{3.298}$$

This is the number of different off-diagonal elements in the $N \times N$ correlation matrix $\mathbf{R_x}$. Unfortunately, such arrays (sometimes called perfect arrays) do not exist for $N > 4$. For larger arrays, we consider two options. In the first option, we construct the array so that $c(\gamma)$ is either zero or one except at the origin. These are called non-redundant arrays and a representative set is shown in Table 3.7.[33]

Table 3.7 Non-redundant Arrays

N	Sensor Separation	D
2	·1·	1
3	·1 · 2·	1
4	·1 · 3 · 2·	1
5	·1 · 3 · 5 · 2·	1.10
6	·1 · 3 · 6 · 2 · 5·	1.13
7	·1 · 3 · 6 · 8 · 5 · 2·	1.19
8	·1 · 3 · 5 · 6 · 7 · 10 · 2·	1.21
9	·1 · 4 · 7 · 13 · 2 · 8 · 6 · 3·	1.22
10	·1 · 5 · 4 · 13 · 3 · 8 · 7 · 12 · 2·	1.22

The corresponding co-arrays for $N > 4$ have "gaps" or "holes" in their values. The number D is the ratio of the aperture length N_a to aperture length of a hypothetical perfect array ($N(N-1)/2$). We look at the significance of these "gaps" later.

In this second option, we construct arrays that have no gaps and have the largest possible aperture. These are referred to as minimum redundancy arrays. We choose the sensor positions to make N_a as large as possible without having any gaps. We can write N_a as

$$N_a = \frac{N(N-1)}{2} - N_R + N_H, \tag{3.299}$$

where N_R is the number of redundancies and N_H is the number of holes. We require $N_H = 0$ in a minimum redundant array.

There has been a significant amount of research on element spacing to achieve as low a redundancy as possible. For $N \leq 17$, minimum redun-

[33] From [JD93].

dancy arrays have been found through exhaustive search routines. These configurations are shown in Table 3.8.[34]

Table 3.8 Minimum Redundancy Linear Arrays

N	N_R	N_a	$\frac{N^2}{N_a}$	$\frac{N(N-1)}{2N_a}$	Array(s)
3	0	3	3.0	1.0	12
4	0	6	2.67	1.0	132
5	1	9	2.78	1.11	1332 & 3411
6	2	13	2.77	1.15	13162 & 15322 & 11443
7	4	17	2.88	1.24	136232 & 114443 & 111554
					116423 & 173222
8	5	23	2.78	1.22	1366232 & 1194332
9	7	29	2.79	1.24	136^3232 & 12377441 & 11(12)43332
10	9	36	2.78	1.25	1237^3441
11	12	43	2.81	1.28	1237^4441
12	16	50	2.88	1.32	1237^5441 & $111(20)54^433$
13	20	58	2.91	1.34	$111(24)54^533$ & $11671(10)^33423$
					143499995122
14	23	68	2.88	1.34	$11671(10)^43423$ & $11355(11)^366611$
15	26	79	2.85	1.33	$11355(11)^466611$
16	30	90	2.84	1.33	$11355(11)^566611$
17	35	101	2.86	1.35	$11355(11)^666611$

Notation n^m means m repetitions of the spacing n.

Several authors have developed techniques for generating low redundancy arrays. Pearson et al. [PPL90] develop an efficient constructive procedure for near-optimal placement of sensors. Ruf [Ruf93] uses simulated annealing to obtain low redundancy arrays and gives results for $N \leq 30$ ($N_a \leq 287$). Linebarger et al. [LST93] provide algorithms for constructing sparse arrays and develop bounds. Linebarger [Lin92] presents a fast method for computing co-arrays. (See also Abramovich et al. [AGGS98], [ASG99a], [ASG99b].)

We consider two examples to illustrate the behavior of the beam patterns.

Example 3.9.1

Consider the MRLA in Figure 3.44 with $d = \lambda/2$. The beam pattern for uniform weighting is shown in Figure 3.46. The HPBW is 0.666 and the BW_{NOT} is 1.385 in ψ-space. Note that the beam pattern does not have a perfect null so we use BW notch-notch. This compares to 1.429 and 3.1416, respectively, for a standard 4-element array and to 0.801 and 1.795, respectively, for a standard 7-element linear array. Thus, in terms of main-lobe characteristics, the MRLA offers improvement over the standard 4-element array.

[34]This table was taken from Linebarger et al. [LST93], but the result is due to a sequence of earlier papers.

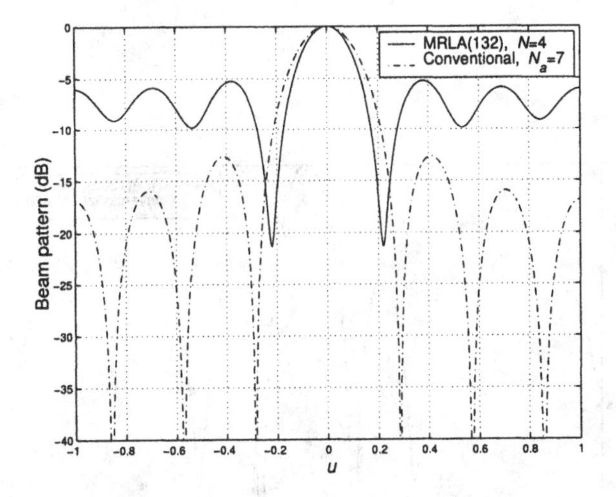

Figure 3.46 Beam pattern for 4-element MRLA with uniform weighting.

The problem with the MRLA is that the sidelobes are significantly higher than the uniform array.

Example 3.9.2 (continuation)

Consider the two 5-element minimum redundancy arrays from Table 3.8. In these cases, $N_a = 9$. The beam patterns are shown in Figure 3.47. For the case 1 (1,3,3,2), the HPBW is 0.464 and the BW_{NOT} is 0.98. For the case 2 (3,4,1,1), the HPBW is 0.473 and the BW_{NOT} is 0.94. This compares to 0.559 and 1.25, respectively, for a standard 10-element linear array.

Just as in the uniform linear case, we can improve the sidelobe behavior by using a non-uniform weighting.

We revisit minimum redundancy arrays at several points in the text and see how their performance compares to standard arrays.

3.9.3 Beam Pattern Design Algorithm

In this section, we derive an algorithm that provides a simple iterative technique for designing desired beam patterns for arbitrary arrays.[35] The algorithm is due to Bell et al. [BVG00] and is based on the techniques developed previously by Olen and Compton [OC90] and Zhou and Ingram [ZI98], [ZI99]. Earlier work using this type of algorithm is contained in Sureau and Keeping [SK82] and Dufort [Duf89]. An alternative approach that uses least squares constraints on sidelobe levels is given in Er [Er92]. Tseng and Griffiths [TG92] also developed an alternative approach to designing beam patterns.

[35]This section is due to Professor Kristine Bell (private communication).

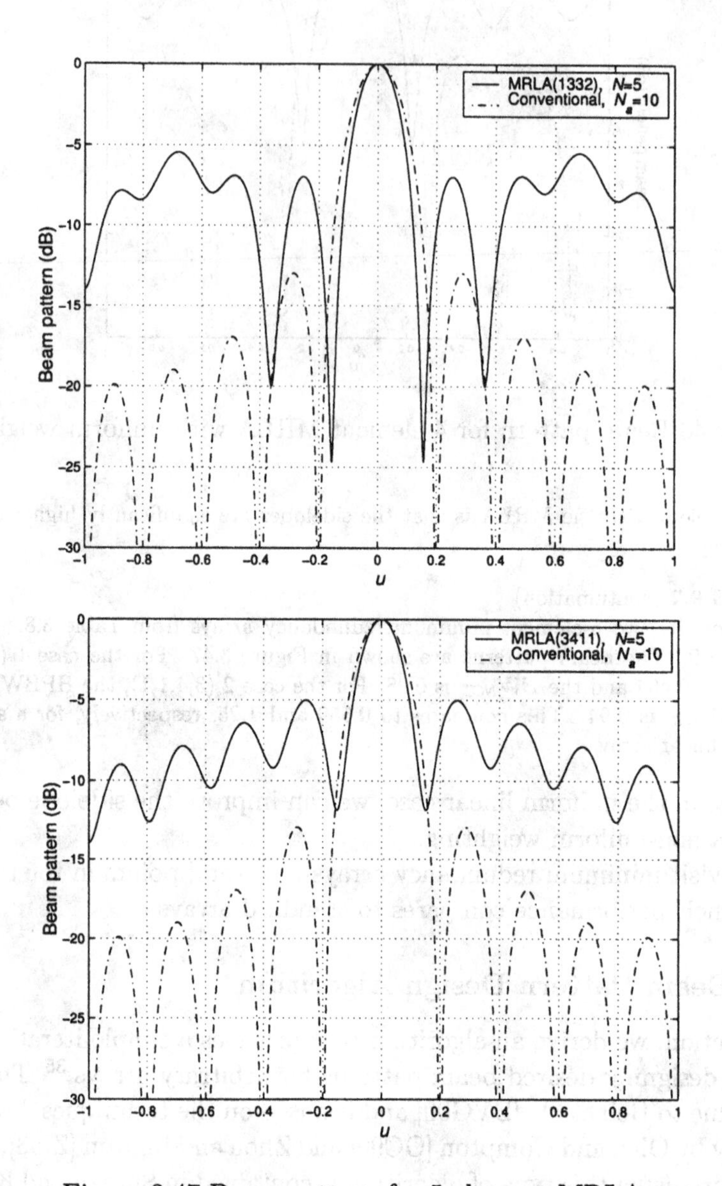

Figure 3.47 Beam pattern for 5-element MRLAs.

The objective is to find weights that maximize the directivity of the array subject to a set of constraints on the beam pattern, which limit the sidelobe levels. We develop the algorithm in the context of linear arrays of isotropic elements, although it applies to arrays of arbitrary geometry, and with non-isotropic elements.

We assume a linear array of isotropic elements on the z-axis with $N \times 1$ array response vector $\mathbf{v}(u)$. When the pattern response at the main response axis or pointing direction is equal to one, the directivity is given by (2.148),

$$\begin{aligned} D &= \left\{\frac{1}{2}\int_{-1}^{1} |B(u)|^2 du\right\}^{-1} \\ &= \left\{\frac{1}{2}\int_{-1}^{1} |\mathbf{w}^H \mathbf{v}(u)|^2 du\right\}^{-1} \\ &= \left\{\mathbf{w}^H \mathbf{A} \mathbf{w}\right\}^{-1}, \end{aligned} \tag{3.300}$$

where

$$\mathbf{A} = \frac{1}{2}\int_{-1}^{1} \mathbf{v}(u)\mathbf{v}^H(u) du. \tag{3.301}$$

The entries in $\mathbf{A}$ are:

$$[\mathbf{A}]_{mn} = \text{sinc}\left(\frac{2\pi}{\lambda}|p_m - p_n|\right), \tag{3.302}$$

where p_n is the position of the nth element.

Let $\mathbf{v}_T = \mathbf{v}(u_T)$ be the array response vector for the steering direction. The basic problem is to maximize the directivity (or equivalently minimize the inverse of the directivity), subject to the unity response constraint at the main response axis, that is,

$$\min \mathbf{w}^H \mathbf{A} \mathbf{w} \quad \text{s.t. } \mathbf{w}^H \mathbf{v}_T = 1. \tag{3.303}$$

The solution is

$$\mathbf{w} = \mathbf{A}^{-1}\mathbf{v}_T\left(\mathbf{v}_T^H \mathbf{A}^{-1}\mathbf{v}_T\right)^{-1}. \tag{3.304}$$

In the special case of a uniform linear array, $\mathbf{A} = \mathbf{I}$, and the maximum directivity weight vector is the uniform weight vector steered to the desired direction, $\mathbf{w} = \frac{1}{N}\mathbf{v}_T$. For both uniformly and non-uniformly spaced arrays, we wish to obtain lower sidelobes by sacrificing some directivity. This can be done by partitioning u-space into r sectors, $\Omega_1, \ldots, \Omega_r$ and defining a desired (although not necessarily realizable) beam pattern in each sector,

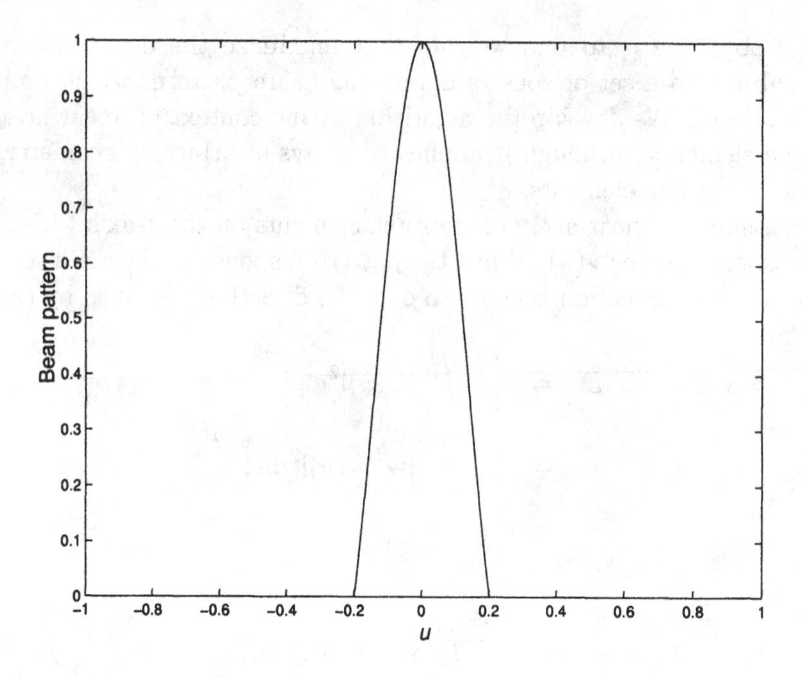

Figure 3.48 Desired three-sector beam pattern.

then limiting deviations between the synthesized and desired beam pattern. A typical desired beam pattern defined in three sectors is shown in Figure 3.48, in which the main beam sector has a beam pattern with some desired main beam shape, and the sidelobe sectors are ideally zero. We assume there is a weight vector $\mathbf{w}_{d,i}$ that generates the desired beam pattern in the ith sector. Let $B_{d,i}(u) = \mathbf{w}_{d,i}^H \mathbf{v}(u)$ be the corresponding beam pattern. The square error between the beam pattern generated by the synthesized weight vector $\mathbf{w}$ and the desired beam pattern over the region Ω_i is given by

$$\begin{aligned}\epsilon_i^2 &= \int_{\Omega_i} \left|B(u) - B_{d,i}(u)\right|^2 du \\ &= \int_{\Omega_i} \left|\mathbf{w}^H \mathbf{v}(u) - \mathbf{w}_{d,i}^H \mathbf{v}(u)\right|^2 du \\ &= (\mathbf{w} - \mathbf{w}_{d,i})^H \mathbf{Q}_i (\mathbf{w} - \mathbf{w}_{d,i}),\end{aligned} \tag{3.305}$$

where

$$\mathbf{Q}_i = \int_{\Omega_i} \mathbf{v}(u)\mathbf{v}(u)^H du. \tag{3.306}$$

Let Ω_i be the region $(u_i - \Delta_i, u_i + \Delta_i)$. The entries in $\mathbf{Q}_i$ are:

$$[\mathbf{Q}_i]_{mn} = e^{j\frac{2\pi}{\lambda}(p_m - p_n)u_i} 2\Delta_i \text{ sinc}\left(\frac{2\pi\Delta_i}{\lambda}|p_m - p_n|\right). \tag{3.307}$$

Now we can maximize directivity subject to constraints on the pattern error as follows:

$$\begin{aligned} \min \quad & \mathbf{w}^H\mathbf{A}\mathbf{w} \text{ s.t. } \mathbf{w}^H\mathbf{v}_T = 1 \\ \text{s.t.} \quad & (\mathbf{w} - \mathbf{w}_{d,i})^H\mathbf{Q}_i(\mathbf{w} - \mathbf{w}_{d,i}) \leq L_i \quad i = 1\ldots r. \end{aligned} \tag{3.308}$$

We define

$$\begin{aligned} F = \quad & \mathbf{w}^H\mathbf{A}\mathbf{w} + \lambda_0(\mathbf{w}^H\mathbf{v}_T - 1) + \lambda_0^*(\mathbf{v}_T^H\mathbf{w} - 1) \\ & + \sum_{i=1}^{r}\lambda_i(\mathbf{w} - \mathbf{w}_{d,i})^H\mathbf{Q}_i(\mathbf{w} - \mathbf{w}_{d,i}). \end{aligned} \tag{3.309}$$

Differentiating with respect to $\mathbf{w}^H$ and setting the result equal to zero gives

$$\mathbf{A}\mathbf{w} + \lambda_0\mathbf{v}_T^H + \sum_{i=1}^{r}\lambda_i[\mathbf{Q}_i(\mathbf{w} - \mathbf{w}_{d,i})] = \mathbf{0}. \tag{3.310}$$

Defining

$$\mathbf{A}_Q = \mathbf{A} + \sum_{i=1}^{r}\lambda_i\mathbf{Q}_i, \tag{3.311}$$

and

$$\mathbf{w}_Q = \sum_{i=1}^{r}\lambda_i\mathbf{Q}_i\mathbf{w}_{d,i}, \tag{3.312}$$

we can write (3.310) as

$$\mathbf{w} = -\lambda_0\mathbf{A}_Q^{-1}\mathbf{v}_T + \mathbf{A}_Q^{-1}\mathbf{w}_Q. \tag{3.313}$$

Solving for λ_0 and substituting the result into (3.313) gives:

$$\begin{aligned} \mathbf{w} = \quad & \mathbf{A}_Q^{-1}\mathbf{v}_T\left(\mathbf{v}_T^H\mathbf{A}_Q^{-1}\mathbf{v}_T\right)^{-1} \\ & + \left[\mathbf{A}_Q^{-1} - \mathbf{A}_Q^{-1}\mathbf{v}_T\left(\mathbf{v}_T^H\mathbf{A}_Q^{-1}\mathbf{v}_T\right)^{-1}\mathbf{v}_T^H\mathbf{A}_Q^{-1}\right]\mathbf{w}_Q. \end{aligned} \tag{3.314}$$

We can obtain tight sidelobe control by defining a set of small sectors in the sidelobe region, as shown in Figure 3.49, and setting the desired beam pattern to zero in these regions. The desired weight vector in each sector

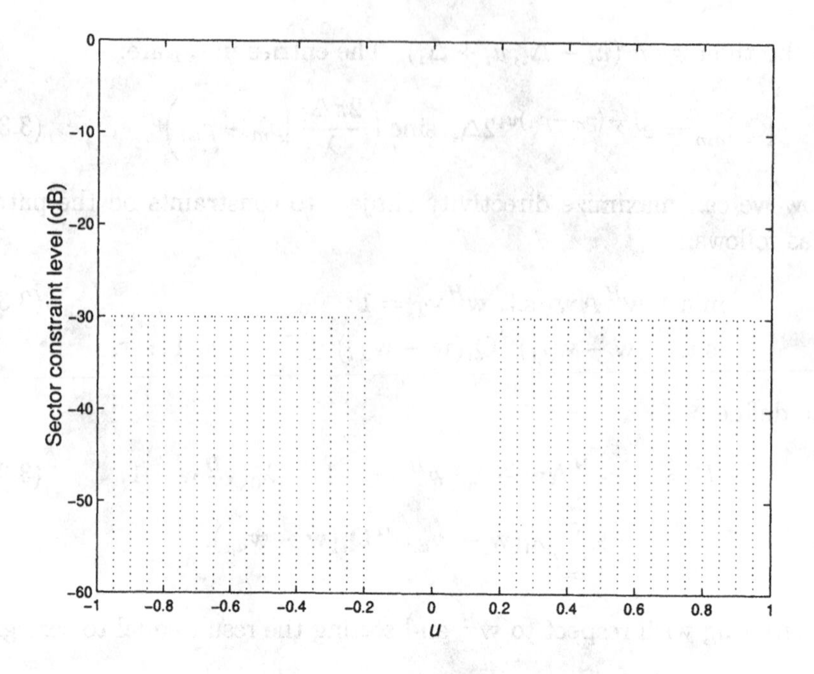

Figure 3.49 Sidelobe sectors.

is just the all-zero vector. In the limit of infinitesimally small sectors, the pattern error constraints become constraints on the magnitude squared of the beam pattern at every point in the sidelobe region. The allowed deviation can be set to the maximum allowable sidelobe level, and the sidelobe levels can be controlled directly. By choosing wider but relatively small sectors, we can still control sidelobe levels fairly accurately. Furthermore, if we choose to constrain pattern "error" only in the sidelobe region and not in the main beam, the desired weight vector in each constrained sector will be zero, and the second term in (3.314) drops out, so the weight vector becomes

$$\mathbf{w} = \mathbf{A}_Q^{-1}\mathbf{v}_T\left(\mathbf{v}_T^H\mathbf{A}_Q^{-1}\mathbf{v}_T\right)^{-1}. \tag{3.315}$$

In this expression, a weighted sum of loading matrices $\mathbf{Q}_i, i = 1 \ldots r$ are added to $\mathbf{A}$. The loading factors balance the maximum directivity pattern with the desired low sidelobe level pattern. There is generally a set of optimum loading levels $\lambda_i, i = 1 \ldots r$ that satisfy the constraints; however, there is no closed-form solution for the loading levels, even when $r = 1$. It can be shown that the mean-square pattern error decreases with increasing λ_i, but at the expense of decreased directivity. An iterative procedure can be used to adjust the loading levels to achieve the sidelobe level constraints. At each iteration, the pattern errors are computed and checked against the constraints. If a constraint is exceeded, the loading for that sector is increased, and the weights are updated.

One way to achieve fast convergence is to let the loading increment at the pth iteration, $\delta_i^{(p)}$, be a fraction of the of the current loading value, that is, $\delta_i^{(p)} = \alpha\lambda_i^{(p)}$. This requires that the initial loading level be non-zero. One possibility is to initialize all of the loading levels to some small value, such as, $\lambda_i^{(0)} = \lambda_0, i = 1 \ldots r$. If the initial loading is small enough, the initial weight vector is essentially the maximum directivity weight vector. The update procedure is:

$$\begin{aligned} &\text{if} \quad \mathbf{w}^{(p-1)H}\mathbf{Q}_i\mathbf{w}^{(p-1)} > L_i, \\ &\text{then} \quad \delta_i^{(p)} = \alpha\lambda_i^{(p-1)}, \\ &\text{else} \quad \delta_i^{(p)} = 0. \end{aligned} \tag{3.316}$$

$$\lambda_i^{(p)} = \lambda_i^{(p)} + \delta_i^{(p)}. \tag{3.317}$$

$$\mathbf{A}_Q^{(p)} = \mathbf{A}_Q^{(p-1)} + \sum_{i=1}^{r} \delta_i^{(p)}\mathbf{Q}_i. \tag{3.318}$$

$$\mathbf{w}^{(p)} = \left(\mathbf{A}_Q^{(p)}\right)^{-1} \mathbf{v}_T \left\{\mathbf{v}_T^H \left(\mathbf{A}_Q^{(p)}\right)^{-1} \mathbf{v}_T\right\}^{-1}. \tag{3.319}$$

The iteration is repeated until a convergence criterion is satisfied.

It is usually necessary to adjust the sectors included in the sidelobe region at each iteration. As the sidelobes are pushed down, the main beam widens, and some sectors previously in the sidelobe region fall in the main beam. The constraints on these sectors must then be dropped.

Example 3.9.3

Consider a standard 10-element linear array. The desired look direction is $u_T = 0$ and the desired sidelobe level is -30 dB. Initially, 80 sidelobe regions are defined as sectors of width $2\Delta_i = 0.02$ in the regions $0 \le u \le -0.2$ and $0.2 \le u \le 1$. The constraint levels are all set to the sidelobe level times the width of each sector $L_i = 2 \times 10^{-5}$. The initial loading level is set to $\lambda_0 = 1$, and $\alpha = 0.3$. In Figure 3.50, we show the beam pattern and sidelobe region evolution. The final beam pattern is obtained after 14 iterations. The final beam pattern is essentially the same as the Dolph Chebychev pattern in Section 3.4.2.

Example 3.9.4

In this example, 10 elements were located along the z-axis. The elements were spaced at a distance of $\lambda/2$ with a random perturbation between $\pm\lambda/4$. Specifically,

$$p_n = \left(-\frac{N-1}{2} + n\right)\frac{\lambda}{2} + \left(d_i - \frac{1}{2}\right)\frac{\lambda}{2}, \tag{3.320}$$

where d_i is a uniform random variable $[0,1]$. The desired look direction is broadside. An equal-ripple beam pattern was designed with -25 dB sidelobes.

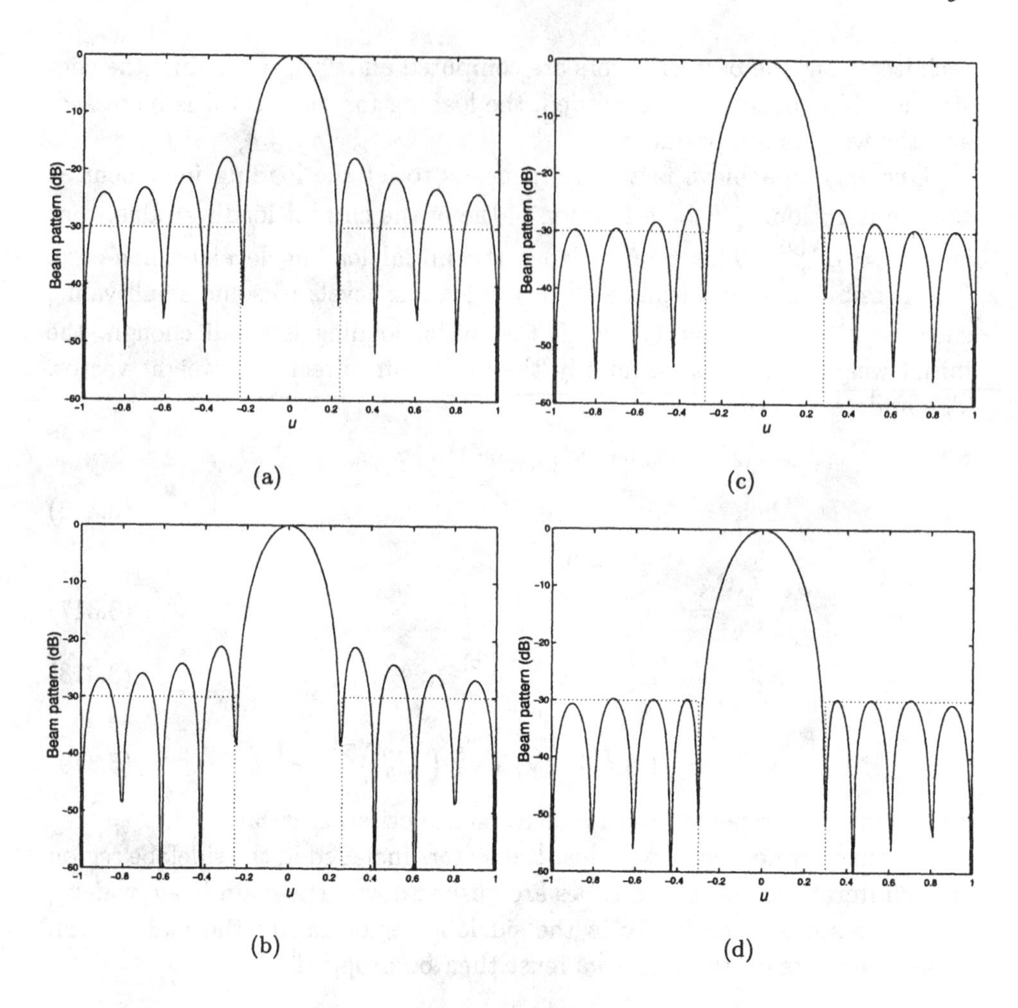

Figure 3.50 Beam pattern evolution for 10-element uniform linear array and −30-dB sidelobes. (a) First iteration; (b) fifth iteration; (c) 10th iteration; (d) 14th iteration.

Table 3.9 Element Locations Used for the Isotropic Linear Random Array

Element No.	Relative Position
0	-2.2509λ
1	-1.6501λ
2	-1.1696λ
3	-0.7138λ
4	-0.1705λ
5	0.2901λ
6	0.7105λ
7	1.1974λ
8	1.7103λ
9	2.2585λ

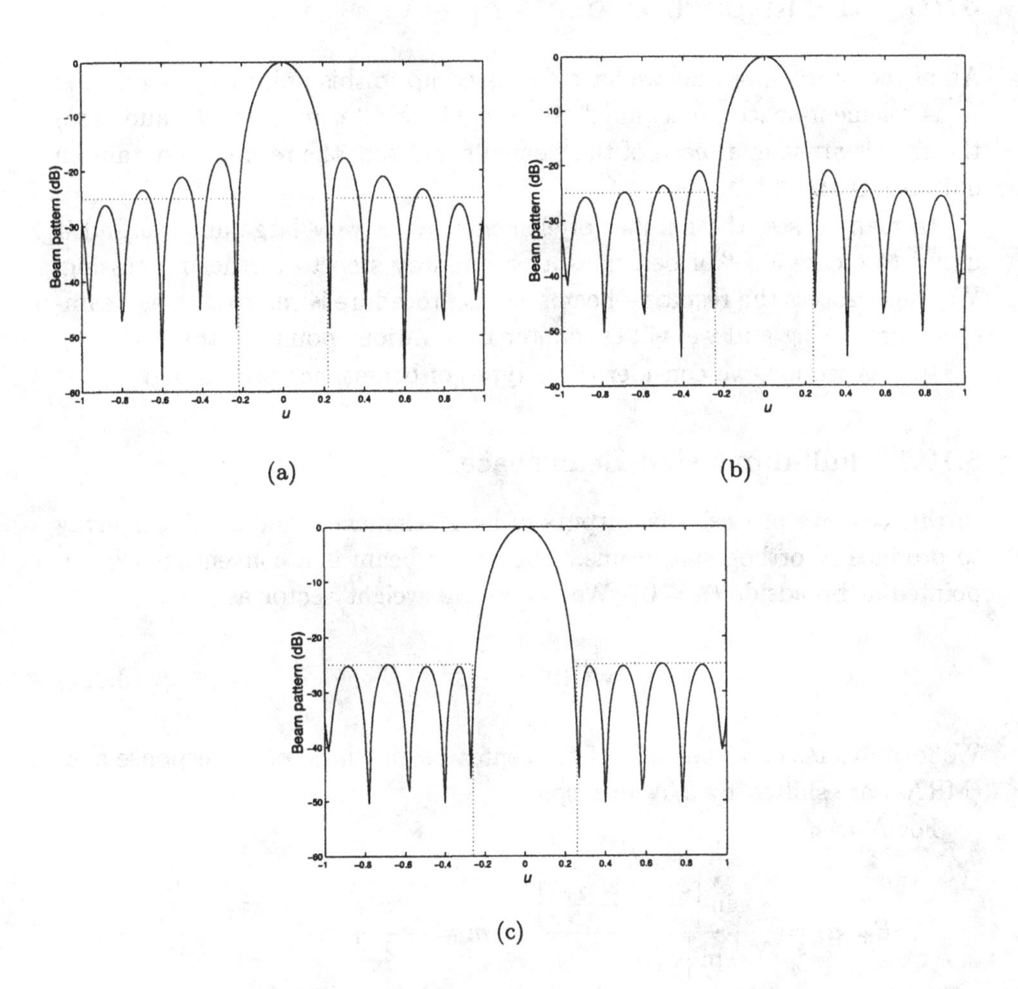

Figure 3.51 Beam pattern evolution for 10-element non-uniform linear array and −25-dB sidelobes. (a) First iteration; (b) fifth iteration; (c) 10th iteration.

The element locations for a particular trial are shown in Table 3.9. The beam pattern evolution is shown in Figure 3.51. Figure 3.51(a) is the initial pattern, Figure 3.51(b) is the fifth iteration, and Figure 3.51(c) is the tenth iteration.

The extension to non-uniform sidelobe control follows easily. We can also force nulls at specific locations. Both of our examples in this section considered linear arrays, but the technique also applies to arbitrary array geometries and non-isotropic sensors. In Chapter 7, we develop an adaptive version of the algorithm.

3.10 Beamspace Processing

All of the processing that we have discussed up to this point can be referred to as "element-space processing." We weight (or filter in the wideband case) the signals arriving at each of the elements and sum the results to obtain an output signal.

In many cases, the number of elements, N, is very large and we find it useful to create a set of beams as a preliminary step to further processing. We then process the resulting beams. This procedure is referred to as beam-space processing and we will encounter it at various points in the text.

In this section, we consider three types of beamspace processing.

3.10.1 Full-dimension Beamspace

In this case we process the outputs of an N-element standard linear array to produce N orthogonal beams. The center beam is a conventional beam pointed at broadside ($u = 0$). We denote the weight vector as

$$\mathbf{w}^H(0) = \frac{1}{N}\mathbf{1}^T. \tag{3.321}$$

We form beams on either side of the center beam whose main response axes (MRAs) are shifted by $2/N$ in u-space.

For N even,

$$B_m(u) = \frac{1}{N}\frac{\sin\left[\frac{\pi N}{2}\left(u - \frac{2m}{N}\right)\right]}{\sin\left[\frac{\pi}{2}\left(u - \frac{2m}{N}\right)\right]}, \quad m = -\frac{N}{2}+1, \cdots \frac{N}{2}. \tag{3.322}$$

There are N beams. The beam corresponding to $m = N/2$ is an endfire beam (it could also have been indexed as $m = -N/2$). In Figure 3.52(a), we show the beams in u-space. In Figure 3.52(b), we show the main lobes of the beams in θ-space. The lobe at $\theta = 180°$ is a grating lobe.

For N odd,

$$B_m(u) = \frac{1}{N}\frac{\sin\left[\frac{\pi N}{2}\left(u - \frac{2m}{N}\right)\right]}{\sin\left[\frac{\pi}{2}\left(u - \frac{2m}{N}\right)\right]}, \quad m = -\frac{N-1}{2}, \cdots \frac{N-1}{2}. \tag{3.323}$$

In this case, there is no endfire beam. In Figure 3.53, we show the beams in u-space and the main lobes in θ-space for $N = 11$.

The MRA for the mth beam occurs at $2m/N$. All of the other beams have nulls at that point. This occurs because the weight vectors are orthogonal

$$\mathbf{w}^H(m)\mathbf{w}(l) = \frac{1}{N}\delta_{ml}. \tag{3.324}$$

The corresponding beams are also orthogonal:

$$B_m(u) = \mathbf{w}^H(m)\mathbf{v}(u), \tag{3.325}$$

and

$$\begin{aligned}\int_{-1}^{1} B_m(u)B_l^*(u)\,du &= \mathbf{w}^H(m)\left(\int_{-1}^{1}\mathbf{v}(u)\mathbf{v}^H(u)\,du\right)\mathbf{w}(l)\\ &= \mathbf{w}^H(m)\,\mathbf{I}\,\mathbf{w}(l) = \frac{1}{N}\delta_{ml}.\end{aligned} \tag{3.326}$$

The result in (3.324) implies that a signal arriving along the MRA of a particular beam will have no output in any other beam. However, a signal that is not along the MRA will appear in the sidelobes of the other beams.

We form an $N \times N$ matrix, $\mathbf{B}_{bs}^H$ whose mth row is $\mathbf{w}^H(m)$. Then

$$\mathbf{x}_{bs} = \mathbf{B}_{bs}^H\mathbf{x}. \tag{3.327}$$

This operation is shown in Figure 3.54.

The matrix, $\mathbf{B}_{bs}^H$, is commonly referred to as Butler matrix [BL61] and is an invertible matrix. From (3.95), we observe that $\mathbf{B}_{bs}^H$ is the DFT matrix. Thus,

$$\mathbf{x} = [\mathbf{B}_{bs}^H]^{-1}\mathbf{x}_{bs}, \tag{3.328}$$

so we have not lost any information by the transformation. In the statistical literature, (3.327) is referred to as the DFT beamformer. Often this transformation makes the implementation of the resulting processing easier. In later chapters, we will develop beamspace adaptive arrays [AC76] and beamspace direction-of-arrival estimators.

In most applications, we work with a reduced-dimension beamspace. We discuss this approach in the next section.

3.10.2 Reduced-dimension Beamspace

Consider the application shown in Figure 3.55. All of the signals of interest are contained in a region ψ_B. We can significantly reduce our subsequent processing if we form a set of beams that span the space.

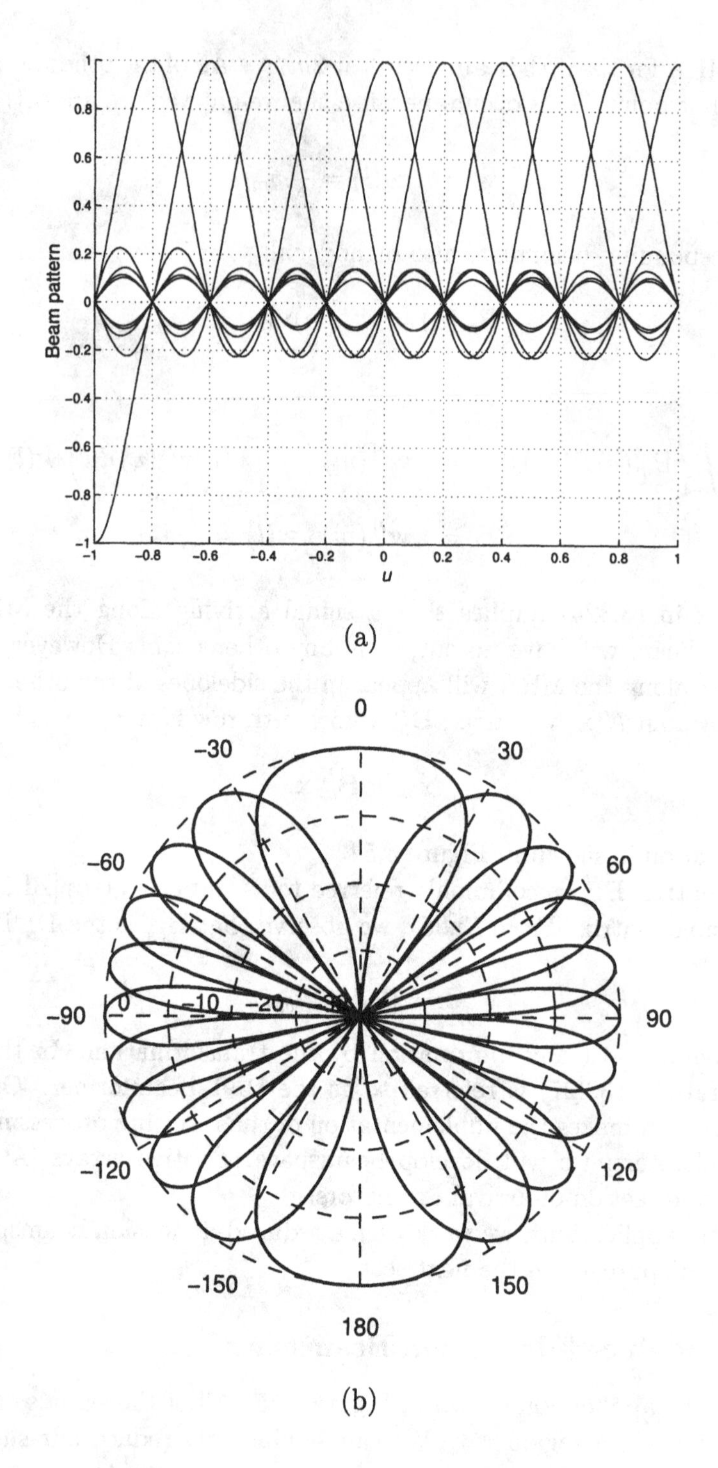

Figure 3.52 Beam patterns: N=10; (a) u-space; (b) θ-space.

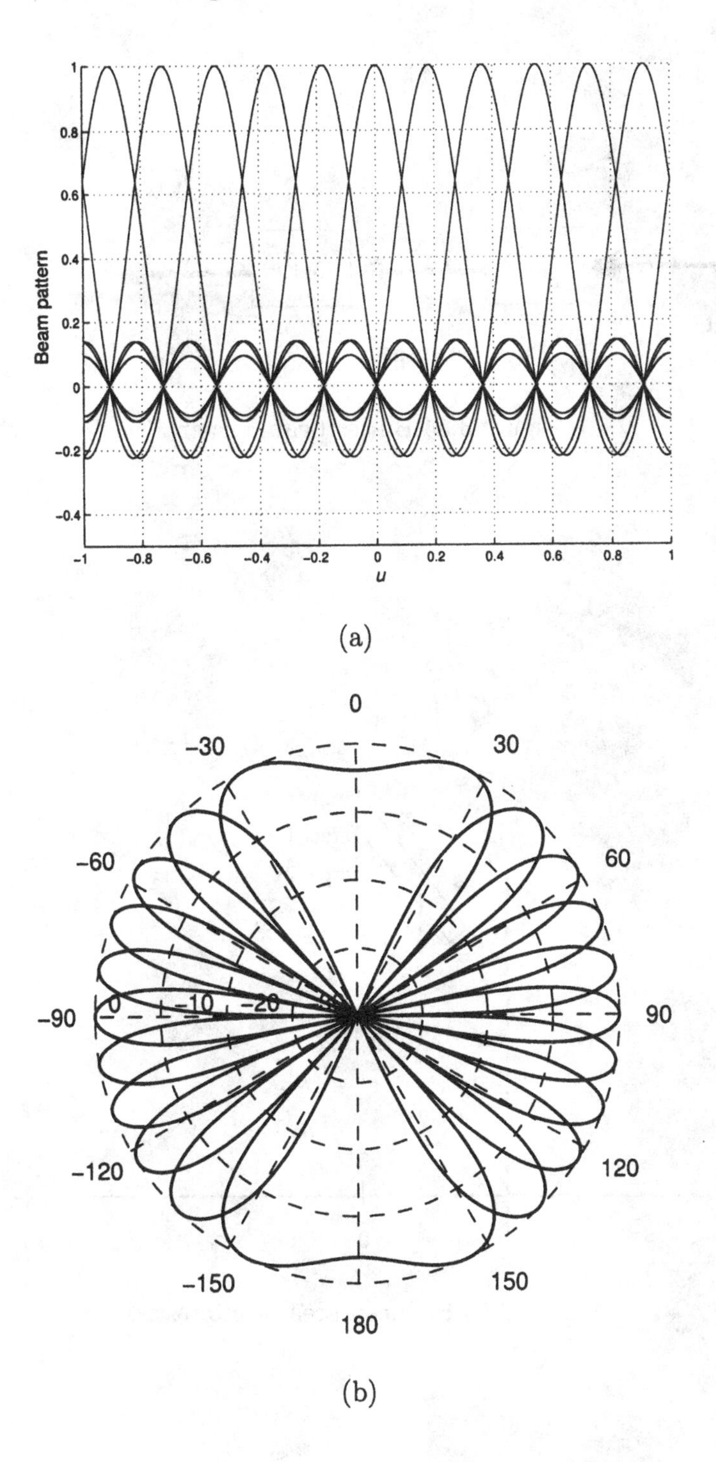

Figure 3.53 Beam patterns: N=11; (a) u-space; (b) θ-space.

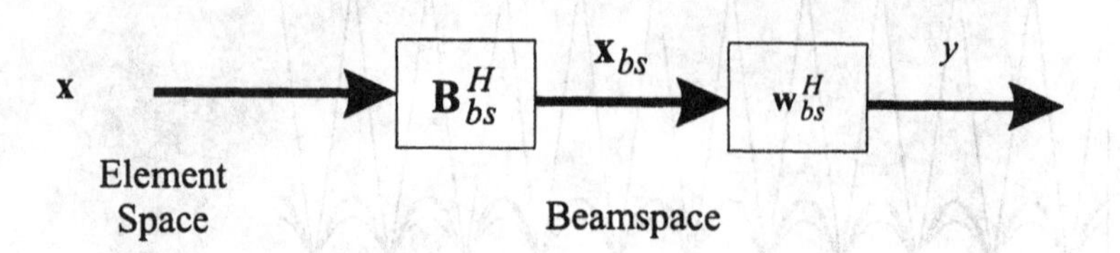

Figure 3.54 Butler beamformer.

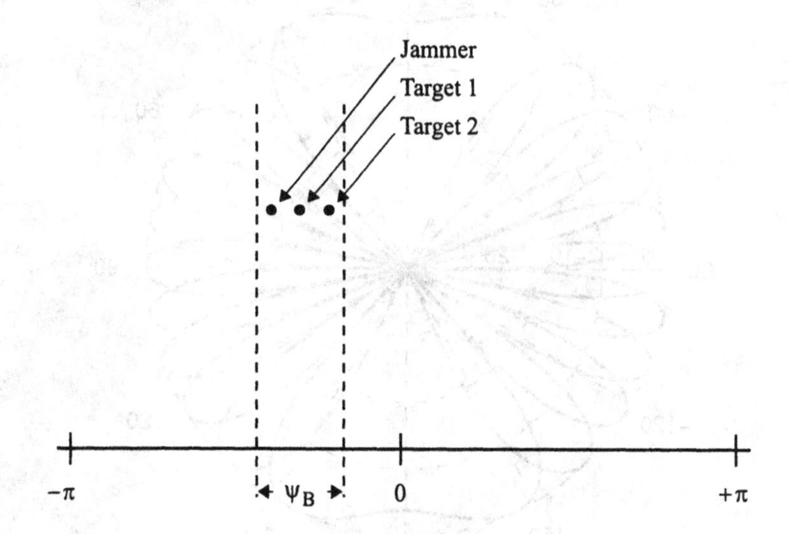

Figure 3.55 Reduced area of interest.

In practice, we normally are scanning the beams over some region in u-space. We form a beam fan consisting of N_{bs} beams and move the MRA of the center beam through u-space in $2/N$ increments. In Figure 3.56, we show several representative positions in u-space for a 7-beam fan and a 32-element standard linear array.

In most applications, we would like to use as few beams as possible without degrading the performance of the system. Later in the text we introduce appropriate statistical measures to quantify the degradation. In constructing the reduced-dimension beamspace we will consider different beamforming matrices in order to maximize performance. Several techniques are logical:

(i) The conventional beamformer (or DFT) set as shown in Figure 3.56 is appropriate.

(ii) In order to reduce out-of-sector interference, we can use low sidelobe beams such as Dolph-Chebychev or Taylor as the component beams. We would normally space the centers of the beams at equal intervals in ψ-space. The columns in the beamspace matrix are not orthogonal. In many applications, we need orthogonal columns. If we denote the original beamspace matrix as $\mathbf{B}_{no}$ where the subscript "no" denotes that the columns are not orthogonal, then if we define the beamspace matrix as,

$$\mathbf{B}_{bs} = \mathbf{B}_{no}\left[\mathbf{B}_{no}^{H}\mathbf{B}_{no}\right]^{-\frac{1}{2}}, \tag{3.329}$$

the columns will be orthogonal. The transformation in (3.329) will increase the sidelobe level of the component beams by several dB.

(iii) Another choice of component beams that have lower sidelobes than conventional beams are the weightings from Section 3.1 that generated beam patterns corresponding to weighted sums of shifted sinc functions. Specifically:

(a) $\cos^m(\pi\tilde{n}/N)$ (3.18)

(b) $\cos^2(\pi\tilde{n}/N)$ (3.18)

(c) Hamming (3.21)

(d) Blackman-Harris (3.23)

These weightings have the property that the out-of-sector zeros of each beam in the beam fan are the same. This property will be useful when we study parameter estimation in Chapters 8 and 9.

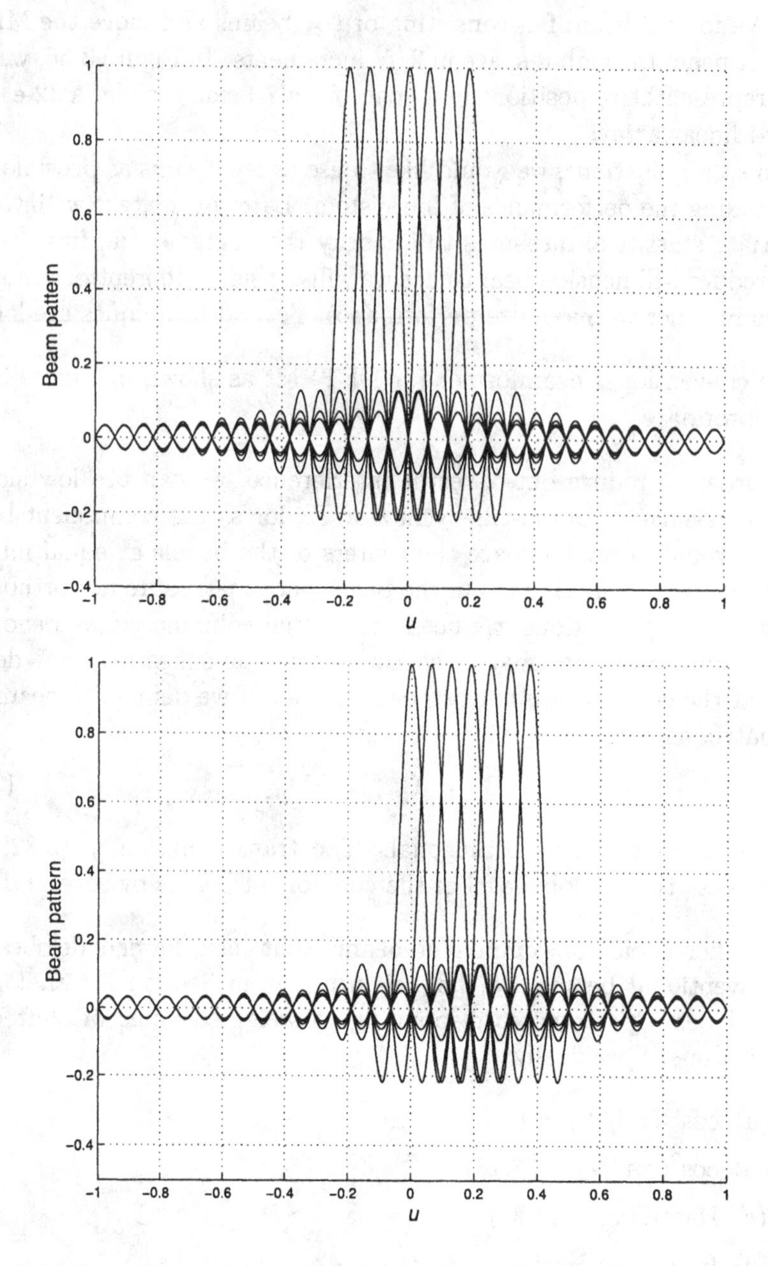

Figure 3.56 Scanning u-space with beam fans: $N = 32, N_{bs} = 7$; (a) center beam at $u = 0$; (b) center beam at $u = 6/N$.

(iv) The first M discrete prolate spheroidal sequences developed in Section 3.1 ((3.25)–(3.35)) are logical candidates because of their power maximization characteristic.

(v) We can develop a set of component beams by using the array manifold vector $\mathbf{v}(\psi)$ and its derivatives with respect to ψ, $\dot{\mathbf{v}}(\psi)$, $\ddot{\mathbf{v}}(\psi)$ evaluated at ψ_c, the location of the center of the beam fan.

Other choices appear in later discussions. The best choice depends on the environment and the specific objective of our array processor.

The final step is to process $\mathbf{x}_{bs}$, the output of the beamspace matrix to produce the array output y. We show the processing in Figure 3.54.

We define the beamspace array manifold vector as

$$\mathbf{v}_{bs}(\psi) = \mathbf{B}_{bs}^H \mathbf{v}_{\psi}(\psi). \tag{3.330}$$

If we use a 7×32 beamspace matrix whose rows correspond to shifted conventional beams, and the center beam is steered to ψ_c, then

$$\left[\mathbf{b}_{bs,m}^H\right]_n = \frac{1}{\sqrt{N}}\left[e^{-j\left(n-\frac{N-1}{2}\right)\left(\psi_c-(m-4)\frac{2\pi}{N}\right)}\right], \quad \begin{aligned} n &= 0,\cdots,N-1, \\ m &= 1,\cdots,7, \end{aligned} \tag{3.331}$$

and

$$[\mathbf{v}_{\psi}(\psi)]_n = e^{j\left(n-\frac{N-1}{2}\right)\psi}, \quad n = 0,\cdots,N-1. \tag{3.332}$$

Using (3.331) and (3.332) in (3.330) and, letting $\psi_c = 0$, gives

$$\mathbf{v}_{bs}(\psi) = \sqrt{N}\begin{bmatrix} \frac{sin\left(\frac{N}{2}\left(\psi+3\frac{2\pi}{N}\right)\right)}{sin\left(\frac{1}{2}\left(\psi+3\frac{2\pi}{N}\right)\right)} \\ \vdots \\ \frac{sin\left(\frac{N}{2}\left(\psi-(m-4)\frac{2\pi}{N}\right)\right)}{sin\left(\frac{1}{2}\left(\psi-(m-4)\frac{2\pi}{N}\right)\right)} \\ \vdots \\ \frac{sin\left(\frac{N}{2}\left(\psi-3\frac{2\pi}{N}\right)\right)}{sin\left(\frac{1}{2}\left(\psi-3\frac{2\pi}{N}\right)\right)} \end{bmatrix} = \sqrt{N}\begin{bmatrix} B_c\left(\psi+3\frac{2\pi}{N}\right) \\ \vdots \\ B_c\left(\psi-(m-4)\frac{2\pi}{N}\right) \\ \vdots \\ B_c\left(\psi-3\frac{2\pi}{N}\right) \end{bmatrix}. \tag{3.333}$$

Note that the beamspace array manifold vector is real. This result allows us to use real computation in many of our subsequent algorithms.

We process the output of the beamspace matrix, $\mathbf{x}_{bs}$, with a $1 \times N_{bs}$ matrix, $\mathbf{w}_{bs}^H$, to obtain a scalar output, y. The resulting beam pattern is

$$B_{\psi}(\psi) = \mathbf{w}_{bs}^H \mathbf{v}_{bs}(\psi). \tag{3.334}$$

In Chapters 6 through 9, we develop optimum beamspace processors for various applications and show the advantages of operating in a reduced-dimension beamspace.

3.10.3 Multiple Beam Antennas

In this case, the physical implementation of the antenna provides a set of multiple beams that span the wavenumber region of interest. Important examples of this case are the transmit and receive antennas on various military satellites (e.g., Mayhan [May76] or Ricardi [Ric76]). In those cases, we start the problem in beamspace.

3.10.4 Summary

We have introduced beamspace processing in this section. It is widely used in practical applications. One advantage is that the complexity of the subsequent processing is reduced significantly. By choosing a suitable beam fan we can usually minimize any loss in performance. We investigate those issues in later sections.

3.11 Broadband Arrays

In many applications, the array is required to process signals over a broad frequency band. One approach is to utilize an array with uniform spacing,

$$d = \frac{\lambda_u}{2}, \tag{3.335}$$

where λ_u is the wavelength of the highest frequency. We then use frequency-dependent weightings to process the beams. The difficulty with this approach is that the required number of sensors may be prohibitive. The spacing in (3.335) is required to avoid grating lobes. If we require that the width of the main lobe be constant across the frequency band of interest, then the total length must be proportional to the λ_l. From (2.109),

$$BW_{NN} = \alpha \frac{\lambda_l}{Nd}, \quad u\text{-space}, \tag{3.336}$$

where α is a constant dependent on the shading. Using (3.335) in (3.336) gives

$$N = \left(\frac{2\alpha}{BW_{NN}}\right) \frac{\lambda_l}{\lambda_u}, \tag{3.337}$$

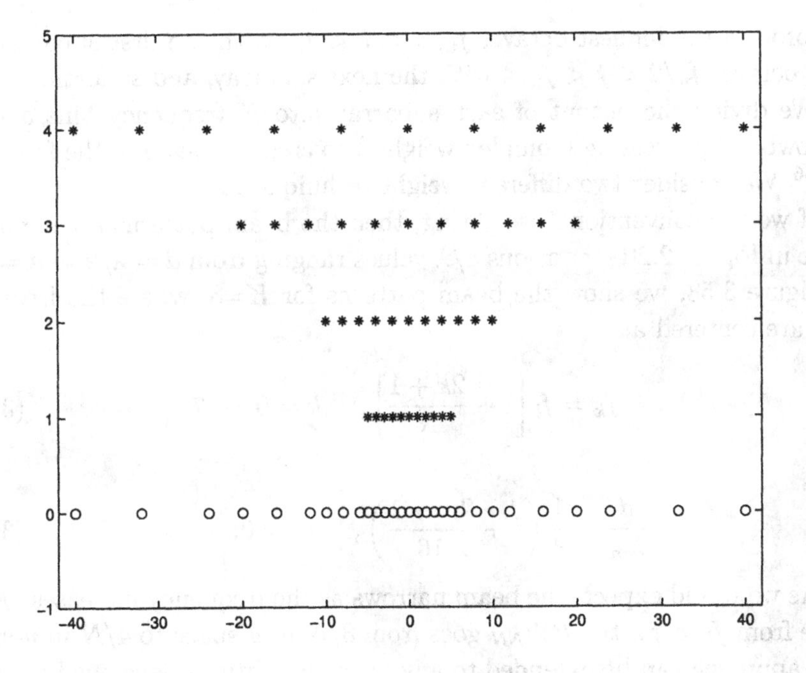

Figure 3.57 Nested arrays.

so the required number of elements is proportional to the band ratio. In this section, we discuss two techniques that utilize arrays with non-uniform element spacing.

The first technique uses a compound array consisting of a nested set of subarrays, each of which is designed for single frequency. The most common type of nested array is shown in Figure 3.57. The bottom subarray has N elements spaced at $d_1 = \lambda_u/2$, where λ_u is the wavelength at the highest frequency. In the figure, $N = 11$. The next subarray has N elements spaced at $d_2 = \lambda_u$, which corresponds to half-wavelength spacing at $f_u/2$. The third subarray has N elements spaced at $d_3 = 2\lambda_u$, which corresponds to half-wavelength spacing at $f_u/4$. The fourth subarray has N element spaced at $d_4 = 4\lambda_u$, which corresponds to half-wavelength spacing at $f_u/8$.

The total number of elements is

$$N_T = N + 3\left(\frac{N+1}{2}\right), \quad N \text{ odd}, \tag{3.338}$$

and

$$N_T = N + 3\frac{N}{2}, \quad N \text{ even}. \tag{3.339}$$

We process the highest octave, $f_u/2 < f < f_u$, with the first subarray, the next octave, $f_u/2 < f < f_u/4$ with the next subarray, and so forth.

We divide the output of each subarray into K frequency bins and use narrowband processing (complex weights) to create a beam in the frequency bin.[36] We consider two different weight techniques.

If we use conventional weighting, then the beam patterns are similar to those in Figure 2.20 for various d/λ values ranging from $d = \lambda/2$ to $d = \lambda/4$. In Figure 3.58, we show the beam patterns for K=8, where the frequency bins are centered at

$$f_k = f_l \left[1 + \frac{2k+1}{16}\right], \quad k = 0, \cdots 7. \tag{3.340}$$

Thus

$$\frac{d}{\lambda_m} = \frac{1}{4}\left(1 + \frac{2m+1}{16}\right), \quad m = 0, \cdots, 7. \tag{3.341}$$

As we would expect, the beam narrows as the frequency increases. As we move from f_l to f_u, the BW_{NN} goes from $8/N$ in u-space to $4/N$ in u-space. This approach can be extended to any of the weightings developed earlier in the chapter. In all cases, the main-lobe width will decrease as we go up in frequency. Note that the beam patterns are repetitive in each octave because of the choice of the subarray.

A different technique is used if we want a constant beam pattern over each octave. The technique is described in Chou [Cho95] (see also [GE93] and [Tuc57]). We first define a desired beam pattern at f_l for a uniform linear array whose spacing is

$$d = \frac{\lambda_u}{2}, \tag{3.342}$$

and

$$\lambda_l = 2\lambda_u. \tag{3.343}$$

For example, if we use uniform weighting,

$$B_u(u) = \frac{1}{N}\frac{\sin\left(\frac{\pi N d}{\lambda_l}u\right)}{\sin\left(\frac{\pi d}{\lambda_l}u\right)} = \frac{1}{N}\frac{\sin(\pi N u)}{\sin \pi u}. \tag{3.344}$$

The beam pattern is shown in Figure 3.59 over the interval $-2 \leq u \leq 2$. Because $d = \lambda_l/4$, we need to consider the beam pattern over twice the

[36] An important issue is the technique for combining the outputs of the different frequency bins to synthesize the output signal. This is discussed in Section 6.13.

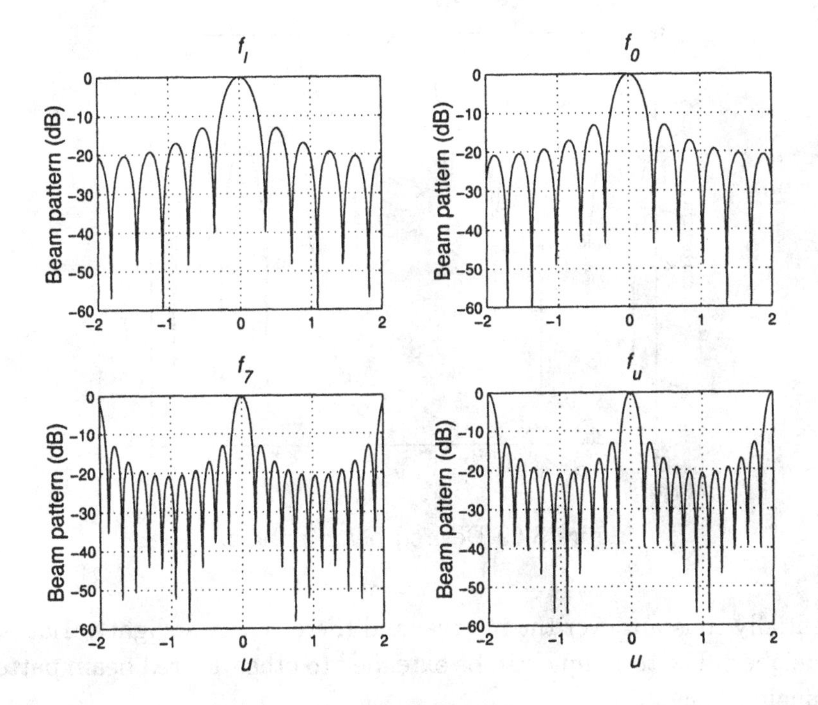

Figure 3.58 Beam patterns: N=11, K=8; f_0 and f_7 plus endpoint references f_l and f_u.

visible region. For each of the frequencies in (3.340), we sample the beam pattern at

$$u_n = \frac{c}{Nd \cdot f_k} n, \quad -\frac{N-1}{2} \leq n \leq \frac{N-1}{2}, \quad N \text{ odd}, \quad 1 \leq k \leq K. \tag{3.345}$$

We perform an IDFT on the samples to obtain the weightings for that frequency bin f_k. We then take the DFT to find the beam pattern for frequency bin f_k. In Figure 3.60, we show the samples and resulting weightings for

$$f_0 = f_l(1 + 1/16), \tag{3.346}$$

and

$$f_7 = f_l(1 + 15/16). \tag{3.347}$$

We recall from Example 3.3.4 that we need samples outside the visible region to specify the weight vector when $d < \lambda/2$. In Figure 3.61, we show the resulting beam pattern over four octaves from 500 to 8000. The main lobe

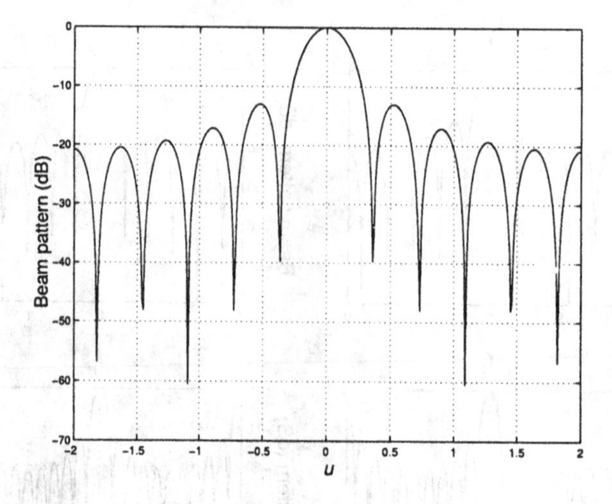

Figure 3.59 Desired beam pattern.

is essentially constant over the interval and there is only a slight variation in the sidelobes. The technique can be extended to other desired beam patterns by changing (3.344).

A possible disadvantage of the nested approach is that only a limited set of band ratios; 2, 4, 8, $\cdots$ are achievable.

An alternate approach that allows more flexibility in broadband array design is described in Doles and Benedict [DB88]. They use the asymptotic theory of unequally spaced arrays that was developed by Ishimaro (e.g., Ishimaru [Ish62], Ishimaru and Chen [IC65], and Chow [Cho65]). The resulting arrays have exponential spacing. The reader is referred to these references for discussions of the technique.

3.12 Summary

This completes our discussion of analysis and synthesis of the weight vectors for linear arrays. We have developed an extensive set of tools to carry out the process.

The majority of the chapter was devoted to two problems. In the first problem, we wanted to generate a pattern that had a narrow main lobe and low sidelobes. Sections 3.1 and 3.4 developed techniques for designing a weight vector that provided a suitable compromise between these conflicting objectives. The Dolph-Chebychev and Taylor weightings are widely used in classical array design and play an important role in the optimum array

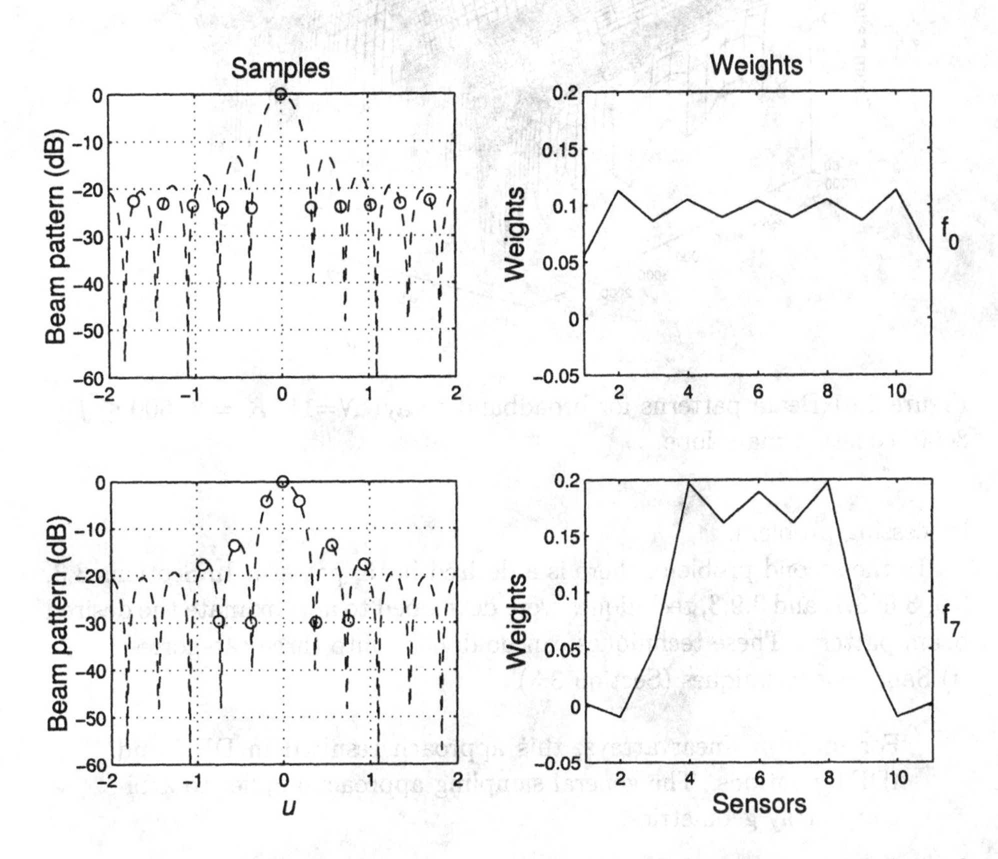

Figure 3.60 Samples and array weights; N=11, K=8, conventional desired pattern.

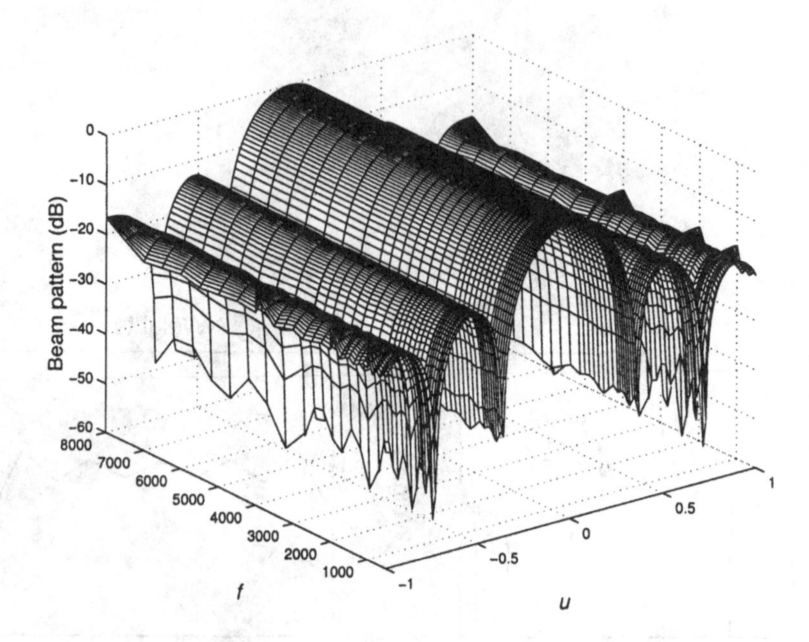

Figure 3.61 Beam patterns for broadband array: N=11, $K = 8$, $500 \leq f \leq 8000$, constant main lobe.

processing problem.

In the second problem, there is a desired beam pattern. In Sections 3.3, 3.5, 3.6, 3.7, and 3.9.3, techniques were developed to approximate the desired beam pattern. These techniques can be divided into three categories:

(i) Sampling techniques (Section 3.3)

> For uniform linear arrays, this approach resulted in DFT and IDFT relations. The general sampling approach applies to arbitrary array geometries.

(ii) Minimax techniques (Sections 3.6 and 3.9.3)

> These techniques impose hard constraints on the allowable deviation of the synthesized beam pattern from the desired beam pattern. They utilize an iterative technique to obtain the weight vector. The technique in Section 3.9.3 extends the hard constraint criterion of Sections 3.4 and 3.6 to arbitrary array geometries.

(iii) Least squares techniques (Sections 3.5 and 3.7)

Least squares techniques (generally with some type of constraints) are widely used in array design. The techniques are applicable to arbitrary array geometries. The least squares criterion leads to a quadratic minimization problem. In Sections 3.5 and 3.7, an analytic solution is available. The quadratic minimization approach applies to arbitrary arrays and is encountered frequently in subsequent discussions.

The remainder of the chapter discussed four other topics: difference beams, non-uniform linear arrays, beamspace processing, and broadband arrays.

Section 3.8 discussed the synthesis of difference beams with low sidelobes; these beams are widely used in applications where we want to estimate the angle of arrival of a plane-wave signal from a target and aim the array at the target. Section 3.9 discussed linear arrays with non-uniform spacing. The concept of a minimum redundancy linear array was introduced. Synthesis techniques for linear arrays with non-uniform spacing were developed. Section 3.10 developed the basic ideas of beamspace processing. Various beamspace applications will be discussed in subsequent chapters. Section 3.11 introduced some of the techniques for broadband array design. This topic is revisited in subsequent chapters.

Many of these techniques can be extended to planar arrays. Chapter 4 discusses planar arrays.

3.13 Problems

P3.1 Spectral Weighting

Problem 3.1.1

Consider a linear aperture of length L:

$$w(z) = 1 - \left(\frac{2z}{L}\right)^2, \quad -\frac{L}{2} \leq z \leq \frac{L}{2}. \tag{3.348}$$

Find the beam pattern, the HPBW, and height of the first sidelobe.

Problem 3.1.2

Consider a linear aperture of length L:

$$w(z) = \left(1 - \left(\frac{2z}{L}\right)^2\right)^2, \quad -\frac{L}{2} \leq z \leq \frac{L}{2}. \tag{3.349}$$

Find the beam pattern, the HPBW, and height of the first sidelobe.

Problem 3.1.3

Consider a linear aperture of length L:

$$w(z) = \left(1 - \left(\frac{2z}{L}\right)^2\right)^{\frac{1}{2}}, \quad -\frac{L}{2} \leq z \leq \frac{L}{2}. \tag{3.350}$$

Find the beam pattern, the HPBW, and height of the first sidelobe.

Problem 3.1.4

Consider a standard linear array with 9 elements.

(a) Compute the DPSS weighting for $\psi_0 = 0.1\pi, 0.2\pi, 0.3\pi, 0.4\pi$.

(b) Plot the resulting beam patterns.

(c) Compute the HPBW, BW_{NN}, and the height of highest sidelobe.

Problem 3.1.5

Show that the beam pattern for the Kaiser weighting is proportional to

$$\frac{\sin\left[\beta\sqrt{\left(\frac{\psi}{\psi_\beta}\right)^2 - 1}\right]}{\sqrt{\left(\frac{\psi}{\psi_\beta}\right)^2 - 1}}, \tag{3.351}$$

where ψ_β is the approximate beamwidth of the main lobe.

Problem 3.1.6

The Lanczos weighting for a standard linear array is defined as

$$w_L(n) = \left\{\frac{\sin\left[\frac{2\pi n}{(N-1)^2}\right]}{\frac{4\pi n}{(N-1)^2}}\right\}^L, \quad L > 0, \tag{3.352}$$

where the indexing is symmetric.

Plot the array weighting and corresponding beam pattern for several values of L.

Problem 3.1.7

Compute Δu_4 (see Problem 2.4.8) for:

(a) Hann weighting

(b) Hamming weighting

(c) Blackman-Harris weighting

and compare to the HPBW.

Problem 3.1.8

Compute Δu_4 for:

(a) Gaussian weighting

(b) DPSS weighting

(c) Kaiser weighting

and compare to the HPBW.

Problem 3.1.9

Consider an 11-element standard linear array with symmetric weighting

$$w_0 = 0.5, \tag{3.353}$$

$$w_n = w_{-n} = 1.0 + 0.5(n-1), \quad n = 1, 2, \cdots, 5. \tag{3.354}$$

(a) Plot $B_\psi(\psi)$.

(b) Where might this beam pattern be useful?

Problem 3.1.10

Consider an 11-element standard linear array with symmetric weighting,

$$w_n = 1.0, \quad n = 0, 1, 2, 3, \tag{3.355}$$

$$w_n = 0.6, \quad n = 4, \tag{3.356}$$

$$w_n = 0.2, \quad n = 5. \tag{3.357}$$

Compute and plot $B_\psi(\psi)$.

Problem 3.1.11

The Riesz (Bochner, Parzen) window [Har78] is defined as

$$w_n = 1.0 - \left|\frac{n}{N/2}\right|^2, \quad 0 \le |n| \le \frac{N}{2}. \tag{3.358}$$

(a) Find and plot the beam pattern for $N = 10$.

(b) Compute the HPBW, BW_{NN}, the height of the first sidelobe, the rate of sidelobe decrease, and the directivity.

Problem 3.1.12

The Riemann window is [Har78]

$$w_n = \frac{\sin\left[\frac{n}{N}2\pi\right]}{\left[\frac{n}{N}2\pi\right]}, \quad 0 \le |n| \le \frac{N}{2}, \tag{3.359}$$

and corresponds to the main lobe of a sinc function. Repeat Problem 3.1.11.

Problem 3.1.13

The de la Vallé-Poussin (Jackson, Parzen) window is a piecewise cubic curve obtained by self-convolving two triangles of half-extent or four rectangles of one-fourth extent. It is defined as

$$w_n = \begin{cases} 1.0 - 6\left[\frac{n}{N/2}\right]^2\left[1.0 - \frac{|n|}{N/2}\right], & 0 \le |n| \le \frac{N}{4}, \\ 2\left[1.0 - \frac{|n|}{N/2}\right]^3, & \frac{N}{4} \le |n| \le \frac{N}{2}. \end{cases} \tag{3.360}$$

Repeat Problem 3.1.11.

Problem 3.1.14

The Tukey window [Har78], [Tuk67] is a cosine lobe of width $(\alpha/2)N$ convolved with a rectangle window of width $(1.0 - \alpha/2)N$. The window evolves from the rectangle to the

Hann window as the parameter α varies from unity to zero. The Tukey window is defined as

$$w_n = \begin{cases} 1.0, & 0 \le |n| \le \alpha\frac{N}{2}, \\ 0.5\left[1.0 + \cos\left(\pi\frac{|n|-\alpha\frac{N}{2}}{(1-\alpha)\frac{N}{2}}\right)\right], & \alpha\frac{N}{2} \le |n| \le \frac{N}{2}. \end{cases} \tag{3.361}$$

(a) Plot w_n and the corresponding beam pattern for $\alpha = 0.25, 0.50, 0.75$.

(b) Compute the HPBW, BW_{NN}, the height of the first sidelobe, the rate of sidelobe decrease, and the directivity.

Problem 3.1.15

The Bohman window [Har78], [Boh60] is obtained by the convolution of two half-duration cosine lobes; thus its transform is the square of the cosine lobe transform. The window is defined as

$$w_n = \left[1.0 - \frac{|n|}{N/2}\right]\cos\left[\pi\frac{|n|}{N/2}\right] + \frac{1}{\pi}\sin\left[\pi\frac{|n|}{N/2}\right], \quad 0 \le |n| \le \frac{N}{2}. \tag{3.362}$$

Repeat Problem 3.1.11.

Problem 3.1.16

The Poisson window [Har78], [Bar64] is a two-sided exponential defined by

$$w_n = \exp\left(-\alpha\frac{|n|}{N/2}\right), \quad 0 \le |n| \le \frac{N}{2}. \tag{3.363}$$

Repeat Problem 3.1.11 for $\alpha = 2.0, 3.0$, and 4.0.

Problem 3.1.17

The Hann-Poisson window is constructed as the product of the Hann and the Poisson windows. The family is defined by

$$w_n = 0.5\left[1.0 + \cos\left(\pi\frac{n}{N/2}\right)\right]\exp\left(-\alpha\frac{|n|}{N/2}\right), \quad 0 \le |n| \le \frac{N}{2}. \tag{3.364}$$

Repeat Problem 3.1.11 for $\alpha = 0.5, 1.0$, and 2.0.

Problem 3.1.18

The Cauchy (Abel, Poisson) window [Har78], [Akh56] is a family of windows parameterized by α and is defined as

$$w_n = \frac{1}{1.0 + \left(\alpha\frac{n}{N/2}\right)^2}, \quad 0 \le |n| \le \frac{N}{2}. \tag{3.365}$$

Repeat Problem 3.1.11 for $\alpha = 3.0, 4.0$, and 5.0.

Problem 3.1.19

The Hamming, Hann, and Blackman windows are constructed from shifted conventional patterns. More generally, we can write

$$w_n = \sum_{m=0}^{N/2} a_m \cos\left(\frac{2\pi}{N}m(|n| - \frac{N-1}{2})\right), \quad 0 \le |n| \le \frac{N}{2} \tag{3.366}$$

and

$$B_\psi(\psi) = \sum_{m=0}^{N/2} (-1)^m \, a_m \left[\Upsilon_C \left(\psi - \frac{2\pi}{N} m \right) + \Upsilon_C \left(\psi + \frac{2\pi}{N} m \right) \right]. \tag{3.367}$$

The Blackman window in the text uses $a_0 = 0.42$, $a_1 = 0.50$, $a_2 = 0.08$.

The "exact Blackman" window uses coefficients that place a zero exactly at $\psi = 3.5(2\pi/N)$ and $\psi = 4.5(2\pi/N)$. The coefficients are:

$$a_0 = \frac{7938}{18608} \doteq 0.42650971, \tag{3.368}$$

$$a_1 = \frac{9240}{18608} \doteq 0.49656062, \tag{3.369}$$

$$a_2 = \frac{1430}{18608} \doteq 0.07684867. \tag{3.370}$$

(a) Repeat Problem 3.1.11 for these coefficients.

(b) Compare to the Blackman window in the text and observe how a small change in w_n gives a significant change in the sidelobes.

Problem 3.1.20

Harris [Har76], [Har78] developed 4-term windows to achieve minimum sidelobe levels. The weighting is defined

$$w_m = a_0 - a_1 \cos\left(\frac{2\pi}{N} m\right) + a_2 \cos\left(\frac{2\pi}{N} 2m\right) - a_3 \cos\left(\frac{2\pi}{N} 3m\right), \quad m = 0, 1, 2, \cdots, N-1. \tag{3.371}$$

Four examples are listed in the following table:

	3-Term (−67 dB)	3-Term (−61 dB)	4-Term (−92 dB)	4-Term (−74 dB)
a_0	0.42323	0.44959	0.35875	0.40217
a_1	0.49755	0.49364	0.48829	0.49703
a_2	0.07922	0.05677	0.14128	0.09392
a_3	—	—	0.01168	0.00183

Problem 3.1.21: Gaussian weightings

The Fourier uncertainty principle suggests that the optimum weighting for an infinitely long aperture has a Gaussian form. More precisely, for a given root-mean-square aperture extent, the weighting that generates the narrowest root-mean-square wavenumber response has a Gaussian function form. Moreover, the Gaussian weighting leads to a Gaussian array response that implies no sidelobes. The limiting factor for any implementation is the finite aperture extent that mitigates the consequences of the Fourier uncertainty principle, since the weighting must be terminated with a step discontinuity, however small it may be. The Gaussian weighting is a family of weightings parameterized by its width relative to the aperture length L. It is given by

$$w_\sigma(z) = \begin{cases} \frac{c(\sigma)}{L} e^{-\frac{1}{2}\left(\frac{z}{\sigma L}\right)^2}, & |z| \leq \frac{L}{2}, \\ 0, & |z| > \frac{L}{2}, \end{cases} \tag{3.372}$$

where $c(\sigma)$ is a normalization constant such that response on the main response axis is unity (or equivalently, the area under the weighting is unity). The constant is given by

$$c(\sigma) = \left[2\sqrt{2\pi}\sigma(1 - Q\left(\frac{1}{2\sigma}\right)\right]^{-1}, \tag{3.373}$$

where $Q(x)$ is the complementary error function for the Gaussian probability density.[37]

The array weightings are

$$w_N(n) = w_N(z)|_{z=\frac{n}{N}}, \quad n = -\frac{N-1}{2}, \cdots, \frac{N-1}{2}. \tag{3.374}$$

(a) Plot the beam pattern for a standard 11-element linear array for $\sigma = 0.25$, 0.177, and 0.125.

(b) Compute the HPBW, BW_{NN}, and first sidelobe height for the three values of σ in part (a).

(a) Discuss your results.

Problem 3.1.22: Binomial[38]

Consider a standard N-element linear array steered to broadside. We want to synthesize a pattern whose zeros are all at $z = -1$. Thus,

$$\begin{aligned} B(z) &= (z+1)^{N-1} \\ &= z^{N-1} + \cdots + a_n z^n + \cdots + 1, \end{aligned} \tag{3.375}$$

where

$$a_n = \frac{(N-1)!}{n!\,(N-1-n)!} \tag{3.376}$$

is the binomial coefficient.

The resulting coefficients are given by Pascal's triangle.

$N=1$	1									
$N=2$	1	1								
$N=3$	1	2	1							
$N=4$	1	3	3	1						
$N=5$	1	4	6	4	1					
$N=6$	1	5	10	10	5	1				
$N=7$	1	6	15	20	15	6	1			
$N=8$	1	7	21	35	35	21	7	1		
$N=9$	1	8	28	56	70	56	28	8	1	
$N=10$	1	9	36	84	126	126	84	36	9	1

(3.377)

(a) Plot the beam pattern for $N = 10$. Compute the HPBW and the tolerance factor.

(b) Discuss the behavior for large N.

P3.2 Array Polynomials and the z-Transform

Problem 3.2.1 [Bal82]

A 3-element array is placed along the z-axis. Assuming the spacing between the elements is $d = \lambda/4$ and the relative amplitude excitation is equal to $a_1 = 1, a_2 = 2, a_3 = 1$:

---

[37] $Q(x) = \frac{1}{\sqrt{2\pi}} \int_x^\infty e^{-\frac{x^2}{2}}\,dx$, which was defined as $erfc_*(x)$ in [DEMT I], [VT68], [VT01a].

[38] The binomial distribution was originally suggested by Stone [Sto]. He proposed that the amplitudes be proportional to the coefficients of a binomial series of the form of (3.375).

(a) Find the nulls of the beam pattern where the interelement phase shift ψ_T (2.127) is 0, $\pi/2$, and $3\pi/2$.

(b) Plot the beam patterns.

Problem 3.2.2

Design a linear array of isotropic elements placed along the z-axis such that the zeros of the array factor occur at $\theta = 0°, 60°$, and $120°$. Assume that the elements are spaced $d = \lambda/4$ apart and that the interelement phase shift between them is $0°$.

(a) Find the required number of elements.

(b) Determine their excitation coefficients.

(c) Write the array factor.

(d) Plot the array factor pattern.

Problem 3.2.3

Consider a linear array along the z-axis. Assume that $d = \lambda/2$. The nulls in the array factor are specified to be $0°, 60°$, and $120°$.

(a) Find the minimum number of array elements.

(b) Specify the array weighting.

(c) Find and plot the array factor.

Problem 3.2.4

Consider a standard 21-element linear array. Plot the zeros in the z-plane for the following array weighting functions:

(a) Triangular

(b) Cosine

(c) Raised cosine

(d) Cosine^2

(e) Cosine^3

(f) Cosine^4

Problem 3.2.5

Consider a standard 21-element linear array. Plot the zeros in the z-plane for the following array weighting functions:

(a) Raised cosine-squared

(b) Hamming

(c) Blackman-Harris

Problem 3.2.6

Consider a standard 15-element linear array. Plot the zeros in the z-plane for the following array weighting functions:

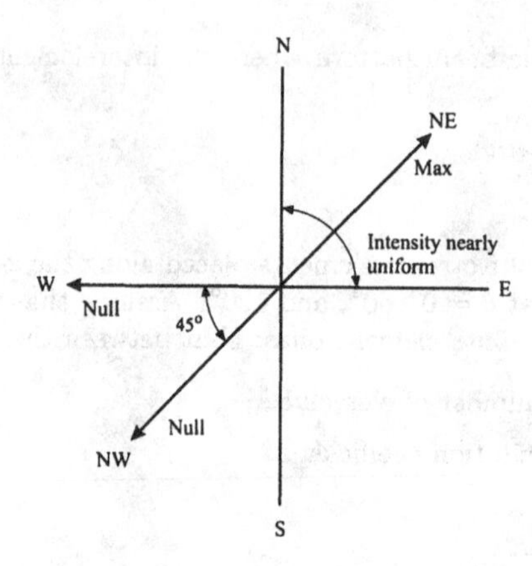

Figure 3.62 Four-tower broadcast array pattern requirements. [Adapted from [Kra88]]

(a) DPSS ($\psi_0 = 0.15\pi, 0.25\pi$, and 0.5π)

(b) Kaiser ($\beta = 2$ and 8)

Problem 3.2.7 [Bal82]

The z-plane array factor of an array of isotropic elements placed along the z-axis is given by

$$AF = z(z^4 - 1). \tag{3.378}$$

Determine the

(a) Number of elements of the array. Indicate any elements with zero weighting (null elements).

(b) Position of each element (including that of null elements) along the z-axis.

(c) Magnitude and phase (in degree) of each element.

(d) Angles where the pattern will equal zero when the total array length (including null elements) is 2λ.

Problem 3.2.8 [Kra88]

Four-tower BC array. A broadcasting station requires the horizontal plane pattern indicated in Figure 3.62. The maximum field intensity is to be radiated northeast with as little decrease as possible in field intensity in the 90° sector between north and east. No nulls are permitted in this sector. Nulls may occur in any direction in the complementary 270° sector. However, it is required that nulls must be present for the directions of due west and due southwest, in order to prevent interference with other stations in these directions.

Design a four-vertical-tower array to fulfill these requirements. The currents are to be equal in magnitude in all towers, but the phase may be adjusted to any relationship. There is also no restriction on the spacing or geometrical arrangements of the towers. Plot the beam pattern.

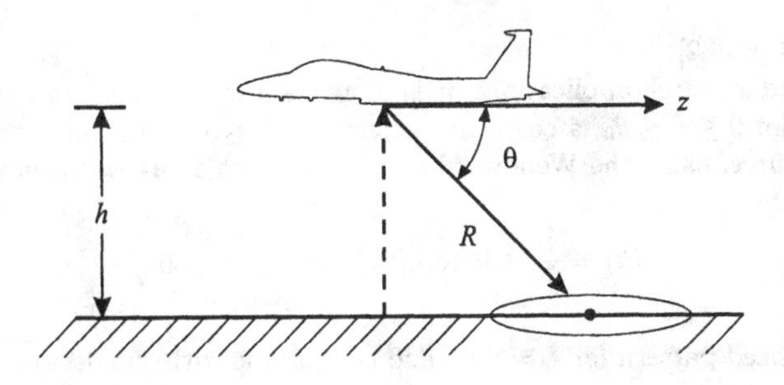

Figure 3.63 Cosecant pattern.

P3.3 Pattern Sampling in Wavenumber Space

The cosecant beam pattern is encountered in a number of applications. The next four problems, which are taken from [Bal82], use the Woodward procedure to synthesize these patterns.

Problem 3.3.1 [Bal82]

In target-search, grounding-mapping radars, and in airport beacons it is desirable to have the echo power received from a target, of constant cross section, to be independent of its range R.

Generally, the far-zone field radiated by an antenna is given by

$$|E(R,\theta,\phi)| = C_0 \frac{|F(\theta,\phi)|}{R}, \tag{3.379}$$

where C_0 is a constant. According to the geometry of Figure 3.63,

$$R = \frac{h}{\sin\theta} = h\csc\theta. \tag{3.380}$$

For a constant value of ϕ, the radiated field expression reduces to

$$|E(R,\theta,\phi=\phi_0)| = C_0 \frac{|F(\theta,\phi=\phi_0)|}{R} = C_1 \frac{|f(\theta)|}{R}. \tag{3.381}$$

A constant value of field strength can be maintained provided the radar is flying at a constant altitude h and the far-field antenna pattern is equal to

$$f(\theta) = C_2 \csc(\theta).$$

This is referred to as a cosecant pattern, and it is used to compensate for the range variations. For very narrow beam antennas, the total pattern is approximately equal to the space or array factor. Design a line source, using the Woodward method, whose array factor is given by

$$AF(\theta) = \begin{cases} 0.342\csc(\theta), & 20^\circ \le \theta \le 60^\circ, \\ 0, & \text{elsewhere}. \end{cases} \tag{3.382}$$

Plot the synthesized pattern for $L = 20\lambda$, and compare it with the desired pattern.

Problem 3.3.2 [Bal82]

Repeat the design of Problem 3.3.1 for a linear array of $N = 21$ elements with a spacing of $d = \lambda/2$ between them.

Problem 3.3.3 [Bal82]

For some radar search applications, it is more desirable to have an antenna that has a square beam for $0 \le \theta \le \theta_0$, a cosecant pattern for $\theta_0 \le \theta \le \theta_m$, and is zero elsewhere. Design a line source, using the Woodward method with an array factor of

$$AF(\theta) = \begin{cases} 1, & 15^\circ \le \theta \le 20^\circ, \\ 0.342\csc(\theta), & 20^\circ \le \theta \le 60^\circ, \\ 0, & \text{elsewhere.} \end{cases} \tag{3.383}$$

Plot the synthesized pattern for $L = 20\lambda$, and compare it with the desired pattern.

Problem 3.3.4 [Bal82]

Repeat the design of Problem 3.3.3 using the Woodward method, for a standard linear array of 41 elements.

Problem 3.3.5

Repeat the design problem in Problem 3.3.1 using the Woodward method.

Problem 3.3.6

Repeat the design problem in Problem 3.3.2 using the Woodward method.

Problem 3.3.7

Design a standard linear array with $N = 21$ when the desired array factor is

$$AF_d(\theta) = \sin^3\theta, \quad 20^\circ \le \theta \le 60^\circ. \tag{3.384}$$

Use the Woodward procedure.

P3.4 Minimum Beamwidth for Specified Sidelobe Level

Problem 3.4.1

Show that the Dolph-Chebychev procedure and the Riblet-Chebychev procedure lead to the same beam pattern for $d = \lambda/2$.

Problem 3.4.2

Consider a standard 10-element linear array pointed at broadside.

(a) Find the Dolph-Chebychev weightings for sidelobes of −20 dB, −30 dB, and −40 dB.

(b) Plot the resulting beam pattern and compute the HPBW, BW_{NN}, and the directivity.

(c) Plot the roots in the z-plane.

Problem 3.4.3

Repeat Problem 3.4.2 for a 20-element linear array with $d = \lambda/4$. Repeat parts (a)–(b).

(c) Find the Riblet-Chebychev weighting for the same sidelobe requirements and compare the BW_{NN}.

Problem 3.4.4

Compute the HPBW of the Chebychev array in Problem 3.4.2 when it is scanned to $\theta = 60^\circ$.

Problem 3.4.5

Consider a standard 7-element linear array pointed at broadside.

(a) Find the Dolph-Chebychev weightings for sidelobe of −20 dB, −30 dB, and −40 dB.

(b) Plot the resulting beam pattern and compute the HPBW, BW_{NN}, and the directivity.

Problem 3.4.6

[Kra88] Calculate the Dolph-Chebychev distribution of a six-source broadside standard linear array for $R = 5, 7$, and 10.

Problem 3.4.7

Consider a standard 5-element linear array. The array is aimed at broadside.

(a) Find the Dolph-Chebychev weightings for −20 dB sidelobes.

(b) Plot the resulting beam pattern.

(c) Find the HPBW and BW_{NN}.

Problem 3.4.8

Consider an 8-element linear array with $d = 3\lambda/4$. The array is aimed at broadside.

(a) Find the Dolph-Chebychev weightings for −40 dB sidelobes.

(b) Plot the resulting beam pattern.

(c) Find the HPBW and BW_{NN}.

Problem 3.4.9

Consider a standard linear array with $N = 15$.

(a) Repeat Problem 3.4.2.

(b) Repeat part (a) for Taylor weighting with $\bar{n} = 6$.

Problem 3.4.10

Consider the model in Example 3.4.2. Find and plot the aperture weighting function and the corresponding beam pattern.

Problem 3.4.11 (continuation)

(a) Plot the beam pattern for the corresponding linear array with $d = \lambda/4$ and $N = 17$.

(b) Plot the beam pattern for the corresponding linear array with $d = \lambda/2$ and $N = 9$.

Problem 3.4.12 (continuation)

Instead of finding the weighting of the discrete array by sampling the continuous weighting, find a discrete weighting so that the nulls of the two patterns are matched (e.g., [Ell81]).

Problem 3.4.13

Consider a 41-element standard linear array. Design a beam pattern using a Villeneuve $\bar{n}$ weighting with a maximum −30-dB sidelobe.

(a) Plot the beam pattern for $\bar{n} = 6$.

(b) How does the pattern change as a function of $\bar{n}$?

(c) Find the array weights for the pattern in part (a).

Problem 3.4.14

Repeat Problem 3.4.13 for −35-dB maximum sidelobes.

Problem 3.4.15

(a) Calculate the directivity of a standard N-element linear array pointed at broadside using a Dolph-Chebychev weighting.

(b) Plot the directivity index DI_B versus N for sidelobe levels of −15, −20, −30, −40, and −60 dB. Let N vary from 10 to 1000.

Problem 3.4.16

Repeat Problem 3.4.15 for a Villeneuve $\bar{n}$ weighting with $\bar{n} = 6$.

Problem 3.4.17

Derive a simple expression for the BW_{NN} of a SLA using Dolph-Chebychev weighting. Plot BW_{NN} versus SLL in dB for $N = 10$.

Problem 3.4.18

In order to find the Chebychev weights, Stegen [Ste53] used the following technique. The array coefficients are represented by an inverse DFT,

$$a_n = \frac{1}{N} \sum_{m=-\frac{N-1}{2}}^{\frac{N-1}{2}} p_m e^{j\frac{2\pi}{N}nm} \tag{3.385}$$

for N odd. The p_m correspond to equally spaced samples of the beam pattern in ψ-space. If the pattern corresponds to a Chebychev polynomial, then

$$p_m = T_{N-1}(x_0 \cos\frac{m\pi}{N}). \tag{3.386}$$

Using (3.386) in (3.385) gives an expression for the coefficients,

$$\begin{aligned} a_n &= \frac{1}{RN}\left\{R + 2\sum_{m=1}^{\frac{N-1}{2}} T_{N-1}\left(x_0 \cos\frac{m\pi}{N}\right)\cdot\left(\cos\frac{2\pi}{N}mn\right)\right\}, \\ n &= 0, 1, \ldots, \frac{N-1}{2}, \end{aligned} \tag{3.387}$$

for N odd. Similarly,

$$\begin{aligned} a_n &= \frac{2}{RN} + \left\{\frac{R}{2}\sum_{m=1}^{\frac{N}{2}} T_{N-1}\left(x_0 \cos\frac{m\pi}{N}\right)\cdot\left(\cos\frac{\pi}{N}(2n-1)m\right)\right\}, \\ n &= 0, 1, \ldots, \frac{N}{2}, \end{aligned} \tag{3.388}$$

for N even.

Use this approach to find the weights in Example 3.4.1.

P3.5 Least Squares Error Pattern Synthesis

Problem 3.5.1

The desired $B_D(\theta)$ is uniform on $-\pi/4 \leq \theta \leq 3\pi/4$. Use the Fourier transform method to find the aperture weighting function for a linear aperture of length $L = 5\lambda$ and $L = 10\lambda$.

(a) Plot the resulting beam pattern.

(b) Calculate the resulting maximum sidelobe.

(c) Calculate BW_{NN}.

(d) Calculate the peak overshoot in the main-lobe region.

Problem 3.5.2

Repeat Problem 3.5.1 for a standard linear array with $N = 11$ and $N = 21$.

Problem 3.5.3

Assume that $B_d(\theta)$ is uniform on $-30° \leq \theta \leq 60°$. Repeat Problem 3.5.1.

Problem 3.5.4

Assume that $B_d(\theta)$ is uniform on $30° \leq \theta \leq 60°$. Repeat Problem 3.5.2.

Problem 3.5.5

The desired beam pattern is uniform on $-\pi/\sqrt{2} \leq \psi \leq \pi/\sqrt{2}$ and zero elsewhere.

(a) Find the Fourier coefficients for $N = 11$ and $N = 21$.

(b) Use a Hann window. Plot the resulting beam pattern. Compare δ, the main-lobe overshoot, and $\Delta\psi$, the transition distance.

(c) Repeat part (b) using a Hamming window.

(d) Assume $N = 21$. Use a Kaiser window. Find β to match the overshoot in part (b). Plot the resulting beam pattern. What is $\Delta\psi$?

(e) Repeat part (d) and match the overshoot in part (c).

Problem 3.5.6

Consider the model in Example 3.5.2.

(a) Use a DPSS window with $\psi_0 = \pi/6$. Plot the resulting beam pattern and compare to Figure 3.33(a).

(b) What is the effect of varying ψ_0 on the window?

P3.6 Minimax Design

Problem 3.6.1

Consider the model in Example 3.6.1 ($N = 11$) with $\delta_p/\delta_s = 10$. Assume δ_p is chosen to match the Hamming window in Example 3.5.2.

(a) Use the Parks-McClellan-Rabiner algorithm to design the optimum weighting and plot the resulting beam pattern.

(b) What $\Delta\psi = \psi_p - \psi_s$. Compare this result with $\Delta\psi$ found in Example 3.6.1.

Problem 3.6.2

Utilize the techniques in Section 3.6.2 to repeat Problem 3.6.1 for the case when $N = 10$.

Problem 3.6.3

Consider a standard linear array. Assume $20\log_{10}\delta_p = -53$ dB and $\delta_p/\delta_s = 10$. Assume the Parks-McClellan-Rabiner algorithm is used to design the array weighting.

Plot $\Delta\psi$ as a function of N, the number of array elements.

P3.7 Null Steering

Problem 3.7.1

Consider a standard 41-element linear array with a Chebychev pattern (−40 dB sidelobe). In Example 3.7.3, we placed four nulls at $u = 0.22, 0.24, 0.26$, and 0.28. An alternative strategy is to place a first- and second-order null at $u = 0.235$ and 0.265. Find the resulting beam pattern and compare to Example 3.7.3.

Problem 3.7.2

Consider a standard 21-element linear array. We require a null at ψ_0. $B_d(\psi)$ corresponds to uniform weighting.

(a) Find the least squares approximation to $B_d(\psi)$ subject to this null constraint.

(b) Plot the beam pattern for $\psi_0 = 3\pi/N$, $\psi_0 = 2\pi/N$, $\psi_0 = \pi/N$, and $\psi_0 = 0.5\pi/N$. Discuss your results.

Problem 3.7.3

Consider a linear array that consists of two segments: (i) an 11-element standard linear array centered at $z = 0$; and (ii) two 5-element standard linear arrays centered at $z = \pm 7.5\lambda$.

(a) Assume that the 21 elements are weighted uniformly. Plot the beam pattern and discuss the grating structure.

(b) We denote the beam pattern in part (a) as $B_d(\psi)$. We require a null at $\psi = 0.5\pi/N$. Find the least squares approximation to $B_d(\psi)$ subject to the null constraint.

P3.8 Asymmetric Beams

Problem 3.8.1

Assume N is even. Let the $w_n, n = 1, 2, \cdots, N/2$ correspond to Hamming weighting, and

$$w_{-n} = -w_n. \tag{3.389}$$

(a) Find and plot the resulting beam pattern.

(b) Find the slope at the origin and the height of the first sidelobe.

Problem 3.8.2

We can divide an asymmetric (N even) array polynomial into two components:

$$B(z) = B_1(z)\, B_2(z), \tag{3.390}$$

where

$$B_1(z) = \frac{1 - z^{-1}}{2}, \tag{3.391}$$

and

$$B_2(z) = \sum_{n=0}^{N-2} b_2(n) z^{-n}, \tag{3.392}$$

and thus

$$B_1(\psi) = \sin(\frac{\psi}{2}). \tag{3.393}$$

(a) Use this approach to find the beam pattern for uniform weighting.

(b) Use this approach to design an asymmetric Hann beam pattern.

Problem 3.8.3 (continuation)

Assume $N = 10$. Then $B_2(z)$ is an eighth-order polynomial.

(a) Choose the coefficients to achieve a Dolph-Chebychev weighting with -40 dB sidelobes.

(b) Plot the resulting beam pattern.

(c) Calculate the slope at the origin.

Problem 3.8.4 (continuation)

Assume that the beam pattern in Problem 3.8.3 is $B_d(\psi)$. Find the least squares approximation to $B_d(\psi)$ with a zero-order null constraint at $\psi_0 = 5\pi/N$.

P3.9 Spatially Non-uniform Linear Arrays

Problem 3.9.1

Consider a 4-element non-uniform linear array whose element spacings are $d, 3d, 2d$.

(a) Plot the beam pattern for uniform weighting.

(b) Uniform weighting corresponds to $w_1 = w_2 = w_5 = w_7 = 1/4$ and $w_3 = w_4 = w_6 = 0$ in a standard linear array. Plot the autocorrelation of $\mathbf{w}$ with itself.

Problem 3.9.2

Consider a 5-element non-uniform linear array whose element spacings are $d, 3d, 5d, 2d$. Repeat Problem 3.9.1.

Problem 3.9.3 (continuation)

Consider the array in Problem 3.9.2. Use the algorithm in Section 3.9.3 to design a beam pattern with -20 dB sidelobes.

Problem 3.9.4

(a) Consider a standard linear array with 21 elements. Use the technique in Section 3.9.3 to design a beam pattern that maximizes the directivity subject to a −35-dB constraint of the sidelobe. Describe your algorithm.

(b) Compare the resulting BW_{NN} with that of a 21-element SLA using Dolph-Chebychev weighting.

P3.10 Beamspace Processing

Problem 3.10.1

Consider a 32-element standard linear array and a 7×32 beamspace matrix whose rows use conventional weighting. We want to implement the null steering techniques in beamspace. Assume

$$B_d(\psi) = \mathbf{v}_{bs}^H(0)\mathbf{v}_{bs}(\psi). \tag{3.394}$$

We want to place a null at ψ_I while minimizing the least squares error between the desired beam pattern and the beam pattern containing the null.

(a) Let $\psi_I = 3\pi/32$. Find $\mathbf{w}_{bs}^H$ and plot the resulting beam pattern.

(b) Repeat part (a) for $\psi_I = 7\pi/32$ and $\psi_I = 13\pi/32$.

(c) Let

$$B_d(\psi) = \mathbf{v}_{bs}^H(\psi_T)\,\mathbf{v}_{bs}(\psi). \tag{3.395}$$

Repeat part (a) for $\psi_T = \pi/32$ and $\psi_I = 5\pi/32$.

Problem 3.10.2

In many applications, we require

$$\mathbf{B}_{bs}^H\mathbf{B}_{bs} = \mathbf{I}. \tag{3.396}$$

If we use a Taylor weighting, the rows of the beamspace matrix are not orthonormal. We denote this matrix as $\mathbf{B}_{no}^H$. We pass the output of $\mathbf{B}_{no}^H$ through an $N_{bs} \times N_{bs}$ matrix $\mathbf{H}_w$. We denote the cascade of the two processors as

$$\mathbf{B}_{bs}^H = \mathbf{H}_w\,\mathbf{B}_{no}^H. \tag{3.397}$$

(a) Show that if

$$\mathbf{H}_w = [\mathbf{B}_{no}^H\,\mathbf{B}_{no}]^{-1/2}, \tag{3.398}$$

then $\mathbf{B}_{bs}^H$ satisfies the orthonormality condition. Is $\mathbf{H}_w$ unique?

(b) Consider the 7×32 beamspace matrix whose rows have Taylor weighting. Verify the above result numerically.

(c) Plot the beam patterns for the orthogonal beams. Discuss the sidelobe behavior.

Problem 3.10.3

Consider an 32-element standard linear array and a $N_{bs} \times N$ beamspace processor where rows are orthonormal conventional beams.

We want to generate a beam pattern corresponding to the $\cos^{N_{bs}}$-element-space weighting in (3.18).

(a) Consider the case when N_{bs} is even. Assume the beam sector is centered at $u_s = 0$. Find the required beam steering directions and beamspace weights for N_{bs} = 2,4, and 6. Plot the resulting beam patterns and calculate the HPBW, BW_{NN}, D_N, and the height of the first sidelobe.

(b) Repeat part (a) for N_{bs} = 3, 5, and 7.

Problem 3.10.4 (continuation Problem 3.10.3)

Repeat Problem 3.10.3 for the case in which the desired beam pattern corresponds to the Hamming beam pattern in (3.21). In this case, $N_{bs} = 3$.

Problem 3.10.5 (continuation Problem 3.10.3)

Repeat Problem 3.10.3 for the case in which the desired beam pattern corresponds to the Blackman-Harris beam pattern in (3.23). In this case, $N_{bs} = 5$.

Problem 3.10.6 (continuation Problem 3.10.5)

Extend the results of Problem 3.10.5 to the case when $N_{bs} = 7$.

Problem 3.10.7 (continuation, Example 3.10.3)

Consider a standard 32-element linear array and the 6 × 32 beamspace processor in Example 3.10.3. We use an asymmetric beamspace weighting,

$$[\mathbf{w}_{bs}]_m = -[\mathbf{w}_{bs}]_{M-m}, \quad m = 1, 2, 3. \tag{3.399}$$

(a) Plot the beam pattern for real constant weights. Plot both $B_{bs}(u)$ on a linear scale and $|B_{bs}(u)|$ in dB.

(b) Consider various other weightings that trade-off slope at the origin versus sidelobe behavior.

P3.11 Broadband Arrays

Problem 3.11.1

Consider the compound array with nested subarrays shown in Figure 3.57. Each subarray has $N = 11$ elements. In some applications, we want to place a null at a specific point in u-space over the entire octave. Assume that we use $K = 8$ frequency bins.

The desired beam pattern corresponds to conventional weighting in each bin (Figure 2.20). Design a beamformer that provides a least squares approximation to the desired beam pattern (e.g., Section 3.7) with a null at $u = 0.30$ in each bin.

Problem 3.11.2

Consider the compound array with nested subarrays shown in Figure 3.57. Assume that we use $K = 8$ frequency bins.

Design a beamformer using the Riblet-Chebychev algorithm in each frequency bin. Plot the resulting beam pattern in each bin.

Bibliography

[AC76] S. P. Applebaum and D. J. Chapman. Adaptive arrays with main beam constraints. *IEEE Trans. Antennas Propag.*, vol.AP-24, pp. 650–662, September 1976.

[AGGS98] Y. I. Abramovich, D. A. Gray, A. Y. Gorokhov, and N. K. Spencer. Positive-definite Toeplitz completion in DOA estimation for nonuniform linear antenna arrays–Part I: Fully augmentable arrays. *IEEE Trans. Signal Process.*, vol.SP-46, pp. 2458–2471, September 1998.

[Akh56] N. I. Akhiezer. *Theory of Approximation.* Ungar, New York, 1956.

[AL69] V. D. Agrawal and Y. T. Lo. Distribution of sidelobe level in random arrays. *Proc. IEEE*, vol.57, pp. 1764–1765, October 1969.

[All61] J. L. Allen. Some extensions of the theory of random error effects on array patterns. Technical Report 36, M.I.T. Lincoln Laboratory, Lexington, Massachusetts, November 1961.

[AS65] M. Abramowitz and I. A. Stegun. *Handbook of Mathematical Functions.* Dover Publications, New York, 1965.

[ASG99a] Y. I. Abramovich, N. K. Spencer, and A. Y. Gorokhov. Positive-definite Toeplitz completion in DOA estimation for nonuniform linear antenna arrays–Part II: Partially augmentable arrays. *IEEE Trans. Signal Process.*, vol.SP-47, pp. 1502–1521, June 1999.

[ASG99b] Y. I. Abramovich, N. K. Spencer, and A. Y. Gorokhov. Resolving manifold ambiguities in direction-of-arrival estimation for nonuniform linear antenna arrays. *IEEE Trans. Signal Process.*, vol.SP-47, pp. 2619–2643, October 1999.

[Bag76] A. B. Baggeroer. Space-time processes and optimal array processing. Technical Report 506, Navy Undersea Center, San Diego, California, December 1976.

[Bal82] C. A. Balanis. *Antenna Theory Analysis and Design.* Wiley, New York, 1982.

[Bar64] N. K. Bary. *A Treatise on Trigonometric Series*, volume 1. Macmillan, New York, 1964.

[Bar89] D. K. Barton. *Modern Radar System Analysis.* Artech House, Boston, 1989.

[Bay68] E. T. Bayliss. Design of monopulse antenna difference patterns with low side lobes. *Bell Syst. Tech. J.*, vol.47, pp. 623–640, 1968.

[BL61] J. Butler and R. Lowe. Beam forming matrix simplifies design of electronically scanned antennas. *Electron. Des.*, vol.9, pp. 120–133, April 1961.

[Bla65] R. B. Blackman. *Linear Data-Smoothing and Prediction in Theory and Practice.* Addison-Wesley, Reading, Massachusetts, 1965.

[Boh60] H. Bohman. Approximate Fourier analysis of distribution functions. *Ark. Mat.*, vol.4, pp. 99–157, 1960.

[Bro88] E. Brookner, editor. *Aspects of Modern Radar.* Artech House, Boston, 1988.

[BVG00] K. L. Bell, H. L. Van Trees, and L. J. Griffiths, Adaptive beampattern control using quadratic constraints for circular array STAP. *8th Annaul Workshop on Adaptive Sensor Array Processing (ASAP 2000)*, M.I.T. Lincoln Laboratory, Lexington, Massachusetts, pp. 43–48, March 2000.

[Che66] E. W. Cheney. *Introduction to Approximation Theory.* McGraw-Hill, New York, 1966.

[Cho65] Y. L. Chow. On grating plateaux of nonuniformly spaced arrays. *IEEE Trans. Antennas Propag.*, vol.13, pp. 208–215, 1965.

[Cho95] T. Chou. Frequency-independent beamformer with low response error. *Proc. ICASSP*, Detroit, Michigan, vol.5, pp. 2995–2998, 1995.

[DB88] J. H. Doles and F. D. Benedict. Broad-band array design using the asymptotic theory of unequally spaced arrays. *IEEE Trans. Antennas Propag.*, vol.AP-36, pp. 27–33, January 1988.

[Dol46] C.L. Dolph. A current distribution for broadside arrays which optimizes the relationship between beamwidth and sidelobe level. *Proc. IRE*, vol.34, pp. 335–348, June 1946.

[Duf89] E. C. Dufort. Pattern synthesis based on adaptive array theory. *IEEE Trans. Antennas Propag.*, vol.AP-37, pp. 1017–1018, 1989.

[Ell63] R. S. Elliott. Beamwidth and directivity of large scanning arrays. *Microwave J., Appendix B*, vol.6, pp. 53–60, 1963.

[Ell75] R. S. Elliott. Design of line source antennas for narrow beamwidth and asymmetric low sidelobes. *IEEE Trans. Antennas Propag.*, vol.AP-23, pp. 100–107, January 1975.

[Ell81] R. S. Elliott. *Antenna Theory and Design.* Prentice-Hall, Englewood Cliffs, New Jersey, 1981.

[Er92] M. H. Er. Array pattern synthesis with a controlled mean-square sidelobe level. *IEEE Trans. Signal Process.*, vol.SP-40, pp. 977–981, April 1992.

[FS74] D. C. Farden and L. L. Scharf. Statistical design of nonrecursive digital filters. *IEEE Trans. Acoust., Speech, Signal Processing*, vol.ASSP-22, pp. 188–196, June 1974.

[GE93] M. M. Goodwin and G. W. Elko. Constant beamwidth beamforming. *Proc. ICASSP*, Minneapolis, Minnesota, vol.1, pp. 169–172, 1993.

[GM55] E.N. Gilbert and S.P. Morgan. Optimum design of directive antenna arrays subject to random variations. *Bell Syst. Tech. J.*, vol.34, pp. 637–663, May 1955.

[HA70] E. L. Hixson and K. T. Au. Wide-bandwidth constant beamwidth acoustic array. *J. Acoust. Soc. Am.*, vol.48, p.117, July 1970.

[Han66] R.C. Hansen, Editor. *Microwave Scanning Antennas (Array Theory and Practice)*, vol.II, Chapter 1, p. 27. Academic Press, New York, 1966.

[Han85] R. C. Hansen. Aperture efficiency of Villeneuve n̄ arrays. *IEEE Trans. Antennas Propag.*, vol.AP-33, pp. 666–669, June 1985.

[Han92] R. C. Hansen. Array pattern control and synthesis. *Proc. IEEE*, vol.80, pp. 141–151, January 1992.

[Han98] R.C. Hansen, Ed. *Phased Array Antennas.* Wiley, New York, 1998.

[Har61] R. F. Harrington. Sidelobe reduction by nonuniform element spacing. *IEEE Trans. Antennas Propag.*, vol.AP-9, pp. 187–192, March 1961.

[Har76] F. J. Harris. High-resolution spectral analysis with arbitrary spectral centers and adjustable spectral resolutions. *J. Comput. Elec. Eng.*, vol.3, pp. 171–191, 1976.

[Har78] F. J. Harris. On the use of windows for harmonic analysis with the discrete fourier transform. *Proc. IEEE*, vol.66, pp. 51–83, January 1978.

[Hau68] R. A. Haubrich. Array design. *Bull. Seismol. Soc. Am.*, vol.58, pp. 977–991, June 1968.

[HTL72] H. S. Hersey, D. W. Tufts, and J. T. Lewis. Interactive minimax design of linear phase nonrecursive digital filters subject to upper and lower function constraints. *IEEE Trans. Audio Elecroacoust.*, vol.AU-20, pp.171–173, June 1972.

[IC65] A. Ishimaru and Y. S. Chen. Thinning and broadbanding antenna arrays by unequal spacings. *IEEE Trans. Antennas Propag.*, vol.AP-13, pp. 34–42, January 1965.

[Ish62] A. Ishimaru. Theory of unequally spaced arrays. *IRE Trans. Antennas Propag.*, vol.AP-10, pp. 691–702, November 1962.

[JD93] D. H. Johnson and D. E. Dudgeon. *Array Signal Processing.* Prentice-Hall, Englewood Cliffs, New Jersey, 1993.

[Joh93] R. C. Johnson. *Antenna Engineering Handbook.* McGraw-Hill, New York, 3rd edition, 1993.

[Kai74] J. F. Kaiser. Nonrecursive digital filter design using the I_0-Sinh window function. In *Proc. IEEE Int. Symp. Circuit Syst.*, San Francisco, California, pp. 20–23, 1974.

[KK66] F. F. Kuo and J. F. Kaiser, editors. *System Analysis by Digital Computer.* Wiley, New York, 1966.

[KPT60] D. D. King, R. F. Packard, and R. K. Thomas. Unequally spaced broad-band antenna arrays. *IRE Trans. Antennas Propag.*, vol.AP-8, pp. 380–385, July 1960.

[Kra88] J. D. Kraus. *Antennas.* McGraw-Hill, New York, 2nd edition, 1988.

[Kre71] J. L. Kreuzer. A synthetic aperture coherent imaging technique. *Acoust. Hologr.*, vol.3, pp. 287–315, 1971.

[Kum92] W. H. Kummer. Basic array theory. *Proc. IEEE*, vol.80, pp. 127–139, January 1992.

[Lin92] D. A. Linebarger. A fast method for computing the coarray of sparse linear arrays. *IEEE Trans. Antennas Propag.*, vol.AP-40, pp. 1109–1112, September 1992.

[LL59] Y. T. Lo and S. W. Lee, editors. *Antenna Handbook Theory, Applications and Design.* Van Nostrand Reinhold, New York, 1959.

[LL93a]. Y. T. Lo and S. W. Lee, editors. *Antenna Handbook (Antenna Theory)*, vol.2, Chapman & Hall, New York, 1993.

[LL93b] Y. T. Lo and S. W. Lee, editors. *Antenna Handbook (Applications)*, vol.3, Chapman & Hall, New York, 1993.

[LL93c] Y. T. Lo and S. W. Lee, editors. *Antenna Handbook (Fundamentals and Mathematical Techniques)*, vol.1, Chapman & Hall, New York, 1993.

[LL93d] Y. T. Lo and S. W. Lee, editors. *Antenna Handbook (Related Issues)*, vol.4, Chapman & Hall, New York, 1993.

[LMT76] J. T. Lewis, R. Murphy, and D. W. Tufts. Design of minimum noise digital filters subject to inequality constraints using quadratic programming. *IEEE Trans. Acoust., Speech, Signal Processing*, vol.ASSP-24, pp.434–436, October 1976.

[Lo63] Y. T. Lo. Sidelobe level in non-uniformly spaced antenna arrays. *IEEE Trans. Antennas Propag.*, vol.AP-11, p. 511, July 1963.

[Lo64a] Y. T. Lo. A mathematical theory of antenna arrays with randomly spaced elements. *IRE Trans. Antennas Propag.*, vol.AP-12, pp.257–268, May 1964.

[Lo64b] Y. T. Lo. A probabilistic approach to the problem of large antenna arrays. *Radio Sci.*, vol.68D, pp. 1011–1019, September 1964.

[Lo68] Y. T. Lo. Random periodic arrays. *Radio Sci.*, vol.3, pp. 425–436, May 1968.

[LP61] H. J. Landau and H. O. Pollak. Prolate spheroidal wave functions, Fourier analysis, and uncertainty: II. *Bell Syst. Tech. J.*, vol.40, pp. 65–84, January 1961.

[LP62] H. J. Landau and H. O. Pollak. Prolate spheroidal wave functions, Fourier analysis, and uncertainty: III. *Bell Syst. Tech. J.*, vol.41, pp. 1295–1336, July 1962.

[LST93] D. A. Linebarger, I. H. Sudborough, and I. G. Tollis. Difference bases and sparse sensor arrays. *IEEE Trans. Inform. Theory*, vol.IT-39, pp. 716–721, March 1993.

[Ma74] M. T. Ma. *Theory and Applications of Antenna Arrays.* Wiley, New York, 1974.

[Maf62] A. L. Maffet. Array factors with nonuniform spacing parameters. *IRE Trans. Antennas Propag.*, vol.AP-10, pp. 131–136, March 1962.

[Mai94] R. J. Mailloux. *Phased Array Antenna Handbook.* Artech House, Boston, 1994.

[May76] J. T. Mayhan. Nulling limitations for a multiple-beam antenna. *IEEE Trans. Antennas and Propag.*, vol.AP-24, pp. 769–779, November 1976.

[MC63] T. M. Maher and D. K. Cheng. Random removal of radiators from large linear arrays. *IEEE Trans. Antennas Propag.*, vol.AP-11, pp. 106–112, March 1963.

[McN85] D. A. McNamara. Optimum monopulse linear array excitations using Zolotarev polynomials. *Electron. Lett.*, vol.21, pp. 681–682, 1985.

[McN93] D. A. McNamara. Direct synthesis of optimum difference patterns for discrete linear arrays using Zolotarev distributions. *Proc. IEE*, vol.MAP-140, pt. H, pp. 495–500, December 1993.

[McN94] D. A. McNamara. Performance of Zolotarev and modified-Zolotarev difference pattern array distributions. *Proc. IEE*, vol.MAP-141, pp. 37–44, February 1994.

[Mil85] T. A. Milligan. *Modern Antenna Design.* McGraw-Hill, New York, 1985.

[MK93] S. K. Mitra and J. F. Kaiser, editors. *Handbook for Digital Signal Processing.* Wiley, New York, 1993.

[MP73] J. H. McClellan and T. W. Parks. A unified approach to the design of optimum FIR linear phase digital filters. *IEEE Trans. Circuit Theory*, vol.CT-20, pp. 697–701, 1973.

[MPR73] J. H. McClellan, T. W. Parks, and L. R. Rabiner. A computer program for designing optimum FIR linear phase digital filters. *IEEE Trans. Audio Electroacoust.*, vol.AU-21, pp. 506–526, 1973.

[MR79] J. H. McClellan and C. M. Rader. *Number Theory in Digital Signal Processing.* Prentice-Hall, Englewood Cliffs, New Jersey, 1979.

[MTL75] R. A. Mucci, D. W. Tufts, and J. T. Lewis. Beam pattern synthesis for line arrays subject to upper and lower constraining bounds. *IEEE Trans. Antennas Propag.*, pp. 732–734, September 1975.

[MTL76] R. A. Mucci, D. W. Tufts, and J. T. Lewis. Constrained least-square synthesis of coefficients for arrays of sensors and FIR digital filters. *IEEE Trans. Aerosp. Electron. Syst.*, vol.AES-12, pp. 195–202, March 1976.

[OC90] C. A. Olen and R. T. Compton, Jr. A numerical pattern synthesis algorithm for arrays. *IEEE Trans. Antennas Propag.*, vol.AP-38, pp. 1666–1676, October 1990.

[OS89] A. V. Oppenheim and R. W. Schafer. *Discrete-Time Signal Processing.* Prentice-Hall, Englewood Cliffs, New Jersey, 1989.

[PB72] A. Papoulis and M. S. Bertran. Digital filtering and prolate functions. *IEEE Trans. Circuit Theory*, vol.CT-19, pp. 674–681, November 1972.

[PL69] A. R. Panicali and Y. T. Lo. A probabilistic approach to large circular and spherical arrays. *IEEE Trans. Antennas Propag.*, vol.AP-17, pp. 514–522, July 1969.

[PM72a] T. W. Parks and J. H. McClellan. Chebyshev approximation for nonrecursive digital filters with linear phase. *IEEE Trans. Circuit Theory*, vol.CT-19, pp. 189–194, 1972.

[PM72b] T. W. Parks and J. H. McClellan. A program for the design of linear phase finite impulse response digital filters. *IEEE Trans. Audio Electroacoust.*, vol.AU-20, pp. 195–199, 1972.

[PPL90] D. Pearson, S. U. Pillai, and Y. Lee. An algorithm for near-optimal placement of sensor elements. *IEEE Trans. Inform. Theory*, vol.IT-36, pp. 1280–1284, November 1990.

[Pra82] S. Prasad. On an index for array optimization and the discrete prolate spheroidal functions. *IEEE Trans. Antennas Propag.*, vol.AP-30, pp. 1021–1023, September 1982.

[PRLN92] J. G. Proakis, C. M. Rader, F. Ling, and C. L. Nikias. *Advanced Digital Signal Processing.* Macmillan Publishing Company, New York, 1992.

[RG75] L. R. Rabiner and B. Gold. *Theory and Application of Digital Signal Processing.* Prentice-Hall, Englewood Cliffs, New Jersey, 1975.

[Rho63] D. R. Rhodes. The optimum line source for the best mean-square approximation to a given radiation pattern. *IEEE Trans. Antennas Propag.*, vol.AP-11, pp. 440–446, July 1963.

[Rib47] H. J. Riblet. A current distribution for broadside arrays which optimizes the relationship between beamwidth and side-lobe level. *Proc. IRE*, vol.35, pp. 489–492, May 1947.

[Ric64] J. R. Rice. *The Approximation of Functions, volume 1. Antenna Handbook.* Addison-Wesley, Reading, Massachusetts, 1964.

[Ric76] L. J. Ricardi. A summary of methods for producing nulls in an antenna radiation pattern. Technical Report, M.I.T. Lincoln Laboratory, Lexington, Massachusetts, August 1976.

[RMP75] L. R. Rabiner, J. H. McClellan, and T. W. Parks. FIR digital filter design techniques using weighted Chebychev approximation. *Proc. IEEE*, vol.63, pp. 595–610, April 1975.

[Ruf93] C. S. Ruf. Numerical annealing of low-redundancy linear arrays. *IEEE Trans. Antennas Propag.*, vol.AP-41, pp. 85–90, January 1993.

[Ruz52] J. Ruze. The effect of aperture errors on the antenna pattern. *Nouvo Cimento*, Suppl. 3(9), pp. 364–380, 1952.

[San60] S. S. Sandler. Some equivalence between equally and unequally spaced arrays. *IRE Trans. Antennas Propag.*, vol.AP-8, pp. 498–500, September 1960.

[Sch43] S. A. Schelkunoff. A mathematical theory of linear arrays. *Bell Syst. Tech. J.*, vol.22, pp. 80–107, 1943.

[SK82] J. C. Sureau and K. J. Keeping. Sidelobe control in cylindrical arrays. *IEEE Trans. Antennas Propag.*, vol.AP-30, pp. 1027–1031, September 1982.

[Sko69] M. I. Skolnik. Nonuniform arrays. In Collin and Zucke, editors, *Antenna Theory*, Part I, chapter 6. McGraw-Hill, New York, 1969.

[Sko80] M. I. Skolnik. *Introduction to Radar Systems*. McGraw-Hill, New York, 1980.

[Sko90] M. I. Skolnik, editor. *Radar Handbook*. McGraw-Hill, New York, 2nd edition, 1990.

[Sle64] D. Slepian. Prolate spheroidal wave functions, Fourier analysis, and uncertainty: IV. *Bell Syst. Tech. J.*, vol.46, pp. 3009–3058, November 1964.

[Sle65] D. Slepian. Some asymptotic expansions for prolate spheroidal wave functions. *J. Math. Phys.*, vol.44, pp. 99–140, June 1965.

[Sle78] D. Slepian. Prolate spheroidal wave functions, Fourier analysis, and uncertainty, V: The discrete case. *Bell Syst. Tech. J.*, vol.57, pp. 1371–1430, May-June 1978.

[SP61] D. Slepian and H. O. Pollak. Prolate spheroidal wave functions, Fourier analysis, and uncertainty: I. *Bell Syst. Tech. J.*, vol.40, pp. 43–64, January 1961.

[SS65] D. Slepian and E. Sonnenblick. Eigenvalues associated with prolate spheroidal wave functions of zero order. *Bell Syst. Tech. J.*, vol.44, pp. 1745–1760, October 1965.

[ST81] W. L. Stutzman and G. A. Thiele. *Antenna Theory and Design*. Wiley, New York, 1981.

[Ste53] R. J. Stegen. Excitation coefficients and beamwidths of Tschebyscheff arrays. *Proc. IRE*, vol.41, pp. 1671–1674, November 1953.

[Ste72] B. D. Steinberg. The peak sidelobe of the phased array having randomly located elements. *IEEE Trans. Antennas Propag.*, vol.AP-20, pp. 312–320, March 1972.

[Ste76] B.D. Steinberg. *Principles of Aperture and Array System Design*. Wiley, New York, 1976.

[Ste82] H. Steyskal. Synthesis of antenna patterns with prescribed nulls. *IEEE Trans. Antennas Propag.*, vol.AP-30, pp. 273–279, March 1982.

[Sto] J. S. Stone. United States Patents No. 1,643,323 and No. 1,715,433.

[Tay53] T. T. Taylor. One parameter family of line sources producing modified Sin U/U patterns. Technical Report 324, Hughes Aircraft Co., Culver City, California, 1953.

[Tay55] T. T. Taylor. Design of line-source antennas for narrow beamwidth and low sidelobes. *IRE Trans. Antennas Propag.*, vol.AP-3, pp. 16–28, January 1955.

[Tay69] G. D. Taylor. Approximation by Functions Having Restricted Ranges III. *J. Math. Anal. Appl.*, vol.27, pp.241–248, 1969.

[TF70] D. W. Tufts and J. T. Francis. Designing digital low-pass filters: comparison of some methods and criteria. *IEEE Trans. Audio Electroacoust.*, vol.AU-18, pp. 487–494, December 1970.

[TG92] C-Y. Tseng and L. J. Griffiths. A simple algorithm to achieve desired patterns for arbitrary arrays. *IEEE Trans. Signal Process.*, vol.SP-40, pp. 2737–2746, November 1992.

[TW70] G. D. Taylor and M. J. Winter. Calculation of best restricted approximations. *SIAM J. Numer. Anal.*, vol.2, pp. 248–255, 1970.

[Tuc57] D. G. Tucker. Arrays with constant beam-width over a wide frequency-range. *Nature*, vol.180, pp. 496-497, September 1957.

[Tuf75] D. W. Tufts. Comments on FIR digital filter design techniques using weighted Chebyshev approximation. *Proc. IEEE*, vol.63, p.1618, November 1975.

[Tuk67] J. W. Tukey. An introduction to the calculations of numerical spectrum analysis. *Spectral Analysis of Time Series*, pp. 25–46, 1967.

[Unz60] H. Unz. Linear arrays with arbitrarily distributed elements. *IRE Trans. Antennas Propag.*, vol.AP-8, pp. 222–223, March 1960.

[Vil84] A. T. Villeneuve. Taylor patterns for discrete arrays. *IEEE Trans. Antennas Propag.*, vol.AP-32, pp. 1089–1093, October 1984.

[VT68] H. L. Van Trees. *Detection, Estimation, and Modulation Theory, Part I.* Wiley, New York, 1968.

[VT01a] H. L. Van Trees. *Detection, Estimation, and Modulation Theory, Part I.* Wiley Interscience, New York, 2001.

[Wee68] W. L. Weeks. *Antenna Engineering.* McGraw-Hill, New York, 1968.

[Wil62] R. E. Willey. Space tapering of linear and planar arrays. *IRE Trans. Antennas Propag.*, vol.AP-10, pp. 1369–1377, July 1962.

[Wol37] I. Wolf. Determination of the radiating system which will produce a specified directional characteristic. *Proc. IRE*, vol.25, pp. 630–643, May 1937.

[Woo46] P. M. Woodward. A method for calculating the field over a plane aperture required to produce a given polar diagram. *J. IEE*, vol.93, pt. IIIA, pp. 1554–1558, 1946.

[ZI98] P. Y. Zhou and M. A. Ingram, A new synthesis algorithm with application to adaptive beamforming, *9th IEEE Workshop on Stat. Signal and Array Process.*, Portland, Oregon, September 1998.

[ZI99] P. Y. Zhou and M. A. Ingram, Pattern synthesis for arbitrary arrays using an adaptive array method, *IEEE Trans. Antennas Propag.*, vol.AP-47, pp. 862–869, May 1999.

[Zio95] L. J. Ziomek. *Fundamentals of Acoustic Field Theory and Space-Time Signal Processing.* CRC Press, Boca Raton, Florida, 1995.

Chapter 4

Planar Arrays and Apertures

In this chapter we discuss analysis and synthesis techniques for planar arrays. A planar array is an array whose elements all lie in the xy-plane.

In the array case, we consider three types of element geometries, as shown in Figures 4.1, 4.2, and 4.3. We find that both the element topology and the boundaries are important. In the aperture case, we consider two cases, as shown in Figure 4.4.

Many of the ideas that we developed for linear arrays and apertures carry over to the planar case. In other cases, extensions are necessary. As in the linear array case, our development is a combination of classical antenna theory and finite impulse response filter theory. Classical antenna references that discuss planar arrays include [Ell81], [Bal82], [Mai94], [Ma74], [Ste81], and [Zio95]. Two-dimensional FIR filter references include [RG75] and [DM84].

In Section 4.1, we consider array geometries utilizing a rectangular element grid. We extend the techniques in Chapter 3 to the analysis and synthesis of rectangular arrays.

In Section 4.2, we develop analysis and synthesis procedures for circular arrays and ring apertures. We show that the Bessel function decomposition replaces the Fourier series decomposition for linear arrays.

In Section 4.3, we develop analysis and synthesis procedures for circular apertures. These apertures correspond to the limiting case for filled circular arrays. The circular aperture also occurs in parabolic reflector antennas.

In Section 4.4, we consider arrays using a hexagonal (also called triangular) element grid. Sampling theory indicates that a hexagonal grid is the most efficient grid and hexagonal grids are widely used in various applications. Hexagonal grid arrays are closely related to rectangular grid arrays.

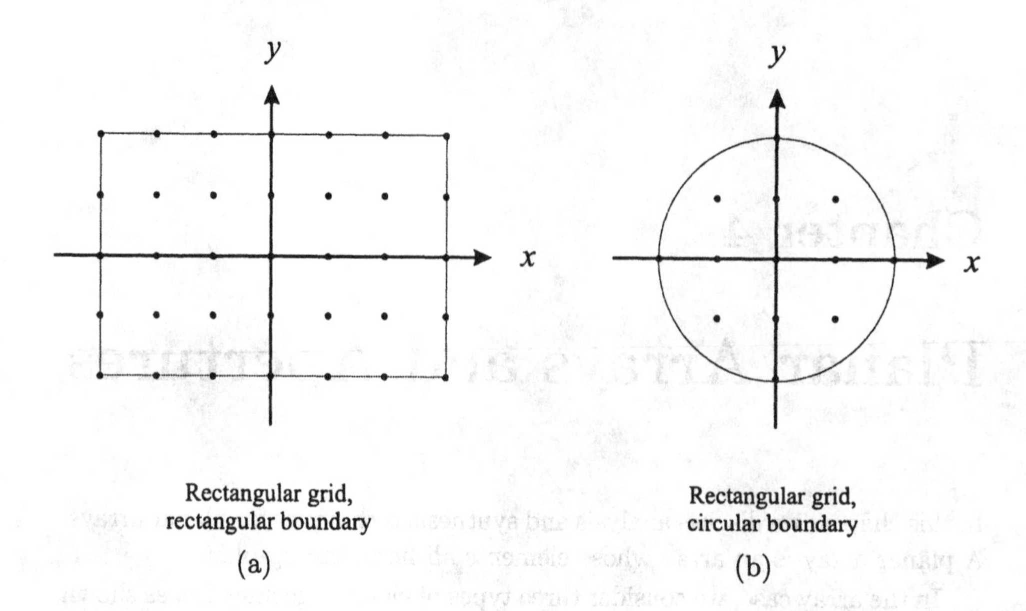

Figure 4.1 (a) Rectangular grid, rectangular boundary; (b) rectangular grid, circular boundary.

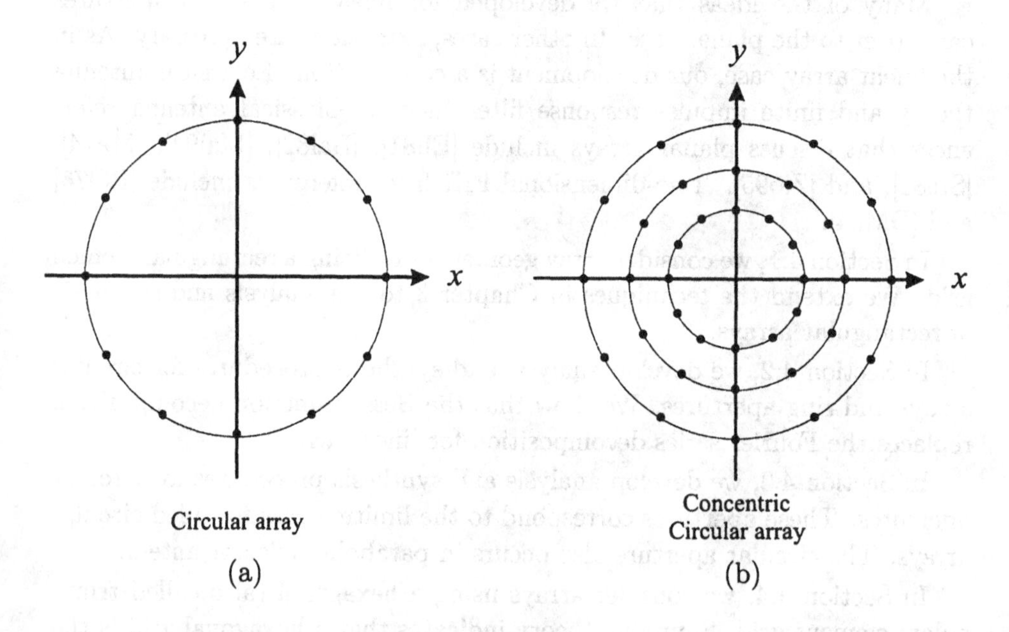

Figure 4.2 (a) Circular array; (b) concentric circular array.

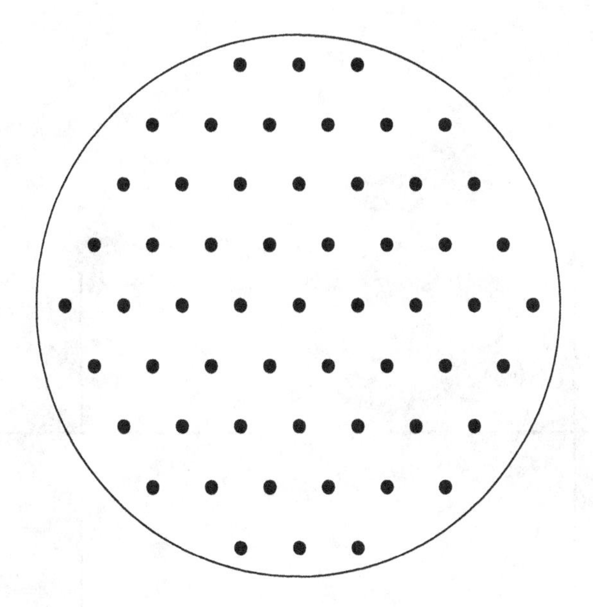

Figure 4.3 Hexagonal arrays; circular boundary.

Our development utilizes the circular aperture results in Section 4.3 which, in turn, uses the circular array results in Section 4.2. This path is the reason the two sections are separated.

In Section 4.5, we discuss nonplanar arrays briefly. In Section 4.6, we briefly summarize our results. The structure of Chapter 4 is shown in Table 4.1.

4.1 Rectangular Arrays

Our discussion of rectangular arrays parallels the development of synthesis techniques for linear arrays in Chapter 3.

4.1.1 Uniform Rectangular Arrays

The geometry for a planar array with a uniform rectangular grid and rectangular boundary is shown in Figure 4.5. We refer to these arrays as **uniform rectangular arrays** (URAs). Utilizing the relations in Chapter 3 we can write the beam pattern as the 2-D Fourier transform of the weighting function

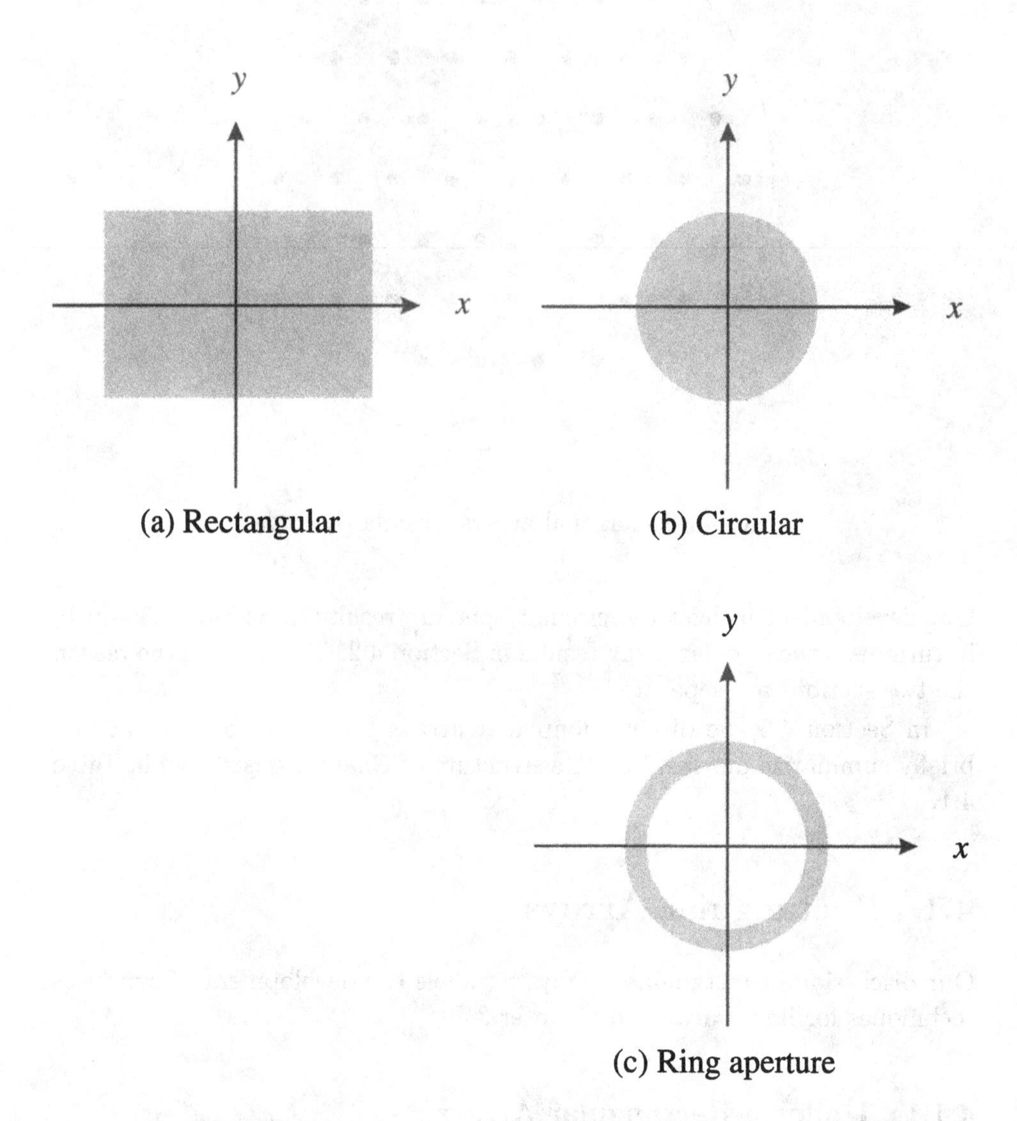

Figure 4.4 Apertures: (a) rectangular; (b) circular; (c) ring.

Table 4.1 Structure of Chapter 4.

4.1 Rectangular	4.4 Hexagonal
Rectangular grid URA, SRA Separable weighting 2-D Z-Transform Circular symmetry 2-D DFT Chebychev	Triangular grid UHA SHA H-R Transformations Beamspace processing

4.2 Circular Arrays	4.3 Circular Aperture
Ring aperture Circular arrays Phase mode exicitation	Separable weighting Taylor synthesis Sampling Difference beams

4.5 Nonplanar Arrays
Cylindrical Spherical

$$B(\psi_x, \psi_y) = e^{-j\left(\frac{N-1}{2}\psi_x + \frac{M-1}{2}\psi_y\right)} \sum_{n=0}^{N-1} \sum_{m=0}^{M-1} w_{nm}^* \, e^{j(n\psi_x + m\psi_y)}, \tag{4.1}$$

where

$$\psi_x = \frac{2\pi}{\lambda} d_x \sin\theta \cos\phi, \tag{4.2}$$

$$\psi_y = \frac{2\pi}{\lambda} d_y \sin\theta \sin\phi. \tag{4.3}$$

We can also express (4.2) and (4.3) in terms of the directional cosines,

$$u_x = \sin\theta \cos\phi, \tag{4.4}$$

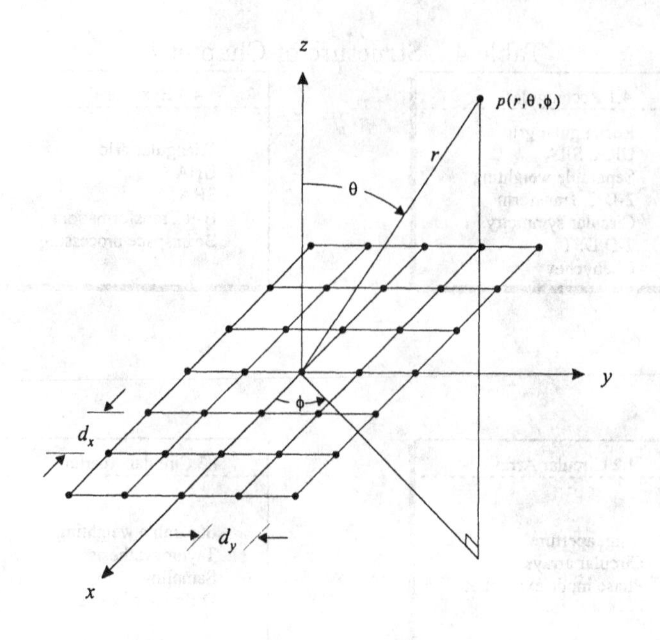

Figure 4.5 Planar array geometry.

$$u_y = \sin\theta \sin\phi. \tag{4.5}$$

The visible region is

$$u_r \triangleq \sqrt{u_x^2 + u_y^2} \le 1. \tag{4.6}$$

In terms of ψ_x, ψ_y, the visible region is

$$\sqrt{\left(\frac{\psi_x}{d_x}\right)^2 + \left(\frac{\psi_y}{d_y}\right)^2} \le \frac{2\pi}{\lambda}. \tag{4.7}$$

For the case in which

$$d_x = d_y = \frac{\lambda}{2}, \tag{4.8}$$

(4.2) and (4.3) reduce to

$$\psi_x = \pi u_x \tag{4.9}$$

and

$$\psi_y = \pi u_y. \tag{4.10}$$

We refer to uniform rectangular arrays that satisfy (4.8) as **standard rectangular arrays** (SRAs). We normally work in (ψ_x, ψ_y) space or (u_x, u_y) space because of the Fourier transform relationship. However, it is important

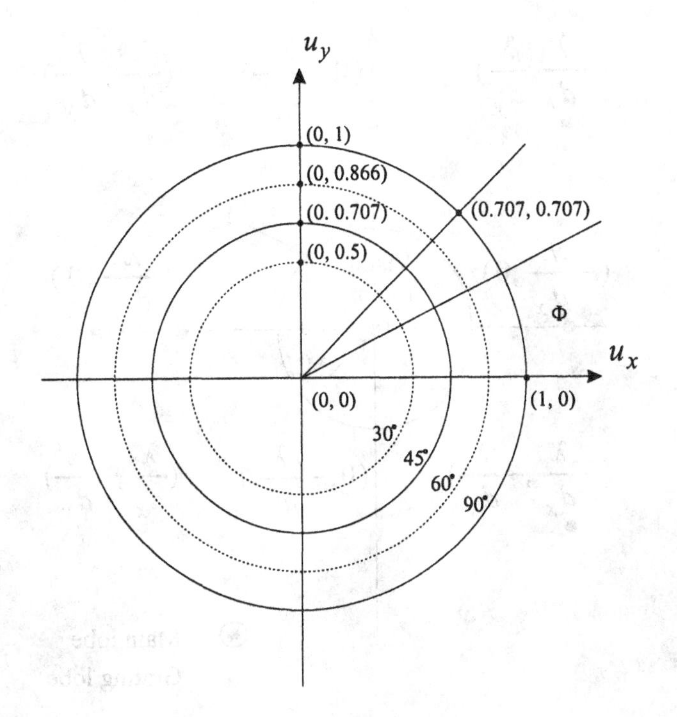

Figure 4.6 Mapping between (θ, ϕ) space and (u_x, u_y) space.

to remember the mapping into (θ, ϕ) space. This is shown in Figure 4.6 for the $d = \lambda/2$ case. The pairwise mapping $(u_x, u_y) \leftrightarrow (\theta, \phi)$ is unique.

As in the linear array case, we must consider the grating lobe structure. To do this, we rewrite (4.1) as

$$B(u_x, u_y) = e^{-j\left(\frac{N-1}{2}\pi u_x + \frac{M-1}{2}\pi u_y\right)} \sum_{m=0}^{M-1} \sum_{n=0}^{N-1} w_{nm}^* \cdot e^{j[nk_0 d_x u_x + mk_0 d_y u_y]}. \tag{4.11}$$

First, consider the case when the array is steered to broadside. Expanding (4.11) and recalling that $k_0 \triangleq |k| = 2\pi/\lambda$, we see that grating lobes will occur at

$$u_x = p\frac{\lambda}{d_x}, \quad p = 1, 2, \cdots, \tag{4.12}$$

$$u_y = q\frac{\lambda}{d_y}, \quad q = 1, 2, \cdots. \tag{4.13}$$

This periodicity is shown in the (u_x, u_y) plane in Figure 4.7.

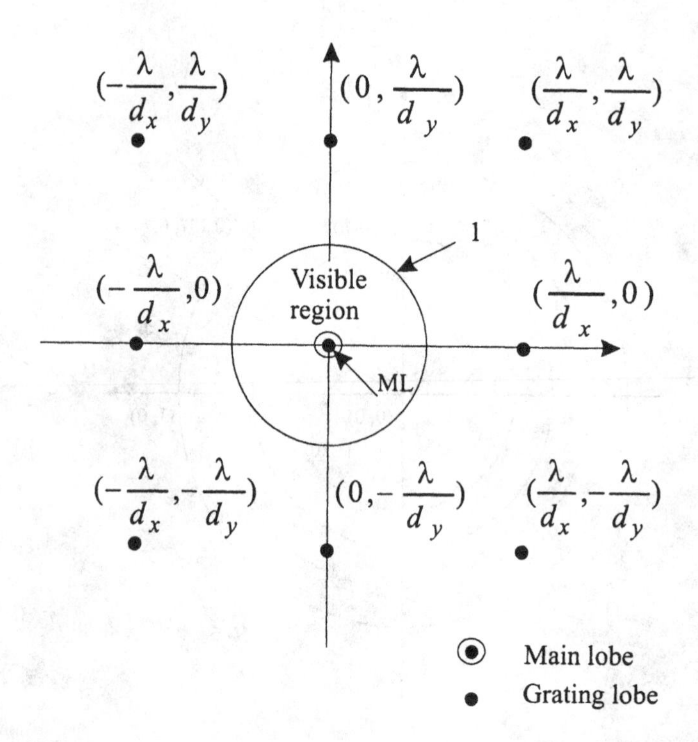

Figure 4.7 Planar array–grating lobe grid (rectangular lattice–rectangular grid).

Now assume the array is steered to (θ_0, ϕ_0).[1] Then,

$$u_{x0} = \sin\theta_0 \cos\phi_0 \tag{4.14}$$

and

$$u_{y0} = \sin\theta_0 \sin\phi_0. \tag{4.15}$$

Also,

$$\frac{u_{y0}}{u_{x0}} = \tan\phi_0 \tag{4.16}$$

and

$$\sqrt{u_{x0}^2 + u_{y0}^2} = \sin\theta_0. \tag{4.17}$$

The location of the main beam in (u_x, u_y) space is given by (4.14) and (4.15) and is shown in Figure 4.8. The grating lobes shift along exactly the same vector as shown in Figure 4.8. We can now determine the values of d_x and d_y required to avoid grating lobes in the visible region. We illustrate the process with a simple example.

[1]The subscript "0" denotes the steering direction.

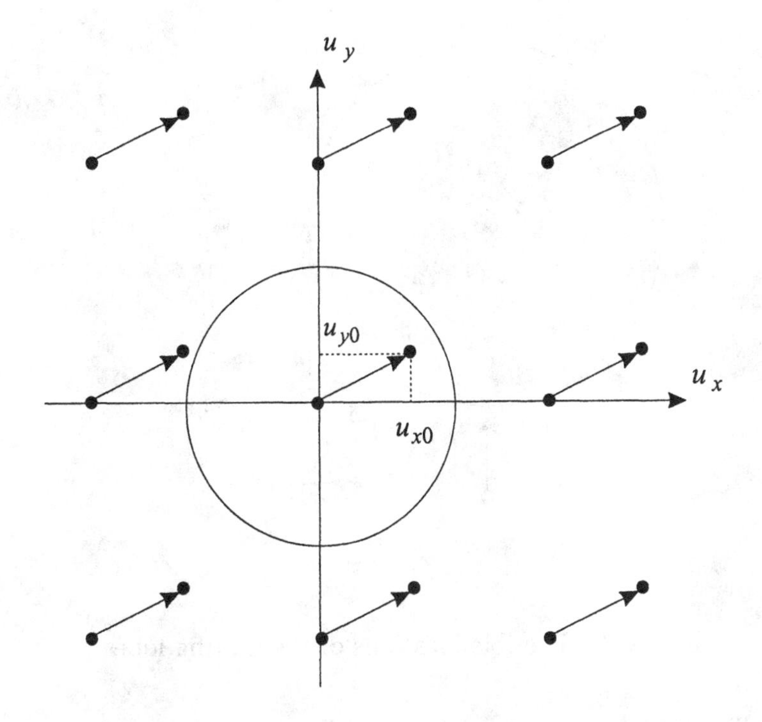

Figure 4.8 Planar array—scanned.

Example 4.1.1

Consider a rectangular grid array with uniform spacing d_x and d_y. We are required to scan from $0° \leq \theta \leq 60°$ and over all values of ϕ. Some possible locations of the grating lobes are shown in Figure 4.9. In the figure, we have chosen d_x and d_y so that the grating lobe is on the border of the visible region in the worst case.

Thus,

$$\frac{\lambda}{d_x} = \frac{\lambda}{d_y} = 1.866, \tag{4.18}$$

and we require,

$$d_x \leq \frac{\lambda}{1.866}, \tag{4.19}$$

and

$$d_y \leq \frac{\lambda}{1.866}. \tag{4.20}$$

If we are required to scan $0° \leq \theta \leq 90°$, then we require

$$d_x \leq \frac{\lambda}{2} \tag{4.21}$$

and

$$d_y \leq \frac{\lambda}{2}. \tag{4.22}$$

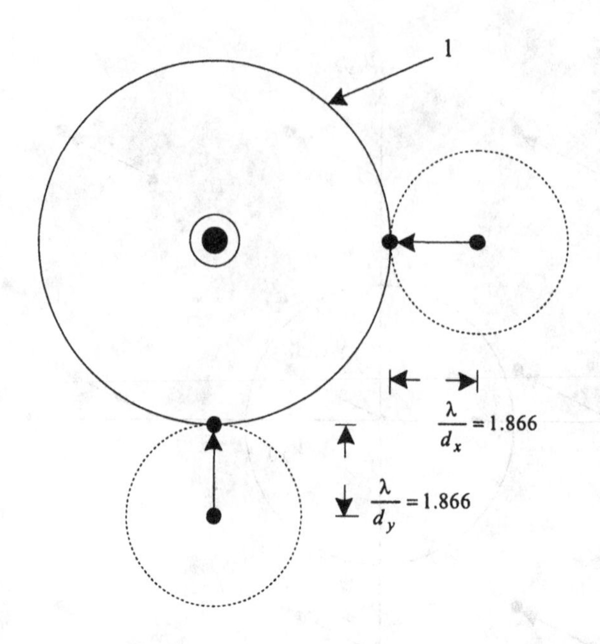

Figure 4.9 Possible locations of the grating lobes.

We will refer to a rectangular array with

$$d_x = d_y = \frac{\lambda}{2}, \tag{4.23}$$

as a standard rectangular array.

For the special case in which we have separable weightings

$$w_{nm} = w_n \, w_m, \tag{4.24}$$

the beam pattern is the product of the two individual array factors,

$$B(\psi_x, \psi_y) = B_x(\psi_x) \cdot B_y(\psi_y). \tag{4.25}$$

If the weighting is uniform in both directions, then

$$B(\psi_x, \psi_y) = \left[\frac{1}{N}\frac{\sin\left(\frac{N}{2}\psi_x\right)}{\sin\left(\frac{\psi_x}{2}\right)}\right]\left[\frac{1}{M}\frac{\sin\left(\frac{M}{2}\psi_y\right)}{\sin\left(\frac{\psi_y}{2}\right)}\right]. \tag{4.26}$$

There are several ways to plot beam patterns for the 2-D case. We illustrate them for this case. In subsequent cases, we use the plot that is most appropriate for the particular problem. We use $M = N = 10$ in the plots.

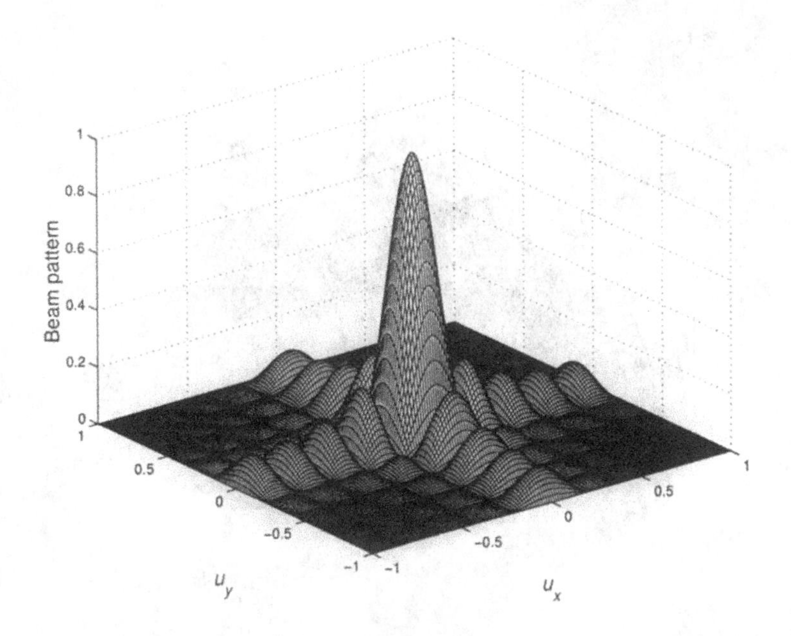

Figure 4.10 Magnitude of beam pattern of standard rectangular array: $N = M = 10$, uniform weighting, linear vertical scale.

(i) In Figure 4.10, we show the amplitude of the beam pattern versus u_x and u_y with a linear vertical scale. In Figure 4.11, we show the amplitude of the beam pattern versus u_x and u_y on a logarithmic scale. Both of these give a good overall view of the pattern but are not convenient for examining the details of the main lobe and sidelobe behavior.

(ii) In Figure 4.12, we show polar plots of the amplitude of the beam pattern versus θ for various values of ϕ. These plots are vertical cuts through the 3-D beam pattern. Note that the right side of the polar plot corresponds to the indicated value of ϕ. The left side corresponds to $\phi + 180°$. In Figure 4.13, we plot the amplitude on a logarithmic scale versus $u_r = \sin\theta$ for various values of ϕ. These plots are referred to as pattern cuts and are generally the most useful for a detailed pattern analysis.

(iii) In Figure 4.14, we show a contour plot of the amplitude in dB versus u_x and u_y. This plot gives a good view of the pattern symmetries and is particularly useful when plotted in color.

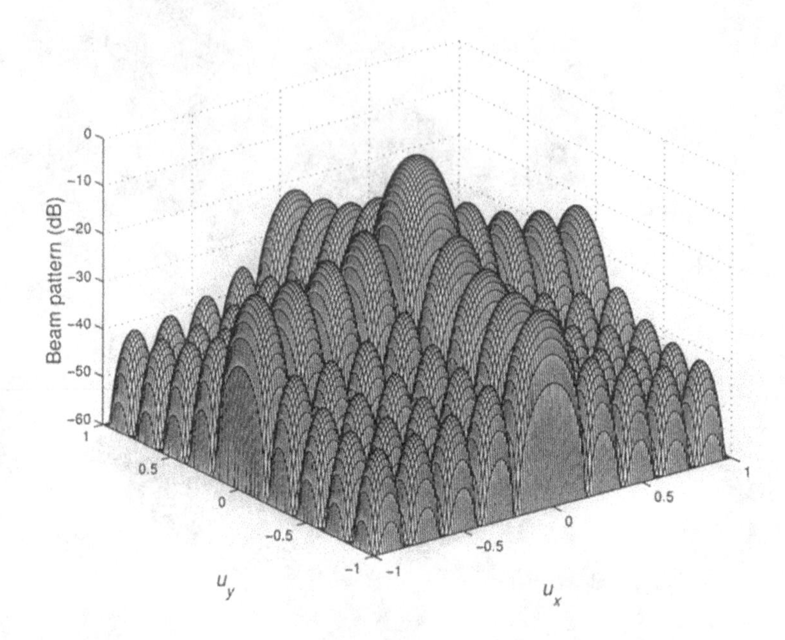

Figure 4.11 Magnitude of beam pattern of standard rectangular array: $N = M = 10$, uniform weighting, vertical scale is $20\log|B_u(u)|$.

4.1.1.1 Beamwidth

The half-power beamwidth for a pattern array is a contour in (u_x, u_y) space or (θ, ϕ) space where the magnitude-squared of the beam pattern is 0.5. For general weights, this contour must be evaluated analytically.

In some cases, the weights are chosen so that when the array is steered to broadside, the 3-dB contour can be approximated by a circle if $M = N$ and an ellipse if $M \neq N$. In these cases, an approximate expression for the axes of the half-power ellipse can be used.

The beamshape versus scan angle is shown in Figure 4.15. As the MRA of the beam moves away from $\theta = 0$, the circular contour becomes elliptical and the beamwidth in the θ-direction increases.

Two planes are chosen to characterize the beamwidth. The first is the elevation plane corresponding to $\phi = \phi_0$ and the second is a plane perpendicular to it. The half-power beamwidths in the two planes are designated by θ_H and Ψ_H. For a large array steered near broadside, θ_H is given approximately by

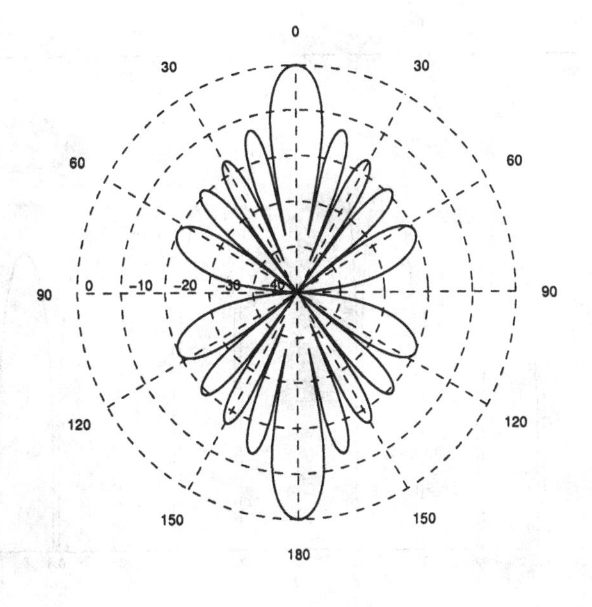

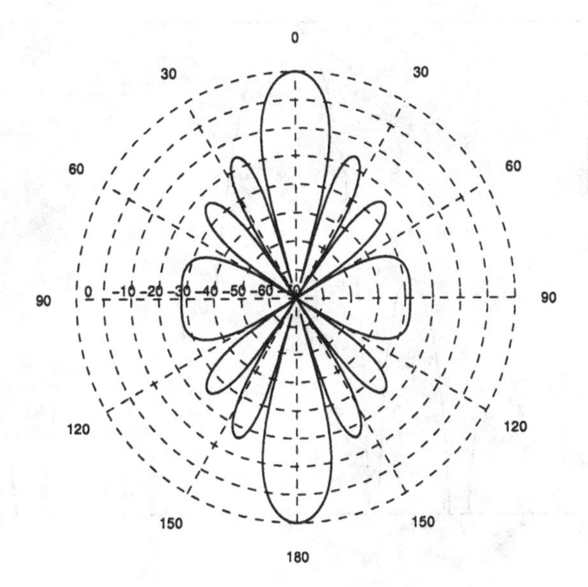

Figure 4.12 Polar plot of beam pattern cut of standard rectangular array: $N = M = 10$, uniform weighting; (a) $\phi = 0°$ or $90°$; (b) $\phi = 45°$.

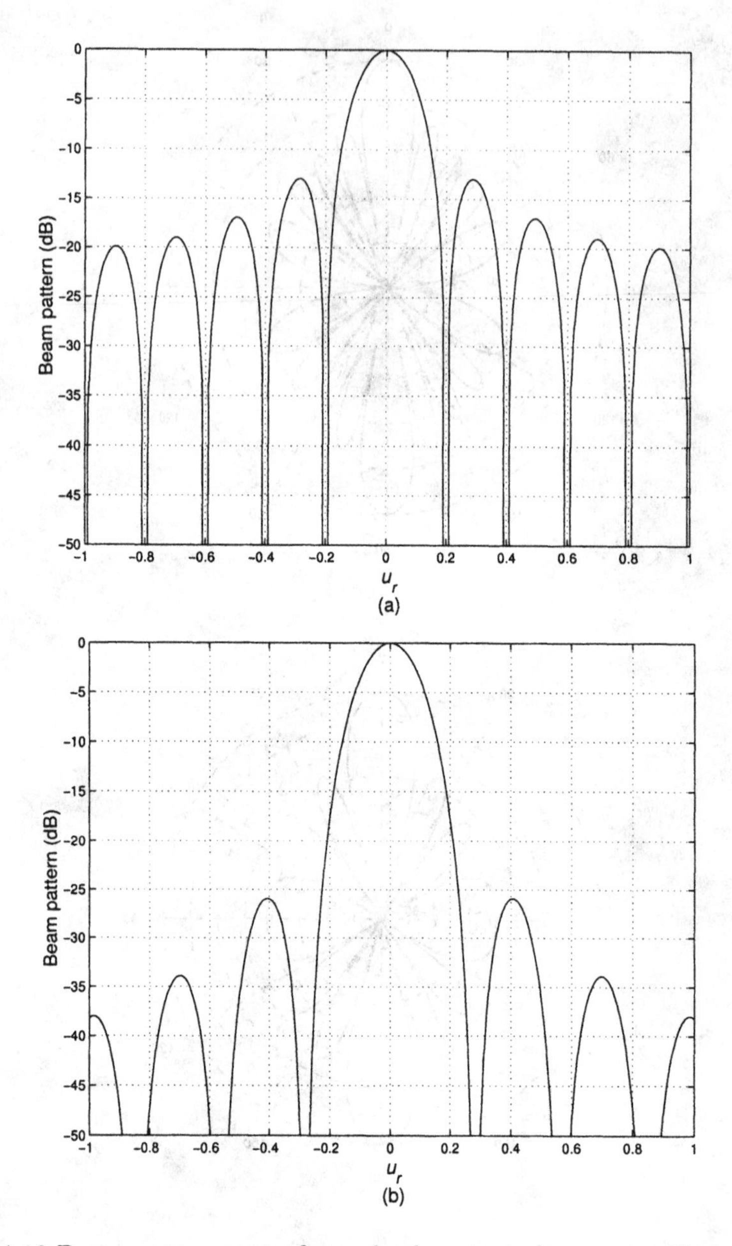

Figure 4.13 Beam pattern cut of standard rectangular array: $N = M = 10$, uniform weighting, plotted versus u_r; (a) $\phi = 0°$ or $90°$; (b) $\phi = 45°$.

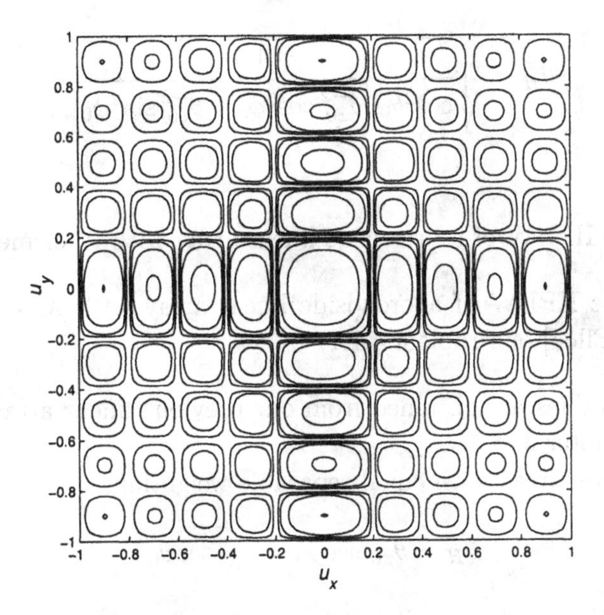

Figure 4.14 Contour plot of magnitude of beam pattern for standard rectangular array: $N = M = 10$, uniform weighting.

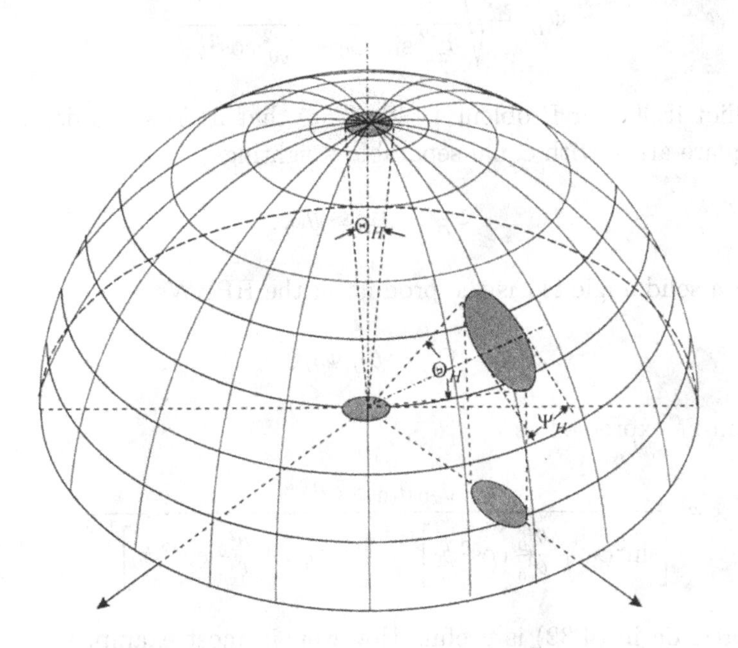

Figure 4.15 Beam shape versus scan angle.

$$\theta_H = \sqrt{\frac{1}{\cos^2\theta_0 \left[\theta_{x0}^{-2}\cos^2\phi_0 + \theta_{y0}^{-2}\sin^2\phi_0\right]}}, \tag{4.27}$$

where

(i) θ_{x0} is the HPBW of a broadside linear array with N elements;

(ii) θ_{y0} is the HPBW of a broadside linear array with M elements (e.g., Elliot [Ell64] or Problem 4.1.1).

The HPBW values are obtained from our previous linear array beamwidth results in Chapter 3.

For a square array with equal separable weightings,

$$\theta_H = \theta_{x0} \sec\theta_0 = \theta_{y0} \sec\theta_0\,, \tag{4.28}$$

which is identical to our result in Chapter 2.

The HPBW in the perpendicular plane is given by

$$\Psi_H = \sqrt{\frac{1}{\theta_{x0}^{-2}\sin^2\phi_0 + \theta_{y0}^{-2}\cos^2\phi_0}} \tag{4.29}$$

(e.g., Elliot [Ell64] or Problem 4.1.2). Note that it does not depend on θ_0. For a square array with equal separable weightings

$$\Psi_H = \theta_{x0} = \theta_{y0}\,. \tag{4.30}$$

The beam solid angle Ω_A is the product of the HPBWs,

$$\Omega_A = \theta_H\,\Psi_H\,, \tag{4.31}$$

which can be expressed as

$$\Omega_A = \frac{\theta_{x0}\,\theta_{y0}\sec\theta_0}{\left[\sin^2\phi_0 + \frac{\theta_{y0}^2}{\theta_{x0}^2}\cos^2\phi_0\right]^{1/2}\left[\sin^2\phi_0 + \frac{\theta_{x0}^2}{\theta_{y0}^2}\cos^2\phi_0\right]^{1/2}}. \tag{4.32}$$

The expression in (4.32) is useful. However, in most examples we plot the actual half-power contour of the beam pattern in the u_x, u_y plane or examine it for various pattern cuts.

4.1.1.2 Directivity of a planar array

The directivity of an array is given by (2.144) as

$$D = \frac{|B(\theta_0, \phi_0)|^2}{\frac{1}{4\pi}\int_0^{2\pi}\int_0^{\pi}|B(\theta,\phi)|^2 \sin\theta d\theta\, d\phi}, \tag{4.33}$$

where (θ_0, ϕ_0) is the MRA. We desire an expression for the directivity that is valid for any planar array in the xy-plane. The directivity does not depend on the choice of the coordinate system, so the result is valid for any planar array.

For an N-element planar array in the xy-plane, we denote the position of the nth element by (p_{x_n}, p_{y_n}). Note that the array can have arbitrary configuration in the xy-plane and that we are indexing the elements with a single index n. Note that N is the total number of elements and would equal NM for the array in Figure 4.5.

The beam pattern can be written as,

$$B(\theta,\phi) = \sum_{n=0}^{N-1} w_n^* \exp\left(j\frac{2\pi}{\lambda}\left(p_{x_n}\sin\theta\cos\phi + p_{y_n}\sin\theta\sin\phi\right)\right). \tag{4.34}$$

The denominator in (4.33) can be written as

$$\begin{aligned} DEN &\triangleq \frac{1}{4\pi}\int_0^{2\pi}\int_0^{\pi}|B(\theta,\phi)|^2 \sin\theta\, d\theta\, d\phi \\ &= \frac{1}{4\pi}\int_0^{2\pi}\int_0^{\pi}\sum_{n=0}^{N-1}\sum_{m=0}^{N-1} w_n^* w_m e^{j\frac{2\pi}{\lambda}\sin\theta[(p_{x_n}-p_{x_m})\cos\phi+(p_{y_n}-p_{y_m})\sin\phi]} \\ &\quad \cdot \sin\theta d\theta d\phi, \end{aligned} \tag{4.35}$$

which reduces to

$$\begin{aligned} DEN &= \sum_{n=0}^{N-1}\sum_{m=0}^{N-1} w_n^* w_m \int_0^{\pi}\frac{1}{2}\sin\theta \cdot d\theta \int_0^{2\pi}\frac{1}{2\pi} \\ &\quad \cdot \exp\left(j\frac{2\pi}{\lambda}\sin\theta\left(\Delta p_{x_{nm}}\cos\phi + \Delta p_{y_{nm}}\sin\phi\right)\right) d\phi, \end{aligned} \tag{4.36}$$

where

$$\Delta p_{x_{nm}} \triangleq p_{x_n} - p_{x_m}, \tag{4.37}$$

and

$$\Delta p_{y_{nm}} \triangleq p_{y_n} - p_{y_m}. \tag{4.38}$$

Now define

$$\rho_{nm} = \left((\Delta p_{x_{nm}})^2 + (\Delta p_{y_{nm}})^2 \right)^{\frac{1}{2}} = \|\Delta \mathbf{p}_{nm}\|, \tag{4.39}$$

and

$$\phi_{nm} = \arctan\left(\frac{\Delta p_{y_{nm}}}{\Delta p_{x_{nm}}} \right). \tag{4.40}$$

Then

$$\Delta p_{x_{nm}} = \rho_{nm} \cos \phi_{nm}, \tag{4.41}$$

and

$$\Delta p_{y_{nm}} = \rho_{nm} \sin \phi_{nm}. \tag{4.42}$$

Using (4.41) and (4.42) in the inner integral in (4.36) gives

$$\begin{aligned} & \int_0^{2\pi} \frac{1}{2\pi} \exp\left(j\frac{2\pi}{\lambda}\rho_{nm} \sin\theta \left(\cos\phi \cos\phi_{nm} + \sin\phi \sin\phi_{nm} \right) \right) d\phi \\ = & \int_0^{2\pi} \frac{1}{2\pi} \exp\left(j\frac{2\pi}{\lambda}\rho_{nm} \sin\theta \left(\cos\left(\phi - \phi_{nm}\right) \right) \right) d\phi \\ = & \; J_0\left(\frac{2\pi}{\lambda}\rho_{nm} \sin\theta \right), \end{aligned} \tag{4.43}$$

where $J_0(\cdot)$ is a Bessel function of order zero. Substituting (4.43) into (4.36) gives

$$\begin{aligned} DEN &= \sum_{n=0}^{N-1}\sum_{m=0}^{N-1} w_n w_m^* \int_0^{\pi} \frac{1}{2} \sin\theta \cdot J_0\left(\frac{2\pi}{\lambda}\rho_{nm} \sin\theta \right) d\theta \\ &= \sum_{n=0}^{N-1}\sum_{m=0}^{N-1} w_n w_m^* \text{sinc}\left(\frac{2\pi}{\lambda}\rho_{nm} \right). \end{aligned} \tag{4.44}$$

Defining a matrix $\mathbf{B}$ with elements

$$[\mathbf{B}]_{nm} = \text{sinc}\left(\frac{2\pi}{\lambda}\rho_{nm} \right), \tag{4.45}$$

the directivity can be written as

$$\boxed{D = \frac{|\mathbf{w}^H \mathbf{v}_0|^2}{\mathbf{w}^H \mathbf{B} \mathbf{w}},} \tag{4.46}$$

where

$$\mathbf{v}_0 = \mathbf{v}(\theta_0, \phi_0), \tag{4.47}$$

is the array manifold vector steered at (θ_0, ϕ_0).

If we normalize the weights so that $B(\theta_0, \phi_0) = 1$, then (4.46) reduces to

$$D = \left[\mathbf{w}^H \mathbf{B} \mathbf{w}\right]^{-1}. \tag{4.48}$$

Note that $\mathbf{B}$ does not reduce to an identity matrix for a standard rectangular array.

For large rectangular grid planar arrays Elliot [Ell64] has shown that

$$D_0 = \pi \cos\theta_0 \, D_x \, D_y. \tag{4.49}$$

The reader is referred to that reference for a discussion of this result. We will use the result in (4.46) to calculate the directivity.

In the linear array we developed a number of techniques for synthesizing desireable beam patterns. In the next several sections, we look at the extension of these techniques to planar arrays and the introduction of modified techniques where appropriate. Before doing that development, it is useful to consider the array manifold vector in more detail.

4.1.2 Array Manifold Vector

The array manifold vector was defined for an arbitrary array in (2.28). For rectangular planar arrays it is convenient to define an array manifold matrix as an intermediate step.

We define

$$\mathbf{v}_m(\boldsymbol{\psi}) = \begin{bmatrix} e^{jm\psi_y} \\ e^{j(\psi_x + m\psi_y)} \\ \vdots \\ e^{j((N-1)\psi_x + m\psi_y)} \end{bmatrix}, \tag{4.50}$$

as the array manifold vector corresponding to the mth line of sensors in the y-direction. The vector $\boldsymbol{\psi}$ is

$$\boldsymbol{\psi} = \begin{bmatrix} \psi_x \\ \psi_y \end{bmatrix}. \tag{4.51}$$

Then,

$$\mathbf{V}_{\boldsymbol{\psi}}(\boldsymbol{\psi}) = \left[\, \mathbf{v}_0(\boldsymbol{\psi}) \,\vdots\, \cdots \,\vdots\, \mathbf{v}_{M-1}(\boldsymbol{\psi}) \,\right], \tag{4.52}$$

is an $N \times M$ array manifold matrix. Then, we can define a vector that is obtained by stacking the vectors to obtain an $NM \times 1$ vector (see (A.106)),

$$vec\left[\mathbf{V}_{\boldsymbol{\psi}}(\boldsymbol{\psi})\right] = \begin{bmatrix} \mathbf{v}_0(\boldsymbol{\psi}) \\ \vdots \\ \mathbf{v}_{M-1}(\boldsymbol{\psi}) \end{bmatrix}. \tag{4.53}$$

It is easy to verify that $vec\left[\mathbf{v}_{\psi}(\psi)\right]$ is conjugate symmetric.

The expression in 4.53 can also be written as a Kroneker product (A.79):

$$vec\left[\mathbf{V}_{\psi}(\psi)\right] = \mathbf{v}(\psi) \otimes \mathbf{v}(\psi). \tag{4.54}$$

We can describe the weight vector using this notation. The weight vector down the mth line is

$$\mathbf{w}_m = \begin{bmatrix} w_{0,m} \\ \vdots \\ w_{n,m} \\ \vdots \\ w_{N-1,m} \end{bmatrix}, \tag{4.55}$$

and the weight matrix is

$$\mathbf{W} = \begin{bmatrix} \mathbf{w}_0 & \cdots & \mathbf{w}_m & \cdots & \mathbf{w}_{M-1} \end{bmatrix}, \tag{4.56}$$

and

$$vec\left[\mathbf{W}\right] = \begin{bmatrix} \mathbf{w}_0 \\ \vdots \\ \mathbf{w}_m \\ \vdots \\ \mathbf{w}_{M-1} \end{bmatrix}. \tag{4.57}$$

Then,

$$B(\psi) = B(\psi_x, \psi_y) = vec^H\left[\mathbf{W}\right] vec\left[\mathbf{V}_{\psi}(\psi)\right]. \tag{4.58}$$

If $vec\left[\mathbf{W}\right]$ is conjugate symmetric (or real and symmetric), then $B(\psi)$ will be real.

For a conventional delay-and-sum beamformer,

$$vec^H\left[\mathbf{W}\right] = vec^H\left[\mathbf{V}_{\psi_s}(\psi_s)\right], \tag{4.59}$$

and

$$B_c(\psi) = vec^H\left[\mathbf{V}_{\psi_s}(\psi_s)\right] vec\left[\mathbf{V}_{\psi}(\psi)\right] \tag{4.60}$$

is real.

We will find the $vec(\cdot)$ notation useful in a number of cases later in the text.

4.1.3 Separable Spectral Weightings

In Section 3.1, we introduced various spectral weightings to obtain desirable sidelobe patterns and acceptable main-lobe beamwidths.

A straightforward approach is to use the linear array spectral weighting along each axis:

$$w_{nm} = w_n \, w_m. \tag{4.61}$$

Since the weighting is separable, the beam pattern will be a product of the two linear array beam patterns. Then,

$$B(\psi_x, \psi_y) = B(\psi_x) \, B(\psi_y). \tag{4.62}$$

In many cases, this will lead to satisfactory patterns. The problem is that it is not clear how the two-dimensional pattern will behave when $\phi \neq 0$ or $\pi/2$. For example, we saw in Figure 4.13 that the sidelobe structure for $\phi = \pi/4$ was significantly different than that for $\phi = 0$ or $\pi/2$.

We consider an example to illustrate the behavior.

Example 4.1.2: Hamming window

We assume that $M = N = 11$ and use the Hamming weighting from (3.21):

$$w_N(n) = \begin{cases} 0.54 + 0.46 \cos\left(\frac{2\pi|n|}{N}\right), & |n| \leq 5, \\ 0, & \text{elsewhere.} \end{cases} \tag{4.63}$$

The resulting pattern is shown in Figure 4.16. In Figure 4.17, we show pattern cuts at $\phi = 0°$ (or $90°$) and $\phi = 45°$. We see that the main lobe is wider for $\phi = 0°$ (or $90°$). The sidelobes are lower for $\phi = 45°$ than for $\phi = 0°$ (or $90°$).

This behavior is typical for separable weightings. In order to obtain beam patterns whose pattern cuts are similar for all ϕ, non-separable weightings are required.

4.1.4 2-D z-Transforms

Just as in the 1-D case, we can write the z-transform of the weighting matrix as

$$B_{\mathbf{z}}(z_1, z_2) = \sum_{n=0}^{N-1} \sum_{m=0}^{M-1} w_{nm} \, z_1^{-n} \, z_2^{-m} \, . \tag{4.64}$$

We define

$$z_1 = e^{j\psi_x} \, , \tag{4.65}$$

and

$$z_2 = e^{j\psi_y} \, . \tag{4.66}$$

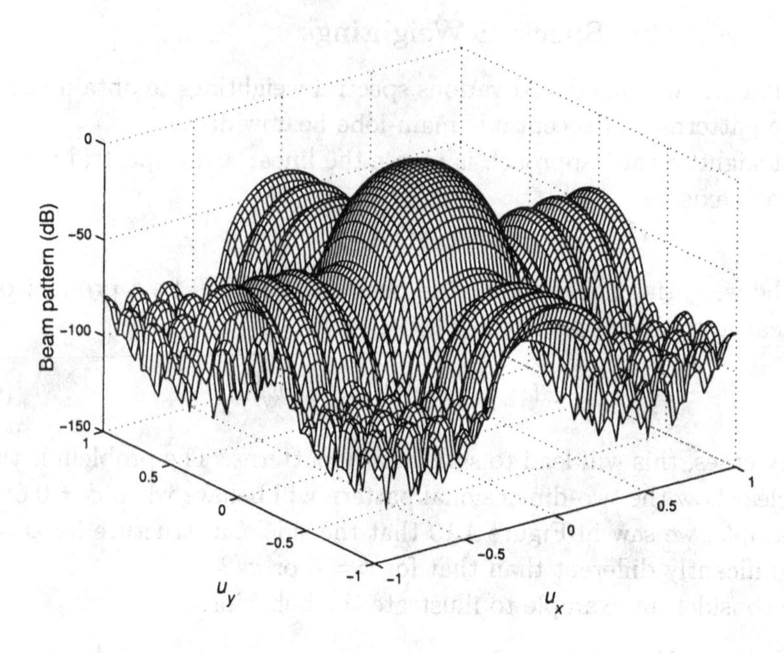

Figure 4.16 Beam pattern: separable Hamming weighting.

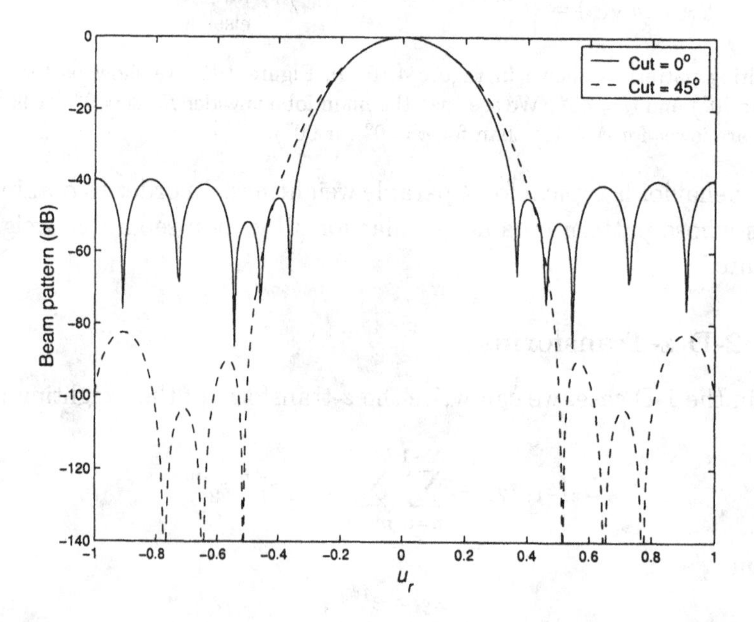

Figure 4.17 Beam pattern cuts: separable Hamming weighting.

The beam pattern can be written as

$$B_{\psi}(\psi_x,\psi_y) = \left[z_1^{-\frac{N-1}{2}} z_2^{-\frac{M-1}{2}} B_{\mathbf{z}}^*(z_1,z_2)\right]_{z_1=e^{j\psi_x},z_2=e^{j\psi_y}}. \tag{4.67}$$

4.1.5 Least Squares Synthesis

In this section, we apply the least squares results in Section 3.5 to planar arrays. The square error can be written as

$$\xi = \frac{1}{4\pi}\int_0^{2\pi}\int_0^{\pi} |B_d(\theta,\phi) - \sum_{m=0}^{M-1} \mathbf{w}_m^H \mathbf{v}_m(\theta,\phi)|^2 \sin\theta \, d\theta \, d\phi \tag{4.68}$$

Differentiating with respect to $\mathbf{w}_m^H$ and setting the result equal to $\mathbf{0}$ gives,[2]

$$\mathbf{w}_{m,o} = \frac{1}{4\pi}\mathbf{A}_m^{-1}\int_0^{2\pi}\int_0^{\pi} \mathbf{v}_m(\theta,\phi)B_d^*(\theta,\phi)\sin\theta\, d\theta\, d\phi,$$
$$m = 0,\cdots,M-1, \tag{4.69}$$

where

$$\mathbf{A}_m = \frac{1}{4\pi}\int_0^{2\pi}\int_0^{\pi} \mathbf{v}_m(\theta,\phi)\mathbf{v}_m^H(\theta,\phi)\sin\theta\, d\theta\, d\phi,$$
$$m = 0,\cdots,M-1 \tag{4.70}$$

is an $N \times N$ matrix. From (4.50), we observe that

$$\mathbf{v}_m(\psi_x,\psi_y)\mathbf{v}_m^H(\psi_x,\psi_y) = \mathbf{v}_m(\theta,\phi)\mathbf{v}_m^H(\theta,\phi) \tag{4.71}$$

is not a function of m so the subscript can be removed from the left side of (4.70). Evaluating (4.70) gives:

$$[\mathbf{A}]_{nl} = \text{sinc}\left[(n-l)\right] = \pi\delta_{nl} \tag{4.72}$$

for a standard rectangular grid. Thus,

$$\mathbf{w}_{m,o} = \frac{1}{4\pi^2}\int_0^{2\pi}\int_0^{\pi} \mathbf{v}_m(\theta,\phi)B_d^*(\theta,\phi)\sin\theta\, d\theta\, d\phi. \tag{4.73}$$

We consider two examples to illustrate typical behavior.

Example 4.1.3

The desired beam pattern is shown in Figure 4.18.

---

[2]This approach is the same as Section 3.5.

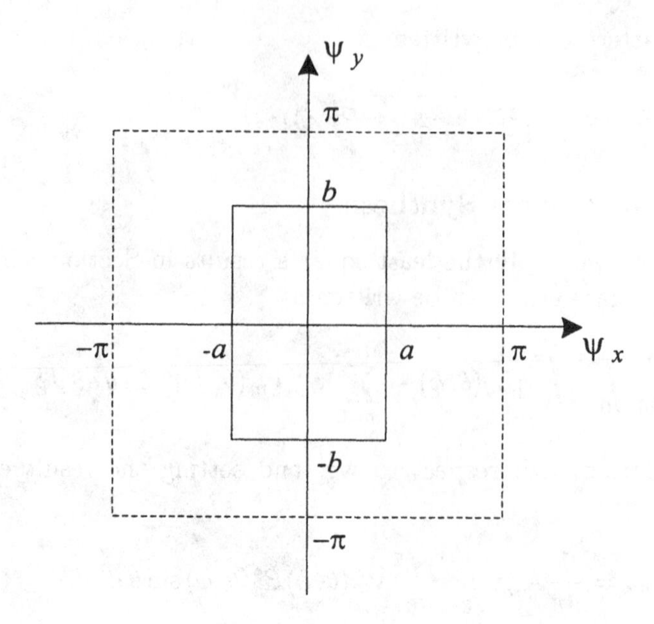

Figure 4.18 Ideal rectangular wavenumber beam pattern.

$$B_{\psi}(\psi_x, \psi_y) = \begin{cases} 1, & -\psi_a \leq \psi_x \leq \psi_a; -\psi_b \leq \psi_y \leq \psi_b, \\ 0, & \text{elsewhere.} \end{cases} \tag{4.74}$$

The limits of the integrand are completely inside the visible region so that we can use rectangular coordinates. The nth component of $\mathbf{w}_{m,o}$ is

$$[\mathbf{w}_{m,o}]_n = \frac{1}{4\pi^2}\int_{-\psi_a}^{\psi_a}\int_{-\psi_b}^{\psi_b} \exp\left\{j\left[\left(n-\frac{N-1}{2}\right)\psi_x + \left(m-\frac{M-1}{2}\right)\psi_y\right]\right\} d\psi_x\, d\psi_y$$

$$n = 0, \cdots, N-1,$$
$$m = 0, \cdots, M-1. \tag{4.75}$$

Integrating gives

$$[\mathbf{w}_{m,o}]_n = \left(\frac{\sin\left(\left(n-\frac{N-1}{2}\right)\psi_a\right)}{\left(n-\frac{N-1}{2}\right)\pi}\right)\left(\frac{\sin\left(\left(m-\frac{M-1}{2}\right)\psi_b\right)}{\left(m-\frac{N-1}{2}\right)\pi}\right)$$

$$n = 0, \cdots, N-1,$$
$$m = 0, \cdots, M-1. \tag{4.76}$$

which is a separable weighting. The weight is real, so $[\mathbf{w}_{m,o}]_n = [\mathbf{w}^*_{m,o}]_n$.

The resulting beam pattern for $N = M = 11$ is a product of the uniform beam pattern in Figure 3.32 with a uniform beam pattern of identical shape (with $\psi_a \neq \psi_b$). As expected, we have an overshoot because of the Gibbs phenomenon.

Just as in the 1-D case, we can alleviate the overshoot problem by using one of the windows discussed in Section 3.1. For example, if we use the

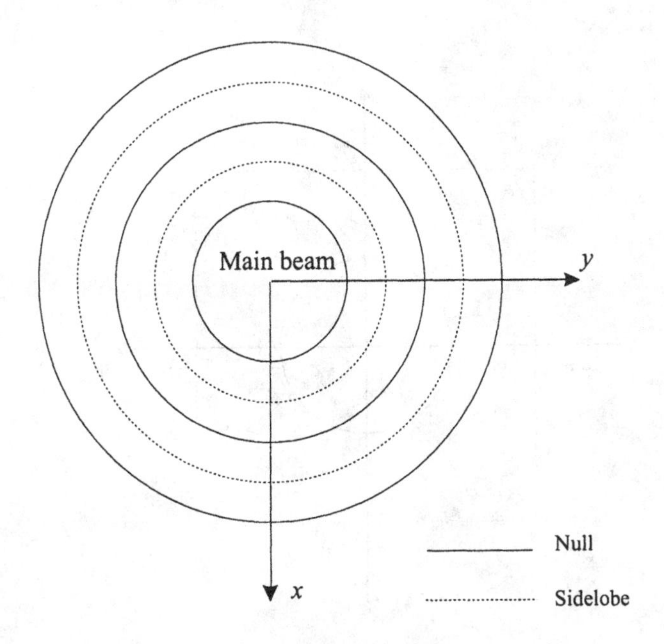

Figure 4.19 Ring-type sidelobe structure of ϕ-symmetric pattern.

Kaiser window from Example 3.4.3, the resulting pattern will be a product of the beam pattern in Figure 3.33 with itself when $\psi_a = \psi_b$.

In a large number of applications, we would like the beam pattern to be uniform in ϕ. This corresponds to circular symmetry in (ψ_x, ψ_y) space. This implies,

$$B_{\psi}(\psi_x, \psi_y) = B_{\psi_r}\left(\sqrt{\psi_x^2 + \psi_y^2}\right). \tag{4.77}$$

The argument of the function on the right side corresponds to the radial wavenumber,

$$\psi_r = \sqrt{\psi_x^2 + \psi_y^2}. \tag{4.78}$$

Beam patterns satisfying (4.77) will have a ring sidelobe structure, as shown in Figure 4.19, and will lead to a non-separable weighting function.

Example 4.1.4[3]

The simplest case corresponds to a desired beam pattern that is constant over a circular region, as shown in Figure 4.20.

In this case, it is useful to use symmetric indexing. Then, for N odd,

$$B_{\psi}(\psi_x, \psi_y) = \sum_{n=-\frac{N-1}{2}}^{\frac{N-1}{2}} \sum_{m=-\frac{M-1}{2}}^{\frac{M-1}{2}} a_{nm}^* \, e^{j\left(n\psi_x + m\psi_y\right)}, \tag{4.79}$$

[3]This example is on p. 446 of [RG75].

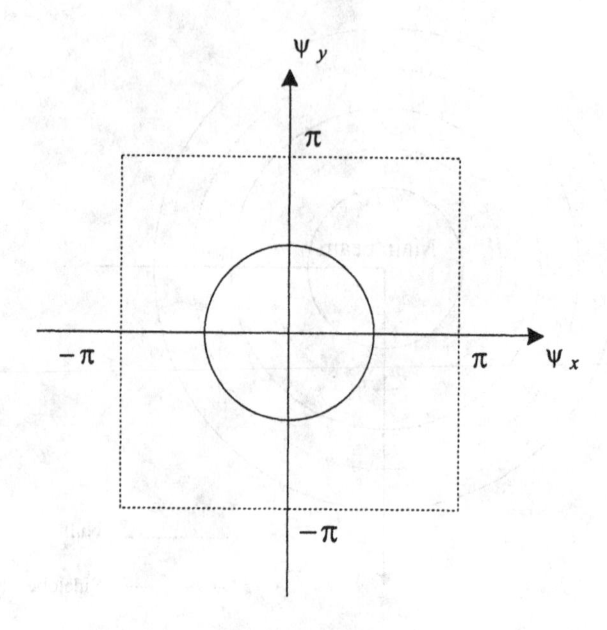

Figure 4.20 Desired beam pattern: circular symmetry.

and, for N even,

$$B_{\psi}(\psi_x, \psi_y) = \sum_{\substack{n=-\frac{N}{2} \\ n\neq 0}}^{\frac{N}{2}} \sum_{\substack{m=-\frac{M}{2} \\ m\neq 0}}^{\frac{M}{2}} a_{nm}^{*}\, e^{j(n\psi_x + m\psi_y)}. \tag{4.80}$$

The weight is denoted by a_{nm}^* when symmetric indexing is used.

Since the beam pattern has circular symmetry, the weighting function will have circular symmetry.

$$a^*(n,m) = a_c^*\left(\sqrt{n^2+m^2}\right). \tag{4.81}$$

A simple way to find $a^*(n,m)$ is to first find $a^*(n,0)$ and then replace n by $\sqrt{n^2+m^2}$:

$$\begin{aligned} a^*(n,0) &= \frac{1}{4\pi^2}\int_{-\psi_R}^{\psi_R} e^{jn\psi_x}\, d\psi_x \int_{-\sqrt{\psi_R^2-\psi_x^2}}^{\sqrt{\psi_R^2-\psi_x^2}} d\psi_y \\ &= \frac{1}{4\pi^2}\int_{-\psi_R}^{\psi_R} e^{jn\psi_x}\left[2\sqrt{\psi_R^2-\psi_x^2}\right] d\psi_x. \end{aligned} \tag{4.82}$$

Letting,

$$\psi_x = \psi_R \sin\varphi, \tag{4.83}$$

where

$$\varphi = \tan^{-1}\left(\frac{\psi_x}{\psi_y}\right). \tag{4.84}$$

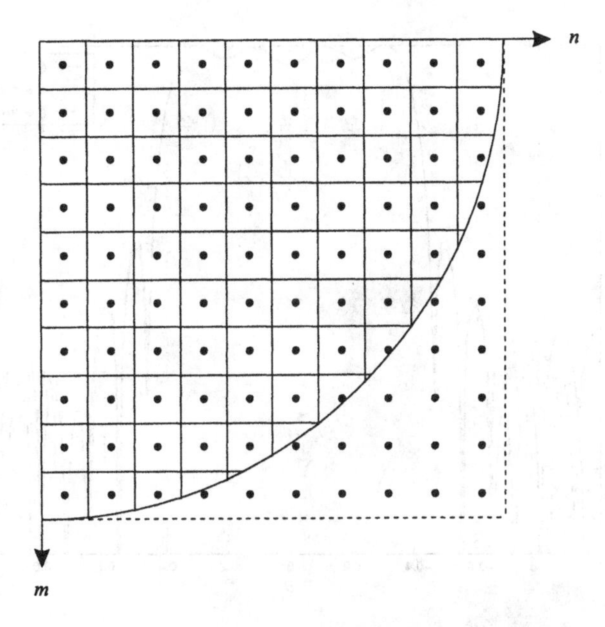

Figure 4.21 One quadrant of a rectangular array: $d_x = d_y = 0.5\lambda$.

Then

$$d\psi_x = \psi_R \cos\varphi \, d\varphi, \tag{4.85}$$

and

$$\begin{aligned} a^*(n,0) &= \frac{1}{4\pi^2}\int_{-\frac{\pi}{2}}^{\frac{\pi}{2}} 2\psi_R^2 \left(\cos^2\varphi\right) e^{j\psi_R n \sin\varphi}\, d\varphi \\ &= \frac{\psi_R J_1(\psi_R n)}{2\pi n}, \end{aligned} \tag{4.86}$$

where $J_1(\cdot)$ is the first-order Bessel function. In (4.86), n is treated as a continuous variable. Then

$$a^*(n,m) = \frac{\psi_R J_1\left(\psi_R \sqrt{n^2+m^2}\right)}{2\pi\sqrt{n^2+m^2}}. \tag{4.87}$$

Now consider a square array with N elements in each direction. One quadrant of a square array with $N = 20$ is shown in Figure 4.21. We consider two ways of truncating $a^*(n,m)$ in (4.86).

The first choice is to let

$$a^*(n,m) = \begin{cases} a^*(\sqrt{n^2+m^2}), & 0 \le \sqrt{n^2+m^2} \le 10, \\ 0, & \text{elsewhere.} \end{cases} \tag{4.88}$$

The effect of (4.88) is to give the square array a circular boundary as shown in Figure 4.21. All of the elements outside of the circular boundary have zero weights.

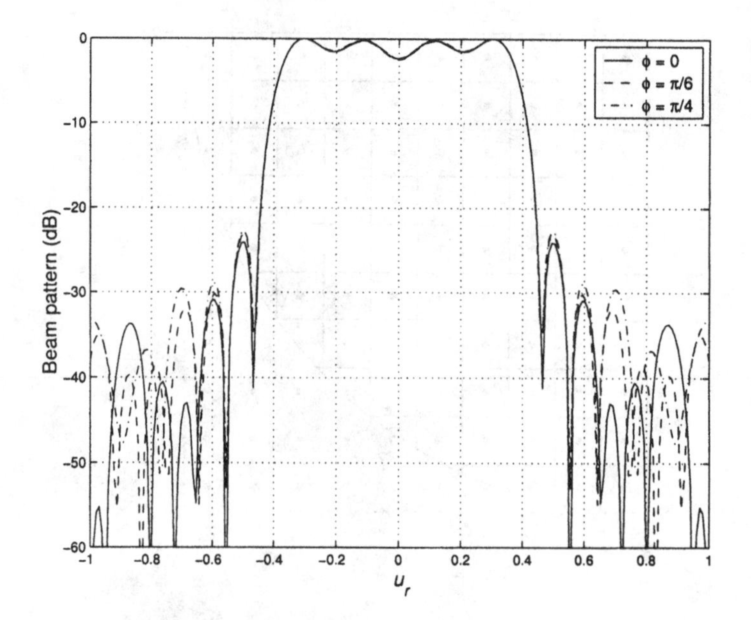

Figure 4.22 Beam pattern cut: standard rectangular grid array with circular boundary: $N = M = 10, \psi_R = 0.4\pi$, $\sqrt{n^2+m^2} \leq 10$, $\phi = 0°, 30°$, and $45°$.

The second choice is to let

$$a^*(n,m) = \begin{cases} a^*(\sqrt{n^2+m^2}), & 0 \leq |n| \leq 10, 0 \leq |m| \leq 10, \\ 0, & \text{elsewhere.} \end{cases} \tag{4.89}$$

In this case, all of the elements have non-zero weights. To illustrate the effect of the two boundaries, we consider the case in which $\psi_R^2 = (0.4\pi)^2$. Then,

$$a^*(n,m) = \frac{0.4\pi J_1\left(0.4\pi\sqrt{n^2+m^2}\right)}{2\pi\sqrt{n^2+m^2}}. \tag{4.90}$$

Beam pattern cuts at $\phi = 0, 30°$, and $45°$ are shown in Figure 4.22 for case 1 and in Figure 4.23 for case 2. We see that using non-zero weights for all of the elements provides better sidelobe behavior. It gives a closer approximation to circular symmetry for the sidelobes but slightly less main-lobe symmetry.

The cuts show the same Gibbs phenomenon as in the one-dimensional case. This leads us to the topic of circularly symmetric windows and circularly symmetric weightings.

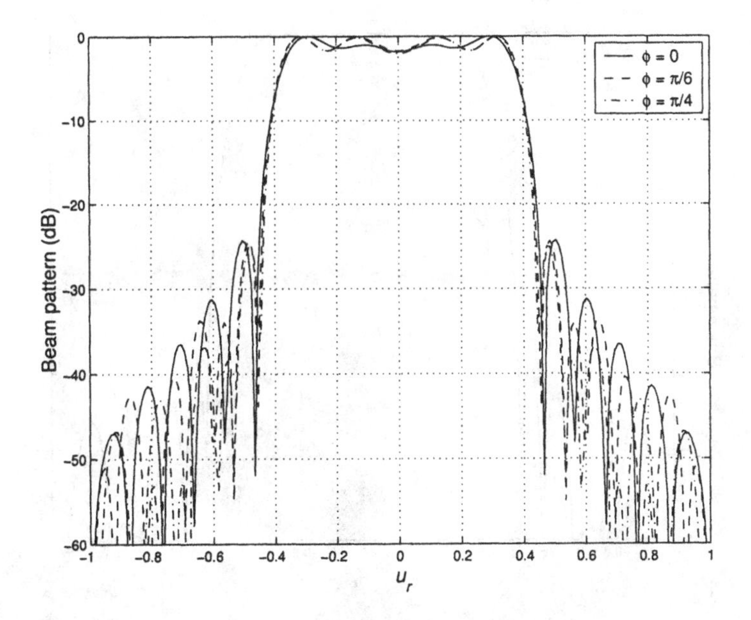

Figure 4.23 Beam pattern cut: standard rectangular grid array with square boundary: $N = M = 10, \psi_R = 0.4\pi, n \leq 10, m \leq 10$, $\phi = 0°, 30°$, and $45°$.

4.1.6 Circularly Symmetric Weighting and Windows

In Section 3.1, we developed a set of spectral weightings and then in Section 3.5 discussed their application as windows. We can extend these ideas into two dimensions in a straightforward manner.

We want the beam pattern associated with the weighting (or window) to approximate a circularly symmetric function. Huang [Hua81] showed that weightings (windows) of the form

$$w_{mn} = w_1\left(\sqrt{n^2+m^2}\right), \tag{4.91}$$

where $w_1(\cdot)$ is a continuous one-dimensional weighting (window), provide good results. Thus all of the weightings in Section 3.1 can be used for the 2-D case. As a simple example, we consider the example in Section 4.1.4.

Example 4.1.5 (continuation)

Assume that we use a Kaiser window

$$w_{K,nm} = \frac{I_0\left[\beta\sqrt{1-\left(\frac{\sqrt{n^2+m^2}}{14}\right)^2}\right]}{I_0(\beta)}, \quad 0 \leq n \leq 10, 0 \leq m \leq 10, \tag{4.92}$$

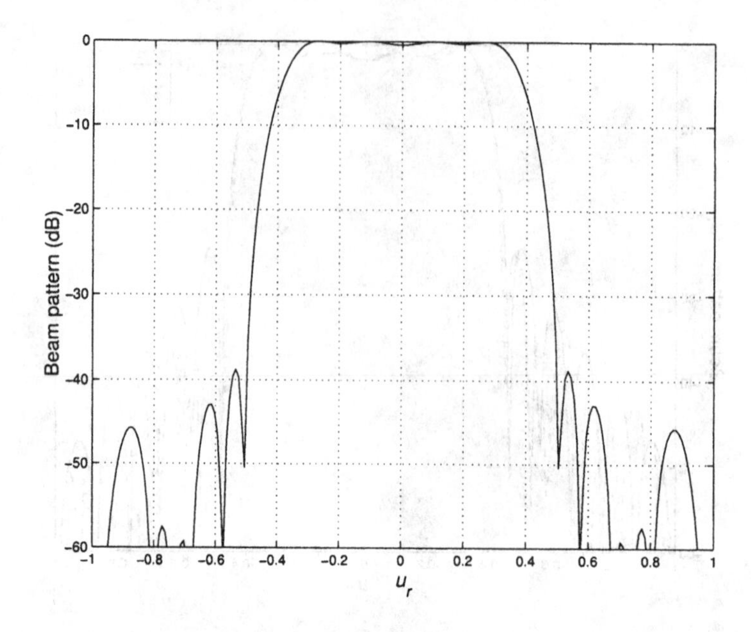

Figure 4.24 Beam pattern cut: standard rectangular grid array with square boundary: $N = M = 10, \psi_R = 0.4\pi$, Kaiser window, $\phi = 0°, 30°$, and $45°$.

where $\beta = 5.0$ and I_0 is the modified Bessel function of zero-order. The constant 14 comes from $\sqrt{200}$, corresponding to the corner element of the array. We apply the Kaiser window to $a^*(n,m)$ in (4.87). Figure 4.24 shows several beam pattern cuts. All three cuts are identical. Figure 4.25 shows a plot of the beam pattern versus u_x and u_y. We see that we have obtained a good approximation to circular symmetry over a significant part of the ψ_1, ψ_2 space.

Other weightings and windows are analyzed in the problems. In Section 4.3, we analyze circular apertures and develop some desirable beam patterns. We will revisit rectangular grid arrays at that point and see how well they can approximate these beam patterns.

4.1.7 Wavenumber Sampling and 2-D DFT

In Section 3.3.3, we saw that Woodward's approach to finding the weighting function for a linear array corresponded to a DFT relation. These ideas are readily extendible to two dimensions. The problem has been studied extensively in the FIR context and is referred to as the frequency sampling problem (e.g., Rabiner and Gold [RG75]).

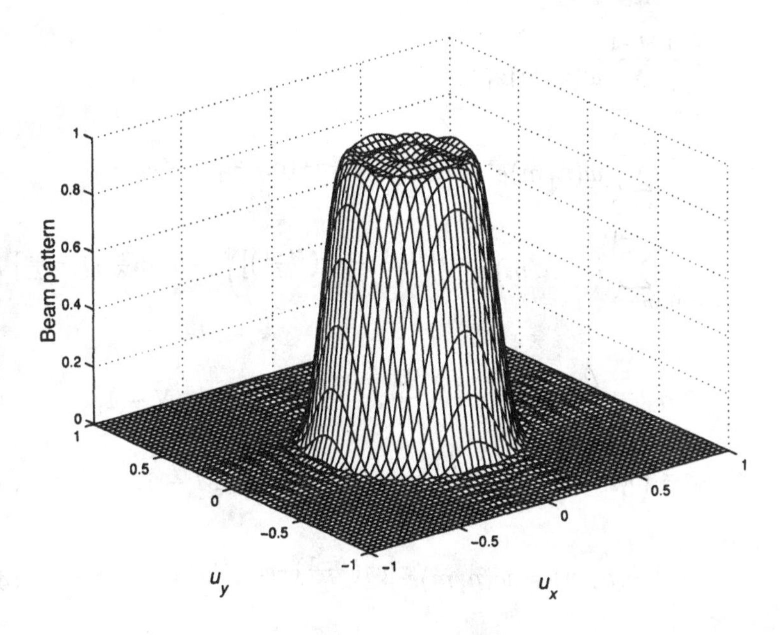

Figure 4.25 Magnitude of beam pattern versus u_x and u_y: standard rectangular grid array with square boundary, Kaiser window.

The z-transform of the array weighting function is

$$B_z(z_1, z_2) = \sum_{n=0}^{N-1} \sum_{m=0}^{M-1} w(n,m) z_1^{-n} z_2^{-m}. \tag{4.93}$$

The beam pattern is

$$B_{\psi}(\psi_x, \psi_y) = e^{-j\left(\frac{N-1}{2}\psi_x + \frac{M-1}{2}\psi_y\right)} \, B_z^*(z_1, z_2)\big|_{z_1=e^{j\psi_x}, z_2=e^{j\psi_y}}, \tag{4.94}$$

and

$$B_z(z_1, z_2)\big|_{z_1=e^{j\psi_x}, z_2=e^{j\psi_y}} = B_{\psi}^*(\psi_x, \psi_y) e^{-j\left(\frac{N-1}{2}\psi_x + \frac{M-1}{2}\psi_y\right)}. \tag{4.95}$$

We sample at

$$z_1 = e^{j\left(k_1 - \frac{N-1}{2}\right)\frac{2\pi}{N}}, \quad k_1 = 0, \cdots, N-1, \tag{4.96}$$

and

$$z_2 = e^{j\left(k_2 - \frac{M-1}{2}\right)\frac{2\pi}{M}}, \quad k_2 = 0, \cdots, M-1. \tag{4.97}$$

This corresponds to samples that are symmetric about the origin. Then,

$$B_{\psi}^*(\psi_{xk_1}, \psi_{yk_2}) e^{-j\left(\frac{N-1}{2}\psi_{xk_1} + \frac{M-1}{2}\psi_{yk_2}\right)}$$

$$= \sum_{n=0}^{N-1}\sum_{m=0}^{M-1} w(n,m) z_1^{-n} z_2^{-m}$$

$$= \sum_{n=0}^{N-1}\sum_{m=0}^{M-1} w(n,m) e^{-j[(k_1-\frac{N-1}{2})\frac{2\pi}{N}n+(k_2-\frac{M-1}{2})\frac{2\pi}{M}m]}$$

$$= \sum_{n=0}^{N-1}\sum_{m=0}^{M-1} \left(w(n,m)e^{j[n\pi(\frac{N-1}{N})+m\pi(\frac{M-1}{M})]}\right) e^{-j(k_1 n\frac{2\pi}{N}+k_2 m\frac{2\pi}{M}]}, \quad (4.98)$$

where

$$\psi_{xk_1} = \left(k_1 - \frac{N-1}{2}\right)\frac{2\pi}{N}, \quad k_1 = 0, 1, \cdots, N-1, \tag{4.99}$$

$$\psi_{yk_2} = \left(k_2 - \frac{M-1}{2}\right)\frac{2\pi}{M}, \quad k_2 = 0, 1, \cdots, M-1. \tag{4.100}$$

Now define

$$b(n,m) = w(n,m)e^{j[n\pi(\frac{N-1}{N})+m\pi(\frac{M-1}{M})]}, \tag{4.101}$$

and

$$B(k_1,k_2) = B_\psi^*(\psi_{xk_1},\psi_{yk_2})e^{-j(\frac{N-1}{2}\psi_{xk_1}+\frac{M-1}{2}\psi_{yk_2})}, \quad \begin{array}{l} k_1 = 0, 1, \cdots, N-1, \\ k_2 = 0, 1, \cdots, M-1. \end{array} \tag{4.102}$$

Then,

$$B(k_1,k_2) = \sum_{n=0}^{N-1}\sum_{m=0}^{M-1} b(n,m)e^{-j(k_1 n\frac{2\pi}{N}+k_2 m\frac{2\pi}{M})}, \quad \begin{array}{l} k_1 = 0, 1, \cdots, N-1, \\ k_2 = 0, 1, \cdots, M-1, \end{array} \tag{4.103}$$

which is the 2-D DFT. The IDFT is

$$b(n,m) = \frac{1}{NM}\sum_{k_1=0}^{N-1}\sum_{k_2=0}^{M-1} B(k_1,k_2)e^{j(k_1 n\frac{2\pi}{N}+k_2 m\frac{2\pi}{M})}. \tag{4.104}$$

To find the weighting function, we:

(i) Sample the desired beam pattern to obtain

$$B_\psi^*\left(\left(k_1 - \frac{N-1}{2}\right)\frac{2\pi}{N}, \left(k_2 - \frac{M-1}{2}\right)\frac{2\pi}{M}\right),$$

$$k_1 = 0, \cdots, N-1, k_2 = 0, \cdots, M-1$$

(ii) Use (4.102) to find $B(k_1, k_2)$

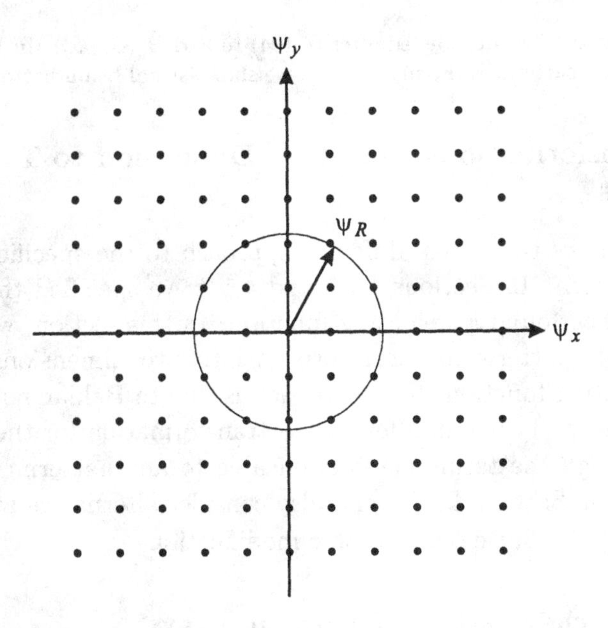

Figure 4.26 Sampling grid in wavenumber space.

(iii) Use (4.104) to find $b(n,m)$

(iv) Use (4.101) to find $w(n,m)$

(v) Use (4.1) to find $\hat{B}_d(\psi_1,\psi_2)$

If the desired beam pattern was synthesizable by a standard planar array, then $\hat{B}_d(\psi_1,\psi_2)$ will equal $B_d(\psi_1,\psi_2)$. In this case, the choice of the initial sampling point is unimportant.

For other desired beam patterns, the choice of the initial sampling point will affect the accuracy of $\hat{B}_d(\psi_1,\psi_2)$ and it may be useful to investigate several options (recall the discussion in Section 3.3.3).

We consider a simple example to demonstrate the technique.

Example 4.1.6

Consider the 11 × 11 sampling grid in ψ-space shown in Figure 4.26. The samples are at $\left(\left(k_1 - \frac{N-1}{2}\right)\frac{2\pi}{N}, \left(k_2 - \frac{M-1}{2}\right)\frac{2\pi}{M}\right)$, $0 \le k_1 \le N-1$, $0 \le k_2 \le M-1$. The desired beam pattern is constant over a circle of radius

$$\psi_R = \frac{2\pi}{N}\sqrt{5}. \tag{4.105}$$

We set

$$B^*_{\psi}(k_1-5,k_2-5) = \begin{cases} 1, & 13 \text{ interior samples,} \\ \frac{1}{2}, & 8 \text{ samples on edge,} \\ 0, & \text{all remaining samples.} \end{cases} \tag{4.106}$$

We carry out the five steps listed after (4.104) to find $\hat{B}_d(\psi_x, \psi_y)$. In Figure 4.27(a), we show the beam pattern. In Figure 4.27(b), we show several beam pattern cuts.

4.1.8 Transformations from One Dimension to Two Dimensions

In this section, we consider a different approach to the specification of the 2-D beam pattern. In Sections 4.1.5 and 4.1.7, we specified the 2-D beam pattern and then found a weighting function. In this section, we start with a desirable 1-D pattern and transform it into two dimensions. We then find the weighting function. The approach is due to Baklanov [Bak66] and Tseng and Cheng [TC68] and develops a transformation for the Chebychev pattern, although the technique is applicable to any pattern. We develop this technique in Section 4.1.8.1 and also consider alternative approaches.

In Section 4.1.8.2, we discuss some modifications to the technique.

4.1.8.1 Chebychev patterns for planar arrays[4]

In this section we consider the planar array shown in Figure 4.5. There are N^2 identical elements. The interelement spacing is d_x in the x-direction and d_y in the y-direction, so that the array is not necessarily square.

We assume that the weights are real and symmetrical about the x and y axes and the elements are cophasal in the direction of scan (θ_0, ϕ_0).

The beam pattern is

$$\begin{aligned} B_e(\theta, \phi) &= B_{\psi}(\psi_x, \psi_y) \\ &= 4 \sum_{m=1}^{N} \sum_{n=1}^{N} a_{mn} \cos\left[(2m-1)\frac{\psi_x}{2}\right] \cos\left[(2n-1)\frac{\psi_y}{2}\right] \end{aligned} \tag{4.107}$$

for an even number of elements in each row and column, and

$$\begin{aligned} B_o(\theta, \phi) &= B_{\psi}(\psi_x, \psi_y) \\ &= \sum_{m=1}^{N+1} \sum_{n=1}^{N+1} \epsilon_m \, \epsilon_n \, a_{mn} \cos\left[(2m-1)\frac{\psi_x}{2}\right] \\ &\quad \cdot \cos\left[(2n-1)\frac{\psi_y}{2}\right] \end{aligned} \tag{4.108}$$

[4] Our discussion follows Tseng and Cheng [TC68].

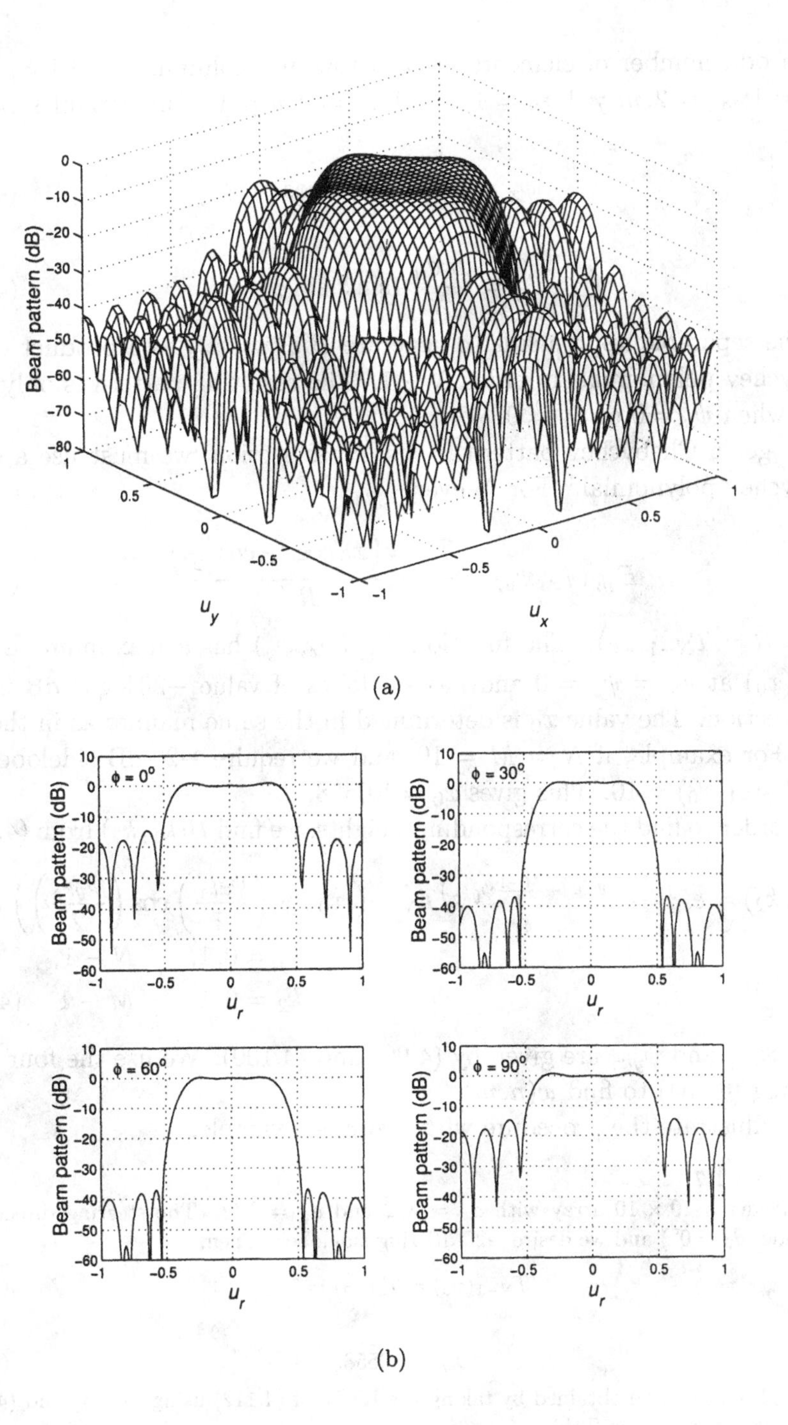

Figure 4.27 Beam pattern synthesized by pattern sampling.

for an odd number of elements in each row and column. In (4.108) $\epsilon_m = 1, m = 1; \epsilon_m = 2, m \neq 1; \epsilon_n = 1, n = 1; \epsilon_n = 2, n \neq 1$. The variables ψ_x and ψ_y are

$$\psi_x = \frac{2\pi d_x}{\lambda} (\sin\theta \cos\phi) \tag{4.109}$$

and

$$\psi_y = \frac{2\pi d_y}{\lambda} (\sin\theta \sin\phi). \tag{4.110}$$

The separable approach is to generate $B_{\psi}(\psi_x, \psi_y)$ as a product of two Chebychev polynomials of order $N-1$. The resulting pattern is only optimum when $\psi_x = 0$ or $\psi_y = 0$.

To get a Chebychev pattern in any cross section, we must use a single Chebychev polynomial.[5] For N even,

$$B_{\psi}(\psi_x, \psi_y) = \frac{T_{N-1}\left(x_0 \cos\frac{\psi_x}{2} \cos\frac{\psi_y}{2}\right)}{R}, \tag{4.111}$$

where $R = T_{N-1}(x_0)$. The function $B_{\psi}(\psi_x, \psi_y)$ has a maximum value of $T_{N-1}(x_0)$ at $\psi_x = \psi_y = 0$ and has sidelobes of value $-20\log R$ dB in any cross section. The value x_0 is determined in the same manner as in the 1-D case. For example, if $N = M = 10$, and we require -20-dB sidelobes, we solve $T_{N-1}(x_0) = 10$. This gives $x_0 = 1.0558$.

In order to find the corresponding weights, we find $B(k_1, k_2)$ from (4.102),

$$B(k_1, k_2) = e^{-j\left(\frac{N-1}{2}\psi_{xk_1} + \frac{M-1}{2}\psi_{yk_2}\right)} T_{N-1}\left(x_0 \cos\left(\frac{\psi_{xk_1}}{2}\right) \cos\left(\frac{\psi_{yk_2}}{2}\right)\right) R^{-1},$$
$$k_1 = 0, 1, \cdots, N-1,$$
$$k_2 = 0, 1, \cdots, M-1. \tag{4.112}$$

where ψ_{xk_1} and ψ_{yk_2} are given by (4.99) and (4.100). We use the four steps following (4.104) to find $w(n, m)$.

We illustrate the procedure with a simple example.

Example 4.1.7

Consider a 10×10 array with $d_x = \lambda/2$ and $d_y = \lambda/2$. The steering direction is broadside ($\theta_0 = 0°$) and we desire -20-dB ring sidelobes. Then,

$$T_{N-1}(x_0) = T_9(x_0) = 10, \tag{4.113}$$

and

$$x_0 = 1.0558. \tag{4.114}$$

The weightings can be obtained by taking the IDFT of (4.112) using (4.104) and (4.102). The results are shown in Table 4.2.

---

[5]This result is due to Baklanov [Bak66].

Table 4.2 Weightings Obtained from (4.112): $w_{mn}, m = 6, \cdots, 10, n = 6, \cdots, 10$

w_{mn}	6	7	8	9	10
6	0.7725	0.5686	0.7961	0.0294	1.0000
7	0.5686	0.9461	0.1186	0.6176	0.6667
8	0.7961	0.1186	0.4859	0.7773	0.2857
9	0.0294	0.6176	0.7773	0.3866	0.0714
10	1.0000	0.6667	0.2857	0.0714	0.0079

The 2-D Dolph-Chebychev pattern is shown in Figure 4.28(a). Several pattern cuts are shown in Figure 4.28(b). We see that they exhibit the desired Dolph-Chebychev behavior. Similar results are available when N is odd.

4.1.8.2 Modified transformations

The transformation in Section 4.1.8.1 used a $\cos(m\psi/2)$ term as a starting point because of the Chebychev beam pattern.

In many other cases when the 1-D beam pattern is real and symmetric we can write it as,

$$B_\psi(\psi) = \sum_{m=0}^{\frac{N-1}{2}} \alpha_m \cos(m\psi), \quad N \text{ odd}, \tag{4.115}$$

where, from Figure 3.22 and Table 3.2,

$$\alpha_m = \begin{cases} a_0, & \\ 2a_m, & 1 \leq m \leq \frac{N-1}{2}. \end{cases} \tag{4.116}$$

From our discussion in Section 3.5, we can write

$$B_\psi(\psi) = \sum_{m=0}^{\frac{N-1}{2}} \tilde{\alpha}_m (\cos\psi)^m, \tag{4.117}$$

where $\tilde{\alpha}_m$ and α_m are related by the Chebychev polynomials (3.133). We create a 2-D beam pattern by using the transformation

$$\cos\psi = \cos\psi_x \cos\psi_y. \tag{4.118}$$

This transformation is a special case of transformation due to McClellan [McC82] and is a generalization of the transformation used by Baklanov [Bak66] and Tseng and Cheng [TC68].

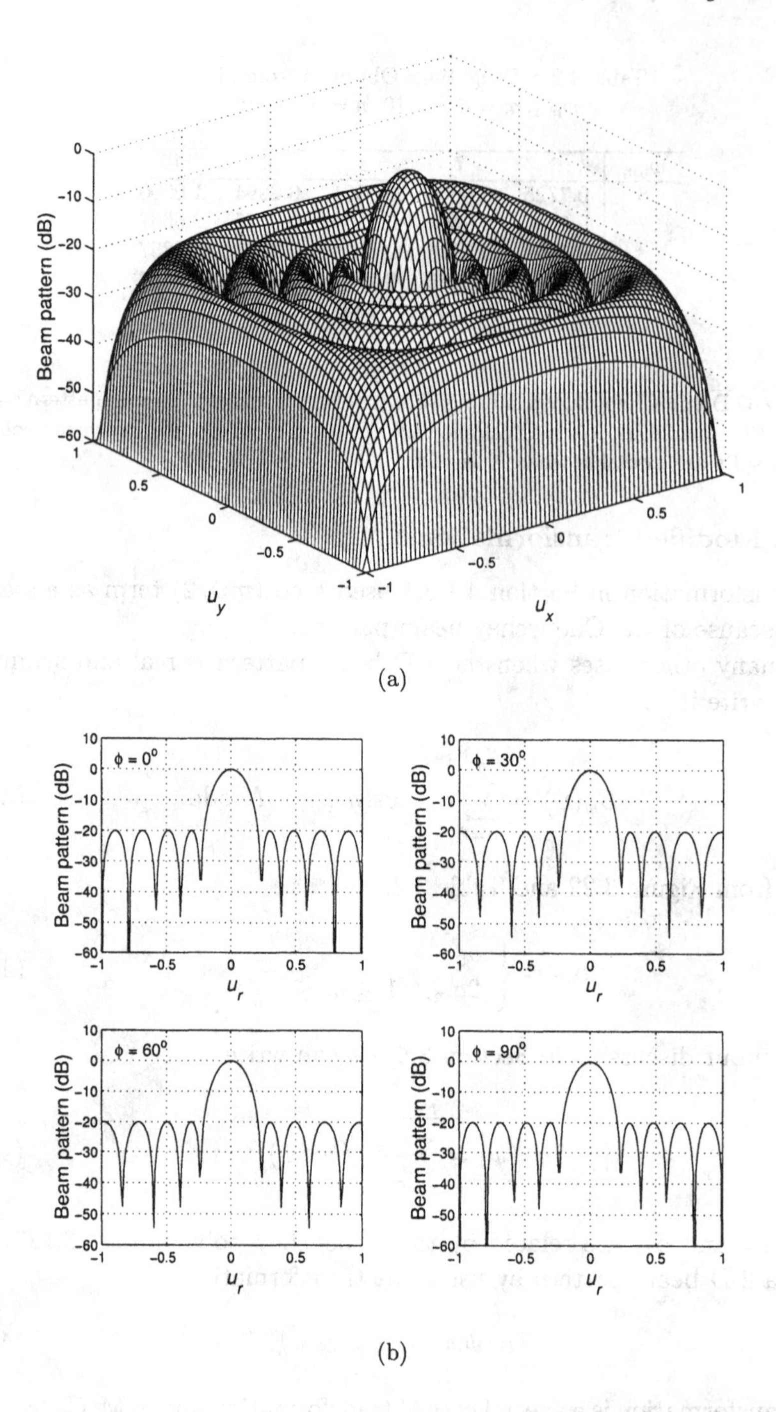

Figure 4.28 Four ϕ-cuts of the beam pattern of a standard rectangular grid array with rectangular boundary; $N = M = 10$; Tseng-Cheng distribution, -20-dB sidelobe level.

The resulting 2-D beam pattern is

$$B_{\psi}(\psi_x, \psi_y) = \sum_{m=0}^{\frac{N-1}{2}} \tilde{\alpha}_m (\cos\psi_x \cos\psi_y)^m . \tag{4.119}$$

Example 4.1.8

Consider an 11-element rectangular array with $d_x = d_y = \lambda/4$. We use a Riblet weighting with -20-dB sidelobes (e.g., (3.162)–(3.168) and Figure 3.28) and the transformation in (4.117)–(4.119). The resulting beam pattern is shown in Figure 4.29(a). Several pattern cuts are shown in Figure 4.29(b). We see that the transformation provides the desired 2-D beam pattern.

These two transformations, (4.111) and (4.119), provide an effective procedure for generating desirable 2-D beam patterns. We have demonstrated them for the Dolph-Chebychev and Riblet-Chebychev patterns, but the general technique is applicable to a large class of 1-D patterns.

4.1.9 Null Steering

Our basic discussion on null constraints in Section 3.7 was valid for arbitrary array geometries. We then considered several examples using linear arrays. We now consider the application to planar arrays. For notational simplicity, we will only consider pattern nulls and not derivative nulls. The derivative null case follows in a straightforward manner (see Problem 4.1.22).

As in (4.50), we can write the array manifold vector for the mth column of the array in Figure 4.30 as

$$\mathbf{v}_m(\boldsymbol{\psi}) = e^{-j\left(\frac{N-1}{2}\psi_x + \frac{M-1}{2}\psi_y\right)} \begin{bmatrix} e^{jm\psi_y} & e^{j(\psi_x + m\psi_y} & \cdots & e^{j((N-1)\psi_x + m\psi_y)} \end{bmatrix}^T . \tag{4.120}$$

We then create an $NM \times 1$ array manifold vector, as in (4.54),

$$vec\left[\mathbf{V}_{\boldsymbol{\psi}}(\boldsymbol{\psi})\right] = \begin{bmatrix} \mathbf{v}_0(\boldsymbol{\psi}) \\ \hline \vdots \\ \hline \mathbf{v}_{M-1}(\boldsymbol{\psi}) \end{bmatrix}, \tag{4.121}$$

where $\mathbf{v}_m(\boldsymbol{\psi})$ is the array manifold for the mth column, as shown in Figure 4.30 for M odd. The vector $\boldsymbol{\psi}$ is

$$\boldsymbol{\psi} = \begin{bmatrix} \psi_x \\ \psi_y \end{bmatrix}, \tag{4.122}$$

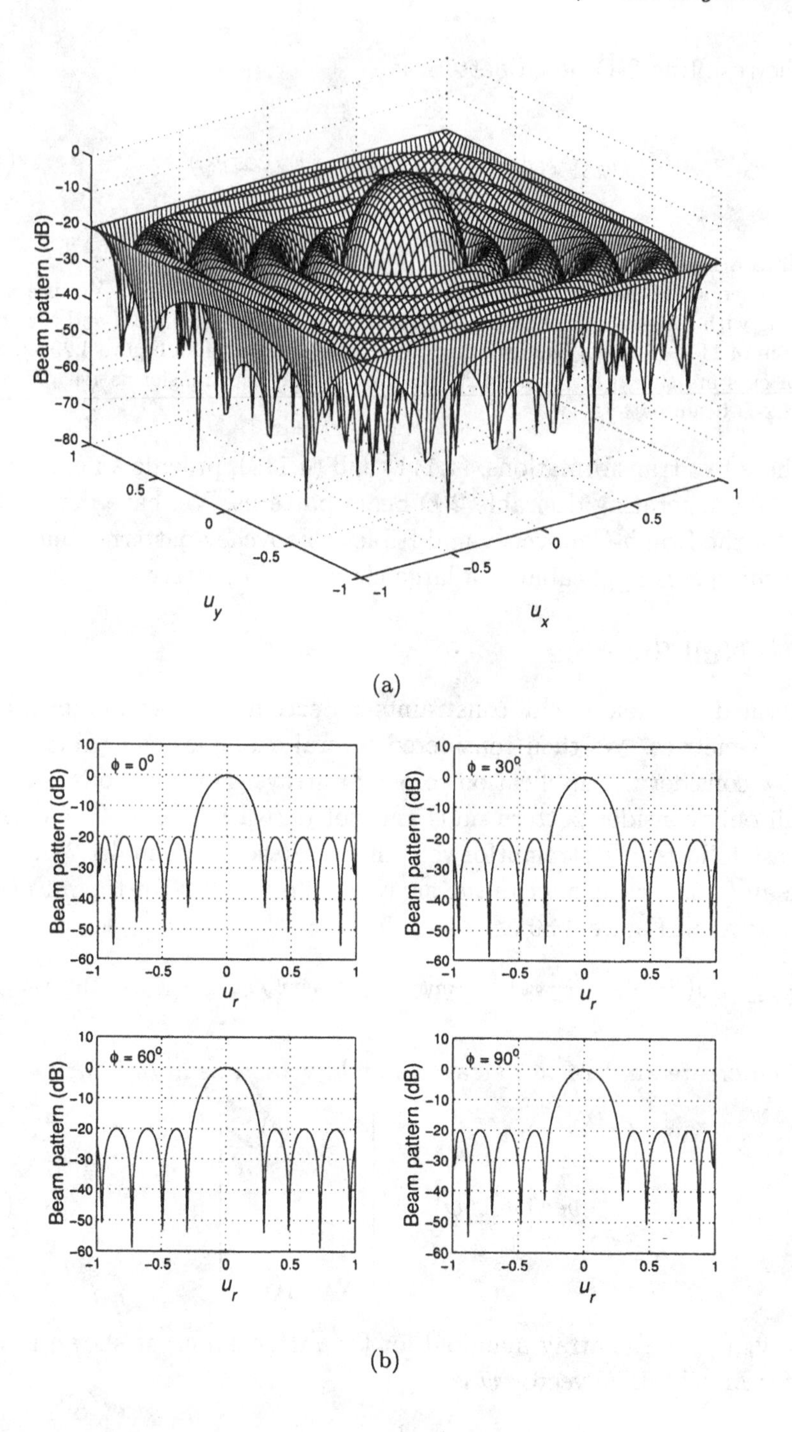

Figure 4.29 Beam pattern with 2-D Riblet-Chebychev weighting: (a) 2-D beam pattern; (b) pattern cuts: $d_x = d_y = \lambda/4$, sidelobes, -20 dB.

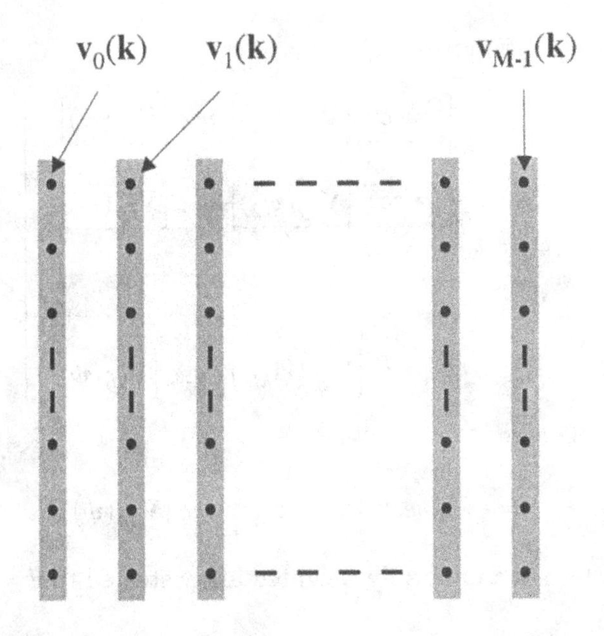

Figure 4.30 Array manifold for rectangular array.

and ψ_x and ψ_y are given in (4.2) and (4.3).

The null constraint matrix is (from (3.251)),

$$\mathbf{C}_0 = \left[\; vec\left[\mathbf{V}_{\psi}(\psi_1)\right] \;\vdots\; vec\left[\mathbf{V}_{\psi}(\psi_2)\right] \;\vdots\; \cdots \;\vdots\; vec\left[\mathbf{V}_{\psi}(\psi_{M_0})\right] \; \right], \tag{4.123}$$

and has dimensions $NM \times M_0$ where M_0 is the number of null constraints.

We assume that the desired weighting is given by an $NM \times 1$ weight vector $\mathbf{w}_d$. Then (3.268) applies directly,

$$\mathbf{w}_0^H = \mathbf{w}_d^H - \left[\mathbf{w}_d^H \, \mathbf{C}_0 \left[\mathbf{C}_0^H \, \mathbf{C}_0\right]^{-1} \mathbf{C}_0^H\right]. \tag{4.124}$$

We use (4.124) to compute the weight vector. The resulting beam pattern is,

$$B_0(\psi) = B_d(\psi) - \mathbf{a}\,\mathbf{C}_0^H \, vec\left[\mathbf{V}_{\psi}(\psi)\right]. \tag{4.125}$$

Using (4.123) in (4.125) gives,

$$B_0(\psi) = B_d(\psi) - \mathbf{a} \begin{bmatrix} vec^H\left[\mathbf{V}_{\psi}(\psi_1)\right] vec\left[\mathbf{V}_{\psi}(\psi)\right] \\ \hdashline vec^H\left[\mathbf{V}_{\psi}(\psi_2)\right] vec\left[\mathbf{V}_{\psi}(\psi)\right] \\ \hdashline \vdots \\ \hdashline vec^H\left[\mathbf{V}_{\psi}(\psi_{M_0})\right] vec\left[\mathbf{V}_{\psi}(\psi)\right] \end{bmatrix}. \tag{4.126}$$

Just as in the 1-D case,

$$B_c(\psi : \psi_m) = \frac{1}{NM} vec^H\left[\mathbf{V}_{\psi}(\psi_m)\right] vec\left[\mathbf{V}_{\psi}(\psi)\right], \tag{4.127}$$

is the beam pattern of a uniformly weighted array steered to $\psi = \psi_m$. Thus,

$$B_0(\psi) = B_d(\psi) - \sum_{m=1}^{M_0} a_m NM B_c(\psi : \psi_m). \tag{4.128}$$

The results in (4.126)–(4.128) are useful to understand the behavior, but (4.124) is used to compute $\mathbf{w}_0^H$.

We consider a simple example to illustrate the result.

Example 4.1.9

Consider a standard 10×10 array with Dolph-Chebychev weighting (−20-dB sidelobes) aimed at broadside. The spacing between elements is $d_x = \frac{\lambda}{2}, d_x = \frac{3\lambda}{4}$. We put a null at $\theta = -30°$ and $\phi = -60°$.

Then

$$\mathbf{C}_0 = vec\left[\mathbf{v}_{\psi}(\psi_m)\right], \tag{4.129}$$

with $\psi_{mx} = -0.25\pi$ and $\psi_{my} = -0.433\pi$. The resulting beam pattern is shown in Figure 4.31.

4.1.10 Related Topics

In this section we have developed the major results for rectangular arrays. There are several topics of interest that have been omitted:

(i) Beamspace processing: the techniques in Section 3.10 can be extended to rectangular arrays in a straightforward manner.

(ii) The beam pattern design algorithms in Section 3.9.3 can be extended to rectangular arrays.

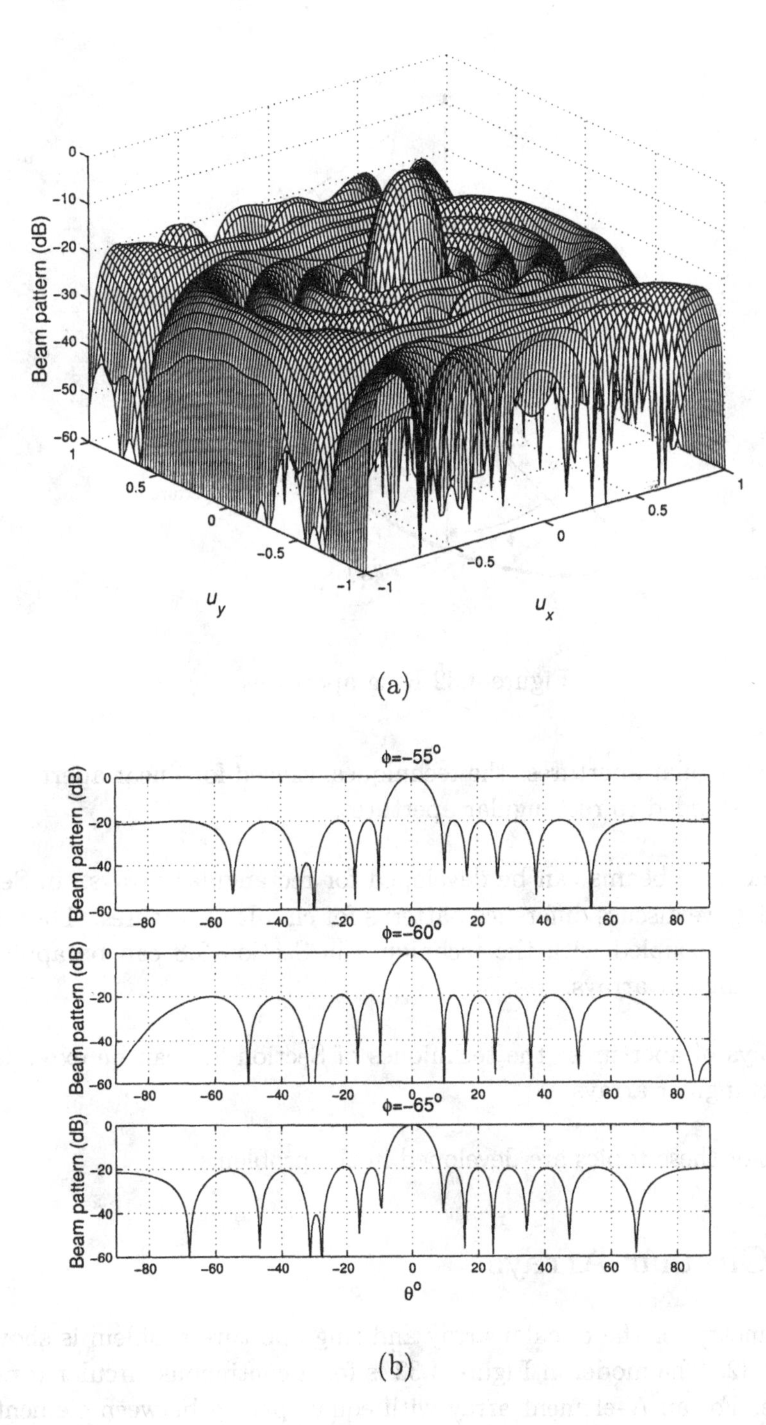

Figure 4.31 Beam pattern of Chebychev array with null.

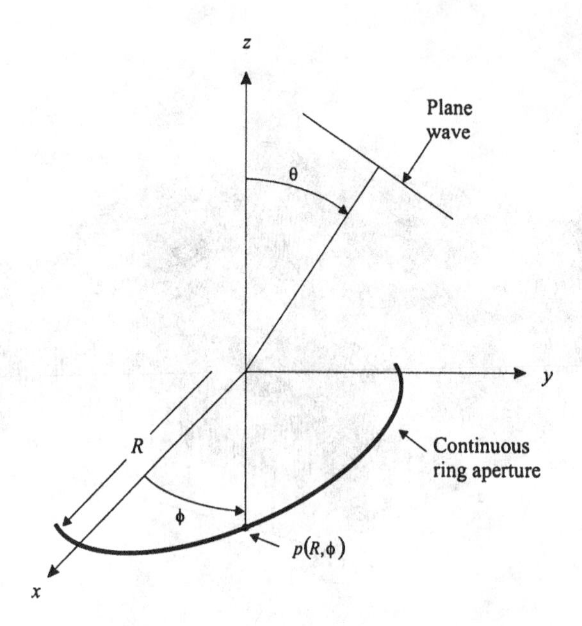

Figure 4.32 Ring apertures.

(iii) Rectangular apertures: the techniques utilized for linear apertures can be extended to rectangular apertures.

(iv) Difference beams can be developed for rectangular arrays. In Section 4.3.4, we discuss difference patterns for circular apertures. These techniques coupled with the techniques in Section 3.8 can be applied to rectangular arrays.

(v) Arrays of apertures: the techniques of Section 2.8 can be extended to rectangular arrays.

Most of these topics are developed in the problems.

4.2 Circular Arrays

The geometry for the circular array and ring aperture problem is shown in Figure 4.32. The model in Figure 4.32 is for a continuous circular (or ring) aperture. For an N-element array with equal spacing between elements we sample the ring around the circumference. We begin our discussion with the continuous ring aperture.

4.2.1 Continuous Circular Arrays (Ring Apertures)

In this section we consider continuous circular arrays. These are also referred to as ring apertures and are shown in Figure 4.4(c)[6] and Figure 4.32.

The first step is to find $\mathbf{v}(\mathbf{k})$ as defined in (2.28). The wavenumber is

$$\mathbf{k} = -\frac{2\pi}{\lambda} \begin{bmatrix} \sin\theta\,\cos\phi \\ \sin\theta\,\sin\phi \\ \cos\theta \end{bmatrix}, \tag{4.130}$$

for a plane wave arriving from $[\theta, \phi]$. The position vector is

$$\mathbf{p}_{\phi_1} = R \begin{bmatrix} \cos\phi_1 \\ \sin\phi_1 \\ 0 \end{bmatrix}, \tag{4.131}$$

for an element at $[R, \phi_1]$. Thus,

$$\begin{aligned} \mathbf{k}^T \mathbf{p}_{\phi_1} &= -\frac{2\pi}{\lambda} R \sin\theta\,[\cos\phi\,\cos\phi_1 + \sin\phi\,\sin\phi_1] \\ &= -\frac{2\pi}{\lambda} R \sin\theta\,[\cos(\phi - \phi_1)]. \end{aligned} \tag{4.132}$$

The frequency wavenumber response is

$$\Upsilon(\omega : \mathbf{k}) = \int_0^{2\pi} w(\phi_1)\, e^{j\frac{2\pi}{\lambda} R \sin\theta[\cos(\phi-\phi_1)]}\, R\, d\phi_1. \tag{4.133}$$

Writing (4.133) as a beam pattern

$$\begin{aligned} B(\theta, \phi) &= \int_0^{2\pi} w(\phi_1) e^{j\frac{2\pi}{\lambda} R \sin\theta[\cos(\phi-\phi_1)]}\, R\, d\phi_1 \\ &= \int_0^{2\pi} w(\phi_1) e^{jk_0 R \sin\theta[\cos(\phi-\phi_1)]} R d\phi_1, \end{aligned} \tag{4.134}$$

where $k_0 = |k| = 2\pi/\lambda$. Since any weighting function will be periodic in ϕ, we expand the aperture weighting function in a Fourier series.

$$w(\phi) = \sum_{m=-\infty}^{\infty} w'_m e^{jm\phi}, \tag{4.135}$$

where

$$w'_m = \frac{1}{2\pi} \int_0^{2\pi} w(\phi)\, e^{-jm\phi}\, d\phi. \tag{4.136}$$

---

[6] [Ma74] has a complete discussion of circular geometries. [Bag76] has this example.

Each term is called a **phase mode excitation** of the aperture.

Then, (4.134) can be written as

$$\begin{aligned} B(\theta,\phi) &= \sum_{m=-\infty}^{\infty} w'_m R \int_0^{2\pi} e^{j[k_0 R \sin\theta \cos(\phi-\phi_1)+m\phi_1]}\, d\phi_1 \\ &= 2\pi R \sum_{m=-\infty}^{\infty} w'_m j^m J_m\left(k_0 R \sin\theta\right) e^{jm\phi}, \end{aligned} \tag{4.137}$$

where $J_m(x)$ is the Bessel function of the first kind of order m.

It is convenient to normalize the weights so that when $w_0 = 1$, $B(0,\phi)$ will equal 1.

$$w'_m = \frac{1}{2\pi R} w_m. \tag{4.138}$$

We denote the component of the beam pattern due to the mth term as $B_m(\theta,\phi)$:

$$B_m(\theta,\phi) = w_m j^m J_m\left(k_0 R \sin\theta\right) e^{jm\phi}. \tag{4.139}$$

We see that each phase mode excitation term gives rise to a spatial harmonic in the beam pattern. This correspondence means that we can take a desired beam pattern in the ϕ-dimension, decompose it into its Fourier components, and separately excite each of the Fourier components with a phase mode term. The weighting must take into account the appropriate Bessel function term. The observation was made by Davies [Dav65], [R+83] and can be used to develop effective pattern synthesis techniques.

First, consider the case of uniform weighting,

$$w_0 = 1 \tag{4.140}$$

and

$$w_m = 0, \quad m \neq 0. \tag{4.141}$$

The pattern is uniform in ϕ and the main response axis is perpendicular to the plane of the aperture.

$$B(\theta,\phi) = J_0\left(k_0 R \sin\theta\right) = J_0(\psi_R), \tag{4.142}$$

where

$$\psi_R = \frac{2\pi}{\lambda} R \sin\theta = 2\pi R_\lambda \sin\theta, \tag{4.143}$$

and R_λ is the radius measured in wavelengths.

For a ring aperture, $J_0(\cdot)$ plays the same role as sinc$(\cdot)$ did for linear aperture. The two functions are shown in Figure 4.33. The first zero of

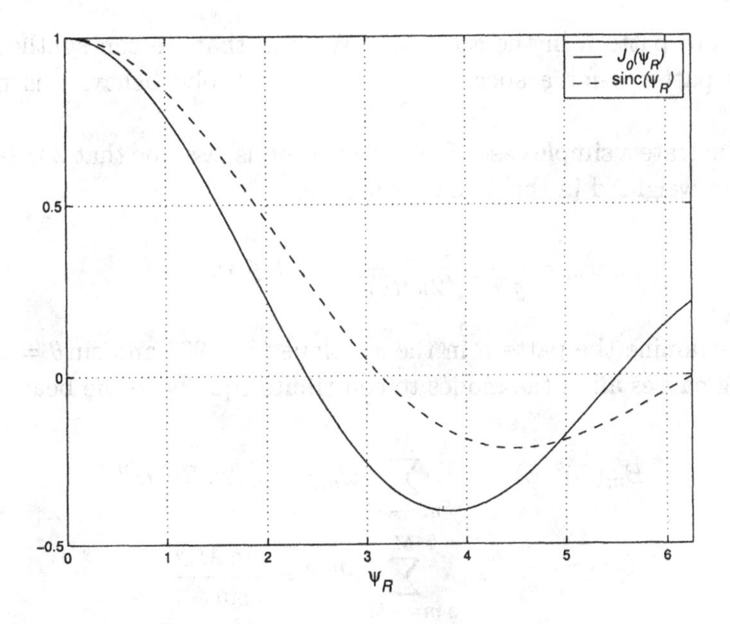

Figure 4.33 Comparison of $J_0(\psi_R)$ and sinc (ψ_R).

$J_0(\psi_R)$ is at $\psi_R = 2.4$. The first sidelobe level (SLL) occurs at $\psi_R = 3.8$ and its height is -7.9 dB. These values can be compared to a linear aperture. The function sinc(ψ_R) has its first zero at $\psi_R = \pi$ and its first sidelobe occurs at $\psi_R = 4.7$ with a height of -13.6 dB.

From (4.143),

$$\theta = \sin^{-1}\left(\frac{\psi_R}{2\pi R_\lambda}\right), \tag{4.144}$$

and the first null is at

$$\theta_{NULL} = \sin^{-1}\left(\frac{2.4}{2\pi R_\lambda}\right). \tag{4.145}$$

As expected, the beamwidth decreases as the radius increases.

The visible region corresponds to

$$0 \leq |\sin\theta| \leq 1, \tag{4.146}$$

or

$$0 \leq \psi_R \leq 2\pi R_\lambda. \tag{4.147}$$

In order to generate a beam pattern with ϕ-dependence we need to utilize phase modes with $m \neq 0$. In many applications, we are primarily interested in the beam pattern in the xy-plane. We find that we can synthesize satisfactory patterns for a specific θ_0, but the sidelobe behavior is poor for $\theta \neq \theta_0$.

To illustrate a simple case of pattern synthesis, assume that $2M+1$ phase modes are weighted in the following manner:

$$w_m = \frac{1}{j^m J_m(2\pi R_\lambda)}, \quad -M \leq m \leq M. \tag{4.148}$$

We first examine the pattern in the xy-plane ($\theta = 90°$ and $\sin\theta = 1$). This weighting causes all of the modes to contribute equally to the beam pattern

$$\begin{aligned} B_m(90°, \phi) &= \sum_{m=-M}^{M} w_m j^m J_m\left(2\pi R_\lambda\right) e^{jm\phi} \\ &= \sum_{m=-M}^{M} e^{jm\phi} = \frac{\sin M\phi}{\sin\phi}. \end{aligned} \tag{4.149}$$

The resulting pattern is the conventional $\sin M\phi/\sin\phi$ pattern in ϕ-space. Note that the pattern extends over 360° and the argument is ϕ. This is in contrast to the ULA, which extends over 180° and the argument is $\psi = \pi\cos\theta$ (for an array along the z-axis).

The amplitude and elevation dependence of each phase mode is governed by the corresponding Bessel function.

For the weighting in (4.148), the pattern for other values of θ is

$$B(\theta, \phi) = \sum_{m=-M}^{M} \frac{J_m\left(2\pi R_\lambda \sin\theta\right)}{J_m\left(2\pi R_\lambda\right)} e^{jm\phi}. \tag{4.150}$$

Using the relation,

$$J_{-m}(x) = (-1)^m J_m(x), \tag{4.151}$$

we can write (4.150) as

$$B(\theta, \phi) = \frac{J_0\left(2\pi R_\lambda \sin\theta\right)}{J_0\left(2\pi R_\lambda\right)} + \sum_{m=1}^{M} 2\frac{J_m\left(2\pi R_\lambda \sin\theta\right)}{J_m\left(2\pi R_\lambda\right)} \cos m\phi. \tag{4.152}$$

For small $|\theta|$, the beam pattern is still acceptable, but it degenerates rapidly as θ increases.

To determine how many modes can be excited for a ring aperture of radius R, we examine the behavior of the Bessel functions in the visible

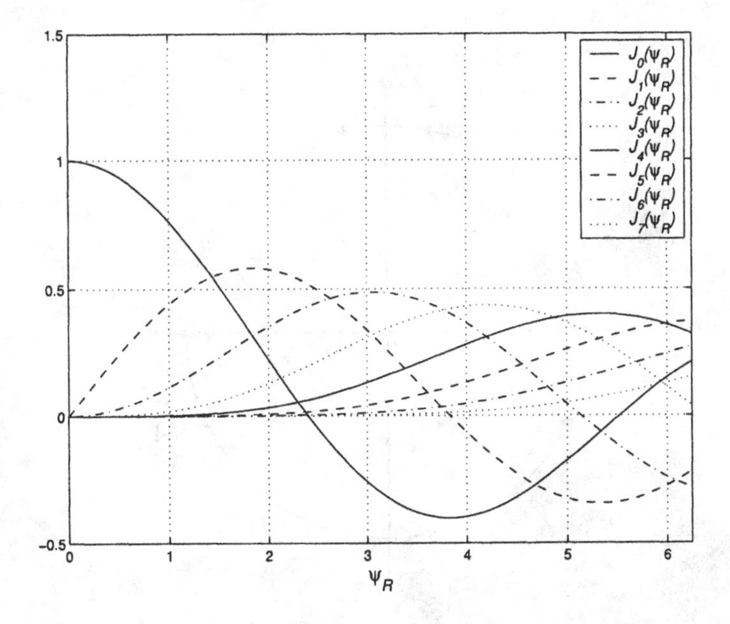

Figure 4.34 $J_0(\psi_R)$ through $J_7(\psi_R)$ versus ψ_R; $\psi_R = 2\pi R_\lambda \sin\theta$: $R_\lambda = 1$.

region. The first seven Bessel functions are plotted versus $x = 2\pi R_\lambda \sin\theta$ in Figure 4.34 for the case when

$$R_\lambda = 1. \tag{4.153}$$

In this case the visible region is $0 < x < 2\pi$. We see that the first six Bessel functions have a non-zero value in the visible region, and that J_7 has a slight non-zero value toward the upper end of the region.

In general, the amplitude is small when the order m exceeds the argument. Thus,

$$M \simeq 2\pi R_\lambda \tag{4.154}$$

is a useful limit. We can use $2M + 1$ phase modes.

In order to have each mode available, we need to choose R so that

$$J_m(2\pi R_\lambda) \neq 0. \tag{4.155}$$

One can show (e.g., p. 307 of [R$^+$83]) that if the diameter, $2R$, is an integral number of half-wavelengths, then (4.155) is satisfied.

We see that by using the phase mode excitation technique, for any given elevation angle θ, we can synthesize a desired pattern in ϕ corresponding to a $(2M+1)$-element linear array. Thus all of the design techniques in Chapter

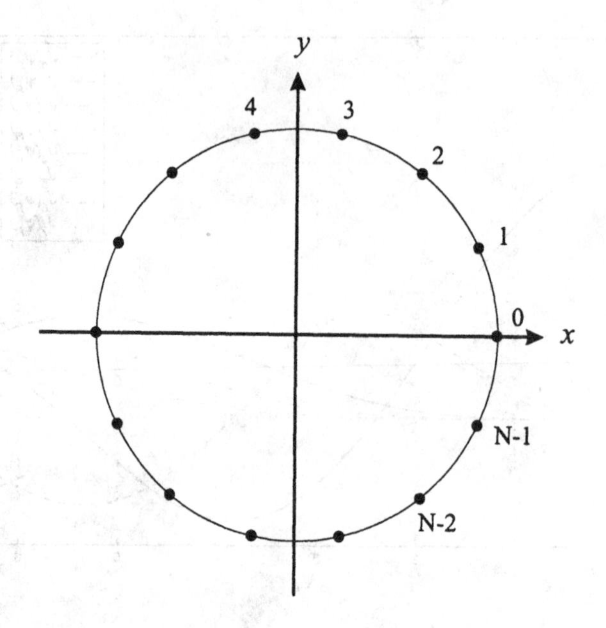

Figure 4.35 Geometry for circular array.

3 can be applied to the ring aperture. In Section 4.2.3, we discuss methods for implementing phase mode excitation beam patterns.

In practice, we normally use a circular array rather than a continuous aperture. We analyze its behavior in the next section.

4.2.2 Circular Arrays

The geometry of interest for the discrete circular array is shown in Figure 4.35. For isotropic elements, the array weighting will correspond to sampling the continuous aperture weighting. If we assume the aperture weighting generated M spatial harmonics, then a sampling theorem argument indicates that we need $(2M+1)$ elements to reproduce these spatial harmonics. Using (4.154), this condition implies

$$N \geq 2\left(\frac{2\pi R}{\lambda}\right) + 1, \tag{4.156}$$

which implies the spacing on the arc is

$$d_{cir} \leq \frac{\lambda}{2}. \tag{4.157}$$

To find the beam pattern, we use a sampling function. We assume that the highest spatial frequency that affects the beam pattern is,

$$M = 2\pi R_{\lambda}. \tag{4.158}$$

Then, the minimum sampling frequency is $4\pi R_{\lambda}$ and the sampling interval around the circle is

$$\phi_T = \frac{2\pi}{2\frac{2\pi}{\lambda}R} = \frac{\lambda}{2R}. \tag{4.159}$$

The sampling function is

$$S_{\phi}(\phi) = \sum_{n=-\infty}^{\infty} \delta(\phi - n\phi_T). \tag{4.160}$$

We can use the expression,

$$\sum_{n=-\infty}^{\infty} \delta(\phi - n\phi_T) = \frac{1}{\phi_T} \sum_{q=-\infty}^{\infty} e^{jqN\phi}, \tag{4.161}$$

and rewrite that sampling function as

$$\begin{aligned} S_{\phi}(\phi) &= \sum_{q=-\infty}^{\infty} e^{jNq\phi} \\ &= 1 + \sum_{q=1}^{\infty} e^{jNq\phi} + \sum_{q=1}^{\infty} e^{-jNq\phi}. \end{aligned} \tag{4.162}$$

Then, the weighting for the mth-order phase mode is,

$$w_m(\phi) = w_m e^{jm\phi} S_{\phi}(\phi). \tag{4.163}$$

Using (4.162) in (4.163) gives

$$w_m(\phi) = w_m e^{jm\phi} + w_m \sum_{q=1}^{\infty} e^{j(Nq+m)\phi} + w_m \sum_{q=1}^{\infty} e^{-j(Nq-m)\phi}. \tag{4.164}$$

The corresponding beam pattern for the mth mode is,

$$\begin{aligned} B_m(\theta, \phi) &= w_m j^m J_m(2\pi R_{\lambda} \sin\theta) e^{jm\phi} \\ &\quad + \sum_{g=1}^{\infty} w_m j^{-g} J_g(2\pi R_{\lambda} \sin\theta) e^{-jg\phi} \\ &\quad + \sum_{h=1}^{\infty} w_m j^h J_h(2\pi R_{\lambda} \sin\theta) e^{jh\phi}, \end{aligned} \tag{4.165}$$

where $g = (Nq - m)$ and $h = (Nq + m)$.

The first term is the desired beam pattern corresponding to the ring aperture and the remaining terms are residual distortion modes due to sampling. The first distortion mode has a Bessel function of order $(N - m)$. It will have negligible amplitude in the visible region if

$$N - m > \frac{2\pi R}{\lambda}. \tag{4.166}$$

However,

$$m \leq M \leq \frac{2\pi R}{\lambda}. \tag{4.167}$$

Thus, (4.166) implies that the distortion modes will be negligible if

$$N \geq \frac{4\pi R}{\lambda} \geq 2M. \tag{4.168}$$

The condition in (4.168) is satisfied if

$$d_{cir} \leq \frac{\lambda}{2}. \tag{4.169}$$

We can make the distortion modes as small as desired by increasing N. In Table 4.3,[7] we show the maximum residual contribution as a function of $(N - M)$ for $R_\lambda = 1$.

Table 4.3 Maximum Residual Contribution as a Function of N

N	13	14	15	16	17	18	19
$J_{N-M}(k_0 r)$	0.158	0.073	0.029	0.010	0.003	8.8e-4	2.3e-4

We see that for $N > 15$, the residual contribution would be approximately 0.01 ($d_{cir} = 0.42$). We will focus our attention in the text on arrays that satisfy (4.168) and (4.169).

The total pattern is

$$B(\theta, \phi) = \sum_{m=-\infty}^{\infty} B_m(\theta, \phi), \tag{4.170}$$

where $B_m(\theta, \phi)$ is given by (4.165). When the main response axis of the array is steered, it is convenient to define a new set of variables.[8] The beam pattern is

---

[7] From [MZ94].

[8] This is a reasonably standard derivation (e.g., [Ma74], p.192).

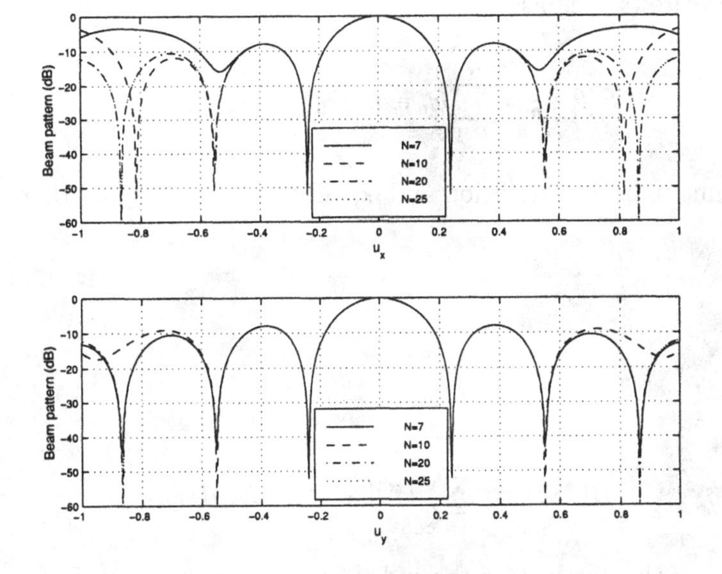

Figure 4.36 Beam patterns for a uniform circular array steered to broadside: (a) $\phi = 0$ and π; (b) $\phi = \pi/2$ and $3\pi/2$.

$$B(\theta, \phi) = \sum_{n=0}^{N-1} w_n \exp\left[jk_0 R \sin\theta \, \cos\left(\phi - \phi_n\right) + j\beta_n\right], \tag{4.171}$$

where β_n is the phase factor with respect to the origin. To align the MRA along (θ_0, ϕ_0),

$$\beta_n = -k_0 R \sin\theta_0 \, \cos\left(\phi_0 - \phi_n\right). \tag{4.172}$$

We now define a new set of variables,

$$\rho = R\left\{\left[\left(\sin\theta \, \cos\phi - \sin\theta_0 \, \cos\phi_0\right)^2 + \right.\right. \\ \left.\left. + \left(\sin\theta \, \sin\phi - \sin\theta_0 \, \sin\phi_0\right)^2\right]^{\frac{1}{2}}\right\}, \tag{4.173}$$

and

$$\cos\xi = \frac{\sin\theta \, \cos\phi - \sin\theta_0 \, \cos\phi_0}{\left[\left(\sin\theta \, \cos\phi - \sin\theta_0 \, \cos\phi_0\right)^2 + \left(\sin\theta \, \sin\phi - \sin\theta_0 \, \sin\phi_0\right)^2\right]^{\frac{1}{2}}}. \tag{4.174}$$

We can rewrite (4.171) as

$$B(\theta,\phi) = \sum_{n=1}^{N} w_n \exp\left[jk_0\rho\cos\left(\xi - \phi_n\right)\right]. \tag{4.175}$$

If we assume uniform excitation and equally spaced elements, then

$$w_n = \frac{1}{N}, \tag{4.176}$$

and

$$\phi_n = \frac{2\pi n}{N}, \tag{4.177}$$

$$B(\theta,\phi) = \sum_{m=-\infty}^{\infty} j^{mN} e^{-jmN\xi} J_{mN}(k_0\rho), \tag{4.178}$$

where mN is the product of the index m and the number of elements N. The term corresponding to $m = 0$, $J_0(k\rho)$ is the principal term and the other terms are the residuals. This expression has the same form as (4.165) and our previous discussion concerning residuals carries over. The behavior of the pattern can be examined by plotting (4.178).

The beam pattern of a uniform circular array with $2\pi R_\lambda = 10$ for the $\theta_0 = 0$ case is plotted in Figure 4.36 for $\phi = 0$ and $\phi = \frac{\pi}{2}$ and several values of N. In this case, (4.178) reduces to

$$B(\theta,\phi) = \sum_{m=-\infty}^{\infty} j^{mN} e^{-jmN\xi} J_{mN}\left(k_0 R\sin\theta\right). \tag{4.179}$$

The main-lobe behavior is adequately described by the J_0 term. The effect of finite N appears in the sidelobe structure.

In Figure 4.37, we show the beam pattern for a 20-element array with $2\pi R_\lambda = 2\pi R/\lambda = 10$ ($d_{cir} = \lambda/2$). In Figure 4.38 we show the vertical patterns along the planes $\phi = 0°$ (the left side of the plot in $\phi = 180°$). As N is increased for a fixed $2\pi R/\lambda$, the beam pattern approaches the beam pattern of the ring aperture.

4.2.3 Phase Mode Excitation Beamformers

Davies [R+83], [Dav65] showed how to excite the phase modes using a Butler beamforming matrix. Our approach is similar to his original work. In order

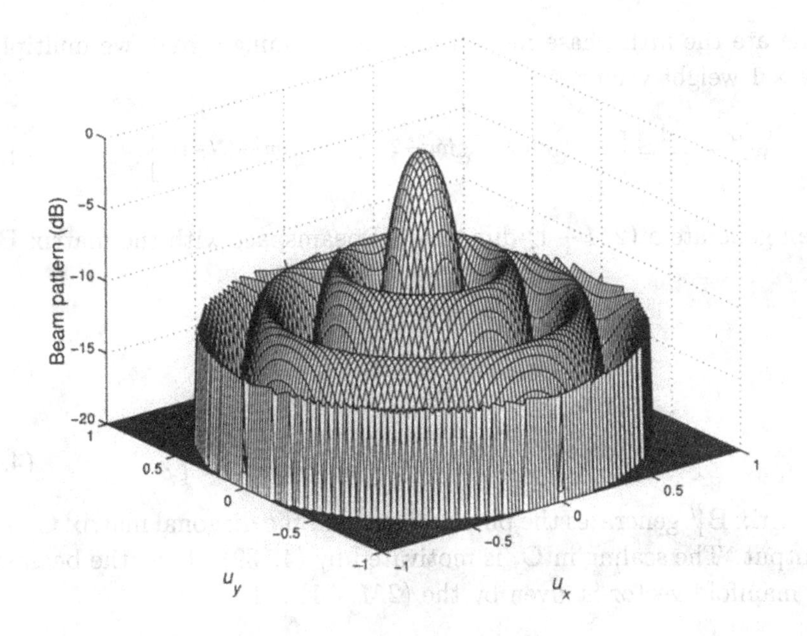

Figure 4.37 Beam pattern for a uniform circular array of 20 elements ($2\pi R/\lambda = 10$).

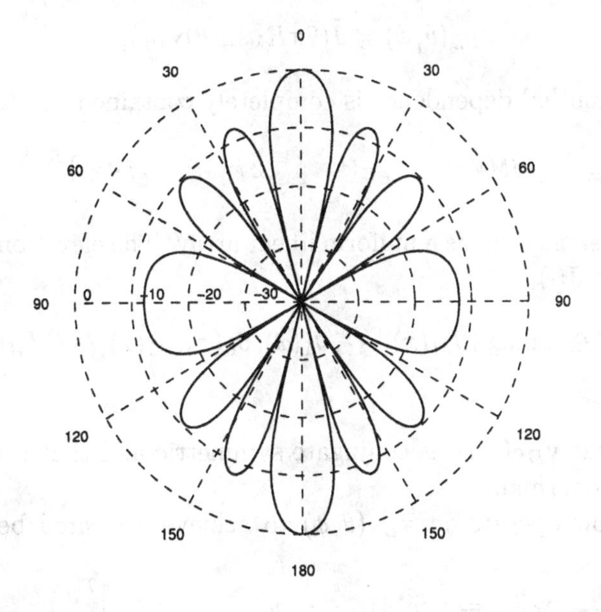

Figure 4.38 Beam patterns as a function of θ for a uniform circular array of 20 elements ($2\pi R/\lambda = 10$) with uniform weighting: $\phi = 0°$ (and $\phi = 180°$).

to generate the mth phase mode from the incoming signal, we multiply by the $N \times 1$ weight vector,

$$\mathbf{w}_m^H = \frac{1}{\sqrt{N}} \left[\begin{array}{ccccc} 1 & e^{jm\frac{2\pi}{N}} & e^{jm\frac{2\pi}{N}2} & \cdots & e^{jm\frac{2\pi}{N}(N-1)} \end{array} \right]. \tag{4.180}$$

We can generate a $(2M+1)$-dimensional beamspace with the matrix $\mathbf{B}_{PM}^H$, where

$$\mathbf{B}_{PM}^H = \mathbf{C}_j \mathbf{B}_1^H, \tag{4.181}$$

$$\mathbf{B}_1 = \left[\begin{array}{ccccc} \mathbf{w}_{-M} & \cdots & \mathbf{w}_0 & \cdots & \mathbf{w}_M \end{array} \right], \tag{4.182}$$

and

$$\mathbf{C}_j \triangleq \operatorname{diag}\left\{ j^{-M}, \cdots, j^{-1}, 1, j^1, \cdots, j^M \right\}. \tag{4.183}$$

The matrix $\mathbf{B}_1^H$ generates the phase modes and the diagonal matrix $\mathbf{C}_j$ scales the output. The scaling in $\mathbf{C}_j$ is motivated by (4.139). Then the beamspace array manifold vector is given by the $(2M+1) \times 1$ vector

$$\mathbf{v}_{BS}(\theta, \phi) = \mathbf{B}_{PM}^H \mathbf{v}(\theta, \phi). \tag{4.184}$$

Using (4.139), we can write (4.184) as

$$\mathbf{v}_{BS}(\theta, \phi) = \tilde{\mathbf{J}}(2\pi R_\lambda \sin\theta) \mathbf{v}(\phi), \tag{4.185}$$

where the azimuthal dependence is completely contained in $\mathbf{v}(\phi)$:

$$\mathbf{v}(\phi) = \left[\begin{array}{ccccccc} e^{-jM\phi} & \cdots & e^{-j\phi} & 1 & e^{j\phi} & \cdots & e^{jM\phi} \end{array} \right]^T, \tag{4.186}$$

which has the same form as a uniform linear array. The elevation dependence is contained in $\tilde{\mathbf{J}}(\cdot)$,

$$\tilde{\mathbf{J}}(x) \quad \triangleq \quad \operatorname{diag}\left\{ J_M(x), \cdots, J_1(x), J_0(x), J_1(x), \cdots, J_M(x) \right\}. \tag{4.187}$$

We observe that $\mathbf{v}_{BS}(\theta, \phi)$ is conjugate symmetric and that the columns of $\mathbf{B}_{PM}$ are orthonormal.

We can now operate on $\mathbf{v}_{BS}(\theta, \phi)$ to achieve a desired beam pattern. Defining,

$$\mathbf{w}_{PM}^H = \left[\begin{array}{ccccc} w_{-M}^* & \cdots & w_0^* & \cdots & w_M^* \end{array} \right]^T, \tag{4.188}$$

the output $y(k)$ is

$$y(k) = \mathbf{w}_{PM}^H \mathbf{B}_{PM}^H \mathbf{x}(k), \tag{4.189}$$

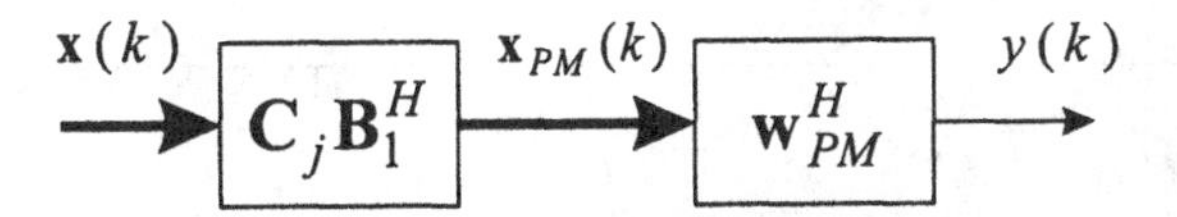

Figure 4.39 Phase mode excitation beamformer.

and the beam pattern is

$$B(\theta,\phi) = \mathbf{w}_{PM}^H \tilde{\mathbf{J}}(2\pi R_\lambda \sin\theta)\mathbf{v}(\phi). \tag{4.190}$$

The beamspace processor is shown in Figure 4.39.

We illustrate the design procedure with an example.

Example 4.2.1

Consider a 25-element uniform circular array with radius $2\pi R = 10\lambda$. Using (4.154), we see that we can excite 21 phase modes.

We design the beam pattern for the case when $\theta = 90°$. The desired pattern corresponds to the Villeneuve pattern in Figure 3.31(a) in ϕ-space instead of u-space.

To achieve the desired pattern, we use

$$\left[\mathbf{w}_{PM}^H\right]_m = \frac{\left[\mathbf{w}_{VIL}^H\right]_m}{j^m J_m(2\pi R_\lambda)}, \tag{4.191}$$

where $\mathbf{w}_{VIL}$ was derived in Section 3.5.4. The resulting beam pattern is plotted versus ϕ for $\theta = 90°$ in Figure 4.40. The pattern exhibits the desired behavior for $\theta = 90°$, but deteriorates rapidly as θ moves away from 90° (these patterns are not shown). To achieve satisfactory behavior we need to introduce vertical directivity in the sensor elements. We do this in the next example.

Example 4.2.2 (continuation)

Consider a 25-element uniform circular array with radius $2\pi R = 10\lambda$. Each element is a linear aperture that is parallel to the z-axis (perpendicular to the xy-plane) with length $L = 10\lambda$. We use the same phase mode weighting in (4.191). In Figure 4.41, the resulting beam pattern is plotted versus ϕ for $\theta = 0.5137\pi$ radians. (This corresponds to the first sidelobe of the element pattern.) The resulting pattern is well-behaved.

The argument of the Bessel function restricts both the elevation beam pattern and the array bandwidth.[9] This limitation is due to cancellation effects between elements at opposite sides of the circle. Therefore, most circular arrays use elements with an element beam pattern whose main response axis is in the radial direction. Synthesis of desirable array beam patterns is more difficult because the pattern is not a product of the element pattern and the array factor. Discussions of this topic are contained in Mailloux [Mai94], Davies [Dav87], and Rahim and Davies [RD82]. The

[9]This discussion follows Chapter 4 of Mailloux [Mai94].

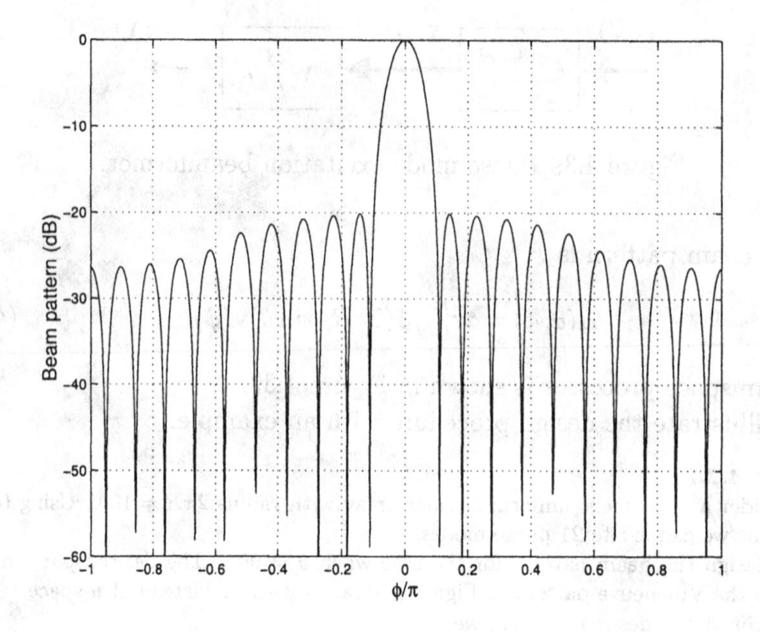

Figure 4.40 Beam pattern versus ϕ: Villenueve phase excitation, $\theta = 90°$.

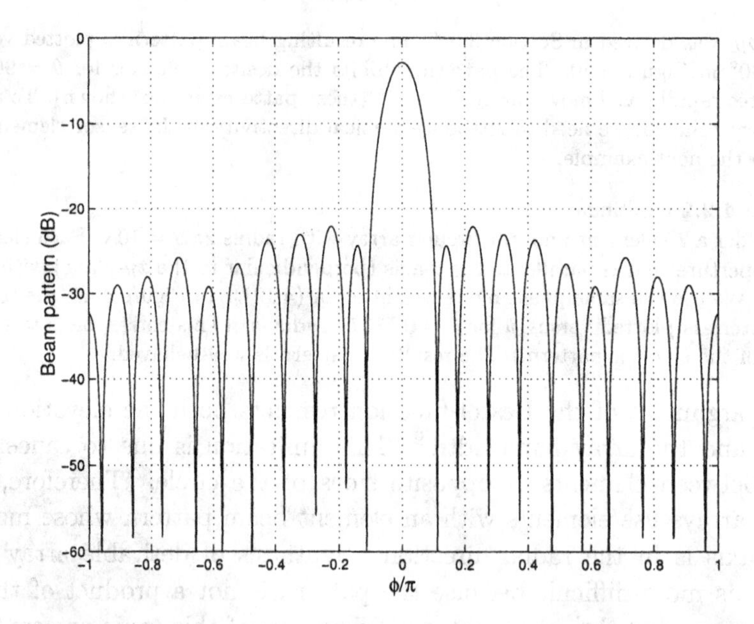

Figure 4.41 Beam pattern versus ϕ for various θ: Villenueve phase excitation, $\theta = 0.514\pi$.

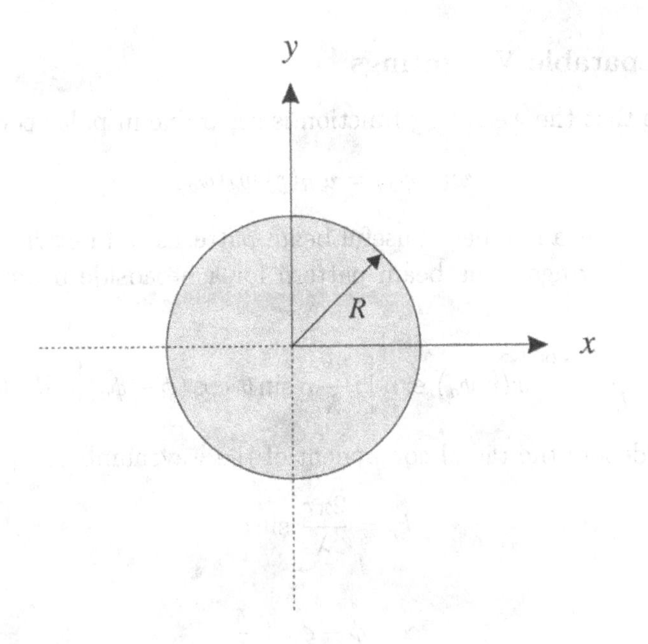

Figure 4.42 Circular aperture.

synthesis techniques that were developed in Section 3.9.3 are applicable to this problem and have been applied to it by Sureau and Keeping [SK82] and Bell et al. [BVT99].

In many applications, the main response axis is scanned in azimuth and only the elements in a 90° sector centered on the MRA are utilized. The techniques in Section 3.9.3 are useful in this case also.

Other references that discuss various aspects of circular arrays include Hansen [Han98], which contains a history of the various advances, and the book by Ma [Ma74].

4.3 Circular Apertures

In this section we consider the characteristics of circular apertures lying in the xy-plane, as shown in Figure 4.42. These apertures are important in many applications (perhaps the most common being the aperture of a parabolic antenna). In addition, they provide the limiting case for several array configurations such as the concentric circular array shown in Figure 4.2(b) and the hexagon array shown in Figure 4.3.

4.3.1 Separable Weightings

We assume that the weighting function is separable in polar coordinates,

$$w(r, \phi_a) = w_R(r)\, w_\phi(\phi_a). \tag{4.192}$$

We will find that a number of useful beam patterns can be synthesized with separable weightings. The beam pattern for a broadside beam ($\theta_0 = 0$) is given by

$$B(\theta, \phi) = \int_0^R \int_0^{2\pi} w(r, \phi_a) \exp\left[j\frac{2\pi}{\lambda} r \sin\theta \cos(\phi - \phi_a)\right] r\, dr\, d\phi_a. \tag{4.193}$$

Letting k_r denote the radial component of the wavenumber,

$$k_r = \frac{2\pi r}{\lambda} \sin\theta, \tag{4.194}$$

and defining

$$\alpha = \phi - \phi_a + \frac{\pi}{2}, \tag{4.195}$$

we can write

$$\sin\alpha = \sin\left(\phi - \phi_a + \frac{\pi}{2}\right) = \cos(\phi - \phi_a). \tag{4.196}$$

Using (4.194)–(4.196), the beam pattern can be written as

$$B(\theta, \phi) = \int_0^{2\pi} d\phi_a \int_0^R w(r, \phi_a) \exp(jk_r \sin\alpha)\, r\, dr. \tag{4.197}$$

We can write the exponential term in (4.197) as a Fourier series,

$$\exp(jk_r \sin\alpha) = \sum_{n=-\infty}^{\infty} J_n(k_r)\, e^{jn\alpha}, \tag{4.198}$$

where

$$J_n(k_r) = \frac{1}{2\pi} \int_{-\pi}^{\pi} \exp\left[\pm j\,(k_r \sin x - nx)\right] dx. \tag{4.199}$$

Using (4.192), (4.195), and (4.198) in (4.197), we have[10]

$$\begin{aligned} B(\theta, \phi) &= \sum_{n=-\infty}^{\infty} j^n e^{jn\phi} \left[\int_0^{2\pi} w_\phi(\phi_a)\, e^{-jn\phi_a}\, d\phi_a\right] \\ &\quad \cdot \int_0^R w_R(r)\, J_n\left(\frac{2\pi r}{\lambda} \sin\theta\right) r\, dr, \end{aligned} \tag{4.200}$$

[10] We use $\exp\left(j\frac{n\pi}{2}\right) = j^n$ and $J_{-n}(x) = (-1)^n J_n(x)$.

which is the desired result. We now consider some simple cases. Assume

$$w_\phi(\phi_a) = e^{jm\phi_a}. \tag{4.201}$$

This is equivalent to a sinusoidal variation in ϕ. Substituting (4.201) into the first integral in (4.200) gives,

$$\int_0^{2\pi} e^{j(m-n)\phi_a}\, d\phi_a = 2\pi\delta_{mn}, \tag{4.202}$$

so (4.200) becomes

$$B(\theta,\phi) = 2\pi j^m\, e^{jm\phi} \int_0^R w_R(r)\, J_m\left(\frac{2\pi r}{\lambda}\sin\theta\right) r\, dr. \tag{4.203}$$

For the special case in which $m = 0$, the weighting in the ϕ-direction is uniform and

$$B(\theta,\phi) = B(\theta) = 2\pi \int_0^R w_R(r)\, J_0\left(\frac{2\pi r}{\lambda}\sin\theta\right) r\, dr. \tag{4.204}$$

If we substitute[11]

$$u_R = \frac{2R}{\lambda}\sin\theta, \tag{4.205}$$

$$p = \frac{\pi r}{R}, \tag{4.206}$$

and

$$g_0(p) = \frac{2R^2}{\pi} w_R\left(\frac{Rp}{\pi}\right), \tag{4.207}$$

then (4.204) can be written as,

$$B(\theta) = \int_0^\pi p\, g_0(p)\, J_0(u_R p)\, dp. \tag{4.208}$$

The reason for this choice of variables is to obtain an integral with $[0,\pi]$ limits.

The final simplification is to assume that $w_R(r)$ is constant ($w_R(r) = c$) from 0 to R. In this case, we can use (4.204) without the change of variables. Then (4.204) becomes

$$B(\theta) = 2\pi c \int_0^R J_0\left(\frac{2\pi r}{\lambda}\sin\theta\right) r\, dr. \tag{4.209}$$

[11]Note that u_R is not the radial component in (ψ_x, ψ_y) space. It contains a factor of R so the visible region is $0 \le u_R \le 2R/\lambda$. This notation is used to be consistent with the antenna literature.

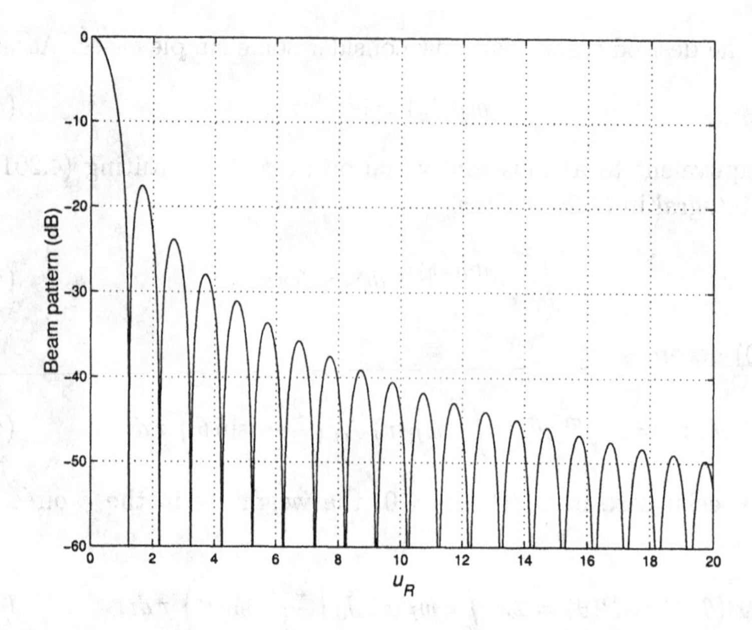

Figure 4.43 Beam pattern of a circular aperture lying in the xy-plane with $R/\lambda = 10$ and uniform weighting; the horizontal axis is $u_R = \psi_R/\pi$ and the visible region is $0 \leq u \leq 2R/\lambda$.

Using

$$\int_0^R x\, J_0(\alpha x)\, dx = \frac{R}{\alpha} J_1(\alpha R), \tag{4.210}$$

we obtain

$$B(\theta) = cR \frac{J_1\left(\frac{2\pi R}{\lambda} \sin\theta\right)}{\frac{\sin\theta}{\lambda}}. \tag{4.211}$$

One can show that

$$B(0) = c\pi R^2, \tag{4.212}$$

so the normalized pattern is

$$B(\theta) = 2\frac{J_1\left(\frac{2\pi R}{\lambda} \sin\theta\right)}{\frac{2\pi R}{\lambda} \sin\theta} = 2\frac{J_1(\pi u_R)}{\pi u_R}. \tag{4.213}$$

The visible region is $0 \leq \psi_R \leq 2\pi R/\lambda$. In Figure 4.43 we plot the normalized beam pattern versus u_R.

The first null of $J_1(\pi u_R)$ is at

$$u_R = 1.22, \tag{4.214}$$

or

$$\theta = \sin^{-1}\left(1.22\,\frac{\lambda}{2R}\right). \tag{4.215}$$

The height of the first sidelobe is -17.6 dB.

One can show that the HPBW is

$$\begin{aligned} \Delta\theta &= 2\sin^{-1}\left(0.257\frac{\lambda}{R}\right) \simeq 0.514\frac{\lambda}{R}\text{rad} \\ &= 29.2\frac{\lambda}{R}\text{degrees}. \end{aligned} \tag{4.216}$$

This compares to the HPBW for a square aperture of the same area

$$\begin{aligned} \Delta\theta_{sq} &= 2\sin^{-1}\left(0.250\frac{\lambda}{R}\right) \simeq 0.50\frac{\lambda}{R}\text{rad} \\ &= 28.65\frac{\lambda}{R}\text{degrees}. \end{aligned} \tag{4.217}$$

Thus, the circular aperture has a slightly larger 3-dB beamwidth, but the sidelobes are significantly lower (-17.6 dB vs. -13.4 dB).

In order to improve the sidelobe characteristics, a nonuniform radial weighting is used. A family of weightings that is used in practice (e.g., [Bal82], [Jas61]) is

$$w_R(r) = \begin{cases} \left[1 - \left(\frac{r}{R}\right)^2\right]^n, & 0 \leq r \leq R, n = 0, 1, 2, 3, \cdots, \\ 0, & \text{elsewhere}. \end{cases} \tag{4.218}$$

For $n = 0$, we have the uniform distribution that we just analyzed. The characteristics for $n = 0, 1$, and 2 are shown in Table 4.4. As n increases, the HPBW increases and the height of the first sidelobe decreases.

Table 4.4 Characteristics of Circular Apertures with Circular Symmetric Weightings[12]

Radial Weighting	Uniform	Radial Taper	Radial Taper Squared
$w_R(r)$	$\left[1-\left(\frac{r}{R}\right)^2\right]^0$	$\left[1-\left(\frac{r}{R}\right)^2\right]^1$	$\left[1-\left(\frac{r}{R}\right)^2\right]^2$
Beam pattern	$2\frac{J_1(\psi_R)}{\psi_R}$	$8\frac{J_2(\psi_R)}{\psi_R^2}$	$48\frac{J_3(\psi_R)}{\psi_R^3}$
Half-power beamwidth (degrees) $R \gg \lambda$	$\frac{29.2}{\frac{R}{\lambda}}$	$\frac{36.4}{\frac{R}{\lambda}}$	$\frac{42.1}{\frac{R}{\lambda}}$
BW_{NN} (degrees) $R \gg \lambda$	$\frac{69.9}{\frac{R}{\lambda}}$	$\frac{93.4}{\frac{R}{\lambda}}$	$\frac{116.3}{\frac{R}{\lambda}}$
First sidelobe (dB)	−17.6	−24.6	−30.6
Directivity	$0.5\left(\frac{2\pi R}{\lambda}\right)^2$	$0.375\left(\frac{2\pi R}{\lambda}\right)^2$	$0.28\left(\frac{2\pi R}{\lambda}\right)^2$

4.3.2 Taylor Synthesis for Circular Apertures

Taylor [Tay60] also developed a synthesis procedure for circular apertures that is a modification of the procedure developed for line sources in Section 3.5.3. He starts with the pattern for a uniformly illuminated circular aperture $J_1(\pi u)/(\pi u)$ and removes a set of zeros from it and adds new zeros.

The roots of $J_1(\pi u)$ are the values of u_m such that

$$J_1(\pi u_m) = 0, \quad m = 1, 2, \cdots. \tag{4.219}$$

The Taylor pattern is obtained by removing first $(\bar{n}-1)$ root pairs and replacing them with $(\bar{n}-1)$ new root pairs

$$B_{TAY}(u) = \frac{J_1(\pi u)}{\pi u} \frac{\prod_{n=1}^{\bar{n}-1}\left(1-\frac{u^2}{z_n^2}\right)}{\prod_{m=1}^{\bar{n}-1}\left(1-\frac{u^2}{u_m^2}\right)}. \tag{4.220}$$

[12] Table follows [Jas61], [Bal82], and [Mai94].

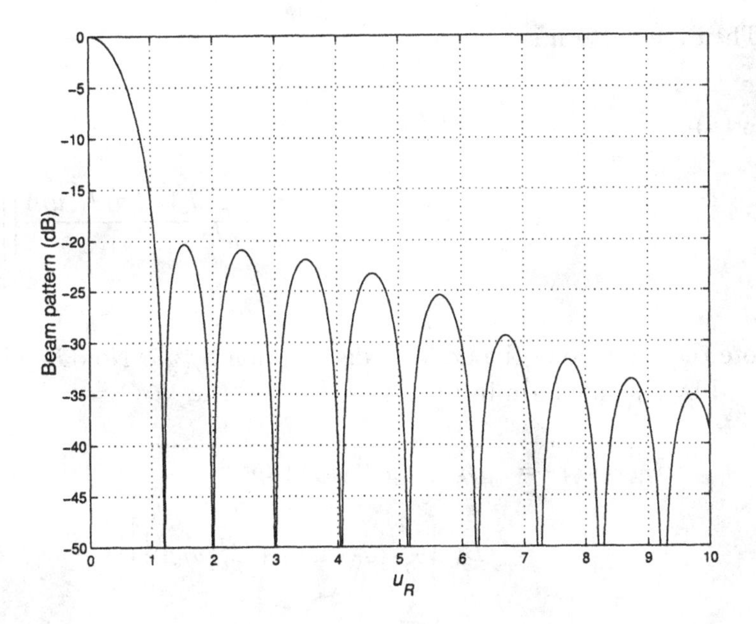

Figure 4.44 Taylor pattern for a circular aperture, $\bar{n} = 6$, -20 dB SLL; $R = 5\lambda$.

Taylor showed that the location of the new roots should be

$$z_n^2 = u_{\bar{n}}^2 \frac{A^2 + \left(n - \frac{1}{2}\right)^2}{A^2 + \left(\bar{n} - \frac{1}{2}\right)^2}, \tag{4.221}$$

where

$$A = \frac{1}{\pi} \cosh^{-1}(R_0), \tag{4.222}$$

or, in other words, $-20 \log_{10} \cosh(\pi A)$ is the desired SLL.

A representative Taylor pattern is shown in Figure 4.44 for $\bar{n} = 6$ and -20-dB sidelobes. The radius R equals 5λ so, from (4.205), the visible region is $|u_R| \leq 10$.

To find the aperture weighting $g_0(p)$ that will produce this pattern we write $g_0(p)$ as a series

$$g_0(p) = \sum_{m=0}^{\infty} B_m \, J_0(u_m p), \tag{4.223}$$

where the u_m are roots of $J_1(\pi u)$ and the B_m are the coefficients in the series. Then the pattern is

$$\begin{aligned} B_{TAY}(u) &= \sum_{m=0}^{\infty} B_m \int_0^{\pi} p\, J_0(u_m p)\, J_0(up)\, dp \\ &= \sum_{m=0}^{\infty} B_m \left[\frac{u_m p J_1(u_m p) J_0(up) - up J_0(u_m p) J_1(up)}{u_m^2 - u^2} \right] \Bigg|_0^{\pi}. \end{aligned} \tag{4.224}$$

We denote the kth zero of the uniform distribution by u_k. Now $B_{TAY}(u_k)$ is determined by a single term in the series,

$$\begin{aligned} B_{TAY}(u_k) &= B_k \int_0^{\pi} p\, J_0^2(u_k p)\, dp \\ &= B_k \left[\frac{p^2}{2} \left[J_0^2(u_k p) + J_1^2(u_k p) \right] \right] \Bigg|_0^{\pi} \\ &= B_k \left[\frac{\pi^2}{2}\, J_0^2(u_k \pi) \right]. \end{aligned} \tag{4.225}$$

Thus,

$$B_k = \frac{2}{\pi^2} \frac{B_{TAY}(u_k)}{J_0^2(u_k \pi)}. \tag{4.226}$$

Since $B_{TAY}(u_k) = 0$ for $k \geq \bar{n}$, the sum in (4.224) is finite and

$$g_0(p) = \frac{2}{\pi^2} \sum_{m=0}^{\bar{n}-1} \frac{B_{TAY}(u_m)}{J_0^2(u_m \pi)} J_0(u_m p), \tag{4.227}$$

where $B_{TAY}(u_m)$ is obtained from (4.220). The zero locations of $J_1(u_m \pi)$ are shown in Table 4.5.

Table 4.5 Zero Locations u_m for $J_1(\pi u_m)$[13]

m	u_m	m	u_m	m	u_m	m	u_m
1	1.2196699	6	6.2439216	11	11.2466228	16	16.2476619
2	2.2331306	7	7.2447598	12	12.2468985	17	17.2477974
3	3.2383155	8	8.2453948	13	13.2471325	18	18.2479181
4	4.2410629	9	9.2458927	14	14.2473337	19	19.2480262
5	5.2439216	10	10.2462933	15	15.2475086	20	20.2481237

[13]From [Ell81].

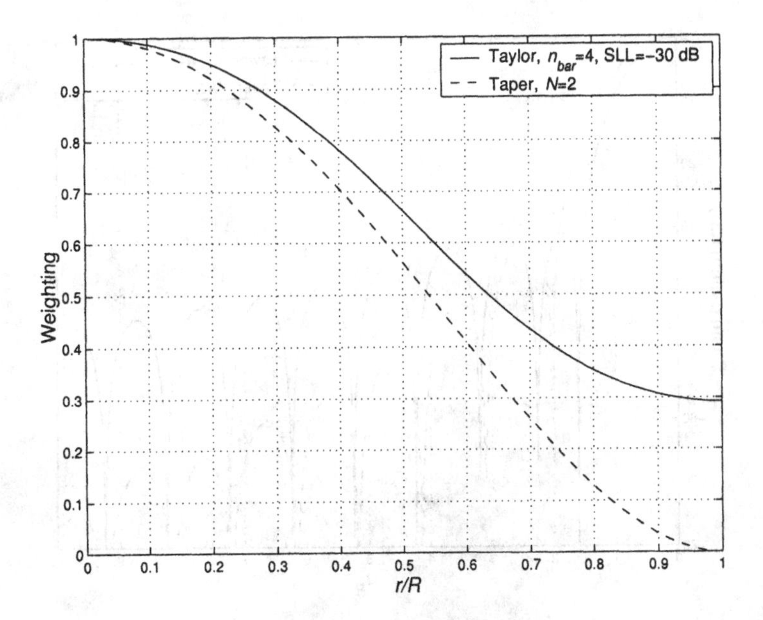

Figure 4.45 Weight function: Taylor weighting and radial taper squared weighting.

Tables of the roots of circular Taylor patterns and the corresponding aperture distributions are given in Hansen [Han59], [Han60]. Hansen[Han60] also compares the characteristics of Taylor weightings and the second-order weighting from Table 4.4. In Figure 4.45, we show the Taylor weighting function for the 30-dB sidelobe case and the second-order weighting from Table 4.3. The beam patterns for the two weightings are shown in Figure 4.46. The main-lobe beamwidth is larger using the radial taper squared, but the sidelobes decay more quickly. In Table 4.6 (from [Han60]), we show the comparative beamwidths for the Taylor weighting and the $(1-(r/R)^2)^N$ weighting. We see that for similar first sidelobe levels there is a significant decrease in the beamwidth obtained by using the Taylor weighting. However the remaining sidelobes decrease more rapidly with the radial taper.

Table 4.6 Comparative Beamwidths

Taylor			$(1-(r/R)^2)^N$		
Sidelobe Level, dB	$\bar{n}$	Beamwidth	Sidelobe Level, dB	N	Beamwidth
25	4	$1.13\lambda/2R$	24.6	1	$1.27\lambda/2R$
30	4	$1.20\lambda/2R$	30.6	2	$1.47\lambda/2R$
35	5	$1.25\lambda/2R$	36.0	3	$1.65\lambda/2R$
40	5	$1.31\lambda/2R$	40.9	4	$1.81\lambda/2R$

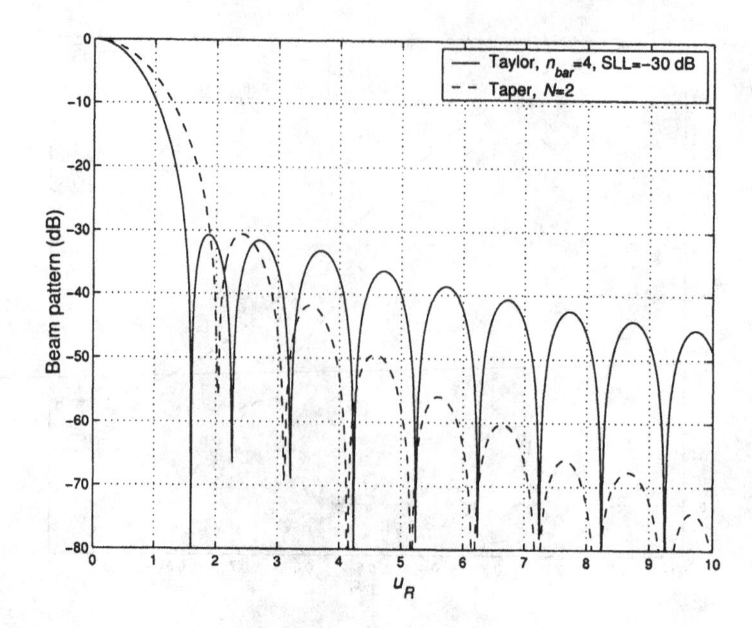

Figure 4.46 Beam patterns for Taylor weighting and radial taper squared weighting.

4.3.3 Sampling the Continuous Distribution

In many applications we approximate the circular aperture by a discrete array. In this section,[14] we discuss the effect that sampling the continuous aperture function has on the beam pattern.

We first consider a rectangular array with a circular boundary. One quadrant of a 20×20 array is shown in Figure 4.47. We want to approximate the Taylor pattern in Figure 4.44.

For the nmth element, the distance from the origin is

$$d_{nm} = \left\{ \left[\frac{(2|n|-1)d_x}{2} \right]^2 + \left[\frac{(2|m|-1)d_y}{2} \right]^2 \right\}^{\frac{1}{2}}, \tag{4.228}$$

where symmetric indexing is used. Then,

$$p_{nm} = \frac{d_{nm}}{R} \tag{4.229}$$

and

$$w_{nm} = g_0(p_{nm}). \tag{4.230}$$

[14]This section follows Elliott [Ell81], pp. 225–230.

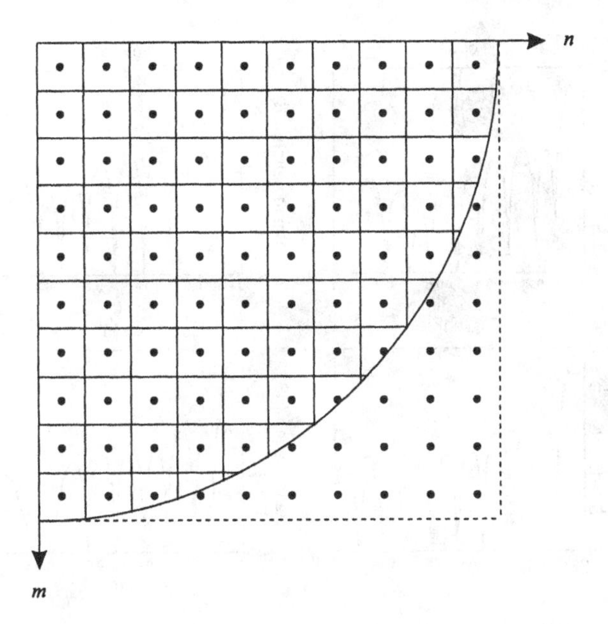

Figure 4.47 One quadrant of a rectangular grid array with circular boundary: $d_x = d_y = 0.5\lambda, R = 5\lambda$.

The beam pattern can be written as

$$B(\theta, \phi) = 4 \sum_{n=1}^{10} \sum_{m=1}^{10} w_{nm} \cos\left[\frac{(2n-1)\psi_x}{2}\right] \cos\left[\frac{(2m-1)\psi_y}{2}\right], \tag{4.231}$$

where

$$\psi_x = \pi \sin\theta \cos\phi, \tag{4.232}$$

$$\psi_y = \pi \sin\theta \sin\phi. \tag{4.233}$$

In Figure 4.48, we show four cuts of the pattern for $\phi = 0°, 15°, 30°, 45°$. We see that there is reasonable agreement with the continuous pattern in Figure 4.44.

In Section 4.4, we discuss hexagonal grids and their ability to generate various circular patterns.

4.3.4 Difference Beams

In Section 3.8, we explored the usage of difference patterns in linear arrays. They play a similar role in circular apertures (and arrays). In this section, we develop a general structure for difference beams.[15]

[15] Our discussion follows Section 6.11 of [Ell81].

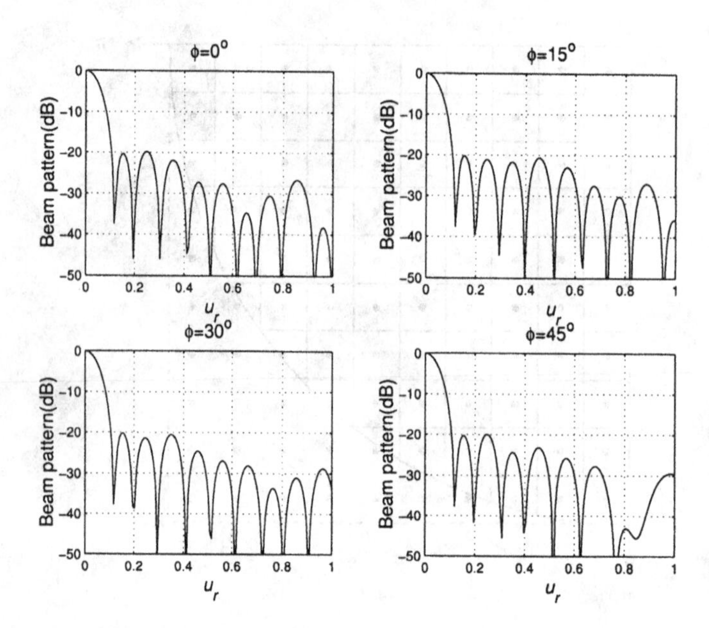

Figure 4.48 Beam pattern cuts: rectangular grid array with circular boundary; $N = 20$, SLL $= -20$ dB, $\bar{n} = 6, R = 5\lambda$, weights obtained by sampling continuous Taylor distribution.

In order to find the beam pattern we write $w(r, \phi)$ as

$$w(r,\phi) = \sum_{n=-\infty}^{\infty} w_n(r)\, e^{jn\phi}. \tag{4.234}$$

Then, from our discussion in Section 4.3.1 (specifically (4.192)–(4.202)), we can write the beam pattern as

$$B(\theta,\phi) = 2\pi \sum_{n=-\infty}^{\infty} (j)^n\, e^{jn\phi} \int_0^R w_n(r)\, J_n(k_r \sin\theta)\, r\, dr. \tag{4.235}$$

In Section 4.3.1 we focused on the case in which only w_0 was non-zero.

In order to construct a difference beam we look at the special case in which only the $n = \pm 1$ weights are non-zero. We denote this difference beam as $D(\theta, \phi)$. Letting $n = \pm 1$ in (4.235) gives

$$D(\theta,\phi) = 2\pi j \int_0^R \Big[e^{j\phi}\, w_1(r)\, J_1(k_r \sin\theta) - e^{-j\phi}\, w_{-1}(r)\, J_{-1}(k_r \sin\theta)\Big]\, r\, dr. \tag{4.236}$$

Since $J_{-1}(x) = -J_1(x)$, if we set $w_{-1}(r) = w_1(r)$, then (4.234) becomes

$$w(r, \phi) = 2w_1(r)\cos\phi, \tag{4.237}$$

and (4.236) becomes

$$D(\theta, \phi) = 4\pi j \cos\phi \int_0^R w_1(r)\, J_1(k_r \sin\theta)\, r\, dr. \tag{4.238}$$

We see that the resulting pattern is of the form

$$D_\alpha(\theta, \phi) = 4\pi j \cos\phi\, D_\theta(\theta), \tag{4.239}$$

where $D_\theta(\theta)$ is the integral in (4.238). Now consider a cut of the pattern through any plane containing the z-axis. Then,

$$D_\alpha(\theta, \phi) = 4\pi j \cos\phi\, D_\theta(\theta), \tag{4.240}$$

and

$$D_\alpha(\theta, \phi + \pi) = -4\pi j \cos\phi\, D_\theta(\theta), \tag{4.241}$$

which give a difference pattern in the various planes containing the z-axis. The function $D_\theta(\theta)$ is weighted with $\cos\phi$ so the maximum slope is in the xz-plane and the function is zero in the yz-plane.

To develop a useful difference pattern for the yz-plane, we set $w_{-1}(r) = -w_1(r)$. Then,

$$w(r, \phi) = 2jw_1(r)\sin\phi, \tag{4.242}$$

$$D_\beta(\theta, \phi) = -4\pi \sin\phi\, D_\theta(\theta). \tag{4.243}$$

This difference pattern provides maximum slope in the yz-plane. Thus, we utilize a beamspace consisting of a sum beam and two difference beams $D_\alpha(\theta, \phi)$ and $D_\beta(\theta, \phi)$. The shape of the difference patterns is determined by $D_\theta(\theta)$. We now focus on how the aperture weighting function affects $D_\theta(\theta)$.

Defining

$$u = \frac{2R}{\lambda}\sin\theta, \quad p = \frac{\pi r}{R}, \tag{4.244}$$

we can write $D_\theta(\theta)$ as,

$$D_\theta(u) = \left(\frac{R}{\pi}\right)^2 \int_0^\pi w_1(p)\, J_1(up)\, p\, dp. \tag{4.245}$$

In order to perform the integration, we expand $w_1(p)$ in an orthogonal expansion of Bessel functions,

$$w_1(p) = \sum_{m=0}^{\infty} A_m \, J_1(u_m p), \tag{4.246}$$

where the u_m are eigenvalues that we will define shortly. Using (4.246) in (4.245) gives

$$\begin{aligned} D_\theta(u) &= \left(\frac{R}{\pi}\right)^2 \sum_{m=0}^{\infty} A_m \int_0^\pi J_1(u_m p)\, J_1(up)\, p\, dp \\ &= \left(\frac{R}{\pi}\right)^2 \sum_{m=0}^{\infty} A_m \left[\frac{u_m p J_1(up)\, J_0(u_m p) - up J_o(up)\, J_1(u_m p)}{u^2 - u_m^2}\right]\Bigg|_0^\pi . \end{aligned} \tag{4.247}$$

Since $vJ_0(v) = J_1(v) + vJ_1'(v)$, where the prime denotes differentiation, (4.247) can be rewritten as

$$\begin{aligned} D_\theta(u) &= \left(\frac{R}{\pi}\right)^2 \sum_{m=0}^{\infty} A_m \left[\frac{u_m p J_1'(u_m p)\, J_1(up) - up J_1'(up)\, J_1(u_m p)}{u^2 - u_m^2}\right]\Bigg|_0^\pi \\ &= \left(\frac{R}{\pi}\right)^2 \sum_{m=0}^{\infty} A_m \frac{\pi u_m J_1'(\pi u_m)\, J_1(\pi u) - \pi u J_1'(\pi u)\, J_1(\pi u_m)}{u^2 - u_m^2}. \end{aligned} \tag{4.248}$$

To get $D_\theta(u_n)$ equal to zero for $n \neq m$, we require either $J_1(\pi u_m) = 0$ or $J_1'(\pi u_m) = 0$. The first choice requires $w_1(\pi) = 0$, which is undesirable, so we use the second option. The u_m are the zeros of $J'(\pi u)$, so

$$J'(\pi u_m) = 0; \tag{4.249}$$

then (4.248) becomes

$$D_\theta(u) = \left(\frac{R}{\pi}\right)^2 \sum_{m=0}^{\infty} A_m \, J_1(\pi u_m) \frac{\pi u J_1'(\pi u)}{u_m^2 - u^2}. \tag{4.250}$$

The zeros of $J_1'(x)$ are tabulated (e.g., p. 252 of [Ell81]). The first 10 zeros are shown in Table 4.7.

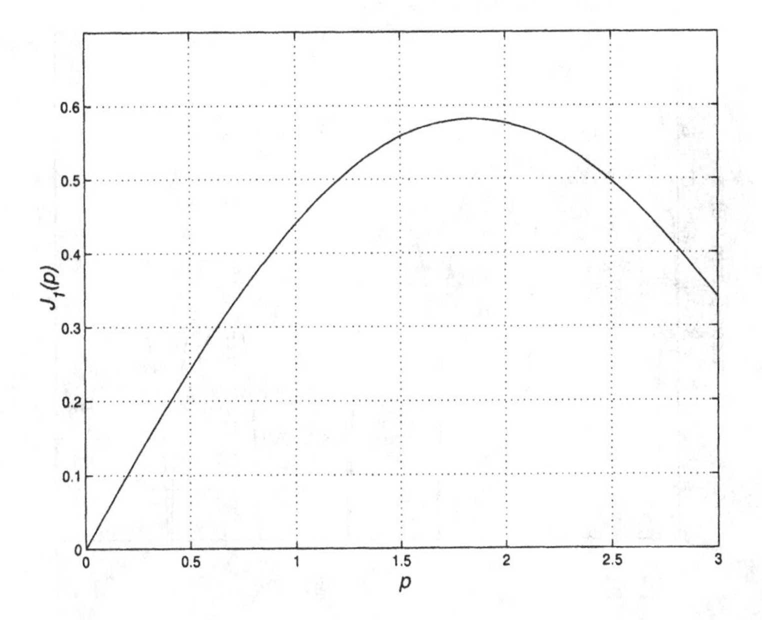

Figure 4.49 Aperture distribution for generic difference pattern: $v = 0.586p$.

Table 4.7 Bessel Function Zeros, $J_1'(\pi u_m) = 0$

m	u_m	m	u_m
0	0.5860670	5	5.7345205
1	1.6970509	6	6.7368281
2	2.7171939	7	7.7385356
3	3.7261370	8	8.7398505
4	4.7312271	9	9.7408945

We now consider the simple case in which there is a single term corresponding to $m = 0$. In this case,

$$D_\theta(u) = \left(\frac{R}{\pi}\right)^2 \frac{A_0 J_1(0.586\pi)}{(0.586)^2} \left[\frac{\pi u J_1'(\pi u)}{1 - \left(\frac{u}{0.586}\right)^2}\right]. \tag{4.251}$$

The aperture weighting $J_1(0.586p)$ is shown in Figure 4.49 and the beam pattern is shown in Figure 4.50. We see that the first sidelobe is at about -14 dB. We would like to preserve an adequate slope at the origin and decrease the height of the sidelobes.

This problem is analogous to the linear array problem that led to a Bayliss difference pattern. Bayliss [Bay68] also derived the difference pattern for a circular aperture.

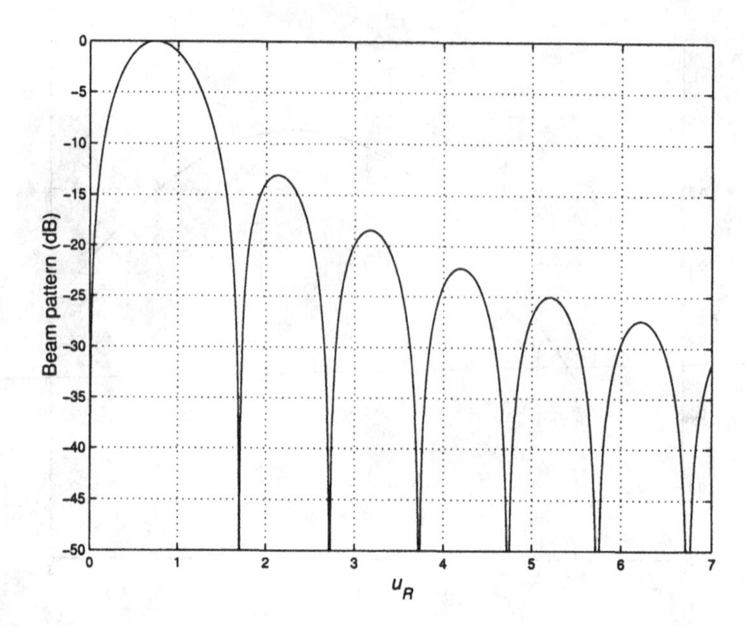

Figure 4.50 Beam pattern for generic aperture distribution: $u_R = 2R\sin\theta/\lambda$.

The reader should consult the above reference for a discussion of Bayliss difference patterns for circular apertures.

A sum beam and two difference beams are often used to estimate (θ, ϕ) of a signal (or target). In later chapters, we use these beams to do beamspace processing.

4.3.5 Summary

In Sections 4.3.1 and 4.3.2 we developed techniques for synthesizing beam patterns for circular apertures. Our emphasis was on patterns in which the weighting in the ϕ-direction was uniform. However, the relationships in (4.200)–(4.203) can be extended to include non-uniform weightings. We found that uniform weighting leads to a $J_1(\psi_R)/\psi_R$ beam pattern. We developed a family of radial taper weightings and Taylor weightings that resulted in lower sidelobes.

In Section 4.3.3, we discussed sampling the continuous distribution using a rectangular grid. Although the performance was satisfactory, we will find that the hexagonal grid in Section 4.4 provides a better approach. In Section 4.3.4, we developed techniques for synthesizing difference beams.

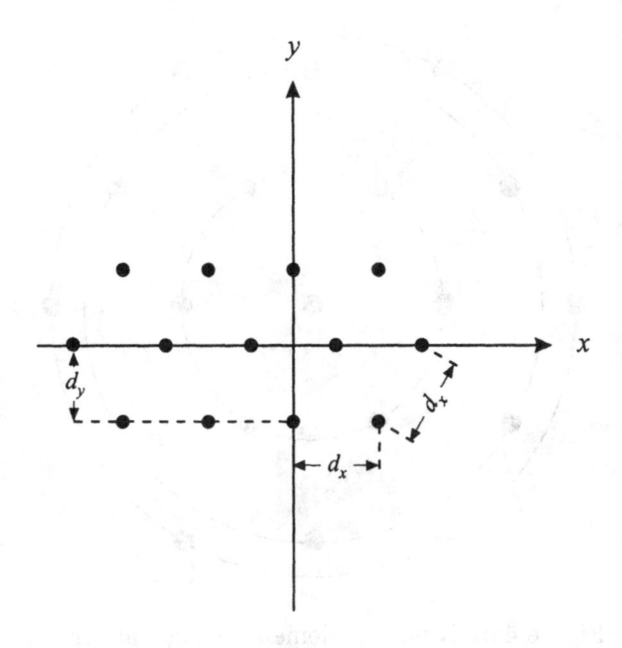

Figure 4.51 Configuration of a typical hexagonal array.

4.4 Hexagonal Arrays

4.4.1 Introduction

In this section, we discuss arrays whose elements are located on a hexagonal (or equilateral triangular) grid, as shown in Figure 4.51. The horizontal interelement spacing is d_x and the vertical spacing between rows is

$$d_y = \frac{\sqrt{3}}{2} d_x. \tag{4.252}$$

The motivation for using hexagonal-grid arrays can be approached from three viewpoints. The first viewpoint emphasizes that the array is sampling a spatial field. Peterson and Middleton [PM62] showed that hexagonal sampling is the optimum sampling strategy for signals that are bandlimited over a circular region of the Fourier plane. In the array case, this corresponds to the visible region,

$$u_x^2 + u_y^2 \le 1. \tag{4.253}$$

We revisit this viewpoint in Chapter 5.

The second viewpoint emphasizes the grating lobe viewpoint. Sharp [Sha61] shows that if the main beam is required to scan inside a cone whose

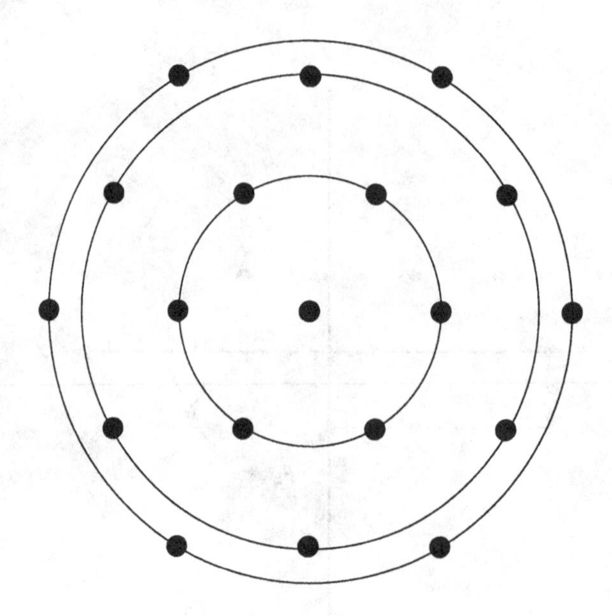

Figure 4.52 Nineteen-element hexagonal arrays.

axis is normal to the array, then the number of elements can be reduced by 13.4%.

A third viewpoint emphasizes the circular symmetry of many desired beam patterns. To illustrate this point, consider the 19-element hexagonal array in Figure 4.52. The hexagonal grid causes the elements to be arranged in concentric circles. From the discussion in Section 4.1, we would anticipate that the hexagonal array would be an efficient arrangement.

Our discussion will focus on triangular-grid arrays that have six equal sides with

$$d_x = \frac{\lambda}{2}, \tag{4.254}$$

and

$$d_y = \frac{\sqrt{3}}{4}\lambda. \tag{4.255}$$

We refer to these as **standard hexagonal arrays** (SHA). The total number of elements will be 7, 19, 37, 61, $\cdots$,

$$N_H = 1 + \sum_{n=1}^{\frac{N_x-1}{2}} 6n, \tag{4.256}$$

where N_x is the number of elements in the horizontal row through the origin, and N_x is odd in order to get a symmetric array. Standard hexagonal arrays with $N_H = 7$ and 19 are shown in Figure 4.53.

This configuration is useful whenever the desired beam pattern is circularly symmetric. An application that uses a similar configuration is a geostationary military satellite operating at 8 GHz that uses a 19-beam multiple beam antenna for reception and 61-beam multiple beam antenna for transmission.

Our discussion of hexagonal arrays is reasonably short because many of the rectangular-grid techniques can be applied directly to hexagonal arrays by using an appropriate coordinate system.

In Section 4.4.2, we discuss several beam pattern design techniques that are representative of possible approaches.

In Section 4.4.3, we discuss a hexagonal-grid to rectangular-grid transformation that will be useful for several applications.

In Section 4.4.4, we summarize our results.

4.4.2 Beam Pattern Design

In this section we discuss techniques for beam pattern design for standard hexagonal arrays. To illustrate the notation, consider the standard 19-element hexagonal array shown in Figure 4.53(b). Each horizontal row of sensors is indexed with the variable n, which ranges from $n = 0$ to $n = N_r - 1$, where N_r is the number of elements in the row. The rows are indexed by the variable m, which ranges from $-(N_x - 1)/2$ to $(N_x - 1)/2$. We see that $N_r = N_x - |m|$.

We define a 19-element array manifold vector as,

$$vec_H(u_x, u_y) = \begin{bmatrix} \mathbf{v}_2^T & \mathbf{v}_1^T & \mathbf{v}_0^T & \mathbf{v}_{-1}^T & \mathbf{v}_{-2}^T \end{bmatrix}^T, \tag{4.257}$$

where

$$\mathbf{v}_2 = e^{j\pi\sqrt{3}u_y} \begin{bmatrix} e^{-j\pi u_x} & 1 & e^{j\pi u_x} \end{bmatrix}^T, \tag{4.258}$$

$$\mathbf{v}_1 = e^{j\pi\frac{\sqrt{3}}{2}u_y} \begin{bmatrix} e^{-j\pi\frac{3u_x}{2}} & e^{-j\pi\frac{u_x}{2}} & e^{j\pi\frac{u_x}{2}} & e^{j\pi\frac{3u_x}{2}} \end{bmatrix}^T, \tag{4.259}$$

$$\mathbf{v}_0 = \begin{bmatrix} e^{-j2\pi u_x} & e^{-j\pi u_x} & 1 & e^{j\pi u_x} & e^{j2\pi u_x} \end{bmatrix}^T, \tag{4.260}$$

$$\mathbf{v}_{-1} = e^{-j\pi\frac{\sqrt{3}}{2}u_y} \begin{bmatrix} e^{-j\pi\frac{3u_x}{2}} & e^{-j\pi\frac{u_x}{2}} & e^{j\pi\frac{u_x}{2}} & e^{j\pi\frac{3u_x}{2}} \end{bmatrix}^T, \tag{4.261}$$

$$\mathbf{v}_{-2} = e^{-j\pi\sqrt{3}u_y} \begin{bmatrix} e^{-j\pi u_x} & 1 & e^{j\pi u_x} \end{bmatrix}^T. \tag{4.262}$$

We see that $vec_H(u_x, u_y)$ is conjugate symmetric.

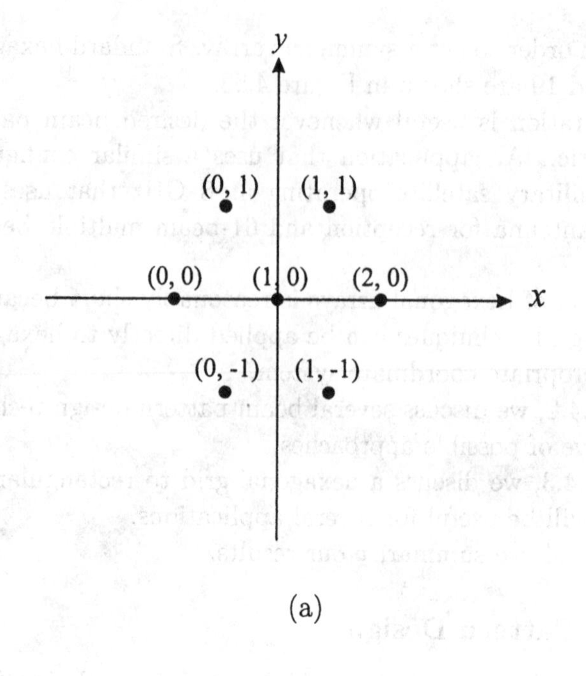

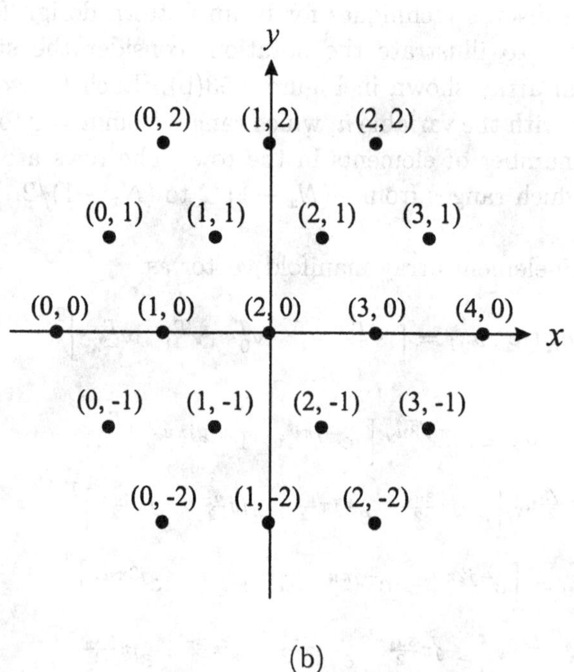

Figure 4.53 Configurations of standard hexagonal arrays: (a) $N_H = 7$; (b) $N_H = 19$.

For a standard hexagonal array with N_x elements along the x-axis, the nmth term in the array manifold vector is

$$[vec_H(u_x, u_y)]_{nm} = \exp\left\{j\pi\left[m\frac{\sqrt{3}}{2}u_y + nu_x - \frac{N_x - |m| - 1}{2}u_x\right]\right\}. \tag{4.263}$$

The beam pattern is

$$B_{\mathbf{u}}(u_x, u_y) = \sum_{m=-\frac{N_x-1}{2}}^{m=\frac{N_x-1}{2}} w_{nm}^* \exp\left\{j\pi\left[m\frac{\sqrt{3}}{2}u_y + \frac{N_x - |m| - 1}{2}u_x\right]\right\} \cdot \sum_{n=0}^{N_x-|m|-1} \exp\{j\pi nu_x\}. \tag{4.264}$$

For uniform weighting,

$$w_{nm} = \frac{1}{N_H}, \tag{4.265}$$

and

$$B_{\mathbf{u}}(u_x, u_y) = \frac{1}{N_H}\sum_{m=-\frac{N_x}{2}}^{\frac{N_x}{2}} \exp\left\{j\pi\left[m\frac{\sqrt{3}}{2}u_y - \frac{N_x - |m| - 1}{2}u_x\right]\right\} \cdot \sum_{n=0}^{N_x-|m|-1} \exp\{j\pi nu_x\}. \tag{4.266}$$

We first consider a design approach in which we construct the desired weighting for the circular aperture using the techniques in Section 4.3. We then sample the aperture weighting on a hexagonal grid to find the element weightings.

To match the hexagonal array to the continuous aperture, we observe that, if R is the radius of the continuous aperture and $2R/\lambda$ is an integer, then

$$R = \frac{N_x}{4}\lambda. \tag{4.267}$$

In our examples, we use $N_x = 11$, which corresponds to a standard hexagonal array with 91 elements.

Example 4.4.1

Consider a SHA with 91 elements with uniform weighting. Then $B_u(u_x, u_y)$ is given by (4.266), with $N_x = 11$ and $N_H = 91$.

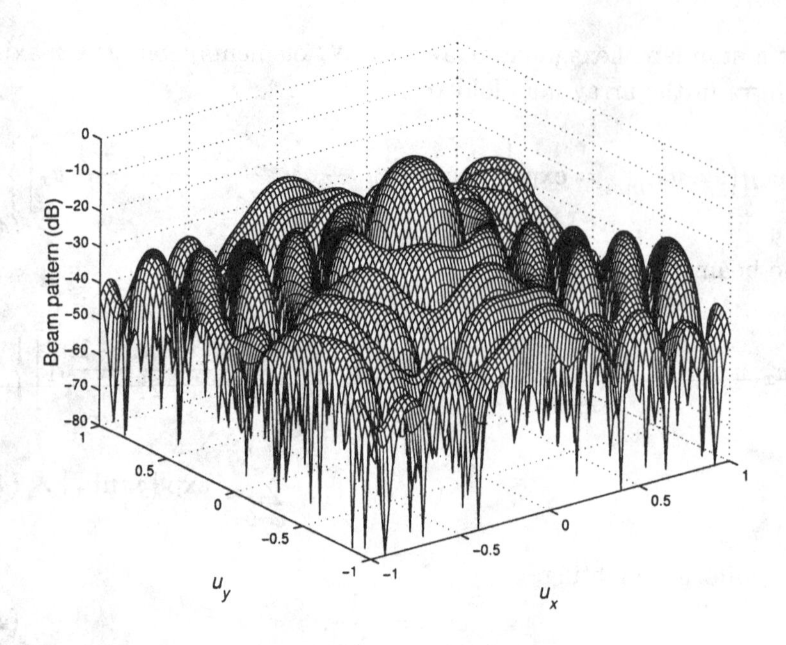

Figure 4.54 Beam pattern of a standard 91-element hexagonal array with uniform weighting.

The beam pattern is shown in Figure 4.54. A contour plot in Figure 4.55 shows the 6-fold symmetry inside the visible region. Beam pattern cuts at $\phi = 0°, 10°, 20°$, and $30°$ are shown in Figure 4.56. The horizontal axis is $u_r = \left(u_x^2 + u_y^2\right)^{\frac{1}{2}}$. These pattern cuts can be compared to the beam pattern of the circular aperture in Figure 4.43. The main lobe and the first sidelobe are almost identical, but the other sidelobes vary from those of the aperture.

In order to reduce the sidelobes we can use a radial taper from Table 4.4.

Example 4.4.2

Consider a standard 91-element hexagonal array. We start with the radial taper in column 2 of Table 4.4.

$$w_R(r) = 1 - \left(\frac{r}{R}\right)^2, \quad 0 \leq r \leq R, \tag{4.268}$$

where

$$R = 2.75\lambda. \tag{4.269}$$

We choose the hexagonal array weights using

$$w_{nm} = 1 - \frac{\left(\left(n - \frac{N_x - |m| - 1}{2}\right)\frac{\lambda}{2}\right)^2 + \left(m\frac{\sqrt{3}}{4}\lambda\right)^2}{R^2}. \tag{4.270}$$

The hexagonal array beam pattern is shown in Figure 4.57 with pattern cuts in Figure 4.58.

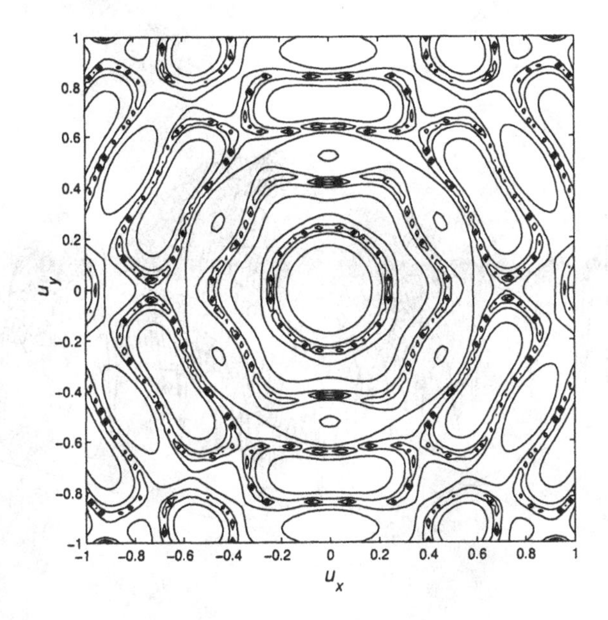

Figure 4.55 Contour plot of a standard 91-element hexagonal array with uniform weighting.

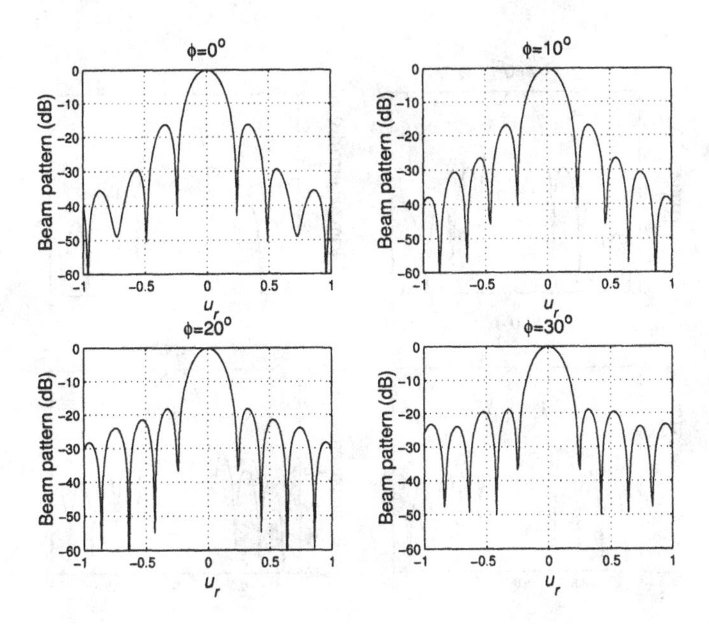

Figure 4.56 Beam pattern cuts of a standard 91-element hexagonal array with uniform weighting: (a)$\phi = 0°$; (b) $\phi = 10°$; (c) $\phi = 20°$; (d) $\phi = 30°$.

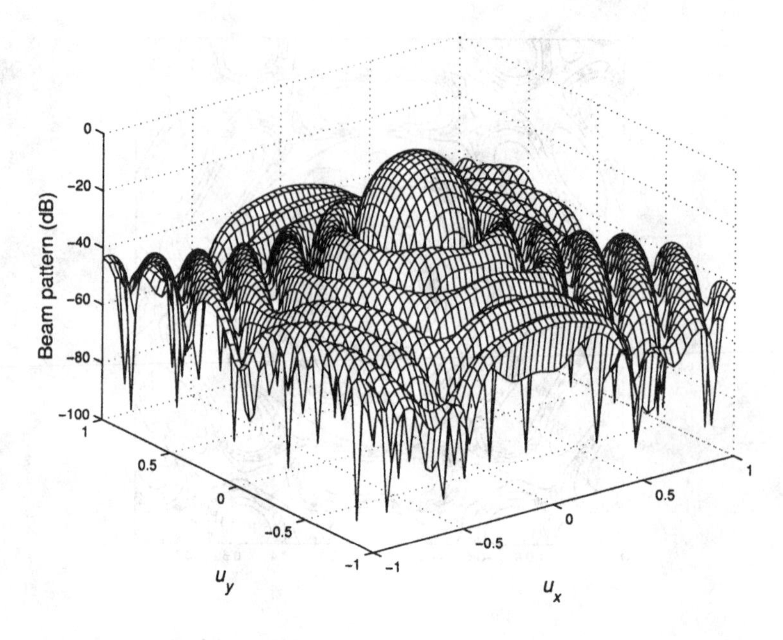

Figure 4.57 Beam pattern of a standard 91-element hexagonal array: radial taper, $R = 2.75\lambda$.

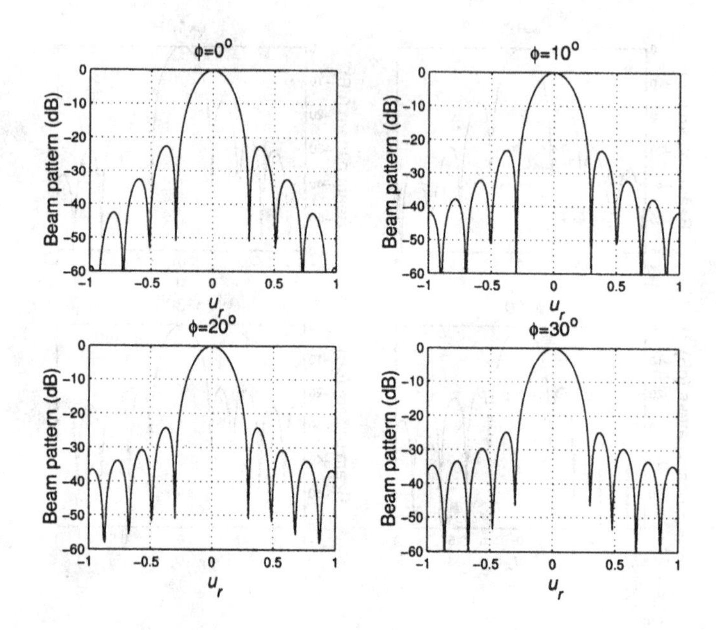

Figure 4.58 Cuts through beam pattern of a standard 91-element hexagonal array with radial taper: $R = 2.75\lambda$.

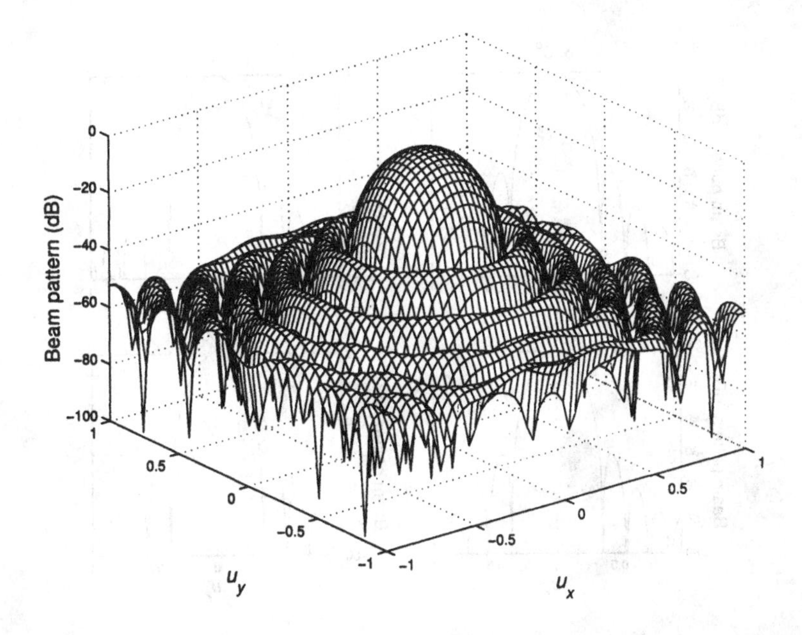

Figure 4.59 Beam pattern of a standard 91-element hexagonal array with radial taper squared weighting: $R = 2.75\lambda$.

The corresponding beam patterns for the weighting in column three of Table 4.4 are shown in Figures 4.59 and 4.60.

As expected, the sidelobe levels are much lower and the main lobe is wider for both radial tapers in comparison to the uniform weightings.

In Example 4.4.2, we have assumed that the desired radial weighting function is known. An alternative approach is to assume that the desired beam pattern is known and use (2.85) to find the weight vector.

If we desire a Dolph-Chebychev beam pattern in each cross section, we use (4.111) :

$$B_\psi(\psi_x, \psi_y) = \frac{T_{N-1}(x_o \cos\frac{\psi_x}{2} \cos\frac{\psi_y}{2})}{R}, \tag{4.271}$$

to generate the desired beam pattern. We sample the resulting beam pattern at N_H points and use (2.85) to find w_{nm}.

If we desire a Taylor beam pattern, we use (4.220) and (4.221) to generate the desired beam pattern. We sample the resulting beam pattern at N_H points and use (2.85) to find w_{nm}.

In both cases, it is important to choose a sampling grid that gives a well-conditioned $\mathbf{V}^H(\psi)$. Several examples are developed in the problems. A Dolph-Chebychev example is done in Problem 4.4.6. The results indicate that the resulting beam pattern is sensitive to the sampling grid and the choice of N_H.

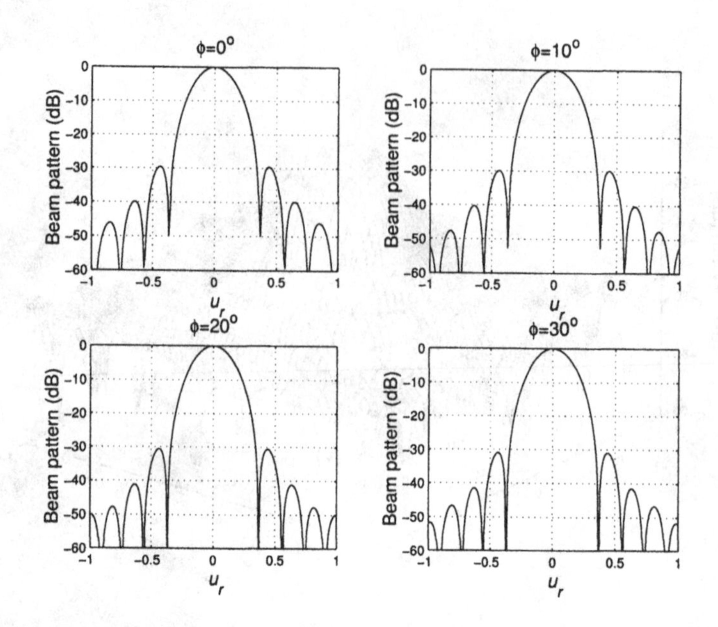

Figure 4.60 Cuts through beam pattern for a standard 91-element hexagonal array with radial taper squared weighting: $R = 2.75\lambda$; (a) $\phi = 0°$; (b) $\phi = 10°$; (c) $\phi = 20°$; (d) $\phi = 30°$.

4.4.3 Hexagonal Grid to Rectangular Grid Transformation

Most of the Fourier transform relationships that we used for beam pattern design and analysis can be utilized with hexagonal arrays by defining a suitable hexagonal Fourier transform or hexagonal DFT. An excellent discussion of hexagonal signal processing is contained in Mersereau [Mer79]

We use an approach due to Lo and Lee [LL83] that is similar and adequate for our purposes. The basic idea is to map the hexagonal array into an equivalent rectangular grid array and formulate the processing using the equivalent array.

In Figure 4.61(b), we show a 19-element standard hexagonal array. In Figure 4.61(a) we show a 19-element array on a standard rectangular grid ($d = \lambda/2$). To obtain the rectangular grid array, we have rotated and stretched the hexagonal grid array.

We write the beam pattern of the rectangular grid array in **v**-space,

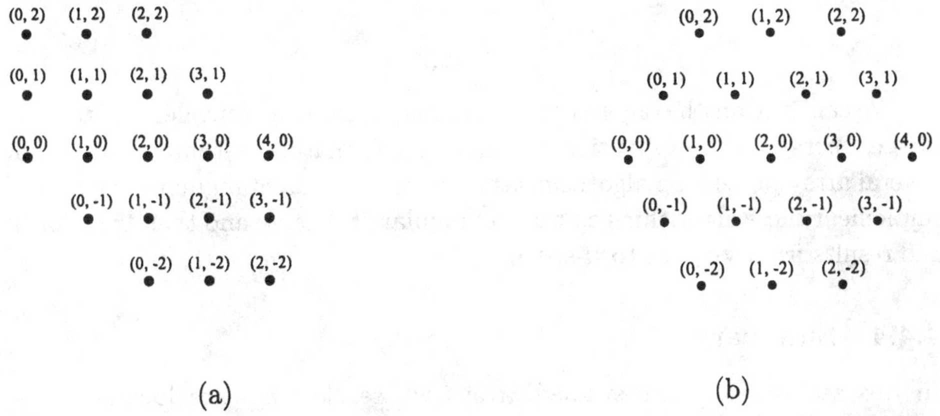

(a) (b)

Figure 4.61 Hexagonal-rectangular transformation.

$$B_{\mathbf{v}}(v_x, v_y) = \sum_{m=-\frac{N_x-1}{2}}^{\frac{N_x-1}{2}} w_{nm}^* \exp\left\{j\pi\left[mv_y - \frac{N_x - |m| - 1}{2}v_x - \frac{m}{2}v_x\right]\right\} \cdot \sum_{n=0}^{N_x-|m|-1} \exp j\pi n v_x, \quad (4.272)$$

where

$$v_x = \sin\theta_r \cos\phi_r, \quad (4.273)$$

and

$$v_y = \sin\theta_r \sin\phi_r, \quad (4.274)$$

and the subscript "r" denotes rectangular.

Comparing the expressions in (4.272) and (4.264), we obtain the relation between **u**-space and **v**-space,

$$v_x = u_x \quad (4.275)$$

and

$$v_y = \frac{u_x}{2} + \frac{\sqrt{3}}{2}u_y, \quad (4.276)$$

or

$$\mathbf{v} = \begin{bmatrix} 1 & 0 \\ \frac{1}{2} & \frac{\sqrt{3}}{2} \end{bmatrix} \mathbf{u} \quad (4.277)$$

and

$$\mathbf{u} = \begin{bmatrix} 1 & 0 \\ -\frac{1}{\sqrt{3}} & \frac{2}{\sqrt{3}} \end{bmatrix} \mathbf{v}. \tag{4.278}$$

We can take any hexagonal array and map it into an equivalent rectangular-grid array. The motivation for this transformation is that we will find several array processing algorithms that work well on rectangular arrays. We implement these algorithms on the rectangular grid array and then transform the results from $\mathbf{v}$-space to $\mathbf{u}$-space.

4.4.4 Summary

In this section, we have studied arrays whose elements are located on a hexagonal (or equilateral-triangular) grid. The grid provides an efficient spatial sampling strategy and is widely used in practice.

In Section 4.4.2, we considered standard hexagonal arrays and assumed that we had synthesized the desired beam pattern using a continuous circular aperture. We then sampled the continuous aperture weighting on a hexagonal grid to obtain the array beam pattern.

In Section 4.4.3, we introduced a hexagonal-grid to rectangular-grid transformation that will enable us to use techniques developed for rectangular grids to solve hexagonal-grid problems. We find this transformation to be useful in several optimal array processing algorithms.

A number of other hexagonal-grid array issues are developed in the problems.

4.5 Nonplanar Arrays

In many applications of interest the physical location of the sensors must conform to the shape of the curved surface that they are mounted on. Examples include arrays mounted on submarines, aircraft, or missiles. These arrays are referred to as **conformal arrays**, and their properties are discussed in numerous references.

References that discuss various aspects of conformal arrays include Borgiotti [Bor87], Hansen [Han81], Mailloux [Mai84], Antonucci and Franchi [AF85], Hansen [Han98], and Kummer et al. [KSV73].

A complete discussion of the issues associated with the various conformal array geometries would take us too far afield, so we will restrict our attention to two commonly used geometries: cylindrical and spherical. We discuss cylindrical arrays in Section 4.5.1 and spherical arrays in Section 4.5.2.

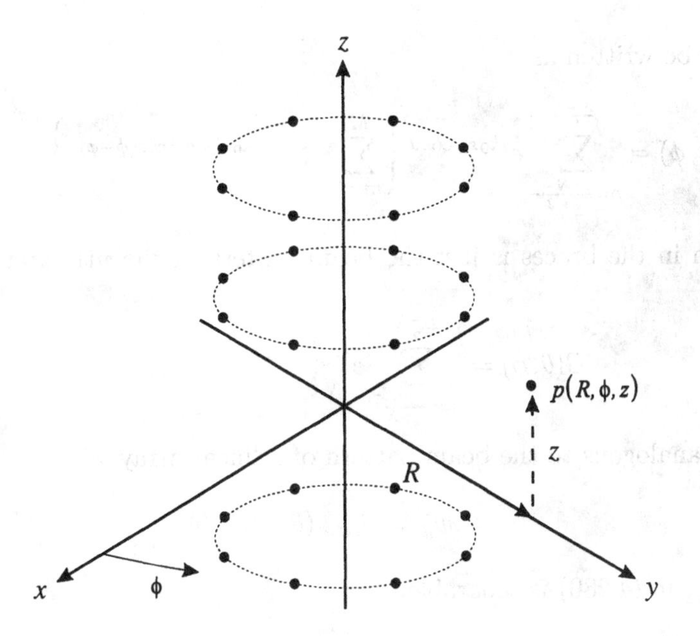

Figure 4.62 Cylindrical array geometry and the cylindrical coordinate system.

4.5.1 Cylindrical Arrays

The cylindrical array geometry and the cylindrical coordinate system are shown in Figure 4.62.

The element beam pattern are assumed to be isotropic. The beam pattern is given by

$$B(\theta,\phi) = \sum_{n=-\frac{N-1}{2}}^{\frac{N-1}{2}} \sum_{m=1}^{M} w_{nm}^* e^{-j\mathbf{k}^T\mathbf{p}_{nm}}, \quad N \text{ odd}, \tag{4.279}$$

where N is the number of circular arrays and M is the number of elements in each circular array. The center of the circular arrays is the z-axis and the array is symmetric in the z-direction about the origin.

We can use the discrete version of (4.134) to write

$$B(\theta,\phi) = \sum_{n=-\frac{N-1}{2}}^{\frac{N-1}{2}} \sum_{m=1}^{M} w_{nm}^* e^{jk_0[R\sin\theta\cos(\phi-\phi_1)+z_n\cos\theta]}. \tag{4.280}$$

This can be written as

$$B(\theta,\phi) = \sum_{n=-\frac{N-1}{2}}^{\frac{N-1}{2}} e^{jk_0 z_n \cos\theta} \left\{ \sum_{m=1}^{M} w_{nm}^* e^{jk_0 R \sin\theta \cos(\phi-\phi_1)} \right\}. \tag{4.281}$$

The term in the braces is just the beam pattern of the nth circular array. Thus,

$$B(\theta,\phi) = \sum_{n=-\frac{N-1}{2}}^{\frac{N-1}{2}} e^{jk_0 z_n \cos\theta} B_{cir,n}(\theta,\phi), \tag{4.282}$$

which is analogous to the beam pattern of a linear array with

$$w_n^* = B_{cir,n}(\theta,\phi). \tag{4.283}$$

If w_{nm}^* in (4.280) is separable,

$$w_{nm}^* = w_n^* w_m^*, \tag{4.284}$$

then (4.281) reduces to

$$\begin{aligned} B(\theta,\phi) &= \sum_{n=-\frac{N-1}{2}}^{\frac{N-1}{2}} w_n^* e^{jk_0 z_n \cos\theta} B_{cir}(\theta,\phi) \\ &= B_{lin}(\theta,\phi) B_{cir}(\theta,\phi), \end{aligned} \tag{4.285}$$

which is the pattern multiplication result from Chapter 2 (2.235).

We consider an example to illustrate a typical pattern.

Example 4.5.1

We consider a cylindrical array consisting of 11 circular arrays with radius $2\pi R = 10\lambda$. We assume that $M = 25$ so that we can utilize 21 phase modes.

We want to create a beam with a narrow beamwidth pointed at $\theta = 0$, $\phi = 0$. We use Dolph-Chebychev weighting in the z-direction and the Villenueve uniform phase mode excitation from Example 4.2.1 for each circular array.

The resulting beam pattern is shown in Figure 4.63. We see that it has acceptable sidelobe behavior.

In many, if not most, applications the elements will have a non-isotropic beam pattern. As in the ring array case, the main response axis of the element pattern will point in a radial direction. Synthesis of a desirable beam pattern is more complicated, but the techniques in Section 3.9.3 are applicable.

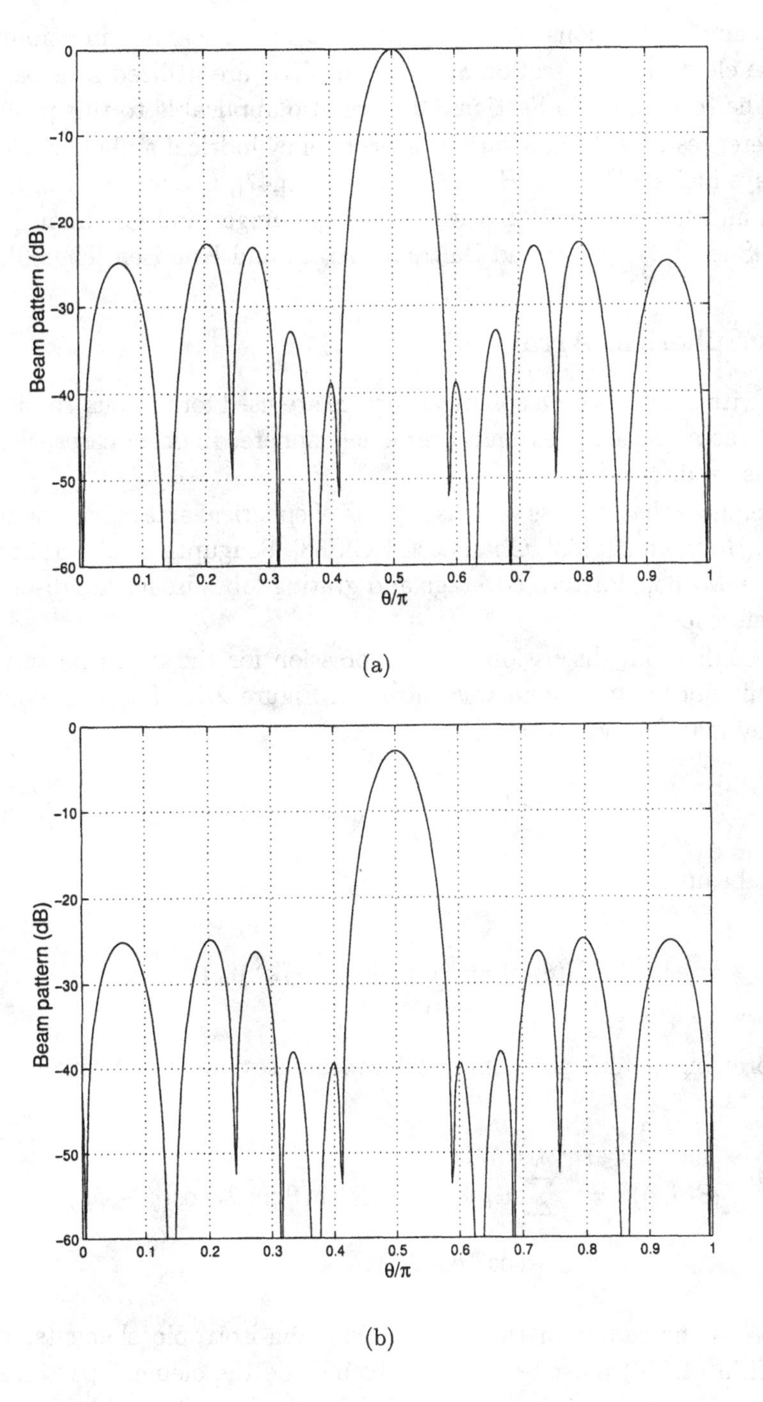

Figure 4.63 Beam pattern cuts at $\phi = 0$ and $\phi = 0.045\pi$.

In many applications, the main response axis is scanned in azimuth and only the elements in a section around the MRA are utilized at a particular time. The techniques in Section 3.9.3 are also applicable to this problem.

References that discuss various aspects of cylindrical and cylindrical sector arrays include Hansen [Han98], James [Jam47], Lee and Lo [LL65], Harrington and Lepage [HL52], Hessel [Hes70], Munger and Gladman [MG70], Sheleg [She75], Borgiotti and Balzano [BB72], and Knudsen [Knu59].

4.5.2 Spherical Arrays

Arrays with elements on a spherical surface are used for various applications. In most cases, the array extends over a hemisphere. In other cases, the entire sphere is used.

References that discuss various aspects of spherical arrays include Schrank [Sch72], Hoffman [Hof63], Chan et al. [CIS68], Sengupta et al. [SSL68], and MacPhie [MP68]. Pattern coverage and grating lobes issues are discussed in these references.

We confine our discussion to an expression for the beam pattern. The spherical coordinate system was shown in Figure 2.1. The nth element of the array manifold vector is

$$[\mathbf{v_k}(\mathbf{k})]_n = \exp\left(-j\mathbf{k}^T\mathbf{p}_n\right), \tag{4.286}$$

and the beam pattern is

$$B(\theta,\phi) = \sum_{n=1}^{N} w_n \exp\left(-j\mathbf{k}^T\mathbf{p}_n\right). \tag{4.287}$$

Expressing $\mathbf{k}$ and $\mathbf{p}$ in spherical coordinates, we can write the beam pattern as

$$\begin{aligned} B(\theta,\phi) &= \sum_{n=1}^{N} w_n \exp\left\{jk_0R\left[\sin\theta\sin\theta_n\cos(\phi-\phi_n)\right.\right.\\ &\qquad \left.\left.+\cos\theta\cos\theta_n\right]\right\}, \end{aligned} \tag{4.288}$$

where R is the radius of the sphere. For non-isotropic elements, the expression in (4.288) must be modified to include the element patterns. The technique in Section 3.9.3 can be used to synthesize a desired beam pattern. Several examples are developed in the problems.

4.6 Summary

In this chapter, techniques for analyzing and synthesizing planar arrays have been developed. The chapter focused on the array geometries: rectangular, rings, circular apertures, and hexagonal arrays that are commonly used in practice.

Section 4.1 considered arrays with sensors on a uniformly spaced rectangular grid. Many of the techniques developed for linear arrays carried over to rectangular grid arrays. Separable weightings produced sidelobe behavior that varied significantly as ϕ changed. Circularly symmetric weightings were developed to improve the behavior.

Section 4.2 considered ring apertures and ring arrays. The function $J_o(\cdot)$ played the same role in a ring aperture that sinc($\cdot$) did for a linear aperture. Phase mode excitation beamformers were developed that allowed the use of linear array weightings to synthesize desirable beam patterns.

Section 4.3 consider circular apertures. The discussion focussed on weighting functions that were separable in polar coordinates. Most of the examples consider desired beam patterns that were uniform in ϕ-space (a circularly symmetric pattern). A family of radial tapers was developed that were effective in controlling the sidelobe levels (Table 4.4). Synthesis of Taylor beam patterns was also developed. Techniques for synthesizing difference beams were developed.

Section 4.4 considered arrays whose sensors were located on an equilateral triangular grid that produced a hexagonal array. The elements lie on a set of concentric circles so the array is particularly suited to cases in which the desired beam pattern has circular symmetry. Beam pattern design techniques were developed. A hexagonal-to-rectangular transformation was developed that is useful in various applications.

Section 4.5 provided a brief discussion of nonplanar arrays.

There are several issues that we have not discussed in our development of classical array theory:

(i) In some applications, there is mutual coupling between the sensor elements. An input to sensor n will cause an output at sensor m.

(ii) In some applications, the incoming signals are polarized (either horizontal and vertical or circular).

(iii) In some applications, the wavefront impinging on the array has curvature.

Although these topics are not discussed in detail, in the discussion of optimum array processing algorithms later in the text, we occasionally point out how these issues impact the performance of the algorithm.

4.7 Problems

P4.1 Rectangular Arrays

Problem 4.1.1

Derive the expressions for θ_H given in (4.27).

Problem 4.1.2

Derive the expressions for ψ_H given in (4.29).

Problem 4.1.3

Assume $N = M = 11$ and $d_x = d_y = \lambda/2$. We use separable Hamming weighting in both the x and y direction: $\theta_0 = 0°$, $\phi_0 = 0°$. Plot the beam pattern and find the directivity D_0.

Problem 4.1.4

Repeat Problem 4.1.3 for separable Dolph-Chebychev weighting with −30-dB sidelobes.

Problem 4.1.5

Consider an 11×11 standard square array. (a) The array MRA is broadside. Assume we use a separable Kaiser weighting with $\beta = 6$ in the x and y directions. Find and plot the resulting beam pattern. Compute the directivity and beamwidth. (b) Repeat part (a) for $\theta_0 = 45°$, $\phi_0 = 30°$.

Problem 4.1.6

Consider a 17×17 square array with $d_x = d_y = \lambda/4$. Use the separable Taylor weighting with −30-dB SLL and $\bar{n} = 6$. Plot the beam pattern and compute the directivity and HPBW.

Problem 4.1.7

Use the $vec(\cdot)$ notation in Section 4.1.2 to derive the beam pattern for a delay-and-sum (conventional) beamformer for a signal arriving from ψ_{xT}, ψ_{yT}. Is the resulting beam pattern real?

Problem 4.1.8

In many cases, we are interested in beam pattern cuts at $\phi = \phi_c$. Modify the results in Section 4.1.2 to take into account that the beam pattern is a function of a single variable.

Problem 4.1.9

Verify that the results in Section 4.1.2 are valid for N and/or M odd. Are there any restrictions?

Problem 4.1.10

Assume N and M are even. Write $vec[\mathbf{w}]$ as

$$vec[\mathbf{w}] = \begin{bmatrix} \mathbf{w}_d \\ \hdashline -\mathbf{J}\,\mathbf{w}_d^* \end{bmatrix}, \tag{4.289}$$

where $\mathbf{w}_d$ is a $NM/2 \times 1$ vector corresponding to the upper half of $vec[\mathbf{w}]$. The total $vec[\mathbf{w}]$ is conjugate asymmetric. Find the resulting beam pattern.

Problem 4.1.11

Repeat Problem 4.1.5 for the circularly symmetric Kaiser weighting in (4.92).

Problem 4.1.12

Repeat Problem 4.1.5 for a circularly symmetric Hamming weighting. (a) Compare your results to the separable Hamming weightings in Example 4.1.1. (b) Compare the directivities of the two beam patterns.

Problem 4.1.13 [Ell81]

Find the separable weightings for a rectangular grid array with a rectangular boundary if $d_x = 5\lambda/8$, $d_y = 3\lambda/4$, $N = 8$, and $M = 12$, and if 25-dB and 35-dB Dolph-Chebychev patterns are desired in the xz and yz planes, respectively. Assume the main beam points at $\theta_0 = 0°$ and plot the -3-dB contour of the main beam. What are the heights of the off-axis sidelobes?

Problem 4.1.14 [Ell81]

In Problem 4.1.13, if the element pattern is hemispherically isotropic in $z > 0$ and is zero in $z < 0$, find the peak directivity. What is the areal beamwidth (defined as the area inside the -3-dB contour)? Find the changes in directivity and areal beamwidth if the beam is scanned to the position $\theta = 30°$, $\phi = 45°$.

Problem 4.1.15 [Ell81]

Design an equispaced planar array under the following specifications.

(a) Rectangular grid, rectangular boundary, separable distribution.

(b) Sum and difference pattern capability.

(c) Sum pattern scannable out to $\theta = 30°$ in any ϕ-cut.

(d) $\theta_{x0} = 14°$ and $\theta_{y0} = 20°$

(e) Both principal cuts are Dolph-Chebychev, -20 dB in the xz-plane and -15 dB in the yz-plane.

Problem 4.1.16 [Ell81]

Assume that the sum pattern weighting found for the array of Problem 4.1.13 is retained, except that the sign of the weighting is reversed for the two quadrants in which $x < 0$. Write an expression for the resulting difference pattern. Plot this difference pattern for $\phi = 0°/180°$ and discuss the SLL.

Problem 4.1.17 [Ell81]

A rectangular grid array with $d_x = d_y = 0.7\lambda$ has a circular boundary for which $R = 3\lambda$. Because of the cutoff corners, there are only 13 elements per quadrant. Find the weighting of this array if one uses Dolph-Chebychev separable weighting with −20-dB sidelobes and sets the weighting of the three cutoff elements equal to zero. Plot the beam pattern in the cuts $\phi = 0°, 15°, 30°$, and $45°$. Find the beamwidth and directivity.

Problem 4.1.18 [Ell81]

Repeat Problem 4.1.17, except use a Tseng-Cheng nonseparable weighting with −20-dB sidelobes. Find the beamwidth and directivity.

Problem 4.1.19

Repeat Problem 4.1.18 using the transformation in (4.119).

Problem 4.1.20

Consider a 17 × 17 square array with $d_x = d_y = \lambda/2$. Use the discretized Taylor pattern derived in Example 3.4.3 as a starting point.

(a) Use the Tseng-Cheng rotation to generate a 2-D beam pattern.

(b) Sample the pattern in wavenumber space. Use the inverse 2-D DFT to find the array weighting function.

Problem 4.1.21

Consider a standard 17 × 17 square array. Use the Villenueve $\bar{n}$ weighting discussed in Section 3.4.4 with $\bar{n} = 6$ and −25-dB sidelobes as a starting point. Repeat Problem 4.1.20.

Problem 4.1.22

Extend the results in Section 4.1.9 to the case of first derivative and second derivative nulls.

Problem 4.1.23

Repeat Example 4.1.9 with zero-, first- and second-order nulls at

$$\psi_{mx} = 0.25\pi, \quad \psi_{my} = 0.433\pi. \tag{4.290}$$

Problem 4.1.24

Repeat Example 4.1.9 with zero-order nulls at the following nine locations:

$$\psi_{mx} = 0.25\pi, \quad \psi_{my} = 0.41\pi, 0.43\pi, 0.45\pi, \tag{4.291}$$

$$\psi_{mx} = 0.27\pi, \quad \psi_{my} = 0.41\pi, 0.43\pi, 0.45\pi, \tag{4.292}$$

$$\psi_{mx} = 0.23\pi, \quad \psi_{my} = 0.41\pi, 0.43\pi, 0.45\pi. \tag{4.293}$$

Problem 4.1.25

Consider the model in Example 4.1.9. The nominal pattern is a symmetrical Dolph-Chebychev with −20-dB sidelobes.

We want to design an array weighting so that in the region $(20° \leq \theta \leq 50°) \cap (40° < \phi \leq 70°)$ the beam pattern is lower than −50 dB.

Design a nulling scheme to achieve this result. Use as few degrees of freedom as possible. Plot a contour plot of your result.

P4.2 Circular Arrays

Problem 4.2.1

Consider an 8-element circular array with equal spacing between elements. It could also be viewed as two 4-element rectangular arrays with the first array oriented along the x-y axes and the second array rotated by 45°. (a) Using this model, find the beam pattern and compare your result to the result in the text. (b) Assume $d = \lambda$ in the rectangular arrays. Plot the resulting beam pattern.

Problem 4.2.2

Extend the approach in Problem 4.2.1 to other values of N.

Problem 4.2.3

Consider a 10-element circular array with uniform weighting whose radius is $R = \lambda$. Assume that an element is added at the origin with weight w_0.

Choose w_0 to reduce the height of the first sidelobe.

Problem 4.2.4

Consider the cylindrical array in Figure 4.62. Assume the circular component has 10 isotropic elements in each circle separated by $d = \lambda/2$. There are 10 circular segments separated by $\lambda/2$. Assume uniform weighting of the 100 elements.

Plot the beam pattern.

Problem 4.2.5

Show that a uniform circular array can be modeled as a non-uniform linear array (e.g., pp. 205 of [Ma74]).

Problem 4.2.6

Show that a uniform circular array can be modeled as a set of uniform linear arrays where each ULA corresponds to a fixed θ (e.g., [TH92]).

Problem 4.2.7 [Ma74]

We want to derive the directivity of a single-ring circular array with isotropic elements.

$$D = \frac{4\pi \left|B_{max}\right|^2}{\int_0^{2\pi}\int_0^{\pi} \left|B(\theta,\phi)\right|^2 \sin\theta d\theta\, d\phi}. \tag{4.294}$$

(a) Show that a typical term in $|B(\theta,\phi)|^2$ can be written as,

$$w_m w_n^* \exp\left[jk\rho_{mn}\sin\theta\cos(\phi - \phi_{mn})\right], \tag{4.295}$$

$$\rho_{mn} = \begin{cases} 2R\sin\frac{|\phi_m - \phi_n|}{2}, & m \neq n, \\ 0, & m = n, \end{cases} \tag{4.296}$$

$$\phi_{mn} = \tan^{-1}\left[\frac{\sin\phi_m - \sin\phi_n}{\cos\phi_m - \cos\phi_n}\right], \quad m \neq n. \tag{4.297}$$

(b) Show the denominator can be written as $4\pi W$ where

$$W = \sum_{m=0}^{N-1}\sum_{n=0}^{N-1} w_m w_n^* \left(\frac{\sin k\rho_{mn}}{k\rho_{mn}}\right). \tag{4.298}$$

Hint: Utilize the following relation [AS65]:

$$\int_0^{\pi/2} J_0(x\sin\theta)\sin\theta\, d\theta = \left(\frac{\pi}{2}\right)^{1/2} \frac{J_{1/2}(x)}{(x)^{1/2}} = \frac{\sin x}{x}. \tag{4.299}$$

(c) Then,

$$D = \frac{|B(\theta_0, \phi_0)|^2}{W}. \tag{4.300}$$

Problem 4.2.8

Consider a uniform circular array with radius equal to 4λ. Assume N is chosen so $d = 0.4\lambda$.

(a) Using phase mode excitation, construct a Hamming pattern in ϕ-space in the xy-plane.

(b) Plot the beam pattern for $\theta = 30°$, $60°$, and $90°$.

P4.3 Circular Apertures

Problem 4.3.1

Consider a circular aperture with a separable weighting,

$$w_R(r) = \begin{cases} 1 - \left(\frac{r}{R}\right)^2, & 0 \le r \le R, \\ 0, & \text{elsewhere}, \end{cases} \tag{4.301}$$

and $w_\phi(\phi)$ is uniform.

Derive the results in the appropriate column of Table 4.4.

Problem 4.3.2 Repeat Problem 4.3.1 with

$$w_r(r) = \begin{cases} \left[1 - \left(\frac{r}{R}\right)^2\right]^2, & 0 \le r \le R, \\ 0, & \text{elsewhere}. \end{cases} \tag{4.302}$$

Problem 4.3.3

Consider the circular aperture in Problem 4.3.1 with the radial taper. Design an aperture weighting that is a least squares approximation to $w_r(r)$ with a constraint that there is a null at $\theta = \frac{110°}{R/\lambda}$ and $\phi = 0°$.

Problem 4.3.4 [Ell81]

A circular Taylor pattern, -20-dB SLL, $\bar{n} = 3$, is desired from a continuous circular aperture for which $R = 3\lambda$. Find A^2, σ, and the modified root position u_1 and u_2. Write the explicit expression for this Taylor pattern. Plot the pattern in $0° \le \theta \le 90°$ and the aperture distribution in $0 \le \rho \le 3\lambda$.

Problem 4.3.5 [Ell81]

A circular grid array with four concentric rings at radii $\rho/\lambda = 0.7, 1.4, 2.1$, and 2.8 is to be weighted to give a pattern approximating a circular Taylor -20-dB SLL, $\bar{n} = 3$. Determine the weightings (a) by conventional sampling of the Taylor distribution; (b) by matching to the nulls of the pattern found in Problem 4.3.4.

Problem 4.3.6 [Ell81]

Read the discussion in [Ell81] on Bayliss difference patterns. For a 6λ-diameter continuous circular aperture, find the weighting that will produce -20-dB SLL, $\bar{n} = 3$, Bayliss difference pattern. Write the explicit expressions for pattern function and aperture distribution. Plot both the pattern and the distribution.

Problem 4.3.7

Consider the following alternative to the Taylor circular aperture design procedure.

(a) Design a linear aperture using the techniques in Section 3.4.3.

(b) Transform the resulting pattern into two dimensions using the Tseng-Cheng transformation.

Utilize this approach for the model in Example 3.4.3 ($L = 10.5\lambda$) and compare the results to the Taylor circular procedure with $R = 5.25\lambda$.

P4.4 Hexagonal Arrays

Problem 4.4.1

Consider a standard 61-element hexagonal array. Plot the beam pattern for the following circular aperture ($R = 2.25\lambda$) weightings:

(a) Uniform

(b) $w_R(r) = 1 - \left(\frac{r}{R}\right)^2$

(c) $w_R(r) = \left[1 - \left(\frac{r}{R}\right)^2\right]^2$

Problem 4.4.2 (continuation)

Repeat Problem 4.4.1 for a standard 127-element hexagonal array.

Problem 4.4.3

(a) Plot the co-array for a standard 61-element hexagonal array.

(b) Discuss how to reduce the redundancy. Find several lower redundancy arrays.

Problem 4.4.4

Consider a standard 19-element hexagonal array.

(a) Develop a full-dimension (19) beamspace procedure using orthogonal beams.

(b) Plot the MRAs of the 19 beams in (u_x, u_y) space.

Problem 4.4.5

Show that (4.266) can be written as

$$B_{\mathbf{u}}(u_x, u_y) = \frac{1}{N_H} \sum_{m=-\frac{N_x-1}{4}, m \text{ odd}}^{\frac{N_x-1}{4}} \left\{ \left[\sum_{n=1}^{\frac{N_x-|m|}{2}} e^{j\pi(n-\frac{1}{2})u_x} + \sum_{n=-\frac{N_x-|m|}{2}}^{-1} e^{j\pi(n+\frac{1}{2})u_x} \right] e^{jm\frac{\sqrt{3}}{2}u_y} \right\} + \sum_{m=-\frac{N_x-1}{2}}^{\frac{N_x-1}{2}} \left\{ \left[\sum_{n=-\frac{N_x-1-|m|}{2}}^{\frac{N_x-1-|m|}{2}} e^{j\pi n u_x} \right] e^{jm\frac{\sqrt{3}}{2}u_y} \right\}. \tag{4.303}$$

Problem 4.4.6

Consider a standard 91-element hexagonal array. Design a beam pattern that has uniform sidelobes that are −40 dB lower than the main lobe. Assume the MRA is broadside and try to minimize the diameter of the main lobe.

P4.5 Nonplanar Arrays

Problem 4.5.1

Consider two parallel standard N-element linear arrays in the xy-plane. The first array lies on the x-axis and is centered at the origin. The second array is parallel to the first array at $y = d_{sep}$ and is centered on the y-axis.

Assume $N = 40$. Plot the beam pattern for various d_{sep}.

Problem 4.5.2

Consider the array in Example 4.5.1. We want to steer the array to θ_0, ϕ_0. Use Hamming weighting in the z-direction and Hamming weighting of the phase modes.

(a) Find the appropriate weights and plot the beam pattern.

(b) What is the HPBW?

Bibliography

[AF85] J. Antonucci and P. Franchi. A simple technique to correct for curvature effects on conformed phase arrays. *Proc. 1985 Antenna Applications Symposium.* RADC/TR-85-743, Vol. 2, December 1985.

[AGW72] N. Amitay, V. Galindo, and C. P. Wu. *Theory and Analysis of Phased Array Antennas.* Wiley-Interscience, New York, 1972.

[AS65] M. Abramowitz and I. A. Stegun. *Handbook of Mathematical Functions.* Dover Publications, New York, 1965.

[Bag76] A. B. Baggeroer. Space-time processes and optimal array processing. Technical Report 506, Navy Undersea Center, San Diego, California, December 1976.

[Bak66] Y. V. Baklanov. Chebyshev distribution of currents for a plane array of radiators. *Radio Eng. Electron. Phys.*, vol.11, pp. 640–642, April 1966.

[Bal82] C. A. Balanis. *Antenna Theory Analysis and Design.* Wiley, New York, 1982.

[Bay68] E. T. Bayliss. Design of monopulse antenna difference patterns with low side lobes. *Bell System Tech. J.*, vol.47, pp. 623–640, May–June 1968.

[BB72] G. V. Borgotti and Q. Balzano. Analysis and element pattern design of periodic arrays of circular aperture on conducting cylinders. *IEEE. Trans. Antennas Propagat.*, vol.AP–20, pp. 547–555, September 1972.

[BVT99] K. L. Bell and H. L. Van Trees. Adaptive and non-adaptive beampattern control using quadratic beampattern constraints. *Proc. 33rd Asilomar Conference on Signals, Systems, and Computers*, Pacific Grove, California, pp. 486–490, October 1999.

[Bor87] G. V. Borgiotti. Conformal arrays. In A. Rudge et al., editors, *The Handbook of Antenna Design*, vol.2, chapter 11, Peter Peregrinus, London, 1987.

[CIS68] A. K. Chan, A. Ishimaru, and R. A. Sigelmann. Equally spaced spherical arrays. *Radio Sci.*, vol.3, pp. 401–404, May 1968.

[Dav65] D. E. N. Davies. A transformation between the phasing techniques required for linear and circular aerial arrays. *Proc. IEE*, vo.112, pp. 2041–2045, November 1965.

[Dav87] D. E. N. Davies. Circular arrays. In A. Rudge et al., editors, *The Handbook of Antenna Design*, vol.2, chapter 12, Peter Peregrinus, London, 1987.

[DM84] D. E. Dudgeon and R. M. Mersereau. *Multidimensional Digital Signal Processing.* Prentice-Hall, Englewood Cliffs, New Jersey, 1984.

[Ell64] R. S. Elliott. Beamwidth and directivity of large scanning arrays. *Microwave J.*, pp. 74–82, January 1964.

[Ell81] R. S. Elliott. *Antenna Theory and Design.* Prentice-Hall, Englewood Cliffs, New Jersey, 1981.

[Han59] R. C. Hansen. Tables of Taylor distributions for circular aperture antennas. Technical Report 587, Hughes Aircraft Co., Culver City, California, February 1959.

[Han60] R. C. Hansen. Tables of Taylor distributions for circular aperture antennas. *IRE Trans. Antennas Propagat.*, vol.AP-8, pp.22–26, January 1960.

[Han81] R. C. Hansen. *Conformal Antenna Array Design Handbook.* Dept. of the Navy, Air Systems Command, AD A11091, September 1981.

[Han98] R.C. Hansen, editor. *Phased Array Antennas.* Wiley, New York, 1998.

[Hes70] A. Hessel. Mutual coupling effects in circular arrays on cylindrical surfaces–aperture design implications and analysis. In A. A. Oliner and G. H. Knittel, editors, *Phased Array Antennas*, pp. 273–291, Artech House, Boston, 1972.

[HL52] R. F. Harrington and W. R. Lepage. Directional antenna arrays of elements circularly disposed about a cylindrical reflector. *Proc. IRE*, vol.40, pp. 83–86, January 1952.

[Hof63] M. Hoffman. Conventions for the analysis of spherical arrays. *IEEE Trans. Antennas Propagat.*, vol.AP-11, pp. 390–393, July 1963.

[Hua81] T. S. Huang. Two-dimensional digital signal processing I. In *Topics in Applied Physics*, vol.42, Springer-Verlag, New York, 1981.

[Jam47] P. W. James. Polar patterns of phase-connected circular arrays. *Proc. IEE*, vol.112, pp. 1839–1847, 1965.

[Jas61] H. Jasik, editor. *Antenna Engineering Handbook*, pp. 2-25–2-26. McGraw-Hill, New York, 1961.

[Knu59] H. Knudsen. Antennas on circular cylinders. *IRE Trans. Antennas Propagat.*, vol.AP–7, pp. S361–S370, December 1959.

[KSV73] W. H. Kummer, A. F. Seaton and A. T. Villeneuve. Conformal antenna arrays study. Final Report, AD–909220, Hughes Aircraft Co., Culver City, California, January 1973.

[LL65] S. W. Lee and Y. T. Lo. Pattern function of circular arc arrays. *IEEE Trans. Antennas Propagat.*, vol.AP–13, pp. 649–650, July 1965.

[LL83] Y. T. Lo and S. W. Lee. *Antenna Handbook, Volume II (Antenna Theory).* Van Nostrand Reinhold, New York, 1983.

[Ma74] M. T. Ma. *Theory and Applications of Antenna Arrays.* Wiley, New York, 1974.

[Mai84] R. J. Mailloux. Conformal and Low-Profile Arrays. In R. C. Johnson and H. Jasik, editors, *Antenna Engineering Handbook*, chapter 21, McGraw-Hill, New York, 1984.

[Mai94] R. J. Mailloux. *Phased Array Antenna Handbook.* Artech House, Boston, 1994.

[McC82] J. H. McClellan. Multidimensional spectral estimation. *Proc. IEEE*, vol.70, no.9, pp. 1029–1039, September 1982.

[Mer79] R. M. Mersereau. The processing of hexagonally sampled two-dimensional signals. *Proc. IEEE*, vol.67, no.6, pp. 930–949, July 1979.

[MG70] A. D. Munger and B. R. Gladman. Pattern analysis for cylindrical and conical arrays. *Proc. Conformed Array Conference*, AD–875 378, Naval Electronics Lab. Center, San Diego, California, January 1970.

[MP68] R. H. MacPhie. The element density of a spherical antenna array. *IEEE Trans. Antennas Propagat.*, vol.AP–16, pp. 125–127, January 1968.

[MZ94] C. P. Mathews and M. D. Zoltowski. Eigenstructure techniques for 2-D angle estimation with uniform circular arrays. *IEEE Trans. Signal Process.*, vol.42, no.9, pp. 2395–2407, September 1994.

[PM62] D. P. Peterson and D. Middleton. Sampling and reconstruction of wave-number limited functions in n-dimensional Euclidean spaces. *Inf. Control*, vol.5, pp. 279–323, April 1962.

[R+83] A. W. Rudge et al., editors. *The Handbook of Antenna Design*, chapter 12. Peregrinus, London, 1983.

[RD82] T. Rahim and D. E. N. Davies. Effect of directional elements on the directional response of circular arrays *Proc. IEE*, vol.129, Part 11, no.1, pp. 18–22, February 1982.

[RG75] L. R. Rabiner and B. Gold. *Theory and Application of Digital Signal Processing.* Prentice-Hall, Englewood Cliffs, New Jersey, 1975.

[Sch72] H. E. Schrank. Basic theoretical aspects of spherical phased arrays. In A. A. Oliner and G. H. Knittel, editors, *Phased Array Antennas.*, pp. 323–327, Artech House, Boston, 1972.

[SSL68] D. L. Sengupta, T. M. Smith and R. W. Larson. Radiation characteristics of a spherical array of circularly polarized elements. *IEEE Trans. Antennas Propagat.*, vol.AP-16, pp. 2–7, January 1968.

[Sha61] E. D. Sharp. A triangular arrangement of planar-array elements that reduces the number needed. *IRE Trans. Antennas Propagat.*, vol.AP-3, pp. 126–129, January 1961.

[She75] B. Sheleg. Circular and cylindrical arrays. *Workshop on Conformal Antennas*, AD-A015 630, Naval Air Systems Command, pp. 107–138, April 1975.

[SK82] J. C. Sureau and K. J. Keeping. Sidelobe control in cylindrical arrays. *IEEE Trans. Antennas Propagat.*, vol.AP-37, pp. 1017–1018, 1989.

[Ste81] S. Stein. Algorithms for ambiguity function processing. *IEEE Trans. Acoust., Speech, Signal Processing*, vol.ASSP-29, no.3, pp. 588–599, June 1981.

[Tay60] T. T. Taylor. Design of circular apertures for narrow beamwidth and low sidelobes. *IRE Trans. Antennas Propagat.*, vol.AP-8, pp. 17–22, 1960.

[TC68] F. I. Tseng and D. K. Cheng. Optimum scannable planar arrays with an invariant side-lobe level. *Proc. IEEE*, vol.56, pp. 1771–1778, 1968.

[TH92] A. H. Tewfik and W. Hong. On the application of uniform linear array bearing estimation techniques to uniform circular arrays. *IEEE Trans. Signal Process.*, vol.40, no.4, pp. 1008–1011, April 1992.

[Zio95] L. J. Ziomek. *Fundamentals of Acoustic Field Theory and Space-Time Signal Processing.* CRC Press, Boca Raton, Florida, 1995.

Chapter 5

Characterization of Space-time Processes

5.1 Introduction

In this chapter we will develop techniques for characterizing space-time random processes and their interaction with arrays and apertures. It produces the statistical basis for the remainder of the text, where we emphasize a statistical approach rather than a deterministic approach to array analysis and synthesis.

In the discussion of classical array processing the input to the array was assumed to be a sinusoidal plane-wave signal. It was characterized by its frequency and wavenumber or by its frequency and direction of arrival. In most of our discussions, we focused on a single frequency, and the beamformer weights were fixed complex numbers.

In the statistical approach, the input to the array consists of desired signals, interfering signals, and noise. Some or all of these inputs are modeled as sample functions of space-time random processes.

These processes may be spread over regions of ω-$\mathbf{k}$ space. The array processor operates on the array output to estimate a waveform, detect a signal, or estimate parameters of a signal. These operations are the vector extensions of the scalar detection, estimation, and modulation problems that were solved in DEMT I [VT68], [VT01a] and DEMT III [VT71], [VT01b]. The first step was to operate on the continuous time functions to obtain a set of random variables that were used to solve the detection or estimation problem. We used a Karhunen-Loève (KL) expansion that was applicable to non-stationary processes and finite time intervals and generated statistically independent Gaussian random variables when the input was a Gaussian random process. This approach can be extended to the vector case.

In most applications and in the majority of the research literature, an alternative approach using sampling is employed. These samples are normally referred to in the literature as **snapshots**. In Section 5.2, snapshot models in the frequency domain and time domain are developed. The objective of the snapshot generation process is to operate on the sensor output and generate a sequence of vectors. Appropriate processing is performed on this vector sequence.

Section 5.3 develops models for spatially spread random processes. A second-moment characterization of the random fields is developed. If processing consists of linear filters and quadratic error criteria, then the second-moment characterization is adequate. However, if we want to develop optimum detectors and estimators, then a complete statistical characterization is necessary. Gaussian space-time processes are defined that are completely characterized by the second moments.

In Section 5.4, the output statistics of arrays and apertures are developed for the case in which the input is a space-time random process. For an array the primary statistic of interest is the **spatial spectral matrix,** $\mathbf{S_x}$, of the snapshot vectors. We will show how the classical beam patterns developed in Chapters 2–4 interact with the frequency-wavenumber characterization of space-time processes.

In Section 5.5, an eigendecomposition of the $N \times N$ spatial spectral matrix $\mathbf{S_x}$ is performed. The resulting eigenvalues and eigenvectors play a key role in many of the subsequent discussions. The concept of signal and noise subspaces is introduced. The subspace concept also plays a key role in subsequent discussions.

Section 5.6 develops parametric models for the frequency-wavenumber spectrum. Various rational transfer function models such as auto-regressive (AR) and auto-regressive moving average (ARMA) models are developed.

Section 5.7 provides a brief summary of the chapter. The structure of Chapter 5 is shown in Table 5.1.

5.2 Snapshot Models

In this section, the snapshot models that play a key role in the subsequent discussion are developed. In Section 5.2.1, frequency-domain snapshot models are developed. In Section 5.2.2, time-domain snapshot models are developed. In Section 5.2.3, the results are summarized.

Table 5.1 Structure of Chapter 5.

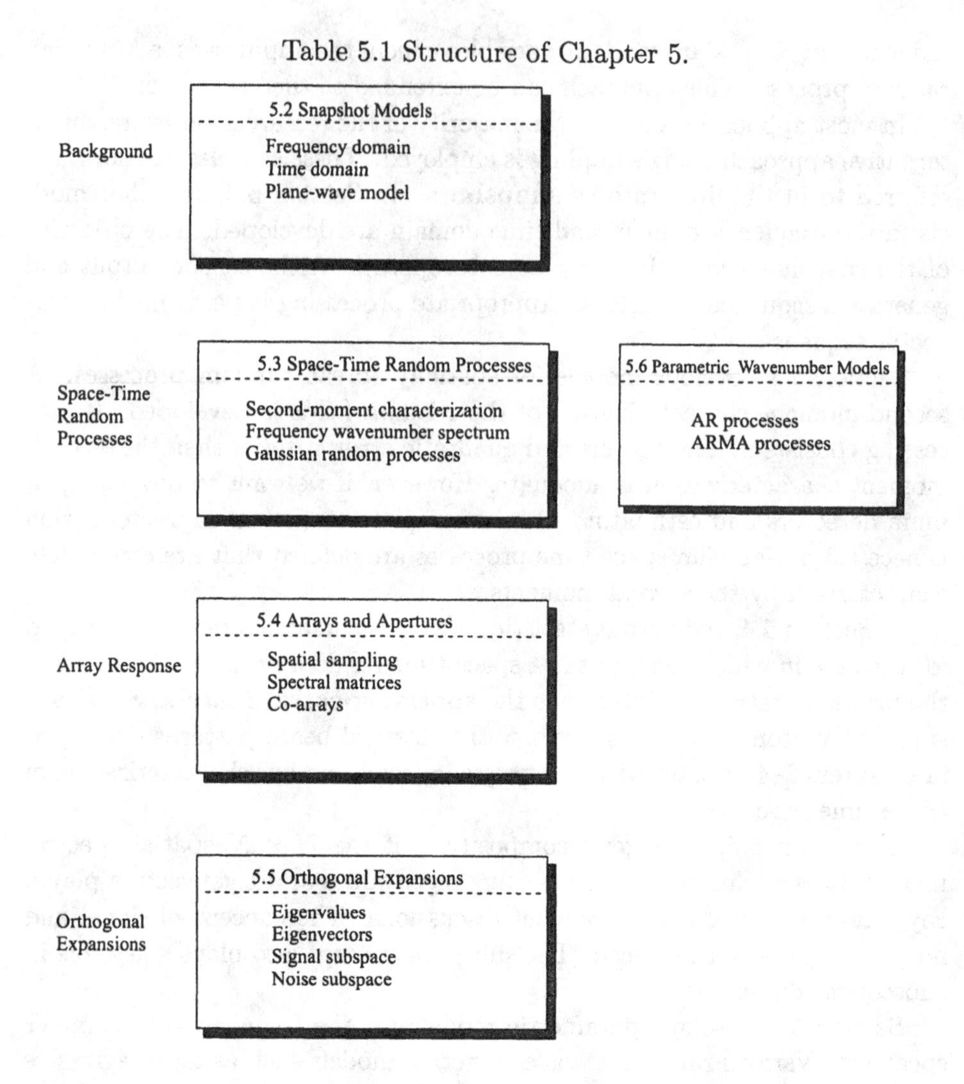

5.2.1 Frequency-domain Snapshot Models

In many applications, we implement the beamforming in the frequency domain. The model is shown in Figure 5.1. The first box converts the sensor input vector from the time domain to the frequency domain. The second box processes the frequency-domain vector to obtain a scalar frequency-domain function. The third box converts the frequency-domain function into a scalar time-domain waveform. The functions in each box are now described.

The objective of the first processor is to generate a set of complex vectors that can be processed to form a beam. We refer to these complex vectors as

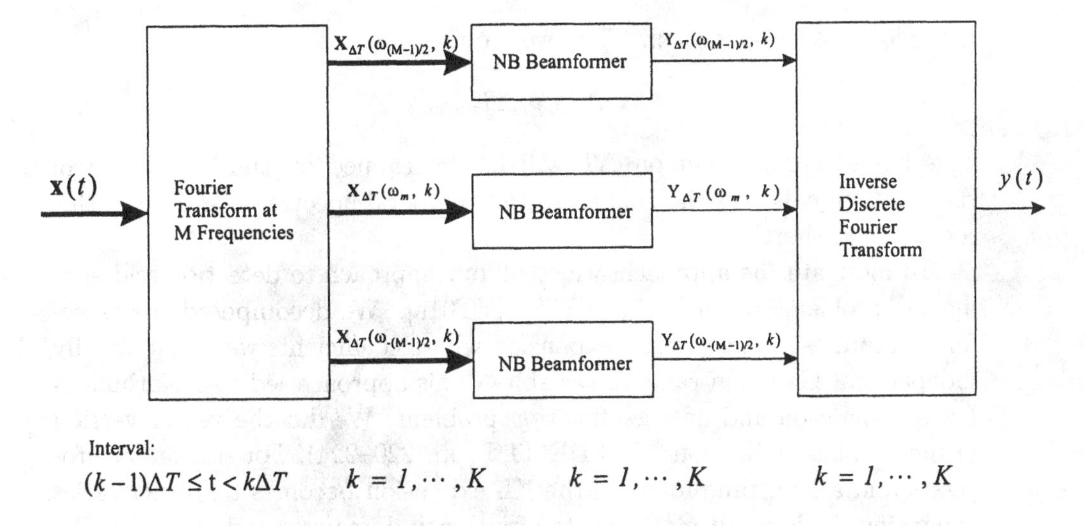

Figure 5.1 Frequency-domain beamformer.

frequency-domain snapshots.

In order to generate these vectors, first divide the total observation interval T into K disjoint intervals of length ΔT. These intervals are indexed with k,

$$\begin{aligned}
k &= 1 \qquad 0 \leq t < \Delta T \\
k &= 2 \qquad \Delta T \leq t < 2\Delta T \\
&\vdots \\
k &= k \qquad (k-1)\Delta T \leq t < k\Delta T \\
&\vdots \\
k &= K \qquad (K-1)\Delta T \leq t < K\Delta T.
\end{aligned} \tag{5.1}$$

As part of the development, we will develop criteria for choosing ΔT.

The first requirement is that ΔT must be significantly greater than the propagation time across the array. Define

$$\Delta T_{\max} \triangleq \max_{d=1,\cdots,D} \left\{ \max_{i,j} \left[\Delta T_{ij}^{(d)} \right] \right\}, \tag{5.2}$$

where $\Delta T_{ij}^{(d)}$ is the travel time between the i and j elements for the dth signal. In a linear array, the maximum propagation time corresponds to a signal arriving from endfire and equals the length of the array divided by the velocity of propagation. Then we require

$$\Delta T \gg \Delta T_{max}. \tag{5.3}$$

The second requirement on ΔT will be determined by the bandwidth of the input signals and the shape of their temporal spectra. We discuss this requirement shortly.

To motivate the approach, we recall our approach to detection and estimation problems in DEMT I [VT68], [VT01a]. We decomposed the waveform of interest using a series expansion whose coefficients were statistically independent Gaussian random variables. This approach led to a Karhunen-Loève expansion and an eigenfunction problem. We did the vector version of the problem in Section 3.7 of DEMT I (pp. 220–224). For stationary processes and a long time interval, the KL expansion becomes a Fourier series expansion. This result motivates the frequency-domain snapshot model. We define this model and analyze the covariance between its components as a function of ΔT.

We assume that the input is a zero-mean bandpass process centered at ω_c. The signal at the origin is $x(t)$. The bandpass spectrum of $x(t)$ and its lowpass equivalent are shown in Figure 5.2.

We first consider the interval $(0, \Delta T)$ and define[1]

$$\begin{aligned} \mathbf{X}_{\Delta T}(\omega_m) &= \frac{1}{\sqrt{\Delta T}} \int_0^{\Delta T} \mathbf{x}(t) e^{-j(\omega_c + m\omega_\Delta)t} dt, \\ m &= -\frac{M-1}{2}, \cdots, 0, \cdots, \frac{M-1}{2}, \end{aligned} \tag{5.4}$$

where the mth Fourier series frequency is

$$\omega_m = \omega_c + m\omega_\Delta, \tag{5.5}$$

and the resolution of the transform is

$$\omega_\Delta = \frac{2\pi}{\Delta T}. \tag{5.6}$$

The expression in (5.4) assumes M is odd. For M even, the limits are $m = -M/2, \cdots, 0, \cdots, \frac{M}{2} - 1$. The value of M will depend on the bandwidth of the bandpass process. For M odd,

$$M = \lfloor B_s \cdot \Delta T \rfloor + 1, \tag{5.7}$$

[1] This discussion is based on Hodgkiss and Nolte [HN76]. Several of their key formulas are due to Blackman [Bla57].

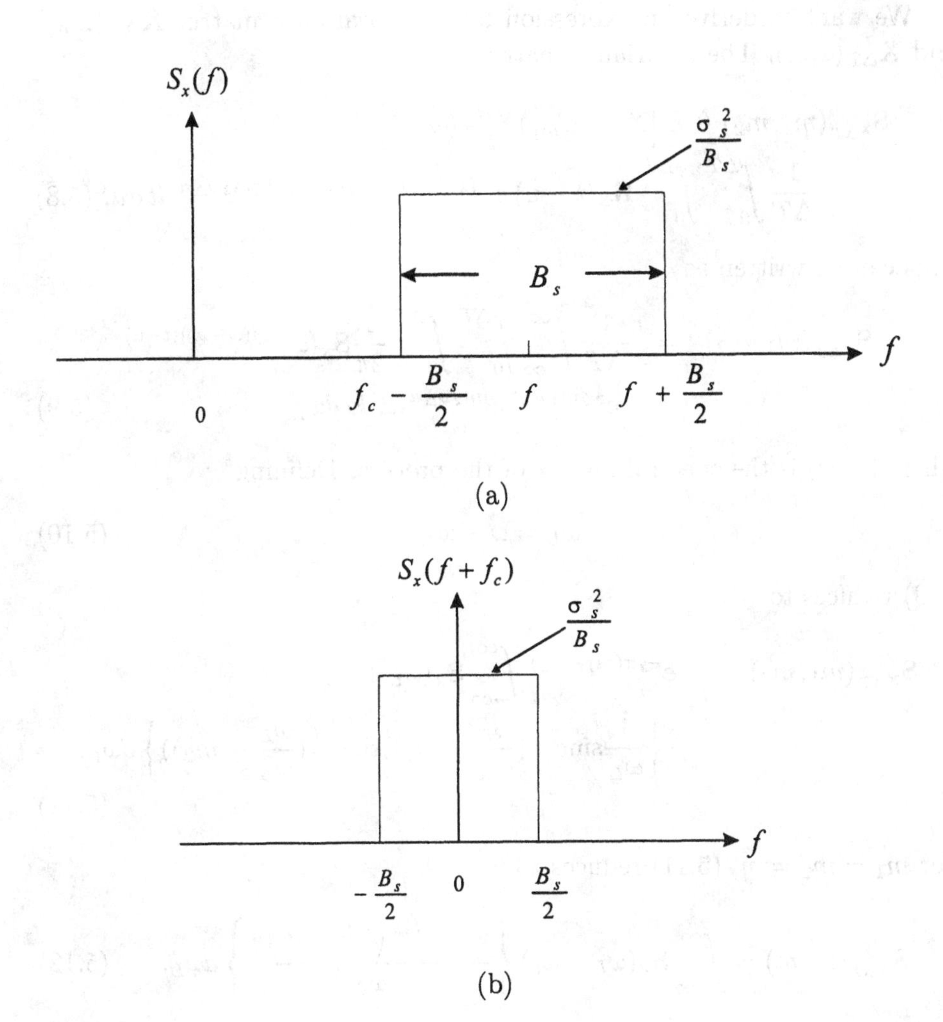

Figure 5.2 Bandpass signal spectrum and corresponding lowpass spectrum.

where $\lfloor B_s \cdot \Delta T \rfloor$ denote the largest integer less that $B_s \cdot \Delta T$. Each of the $\mathbf{X}_{\Delta T}(\omega_m)$ is an N-dimensional complex vector corresponding to the Fourier series coefficient at ω_m. We refer to each of the ω_m as the center frequency of a **frequency bin** whose width is ω_Δ.

We want to derive an expression for the covariance matrix $\mathbf{X}_{\Delta T}(\omega_{m_1})$ and $\mathbf{X}_{\Delta T}(\omega_{m_2})$. The covariance matrix is

$$\begin{aligned}\mathbf{S}_{\mathbf{x}_{\Delta T}}(m_1, m_2) &\triangleq E\left[\mathbf{X}_{\Delta T}(\omega_{m_1})\mathbf{X}_{\Delta T}^H(\omega_{m_2})\right] \\ &= \frac{1}{\Delta T}\int_0^{\Delta T}\int_0^{\Delta T} \mathbf{R}_{\mathbf{x}}(t-u)\, e^{-j\omega_c(t-u)}\, e^{-jm_1\omega_\Delta t + jm_2\omega_\Delta u}\, dt\, du. \end{aligned} \tag{5.8}$$

This can be written as

$$\begin{aligned}\mathbf{S}_{\mathbf{x}_{\Delta T}}(m_1, m_2) &= \frac{1}{\Delta T}\int_{-\infty}^{\infty}\int_0^{\Delta T}\int_0^{\Delta T} \frac{1}{2\pi}\mathbf{S}_{\mathbf{x}}(\omega)\, e^{j(\omega-\omega_c)(t-u)} \\ &\quad \cdot e^{-jm_1\omega_\Delta t}\, e^{jm_2\omega_\Delta u}\, dt\, du\, d\omega, \end{aligned} \tag{5.9}$$

where $\mathbf{S}_{\mathbf{x}}(\omega)$ is the spectral matrix of the process. Defining

$$\omega_L = \omega - \omega_c, \tag{5.10}$$

(5.9) reduces to

$$\begin{aligned}\mathbf{S}_{\mathbf{x}_{\Delta T}}(m_1, m_2) &= e^{-j\pi(m_1-m_2)}\int_{-\infty}^{\infty} \mathbf{S}_{\mathbf{x}}(\omega_L + \omega_c) \\ &\quad \cdot \left\{\frac{1}{\omega_\Delta}\text{sinc}(\pi(\frac{\omega_L}{\omega_\Delta} - m_1))\text{sinc}(\pi(\frac{\omega_L}{\omega_\Delta} - m_2))\right\} d\omega_L. \end{aligned} \tag{5.11}$$

For $m_1 = m_2 = m$, (5.11) reduces to

$$\mathbf{S}_{\mathbf{x}_{\Delta T}}(m, m) = \int_{-\infty}^{\infty} \mathbf{S}_{\mathbf{x}}(\omega_L + \omega_c)\left\{\frac{\text{sinc}^2\left(\pi\left(\frac{\omega_L}{\omega_\Delta} - m\right)\right)}{\omega_\Delta}\right\} d\omega_L. \tag{5.12}$$

We first consider the diagonal terms that correspond to the covariance at a single sensor. For the nth sensor,

$$[\mathbf{S}_{\mathbf{x}_{\Delta T}}(m, m)]_{nn} = \int_{-\infty}^{\infty} [\mathbf{S}_{\mathbf{x}}(\omega_L + \omega_c)]_{nn}\left\{\frac{\text{sinc}^2\left(\pi\left(\frac{\omega_L}{\omega_\Delta} - m\right)\right)}{\omega_\Delta}\right\} d\omega_L. \tag{5.13}$$

In Figure 5.3, we show the components of the integrand for various values of ω_Δ. Figure 5.3(a) shows the $\text{sinc}^2(\cdot)$ function for several values of m. In Figure 5.3(b), we plot $[\mathbf{S_x}(\omega_L + \omega_c)]_{nn}$ versus (ω_L/ω_Δ) for the case in which

$$B_s \cdot \Delta T = 16. \tag{5.14}$$

We typically use the frequency-domain snapshot approach when the time-bandwidth product $B_s \cdot \Delta T$ is large. In practice, $B_s \cdot \Delta T$ products from 16 to 512 are common. The value of ω_Δ determines the frequency resolution of the model and $2\pi B_s/w_\Delta$ determines the number of terms in the Fourier series expression.

In the limit, as $\Delta T \to \infty$, the term in braces in (5.13) approaches an impulse and

$$\lim_{\Delta T \to \infty} \left\{ [\mathbf{S}_{\mathbf{x}_{\Delta T}}(m, m)]_{nn} \right\} = [\mathbf{S_x}(\omega_c + m\omega_\Delta)]_{nn} \,. \tag{5.15}$$

This result is the Wiener-Khinchin theorem (e.g., [Hel91]). A similar result holds for the off-diagonal terms, so the matrix version of (5.15) is

$$\lim_{\Delta T \to \infty} \left\{ \mathbf{S}_{\mathbf{x}_{\Delta T}}(m, m) \right\} = \mathbf{S_x}(\omega_c + m\omega_\Delta). \tag{5.16}$$

We now consider a simple example to illustrate the behavior as a function of $B_s \cdot \Delta T$.

Example 5.2.1

The spectrum of the signal is shown in Figure 5.2. We evaluate the integral in (5.12) for

$$B_s \cdot \Delta T = 2^l, \quad l = 0, \cdots, 8. \tag{5.17}$$

In Table 5.2, we show the normalized value, $[\mathbf{S}_{\mathbf{x}_{\Delta T}}(m, m)]_{nn} / (\sigma_s^2/B_s)$ for the given values of l and selected values of $|m|$.

Table 5.2

$\|m\|$ / l	0	1	2	4	8	16	32	64	128
0	0.7737	0.0787	0.0140	0.0032	0.0008	0.0002	0.0000	0.0000	0.0000
1	0.9028	0.4750	0.0318	0.0067	0.0016	0.0004	0.0001	0.0000	0.0000
2	0.9499	0.9346	0.4874	0.0166	0.0034	0.0008	0.0002	0.0000	0.0000
3	0.9747	0.9731	0.9665	0.4937	0.0084	0.0017	0.0004	0.0001	0.0000
4	0.9873	0.9871	0.9865	0.9832	0.4968	0.0042	0.0008	0.0002	0.0000
5	0.9937	0.9936	0.9936	0.9932	0.9916	0.4984	0.0021	0.0004	0.0001
6	0.9968	0.9968	0.9968	0.9968	0.9966	0.9958	0.4992	0.0011	0.0002
7	0.9984	0.9984	0.9984	0.9984	0.9984	0.9983	0.9979	0.4996	0.0005
8	0.9992	0.9992	0.9992	0.9992	0.9992	0.9992	0.9992	0.9989	0.4998

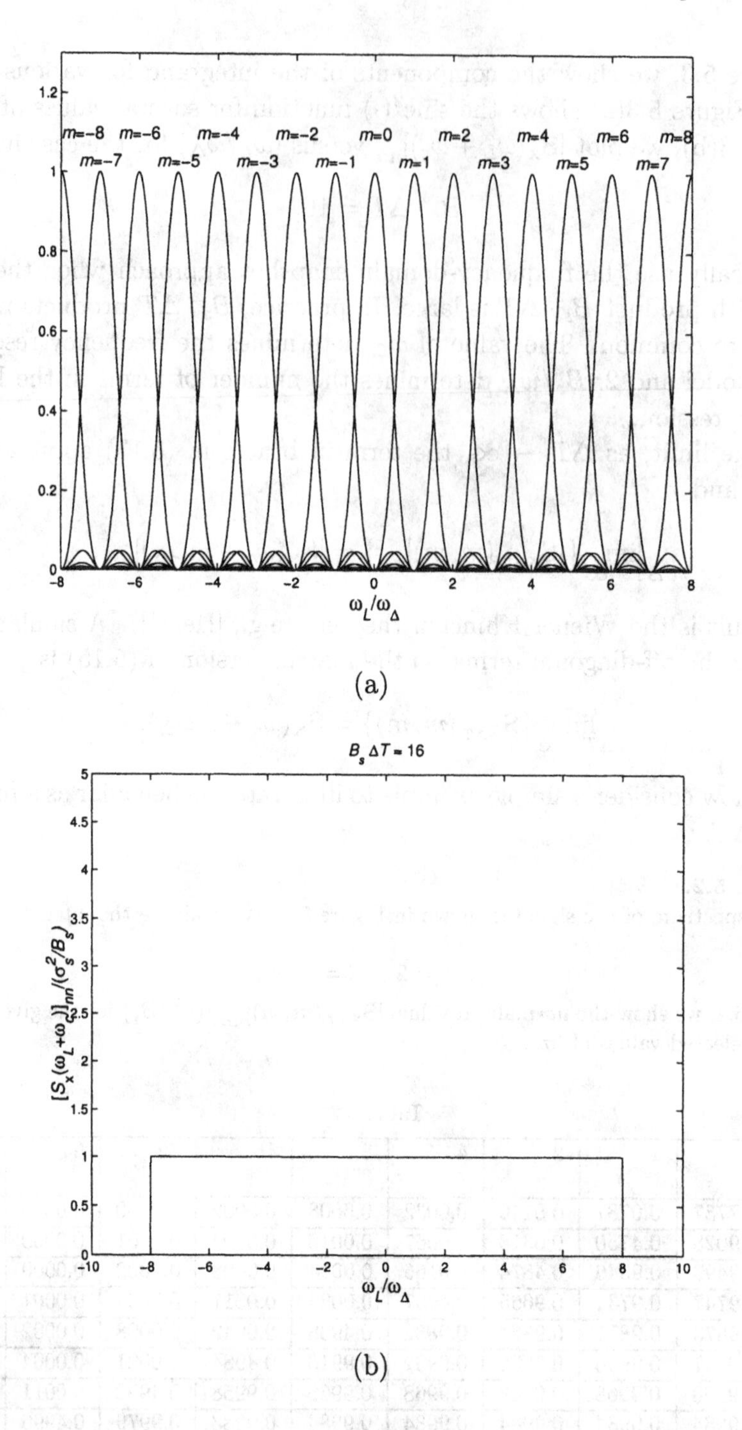

Figure 5.3 Functions in integrand: (a) $\mathrm{sinc}^2(\cdot)$ functions for various m; (b) spectrum of bandpass process ($B_s \cdot \Delta T = 16$).

We see that, for $B_s \cdot \Delta T \geq 16$, the approximation in (5.15) is accurate.

In practice, many of the spectra of interest are not flat. However, if the spectrum is constant over an interval $\pm 2\omega_\Delta - \pm 3\omega_\Delta$ around $m\omega_\Delta$, then we can use the approximation,

$$[\mathbf{S}_{\mathbf{x}_{\Delta T}}(m,m)]_{nn} \simeq [\mathbf{S}_{\mathbf{x}}(\omega_c + m\omega_\Delta)]_{nn} = [\mathbf{S}_{\mathbf{x}}(\omega_m)]_{nn}. \tag{5.18}$$

Example 5.2.1 and the approximation in (5.18) deal with the diagonal terms in $\mathbf{S}_{\mathbf{x}_{\Delta T}}(m,m)$. To investigate the off-diagonal terms we assume a plane-wave model. Carrying out a similar analysis, we find that if the spectrum is constant over $\pm 2\omega_\Delta - \pm 3\omega_\Delta$, then we can use the approximation

$$\mathbf{S}_{\mathbf{x}_{\Delta T}}(m,m) \simeq \mathbf{S}_{\mathbf{x}}(\omega_c + m\omega_\Delta) = \mathbf{S}_{\mathbf{x}}(\omega_m). \tag{5.19}$$

The frequency-domain snapshot model is most appropriate for large $B_s \cdot \Delta T$ products. However, it is useful to consider the case in which

$$B_s \cdot \Delta T \ll 1. \tag{5.20}$$

Now the $\text{sinc}^2(\cdot)$ is approximately unity over the interval where the spectrum is non-zero and we can approximate the integral for $m = 0$ by

$$\begin{aligned} [\mathbf{S}_{\mathbf{x}_{\Delta T}}(0,0)]_{nn} &\simeq \int_{-\pi B_s}^{\pi B_s} \frac{\sigma_s^2}{B_s} \cdot \frac{1}{\omega_\Delta} d\omega \\ &= \sigma_s^2 \Delta T = [\mathbf{S}_{\mathbf{x}}(\omega_c)]_{nn} \cdot \Delta T. \end{aligned} \tag{5.21}$$

We also use the approximation,

$$[\mathbf{S}_{\mathbf{x}_{\Delta T}}(m,m)]_{nn} \simeq 0, \quad m \neq 0. \tag{5.22}$$

If we use the approximations in (5.21) and (5.22), then the beamformer only processes one frequency bin. This case is the narrowband frequency-domain snapshot model.

The frequency-domain snapshot model is occasionally used for the case when

$$B_s \cdot \Delta T = 1. \tag{5.23}$$

In Example 5.2.1, the actual values of $[\mathbf{S}_{\mathbf{x}_{\Delta T}}(m,m)]_{nn}$ for various $B_s \cdot \Delta T$ are calculated. For $B_s \cdot \Delta T = 1$,

$$[\mathbf{S}_{\mathbf{x}_{\Delta T}}(m,m)]_{nn} = \begin{cases} 0.773 & m = 0 \\ 0.079 & |m| = 1 \\ 0.014 & |m| = 2. \end{cases} \tag{5.24}$$

Thus, the approximation in (5.21) is not precise. However, if the desired signal, the interfering signals, and the sensor noise all have flat spectrum over the bandwidth, then the scaling factor (0.773) will be the same. All of the beamformer weights will depend on ratios of the various spectra so the use of (5.21) will not affect the beamformer design or its performance.

The approximation in (5.22) means that the narrowband beamformer is not utilizing the information in the $m \neq 0$ frequency bins. From (5.4), we note that $\mathbf{X}_{\Delta T}(\omega_m)$ for $m = 0$ is just the average value of the envelope over the interval ΔT (multiplied by $\sqrt{\Delta T}$ for normalization purposes). In our subsequent work we assume that we utilize

$$M = \begin{cases} 1 & , \quad B_s \cdot \Delta T \leq 1 \\ B_s \cdot \Delta T + 1, & \quad B_s \cdot \Delta T = 2, 3, \cdots \end{cases} \tag{5.25}$$

and that $B_s \cdot \Delta T$ is large enough that the approximation in (5.19) is valid.[2]

If we are going to process multiple frequency bins, we need to analyze the statistical behavior for $m_1 \neq m_2$. The expression for $\mathbf{S}_{\mathbf{x}_{\Delta T}}(m_1, m_2)$ is given in (5.11). The relevant $\text{sinc}^2(\cdot)$ functions are shown in Figure 5.3(a). For $m_1 \neq m_2$, the two $\text{sinc}(\cdot)$ functions are orthogonal so that, if the elements in $[\mathbf{S}_{\mathbf{x}}(\omega)]$ are constant over an interval $\pm 2\omega_\Delta$ or $\pm 3\omega_\Delta$ on either side of $(\omega_c + m\omega_\Delta)$, the magnitude will approach zero. Hodgkiss and Nolte [HN76] have calculated the values for a flat spectrum over the frequency range $(-B_s/2 \leq f \leq B_s/2)$ and find that if

$$B_s \cdot \Delta T \geq 16, \tag{5.26}$$

the values are essentially zero (see Problem 5.2.3).

We implement a similar expansion in the kth interval to obtain the vector $\mathbf{X}_{\Delta T}(\omega_m, k)$. Due to the stationarity of the process, the spatial covariance matrix is not a function of k. In addition,

$$E\left[\mathbf{X}_{\Delta T}(\omega_{m_1}, k)\mathbf{X}_{\Delta T}^H(\omega_{m_2}, l)\right] \cong 0, \quad k \neq l$$
$$m_1, m_2 = -\frac{M-1}{2}, \cdots, \frac{M-1}{2}, \tag{5.27}$$

for large $B_s \cdot \Delta T$ (see Problem 5.2.4).

Thus, the frequency-domain snapshot model generates a sequence of N-dimensional complex vectors at M discrete frequencies, ω_m.

[2] The $B_s \cdot \Delta T = 1$ case is really not a good fit in the frequency-domain snapshot model. The discussion after (5.24) partially justifies its use. In practice, we would normally use the time-domain model in Section 5.2.2 if all of the signals and interferers are narrowband around ω_c.

In our subsequent discussions, we assume that the approximation in (5.19) is valid and that the snapshots for different m and/or different k are uncorrelated. In the next section, we show that if $\mathbf{x}(t)$ is a real Gaussian random process, then the snapshots for m_i, $i = -(M-1)/2, \cdots, (M-1)/2$, and k, $k = 1, \cdots, K$ are joint circular complex Gaussian random vectors. Thus, the snapshots for different m and/or different k are statistically independent.

5.2.1.1 Gaussian model

In many of our subsequent developments, $\mathbf{x}(t)$ is modeled as a real vector Gaussian random process. We want to show that the $\mathbf{X}_{\Delta T}(\omega_m, k)$ are joint circular complex Gaussian random vectors.[3] First consider the case in which $\mathbf{x}(t)$ is a zero-mean process.

Then, the real part and the imaginary part of $\mathbf{X}_{\Delta T}(\omega_m, k)$ are real zero-mean Gaussian random vectors:

$$\begin{aligned} \mathbf{X}_{\Delta T,C}(\omega_m, k) &\triangleq Re\left(\mathbf{X}_{\Delta T}(\omega_m, k)\right) = \frac{1}{\sqrt{\Delta T}} \int_0^{\Delta T} \mathbf{x}(t) \cos \omega_m t \, dt, \\ m &= -\frac{M-1}{2}, \cdots, 0, \cdots \frac{M-1}{2}, \end{aligned} \tag{5.28}$$

and

$$\begin{aligned} \mathbf{X}_{\Delta T,S} \triangleq Im\left(\mathbf{X}_{\Delta T}(\omega_m, k)\right) &= -\frac{1}{\sqrt{\Delta T}} \int_0^{\Delta T} \mathbf{x}(t) \sin \omega_m t \, dt, \\ m &= -\frac{M-1}{2}, \cdots, 0, \cdots \frac{M-1}{2}. \end{aligned} \tag{5.29}$$

However, in order for $\mathbf{X}_{\Delta T}(\omega_m, k)$ to be a circular complex Gaussian random vector, we require

$$E\left[\mathbf{X}_{\Delta T,C}(\omega_m, k)\mathbf{X}_{\Delta T,C}^T(\omega_m, k)\right] = E\left[\mathbf{X}_{\Delta T,S}(\omega_m, k)\mathbf{X}_{\Delta T,S}^T(\omega_m, k)\right], \tag{5.30}$$

and

$$E\left[\mathbf{X}_{\Delta T,C}(\omega_m, k)\mathbf{X}_{\Delta T,S}^T(\omega_m, k)\right] = \mathbf{0}. \tag{5.31}$$

The conditions in (5.30) and (5.31) can also be written as,

$$E\left[\mathbf{X}_{\Delta T}(\omega_m, k)\mathbf{X}_{\Delta T}^T(\omega_m, k)\right] = \mathbf{0}. \tag{5.32}$$

[3] Complex Gaussian random processes are discussed in Appendix A of DEMT III [VT71], [VT01b]. Other discussions are available in Fuhrman [Fuh98], Miller [Mil74], and Neeser and Massey [NM93].

The equalities in (5.30) and (5.31) can be verified by using the same steps as in (5.4)–(5.12) to compute the various terms. Expressing $\cos\omega_m t$ and $\sin\omega_m t$ in exponential form gives integrals similar to those in (5.9).

Therefore, for a circular complex Gaussian random vector, the probability density can be expressed in terms of a single correlation matrix,

$$\mathbf{S}_{\mathbf{x}_{\Delta T}}(\omega_m) = E\left[\mathbf{X}_{\Delta T}(\omega_m,k)\mathbf{X}_{\Delta T}^H(\omega_m,k)\right]. \tag{5.33}$$

The probability density for the zero-mean case is

$$p_{\mathbf{x}_{\Delta T}}(\mathbf{X}_{\Delta T}(\omega_m,k)) \triangleq \frac{1}{\pi^N|\mathbf{S}_{\mathbf{x}_{\Delta T}}(\omega_m)|} \exp\left\{-\left[\mathbf{X}_{\Delta T}(\omega_m,k)\mathbf{S}_{\mathbf{x}_{\Delta T}}^{-1}\mathbf{X}_{\Delta T}^H(\omega_m,k)\right]\right\}. \tag{5.34}$$

For the non-zero mean case, define the mean vector as

$$\mathbf{m}_{\Delta T}(\omega_m,k) = E\left[\mathbf{X}_{\Delta T}(\omega_m,k)\right], \tag{5.35}$$

and the covariance matrix as

$$\mathbf{K}_{\mathbf{x}_{\Delta T}}(\omega_m) \triangleq E\left[\left[\mathbf{X}_{\Delta T}(\omega_m,k)-\mathbf{m}_{\Delta T}(\omega_m,k)\right]\left[\mathbf{X}_{\Delta T}^H(\omega_m,k)-\mathbf{m}_{\Delta T}^H(\omega_m,k)\right]\right]. \tag{5.36}$$

Note that the mean vector is usually a function of k (time), while the covariance matrix is constant due to the stationary assumption. The probability density is

$$p_{\mathbf{x}_{\Delta T}}(\mathbf{x}_{\Delta T}(\omega_m,k)) \triangleq \frac{1}{\pi^N|\mathbf{K}_{\mathbf{x}_{\Delta T}}(\omega_m)|} \exp\left\{-\left[\mathbf{X}_{\Delta T}(\omega_m,k) - \mathbf{m}_{\Delta T}(\omega_m,k)\right]\cdot \mathbf{K}_{\mathbf{x}_{\Delta T}}^{-1}(\omega_m)\left[\mathbf{X}_{\Delta T}^H(\omega_m,k)-\mathbf{m}_{\Delta T}^H(\omega_m,k)\right]\right\}. \tag{5.37}$$

The probability density in (5.37) will play a central role in many of the subsequent discussions. The condition in (5.30) and (5.31) considered a single snapshot vector, $\mathbf{X}_{\Delta T}(\omega_m,k)$. In addition, the snapshots at different frequencies, $\mathbf{X}_{\Delta T}(\omega_{m_1},k)$ and $\mathbf{X}_{\Delta T}(\omega_{m_2},k)$ are jointly circular complex Gaussian random vectors for $-(M-1)/2 \leq m_1 \leq (M-1)/2, -(M-1)/2 \leq m_2 \leq (M-1)/2, m_1 \neq m_2$, and $k=1,\cdots,K$. The snapshots in different time intervals, $\mathbf{X}_{\Delta T}(\omega_{m_1},k_1)$ and $\mathbf{X}_{\Delta T}(\omega_{m_2},k_2)$ are jointly circular complex Gaussian random vectors for $-(M-1)/2 \leq m_1 \leq (M-1)/2, -(M-1)/2 \leq m_2 \leq (M-1)/2, k_1=1,\cdots,K, k_2=1,\cdots,K, k_1\neq k_2$. These results can be verified in the same manner as the single snapshot result.

Therefore, whenever (5.26) and (5.27) are satisfied, the frequency-domain snapshots can be modeled as statistically independent circular complex Gaussian random vectors. When (5.26) is valid,

$$\mathbf{m}_{\Delta T}(\omega_m, k) \simeq \mathbf{m}_{\mathbf{x}}(\omega_m, k), \tag{5.38}$$

$$\mathbf{K}_{\mathbf{x}_{\Delta T}}(\omega_m) \simeq \mathbf{K}_{\mathbf{x}}(\omega_m), \tag{5.39}$$

and

$$\mathbf{S}_{\mathbf{x}_{\Delta T}}(\omega_m) \simeq \mathbf{S}_{\mathbf{x}}(\omega_m), \tag{5.40}$$

In this case, (5.34) becomes,

$$p_{\mathbf{x}_{\Delta T}}\left(\mathbf{X}_{\Delta T}(\omega_m, k)\right) = \frac{1}{\pi^N |\mathbf{S}_{\mathbf{x}}(\omega_m)|} \exp \left\{ - \left[\mathbf{X}_{\Delta T}(\omega_m, k) \mathbf{S}_{\mathbf{x}}^{-1}(\omega_m) \mathbf{X}_{\Delta T}^H(\omega_m, k) \right] \right\} \tag{5.41}$$

for the zero-mean case. Similarly, (5.37) becomes,

$$p_{\mathbf{x}_{\Delta T}}\left(\mathbf{X}_{\Delta T}(\omega_m, k)\right) = \frac{1}{\pi^N |\mathbf{K}_{\mathbf{x}}(\omega_m)|} \exp \left\{ - \left[\mathbf{X}_{\Delta T}(\omega_m, k) - \mathbf{m}_{\mathbf{x}}(\omega_m, k) \right] \mathbf{K}_{\mathbf{x}}^{-1}(\omega_m) \left[\mathbf{X}_{\Delta T}^H(\omega_m, k) - \mathbf{m}_{\mathbf{x}}^H(\omega_m, k) \right] \right\}. \tag{5.42}$$

The statistical independence result means that the joint densities of the snapshots for different m and k factor into a product of the individual densities.

5.2.1.2 Plane-wave snapshot model

In this section we develop the frequency-domain snapshot model for the case in which the desired signals and interfering signals can be modeled as plane waves. A family of models is developed that plays a central role in many of the subsequent developments. The models can be divided into two cases. In the first case, the desired signals are either deterministic signals or unknown nonrandom signals. In the second case, the desired signals are sample functions of a Gaussian random process. For each case, several examples are discussed.

In case 1, the output of the array is an $N \times 1$ complex vector,

$$\mathbf{x}(t) = \mathbf{x}_s(t) + \mathbf{n}(t), \tag{5.43}$$

where $\mathbf{x}_s(t)$ is either a deterministic signal or an unknown nonrandom signal[4] and $\mathbf{n}(t)$ is a sample function of a zero-mean Gaussian random process. The frequency-domain snapshot model is

$$\begin{aligned}\mathbf{X}_{\Delta T}(\omega_m, k) &= \mathbf{X}_{s,\Delta T}(\omega_m, k) + \mathbf{N}_{\Delta T}(\omega_m, k), \\ m &= 0, \cdots, M-1, \quad k = 1, \cdots, K. \end{aligned} \tag{5.44}$$

In subsequent equations the range of m and k is the same as in (5.44) and is omitted.

Example 5.2.2

In this example, there is a single plane-wave desired signal,

$$\mathbf{X}_{s,\Delta T}(\omega_m, k) = \mathbf{v}(\omega_m, \mathbf{k}_s) F_{s,\Delta T}(\omega_m, k). \tag{5.45}$$

The source-signal snapshot, $F_{s,\Delta T}(\omega_m, k)$, is modeled in one of two ways. In the first way, $F_{s,\Delta T}(\omega_m, k)$ is assumed to be the snapshot of a known (or deterministic) signal. This model is appropriate in many communications or radar problems. In the second way, $F_{s,\Delta T}(\omega_m, k)$ is assumed to be the snapshot of an unknown, but nonrandom waveform. This model is appropriate in many passive sonar or direction-finding applications.

The probability density is a non-zero mean circular complex density,

$$p_{\mathbf{x}_{\Delta T}}\left(\mathbf{X}_{\Delta T}(\omega_m, k)\right) = \frac{1}{\pi^N |\mathbf{S}_{\mathbf{n}_{\Delta T}}|} \exp \left\{-\left[\mathbf{X}_{\Delta T} - \mathbf{v}(\mathbf{k}_s) F_{s,\Delta T}\right] \mathbf{S}_{\mathbf{n}_{\Delta T}}^{-1} \left[\mathbf{X}_{\Delta T} - \mathbf{v}(\mathbf{k}_s) F_{s,\Delta T}\right]^H \right\}, \tag{5.46}$$

where the ω_m and k arguments are suppressed on the right side of the equation. The covariance matrix is the spatial spectral matrix of the noise. Initially, $\mathbf{S}_\mathbf{n}$ is assumed to be known. The mean of the density is either known (deterministic signal) or unknown (unknown nonrandom signal). Using the approximations in Section 5.2.1.1, (5.46) reduces to

$$p_{\mathbf{x}_{\Delta T}}\left(\mathbf{X}_{\Delta T}(\omega_m, k)\right) = \frac{1}{\pi^N |\mathbf{S}_\mathbf{n}|} \exp \left\{-\left[\mathbf{X}_{\Delta T} - \mathbf{v}(\mathbf{k}_s) F_{s,\Delta T}\right] \mathbf{S}_\mathbf{n}^{-1} \left[\mathbf{X}_{\Delta T} - \mathbf{v}(\mathbf{k}_s) F_{s,\Delta T}\right]^H \right\}. \tag{5.47}$$

For the spatial case of spatial white noise,

$$\mathbf{S}_\mathbf{n} = \sigma_\omega^2 \mathbf{I}, \tag{5.48}$$

and (5.47) reduces to

$$p_{\mathbf{x}_{\Delta T}}\left(\mathbf{X}_{\Delta T}(\omega_m, k)\right) = \left(\pi^N \sigma_\omega^2 N\right)^{-1} \exp \left\{-\frac{1}{\sigma_\omega^2} \left[\mathbf{X}_{\Delta T} - \mathbf{v}(\mathbf{k}_s) F_{s,\Delta T}\right] \left[\mathbf{X}_{\Delta T} - \mathbf{v}(\mathbf{k}_s) F_{s,\Delta T}\right]^H \right\}. \tag{5.49}$$

Example 5.2.3

In this example, there is a single plane-wave desired signal, and $(D-1)$ plane-wave interfering signals, and additive noise. All of the source signals are either deterministic

---

[4]These two terms are defined in Example 5.2.2.

signals or unknown nonrandom signals. Then,

$$\mathbf{X}_{\Delta T}(\omega_m, k) = \mathbf{v}(\omega_m, \mathbf{k}_s)F_{s,\Delta T}(\omega_m, k) + \sum_{i=1}^{D-1} \mathbf{v}(\omega_m, \mathbf{k}_i)F_{i,\Delta T}(\omega_m, k) + \mathbf{N}_{\Delta T}(\omega_m, k). \tag{5.50}$$

A composite $N \times D$ array manifold matrix is defined as

$$\mathbf{V}(\omega_m, \mathbf{k}) = [\mathbf{v}(\omega_m, \mathbf{k}_s)\ \mathbf{v}(\omega_m, \mathbf{k}_1)\ \cdots\ \mathbf{v}(\omega_m, \mathbf{k}_{D-1})], \tag{5.51}$$

and a composite source signal vector is defined as

$$\mathbf{F}_{\Delta T}(\omega_m, k) = [F_{s,\Delta T}\ F_{1,\Delta T}\ \cdots\ F_{D-1,\Delta T}]^T. \tag{5.52}$$

Then the snapshot vector can be written as

$$\mathbf{X}_{\Delta T}(\omega_m, \mathbf{k}) = \mathbf{V}(\omega_m, \mathbf{k})\mathbf{F}_{\Delta T}(\omega_m, k) + \mathbf{N}_{\Delta T}(\omega_m, k). \tag{5.53}$$

Using the approximations in Section 5.2.1.1, the probability density is

$$p_{\mathbf{X}_{\Delta T}}(\mathbf{X}_{\Delta T}(\omega_m, k)) = \frac{1}{\pi^N|\mathbf{S_n}|}\exp\left\{-[\mathbf{X}_{\Delta T} - \mathbf{V}(\mathbf{k})\mathbf{F}_{\Delta T}]\,\mathbf{S_n}^{-1}\,[\mathbf{X}_{\Delta T} - \mathbf{V}(\mathbf{k})\mathbf{F}_{\Delta T}]^H\right\}, \tag{5.54}$$

where the ω_m and k arguments are suppressed on the right side of (5.54).

Example 5.2.4

In this example, there is a single plane-wave desired signal, $(D-1)$ plane-wave interfering signals, and additive noise. The desired signal is modeled as a deterministic or unknown nonrandom signal. The $(D-1)$ plane-wave interfering signals are modeled as sample functions of Gaussian random processes. The snapshot vector is the same as in (5.50),

$$\mathbf{X}_{\Delta T}(\omega_m, k) = \mathbf{v}(\omega_m, \mathbf{k}_s)F_{s,\Delta T}(\omega_m, k) + \sum_{i=1}^{D-1} \mathbf{v}(\omega_m, \mathbf{k}_i)F_{i,\Delta T}(\omega_m, k) + \mathbf{N}_{\Delta T}(\omega_m, k) \tag{5.55}$$

A composite $N \times (D-1)$ interference array manifold matrix is defined as

$$\mathbf{V}_I(\omega_m, \mathbf{k}_1) = [\mathbf{v}(\omega_m, \mathbf{k}_1)\ \cdots\ \mathbf{v}(\omega_m, \mathbf{k}_{D-1})], \tag{5.56}$$

and a composite source interference vector is defined as

$$\mathbf{F}_{I,\Delta T}(\omega_m, k) = [F_{1,\Delta T}\ \cdots\ F_{D-1,\Delta T}(\omega_m, k)]^T. \tag{5.57}$$

The interference source spectral matrix is defined as

$$\mathbf{S}_{I,\Delta T}(\omega_m) \triangleq E\left[\mathbf{F}_{I,\Delta T}(\omega_m, k)\mathbf{F}_{I,\Delta T}^H(\omega_m, k)\right], \tag{5.58}$$

which is approximated by

$$\mathbf{S}_{I,\Delta T}(\omega_m) \simeq \mathbf{S}_I(\omega_m). \tag{5.59}$$

The noise-plus-interference spectral matrix is defined as

$$\mathbf{S}_{IN}(\omega_m) = \mathbf{V}_I(\omega_m, \mathbf{k}_I)\mathbf{S}_I(\omega_m)\mathbf{V}_I^H(\omega_m, \mathbf{k}_I) + \mathbf{S_n}(\omega_m). \tag{5.60}$$

The probability density is

$$p_{\mathbf{x}_{\Delta T}}(\mathbf{X}_{\Delta T}(\omega_m, k)) =$$

$$\frac{1}{\pi^N |\mathbf{S}_{IN}(\omega_m)|} \exp\left\{-\left[\mathbf{X}_{\Delta T} - \mathbf{v}(\mathbf{k}_s)F_{\Delta T}\right] \mathbf{S}_{IN}^{-1}(\omega_m) \left[\mathbf{X}_{\Delta T} - \mathbf{v}(\mathbf{k}_s)F_{\Delta T}\right]^H\right\}. \tag{5.61}$$

These three models are used for applications in which the source signals are modeled as nonrandom or deterministic waveforms.

In case 2, the output of the array is an $N \times 1$ complex vector,

$$\mathbf{x}(t) = \mathbf{x}_s(t) + \mathbf{n}(t), \tag{5.62}$$

and both $\mathbf{x}_s(t)$ and $\mathbf{n}(t)$ are zero-mean Gaussian random processes. The frequency-domain snapshot model is

$$\mathbf{X}_{\Delta T}(\omega_m, k) = \mathbf{X}_{s,\Delta T}(\omega_m, k) + \mathbf{N}_{\Delta T}(\omega_m, k). \tag{5.63}$$

Example 5.2.5

In this example, there is a single plane-wave desired signal, $(D-1)$ plane-wave interfering signals, and additive noise. All of the source signals are zero-mean Gaussian random processes. The expressions in (5.50)–(5.53) all apply to this case. Now, the $D \times D$ signal-plus-interference source spectral matrix is defined as

$$\mathbf{S}_{SI,\Delta T}(\omega_m) = E\left[\mathbf{F}_{\Delta T}(\omega_m, k)\mathbf{F}_{\Delta T}^H(\omega_m, k)\right], \tag{5.64}$$

where $\mathbf{F}_{\Delta T}(\omega_m, k)$ is defined in (5.52). The spectral matrix in (5.64) is approximated by

$$\mathbf{S}_{SI}(\omega_m) \simeq \mathbf{S}_{\mathbf{f}}(\omega_m). \tag{5.65}$$

The total snapshot spectral matrix is

$$\begin{aligned} \mathbf{S}_{\mathbf{X}_{\Delta T}} &= E\left[\mathbf{X}_{\Delta T}(\omega_m)\mathbf{X}_{\Delta T}^H(\omega_m)\right] \\ &= \mathbf{V}(\omega_m, \mathbf{k})\mathbf{S}_{SI,\Delta T}(\omega_m)\mathbf{V}^H(\omega_m, \mathbf{k}) + \mathbf{S}_{N,\Delta T}(\omega_m), \end{aligned} \tag{5.66}$$

which is approximated by

$$\mathbf{S}_{\mathbf{x}}(\omega_m) = \mathbf{V}(\omega_m, \mathbf{k})\mathbf{S}_{\mathbf{f}}(\omega_m)\mathbf{V}^H(\omega_m, \mathbf{k}) + \mathbf{S}_{\mathbf{n}}(\omega_m). \tag{5.67}$$

The probability density is

$$p_{\mathbf{X}_{\Delta T}}\left(\mathbf{X}_{\Delta T}(\omega_m, k)\right) = \frac{1}{\pi^N |\mathbf{S}_{\mathbf{x}}|} \exp\left\{-\mathbf{X}_{\Delta T}\mathbf{S}_{\mathbf{x}}^{-1}\mathbf{X}_{\Delta T}^H\right\}, \tag{5.68}$$

where the ω_m and k are suppressed on the right side of (5.68).

The four models in these examples will have a central role in most of our subsequent discussions.

5.2.1.3 Beamforming

We now discuss the other two boxes in Figure 5.1. In the second box, each frequency bin is processed with a narrowband beamformer centered at ω_m. The output in each bin is a complex scalar Gaussian random variable,

$$Y_{\Delta T}(m\omega_\Delta), \quad m = -\frac{M-1}{2}, \cdots, 0, \cdots, \frac{M-1}{2}. \tag{5.69}$$

Chapter 6 develops optimum narrowband beamformers assuming that the appropriate spatial spectral matrices (e.g., $\mathbf{S_x}(\omega_m)$ or $\mathbf{S_n}(\omega_m)$) are known. Chapter 7 develops the adaptive versions.

The final step is to construct the beamformer output that is a scalar complex Gaussian random process,

$$y(t) = \frac{1}{\sqrt{\Delta T}} \sum_{-\frac{M-1}{2}}^{\frac{M-1}{2}} Y_{\Delta T}(m\omega_\Delta) e^{jm\omega_\Delta t}. \tag{5.70}$$

5.2.2 Narrowband Time-domain Snapshot Models

In this section, a time-domain model that is appropriate for narrowband waveforms is developed. In Section 6.13, time-domain models for the broadband case are developed.

First consider the case of a single plane-wave input. The input at the reference sensor located at the origin is a real bandpass signal,

$$f(t) = \sqrt{2} Re\left\{\tilde{f}(t)\, e^{j\omega_c t}\right\}. \tag{5.71}$$

The input at the nth sensor is

$$\begin{aligned} f_n(t) &= f(t-\tau_n) \\ &= \sqrt{2} Re\left\{\tilde{f}(t-\tau_n)\, e^{j\omega_c (t-\tau_n)}\right\}, \end{aligned} \tag{5.72}$$

where τ_n is the time delay from the origin to the nth sensor. The narrowband assumption implies

$$\tilde{f}(t-\tau_n) \simeq \tilde{f}(t), \quad n = 0, \cdots, N-1. \tag{5.73}$$

Using (5.73) in (5.72) gives

$$f_n(t) = \sqrt{2} Re\left\{\tilde{f}(t)\, e^{-j\omega_c \tau_n}\, e^{j\omega_c t}\right\}. \tag{5.74}$$

We normally perform a quadrature demodulation of the sensor outputs prior to time-domain processing. The quadrature demodulation process is shown in Figure 5.4. Figure 5.4(a) shows the actual demodulation process. Figure 5.4(b) shows its complex representation in the time domain. Figure 5.4(c) shows its complex representation in the frequency domain. The lowpass filter has a bandwidth of B_s Hz.

The complex output of the quadrature demodulator at the nth sensor is

$$\tilde{f}_n(t) = \tilde{f}(t)e^{-j\omega_c \tau_n}. \tag{5.75}$$

We recognize the exponential term as the nth element of the array manifold vector $\mathbf{v}(\mathbf{k})$. Thus, the complex vector output of the array (after quadrature demodulation) is

$$\tilde{\mathbf{f}}(t) = \tilde{f}(t)\mathbf{v}(\mathbf{k}), \tag{5.76}$$

where $\tilde{f}(t)$ is a scalar zero-mean complex Gaussian random process. The covariance matrix at time t is

$$\mathbf{R}_{\tilde{\mathbf{f}}}(0) = E\left[\tilde{\mathbf{f}}(t)\tilde{\mathbf{f}}^H(t)\right] = \mathbf{v}(\mathbf{k})R_f(0)\mathbf{v}^H(\mathbf{k}). \tag{5.77}$$

However, $R_f(0)$ is just the power in the signal process, so (5.77) can be written as

$$\mathbf{R}_{\tilde{f}}(0) = \sigma_s^2\mathbf{v}(\mathbf{k})\mathbf{v}^H(\mathbf{k}), \tag{5.78}$$

where σ_s^2 is the signal power.

Generalizing this result to the case of D signals and adding a white noise component gives

$$\mathbf{R}_{\tilde{\mathbf{x}}}(0) = \mathbf{V}(\mathbf{k})\,\mathbf{R}_{\tilde{\mathbf{f}}}(0)\,\mathbf{V}^H(\mathbf{k}) + \sigma_w^2 B_s\mathbf{I}. \tag{5.79}$$

The B_s factor in the second term results from passing the white noise through the filter in the quadrature demodulator. If we assume that $\tilde{f}(t)$ is bandlimited to $-B_s/2 \le f \le B_s/2$ Hz, then we sample $\mathbf{x}(t)$ every $1/B_s$ seconds to obtain a time domain snapshot model. We denote the snapshots of the complex envelope as

$$\tilde{\mathbf{x}}(k), \quad k = 1, 2, \cdots, K. \tag{5.80}$$

Then

$$\mathbf{R}_{\tilde{\mathbf{x}}}(k) \triangleq E\left[\tilde{\mathbf{x}}(k)\tilde{\mathbf{x}}^H(k)\right] = \mathbf{V}(\mathbf{k})\,\mathbf{R}_{\tilde{\mathbf{f}}}(0)\,\mathbf{V}^H(\mathbf{k}) + \sigma_w^2\,B_s\,\mathbf{I}, \quad k = 1, 2, \cdots, K. \tag{5.81}$$

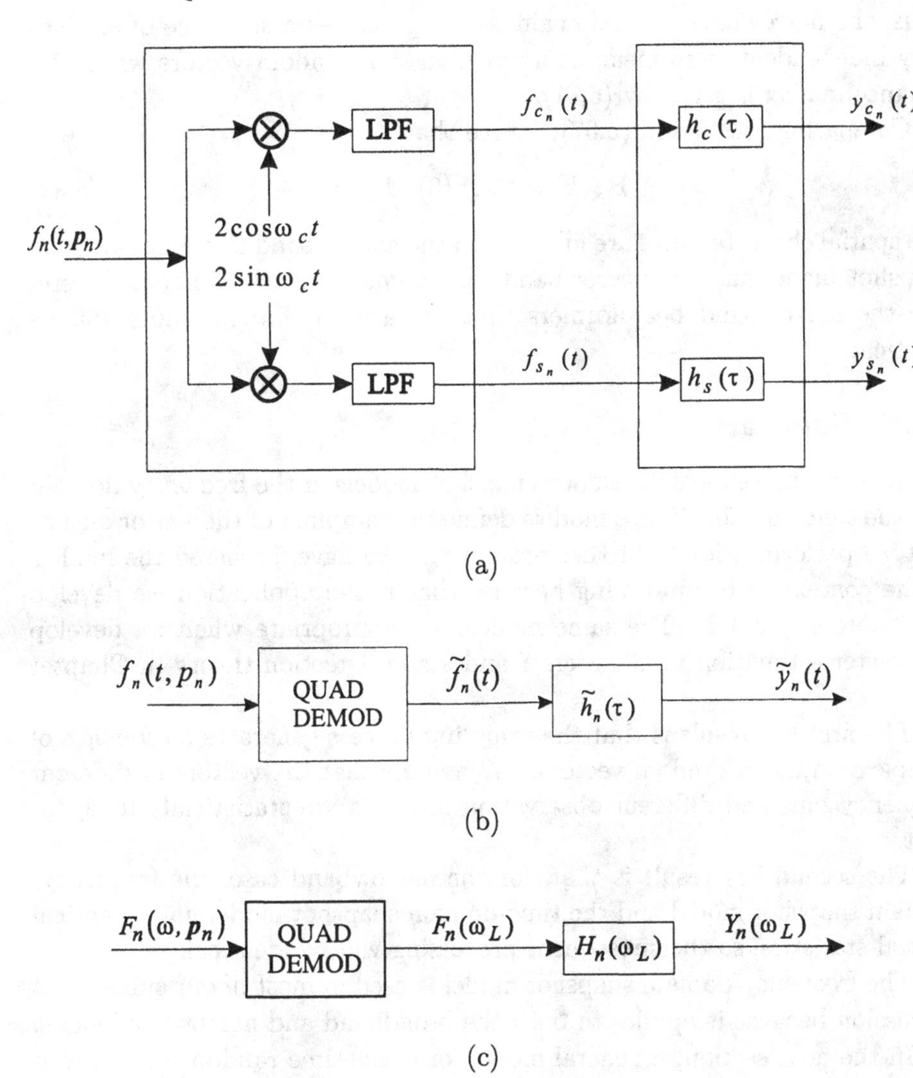

Figure 5.4 Quadrature demodulation.

If we further assume that the components of $\tilde{\mathbf{f}}(t)$ have a flat spectrum, as shown in Figure 5.2(b), then

$$E\left[\tilde{\mathbf{x}}(k_1)\tilde{\mathbf{x}}^H(k_2)\right] = 0, \quad k_1 \neq k_2. \tag{5.82}$$

Thus, the narrowband time-domain model generates a sequence of statistically independent, zero-mean complex Gaussian random vectors whose covariance matrix is given by (5.81).

Comparing (5.81) and (5.67), we see that

$$\mathbf{R}_{\tilde{\mathbf{x}}}(k) = \mathbf{S}_{\mathbf{x}_{\Delta T}}(0) \cdot B_s. \tag{5.83}$$

The spatial characteristics are identical in the narrowband frequency-domain snapshot model and the narrowband time-domain snapshot model. Therefore the narrowband beamformers that operate on the snapshots will be identical.

5.2.3 Summary

In this section, we have developed snapshot models in the frequency domain and the time domain. These models define the sampling of the sensor output that we perform prior to further processing. We have discussed the models in the context of beamforming because that is the application we develop in Chapters 6 and 7. The same models are appropriate when we develop parameter estimation in Chapters 8 and 9 and detection theory in Chapter 10.

The first key result is that the sampling process generates a sequence of complex Gaussian random vectors. We assume that the vectors in different frequency bins and different observation intervals are statistically independent.

The second key result is that, for the narrowband case, the frequency-domain snapshot model and the time-domain snapshot model have identical spatial statistics, so the subsequent processing will be identical.

The frequency-domain snapshot model is used in most of our subsequent discussion because it applies to both the broadband and narrowband cases.

In the next section, a general model for space-time random processes is considered. This model allows us to model spatially spread random processes as well as the finite plane-wave model in Section 5.2.1.2 when the source signals are statistically independent. It does not allow us to model correlated plane waves. For readers whose primary interest is in applications that can be modeled with D plane waves, the next section can be skipped at first reading.

5.3 Space-time Random Processes

In many applications, the signal or noise source has a continuous distribution in space. For these cases, a space-time model of the source process must be developed.

Assume that the complex signal $f(t, \mathbf{p})$ is a scalar waveform that is defined over time and space.[5] We model $f(t, \mathbf{p})$ as a complex random process over the time interval (T_i, T_f) and the spatial region $\mathbf{p} \in \Re^N$. Normally, $\Re$ will be a 3-D space.

We are interested in both a second-moment characterization of the process and a complete characterization of the process.[6] The second-moment characterization of the process is adequate to deal with linear operations on the process. A complete characterization is necessary to formulate optimum detection and estimation problems. We restrict our complete characterization to Gaussian space-time processes. We develop the second-moment characterization in Section 5.3.1 and define the Gaussian process model in Section 5.3.2.

5.3.1 Second-moment Characterization

The mean of the space-time field, $f(t, \mathbf{p})$, is defined as

$$m_f(t, \mathbf{p}) = E\left[f(t, \mathbf{p})\right], \quad T_i \leq t \leq T_f, \quad \mathbf{p} \in \Re^N, \tag{5.84}$$

and the space-time covariance as

$$\begin{aligned} K_f(t_1, t_2 : \mathbf{p}_1, \mathbf{p}_2) &= E\left\{\left[f(t_1, \mathbf{p}_1) - m_f(t_1, \mathbf{p}_1)\right]\left[f^*(t_2, \mathbf{p}_2) - m_f^*(t_2, \mathbf{p}_2)\right]\right\} \\ &\quad T_i \leq t_1, t_2 \leq T_f, \quad \mathbf{p}_1, \mathbf{p}_2 \in \Re^N. \end{aligned} \tag{5.85}$$

In the subsequent discussion we assume that $m_f(t, \mathbf{p}) = 0$.

In the general case the random process can be non-stationary and non-homogeneous. In Part I of DEMT [VT68], [VT01a], we analyzed scalar non-stationary processes and saw that mathematics were more complicated than the stationary case, but the concepts were unchanged. The same results carry over to space-time processes.

In many cases of interest, the process is temporally wide-sense stationary. Then,

$$K_f(t_1, t_2 : \mathbf{p}_1, \mathbf{p}_2) = K_f(\tau : \mathbf{p}_1, \mathbf{p}_2), \tag{5.86}$$

[5] See Appendix A of DEMT, Part III of [VT71], [VT01b] for a discussion of complex processes.

[6] See Sections 5.3.1 and 5.3.3 of DEMT, Part I [VT68], [VT01a] for the corresponding scalar process discussion.

where

$$\tau = t_1 - t_2 \,. \tag{5.87}$$

Note that, in a practical application, the stationarity only has to hold over a time interval that is long compared to the processing time.

In many cases of interest, the process is homogeneous or spatially stationary over a region that is large compared to the spatial dimensions of the aperture or array. Then,

$$K_f(t_1, t_2 : \mathbf{p}_1, \mathbf{p}_2) = K_f(t_1, t_2 : \Delta\mathbf{p}) \,, \tag{5.88}$$

where

$$\Delta\mathbf{p} = \mathbf{p}_1 - \mathbf{p}_2 \,. \tag{5.89}$$

For cases that satisfy both conditions, the process is time-stationary and space-homogeneous,

$$K_f(t_1, t_2 : \mathbf{p}_1, \mathbf{p}_2) = K_f(\tau : \Delta\mathbf{p}) \,. \tag{5.90}$$

We focus most of our attention on random processes that satisfy (5.90).

We define the *temporal frequency spectrum-spatial correlation function*,

$$S_f(\omega : \Delta\mathbf{p}) = \int_{-\infty}^{\infty} K_f(\tau : \Delta\mathbf{p}) e^{-j\omega\tau} d\tau, \tag{5.91}$$

which is the Fourier transform with respect to τ of the space-time covariance function. When $\Delta\mathbf{p} = 0, S_f(\omega : 0)$ is the frequency spectrum at a point in space.

Similarly, we can define a *temporal correlation-spatial wavenumber spectrum* as

$$F_f(\tau : \mathbf{k}) = \int \cdot \int_{\Re^N} d\Delta\mathbf{p} K_f(\tau : \Delta\mathbf{p}) e^{j\mathbf{k}^T \Delta\mathbf{p}} \,. \tag{5.92}$$

Note that the region of integration depends on the dimension of spatial domain (linear, planar, or volumetric). Although we define this function for completeness, we will not use it very often in the subsequent discussion.

A function of major interest is the *frequency-wavenumber spectrum*,

$$P_f(\omega : \mathbf{k}) = \int \cdot \int_{\Re^N} d\Delta\mathbf{p} S_f(\omega : \Delta\mathbf{p}) e^{j\mathbf{k}^T \Delta\mathbf{p}} \,. \tag{5.93}$$

If we assume a three-dimensional space, then

$$P_f(\omega : \mathbf{k}) = \int_{-\infty}^{\infty} \int_{-\infty}^{\infty} \int_{-\infty}^{\infty} d\Delta\mathbf{p} S_f(\omega : \Delta\mathbf{p}) e^{j\mathbf{k}^T \Delta\mathbf{p}} \,, \tag{5.94}$$

and the inverse transform is

$$S_f(\omega : \Delta\mathbf{p}) = \int_{-\infty}^{\infty}\int_{-\infty}^{\infty}\int_{-\infty}^{\infty} \frac{d\mathbf{k}}{(2\pi)^3} P_f(\omega : \mathbf{k}) e^{-j\mathbf{k}^T\Delta\mathbf{p}} . \tag{5.95}$$

$P_f(\omega : \mathbf{k})$ can also be expressed in terms of $K_f(\tau : \Delta\mathbf{p})$ as

$$P_f(\omega : \mathbf{k}) = \int_{-\infty}^{\infty} d\tau \int \cdot \int_{\Re^N} d\Delta\mathbf{p} K_f(\tau : \Delta\mathbf{p}) e^{-j(\omega\tau - \mathbf{k}^T\Delta\mathbf{p})} . \tag{5.96}$$

If we assume a 3-D space, then

$$P_f(\omega : \mathbf{k}) = \int_{-\infty}^{\infty} d\tau \int_{-\infty}^{\infty}\int_{-\infty}^{\infty}\int_{-\infty}^{\infty} d\Delta\mathbf{p} K_f(\tau : \Delta\mathbf{p}) e^{-j(\omega\tau - \mathbf{k}^T\Delta\mathbf{p})} . \tag{5.97}$$

$P_f(\omega : \mathbf{k})$ has properties analogous to the frequency spectrum for scalar processes.

In particular, $f(t, \mathbf{p})$ can be represented as a Stieltjes integral,

$$f(t, \mathbf{p}) = \int_{-\infty}^{\infty} \cdot \cdot \int_{-\infty}^{\infty} e^{j(\omega t - \mathbf{k}^T\mathbf{p})} dF(\omega; \mathbf{k}) , \tag{5.98}$$

where $\mathbf{k}$ has a definition consistent with the space domain (e.g., [Doo53], [Yag57], [Yag60]).

The following properties hold:

$$E\left[|dF(\omega_2 : \mathbf{k}_2) - dF(\omega_1 : \mathbf{k}_1)|^2 \right] = \int_{\omega_1}^{\omega_2} \frac{d\omega}{2\pi} \int_{\mathbf{k}_1}^{\mathbf{k}_2} \frac{d\mathbf{k}}{(2\pi)^N} P_f(\omega : \mathbf{k}) , \tag{5.99}$$

and

$$E\left[(dF(\omega_2 : \mathbf{k}_2) - dF(\omega_1 : \mathbf{k}_1)) (dF(\omega_4 : \mathbf{k}_4) - dF(\omega_3 : \mathbf{k}_3))^* \right] = 0, \tag{5.100}$$

when

$$(\omega_1, \omega_2) \times (\mathbf{k}_1, \mathbf{k}_2) \cap (\omega_3, \omega_4) \times (\mathbf{k}_3, \mathbf{k}_4) = 0 . \tag{5.101}$$

In other words, disjoint frequency-wavenumber bands are uncorrelated.

This is the spatial generalization of a spectral decomposition of a scalar random process. Therefore, we can consider random processes that satisfy (5.99) to be composed of a superposition of uncorrelated plane waves with temporal frequency ω and wavenumber k.

If, in addition, we impose the condition that the process $f(t, \mathbf{p})$ satisfies the homogeneous wave equation, then we have the constraint

$$\frac{\omega}{c} = |\mathbf{k}| . \tag{5.102}$$

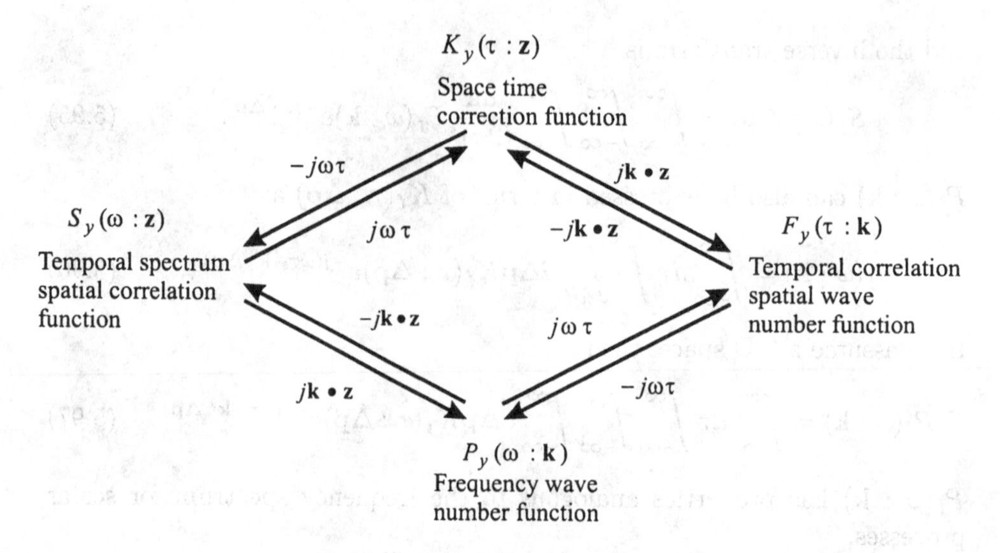

Figure 5.5 Second-moment representations for stationary homogeneous space-time random processes. [From [Bag76]]

This defines a structure that signals propagating in a medium must satisfy. Further, this implies that at any particular ω the frequency-wavenumber function is nonzero only on a sphere of radius $\frac{2\pi}{\lambda} = |\mathbf{k}| = \frac{\omega}{c}$ and is concentrated in that sphere. This constraint is not necessary for $f(t : \mathbf{p})$ to correspond to a propagating process. For example, one may have propagation in three dimensions and yet consider a representation over a 2-D surface, since the 2-D wavenumber value corresponds unambiguously to a 3-D surface except for its sign.

The advantage of this representation for homogeneous fields is that it allows us to make many statements regarding the processing for our signals, which are completely dual to analogous frequency-domain operations when processing temporal waveforms. Naturally, there is a close relationship among the various transforms discussed, which are summarized in Figure 5.5.

When we consider frequency-wavenumber spectra in different dimensions, it is important to understand their relationship. From (5.95), we have the 3-D transform,

$$S_{3f}(\omega : \Delta p_x, \Delta p_y, \Delta p_z) = \int_{-\infty}^{\infty}\int_{-\infty}^{\infty}\int_{-\infty}^{\infty} \frac{d\mathbf{k}_3}{(2\pi)^3} P_{3f}(\omega : \mathbf{k}_3) \, e^{-j[k_x \cdot \Delta p_x + k_y \cdot \Delta p_y + k_z \cdot \Delta p_z]}, \tag{5.103}$$

where we have added the subscript "3" to denote the dimension. Similarly, from (5.95), we have the 2-D transform,

$$S_{2f}(\omega : \Delta p_x, \Delta p_y) = \int_{-\infty}^{\infty}\int_{-\infty}^{\infty} \frac{d\mathbf{k}_2}{(2\pi)^2} P_{2f}(\omega : \mathbf{k}_2) e^{-j[k_x \cdot \Delta p_x + k_y \cdot \Delta p_y]}, \quad (5.104)$$

for a representation in the xy-plane. We observe that

$$S_{2f}(\omega : \Delta p_x, \Delta p_y) = S_{3f}(\omega : \Delta p_x, \Delta p_y, 0). \quad (5.105)$$

Using (5.105) in (5.103),

$$S_{2f}(\omega : \Delta p_x, \Delta p_y) = \int_{-\infty}^{\infty}\int_{-\infty}^{\infty}\int_{-\infty}^{\infty} \frac{d\mathbf{k}_3}{(2\pi)^3} P_{3f}(\omega : \mathbf{k}_3) e^{-j[k_x \cdot \Delta p_x + k_y \cdot \Delta p_y]}$$

$$= \int_{-\infty}^{\infty}\int_{-\infty}^{\infty} \frac{dk_x dk_y}{(2\pi)^2} \left[\int_{-\infty}^{\infty} \frac{dk_z}{2\pi} P_{3f}(\omega : \mathbf{k}_3)\right] e^{-j[k_x \cdot \Delta p_x + k_y \cdot \Delta p_y]}. \quad (5.106)$$

Equating the right sides of (5.106) and (5.104) gives

$$P_{2f}(\omega : \mathbf{k}_2) = \int_{-\infty}^{\infty} \frac{dk_z}{2\pi} P_{3f}(\omega : \mathbf{k}_3). \quad (5.107)$$

Similarly,

$$P_{1f}(\omega : k_x) = \int_{-\infty}^{\infty} \frac{dk_y}{2\pi} P_{2f}(\omega : \mathbf{k}_2). \quad (5.108)$$

In Chapters 2 and 3, we normally used the z-axis for 1-D representations. Then,

$$P_{1f}(\omega : k_z) = \int_{-\infty}^{\infty}\int_{-\infty}^{\infty} \frac{dk_x dk_y}{(2\pi)^2} P_{3f}(\omega : \mathbf{k}). \quad (5.109)$$

We consider a simple example.

Example 5.3.1: Directional Signal

The simplest signal of interest is a plane wave propagating in a direction with speed c as illustrated in Figure 5.6. The space-time process has the form

$$f(t, \mathbf{p}) = f\left[t - \frac{\mathbf{a}^T\mathbf{p}}{c}\right], \quad (5.110)$$

where $f(t)$ is a zero-mean stationary random process with spectrum $S_f(\omega)$.

Then, the space-time correlation function is

$$\begin{aligned} K_f(\tau : \Delta\mathbf{p}) &= E\left(f\left(t - \frac{\mathbf{a}^T\mathbf{p}}{c}\right) f^*\left(t - \tau - \frac{\mathbf{a}^T(\mathbf{p} - \Delta\mathbf{p})}{c}\right)\right] \\ &= K_f\left(\tau - \frac{\mathbf{a}^T\Delta\mathbf{p}}{c}\right). \end{aligned} \quad (5.111)$$

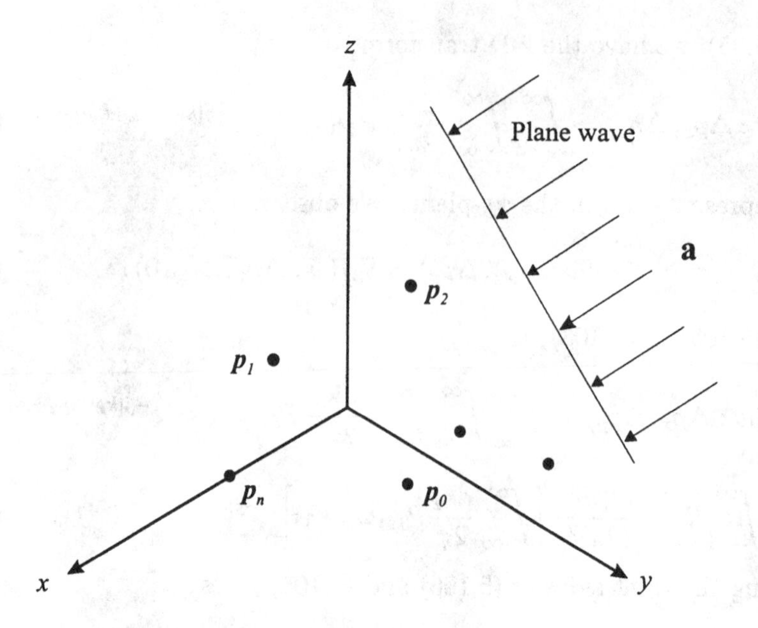

Figure 5.6 Model of plane wave propagation.

The temporal frequency spectrum spatial correlation function is

$$S_{3f}(\omega : \Delta\mathbf{p}) = S_f(\omega)e^{-j\left(\frac{\omega}{c}\right)\left(\mathbf{a}^T\Delta\mathbf{p}\right)} = S_f(\omega)e^{j\frac{2\pi}{\lambda}\mathbf{u}^T\Delta\mathbf{p}}. \tag{5.112}$$

Consequently, we have the cross-spectrum of a plane wave at two different locations that is related to the spectrum of the process at a single point by a simple linear phase shift that reflects the propagation delay between the two points.

Transforming with respect to the spatial variable, we obtain the frequency-wavenumber function

$$P_{3f}(\omega : \mathbf{k}) = \int\int_{\Re^N} d\Delta\mathbf{p} S_f(\omega : \Delta\mathbf{p})e^{j\mathbf{k}^T\Delta\mathbf{p}} = S_{f_o}(\omega)\delta\left(\mathbf{k} - \frac{\omega}{c}\mathbf{a}\right), \tag{5.113}$$

where

$$\delta\left(\mathbf{k} - \frac{\omega}{c}\mathbf{a}\right) = \delta\left(k_x - \frac{\omega}{c}a_x\right)\delta\left(k_y - \frac{\omega}{c}a_y\right)\delta\left(k_z - \frac{\omega}{c}a_z\right), \tag{5.114}$$

is a 3-D impulse function, which in Cartesian coordinates can be written as a product of impulse functions of a single variable.

For a planar representation in the xy-plane,

$$P_{2f}(\omega : k_x, k_y) = \delta(k_x - k_{sx})\delta(k_y - k_{sy})$$

$$= \delta\left(k_x - \frac{2\pi}{\lambda}\sin\theta\cos\phi\right)\delta\left(k_y - \frac{2\pi}{\lambda}\sin\theta\sin\phi\right), \tag{5.115}$$

and for a 1-D representation along the z-axis,

$$P_{1f}(\omega : k_z) = \delta(k_z - k_{sz}) = \delta\left(k_z - \frac{2\pi}{\lambda}\cos\theta\right). \tag{5.116}$$

We find that a plane wave propagating with speed c and direction $\mathbf{a}$ has an impulsive frequency-wavenumber spectrum located at

$$\mathbf{k}_a = \frac{\omega}{c}\mathbf{a}, \tag{5.117}$$

in wavenumber space. We again point out that the space over which this transform is taken must be specified. Theoretically, in a homogeneous isotropic medium the magnitude of the wavenumber $\mathbf{k}$ at a particular frequency should be a constant, indicating that, for three dimensions, the function $P_f(\omega : \mathbf{k})$ is concentrated on a sphere; concentrated on a circle in two dimensions; or concentrated at two points opposite in sign in one dimension.

However, in many cases one considers representing the waves in a 3-D space as projected on a 2-D geometry. This leads to an analysis in which the magnitude of the wavenumber is not a constant at a particular temporal frequency. It should be obvious that any set of statistically independent directional signals propagating as plane waves can be represented with an impulsive frequency-wavenumber function. However, such a representation is not possible if the components are correlated as could possibly be envisioned in some multipath situations. The fundamental difficulty stems from the requirement that components from disjoint regions of wavenumber space must be uncorrelated as specified by (5.98); therefore such a process would not be homogeneous.

We discuss other interesting examples in Sections 5.3.3 and 5.3.4. Before doing that development, we want to define Gaussian space-time processes.

5.3.2 Gaussian Space-time Processes

In Section 2.6 of DEMT I [VT68], [VT01a], we discuss Gaussian random vectors in detail. In Section 3.3.3 of DEMT I [VT68], [VT01a], we discuss scalar Gaussian random processes in detail. We now extend those ideas to vector space-time random processes.

Definition: Let $f(t, \mathbf{p})$ be a space-time random process defined over some time interval $[T_\alpha, T_\beta]$ and spatial region $\Re^N$ with a mean value function $m_f(t, \mathbf{p})$ and covariance function $K_f(t_1, t_2 : \mathbf{p}_1, \mathbf{p}_2)$. If every linear functional of $f(t, \mathbf{p})$ is a Gaussian random variable, then $f(t, \mathbf{p})$ is a Gaussian space-time random process. In other words, if

$$y = \int_{T_\alpha}^{T_\beta} \int_{\Re^N} g(u : \mathbf{p}) f(u : \mathbf{p})\, du\, d\mathbf{p}, \tag{5.118}$$

and $g(u : \mathbf{p})$ is any linear function that $E\left[y^2\right] < \infty$, then in order for $f(u : \mathbf{p})$ to be a Gaussian space-time random process, y must be a Gaussian random variable for every $g(u : \mathbf{p})$ in the above class (denoted by $\mathcal{G}$).[7] Several properties follow immediately from this definition.

<u>**Property 1**</u>: The output of a linear system is a particular linear functional of interest. We denote the impulse response (which is the output at time t and position $\mathbf{p}_1$ due to a unit impulse input at time u and position $\mathbf{p}_2$) as $h(t, u : \mathbf{p}_1, \mathbf{p}_2)$. If the input is $f(t, \mathbf{p})$, which is a sample function from a Gaussian space-time random process, the output $y(t, \mathbf{p})$ is also;

$$\int_{T_\alpha}^{T_\beta} \int_{\Re^N} du d\mathbf{p}_2 h(t, u : \mathbf{p}_1, \mathbf{p}_2) f(u : \mathbf{p}_2) = y(t, \mathbf{p}_1). \tag{5.119}$$

The proof is identical to the proof on page 183 of DEMT I [VT68], [VT01a], with appropriate changes in notation.

<u>**Property 2**</u>: If

$$y_1 = \int_{T_\alpha}^{T_\beta} \int_{\Re^N} g_1(u : \mathbf{p}) f(u : \mathbf{p})\, du\, d\mathbf{p}, \tag{5.120}$$

and

$$y_2 = \int_{T_\alpha}^{T_\beta} \int_{\Re^N} g_2(u : \mathbf{p}) f(u : \mathbf{p})\, du\, d\mathbf{p}, \tag{5.121}$$

where $f(u : \mathbf{p})$ is a Gaussian random process and g_1 and g_2 are in $\mathcal{G}$, then y_1 and y_2 are jointly Gaussian.

<u>**Property 3**</u>: If $f(t, \mathbf{p})$ is a Gaussian random process, for any set of times $t_1, t_2, t_3, \cdots, t_n$ in the time interval $[T_\alpha, T_\beta]$ and locations $\mathbf{p}_1, \mathbf{p}_2, \mathbf{p}_3, \cdots, \mathbf{p}_m$ in the set $\Re^N$, the mn random variables $f_{t_1\mathbf{p}_1}, f_{t_2\mathbf{p}_1}, f_{t_3\mathbf{p}_1}, \cdots, f_{t_n\mathbf{p}_1}, f_{t_1\mathbf{p}_2}, \cdots, f_{t_1\mathbf{p}_m}, \cdots, f_{t_n\mathbf{p}_m}$ are jointly Gaussian random variables. To verify this, use a $g(u : \mathbf{p})$ in (5.118), which is a set of impulses,

$$g(u : \mathbf{p}) = \sum_{i=1}^{n} \sum_{j=1}^{m} g_{ij} \delta(u - t_i) \delta(\mathbf{p} - \mathbf{p}_j). \tag{5.122}$$

[7] Gaussian space-time random processes are sometimes referred to as Gaussian random fields.

We see that our definition has the desirable property that it uniquely specifies the joint density at any set of times. Frequently, Property 3 is used as the basic definition. The disadvantage of this approach is that it is more difficult to prove that our definition and properties 1 and 2 hold.

The Gaussian process we have defined has two main virtues:

1. The physical mechanisms that produce many processes are such that a Gaussian model is appropriate.

2. The Gaussian process has many properties that make analytic results feasible.

Discussions of physical mechanisms that lead logically to Gaussian processes are available in [DR58] and [DR87].

We encounter multiple processes that are jointly Gaussian. The definition is a straightforward extension of the preceding one.

Definition: Let $f_1(t, \mathbf{p})$, $f_2(t, \mathbf{p})$, $\cdots$, $f_N(t, \mathbf{p})$ be a set of random processes defined over the intervals $(T_{\alpha_1}, T_{\beta_1})$, $(T_{\alpha_2}, T_{\beta_2})$, $\cdots$, $(T_{\alpha_N}, T_{\beta_N})$, respectively. If every sum of arbitrary functional of $f_i(t, \mathbf{p})$, $i = 1, \cdots, N$ is a Gaussian random variable, then the processes $f_1(t, \mathbf{p})$, $f_2(t, \mathbf{p})$, $\cdots$, $f_N(t, \mathbf{p})$ are defined to be jointly Gaussian random processes. In other words,

$$y = \sum_{i=1}^{N} \int_{T_{\alpha_i}}^{T_{\beta_i}} \int_{\Re^N} g_i(u : \mathbf{p}) f_i(u, \mathbf{p}) du d\mathbf{p}, \tag{5.123}$$

is a Gaussian random variable.

Most of our random process models will be Gaussian when we study optimum array processing. This assumption is justified in a wide range of physical problems.

In the next section, more detailed models of signal and noise random processes are developed.

5.3.3 Plane Waves Propagating in Three Dimensions

In this section[8] we describe a model that allows us to generate the temporal frequency spatial correlation function for a large class of random processes. This enables us to find the frequency-wavenumber spectrum. Isotropic noise is included as a special case.

[8] This section follows [Bag76].

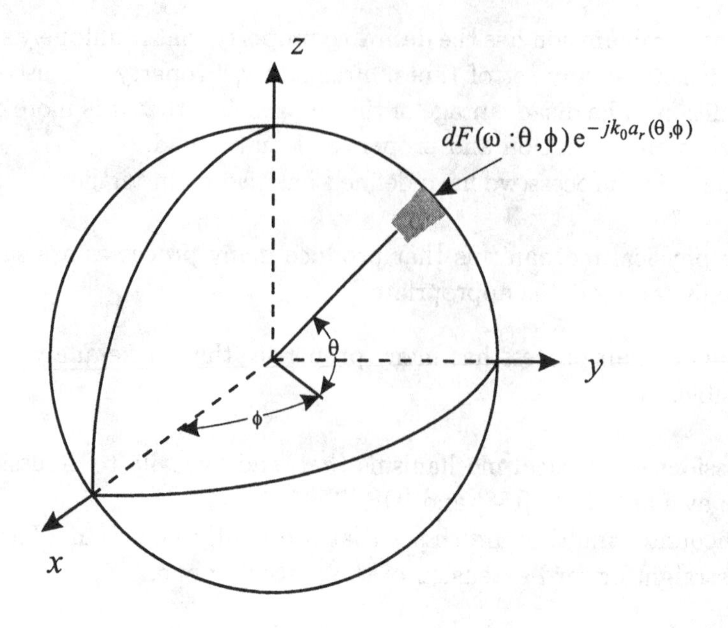

Figure 5.7 Incremental surface area contributing to a plane-wave spatial process.

At temporal frequency ω_o, the noise processes of interest are modeled in three dimensions as positions of infinitesimal plane-wave processes, all radiating towards a common point. These waves may be considered to be generated on the surface of a sphere whose radius is large compared to any geometries or wavelengths of interest. Using the integrated transform representation we have

$$dF(\omega_o, \mathbf{p}) = \int_0^{\pi} d\theta \int_0^{2\pi} \frac{\sin(\theta)d\phi}{4\pi} dF_0(\omega_o : \theta, \phi) e^{-jk_o \mathbf{a}_r^T(\theta,\phi)\mathbf{p}}, \tag{5.124}$$

where

$$f(t, \mathbf{p}) = \int_{-\infty}^{\infty} dF(\omega_o : \mathbf{p}) e^{j\omega t}, \tag{5.125}$$

$k_o = \frac{2\pi}{\lambda}$, and $\mathbf{a}_r(\theta, \phi)$ is a unit vector in the radial direction. Consequently, $-k_o\mathbf{a}_r(\theta, \phi)$ forms a propagation vector $\mathbf{k}(\theta, \phi)$ at a temporal frequency ω_o radiating towards the center of the sphere, or

$$\mathbf{k}(\theta, \phi) = -k_o \mathbf{a}_r(\theta, \phi) = -\frac{2\pi}{\lambda}\mathbf{a}_r(\theta, \phi) = -\frac{\omega_o}{c}\mathbf{a}_r(\theta, \phi). \tag{5.126}$$

The model is shown in Figure 5.7.

We assume that disjoint regions of the sphere radiate uncorrelated components so that

$$E\left[dF_0(\omega_o:\theta_1,\phi_1)dF_0^*(\omega_o:\theta_2,\phi_2)\right]=$$

$$S_0(\omega_o:\theta_1,\phi_1)\left(\frac{\delta(\theta_1-\theta_2)\delta(\phi_1-\phi_2)}{\frac{\sin\theta_1}{4\pi}}\right)\left(\frac{d\omega}{2\pi}\right). \tag{5.127}$$

(The impulse terms should be interpreted formally and should operate simultaneously. The factor of $4\pi/\sin\theta_1$ needs to be introduced because of the use of spherical coordinates.) The temporal spectrum-spatial correlation function is given by

$$E\left[dF(\omega_o:\mathbf{p})dF^*(\omega_o:\mathbf{p}-\Delta\mathbf{p})\right]=S_f(\omega_o:\Delta\mathbf{p})\frac{d\omega}{2\pi}. \tag{5.128}$$

Using (5.124), we have

$$S_f(\omega_o:\Delta\mathbf{p})\frac{d\omega}{2\pi}=\int_0^{\pi}d\theta_1\int_0^{2\pi}\frac{\sin(\theta_1)}{4\pi}d\phi_1\int_0^{\pi}d\theta_2\int_0^{2\pi}\frac{\sin(\theta_2)}{4\pi}$$

$$\cdot E\left[dF_0(\omega_o:\theta_1,\phi_1)dF_0^*(\omega_o:\theta_2,\phi_2)\right]\cdot e^{-jk_o\mathbf{a}_r^T(\theta_1,\phi_1)\mathbf{p}+jk_o\mathbf{a}_r^T(\theta_2,\phi_2)(\mathbf{p}-\Delta\mathbf{p})}d\phi_2. \tag{5.129}$$

Equation (5.127) implies that disjoint (θ,ϕ) are uncorrelated, so (5.129) reduces to

$$\boxed{S_f(\omega_o:\Delta\mathbf{p})=\int_0^{\pi}d\theta\int_0^{2\pi}\frac{\sin(\theta)}{4\pi}d\phi S_o(\omega_o:\theta,\phi)e^{-jk_o\mathbf{a}_r^T(\theta,\phi)\Delta\mathbf{p}}}, \tag{5.130}$$

which is the first desired result. It enables us to find the temporal frequency spectrum-spatial correlation function for a large class of random processes as a function of $S_o(\omega_o:\theta,\phi)$, which is their temporal frequency function at locations (θ,ϕ) on the sphere.

The form in (5.130) is most useful in cases where $S_o(\omega_o:\theta,\phi)$ is known and we want to find the spatial spectral matrix $\mathbf{S_f}(\omega_o)$ at the array.

The frequency-wavenumber function follows from the Fourier transform relationship. It is useful to define the wavenumber $\mathbf{k}$ in both spherical and Cartesian coordinates:

$$\mathbf{k}=k_x\mathbf{i}_x+k_y\mathbf{i}_y+k_z\mathbf{i}_z=k_r\mathbf{a}_r(\theta_k,\phi_k), \tag{5.131}$$

where $\mathbf{a}_r(\theta_k,\phi_k)$ is a unit radial vector in the same direction as $\mathbf{k}$. We have

$$\begin{aligned} P_f(\omega_o : \mathbf{k}) &= \int_{-\infty}^{\infty}\int_{-\infty}^{\infty}\int_{-\infty}^{\infty} d\Delta\mathbf{p} e^{j\mathbf{k}^T\Delta\mathbf{p}} \int_0^{\pi} d\theta \int_0^{2\pi} \frac{\sin(\theta)}{4\pi} d\phi \\ &\quad S_o(\omega_o : \theta, \phi) e^{-jk_o\mathbf{a}_r^T(\theta,\phi)\Delta\mathbf{p}} \\ &= \int_0^{\pi} d\theta \int_0^{2\pi} \frac{\sin(\theta)}{4\pi} \int_{-\infty}^{\infty}\int_{-\infty}^{\infty}\int_{-\infty}^{\infty} d\Delta\mathbf{p} S_o(\omega_o : \theta, \phi) \\ &\quad e^{j(k_r\mathbf{a}_r(\theta_k,\phi_k) - k_o\mathbf{a}_r(\theta,\phi))^T\Delta\mathbf{p}} . \end{aligned} \tag{5.132}$$

The evaluation of the last integral leads to an impulse, or delta function, in wavenumber space, that is,

$$\begin{aligned} &\int_{-\infty}^{\infty}\int_{-\infty}^{\infty}\int_{-\infty}^{\infty} d\Delta\mathbf{p} e^{j(k_r\mathbf{a}_r(\theta_k,\phi_k) - k_o\mathbf{a}_r(\theta,\phi))^T\Delta\mathbf{p}} \\ &= (2\pi)^3 \frac{\delta(k_r - k_o)\delta(\theta_k - \theta)\delta(\phi_k - \phi)}{k_o^2\sin(\theta)} . \end{aligned} \tag{5.133}$$

When we substitute (5.133) into (5.132), we obtain

$$\boxed{P_f(\omega_o : \mathbf{k}) = (2\pi)^3 S_o(\omega_o : \theta_k, \phi_k) \frac{\delta(k_r - k_o)}{4\pi k_o^2} .} \tag{5.134}$$

We have the intuitive interpretation that the resulting frequency-wavenumber function has the same distribution as the spectra of the plane waves at the various locations on the sphere. The delta function arises because these waves are modeled as pure plane waves analogous to pure tones temporally. By starting with this specification of the noise field we can model a large number of ambient fields that may be encountered.

In many applications it is useful to express the temporal frequency spectrum-spatial correlation function in spherical coordinates. To do this, we write $\mathbf{a}_r(\theta, \phi)$ and $\Delta\mathbf{p}$ in terms of rectangular coordinates,

$$\mathbf{a}_r(\theta, \phi) = -\left[\sin\theta\cos\phi \quad \sin\theta\sin\phi \quad \cos\theta\right]^T , \tag{5.135}$$

and

$$\Delta\mathbf{p} = r_p\left[\sin\theta_p\cos\phi_p \quad \sin\theta_p\sin\phi_p \quad \cos\theta_p\right]^T , \tag{5.136}$$

where

$$r_p = |\Delta\mathbf{p}| . \tag{5.137}$$

Using (5.135) and (5.136) in (5.130) gives

$$\boxed{\begin{aligned} S_f(\omega_o : \Delta\mathbf{p}) = &\int_0^{\pi} d\theta \int_0^{2\pi} \frac{\sin\theta}{4\pi} d\phi S_o(\omega_o : \theta, \phi) \cdot \\ &\exp\left\{jk_o r_p\left[\sin\theta\sin\theta_p(\cos(\phi - \phi_p)) + \cos\theta\cos\theta_p\right]\right\} . \end{aligned}} \tag{5.138}$$

We now consider the special case of isotropic noise. For isotropic noise,

$$S_o(\omega : \theta, \phi) = S_o(\omega). \tag{5.139}$$

Using (5.139) in (5.138) gives

$$\begin{aligned} S_f(\omega_o : \mathbf{\Delta p}) &= S_o(\omega_o)\left(\int_0^{\pi} d\theta \frac{\sin\theta}{4\pi} e^{jk_o|\mathbf{\Delta p}|\cos\theta\cos\theta_p}\right) \\ &\cdot \left(\int_0^{2\pi} e^{jk_o|\mathbf{\Delta p}|\sin\theta\sin\theta_p\cos(\phi-\phi_p)} d\phi\right). \end{aligned} \tag{5.140}$$

We know that the result cannot be a function of (θ_p, ϕ_p) due to the spherical symmetry of the field. Therefore, we can let $\theta_p = 0$. Now (5.140) reduces to

$$S_f(\omega_o : \mathbf{\Delta p}) = S_o(\omega_o)\int_0^{\pi} d\theta \frac{\sin\theta}{2} e^{jk_o\mathbf{\Delta p}\cos\theta}. \tag{5.141}$$

Letting $\alpha = \cos\theta$, and integrating , we obtain

$$\begin{aligned} S_f(\omega_o : \mathbf{\Delta p}) &= S_o(\omega_o)\frac{\sin k_o\mathbf{\Delta p}}{k_o\mathbf{\Delta p}} \\ &= S_o(\omega_o)\text{sinc}\left(\frac{2\pi\mathbf{\Delta p}}{\lambda}\right). \end{aligned} \tag{5.142}$$

Thus, the temporal frequency spectrum spatial correlation function is a sinc function whose argument is $\frac{2\pi\mathbf{\Delta p}}{\lambda}$.

In [Bag76], Baggeroer develops expansions of $S_o(\omega_o : \theta, \phi)$ in terms of spherical harmonics that were introduced by Stratton [Str41] to study electromagnetic fields. We normally use numerical techniques for the non-isotropic cases.

5.3.4 1-D and 2-D Projections

In many of our applications, the array will be 1-D or 2-D. In this case we are interested in the projection onto the appropriate dimension. We consider the one-dimensional case first.

5.3.4.1 Projection onto one dimension

If we project onto one dimension (the z-axis), then $\theta_p = 0$ or π in (5.138). Then,

$$S_f(\omega : \Delta p_z) = \int_0^{\pi} d\theta \frac{\sin\theta}{4\pi} e^{jk_o\Delta p_z\cos\theta} \int_0^{2\pi} S_o(\omega : \theta, \phi) d\phi. \tag{5.143}$$

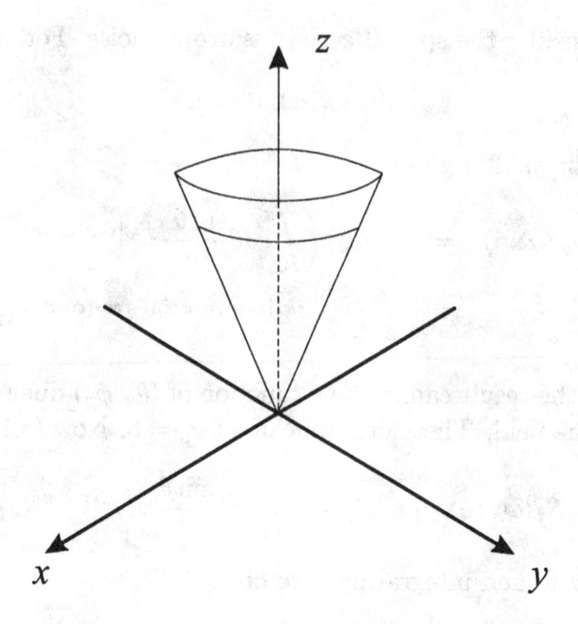

Figure 5.8 Sector signal field.

Since the linear array cannot resolve in the ϕ-direction, define

$$\bar{S}_o(\omega : \theta) = \frac{1}{2\pi}\int_0^{2\pi} S_o(\omega : \theta, \phi)d\phi, \tag{5.144}$$

and (5.143) becomes

$$\boxed{S_f(\omega : \Delta p_z) = \int_0^{\pi} d\theta \frac{\sin\theta}{2} e^{jk_o \Delta p_z \cos\theta}\bar{S}_o(\omega : \theta).} \tag{5.145}$$

This can be rewritten in terms of the directional cosine u,

$$u = \cos\theta, \tag{5.146}$$

as

$$\boxed{S_f(\omega : \Delta p_z) = \frac{1}{2}\int_{-1}^{1} du\, e^{jk_o \Delta p_z u}\bar{S}_{ou}(\omega : u),} \tag{5.147}$$

where $\bar{S}_{ou}(\omega : u) = \bar{S}_o(\omega : \cos^{-1}(u))$.

We consider a simple example to illustrate a typical result.

Example 5.3.2

Consider the spatial distribution shown in Figure 5.8.

$$S_o(\omega : \theta, \phi) = \begin{cases} \sigma_f^2, & \theta \leq \theta_0 \\ 0, & \text{elsewhere.} \end{cases} \tag{5.148}$$

The spectrum is constant over a segment of the sphere and is zero elsewhere. It is uniform in the ϕ-variable. Defining,

$$u_0 = \cos\theta_0, \tag{5.149}$$

(5.147) reduces to

$$\begin{aligned} S_f(\omega : \Delta p_z) &= \frac{\sigma_f^2}{2}\int_{-u_0}^{u_0} e^{jk_o \Delta p_z u} du \\ &= \sigma_f^2 \frac{\sin(k_o \Delta p_z u_0)}{k_o \Delta p_z} = \sigma_f^2 u_0 \text{sinc}(k_o \Delta p_z u_0), \end{aligned} \tag{5.150}$$

which is the desired result.

A second example corresponds to a spatial distribution spread around the broadside direction.

Example 5.3.3

We assume

$$\bar{S}_{ou}(\omega : u) = \frac{\sigma_f^2}{\sqrt{2\pi}\sigma_u} e^{-\frac{u^2}{2\sigma_u^2}}, \quad -1 \leq u \leq 1, \tag{5.151}$$

and σ_u is small compared to unity. Then (5.147) reduces to

$$S_f(\omega : \Delta p_z) = \sigma_f^2 e^{-\frac{k_o^2 \Delta p_z^2 \sigma_u^2}{2}} \int_{-1}^{1} \frac{1}{\sqrt{2\pi}\sigma_u} e^{-\frac{1}{2\sigma_u^2}\left(u - jk_o \Delta p_z \sigma_u^2\right)^2} du. \tag{5.152}$$

For small σ_u, the limits on the integral can be approximated as $-\infty$ to ∞, so the integral equals unity and

$$S_f(\omega : \Delta p_z) = \sigma_f^2 \exp\left\{-\frac{k_o^2 \Delta p_z^2 \sigma_u^2}{2}\right\}. \tag{5.153}$$

Other examples are developed in the problems. We now consider 2-D projections.

5.3.4.2 Projection onto two dimensions

In this case, we assume the 2-D surface of interest is the xy-plane. Define

$$\Delta\mathbf{p}_s = \Delta\mathbf{p}|_{p_z=0} = \left[\Delta p_x \; \Delta p_y\right]^T, \tag{5.154}$$

and

$$\mathbf{k}_s = \mathbf{k}|_{k_z=0} = \left[k_x \; k_y\right]^T. \tag{5.155}$$

We first find $S_f(\omega : \Delta\mathbf{p}_s)$. Substituting $\theta_p = \frac{\pi}{2}$ into (5.138), we obtain

$$\boxed{S_f(\omega_0 : \Delta\mathbf{p}_s) = \int_0^{\pi} d\theta \int_0^{2\pi} d\phi \frac{\sin\theta}{4\pi} S_o(\omega_0 : \theta, \phi) e^{jk_o r_p \sin\theta \cos(\phi - \phi_p)}.} \tag{5.156}$$

We first perform the integration with respect to θ over the regions $[0, \frac{\pi}{2}]$ and $[\frac{\pi}{2}, \pi]$. (This is necessary since two points on the surface of the sphere for the plane-wave model project to the same point on the 2-D surface.) Now change variables in each region. Define

$$\sin\theta = v, \quad 0 \le \theta \le \frac{\pi}{2}, \tag{5.157}$$

and

$$\sin(\pi - \theta) = v, \quad \frac{\pi}{2} < \theta \le \pi. \tag{5.158}$$

Then (5.156) reduces to

$$\begin{aligned} S_f(\omega_0 : \mathbf{p}_s) &= \frac{1}{4\pi}\int_0^1 \frac{v dv}{\sqrt{1-v^2}} \int_0^{2\pi} d\phi e^{jk_o r_p v\cos(\phi-\phi_p)} \left\{ S_o\left(\omega_0 : \sin^{-1}v, \phi\right) \right. \\ & \left. + S_o\left(\omega_0 : \pi - \sin^{-1}v, \phi\right)\right\}, \end{aligned} \tag{5.159}$$

which is the general result.

If $S_o(\omega_0 : \theta, \phi)$ does not have an azimuthal (ϕ) dependence, then we can perform the integration over ϕ. Rewriting (5.156) for this case,

$$\begin{aligned} S_f(\omega_0 : \mathbf{p}_s) &= \frac{1}{4\pi}\int_0^1 \frac{v dv}{\sqrt{1-v^2}} \left\{ S_o\left(\omega_0 : \sin^{-1}(v)\right) \right. \\ & \left. + S_o\left(\omega_0 : \pi - \sin^{-1}v\right)\right\} \\ & \cdot \int_0^{2\pi} d\phi e^{jk_o r_p v\cos(\phi-\phi_p)}. \end{aligned} \tag{5.160}$$

The inner integral corresponds to a zero-order Bessel function. Thus,

$$\begin{aligned} S_f(\omega_0 : \mathbf{p}_s) = \frac{1}{2}\int_0^1 \frac{dv J_0(k_o r_p v)}{\sqrt{1-v^2}} & \left\{ S_o\left(\omega_0 : \sin^{-1}(v)\right) \right. \\ & \left. + S_o\left(\omega_0 : \pi - \sin^{-1}v\right)\right\}. \end{aligned} \tag{5.161}$$

At this point, we must specify $S_f(\omega_0 : \theta)$ in order to proceed. We consider the sector distribution in Figure 5.8.

Example 5.3.4

We define

$$v_0 = \sin\theta_0. \tag{5.162}$$

Then, (5.161) reduces to

$$S_f(\omega_0 : \Delta\mathbf{p}_s) = \frac{S_o(\omega_0)}{2}\int_0^{v_0} \frac{v dv J_0(k_o r_p v)}{\sqrt{1-v^2}}. \tag{5.163}$$

The integral in (5.163) can be evaluated numerically.

Several other spatial distributions corresponding to various ambient noise or signal models are developed in the problems.

5.3.4.3 Summary

In this section, we have developed both a second-moment representation and a Gaussian process representation for space-time random processes. We now study how arrays and apertures respond to these space-time processes.

5.4 Arrays and Apertures

In this section, we discuss how arrays and apertures respond to space-time random processes. In Chapter 2, we showed that a plane wave was the basis function for a linear time-invariant system. In Section 5.2, we modeled the space-time processes as collections of statistically independent plane waves. We now combine these two models. We begin our discussion with arrays.

5.4.1 Arrays

5.4.1.1 Sampling

We first consider a uniformly spaced linear array. The linear array samples the space-time process at spatial intervals d, where d is the interelement spacing. The space-time field is bandlimited in wavenumber space:

$$|k| = \frac{2\pi}{\lambda} \text{ radians/meter } = \frac{1}{\lambda} \text{ cycles/meter }, \tag{5.164}$$

and

$$|k_z| \leq \frac{2\pi}{\lambda} \text{ radians/meter } = \frac{1}{\lambda} \text{ cycles/meter }. \tag{5.165}$$

Since wavenumber and position are conjugate Fourier transform variables, we know that we must sample at

$$d \leq \frac{1}{2\left(\frac{1}{\lambda}\right)} = \frac{\lambda}{2}, \tag{5.166}$$

in order to be able to reconstruct the space-time process without spatial aliasing. This is the same interelement spacing that we found in Chapter 2 using a grating lobe argument.

In two dimensions,

$$|k_x| \leq \frac{2\pi}{\lambda}, \tag{5.167}$$

and

$$|k_y| \leq \frac{2\pi}{\lambda}, \tag{5.168}$$

so that, if we use a rectangular sampling grid, we require

$$d_x \leq \frac{\lambda}{2}, \tag{5.169}$$

and

$$d_y \leq \frac{\lambda}{2}. \tag{5.170}$$

In [PM62], Peterson and Middleton discussed alternate sampling strategies and showed that using the hexagonal array discussed in Section 4.4 with spacing,

$$d_x = \frac{\lambda}{2}, \tag{5.171}$$

$$d_y = \frac{\sqrt{3}}{4}\lambda, \tag{5.172}$$

is the sampling strategy that completely specifies the space-time field with the smallest number of samples (e.g., [PM62], [Mer79] or [DM84]).

5.4.1.2 Spectral matrices

We now consider the output of an N-element array when the input is a sample function of a random space-time process. If the input to the array is the space-time process $x(t, \mathbf{p})$, then the output of the array, in the absence of sensor noise, is

$$\mathbf{x}(t, \mathbf{p}) = \begin{bmatrix} x(t, \mathbf{p}_0) \\ x(t, \mathbf{p}_1) \\ \vdots \\ x(t, \mathbf{p}_{N-1}) \end{bmatrix}, \tag{5.173}$$

where $\mathbf{p}_i$ denotes the position of the ith element.

The spatial spectral matrix is $\mathbf{S}_{\mathbf{x}}(\omega)$ whose ij element is

$$[\mathbf{S}_{\mathbf{x}}(\omega)]_{ij} = \mathbf{S}_x(\omega : \mathbf{p}_i - \mathbf{p}_j), \tag{5.174}$$

where $\mathbf{S}_x(\omega : \mathbf{p}_i - \mathbf{p}_j)$ is the temporal frequency spectrum-spatial correlation function defined in (5.91).

We use the frequency-domain snapshot model of Section 5.2 to obtain $\mathbf{X}_{\Delta T}(\omega_m, k)$. We use the approximation discussed in Section 5.2:

$$E\left[\mathbf{X}_{\Delta T}(\omega_m, k)\mathbf{X}_{\Delta T}^H(\omega_m, k)\right] \simeq \mathbf{S}_{\mathbf{x}}(\omega_m). \tag{5.175}$$

In the beamforming application, $\mathbf{X}(\omega_m, k)$ is processed with a matrix filter $\mathbf{W}^H(\omega_m)$ as shown in Figure 5.9. Then,

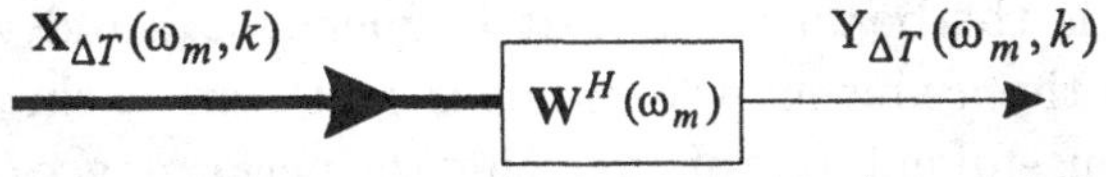

$\mathbf{X}_{\Delta T}(\omega_m, k)$ $\mathbf{W}^H(\omega_m)$ $Y_{\Delta T}(\omega_m, k)$

Figure 5.9 Matrix filter.

$$Y_{\Delta T}(\omega_m, k) = \mathbf{W}^H(\omega_m)\mathbf{X}_{\Delta T}(\omega_m, k)\,, \tag{5.176}$$

and the spectrum of $Y_{\Delta T}(\omega_m, k)$ is

$$\begin{aligned} S_y(\omega_m) &= E\left[Y_{\Delta T}(\omega_m, k)Y^*_{\Delta T}(\omega_m, k)\right] \\ &= \mathbf{W}^H(\omega_m, k)\left[E\left[\mathbf{X}(\omega_m, k)\mathbf{X}^H(\omega_m, k)\right]\mathbf{W}(\omega_m)\right], \end{aligned} \tag{5.177}$$

or

$$S_y(\omega_m) = \mathbf{W}^H(\omega_m)\mathbf{S}_\mathbf{x}(\omega_m)\mathbf{W}(\omega_m)\,. \tag{5.178}$$

5.4.1.3 Beam patterns in frequency-wavenumber space

An alternative expression for $S_y(\omega_m)$ is useful in order to interpret the result in terms of beam patterns and the frequency-wavenumber spectrum.

We can write

$$Y_{\Delta T}(\omega_m, k) = \sum_{n=0}^{N-1} W^*_n(\omega_m)X(\omega_m, k : \mathbf{p}_n)\,. \tag{5.179}$$

The output spectrum is given by (5.177). Using (5.179) in (5.177) and (5.174) in the result gives

$$S_y(\omega_m) = \sum_{n=1}^{N}\sum_{l=1}^{N} W^*_n(\omega_m)S_x(\omega : \mathbf{p}_n - \mathbf{p}_l)W_l(\omega_m)\,. \tag{5.180}$$

Using (5.95) in (5.180) gives

$$\begin{aligned} S_y(\omega_m) &= \int_{-\infty}^{\infty}\int_{-\infty}^{\infty}\int_{-\infty}^{\infty} \frac{d\mathbf{k}}{(2\pi)^3}\sum_{n=1}^{N}\sum_{l=1}^{N} W^*_n(\omega_m)W_l(\omega_m) \\ &\quad \cdot P_x(\omega : \mathbf{k})e^{-j\mathbf{k}(\mathbf{p}_n - \mathbf{p}_l)}\,. \end{aligned} \tag{5.181}$$

Using the definition of the beam pattern, (5.181) can be written as

$$\boxed{S_y(\omega_m) = \int_{-\infty}^{\infty}\int_{-\infty}^{\infty}\int_{-\infty}^{\infty} \left|B(\omega_m, \mathbf{k})\right|^2 P_x(\omega_m, \mathbf{k})\frac{d\mathbf{k}}{(2\pi)^3}\,,} \tag{5.182}$$

where $B(\omega_m, \mathbf{k})$ is the beam pattern at frequency ω_m and wavenumber $\mathbf{k}$ and $P_x(\omega_m, \mathbf{k})$ is the frequency-wavenumber spectrum at the array.

This result is useful in terms of analyzing the performance of an array and in designing suitable beam patterns. It is analogous to the scalar spectral result,

$$S_{out}(\omega_m) = |H(\omega_m)|^2 S_{in}(\omega_m). \tag{5.183}$$

5.4.1.4 Co-arrays

It is useful to explore the relationship in (5.182) further. The term $|B(\omega, \mathbf{k})|^2$ represents the power pattern of the array. We can define an aperture autocorrelation function corresponding to the power pattern. Thus,[9]

$$R_W(\omega : \mathbf{p}) = \int_{-\infty}^{\infty} \cdot \int_{-\infty}^{\infty} |B(\omega, \mathbf{k})|^2 \, e^{j\mathbf{k}^T\mathbf{p}} \frac{d\mathbf{k}}{(2\pi)^3}, \tag{5.184}$$

where the integral is over the appropriate dimension and $\mathbf{p}$ takes on discrete values corresponding to the element positions. For one dimension,

$$R_W(\omega : p_x) = \int_{-\infty}^{\infty} |B(\omega, k_x)|^2 \, e^{jk_x p_x} \frac{dk_x}{2\pi}, \tag{5.185}$$

and for two dimensions,

$$R_W(\omega : p_x, p_y) = \int_{-\infty}^{\infty} \int_{-\infty}^{\infty} |B(\omega, \mathbf{k})|^2 \, e^{j[k_x p_x + k_y p_y]} \frac{dk_x dk_y}{(2\pi)^2}. \tag{5.186}$$

From the transform relationship between the aperture weighting function and the beam pattern, we see $R_W(\omega : \mathbf{p})$ is just the convolution of $\mathbf{W}(\omega)$ with itself. For an N-element linear array (not necessarily equally spaced)

$$R_W(\omega : |m|) = \sum_{n=0}^{N-1} W(n) W^*(n+m), \quad m = 0, \pm 1, \cdots, \pm N, \tag{5.187}$$

where the symmetry of $R_W(\omega : |m|)$ follows from the symmetry of $P_B(\omega : k_x)$ in k_x.

We encountered co-arrays previously in Section 3.9.2 in our discussion of minimum redundancy arrays. This discussion develops the idea from a more fundamental viewpoint of sampling $P_x(\omega : \mathbf{k})$ and showing how the samples of $P_x(\omega : \mathbf{k})$ can be used to reconstruct $P_x(\omega : \mathbf{k})$.

[9]This function was defined in [Bag76]. The application of co-arrays is discussed in [Hau68] and [Kre71].

For simplicity we consider the 1-D case. The function $P_x(\omega : k_z)$ is bandlimited in k_z-space,

$$|k_z| \leq \frac{2\pi}{\lambda}. \tag{5.188}$$

Therefore, from the sampling theorem, if the co-array has elements at

$$md = m\left(\frac{\lambda}{2}\right), \quad -\infty < m < \infty, \tag{5.189}$$

we could reconstruct $P_x(\omega : k_z)$ exactly. In practice, m will be limited by the physical extent of the array. However, we want to have the value at md for all m within the range allowed by the aperture size.

The reason that we use the co-array in the sampling theorem is the relationship to $P_x(\omega : \mathbf{k})$ in (5.182). Thus, as long as the co-array does not have gaps, the sampling theorem will hold. As we saw in Section 3.9.2, the array itself may have significant gaps.

5.4.1.5 Sensor noise

In all physical systems there is noise associated with the observation process. At a minimum, there is noise generated internally by the electronics in the system. We model this as a zero-mean spatially uncorrelated Gaussian random process that we denote by $\mathbf{n}_{su}(t)$. This is the noise introduced in Section 5.2. The contribution to the spatial spectral matrix is

$$\mathbf{S}_{\mathbf{n}_{su}}(\omega_m) = S_n(\omega_m)\mathbf{I}. \tag{5.190}$$

In most cases, we also assume that the temporal spectrum is flat. Then,

$$\mathbf{S}_{\mathbf{n}_{su}}(\omega_m) = \sigma_w^2 \mathbf{I}, \tag{5.191}$$

where σ_w^2 is height of the white noise spectrum.[10] We denote noise vectors satisfying (5.191) as $\mathbf{w}(t)$.

The spectrum at the output of the array processor due to a white noise input is given by substituting (5.191) into (5.178), which results in,

$$S_{yw}(\omega_m) = \sigma_w^2 \left|\mathbf{W}(\omega_m)\right|^2, \tag{5.192}$$

or

$$S_{yw}(\omega_m) = \sigma_w^2 \sum_{i=1}^{N} \left|W_i(\omega_m)\right|^2. \tag{5.193}$$

[10] We use N_0 as the spectral height for the continuous spectral case. When we use the frequency-domain snapshot model we label the snapshot variance σ_w^2 to be consistent with the literature ($\sigma_w^2 = N_0$).

5.4.2 Apertures

All of the results developed for arrays carry over to apertures in a straightforward manner.

The input to the aperture is the space-time process $f(t,p)$. The aperture weighting function is $w(t,p)$. The output of the system is

$$y(t) = \int_{-\infty}^{\infty} d\tau \int_{\mathcal{R}^A} \cdot \int w^*(\tau, \mathbf{p}) f(t-\tau, \mathbf{p}) d\mathbf{p}, \tag{5.194}$$

where $\mathcal{R}^A$ denotes the aperture region. For a linear aperture,

$$y(t) = \int_{-\infty}^{\infty} d\tau \int_{-\frac{L}{2}}^{\frac{L}{2}} w^*(\tau, p_x) f_1(t-\tau, p_x) dp_x, \tag{5.195}$$

where the subscript "1" on $f(\cdot)$ denotes that it is a 1-D representation of the input. If we started with a 2-D representation of the input, then

$$y(t) = \int_{-\infty}^{\infty} d\tau \int_{-\frac{L}{2}}^{\frac{L}{2}} \int_{L_y} w^*(\tau, p_x) \delta(p_y) f_2(t-\tau, p_x, p_y) dp_x dp_y, \tag{5.196}$$

for a linear array on the x-axis.

If we use the frequency domain snapshot model and let $\omega_m = \omega$, then we can also write

$$Y(\omega) = \int_{\mathcal{R}^A} \cdot \int W^*(\omega, \mathbf{p}) F(\omega, \mathbf{p}) d\mathbf{p}. \tag{5.197}$$

The output spectrum is

$$\begin{aligned} S_y(\omega) &= \int_{\mathcal{R}^A} \cdot \int \int_{\mathcal{R}^A} \cdot \int W^*(\omega, \mathbf{p}_\alpha) \Delta T^{-1} E\left[F(\omega, \mathbf{p}_\alpha) F^*(\omega, \mathbf{p}_\beta) \right] \\ & \quad W(\omega, \mathbf{p}_\beta) d\mathbf{p}_\alpha d\mathbf{p}_\beta, \end{aligned} \tag{5.198}$$

or

$$S_y(\omega) = \int_{\mathcal{R}^A} \cdot \int \int_{\mathcal{R}^A} \cdot \int W^*(\omega, \mathbf{p}_\alpha) S_f(\omega : \mathbf{p}_\alpha, \mathbf{p}_\beta) W(\omega, \mathbf{p}_\beta) d\mathbf{p}_\alpha d\mathbf{p}_\beta, \tag{5.199}$$

where $S_f(\omega : \mathbf{p}_\alpha, \mathbf{p}_\beta)$ is the temporal frequency spectrum spatial correlation function of the input space-time process defined in (5.91). Using (5.95) in (5.199), and performing the integration over $\mathbf{p}_\alpha$ and $\mathbf{p}_\beta$ gives

$$S_y(\omega) = \int_{-\infty}^{\infty} \int_{-\infty}^{\infty} \int_{-\infty}^{\infty} |B(\omega, \mathbf{k}|^2 P_f(\omega, \mathbf{k}) \frac{d\mathbf{k}}{(2\pi)^N}, \tag{5.200}$$

which is identical to (5.182).

To define the sensor noise we specify the dimension of the aperture first. For a linear aperture,

$$K_W(t, u : p_1, p_2) = N_o \delta(t-u)\delta(p_1 - p_2) . \quad (5.201)$$

For a planar aperture,

$$K_W(t, u : p_{x_1}, p_{y_1}, p_{x_2}, p_{y_2}) = N_o \delta(t-u)\delta(p_{x_1} - p_{x_2})\delta(p_{y_1} - p_{y_2}) . \quad (5.202)$$

The results in (5.197) and (5.200) are important when the effect of non-isotropic sensors are included in the model. Then the frequency domain snapshot at the output of the nth sensor is

$$X_n(\omega) = \int_{R^A} \cdot \int W^*_{el}(w, \mathbf{p}) F(w, \mathbf{p}) d\mathbf{p} , \quad (5.203)$$

and

$$S_{x_n}(\omega) = \int_{-\infty}^{\infty} \int_{-\infty}^{\infty} \int_{-\infty}^{\infty} |B_{el}(\omega, \mathbf{k})|^2 P_f(\omega, \mathbf{k}) \frac{d\mathbf{k}}{(2\pi)^N} . \quad (5.204)$$

In most of the subsequent discussions (and in most of the array processing literature), the sensor output is assumed to be the value of the space-time field at the sensor location. However, in an actual system design both the temporal and spatial response characteristics should be taken into account.

5.5 Orthogonal Expansions

In Chapter 3 of DEMT I [VT68], [VT01a], we saw the utility of Karhunen-Loève expansion for scalar Gaussian processes. It provided an orthogonal series expansion whose coefficients are statistically independent Gaussian random variables.[11] In Section 3.7 of [VT68], [VT01a], we introduced the corresponding expansion for vector random processes.

By using a frequency-domain snapshot model we have gone directly to a sequence of complex Gaussian random variables whose spatial spectral matrix is $\mathbf{S}_{\mathbf{x}}(\omega_m)$. The model assumes that the snapshots are statistically independent so we can focus on a single snapshot $\mathbf{X}(\omega_m, k)$.[12] Writing $\mathbf{X}(\omega_m, k)$ as a weighted sum of orthonormal vectors gives

$$\mathbf{X}(\omega_m, k) = \sum_{i=1}^{N} x_i \mathbf{\Phi}_i , \quad (5.205)$$

[11]The Karhunen-Loève expansion deals with second moments and produces *uncorrelated* coefficients if we do not impose the Gaussian assumption.

[12]For notational simplicity, the "ΔT" subscript is omitted.

where the ω_m and k dependence are suppressed on the right side of (5.205). The $\mathbf{\Phi}_i$ are orthonormal vectors,

$$\mathbf{\Phi}_i^H \mathbf{\Phi}_j = \delta_{ij}, \quad i,j = 1,\cdots,N, \tag{5.206}$$

and the coefficients are

$$x_i = \mathbf{\Phi}_i^H \mathbf{X}(\omega_m, k), \quad i = 1,\cdots,N. \tag{5.207}$$

The $\mathbf{\Phi}_i$ are chosen so that

$$E\left[x_i x_j^*\right] = \lambda_i \delta_{ij}, \quad i,j = 1,\cdots,N. \tag{5.208}$$

Substituting (5.207) into (5.208) gives

$$E\left[\mathbf{\Phi}_i^H \mathbf{X}(\omega_m, k) \mathbf{X}^H(\omega_m, k) \mathbf{\Phi}_j\right] = \lambda_i \delta_{ij}, \quad i,j = 1,\cdots,N, \tag{5.209}$$

or

$$\mathbf{\Phi}_i^H \mathbf{S}_{\mathbf{x}}(\omega_m) \mathbf{\Phi}_j = \lambda_i \delta_{ij}, \quad i,j = 1,\cdots,N. \tag{5.210}$$

The solution to (5.210) is

$$\lambda_i \mathbf{\Phi}_i = \mathbf{S}_{\mathbf{x}}(\omega_m) \mathbf{\Phi}_i, \quad i = 1,\cdots,N. \tag{5.211}$$

The result in (5.211) is familiar as the matrix eigendecomposition problem that is reviewed in Section A.4.[13]

For simplicity we can suppress ω_m and i in the notation to obtain

$$\lambda \mathbf{\Phi} = \mathbf{S}_{\mathbf{x}} \mathbf{\Phi}. \tag{5.212}$$

The λ_i, $i = 1,\cdots,N$ are the eigenvalues and the $\mathbf{\Phi}_i$, $i = 1,\cdots,N$ are eigenvectors. The spatial spectral matrix is a non-negative definite Hermitian matrix whose eigenvalues are real non-negative numbers. We order the eigenvalues in decreasing size,

$$\lambda_{max} \triangleq \lambda_1 \geq \lambda_2 \cdots \lambda_N = \lambda_{min}. \tag{5.213}$$

The coefficients in the expansion, x_i, $i = 1,\cdots,N$, are zero-mean circular complex Gaussian random variables with variance λ_i.

In many cases, it is useful to expand $\mathbf{S}_{\mathbf{x}}$ using the eigenvalue and eigenvectors,

$$\mathbf{S}_{\mathbf{x}} = \sum_{i=1}^{N} \lambda_i \mathbf{\Phi}_i \mathbf{\Phi}_i^H. \tag{5.214}$$

[13]A brief review of Section A.4 at this point may be useful.

Recall that spatially white noise had a diagonal spectral matrix. From (5.191),

$$\mathbf{S_w} = \sigma_w^2 \mathbf{I}. \tag{5.215}$$

It can be expanded using any set of orthogonal vectors. In particular, if the input spectrum is

$$\mathbf{S_x} = \mathbf{S}_c + \sigma_w^2 \mathbf{I}, \tag{5.216}$$

then $\mathbf{S_x}(\omega)$ can be expanded as

$$\mathbf{S_x} = \sum_{i=1}^{N} \left[\lambda_i^c + \sigma_w^2\right] \mathbf{\Phi}_i \mathbf{\Phi}_i^H, \tag{5.217}$$

where the λ_i^c are the eigenvalues of $\mathbf{S}_c$ and the $\mathbf{\Phi}_i$ are the eigenvectors of $\mathbf{S}_c$.

In many cases of interest, the rank of $\mathbf{S}_c$ is less than N so that some of the λ_i^c will equal zero. For example, if the rank is $D < N$, then

$$\lambda_i = \begin{cases} \lambda_i^c + \sigma_w^2, & i = 1, 2, \cdots, D, \\ \sigma_w^2, & i = D+1, \cdots, N, \end{cases} \tag{5.218}$$

so that there are $N - D$ equal eigenvalues due to the white noise. The inverse of $\mathbf{S_x}$ can be written as

$$\begin{aligned} \mathbf{S_x}^{-1} &= \sum_{i=1}^{N} \left[\frac{1}{\lambda_i^c + \sigma_w^2}\right] \mathbf{\Phi}_i \mathbf{\Phi}_i^H \\ &= \frac{1}{\sigma_w^2}\mathbf{I} - \sum_{i=1}^{N} \frac{\lambda_i^c}{\sigma_w^2(\lambda_i^c + \sigma_w^2)} \mathbf{\Phi}_i \mathbf{\Phi}_i^H. \end{aligned} \tag{5.219}$$

We study the behavior of the eigenvalues and eigenvector for various signal configurations. We first consider plane-wave signals.

5.5.1 Plane-wave Signals

In this section, we look at the special case in which $\mathbf{x}(t)$ consists of D directional plane-wave processes plus white sensor noise. The frequency-domain snapshots are

$$\mathbf{X} = \sum_{d=1}^{D} \mathbf{v_k}(\mathbf{k}_d) f_d + \mathbf{W}, \tag{5.220}$$

where we have suppressed both ω_m and k. We define $\mathbf{V}$ as an $N \times D$-dimensional array manifold matrix,[14]

$$\mathbf{V} \triangleq \left[\begin{array}{c|c|c|c} \mathbf{v}(\mathbf{k}_1) & \mathbf{v}(\mathbf{k}_2) & \cdots & \mathbf{v}(\mathbf{k}_D) \end{array} \right]. \tag{5.221}$$

From (5.67),

$$\mathbf{S_x} = \mathbf{V}\mathbf{S_f}\mathbf{V}^H + \sigma_w^2 \mathbf{I}. \tag{5.222}$$

We assume that no two signals are coherent. This implies that $\mathbf{S_f}$ is positive definite. We will discuss the coherent signal case later. We examine the eigendecomposition of the first term in (5.222). The columns of $\mathbf{V}$ define a D-dimensional subspace that contains all of the signal energy. The eigenvectors provide an orthogonal basis for that subspace. Therefore, each $\mathbf{\Phi}_i$ can be obtained from $\mathbf{V}$ by a linear transformation:

$$\mathbf{\Phi}_i = c_{1i}\mathbf{v}(\mathbf{k}_1) + c_{2i}\mathbf{v}(\mathbf{k}_2) + \cdots + c_{Di}\mathbf{v}(\mathbf{k}_D), \tag{5.223}$$

or

$$\mathbf{\Phi}_i = \mathbf{V}\mathbf{c}_i, \quad i = 1, 2, \cdots, D, \tag{5.224}$$

where $\mathbf{c}_i$ is a $D \times 1$ vector. The eigendecomposition equation is

$$\lambda \mathbf{\Phi} = \mathbf{V}\mathbf{S_f}\mathbf{V}^H \mathbf{\Phi}. \tag{5.225}$$

Using (5.224) in (5.225) gives

$$\lambda \mathbf{V}\mathbf{c}_i = \mathbf{V}\mathbf{S_f}\mathbf{V}^H \mathbf{V}\mathbf{c}_i. \tag{5.226}$$

Equation (5.226) can be written as

$$\mathbf{V}\left[\lambda \mathbf{I} - \mathbf{S_f}\mathbf{V}^H\mathbf{V}\right]\mathbf{c}_i = \mathbf{0}. \tag{5.227}$$

In order for (5.227) to be satisfied, we require

$$\left[\lambda \mathbf{I} - \mathbf{S_f}\mathbf{V}^H\mathbf{V}\right]\mathbf{c}_i = \mathbf{0}. \tag{5.228}$$

This is a homogeneous equation that has a solution for D values of $\lambda_i, i = 1, 2, \cdots, D$. The $\mathbf{c}_i$ are the eigenvectors of $\mathbf{S_f}\mathbf{V}^H\mathbf{V}$. In order for (5.228) to have a non-trivial solution, we require

$$\det\left[\lambda \mathbf{I} - \mathbf{S_f}\mathbf{V}^H\mathbf{V}\right] = 0, \tag{5.229}$$

[14] We have deleted the subscript "$\mathbf{k}$" for simplicity.

which will have D solutions, which we denote as $\lambda_1^s \geq \lambda_2^s \geq \cdots \lambda_D^s$. The superscript "$s$" emphasizes that these eigenvalues are associated with the signal component of $\mathbf{X}$.

For each λ_i^s we can find the corresponding $\mathbf{c}_i$, which determines the corresponding $\mathbf{\Phi}_i, i = 1, 2, \cdots, D$. The eigenvectors corresponding to distinct eigenvalues are orthogonal and we normalize them so that

$$\mathbf{\Phi}_i^H \mathbf{\Phi}_j = \delta_{ij}. \tag{5.230}$$

The eigenvectors corresponding to equal eigenvalues can be chosen to be orthonormal. The $\mathbf{c}_i$ specified by (5.228) contain an arbitrary scale factor because (5.228) is a homogeneous equation. The normalization in (5.230) determines the magnitude of the scale factor. We can collect the eigenvectors into an $N \times D$ matrix,

$$\mathbf{U}_S \triangleq \left[\begin{array}{c:c:c:c} \mathbf{\Phi}_1 & \mathbf{\Phi}_2 & \cdots & \mathbf{\Phi}_D \end{array} \right]. \tag{5.231}$$

The range space of $\mathbf{U}_S$ is referred to as the signal subspace. We can write

$$\mathbf{U}_S = \mathbf{VC}, \tag{5.232}$$

where $\mathbf{C}$ is a non-singular $D \times D$ matrix whose columns are $\mathbf{c}_i, i = 1, 2, \cdots, D$.

We also observe that the ijth element of $\mathbf{V}^H\mathbf{V}$ is

$$\left[\mathbf{V}^H\mathbf{V}\right]_{ij} = \mathbf{v}^H(\mathbf{k}_i)\mathbf{v}(\mathbf{k}_j) \triangleq N B_c\left(\mathbf{k}_j : \mathbf{k}_i\right), \tag{5.233}$$

which is just the conventional beam pattern of the array steered to $\mathbf{k}_i$ and evaluated at $\mathbf{k}_j$. We will use this identification shortly.

Returning to (5.222), we observe that the noise component spans the N-dimensional space. Thus, to represent it, we must add $(N - D)$ more orthonormal vectors. Any set of vectors that are orthogonal with each other and orthogonal to $\mathbf{U}_S$ are satisfactory. We denote this set by the $N \times (N - D)$ matrix,

$$\mathbf{U}_N \triangleq \left[\begin{array}{c:c:c:c} \mathbf{\Phi}_{D+1} & \mathbf{\Phi}_{D+2} & \cdots & \mathbf{\Phi}_N \end{array} \right]. \tag{5.234}$$

The range space of $\mathbf{U}_N$ is referred to as the orthogonal subspace or the noise subspace. The corresponding eigenvalues are σ_w^2. Notice that when we expand $\mathbf{S}_\mathbf{x}$, the $\sigma_w^2\mathbf{I}$ term adds σ_w^2 to each signal eigenvalue. Thus, the total expansion is

$$\mathbf{S}_\mathbf{x} = \sum_{k=1}^{N} \lambda_k \mathbf{\Phi}_k \mathbf{\Phi}_k^H, \tag{5.235}$$

where

$$\lambda_k = \begin{cases} \lambda_k^s + \sigma_w^2, & k = 1, 2, \ldots, D, \\ \sigma_w^2, & k = D+1, \cdots, N. \end{cases} \tag{5.236}$$

Thus, although the noise subspace contains only noise, the signal subspace contains both signal and noise.

We can also write $\mathbf{S_x}$ as

$$\mathbf{S_x} = \mathbf{U}_S \mathbf{\Lambda}_S \mathbf{U}_S^H + \mathbf{U}_N \mathbf{\Lambda}_N \mathbf{U}_N^H, \tag{5.237}$$

where

$$\mathbf{\Lambda}_S = \text{diag}\ [\lambda_1, \lambda_2, \cdots, \lambda_D], \quad D \times D, \tag{5.238}$$

and

$$\mathbf{\Lambda}_N = \text{diag}\ \left[\sigma_w^2, \cdots, \sigma_w^2\right] = \sigma_w^2 \mathbf{I}, \quad (N-D) \times (N-D). \tag{5.239}$$

The notion of dividing the observation space into a signal subspace is fundamental to many of the optimum processing schemes that we develop in the text. We observe that this idea of a signal subspace is just the vector version of our signal subspace derivation in Chapter 4 of DEMT I [VT68], [VT01a].

We now discuss the following example to illustrate the concepts.

Example 5.5.1[15]

The example of interest is the case of two directional signals $F_1(\omega)$ and $F_2(\omega)$, which are frequency-domain snapshots of sample functions of a zero-mean stationary vector random process with spectral matrix,

$$\mathbf{S_f} = \begin{bmatrix} S_1 & S_{12} \\ S_{21} & S_2 \end{bmatrix}. \tag{5.240}$$

The covariance matrix of $\mathbf{X}$ is

$$\mathbf{S_x} = \mathbf{V}\mathbf{S_f}\mathbf{V}^H. \tag{5.241}$$

The wavenumbers of the two signals are $\mathbf{k}_1$ and $\mathbf{k}_2$. So the array manifold vectors are $\mathbf{v}(\mathbf{k}_1)$ and $\mathbf{v}(\mathbf{k}_2)$. Thus,

$$\begin{aligned} \mathbf{v}_1 &= \mathbf{v}(\mathbf{k}_1) \\ \mathbf{v}_2 &= \mathbf{v}(\mathbf{k}_2). \end{aligned} \tag{5.242}$$

[15]The solution for the eigenvalues for two plane-wave sources was published by Kwok and Brandon [KB79a], [KB79b] and Morgan [Mor78] and is utilized by Hudson [Hud81] and Kaveh and Barabell ([KB86a] and [KB86b]).

Inserting (5.240) and **V** into (5.229) gives,

$$\det\left[\lambda\mathbf{I} - N\begin{bmatrix} S_1 & S_{12} \\ S_{21} & S_2 \end{bmatrix}\begin{bmatrix} 1 & B_c^{12} \\ B_c^{21} & 1 \end{bmatrix}\right] = 0, \tag{5.243}$$

where we have defined

$$B_c^{mn} = B_c(\mathbf{k}_n : \mathbf{k}_m) = \frac{\mathbf{v}^H(\mathbf{k}_m)\mathbf{v}(\mathbf{k}_n)}{N}, \quad m,n = 1,2, \tag{5.244}$$

and we observe that

$$B_c^{21} = \left[B_c^{12}\right]^*, \tag{5.245}$$

and

$$S_{21} = S_{12}^*. \tag{5.246}$$

The B_c^{mn} terms represent the spatial correlation between the signals, and the S_{mn} terms represent the temporal correlation.

The determinant in (5.243) is a quadratic equation in λ,

$$\lambda^2 - b\lambda + c = 0, \tag{5.247}$$

where

$$b = N\left[S_1 + S_2 + 2Re\left[S_{12}B_c^{21}\right]\right], \tag{5.248}$$

and

$$c = N^2 S_1 S_2 \left[(1 - |\rho|^2)(1 - |B_c^{12}|^2)\right], \tag{5.249}$$

where

$$\rho = \frac{S_{12}}{(S_1 S_2)^{\frac{1}{2}}}, \tag{5.250}$$

is the temporal signal correlation. Then,

$$\lambda = \frac{b}{2}\left[1 \pm \sqrt{1 - \frac{4c}{b^2}}\right]. \tag{5.251}$$

Using (5.248) and (5.249) in (5.251) gives

$$\boxed{\begin{aligned}\lambda_{1(2)} &= \tfrac{1}{2}N\left[S_1 + S_2 + 2(S_1S_2)^{\frac{1}{2}} Re\left[\rho^* B_c^{12}\right]\right] \\ &\cdot\left[1 \pm \sqrt{1 - \frac{4S_1S_2(1-|B_c^{12}|^2)(1-|\rho|^2)}{\left[S_1+S_2+2(S_1S_2)^{\frac{1}{2}} Re\left[\rho^* B_c^{12}\right]\right]^2}}\right],\end{aligned}} \tag{5.252}$$

where λ_1 corresponds to the plus sign and λ_2 corresponds to the minus sign.

The two eigenvectors are given by (5.224),

$$\mathbf{\Phi}_1 = c_{11}\mathbf{v}(\mathbf{k}_1) + c_{12}\mathbf{v}(\mathbf{k}_2), \tag{5.253}$$

$$\mathbf{\Phi}_2 = c_{21}\mathbf{v}(\mathbf{k}_1) + c_{22}\mathbf{v}(\mathbf{k}_2), \tag{5.254}$$

where $\mathbf{c}_i, i = 1, 2$, is obtained from (5.228) using the eigenvalues obtained from (5.247).

Note that all of the results to this point are for an arbitrary array and the array geometry enters through the conventional beam pattern.

For the case of uncorrelated signals

$$\lambda_{1(2)} = \frac{1}{2} N \left[(S_1 + S_2)\right] \left[1 \pm \sqrt{1 - \frac{4S_1 S_2 (1 - |B_c^{12}|^2)}{(S_1 + S_2)^2}}\right]. \tag{5.255}$$

For equal-power uncorrelated signals, (5.255) reduces to

$$\lambda_{1(2)} = N S_1 \left[1 \pm |B_c^{12}|\right], \tag{5.256}$$

and

$$\mathbf{c}_1 = \frac{1}{\sqrt{2N[1 + |B_c^{12}|]}} \begin{bmatrix} 1 \\ e^{-j\phi(B_c^{12})} \end{bmatrix}, \tag{5.257}$$

$$\mathbf{c}_2 = \frac{1}{\sqrt{2N[1 - |B_c^{12}|]}} \begin{bmatrix} 1 \\ -e^{-j\phi(B_c^{12})} \end{bmatrix}, \tag{5.258}$$

where $\phi(B_c^{12})$ denotes the phase of B_c^{12}. The corresponding normalized eigenvectors are

$$\mathbf{\Phi}_{1(2)} = \frac{\mathbf{v}(\mathbf{k}_1) \pm e^{-j\phi(B_c^{12})} \mathbf{v}(\mathbf{k}_2)}{\sqrt{2N\left[1 \pm |B_c^{12}|\right]}}. \tag{5.259}$$

For a standard linear array,

$$B_c^{12} = \frac{1}{N} \frac{\sin\left(\frac{N}{2}\pi \Delta u\right)}{\sin\left(\frac{1}{2}\pi \Delta u\right)}. \tag{5.260}$$

In Figure 5.10, we plot $\frac{\lambda_1}{NS_1}$ and $\frac{\lambda_2}{NS_1}$ in dB versus $\frac{\Delta u}{BW_{NN}}$ on a log scale. The behavior is what we would expect. For small Δu, the largest eigenvalue approaches 2 and the smallest approaches zero. We see that the plot is linear for $\frac{\Delta u}{BW_{NN}} \leq 0.2$ and that λ_2 has a slope of $\frac{20 \text{ dB}}{\text{decade}}$. When B_c^{12} has a null, the two eigenvalues are equal because the two directional signals already defined a 2-D subspace with an orthogonal base. Once Δu goes beyond the first null in the beam pattern ($\Delta u = \frac{2}{N}$), the value of B_c^{12} is low ($\lesssim 13.4$ dB) so that the two eigenvalues are approximately equal.

In Figure 5.11, we plot the eigenbeams[16] for two signals placed symmetric around $u = 0$ for various values of $\frac{\Delta u}{BW_{NN}}$.

A second case of interest is the unequal signal case when one signal is much larger (≥ 10 dB) than the other. In Figure 5.12, we show the eigenvalue behavior for the case when $\frac{S_1}{S_2} = 0$ dB, 3 dB, 10 dB, and 20 dB.

From (5.252), we see that temporal correlation has the same effect as spatial correlation. In Figure 5.13, we plot the two eigenvalues versus ρ, for various Δu. Note the importance of the phase of ρ; ϕ_ρ, in the eigenvalue behavior.

If $|\rho| = 1$, then (5.252) shows that $\lambda_2 = 0$. Thus, the signal subspace is 1-D instead of 2-D. This situation occurs when the two signals are coherent.

---

[16]An eigenbeam is a beam obtained when the Hermitian transpose of an eigenvector is used as the weight vector.

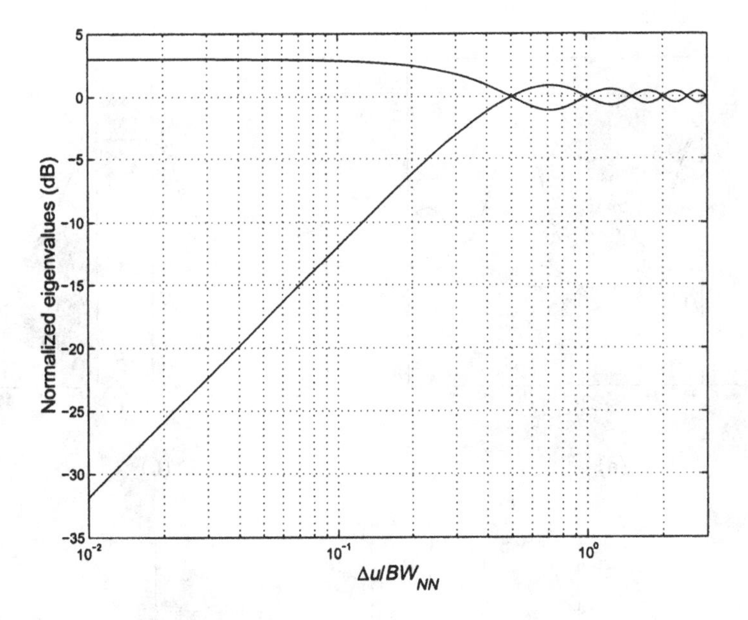

Figure 5.10 Normalized eigenvalues versus $\Delta u / BW_{NN}$.

One of the difficult cases that we encounter in various applications is the case when Δu is small (i.e., the plane waves are arriving from almost the same direction). For small Δu, we define[17]

$$\varepsilon^2 = 1 - |B_c^{12}|^2 . \tag{5.261}$$

Then for small ε^2,

$$\lambda_1 \simeq N(S_1 + S_2), \tag{5.262}$$

and

$$\lambda_2 \simeq N \frac{S_1 S_2 \varepsilon^2}{S_1 + S_2} . \tag{5.263}$$

For small Δu and a standard linear array,

$$B_c^{12} \simeq 1 - \frac{N^2 - 1}{24} (\pi \Delta u)^2 = 1 - \frac{N^2 - 1}{24} (\Delta \psi)^2 . \tag{5.264}$$

For closely spaced sources with high correlation,

$$\lambda_1 \simeq N(S_1^{\frac{1}{2}} + S_2^{\frac{1}{2}})^2 , \tag{5.265}$$

and

$$\lambda_2 \simeq \frac{2N S_1 S_2 \varepsilon_u^2 \varepsilon_\rho^2}{(S_1^{\frac{1}{2}} + S_2^{\frac{1}{2}})^2} . \tag{5.266}$$

[17]The results are contained in [Hud81].

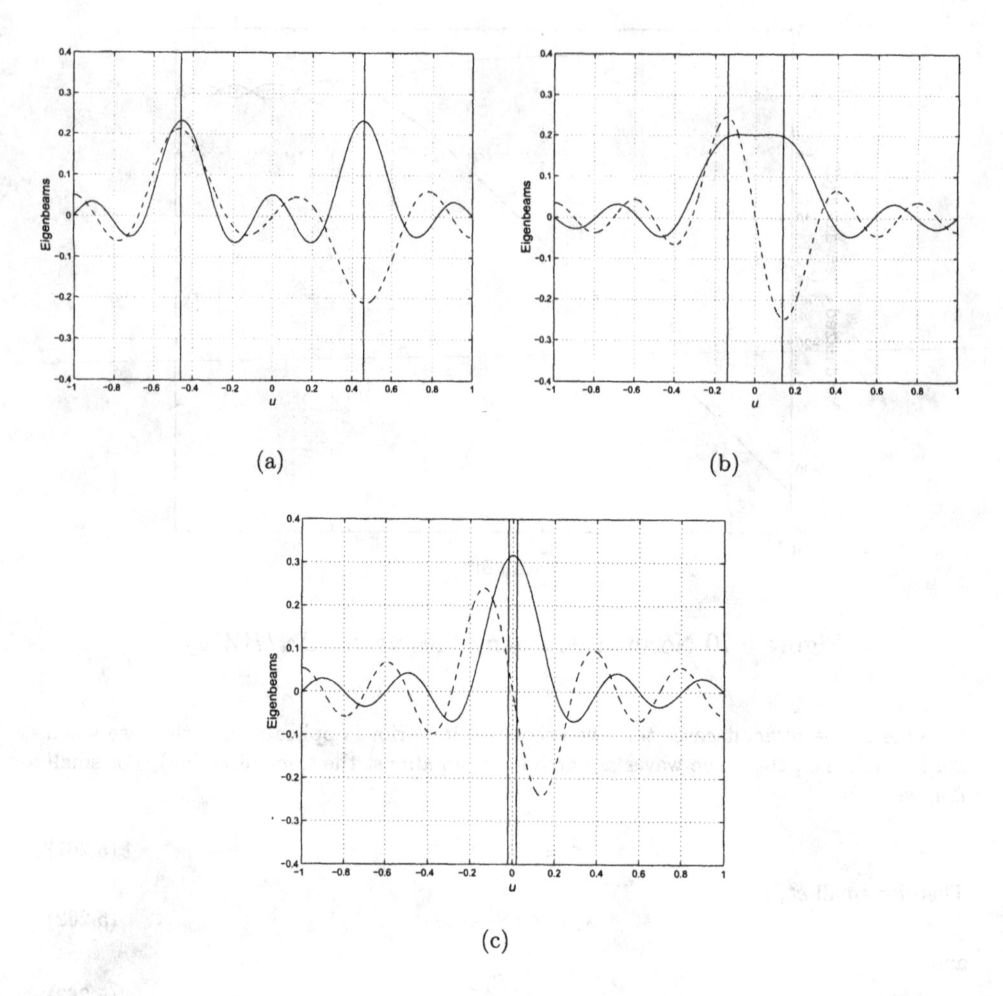

Figure 5.11 Eigenbeams for the two directional signals case: 10-element standard linear array (a) $\Delta u = 2.3BW_{NN}$; (b) $\Delta u = 0.68BW_{NN}$; (c) $\Delta u = 0.1BW_{NN}$.

For more than two directional signals at arbitrary angles, (5.228) must be solved numerically. There are various algorithms that enable this to be accomplished in a straightforward manner (e.g., [KSL94]). For certain geometries and array configurations, we can solve the 3- and 4-directional signal cases analytically by exploiting the centrosymmetric and persymmetric nature of $\mathbf{S_f V}^H\mathbf{V}$. Various other signal and array geometries are studied in the problems.

For temporally uncorrelated signals, the conventional beam pattern matrix $[\mathbf{V}^H\mathbf{V}]$ is the key factor in the eigenvalue calculations. If the directional

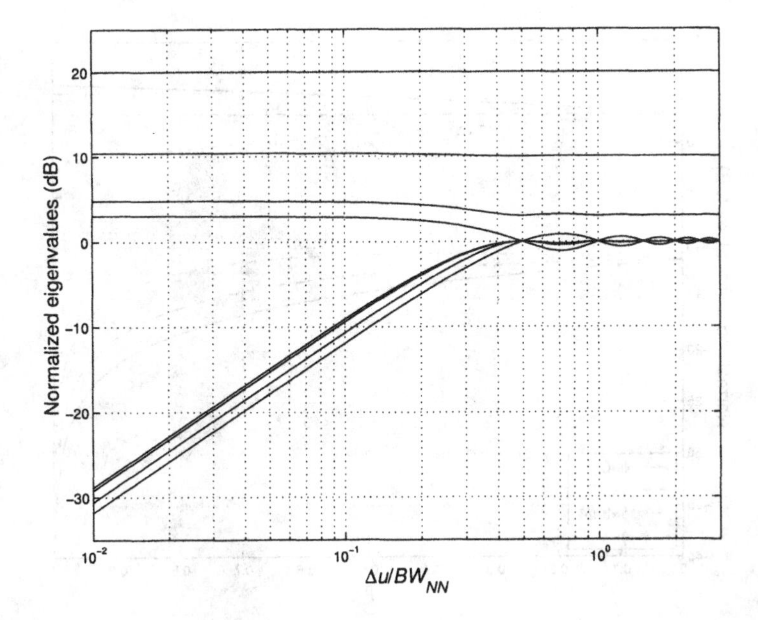

Figure 5.12 Eigenvalue behavior for unequal signals: $S_1/S_2 = 0$ dB, 3 dB, 10 dB, and 20 dB.

signals have the same spectrum and are outside the main lobes of the beams formed by the other directional signals, then the eigenvalues will approach N times the spectral heights of the various signals and the signals will have only modest impact on each other.

5.5.2 Spatially Spread Signals

We now consider an example of spatially spread signals to illustrate the eigendecomposition process.

Example 5.5.2

Consider an N-element standard linear array. The array manifold vector for a signal arriving from direction u is,

$$[\mathbf{v}(u)]_n = e^{j\left(n-\frac{N-1}{2}\right)\pi u}, \quad n = 0, 1, \cdots, N-1. \tag{5.267}$$

The resulting spectral matrix for a signal with unity power is

$$\begin{aligned} [\mathbf{S_x}]_{nm} &= e^{j\left(n-\frac{N-1}{2}\right)\pi u} e^{-j\left(m-\frac{N-1}{2}\right)\pi u} \\ &= e^{j(n-m)\pi u}. \end{aligned} \tag{5.268}$$

Now assume that the signal direction is centered at broadside and is spread uniformly in

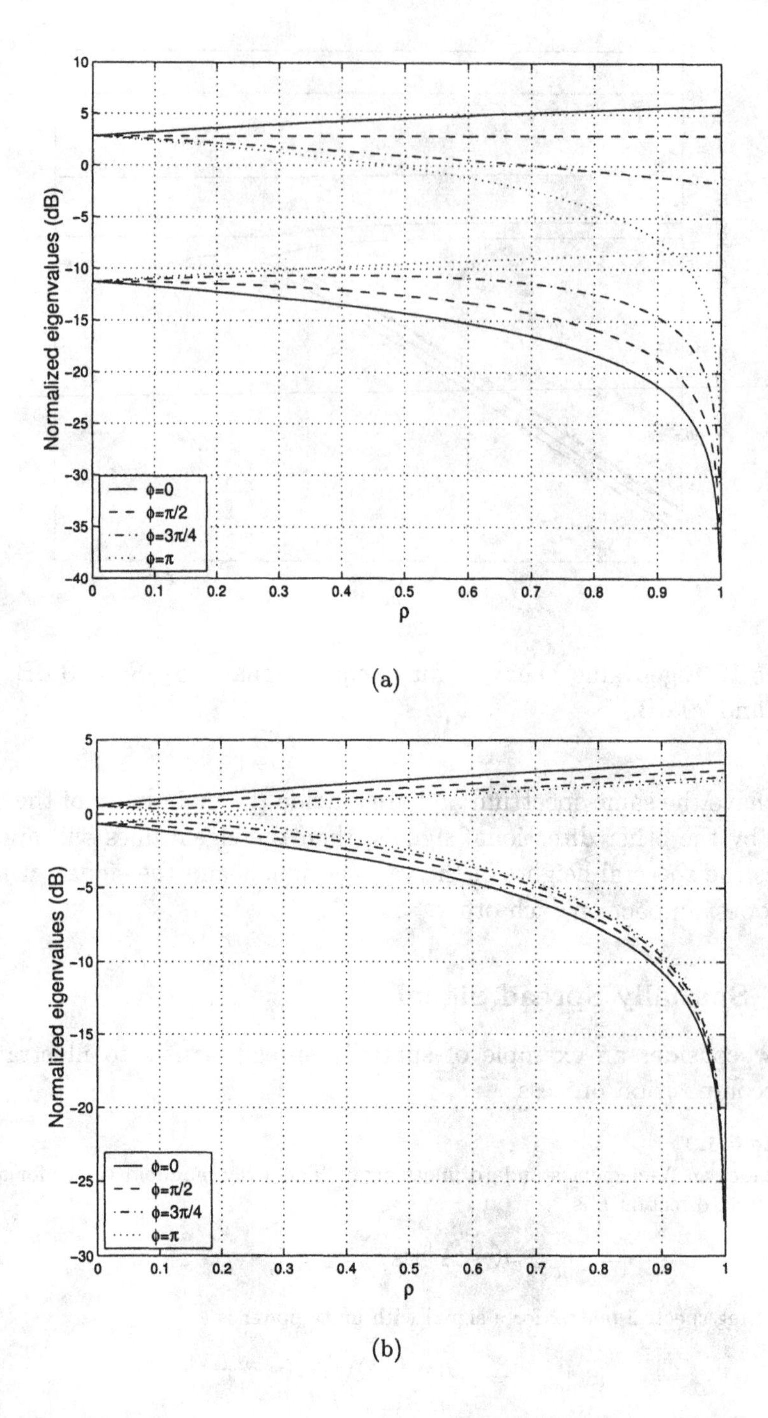

Figure 5.13 Eigenvalues for two correlated signals: $SNR_1 = SNR_2 = 0$ dB (a) $\Delta u/BW_{NN} = 0.0433$; (b) $\Delta u/BW_{NN} = 0.5$.

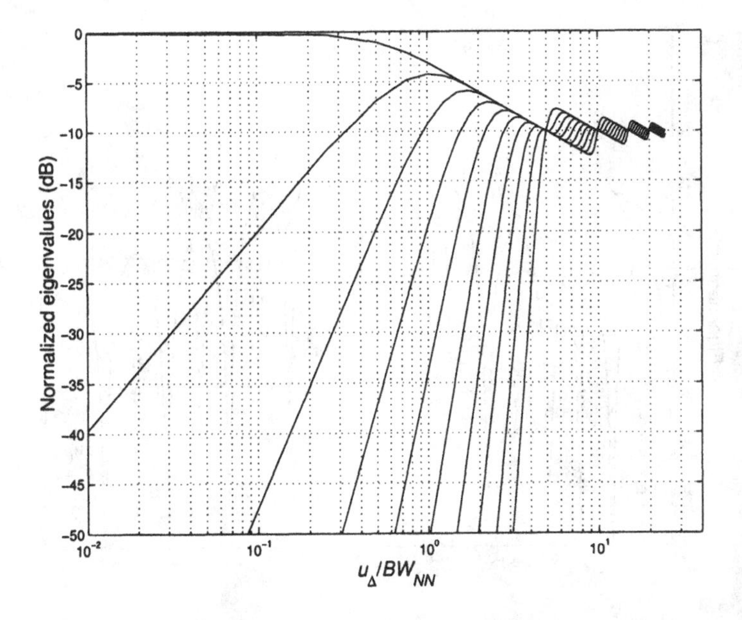

Figure 5.14 Eigenvalues as a function of u_Δ/BW_{NN}.

u-space,

$$P_1(\omega : u) = \begin{cases} \frac{\sigma_s^2}{u_\Delta}, & |u| \le \frac{u_\Delta}{2}, \\ 0, & \text{elsewhere} . \end{cases} \tag{5.269}$$

Then,

$$\begin{aligned} [\mathbf{S_x}]_{nm} &= \int_{-u_\Delta/2}^{u_\Delta/2} \frac{\sigma_s^2}{u_\Delta} e^{-j(n-m)\pi u} \\ &= \sigma_s^2 \operatorname{sinc}\left((n-m)\frac{\pi u_\Delta}{2}\right). \end{aligned} \tag{5.270}$$

Using (5.270) in (5.214) and writing out the matrix multiplication gives

$$\lambda \mathbf{\Phi}_n = \sigma_s^2 \sum_{m=1}^{N} \operatorname{sinc}\left((n-m)\frac{\pi u_\Delta}{2}\right) \mathbf{\Phi}_m, \qquad n = 1, \cdots, N. \tag{5.271}$$

We see that this is exactly the same problem as in Section 3.1 (see Figure 3.5), so the eigenvectors will consist of the *discrete prolate spheroidal sequences.*

In Figure 5.14, we plot the normalized eigenvalues for a 10-element array and various values of u_Δ/BW_{NN}. As u_Δ goes to zero, the wavenumber spectrum approaches that of a single directional signal. As u_Δ goes to 2, the wavenumber spectrum corresponds to "one-dimensional" isotropic noise.

The number of significant eigenvalues is

$$N_{SE} = \frac{u_\Delta}{2} N + 1. \tag{5.272}$$

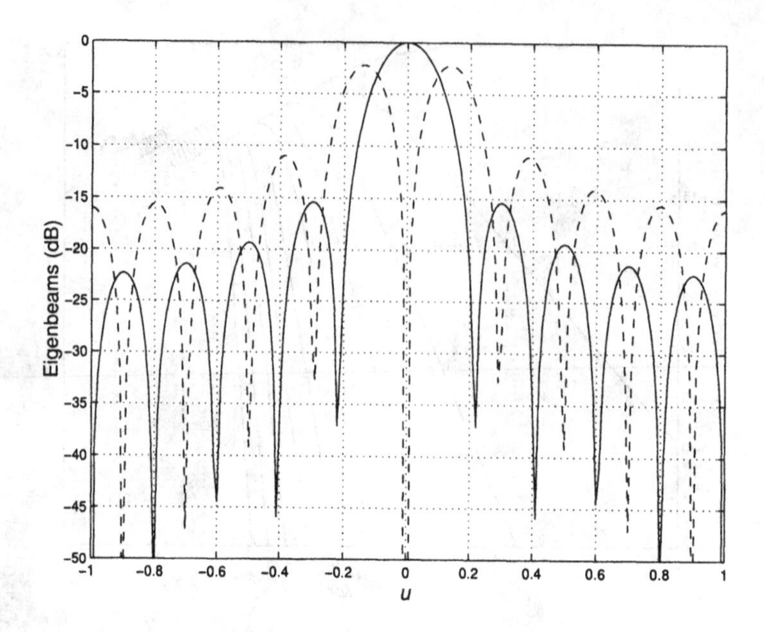

Figure 5.15 Eigenbeams for $u_\Delta = 0.2$.

This is the number we would expect from our sampling discussion in Section 2.4.6 and Section 5.2.4. In Figure 5.15, we show the eigenvectors and eigenbeams corresponding to $u_\Delta = 0.2$. Note that the second eigenvector is asymmetric so the corresponding beam pattern is imaginary and asymmetric. Thus, the eigenbeams correspond to sum and difference beams.

From Figure 5.14, we see that for very small u_Δ there is only one significant eigenvalue. As u_Δ increases, the second eigenvalue increases to a significant level. In later sections, we design an optimum processing algorithms under the assumption that the signal is not spread. We would like to be able to test how large u_Δ can be without violating the assumption. We hypothesize that we can test the behavior by examining the behavior of the second eigenvalue.[18]

In order to get a simple expression, we replace the spatially spread signal by two point sources placed symmetrically about $u = 0$. We choose the spacing so that the variances match. The variance in (5.269) is

$$\sigma_u^2 = \sigma_s^2 \frac{u_\Delta^2}{12}. \tag{5.273}$$

Therefore, we use two uncorrelated point source plane-wave signals with power $\sigma_s^2/2$ located at

$$u_s = \pm \frac{u_\Delta}{2\sqrt{3}}. \tag{5.274}$$

[18]This approach is developed in the context of finite bandwidth signals by Zatman [Zat98]. We consider that model in Section 5.5.3.

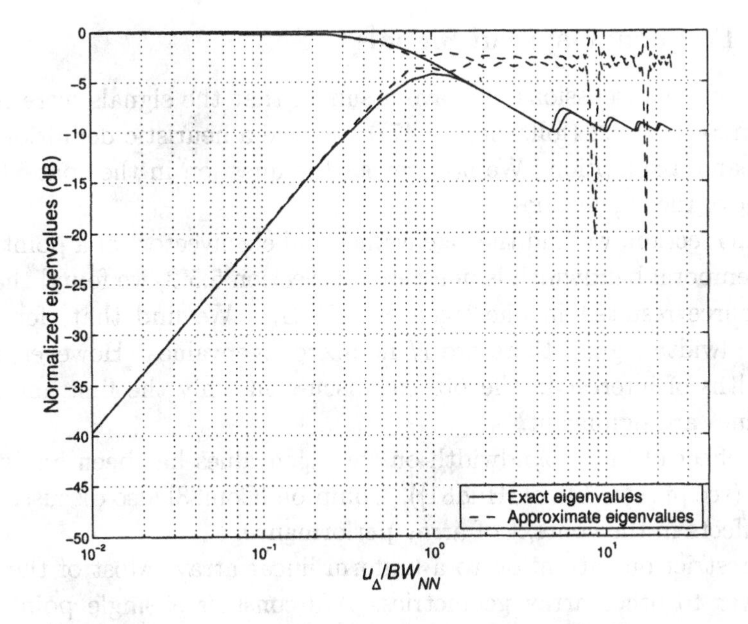

Figure 5.16 Behavior of two largest eigenvalues of a spatially spread signal.

Using (5.274) in (5.256), we obtain

$$\lambda_{1(2)} = \frac{N\sigma_s^2}{2}\left[1 \pm |B_c^{12}|\right], \tag{5.275}$$

where, from (2.97),

$$B_c^{12} = \frac{\sin(N\pi u_s)}{N\sin(\pi u_s)} = \frac{\sin(\frac{N\pi u_\Delta}{2\sqrt{3}})}{N\sin(\frac{\pi u_\Delta}{2\sqrt{3}})}. \tag{5.276}$$

In Figure 5.16, we plot the exact value of $\lambda_{1(2)}$ from (5.271) and the approximate value of $\lambda_{1(2)}$ from (5.275). We see that the approximation is accurate for $u \leq 0.5BW_{NN}$.

In most cases of interest we observe the signals of interest in the presence of additive white noise of spectral height σ_w^2 being added to each eigenvalue. We hypothesize that if the second eigenvalue is less than the noise eigenvalue that the performance of algorithms designed for point source signals will not be degraded significantly when the signal is spatially spread. Thus,

$$\frac{N\sigma_s^2}{2}\left(1 - \left|\frac{\sin(\frac{N\pi u_\Delta}{2\sqrt{3}})}{N\sin(\frac{\pi u_\Delta}{2\sqrt{3}})}\right|\right) \leq \sigma_w^2, \tag{5.277}$$

is a tentative criterion. We will revisit it later in the text.

5.5.3 Frequency-spread Signals

In most of our discussions we have assumed that the signals were narrowband. In Section 2.2 (following (2.47)) we gave a heuristic definition of the narrowband assumption. We now revisit the question in the context of the behavior of the eigenvalues.

In this section, we find the eigenvalues and eigenvectors of a point source whose temporal bandwidth is non-zero. In Section 5.5.2, we found that each point source resulted in one eigenvalue in $\mathbf{S_x}$. We find that, for a non-zero bandwidth signal, there are N non-zero eigenvalues. However, for the bandwidths of interest in the present discussion, only the first and second eigenvalues are significant.

The effect of finite bandwidth on the eigenvalues has been analyzed by Hudson (see pp.144–148 of [Hud81]). Compton [Com88] also discusses bandwidth effects in the context of array performance.

We restrict our attention to a uniform linear array. Most of the results carry over to other array geometries. We consider a single point source arriving from $u_s = \cos\theta_s$ in u-space.

The array manifold vector is

$$[\mathbf{v}(\psi)]_n = e^{j\left(n-\frac{N-1}{2}\right)\psi}, \quad n = 0, \cdots, N-1, \tag{5.278}$$

where

$$\psi = \frac{2\pi d}{\lambda} u_s = \frac{2\pi d f}{c} u_s. \tag{5.279}$$

The interelement time delay is

$$\tau_{nm} = \frac{(n-m)d}{c} u_s. \tag{5.280}$$

If we assume the array design frequency is f_c, then an array with $\lambda_c/2$ spacing has[19]

$$d = \frac{\lambda_c}{2} = \frac{c}{2f_c}. \tag{5.281}$$

Using (5.281) in (5.280) and (5.279) gives

$$\tau_{nm} = \frac{(n-m)}{2f_c} u_s \tag{5.282}$$

and

$$\psi = \frac{\pi f}{f_c} u_s. \tag{5.283}$$

[19] In many cases, the design frequency of the array will be the highest frequency, $f_u = f_c + B_s/2$. A similar result to (5.290) can be derived.

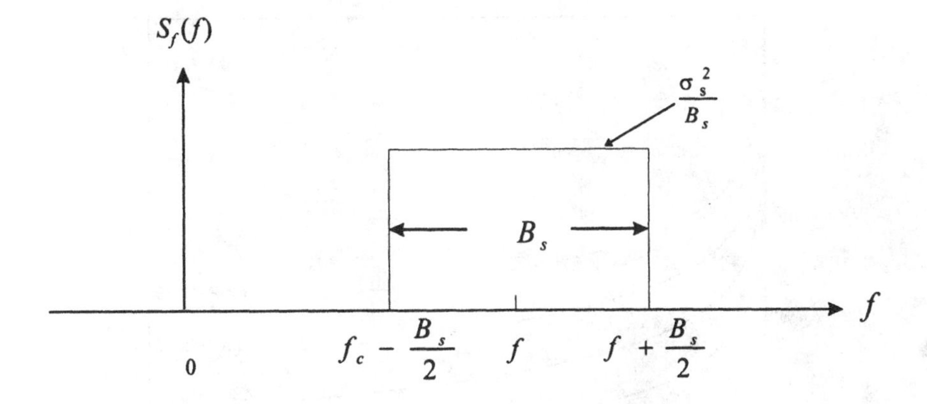

Figure 5.17 Frequency-spread source signal.

For simplicity, we assume the source has the temporal spectrum shown in Figure 5.17. We have assumed narrowband processing, so the spatial spectral matrix for $\omega_\Delta = 0$ is the matrix of interest:

$$[\mathbf{S}_\mathbf{x}(f_c)]_{nm} = \int_{f_c-B_s/2}^{f_c+B_s/2} \frac{\sigma_s^2}{B_s} e^{j\left(n-\frac{N-1}{2}\right)\psi} e^{-j\left(m-\frac{N-1}{2}\right)\psi} df\,. \tag{5.284}$$

Using (5.283) and (5.282) in (5.284) gives

$$[\mathbf{S}_\mathbf{x}(f_c)]_{nm} = \frac{\sigma_s^2}{B_s} \int_{f_c-B_s/2}^{f_c+B_s/2} e^{j2\pi\tau_{nm}f} df\,, \tag{5.285}$$

which reduces to

$$[\mathbf{S}_\mathbf{x}(f_c)]_{nm} = \sigma_s^2 sinc\left(\pi\tau_{nm}B_s\right)\,, \tag{5.286}$$

or

$$[\mathbf{S}_\mathbf{x}(f_c)]_{nm} = \sigma_s^2 sinc\left(\frac{\pi}{2}\left(n-m\right)B_f u_s\right)\,, \tag{5.287}$$

where

$$B_f \triangleq \frac{B_s}{f_c} \tag{5.288}$$

is the ratio of the signal bandwidth to the center frequency.

Comparing (5.287) and (5.270), we see that if we let

$$B_f u_s = u_\Delta\,, \tag{5.289}$$

the two expressions are identical. Thus the eigenvalue plot in Figure 5.14 applies to the frequency-spread case if we relabel the axis.

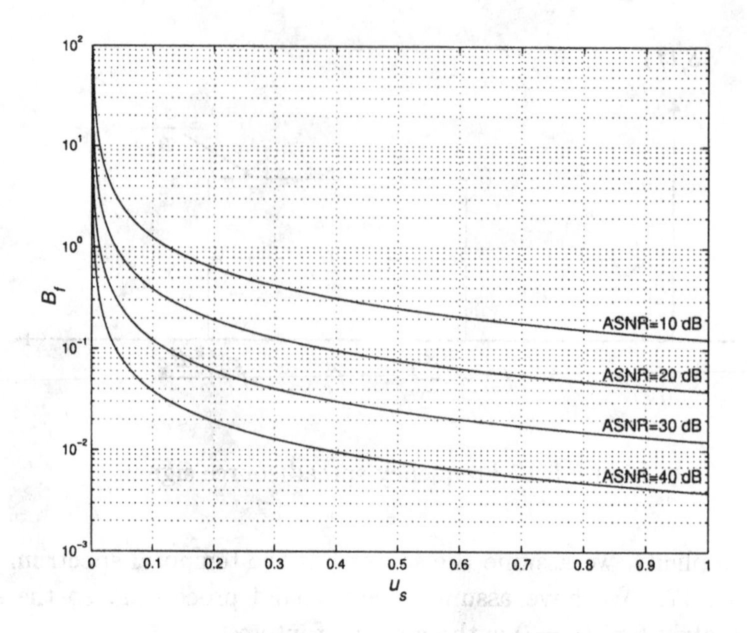

Figure 5.18 Threshold *SNR* for narrowband assumption: standard linear array.

The subsequent discussion leading to (5.277) also applies. Thus, we can hypothesize a test for the narrowband assumption as[20]

$$\frac{N\sigma_s^2}{2}\left(1-\left|\frac{\sin(\frac{N\pi B_f u_s}{2\sqrt{3}})}{N\sin(\frac{\pi B_f u_s}{2\sqrt{3}})}\right|\right) \leq \sigma_w^2 . \tag{5.290}$$

We investigate the validity of this criterion later in the text. In Figure 5.18, we plot (5.290) with an equality sign for various array signal-to-noise ratios (*ASNR*s), where

$$ASNR = \frac{N\sigma_s^2}{\sigma_w^2} . \tag{5.291}$$

The duality between spatial spreading and frequency spreading holds for linear arrays. For other array geometries, the effects must be analyzed individually.

In our discussion, we used a uniform temporal frequency spectrum. The results for other spectral shapes follow in a straightforward manner.

[20]This criterion is due to Zatman [Zat98].

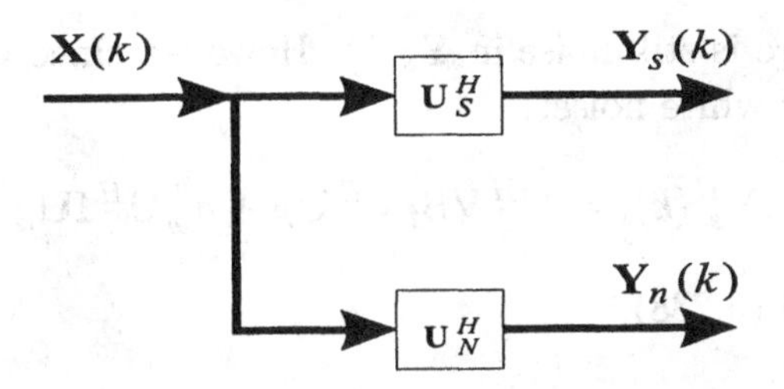

Figure 5.19 Eigenvector beamspace processor.

5.5.4 Closely Spaced Signals

One of the important cases that we encounter in subsequent chapters is the one in which we have D signals that are closely spaced (or clustered) in ψ-space.

Lee [Lee92] has developed approximate expressions for the eigenvalues and eigenvectors that are useful in some analyses. The derivation is lengthy, so we refer the reader to [Lee92] for the derivation and discussion.

5.5.5 Beamspace Processors

In many applications, we will find it useful to interpret the eigenvector decomposition as a beamspace processor. The model is shown in Figure 5.19. The input consists of a Gaussian signal plus additive independent Gaussian sensor noise. In the snapshot model,

$$\mathbf{X}(k) = \mathbf{V}\mathbf{F}(k) + \mathbf{W}(k), \qquad k = 1, 2, \cdots, K. \tag{5.292}$$

The signal subspace processor is $\mathbf{U}_S^H$, where

$$\mathbf{U}_S = \left[\ \mathbf{\Phi}_1 \,\vdots\, \mathbf{\Phi}_2 \,\vdots\, \cdots \,\vdots\, \mathbf{\Phi}_D\ \right], \tag{5.293}$$

is an $N \times D$ matrix of the normalized eigenvectors of $\mathbf{V}\mathbf{S}_\mathbf{f}\mathbf{V}^H$.

The noise subspace processor is $\mathbf{U}_N^H$, where

$$\mathbf{U}_N = \left[\ \mathbf{\Phi}_{D+1} \,\vdots\, \mathbf{\Phi}_{D+2} \,\vdots\, \cdots \,\vdots\, \mathbf{\Phi}_N\ \right], \tag{5.294}$$

is an $N \times (N - D)$ matrix whose columns are orthogonal to $\mathbf{U}_S$.

The advantage of this interpretation is that $\mathbf{Y}_s(k)$, the output of the signal subspace processor, is a sufficient statistic and the output $\mathbf{Y}_n(k)$ can be discarded. All of our subsequent processing deals with the $D \times 1$ vector $\mathbf{Y}_s(k)$.

Note that there is still noise in $\mathbf{Y}_s(k)$. However, since $\mathbf{U}_S$ has orthonormal columns, it is white noise:

$$E\left[\mathbf{Y}_s(k)\mathbf{Y}_s^H(k)\right] = \mathbf{U}_S^H\mathbf{V}\mathbf{S}_\mathbf{f}\mathbf{V}^H\mathbf{U}_S + \sigma_w^2\mathbf{U}_S^H\mathbf{I}\mathbf{U}_S. \tag{5.295}$$

Using (5.237) and (5.238),

$$\mathbf{S}_{\mathbf{Y}_s} = \mathbf{\Lambda}_S + \sigma_w^2\mathbf{I}. \tag{5.296}$$

Thus, our eigenvector beamspace processor has uncorrelated (statistically independent in the Gaussian model) outputs.

5.5.6 Subspaces for Spatially Spread Signals

For directional signal models in the presence of white noise, the concept and construction of a D-dimensional subspace is straightforward. The columns of the $\mathbf{V}$ matrix define a D-dimensional subspace and the corresponding eigenvectors form an orthonormal basis for the subspace.

The subspace concept is also useful for spatially spread signals in which the number of significant eigenvalues is less than N. We can illustrate this idea with a simple example.

Example 5.5.3

Consider the standard linear array in Example 5.5.1 and the wavenumber function in (5.269).

We define $D_s(M)$ as the dimension of the subspace. M is $-10\log$ (percentage of the signal power outside the subspace). In Figure 5.20, we plot $D_s(M)$ for several M for a standard 11-element array versus u_Δ/BW_{NN}.

We see that, for $u_\Delta < BW_{NN}$, the number of eigenvalues required is not sensitive to the specific value of $D_s(M)$. In many applications, we use $M = 20$ dB (or 0.01 signal power outside the subspace).

5.6 Parametric Wavenumber Models

In Section 5.3, we modeled space-time fields by starting with a distribution in space that was based on some knowledge of the environment. Once the statistical properties of the space-time field were specified we could find the statistical properties of the array input.[21]

In this section we take a different approach. We create a parametric model of the mechanism that generates a random field whose spatial spectrum matches observed data at the array. The process generation model is

[21]This section may be omitted at first reading by readers who are primarily interested in plane-wave models.

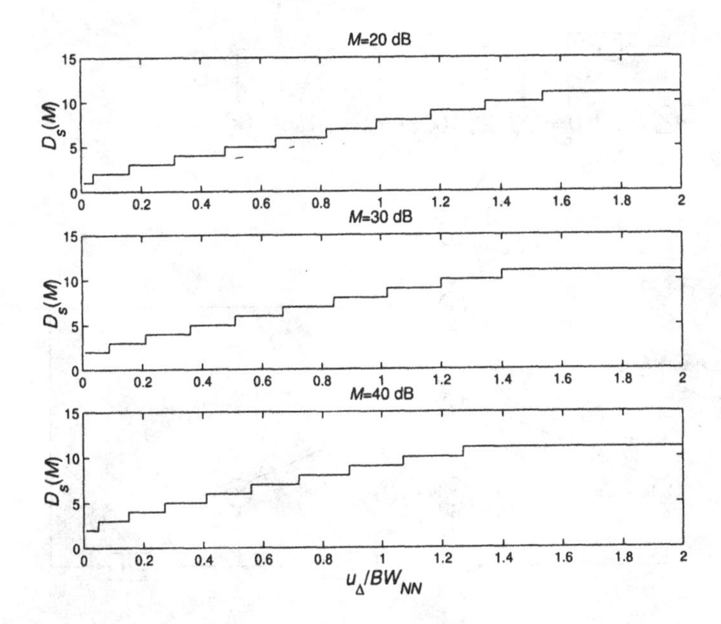

Figure 5.20 Dimension of approximate signal subspace: Number of eigenvalues required for various $D_s(M)$ versus u_Δ/BW_{NN}: $M = 20$ dB, 30 dB, and 40 dB.

not necessarily related to the actual physical mechanism that generates the observed spatial spectra.

The technique was originally used in the estimation of temporal power spectrum.[22] We restrict our attention to a standard N element linear array.

We refer to the model that we develop as a rational transfer function model. These models have been widely used in both continuous and discrete time problems for a number of years.

5.6.1 Rational Transfer Function Models

For the purpose of developing the model, we assume the input sequence to the digital system shown in Figure 5.21 is a white noise sequence $u(n)$ and the output sequence is $x(n)$. Note that $u(n)$ is used to develop the model and is not an actual input to the array. In the array context, $x(n)$ is a complex variable that represents the frequency-domain snapshot from the

[22]There are number of good discussions of parametric models (for example, [Kay88], [Mar87], [PRLN92], [Hay91]). Our discussion is similar to Chapter 5 of [Kay88] and Chapter 6 of [Mar87]. A discussion of various models for spatially spread sources is contained in Wakefield and Kaveh [WK85].

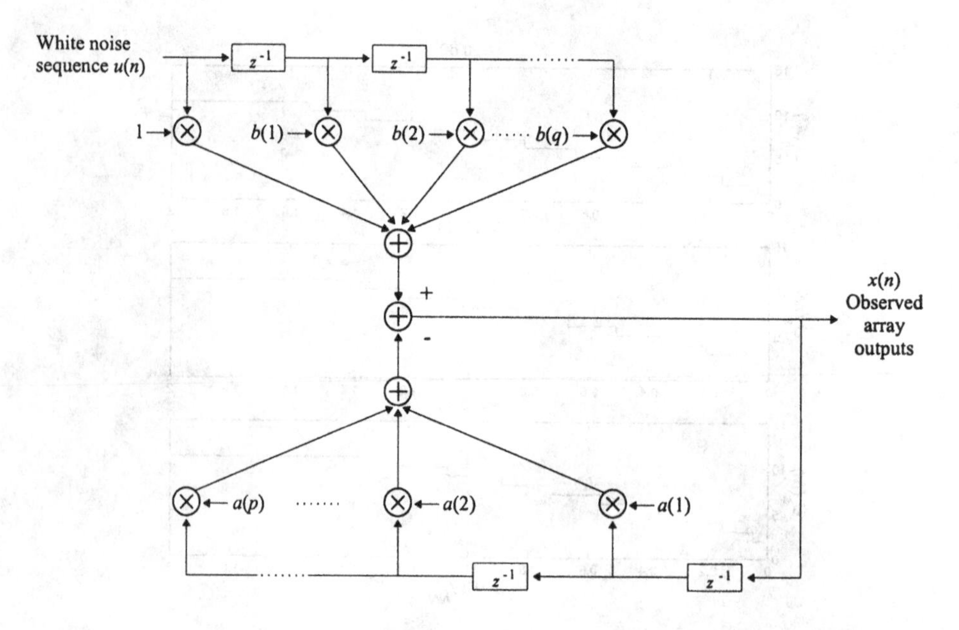

Figure 5.21 Auto-regressive moving average model of random field.

nth element of the array at snapshot time k. Our development does not depend on k, so we suppress k in our notation. We also use $x(n)$ to denote the snapshot instead of $X(n)$ to simplify the subsequent notation.

Our objective is to develop a model for the spatial correlation between sensor outputs. The input and output are related by a linear difference equation

$$x(n) = -\sum_{m=1}^{p} a(m)x(n-m) + \sum_{m=0}^{q} b(m)u(n-m), \quad 1 \leq n \leq N. \tag{5.297}$$

The general model is referred to as the ARMA model. It is shown in Figure 5.21. Note that in (5.297), the sequence is across the array at a single snapshot.

The z-transform of the system is

$$H(z) = \frac{B(z)}{A(z)}, \tag{5.298}$$

where

$$A(z) = \sum_{m=0}^{p} a(m)z^{-m}, \tag{5.299}$$

and

$$B(z) = \sum_{m=0}^{q} b(m) z^{-m}, \tag{5.300}$$

and we require that all the zeros of $A(z)$ be inside the unit circle. This guarantees that $H(z)$ is a stable, realizable system (e.g., [OS89]).

The output spectrum in the z-domain is

$$P_{xx}(z) = H(z)H^*\left(\frac{1}{z^*}\right)P_{uu}(z) \tag{5.301}$$

$$= \frac{B(z)B^*\left(\frac{1}{z^*}\right)}{A(z)A^*\left(\frac{1}{z^*}\right)}P_{uu}(z). \tag{5.302}$$

Evaluating at $z = e^{j\psi}$ gives the wavenumber spectrum of $x(n)$ in ψ-space.[23] Thus,

$$P_{xx}(z)\big|_{z=e^{j\psi}} \triangleq \tilde{P}_{xx}(\psi) = \sigma_u^2 \left|\frac{\tilde{B}(\psi)}{\tilde{A}(\psi)}\right|^2, \tag{5.303}$$

where σ_u^2 is the variance of the input process. We can assume $b(0)$ and $a(0)$ equal 1 and incorporate the filter gain in σ_u^2. This model is referred to as an ARMA(p, q) process.

The corresponding spatial correlation values are obtained by taking the inverse transform of (5.302). We write

$$P_{xx}(z)A(z) = D^*\left(\frac{1}{z^*}\right)B(z)\sigma_u^2, \tag{5.304}$$

where

$$D^*\left(\frac{1}{z^*}\right) = \frac{B^*\left(\frac{1}{z^*}\right)}{A^*\left(\frac{1}{z^*}\right)}. \tag{5.305}$$

The inverse z-transform of (5.304) is

$$\sum_{l=0}^{p} a(l) r_{xx}(m-l) = \sigma_u^2 \sum_{l=0}^{q} b(l) d^*(l-m). \tag{5.306}$$

Since $d(m) = 0$ for $m < 0$, (5.306) reduces to

$$r_{xx}(m) = \begin{cases} -\sum_{l=1}^{p} a(l) r_{xx}(m-l) + \sigma_u^2 \sum_{l=0}^{q-m} d^*(l) b(l+m), & 0 \le m \le q, \\ -\sum_{l=1}^{p} a(l) r_{xx}(m-l), & m \ge q+1. \end{cases} \tag{5.307}$$

[23]Recall that $\psi = k_z d = \pi \cos\theta$ so that there is an exact correspondence between ψ in the space domain and ω in time series analysis.

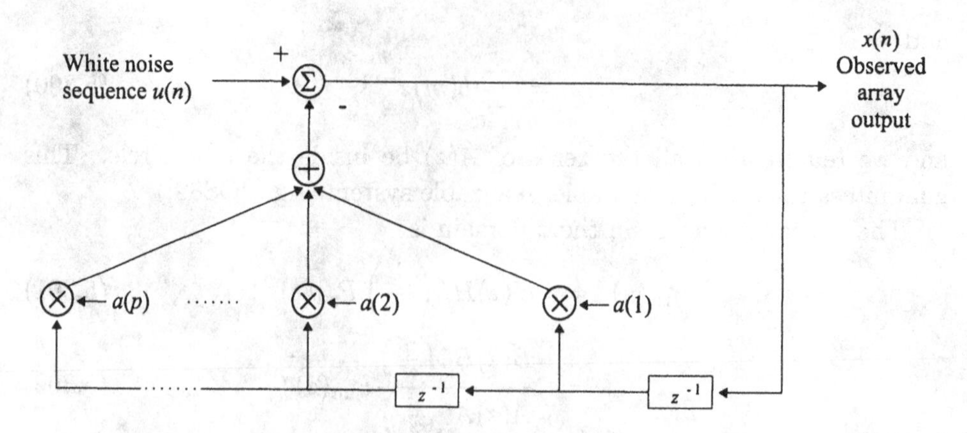

Figure 5.22 Auto-regressive model of random field.

Note that $r_{xx}(m)$ denotes the spatial correlation between the snapshot from sensor element 0 and sensor element m. We focus most of our attention on processes where all of the $b(m)$ except $b(0)$ are equal to zero. In this case,

$$x(n) = -\sum_{m=1}^{p} a(m)x(n-m) + u(n)\,. \tag{5.308}$$

This model is referred to as an auto-regressive process of order p, AR(p). It is auto-regressive because it is a linear regression model acting on itself. If we define

$$\hat{x}(n) = -\sum_{m=1}^{p} a(m)x(n-m), \tag{5.309}$$

then

$$x(n) = \hat{x}(n) + u(n), \tag{5.310}$$

so that $u(n)$ corresponds to the error between the weighted sum of past values and the value $x(n)$.

The power spectral density is

$$\tilde{P}_{xx}(\psi) = \frac{\sigma_u^2}{\left|A(e^{j\psi})\right|^2}\,. \tag{5.311}$$

The AR process model is shown in Figure 5.22.

The spatial correlation values follow directly from the process generation model. From (5.311), we have

$$\tilde{P}_{xx}(\psi)A(e^{j\psi}) = \frac{\sigma_u^2}{A^*\left(e^{j\psi}\right)}\,. \tag{5.312}$$

Taking the inverse transform of (5.312) gives

$$r_{xx}(m) = \begin{cases} -\sum_{l=1}^{p} a(l) r_{xx}(-l) + \sigma_u^2, & m = 0, \\ -\sum_{l=1}^{p} a(l) r_{xx}(m-l), & m \geq 1, \\ r_{xx}^*(-m), & m \leq -1. \end{cases} \tag{5.313}$$

In order to write (5.313) in matrix form, we define a $p \times 1$ vector $\mathbf{y}$ as

$$\mathbf{y} = \begin{bmatrix} x(0) & x(1) & \cdots & x(p) \end{bmatrix}^T. \tag{5.314}$$

The result for $m \geq 1$ can be written in matrix form as

$$\tilde{\mathbf{R}}_{\mathbf{y}} \mathbf{a} = -\mathbf{r}_{\mathbf{y}}, \tag{5.315}$$

where $\tilde{\mathbf{R}}_{\mathbf{y}}$ is a $p \times p$ matrix,

$$\tilde{\mathbf{R}}_{\mathbf{y}} \triangleq \begin{bmatrix} r_{xx}[0] & r_{xx}[-1] & \cdots & r_{xx}[-(p-1)] \\ r_{xx}[1] & r_{xx}[0] & \cdots & r_{xx}[-(p-2)] \\ \vdots & \vdots & \ddots & \vdots \\ r_{xx}[p-1] & r_{xx}[p-2] & \cdots & r_{xx}[0] \end{bmatrix}, \tag{5.316}$$

which is Hermitian because $r_{xx}(-1) = r_{xx}^*(1)$. $\tilde{\mathbf{R}}_{\mathbf{y}}$ is also Toeplitz.

The matrix $\mathbf{a}$ is a $p \times 1$ matrix,

$$\mathbf{a} = \begin{bmatrix} a(1) \,\vdots\, a(2) \,\vdots\, \cdots \,\vdots\, a(p) \end{bmatrix}^T, \tag{5.317}$$

and $\mathbf{r}_{\mathbf{y}}$ is a $p \times 1$ matrix,

$$\mathbf{r}_{\mathbf{y}} = \begin{bmatrix} r_{xx}(1) \,\vdots\, r_{xx}(2) \,\vdots\, \cdots \,\vdots\, r_{xx}(p) \end{bmatrix}^T. \tag{5.318}$$

The equations in (5.313) are referred to as the Yule-Walker equations.[24] The result can also be written in augmented form by incorporating the $m = 0$ part of (5.313):

$$\tilde{\mathbf{R}}_{\mathbf{y}}^A \mathbf{a}^A = \sigma_u^2 \mathbf{e}_1, \tag{5.319}$$

[24]Normally (5.315)–(5.318) are written with $\mathbf{x}$ as the vector. However, in our notation, $\mathbf{x}$ denotes an $N \times 1$ vector of the entire array output.

where

$$\tilde{\mathbf{R}}_{\mathbf{y}}^{A} = \begin{bmatrix} r_{xx}[0] & r_{xx}[-1] & \cdots & r_{xx}[-p] \\ r_{xx}[1] & r_{xx}[0] & \cdots & r_{xx}[-(p-1)] \\ \vdots & \vdots & \ddots & \vdots \\ r_{xx}[p] & r_{xx}[p-1] & \cdots & r_{xx}[0] \end{bmatrix}, \tag{5.320}$$

$$\mathbf{a}^{A} = \left[\, 1 \;\vdots\; a(1) \;\vdots\; \cdots \;\vdots\; a(p) \,\right]^{T}, \tag{5.321}$$

and

$$\sigma_u^2 \mathbf{e}_1 = \left[\, \sigma_u^2 \;\vdots\; 0 \;\vdots\; \cdots \;\vdots\; 0 \,\right]^{T}. \tag{5.322}$$

We consider several examples to illustrate the type of wavenumber spectra that can be generated using the AR process model.

Example 5.6.1

We consider a complex AR (1) process. It follows from (5.313) that

$$r_{xx}(m) = \begin{cases} -a(1) r_{xx}(m-1), & m \geq 1, \\ r_{xx}^{*}(-m), & m \leq -1, \end{cases} \tag{5.323}$$

or

$$r_{xx}(m) = \begin{cases} r_{xx}(0)\,(-a(1))^{m}, & m \geq 1, \\ r_{xx}^{*}(-m), & m \leq -1, \end{cases} \tag{5.324}$$

and

$$r_{xx}(0) = \frac{\sigma_u^2}{1 - |a(1)|^2}. \tag{5.325}$$

Note that the array spatial spectral matrix $\mathbf{S_x}$ is a Hermitian, Toeplitz matrix whose ij element is a function of $(i-j)$ where $(i-j) = k$ in (5.323) and (5.324). For example, if $N = 4$,

$$\mathbf{S_x} = \begin{bmatrix} r_{xx}(0) & r_{xx}^{*}(1) & r_{xx}^{*}(2) & r_{xx}^{*}(3) \\ r_{xx}(1) & r_{xx}(0) & r_{xx}^{*}(1) & r_{xx}^{*}(2) \\ r_{xx}(2) & r_{xx}(1) & r_{xx}(0) & r_{xx}^{*}(1) \\ r_{xx}(3) & r_{xx}(2) & r_{xx}(1) & r_{xx}(0) \end{bmatrix}. \tag{5.326}$$

The input signal power equals $r_{xx}(0)$. The wavenumber spectrum in ψ-space is

$$\tilde{P}_{xx}(\psi) = \frac{\sigma_u^2}{|1 - z_1 e^{-j\psi}|^2}. \tag{5.327}$$

To examine the behavior, we write

$$z_1 = |z_1|\, e^{j\pi\phi_1}. \tag{5.328}$$

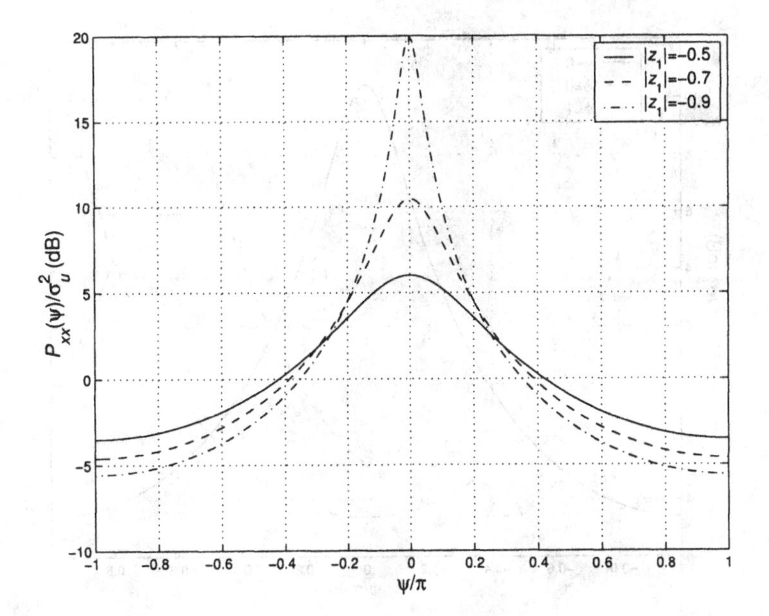

Figure 5.23 Normalized wavenumber spectra for complex AR(1) process: $|z_1| = 0.5, 0.7,$ and 0.9; $\phi_1 = 0$.

Note that $a(1) = -z_1$. In Figure 5.23, we show the wavenumber spectrum for $|z_1| = 0.5, 0.7,$ and 0.9 and $\phi_1 = 0$. We see that it corresponds to a spatially spread signal centered at $\psi = 0$. As $|z_1|$ approaches unity, it approaches the single plane-wave model. In Figure 5.24, we show the spectrum for $|z_1| = 0.7$ and $\phi_1 = 0, 0.6,$ and 1.0. We see that the effect of ϕ_1 is to shift the center of wavenumber spectrum to

$$\psi_c = \pi\phi_1 . \tag{5.329}$$

In order to generate multiple peaks in the wavenumber spectrum, we need to use a higher order model.

Example 5.6.2

In this example, we consider a complex AR(2) process. Since we are trying to match an observed wavenumber spectrum, we start with the wavenumber spectrum,

$$\begin{aligned} \tilde{P}_{xx}(\psi) &= \frac{\sigma_u^2}{|(1 - z_1 e^{-j\psi})(1 - z_2 e^{-j\psi})|^2} \\ &= \frac{\sigma_u^2}{|1 - (z_1 + z_2)e^{-j\psi} + z_1 z_2 e^{-j2\psi}|^2} \\ &= \frac{\sigma_u^2}{|1 + a(1)e^{-j\psi} + a(2)e^{-j2\psi}|^2} . \end{aligned} \tag{5.330}$$

In Figure 5.25, we show the wavenumber spectra for $z_1 = 0.9$ and $z_2 = 0.7e^{-j(0.8\pi)}$. In Figure 5.26, we show the wavenumber spectra for $z_1 = 0.9e^{j(0.5\pi)}, z_2 = 0.9e^{j(0.7\pi)}$. We

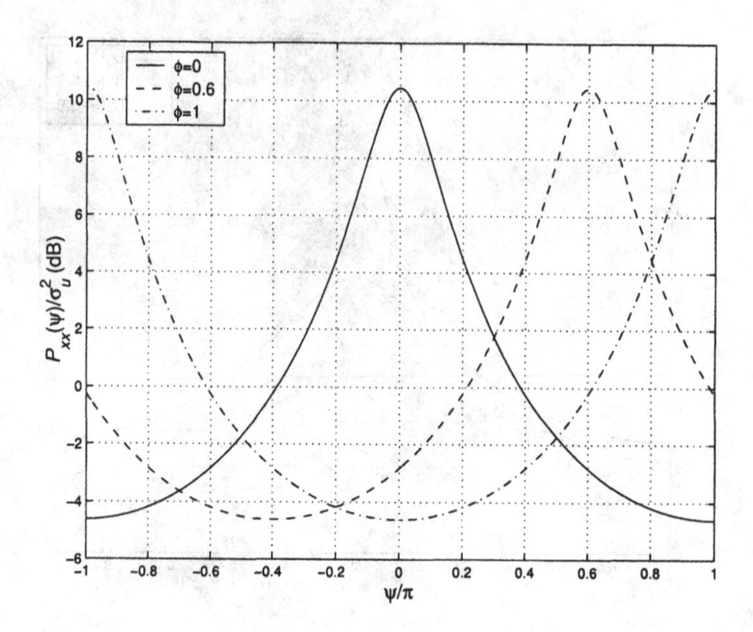

Figure 5.24 Normalized wavenumber spectra for complex AR(1) process: $|z_1| = 0.7$; $\phi_1 = 0, 0.6$, and 1.0.

see that we can achieve reasonable flexibility in the wavenumber spectra by changing the location of the poles.

The correlation matrix can be found by taking the inverse z-transform of $P_2(\psi)$ or by substituting $a(1)$ and $a(2)$ into (5.313) and solving.

Example 5.6.3

In this example, we consider a complex AR(3) process:

$$\tilde{P}_3(\psi) = \frac{\sigma_u^2}{|(1 - z_1 e^{-j\psi})(1 - z_2 e^{-j\psi})(1 - z_3 e^{-j\psi})|^2}. \tag{5.331}$$

In Figure 5.27, we show the wavenumber spectrum for $z_1 = 0.7e^{-j(0.6\pi)}$, $z_2 = 0.7$, $z_3 = 0.7e^{j(0.6\pi)}$. In Figure 5.28, we show the wavenumber spectrum for $z_1 = 0.9e^{-j(0.7\pi)}$, $z_2 = 0.9e^{-j(0.5\pi)}$, and $z_3 = 0.7e^{j(0.4\pi)}$.

The third process of interest corresponds to the case where all of the $a(n)$ coefficients are zero except for $a[0] = 1$. Then,

$$x(n) = \sum_{m=0}^{q} b(m)u(n-m), \tag{5.332}$$

and the process is referred to as a moving average (MA) process.

$$\tilde{P}_{MA}(\psi) = \sigma_u^2 \left|\tilde{B}(\psi)\right|^2. \tag{5.333}$$

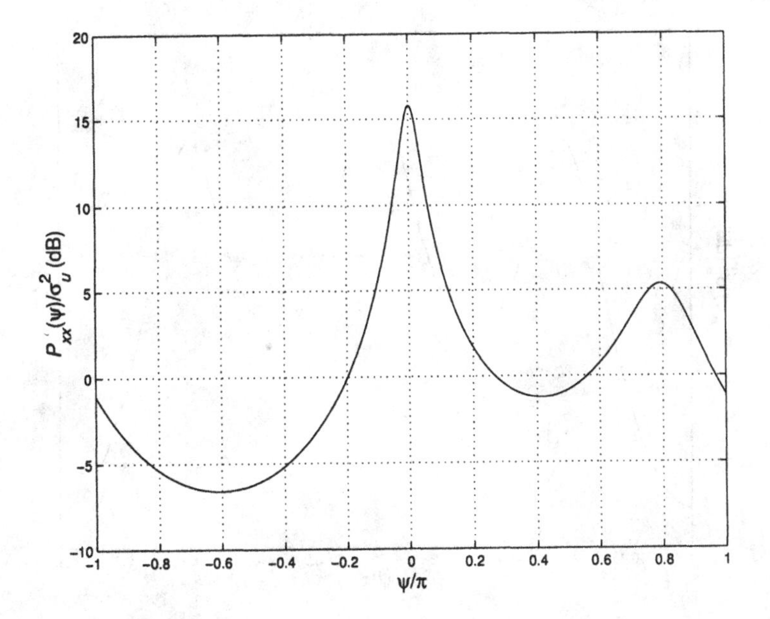

Figure 5.25 Normalized wavenumber spectrum for complex AR(2) process: $z_1 = 0.9, z_2 = 0.7\exp(j(0.8\pi))$.

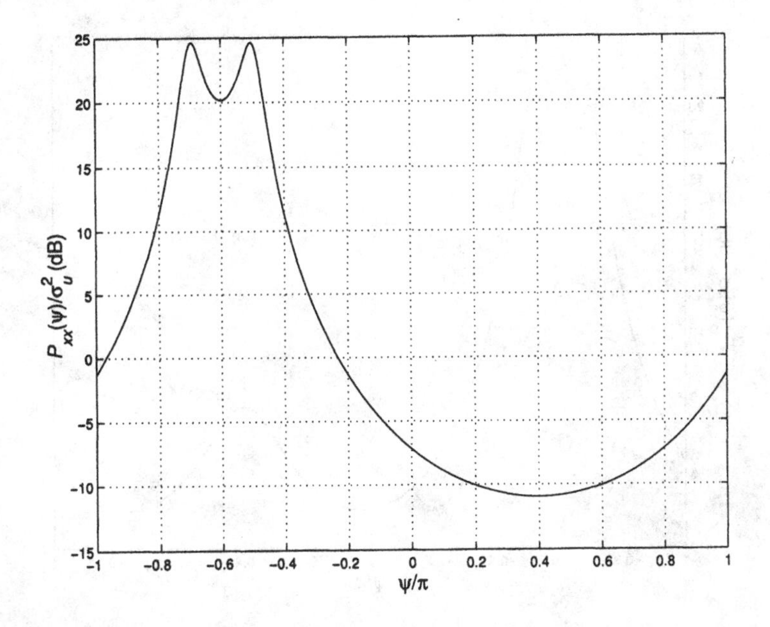

Figure 5.26 Wavenumber spectrum for complex AR(2) process: $z_1 = 0.9\exp(-j0.5\pi), z_2 = 0.9\exp(-j0.7\pi)$.

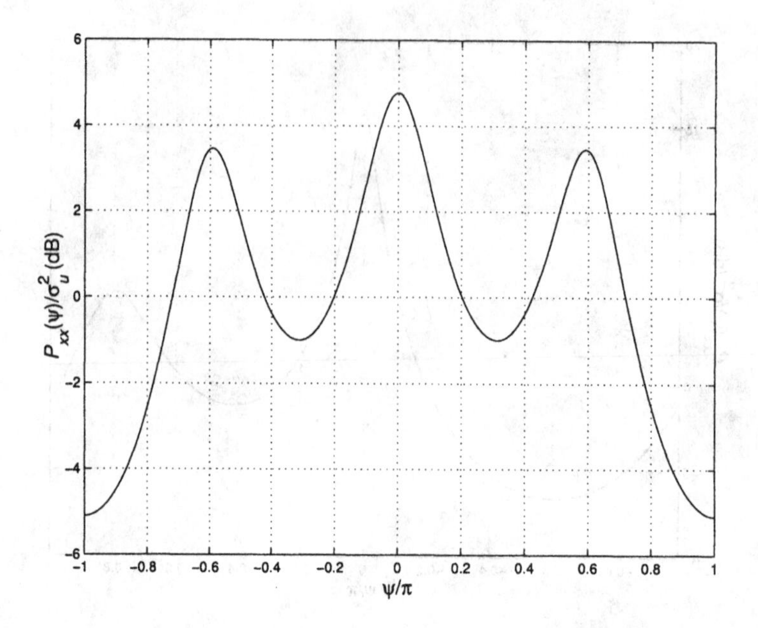

Figure 5.27 Wavenumber spectrum for complex AR(3) process: $z_1 = 0.7\exp(-j0.6\pi)$, $z_2 = 0.7$, $z_3 = 0.7\exp(j0.6\pi)$.

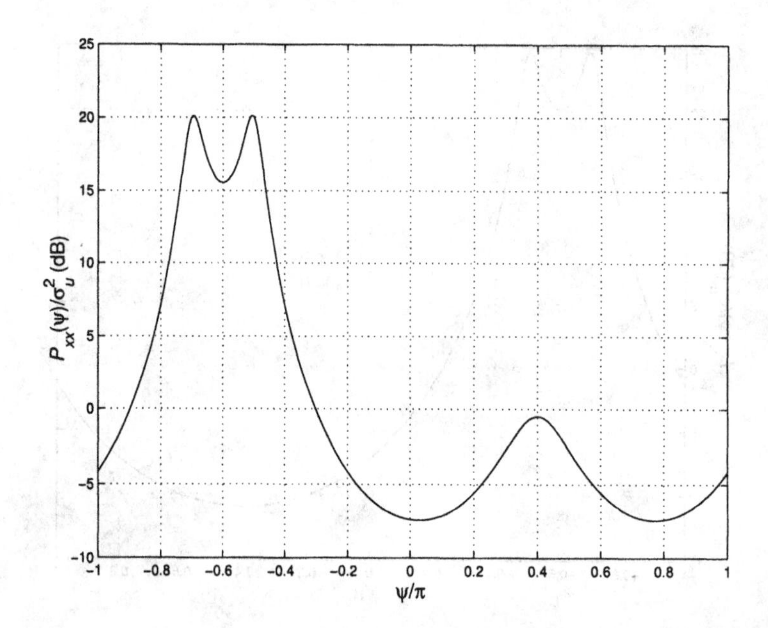

Figure 5.28 Wavenumber spectrum for complex AR(3) process: $z_1 = 0.9\exp(-j0.7\pi)$, $z_2 = 0.9\exp(-j0.5\pi)$, $z_3 = 0.7\exp(j0.4\pi)$.

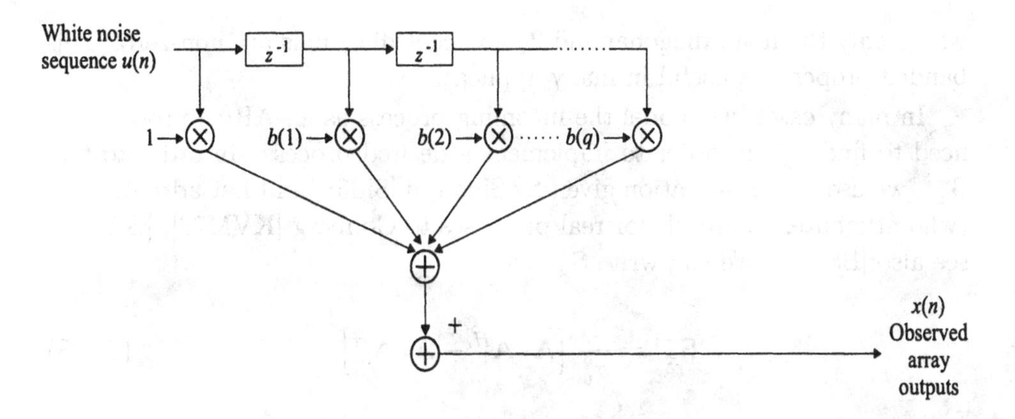

Figure 5.29 Moving average model of random field.

The model is shown in Figure 5.29. The corresponding correlation values can be obtained by letting $a(l) = \delta(l)$ and $d(l) = b(l)$ in (5.307). Then

$$r_{xx}(m) = \begin{cases} \sigma_u^2 \sum_{l=0}^{q-m} b^*(l) b(m+l)\,, & m = 0, 1, \cdots, q, \\ 0\,, & m \geq q+1. \end{cases} \tag{5.334}$$

For a complex MA(1) process,

$$r_{xx}(m) = \begin{cases} \sigma_u^2 \left[1 + |b(1)|^2\right], & m = 0\,, \\ \sigma_u^2 \left[b(1)\right], & m = 1\,, \\ 0\,, & m \geq 2\,. \end{cases} \tag{5.335}$$

For a complex MA(2) process,

$$r_{xx}(m) = \begin{cases} \sigma_u^2 \left[1 + |b(1)|^2 + |b(2)|^2\right], & m = 0\,, \\ \sigma_u^2 \left[b(1) + b^*(1) b(2)\right], & m = 1\,, \\ \sigma_u^2 \left[b(2)\right], & m = 2\,, \\ 0\,, & m \geq 3\,. \end{cases} \tag{5.336}$$

Thus, the spatial spectral matrix for an N-element array is a banded Hermitian, Toeplitz matrix,

$$\mathbf{S_x} = \begin{bmatrix} r_{xx}(0) & r_{xx}^*(1) & r_{xx}^*(2) & 0 & \cdots & 0 \\ r_{xx}(1) & r_{xx}(0) & r_{xx}^*(1) & r_{xx}^*(2) & \cdots & \cdots \\ r_{xx}(2) & r_{xx}(1) & r_{xx}(0) & r_{xx}^*(1) & \cdots & \cdots \\ \vdots & \vdots & \vdots & \vdots & \ddots & \vdots \\ 0 & \cdots & \cdots & \cdots & r_{xx}(1) & r_{xx}(0) \end{bmatrix}, \tag{5.337}$$

where only the main diagonal and $2q$ adjacent diagonals are non-zero. This banded property is useful in many applications.

In many cases, we model the incoming process as an AR(p) process and need to find $\mathbf{S}_{\mathbf{x}}^{-1}$ in order to implement a desired process. In order to find $\mathbf{S}_{\mathbf{x}}^{-1}$, we use a representation given by Siddiqui [Sid58] and LeCadre [LeC89] (who attributes the result for real processes to Gohberg [KVM78], [BAS86], see also [BS83]). We can write $\mathbf{S}_{\mathbf{x}}^{-1}$ as

$$\mathbf{S}_{\mathbf{x}}^{-1} = \frac{1}{\sigma_u^2}\left[\mathbf{A}_1 \mathbf{A}_1^H - \mathbf{A}_3 \mathbf{A}_3^H\right], \tag{5.338}$$

where $\mathbf{A}_1$ and $\mathbf{A}_3$ are two triangular $N \times N$ Toeplitz matrices defined as,

$$\mathbf{A}_1 = \begin{bmatrix} 1 & & & & & \\ a_1 & 1 & & \mathbf{0} & & \\ a_2 & a_1 & 1 & & & \\ \vdots & a_2 & \ddots & 1 & & \\ a_p & \cdots & \ddots & \ddots & 1 & \\ & \ddots & \ddots & \ddots & \ddots & \ddots \\ \mathbf{0} & & a_p & \cdots & \cdots & a_1 \quad 1 \end{bmatrix}, \tag{5.339}$$

$$\mathbf{A}_3 = \begin{bmatrix} & & \mathbf{0} & & \\ \hline a_p^* & & & & \\ \vdots & a_p^* & & \mathbf{0} & \\ a_3^* & \vdots & \ddots & & \\ a_2^* & a_3^* & \cdots & \ddots & \\ a_1^* & a_2^* & a_3^* & \cdots & a_p^* \end{bmatrix}. \tag{5.340}$$

We present a simple example to illustrate the result in (5.338).

Example 5.6.4

Consider a complex AR(1) process and an N-element array. Then,

$$\mathbf{A}_1 = \begin{bmatrix} 1 & & & & \\ a_1 & 1 & & \mathbf{0} & \\ 0 & a_1 & \ddots & & \\ \vdots & \cdots & \ddots & \ddots & \\ 0 & \cdots & \cdots & a_1 & 1 \end{bmatrix}, \tag{5.341}$$

and

$$\mathbf{A}_3 = \begin{bmatrix} 0 & \cdots & \cdots & 0 \\ \vdots & \cdots & \cdots & \vdots \\ 0 & \cdots & \cdots & 0 \\ a_1^* & 0 & \cdots & 0 \end{bmatrix}. \tag{5.342}$$

Substituting (5.341) and (5.342) into (5.338), we can obtain $\mathbf{S}_{\mathbf{x}}^{-1}$.

We see that we can write $\mathbf{S}_{\mathbf{x}}^{-1}$ using triangular matrices that only contain the parameters in the AR(p) model. We find this representation useful in many applications.

5.6.2 Model Relationships

In order to relate the various models, we use results due to Wold [Wol54] and Kolmogorov [Kol41].[25] The specific result of interest is a theorem in [Kol41] that says that any ARMA or MA process can be represented by an infinite order AR process. We develop the procedure to find this representation and the corresponding finite order approximation.

We first consider an ARMA (1,1) process. Then,

$$H(z) = \frac{1 + b(1)z^{-1}}{1 + a(1)z^{-1}}. \tag{5.343}$$

The corresponding infinite order AR(∞) process is

$$H(z) = \frac{1}{C(z)} = \frac{1}{1 + c(1)z^{-1} + c(2)z^{-2} + \cdots}, \tag{5.344}$$

where

$$C(z) = H^{-1}(z) = \frac{1 + a(1)z^{-1}}{1 + b(1)z^{-1}}. \tag{5.345}$$

One can show that the inverse z-transform of $C(z)$ is

$$c(m) = \begin{cases} 1 & m = 0, \\ \\ [a(1) - b(1)]\,(-b(1))^{m-1} & m \geq 1. \end{cases} \tag{5.346}$$

The $c(m)$ are the coefficients that are used in the AR model (they correspond to the $a(m)$ in (5.299)). In order to approximate this with a finite AR model AR(K), we choose K such that

$$c(K+1) \simeq 0, \tag{5.347}$$

[25]This discussion follows [Kay88].

or equivalently

$$b(1)^K \simeq 0. \tag{5.348}$$

Thus, as the zero in the ARMA process gets closer to the unit circle, we require a higher order process to get a good approximation.

For a general ARMA (p, q) process the AR (∞) process parameters can be obtained as the inverse z-transform of $H^{-1}(z)$ or by the recursive difference equation

$$c(l) = -\sum_{m=1}^{q} b(m)c(l-m) + \sum_{m=0}^{p} a(m)\delta(l-m), \quad l \geq 0, \tag{5.349}$$

with initial conditions; $c(-q), c(-q+1), \cdots, c(1)$ set equal to zero.

In our application, we are interested in a particular type of ARMA process, so we will defer an example until we have developed the model.

5.6.3 Observation Noise

It is important to note that our discussion up to this point refers to the signal process of interest (note that the "signal" process might contain various interfering signals that we have sometimes called noise). What we actually observe at the array output is

$$x(n) = f(n) + w(n), \tag{5.350}$$

where $w(n)$ is a zero-mean white observation noise with variance σ_w^2. This observation noise is statistically independent of the white input process $u(n)$. Then

$$\tilde{P}_{xx}(\psi) = \frac{\sigma_u^2 \left[1 + \frac{\sigma_w^2}{\sigma_u^2} \left|\tilde{A}(\psi)\right|^2\right]}{\left|\tilde{A}(\psi)\right|^2}. \tag{5.351}$$

Thus, $x(n)$ corresponds to an ARMA (p, p) process where the transfer function of the MA branch satisfies,

$$\left|B(z)B^*\left(\frac{1}{z^*}\right)\right|^2 = \left[1 + \frac{\sigma_w^2}{\sigma_u^2} A(z)A^*\left(\frac{1}{z^*}\right)\right]. \tag{5.352}$$

Note that we can view this model as one in which a single white noise process generates the composite output spectrum. The presence of this observation noise will cause a problem, both in the development of the algorithms and in the application to actual array processing problems.

In many applications we will approximate the ARMA(p,p) model with a higher order AR model. We consider two examples to illustrate this technique.

Example 5.6.5

Consider a complex AR(1) process in the presence of white noise with spectral height σ_w^2. Using (5.327) in (5.351), we have

$$\tilde{P}_{xx}(\psi) = \frac{\sigma_u^2 + \sigma_w^2 \left[1 + a(1)e^{-j\psi}\right]\left[1 + a^*(1)e^{j\psi}\right]}{|1 + a(1)e^{-j\psi}|^2}. \tag{5.353}$$

To find $b(1)$, we equate the MA portion of the wavenumber spectrum to the numerator in (5.353):

$$\sigma_b^2 \left[1 + b(1)e^{-j\psi}\right]\left[1 + b^*(1)e^{j\psi}\right] = \sigma_u^2 + \sigma_w^2 \left[1 + a(1)e^{-j\psi} + a^*(1)e^{j\psi} + |a(1)|^2\right], \tag{5.354}$$

which implies,

$$\sigma_b^2 \left[1 + |b(1)|^2\right] = \sigma_u^2 + \sigma_w^2 \left[1 + |a(1)|^2\right], \tag{5.355}$$

and

$$\sigma_b^2 b(1) = \sigma_w^2 a(1). \tag{5.356}$$

From (5.356)

$$b(1) = \frac{\sigma_w^2}{\sigma_b^2} a(1). \tag{5.357}$$

Substituting (5.357) into (5.355) gives us a quadratic equation in σ_b^2, which can be solved. This gives the value of $b(1)$. We substitute $b(1)$ into (5.346) to find $c(k)$.

We have shown the results in Figures 5.30, 5.31, and 5.32 for three values of SNR ($\sigma_u^2/\sigma_w^2 = 10$ dB, 0 dB, and -10 dB), two values of $|z_1|$, and various values of K, the order of AR (K) process. In all cases, $|c(k)| \leq 0.001$ for the highest order process.

As the *SNR* increases, the model order K that is required to get a good fit decreases. As $|z_1|$ increases, the required model order increases.

Example 5.6.6

In this example, we consider the complex AR(2) process discussed in Example 5.6.2,

$$\tilde{P}_{xx}(\psi) = \frac{\sigma_u^2 + \sigma_w^2 \left|1 + a(1)e^{-j\psi} + a(2)e^{-j2\psi}\right|^2}{|1 + a(1)e^{-j\psi} + a(2)e^{-j2\psi}|^2}. \tag{5.358}$$

Proceeding as in Example 5.6.5, we obtain three equations,

$$\sigma_b^2 \left[1 + |b(1)|^2 + |b(2)|^2\right] = \sigma_u^2 + \sigma_w^2 \left[1 + |a(1)|^2 + |a(2)|^2\right], \tag{5.359}$$

$$\sigma_b^2 \left[b(1) + b(2)b^*(1)\right] = \sigma_w^2 \left[a(1) + a(2)a^*(1)\right], \tag{5.360}$$

and

$$\sigma_b^2 \left[b(2)\right] = \sigma_w^2 a(2). \tag{5.361}$$

We solve these three equations and use the results in (5.349) to find the $c(m)$.

We have shown the results in Figures 5.33, 5.34, 5.35 for several of the models in Example 5.6.2, for three values ($\sigma_u^2/\sigma_w^2 = 10$ dB, 0 dB, and -10 dB) and various values of K, the order of AR(K) process.

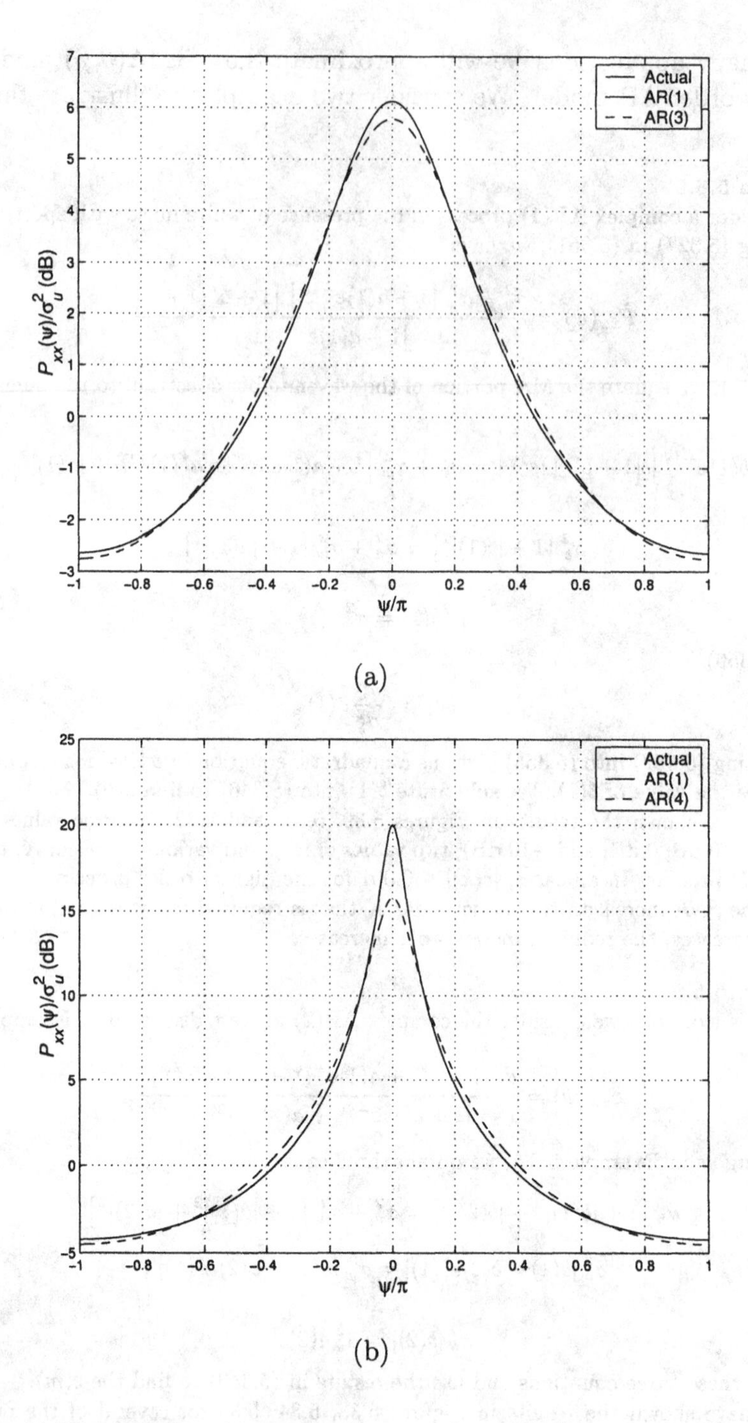

Figure 5.30 Wavenumber spectrum of a complex AR(1) process with additive white noise: $\sigma_u^2/\sigma_w^2 = 10$ dB; (a) $|z_1| = 0.5$; (b) $|z_1| = 0.9$.

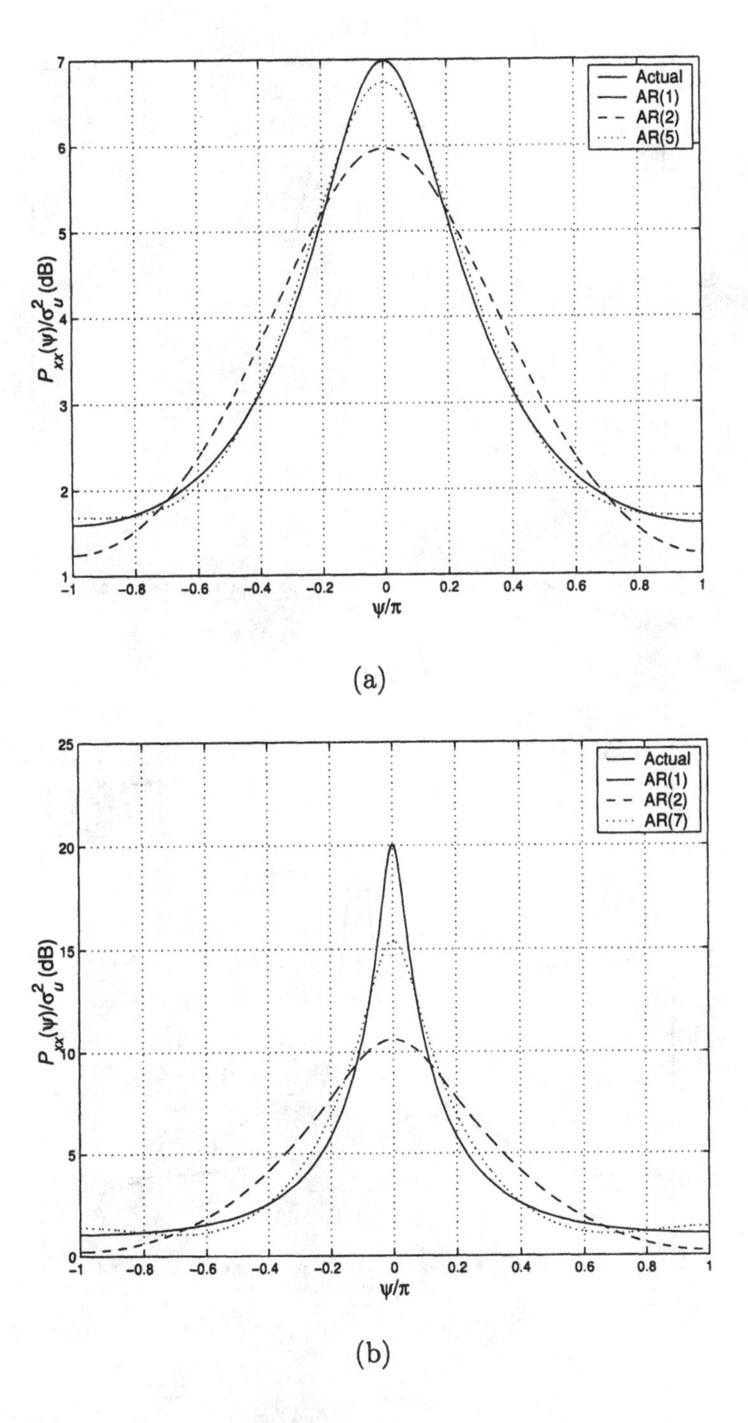

Figure 5.31 Wavenumber spectrum of a complex AR(1) process with additive white noise: $\sigma_u^2/\sigma_w^2 = 0$ dB; (a) $|z_1| = 0.5$; (b) $|z_1| = 0.9$.

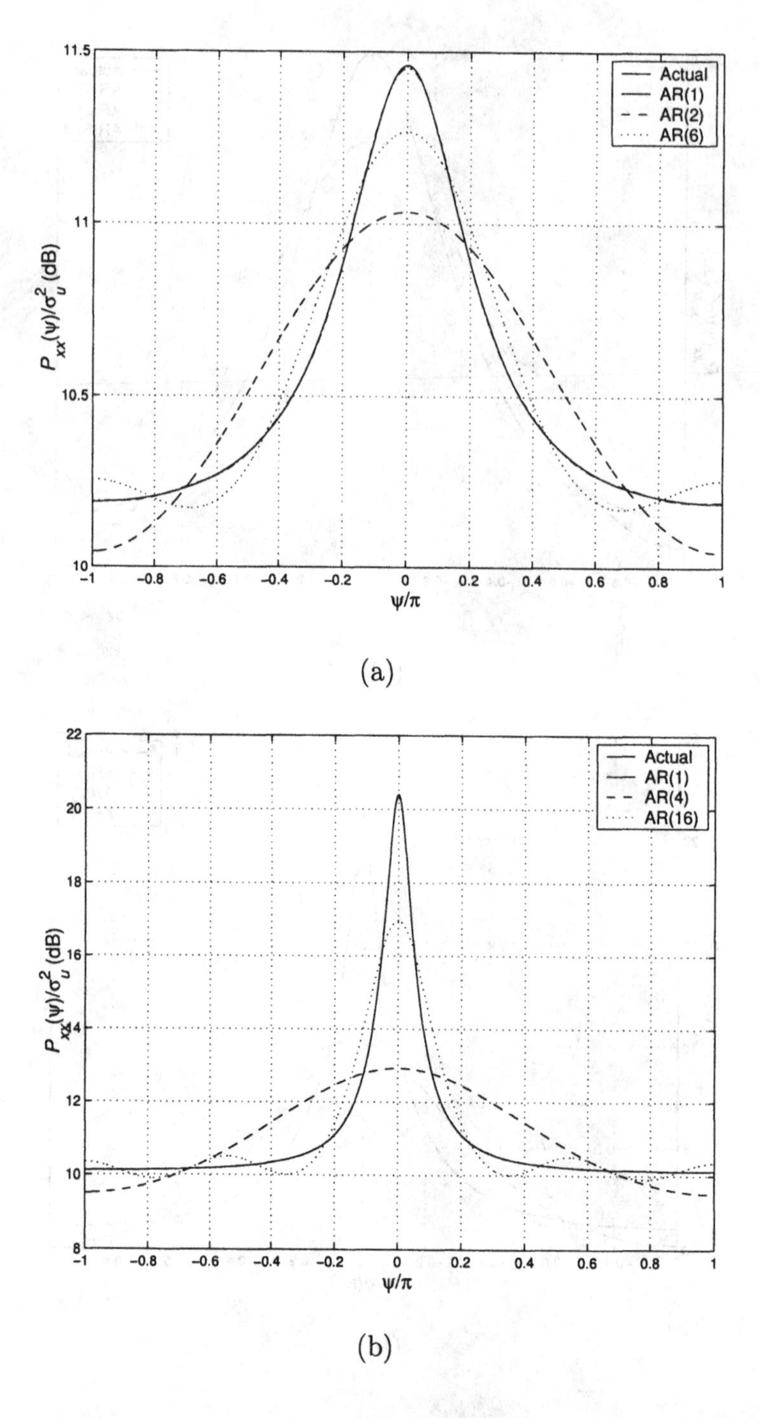

Figure 5.32 Wavenumber spectrum of a complex AR(1) process with additive white noise: $\sigma_u^2/\sigma_w^2 = -10$ dB; (a) $|z_1| = 0.5$; (b) $|z_1| = 0.9$.

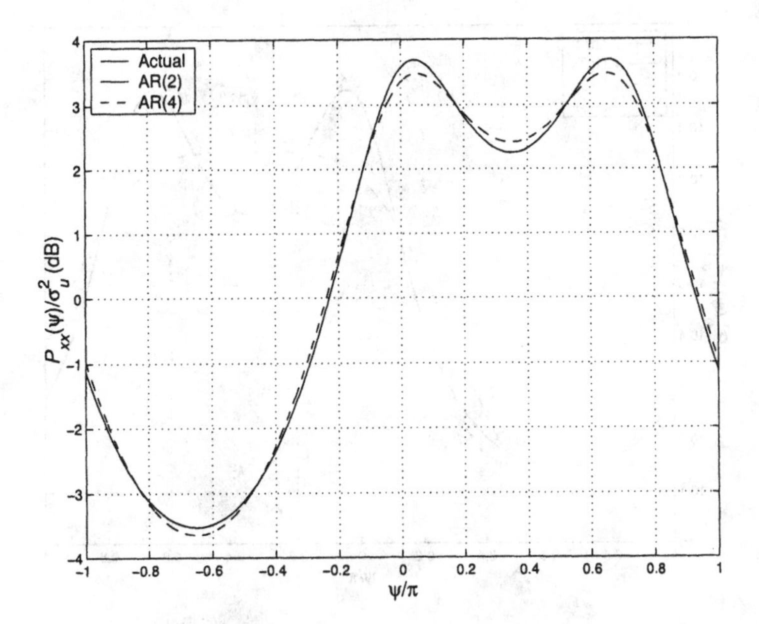

Figure 5.33 Wavenumber spectrum of a complex AR(2) process with additive white noise: $\sigma_u^2/\sigma_w^2 = 10$ dB, $z_1 = 0.5, z_2 = 0.5e^{j\,(0.7\pi)}$.

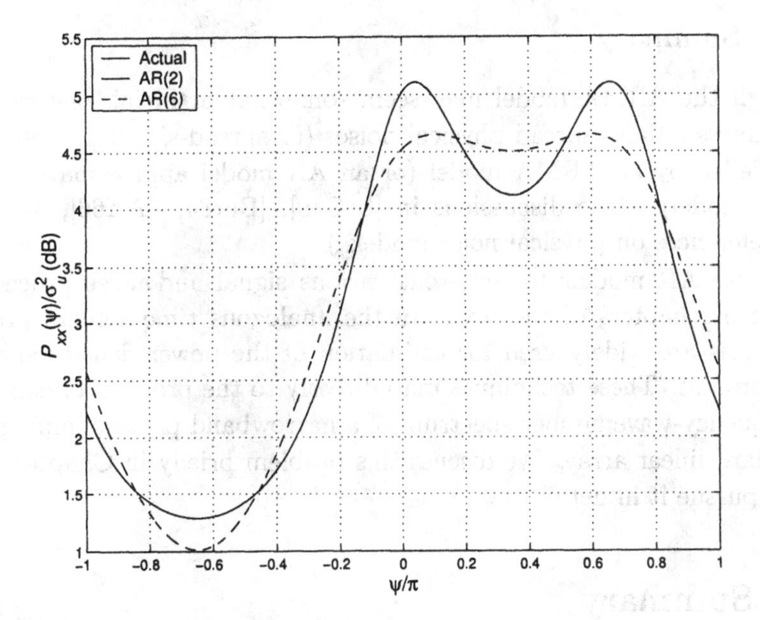

Figure 5.34 Wavenumber spectrum of a complex AR(2) process with additive white noise: $\sigma_u^2/\sigma_w^2 = 0$ dB, $z_1 = 0.5, z_2 = 0.5e^{j\,(0.7\pi)}$.

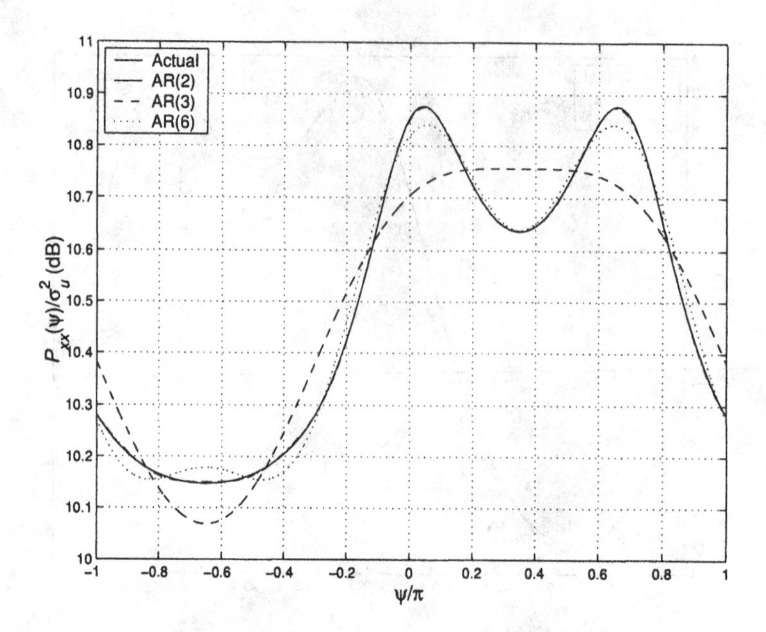

Figure 5.35 Wavenumber spectrum of a complex AR(2) process with additive white noise: $\sigma_u^2/\sigma_w^2 = -10$ dB, $z_1 = 0.5$, $z_2 = 0.5e^{j\,(0.7\pi)}$.

5.6.4 Summary

Although the ARMA model may seem somewhat artificial, one can show that many spatially spread physical noises (or spread-signal processes) can be modelled by an ARMA model (or an AR model approximation to the ARMA model). (See discussions in [LeC89], [Bur84], [AA66], as well as other references on physical noise models.)

We use AR models to represent various signal and noise processes in subsequent chapters of the text. In the analogous time-domain problem, AR models are widely used for estimation of the power density spectrum of the process. These techniques map directly to the problem of estimating the frequency-wavenumber spectrum of a narrowband process impinging on a standard linear array. We discuss this problem briefly in Chapter 10 but do not pursue it in detail.

5.7 Summary

In this chapter we have developed characterizations for random space-time processes that will be used in our study of optimum array processing.

In Section 5.2, we developed the frequency-domain snapshot model that we use throughout most of the text. We also developed the time-domain snapshot model and saw that, for narrowband processes, the two models give the same results.

In Section 5.3, we introduced a more general model of space-time processes and defined Gaussian space-time processes. This general model could accommodate spatially distributed signal and noise sources that are encountered in many applications.

In Section 5.4, we saw how arrays and apertures respond to these space-time processes. We developed an expression for $\mathbf{S_x}$, the spatial spectral matrix. This matrix plays the key role in most of our subsequent discussions.

In Section 5.5, we developed an eigendecomposition of the spatial spectral matrix $\mathbf{S_x}$. The resulting eigenvectors and eigenvalues are the important elements in many of the optimum array processing schemes that we develop in subsequent chapters.

In Section 5.6, we restricted our attention to uniform linear arrays and developed a rational transfer function process model that can be used to model spatially spread signal and noise processes. These AR and ARMA models are widely used to model various physical random processes.

This chapter provides the first part of the background that is needed to develop optimum array processors. In this discussion, we have used ensemble averages and assumed that the various quantities such as $\mathbf{S_f}$, $\mathbf{S_x}$, and σ_w^2 were known. In practice, we must estimate these quantities from a finite amount of data. We discuss these issues in Section 7.2.

In Chapter 6, we develop optimum waveform estimation algorithms using the models developed in this chapter.

5.8 Problems

P5.2 Snapshot Models

Problem 5.2.1

Define

$$\mathbf{X}(\omega_m) = \int_0^{\Delta T} \mathbf{x}(t) e^{-j\omega t}\, dt. \tag{5.362}$$

(a) Show that

$$E\left[\mathbf{X}(\omega_m)\mathbf{X}^H(\omega_m)\right] = \int_0^{\Delta T} (\Delta T - |v|)\mathbf{R_x}(v) \cos\omega_m v\, dv. \tag{5.363}$$

(b) Show that

$$\lim_{\Delta T\to\infty}\left\{\frac{1}{\Delta T}E\left[\mathbf{X}(\omega_m)\mathbf{X}^H(\omega_m)\right]\right\}=\int_{-\infty}^{\infty}\mathbf{R_x}(v)\cos\omega_m v\,dv=\mathbf{S_x}(\omega_m). \quad (5.364)$$

(c) Discuss how large ΔT must be in order for

$$\frac{1}{\Delta T}E\left[\mathbf{X}(\omega_m)\mathbf{X}^H(\omega_m)\right]\simeq\mathbf{S_x}(\omega_m). \quad (5.365)$$

Problem 5.2.2

Consider a standard 10-element linear array. Assume that the signal is a plane wave whose direction cosine is u_s and whose spectrum is shown in Figure 5.2(a). The expression for $\mathbf{S}_{\mathbf{x}_{\Delta T}}(m,m)$ is given in Section 5.2.1.3.

(a) Plot the matrix, $\mathbf{S}_{\mathbf{x}_{\Delta T}}(m,m)$, for $B_s\cdot\Delta T=2^l, l=0,\cdots,7$, $|m|=2^l$, for $u=0$. Discuss your results.

(b) Repeat for $u=0.5$

(c) Repeat for $u=0.9$

Problem 5.2.3

Repeat Problem 5.2.2 for $\mathbf{S}_{\mathbf{x}_{\Delta T}}(m_1,m_2)$. Consider the cases

$$\begin{aligned}m_1=0, \quad m_2=1\\ m_1=0, \quad m_2=2.\end{aligned} \quad (5.366)$$

Problem 5.2.4

Consider the expression in (5.27). Evaluate it for the plane-wave model in Problem 5.2.2, $l=k+1$, and $m_1=m_2$.

P5.3 Space-time Random Processes

Problem 5.3.1 [Bag76]

In an ocean environment, surface or bottom noise can be characterized by

$$S_o(\omega_0:\theta,\phi)=S_o(\omega_0)\left[1+\alpha\cos\theta\right], \quad (5.367)$$

and has azimuthal (ϕ) symmetry. An intensity plot is shown in Figure 5.36.

(a) Plot the real and imaginary part of $S_x(\omega_0:\Delta\mathbf{p})$ versus $\Delta\mathbf{p}$ for various α and $\cos\theta_p$.

(b) Check your result against the analytic formula

$$S_x(\omega_0:\Delta\mathbf{p})=S_o(\omega_o)\left\{\text{sinc}\,(k_o\Delta\mathbf{p})+j\alpha\frac{1}{k_o\Delta\mathbf{p}}\left[\text{sinc}\,(k_o\Delta\mathbf{p})-\cos(k_o\Delta\mathbf{p})\right]\cos\theta_p\right\}. \quad (5.368)$$

Problem 5.3.2 [Bag76]

In an ocean environment, we can model layer noise as

$$S_o(\omega_0:\theta,\phi)=S_o(\omega_0)\left[1-\frac{1}{4}\alpha-\frac{3}{4}\alpha\cos(2\theta)\right], \quad (5.369)$$

with azimuthal symmetry. An intensity plot is shown in Figure 5.37.

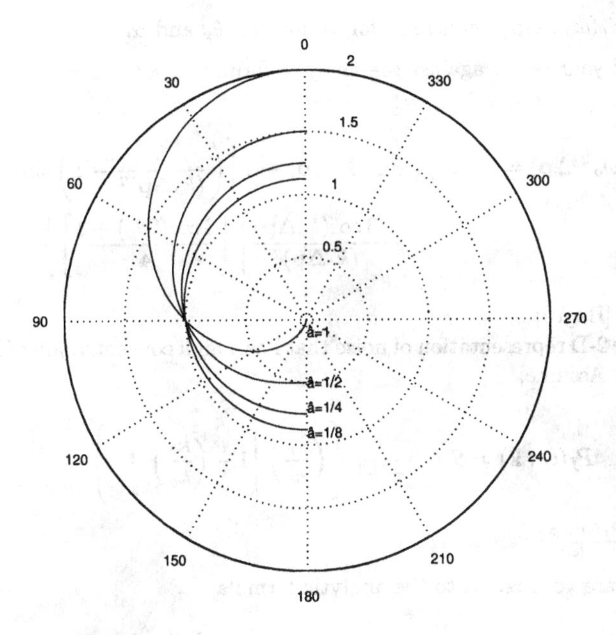

Figure 5.36 Power distribution: surface noise. [From [Bag76]]

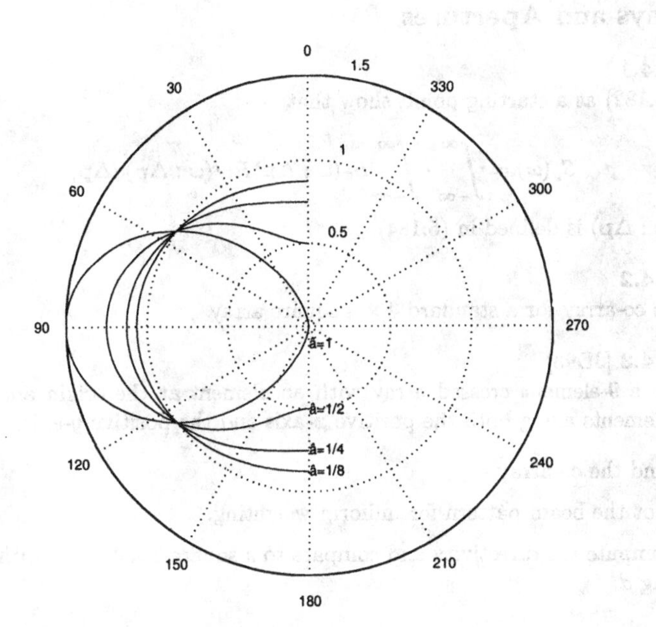

Figure 5.37 Power distribution: layer noise. [From [Bag76]]

(a) Plot $S_f(\omega_0 : \Delta\mathbf{p})$ versus $\Delta\mathbf{p}$ for various $\cos\theta_p$ and α.

(b) Check your result against the analytic formula

$$S_x(\omega_o : \Delta\mathbf{p}) = S_o(\omega_o)\left\{ \text{sinc}(k_o\Delta\mathbf{p}) + \alpha \left[\left(\frac{3}{(k_o\Delta\mathbf{p})^2} - 1 \right) \text{sinc}\,(k_o\Delta\mathbf{p}) - \frac{3\cos(k_o\Delta\mathbf{p})}{(k_o\Delta\mathbf{p})^2} \right] \left[\frac{3\cos(2\theta_p)+1}{4} \right] \right\}. \tag{5.370}$$

Problem 5.3.3 [Bag76]

Consider the 2-D representation of noise that has a high concentration of low wavenumber components. Assume,

$$P_f(\omega : \mathbf{k}) = S_o(\omega)\frac{1}{\pi^2 k_o^2}\left\{ \left(\frac{k_r}{k_o}\right)\left[1 - \left(\frac{k_r}{k_o}\right)^2\right]^{\frac{1}{2}} \right\}^{-1}. \tag{5.371}$$

(a) Plot $\frac{S_f(\omega:\Delta\mathbf{p})}{S_o(\omega)}$ versus $\Delta\mathbf{p}$.

(b) Compare your result to the analytic formula,

$$S_f(\omega : \Delta\mathbf{p}) = S_o(\omega) J_o^2\left(\frac{k_r\Delta\mathbf{p}}{2}\right). \tag{5.372}$$

P5.4 Arrays and Apertures

Problem 5.4.1

Using (5.182) as a starting point, show that

$$S_y(\omega) = \int_{-\infty}^{\infty} \cdot \int_{-\infty}^{\infty} S_x(\omega : \Delta\mathbf{p}) R_W(\omega : \Delta\mathbf{p}) d\Delta\mathbf{p}, \tag{5.373}$$

where $R_W(\omega : \Delta\mathbf{p})$ is defined in (5.184).

Problem 5.4.2

Find the co-array for a standard 4×4 square array.

Problem 5.4.3 [JD93]

Consider a 9-element crossed array with an element at the origin and four equally spaced (d) elements along both the positive x-axis and the positive y-axis.

(a) Find the co-array.

(b) Plot the beam pattern for uniform weighting.

(c) Compute the directivity and compare to a square 3×3 array with interelement spacing d.

Problem 5.4.4

Repeat Problem 5.4.3 for a symmetric 9-element crossed array along the x- and y-axes with interelement spacing d.

Problem 5.4.5 [JD93]

Consider an equilateral triangle array with four equally spaced (d) elements on each side (nine sensors total).

(a) Find the co-array.

(b) Find the beam pattern if the array lies in the xy-plane and the sensors are short dipoles oriented in the z-direction.

P5.5 Orthogonal Expansions

In Problems 5.5.1–5.5.23, the objective is to find the eigenvalues and eigenvectors. The spatial signal (or noise) environment and the array (or aperture) geometry is specified in each problem.

Problem 5.5.1

The input is two equal-power uncorrelated plane waves. The array is a standard 10×10 square array.

(a) Find the normalized eigenvalues.

(b) Plot the eigenvectors for several ψ_1 and ψ_2.

Problem 5.5.2

Repeat Problem 5.5.1 for four equal-power uncorrelated signals that are uniformly spaced in ψ-space.

Problem 5.5.3

The input is three equal-power uncorrelated plane waves that are equally spaced in ψ-space. The array is a standard 10-element linear array.

(a) Plot the eigenvalues for several $\Delta\psi$.

(b) Plot the eigenvectors for several $\Delta\psi$.

Problem 5.5.4.

The input is two equal-power uncorrelated plane waves. The array is a standard 19-element hexagonal array (e.g., Section 4.4).

(a) Find the normalized eigenvalues.

(b) Plot the eigenvectors for several $\Delta\psi$.

Problem 5.5.5

Repeat Problem 5.5.4 for four equal-power uncorrelated signals that are uniformly spaced in ψ-space.

(a) Find the normalized eigenvalues.

(b) Plot the eigenvectors for several $\Delta\psi$.

Problem 5.5.6

Assume that the signal has the spatial correlation function, given by (5.148).
Consider a 20-element linear array along the x-axis with $d = \lambda/2$.
Find the eigenvalues and eigenvectors.

Problem 5.5.7

Assume the signal has the spatial correlation function given by (5.148).
Consider a standard 10-element linear array along the x-axis.
Find the eigenvalues and eigenvectors for various θ_0.

Problem 5.5.8

Repeat Problem 5.5.7 for the sector noise model with the center of the sector moved to $\theta_1 = 30^o, \phi_1 = 0^o$.

Problem 5.5.9

Repeat Problem 5.5.7 for a standard 10×10 square array located in the xy-plane.

Problem 5.5.10

Repeat Problem 5.5.8 for a standard 10×10 square array located in the xy-plane.

Problem 5.5.11

The input is two equal-power uncorrelated plane waves. The aperture is a ring aperture with $R = 3\lambda$.

(a) Find the normalized eigenvalues.

(b) Plot the eigenvectors for several $\Delta\psi$.

Problem 5.5.12

The input is two equal-power uncorrelated plane waves. The array is 12-element circular array with $R = 3\lambda$.

(a) Find the normalized eigenvalues.

(b) Plot the eigenvectors for several $\Delta\psi$.

Problem 5.5.13

Repeat Problem 5.5.11 for a circular aperture with $R = \sqrt{3}\lambda$. Compare the results to the results in Problem 5.5.4.

Problem 5.5.14

Repeat Problem 5.5.13 for a square aperture whose area is the same as the circular aperture in Problem 5.5.13.

Problem 5.5.15

Consider a standard 11-element linear array. The input consists of the spatially spread spectrum described in Example 5.5.1 (with $\psi_\Delta = \pi \cos 15°$) and two equal-power ($P$) uncorrelated plane waves at $\psi_1 = \pi \cos 15°$ and $\psi_2 = \pi \cos 20°$. Define $\beta = \frac{S_x(\omega)\cdot 2\psi_\Delta}{P}$.

(a) Find the normalized eigenvalues.

(b) Check the behavior of the eigenvalues for $\beta = 0$ and $\beta = \infty$.

(c) Plot the eigenvectors for $\beta = 1$

Problem 5.5.16

Consider a standard 10-element linear array and four uncorrelated equal-power plane-wave signals at $\psi = \pm\Delta\psi$ and $\psi = \pm 2\Delta\psi$. Read [Lee92].

(a) Use Lee's procedure to find the eigenvalues and eigenvectors.

(b) Plot the resulting eigenvalues versus $\Delta\psi/BW_{NN}$ using Lee's procedure and compare the results to an exact numerical calculation for several $\Delta\psi$.

(c) Plot the resulting eigenvectors and plot $|\mathbf{e}_i|$ versus $\Delta\psi/BW_{NN}$.

Problem 5.5.17

(a) Repeat Problem 5.5.16(a) for a standard 21-element linear array and six uncorrelated equal-power plane-wave signals at $\psi = \pm\Delta\psi, \pm 2\Delta\psi, \pm 3\Delta\psi$.

(b) Plot the resulting eigenvalues versus $\frac{\Delta\psi}{BW_{NN}}$.

(c) Plot the resulting eigenvectors.

Problem 5.5.18 (continuation: Problem 5.5.16)

Consider a standard 10-element linear array and the frequency-wavenumber function in (5.269).

(a) Use Lee's [Lee92] procedure to find the eigenvalues and eigenvectors.

(b) Compare the results to the results in Example 5.5.1

Problem 5.5.19

Consider a standard 11-element linear array. The input signal is a complex AR(1) process whose frequency-wavenumber spectrum is given by (5.327). Find the eigenvalues and eigenvectors for different values of $a(1)$. Plot the eigenvalues versus $|a(1)|$ and $\angle a(1)$. Plot selected eigenbeams.

Problem 5.5.20

Repeat Problem 5.5.19 for the complex AR(2) process whose frequency-wavenumber spectrum is given by (5.330).

Problem 5.5.21

Repeat Problem 5.5.19 for the complex ARMA (1,1) process whose frequency-wavenumber spectrum is given by (5.353). Does the structure of $\tilde{P}_x(\psi)$ allow any simplification?

Problem 5.5.22

In Section 5.3.4.1, the function $\overline{S}_o(\omega, u)$ is the 1-D frequency-wavenumber response function evaluated for $k_z = -\frac{2\pi}{\lambda}u$. Let

$$\overline{S}_o(\omega, u) = P_x(\omega, k_z)|_{k_z = -\frac{2\pi}{\lambda}u} = \begin{cases} S_x(\omega)\frac{1}{2u_o}, & |u - u_s| \le u_o, \\ 0, & \text{otherwise.} \end{cases} \tag{5.374}$$

The signal is spatially spread around the direction-of-arrival u_s with a uniform distribution with spreading factor u_o.

(a) Show that

$$S_x(\omega, \Delta p_z) = S_x(\omega) e^{j\frac{2\pi}{\lambda}u_s \Delta p_z} \text{sinc}\left(\frac{2\pi}{\lambda} u_o \Delta p_z\right). \tag{5.375}$$

(b) Assume an N-element standard linear array along the z-axis. Find the expression for $S_x(\omega)$.

(c) Plot the eigenvalues and eigenbeams for several directions-of-arrival u_s, and spreading factors u_o.

Problem 5.5.23 (continuation)
Repeat Problem 5.5.22 with

$$\overline{S}_o(\omega, u) = P_x(\omega, k_z)|_{k_z = -\frac{2\pi}{\lambda}u} = S_x(\omega) \frac{1}{\sqrt{2\pi\sigma_o^2}} \exp\left\{-\frac{(u - u_s)^2}{2\sigma_o^2}\right\}, \tag{5.376}$$

and

$$S_x(\omega, \Delta p_z) = S_x(\omega) e^{j\frac{2\pi}{\lambda}u_s \Delta p_z} \exp\left\{-\frac{1}{2}\left(\frac{2\pi}{\lambda}\sigma_o \Delta p_z\right)^2\right\}. \tag{5.377}$$

The signal is spatially spread around the direction-of-arrival u_s with a Gaussian distribution with spreading factor σ_o. Assume $\sigma_o \ll 1$.

Problem 5.5.24
Derive the results in (5.255) and (5.259).

Problem 5.5.25
Consider a standard 10-element linear array, and two uncorrelated plane waves. Assume $S_1 = 1$. Plot the eigenvalues (normalized by$N$) and eigenvectors for various separations for (i)$S_2 = S_1$, (ii)$S_2 = 10S_1$, (iii)$S_2 = 100S_1$, (iv) $S_2 = 1000S_1$. Find the eigenvalues and eigenvectors numerically by constructing $\mathbf{S_x} = \mathbf{V}\mathbf{S_f}\mathbf{V}^H$ and verify that eigenvalues are the same as the analytical expression in (5.255).

Problem 5.5.26
Consider a standard 10-element linear array. There are three uncorrelated plane waves impinging on the array:

Signal: $u_s = 0$, $\sigma_s^2 = 1$

Interferer: $u_1 = 0.3$, $\sigma_1^2 = 10, 1000$

Interferer: $u_1 = -0.5$, $\sigma_2^2 = 10, 1000$

(a) Find the eigenvalues and eigenvectors for the four combinations of interferer powers. Plot the eigenbeams.

(b) Assume that the signal actually arrives from u_a; $-0.1 \le u_a \le 0.1$. Compute the correlation between $\mathbf{v}(u_a)$ and each of the three eigenvectors in part (a). Plot the results versus u_a for the four cases.

(c) Define

$$\mathbf{v}_p(u_a) = \mathbf{v}^H(u_a)\mathbf{U}_{S+I}, \tag{5.378}$$

where $\mathbf{U}_{S+I}$ is the $N \times 3$ matrix defining the "signal-plus-interference" subspace. Then

$$C(u_a) = \frac{1}{N}\|\mathbf{v}_p(u_a)\|^2, \tag{5.379}$$

measures the amount of the signal power contained in the subspace. Plot $C(u_a)$ versus u_a.

Problem 5.5.27

Consider the same nominal model as in Problem 5.5.26(a). Assume that the sensor positions are perturbed using the model in Section 2.6.3.

(a) Express $\mathbf{S_x}$ as a function of $\Delta\mathbf{p}$.

(b) Define

$$\mathbf{v}_p(\Delta\mathbf{p}) = \mathbf{v}^H(u_s)\mathbf{U}_{S+I}(\Delta\mathbf{p}), \tag{5.380}$$

where $\mathbf{U}_{S+I}(\Delta\mathbf{p})$ is the "signal-plus-interference" subspace calculated using the actual array location. The array manifold vector $\mathbf{v}(u_s)$ is calculated using the nominal location with $u_s = 0$. Define

$$C(\Delta\mathbf{p}) = \frac{1}{N}\| \mathbf{v}_p(\Delta\mathbf{p})\|^2. \tag{5.381}$$

Plot $E[C(\Delta\mathbf{p})]$ versus σ_λ^2 for the case of equal power interferers ($\sigma_I^2 = 10$ dB and 30 dB). Discuss your result.

Problem 5.5.28

Consider an N-element standard linear array. The signal is spatially spread and frequency spread. The signal direction is centered at u_s,

$$P_1(f,u) = \begin{cases} \frac{\sigma_s^2}{B_s \Delta u}, & |u-u_s| \leq \frac{\Delta u}{2}, |f-f_c| \leq \frac{B_s}{2}, \\ 0, & \text{elsewhere.} \end{cases} \tag{5.382}$$

(a) Find $\mathbf{S_x}$.

(b) Find the eigenvalues and eigenvectors for a standard 10-element linear array as a function of $u_\Delta / B\omega_{NN}$ and $B_f u_s$.

(c) Develop a test similar to those in (5.277) and (5.290).

P5.6 Parametric Wavenumber Models

Problem 5.6.1

Consider the surface noise model in Problem 5.3.1. Assume we are observing with a standard linear array along the z-axis. Develop a rational transfer model that approximates the frequency-wavenumber function.

Problem 5.6.2

Find the correlation matrix $\mathbf{S_x}$ for the input process whose wavenumber spectrum is given by (5.330). Use the parameter values in Figure 5.25.

Problem 5.6.3

Assume that

$$\mathbf{X}(k) = \mathbf{F}(k) + \mathbf{N}(k), \tag{5.383}$$

where $\mathbf{F}(k)$ is a plane wave from broadside and the $\mathbf{N}(k)$ have the wavenumber spectrum shown in Figure 5.24, with $a(1) = 0.7e^{-j(0.6\pi)}$. We process the input with a standard 10-element linear array. Find the array weighting to maximize the array gain.

Problem 5.6.4

(a) Find $\mathbf{S_x}$ for the complex AR(2) process in (5.330).

(b) Indicate how you would implement the spatial whitening filters.

Problem 5.6.5

Consider an $N \times N$ symmetric tridiagonal Toeplitz matrix whose first row is $[a, b, 0, \cdots, 0]$. The eigenvalues and eigenvectors are given by

$$\lambda^{(k)} = a + 2b\cos\left(\frac{k\pi}{N} + 1\right), \quad k = 1, \cdots, N, \tag{5.384}$$

and

$$\Phi_j^{(k)} = \left(\frac{2}{N} + 1\right)^{1/2} \sin\left(\frac{kj\pi}{N} + 1\right), \quad j, k = 1, \cdots, N. \tag{5.385}$$

(e.g., [RB78] or [GK69]).

(a) Use this result to find an analytic expression for the eigenvalues and eigenvectors of a real AR(1) process.

(b) Plot the eigenbeams for $N = 10$ and several values of $a(1)$.

(c) Check the behavior as $a(1) \rightarrow 1$.

Problem 5.6.6

Repeat Problem 5.6.5 for a complex AR(1) process.

Problem 5.6.7

(a) Find the whitening filter for a complex AR(2) process.

(b) Write $\mathbf{S_x^{-1}}$ in the form of (5.338).

Problem 5.6.8

Repeat Problem 5.6.7 for a complex AR(3) process.

Problem 5.6.9

Consider the case in which

$$\mathbf{X}(k) = \mathbf{F}(k) + \mathbf{W}(k), \tag{5.386}$$

and $\mathbf{F}(k)$ is a complex AR(1) process.

(a) Approximate $\mathbf{x}(k)$ by a higher order AR process. Use this result to find an approximate spatial whitening filter. Define a whiteness measure as

$$\varepsilon_w = \left\| E\left[\mathbf{Y}\mathbf{Y}^H\right] - k\mathbf{I} \right\|_F, \tag{5.387}$$

where $\mathbf{Y}(n)$ is the output of the whitening filter. Show how to compute the whiteness measure.

(b) Consider a standard 10-element linear array. Apply the result in part (a) to the model with $\sigma_u^2/\sigma_n^2 = -10$ dB, 0 dB, 10 dB, and $|a(1)| = 0.5, 0.7, 0.9$ and 0.99 for several values of ε_w.

Problem 5.6.10

Repeat Problem 5.6.9 for a complex AR(2) process.

Bibliography

[AA66] E. M. Arase and T. Arase. Correlations of ambient sea noise. *J. Acoust. Soc. Am.*, vol.40, pp. 205–210, February 1966.

[Bag76] A. B. Baggeroer. Space-time processes and optimal array processing. Technical Report 506, Navy Undersea Center, San Diego, California, December 1976.

[BAS86] A. Ben Artiz and T. Shalom. On inversion of Toeplitz and close to Toeplitz matrices. *Linear Algebra Appl.*, vol.75, pp. 173–192, March 1986.

[Bla57] N. M. Blachman. On Fourier series for Gaussian noise. *Inf. Control* vol.1, pp. 56–63, 1957.

[BS83] A. Botcher and B. Silbermann. *Invertibility and Asymptotics of Toeplitz Matrices.* Akademie Verlag, Berlin, 1983.

[Bur84] W. S. Burdic. *Underwater Acoustic System Analysis.* Prentice-Hall, Englewood Cliffs, New Jersey, 1984.

[Com88] R. T. Compton, Jr. *Adaptive Antennas (Concepts and Performance).* Prentice-Hall, Englewood Cliffs, New Jersey, 1988.

[DM84] D. E. Dudgeon and R. M. Mersereau. *Multidimensional Digital Signal Processing.* Prentice-Hall, Inc., Englewood Cliffs, New Jersey, 1984.

[Doo53] J. Doob. *Stochastic Processes.* Wiley, New York, 1953.

[DR58] W. B. Davenport and W. L Root. *An Introduction to the Theory of Random Signals and Noise.* McGraw-Hill, New York, 1958.

[DR87] W. B. Davenport and W. L Root. *An Introduction to the Theory of Random Signals and Noise.* IEEE Press, New York, 1987.

[Fuh98] D. Fuhrmann. Complex random variables and stochastic processes, In V. Madisetti and D. Williams, editors., Digital Signal Processing Handbook, CRC Press, Boca Raton, Florida, 1998.

[GK69] R. T. Gregory and D. L. Karney. *A Collection of Matrices for Testing Computational Algorithms.* Wiley-Interscience, New York, 1969.

[Hau68] R. A. Haubrich. Array design. *Bull. Seismol. Soc. Am.*, vol.58, pp. 977–991, June 1968.

[Hay91] S. Haykin. *Adaptive Filter Theory.* Prentice-Hall, Englewood Cliffs, New Jersey, second edition, 1991.

[Hel91] C. W. Helstrom. *Probability and Stochastic Processes for Engineers.* Prentice-Hall, Englewood Cliffs, New Jersey, second edition, 1991.

[HN76] W. S. Hodgkiss and L. W. Nolte. Covariance between Fourier coeffcients representing the time waveforms observed from an array of sensors. *J. Acoust. Soc. Am.*, vol.59, pp. 582–590, March 1976.

[Hud81] J.E. Hudson. *Adaptive Array Principles.* Peter Peregrinus, New York and London, 1981.

[JD93] D. H. Johnson and D. E. Dudgeon. *Array Signal Processing.* Prentice-Hall, Englewood Cliffs, New Jersey, 1993.

[Kay88] S. M. Kay. *Modern Spectral Estimation: Theory and Application.* Prentice-Hall, Englewood Cliffs, New Jersey, 1988.

[KB79a] P. C. K. Kwok and P. S. Brandon. Eigenvalues of the noise covariance matrix of a linear array in the presence of 2 directional interferences. *Electron. Lett.*, vol.15, pp. 50–51, January 1979.

[KB79b] P. C. K. Kwok and P. S. Brandon. The optimal radiation pattern of an array in the presence of 2 directional interferences. *Electron. Lett.*, vol.15, pp. 251–252, March 1979.

[KB86a] M. Kaveh and A. J. Barabell. The statistical performance of the MUSIC and the Minimum-Norm algorithms in resolving plane waves in noise. *IEEE Trans. Acoust., Speech, Signal Process.*, vol.ASSP-34, pp. 331–341, April 1986.

[KB86b] M. Kaveh and A. J. Barabell. Corrections to 'The statistical performance of the MUSIC and the Minimum-Norm algorithms in resolving plane waves in noise.' *IEEE Trans. Acoust., Speech, Signal Process.*, vol. ASSP-34, p. 633, June 1986.

[Kol41] A. N. Kolmogorov. Interpolation and extrapolation von stationären zufälligen folgen. *Bull. Acad. Sci. USSR Ser. Math.*, vol.5, pp. 3–14, January 1941.

[Kre71] J. L. Kreuzer. A synthetic aperture coherent imaging technique. *Acoust. Hologr.*, vol.3, pp. 287–315, March 1971.

[KSL94] T. P. Krauss, L. Shure, and J. N. Little. *Signal Processing Toolbox.* The Math Works, Natick, Massachusetts, 1994.

[KVM78] T. Kailath, A. Vieira, and M. Morf. Inverse of Toeplitz operators, innovations and orthogonal polynomials. *SIAM Rev.*, vol.20, pp. 106–119, January 1978.

[LeC89] J.-P. LeCadre. Parametric methods for spatial signal processing in the presence of unknown colored noise fields. *IEEE Trans. Acoust., Speech, Signal Process.*, vol.ASSP-37, pp. 965–983, July 1994.

[Lee92] H. B. Lee. Eigenvalues and Eigenvectors of covariance matrices for signals closely spaced in frequency. *IEEE Trans. Signal Process.*, vol.SP-40, pp. 2518–2535, October 1992.

[Mar87] S. L. Marple, Jr. *Digital Spectral Analysis.* Prentice-Hall, Englewood Cliffs, New Jersey, 1987.

[Mer79] R. M. Mersereau. The processing of hexagonally sampled two-dimensional signals. *Proc. IEEE*, vol.67, pp. 930–949, August 1979.

[Mil74] K. Miller. *Complex Stochastic Processes.* Addison-Wesley, Reading, Massachusetts, 1974.

[Mor78] D. R. Morgan. Partially adaptive array techniques. *IEEE Trans. Antennas Propag.*, vol.AP-26, pp. 823–833, November 1978.

[NM93] F. Neeser and J. Massey. The complex random processes with applications to information theory. *IEEE Trans. Inf. Theory*, vol.IT-39, pp. 1293-1302, July 1993.

[OS89] A. V. Oppenheim and R. W. Schafer. *Discrete-Time Signal Processing.* Prentice-Hall, Englewood Cliffs, New Jersey, 1989.

[Ows85] N. L. Owsley. Overview of adaptive array processing. In *Adaptive Methods in Underwater Acoustics*, pp. 355–374, H.G. Urban, editor, Reidel, Dordrecht, The Netherlands, 1985.

[PM62] D. P. Peterson and D. Middleton. Sampling and reconstruction of wave-number limited functions in n-dimensional Euclidean spaces. *Inf. Control*, vol.5, pp. 279–323, April 1962.

[PRLN92] J. G. Proakis, C. M. Rader, F. Ling, and C. L. Nikias. *Advanced Digital Signal Processing.* Macmillan, New York, 1992.

[RB78] P. A. Roebuck and S. Barnett. A survey of Toeplitz and related matrices. *Int. J. Syst. Sci.*, vol.9, pp. 921–934, 1978.

[Sid58] M. M. Siddiqui. On the inversion of the sample covariance matrix in a stationary autoregressive process. *Ann. Math. Statist.*, vol.58, pp. 585–588, June 1958.

[Str41] J. Stratton. *Electromagnetic Theory.* McGraw-Hill, New York, 1941.

[VT68] H. L. Van Trees. *Detection, Estimation, and Modulation Theory, Part I.* Wiley, New York, 1968.

[VT71] H. L. Van Trees. *Detection, Estimation, and Modulation Theory, Part III.* Wiley, New York, 1971.

[VT01a] H. L. Van Trees. *Detection, Estimation, and Modulation Theory, Part I.* Wiley Interscience, New York, 2001.

[VT01b] H. L. Van Trees. *Detection, Estimation, and Modulation Theory, Part III.* Wiley Interscience, New York, 2001.

[WK85] G. Wakefield and M. Kaveh. Frequency-wavenumber spectral estimation of the pressure field beneath a turbulent boundary layer. *Proc. Winter Annual Meeting of the American Society of Mechanical Engineers*, NCA, vol.1, Miami Beach, Florida November 1985.

[Wol54] H. Wold. *A Study in the Analysis of Stationary Time Series.* Almqvist & Wiksell, Stockholm, 1954.

[Yag57] A. Yaglom. Some classes of random fields m in dimensional space, related to stationary random process. *Theory Prob. Appl.*, p. 273, November 1957.

[Yag60] A. Yaglom. *An Introduction to the Theory of Stationary Random Functions.* Prentice-Hall, Englewood, Cliffs, New Jersey, 1960.

[Zat98] M. Zatman. How narrow is narrowband? *Conference Record of the Thirty-First Asilomar Conference on Signals, Systems and Computers, 1997*, vol.2, pp. 1341–1345, October 1998.

Chapter 6

Optimum Waveform Estimation

6.1 Introduction

In Chapters 2, 3, and 4 we discussed deterministic design techniques and developed the basic ideas of beam patterns, array gain, sidelobe control, and null placement. In Chapter 5 we developed statistical representations of space-time random processes.

In Chapters 6–10, we utilize the statistical representation of the signal and noise processes to design array processors that are optimum in a statistical sense. The first step is to define the objectives of the array processor. We consider the following objectives in the remainder of the text:

(i) The first objective of interest is to **estimate the waveform** of a plane-wave signal impinging on the array in the presence of noise and interfering signals. More generally, we want to estimate the waveform of D plane-wave signals impinging on the array in a similar environment. This problem is sometimes called the "signal copy" problem in the literature. This objective will lead us to beamformers that are optimum in a statistical sense. They are the statistical analog to the deterministic beamformers in Chapters 2–4. In Chapter 6, we assume that the appropriate statistics are known. In Chapter 7, we assume that the necessary statistics must be measured from the data and are led to adaptive beamformers.

(ii) The second objective of interest is to **detect** the presence or absence of a signal that impinges on the array as a plane-wave or a spatially spread signal. More generally, we want to detect which signal belonging to a finite alphabet is present.

This problem is the spatial generalization of the optimum detection problem that we studied in Parts 1 and 3 of DEMT ([VT68], [VT01a]). We study this objective in Chapter 10. In most cases, the spatial aspects of the problem lead to the same beamformers as in Chapters 6 and 7.

(iii) The third objective of interest is to **estimate the direction of arrival** of D plane waves that are impinging on the array in the presence of interference and noise. In most cases of interest, the spatial spectral matrix $\mathbf{S_x}$ and the noise level σ_w^2 are unknown. Thus, we formulate the problem as a more general **parameter estimation problem**, in which there are D desired parameters plus a set of unwanted (or nuisance) parameters.

This problem is the spatial generalization of the parameter estimation problem that we studied in Parts 1 and 3 of DEMT ([VT68] and [VT01a]). We study this problem in Chapters 8 and 9.

There is a problem contained in this third objective that deserves special mention. This is the case in which the array manifold vector is either unknown or is perturbed from its nominal position. In some application our focus is on the signal DOAs and we treat array parameters as nuisance parameters. In another application, the value of the array parameters is also important. This latter case is known as the **array calibration** problem. We treat it briefly in Chapter 9, but do not discuss it in detail.

All of the problems within this objective are characterized by a **finite** parameter set. We want to estimate some (or all) of the parameters in this set.

(iv) The fourth objective of interest is to **estimate the spatial spectrum** of the space-time process that is impinging on the array. If we model the space-time field parametrically as in Section 5.6, then the basic ideas (but not the details) carry over from the parameter estimation problem. However, in many applications, a parametric model is not appropriate and we utilize non-parametric spatial spectrum estimation. We do not cover non-parametric estimation in the text. In Chapter 9, we cite several references on spectrum estimation.

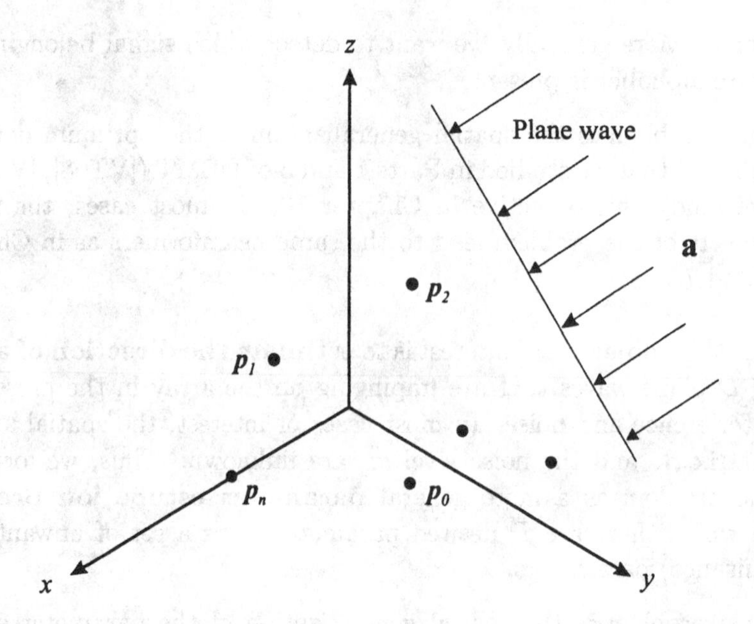

Figure 6.1 Sensor array with single plane-wave signal.

In the course of the remainder of the text we amplify on these objectives and suggest excursions. It is important to always keep the desired objective in mind when evaluating the array processor, because a particular processor may appear under several different objectives and its performance will be evaluated according to different criteria.

For example, in Section 6.2.1, we develop a beamformer referred to as the minimum variance distortionless response (MVDR) or Capon beamformer. It also appears in the parameter estimation context in Chapter 9 and would appear in the non-parametric spatial spectrum estimation problem if we developed it. We must evaluate its performance in the context of the specific objective.

The basic models that we utilize in this chapter are shown in Figures 6.1 through 6.5. The array consists of N sensors located in a 3-D space. Normally we assume isotropic sensors, but we show how to modify the results to take element patterns into account. In Figure 6.1, we show a single plane-wave signal propagating along $\mathbf{a}_s$. The temporal characteristics of the signal of interest include known signals, signals with random parameters, random processes, and unknown nonrandom signals. This includes a large number of communications, radar, sonar, and seismic problems. In Figure 6.2(a), we show multiple plane-wave signals propagating along $\mathbf{a}_i; i = 1, \cdots, L_s$.

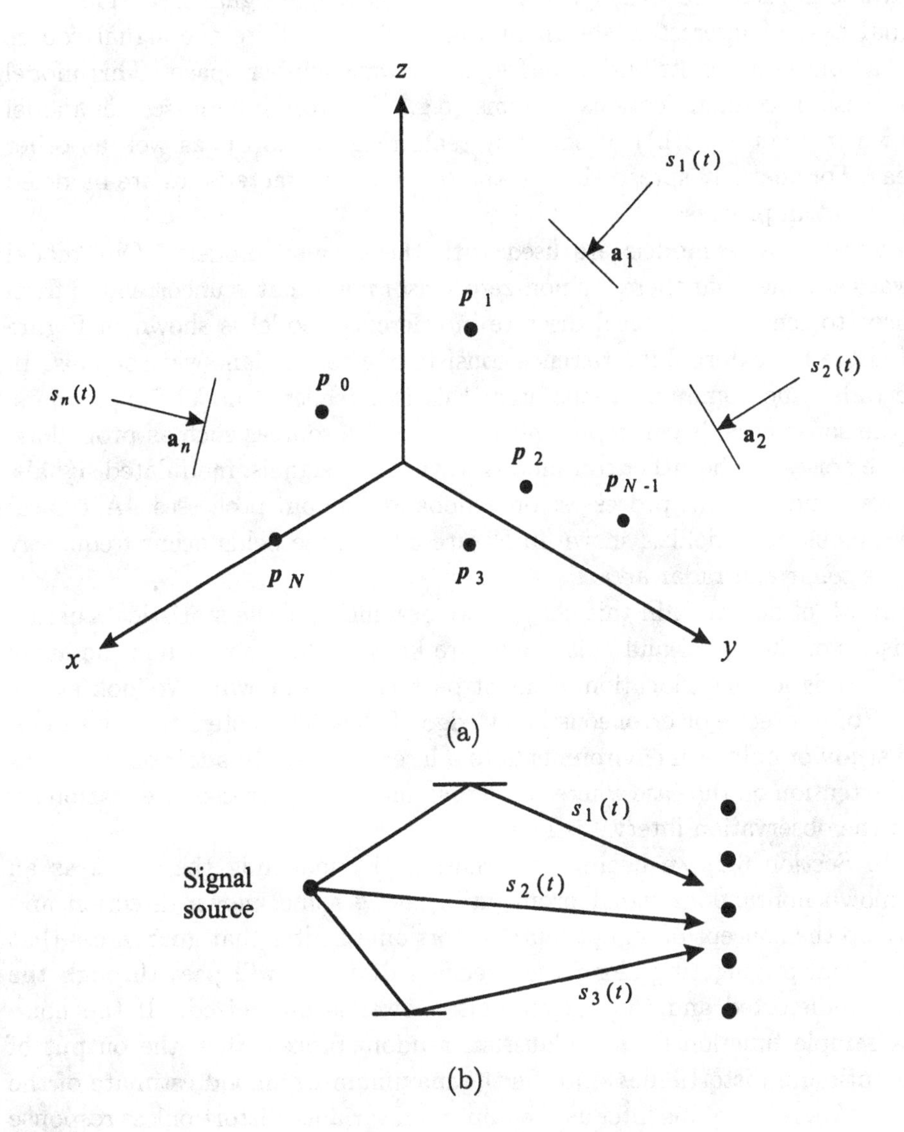

Figure 6.2 (a) Multiple plane-wave signals; (b) multipath environment.

The model arises in the multipath environments of communications, radar, sonar, and seismic environments, as shown in Figure 6.2(b). The temporal characteristics of the signals can vary as in the single-signal case. The third signal case of interest is shown in Figure 6.3(a). Here the signal source has a continuous distribution in frequency-wavenumber space. This model can arise in communications systems (e.g., the tropospheric scatter model shown in Figure 6.3(b)) or sonar systems (Figure 6.3(c)) as well as other areas. For spatially spread signals, the temporal characteristics are modeled as a random processes.

Various noise models are used with these signal models. Our model always assumes that there is a non-zero sensor noise that is uncorrelated from sensor to sensor. A typical discrete interference model is shown in Figure 6.4. Here the external interference consists of a set of plane-wave signals. In the radar and communications area, this is a realistic model for jammers. In the sonar case, it can represent discrete noise sources such as propellors. These noises can be either continuous-wave (CW) signals, modulated signals, narrowband random processes, or wideband random processes. A typical continuous noise field is shown in Figure 6.5. These fields occur frequently in the sonar and radar area.

In all of our work in this chapter we assume that the statistical characteristics of the signal and noise fields are known. We also assume the array characteristics (e.g., location, element patterns) are known. We look at the effect of imprecise or erroneous knowledge of these characteristics, but defer the study of unknown environments to a later chapter. In addition, we focus our attention on the case where the signal and noise processes are stationary and the observation interval is long.

In Section 6.2, we begin our discussion by modeling the signal as an unknown nonrandom signal propagating along some known direction and develop the concept of an optimum distortionless filter that guarantees that any signal propagating along the specified direction will pass through the filter undistorted and the output noise power is minimized. If the noise is a sample function from a Gaussian random process then the output of the optimum distortionless filter is the maximum likelihood estimate of the signal. We refer to this filter as the minimum variance distortionless response (MVDR) filter.

We then consider the case in which the signal is a sample function from a random process and design the optimum linear processor to generate the minimum mean-square error (MMSE) estimate of the signal. The result is the vector version of the Wiener filter that we studied in Chapter 6 of DEMT I [VT68], [VT01a]. We demonstrate that the optimum processor consists of

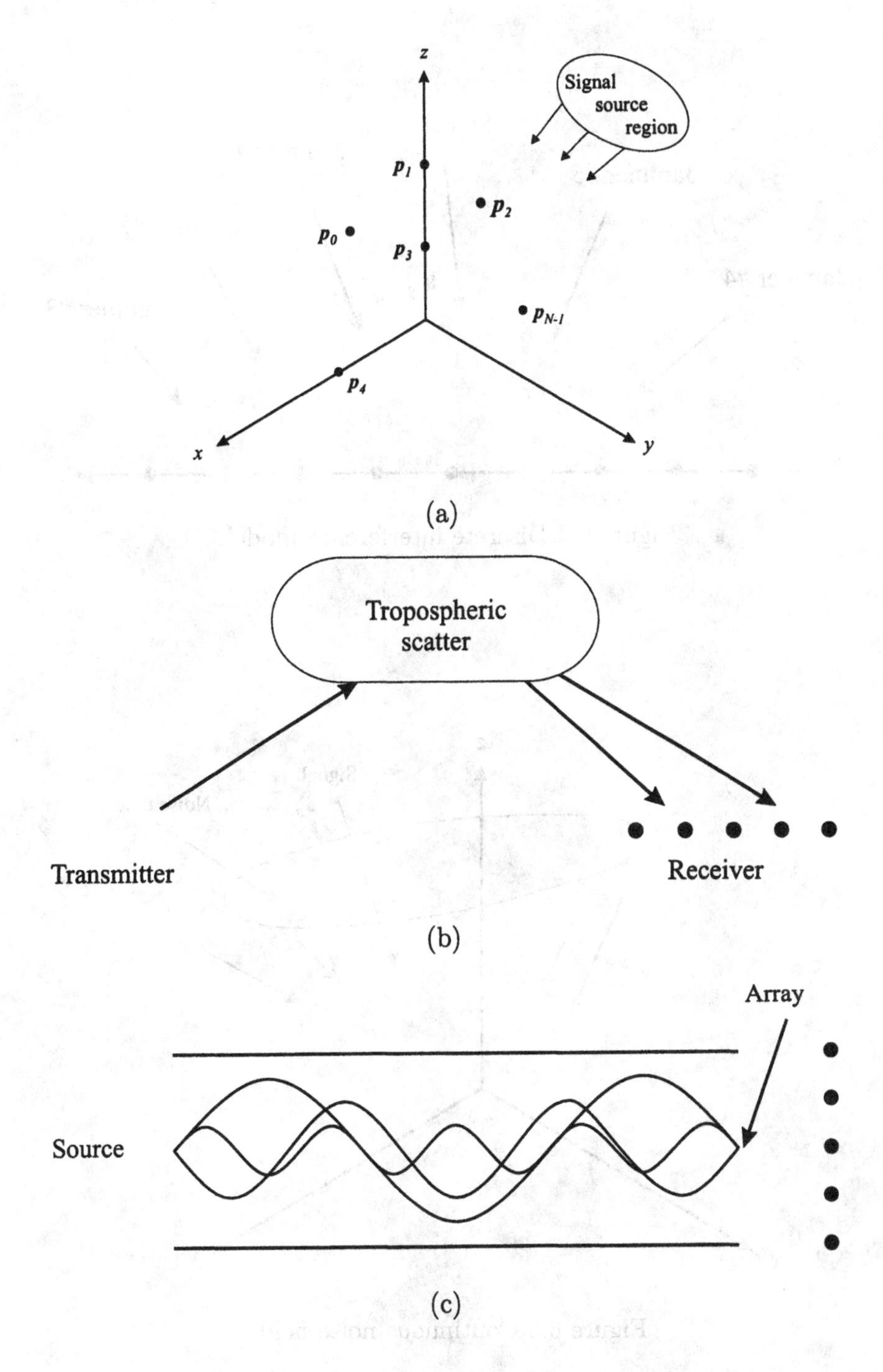

Figure 6.3 (a) Spatially spread signals; (b) troposphere; (c) sonar.

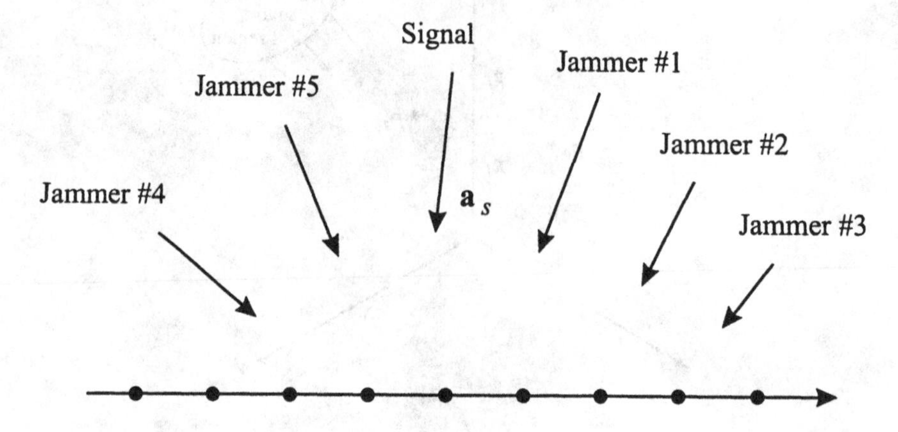

Figure 6.4 Discrete interference model.

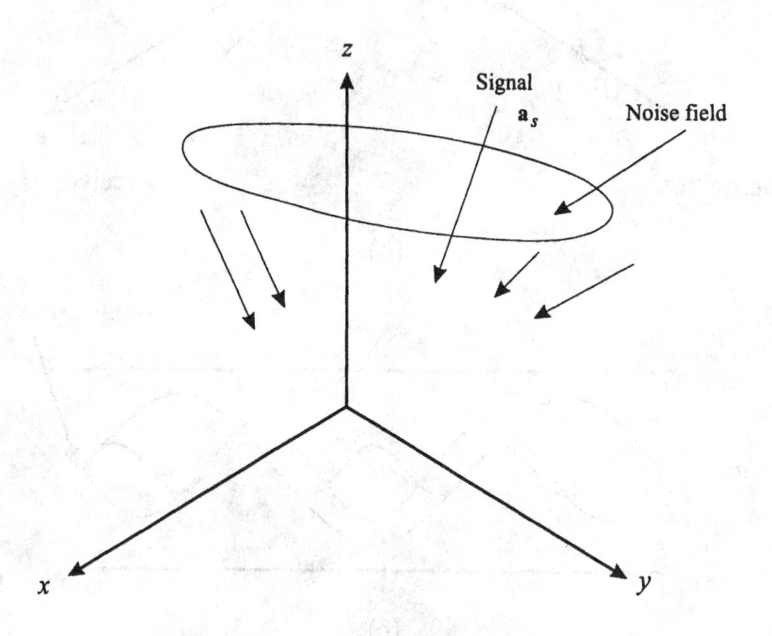

Figure 6.5 Continuous noise field.

an optimum beamformer followed by an optimum scalar filter. We then show that if the signal and noise processes are Gaussian, the optimum linear processor is generating the conditional mean (thus, no nonlinear processor could do better).

We show that if the signal is a single plane wave, then the optimum matrix processor (i.e., the optimum beamformer) is the same for either signal model.

We then examine two other beamformers. The first maximizes the output *SNR*. For a single plane-wave signal, this criterion leads to the same optimum beamformer as the MMSE beamformer. Thus for single plane-wave signals we have a criterion-invariant receiver. The second beamformer assumes that we know (or can measure) $\mathbf{S}_{\mathbf{x}}(\omega)$ but do not know the spectral matrix of the noise component ($\mathbf{S}_{\mathbf{n}}(\omega)$). We choose a steering direction and find the optimum distortionless filter for that direction. We refer to this as the minimum power distortionless response (MPDR) filter. If the steering direction corresponds to the actual signal direction, then the MPDR beamformer reduces to the MVDR beamformer. Later we examine the effects of a mismatch.

All of the results in Section 6.2 apply to arbitrary noise fields. In Section 6.3, we examine the case in which the interference consists of a set of discrete plane waves and find explicit solutions for the optimum beamformers and their performance. We introduce the idea of an eigenbeam receiver as a realization of the optimum processor.

In Section 6.4, we study spatially spread interference. We use the physical noise models from Section 5.3 and the ARMA models from Section 5.6 to model the interference and study the performance of the MVDR and MPDR beamformers with this type of interference.

In Section 6.5, we extend the results to the case in which there are multiple desired plane-wave signals. The results are a logical extension of single plane-wave signals. However, the resulting beamformer has some interesting properties.

A major problem in applying optimum beamformers in operational systems is the potential sensitivity to mismatches between the actual environment and the model used to derive the optimum beamformer.

In Section 6.6, we analyze how the performance of the MVDR beamformer is affected by various types of mismatch. Typical mismatches include DOA mismatch, sensor gain and phase perturbations, and sensor position perturbations. The MVDR beamformer degrades gracefully. However, we find that the performance of the MPDR beamformer degrades rapidly under mismatch. We introduce a technique called diagonal loading to improve robustness.

In Section 6.7, we incorporate additional linear constraints in order to further improve the robustness of the beamformer. We derive the linear constrained minimum variance (LCMV) and the linear constrained minimum power (LCMP) beamformer. We find that, with the addition of diagonal loading, these beamformers are reasonably robust. We also introduce a generalized sidelobe canceller realization of the beamformer.

In Section 5.5, we showed how the use of an orthogonal expansion in terms of eigenvalues and eigenvectors enabled us to represent the set of plane-wave signals and interference in a subspace. In Section 6.8, we develop beamformers that perform an eigendecomposition of $\mathbf{S_x}$. A subset of the eigenvalues and eigenvectors is used to construct the beamformer. Two types of eigenvector beamformers, eigenspace beamformers and dominant-mode rejection beamformers, are developed and analyzed. In the known spectrum case, they have the potential to provide improved performance in the presence of mismatch. In Chapter 7, we see that, in the real-world case, where we have to estimate $\mathbf{S_x}$ or an equivalent statistic, they offer additional advantages.

In Section 6.9, we consider the case in which we first form a reduced-dimension beamspace by processing the data with a small set of non-adaptive beams that span the sector of interest. We then process the beam outputs using an adaptive processor. We find that there are a number of advantages to this technique.

In Section 6.10, we consider the use of quadratic constraints to improve beamformer robustness. We introduced it in Section 6.6 to motivate diagonal loading. We revisit it to more fully exploit its potentials and to set the stage for variable diagonal loading in Chapter 7.

In Section 6.11, we develop an algorithm to minimize the squared error between a desired beam pattern and the actual beam pattern over a region of w-$\mathbf{k}$ space. We minimize the output power subject to a constraint on the squared error. This approach leads to a beamformer that is referred to as a soft-constraint beamformer. It is a generalization of diagonal loading that is useful in some applications.

In Section 6.12, we consider the case in which the desired signal and the interference are coherent ($|\ \rho\ | = 1$) or correlated ($|\ \rho\ | \neq 0$). We find that, as $|\ \rho\ |$ approaches one, there is significant performance degradation because the desired signal is cancelled. We introduce a technique called spatial smoothing to obtain beamformers that will operate in a coherent environment.

In Section 6.13, we consider broadband beamformers. The frequency-domain implementation is a straightforward extension of our narrowband results because of our use of the frequency-domain snapshot model. The time-domain implementation requires an extension of the narrowband time-domain model in Section 5.2.

In Section 6.14, we summarize our results and discuss some of the issues that remain to be explored.

The structure of Chapter 6 is shown in Table 6.1. Most of the material in the first part of this chapter appeared in "Chapter II-5: Multi-dimensional and Multi-variable Processes" [VT69], which was part of the course notes for my course in Advanced Topics in Detection and Estimation Theory taught at M.I.T. in the Spring term of 1969. Although these notes were unpublished, they had a significant underground distribution. Some of the material had been published earlier in [VT66b] and [VT66a] with the grandiose title, "A Unified Theory for Optimum Array Processing." Other references include [VT64], [VT65], and [VT68]. The material was intended to be the last chapter in DEMT III [VT71] and [VT01b], but was omitted because of space limitations. Some of it appeared subsequently in [Bag76]. Other early work in the area covered by this chapter includes Bryn [Bry62], Middleton and Groginsky [MG64], Schweppe [Sch68], Gaardner [Gaa66] [Gaa67], Burg [Bur64], Pritchard [Pri63], [Pri51], Vanderkulk [Van63], Childers and Reed [CR65], Becker and Cron [BC65], Faran and Hills [FH53], Kelly [Kel65], Price [Pri53], [Pri54], [Pri56], Wolf [Wol59], Capon et al. [CGK67], Edelbute et al. [EFK67], Mermoz [Mer64] and Middleton [Mid60]. The material starting in Section 6.6 is new and is based on a large number of journal articles, conference papers, and books that have appeared in this area over the last 30 years.

It is worthwhile re-emphasizing that throughout this chapter we assume that the signal and the interference statistics are known. In Chapter 7, we develop adaptive beamformers that estimate the appropriate statistics from the incoming data. We would expect that these adaptive beamformers will converge to one of the beamformers in this chapter if the processes are stationary and the observation time approaches infinity. Thus, we might consider the beamformers in this chapter to be the "steady-state" version of the corresponding adaptive beamformer.

Even if a particular beamformer developed in this chapter has good performance, it does not guarantee that its adaptive version will have good performance (or be computationally feasible). However, if the beamformer's performance assuming known statistics is poor, then it is unlikely that the adaptive version will be useful.

Table 6.1 Structure of Chapter 6.

Optimum Beamformers	**6.2 Single PW Signal** MVDR, MPDR MMSE	**6.5 Multiple PW Signals** MVDR, MPDR MMSE	
Interference Models	**6.3 Discrete** SPW, MPW Sensitivity Eigenbeams	**6.4 Spatially Spread** Acoustic noise models ARMA models	
Model Mismatch	**6.6 Mismatch** Steering vector Array perturbations Noise models		
Constraints	**6.7 Linear Constraints** Directional Derivative Eigenvector Quiescent pattern	**6.10 Quadratic Constraints** Fixed diagonal loading Variable DL	**6.11 Soft Constraints** Generalized diagonal loading
Reduced Dimension	**6.8 Eigenspace** Principal components Cross spectral	**6.9 Beamspace** BS-MPDR BS-LCMP	
Coherent Signals	**6.12 Coherent Signals** Signal nulling FB averaging Spatial smoothing		
Broadband Beamformers	**6.13 Broadband** Frequency domain Time domain		

Therefore, the purpose of this chapter is to develop a collection of beamformers that have good properties in a known statistics environment. These beamformers serve as candidates for adaptive implementations.

It is important to understand the notation we use in the chapter. We use the frequency-domain snapshot model developed in Section 5.2.1. We assume that the approximations developed in that section are valid. Therefore, the input to the array processor is a sequence of statistically independent circular complex Gaussian random vectors, $\mathbf{X}_{\Delta T}(\omega_m, k), m = -(M-1)/2 \leq m \leq (M-1)/2, k = 1, \cdots, K$. Sections 6.2–6.12 consider narrowband beamformers. Because $\mathbf{X}_{\Delta T}(\omega_{m_1}, k)$ and $\mathbf{X}_{\Delta T}(\omega_{m_2}, k)$ are statistically independent, the narrowband beamformers for each frequency bin are uncoupled. Therefore, it is convenient to suppress the m subscript of ω. The discussion assumes that the spatial spectral matrix is known and is the same for each snapshot. Therefore it is convenient to suppress the k-dependence. Chapter 7 focusses on the behavior as a function of k. The approximations in Section 5.2.1 replace the mean and covariance of $\mathbf{X}_{\Delta T}(\omega_m, k)$ with $\mathbf{m}_{\mathbf{x}}(\omega_m, k)$ and $\mathbf{S}_{\mathbf{x}}(\omega_m)$, respectively. Therefore it is convenient to suppress the ΔT subscript. Thus, to keep the notation as simple as possible, $\mathbf{X}(\omega)$ will be used in Sections 6.2–6.12 to represent $\mathbf{X}_{\Delta T}(\omega_m, k)$.

In many cases, the ω-dependence will also be suppressed and $\mathbf{X}$ will be used. Occasionally, one or more of the subscripts or indices will be used to make a specific point.

It is also useful to note that, since Sections 6.2–6.12 deal with narrowband beamformers, the entire development could have been done in the time domain. For the same spatial characteristics of the signals and noise, the resulting weight vectors in the beamformers are identical.

6.2 Optimum Beamformers

In this section, we derive optimum beamformers for plane-wave signals being received in the presence of a noise field and analyze their performance. We consider four models. The objective of the optimum beamformer is to estimate the signal waveform.

In Section 6.2.1, we assume that the noise is a sample function of a random process with known second-order statistics, but the signal is an unknown nonrandom plane-wave signal arriving from a known direction. We derive the optimum linear array processor to provide the minimum variance unbiased estimate of the signal waveform. We show that this criterion is the same as requiring the array processor to be a distortionless filter (i.e., any waveform coming from the specified signal direction will pass through the filter undistorted) and find the distortionless filter that minimizes the output variance. Finally, we show that if the noise process is Gaussian, then the output of the optimum distortionless filter is the maximum likelihood estimate of the signal waveform.

In Section 6.2.2, we assume that the signals and noises are sample functions of stationary random processes and that their second-order statistics are known. Initially, we consider a single plane-wave signal arriving at the array from a known direction. We derive the optimum linear array processor to estimate the signal waveform using a MMSE criterion. Finally, we observe that if the signal and noise processes are Gaussian, then the optimum linear array processor is optimum over the class of all processors.

In Section 6.2.3, we utilize the signal and noise model of Section 6.2.2. However, now we assume that the criterion is maximum output *SNR*. We show that the optimum processor consists of the same matrix operation as in the second model followed by a different scalar filter.

In Section 6.2.4, we consider a fourth model. This model is closely related to the first model. In this model we select a steering direction that we believe the signal is arriving from. However, the actual signal may be arriving from a different direction. In addition, we assume that we know (or can measure) the statistics of the total received waveform $\mathbf{x}(t)$, but do not know the statistics of the signal and noise components. We derive the distortionless filter for the specified steering direction that minimizes the mean square output power. We refer to the processor derived using this model as the minimum power distortionless response (MPDR) beamformer. When the signal direction and the steering direction coincide, the MPDR beamformer reduces to the MVDR beamformer. In other cases, there may be a significant difference in performance.

In Section 6.2.5, we extend the results to multiple plane-wave signals.

6.2.1 Minimum Variance Distortionless Response (MVDR) Beamformers

In this section, we consider the first model that was described at the beginning of the section. This is the case in which the signal is nonrandom but unknown. In the initial discussion, we consider the case of single plane-wave signal.

The frequency-domain snapshot consists of signal plus noise,

$$\mathbf{X}(\omega) = \mathbf{X}_s(\omega) + \mathbf{N}(\omega). \tag{6.1}$$

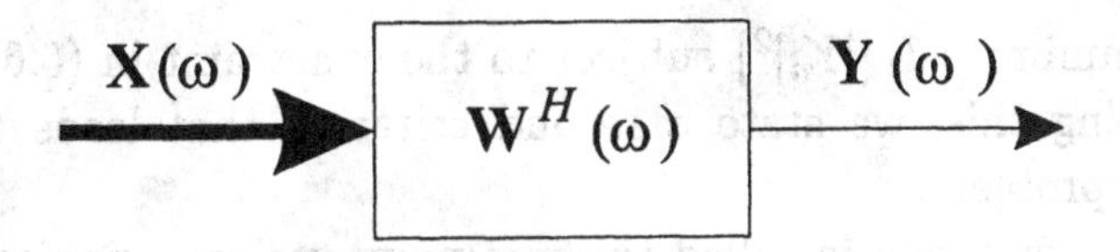

Figure 6.6 Matrix processor.

The signal vector can be written as

$$\mathbf{X}_s(\omega) = F(\omega)\mathbf{v}(\omega : \mathbf{k}_s), \tag{6.2}$$

where $F(\omega)$ is the frequency-domain snapshot of the source signal and $\mathbf{v}(\omega : \mathbf{k}_s)$ is the array manifold vector for a plane wave with wavenumber $\mathbf{k}_s$. The noise snapshot, $\mathbf{N}(\omega)$ is a zero-mean random vector with spectral matrix,

$$\mathbf{S}_\mathbf{n}(\omega) = \mathbf{S}_\mathbf{c}(\omega) + \sigma_w^2\mathbf{I}. \tag{6.3}$$

Later we will add the assumption that $\mathbf{N}(\omega)$ is a zero-mean circular complex Gaussian random vector.

6.2.1.1 Minimum variance distortionless response (MVDR) beamformer

We process $\mathbf{X}(\omega)$ with a matrix operation $\mathbf{W}^H(\omega)$ as shown in Figure 6.6. The dimension of $\mathbf{W}^H(w)$ is $1 \times N$.

The first criterion of interest is called the distortionless criterion. It is required that, in the absence of noise,

$$Y(\omega) = F(\omega), \tag{6.4}$$

for any $F(\omega)$. Under this constraint, we wish to minimize the variance of $Y(\omega)$ in the presence of noise. Thus, we write

$$Y(\omega) = F(\omega) + Y_n(\omega), \tag{6.5}$$

and minimize $E\left[|Y_n(\omega)|^2\right]$.

The constraint of no distortion implies

$$\mathbf{W}^H(\omega)\mathbf{v}(\omega : \mathbf{k}_s) = 1. \tag{6.6}$$

The mean square of the output noise is,

$$E\left[|Y_n|^2\right] = \mathbf{W}^H(\omega)\mathbf{S}_\mathbf{n}(\omega)\mathbf{W}(\omega). \tag{6.7}$$

We want to minimize $E\left[|Y_n|^2\right]$ subject to the constraint in (6.6).[1]

Before doing this, we state a second criterion that leads to the same minimization problem.

This second criterion is called the minimum variance unbiased estimate criterion.[2] Here we require that $Y(\omega)$ be the minimum variance unbiased estimate of $F(\omega)$. This implies

$$E[Y(\omega)] = F(\omega). \tag{6.8}$$

Thus, using (6.5) in (6.8),

$$\begin{aligned} E[Y(\omega)] &= E[F(\omega)] + E[N(\omega)] \\ &= E[F(\omega)], \end{aligned} \tag{6.9}$$

for any $F(\omega)$. This equality implies

$$\mathbf{W}^H(\omega)\mathbf{v}(\omega : \mathbf{k}_s) = 1, \tag{6.10}$$

which is identical to the constraint in (6.6).

Thus we have the same minimization problem as before. We now solve this problem by imposing the constraint in (6.6) by using a Lagrange multiplier. The function that we minimize is

$$\begin{aligned} F \triangleq\; & \mathbf{W}^H(\omega)\mathbf{S}_\mathbf{n}(\omega)\mathbf{W}(\omega) + \lambda(\omega)\left[\mathbf{W}^H(\omega)\mathbf{v}(\omega : \mathbf{k}_s) - 1\right] \\ & + \lambda^*(\omega)\left[\mathbf{v}^H(\omega : \mathbf{k}_s)\mathbf{W}(\omega) - 1\right]. \end{aligned} \tag{6.11}$$

Taking the complex gradient with respect to $\mathbf{W}(\omega)$ and solving gives

$$\mathbf{W}_o^H(\omega) = -\lambda(\omega)\mathbf{v}^H(\omega : \mathbf{k}_s)\mathbf{S}_\mathbf{n}^{-1}(\omega). \tag{6.12}$$

To evaluate $\lambda(\omega)$, we use the constraint in (6.6), which gives

$$\lambda(\omega) = -\left[\mathbf{v}^H(\omega : \mathbf{k}_s)\mathbf{S}_\mathbf{n}^{-1}(\omega)\mathbf{v}(\omega : \mathbf{k}_s)\right]^{-1}. \tag{6.13}$$

Thus,

$$\boxed{\mathbf{W}_o^H(\omega) = \Lambda(\omega : \mathbf{k}_s)\mathbf{v}^H(\omega : \mathbf{k}_s)\mathbf{S}_\mathbf{n}^{-1}(\omega),} \tag{6.14}$$

[1] The idea of combining multiple inputs in a statistically optimum manner under the constraint of no signal distortion is due to Darlington [Dar58]. An interesting discussion of the method is contained in Brown and Nilsson's text [BN62].

[2] This criterion was first considered for this application by Levin [Lev64].

where

$$\boxed{\Lambda(\omega : \mathbf{k}_s) \triangleq \left[\mathbf{v}^H(\omega : \mathbf{k}_s)\mathbf{S}_{\mathbf{n}}^{-1}(\omega)\mathbf{v}(\omega : \mathbf{k}_s)\right]^{-1}.} \tag{6.15}$$

The matrix processor in (6.14) and (6.15) is referred to as the MVDR beamformer and was first derived by Capon [Cap69]. It is sometimes referred to as the Capon beamformer.

For notational simplicity, it is convenient to suppress ω and $\mathbf{k}_s$ in these formulas.

$$\mathbf{v}_s \triangleq \mathbf{v}(\omega_c : \mathbf{k}_s), \tag{6.16}$$

$$\mathbf{S}_{\mathbf{n}}(\omega) \triangleq \mathbf{S}_{\mathbf{n}}, \tag{6.17}$$

$$\Lambda_s \triangleq \left[\mathbf{v}_s^H \mathbf{S}_{\mathbf{n}}^{-1} \mathbf{v}_s\right]^{-1}. \tag{6.18}$$

Then,

$$\mathbf{w}_o^H = \Lambda_s \mathbf{v}_s^H \mathbf{S}_{\mathbf{n}}^{-1} \tag{6.19}$$

or

$$\mathbf{w}_{mvdr}^H = \Lambda_s \mathbf{v}_s^H \mathbf{S}_{\mathbf{n}}^{-1}, \tag{6.20}$$

where the subscript emphasizes the optimization criterion.

6.2.1.2 Maximum likelihood estimators

We consider the same model as in Section 6.2.1.1 except the assumption that $\mathbf{N}(\omega)$ is a circular complex Gaussian random vector is added. The likelihood function at ω_m is

$$l_R(\omega_m) = \left[\mathbf{X}^H(\omega_m) - F^*(\omega_m)\mathbf{v}^H(\omega_m{:}\mathbf{k}_s)\right]\mathbf{S}_{\mathbf{n}}^{-1}(\omega_m)\left[\mathbf{X}(\omega_m) - F(\omega_m)\mathbf{v}(\omega_m{:}\mathbf{k}_s)\right], \tag{6.21}$$

where constant multipliers and additive terms that do not depend on $F(\omega_m)$ have been dropped.

Taking the complex gradient with respect to $F^*(\omega_m)$, setting the result equal to zero and solving the resulting equation gives

$$\hat{\mathbf{F}}(\omega_m)|_{ml} = \frac{\mathbf{v}^H(\omega_m : \mathbf{k}_s)\mathbf{S}_{\mathbf{n}}^{-1}(\omega_m)\mathbf{X}(\omega_m)}{\mathbf{v}^H(\omega_m : \mathbf{k}_s)\mathbf{S}_{\mathbf{n}}^{-1}(\omega_m)\mathbf{v}^H(\omega_m : \mathbf{k}_s)}. \tag{6.22}$$

The estimate in (6.22) is identical to the output of the MVDR processor in (6.14) with $\omega = \omega_m$.

The reason for introducing ω_m in (6.21) is that, in order to find the ML estimate of $f(t)$, the joint ML estimate of $F(\omega_m), m = -(M-1)/2 \leq m \leq$

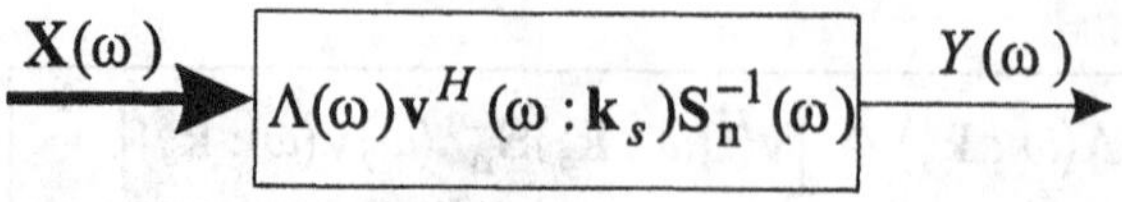

$$\mathbf{X}(\omega) \rightarrow \boxed{\Lambda(\omega)\mathbf{v}^H(\omega:\mathbf{k}_s)\mathbf{S}_{\mathbf{n}}^{-1}(\omega)} \rightarrow Y(\omega)$$

Figure 6.7: Minimum variance distortionless response (MVDR) processor.

$(M-1)/2$ is required. Then $f(t)|_{ml}$ is reconstructed using (5.36). However, because the frequency-domain snapshots for $\omega_{m_1} \neq \omega_{m_2}$ are statistically independent, the joint likelihood function is a sum of the likelihood functions in (6.21) over m. The resulting $F(\omega_m)|_{ml}$ are independent.

The output of the optimum distortionless (or MVDR) matrix is the ML estimate of $F(\omega_m)$ and can be used to generate $f(t)|_{ml}$.[3] The important point is that the ML derivation did not assume a linear operation (matrix multiplication) on $\mathbf{X}(\omega_m)$. As in the scalar case, the Gaussian model leads to a linear processor.

This ML result has occasionally caused confusion in the literature. The MVDR filter provides the ML estimate of the signal $f(t)$ when the signal wavenumber (direction of arrival) is known and $\mathbf{S}_{\mathbf{n}}(\omega)$ is known. In Chapters 8 and 9, we will estimate the DOA of the signal. In most cases, the MVDR filter scanned in wavenumber space does not provide the ML estimate of the signal's DOA.

6.2.1.3 Array gain

We denote the matrix filter in Figure 6.7 by $\mathbf{W}_o^H(\omega)$ where subscript "o" denotes optimum:

$$\mathbf{W}_{mvdr}^H \triangleq \mathbf{W}_o^H(\omega) = \Lambda(\omega)\mathbf{v}^H(\omega:\mathbf{k}_s)\mathbf{S}_{\mathbf{n}}^{-1}(\omega). \tag{6.23}$$

The array gain at a particular frequency ω is given by the ratio of the signal spectrum to the noise spectrum at the output of the distortionless filter compared to the ratio at the input. The spectrum of the noise component of $Y(\omega)$ is

$$\begin{aligned} S_{y_n}(\omega) &= \mathbf{W}_o^H(\omega)\mathbf{S}_{\mathbf{n}}(\omega)\mathbf{W}_o(\omega) \\ &= \mathbf{v}^H(\omega:\mathbf{k}_s)\mathbf{S}_{\mathbf{n}}^{-1}(\omega)\mathbf{S}_{\mathbf{n}}(\omega)\mathbf{S}_{\mathbf{n}}^{-1}(\omega)\mathbf{v}(\omega:\mathbf{k}_s)\cdot\Lambda^2(\omega) \\ &= \Lambda(\omega). \end{aligned} \tag{6.24}$$

Since the $\mathbf{W}_o^H(\omega)$ is distortionless,

$$S_{y_s}(\omega) = S_f(\omega). \tag{6.25}$$

[3] This result was first obtained by Kelly and Levin [KL64].

If we assume that the noise spectrum at each sensor is the same, then the input *SNR* at each sensor is $\frac{S_f(\omega)}{S_n(\omega)}$. Therefore,

$$\begin{aligned} A_o(\omega : \mathbf{k}_s) &= \frac{S_f(\omega)}{\Lambda(\omega)} \Big/ \frac{S_f(\omega)}{S_n(\omega)} = \frac{S_n(\omega)}{\Lambda(\omega)} \\ &= S_n(\omega)\mathbf{v}^H(\omega : \mathbf{k}_s)\mathbf{S}_\mathbf{n}^{-1}(\omega)\mathbf{v}(\omega : \mathbf{k}_s). \end{aligned} \tag{6.26}$$

We define a normalized spectral matrix $\boldsymbol{\rho}_\mathbf{n}(\omega)$ by the relation,

$$\mathbf{S}_\mathbf{n}(\omega) = S_n(\omega)\boldsymbol{\rho}_\mathbf{n}(\omega). \tag{6.27}$$

Then,

$$\boxed{A_o(\omega : \mathbf{k}_s) = \mathbf{v}^H(\omega : \mathbf{k}_s)\boldsymbol{\rho}_\mathbf{n}^{-1}(\omega)\mathbf{v}(\omega : \mathbf{k}_s).} \tag{6.28}$$

Suppressing ω and $\mathbf{k}_s$, (6.28) can be written as

$$A_o = \mathbf{v}_s^H \boldsymbol{\rho}_n^{-1} \mathbf{v}_s. \tag{6.29}$$

Recall from Chapter 2 that the conventional delay-and-sum beamformer is

$$\mathbf{W}_c^H(\omega) = \frac{1}{N}\mathbf{v}^H(\omega : \mathbf{k}_s). \tag{6.30}$$

The output noise is

$$S_{y_{nc}}(\omega) = \frac{1}{N^2}\mathbf{v}^H(\omega : \mathbf{k}_s)\mathbf{S}_\mathbf{n}(\omega)\mathbf{v}(\omega : \mathbf{k}_s), \tag{6.31}$$

and the resulting array gain is

$$\boxed{A_c(\omega : \mathbf{k}_s) = \frac{N^2}{\mathbf{v}^H(\omega{:}\mathbf{k}_s)\boldsymbol{\rho}_\mathbf{n}(\omega)\mathbf{v}(\omega{:}\mathbf{k}_s)},} \tag{6.32}$$

or, suppressing ω and $\mathbf{k}_s$,

$$A_c = \frac{N^2}{\mathbf{v}_s^H \boldsymbol{\rho}_n \mathbf{v}_s}. \tag{6.33}$$

Whenever the $\boldsymbol{\rho}(\omega)$ matrix is diagonal, (i.e., the noises at the different sensors are uncorrelated) the conventional and optimum receivers are identical and the two array gains are identical:

$$A_o(\omega : \mathbf{k}_s) = A_c(\omega : \mathbf{k}_s) = N. \tag{6.34}$$

In other cases,

$$A_o(\omega : \mathbf{k}_s) \geq A_c(\omega : \mathbf{k}_s). \tag{6.35}$$

6.2.2 Minimum Mean-Square Error (MMSE) Estimators

In this section, we find the optimum linear matrix processor whose output is the MMSE estimate of a desired signal snapshot $D(\omega)$.

6.2.2.1 Single plane-wave signal

We first consider the problem of a single plane-wave signal in noise. Thus,

$$\mathbf{X}(\omega) = F(\omega)\mathbf{v}(\omega : \mathbf{k}_s) + \mathbf{N}(\omega), \tag{6.36}$$

where the source signal snapshot, $F(\omega)$, is a scalar zero-mean random variable with variance $S_f(\omega)$. The signal and noise snapshots are uncorrelated. We assume that $\mathbf{N}(\omega)$ contains a spatial white noise component so that

$$\mathbf{S_n}(\omega) = \mathbf{S_c}(\omega) + \sigma_w^2\mathbf{I}. \tag{6.37}$$

The spectral matrix of $\mathbf{X}(\omega)$ is

$$\mathbf{S_x}(\omega) = S_f(\omega)\mathbf{v}(\omega : \mathbf{k}_s)\mathbf{v}^H(\omega : \mathbf{k}_s) + \mathbf{S_n}(\omega). \tag{6.38}$$

For convenience we have assumed the noise snapshot has the same statistics at each sensor. The case of unequal values follows directly.

The desired snapshot $D(\omega)$ is equal to the source snapshot, $F(\omega)$. The matrix processor is an $N \times 1$ matrix denoted by $\mathbf{H}(\omega)$ whose output is $\hat{D}(\omega)$. The mean-square error is

$$\begin{aligned} \xi &= E\left\{|D(\omega) - \mathbf{H}(\omega)\mathbf{X}(\omega)|^2\right\} \\ &= E\left\{(D(\omega) - \mathbf{H}(\omega)\mathbf{X}(\omega))\left(D^*(\omega) - \mathbf{X}^H(\omega)\mathbf{H}^H(\omega)\right)\right\}. \end{aligned} \tag{6.39}$$

Taking the complex gradient with respect to $\mathbf{H}^H(\omega)$ and setting the result equal to zero gives,

$$E\left[D(\omega)\mathbf{X}^H(\omega)\right] - \mathbf{H}(\omega)E\left[\mathbf{X}(\omega)\mathbf{X}^H(\omega)\right] = \mathbf{0}, \tag{6.40}$$

or

$$\mathbf{S}_{d\mathbf{x}^H}(\omega) = \mathbf{H}_o(\omega)\mathbf{S_x}(\omega). \tag{6.41}$$

Thus,

$$\boxed{\mathbf{H}_o(\omega) = \mathbf{S}_{d\mathbf{x}^H}(\omega)\mathbf{S}_\mathbf{x}^{-1}(\omega).} \tag{6.42}$$

From (6.36) and the uncorrelated signal and noise assumption we obtain

$$S_{d\mathbf{x}^H}(\omega) = S_f(\omega)\mathbf{v}^H(\omega : \mathbf{k}_s). \tag{6.43}$$

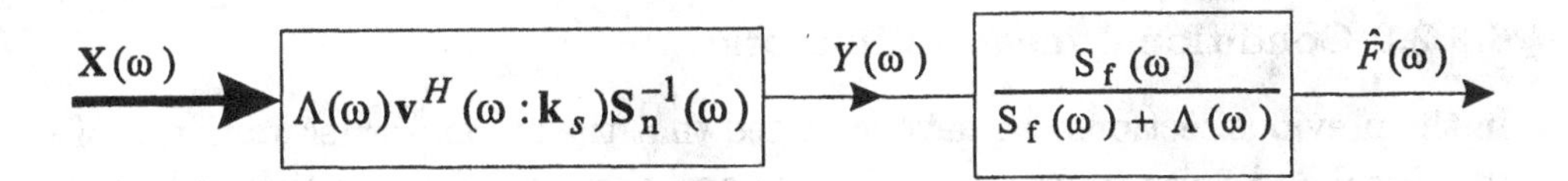

Figure 6.8: MMSE processor.

Then,

$$\mathbf{H}_o(\omega) = S_f(\omega)\mathbf{v}^H(\omega : \mathbf{k}_s)\mathbf{S}_{\mathbf{x}}^{-1}(\omega). \tag{6.44}$$

Inverting (6.38) using the matrix inversion formula, we obtain

$$\mathbf{S}_{\mathbf{x}}^{-1}(\omega) = \mathbf{S}_{\mathbf{n}}^{-1} - \mathbf{S}_{\mathbf{n}}^{-1}S_f\mathbf{v}_s(1 + \mathbf{v}_s^H\mathbf{S}_{\mathbf{n}}^{-1}S_f\mathbf{v}_s)^{-1}\mathbf{v}_s^H\mathbf{S}_{\mathbf{n}}^{-1}, \tag{6.45}$$

where we have suppressed the ω and $\mathbf{k}_s$ arguments on the right side of the equation. Defining

$$\Lambda^{-1}(\omega) = \mathbf{v}^H(\omega : \mathbf{k}_s)\mathbf{S}_{\mathbf{n}}^{-1}(\omega)\mathbf{v}(\omega : \mathbf{k}_s), \tag{6.46}$$

and substituting (6.45) into (6.44), we obtain,

$$\mathbf{H}_o(\omega) = \frac{S_f(\omega)}{S_f(\omega) + \Lambda(\omega)} \cdot \Lambda(\omega)\mathbf{v}^H(\omega : \mathbf{k}_s)\mathbf{S}_{\mathbf{n}}^{-1}(\omega). \tag{6.47}$$

The MMSE processor is shown in Figure 6.8. We see that it consists of the optimum distortionless matrix followed by a scalar multiplier.

Substituting (6.47) into (6.39) gives the optimum mean-square error:

$$\xi_o(\omega) = \frac{S_f(\omega)\Lambda(\omega)}{S_f(\omega) + \Lambda(\omega)}, \tag{6.48}$$

which can also be written as

$$\xi_o(\omega) = \frac{S_f(\omega)\left[\frac{S_n(\omega)}{A_o(\omega)}\right]}{S_f(\omega) + \left[\frac{S_n(\omega)}{A_o(\omega)}\right]}. \tag{6.49}$$

The effect of the array is completely characterized by the optimum array gain $A_o(\omega : \mathbf{k}_s)$.

We look at optimum array gains for various signal and noise environments in Sections 6.3–6.5.

6.2.2.2 Conditional mean estimators

In the previous sections, we have assumed that the second-order statistics of the signal and noise processes are known and then found the optimum linear MMSE matrix processor. From our results with scalar waveforms in DEMT I ([VT68], [VT01a], e.g., pp. 476–479) we would expect that if we assume that the signal and noise processes are vector Gaussian random processes, then the optimum processor will be a linear processor. The proof of this conjecture follows in a straightforward manner.

The frequency-domain snapshot is

$$\mathbf{X}(\omega_m) = F(\omega_m)\mathbf{v}(\omega_m : \mathbf{k}_s) + \mathbf{N}(\omega_m). \tag{6.50}$$

The MMSE estimator is the mean of the *a posteriori* probability density of $F(\omega_m)$, given $\mathbf{X}(\omega_m)$. This mean is referred to as the conditional mean.

The probability density of $\mathbf{X}(\omega_m)$, given $F(\omega_m)$, is

$$p_{\mathbf{X}(\omega_m)|F(\omega_m)}(\cdot) = c_1 \exp\left\{-\left[\mathbf{X}^H(\omega_m) - F^*(\omega_m)\mathbf{v}^H(\omega_m : \mathbf{k}_s)\right]\right.$$
$$\left.\mathbf{S}_{\mathbf{n}}^{-1}(\omega_m)\left[\mathbf{X}(\omega_m) - F(\omega_m)\mathbf{v}(\omega_m : \mathbf{k}_s)\right]\right\}. \tag{6.51}$$

The probability density of $F(\omega_m)$ is

$$p_{F(\omega_m)}(\cdot) = c_2 \exp\left\{-F^*(\omega_m)S_f^{-1}(\omega_m)F(\omega_m)\right\}. \tag{6.52}$$

Using Bayes rule, the *a posteriori* density at ω_m is,

$$p_{F(\omega_m)|\mathbf{X}(\omega_m)}(\cdot) = c_3 \exp\left[-F^*(\omega_m)S_f^{-1}(\omega_m)F(\omega_m)\right]$$
$$\cdot \exp\left[\ \mathbf{X}^H(\omega_m)\mathbf{S}_{\mathbf{n}}^{-1}(\omega_m)\mathbf{X}(\omega_m) - F^*(\omega_m)\mathbf{v}^H(\omega_m : \mathbf{k}_s)\mathbf{S}_{\mathbf{n}}^{-1}(\omega_m)\mathbf{X}(\omega_m)\right.$$
$$-\mathbf{X}^H(\omega_m)\mathbf{S}_{\mathbf{n}}^{-1}(\omega_m)F(\omega_m)\mathbf{v}(\omega_m : \mathbf{k}_s) + F^*(\omega_m)\mathbf{v}^H(\omega_m : \mathbf{k}_s)\mathbf{S}_{\mathbf{n}}^{-1}(\omega_m)$$
$$\left.\mathbf{v}(\omega_m : \mathbf{k}_s)F(\omega_m)\right]. \tag{6.53}$$

Defining

$$H_S^{-1}(\omega_m) = \frac{1}{\Lambda(\omega_m)} + \frac{1}{S_f(\omega_m)}, \tag{6.54}$$

where $\Lambda(\omega_m)$ was defined in (6.15), (6.53) can be written as

$$p_{F(\omega_m)|\mathbf{X}(\omega_m)}(\cdot) = c_4 \exp\left\{\left[F^*(\omega_m) - \mathbf{X}^H(\omega_m)\mathbf{S}_{\mathbf{n}}^{-1}(\omega_m)\mathbf{v}(\omega_m : \mathbf{k}_s)H_S^*(\omega_m)\right]\right.$$
$$H_S^{-1}(\omega_m)\left[F(\omega_m) - H_S(\omega_m)\mathbf{v}^H(\omega_m : \mathbf{k}_s)\mathbf{S}_{\mathbf{n}}^{-1}(\omega_m)\right.$$
$$\left.\left.\mathbf{X}(\omega_m)\right]\right\}. \tag{6.55}$$

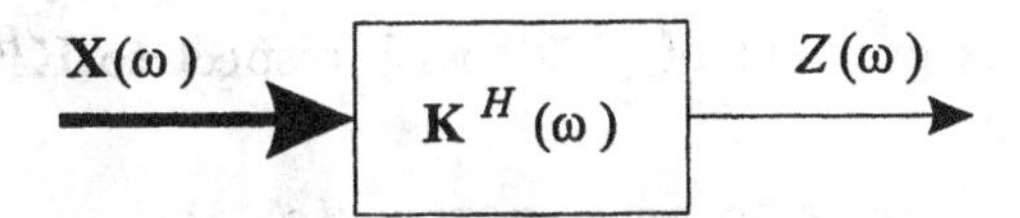

Figure 6.9: Maximum *SNR* processor.

Thus, the conditional mean is,

$$\begin{aligned}\hat{F}(\omega_m) &= H_S(\omega_m)\mathbf{v}^H(\omega_m : \mathbf{k}_s)\mathbf{S}_{\mathbf{n}}^{-1}(\omega_m)\mathbf{X}(\omega_m) \\ &= \frac{S_f(\omega_m)}{S_f(\omega_m) + \Lambda(\omega_m)} \cdot \Lambda(\omega_m)\mathbf{v}^H(\omega_m : \mathbf{k}_s)\mathbf{S}_{\mathbf{n}}^{-1}(\omega_m)\mathbf{X}(\omega_m),\end{aligned} \tag{6.56}$$

which corresponds to the operation in (6.47) with $\omega = \omega_m$. The resulting mean-square error at ω_m is

$$\xi(\omega_m) = H_S(\omega_m) = \frac{S_f(\omega_m)\Lambda(\omega_m)}{S_f(\omega_m) + \Lambda(\omega_m)}. \tag{6.57}$$

Therefore, whenever the Gaussian model is utilized, the optimum MMSE processor will be the linear processor derived in Section 6.2.1.1.

6.2.3 Maximum Signal-to-Noise Ratio (*SNR*)

In the two preceding sections we found the optimum receiver for different criteria. For the single plane-wave signal case, both criteria and signal models led to the same matrix operation. The effect of the array was contained in the array gain. A logical question is, "if we maximized the array gain, would the optimum receiver be different?" It is straightforward to demonstrate that the resulting matrix processor is identical to the processor in the previous sections. Note that, since the input *SNR* is fixed, maximizing the output *SNR* maximizes the array gain.

The system of interest is shown in Figure 6.9. The *SNR* at frequency ω is,

$$\frac{S}{N} \triangleq \frac{\mathbf{K}^H(\omega)\mathbf{S}_{\mathbf{x}_s}(\omega)\mathbf{K}(\omega)}{\mathbf{K}^H(\omega)\mathbf{S}_{\mathbf{n}}(\omega)\mathbf{K}(\omega)}, \tag{6.58}$$

which is a ratio of two quadratic forms.

For notational ease and in anticipation of the final answer, we define[4]

$$\beta \triangleq S/N. \tag{6.59}$$

[4]This derivation is due to N. Owsley (private communication).

Taking the complex gradient of (6.58) with respect to $\mathbf{K}^H$ and setting the result to zero gives

$$\frac{S_{\mathbf{x}_s}\mathbf{K}(\mathbf{K}^H\mathbf{S_n}\mathbf{K}) - \mathbf{S_n}\mathbf{K}(\mathbf{K}^H\mathbf{S}_{\mathbf{x}_s}\mathbf{K})}{(\mathbf{K}^H\mathbf{S_n}\mathbf{K})^2} = 0, \tag{6.60}$$

where we have suppressed the ω dependence. This implies

$$\beta\,\mathbf{S_n}\,\mathbf{K} = \mathbf{S}_{\mathbf{x}_s}\mathbf{K}, \tag{6.61}$$

or

$$\beta\,\mathbf{K} = \left[\mathbf{S_n}^{-1}\mathbf{S}_{\mathbf{x}_s}\right]\mathbf{K}. \tag{6.62}$$

This result is familiar as the eigenvalue problem. Setting $\mathbf{K}$ equal to any eigenvector of $[\mathbf{S_n}^{-1}\mathbf{S}_{\mathbf{x}_s}]$ will satisfy (6.62). However, we want to maximize $\beta = S/N$, so we choose the eigenvector corresponding to the largest eigenvalue. This result is valid for an arbitrary spectral matrix $\mathbf{S}_{\mathbf{x}_s}(\omega)$.

For the special case of a plane-wave signal,

$$\mathbf{S}_{\mathbf{x}_s}(\omega) = S_f(\omega)\mathbf{v}(\omega : \mathbf{k}_s)\mathbf{v}^H(\omega : \mathbf{k}_s). \tag{6.63}$$

Substituting (6.63) into (6.62) gives

$$\beta\,\mathbf{K} = \mathbf{S_n}^{-1}\,S_f\,\mathbf{v}_s\,\mathbf{v}_s^H\,\mathbf{K}, \tag{6.64}$$

where the ω dependence is suppressed. The solution to (6.64) is

$$\mathbf{K} = \mathbf{S_n}^{-1}\mathbf{v}_s, \tag{6.65}$$

and

$$\beta = \left(\frac{S}{N}\right) = S_f(\mathbf{v}_s^H\mathbf{S_n}^{-1}\mathbf{v}_s). \tag{6.66}$$

The matrix operation for maximizing the *SNR* is identical to the MVDR beamformer and the MMSE beamformer. (The constant $\Lambda(\omega : \mathbf{k}_s)$ in (6.24) does not appear but can be included.) For homogeneous noise, (6.66) can be written as

$$\begin{aligned}\frac{S}{N} &= \frac{S_f(\omega)}{S_n(\omega)}\mathbf{v}^H(\omega : \mathbf{k}_s)\boldsymbol{\rho}^{-1}(\omega)\mathbf{v}(\omega : \mathbf{k}_s) \\ &= \frac{S_f(\omega)}{S_n(\omega)}A_o(\omega : \mathbf{k}_s).\end{aligned} \tag{6.67}$$

We see that, for a wide class of criteria, the optimum processor is an MVDR beamformer followed by a scalar filter that depends on the criterion. In [VT66a], we termed this a "criterion-invariant receiver." The important implication is that the MVDR beamformer creates a signal subspace (in this case, 1-D) where all subsequent processing takes place.

6.2.4 Minimum Power Distortionless Response (MPDR) Beamformers

The final beamformer that we develop is referred to as the minimum power distortionless response (MPDR) beamformer. It is closely related to the MVDR beamformer of Section 6.2.1.

The model has two new features. The first is that we will match the distortionless filter $\mathbf{w}^H$ to a plane wave arriving from the direction $\mathbf{a}_m$. Defining

$$\mathbf{v}_m \triangleq \mathbf{v}(\omega : \mathbf{a}_m), \tag{6.68}$$

we require

$$\mathbf{w}^H \mathbf{v}_m = \mathbf{W}^H(\omega)\mathbf{v}(\omega : \mathbf{a}_m) = 1. \tag{6.69}$$

We refer to $\mathbf{v}_m$ as the **steering vector**. Ideally, we would like the steering vector to correspond exactly to the signal vector,

$$\mathbf{v}_m = \mathbf{v}_s, \tag{6.70}$$

but in many cases this may not be true.

The second feature of the model is that we assume that $\mathbf{S}_\mathbf{x}(\omega)$, the spectral matrix of the entire input, is available to design the beamformer. By contrast, our previous models assumed that $\mathbf{S}_\mathbf{n}(\omega)$ and $\mathbf{v}(\omega : \mathbf{k}_s)$ were available (the MMSE model also assumed knowledge of $\mathbf{S}_\mathbf{f}(\omega)$). We want to minimize the total output power subject to the constraint in (6.69). The derivation is identical to that in (6.11)–(6.14). The result is

$$\boxed{\mathbf{w}_{mpdr}^H = \frac{\mathbf{v}_m^H \mathbf{S}_\mathbf{x}^{-1}}{\mathbf{v}_m^H \mathbf{S}_\mathbf{x}^{-1} \mathbf{v}_m}.} \tag{6.71}$$

In the majority of the literature, this beamformer is also referred to as an MVDR beamformer. We use a different name to emphasize that we are using $\mathbf{S}_\mathbf{x}^{-1}$ instead of $\mathbf{S}_\mathbf{n}^{-1}$.

It is straightforward to verify that, if $\mathbf{v}_m = \mathbf{v}_s$, then

$$\mathbf{w}_{mpdr}^H = \mathbf{w}_{mvdr}^H \tag{6.72}$$

(see Problem 6.2.2).

When $\mathbf{v}_s \neq \mathbf{v}_m$, the beamformers are different. In this chapter, $\mathbf{S}_\mathbf{x}$ is assumed to be known. However, in actual applications, $\mathbf{S}_\mathbf{x}$ is estimated. The estimate $\hat{\mathbf{S}}_\mathbf{x}$ will be constructed with the actual $\mathbf{v}_s$, not the model $\mathbf{v}_m$. This is the source of the problem with mismatched signals. In Sections 6.6 and 6.7 we will study the effects of steering vector mismatch on the performance of MVDR and MPDR beamformers. This steering vector mismatch can be caused by signal direction mismatch, array perturbations, or frequency mismatch. We will find that the presence of the signal in $\mathbf{S_x}(\omega)$ can cause significant degradation in the receiver performance when $\mathbf{v}_s \neq \mathbf{v}_m$. We will devote a significant amount of effort to the design of beamformers that are robust to mismatch. The array gain for the MPDR beamformer is

$$\boxed{A_{mpdr} = \frac{\left|\mathbf{v}_m^H \mathbf{S_x}^{-1} \mathbf{v}_s\right|^2}{\mathbf{v}_m^H \mathbf{S_x}^{-1} \boldsymbol{\rho}_\mathbf{n} \mathbf{S_x}^{-1} \mathbf{v}_m}.} \tag{6.73}$$

6.2.5 Summary

In this section, we have considered a single plane-wave signal in the presence of interference and noise. We have shown that, for several important criteria, the optimum beamformer is the MVDR (or Capon) beamformer specified by (6.14) and (6.15). When ω and $\mathbf{k}_s$ are suppressed, it can be written as

$$\boxed{\mathbf{w}_{mvdr}^H = \frac{\mathbf{v}_s^H \mathbf{S_n}^{-1}}{\mathbf{v}_s^H \mathbf{S_n}^{-1} \mathbf{v}_s}.} \tag{6.74}$$

This beamformer is widely used in applications where it is possible to measure or estimate $\mathbf{S_n}(w)$.

In applications where the signal is always present we use the MPDR beamformer specified by (6.71).

In Section 6.3, we examine the performance of the MVDR and MPDR beamformers for discrete interference. In Section 6.4, we examine their performance for spatially spread interference.

6.3 Discrete Interference

In this section, we consider an important special case in which the noise consists of a set of D interfering plane-wave signals plus uncorrelated noise. We assume that $D+1$ is less than N, the number of array elements.

The key result is that in this model the optimum beamformer generates $D+1$ sufficient statistics and then combines them. Thus the problem is reduced to a $(D+1)$-dimensional "signal-plus-interference" subspace rather than the original N-dimensional element space. This leads to significant simplification in the processor.

In Section 6.3.1, we consider a single interfering signal to illustrate the basic ideas. In Section 6.3.2, we consider D interfering signals and develop a "directional interference beamformer." In Section 6.3.3, we summarize our results.

6.3.1 Single Plane-wave Interfering Signal

In this case, we consider a single plane-wave desired signal with array manifold $\mathbf{v}(\omega : \mathbf{k}_s)$ and a noise process consisting of a single plane-wave interfering signal with array manifold $\mathbf{v}(\omega : \mathbf{k}_1)$ plus a white noise component at each sensor with spectral height σ_w^2. We will find the optimum distortionless filter, the array gain of the optimum and conventional beamformer, and the beam pattern of the optimum beamformer. The single plane-wave interferer case illustrates many results that can be generalized to multiple interfering signals.

For a single directional noise,

$$\mathbf{S_n}(\omega) = \sigma_w^2 \mathbf{I} + M_1(\omega)\mathbf{v}(\omega : \mathbf{k}_1)\mathbf{v}^H(\omega : \mathbf{k}_1), \tag{6.75}$$

where $M_1(\omega)$ is the spectrum of the interfering signal. Suppressing ω and $\mathbf{k}$, (6.75) becomes

$$\mathbf{S_n} = \sigma_w^2 \mathbf{I} + M_1 \mathbf{v}_1 \mathbf{v}_1^H. \tag{6.76}$$

Using the matrix inversion lemma,

$$\mathbf{S_n}^{-1} = \frac{1}{\sigma_w^2}\left[\mathbf{I} - \frac{M_1}{\sigma_w^2 + NM_1}\mathbf{v}_1\mathbf{v}_1^H\right]. \tag{6.77}$$

The noise spectrum at each sensor is

$$S_n = (\sigma_w^2 + M_1). \tag{6.78}$$

Using (6.77) in (6.23) gives

$$\mathbf{w}_{mvdr}^H = \mathbf{w}_o^H = \frac{\Lambda}{\sigma_w^2}\mathbf{v}_s^H\left[\mathbf{I} - \frac{M_1}{\sigma_w^2 + NM_1}\mathbf{v}_1\mathbf{v}_1^H\right]. \tag{6.79}$$

Now define

$$\rho_{s1} \triangleq \frac{\mathbf{v}_s^H \mathbf{v}_1}{N}, \tag{6.80}$$

which is the **spatial correlation coefficient** between the desired signal and the interference. Note that

$$\rho_{s1} = B_c(\mathbf{k}_1 : \mathbf{k}_s), \tag{6.81}$$

is the conventional beam pattern aimed as $\mathbf{k}_s$ (the signal wavenumber) and evaluated at $\mathbf{k}_1$ (the interferer wavenumber).

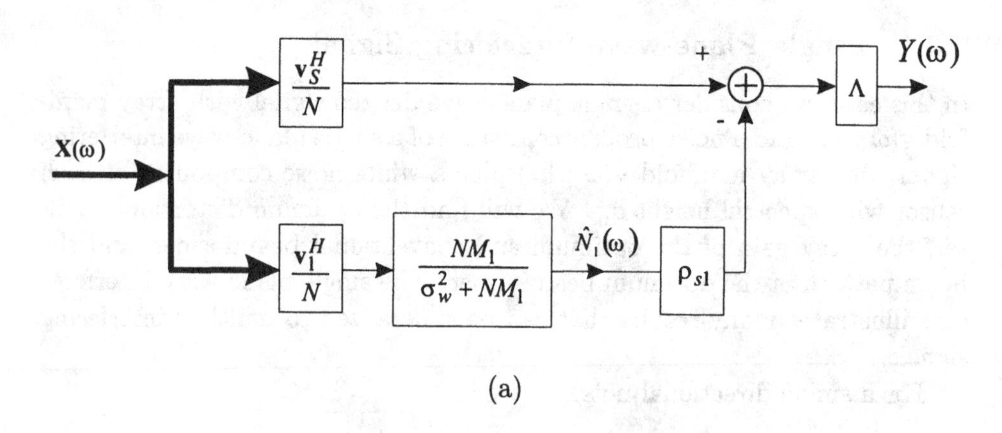

(a)

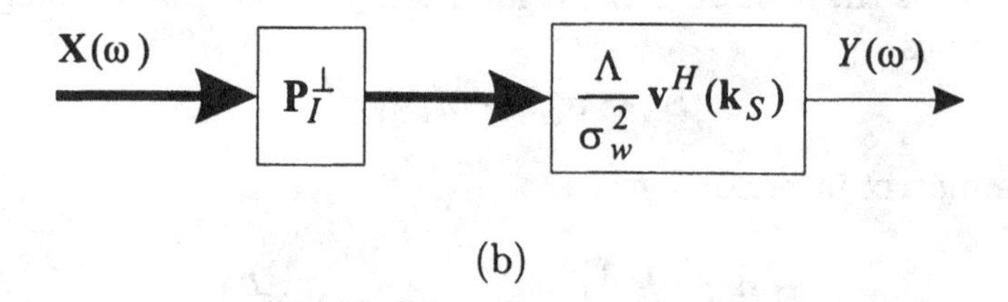

(b)

Figure 6.10 Optimum MVDR processor: single plane-wave interferer; (a) general case; (b) high *INR*.

Then (6.79) reduces to

$$\mathbf{w}_o^H = \frac{\Lambda N}{\sigma_w^2}\left[\frac{\mathbf{v}_s^H}{N} - \rho_{s1}\frac{NM_1}{\sigma_w^2 + NM_1}\cdot\frac{\mathbf{v}_1^H}{N}\right]. \tag{6.82}$$

The normalizing coefficient Λ is given by

$$\begin{aligned}\Lambda &= \left\{\frac{1}{\sigma_w^2}\mathbf{v}_s^H\left[\mathbf{I} - \frac{M_1}{\sigma_w^2 + NM_1}\mathbf{v}_1\mathbf{v}_1^H\right]\mathbf{v}_s\right\}^{-1} \\ &= \left\{\frac{1}{\sigma_w^2}N\left[1 - \frac{NM_1}{\sigma_w^2 + NM_1}|\rho_{s1}|^2\right]\right\}^{-1}. \end{aligned} \tag{6.83}$$

The resulting beamformer is shown in Figure 6.10(a). We see that the beamformer is generating two "spatial" sufficient statistics, one corresponding to the desired signal and one to the interfering signal, and then combining them optimally. Note that each of the spatial processors is precisely a conventional beamformer pointed at the signal and interferer, respectively.

The bottom path has a logical interpretation. It is easy to verify that $\hat{n}_1(t)$ is the minimum mean-square estimate of $n(t)$ (use (6.47) with $M_1 = S_f$ and $\mathbf{S_n} = \sigma_w^2\mathbf{I}$) in the absence of the desired signal. We then subtract a fraction of the estimate corresponding to the spatial correlation of the two signals.

We observe that, if $NM_1 \gg \sigma_w^2$, then (6.79) can be written as

$$\begin{aligned} \mathbf{w}_o^H &= \frac{\Lambda}{\sigma_w^2}\mathbf{v}_s^H\left[\mathbf{I} - \mathbf{v}_1\left[\mathbf{v}_1^H\mathbf{v}_1\right]^{-1}\mathbf{v}_1^H\right] \\ &= \frac{\Lambda}{\sigma_w^2}\mathbf{v}_s^H\mathbf{P}_I^\perp, \end{aligned} \tag{6.84}$$

where $\mathbf{P}_I^\perp$ is the projection matrix onto the subspace orthogonal to the interference. The resulting processor is shown in Figure 6.10(b). The beamformer is placing a perfect null on the interfering signal.

The optimum array gain is obtained by substituting (6.78) and (6.83) into (6.26) to obtain

$$A_o = N\left(1 + \frac{M_1}{\sigma_w^2}\right)\left(1 - \frac{NM_1}{\sigma_w^2 + NM_1}|\rho_{s1}|^2\right), \tag{6.85}$$

$$A_o = N(1+\sigma_I^2)\left[\frac{1 + N\sigma_I^2[1 - |\rho_{s1}|^2]}{1 + N\sigma_I^2}\right], \tag{6.86}$$

where

$$\sigma_I^2 \triangleq \frac{M_1}{\sigma_w^2} \triangleq INR. \tag{6.87}$$

Notice that optimum array gain depends on both σ_I^2, which is an interference-to-white noise ratio (*INR*), and ρ_{s1}, which is the spatial correlation between the signal and the interference.

The limiting cases of (6.85) are of interest. For large $N\sigma_I^2$,

$$A_o(\omega : \mathbf{k}_s) \simeq N(1+\sigma_I^2)[1 - |\rho_{s1}|^2], \tag{6.88}$$

so, for large *INR*s, the array gain can be quite large if $|\rho_{s1}| \neq 1$.

For $|\rho_{s1}| = 1$, (i.e., collinear signal and interference),

$$A_o(\omega : \mathbf{k}_s) = N\frac{1+\sigma_I^2}{1+N\sigma_I^2}, \tag{6.89}$$

which approaches 1 for large σ_I^2.

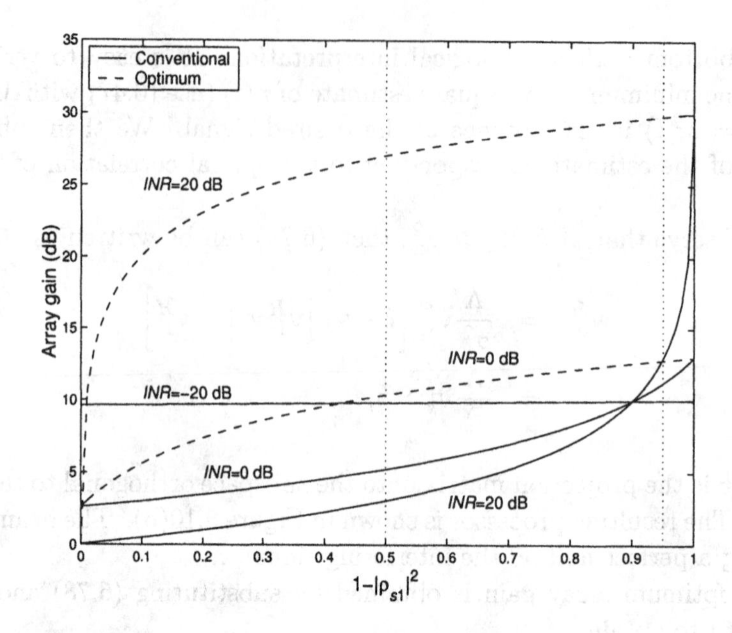

Figure 6.11 Array gain versus $(1 - |\rho_{s1}|^2)$: 10-element array: (a) *INR* = 0 dB; (b) *INR* = 10 dB; (c) *INR* = 20 dB.

The conventional array gain is obtained by substituting (6.76) into (6.32) to obtain

$$A_c(\omega : \mathbf{k}_s) = \frac{N(1+\sigma_I^2)}{1+\sigma_I^2 N|\rho_{s1}|^2}. \tag{6.90}$$

The optimum and conventional array gains are plotted in Figure 6.11 for $N = 10$. Note that the array geometry is embedded in ρ_{s1} through the conventional beam pattern (6.81) so that these results apply to an arbitrary array. We have also shown the HPBW line at $|\rho_{s1}|^2 = 0.5$. For $|\rho_{s1}|^2$ to the left of this line, the interferer is inside the HPBW region of the main lobe. When we examine beam patterns in the examples, we find that this may cause pattern problems.

For large N, and

$$N\sigma_I^2\left[1-|\rho_{s1}|^2\right] \gg 1, \tag{6.91}$$

(6.86) becomes

$$A_o \cong N(1+\sigma_I^2)(1-|\rho_{s1}|^2). \tag{6.92}$$

Therefore, it is convenient to plot A_o/N as shown in Figure 6.12.

The beam pattern for the optimum array can be obtained by substituting

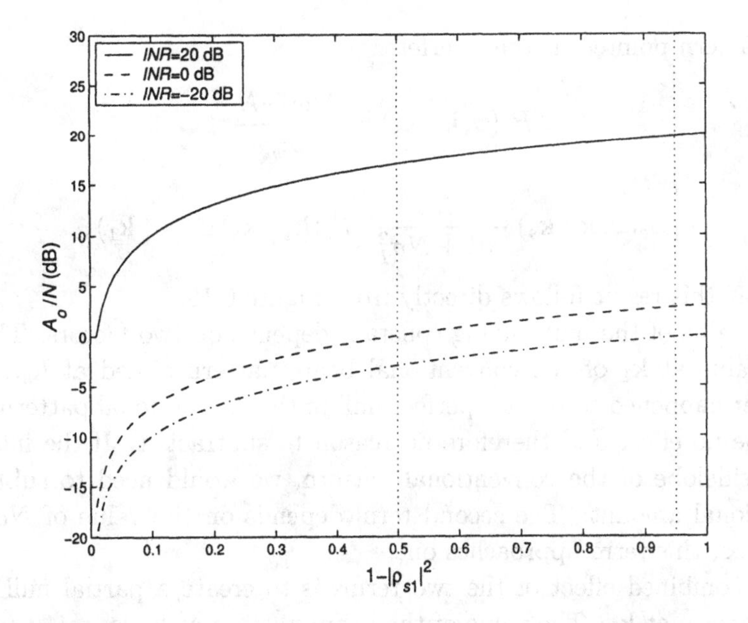

Figure 6.12 Normalized optimum array gain versus $(1 - |\rho_{s1}|^2)$.

(6.82) into (2.51). The result is

$$B_o(\omega : \mathbf{k}_s) = \frac{\Lambda N}{\sigma_w^2}\left[\rho_{sa} - \frac{N\sigma_I^2\rho_{s1}}{1+N\sigma_I^2}\rho_{1a}\right], \tag{6.93}$$

where Λ is the normalizing function in (6.83) and we have suppressed the ω dependence on the right side of (6.93). We observe that

$$\begin{aligned}\rho_{sa} &= \frac{1}{N}\mathbf{v}^H(\omega : \mathbf{k}_s)\mathbf{v}(\omega : \mathbf{k}) \\ &= B_c(\mathbf{k} : \mathbf{k}_s),\end{aligned} \tag{6.94}$$

which is just the conventional beam pattern for a delay-and-sum array beamformer pointed at $\mathbf{k}_s$, and that

$$\rho_{1a} = \frac{1}{N}\mathbf{v}^H(\omega : \mathbf{k}_1)\mathbf{v}(\omega : \mathbf{k}) = B_c(\mathbf{k} : \mathbf{k}_1) \tag{6.95}$$

is just the conventional beam pattern for a delay-and-sum array beamformer pointed at $\mathbf{k}_1$.

Thus, we can interpret the optimum array pattern as the conventional beam pattern pointed at the signal minus a constant times the conventional beam pattern pointed at the interferer

$$B_o(\omega, \mathbf{k} : \mathbf{k}_s) = \frac{\Lambda(\omega) \cdot N}{\sigma_w^2}$$

$$\cdot \left[B_c(\omega, \mathbf{k} : \mathbf{k}_s) - \frac{N\sigma_I^2}{1 + N\sigma_I^2} \cdot B_c(\mathbf{k}_1 : \mathbf{k}_s) B_c(\mathbf{k} : \mathbf{k}_1) \right]. \tag{6.96}$$

Note that this result follows directly from Figure 6.10.

The value of the multiplying constant depends on two factors. The first is the value at $\mathbf{k}_1$ of the conventional beam pattern aimed at $\mathbf{k}_s$. If the interferer happened to be at a perfect null in the conventional pattern, there would be no effect and therefore no reason to subtract it. If the interferer is at a sidelobe of the conventional pattern, we would need to subtract a proportional amount. The second term depends on the value of $N\sigma_I^2$. As $N\sigma_I^2 \to \infty$, this term approaches one.

The combined effect of the two terms is to create a partial null in the beam pattern at $\mathbf{k}_1$. The value of the beam pattern at $\mathbf{k}_1$ is

$$B_o(\mathbf{k}_1 : \mathbf{k}_s) = \frac{\Lambda(\omega)}{\sigma_w^2} \cdot N B_c(\mathbf{k}_1 : \mathbf{k}_s) \left(\frac{1}{1 + N\sigma_I^2} \right), \tag{6.97}$$

as $N\sigma_I^2$ goes to infinity the optimum array pattern has a perfect null at the direction of the interferer. For finite *INR*, it creates a partial null (or notch) whose depth is adjusted to minimize the output variance.

All of the discussion up to this point applies to an arbitrary array. We now specialize the results to a linear uniformly spaced array.

Consider the standard 10-element linear array along the z-axis. From (2.92), the conventional beam pattern for an array pointed at broadside ($u_s = 0$) is

$$B_c(u) = \frac{1}{N} \frac{\sin\left(\frac{N\pi u}{2}\right)}{\sin\left(\frac{\pi u}{2}\right)}, \tag{6.98}$$

where $u = \cos\theta$.

It is useful to divide u-space into three regions in order to analyze the MVDR processor:

SIDELOBE REGION: $0.2 \leq |u| \leq 1.0$,

OUTER MAIN LOBE: $0.045 \leq |u| < 0.2$,

HPBW REGION: $0 < |u| < 0.045$.

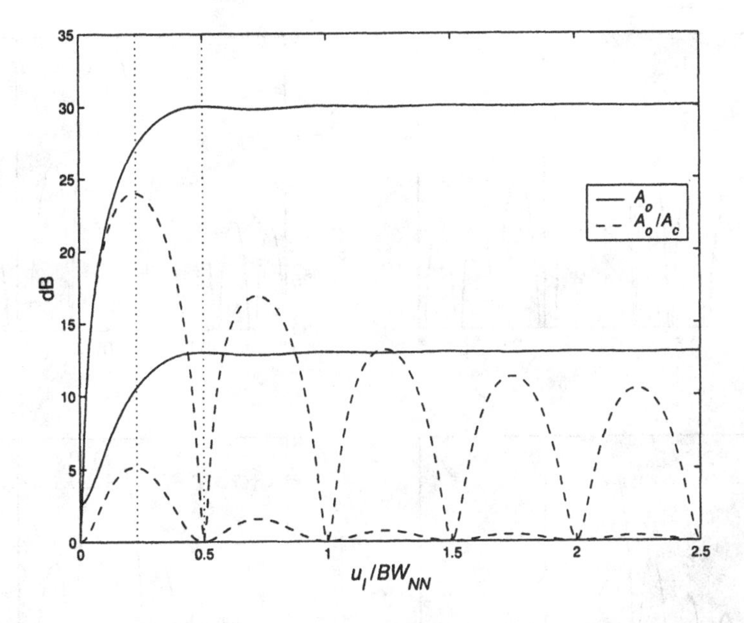

Figure 6.13 A_o and A_o/A_c versus u_I/BW_{NN}; *INR* = 0 dB and 20 dB.

We assume the interferer arrives from u_I. In Figure 6.13, we plot A_o as a function of u_I/BW_{NN}. We also plot A_o/A_c as a function of u_I and observe that its maximum is at the HPBW point. We see that as soon as the interferer is outside the main lobe ($u_I > 0.5BW_{NN}$) the optimum array gain is ($N \cdot INR$) dB. The conventional array gain has the same behavior as the beam pattern in u-space.

In Figure 6.14(a) and 6.14(b), we show the optimum beam pattern for two values of u_I in the sidelobe region for an *INR* (σ_I^2) of 10 dB. In Figure 6.14(c) and (d), we show the optimum beam pattern for the same values of u_I in the sidelobe region for an *INR* (σ_I^2) of 20 dB. We see that the beam pattern is essentially the same as a conventional beam pattern except for the region near the null.

In Figure 6.15, we repeat Figure 6.14 for two values of u_I in the outer main-lobe region; $u_I = 0.18$ and $u_I = 0.09$. For $u_I = 0.18$, the behavior is similar to the sidelobe region except for the growth of the sidelobe nearest to the null. However, at $u_I = 0.09$, two effects are apparent: (1) the main lobe has shifted to the left (away from the null) and its height is larger than unity; (2) the height of the right sidelobe is about -3 dB.

In Figure 6.16, we consider two values of u_I in the HPBW region; $u_I = 0.0433$ and $u_I = 0.02$. We only show the beam pattern inside the BW_{NN}.

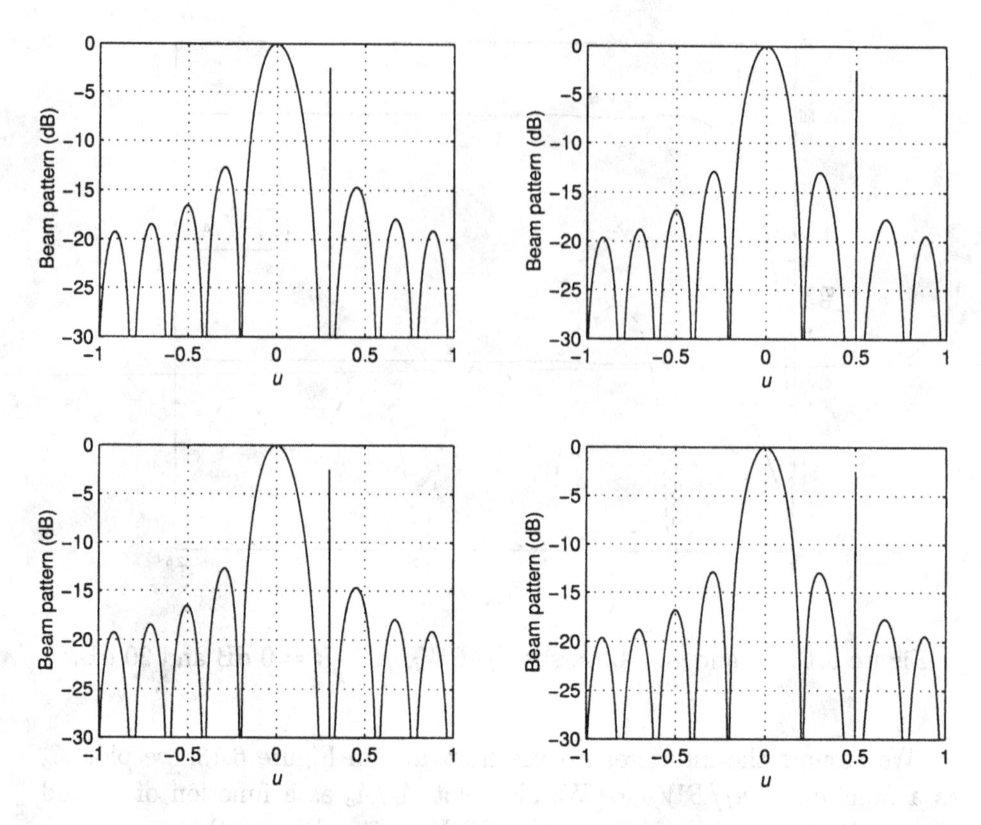

Figure 6.14 Optimum beam pattern for sidelobe interference: (a) $\sigma_I^2 = 10$ dB, $u_I = 0.30$; (b) $\sigma_I^2 = 10$ dB, $u_I = 0.50$; (c) $\sigma_I^2 = 20$ dB, $u_I = 0.30$; (d) $\sigma_I^2 = 20$ dB, $u_I = 0.50$.

We see that the main lobe has split into two "sidelobes" whose heights are significantly larger (about 5 dB at the HPBW point) than the gain in the signal direction.

Thus, as the interferer moves inside the main lobe the peak of the optimum pattern is no longer pointed at the target and the "off-target" gain is significantly larger than the "on-target" gain, which is constrained to be one.

Due to the slope of the beam pattern at the MRA (in this case, when $u_s = 0$), we would anticipate that the beamformer will be sensitive to signal DOA mismatch. In Section 6.6.2, we analyze DOA mismatch. We assume the signal actually arrives from u_a and derive an expression for the normalized array gain $A_{mvdr}(\mathbf{v}_a : \mathbf{v}_m)/A_o(\mathbf{v}_m)$. The result for the scenario in Figure 6.16 is shown in Figure 6.32. We see that the array gain is very sensitive to

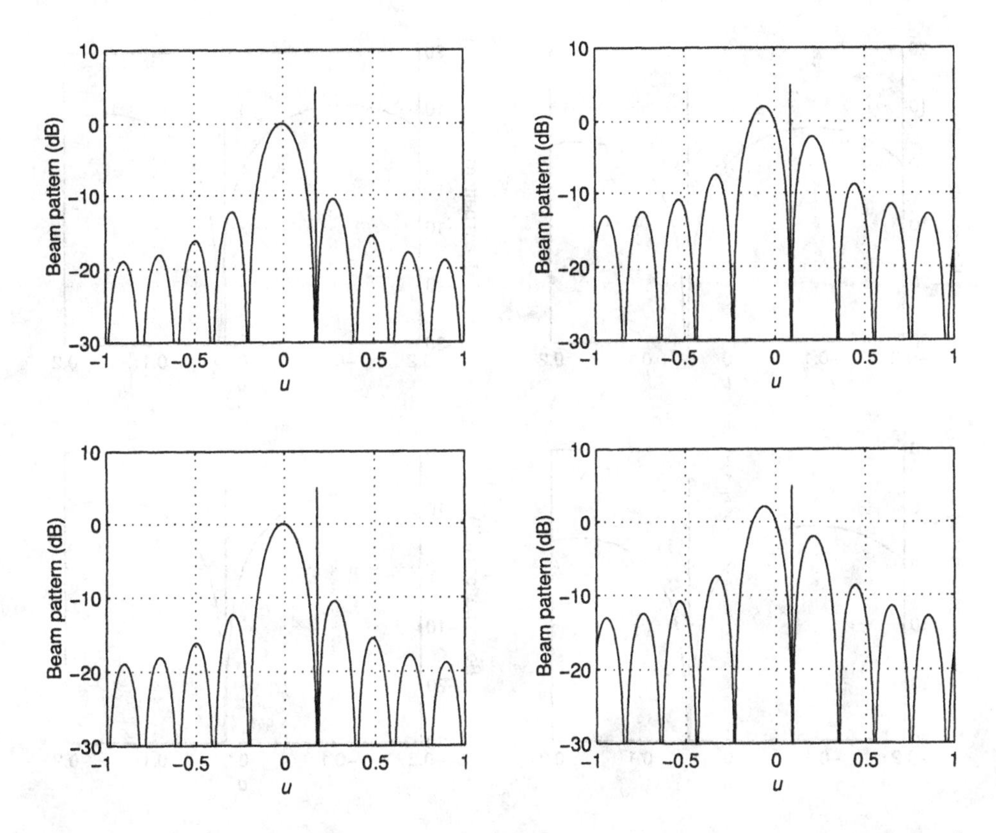

Figure 6.15 Optimum beam pattern for outer main-lobe interference: (a) σ_I^2 = 10 dB, $u_I = 0.18$; (b) σ_I^2 = 10 dB, $u_I = 0.09$; (c) σ_I^2 = 20 dB, $u_I = 0.18$; (d) σ_I^2 = 20 dB, $u_I = 0.09$.

mismatch. In some applications, the variation may be acceptable. However, in most cases, we will want to impose additional constraints in order to maintain reasonably constant performance in the presence of small DOA mismatch.

If the mismatch between the model and actual environment is due to array perturbations, then the sensitivity functions that we encountered in Chapter 2 (2.209) is an appropriate measure.

The sensitivity function is the inverse of the white noise gain,

$$T(\psi_s) = A_w^{-1}(\psi_s) = \left|\mathbf{w}_{do}^H(\psi_s)\right|^2. \tag{6.99}$$

For the single interferer case, the sensitivity function is obtained by substi-

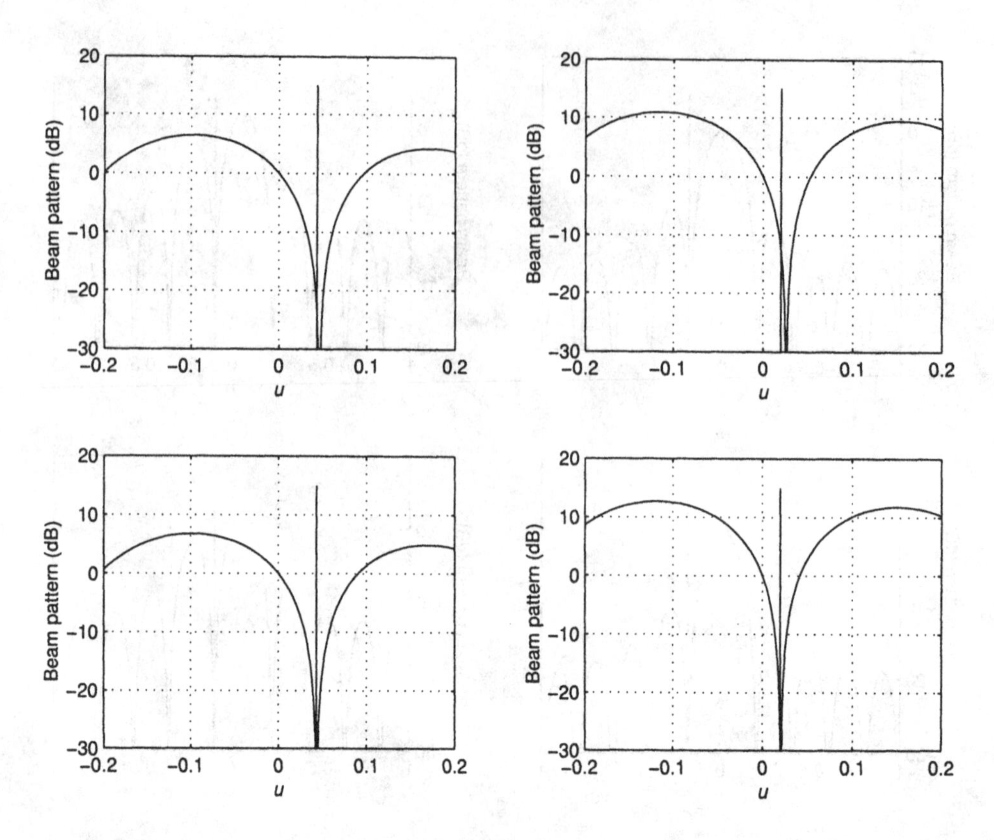

Figure 6.16 Optimum beam pattern for interference inside HPBW: (a) $\sigma_I^2 = 10$ dB, $u_I = 0.0433$; (b) $\sigma_I^2 = 10$ dB, $u_I = 0.02$; (c) $\sigma_I^2 = 20$ dB, $u_I = 0.0433$; (d) $\sigma_I^2 = 20$ dB, $u_I = 0.02$.

tuting (6.82) and (6.83) into (6.99) to give

$$T = \frac{1}{N}\,\frac{1 + 2N\sigma_I^2\alpha^2 + N^2\sigma_I^4\alpha^2}{(1 + N\sigma_I^2\alpha^2)^2}, \tag{6.100}$$

where

$$\alpha^2 = 1 - |\rho_{s1}|^2. \tag{6.101}$$

The sensitivity function is plotted versus α^2 in Figure 6.17(a). This plot is valid for an arbitrary array. In Figure 6.17(b), we plot T versus u_I/BW_{NN} for a standard linear array.

To compare A_o and A_c is interesting, but not particularly relevant, as we argued in Chapter 3 that we would seldom use uniform weighting. A more realistic comparison is to compare A_o to some of the patterns we designed in Chapter 3 (or Chapter 4 for planar arrays).

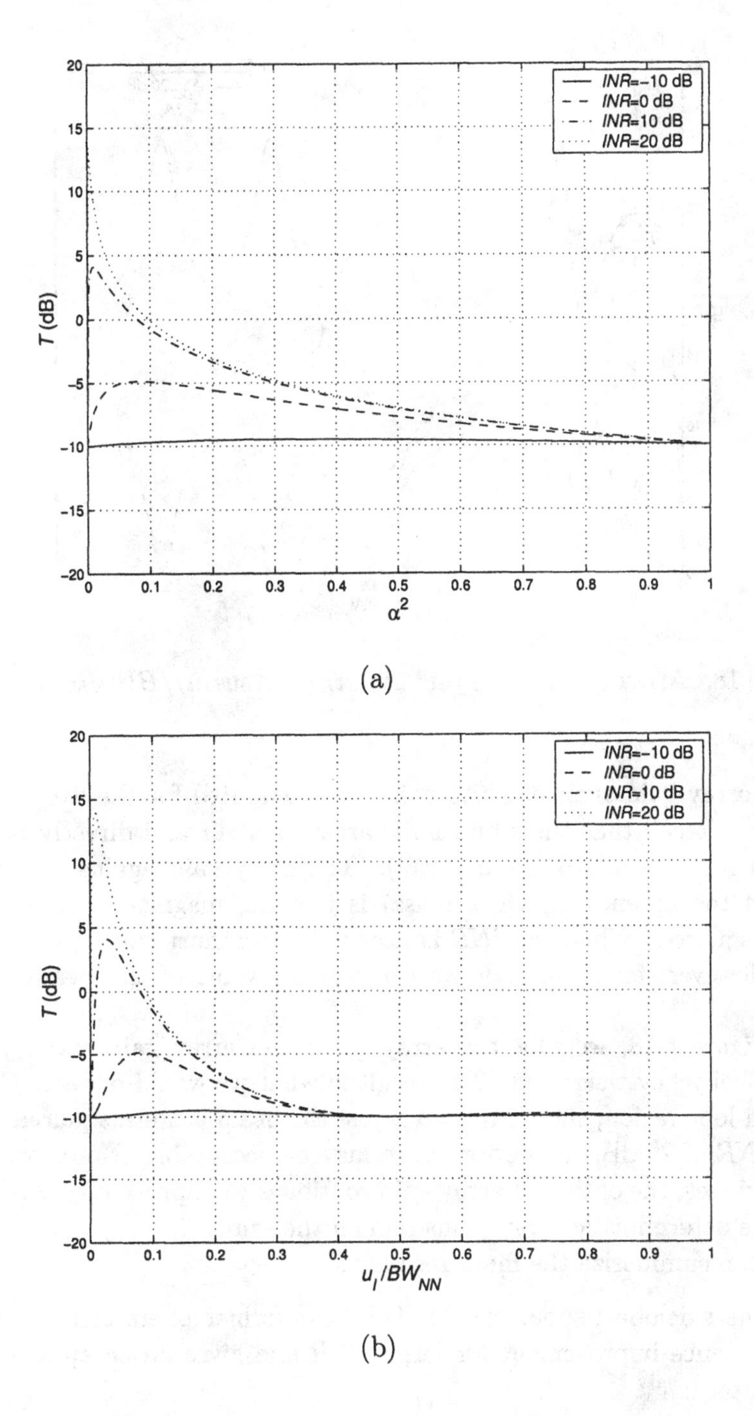

Figure 6.17 Sensitivity function, $T(\alpha)$: (a) $T(\alpha)$ versus α for arbitrary array; (b) $T(\alpha)$ versus u_I/BW_{NN} for standard linear array.

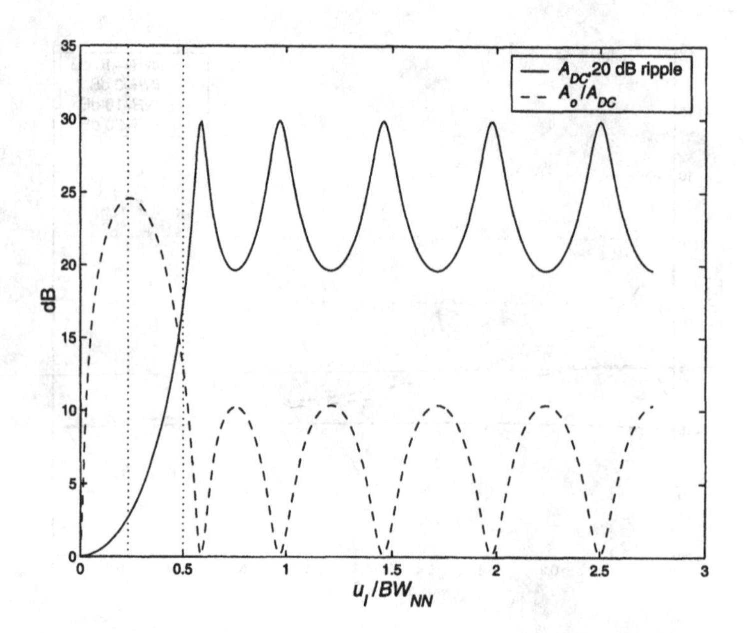

Figure 6.18: Array gain: A_{DC} and A_o/A_{DC} versus u_I/BW_{NN}; *INR* = 20 dB.

The array gain against white noise was computed for the weightings in Chapter 3 (recall that the white noise array gain A_w and directivity D are identical for a standard linear array). The array gain against directional noise (in the absence of white noise) is just the magnitude of the beam pattern squared. When the *INR* is near unity, we must calculate the array gain. However, for large *INR* we can use the value of the beam pattern squared.

In Figure 6.18, we plot the array gains and array gain ratios for the Dolph-Chebychev weighting. The result is what we would expect. Outside the main-lobe region, the array gain is just the beam pattern squared. Thus, for an *INR* of 20 dB, the performance may be acceptable. However, as the *INR* increases, the optimum array gain continues to improve ($A_o \simeq N \cdot INR$) while the deterministic array gains remain the same.

We can summarize the main results as:

(i) In the sidelobe region, the MVDR beamformer offers significant performance improvement for larger *INR* and gives an acceptable beam pattern.

(ii) In the outer mainlobe region, the beam pattern of the MVDR beamformer starts to degenerate.

(iii) In the HPBW region, the MVDR beamformer generates a solution that distorts the main lobe and is sensitive to model mismatch. One of the major issues that we discuss in later sections of this chapter is techniques to provide "main-lobe protection" when a strong interferer arrives in the HPBW region.

It is important to note that our comments about the MVDR beamformer are in the context of the objective of the beamformer. We are steering the the beam in a specific direction u_m. Our objective is to reproduce a signal coming from that direction as accurately as possible. If the signal is a digital signal, we want to detect the bits with a minimum probability of error. We do not want the performance to degrade significantly for signal directions close to u_m. The allowable variation in u_a will depend on how wide an area in u-space is covered by a given steering direction.

In Chapters 8 and 9, we study DOA estimation. In this problem, and the spatial spectral estimation problem, we want the array processor to be very sensitive to the direction of arrival of the signal. We discuss the performance of the MVDR beamformer in that context in Section 9.2.

6.3.2 Multiple Plane-wave Interferers

In this section, we generalize the single interferer results from Section 6.3.1 to the case of D plane-wave interferers.

The realization of interest is shown in Figure 6.19. There are D paths, which are normalized steering and sum devices corresponding to the D interfering noises. Notice that they are not orthogonal. The other path is a normalized steer-and-sum device corresponding to the desired signal. The transform of the interfering noise can be written as

$$\mathbf{N}_c(\omega) = \sum_{i=1}^{D} N_i(\omega)\mathbf{v}(\omega : \mathbf{k}_i) \tag{6.102}$$

and

$$\mathbf{S}_{\mathbf{n}_c}(\omega) = \sum_{i=1}^{D}\sum_{j=1}^{D} S_{n_i n_j}(\omega)\mathbf{v}(\omega : \mathbf{k}_i)\mathbf{v}^H(\omega : \mathbf{k}_j). \tag{6.103}$$

We define an $N \times D$ dimensional matrix,

$$\mathbf{V}_I = \left[\begin{array}{c:c:c:c} \mathbf{v}(\mathbf{k}_1) & \mathbf{v}(\mathbf{k}_2) & \cdots & \mathbf{v}(\mathbf{k}_D) \end{array}\right], \tag{6.104}$$

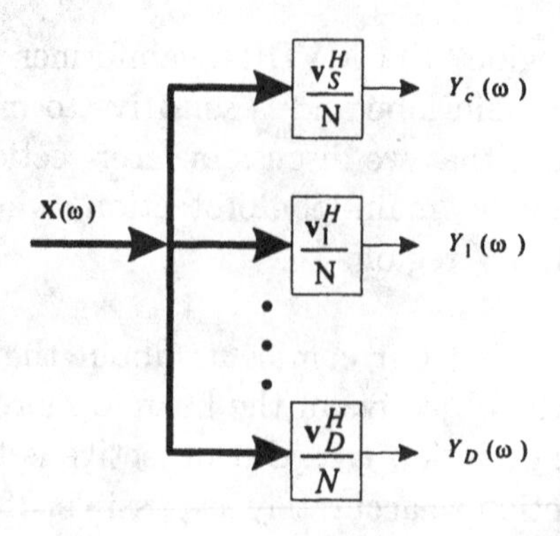

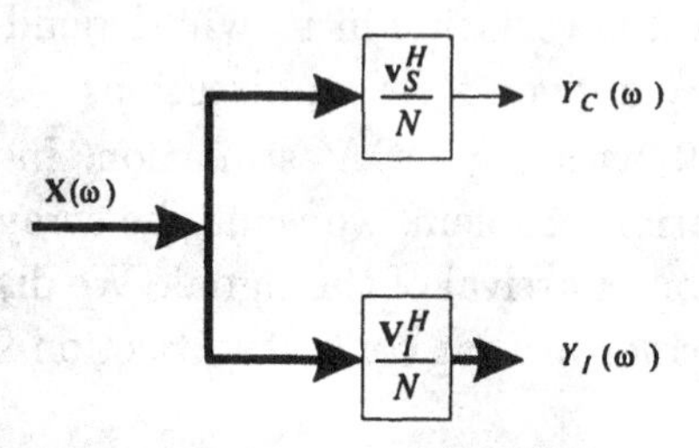

Figure 6.19 Directional noise receiver.

which represents the composite array manifold, and

$$\mathbf{S}_I = \begin{bmatrix} S_{11}(\omega) & S_{21}(\omega) & \cdots \\ S_{12}(\omega) & S_{22}(\omega) & \cdots \\ \vdots & \vdots & S_{DD}(\omega) \end{bmatrix}, \tag{6.105}$$

is the interference spectral matrix. We assume that the signal and interference are statistically independent.

Suppressing the ω and $\mathbf{k}_i$ dependence, the total interference plus white noise spectral matrix is

$$\mathbf{S_n} = \sigma_w^2 \mathbf{I} + \mathbf{V}_I \, \mathbf{S}_I \, \mathbf{V}_I^H. \tag{6.106}$$

Using the matrix inversion lemma

$$\mathbf{S_n}^{-1} = \frac{1}{\sigma_w^2}\mathbf{I} - \frac{1}{\sigma_w^2}\mathbf{V}_I(\mathbf{I} + \mathbf{S}_I\mathbf{V}_I^H \cdot \frac{1}{\sigma_w^2}\mathbf{V}_I)^{-1}\mathbf{S}_I\mathbf{V}_I^H \cdot \frac{1}{\sigma_w^2}. \tag{6.107}$$

Then, (6.107) can be written as

$$\mathbf{S_n}^{-1} = \frac{1}{\sigma_w^2}\left[\mathbf{I} - \mathbf{V}_I(\mathbf{I} + \frac{\mathbf{S}_I}{\sigma_w^2}\mathbf{V}_I^H\mathbf{V}_I)^{-1}\frac{\mathbf{S}_I}{\sigma_w^2}\mathbf{V}_I^H\right]. \tag{6.108}$$

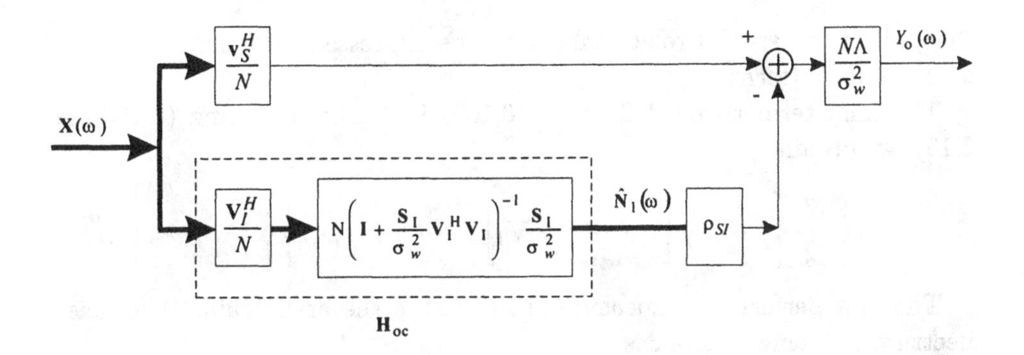

Figure 6.20 Optimal directional noise receiver.

From (6.23), the MVDR beamformer is

$$\begin{aligned} \mathbf{w}_o^H &= \Lambda \mathbf{v}_s^H \mathbf{S}_\mathbf{n}^{-1} \\ &= \frac{N\Lambda}{\sigma_w^2}\left[\frac{\mathbf{v}_s^H}{N} - \boldsymbol{\rho}_{sI}\mathbf{H}_{oc}\right], \end{aligned} \tag{6.109}$$

where

$$\mathbf{H}_{oc} \triangleq \left[\mathbf{I} + \frac{\mathbf{S}_I}{\sigma_w^2}\mathbf{V}_I^H\mathbf{V}_I\right]^{-1}\frac{\mathbf{S}_I}{\sigma_w^2}\mathbf{V}_I^H, \tag{6.110}$$

is a $D \times N$ matrix. We shall see that $\mathbf{H}_{oc}$ is the matrix filter whose output is the minimum mean-square error estimate of $\mathbf{n}_I(t)$ in the absence of the desired signal.

The matrix $\boldsymbol{\rho}_{sI}$ is the $1 \times D$ spatial correlation matrix,

$$\boldsymbol{\rho}_{sI} \triangleq \frac{\mathbf{v}_s^H \mathbf{V}_I}{N}. \tag{6.111}$$

The resulting receiver is shown in Figure 6.20. Thus, the optimum receiver is estimating the directional noise vectors and then subtracting the component of the estimate that is correlated with the desired signal. The ith component of the spatial correlation matrix is

$$[\boldsymbol{\rho}_{sI}]_i = \frac{\mathbf{v}_s^H \mathbf{v}_i}{N} = B_c(\mathbf{k}_i : \mathbf{k}_s). \tag{6.112}$$

Similarly the ijth element of $\mathbf{V}_I^H\mathbf{V}_I$ can be expressed as a conventional beam pattern

$$[\mathbf{V}_I^H\mathbf{V}_I]_{ij} = \mathbf{v}_i^H\mathbf{v}_j = NB_c(\mathbf{k}_j : \mathbf{k}_i). \tag{6.113}$$

Thus, all of the spatial relationships can be expressed in terms of conventional beam patterns.

The final term to be defined in (6.109) is Λ. Substituting (6.108) in (6.15), we obtain

$$\Lambda = \left\{ \frac{N}{\sigma_w^2} \left\{ 1 - \boldsymbol{\rho}_{sI} \left[\mathbf{I} + \frac{\mathbf{S}_I}{\sigma_w^2} \mathbf{V}_I^H \mathbf{V}_I \right]^{-1} \frac{\mathbf{S}_I}{\sigma_w^2} \cdot N \cdot \boldsymbol{\rho}_{sI}^H \right\} \right\}^{-1}. \tag{6.114}$$

The first performance measure of interest is the array gain. The noise spectrum at a single sensor is

$$S_n = \mathbf{1}^T \mathbf{S}_I \mathbf{1} + \sigma_w^2. \tag{6.115}$$

Using (6.108) and (6.115) in (6.28) gives

$$\boxed{A_o = N \left(1 + \frac{\mathbf{1}^T \mathbf{S}_I \mathbf{1}}{\sigma_w^2} \right) (1 - N\alpha),} \tag{6.116}$$

where

$$\alpha = \boldsymbol{\rho}_{sI} \left[\mathbf{I} + \frac{\mathbf{S}_I}{\sigma_w^2} \mathbf{V}_I^H \mathbf{V}_I \right]^{-1} \frac{\mathbf{S}_I}{\sigma_w^2} \boldsymbol{\rho}_{sI}^H. \tag{6.117}$$

The α can also be written as

$$\begin{aligned} \alpha &= tr \left[\boldsymbol{\rho}_{sI} [\mathbf{I} + \frac{\mathbf{S}_I}{\sigma_w^2} \mathbf{V}_I^H \mathbf{V}_I]^{-1} \frac{\mathbf{S}_I}{\sigma_w^2} \boldsymbol{\rho}_{sI}^H \right] \\ &= tr \left[\boldsymbol{\rho}_{sI}^H \boldsymbol{\rho}_{sI} [\mathbf{I} + \frac{\mathbf{S}_I}{\sigma_w^2} \mathbf{V}_I^H \mathbf{V}_I]^{-1} \frac{\mathbf{S}_I}{\sigma_w^2} \right]. \end{aligned} \tag{6.118}$$

The second measure of interest is the optimal beam pattern

$$B_o(\mathbf{k} : \mathbf{k}_s) = \frac{N\Lambda}{\sigma_w^2} \left[\frac{\mathbf{v}_s^H}{N} - \boldsymbol{\rho}_{sI} \left[\mathbf{I} + \frac{\mathbf{S}_I}{\sigma_w^2} \mathbf{V}_I^H \mathbf{V}_I \right]^{-1} \frac{\mathbf{S}_I}{\sigma_w^2} \mathbf{V}_I^H \right] \mathbf{v}(\mathbf{k}). \tag{6.119}$$

Note that

$$\frac{\mathbf{V}_I^H \mathbf{v}_{\mathbf{k}}}{N} = \begin{bmatrix} B_c(\mathbf{k} : \mathbf{k}_1) \\ B_c(\mathbf{k} : \mathbf{k}_2) \\ \vdots \\ B_c(\mathbf{k} : \mathbf{k}_D) \end{bmatrix} \triangleq \mathbf{B}_c(\mathbf{k} : \mathbf{k}_I) \tag{6.120}$$

is a $D \times 1$ matrix of conventional beam patterns. This is what we would expect from the receiver configuration in Figure 6.20. Then (6.119) can be written as

$$\boxed{B_o(\mathbf{k} : \mathbf{k}_s) = \frac{N\Lambda}{\sigma_w^2} \left[B_c(\mathbf{k} : \mathbf{k}_s) - N \boldsymbol{\rho}_{sI} \left[\mathbf{I} + \frac{\mathbf{S}_I}{\sigma_w^2} \mathbf{V}_I^H \mathbf{V}_I \right]^{-1} \frac{\mathbf{S}_I}{\sigma_w^2} \mathbf{B}_c(\mathbf{k} : \mathbf{k}_I) \right].} \tag{6.121}$$

Thus, the optimal beam pattern is the weighted sum of $D+1$ conventional beam patterns corresponding to the desired and interfering signals.

If the interfering signals are uncorrelated, we can write

$$\frac{\mathbf{S}_I}{\sigma_w^2} \triangleq \begin{bmatrix} \sigma_1^2 & & 0 & \\ & \sigma_2^2 & & \\ & 0 & \ddots & \\ & & & \sigma_D^2 \end{bmatrix} \triangleq \boldsymbol{\sigma}_I^2. \tag{6.122}$$

Then, (6.109) can be written as

$$\boxed{\mathbf{w}_o^H = \frac{\Lambda \mathbf{v}_s^H}{\sigma_w^2}\left[\mathbf{I} - \mathbf{V}_I\left[[\boldsymbol{\sigma}_I^2]^{-1} + \mathbf{V}_I^H\mathbf{V}_I\right]^{-1}\mathbf{V}_I^H\right].} \tag{6.123}$$

For large $\boldsymbol{\sigma}_I^2$, this reduces to

$$\boxed{\mathbf{w}_o^H = \frac{\Lambda \mathbf{v}_s^H}{\sigma_w^2}\left[\mathbf{I} - \mathbf{V}_I\left[\mathbf{V}_I^H\mathbf{V}_I\right]^{-1}\mathbf{V}_I^H\right],} \tag{6.124}$$

which can be written as

$$\mathbf{w}_o^H = \frac{\Lambda}{\sigma_w^2}\mathbf{v}_s^H\mathbf{P}_I^{\perp}, \tag{6.125}$$

where $\mathbf{P}_I$ is the projection matrix onto the interference subspace and $\mathbf{P}_I^{\perp}$ is the projection onto a subspace orthogonal to the interference subspace. Thus, we are putting a null on each of the interfering signals.

The single plane-wave case illustrated many of the properties of the multiple plane-wave interference case. In the sidelobe region with multiple separated interferers, the effects are similar to the single interferer case.

We consider two examples to illustrate some interesting behavior.

Example 6.3.1

Consider a standard 10-element linear array. The desired signal is at $\psi = 0$. There are two uncorrelated interfering signals located symmetrically about $\psi = 0$. In Figure 6.21, we show the optimum beam pattern for $\sigma_I^2 = 20$ dB. We see that in the sidelobe region the optimum pattern has a deep notch at the locations of the interferers and the other lobes are well behaved. As the interferers move inside the main lobe, the optimum pattern has two large symmetric sidelobes plus three other sidelobes that are larger than the main lobe.

In Figure 6.22, we show the array gain as a function of u_I/BW_{NN} for various values of σ_I^2 (for $N \geq 10$).

Example 6.3.2

Consider a standard linear array with 21 elements. (This is the same array as in Example 3.7.2.) We assume there are three interferers at $u_1 = 0.21$, $u_2 = 0.22$, and $u_3 = 0.23$.

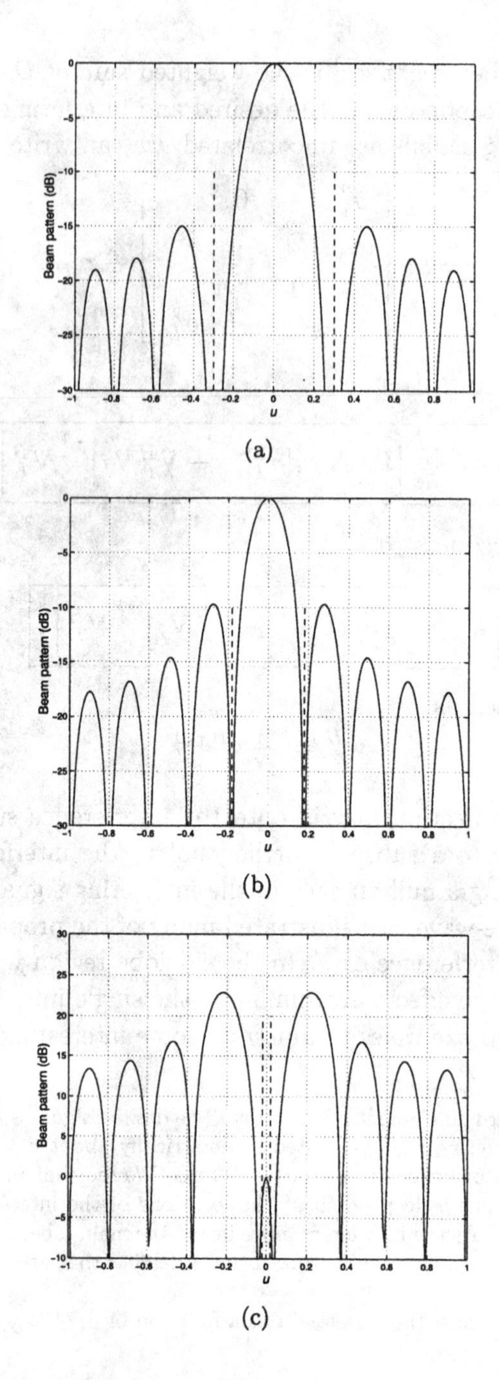

Figure 6.21 MVDR beam pattern: symmetric interference: $INR = 20$ dB; (a) sidelobe region, $u = 0.3$; (b) outer main-lobe region, $u = 0.18$; (c) HPBW region, $u = 0.02$.

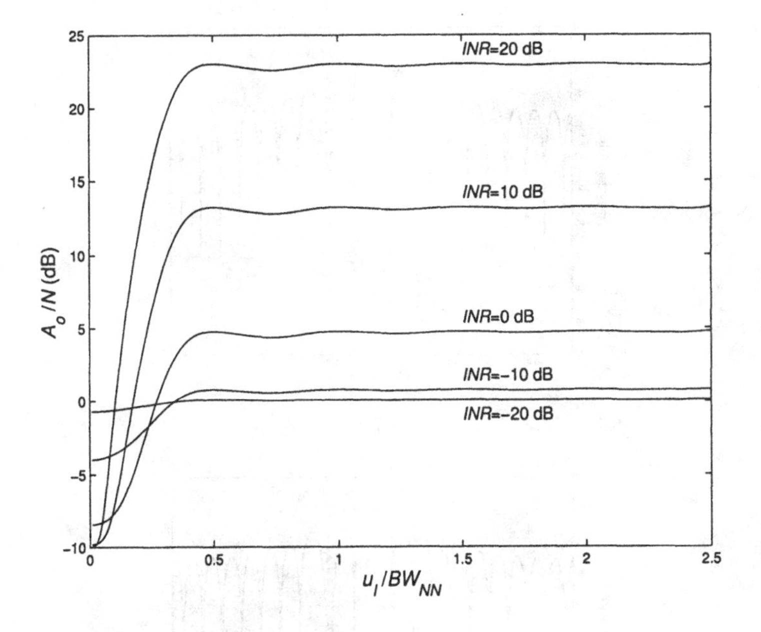

Figure 6.22 Array gain/N versus u_I/BW_{NN}.

The resulting beam pattern is shown in Figure 6.23 for several values of *INR*. The corresponding array gains are indicated.

We see that for $\sigma_I^2 = 20$ dB; the pattern is still different from the result in Example 3.7.2 (Figure 3.40). (The MVDR preserves the main lobe and the first sidelobe.) This is because the MVDR beamformer does not need to create three nulls.

6.3.3 Summary: Discrete Interference

For discrete interference, the optimum receiver forms a signal plus interference subspace and then performs optimal processing in that subspace. We could have derived the optimum processor by arguing that the signal plus interference subspace forms a sufficient statistic for the optimization problem.

If the *INR* is modest, then the optimum processor forms notches (partial nulls) in the direction of the interfering signals. As the *INR* increases, these notches approach perfect nulls.

If the interferers are outside the main lobe, then the beam pattern of the optimum processor is well-behaved and we can obtain a significant improvement in array gain. If the interferers are inside the main lobe, the optimal solution creates large spurious lobes away from the desired MRA. We will look at several alternative approaches later, but will find that main-lobe

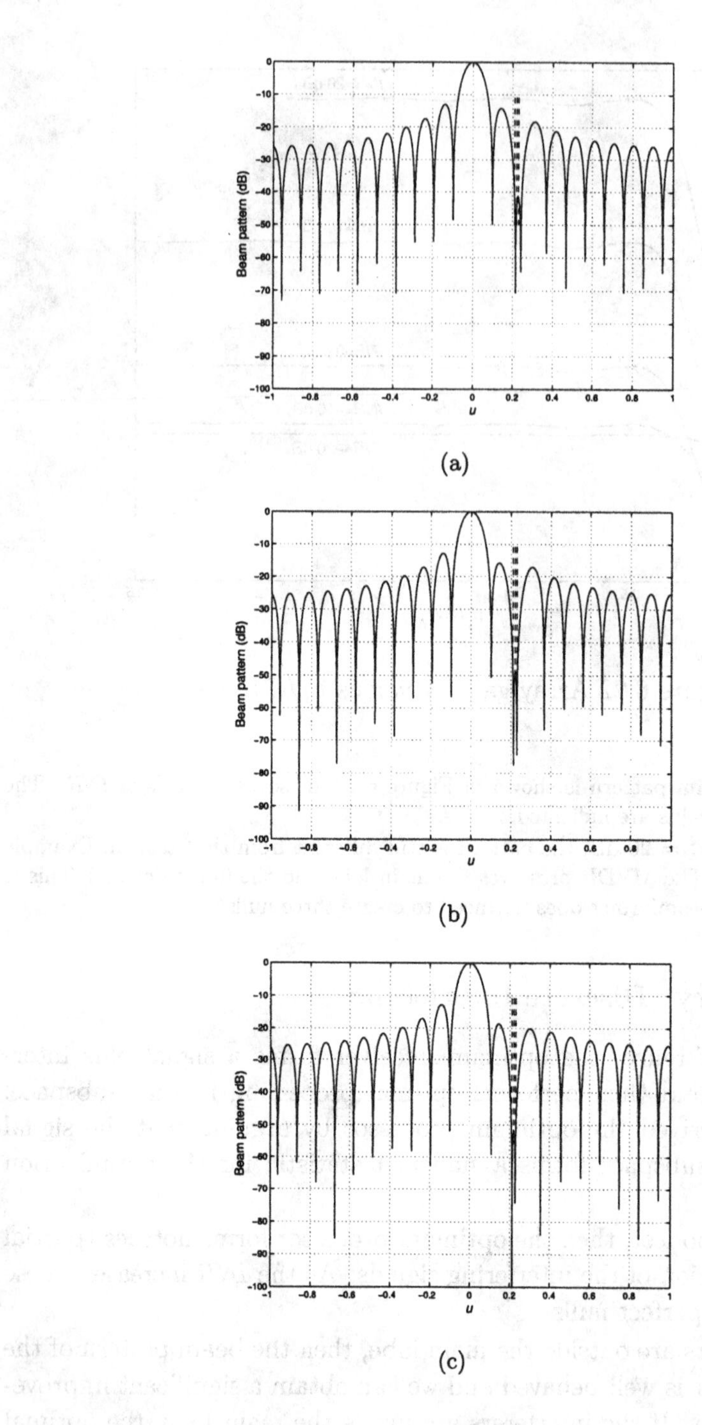

Figure 6.23 Beam pattern for three interferers: N=21, u_s=0, u_I=0.21, 0.22, 0.23; (a) σ_I^2 = 0 dB; (b) σ_I^2 = 10 dB; (c) σ_I^2 = 20 dB.

interference is a difficult problem.

6.4 Spatially Spread Interference

In this section, we consider the performance of the MVDR beamformer in the presence of spatially spread interference plus white noise. In Section 6.4.1, we consider the physical noise models that were developed in Section 5.3.3. In Section 6.4.2, we consider the ARMA models that were developed in Section 5.6.

6.4.1 Physical Noise Models

In Section 5.3.3 and the associated problems (e.g., Problems 5.3.1–5.3.3), we developed several noise models that corresponded to noise fields encountered in acoustic environments.

We consider two examples to illustrate the performance of an optimum processor in that environment.

Example 6.4.1

Consider a standard 10-element linear array oriented in the vertical direction (along the z-axis). The frequency-domain snapshots are

$$\mathbf{X}(\omega) = \mathbf{v}(\psi_T)F(\omega) + \mathbf{N}_c(\omega) + \mathbf{W}(\omega). \tag{6.126}$$

The noise contains a white component with spectral matrix $\sigma_w^2\mathbf{I}$ and a high-surface noise component whose spatial spectral matrix is

$$\mathbf{S}_{\mathbf{n}_c}(\omega) = \mathbf{S}_o(\omega)\left\{\text{sinc}(k_o\mathbf{\Delta p}) + j\alpha\frac{1}{k_o\mathbf{\Delta p}}\left[\text{sinc}(k_o\mathbf{\Delta p}) - \cos(k_o\mathbf{\Delta p})\right]\cos\theta_p\right\}, \tag{6.127}$$

and $\alpha = 1$ (see Problem 5.3.1). The surface noise and the white noise are uncorrelated. The optimum array gain is obtained by:

(i) Evaluating $S_o(\omega : \theta, \phi)$ at the sensor locations to obtain the spectral matrix $\mathbf{S}_{\mathbf{n}_c}(\omega)$.

(ii) Find $\mathbf{S}_n^{-1}(\omega) = \left[\mathbf{S}_{\mathbf{n}_c}(\omega) + \sigma_w^2\mathbf{I}\right]^{-1}$ numerically.

(iii) Substituting into (6.28) to find A_o.

(iv) Substituting into (6.32) gives A_c for comparison.

In Figure 6.24 we show the results as a function of θ_s for several values of $\mathbf{S}_o(\omega)/\sigma_w^2$.

Example 6.4.2 [Bag76]

We consider the same model as in Example 6.4.1 except $\mathbf{S}_{\mathbf{n}_c}(\omega)$ corresponds to a high-layer noise environment. The spatial spectral matrix is

$$\begin{aligned}\mathbf{S}_{\mathbf{n}_c}(\omega) \quad = \quad & S_o(\omega)\left\{\text{sinc}(k_o\mathbf{\Delta p}) + \alpha\left[\left(\frac{3}{(k_o\mathbf{\Delta p})^2} - 1\right)\text{sinc}(k_o\mathbf{\Delta p})\right.\right. \\ & \left.\left. - \frac{3\cos(k_o\mathbf{\Delta p})}{(k_o\mathbf{\Delta p})^2}\right]\left[\frac{3\cos(2\theta_p) + 1}{4}\right]\right\},\end{aligned} \tag{6.128}$$

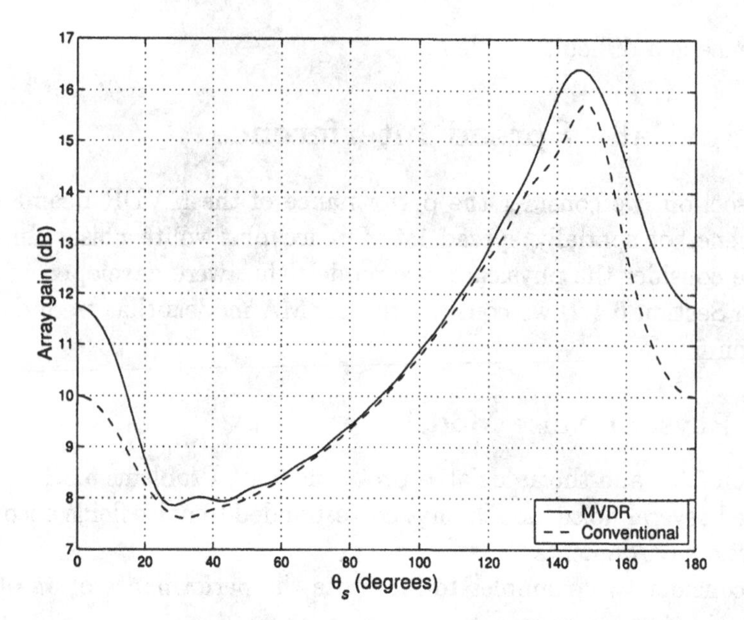

Figure 6.24 MVDR beamformer: array gain versus θ_s for various $\mathbf{S}_n(\omega_c)/\sigma_w^2$ levels; high surface noise.

with $\alpha = 1$ (see Problem 5.3.2). We repeat the steps in Example 6.4.1 to obtain the optimum array gain and the conventional array gain. We show the results in Figure 6.25 as a function of θ_s for several values of $S_o(\omega)/\sigma_w^2$.

6.4.2 ARMA Models

In this case, we assume that the spatially spread noise process can be modeled as an autoregressive process of order p where N, the number of sensor elements in the standard linear array, satisfies

$$N \geq 2p + 1. \tag{6.129}$$

In order to implement the optimum beamformer we need to evaluate $\mathbf{S}_\mathbf{n}^{-1}$, where $\mathbf{n}(k)$ is an AR(p) process (see (5.308) and Figure 5.22). We first consider the case in which the sensor noise is negligible. Subsequently we consider the case of small sensor noise and then the general case.

A formula for $\mathbf{S}_\mathbf{n}^{-1}$ in terms of the AR parameters was given in (5.338). We give two examples.

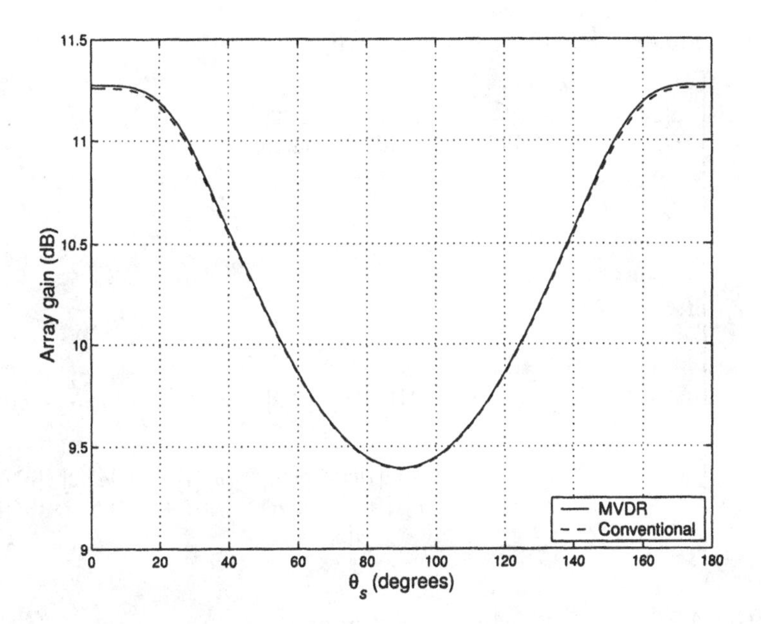

Figure 6.25 MVDR beamformer: array gain versus θ_s for various $\mathbf{S}_n(\omega_c)/\sigma_w^2$ levels; high layer noise.

For a complex AR(1) process,

$$\mathbf{S_n}^{-1} = \frac{1}{\sigma_u^2}\begin{bmatrix} 1 & a^*[1] & 0 & 0 & \cdots & 0 \\ a[1] & 1+|a[1]|^2 & a^*[1] & 0 & \cdots & 0 \\ 0 & a[1] & 1+|a[1]|^2 & a^*[1] & \cdots & 0 \\ \vdots & \vdots & \ddots & \ddots & \ddots & \vdots \\ 0 & 0 & \cdots & a[1] & 1+|a[1]|^2 & a^*[1] \\ 0 & 0 & \cdots & 0 & a[1] & 1 \end{bmatrix}. \tag{6.130}$$

We observe that $\mathbf{S_n}^{-1}$ is a Hermitian, persymmetric matrix, but it is not Toeplitz. We also observe that we can partition $\mathbf{S_n}^{-1}$ as

$$\mathbf{S_n}^{-1} = \begin{bmatrix} \mathbf{a}^* \;\vdots\; \mathbf{0} \\ \mathbf{B} \\ \mathbf{0} \;\vdots\; \mathbf{J\,a} \end{bmatrix}, \tag{6.131}$$

and the middle $(N-2) \times N$ matrix $\mathbf{B}$ is a right circulant matrix as well as a banded Toeplitz matrix.

For a complex AR(2) process, we have,

$$
\mathbf{S}_{\mathbf{n}}^{-1} = \frac{1}{\sigma_u^2}
\left[\begin{array}{cccccc}
1 & a^*[1] & a^*[2] & & & \\
a[1] & 1+|a[1]|^2 & a^*[1]+a[1]a^*[2] & & & \\
a[2] & a[1]+a^*[1]a[2] & 1+|a[1]|^2+|a[2]|^2 & & & \\
\vdots & \ddots & \ddots & & & \\
0 & \cdots & a[1]+a^*[1]a[2] & & & \\
0 & \cdots & a[2] & & & \\
0 & \cdots & 0 & & & \\
 & & 0 & & \cdots & 0 \\
 & & a^*[2] & & \cdots & 0 \\
 & & a^*[1]+a[1]a^*[2] & & \cdots & 0 \\
 & & \ddots & & \ddots & \vdots \\
 & & 1+|a[1]|^2+|a[2]|^2 & & a^*[1]+a[1]a^*[2] & a^*[2] \\
 & & a[1]+a^*[1]a[2] & & 1+|a[1]|^2 & a^*[1] \\
 & & a[2] & & a[1] & 1
\end{array}\right].
\tag{6.132}
$$

For the AR(p), (5.338) can be written as

$$
\left[\mathbf{S}_{\mathbf{n}}^{-1}\right]_{ij} = \frac{1}{\sigma^2}\sum_{k=1}^{N}\left(a[i-k]a^*[j-k] - a^*[N-i+k]a[N-j+k]\right)
$$

$$
i = 1, 2, \cdots, N; j = 1, 2, \cdots, N, \tag{6.133}
$$

where $a[k] = 0$ for $k < 0$ and $k > p$.

We consider a simple example to illustrate the result.

Example 6.4.3

Consider an N-element standard linear array with $\mathbf{n}(k)$ consisting of a complex AR(1) process and assume there is no observation noise. We assume N is odd.

Then, the MVDR filter is

$$
\mathbf{w}_o^H = \Lambda \mathbf{v}^H(\psi_s)\mathbf{S}_{\mathbf{n}}^{-1}, \tag{6.134}
$$

and the optimum array gain is

$$
A_o = \left[E\left[|n_i(k)|^2\right]\right]\Lambda^{-1} = \frac{\sigma_u^2}{(1-|a[1]|^2)}\Lambda^{-1}, \tag{6.135}
$$

where

$$
\Lambda = \left[\mathbf{v}^H(\psi_s)\mathbf{S}_{\mathbf{n}}^{-1}\mathbf{v}(\psi_s)\right]^{-1}. \tag{6.136}
$$

The elements of the MVDR filter (omitting the Λ term temporarily) are obtained by using (6.130) in (6.14)

$$
w'_{o1} = e^{j\frac{N-1}{2}\psi_s}\left[1 - |a_1|e^{-j(\psi_s-\psi_n)}\right], \tag{6.137}
$$

$$
\begin{aligned}
w'_{on} &= e^{j\left[\frac{N-2}{2}-(n-1)\right]\psi_s}\left[-|a_1|e^{j(\psi_s-\psi_n)} + \left(1+|a_1|^2\right) - |a_1|e^{-j(\psi_s-\psi_n)}\right] \\
&= e^{j\left[\frac{N-2}{2}-(n-1)\right]\psi_s}\left[\left(1+|a_1|^2\right) - 2|a_1|\cos(\psi_s-\psi_n)\right] \\
& \qquad n = 2, 3, \cdots, N-1,
\end{aligned} \tag{6.138}
$$

$$w'_{oN} = e^{-j\frac{N-1}{2}\psi_s}\left[1 - |a_1|e^{j(\psi_s - \psi_n)}\right] = w'^*_{o1}, \tag{6.139}$$

where the prime denotes the absence of Λ.

The beam pattern for a standard 11-element linear array is shown in Figure 6.26. We see that as the peak of a_1 moves inside the main lobe, the MVDR beamformer narrows the main lobe, causing a significant increase in the sidelobes.

To find the array gain, we substitute (6.130) into (6.28) and obtain

$$\begin{aligned} A_o &= \frac{1}{1 - |a_1|^2}\left[N - 2(N-1)|a_1|\cos(\psi_s - \psi_n) + (N-2)|a_1|^2\right] \\ &= \frac{1}{1 - |a_1|^2}\left[N(1 - |a_1|^2) - 2(N-1)|a_1|\left(|a_1| - \cos(\psi_s - \psi_n)\right)\right] \\ &= N\left\{1 + \frac{2(N-1)}{N}\frac{|a_1|}{1 - |a_1|^2}\left(|a_1| - \cos(\psi_s - \psi_n)\right)\right\}. \end{aligned} \tag{6.140}$$

The array gain is plotted versus $\Delta u/BW_{NN} = (u_s - u_n)/BW_{NN}$ in Figure 6.27 for several values of $|a_1|$.

As $|a_1|$ approaches unity, the required *INR* for the no white noise assumption to be valid becomes higher. In these cases, the array gain is calculated numerically.

In Figure 6.28, the array gain is plotted versus $\Delta u/BW_{NN}$ for various values of $|a_1|$ and σ_x^2/σ_w^2. We also show the result for a single plane wave from Figure 6.13. We see that, as $|a_1|$ increases, the array gain approaches the single plane-wave result. For small $|a_1|$, the interference is spread over more of u-space and the optimum beamformer is less effective. In the limit, as $|a_1|$ goes to zero, the AR process is identical to white noise and $A_o = 10.41$ dB.

6.5 Multiple Plane-wave Signals

In this section, we consider the case in which there are multiple desired plane-wave signals. In Section 6.5.1, we consider MVDR beamformers. In Section 6.5.2, we consider MMSE beamformers.

6.5.1 MVDR Beamformer

In this section, we find the optimum distortionless matrix processor for D plane-wave signals. Using the frequency-domain snapshot model,

$$\mathbf{X}(\omega) = \mathbf{V}_s\mathbf{F}(\omega) + \mathbf{N}(\omega), \tag{6.141}$$

where $\mathbf{V}_s$ is an $N \times D$ array manifold matrix,

$$\mathbf{V}_s = \left[\begin{array}{c:c:c:c} \mathbf{v}(\omega : \mathbf{k}_1) & \mathbf{v}(\omega : \mathbf{k}_2) & \cdots & \mathbf{v}(\omega : \mathbf{k}_D) \end{array}\right], \tag{6.142}$$

and

$$\mathbf{F}(\omega) = \begin{bmatrix} F_1(\omega) \\ F_2(\omega) \\ \vdots \\ F_D(\omega) \end{bmatrix} \tag{6.143}$$

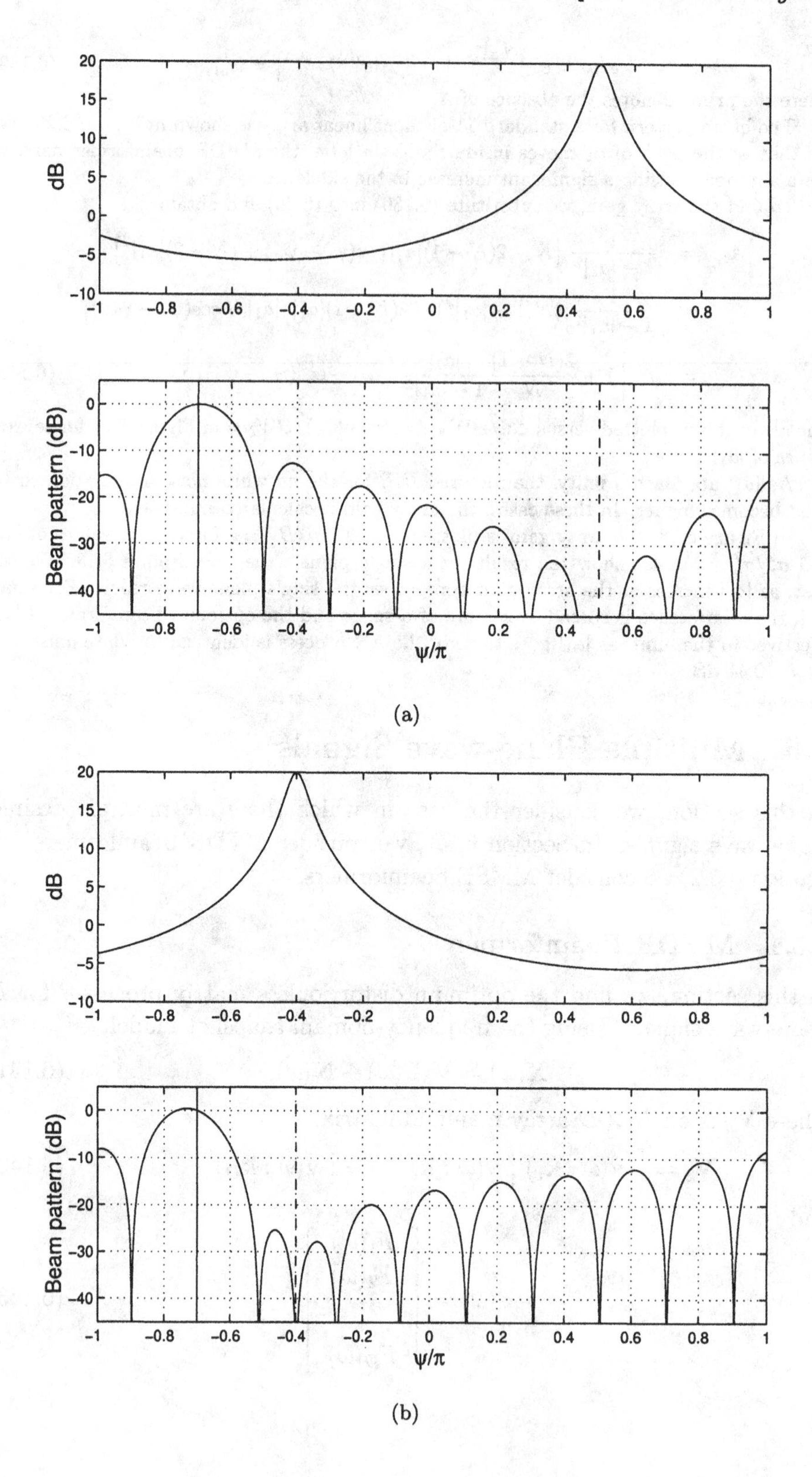

(a)

(b)

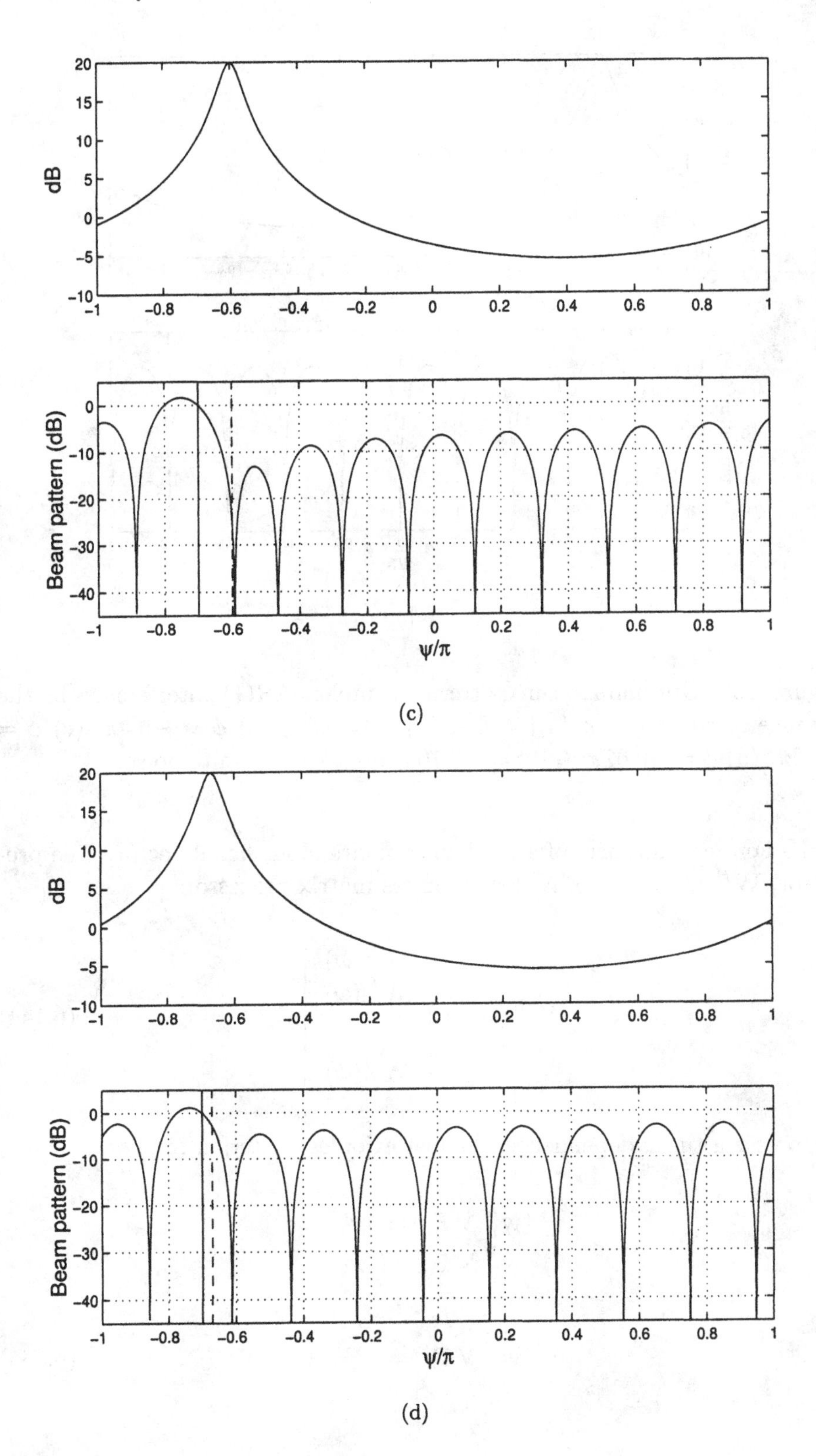
20
15
10
5
0
-5
-10
dB
-1
-0.8
-0.6
-0.4
-0.2
0
0.2
0.4
0.6
0.8
1
Beam pattern (dB)
-40
-30
-20
-10
0
ψ/π

(c)

(d)

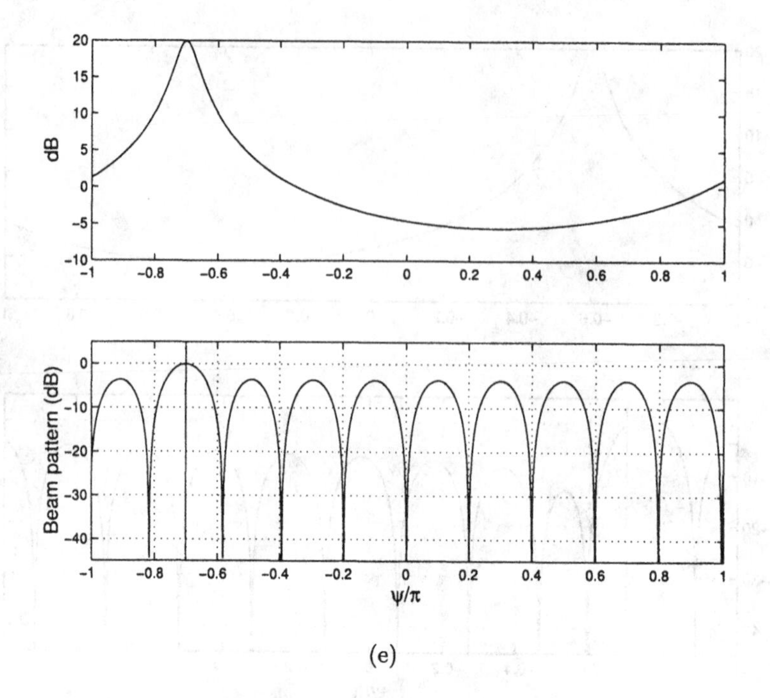

Figure 6.26 Optimum beam pattern: complex AR(1) interference in the absence of white noise: $|a_1| = 0.9$; (a) $\phi = 0.5\pi$; (b) $\phi = -0.4\pi$; (c) $\phi = -0.6\pi$; (d) $\phi = -0.67\pi$; (e) $\phi = -0.70\pi$; no additive white noise.

is the Fourier transform of an unknown nonrandom signal vector. The processor, $\mathbf{W}^H(\omega)$, is a $D \times N$ distortionless matrix processor:

$$\mathbf{W}^H(\omega) = \begin{bmatrix} \mathbf{W}_1^H(\omega) \\ \mathbf{W}_2^H(\omega) \\ \vdots \\ \mathbf{W}_D^H(\omega) \end{bmatrix}. \tag{6.144}$$

Suppressing the ω dependence, the distortionless criterion implies,

$$\begin{aligned} \mathbf{w}_1^H \mathbf{V}\mathbf{F} &= F_1, \\ \mathbf{w}_2^H \mathbf{V}\mathbf{F} &= F_2, \\ &\vdots \\ \mathbf{w}_D^H \mathbf{V}\mathbf{F} &= F_D. \end{aligned} \tag{6.145}$$

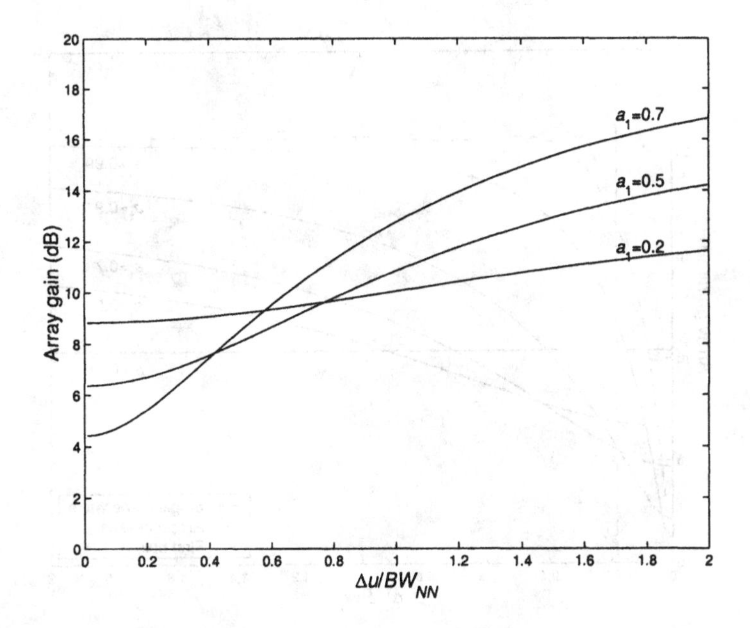

Figure 6.27 Array gain: complex AR(1) interference; A_o versus $\Delta u/BW_{NN}$; $|a_1| = 0.2, 0.5$ and 0.7.

Because (6.145) must be satisfied for arbitrary $\mathbf{F}$, it places D constraints on each $\mathbf{w}_i$. For $\mathbf{w}_1$,

$$\begin{aligned}\mathbf{w}_1^H\,\mathbf{v}_1\,F_1 &= \quad F_1,\\ \mathbf{w}_1^H\,\mathbf{v}_2\,F_2 &= \quad 0,\\ &\vdots\\ \mathbf{w}_1^H\,\mathbf{v}_D\,F_D &= \quad 0.\end{aligned} \tag{6.146}$$

These constraints imply,

$$\begin{aligned}\mathbf{w}_1^H\,\mathbf{v}_1 &= \quad 1,\\ \mathbf{w}_1^H\,\mathbf{v}_2 &= \quad 0,\\ &\vdots\\ \mathbf{w}_1^H\,\mathbf{v}_D &= \quad 0.\end{aligned} \tag{6.147}$$

or, for the ith beamformer,

$$\mathbf{w}_i^H\,\mathbf{v}_j = \delta_{ij} \quad i, j = 1, \cdots, D. \tag{6.148}$$

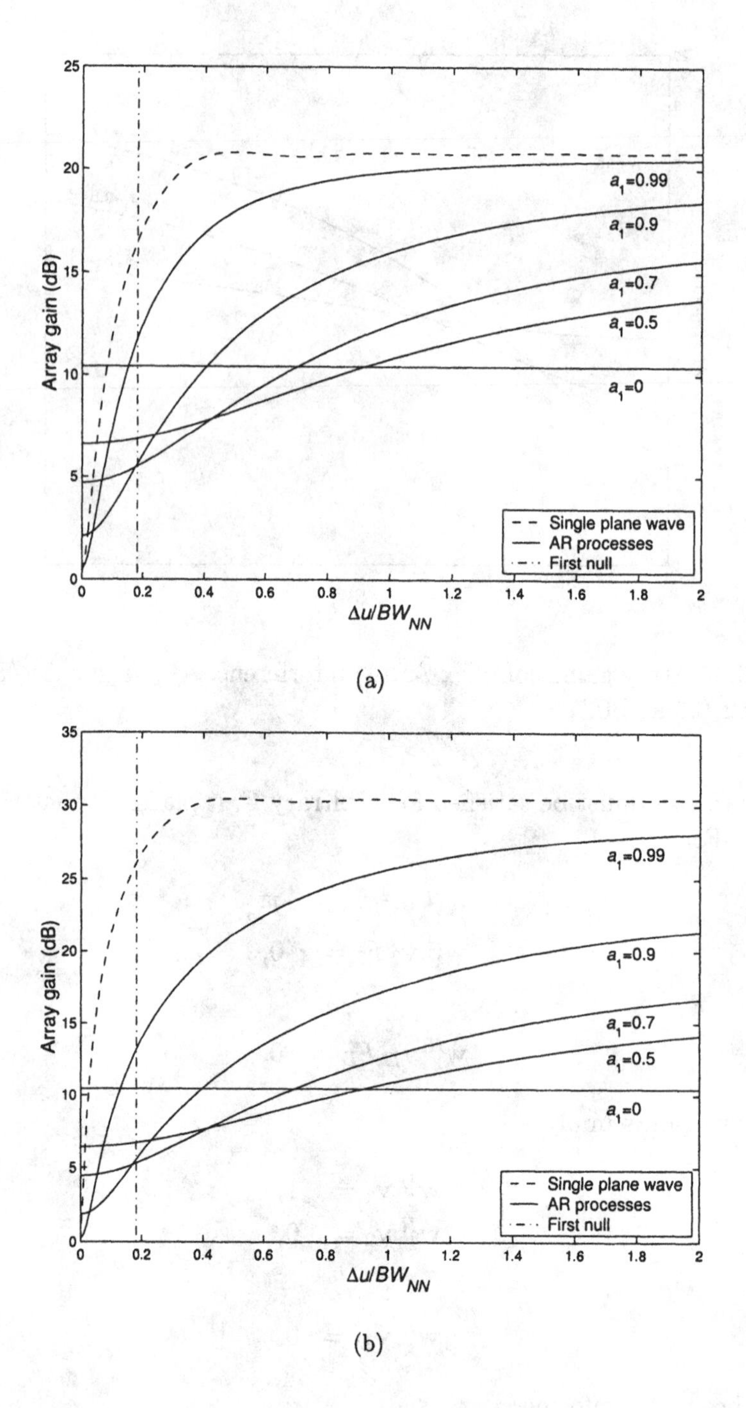

Figure 6.28 Array gain: complex AR(1) interference; A_o versus $\Delta u/BW_{NN}$; $|a_1| = 0.5, 0.7, 0.9$ and 0.99; (a) $\sigma_x^2/\sigma_w^2 = 10$ dB; (b) $\sigma_x^2/\sigma_w^2 = 20$ dB.

Therefore, the ith beamformer is distortionless with respect to the ith signal and puts a perfect null of the other $D-1$ signals. The output noise of the ith beamformer is

$$\sigma_{ni}^2 = \mathbf{w}_i^H \, \mathbf{S_n} \, \mathbf{w}_i. \tag{6.149}$$

We minimize σ_{ni}^2 subject to the D constraints on $\mathbf{w}_i$. We define

$$G_i = \mathbf{w}_i^H \, \mathbf{S_n} \, \mathbf{w}_i + \sum_{j=1}^{D} \lambda_{ij}^* \left(\mathbf{w}_i^H \, \mathbf{v}_j - \delta_{ij} \right) + \sum_{j=1}^{D} \lambda_{ij} \left(\mathbf{v}_j^H \, \mathbf{w}_i - \delta_{ij} \right), \quad i = 1, \cdots, D. \tag{6.150}$$

Minimizing gives

$$\mathbf{w}_i^H \mathbf{S_n} + \lambda_{ij} \mathbf{v}_j^H = 0, \quad i, j = 1, \cdots, D, \tag{6.151}$$

or

$$\mathbf{w}_i^H = -\lambda_{ij} \, \mathbf{v}_j^H \, \mathbf{S}_\mathbf{n}^{-1}, \quad i, j = 1, \cdots, D. \tag{6.152}$$

Substituting (6.152) into (6.148) gives

$$-\lambda_{ii} \, \mathbf{v}_i^H \, \mathbf{S}_\mathbf{n}^{-1} \, \mathbf{v}_i = 1, \quad i = 1, \cdots, D. \tag{6.153}$$

$$-\lambda_{ij} \, \mathbf{v}_i^H \, \mathbf{S}_\mathbf{n}^{-1} \, \mathbf{v}_j = 0, \quad i, j = 1, \cdots, D, \tag{6.154}$$

or

$$-\lambda_{ij} \, \mathbf{v}_i^H \, \mathbf{S}_\mathbf{n}^{-1} \, \mathbf{v}_j = \delta_{ij}, \quad i, j = 1, \cdots, D. \tag{6.155}$$

We define a $D \times D$ matrix $\mathbf{\Lambda}$ whose ijth element is λ_{ij}. Then (6.155) can be written in matrix notation as

$$-\mathbf{\Lambda} [\mathbf{V}^H \, \mathbf{S}_\mathbf{n}^{-1} \, \mathbf{V}] = \mathbf{I}, \tag{6.156}$$

or

$$\mathbf{\Lambda} = -[\mathbf{V}^H \, \mathbf{S}_\mathbf{n}^{-1} \, \mathbf{V}]^{-1}. \tag{6.157}$$

Then, the optimum distortionless beamformer can be written as

$$\boxed{\mathbf{W}_{do}^H = [\mathbf{V}^H \, \mathbf{S}_\mathbf{n}^{-1} \, \mathbf{V}]^{-1} \, \mathbf{V}^H \, \mathbf{S}_\mathbf{n}^{-1}.} \tag{6.158}$$

The output of $\mathbf{W}_{do}^H$ is the $D \times 1$ vector,

$$\hat{\mathbf{F}}_o = \left[\mathbf{V}^H \mathbf{S}_\mathbf{n}^{-1} \mathbf{V} \right]^{-1} \mathbf{V}^H \mathbf{S}_\mathbf{n}^{-1} \mathbf{X}, \tag{6.159}$$

which is also the minimum variance unbiased estimate of $\mathbf{F}$.

It is also straightforward to show that

$$\hat{\mathbf{F}}_{ml} = \hat{\mathbf{F}}_o, \tag{6.160}$$

if the Gaussian assumption is added.

For the special case of white homogeneous noise,

$$\mathbf{S}_\mathbf{n} = \sigma_w^2 \mathbf{I}, \tag{6.161}$$

the optimum processor is

$$\boxed{\mathbf{W}_{do}^H = [\mathbf{V}^H\mathbf{V}]^{-1}\mathbf{V}^H}. \tag{6.162}$$

Thus,

$$\hat{\mathbf{F}}_o = [\mathbf{V}^H\mathbf{V}]^{-1}\mathbf{V}^H\mathbf{X}. \tag{6.163}$$

This result in (6.163) was first pointed out by Schweppe [Sch68]. As he did not include the derivation in the paper, it has been rediscovered in several subsequent papers.

It is useful to rewrite the result in (6.148) in a different manner. Consider the first signal and partition $\mathbf{V}$ as,

$$\mathbf{V} = [\mathbf{v}_1 | \mathbf{V}_I]. \tag{6.164}$$

Then

$$\mathbf{V}^H\mathbf{V} = \left[\begin{array}{c|c} \mathbf{v}_1^H\mathbf{v}_1 & \mathbf{v}_1^H\mathbf{V}_I \\ \hline \mathbf{V}_I^H\mathbf{v}_1 & \mathbf{V}_I^H\mathbf{V}_I \end{array}\right]. \tag{6.165}$$

Using (6.165) in (6.163) gives

$$\hat{F}_{ml1} = c\mathbf{v}_1^H \left\{\mathbf{I} - \mathbf{V}_I(\mathbf{V}_I^H\mathbf{V}_I)^{-1}\mathbf{V}_I^H\right\}\mathbf{X}, \tag{6.166}$$

where c is the normalizing constant,

$$c = \left\{\mathbf{v}_1^H[\mathbf{I} - \mathbf{V}_I(\mathbf{V}_I^H\mathbf{V}_I)^{-1}\mathbf{V}_I^H]\mathbf{v}_1\right\}^{-1}. \tag{6.167}$$

The matrix in the brace is just a projection matrix onto the subspace defined by the columns of $\mathbf{V}_I$ (the interference subspace):

$$\mathbf{P}_I \triangleq \mathbf{V}_I(\mathbf{V}_I^H\mathbf{V}_I)^{-1}\mathbf{V}_I^H. \tag{6.168}$$

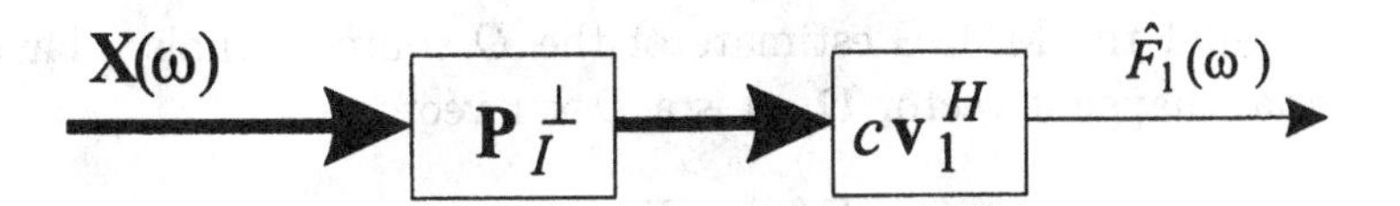

Figure 6.29 Optimum beamformer: multiple plane waves.

The complement of the projection matrix is

$$\mathbf{P}_I^\perp = \mathbf{I} - \mathbf{P}_I. \tag{6.169}$$

Then we can write

$$\hat{F}_{ml1} = c\mathbf{v}_1^H \mathbf{P}_I^\perp \mathbf{X}. \tag{6.170}$$

The representation in (6.169) and (6.170) emphasizes that when we want to simultaneously find the ML estimate of D plane-wave signal waveforms or construct a $D \times N$ MVDR processor, the optimum receiver treats the ith signal ($i = 1, 2, \cdots, D$ as the desired signal and the other $(D-1)$ signals as interferers and puts "perfect" nulls on them. Note that this result only corresponds to (6.124) in the limiting case of large σ_I^2. The optimum beamformer is shown in Figure 6.29.

6.5.2 MMSE Processors

In this case the signal component of the received waveform consists of D plane-wave signals arriving from different directions. The source signals are sample functions from stationary zero-mean random processes that are uncorrelated with the noise field but may be correlated with each other. We define the **source signal snapshot vector** as a $D \times 1$ complex vector,

$$\mathbf{F}(\omega) = \begin{bmatrix} F_1(\omega) \\ F_2(\omega) \\ \vdots \\ F_D(\omega) \end{bmatrix}. \tag{6.171}$$

The source spectral matrix is $\mathbf{S_f}$ and is assumed to be of rank D. The array manifold matrix, $\mathbf{V}_s$, is an $N \times D$ matrix,

$$\mathbf{V}_s = \left[\; \mathbf{v}(\omega : \mathbf{k}_1) \; \vdots \; \mathbf{v}(\omega : \mathbf{k}_2) \; \vdots \; \cdots \; \vdots \; \mathbf{v}(\omega : \mathbf{k}_D) \; \right], \tag{6.172}$$

where we have suppressed the ω dependence on the left side of (6.172). The spectral matrix of the input is

$$\mathbf{S_x} = \mathbf{V}_s \mathbf{S_f} \mathbf{V}_s^H + \mathbf{S_n}. \tag{6.173}$$

We want to find the MMSE estimate of the D source signals. Thus, the desired signal snapshot vector $\mathbf{D}(\omega)$ is a $D \times 1$ vector,

$$\mathbf{D}(\omega) = \mathbf{F}(\omega), \tag{6.174}$$

and $\mathbf{H}_o$ will be a $D \times N$ matrix processor whose output is the desired estimate $\hat{\mathbf{D}}(\omega)$. The mean square error is defined as

$$\xi = E\left\{\| \mathbf{D}(\omega) - \hat{\mathbf{D}}(\omega) \|^2\right\}. \tag{6.175}$$

Proceeding as in the single plane-wave case, we obtain

$$\mathbf{H}_o = \mathbf{S}_{\mathbf{dx}^H}\mathbf{S}_{\mathbf{x}}^{-1}, \tag{6.176}$$

where

$$\mathbf{S}_{\mathbf{dx}^H} = \mathbf{S}_{\mathbf{f}}\mathbf{V}_s^H. \tag{6.177}$$

Taking the inverse of $\mathbf{S}_{\mathbf{x}}(\omega)$ gives

$$\mathbf{S}_{\mathbf{x}}^{-1} = \mathbf{S}_{\mathbf{n}}^{-1}\left[\mathbf{I} - \mathbf{V}_s(\mathbf{I} + \mathbf{S}_{\mathbf{f}}\mathbf{V}_s^H\mathbf{S}_{\mathbf{n}}^{-1}\mathbf{V}_s)^{-1}\mathbf{S}_{\mathbf{f}}\mathbf{V}_s^H\mathbf{S}_{\mathbf{n}}^{-1}\right], \tag{6.178}$$

where the ω and $\mathbf{k}_i (i = 1, \cdots, D)$ dependence are suppressed. Then using (6.177) and (6.178) in (6.176), we have

$$\begin{aligned}
\mathbf{H}_o &= \mathbf{S}_{\mathbf{f}}\mathbf{V}^H\mathbf{S}_{\mathbf{n}}^{-1}\left[\mathbf{I} - \mathbf{V}(\mathbf{I} + \mathbf{S}_{\mathbf{f}}\mathbf{V}^H\mathbf{S}_{\mathbf{n}}^{-1}\mathbf{V})^{-1}\mathbf{S}_{\mathbf{f}}\mathbf{V}^H\mathbf{S}_{\mathbf{n}}^{-1}\right] \\
&= \mathbf{S}_{\mathbf{f}}\mathbf{V}^H\mathbf{S}_{\mathbf{n}}^{-1} - \mathbf{S}_{\mathbf{f}}\mathbf{V}^H\mathbf{S}_{\mathbf{n}}^{-1}\mathbf{V}(\mathbf{I} + \mathbf{S}_{\mathbf{f}}\mathbf{V}^H\mathbf{S}_{\mathbf{n}}^{-1}\mathbf{V})^{-1} \times \\
&\quad \mathbf{S}_{\mathbf{f}}\mathbf{V}^H\mathbf{S}_{\mathbf{n}}^{-1}.
\end{aligned} \tag{6.179}$$

This can be put in more convenient form by rewriting the first term. Then

$$\begin{aligned}
\mathbf{H}_o &= \Big\{(\mathbf{I} + \mathbf{S}_{\mathbf{f}}\mathbf{V}^H\mathbf{S}_{\mathbf{n}}^{-1}\mathbf{V})(\mathbf{I} + \mathbf{S}_{\mathbf{f}}\mathbf{V}^H\mathbf{S}_{\mathbf{n}}^{-1}\mathbf{V})^{-1}\mathbf{S}_{\mathbf{f}}\mathbf{V}^H\mathbf{S}_{\mathbf{n}}^{-1} \\
&\quad -\mathbf{S}_{\mathbf{f}}\mathbf{V}^H\mathbf{S}_{\mathbf{n}}^{-1}\mathbf{V}(\mathbf{I} + \mathbf{S}_{\mathbf{f}}\mathbf{V}^H\mathbf{S}_{\mathbf{n}}^{-1}\mathbf{V})^{-1}\mathbf{S}_{\mathbf{f}}\mathbf{V}^H\mathbf{S}_{\mathbf{n}}^{-1}\Big\},
\end{aligned} \tag{6.180}$$

which reduces to,

$$\boxed{\mathbf{H}_o = (\mathbf{I} + \mathbf{S}_{\mathbf{f}}\mathbf{V}^H\mathbf{S}_{\mathbf{n}}^{-1}\mathbf{V})^{-1}\mathbf{S}_{\mathbf{f}}\mathbf{V}^H\mathbf{S}_{\mathbf{n}}^{-1}.} \tag{6.181}$$

This result is the multiple plane-wave generalization of (6.47). The optimum processor is shown in Figure 6.30. Note that the first matrix operation

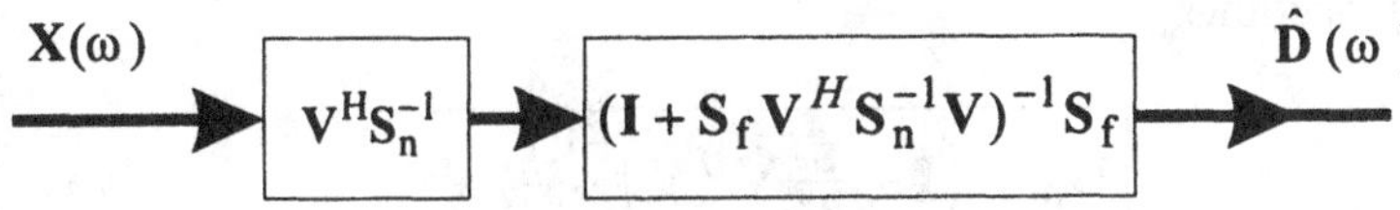

$\mathbf{X}(\omega)$ → $\mathbf{V}^H\mathbf{S}_n^{-1}$ → $(\mathbf{I}+\mathbf{S}_f\mathbf{V}^H\mathbf{S}_n^{-1}\mathbf{V})^{-1}\mathbf{S}_f$ → $\hat{\mathbf{D}}(\omega)$

Figure 6.30 MMSE processor, multiple plane-wave signals.

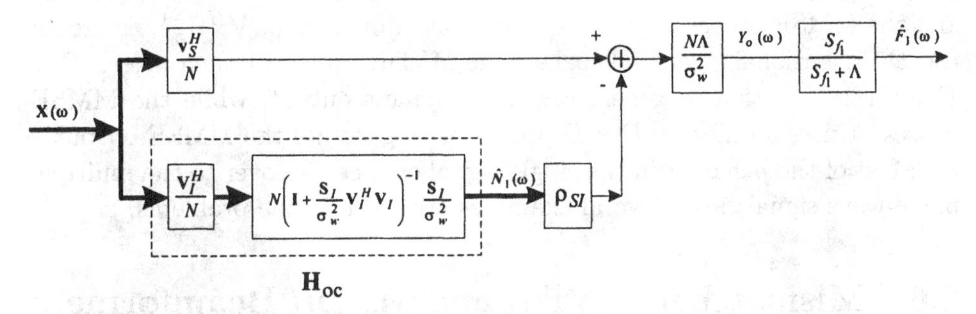

Figure 6.31 MMSE filter for single signal (temporally uncorrelated).

is generating a set of D sufficient statistics that are then multiplied by a matrix that takes into account the signal spectral matrix and generates $\hat{\mathbf{D}}(\omega)$. Note also that the ith output is the MMSE estimate of the corresponding source signal $F_i(\omega)$. In other words, minimizing ξ in (6.175) is equivalent to minimizing each ξ_i,

$$\xi_i = E\left\{\left|D_i(\omega) - \hat{D}_i(\omega)\right|^2\right\}. \tag{6.182}$$

Therefore, we would expect that the multiple plane-wave signal problem is equivalent to D single plane-wave signal problems in which the other $D-1$ signals are treated exactly as if they were noise (see Problem 6.5.5). If the signal of interest (e.g., $F_1(\omega)$) is uncorrelated from the other $D-1$ signals, then,

$$\mathbf{S_f} = \left[\begin{array}{c:c} S_{f_1} & 0 \\ \hdashline 0 & \mathbf{S}_I \end{array}\right]. \tag{6.183}$$

In this case the processor in Figure 6.30 can be partitioned into the configuration shown in Figure 6.31. We observe that this is exactly the distortionless processor shown in Figure 6.10 followed by a scalar gain.

A special case of interest is when the noise is temporally and spatially uncorrelated. If, in addition, the spectral height is the same at each sensor, then

$$\mathbf{S_n}(\omega) = \sigma_w^2\mathbf{I} \tag{6.184}$$

and $\mathbf{H}_o(\omega)$ becomes,

$$\boxed{\mathbf{H}_o = \left(\mathbf{I} + \frac{\mathbf{S_f}}{\sigma_w^2}\mathbf{V}^H\mathbf{V}\right)^{-1}\frac{\mathbf{S_f}}{\sigma_w^2}\mathbf{V}^H.} \tag{6.185}$$

Comparing (6.181) and (6.158), we see that both the MVDR processor and the MMSE processor use the same matrix operation, $\mathbf{V}^H\mathbf{S}_\mathbf{n}^{-1}$, to create the D-dimensional signal subspace. The MVDR processor then uses a $D \times D$ matrix operation to generate a distortionless output, while the MMSE processor uses a different $D \times D$ processor to generate the MMSE output.

Most of the behavior in the single-signal case carries over to the multiple plane-wave signal case. Several examples are given in the problems.

6.6 Mismatched MVDR and MPDR Beamformers

6.6.1 Introduction

In Section 6.2, MVDR and MPDR beamformers were derived and their performance was analyzed. In this section, their behavior is analyzed when they are mismatched. As a starting point, we repeat the formulas for the ideal MVDR and MPDR beamformers.

From (6.14),

$$\mathbf{w}_{mvdr}^H = \Lambda\mathbf{v}_m^H\mathbf{S}_\mathbf{n}^{-1} \qquad \text{(ideal MVDR).} \tag{6.186}$$

From (6.71),

$$\mathbf{w}_{mpdr}^H = \Lambda_1\mathbf{v}_m^H\mathbf{S}_\mathbf{x}^{-1} \qquad \text{(ideal MPDR),} \tag{6.187}$$

where Λ_1 is the reciprocal of the denominator in (6.71). In (6.186) and (6.187), we assume the beamformer is designed assuming the array manifold vector of the desired signal is $\mathbf{v}_m$. The subscript m denotes model. The ω and $\mathbf{k}_s$ are suppressed.

In a typical application the beamformer is scanned across the region of (θ, ϕ) space of interest by changing $\mathbf{v}_m$ in discrete steps. For example, for a standard linear array, one might scan across u-space by changing the steering direction in $u_\Delta = 2/N$ steps. With this step size, conventional beams pointed at different steps are orthogonal. At each step, the MVDR or MPDR beamformer is constructed, assuming the signal of interest arrives along the steering direction $\mathbf{v}_m$. However, all signals in the interval, $-1/N \leq u \leq 1/N$, must be treated for the beamformer pointed at $u = 0$. Therefore, the effect of this DOA mismatch must be considered.

In many applications, the step size can be reduced. This reduces the signal mismatch problem but increases the time required to scan a given area.

There are other sources of mismatch. The sources of mismatch can be divided into three cases.

Case 1

The steering vector in (6.186) or (6.187) is unequal to $\mathbf{v}_s$. This can happen in two ways.

Case 1a

The steering vector is pointed at the wrong point in frequency-wavenumber space because

$$\mathbf{a}_s \neq \mathbf{a}_m, \tag{6.188}$$

$$\omega_s \neq \omega_m, \tag{6.189}$$

or the velocity of propagation is different from that used in the model (this difference may occur in sonar applications).

In (6.188), the steering vector is pointed at the wrong coordinates in azimuth and elevation, but the frequency is correct. In (6.189), the steering vector is pointed at the correct coordinate in azimuth and elevation, but the frequency is incorrect, so the wavenumber is incorrect.

Case 1b

The steering vector is pointed at the correct point in frequency-wavenumber space, but the array has been perturbed. Array perturbations could include errors in sensor gain and phase or location of sensors. Then,

$$\mathbf{v}_m(\mathbf{k}_d) \neq \mathbf{v}_s(\mathbf{k}_d). \tag{6.190}$$

Recall that we studied this problem in Chapter 2 in the context of the change in beam pattern and array gain.

A combination of Cases 1a and 1b is also possible.

Case 2

The spatial spectral estimates will not be exact if there is a finite amount of data. Thus,

$$\hat{\mathbf{S}}_{\mathbf{n}} \neq \mathbf{S}_{\mathbf{n}} \tag{6.191}$$

and

$$\hat{\mathbf{S}}_{\mathbf{x}} \neq \mathbf{S}_{\mathbf{x}}. \tag{6.192}$$

We might expect that, depending on how the various errors are modeled, that Cases 1a, 1b, and 2 would all have a similar effect on the beamformer performance. The advantage of recognizing this similarity is that, when the MVDR or MPDR beamformer is modified to improve performance for one kind of mismatch, the modification will generally improve performance for the other types of mismatch.

In Section 6.6.2, we study the case in which the signal mismatch is due to the desired signal arriving from a different direction than the direction that was used to design the beamformer. This is referred to as DOA mismatch.

In Section 6.6.3, we consider the case in which the steering vector mismatch is due to array perturbations.

In Section 6.6.4, we introduce a technique referred to as **diagonal loading** to make the beamformer more robust in the presence of mismatch.

In Section 6.6.5, we summarize our results and discuss other techniques for improving robustness.

We defer our analysis of Case 2, the finite data problem, until Chapter 7. We will find that the techniques that we use to improve robustness in the presence of signal mismatch will usually improve robustness to spatial spectral matrix mismatch.

6.6.2 DOA Mismatch

In this section, we discuss the problem of signal mismatch in MVDR and MPDR beamformers. We also consider the conventional beamformer for comparison.[5]

The weight vectors for the three beamformers are:

$$\mathbf{w}_c^H = \frac{\mathbf{v}_m^H}{N}, \tag{6.193}$$

$$\mathbf{w}_{mvdr}^H = \frac{\mathbf{v}_m^H \mathbf{S}_\mathbf{n}^{-1}}{\mathbf{v}_m^H \mathbf{S}_\mathbf{n}^{-1} \mathbf{v}_m}, \tag{6.194}$$

and

$$\mathbf{w}_{mpdr}^H = \frac{\mathbf{v}_m^H \mathbf{S}_\mathbf{x}^{-1}}{\mathbf{v}_m^H \mathbf{S}_\mathbf{x}^{-1} \mathbf{v}_m}, \tag{6.195}$$

where

$$\mathbf{v}_m \triangleq \mathbf{v}(\omega : \mathbf{k}_m). \tag{6.196}$$

[5] Our discussion follows Cox [Cox73].

Using the frequency-domain snapshot model, the input is

$$\mathbf{X}(\omega) = \mathbf{v}_a F(\omega) + \mathbf{N}(\omega), \tag{6.197}$$

where

$$\mathbf{v}_a \triangleq \mathbf{v}(\omega : \mathbf{k}_a) \tag{6.198}$$

is the actual array manifold vector. The vector $\mathbf{v}_a$ can be different from $\mathbf{v}_m$ because the signal is actually arriving from a different direction than we assumed, or it could be different because the array geometry is different than our model. In this section, we initially assume that $\mathbf{v}_a$ is deterministic.

If there are interfering plane waves, they are included in $\mathbf{N}(\omega)$. For simplicity, we assume that the noise power is the same at each sensor. We can define a normalized noise covariance matrix as

$$\mathbf{S}_\mathbf{n} = S_n \boldsymbol{\rho}_\mathbf{n}. \tag{6.199}$$

The output power is,

$$\begin{aligned} P_o &= \sigma_s^2 \left| \mathbf{w}^H \mathbf{v}_a \right|^2 + \mathbf{w}^H \mathbf{S}_\mathbf{n} \mathbf{w} \\ &= P_s + P_n. \end{aligned} \tag{6.200}$$

Therefore, the output *SNR* is

$$SNR_o \triangleq \frac{P_s}{P_n} = \frac{\sigma_s^2 \left| \mathbf{w}^H \mathbf{v}_a \right|^2}{\mathbf{w}^H \mathbf{S}_\mathbf{n} \mathbf{w}}. \tag{6.201}$$

The signal mismatch has two negative effects: it can lower the output signal power P_s, and it can raise the output noise power P_n.

The array gain is

$$A = \frac{\left| \mathbf{w}^H \mathbf{v}_a \right|^2}{\mathbf{w}^H \boldsymbol{\rho}_\mathbf{n} \mathbf{w}}. \tag{6.202}$$

The array gain for the three beamformers in (6.193)–(6.195) in the presence of mismatch will be computed. First, consider the conventional beamformer defined by (6.193).

6.6.2.1 Conventional beamformer

The output SNR_o is[6]

$$SNR_o = \frac{\sigma_s^2 \left| \mathbf{v}_m^H \mathbf{v}_a \right|^2}{\mathbf{v}_m^H \mathbf{S}_\mathbf{n} \mathbf{v}_m} = \frac{\sigma_s^2 N^2 \left| B_c(\mathbf{v}_a : \mathbf{v}_m) \right|^2}{\mathbf{v}_m^H \mathbf{S}_\mathbf{n} \mathbf{v}_m}. \tag{6.203}$$

[6]The notation for $B_c(\mathbf{v}_a : \mathbf{v}_m)$ is used to emphasize the dependence on $\mathbf{v}$.

As expected, the output signal power is reduced as a function of the beam pattern. The denominator is unchanged.

The array gain is

$$\boxed{A_c = \frac{N^2|B_c(\mathbf{v}_a:\mathbf{v}_m)|^2}{\mathbf{v}_m^H \boldsymbol{\rho}_\mathbf{n} \mathbf{v}_m}.} \tag{6.204}$$

The ratio of the mismatched array gain to matched array gain is just the power pattern of the array

$$\frac{A_c(\mathbf{v}_a)}{A_c(\mathbf{v}_m)} = |B_c(\mathbf{v}_a : \mathbf{v}_m)|^2 . \tag{6.205}$$

We now consider the MVDR beamformer.

6.6.2.2 MVDR beamformer

In this discussion, it is useful to introduce the generalized cosine notation,

$$\begin{aligned} \cos^2\left(\mathbf{v}_m, \mathbf{v}_a : \boldsymbol{\rho}_\mathbf{n}^{-1}\right) &= \cos^2\left(\mathbf{v}_a, \mathbf{v}_m : \boldsymbol{\rho}_\mathbf{n}^{-1}\right) \\ &\triangleq \frac{\left|\mathbf{v}_m^H \boldsymbol{\rho}_\mathbf{n}^{-1} \mathbf{v}_a\right|^2}{\left(\mathbf{v}_m^H \boldsymbol{\rho}_\mathbf{n}^{-1} \mathbf{v}_m\right)\left(\mathbf{v}_a^H \boldsymbol{\rho}_\mathbf{n}^{-1} \mathbf{v}_a\right)}, \end{aligned} \tag{6.206}$$

which is an inner product in a $\boldsymbol{\rho}_\mathbf{n}^{-1}$ space.

Substituting (6.194) into (6.200) gives the output signal power,

$$\begin{aligned} P_s &= \sigma_s^2 \left|\frac{\mathbf{v}_m^H \mathbf{S}_\mathbf{n}^{-1} \mathbf{v}_a}{\mathbf{v}_m^H \mathbf{S}_\mathbf{n}^{-1} \mathbf{v}_m}\right|^2 = \sigma_s^2 \left|\frac{\mathbf{v}_m^H \boldsymbol{\rho}_\mathbf{n}^{-1} \mathbf{v}_a}{\mathbf{v}_m^H \boldsymbol{\rho}_\mathbf{n}^{-1} \mathbf{v}_m}\right|^2 \\ &= \sigma_s^2 \left(\frac{\mathbf{v}_a^H \boldsymbol{\rho}_\mathbf{n}^{-1} \mathbf{v}_a}{\mathbf{v}_m^H \boldsymbol{\rho}_\mathbf{n}^{-1} \mathbf{v}_m}\right) \cos^2\left(\mathbf{v}_m, \mathbf{v}_a : \boldsymbol{\rho}_\mathbf{n}^{-1}\right). \end{aligned} \tag{6.207}$$

This can also be written in terms of the beam pattern of the MVDR beamformer.

$$P_s = \sigma_s^2 \left|B_{mvdr}(\mathbf{v}_a : \mathbf{v}_m)\right|^2 . \tag{6.208}$$

The output noise power is

$$P_n = \Lambda = \left[\mathbf{v}_m^H \mathbf{S}_\mathbf{n}^{-1} \mathbf{v}_m\right]^{-1}, \tag{6.209}$$

which is unchanged from the nominal model. The output SNR_o is

$$SNR_o = \frac{\sigma_s^2 \left|\mathbf{v}_m^H \mathbf{S}_\mathbf{n}^{-1} \mathbf{v}_a\right|^2}{\mathbf{v}_m^H \mathbf{S}_\mathbf{n}^{-1} \mathbf{v}_m} = \frac{\sigma_s^2}{\sigma_n^2} \frac{\left|\mathbf{v}_m^H \boldsymbol{\rho}_\mathbf{n}^{-1} \mathbf{v}_a\right|^2}{\mathbf{v}_m^H \boldsymbol{\rho}_\mathbf{n}^{-1} \mathbf{v}_m} \tag{6.210}$$

and the mismatched array gain is

$$A_{mvdr} = \frac{\left|\mathbf{v}_m^H \boldsymbol{\rho}_{\mathbf{n}}^{-1} \mathbf{v}_a\right|^2}{\mathbf{v}_m^H \boldsymbol{\rho}_{\mathbf{n}}^{-1} \mathbf{v}_m}. \tag{6.211}$$

The array gain can also be written as

$$A_{mvdr}(\mathbf{v}_a : \mathbf{v}_m) = \frac{|B_{mvdr}(\mathbf{v}_a : \mathbf{v}_m)|^2}{(\mathbf{v}_m^H \boldsymbol{\rho}_{\mathbf{n}}^{-1} \mathbf{v}_m)^{-1}} = |B_{mvdr}(\mathbf{v}_a : \mathbf{v}_m)|^2 A_o(\mathbf{v}_m). \tag{6.212}$$

We can normalize this expression by dividing by $A_o(\mathbf{v}_m)$,

$$\frac{A_{mvdr}(\mathbf{v}_a : \mathbf{v}_m)}{A_o(\mathbf{v}_m)} = |B_{mvdr}(\mathbf{v}_a : \mathbf{v}_m)|^2. \tag{6.213}$$

Thus, the mismatch degradation is completely characterized by the optimum MVDR beam pattern steered to $\mathbf{v}_m$. As long as the interference is in the sidelobe region, the main lobe of the beam pattern is well-behaved, so we would anticipate a gradual degradation due to mismatch.

The expression in (6.212) includes two effects: the beamformer mismatch and the different matched array gain because $\mathbf{v}_a \neq \mathbf{v}_m$. We can isolate the mismatch effect by dividing by $A_o(\mathbf{v}_a)$,

$$\frac{A_{mvdr}(\mathbf{v}_a : \mathbf{v}_m)}{A_o(\mathbf{v}_a)} = |B_{mvdr}(\mathbf{v}_a : \mathbf{v}_m)|^2 \frac{A_o(\mathbf{v}_m)}{A_o(\mathbf{v}_a)}. \tag{6.214}$$

We can also write the expression in (6.214) as

$$\frac{A_{mvdr}(\mathbf{v}_a : \mathbf{v}_m)}{A_o(\mathbf{v}_a)} = \frac{\left|\mathbf{v}_m^H \boldsymbol{\rho}_{\mathbf{n}}^{-1} \mathbf{v}_a\right|^2}{\left(\mathbf{v}_m^H \boldsymbol{\rho}_{\mathbf{n}}^{-1} \mathbf{v}_m\right)\left(\mathbf{v}_a^H \boldsymbol{\rho}_{\mathbf{n}}^{-1} \mathbf{v}_a\right)}. \tag{6.215}$$

Using (6.206), the right side can be written as the generalized cosine,

$$\frac{A_{mvdr}\big|_{\mathbf{v}_m \neq \mathbf{v}_a}}{A_{mvdr}\big|_{\mathbf{v}_m = \mathbf{v}_a}} = \cos^2\left(\mathbf{v}_m, \mathbf{v}_a : \boldsymbol{\rho}_{\mathbf{n}}^{-1}\right), \tag{6.216}$$

which is always less than one. Its value depends on the angular separation of $\mathbf{v}_m$ and $\mathbf{v}_a$ and the noise field. The effect of the noise field enters through its eigenvalues and eigenvectors.

Denoting the eigenvalues and eigenvectors of $\boldsymbol{\rho}_{\mathbf{n}}$ by $(\lambda_1, \cdots, \lambda_N)$ and $(\boldsymbol{\Phi}_1, \cdots \boldsymbol{\Phi}_N)$, respectively, we have

$$\cos^2(\mathbf{v}_m, \mathbf{v}_a : \boldsymbol{\rho}_{\mathbf{n}}^{-1}) = \frac{\left|\mathbf{v}_m^H \boldsymbol{\Phi}\boldsymbol{\Lambda}^{-1}\boldsymbol{\Phi}^H \mathbf{v}_a\right|^2}{(\mathbf{v}_m \boldsymbol{\Phi}\boldsymbol{\Lambda}^{-1}\boldsymbol{\Phi}^H \mathbf{v}_m)(\mathbf{v}_a \boldsymbol{\Phi}\boldsymbol{\Lambda}^{-1}\boldsymbol{\Phi}^H \mathbf{v}_a)} = \frac{\left|\sum_{i=1}^{N} \frac{\rho_{mi}^* \rho_{si}}{\lambda_i}\right|^2}{\sum_{i=1}^{N} \frac{|\rho_{mi}|^2}{\lambda_i} \sum_{i=1}^{N} \frac{|\rho_{si}|^2}{\lambda_i}}, \tag{6.217}$$

where ρ_{mi} is the correlation between $\mathbf{v}_m$ and $\boldsymbol{\Phi}_i$ and ρ_{si} is the correlation between $\mathbf{v}_a$ and $\boldsymbol{\Phi}_i$. (Recall that $\sum \lambda_i = N$ because of the normalization.) Thus, the $\cos^2$ function emphasizes the components of $\mathbf{v}_m$ and $\mathbf{v}_a$ corresponding to small eigenvalues and de-emphasizes those corresponding to large eigenvalues.

Example 6.6.1

Consider a standard 10-element linear array. The interference consists of a single plane-wave interferer plus white noise. We assume the interferer is at $u_I = 0.30$. We assume $|u_a| \leq 0.1$. In Figure 6.32, we plot the expressions in (6.213) for an $INR = 10$ dB ($INR \triangleq \frac{\sigma_I^2}{\sigma_w^2}$). The plot is insensitive to the *INR* value.

Example 6.6.2 (continuation)

Consider the same model as in Example 6.6.1 except u_I is inside the main lobe. In Figure 6.33(a), we assume $u_I = 0.0433$ and plot the expression in (6.213) for several values of *INR*. In Figure 6.33(b), we assume $u_I = 0.02$ and plot the same results.

We see that the array gain is very sensitive to mismatch for both values of u_I. If the actual signal DOA moves away from the interferers location, u_I, the normalized array gain increases dramatically. However, if the actual signal DOA moves toward u_I, the normalized array gain approaches zero. In some applications, this variation may be acceptable. However, in most cases, we will want to impose additional constraints in order to maintain reasonably constant performance in the presence of small DOA mismatch.

6.6.2.3 MPDR beamformer

In this beamformer, the spectral matrix of the *entire* input is inverted instead of the noise-only input. From the results with discrete interference in Section 6.2, we can anticipate what will happen when $\mathbf{v}_m \neq \mathbf{v}_a$. The processor will treat the signal along $\mathbf{v}_a$ as a discrete interfering signal and will put a partial null in its direction. In the limit, as σ_s^2 approaches infinity, it will become a perfect null and the signal will be completely eliminated.

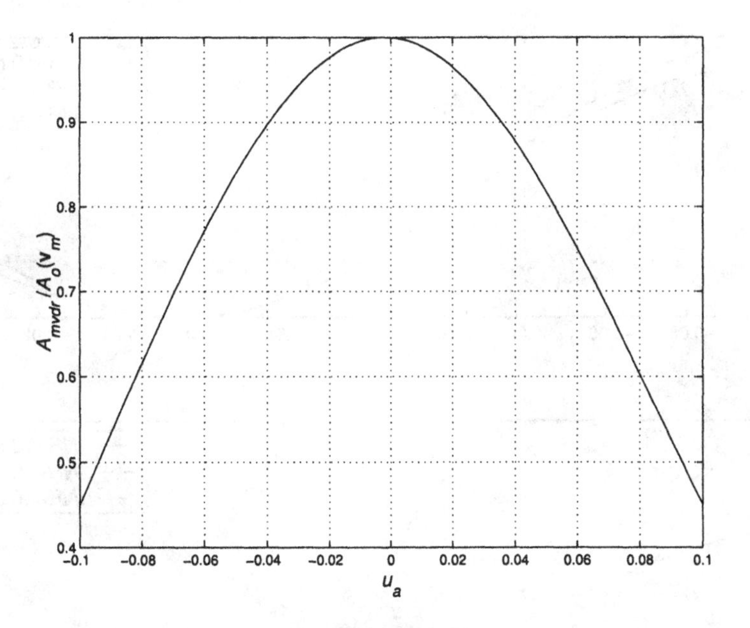

Figure 6.32 Normalized array gain of MVDR beamformer in the presence of mismatch (normalized by $A_o(\mathbf{v}_m)$) single plane-wave interferer with $INR =$ 10 dB at $u_I = 0.3$, $u_m = 0$.

The array gain is obtained from (6.73),

$$A_{mpdr} = \frac{\left|\mathbf{v}_m^H \mathbf{S}_\mathbf{x}^{-1} \mathbf{v}_a\right|^2}{\mathbf{v}_m^H \mathbf{S}_\mathbf{x}^{-1} \boldsymbol{\rho}_\mathbf{n} \mathbf{S}_\mathbf{x}^{-1} \mathbf{v}_m}. \tag{6.218}$$

The spatial spectral matrix is

$$\mathbf{S}_\mathbf{x} = \sigma_s^2 \mathbf{v}_a \mathbf{v}_a^H + \sigma_n^2 \boldsymbol{\rho}_\mathbf{n}, \tag{6.219}$$

where

$$\sigma_n^2 = \sigma_w^2 + \frac{1}{N} tr\left[\mathbf{S}_\mathbf{c}\right]. \tag{6.220}$$

In (6.220), $\mathbf{S}_\mathbf{c}$ denotes the non-white noise component. We use the matrix inversion lemma to obtain

$$\mathbf{S}_\mathbf{x}^{-1} = \frac{1}{\sigma_n^2} \boldsymbol{\rho}_\mathbf{n}^{-1} \left\{ \mathbf{I} - \mathbf{v}_a \mathbf{v}_a^H \boldsymbol{\rho}_\mathbf{n}^{-1} \frac{\sigma_s^2}{\sigma_n^2} \left(1 + \frac{\sigma_s^2}{\sigma_n^2} \mathbf{v}_a^H \boldsymbol{\rho}_\mathbf{n}^{-1} \mathbf{v}_a \right)^{-1} \right\}. \tag{6.221}$$

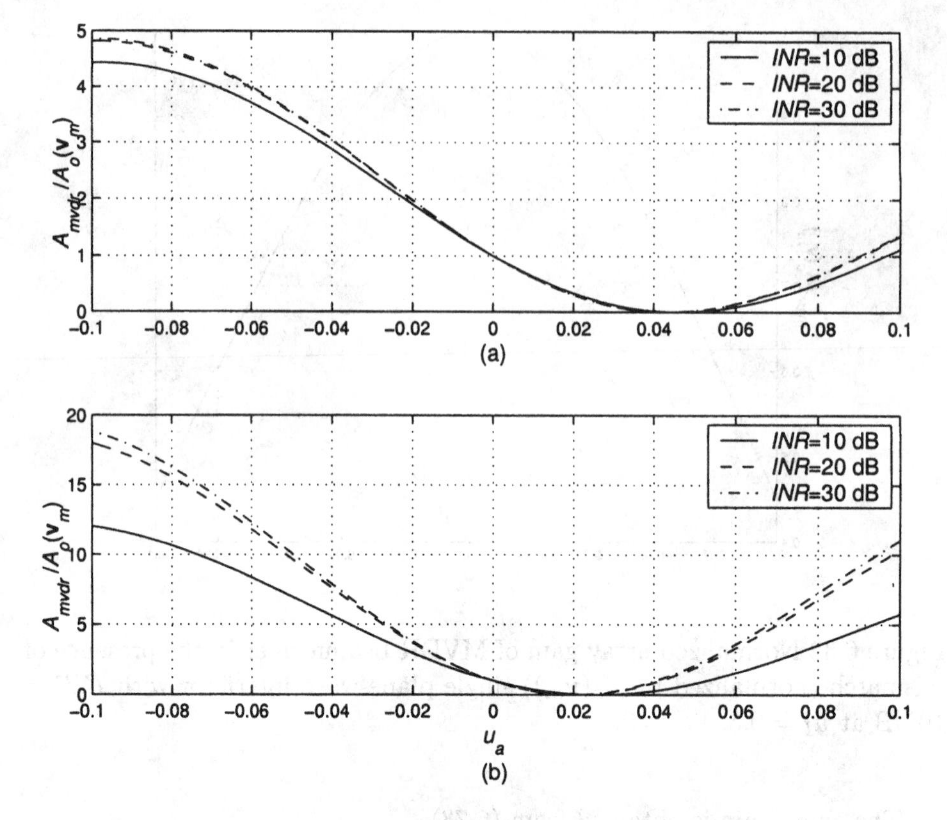

Figure 6.33 Normalized array gain, $A_{mvdr}(\mathbf{v}_a : \mathbf{v}_m)/A_o(\mathbf{v}_m)$, of MVDR beamformer in the presence of mismatch: (a) $u_I = 0.0433$; (b) $u_I = 0.02$.

To evaluate the numerator in (6.218), we write

$$\mathbf{v}_m^H \mathbf{S}_{\mathbf{x}}^{-1} \mathbf{v}_a = \frac{1}{\sigma_n^2} \left\{ \mathbf{v}_m^H \boldsymbol{\rho}_{\mathbf{n}}^{-1} \mathbf{v}_a \left[1 - \mathbf{v}_a^H \boldsymbol{\rho}_{\mathbf{n}}^{-1} \mathbf{v}_a \frac{\sigma_s^2}{\sigma_n^2} \left(1 + \frac{\sigma_s^2}{\sigma_n^2} \mathbf{v}_a^H \boldsymbol{\rho}_{\mathbf{n}}^{-1} \mathbf{v}_a \right)^{-1} \right] \right\}. \tag{6.222}$$

We denote the output *SNR* of an MVDR processor that is matched to $\mathbf{v}_a$ as M,

$$M = \frac{\sigma_s^2}{\sigma_n^2} \mathbf{v}_a^H \boldsymbol{\rho}_{\mathbf{n}}^{-1} \mathbf{v}_a. \tag{6.223}$$

Then (6.222) can be written as

$$\mathbf{v}_m^H \mathbf{S}_{\mathbf{x}}^{-1} \mathbf{v}_a = \frac{\mathbf{v}_m^H \boldsymbol{\rho}_{\mathbf{n}}^{-1} \mathbf{v}_a}{\sigma_n^2 (1 + M)}. \tag{6.224}$$

To evaluate the denominator in (6.218), we write

$$\mathbf{v}_m^H \mathbf{S}_{\mathbf{x}}^{-1} \boldsymbol{\rho}_{\mathbf{n}} \mathbf{S}_{\mathbf{x}}^{-1} \mathbf{v}_m = \mathbf{v}_m^H \frac{1}{\sigma_n^2} \boldsymbol{\rho}_{\mathbf{n}}^{-1} \left[\mathbf{I} - \mathbf{v}_a \mathbf{v}_a^H \boldsymbol{\rho}_{\mathbf{n}}^{-1} \frac{\sigma_s^2}{\sigma_n^2} (1+M)^{-1} \right] \cdot \boldsymbol{\rho}_{\mathbf{n}} \frac{1}{\sigma_n^2} \boldsymbol{\rho}_{\mathbf{n}}^{-1} \left[\mathbf{I} - \mathbf{v}_a \mathbf{v}_a^H \boldsymbol{\rho}_{\mathbf{n}}^{-1} \frac{\sigma_s^2}{\sigma_n^2} (1+M)^{-1} \right] \mathbf{v}_m. \tag{6.225}$$

This reduces to

$$\mathbf{v}_m^H \mathbf{S}_{\mathbf{x}}^{-1} \boldsymbol{\rho}_{\mathbf{n}} \mathbf{S}_{\mathbf{x}}^{-1} \mathbf{v}_m = \frac{1}{\sigma_n^4} \mathbf{v}_m^H \boldsymbol{\rho}_{\mathbf{n}}^{-1} \mathbf{v}_m \left[1 - 2|B_{mvdr}|^2 \mathbf{v}_m^H \boldsymbol{\rho}_{\mathbf{n}}^{-1} \mathbf{v}_m \frac{\sigma_s^2}{\sigma_n^2} (1+M)^{-1} + |B_{mvdr}|^2 \mathbf{v}_m^H \boldsymbol{\rho}_{\mathbf{n}}^{-1} \mathbf{v}_m \frac{M}{(1+M)^2} \frac{\sigma_s^2}{\sigma_n^2} \right], \tag{6.226}$$

where the argument of B_{mvdr} is suppressed. The form of this expression can be simplified if we use the generalized cosine notation and the relationship

$$\sin^2(\mathbf{v}_m, \mathbf{v}_a; \boldsymbol{\rho}_{\mathbf{n}}^{-1}) = 1 - \cos^2(\mathbf{v}_m, \mathbf{v}_a; \boldsymbol{\rho}_{\mathbf{n}}^{-1}). \tag{6.227}$$

Using these definitions in (6.226), the denominator becomes

$$DEN = \frac{1}{\sigma_n^4} \mathbf{v}_m^H \boldsymbol{\rho}_n^{-1} \mathbf{v}_m (1+M)^2 \left[1 + (2M + M^2) \sin^2(\mathbf{v}_m, \mathbf{v}_a; \boldsymbol{\rho}_{\mathbf{n}}^{-1}) \right]. \tag{6.228}$$

Using (6.228) and the magnitude squared of (6.224) in (6.218) gives

$$A_{mpdr}(\mathbf{v}_a : \mathbf{v}_m) = \frac{\mathbf{v}_m^H \boldsymbol{\rho}_{\mathbf{n}}^{-1} \mathbf{v}_m \, |B_{mvdr}(\mathbf{v}_a : \mathbf{v}_m)|^2}{1 + [2M + M^2] \sin^2(\mathbf{v}_m, \mathbf{v}_a; \boldsymbol{\rho}_{\mathbf{n}}^{-1})}, \tag{6.229}$$

or equivalently,

$$\boxed{A_{mpdr} = \frac{\mathbf{v}_a^H \boldsymbol{\rho}_{\mathbf{n}}^{-1} \mathbf{v}_a \cos^2(\mathbf{v}_m, \mathbf{v}_a; \boldsymbol{\rho}_{\mathbf{n}}^{-1})}{1 + [2M + M^2] \sin^2(\mathbf{v}_m, \mathbf{v}_a; \boldsymbol{\rho}_{\mathbf{n}}^{-1})}}. \tag{6.230}$$

Note that either (6.229) or (6.230) can be written in terms of MVDR array gain in a perfectly matched environment,

$$A_{mpdr}(\mathbf{v}_a : \mathbf{v}_m) = \left\{ \frac{A_o(\mathbf{v}_m) |B_o\,(\mathbf{v}_a : \mathbf{v}_m)|^2}{1 + [2M + M^2] \sin^2(\mathbf{v}_m, \mathbf{v}_a; \boldsymbol{\rho}_{\mathbf{n}}^{-1})} \right\} = \left\{ \frac{A_o(\mathbf{v}_a) \cos^2(\mathbf{v}_m, \mathbf{v}_a; \boldsymbol{\rho}_{\mathbf{n}}^{-1})}{1 + [2M + M^2] \sin^2(\mathbf{v}_m, \mathbf{v}_a; \boldsymbol{\rho}_{\mathbf{n}}^{-1})} \right\}. \tag{6.231}$$

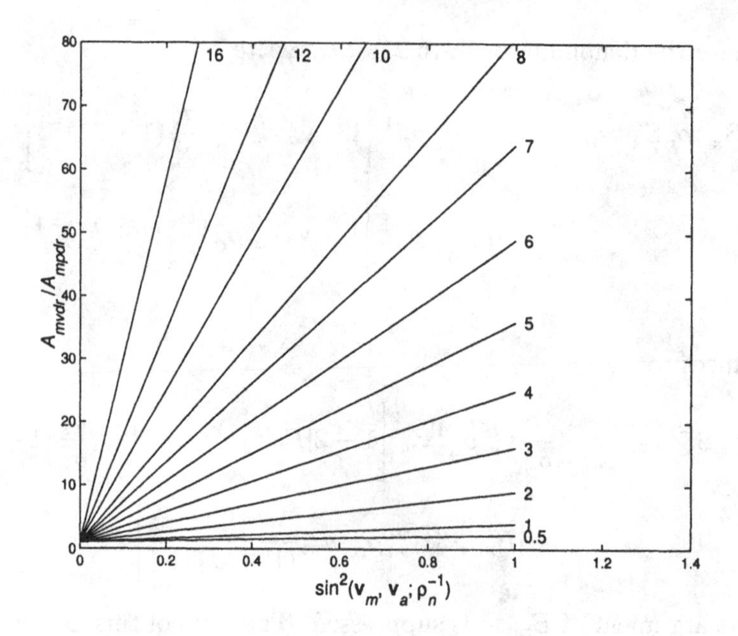

Figure 6.34 Ratio of A_{mvdr} to A_{mpdr} versus $\sin^2(\mathbf{v}_m, \mathbf{v}_a; \boldsymbol{\rho}_\mathbf{n}^{-1})$ for various M.

The numerator of (6.231) is just the array gain of the MVDR filter in the presence of the mismatch. Thus, the denominator of (6.231) indicates the effect of including the signal in the matrix inversion. The ratio of the two array gains is

$$\frac{A_{mvdr}(\mathbf{v}_a : \mathbf{v}_m)}{A_{mpdr}(\mathbf{v}_a : \mathbf{v}_m)} = 1 + [2M + M^2] \sin^2(\mathbf{v}_m, \mathbf{v}_a; \boldsymbol{\rho}_\mathbf{n}^{-1}). \tag{6.232}$$

The result in (6.232) is due to Cox [Cox73].

In Figure 6.34, the array gain ratio is plotted versus $\sin^2(\mathbf{v}_m, \mathbf{v}_a; \boldsymbol{\rho}_\mathbf{n}^{-1})$ for various values of the maximum *SNR*. Since the gain ratio depends on $M^2 \sin^2(\mathbf{v}_m, \mathbf{v}_a; \boldsymbol{\rho}_\mathbf{n}^{-1})$, large values of M can cause significant signal suppression. For example, if $M = 10$ and $\sin^2(\mathbf{v}_m, \mathbf{v}_a; \boldsymbol{\rho}_\mathbf{n}^{-1}) = 0.5$, then the array gain ratio is 61. The output *SNR* of $\mathbf{w}_o^H$ is 5.0 and the output *SNR* of $\mathbf{w}_{mpdr}^H$ is 0.082. As we would expect from our directional noise examples, weak signals suffer less suppression than strong signals.

The results in (6.230) and (6.232) are useful general expressions. However, examining some specific array geometries will give additional insight.

Note that, for white noise, (6.230) reduces to

$$A_{mpdr}(\mathbf{v}_a : \mathbf{v}_m) = \frac{N\, |\rho_{ma}|^2}{1 + [2M + M^2][1 - |\rho_{ma}|^2]}, \tag{6.233}$$

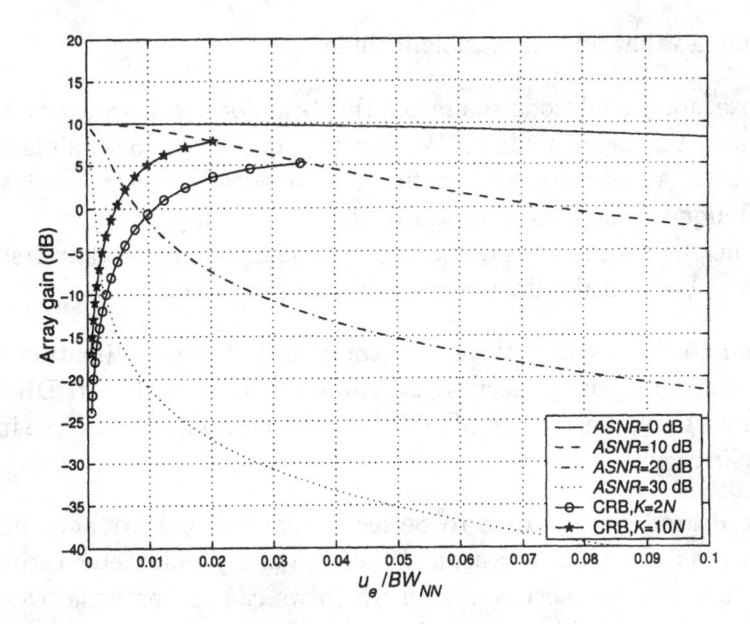

Figure 6.35 Array gain of MPDR beamformer versus $\frac{u_e}{BW_{NN}}$:$N = 10$, *ASNR* $= -10$ dB, $\cdots$, 30 dB.

where

$$|\rho_{ma}|^2 = \cos^2(\mathbf{v}_m, \mathbf{v}_a; \mathbf{I}) = |B_c(\mathbf{v}_a : \mathbf{v}_m)|^2, \tag{6.234}$$

and

$$M = N\frac{\sigma_s^2}{\sigma_w^2} \triangleq ASNR \tag{6.235}$$

is the optimum array output *SNR* in the presence of white noise. The actual output *SNR* is

$$SNR_o = \frac{\sigma_s^2}{\sigma_w^2} A_{mpdr}\left(\mathbf{v}_a : \mathbf{v}_m\right). \tag{6.236}$$

We consider an example to illustrate the behavior.

Example 6.6.3

Consider a standard N-element linear array where the interference is white noise. In Figure 6.35, we plot the array gain in (6.233) versus $\Delta u/BW_{NN}$ for various values of M. The result in Figure 6.35 is valid for $N \geq 10$. We see that if the *ASNR* is high (e.g., ≥ 10 dB) there is significant degradation for small $\Delta u/BW_{NN}$.

We see that as the *SNR* increases the performance degrades significantly with very small DOA mismatch. Therefore, if DOA mismatch is anticipated in the application of interest, then we need to find techniques to avoid the degradation illustrated in Figures 6.34 and 6.35.

Techniques that appear feasible include:

(i) Impose additional constraints on the beamformer to prevent (or decrease) the signal nulling. We consider quadratic constraints in Sections 6.6.4 and 6.10. We will find that quadratic constraints lead to a technique called **diagonal loading**. This technique designs the beamformer by assuming that the white noise σ_w^2 is higher than its actual value. We consider linear constraints in Section 6.7.

(ii) Eliminate (or reduce) the signal component before estimating $\mathbf{S_x}$. If we could achieve perfect signal elimination, then the MPDR beamformer would become an MVDR beamformer. This is possible in some applications.

(iii) The degradation relative to perfectly matched performance becomes worse as the *SNR* increases. When we study parameter estimation, we find that the accuracy with which we can estimate the DOA of a plane-wave signal increases with the *ASNR*. We show that for a single plane-wave signal impinging on a standard linear array in the presence of white Gaussian noise the root-mean-square error in estimating u_a is

$$\frac{(E[|\hat{u}_a - u_a|^2])^{\frac{1}{2}}}{BW_{NN}} = \left(\frac{3}{8K}\left[\frac{1}{ASNR} + \frac{1}{ASNR^2}\right]\frac{1}{1-\frac{1}{N^2}}\right)^{\frac{1}{2}} \tag{6.237}$$

for $ASNR \geq 10$ dB. In (6.237), K denotes the number of snapshots. The estimator for this case is simple to implement. In Figure 6.35, we have superimposed the normalized version of (6.237) onto the plot in Figure 6.35 for $ASNR \geq 10$ dB and relabeled the axis as $u_e/BW_{NN} = (\hat{u}_a - u_a)/BW_{NN}$. For our current purposes, it provides a guideline as to the maximum amount of mismatch that must be considered in a beamformer that includes a preliminary estimation step.

We refer to beamformers that attempt to preserve good performance in the presence of mismatch as **robust beamformers**. A minimal goal for robust beamformers is that their performance should never degrade below the performance of a conventional beamformer (or some other classical beamformer from Chapter 3 or 4) due to allowable mismatch scenarios. In the course of our development, we will see if this is a realistic goal. Note that this goal is not a useful mathematical statement because we have not identified any positive performance objective (such as maximum SNR_o in the absence of mismatch).

Before considering solutions to the DOA mismatch problem, we consider the array perturbation problem and show that it causes a similar degradation.

6.6.3 Array Perturbations

In this section, we discuss how array perturbations affect the performance of the MVDR beamformer and the MPDR beamformer. We recall from Chapter 2, that the effect of random errors in sensor gain and phase and sensor position was to add a constant floor to the beam pattern whose height was inversely proportional to the array gain in the presence of white noise, A_w. If the weight vector was normalized such that

$$\mathbf{w}^H \mathbf{v}_s = 1, \tag{6.238}$$

then

$$A_w = \left\{ \| \mathbf{w} \|^2 \right\}^{-1}, \tag{6.239}$$

so that, as the norm of the weight vector increased, the sensitivity to array perturbations increased.

In this section, we investigate the behavior of the output signal-to-noise ratio (SNR_o) of MVDR and MPDR beamformers in the presence of array perturbations.

This problem has been analyzed using various perturbation models in a number of references (e.g., Nitzberg [Nit76], Kleinberg [Kle80], Mucci and Pridham [MP81], Farrier [Far83], Quazi [Qua82], Godara [God85], [God86], and Youn and Un [YU94]).

The general frequency-domain snapshot model can be written as

$$\mathbf{X}(\omega) = \mathbf{V}(\boldsymbol{\psi}, \boldsymbol{\rho})\mathbf{F}(\omega) + \mathbf{W}(\omega), \tag{6.240}$$

where $\mathbf{F}(\omega)$ is a composite source signal matrix that contains both the desired signal $F_d(\omega)$ and the D plane-wave interfering signals $\mathbf{F}_I(\omega)$. The vector $\boldsymbol{\psi}$ describes the angles of arrival of the $D+1$ signals in $\boldsymbol{\psi}$-space. The vector $\boldsymbol{\rho}$ is a real $M \times 1$ vector that represents the perturbations in the array parameters. The additive white noise has spectral height σ_w^2. For notational simplicity, we assume that the array lies along the x-axis and assume the signals lie in the xy-plane. Thus, the direction of arrival of signals can be described by a scalar variable. We return to the 3-D case later.

We first assume that $\boldsymbol{\rho}$ is a real Gaussian random vector

$$p_{\boldsymbol{\rho}}(\boldsymbol{\rho}) = \frac{1}{(2\pi)^{\frac{M}{2}} \left|\boldsymbol{\Lambda}_{\boldsymbol{\rho}}\right|^{\frac{1}{2}}} \exp\left\{ -\frac{1}{2}(\boldsymbol{\rho} - \boldsymbol{\rho}_0)^T \boldsymbol{\Lambda}_{\boldsymbol{\rho}}^{-1} (\boldsymbol{\rho} - \boldsymbol{\rho}_0) \right\}, \tag{6.241}$$

where ρ_0 represents the nominal value of the parameter vector $\boldsymbol{\rho}$ and $\boldsymbol{\Lambda}_{\boldsymbol{\rho}}$ is the covariance matrix. Later, we consider the case when $\boldsymbol{\rho}$ is a real unknown nonrandom vector.

We rewrite the array manifold vector to include the gain and phase of each sensor explicitly.

$$\mathbf{v}(\psi_i, \boldsymbol{\rho}) = \left[\begin{array}{c:c:c} a_o\, e^{j\phi_0}\, e^{j\frac{2\pi}{\lambda}\mathbf{u}_i^T\mathbf{p}_0} & \cdots & a_{N-1}\, e^{j\phi_{N-1}}\, e^{j\frac{2\pi}{\lambda}\mathbf{u}_i^T\mathbf{p}_{N-1}} \end{array} \right]^T. \tag{6.242}$$

In the general case, the $\boldsymbol{\rho}$ vector is a $5N \times 1$ vector containing gain, phase, and position vectors. Define

$$[\mathbf{a}]_n = a_n^n(1+\Delta a_n), \quad n = 0, \cdots, N-1, \tag{6.243}$$

$$[\boldsymbol{\phi}]_n = \phi_n^n + \Delta\phi_n, \quad n = 0, \cdots, N-1, \tag{6.244}$$

$$[\mathbf{p_x}]_n = p_{xn}^n + \Delta p_{xn}, \quad n = 0, \cdots, N-1, \tag{6.245}$$

$$[\mathbf{p_y}]_n = p_{yn}^n + \Delta p_{yn}, \quad n = 0, \cdots, N-1, \tag{6.246}$$

$$[\mathbf{p_z}]_n = p_{zn}^n + \Delta p_{zn}, \quad n = 0, \cdots, N-1. \tag{6.247}$$

The superscript "n" denotes the nominal value. This model is similar to the model in Section 2.6.3. Then, $\boldsymbol{\rho}$ is

$$\boldsymbol{\rho} = \left[\begin{array}{ccccc} \mathbf{a}^T & \boldsymbol{\phi}^T & \mathbf{p}_x^T & \mathbf{p}_y^T & \mathbf{p}_z^T \end{array} \right]^T. \tag{6.248}$$

In most cases we study a subset of the possible variations and $\boldsymbol{\rho}$ has a smaller dimension. The probability density of $\boldsymbol{\rho}$ is given by (6.241).

In the second perturbation model, we combine $\mathbf{a}$ and $\boldsymbol{\phi}$ into a complex vector $\mathbf{g}$,

$$g_n = a_n\, e^{j\phi_n}, n = 0, \cdots, N-1, \tag{6.249}$$

where

$$g_n = a_n^n\,(1+\Delta a_n)\, e^{j(\phi_n^n + \Delta\phi_n)} = g_n^n\,(1+\Delta a_n)\, e^{j\Delta\phi_n}, \tag{6.250}$$

where Δa_n and $\Delta\phi_n$ are the amplitude and phase errors. For small errors,

$$g_n \simeq g_n^n\,(1+\Delta a_n + j\Delta\phi_n) \triangleq g_n^n\,(1+\Delta g_n). \tag{6.251}$$

If the gain and phase errors are independent with equal variances $\sigma_a^2 = \sigma_\phi^2$, then Δg_n is a zero-mean circular complex random variable. Then,

$$E\left[\Delta g_n \Delta g_n^*\right] = \sigma_g^2 = 2\,\sigma_\phi^2. \tag{6.252}$$

Thus, for small perturbations and equal variance for the gain and phase errors, the two models are the same.

This second model is more restrictive than the first model. The advantage of the second model is that it leads to analytic results in some cases where the first model is analytically difficult.

We consider two examples to illustrate the effect of array perturbations.

Example 6.6.4

Consider a standard 10-element linear array along the x-axis. The signal arrives from broadside and the interferer arrives at $u_x = 0.30$, $u_y = 0$ with an $INR = 20$ dB. Assume that the sensor positions are subject to random perturbations in the x and y directions. The perturbations are statistically independent zero-mean Gaussian random variables with standard deviation σ_p, where $\sigma_p = 0, 0.05\lambda, 0.1\lambda$, and 0.2λ. The perturbations are constant during the snapshot sequence. Thus, the actual array manifold vector, $\mathbf{v}_p$, is the random vector defined by (6.242). We assume that the actual spatial spectral matrix $\mathbf{S}_{\mathbf{x},p}$ is available to design the beamformer. In an actual adaptive beamformer, $\mathbf{S}_{\mathbf{x},p}$ will be the matrix we estimate. We design the beamformer using $\mathbf{v}_m$, the nominal array manifold vector. Thus, using (6.71),

$$\mathbf{w}_{mpdr,p}^H = \frac{\mathbf{v}_m^H \mathbf{S}_{\mathbf{x},p}^{-1}}{\mathbf{v}_m^H \mathbf{S}_{\mathbf{x},p}^{-1} \mathbf{v}_m}. \tag{6.253}$$

In Figure 6.36, we plot the expected value of the array gain versus input SNR for various σ_p^2. We see that the array perturbations cause the array manifold vector $\mathbf{v}_m$ to be mismatched. As the SNR increases, the beamformer attempts to null out the mismatched signal and the array gain decreases.

Example 6.6.5

Consider a standard 10-element linear array. We consider only gain errors and use the first perturbation model. The interference consists of white sensor noise.

The actual received spectral matrix is

$$\mathbf{S}_{\mathbf{x}a} = \sigma_d^2 \tilde{\mathbf{v}}_d \tilde{\mathbf{v}}_d^H + \sigma_w^2 \mathbf{I}, \tag{6.254}$$

where

$$\tilde{\mathbf{v}}_d \triangleq \mathbf{v}_s(\psi, \mathbf{a}) \tag{6.255}$$

and

$$a_n = 1 + \Delta a_n, n = 0, 1, \cdots, N-1. \tag{6.256}$$

The Δa_n are independent zero mean, and

$$\sigma_g^2 \triangleq E\left[|\Delta a_n|^2\right], \quad n = 0, 1, \cdots, N-1. \tag{6.257}$$

The beamformer uses $\mathbf{S}_{\mathbf{x}a}$ to find the optimum MPDR weight vector

$$\mathbf{w}^H = \frac{\mathbf{v}_d^H \mathbf{S}_{\mathbf{x}a}^{-1}}{\mathbf{v}_d^H \mathbf{S}_{\mathbf{x}a}^{-1} \mathbf{v}_d}. \tag{6.258}$$

Note that (6.258) would imply perfect knowledge of $\mathbf{S}_{\mathbf{x}a}$. In practice, we use $\hat{\mathbf{S}}_{\mathbf{x}a}$, and (6.258) represents the limiting behavior of the weight vector. In this case, one can find the SNR_o analytically (e.g., [YU94]). The result is

$$SNR_o = SNR_{in} \left\{ \frac{N + \sigma_g^2}{(1 + (N-1)\sigma_g^2 SNR_{in})^2 + (N-1)(N + \sigma_g^2)\sigma_g^2 SNR_{in}^2} \right\}. \tag{6.259}$$

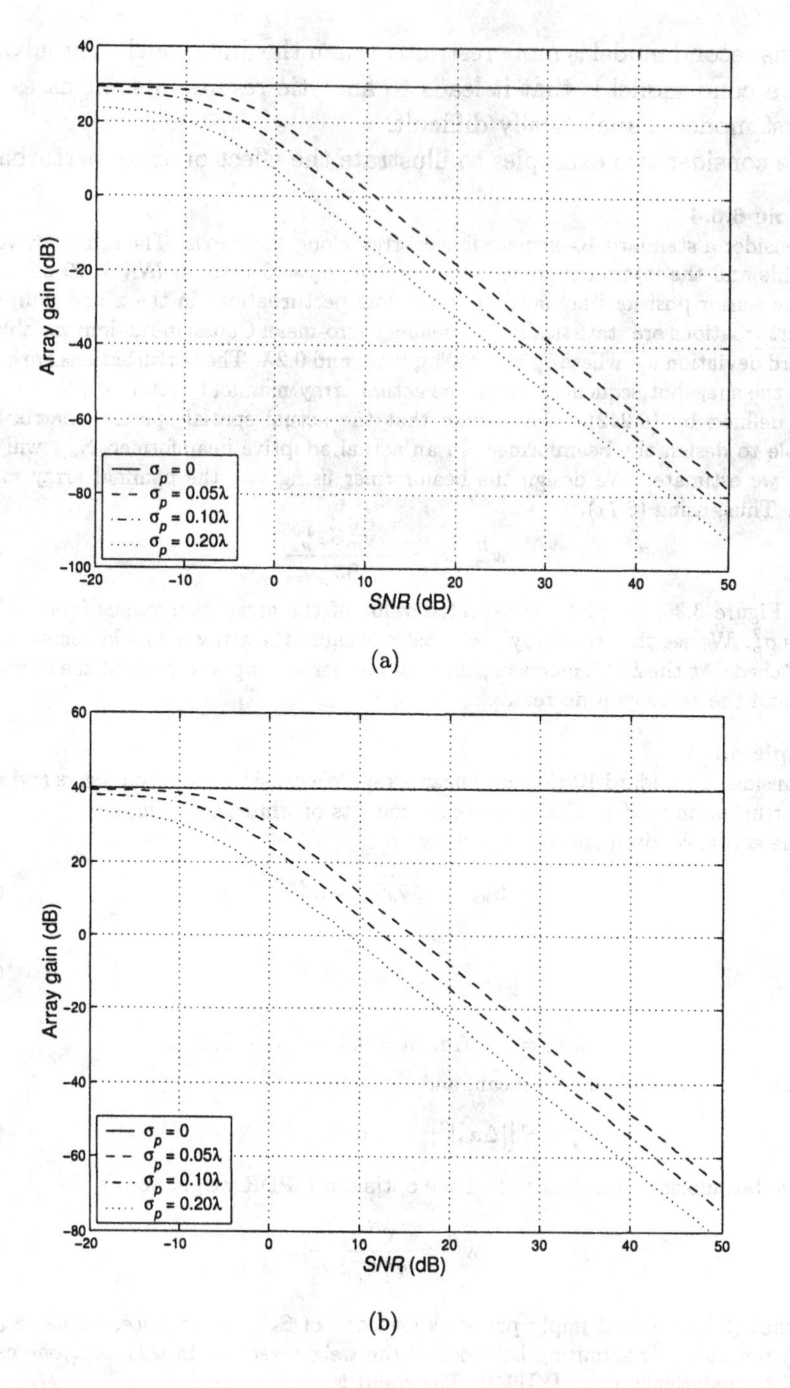

Figure 6.36 Array gain of MPDR beamformer versus SNR for various σ_p: $N = 10, u_s = 0, u_I = 0.30$; 500 trials; (a) *INR* = 20 dB; (b) *INR* = 30 dB.

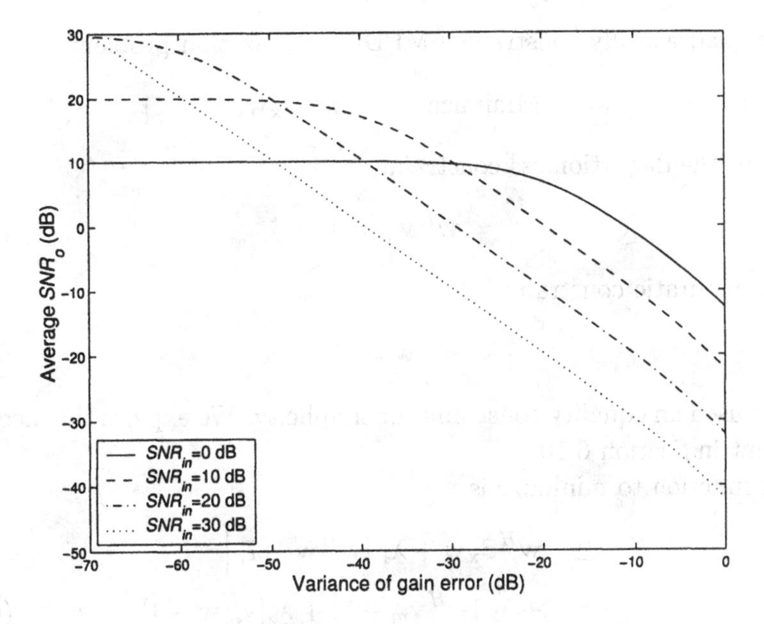

Figure 6.37 SNR_o versus σ_g^2.

For $N \gg 1$, $N \gg \sigma_g^2$ and large SNR_{in},

$$SNR_o \cong \frac{1}{N\sigma_g^2(1+\sigma_g^2)SNR_{in}}. \tag{6.260}$$

The output SNR_o is plotted versus σ_g^2 in Figure 6.37 for $N = 10$. We see that if the SNR_{in} is large, there is significant degradation for small σ_g^2 and the slope is 10 dB per decade, as indicated by (6.260).

6.6.4 Diagonal Loading

We have observed that the sensitivity of the MPDR beamformer to signal DOA mismatch and array perturbations increases as $\|\mathbf{w}\|^2$ increases. This suggests the use of a quadratic constraint,

$$\| \mathbf{w} \|^2 \leq T_o, \tag{6.261}$$

where T_o is a design parameter. From the discussion in Section 2.6.2, we know that

$$T_o \geq \frac{1}{N}. \tag{6.262}$$

We explore the choice of T_o in Section 6.10. In this section, we show how the imposition of a quadratic constraint leads to a procedure that we refer to as **diagonal loading** (DL).

The quadratically constrained MPDR optimization problem is:

$$\text{Minimize} \qquad \mathbf{w}^H \mathbf{S_x} \mathbf{w}, \tag{6.263}$$

subject to the distortionless constraint,

$$\mathbf{w}^H \mathbf{v}_m = 1, \tag{6.264}$$

and the quadratic constraint,

$$\mathbf{w}^H \mathbf{w} = T_o. \tag{6.265}$$

We have used an equality constraint for simplicity. We explore the inequality constraint in Section 6.10.

The function to minimize is

$$\begin{aligned} F \triangleq\ & \mathbf{w}^H \mathbf{S_x} \mathbf{w} + \lambda_1 \left[\mathbf{w}^H \mathbf{w} - T_o\right] \\ & + \lambda_2 \left[\mathbf{w}^H \mathbf{v}_m - 1\right] + \lambda_2^* [\mathbf{v}_m^H \mathbf{w} - 1]. \end{aligned} \tag{6.266}$$

Differentiating with respect to $\mathbf{w}$ and setting the result to zero gives

$$\mathbf{w}^H \mathbf{S_x} + \lambda_1 \mathbf{w}^H + \lambda_2^* \mathbf{v}_m^H = 0. \tag{6.267}$$

Solving for $\mathbf{w}^H$ and solving for λ_2^* by imposing the constraint in (6.264), gives

$$\mathbf{w}^H = \frac{\mathbf{v}_m^H [\mathbf{S_x} + \lambda_1 \mathbf{I}]^{-1}}{\mathbf{v}_m^H [\mathbf{S_x} + \lambda_1 \mathbf{I}]^{-1} \mathbf{v}_m}. \tag{6.268}$$

We see that the effect of the quadratic constraint (QC) is to add a diagonal matrix to $\mathbf{S_x}$ in the formula for $\mathbf{w}^H$. In effect, the MPDR-QC weight vector is designing for a higher white noise level than is actually present.

The value of λ_1 depends on the choice of T_o. In this section, we use a simpler approach and specify λ_1 directly. In Section 6.10, we see that this approach is not optimum. However, it is adequate in many cases and is widely used in practice. To emphasize the decoupling from T_o, we rewrite (6.268) as

$$\mathbf{w}^H = \frac{\mathbf{v}_m^H [\mathbf{S_x} + \sigma_L^2 \mathbf{I}]^{-1}}{\mathbf{v}_m^H [\mathbf{S_x} + \sigma_L^2 \mathbf{I}]^{-1} \mathbf{v}_m}. \tag{6.269}$$

Rewriting (6.269) as

$$\mathbf{w}_{mpdr,dl}^H = \frac{\mathbf{v}_m^H [\mathbf{I} + \frac{\mathbf{S_x}}{\sigma_L^2}]^{-1}}{\mathbf{v}_m^H [\mathbf{I} + \frac{\mathbf{S_x}}{\sigma_L^2}]^{-1} \mathbf{v}_m}, \tag{6.270}$$

we see that, as $\sigma_L^2 \to \infty$, the MPDR-DL beamformer approaches the conventional beamformer.

We can write $\mathbf{S_x}$ as

$$\mathbf{S_x} = \sigma_s^2 \mathbf{v}_a \mathbf{v}_a^H + \mathbf{S}_c + \sigma_w^2 \mathbf{I}, \tag{6.271}$$

where $\mathbf{v}_a$ is the actual array manifold vector, $\mathbf{S_c}$ is the non-white interference, and $\sigma_w^2 \mathbf{I}$ is the uncorrelated noise component. We define the load-to-white noise level (*LNR*) as

$$LNR = \frac{\sigma_L^2}{\sigma_w^2}. \tag{6.272}$$

We consider two examples to illustrate how diagonal loading affects the performance.

Example 6.6.6: MPDR-DL beamformer; mismatched signal

Consider a standard 10-element linear array. The array is steered to broadside ($\mathbf{v}_m = \mathbf{1}$). The interference consists of two equal-power uncorrelated plane-wave signals located at $u_I = \pm 0.30$. Thus,

$$\mathbf{S_x} = \sigma_s^2 \mathbf{v}_a \mathbf{v}_a^H + \sum_{i=1}^{2} \sigma_i^2 \mathbf{v}_i \mathbf{v}_i^H + \sigma_w^2 \mathbf{I}. \tag{6.273}$$

We utilize the beamformer in (6.269) and plot the array gain for various values of *SNR*, *INR*, and *LNR* versus $\frac{u_a}{BW_{NN}}$. Figure 6.38 corresponds to an *INR* =30 dB.

In Figure 6.38(a), we plot the array gain versus $\frac{u_a}{BW_{NN}}$ for various input *SNR* with no loading. We see that there is significant degradation for *SNR* > -10 dB. In Figure 6.38(b), we plot the array gain versus $\frac{u_a}{BW_{NN}}$ with 20-dB *LNR*. We see that loading provides adequate performance for *SNR* ≤ 10 dB. In Figure 6.38(c), we plot the array gain versus $\frac{u_a}{BW_{NN}}$ with 30-dB *LNR*. The curve for high input *SNR* has improved somewhat because we are not nulling the signal as deeply. However, the *LNR* equals the *INR*, so we are not nulling the interferers adequately.

In order to compare results in a more compact manner, we model u_a as a uniform random variable between ± 0.1,

$$p_{u_a} = \begin{cases} 5 & |u_a| \leq 0.1 \\ 0 & \text{elsewhere.} \end{cases} \tag{6.274}$$

The motivation for choosing an uncertainty range of ± 0.1 ($= \frac{1}{N}$ for a 10-element array) is the following scenario. We want to scan over u-space ($-1 \leq u \leq 1$) by forming optimal beams whose steering directions are spaced by $\frac{2}{N}$. This spacing leads to orthogonal conventional beams whose main lobes intersect at $\pm\frac{1}{N}, \pm\frac{3}{N}, \cdots$. Therefore, any signal with $|u_a - u_m| \leq \frac{1}{N}$ would be processed by the beamformer. Signals with $\frac{1}{N} < |u_a - u_m| < \frac{3}{N}$ would be processed by the adjacent beamformers.

There are cases in which the uncertainty in u_a may be much less than this. We could use closer beam spacing or we could do preliminary processing to estimate $\hat{u_a}$. If we estimate u_a, we would use a Gaussian probability density for u_e instead of the uniform density in (6.274). We discuss these cases later.

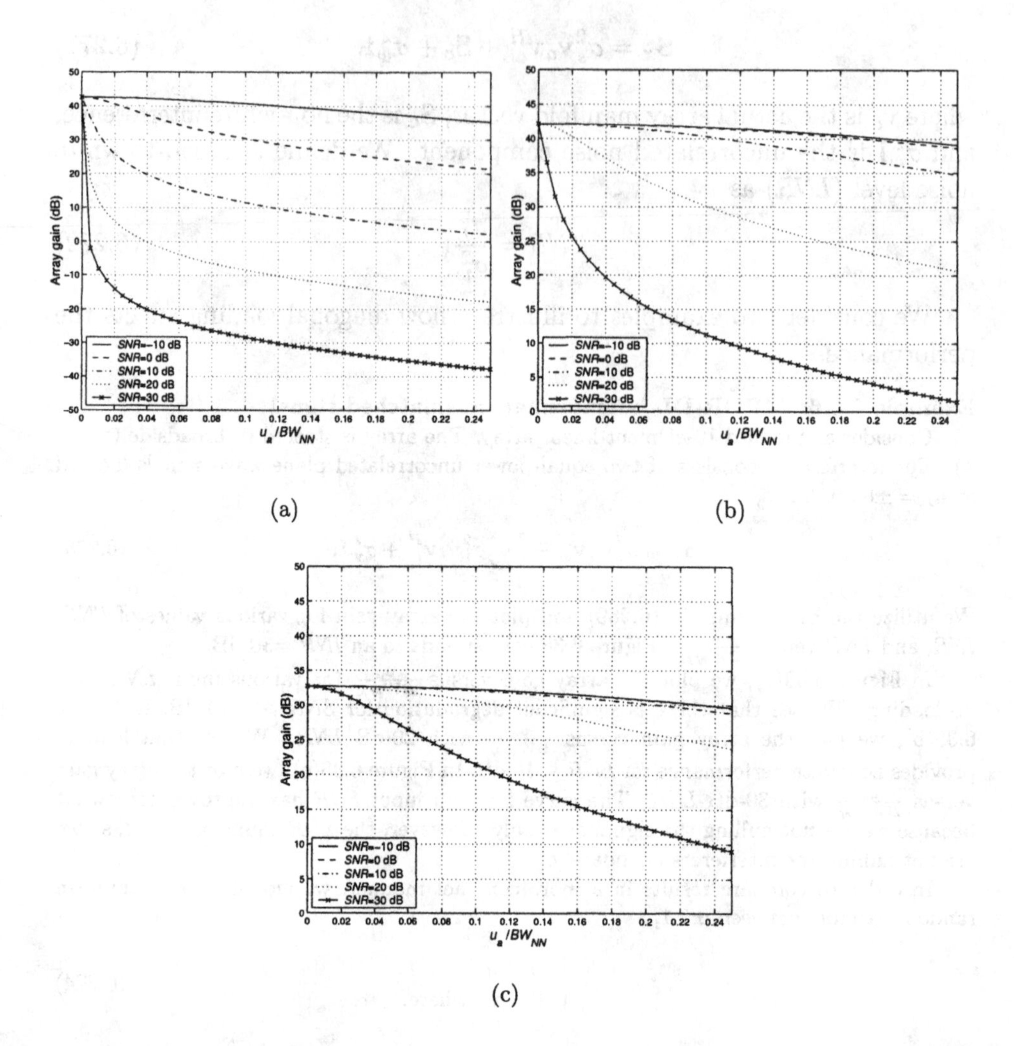

Figure 6.38 MPDR-DL beamformer, array gain versus $\frac{u_a}{BW_{NN}}$ for various *SNR*: $N = 10$, $u_m = 0$, $u_I = \pm 0.30$, $INR = 30$ dB. (a) $\sigma_L^2 = 0$; (b) $LNR = 20$ dB; (c) $LNR = 30$ dB.

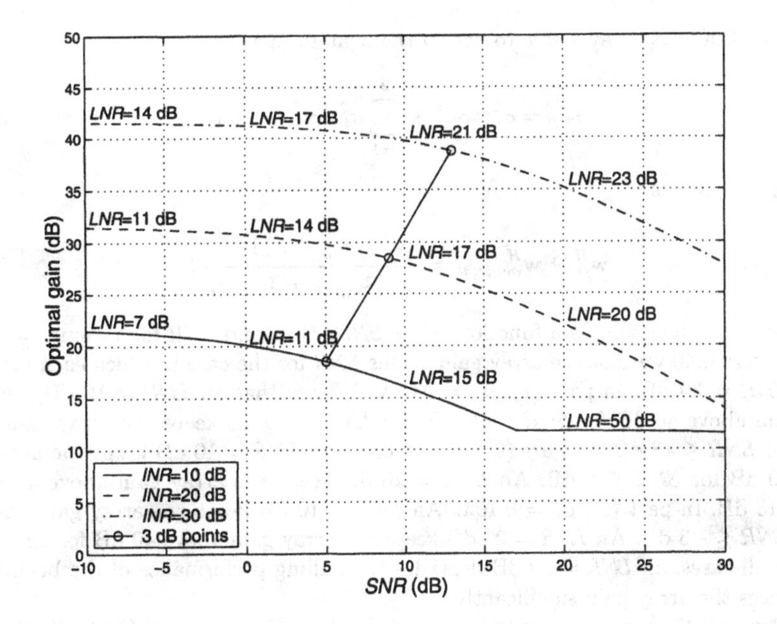

Figure 6.39: MPDR-DL beamformer, array gain, and optimal loading versus *SNR* for various *INR*; N=10, u_m=0, $u_I = \pm 0.3$.

We then compute $E\left[A_{mpdr,dl}\right]$ as a function of *INR* and *SNR* and choose the *LNR* to maximize $E\left[A_{mpdr,dl}\right]$. In Figure 6.39, we plot $E\left[A_{mpdr,dl}\right]$ versus *SNR* for various *INR*. We also show the optimum *LNR* values. Comparing the results in Figure 6.39 with the results in Figure 6.35, we see that diagonal loading provides a significant gain in performance.

This leads to our approximate rule of thumb,

$$SNR + 10 \text{ dB} \leq LNR \leq INR, \tag{6.275}$$

if the *SNR* and *INR* allow these inequalities. The worst case for this technique is when the *SNR* $\geq$ *INR* and both are large ($\geq$ 10 dB). We shall find that many of our beamforming algorithms have difficulty in this region. Although a particular algorithm may have trouble, the *SNR* > *INR* case is a desirable one.

In the next example, we consider the use of diagonal loading to provide robustness in the presence of array perturbations.

Example 6.6.7: MPDR-DL beamformer; array perturbations (continuation)

Consider a standard 10-element linear array along the x-axis. The array is steered to broadside. The interference model is given in (6.273). In this example, the signal is matched ($\mathbf{v}_a = \mathbf{v}_m$) but the sensor positions are perturbed in the x and y directions. The perturbations are statistically independent zero-mean Gaussian random variables with standard deviation σ_p, where $\sigma_p = 0.02\lambda, 0.05\lambda$, and 0.1λ. The perturbations are constant during the snapshot sequence. The actual array manifold vector $\mathbf{v}_p$ is given by (6.242).

We assume that $\mathbf{S}_{\mathbf{x},p}$ is available to design the beamformer,

$$\mathbf{S}_{\mathbf{x},p} = \sigma_s^2 \mathbf{v}_p \mathbf{v}_p^H + \sum_{i=1}^{2} \sigma_i^2 \mathbf{v}_i \mathbf{v}_i^H + \sigma_w^2 \mathbf{I}. \tag{6.276}$$

The weight vector is

$$\mathbf{w}^H \triangleq \mathbf{w}_{mpdr,dl}^H = \frac{\mathbf{v}_m^H \left[\mathbf{S}_{\mathbf{x},p} + \sigma_L^2 \mathbf{I}\right]^{-1}}{\mathbf{v}_m^H \left[\mathbf{S}_{\mathbf{x},p} + \sigma_L^2 \mathbf{I}\right]^{-1} \mathbf{v}_m}. \tag{6.277}$$

We examine the behavior as a function of the *SNR*, *INR*, and *LNR* for various σ_p.

In Figure 6.40 we plot the array gain versus *SNR* for the case in which each interferer has an *INR* = 30 dB. In part (a), $\sigma_p = 0.02\lambda$. We see that an *LNR* = 10 dB keeps the array gain above 40 dB for *SNR* $\leq$ 10 dB. An *LNR* = 20 dB keeps the array gain above 40 dB for *SNR* $\leq$ 18 dB. In Part (b), $\sigma_p = 0.05\lambda$. An *LNR* = 10 dB keeps the array gain above 40 dB for *SNR* $\leq$ 5 dB. An *LNR* = 20 dB keeps the array gain above 40 dB for *SNR* $\leq$ 13 dB. In part (c), $\sigma_p = 0.10\lambda$. An *LNR* = 10 dB keeps the array gain above 40 dB for *SNR* ≤ -3 dB. An *LNR* = 20 dB keeps the array gain above 40 dB for an *SNR* $\leq$ 3 dB. In all cases, an *LNR* = 30 dB degrades the nulling performance of the beamformer and reduces the array gain significantly.

In Figure 6.41, we plot the optimum loading versus $\frac{\sigma_p}{\lambda}$ for various *SNR*. We also show the corresponding array gain. In this case we assume that σ_p/λ and *SNR* are known and we choose the optimum *LNR* based on that knowledge.

In this section, we have seen that diagonal loading offers a significant improvement in performance in the presence of mismatch. If we have reasonably good information on the expected *SNR* and *INR* levels we can select the appropriate amount of fixed diagonal loading.

The disadvantage of this approach is that we may not have enough prior information to determine the correct fixed loading level or the environment may change over time. In Section 6.10, we re-examine the MPDR-QC beamformer in (6.268) and study the choice of T_o. In Chapter 7, we study the finite data problem and develop variable loading algorithms that determine the loading level from the observed data.

We will find that diagonal loading plays a central role in most robust beamformers.

6.6.5 Summary

In this section, we have studied the behavior of MVDR and MPDR beamformers in the presence of various types of differences between the model and the actual environment.

In Section 6.6.2, we studied the case of DOA mismatch. We found that the behavior of the conventional beamformer and the MVDR beamformer was characterized by their beam patterns in the absence of mismatch (see

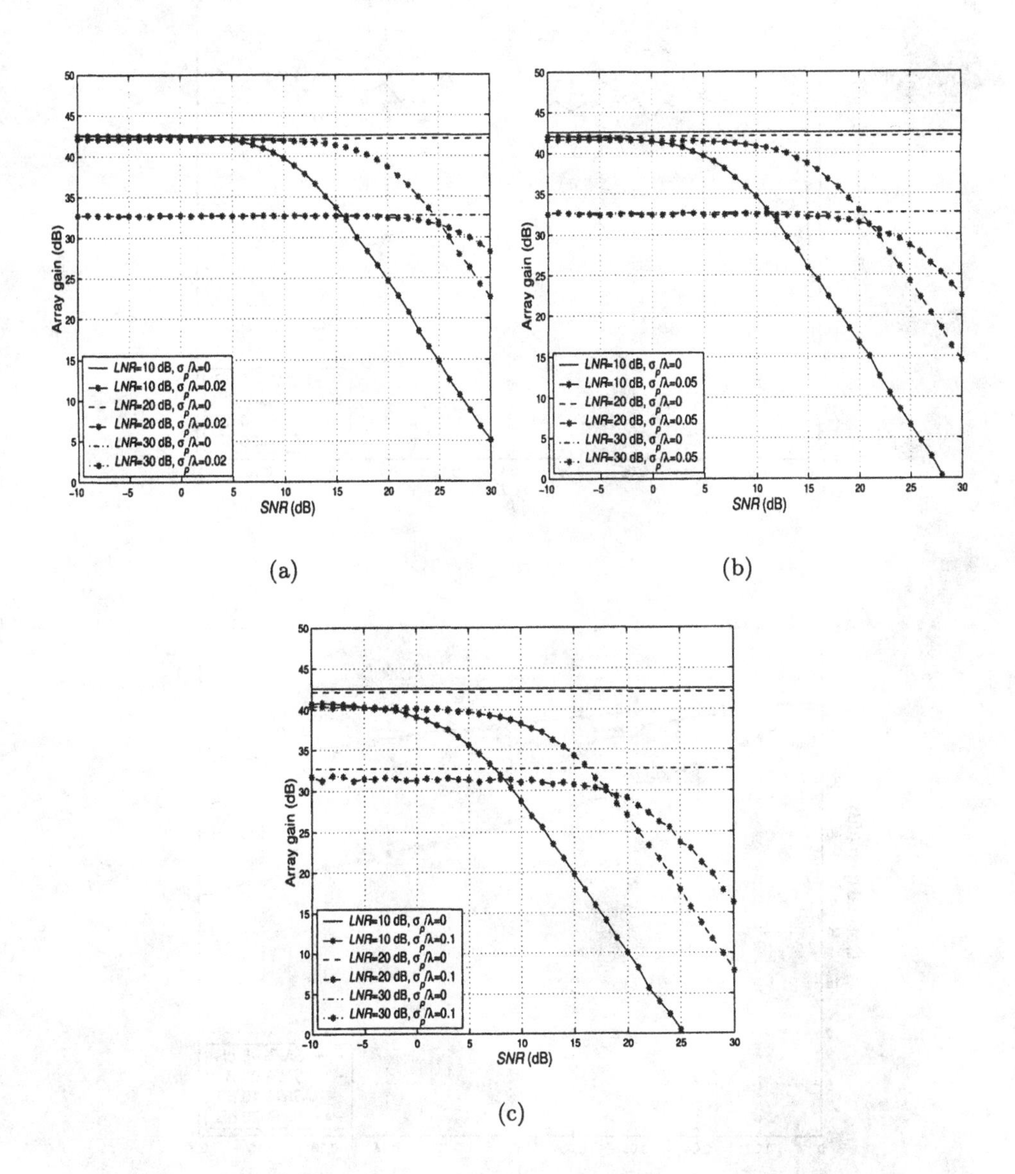

Figure 6.40 MPDR-DL beamformer with array perturbations, array gain versus *SNR* for various *LNR*; $u_I = \pm 0.3$, *INR* =30 dB, 500 trials (a) $\sigma_p = 0.02\lambda$; (b) $\sigma_p = 0.05\lambda$; (c) $\sigma_p = 0.1\lambda\cdot$.

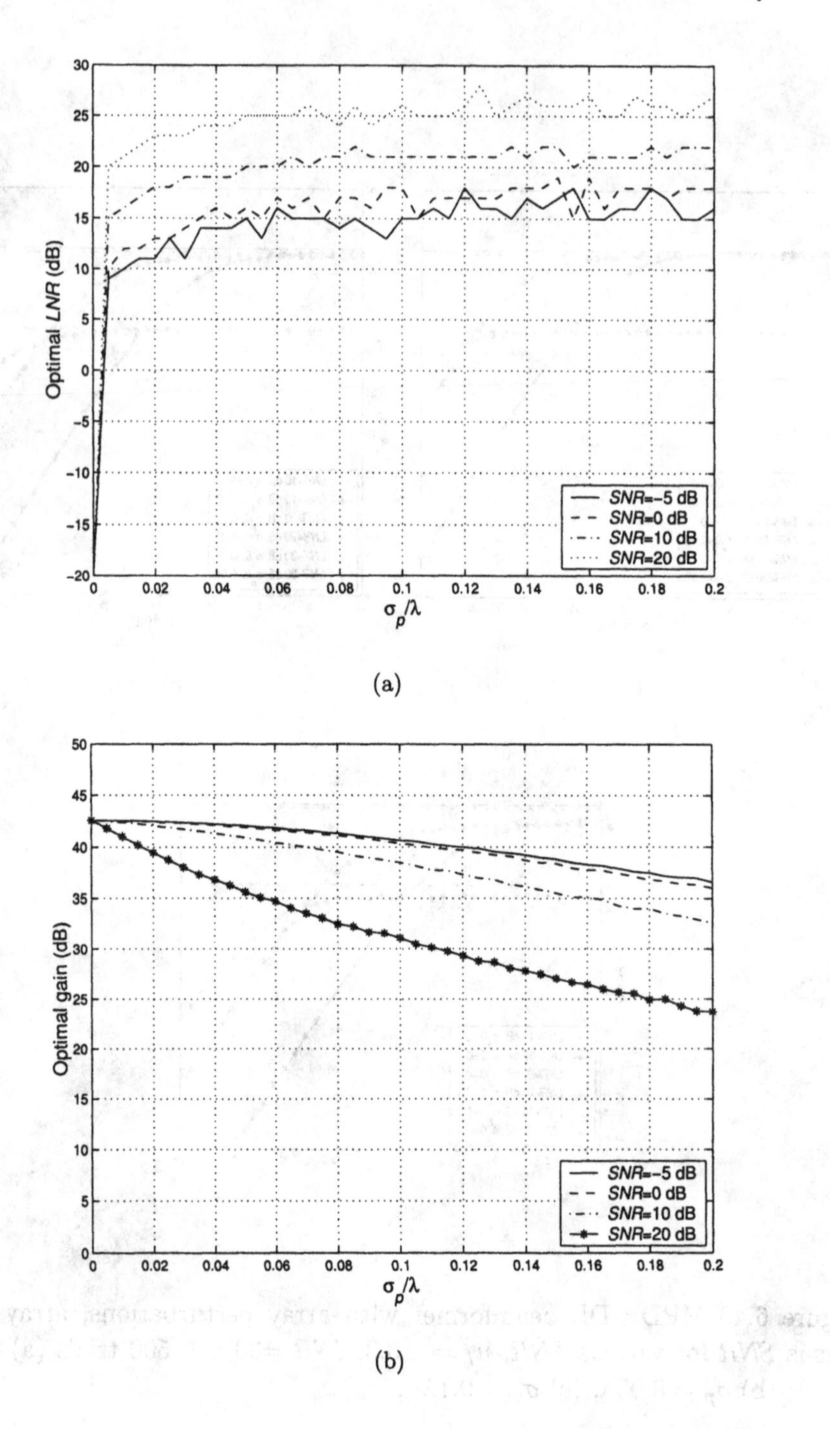

Figure 6.41 MPDR-DL beamformer with array perturbations; 500 trials: (a) optimal *LNR* versus σ_p/λ; (b) array gain versus σ_p/λ.

(6.205) and (6.213)). Therefore, as long as the MVDR beam pattern is well-behaved, the performance degrades gradually in the presence of DOA mismatch. However, the MPDR beamformer treats the mismatched signal as an interferer and attempts to null it. As the *SNR* increases, the degradation due to mismatch increases and small mismatches lead to unacceptable performance.

In Section 6.6.3, we studied the case of array perturbations. We found that the array perturbations caused a performance degradation similar to that in the DOA mismatch case.

In Section 6.6.4, we introduced a quadratic constraint in order to inhibit main-lobe nulling. The quadratic constraint leads to diagonal loading. We found that, if the *INR*s were 15–20 dB larger than the *SNR*, we could use DL and obtain adequate performance for both the DOA mismatch and the array perturbation case.

In the next section, we develop other constraint techniques in order to provide main-lobe protection and increased robustness in the presence of mismatch.

6.7 LCMV and LCMP Beamformers

In Section 6.2, we derived the minimum variance distortionless beamformer by imposing a linear constraint,

$$\mathbf{w}^H \mathbf{v}_m = 1. \tag{6.278}$$

In Section 6.6, we saw that the performance of the resulting MPDR beamformers was not particularly robust to various changes in the environment. In this section, we develop beamformers in which additional linear constraints are imposed to make the beamformer more robust. We refer to beamformers using $\mathbf{S}_\mathbf{n}$ as linear constrained minimum variance (LCMV) beamformers. In many cases, we must use $\mathbf{S}_\mathbf{x}$ instead of $\mathbf{S}_\mathbf{n}$. We refer to these beamformers as linear constrained minimum power (LCMP) beamformers. The majority of the literature uses the LCMV descriptor for both types of beamformers.

In this section, we develop the two beamforming algorithms. We define a set of linear constraints by the $N \times M_c$ constraint matrix, $\mathbf{C}$, whose columns are linearly independent. We require that

$$\mathbf{w}^H \mathbf{C} = \mathbf{g}^H \tag{6.279}$$

or equivalently,

$$\mathbf{C}^H \mathbf{w} = \mathbf{g}. \tag{6.280}$$

Subject to the constraint in (6.279) we derive two beamformers. The LCMV beamformer minimizes

$$\mathbf{P}_n = \mathbf{w}^H \, \mathbf{S_n} \, \mathbf{w}, \tag{6.281}$$

subject to the constraint in (6.279). The LCMP beamformer minimizes

$$\mathbf{P}_o = \mathbf{w}^H \, \mathbf{S_x} \, \mathbf{w}, \tag{6.282}$$

subject to the constraint in (6.279).

In Section 6.7.1, we discuss typical linear constraints that are useful and discuss various choices of $\mathbf{C}$ and $\mathbf{g}$. In Section 6.7.2, we derive the optimum LCMV and LCMP beamformers. In Section 6.7.3, we develop the generalized sidelobe canceller implementation of the LCMV and LCMP beamformers.

In Section 6.7.4, we study the performance of the LCMV and LCMP beamformers. We first consider their performance in a nominal environment. We then consider the various mismatch problems such as steering vector mismatch and array imperfections, and see how the constraints improve robustness.

The constraints in Sections 6.7.1–6.7.4 focus on constraining the beam pattern at specific points in frequency-wavenumber space and impose a hard constraint at those points. In Section 6.7.5, we study a different type of linear constraint called **quiescent pattern constraints**. We use the adjective "quiescent" to describe the beam pattern of the optimum beamformer in the presence of white noise. This beamformer utilizes the classical beam patterns that we developed in Chapters 3 and 4 as a basis for the design. By using these constraints (such as mean-square deviation from the quiescent pattern) we can obtain improved performance in some cases.

In Section 6.7.6, we develop a technique called covariance augmentation to combat moving or broadband interferers. In Section 6.7.7, we summarize our results.

6.7.1 Typical Constraints

We use the frequency domain snapshot model. The input signal is modeled as,

$$\mathbf{X}(\omega) = \mathbf{V}\mathbf{F}(\omega) + \mathbf{N}(\omega). \tag{6.283}$$

The output is

$$Y(\omega) = \mathbf{w}^H \mathbf{X}(\omega). \tag{6.284}$$

We now discuss typical constraints on $\mathbf{w}$.

6.7.1.1 Distortionless constraint

We introduced this constraint in Section 6.2. If the signal of interest corresponds to an array manifold vector $\mathbf{v}_m$, then we require

$$\mathbf{w}^H \mathbf{v}_m = 1, \tag{6.285}$$

which guarantees that any signal propagating along $\mathbf{k}_m$ will pass through the filter undistorted. The subscript "m" denotes model. We would like

$$\mathbf{v}_m = \mathbf{v}_s,$$

where $\mathbf{v}_s$ is the desired signal's array manifold vector. We use $\mathbf{v}_m$ to allow for mismatch.

We use an $N \times 1$ vector $\mathbf{c}_i$ to denote the ith constraint. Unless noted, we always assume that the first constraint is the distortionless constraint. Thus,

$$\mathbf{c}_1 = \mathbf{v}_m. \tag{6.286}$$

We have seen that this constraint will be inadequate to ensure adequate performance in the presence of various mismatches. One possible improvement is to use multiple directional constraints.

6.7.1.2 Directional constraints

The general directional constraint is

$$\mathbf{w}^H \mathbf{v}(\mathbf{k}_i) = g_i, \qquad i = 1, 2, \cdots, M_0, \tag{6.287}$$

where $\mathbf{k}_i$ denotes the wave number along which we want to impose the constraint and g_i is a complex number denoting the value of the constraint.

To illustrate a special case of these constraints that could be used to make the beamformer robust against steering errors, we consider a uniform linear array. Then, the distortionless constraint can be written as

$$\mathbf{w}^H \mathbf{v}(\psi_m) = 1. \tag{6.288}$$

We can try to force a flatter beam pattern near the signal direction by adding two constraints,

$$\mathbf{w}^H \mathbf{v}(\psi_m + \Delta\psi) = 1, \tag{6.289}$$

$$\mathbf{w}^H \mathbf{v}(\psi_m - \Delta\psi) = 1. \tag{6.290}$$

In matrix notation, the constraint matrix is an $N \times 3$ matrix,

$$\mathbf{C} = \left[\mathbf{v}(\psi_m) \vdots \mathbf{v}(\psi_m + \Delta\psi) \vdots \mathbf{v}(\psi_m - \Delta\psi) \right], \tag{6.291}$$

$$\mathbf{g} = \begin{bmatrix} 1 \\ 1 \\ 1 \end{bmatrix}, \tag{6.292}$$

and

$$\mathbf{w}^H \mathbf{C} = \mathbf{g}^H. \tag{6.293}$$

In later examples, we see that this type of constraint can degrade the noise rejection performance significantly. A second possibility is to choose the g_i to match the conventional beamformer shape. Thus,

$$\mathbf{g} = \begin{bmatrix} 1 \\ B_c(\psi_m + \Delta\psi : \psi_m) \\ B_c(\psi_m - \Delta\psi : \psi_m) \end{bmatrix}. \tag{6.294}$$

Depending on the extent of signal mismatch, other values of g_2 and g_3 may be appropriate. For symmetric mismatch, we use

$$\mathbf{g} = \begin{bmatrix} 1 \\ g_c \\ g_c \end{bmatrix}, \tag{6.295}$$

where

$$B_c(\psi_m + \Delta\psi : \psi_m) \leq g_c \leq 1. \tag{6.296}$$

Example 6.7.1

Consider a standard N-element linear array. Assume we impose the distortionless constraint plus a constraint at the HPBW points of the conventional beam pattern, $u = \pm 0.891/N$. Assume the array is pointed at broadside, $\mathbf{C}$ is an $N \times 3$ matrix,

$$\mathbf{C} = \left[\mathbf{1} \vdots \mathbf{v}_u\left(\tfrac{0.891}{N}\right) \vdots \mathbf{v}_u\left(\tfrac{-0.891}{N}\right) \right], \tag{6.297}$$

where $\mathbf{v}_u(\cdot)$ is the array manifold vector in u-space.

The value of $\mathbf{g}$ is given by either (6.292), (6.294), or (6.295).

When we impose the constraints, we assume that the columns of $\mathbf{C}$ are linearly independent so that its rank is M_0, the number of constraints. However, we will find that if the wavenumbers are too close together, the matrix that we must invert will be ill-conditioned.

6.7.1.3 Null constraints

We encountered null constraints in Section 3.7. This type of constraint is appropriate if there is an interfering signal (jammer) coming from a known direction. In this case,

$$\mathbf{w}^H \mathbf{v}(\mathbf{k}_i) = 0\,, \quad i = 2, \cdots, M_0. \tag{6.298}$$

Thus,

$$\mathbf{C} = \left[\begin{array}{c:c:c:c:c} \mathbf{v}_m & \mathbf{v}_2 & \mathbf{v}_3 & \cdots & \mathbf{v}_{M_0} \end{array}\right] \tag{6.299}$$

and

$$\mathbf{g}^T = \left[\begin{array}{ccccc} 1 & 0 & 0 & \cdots & 0 \end{array}\right]. \tag{6.300}$$

We recall from Section 6.3 that the MVDR beamformer only puts a perfect null on a directional noise when σ_I^2/σ_w^2 is infinite.

Thus, a constraint such as

$$\mathbf{w}^H \mathbf{v}(\mathbf{k}_i) = \varepsilon_i\,, \quad i = 2, \cdots, M_0, \tag{6.301}$$

where ε_i is related to σ_I^2/σ_w^2, may be used.

We also recall that the performance of the beamformer was sensitive to interferer mismatch. Thus, we may want to use multiple point constraints in the vicinity of the interferer. For example, in the single interferer and a linear array case, we might impose three constraints

$$\mathbf{C}_I = \left[\begin{array}{c:c:c} \mathbf{v}(\psi_I) & \mathbf{v}(\psi_I + \Delta\psi) & \mathbf{v}(\psi_I - \Delta\psi) \end{array}\right], \tag{6.302}$$

where $\mathbf{C}_I$ is the null part of the constraint matrix, and

$$\mathbf{g}^T = \left[\begin{array}{ccc} 0 & B & B \end{array}\right]. \tag{6.303}$$

Once again, we do not want to put the columns of $\mathbf{C}_I$ too close together. This not only gives an ill-conditioned matrix but it also unnecessarily uses degrees of freedom.

6.7.1.4 Derivative constraints

Another method of controlling the shape of the beam pattern near the peak or a null is to utilize derivative constraints. The complex beam pattern is

$$B(\mathbf{k} : \mathbf{k}_m) = \mathbf{w}^H \mathbf{v}(\mathbf{k}), \tag{6.304}$$

and the corresponding power pattern is

$$P(\mathbf{k}:\mathbf{k}_m) = |B(\mathbf{k}:\mathbf{k}_m)|^2. \tag{6.305}$$

We can also write the above pattern using θ and ϕ as variables. For example,

$$P(\omega,\theta,\phi) = |B(\omega,\theta,\phi)|^2. \tag{6.306}$$

The reason for writing the alternative forms is that we impose derivative constraints in a different manner in the various forms. Three representative choices are:

(i) Beam pattern derivatives

Set the derivatives of $B(\mathbf{k}:\mathbf{k}_s)$ with respect to k_x, k_y, k_z equal to a specified value at some point $\mathbf{k}_c$. Alternatively, we can use derivatives with respect to θ and ϕ.

(ii) Power pattern derivatives

Set the derivatives of $P(\omega,\theta,\phi)$ with respect to θ and ϕ equal to a specified value at some point (θ_c,ϕ_c). Alternatively, we can use derivatives with respect to $\mathbf{k}$.

(iii) Frequency derivatives

Set the derivatives of either $B(\omega,\theta,\phi)$ or $P(\omega,\theta,\phi)$ with respect to ω equal to a specified value at some point ω_L.

We explore these cases briefly. First, consider a standard linear array. Then,

$$B(\psi) = \mathbf{w}^H\mathbf{v}(\psi) \tag{6.307}$$

and

$$\frac{dB(\psi)}{d\psi} = \mathbf{w}^H\frac{d\mathbf{v}(\psi)}{d\psi} \triangleq \mathbf{w}^H\mathbf{d}(\psi), \tag{6.308}$$

where, using symmetric indexing,

$$[\mathbf{d}(\psi)]_n \triangleq \begin{cases} jne^{jn\psi}, \quad -\frac{N-1}{2} \le n \le \frac{N-1}{2}, & N\,\text{odd}, \\[2ex] \begin{cases} j(n-\frac{1}{2})e^{j(n-\frac{1}{2})\psi}, \quad n = 1,\cdots,\frac{N}{2}, \\ j(n+\frac{1}{2})e^{j(n+\frac{1}{2})\psi}, \quad n = -1,\cdots,-\frac{N}{2}, \end{cases} & N\,\text{even}. \end{cases} \tag{6.309}$$

Note that $\mathbf{d}(\psi)$ is conjugate asymmetric.

The second derivative is

$$\frac{d^2B(\psi)}{d\psi^2} = \mathbf{w}^H \frac{\partial}{\partial\psi}\mathbf{d}(\psi) = \mathbf{w}^H \dot{\mathbf{d}}(\psi), \tag{6.310}$$

where, using symmetric indexing,

$$\left[\dot{\mathbf{d}}(\psi)\right]_n = \begin{cases} -n^2 e^{jn\psi}, \quad -\frac{N-1}{2} \le n \le \frac{N-1}{2}, \quad N\,\text{odd}, \\ \\ \begin{cases} -(n-\frac{1}{2})^2 e^{j(n-\frac{1}{2})\psi}, \quad n = 1,\cdots,\frac{N}{2}, \\ -(n+\frac{1}{2})^2 e^{j(n+\frac{1}{2})\psi}, \quad n = -1,\cdots,-\frac{N}{2}, \end{cases} \quad N\,\text{even}, \end{cases} \tag{6.311}$$

is conjugate symmetric.

A typical application is to impose constraints on the derivatives of the main lobe evaluated at its steering direction. A simple example illustrates this application.

Example 6.7.2

We utilize an $N \times 3$ constraint matrix. We assume the array is pointed at broadside.

$$\mathbf{C} = \left[\begin{array}{c:c:c} \mathbf{1} & \mathbf{d}(0) & \dot{\mathbf{d}}(0) \end{array} \right]. \tag{6.312}$$

There are several logical choices for the $\mathbf{g}$ vector. The first choice sets the first and second derivatives equal to zero. Thus,

$$\mathbf{g}^H = \left[\begin{array}{c:c:c} 1 & 0 & 0 \end{array} \right]. \tag{6.313}$$

The second choice matches the behavior of the conventional beam pattern. Thus,

$$\mathbf{g}^H = \left[\begin{array}{c:c:c} 1 & 0 & \ddot{B}_C(\psi)\big|_{\psi=0} \end{array} \right]. \tag{6.314}$$

The third choice uses an intermediate value for g_3,

$$\mathbf{g}^T = \left[\begin{array}{c:c:c} 1 & 0 & g_3 \end{array} \right], \tag{6.315}$$

where

$$\ddot{B}_C(\psi)\big|_{\psi=0} \le g_3 \le 0. \tag{6.316}$$

We could also express the beam pattern in θ-space. Then

$$B(\theta) = \mathbf{w}^H \mathbf{v}_\theta(\theta) \tag{6.317}$$

and

$$\frac{dB(\theta)}{d\theta} = \mathbf{w}^H \frac{d\mathbf{v}_\theta(\theta)}{d\theta} = \mathbf{w}^H \mathbf{d}_\theta(\theta), \tag{6.318}$$

where, using symmetric indexing,

$$[\mathbf{d}_\theta(\theta)]_n \triangleq \begin{cases} jn\pi \sin\theta e^{jn\pi\cos\theta}, & -\frac{N-1}{2} \leq n \leq \frac{N-1}{2}, \quad N \text{ odd}, \\ \begin{cases} j(n-\frac{1}{2})\sin\theta e^{j(n-\frac{1}{2})\pi\cos\theta}, & n = 1, \cdots, \frac{N}{2}, \\ j(n+\frac{1}{2})\sin\theta e^{j(n+\frac{1}{2})\pi\cos\theta}, & n = -1, \cdots, -\frac{N}{2}, \end{cases} & N \text{ even}, \end{cases} \tag{6.319}$$

with similar expressions for the second derivative.

The constraints in (6.312)–(6.314) have been used by a number of authors (e.g., Cox [Cox73], Applebaum and Chapman [AC76], Vural [Vur77], [Vur79], Er and Cantoni [EC83], Steele [Ste83], Buckley and Griffiths [BG86]).

In [BG86], it is pointed out that if one constrains the phase of the derivative by setting (6.319) or (6.309) equal to zero, then the resulting beam pattern will be affected by the choice of the spatial reference point (the origin of coordinates).

Tseng [Tse92] shows this dependency in a straightforward manner. Assume that $\bar{\mathbf{v}}$ is a steering vector defined at a reference point other than $\mathbf{v}$. Then,

$$\bar{\mathbf{v}} = \mathbf{v}e^{j\omega\Delta\tau(\theta)}, \tag{6.320}$$

where $\Delta\tau(\theta)$ is the time delay corresponding to the difference in the two reference points. Thus, if we impose a gain constraint, changing the reference point changes the phase of the beam pattern by a fixed amount, but does not affect the magnitude response.

Imposing a derivative constraint has a different effect. Differentiating (6.320) with respect to θ gives

$$\bar{\mathbf{d}}_\theta \triangleq \frac{\partial \bar{\mathbf{v}}}{\partial \theta} = \mathbf{v}_\theta e^{j\omega\Delta\tau(\theta)} + \mathbf{v}\cdot j\omega\frac{\partial\Delta\tau(\theta)}{\partial\theta}e^{j\omega\Delta\tau(\theta)}. \tag{6.321}$$

Because of the second term, $\bar{\mathbf{v}}_\theta$ is not just a phase-shifted replica of $\mathbf{v}_\theta$. Therefore, both the magnitude and phase responses will be affected by the choice of spatial reference point.

If we differentiate the power pattern, the problem of reference point sensitivity is eliminated. This approach is utilized by Tseng [Tse92], Tseng and Griffiths [TG92a], Thng et al. [TCL95], and others. This leads to a set of nonlinear constraints. However, [Tse92] shows how to transform them to a set of linear constraints.

For a planar array, the best notation will depend on the array geometry.

For a standard rectangular array, using symmetric indexing,

$$B(\psi_x, \psi_y) = \sum_{n=-\frac{N-1}{2}}^{\frac{N-1}{2}} \sum_{m=-\frac{M-1}{2}}^{\frac{M-1}{2}} w_{nm} e^{j[n\psi_x + m\psi_y]}, \quad N \text{ odd}. \tag{6.322}$$

Then,

$$\frac{dB(\psi_x, \psi_y)}{d\psi_x} = \sum_{n=-\frac{N-1}{2}}^{\frac{N-1}{2}} \sum_{m=-\frac{M-1}{2}}^{\frac{M-1}{2}} w_{nm} jn e^{j[n\psi_x + m\psi_y]}, \quad N \text{ odd}, \tag{6.323}$$

$$\frac{dB(\psi_x, \psi_y)}{d\psi_y} = \sum_{n=-\frac{N-1}{2}}^{\frac{N-1}{2}} \sum_{m=-\frac{M-1}{2}}^{\frac{M-1}{2}} w_{nm} jm e^{j[n\psi_x + m\psi_y]}, \quad N \text{ odd}, \tag{6.324}$$

are the first derivatives. The second derivatives are

$$\frac{d^2B(\psi_x, \psi_y)}{d\psi_x^2} = \sum_{n=-\frac{N-1}{2}}^{\frac{N-1}{2}} \sum_{m=-\frac{M-1}{2}}^{\frac{M-1}{2}} w_{nm}(-n^2) e^{j[n\psi_x + m\psi_y]}, \quad N \text{ odd}, \tag{6.325}$$

$$\frac{d^2B(\psi_x, \psi_y)}{d\psi_y^2} = \sum_{n=-\frac{N-1}{2}}^{\frac{N-1}{2}} \sum_{m=-\frac{M-1}{2}}^{\frac{M-1}{2}} w_{nm}(-m^2) e^{j[n\psi_x + m\psi_y]}, \quad N \text{ odd}, \tag{6.326}$$

$$\frac{d^2B(\psi_x, \psi_y)}{d\psi_x d\psi_x} = \sum_{n=-\frac{N-1}{2}}^{\frac{N-1}{2}} \sum_{m=-\frac{M-1}{2}}^{\frac{M-1}{2}} w_{nm}(-mn) e^{j[n\psi_x + m\psi_y]}, \quad N \text{ odd}. \tag{6.327}$$

Thus, to control the first and second derivatives in addition to the distortionless constraint will require six constraints.

When dealing with derivative constraints in either linear or planar arrays it is important to check the columns of $\mathbf{C}$ for linear independence and remove any dependent constraint vectors (or use the SVD approach discussed in Chapter 3, e.g., discussion after (3.296)).

The third case corresponds to frequency derivatives and can be illustrated with a simple example.

Example 6.7.3

Consider a uniform linear array. Then,

$$B(\omega, u) = \sum_{n=-\frac{N-1}{2}}^{\frac{N-1}{2}} w_n e^{j\frac{\omega}{c}und}, \tag{6.328}$$

$$\frac{dB(\omega,u)}{d\omega} = \sum_{n=-\frac{N-1}{2}}^{\frac{N-1}{2}} w_n jn \frac{ud}{c} e^{j\frac{\omega}{c}und}, \tag{6.329}$$

$$\frac{d^2B(\omega,u)}{d\omega^2} = \sum_{n=-\frac{N-1}{2}}^{\frac{N-1}{2}} w_n \left(-n^2\frac{u^2d^2}{c^2}\right) e^{j\frac{\omega}{c}und}. \tag{6.330}$$

For a uniform linear array, the behavior versus frequency is the dual of the behavior versus u. This duality does not necessarily apply to other array geometries.

Before leaving derivative constraints, it is useful to revisit null constraints. As discussed in Section 6.7.1.3, a null constraint can be used if there is an interferer coming from a specific direction. The performance is very sensitive to mismatch in the interferer direction. In order to improve robustness, derivative constraints in the vicinity of the null can be imposed. For a single interferer arriving from ψ_I, a constraint matrix whose null component is

$$\mathbf{C}_I = \left[\begin{array}{ccc} \mathbf{v}(\psi_I) & \dot{\mathbf{v}}(\psi_I) & \ddot{\mathbf{v}}(\psi_I) \end{array}\right], \tag{6.331}$$

where

$$\mathbf{g}_I^T = \left[\begin{array}{ccc} 0 & 0 & g_3 \end{array}\right], \tag{6.332}$$

can be used. These constraints will create a broader null. It is an alternative to (6.301) and (6.302) that may be more useful.

We see from this brief discussion that derivative constraints offer a wide range of choices. After the optimum LCMV and LCMP beamformers are derived, we will explore how the derivative constraints affect the beam pattern and array gain.

6.7.1.5 Eigenvector constraints

A different type of constraint was introduced by Er and Cantoni [EC85] in order to achieve effective control of the beam pattern over a region of ω-$\mathbf{k}$ space. The technique was further developed by Buckley [Buc87] and Van Veen [VV91]. (e.g., Chapter 4 of [HS92]).[7] These constraints are referred to as eigenvector constraints for reasons that will be clear when they are derived.

The basic model assumes that there is a desired response over a specified region in the ω-$\mathbf{k}$ space. We want to define a set of constraints that will minimize the total squared error between the desired response and the actual response. This is analogous to the least squared error approach in Chapter

[7] Our discussion is similar to [EC85].

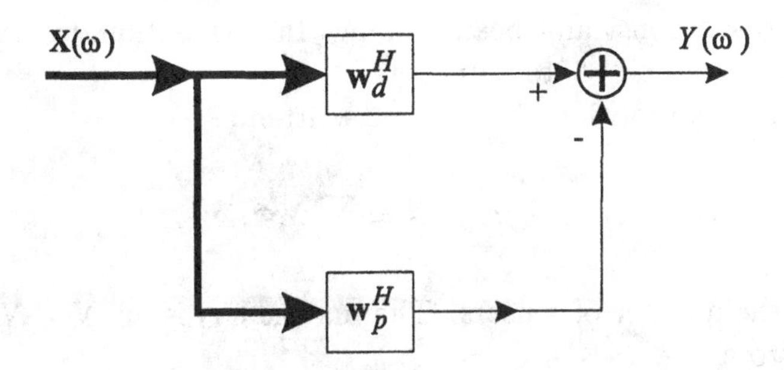

Figure 6.42: Partition of beamformer.

3, except we only specify the desired response over a segment of the ω-**k** space.

The desired beam pattern can be written as

$$B_d(w, \mathbf{k}) = \mathbf{w}_d^H \, \mathbf{v}(w : \mathbf{k}), \tag{6.333}$$

where we have suppressed the w dependence in the weight vector. The squared error is

$$e^2 = \int_{\mathbf{k}\in\mathbf{K}} \int_{\omega\in\Omega} \left| \mathbf{w}_d^H \, \mathbf{v}(\omega, \mathbf{k}) - \mathbf{w}^H \mathbf{v}(\omega : \mathbf{k}) \right|^2 d\mathbf{k}\, d\omega. \tag{6.334}$$

It is convenient to partition the beamformers as shown in Figure 6.42. Then,

$$\mathbf{w} = \mathbf{w}_d - \mathbf{w}_p. \tag{6.335}$$

Substituting (6.335) into (6.334) gives

$$e^2 = \mathbf{w}_p^H \, \mathbf{Q} \, \mathbf{w}_p \tag{6.336}$$

where

$$\mathbf{Q} = \int_{\mathbf{k}\in\mathbf{K}} \int_{\omega\in\Omega} \mathbf{v}(\omega : \mathbf{k}) \mathbf{v}^H(\omega : \mathbf{k}) d\mathbf{k}\, d\omega. \tag{6.337}$$

Note that $\mathbf{Q}$ is just the correlation matrix at the array from a distributed source $S_o(\omega : \mathbf{k}) = 2\pi$ over $\mathbf{K}$ and Ω. It does not depend on the desired pattern shape, but only the constraint region and the array manifold vector. It is a non-negative definite Hermitian matrix.

We use the error expression in (6.336) in two different ways. In this section, we use the principal eigenvectors of $\mathbf{Q}$ to form the linear constraint matrix $\mathbf{C}$. In Section 6.11, we impose a constraint on e^2 (i.e., $e^2 \leq e_0$) and find the resultant beamformer. This leads to a generalized form of diagonal loading called soft constraint beamforming. In this section, we impose a set of linear constraints on $\mathbf{w}_p$ to reduce e^2.

The eigendecomposition of $\mathbf{Q}$ can be written as

$$\mathbf{Q} = \mathbf{U}_Q \mathbf{\Lambda}_Q \mathbf{U}_Q^H = \sum_{i=1}^{N} \lambda_i \mathbf{\Phi}_i \mathbf{\Phi}_i^H, \tag{6.338}$$

where N is the number of sensors. The matrix $\mathbf{U}_\mathbf{Q}$ is an $N \times N$ matrix of the eigenvectors,

$$\mathbf{U}_\mathbf{Q} = \begin{bmatrix} \mathbf{\Phi}_1 & \mathbf{\Phi}_2 & \cdots & \mathbf{\Phi}_N \end{bmatrix}, \tag{6.339}$$

and

$$\mathbf{\Lambda}_\mathbf{Q} = \text{diag}\left\{ \lambda_1 \quad \lambda_2 \quad \cdots \quad \lambda_N \right\}, \tag{6.340}$$

is a diagonal matrix of the ordered eigenvalues.

We impose N_e linear constraints on $\mathbf{w}_p$,

$$\mathbf{\Phi}_i^H \mathbf{w}_p = 0, \quad i = 1, \cdots, N_e. \tag{6.341}$$

We normally choose N_e to correspond to the number of significant eigenvalues of $\mathbf{Q}$. The eigenvalue behavior in Figure 5.14 is typical of many problems of interest. As we increase N_e, we reduce the number of degrees of freedom available for adaptation. However, we reduce the error in (6.336). Using (6.338) and (6.341) in (6.336) gives,

$$e^2 = \sum_{i=N_e+1}^{N} \lambda_i |\mathbf{\Phi}_i^H \mathbf{w}_p|^2. \tag{6.342}$$

The choice of N_e is a trade-off between generating a beam pattern that is close to B_d over a desired region and retaining adequate degrees of freedom. For most $\mathbf{Q}$ matrices, the eigenvalues drop off sharply, so the choice of N_e is clear.

Note that the constraint in (6.341) is on the weight vector in the bottom path in Figure 6.42. In some cases, this model is the easiest to work with. In other cases, we want to impose the constraint directly on $\mathbf{w}$.

To rewrite the constraints, we expand the three weight vectors in terms of their eigenvectors and match the coefficients. The resulting constraint on $\mathbf{w}$ is

$$\mathbf{\Phi}_i^H \mathbf{w} = \mathbf{\Phi}_i^H \mathbf{w}_d, \quad i = 1, \cdots, N_e. \tag{6.343}$$

To put (6.343) in the same notation as (6.279)–(6.280), define

$$\mathbf{C} = \begin{bmatrix} \mathbf{\Phi}_1 & \mathbf{\Phi}_2 & \cdots & \mathbf{\Phi}_{N_e} \end{bmatrix} \tag{6.344}$$

and

$$[\mathbf{g}^H]_i = \mathbf{w}_d^H \, \mathbf{\Phi}_i, \quad i = 1, 2, \cdots, N_e. \tag{6.345}$$

The resulting constraint equation is

$$\mathbf{w}^H \, \mathbf{C} = \mathbf{g}^H. \tag{6.346}$$

Note that we do not include a distortionless constraint because it will be almost linearly dependent on $\mathbf{\Phi}_1$ in most cases.

We consider a simple example to illustrate the behavior.

Example 6.7.4: Narrowband main-lobe constraint

Consider a standard 10-element narrowband linear array. We assume that the desired beam pattern is the conventional beam pattern

$$B_d(u) = B_c(u), \quad -u_d \leq u \leq u_d. \tag{6.347}$$

Then,

$$\mathbf{Q} = \int_{-u_d}^{u_d} \mathbf{v}(u)\mathbf{v}^H(u) \, du. \tag{6.348}$$

The mn element of $\mathbf{Q}$ is

$$\begin{aligned} [\mathbf{Q}]_{mn} &= \int_{-u_d}^{u_d} e^{j\pi u[m-n]} du \\ &= \frac{2\sin[\pi u_d(m-n)]}{\pi(m-n)}, \end{aligned} \tag{6.349}$$

so $\mathbf{Q}$ is a Toeplitz matrix. The eigenvectors are the discrete prolate spheroidal sequences. Note that $\mathbf{Q}$ is completely determined by the array manifold vectors and does not depend on $B_d(u)$. This $\mathbf{Q}$ matrix is the same as the matrix we encountered in Example 5.5.1. The number of significant eigenvalues will depend on u_d. In many cases, we choose $u_d = 1/N$ (or $u_d = 0.1$ for N = 10). From Figure 5.14, we see that there are three significant eigenvalues.

This simple example does not illustrate the real value of eigenvector constraints. We encounter more interesting examples when we study broadband beamforming.

6.7.1.6 Quiescent pattern constraints

There are other types of constraints that will prove useful. The eigenvector constraint matched a desired response over a certain region. An alternative approach would be to specify the pattern over the entire ω-$\mathbf{k}$ plane and try to approximate it in a least squares sense while minimizing the output power (or variance). We explore this approach in Section 6.7.5.

6.7.1.7 Summary

At this point, we have developed a large menu of possible constraints. The next step is to develop the optimum LCMV and LCMP beamformers and then see how the various constraints affect the array gain, the beam pattern, and the robustness to model variations.

6.7.2 Optimum LCMV and LCMP Beamformers

In this section, we find the optimum LCMV and LCMP beamformers for the narrowband case. For M_c linear constraints, the constraint equation can be written as,

$$\mathbf{w}^H \mathbf{C} = \mathbf{g}^H, \tag{6.350}$$

where $\mathbf{w}^H$ is $1 \times N$, $\mathbf{C}$ is $N \times M_c$, $\mathbf{g}^H$ is $1 \times M_c$, and N is the number of sensors. We require that the columns of $\mathbf{C}$ be linearly independent.

We will assume that the first column of $\mathbf{C}$ is $\mathbf{v}_m$ and that the first element of $\mathbf{g}$ is 1 so that the processor is distortionless.

We consider two related optimization problems. In the first case, we assume that $\mathbf{S_n}$ is known or that we will be able to estimate it. In the second case, we assume that $\mathbf{S_x}$ is known or that we will be able to estimate it. These two cases correspond to the MVDR and MPDR beamformers in Section 6.2.

In the first case, we minimize the output due to noise subject to the constraint in (6.350)

$$\sigma_{no}^2 = \mathbf{w}^H \mathbf{S_n} \mathbf{w}. \tag{6.351}$$

We refer to this case as the linear constraint minimum variance (LCMV) beamformer.

In the second case, we minimize the output power subject to the constraint in (6.350)

$$E\left[y^2\right] = \mathbf{w}^H \mathbf{S_x} \mathbf{w}. \tag{6.352}$$

We refer to this case as the linear constraint minimum power beamformer (LCMP).

The minimization procedure is identical for the two cases, so we will solve the first case and indicate the answer for the second case.

We minimize the function

$$J \triangleq \mathbf{w}^H \mathbf{S_n} \mathbf{w} + \left[\mathbf{w}^H \mathbf{C} - \mathbf{g}^H\right] \boldsymbol{\lambda} + \boldsymbol{\lambda}^H \left[\mathbf{C}^H \mathbf{w} - \mathbf{g}\right]. \tag{6.353}$$

The Lagrange multiplier $\boldsymbol{\lambda}$ is an $M_c \times 1$ vector because of the M_c constraints. Taking the complex gradient of $\mathbf{J}$ with respect to $\mathbf{w}^H$ and setting it equal to zero gives

$$\mathbf{S_n w} + \mathbf{C}\boldsymbol{\lambda} = \mathbf{0}, \tag{6.354}$$

or

$$\mathbf{w} = -\mathbf{S_n^{-1} C}\boldsymbol{\lambda}. \tag{6.355}$$

Substituting (6.355) into (6.350) gives

$$-\boldsymbol{\lambda}^H \mathbf{C}^H \mathbf{S_n^{-1} C} = \mathbf{g}^H. \tag{6.356}$$

Solving for $\boldsymbol{\lambda}^H$ and substituting into (6.355) gives

$$\boxed{\mathbf{w}_{lcmv}^H = \mathbf{g}^H \left[\mathbf{C}^H \mathbf{S_n^{-1} C}\right]^{-1} \mathbf{C}^H \mathbf{S_n^{-1}},} \tag{6.357}$$

as the optimum constrained processor for Case 1. The existence of the inverse is guaranteed because $\mathbf{S_n}$ is full rank and the columns of $\mathbf{C}$ are linearly independent.

Similarly, for Case 2,

$$\boxed{\mathbf{w}_{lcmp}^H = \mathbf{g}^H \left[\mathbf{C}^H \mathbf{S_x^{-1} C}\right]^{-1} \mathbf{C}^H \mathbf{S_x^{-1}}} \tag{6.358}$$

is the optimum processor.[8]

One interpretation of the optimum processor in (6.357) is shown in Figure 6.43. The beamformer first forms a set of M_c constraint beams and then combines them to form $Y(\omega)$. Note that the processor can be viewed as operating in a M_c-dimensional constraint subspace.

The combiner utilizes the inverse of the matrix $\left[\mathbf{C}^H \mathbf{S_n^{-1} C}\right]$. One of the issues in the practical implementation of constrained processors is the condition number of this matrix,

$$r = \frac{\lambda_{\max}}{\lambda_{\min}}. \tag{6.359}$$

As this number increases there may be a problem with the numerical accuracy of the results. The r will increase as the columns of $\mathbf{C}$ become more correlated in a space defined by $\mathbf{S_n^{-1}}$ (or $\mathbf{S_x^{-1}}$).

[8]This result is contained in Frost [Fro72] in the context of a tapped delay line implementation of a closed–loop adaptive array processor.

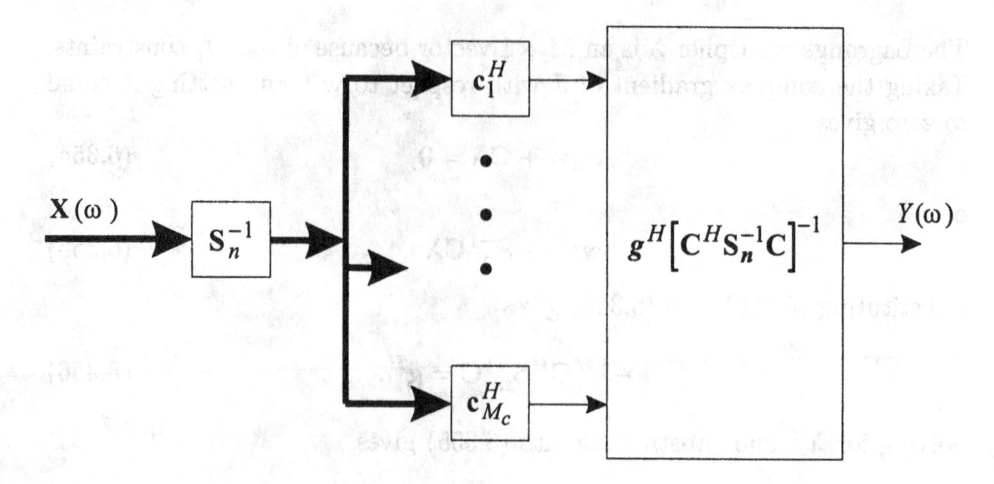

Figure 6.43 Optimum constrained receiver.

If the noise is white, then (6.357) reduces to

$$\boxed{\mathbf{w}_q^H = \mathbf{g}^H \left[\mathbf{C}^H\mathbf{C}\right]^{-1}\mathbf{C}^H}, \tag{6.360}$$

which is referred to as the **quiescent** weight vector (the idea of $\mathbf{w}_q$ was introduced by Applebaum and Chapman [AC76] and utilized by Gabriel [Gab76a], [Gab76b]). We will find that it plays an important role in some of our subsequent discussions.

In the next section, we interpret the optimum LCMV beamformer as a generalized sidelobe canceller.

6.7.3 Generalized Sidelobe Cancellers

A useful implementation of the LCMV (or LCMP) beamformer is obtained by dividing the N-dimensional space into two subspaces, a **constraint** subspace and an **orthogonal** subspace. The constraint subspace is defined by the columns of $\mathbf{C}$, an $N \times M_c$ matrix. The orthogonal subspace is defined by the columns of $\mathbf{B}$, an $N \times (N - M_c)$ matrix. The columns in $\mathbf{B}$ are linearly independent and each column is orthogonal to each column in $\mathbf{C}$. Thus,

$$\mathbf{C}^H\mathbf{B} = \mathbf{0}, \tag{6.361}$$

where $\mathbf{0}$ is a $M_c \times (N - M_c)$ matrix of zeros.

First, we assume that $\mathbf{w}_o$ is known. From (6.358)

$$\mathbf{w}_o^H \triangleq \mathbf{w}_{lcmp}^H = \mathbf{g}^H \left[\mathbf{C}^H\mathbf{S}_{\mathbf{x}}^{-1}\mathbf{C}\right]^{-1}\mathbf{C}^H\mathbf{S}_{\mathbf{x}}^{-1}. \tag{6.362}$$

We partition the processor into two orthogonal components,

$$\mathbf{w}_o^H = \mathbf{w}_c^H - \mathbf{w}_p^H, \tag{6.363}$$

where $\mathbf{w}_c^H$ is defined to be the projection of $\mathbf{w}_o^H$ onto the constraint subspace and $\mathbf{w}_p^H$ is the projection of $\mathbf{w}_o^H$ onto $\mathbf{B}$. The projection matrix onto the constraint subspace is

$$\mathbf{P_C} = \mathbf{C}\left[\mathbf{C}^H\mathbf{C}\right]^{-1}\mathbf{C}^H, \tag{6.364}$$

and

$$\mathbf{w}_c^H = \mathbf{w}_o^H\mathbf{P_C}. \tag{6.365}$$

Substituting (6.362)) and (6.364) into (6.365),

$$\mathbf{w}_c^H = \mathbf{w}_o^H\mathbf{P_C} = \mathbf{g}^H\left[\mathbf{C}^H\mathbf{S}_\mathbf{x}^{-1}\mathbf{C}\right]^{-1}\mathbf{C}^H\mathbf{S}_\mathbf{x}^{-1}\left\{\mathbf{C}\left[\mathbf{C}^H\mathbf{C}\right]^{-1}\mathbf{C}^H\right\} \tag{6.366}$$

or

$$\boxed{\mathbf{w}_c^H = \mathbf{g}^H\left[\mathbf{C}^H\mathbf{C}\right]^{-1}\mathbf{C}^H = \mathbf{w}_q^H,} \tag{6.367}$$

which does not depend on $\mathbf{S_x}$. The weight vector in the upper path, $\mathbf{w}_q^H$, is the quiescent component of $\mathbf{w}_o^H$. If the input is white noise,

$$\mathbf{w}_o^H = \mathbf{w}_q^H. \tag{6.368}$$

The second component of $\mathbf{w}_o^H$ is $\mathbf{w}_p^H$, which can be written as

$$\mathbf{w}_p = -\mathbf{B}\left[\mathbf{B}^H\mathbf{B}\right]^{-1}\mathbf{B}^H\mathbf{w}_o = -\mathbf{P}_\mathbf{C}^\perp\mathbf{w}_o. \tag{6.369}$$

Thus,

$$\mathbf{w}_p^H = \mathbf{g}^H\left[\mathbf{C}^H\mathbf{S}_\mathbf{x}^{-1}\mathbf{C}\right]^{-1}\mathbf{C}^H\mathbf{S}_\mathbf{x}^{-1}\cdot\mathbf{B}\left[\mathbf{B}^H\mathbf{B}\right]^{-1}\mathbf{B}^H. \tag{6.370}$$

This leads to the processor shown in Figure 6.44. This is a correct, but not particularly useful, implementation.

Note that

$$Y_b(\omega) = \mathbf{w}_p^H\mathbf{X}(\omega) \tag{6.371}$$

is obtained by multiplying $\mathbf{X}(\omega)$ by a matrix completely contained in the $\mathbf{B}$ subspace. This suggests dividing the processor in the lower path into two parts, as shown in Figure 6.45. The first processor operates on $\mathbf{X}(\omega)$ to obtain an $(N - M_c) \times 1$ vector. Note that the output of $\mathbf{B}$ cannot contain

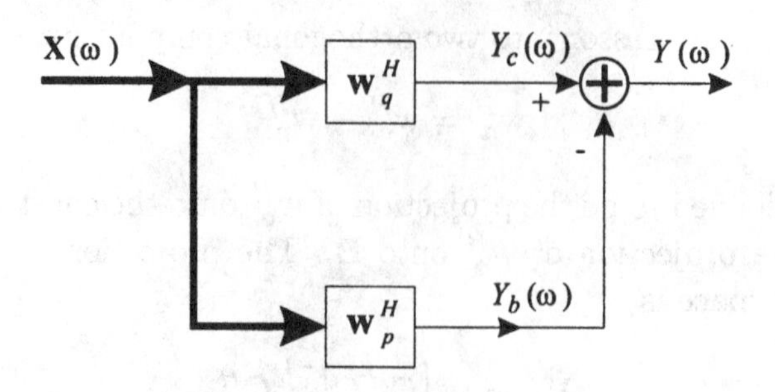

Figure 6.44 Partitioned processor.

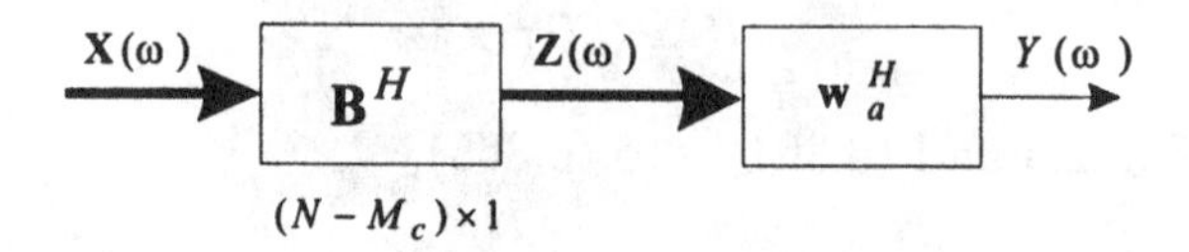

Figure 6.45 Cascade of blocking matrix and adaptive matrix.

any components in the $\mathbf{C}$ space because of (6.361). We refer to $\mathbf{B}$ as the **blocking matrix**. The second matrix operates on the $(N - M_c) \times 1$ vector to produce $Y_b(\omega)$. The formula for $\mathbf{w}_a^H$ follows directly from $\mathbf{w}_p^H$. It is the matrix that will adapt as a function of the data when adaptive beamformers are discussed in Chapter 7. Combining Figures 6.44 and 6.45 gives the realization shown in Figure 6.46. This realization is referred to as a generalized sidelobe canceller because of its similarity to classical sidelobe cancellers. Generalized sidelobe cancellers were utilized by Owsley [Ows71] and Applebaum and Chapman [AC76]. Griffiths and Jim [GJ82] analyzed them and introduced the "generalized sidelobe canceller" name. Other references include Er and Cantoni [EC83] and Cox et al. [CZO87].

We now recall the criterion by which the overall filter was designed. It was to minimize the output power subject to the constraint

$$\mathbf{w}^H \mathbf{C} = \mathbf{g}^H. \tag{6.372}$$

The top processor satisfies the constraint exactly and the bottom path is orthogonal to $\mathbf{C}$ so it cannot impact the constraint. The output power is

$$P_o = [\mathbf{w}_q - \mathbf{B}\mathbf{w}_a]^H \, \mathbf{S}_\mathbf{x} \, [\mathbf{w}_q - \mathbf{B}\mathbf{w}_a]. \tag{6.373}$$

Taking the gradient with respect to $\mathbf{w}_a$ and setting the result equal to zero gives

$$[\mathbf{w}_q^H - \mathbf{w}_a^H \, \mathbf{B}^H] \, \mathbf{S}_\mathbf{x} \, \mathbf{B} = \mathbf{0}, \tag{6.374}$$

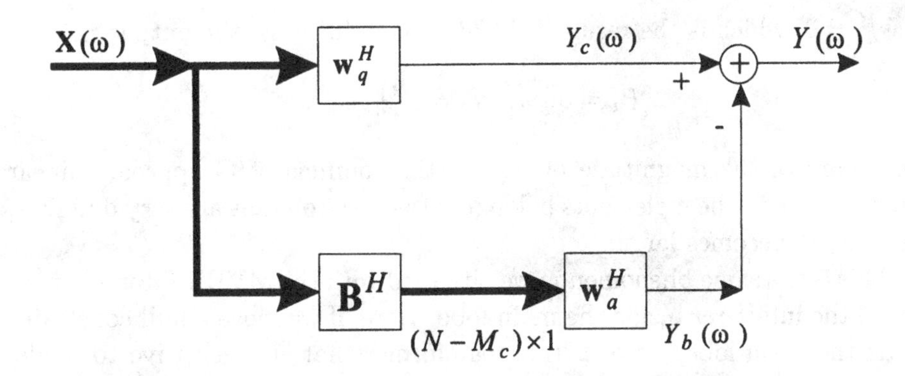

Figure 6.46 Generalized sidelobe canceller.

or[9]

$$\boxed{\hat{\mathbf{w}}_a^H = \mathbf{w}_q^H\,\mathbf{S}_\mathbf{x}\,\mathbf{B}[\mathbf{B}^H\,\mathbf{S}_\mathbf{x}\,\mathbf{B}]^{-1}.} \tag{6.375}$$

For a single plane-wave signal input,

$$\mathbf{S}_\mathbf{x} = \sigma_s^2 \mathbf{v}_m \mathbf{v}_m^H + \mathbf{S}_\mathbf{n}. \tag{6.376}$$

The first column of $\mathbf{C}$ is $\mathbf{v}_m$, so

$$\mathbf{v}_m^H \mathbf{B} = 0, \tag{6.377}$$

and

$$\mathbf{S}_\mathbf{x}\mathbf{B} = \mathbf{S}_\mathbf{n}\mathbf{B}. \tag{6.378}$$

Then, (6.375) reduces to

$$\hat{\mathbf{w}}_a^H = \mathbf{w}_q^H \mathbf{S}_\mathbf{n}\mathbf{B}\left[\mathbf{B}^H\mathbf{S}_\mathbf{n}\mathbf{B}\right]^{-1}. \tag{6.379}$$

Note that (6.377)–(6.379) assume that there is no signal mismatch ($\mathbf{v}_a = \mathbf{v}_m$) and that $\mathbf{S}_\mathbf{x}$ and $\mathbf{S}_\mathbf{n}$ are known. In this case, the LCMV and LCMP beamformers are identical. More generally, as long as the signal is completely contained in the constraint subspace, the two beamformers will be the same. The formula in (6.375) is used to implement the beamformer because the model assumes that only $\mathbf{S}_\mathbf{x}$ is available. In practice, an estimate $\hat{\mathbf{S}}_\mathbf{x}$ is used.

When the noise is white, the left matrix product in (6.379),

$$\mathbf{w}_q^H \left[\sigma_\omega^2 \mathbf{I}\right] \mathbf{B} = \mathbf{0}, \tag{6.380}$$

[9]The result in (6.375) can also be written as $\hat{\mathbf{w}}_a^H = \mathbf{S}_{d\mathbf{z}^H}\mathbf{S}_\mathbf{z}^{-1}$. We exploit this interpretation in Chapter 7.

so $\hat{\mathbf{w}}_a^H = \mathbf{0}$, which is the result in (6.368). For this case, the output noise is

$$P_n = \sigma_w^2 \mathbf{w}_q^H \mathbf{w}_q = \sigma_w^2 \|\mathbf{w}_q\|^2, \tag{6.381}$$

the square of the magnitude of $\mathbf{w}_q$. If the columns of $\mathbf{C}$ approach linear dependency and the $\mathbf{g}$ elements belonging to those columns are very different, then $\|\mathbf{w}_q\|^2$ becomes large.

This is the same phenomenon we observed with the MVDR filter when we moved the interferer inside the main lobe. Here, if we move a null constraint inside the main lobe, we obtain a beamformer that is is sensitive to model mismatch.

Note that $\mathbf{B}$ is not unique. It must satisfy (6.361) and have rank equal to $(N - M_c)$. One method for constructing $\mathbf{B}$ is to find $\mathbf{P}_{\mathbf{C}}^{\perp}$,

$$\mathbf{P}_{\mathbf{C}}^{\perp} = \mathbf{I} - \mathbf{C}\left(\mathbf{C}^H\mathbf{C}\right)^{-1}\mathbf{C}^H, \tag{6.382}$$

which is an $N \times N$ matrix. Then orthonormalize $\mathbf{P}_{\mathbf{C}}^{\perp}$ and choose the first $(N - M_c)$ columns of the orthonormalized matrix. The resulting $\mathbf{B}$ matrix has the property that

$$\mathbf{B}^H\mathbf{B} = \mathbf{I}. \tag{6.383}$$

Note that (6.383) implies that the component of $\mathbf{S}_{\mathbf{z}}$ due to the white noise component of $\mathbf{S}_{\mathbf{x}}$ is white noise with variance σ_w^2. In subsequent discussions, we will always choose $\mathbf{B}$ to satisfy (6.383).

6.7.4 Performance of LCMV and LCMP Beamformers

In this section, we analyze the performance of LCMV and LCMP beamformers with directional, derivative, or eigenvector constraints. In this chapter, we assume the ensemble spatial spectral matrix is available. In Chapter 7, we analyze the behavior when we estimate $\mathbf{S}_{\mathbf{n}}$ or $\mathbf{S}_{\mathbf{x}}$ from the input data.

The purpose of introducing the linear constraints is to provide main-lobe protection in the case of a main-lobe interferer and to prevent performance degradation in the presence of DOA mismatch and/or array perturbations. Our performance discussion in the text emphasizes the DOA mismatch problem. The results for the array perturbation problem are similar and are developed in the problems. Diagonal loading is used in most cases in order to obtain satisfactory performance.

A large number of examples are included in order to adequately explore the utility of the various constraints.

6.7.4.1 Array gain

The expressions for the array gain of LCMV and LCMP beamformers are straightforward. For the LCMV beamformer, $\mathbf{w}$ is given by (6.357). For the LCMP beamformer, $\mathbf{w}$ is given by (6.358).

The output signal power is

$$P_s = \sigma_s^2 \left|\mathbf{w}^H \mathbf{v}_s\right|^2. \tag{6.384}$$

The output noise power is,

$$P_n = \sigma_n^2 \left|\mathbf{w}^H \boldsymbol{\rho}_{\mathbf{n}} \mathbf{w}\right|, \tag{6.385}$$

where $\boldsymbol{\rho}_{\mathbf{n}}$ is the normalized spatial spectral matrix and includes both the white noise and any additional interference (see (6.106)). The array gain is,

$$A_o = \frac{\left|\mathbf{w}^H \mathbf{v}_s\right|^2}{\left|\mathbf{w}^H \boldsymbol{\rho}_{\mathbf{n}} \mathbf{w}\right|}, \tag{6.386}$$

where $\mathbf{w}$ is given by either (6.357) or (6.358). If we include a distortionless constraint, then the numerator in (6.385) is unity and the array gain is

$$A_o = \left[\mathbf{w}^H \boldsymbol{\rho}_{\mathbf{n}} \mathbf{w}\right]^{-1}. \tag{6.387}$$

The output SNR_o is[10]

$$SNR_o = \left(\frac{\sigma_s^2}{\sigma_n^2}\right) A_o. \tag{6.388}$$

The array gain for the LCMV beamformer is obtained by substituting (6.357) into (6.387). The result is

$$\boxed{A_{lcmv} = \frac{1}{\mathbf{g}^H \left[\mathbf{C}^H \boldsymbol{\rho}_{\mathbf{n}}^{-1} \mathbf{C}\right]^{-1} \mathbf{g}}.} \tag{6.389}$$

In order to find the array gain for the LCMP beamformer, we substitute (6.358) into (6.387) and obtain

$$\boxed{A_{lcmp} = \left\{\mathbf{g}^H \left[\mathbf{C}^H \mathbf{S}_{\mathbf{x}}^{-1} \mathbf{C}\right]^{-1} \mathbf{C}^H \mathbf{S}_{\mathbf{x}}^{-1} \boldsymbol{\rho}_{\mathbf{n}} \mathbf{S}_{\mathbf{x}}^{-1} \mathbf{C} \left[\mathbf{C}^H \mathbf{S}_{\mathbf{x}}^{-1} \mathbf{C}\right]^{-1} \mathbf{g}\right\}^{-1}.} \tag{6.390}$$

[10] The subscript "o" on the left side of (6.387) denotes output. On the right-hand side it denotes optimum. The meaning should be clear from the context.

From (6.379), the array gain for the LCMP case is equal to the array gain for the LCMV case when the distortionless constraint is imposed and the signal is perfectly matched. Thus,

$$A_{lcmp} = \frac{1}{\mathbf{g}^H \left[\mathbf{C}^H \rho_{\mathbf{n}}^{-1} \mathbf{C}\right]^{-1} \mathbf{g}}. \tag{6.391}$$

In order to study the effects of mismatch, (6.390) must be used.

The array gain or SNR_o when there is no model mismatch is one performance measure of interest. The second measure is the array gain or SNR_o under mismatched conditions, which was one of the primary reasons for introducing constraints.

In order to analyze the effect of signal mismatch, use $\mathbf{C}_m$ and $\mathbf{v}_m$ in (6.358),

$$\mathbf{w}^H = \mathbf{g}^H \left[\mathbf{C}_m^H \mathbf{S}_{\mathbf{x}}^{-1} \mathbf{C}_m\right]^{-1} \mathbf{C}_m^H \mathbf{S}_{\mathbf{x}}^{-1}, \tag{6.392}$$

where $\mathbf{C}_m$ and $\mathbf{g}_m$ are constructed assuming $\mathbf{v}_m$ is correct. Substitute $\mathbf{w}^H$ from (6.392) and $\mathbf{v}_a$ into (6.386). Carrying out the substitution and simplifying gives

$$A_{lcmp}(\mathbf{v}_a : \mathbf{v}_m) =$$

$$\frac{\left|\mathbf{g}^H \left(\mathbf{C}_m^H \rho_{\mathbf{n}}^{-1} \mathbf{C}_m\right)^{-1} \mathbf{C}_m^H \rho_{\mathbf{n}}^{-1} \mathbf{v}_a\right|^2}{\left(1 + \frac{\sigma_s^2}{\sigma_w^2}\gamma\right)^2 \mathbf{g}^H \left(\mathbf{C}_m^H \rho_{\mathbf{n}}^{-1} \mathbf{C}_m\right) \mathbf{g} + \frac{\sigma_s^4}{\sigma_w^4}\gamma \left|\mathbf{g}^H \left(\mathbf{C}_m^H \rho_{\mathbf{n}}^{-1} \mathbf{C}_m\right)^{-1} \mathbf{C}_m^H \rho_{\mathbf{n}}^{-1} \mathbf{v}_a\right|^2}, \tag{6.393}$$

where

$$\gamma = \mathbf{v}_a^H \rho_{\mathbf{n}}^{-1} \mathbf{v}_a - \mathbf{v}_a^H \rho_{\mathbf{n}}^{-1} \mathbf{C}_m \left(\left(\mathbf{C}_m^H \rho_{\mathbf{n}}^{-1} \mathbf{C}_m\right)^{-1}\right) \mathbf{C}_m^H \rho_{\mathbf{n}}^{-1} \mathbf{v}_a, \tag{6.394}$$

and $\mathbf{v}_a$ is the actual signal manifold vector. The results in (6.393) and (6.394) are due to Steele [Ste83].

We consider a series of examples to illustrate the behavior using various constraints. In all of the examples, we consider a standard linear array and assume the array is steered to broadside. We assume u_a, the actual arrival direction varies between $-0.25BW_{NN}$ and $0.25BW_{NN}$. This assumption corresponds to a model in which we search u-space with beams spaced at $0.5BW_{NN}$.

We can also search u-space using a finer grid. This technique would reduce the mismatch problem at the cost of increased computational complexity.

6.7.4.2 Directional constraints

In this section, we study the behavior using directional constraints.

Example 6.7.5: Directional main-lobe constraints; LCMV beamformer
Consider a standard 10-element linear array. We use the three main-lobe constraints in (6.291),

$$\mathbf{C} = \left[\begin{array}{c:c:c} \mathbf{v}_u(0) & \mathbf{v}_u(-0.0866) & \mathbf{v}_u(0.0866) \end{array} \right], \tag{6.395}$$

where the array manifold vector is specified in u-space. We consider four $\mathbf{g}$ matrices,

$$\mathbf{g} = \left[\begin{array}{ccc} 1 & g_{ci} & g_{ci} \end{array} \right], \quad i = 1, \cdots, 4, \tag{6.396}$$

where

$$g_{c1} = 1, \tag{6.397}$$

$$g_{c2} = B_c(0.0866), \tag{6.398}$$

$$g_{c3} = \frac{1 + B_c(0.0866)}{2}, \tag{6.399}$$

$$g_{c4} = \frac{2 + B_c(0.0866)}{3}. \tag{6.400}$$

We assume the array is pointed at broadside ($u_m = 0$) and the only interference is white noise. The actual signal arrival angle is u_a, and we examine the output SNR_o as a function of $\frac{u_a}{BW_{NN}}$. We assume that the actual value of $\mathbf{S}_\mathbf{n}$ (in this case $\sigma_w^2 \mathbf{I}$) is used to design the beamformer.

In Figure 6.47, we plot the array gain versus $\frac{u_a}{BW_{NN}}$ for an input *SNR* of 20 dB for the four choices of g_c. We also show the MVDR beamformer as a reference. For LCMV beamformers, the array gain is not a function of the input *SNR*. We use this input *SNR* as a reference level in the next example.

We see that the best choice for g_c depends on the range of mismatch that we expect in a particular application. As g_c is increased from $B_c(0.0866)$ to 1, the array gain at $u_a = 0$ decreases, but the slope of the array gain curve versus $\frac{u_a}{BW_{NN}}$ also decreases. For our case in which the maximum mismatch is $0.25 BW_{NN}$, we would use g_{c2}.

In the next example, we use the same model but assume that the actual $\mathbf{S}_\mathbf{x}$, rather than $\mathbf{S}_\mathbf{n}$, is available to design the LCMP beamformer,

$$\mathbf{S}_\mathbf{x} = \sigma_s^2 \mathbf{v}_a \mathbf{v}_a^H + \sigma_w^2 \mathbf{I}. \tag{6.401}$$

We will see that there is a dramatic difference in performance.

Example 6.7.6: Directional main-lobe constraints; LCMP beamformers (continuation)
Consider the same model as in Example 6.7.5. We use the actual $\mathbf{S}_\mathbf{x}$ to design the beamformer. We use the $\mathbf{C}$ matrix in (6.395) and use g_{c2} (6.398) in the $\mathbf{g}$ matrix. We plot the array gain versus $\frac{u_a}{BW_{NN}}$ for various input *SNR* (from -20 dB to 30 dB in 10-dB increments). The results are in Figure 6.48.

We see that there is a significant decrease in array gain for $SNR \geq 10$ dB. We could reduce this loss by adding more directional constraints, but this

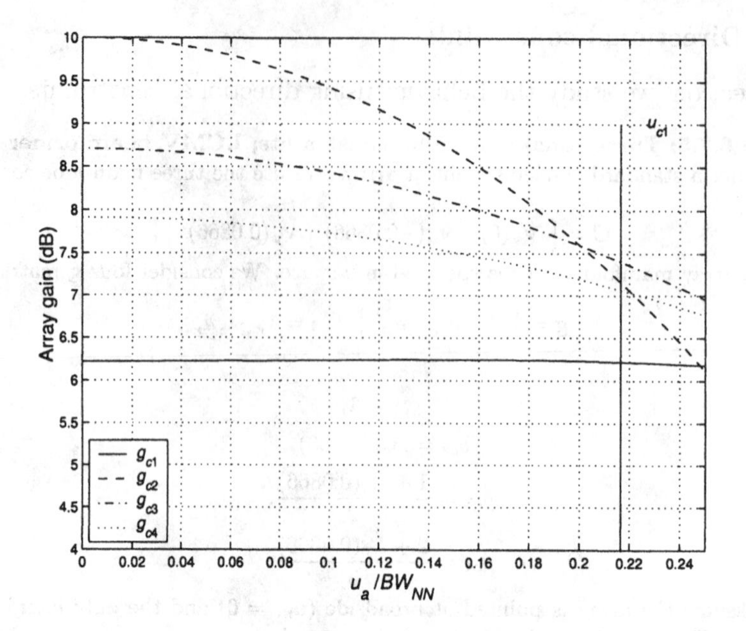

Figure 6.47 LCMV beamformer using directional constraints in white noise environment, SNR = 20 dB, $u_c = \pm 0.866$; array gain versus u_a/BW_{NN}.

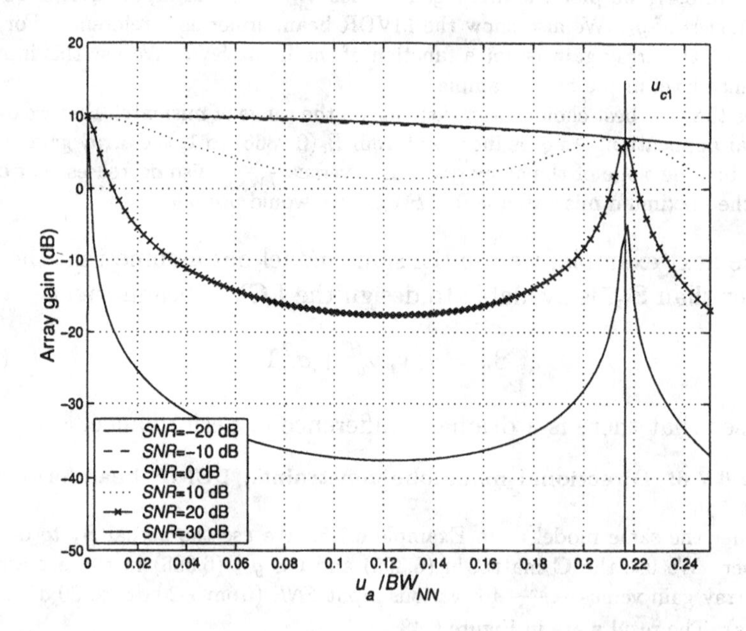

Figure 6.48 LCMP beamformer using directional constraints in white noise environment; $u_c = \pm 0.0866$, $g_c = B_c$; array gain versus u_a/BW_{NN}.

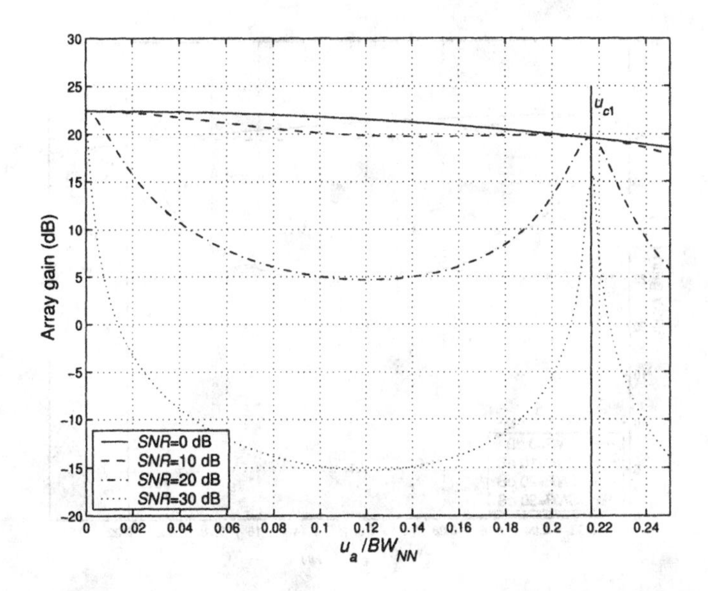

Figure 6.49 LCMP beamformer using directional constraints with two plane-wave interferers (±0.30) and white noise. INR = 10 dB for each interferer; array gain versus u_a/BW_{NN}.

would reduce the degrees of freedom available to null interferers or reduce spatially spread noise.

In order to avoid the signal nulling behavior, we utilize diagonal loading as developed in Section 6.6.4. The appropriate level of diagonal loading depends on the interference environment. Before introducing diagonal loading, we consider an example with multiple plane-wave interferers to provide a baseline case.

Example 6.7.7: Directional main-lobe constraints; LCMP beamformers (continuation)

Consider the same model as in Example 6.7.5. In addition to the white noise interference, we assume that there are two equal-power, uncorrelated plane-wave interferers located at $u_I = \pm 0.30$. Thus,

$$\mathbf{S_x} = \sigma_s^2 \mathbf{v}_a \mathbf{v}_a^H + \sum_{i=1}^{2} \sigma_I^2 \mathbf{v}_i \mathbf{v}_i^H + \sigma_w^2 \mathbf{I}. \tag{6.402}$$

We use the $\mathbf{C}$ matrix in (6.395) and let $g_c = B_c(0.0866)$. We plot the array gain versus u_a/BW_{NN} for various SNR and INR. In Figure 6.49, the INR of each interferer is 10 dB. In Figure 6.50, the INR of each interferer is 30 dB. As in the white noise case, there is significant degradation for higher SNR.

We now introduce diagonal loading.

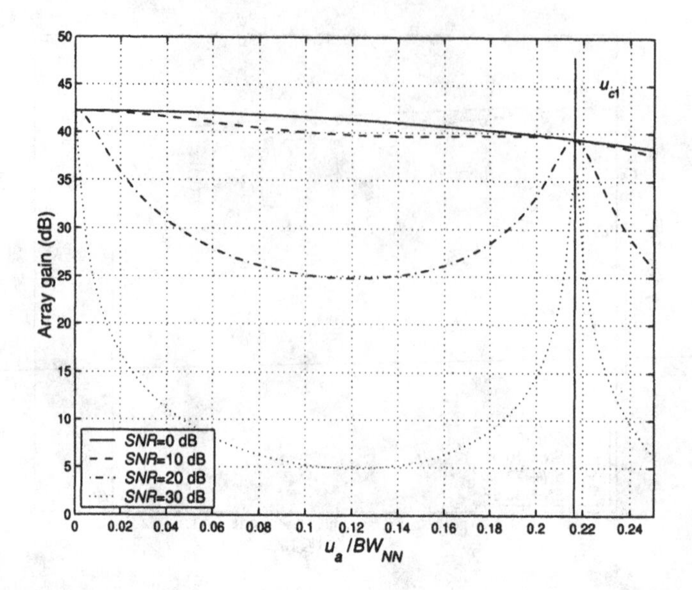

Figure 6.50 LCMP beamformer using directional constraints with two plane-wave interferers (±0.30) and white noise, *INR* = 30 dB for each interferer; array gain versus u_a/BW_{NN}.

Example 6.7.8: Directional main-lobe constraints; LCMP beamformers with diagonal loading (continuation)

Consider the same model as in Example 6.7.7. The actual $\mathbf{S_x}$ is given by (6.402). The weight vector is given by (6.358) modified to include diagonal loading,

$$\mathbf{w}_o^H \triangleq \mathbf{w}_{lcmp,dl}^H = \mathbf{g}^H \left[\mathbf{C}^H \left[\mathbf{S_x} + \sigma_L^2 \mathbf{I}\right]^{-1} \mathbf{C}\right]^{-1} \mathbf{C}^H \left[\mathbf{S_x} + \sigma_L^2 \mathbf{I}\right]^{-1}. \tag{6.403}$$

The appropriate value of σ_L^2 will depend on the *SNR* and *INR*. We define

$$LNR = \frac{\sigma_L^2}{\sigma_w^2}.$$

In order to decrease the signal nulling effect, we would like

$$LNR \geq SNR + 10 \text{ dB}. \tag{6.404}$$

However, to obtain adequate nulling of the interferers, we would like

$$INR(\text{dB}) - (LNR)(\text{dB}) \geq 10 \text{ dB}. \tag{6.405}$$

In some scenarios the two requirements conflict. In Figures 6.51–6.54, we plot the array gain versus u_a/BW_{NN} for various *SNR*, *INR*, and *LNR* combinations.

In Figure 6.51, the *INR* is 30 dB. We plot the array gain with 15 dB loading versus u_a/BW_{NN}. We see that for $SNR \leq 20$ dB, the array gain decreases monotonically to about 38 dB at $u_a = 0.25BW_{NN}$. In Figure 6.52, the *INR* is 30 dB. We plot the array gain with 20-dB loading versus u_a/BW_{NN}. The higher loading moves the curve for low *SNR* down

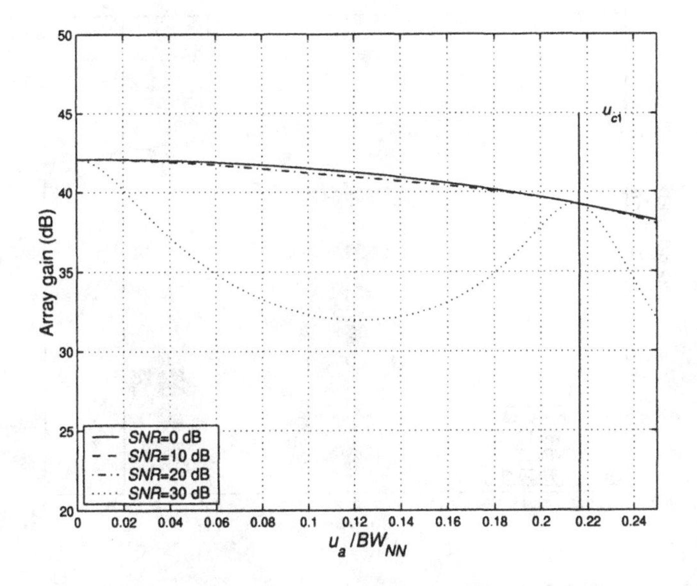

Figure 6.51 LCMP-DL beamformer using directional constraints with two plane-wave interferers (±0.30) and white noise, INR = 30 dB, LNR = 15 dB; array gain versus u_a/BW_{NN}.

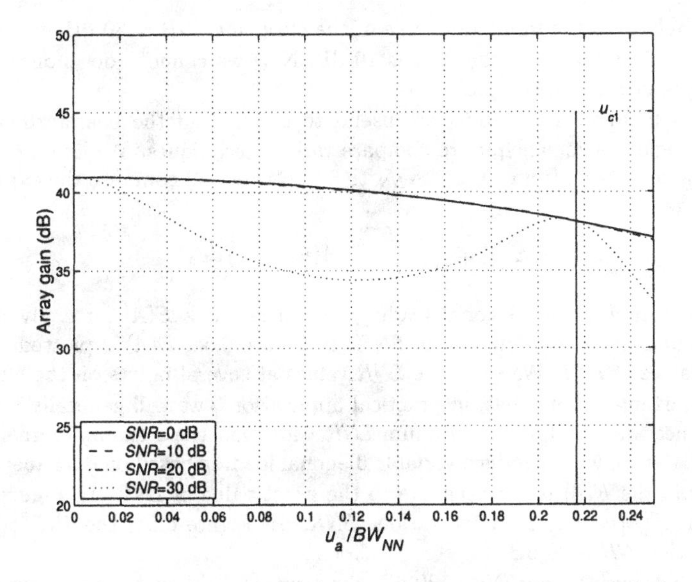

Figure 6.52 LCMP-DL beamformer using directional constraints with two plane-wave interferers (±0.30) and white noise, INR = 30 dB, LNR = 20 dB, array gain versus u_a/BW_{NN}.

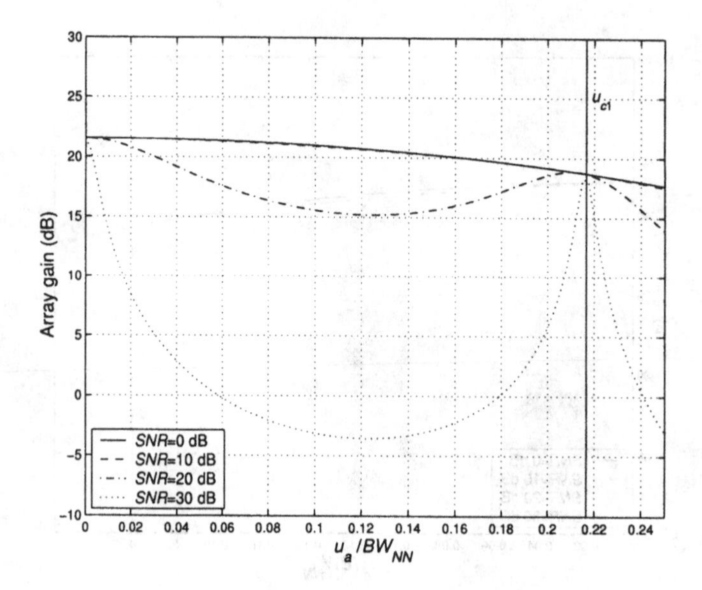

Figure 6.53 LCMP-DL beamformer using directional constraints with two plane-wave interferers (± 0.30) and white noise, *INR* = 10 dB, *LNR* = 10 dB, array gain versus u_a/BW_{NN}.

about 1 dB but raises the minimum value of the curve for *SNR* = 30 dB.

In Figures 6.53 and 6.54, the *INR* is 10 dB. Now we cannot add enough loading to prevent nulling of the 30-dB signal.

The plots in Figures 6.51–6.54 are useful to understand the beamformer behavior. However, a single metric is helpful to compare various techniques. We let u_a be a uniform random variable ranging from $-0.25BW_{NN}$ to $0.25BW_{NN}$ and compute the expected value of the array gain:

$$E[A] = \frac{1}{0.2}\int_{-0.1}^{+0.1} A(u_a : u_m)\, du_a. \tag{6.406}$$

We can then compute the diagonal loading that maximizes $E[A]$ for a given interferer scenario (u_I and *INR*) as a function of *SNR*. In Figure 6.55, $E[A]$ is plotted versus *SNR* for several values of *INR*. We show the *LNR* values at several points on the curves.

It is important to note that, in practical applications, we will generally not have the information necessary to find an optimum *LNR*. However, these examples give us general guidelines. Later we will introduce variable diagonal loading based on data measurements.

In general, the $E[A]$ is not sensitive to the exact value of *LNR*. In Figures 6.56 and 6.57, we plot $E[A]$ versus *LNR* for various *SNR*. In Figure 6.56, the *INR* is 30 dB. In Figure 6.57, the *INR* is 20 dB.

We see that, even for an *SNR* = 30 dB, the range of *LNR*, which results in a 3-dB loss from the optimum *LNR* exceeds 10 dB.

This sequence of examples illustrates typical performance for LCMP-DL algorithms using directional constraints. As long as the *INR* is sufficiently larger than the *SNR*, we can achieve satisfactory performance by adding appropriate diagonal loading. For higher

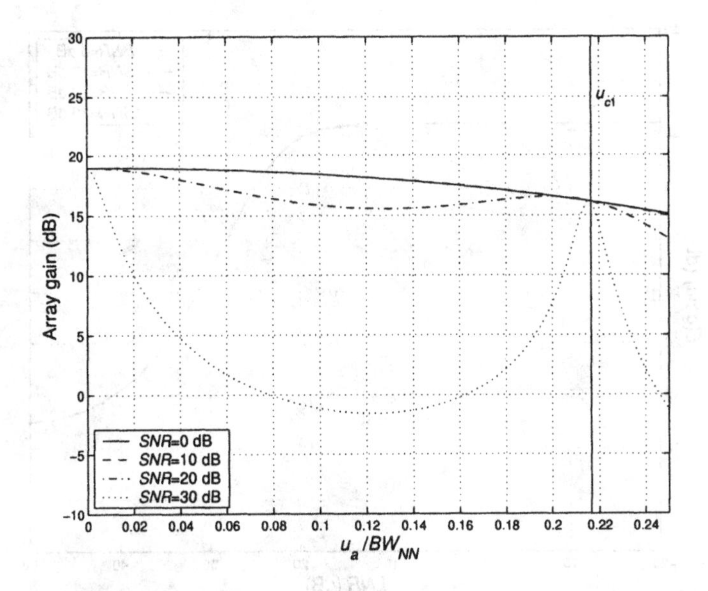

Figure 6.54 LCMP-DL beamformer using directional constraints with two plane-wave interferers (±0.30) and white noise, INR = 10 dB, LNR = 15 dB, array gain versus u_a/BW_{NN}.

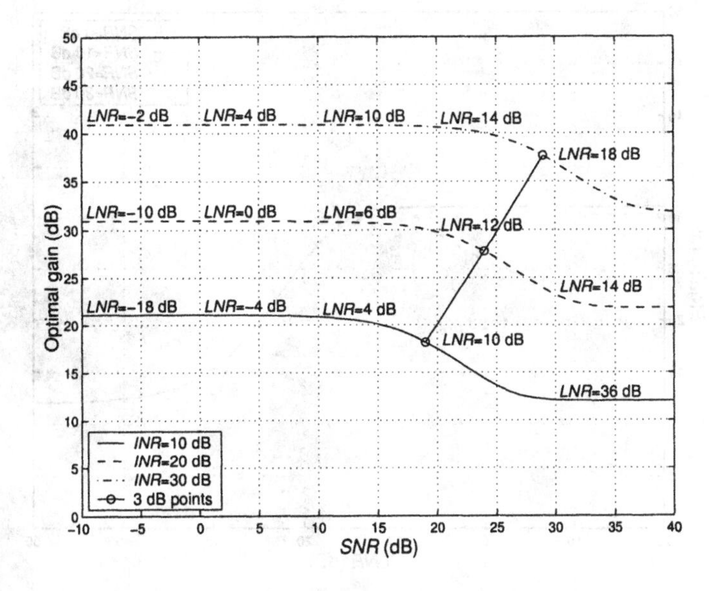

Figure 6.55 LCMP-DL beamformer using directional constraints, various INR and LNR, E[array gain] versus SNR.

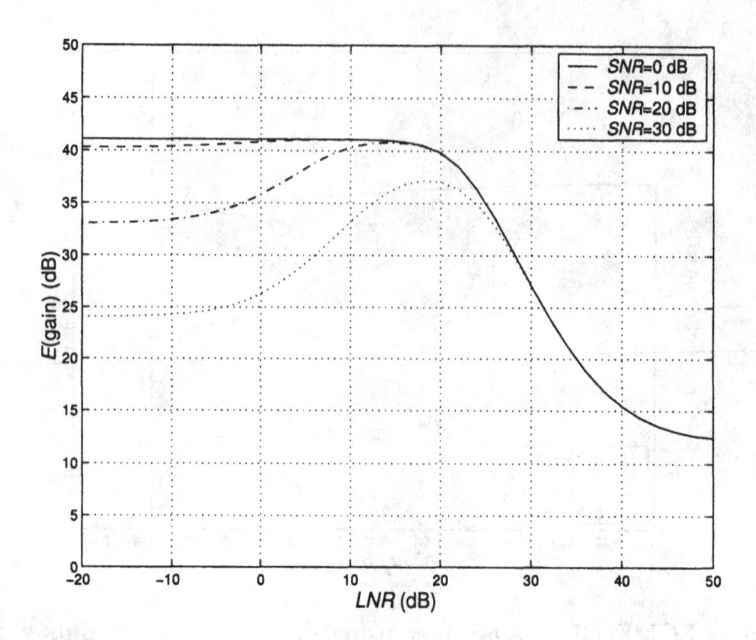

Figure 6.56 LCMP-DL beamformer using directional constraints, $INR = 30$ dB, various SNR, E[array gain] versus LNR.

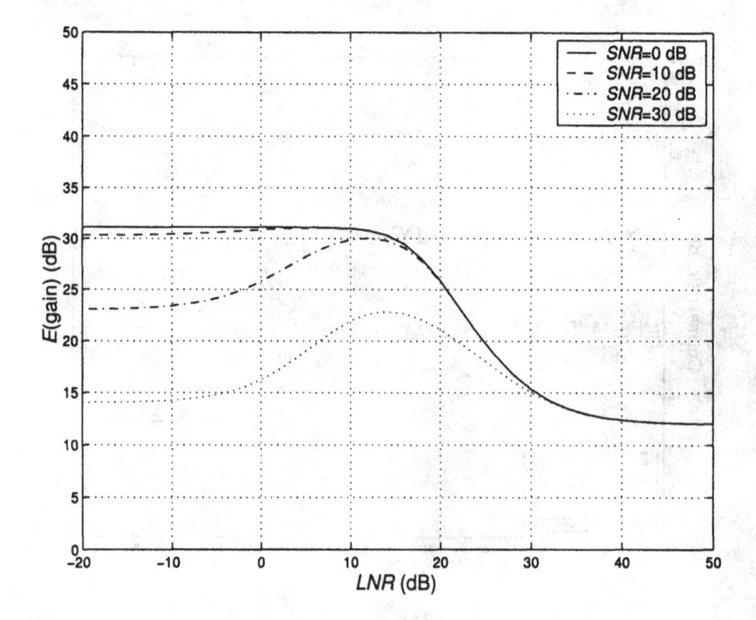

Figure 6.57 LCMP-DL beamformer using directional constraints, $INR = 20$ dB, various SNR, E[array gain] versus LNR.

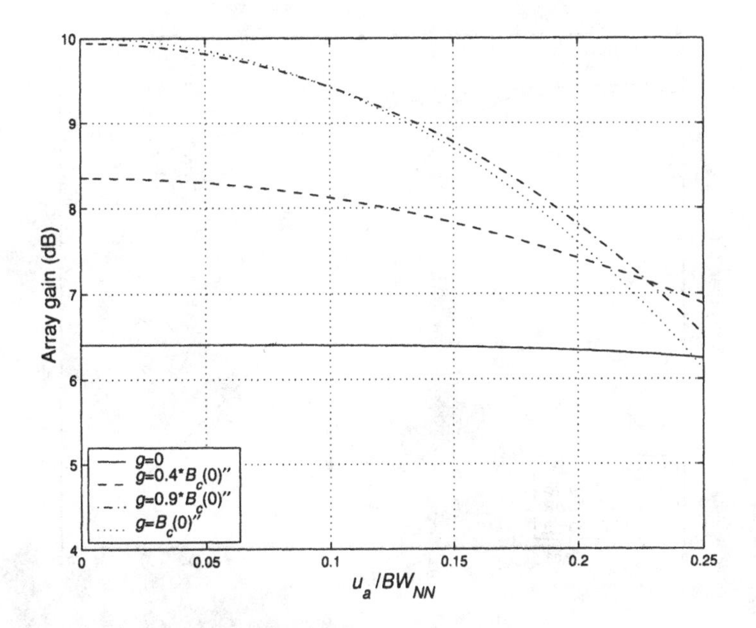

Figure 6.58 LCMP beamformer with second derivative constraints: $SNR =$ 20 dB, white noise environment, array gain versus u_a/BW_{NN}.

SNR, we can reduce the range of mismatch by estimating u_a. We can then move the directional constraints to the steering direction.

6.7.4.3 Derivative constraints

In this section, we consider derivative constraints.

Example 6.7.9: Derivative main-lobe constraints; LCMV beamformer

Consider a standard 10-element linear array. We use distortionless, first derivative, and second derivative constraints evaluated at the origin. From (6.312),

$$\mathbf{C} = \left[\begin{array}{ccc} 1 & d(0) & \dot{d}(0) \end{array} \right]. \tag{6.407}$$

We consider four $\mathbf{g}$ matrices: (6.313), (6.314), and (6.315) with $g_3 = 0.4\left(\frac{1-N^2}{12}\right)$ and $0.9\left(\frac{1-N^2}{12}\right)$.

We assume that the array is pointed at broadside ($u_m = 0$) and that the interference consists of white noise. The actual signal arrival angle is u_a, and we examine the array gain as a function of u_a/BW_{NN}. We assume that the actual value of $\mathbf{S_x}$ (in this case, $\sigma_w^2\mathbf{I}$) is used to design the beamformer.

In Figure 6.58, we plot the array gain versus u_a/BW_{NN} for the LCMV beamformer. Using g_3 equal to $0.9\ddot{B}_c(o)$ gives the best results for this scenario.

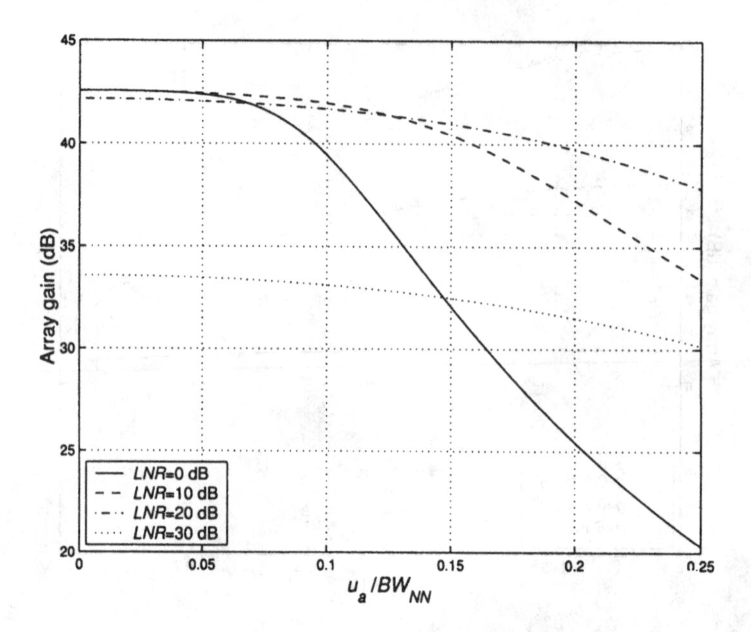

Figure 6.59 LCMV beamformer with second derivative constraints, $SNR =$ 20 dB, $INR = 30$ dB, $u_I = \pm 0.3$, $g_3 = 0.8\ddot{B}_c(0)$, array gain versus $\frac{u_a}{BW_{NN}}$.

We now consider the LCMP case. Based on the results of the directional constraints, we go directly to the diagonally loaded model and include $\sigma_L^2 = 0$ as a special case.

Example 6.7.10: Derivative main-lobe constraints; diagonally loaded LCMP beamformer (continuation)

We utilize the interference plus noise model given in (6.402), and the **C** matrix in (6.407), and the **g** matrix in (6.315) with $g_3 = 0.8\ddot{B}_c(0)$. We consider various combinations of *SNR*, *INR*, and *LNR*.

In Figure 6.59, we plot the array gain versus u_a/BW_{NN} for an $INR = 30$ dB and an $SNR = 20$ dB. We see that, with an $LNR = 20$ dB, the array gain goes from 42 dB at $u_a = 0$ to 38 dB at $u_a = 0.25BW_{NN}$. In Figure 6.60, we plot $E[A]$ versus *SNR* for various values of *INR*.

Comparing Figures 6.55 and 6.60, we see that directional and derivative constraints provide essentially the same performance when optimal DL is used. Comparing these results to Figure 6.39, we see that the two additional constraints allow us to increase the *SNR* by about 15 dB and maintain the same E[array gain].

6.7.4.4 Eigenvector constraints

In this section, we consider eigenvector constraints.

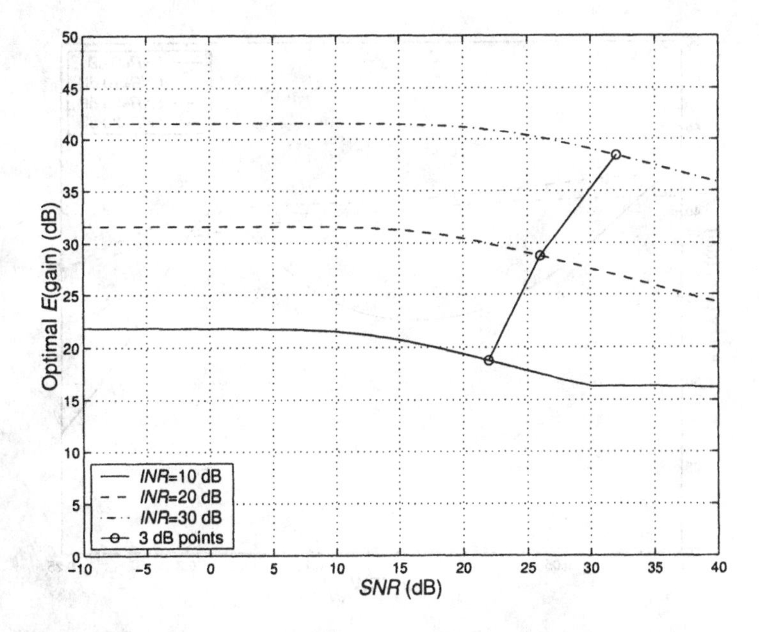

Figure 6.60 LCMP-DL beamformer with second derivative constraints and optimal loading, u_a is uniform $[-0.1, 0.1]$, $u_I = \pm 0.3$, $g_3 = 0.8\ddot{B}_c(0)$, $INR =$ 10, 20, and 30 dB, E[array gain] versus *SNR*.

Example 6.7.11: Eigenvector constraints; LCMP beamformer

Consider a standard 10-element linear array. In (6.347), we assumed that

$$B_d(u) = B_c(u), \quad -u_d \le u \le u_d.$$

We let $u_d = 0.1$. Since $\mathbf{Q}$ in (6.348) does not depend on $B_d(u)$, the eigenvectors and columns of $\mathbf{C}$ are the discrete prolate spheroidal functions. The vector $\mathbf{g}^H$ is given by (6.345).

In Figure 6.61, we plot the array gain versus u_a/BW_{NN} for the case in which the *SNR* is 20 dB and the *INR* of each interferer is 30 dB. As we would expect, the behavior is very similar to the behavior using directional constraints because, in the narrowband case, they are essentially equivalent.

In Figure 6.62, we plot the expected value of the array gain versus *SNR* for various *INR*.

6.7.4.5 Summary

In Section 6.7.4, we have studied the performance of LCMV and LCMP beamformers in the presence of a DOA mismatch. This range of uncertainty $|u_a| \le \frac{1}{N}$ is among the most stressful for testing array robustness. In the problems, we consider array perturbations in the form of sensor position perturbations.

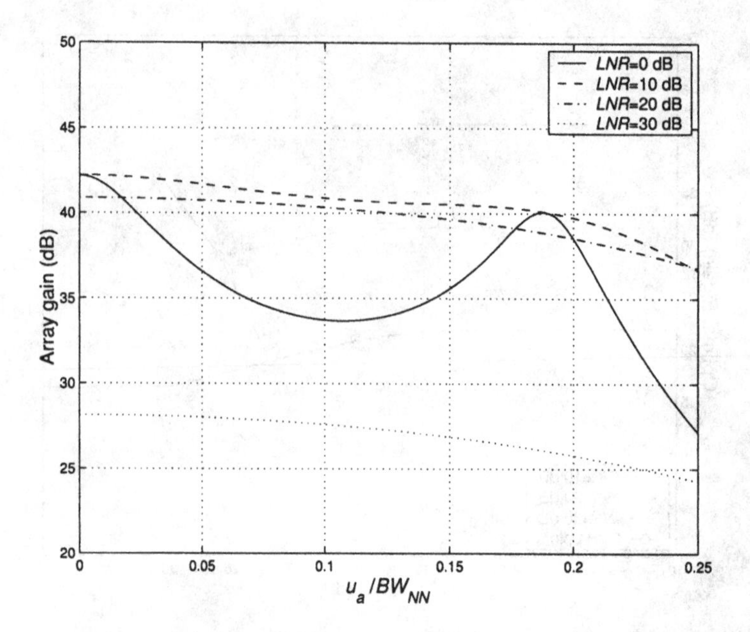

Figure 6.61 LCMP-DL beamformer with three eigenvector constraints, $SNR = 20$ dB, $INR = 30$ dB, $u_I = \pm 0.3$, $B_d(u) = B_c(u)$, $u_d = 0.1$; array gain versus u_a/BW_{NN}.

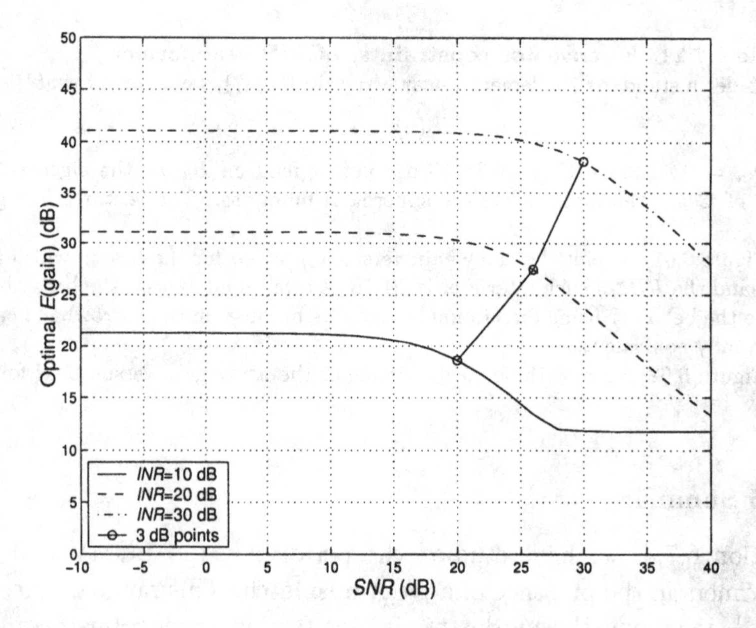

Figure 6.62 LCMP-DL beamformer with eigenvector constraints; E[array gain] versus SNR for various INR.

In the problems, we analyze a wide range of array configurations and interference scenarios. We find the results for the linear array with discrete interference are characteristic of the general case. The key element in the beamformer performance is an appropriate level of diagonal loading. In Section 6.10, we revisit the diagonal loading question. In Chapter 7, we develop variable loading algorithms that depend on the data input.

We found that the case of high *SNR* compared to *INR* led to significantly degraded performance in the presence of DOA mismatch. A logical step would be to perform a preprocessing to estimate the DOA prior to beamforming. We look at this approach briefly in Chapter 7 and in more detail in Chapter 9 when we study parameter estimation.

6.7.5 Quiescent Pattern (QP) Constraints

The constraints that we have discussed up to this point have focused on constraining the beam pattern at specific points in frequency-wavenumber space or in small regions of the space. The eigenvector constraints in Section 6.7.1.5 can control arbitrary regions of the frequency-wavenumber space, but as the region grows more degrees of freedom are required.

In this section, an algorithm is developed that first specifies a desired quiescent pattern (QP). We then specify a region in ω-$\mathbf{k}$ space in which we expect the desired signal to arrive. Constraints are introduced to avoid signal degradation. These constraints are in a space that is orthogonal to the space defined by the quiescent pattern.

Applebaum [AC76], in his early work on adaptive arrays recognized the importance of including desirable quiescent patterns in his adaption procedure. In [AC76], examples are given for a 32-element linear array in which the quiescent pattern utilizes Dolph-Chebyshev weighting with −30-dB sidelobes. Distortionless, first-order derivative, and second-order derivative constraints are imposed and the optimum pattern is found for a single directional interferer and white noise.

Gabriel [Gab87] also discusses the importance of a good quiescent pattern and gives examples. Subsequently, Griffiths and Buckley [GB87] developed a solution in the context of the generalized sidelobe canceller in Figure 6.46. A different approach was developed by Tseng and Griffiths [TG92b]. Our discussion follows [TG92b].

We denote the weight vector of desired quiescent pattern by $\mathbf{w}_{dq}$. In order to specify the constraint, we recall that the quiescent pattern is

$$\mathbf{w}_q = \mathbf{C}(\mathbf{C}^H\mathbf{C})^{-1}\mathbf{g}. \tag{6.408}$$

Therefore, if we define

$$\bar{\mathbf{w}}_{dq} = \frac{\mathbf{w}_{dq}}{\|\mathbf{w}_{dq}\|^2}, \tag{6.409}$$

and use the constraint

$$\bar{\mathbf{w}}_{dq}^H \mathbf{w} = 1, \tag{6.410}$$

then the resulting quiescent pattern is, from (6.408),

$$\mathbf{w}_q = \frac{\mathbf{w}_{dq}}{\|\mathbf{w}_{dq}\|^2} \left(\frac{\mathbf{w}_{dq}^H \mathbf{w}_{dq}}{\|\mathbf{w}_{dq}\|^4} \right)^{-1} = \mathbf{w}_{dq}. \tag{6.411}$$

However, we must impose other constraints to prevent the desired signal from being cancelled by the adaptive weights. We assume there is a response region from which the signal will arrive. We specify that region in a (θ, ϕ, f) space and define

$$\mathbf{Q} = \int_F \int_\Theta \int_\Phi \mathbf{v}(\theta, \phi, f) \mathbf{v}^H(\theta, \phi, f) d\theta \, d\phi \, df. \tag{6.412}$$

Note that this is same $\mathbf{Q}$ that we defined in the eigenvector constraint discussion in (6.337).

Instead of using this matrix directly, we define a modified matrix,

$$\bar{\mathbf{R}}_s = \mathbf{P}_{\bar{w}_{dq}}^{\perp} \mathbf{Q} \mathbf{P}_{\bar{w}_{dq}}^{\perp}, \tag{6.413}$$

where

$$\mathbf{P}_{\bar{w}_{dq}} = \bar{\mathbf{w}}_{dq} \left(\bar{\mathbf{w}}_{dq}^H \bar{\mathbf{w}}_{dq} \right)^{-1} \bar{\mathbf{w}}_{dq}^H \tag{6.414}$$

is the projection matrix with respect to $\bar{\mathbf{w}}_{dq}$. We then construct a matrix $\bar{\mathbf{C}}_s$ whose columns are the principal eigenvectors of $\bar{\mathbf{R}}_s$ and require, in addition to the constraint in (6.410), that

$$\bar{\mathbf{C}}_s^H \mathbf{w} = \mathbf{0}. \tag{6.415}$$

Thus, the quiescent response is unchanged because $\bar{\mathbf{C}}_s$ is orthogonal to $\bar{\mathbf{w}}_{dq}$ and the corresponding constraint values are zero. For example, if we use two eigenvectors:

$$\mathbf{C} = \begin{bmatrix} \bar{\mathbf{w}}_{dq} & \bar{\mathbf{C}}_s \end{bmatrix} = \begin{bmatrix} \bar{\mathbf{w}}_{dq} & \boldsymbol{\psi}_{s1} & \boldsymbol{\psi}_{s2} \end{bmatrix}, \tag{6.416}$$

where $\boldsymbol{\psi}_{s1}$ and $\boldsymbol{\psi}_{s2}$ are the principal eigenvectors of $\bar{\mathbf{R}}_s$. The algorithm relies on the eigenvectors in (6.416) for main-lobe protection.

In a similar manner, we can specify behavior in other regions of interest. For example, consider a particular area in ω-$\mathbf{k}$ space where we expect interference (or jamming). Denoting this region as $\mathbf{R}_J$, we add additional constraints,

$$\bar{\mathbf{C}}_J \mathbf{w}^H = \mathbf{0}, \tag{6.417}$$

where $\bar{\mathbf{C}}_J$ are the principal eigenvectors of $\bar{\mathbf{R}}_J$, which is defined as

$$\bar{\mathbf{R}}_J = \mathbf{P}^{\perp}_{\bar{\mathbf{C}}_s} \mathbf{P}^{\perp}_{\bar{\mathbf{w}}_{dq}} \mathbf{R}_J \mathbf{P}^{\perp}_{\bar{\mathbf{w}}_{dq}} \mathbf{P}^{\perp}_{\bar{\mathbf{C}}_s}. \tag{6.418}$$

Once again, the additional constraints do not change the quiescent response.

The total constraint matrix is

$$\begin{aligned} \mathbf{C} &= \begin{bmatrix} \bar{\mathbf{w}}_{dq} & \bar{\mathbf{C}}_s & \bar{\mathbf{C}}_J \end{bmatrix} \\ &= \begin{bmatrix} \bar{\mathbf{w}}_{dq} & \boldsymbol{\psi}_{s1} & \boldsymbol{\psi}_{s2} & \boldsymbol{\psi}_{J1} & \boldsymbol{\psi}_{J2} \end{bmatrix}, \end{aligned} \tag{6.419}$$

and

$$\mathbf{g}^H = \begin{bmatrix} 1 & \vdots & 0 & \vdots & \cdots & \vdots & 0 \end{bmatrix}. \tag{6.420}$$

Then, the optimum beamformer is given by (6.358),

$$\mathbf{w}_o^H = \mathbf{g}^H \left[\mathbf{C}^H \mathbf{S}_{\mathbf{x}}^{-1} \mathbf{C}\right]^{-1} \mathbf{C}^H \mathbf{S}_{\mathbf{x}}^{-1}. \tag{6.421}$$

To illustrate the algorithm, we first consider an example with the same signal and interference model that we have used in previous examples.

Example 6.7.12

Consider a standard 10-element linear array. The nominal signal arrival angle is $u_s = 0$. The actual signal arrival angle is u_a, where $0 \leq |u_a| \leq 0.1$. We first consider the case in which $\mathbf{w}_{dq}$ corresponds to the conventional beam pattern.

Then

$$\bar{\mathbf{w}}_{dq} = \mathbf{1}/N, \tag{6.422}$$

and

$$\mathbf{P}^{\perp}_{\bar{\mathbf{w}}_{dq}} = \mathbf{I} - \mathbf{1}\mathbf{1}^H/N. \tag{6.423}$$

The region of interest is $|u_a| \leq 0.1$. Then

$$\mathbf{Q} = \int_{-u_d}^{u_d} \mathbf{v}(u)\mathbf{v}^H(u)\,du, \tag{6.424}$$

which is the same $\mathbf{Q}$ as in (6.348) and $u_d = 0.1$. Thus,

$$[\mathbf{Q}]_{mn} = \frac{2\sin[\pi u_d(m-n)]}{\pi(m-n)}. \tag{6.425}$$

We use (6.423) and (6.424) in (6.413) to find $\bar{\mathbf{R}}_s$. We find the eigenvectors of $\bar{\mathbf{R}}_s$ and use the N_e eigenvectors with the largest eigenvalues to construct $\bar{\mathbf{C}}_s$.

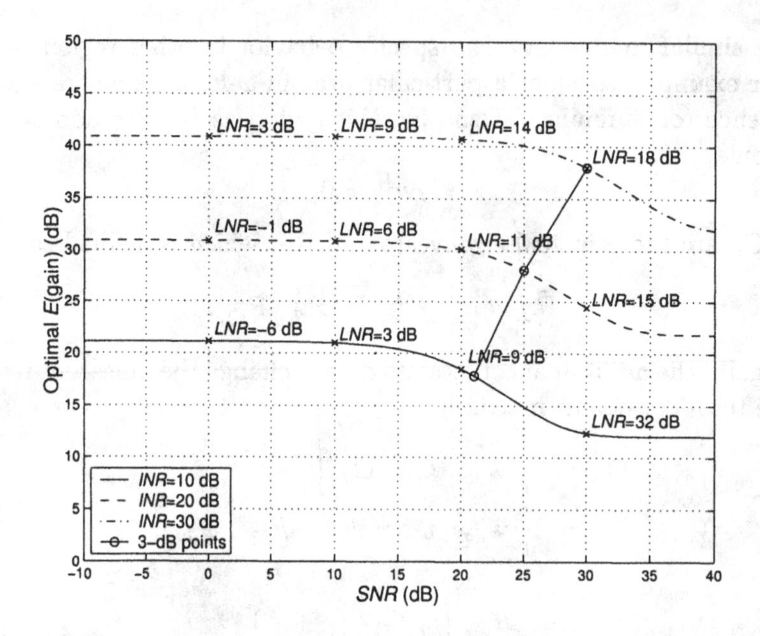

Figure 6.63 LCMP beamformer with quiescent pattern constraints; conventional QP, $N_e = 2$, $u_d = 0.1$, E[array gain] versus *SNR* for various *INR*.

The interference model is given by (6.402). There are two plane-wave interferers at ± 0.30.

$$\mathbf{S_x} = \sigma_s^2 \mathbf{v}_a \mathbf{v}_a^H + \sum_{i=1}^{2} \sigma_I^2 \mathbf{v}_i \mathbf{v}_i^H + \sigma_w^2 \mathbf{I}. \tag{6.426}$$

The weight vector is given by (6.358) with the $N \times (N_e + 1)$ constraint matrix,

$$\mathbf{C} = \left[\begin{array}{c:c:c:c} \bar{\mathbf{w}}_{dq} & \boldsymbol{\psi}_1 & \cdots & \boldsymbol{\psi}_{N_e} \end{array} \right], \tag{6.427}$$

and the $\mathbf{g}$ matrix,

$$\mathbf{g}^H = \left[\begin{array}{cccc} 1 & 0 & \cdots & 0 \end{array} \right]. \tag{6.428}$$

In order to provide a comparison with our previous examples, we let $N_e = 2$ so that there are three constraints. We also add diagonal loading.

In Figure 6.63, we plot the expected value of array gain versus *SNR* for various *INR*. We see that with appropriate *LNR*, we can maintain the array gain over values of *SNR* $\leq$ *INR*.

Example 6.7.13 (continuation)

The model is identical to Example 6.7.12. We use a desired quiescent weighting corresponding to a Taylor-Villenueve weighting with -20 dB maximum sidelobe and $\bar{n} = 2$.

In Figure 6.64, we plot the expected value of the array gain versus *SNR* for various *INR*. We also show the results from Example 6.7.12.

We see that the results for the two choices of quiescent beam patterns

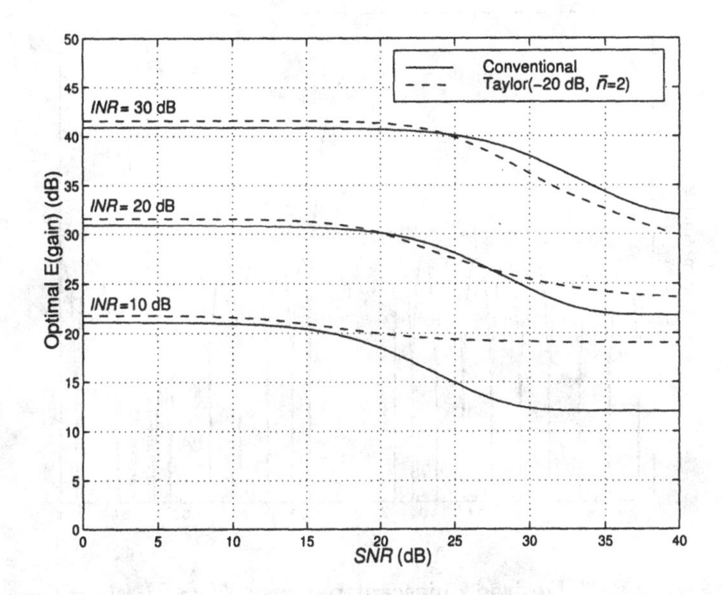

Figure 6.64 LCMP beamformer with quiescent pattern constraints, Taylor QP, $N_e = 2$, $u_d = 0.1$, E[array gain] versus *SNR* for various *INR*.

are similar for $SNR \leq 20 - 25$ dB. For higher *SNR*, a higher *LNR* is required to prevent signal cancellation, and the quiescent sidelobe structure provides the array gain.

To illustrate a different application of the QP technique, we consider a simple example from [TG92b].

Example 6.7.14 [TG92b]

Consider a standard 21-element linear array. The desired signal direction is broadside and the anticipated jammer region is $40° \leq \overline{\theta} \leq 50°$, where $\overline{\theta}$ is the broadside angle.

The desired quiescent pattern is shown in Figure 6.65. It has unit gain in signal direction, -50-dB sidelobes in the anticipated jamming region ($40° \leq \overline{\theta} \leq 50°$) and -30-dB sidelobes elsewhere. It is designed using the techniques in Section 3.9.3. By having lower sidelobes in the area of the anticipated jammer, the adaptive beamformer will converge more rapidly if the jammer arrives in that area.

The first two constraints are

$$\bar{\mathbf{w}}_{dq}^H \mathbf{w} = 1, \tag{6.429}$$

and

$$\mathbf{c}_2^H \mathbf{w} = 0, \tag{6.430}$$

where

$$\mathbf{c}_2 = \mathbf{P}_{\bar{\mathbf{w}}_{dq}}^{\perp} \mathbf{v}(0). \tag{6.431}$$

We refer to (6.429) and (6.430) as constraint set 1.

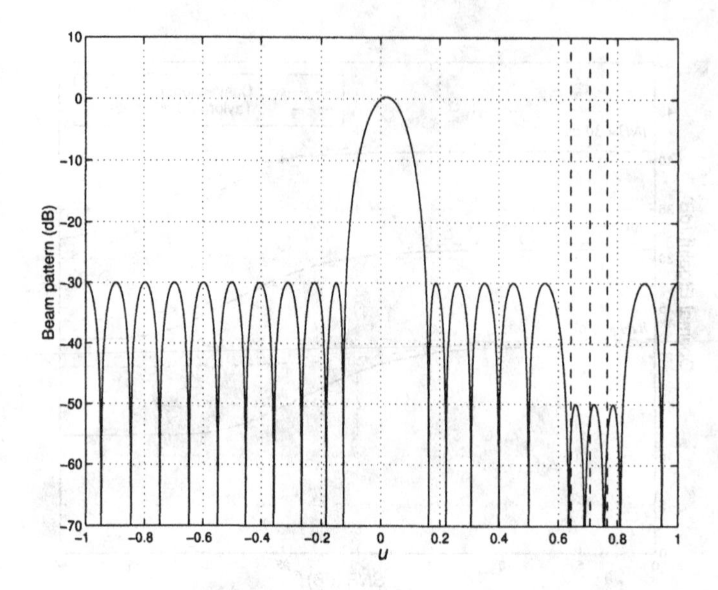

Figure 6.65 Desired quiescent pattern: $N = 21$, $u_s = 0$.

A second constraint set is constructed by constraining the four nulls in the −50-dB sidelobe region (at 38.1°, 42.1°, 47.5°, and 52.3°) to remain nulls. Thus,

$$\mathbf{R}_J = \frac{1}{4}\sum_{i=1}^{4} \mathbf{v}(\bar{\theta}_i)\mathbf{v}^H(\bar{\theta}_i). \tag{6.432}$$

This step leads to four additional constraints, $\boldsymbol{\psi}_i$, where the $\boldsymbol{\psi}_i$ are the eigenvectors of

$$\bar{\mathbf{R}}_J = \mathbf{P}^{\perp}_{\mathbf{C}_2}\mathbf{P}^{\perp}_{\bar{\mathbf{w}}_{dq}}\mathbf{R}_J\mathbf{P}^{\perp}_{\bar{\mathbf{w}}_{dq}}\mathbf{P}^{\perp}_{\mathbf{C}_2}. \tag{6.433}$$

There are a total of six constraints,

$$\mathbf{C} = \left[\begin{array}{cccccc} \bar{\mathbf{w}}_{dq} & \mathbf{c}_2 & \boldsymbol{\psi}_1 & \boldsymbol{\psi}_2 & \boldsymbol{\psi}_3 & \boldsymbol{\psi}_4 \end{array}\right], \tag{6.434}$$

where the $\boldsymbol{\psi}_i$ are the four principal eigenvectors of $\bar{\mathbf{R}}_J$.

We now assume that a jammer with a 30-dB *INR* arrives at 45°. The resulting adapted beam pattern is shown in Figure 6.66. We see that the constraint set preserves the desired quiescent pattern reasonably well and creates a null in the direction of the jammer. The array gain is 42.19 dB for this case.

We also consider the case in which only the first two constraints are used. The resulting adapted beam pattern is shown in Figure 6.67. The array gain is 42.21 dB.

We see that the QP approach provides a great deal of design flexibility and allows us to tailor our optimum beamformer to different anticipated scenarios.

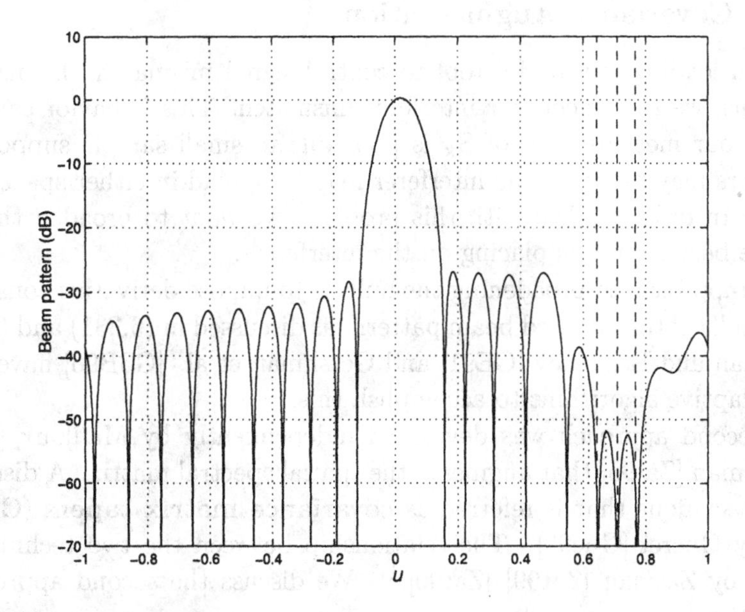

Figure 6.66 Optimum beam pattern: six constraints; $N = 21$, $u_s = 0$, $u_I = 0.707$, $SNR = 0$ dB, $INR = 30$ dB.

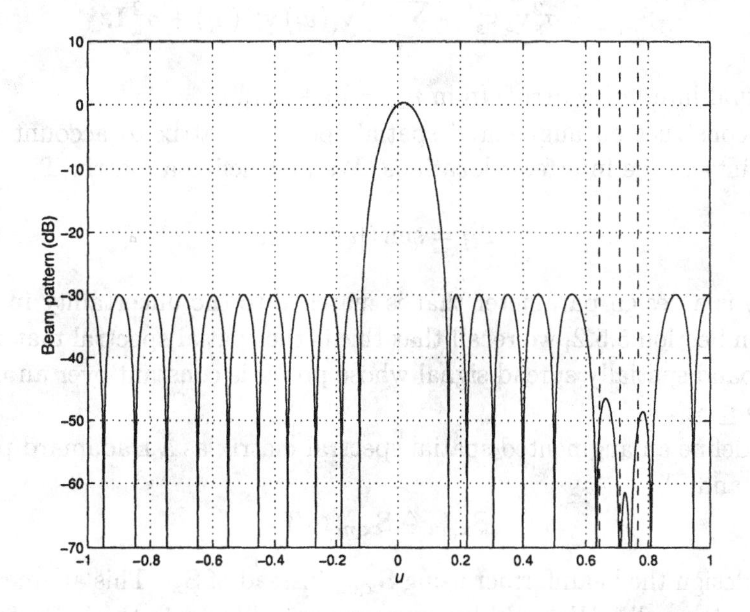

Figure 6.67 Optimum beam pattern: constraint set 1 (2 constraints); $N = 21$, $u_s = 0$, $u_I = 0.707$, $SNR = 0$ dB, $INR = 30$ dB.

6.7.6 Covariance Augmentation

Diagonal loading is a useful tool to control signal mismatch. In many applications, we also encounter interferer mismatch. This behavior can occur because our measurement of $\mathbf{S}_{\mathbf{x}}$ is poor due to small sample support, the interferers may move, or the interferer may be spread in either space or frequency. In order to deal with this problem, we want to broaden the nulls that the beamformer is placing on the interferers.

One approach to broadening the nulls is to impose derivative constraints on the nulls of the adapted beam pattern, as discussed in (6.331) and (6.332). Gershman and Ermolaev [GE91] and Gershman et al. [GSB96] have developed adaptive algorithms to accomplish this.

A second approach was developed independently by Mailloux [Mai95] and Zatman [Zat95] that augments the spatial spectral matrix. A discussion of this technique that is referred as **covariance matrix tapers** (CMT) is given by Guerci [Gue99]. The relationship between the two techniques is derived by Zatman [Zat99] [Zat00].[11] We discuss the second approach in this section.

Assume that the MPDR beamformer is designed under the assumption that

$$\mathbf{S}_{\mathbf{x},m} = \sigma_s^2 \mathbf{v}_s \mathbf{v}_s^H + \sum_{i=1}^{D} \sigma_i^2\, \mathbf{v}_i(u_i)\, \mathbf{v}_i^H(u_i) + \sigma_w^2 \mathbf{I}. \tag{6.435}$$

The actual interferers arrive from $u_{ai} = u_i + \Delta u_i$.

We construct an augmented spatial spectral matrix to account for the uncertainty in the interferer locations. We first define a matrix $\mathbf{T}$,

$$\mathbf{T}_{ij} = \mathrm{sinc}(|i - j|\gamma), \tag{6.436}$$

where γ is a design parameter that is matched to the uncertainty in the u_i.

From Section 5.5.2, we recall that this is the spatial spectral matrix for a narrowband spatially spread signal whose power is constant over an interval $-\gamma \le \psi \le \gamma$.

We define an augmented spatial spectral matrix as a Hadamard product of $\mathbf{S}_{\mathbf{x},m}$ and $\mathbf{T}$,

$$\mathbf{S}_{\mathbf{x},au} \triangleq \mathbf{S}_{\mathbf{x},m} \odot \mathbf{T}. \tag{6.437}$$

We design the beamformer using $\mathbf{S}_{\mathbf{x},au}$ instead of $\mathbf{S}_{\mathbf{x}}$. This augmentation broadens the nulls. We consider an example to illustrate the behavior.

---

[11] We are using $\mathbf{S}_{\mathbf{x}}$, so a more appropriate title is spatial matrix tapers. We use CMT to be consistent with the literature.

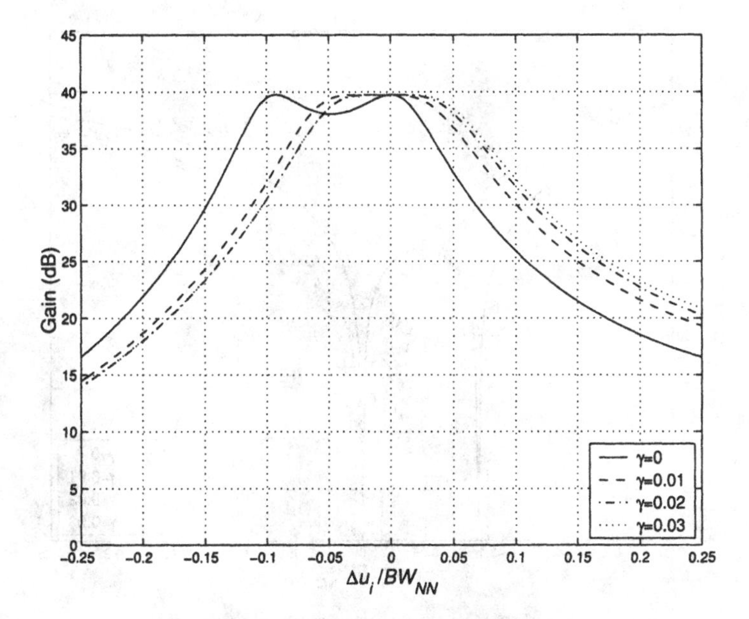

Figure 6.68 Array gain versus $\Delta u_i/BW_{NN}$: N=10, u_a=0, u_I=0.3, $INR = 30$ dB, covariance matrix taper with various γ.

Example 6.7.15

Consider a standard 10-element array. The MPDR beamformer is designed under the assumption that

$$\mathbf{S}_{\mathbf{x},m} = \sigma_s^2 \mathbf{v}_s \mathbf{v}_s^H + \sigma_I^2 v_I(u_I) v_I^H(u_I) + \sigma_w^2 \mathbf{I}. \tag{6.438}$$

The array is steered to broadside. The actual interferer arrives from $u_{Ia} = u_I + \Delta u$.

We consider two beamformers. The first is the MPDR beamformer. The second is a beamformer that is constructed using the augmented $\mathbf{S}_\mathbf{x}$ matrix defined in (6.437). We first consider the case with no signal mismatch and $u_I = 0.30$ with an $INR = 30$ dB. We vary γ from 0 to 0.05 in 0.01 steps.

In Figure 6.68, we plot the array gain versus $\Delta u_I/BW_{NN}$ for the various γ. In Figure 6.69, we plot the beam pattern in the vicinity of $u_I = 0.3$ for $\gamma = 0$, 0.01, 0.02 and 0.03. We see that the null has been broadened with negligible loss in array gain.

6.7.7 Summary

We have explored five types of linear constraints in Section 6.7.4 and 6.7.5. In Figure 6.70, we plot their performance for the signal and interference model that we analyzed in a sequence of examples. We see that the five LCMP-DL beamformers have similar performance for *SNR*s that are less than the 3-dB loss value. They are all significantly better than MPDR-DL. It is important to recognize the contribution of diagonal loading to the

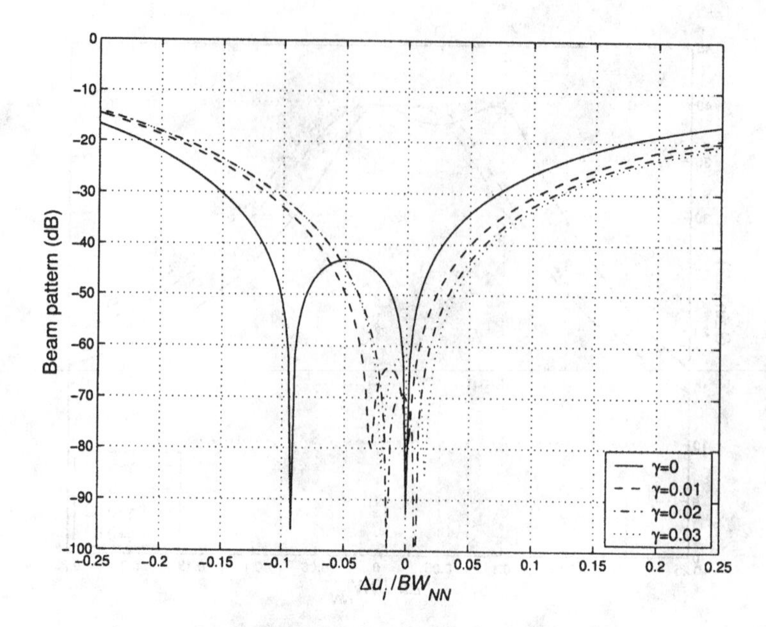

Figure 6.69 Beam pattern around $u_I = 0.30$: covariance matrix taper with various γ.

robust behavior. In most scenarios, it will be a necessary component of the optimum beamformer. Note that, although diagonal loading was introduced as the solution to a well-posed optimization problem (6.261)–(6.268), it is usually applied in a more ad hoc manner. When diagonal loading is used in this manner, it can provide good performance, but it is not optimum in any formal sense.

In the problems, we discuss a number of signal and interference scenarios for DOA mismatch. We also study the array perturbation problem. We find similar results to the above examples.

All of the optimum beamformers that we have discussed up to this point are operating in an N-dimensional space. In the next two sections, we develop algorithms that operate in a reduced-dimension space. We find that this reduced-dimension operation offers significant advantages.

6.8 Eigenvector Beamformers

In the first part of this chapter we found that MVDR and MPDR beamformers are the optimum processors for waveform estimation under a variety

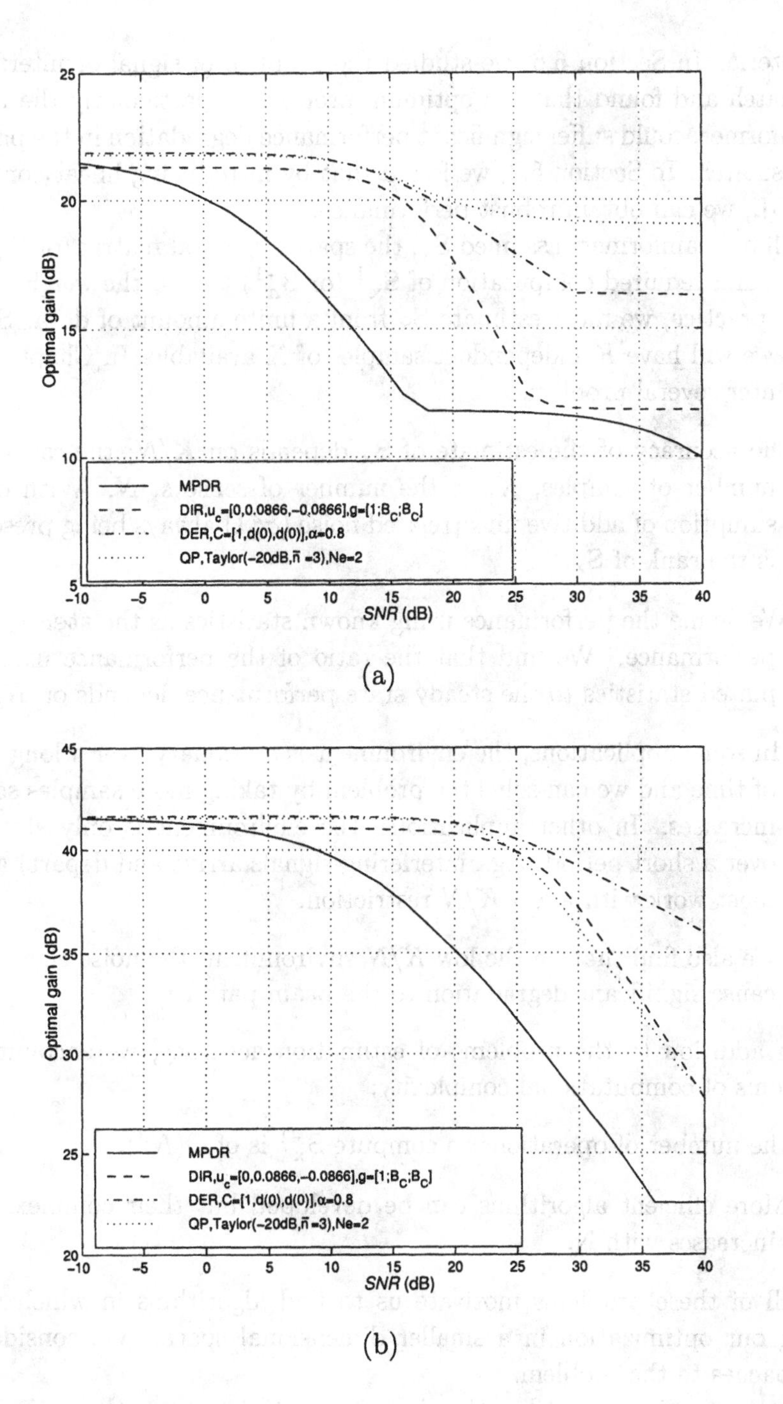

Figure 6.70 Performance comparison using various constraints and optimum DL: $N = 10$, $u_I = \pm 0.3$. (a) $INR = 10$ dB; (b) $INR = 30$ dB.

of criteria. In Section 6.6, we studied the problem of signal or interference mismatch and found that the optimum processors, particularly the MPDR beamformer, could suffer significant performance degradation in the presence of mismatch. In Section 6.7, we found that by introducing linear contraints and DL, we can obtain robust performance.

All of beamformers assumed $\mathbf{S_x}$, the spatial spectral matrix (or $\mathbf{S_n}$), was known and required computation of $\mathbf{S_x}^{-1}$ (or $\mathbf{S_n}^{-1}$) to find the weight vector.

In practice, we must estimate $\mathbf{S_x}$ from a finite amount of data. Specifically, we will have K independent samples of $\mathbf{X}$ available. In Chapter 7, we encounter several problems:

(i) The accuracy of the estimate of $\mathbf{S_x}$ depends on K/N; the ratio of the number of samples, K, to the number of sensors, N. With our assumption of additive uncorrelated noise ($\sigma_w^2\mathbf{I}$) always being present, N is the rank of $\mathbf{S_x}$.

(ii) We define the performance using known statistics as the **steady state** performance. We find that the ratio of the performance using estimated statistics to the steady state performance depends on K/N.

(iii) In some applications, the environment is stationary over a long period of time and we can solve the problem by taking more samples so K/N increases. In other applications, the environment is only stationary over a short period (e.g., interfering signals arrive and depart) and we must work with a low K/N restriction.

(iv) We also find that, in the low K/N environment, the noise eigenvectors cause significant degradation to the beam pattern.

In addition to the problems of estimation accuracy, we encounter the problems of computational complexity:

(i) The number of operations to compute $\mathbf{S_x}^{-1}$ is of $O(N^3)$.

(ii) More efficient algorithms can be developed but their complexity still increases with N.

All of these problems motivate us to find algorithms in which we are doing our optimization in a smaller dimensional space. We consider two approaches to the problem.

In this section, we utilize the eigenvectors to construct the optimization space. Since the eigenvectors and eigenvalues depend on $\mathbf{S_x}$ (in practice, they depend on $\widehat{\mathbf{S}}_\mathbf{x}$) the subspace construction will be data dependent (or adaptive) in actual applications. In Section 6.9, we construct the subspace by preprocessing the input with a set of conventional beams that span the region of (ω-$\mathbf{k}$) space of interest.

Many of the advantages of eigenspace (ES) and beamspace (BS) processing will not become apparent until we study their adaptive realizations. However, the steady state performance provides the foundation for these analyses.

In this section, we develop beamformers that compute the eigenvalues and eigenvectors of $\mathbf{S_x}$ and use them to construct the beamformer. In this chapter, we assume $\mathbf{S_x}$ is available.

In many of the eigenvector beamformers that we develop, we only need to compute the eigenvectors corresponding to the largest eigenvalues. This requirement means that we can exploit various efficient computation techniques in lieu of performing a complete eigendecomposition.

We consider two types of beamformers. They both utilize an eigendecomposition, but they use the results in algorithms that are significantly different. The first type of beamformer projects the input onto a reduced-rank subspace called the eigenspace containing the signal and interference.[12] They then process the projected input in the subspace to form the beam. In the presence of signal mismatch, this technique provides significant performance improvement even in the case of known statistics. It will also provide a reduction in computational complexity. We refer to these beamformers as **eigenspace (ES) beamformers**.

In Section 6.8.1, we develop an algorithm that uses the principal components of $\mathbf{S_x}$ as a basis for a subspace. In Section 6.8.2, we develop an algorithm that uses the cross-correlation between the eigenvectors and $\mathbf{v}_m$ as a criterion for selecting the subspace.

The second type of beamformer uses the eigendecomposition to construct a subspace called the **dominant-mode** (DM) subspace. The algorithm is designed to reject the modes in the DM subspace and is referred to as a dominant-mode rejection (DMR) beamformer. When there is no mismatch and the statistics are known, the DMR and ES beamformers are identical. However, in the presence of signal mismatch, array mismatch, or estimated statistics, their performance can be dramatically different. We develop and analyze this beamformer in Section 6.8.3. We summarize our results in Section 6.8.4.

[12]This statement is true for the cases of discrete interference, known statistics, and no model mismatch. We discuss the general case in Sections 6.8.1 and 6.8.2.

6.8.1 Principal–component (PC) Beamformers

The general approach has been studied under a number of different names. Under the eigendecomposition label, there are algorithms proposed by Hung and Turner [HT83], Citron and Kailath [CK84], Owsley [Ows85], Gabriel [Gab86], Friedlander [Fri88], Haimovich and Bar-Ness [HBN88], [HBN91], Van Veen [VV88], Chang and Yeh [CY92], Youn and Un [YU94], and Yu and Yeh [YY95].

Under the reduced covariance matrix or principal–components (PC) label, the problem has been studied by Kirstein and Tufts [KT85].

Under the projection label, algorithms have been developed by Feldman and Griffiths [FG91], [FG94].

Related work includes that of Er and Cantoni [EC85], Zunich and Griffiths [ZG91], and Bull et al. [BAB90].

The first step in the design is to perform an eigendecomposition of $\mathbf{S_x}$,

$$\mathbf{S_x} = \sum_{i=1}^{N} \lambda_i \mathbf{\Phi}_i \mathbf{\Phi}_i^H = \mathbf{U \Lambda U}^H, \tag{6.439}$$

where $\mathbf{U}$ is an $N \times N$ matrix of eigenvectors,

$$\mathbf{U} \triangleq \begin{bmatrix} \mathbf{\Phi}_1 & \mathbf{\Phi}_2 & \cdots & \mathbf{\Phi}_N \end{bmatrix}, \tag{6.440}$$

and $\mathbf{\Lambda}$ is a diagonal matrix of the ordered eigenvalues,

$$\mathbf{\Lambda} = \text{diag} \begin{bmatrix} \lambda_1 & \lambda_2 & \cdots & \lambda_N \end{bmatrix}. \tag{6.441}$$

In practice, we would use $\hat{\mathbf{S}}_\mathbf{x}$, the sample covariance matrix, to generate the eigendecomposition. We now select D_r of these eigenvectors to form a $N \times D_r$ matrix $\mathbf{U}_r$. The subscript r denotes "reduced."

There are several ways to select the eigenvectors to be used in $\mathbf{U}_r$. The first model assumes there is a single plane-wave signal, D plane-wave interferers, and additive white noise. We select the eigenvectors corresponding to the $D+1$ largest eigenvalues. We refer to these as the principal components. We define

$$\mathbf{U}_{S+I} = \begin{bmatrix} \mathbf{\Phi}_1 & \mathbf{\Phi}_2 & \cdots & \mathbf{\Phi}_{D+1} \end{bmatrix} \tag{6.442}$$

to be the signal-plus-interference subspace. In this model, we must estimate the number of plane waves present. We develop techniques for estimating the value of $D+1$ in Chapter 7 (e.g., the discussion of AIC and MDL in Section 7.9.3).

The second model does not assume any structure for the interference (e.g., it could include a spatially spread component). We perform the eigendecomposition in (6.439) and select the D_r largest eigenvalues. There are various tests for determining if an eigenvalue is large:

$$(i) \quad \frac{\lambda_i}{tr[\mathbf{S_x}]} > n_1, \tag{6.443}$$

the percent of the total power is contained in λ_i,

$$(ii) \quad \frac{\lambda_{i+1}}{\lambda_i} > n_2, \tag{6.444}$$

the ratio of the $(i+1)$th eigenvalue to the ith eigenvalue which has been declared to be large.

After we have selected an appropriate value of D_r, we define a $N \times D_r$ matrix,

$$\mathbf{U}_r = \begin{bmatrix} \mathbf{\Phi}_1 & \mathbf{\Phi}_2 & \cdots & \mathbf{\Phi}_{D_r} \end{bmatrix}, \tag{6.445}$$

which defines the reduced subspace and the $D_r \times D_r$ diagonal matrix,

$$\mathbf{\Lambda}_r = \text{diag}\left\{ \lambda_1 \quad \lambda_2 \quad \cdots \quad \lambda_{D_r} \right\}. \tag{6.446}$$

Using the model in (6.442), we can write

$$\mathbf{S_x} = \mathbf{U}_{S+I}\, \mathbf{\Lambda}_{S+I}\, \mathbf{U}_{S+I}^H + \mathbf{U}_N\, \mathbf{\Lambda}_N\, \mathbf{U}_N^H, \tag{6.447}$$

where $\mathbf{U}_N$ is orthogonal to $\mathbf{U}_{S+I}$. We can write $\mathbf{S_x}^{-1}$ as

$$\mathbf{S_x}^{-1} = \mathbf{U}_{S+I}\, \mathbf{\Lambda}_{S+I}^{-1}\, \mathbf{U}_{S+I}^H + \mathbf{U}_N\, \mathbf{\Lambda}_N^{-1}\, \mathbf{U}_N^H. \tag{6.448}$$

Because $\mathbf{v}_m$ is in $\mathbf{U}_{S+I}$, it is orthogonal to $\mathbf{U}_N$. Therefore the MPDR beamformer reduces to

$$\boxed{\mathbf{w}_{mpdr,es}^H = \gamma_{es}\, \mathbf{v}_m^H\, \mathbf{U}_{S+I}\, \mathbf{\Lambda}_{S+I}^{-1}\, \mathbf{U}_{S+I}^H,} \tag{6.449}$$

where

$$\boxed{\gamma_{es} = \left(\mathbf{v}_m^H \mathbf{U}_{S+I} \mathbf{\Lambda}_{S+I}^{-1} \mathbf{U}_{S+I}^H \mathbf{v}_m \right)^{-1}.} \tag{6.450}$$

We use the symbol γ_{es} instead of our previous symbol Λ to avoid confusion with the eigenvalue matrix. The resulting beamformer is shown in Figure 6.71. We refer to it as a principal component or eigenspace beamformer.

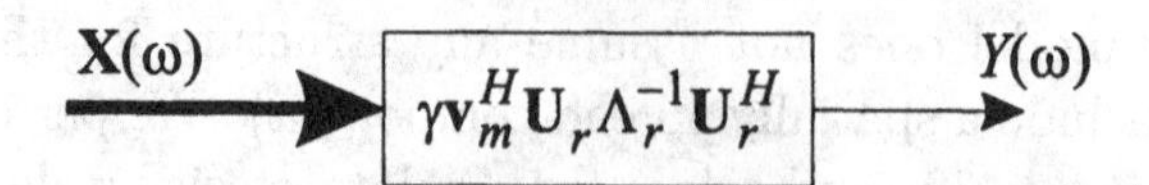

Figure 6.71 Principal component beamformer.

The result in (6.449) can also be written in terms of the individual eigenvalues and eigenvectors,

$$\mathbf{w}_{mpdr,es}^H = \frac{\sum_{i=1}^{D_r} \frac{1}{\lambda_i} (\mathbf{v}_m^H \mathbf{\Phi}_i) \mathbf{\Phi}_i^H}{\sum_{i=1}^{D_r} \frac{1}{\lambda_i} |\mathbf{v}_m^H \mathbf{\Phi}_i|^2}. \tag{6.451}$$

If we use the model in (6.445), then

$$\mathbf{w}_{mpdr,es}^H = \gamma_{es}\, \mathbf{v}_m^H\, \mathbf{U}_r\, \mathbf{\Lambda}_r^{-1}\, \mathbf{U}_r^H. \tag{6.452}$$

The algorithm in this form is referred to in the literature as the eigenspace beamformer (e.g., [HBN91],[CY92]) or the **signal-plus-interference** subspace beamformer (e.g., [Fri88]).

We can also interpret (6.452) as a projection beamformer (e.g., Feldman and Griffiths [FG91], [FG94]).

We define a projected steering vector as

$$\mathbf{v}_p \triangleq \mathbf{P}_{\mathbf{U}_r} \mathbf{v}_m = \mathbf{U}_r \mathbf{U}_r^H \mathbf{v}_m, \tag{6.453}$$

because

$$\mathbf{U}_r^H \mathbf{U}_r = \mathbf{I}. \tag{6.454}$$

The projection beamformer is

$$\mathbf{w}_p = \gamma_p \mathbf{S}_{\mathbf{x}}^{-1} \mathbf{v}_p = \gamma_p \mathbf{S}_{\mathbf{x}}^{-1} \mathbf{U}_r \mathbf{U}_r^H \mathbf{v}_m. \tag{6.455}$$

Expanding $\mathbf{S}_{\mathbf{x}}^{-1}$ gives

$$\mathbf{w}_p = \gamma_p \left[\mathbf{U}_r \mathbf{\Lambda}_r^{-1} \mathbf{U}_r^H + \mathbf{U}_N \mathbf{\Lambda}_N^{-1} \mathbf{U}_N^H \right] \mathbf{U}_r \mathbf{U}_r^H \mathbf{v}_m, \tag{6.456}$$

where $\mathbf{U}_N$ is the $N \times (N - D_r)$ noise subspace that is orthogonal to $\mathbf{U}_r$. Then (6.456) reduces to

$$\mathbf{w}_p = \gamma_p \left[\mathbf{U}_r \mathbf{\Lambda}_r^{-1} \mathbf{U}_r^H \right] \mathbf{v}_m, \tag{6.457}$$

which is the same as (6.452) if we choose $\gamma_p = \gamma_{es}$.

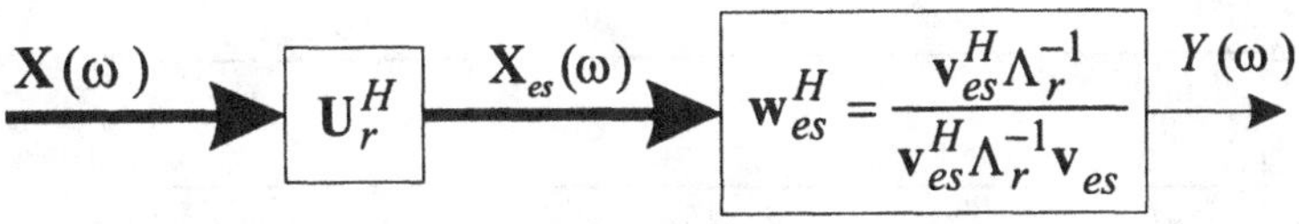

Figure 6.72 Eigenspace beamformer.

We see that we can interpret the PC beamformer as a beamformer that uses the projection of the model array manifold vector onto the principal component subspace in $\mathbf{w}_{mpdr,es}$. We would expect that this projection approach should provide robustness against model signal mismatch.

A third interpretation of the beamformer is also useful. We can define

$$\mathbf{X}_{es} = \mathbf{U}_r^H \mathbf{X}, \tag{6.458}$$

and

$$\mathbf{v}_{es} = \mathbf{U}_r^H \mathbf{v}_m. \tag{6.459}$$

This leads to the processor shown in Figure 6.72. Here

$$\mathbf{w}_{mpdr,es}^H = \frac{\mathbf{v}_{es}^H \mathbf{\Lambda}_r^{-1}}{\mathbf{v}_{es}^H \mathbf{\Lambda}_r^{-1} \mathbf{v}_{es}}. \tag{6.460}$$

The advantage of this representation is that it highlights that the beamformer is operating in the eigenspace.

If we include diagonal loading,

$$\mathbf{w}_{es,mpdr,dl} = \frac{\mathbf{v}_{es}^H \left[\mathbf{\Lambda}_r + \sigma_L^2 \mathbf{I}_{D_r}\right]^{-1}}{\mathbf{v}_{es}^H \left[\mathbf{\Lambda}_r + \sigma_L^2 \mathbf{I}_{D_r}\right]^{-1} \mathbf{v}_{es}}. \tag{6.461}$$

Note that the diagonal loading is in the eigenspace. When $\mathbf{S}_\mathbf{x}$ is known, diagonal loading is not useful to combat DOA mismatch. However, it is useful to protect the main lobe against main-lobe interferers.

For an LCMP beamformer, we define the constraints in the original N–dimensional space.

$$\mathbf{C}^H \mathbf{w} = \mathbf{C}^H \left[\mathbf{U}_r \mathbf{w}_{es}\right] = \mathbf{C}_{es}^H \mathbf{w}_{es} = \mathbf{g}^H, \tag{6.462}$$

where

$$\mathbf{C}_{es}^H \triangleq \mathbf{C}^H \mathbf{U}_r. \tag{6.463}$$

Then, using (6.358),

$$\boxed{\mathbf{w}_{lcmp,es}^H = \mathbf{g}^H \left[\mathbf{C}_{es}^H \mathbf{\Lambda}_r^{-1} \mathbf{C}_{es}\right]^{-1} \mathbf{C}_{es}^H \mathbf{\Lambda}_r^{-1}.} \tag{6.464}$$

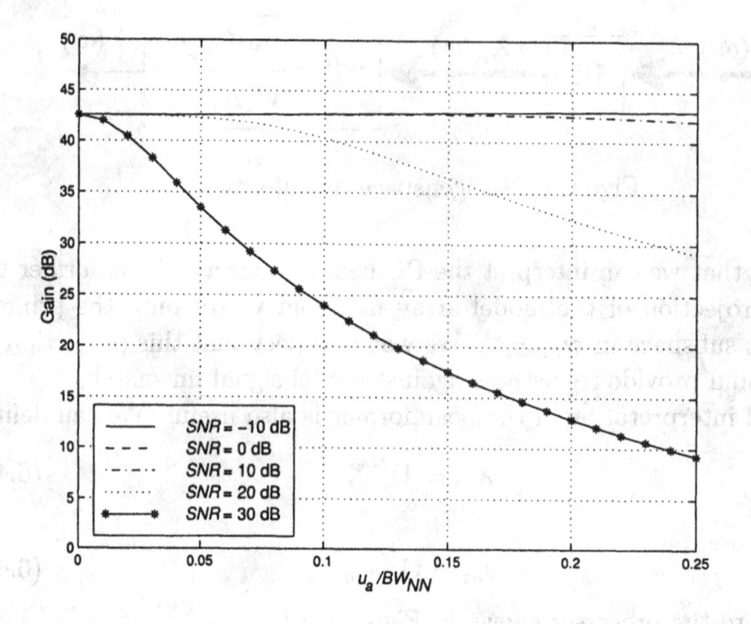

Figure 6.73 MPDR eigenspace beamformer with $D_r = 3$, $u_I = \pm 0.3$, *INR* = 30 dB, various *SNR*; array gain versus u_a/BW_{NN}.

We consider an example to illustrate a typical result.

Example 6.8.1

Consider a standard 10-element linear array. The nominal signal arrival angle is $u_m = 0$. The input spatial spectral matrix is

$$\mathbf{S_x} = \sigma_s^2 \mathbf{v}_a \mathbf{v}_a^H + \sum_{i=1}^{2} \sigma_i^2 \mathbf{v}_i \mathbf{v}_i^H + \sigma_w^2 \mathbf{I}. \tag{6.465}$$

We assume the interfering plane waves are at ±0.30. We perform an eigendecomposition of $\mathbf{S_x}$ and construct a 3-D subspace using the eigenvectors corresponding to the three largest eigenvalues. We use the eigenspace beamformer in (6.460)

In Figure 6.73, we plot the array gain versus u_a/BW_{NN} for an $INR = 30$ dB. Comparing these results to the MPDR-DL results in Figure 6.38(b), we see that the eigenspace beamformer is better for all of the *SNR* considered. The eigenspace beamformer partially compensates for the DOA mismatch by projecting $\mathbf{v}_m$ into the actual subspace.

The result in Example 6.8.1 for low *SNR* is correct but misleading. In practice, we have to estimate D_r and use $\hat{\mathbf{S}}_\mathbf{x}$ for the eigendecomposition. In Section 7.9.2, we find that for low *SNR* we frequently underestimate D_r and lose the eigenvalue that is most closely correlated with the signal. To investigate this effect we repeat Example 6.8.1 with $D_r = 2$ instead of 3.

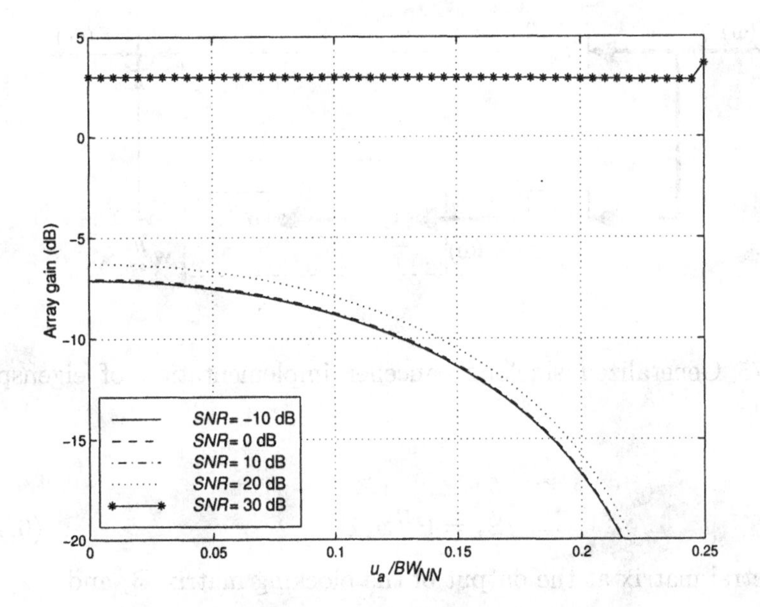

Figure 6.74 MPDR eigenspace beamformer with $D_r = 2$, $u_I = \pm 0.3$, $INR =$ 30 dB, various *SNR*; array gain versus u_a/BW_{NN}.

Example 6.8.2 (continuation)
Consider the same model as in Example 6.8.1. We use the beamformer in (6.451) with $D_r = 2$. In Figure 6.74, we plot the array gain versus u_a/BW_{NN}.

We see that the performance is significantly degraded. This is because there is very little signal content in the eigenspace. Thus, it is important that the dimension of the eigenspace is chosen properly.

We can also implement the eigenspace beamformer in a generalized sidelobe canceller (GSC) format, as shown in Figure 6.75. The adaptive matrix, $\mathbf{w}_a^H$, in Figure 6.46 has been divided into a cascade of two matrices. These matrices are defined in the next several equations. In this case, the eigendecomposition of the spatial spectral matrix of $\mathbf{X}_b(\omega)$ is used (e.g., [VV88]).

The overall weight vector is

$$\mathbf{w} = \mathbf{w}_q - \mathbf{B}\mathbf{w}_a, \tag{6.466}$$

where

$$\mathbf{w}_q = \mathbf{C}[(\mathbf{C}^H\mathbf{C})]^{-1}\mathbf{g}. \tag{6.467}$$

From Section 6.7.3, the optimum adaptive component in Figure 6.46 is

$$\mathbf{w}_{ao}^H = \mathbf{S}_{d\mathbf{z}^H}\mathbf{S}_{\mathbf{z}}^{-1}, \tag{6.468}$$

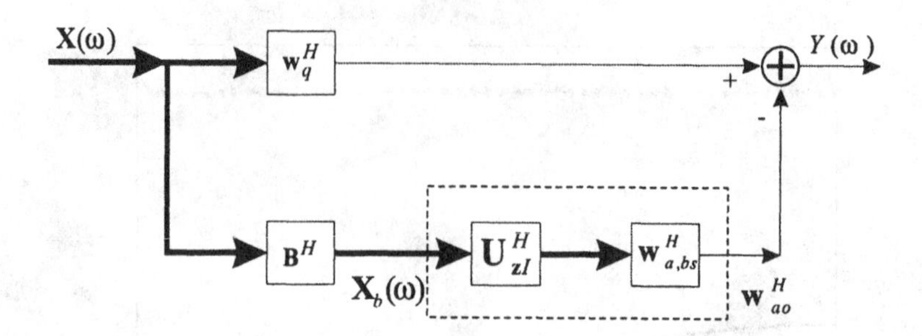

Figure 6.75 Generalized sidelobe canceller implementation of eigenspace beamformer.

where

$$\mathbf{S}_{\mathbf{z}} = \mathbf{B}^H \mathbf{S}_{\mathbf{x}} \mathbf{B} \tag{6.469}$$

is the spectral matrix at the output of the blocking matrix $\mathbf{B}$, and

$$\mathbf{S}_{d\mathbf{z}^H} = \mathbf{w}_q^H \mathbf{S}_{\mathbf{x}} \mathbf{B}. \tag{6.470}$$

We now utilize an eigendecomposition of $\mathbf{S}_{\mathbf{z}}$:

$$\mathbf{S}_{\mathbf{z}} = \mathbf{U}_{\mathbf{z}I} \boldsymbol{\Lambda}_{\mathbf{z}I} \mathbf{U}_{\mathbf{z}I}^H + \mathbf{U}_{\mathbf{z}N} \boldsymbol{\Lambda}_{\mathbf{z}N} \mathbf{U}_{\mathbf{z}N}^H. \tag{6.471}$$

If the signal arrives from the direction that the blocking matrix assumed, then the first subspace in (6.471) is an interference subspace, because the blocking matrix $\mathbf{B}$ has removed all signal components. If there is signal mismatch, then some signal component will leak into the lower path. We refer to $\mathbf{U}_{\mathbf{z}I}$ as an interference subspace even though it may contain leaked signal. $\mathbf{U}_{\mathbf{z}N}$ is orthogonal to the interference subspace. We can also show that $\mathbf{S}_{d\mathbf{z}^H}$ contains the component due to interference only (e.g., [Jab86a]) and $\mathbf{U}_{\mathbf{z}N}$ is orthogonal to $\mathbf{S}_{d\mathbf{z}^H}$.

Therefore,

$$\mathbf{w}_{ao}^H = \mathbf{w}_q^H \mathbf{S}_x \mathbf{B} [\mathbf{U}_{\mathbf{z}I} \boldsymbol{\Lambda}_{\mathbf{z}I}^{-1} \mathbf{U}_{\mathbf{z}I}^H]. \tag{6.472}$$

and

$$\mathbf{w}_{a,es}^H = \mathbf{w}_q^H \mathbf{S}_{\mathbf{x}} \mathbf{B} \mathbf{u}_{\mathbf{z}I} \boldsymbol{\Lambda}_{\mathbf{z}I}^{-1}. \tag{6.473}$$

The results in (6.468) and (6.473) specify the GSC version of the eigenspace beamformer.

We should observe that some of the important issues concerning eigenspace beamformers will not appear until we study the finite data problem and adaptive implementations in Chapter 7. Specifically:

(i) We have to estimate D_r and compute the eigenvalues and eigenvectors. It is important that a significant part of the signal energy is contained in the resulting eigenspace. When the *SNR* is low, we may include a "noise" eigenvector instead of a "signal" eigenvector. This error causes a dramatic decrease in performance.

(ii) It is important to observe that we do not have to compute all of the eigenvalues. We can use the techniques discussed in Golub and Van Loan [GVL89] to compute the eigenvalues in descending order.

6.8.2 Cross-spectral Eigenspace Beamformers

In Section 6.8.1, we performed an eigendecomposition of $\mathbf{S_x}$ and formed a reduced–rank beamformer by choosing the eigenvectors with the D_r largest eigenvalues. In this section, we use a different choice of eigenvectors to form the subspace.[13]

The MPDR beamformer was designed to minimize the output power subject to the distortionless constraint,

$$\mathbf{w}^H \mathbf{v}_m = 1. \tag{6.474}$$

The resulting weight vector is

$$\mathbf{w}^H = \frac{\mathbf{v}_m^H \mathbf{S}_\mathbf{x}^{-1}}{\mathbf{v}_m^H \mathbf{S}_\mathbf{x}^{-1} \mathbf{v}_m}, \tag{6.475}$$

and the output power is

$$P_o = \left(\mathbf{v}_m^H \mathbf{S}_\mathbf{x}^{-1} \mathbf{v}_m \right)^{-1}. \tag{6.476}$$

The eigendecomposition of $\mathbf{S_x}$ is

$$\mathbf{S}_\mathbf{x} = \sum_{i=1}^{N} \lambda_i \mathbf{\Phi}_i \mathbf{\Phi}_i^H \tag{6.477}$$

and

$$\mathbf{S}_\mathbf{x}^{-1} = \sum_{i=1}^{N} \frac{\mathbf{\Phi}_i \mathbf{\Phi}_i^H}{\lambda_i}. \tag{6.478}$$

[13] This approach is discussed in Goldstein and Reed (e.g., [GR97a],[GR97b],[GR97c]). The ideas were discussed earlier in Byerly and Roberts [BR89].

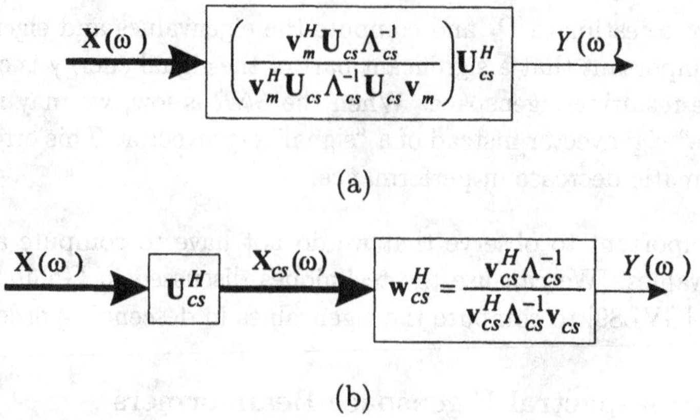

$\mathbf{X}(\omega) \rightarrow \left(\frac{\mathbf{v}_m^H \mathbf{U}_{cs} \mathbf{\Lambda}_{cs}^{-1}}{\mathbf{v}_m^H \mathbf{U}_{cs} \mathbf{\Lambda}_{cs}^{-1} \mathbf{U}_{cs} \mathbf{v}_m} \right) \mathbf{U}_{cs}^H \rightarrow Y(\omega)$

(a)

$\mathbf{X}(\omega) \rightarrow \mathbf{U}_{cs}^H \rightarrow \mathbf{X}_{cs}(\omega) \rightarrow \mathbf{w}_{cs}^H = \frac{\mathbf{v}_{cs}^H \mathbf{\Lambda}_{cs}^{-1}}{\mathbf{v}_{cs}^H \mathbf{\Lambda}_{cs}^{-1} \mathbf{v}_{cs}} \rightarrow Y(\omega)$

(b)

Figure 6.76 Cross-spectral beamformer.

Using (6.478) in (6.476), we have

$$P_o = \left(\mathbf{v}_m^H \sum_{i=1}^{N} \frac{\mathbf{\Phi}_i \mathbf{\Phi}_i^H}{\lambda_i} \mathbf{v}_m \right)^{-1} = \left(\sum_{i=1}^{N} \frac{\left| \mathbf{v}_m^H \mathbf{\Phi}_i \right|^2}{\lambda_i} \right)^{-1}. \tag{6.479}$$

To retain D_c eigenvectors as the basis for the reduced–rank subspace, we choose the D_c largest terms in the summation,

$$\lambda_{(n)}, \mathbf{\Phi}_{(n)} = \underset{\lambda_i, \mathbf{\Phi}_i}{\arg\max} \left\{ \frac{\left| \mathbf{v}_m^H \mathbf{\Phi}_i \right|^2}{\lambda_i} \right\}, \quad (n) = 1, \cdots, D_c. \tag{6.480}$$

We use the subscript "(n)" on this set of eigenvectors because the subscript n has been associated with eigenvectors ordered according to the size of the corresponding eigenvalue.

We define an $N \times D_c$ matrix, $\mathbf{U}_{cs}$, whose columns are the D_c eigenvectors chosen in (6.480). We then use the MPDR beamformer in the reduced-rank space

$$\mathbf{w}_{cs}^H = \frac{\mathbf{v}_m^H \mathbf{U}_{cs} \mathbf{\Lambda}_{cs}^{-1} \mathbf{U}_{cs}^H}{\mathbf{v}_m^H \mathbf{U}_{cs} \mathbf{\Lambda}_{cs}^{-1} \mathbf{U}_{cs}^H \mathbf{v}_m}. \tag{6.481}$$

The **cross-spectral** (CS) beamformer is shown in Figure 6.76(a). An equivalent implementation is shown in Figure 6.76(b). A generalized sidelobe canceller version is shown in Figure 6.77.

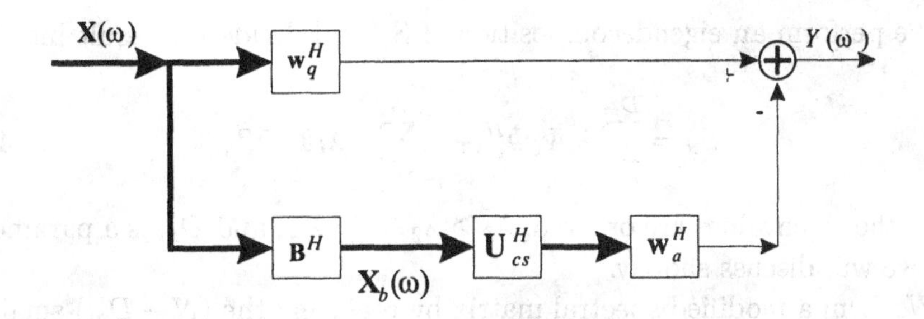

Figure 6.77 Generalized sidelobe canceller implementation of cross-spectral beamformer.

If we try to implement the cross-spectral beamformer for the model in Example 6.8.1, we find that it does not work in the presence of DOA mismatch. For $u_a = 0$, the steering vector is matched to the model steering vector $\mathbf{v}_m$ and the eigenvectors that we pick correspond to the ones with the three largest eigenvalues. The resulting beamformer is the same as in Example 6.8.1. However, when $u_a \neq 0$, we start picking eigenvectors corresponding to the noise subspace and the performance deteriorates rapidly. Adding diagonal loading does not help.

We would anticipate that the cross-spectral beamformer would offer a performance advantage when the dimension of eigenspace was less than the rank of the signal-plus-interference subspace. Goldstein and Reed [GR97b] have an example that demonstrates a performance improvement, but the interference model is tailored to the algorithm. We were able to construct scenarios in which the CS beamformer was better than the PC beamformer under nominal conditions, but it degraded rapidly with mismatch. Another scenario that should offer an advantage is spatially spread interference where the eigenvalues roll off gradually. We consider this case in the problems and find very little performance improvement over the PC beamformer. The computational disadvantage of the CS beamformer is that we must compute all of the eigenvalues in order to find the terms in (6.480).

6.8.3 Dominant–mode Rejection Beamformers

In this section, we develop a class of beamformers referred to as **dominant mode rejection** (DMR) beamformers. The reason for the name will be clear after we derive the beamformer. The DMR beamformer is a special case of an enhanced MPDR beamformer that was derived by Abraham and Owsley [AO90].

We perform an eigendecomposition of $\mathbf{S_x}$ and divide the result into two terms,

$$\mathbf{S_x} = \sum_{i=1}^{D_m} \lambda_i \mathbf{\Phi}_i \mathbf{\Phi}_i^H + \sum_{i=D_{m+1}}^{N} \lambda_i \mathbf{\Phi}_i \mathbf{\Phi}_i^H, \tag{6.482}$$

where the eigenvalues are ordered, $\lambda_1 \geq \lambda_2 \cdots \geq \lambda_N$, and D_m is a parameter that we will discuss shortly.

We form a modified spectral matrix by replacing the $(N - D_m)$ smallest eigenvalues by their average,

$$\alpha \triangleq \frac{1}{N - D_m} \sum_{i=D_m+1}^{N} \lambda_i. \tag{6.483}$$

We can also find α from the expression,

$$\alpha = \frac{1}{N - D_m} \left\{ tr\left[\mathbf{S_x}\right] - \sum_{i=1}^{D_m} \lambda_i \right\}. \tag{6.484}$$

The advantage of (6.484) is that we only have to compute the D_m largest eigenvalues.

Note that, if the model consists of a plane-wave signal plus $(D_m - 1)$ plane-wave interferers in the presence of spatially uncorrelated noise $(\sigma_w^2 \mathbf{I})$, then

$$\alpha = \sigma_w^2, \tag{6.485}$$

when the statistics are known. However when we estimate $\mathbf{S_x}$ using finite data or have spatially correlated background noise, then $\alpha \neq \sigma_w^2$.

The modified spectral matrix is

$$\tilde{\mathbf{S}}_{\mathbf{x}} \triangleq \sum_{i=1}^{D_m} \lambda_i \mathbf{\Phi}_i \mathbf{\Phi}_i^H + \alpha \sum_{i=D_m+1}^{N} \mathbf{\Phi}_i \mathbf{\Phi}_i^H. \tag{6.486}$$

We can denote the dominant mode subspace by the $N \times D_m$ matrix

$$\mathbf{U}_{dm} \triangleq \begin{bmatrix} \mathbf{\Phi}_1 & \mathbf{\Phi}_2 & \cdots & \mathbf{\Phi}_{dm} \end{bmatrix}. \tag{6.487}$$

The orthogonal noise subspace is

$$\mathbf{U}_{dm}^{\perp} \triangleq \begin{bmatrix} \mathbf{\Phi}_{dm+1} & \mathbf{\Phi}_{dm+2} & \cdots & \mathbf{\Phi}_N \end{bmatrix}. \tag{6.488}$$

Then (6.486) can be written as

$$\tilde{\mathbf{S}}_{\mathbf{x}} = \sum_{i=1}^{D_m} \lambda_i \mathbf{\Phi}_i \mathbf{\Phi}_i^H + \alpha \mathbf{U}_{dm}^{\perp} [\mathbf{U}_{dm}^{\perp}]^H. \tag{6.489}$$

We refer to the D_m eigenvectors with the largest eigenvalues as the **dominant modes**. They are due to the signal, plane-wave interferers, and spatially spread interference.

Note that $\tilde{\mathbf{S}}_{\mathbf{x}}$ is a full-rank matrix in contrast to the reduced-rank matrices in Sections 6.8.1 and 6.8.2.

The inverse of $\tilde{\mathbf{S}}_{\mathbf{x}}$ is,

$$\begin{aligned} \tilde{\mathbf{S}}_{\mathbf{x}}^{-1} &= \sum_{i=1}^{D_m} \frac{1}{\lambda_i} \mathbf{\Phi}_i \mathbf{\Phi}_i^H + \frac{1}{\alpha} \mathbf{U}_{dm}^{\perp} [\mathbf{U}_{dm}^{\perp}]^H \\ &= \frac{1}{\alpha} \left[\sum_{i=1}^{D_m} \frac{\alpha}{\lambda_i} \mathbf{\Phi}_i \mathbf{\Phi}_i^H + \mathbf{P}_{dm}^{\perp} \right], \end{aligned} \tag{6.490}$$

where $\mathbf{P}_{dm}^{\perp}$ is the projection matrix onto the subspace orthogonal to $\mathbf{U}_{dm}$. The weight vector is

$$\mathbf{w}_{dm}^H = \frac{\mathbf{v}_m^H \tilde{\mathbf{S}}_{\mathbf{x}}^{-1}}{\mathbf{v}_m^H \tilde{\mathbf{S}}_{\mathbf{x}}^{-1} \mathbf{v}_m} = \frac{\alpha \sum_{i=1}^{D_m} \frac{1}{\lambda_i} (\mathbf{v}_m^H \mathbf{\Phi}_i) \mathbf{\Phi}_i^H + \mathbf{v}_m^H \mathbf{P}_{dm}^{\perp}}{\alpha \sum_{i=1}^{D_m} \frac{1}{\lambda_i} |\mathbf{v}_m^H \mathbf{\Phi}_i|^2 + \mathbf{v}_m^H \mathbf{P}_{dm}^{\perp} \mathbf{v}_m}. \tag{6.491}$$

In the first case of interest, $\mathbf{v}_m$ is completed contained in the DM subspace. For the case of known statistics with no signal mismatch this will occur if D_m is chosen correctly. Then,

$$\mathbf{v}_m^H \mathbf{P}_{dm}^{\perp} = 0, \tag{6.492}$$

and (6.491) reduces to

$$\mathbf{w}_{dm}^H = \frac{\sum_{i=1}^{D_m} \frac{1}{\lambda_i} (\mathbf{v}_m^H \mathbf{\Phi}_i) \mathbf{\Phi}_i^H}{\sum_{i=1}^{D_m} \frac{1}{\lambda_i} |\mathbf{v}_m^H \mathbf{\Phi}_i|^2}. \tag{6.493}$$

Comparing (6.493) and (6.451), we see that the eigenspace beamformer and the DMR beamformer are identical for this case. The DMR beamformer in (6.493) is distortionless in the absence of mismatch.

To investigate the response due to an eigenvector (or mode), we first divide λ_i into a signal and noise component,

$$\lambda_i = \lambda_i^s + \alpha, \quad i = 1, 2, \cdots, D_m. \tag{6.494}$$

The power output when the input spectral matrix is $\lambda_k \mathbf{\Phi}_k \mathbf{\Phi}_k^H$ is

$$P_{\mathbf{\Phi}_k} = \frac{\frac{1}{(\lambda_k^s/\alpha+1)} |\mathbf{v}_m^H \mathbf{\Phi}_k|^2}{\left[\sum_{i=1}^{D_m} \frac{1}{(\lambda_i^s/\alpha+1)} |\mathbf{v}_m^H \mathbf{\Phi}_i|^2\right]^2}. \tag{6.495}$$

If $\lambda_k^s/\alpha \gg 1$, then the output due to the kth mode will be small and that particular mode will be rejected. Hence, the dominant mode rejection (DMR) beamformer name. The second term is a measure of the correlation between the kth eigenvector and $\mathbf{v}_m$. As this correlation increases, the depth of the null on the kth mode decreases. Note that $\mathbf{v}_m$ is not normalized, so that the maximum value of the second term is N^2.

In practice there will be some mismatch in the model. We will have to estimate $\mathbf{S_x}$. The eigendecomposition of $\hat{\mathbf{S}}_\mathbf{x}$ will lead to mismatch in the eigenvalues and eigenvectors. In addition, we will have to estimate D_{dm}, the dimension of the dominant–mode eigenspace. This estimate, $\hat{D}_{dm}$, may be in error. There can also be mismatch in the signal DOA, so that $\mathbf{v}_a \neq \mathbf{v}_m$.

In the next two examples, we discuss the behavior of the DMR beamformer under two types of mismatch. In Example 6.8.3, we consider the case of DOA mismatch. In Example 6.8.4, we choose $\hat{D}_{dm} = D_{dm} - 1$ in order to simulate underestimating the dimension of the dominant–mode eigenspace. Both examples use the same scenarios as in Examples 6.8.1 and 6.8.2.

Example 6.8.3

Consider a standard 10-element linear array. The nominal signal arrival direction is $u_m = 0$. The actual input spectral matrix is

$$\mathbf{S_x} = \sigma_s^2 \mathbf{v}_a \mathbf{v}_a^H + \sum_{i=1}^{2} \sigma_i^2 \mathbf{v}_i \mathbf{v}_i^H + \sigma_w^2 \mathbf{I}. \tag{6.496}$$

We assume the interfering plane waves are at ± 0.30. We perform an eigendecomposition[14] of $\mathbf{S_x}$ and construct a 3-D subspace using the eigenvectors corresponding to the three largest eigenvalues. We use the weight vector in (6.491) with α replaced by σ_w^2,

$$\mathbf{w}_{dm}^H = \frac{\sigma_w^2 \sum_{i=1}^{D_m} \frac{1}{\lambda_i} (\mathbf{v}_m^H \mathbf{\Phi}_i) \mathbf{\Phi}_i^H + \mathbf{v}_m^H \mathbf{P}_{dm}^\perp}{\sigma_w^2 \sum_{i=1}^{D_m} \frac{1}{\lambda_i} |\mathbf{v}_m^H \mathbf{\Phi}_i|^2 + \mathbf{v}_m^H \mathbf{P}_{dm}^\perp \mathbf{v}_m}. \tag{6.497}$$

In Figure 6.78, we plot the array gain versus u_a/BW_{NN}. When the *SNR* $= -10$ dB, there is very gradual degradation. When the *SNR* $= 0$ dB, the degradation is still gradual; by $u_a = 0.1/BW_{NN} = 0.25$, the array gain has decreased by 13 dB. For higher *SNR*, the decrease in array gain is significant.

Other scenarios with different u_i and σ_i^2 give similar results. It appears that, when $D_m = D + 1$, the *SNR* must be at least 10 dB below the weaker interferer in the two interferer case.

Example 6.8.4 (continuation)

Consider the same model as in Example 6.8.3. We use (6.491) with $D_m = 2$ to test the effects of underestimation. Now α includes the third eigenvalue,

$$\alpha = \frac{1}{8}(\lambda_3 + 7\sigma_w^2). \tag{6.498}$$

[14] Note that this is exactly the same eigendecomposition as in Example 6.8.1.

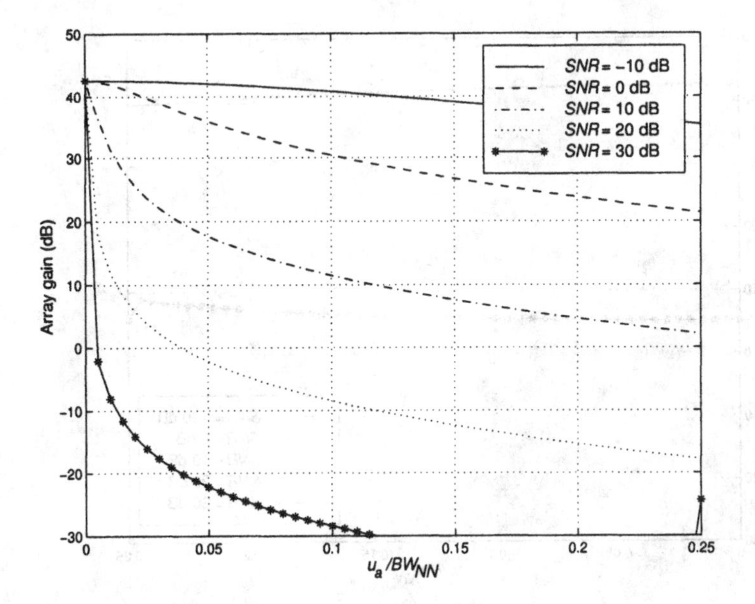

Figure 6.78 DMR beamformer with $D_m = 3$, $u_I = \pm 0.3$, $INR = 30$ dB, various SNR; array gain versus u_a/BW_{NN}.

In Figure 6.79, we plot the array gain versus u_a/BW_{NN}. We see that for an $SNR = -10$ dB, 0 dB, and 10 dB, there is minimal degradation. This is because there is almost no signal content in the two DM eigenvectors. For $SNR = 20$ dB, there is some signal component in the two DM eigenvectors. As u_a increases the geometry causes this signal content to decrease and the array gain goes up. For $SNR = 30$ dB, there is significant signal content and the array gain is very low. Other scenarios give similar results.

This result suggests that the dimension of the dominant mode subspace should equal D, the number of interferers when the $SNR \ll INR$.

The performance of DMR beamformer degrades when $\mathbf{v}_a$ is in the dominant mode subspace. Cox and Pitre [CP97] (e.g., [CPL98] or [Cox00]) have developed a robust DMR beamformer that improves performance under certain scenarios. The reader is referred to those references for a description of the algorithm.

6.8.4 Summary

In this section we have developed two beamformers that utilize the eigendecomposition in a different manner.

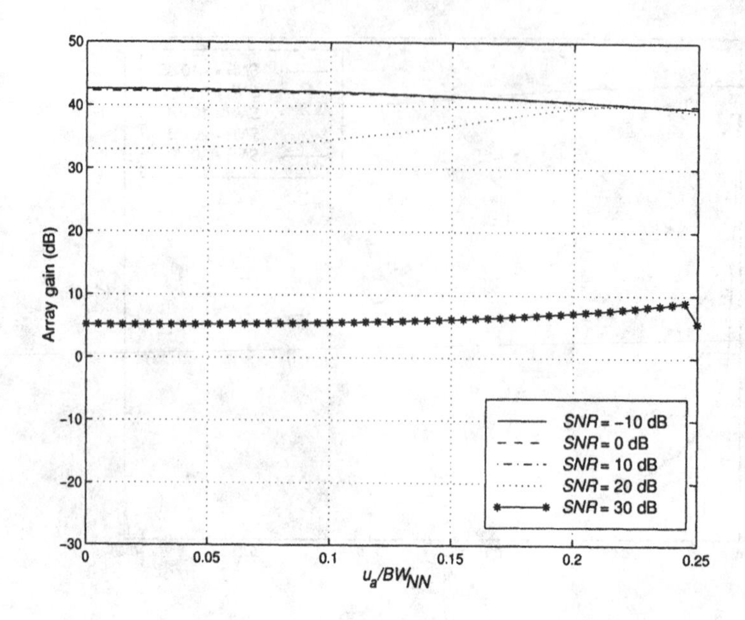

Figure 6.79 DMR beamformer with $D_m = 2$, $u_I = \pm 0.3$, *INR* = 30 dB, various *SNR*; array gain versus u_a/BW_{NN}.

The principal component (or eigenspace) beamformer forms an eigenspace using the eigenvectors with largest eigenvalues. All of the subsequent processing is done in this reduced-dimension eigenspace. In order to have good performance, the desired signal must have a significant component in the eigenspace. In order to accomplish this, the *SNR* must be large enough to cause this inclusion. The exact value required will depend on the signal and interference geometry, the accuracy of the $\hat{\mathbf{S}}_{\mathbf{x}}$ estimator, and other mismatches. In practice, we must also estimate D_r from the data and we will find that it is important not to underestimate D_r.

The dominant mode rejection beamformer performs the same eigendecomposition, but uses it in a different manner. The DMR beamformer averages the smallest eigenvalues and uses the average value in place of λ_i, $i = D_{dm} + 1, \cdots, N$. The result is a modified version of the MPDR beamformer. The eigenvectors that are not in DM subspace only appear in $\mathbf{P}_{dm}^{\perp}$, which can be calculated using the DM eigenvectors. If the model is perfectly matched, then the DMR and PC beamformers are the same. In the presence of mismatch, their behavior is different. In the DMR beamformer we would like to remove the signal from the DM subspace. This result will happen in a low *SNR* environment. The exact value required will depend on

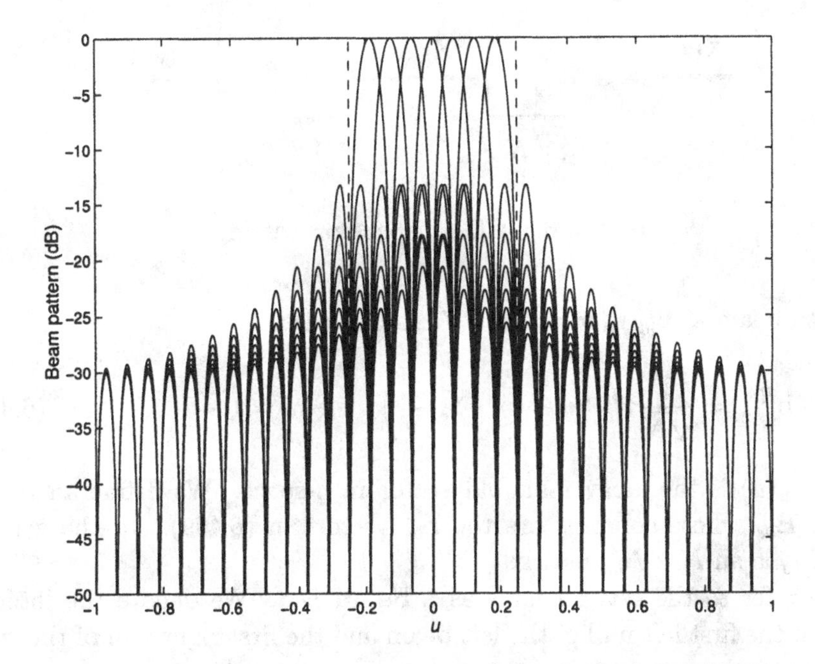

Figure 6.80 Orthogonal beams for beamspace processor.

the signal and interference geometry, the accuracy of the $\hat{\mathbf{S}}_{\mathbf{x}}$ estimator, and other mismatches. When we estimate D_r from the data, an underestimate $\hat{D}_r = D_r - 1$ may improve the performance.

In Chapter 7 (Section 7.9) we look at the behavior when the necessary quantities must be estimated from a data set. Our tentative conclusion is that we can improve on the MPDR performance by using eigendecomposition. For high *SNR*, we would use the eigenspace beamformer. For low *SNR*, we would use the DMR beamformer.

6.9 Beamspace Beamformers

A second approach to reducing the rank of the processor is to perform a preliminary preprocessing with a set of non-adaptive beamformers. We then process the beamformer outputs in an optimum manner. We refer to the space spanned by the output of the beamformers as the beamspace and the overall processor as a **beamspace** beamformer.

A typical example is shown in Figure 6.80. The array is a standard 32-element linear array. We use seven orthogonal conventional beams spaced symmetrically about $u_s = 0$. We denote the weight vector associated with

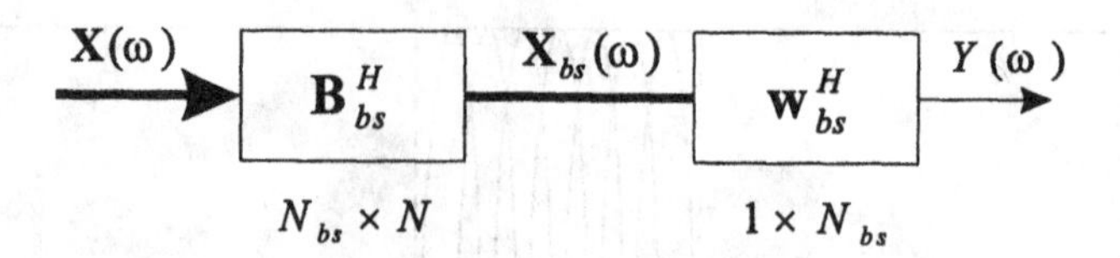

Figure 6.81: Beamspace processing.

the mth beam as $\mathbf{b}_{bs,m}^H, m = 1, \cdots, 7$. In this case,

$$\mathbf{b}_{bs,m}^H = \frac{1}{\sqrt{N}} \mathbf{v}_u^H \left(u - (m-4)\frac{2}{N} \right), \qquad m = 1, \cdots, 7, \tag{6.499}$$

where $\mathbf{v}_u(u)$ is the array manifold vector in u-space. We define an $N \times 7$ matrix $\mathbf{B}_{bs}$ whose columns are the $\mathbf{b}_{bs}$ vectors in (6.499). In the general case, $\mathbf{B}_{bs}$ is an $N \times N_{bs}$ matrix.

We refer to the seven beams as a **beam fan**. We denote the interval between the first left null of the left beam and the first right null of the right beam as the **beamspace sector**.

In Section 6.9.1, we consider beamspace MPDR beamformers with diagonal loading. In Section 6.9.2, we consider beamspace LCMP with diagonal loading. In Section 6.9.3, we summarize our results.

6.9.1 Beamspace MPDR

The model is shown in Figure 6.81. The output of the beamspace matrix is,

$$\mathbf{X}_{bs} = \mathbf{B}_{bs}^H \mathbf{X}, \tag{6.500}$$

and we require

$$\mathbf{B}_{bs}^H \mathbf{B}_{bs} = \mathbf{I}. \tag{6.501}$$

The beamspace array manifold vector is defined as

$$\mathbf{v}_{bs} = \mathbf{B}_{bs}^H \mathbf{v}_s. \tag{6.502}$$

If we use orthogonal conventional beams as in Figure 6.80, then, as discussed in Section 3.10,

$$\left[\mathbf{b}_{bs,m}^H\right]_n = \frac{1}{\sqrt{N}} \left[e^{-j\left(n-\frac{N-1}{2}\right)\left(\psi_c-(m-4)\frac{2\pi}{N}\right)} \right], \quad n = 0, \cdots, N-1, \quad m = 1, \cdots, 7, \tag{6.503}$$

where ψ_c is the MRA of the center beam, and

$$[\mathbf{v}_s]_n = e^{j(n-\frac{N-1}{2})\psi}, \quad n = 0, \cdots, N-1, \tag{6.504}$$

so, for $\psi_c = 0$,

$$\mathbf{v}_{bs}(\psi) = \sqrt{N} \begin{bmatrix} \frac{sin\left(\frac{N}{2}(\psi+3\frac{2\pi}{N})\right)}{sin\left(\frac{1}{2}(\psi+3\frac{2\pi}{N})\right)} \\ \vdots \\ \frac{sin\left(\frac{N}{2}(\psi-(m-4)\frac{2\pi}{N})\right)}{sin\left(\frac{1}{2}(\psi-(m-4)\frac{2\pi}{N})\right)} \\ \vdots \\ \frac{sin\left(\frac{N}{2}(\psi-3\frac{2\pi}{N})\right)}{sin\left(\frac{1}{2}(\psi-3\frac{2\pi}{N})\right)} \end{bmatrix}. \tag{6.505}$$

If we use other weightings such as Chebychev or Taylor to form the beamspace, then

$$\left[\mathbf{b}_{bs,m}^H\right]_n = w_n^* \, e^{-j(n-\frac{N-1}{2})(\psi_c-(m-4)\frac{2\pi}{N})}, \quad n = 0, \cdots, N-1, \quad m = 1, \cdots, 7. \tag{6.506}$$

The output of the beamspace matrix, $\mathbf{X}_{bs}$, is processed with a $1 \times N_{bs}$ matrix, $\mathbf{w}_{bs}^H$ to obtain a scalar output Y. Thus,

$$Y = \mathbf{w}_{bs}^H \mathbf{X}_{bs} = \mathbf{w}_{bs}^H \mathbf{B}_{bs}^H \mathbf{X}. \tag{6.507}$$

The resulting beam pattern is

$$B_\psi(\psi) = \mathbf{w}_{bs}^H \mathbf{v}_{bs}(\psi). \tag{6.508}$$

The spatial spectral matrix in beamspace is

$$\mathbf{S}_{\mathbf{x}_{bs}} = \mathbf{B}_{bs}^H \mathbf{S}_\mathbf{x} \mathbf{B}_{bs}. \tag{6.509}$$

For a single plane-wave signal plus spatially correlated interference and white noise,

$$\mathbf{S}_\mathbf{x} = \sigma_s^2 \mathbf{v}_s \mathbf{v}_s^H + \mathbf{S}_c + \sigma_w^2 \mathbf{I}. \tag{6.510}$$

Then,

$$\mathbf{S}_{\mathbf{x}_{bs}} = \sigma_s^2 \mathbf{B}_{bs}^H \mathbf{v}_s \mathbf{v}_s^H \mathbf{B}_{bs} + \mathbf{B}_{bs}^H \mathbf{S}_c \mathbf{B}_{bs} + \sigma_w^2 \mathbf{B}_{bs}^H \mathbf{B}_{bs}. \tag{6.511}$$

Using (6.501) and (6.502) gives

$$\mathbf{S}_{\mathbf{x}_{bs}} = \sigma_s^2 \mathbf{v}_{bs} \mathbf{v}_{bs}^H + \mathbf{B}_{bs}^H \mathbf{S}_c \mathbf{B}_{bs} + \sigma_w^2 \mathbf{I}. \tag{6.512}$$

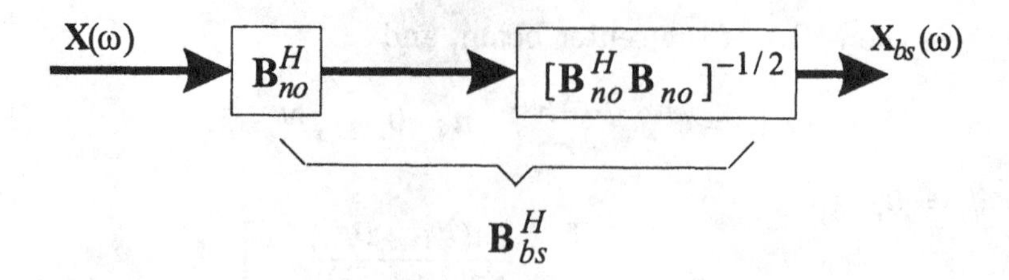

Figure 6.82 Beamspace processing using non-orthogonal beams.

For the special case in which the spatially correlated interference consists of D plane waves

$$\mathbf{S}_c = \mathbf{V}\mathbf{S}_I\mathbf{V}^H, \tag{6.513}$$

and the second term can be written as

$$\mathbf{B}_{bs}^H\mathbf{S}_c\mathbf{B}_{bs} = \mathbf{V}_{bs}\mathbf{S}_I\mathbf{V}_{bs}^H, \tag{6.514}$$

where

$$\mathbf{V}_{bs} \triangleq \mathbf{B}_{bs}^H\mathbf{V}. \tag{6.515}$$

We can now use one of our optimum beamforming algorithms on $\mathbf{X}_{bs}$. Note that the beamspace beamformer has a reduced number of degrees of freedom so that the number of interferers or eigenvectors that we can suppress is reduced.

One way to compensate for this reduction is to use shaped beams to form the beamspace. For example, we can use a set of shifted Dolph-Chebychev or Taylor-Villenueve beams to reduce the sidelobes significantly. Then the beamspace processor can focus on interference within the beamspace sector.

In the shaped beam case, the original beam set may not satisfy (6.501). We denote a non-orthogonal $\mathbf{B}$ as $\mathbf{B}_{no}$. Then (6.512) becomes

$$\mathbf{S}_{\mathbf{x}_{bs}} = \sigma_s^2\mathbf{v}_{bs}\mathbf{v}_{bs} + \mathbf{B}_{no}^H\mathbf{S}_c\mathbf{B}_{no} + \sigma_w^2\mathbf{B}_{no}^H\mathbf{B}_{no}. \tag{6.516}$$

Most of our beamforming algorithms assume a white noise component so we pass $\mathbf{X}_{bs}$ through a whitening matrix, as shown in Figure 6.82:

$$\mathbf{H}_w = \left[\mathbf{B}_{no}^H\mathbf{B}_{no}\right]^{-\frac{1}{2}}. \tag{6.517}$$

The overall matrix filter is

$$\mathbf{B}_{bs}^H = \left[\mathbf{B}_{no}^H\mathbf{B}_{no}\right]^{-\frac{1}{2}}\mathbf{B}_{no}^H. \tag{6.518}$$

We then use $\mathbf{B}_{bs}^H$ as the matrix to form the beamspace. The sidelobes in the component beams of $\mathbf{B}_{bs}$ are several dB higher than the original Dolph-Chebychev or Taylor beams.

We consider several examples to illustrate beamspace beamforming. In Examples 6.9.1 through 6.9.4, we consider a standard 32-element linear array. The beamspace sector is centered around $u = 0$.

Example 6.9.1

In this example, we use the seven conventional beams shown in Figure 6.80 to form the beamspace. We then form the beamspace MPDR beamformer,

$$\mathbf{w}_{mpdr,bs}^H = \frac{\mathbf{v}_{bs}^H \mathbf{S}_{\mathbf{x}_{bs}}^{-1}}{\mathbf{v}_{bs}^H \mathbf{S}_{\mathbf{x}_{bs}}^{-1} \mathbf{v}_{bs}}. \tag{6.519}$$

We steer the array to broadside. Using (6.499) in (6.502), we have

$$\mathbf{v}_{bs}^H = \begin{bmatrix} 0 & 0 & 0 & \sqrt{N} & 0 & 0 & 0 \end{bmatrix}, \tag{6.520}$$

so that

$$\mathbf{w}_{mpdr,bs}^H = \frac{\mathbf{e}_4^T \mathbf{S}_{\mathbf{x}_{bs}}^{-1}}{\sqrt{N}\mathbf{e}_4^T \mathbf{S}_{\mathbf{x}_{bs}}^{-1} \mathbf{e}_4}. \tag{6.521}$$

We assume that the interference consists of two equal-power uncorrelated plane-wave interferers at $\frac{3}{32}$ and $\frac{5}{32}$. We assume that the signal direction is perfectly matched so the *SNR* does not affect the array gain or beam pattern.

For $INR \geq 10$ dB the beam patterns have essentially the same shape. In Figure 6.83, we show the beam patterns for an $INR = 20$ dB (each interferer) for the beamspace MPDR. The two array gains are essentially the same: for element-space MPDR, $A = 37.806$ and for beamspace MPDR, $A = 37.753$.

The advantage of beamspace is that the dimension of the optimum beamformers is 7 instead of the 32-dimensional element-space beamformer, so the computational complexity is reduced. In Chapter 7, we find that the number of snapshots needed to obtain a required accuracy will decrease as the dimension of the space decreases.

In the next example, we investigate the performance in the case of DOA mismatch. As in the element-space beamformer, we use diagonal loading to improve robustness.

Example 6.9.2 (continuation)

We use the same model as in Example 6.9.1 and examine the behavior when $u_a \neq 0$. There are two plane-wave interferers at $u_I = 3/32$ and $u_I = 5/32$. Each interferer has an $INR = 20$ dB. We consider $SNR = 0$ dB and 10 dB and use fixed loading of 10 dB and 15 dB. In Figure 6.84, we plot the array gain versus u_a/BW_{NN} for beamspace MPDR.

The loading provides some robustness against signal mismatch, but an LCMP beamformer may be necessary.

In the next example, we investigate the performance in the case of array perturbations.

Example 6.9.3 (continuation)

We use the same model as in Example 6.9.2 and assume $u_a = 0$. We use the array perturbation model in Example 6.6.4. We use a beamspace MPDR algorithm with diagonal

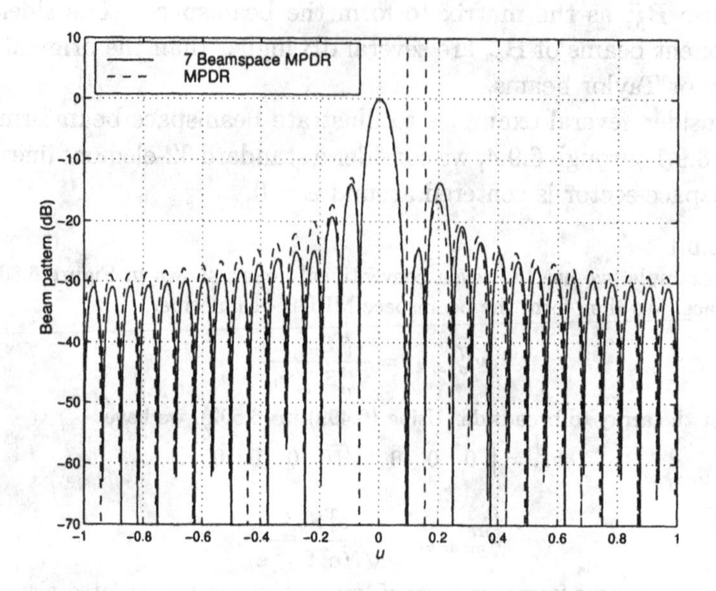

Figure 6.83 Beamspace MPDR; N=32, N_{bs} = 7, u_s = 0, *SNR* = 0 dB, u_I = 0.094, 0.16, *INR* = 20 dB: beam patterns.

loading. In Figure 6.85, we plot the array gain versus $\frac{\sigma_p}{\lambda}$ when appropriate diagonal loading is used. We also include the results for element-space MPDR with diagonal loading.

We use the same *INR* and *LNR* values as in Example 6.9.2. In Figure 6.85(a), the *SNR* is 0 dB. In Figure 6.85(b), the *SNR* is 10 dB.

We see that the beamspace algorithm is always better than the element-space algorithm. For higher *SNR*, the performance difference is significant.

In the next example, we modify the beam patterns of the fixed beams in order to further suppress out-of-sector interference.

Example 6.9.4

Consider a standard 32-element linear array. We form a 7-dimensional beamspace using beams whose steering direction is shifted by $\frac{2}{N}$. Each beam utilizes Taylor (Villenueve) weighting with -30 dB sidelobes and $\bar{n} = 6$. We use (6.518) to generate the orthonormal rows in $\mathbf{B}_{bs}^H$. The interference consists of eight statistically independent plane-wave interferers with DOA in u-space of $\pm\frac{3}{32}, \pm\frac{9}{32}, \pm\frac{11}{32}$, and $\pm\frac{13}{32}$. Note that six of the interferers are outside the sector ($\pm 7/32$) of the beamspace and there are more interferers than beamspace degrees-of-freedom. Each interferer has the same *INR*. We use an MPDR beamformer in beamspace.

In Figure 6.86, we show the resulting beam pattern for the case when the *INR* =30 dB. The array gain is the same as the element-space MPDR beamformer. We see that the beamspace processor has put a "perfect" null on the interferer inside the beamspace sector and reduces the beam pattern to less than -59 dB in the direction of the out-of-sector interferers.

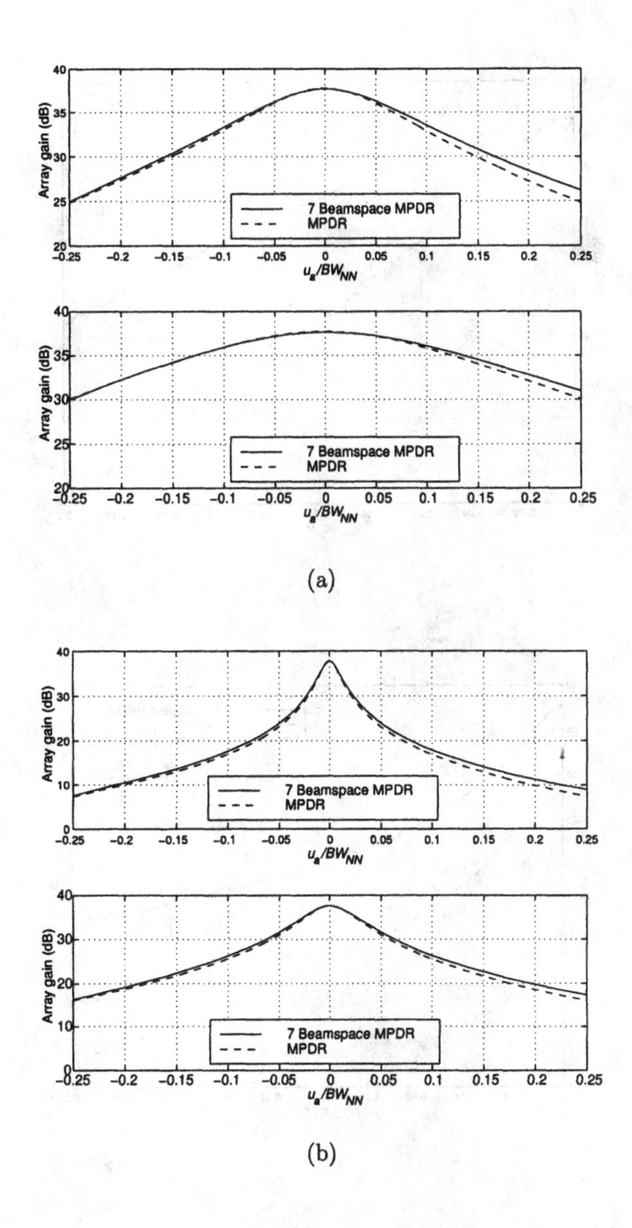

Figure 6.84 Signal mismatch, beamspace MPDR with fixed DL; N=32, N_{bs}=7, u_m =0, *SNR* =0 dB, u_{I1}=3/32, u_{I2}=5/32, INR_1=INR_2=20 dB, array gain versus u_a/BW_{NN}: (a) *SNR* = 0 dB, *LNR* = 10 dB and 15 dB; (b) *SNR* = 10 dB, *LNR* = 15 dB and 20 dB.

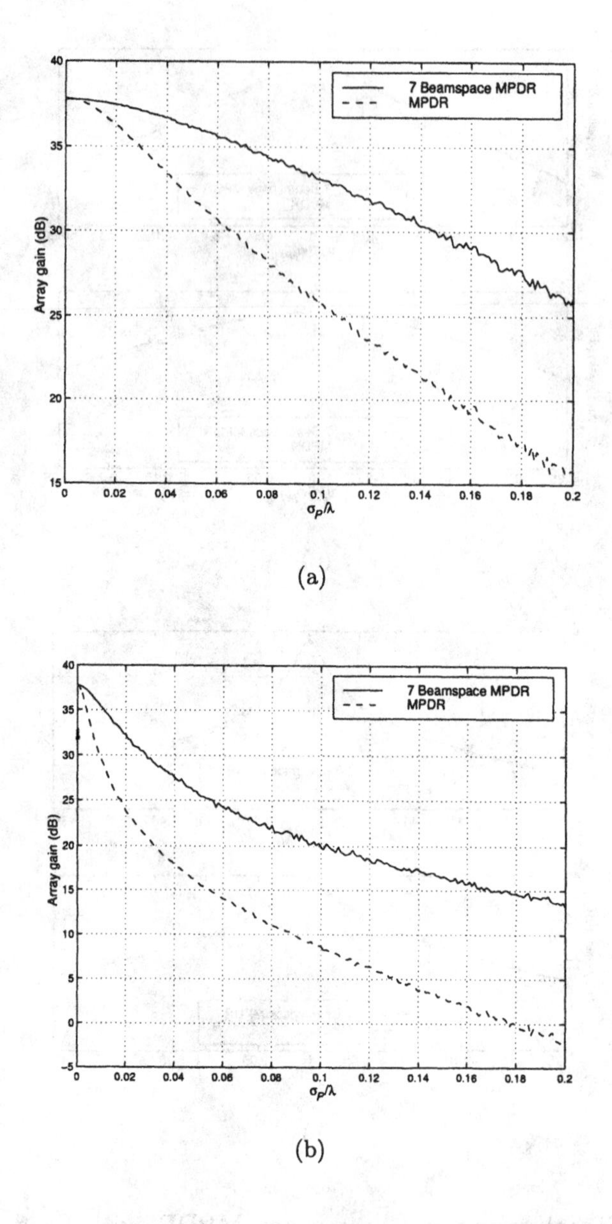

Figure 6.85 Array position perturbations, beamspace MPDR with fixed DL; N=32, N_{bs}=7, u_m = 0, u_{I1}=3/32, u_{I2}=5/32, INR_1=INR_2= 20 dB, array gain versus σ_p/λ, 200 trials: (a) SNR =0 dB, LNR = 10 dB; (b)SNR = 10 dB, LNR = 20 dB.

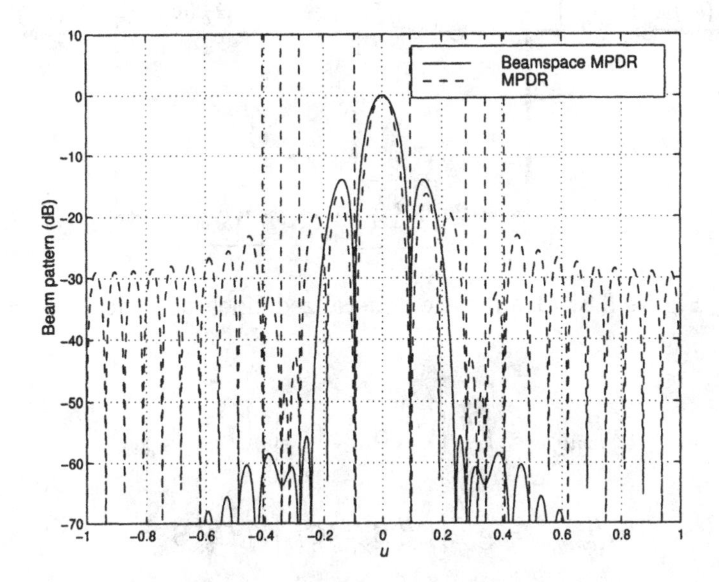

Figure 6.86 Beamspace MPDR beamformer with Taylor weighted beams; N=32, N_{bs}=7, $u_s = 0$, $SNR = 20$ dB, eight interferers with $INR = 30$ dB; beam pattern.

6.9.2 Beamspace LCMP

We can also implement LCMV and LCMP beamformers in beamspace. We first formulate the constraints in element space. From (6.350),

$$\mathbf{w}^H\mathbf{C} = \mathbf{g}^H, \tag{6.522}$$

where $\mathbf{w}^H$ is $1 \times N$, $\mathbf{C}$ is $N \times M_c$, and $\mathbf{g}^H$ is $1 \times M_c$. From Figure 6.81,

$$\mathbf{w}^H = \mathbf{w}_{bs}^H \mathbf{B}_{bs}^H. \tag{6.523}$$

Using (6.523) in (6.522) gives

$$\mathbf{w}_{bs}^H (\mathbf{B}_{bs}^H \mathbf{C}) = \mathbf{g}^H. \tag{6.524}$$

Defining

$$\mathbf{C}_{bs} \triangleq \mathbf{B}_{bs}^H \mathbf{C}, \tag{6.525}$$

(6.524) can be written as

$$\mathbf{w}_{bs}^H \mathbf{C}_{bs} = \mathbf{g}^H. \tag{6.526}$$

Following the steps as in (6.350)–(6.358), the optimum beamspace LCMV and LCMP processors are:

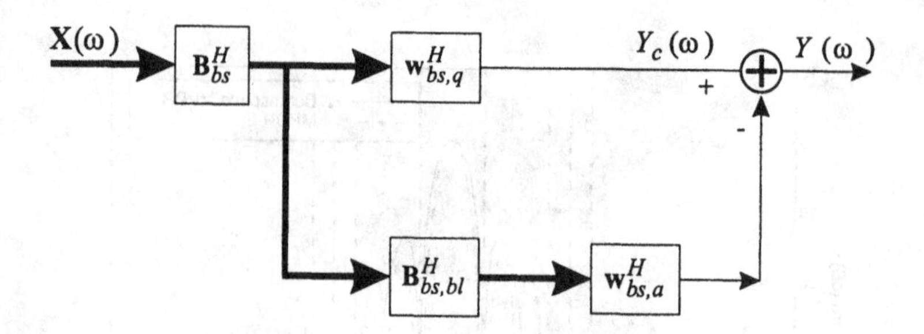

Figure 6.87 Beamspace generalized sidelobe canceller.

$$\mathbf{w}_{lcmv,bs}^H = \mathbf{g}^H \left[\mathbf{C}_{bs}^H \, \mathbf{S}_{\mathbf{n},bs}^{-1} \, \mathbf{C}_{bs}\right]^{-1} \mathbf{C}_{bs}^H \, \mathbf{S}_{\mathbf{n},bs}^{-1}, \tag{6.527}$$

and

$$\mathbf{w}_{lcmp,bs}^H = \mathbf{g}^H \left[\mathbf{C}_{bs}^H \, \mathbf{S}_{\mathbf{x},bs}^{-1} \, \mathbf{C}_{bs}\right]^{-1} \mathbf{C}_{bs}^H \, \mathbf{S}_{\mathbf{x},bs}^{-1}. \tag{6.528}$$

Note that, because the beamspace processor has fewer degrees of freedom, the impact of M_c constraints on the ability to null multiple interferers is increased.

The generalized sidelobe canceller implementation follows in the same manner as Section 6.7.3 and is shown in Figure 6.87.

The quiescent weight vector is

$$\mathbf{w}_{bs,q}^H = \mathbf{g}^H \left[\mathbf{C}_{bs}^H \, \mathbf{C}_{bs}\right]^{-1} \mathbf{C}_{bs}^H. \tag{6.529}$$

The beamspace blocking matrix is defined by the constraint,

$$\mathbf{C}_{bs}^H \, \mathbf{B}_{bsbl} = \mathbf{0}, \tag{6.530}$$

and

$$\mathbf{B}_{bsbl}^H \, \mathbf{B}_{bsbl} = \mathbf{I}. \tag{6.531}$$

The second processor in the lower path is

$$\widehat{\mathbf{w}}_{bs,a}^H = \mathbf{w}_{bs,q}^H \, \mathbf{S}_{\mathbf{x},bs} \, \mathbf{B}_{bsbl} \left[\mathbf{B}_{bsbl}^H \, \mathbf{S}_{\mathbf{x},bs} \, \mathbf{B}_{bsbl}\right]^{-1}. \tag{6.532}$$

The various constraints that were developed in element space can be used in these two configurations. A summary of appropriate constraints is given in Table 6.2.

Table 6.2 Constraint References

Constraint	Section
Directional	6.7.4.1
Derivative	6.7.4.2
Eigenvector	6.7.4.3
Quiescent pattern	6.7.5

In most applications, diagonal loading is also added. In general, beamspace processing works well when the rank of the interference is less than the degrees of freedom available in beamspace. When, the interference rank is too large, the performance of the beamspace algorithm can be much poorer than an element-space algorithm.

6.9.3 Summary: Beamspace Optimum Processors

In this section, we have observed several advantages of beamspace processing:

(i) The computational complexity is reduced from $O(N^3)$ to $O(N_{bs}^3)$.

(ii) The beamspace processor is more robust to array perturbations.

(iii) By using low sidelobe beams to construct the beamspace, the subsequent beamspace beamformer emphasizes the sector of interest in its optimization.

In Chapter 7, we see another important advantage when we estimate $\mathbf{S}_{\mathbf{x},bs}$. The number of snapshots (samples), K, required to obtain an adequate estimate of $\mathbf{S}_{\mathbf{x}}$ is a function of K/N. However, the number of snapshots required to get a comparable estimate of $\mathbf{S}_{\mathbf{x},bs}$ is a function of K/N_{bs}, which may be significantly larger.

One disadvantage of beamspace processing is its loss of available degrees of freedom to combat interference.

In the next section, we revisit the issue of diagonal loading.

6.10 Quadratically Constrained Beamformers

In Section 6.6.4, the technique of constraining $\|\mathbf{w}\|^2$ in order to improve the robustness of MPDR beamformers to signal mismatch and array perturbations was introduced. It resulted in diagonal loading. The level of loading was chosen without coupling it to the original constraint on the norm of the weight vector. Diagonal loading was also used in Section 6.7 to improve the robustness of LCMV and LCMP to mismatch.

In this section, we revisit the quadratic constraint problem and solve for the optimum loading as a function of T_o. First, the derivation in Section 6.6.4 is modified to include linear constraints. Then, a LCMV (or LCMP) beamformer is designed that minimizes the output noise variance subject to the constraint that[15]

$$\mathbf{w}^H\mathbf{w} = T \leq T_o. \tag{6.533}$$

The constraints T_o can only be set within a certain range of values. This range will be discussed shortly.

The LCMP optimization problem is:

$$\text{Minimize} \qquad P_o = \mathbf{w}^H\mathbf{S_x}\mathbf{w}, \tag{6.534}$$

subject to the linear constraints,

$$\mathbf{C}^H\mathbf{w} = \mathbf{g}, \tag{6.535}$$

and the quadratic constraint,

$$\mathbf{w}^H\mathbf{w} \leq T_o. \tag{6.536}$$

Note that, when the signal is mismatched, minimizing P_o in (6.534) is not equivalent to maximizing the array gain. To maximize the array gain we would need $\mathbf{S_n}$ (or an estimate of it). In our model we have assumed that only $\mathbf{S_x}$ or its estimate is available.

We first solve the problem for the equality constraint,

$$\mathbf{w}^H\mathbf{w} = T_o. \tag{6.537}$$

In order to explore the allowable range of T_o, we denote the optimum solution by $\tilde{\mathbf{w}}_o$. We write $\tilde{\mathbf{w}}_o$ as a generalized sidelobe canceller,

$$\begin{aligned}\tilde{\mathbf{w}}_o &= \mathbf{P_C}\tilde{\mathbf{w}}_o + \mathbf{P_C^{\perp}}\tilde{\mathbf{w}}_o \\ &= \mathbf{w}_q + \mathbf{P_C^{\perp}}\tilde{\mathbf{w}}_o.\end{aligned} \tag{6.538}$$

Therefore

$$\tilde{\mathbf{w}}_o^H\tilde{\mathbf{w}}_o = \mathbf{w}_q^H\mathbf{w}_q + \tilde{\mathbf{w}}_o^H\mathbf{P_C^{\perp}}\tilde{\mathbf{w}}_o \geq \mathbf{w}_q^H\mathbf{w}_q. \tag{6.539}$$

So,

$$\min\left\{\tilde{\mathbf{w}}_o^H\tilde{\mathbf{w}}_o\right\} = \mathbf{w}_q^H\mathbf{w}_q. \tag{6.540}$$

---

[15]The first approach to the design of optimum array with a robustness constraint was published in 1955 (Gilbert and Morgan [GM55]) and 1956 (Uzsoky and Solymár [US56]). The specific problem formulation used here was published by Cox et al. [CZO87] and Owsley [Ows73].

Therefore

$$T_o \geq \mathbf{w}_q^H \mathbf{w}_q = \mathbf{g}^H \left(\mathbf{C}^H\mathbf{C}\right)^{-1} \mathbf{C}^H\mathbf{C} \left(\mathbf{C}^H\mathbf{C}\right)^{-1} \mathbf{g}, \tag{6.541}$$

or

$$T_o \geq \mathbf{g}^H \left(\mathbf{C}^H\mathbf{C}\right)^{-1} \mathbf{g}. \tag{6.542}$$

So we cannot set the constraint to be less than $\mathbf{g}^H \left(\mathbf{C}^H\mathbf{C}\right)^{-1} \mathbf{g}$.

We now solve the optimization problem. Initially, we use an equality constraint. The function to minimize is

$$\begin{aligned} F \triangleq\; & \mathbf{w}^H \mathbf{S}_{\mathbf{x}} \mathbf{w} + \lambda_1 \left[\mathbf{w}^H\mathbf{w} - T_o\right] \\ & + \boldsymbol{\lambda}_2 \left[\mathbf{C}^H\mathbf{w} - \mathbf{g}\right] + \left[\mathbf{w}^H\mathbf{C} - \mathbf{g}^H\right] \boldsymbol{\lambda}_2^H. \end{aligned} \tag{6.543}$$

Taking the gradient with respect to $\mathbf{w}$ and setting the result to zero gives

$$\mathbf{w}^H \mathbf{S}_{\mathbf{x}} + \lambda_1 \mathbf{w}^H + \boldsymbol{\lambda}_2 \mathbf{C}^H = \mathbf{0}, \tag{6.544}$$

or

$$\mathbf{w}^H = -\boldsymbol{\lambda}_2 \mathbf{C}^H \left[\lambda_1 \mathbf{I} + \mathbf{S}_{\mathbf{x}}\right]^{-1}. \tag{6.545}$$

Solving for $\boldsymbol{\lambda}_2$ gives

$$\tilde{\mathbf{w}}_o^H = \mathbf{g}^H \left[\mathbf{C}^H \left[\mathbf{S}_{\mathbf{x}} + \lambda_1 \mathbf{I}\right]^{-1} \mathbf{C}\right]^{-1} \mathbf{C}^H \left[\mathbf{S}_{\mathbf{x}} + \lambda_1 \mathbf{I}\right]^{-1}. \tag{6.546}$$

To emphasize that we are really using an inequality constraint, we change notation, letting

$$\tilde{\mathbf{w}}_o = \tilde{\mathbf{w}}_{oqc} \tag{6.547}$$

and

$$\beta = \lambda_1. \tag{6.548}$$

The solution in (6.546) can be written as

$$\tilde{\mathbf{w}}_{oqc} = \left(\mathbf{S}_{\mathbf{x}} + \beta \mathbf{I}\right)^{-1} \mathbf{C} \left[\mathbf{C}^H \left(\mathbf{S}_{\mathbf{x}} + \beta \mathbf{I}\right)^{-1} \mathbf{C}\right]^{-1} \mathbf{g}. \tag{6.549}$$

We can solve for β numerically using (6.549). As in Section 6.6.4, the effect of a positive β is to add a diagonal term to the actual $\mathbf{S}_{\mathbf{x}}$ in the expression defining the optimum beamformer. This technique is the diagonal loading used in Sections 6.6 and 6.7. In those cases, we picked a value of β based on the *SNR* and *INR*. We now want to find the optimum value of β. The beamformer is implemented in a generalized sidelobe canceller configuration.

Assume

$$\mathbf{B}^H\mathbf{B} = \mathbf{I}, \tag{6.550}$$

and write the LCMP as

$$\tilde{\mathbf{w}}_{oqc} = \tilde{\mathbf{S}}_{\mathbf{x}}^{-1}\mathbf{C}\left(\mathbf{C}^H\tilde{\mathbf{S}}_{\mathbf{x}}^{-1}\mathbf{C}\right)^{-1}\mathbf{g}, \tag{6.551}$$

where

$$\tilde{\mathbf{S}}_{\mathbf{x}} = \mathbf{S}_{\mathbf{x}} + \beta\mathbf{I}. \tag{6.552}$$

Then

$$\tilde{\mathbf{w}}_{gsc} = \mathbf{w}_q - \mathbf{B}\tilde{\mathbf{w}}_a \tag{6.553}$$

and

$$\begin{aligned}\tilde{\mathbf{w}}_a &= \left[\mathbf{B}^H\tilde{\mathbf{S}}_{\mathbf{x}}\mathbf{B}\right]^{-1}\mathbf{B}^H\tilde{\mathbf{S}}_{\mathbf{x}}\mathbf{w}_q \\ &= \left[\mathbf{B}^H\mathbf{S}_{\mathbf{x}}\mathbf{B} + \beta\mathbf{B}^H\mathbf{B}\right]^{-1}\left(\mathbf{B}^H\mathbf{S}_{\mathbf{x}}\mathbf{w}_q + \beta\mathbf{B}^H\mathbf{w}_q\right),\end{aligned} \tag{6.554}$$

which reduces to

$$\boxed{\tilde{\mathbf{w}}_a = \left[\mathbf{B}^H\mathbf{S}_{\mathbf{x}}\mathbf{B} + \beta\mathbf{I}\right]^{-1}\mathbf{B}^H\mathbf{S}_{\mathbf{x}}\mathbf{w}_q.} \tag{6.555}$$

Now find β to satisfy

$$\begin{aligned}T_o &\geq \tilde{\mathbf{w}}_{gsc}^H\tilde{\mathbf{w}}_{gsc} \\ &= \mathbf{w}_q^H\mathbf{w}_q + \tilde{\mathbf{w}}_a^H\mathbf{B}^H\mathbf{B}\tilde{\mathbf{w}}_a \\ &= \mathbf{w}_q^H\mathbf{w}_q + \tilde{\mathbf{w}}_a^H\tilde{\mathbf{w}}_a,\end{aligned} \tag{6.556}$$

or

$$\boxed{T_o - \mathbf{w}_q^H\mathbf{w}_q \geq \tilde{\mathbf{w}}_a^H\tilde{\mathbf{w}}_a.} \tag{6.557}$$

Replacing the left side by α^2 gives

$$\boxed{\alpha^2 \geq \tilde{\mathbf{w}}_a^H\tilde{\mathbf{w}}_a,} \tag{6.558}$$

where $\alpha^2 \geq 0$.

When $\beta = 0$, the standard LCMP solution is obtained. When $\beta \to \infty$, the quiescent beamformer is obtained.

As β increases, the norm of $\tilde{\mathbf{w}}_a$ decreases. To see this, we first rewrite (6.555) as

$$\tilde{\mathbf{w}}_a = \left[\mathbf{S}_{\mathbf{z}} + \beta\mathbf{I}\right]^{-1}\mathbf{p}_{\mathbf{z}}, \tag{6.559}$$

where

$$\mathbf{S_z} = \mathbf{B}^H \mathbf{S_x} \mathbf{B}, \tag{6.560}$$

is the spectral matrix of $\mathbf{z}$, the output from the blocking matrix, and

$$\mathbf{p_z} = \mathbf{B}^H \mathbf{S_x} \mathbf{w}_q, \tag{6.561}$$

is the cross-spectral term.

The squared norm of the weight vector $\tilde{\mathbf{w}}_a$ can be written as

$$\tilde{\mathbf{w}}_a^H \tilde{\mathbf{w}}_a = \mathbf{p_z}^H (\mathbf{S_z} + \beta \mathbf{I})^{-2} \mathbf{p_z}. \tag{6.562}$$

Taking the derivative with respect to β gives

$$\frac{d}{d\beta} \tilde{\mathbf{w}}_a^H \tilde{\mathbf{w}}_a = -2\mathbf{p_z}^H (\mathbf{S_z} + \beta \mathbf{I})^{-3} \mathbf{p_z}. \tag{6.563}$$

Since the diagonally loaded data matrix $[\mathbf{S_z} + \beta \mathbf{I}]$ is positive definite when $\beta \geq 0$, the derivative value in (6.563) is negative. Therefore, the weight vector norm is decreasing in β for $\beta > 0$.

The amount of diagonal loading is adjusted to satisfy the quadratic constraint, however, the optimal loading level cannot be directly expressed as a function of the constraint and has to be solved for numerically (Tian et al. [TBV98]).

Assuming $\tilde{\mathbf{w}}_a^H \tilde{\mathbf{w}}_a > \alpha^2$, when β=0, an iterative procedure can be used to solve the constraint equation,

$$\tilde{\mathbf{w}}_a^H \tilde{\mathbf{w}}_a = \mathbf{w}_q^H \mathbf{S_x} \mathbf{B} \left(\mathbf{B}^H \mathbf{S_x} \mathbf{B} + \beta \mathbf{I}\right)^{-2} \mathbf{B}^H \mathbf{S_x} \mathbf{w}_q = \alpha^2. \tag{6.564}$$

Starting with $\beta^{(0)} = 0$, let

$$\beta^{(1)} = \beta^{(0)} + \Delta\beta, \tag{6.565}$$

and

$$\beta^{(k)} = \beta^{(k-1)} + \Delta\beta. \tag{6.566}$$

At each step, compute $\tilde{\mathbf{w}}_a^H \tilde{\mathbf{w}}_a$. Continue the iteration until (6.564) is satisfied.

We consider two examples to illustrate typical behavior.

Example 6.10.1

Consider a standard 10-element linear array. The array is steered to broadside. The interference consists of two uncorrelated equal-power interferers each with an $INR = 20$ dB. The interferers are located at $u = 0.29$ and $u = 0.45$. The signal arrives u_a with an $SNR = 10$ dB. An MPDR-QC beamformer is implemented with $T_o = 0.12$. Note that, whenever a

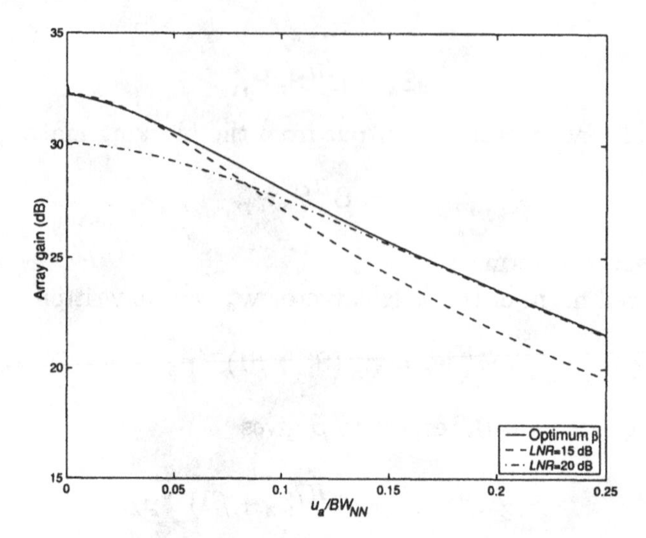

Figure 6.88 MPDR-QC beamformer with $T_o = 0.12$; MPDR-DL beamformer with *LNR* = 15 dB and 20 dB; *SNR* =10 dB, *INR* =20 dB, u_I=0.29 and 0.45; array gain versus u_a/BW_{NN}, 200 trials.

distortionless constraint is included, $T_o \geq 1/N = 0.1$. An MPDR-DL beamformer is also implemented with fixed loading of 15 dB and 20 dB.

In Figure 6.88, the array gain for the three beamformers is plotted versus u_a/BW_{NN}. In Figure 6.89, the value of β is plotted versus u_a/BW_{NN}. In Figure 6.90, the value of $\|\mathbf{w}\|^2$ is plotted versus u_a/BW_{NN}.

In terms of array gain, there is only a moderate advantage in using the optimum β. However, the fixed loading levels were chosen based on knowledge of the *SNR* and *INR*. The constraint T_o was chosen based on a desired white noise gain level and the loading was derived to meet the constraint. This approach is more flexible and will lead to a useful adaptive loading technique.

Example 6.10.2 (continuation)

Consider the same model as in Example 6.10.1. In this example, there is no signal mismatch, but there are array perturbations. The signal arrives at $u_s = 0$. The array lies along the x-axis and the sensor positions are perturbed randomly in the x and y directions. The perturbations of each sensor in the x and y directions are statistically independent zero-mean Gaussian random variables with standard derivation σ_p. The *SNR* is 10 dB and the *INR* of each interferer is 20 dB.

In Figure 6.91, the expected value of the array gain versus $\frac{\sigma_p}{\lambda}$ is plotted for two cases:
(i) MPDR-QC with $T_o = 0.12$.
(ii) MPDR-DL with *LNR* = 15 dB and 20 dB.
In Figure 6.92, the value of β is plotted versus σ_p/λ.

The use of the optimum β provides a 1–2 dB improvement over the fixed loading performance.

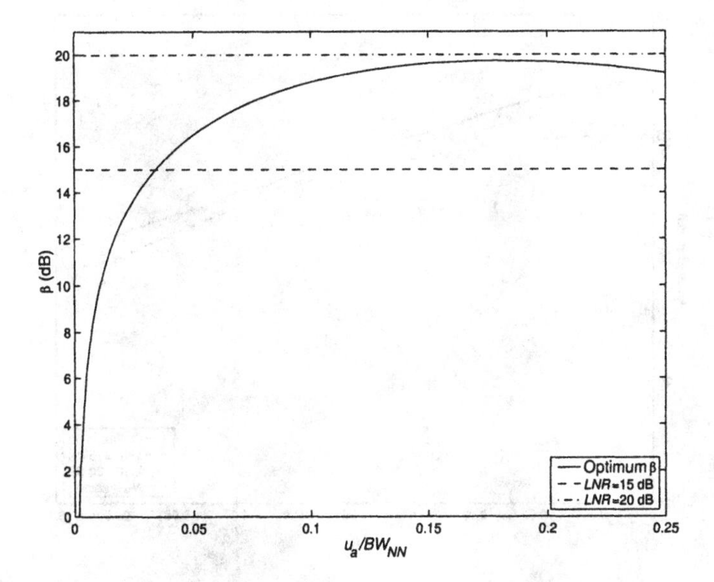

Figure 6.89 β versus u_a/BW_{NN}.

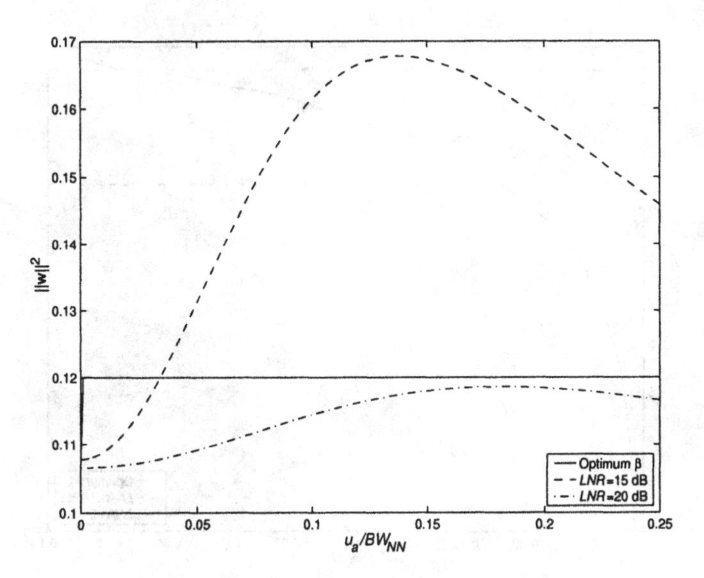

Figure 6.90 $\|\mathbf{w}\|^2$ versus u_a/BW_{NN}.

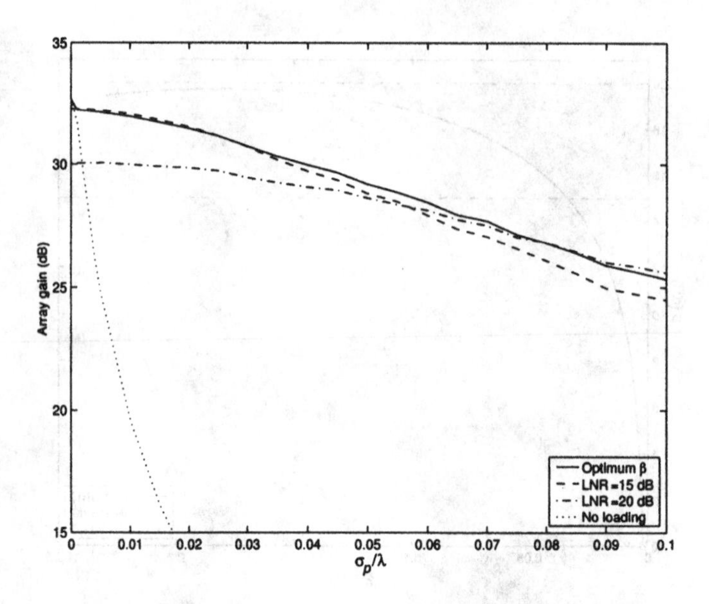

Figure 6.91 Array perturbations: MPDR-QC beamformer with $T_o = 0.12$ and MPDR-DL beamformer with *LNR* = 15 dB and 20 dB, average array gain versus σ_p/λ, 100 trials.

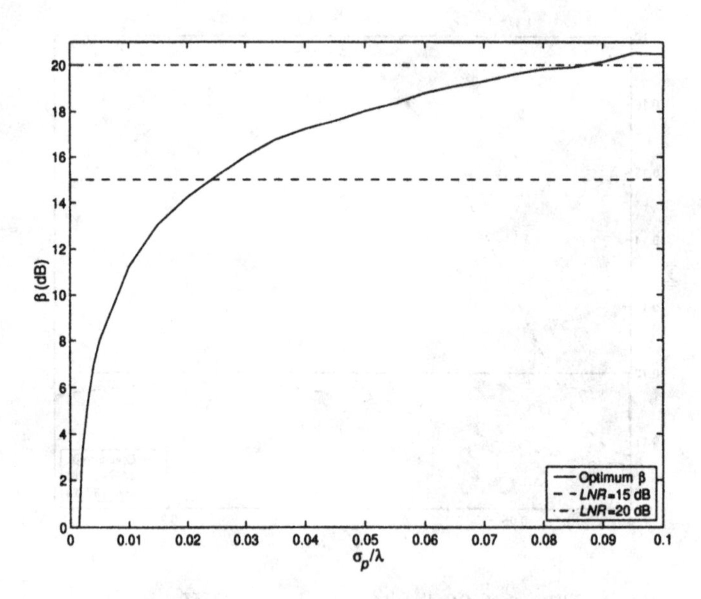

Figure 6.92 β versus σ_p/λ, 100 trials.

In this section, quadratic constraints were introduced as a technique to make the optimum algorithm more robust in the presence of signal mismatch or array perturbations. It also provides main-lobe protection when a strong interferer moves into the main-lobe region. In Chapter 7, we find that it also provides robust behavior when the spatial spectral matrix is estimated from a limited amount of data.

An iterative technique was used to find the value of β required to satisfy the constraint. This technique has an advantage over fixed loading, because it only employs enough loading to meet the constraint. In the examples in this section, an appropriate level of fixed loading was chosen using knowledge of the signal and interference environment. In practice, this may be an unrealistic assumption. However, a suitable value of T_o can be chosen based on the entire range of signal and interference scenarios that are expected.

Quadratic constraints and diagonal loading will be revisited in Chapter 7. They will be a necessary component of most practical algorithms. In the next section, an algorithm is developed that turns out to be a generalization of diagonal loading.

6.11 Soft-constraint Beamformers

In this section, a quadratic constraint technique is developed that is referred to in the literature as a **soft-constraint beamformer**. This type of constraint was introduced by Er and Cantoni [EC85] and developed in detail by Van Veen [VV91]. Er and Cantoni consider the narrowband case and use a planar array as an example. Van Veen considers a broadband case using a tapped delay-line implementation. To simplify the notation, we consider a narrowband standard linear array in the text and develop the generalizations in the problems. The initial discussion is the same as Section 6.7.1.5.

The first step is to specify a desired beam pattern $B_d(\psi)$. We want the beam pattern to approximate it over some region in ψ-space. The desired beam pattern is

$$B_d(\psi) = \mathbf{w}_d^H \, \mathbf{v}(\psi). \tag{6.567}$$

In some cases $B_d(\psi)$ might not be realizable with an N-element array. In that case, $\mathbf{w}_d$ is the weight vector that minimizes the squared error (see Section 3.7.2).

We define a weighted least squares response error between the desired beam pattern and the actual beam pattern over a region in ψ-space as

$$e^2 = \int_{\Psi_0} f(\psi) \left| \mathbf{w}_d^H \, \mathbf{v}(\psi) - \mathbf{w}^H \, \mathbf{v}(\psi) \right|^2 d\psi, \tag{6.568}$$

where Ψ_0 is the region of interest and $f(\psi)$ is a weighting function that allows for emphasis of different regions in ψ-space. The error weight vector is defined as

$$\mathbf{w}_p = \mathbf{w}_d - \mathbf{w}. \tag{6.569}$$

Using (6.569) in (6.568) gives

$$e^2 = \mathbf{w}_p^H \mathbf{Q} \mathbf{w}_p, \tag{6.570}$$

where

$$\mathbf{Q} \triangleq \int_{\Psi_0} f(\psi)\, \mathbf{v}(\psi)\, \mathbf{v}^H(\psi)\, d\psi. \tag{6.571}$$

For the special case of a standard linear array and a constant $f(\psi)$,

$$[\mathbf{Q}]_{mn} = \int_{\Psi_0} e^{j\psi(m-n)}\, d\psi. \tag{6.572}$$

If Ψ_0 is a symmetric interval $[-\psi_0, \psi_0]$, then (6.572) reduces to

$$[\mathbf{Q}]_{mn} = \int_{-\psi_0}^{\psi_0} e^{j\psi(m-n)}\, d\psi = \frac{2\sin(\psi_0(m-n))}{(m-n)}. \tag{6.573}$$

Note that, if $\psi_0 = \pi$, then we are attempting to approximate the desired beam pattern over the entire visible region and $\mathbf{Q}$ is a diagonal matrix. As ψ_0 decreases, the off-diagonal terms increase.

There are several approaches to introducing the constraint. The first ([EC85] or [VV91]) is to minimize the output P_o,

$$P_o = \mathbf{w}^H \mathbf{S}_\mathbf{x} \mathbf{w}, \tag{6.574}$$

subject to the constraint

$$e^2 \leq e_0^2, \tag{6.575}$$

where the value of e_0^2 depends on the application.

Using a Lagrange multiplier, we define

$$F = (\mathbf{w}_d^H - \mathbf{w}_p^H)\, \mathbf{S}_\mathbf{x} (\mathbf{w}_d - \mathbf{w}_p) + \lambda(\mathbf{w}_p^H \mathbf{Q} \mathbf{w}_p - e_0^2). \tag{6.576}$$

Minimizing gives

$$\boxed{\widehat{\mathbf{w}}_p = [\mathbf{S}_\mathbf{x} + \lambda \mathbf{Q}]^{-1} \mathbf{S}_\mathbf{x} \mathbf{w}_d,} \tag{6.577}$$

and λ is chosen as large as possible subject to the constraint,

$$\boxed{\widehat{\mathbf{w}}_p^H \mathbf{Q} \widehat{\mathbf{w}}_p \leq e_0^2.} \tag{6.578}$$

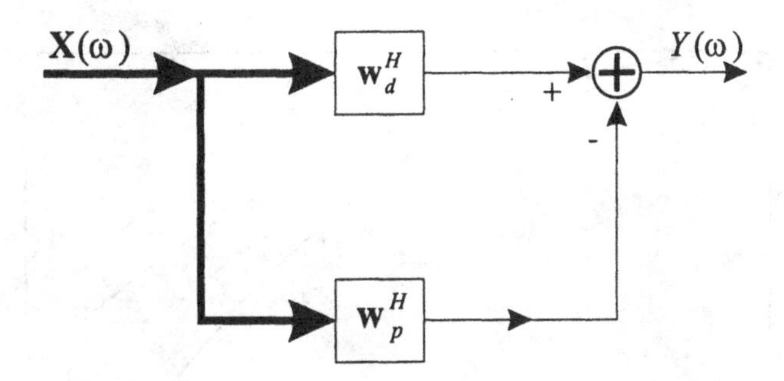

Figure 6.93 Partitioned processor.

We refer to the resulting beamformer as the MPSC beamformer. Van Veen [VV91] refers to it as soft-constrained minimum variance beamformer (SCMV). Note that the result in (6.577) corresponds to a partitioned processor as shown in Figure 6.93. This is the same partitioned processor that we saw in Figure 6.42. We see that MPSC is a generalization of diagonal loading that provides freedom to emphasize different areas of the beam pattern.

Just as in the diagonal loading case, we must solve for λ as a function of e_0^2 using numerical procedures. As $e_0^2 \to \infty$, $\lambda \to 0$, and we have an unconstrained beamformer. In this case, $\mathbf{w}_p \to \mathbf{w}_d$, and the output goes to zero. This result happens because of the absence of a distortionless constraint. As $e_0^2 \to 0$, $\mathbf{w}_p \to 0$, and the beamformer is $\mathbf{w}_d^H$.

To obtain useful results, e_0^2 must be small enough to avoid a zero output, but large enough to allow some adaptation. It is convenient to rewrite e_0 as

$$e_0^2 = \alpha\, \mathbf{w}_d^H \mathbf{Q}\, \mathbf{w}_d, \quad 0 \le \alpha \le 1. \tag{6.579}$$

If $\alpha = 1$, $\widehat{\mathbf{w}}_p = \mathbf{w}_d$ satisfies (6.577) and (6.578) with $\lambda = 0$, so we do not need to consider larger values of e_0^2. To avoid notational confusion between λ and eigenvalues, we replace λ with β and rewrite (6.577) as

$$\boxed{\widehat{\mathbf{w}}_{mpsc,p} = \left[\mathbf{S}_{\mathbf{x}} + \beta\mathbf{Q}\right]^{-1} \mathbf{S}_{\mathbf{x}}\, \mathbf{w}_d,} \tag{6.580}$$

subject to the constraint

$$\boxed{\widehat{\mathbf{w}}_{mpsc,p}^H\, \mathbf{Q}\, \widehat{\mathbf{w}}_{mpsc,p} \le \alpha\, \widehat{\mathbf{w}}_d^H\, \mathbf{Q}\, \mathbf{w}_d, \quad 0 \le \alpha \le 1.} \tag{6.581}$$

The parameter α plays a similar role to T_o in the diagonal loading model. As an alternative to solving for β as a function of α, we can choose β based on expected scenarios. This is analogous to fixed diagonal loading.

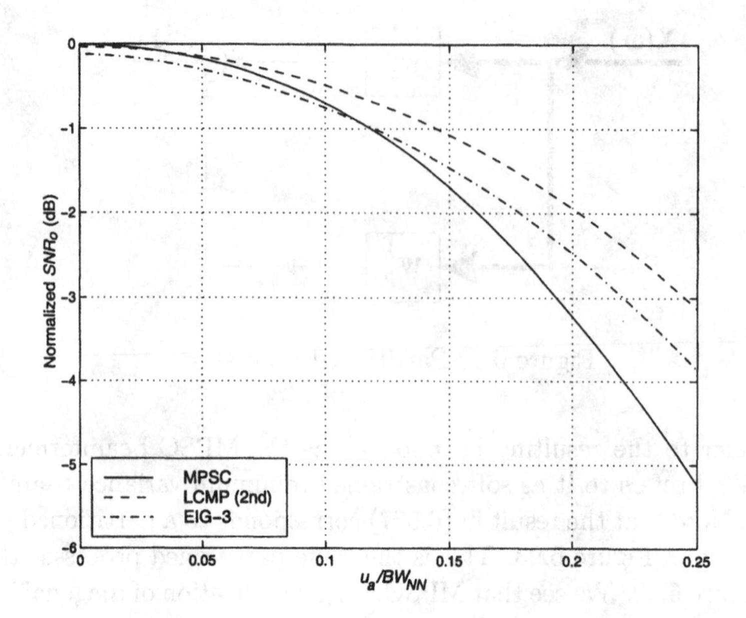

Figure 6.94: MPSC beamformer: normalized SNR_o versus u_a/BW_{NN}, $\mathbf{w}_d = \mathbf{1}/N$, $\psi_0 = 0.1\pi$, $\alpha = 0.001$, $u_I = 0.3$ and 0.5, $INR = 20$ dB, $SNR = 0$ dB.

We consider an example to illustrate typical performance of soft constraint beamformers.

Example 6.11.1

Consider a standard 10-element linear array. The desired beam pattern corresponds to a conventional beam and $\psi_0 = 0.1\pi$. There are two uncorrelated plane-wave interferers at $u = 0.3$ and 0.5 with an INR =20 dB each. The performance of the MPSC beamformer specified by (6.580) with $\alpha = 0.001$ is analyzed.

For comparison, the LCMP beamformer using three eigenvector constraints (Example 6.7.11), and an LCMP beamformer with second derivative constraints (Example 6.7.10) is also analyzed.

In Figure 6.94, the normalized SNR_o is plotted versus u_a/BW_{NN} for an $SNR = 0$ dB. In Figure 6.95, the normalized SNR_o is plotted for an $SNR = 10$ dB. Note that the vertical scale is different in the two figures.

For the $SNR = 0$-dB case, the eigenvector-constraint beamformer is slightly worse for $u_a/BW_{NN} \leq 0.1$ and better for larger u_a. The LCMP beamformer is better over the entire range, but the difference is small for $u_a/BW_{NN} \leq 0.1$.

For the $SNR = 10$-dB case, there is a wider difference in performance. The eigenvector-constraint algorithm is slightly worse for small u_a and slightly better for larger u_a. The LCMP algorithm drops off for u_a/BW_{NN}, but this could be improved by diagonal loading.

Small values of α are used so the soft constraint is very tight. As α is decreased, the MPSC beamformer acts very much like an eigenvector-constraint algorithm that uses all of the principal eigenvalues. Thus, as α decreases, effective degrees of freedom are being lost.

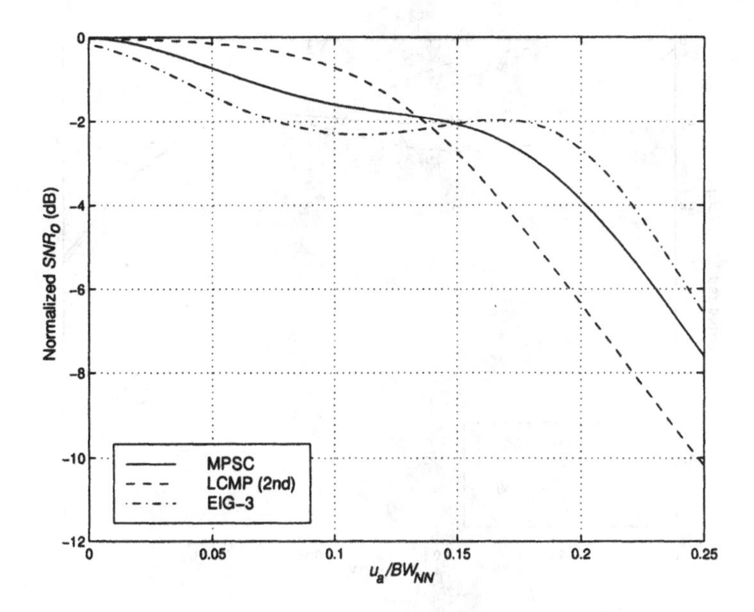

Figure 6.95 MPSC beamformer: normalized SNR_o versus u_a/BW_{NN}, $\mathbf{w}_d = \mathbf{1}/N$, $\psi_0 = 0.1\pi$, $\alpha = 0.001$, $u_I = 0.3$ and 0.5, $INR = 20$ dB, $SNR = 10$ dB.

In order to improve the soft-constraint performance at $u_a = 0$, a linear distortionless constraint can be added. In the general case, a set of linear constraints defined by $\mathbf{C}$ and $\mathbf{g}$ can be added. If we try to implement the beamformer in a direct form, there are two Lagrange multipliers and the solution is complicated. However, a GSC solution is straightforward.

The GSC configuration in Figure 6.46 is used. We minimize

$$[\mathbf{w}_q - \mathbf{B}\mathbf{w}_a]^H \mathbf{S}_\mathbf{x} [\mathbf{w}_q - \mathbf{B}\mathbf{w}_a], \tag{6.582}$$

subject to the constraint

$$[\mathbf{w}_q - \mathbf{B}\mathbf{w}_a - \mathbf{w}_d]^H \mathbf{Q} [\mathbf{w}_q - \mathbf{B}\mathbf{w}_a - \mathbf{w}_d] \leq e_0^2 = \alpha \mathbf{w}_d^H \mathbf{Q} \mathbf{w}_d. \tag{6.583}$$

The constraint in (6.583) can also be written as

$$[(\mathbf{w}_q - \mathbf{w}_d) - \mathbf{B}\mathbf{w}_a]^H \mathbf{Q} [(\mathbf{w}_q - \mathbf{w}_d) - \mathbf{B}\mathbf{w}_a] \leq e_0^2. \tag{6.584}$$

Minimizing using a Lagrange multiplier gives

$$\hat{\mathbf{w}}_a = \left[\mathbf{B}^H (\mathbf{S}_\mathbf{x} + \beta\mathbf{Q})\mathbf{B}\right]^{-1} \mathbf{B}^H \left[\mathbf{S}_\mathbf{x}\mathbf{w}_q + \beta\mathbf{Q}(\mathbf{w}_q - \mathbf{w}_d)\right], \tag{6.585}$$

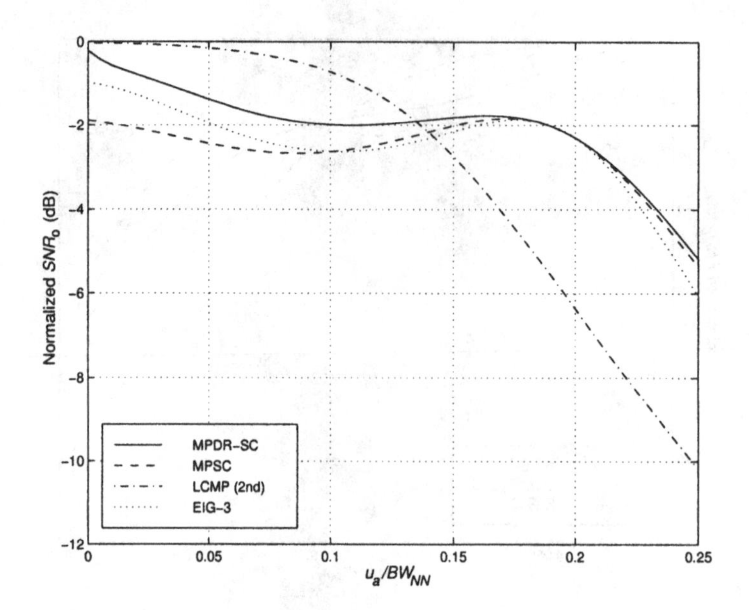

Figure 6.96 MPSC beamformer: normalized SNR_o versus u_a/BW_{NN}, desired pattern is Dolph-Chebychev (−40-dB SLL), $u_I = 0.3$ and 0.5 (20-dB), distortionless constraint plus soft constraint with $\alpha = 10^{-3}$.

where β is chosen to satisfy (6.584). To find β, an iterative procedure similar to the procedure in (6.564)–(6.566) is used.

We consider a simple example to illustrate the behavior.

Example 6.11.2 (continuation)

Consider the same model as in Example 6.11.1 except the desired beam pattern is Dolph-Chebychev with −40-dB SLL. A single distortionless constraint is added. In Figure 6.96, the normalized SNR_o versus u_a/BW_{NN} for four beamformers is plotted. The $SNR =$ 10 dB and $\alpha = 10^{-3}$. As expected, the distortionless constraint improves the performance for small u_a without degrading the performance for large u_a. In Figure 6.97, α is decreased to 10^{-4}. This decrease degrades the performances of MPSC and MPDR-SC for small u_a, but improves them for larger u_a.

The LCMP-SC beamformer provides another technique for improving the robustness of an LCMP beamformer. In some signal, interference, and mismatch scenarios it can provide improved performance over other techniques.

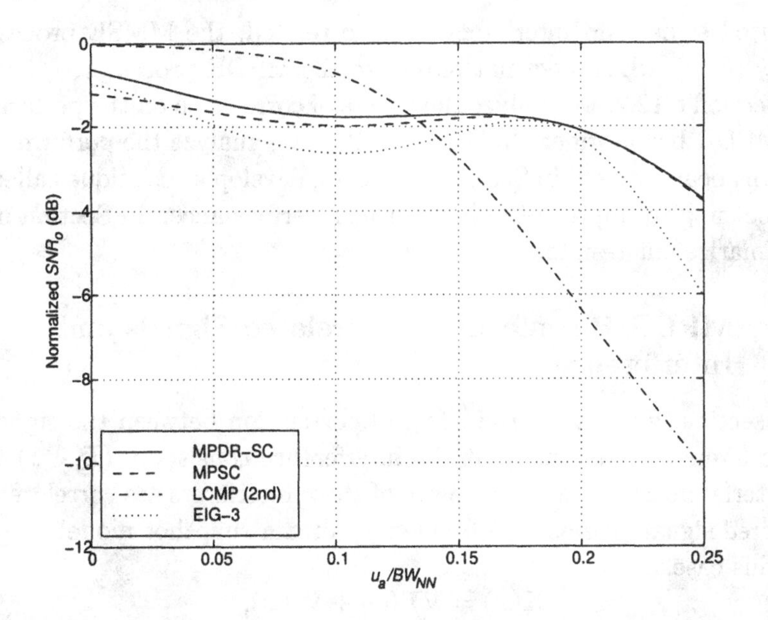

Figure 6.97 MPDR-SC beamformer: normalized SNR_o versus u_a/BW_{NN}, desired pattern is Dolph-Chebychev (−40-dB SLL), $u_I = 0.3$ and 0.5 (20-dB) distortionless constraint plus soft constraint with $\alpha = 10^{-4}$.

6.12 Beamforming for Correlated Signal and Interferences

6.12.1 Introduction

In our discussion up to this point, we have assumed that the desired signal and the interference were uncorrelated. In many applications, the desired signal and interfering signals are correlated ($|\rho| \neq 0$) or coherent ($|\rho| = 1$). This situation can occur in a multipath environment or in a "smart" jamming scenario (i.e., the jamming signal is derived from the transmitted signal).

We consider two approaches to the problem. In the first approach, we implement an MPDR or LCMP beamformer and analyze the effects of correlation between the desired signal and the interference. We then develop a pre-processing technique called **spatial smoothing** that reduces the correlation. The output of this pre-processing step is the input to the MPDR or LCMP beamformer.

In the second approach, we implement a MMSE beamformer. When the desired signal and the interference are uncorrelated, the MMSE and the MPDR processor have the same matrix processing (6.44). However, when the desired signal and interference are correlated, the MMSE processor is given by (6.42), which does not reduce to the MPDR processor.

In Section 6.12.2, we analyze the effects of correlation on the performance on an MPDR beamformer. In Section 6.12.3, we analyze the performance of an MMSE beamformer. In Section 6.12.4, we develop a technique called spatial smoothing to improve the beamformer performance. In Section 6.12.5, we summarize our results.

6.12.2 MPDR Beamformer: Correlated Signals and Interference

In this section, we analyze the effect of correlation between the signal and the interferers. We assume that the interference consists of $(D-1)$ plane-wave interferers and that one or more of these interferers are correlated with the desired signal. We use the frequency-domain snapshot model.

In this case,

$$\mathbf{X}(\omega) = \mathbf{V}\mathbf{F}(\omega) + \mathbf{W}(\omega), \tag{6.586}$$

and the vector $\mathbf{F}(\omega)$ is a D-dimensional zero-mean random vector that contains both the desired signal and the interfering signals,

$$\mathbf{S}_{\mathbf{x}} = \mathbf{V}\mathbf{S}_{\mathbf{f}}\mathbf{V}^H + \sigma_w^2\mathbf{I}. \tag{6.587}$$

The source spectral matrix, $\mathbf{S}_{\mathbf{f}}$, is not diagonal. The $N \times D$ matrix, $\mathbf{V}$, is the array manifold matrix. The additive spatially white noise $\mathbf{W}(\omega)$ is uncorrelated with the signal and interference. If there is a single interfering signal, $\mathbf{S}_{\mathbf{f}}$ can be written as

$$\mathbf{S}_{\mathbf{f}} = \begin{bmatrix} \sigma_s^2 & \sigma_s\sigma_I\rho \\ \sigma_s\sigma_I\rho^* & \sigma_I^2 \end{bmatrix}. \tag{6.588}$$

For $|\rho| = 1$, the source covariance matrix is singular. We want to examine the behavior of the MPDR beamformer as a function of ρ. This problem (or models with similar beamformers) has been analyzed by a number of references. References include: Owsley [Ows74], [Ows80], Cantoni and Godara [CG80], Widrow et al. [WDGN82], [Ows85], Kesler et al. [KBK85], Paulraj and Kailath [PK85], Shahmirian and Kesler [SK85b], Shan and Kailath [SK85a], Su et al. [SSW86], Luthra [Lut86], Takao et al. [TKY86], Reddy et al. [RPK87a], [RPK87b], Paulraj et al. [PRK87], Bresler et al. [BRK88], Zoltowoski [Zol88], Godara [God90], Raghunath and Reddy [RR92b], [RR92a], and Tsai et al. [TYS95]. We utilize results from [TYS95]. In this section, we analyze the performance.

To understand the behavior, it is useful to write $\mathbf{X}(\omega)$ as

$$\mathbf{X}(\omega) = \sigma_s F_d(\omega)\mathbf{v}_s + \sigma_I \left[\rho^* F_d(\omega) + \sqrt{1-|\rho|^2} F_u(\omega)\right] \mathbf{v}_I + \mathbf{W}(\omega). \quad (6.589)$$

The form in (6.589) separates $F_I(\omega)$ into a component that is correlated with the desired signal $F_d(\omega)$ and the remaining uncorrelated term $F_u(\omega)$.

The MPDR beamformer $\mathbf{w}_{mpdr}^H$ uses a distortionless constraint,

$$\mathbf{w}^H \mathbf{v}_s = 1, \quad (6.590)$$

and minimizes the total output power P_o. Intuitively, we would anticipate that the optimum $\mathbf{w}$ would constrain the output from $\mathbf{v}_s$ as required by (6.590), but would add in enough of $F_d(\omega)$ arriving from $\mathbf{v}_I$ to partially cancel the desired signal. The output is

$$\begin{aligned} Y(\omega) &= \mathbf{w}^H \sigma_s F_d(\omega)\mathbf{v}_s + \mathbf{w}^H \sigma_I \left[\rho^* F_d(\omega)\right. \\ &\quad \left. + \sqrt{1-|\rho|^2} F_u(\omega)\right] \mathbf{v}_I + \mathbf{w}^H \mathbf{W}(\omega), \end{aligned} \quad (6.591)$$

or

$$\begin{aligned} Y(\omega) &= \left[\mathbf{w}^H \mathbf{v}_s \sigma_s + \mathbf{w}^H \mathbf{v}_I \sigma_I \rho^*\right] F_d(\omega) \\ &\quad + \mathbf{w}^H \mathbf{v}_I \left[\sigma_I \sqrt{1-|\rho|^2}\right] F_u(\omega) + \mathbf{w}^H \mathbf{W}(\omega) \\ &\triangleq Y_d(\omega) + Y_I(\omega) + Y_n(\omega), \end{aligned} \quad (6.592)$$

where the three terms are uncorrelated.

The output power in the desired signal is

$$\begin{aligned} P_{do} &= E\left[Y_d(\omega) Y_d^*(\omega)\right] \\ &= \left[\mathbf{w}^H \mathbf{v}_s \sigma_s + \mathbf{w}^H \mathbf{v}_I \sigma_I \rho^*\right] \left[\mathbf{v}_s^H \mathbf{w} \sigma_s + \mathbf{v}_I^H \mathbf{w} \sigma_I \rho\right] \\ &= \sigma_s^2 \left|\alpha_s + \alpha_I \frac{\sigma_I \rho^*}{\sigma_s}\right|^2, \end{aligned} \quad (6.593)$$

where

$$\alpha_s \triangleq \mathbf{w}^H \mathbf{v}_s, \quad (6.594)$$

and

$$\alpha_I \triangleq \mathbf{w}^H \mathbf{v}_I. \quad (6.595)$$

For an MPDR beamformer, $\alpha_s = 1$ and (6.593) reduces to

$$P_{do} = \sigma_s^2 \left| 1 + \alpha_I \frac{\sigma_I \rho^*}{\sigma_s} \right|^2. \tag{6.596}$$

The output power of the interference is

$$P_{Io} = E\left[Y_I(\omega) Y_I^H(\omega) \right] = |\alpha_I|^2 (1 - |\rho|^2) \sigma_I^2, \tag{6.597}$$

and the output power due to noise is

$$P_{no} = \sigma_w^2 \parallel \mathbf{w} \|^2. \tag{6.598}$$

The output SNR_o is

$$SNR_o = \frac{P_{do}}{P_{Io} + P_{no}}. \tag{6.599}$$

To evaluate the SNR_o, we use $\mathbf{w}^H$ from (6.14) in (6.593), (6.597), and (6.598). The result is (e.g., [TYS95])

$$SNR_o = \frac{A}{B}, \tag{6.600}$$

where

$$A = \left| \sigma_s \sigma_I^2 \Gamma \Delta + \sigma_w^2 \left(N \sigma_s + \rho^* N B_{dI}^* \sigma_I \right) \right|^2, \tag{6.601}$$

and

$$\begin{aligned} B \ = \ & \sigma_s^2 \sigma_I^4 \Gamma |\rho|^2 \Delta^2 + \sigma_I^2 \sigma_w^2 \Delta \left[N \sigma_s^2 |\rho|^2 \right. \\ & \left. + N \sigma_I^2 + |\rho|^2 \sigma_s \sigma_I N \left(\rho^* B_{dI}^* + \rho B_{dI} \right) + 2\sigma_w^2 \right] \\ & + N^2 \sigma_I^2 \sigma_w^4 |B_{dI}|^2 \Gamma + N \sigma_w^6. \end{aligned} \tag{6.602}$$

Parameters in (6.601) and (6.602) include:

$$\Gamma = 1 - |\rho|^2, \tag{6.603}$$

and

$$B_{dI} = \frac{1}{N} \mathbf{v}_I^H \mathbf{v}_s, \tag{6.604}$$

and

$$\Delta = N^2 \left(1 - |B_{dI}|^2 \right). \tag{6.605}$$

Several limiting cases are of interest:

(i) $\sigma_w^2 = 0$, $|\rho| = 1$. In this case, we use L'Hopital's rule on (6.600) to obtain $SNR_o = 0$.

(ii) High *SNR* and *INR* and a sidelobe interferer. Then (6.600) reduces to

$$SNR_o \simeq \frac{\Gamma^2 \sigma_s^2}{\Gamma |\rho|^2 \sigma_s^2 + \frac{\sigma_w^2}{N}}, \tag{6.606}$$

which indicates the importance of $|\rho|$.

(iii) For main-lobe interference, $|B_{dI}| \simeq 1$,

$$SNR_o \simeq \frac{\sigma_d^2 \left|1 + \rho^* \frac{\sigma_I}{\sigma_s}\right|^2}{\sigma_I^2 (1 - |\rho|^2) + \frac{\sigma_w^2}{N}}. \tag{6.607}$$

The numerator depends on ρ, and the interfering signal can have either a constructive or destructive effect.

We consider an example to illustrate the behavior.

Example 6.12.1

Consider a standard 10-element linear array. In Figure 6.98, the SNR_o versus $|\rho|$ for $u_I = 0.30$ is plotted. The phase, ϕ_ρ, equals 0. The *SNR* is 0 dB and the *INR*/*SNR* ratio is varied from 0.1 to 10.0. In Figure 6.99, the SNR_o for an $SNR = 10$ dB is plotted.

For $|\rho| > 0.5$, there is significant degradation in the performance. This result is typical for a correlated interferer arriving in the sidelobe region.

An alternative approach to the MPDR beamformer is an MMSE beamformer. When the signal and interference are uncorrelated the matrix processing is the same. However, for correlated signals and interference, they are different.

6.12.3 MMSE Beamformer: Correlated Signals and Interference

In this section, we analyze the performance of a MMSE beamformer for the correlated signal and interference model in (6.586)–(6.588).

From (6.42),

$$\mathbf{W}_{mmse}^H = \mathbf{S}_{d\mathbf{x}^H} \mathbf{S}_{\mathbf{x}}^{-1}. \tag{6.608}$$

For the case of a single interferer,

$$S_{d\mathbf{x}^H} = \left[\sigma_s^2 \mathbf{v}_s^H + \sigma_s \sigma_I \rho \mathbf{v}_I^H\right] \tag{6.609}$$

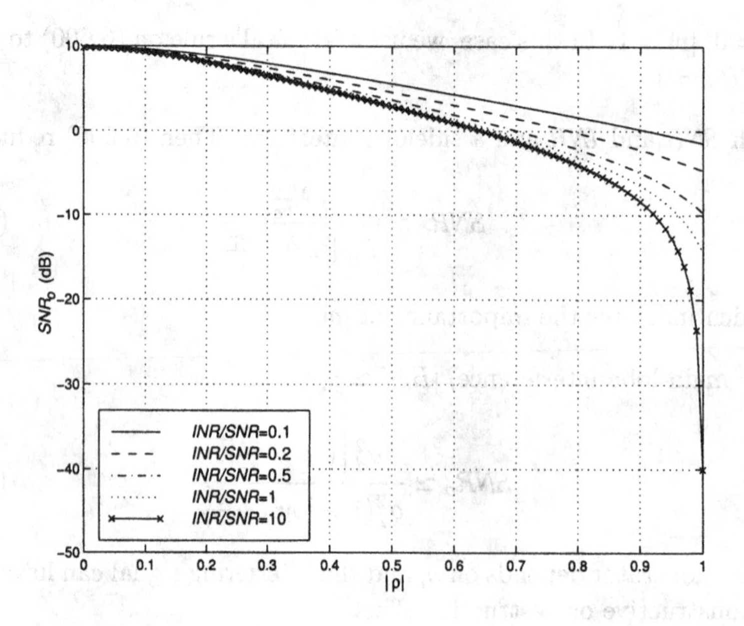

Figure 6.98 MPDR beamformer with correlated signal and interferer, SNR_o versus $|\rho|$; $u_s = 0$, $SNR = 0$ dB, single plane-wave interferer at $u_I = 0.3$, various INR/SNR ratios.

and

$$\mathbf{S}_{\mathbf{x}}^{-1} = \left[\sigma_s^2 \mathbf{v}_s \mathbf{v}_s^H + \sigma_s \sigma_I \rho \mathbf{v}_s \mathbf{v}_I^H + \sigma_s \sigma_I \rho^* \mathbf{v}_I \mathbf{v}_s^H + \sigma_I^2 \mathbf{v}_I \mathbf{v}_I^H + \sigma_\omega^2 \mathbf{I}\right]^{-1}. \quad (6.610)$$

The expressions in (6.593) and (6.597)–(6.599) are valid with

$$\mathbf{w}^H = \mathbf{w}_{mmse}^H. \quad (6.611)$$

Note that $\alpha_s \neq 1$ for $\mathbf{w}_{mmse}^H$.

Example 6.12.2 (continuation)

Consider the same model as in Example 6.12.1. The MMSE weight vector is specified by (6.608)–(6.610). Substituting into (6.593) and (6.597)–(6.599) gives the output SNR_o.

In Figure 6.100, the SNR_o is plotted versus $|\rho|$ for the case when $\rho = |\rho| \exp(j\pi/4)$ and $\sigma_s^2 = 0$ dB. The range $0.7 \leq |\rho| \leq 1.0$ is plotted. For $|\rho| \leq 0.7$, the curve is almost flat. As $|\rho|$ increases, the correlated component of this signal from $\mathbf{v}_I$ adds constructively. In Figure 6.101, the SNR_o is plotted for the case when $\rho = |\rho| \exp(j\pi/4)$ and $\sigma_s^2 = 10$ dB. The range $0.9 \leq |\rho| \leq 1.0$ is plotted. For $|\rho| \leq 0.9$, the curve is almost flat. Once again, the correlated component of this signal from $\mathbf{v}_I$ adds constructively.

In order to achieve this performance, the values of $\mathbf{S}_{d\mathbf{x}^H}$ and $\mathbf{S}_{\mathbf{x}}$ must be known or the beamformer must estimate them. Alternatively, the beamformer can set $\mathbf{v}_s = \mathbf{v}_m$ and estimate $\sigma_s^2, \sigma_I^2, \rho, \mathbf{v}_I$, and $\mathbf{S}_{\mathbf{x}}$. We find that

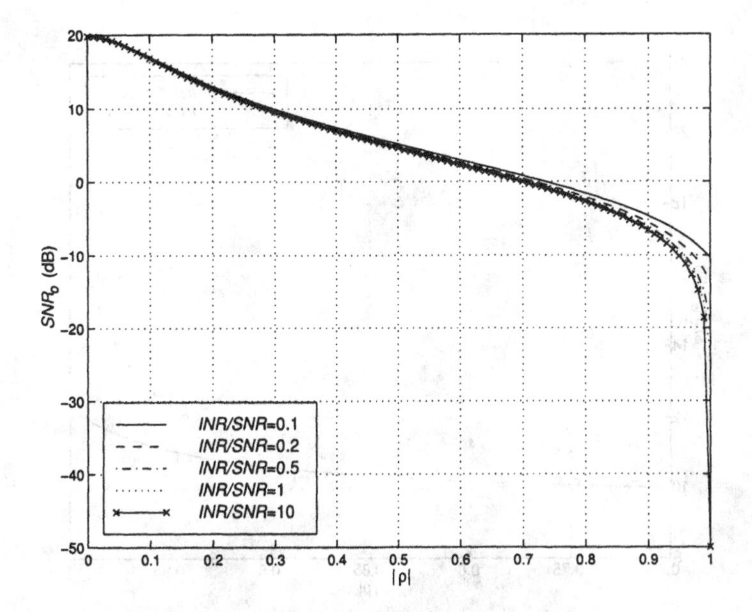

Figure 6.99 MPDR beamformer with correlated signal and interferer, SNR_o versus $|\rho|$; $u_s = 0$, $SNR = 10$ dB, single plane-wave interferer at $u_I = 0.3$, various INR/SNR ratios.

this estimation is more difficult than the estimation of $\mathbf{S_x}$ required for the MPDR beamformer.

One can think of the MPDR beamformer as a matched spatial processor where the beamformer is matched to $\mathbf{v}_m$. The MMSE beamformer is matched to the multipath environment.

6.12.4 Spatial Smoothing and Forward–Backward Averaging

In this section, we introduce a technique referred to as **spatial smoothing (SS)** to remove the singularity in the input signal and interference correlation matrix, which is caused by the coherence between the desired signal and the interference. The original use of spatial smoothing was by Evans et al. [EJS82] in the direction-finding problem. We discuss this application in Chapter 9. The modification to the beamforming problem was done by Shan and Kailath ([SK85b]).

Our discussion is an adaption of [EJS82] and [PK89] to the beamformer problem. If we do not incorporate forward–backward (FB), averaging it reduces to the result in [SK85b], which was analyzed in [RPK87a] and [RR92b].

Spatial smoothing requires a regular array geometry, such as a uniform

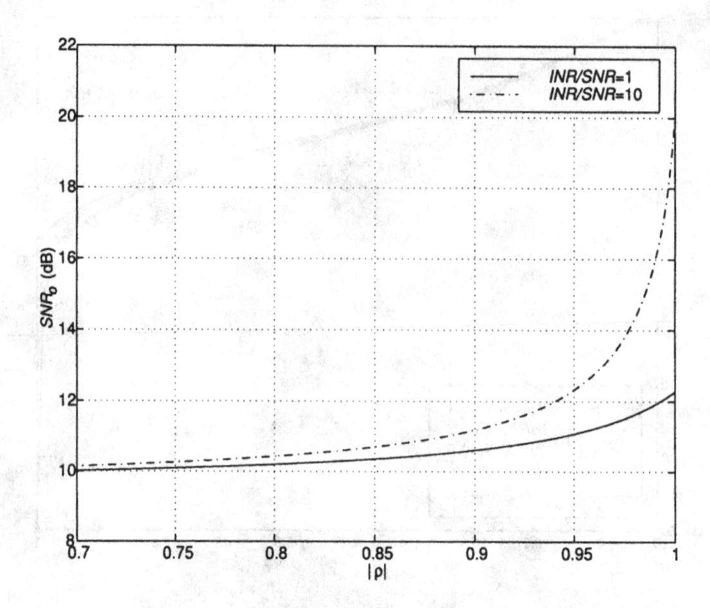

Figure 6.100 MMSE beamformer: SNR_o versus $|\rho|$; $u_a = 0$, $SNR = 0$ dB, single correlated plane-wave interferer at $u_I = 0.0866$, $\rho = |\rho|\exp(j\pi/4)$, $SNR = 0$ dB, various INR/SNR ratios.

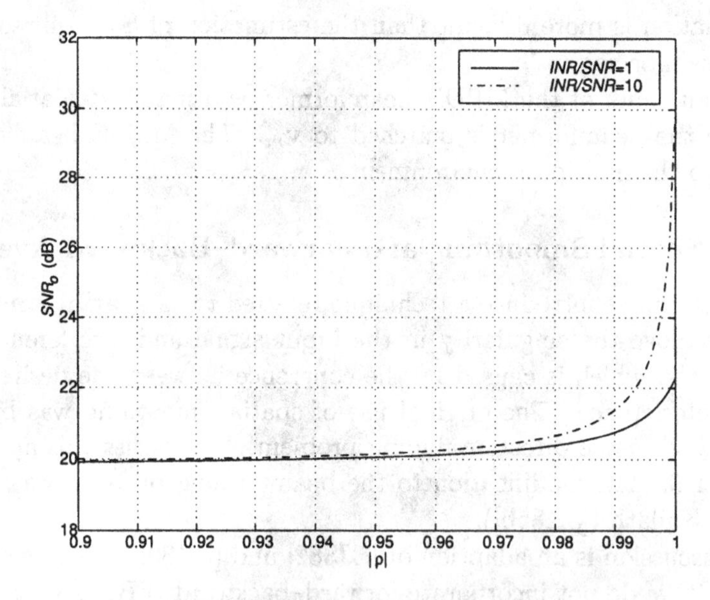

Figure 6.101 MMSE beamformer: SNR_o versus $|\rho|$; $u_a = 0$, $SNR = 0$ dB, single correlated plane-wave interferer at $u_I = 0.02$, $\rho = |\rho|\exp(j\pi/4)$.

linear array or a uniform rectangular array. We restrict our attention to a standard linear array in this section. We put the center of the array at the origin in order to eliminate some artifacts in FB smoothing. Because of this choice, the equations will be different from some of the above references.

We restrict our attention to a standard linear array. The received signal is

$$\mathbf{X}(\omega) = \mathbf{V}\mathbf{F}(\omega) + \mathbf{W}(\omega), \tag{6.612}$$

and the covariance matrix is

$$\mathbf{S_X} = \mathbf{V}\mathbf{S_f}\mathbf{V}^H + \sigma_w^2\mathbf{I}, \tag{6.613}$$

where $\mathbf{S_f}$ is the source-signal spectral matrix and is assumed to be singular. We assume that D, the number of signals, is known or has been estimated. We use a frequency-domain snapshot model.

The linear array of interest is shown in Figure 6.102(a). Note that the array elements are indexed from 1 to N. We construct a set of L subarrays of length $M \geq D+1$, as shown in Figure 6.102(b). Each subarray is shifted by one from the preceding subarray. The ith subarray has the ith element as its initial element. We use an M-element subarray centered at the origin as the reference subarray.

Let

$$\mathbf{V}_M = \left[\, \mathbf{v}_M(\psi_1) \;\vdots\; \mathbf{v}_M(\psi_2) \;\vdots\; \cdots \;\vdots\; \mathbf{v}_M(\psi_D) \,\right] \tag{6.614}$$

denote the set of steering vectors for the reference subarray of length M, where

$$\mathbf{v}_M(\psi) = \left[\, e^{-j\frac{M-1}{2}\psi} \quad e^{-j\frac{M-3}{2}\psi} \quad \cdots \quad e^{j\frac{M-1}{2}\psi} \,\right]. \tag{6.615}$$

We define

$$\mathbf{D} = \text{diag}\left[\, e^{j\psi_1} \;\vdots\; e^{j\psi_2} \;\vdots\; \cdots \;\vdots\; e^{j\psi_D} \,\right]. \tag{6.616}$$

Then the spectral matrix of the received signal at the ith subarray can be written as

$$\mathbf{S}_M^{(i)} = \mathbf{V}_M\,\mathbf{D}^{\frac{M-N}{2}+(i-1)}\,\mathbf{S_f}\left[\mathbf{D}^{\frac{M-N}{2}+(i-1)}\right]^H \mathbf{V}_M^H + \sigma_w^2\,\mathbf{I}. \tag{6.617}$$

We refer to $\mathbf{S}_M^{(i)}$ as the forward spectral matrix of the ith subarray.

The backward spectral matrix is

$$\mathbf{S}_{MB}^{(i)} = \mathbf{J}\left[\mathbf{S}_M^{(i)}\right]^* \mathbf{J}, \tag{6.618}$$

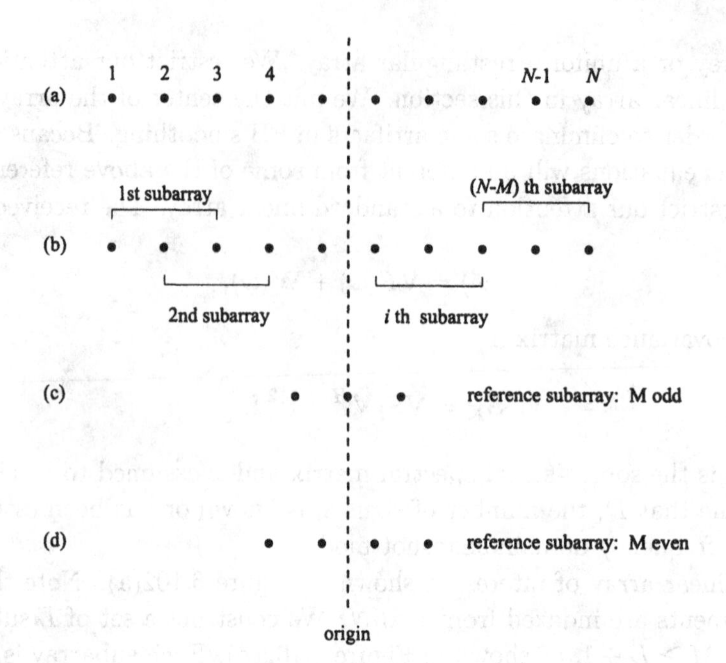

Figure 6.102 (a) Linear array; (b) subarrays; (c) reference subarray: M odd; (d) reference subarray: M even.

where $\mathbf{J}$ is the exchange matrix. The backward spectral matrix can be written as[16]

$$\mathbf{S}_{MB}^{(i)} = \mathbf{J}\,\mathbf{V}_M^* \left[\mathbf{D}^{\frac{M-N}{2}+(i-1)}\right]^* \mathbf{S}_{\mathbf{f}}^* \left[\mathbf{D}^{\frac{M-N}{2}+(i-1)}\right]^{*H} \mathbf{V}_M^{*H}\,\mathbf{J} + \sigma_w^2\,\mathbf{I}. \quad (6.619)$$

Observing that

$$\mathbf{J}\,\mathbf{V}_M^* = \mathbf{V}_M, \quad (6.620)$$

(6.619) reduces to

$$\mathbf{S}_{MB}^{(i)} = \mathbf{V}_M(\mathbf{D}^{\frac{M-N}{2}+(i-1)})^*\mathbf{S}_{\mathbf{f}}^* \left[(\mathbf{D}^{\frac{M-N}{2}+(i-1)})\right]^{*H} \mathbf{V}_M^H + \sigma_w^2\,\mathbf{I}. \quad (6.621)$$

We now perform a weighted average across the subarrays and obtain a smoothed spectral matrix $\mathbf{S}_{SSFB}$:

$$\mathbf{S}_{SSFB} = \sum_{i=1}^{L} w_i \left\{\frac{1}{2}\left(\mathbf{S}_M^{(i)} + \mathbf{S}_{MB}^{(i)}\right)\right\}. \quad (6.622)$$

[16] In Section 7.2.3, we develop FB averaging in more detail. Readers who have not encountered it previously may want to read that section at this point.

In most references (e.g., [RR92a], [SK85b]) the weighting is uniform. Takao et al. [TKY86] introduced the idea of non-uniform weighting.

For uniform weighting,

$$\mathbf{S}_{SSFB} = \frac{1}{2L}\sum_{i=1}^{L}\left(\mathbf{S}_M^{(i)} + \mathbf{S}_{MB}^{(i)}\right). \tag{6.623}$$

We derive the basic result using uniform weighting and then examine the possible advantages of non-uniform weighting.

Since $\mathbf{D}^* = \mathbf{D}^{-1}$, we can write $\mathbf{S}_{SSFB}$ as

$$\begin{aligned}\mathbf{S}_{SSFB} = \mathbf{V}_M & \left[\frac{1}{2L}\sum_{i=1}^{L}\left\{\mathbf{D}^{\frac{M-N}{2}+(i-1)}\,\mathbf{S}_\mathbf{f}\left[\mathbf{D}^{\frac{M-N}{2}+(i-1)}\right]^H\right.\right.\\ & \left.\left.+\left[\mathbf{D}^{\frac{M-N}{2}+(i-1)}\right]^*\mathbf{S}_\mathbf{f}^*\left[\mathbf{D}^{\frac{M-N}{2}+(i-1)}\right]^{*H}\right\}\right]\mathbf{V}_M^H + \sigma_w^2\mathbf{I}.\end{aligned} \tag{6.624}$$

We define the smoothed signal as the term in brackets,

$$\begin{aligned}\mathbf{S}_{\mathbf{f},SSFB} &= \frac{1}{2L}\sum_{i=1}^{L}\left(\mathbf{D}^{\frac{M-N}{2}+(i-1)}\,\mathbf{S}_\mathbf{f}\left[\mathbf{D}^{\frac{M-N}{2}+(i-1)}\right]^H\right.\\ &\qquad\left.+\left[\mathbf{D}^{\frac{M-N}{2}+(i-1)}\right]^*\mathbf{S}_\mathbf{f}^*\left[\mathbf{D}^{\frac{M-N}{2}+(i-1)}\right]^{*H}\right)\\ &= \frac{1}{L}\sum_{i=1}^{L} Re\left[\mathbf{D}^{\frac{M-N}{2}+(i-1)}\,\mathbf{S}_\mathbf{f}\left[\mathbf{D}^{\frac{M-N}{2}+(i-1)}\right]^H\right].\end{aligned} \tag{6.625}$$

Then we can write

$$\mathbf{S}_{SSFB} = \mathbf{V}_M\,\mathbf{S}_{\mathbf{f},SSFB}\,\mathbf{V}_M^H + \sigma_w^2\,\mathbf{I}. \tag{6.626}$$

Note that , if we did not use FB averaging,[17]

$$\mathbf{S}_{\mathbf{f},SS} = \frac{1}{L}\sum_{i=1}^{L}\mathbf{D}^{\frac{M-N}{2}+(i-1)}\,\mathbf{S}_\mathbf{f}\left[\mathbf{D}^{\frac{M-N}{2}+(i-1)}\right]^H, \tag{6.627}$$

and

$$\mathbf{S}_{SS} = \mathbf{V}_M\mathbf{S}_{\mathbf{f},SS}\mathbf{V}_M^H + \sigma_w^2\mathbf{I}. \tag{6.628}$$

The first issue is to determine the number of subarrays need in order for the smoothed-signal covariance matrix to be non-singular. Evans et al.

[17]This is the model used in [SK85a], [SK85b], and [RPK87a].

[EJS82] indicated that L must be greater than $D/2$, but did not prove it. Williams et al. [WPMS88] and Pillai and Kwon [PK89] (apparently independently) have shown that, under mild conditions, that $L \geq D/2$ guarantees that $\mathbf{S}_{\mathbf{f},SSFB}$ will be non-singular. The reader is referred to either of these references for the proof. For the case when $D = 2$, the proof requires that the phase of $[\mathbf{S}_{\mathbf{f}}]_{12} \neq 0$. We will use this criterion. However, if the phase of $[\mathbf{S}_{\mathbf{f}}]_{12}$ is near zero, the FB averaging will provide a matrix that is close to singular. As $M \geq D+1$, and

$$N = L + M - 1, \tag{6.629}$$

we require

$$N \geq \frac{D}{2} + D, \tag{6.630}$$

or

$$N \geq \frac{3D}{2}, \tag{6.631}$$

in order to have a nonsingular $\mathbf{S}_{\mathbf{f},SSFB}$.

Note that the nonsingularity of $\mathbf{S}_{\mathbf{f},SSFB}$ allows us to use various algorithms, such as the MPDR beamformer. However, if $|\rho|$ is close to unity, the resulting performance will still be degraded.

The second issue is how rapidly do $\mathbf{S}_{\mathbf{f},SSFB}$ and $\mathbf{S}_{\mathbf{f},SS}$ approach diagonal matrices as a function of L. This rate will depend on the phase of ρ, the separation in wavenumber space of the correlated plane waves, and the *SNR*.

First, consider forward-only SS with uniform weighting. From (6.627),

$$\left[\mathbf{S}_{\mathbf{f},SS}^{(L)}\right]_{ij} = \mathbf{S}_{ij}\frac{1}{L}\sum_{l=1}^{L}\mathbf{D}_{ii}^{\frac{M-N}{2}+(l-1)}\left[\mathbf{D}_{jj}^{\frac{M-N}{2}+(l-1)}\right]^{*}. \tag{6.632}$$

For $i = j$,

$$\left[\mathbf{S}_{\mathbf{f},SS}^{(L)}\right]_{ij} = \mathbf{S}_{ii}. \tag{6.633}$$

However, for $i \neq j$,

$$\begin{aligned}\frac{1}{L}\sum_{l=1}^{L}\mathbf{D}_{ii}^{\frac{M-N}{2}+(l-1)}\left[\mathbf{D}_{jj}^{\frac{M-N}{2}+(l-1)}\right]^{*} &= \frac{1}{L}\sum_{l=1}^{L}e^{j\frac{M-N}{2}(\psi_i-\psi_j)}\cdot e^{-j(l-1)(\psi_i-\psi_j)} \\ &= e^{j\frac{M-N}{2}\Delta\psi_{ij}}\,\frac{1}{L}\sum_{m=0}^{L-1}e^{jm\Delta\psi_{ij}}.\end{aligned} \tag{6.634}$$

The sum on the right side of (6.634) is familiar as the conventional beam pattern for a standard L-element linear array.

Although it is convenient to recognize the right side of (6.634) as a beam pattern, there is no actual L-element array involved in the beamformer. The parameter L is the number of M-element subarrays used to construct the smoothed spectral matrix.

For the case of two sources,

$$\rho_F^{(L)} = \frac{\left[\mathbf{S}_{\mathbf{f},SS}^{(L)}\right]_{12}}{\sqrt{\mathbf{S}_{11}\mathbf{S}_{22}}}. \tag{6.635}$$

Using (6.634) for the two-source case gives

$$\begin{aligned}\rho_F^{(L)} &= \frac{\rho}{L}\frac{\sin\left(\frac{L\Delta\psi}{2}\right)}{\sin\left(\frac{\Delta\psi}{2}\right)}e^{-j\frac{L-1}{2}\Delta\psi}e^{-j\frac{M-N}{2}\Delta\psi} \\ &= \frac{\rho}{L}\frac{\sin\left(\frac{L\Delta\psi}{2}\right)}{\sin\left(\frac{\Delta\psi}{2}\right)},\end{aligned} \tag{6.636}$$

where $\Delta\psi = \Delta\psi_{12}$.

In Figure 6.103(a), the magnitude of $\rho_F^{(L)}$ is plotted versus L for various values of $\Delta\psi$. In practice, the value of L is constrained by N and D ($N = M + L - 1$ and $M \geq D + 1$). Thus, for $N = 10$ and $D = 2$, the maximum value of L is 8. In Figure 6.103(b), the magnitude of $\rho_F^{(L)}$ is plotted versus M for $N = 10$ and $D = 2$.

A similar result can be obtained for $\mathbf{S}_{\mathbf{f},SSFB}$ in (6.625). For the two-signal case and $i \neq j$, (6.625) can be written as,

$$\left[\mathbf{S}_{\mathbf{f},SSFB}\right]_{ij} = \frac{1}{L}\sum_{l=1}^{L} Re\left[\left[\mathbf{S}_{\mathbf{f}}\right]_{ij} e^{j\left(\frac{M+N}{2}+(l-1)\right)\Delta\psi_{ij}}\right], \tag{6.637}$$

where

$$\Delta\psi_{ij} = \psi_i - \psi_j. \tag{6.638}$$

For the two-signal case, (6.637) reduces to

$$\rho_{FB}^{(L)} = \frac{Re[\rho]}{L}\frac{(\sin\frac{L\Delta\psi}{2})}{\sin(\frac{\Delta\psi}{2})}. \tag{6.639}$$

Thus, the effect of FB smoothing is to replace ρ with $Re[\rho]$ in (6.636). Writing

$$\rho = |\rho|e^{j\phi}, \tag{6.640}$$

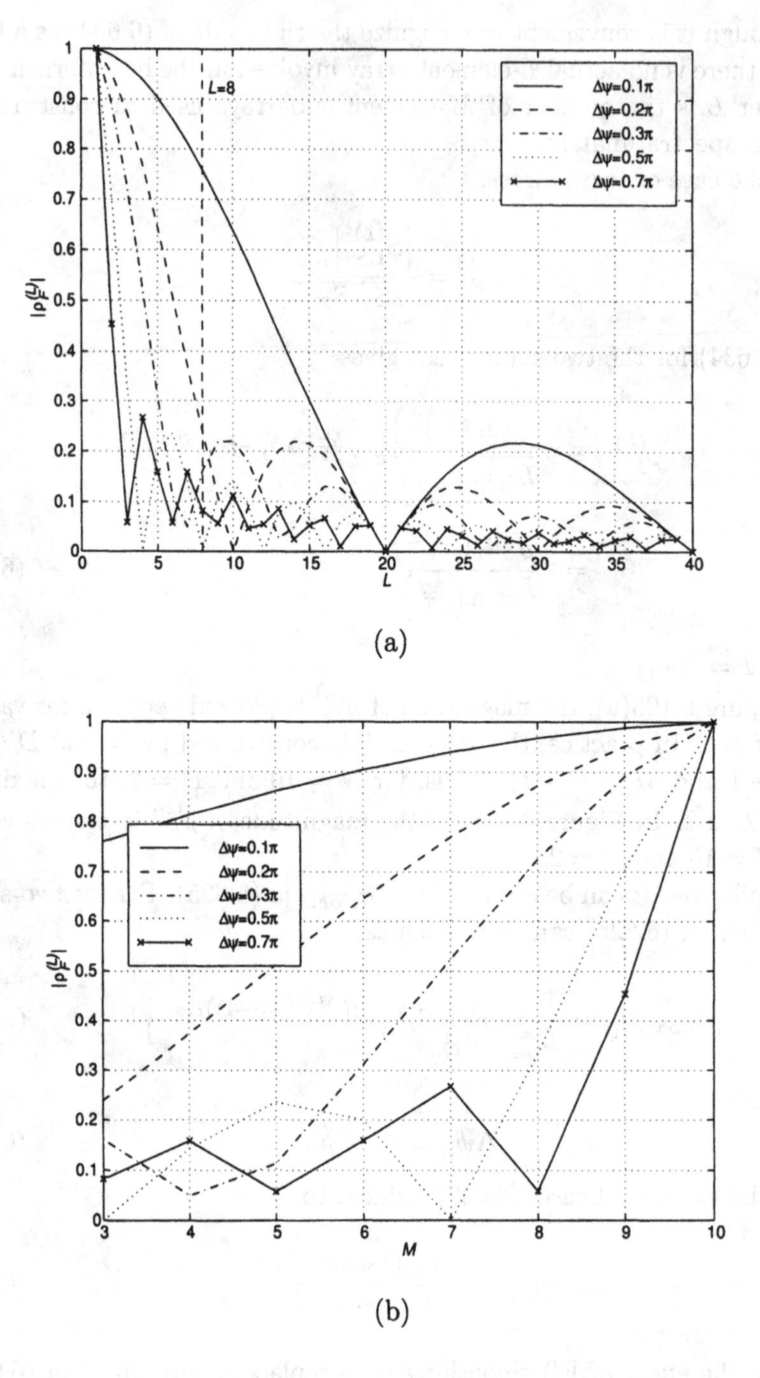

Figure 6.103 Magnitude of smoothed correlation coefficient for various $\Delta\psi$, forward-only smoothing; (a) $|\rho_F^{(L)}|$ versus L; (b) $|\rho_F^{(L)}|$ versus M for N=10.

we see that the phase of ϕ determines the difference between the effect of FB smoothing and forward-only smoothing. If $\phi = \pi/2$ or $3\pi/2$, the correlation is completely removed by the FB averaging and no smoothing is needed. If $\phi = 0$ or π, these is no difference between FB and forward-only smoothing.

The output of the spatial smoothing processing is the $M \times M$ spectral matrix in (6.626) or (6.628). The array manifold vector for the reference subarray is given by (6.614). The weight vector for the MPDR algorithm is

$$\mathbf{w}_{SFB}^H = \frac{\mathbf{v}_M^H \mathbf{S}_{SSFB}^{-1}}{\mathbf{v}_M^H \mathbf{S}_{SSFB}^{-1} \mathbf{v}_M}, \tag{6.641}$$

if (6.626) is used, or

$$\mathbf{w}_S^H = \frac{\mathbf{v}_M^H \mathbf{S}_{SS}^{-1}}{\mathbf{v}_M^H \mathbf{S}_{SS}^{-1} \mathbf{v}_M}, \tag{6.642}$$

if (6.628) is used.

The behavior of the beamformer in (6.642) has been analyzed for the case of two correlated sources by Reddy et al. [RPK87a]. The result is identical to (6.600)–(6.605) with ρ replaced with $\rho_F^{(L)}$ or $\rho_{FB}^{(L)}$, N replaced by M, and B_{dI} defined for an M-element array. The modified equations are:

$$SNR_o = \frac{A_{SS}}{B_{SS}}, \tag{6.643}$$

where

$$A_{SS} = \left|\sigma_1 \sigma_2^2 \Gamma_s \Delta_s + \sigma_w^2 \left(M\sigma_1 + \rho_s^* M B_{12}^* \sigma_2\right)\right|^2, \tag{6.644}$$

$$\begin{aligned} B_{SS} = {} & \sigma_1^2 \sigma_2^2 \Gamma_s |\rho_s|^2 \Delta_s^2 + \sigma_2^2 \sigma_w^2 \Delta_s \\ & \left[M\sigma_1^2 |\rho_s|^2 + M\sigma_2^2 + |\rho_s|^2 \sigma_1 \sigma_2 M \left(\rho_s B_{12} + \rho_s^* B_{12}^*\right) + 2\sigma_w^2\right] \\ & + M^2 \sigma_2^2 \sigma_w^4 |B_{12}|^2 \Gamma_s + M\sigma_w^6. \end{aligned} \tag{6.645}$$

Parameters in (6.644) and (6.645) include

$$\Gamma_s = 1 - |\rho_s|^2, \tag{6.646}$$

$$B_{12} = \frac{1}{M} \mathbf{v}_M^H(\psi_d) \mathbf{v}_M(\psi_I), \tag{6.647}$$

and

$$\Delta_s = M^2 \left(1 - |B_{12}|^2\right). \tag{6.648}$$

The correlation coefficient, ρ_s, denotes either $\rho_F^{(L)}$ or $\rho_{FB}^{(L)}$.

If N is fixed, there are two opposing effects. As M decreases, L increases and the correlation coefficient decreases. However, as M decreases, the array gain for the subarray decreases and its resolution capability decreases. Thus, for a given value of ρ, there will a value of M that maximizes the output *SNR*.

We consider an example to illustrate the behavior.

Example 6.12.3

Consider a standard 10-element linear array. The input consists of two equal-power correlated plane-wave signals with $\mathbf{S_f}$ given by (6.588) with

$$\rho = |\rho|e^{j\phi}. \tag{6.649}$$

The desired signal arrives from $u_s = 0$. The interfering signal arrives from $u_I = 0.30$. The *SNR* and *INR* are both 20 dB. We let

$$L = 11 - M. \tag{6.650}$$

We construct a smoothed spectral matrix using (6.626) or (6.628). We then use an MPDR beamformer with a symmetric M-element array manifold vector. We plot the array gain versus M.

In Figure 6.104, we let $\phi = 0$ and plot the array gain versus M for $|\rho| = 0, 0.1, 0.5, 0.95$, and 1.0. We use forward SS.

We see that the optimum M is 4 for $\rho \geq 0.50$. For $\rho = 0.10, M_o$ is 5. For $\rho = 0$, we do not need SS, so $M_o = 10$ and $L = 1$. For the case when $\phi = 0$, FB smoothing produces the same result.

In Figure 6.105, we let $\phi = \frac{\pi}{4}$ and plot the array gain versus M for $|\rho| = 0, 0.1, 0.5, 0.95$, and 1.0 for FBSS.

For forward SS, the performance does not depend on ϕ, so Figure 6.104 applies. For $\rho \geq 0.5$, the optimum M is 5 and the array gain is 26 dB, compared to 30 dB when $\rho = 0$.

In Figure 6.106, we let $\phi = \frac{\pi}{2}$ and plot the array gain versus M for the same values as in Figure 6.98. FB smoothing decorrelates the signal for $L = 1$, so the optimum M for any value of $|\rho|$ is 10.

The smoothed correlation coefficient is sensitive to the value of $\Delta\psi$ (see Figure 6.103). In Figure 6.107, $\Delta\psi$ is treated as a uniform random variable of $0.2\pi \leq \Delta\psi \leq \pi$. For each realization, the array gain using FB smoothing is plotted versus M. Then the result is averaged over 10,000 trials. For $M = 4$, 5, or 6, the array gain is slightly above 21 dB.

We now revisit the case of non-uniform weighting in the subarray averaging and consider two possible techniques. First consider the case in which we do not use FB averaging. Repeating the steps in (6.632)–(6.636), we obtain

$$\begin{aligned}\left[\mathbf{S}_{\mathbf{f},ss}^{(L)}\right]_{ij} &= \mathbf{S}_{ij}\left\{e^{j\frac{M-N}{2}\Delta\psi_{ij}}\sum_{m=0}^{L-1} w_m e^{jm\Delta\psi_{ij}}\right\}\\ &= \mathbf{S}_{ij}\left\{e^{j\frac{M-N}{2}\Delta\psi_{ij}}e^{j\frac{L-1}{2}\Delta\psi_{ij}}\left(e^{-j\frac{L-1}{2}\Delta\psi_{ij}}\sum_{m=0}^{L-1} w_m e^{jm\Delta\psi_{ij}}\right)\right\}\end{aligned}$$

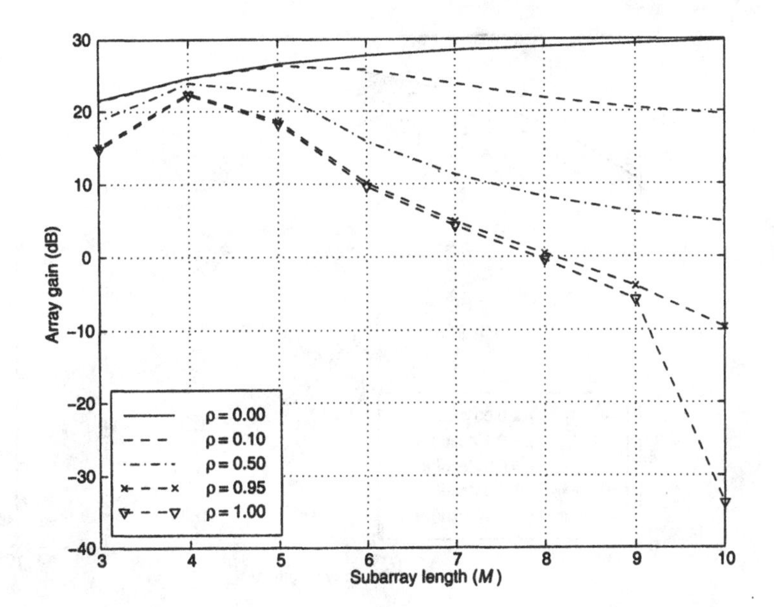

Figure 6.104: MPDR beamformer with forward spatial smoothing, $u_s = 0$, $SNR = 20$ dB, $u_I = 0.3$, $INR = 20$ dB, $\phi = 0$; array gain versus M (subarray length).

$$= \quad \mathbf{S}_{ij} \left\{ e^{-j\frac{L-1}{2}\Delta\psi_{ij}} \sum_{m=0}^{L-1} w_m^* e^{jm\Delta\psi_{ij}} \right\}. \tag{6.651}$$

The sum is just the beam pattern of a standard L-element linear array with non-uniform weighting. Then, for the two-source case

$$\rho^{(L)} = \rho B_w(\Delta\psi), \tag{6.652}$$

where $B_w(\Delta\psi)$ is the beam pattern associated with the weights. For example, we might use a Hamming or Dolph-Chebychev weighting to lower the sidelobes in the function in Figure 6.103. A similar result can be obtained for FBSS.

This technique will be most effective when $\Delta\psi$ and L are such that the interferer in the sidelobe region of an L-element standard linear array.

Example 6.12.4 (continuation)

Consider a standard 10-element array. The input consists of two equal-power coherent plane-wave signals with $\mathbf{S}_\mathbf{f}$ given by (6.588) with

$$\rho = \exp(j\pi/4).$$

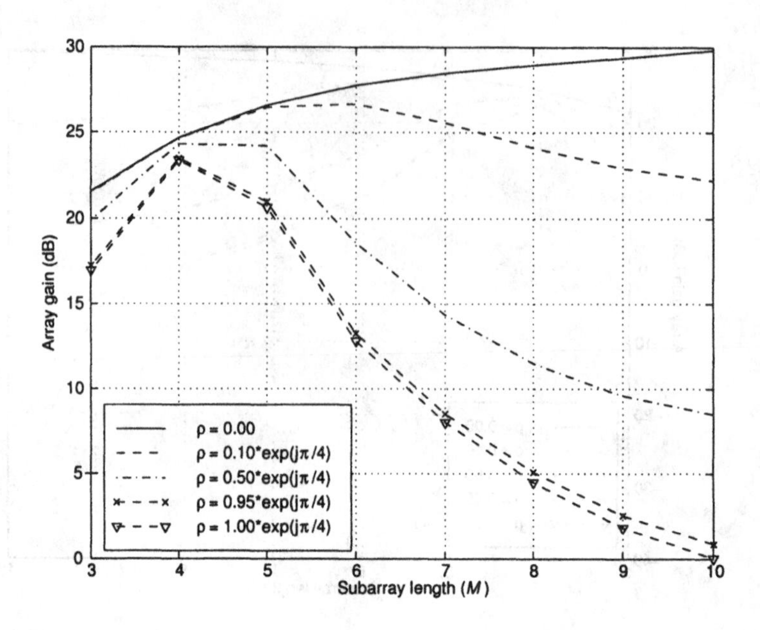

Figure 6.105 MPDR beamformer with forward spatial smoothing, $u_s = 0$, $SNR = 20$ dB, $u_I = 0.3$, $INR = 20$ dB, $\phi = \frac{\pi}{4}$; array gain versus M (subarray length).

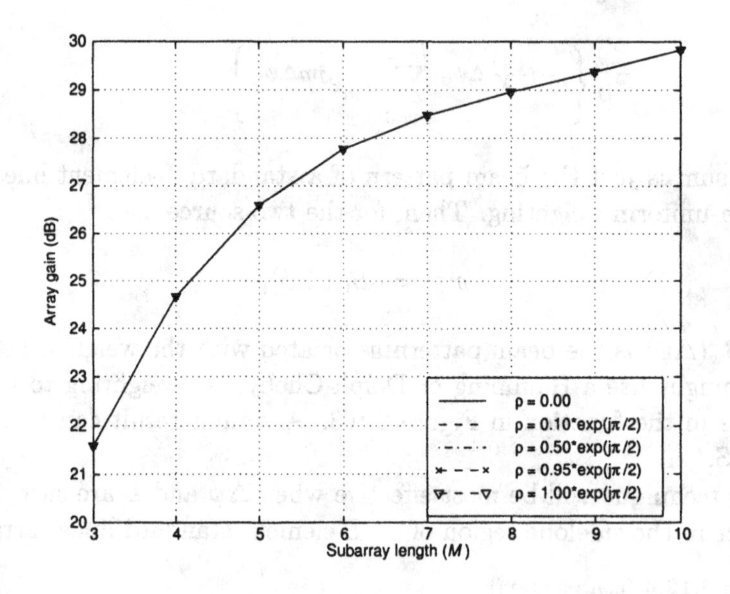

Figure 6.106 MPDR beamformer with FBSS, $u_s = 0$, $SNR = 20$ dB, $u_I = 0.3$, $INR = 20$ dB, $\phi = \frac{\pi}{2}$; array gain versus M (subarray length).

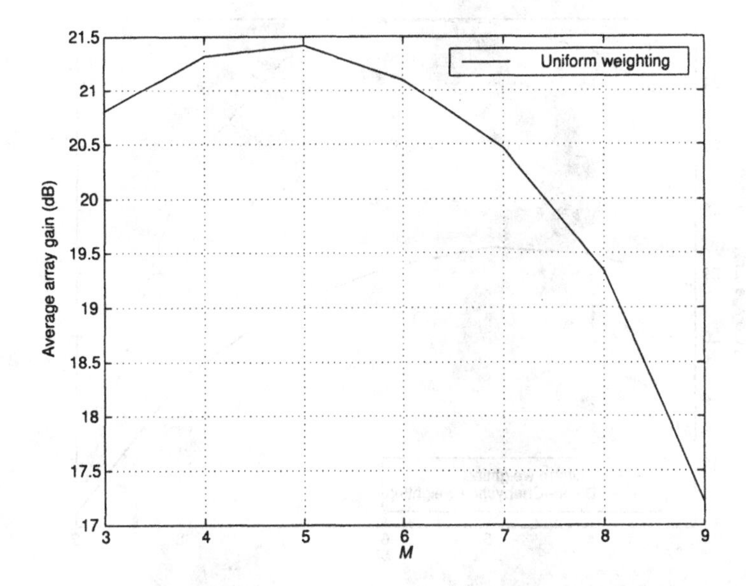

Figure 6.107 MPDR-FBSS beamformer: average array gain versus M, $\rho = \exp(j\pi/4)$, $u_s = 0, u_I$ is a uniform random variable (0.2, 1), $SNR = INR =$ 20 dB, 10,000 trials.

The desired signal arrives from $u_s = 0$. The interfering signal arrives from u_I, where u_I is a uniform random variable (0.2, 1.0). The $SNR = INR = 20$ dB. A FB-smoothed spectral matrix is constructed using a Dolph-Chebychev weighting with −30-dB sidelobe level. The smoothed output is processed using an MPDR beamformer with a symmetric M-element array manifold vector. In Figure 6.108, the average array gain is plotted versus M. The result from Figure 6.107 is shown for comparison. The Dolph-Chebychev weighting provides about a 3-dB improvement for $M = 4$, 5, 6, and 7 and gives an average array gain about 24.5–25 dB.

A different approach suggested by Takao and Kikuma [TK87] is to choose the w_m to make the resulting smoothed matrix as close to Toeplitz as possible. We rewrite (6.622) as

$$\mathbf{S}_{SSFB} = \sum_{k=1}^{L} w_k \mathbf{H}_k \triangleq \mathbf{B}, \tag{6.653}$$

where

$$\mathbf{H}_k \triangleq \frac{1}{2}\left[\mathbf{S}_M^{(k)} + \mathbf{S}_{MB}^{(k)}\right]. \tag{6.654}$$

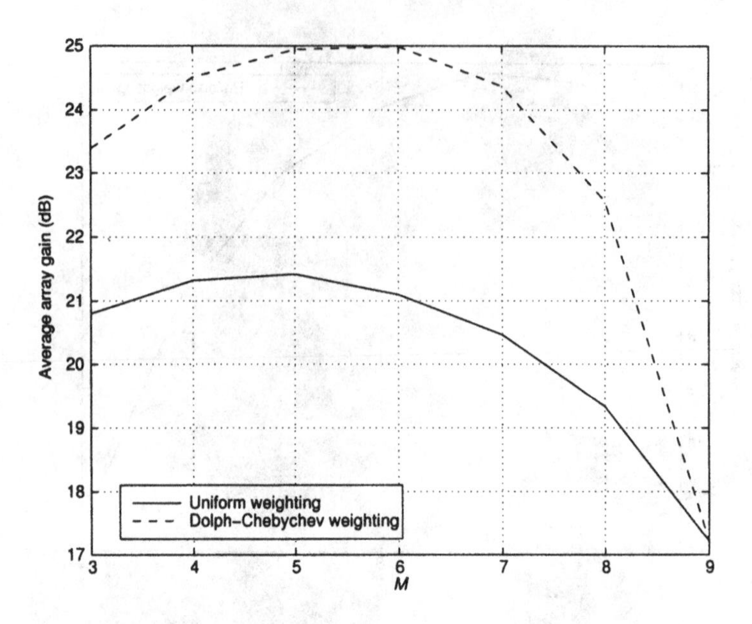

Figure 6.108 MPDR-FBSS beamformer with Dolph-Chebychev and uniform weighting: average array gain versus M, $\rho = \exp(j\pi/4)$, $u_s = 0, u_I$ is a uniform random variable (0.2, 1), *SNR* = *INR* = 20 dB, 10,000 trials.

We test how close $\mathbf{B}$ is to a Toeplitz matrix by defining

$$F = \sum_{i=0}^{M-2} \sum_{k=1}^{M-1} \left| b_{k+i,k} - b_{(i)} \right|^2, \tag{6.655}$$

where

$$b_{(i)} \triangleq \frac{1}{M-i} \sum_{k=1}^{M-i} b_{k+i,k}. \tag{6.656}$$

Then,

$$\begin{aligned} b_{k+i,k} - b_{(i)} &= b_{k+i,k} - \frac{1}{M-i} \sum_{k-1}^{M-i} b_{k+i,k} \\ &= \left[\sum_{l=1}^{L} w_l \mathbf{H}_l \right]_{k+i,k} - \frac{1}{M-i} \sum_{k=1}^{M-i} \left[\sum_{l=1}^{L} w_l \mathbf{H}_l \right]_{k+i,k} \\ &= \sum_{l=1}^{L} w_l \left[[\mathbf{H}_l]_{k+i,k} - \frac{1}{M-i} \sum_{k=1}^{M-i} [\mathbf{H}_l]_{k+i,k} \right] \end{aligned}$$

$$= \sum_{l=1}^{L} w_l \left[\mathbf{e}_l\right]_{k+i,k}, \tag{6.657}$$

where

$$\left[\mathbf{e}_l\right]_{k+i,k} \triangleq \left[\mathbf{H}_l\right]_{k+i,k} - \frac{1}{M-i} \sum_{k=1}^{M-i} \left[\mathbf{H}_l\right]_{k+i,k}. \tag{6.658}$$

Now define the vectors,

$$\mathbf{w} = \begin{bmatrix} w_1 & w_2 & \cdots & w_L \end{bmatrix}^T \tag{6.659}$$

$$\mathbf{e}_{k+i,k} = \begin{bmatrix} \left[\mathbf{e}_1\right]_{k+i,k} & \cdots & \left[\mathbf{e}_L\right]_{k+i,k} \end{bmatrix}^T. \tag{6.660}$$

Then,

$$b_{k+i,k} - b_{(i)} = \mathbf{w}^T \mathbf{e}_{k+i,k}, \tag{6.661}$$

and

$$F = \sum_{i=0}^{M-2} \sum_{k=1}^{M-i} \mathbf{w}^T \mathbf{e}_{k+i,k} \mathbf{e}_{k+i,k}^T \mathbf{w}. \tag{6.662}$$

Defining,

$$\begin{aligned} \mathbf{R}_{\mathbf{ee}} &= \sum_{i=0}^{M-2} \sum_{k=1}^{M-i} \mathbf{e}_{k+i,k} \mathbf{e}_{k+i,k}^T \\ &= Re\left\{ \sum_{i=0}^{M-2} \sum_{k=1}^{M-i} \mathbf{e}_{k+i,k} \mathbf{e}_{k+i,k}^T \right\}, \end{aligned} \tag{6.663}$$

we can write (6.662) as

$$F = \mathbf{w}^T \mathbf{R}_{\mathbf{ee}} \mathbf{w}. \tag{6.664}$$

Minimizing F subject to the constraint

$$\mathbf{w}^T \mathbf{1} = 1, \tag{6.665}$$

gives

$$\hat{\mathbf{w}} = \frac{\mathbf{R}_{\mathbf{ee}}^{-1} \mathbf{1}}{\mathbf{1}^T \mathbf{R}_{\mathbf{ee}}^{-1} \mathbf{1}}. \tag{6.666}$$

In an actual application, $\mathbf{S}_M^{(k)}$ and $\mathbf{S}_{MB}^{(k)}$ will be estimated from the data so the weight in (6.666) will be data dependent. We refer to this technique as **Toeplitz weighting.** It is called **adaptive spatial averaging** in the literature [TK87].

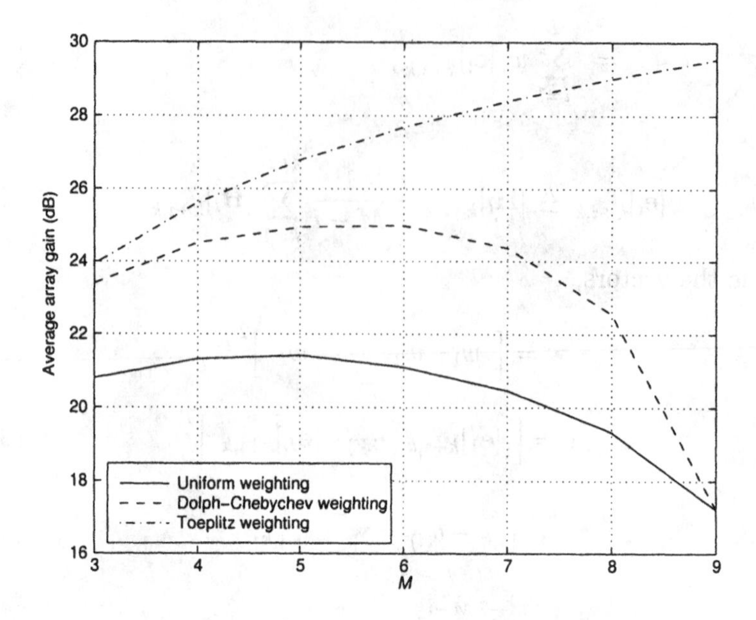

Figure 6.109 Average array gain using FBSS with Toeplitz weighting, $\rho = \exp(j\pi/4)$, $u_s = 0$, u_I uniform (0.2-1.0), $SNR = INR =$ 20 dB, $N = 10$, 10,000 trials.

We consider the models in examples 6.12.3 and 6.12.4 to see the impact on this weighting.

Example 6.12.5 (continuation; Example 6.12.4)

We use the same model as in Example 6.12.4 and consider the case when $\phi = \pi/4$. The result is shown in Figure 6.109. We see that, in the case of known spectral matrices, there is significant improvement.

6.12.5 Summary

In this section, we have considered the case in which the signal and interference are correlated ($|\rho_{ij}| \neq 0$) or coherent ($|\rho_{ij}| = 1$). We utilized a technique called spatial smoothing to reduce the correlation. After constructing a smoothed spectral matrix, we could use any of the beamforming algorithms developed in earlier sections of the chapter. We used MPDR beamformers as examples, but similar results follow for other beamformers.

We used a standard linear array as an example. Similar results hold for planar arrays that exhibit regularity under 2-D shifts.

6.13 Broadband Beamformers

6.13.1 Introduction

In this section we consider beamformers in which either the signal or the interference do not satisfy the narrowband assumption. The initial model of interest is shown in Figure 6.110(a) and (c). The signals and interferers are bandlimited around the carrier frequency ω_c (or $2\pi f_c$), which we refer to as the **center frequency** of the array. The dotted lines denote the spectral boundaries.

As indicated in Figure 6.110(c), the spectrum does not have to be symmetric around ω_c. In some applications, ω_c corresponds to an actual carrier frequency generated at some place in the environment. In other applications, it is chosen for analysis purposes.

We use a quadrature demodulator at each sensor output so the input to the beamformer is a complex waveform that is bandlimited to $|\omega| \leq 2\pi B_s/2$, as shown in Figure 6.110(b) and (d). We process the sensor outputs with a set of linear filters, as shown in Figure 6.111. The notation $W_{f,n}(\omega)$ denotes the frequency-domain transfer function of the nth linear filter. The model in Figure 6.111 is the same model that we introduced in Chapter 2. However, in most of our subsequent discussion, we assumed $\omega = \omega_c$ and suppressed the frequency variable in our discussion. We now reexamine our results for frequencies in the region shown in Figure 6.110(b).

We should observe at this point that we will not actually implement the beamformer as shown in Figure 6.111. We will either use frequency-domain snapshots and operate on discrete frequency bins or we will approximate the $W_{f,n}(\omega)$ using FIR filters. However, the structure in Figure 6.111 provide a useful reference.

In the next several figures, we illustrate some cases in which a broadband beamformer may be useful. For simplicity, we assume a linear array so that we can represent the spatial dependence in u-space. We define

$$f = f_c + f_\Delta \tag{6.667}$$

and represent the frequency dependence in f_Δ-space.

In Figure 6.112, we consider the case of a single plane-wave signal whose frequency spectrum is flat over $\pm B_s/2$. The interference consists of white noise. From Section 5.5.3, we know that, as $B_f = B_s/f_c$ increases, the signal will have multiple eigenvalues and we will need a frequency-dependent beamformer to achieve optimum performance. For this simple model, we need to approximate pure delays over the frequency band of interest.

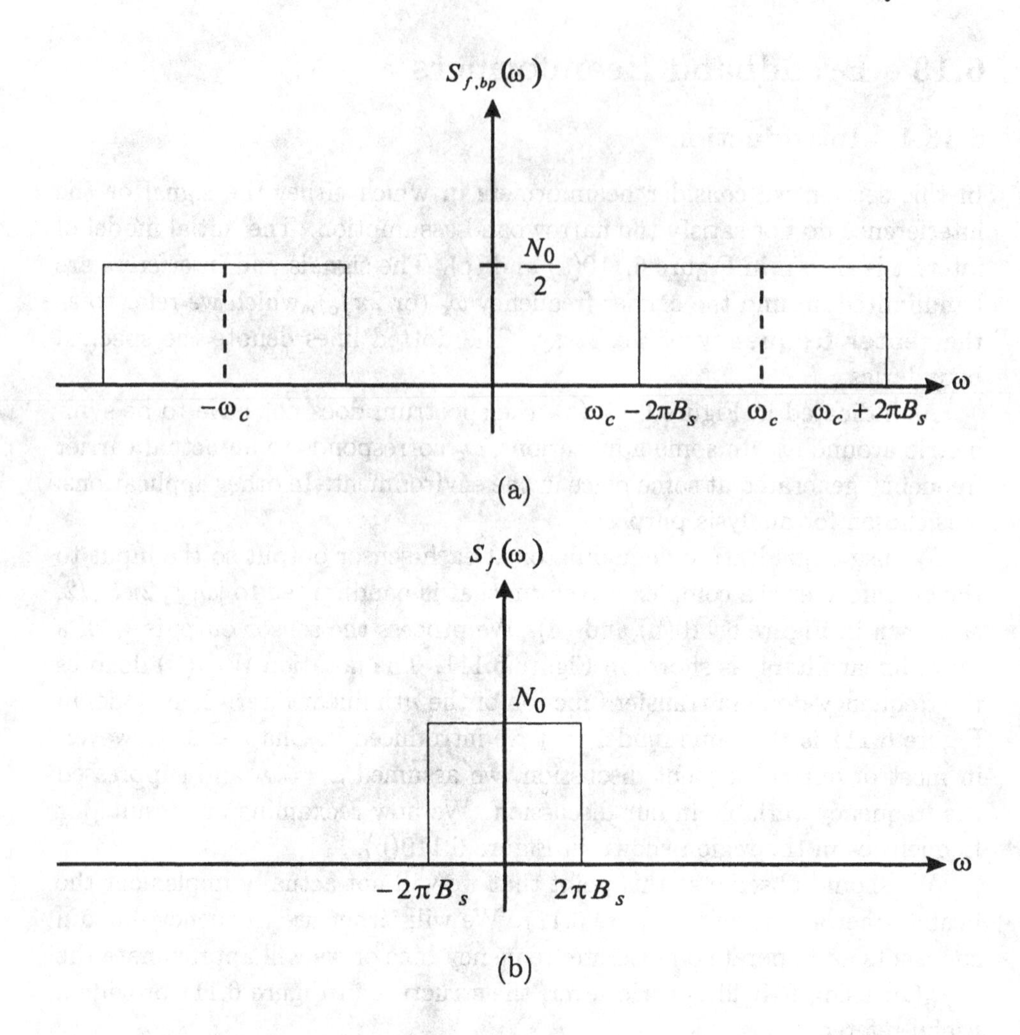

In Figure 6.113, we consider the same signal model as in Figure 6.112. The interference consists of a set of plane-wave interferers whose frequency spectra are identical to the signal spectrum. In this case, our processing gain will come from the spatial filtering and a frequency-dependent MVDR or MPDR beamformer will be appropriate.

From (6.14), the MVDR beamformer is

$$\mathbf{W}_{mvdr}^H(\omega) = \frac{\mathbf{v}_{\omega,u}^H(\omega, u_s)\,\mathbf{S}_{\mathbf{n}}^{-1}(\omega)}{\mathbf{v}_{\omega,u}^H(\omega, u_s)\,\mathbf{S}_{\mathbf{n}}^{-1}(\omega)\,\mathbf{v}_{\omega,u}(\omega, u_s)}.$$

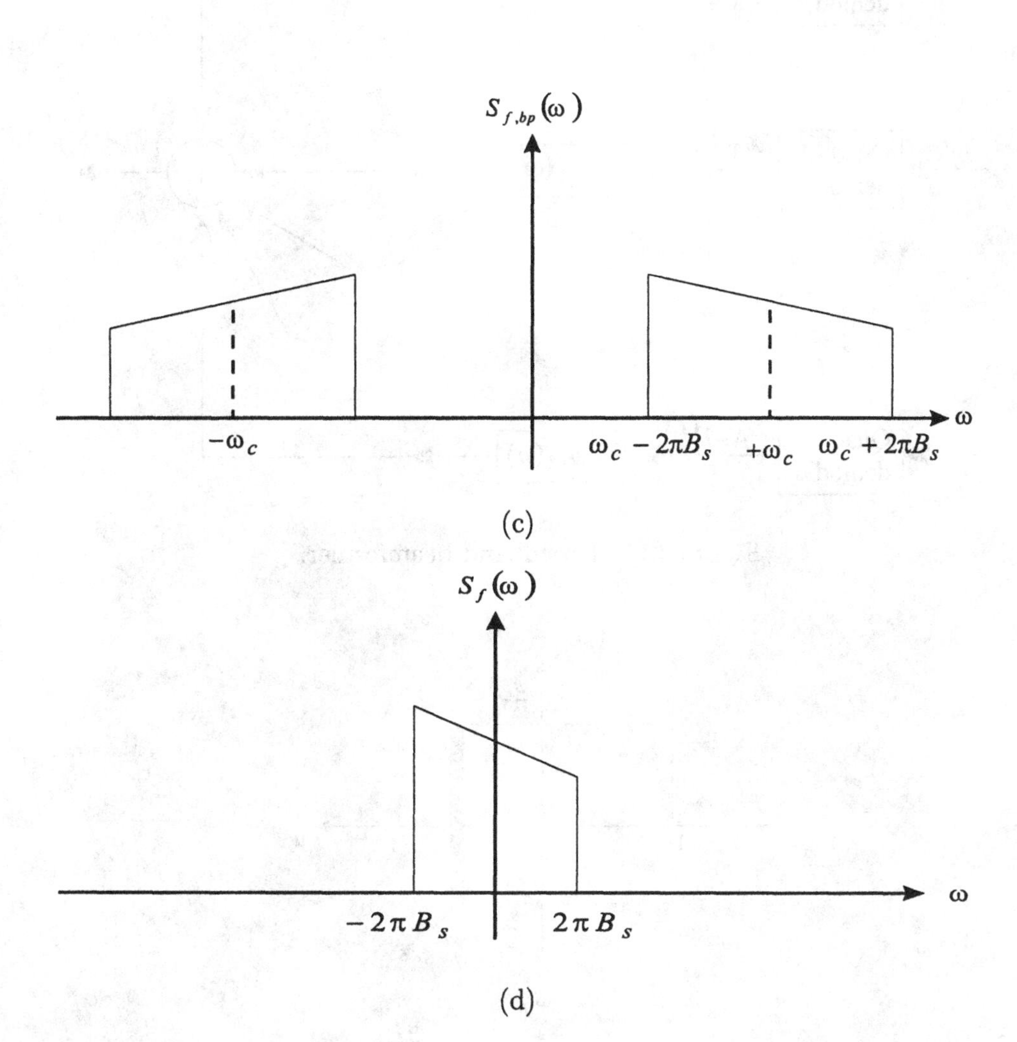

Figure 6.110 Signal spectra: (a) bandpass; (b) complex baseband; (c) bandpass; (d) complex baseband.

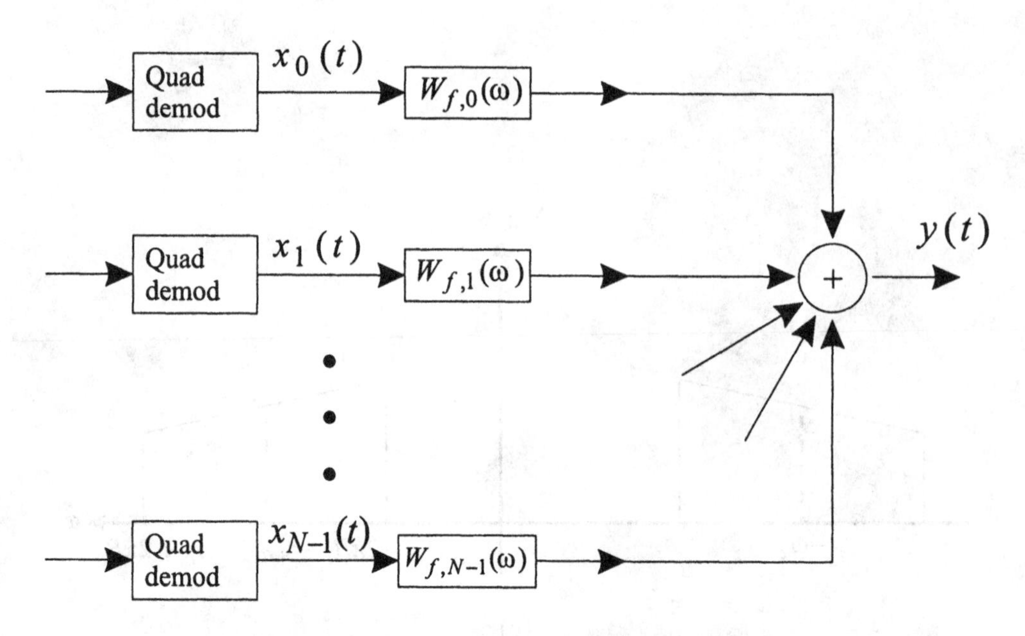

Figure 6.111 Broadband beamformer.

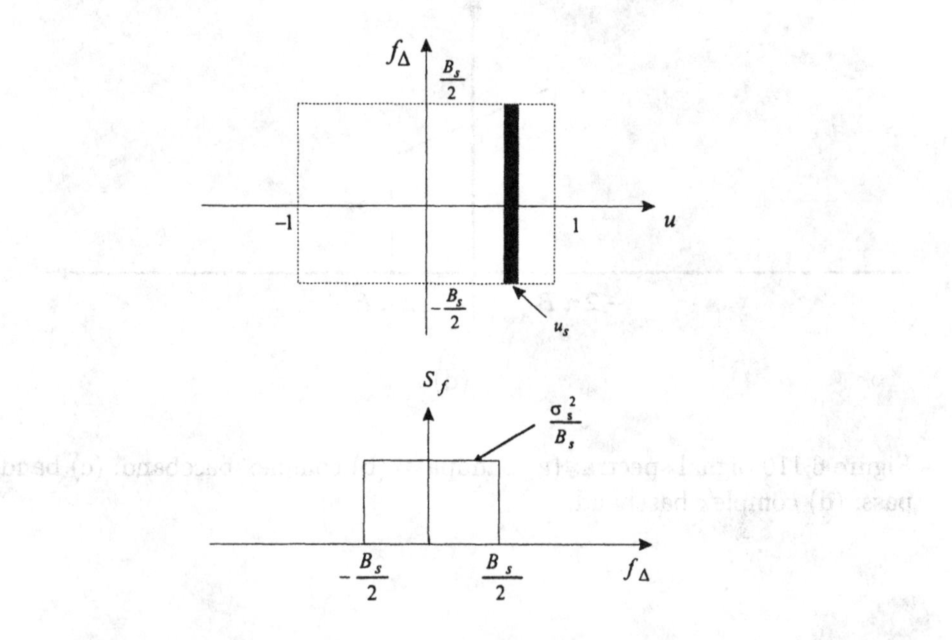

Figure 6.112 Plane-wave signal with flat spectrum located at u_s.

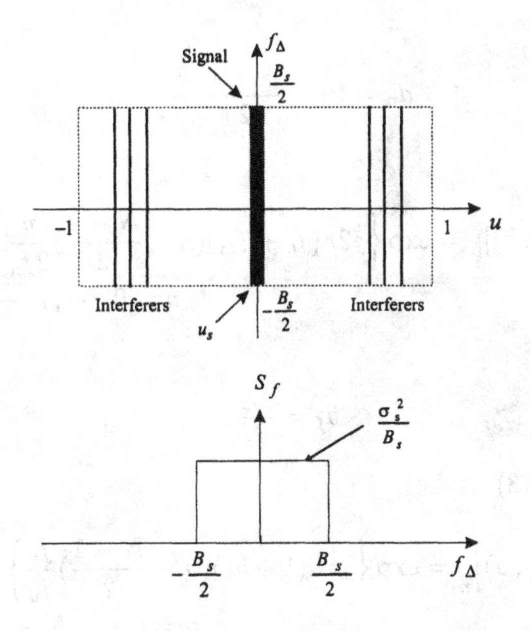

Figure 6.113 Broadband signal and interferers in u-f_Δ space.

From (6.71), the MPDR beamformer is

$$\mathbf{W}_{mpdr}^H(\omega) = \frac{\mathbf{v}_{\omega,u}^H(\omega, u_s)\,\mathbf{S}_{\mathbf{x}}^{-1}(\omega)}{\mathbf{v}_{\omega,u}^H(\omega, u_s)\,\mathbf{S}_{\mathbf{x}}^{-1}(\omega)\,\mathbf{v}_{\omega,u}(\omega, u_s)} \triangleq \Lambda_{\mathbf{x}}(\omega)\,\mathbf{v}_{\omega,u}^H(\omega, u_s)\,\mathbf{S}_{\mathbf{x}}^{-1}(\omega). \tag{6.668}$$

The array manifold vector is

$$\mathbf{v}_{\omega,\mathbf{k}}(\omega, \mathbf{k}) = \left[\; e^{-j\omega\tau_0} \quad e^{-j\omega\tau_1} \cdots \quad e^{-j\omega\tau_{N-1}} \;\right]^T, \tag{6.669}$$

where τ_n is given by (2.21). We assume the highest frequency of interest is $\omega_u (= 2\pi f_u)$ and refer to this as the **design frequency** of the array,[18]

$$f_u = f_c + B_s/2. \tag{6.670}$$

The corresponding **design wavelength** is λ_u. For a standard linear array with design frequency ω_u,

$$\tau_n = \frac{-d_n u}{c}, \tag{6.671}$$

[18]Note that the subscript "u" denotes upper and should not be confused with the directional cosine, $u = \cos\theta$.

and

$$d_n = (n - \frac{N-1}{2})\frac{\lambda_u}{2}. \tag{6.672}$$

(see (2.53)). Then,

$$[\mathbf{v}_{\omega,u}(2\pi f, u)]_n = \exp\left\{j2\pi(f_c + f_\Delta)(n - \frac{N-1}{2})\frac{u}{2f_u}\right\}, \quad n = 0, \cdots, N-1. \tag{6.673}$$

Defining

$$b_f = \frac{f_\Delta}{f_c}, \tag{6.674}$$

we can write (6.673) as

$$\left[\mathbf{v}_{b_f,u}(b_f, u)\right]_n = \exp\left\{j\pi u(1 + b_f)(n - \frac{N-1}{2})\frac{f_c}{f_u}\right\}, \quad n = 0, \cdots, N-1. \tag{6.675}$$

Defining

$$B_f = \frac{B_s}{f_c}, \tag{6.676}$$

(6.675) reduces to

$$\left[\mathbf{v}_{b_f,u}(b_f, u)\right]_n = \exp\left\{j\pi u(\frac{1 + b_f}{1 + \frac{B_f}{2}})(n - \frac{N-1}{2})\right\}, \quad n = 0, \cdots, N-1. \tag{6.677}$$

The spectral matrix can also be written in terms of b_f.

In the cases in Figures 6.112 and 6.113, the broadside beamformer achieves gain through spatial discrimination.

In Figure 6.114, we show a case in which the signal and interferers have different frequency spectra. The desired signal is a plane wave arriving from $u_s = 0$. The temporal waveform of the complex envelope is modeled as a complex AR(1) process centered at $\omega = 0$. The interfering signal is a plane wave arriving from $u_s = 0.2$. The temporal waveform of the complex envelope is modeled as a complex AR(1) process centered at 0.2. In this case, we want to exploit both spatial and frequency filtering, so we use the MMSE beamformer developed in Section 6.2.2.1. From (6.42),

$$\mathbf{W}_{mmse}^H(\omega) = \mathbf{S}_{d\mathbf{x}}^H(\omega)\,\mathbf{S}_{\mathbf{x}}^{-1}(\omega). \tag{6.678}$$

For the plane-wave signal case in which the desired signal and the interference are uncorrelated, (6.43) applies and (6.678) reduces to

$$\mathbf{W}_{mmse}^{H}(\omega) = S_f(\omega)\,\mathbf{v}_{\omega,\mathbf{k}}^{H}(\omega,\mathbf{k}_s)\,\mathbf{S}_{\mathbf{x}}^{-1}(\omega). \tag{6.679}$$

Using (6.668), we can partition $\mathbf{W}_{mmse}^{H}(\omega)$, as shown in Figure 6.115, although this may not be the best implementation in practice.

The purpose of the rest of Section 6.13 is to develop efficient techniques for implementing the various beamformers: $\mathbf{W}_{mpdr}^{H}(\omega)$, $\mathbf{W}_{lcmp}^{H}(\omega)$, $\mathbf{W}_{mmse}^{H}(\omega)$, and $\mathbf{W}_{lcmp,dl}^{H}(\omega)$.

We must develop structures that can be implemented adaptively based on the data inputs. In this chapter, we consider the solution assuming the necessary statistics are known.

In the problems for Section 6.6, we discussed how frequency mismatch (Problem 6.6.11), broadband signals (Problem 6.6.12), and broadband interference (Problem 6.6.13) degrade the performance of a narrowband beamformer. The reader may want to review those problems at this point. In this section, we are concerned with scenarios in which the degradation requires us to implement some form of broadband beamformer.

In Section 6.13.2 we consider a DFT beamformer that obtains the frequency-dependent weighting by taking an FFT and weighting each frequency component separately. This FFT beamformer is the discrete realization of our frequency-domain snapshot model.

In Section 6.13.3, we consider a broadband beamformer that utilizes a tapped delay line or finite impulse response (FIR) filter at the output of each sensor to achieve a frequency-dependent weighting.

In Section 6.13.4, we summarize our results.

6.13.2 DFT Beamformers

In this section, we develop the DFT implementation of a frequency-domain beamformer. Because we have used a frequency-domain snapshot model in our earlier discussion, the structure of the frequency-domain beamformer follows easily from Figure 6.111. We take the finite-time Fourier transform of $\mathbf{x}(t)$ over the time interval $(k-1)\Delta T \leq t < k\Delta T$ at a set of frequencies ω_m that are separated by $\omega_\Delta = 2\pi/\Delta T$. We assume that ΔT is large enough that the snapshots are statistically independent. We then implement the narrowband beamformers at each of the M frequencies.

In practice, we implement the frequency-domain beamformer by sampling $\mathbf{x}(t)$ every $1/B_s$ seconds and obtain M samples. We then perform a

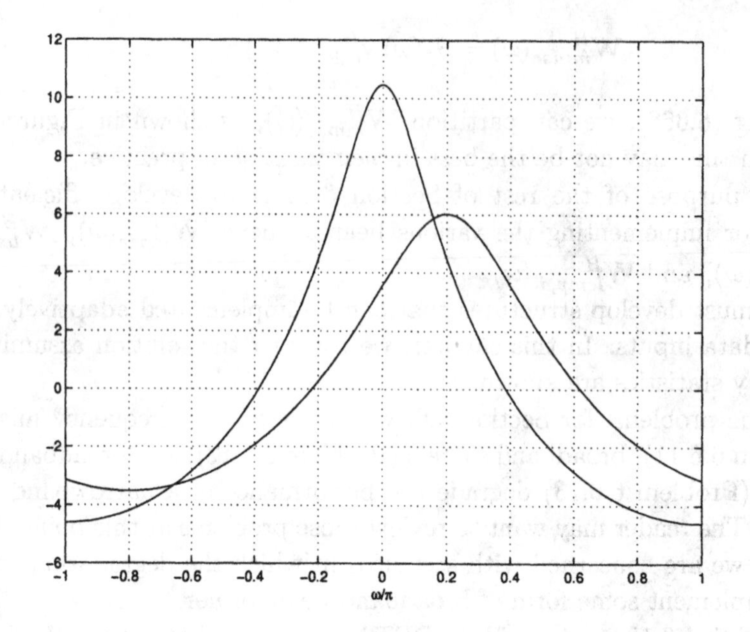

Figure 6.114 Broadband signal and interferers in $u - f_{\triangle}$ space; different frequency spectra.

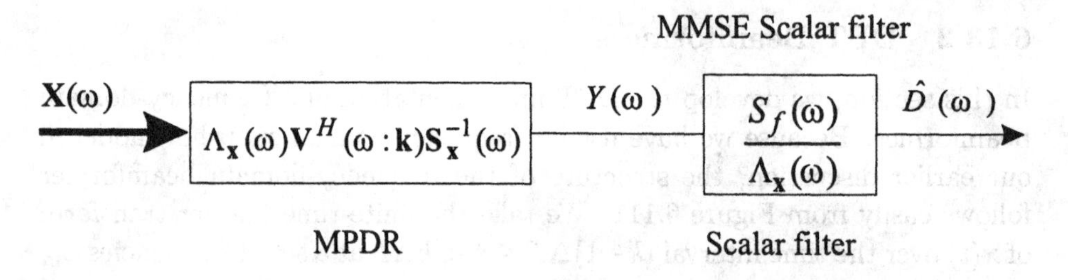

Figure 6.115 Partitioned MMSE beamformer.

DFT to obtain M frequency samples. We accomplish the DFT operation with a FFT. The concept is straightforward, but there is a fair amount of detail that must be developed to do the implementation. The bulk of this section is devoted to developing some of the necessary detail.

Frequency-domain beamformers have been discussed in a number of references. Most of these references focus on the adaptive behavior.

We assume that the signals of interest are bandlimited, as shown in Figure 6.110. The signal is a bandpass signal and the first step in the processing is to perform the quadrature demodulation to obtain a complex low-pass signal that is bandlimited to $|f| \leq B_s/2$. We divide our discussion into several steps.

Step 1: Generation of frequency-domain vectors: An array with DFT processing is shown in Figure 6.116. The output of each sensor is passed through a quadrature demodulator and is sampled every T_s seconds where

$$T_s = \frac{1}{B_s}. \tag{6.680}$$

We buffer M samples and take the DFT of the M samples. For the present discussion, we assume M is chosen to be a power of 2. The total observation interval on each snapshot is

$$\Delta T = M T_s\,. \tag{6.681}$$

We discuss the case in which we have to add zeros to the actual samples to obtain a power of 2 later. There are other reasons for zero padding.

The output of the DFT is a set of M frequency values (or bins) that are separated in frequency by $1/MT_s$. Thus, the frequency resolution is

$$\Delta f = 1/M T_s = B_s/M. \tag{6.682}$$

Many algorithms require high-frequency resolution, so M may be a large number.

In the mth frequency bin there is an $N \times 1$ complex vector, $\mathbf{X}\{m\}$, $m = 0, \cdots, M-1$.

$$\mathbf{X}\{m\} \triangleq [\; X_0\{m\} \quad X_1\{m\} \quad \cdots \quad X_{N-1}\{m\} \;]^T, \qquad m = 0, \cdots, M-1. \tag{6.683}$$

The argument m represents the number of the bin. We will relate it to the corresponding frequency shortly. Recall that the continuous Fourier transform at frequency ω_m is denoted $\mathbf{X}_{\Delta T}(\omega_m)$ for a finite-time interval.

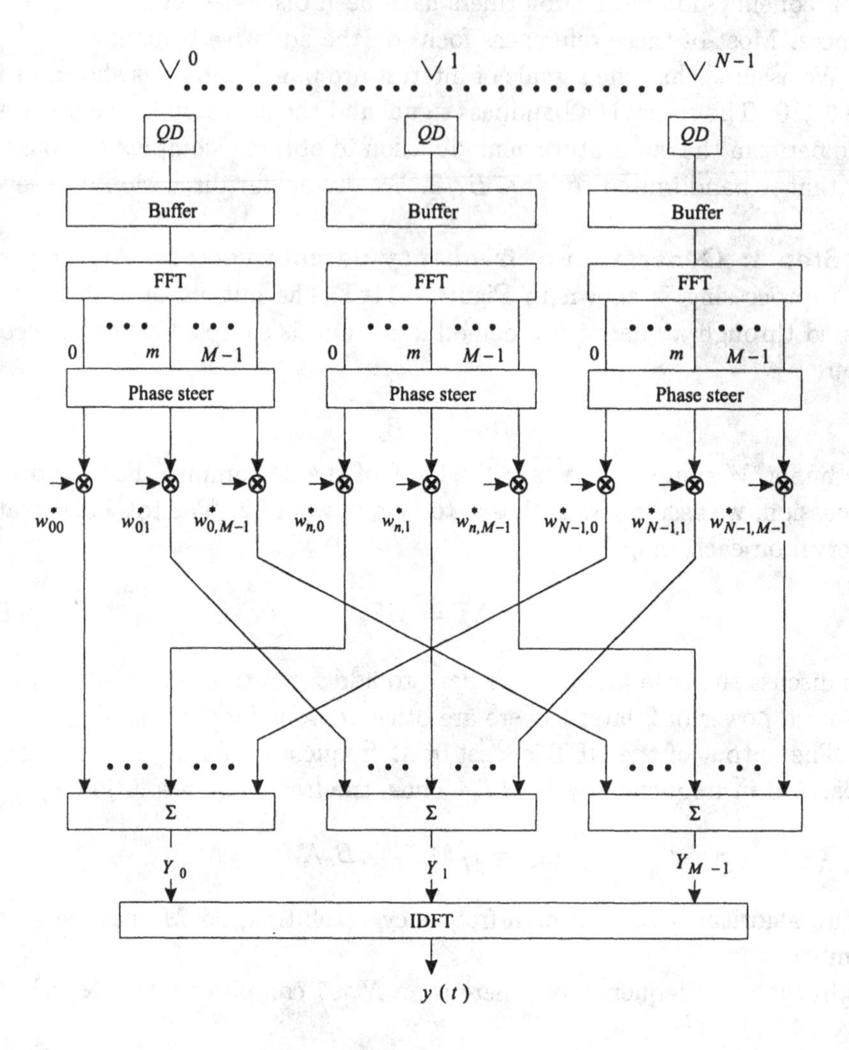

Figure 6.116 FFT processing.

We use $\mathbf{X}\{m\}$ to denote the DFT in the mth bin. Using a different shape of the brackets is an abuse of notation but should not cause confusion due to the context.

Normally, we process each frequency bin independently using one of the narrowband algorithms we have developed in this chapter.[19] The result is a set of complex scalars,

$$Y\{m\} = \mathbf{w}^H\{m\}\,\mathbf{X}\{m\}, \quad m = 0, \cdots, M-1. \tag{6.684}$$

The final step is to take an IDFT to obtain M output samples in the time domain. We refer to this procedure as **block processing**.

We next develop the equations for the DFT beamformer with block processing. There is direct correspondence with the discussion of DFT in Section 3.3.3.

We denote the time of the most recent sample as t_0. The current sample from the nth sensor is $x_n(t_0)$. The previous sample is $x_n(t_0 - T_s)$. We introduce the notation

$$x_n(t_0) = x_n(0), \tag{6.685}$$

$$x_n(t_0 - T_s) = x_n(1), \tag{6.686}$$

$$x_n(t_0 - (M-1)T_s) = x_n(M-1). \tag{6.687}$$

The subscript "n" denotes the sensor whose output is being sampled. The argument specifies the sample.

We define a complex vector, $\widetilde{\mathbf{x}}_n$, consisting of the time samples. To be consistent with conventional DFT notation, we reorder the time samples so that the most recent sample is at the bottom of the vector,

$$\widetilde{\mathbf{x}}_n \triangleq \begin{bmatrix} x_n(t_0 - (M-1)T_s) \\ \vdots \\ x_n(t_0) \end{bmatrix} = \begin{bmatrix} x_n(M-1) \\ x_n(M-2) \\ \vdots \\ x_n(0) \end{bmatrix}, \quad n = 0, \cdots, N-1. \tag{6.688}$$

This is a vector of the time samples. We use a superscript "$\sim$" to distinguish it from the vector down the array.

The DFT of $\widetilde{\mathbf{x}}_n$ is

$$X_n\{m\} = \sum_{k=0}^{M-1} x_n(M-1-k)\,(e^{-j\frac{2\pi}{M}})^{km}$$

[19]We discuss whether this technique is optimum later in this section.

$$= \sum_{k=0}^{M-1} x_n(M-1-k)\,(F_M)^{km}, \quad m = 0, 1, \cdots, M-1, \tag{6.689}$$

where

$$F_M \triangleq e^{-j\frac{2\pi}{M}}. \tag{6.690}$$

In matrix notation,

$$\widetilde{\mathbf{X}}_n = \mathbf{F}_M\, \widetilde{\mathbf{x}}_n, \tag{6.691}$$

where

$$[\mathbf{F}_M]_{kl} = (F_M)^{kl} = \left(e^{-j\frac{2\pi}{M}}\right)^{kl}. \tag{6.692}$$

The vector $\widetilde{\mathbf{X}}_n$ corresponds to the M frequency components at the output of sensor n,

$$\widetilde{\mathbf{X}}_n \triangleq [\; X_n\{0\} \;\; \cdots \;\; X_n\{m\} \;\; \cdots \;\; X_n\{M-1\} \;]^T. \tag{6.693}$$

Recall that the frequency samples are arranged in the following order,[20]

m	f
0	0
1	$\triangle f$
$\vdots$	
$\frac{M}{2}-1$	$(\frac{M}{2}-1)\triangle f$
$\frac{M}{2}$	$\frac{M}{2}\triangle f = -\frac{M}{2}\triangle f$
$\frac{M}{2}+1$	$(\frac{M}{2}+1)\triangle f = (-\frac{M}{2}+1)\triangle f$
$\vdots$	
$M-1$	$(M-1)\triangle f = -\triangle f$

The weight vector is constructed in a similar manner. We define

$$\widetilde{\mathbf{w}}_n \triangleq \begin{bmatrix} w_{n,M-1} \\ w_{n,M-2} \\ \vdots \\ w_{n,0} \end{bmatrix}. \tag{6.694}$$

Then, proceeding as in (6.690)–(6.693),

$$\widetilde{\mathbf{W}}_n = \mathbf{F}_M\, \widetilde{\mathbf{w}}_n, \tag{6.695}$$

[20] For example, [OS89].

where

$$\widetilde{\mathbf{W}}_n = \left[\begin{array}{cccc} W_n\{0\} & W_n\{1\} & \cdots & W_n\{M-1\} \end{array} \right]^T. \tag{6.696}$$

In each frequency bin we construct a vector down the array

$$\mathbf{X}\{m\} \triangleq \left[\begin{array}{cccc} X_0\{m\} & X_1\{m\} & \cdots & X_{N-1}\{m\} \end{array} \right]^T, \quad m = 0, \cdots, M-1. \tag{6.697}$$

This vector consists of the frequency components of the array in frequency bin m. We process each of these vectors with a **narrowband beamforming** algorithm to obtain a set of M scalar outputs

$$Y\{m\} = \mathbf{w}^H\{m\}\,\mathbf{X}\{m\}, \quad m = 0, \cdots M-1. \tag{6.698}$$

We define the vector across frequencies as

$$\widetilde{\mathbf{Y}} = \left[\begin{array}{cccc} Y\{0\} & Y\{1\} & \cdots & Y\{M-1\} \end{array} \right]^T. \tag{6.699}$$

We take inverse DFT (IDFT) to obtain M output time samples,

$$y(M-1-k) = \frac{1}{M}\sum_{m=0}^{M-1} Y\{m\}\left(e^{j\frac{2\pi}{M}}\right)^{km}, \quad k = 0, \cdots M-1. \tag{6.700}$$

In matrix form,

$$\widetilde{\mathbf{y}} = \mathbf{F}_M^{-1}\widetilde{\mathbf{Y}} = \frac{1}{M}\mathbf{F}_M^H\widetilde{\mathbf{Y}}, \tag{6.701}$$

where

$$\begin{aligned} \widetilde{\mathbf{y}} &\triangleq \left[\begin{array}{cccc} y(M-1) & y(M-2) & \cdots & y(0) \end{array} \right]^T \\ &= \left[\begin{array}{ccc} y(t_0-(M-1)T_s) & \cdots & y(t_0) \end{array} \right]^T. \end{aligned} \tag{6.702}$$

By using the weighting techniques in (6.698) we are treating each frequency component independently. Most DFT beamformers used in practice are implemented in this manner.

We can also use a **sliding window technique**, in which the DFT is computed each time a new sample enters the buffer. In this case, we write (6.701) as

$$y(M-1-k) = \frac{1}{M}\sum_{m=0}^{M-1} Y\{m\}\left(e^{j\frac{2\pi}{M}}\right)^{km}, \quad k = 0, \cdots M-1, \tag{6.703}$$

and observe that the first sample is

$$y\left(t_0-(M-1)T_s\right)=\frac{1}{M}\sum_{m=0}^{M-1}Y\{m\}, \tag{6.704}$$

so that the output is obtained by summing the frequency samples. The result in (6.704) corresponds to the IDFT at $t=t_0-(M-1)T_s$. We avoid doing the entire IDFT because we only want a single time sample.

If we want to consider more general processing, we would define a composite $NM\times 1$ data vector

$$\widetilde{\mathbf{X}}\triangleq\begin{bmatrix}\widetilde{\mathbf{X}}\{0\}\\ \widetilde{\mathbf{X}}\{1\}\\ \vdots\\ \widetilde{\mathbf{X}}\{M-1\}\end{bmatrix}, \tag{6.705}$$

and a composite weight vector,

$$\widetilde{\mathbf{W}}\triangleq\begin{bmatrix}\widetilde{\mathbf{W}}\{0\}\\ \widetilde{\mathbf{W}}\{1\}\\ \vdots\\ \widetilde{\mathbf{W}}\{M-1\}\end{bmatrix}. \tag{6.706}$$

Then

$$Y=\widetilde{\mathbf{W}}^H\widetilde{\mathbf{X}}. \tag{6.707}$$

The model in (6.705)–(6.707) allows us to use a wider class of constraints on the frequency domain beamformers.[21] For the present, we utilize independent narrowband beamformers in each frequency bin and sum the output using (6.704) to obtain the output.

Step 2: Array manifold vectors: In order to construct the weighting for the mth frequency bin, we need the steering vector and the spatial spectral matrix.

The array manifold vector as a function of frequency is

$$\mathbf{v}_\omega=\begin{bmatrix}e^{-j2\pi f\tau_0} & e^{-j2\pi f\tau_1} & \dots & e^{-j2\pi f\tau_{N-1}}\end{bmatrix}^T,$$

[21] The more general processing in (6.705)–(6.707) is not used in any actual systems that we are aware of.

where τ_n is the delay to the nth sensor relative to the origin of coordinates. The subscript "ω" denotes that we retain the frequency dependence explicitly.

The frequency at the mth frequency bin is

$$f_m = \begin{cases} f_c + m\Delta f & m = 0, \cdots, \frac{M}{2} - 1, \\ f_c + (m - M)\Delta f & m = \frac{M}{2}, \cdots, M - 1, \end{cases} \tag{6.708}$$

where

$$\Delta f = \frac{1}{MT_s} = \frac{B_s}{M}. \tag{6.709}$$

For simplicity, we consider a standard linear array. Then

$$\tau_n = -\frac{d_n u}{c}, \tag{6.710}$$

and

$$d_n = (n - \frac{N-1}{2})d = (n - \frac{N-1}{2})\frac{\lambda_u}{2}, \quad n = 0, \cdots, N-1, \tag{6.711}$$

where λ_u is the wavelength corresponding to the highest (upper) frequency.

Using (6.708), (6.710), and (6.711) in (6.705) gives an expression for nth element of the array manifold vector for the mth frequency bin,

$$[\mathbf{v}_m(u)]_n = \begin{cases} \exp\left\{j2\pi(f_c + m\Delta f)(n - \frac{N-1}{2})\frac{\lambda_u u}{2c}\right\}, \\ \qquad m = 0, \cdots, \frac{M}{2} - 1, \quad n = 0, \cdots, N-1, \\ \exp\left\{j2\pi\left(f_c + (m - M)\Delta f\right)(n - \frac{N-1}{2})\frac{\lambda_u u}{2c}\right\}, \\ \qquad m = \frac{M}{2}, \cdots, M-1, \quad n = 0, \cdots, N-1. \end{cases} \tag{6.712}$$

Using

$$f_c = \frac{c}{\lambda_c}, \tag{6.713}$$

and

$$B_f = \frac{B_s}{f_c}, \tag{6.714}$$

we have

$$\boxed{[\mathbf{v}_m(u)]_n = \begin{cases} \exp\{j\pi u \frac{1}{1+\frac{B_f}{2}}\left[(n - \frac{N-1}{2}) + (\frac{m}{M})(n - \frac{N-1}{2})B_f\right]\}, \\ \qquad m = 0, \cdots, \frac{M}{2} - 1, \quad n = 0, \cdots, N-1. \\ \exp\{j\pi u \frac{1}{1+\frac{B_f}{2}}\left[(n - \frac{N-1}{2}) + (\frac{m}{M} - 1)(n - \frac{N-1}{2})B_f\right]\}, \\ \qquad m = \frac{M}{2}, \cdots, M-1, \quad n = 0, \cdots, N-1. \end{cases}} \tag{6.715}$$

as the desired result. The first term in the exponent is the familiar narrow-band result and the second term models the broadband effect.

Step 3: Spatial spectral matrix: The next step is to compute the spatial spectral matrix in frequency bin m. We use a derivation similar to that given in Godard [God95]. The spectral spatial matrix in the mth bin is

$$\mathbf{S}_{\mathbf{X}}\{m\} \triangleq E\left[\mathbf{X}\{m\}\mathbf{X}^H\{m\}\right]. \tag{6.716}$$

We use the brace to distinguish this matrix from the usual (continuous in ω) spectral matrix $\mathbf{S}_{\mathbf{X}}(\omega)$.

The np-element of this matrix is

$$[\mathbf{S}_{\mathbf{X}}\{m\}]_{np} = E\left[X_n\{m\}X_p^*\{m\}\right]. \tag{6.717}$$

From the DFT relationship,

$$X_n\{m\} = \sum_{k=0}^{M-1} x_n(M-1-k)\left(e^{-j\frac{2\pi}{M}}\right)^{km}, \tag{6.718}$$

where $x_n(k)$ is the time sample from the kth tap of the nth sensor. Using (6.718) in (6.717) gives

$$[\mathbf{S}_{\mathbf{X}}\{m\}]_{np} = E\left\{\sum_{k=0}^{M-1} x_n(M-1-k)\left(e^{-j\frac{2\pi}{M}}\right)^{km}\sum_{q=0}^{M-1} x_p^*(M-1-q)\left(e^{j\frac{2\pi}{M}}\right)^{qm}\right\}. \tag{6.719}$$

Now recall from (6.688) that

$$\tilde{\mathbf{x}}_n = \left[\begin{array}{cccc} x_n(M-1) & \cdots & x_n(1) & x_n(0) \end{array}\right]^T. \tag{6.720}$$

We also define a vector corresponding to the kth row of the DFT matrix,

$$\mathbf{e}(m) = \begin{bmatrix} 1 \\ e^{\left(j\frac{2\pi}{M}\right)m} \\ \vdots \\ e^{\left(j\frac{2\pi}{M}\right)km} \\ \vdots \\ e^{\left(j\frac{2\pi}{M}\right)(M-1)m} \end{bmatrix}, \quad m = 0, \cdots, M-1. \tag{6.721}$$

Then, (6.719) can be written as

$$\begin{aligned}
[\mathbf{S_X}\{m\}]_{np} &= \mathbf{e}^H(m)\, E\left[\tilde{\mathbf{x}}_n \tilde{\mathbf{x}}_p^H\right] \mathbf{e}(m) \\
&= \mathbf{e}^H(m) \left\{ [\mathbf{R_x}]_{np} \right\} \mathbf{e}(m),
\end{aligned} \tag{6.722}$$

where $[\mathbf{R_x}]_{np}$ is the temporal correlation between the tap vector at sensor n and the tap vector at sensor p.

The next step is to derive an expression for $[\mathbf{R_x}]_{np}$. The correlation function matrix of the complex baseband vector $\mathbf{x}(t)$ can be written as

$$\mathbf{R}_{\mathbf{x}(t)}(\tau) = \int_{-\infty}^{\infty} \mathbf{S_x}(2\pi f_\Delta)\, e^{j2\pi f_\Delta \tau}\, df_\Delta, \tag{6.723}$$

where $\mathbf{S_x}(2\pi f)$ is the spectral matrix of the complex baseband vector waveform

$$\mathbf{S_x}(2\pi f_\Delta) = 2\left[\mathbf{S}_{\mathbf{x},bp}(2\pi f_\Delta + 2\pi f_c)\right]_{LP}. \tag{6.724}$$

Then

$$\mathbf{x}_k(t) = \mathbf{x}\left(t - (M-1-k)T_s\right) \tag{6.725}$$

and

$$\mathbf{x}_q^*(t) = \mathbf{x}^*\left(t - (M-1-q)T_s\right). \tag{6.726}$$

The correlation matrix is,

$$[\mathbf{R_x}]_{kq} = \mathbf{R}_{\mathbf{x}(t)}\left((k-q)T_s\right). \tag{6.727}$$

Using (6.723),

$$[\mathbf{R_x}]_{kq} = \int_{-\infty}^{\infty} \mathbf{S_x}(2\pi f_\Delta)\, e^{j2\pi f_\Delta [(k-q)T_s]} df_\Delta. \tag{6.728}$$

Now define

$$\mathbf{e}(f_\Delta) = \left[\begin{array}{ccccccc} 1 & e^{j2\pi f_\Delta T_s} & \cdots & e^{j2\pi f_\Delta T_s q} & \cdots & e^{j2\pi f_\Delta T_s (M-1)} \end{array}\right]^T. \tag{6.729}$$

Using (6.729), we can write (6.728) in matrix form as

$$[\mathbf{R_x}] = \int_{-\infty}^{\infty} \mathbf{S_x}(2\pi f_\Delta)\mathbf{e}(f_\Delta)\mathbf{e}^H(f_\Delta) df_\Delta. \tag{6.730}$$

Using (6.730) in (6.722) gives

$$[\mathbf{S_X}\{m\}]_{np} = \int_{-\infty}^{\infty} \mathbf{S_x}(2\pi f_\Delta)\mathbf{e}^H\{m\}\mathbf{e}(f_\Delta)\mathbf{e}^H(f_\Delta)\mathbf{e}\{m\} df_\Delta. \tag{6.731}$$

Now define

$$a(f_\Delta, m) \triangleq \left|\mathbf{e}^H(f_\Delta)\mathbf{e}\{m\}\right|^2. \tag{6.732}$$

Using (6.732) in (6.731) gives

$$\mathbf{S_x}\{m\} = \int_{-\infty}^{\infty} \mathbf{S_x}(2\pi f_\Delta)\, a(f_\Delta, m))df_\Delta. \tag{6.733}$$

Then,

$$\begin{aligned} \mathbf{e}^H(f_\Delta)\,\mathbf{e}\{m\} &= \sum_{q=0}^{M-1} e^{-j2\pi(f_\Delta T_s - \frac{m}{M})q} \\ &= e^{-j\pi(M-1)(f_\Delta T_s - \frac{m}{M})} \frac{\sin\pi(f_\Delta T_s M - m)}{\sin\frac{\pi}{M}(f_\Delta T_s M - m)}, \end{aligned} \tag{6.734}$$

or

$$\begin{aligned} a(f_\Delta, m) &= \frac{\sin^2\left(\pi(f_\Delta T_s M - m)\right)}{\sin^2\left(\frac{\pi}{M}(f_\Delta T_s M - m)\right)} = \frac{\sin^2\left(\pi(\frac{f_\Delta}{B_s}M - m)\right)}{\sin^2\left(\frac{\pi}{M}(\frac{f_\Delta}{B_s}M - m)\right)} \\ &= \frac{\sin^2\left(\frac{\pi M}{B_s}(f_\Delta - m\Delta f)\right)}{\sin^2\left(\frac{\pi}{B_s}(f_\Delta - m\Delta f)\right)}, \end{aligned} \tag{6.735}$$

where

$$B_s = T_s^{-1} \tag{6.736}$$

is the sampling frequency. The function $a(f_\Delta, m)$ provides a window on $\mathbf{S_x}(2\pi f_\Delta)$ and reflects how power at frequencies unequal to f_m (the frequency of the mth bin) affects $\mathbf{S_x}\{m\}$. A normalized plot of $a(f_\Delta, m)$ with $m = 0$ for $M = 16$ and $M = 64$ is shown in Figure 6.117 . The normalization constant is M^2.

For $m = 0$, the first zeros are at

$$\frac{f_\Delta}{B_s} = \pm\frac{1}{M}. \tag{6.737}$$

The bin spacing is

$$\frac{\Delta f}{B_s} = \frac{1}{M}. \tag{6.738}$$

If $\mathbf{S_x}(2\pi f)$ is essentially constant over the interval,

$$-\frac{1}{2M} \leq \frac{f_\Delta}{B_s} \leq \frac{1}{2M}, \tag{6.739}$$

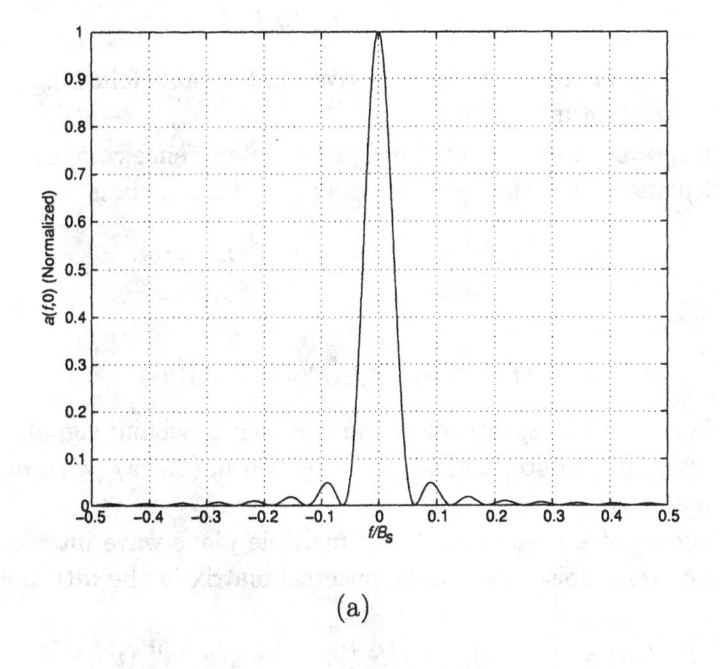

(a)

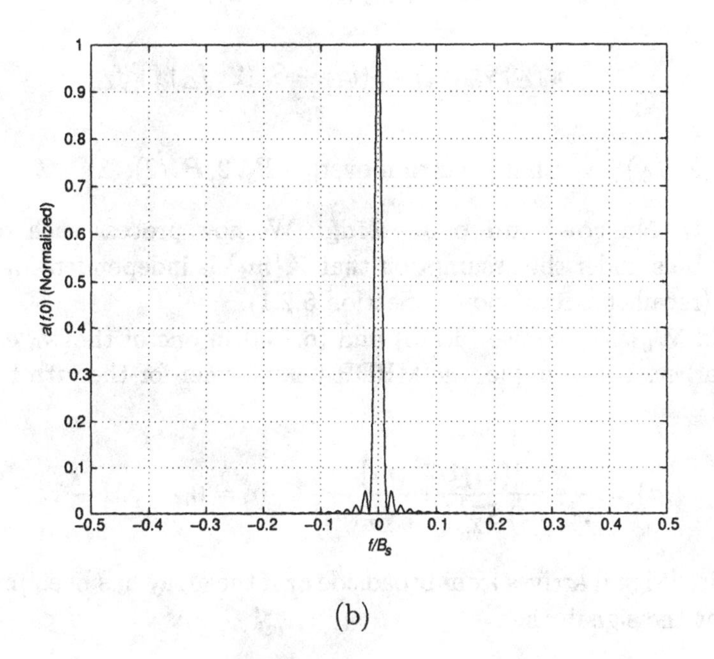

(b)

Figure 6.117 Normalized $a(f,k)$ versus f/B_s: (a)$K = 16$; (b)$K = 64$.

then

$$\mathbf{S_X}\{m\} \simeq M^2\,\mathbf{S_x}(2\pi f_\Delta), \tag{6.740}$$

where the Δf-to-m correspondence is given in the table following (6.693) or through the relation in (6.708).

For the special case in which the signal or interference component of $\mathbf{x}(t)$ consists of plane waves, the corresponding spectral matrix is

$$\left[\mathbf{S_X}(2\pi f_\Delta)\right]_{np} = \mathbf{S}_f(2\pi f_\Delta)\,e^{j2\pi f_\Delta(\tau_p-\tau_n)}, \tag{6.741}$$

or

$$\mathbf{S_X}(2\pi f_\Delta) = \mathbf{S}_f(2\pi f_\Delta)\,\mathbf{v}_m(u_s)\,\mathbf{v}_m^H(u_s), \tag{6.742}$$

where $\mathbf{S}_f(2\pi f_\Delta)$ is the spectrum of the complex baseband signal, u_s is the signal direction in u-space, and $\mathbf{v}_m(u)$ is defined in (6.715) (with m and Δf correctly matched).

For the case of a single signal and multiple plane-wave interferers with independent white noise, the spatial spectral matrix in the mth bin is,

$$\mathbf{S_X}\{m\} = \int_{-\infty}^{\infty} a(f_\Delta, m)\,[S_f(2\pi f_\Delta)\,\mathbf{v}_m(u_s)\,\mathbf{v}_m^H(u_s)$$

$$+\sum_{i=1}^{J} S_{f_i}(2\pi f_\Delta)\,\mathbf{v}_m(u_i)\,\mathbf{v}_m^H(u_i) + S_w(2\pi f_\Delta)\,\mathbf{I}\,]df_\Delta, \tag{6.743}$$

where $S_w(2\pi f_\Delta)$ has a flat spectrum over $(-B_s/2, B_s/2)$,

Step 4: Narrowband processing: We now process each of the m frequency bins under the assumption that $\mathbf{X}\{m_i\}$ is independent of $\mathbf{X}\{m_j\}$, $m_i \neq m_j$ (recall the discussion in Section 5.2.1).

To find $\mathbf{W}_0\{m\}$, we use (6.715) and (6.733) in one of the $\mathbf{w}_0$ equations defined earlier. For example, the MPDR beamformer for the mth frequency bin is

$$\mathbf{W}_{mpdr}^H\{m\} = \frac{\mathbf{v}_m^H(u_s)\,\mathbf{S_X}^{-1}\{m\}}{\mathbf{v}_m^H(u_s)\,\mathbf{S_X}^{-1}\{m\}\,\mathbf{v}_m(u_s)}, \quad m = 0, \cdots, M-1. \tag{6.744}$$

If the desired signal arrives from broadside or if the array has been presteered to point at the signal, then

$$\mathbf{W}_{mpdr}^H\{m\} = \frac{\mathbf{1}^T\,\mathbf{S_X}^{-1}\{m\}}{\mathbf{1}^T\,\mathbf{S_X}^{-1}\{m\}\,\mathbf{1}}, \quad m = 0, \cdots, M-1. \tag{6.745}$$

The LCMP beamformer follows from (6.358) as

$$\mathbf{W}_{lcmp}^{H}\{m\} = \mathbf{g}_m^H \left[\mathbf{C}_m^H \mathbf{S}_{\mathbf{X}}^{-1}\{m\}\mathbf{C}_m\right]^{-1} \mathbf{C}_m^H \mathbf{S}_{\mathbf{X}}^{-1}\{m\}, \tag{6.746}$$

corresponding to the constraint equation,

$$\mathbf{W}^H\{m\}\,\mathbf{C}_m = \mathbf{g}_m^H.$$

To evaluate the output SNR_0 or the array gain, we have

$$P_o = E\left[|y(t_0)|^2\right] = E\left[|y(0)|^2\right]. \tag{6.747}$$

Substituting (6.704) into (6.719) gives

$$P_o = \frac{1}{M^2}\,E\left[\sum_{m=0}^{M-1} Y\{m\} \sum_{l=0}^{M-1} Y^*\{l\}\right]. \tag{6.748}$$

If we assume the $Y\{m\}$ are uncorrelated, this reduces to

$$P_o = \frac{1}{M^2}\,E\left[\sum_{m=0}^{M-1} |Y\{m\}|^2\right]. \tag{6.749}$$

The signal component output of the mth bin is,

$$Y\{m\} = \mathbf{W}_0^H\{m\}\,\mathbf{F}\{m\}, \tag{6.750}$$

where $\mathbf{F}\{m\}$ is the DFT of the signal in the mth bin. Using (6.750) in (6.749) gives

$$P_{so} = \frac{1}{M^2}\sum_{m=0}^{M-1} \mathbf{W}_0^H\{m\}\,\mathbf{S}_\mathbf{f}\{m\}\,\mathbf{W}_0\{m\}, \tag{6.751}$$

where $\mathbf{S}_\mathbf{f}\{m\}$ is given by the first term on the right side of (6.743).

If each weight vector contains a distortionless constraint, then

$$P_{so} = \sigma_s^2. \tag{6.752}$$

Similarly, the output due to interference and noise is

$$P_{in,o} = \frac{1}{M^2}\sum_{m=0}^{M-1} \mathbf{W}_0^H\{m\}\,\mathbf{S}_\mathbf{n}\{m\}\,\mathbf{W}_0\{m\}, \tag{6.753}$$

where $\mathbf{S}_\mathbf{n}\{m\}$ is given by the second and third term on the right side of (6.743).

Thus, the SNR_0 is

$$SNR_0 = \frac{P_o}{P_{in,o}} \tag{6.754}$$

and the array gain is

$$A_0 = \frac{\sigma_n^2 \, P_o}{\sigma_s^2 \, P_{in,o}}. \tag{6.755}$$

This completes our derivation. We now consider several examples.

Example 6.13.1

Consider a 10-element uniform linear array with element spacing equal to $\lambda_u/2$. The desired signal is a plane-wave signal arriving from u_s with fractional bandwidth B_f. The signal spectrum is flat, as shown in Figure 6.112, with total power σ_s^2. The interference is white noise. We use a conventional beamformer in each frequency bin. For white noise, the conventional beamformer is also the MPDR beamformer. The array gain would be 10 if the signal were narrowband or if the signal arrived from broadside.

In Figure 6.118, we plot the array gain versus u_s/BW_{NN} for $M = 1, 2, 4, 8$, and 16. In part (a), we consider $B_f = 0.2$. In part (b), we consider $B_f = 0.4$. In part (c), we consider $B_f = 0.6$. In part (d), we consider $B_f = 0.8$.

We see that, by increasing the number of frequency bins, we can approach the same performance as in the narrowband case.

Example 6.13.2

Consider a 10-element uniform linear array with element spacing equal to $\lambda_u/2$. The desired signal arrives from broadside. It has a flat spectrum with bandwidth B_s and fractional bandwidth B_f. A single interfering plane-wave signal with the same spectrum arrives from u_I. The *SNR* is 0 dB and the *INR* is 30 dB. We use an MPDR beamformer in each frequency bin.

In Figure 6.119, we plot the array gain versus u_I/BW_{NN} for $M = 1, 2, 4, 8$, and 16. In part (a), we consider $B_f = 0.4$. In part (b), we consider $B_f = 0.8$.

We see that there is marginal improvement with increasing M. However, this example is misleading because the array processor is not stressed by the single interferer and uses its temporal degrees of freedom to suppress the interferer. In the next example we consider a more complicated environment.

Example 6.13.3

Consider the same model as in Example 6.13.1, except we assume that there are $N-1$ uncorrelated plane-wave interferers. $N-2$ narrowband interferers are fixed in orthogonal locations in u-space,

$$u_i = \pm 0.3, \pm 0.5, \pm 0.7, \pm 0.9 \quad i = 1, \cdots, 4. \tag{6.756}$$

The broadband interferer arrives from u_I and we vary u_I from $-1 \leq u_I \leq 1$. Each interferer has a 30 dB *INR*.

In Figure 6.120, we plot the array gain versus u_I/BW_{NN} for $M = 1, 2, 4, 8$, and 16. In part (a), we consider $B_f = 0.4$. In part (b), we consider $B_f = 0.8$.

If we compare Figure 6.119(a) with Figure 6.120(a), we see that the array gain is larger with multiple interferers. The reason is that the total input of the interferers is larger by 9 dB and the beamformer can place significant nulls of the eight narrowband interferers

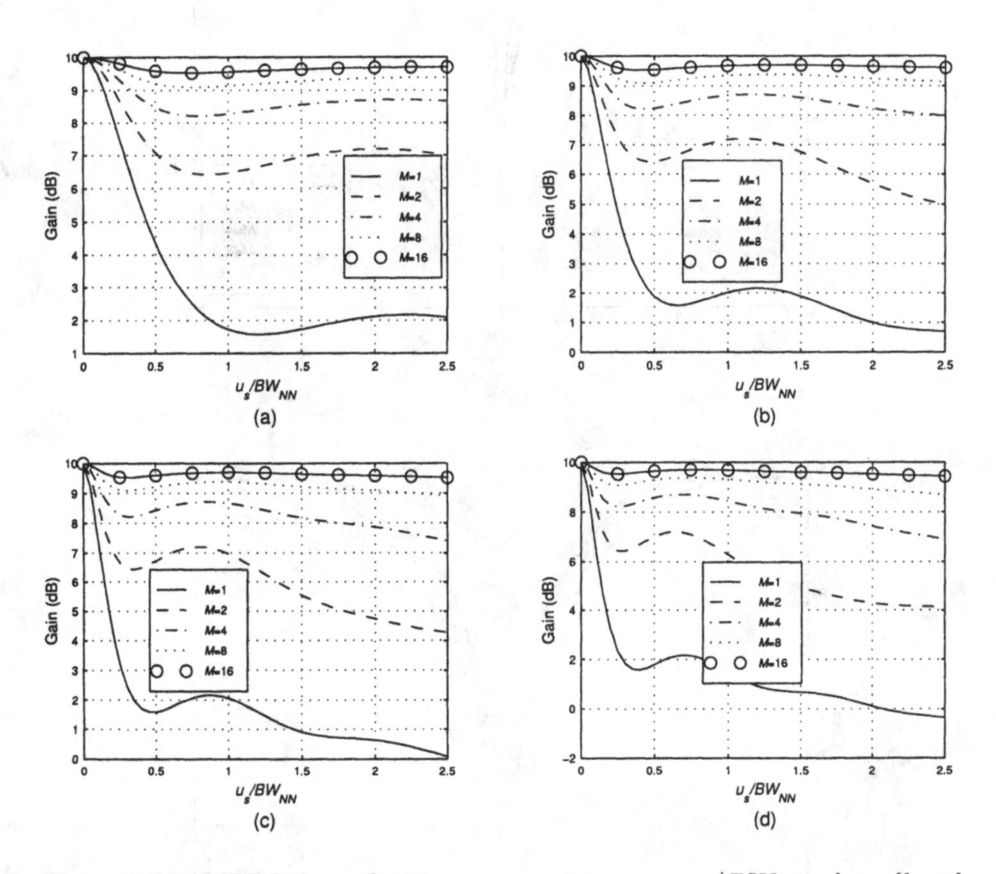

Figure 6.118 MPDR beamformer; array gain versus u_s/BW_{NN}, broadband signal in white noise, $M = 1, 2, 4, 8$, and 16; (a) $B_f = 0.2$, (b) $B_f = 0.4$, (c) $B_f = 0.6$, (d) $B_f = 0.8$.

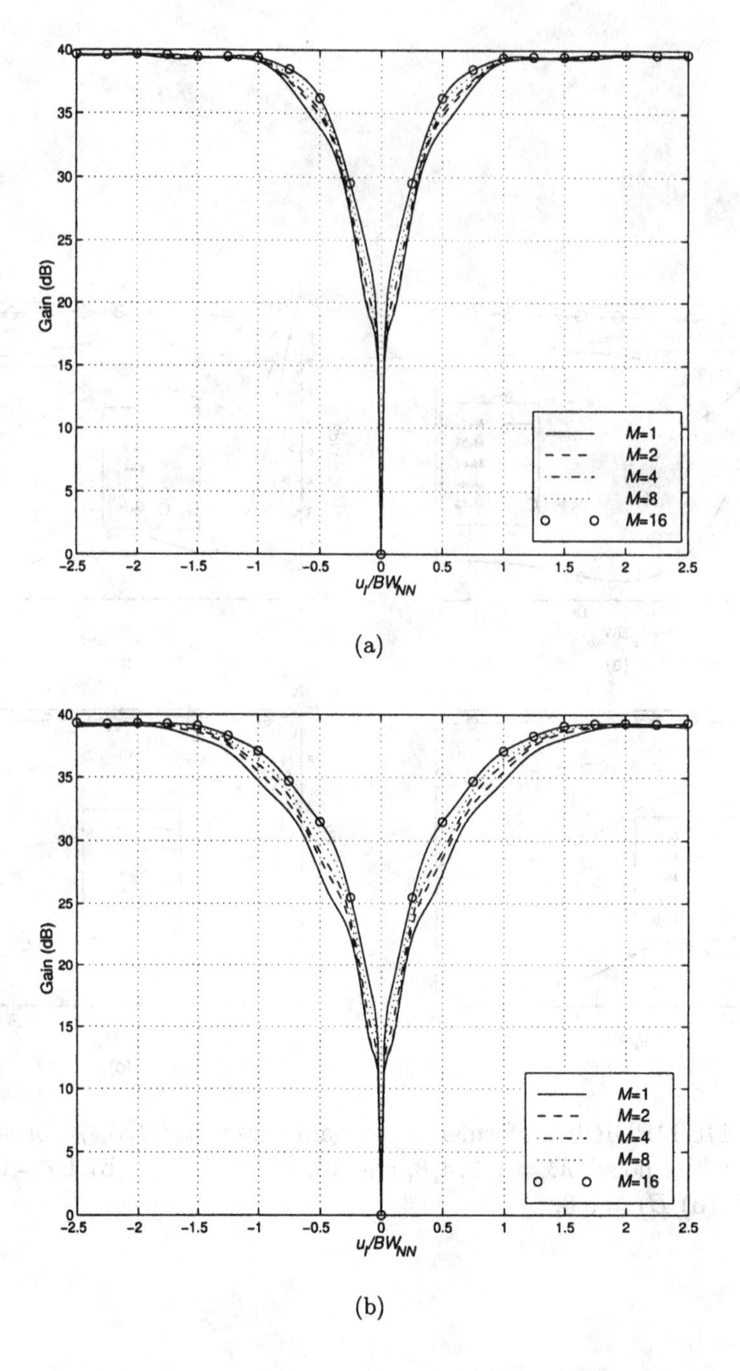

Figure 6.119 Array gain versus u_I/BW_{NN}: broadband signal in white noise; $N = 10$, *SNR* = 0 dB, *INR* = 30 dB, $M = 1, 2, 4, 8$, and 16: (a)$B_f = 0.4$; (b)$B_f = 0.8$.

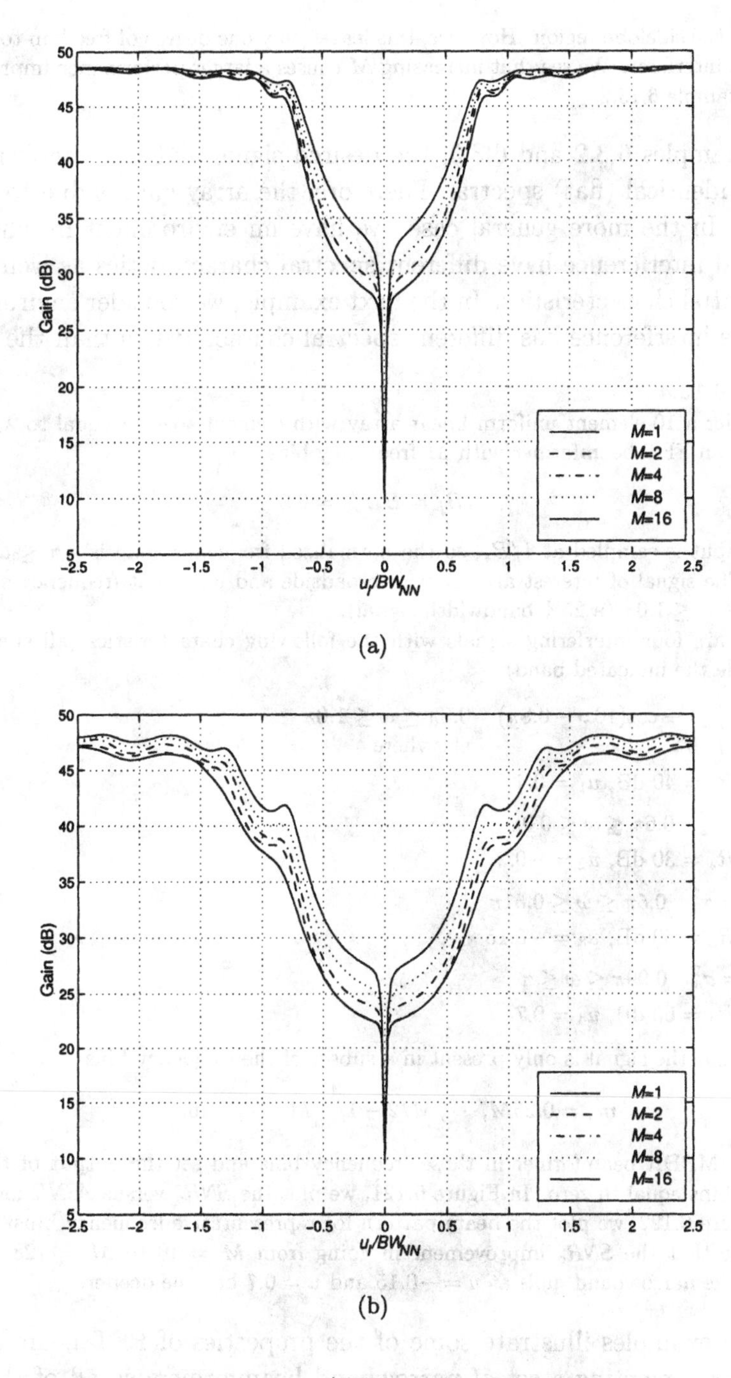

Figure 6.120 Array gain versus u_I/BW_{NN}; broadband signal in white noise: (a)$B_f = 0.4$; (b)$B_f = 0.8$.

that are in the sidelobe region. However, this leaves only one degree of freedom to null the broadband interferer. We see that increasing M causes a larger performance improvement than in Example 6.13.2.

In Examples 6.3.2 and 6.3.3, the desired signal and the interfering signal have identical (flat) spectra. Therefore, the array gain is due to spatial filtering. In the more general case, we have an environment in which the signal and interference have different spectral characteristics as well as different spatial characteristics. In the next example, we consider environments where the interference has different spectral characteristics than the signal.

Example 6.13.4

Consider a 10-element uniform linear array with element spacing equal to $\lambda_u/2$. We implement an FFT beamformer with M frequency bins.

$$B_f = B_s/f_c = 0.8.$$

The input is sampled at $1/B_s$, so the normalized frequency span is $-\pi \leq \omega \leq \pi$ in radians. The signal of interest arrives from broadside and has a flat frequency spectrum over $0.5\pi \leq \omega \leq 1.0\pi$ (a 25% bandwidth signal).

There are four interfering signals with the following characteristics (all spectra are zero outside the indicated band):

(i) $S_{I1} = \begin{cases} 1 + \cos(10\omega - 0.8\pi) & 0.6\pi \leq \omega \leq 1.0\pi \\ 0 & \text{elsewhere} \end{cases}$

$INR_1 = 40$ dB, $u_1 = 0.4$

(ii) $S_{I2} = \sigma_2^2 \quad 0.6\pi \leq \omega \leq 0.9\pi$

$INR_2 = 30$ dB, $u_2 = -0.2$

(iii) $S_{I3} = \sigma_3^2 \quad 0.5\pi \leq \omega \leq 0.51\pi$

$INR_3 = 60$ dB, $u_3 = -0.15$

(iv) $S_{I4} = \sigma_4^2 \quad 0.99\pi \leq \omega \leq \pi$

$INR_4 = 60$ dB, $u_4 = 0.7$

Note that the signal is only present in a subset of the frequency bins

$$m_s = 0.25M, \cdots, M/2 - 1, \quad M = 4, 8, 16, \cdots. \tag{6.757}$$

We use an MPDR beamformer in those frequency bins and set the output of the other frequency bins equal to zero. In Figure 6.121, we plot the SNR_o versus $ASNR$ for various M. In Figure 6.122, we plot the beam pattern for representative frequency bins.

We see that the SNR_o improvement in going from $M = 16$ to $M = 128$ is small. However, the narrowband nulls at $u = -0.15$ and $u = 0.7$ become deeper.

These examples illustrate some of the properties of FFT beamformers. We are implementing a set of narrowband beamformers so all of the techniques that we developed in Sections 6.2–6.12 carry over in a straightforward manner.

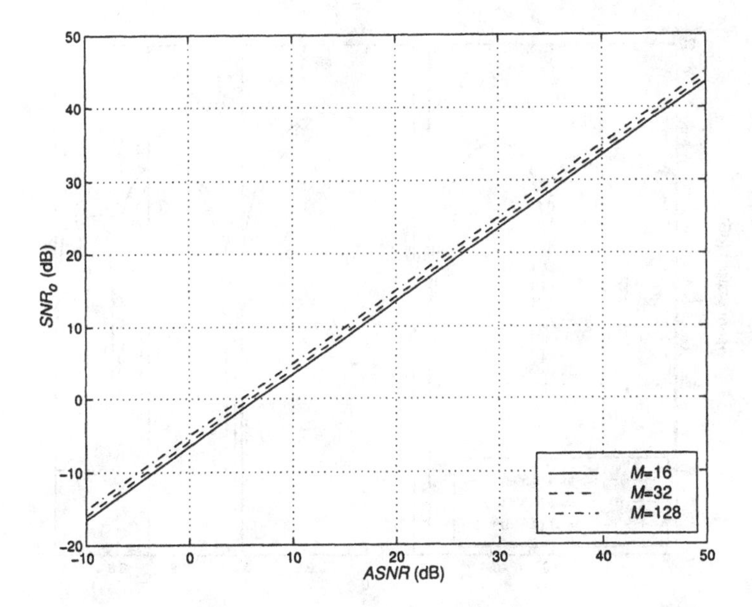

Figure 6.121 MPDR FFT beamformers: SNR_o versus $ASNR$, four interferers.

In this section we have analyzed the performance of DFT/FFT beamformer under the assumption that the relevant statistics about the signal and noise processes are known.

In the next section, we develop beamformers in which the filter behind each sensor is implemented in the time domain.

6.13.3 Finite impulse response (FIR) Beamformers

In this section, we study broadband beamformers that are implemented by placing a tapped delay line or finite impulse response(FIR) filter at the output of each sensor. We choose the impulse response (or tap weights) to achieve the desired frequency-wavenumber response. In practice, these weights are determined adaptively from the data. In this section, we assume that the statistics are known so the resulting beamformers correspond to the steady-state behavior. In Chapter 7, we develop the adaptive version of the FIR beamformer. An important advantage of the time domain implementation is that we can update the beamformer when each new snapshot arrives, in contrast to the FFT beamformer, which requires a block of snapshots to perform the FFT.

In Section 6.13.3.1, we develop the FIR model for a broadband beamformer. In Section 6.13.3.2, we analyze the performance.

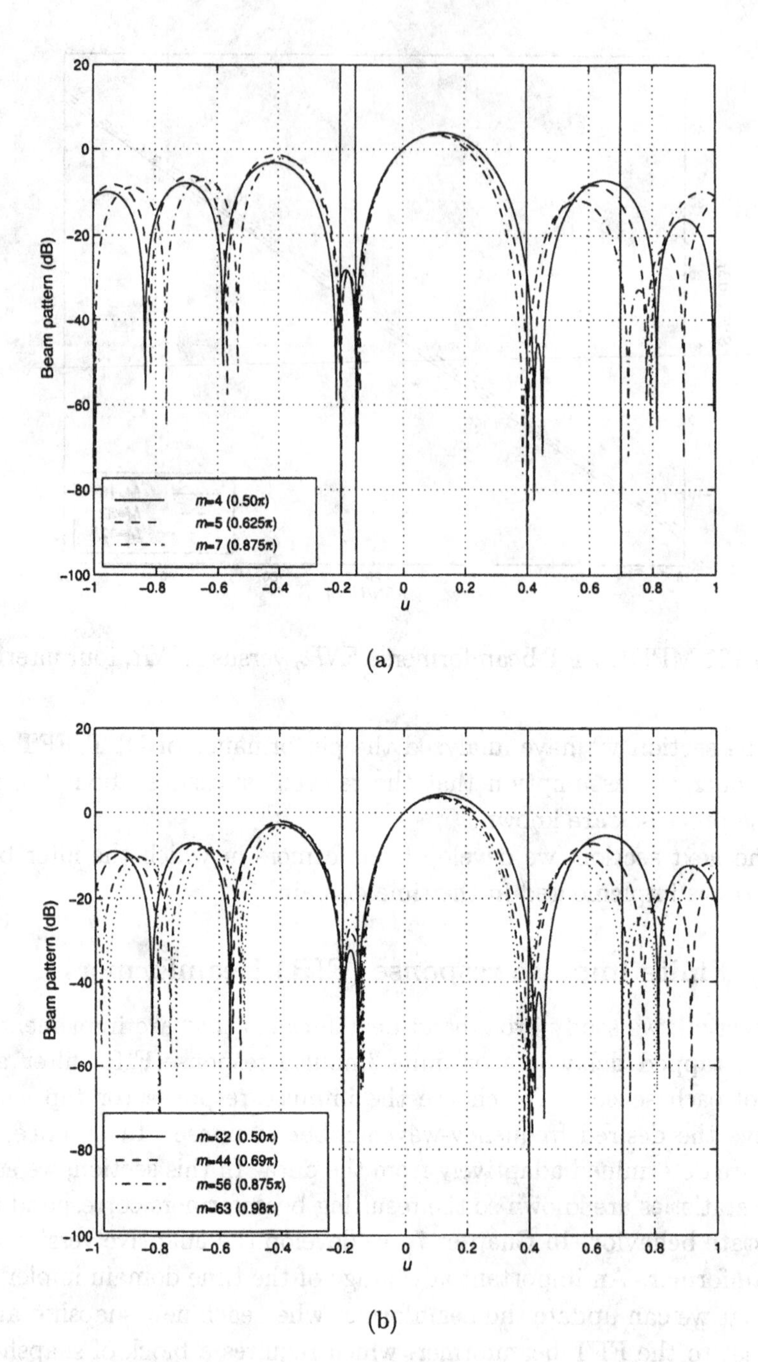

Figure 6.122 Beam patterns at selected frequencies: (a) $M = 16$; (b) $M = 128$.

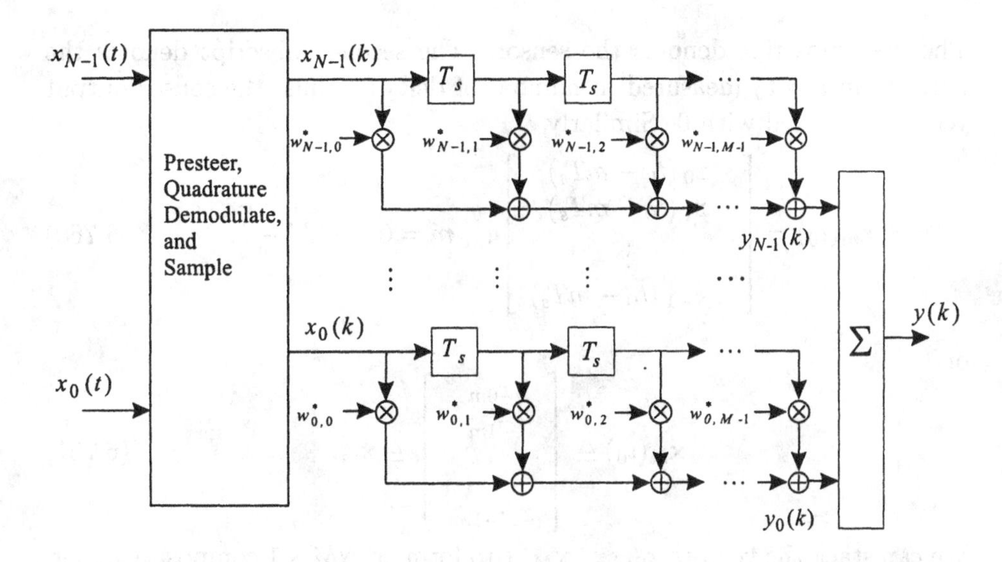

Figure 6.123 FIR beamformer.

6.13.3.1 FIR model

The FIR model of interest is shown in Figure 6.123. We assume that the output of each sensor has been quadrature demodulated so that the input to each of the FIR filter is a complex waveform. We are implementing the FIR at baseband.[22]

In order to simplify our initial discussion, we assume that the array has been pre-steered so the desired signal is aligned at the various sensors in the array.

We assume that the maximum frequency in the baseband signal is $B_s/2$. The tap spacing corresponds to the Nyquist sampling rate,

$$T_s = \frac{1}{B_s}. \tag{6.758}$$

The input to the first set of weights at time t_0 is

$$\mathbf{x}_0(t_0) = \begin{bmatrix} x_0(t_0) \\ x_1(t_0) \\ x_2(t_0) \\ \vdots \\ x_{N-1}(t_0) \end{bmatrix} \triangleq \begin{bmatrix} x_{00} \\ x_{10} \\ x_{20} \\ \vdots \\ x_{N-1,0} \end{bmatrix} \triangleq \mathbf{x}_0. \tag{6.759}$$

[22] A number of discussions in the literature (e.g., [Com88a]) utilize a bandpass implementation and a tapped delay line.

The first subscript denotes the sensor. The second subscript denotes the delay from $t = t_0$ measured in number of delays. Thus, the sensor output vector is indexed with 0. Similarly,

$$\mathbf{x}_m(t_0) = \begin{bmatrix} x_0(t_0 - mT_s) \\ x_1(t_0 - mT_s) \\ \vdots \\ x_{N-1}(t_0 - mT_s) \end{bmatrix}, \quad m = 0, \cdots, M-1, \tag{6.760}$$

or

$$\mathbf{x}_m(t_0) \triangleq \begin{bmatrix} x_{0m} \\ x_{1m} \\ \vdots \\ x_{N-1,m} \end{bmatrix} \triangleq \mathbf{x}_m. \tag{6.761}$$

We can stack the vectors $\mathbf{x}_0, \cdots, \mathbf{x}_{M-1}$ to form an $NM \times 1$ composite vector,

$$\mathbf{x} \triangleq \begin{bmatrix} \mathbf{x}_0 \\ \mathbf{x}_1 \\ \vdots \\ \mathbf{x}_{M-1} \end{bmatrix}. \tag{6.762}$$

We refer to this vector as a **tap-stacked** vector because the component vectors are indexed across the taps.

We can also specify the vector $\mathbf{x}$ with the double index, $[\mathbf{x}]_{nm}$, $n = 0, \cdots, N-1$, $m = 0, \cdots, M-1$, where the ordering is given by (6.761) and (6.762).

We can define an $NM \times 1$ array manifold vector corresponding to the tap-stacked vector $\mathbf{x}$. We define the spatial origin to be at the center of the array and the temporal origin to be at the sensor output. Then, the delay to a specific tap (for a linear array) is

$$\tau_{nm} = -\frac{d_n u}{c} + mT_s. \tag{6.763}$$

We first consider the case in which the quadrature demodulators are omitted. The $NM \times 1$ array manifold vector is given by

$$[\mathbf{v}_{f,u}]_{nm} = e^{-j2\pi f \tau_{nm}}. \tag{6.764}$$

The array manifold vector down the mth tap is just the mth component vector,

$$[\mathbf{v}_{f,u}]_m = \left[\begin{array}{c:c:c:c} e^{-j2\pi f(\tau_0 + mT_s)} & e^{-j2\pi f(\tau_1 + mT_s)} & \cdots & e^{-j2\pi f(\tau_{N-1} + mT_s)} \end{array}\right]^T. \tag{6.765}$$

If the linear array is uniformly spaced at

$$d = \frac{\lambda_u}{2}, \tag{6.766}$$

where λ_u is the wavelength corresponding to the highest frequency, then

$$[\mathbf{v}_{f,u}]_{nm} = \exp\left\{j2\pi f\left[(n - \frac{N-1}{2})\frac{u}{2f_u} - mT_s\right]\right\}. \tag{6.767}$$

If we include the quadrature demodulators, then the appropriate frequency for the taps is f_Δ. The array manifold vector is

$$[\mathbf{v}_{f_\Delta,u}(f_\Delta, u)]_{nm} = \exp\left\{j2\pi\left[f(n - \frac{N-1}{2})\frac{u}{2f_u} - f_\Delta(mT_s)\right]\right\}, \tag{6.768}$$

where

$$f_u = f_c + \frac{B_s}{2}. \tag{6.769}$$

The result in (6.768) can be rewritten as

$$[\mathbf{v}_{f_\Delta,u}(f_\Delta, u)]_{nm} = \exp\left\{j\pi\left[\frac{\left(1+\frac{f_\Delta}{f_c}\right)}{\left(1+\frac{B_f}{2}\right)}(n - \frac{N-1}{2})u - 2\frac{f_\Delta}{B_s}m\right]\right\}, \tag{6.770}$$

where

$$B_f \triangleq \frac{B_s}{f_c}. \tag{6.771}$$

The conjugate of the complex weights at the mth tap are denoted by the $N \times 1$ vector

$$\mathbf{w}_m = \begin{bmatrix} w_{0m} \\ w_{1m} \\ \vdots \\ w_{N-1,m} \end{bmatrix}. \tag{6.772}$$

The composite tap-stacked weight vector is denoted by the $NM \times 1$ vector,

$$\mathbf{w} = \begin{bmatrix} \mathbf{w}_0 \\ \mathbf{w}_1 \\ \mathbf{w}_2 \\ \vdots \\ \mathbf{w}_{M-1} \end{bmatrix}. \tag{6.773}$$

The array output at t_0 can be written as

$$y(t_0) = \sum_{m=0}^{M-1} \mathbf{w}_m^H \mathbf{x}_m, \tag{6.774}$$

or, equivalently,

$$y(t_0) = \mathbf{w}^H \mathbf{x}. \tag{6.775}$$

The beam pattern of the TDL beamformer is obtained by replacing $\mathbf{x}$ in (6.775) by the array manifold vector in (6.770),

$$B(f_\Delta, u) = \mathbf{w}^H \mathbf{v}_{f_\Delta,u}(f_\Delta, u). \tag{6.776}$$

For a plane-wave signal coming from the look direction, the values down the sensor at each tap are identical,

$$\begin{aligned} \mathbf{f}_0(t_0) &= f(t_0)\,\mathbf{1}, \\ &\vdots \\ \mathbf{f}_m(t_0) &= f(t_0 - mT_s)\,\mathbf{1}, \quad m = 0, \cdots, M-1. \end{aligned} \tag{6.777}$$

Thus, for "look-direction" signals, we can use the equivalent tapped delay shown in Figure 6.124. The equivalent weights are

$$h_m^* = \sum_{n=0}^{N-1} w_{nm}^*, \quad m = 0, \cdots, M-1, \tag{6.778}$$

and the output is

$$y(t_0) = \sum_{n=0}^{M-1} h_m^* f(t_0 - mT_s). \tag{6.779}$$

In order to implement the beamformer, we need the covariance matrix of $\mathbf{x}$ in (6.762). In Section 6.13.2, we derived the covariance matrix of $\widetilde{\mathbf{x}}$, which contained the same elements arranged in a different order.

Using the array manifold vector in (6.768), the $NM \times NM$ correlation matrix is given by

$$\mathbf{R}_\mathbf{x} = \int \mathbf{S}_\mathbf{x}(2\pi f_\Delta)\, df_\Delta = \int\int S_x(f_\Delta, u) \mathbf{v}_{f_\Delta,u}(f_\Delta, u) \mathbf{v}_{f_\Delta,u}^H(f_\Delta, u)\, df du, \tag{6.780}$$

where $S_x(f_\Delta, u)$ is the source spectrum in (f_Δ, u) space. For a single plane-wave signal arriving from u_s, (6.780) reduces to

$$\mathbf{R}_\mathbf{x} = \int_{-B_s/2}^{B_s/2} S_f(2\pi f_\Delta) \mathbf{v}_{f_\Delta,u}(f_\Delta, u) \mathbf{v}_{f_\Delta,u}^H(f_\Delta, u) df_\Delta. \tag{6.781}$$

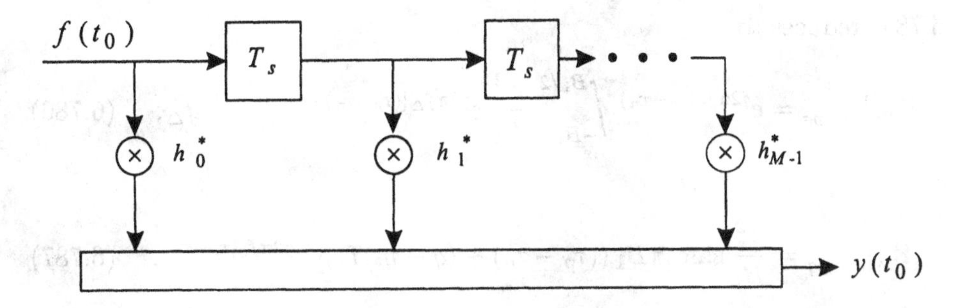

Figure 6.124 Equivalent "look-direction" FIR filter.

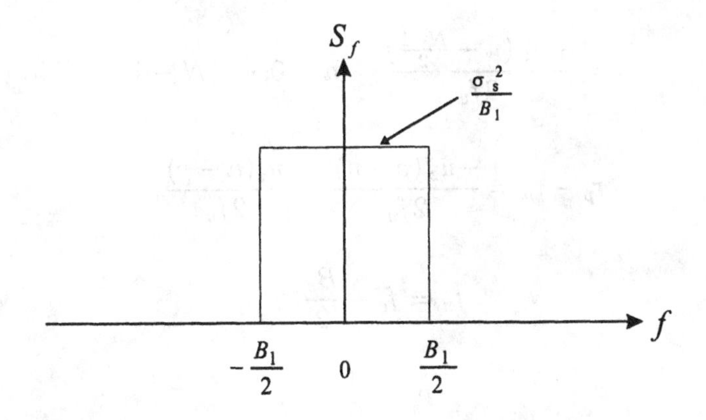

Figure 6.125 Flat signal spectrum $(-B_1/2, +B_1/2)$.

Using (6.768), the nm, pq element is

$$[\mathbf{R_x}]_{nm,pq} = \int_{-B_s/2}^{B_s/2} S_f(2\pi f_\Delta) e^{j2\pi f(\tau_p - \tau_n)} e^{j2\pi f_\Delta (q-m)T_s} df_\Delta. \tag{6.782}$$

In many cases, the signals or interferers will not occupy the entire bandwidth $[-B_s/2, B_s/2]$. We consider a signal with the flat spectrum shown in Figure 6.125, where

$$B_1 \leq B_s. \tag{6.783}$$

Then,

$$[\mathbf{R_x}]_{nm,pq} = \int_{-B_1/2}^{B_1/2} \frac{\sigma_s^2}{B_1} e^{j2\pi f(\tau_p - \tau_n)} e^{j2\pi f_\Delta (q-m)T_s} df_\Delta. \tag{6.784}$$

Writing

$$f = f_c + f_\Delta, \tag{6.785}$$

(6.784) reduces to

$$[\mathbf{R_x}]_{nm,pq} = e^{j2\pi f_c(\tau_p-\tau_n)} \int_{-B_1/2}^{B_1/2} \frac{\sigma_s^2}{B_1} e^{j2\pi f_\triangle[(\tau_p-\tau_n)+(q-m)T_s]} \, df_\triangle, \tag{6.786}$$

or

$$[\mathbf{R_x}]_{nm,pq} = \frac{\sigma_s^2}{B_1} \operatorname{sinc}\left[\pi B_1 \left((\tau_p - \tau_n) + (q-m)T_s\right)\right] e^{j2\pi f_c(\tau_p-\tau_n)}. \tag{6.787}$$

The result in (6.787) is valid for an arbitrary array configuration.

If we restrict our attention to a standard linear array along the z-axis, then

$$\tau_n = \frac{-u_z\left(n - \frac{N-1}{2}\right)}{2f_u}, \quad n = 0, \cdots, N-1. \tag{6.788}$$

Then,

$$\tau_p - \tau_n = \frac{-u_z\,(p-n)}{2f_u} = \frac{u_z(n-p)}{2f_u}, \tag{6.789}$$

where

$$f_u = f_c + \frac{B_s}{2}. \tag{6.790}$$

If we let

$$T_s = \frac{1}{B_s}, \tag{6.791}$$

and define two fractional bandwidths,

$$B_{fs} = \frac{B_s}{f_c}, \tag{6.792}$$

and

$$B_{f1} = \frac{B_1}{f_c}, \tag{6.793}$$

we can write (6.787) as

$$\boxed{[\mathbf{R_x}]_{nm,pq} = \frac{\sigma_s^2}{B_1} \operatorname{sinc}\left[\pi B_{f1} \frac{u_z(n-p)}{2(1+\frac{1}{2}B_{fs})} + \pi \frac{B_1}{B_s}(q-m)\right] e^{j\pi \frac{u_z(n-p)}{1+\frac{1}{2}B_{fs}}}.} \tag{6.794}$$

The first term in the sinc function is due to the delay line and the second term is due to sensor location.

We now have all of the quantities necessary to specify the optimum beamformer. All of our narrowband results can be extended by suitably defining the constraint matrices.

6.13.3.2 Linearly constrained beamformers

In this section, we consider several types of broadband beamformers with linear constraints. In Section 6.13.3.2.1, we develop MVDR and MPDR beamformers. In Section 6.13.3.2.2, we develop LCMV and LCMP beamformers. In Section 6.13.3.2.3, we develop the generalized sidelobe canceller implementation. In Section 6.13.3.2.4, we summarize our results.

6.13.3.2.1 MVDR and MPDR beamformers

We first consider MVDR and MPDR beamformers. In order to introduce a constraint on the look-direction response, we define the $NM \times 1$ vector $\mathbf{c}_m, m = 0, \cdots, M-1$ as

$$\mathbf{c}_m = \begin{bmatrix} \mathbf{0}_N \\ \mathbf{0}_N \\ \vdots \\ \mathbf{1}_N \\ \vdots \\ \mathbf{0}_N \end{bmatrix}, \tag{6.795}$$

where the $\mathbf{1}_N$ vector is in the mth position. We next define an $NM \times M$ matrix, $\mathbf{C}$,

$$\mathbf{C} \triangleq \begin{bmatrix} \mathbf{c}_0 & \mathbf{c}_1 & \cdots & \mathbf{c}_{M-1} \end{bmatrix}. \tag{6.796}$$

Thus, $\mathbf{C}$ is a sparse matrix,

$$\mathbf{C} = \begin{bmatrix} \mathbf{1} & \mathbf{0} & \cdots & \mathbf{0} \\ \mathbf{0} & \mathbf{1} & & \vdots \\ \vdots & \mathbf{0} & \ddots & \mathbf{0} \\ \mathbf{0} & \mathbf{0} & \cdots & \mathbf{1} \end{bmatrix} = \mathbf{I}_M \otimes \mathbf{1}_N. \tag{6.797}$$

The constraint equation is

$$\mathbf{w}^H \mathbf{C} = \mathbf{g}^H, \tag{6.798}$$

where

$$\mathbf{g} \triangleq \begin{bmatrix} g_0 & g_1 & \cdots & g_{M-1} \end{bmatrix}^T, \tag{6.799}$$

is an $M \times 1$ vector.

If we impose the constraint in (6.797) on the weights, then the weight vector $\mathbf{h}$ in Figure 6.124 is

$$\mathbf{h} = \mathbf{g}. \tag{6.800}$$

The values of $\mathbf{g}$ will determine the temporal frequency response in the look direction. Consider a uniform linear array. If we denote the frequency response for the signal arriving from u as $H(\omega, u)$, then the look-direction frequency response is

$$H(\omega,0) = \sum_{m=0}^{M-1} \int_{-\infty}^{\infty} e^{-j\omega t}\, g_m\, \delta(t - mT_s)\, dt. \tag{6.801}$$

Performing the integration gives

$$H(\omega,0) = \sum_{m=0}^{M-1} g_m\, e^{-jm\omega T_s}. \tag{6.802}$$

The MVDR and MPDR beamformers require that the look-direction response be distortionless. In this case,

$$\mathbf{g}^H = \begin{bmatrix} 1 & 0 & \cdots & 0 \end{bmatrix}, \tag{6.803}$$

and

$$H(\omega,0) = 1. \tag{6.804}$$

We minimize

$$P_o = \mathbf{w}^H\, \mathbf{R}_{\mathbf{x}}\, \mathbf{w}, \tag{6.805}$$

subject to the constraint in (6.797). The result is

$$\boxed{\mathbf{w}_{mpdr}^H = \mathbf{g}^H \left[\mathbf{C}^H \mathbf{R}_{\mathbf{x}}^{-1} \mathbf{C}\right]^{-1} \mathbf{C}^H \mathbf{R}_{\mathbf{x}}^{-1}.} \tag{6.806}$$

Similarly, if $\mathbf{R}_n$ is available, then

$$\boxed{\mathbf{w}_{mvdr}^H = \mathbf{g}^H \left[\mathbf{C}^H \mathbf{R}_{\mathbf{n}}^{-1} \mathbf{C}\right]^{-1} \mathbf{C}^H \mathbf{R}_{\mathbf{n}}^{-1}.} \tag{6.807}$$

Note that $\mathbf{w}_{lcmp}^H$ and $\mathbf{w}_{lcmv}^H$ are $NM \times 1$ complex vectors. We use M constraints in both (6.806) and (6.807) so that there are $(NM - M)$ degrees of freedom remaining.

The results in (6.806) and (6.807) are due to Frost [Fro72], and the resulting beamformer is sometimes referred to in the literature as the Frost beamformer. In Chapter 7, we discuss adaptive implementations of this beamformer. Frost only considered the distortionless constraint in [Fro72], but he indicated that the addition of other linear constraints was straightforward. We next consider broadband LCMV and LCMP beamformers.

6.13.3.2.2 LCMV and LCMP beamformers

The extension to LCMV and LCMP beamformers follows easily. The various constraint sets that we discussed for narrowband processors can be extended to the broadband TDL case. For N_d derivative constraints, we define an $NM \times N_d$ matrix,

$$\mathbf{c}_m = \begin{bmatrix} \mathbf{0}_N \\ \mathbf{0}_N \\ \vdots \\ \mathbf{c}_{N_d} \\ \vdots \\ \mathbf{0}_N \end{bmatrix}, \tag{6.808}$$

where $\mathbf{c}_{N_d}$ is the basic $N \times N_d$ constraint matrix, which is in the mth position.

For three spatial derivative constraints,

$$\mathbf{c}_{N_d} = \begin{bmatrix} \mathbf{1} & \dot{\mathbf{v}}(0) & \ddot{\mathbf{v}}(0) \end{bmatrix}. \tag{6.809}$$

Then $\mathbf{C}$ is an $NM \times N_dM$ constraint matrix,

$$\mathbf{C} = \begin{bmatrix} \mathbf{c}_{N_d} & \mathbf{0}_{NN_d} & \cdots & \mathbf{0}_{NN_d} \\ \mathbf{0}_{NN_d} & \mathbf{c}_{N_d} & & \vdots \\ \vdots & \mathbf{0}_{NN_d} & \mathbf{c}_{N_d} & \\ \vdots & \vdots & \ddots & \mathbf{0}_{NN_d} \\ \mathbf{0}_{NN_d} & \mathbf{0}_{NN_d} & \cdots & \mathbf{c}_{N_d} \end{bmatrix} = \mathbf{I}_M \otimes \mathbf{c}_{N_d}. \tag{6.810}$$

The $\mathbf{g}$ matrix is an $N_dM \times 1$ matrix,

$$\mathbf{g}^H = \begin{bmatrix} \mathbf{g}_1^H & \mathbf{0}_{N_d}^T & \cdots & \mathbf{0}_{N_d}^T \end{bmatrix}, \tag{6.811}$$

where

$$\mathbf{g}_1^H = \begin{bmatrix} 1 & g_2 & g_3 \end{bmatrix}. \tag{6.812}$$

The result for derivative constraints is due to Er and Cantoni [EC85].

Just as in the narrowband case, it is necessary to check that the columns in the constraint matrix are linearly independent.

The resulting weight vectors are

$$\mathbf{w}_{lcmp}^H = \mathbf{g}^H \left[\mathbf{C}^H \mathbf{R}_{\mathbf{x}}^{-1} \mathbf{C}\right]^{-1} \mathbf{C}^H \mathbf{R}_{\mathbf{x}}^{-1}, \tag{6.813}$$

and

$$\mathbf{w}_{lcmv}^H = \mathbf{g}^H \left[\mathbf{C}^H \mathbf{R}_\mathbf{n}^{-1} \mathbf{C}\right]^{-1} \mathbf{C}^H \mathbf{R}_\mathbf{n}^{-1}. \tag{6.814}$$

These are identical in form to (6.806) and (6.807). However, the $\mathbf{g}^H$ and $\mathbf{C}$ matrices are different.

The block structure in (6.797) and (6.810) will lead to important simplifications when we implement the beamformer adaptively.

All of the linear constraints that we studied in the narrowband case can be extended to the broadband case. However, for constraints such as multiple-direction constraints, eigenvector constraints, or quiescent pattern constraints, the constraint matrix no longer has the block structure in (6.797) or (6.810). In the adaptive implementation, the computational complexity is increased. We do not discuss these constraints in the text. A discussion of eigenvector constraints is available in Van Veen (Chapter 4 in [HS92]).

In order to improve robustness, we can also incorporate diagonal loading. We impose a quadratic constraint on the composite weight vector $\mathbf{w}$ in (6.798),

$$\| \mathbf{w} \|^2 \leq T_0 \tag{6.815}$$

and minimize the output power. Proceeding as in the narrowband case, we obtain

$$\mathbf{w}_{lcmp,dl}^H = \mathbf{g}^H \left[\mathbf{C}^H \left[\mathbf{R}_\mathbf{x} + \lambda \mathbf{I}\right]^{-1} \mathbf{C}\right]^{-1} \mathbf{C}^H \left[\mathbf{R}_\mathbf{x} + \lambda \mathbf{I}\right]^{-1}, \tag{6.816}$$

where λ is chosen to satisfy (6.815). Alternatively, we can choose a fixed loading level based on the expected scenario.

6.13.3.2.3 Generalized sidelobe canceller

In this subsection, we develop the generalized sidelobe canceller form of a broadband MPDR beamformer using tapped delay lines. The GSC form was derived independently by Griffiths and Jim [GJ82] and Byun and Gangi [BG81].

We only consider the case in which the constraint matrix $\mathbf{C}$ has the sparse structure in (6.797) and (6.810). The broadband version of the generalized sidelobe canceller is shown in Figure 6.126. There is a preprocessing operation that steers the array in the desired direction, performs a quadrature demodulation, and samples the demodulated waveform at $T_s = 1/B_s$ intervals. The output is a sequence of complex vectors $\mathbf{x}(k)$.

The upper path consists of an FIR filter of length M with fixed weights that generates the desired quiescent response. The composite $NM \times 1$ input

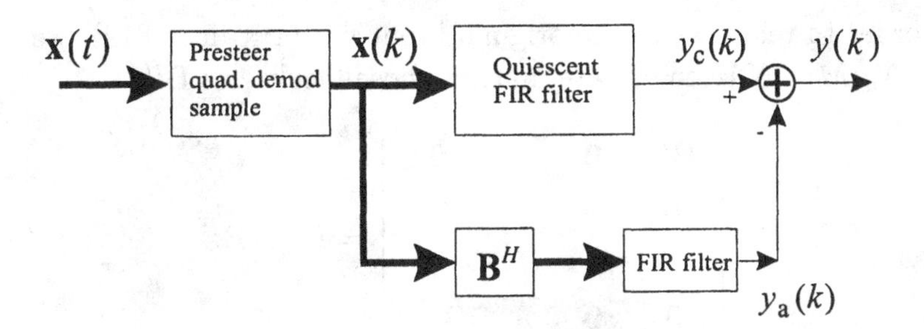

Figure 6.126 Time domain implementation of generalized sidelobe canceller.

vector in the upper filter is $\mathbf{x}$, as in (6.762). The $NM \times 1$ weight vector in the upper path is designed to provide the quiescent response. From (6.813), letting $\mathbf{R_x} = \mathbf{I}$, we have

$$\mathbf{w}_q^H = \mathbf{g}^H [\mathbf{C}^H \mathbf{C}]^{-1} \mathbf{C}^H, \tag{6.817}$$

where $\mathbf{g}^H$ and $\mathbf{C}$ are given by (6.810) and (6.811), respectively. The output of the upper path is denoted by $y_c(k)$.

The blocking matrix $\mathbf{B}^H$ in the lower path is an $(N - N_d) \times N$ matrix that satisfies

$$\mathbf{C}_{N_d}^H \mathbf{B} = \mathbf{0}, \tag{6.818}$$

where $\mathbf{C}_{N_d}$ is the basic $N \times N_d$ constraint matrix in (6.808) and $\mathbf{0}$ is a $N_d \times (N - N_d)$ matrix of zeros. Note that because the constraints are spatial constraints, the blocking matrix does not require an FIR filter. As in the narrowband case, we require

$$\mathbf{B}^H \mathbf{B} = \mathbf{I}. \tag{6.819}$$

We note the output of the blocking matrix by $\mathbf{x}_b(k)$. At time t_0, the vector input to the mth taps in the adaptive FIR filter is

$$\mathbf{x}_{b:m} \triangleq \mathbf{x}_b(t_0 - mT_s), \tag{6.820}$$

and we can define a composite $(N - N_d) \times M$ vector

$$\mathbf{x}_b = \begin{bmatrix} \mathbf{x}_{b:0} \\ \mathbf{x}_{b:1} \\ \vdots \\ \mathbf{x}_{b:M-1} \end{bmatrix}. \tag{6.821}$$

In order to relate $\mathbf{x}_b$ to the original input vector $\mathbf{x}$ in (6.819), we define a $(N - N_d)M \times NM$ composite matrix whose blocks are $\mathbf{B}^H$,

$$\mathbf{B}_{comp}^H \triangleq \begin{bmatrix} \mathbf{B}^H & \mathbf{0} & \cdots & \mathbf{0} \\ \mathbf{0} & \mathbf{B}^H & & \vdots \\ \vdots & & \ddots & \\ \mathbf{0} & \mathbf{0} & \cdots & \mathbf{B}^H \end{bmatrix} = \mathbf{I}_M \otimes \mathbf{B}^H. \tag{6.822}$$

Then,

$$\mathbf{x}_b = \mathbf{B}_{comp}^H \mathbf{x} \tag{6.823}$$

The FIR filter in the lower path is chosen to minimize the output power. Following the same approach as in (6.373)–(6.375), we obtain the $1 \times (N - N_d)M$ weight vector

$$\hat{\mathbf{w}}_a^H = \mathbf{w}_q^H \mathbf{R}_{\mathbf{x}} \mathbf{B}_{comp} \left[\mathbf{B}_{comp}^H \mathbf{R}_{\mathbf{x}} \mathbf{B}_{comp} \right]^{-1}. \tag{6.824}$$

The advantage of the generalized sidelobe canceller structure will become more apparent when we implement the adaptive version in Section 7.13. We have replaced the constrained minimization problem with an unconstrained minimization problem. In this derivation, we have implemented the FIR filter in a direct form. In Section 7.13, we mention other implementations such as lattice filters that may exhibit better adaptive behavior.

In the next subsection, we consider several examples.

6.13.3.2.4 Beamformer performance

We consider several examples to illustrate the behavior of finite impulse response beamformers.

The first example uses the same signal and interference model as in the FFT beamformer discussion so we can compare performance.

Example 6.13.5: MPDR TDL beamformer; no mismatch (continuation; Example 6.13.2)

Consider the same signal and interference model as in Example 6.13.2. The desired signal arrives from broadside and has a flat spectrum with bandwidth B_s and fractional bandwidth B_f. A single interfering plane-wave signal with the same spectrum arrives from u_I. The $SNR = 0$ dB and the $INR = 30$ dB. We sample the input at $T_s = B_s^{-1}$. We implement an MPDR beamformer using an M-tap FIR filter.

In Figure 6.127, we show the array gain versus u_I/BW_{NN} for $B_f = 0.4$ and $M = 2, 4,$ and 8. As we would expect from the FFT beamformer results, $M = 2$ provides all of the available performance improvement.

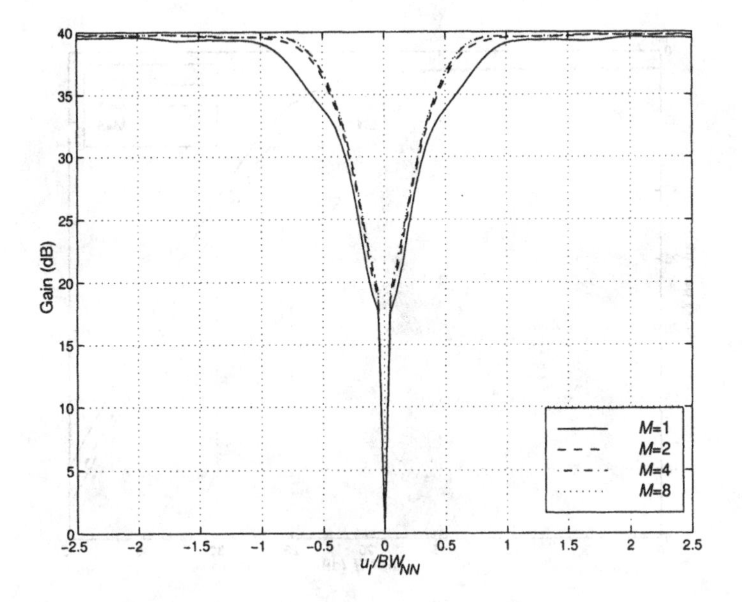

Figure 6.127 Time-domain MPDR beamformer: $u_s = 0$, $INR = 30$ dB, $B_f = 0.4$, array gain versus u_I/BW_{NN}.

Example 6.13.6: LCMP beamformer with derivative constraints; signal mismatch

Consider a uniformly spaced 10-element array. The interelement spacing is $\lambda_u/2$, where λ_u is the wavelength of the highest frequency in the input signal. There is an M-tap FIR filter at the output of each sensor.

The array is steered to broadside. We use the derivative constraint matrix in (6.809)–(6.812) with $g_2 = 0$ and $g_3 = \ddot{B}_c(0)$. Note that we are using $3M$ degrees of freedom.

The signal actually arrives from $u_a = 0.05$ and has a flat spectrum over the frequency band $0 \leq |f_\Delta| \leq 0.5B_s$. $B_f = 0.4$. The *SNR* varies from -10 dB to 30 dB.

A single plane-wave interferer arrives from $u_I = 0.45$ with the same frequency spectrum. The $INR = 30$ dB.

In Figure 6.128(b), we plot the average SNR_o versus *ASNR* for $M = 2, 4$, and 8. In Figure 6.128(a), we plot the array gain versus *ASNR* for $M = 2, 4$, and 8.

In Figure 6.129, we plot the beam pattern at $f_\Delta = 0, 0.2B_s, 0.4B_s$ for $M = 8$ and $SNR = 10$ dB.

For $ASNR < 25$ dB, the derivative constraints prevent signal nulling, and the array puts a null in the interferer direction across the frequency band.

Example 6.13.7 (continuation; Example 6.13.4)

Consider a uniformly spaced 10-element array. The interelement spacing is $\lambda_u/2$, where λ_u is the wavelength of the highest frequency in the input signal. There is an M-tap FIR filter at the output of each sensor.

$$B_f = B_s/f_c = 0.8. \tag{6.825}$$

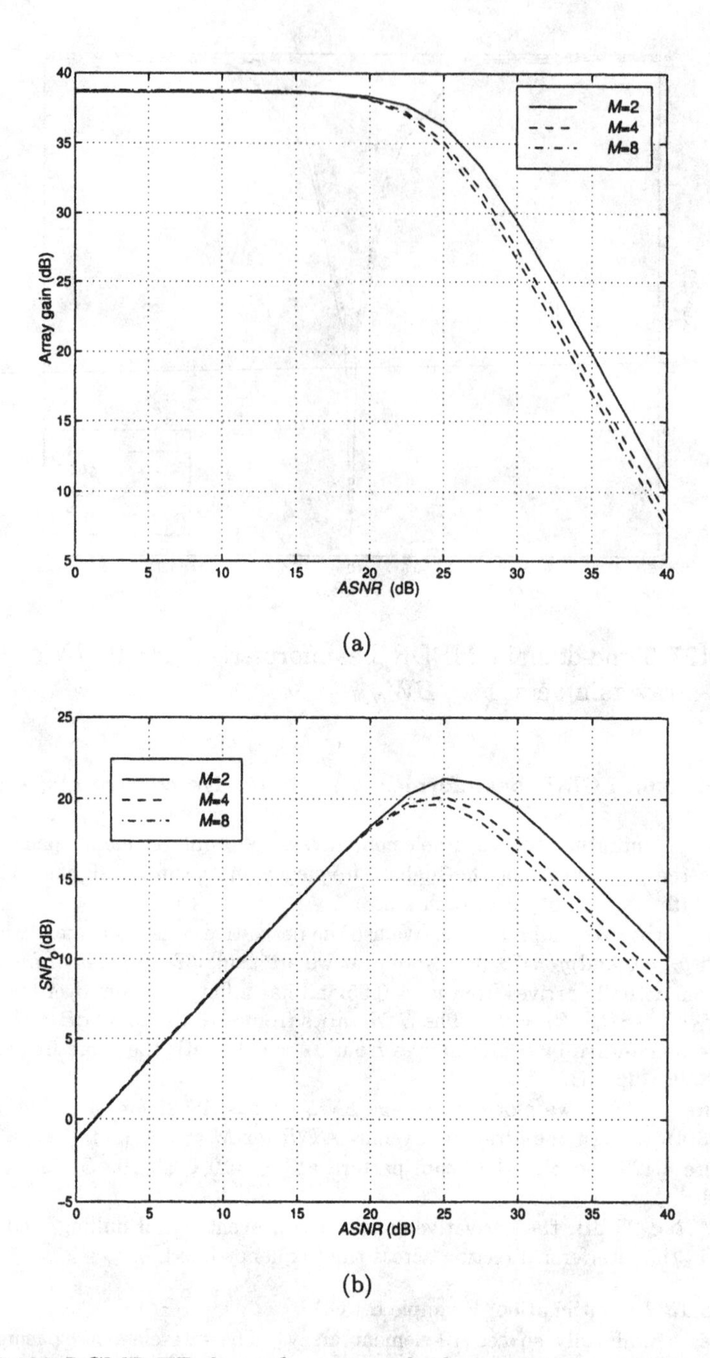

Figure 6.128 LCMP-TD beamformer with derivative constraints; $u_s = 0$, $u_a = 0.05$, $u_I = 0.45$, $INR = 30$ dB, $B_{fs} = 0.4$, $B_1/B_s = 0.4$: (a) SNR_o versus $ASNR$; (b) array gain versus $ASNR$.

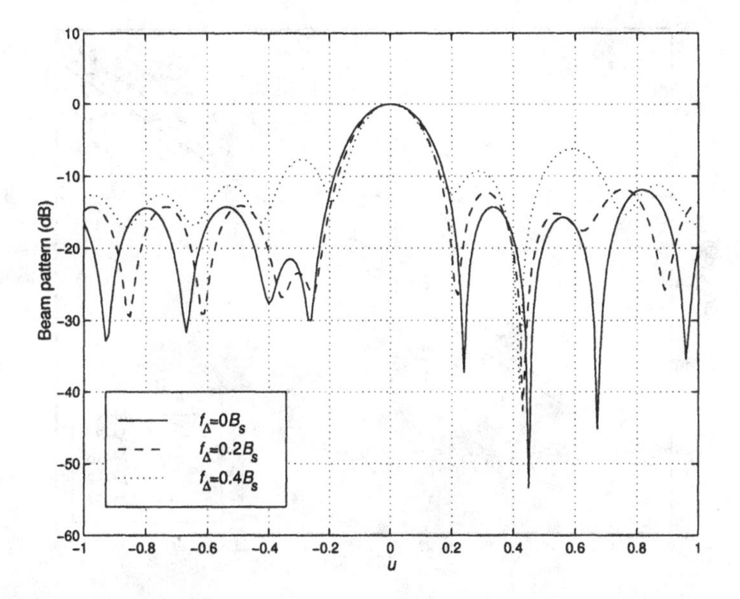

Figure 6.129 Beam pattern at various frequencies; $M = 8$, $f_\Delta = 0, 0.2B_s, 0.4B_s$.

The input is sampled at $1/B_s$, so the normalized frequency span is $-\pi \leq \omega \leq \pi$ in radians. The signal of interest arrives from broadside and has a flat frequency spectrum over $0.5\pi \leq \omega \leq 1.0\pi$ (a 25% bandwidth signal).

There are four interfering signals with the following characteristics (all spectra are zero outside the indicated band):

(i) $S_{I1} = \begin{cases} 1 + \cos(10\omega - 0.8\pi), & 0.6\pi \leq \omega \leq 1.0\pi, \\ 0, & \text{elsewhere}, \end{cases}$

$$INR_1 = 40 \text{ dB}, u_1 = 0.4; \tag{6.826}$$

(ii) $S_{I2} = \sigma_2^2 \quad 0.6\pi \leq \omega \leq 0.9\pi,$

$$INR_2 = 30 \text{ dB}, u_2 = -0.2; \tag{6.827}$$

(iii) $S_{I3} = \sigma_3^2 \quad 0.5\pi \leq \omega \leq 0.51\pi,$

$$INR_3 = 60 \text{ dB}, u_3 = -0.15; \tag{6.828}$$

(iv) $S_{I4} = \sigma_4^2 \quad 0.99\pi \leq \omega \leq \pi,$

$$INR_4 = 60 \text{ dB}, u_4 = 0.7. \tag{6.829}$$

We implement a time-domain MPDR (MPDR-TD) beamformer using M-taps. In Figure 6.130 we plot the SNR_o versus $ASNR$ for various M. In Figure 6.131, we plot the beam pattern at selected frequencies.

Comparing the results in Figures 6.121 and 6.130, we see that the time-domain implementation is about 4 dB better than the FFT implementation.

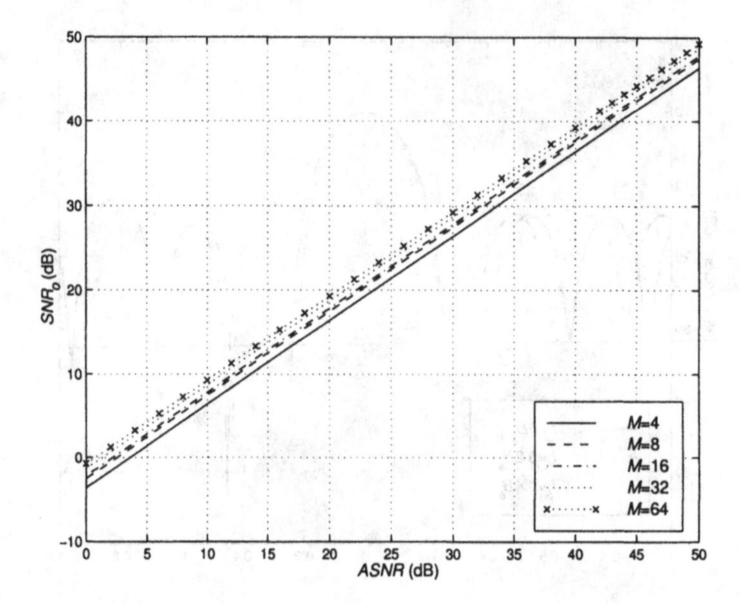

Figure 6.130 Average SNR_o versus $ASNR$, MPDR-TD beamformer with M taps; $B_f = 0.8$; $M = 4, \cdots, 64$.

6.13.3.2.5 Summary: Time-domain beamforming

In this section, we have discussed time-domain beamforming. In Section 7.11, we develop adaptive implementations of the time-domain beamformers that we have developed in this section.

6.13.4 Summary: Broadband Processing

In this section we have discussed two methods of implementing broadband beamformers. In Section 6.13.1, we developed the necessary background by reintroducing the frequency dependence in the array manifold vector. The result in (6.675) was for a uniform linear array, but the extension to an arbitrary array geometry is straightforward. If we replaced the complex weights in a narrowband beamformer with a complex linear filter, then the formulas for the MPDR, LCMP, and MMSE beamformers followed easily.

In practice, we must estimate the appropriate statistics and implement these complex linear filters adaptively. The challenge is to find efficient implementations that have the capability to adapt.

In Section 6.13.2, we developed frequency-domain implementations. We sampled the output of the quadrature demodulators at the sensors to ob-

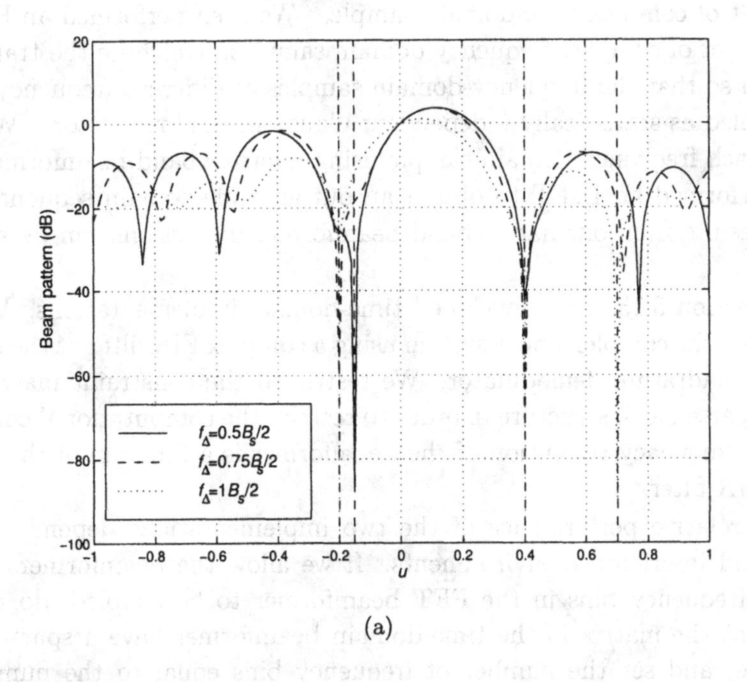

(a)

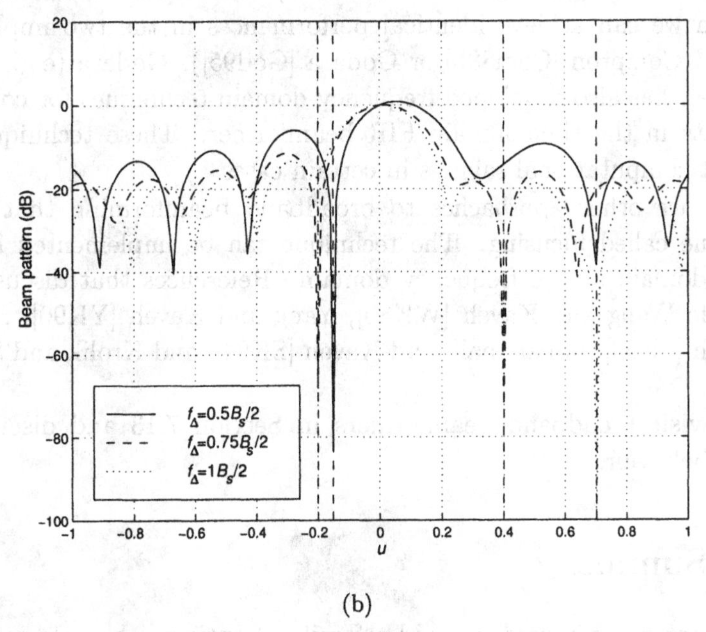

(b)

Figure 6.131 Broadband MPDR-TD beamformer: $SNR = 0$ dB, $B_f = 0.8$; beam patterns at $f_\Delta = 0.5B_s/2$, $0.75B_s/2$, $B_s/2$: (a)$M = 4$; (b)$M = 64$.

tain a set of complex time-domain samples. We then performed an FFT to obtain a set of complex frequency-domain samples. We chose the transform duration so that the frequency-domain samples at different frequency could be modeled as statistically independent Gaussian random vectors. We processed each frequency-domain sample using a narrowband beamformer. We then performed an IDFT to obtain an output time-domain sequence. All of the results from our narrowband beamformer discussions can be applied directly.

In Section 6.13.3, we developed time-domain implementations. We implemented the complex linear system using a complex FIR filter at the output of each quadrature demodulator. We restricted the constraint matrices to have a sparse block structure in order to restrict the computational complexity. The frequency resolution of the beamformer is a function of the length of the FIR filter.

The relative performance of the two implementations depends on the signal and interference environments. If we allow the beamformers in the various frequency bins in the FFT beamformer to be coupled, do not require that the matrix in the time-domain beamformer have a sparse block structure, and set the number of frequency bins equal to the number of taps, then we can achieve identical performances in the two implementations (e.g., Compton [Com88b] or Godara [God95]). Godara (e.g., [God95] and [GJ99]) has also developed frequency-domain techniques for computing the weights in the time-domain FIR beamformer. These techniques offer significant computational savings in certain cases.

There are other approaches to broadband beamforming that rely on a technique called focusing. The technique can be implemented in either the time domain or the frequency domain. References that discuss focusing include Wang and Kaveh [WK85], Yang and Kaveh [YK90], Sivanand and Kaveh [SK90], Simanapalli and Kaveh [SK94], and Krolik and Swingler [KS90].

We revisit broadband beamformers in Section 7.13 and discuss their adaptive behavior.

6.14 Summary

In this chapter, we have developed a family of optimum beamformers. Our development and analysis assumed that the statistics of the environment, either $\mathbf{S_x}$ or $\mathbf{S_n}$, were known.

The discussion in Section 6.2 considered the MVDR and MPDR beamformers and demonstrated their optimality under a number of criteria.

Sections 6.3 and 6.4 analyzed their performance for discrete interference and spatially spread interference, respectively. We found that, when the interference was in the sidelobe region, the beam pattern was well-behaved and the beamformer was effective in reducing the output due to the interference. When the interferer was inside the main-lobe region, the beam pattern degraded and the output SNR_o decreased significantly. The resulting beamformer was also sensitive to the model assumptions. These problems of signal mismatch and main-lobe interference motivated modifications to the MPDR and MVDR beamformers in order to provide main-lobe protection.

Section 6.5 extended the results to multiple plane-wave input signals. We found that the MVDR and MPDR beamformers imposed a distortionless constraint on the signal of interest and imposed perfect nulls on the remaining signals. The MMSE beamformers treated the other signals as interferers and adjusted the gain in their direction as a function of their *SNR*.

Section 6.6 introduced the problems of mismatch between the model and the actual environment that are present in most applications. We studied the effects of DOA mismatch and array perturbations. We found that the MPDR beamformer was sensitive to mismatch and tried to null the signal of interest. We introduced the technique of diagonal loading to combat the problem. We found that, when the *SNR* was less than the individual *INR*s, diagonal loading provided significant improvement. Subsequent discussions in Chapter 7 lead us to the conclusion that some form of diagonal loading is an important component of most beamforming algorithms. The disadvantage of relying completely on diagonal loading to solve the mismatch or main-lobe interferer problem was that the beamformer lost its ability to form nulls on interferers with *INR*s less than the *LNR*.

Section 6.7 introduced linear constraints to improve the performance in the presence of mismatch. Directional and derivative constraints tried to control the behavior of the beam pattern at specific points (or small regions) in frequency-wavenumber space. Eigenvector and quiescent pattern constraints tried to control the behavior over larger regions. The resulting LCMV and LCMP beamformers, coupled with diagonal loading, provided significant performance improvement. We also introduced the generalized sidelobe canceller that separated the constraints and the adaptive component of the beamformer in an effective manner.

Section 6.8 exploited the result that the plane-wave signal and D plane-wave interferers define a $(D+1)$-dimensional subspace that provides a sufficient statistic for the waveform estimation (or detection) problem. By projecting the incoming waveform into this subspace, we can implement eigenspace beamformers, which have improved performance and reduced computational complexity. In Chapter 7, we find that they have good adaptive performance.

Section 6.9 developed beamspace processing. In this technique, we transform the input from element space into a reduced-dimension beamspace using a set of fixed beams (either conventional beams or one of the tapered beams developed in Chapters 3 and 4). We process the beam outputs using one of the optimum processors developed in the earlier section of the chapter. Beamspace processing provides reduced computational complexity without significant loss in performance for interferers in the beamspace sector. In Chapter 7, we find that beamspace processors provide faster adaptation to the optimum processor than the corresponding element-space processors. For conjugate symmetric array manifolds and conjugate symmetric beamspace matrix, we find that there is the additional advantage that we can use real computation to find the beamformer. The disadvantage is the loss of adaptive degrees of freedom.

Section 6.10 revisited quadratically constrained beamformers that were introduced in Section 6.6.4. The norm of the weight vector was constrained and the resulting level of diagonal loading was computed. The QC formulation in this section also leads naturally to a data-dependent variable loading in Chapter 7.

Section 6.11 revisited the problem of approximating a desired beam pattern over some region of frequency-wavenumber space that we first encountered in our discussion of eigenvector constraints in Section 6.7.1.5. We developed an algorithm that is a generalization of diagonal loading. The resulting algorithms are referred to as soft constraint algorithms and provide performance improvements in several application areas.

Section 6.12 considered the problem in which the signal and interference are temporally correlated. This model occurs in a multipath environment and other scenarios. We introduced a technique called spatial smoothing to reduce the correlation prior to beamforming.

Section 6.13 considered broadband beamformers and developed frequency domain and time domain implementations.

In this chapter we have assumed that the necessary statistics are known. In Chapter 7, we develop adaptive beamformers that rely on the observed data to develop the beamformer. If the environment is stationary, we would anticipate that the adaptive beamformers would approach one of the optimum beamformers developed in this chapter. We will develop various adaptive techniques and explore their adaptive behavior, computational complexity, and numerical stability.

6.15 Problems

P6.2 Optimum Beamformers

Problem 6.2.1

Assume that we have a sequence of statistically independent frequency-domain snapshots

$$X(w) = \mathbf{F}(w) + \mathbf{N}(w), \tag{6.830}$$

where $\mathbf{F}(w)$ and $\mathbf{N}(w)$ are statistically independent with spectral matrices $\mathbf{S_f}(\omega)$ and $\mathbf{S_n}(\omega)$.

Find the linear filter that maximizes the *SNR*.

Problem 6.2.2

Show that, if $\mathbf{v}_m(\psi) = \mathbf{v}_s(\psi)$, the MPDR beamformer in (6.71) is identical to the MVDR beamformer in (6.14).

Problem 6.2.3

Consider the case where

$$\mathbf{X}(\omega) = F_1(\omega)\mathbf{v}(\mathbf{k}_1) + F_2(\omega)\mathbf{v}(\mathbf{k}_2) + \mathbf{W}(\omega), \tag{6.831}$$

where

$$S_{F_1}(\omega), \quad S_{F_2}(\omega), \quad S_{F_1F_2}(\omega), \quad \text{and} \quad S_{\mathbf{w}}(\omega) = \sigma_w^2\mathbf{I}, \tag{6.832}$$

are known, and

$$D(\omega) = F_1(\omega). \tag{6.833}$$

(a) Find the optimum MMSE matrix processor to estimate $D(\omega)$.

(b) Assume the signals are narrowband around ω_c. Plot ξ_u as a function of $S_{F_1F_2}(\omega_c)$ and $\rho_{12}(\omega_c)$, the spatial correlation.

Problem 6.2.4

Consider a standard 10-element linear array. The signal is a plane wave arriving from broadside. The interference consists of a plane wave arriving from u_I with an $INR = \sigma_I^2/\sigma_w^2$ and white noise with spectral height σ_w^2.

(a) Plot the array gain versus u_I/BW_{NN} for an MVDR beamformer. Consider *INRs*=0, 10, 20, and 30 dB.

(b) Plot $\| \mathbf{w}_{mvdr}\|^2$ versus u_I/BW_{NN} for the same *INRs*.

Problem 6.2.5 (continuation)

Consider a standard N-element linear array. The rest of the model is the same as in Problem 6.2.4.

Define the normalized array gain as

$$AG_n \triangleq AG/N, \tag{6.834}$$

the array *INR* as

$$AINR = N \cdot INR, \tag{6.835}$$

and

$$u_{In} = u_I / BW_{NN}. \tag{6.836}$$

(a) Plot the normalized array gain versus u_I/BW_{NN} for *AINRs* = 10, 20, 30, and 40 dB, for N = 4, 8, 10, 32, and 64.

(b) Discuss your results. Are the normalized plots essentially the same after some value of N?

Problem 6.2.6

Consider a standard 10-element linear array. Assume $\mathbf{v}_{\psi_m}$ corresponds to broadside. Assume $\mathbf{S_n} = \mathbf{I}$ and σ_s^2 is the signal strength. The signal actually arrives from u_a, where $|u_a| \leq 0.1$. Plot A_{mpdr} vs u_a for the interval $[0, -0.1]$. Indicate the points where

$$A_{mpdr}(u_a) = 0.5 A_{mpdr}(0). \tag{6.837}$$

Problem 6.2.7 (continuation)

Repeat Problem 6.2.7 for the interference model in Problem 6.2.4.

Problem 6.2.8: MVQR Beamformers

In many applications, we want to have a quiescent pattern with low sidelobes. We define the weight vector of the desired quiescent pattern as $\mathbf{w}_{dq}^H$. We define a normalized version of $\mathbf{w}_{dq}^H$ as,

$$\overline{\mathbf{w}}_{dq}^H = \frac{\mathbf{w}_{dq}^H}{\|\mathbf{w}_{dq}\|^2}. \tag{6.838}$$

We impose the constraint

$$\overline{\mathbf{w}}_{dq}^H \mathbf{w} = 1, \tag{6.839}$$

on the weight vector and minimize

$$\mathbf{w}^H \mathbf{S_n} \mathbf{w} \tag{6.840}$$

subject to the constraint.

(a) Show that

$$\mathbf{w}_{mvqr}^H = \Lambda_q \overline{\mathbf{w}}_{dq}^H \mathbf{S_n}^{-1}, \tag{6.841}$$

where

$$\Lambda_q = \left[\overline{\mathbf{w}}_{dq}^H \mathbf{S_n}^{-1} \overline{\mathbf{w}}_{dq}\right]^{-1}. \tag{6.842}$$

The resulting beamformer is called the **minimum variance quiescent response (MVQR)** beamformer.

(b) Derive an expression for the array gain.

(c) What is the array gain when

$$\mathbf{S_n} = \sigma_w^2 \mathbf{I}? \tag{6.843}$$

Problem 6.2.9: MPQR Beamformers (continuation)

Define

$$\overline{\mathbf{w}}_{dq}^H = \frac{\mathbf{w}_{dq}^H}{\|\mathbf{w}_{dq}\|^2} \tag{6.844}$$

and impose the constraint on the weight vector

$$\overline{\mathbf{w}}_{dq}^{H}\mathbf{w} = 1. \tag{6.845}$$

Minimize

$$\mathbf{w}^{H}\mathbf{S}_{\mathbf{x}}\mathbf{w} \tag{6.846}$$

subject to the constraint. The resulting beamformer is called the **minimum power quiescent response (MPQR)** beamformer.

(a) Show that

$$\mathbf{w}_{mpqr}^{H} = \Lambda_{qp}\overline{\mathbf{w}}_{dq}^{H}\mathbf{S}_{\mathbf{x}}^{-1}, \tag{6.847}$$

where

$$\Lambda_{qp} = \left[\overline{\mathbf{w}}_{dq}^{H}\mathbf{S}_{\mathbf{x}}^{-1}\overline{\mathbf{w}}_{dq}\right]^{-1}. \tag{6.848}$$

(b) Derive an expression for the array gain.

P6.3 Discrete Interference

Problem 6.3.1

Consider a standard 10-element linear array. Compare the performance of a least squares Dolph-Chebychev beamformer (−40-dB sidelobe) with a zero-order null at u_I (see Section 3.7) with the MVDR processor with a single plane-wave interferer at u_I plus white noise (σ_w^2).

(a) Compute the array gain of the two beamformers and plot their beam patterns for $u_I = 0.3$ and 0.50 and $\sigma_I^2 = 10$ dB and 50 dB.

(b) Compute the array gain of the two beamformers and plot their beam patterns for $u_I = 0.09$ and 0.18 and $\sigma_I^2 = 10$ dB and 50 dB.

(c) Compute the array gain of the two beamformers and plot their beam patterns for $u_I = 0.02$ and 0.0433 and $\sigma_I^2 = 10$ dB and 50 dB.

Problem 6.3.2

Repeat Problem 6.3.1 using the Villenueve $\bar{n}$ beamformer of Section 3.4.4 as the original weighting.

Problem 6.3.3

Repeat Problem 6.3.1 using the Kaiser beamformer ($\beta = 6$) as the original weighting.

Problem 6.3.4

Repeat Problem 6.3.1 using the Hann beamformer as the original weighting.

Problem 6.3.5

Consider a standard 10-element linear array that uses the MVDR beamformer with a single plane-wave interferer plus white noise. Assume the MVDR beamformer is designed assuming the plane wave is at u_I.

The actual location of the plane wave in u-space is a Gaussian random variable, $N(u_I, \sigma_u^2)$.

(a) Assume u_I is in the sidelobe region and σ_u is small. Find $E[A_o(u_I)]$. Plot your result for $u_I = 0.30, \sigma_u = 0.05, \sigma_I^2 = 10$ dB and 20 dB.

(b) Repeat part (a) for $u_I = 0.18$.

Problem 6.3.6

(a) Repeat Problem 6.3.5 for the Villenueve $\bar{n}$ beamformer in Problem 6.3.2.

(b) Compare your result to the result in Problem 6.3.5.

Problem 6.3.7

Consider a standard 10-element linear array with $d = \lambda_c/2$ where λ_c is the wavelength corresponding to f_c. The frequency spectrum of the single plane-wave interferer is uniform from $f_c - B_s/2 \leq f \leq f_c + B_s/2$. Denote the fractional bandwidth as $B_f = B_s/f_c$

(a) We design an MVDR beamformer assuming that the interferer is at f_c. Assume that

$$INR \triangleq \sigma_I^2/\sigma_w^2 = 30 \text{ dB}. \tag{6.849}$$

Plot the array gain versus u_I/BW_{NN} over the range

$$0.5 \leq u_I/BW_{NN} \leq 2.5, \tag{6.850}$$

for $B_f = 0$, 0.1, 0.2, and 0.4. Plot the beam patterns for several values of u_I.

(b) Repeat part (a) for the case in which the MVDR beamformer is designed using the correct frequency spectrum of the interferer. Plot the beam patterns for several values of u_I.

(c) We want to show that a frequency-spread interferer has the same effect as a cluster of narrowband interferers. Derive a formula that allows us to replace a single frequency spread interferer with M interferers clustered around u_I. Show how the number of interferers and their spacing is related to u_I and B_f.

(d) Discuss the results in part (c) in terms of degrees of freedom available to the beamformer.

Problem 6.3.8

Consider a standard 10-element linear array. We implement the MPDR beamformer in (6.71), assuming the desired signal arrives from broadside. The interference consists of two plane-wave interferers at $u_I = 0.30$ and 0.50 with an $INR = 20$ dB (each). The additive white noise has spectral height σ_w^2. The signal actually arrives from u_a.

(a) Plot the array gain versus u_a/BW_{NN} over the range $0 \leq u_a/BW_{NN} \leq 0.25$ for $SNR = -10$ dB, 0 dB, 10 dB, and 20 dB.

(b) Plot the beam patterns for $u_a = 0.02$, 0.05, and 0.1.

(c) Discuss your results.

Problem 6.3.9 (continuation; Problem 6.2.8)

Consider a standard 10-element linear array. We implement the MVQR beamformer derived in Problem 6.2.8.

The desired quiescent beam pattern is a Dolph-Chebychev beam pattern with −40-dB SLL. The signal arrives from broadside. The interference consists of a single plane-wave interferer with an $INR = 10$ dB or 20 dB.

Plot the array gain versus u_I/BW_{NN} for the two INR levels.

Plot the beam patterns for $u_I = 0.50$, 0.30, 0.18, 0.09, 0.0433, and 0.02. Compare your results to the results in Figures 6.13–6.16.

Problem 6.3.10 (continuation; Problems 6.2.9 and 6.3.8)

Consider the same signal and interference model as in Problem 6.3.8. We use the MPQR beamformer in Problem 6.2.9 with a Dolph-Chebychev quiescent beam pattern with −40-dB SLL. Repeat Problem 6.3.9.

Problem 6.3.11 (continuation; Problem 6.2.10)

Consider the MPQR beamformer in Problem 6.3.10. We add a second constraint,

$$\mathbf{C}_2^H \mathbf{v}_s \triangleq \mathbf{P}_{\overline{\mathbf{w}}_{dq}}^{\perp} \mathbf{v}_s = 1, \tag{6.851}$$

where $\mathbf{P}_{\overline{\mathbf{w}}_{dq}}$ is the projection matrix onto $\overline{\mathbf{w}}_{dq}$.

(a) Derive the optimum beamformer. Verify that the quiescent pattern is unchanged.

(b) Repeat Problem 6.3.10 and compare your results.

Problem 6.3.12

As a modification to the MVDR beamformer, we can require both a distortionless output at $\mathbf{k} = \mathbf{k}_s$ and a zero-order null at $\mathbf{k} = \mathbf{k}_I$. Assume that the noise spectral matrix is $\mathbf{S_n}$.

(a) Find the minimum variance beamformer subject to the two constraints. Find the expression for the array gain A_{co}.

(b) Specialize your result to a 10-element standard linear array. The desired signal arrives from broadside. Assume that the interference consists of a single plane wave at u_I with an *INR* that is a parameter. The additive white noise has a spectral height σ_w^2. Plot A_{co} versus u_I/BW_{NN} for *INR*s = 0 dB, 10 dB, 20 dB, and 30 dB. Compare your result to (6.28). Plot A_{co}/A_o versus u_I/BW_{NN} for the above *INR*s. Discuss your results.

Problem 6.3.13

Consider a standard 19-element hexagonal array in the xy-plane. The desired signal arrives at $\theta_s = 0°$. Two plane-wave interferers arrive at

$$\begin{aligned} u_r &= 0.5, \phi = 0, \\ u_r &= 0.5, \phi = 30°. \end{aligned} \tag{6.852}$$

Each interferer has an *INR* = 20 dB. Design an MPDR beamformer. Compute the array gain. Plot a contour plot of the beam pattern and pattern cuts through $\phi = 0°, 10°, 20°$, and $30°$.

Problem 6.3.14

Repeat Problem 6.3.13 for a standard 37-element hexagonal array.

Problem 6.3.15

Repeat Problem 6.3.14 for a standard 61-element hexagonal array.

Problem 6.3.16

Consider a standard 10-element linear array. The noise consists of three plane-wave interferers located in u-space at u_I, $u_I + \Delta u$, $u_I - \Delta u$ plus additive sensor noise of height σ_w^2.

(a) The three plane waves are uncorrelated and have equal power, $\sigma_I^2/3$. Assume that the three plane waves are in the sidelobe region. Find the MVDR beamformer (signal is at $u_s = 0$) and compute the array gain as a function of u_I and Δu.

(b) Plot the beam pattern for $u_I = 0.3$ and $\Delta u = 0.03$ and $\Delta u = 0.003$.

(c) Discuss the behavior as $\Delta u \to 0$ in terms of the condition number of $\left[\mathbf{v}_I^H \mathbf{v}_I\right]$.

(d) Compare the result in parts (b) and (c) to a constrained MVDR beamformer with zero-order, first-order, and second-order null at u_I.

(e) Repeat parts (a) and (b) for $u_I = 0.2$ and $\Delta = 0.04$.

Problem 6.3.17

Consider a standard 37-element hexagonal array in the xy-plane. The interference consists of 10 plane-wave interferers distributed uniformly on a circle in u-space with radius $u_{RI} = \sqrt{u_{xI}^2 + u_{yI}^2}$. The interferers are uncorrelated equal power signals ($\sigma_i^2 = \sigma_I^2/10$) and the additive noise has variance σ_w^2.

(a) Derive the MVDR beamformer and compute the array gain.

(b) Plot pattern cuts for u_{RI} in the sidelobe region and $\sigma_I^2/\sigma_w^2 = 20$ dB.

(c) Repeat part (b) for the outer main-lobe region.

(d) Assume we have M uncorrelated equal power interferers with spectrum σ_I^2/M. Analyze the MVDR beamformer as $M \to \infty$.

P6.4 Spatially Spread Interference

Problem 6.4.1

Consider a vertical standard 10-element linear array. Assume the noise consists of the surface noise model in Problem 5.3.1 with $\alpha = 0.25$ plus additive white noise with spectrum height σ_ω^2.

Calculate the array gain as a function of θ_T, the target signal's angle of arrival measured from the vertical for *INR* values of 10 dB and 20 dB.

Problem 6.4.2

Repeat Problem 6.4.1 for the case of high surface noise: $\alpha = 1$.

Problem 6.4.3

Repeat Problem 6.4.1 for the layer noise model in Problem 5.3.2. Assume low layer noise: $\alpha = 0.25$.

Problem 6.4.4

Repeat Problem 6.4.3 for the case of high layer noise: $\alpha = 1$.

Problem 6.4.5

Consider the sector noise model in Figure 5.8. Denote the angle of the cone boundary with the xy-plane as $\bar{\theta}_c$, $S_o(\omega) = S_o$. Assume that we have standard 10-element linear array along the x-axis and the signal is a plane wave arriving from $[\theta_s, \psi_s = 0]$. In addition, there is additive white noise (σ_w^2).

(a) Find the MVDR beamformer. Plot the resulting beam pattern for (i) $\theta_s = 30°, \bar{\theta}_c = 75°$, *INR* $= 20$ dB; (ii) $\theta_s = 30°, \bar{\theta}_c = 60°$, *INR* $= 20$ dB; (iii) $\theta_s = 45°, \bar{\theta}_c = 75°$, *INR* $= 30$ dB.

(b) Compute the array gain for the above cases.

Problem 6.4.6

Repeat Problem 6.4.5 for a standard 10×10 rectangular array in the xy-plane.

Problem 6.4.7

Consider a standard 21-element linear array. Assume the noise spectrum corresponds to a complex AR(2) process plus additive white noise with spectral height σ_w^2.

(a) Find the MVDR receiver for a signal located at $u_s = 0.4$.

(b) Find the array gain for AR spectrum in Figure 6.19, assuming $\sigma_w^2 = 0$. Plot the beam pattern.

(c) Repeat part (b) for AR spectrum in Figure 6.20, assuming $\sigma_w^2 = 0$.

(d) Repeat part (b) for $\sigma_{AR}^2/\sigma_w^2 = 20$ dB.

(e) Compare your result in part (d) to the case of two plane-wave interferers located at $u_{I1} = 0, u_{I2} = 0.8$.

Problem 6.4.8

Consider the same linear array and noise model as in Problem 6.4.7. Assume $u_s = 0$ and $z_1 = |z_1| \exp(-0.15\pi)$ and $z_2 = |z_2| \exp(+0.15\pi)$. Assume the *INR* is 20 dB.

(a) Plot the beam pattern of the MVDR receiver for $|z_1| = |z_2| = 0.5, 0.9$ and 0.99.

(b) Find the array gain for the above cases.

Problem 6.4.9

Consider a standard 10-element linear array. The signal is at $u = 0$. The interference has a flat spatial spectrum in one dimension:

$$S_I(u) = \frac{S_I}{2u_\Delta} \qquad u_I - u_\Delta \leq u \leq u_I + u_\Delta. \tag{6.853}$$

The additive white noise has spectral height σ_w^2. Find the optimum beam pattern and compute the array gain for

(a)

$$u_I = 0.3, \quad u_\Delta = 0.1, \quad S_I/\sigma_w^2 = 20 \text{ dB}. \tag{6.854}$$

(b)

$$u_I = 0.2, \quad u_\Delta = 0.1, \quad S_I/\sigma_w^2 = 20 \text{ dB}. \tag{6.855}$$

(c)

$$u_I = 0.5, \quad u_\Delta = 0.2, \quad S_I/\sigma_w^2 = 40 \text{ dB}. \tag{6.856}$$

Problem 6.4.10

Consider a 32-element uniformly spaced linear array with $d = \frac{\lambda_c}{8}$. Assume that

$$\mathbf{S}_\mathbf{n}(\omega_c) = \sigma_I^2(\omega_c)\text{sinc}\left(\frac{\omega_c}{c}|d|\right) + \sigma_w^2\,\mathbf{I}, \tag{6.857}$$

which is a mixture of isotropic and white noise. The signal and noise are narrowband.

(a) Find the array gain as a function of u_s.

(b) Plot the magnitude of the beam pattern for $u_s = 0, 0.5$ and 1.0 and various *INR*.

(c) Plot the sensitivity function for the same values as in part (b).

P6.5 Multiple Plane-wave Signals

Problem 6.5.1

Consider the case of D plane-wave signals impinging on the array from $\mathbf{k}_1, \mathbf{k}_2, \cdots, \mathbf{k}_D$. The noise consists of a single plane wave at $\mathbf{k}_J$ plus white noise with spectral height σ_w^2. Thus,

$$\mathbf{S_n} = \mathbf{v}^H(\mathbf{k}_J)S_J\mathbf{v}(\mathbf{k}_J) + \sigma_w^2\mathbf{I}. \tag{6.858}$$

(a) Find the $D \times N$ MVDR processor. Interpret your result.

(b) Is the result in part (a) the same as the $(D+1) \times N$ MVDR processor for $D+1$ plane-wave signals impinging on the array from $\mathbf{k}_1, \mathbf{k}_2, \cdots, \mathbf{k}_D$, $\mathbf{k}_J$? Justify your answer.

Problem 6.5.2

Assume there are two signals impinging on an array in the presence of noise that is uncorrelated with the signals and has a spectral matrix $\mathbf{S_n}(\omega)$:

$$\mathbf{X}(\omega) = \mathbf{V}\mathbf{F}(\omega) + \mathbf{N}(\omega). \tag{6.859}$$

Create a 2-D signal subspace

$$\mathbf{X}_s(\omega) = \mathbf{V}^H\mathbf{X}(\omega). \tag{6.860}$$

(a) Find the MMSE processor to estimate $\mathbf{F}(\omega)$ using $\mathbf{X}_s(\omega)$ as the input.

(b) Compare your result with (6.181).

Problem 6.5.3

Consider a standard 10-element linear array. The frequency-domain snapshots are

$$\mathbf{X}(\omega) = \mathbf{V}\mathbf{F}(\omega) + \mathbf{W}(\omega), \tag{6.861}$$

where $\mathbf{F}(\omega)$ is 2×1 source-signal vector,

$$\mathbf{F}(\omega) = \begin{bmatrix} F_s(\omega) \\ F_I(\omega) \end{bmatrix}, \tag{6.862}$$

with spectral matrix

$$\mathbf{S_f}(\omega) = \begin{bmatrix} \sigma_s^2 & \sigma_s\sigma_I\rho \\ \sigma_s\sigma_I\rho^* & \sigma_I^2 \end{bmatrix}. \tag{6.863}$$

The array manifold matrix $\mathbf{V}$ is known. The source signal $\mathbf{F}(\omega)$ and additive-noise vector $\mathbf{W}(\omega)$ $\left(\mathbf{S_W}(\omega) = \sigma_w^2\mathbf{I}\right)$ are uncorrelated.

(a) Find the MMSE processor to estimate $F_s(\omega)$.

(b) Assume the signal arrives from broadside with an *SNR* $= 10$ dB. Plot the output SNR_o and the normalized MSE versus ψ_I for *INRs* $= 10$, 20, and 30 dB. Discuss your result.

Problem 6.5.4

Derive a "pure" eigenbeam receiver. In this case the $D+1$ sufficient statistics are the $D+1$ eigenbeams.

Problem 6.5.5

Consider the problem of finding the MMSE estimate for multiple plane-wave signals (D signals)

$$\mathbf{S_x} = \mathbf{V S_f V}^H + \mathbf{S_n}, \tag{6.864}$$

$$\mathbf{V} = \left[\mathbf{v}_1 \ \vdots \ \mathbf{v}_2 \ \vdots \ \cdots \ \vdots \ \mathbf{v}_D \right]. \tag{6.865}$$

The resulting MMSE filter is

$$\mathbf{H}_o = \left(\mathbf{I} + \mathbf{S_f V}^H \mathbf{S_n}^{-1} \mathbf{V}\right)^{-1} \mathbf{S_f V}^H \mathbf{S_n}^{-1}, \tag{6.866}$$

which reduces to

$$\mathbf{H}_o = \left(\mathbf{I} + \frac{\mathbf{S_f}}{\sigma_w^2} \mathbf{V}^H \mathbf{V}\right)^{-1} \frac{\mathbf{S_f}}{\sigma_w^2} \mathbf{V}^H, \tag{6.867}$$

for white noise.

Now consider the problem of finding the MMSE estimate of the first signal in the presence of $D-1$ interfering signals plus white noise. This will result in a $1 \times N$ filter that is denoted as $\mathbf{H}_1$.

(a) Find $\mathbf{H}_1$.

(b) Prove $\hat{f}_1(t) = y_1(t)$.

Problem 6.5.6

The received waveform is an $N \times 1$ vector,

$$\mathbf{X}(\omega) = \sum_{i=1}^{K_1} \mathbf{v}(\omega : \mathbf{k}_i^S) F_i(\omega) + \sum_{j=1}^{K_2} \mathbf{v}(\omega : \mathbf{k}_j^N) N_j(\omega) + \mathbf{W}(\omega), \tag{6.868}$$

where

$$\mathbf{F}(\omega) = \begin{bmatrix} F_1(\omega) \\ \vdots \\ F_{K_1}(\omega) \end{bmatrix}, \tag{6.869}$$

and $\mathbf{S_f}(\omega)$ is the corresponding signal spectral matrix

$$\mathbf{N}(\omega) = \begin{bmatrix} N_1(\omega) \\ \vdots \\ N_{K_2}(\omega) \end{bmatrix}, \tag{6.870}$$

and $\mathbf{S_n}(\omega)$ is the corresponding noise spectral matrix and $\mathbf{W}(\omega)$ corresponds to a white random process with spectral matrix $\sigma_w^2\mathbf{I}$.

Assume $K_1 + K_2 < N$. Define

$$\mathbf{V}_F = \left[\ \mathbf{v}(\omega : \mathbf{k}_1^S) \ \vdots \ \cdots \ \vdots \ \mathbf{v}(\omega : \mathbf{k}_{K_1}^S) \ \right] \tag{6.871}$$

and

$$\mathbf{V}_N = \left[\ \mathbf{v}(\omega : \mathbf{k}_1^N) \ \vdots \ \cdots \ \vdots \ \mathbf{v}(\omega : \mathbf{k}_{K_2}^N) \ \right]. \tag{6.872}$$

(a) Find $\mathbf{H}_W(\omega)$, the whitening filter. $\mathbf{H}_W(\omega)$ whitens the total noise input.

(b) Define $\mathbf{S}_W(\omega)$ as the total output of the whitening filter when the input is signal plus noise. Find the spectrum of $\mathbf{S}_W(\omega)$.

(c) What is rank of signal-component spectral matrix?

(d) Does the result in part (c) generalize to an arbitrary $\mathbf{S_n}(\omega)$?

Problem 6.5.7

Consider the case of a scalar Gaussian process propagating along $\mathbf{v}(\omega : \mathbf{k}_s)$. Then

$$\mathbf{S_f}(\omega) = \mathbf{v}(\omega : \mathbf{k}_s) S_f(\omega) \mathbf{v}^H(\omega : \mathbf{k}_s). \tag{6.873}$$

(a) Show that $H_1(\omega)$, the MMSE processor is

$$\mathbf{H}_1(\omega) = \mathbf{v}_s S_f(\omega) \mathbf{v}_s^H \left[\sigma_w^2 \mathbf{I} + \mathbf{v}_s S_f(\omega) \mathbf{v}_s^H\right]^{-1}. \tag{6.874}$$

(b) Show that $H_1(\omega)$ reduces to

$$\mathbf{H}_1(\omega) = \mathbf{v}_s \left\{ \frac{1}{N} \frac{S_f(\omega)}{\frac{\sigma_w^2}{N} + S_f(\omega)} \right\} \mathbf{v}_s^H. \tag{6.875}$$

Sketch the optimum processor. Note that the optimal processor is an MVDR beamformer followed by the optimum scalar processor.

Problem 6.5.8

The frequency-domain snapshots at a standard 10-element linear array are

$$\mathbf{X}(\omega) = F(\omega)\mathbf{v}_\psi(\psi_1) + \alpha_2 F(\omega)\mathbf{v}_\psi(\psi_2) + \alpha_3 F(\omega)\mathbf{v}_\psi(\psi_3) + \mathbf{W}(\omega), \tag{6.876}$$

where $F(\omega)$ is the Fourier transform of the signal whose spectrum is $S_f(\omega)$, α_2 and α_3 are known complex parameters, and $\mathbf{W}(\omega)$ is spatially white noise with spectral height σ_w^2. The signal is narrowband so ω can be suppressed.

(a) Find the linear processor that provides a MMSE estimate of $F(\omega)$.

(b) The array *SNR* is defined as

$$ASNR = \frac{N\, S_f(\omega)}{\sigma_w^2}. \tag{6.877}$$

Assume $\psi_1 = 0$, $\psi_3 = -\psi_2$, and $\alpha_2 = 0.5\alpha_3$. Plot the the output SNR_o and the MMSE error versus *ASNR* for several α_2 and ψ_2.

P6.6 Mismatched MVDR and MPDR Beamformers

Problem 6.6.1

Consider a standard 10-element linear array with white noise interference. Assume that the optimum MPDR beamformer is steered to broadside ($\mathbf{v}_m = \mathbf{1}$ and $u_m = 0$).

(a) Assume that the desired signal actually arrives from u_a. The input *SNR* is σ_s^2/σ_w^2. Plot A_{mpdr} versus u_a for various *SNR* (−10 dB to 30 dB).

(b) Plot A_{mpdr} versus *SNR* for $u_a = 0.25BW_{NN}$ and $u_a = 0.0433BW_{NN}$.

(c) Assume that u_a is a random variable whose probability density is uniform $[-u_1 \le u_a \le u_1]$. Plot $E[A_{mpdr}]$ versus *SNR* for $u_1 = 0.0433BW_{NN}$, and $0.25BW_{NN}$.

(d) Assume u_a is a zero-mean Gaussian random variable whose standard deviation is given by (6.237). Repeat part (c).

Problem 6.6.2 (continuation; Problem 6.6.1)

Consider a standard 10-element linear array. The interference consists of two plane-wave interferers at $u = 0.29$ and 0.45, each with an *INR* = 20 dB. We design an optimum MPDR beamformer steered to broadside.

(a) Repeat part (a) of Problem 6.6.1.

(b) Repeat part (b) of Problem 6.6.1.

(c) Add a third interferer at $u_I = 0.25BW_{NN}(u_I = 0.1)$. Repeat part (a) for $u_a \le 0.05$.

(d) Add diagonal loading with various *LNR*. Repeat part (c) for representative *LNR*.

Problem 6.6.3

Consider a standard 10-element linear array. Consider the multiple plane-wave MPDR beamformer. (It corresponds to the MVDR beamformer in (6.158), with $\mathbf{S_n}$ replaced by $\mathbf{S_x}$.)

There are four plane-wave signals impinging on the array plus white noise. The beamformer is designed assuming the signals arrive from

$$u_{s1} = 0, \quad u_{s2} = 0.3, \quad u_{s3} = 0.5, \quad u_{s4} = -0.7. \tag{6.878}$$

Each signal has the same *SNR*. The actual signals arrive from

$$u_{a1} = u_a, \quad u_{a2} = 0.3 + u_a, \quad u_{a3} = 0.5 - u_a, \quad u_{a4} = -0.7 + u_a, \tag{6.879}$$

where $|u_a| \le 0.05$.

(a) Plot the output SNR_o versus u_a for signal 1 for *SNR* = −10 dB, ···, 30 dB in 10-dB steps.

(b) Add diagonal loading. Repeat part (a) for representative *LNR*.

Problem 6.6.4 (continuation; Problem 6.3.13)

Consider the same model and environment as Problem 6.3.13. The signal actually arrives from (θ_a, ϕ_a).

(a) Plot the array gain versus $u_{ra} = \sin\theta_a$ for $\phi_a = 0°, 10°, 20°$, and $30°$ and *SNR* = 0, 10, and 20 dB. Assume $|u_{ra}| \le 0.5$

(b) Add diagonal loading. Repeat part (a) for various *LNR*.

Problem 6.6.5 (continuation; Problem 6.3.13)

Consider the same model and environment as Problem 6.3.13. The plane of the array rotates slightly around the y-axis so that the angle between the plane of the array and the x-axis is γ_x degrees.

Plot the array gain versus γ_x for $0 < \gamma_x < 5°$. Consider *SNR* = 0, 10, and 20 dB.

Example 6.6.6 (continuation; Example 6.6.4)

Consider the same model as in Example 6.6.4.

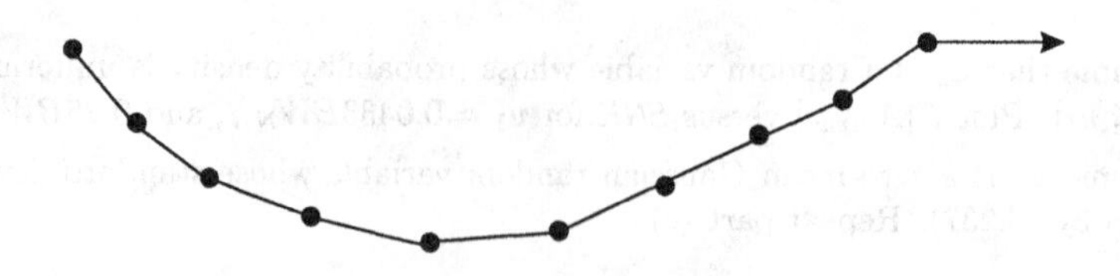

Figure 6.132 Towed array.

(a) Assume the position perturbations are only in the z-direction. Repeat Example 6.6.4.

(b) Assume the position perturbations occur only in the y-direction. Repeat Example 6.6.4.

In both parts, compare your results to the results in Example 6.6.4.

Problem 6.6.7

Consider a standard 10-element linear array. The signal arrives from broadside and two equal-power uncorrelated interferers arrive from $u_I = 0.30$ and 0.50 with an *INR* = 20 dB (each). Design an MPDR beamformer and calculate the array gain.

(a) Assume the nth sensor fails. When the sensor fails, its output is set to zero. Calculate the resulting array gain for two cases:

(i) The remaining $N-1$ weights remain the same.

(ii) The MPDR weights are recomputed for the $N-1$ operational sensors.

Consider different values of n.

(b) Repeat part (a) for two sensor failures. Consider various sensor pairs.

Problem 6.6.8

Consider a standard 48-element towed linear array. The signal is a plane-wave signal arriving from broadside. There are two plane-wave interferers arriving from u_I=3/48 and 5/48 with a 30-dB *INR* (each).

When the vehicle towing the array turns, the array shape can be approximated as a segment of a circle in the xy-plane, as shown in Figure 6.132. The diameter of the circle measured in wavelengths is d_p.

Plot the array gain versus d_p (measured in wavelengths under two assumptions):

(a) The MPDR weights do not change from their values calculated assuming a linear array.

(b) The positions of sensors are known and the MPDR weights are recalculated.

Problem 6.6.9

Consider a standard 10-element linear array along the x-axis. The signal arrives from broadside and two plane-wave interferers arrive from $u_I = 0.30$ and 0.50. We use the array perturbation model in (6.242)–(6.247).

(a) Consider independent gain and phase perturbations with standard derivation σ_g and σ_ϕ, respectively. Plot the array gain versus *SNR* for various σ_g and σ_ϕ. For *SNR* = −10 dB, 0 dB, 10 dB, and 20 dB, plot a contour map showing array gain versus σ_g and σ_p.

(b) Compare your results to those in Example 6.6.5. Discuss the relative importance of gain versus phase variations.

Problem 6.6.10

Consider a standard 10-element linear array with $d = \lambda_c/2$ and white noise interference. Assume that the optimum MPDR beamformer is steered to $\theta_m = 30°$.

(a) Assume that the desired signal arrives from 30°, but its frequency is $f_c + f_a$. The input *SNR* is σ_s^2/σ_w^2. Plot A_{mpdr} versus f_a for various *SNR* (−10 dB to 30 dB).

(b) Plot A_{mpdr} versus *SNR* for $f_a = 0.2f_c$ and $f_a = 0.4f_c$.

(c) Assume that f_a is a random variable whose probability density is uniform $[-f_1 \leq f_a \leq f_1]$. Plot $E[A_{mpdr}]$ versus *SNR* for $f_1 = 0.04f_c, 0.10f_c, 0.20f_c$.

Problem 6.6.11: Frequency mismatch

Consider a standard 10-element linear array designed for frequency f_c. A narrowband plane-wave signal at frequency f arrives from direction u_s. The interference is white noise. We define

$$f = f_c + f_\Delta, \tag{6.880}$$

and

$$b_f = \frac{f_\Delta}{f_c}. \tag{6.881}$$

(a) Show that the array manifold vector $\mathbf{v}_f$ can be written as

$$[\mathbf{v}_f]_n = \exp\left(j\pi(n - \frac{N-1}{2})\right) u_f, \tag{6.882}$$

with

$$u_f \triangleq (1 + b_f)u. \tag{6.883}$$

(b) Design a MVDR beamformer for frequency f_c. Plot the array gain (normalized by the array gain when $f_\Delta = 0$) versus u_s for $b_f = 0.2, 0.4, 0.6$, and 0.8.

(c) Repeat part (b) for a MPDR beamformer.

Problem 6.6.12: Broadband signal (continuation)

Consider a standard 10-element linear array designed for frequency f_c. A plane-wave signal arrives from direction u_s. The signal has a flat frequency spectrum over the band $f_c - B_s/2 \leq f \leq f_c + B_s/2$. The interference is white noise. Define

$$B_f \triangleq \frac{B_s}{f_c}. \tag{6.884}$$

(a) Design a MVDR beamformer for frequency f_c. Plot the array gain versus u_s for B_f=0.1, 0.2, and 0.4.

(b) Repeat part (a) for a MPDR beamformer.

Problem 6.6.13: Broadband interference

Consider a standard 10-element linear array designed for frequency f_c. A narrowband plane-wave signal arrives from direction u_s.

There are two plane-wave interferers arriving from $u_I = 0.30$ and 0.50 with an *INR* = 20 dB (each). Each interferer has a flat frequency spectrum over the band

$$f_c - B_I/2 \le f \le f_c + B_I/2. \tag{6.885}$$

Define

$$B_{fI} \triangleq \frac{B_I}{f_c}. \tag{6.886}$$

(a) Design a MVDR beamformer using the assumption that the interferers are narrowband. Evaluate the array gain for $B_{fI} = 0$, 0.1, 0.2, and 0.4. Plot the beam patterns.

(b) Let the DOA and the *INR* of the second interferer be a variable. Plot the array gain versus u_{I2} for various *INR*.

(c) Design a MVDR beamformer using the actual spectral matrix $\mathbf{S_n}(w)$ corresponding to the actual interferer spectra. The beamformer can only use complex weights (not complex filters). Evaluate the array gain for B_{fI}=0, 0.1, 0.2, and 0.4. Plot the resulting beam patterns. Explain your results.

Problem 6.6.14

Consider a standard 10-element linear array whose input is given by (6.589). The various output powers P_{do}, P_{io}, and P_{no} are given by (6.593), (6.597), and (6.598), respectively. The input *SNR* is σ_s^2/σ_w^2. The input *INR* is σ_I^2/σ_w^2.

(a) Assume $u_I = 0.3$. Plot P_{do}, P_{io}, P_{no}, and SNR_o versus $|\rho|$ for various *SNR* and *INR*. Does the phase of ρ affect any of the results?

(b) Now assume that diagonal loading is used with σ_L^2 added to σ_w^2 for purposes of computing $\mathbf{S_x}^{(L)}$. Use $\mathbf{S_x}^{(L)}$ to determine the $\mathbf{w}_{mpdr,dl}$. Repeat part (a) for various σ_L^2/σ_w^2. Note that $\mathbf{S_x}$ is still given by (6.587).

Problem 6.6.15 (continuation)

Repeat Problem 6.6.14 for $u_I = 0.18$.

P6.7 LCMV and LCMP Beamformers

Problem 6.7.1

Consider a standard 10-element linear array with white noise interference. Compare the performance using directional, derivative, and eigenvector constraints in a LCMP beamformer. In each case, utilize three constraints and add diagonal loading. The *LNR* is a design parameter. Show the MPDR-DL results for comparison.

(a) Consider two sets of directional constraints. The first set is at $u_1 = 0$, $u_2 = 0.0433$, and $u_3 = -0.0433$. The second set is $u_1 = 0$, $u_2 = 0.1$, and $u_3 = -0.1$. For each set we use the $\mathbf{g}$ given by (6.292) and (6.294). Plot SNR_o versus u_a/BW_{NN} for various *SNR* (0 dB, 10 dB, 20 dB). Plot the corresponding beam patterns.

(b) Repeat part (a) for the derivative constraints in (6.312). Utilize the $\mathbf{g}$ given by (6.313) and (6.314).

(c) Repeat part (a) for three eigenvector constraints where we try to match a conventional beam pattern between ± 0.10.

(d) Compare your results and develop some design guidelines.

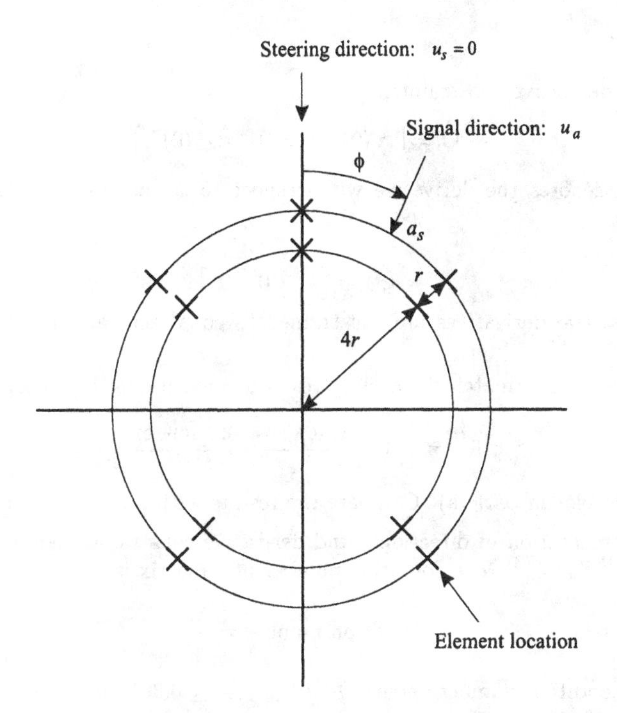

Figure 6.133 Planar dual array.

Problem 6.7.2 (continuation)

Repeat Problem 6.7.1 for the case of a single plane-wave interferer (*INR* = 0, 10, 20 dB) in the sidelobe region ($u_I = 0.3$).

Problem 6.7.3 (continuation)

Repeat Problem 6.7.2 for the case in which the DOA of the interferer is a variable. Study the behavior as u_I moves from the sidelobe region into the main-lobe region.

Problem 6.7.4 (continuation; Problem 6.7.1)

Repeat Problem 6.7.1 for the case case of two uncorrelated plane-wave interferers at u_{I1} =0.30 and $u_{I2} = -0.50$. The interferers are equal power (*INR* = 0, 10, 20 dB each).

Problem 6.7.5 (continuation)

Repeat Problem 6.7.4 for the case of correlated interferers. Consider $|\rho| = 0, 0.5, 0.9, 0.95$, and 1.0.

Problem 6.7.6 [Ste83]

Consider the 10-element planar dual ring array shown in Figure 6.133. Compare the performance using directional and derivative constraints. The array is steered to $\phi = 0$. The signal arrives at ϕ_s. The interference consists of isotropic noise.

(a) Use three directional constraints at $\phi = -5°$, 0, and $+5°$. Use a **g** vector corresponding to a flat response and a conventional response. Plot the SNR_o (in dB) versus ϕ for input *SNR* = −12 dB, −6 dB, 0 dB, 6 dB, and $r/\lambda = 0.2$. Discuss your results.

(b) Use three derivative constraints,

$$\mathbf{C} = \left[\begin{array}{ccc} \mathbf{v}(0) & \mathbf{d}_\phi(0) & \mathbf{d}_{\phi\phi}(0) \end{array} \right], \tag{6.887}$$

where $\mathbf{d}_\phi$ denotes the derivative with respect to ϕ and $\mathbf{d}_{\phi\phi}$ denotes the second derivative.

In the first case,

$$\mathbf{g}^H = \left[\begin{array}{c:c:c} 1 & 0 & 0 \end{array} \right], \tag{6.888}$$

so the first two derivatives are constrained to equal zero at $\phi_s = 0$, the steering direction.

In the second case, match the derivatives of a conventional beamformer.

$$\mathbf{g}^H = \left[\begin{array}{c:c:c} 1 & \frac{\mathbf{d}_\phi^H(0)\mathbf{v}(0)}{N} & \frac{\mathbf{d}_{\phi\phi}^H(0)\mathbf{v}(0)}{N} \end{array} \right], \tag{6.889}$$

Repeat the plot in part (a). Compare the results with part (a).

(c) The implementation of directional and derivative constraints requires the calculation of $[\mathbf{C}^H \mathbf{S_x} \mathbf{C}]^{-1}$. The condition number of a matrix is

$$\text{Condition number} = \frac{\lambda_{max}}{\lambda_{min}}. \tag{6.890}$$

Plot the condition number versus r/λ $(0 \leq r/\lambda \leq 0.30)$ for three directional constraints at 1°, 3°, and 5° and derivative constraints. Discuss the implication of your results.

Problem 6.7.7 [Jab86a],[Jab86b]

For directional and derivative constraints, an easy technique for constructing $\mathbf{B}$ is to treat each column of $\mathbf{C}$ successively. Consider a standard N-element linear array and assume that $\mathbf{C}$ is given by (6.291). First pass $\mathbf{X}(\omega)$ through the bidiagonal $(N-1) \times N$ matrix $\mathbf{B}_1^H$,

$$\mathbf{B}_1^H = \begin{bmatrix} 1 & -1 & & & & \\ & 1 & -1 & & \mathbf{0} & \\ & & 1 & -1 & & \\ & \mathbf{0} & & \ddots & \ddots & \\ & & & & 1 & -1 \end{bmatrix}. \tag{6.891}$$

Then,

$$\mathbf{X}_1(\omega) = \mathbf{B}_1^H \mathbf{X}(\omega) \tag{6.892}$$

will not contain any components along $\mathbf{v}(\psi_m)$.

Then pass $\mathbf{X}_1(\omega)$ through the bidiagonal $(N-2) \times (N-1)$ matrix $\mathbf{B}_2^H$,

$$\mathbf{B}_2^H = \begin{bmatrix} 1 & -e^{-j\psi_1} & & & & \\ & 1 & -e^{-j\psi_1} & & \mathbf{0} & \\ & & 1 & -e^{-j\psi_1} & & \\ & \mathbf{0} & & \ddots & \ddots & \\ & & & & 1 & -e^{-j\psi_1} \end{bmatrix}, \tag{6.893}$$

where $\psi_1 = \psi_m + \Delta\psi$.

Then,

$$\mathbf{X}_2(\omega) = \mathbf{B}_2^H \mathbf{X}_1(\omega) \tag{6.894}$$

will not contain any components along $\mathbf{v}(\psi_1)$.

Finally, pass $\mathbf{X}_2(\omega)$ through the bidiagonal $(N-3) \times (N-2)$ matrix $\mathbf{B}_3^H$,

$$\mathbf{B}_3^H = \begin{bmatrix} 1 & -e^{-j\psi_2} & & & & \\ & 1 & -e^{-j\psi_2} & & \mathbf{0} & \\ & & 1 & -e^{-j\psi_2} & & \\ & & & \ddots & \ddots & \\ & \mathbf{0} & & & 1 & -e^{-j\psi_2} \end{bmatrix}, \tag{6.895}$$

where $\psi_2 = \psi_m - \Delta\psi$.

Then,

$$\mathbf{Z}(\omega) = \mathbf{B}_3^H \mathbf{X}_2(\omega) = \mathbf{B}_3^H \mathbf{B}_2^H \mathbf{B}_1^H \mathbf{X}(\omega) \tag{6.896}$$

will be orthogonal to the constraint subspace.

The advantage of this approach is that the matrices are sparse, which saves computation. A possible disadvantage is that

$$\mathbf{B}^H \mathbf{B} \neq \mathbf{I}. \tag{6.897}$$

(a) Consider a standard 10-element linear array. Implement a LCMP beamformer using three directional constraints with $\Delta\psi = 0.25 BW_{NN}$. Assume there is a single plane-wave interferer at $u_I = 0.3$ (INR = 0, 10, 20 dB). The signal arrives from u_a. (SNR = 0, 10, 20 dB). Plot the array gain versus u_a/BW_{NN} for $0 \leq u_a/BW_{NN} \leq 0.25$.

(b) Discuss the implications of (6.897).

Problem 6.7.8 (continuation)

The same technique can be used for derivative constraints. Pass $\mathbf{X}_1(\omega)$ through the tridiagonal $(N-2) \times N$ matrix $\mathbf{B}_2^H$,

$$\mathbf{B}_2^H = \begin{bmatrix} -1 & 2 & -1 & & & & \\ & -1 & 2 & -1 & & \mathbf{0} & \\ & & -1 & 2 & -1 & & \\ & \mathbf{0} & & \ddots & \ddots & \ddots & \\ & & & & -1 & 2 & -1 \end{bmatrix}. \tag{6.898}$$

Then

$$\mathbf{X}_2(\omega) = \mathbf{B}_2^H \mathbf{B}_1^H \mathbf{X}(\omega), \tag{6.899}$$

where $\mathbf{B}_1^H$ is defined in (6.891), contains no components in the constraint space.

To include the second derivative constraint in (6.310) with $\psi = 0$, use the $(N-3) \times N$ matrix $\mathbf{B}_3^H$,

$$\mathbf{B}_3^H = \begin{bmatrix} -1 & 3 & -3 & 1 & & & & \\ & -1 & 3 & -3 & 1 & & \mathbf{0} & \\ & & -1 & 3 & -3 & 1 & & \\ & \mathbf{0} & & \ddots & \ddots & \ddots & \ddots & \\ & & & & -1 & 3 & -3 & 1 \end{bmatrix}. \tag{6.900}$$

Then,

$$\mathbf{Z}(\omega) = \mathbf{B}_3^H \mathbf{B}_2^H \mathbf{B}_1^H \mathbf{X}(\omega), \tag{6.901}$$

contains no components in the constraint space.

(a) Consider a standard 10-element linear array. Repeat part (b) of Problem 6.7.2.

(b) Discuss the implications of (6.897).

Problem 6.7.9

Analyze the behavior of a LCMP-GSC beamformer in which fixed nulls are placed in the direction of strong interferers whose direction has been previously determined. Implement the beamformer as a GSC.

Consider a standard 20-element linear array. The nominal signal direction is broadside. There are four uncorrelated plane-wave interferers at $u_{I1} = 0.30$ ($INR = 40$ dB), $u_{I2} = 0.50$ ($INR = 30$ dB), $u_{I3} = 0.70$ ($INR = 50$ dB), $u_{I4} = -0.50$ ($INR = 10$ dB).

There is an additional interferer at $u_I = -0.3$ (variable *INR*).

(a) Design a LCMP-GSC beamformer in which the second through fifth columns of $\mathbf{C}$ force perfect nulls on the four interferers. Calculate the array for various *INR* of the additional interferer. Compare your results to a single distortionless constraint MPDR beamformer.

(b) Assume that the location of the interferers has been estimated and that the error in the estimate is a zero-mean Gaussian random variable with standard deviation σ_e (in u-space). The estimate associated with each interferer is independent. Use a Monte Carlo procedure to evaluate the loss in array gain for several σ_e. How can you make the beamformer more robust to estimation errors?

Problem 6.7.10

Consider a standard 21-element linear array. The desired quiescent pattern is a Dolph-Chebychev pattern with −40-dB SLL.

The nominal signal DOA is broadside. The actual signal arrives from u_a where u_a is a uniform random variable, $|u_a| \leq 0.1$. Implement a LCMP-DL beamformer using a quiescent pattern constraint and diagonal loading.

There are two uncorrelated interferers arriving from u_{I1} and u_{I2}. Consider *INRs* of 10, 20, and 30 dB.

(a) Assume $u_{I1} = 3/21$ and $u_{I2} = -3/21$. Plot the array gain versus *SNR* for the three *INR*. Use appropriate loading. Plot the beam patterns. Explain the behavior of the beamformer.

(b) Repeat part (a) for $u_{I1} = 3/21$ and $u_{I2} = -5/21$.

Problem 6.7.11 (continuation; Problem 6.7.10)

Consider a standard 21-element linear array. The desired quiescent pattern is a Dolph-Chebychev pattern with −40-dB SLL. Use the same signal and interference model as in Problem 6.7.10.

Consider the following constraints:

(i) $\mathbf{w}_{dq}/\|\mathbf{w}_{dq}\|^2$ is the only constraint.

(ii) $\mathbf{w}_{dq}/\|\mathbf{w}_{dq}\|^2$ plus first- and second-order derivative constraints. Use an appropriate diagonal loading level.

(a) Repeat part (a) of Problem 6.7.10.

(b) Repeat part (b) of Problem 6.7.10.

(c) Compare your results to those in Problem 6.7.10.

Problem 6.7.12

Consider a standard 21-element linear array. The nominal signal arrival angle is $u_s = 0$. The actual signal arrival angle is u_a, where u_a is a random variable with uniform probability density, $0 \leq |u_a| \leq 0.1$. The desired quiescent beam pattern is a Dolph-Chebychev beam pattern with −40-dB SLL.

Use the constraints in (6.416)and diagonal loading.

(a) Assume there are four plane-wave interferers:

$$\begin{aligned} u_I1 &= 0.3, \quad INR = 50 \text{ dB}, \\ u_I2 &= -0.4, \quad INR = 40 \text{ dB}, \\ u_I3 &= 0.5, \quad INR = 30 \text{ dB}, \\ u_I4 &= -0.6, \quad INR = 20 \text{ dB}. \end{aligned} \tag{6.902}$$

Use a total of three constraints, $\overline{\mathbf{w}}_{dq}$, and two eigenvectors. Choose an appropriate *LNR*. Plot the expected value of the array gain versus *SNR*.

(b) Let the number of eigenvectors constraints be a design parameter. Repeat part (a).

Explain your results. How would they change if the number of interferers or their strength changed.

Problem 6.7.13 (continuation; Problem 6.7.12)

Consider a standard 21-element linear array. Use the quiescent pattern technique and investigate how the choice of the quiescent pattern affects the performance. Consider the following quiescent patterns:

(i) Hann
(ii) Hamming
(iii) Blackman-Harris
(iv) Dolph-Chebychev (−20-dB SLL)
(v) Dolph-Chebychev (−30-dB SLL)
(vi) Taylor (−30-dB SLL, $\overline{n} = 6$)

Repeat Problem 6.7.12.

P6.8 Eigenvector Beamformers

Problem 6.8.1

Consider a standard 10-element array. The nominal signal direction is broadside. The actual signal arrives from u_a, where $|u_a| \leq 0.1$ with an $SNR = -10$, 0, 10, or 20 dB. There are two equal-power uncorrelated interferers at $u_I = 0.29$ and 0.45. Consider *INR*s of 0, 10, 20, and 30 dB.

Use an eigenspace beamformer and plot the array gain versus u_a/BW_{NN} for three scenarios.

(a) The dimension of the $S + I$ subspace is correctly estimated at 3. Plot the array gain versus u_a/BW_{NN}. Note that the eigenspace space is based on the actual u_a. Consider the *SNR* and *INR* values specified above.

(b) The dimension of the $S+I$ subspace is underestimated at 2. The eigenspace consists of the eigenvectors with the two largest eigenvalues. Repeat part (a).

(c) The dimension of the $S+I$ subspace is overestimated at 4. One eigenvector due to the white noise is added to the eigenspace in part (a). Repeat part (a).

(d) Compare your results in parts (a)–(c).

Problem 6.8.2 (continuation; Problem 6.8.1)

Consider the same model as in Problem 6.8.1 except there are four interferers:

$$\begin{aligned} u_{I1} &= 0.29, \quad INR = 30 \text{ dB}, \\ u_{I2} &= 0.45, \quad INR = 20 \text{ dB}, \\ u_{I3} &= 0.55, \quad INR = 10 \text{ dB}, \\ u_{I4} &= -0.50, \quad INR = 10 \text{ dB}, \end{aligned} \tag{6.903}$$

The *SNR* various from -10 dB to 30 dB in 10-dB steps. Consider eigenspace dimensions of 4, 5, and 6. Repeat Problem 6.8.1. Discuss your results.

Problem 6.8.3 (continuation; Problem 6.8.2)

Consider the model as in Problem 6.8.2. Impose second-order derivative constraints in (6.464). Repeat Problem 6.8.2 for the case in which the eigenspace dimension is 5.

Problem 6.8.4

Consider a 10-element uniform linear array with $d = \lambda/4$. The nominal signal direction is broadside. The interference consists of isotropic noise and white noise. Define

$$INR_{ISO} = \sigma_{ISO}^2/\sigma_w^2.$$

(a) Assume the signal arrives from broadside. Consider INR_{ISO} = 0, 10, 20, and 30 dB. Plot the array gain versus the dimension of the eigenspace for the indicated INR_{ISO}.

(b) Now assume the signal arrives from u_a where $|u_a| \leq 0.1$. Plot the array gain versus u_a/BW_{NN} for the indicated INR_{ISO}. Choose the dimension based on the results in part (a).

(c) Repeat parts (a) and (b) for the cross-spectral beamformer in Section 6.8.2.

Problem 6.8.5 (continuation)

Consider the same model and isotropic noise environment as in Problem 6.8.4. There are also two equal-power uncorrelated plane-wave interferers at $u_I = -0.25$ and 0.3 with *INR*s that vary from 0 dB to 30 dB in 10-dB steps.

Repeat Problem 6.8.4.

Problem 6.8.6

Consider a standard 10-element array. The steering direction is broadside ($u_s = 0$). The interference consists of three plane waves,

$$\begin{aligned} u_{I1} &= 0.30, & INR &= 20 \text{ dB}, \\ u_{I2} &= -0.50, & INR &= 25 \text{ dB}, \\ u_{I3} & \text{ is variable}, & INR &= 30 \text{ dB}. \end{aligned} \tag{6.904}$$

The *SNR* is 10 dB. We use $D_m = 4$ and 15 dB of diagonal loading. Read [CP97] or [CPL98]. Implement the robust DMR algorithm.

Plot the output SNR_o versus u_{I3}.

Problem 6.8.7

Consider a standard 32-element linear array.

The signal is a plane wave arriving from $u_a = 0.01$. There are four interferers at:

$$\begin{aligned} u_{I1} &= \quad 5/32, \quad INR = 20 \text{ dB}, \\ u_{I2} &= \quad 8/32, \quad INR = 30 \text{ dB}, \\ u_{I3} &= \quad 13/32, \quad INR = 20 \text{ dB}, \\ u_{I4} &= -17/32, \quad INR = 30 \text{ dB}. \end{aligned} \tag{6.905}$$

(a) Design an eigenspace beamformer and plot the array gain versus *SNR*. Plot the beam pattern for $SNR = 10$ dB.

(b) Assume there are eight interferers. The location in u-space of the ith interferer is a uniform random variable $(-1 \leq u_i \leq 1)$, $i = 1, \cdots, 8$. The location random variables are statistically independent. The *INR* of each interferer is one of five values, 0, 10, 20, 30, 40 dB with equal probability. The *INR* is independent of the location.

Design an eigenspace beamformer. Specify the assumptions you use in the design. Find the array gain and plot the beam pattern for a single trial.

(c) Run 50 trials of part (b) and compute the average array gain.

P6.9 Beamspace (BS) Beamformers

Problem 6.9.1 (continuation; Example 6.9.1)

Consider a standard 32-element linear array and a 7×32 beamspace matrix using conventional beams. The signal is a plane-wave signal arriving from u_s where $|u_s| \leq 0.25 BW_{NN}$. The signal arrival direction u_s is known to the beamformer. The interferers consist of two equal-power uncorrelated plane-wave interferers at $u_I = 3/32$ and $5/32$.

(a) Find the optimum beamspace MPDR beamformer. Compute the array gain and compare it to the result in Example 6.9.1. Discuss your results.

(b) Discuss various ways to improve performance and evaluate one of them.

Problem 6.9.2 (continuation; Example 6.9.2)

Consider the same model and environment as in Example 6.9.2. Design a LCMP beamformer using second-order derivative constraints and diagonal loading.

(a) Plot the array gain versus u_a and compare the results to those in Figure 6.84.

(b) Add three out-of-sector interferers at $u_I = -11/32$, $9/32$, and $15/32$, each with an *INR* of 20 dB. Plot the array gain versus u_a and compare the results to part (a).

Problem 6.9.3 (continuation; Example 6.9.2)

Consider the model in Example 6.9.2 except we use Dolph-Chebychev beams with −40-dB SLL to form $\mathbf{B}_{no}^H$ and orthonormalize them to form $\mathbf{B}_{bs}^H$.

(a) Repeat Example 6.9.2 and compare your results to Figure 6.84.

(b) Repeat Problem 6.9.2 and compare your results.

Problem 6.9.4 (continuation; Example 6.9.6)

Consider a standard 32-element linear array and a 7×32 beamspace matrix. Assume there are 10 uncorrelated interferers at 3/32, 7/32, 11/32, 15/32, 19/32 and −5/32, −9/32, −13/32, −17/32, −21/32. Each interferer has a 20-dB *INR*.

Consider various beamspace matrices such as:

(i) Conventional beams.

(ii) Dolph-Chebychev (−30-dB SLL, −40-dB SLL) beams.

(iii) Taylor beams.

Design an MPDR beamformer for a signal arriving from $u_s = 0$. Evaluate the array gain for the various beamspace matrices. Compare the result to the array gain for element-space MPDR.

Problem 6.9.5 (continuation; Example 6.9.2)

Consider a standard 32-element linear array and the 7×32 beamspace matrix given in Example 6.9.1. The interference environment is the same as in Example 6.9.2. Use an LCMP beamspace beamformer with a quiescent pattern constraint and fixed diagonal loading. The desired quiescent beam corresponds to $\cos^6(\pi\tilde{n}/N)$ weighting.

(a) Plot the array gain versus u_a/BW_{NN} for an *SNR* =0 and 10 dB and several levels of diagonal loading.

(b) Compare your results to those in Figure 6.84. Suggest some design guidelines.

Problem 6.9.6

Consider a standard 32-element linear array and a 6×32 beamspace matrix. The beam sector is centered at broadside. The rows of the beamspace matrix are conventional beams with steering directions,

$$u_s = \pm 1/32, \pm 3/32, \pm 5/32. \tag{6.906}$$

The interference consists of two equal-power uncorrelated plane-wave interferers at 4/32 and 6/32. Consider *INR*s of 10, 20, and 30 dB. Consider *SNR*s of 0, 10 dB, and 20 dB.

(a) Evaluate the performance of an MPDR beamformer. Assuming no signal mismatch. Plot the beam pattern and evaluate the array gain. Compare your result to the results in Example 6.9.1.

(b) Consider the case of signal mismatch where $|u_a| \le 1/32$. Use an appropriate level of fixed diagonal loading. Plot the array gain versus u_a/BW_{NN}. Compare your result to the results in Example 6.9.2.

(c) Assume the signal arrives from u_a where $|u_a| \le 1/32$. We estimate u_a. We assume that the estimate $\hat{u}_a$ is unbiased and its standard deviation is given by (6.237)

$$\left(E\left[|\hat{u}_a - u_a|^2\right]\right)^{\frac{1}{2}} = \frac{4}{N}\left(\frac{3}{8K}\left[\frac{1}{ASNR} + \frac{1}{ASNR^2}\right]\frac{1}{1-1/N^2}\right)^{\frac{1}{2}}. \tag{6.907}$$

We construct a MPDR beamformer steered at $\hat{u}_a$. Evaluate the performance. Plot the array gain versus u_a for representative *SNR*. Plot the array gain versus *SNR* for representative u_a. Use $K/N = 2$ and $K/N = 10$.

Problem 6.9.7 (continuation; Problem 6.9.6)

Consider the same model as in Problem 6.9.5. Define a quiescent beam corresponding to a $\cos^5(\pi\tilde{n}/N)$ weighting. We construct a beamformer that satisfies the constraint

$$\overline{\mathbf{w}}_{dq}^H \mathbf{w} = 1. \tag{6.908}$$

(a) Repeat part (a) of Problem 6.9.6.

(b) Repeat part (b) of Problem 6.9.6.

(c) Design a beamspace LCMP beamformer using the quiescent pattern constraint and appropriate fixed loading. Plot the array gain versus u_a/BW_{NN}. Compare your result to the results in Example 6.9.2 and Problem 6.9.6.

Problem 6.9.8

Consider a standard 32-element linear array and a 5×32 beamspace matrix centered at broadside. The rows of the beamspace matrix are conventional beams with steering directions,

$$u_s = 0, \pm 2/32, \pm 4/32. \tag{6.909}$$

The interference consists of two equal-power uncorrelated plane-wave interferers at 3/32 and 5/32. Consider *INRs* of 10, 20, 30 dB. Consider *SNRs* of 0, 10, and 20 dB.

(a) Evaluate the performance of a MPDR beamformer assuming no signal mismatch. Plot the beam pattern and evaluate the array gain. Compare your result to the result in Example 6.9.1.

(b) Repeat part (a) for signal mismatch.

Problem 6.9.9 (continuation; Problem 6.9.8)

Consider the same model as in Problem 6.9.8. Consider two quiescent beam patterns:
(i) Blackman-Harris
(ii) $\cos^4(\pi n/N)$

(a) Construct a beamformer that satisfies the constraint,

$$\overline{\mathbf{w}}_{dq}^H \mathbf{w} = 1 \tag{6.910}$$

and uses fixed diagonal loading with appropriate *LNR*. Consider the signal mismatch problem where $|u_a| \leq 1/32$. Plot the array gain versus u_a/BW_{NN}. Compare your results to the results in Example 6.9.2.

(b) Design a beamspace LCMP beamformer using the a quiescent pattern constraint. Repeat part (a).

(c) Discuss your results and suggest design guidelines.

Problem 6.9.10

Consider a standard 32-element linear array and a 4×32 beamspace matrix centered at broadside. The rows of the beamspace matrix are conventional beams with steering directions,

$$u_s = \pm 1/32, \pm 3/32. \tag{6.911}$$

The interference consists of two equal power uncorrelated plane-wave interferers at 2/32 and 4/32. We consider *INRs* of 10, 20, 30 dB. We consider *SNRs* of 0, 10, and 20 dB.

Evaluate the performance of a MPDR beamformer assuming no signal mismatch. Plot the beam pattern and evaluate the array gain. Compare your result to the results in Example 6.9.1.

Problem 6.9.11

Consider the same model as in Problem 6.9.10. Consider a quiescent beam pattern of $\cos^3(\pi n/N)$.

(a) Construct a beamformer that satisfies the constraint,

$$\overline{\mathbf{w}}_{dq}^H \mathbf{w} = 1 \tag{6.912}$$

and uses fixed diagonal loading with appropriate *LNR*. Consider the signal mismatch problem where $|u_a| \leq 1/32$. Plot the array gain versus u_a/BW_{NN}. Compare your results to the results in Example 6.9.2.

(b) Discuss your results. What are the limitations of this low-dimensional beamspace?

Problem 6.9.12

As an alternative to eigenspace processing, consider the following beamspace algorithm.

In order to construct the beamspace matrix, scan a conventional beam over u-space with a scan step size of $2/N$. Detect the presence or absence of the a signal (or interferer) at each scan location and estimate its power. If the *SNR* exceeds 10 dB, form a conventional beam steered at it and use that beam as a row in the beamspace matrix. For the beamspace sector centered at $u_s = 0$, always include a beam centered at $u_s = 0$. Then form a MPDR beamformer in beamspace.

Consider a standard 32-element linear array and a $N_{bs} \times N$ beamspace matrix. The signal is a plane wave with an $SNR = 10$ dB arriving from $u_a = 0.01$.

(a) Assume that the scan detects four interferers at:

$$\begin{aligned} u_{I1} &= 5/32, \quad INR = 20 \text{ dB}, \\ u_{I2} &= 8/32, \quad INR = 30 \text{ dB}, \\ u_{I3} &= 13/32, \quad INR = 20 \text{ dB}, \\ u_{I4} &= -17/32, \quad INR = 30 \text{ dB}. \end{aligned} \tag{6.913}$$

Form a 5×32 beamspace matrix and then use an MPDR beamformer in beamspace. Find the array gain and plot the beam pattern. Compare your result to the result in Problem 6.8.7 and to an element-space beamformer.

(b) Assume there are eight interferers. The location in u-space of the ith interferer is a uniform random variable $(-1 \leq u_i \leq 1)$, $i = 1, \cdots, 8$. The location random variables are statistically independent. The *INR* of each interferer is one of five values; 0, 10, 20, 30, 40 dB with equal probability. The *INR* is independent of the location. (Note that a 0-dB interferer will not be detected.) Find the array gain and plot the beam pattern for a single trial. Compare your result to the result in Problem 6.8.7 and to an element-space beamformer. Use the same data set.

(c) Run 50 trials of part (b) and compute the average array gain.

Problem 6.9.13 (continuation; Example 6.9.1)

Consider a standard 32-element linear array and a 7×32 beamspace matrix. The rows of the beamspace matrix are determined adaptively.

Each row of the beamspace processor is an MPDR beamformer with a distortionless constraint in its steering direction. Assume that the beamspace beams can be formed during a period in which there are no signals or interferers arriving in the beamspace sector.

There are eight interferers at:

$$
\begin{aligned}
u &= 11/32, -13/32, 19/32 : INR = 40 \text{ dB}, \\
u &= 15/32, 23/32, -17/32 : INR = 30 \text{ dB}, \\
u &= -21/32, 27/32 : INR = 20 \text{ dB},
\end{aligned}
\tag{6.914}
$$

present during the beamspace matrix construction. After the beamspace matrix is established, two interferers, each with an *INR*=20 dB, arrive from 3/32 and 5/32. The eight interferers remain.

(a) Design a beamspace MPDR processor for a signal arrival direction $u_s = 0$. Evaluate its array gain and plot its beam pattern.

(b) Compare the performance in part (a) to an element-space MPDR beamformer.

Problem 6.9.14 (continuation; Problem 6.9.13)

Consider the same model as in Problem 6.9.13. In order to construct the beamspace matrix, scan a conventional beam over u-space with a scan step size of $2/N$. Detect the presence or absence of a signal (or interferer) at each scan location and estimate its power. If the *SNR* exceeds 10 dB and it is outside the beamspace sector, place a null on it in each of the beamspace rows.

After the beamspace matrix is established, the two interferers appear at 5/32 and 7/32 (*INR* = 20 dB each).

(a) Repeat Problem 6.9.13. Compare your result to the results in Problem 6.9.13.

(b) How does the scan step size affect performance?

Problem 6.9.15

Consider a standard 32-element linear array and a 5×32 beamspace matrix using conventional beams centered at $u = 0$, $\pm 2/32$, and $\pm 4/32$. The desired quiescent beam is the Blackman-Harris beam.

The nominal signal arrival direction is $u_s = 0$. The actual signal arrival angle is u_a, where $|u_a| \leq 1/32$. There are four uncorrelated plane-wave interferers:

$$
\begin{aligned}
u_1 &= 5/32, INR = 20 \text{ dB}, \\
u_2 &= 13/32, INR = 50 \text{ dB}, \\
u_3 &= 19/32, INR = 50 \text{ dB}, \\
u_4 \text{ variable: } 3/32 &\leq u_4 \leq 21/32, INR = 20 \text{ dB}.
\end{aligned}
\tag{6.915}
$$

Design a beamspace LCMP-DL beamformer using a quiescent pattern constraint. Evaluate the performance for various u_a, u_4, and *SNR*. Discuss the trade-offs between low out-of-sector sidelobes and the ability to null in-sector interferers.

P6.10 Quadratically Constrained Beamformers

Problem 6.10.1 (continuation; Example 6.10.1)

Consider a standard 10-element linear array. The array is steered to broadside. The interference consists of two uncorrelated equal-power interferers at $u = 0.29$ and 0.45. The signal arrives at u_a. The *SNR*, *INR*, and T_o are parameters that we want to investigate.

(a) The *SNR* is 0 dB. Consider *INR*s of 0, 10, 20, and 30 dB. Plot the output SNR_o versus u_a/BW_{NN} for various T_o, where $1/N \leq T_o \leq 2/N$. Explain your results. Use the optimum β. Plot β versus u_a/BW_{NN}.

(b) Repeat part (a) for an $SNR = 10$ dB.

(c) Repeat part (a) for an $SNR = 20$ dB.

(d) Based on the results of parts (a)–(c), suggest guidelines for choosing T_o.

Problem 6.10.2 (continuation)

Consider the same model as in Problem 6.10.1 with $u_a = 0$. Add a third interferer with an $INR_3 = 20$ dB whose location u_{I3} is a parameter.

(a) Consider *INR*s of 0, 10, 20, and 30 dB for the original two interferers. Plot the output SNR_o versus u_{I3}/BW_{NN} for various T_o, where $1/N \leq T_o \leq 2/N$. Use the optimum β and the approximate β.

(b) Suggest guidelines for choosing T_o.

Problem 6.10.3 (continuation)

Consider the same model as in Problems 6.10.1 and 6.10.2. Investigate the sensitivity to the choice of β. Use these sensitivity results to develop criteria for how accurate the value of β should be in order to preserve almost optimum performance.

Problem 6.10.4 (continuation; Example 6.10.1)

Consider the same model as in Example 6.10.1. Implement an LCMP-QC beamformer with second order derivative constraints,

$$\mathbf{C} = \begin{bmatrix} \mathbf{1} & \mathbf{d}(0) & \dot{\mathbf{d}}(0) \end{bmatrix}, \tag{6.916}$$

$$\mathbf{g}^H = \begin{bmatrix} 1 & 0 & \ddot{\mathbf{B}}_c(0), \end{bmatrix}, \tag{6.917}$$

and

$$\mathbf{w}^H\mathbf{w} \leq T_o \tag{6.918}$$

(a) The *SNR* is 0 dB. Consider *INR*s of 0, 10, 20, and 30 dB. Plot the output SNR_o versus u_a/BW_{NN} for various T_o. Explain your results. Use the optimum β. Plot β versus u_a/BW_{NN}.

(b) Repeat part (a) for an $SNR = 10$ dB.

(c) Repeat part (a) for an $SNR = 20$ dB.

(d) Based on the results of parts (a)–(c), suggest guidelines for choosing T_o.

(e) In Example 6.10.1 and Problems 6.10.1–6.10.3 the beamformer relied on QC to provide main-lobe protection. In Example 6.10.2 and Problems 6.10.4–6.10.5, the beamformer relied on both linear constraints and QC to provide main-lobe protection. Discuss the advantages and disadvantages of the two approaches in the context of the environment.

Problem 6.10.5 (continuation; Problem 6.10.4)

Consider the same model as in Problem 6.10.4 with $u_a = 0$. Add a third interferer with an $INR_3 = 20$ dB whose location u_{I3} is a parameter.

(a) Consider *INRs* of 0, 10, 20, and 30 dB for the original two interferers. Plot the output SNR_o versus u_{I3}/BW_{NN} for various T_o. Use the optimum β.

(b) Suggest guidelines for choosing T_o.

Problem 6.10.6 (continuation; Problem 6.10.1)

Consider a standard 10-element linear array and the signal and interference environment described in Problem 6.10.1. Use the quiescent pattern constraint technique in Section 6.7.5 to determine the linear constraints. Consider two quiescent patterns: Hamming, and Dolph-Chebychev with −40-dB SLL. Impose a quadratic constraint.

Repeat the analyses in Problem 6.10.1.

Problem 6.10.7 (continuation; Problem 6.7.5)

Consider a standard 21-element linear array. Use the quiescent pattern constraint technique in Section 6.7.5 to determine the linear constraint. In addition, impose a quadratic constraint. Consider the six quiescent patterns in Problem 6.7.13. The nominal signal direction is broadside and u_a is a uniform random variable ($-1/21 \leq u_a \leq 1/21$). There are six interferers in the sidelobe region (3/21, 5/21, 11/21, −7/21, −9/21, −15/21).

(a) The *SNR* is 0 dB. Consider *INRs* of 0, 10, 20, and 30 dB. Plot the output SNR_o versus u_a/BW_{NN} for various T_o. Consider up to five constraints. Do not simply generate a large set of plots. Determine the important factors and illustrate them. Use the optimum β. Plot β versus u_a/BW_{NN}. Explain your results.

(b) Repeat part(a) for an *SNR* = 10 dB.

(c) Repeat part(a) for an *SNR* = 20 dB.

(d) Based on the results of parts (a)–(c), suggest guidelines for choosing the quiescent pattern, the number of constraints, and T_o.

Problem 6.10.8: Beamspace-QC algorithms

Consider a standard 32-element linear array and a 7 × 32 beamspace matrix using conventional beams. Use the signal mismatch and interference model in Example 6.9.2. Consider two quadratically constrained beamformers.

(i) MPDR (see Example 6.9.2).

(ii) LCMP using a quiescent pattern constraint, with $\cos^6(\pi\tilde{n}/N)$ the quiescent pattern.

(a) Plot the array gain versus u_a/BW_{NN} for the two beamformers (consider various values of T_o). Plot the resulting beam patterns. Compare your results to those in Example 6.9.2 and Problem 6.9.5.

(b) Discuss your results and suggest design guidelines.

Problem 6.10.9 (continuation; Problem 6.9.6)

Consider a standard 32-element linear array and the 6×32 beamspace matrix described in Problem 6.9.6. Consider two quadratically constrained beamformers:

(i) MPDR (see Example 6.9.6).

(ii) LCMP-QP-QC with $\cos^5(\pi\tilde{n}/N)$ quiescent pattern (see Problem 6.9.7).

The signal and interference environment is the same as Problem 6.9.6.

(a) Plot the array gain versus u_a/BW_{NN} for the two beamformers (consider various values of T_o). Plot the resulting beam patterns. Compare your results to those in Problem 6.9.6 and 6.9.7.

(b) Discuss your results and suggest design guidelines. What are the advantages and disadvantages of this beamspace processor compared to the 7×32 processor in Problem 6.10.8.

P6.11 Soft Constraint Beamformers

Problem 6.11.1 (continuation; Example 6.11.1)

Consider the same signal and interference model as in Example 6.11.1. Study the effect of the choice of $\mathbf{w}_d$ in (6.567). Consider Dolph-Chebychev beam patterns with SLLs of -20, -30, and -40 dB. Use $\psi_0 = 0.1\pi$ and $\alpha = 0.03$.

(a) Plot the normalized SNR_o versus u_a/BW_{NN} for the MPSC beamformer. Consider $SNR = 0$ dB and 10 dB. Compare your results to the results in Figures 6.94 and 6.95.

(b) Vary α and study the effect on SNR_o.

Problem 6.11.2

Three of the algorithms developed in this chapter use the $\mathbf{Q}$ matrix in (6.571) as a starting point:

(i) Eigenvector constraints in Section 6.7.4.3.

(ii) Quiescent pattern constraints in Section 6.7.5.

(iii) Soft constraints in Section 6.11.

Consider several signal-and-interference scenarios and compare the performance of the three algorithms. Based on these analyses, discuss the trade-offs involved and suggest design guidelines.

Problem 6.11.3

A general formulation of the $\mathbf{Q}$ matrix is given in (6.412). For a linear array, $\mathbf{Q}$ can be written as

$$\mathbf{Q} = \int_F \int_\psi \mathbf{v}(\psi, f)\, \mathbf{v}^H(\psi, f)\, d\psi\, df. \tag{6.919}$$

Consider a 10-element uniform linear array with spacing $d = \lambda_c/2$ where λ_c is the wavelength corresponding to the design frequency f_c. Assume the interval is

$$f_c - 0.2f_c \leq f \leq f_c + 0.2f_c. \tag{6.920}$$

Use the signal and interference model in Example 6.11.1. Consider different frequencies for the signal. Denote the frequency of the signal as f_s.

(a) Assume $u_a = 0.5$. Plot the normalized SNR_o versus $\Delta f/f_c$, for the MPSC beamformer where

$$\Delta f = f_s - f_c.$$

Consider various α.

(b) Assume $\Delta f = 0$. Plot the normalized SNR_o versus u_a/BW_{NN}. Compare your results to those in Figures 6.94 and 6.95 and in part (a). Does the addition of a soft-frequency constraint impact the performance?

(c) Plot a contour plot of the normalized SNR_o versus $(\Delta f/f_c, u_a/BW_{NN})$. Discuss your results.

Problem 6.11.4

In some applications, it is easier to specify $\mathbf{Q}$ as a sum of points in the (f, θ, ϕ) space rather than an integral. Then,

$$\mathbf{Q} = \sum_{m=1}^{M} c_m \, \mathbf{v}(f_m, \theta_m, \phi_m) \, \mathbf{v}^H(f_m, \theta_m, \phi_m) \tag{6.921}$$

where c_m is a weighting function to allow emphasis on certain points.

As a special case, consider a standard 10-element linear array and the signal-and-interference model in Example 6.11.1. There is no frequency mismatch. Thus

$$\mathbf{Q} = \sum_{m=1}^{M_c} c_m \, \mathbf{v}(\psi_m) \, \mathbf{v}^H(\psi_m). \tag{6.922}$$

(a) Choose M_c in order to achieve equivalent performance to Example 6.11.1.

(b) Now assume that $M_c = 3$ and that $\psi_m = 0, \pm 0.1$. Plot the normalized SNR_o versus u_a/BW_{NN}. Compare your results to an LCMP beamformer using three directional constraints and diagonal loading.

Problem 6.11.5

Consider an algorithm in which both a soft constraint and a quadratic constraint are imposed.

(a) Show that the resulting optimum weight vector is

$$\widehat{\mathbf{w}}_p = \left[[\mathbf{S}_\mathbf{x} + \lambda_2 \, \mathbf{I}] + \lambda_1 \, \mathbf{Q}\right]^{-1} [\mathbf{S}_\mathbf{x} + \lambda_2 \, \mathbf{I}]\mathbf{w}_d, \tag{6.923}$$

where λ_1 and λ_2 are chosen to satisfy the two constraints,

$$\widehat{\mathbf{w}}_p^H \mathbf{Q} \widehat{\mathbf{w}}_p \leq e_0^2, \tag{6.924}$$

and

$$(\mathbf{w}_d - \mathbf{w}_p)^H (\mathbf{w}_d - \mathbf{w}_p) \leq T_o. \tag{6.925}$$

(b) Consider various signal-and-interference scenarios and see if there are cases in which the added complexity leads to a significant performance improvement.

Problem 6.11.6 (continuation; Example 6.11.1)

Consider the same model as in Example 6.11.1. As an alternative to the approach in Problem 6.11.5, define

$$\mathbf{Q}_{au} = \gamma \mathbf{Q} + (1 - \gamma)\mathbf{I}, \quad 0 \leq \gamma \leq 1, \tag{6.926}$$

where $\mathbf{Q}$ is defined in (6.571). Plot SNR_o versus u_a/BW_{NN} for various values of γ and α. Compare your results to the results in Example 6.11.1

Problem 6.11.7 (continuation; Example 6.7.10)

Extend the technique of soft constraints to derivative constraints. Derive the optimum $\mathbf{w}_p$.

Consider the signal-and-interference model in Example 6.7.10. Plot the array gain versus u_a/BW_{NN} and compare your results to the results in Figure 6.59.

Problem 6.11.8: Beamspace soft constraints (continuation; Example 6.9.2 and Problem 6.10.15)

Consider a standard 32-element linear array and the 7×32 standard beamspace matrix given in Example 6.9.1. Use the same signal-and-interference model as in Example 6.9.2. Consider two desired beam patterns (conventional and $\cos^6(\pi\tilde{n}/N)$). Utilize soft constraints over the interval $-1/32 \leq u \leq 1/32$.

(a) Plot the array gain versus u_a/BW_{NN} and compare your results to those in Example 6.9.2.

(b) Discuss your comparison. How dependent is it on the specific signal-and-interference model?

P6.12 Beamforming for Correlated Signal and Interferences

Problem 6.12.1

Consider the spatial smoothing approach in which the subarray spectral matrices are averaged with different weights. Then, from (6.622),

$$\mathbf{S}_{SSFB} = \frac{1}{2}\sum_{i=1}^{L} w_i \left(\mathbf{S}_M^{(i)} + \mathbf{S}_{MB}^{(i)}\right), \tag{6.927}$$

where the weights are real and

$$\sum_{i=1}^{L} w_i = 1. \tag{6.928}$$

(a) Derive the equation corresponding to (6.637).

(b) Assume the w_i are chosen to correspond to an L-element Hamming weighting. Plot the results analogous to those in Figure 6.104 for this weighting

(c) Discuss your results. Is the $u_I = 0.30$ case representative?

Problem 6.12.2 (continuation)

Consider a standard 32-element linear array. Repeat Problem 6.12.1 for the following sub-array weightings:

(i) Uniform
(ii) Hamming
(iii) Dolph-Chebychev

Test various u_I in the sidelobe region.

Problem 6.12.3

Consider a standard rectangular array with $N = M$. Extend the technique of spatial smoothing to the rectangular array.

(a) Derive the necessary equations

(b) Develop plots similar to Figure 6.104 for several interference scenarios.

P6.13 Broadband Beamformers

Problem 6.13.1

Consider a uniformly spaced 10-element array. The interelement spacing is $\lambda_u/2$, where λ_u is the wavelength of the highest frequency in the input signal. Implement an FFT beamformer with M frequency bins.

$$B_f = B_s/f_c = 0.6. \tag{6.929}$$

The input is sampled at $1/B_s$ so the normalized frequency span is $-\pi \leq \omega \leq \pi$ in radians. The signal of interest arrives from $u_s = 0.3$ and has a flat frequency spectrum over $-0.25\pi \leq \omega \leq 0.25\pi$ (a 50% bandwidth signal).

There are three interfering signals with the following characteristics (all spectra are zero outside the indicated band):

$$S_{I1} = \sigma_1^2, \quad -0.25\pi < \omega < 0, \quad u_I = 0, \quad INR_1 = 40 \text{ dB}. \tag{6.930}$$

$$S_{I2} = \sigma_2^2, \quad -0.01 \leq \omega \leq 0.01, \quad u_I = 0.5, \quad INR_2 = 60 \text{ dB}. \tag{6.931}$$

$$S_{I3} = \sigma_3^2, \quad 0 < \omega \leq 0.25\pi, \quad u_I = -0.2, \quad INR_2 = 30 \text{ dB}. \tag{6.932}$$

Consider M = 4, 8, 16, 32, and 64. Use an MPDR beamformer in the appropriate bins and set the output of the other frequency bins equal to 0.

Plot the SNR_o versus $ASNR$ for the various M. Discuss your results.

Problem 6.13.2 (continuation)

Consider the same model as in Problem 6.13.1, except the signal spectrum is

$$S_s = \begin{cases} 1 + \cos 10\omega & -0.25\pi \leq \omega \leq 0.25\pi, \\ 0 & \text{elsewhere.} \end{cases} \tag{6.933}$$

Repeat Problem 6.13.1.

Problem 6.13.3 (continuation)

Consider the same model as in Problem 6.13.1, except there is a single narrowband interferer,

$$S_{I1} = \sigma_1^2, \quad \omega_1 - 0.01 \leq \omega \leq \omega_1 + 0.01,$$
$$INR_1 = 50 \text{ dB}. \tag{6.934}$$

Investigate the behavior as a function of u_I, ω_1, and M. Plot the SNR_o versus $ASNR$ for various combinations. Discuss your results.

Bibliography

[AC76] S. P. Applebaum and D. J. Chapman. Adaptive arrays with main beam constraints. *IEEE Trans. Antennas Propag.*, vol.AP-24, pp. 650–662, September 1976.

[AO90] D.A. Abraham and N.L. Owsley. Beamforming with dominant mode rejection. *Proc. IEEE Oceans 90*, pp. 470–475, 1990.

[BAB90] J. F. Bull, M. A. Arnao, and L. R. Burgess. Hypersensitivity effects in adaptive antenna arrays. *Proc. IEEE Antennas Propag. Soc. Symp.*, pp. 396–399, May 1990.

[Bag76] A. B. Baggeroer. Space-time processes and optimal array processing. Technical Report 506, Navy Undersea Center, San Diego, California, December 1976.

[BC65] C. J. Becker and B. F. Cron. Optimum array gain for directional noise. Technical Report 656, Underwater Systems Lab, October 1965.

[BG81] B. S. Byun and A. F. Gangi. A constraint-elimination technique for linearly constrained arrays. *IEEE Trans. Geosci. Remote Sensing*, vol.GE-19, pp. 8–15, January 1981.

[BG86] K. M. Buckley and L. J. Griffiths. An adaptive generalized sidelobe canceller with derivative constraints. *IEEE Trans. Antennas Propag.*, vol.AP-34, pp. 311–319, March 1986.

[BH86] E. Brookner and J.M. Howell. Adaptive-adaptive array processing. *Proc. IEEE*, vol.74, pp. 602–604, April 1986.

[BKS88] A. B. Baggeroer, W. A. Kuperman, and H. Schmidt. Matched field processing: Source localization in correlated noise as an optimum parameter estimation problem. *J. Acoust. Soc. Am.*, vol.83, pp. 571–587, February 1988.

[BN62] R. G. Brown and J. W. Nilsson. *Introduction to Linear Systems Analysis*. Wiley, New York, 1962.

[BR89] K. A. Byerly and R. A. Roberts. Output power based partially adaptive array design. *Proc. 23rd Asilomar Conf. on Signals, Systems and Computers*, Pacific Grove, California, pp. 576–580, November 1989.

[BRK88] Y. Bresler, V. U. Reddy, and T. Kailath. Optimum beamforming for coherent signal and interferences. *IEEE Trans. Acoust., Speech, Signal Process.*, vol.ASSP-36, pp. 833–842, June 1988.

[Bry62] F. Bryn. Optimum signal processing of three-dimensional array operating on Gaussian signals and noise. *J. Acoust. Soc. Am.*, vol.34, pp. 289–297, March 1962.

[Buc87] K. M. Buckley. Spatial/spectral filtering with linearly constrained minimum variance beamformers. *IEEE Trans. Acoust., Speech, Signal Process.*, vol.ASSP-35, pp. 249–266, March 1987.

[Bur64] J. P. Burg. Three-dimensional filtering with an array of seismometers. *Geophysics*, vol.XXIX, no.5, pp. 693–713, October 1964.

[Cap69] J. Capon. High-resolution frequency-wavenumber spectrum analysis. *Proc. IEEE*, vol.57, pp. 1408–1418, August 1969.

[CG80] A. Cantoni and L.C. Godara. Resolving the direction of sources in a correlated field incident on an array. *J. Acoust. Soc. Am.*, vol.67, pp. 1247–1255, April 1980.

[CGK67] J. Capon, R. J. Greenfield, and R. J. Kolker. Multidimensional maximum-likelihood processing of a large aperture seismic array. *Proc. IEEE*, vol.56, pp. 192–211, February 1967.

[Cha76] D. J. Chapman. Partial adaptivity for the large array. *IEEE Trans. Antennas Propag.*, vol.AP-24, pp. 685–696, September 1976.

[CK84] T. K. Citron and T. Kailath. An improved eigenvector beamformer. *Proc. ICASSP*, San Diego, California, 1984.

[Com88a] R. T. Compton, Jr. *Adaptive Antennas (Concepts and Performance)*. Prentice-Hall, Englewood Cliffs, New Jersey, 1988.

[Com88b] R. T. Compton, Jr. The relationship between tapped delay-line and FFT processing in adaptive arrays. *IEEE Trans. Antennas Propag.*, vol.AP36, pp. 15–26, January 1988.

[Cox73] H. Cox. Resolving power and sensitivity to mismatch of optimum array processors. *J. Acoust. Soc. Am.*, vol.ASSP-54, pp. 771–785, September 1973.

[Cox00] H. Cox. Multi-Rate Adaptive Beamforming (MRABF). *Proc. IEEE Sensor Array and Multichannel Signal Processing Workshop*, Cambridge, Massachusetts, pp. 306–309, March 2000.

[CP97] H. Cox. and R. Pitre. Robust DMR and multi-rate adaptive beamforming. *Proc. 31st Asilomar Conf. on Signals, Systems and Computers*, Pacific Grove, California, November 1997.

[CPL98] H. Cox, R. Pitre, and H. Lai. Robust adaptive matched field processing. *Proc. 32nd Asilomar Conf. on Signals, Systems and Computers*, Pacific Grove, California, November 1998.

[CR65] D. Childers and I. Reed. On the theory of continuous array processing. *IEEE Trans. Aerosp. Navig. Electron.*, vol.AES-12, pp. 103–109, June 1965.

[CY92] L. Chang and C. Yeh. Performance of DMI and eigenspace-based beamformers. *IEEE Trans. Antennas Propag.*, vol.AP-40, pp. 1336–1347, November 1992.

[CZO87] H. Cox, R. M. Zeskind, and M. M. Owen. Robust adaptive beamforming. *IEEE Trans. Acoust., Speech, Signal Process.*, vol.-ASSP-35, pp. 1365–1376, October 1987.

[Dar58] S. Darlington. Linear least-squares smoothing and prediction, with applications. *Bell Syst. Tech. J.*, vol.37, pp. 1221–1294, September 1958.

[EC81] M. H. Er and A. Cantoni. An alternative formulation for an optimum beamformer with robustness capability. *Proc. IEE*, vol. 132, pt. F, pp. 447–460, October 1985.

[EC83] M. H. Er and A. Cantoni. Derivative constraints for broad-band element space antenna array processors. *IEEE Trans. Acoust., Speech, Signal Process.*, vol.ASSP-31, pp. 1378–1393, December 1983.

[EC85] M. H. Er and A. Cantoni. An alternative formulation for an optimum beamformer with robustness capability. *Proc. IEE*, vol.132 pt. F, pp. 447–460, October 1985.

[EC90] M. H. Er and A. Cantoni. A unified approach to the design of robust narrowband antenna array processors. *IEEE Trans. Antennas Propag.*, vol.AP-38, pp. 17–23, January 1990.

[EFK67] D.J. Edelblute, J.M. Fisk, and G.L. Kinnison. Criteria for optimum-signal-detection theory for arrays. *J. Acoust. Soc. Am.*, vol.41, pp. 199–205, January 1967.

[EJS82] J.E. Evans, J.R. Johnson, and D.F. Sun. Application of advanced signal processing techniques to angle of arrival estimation in ATC navigation and surveillance systems. Technical Report, M.I.T. Lincoln Laboratory, Lexington, Massachusetts, June 1982.

[Far83] D.R. Farrier. Gain of an array of sensors subjected to processor perturbation. *Proc. IEE*, vol.72, pt. H, pp. 251–254, June 1983.

[FG91] D. D. Feldman and L. J. Griffiths. A constraint projection approach for robust adaptive beamforming. *Proc. ICASSP*, Toronto, Canada, vol.2, pp. 1381–1384, May 1991.

[FG94] D. D. Feldman and L. J. Griffiths. A projection approach for robust adaptive beamforming. *IEEE Trans. Signal Process.*, vol.SP-42, pp. 867–876, April 1994.

[FH53] J. J. Faran and R. Hills. Wide-band directivity of receiving arrays. Technical Report 31, Acoustic Research Laboratory, Harvard University, Cambridge, Massachusetts, May 1953.

[FT98] B.E. Freburger and D.W. Tufts. Case study of principal component inverse and cross spectral metric for low rank interference adaptation. *Proc. ICASSP*, Seattle, Washington, vol.4, pp. 1977–1980, May 1998.

[Fri88] B. Friedlander. A signal subspace method for adaptive interference cancellation. *IEEE Trans. Acoust., Speech, Signal Process.*, vol.ASSP-36, pp. 1835–1845, December 1988.

[Fro72] O. L. Frost III. An algorithm for linearly constrained adaptive array processing. *Proc. IEEE*, vol.60, pp. 926–935, August 1972.

[Gaa66] N.T. Gaarder. The design of point detector arrays: II. *IEEE Trans. Inf. Theory*, vol.IT-12, pp. 112–120, April 1966.

[Gaa67] N.T. Gaarder. The design of point detector arrays: I. *IEEE Trans. Inf. Theory*, vol.IT-13, pp. 42–50, January 1967.

[Gab76a] W. F. Gabriel. Adaptive arrays - An introduction. *Proc. IEEE*, vol.64, pp. 239–271, February 1976.

[Gab76b] W. F. Gabriel. Preface-Special issue on adaptive antennas. *IEEE Trans. Antennas Propag.*, vol.AP-24, pp. 573–574, September 1976.

[Gab86] W. F. Gabriel. Using spectral estimation technique in adaptive processing antenna system. *IEEE Trans. Antennas Propag.*, vol.AP-34, pp. 291–300, March 1986.

[Gab87] W. F. Gabriel. *Adaptive Processing Antenna Systems.* IEEE Antennas and Propagation Short Course. Naval Research Laboratory, Virginia Polytechnic Institute and State University, Blacksburg, Virginia , June 1987.

[GB87] L. J. Griffiths and K. M. Buckley. Quiescent pattern control in linearly constrained adaptive arrays. *IEEE Trans. Acoust., Speech, Signal Process.*, vol.ASSP-35, pp. 917–926, July 1987.

[GE91] A. B. Gershman and V. T. Ermolaev. Synthesis of the weight distribution of an adaptive array with wide dips in the directional pattern. *Radiophys. Quantum Electron.*, vol.34, pp. 720–724, August 1991.

[GJ82] L. J. Griffiths and C. W. Jim. An alternative approach to linearly constrained adaptive beamforming. *IEEE Trans. Antennas Propag.*, vol.AP-30, pp. 27–34, January 1982.

[GJ99] L. C. Godara and M. R. S. Jahromi. Limitations and capabilities of frequency domain broadband constrained beamforming schemes. *IEEE Trans. Signal Process.*, vol.SP-47, pp. 2386–2395, September 1999.

[GM55] E.N. Gilbert and S.P. Morgan. Optimum design of directive antenna arrays subject to random variations. *Bell Syst. Tech. J.*, vol.34, pp. 637–663, May 1955.

[God85] L.C. Godara. The effect of phase-shifter error and the performance of an antenna array beamformer. *IEEE J. Oceanogr. Eng.*, vol.OE-10, pp. 278–284, July 1985.

[God86] L. C. Godara. Error analysis of the optimal antenna array processors. *IEEE Trans. Aerosp. Electron. Syst.*, vol.AES-22, pp. 395–409, July 1986.

[God90] L. C. Godara. Beamforming in the presence of correlated arrivals using structured correlation matrix. *IEEE Trans. Acoust., Speech, Signal Process.*, vol.ASSP-38, pp. 1–15, January 1990.

[God95] L. C. Godara. Application of the fast Fourier transform to broadband beamforming. *J. Acoust. Soc. Am.*, vol.98, pp. 230–239, July 1995.

[GR97a] J. Goldstein and I. Reed. Reduced rank adaptive filtering. *IEEE Trans. Aerosp. Electron. Syst.*, vol.IEEE-AES-45, p. 492, February 1997.

[GR97b] J. Goldstein and I. Reed. Subspace selection for partially adaptive sensor array processing. *IEEE Trans. Aerosp. Electron. Syst.*, vol.IEEE-AES-33, pp. 539-544, April, 1997.

[GR97c] J. Goldstein and I. Reed. Theory of partially adaptive radar. *IEEE Trans. Aerosp. Electron. Syst.*, vol.IEEE-AES-33, pp. 1309-1325, October 1997.

[Gra82] D. A. Gray. Formulation of the maximum signal-to-noise array processor in beam space. *J. Acoust. Soc. Am*, vol.72, pp. 1195-1201, October 1982.

[Gri77] L.J. Griffiths. An adaptive beamformer which implements constraints using an auxiliary array processor. In G. Tacconi Ed.,*Aspects of Signal Processing, PROC. NATO ASI*, Reidel, Dordrecht, The Netherlands, pp. 517–522, 1977.

[GSB96] A. B.Gershman, G. V. Serebryakov, and J. F. Bohme. Constrained Hung-Turner adaptive beamforming algorithm with additional robustness to wideband and moving jammers. *IEEE Trans. Antennas Propag.*, vol.AP-44, pp. 361-367, 1996.

[Gue99] J.R. Guerci. Theory and application of covariance matrix tapers for robust adaptive beamforming. *IEEE Trans. Signal Process.*, vol.SP-47, pp. 997-985, April 1999.

[GVL89] G. H. Golub and C. F. Van Loan. *Matrix Computations.* The Johns Hopkins University Press, Baltimore, Maryland, 1989.

[HBN88] A. M. Haimovich and Y. Bar-Ness. Adaptive antenna arrays using eigenvector methods. *IEEE INt. Conf. Antennas Propagat.*, 1988.

[HBN91] A. M. Haimovich and Y. Bar-Ness. An eigenanalysis interference canceller. *IEEE Trans. Acoust., Speech, Signal Process.*, vol.ASSP-39, pp. 76–84, January 1991.

[HS92] S. Haykin and A. Steinhardt, Editors. *Adaptive Radar Detection and Estimation.* Wiley, New York, 1992.

[HT83] E. K. L. Hung and R. M. Turner. A fast beamforming algorithm for large arrays. *IEEE Trans. Aerosp. Electron. Syst.*, vol.AES-19, pp. 598–607, July 1983.

[Jab86a] N. K. Jablon. Adaptive beamforming with the generalized sidelobe canceller in the presence of array imperfections. *IEEE Trans. Antennas Propag.*, vol.AP-34, pp. 996–1012, August 1986.

[Jab86b] N. K. Jablon. Steady state analysis of the generalized sidelobe canceller by adaptive noise cancelling techniques. *IEEE Trans. Antennas Propag.*, vol.AP-34, pp. 330–337, March 1986.

[KBK85] S. B. Kesler, S. Boodaghians, and J. Kesler. Resolving uncorrelated and correlated sources by linear prediction. *IEEE Trans. Antennas Propag.*, vol.AP-33, pp. 1221–1227, November 1985.

[Kel65] E. J. Kelly. A comparison of seismic array processing schemes. Technical Report 1965-21, M.I.T. Lincoln Laboratory, Lexington, Massachusetts, June 1965.

[KL64] E. J. Kelly, Jr. and M.J. Levin. Signal parameter estimation for seismometer arrays. MIT DDC 435-489 Technical Report 339, M.I.T. Lincoln Laboratory, Lexington, Massachusetts, January 1964.

[Kle80] L. I. Kleinberg. Array gain for signals and noise having amplitude and phase fluctuations. *J. Acoust. Soc. Am.*, vol.67, pp. 572–576, February 1980.

[KS90] J. Krolik and D. Swingler. Focused wideband array processing by spatial resampling. *IEEE Trans. Acoust., Speech, Signal Process.*, vol.ASSP-38, pp. 356–360, February 1990.

[KT85] I. P. Kirsteins and D. W. Tufts. On the probability density of signal-to-noise ratio in an improved adaptive detector. *Proc. ICASSP*, Tampa, Florida, vol.1, pp. 572–575, 1985.

[Lev64] M.J. Levin. Maximum-likelihood array processing. Technical Report DDC 455743, M.I.T. Lincoln Laboratory, Lexington, Massachusetts, December 1964.

[Lut86] A. K. Luthra. A solution to the adaptive nulling problem with a look-direction constraint in the presence of coherent jammers. *IEEE Trans. Antennas Propag.*, vol.AP-34, pp. 702–710, May 1986.

[Mai95] R.J. Mailloux. Covariance matrix augmentation to produce adaptive array pattern troughs. *Electron. Lett.*, vol.31, pp. 771–772, July 1995.

[Mer64] H.F. Mermoz. Filtrage adapté et utilisation optimale d'une antenne. In *NATO Adavanced Study Institute Signal Processing Emphasis Underwater Acoustics*, Grenoble, France, 1964.

[MG64] D. Middleton and H. Groginsky. Detection of random acoustic signals with distributed elements: Optimum receiver structures for normal signal and noise fields. *Proc. Acoustic Society Symposium of 1964*, 1964.

[MG88] T.T. Ma and L.J. Griffiths. A solution space approach to achieving partially adaptive arrays. *Proc. ICASSP*, NewYork, New York, vol.V, pp. 2869–2872, April 1988.

[Mid60] D. Middleton. *An Introduction to Statistical Communication Theory.* McGraw-Hill, New York, 1960.

[Mor78] D.R. Morgan. Partially adaptive array techniques. *IEEE Trans. Antennas Propag.*, vol.AP-26, pp. 823–833, November 1978.

[MP81] R.A. Mucci and R.G. Pridham. Impact of beam steering errors on shifted sideband and phase shift beamforming techniques. *J. Acoust. Soc. Am.*, vol.69, pp. 1360–1368, May 1981.

[MSC81] J. T. Mayhan, A. J. Simmons, and W. C. Cummings. Wide-band adaptive antenna nulling using tapped delay lines. *IEEE Trans. Antennas Propag.*, vol.AP-29, pp. 923–936, November 1981.

[Nit76] R. Nitzberg. Effect of errors in adaptive weights. *IEEE Trans. Aerosp. Electron. Syst.*, vol.AES-12, pp. 369–373, May 1976.

[OS89] A. V. Oppenheim and R. W. Schafer. *Discrete-Time Signal Processing.* Prentice-Hall, Englewood Cliffs, New Jersey, 1989.

[Ows71] N. L. Owsley. Source location with an adaptive antenna array. Technical Report, Naval Underwater Systems Center, National Technical Information Service, Springfield, Virginia, January 1971.

[Ows73] N. L. Owsley. A recent trend in adaptive spatial processing for sensor arrays: Constrained adaptation. In *Proc. NATO Advanced Study Institute on Signal Processing*, London and New York, pp. 591–604, 1973.

[Ows74] N. L. Owsley. Noise cancellation in the presence of correlated signal and noise. Technical Report 4639, Naval Underwater Systems Center, New London, Connecticut, January 1974.

[Ows80] N. L. Owsley. An overview of optimum-adaptive control in sonar array processing. In K. S. Narendra and R. V. Monopoli, editors, *Applications of Adaptive Control*, pp. 131–164. Academic Press, New York, 1980.

[Ows85] N. L. Owsley. Sonar array processing. In S. Haykin, editor, *Array Signal Processing.* Prentice-Hall, Englewood Cliffs, New Jersey, 1985.

[PK85] A. Paulraj and T. Kailath. On beamforming in the presence of multipath. *Proc. ICASSP*, vol.1, pp. 564–567, Tampa, Florida, March 1985.

[PK89] S. U. Pillai and B. H. Kwon. Forward/backward spatial smoothing techniques for coherent signal identification. *IEEE Trans. Acoust., Speech, Signal Process.*, vol.ASSP-37, pp. 8–15, January 1989.

[Pri51] R. L. Pritchard. Directivity of acoustic linear point arrays. Technical Report 21, Acoustic Research Laboratory, Harvard University, Cambridge, Massachusetts, January 1951.

[Pri53] R. Price. Statistical theory applied to communication through multipath disturbances. Technical Report, M.I.T. Research Laboratory of Electronics, Lexington, Massachusetts, September 1953.

[Pri54] R. Price. The detection of signals perturbed by scatter and noise. *IRE Trans.*, vol.PGIT-4, pp. 163–170, September 1954.

[Pri56] R. Price. Optimum detection of random signals in noise with application to scatter-multipath communication. *IRE Trans.*, vol.PGIT, pp. 125–135, December 1956.

[Pri63] R. Pritchard. Optimum directivity patterns for linear point arrays. *J. Acoust. Soc. Am.*, vol.25, pp. 879–891, September 1953.

[PRK87] A. Paulraj, V. U. Reddy, and T. Kailath. Analysis of signal cancellation due to multipath in optimum beamformers for moving arrays. *IEEE J. Oceanogr. Eng.*, vol.OE-12, pp. 163–172, January 1987.

[Qua82] A. H. Quazi. Array beam response in the presence of amplitude and phase fluctuations. *J. Acoust. Soc. Am.*, vol.72, pp. 171–180, July 1982.

[RC79] W.E. Rodgers and R. T. Compton, Jr. Adaptive array bandwidth with tapped delay-line processing. *IEEE Trans. Aerosp. Electron. Syst.*, vol.AES-15, pp. 21–28, January 1979.

[RPK87a] V. U. Reddy, A. Paulraj, and T. Kailath. Performance analysis of the optimum beamformer in the presence of correlated sources and its behavior under spatial smoothing. *IEEE Trans. Acoust., Speech, Signal Process.*, vol.ASSP-35, pp. 927–936, July 1987.

[RPK87b] R. Roy, A. Paulraj, and T. Kailath. Comparative performance of ESPRIT and MUSIC for direction-of-arrival estimation. *Proc. ICASSP*, vol.IV, pp. 2344–2347, Dallas, Texas, April 1987.

[RR92a] K. J. Raghunath and V. U. Reddy. Finite data analysis of MVDR beamformer: With and without spatial smoothing. *IEEE Trans. Antennas Propag.*, vol.ASSP-11, pp. 1226–1236, November 1992.

[RR92b] K. J. Raghunath and V. U. Reddy. A note on spatially weighted subarray covariance averaging schemes. *IEEE Trans. Antennas Propag.*, vol.ASSP-40, pp. 720–723, June 1992.

[Sch68] F. Schweppe. Sensor array data processing for multiple signal sources. *IEEE Trans. Inf. Theory*, vol.IT-4, pp. 294–305, March 1968.

[Sid58] M. M. Siddiqui. On the inversion of the sample covariance matrix in a stationary autoregressive process. *Ann. Math. Statist.*, vol.58, pp. 585–588, 1958.

[SK85a] V. Shahmirian and S. Kesler. Bias and resolution of the vector space methods in the presence of coherent planewaves. In *Proc. ICASSP*, Tampa, Florida, vol.IV, pp. 2320–2323, 1985.

[SK85b] T.-J. Shan and T. Kailath. Adaptive beamforming for coherent signals and interference. *IEEE Trans. Acoustics, Speech, Signal Process.*, vol.ASSP-33, pp. 527–534, June 1985.

[SK90] S. Sivanand and M. Kaveh. New results in broadband adaptive beamforming. *Proc. Asilomar Conference on Communications, Signal Processing and Computers*, Pacific Grove, California, November 1990.

[SK94] S. Simanapalli and M. Kaveh. Broadband focusing for partially adaptive beamforming. *IEEE Trans. Aerosp. Electron. Syst.*, vol.AES-30, pp. 68–79, January 1994.

[SSW86] Y.-L. Su, T.-J. Shan, and B. Widrow. Parallel spatial processing: A cure for signal cancellation in adaptive arrays. *IEEE Trans. Antennas Propag.*, vol.AP-34, pp. 347–355, March 1986.

[Ste83] A.K. Steele. Comparison of directional and derivative constraints for beamformers subject to multiple linear constraints. *Proc. IEE*, vol.130, pts. F and H, pp. 41–45, 1983.

[TBV98] Z. Tian, K. L. Bell, and H. L. Van Trees. A recursive least squares implementation for adaptive beamforming under quadratic constraint. *Proc. ICASSP*, Seattle, Washington, vol.IV, pp. 2053–2056, May 1998.

[TCL95] I. Thng, A. Cantoni, and Y. H. Leung. Constraints for maximally flat optimum broadband antenna arrays. *IEEE Trans. Signal Process.*, vol.SP-43, pp. 1334–1347, June 1995.

[TG92a] C-Y. Tseng and L. J. Griffiths. A simple algorithm to achieve desired pattern for arbitrary arrays. *IEEE Trans. Signal Process.*, vol.SP-40, pp. 2737–2746, November 1992.

[TG92b] C-Y. Tseng and L. J. Griffiths. A unified approach to the design of linear constraints in minimum variance adaptive beamformers. *IEEE Trans. Antennas Propag.*, vol.AP-40, pp. 1533–1542, December 1992.

[TK87] K. Takao and N. Kikuma. Adaptive array utilizing an adaptive spatial averaging technique for multipath environments. *IEEE Trans. Antennas Propag.*, vol.AP-35, pp. 1389–1396, December 1987.

[TKY86] K. Takao, N. Kikuma, and T. Yano. Toeplitzization of correlation matrix in multipath environment. *Proc. ICASSP*, vol.III, pp. 1873–1876, Tokyo, Japan, April 1986.

[Tse92] C.-Y. Tseng. Minimum variance beamforming with phase-independent derivative constraints. *IEEE Trans. Antennas Propag.*, vol.AP-40, pp. 285–294, March 1992.

[TYS95] C.-J. Tsai, J.-F. Yang, and T.-H. Shiu. Performance analysis of beamformers using effective SINR on array parameters. *IEEE Trans. Signal Process.*, vol.SP-43, pp. 300–303, January 1995.

[US56] M. Uzsoky and L. Solymár. Theory of superdirective linear arrays. *Acta Phys. Acad. Sci. Hung.*, vol.6, pp. 185–205, June 1956.

[Van63] V. Vanderkulk. Optimum processing for acoustic arrays. *J. Br. IRE*, vol.26, pp. 286–292, October 1963.

[VT64] H. L. Van Trees. *Optimum Signal Design and Processing for Reverberation-Limited Environments*, Report No. 1501064, pp. 1–115. Arthur D. Little, Cambridge, Massachusetts, October 1964.

[VT65] H. L. Van Trees. Optimum signal design and processing for reverberation-limited environments. *IEEE Trans. Mil. Electron.*, vol.MIL-9, pp. 212–229, July–October 1965.

[VT66a] H. L Van Trees. Optimum processing for passive sonar arrays. In *Proc. IEEE Ocean Electronics Symp.*, pp. 41–65, Honolulu, Hawaii, 1966.

[VT66b] H. L. Van Trees. A unified theory for optimum array processing. Technical Report 4160866, Dept. of the Navy, Naval Ship Systems Command, Arthur D. Little, Cambridge, Massachusetts, August 1966.

[VT68] H. L. Van Trees. *Detection, Estimation, and Modulation Theory, Part I.* Wiley, New York, 1968.

[VT01a] H. L. Van Trees. *Detection, Estimation, and Modulation Theory, Part I.* Wiley Interscience, New York, 2001.

[VT69] H. L. Van Trees. *Multi-Dimensional and Multi-Variable Processes.* unpublished class notes, 1969.

[VT71] H. L. Van Trees. *Detection, Estimation, and Modulation Theory, Part III.* Wiley, New York, 1971.

[VT01b] H. L. Van Trees. *Detection, Estimation, and Modulation Theory, Part III.* Wiley Interscience, New York, 2001.

[Vur77] A.M. Vural. A comparative performance study of adaptive array processors. *Proc. ICASSP*, Hartford, Connecticut, vol.I, pp. 695–700, May 1977.

[Vur79] A.M. Vural. Effects of perturbations on the performance of optimum/adaptive arrays. *IEEE Trans. Aerosp. Electron. Syst.*, vol.AES-15, pp. 76–87, January 1979.

[VV88] B. D. Van Veen. Eigenstructure based partially adaptive array design. *IEEE Trans. Antennas Propag.*, vol.AP-36, pp. 357–362, March 1988.

[VV89] B. D. Van Veen. An analysis of several partially adaptive beamformer designs. *IEEE Trans. Acoust., Speech, Signal Process.*, vol.ASSP-37, pp. 192–203, February 1989.

[VV91] B. D. Van Veen. Minimum variance beamforming with soft response constraints. *IEEE Trans. Signal Process.*, vol.SP-39, pp. 1964–1972, September 1991.

[VVR87] B.D. Van Veen and R.A. Roberts. Partially adaptive beamformer design via output power minimization. *IEEE Trans. Acoust., Speech, Signal Process.*, vol.ASSP-35, pp. 1524–1532, November 1987.

[WAN84] H. Watanabe, R. Azuma, and A. Noguchi. Interference rejection characteristics of spatially averaged learning adaptive antenna. Tech. Group on Antennas Propagat., IECE Japan, No.AP84-94, 1984

[WDGN82] B. Widrow, K. M. Duvall, R. P. Gooch, and W. C. Newman. Signal cancellation phenomena in adaptive antennas: Causes and cures. *IEEE Trans. Antennas Propag.*, vol.AP-30, pp. 469–478, 1982.

[Whi76] W.D. White. Cascade preprocessors for adaptive antennas. *IEEE Trans. Antennas Propag.*, vol.AP-24, pp. 670–684, September 1976.

[WK85] H. Wang and M. Kaveh. Coherent signal subspace processing for the detection and estimation of angles of arrival of multiple wideband sources. *IEEE Trans. Acoust., Speech, Signal Process.*, pp. 823–831, August 1985.

[Wol59] J. K. Wolf. *On the Detection and Estimation Problem for Multiple Non-Stationary Random Processes.* PhD thesis, Department of Electrical Engineering, Princeton University, Princeton, New Jersey, 1959.

[WPMS88] R. T. Williams, S. Prasad, A. K. Mahalanabis, and L. H. Sibul. An improved spatial smoothing technique for bearing estimation in a multipath environment. *IEEE Trans. Acoust., Speech, Signal Process.*, vol.ASSP-36, pp. 425–431, April 1988.

[WS74] L. P. Winkler and M. Schwartz. Constrained array optimization by penalty function techniques. *J. Acoust. Soc. Am.*, vol.55, pp. 1042–1048, May 1974.

[YK90] J. F. Yang and M. Kaveh. Coherent signal-subspace transformation beamformer. *IEE Proc.*, vol. 137, Pt. F, No. 4, pp. 267–275, August 1990.

[YU94] W. S. Youn and C. K. Un. Robust adaptive beamforming based on the eigenstructure method. *IEEE Trans. Signal Process.*, vol.SP-42, pp. 1543–1547, June 1994.

[YY95] J.-L. Yu and C.-C. Yeh. Generalized eigenspace-based beamformers. *IEEE Trans. Signal Process.*, vol.SP-43, pp. 2453–2461, November 1995.

[Zat95] M. Zatman. Production of adaptive array troughs by dispersion synthesis. *Electron. Let.*, vol.31, pp. 2141-2142, 1995.

[Zat99] M. Zatman. The relationship between covariance matrix tapers and multiple constraints. M.I.T. Lincoln Laboratory, Internal Memo. 1-6, Cambridge, Massachusetts, February 1999.

[Zat00] M. Zatman. Comments on "Theory and application of covariance matrix tapers for robust adaptive beamforming." *IEEE Trans. Signal Process.*, vol.SP-48, pp. 1796-1800, June 2000.

[ZG91] G. T. Zunich and L. J. Griffiths. A robust method in adaptive array processing for random phase errors. *Proc. ICASSP*, Toronto, Canada, vol.II, pp. 1357–1360, May 1991.

[ZL91a] M. D. Zoltowski and T. Lee. Maximum likelihood based sensor array signal processing in the beamspace domain for low-angle radar tracking. *IEEE Trans. Acoust., Speech, Signal Process.*, vol.ASSP-39, pp. 656–671, 1991.

[ZL91b] M. D. Zoltowski and T. Lee. Beamspace. ML bearing estimation incorporating low-angle geometry. *IEEE Trans. Acoust., Speech, Signal Process.*, vol.ASSP-27, pp. 441–458, 1991.

[ZL91c] M. D. Zoltowski and T. Lee. Interference cancellation matrix beamforming for 3-D beamspace ML/MUSIC bearing estimation. *IEEE Trans. Acoust., Speech, Signal Process.*, vol.ASSP-39, pp. 1858–1876, 1991.

[Zol88] M. D. Zoltowski. On the performance analysis of the MVDR beamformer in the presence of correlated interference. *IEEE Trans. Acoust., Speech, Signal Process.*, vol.ASSP-36, pp. 945–947, June 1988.

Chapter 7

Adaptive Beamformers

7.1 Introduction

In this chapter we develop techniques for implementing the algorithms developed in Chapter 6 using the incoming data. In Chapter 6, we assumed that we knew the signal directions and the various spatial spectral matrices ($\mathbf{S_x}$ or $\mathbf{S_n}$). In actual applications, we must estimate these quantities (or appropriate surrogates) from the incoming data. The resulting beamformers will adapt to the incoming data and are referred to as **adaptive beamformers**. In this chapter we develop the theory and practice of adaptive beamformers.

The adaptive beamformers that we develop can be divided into three general categories:

(i) Beamformers that estimate the spatial spectral matrix $\mathbf{S_x}$ or $\mathbf{S_n}$ (or the correlation $\mathbf{R_x}$ or $\mathbf{R_n}$ if the implementation is in the time domain) and use the estimate in the appropriate formula from Chapter 6 (e.g., (6.14), (6.71)). This implementation requires the inversion of the sample covariance matrix and is frequently referred to as the sample matrix inversion (SMI) technique. It is also referred to as the direct matrix inversion (DMI) technique (e.g., [MM80]) or the estimate-and-plug technique (e.g., [Hay85]). It is a block data processor.

(ii) Beamformers that implement the inversion recursively are the second category. We reformulate the algorithm as a least squares algorithm and develop a recursive version. These recursive least squares (RLS) implementations potentially have performance that is similar to the SMI beamformer.

(iii) A third approach is to adapt classical steepest descent algorithms to the optimization problem in order to find $\mathbf{w}_{opt}$. This approach leads to the least mean square (LMS) algorithm. These algorithms require less computation, but converge slower to the optimum solution.

In Section 7.2, we discuss various techniques for estimating the covariance matrix (or spectral matrix) of the array output. The eigenvalues and eigenvectors play a key role in several subsequent discussions. We show how they can be derived directly from the data using a singular value decomposition (SVD).

In Section 7.3, we implement the MVDR and MPDR beamformers using a technique called **sample matrix inversion** (SMI). As the name implies, we use $\mathbf{C}_{\mathbf{x}}$, the sample correlation matrix in place of $\mathbf{S}_{\mathbf{x}}$ and invert it to obtain the MVDR or MPDR beamformers. We show how the use of a finite amount of data affects the performance.

In Section 7.4, we reformulate the problem using a least squares formulation and introduce an exponentially weighted sample spectral matrix $\mathbf{\Phi}$. We find that the resulting MPDR beamformer is identical to the MPDR beamformer of Chapter 6, with the ensemble spatial spectral matrix $\mathbf{S}_{\mathbf{x}}$ replaced by $\mathbf{\Phi}$. We then develop a recursive implementation of the algorithm which is denoted as the **recursive least squares** (RLS) algorithm. We compare its performance to the SMI algorithm and develop various diagonal loading methods.

In Section 7.5, we develop more efficient recursive algorithms that have better numerical stability and are computationally simpler than the algorithms in Section 7.4. We show how these algorithms can be put in a structure that can be implemented efficiently in VLSI.

In Section 7.6, we begin discussion of a group of algorithms that rely on the quadratic characteristic of the error surface and utilize gradient techniques to find the optimum weight vector. Section 7.6 discusses steepest descent algorithms. These algorithms are deterministic and provide background for the stochastic gradient algorithms that we actually employ.

In Section 7.7, we develop **least mean-square** (LMS) algorithms and investigate their performance. The LMS algorithms are computationally much simpler than the SMI and RLS algorithms, but they converge much more slowly to the optimum solution.

In Section 7.8, we study the important problem of detecting the number of plane-wave signals (including both desired signals and interfering signals) that are impinging on the array. In this chapter, we need this information in order to implement the adaptive version of the eigenspace beamformers that we developed in Section 6.8. In Chapters 8 and 9, we will need this information for parameter estimation.

In Section 7.9, we study adaptive eigenvector beamformers. We first utilize an SMI implementation and find that we obtain faster convergence to the optimum beamformer because of the reduced degrees of freedom.

In Section 7.10, we study beamspace adaptive beamformers. In Section 6.9, we found that (in most scenarios) we could achieve performance similar to element-space beamformers with reduced computational complexity. In the adaptive case, we find that we can obtain faster convergence to the optimum solution because of the reduced dimension.

In Section 7.11, we study the adaptive implementation of the broadband beamformers that we developed in Section 6.13. We restrict our discussion to time-domain implementations.

In Section 7.12, we summarize our results and discuss some open issues. The structure of the chapter is shown in Table 7.1.

7.2 Estimation of Spatial Spectral Matrices

In the discussion up to this point in the book, we have assumed that the second-order statistics of the input process were known. In practice, we usually have to estimate these statistics from a finite amount of data. We have available a sequence of snapshots, $\mathbf{X}_1, \mathbf{X}_2, \cdots, \mathbf{X}_K$, where $\mathbf{X}_k$ is an N-dimensional vector corresponding to the frequency-domain snapshot at time k. We process these snapshots to obtain an estimate of $\mathbf{S}_\mathbf{x}$, which we denote as $\hat{\mathbf{S}}_\mathbf{x}$.

The two issues of interest are:

(i) What is the appropriate estimator?

(ii) How well does it perform?

We will find that, as we assume more prior information about the process, the estimators become more accurate. However, they also become more computationally complex and dependent on the prior assumptions. A particular problem of interest to us is the case in which $\mathbf{S}_\mathbf{x}$ has the structure in (6.173) but $\mathbf{C}_\mathbf{x}$, the sample spectral matrix defined in (7.3), does not have that structure. Thus, our minimization depends on how much structure we impose on $\hat{\mathbf{S}}_\mathbf{x}$.

A logical measure of estimator performance is the Frobenius norm of the error matrix,

$$\xi_F = \| \hat{\mathbf{S}}_\mathbf{x} - \mathbf{S}_\mathbf{x} \|_F . \tag{7.1}$$

Table 7.1 Structure of Chapter 7

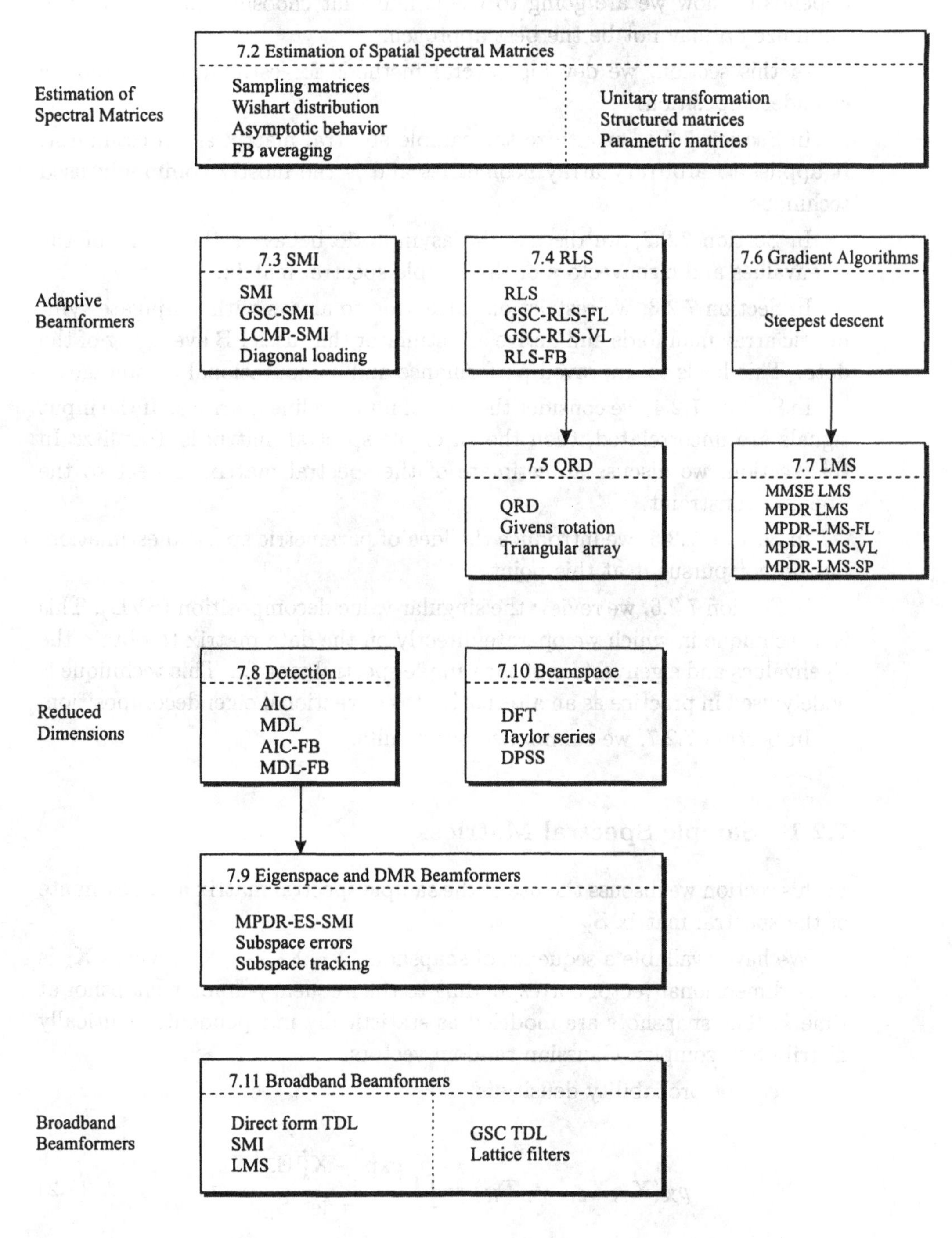

However, we find that the appropriate measure of the quality of the estimate depends on how we are going to use it and that choosing an estimator to minimize ξ_F may not be the best approach.

In this section, we develop several methods for estimating $\mathbf{S_x}$ and its eigendecomposition.

In Section 7.2.1, we utilize the sample spectral matrix as an estimator. It applies to arbitrary array geometries and is the mostly commonly used technique.

In Section 7.2.2, we discuss the asymptotic behavior (large K) of the eigenvalues and eigenvectors of the sample spectral matrix.

In Section 7.2.3, we restrict our attention to arrays with conjugate symmetric array manifolds and derive an estimator that uses FB averaging of the data. This leads to improved performance and computational advantages.

In Section 7.2.4, we consider the case of uniform linear arrays. If the input signals are uncorrelated, then the ensemble spectral matrix is Toeplitz. In this section, we discuss an estimate of the spectral matrix subject to the Toeplitz constraint.

In Section 7.2.5, we introduce the idea of parametric spectral estimation, but do not pursue it at this point.

In Section 7.2.6, we review the singular value decomposition (SVD). This is a technique in which we operate directly on the data matrix to obtain the eigenvalues and eigenvectors of the sample spectral matrix. This technique is widely used in practice as an alternative to conventional eigendecomposition.

In Section 7.2.7, we summarize our results.

7.2.1 Sample Spectral Matrices

In this section we discuss the use of the sample spectral matrix as an estimate of the spectral matrix $\mathbf{S_x}$.

We have available a sequence of snapshots, $\mathbf{X}_1, \mathbf{X}_2, \cdots, \mathbf{X}_K$, where $\mathbf{X}_k$ is an N-dimensional vector corresponding to the frequency-domain snapshot at time k. The snapshots are modeled as statistically independent, identically distributed, complex Gaussian random vectors.

The joint probability density is

$$p_{\mathbf{X}}(\mathbf{X}_1, \mathbf{X}_2, \cdots, \mathbf{X}_K) = \prod_{i=1}^{K} \frac{\exp\left[-\mathbf{X}_k^H \mathbf{S}_{\mathbf{x}}^{-1} \mathbf{X}_k\right]}{\pi \det\left[\mathbf{S}_{\mathbf{x}}\right]}. \tag{7.2}$$

We define the sample spectral matrix $\mathbf{C_x}$ as[1]

$$\mathbf{C_x} = \frac{1}{K}\sum_{k=1}^{K}\mathbf{X}_k\mathbf{X}_k^H = \frac{1}{K}\sum_{k=1}^{K}\mathbf{X}(k)\mathbf{X}^H(k). \tag{7.3}$$

We can also write $\mathbf{C_x}$ in terms of a $N \times K$ data matrix, $\tilde{\mathbf{X}}$,

$$\tilde{\mathbf{X}} = \frac{1}{\sqrt{K}}\left[\ \mathbf{X}(1) \ \vdots\ \mathbf{X}(2) \ \vdots\ \cdots \ \vdots\ \mathbf{X}(K)\ \right], \tag{7.4}$$

or

$$\tilde{\mathbf{X}} = \frac{1}{\sqrt{K}}\left[\begin{array}{c:c:c:c} X_0(1) & X_0(2) & & X_0(K) \\ X_1(1) & X_1(2) & & X_1(K) \\ \vdots & & \cdots & \\ X_{N-1}(1) & X_{N-1}(2) & & X_{N-1}(K) \end{array}\right], \tag{7.5}$$

and

$$\mathbf{C_x} = \tilde{\mathbf{X}}\tilde{\mathbf{X}}^H. \tag{7.6}$$

Substituting $\mathbf{C_x}$ into (7.2), taking the logarithm and dropping constant terms, we have

$$L\left(\mathbf{S_x^{-1}}\right) = \ln\det\left[\mathbf{S_x^{-1}}\right] - \text{tr}\left[\mathbf{S_x^{-1}C_x}\right], \tag{7.7}$$

so $\mathbf{C_x}$ is a sufficient statistic to estimate $\mathbf{S_x^{-1}}$. Taking the matrix gradient of $L\left(\mathbf{S_x^{-1}}\right)$ (using (A.387) and (A.380)) and setting the result equal to zero gives

$$\left[\mathbf{S_x^T} - \mathbf{C_x^T}\right]_{\mathbf{S_x}=\hat{\mathbf{S}}_\mathbf{x}} = \mathbf{0}. \tag{7.8}$$

If we do not impose any structure on $\mathbf{S_x}$, then

$$\hat{\mathbf{S}}_\mathbf{x} = \mathbf{C_x}. \tag{7.9}$$

Thus, the maximum likelihood estimate of the spectral matrix is the sample correlation matrix. We observe that $\mathbf{C_x}$ is Hermitian, and if $K \geq N$, it is positive definite.

The elements of the sample spectral matrix have a probability density given by the complex Wishart density (e.g., Goodman [Goo63], [Hay96], [And63], [And84], or [Mui82]).[2]

---

[1] We use both $\mathbf{X}_k$ and $\mathbf{X}(k)$ to denote the kth snapshot vector. The second version is more appropriate when we write the components of the vector.

[2] Our discussion is based on Kelly and Forsythe [KF89]. The appendices of [KF89] contain an extensive discussion of techniques for multivariate statistical analysis.

Defining

$$\mathbf{C}_K = K\mathbf{C}_{\mathbf{x}}, \tag{7.10}$$

one can show that

$$p_{\mathbf{C}_K}(\mathbf{C}_K) = \frac{|\mathbf{C}_K|^{K-N}}{\Gamma_N(K)|\mathbf{S}_{\mathbf{x}}|^K} etr\left[-\mathbf{S}_{\mathbf{x}}^{-1}\mathbf{C}_K\right], \tag{7.11}$$

where

$$etr[\mathbf{A}] \triangleq \exp[tr\mathbf{A}], \tag{7.12}$$

and $\Gamma_N(K)$ is a generalization of the Gamma function,

$$\Gamma_N(K) = \Pi^{N(N-1)/2} \prod_{j=0}^{N-1} \Gamma(K-j). \tag{7.13}$$

Note that $\Gamma_1(K) = \Gamma(K)$.

The probability density in (7.11) is referred to as the complex Wishart density and denoted by $\mathcal{W}_N(K, \mathbf{S}_{\mathbf{x}})$ and is defined over a space of non-negative definite Hermitian matrices.

It is a generalization of the complex chi-squared density. If N equals one (a single element), then $\mathbf{X}$ is a scalar, X, and

$$C_K = \sum_{k=1}^{K} |X(k)|^2, \tag{7.14}$$

and $\mathbf{S}_{\mathbf{x}} = \sigma_X{}^2$. Defining

$$\chi^2(K) = \frac{C_K}{\sigma_X{}^2}, \tag{7.15}$$

and using (7.11), the probability density of $\chi^2(K)$ is

$$p_{\chi^2}(Y) = \frac{(Y)^{K-1}}{\Gamma(K)} \exp(-Y), \quad Y \geq 0, \tag{7.16}$$

which is familiar as the complex chi-squared probability density with K degrees of freedom. Several properties of the complex chi-squared probability density are derived in the problems.

The complex Wishart density has several properties that we use in the sequel (e.g., [KF89], [Hay96], [Mui82], [And84]).

We assume that $\mathbf{C}_K$ has a complex Wishart density $\mathcal{W}_N(K, \mathbf{S}_{\mathbf{x}})$. Then:

1. Let $\mathbf{a}$ be any $N \times 1$ random vector that is independent of $\mathbf{C}_K$ and the $p(\mathbf{a}) = \mathbf{0} = 0$. Then:

$$y_1 \triangleq \frac{\mathbf{a}^H\, \mathbf{C}_K\, \mathbf{a}}{\mathbf{a}^H\, \mathbf{S}_{\mathbf{x}}\, \mathbf{a}}, \tag{7.17}$$

is chi-square distributed with K degrees of freedom and is independent of $\mathbf{a}$, and

$$y_2 \triangleq \frac{\mathbf{a}^H\, \mathbf{S}_{\mathbf{x}}^{-1}\, \mathbf{a}}{\mathbf{a}^H\, \mathbf{C}_K^{-1}\, \mathbf{a}}, \tag{7.18}$$

is chi-square distributed with $K - N + 1$ degrees of freedom.

2. Let $\mathbf{B}$ be a $N \times M$ matrix of rank M, then:

(a) $\mathbf{B}^H \mathbf{C}_K \mathbf{B}$ is $\mathcal{W}_M(K, \mathbf{B}^H \mathbf{S}_{\mathbf{x}} \mathbf{B})$. (This could correspond to a beamspace processor.)

(b) $\left[\mathbf{B}^H \mathbf{C}_K \mathbf{B}\right]^{-1}$ is $\mathcal{W}_M\left(K - N + M, \left[\mathbf{B}^H \mathbf{C}_K^{-1} \mathbf{B}\right]^{-1}\right)$.

We can also show that, if $\mathbf{b}$ is a fixed non-zero vector, then

$$\begin{aligned} E\left[\mathbf{b}^H \mathbf{C}_K^{-1} \mathbf{b}\right] &= \left(\mathbf{b}^H \mathbf{C}_K^{-1} \mathbf{b}\right) E\left[\frac{1}{\chi^2(K - N + 1)}\right] \\ &= \frac{\mathbf{b}^H \mathbf{C}_K^{-1} \mathbf{b}}{K - N - 1}, \quad K > N + 1. \end{aligned} \tag{7.19}$$

This result implies

$$E\left[\mathbf{C}_K^{-1}\right] = \frac{\mathbf{S}_{\mathbf{x}}^{-1}}{K - N - 1}, \quad K > N + 1. \tag{7.20}$$

The statistical properties of $\mathbf{C}_K$ as an estimator are discussed in various statistics texts (e.g., Anderson [And84]). We are normally interested in the statistical behavior of functions derived from $\mathbf{C}_K$, so we focus our attention on their behavior. In the next section, we consider the statistical behavior of the eigenvectors and eigenvalues.

7.2.2 Asymptotic Behavior

In many applications we do an eigendecomposition of the estimated spectral matrix $\hat{\mathbf{S}}_{\mathbf{x}}$,

$$\hat{\mathbf{S}}_{\mathbf{x}} = \sum_{i=1}^{N} \hat{\lambda}_i\, \hat{\mathbf{\Phi}}_i\, \hat{\mathbf{\Phi}}_i^H. \tag{7.21}$$

We want to investigate the properties of the $\hat{\lambda}_i$ and $\hat{\mathbf{\Phi}}_i$ for large K. The original results are due to Anderson [And63] and were extended to the complex case by Gupta [Gup65] (e.g., [Bri81], [Wil65], [KW91]).

One can show that the eigenvalues $\hat{\lambda}_i$ are asymptotically Gaussian and independent of the eigenvectors $\hat{\mathbf{\Phi}}_i$ for $i = 1, 2, \cdots, N$. In addition,

$$E\left[\hat{\lambda}_i\right] = \lambda_i + O\left(K^{-1}\right), \tag{7.22}$$

and

$$Cov\left[\hat{\lambda}_i, \hat{\lambda}_j\right] = \delta_{ij}\frac{\lambda_i^2}{K} + O\left(K^{-2}\right). \tag{7.23}$$

The following results concerning the estimated eigenvectors can be derived (e.g., [KW91]). Define

$$\hat{\mathbf{\Phi}}_i = \mathbf{\Phi}_i + \boldsymbol{\eta}_i. \tag{7.24}$$

Then

$$E\left[\boldsymbol{\eta}_i\right] = -\frac{\lambda_i}{2K}\sum_{\substack{k=1\\k\neq i}}^{N}\frac{\lambda_k}{\left(\lambda_i-\lambda_k\right)^2}\,\mathbf{\Phi}_i \triangleq c_i\mathbf{\Phi}_i, \tag{7.25}$$

$$E\left[\boldsymbol{\eta}_i\,\boldsymbol{\eta}_j^H\right] = \frac{\lambda_i}{K}\sum_{\substack{k=1\\k\neq i}}^{N}\frac{\lambda_k}{\left(\lambda_i-\lambda_k\right)^2}\,\mathbf{\Phi}_k\,\mathbf{\Phi}_k^H\delta_{ij}, \tag{7.26}$$

$$E\left[\boldsymbol{\eta}_i\,\boldsymbol{\eta}_j^T\right] = -\frac{\lambda_i\lambda_j}{K\left(\lambda_i-\lambda_j\right)^2}\,\mathbf{\Phi}_j\,\mathbf{\Phi}_i^T(1-\delta_{ij}). \tag{7.27}$$

We use these results later in the text to analyze the asymptotic behavior of various processing algorithms.

7.2.3 Forward–Backward Averaging

The sample correlation matrix is widely used as an estimate of $\mathbf{S_x}$. When the array manifold vector is conjugate symmetric,

$$\mathbf{v}_\psi(\psi) = \mathbf{J}\mathbf{v}_\psi^*(\psi), \tag{7.28}$$

then we can introduce constraints in the estimator to improve performance. We discuss this technique in this section.

To illustrate the procedure, we use a linear array with an even number of elements placed symmetrically about the origin. We index the elements from 1 to N. We do not require that they be uniformly spaced. Thus,

$$d_i = -d_{N+1-i}. \tag{7.29}$$

The kth snapshot from the array when the input is a plane wave from DOA θ is,

$$\mathbf{X}(k) = f(k)\left[\; e^{-j\frac{2\pi}{\lambda}d_1\cos\theta} \;\vdots\; \cdots \;\vdots\; e^{-j\frac{2\pi}{\lambda}d_N\cos\theta} \;\right]^T + \mathbf{w}(k), \tag{7.30}$$

where $f(k)$ is a zero-mean complex Gaussian variable and $\mathbf{w}(k)$ is a complex Gaussian noise vector.

The spectral matrix is

$$\mathbf{S_x} = E\left[\mathbf{X}(k)\mathbf{X}^H(k)\right]. \tag{7.31}$$

The ij element is,

$$\left[\mathbf{S_x}\right]_{ij} = \sigma_s^2\left[e^{-j\frac{2\pi}{\lambda}(d_i-d_j)\cos\theta}\right] + \sigma_w^2\delta_{ij}. \tag{7.32}$$

Using (7.29) in (7.32) we have

$$\left[\mathbf{S_x}\right]_{N+1-i,N+1-j} = \left\{\left[\mathbf{S_x}\right]_{ij}\right\}^*. \tag{7.33}$$

Thus, $\mathbf{S_x}$ is a centrohermitian matrix (A.137). It is also a Hermitian matrix,

$$\mathbf{S_x}^H = \mathbf{S_x}. \tag{7.34}$$

Therefore, it is also persymmetric (A.134). As an example, for $N = 4$,

$$\mathbf{S_x} = \begin{bmatrix} s_{11} & s_{12} & s_{13} & s_{14} \\ s_{12}^* & s_{22} & s_{23} & s_{13} \\ s_{13}^* & s_{23}^* & s_{22} & s_{12} \\ s_{14}^* & s_{13}^* & s_{12}^* & s_{11} \end{bmatrix}. \tag{7.35}$$

The centrohermitian-Hermitian property implies that

$$\mathbf{S_x} = \mathbf{J}\mathbf{S_x^*}\mathbf{J}, \tag{7.36}$$

where $\mathbf{J}$ is the exchange matrix (A.125). Similarly,

$$\mathbf{S_x}^{-1} = \mathbf{J}\mathbf{S_x^*}^{-1}\mathbf{J}. \tag{7.37}$$

The left $\mathbf{J}$ reverses the rows and the right $\mathbf{J}$ reverses the columns.

We want to find the maximum likelihood estimate of $\mathbf{S_x}$ subject to the centrohermitian-Hermitian constraint.

The sample spectral matrix $\mathbf{C_x}$ is defined in (7.3). We refer to the construction of $\mathbf{C_x}$ as forward averaging. We now define a technique called backward averaging. The technique is implemented by reversing and conjugating the snapshot vector $\mathbf{X}[k]$. Define

$$\mathbf{X}^J = \mathbf{J}\mathbf{X}^*. \tag{7.38}$$

We define a backward averaged sample spectral matrix as

$$\mathbf{C}_{\mathbf{x},b} \triangleq \frac{1}{K}\sum_{k=1}^{K} \mathbf{J}\mathbf{X}_k^*\mathbf{X}_k^T\mathbf{J} = \mathbf{J}\mathbf{C}_{\mathbf{x}}^*\mathbf{J}, \tag{7.39}$$

and FB averaged sample spectral matrix as

$$\mathbf{C}_{\mathbf{x},fb} = \frac{1}{2K}\sum_{k=1}^{K}\left(\mathbf{X}_k\mathbf{X}_k^H + \mathbf{J}\mathbf{X}_k^*\mathbf{X}_k^T\mathbf{J}\right) = \frac{1}{2}\left(\mathbf{C}_{\mathbf{x}} + \mathbf{J}\mathbf{C}_{\mathbf{x}}^*\mathbf{J}\right). \tag{7.40}$$

The joint probability density function was given in (7.2). We repeat (7.7) as,

$$L\left(\mathbf{S}_{\mathbf{x}}^{-1}\right) = \ln\det\left[\mathbf{S}_{\mathbf{x}}^{-1}\right] - \operatorname{tr}\left[\mathbf{S}_{\mathbf{x}}^{-1}\mathbf{C}_{\mathbf{x}}\right]. \tag{7.41}$$

We maximize $L\left(\mathbf{S}_{\mathbf{x}}^{-1}\right)$ subject to the constraint,

$$\hat{\mathbf{S}}_{\mathbf{x}} = \mathbf{J}\hat{\mathbf{S}}_{\mathbf{x}}^*\mathbf{J}. \tag{7.42}$$

Observe that

$$\begin{aligned}
\operatorname{tr}\left[\mathbf{S}_{\mathbf{x}}^{-1}\mathbf{C}_{\mathbf{x}}\right] &= \operatorname{tr}\left[\left[\mathbf{J}\mathbf{S}_{\mathbf{x}}^{*\,-1}\mathbf{J}\right]\mathbf{C}_{\mathbf{x}}\right] \\
&= \operatorname{tr}\left[\left[\mathbf{S}_{\mathbf{x}}^*\right]^{-1}\mathbf{J}\mathbf{C}_{\mathbf{x}}\mathbf{J}\right] \\
&= \operatorname{tr}\left[\mathbf{S}_{\mathbf{x}}^{-1}\mathbf{J}\mathbf{C}_{\mathbf{x}}^*\mathbf{J}\right],
\end{aligned} \tag{7.43}$$

where the last equality follows because the trace is real. Using (7.43), the ln likelihood function in (7.41) can be written as

$$\begin{aligned}
L\left(\mathbf{S}_{\mathbf{x}}^{-1}\right) &= \ln\det\left[\mathbf{S}_{\mathbf{x}}^{-1}\right] - \operatorname{tr}\left[\frac{1}{2}\mathbf{S}_{\mathbf{x}}^{-1}\left[\mathbf{C}_{\mathbf{x}} + \mathbf{J}\mathbf{C}_{\mathbf{x}}^*\mathbf{J}\right]\right] \\
&= \ln\det\left[\mathbf{S}_{\mathbf{x}}^{-1}\right] - \operatorname{tr}\left[\mathbf{S}_{\mathbf{x}}^{-1}\mathbf{C}_{\mathbf{x},fb}\right].
\end{aligned} \tag{7.44}$$

This equation has the same form as (7.7). Thus, the unconstrained maximum of (7.44) is $\mathbf{C}_{\mathbf{x},fb}$ which is a centrohermitian-Hermitian matrix. Therefore,

$$\hat{\mathbf{S}}_{\mathbf{x},fb} = \mathbf{C}_{\mathbf{x},fb}. \tag{7.45}$$

is the constrained maximum likelihood estimate.[3]

We can also write (7.45) using the data matrix in (7.5) as

$$\hat{\mathbf{S}}_{\mathbf{x},fb} = \frac{1}{2}\left(\tilde{\mathbf{X}}\tilde{\mathbf{X}}^H + \mathbf{J}\tilde{\mathbf{X}}^*\tilde{\mathbf{X}}^T\mathbf{J}\right). \tag{7.46}$$

The result in (7.46) can also be written as

$$\hat{\mathbf{S}}_{\mathbf{x},fb} = \tilde{\mathbf{X}}_{fb}\tilde{\mathbf{X}}_{fb}^H, \tag{7.47}$$

where

$$\tilde{\mathbf{X}}_{fb} = \frac{1}{\sqrt{2}}\left[\ \tilde{\mathbf{X}} \ \vdots\ \mathbf{J}\tilde{\mathbf{X}}^*\ \right] \tag{7.48}$$

is an $N \times 2K$ data matrix.

Although a linear array was used as an example, it is important to note that an array manifold vector for a large class of arrays can be written so that it is conjugate symmetric. The symmetry requirement on the array geometry is that for every element located at $\mathbf{p}_n$, there must be an identical element at $-\mathbf{p}_n$. Arrays that satisfy this condition include:

(i) Standard rectangular array;

(ii) Standard hexagonal array ($N = 7, 19, 37, 61, \cdots$);

(iii) Uniform circular array (N even);

(iv) Uniform cylindrical array (N, M even);

(v) Concentric circles array (N_i even).

The exploitation of centrohermitian-persymmetric matrices for communication problems was introduced by Cantoni and Butler [CB76]. They were used by Nitzberg [Nit80] for adaptive arrays and by Evans et al. [EJS82] for DOA estimation using arrays. Nuttall [Nut76] utilizes FB averaging in the context of linear prediction techniques.

---

[3] An early derivation of the result is due to Nitzberg [Nit80] who references Rao [Rao46] and Cantoni and Butler [CB76]. Our derivation is due to Jansson and Stoica [JS99].

$\hat{\mathbf{S}}_{\mathbf{x},fb}$ has properties that will be important in our signal processing discussion. We discuss three of these properties in this section and develop others at appropriate points in the text.[4]

Property 1: Complex FB spectral matrices can be transformed into real spectral matrices.

One of the operations that we will use frequently is

$$\hat{\mathbf{S}}_{12} = \mathbf{B}_1^H \hat{\mathbf{S}}_{\mathbf{x},fb} \mathbf{B}_2, \tag{7.49}$$

where $\mathbf{B}_1$ and $\mathbf{B}_2$ are $N \times M$ matrices. As an example, $\hat{\mathbf{S}}_{12}$ is the spectral matrix at the output of a beamspace transformation with

$$\mathbf{B}_{bs}^H = \mathbf{B}_1^H = \mathbf{B}_2^H. \tag{7.50}$$

If $\mathbf{B}_1$ and $\mathbf{B}_2$ are both column conjugate symmetric,

$$\mathbf{B}_1 = \mathbf{J}\mathbf{B}_1^*, \quad \mathbf{B}_2 = \mathbf{J}\mathbf{B}_2^*, \tag{7.51}$$

or both column conjugate asymmetric,

$$\mathbf{B}_1 = -\mathbf{J}\mathbf{B}_1^*, \quad \mathbf{B}_2 = -\mathbf{J}\mathbf{B}_2^*, \tag{7.52}$$

then $\hat{\mathbf{S}}_{12}$ is real.

This result follows directly from the definitions,

$$\begin{aligned} \mathbf{S}_{12}^* &= \left(\mathbf{B}_1^H \hat{\mathbf{S}}_{\mathbf{x},fb} \mathbf{B}_2\right)^* = \mathbf{B}_1^T \hat{\mathbf{S}}_{\mathbf{x},fb}^* \mathbf{B}_2^* = \mathbf{B}_1^T \left(\mathbf{J}\hat{\mathbf{S}}_{\mathbf{x},fb}\mathbf{J}\right) \mathbf{B}_2^* \\ &= \left(\mathbf{J}\mathbf{B}_1\right)^T \hat{\mathbf{S}}_{\mathbf{x},fb} \left(\mathbf{J}\mathbf{B}_2^*\right) = \mathbf{B}_1^H \hat{\mathbf{S}}_{\mathbf{x},fb} \mathbf{B}_2 = \mathbf{S}_{12}. \end{aligned} \tag{7.53}$$

As $\hat{\mathbf{S}}_{12}$ and $\hat{\mathbf{S}}_{12}^*$ are equal, $\hat{\mathbf{S}}_{12}$ is real.

Property 1A: If the beamspace matrix $\mathbf{B}_{bs}$ has columns that are either conjugate symmetric or conjugate asymmetric, then

$$Re\{\mathbf{B}_{bs}^H \hat{\mathbf{S}}_{\mathbf{x}} \mathbf{B}_{bs}\} = \mathbf{B}_{bs}^H \hat{\mathbf{S}}_{\mathbf{x},fb} \mathbf{B}_{bs}. \tag{7.54}$$

Therefore, we can map the data $\tilde{\mathbf{X}}$ into beamspace and compute,

$$\hat{\mathbf{S}}_{\mathbf{x},fb} = \tilde{\mathbf{X}}_{bs} \tilde{\mathbf{X}}_{bs}^H, \tag{7.55}$$

---

[4]The first part of our discussion follows Linebarger et al. [LDD94] with somewhat different notation.

and take the real part,

$$\hat{\mathbf{S}}_{\mathbf{x},bs,fb} = Re\{\hat{\mathbf{S}}_{\mathbf{x},bs}\}, \tag{7.56}$$

to obtain FB averaging in beamspace. The result in (7.54) follows by substituting the various matrices into (7.54) and using the conjugate symmetry properties.

The implication of this is that if we process the data using FB averaging and then use beamspace processing (Section 3.10), the resulting spectral matrix in beamspace will be real. Alternatively, we can process the data in beamspace without FB averaging and then take the real part of the estimated spectral matrix. The advantage of this result is that we can use real arithmetic for subsequent operations, which gives a computational saving of up to 75%. Computational advantages for various operations are discussed in detail in [LDD94] (see also [ZKS93]).

A particular beamspace processor of interest is the eigenvector beamspace processor in Figure 5.19. If $\hat{\mathbf{S}}_{\mathbf{x},fb}$ is a centrohermitian-persymmetric matrix (e.g., a linear array), then $\frac{N}{2}$ eigenvectors will be conjugate symmetric and $\frac{N}{2}$ eigenvectors will be conjugate asymmetric (see Section A.4.2.2 in Appendix A). Therefore, the beamspace spectral matrices will be real.

We also observe that the elements of $\hat{\mathbf{S}}_{\mathbf{x},fb}$ will not have a complex Wishart probability density. In [PK89], the probability density and various asymptotic properties are derived.

In many cases, we process the data in element space. Here we use a unitary transformation to obtain the advantage of real processing. The next two properties develop this technique.

Property 2: Eigenvalues and eigenvectors: Consider the product,

$$\mathbf{S}_{\mathbf{Q}} = \mathbf{Q}^H \hat{\mathbf{S}}_{\mathbf{x},fb} \mathbf{Q}, \tag{7.57}$$

where $\mathbf{Q}$ is unitary and column conjugate symmetric. If $\hat{\mathbf{S}}_{\mathbf{x},fb}$ has even dimension, then

$$\mathbf{Q} = \frac{1}{\sqrt{2}} \begin{bmatrix} \mathbf{I} & j\mathbf{I} \\ \mathbf{J} & -j\mathbf{J} \end{bmatrix}, \tag{7.58}$$

where $\mathbf{I}$ and $\mathbf{J}$ have dimension $N/2$, is computationally attractive because of its sparse structure. If $\hat{\mathbf{S}}_{\mathbf{x},fb}$ has odd dimension,

$$\mathbf{Q} = \frac{1}{\sqrt{2}} \begin{bmatrix} \mathbf{I} & \mathbf{0} & j\mathbf{I} \\ \mathbf{0}^{\mathbf{T}} & \sqrt{2} & \mathbf{0}^{\mathbf{T}} \\ \mathbf{J} & \mathbf{0} & -j\mathbf{J} \end{bmatrix} \tag{7.59}$$

is used. Because $\hat{\mathbf{S}}_{\mathbf{x},fb}$ is centrohermitian and Hermitian, $\mathbf{S}_{\mathbf{Q}}$ is symmetric.

We consider the even-dimension case in the text. The eigenvector decomposition of $\mathbf{S}_{\mathbf{Q}}$ can be written as,

$$\mathbf{S}_{\mathbf{Q}} = \mathbf{U}_c \mathbf{\Lambda}_c \mathbf{U}_c^T, \tag{7.60}$$

where $\mathbf{U}_c$ is an $N \times N$ matrix of the eigenvectors

$$\mathbf{U}_c = \begin{bmatrix} \boldsymbol{\phi}_{c1} & \boldsymbol{\phi}_{c2} & \cdots & \boldsymbol{\phi}_{cN} \end{bmatrix}, \tag{7.61}$$

and $\mathbf{\Lambda}_c$ is the diagonal matrix of eigenvalues,

$$\mathbf{\Lambda}_c \triangleq \text{diag}\left\{\lambda_{c1}, \lambda_{c2}, \cdots, \lambda_{cN}\right\}. \tag{7.62}$$

We can partition $\mathbf{U}_c$ into an upper and lower half:

$$\mathbf{U}_c = \begin{bmatrix} \mathbf{U}_{c1} \\ \mathbf{U}_{c2} \end{bmatrix}. \tag{7.63}$$

Then (7.60) can be written as

$$\mathbf{S}_{\mathbf{Q}} = \begin{bmatrix} \mathbf{U}_{c1} \\ \mathbf{U}_{c2} \end{bmatrix} \mathbf{\Lambda}_c \begin{bmatrix} \mathbf{U}_{c1}^T & \mathbf{U}_{c2}^T \end{bmatrix}. \tag{7.64}$$

Since $\mathbf{Q}$ is unitary, the eigenvalues of $\hat{\mathbf{S}}_{\mathbf{x},fb}$ will be the same as the eigenvalues of $\mathbf{S}_{\mathbf{Q}}$. Thus, we can write the eigendecomposition of $\hat{\mathbf{S}}_{\mathbf{x},fb}$ as

$$\hat{\mathbf{S}}_{\mathbf{x},fb} = \mathbf{U}_R \mathbf{\Lambda}_c \mathbf{U}_R^T. \tag{7.65}$$

The eigenvectors of $\hat{\mathbf{S}}_{\mathbf{x},fb}$ follow directly from the unitary property of $\mathbf{Q}$. The relation (7.57) can be written as

$$\mathbf{Q}\mathbf{S}_{\mathbf{Q}}\mathbf{Q}^H = \mathbf{Q}\mathbf{Q}^H \hat{\mathbf{S}}_{\mathbf{x},fb} \mathbf{Q}\mathbf{Q}^H = \hat{\mathbf{S}}_{\mathbf{x},fb}. \tag{7.66}$$

Thus,

$$\hat{\mathbf{S}}_{\mathbf{x},fb} = \mathbf{Q}\mathbf{U}_c \mathbf{\Lambda}_c \mathbf{U}_c^T \mathbf{Q}^H, \tag{7.67}$$

so

$$\mathbf{U}_R = \mathbf{Q}\mathbf{U}_c = \frac{1}{\sqrt{2}} \begin{bmatrix} \mathbf{U}_{c1} + j\mathbf{U}_{c2} \\ \mathbf{J}\left(\mathbf{U}_{c1} - j\mathbf{U}_{c2}\right) \end{bmatrix}. \tag{7.68}$$

Note that the transformation $\mathbf{Q}$ has a block identity-exchange matrix structure. Therefore, both the transformation in (7.57) and the inverse transformation (7.65) involve only additions rather than general matrix multiplication.

Property 3: Efficient construction of $\hat{\mathbf{S}}_{\mathbf{x},fb}$: The FB sample correlation matrix is given by (7.46) as

$$\hat{\mathbf{S}}_{\mathbf{x},fb} = \frac{1}{2}\left(\tilde{\mathbf{X}}\tilde{\mathbf{X}}^H + \mathbf{J}\tilde{\mathbf{X}}^*\tilde{\mathbf{X}}^T\mathbf{J}\right) = \tilde{\mathbf{X}}_{fb}\tilde{\mathbf{X}}_{fb}^H, \tag{7.69}$$

where $\tilde{\mathbf{X}}_{fb}$ is defined in (7.48). We now write,

$$\mathbf{S_Q} = \mathbf{Q}^H\hat{\mathbf{S}}_{\mathbf{x},fb}\mathbf{Q} = \mathbf{Q}^H\tilde{\mathbf{X}}_{fb}\tilde{\mathbf{X}}_{fb}^H\mathbf{Q} \triangleq \mathbf{Z}_{fb}\mathbf{Z}_{fb}^T, \tag{7.70}$$

where

$$\mathbf{Z}_{fb} = \mathbf{Q}^H\tilde{\mathbf{X}}_{fb}\mathbf{L}, \tag{7.71}$$

and $\mathbf{L}$ is a unitary transform defined as

$$\mathbf{L} = \frac{1}{\sqrt{2}}\begin{bmatrix} \mathbf{I} & j\mathbf{I} \\ \mathbf{I} & -j\mathbf{I} \end{bmatrix}, \tag{7.72}$$

and $\mathbf{Z}_{fb}$ is real.

For N even, we partition $\tilde{\mathbf{X}}$ into two $(N/2) \times K$ matrices,

$$\tilde{\mathbf{X}} = \begin{bmatrix} \tilde{\mathbf{X}}_1 \\ \tilde{\mathbf{X}}_2 \end{bmatrix}. \tag{7.73}$$

Using (7.73) along with (7.69)–(7.72), we obtain

$$\mathbf{Z}_{fb} = \begin{bmatrix} Re\left(\tilde{\mathbf{X}}_1 + \mathbf{J}\tilde{\mathbf{X}}_2\right) & -Im\left(\tilde{\mathbf{X}}_1 + \mathbf{J}\tilde{\mathbf{X}}_2\right) \\ Im\left(\tilde{\mathbf{X}}_1 - \mathbf{J}\tilde{\mathbf{X}}_2\right) & Re\left(\tilde{\mathbf{X}}_1 - \mathbf{J}\tilde{\mathbf{X}}_2\right) \end{bmatrix}, \tag{7.74}$$

which is real.

The steps to construct $\hat{\mathbf{S}}_{\mathbf{x},fb}$ can be summarized:

1. Construct $\mathbf{Z}_{fb}$ using the forward-only data matrix in (7.74) [$2NK$ real additions].

2. Construct $\mathbf{S_Q}$ using (7.70) [$2N^2K$ real multiplications and $2N^2K$ real additions].

3. Construct $\hat{\mathbf{S}}_{\mathbf{x},fb}$ using (7.66) [N^2 real additions].

Note that, in this case, we use the unitary transformations to get a real data matrix before performing any operations. We will find these techniques to be useful in many applications.

7.2.4 Structured Spectral Matrix Estimation

The derivation in the previous section exploited the centrosymmetric persymmetric structure of $\mathbf{S_x}$ in order to find a maximum likelihood estimate. For a uniform linear array, $\mathbf{S_x}$ is also Toeplitz. We would like to exploit that structure to find a maximum likelihood estimate of $\mathbf{S_x}$.

This problem was introduced by Burg et al. [BLW82] and has been studied extensively in the literature (e.g., [And73], [Deg87], [Cad88], [WH88], [VVW88], [FM88], [Fuh88], [FTM88], [DMS89], [MFOS91], [Fuh91], [WJ93], [TM94], [FM97], [LSL98]). Li et al. [LSL99] derive a computationally efficient technique that provides an asymptotic ML estimate of a structured spectral matrix. In most of our applications, we use either $\widehat{\mathbf{S}}_\mathbf{x}$ or $\hat{\mathbf{S}}_{\mathbf{x},fb}$ as the estimate, so the reader is referred to this literature for a discussion of the issues.

7.2.5 Parametric Spatial Spectral Matrix Estimation

In many applications, we can construct a parametric model of $\mathbf{S_x}$ and then estimate the parameters in order to construct the estimate $\hat{\mathbf{S}}_\mathbf{x}$. We consider two parametric models that are widely used.

In the first model, we assume the input consists of D uncorrelated plane waves plus additive white noise. Then,

$$\mathbf{S_x} = \mathbf{V}(\psi)\mathbf{S_f}\mathbf{V}^H(\psi) + \sigma_w^2\mathbf{I}, \tag{7.75}$$

where $\mathbf{V}(\psi)$ is an $N \times D$ matrix composed of the array manifold vectors,

$$\mathbf{V}(\psi) = \left[\begin{array}{cccc} \mathbf{v}(\psi_1) & \mathbf{v}(\psi_2) & \cdots & \mathbf{v}(\psi_D) \end{array}\right], \tag{7.76}$$

and $\mathbf{S_f}$ is a diagonal matrix of signal powers,

$$\mathbf{S_f} = \text{diag}\left[\begin{array}{cccc} \sigma_1^2, & \sigma_2^2, & \cdots & \sigma_D^2 \end{array}\right]. \tag{7.77}$$

We find a maximum likelihood estimate of the $2D+1$ parameters,

$$\theta \triangleq \{\ \psi_1,\ \ \psi_2,\ \ \cdots\ \ \psi_D,\ \ \sigma_1^2,\ \ \sigma_2^2,\ \ \cdots\ \ \sigma_D^2,\ \ \sigma_w^2\ \}. \tag{7.78}$$

We use these estimates in (7.75) to obtain

$$\hat{\mathbf{S}}_\mathbf{x} = \mathbf{V}(\hat{\psi})\hat{\mathbf{S}}_\mathbf{f}\mathbf{V}^H(\hat{\psi}) + \hat{\sigma}_w^2\mathbf{I}. \tag{7.79}$$

We discuss this technique in Chapter 8 after we have studied parameter estimation.

The second type of parametric model is the parametric wavenumber model introduced in Section 5.6. It is applicable to standard linear arrays. Here we estimate the parameters in the model and use these estimates to construct $\hat{\mathbf{S}}_{\mathbf{x}}$. This problem is the dual of the temporal spectral estimation problem, and the techniques are discussed extensively in the literature (e.g., books by Kay [Kay88], Marple [Mar87], and Stoica and Moses [SM97]). We do not develop this general approach in the text. A few simple cases are covered in the problem section of Chapter 8.

7.2.6 Singular Value Decomposition

A significant portion of our subsequent development will deal with the spatial spectral matrix of the received waveform at the sensors. In order to estimate the spatial spectral matrix, we form the sample spectral matrix

$$\hat{\mathbf{S}}_{\mathbf{x}} = \mathbf{C}_{\mathbf{x}} = \frac{1}{K}\sum_{k=1}^{K}\mathbf{X}(k)\mathbf{X}^H(k). \tag{7.80}$$

If the array manifold vector is conjugate symmetric, we can use the FB averaging technique in (7.46)–(7.48) to construct the sample spectral matrix.

In many applications, we want to find the eigenvalues and eigenvectors of $\hat{\mathbf{S}}_{\mathbf{x}}$ or $\hat{\mathbf{S}}_{\mathbf{x},fb}$. The SVD technique enables us to find the eigenvalues and eigenvectors directly from the data matrix $\tilde{\mathbf{X}}$ in (7.4) or $\tilde{\mathbf{X}}_{fb}$ in (7.48). The technique is developed in Section A.5.

There are important computational reasons for working directly with the data matrix rather than the sample spectral matrix. The dynamic range required to deal with $\hat{\mathbf{S}}_{\mathbf{x}}$ is doubled. Thus, for a specified numerical accuracy, the required word length is doubled.

There are several efficient computational schemes for computing the SVD. All of the various computational programs such as LINPACK, EISPACK, and MATLAB have SVD algorithms included. SVD is widely used in a number of signal processing applications, and there is extensive literature on the topic (e.g., Deprettere [Dep88], in particular the tutorials in Part I; the discussion in Chapter 11 of Haykin [Hay96] and several of his references; Klema and Laub [KL80]; and Eckhart and Young [EY36]).

7.2.7 Summary

In this section, we have developed techniques for estimating the spectral matrix of the snapshot vector at the sensor output.

In Section 7.2.1, we introduced the (forward-only) sample spectral matrix and showed that, if we did not impose any structure on the spectral matrix, then $\mathbf{C_x}$ is the maximum likelihood estimate of $\mathbf{S_x}$. The statistics of the sample spectral matrix are described by the complex Wishart probability density in (7.11).

In many applications, we use the eigenvectors and/or eigenvalues of the sample spectral matrix. Even when we do not use the eigenvectors and eigenvalues directly, they are often the key to the performance. In Section 7.2.2, expressions for the asymptotic behavior of the eigenvectors and eigenvalues were given. Later these results will be used to evaluate the performance of various array processing algorithms.

In Section 7.2.3, the property that many of the arrays that are used in practice have array manifold vectors that are conjugate symmetric was exploited. This property leads to a centrosymmetric Hermitian spectral matrix. In this case the maximum likelihood estimate is given by a FB averaged sample spectral matrix $\mathbf{C}_{\mathbf{x},fb}$. This property not only provides performance improvement but allows the use of real computation.

In Section 7.2.4, the idea of structured spectral estimation was introduced but not developed.

In Section 7.2.5, the idea of parametric spatial spectral matrix estimation was introduced but not developed.

In Section 7.2.6, we discussed the SVD technique that is developed in detail in Section A.5. This technique allowed us to find the eigenvalues and eigenvectors of the sample covariance matrix directly from the data.

This section provides the foundation for the adaptive beamforming algorithms that are developed in the remainder of the chapter. In some algorithms, we will utilize $\widehat{\mathbf{S}}_\mathbf{x}$ or the eigenvalues and eigenvectors explicitly. In other algorithms, they are used implicitly. However, they always have a major influence on the performance of the algorithm.

7.3 Sample Matrix Inversion (SMI)

In adaptive beamformers, the sample spectral matrix $\hat{\mathbf{S}}_\mathbf{x}$ is estimated from the data samples. We recall from Section 7.2, that the maximum likelihood estimate for an unstructured matrix is just the sample spectral matrix. From (7.9),

$$\hat{\mathbf{S}}_\mathbf{x} = \frac{1}{K}\sum_{k=1}^{K}\mathbf{X}(k)\mathbf{X}^H(k). \tag{7.81}$$

In some applications,we can observe the input without the signal being present. For example, in the interval between returned pulses in an active radar. Then, we can construct an estimate of $\mathbf{S_n}$ in the same manner,

$$\hat{\mathbf{S}}_{\mathbf{n}} = \frac{1}{K}\sum_{k=1}^{K}\mathbf{N}(k)\mathbf{N}^H(k). \tag{7.82}$$

If $\hat{\mathbf{S}}_{\mathbf{n}}$ is available, we use it in place of $\mathbf{S_n}$ in the MVDR beamformer (see (6.14) and (6.15)) to obtain

$$\boxed{\mathbf{w}^H_{mvdr,smi} \triangleq \Lambda_{smi}\mathbf{v}_s^H\hat{\mathbf{S}}_{\mathbf{n}}^{-1},} \tag{7.83}$$

and

$$\Lambda_{smi} = \left(\mathbf{v}_s^H\hat{\mathbf{S}}_{\mathbf{n}}^{-1}\mathbf{v}_s\right)^{-1}. \tag{7.84}$$

The presence of $\hat{\mathbf{S}}_{\mathbf{n}}^{-1}$ gives rise to the name **sample matrix inversion** (SMI) for this approach.

If only $\hat{\mathbf{S}}_{\mathbf{x}}^{-1}$ is available,we use (7.81) in (6.71) to obtain

$$\boxed{\mathbf{w}^H_{mpdr,smi} \triangleq \frac{\mathbf{v}_m^H\hat{\mathbf{S}}_{\mathbf{x}}^{-1}}{\mathbf{v}_m^H\hat{\mathbf{S}}_{\mathbf{x}}^{-1}\mathbf{v}_m},} \tag{7.85}$$

where $\mathbf{v}_m$ is the nominal signal direction.

If the array is conjugate symmetric, we can use $\hat{\mathbf{S}}_{\mathbf{x},fb}$ as an estimate of $\mathbf{S_x}$ (or $\mathbf{S_n}$).

In Figure 7.1, we show a diagram of the SMI beamformer. In this configuration, we are processing the data in blocks of size K. In Figure 7.2, we show a diagram of the SMI beamformer in the generalized sidelobe canceller configuration.

The equations specifying the SMI GSC implementation are

$$\hat{\mathbf{S}}_{\mathbf{z}}(K) = \frac{1}{K}\sum_{k=1}^{K}\mathbf{Z}(k)\mathbf{Z}^H(k) = \mathbf{B}^H\hat{\mathbf{S}}_{\mathbf{x}}(K)\mathbf{B}, \tag{7.86}$$

$$\hat{\mathbf{S}}_{\mathbf{z}y_c^*}(K) = \frac{1}{K}\sum_{k=1}^{K}\mathbf{Z}(k)Y_c^*(k) = \mathbf{B}^H\hat{\mathbf{S}}_{\mathbf{x}}(K)\mathbf{w}_q, \tag{7.87}$$

and

$$\begin{aligned}\hat{\mathbf{w}}_a(K) &= \hat{\mathbf{S}}_{\mathbf{z}}^{-1}(K)\hat{\mathbf{S}}_{\mathbf{z}y_c^*}(K)\\ &= \left[\mathbf{B}^H\hat{\mathbf{S}}_{\mathbf{x}}(K)\mathbf{B}\right]^{-1}\mathbf{B}^H\hat{\mathbf{S}}_{\mathbf{x}}(K)\mathbf{w}_q.\end{aligned} \tag{7.88}$$

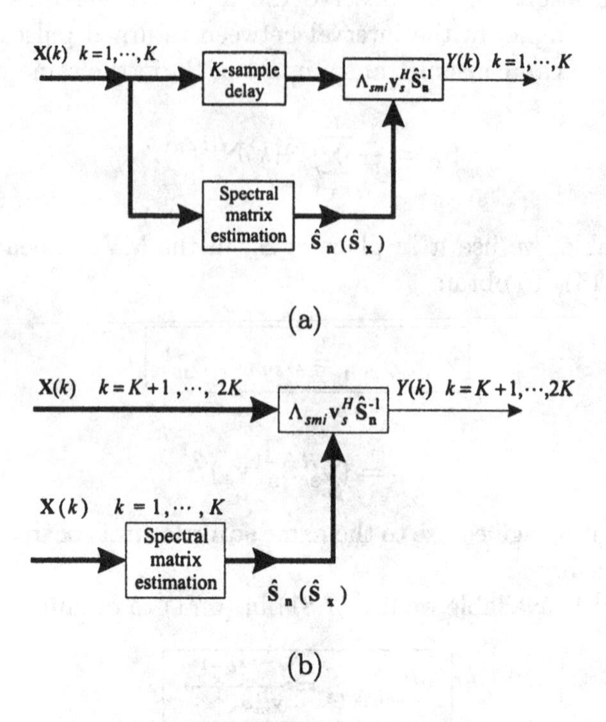

Figure 7.1 Diagram of the SMI beamformer: (a) use of same data vector; (b) use of new data vector.

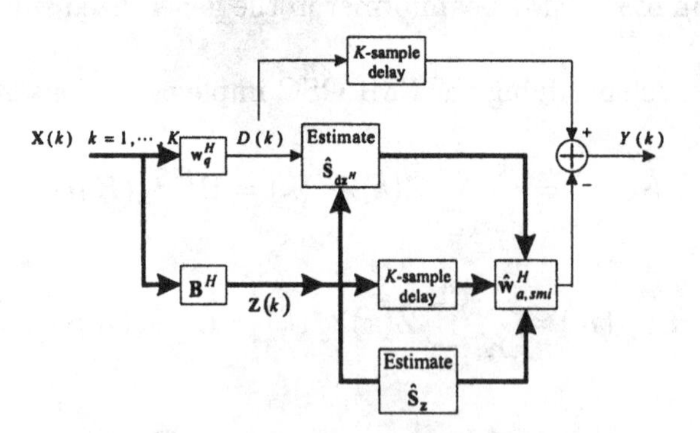

Figure 7.2 Diagram of the SMI beamformer in GSC configuration.

In Section 7.3.1, we discuss the statistical behavior of the *SINR$_{smi}$* as a function of the number of samples K. This behavior will indicate how quickly the performance of MVDR beamformer converges to the results in Chapter 6, which assumed perfect knowledge of either $\mathbf{S_x}$ or $\mathbf{S_n}$.

In Section 7.3.2, we discuss how the SMI MVDR beamformer can be implemented in a recursive manner.

In Section 7.3.3,we introduce the technique of diagonal loading in which we add a constant diagonal matrix to $\hat{\mathbf{S}}_\mathbf{x}$ or $\hat{\mathbf{S}}_\mathbf{n}$ before utilizing them in the weight vector computation. This technique is called **regularization** in the statistical literature.

In Section 7.3.4, we revisit the case of conjugate-symmetric array manifold vectors and show how the implementation can be simplified when we use FB averaging.

7.3.1 *SINR$_{smi}$* Behavior: MVDR and MPDR

In this section we discuss the *SINR* behavior of MVDR and MPDR beamformers as a function of K.

We first consider the beamformer in (7.83). We denote the weight vector using K samples as

$$\hat{\mathbf{w}}^H_{mvdr,smi}(K) \triangleq \Lambda_{smi}(K)\mathbf{v}_s^H\hat{\mathbf{S}}_\mathbf{n}^{-1}(K). \tag{7.89}$$

When the context is clear, we suppress the subscript and use $\hat{\mathbf{w}}(K)$. We want to investigate the *SINR* behavior as a function of K. We assume that $\mathbf{v}_s$ is known. The signal, interference, and noise inputs are sample functions of discrete time random processes so the *SINR$_o$* is a random variable. The *SINR$_o$* on the lth trial at sample K is

$$SINR_o(l) = \frac{\hat{\mathbf{w}}^H(K)\mathbf{v}_s f_l(K) f_l^H(K)\mathbf{v}_s^H\hat{\mathbf{w}}(K)}{\hat{\mathbf{w}}^H(K)\left[\mathbf{n}_l(K)\mathbf{n}_l^H(K)\right]\hat{\mathbf{w}}(K)}, \tag{7.90}$$

where $f_l(K)$ and $\mathbf{n}_l(K)$ are the sample values on the lth trial. The vector $\mathbf{n}_l(K)$ contains the interference and the white sensor noise.

In Chapter 6, the "noise" was assumed to contain both the white sensor noise and the interference. To emphasize that the noise vector $\mathbf{n}$ contains both the white sensor noise and any additional interference, we use *SINR$_o$* in several sections in Chapter 7. This is consistent with the adaptive beamformer literature.

We also define a $SINR_{smi}$ in which the $\mathbf{n}_l(K)\mathbf{n}_l^H(K)$ term in (7.90) is replaced by $\mathbf{S_n}$,

$$SINR_{smi} \triangleq \frac{\sigma_s^2 \left|\hat{\mathbf{w}}^H(K)\mathbf{v}_s\right|^2}{\hat{\mathbf{w}}^H(K)\mathbf{S_n}\hat{\mathbf{w}}(K)}. \tag{7.91}$$

The expression in (7.91) has a smoother behavior and is the definition that is normally used in the literature. We define a random variable $\rho(K)$,

$$\rho(K) \triangleq \frac{SINR_{smi}(K)}{SINR_{mvdr}}, \quad 0 \leq \rho(K) \leq 1, \tag{7.92}$$

which is the ratio of the $SINR$ due to the use of $\hat{\mathbf{S}}_\mathbf{n}$ to the $SINR$ using $\mathbf{S_n}$. For simplicity, we suppress the K dependence and write

$$\rho \triangleq \frac{SINR_{smi}}{SINR_{mvdr}}, \quad 0 \leq \rho \leq 1. \tag{7.93}$$

We want to find the probability density of ρ and its mean and variance. The solution is given by Reed et al. [RMB74]. They indicate that their development is based in part on Capon and Goodman [CG70], Goodman [Goo63], and unpublished work by Goodman.

The development is lengthy, so we quote the result. Using the properties of the complex Wishart density (see Section 7.2.1), they show that ρ has a beta probability density,

$$p(\rho) = \frac{K!}{(N-2)!(K+1-N)!}(1-\rho)^{N-2}\rho^{K+1-N}, \quad 0 \leq \rho \leq 1. \tag{7.94}$$

This probability density is valid for $N \geq 2$ and $K \geq N$. In (7.94), N is the number of elements. The probability density is shown in Figure 7.3 for various values of K.

We observe that the probability density does not depend on the signal strength or the noise and interference environment.

The mean of ρ is

$$E[\rho] = \frac{K+2-N}{K+1}, \tag{7.95}$$

and the variance is

$$Var[\rho] = \frac{(K+2-N)(N-1)}{(K+1)^2(K+2)}. \tag{7.96}$$

Therefore,

$$E[SINR_{smi}] = \frac{K+2-N}{K+1} SINR_{mvdr}. \tag{7.97}$$

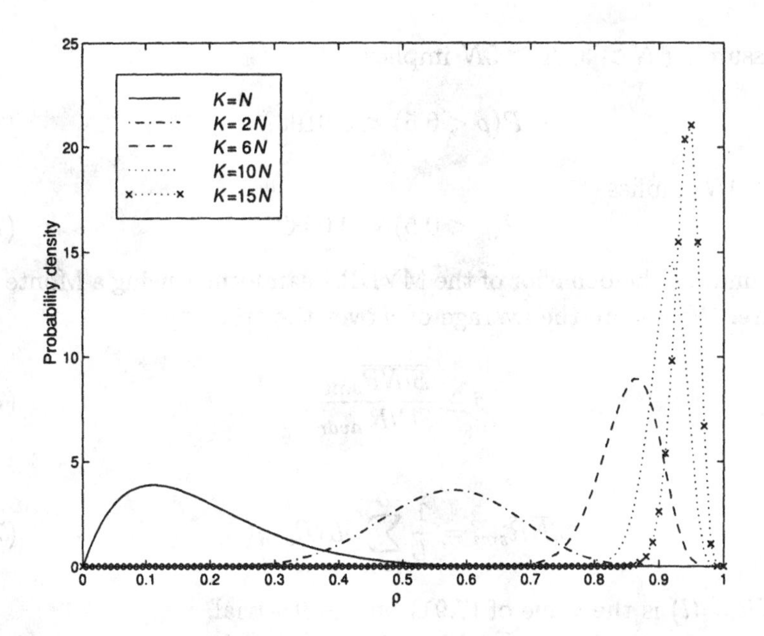

Figure 7.3 Probability density of ρ.

If we desire that $E[SINR_{smi}] = \alpha SINR_{mvdr}$, then we require,

$$K = \frac{1}{1-\alpha}(N-2+\alpha) \simeq \frac{1}{1-\alpha}N. \tag{7.98}$$

Thus, $K = 2N - 3$ obtains an $E[SINR_{smi}]$ that is 3 dB lower. To achieve $\alpha = 0.95$ requires $K = 20N$. In practice, a common rule of thumb is that $K = 2N$ samples are required for "satisfactory" performance.

When $K = 2N - 3$, $p(\rho)$ is symmetric about 0.5, so $Pr[\rho < 0.5]$ equals 0.5. Boronson [Bor80] suggests that a useful measure is to determine the value of K such that

$$P[\rho < 1 - \delta] < \epsilon. \tag{7.99}$$

One can show [AS65] that

$$P[\rho < 1 - \delta] = \sum_{m=0}^{N-2} b(m; K, \delta), \tag{7.100}$$

where

$$b(m; K, \delta) = \binom{K}{m} \delta^m (1-\delta)^{K-m}. \tag{7.101}$$

Then, assuming $N > 3$, $K \geq 3N$ implies

$$P(\rho < 0.5) < 0.0196, \tag{7.102}$$

and $K \geq 4N$ implies

$$P(\rho < 0.5) < 0.0032. \tag{7.103}$$

We simulate the behavior of the MVDR beamformer using a Monte Carlo procedure. We denote the average of ρ over the trials by $\bar{\rho}$,

$$\bar{\rho} \triangleq \frac{\overline{SINR_{smi}}}{SINR_{mvdr}}, \tag{7.104}$$

where

$$\overline{SINR_{smi}} = \frac{1}{L}\sum_{l=1}^{L} SINR_{smi}(l), \tag{7.105}$$

and $SINR_{smi}(l)$ is the value of (7.91) on the lth-trial.

We illustrate the behavior with a simple example.

Example 7.3.1

Consider a standard 10-element linear array. The signal arrives from $u_s = 0$, and there is a single interferer at $u_I = 0.15$ with an $INR = 10$ dB. In Figure 7.4, we plot $\bar{\rho}$ for 200 trials for both $\hat{\mathbf{S}}_{\mathbf{n}} = \mathbf{C}_{\mathbf{n}}$ and $\hat{\mathbf{S}}_{\mathbf{n}} = \mathbf{C}_{\mathbf{n},fb}$. For the $\hat{\mathbf{S}}_{\mathbf{n}} = \mathbf{C}_{\mathbf{n}}$ case we show $E[\rho]$ as given by (7.95). For modest K, the use of $\hat{\mathbf{S}}_{\mathbf{n}} = \mathbf{C}_{\mathbf{n},fb}$ halves the number of required samples to achieve a given value of ρ. Note that, although we specify a signal and interference environment, the result does not depend on it.

Monzingo and Miller [MM80] extended the result to include the case of the signal present, and the beamformer is given by (7.85). We define η to be the ratio of the $SINR$s,

$$\begin{aligned} \eta &= \frac{SINR_{mpdr,smi}}{SINR_{mpdr}} \\ &= \frac{\rho'}{SINR_{mpdr}(1-\rho')+1}, \end{aligned} \tag{7.106}$$

where ρ' has the same probability density as ρ. The expectation of η can be written as an infinite series,

$$\begin{aligned} E[\eta] &= \frac{a}{a+b}\left\{1+\sum_{i=1}^{\infty}(-SINR_{mpdr})^{i}\left(\frac{b}{a+b+1}\right)\right. \\ &\qquad \left. \cdot\left(\frac{b+1}{a+b+2}\right)\cdots\left(\frac{i+b-1}{a+b+i}\right)\right\}, \end{aligned} \tag{7.107}$$

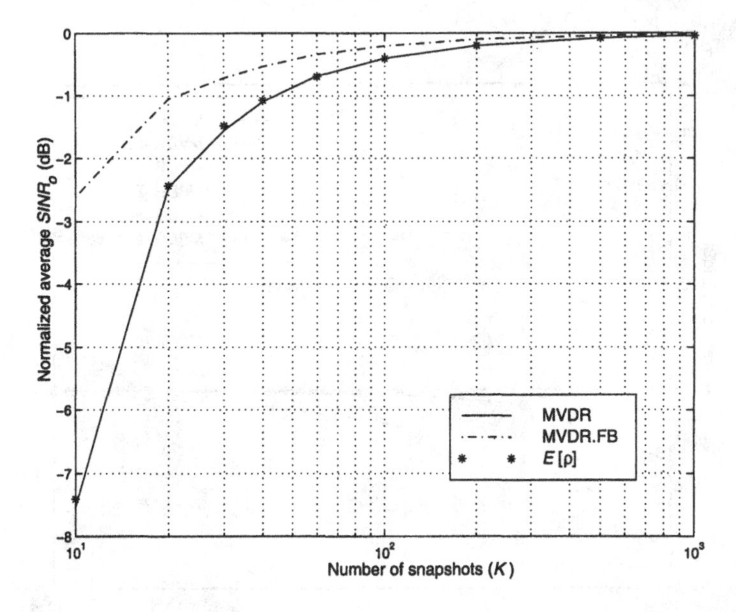

Figure 7.4: MVDR SMI beamformer: $\bar{\rho}$ and $E[\rho]$ versus K.

where

$$a = K - N + 2, \tag{7.108}$$

$$b = N - 1. \tag{7.109}$$

We can approximate the sum in the brace to obtain an approximate expression,[5]

$$E[\eta] \simeq \frac{a}{a+b}\left\{\frac{1}{1 + SINR_{mpdr} \cdot \frac{b}{a+b+1}}\right\}. \tag{7.110}$$

Example 7.3.2

Consider a standard 10-element linear array. The signal arrives from $u_s = 0$. The *SNR* is varied from 0 dB to 30 dB in 10-dB steps. Two equal-power interferers arrive from $u_I = 0.29$ and 0.45, each with an *INR* = 20 dB.

In Figure 7.5 we show the results of a Monte Carlo simulation. We plot the average $SINR_o$ versus K for both $\hat{\mathbf{S}}_{\mathbf{x}} = \mathbf{C}_{\mathbf{x}}$ and $\hat{\mathbf{S}}_{\mathbf{x},fb} = \mathbf{C}_{\mathbf{x},fb}$. We also show the analytic results using (7.97) and (7.110). We see that the presence of the signal in $\hat{\mathbf{S}}_{\mathbf{x}}$ causes a significant degradation in the performance of the SMI beamformer.

In Figure 7.6, we plot $\bar{\eta}$, which is the normalized quantity. This plot shows the effect of the signal level.

[5]Z. Tian, private communication.

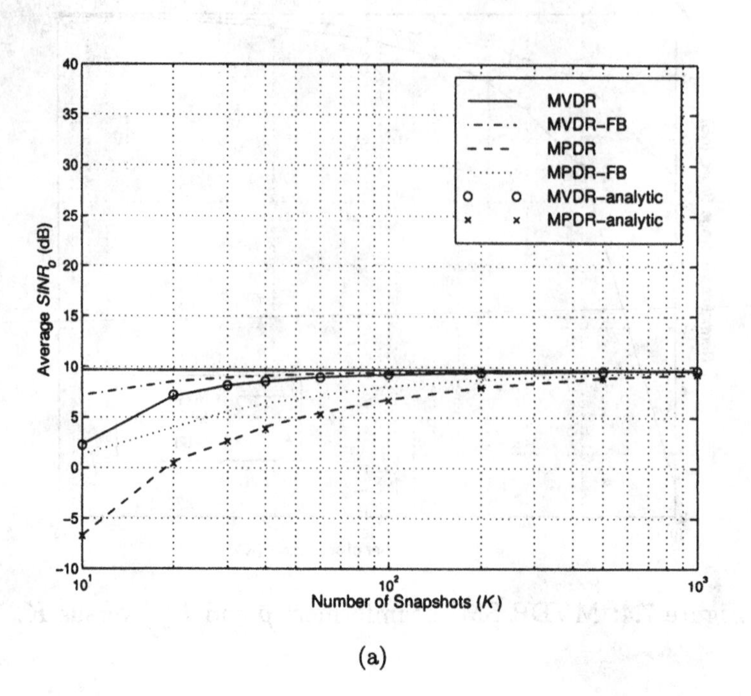
MVDR
MVDR–FB
MPDR
MPDR–FB
MVDR–analytic
MPDR–analytic
Average $SINR_o$ (dB)
Number of Snapshots (K)
40
35
30
25
20
15
10
5
0
−5
−10
10^1
10^2
10^3

(a)

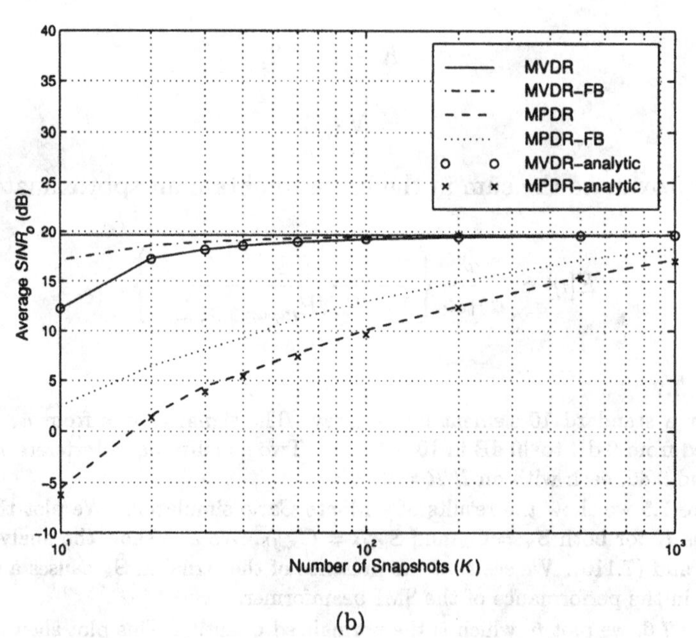
MVDR
MVDR–FB
MPDR
MPDR–FB
MVDR–analytic
MPDR–analytic
Average $SINR_o$ (dB)
Number of Snapshots (K)
40
35
30
25
20
15
10
5
0
−5
−10
10^1
10^2
10^3

(b)

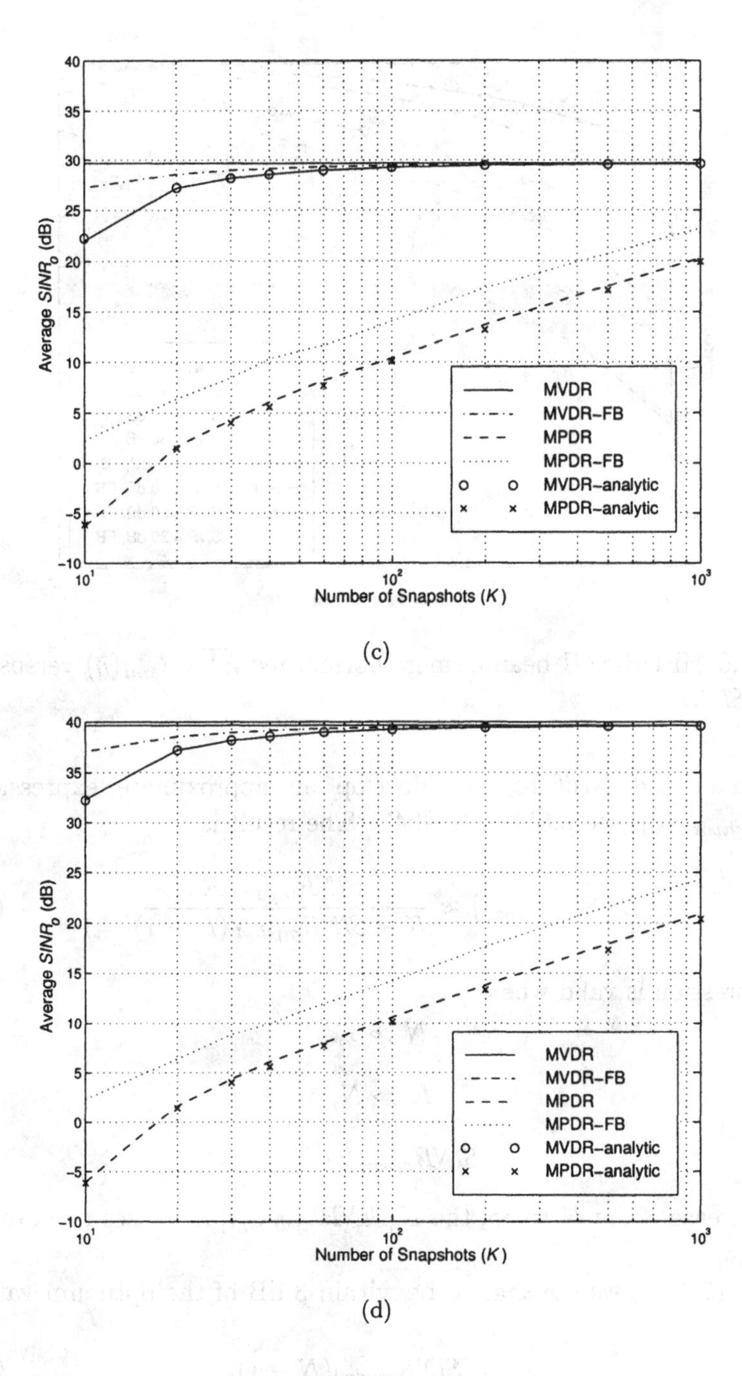

Figure 7.5 MVDR-SMI and MPDR-SMI beamformers: average $SINR_o$ versus K: (a) $SNR = 0$ dB; (b) $SNR = 10$ dB; (c) $SNR = 20$ dB; (d) $SNR = 30$ dB.

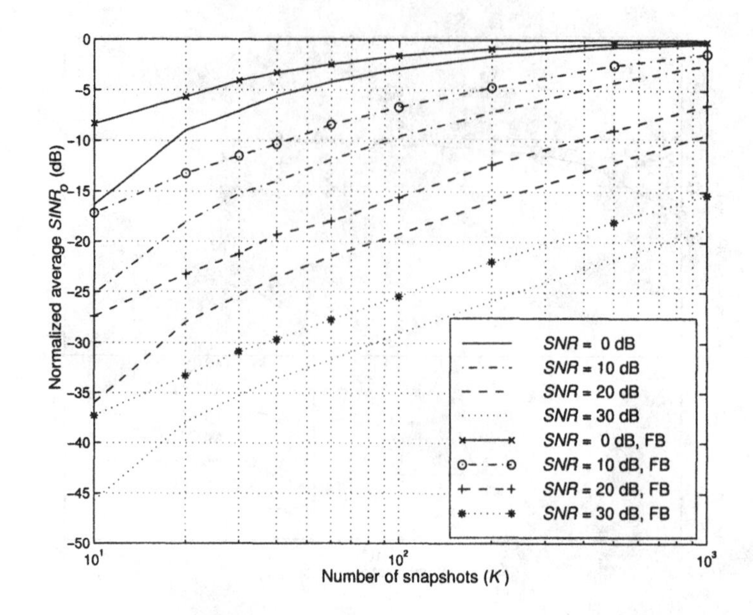

Figure 7.6 MPDR-SMI beamformer: normalized $\overline{SINR_{smi}}(\bar{\eta})$ versus K for various *SNR*.

Feldman and Griffiths also develop an approximate expression for $E\left[SINR_{mpdr,smi}\right]$ (see p.871 of [FG94]). The result is

$$E\left\{SINR_{mpdr}\right\} \approx \frac{SINR_{mpdr} \cdot K}{K + SINR_{mpdr} \cdot (N-1)}. \tag{7.111}$$

This expression is valid when

$$N \gg 1,$$

$$K \gg N,$$

$$SINR_{mpdr} \gg N.$$

As K increases from N to ∞, the $E\left[SINR_{mpdr,smi}\right]$ increases monotonically to $SINR_{mpdr}$.

From (7.111), we see that to be within 3 dB of the optimum value, we require

$$K = (SINR_{mpdr})(N-1), \tag{7.112}$$

which will be significantly larger than the previous $K = 2N$ result in most cases of interest.

7.3.2 LCMV and LCMP Beamformers

In the LCMV and LCMP beamformers developed in Section 6.7.2, we impose additional constraints on the beamformer and reduce the adaptive degrees of freedom. Monzingo and Miller [MM80] argue that we can use the results in (7.94) by defining $(N-1)$ as the number of adaptive degrees of freedom instead of the number of elements minus one. Van Veen ([VV91a], [VV91b], [HS92]) obtains a similar result (e.g., Reed et al. [RMB74], Capon and Goodman [CG70], Monzingo and Miller [MM80], Baggeroer [Bag76], Ganz et al. [GMW90], Kelly [Kel86], and Kelly and Forsythe [KF89]).

We consider a simple example to illustrate the behavior.

Example 7.3.3 (continuation, Example 7.3.2)

Consider a standard 10-element array. The signal-and-interference model are the same as in Example 7.3.2. We use an LCMP beamformer with derivative constraints (see Examples 6.7.2 and 6.7.10). The $\mathbf{C}$ and $\mathbf{g}^H$ matrices are given by (6.312) and (6.314), respectively:

$$\mathbf{C} = \left[\ 1 \ \vdots\ \mathbf{d}(0) \ \vdots\ \dot{\mathbf{d}}(0)\ \right], \tag{7.113}$$

and

$$\mathbf{g}^H = \left[\ 1 \ \vdots\ 0 \ \vdots\ \ddot{B}_c(0)\ \right]. \tag{7.114}$$

In Figure 7.7(a), we show $\bar{\rho}$ obtained from a Monte Carlo simulation for an LCMV beamformer. We also show $E[\rho]$ obtained from (7.95) with $N=8$. Comparing these results to those in Figure 7.4, we see that, by reducing the available adaptive degrees of freedom, we have improved the normalized SMI performance. In Figure 7.7(b), we plot $\bar{\eta}$, the normalized average $SINR_o$ versus K for an LCMP beamformer. Comparing these results to those in Figure 7.6, we see an improvement.

7.3.3 Fixed Diagonal Loading

The concept of diagonal loading is straightforward. We use,

$$\tilde{\mathbf{S}}_{\mathbf{x},L} = \frac{1}{K}\sum_{k=1}^{K}\mathbf{X}(k)\mathbf{X}^H(k) + \sigma_L^2\mathbf{I} \tag{7.115}$$

in place of the estimated spectral matrix in order to design $\hat{\mathbf{w}}$. We use $\tilde{\mathbf{S}}_{\mathbf{x},L}$ instead of $\hat{\mathbf{S}}_{\mathbf{x}}$ because it is not used as an estimate of $\mathbf{S}_{\mathbf{x}}$. We have added a diagonal matrix of level σ_L^2. We encountered diagonal loading in Sections 6.6.4 and 6.10 in the context of quadratically constrained beamformers. In this section, we use it for three purposes:

(i) To improve the $SINR_{smi}$ performance of the MPDR beamformer;

(ii) To implement beamformers when $K < N$;

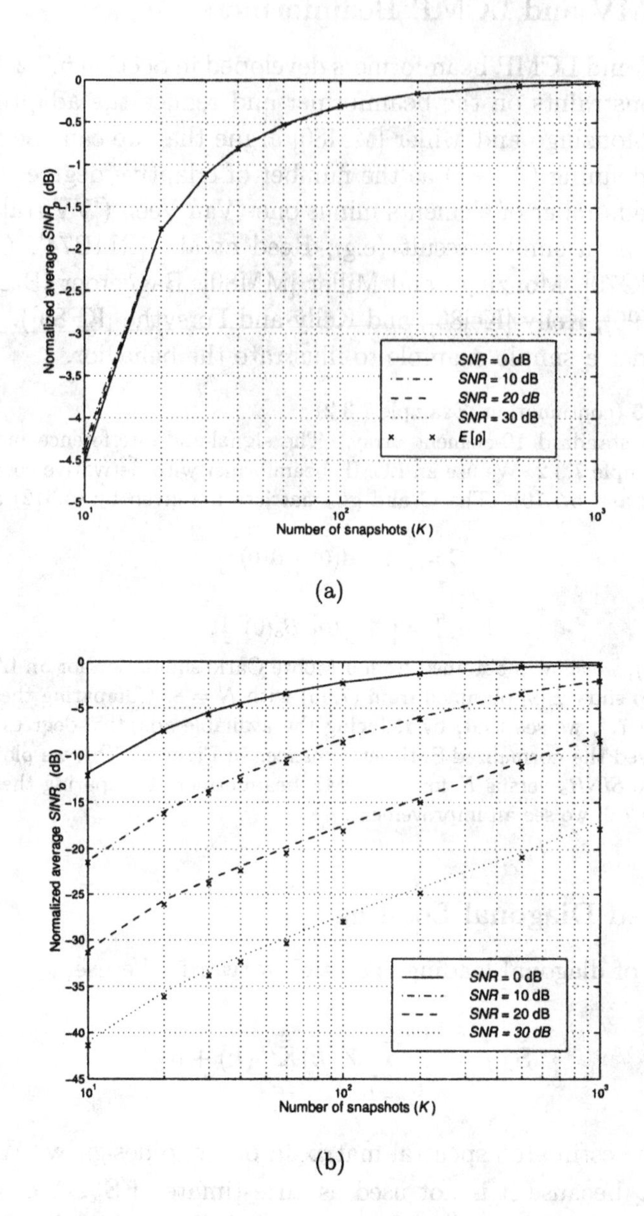

Figure 7.7 LCMV-SMI and LCMP-SMI beamformers: $u_s = 0$, $u_I = 0.29$, and 0.45, $INR = 20$ dB, $SNR = 0$, 10, 20, 30 dB; normalized $\overline{SINR_o}$ versus K: (a) LCMV-SMI; (b) LCMP-SMI beamformers.

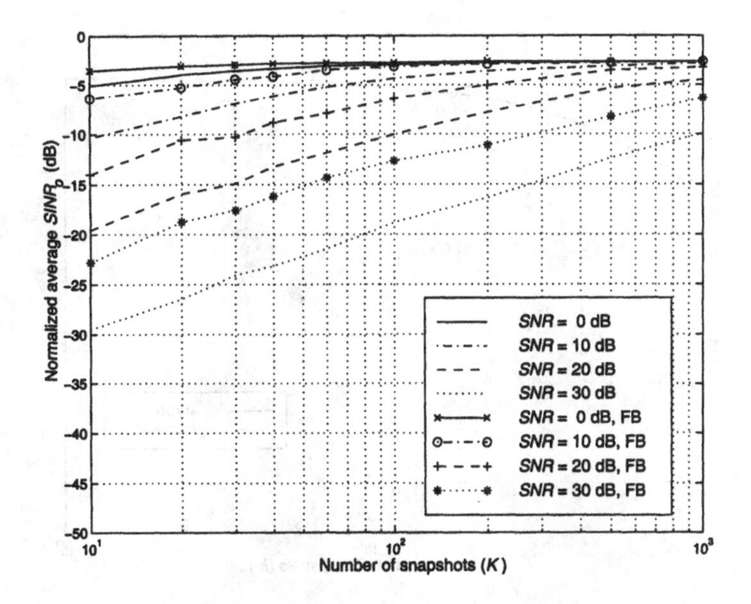

Figure 7.8: MPDR-SMI beamformer: INR = 20 dB, σ_L^2/σ_W^2 = 20 dB; normalized $\overline{SINR_o}$ versus K.

(iii) To achieve better sidelobe control and main-beam shaping in the SMI algorithm.

To demonstrate how diagonal loading improves the $SINR_{smi}$ behavior, we consider the same model as in Section 7.3.1.

Example 7.3.4 (continuation, Example 7.3.2)

Consider a standard 10-element linear array. The desired signal arrives from $u_s = 0$ and two equal-power interferers arrive from $u_I = 0.29$ and 0.45, each with an INR = 20 dB. We use

$$\tilde{\mathbf{S}}_{\mathbf{x},L} = \mathbf{C}_{\mathbf{x}} + \sigma_L^2\mathbf{I}, \tag{7.116}$$

or

$$\tilde{\mathbf{S}}_{\mathbf{x},L} = \mathbf{C}_{\mathbf{x},fb} + \sigma_L^2\mathbf{I}. \tag{7.117}$$

in (7.85). In Figure 7.8, we plot the normalized average $SINR_o$ versus K for σ_L^2 = 20 dB. We see that there is significant improvement for all SNR. In Figure 7.9, we show the MVDR behavior. There is also improvement, but it is less significant.

When we use diagonal loading we can implement the beamformer for $K < N$ because the $\sigma_L^2\mathbf{I}$ term makes $\tilde{\mathbf{S}}_{\mathbf{x},L}$ non-singular.

Several references (e.g., Hudson [Hud81], Gabriel [Gab85], and Brookner and Howell [BH85]) show that if there are D strong interferences and $N \gg D$, the beamformer can achieve effective nulling with $K = 2D$ samples. With

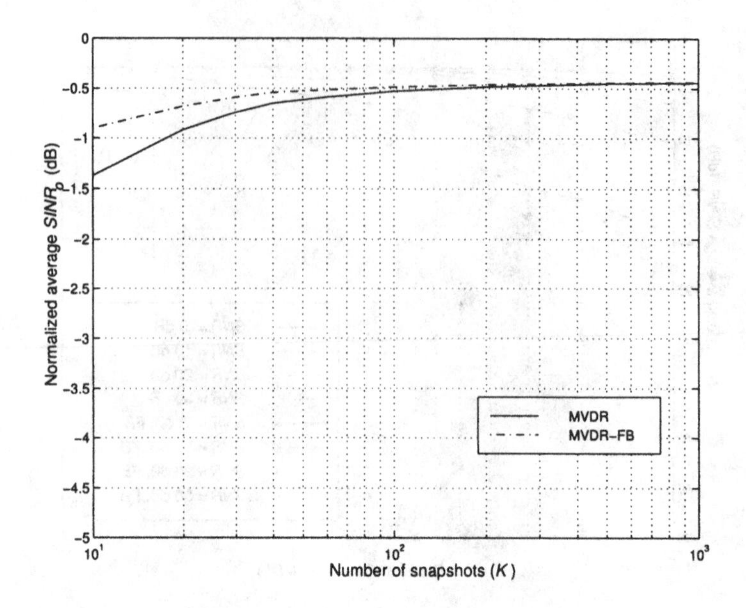

Figure 7.9 MVDR-SMI beamformer: INR = 20 dB, σ_L^2/σ_W^2 = 20 dB; normalized $\overline{SINR_o}$ versus K.

FB averaging, $K = D$. We revisit Example 7.3.4 and explore the behavior for $K = 2, 4$, and 6.

Example 7.3.4 (revisited)

Consider the same model as in Example 7.3.4. In Figure 7.10, we plot the normalized average $SINR_o$ versus K for $K \geq D$ for an MPDR-SMI beamformer with fixed diagonal loading.

We see that there is significant improvement in the $\overline{SINR_o}$ due to diagonal loading with a small number of samples. However, it is also important to examine the beam pattern behavior. In Figure 7.11, we show the beam pattern for the two-interferer case (u_I = 0.29, 0.15) with a high INR (20 dB), a modest SNR (10 dB), for $K = 4$ using FB averaging. We vary σ_L^2/σ_w^2 from -10 dB to 20 dB in 10-dB steps. We observe that, although the nulls are deep and close to the interferer, a typical beam pattern has an undesirable sidelobe structure unless we use diagonal loading.

The same problem continues when $K > N$.

Kelly [Kel87a], [Kel87b] shows that the expected value of the sidelobes of the adapted pattern is

$$E[SLL] = \frac{1}{K+1}. \tag{7.118}$$

We consider a simple example to illustrate this effect.

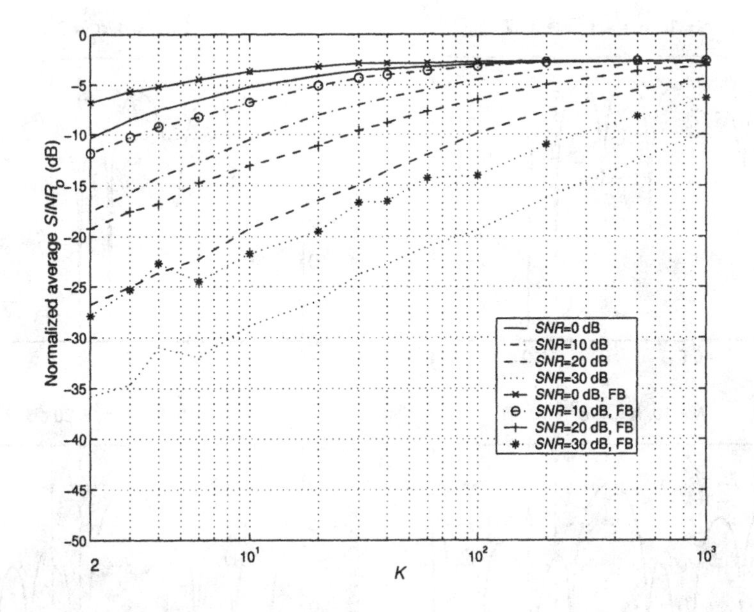

Figure 7.10 MPDR-SMI beamformer with fixed diagonal loading: normalized $\overline{SINR}_o$ versus K, $u_I = 0.29, 0.45$, $INR = 20$ dB, $LNR = \sigma_L^2/\sigma_W^2 = 20$ dB.

Example 7.3.5[6]

Consider a standard 10-element linear array. A Dolph-Chebychev weighting on the steering vector $\mathbf{v}(\psi_s)$ provided a quiescent beam with -30-dB sidelobes. We use the constraint $\mathbf{w}^H\mathbf{w}_q = 1$. The beam is steered to broadside. A single plane-wave source with an *INR* of 30 dB inpinges on the array at $u_I = 0.29$. There is no diagonal loading.

The results of a simulation for $K = 2N$ and $K = 6N$ are shown in Figure 7.12.

To understand the behavior we write the weighting vector using the eigenvector expansion of $\hat{\mathbf{S}}_\mathbf{n}$.[7] We assume that $\sigma_L^2 = 0$ and that there are D plane-wave interferers. The estimated eigenvalues are ordered,

$$\hat{\lambda}_1 \geq \hat{\lambda}_2 \geq \cdots \hat{\lambda}_{min}. \tag{7.119}$$

The eigenvalues can be written as

$$\hat{\lambda}_i = \begin{cases} \hat{\lambda}_i^I + \hat{\lambda}_{min}, & i = 1, 2, \cdots, D \\ \hat{\lambda}_i^\epsilon + \hat{\lambda}_{min}, & i = D+1, D+2, \cdots, N-1 \\ \hat{\lambda}_{min}, & i = N, \end{cases} \tag{7.120}$$

[6]This example is adapted from Carlson [Car88].

[7]We discuss the behavior for$\mathbf{S}_\mathbf{n}$ and MVDR beamformers. A similar discussion follows for $\mathbf{S}_\mathbf{x}$ and MPDR beamformers.

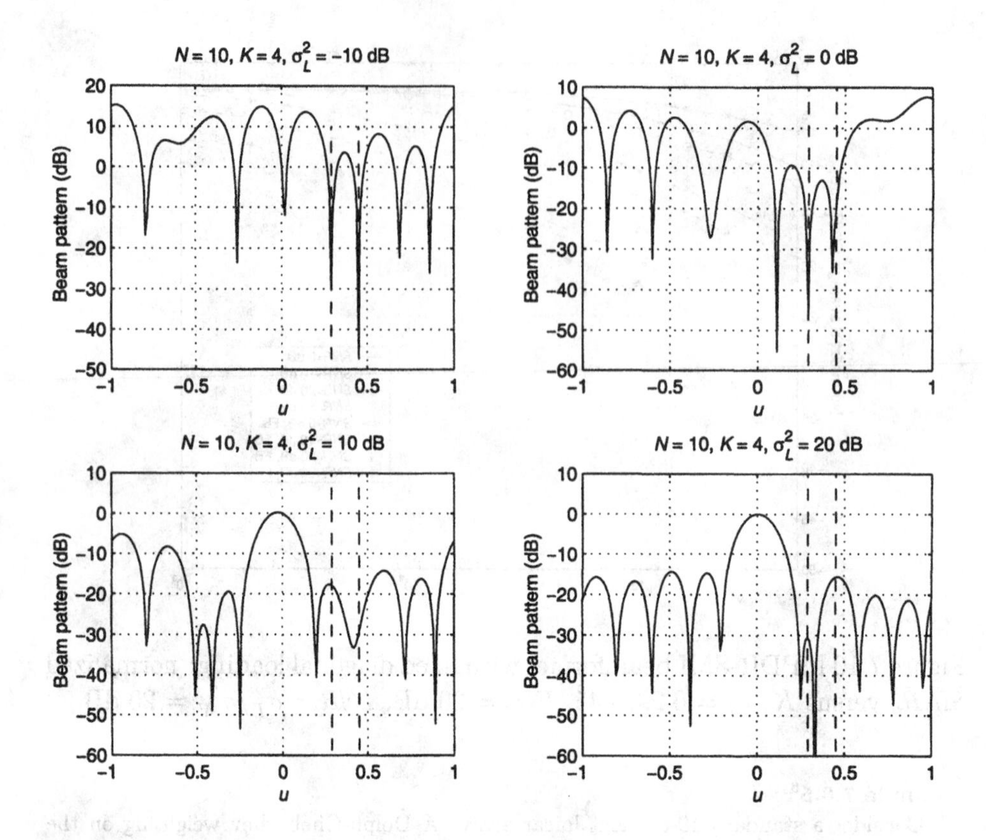

Figure 7.11: MPDR-FL-SMI beamformer: beam pattern for two interferers, $K = 4$: (a) *LNR* = -10 dB; (b) *LNR* = 0 dB; (c) *LNR* = 10 dB; (d) *LNR* = 20 dB.

where $\hat{\lambda}_{min}$ is the smallest (minimum eigenvalue) of $\hat{\mathbf{S}}_{\mathbf{n}}$; $\hat{\lambda}_i^{\epsilon}$ is the difference between the other $N-D-1$ noise eigenvalues and $\hat{\lambda}_{min}$; and $\hat{\lambda}_i^I$ is the estimate of the interference eigenvalues. Note that if $\hat{\mathbf{S}}_{\mathbf{n}}$ equaled $\mathbf{S}_{\mathbf{n}}$, then the $N-D$ noise eigenvalues would be equal and $\hat{\lambda}_i^{\epsilon}$ would be zero.

The inverse of $\hat{\mathbf{S}}_{\mathbf{n}}^{-1}$ can be written as

$$\begin{aligned}\hat{\mathbf{S}}_{\mathbf{n}}^{-1} &= \sum_{i=1}^{D} \frac{1}{\hat{\lambda}_i^I + \hat{\lambda}_{min}} \mathbf{\Phi}_i \mathbf{\Phi}_i^H + \sum_{i=D+1}^{N-1} \frac{1}{\hat{\lambda}_i^{\epsilon} + \hat{\lambda}_{min}} \mathbf{\Phi}_i \mathbf{\Phi}_i^H \\ &\quad + \frac{1}{\hat{\lambda}_{min}} \mathbf{\Phi}_i \mathbf{\Phi}_i^H, \end{aligned} \tag{7.121}$$

or

$$\hat{\mathbf{S}}_{\mathbf{n}}^{-1} = \frac{1}{\hat{\lambda}_{min}} \left\{ \mathbf{I} - \sum_{i=1}^{N} \frac{\hat{\lambda}_i - \hat{\lambda}_{min}}{\hat{\lambda}_i} \mathbf{\Phi}_i \mathbf{\Phi}_i^H \right\}. \tag{7.122}$$

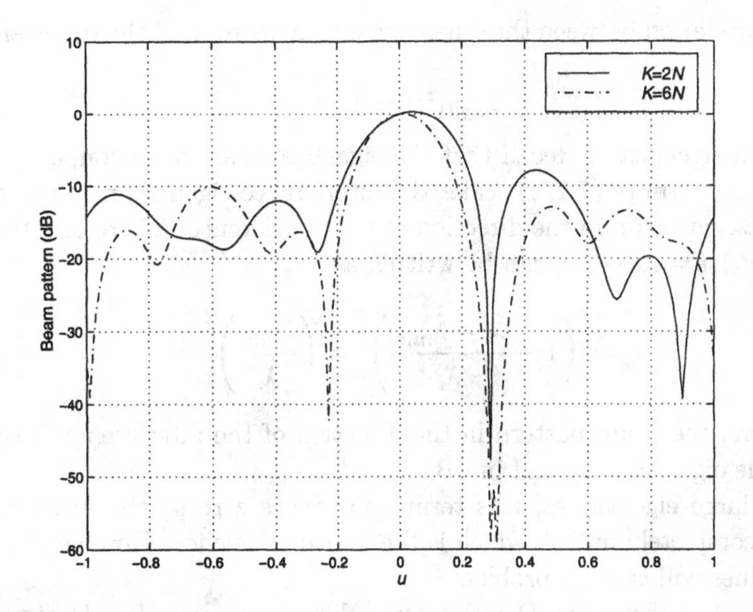

Figure 7.12: MPDR-SMI beamformer with Dolph-Chebychev quiescent beam pattern; adapted beam pattern with $2N$ and $6N$ samples: $u_s = 0$, $u_I = 0.29$, $INR = 30$ dB.

Using (7.122) in (7.83), the optimum weight vector can be written as

$$\mathbf{w}_o^H = \mathbf{w}_q^H - \sum_{i=1}^{N} \frac{\hat{\lambda}_i - \hat{\lambda}_{min}}{\hat{\lambda}_i} \mathbf{w}_q^H \mathbf{\Phi}_i \mathbf{\Phi}_i^H, \tag{7.123}$$

where $\mathbf{w}_q$ is the weight vector in a white noise, or quiescent, environment.[8]

The resulting beam pattern is

$$\begin{aligned} B_o(\psi) &= B_q(\psi) - \sum_{i=1}^{N} \frac{\hat{\lambda}_i - \hat{\lambda}_{min}}{\hat{\lambda}_i} \rho_{qi} B_{eig,i}(\psi) \\ &= B_q(\psi) - \sum_{i=1}^{D} \frac{\hat{\lambda}_i^I}{\hat{\lambda}_i} \rho_{qi} B_{eig,i}(\psi) \\ &\quad - \sum_{i=D+1}^{N-1} \frac{\hat{\lambda}_i^{\epsilon}}{\hat{\lambda}_i^{\epsilon} + \hat{\lambda}_{min}} \rho_{qi} B_{eig,i}(\psi), \end{aligned} \tag{7.124}$$

where

$$\rho_{qi} = \mathbf{w}_q^H \mathbf{\Phi}_i, \tag{7.125}$$

[8] We have not normalized $\mathbf{w}_q$ and $\mathbf{w}_o$, in order to simplify the notation.

is the correlation between the quiescent weight vector and the ith eigenvector and,

$$B_{eig,i}(\psi) = \mathbf{\Phi}_i^H \mathbf{v}(\psi), \tag{7.126}$$

is the ith eigenbeam. Recall that the eigenbeams are orthogonal.

The ρ_{qi} term in (7.124) scales the interference eigenbeam to be equal to the quiescent beam in the direction of the interference. Therefore, the value of $B_o(\psi_{Ii}), i = 1, \cdots, D$, can be written as

$$\left(1 - \frac{\hat{\lambda}_i - \hat{\lambda}_{min}}{\hat{\lambda}_i}\right)^2 = \left(\frac{\hat{\lambda}_{min}}{\hat{\lambda}_i}\right)^2. \tag{7.127}$$

Therefore, the beam pattern in the direction of the interference is equal to twice the eigenvalue spread in dB.

For large eigenvalues, this term approaches zero so the interferers are almost completely nulled, which is the desired behavior. However, the noise eigenvalues will cause a problem.

Assuming there are D interferers, then there are $N - D$ eigenvalues corresponding to noise. If we knew their values exactly, then they would all be equal,

$$\lambda_i = \lambda_{i+1} = \cdots = \lambda_D = \lambda_{min}, \quad i > N - D. \tag{7.128}$$

Then, the spread in (7.127) would be unity and the white noise would have no impact on the quiescent pattern, which is the desired behavior. However, we are estimating the λ_i from the sample spectral matrix and, for small K, there will be a significant spread around the correct value. This means that the beamformer will be subtracting noise eigenbeams (which are random in nature) from the quiescent beam pattern. From (7.124), we see that the noise eigenbeams would not be added if $\hat{\lambda}_i^\epsilon$ were zero. We show this behavior for the model in Example 7.3.5. Note that this is an MPDR example.

Example 7.3.6 (continuation)

In Figure 7.13, we show the noise eigenvalue spread as a function of the number of snapshots. We see that there is significant improvement between $K = N$ and $K = 2N$, but the subsequent improvement is gradual.

In Figure 7.14, we show the highest sidelobe in the adapted pattern as a function of the number of snapshots. The level is similar to that predicted by (7.118) (which corresponded to the average level).

In Figure 7.15, we show the reduction in array gain as a function of the number of snapshots. This behavior is consistent with the result in (7.110).

This behavior leads to the conclusion that the noise eigenvalues and eigenvectors can create a significant problem. Once again, we use diagonal loading.

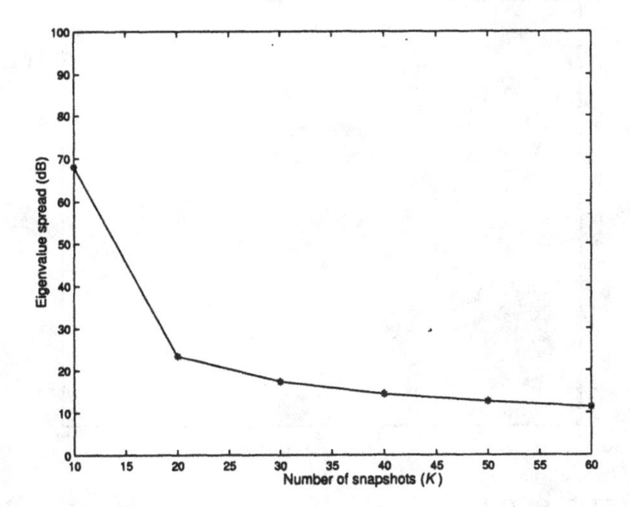

Figure 7.13 Noise eigenvalue spread (ratio of largest noise eigenvalue to smallest) as a function of the number of independent snapshots included in the spectral matrix; SNR = 20 dB, u_I = 0.29, INR = 30 dB, eight noise eigenvalues.

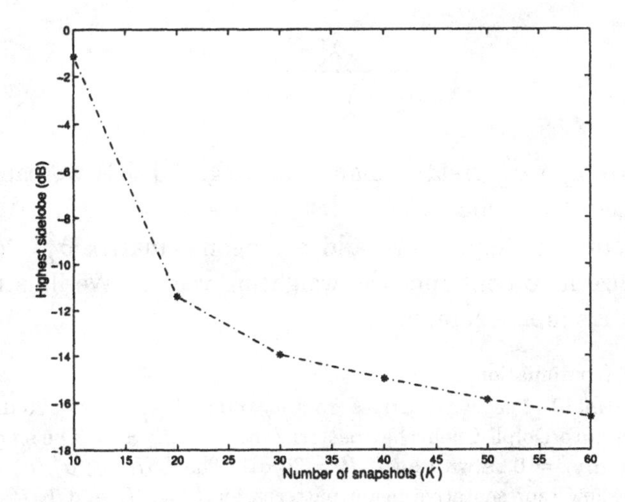

Figure 7.14 MPDR-SMI beamformers: highest sidelobe in adapted patterns as a function of the number of independent snapshots included in the spectral matrix. Dolph-Chebychev (−30 dB SLL) quiescent beam, SNR = 20 dB, INR = 30 dB.

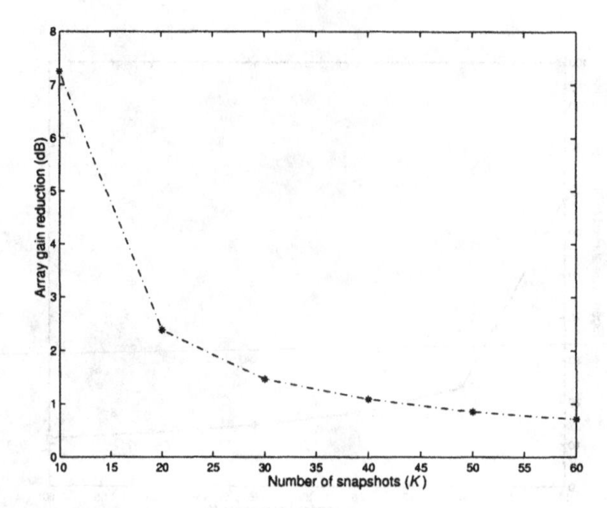

Figure 7.15 MPDR-SMI beamformer: array gain reduction in adapted main beam as a function of the number of independent snapshots included in the covariance matrix.

If we use (7.115) to construct the weight vector, then the coefficient on the noise eigenbeams in (7.124) is

$$\frac{\hat{\lambda}_i^{\epsilon}}{\hat{\lambda}_i^{\epsilon} + \hat{\lambda}_{min} + \sigma_L^2}. \tag{7.129}$$

Thus, by choosing a σ_L^2 greater than $\hat{\lambda}_{min}$ (e.g., 10 dB) we can essentially eliminate the noise eigenbeams.

The technique is simple. We add a diagonal matrix $\sigma_L^2 \mathbf{I}$ to $\hat{\mathbf{S}}_{\mathbf{n}}$ or $\hat{\mathbf{S}}_{\mathbf{x}}$ before we utilize it to construct the weighting vector. We illustrate typical behavior with a simple example.

Example 7.3.7 (continuation)

Consider a SLA 10. The signal arrives from $u_s = 0$ with an $SNR = 20$ dB. The quiescent beam pattern is a Dolph-Chebychev pattern with −30-dB SLL. The single interfering signal arrives from $u_I = 0.29$ with an $INR = 30$ dB. The LNR ($\triangleq \sigma_L^2/\sigma_w^2$) is 10 dB. In Figure 7.16, we show representative beam patterns for $K = 2N$ and $K = 6N$ with and without loading. We see that the sidelobe behavior is improved significantly.

In Figure 7.17, we show the noise eigenvalue spread as a function of σ_L^2/σ_w^2 for various values of K. In Figure 7.18, we show the highest sidelobes in the adapted pattern as a function of σ_L^2/σ_w^2 for various values of K. In Figure 7.19, we show the reduction in array gain as a function of σ_L^2/σ_w^2 for various values of K. We see that, for all of the performance metrics of interest, loading offers significant improvement.

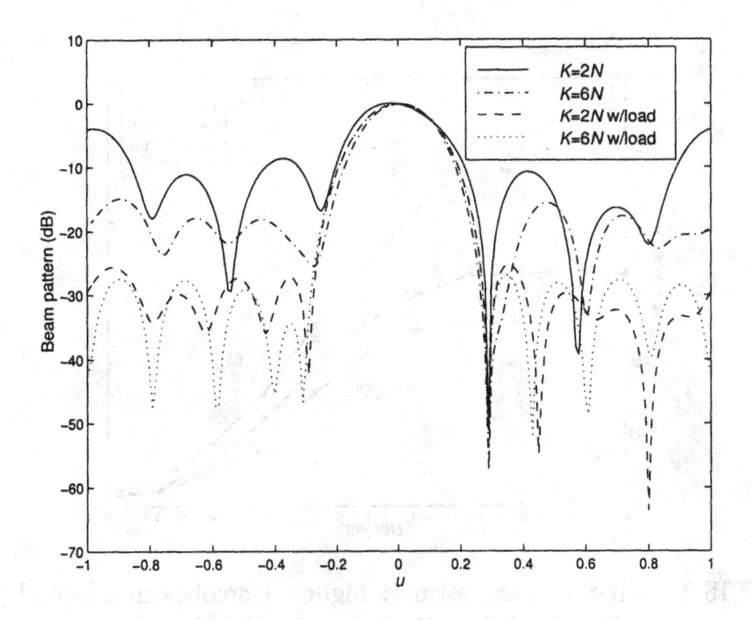

Figure 7.16 MPDR-SMI beamformer: adapted beam pattern with $6N$ snapshots and diagonal loading; $u_s = 0$, $SNR = 20$ dB, $u_I = 0.29$, $INR = 30$ dB, $LNR = 10$ dB.

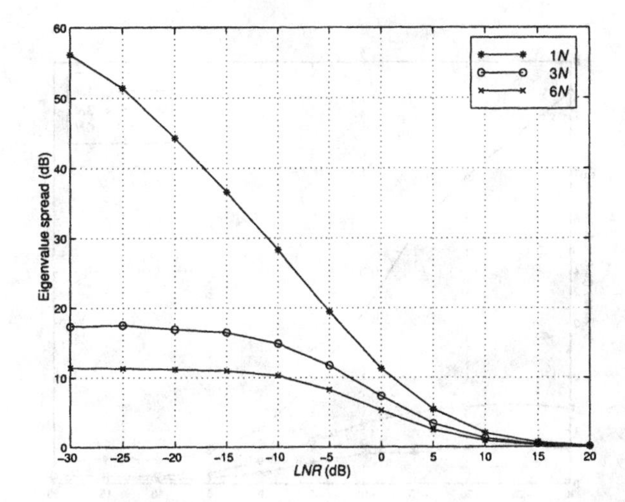

Figure 7.17 MPDR-SMI beamformer: noise eigenvalue spread (ratio of largest noise eigenvalue to smallest) as a function of the loading level for $1N$, $3N$, and $6N$ independent snapshots in spectral matrix; $u_s = 0$, $SNR = 20$ dB, $u_I = 0.29$, $INR = 30$ dB.

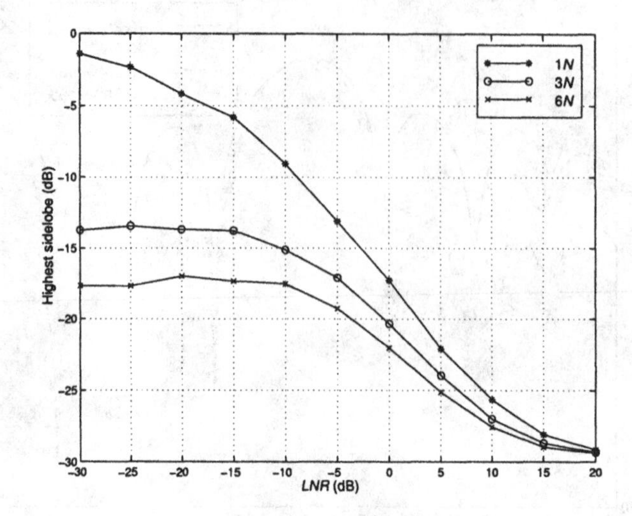

Figure 7.18 MPDR-SMI beamformer: highest sidelobes in adapted pattern as function of loading level for $1N$, $3N$, and $6N$ independent snapshots in spectral matrix.

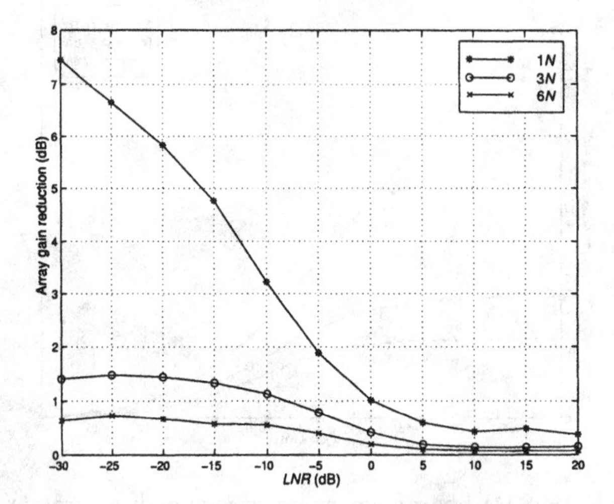

Figure 7.19 MPDR-SMI beamformer: array gain reduction as a function of loading level for $1N$, $3N$, and $6N$ independent snapshots in spectral matrix.

The disadvantage is that the beamformer loses its ability to adapt against small eigenvalues. These can occur if there are small power interferers or if there are two or more large interferers close together. From Figure 5.14, we see that as the interferer separation becomes smaller than the BW_{NN}, the second (and higher) eigenvalues become quite small. Depending on the scenario, this may or may not be a problem.

Because of its simplicity and potential performance improvement, diagonal loading is an attractive modification to the SMI algorithm in most applications. The disadvantage of fixed diagonal loading is that we need to have prior knowledge of the signal and interferer levels in order to select the appropriate value for σ_L^2. In many applications of interest we have enough prior information about the anticipated environment to choose an appropriate loading level.

Another technique for dealing with the errors in estimating the noise eigenvalues is to replace the estimates of the $(N-D)$ smallest eigenvalues by their average value,

$$\alpha \triangleq \frac{1}{N-D} \sum_{i=D+1}^{N} \hat{\lambda}_i. \tag{7.130}$$

This approach is used in the dominant mode rejection (DMR) beamformer that was discussed in Section 6.8.3. We study its adaptive behavior in Section 7.9.

7.3.4 Toeplitz Estimators

In Section 7.2.4, we indicated that there are techniques for estimating $\mathbf{S_x}$ that exploit the Toeplitz structure of $\mathbf{S_x}$. Fuhrmann [Fuh91] has shown that these techniques lead to a significant performance improvement in the SMI beamformer. The reader is referred to this reference for a discussion of the technique.

7.3.5 Summary

In this section, we have studied the SMI implementation of the MVDR and MPDR beamformer.

For MVDR beamformers, we defined a random variable ρ, which is the ratio of the $SINR_{smi}$ to the steady state $SINR_{mvdr}$. The probability density of ρ does not depend on the interference environment. For $\hat{\mathbf{S}}_\mathbf{n} = \mathbf{C_x}$, $E[\rho]$ was -1 dB for $K = 4N$. If we use FB averaging, $E[\rho] = -1$ dB for $K = 2N$ and we effectively double our sample support.

The performance can be improved by using diagonal loading. This technique, which consists of adding a diagonal matrix to either $\hat{\mathbf{S}}_{\mathbf{x}}$ or $\hat{\mathbf{S}}_{\mathbf{n}}$, works best in a strong interferer environment. Diagonal loading reduces the sidelobe level, improves the $SINR_{smi}$ performance, and allows $K < N$ sample support. Diagonal loading plays an important role in most beamformer implementations.

Similar results were obtained for the MPDR beamformer. However, in the presence of a strong signal the convergence was much slower. Diagonal loading provided more dramatic improvements for the MPDR model.

The SMI technique is effective in many applications. However, its computational requirement is a function of N^3, and it is a block algorithm. In the next section, we develop a recursive algorithm.

7.4 Recursive Least Squares (RLS)

In this section, we develop a recursive implementation of the sample matrix inversion algorithm. In order to do this efficiently, we first reformulate the MPDR beamformer and the MMSE beamformer as least squares problems. We do that in Section 7.4.1.

In Section 7.4.2, we develop a recursive algorithm for the MPDR beamformer. In Section 7.4.3 we develop a recursive algorithm for the least squares estimation (LSE) beamformer. Our discussion is adapted from the development in Chapter 13 of [Hay96].[9]

We should observe that the recursive algorithms that we develop in this section are adequate for our present purposes. In Section 7.5, we revisit the implementation problem and develop algorithms that are more computationally efficient and have better numerical stability.

7.4.1 Least Squares Formulation

The method of least squares estimation was invented by Gauss in 1795 in his studies of motion of heavenly bodies.[10] In 1912, Fisher [Fis12] introduced the maximum likelihood method, which leads to the same result if the observation noise is assumed to be Gaussian. Kolmogorov [Kol41a], [Kol41b] in 1941 and Wiener [Wie49] independently invented MMSE filter theory that

[9] An alternative method to develop the RLS algorithm is start with a deterministic Newton-type algorithm and find the stochastic version of it. We discuss this approach briefly in Section 7.7.

[10] Sorenson [Sor70] has provided a good discussion of the history of LSE and its evolution from Gauss to Swerling and Kalman. Our historical discussion is based on this reference.

we have studied in DEMT I [VT68], [VT01a]. In our study of the MMSE filter we utilized ensemble averages. By contrast, LSE utilizes time averages. We find that the least squares beamformers have the same form as the MPDR and MMSE beamformers, with the ensemble averages replaced by time averages.

In the middle 1950s, there was a sequence of papers dealing with recursive implementations of LSEs. These include Follin [CF56], Swerling [Swe58], [Swe59], Kalman and Bucy [KB61]. The culmination was the discrete time Kalman filter [Kal60], which has been applied to a wide range of applications in the last four decades. As Sorenson points out, Swerling's recursive algorithm predates Kalman's work and is essentially the same algorithm. We now derive the least squares estimator.

The output of a distortionless response beamformer is

$$Y(k) = D(k) + N(k), \quad k = 1, 2, \cdots, K, \tag{7.131}$$

where $D(k)$ is the desired signal. The estimation error is $N(k)$. In the least squares approach, we minimize a weighted summation of the squared error

$$\xi_N(K) = \sum_{k=1}^{K} \mu^{K-k} |N(k)|^2, \tag{7.132}$$

where μ is a positive constant less than 1. It provides an exponential weighting factor of the error so that the importance of past errors decrease as their distance from the current sample time K increases. This factor allows the beamformer to accommodate possible non-stationarities in the input. Typically μ is close to unity.

Due to the distortionless constraint, the minimization of $\xi_N(K)$ is equivalent to minimizing

$$\xi_Y(K) = \sum_{k=1}^{K} \mu^{K-k} |Y(k)|^2. \tag{7.133}$$

As in Chapter 6,

$$Y(k) = \mathbf{w}^H(K)\mathbf{X}(k), \tag{7.134}$$

and the distortionless constraint is

$$\mathbf{w}^H(K)\mathbf{v}_s = 1. \tag{7.135}$$

Note that $\mathbf{w}^H(K)$ is a function of K because it will adapt as we receive more data. We minimize $\xi_Y(K)$ subject to the constraint in (7.135). Define

$$F \triangleq \sum_{k=1}^{K} \mu^{K-k} \mathbf{w}^H(K)\mathbf{X}(k)\mathbf{X}^H(k)\mathbf{w}(K)$$

$$+\lambda\left[\mathbf{w}^H(K)\mathbf{v}_s - 1\right] + \lambda^*\left[\mathbf{v}_s^H\mathbf{w}(K) - 1\right], \tag{7.136}$$

or

$$F = \mathbf{w}^H(K)\mathbf{\Phi}(K)\mathbf{w}(K) + 2\mathrm{Re}\left[\lambda\left[\mathbf{w}^H(K) - 1\right]\right], \tag{7.137}$$

where

$$\mathbf{\Phi}(K) = \sum_{k=1}^{K} \mu^{K-k}\mathbf{X}(k)\mathbf{X}^H(k). \tag{7.138}$$

The matrix $\mathbf{\Phi}(K)$ is introduced to denote the exponential weighted sample spectral matrix, in contrast with the conventional sample spectral matrix $\mathbf{C_x}(K)$.

Substituting (7.138) into (7.136) gives

$$F = \mathbf{w}^H(K)\mathbf{\Phi}(K)\mathbf{w}(K) + \lambda\left[\mathbf{w}^H(K)\mathbf{v}_s - 1\right] + \lambda^*\left[\mathbf{v}_s^H\mathbf{w}(K) - 1\right]. \tag{7.139}$$

Taking the complex gradient with respect to $\mathbf{w}^H(K)$, setting the result to zero, and solving for the Lagrange multiplier gives

$$\boxed{\hat{\mathbf{w}}_{mpdr}(K) = \frac{\mathbf{\Phi}^{-1}(K)\mathbf{v}_s}{\mathbf{v}_s^H\mathbf{\Phi}^{-1}(K)\mathbf{v}_s} = \Lambda(K)\mathbf{\Phi}^{-1}(K)\mathbf{v}_s,} \tag{7.140}$$

where

$$\boxed{\Lambda(K) \triangleq \left[\mathbf{v}_s^H\mathbf{\Phi}^{-1}(K)\mathbf{v}_s\right]^{-1}.} \tag{7.141}$$

Note that

$$\xi_Y(K) = \left[\mathbf{v}_s^H\mathbf{\Phi}^{-1}(K)\mathbf{v}_s\right]^{-1} = \Lambda(K). \tag{7.142}$$

We see that the least squares distortionless response beamformer is the MPDR beamformer of Chapter 6 with the ensemble average replaced by a weighted time average.

Note that

$$E\left[\mathbf{\Phi}(K)\right] = \frac{1-\mu^K}{1-\mu}\mathbf{S_x}, \tag{7.143}$$

which is approximately

$$E\left[\mathbf{\Phi}(K)\right] \simeq \frac{1}{1-\mu}\mathbf{S_x}, \tag{7.144}$$

for large K, so $\mathbf{\Phi}(K)$ provides an asymptotically biased estimate of $\mathbf{S_x}$.

for large K, so $\mathbf{\Phi}(K)$ provides an asymptotically biased estimate of $\mathbf{S_x}$.

An alternative definition that provides an unbiased estimate is

$$\mathbf{\Phi}'(K) = \frac{1-\mu}{1-\mu^K} \sum_{k=1}^{K} \mu^{K-k} \mathbf{X}(k)\mathbf{X}^H(k), \tag{7.145}$$

which, for large K, reduces to

$$\mathbf{\Phi}'(K) = (1-\mu) \sum_{k=1}^{K} \mu^{K-k} \mathbf{X}(k)\mathbf{X}^H(k). \tag{7.146}$$

Using (7.145) in (7.140) gives the same result as before because it appears in both the numerator and denominator. However, when we add diagonal loading we have to take the $(1-\mu)$ factor into account.

The least squares solution analogous to the MMSE beamformer follows in a similar manner. We assume there is a desired response $D(k)$. The error at time k is

$$e(k) = D(k) - \mathbf{w}^H(K)\mathbf{X}(k), \quad k = 1, \cdots, K. \tag{7.147}$$

Note $\mathbf{w}^H(K)$ is the weight vector at K. We minimize

$$\begin{aligned} \xi_\mu(K) &= \sum_{k=1}^{K} \mu^{K-k} |e(k)|^2 \\ &= \sum_{k=1}^{K} \mu^{K-k} \left(D(k) - \mathbf{w}^H(K)\mathbf{X}(k)\right) \left(D(k)^* - \mathbf{X}^H(k)\mathbf{w}(K)\right). \end{aligned} \tag{7.148}$$

Taking the gradient with respect to $\mathbf{w}^H(K)$ and setting the result to zero, gives

$$\boxed{\hat{\mathbf{w}}_{lse}(K) = \mathbf{\Phi}^{-1}(K)\mathbf{\Phi}_{\mathbf{x}d^*}(K),} \tag{7.149}$$

where $\mathbf{\Phi}(K)$ is defined in (7.138) and

$$\boxed{\mathbf{\Phi}_{\mathbf{x}d^*}(K) \triangleq \textstyle\sum_{k=1}^{K} \mu^{K-k} \mathbf{X}(k) D^*(k).} \tag{7.150}$$

This result is the MMSE beamformer of Chapter 6, with the ensemble averages replaced by weighted time averages.

The output $Y(K)$ is

$$Y(K) = \hat{\mathbf{w}}_{lse}^H(K)\mathbf{X}(K). \tag{7.151}$$

In order to implement (7.149), we must generate $\mathbf{\Phi}_{\mathbf{x}d^*}(K)$. We discuss techniques for doing this later in this section.

7.4.2 Recursive Implementation

In order to implement $\hat{\mathbf{w}}_{mpdr}(K)$ recursively, we need an algorithm to find $\mathbf{\Phi}^{-1}(K)$ from $\mathbf{\Phi}^{-1}(K-1)$. From (7.138),

$$\mathbf{\Phi}(K) = \mu\mathbf{\Phi}(K-1) + \mathbf{X}(K)\mathbf{X}^H(K). \tag{7.152}$$

The desired iteration follows directly from the matrix inversion formula in (A.50)[11]

$$\mathbf{\Phi}^{-1}(K) = \mu^{-1}\mathbf{\Phi}^{-1}(K-1) - \frac{\mu^{-2}\mathbf{\Phi}^{-1}(K-1)\mathbf{X}(K)\mathbf{X}^H(K)\mathbf{\Phi}^{-1}(K-1)}{1+\mu^{-1}\mathbf{X}^H(K)\mathbf{\Phi}^{-1}(K-1)\mathbf{X}(K)}. \tag{7.153}$$

We now define

$$\mathbf{P}(K) = \mathbf{\Phi}^{-1}(K), \tag{7.154}$$

and

$$\mathbf{g}(K) = \frac{\mu^{-1}\mathbf{P}(K-1)\mathbf{X}(K)}{1+\mu^{-1}\mathbf{X}^H(K)\mathbf{P}(K-1)\mathbf{X}(K)}. \tag{7.155}$$

The choice of notation is deliberate because of the relationship between recursive least squares and Kalman filtering.

Using (7.154) and (7.155) in (7.153) gives

$$\mathbf{P}(K) = \mu^{-1}\mathbf{P}(K-1) - \mu^{-1}\mathbf{g}(K)\mathbf{X}^H(K)\mathbf{P}(K-1), \tag{7.156}$$

which is known as the Riccati equation.

Post-multiplying both sides of (7.156) by $\mathbf{X}(K)$ and using (7.155), one can show that

$$\mathbf{g}(K) = \mathbf{P}(K)\mathbf{X}(K) = \mathbf{\Phi}^{-1}(K)\mathbf{X}(K). \tag{7.157}$$

The vector $\mathbf{g}(K)$ is referred as the gain vector.

We now develop a recursive equation for $\hat{\mathbf{w}}_{mpdr}(K)$. Suppressing the subscript on $\hat{\mathbf{w}}(K)$, (7.140) can be written as

$$\hat{\mathbf{w}}(K) = \Lambda(K)\mathbf{P}(K)\mathbf{v}_s. \tag{7.158}$$

Using (7.156) in (7.158) gives

$$\begin{aligned}\hat{\mathbf{w}}(K) &= \Lambda(K)\left[\mu^{-1}\mathbf{P}(K-1) - \mu^{-1}\mathbf{g}(K)\mathbf{X}^H(K)\mathbf{P}(K-1)\right]\mathbf{v}_s \\ &= \left\{\frac{\Lambda(K)}{\mu\Lambda(K-1)}\left[\mathbf{I} - \mathbf{g}(K)\mathbf{X}^H(K)\right]\right\}\hat{\mathbf{w}}(K-1).\end{aligned} \tag{7.159}$$

---

[11] This approach is due originally to Baird [Bai74]. It is discussed in Hudson's book [Hud81], pp. 124–125, Compton's book [Com88], pp. 318–326, and Haykin's book [Hay96], pp. 566–571.

The term in curly brackets is an $N \times N$ matrix used to update $\hat{\mathbf{w}}(K-1)$.

The last step is to initialize the algorithm. Haykin [Hay96] suggests augmenting $\mathbf{\Phi}(K)$ with a white noise term. Thus,

$$\mathbf{\Phi}(K) = \sum_{k=1}^{K} \mu^{K-k} \mathbf{X}(k)\mathbf{X}^H(k) + \sigma_o^2 \mu^K \mathbf{I}, \tag{7.160}$$

where σ_o^2 is a small positive constant. This augmentation is just exponentially decaying diagonal loading. Then, for $K = 0$ (no snapshots received)

$$\mathbf{\Phi}(0) = \sigma_o^2 \mathbf{I}, \tag{7.161}$$

and

$$\mathbf{P}(0) = \frac{1}{\sigma_o^2}\mathbf{I}. \tag{7.162}$$

We can choose $\hat{\mathbf{w}}(0)$ to be any vector satisfying the distortionless constraint. We refer to this vector as the **quiescent** weight vector, $\mathbf{w}_q$.

In the time-domain case (equalizers), it is normal to make σ_o^2 very small and let $\hat{\mathbf{w}}(0) = \mathbf{0}$. However, a small σ_o^2 gives very poor beam patterns (distorted main lobes and high sidelobes) for small K. If we assume the array is in operation prior to adaption, then $\hat{\mathbf{w}}(0) = \mathbf{0}$ is not a useful initial weighting. We normally initialize the algorithm with

$$\hat{\mathbf{w}}(0) = \frac{\mathbf{v}_s}{N}, \tag{7.163}$$

or a $\mathbf{w}_q$ with a better sidelobe pattern that satisfies the distortionless criterion.

The effect of the diagonal loading term, $\sigma_o^2\mathbf{I}$, decays rapidly. Taking the expectation of (7.160) gives

$$E\left[\mathbf{\Phi}(K)\right] = \left[\frac{1-\mu^K}{1-\mu}\right]\mathbf{S}_{\mathbf{x}} + \mu^K \sigma_o^2 \mathbf{I}. \tag{7.164}$$

Defining,

$$\tilde{\mathbf{\Phi}}(K) = \left[\frac{1-\mu}{1-\mu^K}\right]\mathbf{\Phi}(K), \tag{7.165}$$

(7.164) can be rewritten as

$$E\left[\tilde{\mathbf{\Phi}}(K)\right] = \mathbf{S}_{\mathbf{x}} + \left[\frac{1-\mu}{1-\mu^K}\right]\mu^K \sigma_o^2 \mathbf{I}. \tag{7.166}$$

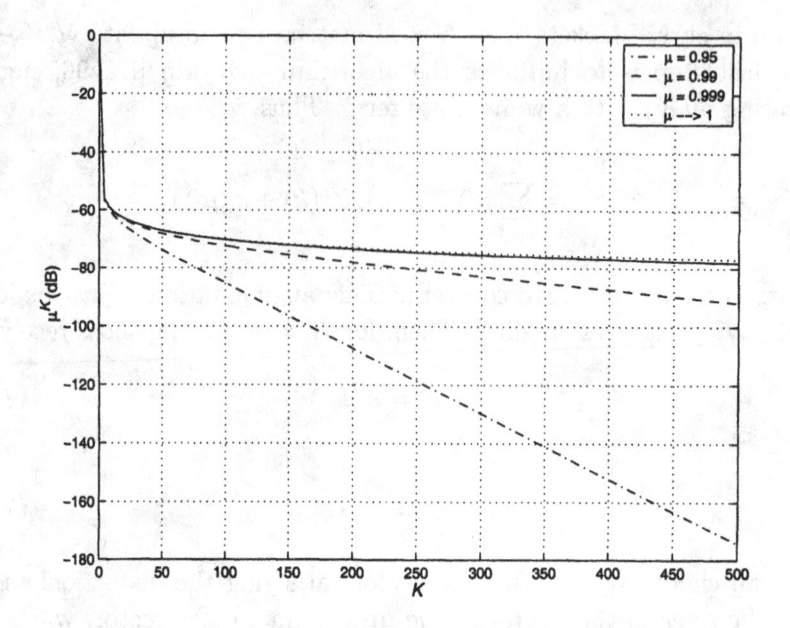

Figure 7.20: Diagonal loading decay in RLS algorithm.

In Figure 7.20, we plot the diagonal loading term as a function of K. Note that as μ approaches unity, we can approximate the fraction as

$$\left[\frac{1-\mu}{1-\mu^K}\right] \simeq \frac{1}{K}. \tag{7.167}$$

Because of this rapid decay, we may have to add additional diagonal loading to maintain robustness. We will revisit that issue later.

Sayed and Kailath [SK94] pointed out that, with this choice of $\mathbf{P}(0)$, we are actually minimizing

$$\min_{\mathbf{w}(K)} \left[\sigma_o^2 \mu^K \parallel \mathbf{w}(K) \parallel^2 + \sum_{k=1}^{K} \mu^{K-k} |Y(k)|^2\right], \tag{7.168}$$

where $Y(k)$ is given by (7.134). Note that the first term is the norm of the weight vector multiplied by an exponentially decreasing constant. We will encounter the first term in various contexts.

The steps can be summarized as:

Initialize the algorithm with

$$\mathbf{P}(0) = \frac{1}{\sigma_o^2}\mathbf{I}, \tag{7.169}$$

$$\hat{\mathbf{w}}(0) = \frac{\mathbf{v}_s}{N}, \tag{7.170}$$

or a $\mathbf{w}_q$ with a better sidelobe pattern that satisfies the distortionless criterion. Note that (7.169) provides the initialization for $\Lambda(0)$ by using (7.169) in (7.173).

At each snapshot, $K = 1, 2, \cdots$, compute

$$\mathbf{g}(K) = \frac{\mu^{-1}\mathbf{P}(K-1)\mathbf{X}(K)}{1 + \mu^{-1}\mathbf{X}^H(K)\mathbf{P}(K-1)\mathbf{X}(K)}, \tag{7.171}$$

$$\mathbf{P}(K) = \mu^{-1}\mathbf{P}(K-1) - \mu^{-1}\mathbf{g}(K)\mathbf{X}^H(K)\mathbf{P}(K-1), \tag{7.172}$$

$$\Lambda(K) = \left[\mathbf{v}_s^H\mathbf{P}(K)\mathbf{v}_s\right]^{-1}, \tag{7.173}$$

and

$$\hat{\mathbf{w}}_{mpdr}(K) = \frac{\Lambda(K)}{\mu\Lambda(K-1)}\left[\mathbf{I} - \mathbf{g}(K)\mathbf{X}^H(K)\right]\hat{\mathbf{w}}_{mpdr}(K-1). \tag{7.174}$$

Then,

$$Y(K) = \hat{\mathbf{w}}_{mpdr}^H(K)\mathbf{X}(K). \tag{7.175}$$

We can also write (7.174) as

$$\hat{\mathbf{w}}_{mpdr}(K) = \frac{\Lambda(K)}{\mu\Lambda(K-1)}\left[\hat{\mathbf{w}}_{mpdr}(K-1) - \mathbf{g}(K)\tilde{Y}^*(K)\right], \tag{7.176}$$

where

$$\tilde{Y}(K) \triangleq \hat{\mathbf{w}}_{mpdr}^H(K-1)\mathbf{X}(K). \tag{7.177}$$

The expressions in (7.174) and (7.176) are identical in the absence of numerical errors. In practice, we normally use (7.176) so that the RLS algorithm is operating as a closed-loop system, as shown in Figure 7.21. This implementation is due to Baird [Bai73] (e.g., [Hud79] and p.127 in [Hud81]).

We consider the following example to illustrate the behavior.

Example 7.4.1 (continuation, Examples 6.3.2–6.3.5, 7.3.2)

Consider a standard 10-element linear array in which the desired signal arrives from $u_s = 0$, and two equal-power interfering signals arrive from $u_I = 0.29$ and 0.45 with an *INR* = 20 dB. We use a recursive LSE to implement the MPDR beamformer ((7.169)–(7.177)).

In Figure 7.22, we plot the average output $SINR_o$ versus K for $\mu = 0.99$ and $\mu = 0.999$. We also show the SMI beamformer performance with no diagonal loading for comparison. In Figure 7.22(a), the initial loading, σ_o^2/σ_w^2, equals −10 dB. For $\mu = 0.999$, the behavior of RLS and SMI is essentially equal. For $\mu = 0.99$, the RLS performance levels off because we are effectively using about $((1-\mu)^{-1} = 100)$ snapshots. In Figure 7.22(b), the initial

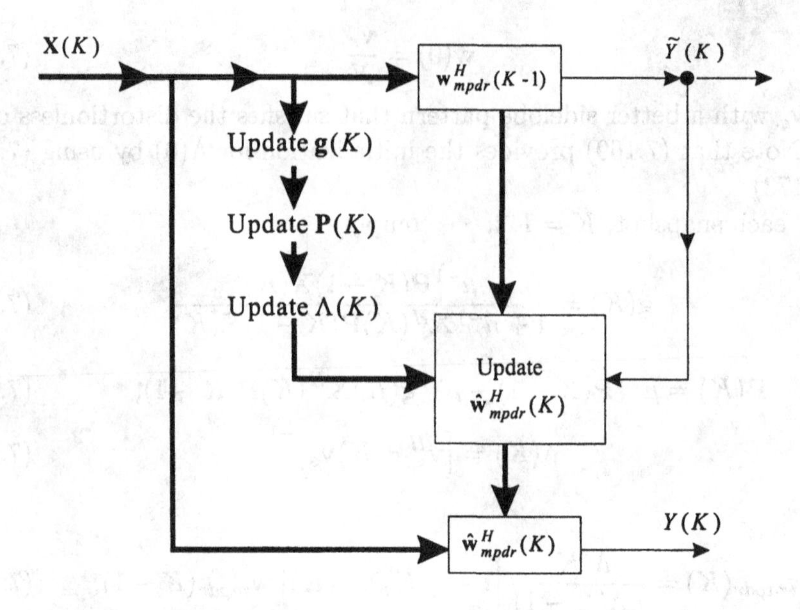

Figure 7.21 MPDR-RLS algorithm: closed-loop implementation.

loading is 0 dB. The increased loading causes RLS to be better for $K < 40$. For larger K, it has decayed and the performance of RLS with $\mu = 0.999$ and SMI are the same. In Figure 7.22(c), the initial loading is 10 dB. Now RLS is better for $K < 200$.

In Figure 7.23, we compare the performance for various *SNR* with $\sigma_o^2/\sigma_w^2 = 10$ dB.

We observe that the RLS algorithm does not perform as well as the diagonally loaded SMI algorithm. The reason is that the diagonal loading in the RLS algorithm decreases exponentially. This suggests defining a different augmented matrix,

$$\mathbf{\Phi}_a(K) = \sum_{k=1}^{K} \mu^{K-k}\mathbf{X}(k)\mathbf{X}^H(k) + \sigma_L^2\mathbf{I}, \tag{7.178}$$

To illustrate the behavior we use an algorithm that accomplishes (7.178), but is not practical to implement in practice.

We replace $\mathbf{\Phi}(K)$ in (7.152) with

$$\mathbf{\Phi}_a(K) = \mu\mathbf{\Phi}_a(K-1) + \mathbf{X}(K)\mathbf{X}^H(K) + (1-\mu)\sigma_L^2\mathbf{I}, \tag{7.179}$$

and write (7.156) as

$$\mathbf{P}_1(K) = \mu^{-1}\mathbf{P}(K-1) - \mu^{-1}\mathbf{g}(K)\mathbf{X}^H(K)\mathbf{P}(K-1). \tag{7.180}$$

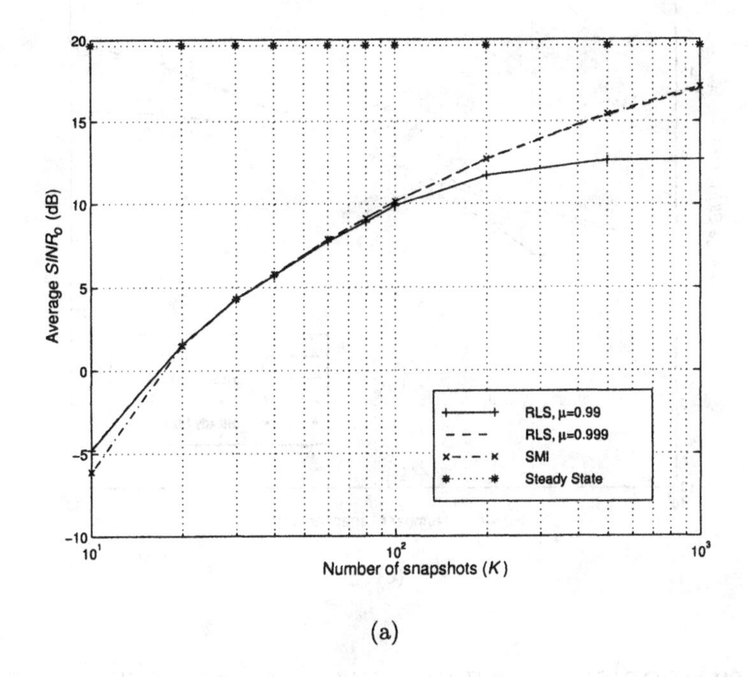

Average $SINR_o$ (dB)
Number of snapshots (K)
RLS, μ=0.99
RLS, μ=0.999
SMI
Steady State

(a)

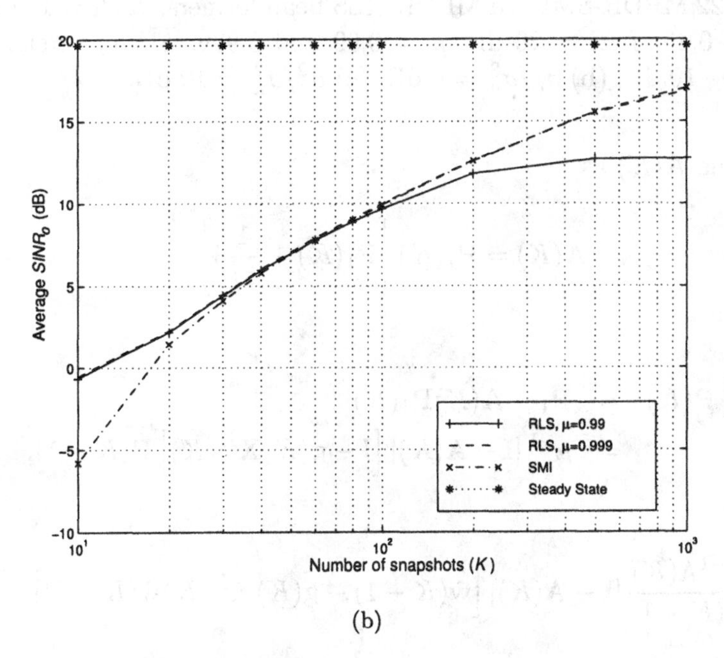

Average $SINR_o$ (dB)
Number of snapshots (K)
RLS, μ=0.99
RLS, μ=0.999
SMI
Steady State

(b)

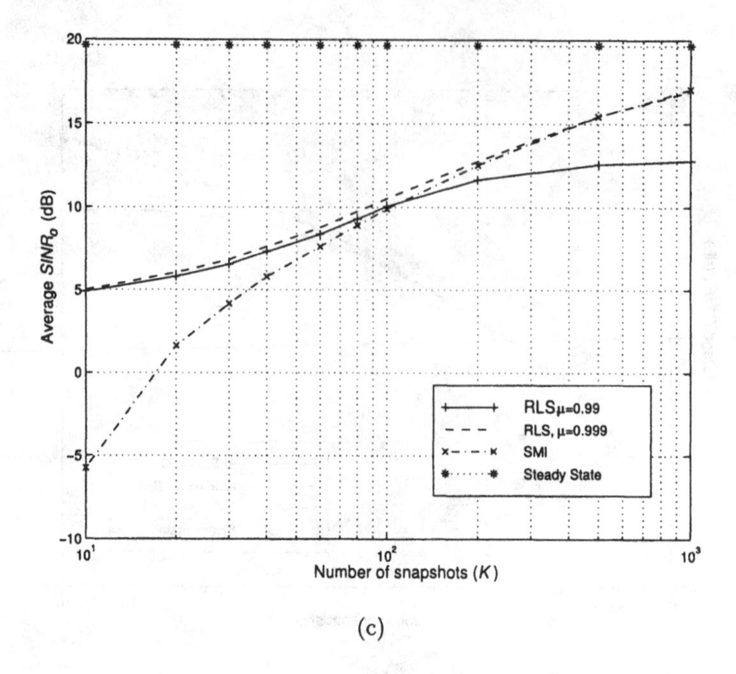

(c)

Figure 7.22 MPDR-SMI and MPDR-RLS beamformers: $SNR = 10$ dB, $u_I =$ 0.29 and 0.45, $INR = 20$ dB, $\mu = 0.99$ and 0.999; $\overline{SINR}_o$ versus K: (a) $\sigma_o^2/\sigma_w^2 = -10$ dB; (b) $\sigma_o^2/\sigma_w^2 = 0$ dB; (c) $\sigma_o^2/\sigma_w^2 = 10$ dB.

Now define $\mathbf{A}(K)$ as

$$\mathbf{A}(K) = \mathbf{P}_1(K)\left[\mathbf{P}_1(K) + \frac{1}{\sigma_L^2}\mathbf{I}\right]^{-1}. \tag{7.181}$$

Then,

$$\begin{aligned}\mathbf{P}(K) &= \mathbf{P}_1 - \mathbf{A}(K)\mathbf{P}_1(K) \\ &= \mu^{-1}\left[\mathbf{I} - \mathbf{A}(K)\right]\left[\mathbf{I} - \mathbf{g}(K)\mathbf{X}^H(K)\right]\mathbf{P}(K-1),\end{aligned} \tag{7.182}$$

with

$$\hat{\mathbf{w}} = \frac{\mu^{-1}\Lambda(K)}{\Lambda(K-1)}\left[\mathbf{I} - \mathbf{A}(K)\right]\left[\hat{\mathbf{w}}(K-1) - \mathbf{g}(K)\mathbf{X}^H(K)\hat{\mathbf{w}}(K-1)\right]. \tag{7.183}$$

The algorithm is not practical because the inverse in (7.181) is required at each iteration. However, it provides a performance reference. An alternative

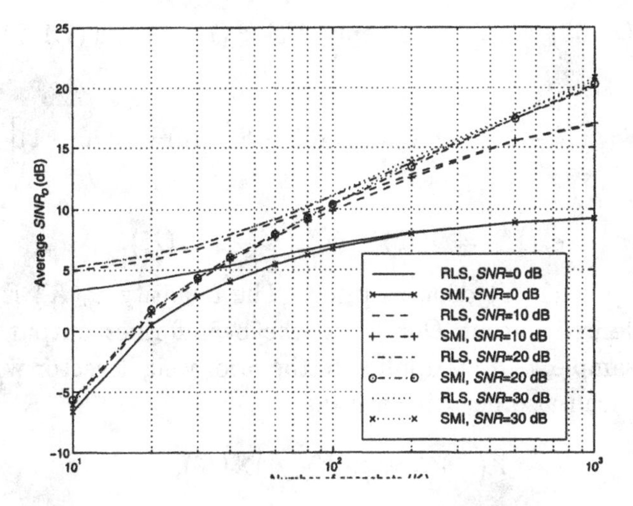

Figure 7.23: MPDR-SMI and MPDR-RLS beamformers: $u_I = 0.29, 0.45$, $INR = 20$ dB, $\mu = 0.999$, $\sigma_o^2/\sigma_w^2 = 10$ dB, $SNR = 0, 10, 20, 30$ dB; $\overline{SINR}_o$ versus K.

approach is to use a moving window of length $1/(1-\mu)$. The moving window counterpart to (7.178) is easy to update[12] (see Problem 7.4.24).

In the GSC structure, we can utilize an approximate procedure. We develop it in Section 7.4.4.

7.4.3 Recursive Implementation of LSE Beamformer

Using a similar approach one can derive a recursive implementation of the LSE beamformer discussed in (7.149) and (7.150). The equations specifying $\hat{\mathbf{w}}_{lse}(K)$ are

$$\hat{\mathbf{w}}_{lse}(K) = \mathbf{\Phi}^{-1}(K)\mathbf{\Phi}_{\mathbf{x}d^*}(K) = \mathbf{P}(K)\mathbf{\Phi}_{\mathbf{x}d^*}(K) \tag{7.184}$$

and

$$\begin{aligned} \mathbf{\Phi}_{\mathbf{x}d^*}(K) &= \sum_{k=1}^{K} \mu^{K-k}\mathbf{X}(k)D^*(k) \\ &= \mathbf{X}(K)D^*(K) + \mu\mathbf{\Phi}_{\mathbf{x}d^*}(K-1). \end{aligned} \tag{7.185}$$

[12]This alternative was suggested by P. Stoica (private communication).

The results in (7.169)–(7.172) still apply. Using (7.185) and (7.172) in (7.184) gives

$$\hat{\mathbf{w}}_{lse}(K) = \hat{\mathbf{w}}_{lse}(K-1) + \mathbf{g}(K)\left[D^*(K) - \mathbf{X}^H(K)\hat{\mathbf{w}}_{lse}(K-1)\right]. \quad (7.186)$$

Now define

$$e_p(K) = D(K) - \hat{\mathbf{w}}_{lse}^H(K-1)\mathbf{X}(K). \quad (7.187)$$

The subscript p in $e_p(K)$ denotes prior. The quantity $e_p(K)$ is the error between the desired output $D(K)$ and the beamformer output when the current input sample $\mathbf{X}(K)$ is applied to the prior weight vector $\hat{\mathbf{w}}_{lse}^H(K-1)$. The prior error can also be written using

$$\tilde{Y}(K) = \hat{\mathbf{w}}_{lse}^H(K-1)\mathbf{X}(K), \quad (7.188)$$

to provide a closed-loop implementation. Using (7.187) in (7.186) gives

$$\hat{\mathbf{w}}_{lse}(K) = \hat{\mathbf{w}}_{lse}(K-1) + \mathbf{g}(K)e_p^*(K), \quad (7.189)$$

and

$$Y(K) = \hat{\mathbf{w}}_{lse}^H(K)\mathbf{X}(K). \quad (7.190)$$

The algorithm can be summarized:

1. Initialize the algorithm with

$$\mathbf{P}(0) = \frac{1}{\sigma_o^2}\mathbf{I}, \quad (7.191)$$

$$\hat{\mathbf{w}}(0) = \frac{\mathbf{v}_s}{N}, \quad (7.192)$$

or a $\mathbf{w}_q$ with a better sidelobe pattern.

2. At each snapshot, $K = 1, 2, \cdots$, compute

$$\mathbf{g}(K) = \frac{\mu^{-1}\mathbf{P}(K-1)\mathbf{X}(K)}{1 + \mu^{-1}\mathbf{X}^H(K)\mathbf{P}(K-1)\mathbf{X}(K)}, \quad (7.193)$$

and

$$\mathbf{P}(K) = \mu^{-1}\mathbf{P}(K-1) - \mu^{-1}\mathbf{g}(K)\mathbf{X}^H(K)\mathbf{P}(K-1). \quad (7.194)$$

3. Compute $e_p(K)$ using (7.187).

4. Compute $\hat{\mathbf{w}}_{lse}(K)$ using (7.189).

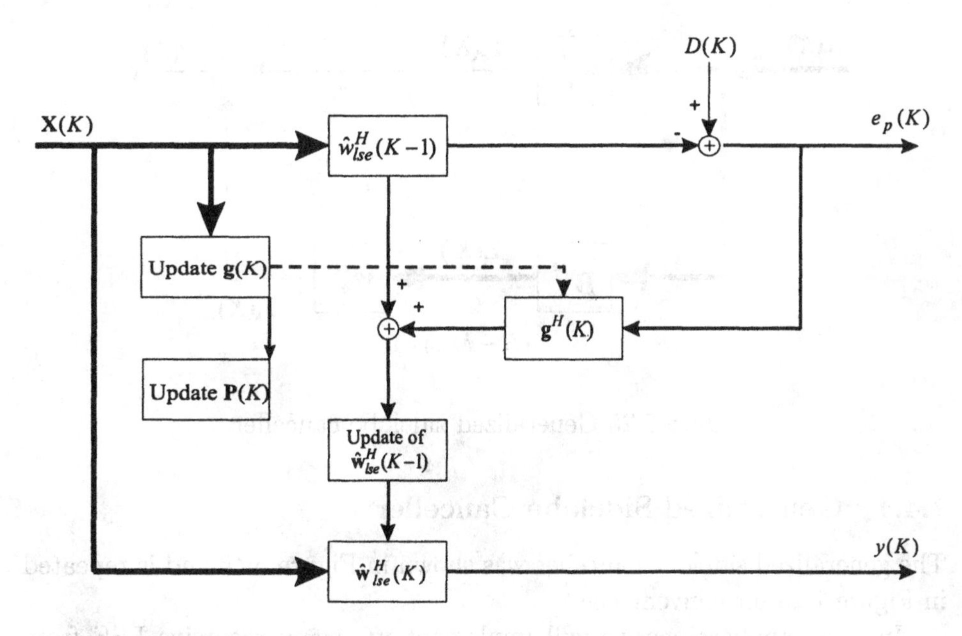

Figure 7.24 Implementation of LSE beamformer.

5. Compute the output $Y(K)$ using (7.190).

Note that in the direct form implementation of the LSE beamformer, the desired signal $D(k), k = 1, 2, \cdots, K$ must be known at the receiving antenna. In a typical communications system implementation, the desired signal $D(k)$ could be supplied by a training sequence transmitted at the beginning of a frame. We can then switch to a decision directed mode for continuing operation.

Note that, in contrast to the MVDR and MPDR beamformers, the receiver does not need to know the direction of arrival of the signal. Figure 7.24 shows the implementation of the algorithm.

In our discussion in the text, the emphasis is on MPDR and MVDR beamformers. When we implement these beamformers as generalized sidelobe cancellers (see Section 6.7.3 and Figure 6.46), the LSE beamformer is the adaptive element in the lower path and the desired signal is the output of the quiescent vector processor in the upper path. We discuss this implementation in detail in the next section.

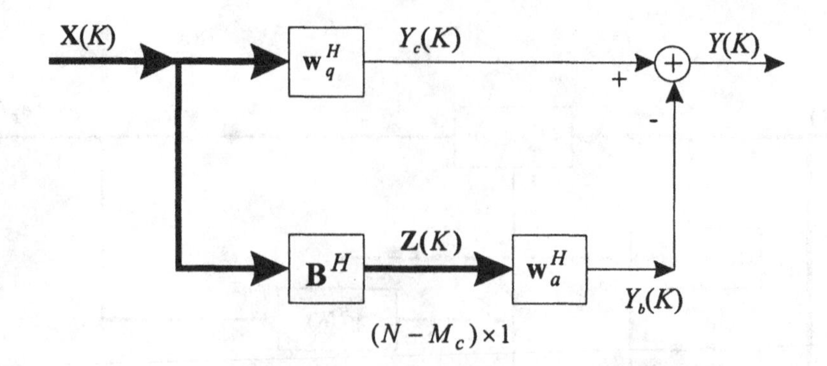

Figure 7.25 Generalized sidelobe canceller.

7.4.4 Generalized Sidelobe Canceller

The generalized sidelobe canceller was shown in Figure 6.46 and is repeated in Figure 7.25 for convenience.

In most applications we will implement $\mathbf{w}_a$ using recursive LSE from Section 7.4.3 or an LMS algorithm that we will discuss in Section 7.7. The adaptive processor $\mathbf{w}_a^H$ is a least squares estimator (Section 7.4.3), with $Y_c(k)$ as the desired signal and $\mathbf{Z}(k)$, the output of the blocking matrix, as the input.

From (7.138),

$$\mathbf{\Phi}_{\mathbf{x}}(K) = \sum_{k=1}^{K} \mu^{K-k}\mathbf{X}(k)\mathbf{X}^H(k). \tag{7.195}$$

Then,

$$\mathbf{\Phi}_{\mathbf{z}}(K) = \sum_{k=1}^{K} \mu^{K-k}\mathbf{Z}(k)\mathbf{Z}^H(k) = \mathbf{B}^H\mathbf{\Phi}_{\mathbf{x}}(K)\mathbf{B}, \tag{7.196}$$

and

$$\mathbf{\Phi}_{\mathbf{z}y_c^*}(K) = \sum_{k=1}^{K} \mu^{K-k}\mathbf{Z}(k)\mathbf{Y}_c^*(k) = \mathbf{B}^H\mathbf{\Phi}_{\mathbf{x}}(K)\mathbf{w}_q. \tag{7.197}$$

We define

$$\mathbf{P}_{\mathbf{z}}(K) = \mathbf{\Phi}_{\mathbf{z}}^{-1}(K). \tag{7.198}$$

For this case the RLS algorithm is adapted from (7.193)–(7.194) and (7.187)–(7.190):

$$\mathbf{g}_{\mathbf{z}}(K) = \frac{\mu^{-1}\mathbf{P}_{\mathbf{z}}(K-1)\mathbf{Z}(K)}{1+\mu^{-1}\mathbf{Z}^H(K)\mathbf{P}_{\mathbf{z}}(K-1)\mathbf{Z}(K)}. \tag{7.199}$$

$$\mathbf{P}_{\mathbf{z}}(K) = \mu^{-1}\mathbf{P}_{\mathbf{z}}(K-1) - \mu^{-1}\mathbf{g}_{\mathbf{z}}(K)\mathbf{Z}^H(K)\mathbf{P}_{\mathbf{z}}(K-1). \tag{7.200}$$

$$e_p(K) = Y_c(K) - \hat{\mathbf{w}}_a^H(K-1)\mathbf{Z}(K). \tag{7.201}$$

$$\hat{\mathbf{w}}_a(K) = \hat{\mathbf{w}}_a(K-1) + \mathbf{g}_{\mathbf{z}}(K)e_p^*(K). \tag{7.202}$$

The result in (7.202) can also be written as

$$\hat{\mathbf{w}}_a(K) = \hat{\mathbf{w}}_a(K-1) + \mathbf{g}_{\mathbf{z}}(K)\left[Y_c^*(K) - \mathbf{Z}^H(K)\hat{\mathbf{w}}_a(K-1)\right], \tag{7.203}$$

or

$$\hat{\mathbf{w}}_a(K) = \hat{\mathbf{w}}_a(K-1) + \mathbf{g}_{\mathbf{z}}(K)\left[Y_c^*(K) - \tilde{Y}_b^*(K)\right]. \tag{7.204}$$

The initial conditions are

$$\mathbf{P}_{\mathbf{z}}(0) = \frac{1}{\sigma_o^2}\left[\mathbf{B}^H\mathbf{B}\right]^{-1} = \frac{1}{\sigma_o^2}\mathbf{I}_{(N-M_c)}. \tag{7.205}$$

$$\hat{\mathbf{w}}_a(0) = \mathbf{0}. \tag{7.206}$$

Note that we are operating on an $N - M_c$ vector which provides a computational advantage.

In order to include fixed loading, note that

$$\mathbf{g}_{\mathbf{z}}(K) = \mathbf{P}_{\mathbf{z}}(K)\mathbf{Z}(K). \tag{7.207}$$

Then we can write

$$\hat{\mathbf{w}}_a(K) = \hat{\mathbf{w}}_a(K-1) + \mathbf{P}_{\mathbf{z}}(K)\mathbf{Z}(K)Y_c^*(K) - \mathbf{P}_{\mathbf{z}}(K)\mathbf{Z}(K)\mathbf{Z}^H(K)\hat{\mathbf{w}}_a(K-1). \tag{7.208}$$

At each iteration, there is a term of the form $\mathbf{Z}(K)\mathbf{Z}^H(K)$ to which a diagonal matrix can be added. The fixed diagonal loading RLS update becomes

$$\begin{aligned}\hat{\mathbf{w}}_a(K) &= \hat{\mathbf{w}}_a(K-1) + \mathbf{P}_{\mathbf{z}}(K)\mathbf{Z}(K)Y_c^*(K) - \mathbf{P}_{\mathbf{z}}(K) \times \\ &\quad \left[\mathbf{Z}(K)\mathbf{Z}^H(K) + \sigma_L^2\mathbf{I}\right]\hat{\mathbf{w}}_a(K-1).\end{aligned} \tag{7.209}$$

This update equation can be written as

$$\hat{\mathbf{w}}_a(K) = \hat{\mathbf{w}}_a(K-1) + \mathbf{g}_{\mathbf{z}}(K)e_p^*(K) - \sigma_L^2\mathbf{P}_{\mathbf{z}}(K)\hat{\mathbf{w}}_a(K-1), \tag{7.210}$$

or

$$\hat{\mathbf{w}}_a(K) = \left[\mathbf{I} - \sigma_L^2\mathbf{P}_{\mathbf{z}}(K)\right]\hat{\mathbf{w}}_a(K-1) + \mathbf{g}_{\mathbf{z}}(K)e_p^*(K). \tag{7.211}$$

Note that this approach is ad hoc in that it does not actually add loading to the sample spectral matrix.

We consider two examples to illustrate the performance.

Example 7.4.2 (continuation, Example 7.4.1)

Consider the same model as in Example 7.4.1. We repeat the simulation using the same data set and parameter values. We incorporate diagonal loading using (7.209). We also show the results using the diagonal loading technique in (7.179)–(7.183) (denoted by RLS, FLx). The results are shown in Figure 7.26.

For $\mu = 0.999$, the (impractical) fixed loading scheme has the same performance as the diagonally loaded SMI algorithm. The GSC implementation in (7.211) is only slightly worse and is straightforward to realize. For $\mu = 0.99$, the results have the same behavior for $K \leq 100$. They diverge for $K > 100$ because of the smaller μ.

Example 7.4.3 (continuation, Example 7.3.3)

Consider a standard 10-element linear array. We use an LCMP beamformer with derivative constraints. Assuming the signal impinges from $u_s = 0$,

$$\mathbf{C} = \left[\begin{array}{c:c:c} \mathbf{1} & \dot{\mathbf{v}}_s(0) & \ddot{\mathbf{v}}_s(0) \end{array}\right] \tag{7.212}$$

and

$$\mathbf{g}^H = \left[\begin{array}{c:c:c} 1 & 0 & \ddot{B}_c(0) \end{array}\right]. \tag{7.213}$$

We use a blocking matrix consisting of the first seven columns of orth $\left[\mathbf{P}_{\mathbf{C}}^{\perp}\right]$. We implement the recursion given in (7.199)–(7.206).

There are two equal-power uncorrelated interferers impinging on the array from $u_I = 0.29, 0.45$. We consider the following parameter values:

(i) $SNR = 10$ dB

(ii) $INR = 20$ dB

(iii) $\sigma_o^2/\sigma_w^2 = 10$.

The results are shown in Figure 7.27. We see that the SMI and RLS implementations have essentially the same performance. The ad hoc fixed loading technique works well in environments where the $SNR < INR$ and $LNR(\sigma_L^2/\sigma_w^2)$ is chosen properly. The disadvantage is that some prior knowledge of the environment is required to choose the appropriate value of σ_L^2. In the next section, we revisit the quadratic constraint approach in Section 6.10 and derive a data-dependent variable loading algorithm.

7.4.5 Quadratically Constrained RLS

The RLS algorithm is implemented using a GSC configuration. A quadratic constraint is imposed. From (6.559)

$$\tilde{\mathbf{w}}_a = \left[\mathbf{S}_{\mathbf{z}} + \beta \mathbf{I}\right]^{-1} \mathbf{p}_{\mathbf{z}}, \tag{7.214}$$

and β is chosen so that

$$\tilde{\mathbf{w}}_a^H \tilde{\mathbf{w}}_a \leq \alpha^2. \tag{7.215}$$

We develop an approximate technique so that the constraint equation (7.215) does not have to be solved at each step.

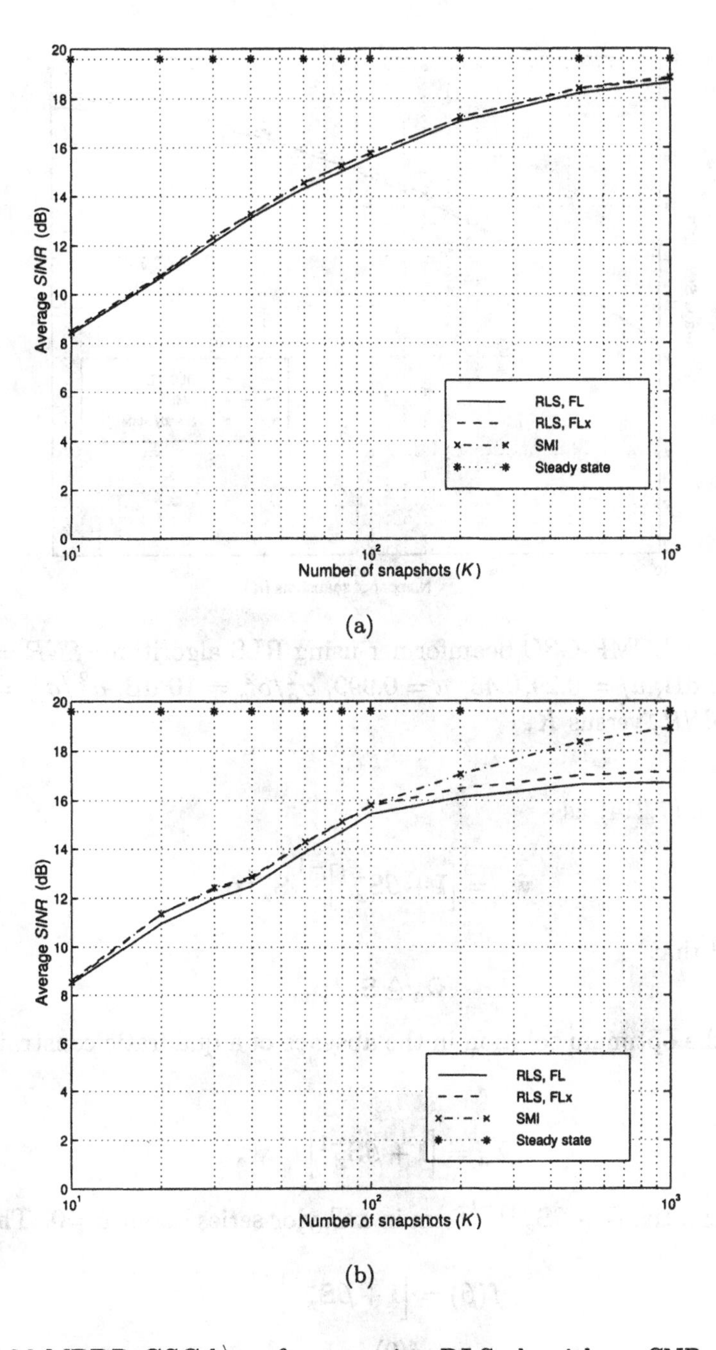

Figure 7.26 MPDR-GSC beamformer using RLS algorithm: $SNR = 10$ dB, $INR = 20$ dB, $u_I = 0.29, 0.45$, $\sigma_o^2/\sigma_w^2 = 10$ dB, $\sigma_L^2/\sigma_w^2 = 10$ dB; average $SINR_o$ versus K. (a) $\mu = 0.999$; (b) $\mu = 0.99$.

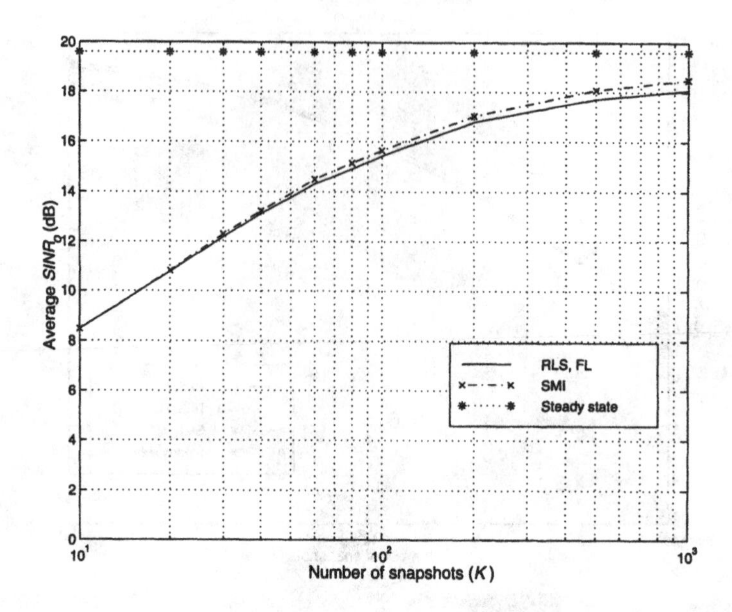

Figure 7.27: LCMP-GSC beamformer using RLS algorithm: $SNR = 10$ dB, $INR = 20$ dB, $u_I = 0.29, 0.45$, $\mu = 0.999$, $\sigma_o^2/\sigma_w^2 = 10$ dB, $\sigma_L^2/\sigma_w^2 = 10$ dB; average $SINR_o$ versus K.

Rewrite (7.214) as

$$\widetilde{\mathbf{w}}_a = \left[\mathbf{I} + \beta \mathbf{S}_\mathbf{z}^{-1}\right]^{-1} \mathbf{S}_\mathbf{z}^{-1}\mathbf{p}_\mathbf{z}, \tag{7.216}$$

and recall that

$$\widehat{\mathbf{w}}_a \triangleq \mathbf{S}_\mathbf{z}^{-1}\mathbf{p}_\mathbf{z}, \tag{7.217}$$

which is the optimum solution in the absence of a quadratic constraint ($\beta = 0$).

Then

$$\widetilde{\mathbf{w}}_a = \left[\mathbf{I} + \beta \mathbf{S}_\mathbf{z}^{-1}\right]^{-1} \widehat{\mathbf{w}}_a. \tag{7.218}$$

Now expand the $\left[\mathbf{I} + \beta \mathbf{S}_\mathbf{z}^{-1}\right]^{-1}$term in a Taylor series about β =0. The terms are

$$f(\beta) = \left[\mathbf{I} + \beta \mathbf{S}_\mathbf{z}^{-1}\right]^{-1}, \tag{7.219}$$

$$f(0) = \mathbf{I}, \tag{7.220}$$

and

$$f'(\beta) = -\left[\mathbf{I} + \beta \mathbf{S}_\mathbf{z}^{-1}\right]^{-2} \mathbf{S}_\mathbf{z}^{-1}, \tag{7.221}$$

$$f'(0) = -\mathbf{S}_{\mathbf{z}}^{-1}. \tag{7.222}$$

Retaining the first two terms,

$$\left[\mathbf{I} + \beta\mathbf{S}_{\mathbf{z}}^{-1}\right]^{-1} \simeq \mathbf{I} - \beta\mathbf{S}_{\mathbf{z}}^{-1}. \tag{7.223}$$

Using (7.223) in (7.218) gives

$$\widetilde{\mathbf{w}}_a \simeq \left[\mathbf{I} - \beta\mathbf{S}_{\mathbf{z}}^{-1}\right]\widehat{\mathbf{w}}_a = \widehat{\mathbf{w}}_a - \beta\mathbf{S}_{\mathbf{z}}^{-1}\widehat{\mathbf{w}}_a. \tag{7.224}$$

Defining

$$\mathbf{v} \triangleq \mathbf{S}_{\mathbf{z}}^{-1}\widehat{\mathbf{w}}_a, \tag{7.225}$$

(7.224) becomes

$$\widetilde{\mathbf{w}}_a = \widehat{\mathbf{w}}_a - \beta\mathbf{v}. \tag{7.226}$$

Using (7.226) in the constraint equation (7.215) gives

$$\widetilde{\mathbf{w}}_a^H\widetilde{\mathbf{w}}_a = (\widehat{\mathbf{w}}_a - \beta\mathbf{v})^H(\widehat{\mathbf{w}}_a - \beta\mathbf{v}) = \alpha^2, \tag{7.227}$$

which is a quadratic equation in β^2,

$$\beta^2(\mathbf{v}^H\mathbf{v}) + \beta\left(-2Re(\mathbf{v}^H\widehat{\mathbf{w}}_a)\right) + (\widehat{\mathbf{w}}_a^H\widehat{\mathbf{w}}_a - \alpha^2) = 0. \tag{7.228}$$

Define

$$a = \mathbf{v}^H\mathbf{v}, \tag{7.229}$$

$$b = -2Re(\mathbf{v}^H\widehat{\mathbf{w}}_a) = -2\widehat{\mathbf{w}}_a^H\mathbf{S}_{\mathbf{z}}^{-1}\widehat{\mathbf{w}}_a, \tag{7.230}$$

and

$$c = \widehat{\mathbf{w}}_a^H\widehat{\mathbf{w}}_a - \alpha^2. \tag{7.231}$$

Then, using the equality sign in (7.228),

$$\beta = \frac{-b \pm \sqrt{b^2 - 4ac}}{2a}. \tag{7.232}$$

we observe the following characteristics:

(i) If $\widehat{\mathbf{w}}_a^H\widehat{\mathbf{w}}_a \leq \alpha^2$, no diagonal loading is needed and we set $\beta = 0$. Therefore, $c > 0$ for the cases where a non-zero β is needed.

(ii) $b \leq 0$.

(iii) $a \geq 0$.

(iv) $(b^2 - 4ac)$ may be positive or negative.

(v) If $b^2 - 4ac > 0$, there are two real positive solutions.

(vi) If $b^2 - 4ac < 0$, there are two complex conjugate solutions whose real part $(-b/2a)$ is positive.

If condition (v) applies, choose the smallest value. This choice causes $\tilde{\mathbf{w}}_a$ to be closest to $\hat{\mathbf{w}}_a$. The resulting solution meets the constraint.

If condition (vi) applies, choose the real part,

$$\beta = \frac{-b}{2a}. \tag{7.233}$$

In this case, the solution does not meet the constraint but, for a vector of the form in (7.226), it is the closest. It is convenient to write β as

$$\beta = \frac{-b - Re(\sqrt{b^2 - 4ac})}{2a}, \tag{7.234}$$

which applies to both conditions (v) and (vi).

We refer to this algorithm as the variable loading algorithm. The next step is to apply the variable loading algorithm at each step in the RLS iteration. Let $\tilde{\mathbf{w}}_a(K)$ denote the quadratically constrained adaptive weight vector. The steps in the variable loading algorithm can be summarized:

1. Compute the standard RLS update in (7.199)-(7.200) and (7.203). Let $\hat{\mathbf{w}}_a(K)$ denote the standard weight vector update:

$$\hat{\mathbf{w}}_a(K) = \tilde{\mathbf{w}}_a(K-1) + \mathbf{g}_{\mathbf{z}}(K)\left[Y_c^*(K) - \mathbf{Z}^H(K)\tilde{\mathbf{w}}_a(K-1)\right]. \tag{7.235}$$

2. Test the norm of $\hat{\mathbf{w}}_a(K)$. If $\| \hat{\mathbf{w}}_a(K) \|^2 \leq \alpha^2$, then

$$\tilde{\mathbf{w}}_a(K) = \hat{\mathbf{w}}_a(K). \tag{7.236}$$

3. If $\| \hat{\mathbf{w}}_a(K) \|^2 > \alpha^2$, define

$$\begin{aligned}
\mathbf{v}_a(K) &= \mathbf{P}_{\mathbf{z}}(K)\hat{\mathbf{w}}_a(K), && (7.237)\\
a &= \| \mathbf{v}_a(K) \|^2, && (7.238)\\
b &= -2\mathrm{Re}\left\{\mathbf{v}_a^H(K)\hat{\mathbf{w}}_a(K)\right\}, && (7.239)\\
c &= \| \hat{\mathbf{w}}_a(K) \|^2 - \alpha^2, && (7.240)\\
\beta(K) &= \frac{-b - Re\left\{\sqrt{b^2 - 4ac}\right\}}{2a}. && (7.241)\\
& && (7.242)
\end{aligned}$$

Then,

$$\tilde{\mathbf{w}}_a(K) = \hat{\mathbf{w}}_a(K) - \beta(K)\mathbf{v}_a(K). \tag{7.243}$$

Note that $\tilde{\mathbf{w}}_a(K-1)$ is used on the right side of (7.235).

A more detailed discussion on the RLS-VL algorithm is given in Tian et al. [TBV01]. A simpler technique would be to simply scale $\hat{\mathbf{w}}_a$,

$$\tilde{\mathbf{w}}_a = \hat{\mathbf{w}}_a \frac{\alpha}{\| \hat{\mathbf{w}}_a \|}. \tag{7.244}$$

This technique was suggested by Cox et al. [CZO87] in conjunction with the LMS algorithm. Note that $\hat{\mathbf{w}}_a$ is in a space that is orthogonal to $\mathbf{v}_s$, so scaling $\hat{\mathbf{w}}_a$ does not affect the distortionless constraint. In Section 7.7, we find it to be effective for that application. It does not appear to be effective with the RLS algorithm.

7.4.6 Conjugate Symmetric Beamformers

All of the discussion to this point applied to arbitrary arrays, although we frequently used linear arrays as examples. Recall that, whenever the array is symmetric about the origin, the array manifold vectors are conjugate symmetric and $\mathbf{S_x}$ (and $\hat{\mathbf{S}}_\mathbf{x}$) are Hermitian persymmetric. In Section 7.2, we observed that the constrained ML estimate of $\mathbf{S_x}$ was obtained by FB averaging of the data.

$$\begin{aligned}\tilde{\mathbf{S}}_\mathbf{x} &= \frac{1}{2}\left[\mathbf{C_x} + \mathbf{J}\mathbf{C}_\mathbf{x}^*\mathbf{J}\right] \\ &= \frac{1}{2K}\sum_{k=1}^{K}\left[\mathbf{X}(k)\mathbf{X}^H(k) + \mathbf{J}\mathbf{X}^*(k)\mathbf{X}^T(k)\mathbf{J}\right].\end{aligned} \tag{7.245}$$

In our discussion of SMI algorithms in Section 7.3.1 (e.g., Example 7.3.2 and Figure 7.5) there was significant improvement obtained by using FB averaging. We demonstrate a similar improvement for RLS beamformers and show how to implement the algorithm.

Note that "forward-backward" was used to be consistent with the temporal literature. We are really averaging across the array in opposite directions. The corresponding weighted average is

$$\tilde{\mathbf{\Phi}}(K) = \frac{1}{2}\sum_{k=1}^{K}\mu^{K-k}\left[\mathbf{X}(k)\mathbf{X}^H(k) + \mathbf{J}\mathbf{X}^*(k)\mathbf{X}^T(k)\mathbf{J}\right]. \tag{7.246}$$

Historically, conjugate symmetry was first utilized in the parameter estimation problem (e.g., Evans et al. [EJS82]). We discuss this application in Chapters 8 and 9. The application to adaptive beamformers is due to Huarng and Yeh [HY91], and our discussion follows that reference.

Consider the direct form implementation of the MPDR beamformer using (7.245). Then,

$$\tilde{\mathbf{w}}_{mpdr} = \Lambda \tilde{\mathbf{\Phi}}^{-1} \mathbf{v}_s, \tag{7.247}$$

and the weight vector $\tilde{\mathbf{w}}_{mpdr}$ is conjugate symmetric.

We manipulate the data into a form where real computations can be utilized. We use the unitary transformation defined in (7.58) and (7.59) to accomplish this goal. $\mathbf{Q}$ is defined as

$$\mathbf{Q} = \begin{cases} \frac{1}{\sqrt{2}} \begin{bmatrix} \mathbf{I} & j\mathbf{I} \\ \mathbf{J} & -j\mathbf{J} \end{bmatrix}, & \text{for even } N, \\ \frac{1}{\sqrt{2}} \begin{bmatrix} \mathbf{I} & \mathbf{0} & j\mathbf{I} \\ \mathbf{0}^T & \sqrt{2} & \mathbf{0}^T \\ \mathbf{J} & \mathbf{0} & -j\mathbf{J} \end{bmatrix}, & \text{for odd } N. \end{cases} \tag{7.248}$$

Note that the $\mathbf{I}$ and $\mathbf{J}$ matrices have dimension $N/2 \times N/2$. $\mathbf{Q}$ has two important features,

$$\mathbf{Q}^H = \mathbf{Q}^{-1}, \tag{7.249}$$

and

$$\mathbf{Q}^* = \mathbf{J}\mathbf{Q}. \tag{7.250}$$

The transformed quantities are given by

$$\bar{\mathbf{v}}_s = \mathbf{Q}^H \mathbf{v}_s, \tag{7.251}$$

$$\bar{\mathbf{\Phi}} = \mathbf{Q}^H \tilde{\mathbf{\Phi}} \mathbf{Q}, \tag{7.252}$$

$$\bar{\mathbf{w}} = \mathbf{Q}^H \tilde{\mathbf{w}} = \bar{\Lambda} \bar{\mathbf{\Phi}}^{-1} \bar{\mathbf{v}}_s, \tag{7.253}$$

and

$$\bar{\Lambda} = \left[\bar{\mathbf{v}}_s^T \bar{\mathbf{\Phi}}^{-1} \bar{\mathbf{v}}_s \right]^{-1}. \tag{7.254}$$

One can show easily that $\bar{\mathbf{v}}_s$ is a real vector and $\bar{\mathbf{\Phi}}$ is a real symmetric matrix. Therefore (7.253) can be solved using real computation. The complex weight vector $\tilde{\mathbf{w}}$ is given by

$$\tilde{\mathbf{w}} = \mathbf{Q}\bar{\mathbf{w}}. \tag{7.255}$$

The weight vector $\tilde{\mathbf{w}}$ is used to process the input data. We now discuss how to compute $\bar{\mathbf{\Phi}}^{-1}$ iteratively using real computation. The approach is analogous to the iterative procedure in Section 7.3. We can write $\bar{\mathbf{\Phi}}(K)$ at the Kth snapshot as

$$\bar{\mathbf{\Phi}}(K) = \mu\bar{\mathbf{\Phi}}(K-1) + \frac{1}{2}\mathbf{Q}^H \Big[\mathbf{X}(K)\mathbf{X}^H(K) + \mathbf{J}\mathbf{X}^*(K)\mathbf{X}^T(K)\mathbf{J}\Big] \mathbf{Q}. \tag{7.256}$$

There are various ways to update $\bar{\mathbf{\Phi}}^{-1}(K)$ from $\bar{\mathbf{\Phi}}^{-1}(K-1)$. We first manipulate $\bar{\mathbf{\Phi}}(K)$ into a form that will only utilize real calculation. We rewrite the term in the brackets of (7.256) as

$$\begin{aligned}\mathbf{X}(K)\mathbf{X}^H(K) + \mathbf{J}\mathbf{X}^*(K)\mathbf{X}^T(K)\mathbf{J} = \frac{1}{2}\Big\{ &[\mathbf{X}(K) + \mathbf{J}\mathbf{X}^*(K)] \\ &\cdot [\mathbf{X}(K) + \mathbf{J}\mathbf{X}^*(K)]^H \\ &+ [-j\mathbf{X}(K) + j\mathbf{J}\mathbf{X}^*(K)] \\ &\cdot [-j\mathbf{X}(K) + j\mathbf{J}\mathbf{X}^*(K)]^H\Big\},\end{aligned} \tag{7.257}$$

and define

$$\bar{\mathbf{X}}_1(K) = \frac{1}{2}\mathbf{Q}^H [\mathbf{X}(K) + \mathbf{J}\mathbf{X}^*(K)] = Re\left[\mathbf{Q}^H\mathbf{X}(K)\right], \tag{7.258}$$

and

$$\bar{\mathbf{X}}_2(K) = \frac{1}{2}\mathbf{Q}^H [-j\mathbf{X}(K) + j\mathbf{J}\mathbf{X}^*(K)] = Im\left[\mathbf{Q}^H\mathbf{X}(K)\right]. \tag{7.259}$$

Now (7.256) can be written as

$$\bar{\mathbf{\Phi}}(K) = \mu\bar{\mathbf{\Phi}}(K-1) + \left[\bar{\mathbf{X}}_1(K)\bar{\mathbf{X}}_1^T(K) + \bar{\mathbf{X}}_2(K)\bar{\mathbf{X}}_2^T(K)\right]. \tag{7.260}$$

We can now update $\bar{\mathbf{\Phi}}^{-1}(K)$ using any convenient recursive formula. A straightforward approach is to do two rank-one updates using the matrix inversion lemma. We can modify the recursion in (7.153)–(7.159). The first update is

$$\bar{\mathbf{P}}_1(K) = \mu^{-1}\bar{\mathbf{P}}(K-1) - \mu^{-1}\bar{\mathbf{g}}_1(K)\overline{\mathbf{X}}_1^T(K)\bar{\mathbf{P}}(K-1), \tag{7.261}$$

$$\bar{\mathbf{g}}_1(K) = \frac{\mu^{-1}\bar{\mathbf{P}}(K-1)\overline{\mathbf{X}}_1(K)}{1+\mu^{-1}\overline{\mathbf{X}}_1^T(K)\bar{\mathbf{P}}(K-1)\overline{\mathbf{X}}_1(K)}, \tag{7.262}$$

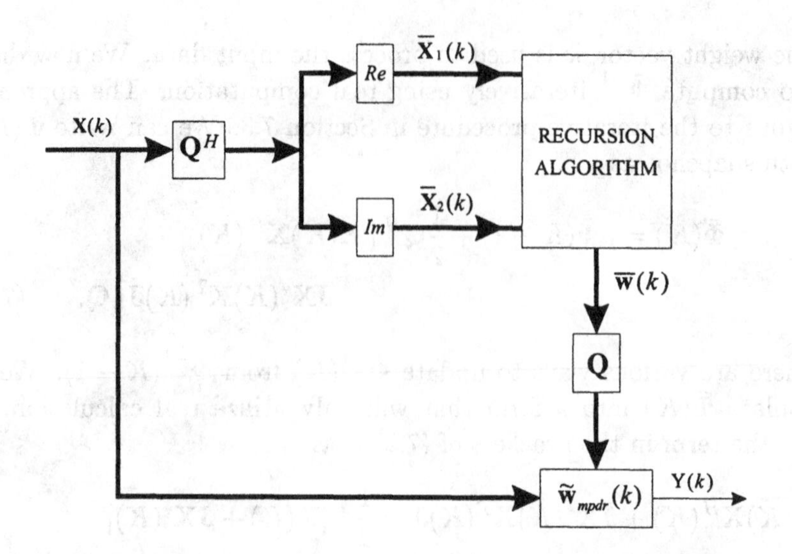

Figure 7.28 Recursive algorithm with FB averaging.

and

$$\bar{\mathbf{w}}_1(K) = \frac{\bar{\mathbf{\Lambda}}_1(K)}{\mu\bar{\mathbf{\Lambda}}(K-1)}\left[\mathbf{I} - \bar{\mathbf{g}}_1(K)\overline{\mathbf{X}}_1^T(K)\right]\bar{\mathbf{w}}(K-1). \tag{7.263}$$

The second update is

$$\bar{\mathbf{P}}(K) = \bar{\mathbf{P}}_1(K) - \bar{\mathbf{g}}(K)\overline{\mathbf{X}}_2^T(K)\bar{\mathbf{P}}_1(K), \tag{7.264}$$

$$\bar{\mathbf{g}}(K) = \frac{\bar{\mathbf{P}}_1(K)\overline{\mathbf{X}}_2(K)}{1 + \overline{\mathbf{X}}_2^T(K)\bar{\mathbf{P}}_1(K)\overline{\mathbf{X}}_2(K)}, \tag{7.265}$$

and

$$\bar{\mathbf{w}}(K) = \frac{\bar{\mathbf{\Lambda}}(K)}{\bar{\mathbf{\Lambda}}_1(K-1)}\left[\mathbf{I} - \bar{\mathbf{g}}(K)\overline{\mathbf{X}}_2^T(K)\right]\bar{\mathbf{w}}_1(K). \tag{7.266}$$

After the recursion in (7.261)–(7.266) at a particular K, we find

$$\tilde{\mathbf{w}}_{mpdr}(K) = \mathbf{Q}\bar{\mathbf{w}}(K), \tag{7.267}$$

and the output is

$$Y(K) = \tilde{\mathbf{w}}_{mpdr}^H(K)\mathbf{X}(K). \tag{7.268}$$

The beamformer is shown in Figure 7.28.

We consider a simple example to indicate the performance improvement that can be achieved.

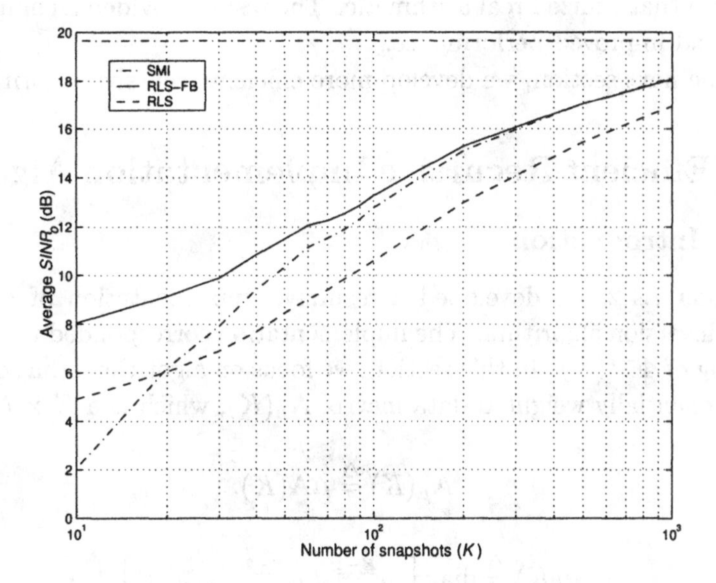

Figure 7.29 MPDR beamformer using RLS algorithm with FB averaging: $u_s = 0, SNR = 10$ dB, $u_I = 0.29, 0.45$, $INR = 20$ dB each, $\sigma_o^2/\sigma_w^2 = 10$ dB, $\mu = 0.999, 200$ trials; average $SINR_o$ versus K.

Example 7.4.4 (continuation, Example 7.4.1)

Consider a standard 10-element linear array. We use the same signal and interference model as in Example 7.4.1. We implement an MPDR beamformer using the RLS algorithm with FB averaging. The results are shown in Figure 7.29.

We see that RLS-FB with K snapshots has the same performance as RLS with $2K$ snapshots. The advantage is somewhat greater at small values of K.

7.4.7 Summary

In this section, we have developed the least squares estimation version of the adaptive beamformer problem. We then developed a recursive least squares implementation of the beamformer and compared its performance to the SMI beamformers in Section 7.3. We found that by the use of appropriate diagonal loading, the RLS and SMI beamformers have comparable performance.

We developed the generalized sidelobe canceller implementation and compared the RLS implementation to the SMI implementation.

We considered the case of conjugate symmetric arrays and developed an algorithm that utilized real arithmetic. The result provided a computational saving and improved performance.

In the next section, we develop more efficient recursive algorithms.

7.5 Efficient Recursive Implementation Algorithms

7.5.1 Introduction

In Section 7.4.2, we developed a recursive implementation of the sample matrix inversion algorithm. The implementation corresponded to a rank-one updating of $\mathbf{\Phi}^{-1}(K)$. In this section, we focus on algorithms that operate on the exponentially weighted data matrix $\mathbf{A}_\mu(K)$, which is a $K \times N$ complex matrix,

$$\mathbf{A}_\mu(K) \triangleq \boldsymbol{\mu}\mathbf{A}(K), \tag{7.269}$$

where

$$\boldsymbol{\mu}(K) \triangleq \text{diag}\left[\begin{array}{cccc} \mu^{\frac{K-1}{2}} & \mu^{\frac{K-2}{2}} & \cdots & 1 \end{array}\right], \tag{7.270}$$

and [13]

$$\mathbf{A}(K) \triangleq \begin{bmatrix} \mathbf{X}^T(1) \\ \mathbf{X}^T(2) \\ \vdots \\ \mathbf{X}^T(K) \end{bmatrix}. \tag{7.271}$$

From (7.138), we observe that

$$\mathbf{\Phi}^*(K) = \mathbf{A}_\mu^H(K)\mathbf{A}_\mu(K). \tag{7.272}$$

The development of efficient numerically stable recursive algorithms has received significant attention over the last several decades because of their widespread application in the areas of adaptive filtering, adaptive beamforming, and system identification.

The topic is treated extensively in most textbooks on adaptive filters; e.g., Haykin ([Hay91], [Hay96]), Proakis et al. [PRLN92], Widrow and Sterns [WS85], Orfandis [Orf88], Kalouptsidis and Theodoridis [KT93], Honig and Messerschmitt [HM84], Alexander [Ale86], and Treichler et al. [TJL87].

There are numerous papers dealing with specific algorithms that we will indicate as we develop the various algorithms. Various other papers (e.g.,

[13]The $\mathbf{A}(K)$ data matrix is related to the data matrix $\tilde{\mathbf{X}}(K)$ defined in (7.4) by $\mathbf{A}^T(K) = \sqrt{K}\tilde{\mathbf{X}}$. The $\mathbf{A}(K)$ notation is more commonly used in the QR decomposition (QRD) literature.

Yuen [Yue91] and Sayed and Kailath [SK94]) show how the various algorithms are related. Godara [God74] showed the relationship to Kalman filtering.

As we discuss the various algorithms, it is useful to remember that, if we had perfect numerical precision and no perturbations in the model, then the performance of the various implementations would be the same. The issues of interest are:

(i) Numerical stability;

(ii) Computational complexity;

(iii) Capability of parallel computation to improve speed and allow real-time implementation.

In Section 7.5.2, we develop a recursive algorithm that is referred to as the **QR decomposition** (QRD) or square-root algorithm. It provides an efficient implementation for both LCMP and LSE narrowband beamformers. In the text, the LSE version is developed. It can be used in the lower branch of the GSC beamformer to implement LCMP beamformers.

7.5.2 QR Decomposition (QRD)

In this section, we show how the QR decomposition in Section A.6 can be utilized in a recursive algorithm. Our discussion follows McWhirter and Prouder's chapter in [KT93] (see Chapter 7, pp. 260–321). We should note that the authors of that chapter developed many of the original results in the application of QRD techniques to adaptive beamforming and the implementation of those techniques.

We formulate the problem as a least squares estimation problem (see Section 7.4.1). Using the transpose of (7.134), we can write the output of the beamformer at time k as[14]

$$Y(k) = \mathbf{X}^T(k)\mathbf{w}^*(K), \tag{7.273}$$

and, from (7.147), the error is

$$e(k) = D(k) - \mathbf{X}^T(k)\mathbf{w}^*(K), \quad k = 1, \cdots, K. \tag{7.274}$$

As in (7.148), we minimize a weighted residual error

$$\xi_\mu(K) = \|\mathbf{e}_\mu(K)\|^2, \tag{7.275}$$

[14] We use this form rather than $\mathbf{w}^H\mathbf{X}(K)$ for convenience in presenting the final processor.

where

$$\mathbf{e}_\mu(K) = \boldsymbol{\mu}(K)\begin{bmatrix} e(1) & e(2) & \cdots & e(K)\end{bmatrix}^T. \tag{7.276}$$

The error vector in (7.276) can be written as

$$\mathbf{e}_\mu(K) = \mathbf{d}_\mu(K) - \mathbf{A}_\mu(K)\mathbf{w}^*(K), \tag{7.277}$$

where

$$\mathbf{A}_\mu(K) = \boldsymbol{\mu}(K)\begin{bmatrix} \mathbf{X}^T(1) \\ \mathbf{X}^T(2) \\ \vdots \\ \mathbf{X}^T(K)\end{bmatrix}, \tag{7.278}$$

and

$$\mathbf{d}_\mu(K) = \boldsymbol{\mu}(K)\begin{bmatrix} D(1) \\ D(2) \\ \vdots \\ D(K)\end{bmatrix}. \tag{7.279}$$

The subscript "μ" denotes the inclusion of the exponential weighting.

The solution to the minimization problem was given in (7.184), which can be rewritten as

$$\boldsymbol{\Phi}^*(K)\hat{\mathbf{w}}_{lse}^*(K) = \boldsymbol{\Phi}_{\mathbf{x}d^*}^*(K). \tag{7.280}$$

Using (7.272) and (7.4) in (7.280) gives

$$\mathbf{A}_\mu^H(K)\mathbf{A}_\mu(K)\hat{\mathbf{w}}_{lse}^*(K) = \mathbf{A}_\mu^H(K)\mathbf{d}_\mu(K). \tag{7.281}$$

We now develop an alternative approach to the recursion algorithm in Section 7.4 based on the QRD, which has good numerical properties. As discussed in Section A.6, we can find a $K \times K$ unitary matrix, $\mathbf{Q}(K)$ such that

$$\mathbf{Q}(K)\mathbf{A}_\mu(K) = \begin{bmatrix} \tilde{\mathbf{R}}(K) \\ \mathbf{0}\end{bmatrix}, \tag{7.282}$$

where $\tilde{\mathbf{R}}(K)$ is an $N \times N$ upper triangular matrix and $\mathbf{0}$ is a $(K-N) \times N$ matrix of zeros. Because $\mathbf{Q}(K)$ is unitary, $\tilde{\mathbf{R}}(K)$ is just the Cholesky factor of the data covariance matrix $\boldsymbol{\Phi}(K)$. Using (7.249), (7.281) can be written as

$$\mathbf{A}_\mu^H(K)\mathbf{Q}^H\mathbf{Q}\mathbf{A}_\mu(K)\hat{\mathbf{w}}_{lse}^*(K) = \mathbf{A}_\mu^H(K)\mathbf{Q}^H\mathbf{Q}\mathbf{d}_\mu(K). \tag{7.283}$$

Substituting (7.282) into (7.283) gives

$$\begin{bmatrix} \tilde{\mathbf{R}}(K) \\ \mathbf{0}\end{bmatrix}^H \begin{bmatrix} \tilde{\mathbf{R}}(K) \\ \mathbf{0}\end{bmatrix} \hat{\mathbf{w}}_{lse}^*(K) = \begin{bmatrix} \tilde{\mathbf{R}}(K) \\ \mathbf{0}\end{bmatrix}^H \mathbf{Q}\mathbf{d}_\mu(K). \tag{7.284}$$

The $K \times 1$ vector, $\mathbf{Qd}_\mu(K)$ can be partitioned as

$$\mathbf{Qd}_\mu(K) = \begin{bmatrix} \mathbf{p}(K) \\ \mathbf{v}(K) \end{bmatrix}, \tag{7.285}$$

where $\mathbf{p}(K)$ is an $N \times 1$ vector. Then, (7.284) reduces to

$$\tilde{\mathbf{R}}(K)\hat{\mathbf{w}}_{lse}^*(K) = \mathbf{p}(K), \tag{7.286}$$

which is straightforward to solve because of the upper triangular structure of $\tilde{\mathbf{R}}(K)$. The minimum weighted error can be written as

$$\xi_o(K) = \|\mathbf{v}(K)\|^2. \tag{7.287}$$

In order to use (7.286), the QRD in (7.282) must be implemented.

As discussed in Section A.6, the triangularization can be accomplished with either a Givens rotation (see Section A.6.2) or a Householder transformation (see Section A.6.3). We shall find that the Givens rotation is particularly suited to adaptive beamforming because it leads to an efficient algorithm for recursive updating.

To develop the recursive algorithm, we assume that $\mathbf{A}_\mu(K-1)$ has already been reduced to a triangular form by the unitary transformation,

$$\mathbf{Q}(K-1)\mathbf{A}_\mu(K-1) = \begin{bmatrix} \tilde{\mathbf{R}}(K-1) \\ \mathbf{0} \end{bmatrix}. \tag{7.288}$$

First, define a $K \times K$ unitary matrix,

$$\bar{\mathbf{Q}}(K-1) = \begin{bmatrix} \mathbf{Q}(K-1) & \mathbf{0} \\ \mathbf{0}^T & 1 \end{bmatrix}, \tag{7.289}$$

and write $\mathbf{A}_\mu(K)$ as,

$$\mathbf{A}_\mu(K) = \begin{bmatrix} \mu^{\frac{1}{2}}\mathbf{A}_\mu(K-1) \\ \mathbf{X}^T(K) \end{bmatrix}. \tag{7.290}$$

Then,

$$\begin{aligned} \bar{\mathbf{Q}}(K-1)\mathbf{A}_\mu(K) &= \bar{\mathbf{Q}}(K-1)\begin{bmatrix} \mu^{\frac{1}{2}}\mathbf{A}_\mu(K-1) \\ \mathbf{X}^T(K) \end{bmatrix} \\ &= \begin{bmatrix} \mu^{\frac{1}{2}}\tilde{\mathbf{R}}(K-1) \\ \mathbf{0} \\ \mathbf{X}^T(K) \end{bmatrix}. \end{aligned} \tag{7.291}$$

The required triangularization can be completed by using a sequence of complex Givens rotations to eliminate $\mathbf{X}^T(K)$. We demonstrate the procedure with a simple example.

Example 7.5.1

We assume $N = 3$ and $K = 5$. Then, the right side of (7.291) can be written as

$$\mathbf{M}_1 = \begin{bmatrix} \mu^{\frac{1}{2}}\tilde{r}_{11} & \mu^{\frac{1}{2}}\tilde{r}_{12} & \mu^{\frac{1}{2}}\tilde{r}_{13} \\ 0 & \mu^{\frac{1}{2}}\tilde{r}_{22} & \mu^{\frac{1}{2}}\tilde{r}_{23} \\ 0 & 0 & \mu^{\frac{1}{2}}\tilde{r}_{33} \\ 0 & 0 & 0 \\ x_1(5) & x_2(5) & x_3(5) \end{bmatrix}. \tag{7.292}$$

The $\tilde{r}_{ii}$ are real and the $\tilde{r}_{ij}$ $(i \neq j)$ and the x_i are complex. The first Givens rotation uses the first and last rows to eliminate $x_1(5)$. Thus,

$$\mathbf{G}_1 = \begin{bmatrix} c_1 & \mathbf{0}^T & s_1^* \\ & 1 & \\ \mathbf{0} & 1 & \mathbf{0} \\ & 1 & \\ -s_1 & \mathbf{0}^T & c_1 \end{bmatrix}, \tag{7.293}$$

where

$$|c_1|^2 + |s_1|^2 = 1, \tag{7.294}$$

and c_1 can be real without loss of generality.

$$c_1 = \frac{\mu^{\frac{1}{2}}\tilde{r}_{11}}{\sqrt{\mu\tilde{r}_{11}^2 + |x_1|^2}}, \tag{7.295}$$

and

$$s_1 = \frac{x_1}{\sqrt{\mu\tilde{r}_{11}^2 + |x_1|^2}}, \tag{7.296}$$

where we have suppressed the $K = 5$ argument. Multiplying $\mathbf{M}_1$ by $\mathbf{G}_1$ and using (7.295) and (7.296) gives

$$\begin{aligned} \mathbf{G}_1\mathbf{M}_1 &= \left[\begin{array}{c:c:c} c_1\mu^{\frac{1}{2}}\tilde{r}_{11} + s_1^*x_1 & c_1\mu^{\frac{1}{2}}\tilde{r}_{12} + s_1^*x_2 & c_1\mu^{\frac{1}{2}}\tilde{r}_{13} + s_1^*x_3 \\ 0 & \mu^{\frac{1}{2}}\tilde{r}_{22} & \mu^{\frac{1}{2}}\tilde{r}_{23} \\ 0 & 0 & \mu^{\frac{1}{2}}\tilde{r}_{33} \\ 0 & 0 & 0 \\ 0 & -s_1\mu^{\frac{1}{2}}\tilde{r}_{12} + c_1x_2 & -s_1\mu^{\frac{1}{2}}\tilde{r}_{13} + c_1x_3 \end{array}\right] \\ &= \left[\begin{array}{ccc} \tilde{r}'_{11} & \tilde{r}'_{12} & \tilde{r}'_{13} \\ 0 & \mu^{\frac{1}{2}}\tilde{r}_{22} & \mu^{\frac{1}{2}}\tilde{r}_{23} \\ 0 & 0 & \mu^{\frac{1}{2}}\tilde{r}_{33} \\ 0 & 0 & 0 \\ \hdashline 0 & x'_2 & x'_3 \end{array}\right]. \end{aligned} \tag{7.297}$$

Note that the first element in the bottom row is eliminated and the elements in both the first row and the last row are changed.

The second Givens rotation uses the second row and the last row to eliminate x_2':

$$\mathbf{G}_2 = \left[\begin{array}{c:cccc} 1 & & \mathbf{0}^T & & \\ \hdashline & c_2 & 0 & 0 & s_2^* \\ \mathbf{0} & 0 & 1 & 0 & 0 \\ & 0 & 0 & 1 & 0 \\ & -s_2 & 0 & 0 & c_2 \end{array}\right]. \tag{7.298}$$

Choosing c_2 and s_2 in a manner analogous to (7.295) and (7.296), we obtain

$$\mathbf{G}_2\mathbf{G}_1\mathbf{M}_1 = \begin{bmatrix} \tilde{r}'_{11} & \tilde{r}'_{12} & \tilde{r}'_{13} \\ 0 & \tilde{r}'_{22} & \tilde{r}'_{23} \\ 0 & 0 & \mu^{\frac{1}{2}}\tilde{r}_{33} \\ 0 & 0 & 0 \\ 0 & 0 & x''_3 \end{bmatrix}. \tag{7.299}$$

Note that the first row is unchanged and the second and bottom rows are changed. Similarly,

$$\mathbf{G}_3 = \begin{bmatrix} 1 & & \mathbf{0} & & \\ & 1 & & & \\ & & c_3 & 0 & s_3^* \\ \mathbf{0} & & 0 & 1 & 0 \\ & & -s_3 & 0 & c_3 \end{bmatrix}, \tag{7.300}$$

and

$$\mathbf{G}_3\mathbf{G}_2\mathbf{G}_1\mathbf{M}_1 = \begin{bmatrix} \tilde{\mathbf{R}}(N) \\ \mathbf{0} \end{bmatrix}, \tag{7.301}$$

which is the desired result. Note that the $\mathbf{G}_3$ operation only changes the third row and the bottom row.

In general, we can write

$$\mathbf{Q}(K) = \hat{\mathbf{Q}}(K)\bar{\mathbf{Q}}(K-1), \tag{7.302}$$

where

$$\hat{\mathbf{Q}}(K) = \mathbf{G}_N \cdots \mathbf{G}_2\mathbf{G}_1, \tag{7.303}$$

and

$$\hat{\mathbf{Q}}(K) \begin{bmatrix} \mu^{\frac{1}{2}}\tilde{\mathbf{R}}(K-1) \\ \mathbf{0} \\ \mathbf{X}^T(K) \end{bmatrix} = \begin{bmatrix} \tilde{\mathbf{R}}(K) \\ \mathbf{0} \\ \mathbf{0}^T \end{bmatrix}, \tag{7.304}$$

which is the desired result. Note that each Givens rotation only changes the row corresponding to its subscript and the bottom row.

The term in (7.285) must also be computed recursively. The same $\mathbf{Q}(K)$ can be used to perform the recursion on $\mathbf{p}(K)$. At the $(K-1)$ iteration,

$$\mathbf{Q}(K-1)d_\mu(K-1) = \begin{bmatrix} \mathbf{p}(K-1) \\ \mathbf{v}(K-1) \end{bmatrix}, \tag{7.305}$$

where $\mathbf{p}(K-1)$ is $N \times 1$ and $\mathbf{v}(K-1)$ is $(K-1-N) \times 1$. The matrix $\bar{\mathbf{Q}}(K-1)$ is defined in (7.289). Write

$$\bar{\mathbf{Q}}(K-1)\mathbf{d}_\mu(K-1) = \bar{\mathbf{Q}}(K-1) \begin{bmatrix} \mu^{\frac{1}{2}}\mathbf{d}_\mu(K-1) \\ D(K) \end{bmatrix}. \tag{7.306}$$

Substituting (7.289) into (7.306) gives

$$\begin{aligned} \bar{\mathbf{Q}}(K-1)\mathbf{d}_\mu(K-1) &= \begin{bmatrix} \mathbf{Q}(K-1) & \mathbf{0} \\ \mathbf{0}^T & 1 \end{bmatrix} \begin{bmatrix} \mu^{\frac{1}{2}}\mathbf{d}_\mu(K-1) \\ D(K) \end{bmatrix} \\ &= \begin{bmatrix} \mu^{\frac{1}{2}}\mathbf{p}(K-1) \\ \mu^{\frac{1}{2}}\mathbf{v}(K-1) \\ D(K) \end{bmatrix}. \end{aligned} \tag{7.307}$$

Now apply $\hat{\mathbf{Q}}(K)$ from (7.303),

$$\hat{\mathbf{Q}}(K) \begin{bmatrix} \mu^{\frac{1}{2}}\mathbf{p}(K-1) \\ \mu^{\frac{1}{2}}\mathbf{v}(K-1) \\ D(K) \end{bmatrix} = \begin{bmatrix} \mathbf{p}(K) \\ \mu^{\frac{1}{2}}\mathbf{v}(K-1) \\ \tilde{e}(K) \end{bmatrix} = \begin{bmatrix} \mathbf{p}(K) \\ \mathbf{v}(K) \end{bmatrix}. \tag{7.308}$$

Multiplication by $\mathbf{G}_1$ updates $[\mathbf{p}(K)]_1$ and changes the last element in the overall vector. Multiplication by $\mathbf{G}_2$ updates $[\mathbf{p}(K)]_2$ and changes the last element in the overall vector. Continuing, multiplication by $\mathbf{G}_N$ updates $[\mathbf{p}(K)]_N$ and generates $\tilde{e}(K)$ (whose significance is discussed shortly). The other elements of $\mathbf{v}(K)$ do not need to be computed.

This discussion shows that by applying $\mathbf{Q}(K)$ recursively to $A_\mu(K)$ and $\mathbf{d}_\mu(K)$, the $\tilde{\mathbf{R}}(K)$ and $\mathbf{p}(K)$ required to solve (7.286) can be generated. However, actually using $\mathbf{Q}(K)$ is inefficient. First, observe that the right side of (7.291) is already available from the $(K-1)$ step, so the $\bar{\mathbf{Q}}(K-1)$ multiplication is not necessary. Now, examine the Givens rotation in Example 7.5.1:

(i) The $\mathbf{G}_1$ operation consists of computing c_1 and s_1 and then computing three new elements in the top row and two new elements in the bottom row. The left element in the bottom row becomes zero, and all other elements are unchanged.

(ii) The $\mathbf{G}_2$ operation consists of computing c_2 and s_2 and then computing two new elements in the second row and one new element in the bottom row.

(iii) The $\mathbf{G}_3$ operation consists of computing c_3 and s_3 and then computing one new element in the third row.

The formulas for each of these calculations are given in Section A.6.3. These formulas are implemented instead of the complete matrix multiplication. A similar discussion applies to $\mathbf{p}(K)$ updates.

At this point, $\hat{\mathbf{w}}_{lse}(K)$ can be obtained by solving (7.286). The solution is straightforward because of the upper triangular structure of $\tilde{\mathbf{R}}(K)$. However, $\hat{d}(K)$ can be generated without finding an explicit expression for $\hat{\mathbf{w}}_{lse}(K)$.

The error at time K is

$$e(K) = D(K) - \mathbf{X}^T(K)\hat{\mathbf{w}}_{lse}^*(K). \tag{7.309}$$

Write $\hat{\mathbf{Q}}(K)$ as a partitioned $K \times K$ matrix

$$\hat{\mathbf{Q}}(K) = \begin{bmatrix} \mathbf{\Sigma}(K) & \mathbf{0}^T & \mathbf{q}(K) \\ \mathbf{0} & \mathbf{I} & \mathbf{0} \\ \boldsymbol{\sigma}^T(K) & \mathbf{0}^T & \tilde{\alpha}(K) \end{bmatrix}, \tag{7.310}$$

where $\mathbf{\Sigma}(K)$ is $N \times N$, $\mathbf{q}(K)$ and $\boldsymbol{\sigma}(K)$ are $N \times 1$, and $\tilde{\alpha}(K)$ is a scalar. The first step is to find $\tilde{\alpha}(K)$.

From (7.303),

$$\hat{\mathbf{Q}}(K) = \prod_{i=1}^{N} \mathbf{G}_i, \tag{7.311}$$

where the $\mathbf{G}_i$ are the Givens rotation matrices. They have two elements on the diagonal equal to c_i, and the remaining diagonal elements are one. They have only two non-zero off-diagonal elements s_i^* and $-s_i$. The model from Example 7.5.1 illustrates their structure.

Example 7.5.2 (continuation)

$$\hat{\mathbf{Q}}(5) = \begin{bmatrix} 1 & 0 & 0 & 0 & 0 \\ 0 & 1 & 0 & 0 & 0 \\ 0 & 0 & c_3 & 0 & s_3^* \\ 0 & 0 & 0 & 1 & 0 \\ 0 & 0 & -s_3 & 0 & c_3 \end{bmatrix} \begin{bmatrix} 1 & 0 & 0 & 0 & 0 \\ 0 & c_2 & 0 & 0 & s_2^* \\ 0 & 0 & 1 & 0 & 0 \\ 0 & 0 & 0 & 1 & 0 \\ 0 & -s_2 & 0 & 0 & c_2 \end{bmatrix} \left[\begin{array}{c|ccc|c} c_1 & 0 & 0 & 0 & s_1^* \\ \hline 0 & & \mathbf{I} & & \mathbf{0} \\ \hline -s_1 & 0 & 0 & 0 & c_1 \end{array}\right]. \tag{7.312}$$

Thus,

$$\tilde{\alpha}(5) = c_1 c_2 c_3. \tag{7.313}$$

In general,

$$\tilde{\alpha}(K) = \prod_{i=1}^{N} c_i, \tag{7.314}$$

where the c_i are the cosine factors in the Givens rotations.

Next, multiply both sides of (7.304) by $\hat{\mathbf{Q}}^H(K)$ and use the unitary character of $\hat{\mathbf{Q}}(K)$ to obtain

$$\mathbf{X}^T(K) = \mathbf{q}^H(K)\tilde{\mathbf{R}}(K). \tag{7.315}$$

Repeating the process with (7.308) gives

$$D(K) = \mathbf{q}^H(K)\mathbf{p}(K) + \tilde{\alpha}(K)\tilde{e}(K). \tag{7.316}$$

Substituting (7.315) and (7.316) into (7.309) gives

$$e(K) = -\mathbf{q}^H(K)\tilde{\mathbf{R}}(K)\mathbf{w}^*(K) + \mathbf{q}^H(K)\mathbf{p}(K) + \tilde{\alpha}(K)\tilde{e}(K). \tag{7.317}$$

However, (7.286) implies that the first two terms on the right side of (7.317) sum to zero. Thus,

$$e(K) = \tilde{\alpha}(K)\tilde{e}(K), \tag{7.318}$$

and

$$\hat{D}(K) = D(K) - e(K). \tag{7.319}$$

The significance of the result in (7.318) is that the quantities $\tilde{\alpha}(K)$ and $\tilde{e}(K)$ are generated by the recursive algorithm and we do not need to find $\hat{\mathbf{w}}_{lse}(K)$.

The last step in the development is to show how the Givens rotation algorithm may be implemented in parallel with a triangular processor array.

The triangular processor array is shown in Figure 7.30, for the case of $N = 4$.

The input snapshots $\mathbf{X}^T(K)$ and the desired signal $D(K)$ enter the array from the top. Each of the internal cells is performing part of a Givens rotation as shown in Figure 7.31 and Table 7.2.

At time $(K-1)$, the left boundary cells and the internal cells have the elements of $\tilde{\mathbf{R}}(K-1)$ stored. The right boundary cells have the elements of $\mathbf{p}(K)$ stored.

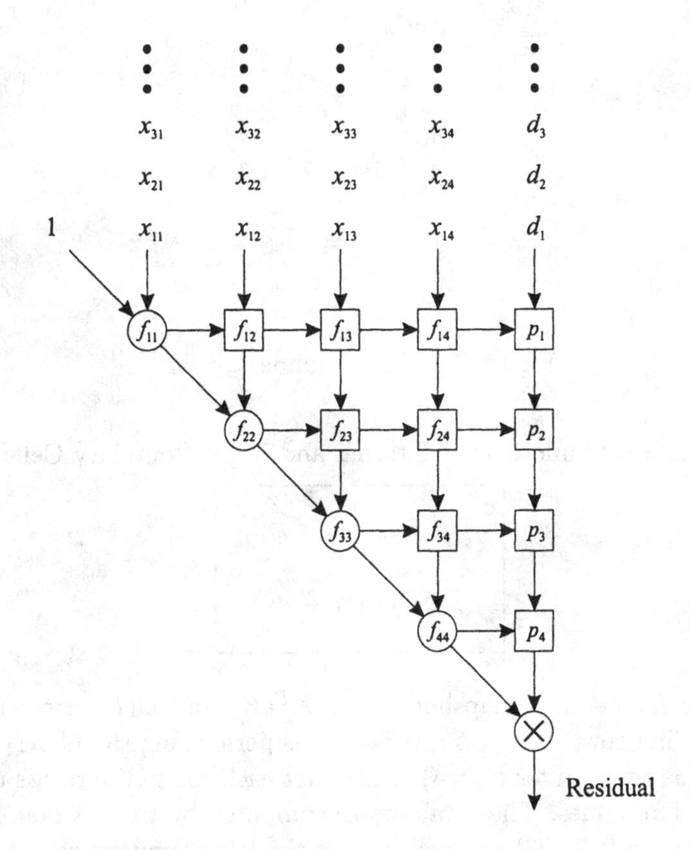

Figure 7.30 Triangular processor array.

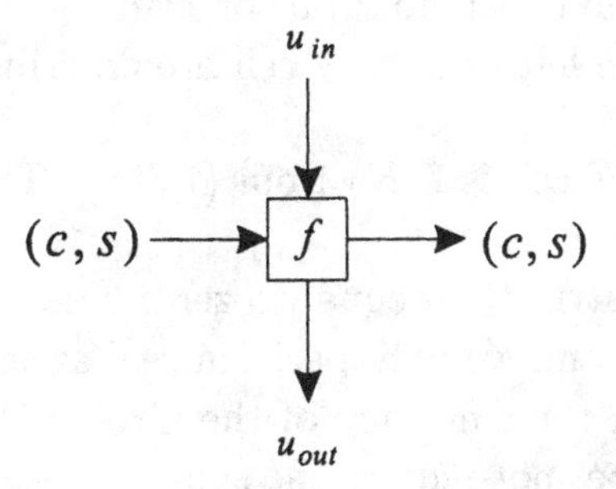

Figure 7.31 Processing elements for triangular QRD array: internal cells and right boundary cells.

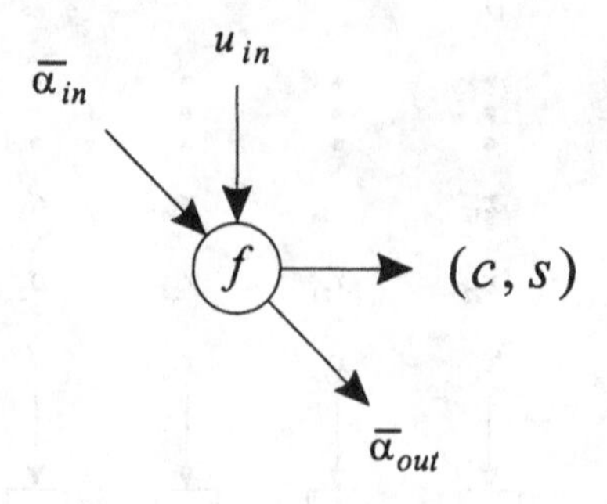

Figure 7.32 Left boundary cells.

Table 7.2 Functions of Internal and Right Boundary Cells

$$u_{out} = -s_1 \mu^{\frac{1}{2}} \tilde{r} + c_1 u_{in}$$

$$\tilde{r}_{new} = c_1 \mu^{\frac{1}{2}} \tilde{r} + s_1^* u_{in}$$

At time K, the new snapshot vector $\mathbf{X}^T(K)$ and $D(K)$ are sent to the cells in the first row. The first row of cells is performing the $\mathbf{G}_1\mathbf{M}_1$ rotation of (7.297) (as adapted for $N = 4$). Note that each cell in the row is using the same c_1 and s_1 values. These values are computed by the left boundary cell using (7.295) and (7.296) (we will discuss the left boundary cell in the next paragraph). As indicated in (7.297), the data output of the left boundary cell is zero. The second row of cells carries out the $\mathbf{G}_2$ rotation in (7.298). The third and fourth rows complete the rotation. The right boundary cell is using the same rotation with an input of $D(K)$.

The functions in the left boundary cell are described in Figure 7.32 and Table 7.3.

The output of the E cell is $\tilde{e}(K)$ from (7.308). Then $\hat{D}(K)$ is obtained from (7.319).

We initialize the matrix $\tilde{\mathbf{R}}(0)$ equal to zero. Therefore all of the $\tilde{r}$ in the boundary cells are real and we only perform real arithmetic.

We can improve on the efficiency of the Givens QRD algorithm by devising a version that does not require the square-root operation. Gentleman [Gen73] and Hammarling [Ham74] have developed efficient algorithms using a modified Givens rotation that does not require square-root operations. The algorithm is discussed in Section 7.2.4 of McWhirter and Prouder's chapter in [KT93] and in Section 7.3.2 of [PRLN92]. The reader is referred to these references for a discussion of the square-root-free Givens algorithm.

Table 7.3 Description of Element Functions in Figure 7.32

If $u_{in} = 0$	If $u_{in} \neq 0$
$c = 1$	$c = \frac{\mu^{\frac{1}{2}} \tilde{r}_{old}}{\sqrt{\mu \tilde{r}^2_{old} + \|u_{in}\|^2}}$
$s = 0$	$s = \frac{x_1}{\sqrt{\mu \tilde{r}^2_{old} + \|u_{in}\|^2}}$
$\tilde{\alpha}_{out} = \tilde{\alpha}_{in}$	$\tilde{\alpha}_{out} = c\tilde{\alpha}_{in}$
$\tilde{r}_{new} = \mu^{\frac{1}{2}} \tilde{r}_{old}$	$\tilde{r}_{new} = \sqrt{\mu \tilde{r}^2_{old} + \|u_{in}\|^2}$

In this section, we have developed the QRD implementation of the least squares estimation problem. It can be used in the lower branch of a GSC beamformer. A direct form LCMP or LCMV beamformer can also be implemented using a QRD. The MVDR version is derived in McWhirter and Shepherd [MS89] (see pp. 619–621 of Haykin [Hay96] for a discussion of MVDR beamformer).

The QRD technique is important because it is numerically stable and can be implemented in triangular processor arrays. A disadvantage is that it is not clear how to efficiently incorporate diagonal loading.

7.6 Gradient Algorithms

7.6.1 Introduction

In Sections 7.6 and 7.7, we discuss a group of algorithms that rely on the quadratic characteristic of the error surface and utilize gradient techniques to find the optimum weight vector. In Section 7.6, we assume that the ensemble statistics are known. We can use deterministic gradient techniques to find the optimum weight vector. In Section 7.7, we consider stochastic gradient algorithms, in which the statistics must be estimated from the data.

The two deterministic gradient techniques that are most widely used are the Newton algorithm (and various quasi-Newton algorithms) and the steepest descent algorithm. These techniques are discussed in a number of texts dealing with classical optimization theory, as well as books on adaptive filtering (e.g., Widrow and Stearns [WS85] and Haykin [Hay96]).

We will discuss the Newton algorithm in Section 8.6 in the context of parameter estimation. In Section 7.6, we only consider the steepest descent method. Our motivation is that the stochastic version of the steepest descent algorithm is the **least-mean-square** (LMS) algorithm, which is widely used in practice.

The major advantage of steepest descent and LMS algorithms is their computational simplicity, which is $O(N)$ in the narrowband model, where N is the number of sensors. For broadband processors, the complexity is $O(NM)$, where M is the number of discrete frequencies or taps that are utilized. Their major disadvantage is that their rate of convergence depends on the eigenvalue spread in $\mathbf{S}_{\mathbf{x}}$ and may be slow in a multiple interference environment.

The LMS algorithms are due to Widrow et al. [WMGG67]. Widrow references earlier papers by Shor [Sho66] and Mermoz [Mer65]. In addition to adaptive antennas, the LMS algorithm has been applied in a number of other areas including adaptive equalization, adaptive deconvolution, adaptive noise cancelling, and adaptive line enhancement (e.g., Widrow and Stearns [WS85] and Haykin [Hay96]).

At the same time, Howells and Applebaum were developing narrowband adaptive antennas that were equivalent to the LMS narrowband arrays, but their work was not publicized to the same extent (e.g., [App66], [App76], and [How76]).

In Section 7.6.2, we develop deterministic gradient algorithms for the MMSE beamformer using the method of steepest descent. In Section 7.6.3, we develop the steepest descent version of the LCMP beamformer. In Section 7.6.4, we summarize our results.

In Section 7.7, we develop various versions of the LMS algorithm, which is a stochastic gradient algorithm. For notational simplicity, Sections 7.6 to 7.7 consider narrowband arrays. In Section 7.11, we show how the LMS algorithm is applied to broadband arrays.

Although we do not discuss it in detail, we should note that if we utilized a Newton-type algorithm for the deterministic gradient algorithm, the stochastic version is called the LMS-Newton algorithm (e.g., Glentis et al. [GBT99] or Diniz et al. [DCA95]). The most common version of this algorithm is identical to the RLS algorithm in Section 7.4.

7.6.2 Steepest Descent: MMSE Beamformers

The simplest way to motivate the stochastic gradient algorithms that we utilize in the next section is to first consider deterministic gradient algorithms.

These deterministic gradient algorithms are discussed in most books or tutorial articles on adaptive filtering or beamforming (e.g., Widrow and Stearns [WS85], Chapter 8 of [Hay96], Monzingo and Miller [MM80], Compton [Com88], Proakis et al. [PRLN92]).

We consider two algorithms. In Section 7.6.2, we consider the MMSE beamformer and develop a steepest descent algorithm to find $\hat{\mathbf{w}}_o$. These results apply to LCMP beamformers if we use the generalized sidelobe canceller implementation of Figure 6.46 and can be adapted easily to accomodate quadratic constraints. In Section 7.6.3, we consider a direct form implementation of the LCMP beamformer and derive a steepest descent algorithm to find $\hat{\mathbf{w}}_{lcmp}$.

From Section 6.2, the MSE is[15]

$$\begin{aligned}\xi(\mathbf{w}) &= E\left[(D-\mathbf{w}^H\mathbf{X})(D^*-\mathbf{X}^H\mathbf{w})\right] \\ &= \sigma_d^2-\mathbf{w}^H\mathbf{p}-\mathbf{p}^H\mathbf{w}+\mathbf{w}^H\mathbf{S_x}\mathbf{w},\end{aligned} \tag{7.320}$$

where

$$\mathbf{p}\triangleq E[\mathbf{X}D^*]=\mathbf{S}_{\mathbf{x}d^*}. \tag{7.321}$$

In the discussion in this section, we assume $\mathbf{p}$ is known.

The gradient with respect to $\mathbf{w}^H$ is

$$\nabla\xi_{\mathbf{w}^H}=-\mathbf{p}+\mathbf{S_x}\mathbf{w}. \tag{7.322}$$

Setting the gradient equal to zero gives the familiar equation for the Wiener-Hopf beamformer,

$$\mathbf{S_x}\mathbf{w}_o=\mathbf{p}. \tag{7.323}$$

The resulting MMSE is

$$\xi_o=\sigma_d^2-\mathbf{w}_o^H\mathbf{S_x}\mathbf{w}_o=\sigma_d^2-\mathbf{p}^H\mathbf{w}_o. \tag{7.324}$$

Instead of solving (7.323) directly by inverting $\mathbf{S_x}$, we solve it by a gradient search technique. The error surface is quadratic so the search procedure should converge to the unique minimum ξ point.

We should observe that there are other search techniques, such as the Newton or quasi-Newton algorithm for minimizing $\xi(\mathbf{w})$. We use the steepest

[15]Our discussion is similar to that in [WS85] and Haykin [Hay96].

descent technique because of its computational simplicity. The cost will be slow convergence.

We define

$$\mathbf{w}(K) = \mathbf{w}(K-1) + \alpha\left(-\nabla\xi_{\mathbf{w}^H}\right), \tag{7.325}$$

where α is a real parameter, which we refer to as the step size parameter. In most cases, we use a variable step size parameter $\alpha(K)$. Using (7.322) in (7.325) gives

$$\mathbf{w}(K) = \mathbf{w}(K-1) + \alpha\left[\mathbf{p} - \mathbf{S_x}\mathbf{w}(K-1)\right], \quad K = 1, 2, \cdots. \tag{7.326}$$

In order to examine its behavior, it is convenient to define the weight-error vector, $\mathbf{w}_e(K)$,

$$\mathbf{w}_e(K) \triangleq \mathbf{w}(K) - \mathbf{w}_o. \tag{7.327}$$

Using (7.323) and (7.327) in (7.326) gives

$$\mathbf{w}_e(K) = \left(\mathbf{I} - \alpha\mathbf{S_x}\right)\mathbf{w}_e(K-1). \tag{7.328}$$

The next step is to rotate (7.328) into an orthonormal coordinate system. Using (5.237), we write

$$\mathbf{S_x} = \mathbf{U}\mathbf{\Lambda}\mathbf{U}^H. \tag{7.329}$$

Substituting (7.329) in (7.328), multiplying by $\mathbf{U}^H$, and utilizing the unitary nature of $\mathbf{U}$, gives

$$\mathbf{U}^H\mathbf{w}_e(K) = \left(\mathbf{I} - \alpha\mathbf{\Lambda}\right)\mathbf{U}^H\mathbf{w}_e(K-1). \tag{7.330}$$

We now define[16]

$$\mathbf{v}(K) \triangleq \mathbf{U}^H\mathbf{w}_e(K). \tag{7.331}$$

Using (7.331) in (7.330) gives

$$\mathbf{v}(K) = \left(\mathbf{I} - \alpha\mathbf{\Lambda}\right)\mathbf{v}(K-1). \tag{7.332}$$

Assuming $\mathbf{w}(0) = \mathbf{0}$, then

$$\mathbf{v}(0) = -\mathbf{U}^H\mathbf{w}_o. \tag{7.333}$$

The coefficient of $\mathbf{v}(K-1)$ in (7.332) is diagonal so the components of $\mathbf{v}(K)$ can be treated independently,

$$v_n(K) = \left(1 - \alpha\lambda_n\right)v_n(K-1), \quad n = 1, 2, \cdots, N. \tag{7.334}$$

[16]Note that $\mathbf{v}(K)$ is not related to the array manifold vector.

The solution to (7.334) is

$$v_n(K) = (1 - \alpha\lambda_n)^K \, v_n(0), \quad n = 1, 2, \cdots, N. \tag{7.335}$$

All of the eigenvalues are real and positive, so $v_n(K)$ is a geometric series. In order for the sequence to converge, we require

$$|1 - \alpha\lambda_n| < 1, \quad n = 1, 2, \cdots, N, \tag{7.336}$$

which implies

$$0 < \alpha < \frac{2}{\lambda_{max}}, \tag{7.337}$$

where λ_{max} is the maximum eigenvalue of $\mathbf{S_x}$. The condition in (7.337) is necessary and sufficient.

We can fit a continuous envelope to the geometric series (e.g., p. 347 of [Hay96]),

$$1 - \alpha\lambda_n = \exp\left(-\frac{1}{\tau_n}\right), \tag{7.338}$$

where τ_n is the time constant,

$$\tau_n = \frac{-1}{\ln(1 - \alpha\lambda_n)} \tag{7.339}$$

or, for very small step sizes,

$$\tau_n \simeq \frac{1}{\alpha\lambda_n}, \quad \alpha \ll 1. \tag{7.340}$$

The transient behavior of the weight vector follows easily. Pre-multiplying (7.331) by $\mathbf{U}$, using the unitary nature of $\mathbf{U}$, and adding $\mathbf{w}(0)$, we obtain

$$\mathbf{w}(K) = \mathbf{w}_o + \mathbf{U}\mathbf{v}(K). \tag{7.341}$$

The beam pattern at iteration K can be written as

$$\begin{aligned} B_K(\psi) &\triangleq \mathbf{w}^H(K)\mathbf{v}_a(\psi) \\ &= \mathbf{w}_o^H\mathbf{v}_a(\psi) + \mathbf{v}^H(K)\mathbf{U}^H\mathbf{v}_a(\psi), \end{aligned} \tag{7.342}$$

where $\mathbf{v}_a(\psi)$ is the array manifold vector. Using (7.335), we can write (7.342) as,

$$B_K(\psi) = B_o(\psi) + \sum_{n=1}^{N} v_n(0) \, (1 - \alpha\lambda_n)^K \, B_{eig,n}(\psi), \tag{7.343}$$

where

$$B_{eig,n}(\psi) = \mathbf{\Phi}_n^H \mathbf{v}_a(\psi), \quad n = 1, 2, \cdots, N, \tag{7.344}$$

is the nth eigenbeam. The steady state solution can also be written as a sum of eigenbeams. Replacing $\mathbf{S_x}$ with its eigenvector expansion in (7.329) gives

$$\mathbf{w}_o = \sum_{n=1}^{N} \frac{1}{\lambda_n} \mathbf{\Phi}_n \mathbf{\Phi}_n^H \mathbf{p}. \tag{7.345}$$

Thus,

$$B_o(\psi) = \sum_{n=1}^{N} \left(\frac{\mathbf{p}^H \mathbf{\Phi}_n}{\lambda_n} \right) B_{eig,n}(\psi). \tag{7.346}$$

The size of α will be limited by the largest eigenvalue. The corresponding eigenbeam will converge with a time constant $\alpha\lambda_{max}$. However the eigenbeam corresponding to the smallest eigenvalue converges with a time constant of $\alpha\lambda_{min}$. If $\lambda_{max}/\lambda_{min}$ is large, then the convergence time to the optimum pattern may be unacceptably long.

All of the discussion up to (7.340) is applicable to an arbitrary MMSE filtering problem. Before proceeding, it is important to show how we use steepest descent for beamforming. We first consider a representative example and then generalize the results.

Example 7.6.1

Consider an N-element array. The desired signal has power σ_s^2 and arrives along a plane wave from $\mathbf{v}_s$. There is a single plane-wave interferer arriving along $\mathbf{v}_I$ with power σ_I^2. There is additive white noise with variance σ_w^2.

Therefore, the input correlation matrix is

$$\mathbf{S_x} = \sigma_s^2 \mathbf{v}_s \mathbf{v}_s^H + \sigma_I^2 \mathbf{v}_I \mathbf{v}_I^H + \sigma_w^2 \mathbf{I}. \tag{7.347}$$

The input is,

$$\mathbf{X}(k) = \mathbf{v}_s F(k) + \mathbf{v}_I S_I(k) + \mathbf{N}(k), \tag{7.348}$$

so

$$\mathbf{p} \triangleq \mathbf{S}_{\mathbf{x}d^*} = E[\mathbf{X}(k) F^*(k)] = \mathbf{v}_s \sigma_s^2, \tag{7.349}$$

because the desired signal is uncorrelated with the interfering signal and the additive noise.

We first expand $\mathbf{S_x}$ in an eigenvector expansion.

$$\mathbf{S_x} = \lambda_1 \mathbf{\Phi}_1 \mathbf{\Phi}_1^H + \lambda_2 \mathbf{\Phi}_2 \mathbf{\Phi}_2^H + \sum_{n=3}^{N} \lambda_n \mathbf{\Phi}_n \mathbf{\Phi}_n^H. \tag{7.350}$$

The first two eigenvectors are linear combinations of $\mathbf{v}_s$ and $\mathbf{v}_I$. The remaining eigenvectors are the noise eigenvectors and are orthogonal to $\mathbf{\Phi}_1$ and $\mathbf{\Phi}_2$ and, therefore, $\mathbf{v}_s$ and $\mathbf{v}_I$. We can rewrite (7.345) as

$$\mathbf{w}_o = \mathbf{S_x}^{-1} \mathbf{p}. \tag{7.351}$$

Substituting (7.349) and the inverse of (7.350) into (7.351) gives

$$\mathbf{w}_o = \frac{\sigma_s^2}{\lambda_1}\mathbf{\Phi}_1\left(\mathbf{\Phi}_1^H\mathbf{v}_s\right) + \frac{\sigma_s^2}{\lambda_2}\mathbf{\Phi}_2\left(\mathbf{\Phi}_2^H\mathbf{v}_s\right), \tag{7.352}$$

where we have used the orthogonality of $\mathbf{\Phi}_n, n = 3, \cdots, N$ and $\mathbf{v}_s$. Assuming $\mathbf{w}(0) = 0$, then, using (7.352) in (7.333) gives,

$$\mathbf{v}(0) = -\mathbf{U}^H\left[\frac{\sigma_s^2}{\lambda_1}\mathbf{\Phi}_1\left(\mathbf{\Phi}_1^H\mathbf{v}_s\right) + \frac{\sigma_s^2}{\lambda_2}\mathbf{\Phi}_2\left(\mathbf{\Phi}_2^H\mathbf{v}_s\right)\right], \tag{7.353}$$

so

$$v_1(0) = -\frac{\sigma_s^2}{\lambda_1}\left(\mathbf{\Phi}_1^H\mathbf{v}_s\right), \tag{7.354}$$

$$v_2(0) = -\frac{\sigma_s^2}{\lambda_2}\left(\mathbf{\Phi}_2^H\mathbf{v}_s\right), \tag{7.355}$$

and

$$v_n(0) = 0, \quad n = 3, \cdots, N. \tag{7.356}$$

Thus, there are two natural modes in the steepest descent algorithm, and the algorithm behavior can be studied in a 2-D space (the signal-plus-interference subspace).

Note that the noise power enters into the beamformer through λ_1 and λ_2,

$$\lambda_1 = \lambda_1' + \sigma_w^2, \tag{7.357}$$

$$\lambda_2 = \lambda_2' + \sigma_w^2, \tag{7.358}$$

where λ_1' and λ_2' are the eigenvalues due to the signal and interference. Therefore the eigenvalue behavior will be a function of both the signal and interference characteristics (geometry and power levels) and the white noise level.

A more realistic initial condition is to assume

$$\mathbf{w}(0) = \frac{\mathbf{v}_s}{N}. \tag{7.359}$$

This weighting corresponds to uniform quiescient weighting. We consider this case in Example 7.6.2

Example 7.6.2 (continuation)

We consider the same model as in Example 7.6.1 with the initial weight-error vector given by (7.359). The optimum weight vector $\mathbf{w}_o$ is still given by (7.352). The initial weight-error vector is

$$\mathbf{w}_e(0) = \frac{\mathbf{v}_s}{N} - \frac{\sigma_s^2}{\lambda_1}\mathbf{\Phi}_1\left(\mathbf{\Phi}_1^H\mathbf{v}_s\right) - \frac{\sigma_s^2}{\lambda_2}\mathbf{\Phi}_2\left(\mathbf{\Phi}_2^H\mathbf{v}_s\right), \tag{7.360}$$

and

$$\mathbf{v}(0) = \mathbf{U}^H\frac{\mathbf{v}_s}{N} - \mathbf{U}^H\left\{\frac{\sigma_s^2}{\lambda_1}\mathbf{\Phi}_1\left(\mathbf{\Phi}_1^H\mathbf{v}_s\right) + \frac{\sigma_s^2}{\lambda_2}\mathbf{\Phi}_2\left(\mathbf{\Phi}_2^H\mathbf{v}_s\right)\right\}. \tag{7.361}$$

Then,

$$v_n(0) = \mathbf{\Phi}_n^H\frac{\mathbf{v}_s}{N} - \mathbf{\Phi}_n^H\left\{\frac{\sigma_s^2}{\lambda_1}\mathbf{\Phi}_1\left(\mathbf{\Phi}_1^H\mathbf{v}_s\right) + \frac{\sigma_s^2}{\lambda_2}\mathbf{\Phi}_2\left(\mathbf{\Phi}_2^H\mathbf{v}_s\right)\right\}. \tag{7.362}$$

Using the orthogonality of the eigenvectors,

$$v_1(0) = \left(\frac{1}{N} - \frac{\sigma_s^2}{\lambda_1}\right) \mathbf{\Phi}_1^H \mathbf{v}_s, \tag{7.363}$$

$$v_2(0) = \left(\frac{1}{N} - \frac{\sigma_s^2}{\lambda_2}\right) \mathbf{\Phi}_2^H \mathbf{v}_s, \tag{7.364}$$

and

$$v_n(0) = 0, \quad n = 3, \cdots, N. \tag{7.365}$$

Thus, once again there are two natural modes in the steepest descent algorithm. We only get a non-zero $v_n(0)$ for $n \geq 3$ when $\mathbf{w}(0)$ has components outside of the $\mathbf{\Phi}_1^H, \mathbf{\Phi}_2^H$ subspace. This behavior occurs if we use a quiescient weighting such as Dolph-Chebychev to obtain lower sidelobes in the quiescient beam pattern.

The results in Examples 7.6.1 and 7.6.2 generalize directly to the general case of a plane-wave desired signal and D interfering plane-wave signals. In this case, the steepest descent algorithm is operating in a $D+1$ subspace.

We next investigate the transient behavior of the MSE. We then return to Examples 7.6.1 and 7.6.2 and consider its behavior for various signal-and-interference environments.

The performance is determined by the transient behavior of the MSE. Using (7.320), (7.324), and (7.331), we have

$$\xi(K) = \xi_0 + \sum_{n=1}^{N} \lambda_n \left|v_n(K)\right|^2, \tag{7.366}$$

where $\xi(K)$ is defined to be the MSE at sample K.[17] Substituting (7.335) into (7.366), we obtain

$$\xi(K) = \xi_0 + \sum_{n=1}^{N} \lambda_n \left(1 - \alpha\lambda_n\right)^{2K} \left|v_n(0)\right|^2, \tag{7.367}$$

and the transient MSE is

$$\begin{aligned} \xi_{sd}(K) &= \xi(K) - \xi_0 \\ &= \sum_{n=1}^{N} \lambda_n \left(1 - \alpha\lambda_n\right)^{2K} \left|v_n(0)\right|^2, \end{aligned} \tag{7.368}$$

where the subscript "*sd*" denotes steepest descent. The transient behavior of the error is a sum of geometric series, each one corresponding to a mode

[17]Note that $\xi(K)$ is an ensemble average at iteration K and is not the same as the weighted residual error $\xi(K)$ in (7.275).

of the algorithm. We illustrate the behavior with the same model as in Example 7.6.1.

Example 7.6.3 (continuation)

We consider the same model as in Example 7.6.1. One can show that

$$\frac{\xi_o}{\sigma_s^2} = \frac{1 + N\,(INR)}{1 + N\,(SNR + INR) + N^2\,(SNR)\,(INR)\left(1 - |\rho_{SI}|^2\right)} \tag{7.369}$$

where

$$INR = \frac{\sigma_I^2}{\sigma_w^2}, \tag{7.370}$$

and

$$SNR = \frac{\sigma_s^2}{\sigma_w^2}. \tag{7.371}$$

Using (7.354) and (7.355) in (7.368), we see that the normalized transient error is

$$\frac{\xi_{sd}(K)}{\sigma_s^2} = \frac{\sigma_s^2}{\lambda_1}\,(1 - \alpha\lambda_1)^{2K}\left|\mathbf{\Phi}_1^H \mathbf{v}_s\right|^2 + \frac{\sigma_s^2}{\lambda_2}\,(1 - \alpha\lambda_2)^{2K}\left|\mathbf{\Phi}_2^H \mathbf{v}_s\right|^2. \tag{7.372}$$

From the above discussion we see that the behavior of the algorithm will be determined by

(i) The step size α;

(ii) The eigenvalue spread $\lambda_{max}/\lambda_{min}$ in $\mathbf{S_x}$.

We now consider two simple examples to illustrate the behavior. From Chapter 5, we know how the plane-wave spatial distribution will affect the eigenvalues. When looking at the results, it is important to remember that we are using ensemble averages. By contrast, in Sections 7.3 to 7.5, the recursions used time averages. In Section 7.7, we will use time averages and obtain a more valid comparison.

In the next two examples, we consider a standard 10-element linear array. We assume that there is a 10 dB *SNR* plane-wave signal arriving from broadside and interfering plane-wave signals arriving from various u_I. By varying the power of the interfering signal and its location we can see how the eigenvalue spread affects the algorithm performance. We use $\mathbf{w}(0) = \mathbf{v}_s/N$ as the initial condition.

In each example we show two figures. The first is a representative beam pattern for various K, and the second is a plot of the transient MSE versus K.

We have chosen the parameters to show the effect of different eigenvalue spreads. We have used a scenario with a single signal to keep the plots reasonably simple.

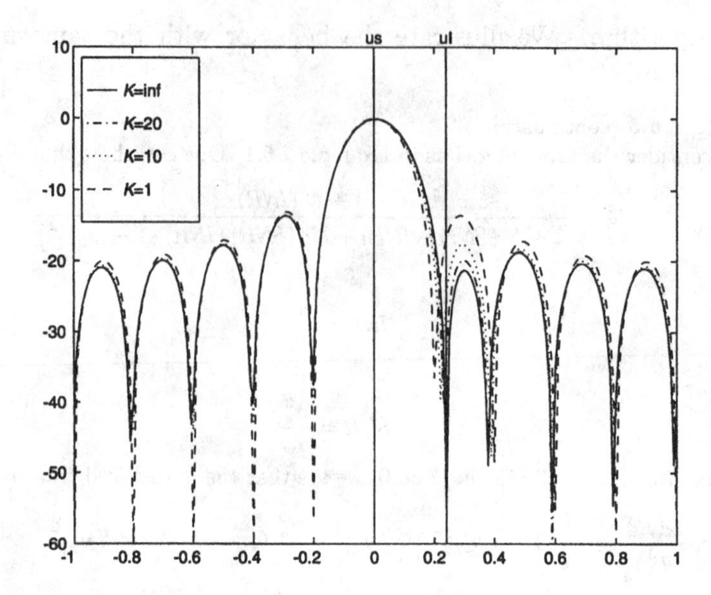

Figure 7.33: Steepest descent beamformer; beam patterns for various K; $u_s = 0, SNR = 10$ dB, $u_I = 0.24, INR = 10$ dB, $\mathbf{w}(0) = \mathbf{v}_s/N, \alpha = 0.001$.

Example 7.6.4

In this example, we use the following parameters:

$$SNR = 10 \text{ dB}, \quad u_s = 0,$$
$$INR = 10 \text{ dB}, \quad u_I = 0.24.$$

The resulting eigenvalues are

$$\lambda_1 = 117$$
$$\lambda_2 = 85.$$

We use $\alpha = 0.001$.

In Figure 7.33, we show the beam patterns for various K for $\mathbf{w}(0) = \mathbf{v}_s/N$. In Figure 7.34, we show the transient mean square error $\xi_{sd}(K)$.

In this case the eigenvalue ratio is 1.376 and the two orthogonal modes behave in a similar manner.

The steepest descent beamformer has reached steady state by $K = 30$. The small eigenvalue spread is a favorable scenario for the SD algorithm.

Example 7.6.5

In this example, we use the following parameters:

$$u_s = 0 \qquad SNR = 10 \text{ dB},$$
$$u_{I1} = 0.29 \qquad INR_1 = 20 \text{ dB},$$
$$u_{I2} = -0.45 \qquad INR_2 = 40 \text{ dB}.$$

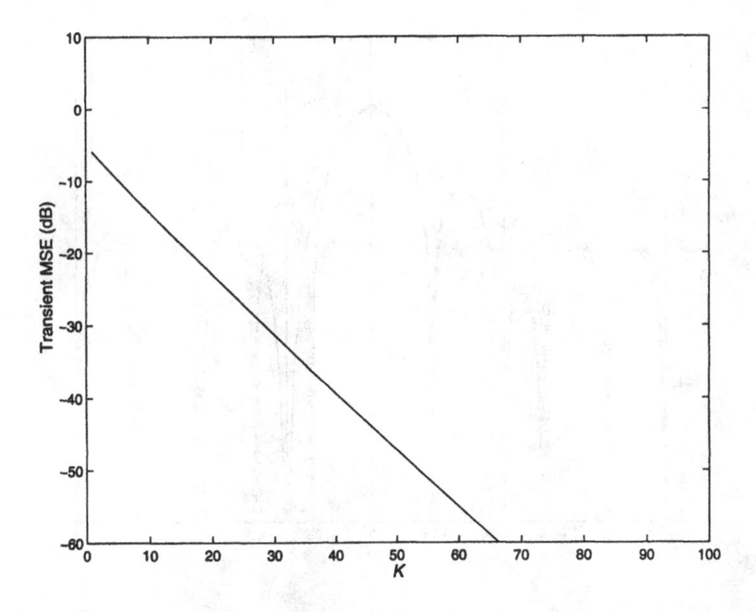

Figure 7.34: Steepest descent beamformer; $\xi_{sd}(K)$ versus K.

The resulting eigenvalues are

$$\lambda_1 = 1.0001 \times 10^5,$$
$$\lambda_2 = 0.01 \times 10^5, \quad \lambda_3 = 0.0009 \times 10^5.$$

We use $\alpha = 4.9995 \times 10^{-6}$.

In Figure 7.35, we show the beam patterns for various K for $\mathbf{w}(0) = \mathbf{v}_s/N$. In Figure 7.36, we show the transient mean square error $\xi_{sd}(K)$.

In this case, the dominant eigenvector is highly correlated with the array manifold vector for the 40 dB interferer at $u_{I2} = -0.45$, so it is nulled quickly. The convergence of the beam pattern at $u_{I1} = 0.29$ is much slower and, by $K = 100$, the beam pattern has not started to form a null.

This completes our initial discussion of the steepest descent algorithm for solving the MMSE problem. We return to it in the next section in the context of generalized sidelobe canceller.

7.6.3 Steepest Decent: LCMP Beamformer

In this section, we derive a steepest descent algorithm to find the LCMP beamformer that we originally studied in Section 6.7. The result is due to Frost[18] [Fro72], and our derivation follows that reference.

[18][Fro72] actually considers a gain-only constraint, but points out that the extension to multiple constraints is straightforward. He also considers the broadband model using a

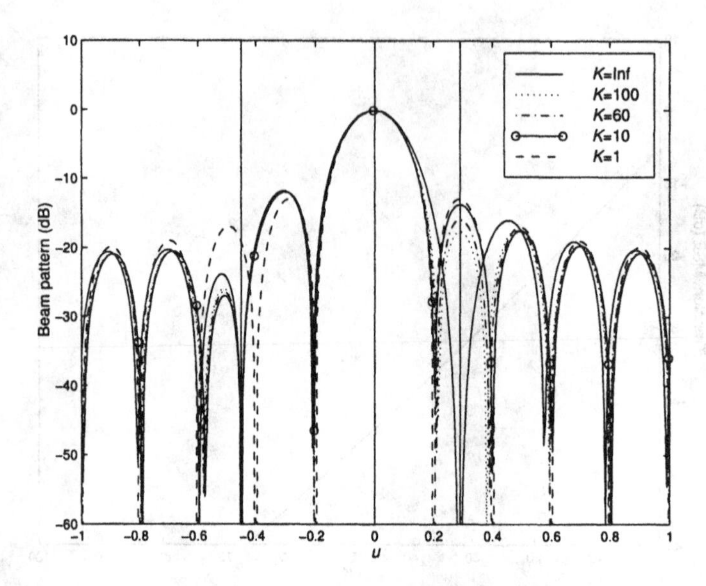

Figure 7.35 Steepest descent beamformer: $u_s = 0, SNR = 10$ dB, $u_{I1} = 0.29, INR_1 = 20$ dB, $u_{I2} = -0.45$, $INR_2 = 40$ dB, $\mathbf{w}(0) = \mathbf{v}_s/N$, $\alpha = 4.9995 \times 10^{-6}$; beam patterns for various K.

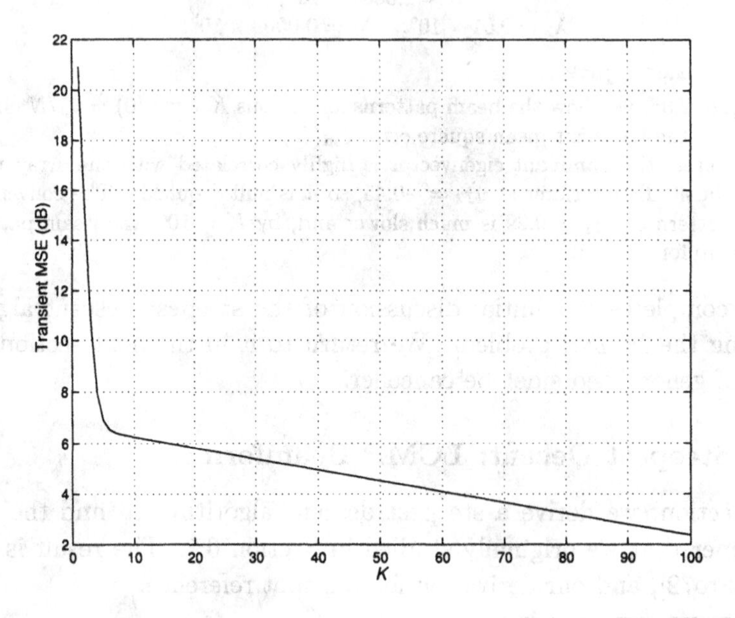

Figure 7.36 Steepest descent beamformer: $\xi_{sd}(K)$ versus K.

From (6.353), we want to minimize

$$J \triangleq \mathbf{w}^H \mathbf{S_x} \mathbf{w} + \left[\mathbf{w}^H \mathbf{C} - \mathbf{g}^H\right] \boldsymbol{\lambda} + \boldsymbol{\lambda}^H \left[\mathbf{C}^H \mathbf{w} - \mathbf{g}\right]. \tag{7.373}$$

The gradient with respect to $\mathbf{w}^H$ is

$$\nabla_{\mathbf{w}^H} = \mathbf{S_x w} + \mathbf{C}\boldsymbol{\lambda}. \tag{7.374}$$

In Section 7.6.2, we set the gradient equal to zero and found that

$$\mathbf{w}_o = \mathbf{S_x}^{-1} \mathbf{C} \left[\mathbf{C}^H \mathbf{S_x} \mathbf{C}\right]^{-1} \mathbf{g} \tag{7.375}$$

(see (6.358)).

To find $\mathbf{w}$ using a steepest descent algorithm, we follow the same procedure as in Section 7.6.2.

$$\begin{aligned} \mathbf{w}(K) &= \mathbf{w}(K-1) - \alpha \nabla_{\mathbf{w}^H} \left(J(K-1)\right) \\ &= \mathbf{w}(K-1) - \alpha \left[\mathbf{S_x w}(K-1) + \mathbf{C}\boldsymbol{\lambda}(K-1)\right]. \end{aligned} \tag{7.376}$$

We require $\mathbf{w}(K)$ to satisfy the constraint.

$$\mathbf{g} = \mathbf{C}^H \mathbf{w}(K) = \mathbf{C}^H \mathbf{w}(K-1) - \alpha \mathbf{C}^H \mathbf{S_x w}(K-1) - \alpha \mathbf{C}^H \mathbf{C} \boldsymbol{\lambda}(K-1). \tag{7.377}$$

Solving for $\boldsymbol{\lambda}(K-1)$ and substituting into (7.376) gives

$$\begin{aligned} \mathbf{w}(K) &= \mathbf{w}(K-1) - \alpha \left(\mathbf{I} - \mathbf{C}\left(\mathbf{C}^H \mathbf{C}\right)^{-1} \mathbf{C}^H\right) \mathbf{S_x w}(K-1) \\ &\quad + \mathbf{C}\left(\mathbf{C}^H \mathbf{C}\right)^{-1} \left[\mathbf{g} - \mathbf{C}^H \mathbf{w}(K-1)\right]. \end{aligned} \tag{7.378}$$

Frost [Fro72] points out that one should not assume that the term in the last brackets is zero. By retaining it we improve the numerical stability compared to previous algorithms ([Ros60], [BOBH69]).

From (6.367),

$$\mathbf{w}_q = \mathbf{C}\left(\mathbf{C}^H \mathbf{C}\right)^{-1} \mathbf{g}, \tag{7.379}$$

and from (6.364),

$$\mathbf{P}_{\mathbf{C}}^{\perp} = \left[\mathbf{I} - \mathbf{C}\left(\mathbf{C}^H \mathbf{C}\right)^{-1} \mathbf{C}^H\right]. \tag{7.380}$$

---

tapped-delay line. We study that case in Section 7.11.

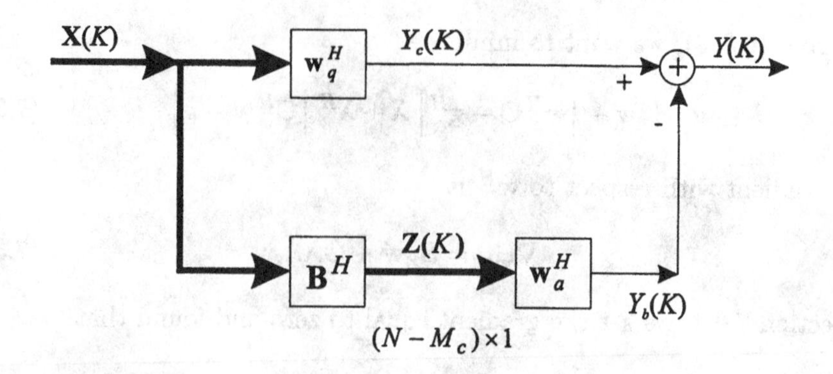

Figure 7.37: Generalized sidelobe canceller.

We can write (7.378) as

$$\mathbf{w}(K) = \mathbf{P}_{\mathbf{C}}^{\perp}\left[\mathbf{w}(K-1) - \alpha \mathbf{S}_{\mathbf{x}} \mathbf{w}(K-1)\right] + \mathbf{w}_q, \tag{7.381}$$

or

$$\boxed{\mathbf{w}(K) = \mathbf{P}_{\mathbf{C}}^{\perp}\left[\mathbf{I} - \alpha \mathbf{S}_{\mathbf{x}}\right] \mathbf{w}(K-1) + \mathbf{w}_q,} \tag{7.382}$$

with

$$\boxed{\mathbf{w}(0) = \mathbf{w}_q.} \tag{7.383}$$

Before we analyze the error behavior of the steepest descent LCMP algorithm we want to develop the generalized sidelobe canceller version of the algorithm.

The GSC implementation was developed in Section 7.4.4 and is shown in Figure 7.37.

Recall that

$$\mathbf{B}^H \mathbf{C} = \mathbf{0}. \tag{7.384}$$

In addition, we assume that the columns of $\mathbf{B}$ are orthonormal,

$$\mathbf{B}^H \mathbf{B} = \mathbf{I}. \tag{7.385}$$

We denote the total weight vector in Figure 7.37 by $\mathbf{w}_{gsc}$,

$$\mathbf{w}_{gsc} = \mathbf{w}_q - \mathbf{B}\mathbf{w}_a. \tag{7.386}$$

The steepest descent algorithm for $\mathbf{w}_a$ is just the MMSE-SD algorithm in (7.326) with the desired signal equal to $Y_c(k)$. The result is

$$\boxed{\mathbf{w}_a(K) = \mathbf{w}_a(K-1) + \alpha\left[\mathbf{p}_{\mathbf{B}} - \mathbf{S}_{\mathbf{z}} \mathbf{w}_a(K-1)\right],} \tag{7.387}$$

where

$$\boxed{\mathbf{S_z} \triangleq \mathbf{B}^H \mathbf{S_x} \mathbf{B},} \tag{7.388}$$

and

$$\boxed{\mathbf{p_B} \triangleq E[\mathbf{Z}D^*] = \mathbf{B}^H \mathbf{S_x} \mathbf{C} \left(\mathbf{C}^H \mathbf{C}\right)^{-1} \mathbf{g}.} \tag{7.389}$$

One can show that if we use identical initial conditions in the direct form LCMP-SD beamformer and the GSC version of the LCMP-SD beamformer,

$$\mathbf{w}_{gsc}(0) = \mathbf{w}_{lcmp}(0) = \mathbf{w}_q, \tag{7.390}$$

which implies

$$\mathbf{w}_a(0) = \mathbf{0}, \tag{7.391}$$

then the behavior of the two SD algorithms will be identical ((7.385) must be satisfied).

We can analyze the weight-error vector of the LCMP-SD beamformers in exactly the same manner as Section 7.6.2. Define

$$\mathbf{w}_{ae}(K) = \mathbf{w}_a(K) - \mathbf{w}_{ao}, \tag{7.392}$$

where $\mathbf{w}_{ao}$ is the optimum weight vector given by

$$\mathbf{p_B} = \mathbf{S_z} \mathbf{w}_{ao}. \tag{7.393}$$

Substituting (7.392) into (7.387) gives

$$\mathbf{w}_{ae}(K) = [\mathbf{I} - \alpha \mathbf{S_z}]\, \mathbf{w}_{ae}(K-1), \tag{7.394}$$

which has the same form as (7.328). To analyze the behavior, rotate into an orthogonal coordinate system,

$$\mathbf{S_z} = \mathbf{B}^H \mathbf{S_x} \mathbf{B} = \mathbf{U}_c \boldsymbol{\lambda}_c \mathbf{U}_c^H. \tag{7.395}$$

Then

$$\mathbf{v}_c(K) = \mathbf{U}_c^H \mathbf{w}_e(K) = [\mathbf{I} - \alpha \boldsymbol{\Lambda}_c]\, \mathbf{v}_c(K-1). \tag{7.396}$$

If there are M constraints, then $\mathbf{S_z}$ will have N–M eigenvalues and eigenvectors. One can show that

$$\lambda_{min} \leq \lambda_{B,min} \leq \lambda_{B,max} \leq \lambda_{max} \tag{7.397}$$

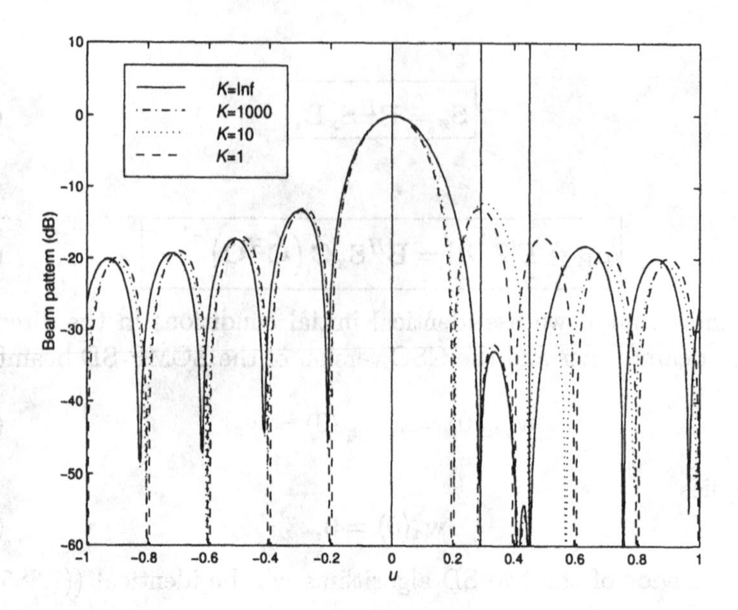

Figure 7.38 MPDR-GSC-SD beamformer, beam patterns for various K; $u_s = 0$, $SNR = 10$ dB, $u_{I1} = 0.29$, $INR_1 = 10$ dB, $u_{I2} = 0.45$, $INR_2 = 30$ dB, $\alpha = 0.5/\lambda_{max}$.

(see [Fro72]). Therefore the convergence depends on fewer modes, and the rates are between the slowest and fastest MMSE modes.

We consider a simple example to illustrate the behavior. The eigenvalues of $\mathbf{S_z}$ determine the convergence rate, so we need a scenario with multiple interferers in order to have an interesting example.

Example 7.6.6

Consider a SLA10 array and the same signal-and-interference model as in Example 7.4.1. The desired signal arrives from $u_s = 0$ with an $SNR = 10$ dB. Two uncorrelated interferers arrive from $u_I = 0.29$ and $u_I = 0.45$. The INR for the interferer at 0.29 is 10 dB. The INR for the interferer at 0.45 is varied from 10 dB to 60 dB in 10 dB steps. We implement an MPDR beamformer with a single distortionless constraint at $u_s = 0$. The initial condition is $\mathbf{w}_a = 0$ and $\mathbf{w}_q = \mathbf{v}_s/N$. We use $\alpha = 0.5/\lambda_{max}$.

In Figure 7.38, we show representative beam patterns for various K with $INR_2 = 30$ dB. We see interferer 2 is nulled by $K = 10$, but interferer 1 is not nulled until $K = 1000$. In Figure 7.39, we plot the $SINR_o$ versus K. As the eigenvalue spread increases, the convergence to steady state slows dramatically.

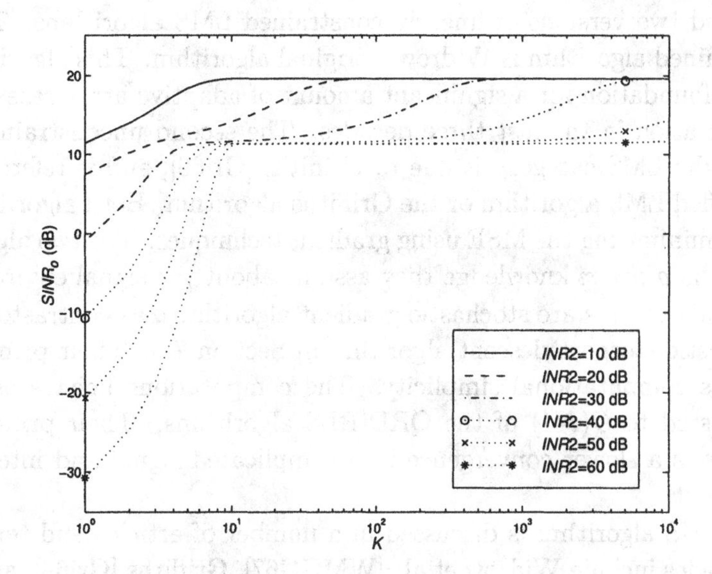

Figure 7.39 MPDR-GSC-SD beamformer, $SINR_o$ versus K; $u_s = 0$, $SNR =$ 10 dB, $u_{I1} = 0.29$, $INR_1 = 10$ dB, $u_{I2} = 0.45$, $INR_2 = 10$ dB, $\cdots$, 60 dB, $\alpha = 0.5/\lambda_{max}$.

7.6.4 Summary

In this section, we have discussed the steepest descent algorithm for the cases of most interest in adaptive beamforming, the MMSE beamformer and the LCMP beamformer. We have demonstrated that they converge to the optimum beamformer if the step size α is chosen appropriately.

As we pointed out in the introduction, there are other deterministic gradient algorithms. Various Newton-type algorithms are most widely used. A discussion of them is available in Chapter 4 of [WS85] in the context of adaptive filtering.

In the next section we develop the stochastic gradient version of these algorithms and analyze their performance.

7.7 LMS Algorithms

In this section we develop the least mean-square (LMS) algorithm. It was originated by Widrow and his colleagues at Stanford University (e.g., [WS85], [WMGG67]). We develop two versions of unconstrained MMSE-LMS algorithms and two versions of linearly constrained LMS algorithms. The first unconstrained algorithm is Widrow's original algorithm. This algorithm has been the foundation for a significant amount of adaptive array research and implementation in the last three decades. The second unconstrained algorithm in the LMS category is due to Griffiths [Gri69], and is referred to as the modified LMS algorithm or the Griffiths algorithm. Both algorithms are based on minimizing the MSE using gradient techniques. The two algorithms differ in the *a priori* knowledge they assume about the signal environment.

Both algorithms are stochastic gradient algorithms, as contrasted to the deterministic steepest descent algorithm in Section 7.6. Their primary advantage is computational simplicity. The computations behave as $O(N)$, as contrasted to $O(N^2)$ of the QRD-RLS algorithms. Their primary disadvantage is a slower convergence in a complicated signal and interference environment.

The LMS algorithm is discussed in a number of articles and textbooks. Early articles include Widrow et al. [WMGG67], Griffiths [Gri69], and Frost [Fro72]. A recent tutorial article is Glentis et al. [GBT99], which has an extensive reference list. Book references include Compton [Com88], Monzingo and Miller [MM80], Widrow and Stearns [WS85], Sibul [Sib87], and Haykin [Hay96].

In Section 7.7.1, we derive the LMS algorithms. In Section 7.7.2, we study the statistical properties of the LMS algorithms. In Section 7.7.3, we demonstrate the algorithm behavior for several interesting cases. In Section 7.7.4, we consider LMS algorithms with quadratic constraints. In Section 7.7.5, we summarize our results.

7.7.1 Derivation of the LMS Algorithms

The LMS algorithm is the stochastic version of the steepest descent algorithm. In this section, we derive several versions of LMS algorithms.

7.7.1.1 Widrow LMS algorithm

The first LMS algorithm of interest is due to Widrow et al. [WH60], [WMGG67], and is the stochastic version of the steepest descent algorithm in (7.322).

The gradient in the SD algorithm was given by (7.322), which we repeat

$$\nabla \xi_{\mathbf{w}^H} = -\mathbf{p} + \mathbf{S}_{\mathbf{x}}\mathbf{w}. \tag{7.398}$$

In the LMS algorithm, we must estimate $\mathbf{p}$ and $\mathbf{S}_{\mathbf{x}}$.

A simple choice for estimates is

$$\hat{\mathbf{p}}(K) = \mathbf{X}(K)D^*(K), \tag{7.399}$$

and

$$\hat{\mathbf{S}}_{\mathbf{x}} = \mathbf{X}(K)\mathbf{X}^H(K). \tag{7.400}$$

In other words, the instantaneous values are used as estimates. Better estimates could be obtained by time averaging,

$$\hat{\mathbf{p}}(K) = \frac{1}{N_I} \sum_{k=K-N_I}^{K} \mathbf{X}(k)D^*(k), \tag{7.401}$$

and

$$\hat{\mathbf{S}}_{\mathbf{x}} = \frac{1}{N_I} \sum_{k=K-N_I}^{K} \mathbf{X}(k)\mathbf{X}^H(k). \tag{7.402}$$

Use of (7.401) and (7.402) leads to an LMS algorithm that is referred to as a sliding window LMS (SW-LMS) algorithm in the literature.

We could also use exponential weighting in the averages, as in Sections 7.3 and 7.4.

We use (7.399) and (7.400) in our present discussion. We find that the LMS algorithm provides an averaging effect. Then, the estimate of the gradient is

$$\hat{\nabla}\xi(K) = -\mathbf{X}(K)D^*(K) + \mathbf{X}(K)\mathbf{X}^H(K)\hat{\mathbf{w}}(K), \tag{7.403}$$

and the LMS algorithm is

$$\hat{\mathbf{w}}(K) = \hat{\mathbf{w}}(K-1) + \alpha(K)\mathbf{X}(K)\left[D^*(K) - \mathbf{X}^H(K)\hat{\mathbf{w}}(K-1)\right]. \tag{7.404}$$

We use the notation $\hat{\mathbf{w}}(K)$ because the algorithm is using an estimate of the gradient instead of the actual gradient. When we write the algorithm as in (7.404), it is an open-loop algorithm and its structure is similar to the RLS algorithm in Section 7.4. We will discuss the similarities in more detail in Section 7.7.1.5.

We can also write (7.404) as

$$\boxed{\begin{aligned} \hat{\mathbf{w}}(K) &= \hat{\mathbf{w}}(K-1) + \alpha(K)\mathbf{X}(K)\left[D^*(K) - \tilde{Y}_p^*(K)\right] \\ &= \hat{\mathbf{w}}(K-1) + \alpha(K)\mathbf{X}(K)e_p^*(K), \end{aligned}} \tag{7.405}$$

where

$$\tilde{Y}_p^*(K) \triangleq \mathbf{X}^H(K)\hat{\mathbf{w}}(K-1). \tag{7.406}$$

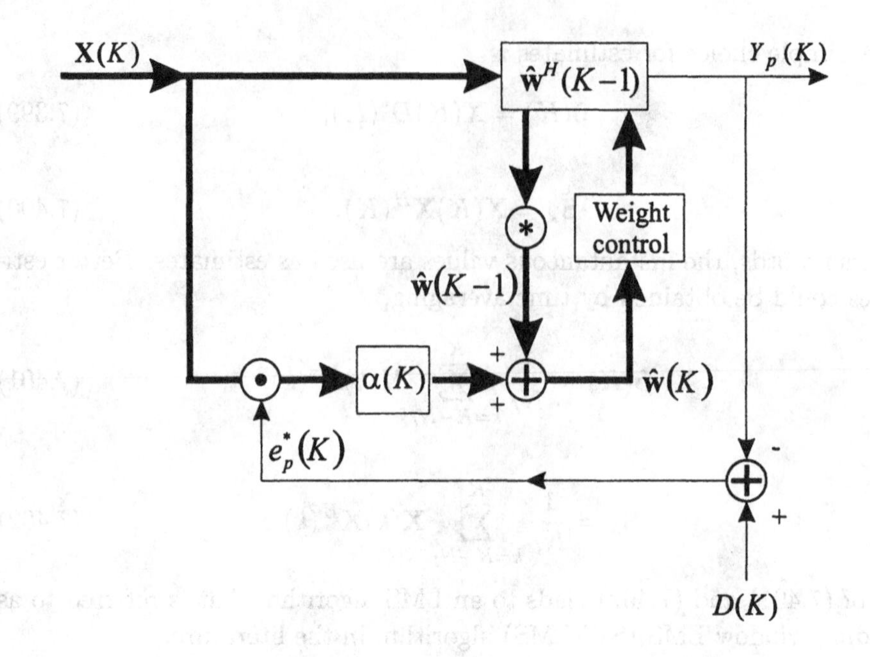

Figure 7.40 Block diagram of LMS algorithm.

A block diagram is shown in Figure 7.40.

If we were dealing with infinite precision arithmetic and perfect components, then (7.404) and (7.405) are identical. However, in an actual system there will be errors, so the LMS algorithm uses the actual output $\tilde{Y}_p^*(K)$ shown in Figure 7.40 and operates as a closed-loop system. The closed-loop operation tends to compensate for errors.

The initial condition, $\hat{\mathbf{w}}(0)$, will depend on the model. Common choices are $\hat{\mathbf{w}}(0) = \mathbf{0}$, $\hat{\mathbf{w}}(0) = \mathbf{v}_s/N$, or $\hat{\mathbf{w}}(0) = \mathbf{w}_q$.

At each iteration, $\mathbf{X}(K)$, $D(K)$, and $\hat{\mathbf{w}}(K)$ are required. The estimates at a particular time may have large variances, and they introduce what is referred to as gradient noise. The result in (7.404) is the complex narrowband version of Widrow's original LMS algorithm (see [WMB75]).

Just as in the RLS case, introducing diagonal loading can provide improved performance in many cases. To include diagonal loading, modify (7.404) to

$$\begin{aligned}\hat{\mathbf{w}}(K) &= \hat{\mathbf{w}}(K-1) + \alpha(K)\mathbf{X}(K)D^*(K) \\ &\quad -\alpha(K)\left\{\left[\sigma_L^2\mathbf{I} + \mathbf{X}(K)\mathbf{X}^H(K)\right]\hat{\mathbf{w}}(K-1)\right\}, \end{aligned} \tag{7.407}$$

which can be rewritten as,

$$\hat{\mathbf{w}}(K) = \left(1 - \alpha(K)\sigma_L^2\right)\hat{\mathbf{w}}(K-1) + \alpha(K)\mathbf{X}(K)\left[D^*(K) - \tilde{Y}_p^*(K)\right], \quad (7.408)$$

or

$$\boxed{\hat{\mathbf{w}}(K) = \beta_L(K)\hat{\mathbf{w}}(K-1) + \alpha(K)\mathbf{X}(K)\left[D^*(K) - \tilde{Y}_p^*(K)\right],} \quad (7.409)$$

where

$$\beta_L(K) = 1 - \alpha(K)\sigma_L^2. \quad (7.410)$$

We can also write (7.409) as

$$\hat{\mathbf{w}}(K) = \beta_L(K)\hat{\mathbf{w}}(K-1) + \alpha(K)\mathbf{X}(K)e_p^*(K). \quad (7.411)$$

We can also let the diagonal loading vary as a function of K by replacing σ_L^2 with $\sigma_L^2(K)$.

Due to its simplicity, the LMS algorithm can also be implemented using analog techniques. Analog implementations are discussed in [Com88].

7.7.1.2 Griffiths LMS algorithm

A second version of the LMS algorithm is due to Griffiths ([Gri68], [Gri69]), and is referred to as the modified LMS algorithm, the Griffiths algorithm, or the steered direction algorithm. It assumes that

$$\mathbf{S}_{\mathbf{x}d^*} = E\left[\mathbf{X}(k)D^*(k)\right], \quad (7.412)$$

is known. In the narrowband case,

$$\mathbf{X}(k) = \mathbf{v}_s F(k) + \mathbf{N}(k), \quad (7.413)$$

and

$$D^*(k) = F^*(k), \quad (7.414)$$

so

$$\mathbf{S}_{\mathbf{x}d^*} = \sigma_s^2 \mathbf{v}_s. \quad (7.415)$$

Thus, the Griffiths algorithm assumes the signal direction of arrival and the signal power are known. However, the desired signal $D(k)$ is not required.

The modified LMS algorithm is

$$\boxed{\hat{\mathbf{w}}(K) = \hat{\mathbf{w}}(K-1) + \alpha(K)\left[\sigma_s^2\mathbf{v}_s - \mathbf{X}(K)\tilde{Y}_p^*(K)\right].} \quad (7.416)$$

The result in (7.416) is due to Griffiths [Gri69]. It is important to note that, although $\mathbf{v}_s$ appears in (7.416), it is not imposing a hard constraint.

If we utilize diagonal loading, (7.416) becomes

$$\hat{\mathbf{w}}(K) = \beta_L(K)\hat{\mathbf{w}}(K-1) + \alpha(K)\left[\sigma_s^2\mathbf{v}_s - \mathbf{X}(K)\tilde{Y}_p^*(K)\right]. \tag{7.417}$$

7.7.1.3 Frost LMS algorithm

The third version of the LMS algorithm imposes a linear constraint. This algorithm is the stochastic gradient version of the steepest descent algorithm in Section 7.6.3. Using (7.400) in (7.382) gives

$$\hat{\mathbf{w}}(K) = \mathbf{P}_{\mathbf{C}}^{\perp}\left[\hat{\mathbf{w}}(K-1) - \alpha(K)\mathbf{X}(K)\tilde{Y}_p^*(K)\right] + \mathbf{w}_q, \tag{7.418}$$

where

$$\tilde{Y}_p(K) = \hat{\mathbf{w}}^H(K-1)\mathbf{X}(K), \tag{7.419}$$

and

$$\hat{\mathbf{w}}(0) = \mathbf{w}_q. \tag{7.420}$$

For the simple case of the MPDR beamformer,

$$\mathbf{w}_q = \frac{1}{N}\mathbf{v}_s, \tag{7.421}$$

and

$$\mathbf{P}_{\mathbf{C}}^{\perp} = \mathbf{I} - \mathbf{v}_s\left(\mathbf{v}_s^H\mathbf{v}_s\right)^{-1}\mathbf{v}_s^H. \tag{7.422}$$

For the general LCMP case,

$$\mathbf{w}_q = \mathbf{C}\left(\mathbf{C}^H\mathbf{C}\right)^{-1}\mathbf{g}, \tag{7.423}$$

and

$$\mathbf{P}_{\mathbf{C}}^{\perp} = \mathbf{I} - \mathbf{C}\left(\mathbf{C}^H\mathbf{C}\right)^{-1}\mathbf{C}^H. \tag{7.424}$$

This is the narrowband complex version of the linearly constrained beamformer which was originally derived by Frost [Fro72]. Frost derived the real wideband version for a gain-only (i.e., distortionless) constraint, but indicated the extension to multiple constraints would be straightforward.

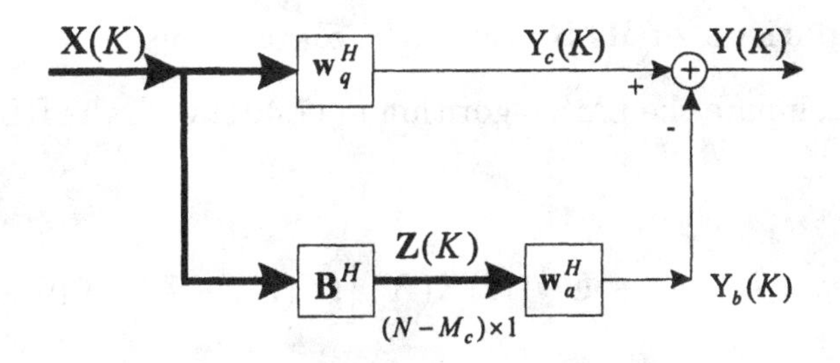

Figure 7.41 Generalized sidelobe canceller.

7.7.1.4 Generalized sidelobe canceller LMS algorithm

The fourth version of the LMS algorithm is the GSC version, which is shown in Figure 7.41. Recall that $Y_c(k)$ corresponds to $D(k)$ in the MMSE algorithm. Adapting (7.405),

$$\hat{\mathbf{w}}_a(K) = \hat{\mathbf{w}}_a(K-1) + \alpha(K)\mathbf{Z}(K)e_{pa}^*(K), \tag{7.425}$$

where

$$\begin{aligned} e_{pa}(K) &= Y_c(K) - \hat{\mathbf{w}}_a^H(K-1)\mathbf{Z}(K) \\ &= Y_c(K) - \tilde{Y}_{bp}(K). \end{aligned} \tag{7.426}$$

The total weight vector is

$$\hat{\mathbf{w}}(K) = \mathbf{w}_q - \mathbf{B}\hat{\mathbf{w}}_a(K), \tag{7.427}$$

and

$$Y(K) = \hat{\mathbf{w}}^H(K)\mathbf{X}(K). \tag{7.428}$$

The initial condition is

$$\hat{\mathbf{w}}_a(0) = \mathbf{0}. \tag{7.429}$$

The result in (7.425)–(7.429) is the narrowband complex version of the GSC beamformer derived by Griffiths and Jim [GJ82].

If we choose $\mathbf{B}$ so that

$$\mathbf{B}^H\mathbf{B} = \mathbf{I}, \tag{7.430}$$

the adaptive performance of GSC implementation will be identical to the adaptive performance of direct form implementation. Diagonal loading can be included.

7.7.1.5 Comparison of RLS and LMS algorithms

It is useful to compare the LMS algorithm in (7.404) with the RLS algorithm of Section 7.4.[19]

From (7.189),

$$\hat{\mathbf{w}}_{rls}(K) = \hat{\mathbf{w}}_{rls}(K-1) + \mathbf{\Phi}^{-1}(K)\mathbf{X}(K)\left[D^*(K) - \mathbf{X}^H(K)\hat{\mathbf{w}}_{rls}(K-1)\right] \tag{7.431}$$

and, repeating (7.404)

$$\hat{\mathbf{w}}_{lms}(K) = \hat{\mathbf{w}}_{lms}(K-1) + \alpha(K)\mathbf{X}(K)\left[D^*(K) - \mathbf{X}^H(K)\hat{\mathbf{w}}_{lms}(K-1)\right]. \tag{7.432}$$

We see that the difference between the two algorithms, as expressed by (7.431) and (7.432), is that the error term is multiplied by $\mathbf{\Phi}^{-1}(K)$ in the RLS algorithm and $\alpha(K)\mathbf{I}$ in the LMS algorithm.

The multiplication by $\mathbf{\Phi}^{-1}(K)$ causes each mode to converge at the same rate in the RLS algorithm so that the eigenvalue spread does not limit the convergence rate. The disadvantage is that computing $\mathbf{\Phi}^{-1}(K)$ requires more computation. Recall that

$$\mathbf{g}(K) = \mathbf{P}(K)\mathbf{X}(K) = \mathbf{\Phi}^{-1}(K)\mathbf{X}(K), \tag{7.433}$$

is the Kalman gain.

Our approach to deriving the LMS algorithm started with the deterministic steepest descent algorithm in Section 7.6. We replaced the deterministic quantities with estimated quantities to obtain the LMS algorithm. If we had started with a deterministic Newton algorithm and replaced the deterministic quantities with estimated quantities, we would have obtained the **LMS/Newton algorithm** or **LMSN algorithm**. This approach is discussed in Chapters 4 and 8 of [WS85]. Their discussion leads to an algorithm referred to as the **sequential regression algorithm** (**SER** algorithm) which they attribute to Graupe [Gra72], Ahmed et al. [ASHP77], Parikh and Ahmed [PA78], Ahmed et al. [AHUS79], and Lee [Lee66]. However, if the various parameters are chosen appropriately, the algorithm is identical to the RLS algorithm. The relation between the two algorithms is also discussed in Diniz et al. [DCA95].

In the next section, we derive the statistical properties of the LMS algorithms derived in Sections 7.7.1.1–7.7.1.4. Then, in Section 7.7.3, we simulate the algorithms for typical scenarios.

---

[19] The relationship between the RLS and LMS algorithms is discussed in a number of texts (e.g., Section 6.2.4 of [PRLN92] or Sections 13.6 and 13.7 of [Hay96]).

7.7.2 Performance of the LMS Algorithms

In order to understand the behavior of the LMS algorithms, we consider the following six questions:

1. Under what condition, and in what sense, does the weight vector converge to the optimum weight vector?

2. Assuming the weight vector converges, what is its transient behavior? How long does it take to converge?

3. As we are dealing with an estimate of the gradient, there may be some residual error in the weight vector. How large is this error and what does it depend on?

4. How complex is the computation required to implement the algorithm?

5. How is the desired signal derived?

6. How does the system behave for some typical examples?

We discuss the first five questions in this section and look at examples in Section 7.7.3. Most of these questions were answered in the original paper by Widrow and his colleagues [WMGG67]. Haykin [Hay96] has added several new arguments that we will utilize. There is a good tutorial discussion of the LMS algorithm and other algorithms in Glentis et al. [GBT99]. The article also contains extensive references.

In our discussion, we assume the input vectors $\mathbf{X}(1)$, $\mathbf{X}(2)$, $\cdots$, $\mathbf{X}(K)$ are statistically independent. We assume that $\mathbf{X}(K)$ is statistically independent of all previous samples of the desired response $D(1), \cdots, D(K-1)$. At time K, the desired response $D(K)$ is dependent on $\mathbf{X}(K)$, but it is statistically independent of all previous samples of the desired response. The input $\mathbf{X}(K)$ and the desired response $D(K)$ are assumed to be jointly Gaussian for all K.

7.7.2.1 Weight-error vector mean

We define the weight-error vector in the LMS algorithm as

$$\hat{\mathbf{w}}_e(K) = \hat{\mathbf{w}}(K) - \mathbf{w}_o. \tag{7.434}$$

Using (7.434) in (7.404) gives

$$\hat{\mathbf{w}}_e(K) = \hat{\mathbf{w}}_e(K-1) + \alpha(K)\mathbf{X}(K)\left[D^*(K) - \mathbf{X}^H(K)\left(\hat{\mathbf{w}}_e(K-1) + \mathbf{w}_o\right)\right]. \tag{7.435}$$

Taking the expectation of both sides, using the independence assumption, and using (7.323), we obtain

$$E\left[\hat{\mathbf{w}}_e(K)\right] = \left[\mathbf{I} - \alpha(K)\mathbf{S}_{\mathbf{x}}\right] E\left[\hat{\mathbf{w}}_e(K-1)\right]. \tag{7.436}$$

We see that (7.436) and (7.328) have identical form. Since $E\left[\hat{\mathbf{w}}_e(0)\right] = \mathbf{w}_e(0)$,

$$E\left[\hat{\mathbf{w}}_e(K)\right] = \mathbf{w}_e(K), \tag{7.437}$$

so that the mean of the LMS weight-error vector is the weight-error vector in the steepest descent algorithm.

We also observe that we could obtain the same result for the Griffiths LMS algorithm (7.416).

7.7.2.2 Weight-error correlation matrix

The correlation matrix of $\hat{\mathbf{w}}_e(K)$ is[20]

$$\mathbf{R}_e(K) = E\left[\hat{\mathbf{w}}_e(K)\hat{\mathbf{w}}_e^H(K)\right]. \tag{7.438}$$

To analyze (7.438), we rewrite (7.435) as

$$\begin{aligned}\hat{\mathbf{w}}_e(K) &= \hat{\mathbf{w}}_e(K-1) + \alpha(K)\mathbf{X}(K)\left[e_o^*(K) - \mathbf{X}^H(K)\hat{\mathbf{w}}_e(K-1)\right] \\ &= \left[\mathbf{I} - \alpha(K)\mathbf{X}(K)\mathbf{X}^H(K)\right]\hat{\mathbf{w}}_e(K-1) + \alpha(K)\mathbf{X}^H(K)e_o(K),\end{aligned} \tag{7.439}$$

where

$$e_o(K) = D(K) - \mathbf{w}_o^H\mathbf{X}(K), \tag{7.440}$$

is the estimation error using the MMSE weight vector. If $\alpha(K)$ is small, then the coefficient of $\hat{\mathbf{w}}_e(K-1)$ in (7.439) is close to $\mathbf{I}$ and the behavior of $\hat{\mathbf{w}}_e(K-1)$ may be approximated by the stochastic difference equation:[21]

$$\hat{\mathbf{w}}_e(K) = \left[\mathbf{I} - \alpha(K)\mathbf{S}_{\mathbf{x}}\right]\hat{\mathbf{w}}_e(K-1) + \alpha(K)\mathbf{X}^H(K)e_o(K). \tag{7.441}$$

Substituting (7.441) into (7.438) gives

$$\begin{aligned}\mathbf{R}_e(K) &= \left[\mathbf{I} - \alpha(K)\mathbf{S}_{\mathbf{x}}\right]\mathbf{R}_e(K-1)\left[\mathbf{I} - \alpha(K)\mathbf{S}_{\mathbf{x}}\right] \\ &\quad + \alpha^2(K)E\left[e_o^*(K)\mathbf{X}(K)\mathbf{X}^H(K)e_o(K)\right].\end{aligned} \tag{7.442}$$

[20]This discussion is similar to Chapter 9 of [Hay96].

[21]This approach is due to Kushner [Kus84] and is described on p. 396 of [Hay96].

Using the moment factorization of jointly Gaussian random variables and the independence assumption,

$$E\left[e_o^*(K)\mathbf{X}(K)\mathbf{X}^H(K)e_o(K)\right] = \xi_o \mathbf{S_x}, \tag{7.443}$$

where

$$\xi_o = E\left[|e_o(K)|^2\right], \tag{7.444}$$

is the MSE using the optimum weight vector. Then, (7.442) can be written as

$$\mathbf{R}_e(K) = [\mathbf{I} - \alpha(K)\mathbf{S_x}]\,\mathbf{R}_e(K-1)\,[\mathbf{I} - \alpha(K)\mathbf{S_x}] + \alpha^2(K)\xi_0\mathbf{S_x}. \tag{7.445}$$

We can use (7.445) to compute the excess MSE. The LMS error can be written as

$$e_{lms}(K) = D(K) - \hat{\mathbf{w}}^H(K)\mathbf{X}(K). \tag{7.446}$$

Using (7.434) and (7.440) in (7.446) gives

$$e_{lms}(K) = e_o(K) - \hat{\mathbf{w}}_e^H(K)\mathbf{X}(K). \tag{7.447}$$

The mean-square LMS error at iteration K is

$$\begin{aligned}
\xi_{lms}(K) &= E\left[\left|e_o(K) - \hat{\mathbf{w}}_e^H(K)\mathbf{X}(K)\right|^2\right] \\
&= \xi_o + E\left[\hat{\mathbf{w}}_e^H(K)\mathbf{X}(K)\mathbf{X}^H(K)\hat{\mathbf{w}}_e(K)\right] \\
&= \xi_o + tr\left\{E\left[\mathbf{X}(K)\mathbf{X}^H(K)\right]E\left[\hat{\mathbf{w}}_e(K)\hat{\mathbf{w}}_e^H(K)\right]\right\} \\
&= \xi_o + tr\left\{\mathbf{S_x}\mathbf{R}_e(K)\right\}.
\end{aligned} \tag{7.448}$$

Now define an excess MSE,

$$\xi_{ex}(K) = \xi_{lms}(K) - \xi_o = tr\left\{\mathbf{S_x}\mathbf{R}_e(K)\right\}. \tag{7.449}$$

The matrix inside the trace is positive definite so, at each value of K, the excess error is positive. The $\mathbf{R}_e(K)$ term in (7.448) can be evaluated using (7.445). Using results by Macchi [Mac95], Haykin (pp. 399–402 of [Hay96]) analyzes the transient behavior. As $K \to \infty$, the excess MSE approaches a constant value if

$$0 < \alpha(K) < \frac{2}{\lambda_{max}}, \tag{7.450}$$

where λ_{max} is the largest eigenvalue of $\mathbf{S_x}$. If (7.450) is satisfied, the steady state excess error is

$$\xi_{ex}(\infty) = \xi_o \sum_{i=1}^{N} \frac{\alpha(K)\lambda_i}{2 - \alpha(K)\lambda_i} \tag{7.451}$$

(e.g., pp. 397–400 of [Hay96]). The misadjustment is defined as the ratio of the excess error $\xi_{ex}(\infty)$ to ξ_0:

$$\mathcal{M} = \frac{\xi_{ex}(\infty)}{\xi_0} = \sum_{i=1}^{N} \frac{\alpha(K)\lambda_i}{2 - \alpha(K)\lambda_i} \tag{7.452}$$

(e.g., p. 400 of [Hay96] or [WMGG67]).

In their original work, Widrow et al. [WMGG67] provide several useful approximations.

Instead of using the condition in (7.450), we use a more conservative upper limit,

$$0 < \alpha(K) < \frac{2}{\sum_{i=1}^{N} \lambda_i} = \frac{2}{tr\left[\mathbf{S_x}\right]} = \frac{2}{E\left[\| \mathbf{X}(K) \|^2\right]} = \frac{2}{E\left[\sum_{n=0}^{N-1} |X(n,K)|^2\right]}. \tag{7.453}$$

The motivation for this step is that $E[\| \mathbf{X}(K) \|^2]$ can be estimated without an eigendecomposition. The denominator on the far right side of (7.453) is the expectation of the sum of the square of the received waveforms at each of the sensors at sample K. If

$$\alpha(K)\lambda_{max} \ll 2, \tag{7.454}$$

then (7.452) can be written as

$$\mathcal{M} \cong \frac{\alpha(K)}{2} \sum_{i=1}^{N} \lambda_i = \frac{\alpha(K)}{2} E\left[\| \mathbf{X}(K) \|^2\right]. \tag{7.455}$$

Defining the average eigenvalue as

$$\lambda_{av} = \frac{1}{N} \sum_{n=1}^{N} \lambda_n, \tag{7.456}$$

and assuming that $\alpha(K)$ is constant, the transient MSE curve can be approximated with a single exponential with time constant $\tau_{mse,av}$,

$$\tau_{mse,av} \simeq \frac{1}{2\alpha\lambda_{av}}, \tag{7.457}$$

and

$$\mathcal{M} \simeq \frac{\alpha N \lambda_{av}}{2} = \frac{N}{4\tau_{mse,av}}. \tag{7.458}$$

Therefore, a small $\mathcal{M}$ requires a long setting time and vice versa.

In practice, the denominator in (7.453) is approximated by an estimate obtained from the data. In most cases, we use a sample dependent $\alpha(K)$:

$$\alpha(K) = \frac{\gamma}{\beta + \mathbf{X}^H(K)\mathbf{X}(K)}, \tag{7.459}$$

with $\beta > 0$ and $0 < \gamma < 2$, is referred to as the normalized LMS algorithm. (e.g., Goodwin and Sin [GS84] or Söderström and Stoica [SS89])

A second version,

$$\boxed{\alpha(K) = \frac{\gamma}{\sigma_x^2(K)},} \tag{7.460}$$

where

$$\boxed{\sigma_x^2(K) = \beta\sigma_x^2(K-1) + (1-\beta)\mathbf{X}^H(K)\mathbf{X}(K),} \tag{7.461}$$

with $0 < \beta < 1$, is referred to as the **power normalized LMS (PNLMS)** algorithm.

Note that β is providing exponential weighting of $\mathbf{X}^H(k)\mathbf{X}(k)$, just as μ did with the sample covariance matrix. Normally, β will be close to one (e.g., $\beta \geq 0.99$). The constant γ in the numerator satisfies $0 < \gamma < 2$. Typical values are $.005 < \gamma < 0.05$. If γ is too small, the step size leads to slow convergence. If γ is too large, the algorithm will have a large excess error or have stability problems.

All of the examples use the PNLMS algorithm.

7.7.2.3 Performance of the linear constrained algorithms

The statistical performance of the linear constrained LMS algorithm is analyzed in a similar manner to the unconstrained LMS algorithm. Frost [Fro72] derives a number of useful results and the reader is referred to that reference for details.

7.7.3 LMS Algorithm Behavior

In this section, we simulate the behavior of the LMS algorithm for various signal and interference scenarios. In Section 7.7.3.1, we study the LMS version of the MMSE beamformer. In Section 7.7.3.2, we study the LMS version of the LCMP beamformer.

7.7.3.1 MMSE-LMS beamformers

We consider two examples to illustrate the behavior of the Griffiths LMS algorithm in (7.416) and (7.417).

Example 7.7.1 (continuation, Example 7.6.4)

Consider the same model as in Example 7.6.4. The signal arrives from $u_s = 0$ with an *SNR* = 10 dB. A single interferer arrives from $u_I = 0.24$ with an *INR* = 10 dB. The trace of $\mathbf{S_x}$ is 210 and the eigenvalues are:

$$\begin{aligned} \lambda_1 &= 117 \\ \lambda_2 &= 85 \\ \lambda_3 &= \cdots = \lambda_{10} = 1. \end{aligned}$$

The average eigenvalue, λ_{av}, equals 21. We use the PNLMS algorithm from (7.460)–(7.461), with $\gamma = 0.01$ and $\beta = 0.99$.

In Figure 7.42, we plot the average squared error and the average $SINR_o$ versus K. We also show the steepest descent results from Example 7.6.4. In Figure 7.43, we plot the beam patterns at $K = 20, 100, 500$, and 1000.

We see that the PNLMS algorithm approaches the steady state result by $K = 100$. The beam pattern has placed a significant null on the interferer by $K = 200$. The single interferer case with no signal mismatch is not a challenging scenario for the LMS algorithm.

Example 7.7.2 (continuation, Example 7.6.5)

Consider the same model as in Example 7.6.5. The signal arrives from $u_s = 0$ with an *SNR* = 10 dB. The first interferer arrives from $u_{I1} = 0.29$ with $INR_1 = 20$ dB. The second interferer arrives from $u_{I2} = -0.45$ with $INR_2 = 40$ dB. The PNLMS algorithm is implemented with $\gamma = 0.01$ and $\beta = 0.99$. In Figure 7.44, the average squared error versus K and the average $SINR_o$ versus K are plotted. In Figure 7.45, representative beam patterns at $K = 20, 100, 500$, and 1000 are plotted.

7.7.3.2 LCMP-LMS beamformers

In this section, we analyze the LMS implementation of the LCMP beamformer behavior for the same signal and interference model that we studied for SMI and RLS beamformers.

We consider a standard 10-element linear array. There is no signal mismatch or array perturbation. We do not use diagonal loading. We consider an example with two interferers in order to have two eigenvalues in the LMS algorithm.

Example 7.7.3

Consider a standard 10-element linear array. The desired signal arrives from $u_s = 0$. The interfering signals arrive from $u_I = 0.29$ and 0.45, each with an *INR* = 20 dB and 40 dB, respectively. The *SNR* = 10 dB. We implement the GSC version of the LMS MPDR beamformer. We use the PNLMS algorithm with $\gamma = 0.01$, $\beta = 0.9$ and $\gamma = 0.01$, $\beta = 0.99$ In Figure 7.46, we plot the average $SINR_o$ versus K. We also show the SMI with no diagonal loading result and the RLS result with $\sigma_o^2/\sigma_w^2 = 10$ dB.

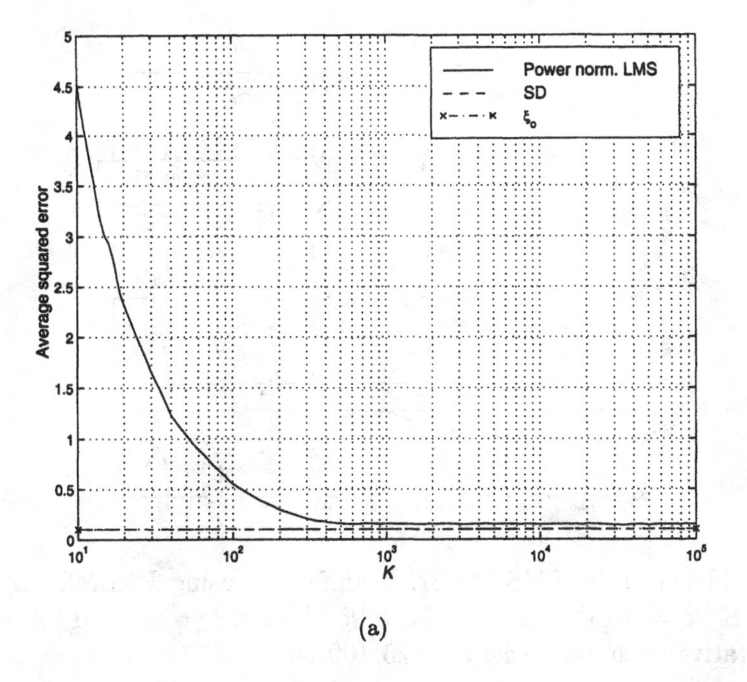

(a)

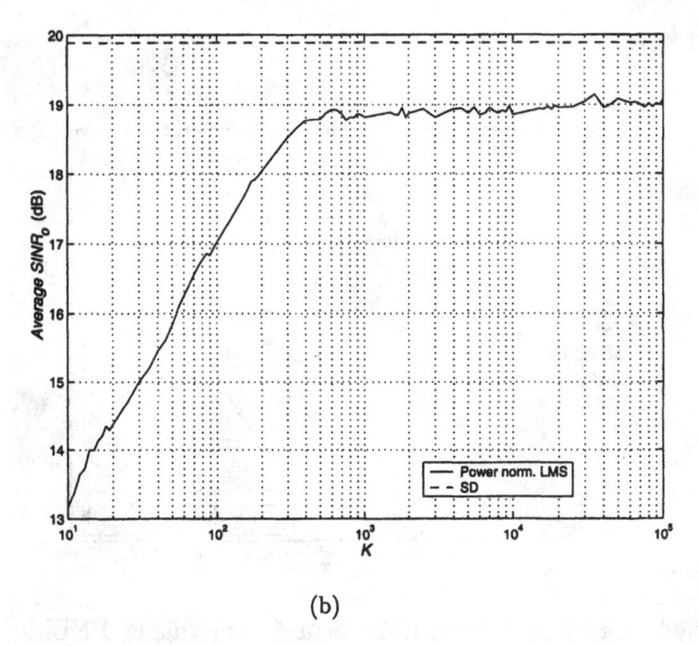

(b)

Figure 7.42: Griffiths MMSE-LMS beamformer using PNLMS algorithm: $u_s = 0, SNR = 10$ dB, $u_I = 0.24, INR = 10$ dB, $\gamma = 0.01$, $\beta = 0.99$, 200 trials: (a) average squared error versus K; (b) average $SINR_o$ versus K.

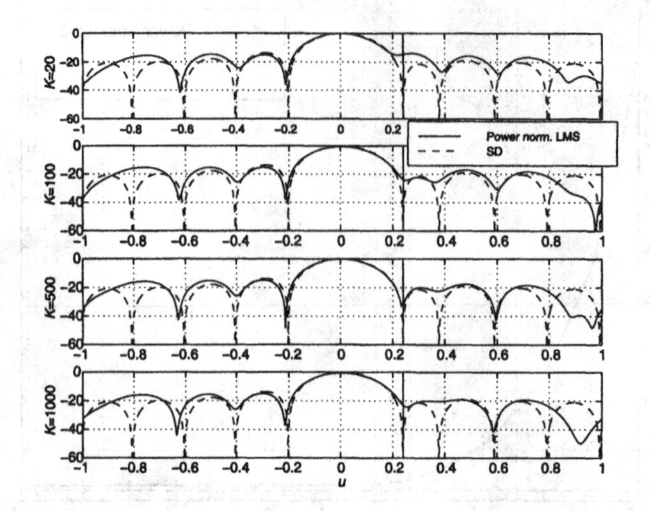

Figure 7.43: Griffiths LMS-MMSE beamformer using PNLMS algorithm: $u_s = 0, SNR = 10$ dB, $u_I = 0.24, INR = 10$ dB, $\gamma = 0.01$, $\beta = 0.99$; representative beam patterns; $K = 20, 100, 500, 1000$.

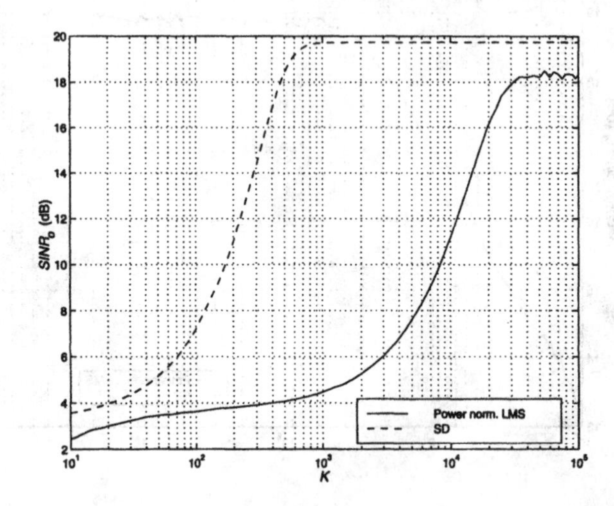

Figure 7.44: Griffiths MMSE-LMS beamformer using PNLMS algorithm: $u_s = 0, SNR = 10$ dB, $u_{I1} = 0.29, INR_1 = 20$ dB, $u_{I2} = -0.45, INR_2 = 40$ dB, $\gamma = 0.01$, $\beta = 0.99$; average $SINR_o$ versus K.

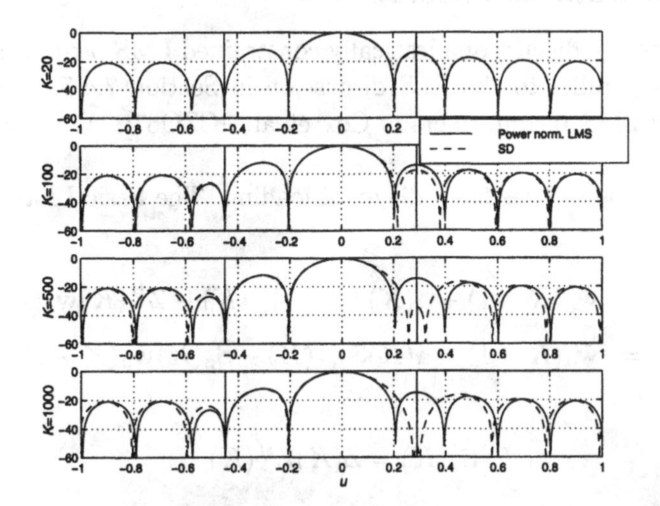

Figure 7.45: Griffiths MMSE-LMS beamformer using PNLMS algorithm: $u_s = 0, SNR = 10$ dB, $u_{I1} = 0.29, INR_1 = 20$ dB, $u_{I2} = -0.45, INR_2 = 40$ dB, $\gamma = 0.01$, $\beta = 0.99$; representative beam patterns; $K = 20, 100, 500, 1000$.

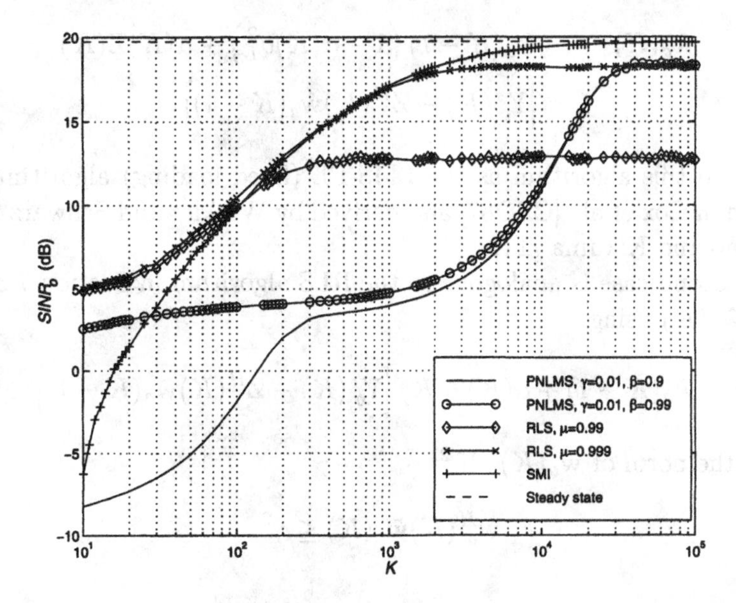

Figure 7.46: MPDR beamformer implemented using PNLMS, RLS, and SMI algorithms: $u_s = 0, SNR = 10$ dB, $u_{I1} = 0.29, INR_1 = 10$ dB, $u_{I2} = 0.45, INR_2 = 40$ dB, $\gamma = 0.01$, $\beta = 0.9$ and 0.99, $\sigma_o^2/\sigma_w^2 = 10$ dB; average $SINR_o$ versus K.

7.7.4 Quadratic Constraints

In this section, we discuss quadratically constrained LMS beamformers. The discussion is parallel to the RLS discussion in Section 7.4.5. The original reference for LMS beamformers is Cox et al. [CZO87]. We develop three algorithms.

The first case utilizes fixed diagonal loading. The basic LMS-GSC algorithm is

$$\begin{aligned}\hat{\mathbf{w}}_a(K) &= \hat{\mathbf{w}}_a(K-1) + \alpha(K)\mathbf{Z}(K)\left[Y_c^*(K) - \mathbf{Z}^H(K)\hat{\mathbf{w}}_a(K-1)\right] \\ &= \hat{\mathbf{w}}_a(K-1) + \alpha(K)\hat{\mathbf{S}}_{\mathbf{z}y_c^*}(K) - \hat{\mathbf{S}}_{\mathbf{z}}(K)\hat{\mathbf{w}}_a(K-1). \qquad (7.462)\end{aligned}$$

We replace

$$\hat{\mathbf{S}}_{\mathbf{z}}(K) = \mathbf{Z}(K)\mathbf{Z}^H(K) \qquad (7.463)$$

with

$$\hat{\mathbf{S}}_{\mathbf{z}}(K) = \mathbf{Z}(K)\mathbf{Z}^H(K) + \sigma_L^2\mathbf{I}. \qquad (7.464)$$

Then, (7.462) becomes

$$\begin{aligned}\hat{\mathbf{w}}_a(K) &= \hat{\mathbf{w}}_a(K-1)\left(1-\alpha(K)\sigma_L^2\right) + \alpha(K)\mathbf{Z}(K) \\ &\quad \left[Y_c^*(K) - \mathbf{Z}^H(K)\hat{\mathbf{w}}_a(K-1)\right]. \qquad (7.465)\end{aligned}$$

We refer to this algorithm as the LMS-FL (fixed loading) algorithm. It is discussed in Cox et al. [CZO87] and derived by Winkler and Schwarz [WS74] and Takao and Kikuma [TK86].

The second case is analogous to the RLS algorithm in Section 7.3.4. We update $\hat{\mathbf{w}}_a(K)$ using

$$\tilde{\mathbf{w}}_a(K) = \hat{\mathbf{w}}_a(K-1) + \alpha(K)\mathbf{Z}(K)\left[Y_c^*(K) - \mathbf{Z}^H(K)\hat{\mathbf{w}}_a(K-1)\right]. \qquad (7.466)$$

We test the norm of $\tilde{\mathbf{w}}_a(K)$. If [22]

$$\tilde{\mathbf{w}}_a^H(K)\tilde{\mathbf{w}}_a(K) \le \gamma^2, \qquad (7.467)$$

we set

$$\hat{\mathbf{w}}_a(K) = \tilde{\mathbf{w}}_a(K). \qquad (7.468)$$

If

$$\tilde{\mathbf{w}}_a^H(K)\tilde{\mathbf{w}}_a(K) > \gamma^2, \qquad (7.469)$$

[22] We use γ^2 instead of α^2 to avoid confusion with the LMS step size parameter $\alpha(K)$.

we add a diagonal loading term to $\tilde{\mathbf{w}}_a(K)$,

$$\begin{aligned}\hat{\mathbf{w}}_a(K) &= \tilde{\mathbf{w}}_a(K) - \alpha(K)\beta(K)\hat{\mathbf{w}}_a(K-1) \\ &= \tilde{\mathbf{w}}_a(K) - \beta(K)\hat{\mathbf{v}}_{lms}(K).\end{aligned} \tag{7.470}$$

where

$$\hat{\mathbf{v}}_{lms}(K) = \alpha(K)\hat{\mathbf{w}}_a(K-1). \tag{7.471}$$

We then solve the quadratic equation

$$\begin{aligned}&\hat{\mathbf{v}}_{lms}^H(K)\hat{\mathbf{v}}_{lms}(K)\beta^2(K) \\ &-2Re\left[\hat{\mathbf{v}}_{lms}^H(K)\tilde{\mathbf{w}}_a(K)\right]\beta(K) \\ &+\tilde{\mathbf{w}}_a^H(K)\tilde{\mathbf{w}}_a(K) - \gamma^2 = 0.\end{aligned} \tag{7.472}$$

We solve (7.472) and use the results in the following manner. The two roots of (7.472) are denoted by $\beta_1(K)$ and $\beta_2(K)$.

If there are two positive real roots, we use the smallest root. In this case we meet the constraint. If there are two negative real roots, we set $\beta_1(K)$ and $\beta_2(K) = 0$ and do not meet the constraint. Simulations indicate that this result happens infrequently. If there are two complex roots, they have the same real part. We use the real part as the diagonal load. We refer this algorithm as the LMS-VL (variable loading) algorithm.

The third case is the LMS-SP (scaled projection) algorithm due to Cox et al. [CZO87].

Once again, we require

$$\| \mathbf{w}_a \|^2 \leq \gamma^2. \tag{7.473}$$

We update $\hat{\mathbf{w}}_a(K)$ recursively using (7.466),

$$\begin{aligned}\tilde{\mathbf{w}}_a(K) &= \hat{\mathbf{w}}_a(K-1) + \alpha(K)\mathbf{Z}(K)\left[Y_c^*(K) - \mathbf{Z}^H(K)\hat{\mathbf{w}}_a(K-1)\right] \\ &= \hat{\mathbf{w}}_a(K-1) + \alpha(K)\mathbf{Z}(K)e_p^*(K) \\ &= \hat{\mathbf{w}}_a(K-1) + \mathbf{g}_\mathbf{z}(K)e_p^*(K).\end{aligned} \tag{7.474}$$

Note that $\hat{\mathbf{w}}_a^H(K-1)$ satisfies the constraint in (7.473). The next step is shown graphically in Figure 7.47.

If $\tilde{\mathbf{w}}_a(K)$ satisfies (7.473), we use it as $\hat{\mathbf{w}}_a(K)$. Otherwise we scale it to satisfy (7.473). Thus,

$$\hat{\mathbf{w}}_a(K) = \begin{cases} \tilde{\mathbf{w}}_a(K), & \| \tilde{\mathbf{w}}_a(K) \|^2 \leq \gamma^2, \\ c(K)\tilde{\mathbf{w}}_a(K), & \| \tilde{\mathbf{w}}_a(K) \|^2 > \gamma^2, \end{cases} \tag{7.475}$$

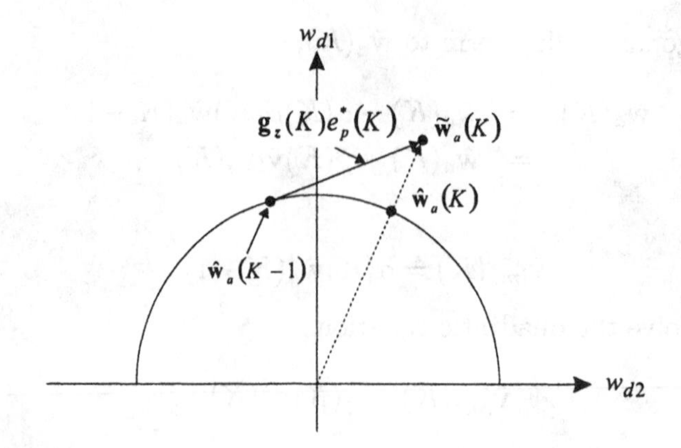

Figure 7.47 Scaling the tentative update vector.

where $c(K)$ scales $\tilde{\mathbf{w}}_a(K)$ so that $\hat{\mathbf{w}}_a(K)$ satisfies (7.473),

$$c(K) = \frac{\gamma}{\| \tilde{\mathbf{w}}_a(K) \|}. \tag{7.476}$$

We see that all of the constraints are satisfied at each step in the recursion.

We consider an example to show the performance of the three diagonal loading algorithms.

Example 7.7.4

Consider a standard 10-element array. The nominal signal direction is $u_s = 0$. We implement an MPDR beamformer using LMS-FL, LMS-VL, and LMS-SP. There is a single interferer at $u_I = 0.29$ with an $INR = 20$ dB.

The signal arrives from $u_a = 0.03$ with a $SNR = 10$ dB. We use $T_o = 0.2$ in the LMS-VL and LMS-SP algorithms. We use $\sigma_L^2/\sigma_w^2 = 10$ dB in the LMS-FL algorithm.

In Figure 7.48, we plot the average $SINR_o$ versus K. The $\overline{SINR_o}$ behavior of the three algorithms is identical. Note that the steady state value is 20 dB, so the LMS is converging slowly.

In Figure 7.49, we plot representative beam patterns. The three algorithms are similar. A significant null does not appear until K = 1,000. Due to the slow convergence and diagonal loading there is no signal nulling.

The general conclusion is:

(i) The LMS-VL and LMS-SP behave in an almost identical manner under most scenarios. In most applications, the LMS-SP algorithm will be used.

(ii) The LMS algorithm convergence will be slow in many cases of interest.

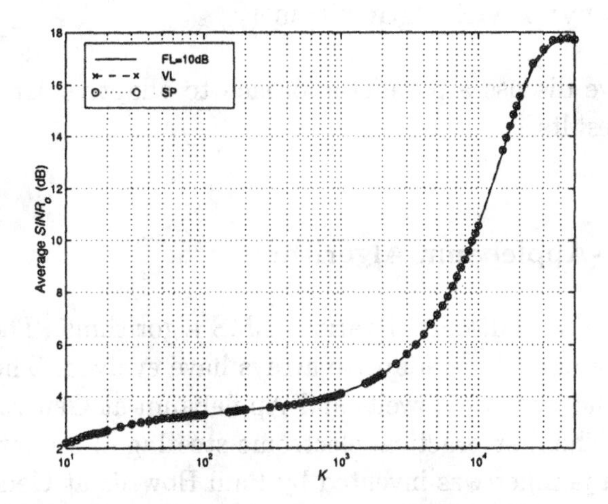

Figure 7.48 MPDR-LMS beamformer with signal mismatch using fixed loading, variable loading, and scaled projection: $u_s = 0, u_a = 0.03, SNR = 10$ dB, $u_I = 0.29, INR = 20$ dB, $T_o = 0.2$, $\sigma_L^2/\sigma_w^2 = 10$ dB; average $SINR_o$ versus K.

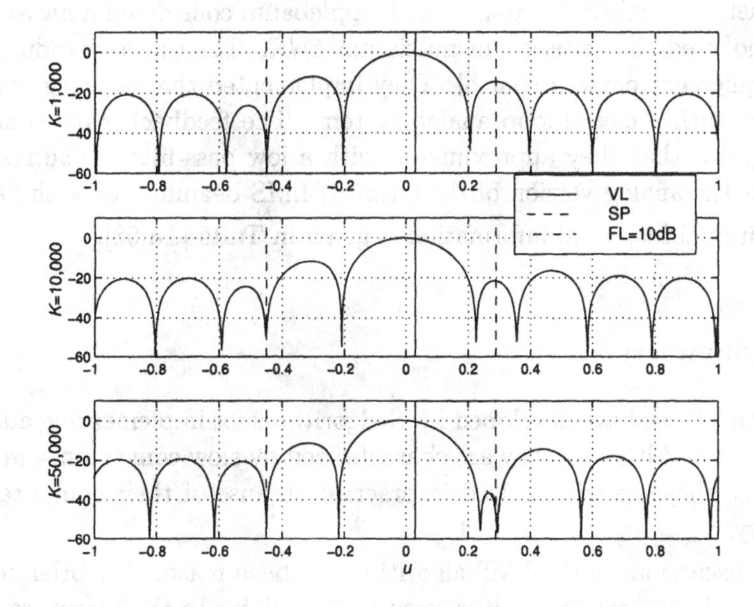

Figure 7.49 MPDR-LMS beamformer with signal mismatch using fixed loading, variable loading, and scaled projection: $u_s = 0, u_a = 0.03, SNR = 10$ dB, $u_I = 0.29, INR = 20$ dB, $T_o = 0.2$, $\sigma_L^2/\sigma_w^2 = 10$ dB; representative beam patterns at $K = 10, 100$, and 1,000.

7.7.5 Summary: LMS algorithms

In this section, we discuss a parallel approach to adaptive arrays and then summarize our results.

7.7.5.1 Howells-Applebaum Algorithms

In this section, we have discussed various LMS algorithms. They represent one of the paths along which adaptive arrays have evolved. The other path originated with the work of Howells and Applebaum at General Electric in the early 1960s. The first practical electronic steering of an antenna null in the direction of a jammer was invented by Paul Howells at General Electric in Syracuse, NY. His 1959 U.S. Patent, number 3,202,990, was titled "Intermediate Frequency Side-Lobe Cancellor." Their original work was published as a Syracuse University Research Corporation report [App66]. Because the circulation was limited, [App66] was republished as [App76]. Several subsequent papers (e.g., [AC76]) discuss the original work and extensions.

In their original work, Howells and Applebaum considered a narrowband array and used the criterion of maximum *SNR*. They also introduced the idea of quiescent pattern control. They implemented the resulting adaptive processor with a closed-loop analog system. The feedback path contained an integrator that they approximated with a low pass-filter. The resulting system is the analog version of the Griffiths LMS beamformer with DL.

Additional historical information is given in Tufts [Tuf98].

7.7.5.2 Summary

In Section 7.7, we have developed LMS algorithms for implementing adaptive beamformers. Although they are characterized by slow convergence in many situations, they are widely used in practice because of their computational simplicity.

Our discussions of the LMS algorithm has been reasonably brief because a number of comprehensive discussions are available in the literature.

For further discussion of LMS algorithms for adaptive beamforming and adaptive filtering, the reader is referred to Widrow and Stearns [WS85], Haykin [Hay96], Compton [Com88], Monzingo and Miller [MM80], and Glentis et al. [GBT99].

7.8 Detection of Signal Subspace Dimension

In Section 6.8, we derived several eigenvector beamformers and saw that they offered an improvement in performance when the statistics are known. In practice, we must estimate the eigenvalues and eigenvectors using the incoming data and then select an appropriate subspace. In the case in which the desired signals and interferers are plane-wave signals we can refer to this subspace as the signal-plus-interference subspace. For simplicity, we consider all of the plane-wave signals to be signals of interest and refer to this subspace as the signal subspace.

The eigenvector beamformers provide motivation for our current discussion of detecting the subspace dimension. However, when we study parameter estimation in Chapters 8 and 9, we find that almost all of the estimation algorithms assume that the number of signals are known. Therefore, we develop the detection problem in a general framework, so that the results are useful in both the adaptive beamforming problem and the parameter estimation problem. We apply the results to the adaptive eigenspace beamformer in Section 7.9.

The techniques that we develop in this section are non-parametric techniques. Although we model the received vector as

$$\mathbf{X}(k) = \mathbf{V}(\psi)\mathbf{F}(k) + \mathbf{W}(k), \quad k = 1, 2, \cdots, K, \tag{7.477}$$

we do not exploit the structure of $\mathbf{V}(\psi)$ in our algorithms. In (7.477), $\mathbf{F}(k)$ is a $d \times 1$ source vector that is a complex Gaussian vector. $\mathbf{V}(\psi)$ is an $N \times d$ array manifold matrix and $\mathbf{W}(k)$ is a complex Gaussian vector with a diagonal spectral matrix $(\sigma_w^2 \mathbf{I})$. The number of signals d is unknown and we want to estimate its value. Because d is an integer, we refer to the problem as a detection problem.

If none of the signals are coherent with each other, then the rank of the subspace equals the number of signals.

However, we recall from (5.252) that if

$$| \rho_{ij} |= 1, \tag{7.478}$$

for some value of ij, then we will have one or more zero eigenvalues and

$$d' < d. \tag{7.479}$$

As $| \rho_{ij} |$ approaches one, the algorithms have a harder time correctly estimating the number of signals.

All of the approaches in this section perform an eigendecomposition of the sample spectral matrix or a singular value decomposition of the data matrix to find a set of estimated eigenvalues. The algorithms test some function of these eigenvalues to determine d or d'.

In Section 7.8.1, we develop the set of detection algorithms that are most widely used in the array processing. All of the tests in Section 7.8.1 utilize some function of the eigenvalues of $\hat{\mathbf{S}}_{\mathbf{x}}$ as the data component of the test. The performance of the various algorithms is analyzed for a sequence of examples.

In Section 7.8.2, we develop a detection algorithm that utilizes the eigenvectors of $\hat{\mathbf{S}}_{\mathbf{x}}$.

7.8.1 Detection Algorithms

We use the familiar frequency-domain snapshot model in (7.477). We assume the columns of $\mathbf{V}$ are linearly independent.

The spectral matrix is

$$\mathbf{S}_{\mathbf{x}} = \mathbf{V}\mathbf{S}_{\mathbf{f}}\mathbf{V}^H + \sigma_w^2\mathbf{I}. \tag{7.480}$$

We estimate d, the rank of $\mathbf{S}_{\mathbf{f}}$.

If $\mathbf{V}\,\mathbf{S}_{\mathbf{f}}\,\mathbf{V}^H$ is of rank d, the eigendecomposition of $\mathbf{S}_{\mathbf{x}}$ can be written as

$$\mathbf{S}_{\mathbf{x}} = \mathbf{U}_S\mathbf{\Lambda}_S\mathbf{U}_S^H + \mathbf{U}_N\mathbf{\Lambda}_N\mathbf{U}_N^H, \tag{7.481}$$

where

$$\mathbf{\Lambda}_N = \text{diag}\left[\begin{array}{ccc} \sigma_w^2 & \cdots & \sigma_w^2 \end{array}\right], \tag{7.482}$$

is a diagonal matrix with $N-d$ elements.

In practice, we have $\hat{\mathbf{S}}_{\mathbf{x}}$ available and we compute

$$\hat{\lambda}_1 \geq \hat{\lambda}_2 \geq \hat{\lambda}_3 \cdots \hat{\lambda}_d \geq \hat{\lambda}_{d+1} \geq \cdots \hat{\lambda}_N, \tag{7.483}$$

via an eigendecomposition of $\hat{\mathbf{S}}_{\mathbf{x}}$ or an SVD of the data matrix. We want to detect the value of d.

The first category of tests are referred to as sequential hypothesis (SH) tests or sphericity tests. They originated in the statistical field (e.g., [Mau40], [And63], or [Mui82]).

We want to find the likelihood ratio between the hypothesis that the $(N-d)$ smallest eigenvalues are equal versus the hypothesis that only the $(N-d-1)$ smallest eigenvalues are equal. We use d as the variable in our test.

Anderson [And63] showed that, for $K \gg N$, if $\mathbf{V}$ is of rank d, the $N-d$ smallest eigenvalues cluster around σ_w^2. Specifically, $\hat{\lambda}_k - \sigma_w^2 = O\left(K^{-\frac{1}{2}}\right)$ for $k = d+1, \cdots, N$.

Anderson [And63] also showed that a sufficient statistic is

$$L_d(d) = K\,(N-d)\,\ln\left\{\frac{\frac{1}{N-d}\sum_{k=d+1}^{N}\hat{\lambda}_k}{\left(\prod_{k=d+1}^{N}\hat{\lambda}_k\right)^{\frac{1}{N-d}}}\right\}. \tag{7.484}$$

The term in braces is the ratio of the arithmetic mean of the $(N-d)$ smallest eigenvalues to the geometric mean of the $(N-d)$ smallest eigenvalues. We now have a 1-D optimization problem to find $\hat{d}$. Note that, if the $N-d$ smallest eigenvalues are equal, then $L_d(d) = 0$.

One can show that asymptotically $(2L(d))$ corresponds to a chi-squared random variable $\chi^2\left((N-d)^2 - 1\right)$ (e.g., Gupta [Gup65]). Thus, if

$$\nu = 2L_d(d), \tag{7.485}$$

then

$$f_\nu(d) = \frac{\nu^{\left(\frac{1}{2}\left(\left((N-d)^2-1\right)-2\right)\right)} \cdot e^{\left(-\frac{1}{2}\nu\right)}}{2^{\left(\frac{(N-d)^2-1}{2}\right)} \cdot \Gamma\left(\frac{(N-d)^2-1}{2}\right)}. \tag{7.486}$$

We choose an arbitrary confidence interval (e.g., 99%). We assume H_0 (i.e., $d=0$) is true and compute $\gamma_{99}^{(0)}\left(\chi^2_{N^2-1}\right)$. If

$$\nu(0) \leq \gamma_{99}^{(0)}, \tag{7.487}$$

we set $\hat{d} = 0$. If $\nu(0) > \gamma_{99}^{(0)}$, we compute $\gamma_{99}^{(1)}\left(\chi^2_{(N-1)^2-1}\right)$. Then, if

$$\nu(1) \leq \gamma_{99}^{(1)}, \tag{7.488}$$

we set $\hat{d} = 1$. If $\nu(1) > \gamma_{99}^{(1)}$, we continue the process until we find

$$\nu(d) \leq \gamma_{99}^{(d)}\left(\chi^2_{(N-d)^2-1}\right). \tag{7.489}$$

Bartlett [Bar54] and Lawley [Law56] developed tests based on (7.484). Simkins [Sim80] applied these results to the direction-of-arrival estimation problem.

One of the problems with the above approaches is the choice of the threshold. In addition, a sequence of tests is required. We do not pursue sequential hypothesis tests further in the text. The reader is referred to the references and the problems for representative results.

The second category of tests utilize $L_d(d)$ and add a penalty function related to the degrees of freedom. The result is a function of d that we then minimize. The two resulting tests were obtained using different approaches.

Akaike [Aka74] introduced an information-theoretic criterion, which is referred to in the literature as the Akaike Information Criterion (AIC). He considered a parameterized family of models with probability densities $p_{\mathbf{x}}(\mathbf{x}|\boldsymbol{\theta})$. He proposed to choose the model that has the minimum AIC where

$$AIC = -2 \ln p_{\mathbf{x}}(\mathbf{x}|\hat{\boldsymbol{\theta}}) + 2k_p, \tag{7.490}$$

where $\hat{\boldsymbol{\theta}}$ is the ML estimate of $\boldsymbol{\theta}$ and k_p is the number of freely adjusted parameters in $\boldsymbol{\theta}$.[23] The first term is the log-likelihood for the ML estimate of $\boldsymbol{\theta}$. The second term is a correction term. Akaike introduced it so that the AIC is an unbiased estimate of Kullback-Liebler distance between $p_{\mathbf{x}}(\mathbf{x}|\boldsymbol{\theta})$ and $p_{\mathbf{x}}(\mathbf{x}|\hat{\boldsymbol{\theta}})$.

Two different approaches were taken by Schwartz [Sch78] and Rissanen [Ris78]. Schwartz [Sch78] utilized a Bayesian approach, assigning a prior probability to each model, and selected the model with the largest *a posteriori* probability. Rissanen used an information theoretic argument. One can think of the model as an encoding of the observation. He proposed choosing the model that gave the minimum code length. In the large sample limit, both approaches lead to the same criterion

$$MDL = -\ln p_{\mathbf{x}}(\mathbf{x}|\hat{\boldsymbol{\theta}}) + \frac{1}{2}\, k_p \ln K, \tag{7.491}$$

where MDL denotes minimum description length.

In [WK85] the AIC and MDL criteria were applied to the array problem.[24] The starting point is to define a family of spectral matrices,

$$\mathbf{S}_{\mathbf{x}}^{(d)} = \boldsymbol{\Psi}^{(d)} + \sigma_w^2 \mathbf{I}, \quad d = 0, 1, \cdots, N-1, \tag{7.492}$$

where $\boldsymbol{\Psi}^{(d)}$ is a non-negative definite matrix of rank d and σ_w^2 is an unknown scalar. We write $\mathbf{S}_{\mathbf{x}}^{(d)}$ as

$$\mathbf{S}_{\mathbf{x}}^{(d)} = \sum_{i=1}^{d} \lambda_i\, \boldsymbol{\Phi}_i\, \boldsymbol{\Phi}_i^H + \sigma_w^2 \sum_{i=d+1}^{N} \boldsymbol{\Phi}_i\, \boldsymbol{\Phi}_i^H. \tag{7.493}$$

[23] $\boldsymbol{\theta}$ will be defined in the context of the array problem in (7.494).

[24] Our discussion in the next part of this section follows Wax and Kailath [WK85].

The parameter vector $\boldsymbol{\theta}^{(d)}$ of the model is

$$\boldsymbol{\theta}^{(d)} = \left[\lambda_1, \cdots, \lambda_d, \sigma_w^2, \boldsymbol{\Phi}_1^T, \cdots, \boldsymbol{\Phi}_d^T\right]^T. \tag{7.494}$$

Note that $\boldsymbol{\theta}$ does not include $\boldsymbol{\psi}$. Also observe that we are estimating the parameters in the eigendecomposition of $\mathbf{S}_{\mathbf{x}}^{(d)}$ without imposing any structure on it.

The ln likelihood function was derived in (7.7),

$$L\left(\boldsymbol{\theta}^{(d)}\right) = -K \ln \det \mathbf{S}_{\mathbf{x}}^{(d)} - K \text{ tr } \left[\left[\mathbf{S}_{\mathbf{x}}^{(d)}\right]^{-1} \mathbf{C}_{\mathbf{x}}\right] \quad d = 0, 1, \cdots, N-1, \tag{7.495}$$

where $\mathbf{C}_{\mathbf{x}}$ is the sample spectral matrix. From (7.9), the maximum likelihood estimate of $\mathbf{S}_{\mathbf{x}}^{(d)}$ is

$$\hat{\mathbf{S}}_{\mathbf{x}}^{(d)} = \mathbf{C}_{\mathbf{x}}, d = 0, 1, \cdots, N-1. \tag{7.496}$$

Now expand $\mathbf{C}_{\mathbf{x}}$ using an eigendecomposition,

$$\mathbf{C}_{\mathbf{x}} = \sum_{i=1}^{N} \hat{\lambda}_i \hat{\boldsymbol{\Phi}}_i \hat{\boldsymbol{\Phi}}_i^H = \sum_{i=1}^{d} \hat{\lambda}_i \hat{\boldsymbol{\Phi}}_i \hat{\boldsymbol{\Phi}}_i^H + \sum_{i=d+1}^{N} \hat{\lambda}_i \hat{\boldsymbol{\Phi}}_i \hat{\boldsymbol{\Phi}}_i^H, \tag{7.497}$$

where $\hat{\lambda}_i$ and $\hat{\boldsymbol{\Phi}}_i$ are the eigenvalues and eigenvectors of the sample covariance matrix $\mathbf{C}_{\mathbf{x}}$. Substituting (7.493) and (7.497) into (7.496) gives the maximum likelihood estimate of the components of $\boldsymbol{\theta}^{(d)}$:

$$(\hat{\lambda}_i)_{ml} = \hat{\lambda}_i, \quad i = 1, \cdots, d, \tag{7.498}$$

$$(\hat{\boldsymbol{\Phi}}_i)_{ml} = \hat{\boldsymbol{\Phi}}_i, \quad i = 1, \cdots, d, \tag{7.499}$$

$$(\hat{\sigma}_w^2)_{ml} = \frac{1}{N-d} \sum_{i=d+1}^{N} \hat{\lambda}_i. \tag{7.500}$$

Thus, the maximum likelihood estimate is

$$\hat{\boldsymbol{\theta}}^{(d)} = \left[\hat{\lambda}_1, \cdots, \hat{\lambda}_d, \hat{\sigma}_w^2, \hat{\boldsymbol{\Phi}}_1^T, \cdots, \hat{\boldsymbol{\Phi}}_d^T\right]^T, \tag{7.501}$$

where the *ml* subscript is dropped for simplicity. Substituting (7.501) and (7.500) in (7.495) gives

$$L_d(d) = K(N-d) \ln \left\{ \frac{\frac{1}{N-d} \sum_{i=d+1}^{N} \hat{\lambda}_i}{\left(\prod_{i=d+1}^{N} \hat{\lambda}_i\right)^{\frac{1}{N-d}}} \right\}, \tag{7.502}$$

where we have dropped terms that do not depend on d. This is the same expression as Anderson's sufficient statistic in (7.484).

It is convenient to write the AIC test as

$$L_d(d) + p(d), \tag{7.503}$$

where $p(d)$ is a penalty function. The penalty function is determined by the number of degrees of freedom in the model (the number of free parameters in $\boldsymbol{\theta}^{(d)}$).

The parameter vector $\boldsymbol{\theta}^{(d)}$ consists of $d+1$ eigenvalues and d eigenvectors.[25] The eigenvalues are real and hence count as $d+1$ real parameters, whereas the eigenvectors are complex, of unit norm and mutually orthogonal. To count the number of degrees-of-freedom (DOF) required to describe the d eigenvectors, observe that because the eigendecomposition of a complex matrix is invariant to multiplication of each eigenvector by a pure phase factor, we can describe an eigenvector with $2N-1$ real parameters, say by fixing the first element to be real. Since the eigenvectors obey certain constraints, we have to deduct d DOF due to their unit norm and $2(d(d-1)/2)$ due to their mutual orthogonalization. Therefore, the total number of degrees of freedom is

$$\begin{aligned} k_p &= d + 1 + 2Nd - 2d - d(d-1) \\ &= d(2N-d) + 1. \end{aligned} \tag{7.504}$$

We can drop the one, so

$$p(d) = d(2N-d), \tag{7.505}$$

and the AIC test is

$$AIC(d) \triangleq \{L_d(d) + [d(2N-d)]\}, \tag{7.506}$$

and

$$\hat{d}_{AIC} = \arg\min_d \{AIC(d)\}. \tag{7.507}$$

For the MDL test,

$$\begin{aligned} MDL(d) &= L_d(d) + \frac{1}{2}p(d)\ln K \\ &= L_d(d) + \frac{1}{2}\left[d(2N-d)+1\right]\ln K, \end{aligned} \tag{7.508}$$

[25] This explanation is due to M. Wax (private communication).

and

$$\hat{d}_{MDL} = \arg\min_{d} \{MDL(d)\}. \tag{7.509}$$

In [WK85], it is shown that the MDL estimate is consistent.[26] In other words, as K goes to infinity, $\hat{d}_{MDL}$ approaches d. It is also shown that the AIC is inconsistent and, asymptotically, tends to overestimate the number of signals. However, we will find that, for a small K, the AIC generally has a higher probability of a correct decision.

We have seen the benefit of FB averaging in a coherent or correlated signal environment. We now discuss how the above detection tests must be modified to account for the use of $\mathbf{C}_{\mathbf{x},fb}$ instead of $\mathbf{C}_{\mathbf{x}}$, the forward-only sample spectral matrix.

The problem is solved by Xu et al. [XRK94]. They derive sequential hypothesis (SH), MDL, and AIC tests that are closely related to the above tests. Their key result is that the tests have the same structure; the only change is in the penalty function.

The FB-MDL and FB-AIC functions are

$$AIC_{FB}(d) = L_d(d) + \frac{1}{2}d\,(2N - d + 1), \tag{7.510}$$

and

$$MDL_{FB}(d) = L_d(d) + \frac{1}{4}d\,(2N - d + 1)\ln K. \tag{7.511}$$

where $L_d(d)$ is given by (7.502) where the $\hat{\lambda}_i$ are the eigenvalues of $\mathbf{C}_{\mathbf{x},fb}$. The effect of the FB averaging is to reduce the free adjustable parameters by a factor of 2. We find the value of d that minimizes (7.510) or (7.511) and denote it by $\hat{d}$. Xu et al. [XRK94] also shows that FB-MDL is consistent.

Wong et al. [WZRY90] derive an alternative approach to detecting the number of signals. They argue that the number of signals is independent of the orientation of the array, so the eigenvectors provide little information. They derive a modified likelihood function consisting of the marginal probability density function of the eigenvalues. They derive two modified criteria using this modified likelihood function and the penalty functions of AIC and MDL. One of the criteria offers improved performance in a number of scenarios and should be considered as a candidate solution. Wu and Tam ([TW96], [WT01]) also discuss this approach. The reader is referred to the above references for details.

Another category of tests is due to Tufts and his colleagues (e.g., Shah and Tufts [ST94]) and focuses on scenarios with short data records. The

[26] See subsequent discussion by Zhao et al. ([ZKB87], [ZKB86]).

method consists of sequential constant false alarm rate tests on the sums of the eigenvalues of $\hat{\mathbf{S}}_{\mathbf{x}}$. In several scenarios, it offers improved performance compared to AIC and MDL. The reader is referred to the above reference for details. An approach that deals with unknown correlated noise fields is derived in Stoica and Cedervall [SC97].

As pointed out in the introduction to this section, all of the algorithms in this section are nonparametric. In Chapter 8, we revisit the detection problem and discuss parametric detection algorithms.

We now consider a sequence of examples in which we simulate the performance of AIC and MDL algorithms.

The examples are chosen to illustrate how the following factors affect the performance:

(i) Number of signals;

(ii) Signal strengths;

(iii) Signal separations;

(iv) Signal correlation;

(v) Number of snapshots.

There are three possible outcomes of the tests:[27]

(i) $\hat{d} = d$, which is the desired result. We denote the probability that $\hat{d} = d$ as P_D (the probability of correct detection).

(ii) $\hat{d} < d$. In this case, we have underestimated the signal subspace. We denote the probability as P_M (probability of miss). In almost all applications, we want to avoid this event because it causes us to lose one or more signals.

(iii) $\hat{d} > d$. In this case, we have overestimated the signal subspace. We denote the probability as P_F (probability of false alarm). In most applications this event causes some degradation in performance. However, the extra eigenvector(s) is due to noise. If we try to track the eigenspace as additional data arrive, the subspace tracker drops it and reduces the dimension. If we do block processing, the next block of data will put the extra eigenvector(s) in a different place (or omit them). We rely on this subsequent processing to correct overestimation.

---

[27] Our comments on the outcomes are in the context of beamforming. We discuss the effects on parameter estimation algorithms later.

We now consider a sequence of examples to explore the behavior. In all of the examples, we consider a standard 10-element linear array.

In the first four examples, we consider a standard 10-element linear array with two equal-power signals impinging on it. The two-signal case has been studied in detail by Kaveh and Wang (see Chapter 5 of [Hay91]). They derive analytic expressions for P_M and P_F based on the asymptotic properties of the eigenvalues and also provide simulation results. We use the simulation approach.

In this case, the outcomes are:

$$P_M : \hat{d} = 0 \text{ or } 1$$

$$P_D : \hat{d} = 2$$

$$P_F : \hat{d} \geq 3$$

We consider four algorithms: AIC, MDL, AIC-FB, and MDL-FB.

Example 7.8.1

In this example, the signal DOAs are symmetric about $u = 0$ and the signal separation corresponds to 0.5 HPBW. The signals are uncorrelated. We assume $K = 100$. We vary the *SNR*.

In Figure 7.50, we plot P_D versus *ASNR*[28] for the four algorithms. In Figure 7.51, we plot P_M versus *ASNR* for the four algorithms. In Figure 7.52, we plot P_F versus *ASNR* for the four algorithms.

We see that the AIC algorithm provides correct detection with an *ASNR* that is 3–4 dB lower than the MDL algorithms in the $0.2 \leq P_D \leq 0.9$ range. However as the *ASNR* increases, the P_D levels off at about $P_D = 0.92$ and the P_{FA} is about 0.08. Above *ASNR* = 6 dB, P_M for AIC is zero. The AIC-FB has a higher P_D for low *ASNR*, but levels off at about 0.83 as the *ASNR* increases.

The P_D for the MDL algorithm is 1.0 for *ASNR* $\geq$ 9 dB (7 dB for MDL-FB).

These results suggest that, for this particular scenario, if we can be certain that the *ASNR* will be greater than 10 dB for the signals of interest, then we would use MDL or MDL-FB. However, if we are interested in signals whose *ASNR* may be as low as 0 dB, then we use AIC or AIC-FB and try to eliminate false alarms by subsequent processing.

Example 7.8.2 (continuation, Example 7.8.1)

Consider the same model as in Example 7.8.1 except K, the number of snapshots is allowed to vary. The signal separation is 0.5 HPBW.

We plot the required *ASNR* versus K to achieve $P_D = 0.2$, 0.5, and 0.8. In Figure 7.53, we consider the AIC algorithm. In Figure 7.54, we consider the MDL algorithm.

---

[28]The *ASNR* is defined as $N(SNR)$. It is useful to plot results versus *ASNR*, because, in most cases, for $N \geq 10$, the result will not depend on N.

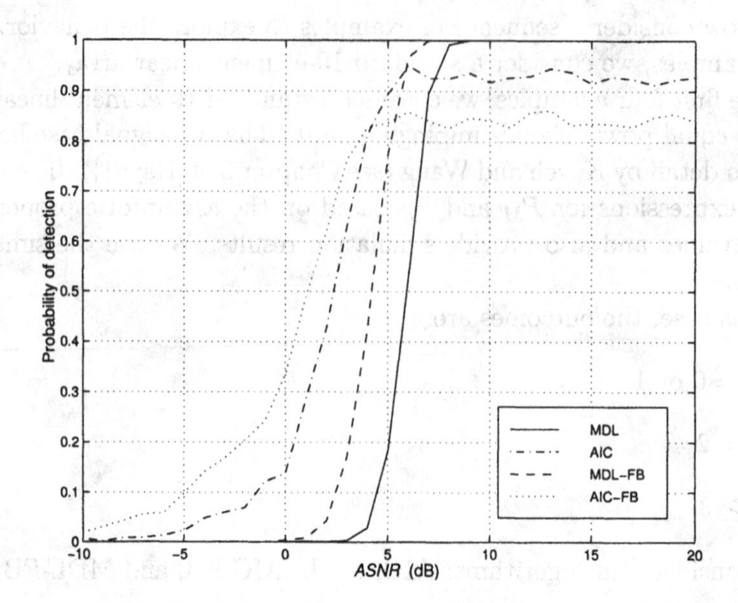

Figure 7.50 Performance of AIC and MDL algorithms versus *ASNR*: two equal-power signals, Δu = 0.5 HPBW, K = 100, 1,000 trials, P_D versus *ASNR*.

We see that the AIC curves are parallel with a decrease in the detection threshold of about 5 dB/decade. The MDL curves are parallel with a similar slope.

Example 7.8.3 (continuation, Example 7.8.1)

Consider a standard 10-element array. The signals have equal power and are uncorrelated. The signals impinge on the array from $\pm\Delta u/2$.

In Figure 7.55, we plot the *SNR* required to achieve P_D = 0.2, 0.5, and 0.8 versus Δu/HPBW for the AIC-FO algorithm. In Figure 7.56, we plot the same results for the MDL-FO algorithm.

We see that the threshold decreases at a rate of about 20 dB/decade. The MDL curves have a similar slope, but are about 4 dB higher.

Example 7.8.4 (continuation, Example 7.8.1)

In this example, we study the behavior as the number of signals, d, is varied. Consider a standard 10-element linear array. We assume that there are d signals placed symmetrically about $u = 0$. We consider two spacings. In case 1, the spacing between the signals is 0.5 HPBW. In case 2, the spacing between the signals is 1.0 HPBW. The number of snapshots is 100. The signals are equal power and uncorrelated.

In Figure 7.57, the signal separation is 0.5 HPBW. We plot P_D versus the *ASNR* for each signal for the AIC-FO and MDL-FO algorithms. If we examine the *ASNR* required for P_D = 0.5, we see that the additional *ASNR* required is about $(d + 4)$ dB for each additional signal up to $d = 7$, and then it is about 11 dB for $d = 8$ and 9.

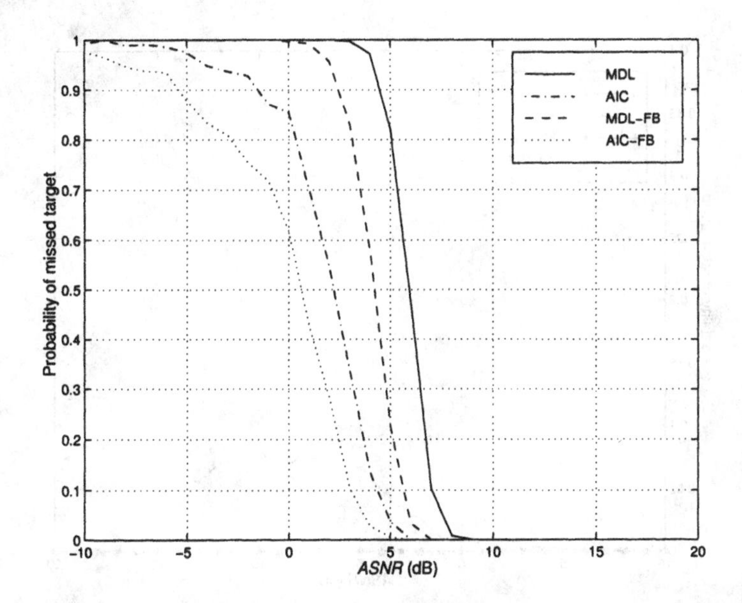

Figure 7.51 Performance of AIC and MDL algorithms versus *ASNR*: two equal-power signals, $\Delta u = 0.5$ HPBW, $K = 100$, 1,000 trials, P_M versus *ASNR*.

In Figure 7.58, the signal separation is 1.0 HPBW. We plot the same results as in Figure 7.57. We see that the effect of increasing d is significantly less.

In this subsection, we have studied AIC and MDL tests for detecting the number of plane-wave signals. We first considered the familiar case of two equal-power uncorrelated signals separated by 0.5 HPBW and $K = 100$. We then considered the effects of:

(i) K, the number of snapshots;

(ii) Δu, the signal separation;

(iii) d, the number of signals.

In all of these case, the AIC algorithm performed better than the MDL algorithm in the $0.2 \leq P_D \leq 0.9$ range. However, as the *ASNR* (or K) increased, the P_D of the AIC algorithm did not approach unity and the AIC algorithm overestimated the number of signals. The P_D of the MDL algorithm approached unity as the *ASNR* (or K) increased.

The choice of the algorithm will depend on the anticipated scenario and the subsequent signal processor that uses the estimate of d. In Section

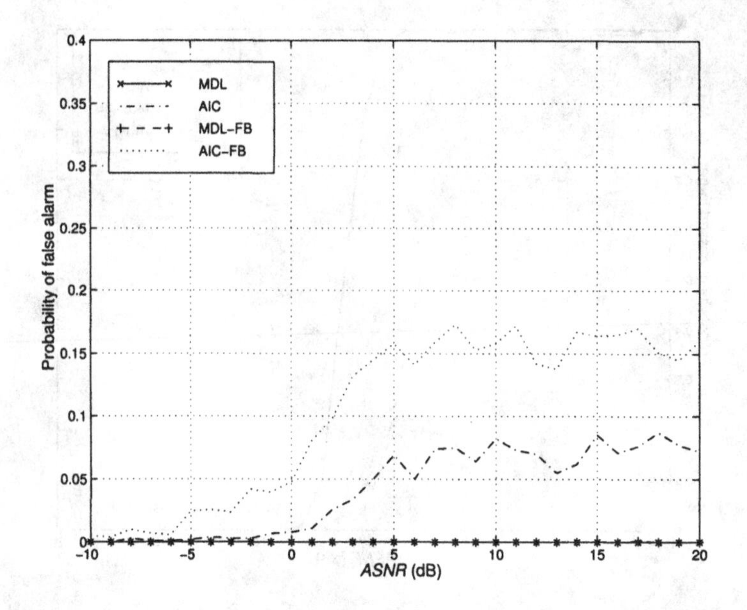

Figure 7.52 Performance of AIC and MDL algorithms versus *ASNR*: two equal-power signals, Δu = 0.5 HPBW, K = 100, 1,000 trials, P_F versus *ASNR*.

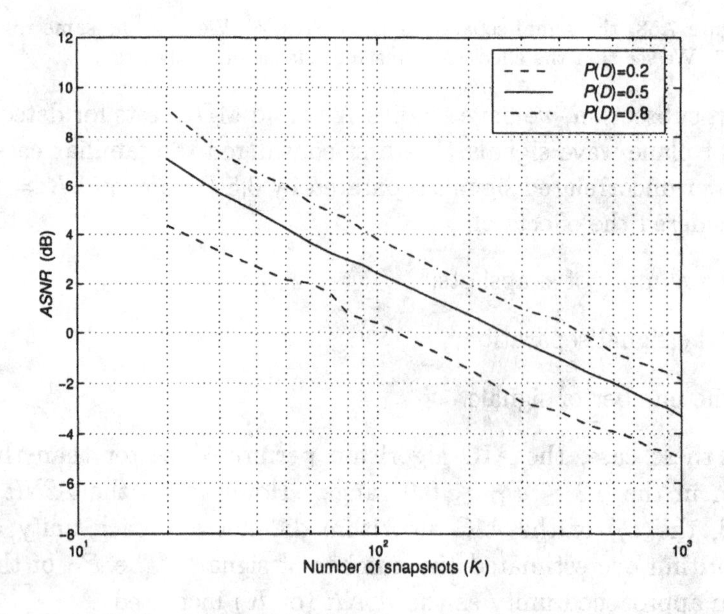

Figure 7.53 AIC algorithm; two equal-power signals, Δu = 0.5 HPBW, $P_D = 0.2$, 0.5, and 0.8; *ASNR* detection threshold versus K.

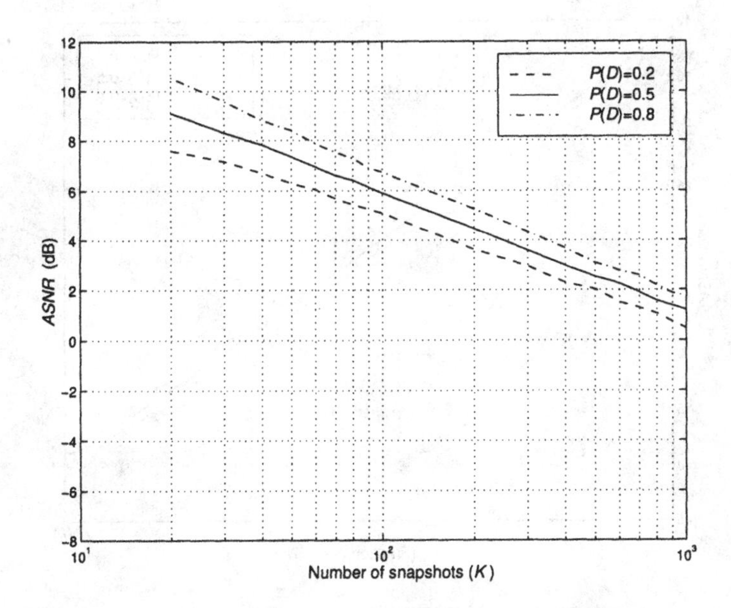

Figure 7.54 MDL algorithm; two equal-power signals; $\Delta u = 0.5$ HPBW, $P_D = 0.2$, 0.5, and 0.8, *ASNR* detection threshold versus K.

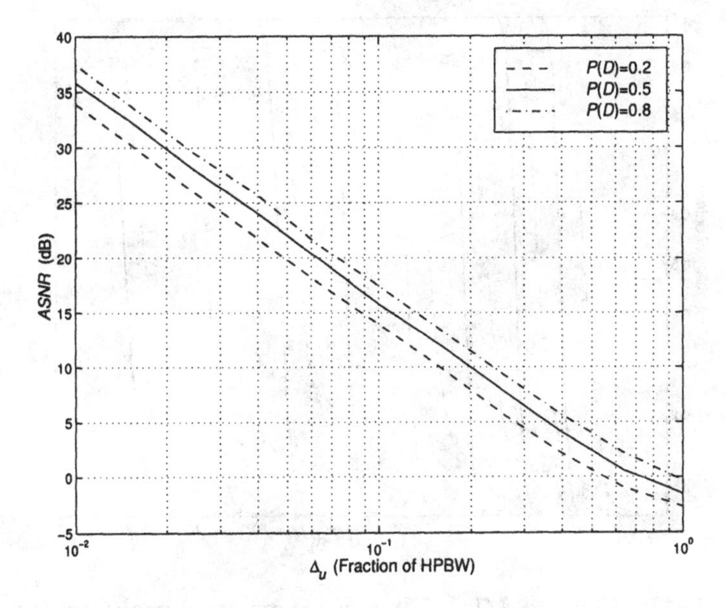

Figure 7.55 AIC-FO algorithm; two equal-power signals, $K = 100$; detection threshold versus $\Delta u/\text{HPBW}$.

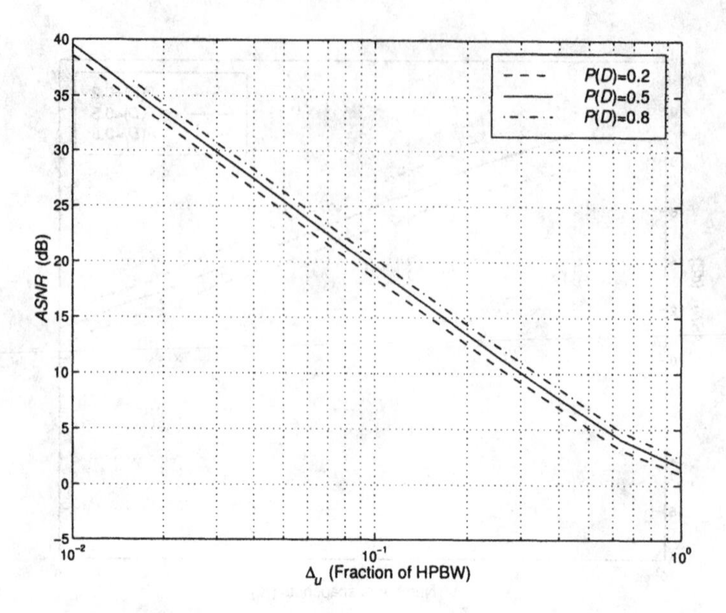

Figure 7.56 MDL-FO algorithm; two equal-power signals, $K = 100$; detection threshold versus Δu/HPBW.

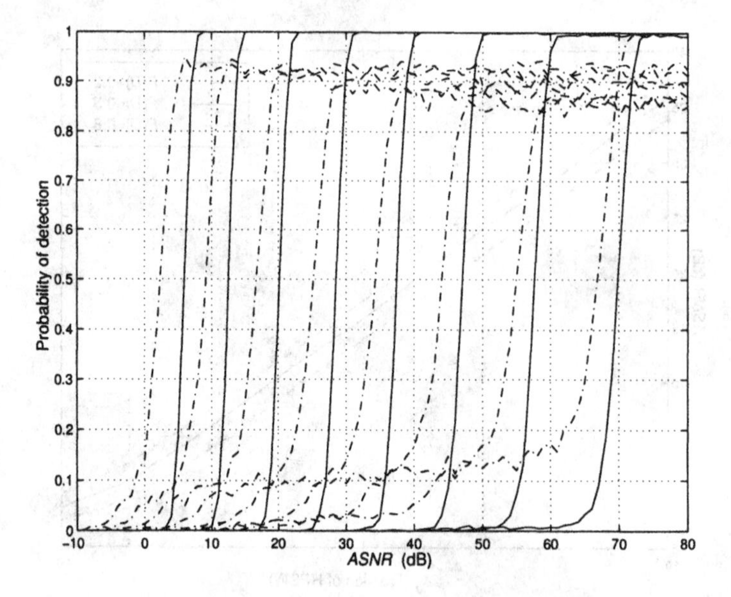

Figure 7.57 AIC-FO and MDL-FO algorithms; uniformly spaced, equal-power signals, $\Delta u = 0.5$ HPBW, P_D versus *ASNR* for each signal, $D = 2, \cdots, 9$.

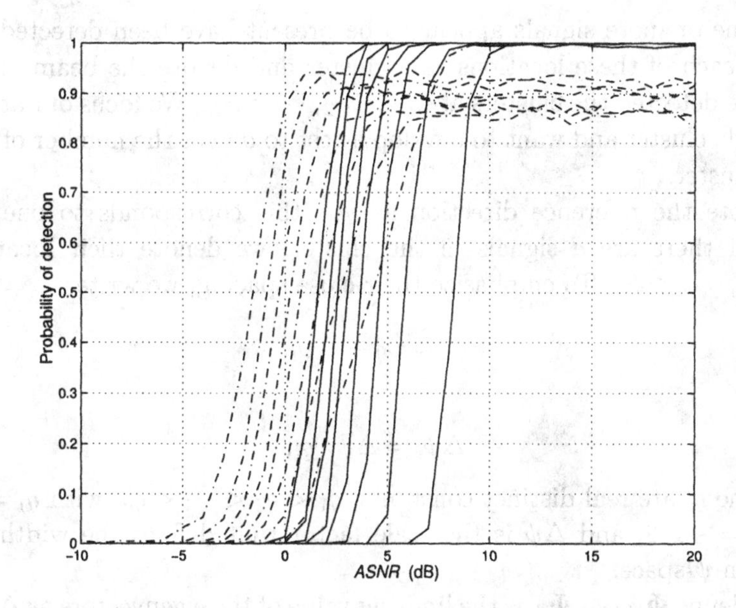

Figure 7.58 AIC-FO and MDL-FO algorithms; uniformly spaced, equal-power signals, Δu = 1.0 HPBW, P_D versus *ASNR* for each signal, $D = 2, \cdots, 9$.

7.9, we study the model in which we use $\hat{d}$ to determine the dimension of an eigenspace beamformer. In Chapters 8 and 9, we use $\hat{d}$ as an input to various parameter estimation algorithms.

In this subsection, we used tests that utilized the estimated eigenvalues. In the next subsection, we develop tests that utilize the estimated eigenvectors.

7.8.2 Eigenvector Detection Tests

The tests in Section 7.8.1 utilized the eigenvalues of $\mathbf{S_x}$ in their detection tests. When the signal separation is small or $|\rho|$ is close to one, some of the signal eigenvalues are small enough that the tests are not reliable. Xu et al. [XPK92] introduced the idea of eigenvector tests. Lee and Li [LL94] developed an efficient technique that is useful for closely spaced signals. It exploits Lee's earlier work on eigenvalues and eigenvectors of closely spaced signals [Lee92] that we discussed in Section 5.5.4. We develop the eigenvector detection technique (EDT) in this section. Our discussion follows [LL94].

We assume that we have done a preliminary beamscan using a conventional beamformer or an MPDR beamformer. Several locations in u-space where one or more signals appear to be present have been detected.[29] We refer to each of these locations as a cluster and denote the beam locations where we detected the clusters at $\psi_{C1}, \psi_{C2}, \cdots, \psi_{CL}$. We focus our attention of a single cluster and want to develop a test to detect the number of signals in the cluster.

Denote the reference direction as ψ_0 (this corresponds to one of the ψ_{Ci}). If there are d signals in the cluster, we denote their location by $\psi_1 < \psi_2, \cdots, < \psi_d$. To emphasize their close spacing, we write

$$\psi_i = \psi_0 + \Delta\psi_i, \tag{7.512}$$

and

$$\Delta\psi_i = \Delta\psi \cdot q_i, \tag{7.513}$$

where the q_i are real distinct constants, $q_1 < q_2 < \cdots < q_d$, with $q_1 = -1/2$ and $q_d = +1/2$, and $\Delta\psi$ is the scale factor that defines the width of the cluster in ψ-space.

We define $\mathbf{\Phi}_1, \cdots, \mathbf{\Phi}_d$ as the limiting value of the eigenvectors as $\Delta\psi \to 0$. We calculate the spatial derivatives of $\mathbf{v}(\psi)$ evaluated at ψ_0,

$$\mathbf{v}^k(\psi_0) \triangleq \frac{d\mathbf{v}^k(\psi)}{d\psi}\Big|_{\psi=\psi_0}, \tag{7.514}$$

and define

$$\dot{\mathbf{V}} = \begin{bmatrix} \mathbf{v}(\psi_0) & \mathbf{v}^1(\psi_0) & \cdots & \mathbf{v}^{N-1}(\psi_0) \end{bmatrix}. \tag{7.515}$$

Then [Lee92] shows that the eigenvectors, $\mathbf{\Phi}_1, \cdots, \mathbf{\Phi}_d$, are the first d columns of the matrix obtained by the left-to-right Gram-Schmidt orthonormalization of $\dot{\mathbf{V}}$.

We next compute the ordered eigenvectors of the sample covariance matrix. We denote these eigenvectors as $\hat{\boldsymbol{\psi}}_1, \hat{\boldsymbol{\psi}}_2, \cdots \hat{\boldsymbol{\psi}}_N$. We then test successive eigenvectors, $\hat{\boldsymbol{\psi}}_i$, for closeness to $\mathbf{\Phi}_i$,

$$\delta_i \triangleq |\hat{\boldsymbol{\psi}}_i^H \mathbf{\Phi}_i|^2. \tag{7.516}$$

If

$$\delta_i \geq T, \tag{7.517}$$

where T ($T < 1$) is a threshold, we say that the eigenvector corresponds to a signal eigenvector. Thus, $\hat{d}$ corresponds to the largest value of i that passes the threshold.

---

[29] We discuss detection algorithms in Chapter 10. However, in this case, the detection algorithms in DEMT III [VT71], [VT01b] are applicable.

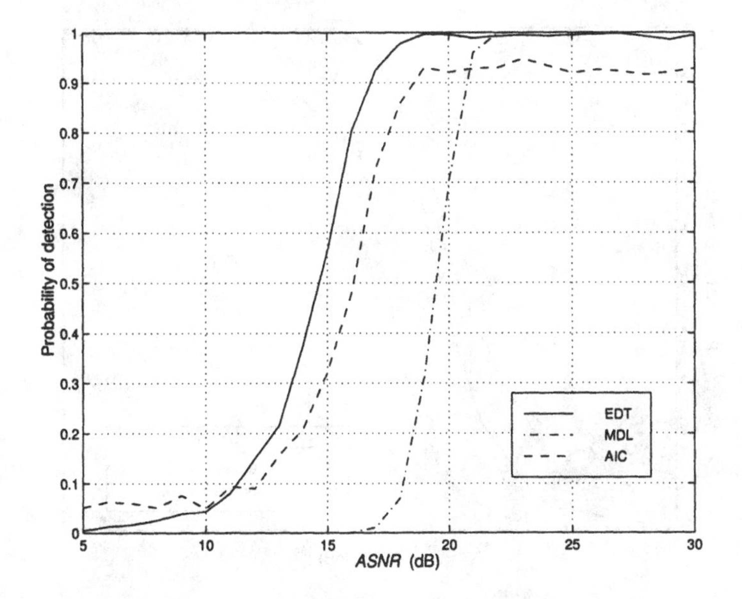

Figure 7.59 EDT, MDL, and AIC algorithms; two equal-power signals, $\Delta u =$ 0.1 HPBW, $K = 100$, 1,000 trials; P_D versus *ASNR*.

We consider a sequence of examples to illustrate the behavior of the EDT algorithm. The examples are similar to those in [LL94].

Example 7.8.5

Consider a standard 10-element linear array. Two equal-power uncorrelated signals impinge on the array from $\pm\Delta\psi/2$, where $\Delta\psi$ is 0.1 HPBW (0.022 radians for this array). We use $\psi_0 = 0$ and $T = 0.5$. We assume $K = 100$.

In Figure 7.59, we show P_D versus *ASNR* for the EDT algorithm. We also show the AIC and MDL results.

We see that the EDT is about 2 dB better then the AIC algorithm and its P_D approaches unity for $ASNR \geq 20$ dB.

However, the EDT algorithm was given additional prior information. In the next example, we utilize that information to modify the AIC and MDL algorithms.

Example 7.8.6 (continuation)

The EDT algorithm assumed that a beamscan provided prior information regarding the centroid of the cluster. With this prior information we could utilize AIC or MDL in beamspace. We first define[30]

$$\mathbf{B}_{no}^H = \dot{\mathbf{V}}_3, \tag{7.518}$$

[30]We motivate this choice when we study parameter estimation in Chapter 8.

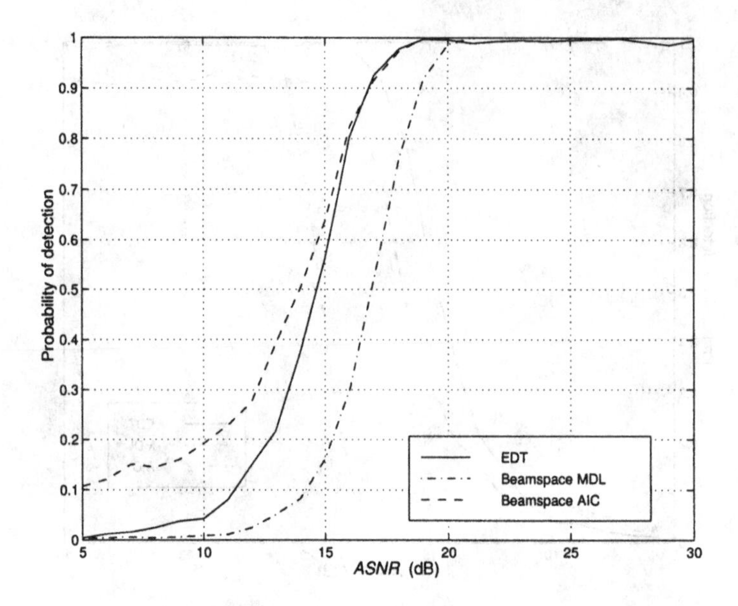

Figure 7.60 EDT, BS-AIC, and BS-MDL algorithms; two equal-power signals, $\Delta u = 0.1$HPBW, $K = 100$, 1,000 trials, P_D versus *ASNR*.

where $\dot{\mathbf{V}}_3$ is an $N \times 3$ matrix containing the first three columns of $\dot{\mathbf{V}}$. Then, using (6.518),

$$\mathbf{B}_{bs}^H = \left[\dot{\mathbf{V}}_3^H \dot{\mathbf{V}}_3\right]^{-\frac{1}{2}} \dot{\mathbf{V}}_3^H. \tag{7.519}$$

In Figure 7.60, we plot P_D versus *ASNR*. We see that beamspace AIC performs better than EDT for $P_D \leq 0.6$ and then the two algorithms have similar performance.

Example 7.8.7 (continuation, Example 7.8.5)

Consider the same model as in Example 7.8.5. We repeat the test for $\Delta u = 0.5$ HPBW. The results are shown in Figure 7.61. For clarity, we show the results for the EDT, BS-AIC, and AIC algorithms. We see that, even with the larger spacing the EDT test performs better than the AIC tests.

The EDT algorithm provides an alternative approach to detecting the number of closely spread plane-wave signals that are present. It requires a preliminary processing step to detect the location of signal clusters. It appears that, if we utilize this preliminary processing, then beamspace AIC will have a similar performance.

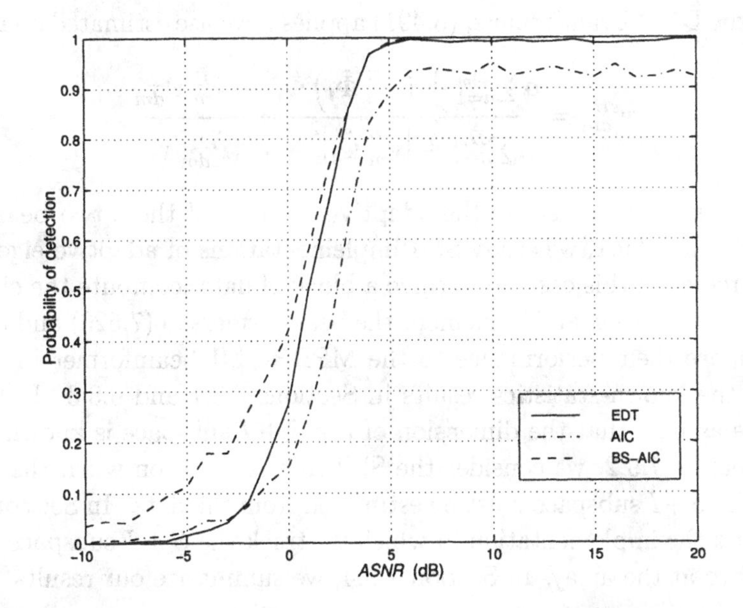

Figure 7.61 EDT, AIC, BS-AIC algorithms, two equal-power signals, $\Delta u =$ 0.5 HPBW, $K = 100$, 1,000 trials, P_D versus *ASNR*.

7.9 Eigenspace and DMR Beamformers

In Section 6.8, we discussed several beamformers that utilized an eigendecomposition to implement the beamformer. We assumed that the spatial spectral matrix, $\mathbf{S_x}$, is known. In this section, we consider the case in which $\mathbf{S_x}$ is estimated from the data. We focus our attention on two beamformers, the principal-component (or eigenspace) beamformer of Section 6.8.1 and the dominant-mode rejection beamformer of Section 6.8.3. We restrict our discussion in the text to the model in which the desired signal and interference are linearly independent plane waves. For the eigenspace beamformers, (6.449) applies if we introduce the estimated quantities in place of the ensemble statistics. The MPDR eigenspace beamformer is

$$\mathbf{w}_{mpdr,es}^H = \gamma_{es} \mathbf{v}_m^H \hat{\mathbf{U}}_{S+I} \hat{\mathbf{\Lambda}}_{S+I}^{-1} \hat{\mathbf{U}}_{S+I}^H, \tag{7.520}$$

where

$$\gamma_{es} = \left(\mathbf{v}_m^H \hat{\mathbf{U}}_{S+I} \hat{\mathbf{\Lambda}}_{S+I}^{-1} \hat{\mathbf{U}}_{S+I}^H \mathbf{v}_m \right)^{-1}. \tag{7.521}$$

For the DMR beamformers, (6.491) applies if we use estimated quantities,

$$\mathbf{w}_{dm}^H = \frac{\alpha \sum_{i=1}^{\hat{D}_m} \frac{1}{\hat{\lambda}_i} \left(\mathbf{v}_m^H \hat{\mathbf{\Phi}}_i \right) \hat{\mathbf{\Phi}}_i^H + \mathbf{v}_m^H \hat{\mathbf{P}}_{dm}^{\perp}}{\alpha \sum_{i=1}^{\hat{D}_m} \frac{1}{\hat{\lambda}_i} \left| \mathbf{v}_m^H \hat{\mathbf{\Phi}}_i \right|^2 + \mathbf{v}_m^H \hat{\mathbf{P}}_{dm}^{\perp} \mathbf{v}_m}. \tag{7.522}$$

In this section, we study the adaptive versions of these two beamformers. In Section 7.9.1, we study SMI implementations of adaptive eigenspace beamformers. In this case, we receive a block of data, compute the eigenvalues and eigenvectors, and implement the beamformers in (7.520) and (7.522). We compare their performance to the MPDR-SMI beamformer in Section 7.3 and the known-statistics results in Sections 6.8.1 and 6.8.3. In Section 7.9.1, we assume that the dimension of the $S+I$ subspace is known.

In Section 7.9.2, we consider the SMI implementation when the dimension of the $S+I$ subspace must be estimated from the data. In Section 7.9.3, we discuss the implementation in which we track the $S+I$ subspace as new data arrive at the array. In Section 7.9.4, we summarize our results.

7.9.1 Performance of SMI Eigenspace Beamformers

There are several performance questions that are pertinent to the eigenspace beamformers:

(i) SMI performance. How quickly do eigenspace beamformers converge to the asymptotic solution?

(ii) How well do the eigenspace beamformers perform with signal mismatch ($\mathbf{v}_a \neq \mathbf{v}_m$)?

(iii) How well do eigenspace beamformers perform in the presence of array perturbations?

(iv) How well do eigenspace beamformers perform when the dimension of the signal subspace must be estimated?

We consider the first two questions in this section under the assumption that the dimension of the signal plus interference subspace is known. We consider the fourth question in Section 7.9.2.

Several analyses for the eigenspace beamformer have been developed using various series expansions and approximations. One approximation, which is derived by Chang and Yeh [CY92] (e.g., [FG94] and [YL96]), is

$$E[SINR_o] \simeq \frac{SINR_o K}{K + SINR_o \cdot (N_{eig} - 1)}$$

$$= \frac{SINR_o K}{K + SINR_o(N_I)}, \tag{7.523}$$

where $SINR_o$ is the output $SINR$ when $\mathbf{S_x}$ is known, N_{eig} is the number of eigenvalues used, and N_I is the number of interferers.

In order to have the $E[SINR_o]$ within 3 dB of $SINR_o$, we require

$$K = N_I \cdot SINR_o. \tag{7.524}$$

This result can be compared to (7.111) from Section 7.2.1,

$$K = (SINR_{mpdr})\,(N-1) \simeq SINR_{mpdr} N. \tag{7.525}$$

Thus, the eigenspace algorithm has reduced the multiplier from N, the number of degrees of freedom to N_I, the number of interferers.

We can also use (7.110) with N replaced with N_{eig},

$$E[\eta] \simeq \frac{K - N_{eig} + 2}{K+1} \left\{ \frac{1}{1 + SINR_{mpdr-es} \frac{N_{eig}-1}{K+2}} \right\}. \tag{7.526}$$

We consider a simple example to illustrate the behavior.

Example 7.9.1: Eigenspace beamformer (continuation, Example 7.3.2)

Consider a standard 10-element linear array. The signal arrives from $u_s = 0$. The *SNR* is 0 dB and 20 dB. Two equal-power uncorrelated interferers arrive at $u_I = 0.29$ and 0.45, each with an $INR = 20$ dB.

In Figure 7.62, we plot the average $SINR_o$ versus K for the SMI-eigenspace beamformer with and without FB averaging. The horizontal line denotes the steady state value. We also show the MPDR results from Figure 7.5. We see that the eigenspace beamformer offers significant improvement. For low *SNR*, the analytic expression is accurate. As the *SNR* increases, the analytic expression underestimates the performance.

In this case, the improved performance is due to the reduced dimensionality of the subspace.

Example 7.9.2: DMR beamformer (continuation)

Consider the same model as in Example 7.9.1. In Figure 7.63, we plot the average $SINR_o$ versus K for the SMI-DMR beamformer using $D_m = 3$ with and without FB averaging. The $SNR = 0$ dB in Figure 7.63(a) and 10 dB in Figure 7.63(b). The performance is several dB worse than the eigenspace beamformer. The estimates of the three dominant mode eigenvectors have some inaccuracy due to the finite data set. This mismatch causes some nulling of the signal.

In the next two examples, we consider the case of signal DOA mismatch.

Example 7.9.3: Eigenspace beamformer with mismatch (continuation)

Consider the same nominal model as in Example 7.9.1. The beamformer assumes that the desired signal is arriving from $u_s = 0$. In practice, the signal arrives from $u_a = -0.05$.

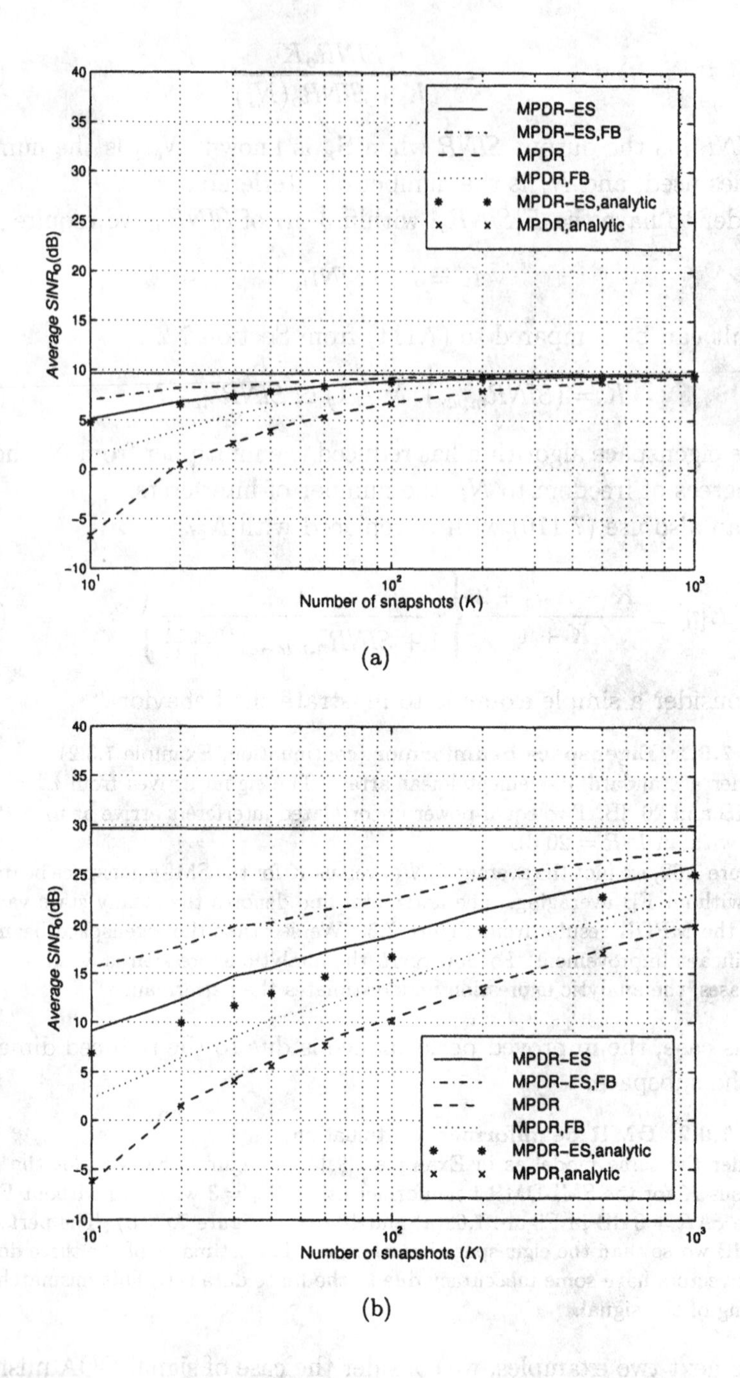

Figure 7.62 Eigenspace and MPDR beamformers using SMI algorithm: $u_s = 0, u_I = 0.29, 0.45, INR = 20$ dB. 200 trials: (a) $SNR = 0$ dB; (b) $SNR = 20$ dB, average $SINR_o$ versus K.

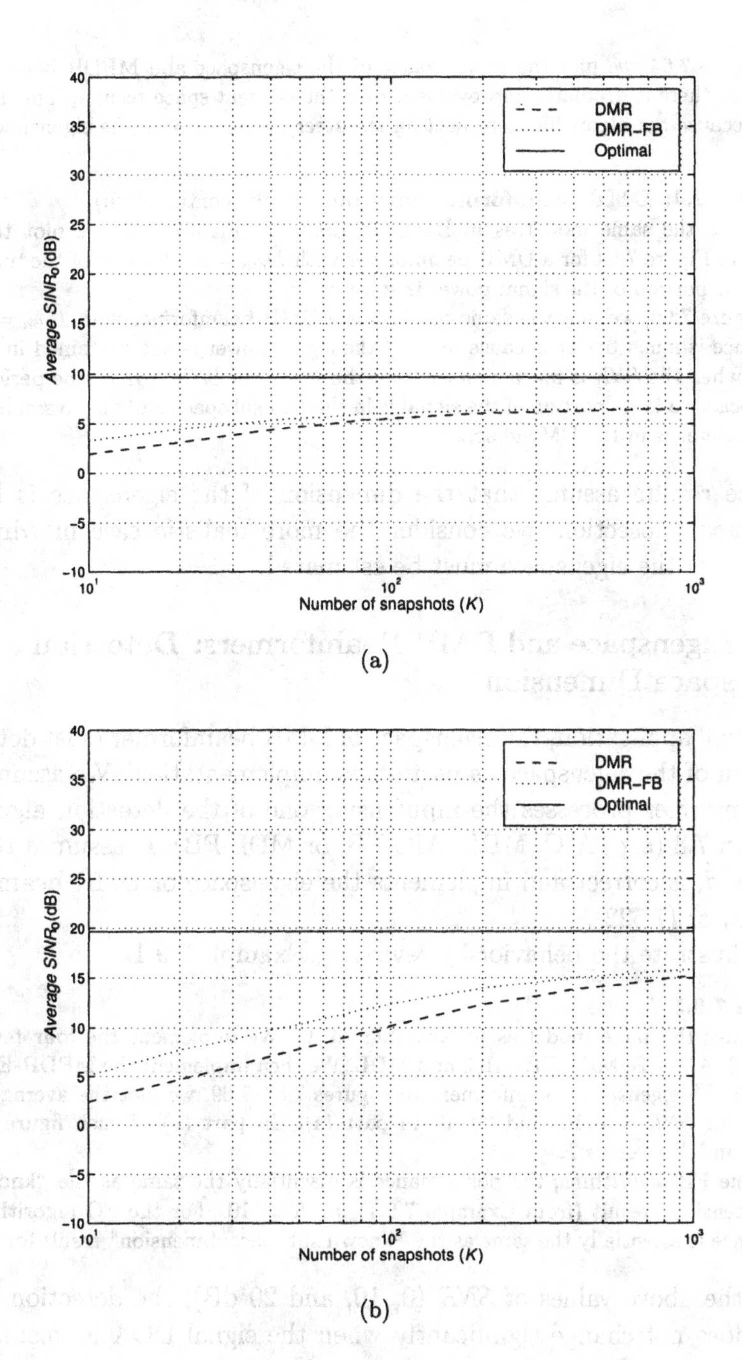

Figure 7.63 DMR beamformer with known subspace dimension using SMI algorithm: $u_s = 0$, $u_I = 0.29$ and 0.45, $INR = 20$ dB: (a) $SNR = 0$ dB; (b) $SNR = 10$ dB, average $SINR_o$ versus K.

In Figure 7.64, we plot the performance of the eigenspace and MPDR beamformers. We see that there is dramatic improvement over the element space results. This improvement is because the algorithm is projecting the steering vector onto the eigenspace.

Example 7.9.4: DMR beamformer with mismatch (continuation)

Consider the same model as in Example 7.9.3. In Figure 7.65, we plot the same results as in Figure 7.64 for a DMR beamformer with $\hat{D}_{dm} = 3$. Because of the mismatch, a significant portion of the signal power is rejected.

In Figure 7.66, we plot the same results for a DMR beamformer with $\hat{D}_{dm} = 2$. The performance is much better because most of the signal power is not contained in the DM subspace when the *SNR* is much smaller than the *INR*. For $SNR \geq INR$, the performance is poor because a large portion of the signal is in the DM subspace and only a small portion of the interferer is in the DM subspace.

These results assume that the dimension of the eigenspace is known. In the next subsection, we consider the more realistic case in which the dimension of the eigenspace must be estimated.

7.9.2 Eigenspace and DMR Beamformers: Detection of Subspace Dimension

In an actual application, the eigenspace or DMR beamformer must detect the dimension of the eigenspace as part of the implementation. We assume that the beamformer processes the input using one of the detection algorithms in Section 7.8 (e.g., AIC, MDL, AIC-FB, or MDL-FB). It assumes that the estimate, $\hat{d}$, is correct and implements the eigenspace or DMR beamformer in (7.520) or (7.522).

We illustrate the behavior by revisiting Example 7.9.1.

Example 7.9.5

Consider the same model as in Example 7.9.1. We implement the four tests from Section 7.8, AIC-FB, MDL-FB, AIC, and MDL. We then implement the MPDR-ES-FB or the MPDR-ES eigenspace beamformer. In Figures 7.67–7.69, we plot the average $SINR_o$ versus K for $SNR = 0, 10$, and 20 dB in part (a). In part (b) of each figure we plot P_D, P_{FA}, and P_M versus K.

For the FB algorithms, the performance is essentially the same as the "known subspace dimension" result (from Example 7.9.1) for $K \geq 10$. For the FO algorithms, the performance is essentially the same as the "known subspace dimension" result for $K \geq 20$.

For the above values of *SNR* (0, 10, and 20 dB), the detection performance does not change significantly when the signal DOA is mismatched. Therefore, the results in Example 7.9.3 apply to the case of signal mismatch when the subspace dimension must be estimated.

In order for the detection algorithm to degrade the eigenspace beamformer, the *SNR* must be low enough to cause P_M to be significant. This

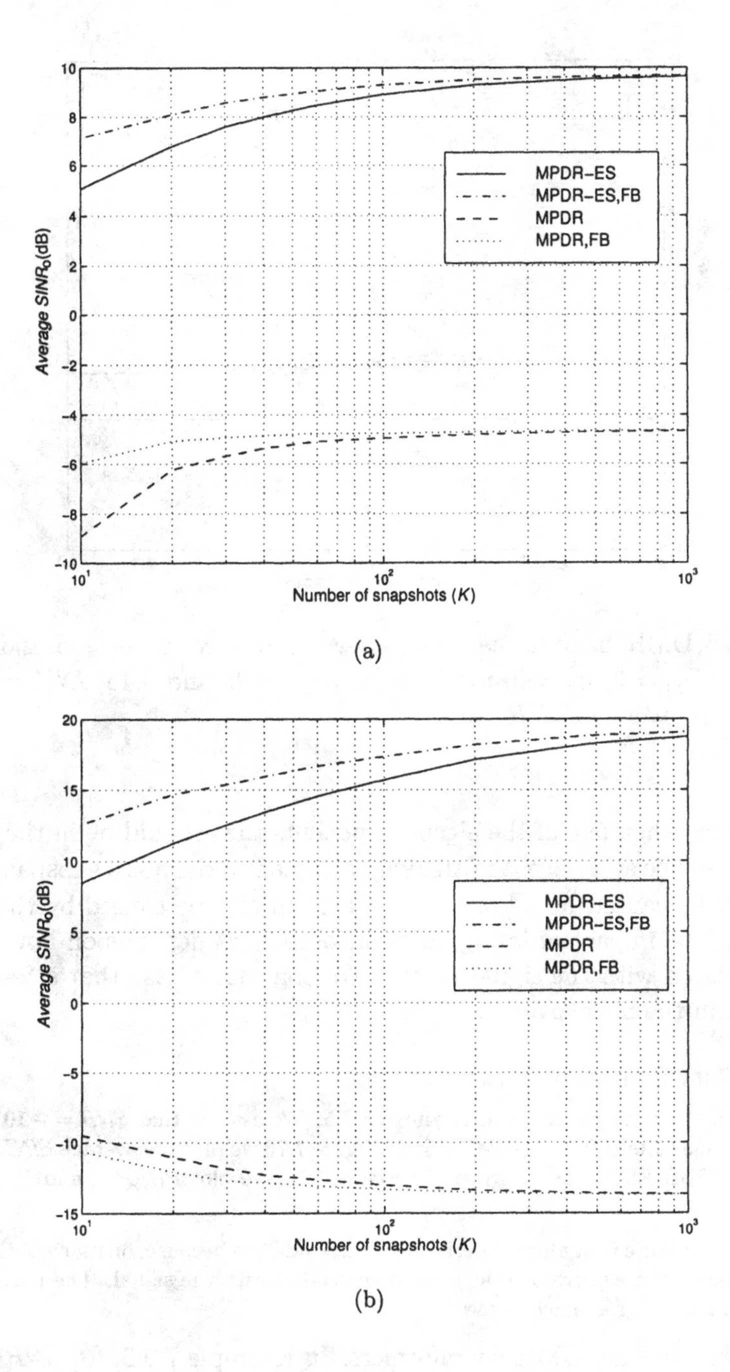

Figure 7.64 Eigenspace beamformer with signal mismatch: $u_s = 0, u_a = -0.05, u_I = 0.29, 0.45, INR = 20$ dB, 200 trials: (a) $SNR = 0$ dB; (b) $SNR = 10$ dB; average $SINR_o$ versus K.

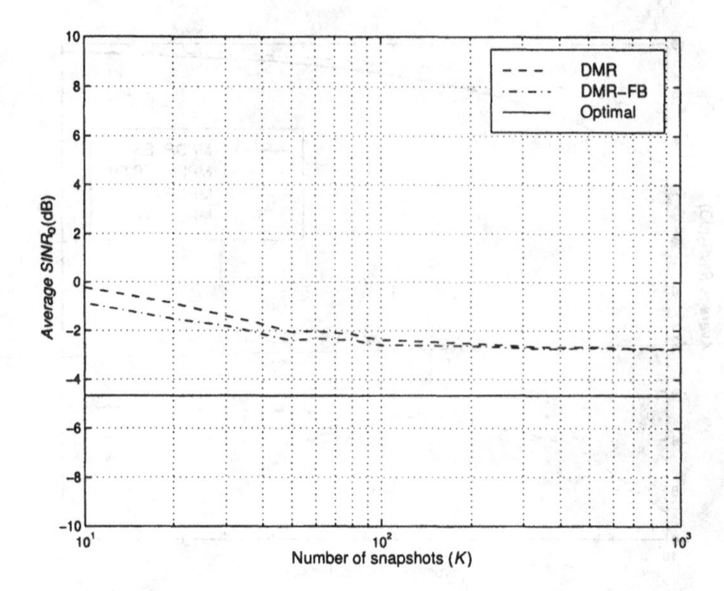

Figure 7.65 DMR beamformer with known subspace dimension and signal mismatch: $\hat{D}_m = 3$, $u_s = 0$, $u_d = -0.05$, $u_I = 0.29$ and 0.45, $INR = 20$ dB, $SNR = 0$ dB; average $SINR_o$ versus K.

result occurs when one of the eigencomponents that should be in the signal-plus-interference subspace is mistakenly assigned to the noise subspace. This result may be caused by a low *SNR*, but it can also be caused by the signal geometry. The impact of losing an eigencomponent depends on how closely it is correlated with the signal vector. We consider a case that stresses the algorithm in the next example.

Example 7.9.6 (continuation, Example 7.9.5)

Consider the same model as in Example 7.9.5. We assume that $SNR = -10$ dB. We repeat the same tests as in Example 7.9.5. In Figure 7.70 we plot the average $SINR_o$ versus K for the MPDR-ES-FB beamformer. In Figure 7.71, we plot P_D, P_M, and P_{FA} versus K.

We see that there is significant performance degradation because, on many of the trials, we fail to detect the eigenvalue that is most correlated with the signal. The performance without FB averaging is much worse.

We next consider DMR beamformers. In Example 7.9.5, for SNR=0, 10, and 20 dB, the performance when the dimension is estimated is essentially the same as the known dimension case. Therefore, the case of most interest is the $SNR = -10$ dB case in Example 7.9.6.

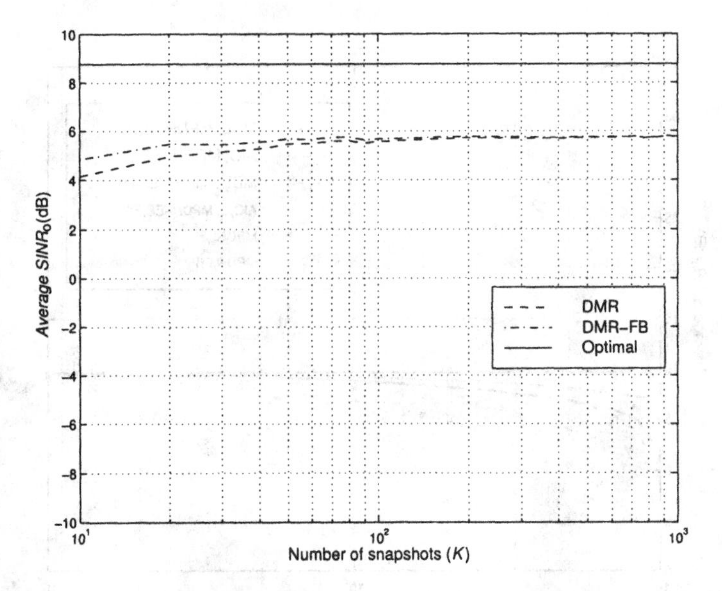

Figure 7.66 DMR beamformer with known subspace dimension and signal mismatch: $\hat{D}_m = 2$, $u_s = 0$, u_d=-0.05, $u_I = 0.29$ and 0.45, *INR* = 20 dB, *SNR* = 0 dB; average output $SINR_o$ versus K.

Example 7.9.7 (continuation)

Consider the same model as in Example 7.9.6. The detection algorithm results in Figure 7.71 also apply to this case. We implement the DMR beamformer with $d_{dm} = \hat{d}$ where $\hat{d}$ is the dimension estimate from one of the four detection algorithms. We use both FB and forward-only averaging. The results in Figure 7.71 show that, for small K, the detection algorithms are underestimating the dominant-mode subspace. Our results in Example 6.8.4 (see Figure 6.79) indicate that DMR should work well in this region. For K>100, AIC-FB either correctly estimates d_{dm} or overestimates d_{dm}. For K>200, MDL-FB correctly estimates d_{dm}. However, there is no signal mismatch, so the degradation due to having the signal in the DM subspace should be minimal.

In Figure 7.72, we plot the average $SINR_o$ versus K for the four detection algorithms. We see that the MDL-DMR-FB algorithms are close to steady state by $K = 20$. The DMR algorithm using MDL-DMR-FO detection is close to steady state by $K = 30$ because of its underestimation behavior: The DMR algorithm using AIC-FO converges much more slowly because of its overestimation behavior. For K<100, the MPDR-DMR-FB algorithms perform better than the MPDR-FB algorithm.

For higher *SNR*, the detection results in Figures 7.67, 7.68, and 7.69 are applicable. We see that, for K>20 ($2N$), the MDL-FB P_D is close to one. Therefore, the results in Example 7.9.2 should be close to the actual performance. Including the signal in the DM subspace causes performance degradation. When the *SNR* is less than the *INR*, we try an approach with

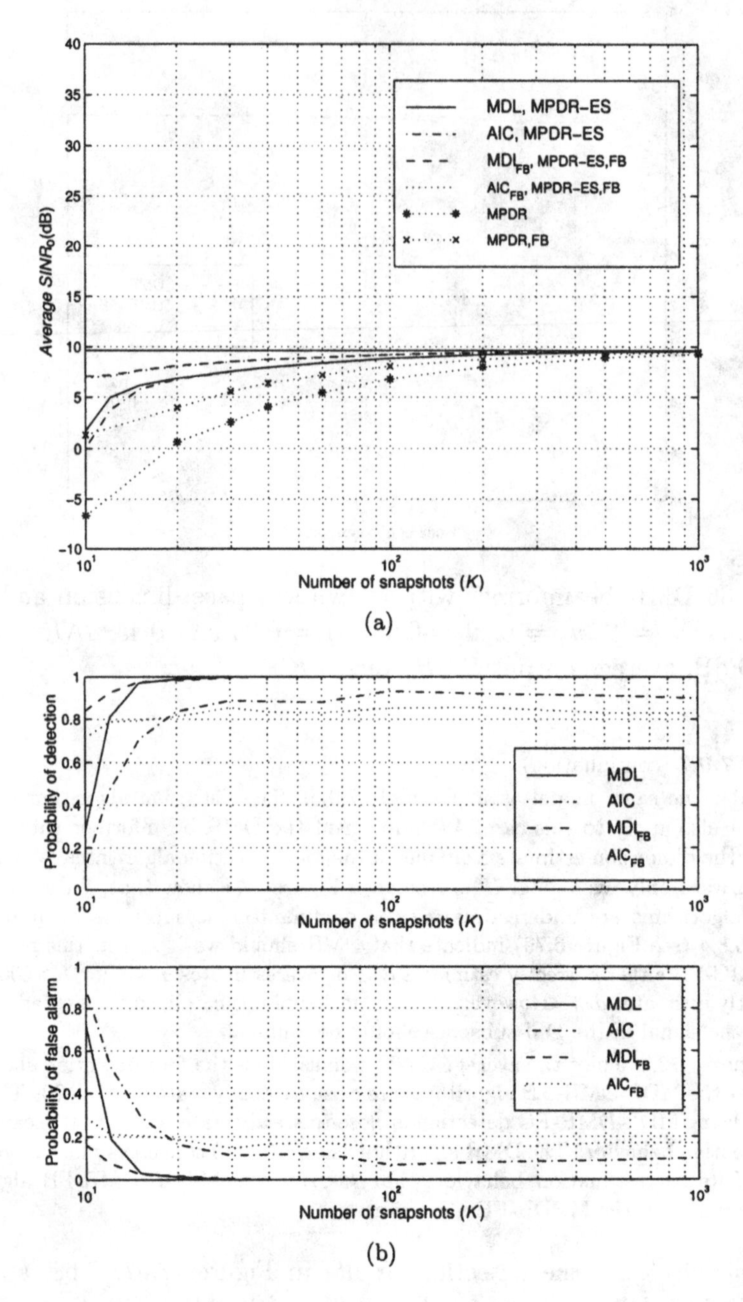

Figure 7.67 Eigenspace Beamformers using AIC and MDL detection algorithms; $u_s = 0$, $SNR = 0$ dB, $u_I = 0.29, 0.45$, $INR = 20$ dB, 200 trials: (a) average $SINR_o$ versus K; (b) P_D and P_{FA} versus K.

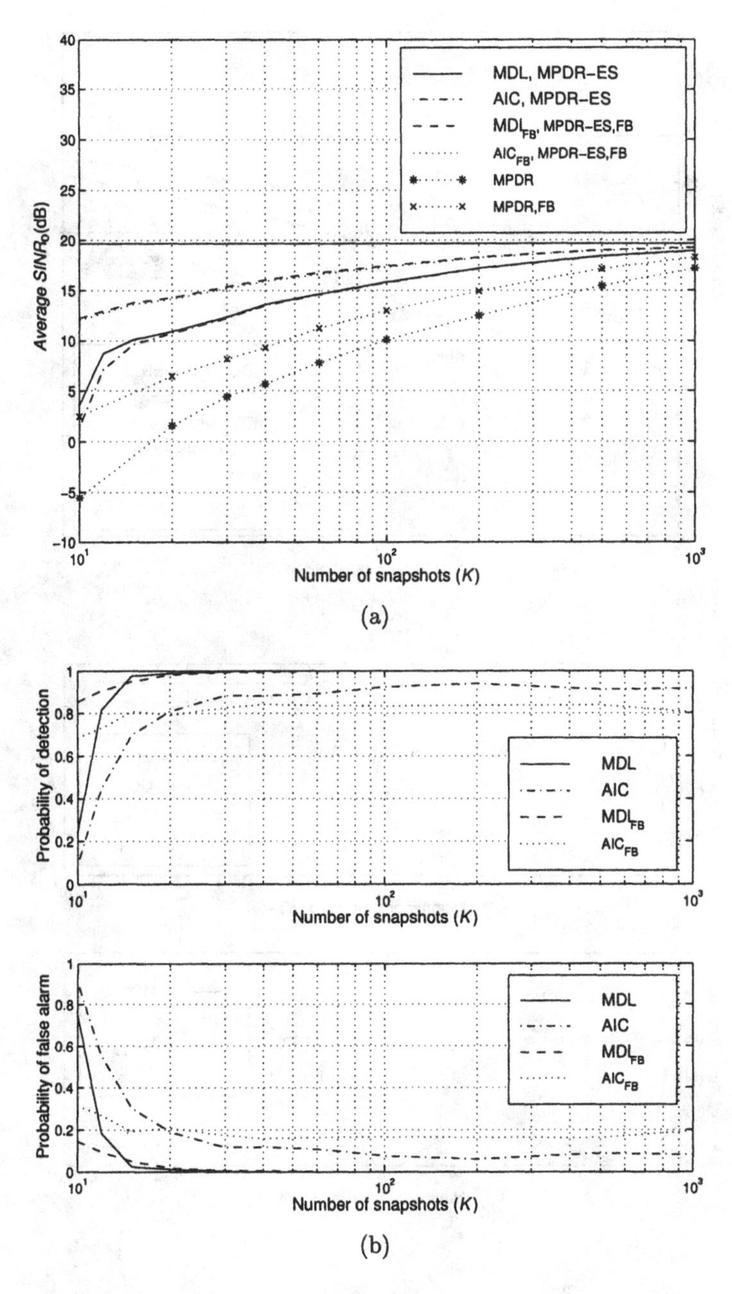

Figure 7.68 Eigenspace Beamformers using AIC and MDL detection algorithms: $u_s = 0$, $SNR = 10$ dB, $u_I = 0.29, 0.45$, $INR = 20$ dB, 200 trials: (a) average $SINR_o$ versus K; (b) P_D and P_{FA} versus K.

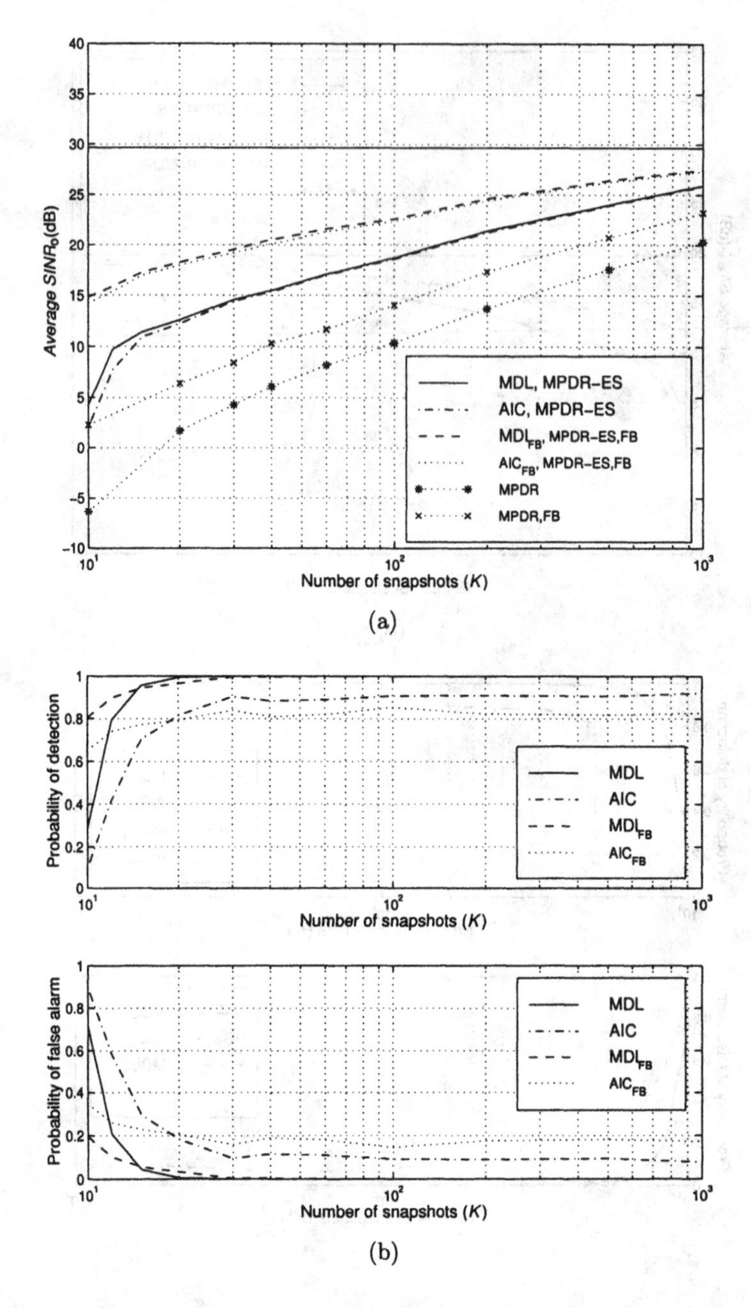

Figure 7.69 Eigenspace beamformers using AIC and MDL detection algorithms: $u_s = 0$, $SNR = 20$ dB, $u_I = 0.29, 0.45$, $INR = 20$ dB, 200 trials: (a) average $SINR_o$ versus K; (b) P_D and P_{FA} versus K.

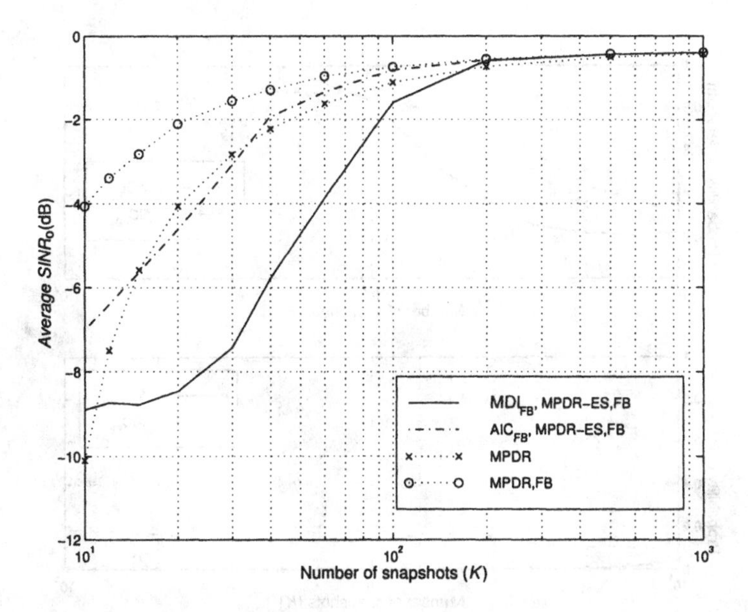

Figure 7.70 Eigenspace beamformers using AIC and MDL detection algorithms: $u_s = 0$, $SNR = -10$ dB, $u_I = 0.29, 0.45$, $INR = 20$ dB, 500 trials: average $SINR_o$ versus K.

$d_{dm} = \hat{d} - 1$ to improve performance.

Example 7.9.8 (continuation)

Consider the same model as in Example 7.9.5 with $SNR = 10$ dB. We implement the DMR algorithm with $d_{dm} = \hat{d} - 1$ (and $d_{dm} = \hat{d}$ for comparison). To simplify the figure, we only show the results for MDL-FB and AIC-FB. In Figure 7.73 we plot the average $SINR_o$ versus K for an $SNR = 10$ dB. We see that, for $K < 500$, deliberately underestimating the dimension of the DMR subspace improves the performance. For $K < 100$, the improvement in this example exceeds 3 dB.

In this section, we have studied the performance of eigenspace and DMR beamformers for the case in which we had to detect the dimension of the signal-plus-interference subspace. We used the AIC and MDL tests that were derived in Section 7.8 to detect the dimension.

For adaptive eigenspace beamformers, it is essential that the signal be included in the eigenspace. Therefore, for low *SNR* and small sample support we use the AIC-FB algorithm because it tends to overestimate, and the degradation in performance due to overestimation (false alarm) is less than the degradation due to underestimation (miss).

For higher *SNR*, AIC-FB still performs better, but the probability of

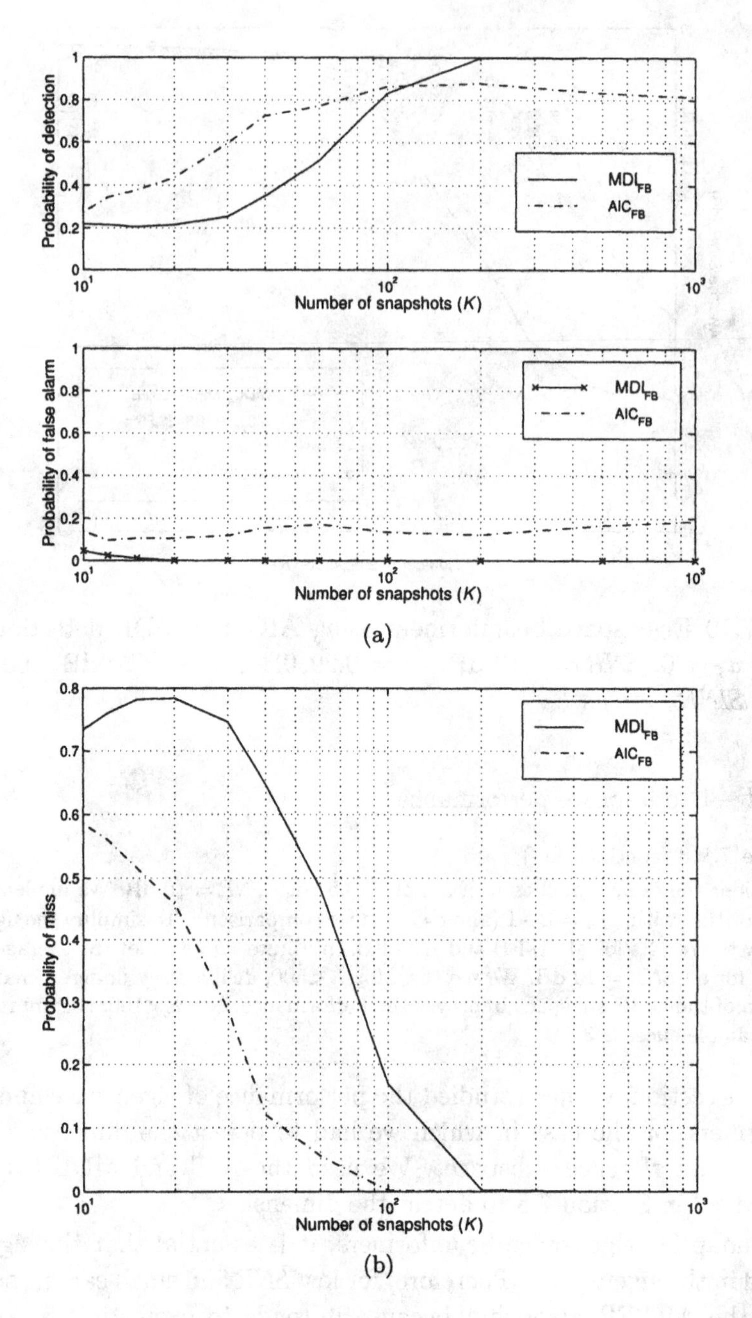

Figure 7.71 Performance of detection algorithms: $u_s = 0$, $SNR = -10$ dB, $u_I = 0.29, 0.45$, $INR = 20$ dB, 500 trials: (a) P_D and P_F versus K; (b) P_M versus K.

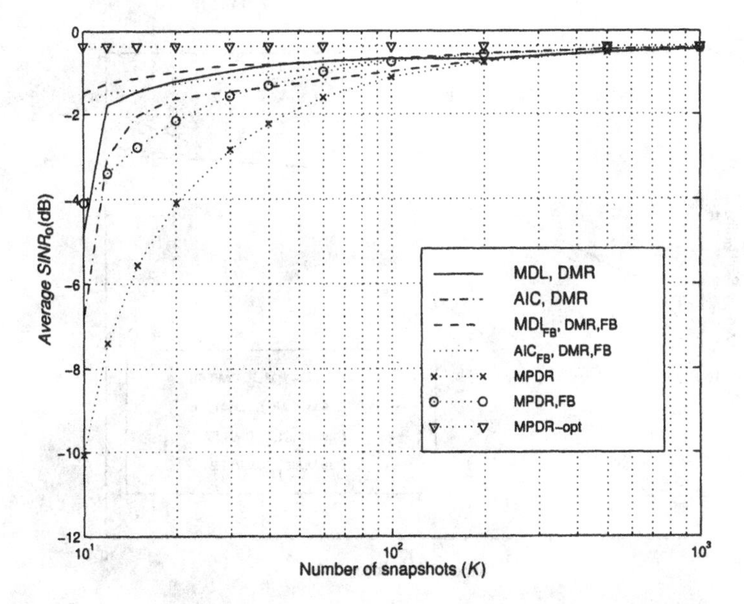

Figure 7.72 DMR beamformer with AIC and MDL detection algorithms, $d_m = \hat{d}$: $u_s = 0$, $SNR = -10$ dB, u_I=0.29 and 0.45, $INR = 20$ dB, 500 trials; average $SINR_o$ versus K.

correct detection increases. As the *SNR* increases, the convergence is slower because of the signal presence, but the performance is always better than MPDR.

For adaptive DMR beamformers, it is better to have the signal excluded from the DM subspace whenever the DM subspace is inexact (e.g., due to the signal mismatch or finite data estimates). For low *SNR*, MDL-FB performs slightly better than AIC-FB because of its tendency to underestimate in the low sample support region. Cox and Pitre [CP97] and Cox et al. [CPL98] have developed a robust version of the DMR beamformer. These use an auxiliary test to decrease signal rejection.

These results indicate that the eigenspace algorithm is the best choice when the *SNR* is high enough that most of the signal power is included in the eigenspace. For low *SNR*, the DMR beamformer performs better. The detection algorithms in Section 7.8 that we use in this section used block processing. In the next section, we consider algorithms that track the subspace as new data arrives.

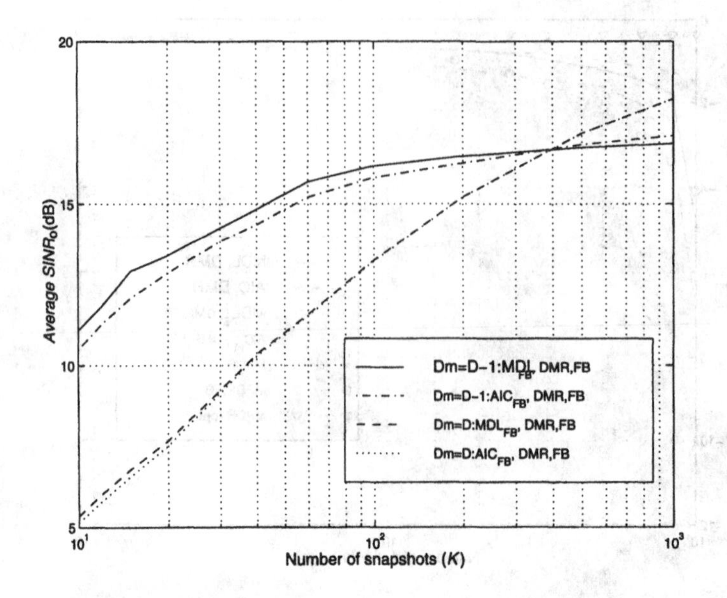

Figure 7.73: DMR beamformer with AIC and MDL detection algorithms, $d_m = \hat{d} - 1$: $u_s = 0$, $SNR = 10$ dB, u_I=0.29 and 0.45, $INR = 20$ dB, 500 trials; average $SINR_o$ versus K.

7.9.3 Subspace tracking

In Sections 7.9.1 and 7.9.2, the signal subspace is computed using a batch eigenvalue decomposition (ED) of the sample covariance matrix or using a SVD of the data matrix.[31] There are two disadvantages to this approach. The first is that the computational complexity is $O(N^3)$, where N is the number of sensors. In most cases of interest to us, we are only interested in the D signal eigenvalues and eigenvectors. Techniques have been developed to reduce the computational complexity. A discussion of these techniques is given in Tufts and Melissinos [TM86].

The second disadvantage is that, in adaptive beamforming (or parameter estimation) we would like to update the signal eigenvalues and eigenvectors of the sample covariance matrix (or the eigenvalues and singular vectors of the data matrix) each time a new data vector arrives. This updating is important because:

[31]This section is based on a research paper by one of my doctoral students, Zhi Tian [Tia99]

(i) Each additional snapshot will improve the accuracy of our subspace estimate;

(ii) In many applications of interest, the signal and/or interferers are moving so the subspace changes;

(iii) New interfering signals may appear.

Any of these effects will cause the estimate of the subspace to change.

For these reasons, we would like to develop algorithms that track or update the eigendecomposition as new data vectors arrive. For adaptive eigenspace beamforming we need to track both the eigenvalues and eigenvectors. In Chapter 9, when we study parameter estimation, we will discuss algorithms (e.g., MUSIC) that utilize the subspace (defined by the eigenvectors), but do not need the eigenvalues. In this section, we restrict our attention to algorithms that track both the eigenvalues and eigenvectors.

We focus our attention on the case in which we have a signal and interference environment that is fixed over the observation interval. However, the algorithms can adapt to an additional interfering signal arriving or an existing interfering signal departing.

A second scenario of interest is the case in which the desired signal and interfering signals are moving and we want to track the subspace.

A number of adaptive algorithms for subspace tracking have been developed in the literature. A survey of work through 1990 is available in a paper by Comon and Golub [CG90]. More recent results are discussed in Dowling et al. [DAD94]. Most techniques can be put into three categories:

(i) Variation of rank-one updating;

Golub [Gol73] developed an eigenvector (EV) updating scheme. Bunch et al. [BNS78] and Bunch and Nielsen [BN78] develop an algorithm that is guaranteed to converge. Karasalo [Kar86] provides an alternative implementation. DeGroat and Roberts [DR90] developed a numerically stable algorithm (ROSE) that provides good performance. MacInnes [MacI98] develops an update algorithm based on operator restriction (OPERA). Schreiber [Sch86] has developed implementations that use real operations. The FAST algorithm of Tufts et al. [TRC97] is another efficient example.

(ii) Variations of QR factorizations;

This category includes various rank revealing decompositions, such as Stewart's URV algorithm [Ste92] and the TQR SVD of Dowling et al.

[DAD94]. Strobach [Str96] developed a low rank adaptive filter (LORAF) based on sequential orthogonal iteration (suggested by Owsley [Ows78]).

(iii) Constrained or unconstrained optimization problem:

The eigenvector decomposition (EVD) or SVD can be formulated as an optimization problem and various techniques, such as gradient-based methods, Gauss-Newton iteration, and conjugate gradient techniques, can be applied to seek the largest or smallest eigenvalues and their corresponding eigenvectors adaptively. Yang and Kaveh [YK88] developed an algorithm using an LMS-type gradient estimator. Yang [Yan95a], [Yan95b] developed an RLS estimator called PAST.

From the computational point of view, we may distinguish among methods requiring $O(N^2d), O(N^2), O(Nd^2), O(Nd)$ arithmetic operations every update, where N is the dimension of the input data vector and d is the dimension of the subspace that we are tracking. The $O(N^2d)$ and $O(N^2)$ techniques are of little practical interest in applications where N is much larger than d. The wide range of the computational complexity is mainly due to the fact that some algorithms update the complete eigenstructure, whereas others track only the signal or noise subspace.

We looked at a number of algorithms, but only studied three algorithms in detail:

(i) Rank-one signal eigenstructure updating (ROSE), [DR90];

(ii) Low rank adaptive filter (LORAF), [Str96];

(iii) Projection approximation subspace tracking (PAST). [Yan95a], [Yan95b]

All of these algorithms are suitable for array processing[32] and are of complexity $O(Nd^2)$ or $O(Nd)$.

In each case, we required the algorithm to estimate the dimension of the subspace and estimate the eigenvalues and eigenvectors. We found that ROSE and a slightly modified version of LORAF had the best performance. The reader is referred to the above references and [Tia99] for a detailed discussion. There are a number of other algorithms that may be useful in a particular application. These include:

(i) Operator restriction algorithm (OPERA): [MacI98], [MV96];

[32]Some adaptive filter algorithms require a serial data input.

(ii) Fast approximate subspace tracking (FAST): Tufts et al. [TRC97] and Real et al. [RTC99];

(iii) Prony-Lanczos algorithm: Tufts and Melissinos [TM86];

(iv) PC method: Champagne [Cha94];

(v) Data-domain signal subspace updating algorithm (DDSSU): Youn and Un [YU94].

The reader is referred to the references for further discussions of these algorithms.[33]

A different approach uses a geometric entity referred to as a Grassmann manifold. Early work in this area is by Bucy [Buc91] and Smith [Smi93]. Further work is contained in Fuhrmann [Fuh98], Fuhrmann et al. [FSM96], and Srivastava [Sri00]. The reader is referred to these references for further discussion.

7.9.4 Summary

In this section, we have discussed adaptive eigenspace and adaptive DMR beamformers. The first two sections utilized block processing. In Section 7.9.1, we assumed that the dimension of the signal-plus-interference subspace was known and demonstrated that the reduced dimension of the subspace resulted in improved adaptive performance compared to a full-rank (N) beamformer. This improvement was significant if the ratio N/N_{eig} was large.

In Section 7.9.2, we considered the more realistic case in which the dimension of the subspace must be estimated. Guidelines for choosing an eigenspace or DMR beamformer were suggested.

The algorithms in Sections 7.9.1 and 7.9.2 are block algorithms. In Section 7.9.3, we identified several subspace tracking algorithms. In these algorithms, the eigendecomposition is updated each time a new data vector arrives. For a stationary environment, some of the algorithms provided the same performance as the block EVD or SVD. More importantly, they enabled us to track changes in the eigenvectors and eigenvalues.

Eigenspace and DMR algorithms provide effective performance in many scenarios and should be considered as a candidate design in many cases. In

[33] We recognize that our discussion of subspace tracking algorithms is too brief for the reader to understand the issues without extensive outside reading. In an earlier draft we derived the three algorithms, but the necessary discussion was long and not comprehensive enough.

the next section, we discuss adaptive beamformers operating in beamspace. This approach provides an alternative way to achieve the benefits of reduced-rank processing.

7.10 Beamspace Beamformers

We introduced beamspace beamformers in Section 3.10 in the context of classical array design. In Section 6.9, we developed optimum beamspace processors. In this section, we develop the beamspace versions of the various adaptive beamformers that we have developed in the earlier sections of the chapter. We use the model in Section 6.9.

The central result of this section is that, because of the reduced degrees of freedom in beamspace (N_{bs} versus N), the adaptation will be faster. The performance will have a K/N_{bs} dependence in beamspace versus a K/N dependence in element space.

The basic relation of interest is

$$\mathbf{X}_{bs} = \mathbf{B}_{bs}^H \mathbf{X}, \tag{7.527}$$

where

$$\mathbf{B}_{bs}^H \mathbf{B}_{bs} = \mathbf{I}. \tag{7.528}$$

The beamspace array manifold vector is

$$\mathbf{v}_{bs}(\psi) = \mathbf{B}_{bs}^H \mathbf{v}_s(\psi). \tag{7.529}$$

The beamspace spatial spectral matrix is

$$\mathbf{S}_{\mathbf{x},bs} = \mathbf{B}_{bs}^H \mathbf{S}_{\mathbf{x}} \mathbf{B}_{bs}. \tag{7.530}$$

If the array manifold vector is conjugate symmetric and the columns of $\mathbf{B}_{bs}$ are conjugate symmetric then, if we take the real part of the estimated beamspace matrix, it is equivalent to FB averaging of the data.

We define an $N_{bs} \times K$ beamspace data matrix $\tilde{\mathbf{X}}_{bs}$, which has the same structure as (7.4). Then

$$\hat{\mathbf{S}}_{\mathbf{x},bs,fb} = Re[\tilde{\mathbf{X}}_{bs}\tilde{\mathbf{X}}_{bs}^H] = \mathbf{B}_{bs}^H \hat{\mathbf{S}}_{\mathbf{x},fb} \mathbf{B}_{bs}. \tag{7.531}$$

In addition, $\mathbf{v}_{bs}$ is real in the conjugate symmetric case.

The requirement for conjugate symmetry of the rows of $\mathbf{B}_{bs}^H$ is satisfied by many beamspace matrices used in practice. Therefore, we can use real arithmetic to implement the adaptive beamspace beamformer, which offers significant computation savings. Note that $\mathbf{X}_{bs}$ is complex so that we have to reformulate our algorithms in a manner similar to the algorithm in Section 7.4.6 in order to actually implement the processors using real arithmetic.

In Section 7.10.1, we consider SMI algorithms. In Section 7.10.2, we consider RLS algorithms. In Section 7.10.3, we consider LMS algorithms. In Section 7.10.4, we summarize our results.

7.10.1 Beamspace SMI

The SMI algorithms apply directly in beamspace with suitable notational changes. The examples parallel those in Section 7.3. In some cases, we include the element-space result for comparision. We first consider an MVDR beamformer with no diagonal loading.

The first example is beamspace SMI and is analogous to Example 7.3.1.

Example 7.10.1: Beamspace MVDR

Consider a standard 32-element linear array. The beamspace matrix is 7×32 and uses conventional beams spaced at $2/N$ intervals in u-space. The array is aimed at broadside. There is a single interferer with an $INR = 10$ dB at $u_I = 3/32$.

We use an MVDR beamformer. In Figure 7.74, we plot the output $SINR_o$ versus K, the number of snapshots. We also plot the analytic result for $E[\rho]$ from (7.95) with N replaced by N_{bs}. In Figure 7.74(a) we plot versus K/N_{bs}.

In Figure 7.74(b), we plot $SINR_o$ versus K/N and show the element-space results for comparison. As expected, the beamspace SMI converges more quickly to the steady state value.

In the next example, we consider the same model and study the MPDR beamformer.

Example 7.10.2: Beamspace MPDR (continuation)

Consider the same model as in Example 7.10.1. In Figure 7.75(a), we plot the beamspace and element-space results for an $SNR = 0$ dB. In Figure 7.75(b), we plot the same results for an $SNR = 10$ dB. In Figure 7.75(c), we plot the same results for an $SNR = 20$ dB.

We see that there is a significant improvement in the performance. For the same level of $SINR_o$ loss the ratio in the number of required snapshots is N_{bs}/N.

We next consider the use of fixed diagonal loading.

Example 7.10.3 (continuation)

Consider the same model as in Example 7.10.1 except the INR is increased to 20 dB. We consider two cases:

(1) $SNR = 0$ dB, $INR = 15$ dB

(2) $SNR = 10$ dB, $INR = 15$ dB

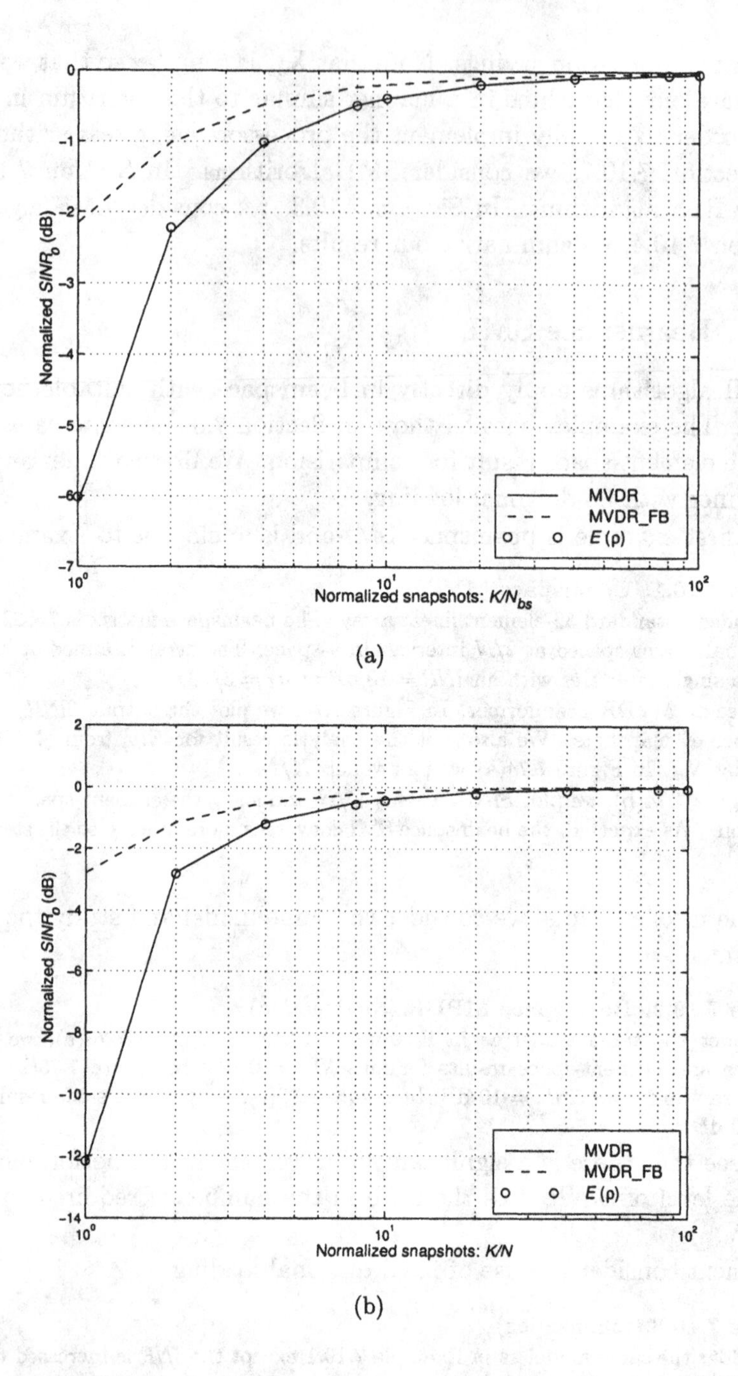

Figure 7.74 Beamspace MVDR with and without FB averaging: $N = 32, N_{bs} = 7, u_s = 0, u_I = 3/N, INR = 10$ dB, 200 trials; normalized SNR: (a) versus K/N_{bs}; (b) versus K/N.

In Figure 7.76(a), we plot the normalized $SINR_o$ versus K/N_{bs} for case 1. In Figure 7.76(b), we plot representative beam patterns for $K = 25$. Figure 7.77 plots the same results for case 2. For both *SNR* levels, the loading provides a useful improvement in performance.

We next consider the signal mismatch problem. In Example 7.10.4, we use diagonal loading to combat signal DOA mismatch. In Example 7.10.5, we also use additional constraints.

Example 7.10.4 (continuation)

Consider the same model as in Example 7.10.3. We use the beamspace MPDR with fixed diagonal loading, $LNR = 20$ dB. There is a single interferer at $u_I = 3/32$ with an $INR = 20$ dB. The array is aimed at broadside but the signal arrives from $u_a = -0.01$ with an $SNR = 10$ dB.

In Figure 7.78, we plot the output $SINR_o$ versus K/N_{bs}.

The next example of an SMI beamspace processor is a GSC implementation. The block diagram of the BS-GSC is shown in Figure 6.87. The equations specifying the BS-GSC are given in Section 6.9 ((6.524)–(6.528)). We consider an example using derivative constraints.

Example 7.10.5: LCMP-BS-GSC-SMI beamformer (continuation)

Consider the same model as in Example 7.10.4. Once again the array is aimed at broadside. The signal arrives from $u_a = -0.01$ with an $SNR = 10$ dB. There is a single interferer at $u_I = 3/32$ with an $INR = 20$ dB. We formulate the second-order derivative constraints in element space,

$$\mathbf{C} = \begin{bmatrix} \mathbf{1} & \dot{\mathbf{v}}(0) & \ddot{\mathbf{v}}(0) \end{bmatrix}, \tag{7.532}$$

and

$$\mathbf{g}^H = \begin{bmatrix} 1 & 0 & \ddot{B}_c(0) \end{bmatrix}. \tag{7.533}$$

Then, from (6.525),

$$\mathbf{C}_{bs} = \mathbf{B}_{bs}^H \mathbf{C}, \tag{7.534}$$

and, from (6.528),

$$\mathbf{w}_{lcmp,bs}^H = \mathbf{g}^H \left[\mathbf{C}_{bs}^H \mathbf{S}_{\mathbf{x},bs}^{-1} \mathbf{C}_{bs}^H\right]^{-1} \mathbf{C}_{bs}^H \mathbf{S}_{\mathbf{x},bs}^{-1}. \tag{7.535}$$

We implement (7.535) using SMI with $\mathbf{S}_{\mathbf{x},bs}$ replaced with $\widehat{\mathbf{S}}_{\mathbf{x},bs,fb}$.

We add fixed diagonal loading of 17 dB. In Figure 7.79, we plot the average $SINR_o$ versus K/N_{bs}. We can compare these results to those in Figure 7.78. With the BS-MPDR beamformer, we needed an $LNR = 20$ dB to prevent signal nulling. With the LCMP-BS beamformer the derivative constraints helped prevent signal nulling, so an $LNR = 17$ dB is adequate. The resulting steady state $SINR_o$ was 3 dB higher and the convergence rate was similar to the MPDR case. We see that the additional constraints improve the mismatch performance.

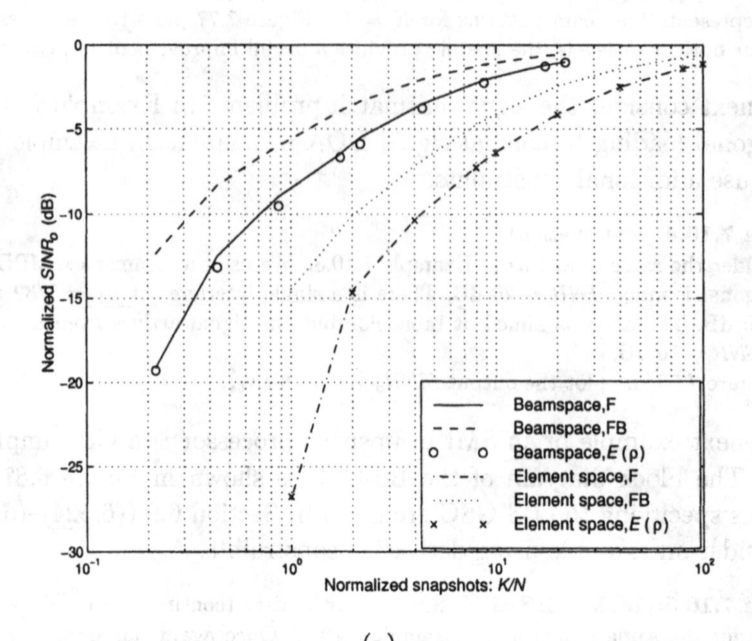
Normalized $SINR_o$ (dB)
Normalized snapshots: K/N
Beamspace,F
Beamspace,FB
Beamspace,$E(\rho)$
Element space,F
Element space,FB
Element space,$E(\rho)$

(a)

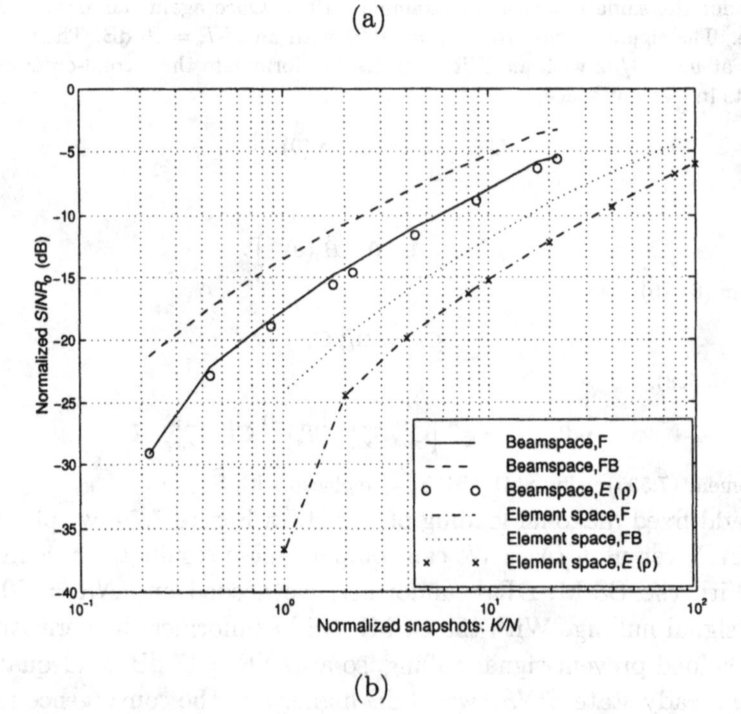
Normalized $SINR_o$ (dB)
Normalized snapshots: K/N
Beamspace,F
Beamspace,FB
Beamspace,$E(\rho)$
Element space,F
Element space,FB
Element space,$E(\rho)$

(b)

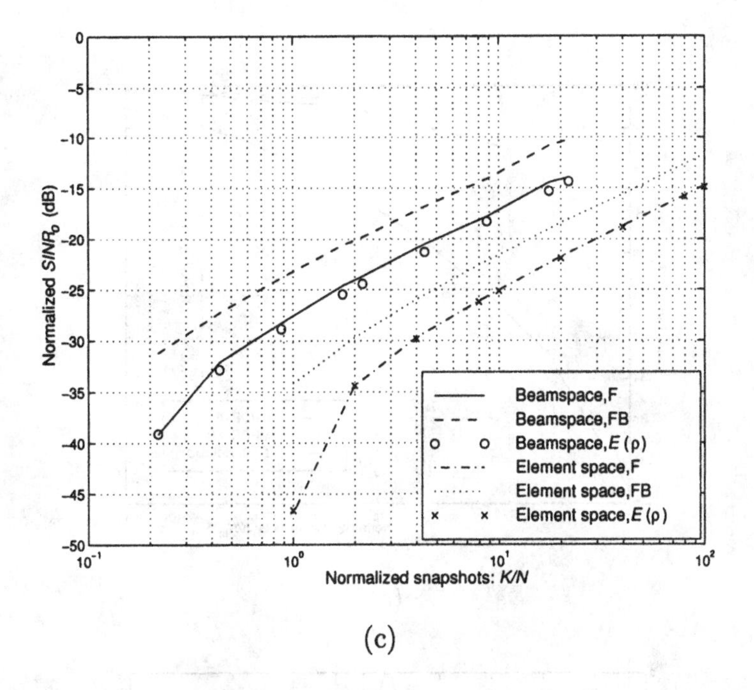

(c)

Figure 7.75 Beamspace MPDR with and without FB averaging: $N = 32, N_{bs} = 7, u_s = 0, u_I = 3/N, INR = 10$ dB, 200 trials; normalized $SINR_o$ versus K/N: (a) $SNR = 0$ dB; (b) $SNR = 10$ dB; (c) $SNR = 20$ dB.

In this section, we have studied SMI implementations of various beamspace beamformers. The key results are:

(i) The convergence to steady state depends on N_{bs}, the dimension of the beamspace instead of N, the number of array elements.

(ii) The computational complexity depends on N_{bs}, instead of N.

Most of our examples have used either $N = 10$ or $N = 32$ because many of the results scale. In many applications, N is much larger so that N_{bs}/N is very small. Thus, the advantages in (i) and (ii) are more dramatic.

In practice, we steer the array across the space of interest in discrete steps and perform beamspace processing at each step.

7.10.2 Beamspace RLS

In this section, we discuss beamspace RLS. All of the algorithms in Sections 7.4 and 7.5 can be modified for beamspace implementation in a straightfor-

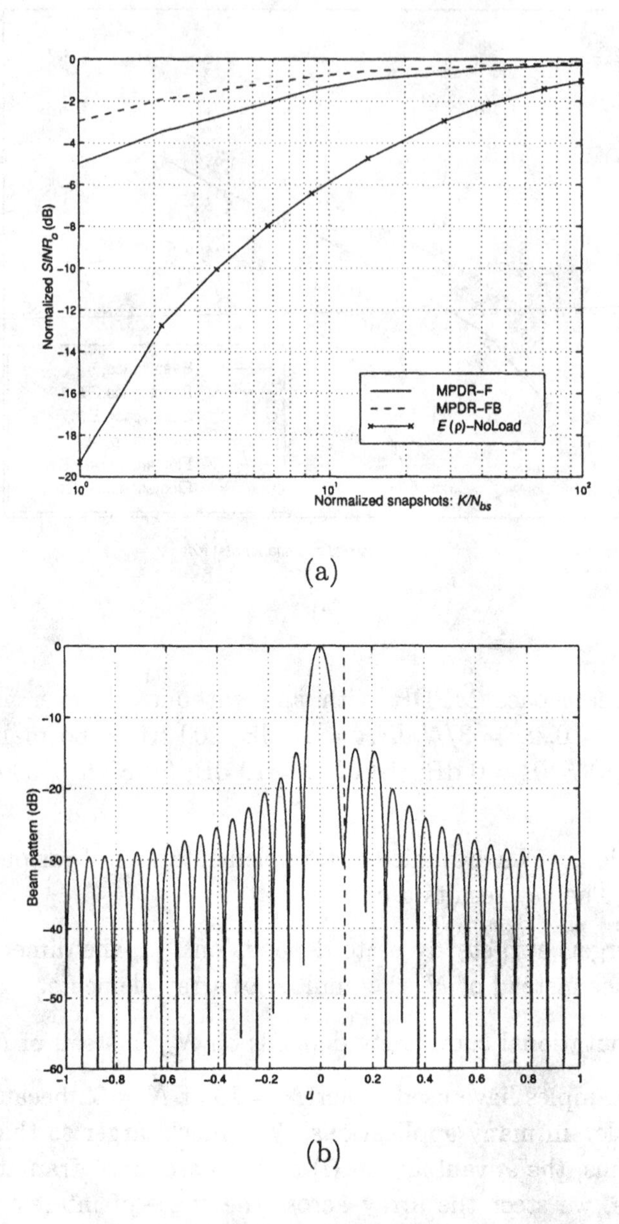

Figure 7.76 Beamspace MPDR with fixed loading: $N = 32, N_{bs} = 7, u_s = 0, SNR = 0$ dB, $u_I = 3/32, INR = 20$ dB, $LNR = 15$ dB: (a) normalized $SINR_o$ versus K/N_{bs}; (b) typical beam patterns, $K = 25$: $SNR = 0$ dB, $INR = 20$ dB, $LNR = 15$ dB.

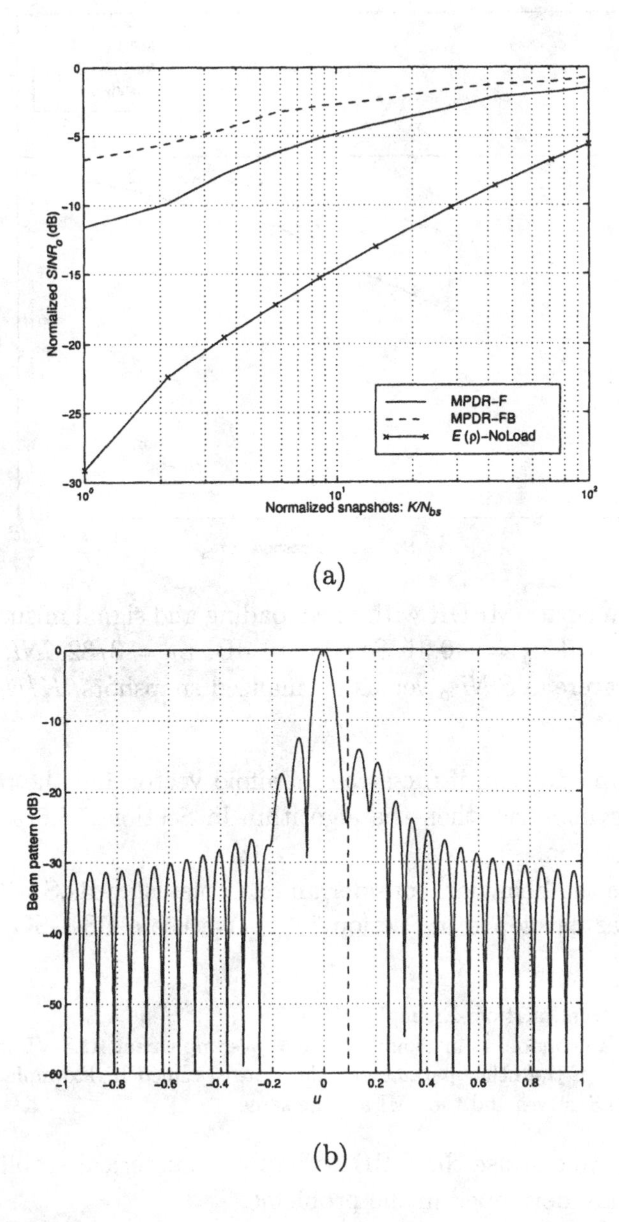

Figure 7.77 Beamspace MPDR with fixed loading: $N = 32, N_{bs} = 7, u_s = 0, SNR = 10$ dB, $u_I = 3/32, INR = 20$ dB, $LNR = 15$ dB: (a) normalized $SINR_o$ versus K/N_{bs}; (b) typical beam patterns, K=25: $SNR = 10$ dB, $INR = 20$ dB, $LNR = 15$ dB.

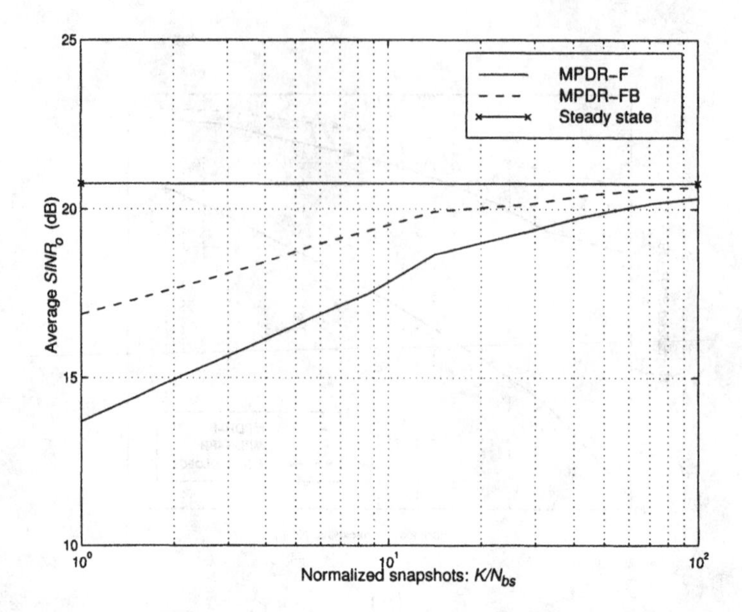

Figure 7.78 Beamspace MPDR with fixed loading and signal mismatch: $N = 32, N_{bs} = 7, u_s = 0, u_a = -0.01, SNR = 10$ dB, $u_I = 3/32, INR = 20$ dB, $LNR = 20$ dB; average $SINR_o$ versus normalized snapshots, K/N_{bs}.

ward manner. In addition, if the array manifold vector and the rows of $\mathbf{B}_{bs}^H$ are conjugate symmetric, then the algorithm in Section 7.4.6 allows us to use real computation.

To illustrate the behavior, consider an example using RLS with variable loading that was developed in Section 7.4.5. We use a BS-GSC implementation.

Example 7.10.6 (continuation, Example 7.10.4)

Consider the same model as in Example 7.10.4. We implement RLS-VL in beamspace with μ=0.999. All of the other parameter values are specified in Example 7.10.4. For $K/N_{bs} > 5$, the RLS curves and the SMI are the same.

In practice, we can use the QRD to improve numerical stability. Other RLS examples are developed in the problems.

7.10.3 Beamspace LMS

In this section, we discuss beamspace LMS. All of the LMS algorithms in Section 7.7 can be modified for beamspace implementation in a straightforward manner. To illustrate the behavior we consider an example using

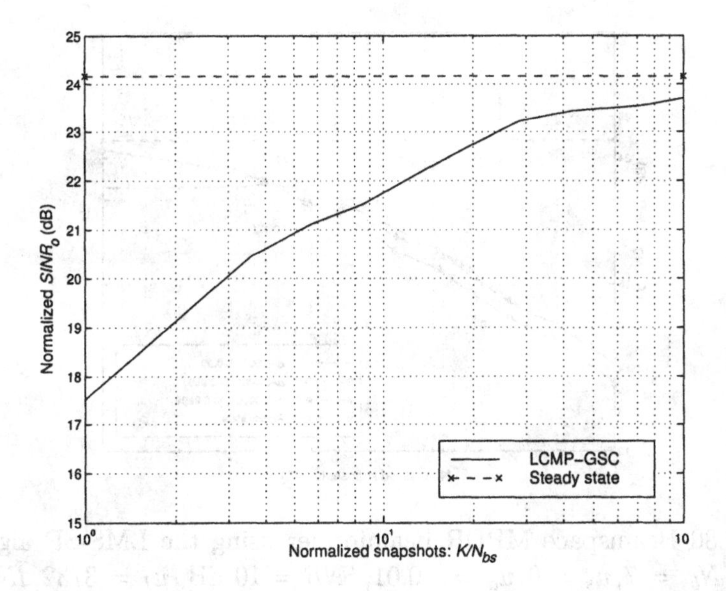

Figure 7.79 Beamspace LCMP beamformer with derivative constraints and fixed loading: $N = 32, N_{bs} = 7, u_s = 0, u_a = -0.01, SNR = 10$ dB, $u_I = 3/32, INR = 20$ dB, $LNR = 17$ dB; average $SINR_o$ versus normalized snapshots, K/N_{bs}.

LMS-SP in beamspace. The algorithm is described in Section 7.7.4.

Example 7.10.7 (continuation, Example 7.10.4)
Consider the same model as in Example 7.10.4. $N = 32$ and $N_{bs} = 7$. We implement LMS-SP in beamspace. We use $T_o = 1.2/32$ and

$$\alpha(K) = \frac{\gamma}{\sigma_x^2(K)} \tag{7.536}$$

where

$$\sigma_x^2(K) = \beta\sigma_x^2(K-1) + (1-\beta)\mathbf{X}_{bs}^H(K)\mathbf{X}_{bs}(K) \tag{7.537}$$

(see (7.474) and (7.475)). In Figure 7.80, we plot the average $SINR_o$ versus K/N_{bs} for several values of γ and β.

We see that the convergence is much faster than in element space.

7.10.4 Summary: Adaptive Beamspace Processing

In this section, we have studied beamspace processing. Several important results are:

(i) The convergence behavior of the algorithms is a function of K/N_{bs} instead of K/N. In most applications, the difference will be significant.

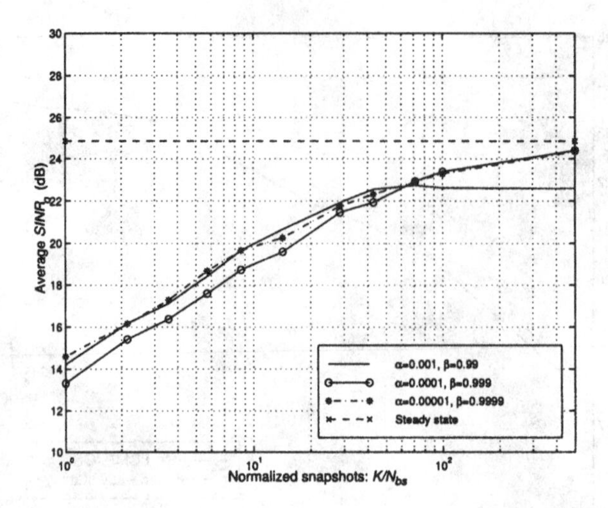

Figure 7.80 Beamspace MPDR beamformer using the LMS-SP algorithm: $N = 32, N_{bs} = 7, u_s = 0, u_a = -0.01, SNR = 10$ dB, $u_I = 3/32, INR = 20$ dB, $LNR = 17$ dB; average $SINR_o$ versus normalized snapshots, K/N_{bs}.

(ii) If the array is conjugate symmetric and the rows in the beamspace matrix are conjugate symmetric, then we can implement FB averaging by taking the real part of $\hat{\mathbf{S}}_{\mathbf{x},bs}$. This technique allows real processing to compute the necessary inverses. The result of (i) and (ii) can provide an order of magnitude improvement in the required number of snapshots to achieve a specified $SINR_o$ loss from the steady state performance in many cases.

(iii) All of the element space algorithms carry over to beamspace. However, the computational complexity is a function of N_{bs} rather than N.

As a result of these advantages, beamspace processors are widely used in practice.

7.11 Broadband Beamformers

In this section, we consider the adaptive behavior of the broadband beamformers that we developed in Section 6.13. We discussed a frequency-domain implementation and a time-domain implementation. The frequency-domain beamformer consisted of M uncoupled narrowband beamformers. Therefore, we can use the narrowband results in the earlier sections in this chapter to

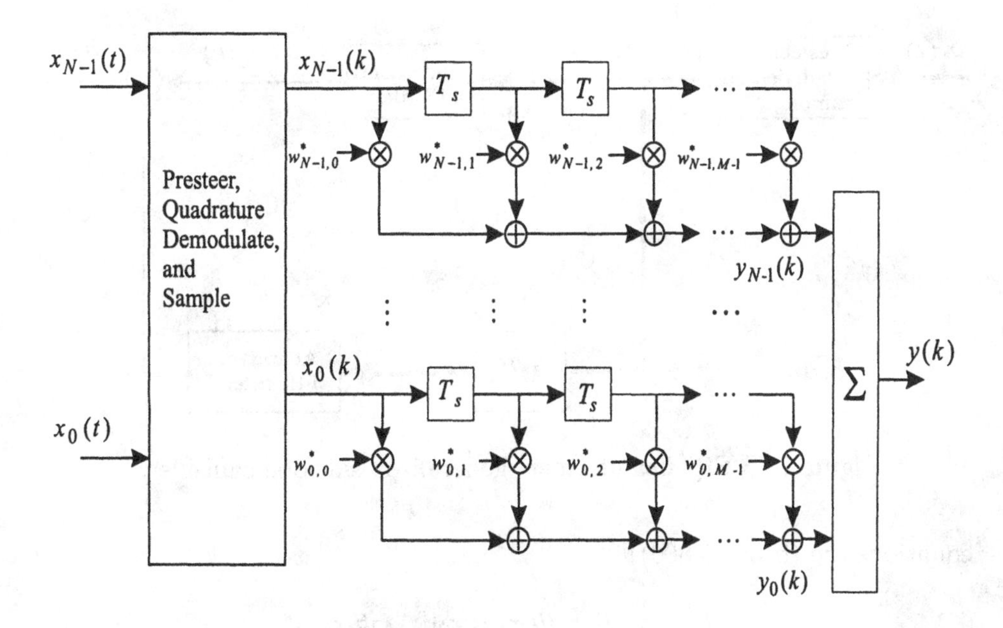

Figure 7.81 FIR beamformer: direct form.

analyze the broadband behavior. We consider several examples in the problems. In this section, we restrict our attention to broadband time-domain beamformers.

The starting point of the discussion is the time-domain model shown in Figure 7.81. This model is the direct form FIR model that we introduced in Figure 6.123.

The second time-domain model of interest is the GSC model in Figure 7.82 that we introduced in Figure 6.126. This GSC model assumes that the constraint matrix has the block structure in (6.797) or (6.810) and therefore, $\mathbf{B}_{comp}^H$ has the sparse structure in (6.822).

In Section 7.11.1, we discuss the SMI implementation of the beamformers. In Section 7.11.2, we discuss the LMS implementation using the FIR filters in Figures 7.81 and 7.82. In Section 7.11.3, we discuss a lattice implementation of the FIR filter in order to improve the LMS convergence. In Section 7.11.4, we summarize our results.

7.11.1 SMI Implementation

In this section, we discuss the SMI implementation of the time-domain beamformer. We use the direct form model in Figure 7.81. The steady state

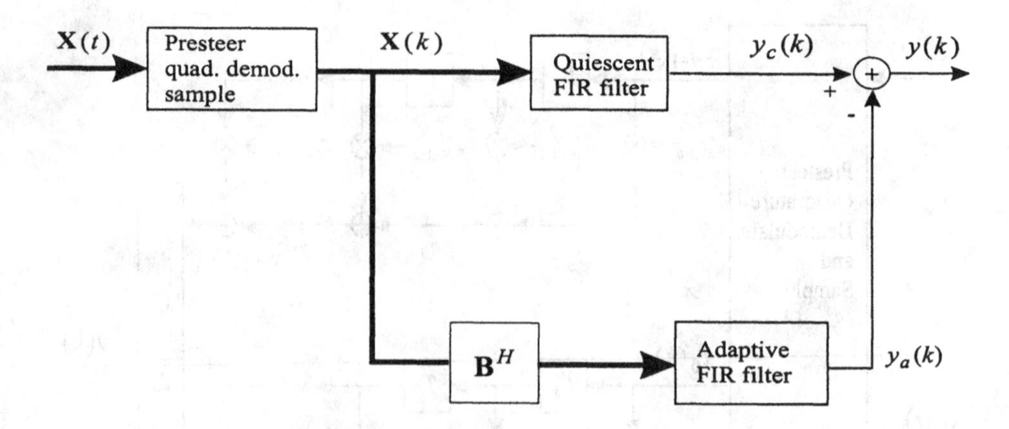

Figure 7.82 FIR beamformer: generalized sidelobe canceller.

equations are (from (6.806))

$$\mathbf{w}_{lcmp}^H = \mathbf{g}^H \left[\mathbf{C}^H \mathbf{R}_{\mathbf{x}}^{-1} \mathbf{C}\right]^{-1} \mathbf{C}^H \mathbf{R}_{\mathbf{x}}^{-1}, \tag{7.538}$$

where $\mathbf{C}$ is an $NM \times M_c M$ constraint matrix that has the block structure shown in (6.810).

For the MPDR case,

$$\mathbf{C} = \begin{bmatrix} \mathbf{1} & \mathbf{0} & \cdots & \mathbf{0} \\ \mathbf{0} & \mathbf{1} & & \vdots \\ \vdots & & \ddots & \mathbf{0} \\ \mathbf{0} & \mathbf{0} & \cdots & \mathbf{1} \end{bmatrix} = \mathbf{I}_M \otimes \mathbf{1}_N \tag{7.539}$$

and

$$\mathbf{g}^H = \begin{bmatrix} 1 & 0 & \cdots & 0 \end{bmatrix}. \tag{7.540}$$

In the SMI implementation, we estimate the $NM \times NM$ matrix $\mathbf{R}_{\mathbf{x}}$ by constructing the sample covariance matrix,

$$\hat{\mathbf{R}}_{\mathbf{x}} \triangleq \frac{1}{K} \sum_{k=1}^{K} \mathbf{x}(k)\mathbf{x}^H(k), \tag{7.541}$$

where $\mathbf{x}(k)$ is $NM \times 1$ composite input vector defined in (6.761). As in the narrowband case, we add diagonal loading,

$$\tilde{\mathbf{R}}_{\mathbf{x}} = \hat{\mathbf{R}}_{\mathbf{x}} + \sigma_L^2 \mathbf{I}_{NM}. \tag{7.542}$$

We then compute the $NM \times 1$ weight vector,

$$\mathbf{w}_{mpdr}^H = \mathbf{g}^H \left[\mathbf{C}^H \left[\hat{\mathbf{R}}_{\mathbf{x}} + \sigma_L^2 \mathbf{I}_{NM}\right]^{-1} \mathbf{C}\right]^{-1} \mathbf{C}^H \left[\hat{\mathbf{R}}_{\mathbf{x}} + \sigma_L^2 \mathbf{I}_{NM}\right]^{-1}, \quad (7.543)$$

and

$$\hat{y}(K) = \mathbf{w}_{mpdr}^H \mathbf{x}(K). \quad (7.544)$$

We consider the same model as in Example 6.13.5.

Example 7.11.1[34] (continuation, Example 6.13.5)

Consider a uniform 10-element array with spacing $d = \lambda_u/2$. There is an M-tap FIR filter at the output of each sensor's quadrature demodulator. We implement a broadbaud MPDR beamformer using the SMI algorithm. The desired signal arrives from broadside ($u_s = 0$) with a flat spectrum with $B_f = 0.4$. The interfering signal arrives from $u_I = 0.29$ with an identical spectrum. The $SNR = 0$ dB and the $INR = 30$ dB.

In Figure 7.83, we plot the average $SINR_o$ versus K for the $M = 4$ case. Four curves are shown:

(i) $LNR = 0$ dB ($SINR_{ss} = 9.24$ dB)

(ii) $LNR = 5$ dB ($SINR_{ss} = 9.13$ dB)

(iii) $LNR = -50$ dB ($SINR_{ss} = 9.26$ dB)

(iv) $\overline{\eta}(SINR_{ss})$; from (7.110), with $LNR = -50$ dB and $N = 40$.

The algorithm with -50-dB loading has a slightly larger steady state $SINR_o$, but significantly poorer performance for $K < 500$. The behavior is accurately described by (7.110). The LNR = 0-dB and LNR = 5-dB cases have essentially the same steady-state but much better transient behavior. The LNR = 5-dB case is almost one dB higher than the LNR = 0-dB case at $K = 100$. LNRs greater than 5 dB lower the steady-state value without improving the transient behavior. Note that an $LNR = 5$ dB is much lower than the LNR for the corresponding narrowband model.

In Figure 7.84, a representative beam pattern is plotted for $K = 1,000$. The nulls are in the correct location and the sidelobes are well-behaved.

In Figure 7.85, we plot the average $SINR_o$ versus K for $M = 2$. The steady state value is 8.874 dB. At $K = 1,000$, the SMI algorithm is very close to the steady state value. In Figure 7.86, a representative beam pattern is plotted for $K = 1,000$. The nulls are in the correct location and the sidelobes are well-behaved.

The $M = 2$ case has almost the same performance as the $M = 4$ case, and the computational complexity is reduced. The reason for this behavior is that the signal and interferer have identical spectra, so the additional frequency resolution obtained by additional taps is not required.

As in the narrowband case, the SMI algorithm with appropriate loading provides good performance. The primary disadvantage is that the matrices that must be inverted are $NM \times NM$, which increases the computational complexity.

[34]The results in Examples 7.11.1 and 7.11.2 are due to J. Hiemstra and R. Jeffers (private communication).

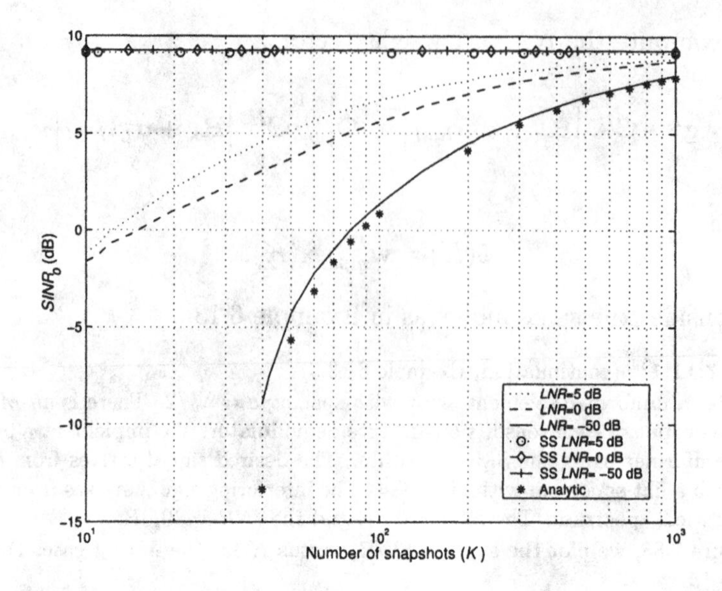

Figure 7.83 Broadband MPDR time-domain beamformer with SMI algorithm: $M = 4, u_s = 0, SNR = 0$ dB, $B_{fs} = 0.4, u_I = 0.29, INR = 30$ dB, $B_{f_I} = 0.4, LNR = 0$ dB, 5 dB, and -50 dB, 100 trials; average $SINR_o$ versus K.

7.11.2 LMS Implementation

In this section, we discuss the LMS implementation of the TDL beamformer. In the first subsection we use the direct form structure in Figure 7.81. In the second subsection we use the GSC structure in Figure 7.82.

7.11.2.1 Direct-form FIR structure

We modify the results in Section 7.7 to accommodate the FIR case. The starting point is (7.418) – (7.420). Recalling that

$$\mathbf{w}_q = \mathbf{C}\left(\mathbf{C}^H\mathbf{C}\right)^{-1}\mathbf{g}, \tag{7.545}$$

and

$$\mathbf{P}_{\mathbf{C}}^{\perp} \triangleq \mathbf{I} - \mathbf{C}\left(\mathbf{C}^H\mathbf{C}\right)^{-1}\mathbf{C}^H, \tag{7.546}$$

the LMS equations are given by (7.418) – (7.420). For the FIR model,

$$\mathbf{w}(0) = \mathbf{w}_q, \tag{7.547}$$

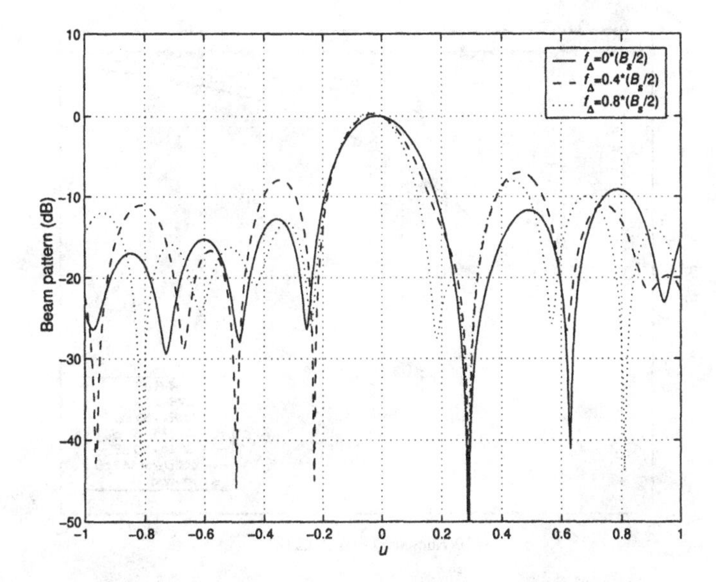

Figure 7.84 Broadband MPDR time-domain beamformer with SMI algorithm: $M = 4, u_s = 0, SNR = 0$ dB, $B_{fs} = 0.4, u_I = 0.29, INR = 30$ dB, $B_{f_I} = 0.4, LNR = 0$ dB, 5 dB, and -50 dB, 100 trials; representative beam patterns at $K = 1,000$ at three frequencies.

and

$$\mathbf{w}(K) = \mathbf{P}_{\mathbf{c}}^{\perp}\left[\hat{\mathbf{w}}(K-1) - \Delta(K)y_p(K)\mathbf{x}(K)\right] + \mathbf{w}_q, \tag{7.548}$$

where

$$y_p(K) = \mathbf{x}^H(K)\mathbf{w}(K-1), \tag{7.549}$$

and

$$\boldsymbol{\Delta}(k) = \text{diag}\left[\Delta_0(k), \Delta_1(k), \cdots, \Delta_m(k)\right] \otimes \mathbf{I}_N. \tag{7.550}$$

We first consider the MPDR case in which (from (6.797))

$$\mathbf{C} = \begin{bmatrix} \mathbf{1} & & & \\ & \mathbf{1} & & \\ & & \ddots & \\ & & & \mathbf{1} \end{bmatrix} = \mathbf{I}_M \otimes \mathbf{1}_N, \tag{7.551}$$

and

$$\mathbf{g} = \begin{bmatrix} 1 & 0 & \cdots & 0 \end{bmatrix}^T. \tag{7.552}$$

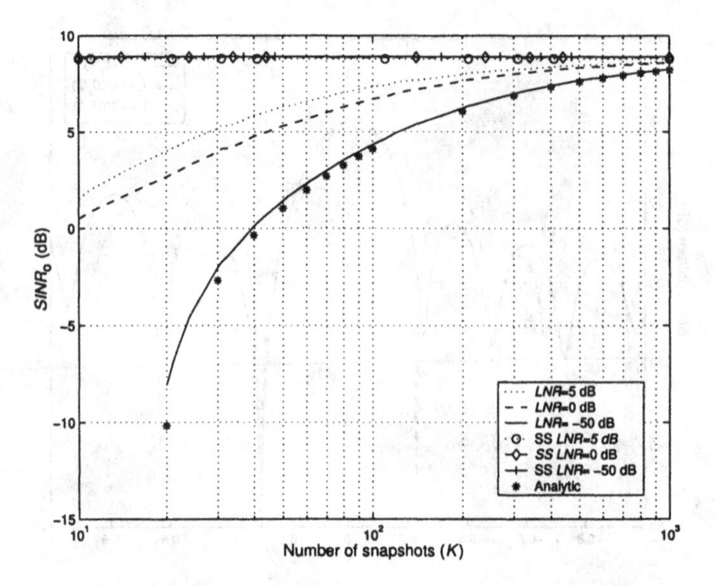

Figure 7.85 Broadband MPDR time-domain beamformer with SMI algorithm: $M = 2, u_s = 0$, $SNR = 0$ dB, $B_{fs} = 0.4$, $u_I = 0.29$, $INR = 30$ dB, $B_{f_I} = 0.4$, $LNR = 0$ dB, and -50 dB, 100 trials; average $SINR_o$ versus K; steady state results are also shown.

Using (7.551) in (7.546), we have

$$\mathbf{P}_{\mathbf{c}}^{\perp} = \mathbf{I}_{NM} - \frac{1}{N}\left(\mathbf{I}_M \otimes \begin{bmatrix} \mathbf{1} & \mathbf{1} & \cdots & \mathbf{1} \\ \mathbf{1} & \mathbf{1} & \cdots & \mathbf{1} \\ \vdots & & \ddots & \vdots \\ \mathbf{1} & & & \mathbf{1} \end{bmatrix}\right). \tag{7.553}$$

Using (7.552) and (7.553) in (7.548) gives the LMS equations,

$$\begin{aligned} \mathbf{w}_0(K) &= \mathbf{w}_0(K-1) - \Delta_0(K) y_p(K) \mathbf{x}_0(K) \\ &\quad -\frac{1}{N}\mathbf{1}\,\mathbf{1}^T \times [\mathbf{w}_0(K-1) - \Delta_0(K) y_p(K) \mathbf{x}_0(K)] + \frac{1}{N}\mathbf{1}, \end{aligned} \tag{7.554}$$

$$\begin{aligned} \mathbf{w}_m(K) &= \mathbf{w}_m(K-1) - \Delta_m(K) y_p(K) \mathbf{x}_m(K) \\ &\quad -\frac{1}{N}\mathbf{1}^T [\mathbf{w}_m(K-1) - \Delta_m(K) y_p(K) \mathbf{x}_m(K)], \\ &\quad m = 1, \cdots, M-1, \end{aligned} \tag{7.555}$$

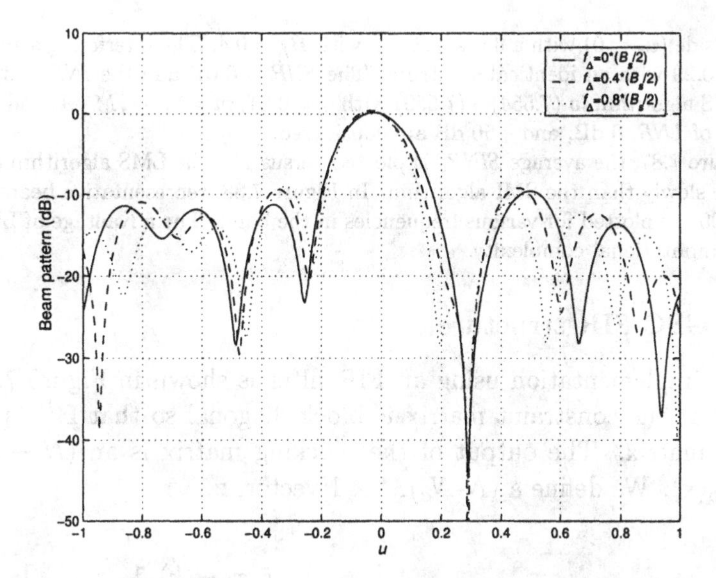

Figure 7.86 Broadband MPDR time-domain beamformer with SMI algorithm: $M = 2$, $u_s = 0$, $SNR = 0$ dB, $B_{fs} = 0.4$, $u_I = 0.29$, $INR = 30$ dB, $B_{f_I} = 0.4$, $LNR = 0$ dB, 100 trials; representative beam patterns at $K = 1,000$ at three frequencies.

and

$$y_p(K) = \sum_{m=0}^{M-1} \mathbf{x}_m^H(K)\mathbf{w}_m(K-1). \tag{7.556}$$

Note that the coupling of the tap weight vectors only occurs in (7.556). This result is because of the diagonal structure of $\boldsymbol{\Delta}(k)$ in (7.550). Therefore, we can use different $\Delta_m(K)$ in (7.554)–(7.555). We used the power normalized version,

$$\Delta_m(K) = \frac{\gamma_m}{\sigma_m^2(K)}, \tag{7.557}$$

where

$$\sigma_m^2(K) = \beta\sigma_m^2(K-1) + (1-\beta) \parallel \mathbf{x}_m(K) \parallel^2 \tag{7.558}$$

with

$$\sigma_m^2(0) = 0, \qquad m = 0, \cdots, M-1. \tag{7.559}$$

The next example uses the same signal and interference model as Example 7.11.2.

Example 7.11.2 (continuation, Example 7.11.1)

Consider a uniform 10-element array with spacing $d = \lambda_u/2$. There is a 4-tap FIR filter at the output of each sensor's quadrature demodulator. The desired signal arrives from broadside ($u_s = 0$) with a flat spectrum with $B_f = 0.4$. The interfering signal arrives from $u_I = 0.29$ with an identical spectrum. The *SNR* = 0 dB and the *INR* = 30 dB. We use the LMS algorithm in (7.554) – (7.559) with $\gamma = 0.01$, $m = 0, \cdots, M-1$, and $\beta = 0.99$. Two levels of *LNR*, 0 dB, and −50 dB are considered.

In Figure 7.87, the average $SINR_o$ is plotted versus K. The LMS algorithm converges much more slowly than the SMI algorithm. In Figure 7.88, representative beam patterns at $K = 1000$ are plotted for various frequencies in the band. The advantage of LMS is the reduced computational complexity.

7.11.2.2 GSC FIR structure

The GSC implementation using an FIR filter is shown in Figure 7.82. We assume that the constraint matrix is block diagonal so that $\mathbf{B}^H$ is a $(N - N_d) \times N$ matrix. The output of the blocking matrix is an $(N - N_d) \times 1$ vector, $\mathbf{x}_b(k)$. We define a $(N\text{-}N_d)M \times 1$ vector, $\mathbf{z}$,

$$
\mathbf{z}(k) \triangleq \begin{bmatrix} \mathbf{x}_b(k) \\ \mathbf{x}_b(k-T_s) \\ \vdots \\ \mathbf{x}_b(k-(M-1)T_s) \end{bmatrix} = \begin{bmatrix} \begin{bmatrix} x_{b,00}(k) \\ x_{b,10}(k) \\ \vdots \end{bmatrix} \\ \begin{bmatrix} x_{b,00}(k-T_s) \\ x_{b,10}(k-T_s) \\ \vdots \end{bmatrix} \\ \vdots \\ \begin{bmatrix} x_{b,00}(k-(M-1)T_s) \\ x_{b,10}(k-(M-1)T_s) \\ \vdots \end{bmatrix} \end{bmatrix}. \tag{7.560}
$$

The LMS equations are (from (7.425) – (7.427)),

$$
e_{pa}(K) = y_c(K) - \hat{\mathbf{w}}_a^H(K-1)\mathbf{z}(K), \tag{7.561}
$$

$$
\hat{\mathbf{w}}_a(K) = \hat{\mathbf{w}}_a^H(K-1) + \boldsymbol{\Delta}(K)\mathbf{z}(K)e_{pa}^*(K), \tag{7.562}
$$

and

$$
\hat{\mathbf{w}}(K) = \mathbf{w}_q - \mathbf{B}\hat{\mathbf{w}}_a(K). \tag{7.563}
$$

The LMS equations are coupled only through (7.561).

$$
\begin{aligned}
\hat{w}_{a,00}(K) &= \hat{w}_{a,00}^H(K-1) + \Delta_0(K)x_{b,00}(K)e_{pa}^*(K), \qquad (7.564) \\
&\vdots \\
\hat{w}_{a,rq}(K) &= \hat{w}_{a,rq}^H(K-1) + \Delta_r(K)x_{b,rq}(K)e_{pa}^*(K)
\end{aligned}
$$

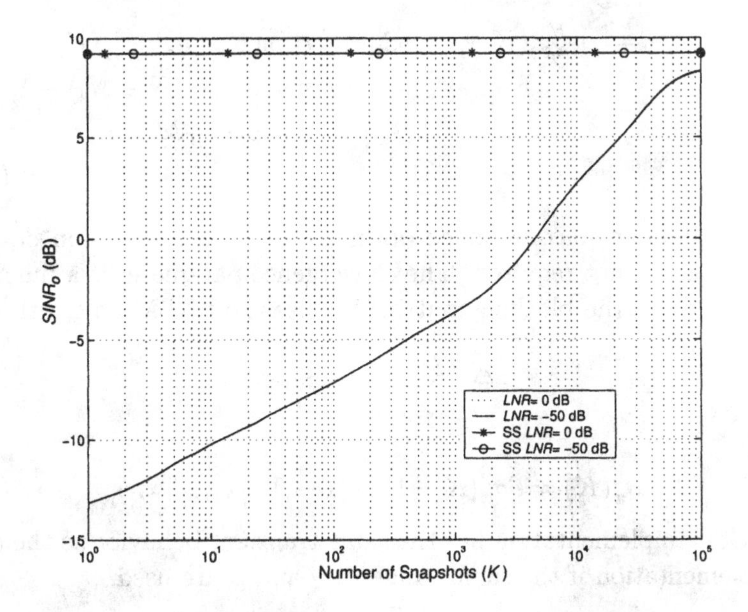

Figure 7.87 Broadband MPDR time-domain beamformer with LMS algorithm: $M = 4, u_s = 0, SNR = 0$ dB, $B_{fs} = 0.4, u_I = 0.29, INR = 30$ dB, $B_{f_I} = 0.4, LNR = 0$ dB, and -50 dB, 100 trials; average $SINR_o$ versus K.

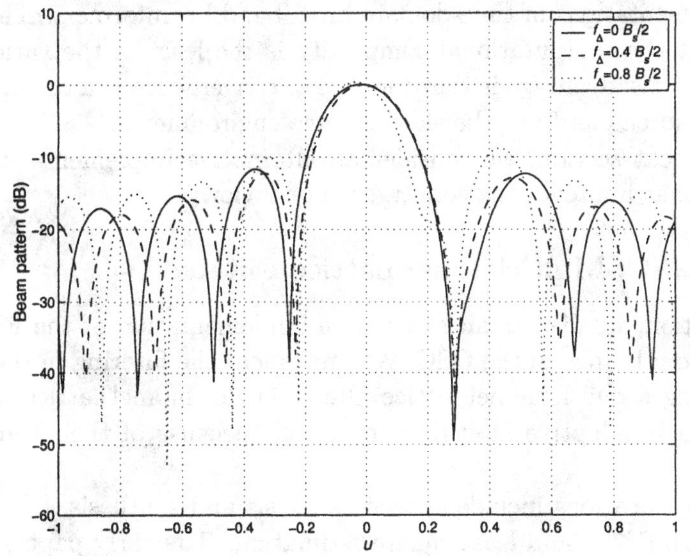

Figure 7.88 Broadband MPDR time-domain beamformer with LMS algorithm: $M = 4, u_s = 0, SNR = 0$ dB, $B_{fs} = 0.4, u_I = 0.29, INR = 30$ dB, $B_{f_I} = 0.4, LNR = 0$ dB, 100 trials; representative beam patterns at $K = 1,000$ at three frequencies.

$$= \hat{w}_{a,rq}^{H}(K-1) + \Delta_r(K) x_{b,r0}(K - qT_s) e_{pa}^{*}(K),$$
$$r = 0, \cdots, (N - N_d) - 1,$$
$$q = 0, \cdots, M - 1. \tag{7.565}$$

Because it is an unconstrained optimization, the summation down the array in (7.555) is not required. The convergence parameter is a function of r, the index down the blocking matrix. We use the PNLMS algorithm,

$$\Delta_r(K) = \frac{\gamma_r}{\sigma_x^2(K)}, \tag{7.566}$$

where

$$\sigma_x^2(K) = \beta \sigma_x^2(K-1) + (1-\beta) \parallel \mathbf{x}_{b,q} \parallel^2 . \tag{7.567}$$

The GSC implementation has the same transient behavior as the direct-form implementation if the same values of γ and β are used.

7.11.2.3 Summary

The FIR implementation using the LMS algorithm is one of the more widely used implementations of time-domain broadband beamformers. The advantage is that the computational complexity is the least of the various algorithms. The disadvantage is that the rate of convergence is a function of the eigenvalue spread and may be slow in some environments.

In the next section, we consider an alternative implementation of the FIR filter that improves the convergence behavior.

7.11.3 GSC: Multichannel Lattice Filters

In this section, we discuss an alternative implementation of the FIR filters in the bottom branch of the GSC. We implement the filtering in the bottom branch using a multichannel lattice filter. Single-channel lattice filters are widely used in adaptive filtering applications because of their convergence properties.

Early applications included speech analysis and synthesis, [IS71], linear prediction, [MV78], and least square estimation. The early papers by Griffiths [Gri77], [Gri78] are particularly relevant to our application. In the past twenty years there have been a large number of papers dealing with adaptive algorithms using lattice structures. Chapter 6 (by F. Ling) in [KT93], Chapters 4 – 7 in Proakis et al. [PRLN92], and Chapters 6 and 15 in [Hay96] provide thorough discussions. Most of the applications have considered single-input systems. For the array processing problem we need a multichannel lattice filter implementation because the input is a $N \times 1$ complex vector. Lattice filters are based on MMSE theory so we implement our processor as a GSC in which the desired scalar signal is $y_c(k)$.

The multichannel lattice filter implementation for the array processing problem is given in Griffiths [Gri77], [Gri78]. Youn and Chang [YC86] have discussed the implementation in more detail. A similar algorithm is given in Table 6.9 on p. 239 of [KT93].

The advantage of the lattice structure is that the equations to update the stages are uncoupled. Therefore, one can choose the adaption parameters in each stage in a manner that improves convergence. An LMS algorithm to update the stages is given in [Gri78] and [YC86].

Just as in the narrowband case, we can improve the adaptive behavior by using an RLS algorithm. QRD multichannel lattice algorithms are discussed by McWhirter and Proudler in Section 7.4.2 of [KT93]. They are also discussed by Ling in Section 6.3.3 of [KT93]. The reference lists at the end of these two chapters provide further material. Both Haykin [Hay96] and Proakis et.al [PRLN92] contain discussions. The interested reader is referred to these references for a complete discussion.[35]

7.11.4 Summary

In this section we have provided a brief discussion of the adaptive behavior of time-domain broadband beamformers. We restricted our discussion to SMI and LMS algorithms.

7.12 Summary

In this chapter, we have studied adaptive beamformers. The topics in the chapter can be divided into four parts.

The first part of the chapter (Section 7.2) developed techniques for estimating the spatial spectral matrix. We emphasized the sample spectral matrix and the FB averaged sample spectral matrix because of their maximum likelihood character, their relative simplicity, and their widespread usage. Other techniques that added more structure to the model were introduced but not developed.

[35]Our discussion of lattice filters requires the reader to consult the suggested references to understand the issues. A more self contained discussion would require a lot of additional background that we have not developed.

The second part of the chapter (Sections 7.3–7.7) developed the three adaptive algorithms that are most widely used.

The SMI algorithm is a block processing algorithm in which the estimated spatial spectral matrix is substituted for the ensemble spectral matrix. We found that we needed to include diagonal loading to obtain satisfactory performance. The convergence to steady state is a function of K/N. The disadvantage of the SMI approach is that the computational complexity is a function of N^3, so that the SMI algorithm is only practical for modest size arrays.

The RLS algorithm can be developed in several different ways. We used a least squares formulation and then developed a recursive implementation that was similar to a Kalman filter. A recursive QRD version was developed that was stable and computationally efficient. Once again, diagonal loading was utilized.

The LMS algorithm required the least amount of computation. We developed it as a stochastic version of the steepest descent algorithm. The disadvantage of the LMS algorithm was that its convergence depended on the eigenvalue spread and may be slow in certain environments.

The third part of the chapter (Sections 7.8–7.10) focused on techniques for reducing the dimension of the adaptive processor. The two techniques were eigenspace processing and beamspace processor. In order to implement eigenspace processors, we must estimate the dimension of the signal subspace. In Section 7.8, we developed the MDL and AIC algorithms to estimate the signal subspace dimension and briefly discussed subspace tracking algorithms. We found that eigenspace beamformers were very effective and reduced the convergence time and computational complexity. The disadvantage of eigenspace beamformers is that, under certain scenarios, we may lose a component of the eigenspace that is highly correlated with the signal. This loss causes a dramatic degradation in performance. BS processors construct the subspace in a deterministic manner, normally with a set of orthogonal beams spaced around the steering direction. We then utilize the SMI, RLS, or LMS algorithm in beamspace. The resulting algorithms are computationally simpler and converge faster than the corresponding element space algorithms.

The fourth part of the chapter (Section 7.11) discussed adaptive time domain implementations of broadband beamformers. The discussion introduced the issues and considered several examples of LMS and SMI implementations. We referred the reader to the literature for a discussion of lattice structures and RLS implementations.

There are several issues that we introduced in Chapter 6 that we have not discussed in the adaptive context:

(i) Array perturbations. In Section 6.6.3, we discussed how array perturbations degrade the steady state performance. Similar effects occur in the adaptive case, but no new issues arise. Wax and Anu [WA96] discuss the closely related problems of steering vector errors and finite data.

(ii) Coherent signals and interferences. In Section 6.12, we discussed beamforming when the desired signals were coherent or correlated. We showed that spatial smoothing provided an effective algorithm in many scenarios. The adaptive version of the algorithm is developed in Shan and Kailath [SK85] and Takao and Kikuma [TK87]. The reader is referred to these references for a discussion of adaptive spatial smoothing.

(iii) Covariance augmentation. In Section 6.7.6, we discussed the use of covariance matrix tapers to broaden the nulls created by the adaptive beamformer. A discussion of the adaptive version is given in [Gue99], [Gue00], [Zat99] and [Zat00]. Another technique for broadening the nulls by imposing derivative constraints is discussed in [GSB96].

This chapter completes our discussion of beamforming. In the next two chapters we discuss the parameter estimation problem.

7.13 Problems

P7.2 Estimation of Spatial Spectral Matrices

Problem 7.2.1

Compute the bias of the sample spectral matrix as an estimator of $\mathbf{S_x}$.

Problem 7.2.2 [Wil62]

(a) Show that the chi-squared density is reproductive with respect to $K/2$. In other words,

$$E\left[\chi^{2r}(K)\right] = \frac{2^r\,\Gamma\left(\frac{K}{2}+r\right)}{\Gamma\left(\frac{K}{2}\right)}. \tag{7.568}$$

(b) Using the result in part (a), show that

$$E\left[\chi^{2}(K)\right] = K, \tag{7.569}$$

$$E\left[\chi^{4}(K)\right] = K(K+2), \tag{7.570}$$

$$var\left[\chi^{2}(K)\right] = K(K+2) - K^2 = 2K. \tag{7.571}$$

Problem 7.2.3

Show that

$$E\left[\frac{1}{\chi^2(K)}\right] = \frac{1}{2}\frac{\Gamma\left(\frac{K}{2}-1\right)}{\Gamma\left(\frac{K}{2}\right)}$$

$$= \frac{1}{2}\frac{\Gamma\left(\frac{K}{2}-1\right)}{\left(\frac{K}{2}-1\right)\Gamma\left(\frac{K}{2}-1\right)} = \frac{1}{K-2}. \tag{7.572}$$

Problem 7.2.4

Extend the technique of FB averaging to a standard linear array with an odd number of elements N.

Problem 7.2.5

Consider a standard 19-element hexagonal array. Show how to construct a conjugate symmetric array manifold vector.

Problem 7.2.6

Consider the correlation matrix $\mathbf{S_x}$ in (7.75). Assume that

$$\mathbf{S_f} = \begin{bmatrix} 1 & e^{j\alpha\pi} \\ e^{-j\alpha\pi} & 1 \end{bmatrix}. \tag{7.573}$$

Assume that FB averaging is used.

(a) Find $\mathbf{S}_{\mathbf{x},fb}$.

(b) Discuss the behavior as a function of α.

Problem 7.2.7

Generate a Gaussian random sequence corresponding to a real AR(1) process acting as an input to a standard 10-element linear array.

(a) Calculate $\hat{\mathbf{S}}_\mathbf{x}$ and $\hat{\mathbf{S}}_{\mathbf{x},fb}$ for $K = 10, 20, 100, 1000$ for one trial.

(b) Repeat for 50 trials and $K = 100$ and analyze your results.

Problem 7.2.8

In this problem we find the least squares estimate of $\mathbf{S_x}$ with a Toeplitz constraint imposed.[36] Specifically, we define

$$e^2 = \text{tr}\left[(\mathbf{C_x} - \mathbf{S_x})^T(\mathbf{C_x} - \mathbf{S_x})\right], \tag{7.574}$$

where $\mathbf{C_x}$ is the sample covariance matrix defined in (7.3). The squared error is the Frobenius norm,

$$e^2 = \sum_{i=1}^{N}\sum_{j=1}^{N}(c_{ij} - s_{ij})^2. \tag{7.575}$$

[36]Our discussion follows Scharf [Sch91], which in turn references Lawson and Hansen [LH74].

(a) We impose the structure on $\mathbf{S_x}$ through the vector parameter $\boldsymbol{\theta}$ and denote it by $\mathbf{S_x}(\boldsymbol{\theta})$. Minimize e^2 with respect to $\boldsymbol{\theta}$ and show that

$$\frac{\partial e^2}{\partial \theta_n} = -2 \text{ tr } \left[(\mathbf{C_x} - \mathbf{S_x})\right]\Big|_{\mathbf{S_x}=\hat{\mathbf{S}}_\mathbf{x}} = 0. \tag{7.576}$$

(b) Show that the second derivative is

$$\text{tr } \left[\left(\frac{\partial \mathbf{S_x}}{\partial \theta_n}\right)^T \frac{\partial \mathbf{S_x}}{\partial \theta_n}\right] + \text{ tr } \left[(\mathbf{C_x} - \mathbf{S_x})^T \frac{\partial^2 \mathbf{S_x}}{\partial \theta_n}\right], \tag{7.577}$$

which will be evaluated to verify that the solution is a minimum.

(c) For a Hermitian Toeplitz structure we parameterize $\mathbf{S_x}(\boldsymbol{\theta})$ as

$$\mathbf{S_x} = s_o\mathbf{Q}_0 + \sum_{n=1}^{N-1} s_n \mathbf{Q}_n + \sum_{n=1}^{N} s_n^* \mathbf{Q}_n^T. \tag{7.578}$$

This corresponds to,

$$\mathbf{S_x} = \begin{bmatrix} s_0 & s_1 & \cdots & s_{N-1} \\ s_1^* & s_0 & \ddots & \\ \vdots & & \ddots & s_1 \\ s_{N-1} & \cdots & s_1^* & s_0 \end{bmatrix}, \tag{7.579}$$

where

$$\mathbf{Q}_0 = \mathbf{I}. \tag{7.580}$$

and

$$\mathbf{Q}_n = \begin{bmatrix} 0 & 0 & 1 & 0 & 0 \\ & 0 & 0 & 1 & 0 \\ & & 0 & & 1 \\ & & & \ddots & \\ & & & & 0 \end{bmatrix}. \tag{7.581}$$

is an upper triangular matrix with 1s along the nth diagonal and zeros everywhere. Differentiate with respect to s_0, $Re[s_n]$, and $Im[s_n]$ and set the result to zero.

$$\text{tr } \left[(\mathbf{C_x} - \mathbf{S_x})^T \mathbf{I}\right] = 0. \tag{7.582}$$

$$\text{tr } \left[(\mathbf{C_x} - \mathbf{S_x})^T \left(\mathbf{Q}_n + \mathbf{Q}_n^T\right)\right] = 0. \tag{7.583}$$

$$\text{tr}\left[(\mathbf{C_x} - \mathbf{S_x})^T\left(\mathbf{Q}_n - \mathbf{Q}_n^T\right)\right] = 0. \tag{7.584}$$

(d) Use the above results and show that,

$$\hat{s}_0 = \frac{1}{N}\sum_{n=1}^{N}\mathbf{C}_{ii}, \tag{7.585}$$

$$Re\left[\hat{s_n}\right] = \frac{1}{2(N-M)}\left[\sum_{j=n+1}^{N}\mathbf{C}_{(j-n)j} + \sum_{i=n+1}^{N}\mathbf{C}_{i(i-n)}\right], \tag{7.586}$$

and

$$Im\left[\hat{s}_n\right] = \frac{j}{2(N-n)}\left[\sum_{j=n+1}^{N}\mathbf{C}_{(j-n)j} - \sum_{i=n+1}^{N}\mathbf{C}_{i(i-n)}\right], \tag{7.587}$$

We see that the solution averages along the appropriate diagonals to obtain a Hermitian Toeplitz matrix. We refer to this estimated matrix as $\hat{\mathbf{S}}_{\mathbf{x},Toe}$. This estimate is intuitively satisfying. However, the resulting estimate is not non-negative definite and gives poor results in some applications.

Problem 7.2.9

Consider the model in which $\mathbf{x}(k)$ is generated by two equal power, uncorrelated plane waves plus white noise.

(a) Find an expression for the asymptotic (large K) behavior of the eigenvalues and eigenvectors.

(b) Plot for a standard 10-element linear array.

P7.3 Sample Matrix Inversion (SMI)

Problem Note 7.3.1: In this section, we simulate the performance of the SMI implementation of adaptive beamformers for various scenarios including those in Table 7.4. The scenarios are chosen to be representative of problems that may be encountered in practical applications. In subsequent sections, we consider different adaptive implementations for the same scenarios. The same data set should be used for the different beamformers. Use 100 trials in your simulation. All of the scenarios include additive white noise.

Problem 7.3.1

Consider a SLA10 and Test Scenario 1.

(a) Simulate the $\mathbf{w}_{mpdr,smi}^H$ beamformer in (7.85). Choose several values of *SNR* and *INR*. Plot $SINR_o$ versus K. Plot representative beam patterns for $K = 2N, 6N, 10N$, and $100N$. Discuss your results.

(b) Repeat part (a) with various levels of fixed diagonal loading. Discuss your results.

Problem 7.3.2 (continuation)

Repeat Problem 7.3.1 for Test Scenario 1m with $u_a = 0.02$ and 0.04.

Problem 7.3.3 (continuation, Problem 7.3.1)

Consider the model in Problems 7.3.1. Implement $\mathbf{w}_{mpdr,smi}^H$ as a GSC with $\mathbf{B}^H\mathbf{B} = \mathbf{I}$. Repeat Problem 7.3.1. Compare the performance on specific sample functions as well as an averaged basis.

Table 7.4

Test Scenario	Signal	Interferers	
		Location	Strength
1	$u_m = 0$ $SNR = 0, 10, \cdots, 30$ dB	Separated Sidelobe $u_1 = 0.30$ $u_2 = 0.50$ $u_3 = 0.70$ $INR = 0, \cdots, 40$ dB	$INR_i = 0, \cdots, 40$ dB i=1, 2, 3
1m	$\lvert u_a \rvert \leq u_m$	TS1	TS1
2	$u_s = 0$ $SNR = 0, \cdots, 30$ dB	Clustered Sidelobe $u_1 = 0.28$ $u_2 = 0.30$ $u_3 = 0.32$	Strength $INR_i = 0, \cdots, 40$ dB i=1, 2, 3
2m	$\lvert u_a \rvert \leq u_m$	TS2	TS2
3	$u_s = 0$ $SNR = 0, \cdots, 30$ dB	Outer Main Lobe $u_1 = 0.18$ $u_2 = -0.30$	Strength $INR_i = 0, \cdots, 40$ dB i=1, 2
3a	$\lvert u_a \rvert \leq u_m$	TS3	TS3
4	$u_s = 0$	Main Lobe $u_1 = 0.08$	Strength $INR_i = 0, \cdots, 40$ dB
5	$u_s = 0$	Random Sidelobe $\lvert u_i \rvert \geq 0.20$ $u_i = 1, \cdots, D$	Strength $INR_i = 0, \cdots, 40$ dB $i = 0, \cdots, D$
6	$u_s = 0$	Correlated Interference $u_1 = 0.30$ $\lvert \rho \rvert \neq 0$	$INR = 0, \cdots, 40$ dB
7	$u_s = 0$	Spatially Spread Section 5.7.1 Complex AR(1) $\phi_a = 0.3, \lvert a(1) \rvert = 0.5$, 0.7, and 0.9	INR=10, 20, and 30 dB

Notes for Table 7.4

1. In TS5, assume that the location of each interferer is a uniform random variable $0.20 \leq |u_i| \leq 0.90$, $i = 1, \cdots, D$ and that the locations are statistically independent. The INR_i takes on one of five values, 0, 10, 20, 30, and 40 dB with equal probability. The locations and *INR*s of the interferers are statistically independent. They are fixed over the snapshots but are different realizations on each trial.
2. In TS7, the additive white noise causes the interference plus noise model to be an ARMA process.
3. In TS3, an extra interferer is added in the sidelobe.

Problem 7.3.4

Repeat Problem 7.3.1 for a SLA10 and Test Scenario 2.

Problem 7.3.5

Repeat Problem 7.3.2 for a SLA10 and Test Scenario 2m.

Problem 7.3.6

Repeat Problem 7.3.1 for a SLA10 and Test Scenario 3.

Problem 7.3.7

Repeat Problem 7.3.2 for a SLA10 and Test Scenario 3m.

Problem 7.3.8

Repeat Problem 7.3.1 for a SLA10 and Test Scenario 4.

Problem 7.3.9

Repeat Problem 7.3.1 for a SLA10 and Test Scenario 5 with $D = 4$.

Problem 7.3.10

Repeat Problem 7.3.1 for a SLA10 and Test Scenario 6 with $\rho = 0.95 \exp(j\pi/4)$.

Problem 7.3.11

Repeat Problem 7.3.1 for a SLA10 and Test Scenario 7.

Problem Note 7.3.2: In several of the previous problems, the scenario included signal mismatch or a main-lobe interferer. We attempted to provide main-lobe protection by using diagonal loading. In Section 6.7, we utilized linear constraints to provide main-lobe protection. The linear constraint sets included:

(a) Directional (6.287).

(b) Derivative (6.308), (6.310).

(c) Eigenvector (6.344).

(d) Quiescent pattern (6.419).

In the next five problems, we revisit some of the previous problems where main-lobe protection was needed. In each problem, choose one or more of the above constraints and implement the LCMP-SMI beamformer. Choose several values of *SNR* and *INR* and use various *LNR*. In each problem,

(a) Plot $SINR_o$ versus K.

(b) Plot representative beam patterns for $K = 2N, 6N, 10N$, and $100N$.

Compare the diagonal loading level needed in the LCMP beamformer with the diagonal loading level needed in the MPDR beamformer.

Problem 7.3.12 (continuation, Problem 7.3.2)

Repeat Problem 7.3.2 for a SLA10 and Test Scenario 1m. Use $u_a = 0.02$ and 0.04.

Problem 7.3.13 (continuation, Problem 7.3.5)

Repeat Problem 7.3.5 for a SLA10 and Test Scenario 2m. Use $u_a = 0.02$ and 0.04.

Problem 7.3.14 (continuation, Problem 7.3.7)

Repeat Problem 7.3.7 for a SLA10 and Test Scenario 3m. Use $u_a = 0.02$ and 0.04.

Problem 7.3.15 (continuation, Problem 7.3.8)
Repeat Problem 7.3.8 for a SLA10 and Test Scenario 4m. Use $u_a = 0.02$ and 0.04.

Problem 7.3.16 (continuation, Problem 7.3.10)
Repeat Problem 7.3.10 for a SLA10 and Test Scenario 6m. Use $u_a = 0.02$ and 0.04.

Problem 7.3.17
Consider a SLA20. Assume $u_s = 0$ and the *SNR* is 10 dB. Assume there are two interfering signals at $u_{I1} = 0.15$ and $u_{I2} = -0.45$. Each interfering signals has an *INR* = 30 dB.

The random variable ρ is defined in (7.92) and the random variable η is defined in (7.106).

(a) Plot $E[\rho]$ and $\bar{\rho}$ for $K = 20, 40, 60, 80$, and 100 for MVDR beamformer using $\widehat{\mathbf{S}}_{\mathbf{n}} = \mathbf{C}_{\mathbf{n}}$: use 200 trials.

(b) Repeat for the case of $\widehat{\mathbf{S}}_{\mathbf{n}} = \mathbf{C}_{\mathbf{n},fb}$. Plot $\bar{\rho}$. Also plot $E[\rho]$ from (7.95) replacing K with $2K$.

(c) Plot a histogram of ρ for $K = 40$. Use 500 trials.

(d) Consider the case when $K = 4, 10$, and 16. Add DL (perhaps *LNR* = 10 dB) and plot $\bar{\rho}$.

(e) Plot $\bar{\eta}$ for the MPDR beamformer for $K = 20, 40, 60, 80$, and 100. Compare to the analytic expression given by (7.110) and (7.111).

(f) Repeat parts (a) and (b) for the GSC configuration. Specify the $\mathbf{B}$ matrix that you use ($\mathbf{B}^H\mathbf{B} = \mathbf{I}$).

P7.4 Recursive Least Squares

Problem Note 7.4.1: In this section, we simulate the performance of the RLS implementation for various scenarios including those in Table 7.4. The parameters that must be specified are μ in (7.138), the initial conditions (σ_o^2 in (7.161) and $\hat{\mathbf{w}}(0)$ in (7.163)), and the diagonal loading σ_L^2 in (7.209). In each problem, consider four RLS implementations:

(a) RLS with no diagonal loading.

(b) GSC-RLS with fixed diagonal loading (Section 7.4.4). σ_L^2/σ_w^2 must be specified. (Include $\sigma_L^2 = 0$ as a reference).

(c) GSC-RLS with variable diagonal loading (Section 7.4.5). T_o must be specified.

(d) Conjugate symmetric RLS (Section 7.4.6) (in order to use diagonal loading, we must derive the GSC implementation of RLS-FB; see Problem 7.4.12).

You must choose the values of these parameters in your solution. Use the same data set as in Section P.7.3. In each problem,

(a) Plot $SINR_o$ versus K.

(b) Plot representative beam patterns for $K = 2N, 6N, 10N$, and $100N$.

Compare your results to SMI results for the same scenario. Discuss the effects of the various parameter values. Discuss any computational issues. Conduct 100 trials.

The test scenario from Table 7.4 that we use with each problem is shown in Table 7.5. Recall that the details of the scenario were given in the SMI problem section. We also show the problem numbers for P7.5, P7.6, and P7.7.

Problem Note 7.4.2: The comments in Problem Note 7.3.2 also apply to the RLS implementation. In the next five problems we consider LCMP beamformers implemented in a GSC structure. Consider one or more of the constraint sets in Problem Note 7.3.2 and plot the results indicated in Problem Note 7.3.2. The problems are listed in Table 7.5.

Table 7.5

Test Scenario	SMI	RLS	QRD	SD	LMS
1	7.3.1	7.4.1	7.5.1	7.6.5	7.7.1
1m	7.3.2	7.4.2	7.5.2	7.6.6	7.7.2
2	7.3.4	7.4.3	7.5.3	7.6.7	7.7.3
2m	7.3.5	7.4.4	7.5.4	7.6.8	7.7.4
3	7.3.6	7.4.5	7.5.5	7.6.9	7.7.5
3m	7.3.7	7.4.6	7.5.6	7.6.10	7.7.6
4	7.3.8	7.4.7	7.5.7	7.6.11	7.7.7
5	7.3.9	7.4.8	7.5.8	7.6.12	7.7.8
6	7.3.10	7.4.9	7.5.9	7.6.13	7.7.9
7	7.3.11	7.4.10	7.5.10	7.6.14	7.7.10
1m	7.3.12	7.4.11	7.5.11	7.6.15	7.7.11
2m	7.3.13	7.4.12	7.5.12	7.6.16	7.7.12
3m	7.3.14	7.4.13	7.5.13	7.6.17	7.7.13
4	7.3.15	7.4.14	7.5.14	7.6.18	7.7.14
6	7.3.16	7.4.15	7.5.15	7.6.19	7.7.15

Problem 7.4.16

(a) Derive the equations that specify the GSC implementation of the conjugate symmetric RLS beamformer in Section 7.4.6. Your equations should use real computation.

(b) Modify your results in part (a) to incorporate diagonal loading.

Problem 7.4.17

Repeat Example 7.4.1 for the case in which there are two interferers at $u_I = \pm 0.29$. All other parameters remain the same. Compare your results to those in Problem 6.3.7.

Problem 7.4.18

Consider the LCMP beamformer in Example 7.3.1. Develop a recursive GSC implementation of the beamformer. Simulate its performance and compare your results to those in Example 7.3.1.

Problem 7.4.19

One of the reasons for using $\mu \neq 1$ is to accommodate non-stationary environments. Consider the model in Problem 7.4.17. Denote the *INR* of the interferer at $u_I = 0.29$ on the kth snapshot as $INR_1(k)$. Denote the *INR* of the interferer at $u_I = -0.29$ on the kth snapshot as $INR_2(k)$. Assume

$$INR_1(k) = \begin{cases} 10^3, & k = 1, \cdots, 50 \\ 0, & k = 51, \cdots, 100 \\ 10^3, & k = 101, \cdots, 150, \end{cases}$$

and continues in the same pattern.

$$INR_2(k) = \begin{cases} 10^3, & k = 1, \cdots, 25 \\ 0, & k = 26, \cdots, 75 \\ 10^3, & k = 76, \cdots, 125, \end{cases}$$

and continues in the same pattern.

Study the behavior of the RLS MPDR beamformer for various choices of μ. Use 10 dB diagonal loading and $SNR = 10$ dB.

Problem 7.4.20

Repeat Example 7.4.1 for the scenario in Problem 7.3.17, that is, $N = 20, u_s = 0, u_{I1} = 0.15$ and $u_{I2} = -0.45$. Consider the following parameters:

(i) $\mu = 0.95, 0.99, 0.999$,

(ii) $\sigma^2 = 10, \sigma^2_{L,smi} = 0$. (i.e., no SMI DL),

(iii) $w(0) = v_s/N$,

(iv) $INR = 20$ dB,

(v) $SNR = 0$, 10, 20 dB.

Plot average *SINR* behavior as well as sample beampatterns.

Problem 7.4.21 (continuation)

Consider the same scenario as in Problem 7.4.20. Implement the FB RLS beamformer in Section 7.4.6. Compare your results to those in Problem 7.4.20. Comment on the difference in average *SINR*, beampatterns, and computational load.

Problem 7.4.22 (continuation)

Consider the same scenario as in Problem 7.4.20. Consider an LCMP beamformer with three mainbeam constraints. You may use any type of constraints you wish (e.g., directional, derivative, quiescent pattern), but specify your choice. Implement the direct form and GSC form of the LCMP beamformer. Assume $u_s = 0.03$. Compare your results to those obtained in Problems 7.4.20 and 7.4.21. Compare/comment on the computational complexity of the direct versus GSC beamformers.

Problem 7.4.23 (continuation)

Implement the GSC LCMP beamformer from Problem 7.4.22 with (1) fixed loading at 10 dB use (7.211), (2) variable loading, (3)scaled projection. Compare your results to those obtained in Problem 7.4.22. Use a constraint level of $T_o = 2/N$.

Problem 7.4.24

Repeat the derivation in (7.178)–(7.183) for the case of a moving window of length $1/1 - \mu$. Simulate its performance for several of the examples in the text.

P7.5 Efficient Recursive Implementation Algorithms

Problem Note 7.5.1: The QRD that we have developed utilizes a GSC implementation of the RLS beamformer and does not incorporate diagonal loading. The objective is to obtain an algorithm that has good numerical stability and reduced computation. The first ten problems repeat implementation (b) in Problem Note 7.4.1, with $\sigma^2_L = 0$ for the scenarios in Problem 7.4.1–7.4.10. In addition to understanding the algorithm, your goal should be to notice the improved numerical stability.

Problem 7.5.1

Consider a SLA10 and Test Scenario 1. Implement the RLS beamformer using a GSC configuration. Implement the adaptive component in the lower path using a QRD.

(a) Plot $SINR_o$ versus K.

(b) Plot representative beam patterns for $K = 2N, 6N, 10N$, and $100N$.

The remaining problem numbers are shown in Table 7.5 (Problems 7.5.2–7.5.10).

Problem Note 7.5.2:

The comments in Problem Notes 7.3.2 and 7.4.2 also apply to the QRD implementation. In the next five problems we consider the same models as in Problems 7.4.11–7.4.15 and use a GSC-RLS-QRD implementation. The problem numbers are shown in Table 7.5 (Problems 7.5.11–7.5.15).

Problem 7.5.16 (continuation, Problem 7.4.20)

Consider the same model as in Problem 7.4.20.

(a) Implement the QRD algorithm for the RLS-GSC beamformer. Compare your results to those in Problem 7.4.20.

(b) Implement the QRD algorithm for the direct form RLS beamformer. Compare your results to those in part (a).

P7.6 Gradient Algorithms

Problem Note 7.6.1

The first set of problems consider direct form MMSE beamformers.

Problem 7.6.1

Consider the model in Example 7.6.5. Assume that the $SNR = 0$ dB, $INR_1 = 10$ dB, $INR_2 = 50$ dB. Repeat the calculations in Example 7.6.5. Discuss your results.

Problem 7.6.2

Consider the model in Example 7.6.5. Assume that $\mathbf{w}(0)$ is a Dolph-Chebychev pattern with -20 dB sidelobes.

(a) Repeat the calculations in Example 7.6.5.

(b) Are $v_n(K), n \geq 3$ equal to zero? If not, plot them as a function of K.

Problem 7.6.3

Consider a SRA10. The desired signal arrives at $u_s = 0$ with an $SNR = 10$ dB. Assume that two interferers arrive at u_1 and u_2. Consider the following parameters:

$$INR_1 = 30 \text{ dB}, u_1 = 0.29,$$

$$INR_2 = 45 \text{ dB}, u_2 = 0.49.$$

(a) Repeat the calculations in Example 7.6.4 (choose an appropriate α). Discuss your results.

(b) Modify the algorithm to include a variable step size $\alpha(K)$. Compare your results to those in part(a).

Problem 7.6.4

Repeat Problem 7.6.3 for the following parameter set:

$$INR_1 = 30 \text{ dB}, u_1 = 0.27,$$

$$INR_2 = 30 \text{ dB}, u_2 = 0.31.$$

Problem Note 7.6.2

The next set of problems utilize the test scenarios in Table 7.4. In all of the problems we use a GSC implementation and use the MMSE steepest descent algorithm to find $\mathbf{w}_a$ in the lower branch ((7.387)–(7.389)). We assume that (7.385) is satisfied. In all problems, plot

(a) Plot $SINR_o$ versus K.

(b) Plot representative beam patterns for $K = 2N, 6N, 10N$, and $100N$.

Compare your results to the RLS results in Problem Section 7.4. Note that we are using known ensemble statistics in this section. In Sections 7.3–7.5, we used measured statistics. A fairer comparison will be used when we simulate the LMS algorithm in Section 7.7.

The problem numbers are shown in Table 7.5.

P7.7 LMS Algorithms

Problem Note 7.7.1

The first set of problems consider the test scenarios in Table 7.4. Consider LMS algorithms:

(a) Griffiths LMS algorithm with diagonal loading (7.417).

(b) Frost MPDR LMS algorithm, direct form with diagonal loading (7.418).

(c) Frost LCMP LMS algorithm, direct form with diagonal loading (7.418).

(d) MPDR-GSC-LMS with diagonal loading (7.426).

(e) LCMP-GSC-LMS with diagonal loading (7.426).

In all cases, use the PNLMS algorithm with $\alpha(K)$ given by (7.460) and (7.461).

The parameters to be selected for the algorithms are γ and β in (7.460) and (7.461) and the *LNR*. The *SNR* and *INR* must also be selected from Table 7.5.

The problem number are shown in Table 7.3. In the first ten problems do parts (a), (b), and (d). In the next five problems do parts (c) and (e). The list in Table 7.5 corresponds to Problems 7.7.1–7.7.15.

Problem 7.7.16

Consider a SLA10. The signal arrives from broadside with an $SNR = 10$ dB. There are two interfering plane wave signals:

$$\begin{aligned} u_{I1} &= 0.29, & INR_1 &= 30 \text{ dB}, \\ u_{I2} &= 0.50, & INR_2 &= 30 \text{ dB}, \\ u_{I3} &= -0.29, & INR_3 &= 20 \text{ dB}, \\ u_{I4} &= -0.50, & INR_4 &= 10 \text{ dB}. \end{aligned}$$

Compare the performance of SMI, RLS, and LMS beamformers.

Assume that we use an LCMV beamformer with a single constraint

$$\mathbf{w}^H \mathbf{v}_s = 1. \tag{7.588}$$

Design an SMI beamformer with appropriate diagonal loading. Design an RLS beamformer with appropriate μ and σ_o^2. Design an LMS beamformer with appropriate $\alpha(K)$ and σ_L^2. Simulate your beamformers and plot the following results:

(i) $\overline{SINR_o}$ versus K ($K = 10, \cdots, 1000$);

(ii) Beam patterns for representative trials at $K = 2N$, $6N$, $10N$, $50N$, and $100N$.

Discuss your results.

Problem 7.7.17 (continuation, Problem 7.4.20)

Consider the same model as in Problem 7.4.20. Implement the power normalized version ((7.460) and (7.461)) of the Griffiths LMS beamformer. Try different values of γ and β and various levels of diagonal loading. Plot the average $SINR_o$ versus K and the beampatterns for various K. Discuss your results.

Problem 7.7.18 (continuation)

Repeat Problem 7.7.17 for an MPDR-LMS beamformer in both the direct form and GSC structure.

Problem 7.7.19 (continuation)

Repeat Problem 7.7.18 for an MPDR-GSC-LMS beamformer using a quadratic constraint, $T_o = 2/N$. Compare fixed loading, variable loading, and scaled projection.

P7.8 Detection of Signal Subspace Dimension

Problem Note 7.8.1

The next ten problems consider an SLA10. In each problem, plot P_D, P_M, and P_{FA} versus *ASNR* for MDL, MDL-FB, AIC, AIC-FB, EDT, and EDT-FB. For the two-signal case, assume that $u_1 = -\Delta u/2$, $u_2 = \Delta u/2$.

Problem 7.8.1

Consider two equal-power uncorrelated signals and four values of Δu; 0.1 HPBW, 0.5 HPBW, 1.0 HPBW, and $\Delta u = 0.3$. Consider the low sample support case, $K = 20$.

Problem 7.8.2

Consider two uncorrelated signals with unequal-power; (i) $ASNR_1 = 10 ASNR_2$, (ii) $ASNR_1 = 100 ASNR_2$. Assume $K = 100$.

Problem 7.8.3

Consider two equal-power correlated signals. The values of ρ of interest are:

(i) $|\rho| = 0.95$, $\phi_\rho = 0, \pi/4, \pi/2$,

(ii) $|\rho| = 0.99$, $\phi_\rho = 0, \pi/4, \pi/2$,

(iii) $|\rho| = 1.0$, $\phi_\rho = 0, \pi/4, \pi/2$.

Assume $K = 100$.

Problem 7.8.4

Repeat Examples 7.8.5–7.8.7 for the case in which FB averaging is used.

Problem 7.8.5

Consider three equal-power uncorrelated signals located at $u_1 = -0.18$, $u_2 = 0$, $u_3 = 0.18$. Assume $K = 100$.

Problem 7.8.6 (continuation)

Consider three uncorrelated signals located at $u_1 = -0.18$, $u_2 = 0$, $u_3 = 0.18$. Assume $K = 100$.

(a) Assume $ASNR_1 = ASNR_3 = 10ASNR_2$.

(b) Assume $ASNR_1 = ASNR_3 = 100ASNR_2$.

Problem 7.8.7 (continuation)

Consider three equal-power correlated signals located at $u_1 = -0.18$, $u_2 = 0$, $u_3 = 0.18$. Assume $K = 100$. Consider various signal correlation matrices, $\boldsymbol{\rho}$.

Problem 7.8.8 (continuation, Example 7.8.5)

The performance of the EDT and EDT-FB algorithms depend on the location of the signals relative to ψ_o (7.514). In this problem, we consider the model in Example 7.8.5 except

$$u_1 = u_c - \Delta u/2 \tag{7.589}$$

and

$$u_2 = u_c + \Delta u/2, \tag{7.590}$$

where $|u_c| \leq 0.1$ is a parameter and $\Delta u = 0.1$ HPBW.

Assume $K = 100$. Plot P_D versus $ASNR$ for $u_c = 0$, 0.2, 0.4, 0.6, 0.8, and 1.0 times $\Delta u/2$. Consider the EDT, BS-AIC, and BS-MDL algorithms with both forward-only (FO) and FB averaging.

Problem 7.8.9

Repeat Problem 7.8.8 for $\Delta u = 0.5$ HPBW.

Problem 7.8.10 (continuation: Problem 7.8.1)

Read the paper by Shah and Tufts [ST94]. Implement their algorithm for the model in Problem 7.8.1. Compare the results to these in Problem 7.8.1.

P7.9 Eigenspace and DMR Beamformers

Problem 7.9.1: The first fifteen problems utilize the test scenarios in Table 7.6. In each problem, we use AIC-FB or MDL-FB to detect the dimension of the signal-plus-interference subspace. We then implement an MPDR or LCMP beamformer in eigenspace. We utilize SMI processing in the eigenspace beamformer. In each problem, plot

(a) $SINR_o$ versus K.

(b) P_D and P_{FA} versus K.

(c) Repeat part (a) for a DMR beamformer.

Discuss your results. The assignment by problems to test scenarios is given in Table 7.4. Problems 7.9.1–7.9.15 are described by this note.

Problem Note 7.9.2: The next several problems consider subspace tracking and require you to read the original references. In each problem, you must develop the subspace tracker and then implement the eigenspace beamformer. For some of the subspace tracking algorithms, (7.520) can be used directly with an RLS or LMS adaptive implementation. In other cases, the eigenspace algorithm must be matched to the subspace tracking algorithm.

Table 7.6

Test Scenario	Eigenspace	Beamspace	Array perturbations
1	7.9.1	7.10.1	7.12.1
1m	7.9.2	7.10.2	7.12.2
2	7.9.3	7.10.3	7.12.3
2m	7.9.4	7.10.4	7.12.4
3	7.9.5	7.10.5	7.12.5
3m	7.9.6	7.10.6	7.12.6
4	7.9.7	7.10.7	7.12.7
5	7.9.8	7.10.8	7.12.8
6	7.9.9	7.10.9	7.12.9
7	7.9.10	7.10.10	7.12.10
1m	7.9.11	7.10.11	7.12.11
2m	7.9.12	7.10.12	7.12.12
3m	7.9.13	7.10.13	7.12.13
4	7.9.14	7.10.14	7.12.14
6	7.9.15	7.10.15	7.12.15

Problem 7.9.16

Read the discussion of the ROSE algorithm in DeGroat and Roberts [DR90]. Implement their algorithm and the corresponding eigenspace beamformer for the signal and interference model in Example 6.9.1. Note that the dimension of the subspace must be detected. Compare your results to the SMI results in Example 6.9.1.

Problem 7.9.17

Read the discussion of the OPERA algorithm in MacInnes [MacI98]. Repeat Problem 7.9.16.

Problem 7.9.18

Read the discussion of the FAST algorithm in Real et. al. [RTC99]. Repeat Problem 7.9.16.

Problem 7.9.19

Read the discussion of the LORAF algorithm in Strobach [Str96]. Repeat Problem 7.9.16.

Problem 7.9.20

Read the discussion of PAST and PAST-D in Yang [Yan95a], [Yan95b]. Repeat Problem 7.9.16.

Problem 7.9.21

A primary motivation for subspace tracking is that it enables the eigenspace beamformer to change dimension when the signal-plus-interference subspace changes.

Derive a scenario in which interferers appear and/or disappear every 100 snapshots. Test one or more of the above subspace trackers and eigenspace beamformers with the above scenario.

P7.10 Beamspace Beamformers

Problem Note 7.10.1

The first sixteen problems utilize the test scenarios in Table 7.6. Use a 5×10 DFT beamspace matrix in each problem. In each problem, implement:

(i) SMI,

(ii) RLS,

(iii) LMS.

Use appropriate diagonal loading. In each problem, plot

(a) $SINR_o$ versus K,

(b) Representative beam patterns for $K = 2N, 6N, 10N$, and $100N$.

The problem assignment to test scenarios is shown in Table 7.6. In all of the beamspace problems, utilize FB averaging of the data.

Problem Note 7.10.2

A 32-element SLA provides more flexibility to study beamspace processing. In Table 7.7, several test scenarios are listed. Use a 7×32 DFT beamspace matrix in each problem. In each problem, implement:

(i) SMI,

(ii) RLS,

(iii) LMS.

Use appropriate diagonal loading. In each problem, plot

(a) $SINR_o$ versus K.

(b) Representative beam patterns for $K = 2N, 6N, 10N$, and $100N$.

Problems 7.10.17–7.10.21 correspond to BS-MPDR implementations for the five test scenarios. Problems 7.10.22 and 7.10.23 correspond to BS-LCMP implementations for test scenarios 1m and 3m, respectively.

Table 7.7

Test Scenario	Signal	Interference Location	Interference Strength
1	$u_m = 0$ $SNR = -10, 0, \cdots, 20$ dB	$u_1 = 3/32$ $u_2 = 5/32$ $u_3 = 7/32$	$INR_i = 0, \cdots, 40$ dB $i = 1, \cdots, 3$
1m	$u_a = \pm 1/128, \pm 1/64$, or $\pm 1/32$	TS1	
2	TS1	TS1 plus 50 dB interferer at $u_i = 27/32$	
3	$u_m = 0$	$u_1 = 1/32$ $u_2 = -3/32$	
3m	$u_m = \pm 1/128$	TS3	

Problem Note 7.10.3

The next set of problems studies the adaptive behavior of the beamformers that we studied in Problems 6.9.4–6.9.15. Problems 7.10.24–7.10.35 correspond to those problems, respectively. Problems 7.10.36 and 7.10.37 correspond to Problems 6.10.8 and 6.10.9, respectively. In each problem, implement:

(i) SMI,

(ii) RLS,

(iii) LMS.

Use appropriate diagonal loading. In each problem, plot

(a) $SINR_o$ versus K.

(b) Representative beam patterns for $K = 2N, 6N, 10N$, and $100N$.

P7.11 Broadband Beamformers

Problem 7.11.1

Consider a ULA10. The interelement spacing is $\lambda_u/2$, where λ_u is the wavelength of the highest frequency in the input signal. We use the signal and interference model in Examples 6.13.2 and 6.13.5.

(a) Implement an FFT beamformer using uncoupled adaptive narrowband beamformers in each bin. Consider M = 4, 8, and 16. Implement the adaptive narrowband beamformers using SMI, RLS, and LMS algorithms. Plot the $SINR_o$ versus K. Discuss your results.

(b) Repeat part (a) using a time-domain beamformer with the same degrees of freedom. Discuss your results.

Problem 7.11.2 (continuation, Example 7.11.1)

Consider the same model as in Example 7.11.1. In each of following parts, implement a time-domain adaptive beamformer, plot $SINR_o$ versus K and discuss your results. Simulate both SMI and LMS implementations.

(a) Denote the number of taps by M. Simulate the beamformer for M = 6, 8, and 10.

(b) Change u_s to -0.10. Simulate the performance for various SNR and M.

(c) Assume the desired signal arrives from -0.10 with a flat spectrum with $B_f = 0.4$. The interfering signal arrives from 0.29 with a flat spectrum over the normalized frequency range $0.6\pi < \omega < 1.0\pi$.

Problem 7.11.3

Repeat Problem 7.11.2 using an FFT beamformer.

Problem 7.11.4 (continuation, Example 6.13.5)

Consider the same model as in Example 6.13.5. Assume $u_I = 0.5BW_{NN}$ and $M = 2$ and 4. Use a GSC configuration and implement an SMI version and an LMS version. The $SNR = 0$ dB and the $INR = 30$ dB. Plot the $SINR_o$ versus K and discuss your results.

Problem 7.11.5 (continuation, Example 6.13.6)

Consider the same model as in Example 6.13.6. Use a GSC configuration and implement an SMI version and an LMS version.

Assume the $ASNR = 20$ dB, $u_s = 0.05$, $INR = 30$ dB, $u_I = 0.45$, and $M = 4$.

Plot the average $SINR_o$ versus K. Discuss your results.

P7.12 Summary

Problem Note 7.12.1

Problems 7.12.1–7.12.15 study the effect of array perturbations on the various algorithms. In each problem assume that the array is on the z-axis and that element locations are perturbed in the x-, y-, and z- directions using the model in (2.174) and (2.185). Let $\sigma_\lambda = 0.02, 0.05$, and 0.1. Repeat the problem in the same rows of Tables 7.4, 7.5, and 7.6 for the following algorithms:

(a) SMI,

(b) RLS,

(c) QRD,

(d) SD,

(e) LMS,

(f) Eigenspace and DMR,

(g) Beamspace.

Bibliography

[AC76] S. P. Applebaum and D. J. Chapman. Adaptive arrays with main beam constraints. *IEEE Trans. Antennas Propag.*, vol.AP-24, pp. 650–662, September 1976.

[AH89] A. H. Abdallah and Y. H. Hu. Parallel VLSI computing array implementation for signal subspace updating algorithm. *IEEE Trans. Acoust., Speech Signal Process*, vol.ASSP-37, pp. 742–748 May 1989.

[Aka74] H. Akaike. A new look at the statistical model identification. *IEEE Trans. Autom. Control*, vol.AC-19, pp. 716–723, June 1974.

[Ale86] S. T. Alexander. *Adaptive Signal Processing - Theory and Applications.* Wiley, New York, 1986.

[And63] T. W. Anderson. Asymptotic theory for principal component analysis. *Ann. Math. Stat.*, vol.34, pp. 122–148, January 1963.

[And73] T. W. Anderson. Asymptotically efficient estimation of covariance matrices with linear structure. *Ann. Stat.*, vol.1, pp. 135–141, January 1973.

[And84] T. W. Anderson. *An Introduction to Multivariate Statistical Analysis.* Wiley, New York, 2nd edition, 1984.

[AO90] D.A. Abraham and N.L. Owsley. Beamforming with dominant mode rejection. *IEEE Oceans 90*, pp. 470–475, 1990.

[App66] S.P. Applebaum. Adaptive arrays. Technical Report, Syracuse University Research Corp., August 1966.

[App76] S.P. Applebaum. Adaptive arrays. *IEEE Trans. Antennas Propag.*, vol.AP-24, pp. 585–598, September 1976.

[AS65] M. Abramowitz and I. A. Stegun. *Handbook of Mathematical Functions.* Dover Publications, New York, 1965.

[ASHP77] N. Ahmed, D. L. Soldan, D. A. Hummels, and D. D. Parikh. Sequential regression considerations of adatptive filtering. *Electron. Lett.*, p. 446, July 1977.

[AHUS79] N. Ahmed, D. A. Hummels, M. Uhl, and D. L. Soldan. A short term sequential regression algorithm. *IEEE Trans. Acoust. Speech, Signal Process*, vol.ASSP-27, p. 453, October 1979.

[Bag76] A. B. Baggeroer. Space-time processes and optimal array processing. Technical Report 506, Navy Undersea Systems Center, San Diego, Calfornia, December 1976.

[Bai73] C. A. Baird. Recursive algorithms for adaptive array antennas. Technical Report Contract F30602-72-C-0499, Rome Air Development Center, Rome, New York, September 1973.

[Bai74] C. A. Baird. Recursive processing for adaptive arrays. In *Proc. Adaptive Antenna Systems Workshop*, Naval Research Lab., Washington, D. C., March 1974; see also NRL Report 7803.

[Bar54] M. S. Barlett. A note on the multiplying factors for various χ^2 approximations. *J. R. Stat. Soc.*, vol.16, pp. 296–298, 1954.

[BG86] K. M. Buckley and L. J. Griffiths. An adaptive generalized sidelobe canceller with derivative constraints. *IEEE Trans. Antennas Propag.*, vol.AP-34, pp. 311–319, March 1986.

[BH85] E. Brookner and J. M. Howell. Adaptive-adaptive array processing. *Proc. Phased Arrays Symp.*, pp. 133–134, 1985.

[BLW82] J. P. Burg, D. G. Luenberger, and D. L. Wenger. Estimation of structured covariance matrices. *Proc. IEEE*, vol.70, pp. 963–974 September 1982.

[BN78] J. R. Bunch and C. P. Nielsen. Updating the singular value decomposition. *Numer. Math.*, vol.31, pp. 111–129, 1978.

[BNS78] J. R. Bunch, C. P. Nielsen, and D. C. Sorensen. Rank-one modification of the symmetric eigenproblem. *Numer. Math.*, vol.31, pp. 31–48, 1978.

[BOBH69] A. H. Booker, C. Y. Ong, J. P. Burg, and G. D. Hair. Multiple-constraint adaptive filtering. Technical Report, Texas Instruments, Science Services Division, Dallas, Texas, April 1969.

[Bor80] D.M. Boroson. Sample size considerations for adaptive arrays. *IEEE Trans. Aerospace and Electron Syst.*, vol.AES-16, pp. 446–451, July 1980.

[Bri81] D. R. Brillinger, Editor. *Time Series: Data Analysis and Theory* (expanded ed.). Holden-Day, San Francisco, Calfornia, 1981.

[BS92] C. H. Bischof and G. M. Shroff. On updating signal subspaces. *IEEE Trans. Signal Process.*, vol.SP-40, pp. 96–105, January 1992.

[Buc91] R. S. Bucy. Geometry and multiple direction estimation. *Inform. Sci.*, vol.57-58, pp. 145–58, 1991.

[Cad88] J. A. Cadzow. Signal enhancement – A composite property mapping algorithm. *IEEE Trans. Acoust., Speech, Signal Process.*, Vol. 36, pp. 49–62, January 1988.

[Car88] B. D. Carlson. Covariance matrix estimation errors and diagonal loading in adaptive arrays. *IEEE Trans. Aerospace and Elec. Syst.*, vol.AES-24, pp. 397–401, July 1988.

[CB76] A. Cantoni and P. Butler. Eigenvalues and eigenvectors of symmetric centrosymmetric matrices. *Linear Algebra and its Applications*, vol.13, pp. 275–288, 1976.

[CF56] A. G. Carlton and J. W. Follin. Recent developments of fixed and adaptive filtering. Technical Report 21, AGAR Dograph, 1956.

[CG70] J. Capon and N. R. Goodman. Probability distributions for estimators of the frequency-wavenumber spectrum. *Proc. IEEE*, vol.58, pp. 1785–1786, October 1970.

[CG90] P. Comon and G. Golub. Tracking a few extreme singular values and vectors in signal processing. *Proc. IEEE*, vol.78, pp. 1327–1343, August 1990.

[Cha87] T. F. Chan. Rank revealing QR factorizations. *Linear Algebra, Appl.*, vol.88/89, pp. 67–82, 1987.

[Cha94] B. Champagne. Adaptive eigendecomposition of data covariance matrices based on first-order perturbations. *IEEE Trans. Signal Process.*, vol.SP-42, pp. 2758–2770, October 1994.

[Com88] R. T. Compton, Jr. *Adaptive Antennas (Concepts and Performance).* Prentice-Hall, Englewood Cliffs, New Jersey, 1988.

[CP97] H. Cox and R. Pitre. Robust DMR and multi-rate adaptive beamforming. *Proc. 31st Asilomar Conf. on Signals, Systems and Computer*, Pacific Grove, California, November 1997.

[CPL98] H. Cox, R. Pitre, and H. Lai. Robust adaptive matched field processing. *Proc. 32nd Asilomar Conf. on Signals, Systems and Computer*, Pacific Grove, California, November 1998.

[CY92] L. Chang and C. Yeh. Performance of DMI and eigenspace-based beamformers. *IEEE Trans. Antennas Propag.*, vol.AP-40, pp. 1336–1347, November 1992.

[CZO87] H. Cox, R. M. Zeskind, and M. M. Owen. Robust adaptive beamforming. *IEEE Trans. Acoust., Speech, Signal Process.*, vol.ASSP-35, pp. 1365–1376, October 1987.

[DAD94] E. M. Dowling, L. P. Ammann, and R. D. DeGroat. A TQR-iteration based adaptive SVD for real-time angle and frequency tracking. *IEEE Trans. Signal Process.*, vol.SP-42, pp. 914–926, April 1994.

[DCA95] P. S. R. Diniz, M. L. R. de Campos, and A. Antoniou. Analysis of LMS-Newton adaptive filtering algorithms with variable convergence factor. *IEEE Trans. Signal Process.*, vol.SP-43, pp. 617–627, March 1995.

[Deg87] S. Degerine. Maximum likelihood estimation of autocovariance matrices from replicated short time series. *J. Time Ser. Anal.*, Vol. 8, pp. 135–146, 1987.

[DeG92] R. D. DeGroat. Noniterative subspace tracking. *IEEE Trans. Signal Process.*, vol.SP-40, pp. 571–577, March 1992.

[DeGR87] R. D. DeGroat and R. A. Roberts. A family of rank-one eigensturcture updating methods. *Proc. 21st Asilomar Conf. on Signals, Systems, and Computers*, Pacific Grove, California, November 1987.

[Dep88] E. F. Deprettere, Editor. *SVD and Signal Processing: Algorithms, Applications and Architectures.* North-Holland, Amsterdam 1988.

[DMS89] A. Dembo, C. L. Mallows, and L. A. Shepp. Embedding nonnegative-definite Toeplitz matrices in nonnegative-definite circulant matrices, with application to covariance estimation. *IEEE Trans. Inf. Theory*, vol.IT-35, pp. 1206–1212, November 1989.

[DR90] R. D. DeGroat and R. A. Roberts. Efficient, numerically stabilized rank-one eigenstructure updating. *IEEE Trans. Acoust., Speech, and Sig. Proc.*, vol.ASSP-38, pp. 301–316, February 1990.

[EC85] M. H. Er and A. Cantoni. An alternative formulation for an optimum beamformer with robustness capability. *Proc. IEE*, vol.132, pp. 447–460, October 1985.

[EC86] M. H. Er and A. Cantoni. An unconstrained partitioned realization for derivative constrained broad-band antenna array processors. *IEEE Trans. Acoust., Speech, Signal Process.*, vol.ASSP-34, pp. 1376–1379, December 1986.

[EJS82] J.E. Evans, J.R. Johnson, and D.F. Sun. Application of advanced signal processing techniques to angle of arrival estimation in ATC navigation and surveillance systems. Technical report, M.I.T. Lincoln Laboratory, Lexington, Massachusetts, June 1982.

[EY36] G. Eckart and G. Young. The approximation of one matrix by another of lower rank. *Psychometrika*, vol.1, pp. 211–218, 1936.

[Far83] D.R. Farrier. Gain of an array of sensors subjected to processor perturbation. *Proc. Inst. Elec. Eng.*, vol.72, pp. 251–254, June 1983.

[FG94] D. D. Feldman and L. J. Griffiths. A projection approach for robust adaptive beamforming. *IEEE Trans. Signal Process.*, vol.SP-42, pp. 867–876, April 1994.

[Fis12] R. A. Fisher. On an absolute criterion for fitting frequency curves. *Mess. of Math.*, vol.41, pp. 155–165, 1912.

[FM88] D. R. Fuhrmann and M. I. Miller. On the existence of positive-definite maximum-likelihood estimates of structured covariance matrices. *IEEE Trans. Inf. Theory*, vol.IT-34, pp. 722–729, July 1988.

[FM97] D. R. Fuhrmann and M. I. Miller. Correction to "On the existence of positive-definite maximum-likelihood estimates of structured covariance matrices." *IEEE Trans. Inf. Theory*, vol.IT-43, pp. 1094–1096, May 1997.

[Fro72] O. L. Frost III. An algorithm for linearly constrained adaptive array processing. *Proc. IEEE*, vol.60, pp. 926–935, August 1972.

[FSM96] D. R. Fuhrmann, A. Srivastava, and H. Moon. Subspace tracking via rigid body dynamics. *Proc. 8th Stat. Signal Array Processing Workshop*, Corfu, Greece, pp. 578–581, June 1996.

[FTM88] D. R. Fuhrmann, M. J. Turmon, and M. I. Miller. Efficient implementation of the EM algorithm for Toeplitz covariance estimation. *Proc. Annu. Conf. Information Science and Systems*, Princeton, New Jersey, March 1988.

[Fuh88] D. R. Fuhrmann. Progress in structured covariance estimation. *Proc. 4th ASSP Workshop on Spectrum Estimation and Modeling*, pp. 158–161, Minneapolis, Minnesota, August 1988.

[Fuh91] D. R. Fuhrmann. Application of Toeplitz covariance estimation to adaptive beamforming and detection. *IEEE Trans. on Signal Process.*, vol. 39, pp. 2194–2198, October 1991.

[Fuh98] D. R. Fuhrmann. A geometric approach to subspace tracking. *Proc. 31st Asilomar Conf. on Signals, Systems, and Computers*, vol.1, pp. 783–787, Pacific Grove, California, November 1998.

[Gab85] W. F. Gabriel. Using spectral estimation techniques in adaptive array systems. *Proc. of the Phased Arrays 1985 Symp.*, pp. 109–113, Rome, New York, 1985.

[GAS87] D. Gray, B. Anderson, and P. Sim. Estimation of structured covariances with application to array beamforming. *Circuits, Syst., Signal Process.*, vol.6, pp. 421–447, June 1987.

[GBT99] G. O. Glentis, K. Berberidis, and S. Theodoridis. Efficient least squares adaptive algorithms for FIR transversal filtering. *IEEE Signal Process. Mag.*, vol.16, pp. 13–41, January 1999.

[Gen73] W. M. Gentleman. Least-squares computation by Givens transformations without square-roots. *J. Inst. Math. Appl.*, vol.12, pp. 329–369, 1973.

[GJ82] L. J. Griffiths and C. W. Jim. An alternative approach to linearly constrained adaptive beamforming. *IEEE Trans. Antennas Propag.*, vol.AP-30, pp. 27–34, January 1982.

[GMW90] M. W. Ganz, R. L. Moses, and S. L. Wilson. Convergence of the SMI and the diagonally loaded SMI algorithms with weak interference. *IEEE Trans. Antennas Propag.*, vol.AP-38, pp. 394–399, March 1990.

[God74] D. Godard. Channel equalization using a Kalman filter for fast data transmission. *IBM J. Res. Dev.*, vol.18, pp. 267–273, May 1974.

[God85] L.C. Godara. The effect of phase-shifter error and the performance of an antenna array beamformer. *IEEE J. Oceanogr. Eng.*, vol.OE-10, pp. 278–284, July 1985.

[God86] L. C. Godara. Error analysis of the optimal antenna array processors. *IEEE Trans. Aerosp. Electron. Syst.*, vol.AES-22, pp. 395–409, July 1986.

[Gol73] G. H. Golub. Some modified matrix eigenvalue problems. *SIAM Rev.*, vol.15, pp. 318–334, 1973.

[Goo63] N. R. Goodman. Statistical analysis based on a certain complex Gaussian distribution. *Ann. Math. Stat.*, vol.34, pp. 152–180, 1963.

[GR80] I. S. Gradshteyn and I. M. Ryzhik. *Table of Integrals, Series, and Products.* Academic Press, New York, 1980.

[Gra72] D. Graupe. *Identification of Systems*, chapter 6, Van Nostrand Reinhold, New York, 1972.

[Gri68] L. J. Griffiths. *Signal extraction using real-time adaptation of a linear multichannel filter.* PhD thesis, Stanford University, Stanford, California, January 1968.

[Gri69] L.J. Griffiths. A simple adaptive algorithm for real-time processing in antenna arrays. *Proc. IEEE*, vol.57, pp. 1696–1704, October 1969.

[Gri77] L.J. Griffiths. A continuously-adaptive filter implemented as a lattice structure. *Proc. ICASSP*, Hartford, Connecticut, May 1977.

[Gri78] L.J. Griffiths. An adaptive lattice structure for noise-cancelling applications. *Proc. ICASSP*, Tulsa, Oklahoma, pp. 87-90, 1978.

[GS84] G. Goodwin and K. Sin. *Adaptive Filtering, Prediction and Control.* Prentice Hall, Eaglewood Cliffs, New Jersey, 1984.

[GSB96] A. B.Gershman, G. V. Serebryakov, and J. F. Bohme. Constrained Hung-turner adaptive beamforming algorithm with additional robustness to wide-band and moving jammers. *IEEE Trans. Antennas Propag.*, vol.AP-44, pp. 361–367, 1996.

[Gue99] J. R. Guerci. Theory and application of covariance matrix tapers for robust adaptive beamforming. *IEEE Trans. Signal Process.*, vol.SP-47, pp. 977–985, April 1999.

[Gue00] J. R. Guerci. Reply to "Comments on 'Theory and application of covariance matrix tapers for robust adaptive beamforming.' " *IEEE Trans. Signal Process.*, vol.SP-48, p. 1800, June 2000.

[Gup65] R. P. Gupta. Asymptotic theory for principal component analysis in the complex case. *J. Indian Stat. Assoc.*, vol.3, pp. 97–106, 1965.

[Ham74] S. Hammarling. A note on modifications to the Givens plane rotation. *J. Inst. Math. Appl.*, vol.3, pp. 215–218, 1974.

[Hay85] S. Haykin. *Array Signal Processing.* Prentice-Hall, Englewood Cliffs, New Jersey, 1985.

[Hay91] S. Haykin. *Adaptive Filter Theory*, Prentice-Hall, Englewood Cliffs, New Jersey, 2nd edition, 1991.

[Hay96] S. Haykin. *Adaptive Filter Theory.* Prentice-Hall, Upper Saddle River, New Jersey, 3rd edition, 1996.

[HM84] M. L. Honig and D. G. Messerschmitt. *Adaptive Filters - Structures, Algorithms and Applications.* Kluwer, Dordrecht, The Netherlands, 1984.

[How76] P. W. Howells. Explorations in fixed and adaptive resolution at GE and SURC. *IEEE Trans. Antennas Propag.*, vol.AP-24, pp. 575–584, September 1976.

[HS92] S. Haykin and A. Steinhardt, Editors. *Adaptive Radar Detection and Estimation.* Wiley, New York, 1992.

[Hud79] J. E. Hudson. A Kalman type algorithm for adaptive radar arrays and modelling of non-stationary weights. In *IEE Conf. Publ.*, p. 180, 1979; also, in "Case Studies in Advanced Signal Processing."

[Hud81] J.E. Hudson. *Adaptive Array Principles.* Peter Peregrinus, New York and London, 1981.

[HY91] K.-C. Huarng and C.-C. Yeh. Adaptive beamforming with conjugate symmetric weights. *IEEE Trans. Antennas Propag.*, vol.AP-39, pp. 926–932, July 1991.

[IS71] F. Itakura and S. Saito. Digital filtering techniques for speech analysis and synthesis. *Proc. 7th Int. Conf. on Acoustics*, vol.3, paper 25C-1, pp. 261–264, 1971.

[Jim77] C. W. Jim. A comparison of two LMS constrained optimal array structures. *Proc. IEEE*, vol.65 pp. 977–978, December 1977.

[JS99] M. Jansson and P. Stoica. Forward-only and forward-backward sample covariances—A comparative study. *Signal Process.*, vol.77, pp. 235–245, 1999.

[Kal60] R. E. Kalman. A new approach to linear filtering and prediction problems. *J. Basic Eng.*, vol.82D, pp. 35–45, March 1960.

[Kar86] I. Karasalo. Estimating the covariance matrix by signal subspace averaging. *IEEE Trans. Acoust., Speech, Signal Process.*, vol.ASSP-34, pp. 8–12, January 1986.

[Kay88] S. M. Kay. *Modern Spectral Estimation.* Prentice-Hall, Englewood Cliffs, New Jersey, 1988.

[KB61] R. E. Kalman and R. S. Bucy. New results in linear filtering and prediction theory. *J. Basic Eng.*, vol.83D, pp. 95–108, 1961.

[Kel86] E. J. Kelly. An adaptive detection algorithm. *IEEE Trans. Aerospace Electron. Syst.*, vol.AES-22, pp. 115–127, March 1986.

[Kel87a] E. J. Kelly. Adaptive detection in non-stationary interference, Part III. Technical Report 761, M.I.T. Lincoln Laboratory, Lexington, Massachusetts, August 1987.

[Kel87b] E. J. Kelly. Performance of an adaptive detection algorithm; rejection of unwanted signals. *IEEE Trans. Aerospace Electron. Syst.*, vol.AES-25, pp. 122–133, March 1987.

[KF89] E. J. Kelly and K. M. Forsythe. Adaptive detection and parameter estimation for multidimensional signal models. Technical Report, M.I.T. Lincoln Laboratory, Lexington, Massachusetts, April 1989.

[KL80] V. C. Klema and A. J. Laub. The singular value decomposition: Its computation and some applications. *IEEE Trans. Autom. Control*, vol.AC-25, pp. 164–176, April 1980.

[Kle80] L. I. Kleinberg. Array gain for signals and noise having amplitude and phase fluctuations. *J. Acoust. Soc. Am.*, vol.67, pp. 572–576, February 1980.

[Kol41a] A. N. Kolmogorov. Interpolation and extrapolation of stationary random sequences. Technical Report RM-3090-PR, RAND Corp., Santa Monica, California, 1941. Translated by W. Doyle and J. Selin in 1962.

[Kol41b] A. N. Kolmogorov. Interpolation and extrapolation von stationären zufälligen folgen. *Bull. Acad. Sci. USSR Ser. Math.*, vol.5, pp. 3–14, 1941.

[KT93] N. Kalouptsidis and S. Theodoridis. *Adaptive System Identification and Signal Processing Algorithms.* Prentice-Hall, Eaglewood Cliffs, New Jersey, 1993.

[Kus84] H. J. Kushner. Approximation and weak convergence methods for random processes with applications to stochastic system theory. MIT Press, Cambridge, Massachusetts, 1984.

[KW91] M. Kaveh and H. Wang. Threshold properties of narrow-band signal-subspace array processing methods. In S. Haykin, editor, *Advances in Spectrum Analysis and Array Processing*, vol.II, pp. 173–220, Prentice-Hall, Englewood Cliffs, New Jersey, 1991.

[Law56] D. N. Lawley. Tests of significance of the latent roots of the covariance and correlation matrices. *Biometrica*, vol.43, pp. 128–136, 1956.

[LDD94] D. A. Linebarger, R. D. DeGroat, and E. M. Dowling. Efficient direction-finding methods employing forward/backward averaging. *IEEE Trans. Signal Process.*, vol.SP-42, pp. 2136–2145, August 1994.

[Lee66] R. C. K. Lee. *Optimal Estimation, Identification, and Control*, MIT Press, Cambridge, Massachusetts, 1966.

[Lee92] H. B. Lee. Eigenvalues and eigenvectors of covariance matrices for signals closely spaced in frequency. *IEEE Trans. Signal Process.*, vol.SP-40, pp. 2518–2535, October 1994.

[LH74] C. L. Lawson and R. J. Hanson. *Solving Least Squares Problems.* Prentice-Hall, Englewood Cliffs, New Jersey, 1974.

[Lig73] W. S. Liggett. Passive sonar: Fitting models to multiple time series. J.W.R. Griffiths and P.L. Stocklin, Editors, *Signal Processing, Proc. NATO ASI Signal Processing*, pp. 327–345, Academic Press, New York, 1973.

[LL94] H. Lee and F. Li. An eigenvector technique for detecting the number of emitters in a cluster. *IEEE Trans. Signal Process.*, vol.SP-42, pp. 2380–2388, September 1994.

[LSL98] H. Li, P. Stoica, and J. Li. Capon estimation of covariance sequences. *Circuits, Syst., Signal Process.*, vol.17, pp. 29–49, January 1999.

[LSL99] H. Li, P. Stoica, and J. Li. Computationally efficient maximum likelihood estimation of structured covariance matrices. *IEEE Trans. Signal Process.*, vol.47, pp. 1314–1323, May 1999.

[Mac95] O. Macchi. *Adaptive Processing: The LMS Approach with Applications in Transmission.* Wiley, New York, 1995.

[MacI98] C. S. MacInnes. Fast, accurate subspace tracking using operator restriction analysis. *Proc. ICASSP*, pp. 1357–1360, Seattle, Washington, 1998.

[Mar87] S. L. Marple, Jr. *Digital Spectral Analysis With Applications.* Prentice-Hall, Englewood Cliffs, New Jersey, 1987.

[Mau40] J. W. Mauchley. Significance test for sphericity of a normal n-variate distribution. *Ann. Math. Stat.*, vol.11, pp. 204–209, 1940.

[Mer65] H. Mermoz. *Adaptive filtering and optimal utilization of an antenna.* Ph.D. thesis, U.S. Navy Bureau of Ships (translation 903 of Ph.D. thesis, Institut Polytechnique, Grenoble, France), Washington D.C., October 1965.

[MFOS91] M. I. Miller, D. R. Fuhrmann, J. A. O'Sullivan, and D. L. Snyder. Maximum-likelihood methods for Toeplitz covariance estimation and radar imaging. In S. Haykin, editor, *Advances in Spectrum Analysis and Array Processing*, vol. II, pp. 145–172. Prentice-Hall, Englewood Cliffs, New Jersey, 1991.

[MM80] R. A. Monzingo and T. W. Miller. *Introduction to Adaptive Arrays.* Wiley, New York, 1980.

[Mos70] J. L. Moschner. Adaptive filtering with clipped input data. Technical Report, Doc. SEL-70-053, Stanford Electronics Laboratories, Stanford, California, June 1970.

[MP81] R.A. Mucci and R.G. Pridham. Impact of beam steering errors on shifted sideband and phase shift beamforming techniques. *J. Acoust. Soc. Am.*, vol.69, pp. 1360–1368, May 1981.

[MS87] H. Messer and P. M. Schultheiss. Estimation of source location and spectral parameters with arrays subject to coherent interference. *11th Colloq. Gretsi*, Nice, France, June 1987.

[MS87b] M. I. Miller and D. L. Snyder. The role of likelihood and entropy in incomplete-data problems: applications to estimating point-process intensities and Toeplitz constrained covariances. *Proc. IEEE*, vol.75, July 1987.

[MS89] J. G. McWhirter and T. J. Shepherd. Systolic array processor for MVDR beamforming. *IEE Proc. (London)* , part F, vol.136, pp. 75–80, 1989.

[Mui82] R. J. Muirhead. *Aspects of Multivariate Statistical Theory.* Wiley, New York, 1982.

[MV78] J. Makhoul and R. Viswanathan. Adaptive lattice methods for linear prediction. *Proc. ICASSP*, vol.1, pp. 87–90, Tulsa, Oklahoma, April 1978.

[MV96] C. S. MacInnes and R. J. Vaccaro. Tracking directions-of-arrival with invariant subspace updating. *Proc. ICASSP*, vol.4, pp. 2896–2899, Atlanta, Georgia, 1996.

[Nit76] R. Nitzberg. Effect of errors in adaptive weights. *IEEE Trans. Aerosp. Electron. Syst.*, vol.AES-12, pp. 369–374, May 1976.

[Nit80] R. Nitzberg. Application of maximum likelihood estimation of persymmetric covariance matrices to adaptive processing. *IEEE Trans. Aerosp. Electron. Syst.*, vol.AES-16, pp. 124–127, January 1980.

[Nut76] A. Nuttall. Spectral analysis of a univariate process with bad data points, via maximum entropy and linear predictive techniques. Technical Report 5303, Naval Undersea Systems Center, San Diego, California, March 1976.

[Orf88] S. J. Orfanidis. *Optimum Signal Processing.* McGraw-Hill, New York, 2nd edition, 1988.

[Ows73] N. L. Owsley, Editor, A recent trend in adaptive spatial processing for sensor arrays: Constrained adaptation. *Proc. of NATO Advanced Study Institute on Signal Processing*, editor, *Signal Processing*, pp. 591–604, Academic Press, London and New York, 1973.

[Ows78] N. L. Owsley. Adaptive data orthogonalization. *Proc. ICASSP*, vol.1, pp. 109–112, Tampa, Florida, 1978.

[PA78] D. Parikh and N. Ahmed. A sequential regression algorithm for recursive filters. *Electron. Lett.*, p. 266, April, 1978.

[PK89] S. U. Pillai and B. H. Kwon. Forward/backward spatial smoothing techniques for coherent signal identification. *IEEE Trans. Acoust., Speech, Signal Process.*, vol.ASSP-37, pp. 8–15, January 1989.

[PRLN92] J. G. Proakis, C. M. Rader, F. Ling, and C. L. Nikias. *Advanced Digital Signal Processing.* Macmillan Publishing Company, New York, 1992.

[Qua82] A. H. Quazi. Array beam response in the presence of amplitude and phase fluctuations. *J. Acoust. Soc. Am.*, vol.72, pp. 171–180, July 1982.

[Rao46] C. R. Rao. *Linear Statistical Inference and Its Applications.* Wiley, New York, 1946.

[Ris78] J. Rissanen. Modeling by shortest data description. *Automatica*, vol.14, pp. 465–471, 1978.

[RMB74] I.S. Reed, J.D. Mallett, and L.E. Brennan. Rapid covergence rate in adaptive arrays. *IEEE Trans. Aerospace and Electron. Syst.*, vol.AES-10, pp. 853–863, November 1974.

[Ros60] J. B. Rosen. The gradient projection method for nonlinear programming, pt. 1: Linear constraints. *J. Soc. Ind. Appl. Math.*, vol.8, p. 181, 1960.

[RTC99] E. C. Real, D. W. Tufts, and J. W. Cooley. Two algorithms for fast approximate subspace tracking. *IEEE Trans. Signal Process.*, vol.47, pp. 1936–1945, July 1999.

[Rut69] H. Rutishauser. Simultaneous iteration method for symmetric matrices. *Numer. Math.*, vol.13, pp. 4–13, 1969.

[SC97] P. Stoica and M. Cedervall. Detection tests for array processing in unknown correlated noise fields. *IEEE Trans. Signal Process.*, vol.45, pp. 2351–2362, September 1997.

[Sch78] G. Schwartz. Estimating the dimension of a model. *Ann. Stat.*, vol.6, pp. 461–464, 1978.

[Sch86] R. Schreiber. Implementation of adaptive array algorithms. *IEEE Trans. Acoust., Speech, Signal Process.*, vol.ASSP-34, pp. 1038–1045, October 1986.

[Sch91] L. L. Scharf. *Statistical Signal Processing: Detection, Estimation, and Time Series Analysis.* Addison-Wesley, Reading, Massachusetts, 1991.

[Sho66] S. W. W. Shor. Adaptive technique to discriminate against coherent noise in a narrow-band system. *J. Acoust. Soc. Am.*, vol.39, pp. 74–78, January 1966.

[Sib87] L. H. Sibul, editor. *Adaptive Signal Processing.* IEEE Press, New York, 1987.

[Sim80] D. N. Simkins. *Multichannel Angle-of-Arrival Estimation.* Ph.D. thesis, Stanford University, Stanford, California, 1980.

[SK85] T. Shan and T. Kailath. Adaptive beamforming for coherent signals and interference. *IEEE Trans. Acoust., Speech, Signal Process.*, vol.ASSP-33, pp. 527–536, 1985.

[SK94] A. H. Sayed and T. Kailath. Constraints for maximally flat optimum broadband antenna arrays. *IEEE Signal Process. Mag.*, vol.11, pp. 18–60, 1994.

[SM97] P. Stoica and R. L. Moses. *Introduction to Spectral Analysis.* Prentice-Hall, Upper Saddle River, New Jersey, 1997.

[Smi93] S. T. Smith. *Geometric Optimization Methods for Adaptive Filtering.* Ph.D. dissertation, Harvard University, Cambridge, Massachusetts, May 1993.

[Sor70] H. W. Sorenson. Least-squares estimation: from Gauss to Kalman. *IEEE Spectrum*, vol.7, pp. 63–68, July 1970.

[Sri00] A. Srivastava. A Bayesian approach to geometric subspace estimation. *IEEE Trans. Signal Process.*, vol.SP-48, pp. 1390–1400, May 2000.

[SS89] T. Söderström and P. Stoica. *System Identification.* Prentice-Hall, Englewood Cliffs, New Jersey, 1989.

[Ste75] G. W. Stewart. Methods of simultaneous iteration for calculating eigenvectors of matrices. *Topics in Numerical Analysis II*, J. H. Miller, Editor, pp.169–185, Academic Press, New York, 1975.

[Ste76] G. W. Stewart. Simultaneous iteraton for computing invariant subspaces of non-Hermitian matrices. *Numer. Math.*, vol.25, pp. 123–136, 1976.

[Ste92] G. W. Stewart. An updating algorithm for subspace tracking. *IEEE Trans. Signal Process.*, vol.SP-40, pp. 1535–1541, June 1992.

[ST94] A. A. Shah and D. W. Tufts. Determination of the dimension of a signal subspace from short data records. *IEEE Trans. Signal Process.*, vol.SP-42, pp. 2531–2535, September 1994.

[Str96] P. Strobach. Low-rank adaptive filters. *IEEE Trans. Signal Processs.*, vol.SP-44, pp. 2932–2947, December 1996.

[Swe58] P. Swerling. A proposed stagewise differential correction procedure for satellite tracking and prediction. Technical Report P-129, RAND Corp., Santa Monica, California, January 1958.

[Swe59] P. Swerling. A proposed stagewise differential correction procedure for satellite tracking and prediction. *J. Astronaut. Sci.*, vol.6, 1959.

[TBV01] Z. Tian, K. L. Bell, and H. L. Van Trees. A recursive least squares implementation for LCMP beamforming under quadratic constraint. *IEEE Trans. Signal Process.*, vol.49, pp. 1138-1145, June 2001.

[Tia99] Z. Tian. On subspace tracking. Internal Report, Array Processing Lab, George Mason University, Fairfax, Virginia, 1999.

[TJL87] J. R. Treichler, C. R. Johnson, Jr., and M. G. Larimore. *Theory and Design of Adaptive Filters.* Wiley, New York, 1987.

[TK86] K. Takao and N. Kikuma. Tamed adaptive antenna arrray. *IEEE Trans. Antennas Propag.*, vol.AP-34, pp. 388–394, March 1986.

[TK87] K. Takao and N. Kikuma. Adaptive array utilizing an adaptive spatial averaging technique for multipath environment. *IEEE Trans. Antennas Propag.*, vol.AP-35, pp. 1388–1396, December 1987.

[TM86] D. W. Tufts and C. D. Melissinos. Simple, effective computation of principal eigenvectors and their eigenvalues and application to high-resolution estimation of frequencies. *IEEE Trans. Acoust., Speech Signal Process.*, vol.ASSP-34, pp. 1046–1053, October 1986.

[TM94] M. J. Turmon and M. I. Miller. Maximum-likelihood estimation of complex sinusoids and Toeplitz covariances. *IEEE Trans. Signal Process.*, vol.SP-42, pp. 1074–1086, May 1994.

[TRC97] D. W. Tufts, E. C. Real, and J. W. Cooley. Fast approximate subspace tracking (FAST). *Proc. ICASSP*, vol.1, pp. 547–560, Munich, Germany, 1997.

[Tuf98] D. W. Tufts. A perspective on the history of underwater acoustic signal processing. *IEEE Signal Process. Mag.*, pp. 23–27, July 1998.

[TW96] K. W. Tam and Y. Wu. On the rate of convergence of some consistent estimates of the number of signals. *8th IEEE Signal Processing Workshop*, pp. 44–47, Corfu, Greece, 1996.

[Ube99] J. Uber. *Detection of Signals.* Ph.D. thesis proposal, George Mason University, Fairfax, Virginia, December 2001.

[VT68] H. L. Van Trees. *Detection, Estimation, and Modulation Theory, Part I.* Wiley, New York, 1968.

[VT71] H. L. Van Trees. *Detection, Estimation, and Modulation Theory, Part III.* Wiley, New York, 1971.

[VT01a] H. L. Van Trees. *Detection, Estimation, and Modulation Theory, Part I.* Wiley Interscience, New York, 2001.

[VT01b] H. L. Van Trees. *Detection, Estimation, and Modulation Theory, Part III.* Wiley Interscience, New York, 2001.

[VV91a] B. D. Van Veen. Adaptive convergence of linearly constrained beamformers based on the sample covariance matrix. *IEEE Trans. Signal Process.*, vol.SP-39, pp. 1470–1473, June 1991.

[VV91b] B. D. Van Veen. Adaptive convergence of linearly constrained beamformers based on the sample covariance matrix. Technical Report, Department of Electrical and Computer Engineering, University of Wisconsin, Madison, Wisconsin, 1991.

[VVW88] B. Van Veen and B. Williams. Structured covariance matrices and dimensional reduction in array processing. *Proc. 4th ASSP Workshop on Spectrum Estimation and Modeling*, pp. 168–171, Minneapolis, Minnesota, August 1988.

[WA96] M. Wax and Y. Anu. Performance analysis of the minimum variance beamformer in the presence of steering vector errors. *IEEE Trans. Signal Process.*, vol.SP-44, pp. 938–947, April 1996.

[WH60] B. Widrow and M. E. Hoff, Jr. Adaptive switching circuits. *IRE WESCON Conv. Rec.*, vol.4, pp. 96–104, 1960.

[WH88] D. M. Wilkes and M. H. Hayes. Iterated Toeplitz approximation of covariance matrices. *Proc. ICASSP*, pp. 1663–1666, New York, April 1988.

[Wie49] N. Wiener. *The Extrapolation , Interpolation and Smoothing of Stationary Time Series.* Wiley, New York, 1949.

[Wil62] S. S. Wilks. *Mathematical Statistics.* Wiley, New York, 1962.

[Wil65] J. H. Wilkinson. *The Algebraic Eigenvalue Problem.* Oxford University Press, New York, 1965.

[Wil86] D. B. Williams. *Eigenvalue analysis for source detection with narrowband passive arrays.* Master's thesis, Rice University, Houston, Texas, September 1986.

[Wil92] D. B. Williams. Comparison of AIC and MDL to the minimum probability of error criterion. *Proc. Sixth SSAP Workshop in Statistical Signal & Array Processing*, Victoria, Canada, pp. 114–117, October 1992.

[Wil94] D. B. Williams. Counting the degrees of freedom when using AIC and MDL to detect signals. *IEEE Trans. Signal Process.*, vol.SP-42, pp. 3282–3284, November 1994.

[WJ90] D. B. Williams and D. H. Johnson. Using the sphericity test for source detection with narrow-band passive arrays. *IEEE Trans. Acoust., Speech, Signal Process.*, vol.ASSP-38, pp. 2008–2014, November 1990.

[WJ93] D. B. Williams and D. H. Johnson. Robust estimation of structured covariance matrices. *IEEE Trans. Signal Process.*, vol.41, pp. 2891–2906, September 1993.

[WK85] M. Wax and T. Kailath. Detection of signals by information theoretic criteria. *IEEE Trans. Acoust., Speech, Signal Process.*, vol.ASSP-33, pp. 387–392, April 1985.

[WMB75] B. Widrow, J. McCool, and M. Ball. The complex LMS algorithm. *Proc. IEEE*, vol.63, pp. 719–720, April 1975.

[WMGG67] B. Widrow, P.E. Mantey, L.J. Griffiths, and B.B. Goode. Adaptive antenna systems. *Proc. IEEE*, vol.55, pp. 2143–2159, December 1967.

[WS74] L. P. Winkler and M. Schwartz. Constrained array optimization by penalty function techniques. *J. Acoust. Soc. Am.*, vol.55, pp. 1042–1048, May 1974.

[WS85] B. Widrow and S.D. Stearns. *Adaptive Signal Processing.* Prentice-Hall, Englewood Cliffs, New Jersey, 1985.

[WT01] Y. Wu and K. W. Tam. On determination of the number of signals. *Proceedings of the International Association of Science and Technology for Development International Conference, Signal and Image Processing*, pp. 113–117, Honolulu, Hawai, August 2001.

[WZRY90] K. M. Wong, Q. Zhang, J .P. Reilly, and P. C. Yip. On information theoretic criteria for determining the number of signals in high resolution array processing. *IEEE Trans. Acoustics, Speech, Signal Process.*, vol.38, pp. 1959–1971, November 1990.

[XPK92] W. Xu, J. Pierre, and M. Kaveh. Practical detection with calibrated arrays. *Proc. Sixth IEEE-SP Workshop Statistical Signal Array Processing*, Victoria, Canada, October 1992.

[XRK94] G. Xu, R. H. Roy, and T. Kailath. Detection of number of sources via exploitation of centro-symmetry property. *IEEE Trans. Signal Process.*, vol.SP-42, pp. 102–112, January 1994.

[Yan95a] B. Yang. Projection approximation subspace tracking. *IEEE Trans. Signal Process.*, vol.SP-43, pp. 95–107, January 1995.

[Yan95b] B. Yang. An extension of the PASTd algorithm to both rank and subspace tracking. *IEEE Signal Process. Lett.*, vol.2, pp. 179–182, September 1995.

[YC86] D. H. Youn and B. Chang. Multichannel lattice filter for an adaptive array processor with linear constraints. *Proc. ICASSP*, Tokyo, Japan, pp. 1829–1832, 1986.

[YK88] J. Yang and M. Kaveh. Adaptive eigensubspace algorithms for direction or frequency estimation and tracking. *IEEE Trans. Acoust., Speech, Signal Process.*, vol.ASSP-36, pp. 241–251, February 1988.

[YL96] S.-J. Yu and J.-H. Lee. The statistical performance of eigenspace-based adaptive array beamformers. *IEEE Trans. Antennas Propag.*, vol.AP-44, pp. 665–671, May 1996.

[YU94] W. S. Youn and C. K. Un. Robust adaptive beamforming based on the eigenstructure method. *IEEE Trans. Signal Process.*, vol.SP-42, pp. 1543–1547, June 1994.

[Yue91] S. M. Yuen. Exact least squares adaptive beamforming using an orthogonalization network. *IEEE Trans. Aerospace and Electron. Syst.*, vol.AES-27, pp. 311–330, March 1991.

[Zat99] M. Zatman. The relationship between covariance matrix tapers and multiple constraints, Technical Report, M.I.T. Lincoln Laboratory, Lexington, Massachusetts, pp. 1–6, February 1999.

[Zat00] M. Zatman. Comments on "Theory and application of covariance matrix tapers for robust adaptive beamforming." IEEE Trans. Signal Process., vol.SP-48, pp. 1796–1800, June 2000.

[ZKB86] L. C. Zhao, P. R. Krishnaiah, and Z. D. Bai. On detection of the number of signals in presence of white noise. *J. Multivariate Anal.*, vol.20, pp. 1–20, 1986.

[ZKB87] L. C. Zhao, P. R. Krishnaiah, and Z. D. Bai. Remarks on certain criteria for detection of number of signals. *IEEE Trans. Acoust., Speech, Signal Process.*, vol.ASSP-35, pp. 129–132, February 1987.

[ZKS93] M. D. Zoltowski, G. M. Kautz, and S. D. Silverstein. Beamspace root-MUSIC. *IEEE Trans. Signal Process.*, vol.SP-41, pp. 344–364, January 1993.

Chapter 8

Parameter Estimation I: Maximum Likelihood

8.1 Introduction

In Chapters 8 and 9, we consider the problem in which the parameter (or parameters) enters into the received signal in a nonlinear manner.

An important problem in array processing where the parameter is embedded in the received waveform in a nonlinear manner is the case of a plane wave with an unknown wavenumber arriving at the array. For example, in the narrowband snapshot model with a linear array, the received snapshots are

$$\mathbf{x}(k) = \mathbf{v}(\psi_s)f(k) + \mathbf{n}(k), \quad k = 1, 2, \cdots, K, \tag{8.1}$$

and the parameter ψ_s is unknown. For D plane waves,

$$\mathbf{x}(k) = \mathbf{V}(\boldsymbol{\psi}_s)\mathbf{f}(k) + \mathbf{n}(k), \quad k = 1, 2, \cdots, K, \tag{8.2}$$

where

$$\boldsymbol{\psi}_s = \begin{bmatrix} \psi_1 & \psi_2 & \cdots & \psi_D \end{bmatrix}^T, \tag{8.3}$$

is an unknown D-dimensional vector. In our initial discussion, we assume D is known. In Section 7.8, techniques for estimating D were discussed. The topic is revisited in Section 8.8 in the context of parameter estimation.

For an arbitrary array with a single plane-wave input, the narrowband snapshot model is

$$\mathbf{x}(k) = \mathbf{v}(\boldsymbol{\psi}_s)f(k) + \mathbf{n}(k), \quad k = 1, 2, \cdots, K, \tag{8.4}$$

where

$$\psi_s = \begin{bmatrix} \psi_x & \psi_y \end{bmatrix}^T . \tag{8.5}$$

is a 2-D vector.

Alternatively, we could write (8.4) as

$$\mathbf{x}(k) = \mathbf{v}(\theta, \phi) f(k) + \mathbf{n}(k), \quad k = 1, 2, \cdots, K, \tag{8.6}$$

and estimate θ and ϕ.

The problem of estimating the wavenumber or angle of arrival of a plane wave (or multiple plane waves) is commonly referred to as the direction finding (DF) or direction of arrival (DOA) estimation problem. It is important in radar, sonar, seismic systems, electronic surveillance, medical diagnosis and treatment, and radio astronomy. Because of its widespread application and the difficulty of obtaining the optimum estimator, the topic has received a significant amount of attention over the last several decades. Many of our results and examples will focus on the direction finding problem.

However, it is important to note that there are many other nonlinear parameter estimation problems of interest in the array processing area. A representative list includes:

(i) Estimating the position of the sensors in array and/or their gain and phase characteristics (the calibration problem);

(ii) Using an AR or ARMA model for the signals' spatial (and/or temporal) characteristics and estimating the model parameters. Alternatively, we may need to estimate the noise/interference parameters;

(iii) Estimating the range of a target in the near field.

Other parameter estimation problems will arise in the course of our development.

Our discussion considers parameters that are constant during the observation period. The extension to the problem of tracking a time-varying parameter (i.e., a moving target) is of obvious interest, but would take us too far afield.

Our discussion of the parameter estimation problem is divided into two chapters. In Chapter 8, we focus on maximum likelihood (ML) and maximum *a posteriori* probability (MAP) estimators and on bounds on the performance of any estimator. In Chapter 9, we develop a number of other estimation procedures that are computationally simpler than the ML estimator and, in many cases, provide adequate performance.

In Section 8.2, we review several classical estimation results that are used in subsequent sections of the chapter. In Section 8.3, we describe the parameter estimation model that we used in the subsequent discussion.

In Section 8.4, we derive the Cramér-Rao bound (CRB) for the multiple-parameter DOA estimation problem. We recall that the ML estimate approaches this bound under certain conditions. However, we observed in the scalar case that, as the *SNR* and/or the number of snapshots is decreased, the estimators exhibit a threshold phenomenon and the variance (or mean square error) increases rapidly above the Cramér-Rao bound.

In Section 8.5, we consider the problem of estimating the direction-of-arrivals (DOAs) or wavenumbers of D plane waves in the presence of additive Gaussian noise. We first develop two maximum likelihood estimates. The first estimate assumes the source signals are sample functions from Gaussian random processes. We refer to this estimate as the unconditional (or stochastic) maximum likelihood (UML) estimate. We show that its performance approaches the Cramér-Rao bound asymptotically. The second estimate assumes the source signals are unknown, but nonrandom, signals. We develop an estimator that is referred to as the conditional (or deterministic) maximum likelihood (CML) estimate, analyze its performance, and compare its performance to that of the UML estimator and the Cramér-Rao bound. The solution for the maximum likelihood estimator is computationally complex. We develop several other multidimensional estimators that have similar asymptotic performance, but are easier to implement.

In Section 8.6, we develop various computational algorithms that enable us to find the estimators in an efficient manner. The techniques in this section are valid for arbitrary array geometries.

In Section 8.7, we restrict our attention to standard linear arrays and develop a polynomial parameterization of the estimation problem. We then develop efficient estimation procedures using this polynomial representation.

The discussion in Sections 8.2 through 8.7 assumes that the dimension of the parameter vector that we are estimating is known. In Section 8.8, we review techniques for estimating the dimension of the parameter vector. The detection algorithms that we developed in Section 7.8 are directly applicable to the estimation problem. We also discuss algorithms that jointly detect the number of signals and estimate their DOAs.

In Section 8.9, we consider spatially spread signals, which can be characterized by a parameter vector of a reasonable dimension. We first consider the model in Section 5.3 in which the source is characterized by a spectral distribution $S_o(\omega_0 : \theta, \phi)$ on a large sphere. We utilize a parametric model for $S_o(\omega_0 : \theta, \phi)$ and find the maximum likelihood estimate of the parameters. We next consider the parametric wavenumber models (AR and ARMA models) of Section 5.6 and discuss maximum likelihood estimates of the parameters.

In Section 8.10, we study parameter estimation in beamspace. Just as in the adaptive beamforming, we find that operating in beamspace provides a reduction in computational complexity and certain performance improvements.

In Section 8.11, we consider the impact on the estimation performance when there are sensor gain and phase errors or errors in the location of the sensors. This is the sensitivity problem that we encountered previously in our classical array discussion in Chapter 2 and our beamformer discussion in Chapter 7. We first develop a hybrid Cramér-Rao bound that indicates how the perturbations affect the variance of the DOA estimates. We then analyze the behavior of the ML estimates in the presence of perturbation. Finally, we develop techniques to jointly estimate the DOA parameters and the array parameters (such as sensor location, gain, and phase). The estimation of the array parameters is referred to as the calibration problem.

In Section 8.12, we summarize our results and discuss some related topics. In Table 8.1, we show the structure of the chapter.

8.2 Maximum Likelihood and Maximum *a posteriori* Estimators

One of the estimators that we will use in many applications is the maximum likelihood estimate of a vector parameter $\boldsymbol{\theta}$. In the discussion in this section, $\boldsymbol{\theta}$ is an arbitrary vector parameter and the notation $\boldsymbol{\theta}$ does not denote an angle. We recall from our discussion in Section 2.4 (pp. 52–86) of DEMT I [VT68] that we can bound the variance of any unbiased estimator by the Cramér-Rao bound. We can also show that, under conditions that are often encountered in practice, the ML estimator is unbiased and, asymptotically (as $K \to \infty$) its variance approaches the Cramér-Rao bound. Thus, the ML estimator is an efficient estimator. This asymptotic behavior is one of the motivations for the use of ML estimator. Moreover, we find that the ML estimator also exhibits good performance in the non-asymptotic region.

When $\boldsymbol{\theta}$ is a random variable we use maximum *a posteriori* probability (MAP) estimators. We can bound the MSE performance of any estimator using the Bayesian version of the Cramér-Rao Bound (see p. 84 of DEMT I [VT68], [VT01a]).

Table 8.1 Structure of Chapter 8

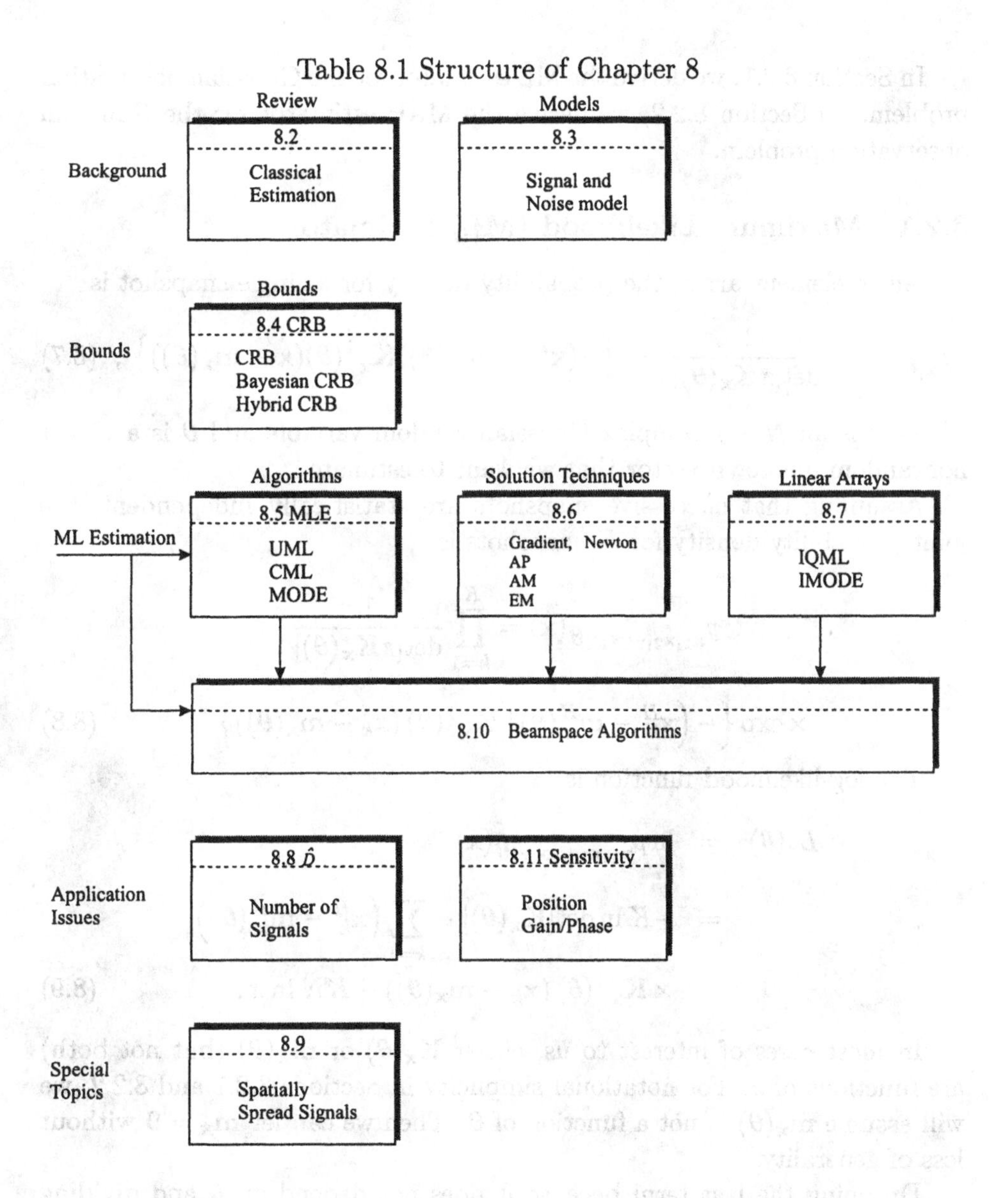

In many of the applications that we consider, the model reduces to the case in which the observation is a complex Gaussian random vector $\mathbf{x}$ whose mean $\mathbf{m_x}(\boldsymbol{\theta})$ and covariance $\mathbf{K_x}(\boldsymbol{\theta})$ depend on the vector parameter $\boldsymbol{\theta}$ that we want to estimate. In view of its widespread usage in subsequent sections, it is worthwhile to review some results from classical estimation theory before proceeding to the physical problems of interest.[1]

[1]The reader may want to review pp. 52–86 of DEMT I [VT68], [VT01a] as background

In Section 8.2.1, we derive the ML estimator for the Gaussian observation problem. In Section 8.2.2, we derive the MAP estimator for the Gaussian observation problem.

8.2.1 Maximum Likelihood (ML) Estimator

For an N-element array, the probability density for a single snapshot is

$$p_{\mathbf{x}|\boldsymbol{\theta}}(\mathbf{x}) = \frac{1}{\det[\pi \mathbf{K}_{\mathbf{x}}(\boldsymbol{\theta})]} \exp\left\{-(\mathbf{x}^H - \mathbf{m}_{\mathbf{x}}^H(\boldsymbol{\theta}))\mathbf{K}_{\mathbf{x}}^{-1}(\boldsymbol{\theta})(\mathbf{x} - \mathbf{m}_{\mathbf{x}}(\boldsymbol{\theta}))\right\}, \tag{8.7}$$

where $\mathbf{x}$ is an $N \times 1$ complex Gaussian random variable and $\boldsymbol{\theta}$ is a $D \times 1$ nonrandom unknown vector that we want to estimate.

Assuming that successive snapshots are statistically independent, the joint probability density for K snapshots is

$$p_{\mathbf{x}_1,\mathbf{x}_2,\cdots,\mathbf{x}_K|\boldsymbol{\theta}}(\mathbf{x}) = \prod_{k=1}^{K} \frac{1}{\det[\pi \mathbf{K}_{\mathbf{x}}(\boldsymbol{\theta})]}$$

$$\times \exp\left\{-\left(\mathbf{x}_k^H - \mathbf{m}_{\mathbf{x}}^H(\boldsymbol{\theta})\right)\mathbf{K}_{\mathbf{x}}^{-1}(\boldsymbol{\theta})\left(\mathbf{x}_k - \mathbf{m}_{\mathbf{x}}(\boldsymbol{\theta})\right)\right\}. \tag{8.8}$$

The log-likelihood function is

$$\begin{aligned} L_{\mathbf{x}}(\boldsymbol{\theta}) &= \ln p_{\mathbf{x}_1,\mathbf{x}_2,\cdots,\mathbf{x}_K|\boldsymbol{\theta}}(\mathbf{x}) \\ &= -K \ln \det[\mathbf{K}_{\mathbf{x}}(\boldsymbol{\theta})] - \sum_{k=1}^{K} \left(\mathbf{x}_k^H - \mathbf{m}_{\mathbf{x}}^H(\boldsymbol{\theta})\right) \\ &\quad \times \mathbf{K}_{\mathbf{x}}^{-1}(\boldsymbol{\theta})\left(\mathbf{x}_k - \mathbf{m}_{\mathbf{x}}(\boldsymbol{\theta})\right) - KN \ln \pi . \end{aligned} \tag{8.9}$$

In most cases of interest to us, either $\mathbf{K}_{\mathbf{x}}(\boldsymbol{\theta})$ or $\mathbf{m}_{\mathbf{x}}(\boldsymbol{\theta})$ (but not both) are functions of $\boldsymbol{\theta}$. For notational simplicity in Sections 8.2.1 and 8.2.2, we will assume $\mathbf{m}_{\mathbf{x}}(\boldsymbol{\theta})$ is not a function of $\boldsymbol{\theta}$. Then we can let $\mathbf{m}_{\mathbf{x}} = \mathbf{0}$ without loss of generality.

Dropping the last term because it does not depend on $\boldsymbol{\theta}$ and dividing through by K, we have

$$L(\boldsymbol{\theta}) = -\left[\ln \det[\mathbf{K}_{\mathbf{x}}(\boldsymbol{\theta})] + \frac{1}{K}\sum_{k=1}^{K} \mathbf{x}_k^H \mathbf{K}_{\mathbf{x}}^{-1}(\boldsymbol{\theta})\mathbf{x}_k\right]. \tag{8.10}$$

(We omit the subscript on $L(\boldsymbol{\theta})$ because we have dropped the constant terms.) We can write the second term as

---

for this discussion.

$$\begin{aligned}\frac{1}{K}\sum_{k=1}^{K}\mathbf{x}_k^H\mathbf{K}_\mathbf{x}^{-1}(\boldsymbol{\theta})\mathbf{x}_k &= \operatorname{tr}\left[\frac{1}{K}\sum_{k=1}^{K}\mathbf{x}_k^H\mathbf{K}_\mathbf{x}^{-1}(\boldsymbol{\theta})\mathbf{x}_k\right] \\ &= \operatorname{tr}\left[\mathbf{K}_\mathbf{x}^{-1}(\boldsymbol{\theta})\cdot\frac{1}{K}\sum_{k=1}^{K}\mathbf{x}_k\mathbf{x}_k^H\right] \\ &= \operatorname{tr}\left[\mathbf{K}_\mathbf{x}^{-1}(\boldsymbol{\theta})\,\mathbf{C}_\mathbf{x}\right], \end{aligned} \tag{8.11}$$

where $\mathbf{C}_\mathbf{x}$ is the sample correlation matrix defined in Section 7.2.1.

$$\mathbf{C}_\mathbf{x} \triangleq \frac{1}{K}\sum_{k=1}^{K}\mathbf{x}_k\mathbf{x}_k^H. \tag{8.12}$$

Then (8.10) can be written as

$$L(\boldsymbol{\theta}) = -\left\{\ln\det[\mathbf{K}_\mathbf{x}(\boldsymbol{\theta})] + \operatorname{tr}\left[\mathbf{K}_\mathbf{x}^{-1}(\boldsymbol{\theta})\,\mathbf{C}_\mathbf{x}\right]\right\}. \tag{8.13}$$

The ML estimate is given by the value of $\boldsymbol{\theta}$ that maximizes $L(\boldsymbol{\theta})$. In general, we must conduct a search procedure to find that value. One of the topics that we study in Section 8.7 is efficient procedures for finding the maximum in the D-dimensional space.

A necessary, but not sufficient, condition is that

$$\frac{\partial}{\partial\theta_i}\left[L(\boldsymbol{\theta})\right]_{\boldsymbol{\theta}=\hat{\boldsymbol{\theta}}_{ml}} = 0, \quad i = 1, 2, \cdots, D. \tag{8.14}$$

This can be written in more compact form as

$$\nabla_{\boldsymbol{\theta}}\left[L(\boldsymbol{\theta})\right]_{\boldsymbol{\theta}=\hat{\boldsymbol{\theta}}_{ml}} = \mathbf{0}, \tag{8.15}$$

where $\nabla_{\boldsymbol{\theta}}$ is the $D\times 1$ derivative matrix defined in (A.362),

$$\nabla_{\boldsymbol{\theta}} = \begin{bmatrix}\frac{\partial}{\partial\theta_1}\\ \frac{\partial}{\partial\theta_2}\\ \vdots\\ \frac{\partial}{\partial\theta_D}\end{bmatrix}. \tag{8.16}$$

Using the derivative formulas in Section A.7 ((A.390) and (A.391)),

$$\begin{aligned}\frac{\partial L(\boldsymbol{\theta})}{\partial\theta_i} &= -\operatorname{tr}\left[\mathbf{K}_\mathbf{x}^{-1}(\boldsymbol{\theta})\frac{\partial\mathbf{K}_\mathbf{x}(\boldsymbol{\theta})}{\partial\theta_i}\right] + \\ &\quad + \operatorname{tr}\left\{\left[\mathbf{K}_\mathbf{x}^{-1}(\boldsymbol{\theta})\,\mathbf{C}_\mathbf{x}\,\mathbf{K}_\mathbf{x}^{-1}(\boldsymbol{\theta})\right]\frac{\partial\mathbf{K}_\mathbf{x}(\boldsymbol{\theta})}{\partial\theta_i}\right\}, \quad i = 1, 2, \cdots, D. \end{aligned} \tag{8.17}$$

Setting (8.17) equal to zero gives a necessary condition on $\hat{\theta}_{ml}$. Thus,

$$\left\{\operatorname{tr}\left[\left[\mathbf{K}_{\mathbf{x}}^{-1}(\boldsymbol{\theta})\,\mathbf{C}_{\mathbf{x}}\,\mathbf{K}_{\mathbf{x}}^{-1}(\boldsymbol{\theta})\right]\frac{\partial\mathbf{K}_{\mathbf{x}}(\boldsymbol{\theta})}{\partial\theta_i}\right] - \operatorname{tr}\left[\mathbf{K}_{\mathbf{x}}^{-1}(\boldsymbol{\theta})\frac{\partial\mathbf{K}_{\mathbf{x}}(\boldsymbol{\theta})}{\partial\theta_i}\right]\right\}\Bigg|_{\hat{\boldsymbol{\theta}}=\hat{\boldsymbol{\theta}}_{ml}} = 0\,, \quad i = 1, 2, \cdots, D\,, \tag{8.18}$$

or

$$\left\{\operatorname{tr}\left[\left[\mathbf{K}_{\mathbf{x}}^{-1}(\boldsymbol{\theta})\,\mathbf{C}_{\mathbf{x}}\,\mathbf{K}_{\mathbf{x}}^{-1}(\boldsymbol{\theta}) - \mathbf{K}_{\mathbf{x}}^{-1}(\boldsymbol{\theta})\right]\frac{\partial\mathbf{K}_{\mathbf{x}}(\boldsymbol{\theta})}{\partial\theta_i}\right]\right\}\Bigg|_{\hat{\boldsymbol{\theta}}=\hat{\boldsymbol{\theta}}_{ml}} = 0\,, \quad i = 1, 2, \cdots, D\,. \tag{8.19}$$

provides the necessary conditions on $\hat{\theta}_{ml}$.

When D is small (e.g., $D \leq 3$), we can use a grid search to find the approximate location of the peak of $L(\boldsymbol{\theta})$ and then use a gradient search technique to solve (8.18) in order to find the exact maximum. When D is large, we have to find more efficient techniques.

8.2.2 Maximum *a posteriori* (MAP) Estimator

In this case we assume that the vector parameter $\boldsymbol{\theta}$ is a random variable with a known probability density $p_{\boldsymbol{\theta}}(\boldsymbol{\theta})$. The MAP estimator is the value of $\boldsymbol{\theta}$ that maximizes the *a posteriori* density. This is equivalent to finding the value of $\boldsymbol{\theta}$ that maximizes

$$L_r(\boldsymbol{\theta}) = -\left\{\ln\det[\mathbf{K}_{\mathbf{x}}(\boldsymbol{\theta})] + \operatorname{tr}\left[\mathbf{K}_{\mathbf{x}}^{-1}(\boldsymbol{\theta})\,\mathbf{C}_{\mathbf{x}}\right]\right\} + \ln p_{\boldsymbol{\theta}}(\boldsymbol{\theta})\,, \tag{8.20}$$

where the subscript "r" denotes we are dealing with a random variable.

In many cases of interest we will model $\boldsymbol{\theta}$ as a zero-mean real Gaussian random vector whose probability density is

$$p_{\boldsymbol{\theta}}(\boldsymbol{\theta}) = \frac{1}{(2\pi)^{\frac{D}{2}}\,|\mathbf{K}_{\boldsymbol{\theta}}|^{\frac{1}{2}}}\exp\left[-\frac{1}{2}\boldsymbol{\theta}^T\,\mathbf{K}_{\boldsymbol{\theta}}^{-1}\,\boldsymbol{\theta}\right]. \tag{8.21}$$

The matrix $\mathbf{K}_{\boldsymbol{\theta}}$ is the covariance matrix of the probability density. It is *not* a function of $\boldsymbol{\theta}$.

To find a necessary condition on the MAP estimate, we differentiate (8.20) with respect to $\boldsymbol{\theta}$ and set the result equal to zero. The derivative of the first term is the left sides of (8.17), (8.18), and (8.19). The derivative of the second term is given by (A.370). Adding the two terms and solving the resulting equation gives a necessary condition on the MAP estimate,

$$\hat{\boldsymbol{\theta}}_{map} = \mathbf{K}_{\boldsymbol{\theta}}\nabla_{\boldsymbol{\theta}}\left[L(\boldsymbol{\theta})\right]\Big|_{\boldsymbol{\theta}=\hat{\boldsymbol{\theta}}_{map}}\,, \tag{8.22}$$

where the ith element of $\nabla_{\boldsymbol{\theta}}\left[L(\boldsymbol{\theta})\right]$ is given by (8.17).

We will look at numerous examples of ML and MAP estimates after we develop our estimation model.

8.2.3 Cramér-Rao Bounds

In order to understand the potential performance of parameter estimation algorithms we develop a set of bounds on their performance. In this section we develop the classical CRB and the Bayesian version of the bound.

These bounds are discussed in Section 2.4 of [VT68] and [VT01a]. The original development of the classic CRB is discussed in [Fis25], [Cra46], and [Rao46]. The Bayesian version of the bound was introduced in [VT68] and [VT01a].

We derive the classic CRB, the Bayesian version of the bound, and a hybrid bound for the case of a Gaussian observation model.

8.2.3.1 Classic Cramér-Rao bound

In this case the log-likelihood function is given by (8.9). The snapshots are assumed to be statistically independent, so we can deal with a single snapshot and combine the results at the end of the discussion. The log-likelihood function for single snapshot is

$$\begin{aligned} L_{\mathbf{x}}(\boldsymbol{\theta}) &\triangleq \ln p_{\mathbf{x}|\boldsymbol{\theta}}(\mathbf{x}) \\ &= -\ln\det[\pi\mathbf{K}_{\mathbf{x}}(\boldsymbol{\theta})] - \left\{\left(\mathbf{x}^H - \mathbf{m}^H(\boldsymbol{\theta})\right)\mathbf{K}_{\mathbf{x}}^{-1}(\boldsymbol{\theta})\left(\mathbf{x}-\mathbf{m}(\boldsymbol{\theta})\right)\right\}. \end{aligned} \tag{8.23}$$

The CRB provides a bound on the covariance matrix of any *unbiased* estimate of $\boldsymbol{\theta}$. We denote the covariance matrix of the estimation errors by $\mathbf{C}(\boldsymbol{\theta})$. Then,

$$\mathbf{C}(\boldsymbol{\theta}) \triangleq E\left[[\hat{\boldsymbol{\theta}}-\boldsymbol{\theta}]\,[\hat{\boldsymbol{\theta}}-\boldsymbol{\theta}]^T\right]. \tag{8.24}$$

The multiple-parameter CRB states that

$$\boxed{\mathbf{C}(\boldsymbol{\theta}) \geq \mathbf{C}_{CR}(\boldsymbol{\theta}) \triangleq \mathbf{J}^{-1},} \tag{8.25}$$

for any unbiased estimate of $\boldsymbol{\theta}$. The matrix inequality means that $\mathbf{C}(\boldsymbol{\theta}) - \mathbf{C}_{CR}(\boldsymbol{\theta})$ is a non-negative definite matrix. The $\mathbf{J}$ matrix is commonly referred to as Fisher's information matrix (or FIM).[2]

[2] The reader may want to review Section 2.4.3 of DEMT I [VT68], [VT01a] at this point.

The elements in $\mathbf{J}$ are

$$\begin{aligned} J_{ij} &\triangleq E\left[\frac{\partial L_{\mathbf{x}}(\boldsymbol{\theta})}{\partial \theta_i} \cdot \frac{\partial L_{\mathbf{x}}(\boldsymbol{\theta})}{\partial \theta_j}\right] \\ &= -E\left[\frac{\partial^2 L_{\mathbf{x}}(\boldsymbol{\theta})}{\partial \theta_i \, \partial \theta_j}\right], \end{aligned} \tag{8.26}$$

or

$$\mathbf{J} = -E\left[\nabla_{\boldsymbol{\theta}}\left(\nabla_{\boldsymbol{\theta}}\left(L_{\mathbf{x}}(\boldsymbol{\theta})\right)^T\right)\right]. \tag{8.27}$$

The result in (8.25) also provides a bound on the variance of any unbiased estimate of θ_i,

$$var\left[\hat{\theta}_i - \theta_i\right] \geq \left[\mathbf{C}_{CR}(\boldsymbol{\theta})\right]_{ii} = \left[\mathbf{J}^{-1}\right]_{ii}. \tag{8.28}$$

The result in (8.28) can also be written as a Hadamard product,

$$var\left[\hat{\theta}_i - \theta_i\right] \geq \left[\mathbf{I} \odot \mathbf{C}_{CR}(\boldsymbol{\theta})\right]_{ii}. \tag{8.29}$$

For the problem of interest, J_{ij} is obtained by differentiating (8.23) and taking the expectation. The second derivative indicated in (8.26) is

$$\begin{aligned} \frac{\partial^2}{\partial \theta_i \, \partial \theta_j}\left[L_{\mathbf{x}}(\boldsymbol{\theta})\right] = \; & \text{tr}\left[-\mathbf{K}_{\mathbf{x}}^{-1}(\boldsymbol{\theta})\frac{\partial \mathbf{K}_{\mathbf{x}}(\boldsymbol{\theta})}{\partial \theta_j}\mathbf{K}_{\mathbf{x}}^{-1}(\boldsymbol{\theta})\frac{\partial \mathbf{K}_{\mathbf{x}}(\boldsymbol{\theta})}{\partial \theta_i} + \mathbf{K}_{\mathbf{x}}^{-1}(\boldsymbol{\theta})\frac{\partial^2 \mathbf{K}_{\mathbf{x}}(\boldsymbol{\theta})}{\partial \theta_i \, \partial \theta_j}\right] \\ & - \left[\left(\mathbf{x}^H - \mathbf{m}^H(\boldsymbol{\theta})\right)\left(-\mathbf{K}_{\mathbf{x}}^{-1}(\boldsymbol{\theta})\frac{\partial \mathbf{K}_{\mathbf{x}}(\boldsymbol{\theta})}{\partial \theta_j}\mathbf{K}_{\mathbf{x}}^{-1}(\boldsymbol{\theta})\frac{\partial \mathbf{K}_{\mathbf{x}}(\boldsymbol{\theta})}{\partial \theta_i}\mathbf{K}_{\mathbf{x}}^{-1}(\boldsymbol{\theta})\right.\right. \\ & \qquad + \mathbf{K}_{\mathbf{x}}^{-1}(\boldsymbol{\theta})\frac{\partial^2 \mathbf{K}_{\mathbf{x}}(\boldsymbol{\theta})}{\partial \theta_i \partial \theta_j}\mathbf{K}_{\mathbf{x}}^{-1}(\boldsymbol{\theta}) \\ & \qquad \left. + \mathbf{K}_{\mathbf{x}}^{-1}(\boldsymbol{\theta})\frac{\partial \mathbf{K}_{\mathbf{x}}(\boldsymbol{\theta})}{\partial \theta_i}\mathbf{K}_{\mathbf{x}}^{-1}(\boldsymbol{\theta})\frac{\partial \mathbf{K}_{\mathbf{x}}(\boldsymbol{\theta})}{\partial \theta_j}\mathbf{K}_{\mathbf{x}}^{-1}(\boldsymbol{\theta})\right) \\ & \left. \cdot\left(\mathbf{x} - \mathbf{m}(\boldsymbol{\theta})\right)\right] - 2Re\left\{-\frac{\partial^2 \mathbf{m}^H(\theta)}{\partial \theta_i \partial \theta_j}\mathbf{K}_{\mathbf{x}}^{-1}(\boldsymbol{\theta})\left[\mathbf{x} - \mathbf{m}(\boldsymbol{\theta})\right]\right. \\ & \qquad + \frac{\partial \mathbf{m}^H(\boldsymbol{\theta})}{\partial \theta_i}\mathbf{K}_{\mathbf{x}}^{-1}(\boldsymbol{\theta})\frac{\partial \mathbf{K}_{\mathbf{x}}(\boldsymbol{\theta})}{\partial \theta_j}\mathbf{K}_{\mathbf{x}}^{-1}(\boldsymbol{\theta})\left[\mathbf{x} - \mathbf{m}(\boldsymbol{\theta})\right] \\ & \qquad + \frac{\partial \mathbf{m}^H(\boldsymbol{\theta})}{\partial \theta_j}\mathbf{K}_{\mathbf{x}}^{-1}(\boldsymbol{\theta})\frac{\partial \mathbf{K}_{\mathbf{x}}(\boldsymbol{\theta})}{\partial \theta_i}\mathbf{K}_{\mathbf{x}}^{-1}(\boldsymbol{\theta})\left[\mathbf{x} - \mathbf{m}(\boldsymbol{\theta})\right] \\ & \qquad \left. + \frac{\partial \mathbf{m}^H(\boldsymbol{\theta})}{\partial \theta_i}\mathbf{K}_{\mathbf{x}}^{-1}(\boldsymbol{\theta})\frac{\partial \mathbf{m}(\boldsymbol{\theta})}{\partial \theta_j}\right\}. \end{aligned} \tag{8.30}$$

To take the expectation, we rewrite the second term as

$$E\left\{-[\mathbf{x}^H - \mathbf{m}(\boldsymbol{\theta})](\cdots)[\mathbf{x} - \mathbf{m}(\boldsymbol{\theta})]\right\} =$$

$$E\left\{-\mathrm{tr}\left\{(\cdots)[\mathbf{x} - \mathbf{m}(\boldsymbol{\theta})][\mathbf{x}^H - \mathbf{m}(\boldsymbol{\theta})]\right\}\right\} = -\mathrm{tr}\left\{(\cdots)\mathbf{K}_{\mathbf{x}}(\boldsymbol{\theta})\right\}. \tag{8.31}$$

Taking the expectation of (8.30), using (8.31), and observing that the expectation of the first three terms in the brackets in (8.30) are zero gives

$$\begin{aligned} E\left[\frac{\partial^2}{\partial\theta_i\,\partial\theta_j}[L_{\mathbf{x}}(\boldsymbol{\theta})]\right] = \mathrm{tr}\Bigg\{&\left[-\mathbf{K}_{\mathbf{x}}^{-1}(\boldsymbol{\theta})\frac{\partial\mathbf{K}_{\mathbf{x}}(\boldsymbol{\theta})}{\partial\theta_j}\mathbf{K}_{\mathbf{x}}^{-1}(\boldsymbol{\theta})\frac{\partial\mathbf{K}_{\mathbf{x}}(\boldsymbol{\theta})}{\partial\theta_i} + \mathbf{K}_{\mathbf{x}}^{-1}(\boldsymbol{\theta})\frac{\partial^2\mathbf{K}_{\mathbf{x}}(\boldsymbol{\theta})}{\partial\theta_i\,\partial\theta_j}\right] \\ &+\left[\mathbf{K}_{\mathbf{x}}^{-1}(\boldsymbol{\theta})\frac{\partial\mathbf{K}_{\mathbf{x}}(\boldsymbol{\theta})}{\partial\theta_j}\mathbf{K}_{\mathbf{x}}^{-1}(\boldsymbol{\theta})\frac{\partial\mathbf{K}_{\mathbf{x}}(\boldsymbol{\theta})}{\partial\theta_i} - \mathbf{K}_{\mathbf{x}}^{-1}(\boldsymbol{\theta})\frac{\partial^2\mathbf{K}_{\mathbf{x}}(\boldsymbol{\theta})}{\partial\theta_i\partial\theta_j}\right. \\ &\left.- \mathbf{K}_{\mathbf{x}}^{-1}(\boldsymbol{\theta})\frac{\partial\mathbf{K}_{\mathbf{x}}(\boldsymbol{\theta})}{\partial\theta_i}\mathbf{K}_{\mathbf{x}}^{-1}(\boldsymbol{\theta})\frac{\partial\mathbf{K}_{\mathbf{x}}(\boldsymbol{\theta})}{\partial\theta_j}\right]\Bigg\} \\ &-2Re\left\{\frac{\partial\mathbf{m}^H(\boldsymbol{\theta})}{\partial\theta_i}\mathbf{K}_{\mathbf{x}}^{-1}(\boldsymbol{\theta})\frac{\partial\mathbf{m}(\boldsymbol{\theta})}{\partial\theta_j}\right\}. \end{aligned} \tag{8.32}$$

The first four terms sum to zero. Thus,

$$J_{ij} = -E\left[\frac{\partial^2}{\partial\theta_i\,\partial\theta_j}[L_{\mathbf{x}}(\boldsymbol{\theta})]\right], \tag{8.33}$$

is

$$\boxed{\begin{aligned} J_{ij} \;=\; &\mathrm{tr}\left[\mathbf{K}_{\mathbf{x}}^{-1}(\boldsymbol{\theta})\frac{\partial\mathbf{K}_{\mathbf{x}}(\boldsymbol{\theta})}{\partial\theta_i}\mathbf{K}_{\mathbf{x}}^{-1}(\boldsymbol{\theta})\frac{\partial\mathbf{K}_{\mathbf{x}}(\boldsymbol{\theta})}{\partial\theta_j}\right] \\ &+2Re\left[\frac{\partial\mathbf{m}^H(\boldsymbol{\theta})}{\partial\theta_i}\mathbf{K}_{\mathbf{x}}^{-1}(\boldsymbol{\theta})\frac{\partial\mathbf{m}(\boldsymbol{\theta})}{\partial\theta_j}\right]. \end{aligned}} \tag{8.34}$$

In some cases, the mean is either zero or not a function of $\boldsymbol{\theta}$ so that (8.34) reduces to[3]

$$\boxed{J_{ij} = \mathrm{tr}\left[\mathbf{K}_{\mathbf{x}}^{-1}(\boldsymbol{\theta})\frac{\partial\mathbf{K}_{\mathbf{x}}(\boldsymbol{\theta})}{\partial\theta_i}\mathbf{K}_{\mathbf{x}}^{-1}(\boldsymbol{\theta})\frac{\partial\mathbf{K}_{\mathbf{x}}(\boldsymbol{\theta})}{\partial\theta_j}\right].} \tag{8.35}$$

Using (A.392), we obtain a second form of (8.34) that is useful in many cases,

$$\begin{aligned} J_{ij} \;=\; &-\mathrm{tr}\left[\frac{\partial\mathbf{K}_{\mathbf{x}}^{-1}(\boldsymbol{\theta})}{\partial\theta_i}\frac{\partial\mathbf{K}_{\mathbf{x}}(\boldsymbol{\theta})}{\partial\theta_j}\right] + \\ &+2Re\left[\frac{\partial\mathbf{m}^H(\boldsymbol{\theta})}{\partial\theta_i}\mathbf{K}_{\mathbf{x}}^{-1}(\boldsymbol{\theta})\frac{\partial\mathbf{m}(\boldsymbol{\theta})}{\partial\theta_j}\right]. \end{aligned} \tag{8.36}$$

[3] This result was first published by Bangs [Ban71].

For the special case in which $\boldsymbol{\theta}$ is a scalar θ,

$$var[\hat{\theta}-\theta] \geq \left\{\left[\operatorname{tr}\left(\left[\mathbf{K}_{\mathbf{x}}^{-1}(\theta)\frac{\partial \mathbf{K}_{\mathbf{x}}(\theta)}{\partial \theta}\right]^2\right)\right] + +2Re\left[\frac{\partial \mathbf{m}^H(\theta)}{\partial \theta}\mathbf{K}_{\mathbf{x}}^{-1}(\theta)\frac{\partial \mathbf{m}(\theta)}{\partial \theta}\right]\right\}^{-1}, \tag{8.37}$$

or equivalently,

$$var[\hat{\theta}-\theta] \geq \left\{\left[-\operatorname{tr}\left(\frac{\partial \mathbf{K}_{\mathbf{x}}^{-1}(\theta)}{\partial \theta}\frac{\partial \mathbf{K}_{\mathbf{x}}(\theta)}{\partial \theta}\right)\right] + +2Re\left[\frac{\partial \mathbf{m}^H(\theta)}{\partial \theta}\mathbf{K}_{\mathbf{x}}^{-1}(\theta)\frac{\partial \mathbf{m}(\theta)}{\partial \theta}\right]\right\}^{-1}. \tag{8.38}$$

Note that the results in (8.34)–(8.38) are quite general and provide a starting point for many subsequent derivations. They apply whenever the observation is a complex Gaussian random process whose mean and covariance matrix are functions of the parameters of interest.

In many applications, the parameter vector $\boldsymbol{\theta}$ contains the parameters of interest to us as well as other unwanted parameters. A typical problem of this type is the case where we observe a plane-wave signal in additive spatially white Gaussian noise. We want to estimate the DOA but the signal power and the noise variance are unknown. In this case, the DOA, is the desired (or wanted) parameter, and signal power and the noise variance are unwanted parameters.

We partition $\boldsymbol{\theta}$ into two vectors,

$$\boldsymbol{\theta} = \begin{bmatrix} \boldsymbol{\theta}_w \\ \boldsymbol{\theta}_u \end{bmatrix}, \tag{8.39}$$

where $\boldsymbol{\theta}_w$ is a $D_1 \times 1$ nonrandom real vector containing the wanted parameters and $\boldsymbol{\theta}_u$ is a $D_2 \times 1$ nonrandom real vector containing the unwanted parameters.

We write $\mathbf{J}$ as a partitioned matrix,

$$\mathbf{J} = \left[\begin{array}{c|c} \mathbf{J}_{\boldsymbol{\theta}_w\boldsymbol{\theta}_w} & \mathbf{J}_{\boldsymbol{\theta}_w\boldsymbol{\theta}_u} \\ \hline \mathbf{J}_{\boldsymbol{\theta}_u\boldsymbol{\theta}_w} & \mathbf{J}_{\boldsymbol{\theta}_u\boldsymbol{\theta}_u} \end{array}\right], \tag{8.40}$$

where $\mathbf{J}_{\boldsymbol{\theta}_w\boldsymbol{\theta}_w}$ is a $D_1 \times D_1$ matrix whose elements are given by (8.34) with $i, j = 1, \cdots, D_1$. $\mathbf{J}_{\boldsymbol{\theta}_w\boldsymbol{\theta}_u}$ is a $D_1 \times D_2$ matrix whose elements are given by (8.34) with $i = 1, \cdots, D_1$; $j = D_1 + 1, \cdots, D$:

$$\mathbf{J}_{\boldsymbol{\theta}_u\boldsymbol{\theta}_w} = \mathbf{J}^H_{\boldsymbol{\theta}_w\boldsymbol{\theta}_u}. \tag{8.41}$$

$\mathbf{J}_{\boldsymbol{\theta}_u\boldsymbol{\theta}_u}$ is a $D_2 \times D_2$ matrix whose elements are given by (8.34) with $i, j = D_1 + 1, \cdots, D$.

We partition the CRB in a similar manner.

$$\mathbf{C}_{CR}(\boldsymbol{\theta}) = \left[\begin{array}{c:c} \mathbf{C}_{CR}(\boldsymbol{\theta}_w) & \mathbf{C}_{CR}(\boldsymbol{\theta}_w, \boldsymbol{\theta}_u) \\ \hdashline \mathbf{C}_{CR}(\boldsymbol{\theta}_u, \boldsymbol{\theta}_w) & \mathbf{C}_{CR}(\boldsymbol{\theta}_u) \end{array}\right]. \tag{8.42}$$

We use the formula for the inverse of a block partitioned matrix to obtain

$$\boxed{\mathbf{C}_{CR}(\boldsymbol{\theta}_w) = \left[\mathbf{J}_{\boldsymbol{\theta}_w\boldsymbol{\theta}_w} - \mathbf{J}_{\boldsymbol{\theta}_w\boldsymbol{\theta}_u}\mathbf{J}^{-1}_{\boldsymbol{\theta}_u\boldsymbol{\theta}_u}\mathbf{J}_{\boldsymbol{\theta}_u\boldsymbol{\theta}_w}\right]^{-1}.} \tag{8.43}$$

The second term represents the effect of the unwanted parameters on the estimation error of the wanted parameters. It is always non-negative and will only be zero when $\mathbf{J}_{\boldsymbol{\theta}_w\boldsymbol{\theta}_u}$ is zero. This corresponds to uncoupled parameters. If $\mathbf{J}_{\boldsymbol{\theta}_w\boldsymbol{\theta}_u}$ is non-zero, the minus sign before the second term causes $C_{CR}(\boldsymbol{\theta}_w)$ to increase.

In array processing, we generally analyze the problem in ψ-space or u-space. In order to translate these results into angle-space, we need to consider the Fisher information matrix for functions of a variable.

We define a new $D_\gamma \times 1$ vector $\boldsymbol{\gamma}$, which is related to $\boldsymbol{\theta}$ by the functional relationship,

$$\boldsymbol{\gamma} = \mathbf{f}(\boldsymbol{\theta}). \tag{8.44}$$

Then,[4]

$$[\mathbf{J}(\boldsymbol{\gamma})]_{ij} = E\left\{\sum_{p=1}^{D_\gamma} \frac{\partial\theta_p}{\partial\gamma_i}\frac{\partial L_{\mathbf{x}}(\boldsymbol{\theta})}{\partial\theta_p}\sum_{q=1}^{D_\gamma}\frac{\partial L_{\mathbf{x}}}{\partial\theta_q}\frac{\partial\theta_q}{\partial\gamma_j}\right\}. \tag{8.45}$$

We define the $D_\gamma \times D_\gamma$ matrix, $\mathbf{G}$, as

$$[\mathbf{G}]_{ip} = \frac{\partial\theta_p}{\partial\gamma_i}. \tag{8.46}$$

[4]See discussion on p. 83 of DEMT I [VT68] [VT01a].

Taking the expectation, (8.45) reduces to

$$\mathbf{J}(\gamma) = \mathbf{G}\mathbf{J}(\boldsymbol{\theta})\mathbf{G}^T, \tag{8.47}$$

and the corresponding Cramér-Rao bound is

$$\mathbf{C}_{CR}(\gamma) = \mathbf{G}^{-T}\mathbf{C}_{CR}(\boldsymbol{\theta})\mathbf{G}^{-1}. \tag{8.48}$$

In most applications that consider, γ_i, is a function of only θ_i so that the $\mathbf{G}$ matrix is diagonal, (see Problems 8.2.1 and 8.2.2).

8.2.3.2 Bayesian Cramér-Rao bounds

In Section 2.4.3 of DEMT I [VT68] [VT01a], we derived a bound on the mean-square error in estimating a random vector parameter $\boldsymbol{\theta}$. The classic Cramér-Rao bound depends on the actual value of the parameter and can be described as a "local bound". The Bayesian bound utilizes the *a priori* probability density of the parameter and provides a "global bound" that does not depend on the value of the parameter on a specific trial.

We denote the information matrix by $\mathbf{J}_B$. The subscript B denotes the Bayesian version of the Fisher information matrix. The $\mathbf{J}_B$ matrix consists of two parts,

$$\mathbf{J}_B = \mathbf{J}_D + \mathbf{J}_P, \tag{8.49}$$

where the subscript "D" denotes the information due to the data and the subscript "P" denotes the information due to prior knowledge.

$$[\mathbf{J}_D]_{ij} \triangleq -E\left[\frac{\partial^2 L_{\mathbf{x}}(\boldsymbol{\theta})}{\partial\theta_i\,\partial\theta_j}\right], \tag{8.50}$$

where the expectation is over both $\mathbf{x}$ and $\boldsymbol{\theta}$, and

$$[\mathbf{J}_P]_{ij} \triangleq -E\left[\frac{\partial^2 \ln p_{\boldsymbol{\theta}}(\boldsymbol{\theta})}{\partial\theta_i\,\partial\theta_j}\right], \tag{8.51}$$

where the expectation is over $\boldsymbol{\theta}$.

The correlation matrix of errors is

$$\mathbf{R}_{\boldsymbol{\epsilon}} = E\left[\left(\hat{\boldsymbol{\theta}} - \boldsymbol{\theta}\right)\left(\hat{\boldsymbol{\theta}} - \boldsymbol{\theta}\right)^T\right]. \tag{8.52}$$

Note that $\mathbf{R}_{\boldsymbol{\epsilon}}$ is not a function of $\boldsymbol{\theta}$ because it is a random vector and we have taken the expected value (unlike $\mathbf{C}(\boldsymbol{\theta})$ in the nonrandom parameter case). From Property 2 on p. 84 of DEMT I [VT68] [VT01a],

$$\boxed{\mathbf{R}_{\boldsymbol{\epsilon}} \geq \mathbf{J}_B^{-1}.} \tag{8.53}$$

The inequality means that the matrix $\left[\mathbf{R}_{\epsilon} - \mathbf{J}_B^{-1}\right]$ is non-negative definite. Note that this is a bound on the mean-square errors.

For the mean-square error of the ith component of $\boldsymbol{\theta}$,

$$E\left[\left|\hat{\theta}_i - \theta_i\right|^2\right] \geq \left[\mathbf{J}_B^{-1}\right]_{ii}. \tag{8.54}$$

For the Gaussian observation model, we use (8.34) in (8.50). Then,

$$\begin{aligned}[\mathbf{J}_D]_{ij} &= E_{\boldsymbol{\theta}}\left[\operatorname{tr}\left[\mathbf{K}^{-1}(\boldsymbol{\theta})\frac{\partial \mathbf{K}_{\mathbf{x}}(\boldsymbol{\theta})}{\partial \theta_i}\mathbf{K}^{-1}(\boldsymbol{\theta})\frac{\partial \mathbf{K}(\boldsymbol{\theta})}{\partial \theta_j}\right]\right. \\ &\quad \left. + 2Re\left[\frac{\partial \mathbf{m}^H(\boldsymbol{\theta})}{\partial \theta_i}\mathbf{K}_{\mathbf{x}}^{-1}(\boldsymbol{\theta})\frac{\partial \mathbf{m}(\boldsymbol{\theta})}{\partial \theta_j}\right]\right],\end{aligned} \tag{8.55}$$

where the subscript "$\boldsymbol{\theta}$" indicates that the expectation is with respect to the random parameter $\boldsymbol{\theta}$.

For the special case of a real parameter vector $\boldsymbol{\theta}$, whose *a priori* density is a multivariate Gaussian density with zero mean,

$$p_{\boldsymbol{\theta}}(\boldsymbol{\theta}) = \frac{1}{(2\pi)^{\frac{D}{2}}\left|\mathbf{K}_{\boldsymbol{\theta}}\right|^{\frac{1}{2}}}\exp\left[-\frac{1}{2}\boldsymbol{\theta}^T\mathbf{K}_{\boldsymbol{\theta}}^{-1}\boldsymbol{\theta}\right], \tag{8.56}$$

and

$$\boxed{\mathbf{J}_P = \mathbf{K}_{\boldsymbol{\theta}}^{-1}.} \tag{8.57}$$

8.2.3.3 Hybrid Cramér-Rao-type bounds

In our study of array model perturbations and array calibration we find it useful to introduce a hybrid version of classic CRB and the Bayesian CRB.

We divide the parameter vector into a nonrandom and random components,

$$\boldsymbol{\theta} = \left[\begin{array}{c}\boldsymbol{\theta}_1 \\ \hline \boldsymbol{\theta}_2\end{array}\right], \tag{8.58}$$

and assume $\boldsymbol{\theta}_1$ is a $D_1 \times 1$ nonrandom vector and $\boldsymbol{\theta}_2$ is a $D_2 \times 1$ random vector. Then $\mathbf{J}_B$ is given by (8.49).

If $\boldsymbol{\theta}_2$ is a zero-mean random variable with a Gaussian probability density, then, $\mathbf{J}_P$ is

$$\mathbf{J}_P = \left[\begin{array}{c|c}\mathbf{0} & \mathbf{0} \\ \hline \mathbf{0} & \mathbf{K}_{\boldsymbol{\theta}_2}^{-1}\end{array}\right]. \tag{8.59}$$

The $\mathbf{J}_D$ matrix contains an expectation over $\boldsymbol{\theta}_2$. For example, in the Gaussian observation model,

$$\begin{aligned}
[\mathbf{J}_D]_{ij} &= E_{\boldsymbol{\theta}_2}\left[\mathrm{tr}\left[\mathbf{K}^{-1}(\boldsymbol{\theta})\frac{\partial \mathbf{K}_{\mathbf{x}}(\boldsymbol{\theta})}{\partial \theta_i}\mathbf{K}^{-1}(\boldsymbol{\theta})\frac{\partial \mathbf{K}(\boldsymbol{\theta})}{\partial \theta_j}\right]\right. \\
&\quad \left. + 2Re\left[\frac{\partial \mathbf{m}^H(\boldsymbol{\theta})}{\partial \theta_i}\mathbf{K}_{\mathbf{x}}^{-1}(\boldsymbol{\theta})\frac{\partial \mathbf{m}(\boldsymbol{\theta})}{\partial \theta_j}\right]\right], \qquad (8.60)
\end{aligned}$$

where the subscript "$\boldsymbol{\theta}_2$" indicates that the expectation is with respect to the random parameter $\boldsymbol{\theta}_2$.

We write the inverse of the $\mathbf{J}_B$ as a partitioned matrix,

$$\mathbf{J}_B^{-1} \triangleq \left[\begin{array}{c|c} \mathbf{C}_{CR}(\boldsymbol{\theta}_1) & \mathbf{C}_1 \\ \hline \mathbf{C}_1^H & \mathbf{C}_{BCR}(\boldsymbol{\theta}_2) \end{array}\right]. \qquad (8.61)$$

The $D_1 \times D_1$ matrix, $\mathbf{C}_{CR}(\boldsymbol{\theta}_1)$, provides a lower bound on the covariance of any unbiased estimate of $\boldsymbol{\theta}_1$. The $D_2 \times D_2$ matrix, $\mathbf{C}_{CR}(\boldsymbol{\theta}_2)$, provides a lower bound of the mean-square error correlation matrix of $\boldsymbol{\theta}_2$. The proof of this assertion is a straightforward modification of the discussion in Section 2.4 of DEMT I [VT68] [VT01a]. The hybrid bound was introduced into the array processing literature by Rockah and Schultheiss [RS87a], [RS87b]. We will utilize it in Section 8.11.

8.2.3.4 Multiple snapshots

All of the expressions up to this point assumed a single snapshot (or sample). If we have K independent snapshots, then the log-likelihood function in (8.23) is multiplied by K and the FIM, $\mathbf{J}$, is multiplied by K. The classic CRB in (8.25) is divided by K.

In the Bayesian bound in Section 8.2.3.2, the data component $\mathbf{J}_D$ is multiplied by K. Similarly, in the hybrid bound in Section 8.2.3.3, $\mathbf{J}_D$ is multiplied by K.

8.2.3.5 Summary

The various forms of the CRB are important because, in many applications, the ML estimate (or the MAP estimate) approach the bound asymptotically as K goes to infinity.

In most of those applications, the value of K where the performance of estimator becomes close to the bound is a function of the signal to *SINR*.

We will quantify these ideas after we develop the model of interest.

8.3 Parameter Estimation Model

In this section, we develop the parameter estimation model that we utilize in our discussions in the remainder of this chapter and in Chapter 9. It is the same model we used in the beamformer problem in Chapters 6 and 7 except the parameters in the model such as the wavenumbers of the impinging plane waves are unknown.

In Section 8.3.1, we consider the case in which we have multiple plane waves impinging on the array. We want to estimate their directions of arrival (DOAs) and various signal parameters.

In Section 8.3.2, we consider the case in which the array configuration is perturbed from its nominal configuration. For example, the sensor positions may be perturbed from their nominal location, or the amplitude and phase of the sensor response may be different from their nominal values. We develop a model that incorporates these perturbations.

In Section 8.3.3, we consider spatially spread signals that can be modelled parametrically.

In Section 8.3.4, we summarize our results.

8.3.1 Multiple Plane Waves

In this section we develop the model that we use to estimate the parameters of multiple plane waves impinging on an array. We develop a frequency-domain model and a time-domain model. We then introduce notation to treat both models with a single generic notation.

The first model of interest is the frequency-domain snapshot model that we encountered previously in Section 5.2 and was used throughout Chapters 6 and 7.

The frequency-domain snapshot model is appropriate for either narrow-band or wideband processes. For the narrowband case, we only use the snapshot corresponding to the carrier frequency.

For the case of a linear array, we can write the snapshots as

$$\mathbf{X}(k) = \mathbf{V}(\boldsymbol{\psi})\mathbf{F}(k) + \mathbf{N}(k), \quad k = 1, \cdots, K, \tag{8.62}$$

where

$$\boldsymbol{\psi} = \left[\begin{array}{ccc} \psi_1 & \cdots & \psi_D \end{array} \right]^T, \tag{8.63}$$

is a $D \times 1$ vector containing the wavenumbers of the D plane waves ($\psi_i = k_i d$) and $\mathbf{V}(\boldsymbol{\psi})$ is the array manifold matrix, $\mathbf{F}(k)$ is a $D \times 1$ vector of the source signals and $\mathbf{N}(k)$ is an $N \times 1$ vector of the noise.

For the arbitrary array case, ψ is a $D \times 2$ matrix,

$$\psi = \begin{bmatrix} \psi_1 & \cdots & \psi_D \end{bmatrix}, \tag{8.64}$$

where

$$\psi_i = \begin{bmatrix} \psi_{x_i} & \psi_{y_i} \end{bmatrix}^T \tag{8.65}$$

contains the two wavenumbers of each source.

The spatial spectral matrix is

$$\mathbf{S_x} = \mathbf{V}(\psi)\mathbf{S_f}\mathbf{V}^H(\psi) + \mathbf{S_n}. \tag{8.66}$$

For the narrowband case, the time-domain snapshot model discussed in Section 5.2 is also appropriate.

In this case,

$$\mathbf{x}(k) = \mathbf{V}(\psi)\,\mathbf{f}(k) + \mathbf{n}(k), \quad k = 1, 2, \cdots, K. \tag{8.67}$$

where the argument "k" denotes the time-domain sample (or snapshot) at $t = k$. We assume the samples are from a stationary zero-mean Gaussian random process. Thus,

$$\begin{aligned} \mathbf{R_x} &= E\left[\mathbf{x}(k)\,\mathbf{x}^H(k)\right] \\ &= \mathbf{V}(\psi)\,\mathbf{R_f}\,\mathbf{V}^H(\psi) + \mathbf{R_n}. \end{aligned} \tag{8.68}$$

The majority of our discussion focuses on the narrowband case. We normally utilize the frequency-domain notation in (8.66), because the extension to the wideband case is straightforward. In all of our snapshot models, we assume that successive snapshots are statistically independent.

We emphasize four different models for the signal. In the narrowband case, these are:

Case S1: The source signals are sample functions from a zero-mean vector stationary complex Gaussian random process and $\mathbf{S_f}$ is known.

Case S2: The source signals are sample functions from a zero-mean vector stationary complex Gaussian random process and $\mathbf{S_f}$ is unknown.

In Case S2, the source signals may be uncorrelated, but we do not know that, so we must consider a general $\mathbf{S_f}$ in our estimators and bounds. Case S2u considers the case where we know *a priori* that the source signals are uncorrelated.

Case S2u: The source signals are sample functions from a zero-mean vector stationary complex Gaussian random process. They are uncorrelated (and therefore, statistically independent). $\mathbf{S_f}$ is a diagonal matrix denoted by $\mathbf{\Lambda_f}$. $\mathbf{\Lambda_f}$ is unknown.

Case S3: The source signals are considered to be unknown complex sequences in order to estimate ψ. In this case, the performance will depend on the values in the complex sequences (just as in classical ML estimation). In order to compare the performance to Case S2, we can consider the unknown complex sequence to be a particular sample function from a zero-mean stationary complex random process. We can then use the statistics of the process to evaluate the performance. ML estimators for this model are referred to as conditional (on a particular sample function) ML estimators. The spectral matrix in (8.66) will have to be revised for this case.

There are corresponding signal models for the wideband case that we will discuss in various sections of the chapter.

In communication systems, we may want to impose more structure in the signal model. One model that we will consider is:

Case S4: The source signal is known. The received signal component from the ith source is $\alpha_i f_i(k), k = 1, \cdots, K$, where α_i is a complex constant. This corresponds to a communication system where a known training sequence is sent.

We could also consider cyclostationary signals (e.g., [Gaa88], [SG92], [XK92]) or [Sch94] and constant modulus (CM) signals (e.g., [GL86]). The reader is referred to these references for a discussion of these models.

There are three noise models for the narrowband case:

Case N1:

$$E\left[\mathbf{N}(k)\,\mathbf{N}^H(k)\right] = \sigma_w^2\,\mathbf{I} \tag{8.69}$$

and σ_w^2 is known.

Case N2:

$$E\left[\mathbf{N}(k)\,\mathbf{N}^H(k)\right] = \sigma_w^2\,\mathbf{I} \tag{8.70}$$

and σ_w^2 is unknown.

Case N3:

$$E\left[\mathbf{N}(k)\,\mathbf{N}^H(k)\right] = \mathbf{S_n} \tag{8.71}$$

and $\mathbf{S_n}$ is unknown. Note that, if $\mathbf{S_n}$ were known, Case N3 could be reduced to Case N1 by a spatial whitening filter.

There are corresponding noise models for the wideband case that we will discuss in later sections.

For each of these models we can identify the parameters that must be estimated. For the one-dimensional case, a representative set is summarized in Table 8.2. For the case of a complex parameter such as $\mathbf{f}_k$, we estimate both real and imaginary parts.

Table 8.2: Parameters to be Estimated

Signal Case	Noise Case	Parameters To Estimate	Dimension of $\boldsymbol{\theta}$
S1	N1	$\psi_i, i = 1, \cdots, D$	D
S1	N2	$\psi_i, i = 1, \cdots, D$ σ_w^2	$D+1$
S1	N3	$\psi_i, i = 1, \cdots, D$ $\mathbf{S_n}$	$D+N^2$
S2	N1	$\psi_i, i = 1, \cdots, D$ $\mathbf{S_f}$	$D+D^2$
S2u	N1	$\psi_i, i = 1, \cdots, D$ $\sigma_{s_i}^2, i = 1, \cdots, D$	$2D$
S3	N1	$\psi_i, i = 1, \cdots, D$ $\mathbf{f}_k, k = 1, \cdots, K$	$D+2DK$
S4	N1	$\psi_i, i = 1, \cdots, D$ $\alpha_i, i = 1, \cdots, D$	$3D$

In the (S1, N3) and (S2, N1) cases the dimension of $\boldsymbol{\theta}$ may be different if constraints are placed on $\mathbf{S_f}$ or $\mathbf{S_n}$. The combinations (S2, N2), (S2, N3), (S3, N2), (S3, N3) are obvious modifications of earlier cases.

8.3.2 Model Perturbations

The nominal model given in (8.62) is

$$\mathbf{X}(k) = \mathbf{V}(\psi)\,\mathbf{F}(k) + \mathbf{N}(k), \quad k = 1, 2, \cdots, K. \tag{8.72}$$

In order to model the perturbations, we write (8.72) as

$$\mathbf{X}(k) = \mathbf{V}(\psi, \rho)\,\mathbf{F}(k) + \mathbf{N}(k), \quad k = 1, 2, \cdots, K, \tag{8.73}$$

where $\boldsymbol{\rho}$ represents the parameters whose perturbations we will investigate. We refer to $\boldsymbol{\rho}$ as the **perturbation vector.**[5]

We rewrite the array manifold vector to include the gain and phase of each sensor explicitly

$$\mathbf{v}(\psi_i, \boldsymbol{\rho}) = \left[\; a_0 e^{j\phi_0} e^{j\frac{2\pi}{\lambda}\mathbf{u}_i^T \mathbf{p}_0} \;\vdots\; \cdots \;\vdots\; a_{N-1} e^{j\phi_{N-1}} e^{j\frac{2\pi}{\lambda}\mathbf{u}_i^T \mathbf{p}_{N-1}} \;\right]^T. \tag{8.74}$$

In the general case, the $\boldsymbol{\rho}$ vector is a $5N \times 1$ vector containing a gain, phase, and position vector. Define

$$[\mathbf{a}]_n = a_n^n (1 + \Delta a_n), \quad n = 0, \cdots, N-1, \tag{8.75}$$

$$[\boldsymbol{\phi}]_n = \phi_n^n + \Delta\phi_n, \quad n = 0, \cdots, N-1, \tag{8.76}$$

$$[\mathbf{p}_x]_n = p_{xn}^n + \Delta p_{xn}, \quad n = 0, \cdots, N-1, \tag{8.77}$$

$$[\mathbf{p}_y]_n = p_{yn}^n + \Delta p_{yn}, \quad n = 0, \cdots, N-1, \tag{8.78}$$

$$[\mathbf{p}_z]_n = p_{zn}^n + \Delta p_{zn}, \quad n = 0, \cdots, N-1. \tag{8.79}$$

The superscript "n" denotes the nominal value. This model is similar to the model in Section 2.6.3 and Section 6.6.3. Then, $\boldsymbol{\rho}$ is

$$\boldsymbol{\rho} = \left[\mathbf{a}^T \; \boldsymbol{\phi}^T \; \mathbf{p}_x^T \; \mathbf{p}_y^T \; \mathbf{p}_z^T\right]^T. \tag{8.80}$$

In most cases we study a subset of the possible variations and $\boldsymbol{\rho}$ has a smaller dimension. Normally we assume that $\boldsymbol{\rho}$ has a multivariate Gaussian density,

$$p_{\boldsymbol{\rho}}(\boldsymbol{\rho}) = \frac{1}{(2\pi)^{\frac{5N}{2}} |\boldsymbol{\Lambda}_{\boldsymbol{\rho}}|^{\frac{1}{2}}} \exp\left[-\frac{1}{2}\left[\boldsymbol{\rho} - \boldsymbol{\rho}_0\right]^T \boldsymbol{\Lambda}_{\boldsymbol{\rho}}^{-1} \left[\boldsymbol{\rho} - \boldsymbol{\rho}_0\right]\right], \tag{8.81}$$

where $\boldsymbol{\rho}_0$ denotes the nominal value and $\boldsymbol{\Lambda}_{\boldsymbol{\rho}}$ is the covariance matrix.

The second kind of variation is in the noise environment. Our nominal model assumes

$$E\left[\mathbf{N}(k)\,\mathbf{N}^H(k)\right] = \sigma_w^2 \mathbf{I}. \tag{8.82}$$

We can examine correlated noise by defining

$$E\left[\mathbf{N}(k)\,\mathbf{N}^H(k)\right] = \sigma_w^2 \mathbf{I} + \tilde{\boldsymbol{\Lambda}}. \tag{8.83}$$

We look at the effect of $\tilde{\boldsymbol{\Lambda}}$ on processing performance.

There are three ways to deal with model perturbations:

[5] Although we refer to $\boldsymbol{\rho}$ as the perturbation vector, it contains the nominal values.

(i) We choose some nominal value for the array parameters and estimate the DOAs as if they were correct. We then investigate how the estimator performance degrades as the parameters vary from their nominal value.

(ii) We can model the variations as random variables, treat them as nuisance parameters, and integrate them out of the likelihood function.

(iii) We can jointly estimate the model parameters and the DOAs.

We will look at examples of these techniques in subsequent sections.

8.3.3 Parametric Spatially Spread Signals

In this case we assume that we have D spatially spread signals impinging on the array and the spatial characteristics of the ith signal can be characterized by an L_i-dimensional vector.

A simple example of this situation would be the case in which each of the D signals could be modeled as a complex AR(1) process (see Figures 5.24 and 5.25). For each signal we would estimate two parameters ϕ_a and $|a(1)|$ (or alternatively $Re[a(1)]$ and $Im[a(1)]$).

We will consider models of this type in Section 8.9.

8.3.4 Summary

There are other propagation models that may be appropriate in certain applications. These include multipath models and near-field models. The reader is referred to the literature for a discussion of these models. We will focus our attention on the three models described above.

8.4 Cramér-Rao Bounds

In Section 8.2.3, we developed the classical CRB for parameter estimation. In this section, we apply those results to the signal and noise models that we developed in Section 8.3.

In Section 8.4.1, we consider the case in which the signal vector is a sample function from a Gaussian random process with unknown signal spectral matrix and the noise is a sample function from a spatially white Gaussian random process. This case includes (S2, N1) and (S2, N2).

In Section 8.4.2, we consider the case in which the signal vector is a sample function from a Gaussian random process and we know *a priori* that the component signals are statistically independent with unknown $\mathbf{\Lambda_f}$. The noise is a sample function from a spatially white Gaussian random process. This case includes (S2u, N1) and (S2u, N2).

In Section 8.4.3, we consider the case in which the signal is a sample function from a Gaussian random process with a known signal spectrum and the noise is a sample function from a spatially white Gaussian random process.

In Section 8.4.4, we consider the case in which the signal is modelled as a nonrandom, unknown waveform and the noise is a sample function from a spatially white Gaussian random process.

In Section 8.4.5, we consider the case in which the signals are modelled as known waveforms with complex amplitude multiplies. The noise is a sample function from a spatially white complex Gaussian random process. This corresponds to case (S4, N1) and (S4, N2).

In Section 8.4.6, we summarize our results.

Before beginning the development, a brief commentary on the steps is useful:

(i) The first step is to derive the Fisher information matrix $\mathbf{J}$ (8.34) or $\mathbf{J}_B$ (8.49). This step is usually straightforward, but may be tedious.

(ii) In most cases, we are primarily interested in estimating $\boldsymbol{\psi}$, the DOAs of the D signals. We partition $\mathbf{J}$ into blocks corresponding to $\boldsymbol{\psi}$ and blocks corresponding to the other (unwanted) parameters and blocks corresponding to cross-terms. We partition $\mathbf{C}_{CR}$ similarly and use the formula for the inverse of block matrices to find the block of $\mathbf{C}_{CR}$ corresponding to $\boldsymbol{\psi}$.

This step is straightforward, but may contain complicated expressions. In many of our examples, we evaluate the expression numerically and plot the result.

(iii) The final step is to obtain a compact expression for $\mathbf{C}_{CR}(\boldsymbol{\psi})$. This step is normally a tour de force of linear algebra. When successful, it offers the ability to compare the bounds for different models. We will quote some of these results, but will normally not derive them.

8.4.1 Gaussian Model: Unknown Signal Spectrum

8.4.1.1 Cramér-Rao bound

The first model of interest corresponds to the signals being a sample function from a zero-mean stationary complex Gaussian random process whose spectral matrix $\mathbf{S_f}$ is unknown and the additive noise is a sample function from a spatially white Gaussian random process (this is Case (S2, N2) in Section 8.2).

We consider an array model in which the wavenumber is 1-D and assume that there are D signals impinging on the array. Thus,

$$\mathbf{X}(k) = \mathbf{V}(\boldsymbol{\psi})\mathbf{F}(k) + \mathbf{N}(k), \tag{8.84}$$

and

$$\mathbf{S_n} = \sigma_w^2 \mathbf{I}, \tag{8.85}$$

where σ_w^2 is unknown, and $\mathbf{S_f}$ is the unknown signal spectral matrix.

The input spectral matrix is

$$\mathbf{S_x} = \mathbf{V}(\boldsymbol{\psi})\mathbf{S_f}\mathbf{V}^H(\boldsymbol{\psi}) + \sigma_w^2 \mathbf{I}. \tag{8.86}$$

In this case, the parameter vector $\boldsymbol{\theta}$ can be written as,

$$\boldsymbol{\theta} \triangleq \begin{bmatrix} \boldsymbol{\psi}, & \boldsymbol{\mu}, & \sigma_w^2 \end{bmatrix}^T, \tag{8.87}$$

where $\boldsymbol{\psi}$ is a $D \times 1$ vector corresponding to signal DOAs ($\psi_i = \pi \cos\theta_i$), $\boldsymbol{\mu}$ is a real vector corresponding to the elements in $\mathbf{S_f}$, and σ_w^2 is a scalar corresponding to the noise spectral height. The matrix $\mathbf{S_f}$ is a complex Hermitian matrix.

$$\dim[\boldsymbol{\mu}] = D + \sum_{k=1}^{D-1} 2k = D^2. \tag{8.88}$$

We construct the $\mathbf{J}$ matrix in (8.35) in a partitioned form,

$$\mathbf{J} = \left[\begin{array}{c|c|c} \mathbf{J}_{\boldsymbol{\psi}\boldsymbol{\psi}} & \mathbf{J}_{\boldsymbol{\psi}\boldsymbol{\mu}} & \mathbf{J}_{\boldsymbol{\psi}\sigma_w^2} \\ \hline \mathbf{J}_{\boldsymbol{\mu}\boldsymbol{\psi}} & \mathbf{J}_{\boldsymbol{\mu}\boldsymbol{\mu}} & \mathbf{J}_{\boldsymbol{\mu}\sigma_w^2} \\ \hline \mathbf{J}_{\sigma_w^2\boldsymbol{\psi}} & \mathbf{J}_{\sigma_w^2\boldsymbol{\mu}} & \mathbf{J}_{\sigma_w^2\sigma_w^2} \end{array}\right]. \tag{8.89}$$

Each element is obtained by substituting $\mathbf{S_x}$ from (8.86) for $\mathbf{K_x}$ in (8.35). We then must find $\mathbf{S_x}^{-1}$, compute the derivative with respect to each parameter, and evaluate the trace in (8.35). Thus,

$$\left[\mathbf{J}_{\boldsymbol{\psi}\boldsymbol{\psi}}\right]_{ij} = \operatorname{tr}\left[\mathbf{S_x}^{-1}\frac{\partial \mathbf{S_x}}{\partial \psi_i}\mathbf{S_x}^{-1}\frac{\partial \mathbf{S_x}}{\partial \psi_j}\right]. \tag{8.90}$$

Using the matrix inversion lemma (A.51) on (8.86) gives

$$\mathbf{S}_{\mathbf{x}}^{-1} = \frac{1}{\sigma_w^2}\left[\mathbf{I} - \mathbf{V}\left(\mathbf{V}^H\mathbf{V} + \sigma_w^2\mathbf{S}_{\mathbf{f}}^{-1}\right)^{-1}\mathbf{V}^H\right]. \tag{8.91}$$

The derivative needed for the $\mathbf{J}_{\psi\psi}$ matrix is

$$\frac{\partial \mathbf{S}_{\mathbf{x}}}{\partial \psi_i} = \frac{\partial \mathbf{V}(\psi)}{\partial \psi_i}\mathbf{S}_{\mathbf{f}}\mathbf{V}^H(\psi) + \mathbf{V}(\psi)\mathbf{S}_{\mathbf{f}}\frac{\partial \mathbf{V}^H(\psi)}{\partial \psi_i}. \tag{8.92}$$

The ith column of the derivative matrix is

$$\left[\frac{\partial \mathbf{V}(\psi)}{\partial \psi_i}\right]_i = \left.\frac{\partial \mathbf{v}(\psi)}{\partial \psi}\right|_{\psi=\psi_i} \triangleq \mathbf{d}(\psi_i), \tag{8.93}$$

and the remaining columns are zero. Using (8.91), (8.92), and (8.93) in (8.90) we obtain $\left[\mathbf{J}_{\psi\psi}\right]_{ij}$. We carry out the details of the derivation in Section 8.4.2.

The terms in the other submatrices follow in a similar manner. This procedure has been carried out by Weiss and Friedlander [WF93] (The evaluation of the $\mathbf{J}_{ij}$ is carried out in [WF90] and summarized in [WF93].) The derivation is complicated and the interested reader should consult those two references.

We next partition the CRB matrix in a similar manner,

$$\mathbf{C}_{CR} = \left[\begin{array}{c:c:c} \mathbf{C}_{CR}(\psi) & \mathbf{C}_{CR}(\psi,\mu) & \mathbf{C}_{CR}(\psi,\sigma_w^2) \\ \hdashline \mathbf{C}_{CR}(\mu,\psi) & \mathbf{C}_{CR}(\mu) & \mathbf{C}_{CR}(\mu,\sigma_w^2) \\ \hdashline \mathbf{C}_{CR}(\sigma_w^2,\psi) & \mathbf{C}_{CR}(\sigma_w^2,\mu) & \mathbf{C}_{CR}(\sigma_w^2) \end{array}\right]. \tag{8.94}$$

We then use the partitioned matrix inversion formula (A.68) to obtain $\mathbf{C}_{CR}(\psi)$.

The result is

$$\boxed{\mathbf{C}_{CR}(\psi) = \frac{\sigma_w^2}{2K}\left\{Re\left[\left[\mathbf{S}_{\mathbf{f}}\mathbf{V}^H\mathbf{S}_{\mathbf{x}}^{-1}\mathbf{V}\mathbf{S}_{\mathbf{f}}\right] \odot \left[\mathbf{D}^H\mathbf{P}_{\mathbf{V}}^{\perp}\mathbf{D}\right]^T\right]\right\}^{-1},} \tag{8.95}$$

where $\odot$ denotes the Hadamard product (A.71). The result includes K to account for the K independent snapshots. Note that

$$\mathbf{P}_{\mathbf{V}}^{\perp} = \left[\mathbf{I} - \mathbf{V}\left(\mathbf{V}^H\mathbf{V}\right)^{-1}\mathbf{V}^H\right] \tag{8.96}$$

is the projection matrix onto the noise subspace, and

$$\mathbf{D} = \dot{\mathbf{V}}_{\psi} = \left[\frac{\partial \mathbf{v}(\psi_1)}{\partial \psi_1} \,\vdots\, \frac{\partial \mathbf{v}(\psi_2)}{\partial \psi_2} \,\vdots\, \cdots \,\vdots\, \frac{\partial \mathbf{v}(\psi_D)}{\partial \psi_D} \right] \tag{8.97}$$

is the derivative matrix. The result in (8.95) was first obtained by Stoica and Nehorai [SN89a] by analyzing the asymptotic behavior of the maximum likelihood estimate (e.g., Stoica and Nehori [SN89b], Viberg [Vib89], Ottersten [Ott89], and Stoica and Nehorai [SN90a], [SN90b]). The first direct derivation is due to Weiss and Friedlander [WF93], [WF90], and [WF91a]. The most efficient derivation is contained in Stoica et al. [SLG01]. The reader is referred to these references for the details of the derivation.

There are two alternative forms of (8.95) that we will encounter in the sequel. The first is an obvious modification,

$$\mathbf{C}_{CR}(\psi) = \frac{\sigma_w^2}{2K} \left\{ Re \left[\left[\mathbf{D}^H \mathbf{P}_{\mathbf{V}}^{\perp} \mathbf{D} \right] \odot \left[\mathbf{S_f} \mathbf{V}^H \mathbf{S_x}^{-1} \mathbf{V} \mathbf{S_f} \right]^T \right] \right\}^{-1}, \tag{8.98}$$

and the second form specifies the ij element of $\mathbf{C}_{CR}^{-1}$,

$$\left[\mathbf{C}_{CR}^{-1} \right]_{ij} = \frac{2K}{\sigma_w^2} \left\{ Re \left[\mathrm{tr} \left[\left[\mathbf{D}_j^H \mathbf{P}_{\mathbf{V}}^{\perp} \mathbf{D}_i \right] \left[\mathbf{S_f} \mathbf{V}^H \mathbf{S_x}^{-1} \mathbf{V} \mathbf{S_f} \right] \right] \right] \right\}. \tag{8.99}$$

Note that if we want to find the CRB with respect to arrival angle θ, we substitute

$$\mathbf{D}_{\theta} = \dot{\mathbf{V}}_{\theta}. \tag{8.100}$$

We define

$$\mathbf{H} \triangleq \mathbf{D}^H \mathbf{P}_{\mathbf{V}}^{\perp} \mathbf{D}, \tag{8.101}$$

and rewrite (8.95) as,

$$\boxed{\mathbf{C}_{CR}(\psi) = \frac{\sigma_w^2}{2K} \left\{ Re \left[\left(\mathbf{S_f} \mathbf{V}^H \mathbf{S_x}^{-1} \mathbf{V} \mathbf{S_f} \right) \odot \mathbf{H}^T \right] \right\}^{-1}.} \tag{8.102}$$

In [WF90], it is shown that the expressions in (8.95) and (8.102) are also valid for the case in which σ_w^2 is known.

We can simplify the first term in (8.102). We define

$$\mathbf{\Sigma} = \frac{\mathbf{S_f}}{\sigma_w^2}, \tag{8.103}$$

and expand $\mathbf{S}_{\mathbf{x}}^{-1}$ using (8.91). Note that $\boldsymbol{\Sigma}$ is a matrix *SNR*:

$$\begin{aligned}
\mathbf{V}^H\mathbf{S}_{\mathbf{x}}^{-1}\mathbf{V}\mathbf{S}_{\mathbf{f}} &= \mathbf{V}^H\left[\mathbf{I}-\mathbf{V}\boldsymbol{\Sigma}\left(\mathbf{V}^H\mathbf{V}\boldsymbol{\Sigma}+\mathbf{I}\right)^{-1}\mathbf{V}^H\right]\mathbf{V}\boldsymbol{\Sigma} \\
&= \mathbf{V}^H\mathbf{V}\boldsymbol{\Sigma}-\mathbf{V}^H\mathbf{V}\boldsymbol{\Sigma}\left(\mathbf{V}^H\mathbf{V}\boldsymbol{\Sigma}+\mathbf{I}\right)^{-1}\mathbf{V}^H\mathbf{V}\boldsymbol{\Sigma} \\
&= \left[\mathbf{I}-\mathbf{V}^H\mathbf{V}\boldsymbol{\Sigma}\left(\mathbf{V}^H\mathbf{V}\boldsymbol{\Sigma}+\mathbf{I}\right)^{-1}\right]\mathbf{V}^H\mathbf{V}\boldsymbol{\Sigma}. \qquad (8.104)
\end{aligned}$$

We can rewrite the identity matrix in the bracket as

$$\begin{aligned}
\mathbf{V}^H\mathbf{S}_{\mathbf{x}}^{-1}\mathbf{V}\mathbf{S}_{\mathbf{f}} &= \left[\left(\mathbf{V}^H\mathbf{V}\boldsymbol{\Sigma}+\mathbf{I}\right)\left(\mathbf{V}^H\mathbf{V}\boldsymbol{\Sigma}+\mathbf{I}\right)^{-1}-\right. \\
&\qquad \left.\mathbf{V}^H\mathbf{V}\boldsymbol{\Sigma}\left(\mathbf{V}^H\mathbf{V}\boldsymbol{\Sigma}+\mathbf{I}\right)^{-1}\right]\mathbf{V}^H\mathbf{V}\boldsymbol{\Sigma} \\
&= \left(\mathbf{V}^H\mathbf{V}\boldsymbol{\Sigma}+\mathbf{I}\right)^{-1}\mathbf{V}^H\mathbf{V}\boldsymbol{\Sigma}. \qquad (8.105)
\end{aligned}$$

Using (8.105) in (8.102) we obtain

$$\boxed{\mathbf{C}_{CR}(\boldsymbol{\psi}) = \frac{\sigma_w^2}{2K}\left\{Re\left\{\left[\mathbf{S}_{\mathbf{f}}\left[\left(\mathbf{I}+\mathbf{V}^H\mathbf{V}\frac{\mathbf{S}_{\mathbf{f}}}{\sigma_w^2}\right)^{-1}\left(\mathbf{V}^H\mathbf{V}\frac{\mathbf{S}_{\mathbf{f}}}{\sigma_w^2}\right)\right]\right]\odot\mathbf{H}^T\right\}\right\}^{-1}.} \qquad (8.106)$$

The result in (8.106) is the formula we shall use for most of our CRB calculations. As pointed out previously, the result is valid for both known and unknown noise variance.

For large values of $\mathbf{V}^H\mathbf{V}\mathbf{S}_{\mathbf{f}}/\sigma_w^2$, the term in the innermost bracket approaches the identity matrix $\mathbf{I}$. To show this relationship, we expand $\mathbf{V}^H\mathbf{V}\boldsymbol{\Sigma}$ in terms of its eigenvalues and eigenvectors,

$$\mathbf{V}^H\mathbf{V}\boldsymbol{\Sigma} = \sum_{i=1}^{N}\lambda_i\,\boldsymbol{\Phi}_i\,\boldsymbol{\Phi}_i^H. \qquad (8.107)$$

Then,

$$\begin{aligned}
\left[\mathbf{I}+\mathbf{V}^H\mathbf{V}\boldsymbol{\Sigma}\right]^{-1}\mathbf{V}^H\mathbf{V}\boldsymbol{\Sigma} &= \sum_{i=1}^{N}\frac{\lambda_i}{1+\lambda_i}\boldsymbol{\Phi}_i\,\boldsymbol{\Phi}_i^H \\
&= \sum_{i=1}^{N}\frac{1}{1+\frac{1}{\lambda_i}}\boldsymbol{\Phi}_i\,\boldsymbol{\Phi}_i^H. \qquad (8.108)
\end{aligned}$$

Assuming that $\lambda_i > 1$, we can expand the denominator, giving

$$\begin{aligned}\left[\mathbf{I}+\mathbf{V}^H\mathbf{V}\mathbf{\Sigma}\right]^{-1}\mathbf{V}^H\mathbf{V}\mathbf{\Sigma} &= \mathbf{I}-\sum_{i=1}^{N}\frac{1}{\lambda_i}\mathbf{\Phi}_i\mathbf{\Phi}_i^H+\cdots \\ &= \mathbf{I}-\left[\mathbf{V}^H\mathbf{V}\mathbf{\Sigma}\right]^{-1}+\cdots. \end{aligned} \tag{8.109}$$

The second and higher-order terms approach zero as λ_i approaches infinity.

These terms will be negligible when $\mathbf{S_f}/\sigma_w^2$ is large and:

(i) $\mathbf{S_f}$ is not close to singular. If the signals are correlated, $\mathbf{S_f}$ approaches singularity as $|\rho_{ij}|$ approaches one for some ij pair.

(ii) If the signals are uncorrelated, the smallest $[\mathbf{S_f}]_{ii}/\sigma_w^2$ must be large.

(iii) The signals are not too closely spaced. Closely spaced signals result in $\mathbf{V}^H\mathbf{V}$ approaching a singular matrix.

When these conditions are satisfied, then we can approximate (8.106) as

$$\boxed{\mathbf{C}_{CR}(\psi) \cong \frac{\sigma_w^2}{2K}\left\{Re\left[\mathbf{S_f}\odot\mathbf{H}^T\right]\right\}^{-1} \triangleq \mathbf{C}_{ACR}(\psi).} \tag{8.110}$$

where $\mathbf{C}_{ACR}$ denotes the approximate CRB.

We consider several simple examples and then return to the CRB in (8.106) and derive some general properties.

8.4.1.2 Single signals

Example 8.4.1

Consider the case of a single signal. In this case J_ψ is a scalar. Then, (8.106) can be written as

$$\begin{aligned} J_\psi &= Re\left[\frac{2K\sigma_s^2}{\sigma_w^2}\cdot\left(1+N\frac{\sigma_s^2}{\sigma_w^2}\right)^{-1}\left(N\frac{\sigma_s^2}{\sigma_w^2}\right)\right]\left[\mathbf{d}^H\left(\mathbf{1}-\frac{\mathbf{v}\mathbf{v}^H}{N}\right)\mathbf{d}\right] \\ &= \frac{2K}{N}\left[\frac{(ASNR)^2}{1+ASNR}\right]\left[\mathbf{d}^H\left(\mathbf{1}-\frac{\mathbf{v}\mathbf{v}^H}{N}\right)\mathbf{d}\right], \end{aligned} \tag{8.111}$$

where the *ASNR* is

$$ASNR = N\frac{\sigma_s^2}{\sigma_w^2}. \tag{8.112}$$

The signal-to-noise behavior is contained in the term in the first bracket and the array geometry is contained in the second bracket.

If the array manifold is conjugate symmetric ($\mathbf{v} = \mathbf{J}\,\mathbf{v}^*$), then

$$\mathbf{d}^H\,\mathbf{v} = 0, \tag{8.113}$$

and

$$J_\psi = \frac{2K}{N}\left[\frac{(ASNR)^2}{1+ASNR}\right] \parallel \mathbf{d} \parallel^2 . \tag{8.114}$$

Then,

$$\begin{aligned} C_{CR}(\psi) &= \frac{N}{2K}\left[\frac{1+ASNR}{(ASNR)^2}\right] \parallel \mathbf{d} \parallel^{-2} \\ &= \frac{N}{2K}\left[\frac{1}{ASNR}+\frac{1}{(ASNR)^2}\right] \parallel \mathbf{d} \parallel^{-2} . \end{aligned} \tag{8.115}$$

We see that for small *ASNR* the variance bound is proportional to $(ASNR)^{-2}$ and for large *ASNR*, it is proportional to $(ASNR)^{-1}$. We also observe that the effect of the array geometry is completely contained in the $\|\mathbf{d}\|^{-2}$ term.

We recall that the conventional beam pattern is

$$B_c(\psi_1,\psi_2) = \mathbf{v}^H(\psi_1)\,\mathbf{v}(\psi_2), \tag{8.116}$$

where ψ_1 is steering direction and ψ_2 is direction of the incoming plane wave. Then,

$$\frac{\partial^2 B_c(\psi_1,\psi_2)}{\partial\psi_1\,\partial\psi_2} = \mathbf{d}^H(\psi_1)\,\mathbf{d}(\psi_2), \tag{8.117}$$

and

$$\parallel \mathbf{d} \parallel^2 = \left.\frac{\partial^2 B_c(\psi_1,\psi_2)}{\partial\psi_1\,\partial\psi_2}\right|_{\psi_1=\psi_2} . \tag{8.118}$$

The beam pattern in wavenumber space is only a function of $\psi_1-\psi_2$, so we can also write (8.118) as

$$\parallel \mathbf{d} \parallel^2 = -\left.\frac{\partial^2 B_c(\psi_1,\psi_2)}{\partial^2\psi_2}\right|_{\psi_2=0} . \tag{8.119}$$

Thus, the effect of the array geometry on the CRB for a single signal is completely specified by the second derivative (or the radius of curvature) of the conventional beam pattern.

If we use the approximation in (8.110), we obtain

$$C_{ACR}(\psi) = \frac{N}{2K}\left[\frac{1}{ASNR}\right]\|\mathbf{d}\|^{-2}. \tag{8.120}$$

We see that $C_{CR}(\psi)$ approaches $C_{ACR}(\psi)$ for large *ASNR*. However, $C_{ACR}(\psi)$ loses the quadratic dependence at low *ASNR*.

We can obtain an explicit expression for the term in the second bracket in (8.111). We do not need to assume linearity or conjugate symmetry.

The array manifold vector is

$$\mathbf{v}(\psi) = \left[\begin{array}{cccc} e^{j\frac{2d_1}{\lambda}\psi} & e^{j\frac{2d_2}{\lambda}\psi} & \cdots & e^{j\frac{2d_N}{\lambda}\psi} \end{array}\right]^T . \tag{8.121}$$

Then,

$$\mathbf{d}^H\mathbf{d} = \left(\frac{2}{\lambda}\right)^2 \sum_{i=1}^{N} d_i^2, \tag{8.122}$$

and the second bracket in (8.111) can be written as

$$\mathbf{d}^H\mathbf{d} - \frac{\left|\mathbf{v}^H(\psi)\mathbf{d}\right|^2}{N} = 4\left\{\sum_{i=1}^{N}\left(\frac{d_i}{\lambda}\right)^2 - \frac{1}{N}\left(\sum_{i=1}^{N}\frac{d_i}{\lambda}\right)^2\right\}. \tag{8.123}$$

Defining,

$$\mathbf{p} = \left[\begin{array}{cccc} \frac{d_1}{\lambda} & \frac{d_2}{\lambda} & \cdots & \frac{d_N}{\lambda} \end{array}\right]^T, \tag{8.124}$$

we can write (8.123) as

$$\mathbf{d}^H\mathbf{d} - \frac{\left|\mathbf{v}^H(\psi)\mathbf{d}\right|^2}{N} = 4\mathbf{p}^T\left[\mathbf{I} - \frac{1}{N}\mathbf{1}\mathbf{1}^T\right]\mathbf{p}. \tag{8.125}$$

We now apply this result to a linear array (similar simplifications follow for conjugate symmetric arrays).

Example 8.4.2 (continuation)

Consider a standard linear array and first assume N is odd for notational convenience. Then

$$\mathbf{p} = \left[\begin{array}{ccc} -\frac{N-1}{2} & \cdots & \frac{N-1}{2} \end{array}\right], \tag{8.126}$$

where we have chosen the origin at the phase center of the array. Then,

$$\mathbf{p}^T\mathbf{1} = 0, \tag{8.127}$$

and

$$\mathbf{p}^T\mathbf{p} = 2\sum_{i=1}^{\frac{N-1}{2}} i^2 = \frac{N(N^2-1)}{12}. \tag{8.128}$$

For N even, (8.127) is still valid and

$$\mathbf{p}^T\mathbf{p} = 2\sum_{i=1}^{\frac{N}{2}}\left(i-\frac{1}{2}\right)^2 = \frac{N(N^2-1)}{12}. \tag{8.129}$$

Then, substituting into (8.115), we obtain

$$C_{CR}(\psi) = \frac{1}{K}\left[\frac{1}{ASNR} + \frac{1}{(ASNR)^2}\right]\frac{6}{(N^2-1)}. \tag{8.130}$$

For large N, it is proportional to N^{-2}. Note that the *ASNR* is proportional to N, so that for a fixed element *SNR*, the leading term in (8.130) is proportional to N^{-3}.

A more general curve is obtained if we normalize the bound by the BW_{NN} (which is $4/N$ for a standard linear array with uniform weighting):

$$\begin{aligned} C_{CRN}(\psi) &\triangleq \frac{C_{CR}(\psi)}{BW_{NN}^2} \\ &= \frac{3}{8K}\left[\frac{1}{ASNR} + \frac{1}{(ASNR)^2}\right]\frac{1}{1-\frac{1}{N^2}}. \end{aligned} \tag{8.131}$$

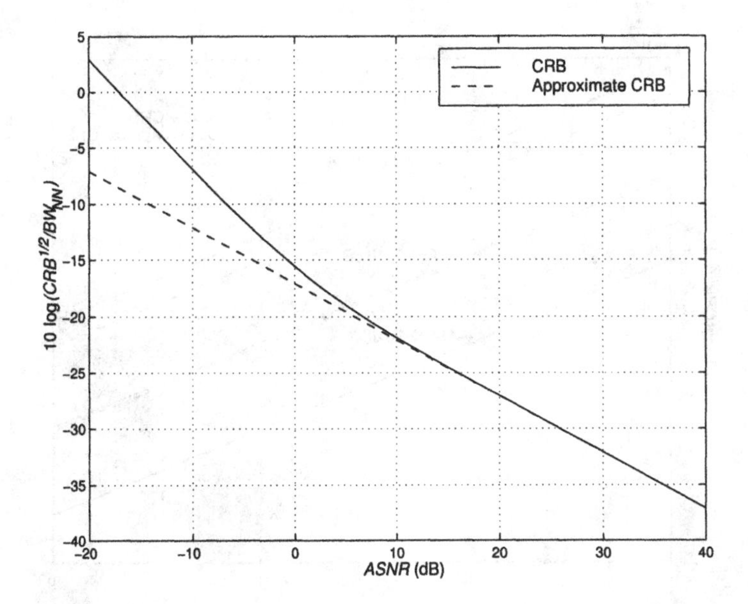

Figure 8.1 Single plane-wave signal: square root of $CRB(\psi)$ vs. *ASNR*, $N = 10$, $K = 100$.

The corresponding result for the normalized version of $C_{ACR}(\psi)$ in (8.120) is the same as in (8.131) except the $(ASNR)^2$ is omitted.

The result for two normalized bounds are plotted versus *ASNR* for a standard 10-element array in Figure 8.1 and versus K for various *ASNR* in Figure 8.2. The scale on the vertical axis is $10\log\left(\sqrt{C_{CR}(\psi)}/BW_{NN}\right)$. Plotted versus *ASNR*, the two bounds are essentially equal above 10 dB. However, plotted versus K, there is a significant difference at lower *ASNR*. The value of plotting our results with this normalization is that the results do not vary significantly (≤ 0.01) with N for $N \geq 10$.

Another view of $C_{CRN}(\psi)$ that is useful is shown in Figure 8.3. Here, we plot the contours of constant $C_{CRN}(\psi)$ versus *ASNR* and K. We will find this presentation useful when we study the performance of various estimators.

It is useful to introduce several ideas with respect to the *ASNR*-K plane at this point. Some of ideas will become clearer when we examine specific estimators.

In Figure 8.3, we have identified intervals on each axis in the *ASNR*-K plane. It is important to note that the exact boundaries vary with the particular application. On the K axis, the three intervals are:

(i) Low sample support

Here $K < 10N$. In this interval, particularly at the lower end, the stability of $\mathbf{C_x}$ is an important issue. In many applications, we are forced

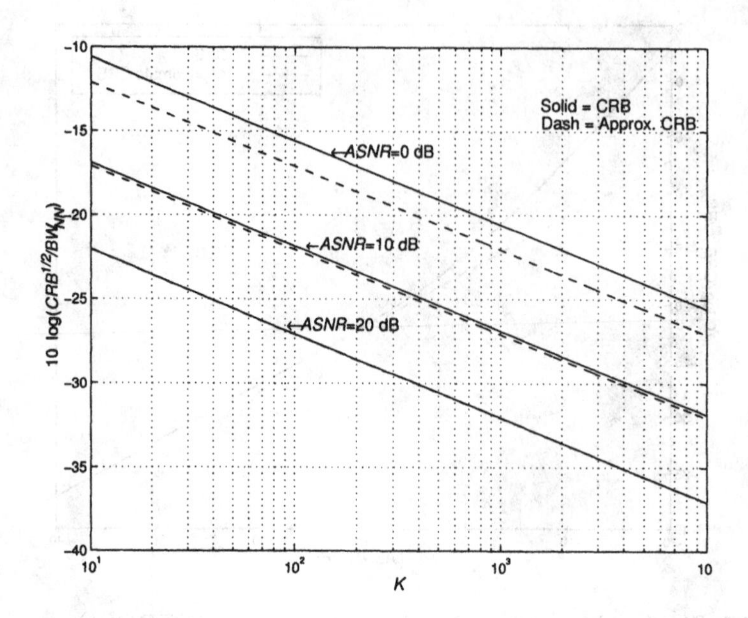

Figure 8.2 Single plane-wave signal: square root of $CRB(\psi)$ vs. K for various *ASNR*.

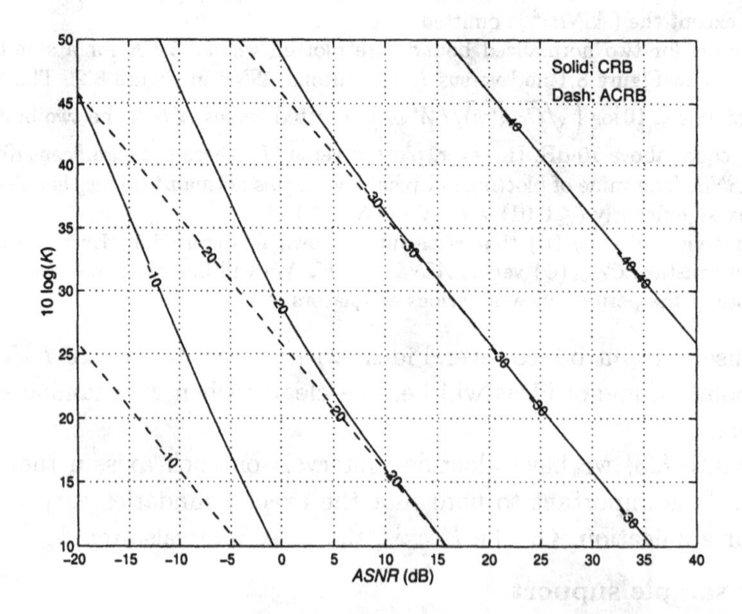

Figure 8.3 Single plane-wave signal: $CRB(\psi)$ in *ASNR*-K plane.

to work in this interval because of a limited number of observations or because of the non-stationarity of the data.

(ii) Asymptotic

Here K is large and we can use analyses that neglect terms of $O(K^{-1})$. Specifically, the asymptotic behavior of the eigenvalues and eigenvectors that was given in Section 7.2.2 can be exploited. When we discuss the asymptotic properties of estimators (particularly ML estimators), we are referring to this region.

(iii) Transition

In this region we transition from one mode of behavior to another mode. Many analyses focus on identifying the boundaries of the transition region.

On the *ASNR* axis, the three intervals are:

(i) Low *ASNR*

Here the *ASNR* is so low that the data do not provide any useful information.

(ii) High *ASNR*

Here the *ASNR* is high enough that, in many cases, the error is small enough that we can analyze it using a Taylor series expansion around the correct value.

(iii) Transition

In this region we transition from the small error behavior mode to a large error behavior mode.

These intervals provide one way to think about the problem that we will find to be useful.

It is important to note that in many cases the transition regions are bounded by iso-MSE curves in the *ASNR*-K plane. We develop this behavior later in this section.

Note that we have considered the parameter in ψ-space. Recall that the corresponding angle is

$$\theta = \cos^{-1}\left(\frac{\psi}{\pi}\right). \tag{8.132}$$

Therefore the bound on angle estimate depends on the actual value of the angle. The $\mathbf{G}$ matrix in (8.46) is derived in Problems 8.2.1 and 8.2.2. It is a diagonal matrix so the mapping of the CRB from ψ-space to θ-space is straightforward.

For the case of two plane-wave signals, the algebraic expressions in the general case are complicated enough that they do not provide much insight into the bound. Most of our results will use numerical evaluation. We note that the expression in (8.130) always lower bounds the multiple-signal case.

In Section 8.4.1.3, we consider uncorrelated signals, and in Section 8.4.1.4 we consider correlated signals.

8.4.1.3 Uncorrelated signals

In this section, we consider the case of two uncorrelated plane-wave signals. However, we do not know *a priori* that they are uncorrelated. We evaluate $C_{CR}(\boldsymbol{\psi})$ using (8.102). The first example is for arbitrary array geometry.

Example 8.4.3

Consider the case of two plane-wave signals. We evaluate the $\mathbf{H}$ matrix first. For two signals,

$$\mathbf{V} = \left[\begin{array}{c:c} \mathbf{v}(\psi_1) & \mathbf{v}(\psi_2) \end{array} \right]. \tag{8.133}$$

To evaluate $\mathbf{H}$ in (8.101), we first find $\mathbf{D}^H\mathbf{D}$,

$$\mathbf{D}^H\mathbf{D} = \left[\begin{array}{c:c} \mathbf{d}_1^H\mathbf{d}_1 & \mathbf{d}_1^H\mathbf{d}_2 \\ \hdashline \mathbf{d}_2^H\mathbf{d}_1 & \mathbf{d}_2^H\mathbf{d}_2 \end{array} \right]. \tag{8.134}$$

Each of the terms in (8.134) can be written in terms of the conventional beam pattern

$$\begin{aligned} \mathbf{d}_i^H \mathbf{d}_j &= \frac{\partial \mathbf{v}_i^H}{\partial \psi_i} \frac{\partial \mathbf{v}_j}{\partial \psi_j} = \frac{\partial^2}{\partial \psi_i \, \partial \psi_j} B_C\left(\psi_i , \psi_j\right) \\ &\triangleq B_{\psi_i \psi_j}\left(\psi_i , \psi_j\right), \quad i = 1,2; j = 1,2. \end{aligned} \tag{8.135}$$

Therefore,

$$\mathbf{D}^H\mathbf{D} = \left[\begin{array}{c:c} B_{\psi_1\psi_1}\left(\psi_1 , \psi_1\right) & B_{\psi_1\psi_2}\left(\psi_1 , \psi_2\right) \\ \hdashline B^*_{\psi_1\psi_2}\left(\psi_1 , \psi_2\right) & B_{\psi_2\psi_2}\left(\psi_2 , \psi_2\right) \end{array} \right]. \tag{8.136}$$

The next term of interest is $\mathbf{P_V}$,

$$\mathbf{V}^H\mathbf{V} = N \left[\begin{array}{cc} 1 & B\left(\psi_1 , \psi_2\right) \\ B^*\left(\psi_1 , \psi_2\right) & 1 \end{array} \right]. \tag{8.137}$$

and

$$\left[\mathbf{V}^H\mathbf{V}\right]^{-1} = \frac{1}{N\left[1 - \left|B\left(\psi_1 , \psi_2\right)\right|^2\right]} \left[\begin{array}{cc} 1 & -B\left(\psi_1 , \psi_2\right) \\ -B^*\left(\psi_1 , \psi_2\right) & 1 \end{array} \right]. \tag{8.138}$$

We use B_{12} as abbreviated notation for $B(\psi_1, \psi_2)$. Then, we can write

$$\begin{aligned}\mathbf{P_V} &= \mathbf{V}\left[\mathbf{V}^H\mathbf{V}\right]^{-1}\mathbf{V}^H \\ &= \frac{1}{N\left[1-|B_{12}|^2\right]}\left[\begin{array}{c:c}\mathbf{v}_1 & \mathbf{v}_2\end{array}\right]\left[\begin{array}{cc}1 & -B_{12} \\ -B_{12}^* & 1\end{array}\right]\left[\begin{array}{c}\mathbf{v}_1^H \\ \hdashline \mathbf{v}_2^H\end{array}\right] \\ &= \frac{1}{N\left[1-|B_{12}|^2\right]}\left\{\mathbf{v}_1\mathbf{v}_1^H - B_{12}\mathbf{v}_1\mathbf{v}_2^H - B_{12}^*\mathbf{v}_2\mathbf{v}_1^H + \mathbf{v}_2\mathbf{v}_2^H\right\}. \end{aligned} \tag{8.139}$$

The ijth element of the second term in $\mathbf{H}$ is,

$$\begin{aligned}\left[\mathbf{D}^H\mathbf{P_V}\mathbf{D}\right]_{ij} &= \frac{1}{N\left[1-|B_{12}|^2\right]}\cdot\left\{\mathbf{d}_i^H\left[\mathbf{v}_1\mathbf{v}_1^H - B_{12}\mathbf{v}_1\mathbf{v}_2^H\right.\right. \\ &\qquad \left.\left. -B_{21}^*\mathbf{v}_2\mathbf{v}_1^H + \mathbf{v}_2\mathbf{v}_2^H\right]\mathbf{d}_j\right\}, \quad i,j=1,2. \end{aligned} \tag{8.140}$$

For conjugate symmetric $\mathbf{v}$,

$$\mathbf{d}_i^H\mathbf{v}_i = \mathbf{v}_i^H\mathbf{d}_i = 0, \quad i=1,2. \tag{8.141}$$

Thus, (8.140) reduces to

$$\mathbf{D}^H\mathbf{P_V}\mathbf{D} = \frac{1}{N\left[1-|B_{12}|^2\right]} \times \left[\begin{array}{c:c}\left|\mathbf{d}_1^H\mathbf{v}_2\right|^2 & -B_{21}^*\left(\mathbf{d}_1^H\mathbf{v}_2\right)\left(\mathbf{v}_1^H\mathbf{d}_2\right) \\ \hdashline -B_{12}\left(\mathbf{d}_2^H\mathbf{v}_1\right)\left(\mathbf{v}_2^H\mathbf{d}_1\right) & \left|\mathbf{d}_2^H\mathbf{v}_1\right|^2\end{array}\right]. \tag{8.142}$$

From (8.116)

$$\mathbf{d}_1^H\mathbf{v}_2 = \frac{\partial B_c(\psi_1,\psi_2)}{\partial\psi_1} = \frac{\partial B_c(\psi_1-\psi_2)}{\partial\psi_1} = -\frac{\partial B_c(\psi_1-\psi_2)}{\partial\psi_2} = -B_{\psi_2}(\Delta\psi), \tag{8.143}$$

$$\mathbf{d}_2^H\mathbf{v}_1 = \frac{\partial B_c(\psi_1-\psi_2)}{\partial\psi_2} = B_{\psi_2}(\Delta\psi), \tag{8.144}$$

where

$$\Delta\psi = \psi_1 - \psi_2. \tag{8.145}$$

Then,

$$H_{11} = B_{\psi_1\psi_1}(\psi_1,\psi_1) - \frac{1}{N\left[1-|B_{12}|^2\right]}\left|B_{\psi_2}(\Delta\psi)\right|^2, \tag{8.146}$$

$$H_{22} = B_{\psi_2\psi_2}(\psi_2,\psi_2) - \frac{1}{N\left[1-|B_{12}|^2\right]}\left|B_{\psi_2}(\Delta\psi)\right|^2, \tag{8.147}$$

$$H_{12} = B_{\psi_1\psi_2}(\psi_1,\psi_2) - \frac{B_{12}}{N\left[1-|B_{12}|^2\right]}\left|B_{\psi_2}(\Delta\psi)\right|^2, \tag{8.148}$$

and

$$H_{21} = H_{12}^*. \tag{8.149}$$

The final step is to evaluate the Hadamard product. For the special case of uncorrelated sources,

$$\left[\mathbf{S_f}\right]_{12} = \left[\mathbf{S_f}\right]_{21} = 0. \tag{8.150}$$

and

$$\mathbf{S_f} = \text{diag}\left[\sigma_{s1}^2, \sigma_{s2}^2\right] = \mathbf{\Lambda_f}, \tag{8.151}$$

and

$$\mathbf{\Lambda_f'} = \frac{\mathbf{\Lambda_f}}{\sigma_w^2}. \tag{8.152}$$

Then (8.106) can be written as

$$\mathbf{C}_{CR}(\boldsymbol{\psi}) = \frac{1}{2K}\left\{Re\left\{\left[\mathbf{\Lambda_f'}\left[\mathbf{I}+\mathbf{V}^H\mathbf{V}\mathbf{\Lambda_f'}\right]^{-1}\mathbf{V}^H\mathbf{V}\mathbf{\Lambda_f'}\right]\odot\mathbf{H}^T\right\}\right\}^{-1}. \tag{8.153}$$

Note that, even with uncorrelated signals, the matrix in the first bracket will not be diagonal unless the conditions following (8.109) are satisfied. If these conditions are satisfied, then we can use the approximation in (8.110). The matrix in the first bracket in (8.153) will be diagonal and the resulting Hadamard product will be diagonal. Thus,

$$C_{ACR}(\psi_i) = \frac{1}{2K}\frac{H_{ii}^{-1}}{SNR_i}, i = 1, 2, \tag{8.154}$$

where H_{ii} is given by (8.146) and (8.147).

In the next example, we specialize the results to the case of a standard linear array.

Example 8.4.4 (continuation)

For a standard linear array the expression in (8.154) reduces to

$$C_{ACR}(\psi_1) = \frac{6}{K\cdot ASNR(N^2-1)}\left\{1-\frac{12\,|B_{\psi_2}(\Delta\psi)|^2}{N^2(N^2-1)\left[1-|B_{12}|^2\right]}\right\}^{-1}. \tag{8.155}$$

The term in braces represents the increase in the bound due to the presence of the second signal. Note that the bound does not depend on the power of the second signal.

This is because the result is asymptotic and as the error becomes small, we find that the ML estimator places a null on the second signal (recall the discussion in Section 6.3). Later we shall see that the result is not true in the non-asymptotic area.

In Figure 8.4, we plot the square root of the normalized CRB versus *ASNR* for $N = 10$, $K = 100$, and various values of $\Delta\psi/BW_{NN}$. We show the bound given by (8.110) and the exact bound given by (8.106). We also show the single-signal bound given by (8.130).

We see that, as Δ decreases, $C_{CR}(\psi)$ separates from $C_{ACR}(\psi)$ at higher *ASNR*. For $\Delta > 0.5BW_{NN}$, the two-signal bound is essentially the same as the single-signal bound for $ASNR > 0$ dB.

In Figure 8.5, we plot the exact CRB (8.106) versus normalized signal separation for various *SNR*. We see that the bound is proportional to $(\Delta u)^{-2}$ over the region $\Delta u/BW_{NN} \leq 1.0$ and is essentially constant for $\Delta u/BW_{NN} > 1.0$. We indicate the HPBW, $0.5BW_{NN}$, and BW_{NN}. The separation, $\Delta u = 0.5BW_{NN}$, is referred to as the Rayleigh resolution value (recall the discussion in Chapter 2).

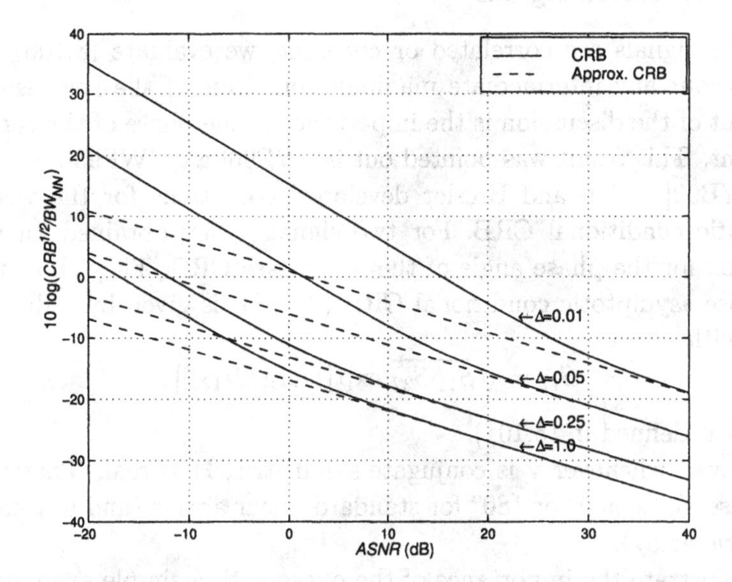

Figure 8.4 Square root of normalized CRB versus *ASNR* : $N = 10$, $K = 100$, two signals, various $\Delta\psi/BW_{NN}$.

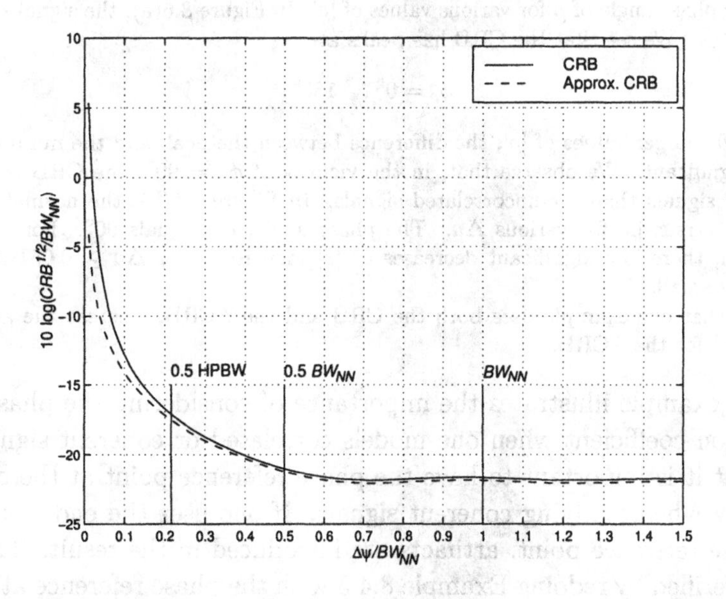

Figure 8.5 Square root of normalized CRB versus $\Delta\psi/BW_{NN}$ (normalized signal separation in ψ-space): two signals, $ASNR = 10$ dB, $N = 10$, $K = 100$.

8.4.1.4 Correlated signals

When the signals are correlated or coherent, we evaluate (8.106) numerically without any intermediate manipulation. One of the key results that comes out of the discussion is the importance of the phase of the correlation coefficient. This result was pointed out in [WF90] and [WF93].

In [YB92] , Yau and Bresler developed conditions for the worst case asymptotic conditional CRB. For two signals, they obtained an analytic expression for the phase angle of the worst case CRB. They show that the worst case asymptotic conditional CRB (ACRB) is given by fully coherent signals with

$$S_{12} = \sqrt{S_{11}\, S_{22}}\, \exp\left[j\left[\arg\left(H_{12}\right)\right]\right], \tag{8.156}$$

where $\mathbf{H}$ is defined in (8.101).

However, whenever $\mathbf{v}$ is conjugate symmetric, $\mathbf{H}$ is real. Therefore, the worst case phase is 0° or 180° for standard linear arrays (and any conjugate symmetric array).

We illustrate the importance of the phase with a simple example.

Example 8.4.5

Consider a standard 10-element linear array. The *ASNR* is 20 dB and $K = 100$. In Figure 8.6, we plot the normalized CRB given by (8.106) and the ACRB given by (8.110) versus the phase angle of ρ for various values of $|\rho|$. In Figure 8.6(a), the signal separation is $0.05BW_{NN}$. We see that the CRB has peaks at

$$\phi_\rho = 0^\circ, \quad 180^\circ, \tag{8.157}$$

and that for larger values of $|\rho|$, the difference between the peak and the minimum value can be significant. We observe that, in the vicinity of $\phi_\rho = 90^\circ$, the CRB is lower for correlated signals than for uncorrelated signals. In Figure 8.6(b), the normalized CRB is plotted versus $|\rho|$ for various Δu. The phase angle, ϕ_ρ, equals 90°. For very small separation, there is a significant decreases as $|\rho|$ increases. For $\Delta u \geq 0.05BW_{NN}$, the decrease is small.

Note that our example uses both the CRB and the ACRB, whereas the analysis in [YB92] was for the ACRB.

This example illustrates the importance of considering the phase of the correlation coefficient when one models correlated or coherent signals. We note that it is important to have the phase reference point at the center of the array when studying coherent signals. If one uses the end element for the phase reference point, artifacts are introduced in the result. This issue can be verified by redoing Example 8.4.5 with the phase reference at the end element.

In Figure 8.7, we show the normalized bounds versus N for a small separation $\Delta\psi = 0.05BW_{NN}$ and an *ASNR* = 20 dB. The phase of ρ is zero.

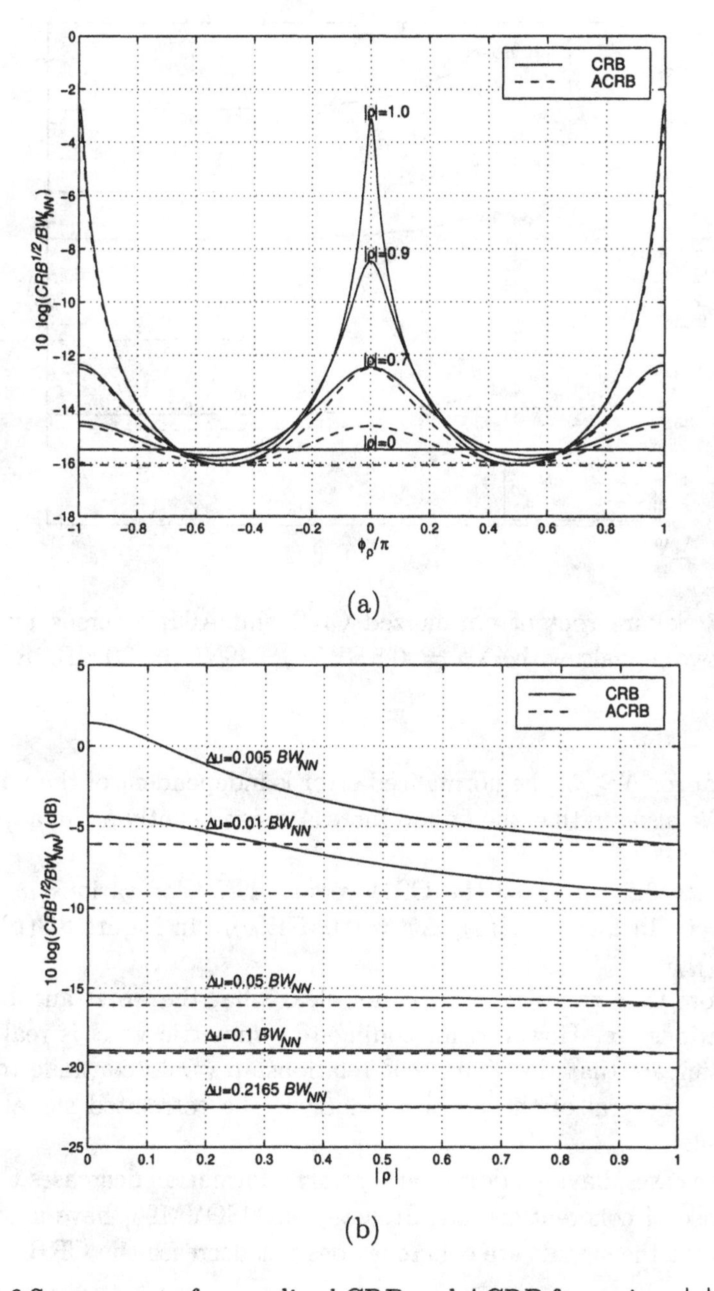

Figure 8.6 Square root of normalized CRB and ACRB for various $|\rho|$: two signals, $N = 10, K = 100$: (a) normalized CRB and ACRB versus $\phi_\rho/\pi, \Delta u = 0.05BW_{NN}$; (b) normalized CRB and ACRB versus $|\rho|, \phi_\rho = \pi/2$, various Δu.

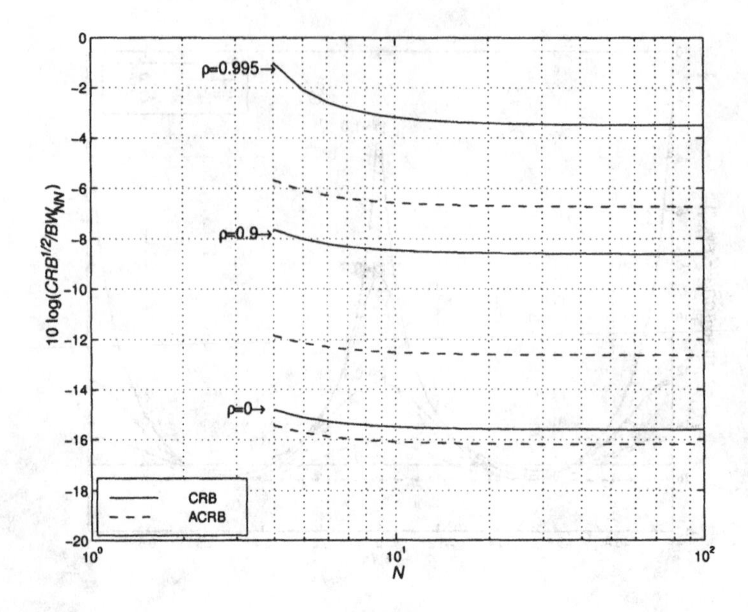

Figure 8.7 Square root of normalized CRB and ACRB versus number of sensors: two signals with $\Delta\psi = 0.5BW_{NN}$, $ASNR = 20$ dB, $K = 100$, various ρ.

We see that for $N \geq 5$, the normalized error is independent of the number of sensors. We also see that the bound increases significantly as $|\rho|$ approaches one.

In Figure 8.8, we show the CRB versus *ASNR* for various ρ. In this case, $\phi_\rho = 0$. In Figure 8.8(a), $\Delta\psi = 0.05BW_{NN}$. In Figure 8.8(b), $\Delta\psi = 0.2165BW_{NN}$.

For more than two signals, it is not clear what the worst and best case phase relations are. However, for conjugate symmetric $\mathbf{v}$, $\mathbf{H}$ is real and we would anticipate that the 0° or 180° relationship would continue to be the worst case. Several of the problems analyze the correlated signal case in more detail.

In some cases, having additional *a priori* information decreases the CRB. For the case of coherent signals, Stoica et al. [SOVM96] have shown that knowing that the signals are coherent does not decrease the CRB.

8.4.1.5 Closely spaced signals

As we have seen from our examples, as the signal separation decreases, the CRB increases. This corresponds to the intuitive result that the parame-

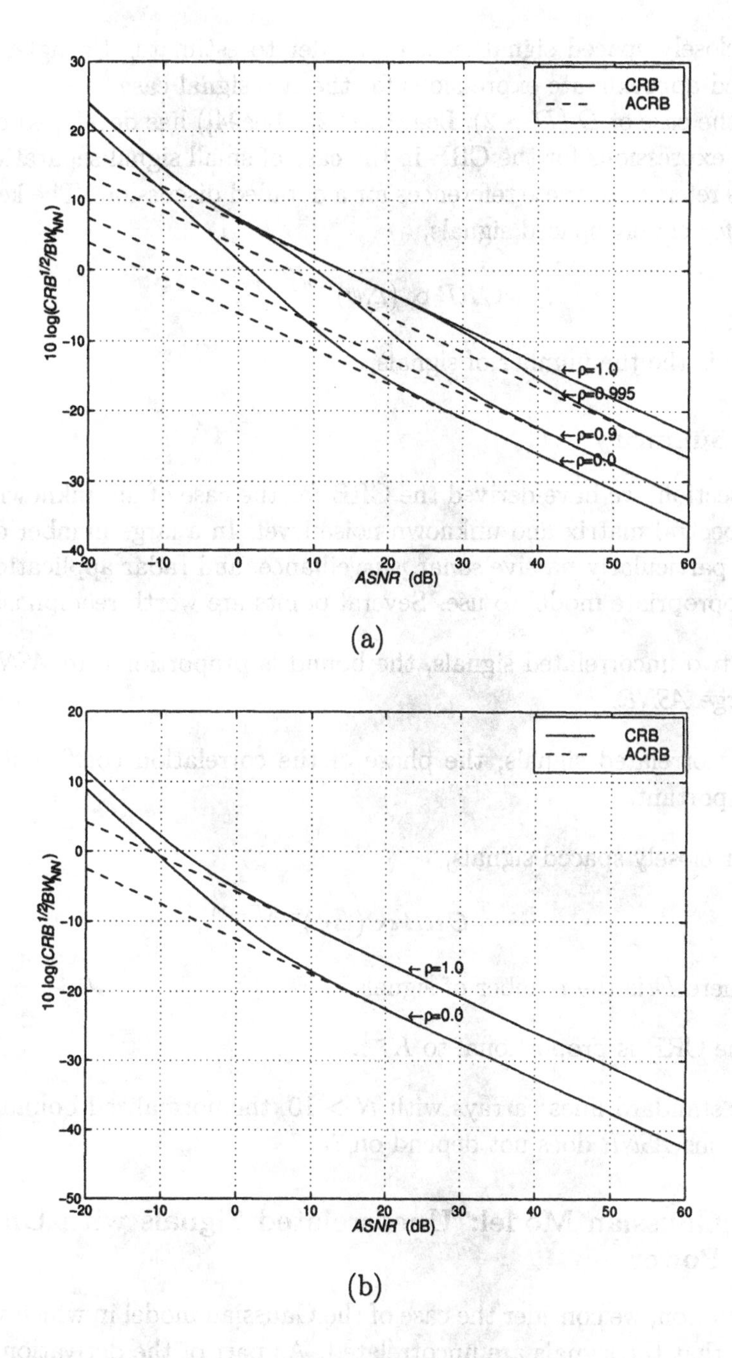

Figure 8.8 Square root of normalized CRB and ACRB versus *ASNR*: various values of ρ, phase of ρ is zero: two signals, $N = 10$, $K = 100$: (a) $\Delta\psi = 0.05BW_{NN}$; (b) $\Delta\psi = 0.2165BW_{NN}$.

ters of closely spaced signals will be harder to estimate. Swingler [Swi93] developed approximate expressions for the two-signal case.

For the case of D $(D > 2)$, Lee ([Lee92], [Lee94]) has developed compact analytic expressions for the CRB in the case of small signal separation. The reader is referred to these references for a detailed discussion. The key result is that, for closely spaced signals,

$$CRB \propto (\Delta\psi)^{-2(D-1)}, \tag{8.158}$$

where D is the the number of signals.

8.4.1.6 Summary

In this section, we have derived the CRB for the case of an unknown source signal spectral matrix and unknown noise level. In a large number of applications, particularly passive sonar, surveillance, and radar applications, this is the appropriate model to use. Several points are worth reemphasizing:

(i) For two uncorrelated signals, the bound is proportional to $ASNR^{-1}$ for large $ASNR$.

(ii) For correlated signals, the phase of the correlation coefficient is very important.

(iii) For closely spaced signals,

$$CRB \propto (\Delta\psi)^{-2(D-1)},$$

where D is the number of signals.

(iv) The CRB is proportional to K^{-1}.

(v) For standard linear arrays with $N > 10$, the normalized bound plotted versus $ASNR$ does not depend on N.

8.4.2 Gaussian Model: Uncorrelated Signals with Unknown Power

In this section, we consider the case of the Gaussian model in which we know *a priori* that the signals are uncorrelated. As part of the derivation we also find the CRB for the case of known source spectrum. The CRB for uncorrelated signals was used in many of the early studies of DOA estimation (e.g., Schmidt [Sch81] or Appendix A of Barabell et al. [BCD+84]). Porat and

Friedlander [PF88] use a high *SNR* approximation to it in their discussion of the relative efficiency of MUSIC. Jansson et al. [JGO99] have derived an expression for the uncorrelated CRB.

The spatial spectral matrix is

$$\mathbf{S_x} = \mathbf{V}\mathbf{S_f}\mathbf{V}^H + \sigma_w^2\mathbf{I}. \tag{8.159}$$

We can write $\mathbf{S_x}$ as

$$\mathbf{S_x} = \sum_{i=1}^{D} \sigma_i^2 \mathbf{v}(\psi_i)\mathbf{v}^H(\psi_i) + \sigma_w^2\mathbf{I}. \tag{8.160}$$

There are $2D+1$ unknown parameters,

$$\boldsymbol{\theta} = \begin{bmatrix} \boldsymbol{\psi}^T & \boldsymbol{\sigma}^{2T} & \sigma_w^2 \end{bmatrix}^T, \tag{8.161}$$

where

$$\boldsymbol{\sigma}^2 \triangleq \begin{bmatrix} \sigma_1^2 & \sigma_2^2 & \cdots & \sigma_D^2 \end{bmatrix}^T. \tag{8.162}$$

We divide $\boldsymbol{\theta}$ into two parts as in (8.39),

$$\boldsymbol{\theta} = \begin{bmatrix} \boldsymbol{\psi}^T & \boldsymbol{\theta}_u^T \end{bmatrix}^T, \tag{8.163}$$

where

$$\boldsymbol{\theta}_u = \begin{bmatrix} \boldsymbol{\sigma}^{2T} & \sigma_w^2 \end{bmatrix}^T \tag{8.164}$$

is a $D+1$ vector containing the unwanted parameters.

The Fisher information can be partitioned as

$$\mathbf{J} = \begin{bmatrix} \mathbf{J}_{\boldsymbol{\psi}\boldsymbol{\psi}} & \mathbf{J}_{\boldsymbol{\psi}\boldsymbol{\theta}_u} \\ \mathbf{J}_{\boldsymbol{\theta}_u\boldsymbol{\psi}} & \mathbf{J}_{\boldsymbol{\theta}_u\boldsymbol{\theta}_u} \end{bmatrix}. \tag{8.165}$$

Then,

$$\mathbf{J}^{-1} = \left[\begin{array}{c:c} (\mathbf{J}_{\boldsymbol{\psi}\boldsymbol{\psi}} - \mathbf{J}_{\boldsymbol{\psi}\boldsymbol{\theta}_u}\mathbf{J}_{\boldsymbol{\theta}_u\boldsymbol{\theta}_u}^{-1}\mathbf{J}_{\boldsymbol{\theta}_u\boldsymbol{\psi}})^{-1} & (\ ) \\ \hdashline (\ ) & (\ ) \end{array}\right], \tag{8.166}$$

where only the upper left block matrix is of interest, and

$$\mathbf{C}_{CR}(\boldsymbol{\psi}) = \left[\mathbf{J}_{\boldsymbol{\psi}\boldsymbol{\psi}} - \mathbf{J}_{\boldsymbol{\psi}\boldsymbol{\theta}_u}\mathbf{J}_{\boldsymbol{\theta}_u\boldsymbol{\theta}_u}^{-1}\mathbf{J}_{\boldsymbol{\theta}_u\boldsymbol{\psi}}\right]^{-1}. \tag{8.167}$$

We next derive the entries in $\mathbf{J}$. In the derivation of $\mathbf{J}_{\boldsymbol{\psi}\boldsymbol{\psi}}$, we will not assume that $\mathbf{S_f}$ is diagonal, so the resulting $\mathbf{J}_{\boldsymbol{\psi}\boldsymbol{\psi}}$ will provide a bound for the general known signal spectrum case.

The ij term in $\mathbf{J}_{\psi\psi}$ is

$$\left[\mathbf{J}_{\psi\psi}\right]_{ij} = K \text{ tr}\left[\mathbf{S}_{\mathbf{x}}^{-1}\frac{\partial \mathbf{S}_{\mathbf{x}}}{\partial \psi_i}\mathbf{S}_{\mathbf{x}}^{-1}\frac{\partial \mathbf{S}_{\mathbf{x}}}{\partial \psi_j}\right]. \tag{8.168}$$

The derivative with respect to ψ_i is

$$\frac{\partial \mathbf{S}_{\mathbf{x}}}{\partial \psi_i} = \frac{\partial \mathbf{V}(\psi)}{\partial \psi_i}\mathbf{S}_{\mathbf{f}}\mathbf{V}^H(\psi) + \mathbf{V}\mathbf{S}_{\mathbf{f}}\frac{\partial \mathbf{V}^H(\psi)}{\partial \psi_i}. \tag{8.169}$$

Now define

$$\mathbf{D} = \sum_{i=1}^{D}\frac{\partial \mathbf{V}(\psi)}{\partial \psi_i} = \left[\begin{array}{ccc} \frac{\partial \mathbf{v}(\psi_1)}{\partial \psi_1} & \cdots & \frac{\partial \mathbf{v}(\psi_D)}{\partial \psi_D}\end{array}\right]. \tag{8.170}$$

Then, we can write

$$\frac{\partial \mathbf{V}(\psi)}{\partial \psi_i} = \mathbf{D}\mathbf{e}_i\mathbf{e}_i^T, \tag{8.171}$$

where $\mathbf{e}_i$ is a $D \times 1$ vector whose ith element is unity and whose remaining elements are zero (see A.104):

$$\begin{aligned}\left[\mathbf{J}_{\psi\psi}\right]_{ij} = K \text{ tr}\Big[\mathbf{S}_{\mathbf{x}}^{-1}\left[\mathbf{D}\mathbf{e}_i\mathbf{e}_i^T\mathbf{S}_{\mathbf{f}}\mathbf{V}^H + \mathbf{V}\mathbf{S}_{\mathbf{f}}\mathbf{e}_i\mathbf{e}_i^T\mathbf{D}^H\right] \\ \times\, \mathbf{S}_{\mathbf{x}}^{-1}\left[\mathbf{D}\mathbf{e}_j\mathbf{e}_j^T\mathbf{S}_{\mathbf{f}}\mathbf{V}^H + \mathbf{V}\mathbf{S}_{\mathbf{f}}\mathbf{e}_j\mathbf{e}_j^T\mathbf{D}^H\right]\Big],\end{aligned} \tag{8.172}$$

or

$$\begin{aligned}\left[\mathbf{J}_{\psi\psi}\right]_{ij} = K \text{ tr}\Big[&\mathbf{S}_{\mathbf{x}}^{-1}\mathbf{D}\mathbf{e}_i\mathbf{e}_i^T\mathbf{S}_{\mathbf{f}}\mathbf{V}^H\mathbf{S}_{\mathbf{x}}^{-1}\mathbf{D}\mathbf{e}_j\mathbf{e}_j^T\mathbf{S}_{\mathbf{f}}\mathbf{V}^H \\ &+\mathbf{S}_{\mathbf{x}}^{-1}\mathbf{V}\mathbf{S}_{\mathbf{f}}\mathbf{e}_i\mathbf{e}_i^T\mathbf{D}^H\mathbf{S}_{\mathbf{x}}^{-1}\mathbf{D}\mathbf{e}_j\mathbf{e}_j^T\mathbf{S}_{\mathbf{f}}\mathbf{V}^H \\ &+\mathbf{S}_{\mathbf{x}}^{-1}\mathbf{D}\mathbf{e}_i\mathbf{e}_i^T\mathbf{S}_{\mathbf{f}}\mathbf{V}^H\mathbf{S}_{\mathbf{x}}^{-1}\mathbf{V}^H\mathbf{S}_{\mathbf{f}}\mathbf{e}_j\mathbf{e}_j^T\mathbf{D}^H \\ &+\mathbf{S}_{\mathbf{x}}^{-1}\mathbf{V}\mathbf{S}_{\mathbf{f}}\mathbf{e}_i\mathbf{e}_i^T\mathbf{D}^H\mathbf{S}_{\mathbf{x}}^{-1}\mathbf{V}^H\mathbf{S}_{\mathbf{f}}\mathbf{e}_j\mathbf{e}_j^T\mathbf{D}^H\Big].\end{aligned} \tag{8.173}$$

Using the trace property in (A.28),

$$\begin{aligned}\left[\mathbf{J}_{\psi\psi}\right]_{ij} = K \text{ tr}\Big[&\left(\mathbf{e}_j^T\mathbf{S}_{\mathbf{f}}\mathbf{V}^H\mathbf{S}_{\mathbf{x}}^{-1}\mathbf{D}\mathbf{e}_i\right)\left(\mathbf{e}_i^T\mathbf{S}_{\mathbf{f}}\mathbf{V}^H\mathbf{S}_{\mathbf{x}}^{-1}\mathbf{D}\mathbf{e}_j\right) \\ &+\left(\mathbf{e}_j^T\mathbf{S}_{\mathbf{f}}\mathbf{V}^H\mathbf{S}_{\mathbf{x}}^{-1}\mathbf{V}\mathbf{S}_{\mathbf{f}}\mathbf{e}_i\right)\left(\mathbf{e}_i^T\mathbf{D}^H\mathbf{S}_{\mathbf{x}}^{-1}\mathbf{D}\mathbf{e}_j\right) \\ &+\left(\mathbf{e}_j^T\mathbf{D}^H\mathbf{S}_{\mathbf{x}}^{-1}\mathbf{D}\mathbf{e}_i\right)\left(\mathbf{e}_i^T\mathbf{S}_{\mathbf{f}}\mathbf{V}^H\mathbf{S}_{\mathbf{x}}^{-1}\mathbf{V}\mathbf{S}_{\mathbf{f}}\mathbf{e}_j\right) \\ &+\left(\mathbf{e}_j^T\mathbf{D}^H\mathbf{S}_{\mathbf{x}}^{-1}\mathbf{V}\mathbf{S}_{\mathbf{f}}\mathbf{e}_i\right)\left(\mathbf{e}_i^T\mathbf{D}^H\mathbf{S}_{\mathbf{x}}^{-1}\mathbf{V}\mathbf{S}_{\mathbf{f}}\mathbf{e}_j\right)\Big].\end{aligned} \tag{8.174}$$

Now all the terms are scalars and the trace can be removed. The fourth term is the conjugate of the first term and the third term is the conjugate of the second term. Therefore, (8.174) can be written as

$$\begin{aligned}\left[\mathbf{J}_{\psi\psi}\right]_{ij} = 2K Re\Big\{&\left(\mathbf{e}_i^T\mathbf{S}_\mathbf{f}\mathbf{V}^H\mathbf{S}_\mathbf{x}^{-1}\mathbf{D}\mathbf{e}_j\right)\left(\mathbf{e}_j^T\mathbf{S}_\mathbf{f}\mathbf{V}^H\mathbf{S}_\mathbf{x}^{-1}\mathbf{D}\mathbf{e}_i\right)\\ &+\left(\mathbf{e}_i^T\mathbf{S}_\mathbf{f}\mathbf{V}^H\mathbf{S}_\mathbf{x}^{-1}\mathbf{V}\mathbf{S}_\mathbf{f}\mathbf{e}_j\right)\left(\mathbf{e}_j^T\mathbf{D}^H\mathbf{S}_\mathbf{x}^{-1}\mathbf{D}\mathbf{e}_i\right)\Big\},\end{aligned} \tag{8.175}$$

or, in matrix form,

$$\boxed{\begin{aligned}\mathbf{J}_{\psi\psi} = 2K Re\ \Big\{&\left[\mathbf{S}_\mathbf{f}\mathbf{V}^H\mathbf{S}_\mathbf{x}^{-1}\mathbf{D}\right]\odot\left[\mathbf{S}_\mathbf{f}\mathbf{V}^H\mathbf{S}_\mathbf{x}^{-1}\mathbf{D}\right]^T\\ &+\left[\mathbf{S}_\mathbf{f}\mathbf{V}^H\mathbf{S}_\mathbf{x}^{-1}\mathbf{V}\mathbf{S}_\mathbf{f}\right]\odot\left[\mathbf{D}^H\mathbf{S}_\mathbf{x}^{-1}\mathbf{D}\right]^T\Big\},\end{aligned}} \tag{8.176}$$

which is the desired result. The expression in (8.176) can also be written as

$$\begin{aligned}\mathbf{J}_{\psi\psi} = 2K Re\Big\{&\left[\mathbf{S}_\mathbf{f}\mathbf{V}^H\mathbf{S}_\mathbf{x}^{-1}\mathbf{D}\right]^T\odot\left[\mathbf{S}_\mathbf{f}\mathbf{V}^H\mathbf{S}_\mathbf{x}^{-1}\mathbf{D}\right]\\ &+\left[\mathbf{S}_\mathbf{f}\mathbf{V}^H\mathbf{S}_\mathbf{x}^{-1}\mathbf{V}\mathbf{S}_\mathbf{f}\right]^T\odot\left[\mathbf{D}^H\mathbf{S}_\mathbf{x}^{-1}\mathbf{D}\right]\Big\}.\end{aligned} \tag{8.177}$$

The expression in (8.176) is for general $\mathbf{S}_\mathbf{f}$. In order to derive the other terms in the $\mathbf{J}$ matrix, we assume that (8.160) is satisfied.

We partition the $\boldsymbol{\theta}_u$ vector as in (8.164), so the cross-matrices can be written as

$$\mathbf{J}_{\psi\boldsymbol{\theta}_u} = \left[\begin{array}{cc}\mathbf{J}_{\psi\mathbf{s}} & \mathbf{J}_{\psi n}\end{array}\right], \tag{8.178}$$

and

$$\mathbf{J}_{\boldsymbol{\theta}_u\boldsymbol{\theta}_u} = \left[\begin{array}{c:c}\mathbf{J}_{\mathbf{ss}} & \mathbf{J}_{\mathbf{s}n}\\ \hdashline \mathbf{J}_{n\mathbf{s}} & \mathbf{J}_{nn}\end{array}\right], \tag{8.179}$$

The first cross-matrix is

$$\begin{aligned}J_{\psi_i\sigma_j^2} &= K\ \mathrm{tr}\left\{\mathbf{S}_\mathbf{x}^{-1}\sigma_i^2(\mathbf{d}_i\mathbf{v}_i^H+\mathbf{v}_i\mathbf{d}_i^H)\mathbf{S}_\mathbf{x}^{-1}\mathbf{v}_j\mathbf{v}_j^H)\right\}\\ &= K\ \mathrm{tr}\left\{\mathbf{S}_\mathbf{x}^{-1}\sigma_i^2\mathbf{d}_i\mathbf{v}_i^H\mathbf{S}_\mathbf{x}^{-1}\mathbf{v}_j\mathbf{v}_j^H+\mathbf{S}_\mathbf{x}^{-1}\sigma_i^2\mathbf{v}_i\mathbf{d}_i^H\mathbf{S}_\mathbf{x}^{-1}\mathbf{v}_j\mathbf{v}_j^H\right\}\\ &= K\left\{(\mathbf{v}_j^H\mathbf{S}_\mathbf{x}^{-1}\mathbf{d}_i)(\sigma_i^2\mathbf{v}_i^H\mathbf{S}_\mathbf{x}^{-1}\mathbf{v}_j)+(\mathbf{v}_j^H\mathbf{S}_\mathbf{x}^{-1}\mathbf{v}_i\sigma_i^2)(\mathbf{d}_j^H\mathbf{S}_\mathbf{x}^{-1}\mathbf{v}_j)\right\},\end{aligned} \tag{8.180}$$

or

$$J_{\psi_i \sigma_j^2} = 2K Re\left\{(\sigma_i^2 \mathbf{v}_i^H \mathbf{S}_\mathbf{x}^{-1} \mathbf{v}_j)(\mathbf{v}_j^H \mathbf{S}_\mathbf{x}^{-1} \mathbf{d}_i)\right\}, \tag{8.181}$$

or

$$J_{\psi_i \sigma_j^2} = 2K Re\left\{[\mathbf{S_f} \mathbf{V}^H \mathbf{S}_\mathbf{x}^{-1} \mathbf{V}]_{ij} [\mathbf{V}^H \mathbf{S}_\mathbf{x}^{-1} \mathbf{D}]_{ji}\right\}. \tag{8.182}$$

In matrix form,

$$\boxed{\mathbf{J}_{\boldsymbol{\psi}\mathbf{s}} = 2K Re\left\{[\mathbf{S_f} \mathbf{V}^H \mathbf{S}_\mathbf{x}^{-1} \mathbf{V}] \odot [\mathbf{V}^H \mathbf{S}_\mathbf{x}^{-1} \mathbf{D}]^T]\right\}.} \tag{8.183}$$

The next cross-matrix is

$$\begin{aligned} J_{\psi_i \sigma_w^2} &= K \operatorname{tr}\left\{\mathbf{S}_\mathbf{x}^{-1} \sigma_i^2 (\mathbf{d}_i \mathbf{v}_i^H + \mathbf{v}_i \mathbf{d}_i^H) \mathbf{S}_\mathbf{x}^{-1}\right\} \\ &= K \operatorname{tr}\left\{\mathbf{S}_\mathbf{x}^{-1} \sigma_i^2 \mathbf{d}_i \mathbf{v}_i^H \mathbf{S}_\mathbf{x}^{-1} + \mathbf{S}_\mathbf{x}^{-1} \sigma_i^2 \mathbf{v}_i \mathbf{d}_i^H \mathbf{S}_\mathbf{x}^{-1}\right\} \\ &= K \left\{\sigma_i^2 \mathbf{v}_i^H \mathbf{S}_\mathbf{x}^{-2} \mathbf{d}_i + \mathbf{d}_i^H \mathbf{S}_\mathbf{x}^{-2} \mathbf{v}_i \sigma_i^2\right\}, \end{aligned} \tag{8.184}$$

which reduces to

$$\begin{aligned} J_{\psi_i \sigma_w^2} &= 2K Re\left\{\sigma_i^2 \mathbf{v}_i^H \mathbf{S}_\mathbf{x}^{-2} \mathbf{d}_i\right\} \\ &= 2K Re\left\{[\mathbf{S_f} \mathbf{V}^H \mathbf{S}_\mathbf{x}^{-2} \mathbf{D}]_{ii}\right\}. \end{aligned} \tag{8.185}$$

In matrix form

$$\boxed{\mathbf{J}_{\boldsymbol{\psi} n} = 2K \mathrm{DIAG}\left[Re(\mathbf{S_f} \mathbf{V}^H \mathbf{S}_\mathbf{x}^{-2} \mathbf{D})\right].} \tag{8.186}$$

The signal power matrix is

$$J_{\sigma_i^2 \sigma_j^2} = K \operatorname{tr}\left\{\mathbf{S}_\mathbf{x}^{-1} \mathbf{v}_i \mathbf{v}_i^H \mathbf{S}_\mathbf{x}^{-1} \mathbf{v}_j \mathbf{v}_j^H\right\}. \tag{8.187}$$

or

$$\begin{aligned} J_{\sigma_i^2 \sigma_j^2} &= K \left\{(\mathbf{v}_i^H \mathbf{S}_\mathbf{x}^{-1} \mathbf{v}_j)(\mathbf{v}_j^H \mathbf{S}_\mathbf{x}^{-1} \mathbf{v}_i)\right\} \\ &= K \left\{[\mathbf{V}^H \mathbf{S}_\mathbf{x}^{-1} \mathbf{V}]_{ij} [\mathbf{V}^H \mathbf{S}_\mathbf{x}^{-1} \mathbf{V}]_{ji}\right\}. \end{aligned} \tag{8.188}$$

In matrix form

$$\boxed{\mathbf{J}_{\mathbf{ss}} = K \left\{\left[\mathbf{V}^H \mathbf{S}_\mathbf{x}^{-1} \mathbf{V}\right] \odot \left[\mathbf{V}^H \mathbf{S}_\mathbf{x}^{-1} \mathbf{V}\right]^T\right\}.} \tag{8.189}$$

The noise-related matrices are

$$J_{\sigma_i^2 \sigma_w^2} = K \operatorname{tr}\left\{\mathbf{S}_\mathbf{x}^{-1} \mathbf{v}_i \mathbf{v}_i^H \mathbf{S}_\mathbf{x}^{-1}\right\}, \tag{8.190}$$

or

$$\begin{aligned} J_{\sigma_i^2 \sigma_w^2} &= K \mathbf{v}_i^H \mathbf{S}_{\mathbf{x}}^{-2} \mathbf{v}_i \\ &= K \left[\mathbf{V}^H \mathbf{S}_{\mathbf{x}}^{-2} \mathbf{V} \right]_{ii}. \end{aligned} \tag{8.191}$$

In matrix form

$$\boxed{\mathbf{J}_{\mathbf{s}n} = K \mathrm{DIAG} \left[\mathbf{V}^H \mathbf{S}_{\mathbf{x}}^{-2} \mathbf{V} \right].} \tag{8.192}$$

Similarly,

$$J_{\sigma_w^2 \sigma_w^2} = K \ \mathrm{tr} \left\{ \mathbf{S}_{\mathbf{x}}^{-1} \mathbf{S}_{\mathbf{x}}^{-1} \right\} = K \ \mathrm{tr} \left\{ \mathbf{S}_{\mathbf{x}}^{-2} \right\}, \tag{8.193}$$

and

$$\boxed{\mathbf{J}_{nn} = K \mathrm{tr} \left\{ \mathbf{S}_{\mathbf{x}}^{-2} \right\}.} \tag{8.194}$$

In addition to the six matrices in boxes ((8.176), (8.183), (8.186), (8.189), (8.192), and (8.194)) there are three Hermitian transposes,

$$\mathbf{J}_{\mathbf{s}\psi} = \mathbf{J}_{\psi \mathbf{s}}^H, \tag{8.195}$$

$$\mathbf{J}_{n\psi} = \mathbf{J}_{\psi n}^H, \tag{8.196}$$

and

$$\mathbf{J}_{n\mathbf{s}} = \mathbf{J}_{\mathbf{s}n}^H. \tag{8.197}$$

We substitute these eight equations into (8.178) and (8.179) and the result along with (8.176) into (8.167). All of the terms in (8.167) are explicitly defined and we can evaluate the uncorrelated CRB. We consider a simple example to illustrate the behavior.

Example 8.4.6

Consider a standard 10-element linear array. We plot the uncorrelated CRB given by (8.167) and the standard CRB given by (8.106) versus $ASNR_1$ for several signal separations; $0.01 BW_{NN}$, $0.05 BW_{NN}$, $0.25 BW_{NN}$, and $0.75 BW_{NN}$. We consider three signal models:

(i) Figure 8.9: $ASNR_2 = ASNR_1$

(ii) Figure 8.10: $ASNR_2 = ASNR_1 + 10$ dB

(iii) Figure 8.11: $ASNR_2 = ASNR_1 + 20$ dB

In each figure, the (a) plot corresponds to signal 1 and the (b) plot corresponds to signal 2. The value of K is 100.

We see that:

(i) For unequal power signals, the bound for the lower power signal is almost the same whether we use the standard or uncorrelated CRB.

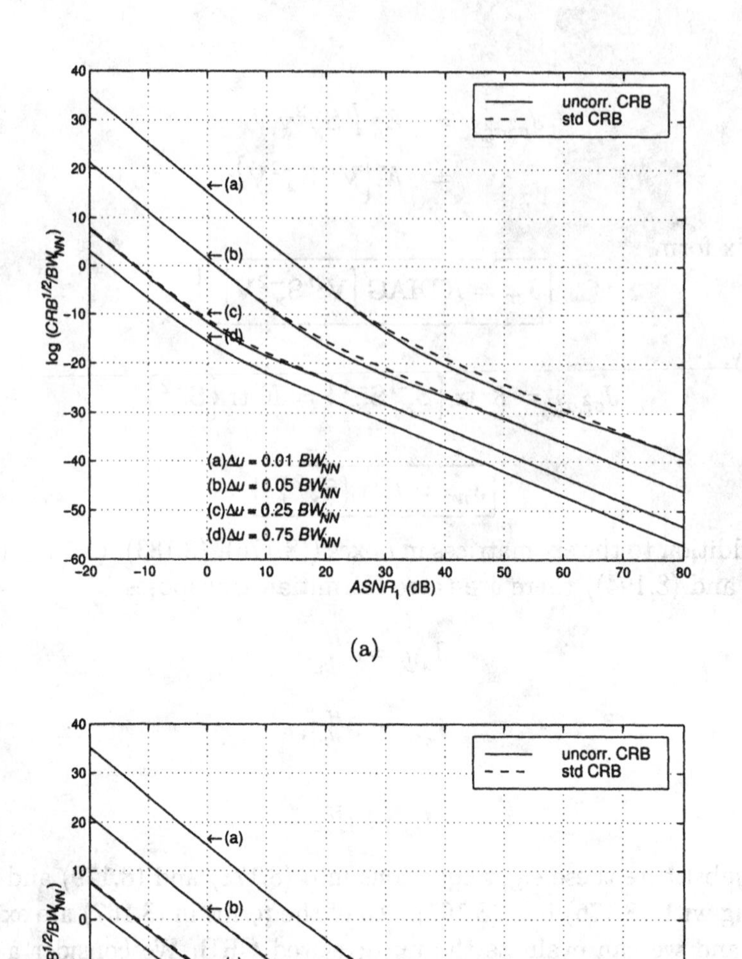

(a)

(b)

Figure 8.9 Normalized CRBs versus *ASNR*: two signals, $N = 10, K = 100$; $ASNR_2 = ASNR_1$: (a) CRBs for signal 1 versus $ASNR_1$; (b) CRBs for signal 2 versus $ASNR_2$.

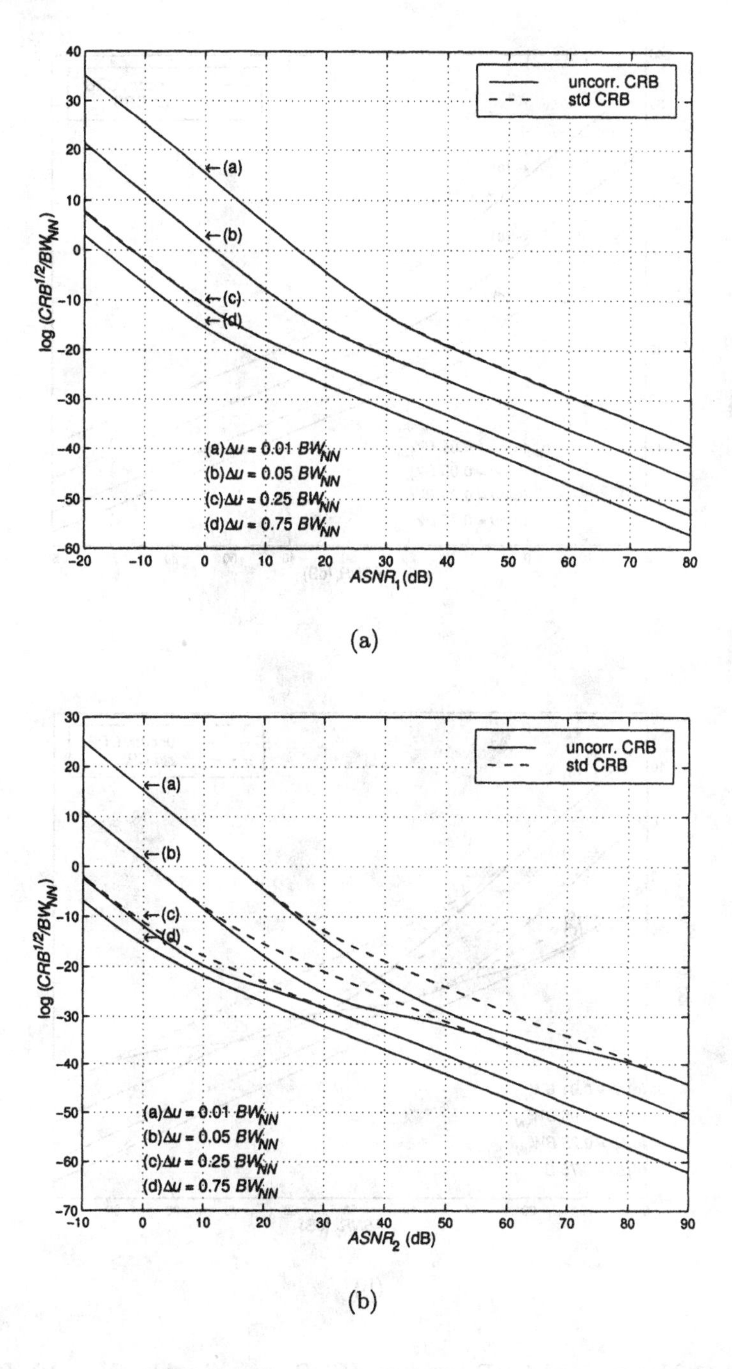

Figure 8.10 Normalized CRBs versus *ASNR*: two signals, $N = 10, K = 100$; $ASNR_2 = ASNR_1 + 10$ dB: (a) CRBs for signal 1 versus $ASNR_1$; (b) CRBs for signal 2 versus $ASNR_2$.

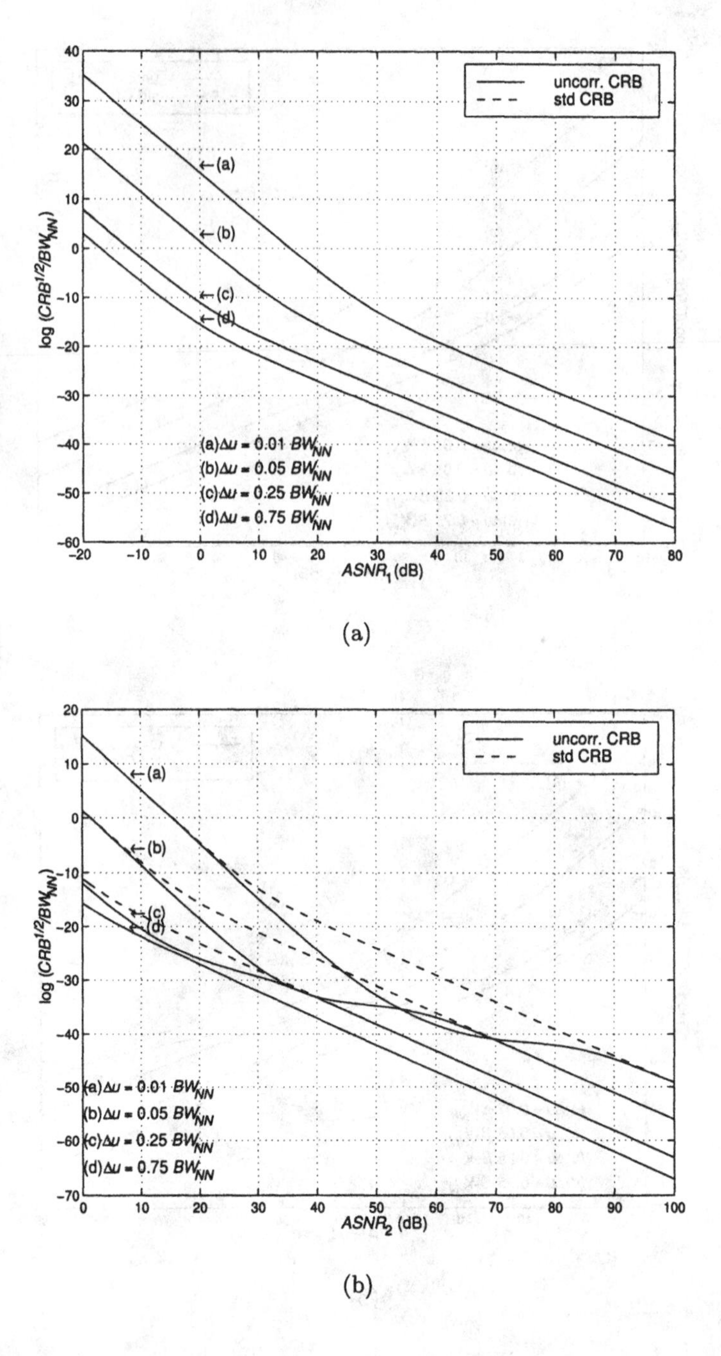

Figure 8.11 Normalized CRBs versus *ASNR*: two signals, $N = 10, K = 100$; $ASNR_2 = ASNR_1 + 20$ dB: (a) CRBs for signal 1 versus $ASNR_1$; (b) CRBs for signal 2 versus $ASNR_2$.

(ii) The bound for the higher power signal can be significantly lower using the uncorrelated signal model rather than the general correlation model, indicating a possible improvement in DOA estimation for schemes that use this knowledge. The difference is most pronounced:

(a) as Δu decreases

(b) as the difference in *ASNR* increases.

These results do not change significantly for $N \geq 6$. There are similar trends for $N = 3$, 4, and 5.

Jansson et al. [JGO99] give several comparisons of the uncorrelated CRB and the general CRB. They also derive an estimator that achieves the uncorrelated CRB asymptotically.

8.4.3 Gaussian Model: Known Signal Spectrum

In this section we consider the case of the Gaussian model with a known signal spectrum and known σ_w^2. The result was derived in Section 8.4.2 (8.176). Now the parameter vector $\boldsymbol{\theta}$ is

$$\boldsymbol{\theta} \stackrel{\Delta}{=} \boldsymbol{\psi}. \tag{8.198}$$

The Fisher information matrix is the term in (8.176). Thus the CRB is[6]

$$\boxed{\begin{aligned} C_{CR}(\boldsymbol{\psi}) &= \mathbf{J}_{\boldsymbol{\psi}\boldsymbol{\psi}}^{-1} = \frac{1}{2K}\left\{Re\left\{\left(\mathbf{S_f}\mathbf{V}^H\mathbf{S_x}^{-1}\mathbf{V}\mathbf{S_f}\right) \odot \left(\mathbf{D}^H\mathbf{S_x}^{-1}\mathbf{D}\right)^T \right.\right. \\ &\qquad \left.\left. + \left(\mathbf{S_f}\mathbf{V}^H\mathbf{S_x}^{-1}\mathbf{D}\right) \odot \left(\mathbf{S_f}\mathbf{V}^H\mathbf{S_x}^{-1}\mathbf{D}\right)^T\right\}\right\}^{-1}, \end{aligned}} \tag{8.199}$$

where $\mathbf{D}$ was defined in (8.170) and $\odot$ denotes the Hadamard product. The second term in the upper left partition in (8.166) is always a non-negative definite matrix. Therefore, if $\mathbf{S_x}$ is diagonal, then the known signal spectrum bound will be less than or equal to the uncorrelated signal bound in (8.167).

Note that the bound in (8.199) applies to correlated signals also. The expression in (8.199) is also the upper left matrix in (8.89), so a similar statement applies to correlated and coherent signals. A good discussion of the coherent signal case is given Stoica et al. [SOVM96].

---

[6]The result in (8.199) was published by Weiss and Friedlander [WF91a] and derived in [WF90]. It is also derived in Appendix A of [BCD+84]. [Wax85] derived the CRB for essentially this problem, but the form of the result was less compact.

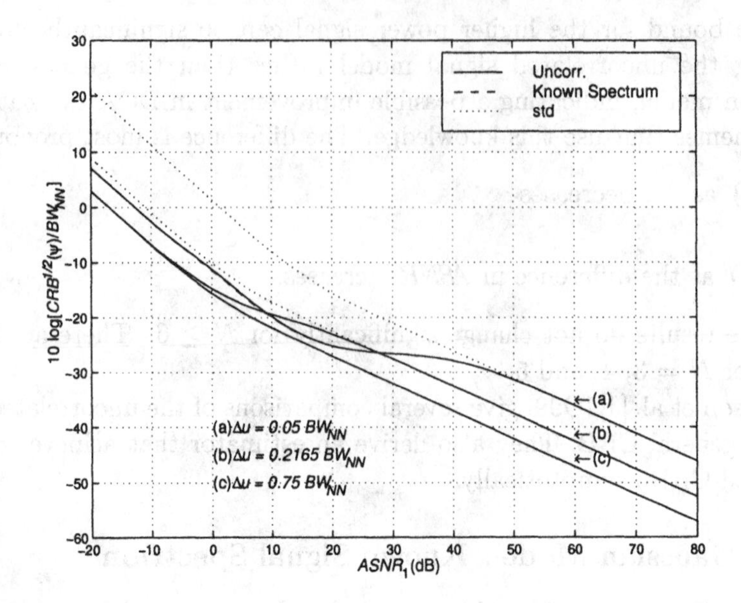

Figure 8.12 Normalized CRBs versus $ASNR_1$: $ASNR_2 = ASNR_1$.

We consider a simple example to show how knowledge of the spectrum affects the CRB.

Example 8.4.7 (continuation)

We consider a standard 10-element linear array. We first consider the case of two uncorrelated signals. We consider the same parameter set as in Example 8.4.6. The results are shown in Figures 8.12, 8.13, and 8.14. In each figure, we plot the normalized CRB using (8.106), (8.167), and (8.199) versus *ASNR*.

We see that the uncorrelated signal bound and the known spectrum bound appear to be the same. There is actually a very small difference. For $\Delta u = 0.2165 BW_{NN}$ and $ASNR_1 = 0$ dB, the difference is 0.02 dB.

Example 8.4.8(continuation)

Consider the same model as in Example 8.4.7 except the two signals are coherent, with $\rho = 1$. Note that the phases of ρ is $0°$, so forward-backward averaging cannot decorrelate the signals. In Figures 8.15 and 8.16, we plot the normalized CRB for the unknown signal spectral matrix case from (8.106) and for the known spectral matrix case from (8.199).

For a closely spaced signals there is a significant difference in the two bounds. For wider spacings, they coincide.

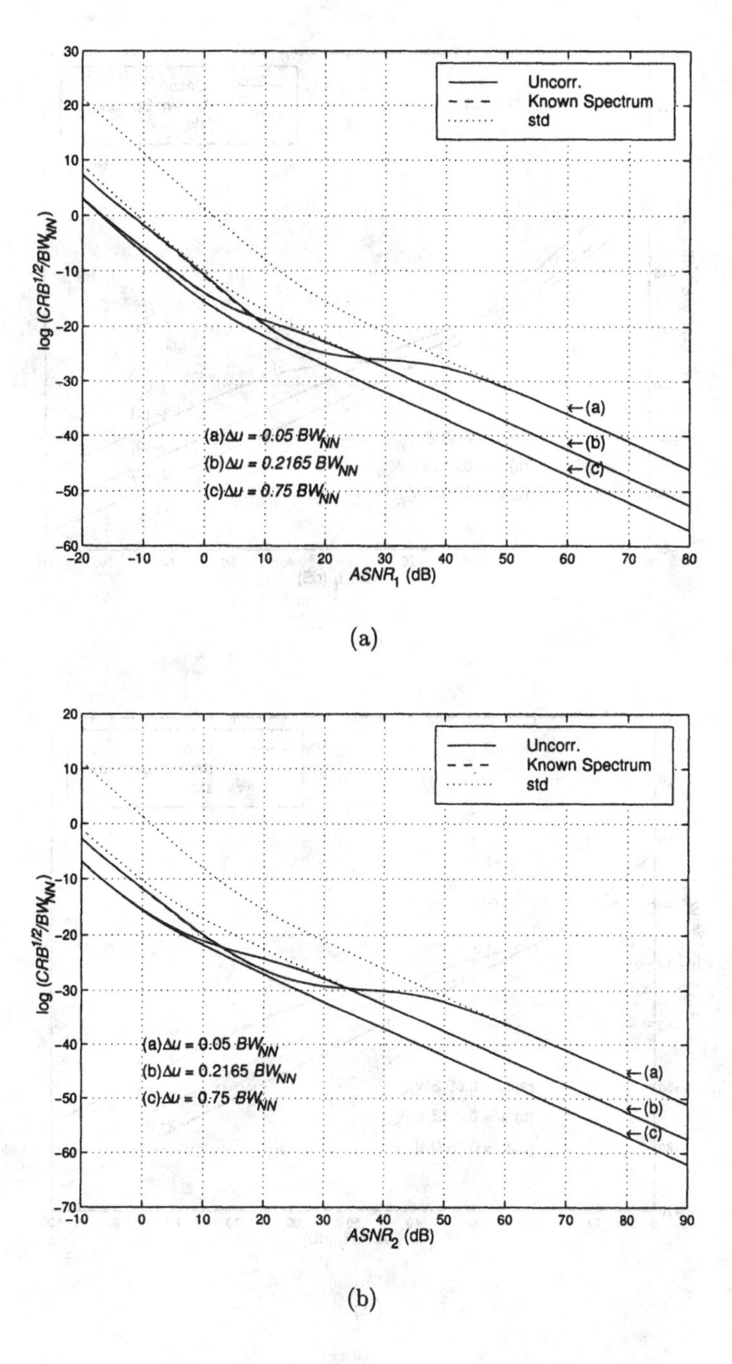

Figure 8.13 Normalized CRBs versus *ASNR*: $ASNR_2 = ASNR_1 + 10$ dB: (a) CRBs for signal 1 versus $ASNR_1$; (b) CRBs for signal 2 versus $ASNR_2$.

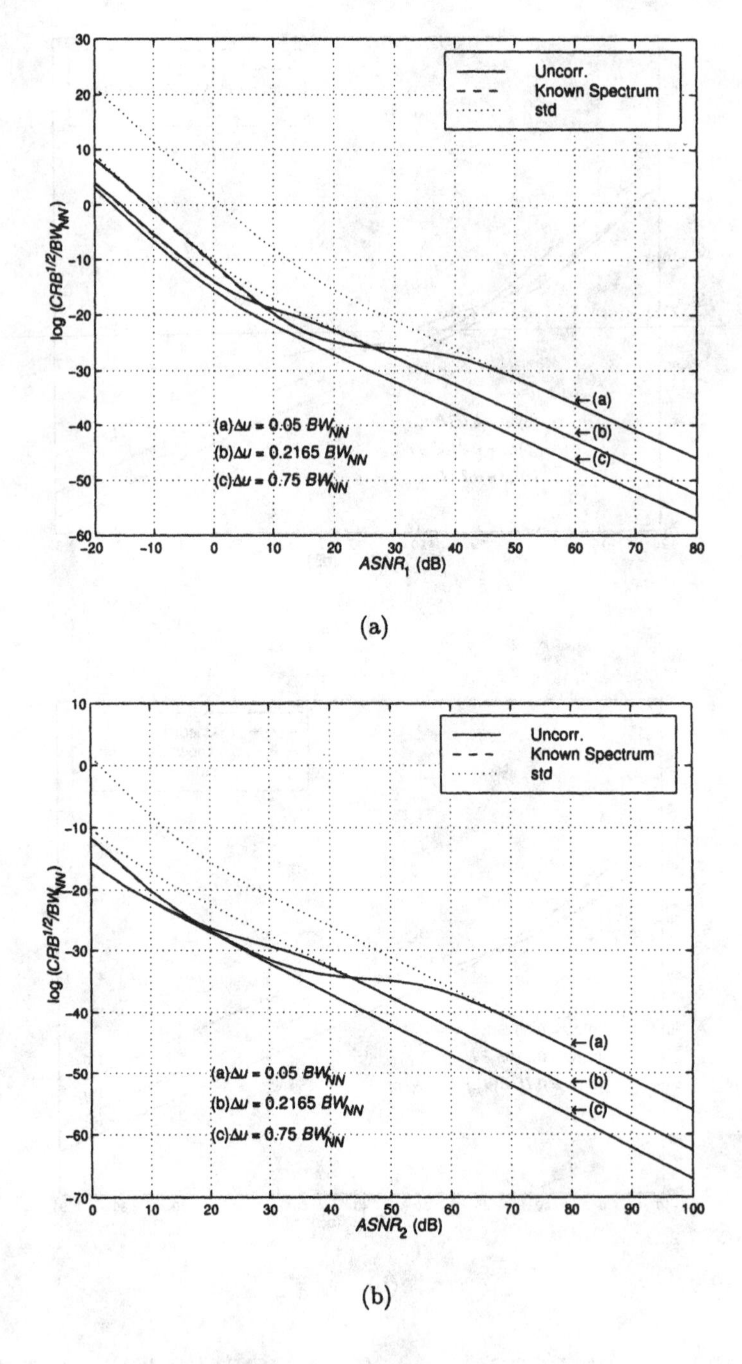

Figure 8.14 Normalized CRBs versus *ASNR*: $ASNR_2 = ASNR_1 + 20$ dB: (a) CRBs for signal 1 versus $ASNR_1$; (b) CRBs for signal 2 versus $ASNR_2$.

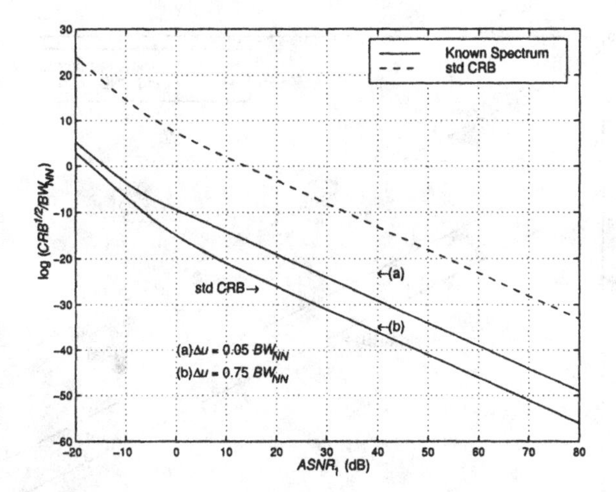

Figure 8.15 Normalized CRBs versus $ASNR_1$: two signals, $N = 10$, $K = 100$, $ASNR_2{=}ASNR_1$, $\rho = 1$.

8.4.4 Nonrandom (Conditional) Signal Model

The nonrandom signal model was introduced in Section 8.3.1. For a narrowband model, the snapshot model is (from (8.67))

$$\mathbf{X}(k) = \mathbf{V}(\boldsymbol{\psi})\,\mathbf{F}(k) + \mathbf{N}(k), \quad k = 1, 2, \cdots, K. \tag{8.200}$$

We assume that $\mathbf{F}(k)$ is an unknown nonrandom complex source-signal vector ($D \times 1$). Alternatively, we can view $\mathbf{F}(k)$ as a specific sample function of a complex random process, and we derive an estimator for this sample function and bound its performance. This second viewpoint leads to the description as a "conditional" (conditioned on $\mathbf{F}(k), k = 1, 2, \cdots, K$) estimator and a conditional Cramér-Rao bound. The two viewpoints lead to identical results.

The nonrandom (conditional) signal model was described in Section 8.3. To evaluate the CRB we utilize (8.35). The unknown parameter vector is

$$\boldsymbol{\theta} \triangleq \begin{bmatrix} \boldsymbol{\psi}, & \mathbf{F}, & \sigma_w^2 \end{bmatrix}. \tag{8.201}$$

The vector $\boldsymbol{\psi}$ and the scalar σ_w^2 have been defined previously (8.87). $\mathbf{F}$ is a real vector that contains the signal values at snapshot times $k = 1, 2, \cdots, K$. Thus,

$$\mathbf{F} \triangleq \begin{bmatrix} Re\,[\mathbf{F}(1)]^T & Im\,[\mathbf{F}(1)]^T & Re\,[\mathbf{F}(2)]^T & \cdots & Im\,[\mathbf{F}(K)]^T \end{bmatrix}^T. \tag{8.202}$$

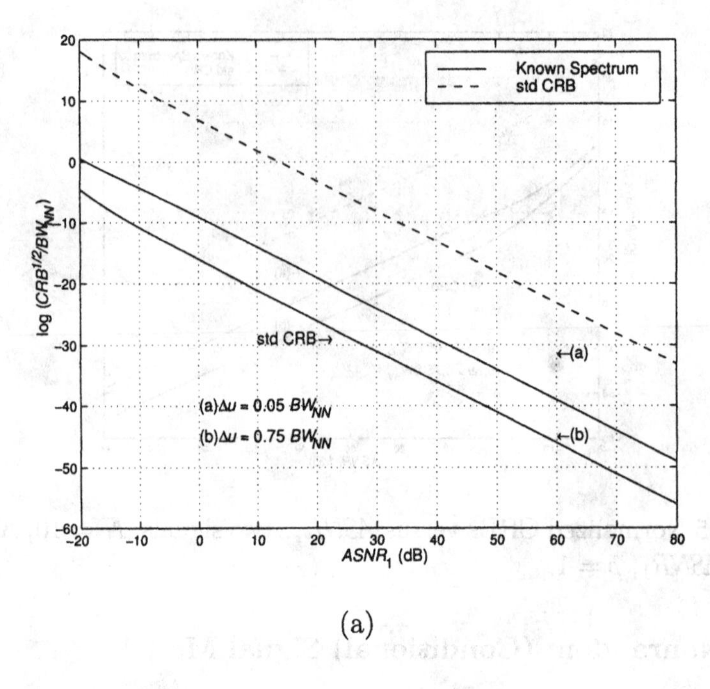

(a)

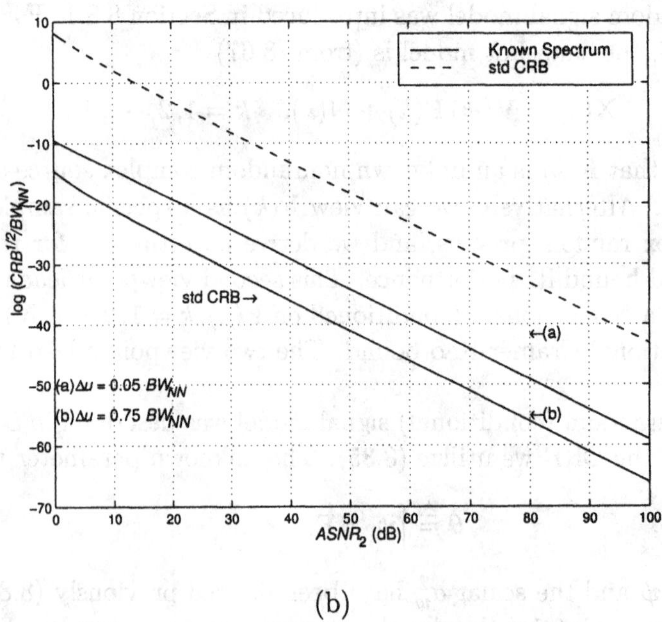

(b)

Figure 8.16 Normalized CRBs versus $ASNR_1$: $ASNR_2 = ASNR_1 + 20$ dB: (a) CRBs for signal 1 versus $ASNR_1$; (b) CRBs for signal 2 versus $ASNR_2$.

It is a $2DK$ dimension vector.

The Fisher information matrix has the same structure as in (8.89).

$$\mathbf{J} = \begin{bmatrix} \mathbf{J}_{\psi\psi} & \mathbf{J}_{\psi\mathbf{F}} & \mathbf{J}_{\psi\sigma_w^2} \\ \mathbf{J}_{\mathbf{F}\psi} & \mathbf{J}_{\mathbf{FF}} & \mathbf{J}_{\mathbf{F}\sigma_w^2} \\ \mathbf{J}_{\sigma_w^2\psi} & \mathbf{J}_{\sigma_w^2\psi} & \mathbf{J}_{\sigma_w^2\sigma_w^2} \end{bmatrix}. \tag{8.203}$$

The three principal submatrices are of dimension $D \times D$, $2KD \times 2KD$, and 1×1, respectively.

To evaluate the terms, we use (8.34) with

$$\mathbf{K_x}(\boldsymbol{\theta}) = \sigma_w^2 \mathbf{I}, \tag{8.204}$$

and

$$\mathbf{m}(\boldsymbol{\theta}, k) = \mathbf{V}(\boldsymbol{\psi})\mathbf{F}(k). \tag{8.205}$$

We go through the steps in the derivation of the bound because it is representative of the technique for deriving a multiple-parameter CRB. Our derivation uses (8.204) and (8.205) and borrows techniques from Appendices C and G of Stoica and Nehorai [SN89a].

Derivation of CCRB[7]

The elements in the upper left sub-matrix are

$$\begin{aligned} \left[\mathbf{J}_{\psi\psi}\right]_{ij} &= \frac{2}{\sigma_w^2} Re \sum_{k=1}^{K} \left[\mathbf{F}^H(k) \frac{\partial \mathbf{V}^H(\boldsymbol{\psi})}{\partial \psi_i} \frac{\partial \mathbf{V}(\boldsymbol{\psi})}{\partial \psi_j} \mathbf{F}(k) \right] \\ &= \frac{2}{\sigma_w^2} Re \sum_{k=1}^{K} \left[F_i^*(k) \frac{\partial \mathbf{v}^H(\psi_i)}{\partial \psi_i} \frac{\partial \mathbf{v}(\psi_j)}{\partial \psi_j} F_j(k) \right] \\ &= \frac{2}{\sigma_w^2} Re \sum_{k=1}^{K} \left[F_i^*(k)\, \mathbf{d}^H(\psi_i)\, \mathbf{d}(\psi_j)\, F_j(k) \right]. \end{aligned} \tag{8.206}$$

Defining

$$\mathbf{D} = \left[\begin{array}{cccc} \mathbf{d}(\psi_1) & \mathbf{d}(\psi_2) & \cdots & \mathbf{d}(\psi_D) \end{array}\right], \tag{8.207}$$

and

$$\mathbf{F}(k) = \text{diag}\left\{F_1(k), F_2(k), \cdots, F_D(k)\right\}, \tag{8.208}$$

we can write $\mathbf{J}_{\psi\psi}$ as

$$\mathbf{J}_{\psi\psi} = \frac{2}{\sigma_w^2} Re \sum_{k=1}^{K} \mathbf{F}^H(k)\, \mathbf{D}^H \mathbf{D}\, \mathbf{F}(k). \tag{8.209}$$

[7] A simple derivation of the CCRB is given in Stoica and Larsson [SL01]. It was motivated by the results in Gu [Gu00]. It reparameterizes the original model to obtain a block-diagonal Fisher information matrix.

The second principal submatrix $\mathbf{J_{FF}}$ is $2KD \times 2KD$ and is block diagonal. Each block is $2D \times 2D$ and has an identical structure. The kth block corresponds to the kth snapshot. It has a structure

$$[\mathbf{J_{FF}}(k)] = \frac{2}{\sigma_w^2}\begin{bmatrix} \mathbf{A}^R & -\mathbf{A}^I \\ \mathbf{A}^I & \mathbf{A}^R \end{bmatrix}, \tag{8.210}$$

where the superscripts "R" and "I" denote the real and imaginary parts, respectively, of $\mathbf{A}$. We now derive $\mathbf{A}^R$ and $\mathbf{A}^I$:

$$\begin{aligned}\frac{2}{\sigma_w^2}\left[\mathbf{A}^R\right]_{ij} &= [[\mathbf{J_{FF}}(k)]]_{ij} = \frac{2}{\sigma_w^2}\sum_{k=1}^{K} Re\frac{\partial \mathbf{m}^H}{\partial F_i^R(k)}\frac{\partial \mathbf{m}}{\partial F_j^R(k)} \\ &= \frac{2}{\sigma_w^2} Re\left\{\mathbf{v}^H(\psi_i)\mathbf{v}(\psi_j)\right\}, \quad i,j = 1,2,\cdots,D \end{aligned}\tag{8.211}$$

Therefore, we can define

$$\mathbf{A} = \left[\mathbf{V}^H\mathbf{V}\right], \tag{8.212}$$

and

$$\mathbf{A}^R = Re\left[\mathbf{V}^H\mathbf{V}\right]. \tag{8.213}$$

To find the matrix in the upper right corner we use

$$\begin{aligned}[[\mathbf{J_{FF}}(k)]]_{ij} &= \frac{2}{\sigma_w^2}\sum_{k=1}^{K} Re\frac{\partial \mathbf{m}^H}{\partial F_i^R(k)}\frac{\partial \mathbf{m}}{\partial F_j^I(k)} \\ &= \frac{2}{\sigma_w^2} Re\left\{j\mathbf{v}^H(\psi_i)\mathbf{v}(\psi_j)\right\}, \quad \begin{matrix} i = 1,2,\cdots,D; \\ j = D+1,\cdots,2D \end{matrix} \\ &= -\frac{2}{\sigma_w^2} Im\left\{\mathbf{v}^H(\psi_i)\mathbf{v}(\psi_j)\right\}, \quad \begin{matrix} i = 1,2,\cdots,D; \\ j = D+1,\cdots,2D. \end{matrix}\end{aligned}\tag{8.214}$$

Therefore, the matrix in the upper right-hand corner can be written as $-\mathbf{A}^I$. Note that the matrices do not depend on k, so each block is identical. The two other sub-matrices follow in a similar manner.

The third principal sub-matrix is a scalar. Using (8.34)

$$\mathbf{J}_{\sigma_w^2\sigma_w^2} = \sum_{k=1}^{K} \mathrm{tr}\left[\frac{1}{\sigma_w^2}\mathbf{I}\cdot\frac{1}{\sigma_w^2}\mathbf{I}\right] = \frac{KN}{\sigma_w^4}. \tag{8.215}$$

We next evaluate the off-diagonal matrices. From (8.34), (8.204), and (8.205), we have

$$\mathbf{J}_{\psi\sigma_w^2} = \mathbf{J}_{\mathbf{F}\sigma_w^2} = \mathbf{0}. \tag{8.216}$$

We partition the $2KD \times D$ matrix, $\mathbf{J}_{\mathbf{F}\psi}$, into K $2D \times D$ matrices:

$$\begin{aligned}\left[\mathbf{J}_{\mathbf{F}\psi}(k)\right]_{ij} &= \frac{2}{\sigma_w^2}\sum_{k=1}^{K} Re\left\{\left[\frac{\partial}{\partial F_i^R(k)}\mathbf{F}^H(k)\mathbf{V}^H(\psi)\right]\left[\frac{\partial}{\partial \psi_j}\mathbf{V}(\psi)\mathbf{F}(k)\right]\right\} \\ &= \frac{2}{\sigma_w^2} Re\left\{\mathbf{V}^H(\psi)\frac{\partial \mathbf{v}(\psi_j)}{\partial \psi_j}\mathbf{F}(k)\right\}, \quad \begin{matrix} i = 1,2,\cdots,D; \\ j = 1,2,\cdots,D. \end{matrix}\end{aligned}\tag{8.217}$$

Similarly,

$$\left[\mathbf{J}_{\mathbf{F}\boldsymbol{\psi}}(k)\right]_{ij} = \frac{2}{\sigma_w^2}\sum_{k=1}^{K} Re\left\{\left[\frac{\partial}{\partial F_i^I(k)}\mathbf{F}^H(k)\mathbf{V}^H(\boldsymbol{\psi})\right]\left[\frac{\partial}{\partial \psi_j}\mathbf{V}(\boldsymbol{\psi})\mathbf{F}(k)\right]\right\}$$

$$= \frac{2}{\sigma_w^2} Im\left\{\mathbf{V}^H(\boldsymbol{\psi})\frac{\partial \mathbf{v}(\psi_j)}{\partial \psi_j}\mathbf{F}(k)\right\}, \quad \begin{array}{l} i = D+1,\cdots,2D; \\ j = 1,2,\cdots,D. \end{array} \tag{8.218}$$

Thus,

$$\left[\mathbf{J}_{\mathbf{F}\boldsymbol{\psi}}(k)\right] = \begin{bmatrix} \boldsymbol{\Delta}^R(k) \\ \hline \boldsymbol{\Delta}^I(k) \end{bmatrix}, \tag{8.219}$$

where

$$\boldsymbol{\Delta}(k) \triangleq \frac{2}{\sigma_w^2}\left[\mathbf{V}^H(\boldsymbol{\psi})\mathbf{D}(\boldsymbol{\psi})\mathbf{F}(k)\right]. \tag{8.220}$$

Similarly,

$$\left[\mathbf{J}_{\boldsymbol{\psi}\mathbf{F}}(k)\right] = \left[\begin{array}{cc} \boldsymbol{\Delta}^{RT}(k) & \boldsymbol{\Delta}^{IT}(k) \end{array}\right]. \tag{8.221}$$

We also define

$$\boldsymbol{\Delta}^T \triangleq \left[\begin{array}{c:c:c:c} \boldsymbol{\Delta}^T(1) & \boldsymbol{\Delta}^T(2) & \cdots & \boldsymbol{\Delta}^T(K) \end{array}\right]. \tag{8.222}$$

The information matrix is,

$$\mathbf{J} = \left[\begin{array}{c:cc:c:cc:c} \mathbf{J}_{\boldsymbol{\psi}\boldsymbol{\psi}} & \boldsymbol{\Delta}^T(1) & & \cdots & \boldsymbol{\Delta}^T(K) & & \mathbf{0} \\ \hdashline \boldsymbol{\Delta}(1) & \begin{array}{c}\mathbf{A}^R \\ \mathbf{A}^I\end{array} & \begin{array}{c}-\mathbf{A}^I \\ \mathbf{A}^R\end{array} & & \mathbf{0} & & \mathbf{0} \\ \hdashline & & & \ddots & & & \\ \hdashline \boldsymbol{\Delta}(K) & \mathbf{0} & & & \begin{array}{c}\mathbf{A}^R \\ \mathbf{A}^I\end{array} & \begin{array}{c}-\mathbf{A}^I \\ \mathbf{A}^R\end{array} & \\ \hdashline \mathbf{0} & \mathbf{0} & & & & & \frac{KN}{\sigma_w^4} \end{array}\right]. \tag{8.223}$$

The bound on σ_w^2 is not coupled. Thus,

$$CRB(\sigma_w^2) = \frac{\sigma_w^4}{KN}. \tag{8.224}$$

Using the formula for the inverse of a partitioned matrix on the $(D+2KD)\times(D+2KD)$ matrix in the upper left-hand corner (A.68), we obtain the conditional CRB (CCR) on $\boldsymbol{\psi}$,

$$C_{CCR}^{-1}(\boldsymbol{\psi}) = \mathbf{J}_{\boldsymbol{\psi}\boldsymbol{\psi}} - \boldsymbol{\Delta}^T\mathbf{J}_{\mathbf{FF}}^{-1}\boldsymbol{\Delta}. \tag{8.225}$$

To evaluate the second term, we write

$$\mathbf{J}_{\mathbf{FF}}^{-1} = \begin{bmatrix} \mathbf{B}^R & -\mathbf{B}^I \\ \mathbf{B}^I & \mathbf{B}^R \end{bmatrix}, \tag{8.226}$$

where

$$\mathbf{B} \triangleq \mathbf{A}^{-1}, \tag{8.227}$$

(see (A.68)). Then, one can show by direct substitution that

$$\mathbf{\Delta}^T(k)\mathbf{B}\mathbf{\Delta}(k) = Re\left[\mathbf{\Delta}^H(k)\mathbf{B}\mathbf{\Delta}(k)\right]. \tag{8.228}$$

Using (8.228) in (8.225) gives

$$C_{CCR}^{-1}(\psi) = \mathbf{J}_{\psi\psi} - \sum_{k=1}^{K} Re\left[\mathbf{\Delta}^H(k)\mathbf{B}\mathbf{\Delta}(k)\right]. \tag{8.229}$$

Using (8.220) and (8.209) in (8.229) gives

$$C_{CCR}^{-1}(\psi) = \frac{2}{\sigma_w^2}\sum_{k=1}^{K} Re\ \Big\{\mathbf{F}^H(k)\mathbf{D}^H\mathbf{D}\mathbf{F}(k) \\ -\mathbf{F}^H(k)\mathbf{D}^H\mathbf{V}\left(\mathbf{V}^H\mathbf{V}\right)^{-1}\mathbf{V}^H\mathbf{D}\mathbf{F}(k)\Big\}. \tag{8.230}$$

This result can be written as

$$\begin{aligned} C_{CCR}^{-1}(\psi) &= \frac{2}{\sigma_w^2}\sum_{k=1}^{K} Re\left\{\mathbf{F}^H(k)\mathbf{D}^H\left[\mathbf{I}-\mathbf{P_V}\right]\mathbf{D}\mathbf{F}(k)\right\} \\ &= \frac{2}{\sigma_w^2}\sum_{k=1}^{K} Re\left\{\mathbf{F}^H(k)\mathbf{D}^H\mathbf{P}_{\mathbf{V}}^{\perp}\mathbf{D}\mathbf{F}(k)\right\}, \end{aligned} \tag{8.231}$$

which is the desired result.

We can rewrite (8.231) more compactly by defining

$$\mathbf{H} \triangleq \mathbf{D}^H\mathbf{P}_{\mathbf{V}}^{\perp}\mathbf{D}. \tag{8.232}$$

Then,

$$\begin{aligned} \left[\mathbf{C}_{CCR}^{-1}(\psi)\right]_{ij} &= \frac{2}{\sigma_w^2} Re\left\{\mathbf{d}^H(\psi_i)\mathbf{P}_{\mathbf{V}}^{\perp}\mathbf{d}(\psi_j)\sum_{k=1}^{K} F_i^*(k)F_j(k)\right\} \\ &= \frac{2K}{\sigma_w^2} Re\left\{\mathbf{H}_{ij}\cdot\frac{1}{K}\sum_{k=1}^{K} F_j(k)F_i^*(k)\right\} \\ &= \frac{2K}{\sigma_w^2} Re\left\{\mathbf{H}_{ij}\cdot\left[\hat{\mathbf{S}}_{\mathbf{f}}\right]_{ji}\right\}, \end{aligned} \tag{8.233}$$

where

$$\hat{\mathbf{S}}_{\mathbf{f}} = \frac{1}{K}\sum_{k=1}^{K}\mathbf{F}(k)\mathbf{F}^H(k). \tag{8.234}$$

The result in (8.233) can be written using the Hadamard product as

$$\mathbf{C}_{CCR}^{-1}(\boldsymbol{\psi}) = \frac{2K}{\sigma_w^2} Re\left[\mathbf{H}\odot\hat{\mathbf{S}}_{\mathbf{f}}^T\right]. \tag{8.235}$$

As we would expect, the bound depends on the actual value of the vector parameter $\boldsymbol{\psi}$ and the actual signal waveform (through $\hat{\mathbf{S}}_{\mathbf{f}}$). However, if we now assume that $\mathbf{x}(k)$ is a sample function from an ergodic random process, then as K goes to infinity, $\hat{\mathbf{S}}_{\mathbf{f}}$ will approach the actual spectral matrix $\mathbf{S}_{\mathbf{f}}$ and we can write the asymptotic conditional Cramér-Rao bound (ACCR) as

$$\boxed{\mathbf{C}_{ACCR}(\boldsymbol{\psi}) = \frac{\sigma_w^2}{2K}\left[Re\left[\mathbf{H}\odot\mathbf{S}_{\mathbf{f}}^T\right]\right]^{-1}.} \tag{8.236}$$

This is the same bound that we encountered in Section 8.4.2.1 as a high *SNR* approximation to the standard CRB which we also refer to as the stochastic or unconditional CRB. We saw several examples of how the two bounds compared in that section. In all of the examples, the $\mathbf{C}_{CR}(\boldsymbol{\psi})$ was above $\mathbf{C}_{ACCR}(\boldsymbol{\psi})$ and they converged as the *SNR* increased. In other words, the stochastic CRB was a better (tighter) bound than the asymptotic conditional CRB (ACCR). We now show that

$$\mathbf{C}_{CR}(\boldsymbol{\psi}) \geq \mathbf{C}_{ACCR}(\boldsymbol{\psi}). \tag{8.237}$$

First, consider $\mathbf{C}_{CR}(\boldsymbol{\psi})$ as given by (8.102). Using the matrix inversion lemma we can write[8]

$$\begin{aligned}\mathbf{S}_{\mathbf{f}}\mathbf{V}^H\mathbf{S}_{\mathbf{x}}^{-1}\mathbf{V}\mathbf{S}_{\mathbf{f}} &= \mathbf{S}_{\mathbf{f}} - \mathbf{S}_{\mathbf{f}}\left[\mathbf{I} - \mathbf{V}^H\left[\mathbf{V}\mathbf{S}_{\mathbf{f}}\mathbf{V}^H + \sigma_w^2\mathbf{I}\right]^{-1}\mathbf{V}\mathbf{S}_{\mathbf{f}}\right] \\ &= \mathbf{S}_{\mathbf{f}} - \mathbf{S}_{\mathbf{f}}\left[\mathbf{I} + \mathbf{V}^H\sigma_w^{-2}\mathbf{V}\mathbf{S}_{\mathbf{f}}\right]^{-1}. \end{aligned} \tag{8.238}$$

Now the second matrix on the right side of (8.238) is Hermitian and non-negative definite. Thus,

$$\mathbf{S}_{\mathbf{f}}\mathbf{V}^H\mathbf{S}_{\mathbf{x}}^{-1}\mathbf{V}\mathbf{S}_{\mathbf{f}} \leq \mathbf{S}_{\mathbf{f}}. \tag{8.239}$$

---

[8]This result is from Ottersten et al. [OVK92]. See also Stoica and Nehorai [SN90b]. Several useful order relationships are discussed in Stoica and Sharman [SS90a] and Stoica and Nehorai [SN89a].

Then, using the properties of the Hadamard product (see (A.77)–(A.79)),

$$\left[Re\left\{\mathbf{H}\odot\left(\mathbf{S_f}\mathbf{V}^H\mathbf{S_x}^{-1}\mathbf{V}\mathbf{S_f}\right)^T\right\}\right]^{-1}\geq\left[Re\left\{\mathbf{H}\odot\mathbf{S_f}^T\right\}\right]^{-1}, \tag{8.240}$$

so that

$$\mathbf{C}_{CR}(\psi)\geq\mathbf{C}_{ACCR}(\psi). \tag{8.241}$$

If $\mathbf{H}$ and $\mathbf{S_f}$ are both positive definite, then the inequality is strict.

We recall from Examples 8.4.1 and 8.4.2 that even though the inequality is strict, the difference is negligible when the eigenvalues of $\mathbf{V}^H\mathbf{V}\mathbf{S_f}/\sigma_w^2$ become large.

8.4.5 Known Signal Waveforms

In this section we consider the case in which the signal waveforms are known. One application of this model is in communication systems in which a training sequence is sent to enable the receiver to synchronize. The CRB for this case is derived in Li and Compton [LC93].

The samples of the signal from ψ_i are denoted by

$$f_i(k)=\alpha_i p_i(k), \quad k=1,2,\cdots,K, \tag{8.242}$$

where $p_i(k)$ are the samples from the known signal and α_i is a complex constant. We consider both known and unknown α_i.

The time-domain samples are

$$\begin{aligned}\mathbf{x}(k) &= \mathbf{V}(\psi)\mathbf{f}(k)+\mathbf{n}(k)\\ &= \mathbf{V}(\psi)\mathbf{p}(k)\boldsymbol{\alpha}+\mathbf{n}(k), \quad k=1,2,\cdots,K,\end{aligned} \tag{8.243}$$

where

$$\mathbf{p}(k)\triangleq\operatorname{diag}\{p_1(k),p_2(k),\cdots,p_D(k)\}, \tag{8.244}$$

and

$$\boldsymbol{\alpha}\triangleq\begin{bmatrix}\alpha_1 & \alpha_2 & \cdots & \alpha_D\end{bmatrix}^T. \tag{8.245}$$

The case for known $\boldsymbol{\alpha}$ is a degenerate version of the model in Section 8.4.4. The CRB is given by the inverse of (8.209)

$$CRB(\psi)=\frac{\sigma_w^2}{2}\left\{Re\sum_{k=1}^{K}\mathbf{F}^H(k)\,\mathbf{D}^H\,\mathbf{D}\,\mathbf{F}(k)\right\}^{-1}, \tag{8.246}$$

where

$$\mathbf{F}(k) = \text{diag}\left\{\alpha_1 p_1(k), \alpha_2 p_2(k), \cdots, \alpha_D p_D(k)\right\}, \tag{8.247}$$

For the case of unknown $\boldsymbol{\alpha}$, we replace the Fisher information matrix in (8.203) with

$$\mathbf{J} = \left[\begin{array}{c|c|c} \mathbf{J}_{\psi\psi} & \mathbf{J}_{\psi\alpha_R} & \mathbf{J}_{\psi\alpha_I} \\ \hline \mathbf{J}_{\alpha_R\psi} & \mathbf{J}_{\alpha_R\alpha_R} & \mathbf{J}_{\alpha_R\alpha_I} \\ \hline \mathbf{J}_{\alpha_I\psi} & \mathbf{J}_{\alpha_I\alpha_R} & \mathbf{J}_{\alpha_I\alpha_I} \end{array}\right], \tag{8.248}$$

where $\boldsymbol{\alpha}_R$ and $\boldsymbol{\alpha}_I$ are the real and imaginary parts, respectively, of $\boldsymbol{\alpha}$. Note that we omit the σ_w^2 term because there is no cross-coupling. The steps for deriving the new submatrices are analogous to those in (8.210)–(8.214). Carrying out those steps, substituting the results into (8.248), inverting $\mathbf{J}$, and retaining the matrix in the upper corner, we obtain

$$CRB(\psi) = \frac{\sigma_w^2}{2}\left[\frac{2}{\sigma_w^2}\mathbf{J}_{\psi\psi} - Re\left[\mathbf{A}^H\mathbf{B}^{-1}\mathbf{A}\right]\right]^{-1}, \tag{8.249}$$

where

$$\mathbf{A} \triangleq \sum_{k=1}^{K} \mathbf{F}^H(k)\mathbf{V}^H\mathbf{D}\mathbf{F}(k), \tag{8.250}$$

and

$$\mathbf{B} \triangleq \sum_{k=1}^{K} \mathbf{F}^H(k)\mathbf{V}^H\mathbf{V}\mathbf{F}(k). \tag{8.251}$$

Using the same technique as in (8.233)–(8.235), we can write,

$$\frac{2}{\sigma_w^2}\mathbf{J}_{\psi\psi} = K Re\left\{\mathbf{D}^H\mathbf{D} \odot \widehat{\mathbf{S}}_{\mathbf{f}}^T\right\}, \tag{8.252}$$

$$\mathbf{A}_K \triangleq \frac{1}{K}\mathbf{A} = \left\{\mathbf{V}^H\mathbf{D} \odot \widehat{\mathbf{S}}_{\mathbf{f}}^T\right\}, \tag{8.253}$$

$$\mathbf{B}_K \triangleq \frac{1}{K}\mathbf{B} = \left\{\mathbf{V}^H\mathbf{V} \odot \widehat{\mathbf{S}}_{\mathbf{f}}^T\right\}, \tag{8.254}$$

where

$$\widehat{\mathbf{S}}_{\mathbf{f}} \triangleq \frac{1}{K}\sum_{k=1}^{K} \mathbf{f}(k)\mathbf{f}^H(k). \tag{8.255}$$

Using (8.252)–(8.254) in (8.249) gives

$$CRB(\psi) = \frac{\sigma_w^2}{2K}\left[Re\left\{\mathbf{D}^H\mathbf{D} \odot \widehat{\mathbf{S}}_{\mathbf{f}}^T - \mathbf{A}_K^H\mathbf{B}_K^{-1}\mathbf{A}_K\right\}\right]^{-1}. \tag{8.256}$$

For the special case in which $\widehat{\mathbf{S}}_{\mathbf{f}}$ is diagonal, the $CRB(\psi)$ is diagonal.

The result in (8.256) is for an arbitrary array geometry. If we consider the case of a standard linear array and a diagonal $\widehat{\mathbf{S}}_{\mathbf{f}}$, then the result can be simplified.

For unknown $\boldsymbol{\alpha}$,

$$CRB(\psi) = \frac{6\sigma_w^2}{K(N^2-1)N}\mathrm{diag}\left\{P_1^{-1}, P_2^{-1}, \cdots, P_D^{-1}\right\}, \tag{8.257}$$

where

$$P_i = \frac{1}{K}\sum_{k=1}^{K}|f_i(k)|^2. \tag{8.258}$$

is the average power in the ith signal and contains the effect of α_i.

The result in (8.257) can also be written as

$$CRB(\psi) = \frac{6}{K(N^2-1)}\mathrm{diag}\left\{ASNR_1^{-1}, ASNR_2^{-1}, \cdots, ASNR_D^{-1}\right\}. \tag{8.259}$$

Comparing (8.259) with (8.115) and (8.131), we see that the CRB for any signal in a known multiple-signal environment is the same as the CRB for a single signal in the unknown signal environment. This is a logical result because we can use the temporal characteristics of the known signals to eliminate all of the signals except the desired signals. For two signals, we can obtain a diagonal $\widehat{\mathbf{S}}_{\mathbf{f}}$ for arbitrary K. However, for $D > 2$, there are residual off-diagonal terms whose magnitude decreases as K increases.

We observe that the CRB is not a function of the angle between the plane waves. We also observe that there is no requirement that D, the number of signals, be less than N, the number of sensors.

8.4.6 Summary

In this section, we have developed CRBs for several signal and noise models that we will study in detail in the remainder of Chapter 8 and Chapter 9. The fundamental result that is the starting point for all of the derivation is

(8.34), which gives the expression for the ij element in the Fisher information matrix,

$$\begin{aligned} J_{ij} &= tr\left[\mathbf{K}_{\mathbf{x}}^{-1}(\boldsymbol{\theta})\frac{\partial\mathbf{K}_{\mathbf{x}}(\boldsymbol{\theta})}{\partial\theta_i}\mathbf{K}_{\mathbf{x}}^{-1}(\boldsymbol{\theta})\frac{\partial\mathbf{K}_{\mathbf{x}}(\boldsymbol{\theta})}{\partial\theta_j}\right] \\ &\quad +2Re\left[\frac{\partial\mathbf{m}^H(\boldsymbol{\theta})}{\partial\theta_i}\mathbf{K}_{\mathbf{x}}^{-1}(\boldsymbol{\theta})\frac{\partial\mathbf{m}(\boldsymbol{\theta})}{\partial\theta_j}\right]. \end{aligned} \tag{8.260}$$

This result is valid whenever $\mathbf{X}(k)$ is a complex Gaussian vector with mean $\mathbf{m}_{\mathbf{x}}(\boldsymbol{\theta})$ and covariance matrix $\mathbf{K}_{\mathbf{x}}(\boldsymbol{\theta})$. The vector $\boldsymbol{\theta}$ includes all parameters that are unknown (both wanted parameters and unwanted, or nuisance, parameters).

In the cases we have developed in this section, $\mathbf{X}(k)$ is the sensor output. Later we look at cases where we pre-process $\mathbf{X}(k)$ with a linear transformation prior to doing the parameter estimation. In those cases, we can use (8.260) on the output of the transformation to compute the CRB. Ideally, we would like the transformation to generate a sufficient statistic for the desired parameter estimation problem so that the CRB would remain the same. In practice, we often have to use transformations that only increase the CRB slightly.

There are a number of other models where we utilize the CRB:

(i) Planar arrays

In this case, (8.200) can be written as

$$\mathbf{X}(k) = \mathbf{V}(\boldsymbol{\psi})\mathbf{F}(k) + \mathbf{N}(k), \tag{8.261}$$

where $\boldsymbol{\psi}$ is defined as

$$\boldsymbol{\psi} = \begin{bmatrix} \boldsymbol{\psi}_1 & \boldsymbol{\psi}_2 & \cdots & \boldsymbol{\psi}_D \end{bmatrix}^T. \tag{8.262}$$

Each component vector is a 2×1 real vector. We will use

$$\boldsymbol{\psi}_i = \begin{bmatrix} \psi_{x_i} \\ \psi_{y_i} \end{bmatrix} \quad i = 1, \cdots, D. \tag{8.263}$$

in the subsequent discussion but the components could also be (θ_i, ϕ_i) if desired.

Yau and Bresler [YB92] have derived the asymptotic conditional Cramér-Rao bound for the case in which $\boldsymbol{\psi}_i$ is a $M \times 1$ vector. We quote their result for the $M = 2$ case.

We define

$$\mathbf{D}_i = \begin{bmatrix} \frac{\partial \mathbf{v}(\boldsymbol{\psi}_i)}{\partial \psi_{x_i}} & \frac{\partial \mathbf{v}(\boldsymbol{\psi}_i)}{\partial \psi_{y_i}} \end{bmatrix} \quad i = 1, \cdots, D, \tag{8.264}$$

and

$$\mathbf{D} = \begin{bmatrix} \mathbf{D}_1 & \mathbf{D}_2 & \cdots & \mathbf{D}_D \end{bmatrix}, \tag{8.265}$$

which is a $N \times 2D$ matrix. As in (8.232), we define

$$\mathbf{H}_2 \triangleq \mathbf{D}^H \mathbf{P}_{\mathbf{V}}^{\perp} \mathbf{D}. \tag{8.266}$$

We use the subscript 2 to denote that $\boldsymbol{\psi}_i$ has two components.

We define a 2×2 matrix of ones as,

$$\mathbf{1}_{2\times 2} = \begin{bmatrix} 1 & 1 \\ 1 & 1 \end{bmatrix}. \tag{8.267}$$

Then, the ACCR is

$$\begin{aligned} C_{ACCR}(\boldsymbol{\psi}) &= C_{ACCR}(\boldsymbol{\psi}_1, \boldsymbol{\psi}_2, \cdots, \boldsymbol{\psi}_D) \\ &= \frac{\sigma_w^2}{2K} \left[Re \left[\mathbf{H}_2 \odot \left[\mathbf{S}_{\mathbf{f}}^T \otimes \mathbf{1}_{2\times 2} \right] \right] \right]^{-1}, \end{aligned} \tag{8.268}$$

where $\odot$ is a Hadamard product (A.71) and $\otimes$ is a Kronecker product (A.80). Yau and Bresler [YB92] derived the conditional CRB, but the modification to obtain the asymptotic bound is clear.

We consider some examples in the problems. Note that this result also can be used for multiple polarization signals.

(ii) Broadband signals

We discuss broadband signals in Section 8.5.6 and discuss the appropriate CRB.

(iii) Spatially spread signals

We discuss parametric models for spatially spread signals in Section 8.9 and derive appropriate CRBs.

(iv) Beamspace processing

In Sections 6.9 and 7.10 we discussed beamspace beamformers. We will also use beamspace processing for parameter estimation. The beamspace matrix is $\mathbf{B}_{bs}^H$ and we require

$$\mathbf{B}_{bs}^H \mathbf{B}_{bs} = \mathbf{I}. \tag{8.269}$$

The beamspace steering vector is

$$\mathbf{v}_{bs} = \mathbf{B}_{bs}^H \mathbf{v}. \tag{8.270}$$

We define a projection matrix,

$$\mathbf{P}_{\mathbf{V}_{bs}} = \mathbf{B}_{bs}^H \mathbf{V} \left[\mathbf{V}^H \mathbf{P}_{\mathbf{B}_{bs}} \mathbf{V} \right]^{-1} \mathbf{V}^H \mathbf{B}_{bs}. \tag{8.271}$$

The beamspace spectral matrix is

$$\begin{aligned} \mathbf{S}_{\mathbf{x}_{bs}} &= \mathbf{B}_{bs}^H \mathbf{S}_{\mathbf{x}} \mathbf{B}_{bs} \\ &= \mathbf{B}_{bs}^H \mathbf{V} \mathbf{S}_{\mathbf{f}} \mathbf{V}^H \mathbf{B}_{bs} + \sigma_w^2 \mathbf{I}. \end{aligned} \tag{8.272}$$

The beamspace noise is white due to (8.269).

Thus, the CRB for beamspace estimation is

$$\begin{aligned} CRB_{bs}(\psi) &= \frac{\sigma_w^2}{2K} \left[Re \left\{ \left[\mathbf{D}^H \mathbf{B}_{bs} \mathbf{P}_{\mathbf{V}_{bs}}^{\perp} \mathbf{B}_{bs}^H \mathbf{D} \right] \right. \right. \\ & \left. \left. \odot \left[\mathbf{S}_{\mathbf{f}} \mathbf{V}^H \mathbf{B}_{bs} \mathbf{S}_{\mathbf{X}_{bs}}^{-1} \mathbf{B}_{bs}^H \mathbf{V} \mathbf{S}_{\mathbf{f}} \right]^T \right\} \right]^{-1}. \end{aligned} \tag{8.273}$$

We discuss several examples in the problems and study beamspace estimators in more detail in Sections 8.11 and 9.7.

(v) Range and bearing estimation

In the case of near-field sources, the parameter vector includes both range and bearing. Several references discuss this model and derive CRB (e.g., Rockah and Schultheiss [RS87a] [RS87b] or Huang and Barkat [HB91]).

(vi) Multipath models

In many applications, multipath is an important factor. Several references discuss parameter estimation in radar and sonar systems. Rendas and Moura [RM91] derive the CRB for a model that is appropriate for the sonar environment.

(vii) Minimally redundant arrays

The CRB results applied to arbitary array geometries, but our examples discussed uniform linear arrays. Another interesting class of arrays is the minimally redundant arrays discussed in Section 3.9. The CRB for this case are discussed in Chambers et al. [CTSD96] and Abramovich et al. [AGGS98]. They also discuss estimation techniques. We develop several examples in the problems.

(viii) Cyclostationary signals

In many communication systems the modulated signals exhibit a cyclostationary (periodic correlation) property that can be exploited to improve the DOA estimation performance. Schell [Sch94] derives the CRB for this model.

(ix) Computation

Computation of the inverse of the Fisher information matrix may be diffcult for large parameter sets. Hero et al. [HUSF97] develop a recusive algorithm to compute the bound.

We encounter other examples as we proceed through Chapters 8 and 9. We now consider maximum likelihood estimation procedures. In many cases of interest, the maximum likelihood estimates achieve the Cramér-Rao bound as K, the number of snapshots, goes to infinity.

8.5 Maximum Likelihood Estimation

In this section, we derive the maximum likelihood estimator for the DOAs of D plane-wave signals. In Section 8.5.1, we consider the model in which the source signals are sample functions from a Gaussian random process with an unknown $\mathbf{S_f}$.

The maximum likelihood estimator that we consider in Section 8.5.1 is sometimes referred to in the literature as the unconditional ML estimate or the stochastic ML estimate (e.g., [SN90b]).

In Section 8.5.2, we consider a model in which we treat the source signals as nonrandom but unknown and derive an ML estimate that is referred to in the literature as the deterministic (or conditional) ML estimate. We compare the performance of the two estimates.

In Section 8.5.3, we discuss the asymptotic performance of the maximum likelihood estimators. In Section 8.5.4, we extend the results to wideband signals. In Section 8.5.5, we summarize our results.

8.5.1 Maximum Likelihood Estimation

In this section, we derive a family of unconditional ML (UML) estimators. The reason that we have several UML estimators rather than a single estimator will become apparent as we proceed.

We use the frequency-domain snapshot model from Section 8.3.1 (8.62). We assume that we have D plane-wave signals arriving at the array from directions $\psi_1, \psi_2, \cdots, \psi_D$. The signals are sample functions from Gaussian random processes whose source spectral matrix is unknown. The signals are corrupted by additive spatially uncorrelated Gaussian noise with spectral height $S_n(\omega)$. Thus,

$$\mathbf{X}(k) = \mathbf{V}(\psi)\mathbf{F}(k) + \mathbf{N}(k), \quad k = 1, 2, \cdots, K. \tag{8.274}$$

$$\mathbf{S_f} = E\left[\mathbf{F}(k)\mathbf{F}^H(k)\right]. \tag{8.275}$$

We assume that

$$\mathbf{S_n} = \sigma_w^2 \mathbf{I}, \tag{8.276}$$

and that σ_w^2 is known. We consider the unknown σ_w^2 case subsequently. The component vector ψ_i summarizes the parameters from the D signals that we want to estimate. The component vector ψ_i is the 1- or 2-D wavenumber. Alternatively, we could estimate the azimuth and elevation angles (θ_i, ϕ_i).

The first estimator will be referred to as the asymptotic ML (AML) estimator.[9]

8.5.1.1 AML estimators

The ML estimators for this model have been derived by Böhme [Boh86] and Jaffer [Jaf88]. Our discussion follows the latter reference. To simplify the notation, we indicate the snapshot number by a subscript.

The likelihood function is

$$L(\psi, \mathbf{S_f}) = -\ln\det \mathbf{S_x} - \frac{1}{K}\sum_{k=1}^{K} \mathbf{X}_k^H\, \mathbf{S_x}^{-1}\, \mathbf{X}_k, \tag{8.277}$$

and

$$\mathbf{S_x} = \mathbf{V}(\psi)\mathbf{S_f}\mathbf{V}^H(\psi) + \sigma_w^2\mathbf{I}, \tag{8.278}$$

where we have dropped unnecessary constants. To avoid confusion between the sample covariance matrix and the ML estimate of $\mathbf{S_x}$, we use $\mathbf{C_x}$ to designate the sample covariance matrix. Recall that the sample covariance matrix is

$$\mathbf{C_x} = \frac{1}{K}\sum_{k=1}^{K} \mathbf{X}_k\, \mathbf{X}_k^H. \tag{8.279}$$

---

[9]The abbreviation "AML" is sometimes used to denote approximate ML estimate in the literature. We do not use it in that manner.

Using (8.279), we can write (8.277) as

$$\begin{aligned} L(\psi, \mathbf{S_f}) &= -\left[\ln\det\mathbf{S_x} + \mathrm{tr}\left[\frac{1}{K}\sum_{k=1}^{K}\mathbf{X}_k^H\,\mathbf{S_x}^{-1}\mathbf{X}_k\right]\right] \\ &= -\left[\ln\det\mathbf{S_x} + \mathrm{tr}\left[\mathbf{S_x}^{-1}\cdot\frac{1}{K}\sum_{k=1}^{K}\mathbf{X}_k\,\mathbf{X}_k^H\right]\right] \\ &= -\left[\ln\det\mathbf{S_x} + \mathrm{tr}\left[\mathbf{S_x}^{-1}\,\mathbf{C_x}\right]\right]. \end{aligned} \tag{8.280}$$

Fortunately, the solution is separable so that we can maximize over $\mathbf{S_x}$ to obtain an explicit function of ψ and then maximize over ψ to get the total solution.

We denote the ijth element of $\mathbf{S_f}$ as S_{ij}. Then $\mathbf{S_x}$ can be written as

$$\mathbf{S_x} = \sum_{i=1}^{D}\sum_{j=1}^{D} S_{ij}\mathbf{v}(\psi_i)\mathbf{v}^H(\psi_j) + \sigma_w^2\mathbf{I}. \tag{8.281}$$

Differentiating the first term in (8.280) using (A.390), we have

$$\frac{\partial\ln|\mathbf{S_x}|}{\partial S_{ij}} = \mathrm{tr}\left[\left[\frac{\partial\ln|\mathbf{S_x}|}{\partial\mathbf{S_x}}\right]^T\frac{\partial\mathbf{S_x}}{\partial S_{ij}}\right], \tag{8.282}$$

and, from (A.387),

$$\frac{\partial\ln|\mathbf{S_x}|}{\partial\mathbf{S_x}} = \left[\mathbf{S_x}^{-1}\right]^T. \tag{8.283}$$

From (8.281),

$$\frac{\partial\mathbf{S_x}}{\partial S_{ij}} = \mathbf{v}(\psi_i)\mathbf{v}^H(\psi_j). \tag{8.284}$$

Using (8.283) and (8.284) in (8.282) gives

$$\frac{\partial\ln|\mathbf{S_x}|}{\partial S_{ij}} = \mathrm{tr}\left[\mathbf{S_x}^{-1}\mathbf{v}(\psi_i)\mathbf{v}^H(\psi_j)\right] = \mathrm{tr}\left[\mathbf{v}^H(\psi_j)\mathbf{S_x}^{-1}\mathbf{v}(\psi_i)\right]. \tag{8.285}$$

Differentiating the second term in (8.280) using (A.391) and (A.383), we have

$$\begin{aligned} \frac{\partial\mathrm{tr}\left[\mathbf{S_x}^{-1}\mathbf{C_x}\right]}{\partial S_{ij}} &= \mathrm{tr}\left\{\left[\frac{\partial\mathrm{tr}\left[\mathbf{S_x}^{-1}\mathbf{C_x}\right]}{\partial\mathbf{S_x}}\right]^T\frac{\partial\mathbf{S_x}}{\partial S_{ij}}\right\} \\ &= \mathrm{tr}\left\{-\mathbf{S_x}^{-1}\mathbf{C_x}\mathbf{S_x}^{-1}\mathbf{v}(\psi_i)\mathbf{v}^H(\psi_j)\right\} \\ &= \mathrm{tr}\left\{-\mathbf{v}^H(\psi_j)\mathbf{S_x}^{-1}\mathbf{C_x}\mathbf{S_x}^{-1}\mathbf{v}(\psi_i)\right\}. \end{aligned} \tag{8.286}$$

A necessary condition is that

$$\frac{\partial L(\psi, \mathbf{S_f})}{\partial S_{ij}} = 0, \quad i,j = 1,2,\cdots,D. \tag{8.287}$$

Substituting (8.285) and (8.286) into (8.287) and noting that both terms are scalars,

$$\mathbf{v}^H(\psi_j)\left[\mathbf{S_x}^{-1}\mathbf{C_x}\mathbf{S_x}^{-1} - \mathbf{S_x}^{-1}\right]\mathbf{v}(\psi_i) = \mathbf{0}, \quad i,j = 1,2,\cdots,D, \tag{8.288}$$

or

$$\mathbf{V}^H(\psi)\left[\mathbf{S_x}^{-1}\mathbf{C_x}\mathbf{S_x}^{-1} - \mathbf{S_x}^{-1}\right]\mathbf{V}(\psi) = \mathbf{0}. \tag{8.289}$$

To get (8.289) into a more usable form, we write,

$$\mathbf{S_x}^{-1} = \frac{1}{\sigma_w^2}\left[\mathbf{I} - \mathbf{V}\left[\mathbf{S_f}\mathbf{V}^H\mathbf{V} + \sigma_w^2\mathbf{I}\right]^{-1}\mathbf{S_f}\mathbf{V}^H\right], \tag{8.290}$$

where we have suppressed the ψ dependence of $\mathbf{V}(\psi)$.

$$\begin{aligned}
\mathbf{S_x}^{-1}\mathbf{V} &= \frac{\mathbf{V}}{\sigma_w^2}\left[\mathbf{I} - \left[\mathbf{S_f}\mathbf{V}^H\mathbf{V} + \sigma_w^2\mathbf{I}\right]^{-1}\mathbf{S_f}\mathbf{V}^H\mathbf{V}\right] \\
&= \frac{\mathbf{V}}{\sigma_w^2}\left[\left[\mathbf{S_f}\mathbf{V}^H\mathbf{V} + \sigma_w^2\mathbf{I}\right]^{-1}\left(\left[\mathbf{S_f}\mathbf{V}^H\mathbf{V} + \sigma_w^2\mathbf{I}\right] - \mathbf{S_f}\mathbf{V}^H\mathbf{V}\right)\right] \\
&= \mathbf{V}\left[\mathbf{S_f}\mathbf{V}^H\mathbf{V} + \sigma_w^2\mathbf{I}\right]^{-1} = \mathbf{V}\mathbf{S_x}^{-1}.
\end{aligned} \tag{8.291}$$

Using (8.291) and its conjugate transpose in (8.289) gives

$$\left[\mathbf{S_f}\mathbf{V}^H\mathbf{V} + \sigma_w^2\mathbf{I}\right]^{-1}\mathbf{V}^H\left[\mathbf{C_x} - \mathbf{S_x}\right]\mathbf{V}\left[\mathbf{S_f}\mathbf{V}^H\mathbf{V} + \sigma_w^2\mathbf{I}\right]^{-1} = \mathbf{0}. \tag{8.292}$$

Note that one cannot let $\widehat{\mathbf{S}}_{\mathbf{x},ml} = \mathbf{C_x}$ in order to solve (8.292) because $\widehat{\mathbf{S}}_{\mathbf{x},ml}$ must have the structure in (8.281) and, with probability 1, $\mathbf{C_x}$ will not have that structure.

The condition in (8.292) implies

$$\left[\mathbf{V}^H\left[\mathbf{C_x} - \mathbf{S_x}\right]\mathbf{V}\right]_{\mathbf{S_x} = \widehat{\mathbf{S}}_{\mathbf{x},ml}} = \mathbf{0}. \tag{8.293}$$

Substituting

$$\mathbf{S_x} = \mathbf{V}\mathbf{S_f}\mathbf{V}^H + \sigma_w^2\mathbf{I}, \tag{8.294}$$

into (8.293) gives

$$\mathbf{V}^H\mathbf{C_x}\mathbf{V} = \mathbf{V}^H\mathbf{V}\mathbf{S_f}\mathbf{V}^H\mathbf{V} + \sigma_w^2\mathbf{V}^H\mathbf{V}. \tag{8.295}$$

The solution to (8.295) is denoted by $\widehat{\mathbf{S}}_{\mathbf{f},ml}(\psi)$:

$$\boxed{\widehat{\mathbf{S}}_{\mathbf{f},ml}(\psi) = \left[\mathbf{V}^H\mathbf{V}\right]^{-1}\mathbf{V}^H\left[\mathbf{C}_{\mathbf{x}} - \sigma_w^2\mathbf{I}\right]\mathbf{V}\left[\mathbf{V}^H\mathbf{V}\right]^{-1}.} \tag{8.296}$$

The result in (8.296) can also be written as

$$\widehat{\mathbf{S}}_{\mathbf{f},ml}(\psi) = \mathbf{V}^{\dagger}\left[\mathbf{C}_{\mathbf{x}} - \sigma_w^2\mathbf{I}\right]\left[\mathbf{V}^{\dagger}\right]^H, \tag{8.297}$$

where

$$\mathbf{V}^{\dagger} = \left[\mathbf{V}^H\mathbf{V}^{-1}\right]\mathbf{V}^H, \tag{8.298}$$

is the Moore-Penrose pseudoinverse. This result was previously given in [Sch79] and [Jaf85].

The result in (8.296) does not guarantee that $\widehat{\mathbf{S}}_{\mathbf{f},ml}(\psi)$ is non-negative definite because the maximization with respect to $\mathbf{S}_{\mathbf{f}}$ was over the set of Hermitian matrices and not over the set of non-negative definite matrices. We discuss the implication of this result after we complete the derivation.

We define

$$\widehat{\mathbf{S}}_{\mathbf{x},ml}(\psi) = \mathbf{V}(\psi)\widehat{\mathbf{S}}_{\mathbf{f},ml}(\psi)\mathbf{V}^H(\psi) + \sigma_w^2\mathbf{I}, \tag{8.299}$$

where $\widehat{\mathbf{S}}_{\mathbf{f},ml}(\psi)$ is given by (8.296).

The relation in (8.299) can also be written as

$$\widehat{\mathbf{S}}_{\mathbf{x},ml}(\psi) = \mathbf{P}_{\mathbf{V}}\left[\mathbf{C}_{\mathbf{x}} - \sigma_w^2\mathbf{I}\right]\mathbf{P}_{\mathbf{V}} + \sigma_w^2\mathbf{I}, \tag{8.300}$$

where $\mathbf{P}_{\mathbf{V}}$ is the projection matrix onto the range of $\mathbf{V}(\psi)$.

To find $\hat{\psi}_{ml}$, we maximize (8.280) with $\widehat{\mathbf{S}}_{\mathbf{x},ml}(\psi)$, as given by (8.300), substituted for $\mathbf{S}_{\mathbf{x}}$. Thus,

$$\hat{\psi}_{ml} = \arg\max_{\psi}\left\{-\left[\ln\det\widehat{\mathbf{S}}_{\mathbf{x}}(\psi) + \operatorname{tr}\left[\widehat{\mathbf{S}}_{\mathbf{x}}^{-1}(\psi)\mathbf{C}_{\mathbf{x}}\right]\right]\right\}, \tag{8.301}$$

and

$$\widehat{\mathbf{S}}_{\mathbf{f}} \triangleq \widehat{\mathbf{S}}_{\mathbf{f},ml} = \widehat{\mathbf{S}}_{\mathbf{f}}(\hat{\psi}_{ml}), \tag{8.302}$$

where we drop the "ml" subscript on $\widehat{\mathbf{S}}_{\mathbf{f}}$ for simplicity.

The last step is to get the right side of (8.301) into a more usable form. Using (8.290),

$$\widehat{\mathbf{S}}_{\mathbf{x}}^{-1}(\psi)\mathbf{C}_{\mathbf{x}} = \frac{1}{\sigma_w^2}\mathbf{C}_{\mathbf{x}} - \frac{1}{\sigma_w^2}\mathbf{V}\left[\widehat{\mathbf{S}}_{\mathbf{f}}\mathbf{V}^H\mathbf{V} + \sigma_w^2\mathbf{I}\right]^{-1}\widehat{\mathbf{S}}_{\mathbf{f}}\mathbf{V}^H\mathbf{C}_{\mathbf{x}}. \tag{8.303}$$

From (8.297),

$$\begin{aligned}\widehat{\mathbf{S}}_{\mathbf{f}}\mathbf{V}^H\mathbf{V}+\sigma_w^2\mathbf{I} &= \left[\mathbf{V}^H\mathbf{V}\right]^{-1}\mathbf{V}^H\left[\mathbf{C}_{\mathbf{x}}-\sigma_w^2\mathbf{I}\right]\mathbf{V}\left[\mathbf{V}^H\mathbf{V}\right]^{-1}\mathbf{V}^H\mathbf{V}+\sigma_w^2\mathbf{I} \\ &= \left[\mathbf{V}^H\mathbf{V}\right]^{-1}\mathbf{V}^H\mathbf{C}_{\mathbf{x}}\mathbf{V}, \end{aligned} \tag{8.304}$$

so

$$\left[\widehat{\mathbf{S}}_{\mathbf{f}}\mathbf{V}^H\mathbf{V}+\sigma_w^2\mathbf{I}\right]^{-1}=\left[\mathbf{V}^H\mathbf{C}_{\mathbf{x}}\mathbf{V}\right]^{-1}\mathbf{V}^H\mathbf{V}. \tag{8.305}$$

Assuming $\mathbf{C}_{\mathbf{x}}$ is positive definite, the inverse exists.[10]

Then, from (8.303),

$$\begin{aligned}\operatorname{tr}\left[\widehat{\mathbf{S}}_{\mathbf{x}}^{-1}(\psi)\mathbf{C}_{\mathbf{x}}\right] &= \frac{\operatorname{tr}\left[\mathbf{C}_{\mathbf{x}}\right]}{\sigma_w^2}-\frac{1}{\sigma_w^2}\operatorname{tr}\left[\mathbf{V}\left[\mathbf{V}^H\mathbf{C}_{\mathbf{x}}\mathbf{V}\right]^{-1}\mathbf{V}^H\mathbf{V}\widehat{\mathbf{S}}_{\mathbf{f}}\mathbf{V}^H\mathbf{C}_{\mathbf{x}}\right] \\ &= \frac{\operatorname{tr}\left[\mathbf{C}_{\mathbf{x}}\right]}{\sigma_w^2}-\frac{1}{\sigma_w^2}\operatorname{tr}\left[\widehat{\mathbf{S}}_{\mathbf{f}}\mathbf{V}^H\mathbf{V}\right].\end{aligned} \tag{8.306}$$

Using (8.304) we have

$$\begin{aligned}\operatorname{tr}\left[\widehat{\mathbf{S}}_{\mathbf{f}}\mathbf{V}^H\mathbf{V}\right] &= \operatorname{tr}\left[\left[\mathbf{V}^H\mathbf{V}\right]^{-1}\mathbf{V}^H\mathbf{C}_{\mathbf{x}}\mathbf{V}-\sigma_w^2\mathbf{I}\right] \\ &= \operatorname{tr}\left[\mathbf{P}_{\mathbf{V}}\mathbf{C}_{\mathbf{x}}\right]-D\sigma_w^2.\end{aligned} \tag{8.307}$$

Using (8.307) in (8.306) gives,

$$\begin{aligned}\operatorname{tr}\left[\widehat{\mathbf{S}}_{\mathbf{x}}^{-1}(\psi)\mathbf{C}_{\mathbf{x}}\right] &= \frac{1}{\sigma_w{}^2}\left\{\operatorname{tr}\left[\left[\mathbf{I}-\mathbf{P}_{\mathbf{V}}\right]\mathbf{C}_{\mathbf{x}}\right]\right\}-D \\ &= \frac{1}{\sigma_w{}^2}\left\{\operatorname{tr}\left[\mathbf{P}_{\mathbf{V}}^{\perp}\mathbf{C}_{\mathbf{x}}\right]\right\}-D.\end{aligned} \tag{8.308}$$

From (8.300),

$$\begin{aligned}\widehat{\mathbf{S}}_{\mathbf{x}}(\psi) &= \mathbf{P}_{\mathbf{V}}\mathbf{C}_{\mathbf{x}}\mathbf{P}_{\mathbf{V}}+\sigma_w^2\left[\mathbf{I}-\mathbf{P}_{\mathbf{V}}\right] \\ &= \mathbf{P}_{\mathbf{V}}\mathbf{C}_{\mathbf{x}}\mathbf{P}_{\mathbf{V}}+\sigma_w^2\mathbf{P}_{\mathbf{V}}^{\perp}.\end{aligned} \tag{8.309}$$

Using (8.308) and (8.309) in (8.301) and dropping terms that do not depend on ψ gives

$$\boxed{\hat{\psi}_{aml} = \underset{\psi}{\arg\max}\left\{-\ln\det\left[\mathbf{P}_{\mathbf{V}}\mathbf{C}_{\mathbf{x}}\mathbf{P}_{\mathbf{V}}+\sigma_w^2\mathbf{P}_{\mathbf{V}}^{\perp}\right]-\frac{1}{\sigma_w^2}\operatorname{tr}\left[\mathbf{P}_{\mathbf{V}}^{\perp}\mathbf{C}_{\mathbf{x}}\right]\right\}.} \tag{8.310}$$

[10] $\mathbf{C}_{\mathbf{x}}$ is positive definite if $K \geq N$.

The estimator defined by (8.310) is referred to in the literature as the stochastic or unconditional ML estimate. We use the subscript "aml" for asymptotic maximum likelihood. A brief discussion of the reason for this description is useful.[11] If $\mathbf{S_f}$ is strictly positive definite, then, since the ML estimates are consistent, $\widehat{\mathbf{S}}_{\mathbf{f},ml}(\widehat{\psi})$ tends to $\mathbf{S_f}$ as $K \to \infty$. Therefore $\widehat{\mathbf{S}}_{\mathbf{f},ml}(\widehat{\psi})$ must be positive definite and therefore is a valid ML estimate. Hence, (8.310) provides a large-sample realization of the ML estimator in the case of a non-singular $\mathbf{S_f}$ (non-coherent sources). The case of coherent sources is discussed in detail in Stoica et al. [SOVM96], and the reader is referred to that reference.[12] The key result (8.310) still provides a large-sample realization of $\widehat{\psi}_{ml}$.

In order to find $\hat{\psi}_{aml}$, we need to perform a maximization over a mD-dimensional space where m is the number of parameters to be estimated in each plane wave (normally one or two). In Section 8.7, we discuss various implementation techniques to perform this maximization. All of these techniques require a significant amount of computation.

Before considering some examples, we indicate the effect of unknown noise variance on the ML estimate.

If the noise variance σ_w^2 is unknown, then the likelihood function in (8.280) becomes,

$$L\left(\psi, \mathbf{S_f}, \sigma_w^2\right) = -\left[\ln\det \mathbf{S_x} + \operatorname{tr}\left[\mathbf{S_x}^{-1}\,\mathbf{C_x}\right]\right], \tag{8.311}$$

where

$$\mathbf{S_x} = \mathbf{V}\,\mathbf{S_f}\,\mathbf{V}^H + \sigma_w^2\,\mathbf{I}. \tag{8.312}$$

Proceeding in exactly the same manner as above, we find

$$\hat{\sigma}_w^2 = \frac{\operatorname{tr}\left[\mathbf{P}_{\mathbf{V}}^{\perp}\,\mathbf{C_x}\right]}{N-D}. \tag{8.313}$$

Substituting (8.313) into (8.310), we obtain

$$\hat{\psi}_{aml} = \arg\max_{\psi}\left\{-\ln\det\left[\mathbf{P_V C_x P_V} + \frac{\operatorname{tr}\left[\mathbf{P}_{\mathbf{V}}^{\perp}\mathbf{C_x}\right]\mathbf{P}_{\mathbf{V}}^{\perp}}{N-D}\right]\right\}, \tag{8.314}$$

because the second term in (8.310) is no longer a function of ψ. We can

[11] This discussion follows [SOVM96].

[12] M. Viberg directed me to this reference (private communication).

maximize the term in braces by minimizing the determinant. Thus,

$$\boxed{\hat{\psi}_{aml} = \underset{\psi}{\arg\min} \left\{ \det \left[\mathbf{P_V C_x P_V} + \frac{\mathrm{tr}[\mathbf{P}_\mathbf{V}^\perp \mathbf{C_x}]\mathbf{P}_\mathbf{V}^\perp}{N-D} \right] \right\}.} \tag{8.315}$$

for the case of unknown noise variance.

When one carries out the derivation leading to (8.314), there is an intermediate step containing

$$\widehat{\mathbf{S}}_{\mathbf{f},ml}(\psi) = \left[\mathbf{V}^H\mathbf{V}\right]^{-1}\mathbf{V}^H\left[\mathbf{C_x} - \hat{\sigma}_w^2\mathbf{I}\right]\mathbf{V}\left[\mathbf{V}^H\mathbf{V}\right]^{-1}, \tag{8.316}$$

with $\hat{\sigma}_w^2$ given by (8.313). The resulting $\widehat{\mathbf{S}}_{\mathbf{f},ml}(\psi)$ is not necessarily positive definite.

We consider a simple example to demonstrate the behavior of the ML estimator.

Example 8.5.1

We consider a standard 10-element linear array. The signal is a single plane wave arriving from $u_s = 0$. We assume the noise spectrum height σ_w^2 is known. We find $\hat{u}_{ml}$ by calculating the term in brackets in (8.310) over a dense grid of points $(-1 \leq u \leq 1)$. We then use a local minimization routine to find the exact minimum.

In Figure 8.17, we plot the normalized root mean-square error (RMSE) versus *ASNR*. In Figure 8.18, we plot the normalized RMSE versus K for several *ASNR*. In both figures we also plot the CRB.

We see that for $ASNR \geq -2$ dB, the RMSE of AML estimator coincides with the CRB. However, as the *ASNR* decreases, the RMSE increases sharply. This behavior is referred as the threshold phenomenon and is a characteristic of most nonlinear parameter estimation problems (e.g., Section 2.4 of DEMT I [VT68], [VT01a] or Wozencraft and Jacobs [WJ65]). In order to understand the behavior, we show a scatter plot of the estimates for various *ASNR* in Figure 8.19 and the corresponding histogram in Figure 8.20. We see that, above threshold, the estimates are clustered around the correct value.

At $ASNR = -3$ dB, most of the estimates (93/100) are clustered around the correct value, but the others correspond to a subsidiary peak that is not close to the correct value. These errors are sometimes referred to as anomalous (or global) errors. Note that even a few anomalous errors cause a dramatic increase in the RMSE, because they are so large compared to the local errors (the errors clustered around the correct value).

As the *ASNR* decreases further, the number of local errors decreases. At an $ASNR = -12$ dB, the histogram shows that the errors are spread across u-space in an almost uniform manner.

It is convenient to write total MSE as

$$MSE = p_{lo}(MSE_{lo}) + (1 - p_{lo})(MSE_{gl}), \tag{8.317}$$

where p_{lo} denotes the probability that the estimation error is local. The local MSE, "MSE_{lo}" is usually characterized by the CRB. It is usually difficult to calculate p_{lo} except in the asymptotic (large K) region. We should observe that, in this example, the threshold occurs just as p_{lo} decreases from unity. We find that this characteristic is true in most cases. Thus, the region of primary interest is usually

$$0.95 \leq p_{lo} \leq 1.0. \tag{8.318}$$

The global MSE, "MSE$_{gl}$" is usually hard to calculate as we enter the threshold region. However, further below threshold, it generally approaches the variance of a uniform random variable (one-third in u-space for $u_s = 0$). If we have *a priori* information about the portion of u-space where the signal can originate, the MSE below threshold may be lower.

We will discuss performance issues further after we consider the multiple plane-wave problem.

Example 8.5.2

Consider a standard 10-element linear array with two equal-power plane-wave signals impinging on it. The signal separation is $\Delta\psi_R = 0.2165 BW_{NN}$. The signals are uncorrelated. In Figure 8.21, we plot the normalized RMSE versus *ASNR* for K = 100 snapshots. We see that for *ASNR* $\geq$ 5 dB, the RMSE is equal to the CRB. At an *ASNR* = 5 dB, a threshold occurs and the RMSE increases rapidly.

In order to understand this behavior we show a sequence of scatter plots for various *ASNR*. In Figure 8.22, the *ASNR* = 6 dB and the errors are local. The likelihood function has a single peak that is close to the correct value. Note that we always assign $\hat{u}_2 \geq \hat{u}_1$ so that only the upper left triangle is needed.

In Figure 8.23, the *ASNR* = 0 dB. We see that the majority of the errors are local. However, there are a number of points in which $\hat{u}_1$ is at the midpoint of the two signal locations,

$$\hat{u}_1 = \frac{u_1 + u_2}{2} = 0, \tag{8.319}$$

and the second estimate $\hat{u}_2$ is scattered from $u = 0$ to $u = 1$.There is similar behavior for $\hat{u}_2$.

When this result occurs we say that the ML estimator cannot resolve the two signals. In order to compute the probability of resolution, we say that the signals are resolved if

$$|\hat{u}_1 - u_1| \leq \min(\frac{u_2 - u_1}{2}, \frac{2}{N}), \tag{8.320}$$

and

$$|\hat{u}_2 - u_2| \leq \min(\frac{u_2 - u_1}{2}, \frac{2}{N}). \tag{8.321}$$

Each of the conditions in (8.320) and (8.321) correspond to a local error for the respective parameter. The probability of resolution, P_R, is the probability that both (8.320) and (8.321) are satisfied. Then, we can write the total MSE as

$$MSE = P_R(MSE_{lo}) + (1 - P_R)(MSE_{gl}). \tag{8.322}$$

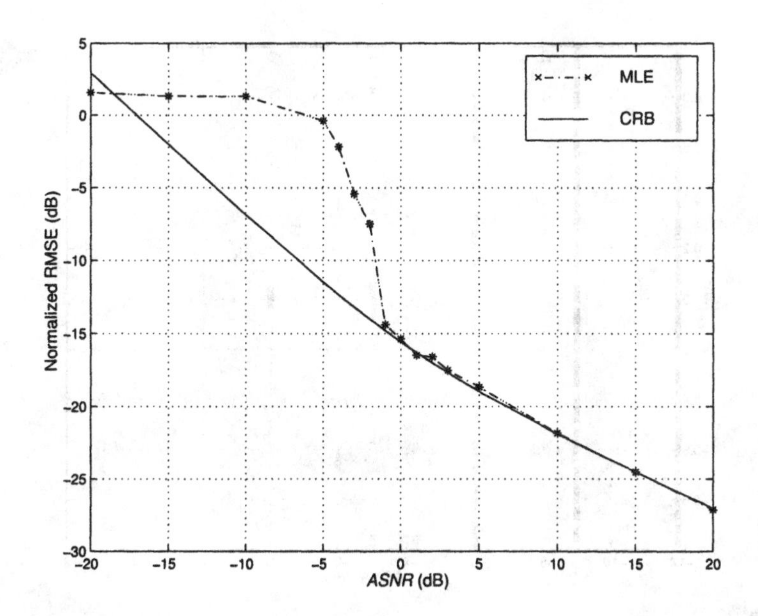

Figure 8.17 AML estimator, single signal, $u_s = 0$, $N = 10$, $K = 100$: normalized RMSE versus *ASNR*.

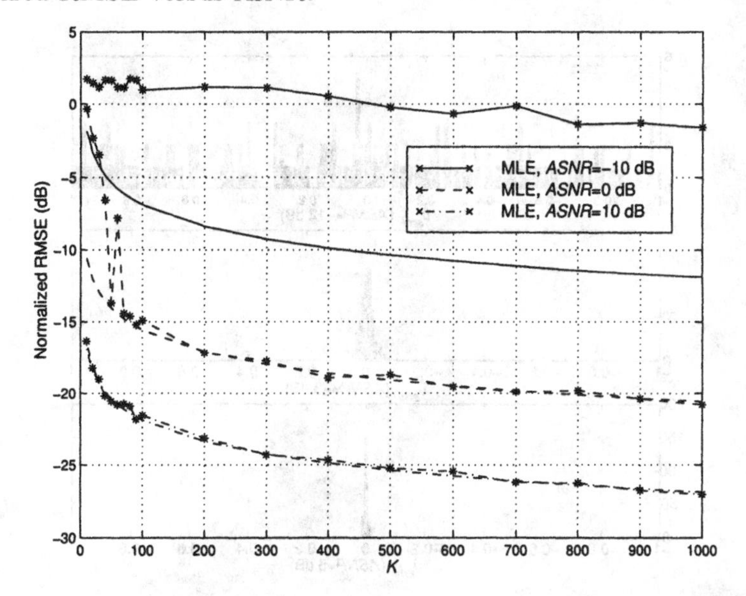

Figure 8.18 AML estimator, single signal, $u_s = 0$, $ASNR = -10$ dB, 0 dB, 10 dB: normalized RMSE versus K.

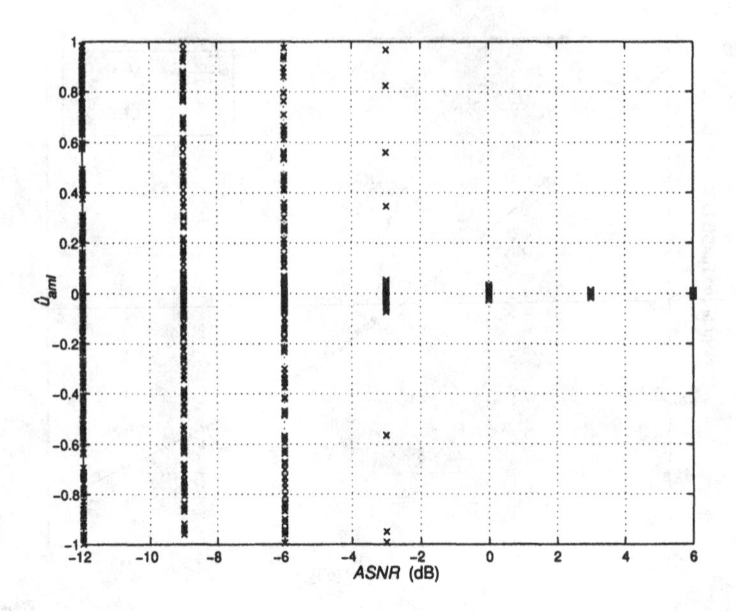

Figure 8.19 AML estimator, single signal, $u_s = 0$, $N = 10$, $K = 100$, 100 trials: scatter plot versus *ASNR*.

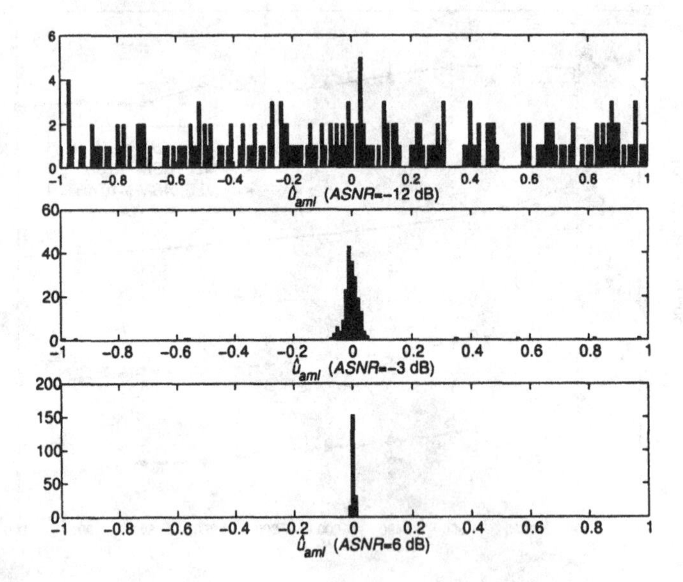

Figure 8.20 AML estimator, single signal, $u_s = 0$, $N = 10$, $K = 100$, 100 trials: histogram for *ASNR* $= -12$ dB, -3 dB, 6 dB.

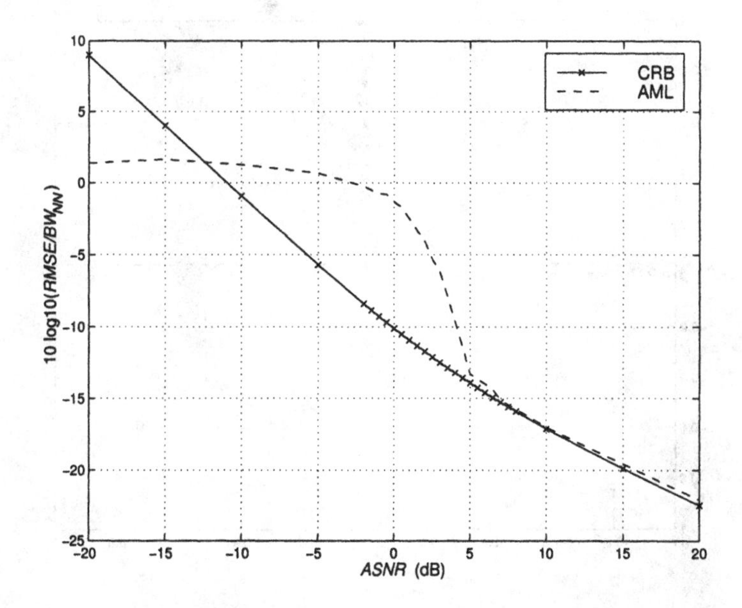

Figure 8.21 AML estimator: two equal-power uncorrelated plane-wave signals at $\pm\Delta u_R/2\,(0.2165BW_{NN}/2)$, $K = 100$: normalized RMSE versus *ASNR*.

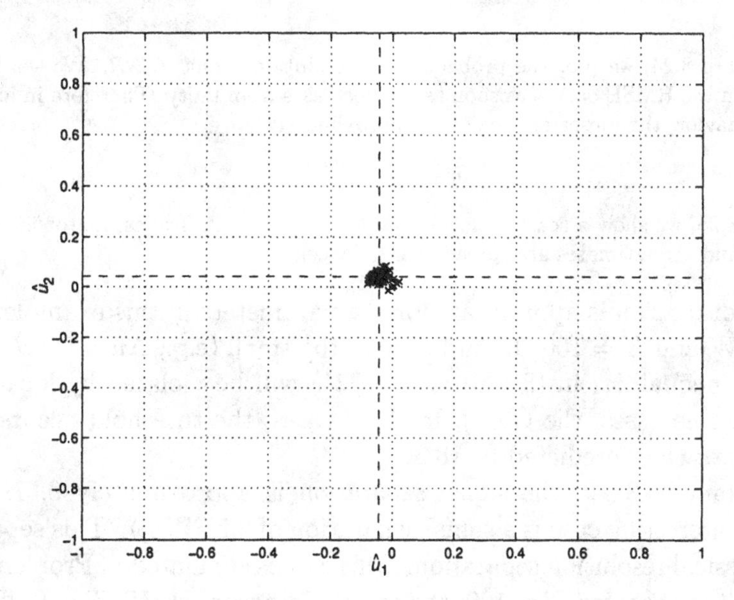

Figure 8.22 AML estimator: two equal-power uncorrelated plane-wave signals at $\pm\Delta u_R/2(0.2165BW_{NN}/2), K = 100$, *ASNR* $= 6$ dB: scatter plot of AML estimates.

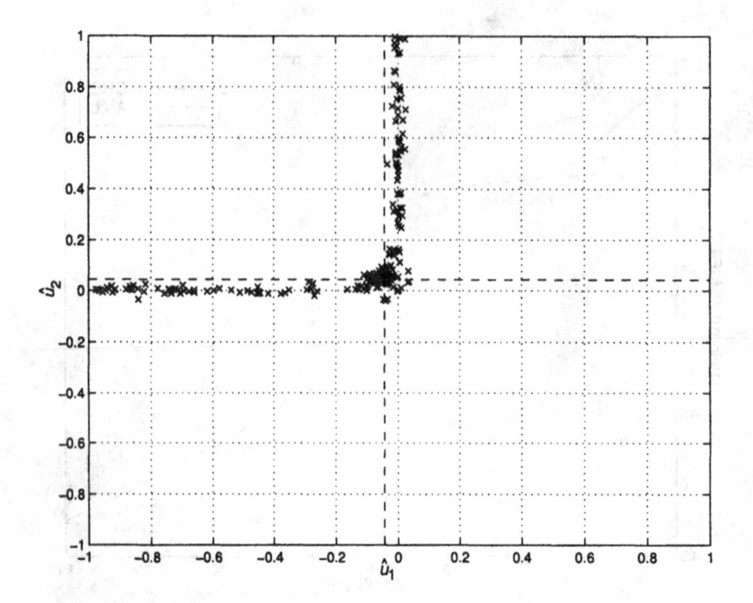

Figure 8.23 AML estimator: two equal-power uncorrelated plane-wave signals at $\pm\Delta u_R/2(0.2165BW_{NN}/2), K = 100$, $ASNR = 0$ dB: scatter plot of AML estimates.

In Figure 8.24, we plot the probability of resolution versus $ASNR$. We see that the threshold in the RMSE occurs as soon as P_R decreases from unity. Therefore in analyzing the P_R behavior, the important region is approximately,

$$0.95 \leq P_R \leq 1.0. \tag{8.323}$$

In Figure 8.25, we show a scatter plot for an $ASNR = -6$ dB. The estimator is far below threshold and the estimates are spread over u-space.

This discussion is appropriate for the parameters in this example, $\Delta u \simeq 0.5HPBW$ and $K = 100$. If Δu becomes too small (e.g., $\Delta u = 0.1HPBW$), then the inequalities in (8.320) and (8.321) will be violated by local errors (and, in some cases, the CRB). In these cases, the threshold occurs at P_R values lower than predicted by (8.323).

In Example 8.5.2, the signal separation is approximately $0.5HPBW$. Another interesting case is a signal separation of $0.5BW_{NN}$. This separation is the classical resolution separation. This case is simulated in Problem 8.5.5. The result is that, for $K = 100$, the threshold occurs at $ASNR = 0$ dB. This result provides a quantitative basis for the classical resolution definition.

In Figure 8.17, the threshold for a single-signal occurs at $ASNR = -1$ dB. Thus, for uncorrelated signals, when the separation exceeds the classical

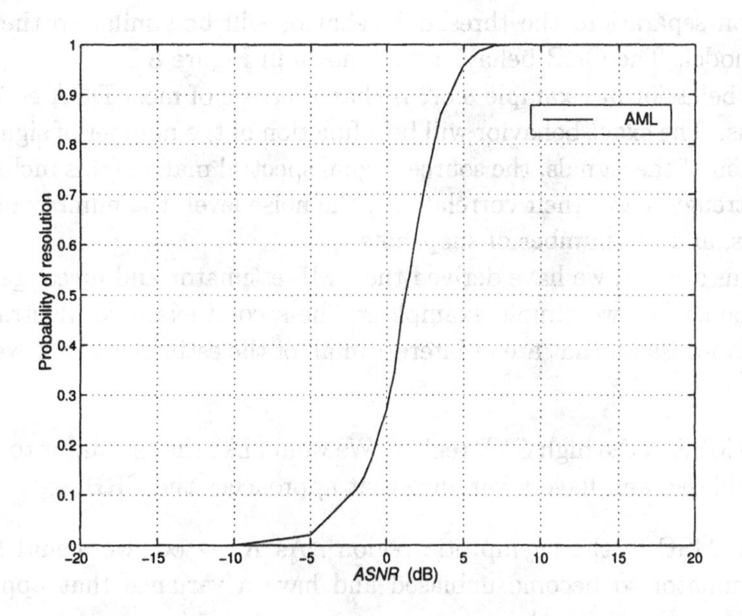

Figure 8.24 AML estimator: two equal-power uncorrelated plane-wave signals at $\pm\Delta u_R/2(0.2165BW_{NN}/2), K = 100$: probability of resolution versus *ASNR*.

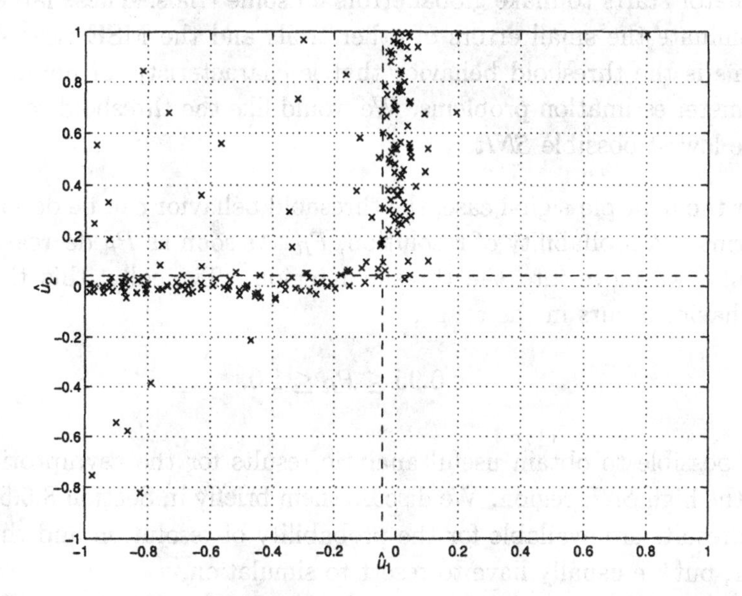

Figure 8.25 AML estimator: two equal-power uncorrelated plane-wave signals at $\pm\Delta u_R/2(0.2165BW_{NN}/2), K = 100$, *ASNR* = −6 dB: scatter plot of AML estimates.

resolution separation, the threshold behavior will be similar to the single-signal model. The CRB behavior was shown in Figure 8.5.

The behavior in Example 8.5.2 is characteristic of most DOA estimation problems. The exact behavior will be a function of the number of signals, the separation of the signals, the source- signal spectral matrix (this includes the signal strengths and their correlation), the noise level, the number of sensor elements, and the number of snapshots.

In this section, we have derived the AML estimator and investigated the performance for two simple examples. The second example illustrates the performance issues that are of interest in all of the estimators that we study. They are:

(i) The MSE in the high *SNR* region. We would like the estimator to become unbiased and have a variance that approaches the CRB.

(ii) The MSE in the asymptotic region. As $K \to \infty$, we would like the estimator to become unbiased and have a variance that approaches the CRB.

(iii) For a given K, as the *SNR* decreases, we reach a point in which the estimator starts to make global errors on some trials. These large errors dominate the small errors of other trials and the MSE rises sharply. This is the threshold behavior that is characteristic of nonlinear parameter estimation problems. We would like the threshold to occur at the lowest possible *SNR*.

(iv) For the multiple-signal case, the threshold behavior can be described in terms of a probability of resolution, P_R. As soon as P_R decreases from unity, we move into the threshold region. Typically, this threshold behavior occurs in the range,

$$0.95 \leq P_R \leq 1.0.$$

It is possible to obtain useful analytic results for the asymptotic (high K) and the high *SNR* region. We discuss them briefly in Section 8.5.5. Some analytic results are available for the probability of resolution and threshold behavior, but we usually have to resort to simulation.

The above four factors relate to performance. The fifth issue is the computational complexity of the estimator. After we discuss the performance capabilities, we look for estimators that are less complex but perform close to optimum.

We will use a set of test scenarios to study the behavior of the parameter estimation algorithms that we develop in this chapter and in Chapter 9. Several of them are designed to stress the estimators. We study other scenarios that exhibit interesting properties, but the test scenarios in Table 8.3 allow a consistent comparison across algorithms. We use a 10-element standard linear array (SLA10) throughout the comparisons. We consider a number of other array geometries in examples and problems, but only do a complete comparison for the SLA. We normalize the RMSE and standard deviation by the BW_{NN} and plot the results versus *ASNR*. Thus the results are valid for any SRA with $N \geq 10$. For smaller arrays, we need to check the result for applicability. The ten test scenarios are shown in Table 8.3.

Table 8.3: Test Scenarios

Test scenario	No. of Signals	Power	Correlation	Separation	K
1	2	Equal	$\rho = 0$	$\Delta\psi = \Delta\psi_R$	$\geq 10N$
2	2	Equal	$\rho = 0$	$\Delta\psi = 0.05BW_{NN}$	$\geq 10N$
3	2	$ASNR_2 = 10ASNR_1$ $ASNR_2 = 100ASNR_1$	$\rho = 0$	$\Delta\psi = \Delta\psi_R$	$\geq 10N$
4	2	Equal	$0 < \lvert\rho\rvert < 1.0$	$\Delta\psi = \Delta\psi_R$	$\geq 10N$
5	2	Equal	$\lvert\rho\rvert = 1$	$\Delta\psi = \Delta\psi_R$	$\geq 10N$
6	2	Equal	$\rho = 0$	$\Delta\psi = \Delta\psi_R$	$\geq 2N$
7	3	Equal	$\boldsymbol{\rho} = \mathbf{I}$	$\psi_1 = -\Delta\psi_R$ $\psi_2 = 0$ $\psi_3 = \Delta\psi_R$	$\geq 10N$
8	3	Equal	$\boldsymbol{\rho} \neq \mathbf{I}$	$\psi_1 = -\Delta\psi_R$ $\psi_2 = 0$ $\psi_3 = \Delta\psi_R$	$\geq 10N$
9	3	$ASNR_1 = ASNR_3$ $= 10\ ASNR_2$ $ASNR_1 = ASNR_3$ $= 100\ ASNR_2$	$\boldsymbol{\rho} = \mathbf{I}$	$\psi_1 = -\Delta\psi_R$ $\psi_2 = 0$ $\psi_3 = \Delta\psi_R$	$\geq 10N$
10	5	Equal	$\boldsymbol{\rho} = \mathbf{I}$	$\psi_1 = -2\Delta\psi_R$ $\psi_2 = -\Delta\psi_R$ $\psi_3 = 0$ $\psi_4 = \Delta\psi_R$ $\psi_5 = 2\Delta\psi_R$	$\geq 10N$

Notes: Table 8.3

1. Recall that $\Delta\psi_R \triangleq 0.2165BW_{NN}$. This is approximately $0.5HPBW$. The subscript "R" denotes reference.
2. Because we are working in ψ-space (or u-space), the estimation results for element space algorithms depend on signal separation, not absolute location. For beamspace algorithms, the location with respect to the beam fan must be specified.
3. Various $\boldsymbol{\rho}$ matrices will be considered in Test Scenario 8.

In the text, we only show representative test scenarios. Before studying the behavior of AML in more detail, we look at other stochastic estimators.

8.5.1.2 Unconditional ML estimators

We now return to the issue that was pointed out after (8.296). The estimate of $\widehat{\mathbf{S}}_{\mathbf{f},ml}$ given by (8.296) may not be non-negative definite and, therefore, it is not necessarily a proper estimate to use in the subsequent part of the algorithm.

Bresler [Bre88] has analyzed this issue in detail and devised an ML algorithm that guarantees a non-negative definite estimate of $\widehat{\mathbf{S}}_{\mathbf{f}}$. Bresler's algorithm assumed that ψ was known, but it is straightforward to develop an iterative version of his algorithm. We implemented this iterative version of the Bresler algorithm for the test scenarios in Table 8.3 and found that the performance in estimating ψ was essentially the same as the AML algorithm over the entire range of *ASNR* (both above and below threshold).

There are two other estimates of $\widehat{\mathbf{S}}_{\mathbf{f}}$ that lead to different estimators for the unknown σ_w^2 case. The first approach uses

$$\widetilde{\sigma}_w^2 = \frac{1}{N-D} \sum_{i=D+1}^{N} \widehat{\lambda}_i, \tag{8.324}$$

where the $\widehat{\lambda}_i$ are the $N-D$ smallest eigenvalues of $\mathbf{C_x}$, as the estimate of σ_w^2. Then, we let

$$\widehat{\mathbf{S}}_{\mathbf{f},I} = \mathbf{V}^{\dagger}\left(\mathbf{C_x} - \widetilde{\sigma}_w^2\mathbf{I}\right)\left(\mathbf{V}^{\dagger}\right)^H, \tag{8.325}$$

where $\mathbf{V}^{\dagger}$ is the Moore-Penrose pseudoinverse of $\mathbf{V}$,

$$\mathbf{V}^{\dagger} \triangleq [\mathbf{V}^H\mathbf{V}]^{-1}\mathbf{V}^H. \tag{8.326}$$

Then, by analogy with (8.300)

$$\widehat{\mathbf{S}}_{\mathbf{x},I}(\psi) = \mathbf{P_V}\left[\mathbf{C_x} - \widetilde{\sigma}_w^2\mathbf{I}\right]\mathbf{P_V} + \widetilde{\sigma}_w^2\mathbf{I}. \tag{8.327}$$

Using (8.327) in (8.301) and proceeding as before, we obtain,

$$\hat{\psi}_{wml} = \arg\min_{\boldsymbol{\psi}}\left\{\left[\ln\det\widehat{\mathbf{S}}_{\mathbf{x},I}(\psi) + \operatorname{tr}\left[\widehat{\mathbf{S}}_{\mathbf{x},I}^{-1}(\psi)\mathbf{C_x}\right]\right]\right\}, \tag{8.328}$$

This estimate is due to Wax et al. [WSK82].

Schmidt [Sch79] suggested using

$$\widehat{\mathbf{S}}_{\mathbf{f},II}(\psi) = \mathbf{V}^{\dagger}\widehat{\mathbf{U}}_S\widehat{\boldsymbol{\Lambda}}_S\widehat{\mathbf{U}}_S^H\left(\mathbf{V}^{\dagger}\right)^H. \tag{8.329}$$

This estimate is unbiased and consistent. Then,

$$\widehat{\mathbf{S}}_{\mathbf{x},II}(\psi) = \mathbf{P}_{\mathbf{V}}\widehat{\mathbf{U}}_S\hat{\mathbf{\Lambda}}_S\widehat{\mathbf{U}}_S^H\mathbf{P}_{\mathbf{V}} + \widetilde{\sigma}_w^2\mathbf{I}. \tag{8.330}$$

where $\widetilde{\sigma}_w^2$ is given by (8.324) and

$$\hat{\psi}_{scml} = \arg\min_{\psi}\left\{\ln\det\widehat{\mathbf{S}}_{\mathbf{x},II}(\psi) + \mathrm{tr}\left[\widehat{\mathbf{S}}_{\mathbf{x},II}^{-1}(\psi)\mathbf{C}_{\mathbf{x}}\right]\right\}, \tag{8.331}$$

where $\widehat{\mathbf{S}}_{\mathbf{x},II}(\psi)$ is given by (8.330).

We implemented the two algorithms for the test scenarios in Table 8.3 and found that their performance was essentially the same as the AML estimator over the entire range of *ASNR*. Stoica et al. [SOVM96] derive a ML estimator for coherent signals that guarantees that $\widehat{\mathbf{S}}_{\mathbf{x}}$ is a positive semi-definite matrix of a given rank. They show that the AML estimate in (8.315) is the large-sample realization of the ML estimate even when the signals are coherent.

We focus our subsequent work on the AML estimator. In the next section, we simulate the AML estimator for some of the test scenarios in Table 8.3.

8.5.1.3 Performance of AML estimators

We consider a sequence of six examples to illustrate the behavior of the AML algorithms. All of the examples utilize a standard 10-element linear array with two signals impinging on it.

Example 8.5.3 (continuation, Example 8.5.2)

We use the same model as in Example 8.5.2, except the separation is reduced to $0.05BW_{NN}$. The signals are uncorrelated ($\rho = 0$). The results are shown in Figure 8.26. We see that the threshold has moved about 13 dB to the right. Later, we explore the threshold behavior as a function of $\Delta\psi$.

Example 8.5.4 (continuation)

We use the same model as in Example 8.5.2, except $|\rho| = 0.95$. The separation $\Delta\psi$ equals $\Delta\psi_R$. We consider two phase angles for ρ. In the first case, $\phi_\rho = 0$, and in the second case $\phi_\rho = \pi/4$. The results for $\phi_\rho = 0$ are shown in Figure 8.27. We see that the threshold is 5 dB higher and occurs at *ASNR* = 10 dB. Above the threshold the AML estimator converges to the CRB. We recall from Figure 8.6 that the CRB is higher for the correlated signal case. The results for $\phi_\rho = \pi/4$ are shown in Figure 8.28. We see that the behavior for this phase angle is very similar to the uncorrelated case. This result illustrates the importance of considering various phase angles when studying the correlated signal model.

We also simulated the AML estimator for $|\rho| = 1$ with $\phi_\rho = 0$ and $\pi/4$. The results are very similar to the $|\rho| = 0.95$ case.

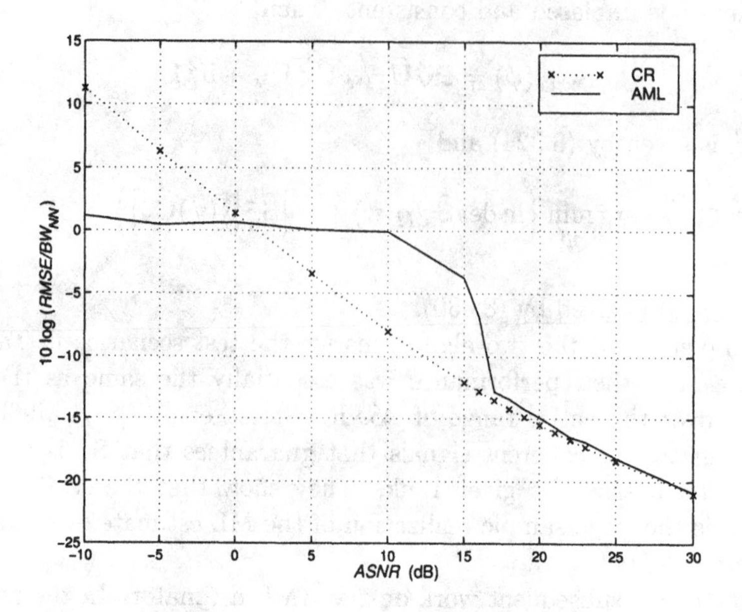

Figure 8.26 AML estimator: two equal-power uncorrelated plane-wave signals, $\Delta u = 0.05BW_{NN}$, $K = 10N$; normalized RMSE versus *ASNR*.

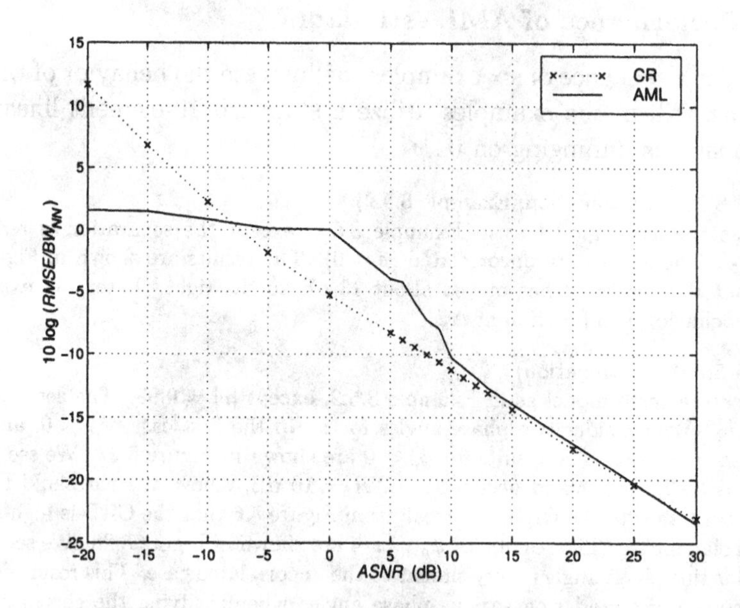

Figure 8.27 AML estimator: two equal-power correlated signals, $\Delta u = 0.2165BW_{NN}$, $K = 10N$, $\rho = 0.95$, $\phi_\rho = 0$; normalized RMSE versus *ASNR*.

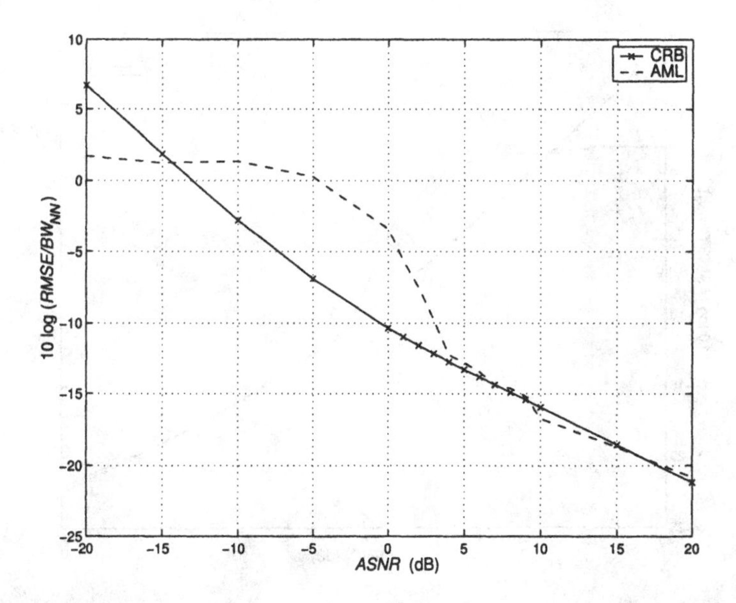

Figure 8.28 AML estimator: two equal-power correlated signals, $\Delta u = 0.2165 BW_{NN}$, $K = 10N$, $\rho = 0.95$, $\phi_\rho = \pi/4$; normalized RMSE versus *ASNR*. (Note that the vertical scale is different from Figure 8.27.)

Example 8.5.5: Unequal signal powers

Consider a standard 10-element linear array with two uncorrelated plane-wave signals impinging on it. The signal separation is $\Delta\psi_R$. The *SNR* of signal 2 is 20 dB higher than the *SNR* of signal 1. The results are shown in Figures 8.29 and 8.30. The RMSE of the weaker signal moves away from the CRB at $ASNR_1 = 1$ dB. The RMSE of the stronger signal moves away from the CRB at $ASNR_2 = 24$ dB.

Example 8.5.6: (Low sample support)

We consider a standard 10-element linear array. We use the same signal model as in Example 8.5.3 except $K = 20$ $(2N)$. The results are plotted in Figure 8.31. We see that the threshold occurs at an $ASNR = 11$ dB. This value is 6 dB higher than the $K = 10N$ case.

We have considered four scenarios in this section. Above threshold, the AML estimator approaches the CRB. However, the location of the threshold varies with the parameters in the scenario.

In our discussion of the CRB, we found that, if we knew that the signals were uncorrelated, the CRB was lower in certain scenarios. The next logical step is to develop a ML estimator that exploits that *a priori* knowledge. An uncorrelated ML has been developed by Bell [Bell99] and shows improved performance in the scenarios where the CRBs differed. The reader is referred

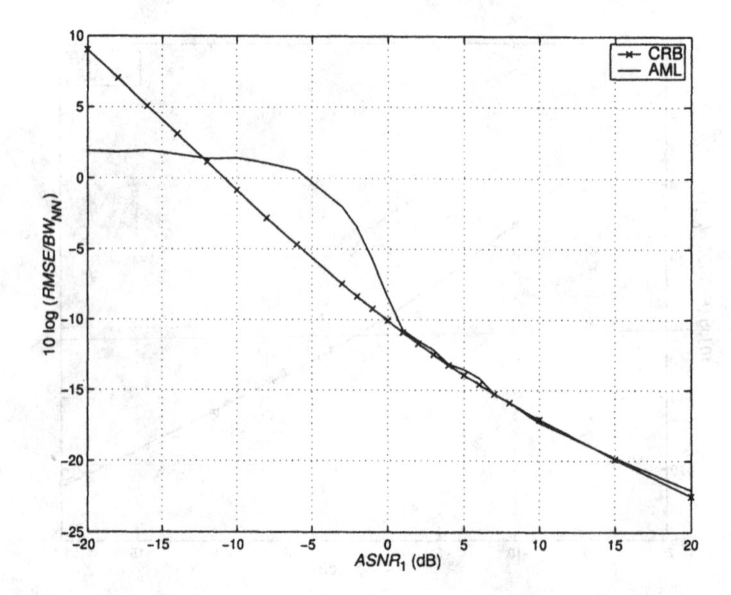

Figure 8.29 AML estimator: two uncorrelated plane-wave signals with unequal power, $\Delta u = 0.2165BW_{NN}$, $ASNR_2 = ASNR_1 + 20$ dB; normalized RMSE of signal 1 versus $ASNR_1$.

to this reference for further discussion.

8.5.2 Conditional Maximum Likelihood Estimators

In this section, we develop a ML estimator that is referred to in the literature as a deterministic or conditional maximum likelihood (CML) estimator. The reason for the name will be clear when we formulate the model.

In this section, we derive the estimator and investigate its performance. We compare its performance to the stochastic maximum likelihood estimator of Section 8.5.1 and to the CRB.

We consider the same snapshot model as in Section 8.5.1,

$$\mathbf{X}(k) = \mathbf{V}\mathbf{F}(k) + \mathbf{N}(k), \quad k = 1, 2, \cdots, K. \tag{8.332}$$

However, we model the source signals as unknown nonrandom signals. Thus, $\mathbf{F}(k)$ is a $D \times 1$ vector

$$\mathbf{F}(k) = \left[\begin{array}{c|c|c|c} F_1(k) & F_2(k) & \cdots & F_D(k) \end{array} \right]^T, \tag{8.333}$$

whose elements are unknown nonrandom complex numbers.

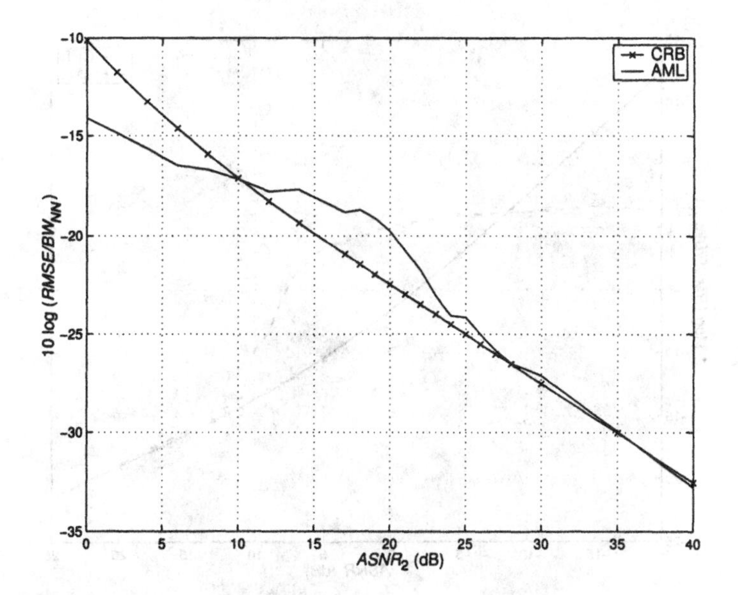

Figure 8.30 AML estimator: two uncorrelated plane-wave signals with unequal power, $\Delta u = 0.2165 BW_{NN}$, $ASNR_2 = ASNR_1 + 20$ dB; normalized RMSE of signal 2 versus $ASNR_2$.

Alternatively, we can consider $\mathbf{F}(k)$ as a specific sample function from a random process and design an estimator based on that sample function. This is sometimes referred to as *conditional maximum likelihood* estimation.

The noise process is a sample function from a Gaussian random process. We assume that it is spatially uncorrelated so that

$$E\left[\mathbf{N}(k)\mathbf{N}^H(k)\right] = \sigma_w^2\mathbf{I}. \tag{8.334}$$

The noise in successive snapshots is assumed to be statistically independent. The spectral height σ_w^2 is known. Later we consider the case of unknown σ_w^2. In that case we find the ML estimate of σ_w^2 as part of the estimation process. We assume the number of signals, D, is known.

The joint probability density is

$$p_{\mathbf{x}}(\mathbf{X}) = \prod_{k=1}^{K} \frac{1}{|\pi\sigma_w^2\mathbf{I}|} \exp\left\{-\frac{1}{\sigma_w^2}|\mathbf{X}(k) - \mathbf{V}(\psi)\mathbf{F}(k)|^2\right\},, \tag{8.335}$$

and the ln likelihood function is

$$L(\psi, \mathbf{F}) = -KN\ln\sigma_w^2 - \frac{1}{\sigma_w^2}\sum_{k=1}^{K}|\mathbf{X}(k) - \mathbf{V}(\psi)\mathbf{F}(k)|^2 \tag{8.336}$$

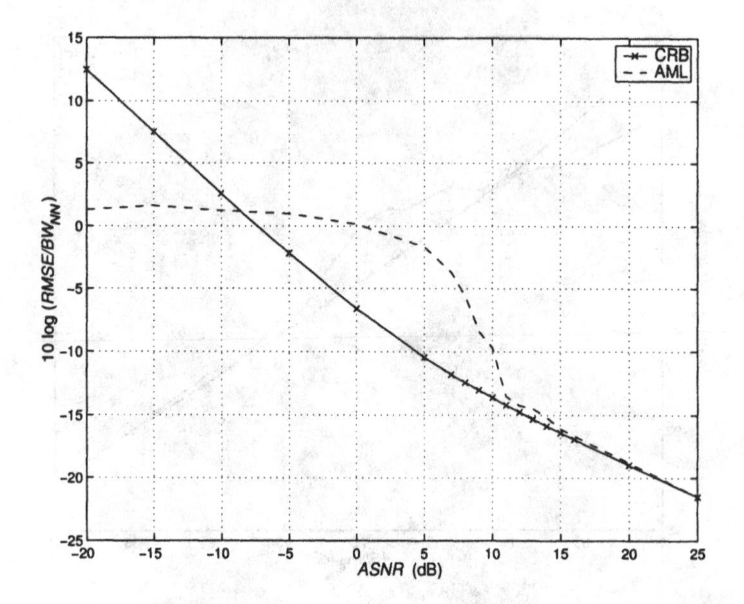

Figure 8.31 AML estimator: two equal-power uncorrelated plane-wave signals with low sample support, $\Delta u = 0.2165BW_{NN}$, $K = 2N$; normalized RMSE versus *ASNR*.

(we have dropped the term containing π), where $\boldsymbol{\psi}$ is an $D \times 1$ vector and $\mathbf{F}$ is a $D \times K$ matrix whose kth column is $\mathbf{F}(k)$. We want to maximize this function over the unknown parameters $(\boldsymbol{\psi}, \mathbf{F})$. This is equivalent to minimizing[13]

$$L_2(\boldsymbol{\psi}, \mathbf{F}) \triangleq \sum_{k=1}^{K} \left|\mathbf{X}(k) - \mathbf{V}(\boldsymbol{\psi})\mathbf{F}(k)\right|^2. \tag{8.337}$$

If we fix $\boldsymbol{\psi}$ and minimize over $\mathbf{F}(k)$, the result is just the matrix MVDR filter of Section 6.5 (see (6.200)). Thus,

$$\widehat{\mathbf{F}}(k) = \mathbf{W}_o^H(\boldsymbol{\psi})\mathbf{X}(k), \tag{8.338}$$

where

$$\mathbf{W}_o^H(\boldsymbol{\psi}) = \left[\mathbf{V}^H(\boldsymbol{\psi})\mathbf{V}(\boldsymbol{\psi})\right]^{-1}\mathbf{V}^H(\boldsymbol{\psi}) = \mathbf{V}^{\dagger}(\boldsymbol{\psi}). \tag{8.339}$$

The result in (8.338) and (8.339) is exactly what we derived in Section 6.5 where $\boldsymbol{\psi}$ was known. Substituting (8.338) and (8.339) into (8.337), we obtain

[13]The subscripts signify that the functions are different although they have the same minimum (or maximum).

the function

$$L_3(\psi) = \sum_{k=1}^{K} \left| \mathbf{X}(k) - \mathbf{V}(\psi) \left[\mathbf{V}^H(\psi)\mathbf{V}(\psi) \right]^{-1} \mathbf{V}^H(\psi)\mathbf{X}(k) \right|^2, \tag{8.340}$$

which is to be minimized over ψ. We recognize the coefficient of the second term as the projection matrix onto the columns of $\mathbf{V}(\psi)$

$$\begin{aligned} L_3(\psi) &= \sum_{k=1}^{K} \left| \mathbf{X}(k) - \mathbf{P}_{\mathbf{V}(\psi)}\mathbf{X}(k) \right|^2 \\ &= \sum_{k=1}^{K} \left| \mathbf{P}^{\perp}_{\mathbf{V}(\psi)}\mathbf{X}(k) \right|^2. \end{aligned} \tag{8.341}$$

Thus we can either minimize $L_3(\psi)$ over ψ or, equivalently, maximize

$$\boxed{L_4(\psi) = \textstyle\sum_{k=1}^{K} \left| \mathbf{P}_{\mathbf{V}(\psi)}\mathbf{X}(k) \right|^2.} \tag{8.342}$$

To get a useful geometric interpretation of (8.342), recall our discussion of signal subspaces in Section 5.5. The signal component of $\mathbf{X}(k)$ always lies in the signal subspace defined by the columns of $\mathbf{V}(\psi)$. The noise has two effects; it adds noise into the signal subspace and it adds noise in an orthogonal space that causes $\mathbf{X}(k)$ to lie outside the signal subspace. The CML estimator finds the D steering vectors that form a signal subspace that is as close as possible to the $\mathbf{X}(k), k = 1, 2, \cdots, K$. Closeness is measured by the magnitude of the projection of $\mathbf{X}(k), k = 1, 2, \cdots, K$ onto the estimated signal subspace.[14]

A second interpretation of (8.342) is also useful. We can rewrite (8.342) as

$$\begin{aligned} L_5(\psi) &= \frac{1}{K}\text{tr}\left[\mathbf{P}_{\mathbf{V}} \sum_{k=1}^{K} \mathbf{X}(k)\mathbf{X}(k)^H \mathbf{P}_{\mathbf{V}} \right] \\ &= \text{tr}\left[\mathbf{P}_{\mathbf{V}}\mathbf{P}_{\mathbf{V}}\mathbf{C}_{\mathbf{x}} \right] = \text{tr}\left[\mathbf{P}_{\mathbf{V}}\mathbf{C}_{\mathbf{x}} \right], \end{aligned} \tag{8.343}$$

where

$$\mathbf{C}_{\mathbf{x}} \triangleq \frac{1}{K} \sum_{k=1}^{K} \mathbf{X}(k)\mathbf{X}(k)^H. \tag{8.344}$$

[14]This result in (8.342) is equivalent to that first obtained by Schweppe [Sch68]; the geometric interpretation is due to Ziskind and Wax [ZW88a], [ZW88b].

Thus, the CML estimate is

$$\hat{\psi}_{cml} = \arg\max_{\psi} \left\{ \operatorname{tr}\left[\mathbf{P_V C_x}\right] \right\}, \tag{8.345}$$

or equivalently,

$$\hat{\psi}_{cml} = \arg\min_{\psi} \left\{ \operatorname{tr}\left[\mathbf{P_V^{\perp} C_x}\right] \right\}. \tag{8.346}$$

Now decompose $\mathbf{C_x}$ using an eigenvector expansion,

$$\mathbf{C_x} = \sum_{i=1}^{N} \hat{\lambda}_i \widehat{\mathbf{\Phi}}_i \widehat{\mathbf{\Phi}}_i^H. \tag{8.347}$$

Then we can write

$$\boxed{L_5(\psi) = \textstyle\sum_{i=1}^{N} \hat{\lambda}_i \left| \mathbf{P_V} \widehat{\mathbf{\Phi}}_i \right|^2.} \tag{8.348}$$

Thus, the CML estimator projects each of the estimated eigenvectors onto the signal subspace, weights the magnitude squared of that projection with the estimated eigenvalue, and sums over all eigenvectors. It does this for each value of ψ and chooses the value of ψ that gives the maximum value.

If we compare the conditional likelihood function in (8.345) with the asymptotic stochastic likelihood function in (8.310), we see that the difference is the first term in (8.310):

$$L_{aml}(\psi) - L_{cml}(\psi) = -\ln\det\left[\mathbf{P_V C_x P_V} + \sigma_w^2 \mathbf{P_V^{\perp}}\right]. \tag{8.349}$$

Thus, the two estimates will be similar when the term in (8.349) is negligible compared to $\operatorname{tr}\left[\mathbf{P_V^{\perp} C_x}\right]/\sigma_w^2$. Conversely, the two estimates will be different when the term in (8.349) is significant compared to $\operatorname{tr}\left[\mathbf{P_V^{\perp} C_x}\right]/\sigma_w^2$ and has its maximum at a different ψ.

By considering several examples, we observe that the estimates are more likely to be different in the following cases:

(i) Small signal separation relative to BW_{NN}.

(ii) Sources with high correlation and coherent sources.

(iii) A mixture of high *SNR* and low *SNR* signals.

If the noise variance σ_w^2 is unknown, then we estimate it. Returning to (8.336), we must retain the first term. Using (8.336) and (8.343), we have

$$L(\psi, \widehat{\mathbf{F}}, \sigma_w^2) = -N \ln \sigma_w^2 - \frac{1}{\sigma_w^2} \operatorname{tr}\left[\mathbf{P_V^{\perp} C_x}\right]. \tag{8.350}$$

Differentiating with respect to σ_w^2 and setting the result equal to zero, we obtain

$$\hat{\sigma}_w^2 = \frac{\operatorname{tr}\left[\mathbf{P}_{\mathbf{V}}^{\perp}\mathbf{C}_{\mathbf{x}}\right]}{N}. \tag{8.351}$$

Note that this estimate is different than the estimate in (8.313). Substituting (8.351) into (8.350) gives

$$L(\psi) = -N \ln\left(\operatorname{tr}\left[\mathbf{P}_{\mathbf{V}}^{\perp}\mathbf{C}_{\mathbf{x}}\right]\right). \tag{8.352}$$

Maximizing (8.352) is equivalent to minimizing the argument of the logarithm, so

$$\hat{\psi}_{cml} = \arg\min_{\psi}\left\{\operatorname{tr}\left[\mathbf{P}_{\mathbf{V}}^{\perp}\mathbf{C}_{\mathbf{x}}\right]\right\}, \tag{8.353}$$

which is identical to (8.346). Thus, knowing the variance of the noise does not affect the CML estimate.

Before doing a set of examples, we derive a closely related estimator.

8.5.3 Weighted Subspace Fitting

We can interpret the CML estimator as an algorithm that fits the subspace spanned by $\mathbf{V}(\psi)$ to the measurements,

$$\tilde{\mathbf{X}}_K \triangleq \left[\begin{array}{cccc} \mathbf{X}(1) & \mathbf{X}(2) & \cdots & \mathbf{X}(K) \end{array}\right], \tag{8.354}$$

in a least squares sense. To obtain this interpretation we rewrite $L_2(\psi, \mathbf{F})$ in (8.337) as a Frobenius norm,

$$\begin{aligned} L_2(\psi, \mathbf{F}) &= \left\|\tilde{\mathbf{X}}_K - \mathbf{V}(\psi)\mathbf{F}\right\|_F^2 \\ &= \operatorname{tr}\left\{\left[\tilde{\mathbf{X}}_K - \mathbf{V}(\psi)\mathbf{F}\right]^H\left[\tilde{\mathbf{X}}_K - \mathbf{V}(\psi)\mathbf{F}\right]\right\}, \end{aligned} \tag{8.355}$$

and

$$\left(\hat{\psi}, \widehat{\mathbf{F}}\right) = \arg\min_{\psi, \mathbf{F}}\left\{L_2(\psi, \mathbf{F})\right\}. \tag{8.356}$$

From (8.338)

$$\widehat{\mathbf{F}} = \mathbf{V}^{\dagger}\tilde{\mathbf{X}}_K, \tag{8.357}$$

where $\mathbf{V}^{\dagger}$ is the Moore-Penrose pseudoinverse,

$$\mathbf{V}^{\dagger} = \left[\mathbf{V}^H\mathbf{V}\right]^{-1}\mathbf{V}^H. \tag{8.358}$$

Then (8.356) reduces to

$$\hat{\psi}_{cml} = \arg\min_{\psi} \left\{ \mathrm{tr} \left[\mathbf{P}_{\mathbf{V}}^{\perp} \tilde{\mathbf{X}}_K \right]^H \left[\mathbf{P}_{\mathbf{V}}^{\perp} \tilde{\mathbf{X}}_K \right] \right\}, \tag{8.359}$$

which can be written as

$$\begin{aligned} \hat{\psi}_{cml} &= \arg\min_{\psi} \left\{ \mathrm{tr} \left[\mathbf{P}_{\mathbf{V}}^{\perp} \tilde{\mathbf{X}}_K \tilde{\mathbf{X}}_K^H \right] \right\} \\ &= \arg\min_{\psi} \left\{ \mathrm{tr} \left[\mathbf{P}_{\mathbf{V}}^{\perp} \mathbf{C}_{\mathbf{x}} \right] \right\}. \end{aligned} \tag{8.360}$$

We now consider more general subspace fitting techniques. The goal is to find an estimator that approaches the CRB asymptotically and has good performance in the threshold region. In addition, we would like the estimator to be computationally simpler than $\hat{\psi}_{cml}$.

The notion of weighted subspace fitting was introduced by Viberg and Ottersten [VO91].

We let $\mathbf{M}$ be an $N \times D'$ matrix representing the data, D' is the rank of $\mathbf{V}(\psi)\mathbf{F}$ (unless the signals are coherent, $D' = D$). An example of an $\mathbf{M}$ matrix that we use is obtained by an eigendecomposition of $\mathbf{C}_{\mathbf{x}}$,

$$\mathbf{C}_{\mathbf{x}} = \widehat{\mathbf{U}}_S \hat{\mathbf{\Lambda}}_S \widehat{\mathbf{U}}_S^H + \widehat{\mathbf{U}}_N \hat{\mathbf{\Lambda}}_N \widehat{\mathbf{U}}_N^H. \tag{8.361}$$

If we only utilize the data in the estimated signal subspace, we can define $\mathbf{M}$ as,

$$\mathbf{M} = \widehat{\mathbf{U}}_S \hat{\mathbf{\Lambda}}_S^{\frac{1}{2}}. \tag{8.362}$$

More generally we can write

$$\mathbf{M} = \widehat{\mathbf{U}}_S \mathbf{W}^{\frac{1}{2}}, \tag{8.363}$$

where $\mathbf{W}$ is a $D' \times D'$ diagonal matrix. This definition leads to the name, **weighted (signal) subspace fitting** (WSF). We define the cost function,

$$\begin{aligned} L(\psi, \mathbf{T}) &\triangleq \|\mathbf{M} - \mathbf{V}(\psi)\mathbf{T}\|_F^2 \\ &= \mathrm{tr} \left\{ [\mathbf{M} - \mathbf{V}(\psi)\mathbf{T}]^H [\mathbf{M} - \mathbf{V}(\psi)\mathbf{T}] \right\}, \end{aligned} \tag{8.364}$$

where $\mathbf{T}$ is a $D \times D'$ matrix. $\mathbf{M}$ is taking the place of $\tilde{\mathbf{X}}_K$ and incorporates the data input through (8.361). $\mathbf{T}$ is taking the place of $\mathbf{F}$.

Minimizing $L(\psi, \mathbf{T})$ with respect to $\mathbf{T}$ gives,

$$\widehat{\mathbf{T}} = \mathbf{V}^{\dagger}\mathbf{M}. \tag{8.365}$$

Substituting (8.365) into (8.364) and using the same steps as in (8.343)–(8.346), we obtain

$$\hat{\psi} = \arg\min_{\psi} \left\{ \text{tr} \left[\mathbf{P}_{\mathbf{V}}^{\perp} \mathbf{M}\mathbf{M}^{H} \right] \right\}. \tag{8.366}$$

Using (8.363) in (8.366) gives

$$\hat{\psi} = \arg\min_{\psi} \left\{ \text{tr} \left[\mathbf{P}_{\mathbf{V}}^{\perp} \widehat{\mathbf{U}}_{S} \mathbf{W} \widehat{\mathbf{U}}_{S}^{H} \right] \right\}. \tag{8.367}$$

The next step is to determine an optimum weighting $\mathbf{W}$. This is difficult to do for arbitrary K. Viberg and Ottersten [VO91] consider the asymptotic case and show that the weighting

$$\mathbf{W}_{ao} \triangleq \hat{\tilde{\mathbf{\Lambda}}}^{2} \hat{\mathbf{\Lambda}}_{S}^{-1}, \tag{8.368}$$

where

$$\hat{\tilde{\mathbf{\Lambda}}} \triangleq \hat{\mathbf{\Lambda}}_{S} - \hat{\sigma}_{w}^{2} \mathbf{I}, \tag{8.369}$$

with

$$\hat{\sigma}_{w}^{2} \triangleq \frac{1}{N-D} \sum_{i=D+1}^{N} \hat{\lambda}_{i}, \tag{8.370}$$

gives an estimator that asymptotically achieves the stochastic CRB. We investigate its performance in the non-asymptotic region by simulation. We refer to the estimator described by (8.367)–(8.370) as the asymptotically optimal WSF estimator (AOWSF) or more compactly, WSF_{ao}. When $D' = D$, the AOWSF cost function is equivalent to the method of direction estimation (MODE) algorithm derived by Stoica and Sharman [SS90a]. The two algorithms of interest are summarized in Table 8.4.

Table 8.4

CML	$\hat{\psi} = \arg\min_{\psi} \left\{ \text{tr} \left[\mathbf{P}_{\mathbf{V}}^{\perp} \mathbf{C}_{\mathbf{x}} \right] \right\}.$
WSF_{ao}/MODE	$\hat{\psi} = \arg\min_{\psi} \left\{ \text{tr} \left[\mathbf{P}_{\mathbf{V}}^{\perp} \hat{\mathbf{U}}_{S} \hat{\tilde{\mathbf{\Lambda}}}^{2} \hat{\mathbf{\Lambda}}_{S}^{-1} \hat{\mathbf{U}}_{S}^{H} \right] \right\}.$

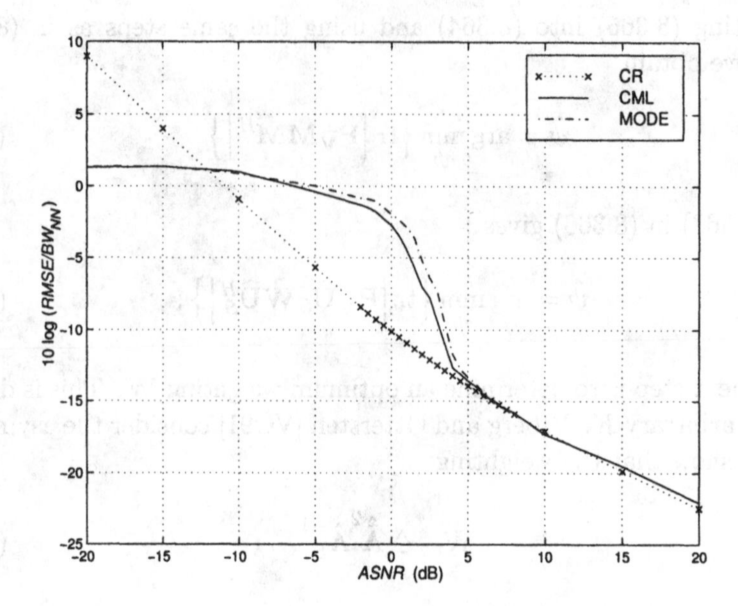

Figure 8.32 Normalized RMSE versus *ASNR*: CML and MODE estimators; $\Delta\psi = \Delta\psi_R$, $N = 10$, $K = 100$, $\rho = 0$, 100 trials.

We consider several of the examples that we discussed in the stochastic maximum likelihood case.

Example 8.5.7 (continuation, Example 8.5.2)

We consider the same model as in Example 8.5.2. The results are shown in Figure 8.32. We have also shown the AML estimator for comparison. We see that the estimators are close to the CRB for *ASNR*s greater than 10 dB. Below the threshold, CML and WSF (MODE) behave in a similar manner. We see that the threshold of all of conditional estimators are slightly to the left of the threshold of the AML estimator.

In Example 8.5.3, we reduced the separation to $\Delta u = 0.05 BW_{NN}$. We implemented the CML and MODE algorithms for this case and found that their threshold was the same as the AML estimator ($\simeq$ 17 dB) and that their RMSE approached the CRB above threshold.

Example 8.5.8 (continuation, Example 8.5.4)

We consider the same model as in Example 8.5.4. $\Delta\psi = \Delta\psi_R$ and $\rho = 0.95 e^{j\phi_\rho}$. In Figure 8.33, we show the results for $\phi_\rho = 0$. We see that MODE performs better than CML. The MODE threshold is slightly higher than the AML threshold in Figure 8.27. Above threshold, the approach to the CRB is slower than the uncorrelated case for both estimators.

In Figure 8.34, we plot the same results for $\phi_\rho = \pi/4$. The difference in performance between MODE and CML is larger. The MODE threshold is about 1 dB higher than the

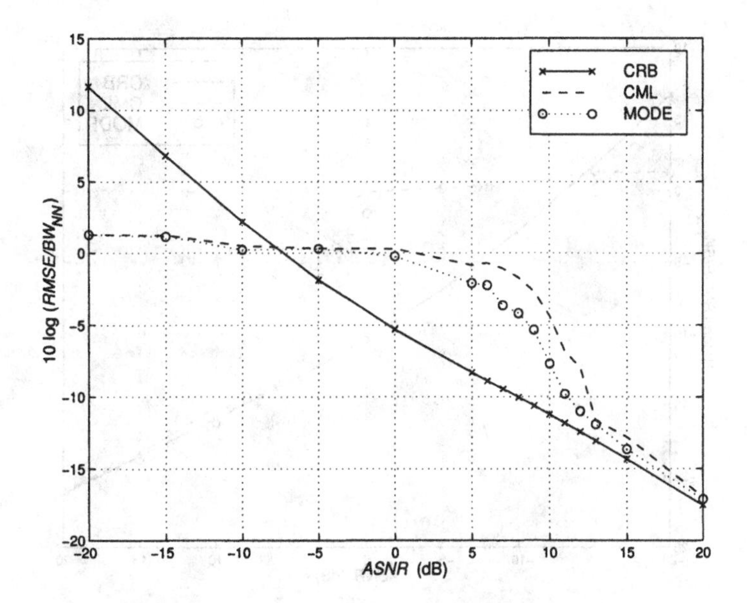

Figure 8.33 Normalized RMSE versus *ASNR*: MODE and CML estimators; $\Delta\psi = \Delta\psi_R$, $N = 10$, $K = 100$, $\rho = 0.95$, 100 trials.

AML threshold.

In Example 8.5.5, we considered the case of unequal signal powers. We implemented the CML and MODE algorithms for this scenario. The CML and MODE algorithms have very similar behavior and their performance is almost the same as the AML algorithm.

In Example 8.5.6, we considered the case of low sample support ($K = 2N$). We implemented the CML and MODE algorithms for this scenario. The CML and MODE algorithms have very similar performance and their performance is almost the same as the AML algorithm.

It is risky to draw general conclusions from a limited set of examples. However, it appears that AML, CML, and MODE all have similar performance in several interesting scenarios. In the correlated signal case, MODE is better than CML and similar to AML.

In the next section we discuss the asymptotic performance of the estimators that we have derived.

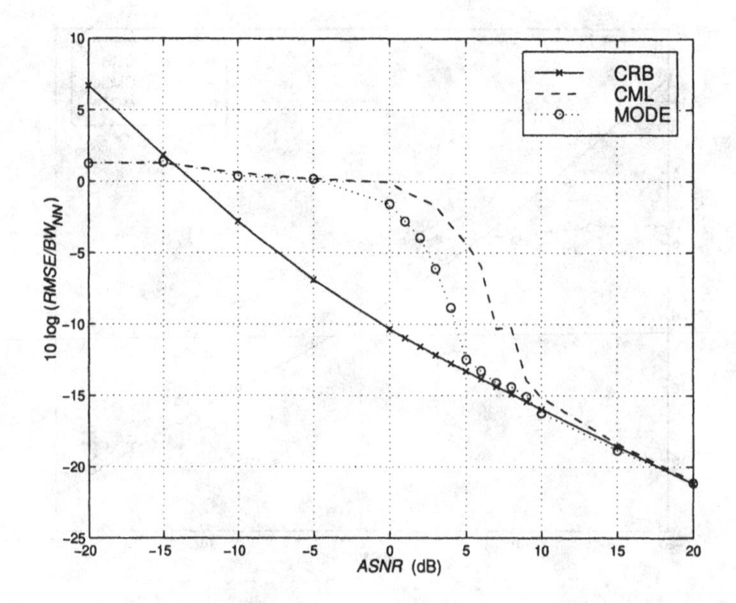

Figure 8.34 Normalized RMSE versus *ASNR*; MODE and CML estimators, $\Delta\psi = \Delta\psi_R$, $K = 10N$, $\rho = 0.95e^{j\pi/4}$, 100 trials.

8.5.4 Asymptotic Performance

The performance of estimators as the number of snapshots, K, goes to infinity is referred as the asymptotic performance and has been extensively studied in classical statistical theory (see, e.g., [And84] or [KS61]). An estimate is consistent if

$$\lim_{K\to\infty} E[\hat{\boldsymbol{\theta}}] = \boldsymbol{\theta}\,, \tag{8.371}$$

and is efficient if the variance of θ_i approaches the CRB.

The asymptotic performance of the UML and CML estimators has been analyzed by Stoica and Nehorai ([SN89a]), [SN90a], and [SN90b]). Their approach to the UML (AML) estimator utilizes a series expansion around the correct parameter value and standard statistical theory of ML estimators (e.g., [KS61]). They show that the AML estimator is consistent and that the variance of each component of $\boldsymbol{\psi}$ approaches the CRB. They reference an earlier paper by Ottersten and Ljung [OL89] that gives a direct derivation.

The CML estimator does not obey the regularity conditions of ML estimator theory because the numbers of parameters approaches infinity as $K \to \infty$. Stoica and Nehorai [SN89a] show that the CML estimator is not consistent as $K \to \infty$ and that it is not efficient. For a single signal, they show that

$$\frac{\sigma^2_{CML}}{\sigma^2_{CRB}} = 1 + \frac{1}{ASNR}. \tag{8.372}$$

In most applications, in order for the estimator to be above threshold, the *ASNR* is such that the second term in (8.372) is very small. We saw this behavior in our simulation examples where, on the scale that we were plotting the MSE, the RMSE of the CML estimator appeared to approach the CRB bound.

Other references that discuss the asymptotic behavior of ML estimates include Sandkühler and Böhme [SB87] and Benitz [Ben93]. We discuss asymptotic performance of other estimators in Section 9.5.

Note that we are frequently interested in the performance of the estimators for a fixed K as the *ASNR* goes to infinity. In many cases, the behavior is similar but a different analysis is required.

Asymptotic analyses play an important role in parameter estimation problems, and we quote results at various points in Chapters 8 and 9. It is important to remember two points when using the results of asymptotic analyses:

(i) Without simulations, it is often difficult to determine the values of *SNR* and K where the asymptotic results are valid.

(ii) It is desireable for an estimator to be consistent and efficient. However, we will encounter a number of useful estimators that are "almost consistent" or "almost efficient." If the bias or excess variance above the CRB is very small, the performance may be satisfactory. In many cases, the error due to the model mismatch will be larger than these terms.

8.5.5 Wideband Signals

A detailed discussion of the extension of ML techniques to wideband signals is contained in Doron and Weiss [DW92], [DWM93]. Our discussion follows [DW92]. We consider the stochastic ML estimate first. We use the frequency-domain snapshot model,

$$\mathbf{X}_k(m\omega_0) = \mathbf{V}(m\omega_0, \psi)\, \mathbf{F}_k(m\omega_0) + \mathbf{N}_k(m\omega_0), \qquad \begin{array}{c} m = 1, 2, \cdots, M \\ k = 1, 2, \cdots, K. \end{array} \tag{8.373}$$

There are three noise cases of interest:

1. $\eta_m \triangleq \sigma_w^2(m\omega_0), \quad m = 1, 2, \cdots, M$, is known.

2. $\eta_m \triangleq \sigma_w^2(m\omega_0), \quad m = 1, 2, \cdots, M$, is unknown.

3. $\eta \triangleq \sigma_w^2(m\omega_0) = \sigma_w^2, \quad m = 1, 2, \cdots, M$, and σ_w^2 is unknown.

We first consider the wideband version of the asymptotic ML estimate developed in Section 8.5.1 (see (8.310) and (8.315)).

The compressed likelihood functions follow in a manner analogous to the narrowband case. For case 1,

$$L_1(\psi) = \sum_{m=1}^{M} \left\{ -\frac{1}{\eta_m} \text{tr} \left[\mathbf{P}_{\mathbf{V}_m}^{\perp} \mathbf{C}_m \right] - \ln \det \left[\mathbf{P}_{\mathbf{V}_m} \mathbf{C}_m \mathbf{P}_{\mathbf{V}_m} + \eta_m \mathbf{P}_{\mathbf{V}_m}^{\perp} \right] \right\}, \tag{8.374}$$

where

$$\mathbf{C}_m \triangleq \frac{1}{K} \sum_{k=1}^{K} \mathbf{X}_k(m\omega_0) \mathbf{X}_k^H(m\omega_0), \tag{8.375}$$

is the sample covariance matrix at $m\omega_0$. Then,

$$\hat{\psi}_{aml} = \arg\max_{\psi} \sum_{m=1}^{M} \left\{ -\ln \det \left[\mathbf{P}_{\mathbf{V}_m} \mathbf{C}_m \mathbf{P}_{\mathbf{V}_m} + \eta_m \mathbf{P}_{\mathbf{V}_m}^{\perp} \right] - \frac{1}{\eta_m} \text{tr} \left[\mathbf{P}_{\mathbf{V}_m}^{\perp} \mathbf{C}_m \right] \right\}. \tag{8.376}$$

Note that $\mathbf{V}_m$, the array manifold at $\omega = m\omega_0$ is a function of m and therefore $\mathbf{P}_{\mathbf{V}_m}$ is a function of m.

For case 2,

$$\hat{\eta}_m = \frac{\text{tr} \left[\mathbf{P}_{\mathbf{V}_m}^{\perp} \mathbf{C}_m \right]}{N - D}, \tag{8.377}$$

and

$$L_2(\psi) = -H_S - (N - D) \sum_{m=1}^{M} \ln \left\{ \text{tr} \left[\mathbf{P}_{\mathbf{V}_m}^{\perp} \mathbf{C}_m \right] \right\}, \tag{8.378}$$

where

$$H_S = \sum_{m=1}^{M} \ln \det \left[\mathbf{P}_{\mathbf{V}_m} \mathbf{C}_m \mathbf{P}_{\mathbf{V}_m} + \mathbf{P}_{\mathbf{V}_m}^{\perp} \right]. \tag{8.379}$$

For case 3,

$$\hat{\eta} = \frac{\sum_{m=1}^{M} \text{tr} \left[\mathbf{P}_{\mathbf{V}_m}^{\perp} \mathbf{C}_m \right]}{(N - D)M}, \tag{8.380}$$

and

$$L_3(\psi) = -H_S - M(N-D)\ln \sum_{m=1}^{M} \text{tr}\left[\mathbf{P}_{\mathbf{V}_m}^{\perp}\mathbf{C}_m\right]. \tag{8.381}$$

The likelihood functions for the CML estimator follow in a similar manner.

For case 1 (known noise variance case),

$$L_1^C(\psi) = \sum_{m=1}^{M} \frac{1}{\eta_m}\text{tr}\left[\mathbf{P}_{\mathbf{V}_n}\mathbf{C}_m\right], \tag{8.382}$$

and

$$\hat{\psi}_{cml} = \arg\max_{\psi} L_1^C(\psi). \tag{8.383}$$

For case 2,

$$L_2^C(\psi) = -\sum_{m=1}^{M} \ln\left(\text{tr}\left[\mathbf{P}_{\mathbf{V}_m}^{\perp}\mathbf{C}_m\right]\right), \tag{8.384}$$

where we have dropped unnecessary constants. Thus,

$$\hat{\psi}_{cml} = \arg\max_{\psi}\left\{\sum_{m=1}^{M} \ln\left(\text{tr}\left[\mathbf{P}_{\mathbf{V}_m}^{\perp}\mathbf{C}_m\right]\right)\right\}. \tag{8.385}$$

For case 3,

$$L_3^C(\psi) = \sum_{m=1}^{M} \text{tr}\left[\mathbf{P}_{\mathbf{V}_m}\mathbf{C}_m\right]. \tag{8.386}$$

Thus,

$$\hat{\psi}_{cml} = \arg\max_{\psi} \sum_{m=1}^{M} \text{tr}\left[\mathbf{P}_{\mathbf{V}_m}\mathbf{C}_m\right]. \tag{8.387}$$

The noise estimate is

$$\hat{\eta}_n = \left.\frac{\sum_{m=1}^{M} \text{tr}\left[\mathbf{P}_{\mathbf{V}_m}^{\perp}\mathbf{C}_m\right]}{MN}\right|_{\psi=\psi_{cml}^{(3)}}. \tag{8.388}$$

The extension of the CRB to the wideband case follows in a straightforward manner. The other ML techniques that were developed for the narrowband case can also be extended to the wideband case.

Böhme and his colleagues have published a number of papers on wideband ML estimation (e.g., [Boh89], [KDB93], [KMB92], and [KB93]). We revisit the wideband case briefly in Section 9.10 and give a few more references. However, we do not study the case in a detailed manner.

8.5.6 Summary

In this section, we have studied ML estimators in detail. We have found the estimates using a grid search over a D-dimensional u-space followed by a local minimization. Due to the computational complexity, this search approach is generally not used in practice. In the next two sections, we develop computational algorithms that provide good results with reduced computational complexity.

8.6 Computational Algorithms

In this section, we discuss computational techniques for finding stochastic maximum likelihood estimators and CML estimators.

In Section 8.6.1, we discuss several optimization techniques. In Section 8.6.2, we develop the alternating projection (AP) and alternating maximization algorithm. In Section 8.6.3, we develop the expectation-maximization algorithm. In Section 8.6.4, we summarize our results and discuss other techniques that are available.

8.6.1 Optimization Techniques

In this section, we give a brief discussion of search techniques that can be used to solve the minimization problem,

$$\hat{\boldsymbol{\psi}} = \arg \min_{\boldsymbol{\psi}}[F(\boldsymbol{\psi})]. \tag{8.389}$$

We are interested in three expressions for $F(\boldsymbol{\psi})$ corresponding to the AML, CML, and WSF estimatiors.

From (8.315),

$$F_{aml} = \det\left[\mathbf{P_V C_x P_V} + \frac{\operatorname{tr}\left[\mathbf{P_V^\perp C_x}\right]\mathbf{P_V^\perp}}{N-D}\right]. \tag{8.390}$$

From Table 8.4,

$$F_{cml} = \operatorname{tr}\left[\mathbf{P_V^\perp C_x}\right], \tag{8.391}$$

and

$$F_{wsf} = \operatorname{tr}\left[\mathbf{P_V^\perp \hat{U}}_S \hat{\tilde{\boldsymbol{\Lambda}}}^2 \hat{\boldsymbol{\Lambda}}_S^{-1} \hat{\mathbf{U}}_S^H\right]. \tag{8.392}$$

Most of the techniques that we discuss in this section originated in the nonlinear optimization area and are adapted to solve our estimation problems. The reader needs to explore some the optimization sources, such as Dennis and Schnabel [DS96], Gill et al. [GMW81], or Nash and Sofer [NS96], to get a comprehensive discussion. These sources have been utilized for ML estimation by Ottersten et al. [OVSN93], and several of our results are from that reference.

We utilize a search technique to find the value of ψ that minimizes $F(\psi)$. The basic idea is to model $F(\psi)$ as a quadratic function in the vicinity of the minimum. We select a starting point ψ_0. We then find a descent direction that will cause $F(\psi)$ to decrease and calculate a step size to determine how far to move in the descent direction. We discuss techniques for choosing $\hat{\psi}_0$ in Section 8.6.2.

At the first iteration,

$$\hat{\psi}_1 = \hat{\psi}_0 + \mathbf{a}_0 \left[\nabla_{\psi} F(\psi)\right]\Big|_{\psi=\hat{\psi}_0}. \tag{8.393}$$

In (8.393), $\hat{\psi}_0$ and $\hat{\psi}_1$ are $D \times 1$ vectors, $\nabla_{\psi} F(\psi)$ is the $D \times 1$ gradient vector,

$$\nabla_{\psi} F(\psi) \triangleq [\frac{\partial F(\psi)}{\partial \psi_1} \cdots \frac{\partial F(\psi)}{\partial \psi_D}]^T \triangleq \mathbf{F}'(\psi), \tag{8.394}$$

and $\mathbf{a}_0$ is a $D \times D$ matrix. The gradient vector will equal $\mathbf{0}$ at a stationary point and

$$\nabla_{\psi} F(\psi) = \mathbf{0}, \tag{8.395}$$

is a necessary but not sufficient condition for a minimum. The matrix $\mathbf{a}_0$ determines both the descent direction and the step size. We will separate $\mathbf{a}_0$ into two terms in subsequent equations. At the kth iteration,

$$\hat{\psi}_{k+1} = \hat{\psi}_k + \mathbf{a}_k \left\{\left[\nabla_{\psi} F(\psi)\right]_{\psi=\hat{\psi}_k}\right\}. \tag{8.396}$$

We continue the iteration until we satisfy a stopping rule of the form,

$$|F(\hat{\psi}_{k+1}) - F(\hat{\psi}_k)| < \delta_1, \tag{8.397}$$

or

$$\| \hat{\psi}_{k+1} - \hat{\psi}_k \| < \delta_2. \tag{8.398}$$

is satisfied.

In the classical Newton method,

$$\mathbf{a}_k = -\left[\nabla^2 F(\psi)\right]^{-1}_{\psi=\hat{\psi}_k}, \tag{8.399}$$

where $\nabla^2 F(\psi)$ is the Hessian matrix.

$$\mathbf{H}(\psi_k) \triangleq \left[\nabla^2 F(\psi)\right]_{ij} \triangleq \frac{\partial^2 F(\psi)}{\partial\psi_i \partial\psi_j}. \tag{8.400}$$

In order for $-\mathbf{H}^{-1}(\psi_k)\mathbf{F}'(\psi_k)$ to be a descent direction, $\mathbf{H}(\psi_k)$ must be positive definite. This may not be true if we are too far from the minimum, because $F(\psi)$ may not be quadratic. There are different techniques for modifying $\mathbf{H}$ to make it positive definite. One approach is to use DL,

$$\tilde{\mathbf{H}}_k = \mathbf{H}_k + \sigma_k^2 \mathbf{I}, \tag{8.401}$$

where σ_k^2 is chosen at each step to make $\tilde{\mathbf{H}}_k$ both positive definite and well-conditioned. This technique is discussed in Chapter 5 of [DS96].

In practice, we want to adjust the step size, so we use a damped Newton algorithm

$$\hat{\psi}_{k+1} = \hat{\psi}_k - \mu_k \mathbf{H}_k^{-1} \mathbf{F}'_k, \tag{8.402}$$

where $\mathbf{F}'_k$ is the gradient defined in (8.394) (e.g., [GMW81], [DS83], [DS96]).

In order to choose the step length μ_k, we choose a $\mu < 1$ and let

$$\mu_k = (\mu)^i, \quad i \geq 0. \tag{8.403}$$

At each step in the iteration, we try successive values of i starting at $i = 0$ and use the smallest i that causes an adequate decrease in $F(\psi_k)$. For example, if $\mu = 0.5$, we would try

$$\mu_k = 1, \frac{1}{2}, \left(\frac{1}{2}\right)^2, \left(\frac{1}{2}\right)^3, \cdots. \tag{8.404}$$

until we obtain a satisfactory decrease in $F(\psi_k)$. Quadratic convergence is obtained if the step length converges to unity.

The advantage of the Newton algorithm is that one can show quadratic convergence. However, in order to implement the Newton method, we must compute a matrix of second derivatives and invert a $D \times D$ matrix. In most applications, we try to find a computationally simpler approximation to the Newton method that still converges at an adequate rate.

One approach is to replace the Hessian by an approximate Hessian that has the same form as the asymptotic form of the Hessian matrix. This approach is referred to in the statistical literature as the **scoring** method. Ottersten [OVSN93] have derived the expression and we use their derivation.

Using (8.296), we can rewrite (8.390) as

$$F_{aml}(\psi) = \ln \det \left[\mathbf{V}\hat{\mathbf{S}}_{\mathbf{f}}(\psi)\mathbf{V}^H + \hat{\sigma}_w^2 \mathbf{I}\right], \tag{8.405}$$

where, from (8.313),

$$\hat{\sigma}_w^2(\psi) = \frac{\text{tr}\left[\mathbf{P}_{\mathbf{V}}^{\perp}\mathbf{C}_{\mathbf{x}}\right]}{N-D}. \tag{8.406}$$

We included the ln in (8.405) to be consistent with the derivation in [OVSN93]. Then, using $|\mathbf{I}+\mathbf{AB}| = |\mathbf{I}+\mathbf{BA}|$, we can write (8.405) as

$$\begin{aligned} F_{aml}(\psi) &= \ln\left\{\hat{\sigma}_w^{2N}(\psi)\left|\hat{\sigma}_w^{-2}(\psi)\hat{\mathbf{S}}_{\mathbf{f}}(\psi)\mathbf{V}^H\mathbf{V}+\mathbf{I}\right|\right\} \\ &= \ln\left\{\hat{\sigma}_w^{2N}(\psi)\left|\hat{\sigma}_w^{-2}(\psi)\left[\mathbf{V}^{\dagger}\left[\mathbf{C}_{\mathbf{x}} - \hat{\sigma}_w^2(\psi)\mathbf{I}\right]\mathbf{V}+\mathbf{I}\right]\right|\right\} \\ &= \ln\left\{\hat{\sigma}_w^{2N}(\psi)\left|\hat{\sigma}_w^{-2}(\psi)\left[\mathbf{V}^{\dagger}\mathbf{C}_{\mathbf{x}}\mathbf{V}\right]\right|\right\} \\ &= \ln\left\{\hat{\sigma}_w^{2(N-D)}(\psi)\left|\mathbf{V}^{\dagger}\mathbf{C}_{\mathbf{x}}\mathbf{V}\right|\right\}. \end{aligned} \tag{8.407}$$

Now define a matrix

$$\mathbf{G} \triangleq \mathbf{V}\left[\left[\mathbf{V}^H\mathbf{C}_{\mathbf{x}}\mathbf{V}\right]^{-1} - \hat{\sigma}_w^{-2}(\psi)\left[\mathbf{V}^H\mathbf{V}\right]^{-1}\right]. \tag{8.408}$$

The gradient matrix is

$$\mathbf{F}'_{aml}(\psi) = 2Re(\text{DIAG}\left[\mathbf{G}^H\mathbf{C}_{\mathbf{x}}\mathbf{P}_{\mathbf{V}}^{\perp}\mathbf{D}\right]), \tag{8.409}$$

where $\mathbf{D}$ is defined in (8.97) and DIAG[$\mathbf{A}$] is a column matrix formed from the diagonal elements of $\mathbf{A}$.

Because the AML estimate is efficient, the approximate Hessian is derived from the CRB in (8.102),

$$\mathbf{H}_{aml,a}(\psi) = \frac{2}{\sigma_w^2} Re\left\{\left[\mathbf{D}^H\mathbf{P}_{\mathbf{V}}^{\perp}\mathbf{D}\right] \odot \left[\mathbf{S}_{\mathbf{f}}\mathbf{V}^H\mathbf{S}_{\mathbf{x}}^{-1}\mathbf{V}\mathbf{S}_{\mathbf{f}}\right]^T\right\}. \tag{8.410}$$

We replace the ensemble statistics $\mathbf{S}_{\mathbf{f}}$ and σ_w^2 by their AML estimates $\hat{\mathbf{S}}_f(\psi)$ and $\hat{\sigma}_w^2(\psi)$ and use (8.299) to replace $\mathbf{S}_{\mathbf{x}}$ by

$$\hat{\mathbf{S}}_{\mathbf{x}}(\psi) = \mathbf{V}\hat{\mathbf{S}}_{\mathbf{f}}(\psi)\mathbf{V}^H + \hat{\sigma}_w^2(\psi)\mathbf{I}. \tag{8.411}$$

Then, the second matrix in (8.410) can be written as

$$\hat{\mathbf{S}}_{\mathbf{f}}(\psi)\mathbf{V}^H\left[\mathbf{V}\hat{\mathbf{S}}_{\mathbf{f}}(\psi)\mathbf{V}^H+\hat{\sigma}_w^2(\psi)\mathbf{I}\right]^{-1}\mathbf{V}\hat{\mathbf{S}}_{\mathbf{f}}(\psi)=\hat{\sigma}_w^4(\psi)\mathbf{G}^H\mathbf{C}_{\mathbf{x}}\mathbf{G}. \quad (8.412)$$

Using (8.412) in (8.410) gives the approximate Hessian matrix,

$$\mathbf{H}_{aml,a}(\psi)=2\hat{\sigma}_w^2(\psi)Re\left\{\left[\mathbf{D}^H\mathbf{P}_{\mathbf{V}}^{\perp}\mathbf{D}\right]\odot\left[\mathbf{G}^H\mathbf{C}_{\mathbf{x}}\mathbf{G}\right]^T\right\}. \quad (8.413)$$

The terms in (8.402) are now defined. At each step, we compute $\hat{\sigma}_w^2(\psi_k)$ using (8.406) and use the result in (8.408), (8.409), and (8.413) to compute $\mathbf{F}'_{aml}(\psi_k)$ and $\mathbf{H}_{aml,a}(\psi_k)$. We substitute the results in (8.402) and use (8.403) and (8.404) to compute the step size. We then compute $\hat{\psi}_{k+1}$. We continue the iteration until the stopping rule is satisfied. We choose $\hat{\psi}_0$ using the AP or AM technique, which we derive in Section 8.6.2.

The approximate Hessians for the CML estimator and the WSF$_{ao}$ estimator are also derived in [OVSN93]. These authors also consider several examples and discuss efficient computational techniques. The reader is referred to this reference for a more detailed discussion.

In the next section, we discuss a different approach to quasi-Newton methods.

8.6.1.1 Quasi-Newton methods

There are a number of other algorithms that attempt to retain some of the good properties of the Newton algorithm, but with a reduced computational cost. In this section, we present one of these algorithms that is effective in array processing problems.

The starting point for the methods in this section is the Newton algorithm.

Once again, we choose an initial estimate $\hat{\psi}_0$. The iteration is

$$\hat{\psi}_{k+1}=\hat{\psi}_k-\mu_k\mathbf{H}_k^{-1}(\psi_k)\mathbf{F}'(\psi). \quad (8.414)$$

We use a modified form of the Newton algorithm which is referred to as the **modified variable projection algorithm** (e.g., [Kau75], [RW80], [GP73]). The algorithm was applied to the array processing problem by Viberg et al. [VOK91], and our discussion follows that reference. We consider the CML and WSF$_{ao}$ estimators. We write $F(\psi)$ as

$$F(\psi)=\mathrm{tr}\left\{\mathbf{P}_{\mathbf{V}}^{\perp}(\psi)\mathbf{M}\mathbf{M}^H\right\}=\left\|\mathbf{P}_{\mathbf{V}}^{\perp}(\psi)\mathbf{M}\right\|_F^2. \quad (8.415)$$

The matrix $\mathbf{M}$ represents the data. In the case of CML,

$$\mathbf{M}\mathbf{M}^H = \mathbf{C}_{\mathbf{x}}, \tag{8.416}$$

so $\mathbf{M}$ equals $\tilde{\mathbf{X}}_K$.

In the case of weighted subspace fitting,

$$\mathbf{M} = \widehat{\mathbf{U}}_s \mathbf{W}^{\frac{1}{2}} \tag{8.417}$$

(see (8.363)).

We now define a vector $\mathbf{r}$ by stacking the columns of $\mathbf{P}_{\mathbf{V}}^{\perp}(\boldsymbol{\psi})\mathbf{M}$,

$$\mathbf{r} = vec\left(\mathbf{P}_{\mathbf{V}}^{\perp}(\boldsymbol{\psi})\mathbf{M}\right). \tag{8.418}$$

Using (8.418) in (8.415) gives

$$F(\boldsymbol{\psi}) = |\mathbf{r}|^2. \tag{8.419}$$

The gradient of $F(\boldsymbol{\psi})$ with respect to ψ_i is

$$\frac{\partial}{\partial \psi_i} F = 2Re\left\{\left(\frac{\partial \mathbf{r}}{\partial \psi_i}\right)^H \mathbf{r}\right\} = 2Re\left\{\mathbf{r}_i^H \mathbf{r}\right\}, \tag{8.420}$$

where

$$\mathbf{r}_i \triangleq \frac{\partial \mathbf{r}}{\partial \psi_i} = \frac{\partial \mathbf{P}_{\mathbf{V}}^{\perp}(\boldsymbol{\psi})}{\partial \psi_i}\mathbf{M}. \tag{8.421}$$

The derivative of the projection matrix is,

$$\frac{\partial}{\partial \psi_i}\mathbf{P}_{\mathbf{V}}^{\perp} = -\frac{\partial}{\partial \psi_i}\mathbf{P}_{\mathbf{V}} = -\mathbf{P}_{\mathbf{V}}^{\perp}\mathbf{V}_i\mathbf{V}^{\dagger} - \left(-\mathbf{P}_{\mathbf{V}}^{\perp}\mathbf{V}_i\mathbf{V}^{\dagger}\right)^H, \tag{8.422}$$

where

$$\mathbf{V}_i \triangleq \frac{\partial \mathbf{V}}{\partial \psi_i}. \tag{8.423}$$

Using (8.422) in (8.420) gives

$$\mathbf{F}'(\boldsymbol{\psi}) = -2Re\left\{\mathrm{DIAG}\left[\mathbf{V}^{\dagger}\mathbf{M}\mathbf{M}^H\mathbf{P}_{\mathbf{V}}^{\perp}\mathbf{D}(\boldsymbol{\psi})\right]\right\}, \tag{8.424}$$

where $\mathbf{D}(\boldsymbol{\psi})$ is defined in (8.97) and DIAG[$\mathbf{A}$] is defined after (8.409).

The next step is to derive the Hessian,

$$\frac{\partial^2}{\partial \psi_i \partial \psi_j} F = 2Re\left\{\mathbf{r}_i^H \mathbf{r}_j + \mathbf{r}_{ij}^H \mathbf{r}\right\}. \tag{8.425}$$

The Gaussian modification of the Newton method assumes that the second term in (8.425) is small and approximates the Hessian by

$$\mathbf{H}_G = 2Re\left\{\mathbf{r}_i^H \mathbf{r}_j\right\}. \tag{8.426}$$

Discarding the second derivative guarantees that (8.414) is a descent method because $\mathbf{H}_G$ is nonnegative definite. The resulting algorithm is the variable projection algorithm of Golub and Pereyra [GP73]. Using (8.418) in (8.422) and observing that $\mathbf{V}^\dagger \mathbf{P}_{\mathbf{V}}^\perp = \mathbf{0}$, we obtain

$$\mathbf{r}_i^H \mathbf{r}_j = \text{tr}\left\{\left[\left[\mathbf{V}^\dagger\right]^H \mathbf{V}_i^H \mathbf{P}_{\mathbf{V}}^\perp \mathbf{V}_j \mathbf{V}^\dagger + \mathbf{P}_{\mathbf{V}}^\perp \mathbf{V}_i \mathbf{V}^\dagger \left[\mathbf{V}^\dagger\right]^H \mathbf{V}_j^H \mathbf{P}_{\mathbf{V}}^\perp\right] \mathbf{M}\mathbf{M}^H\right\}. \tag{8.427}$$

Kaufman [Kau75] modifies the Gauss-Newton algorithm by deleting the second term in (8.427). With this modification,

$$H_{ij} = 2Re\left\{\text{tr}\left[\left[\mathbf{V}^\dagger\right]^H \mathbf{V}_i^H \mathbf{P}_{\mathbf{V}}^\perp \mathbf{V}_j \mathbf{V}^\dagger \mathbf{M}\mathbf{M}^H\right]\right\}, \tag{8.428}$$

which can be expressed in matrix notation as,

$$\mathbf{H} = 2Re\left\{\left[\mathbf{D}^H \mathbf{P}_{\mathbf{V}}^\perp \mathbf{D}\right] \odot \left[\mathbf{V}^\dagger \mathbf{M}\mathbf{M}^H \left[\mathbf{V}^\dagger\right]^H\right]^T\right\}. \tag{8.429}$$

The iteration algorithm is defined by (8.414), (8.424), and (8.429). It is pointed out in [OVSN93] that, for the WSF case, this algorithm is the same as the WSF version of the asymptotic Hessian (scoring) algorithm discussed in Section 8.6.1.1. (We did not derive the WSF version.)

In order to obtain convergence to a global minimum we must initialize the algorithm appropriately. In Section 8.7.2.1, we develop the alternating projection algorithm that we use to initialize the modified Gauss-Newton algorithm. In [OVSN93], examples are given to show the effect of initialization accuracy.

8.6.1.2 Summary

In this section, we have provided a brief discussion of gradient techniques for solving ML estimation problems. The reader is referred to the various references, particularly [OVSN93] for further discussion.

8.6.2 Alternating Maximization Algorithms

In this section, we develop techniques that are sometimes referred to in the optimization literature as the relaxation method. The basic idea is straightforward. We have a function of D variables, $F(\boldsymbol{\psi})$. We want to find the value of $\boldsymbol{\psi}$ that maximizes $F(\boldsymbol{\psi})$. We accomplish this by a sequence of 1-D maximization problems.

The technique was applied to the array processing problem by Ziskind and Wax [ZW88b]. For the CML cases, they utilize a property of projection matrices to simplify the algorithm, and the resulting algorithm is referred to in the array processing literature as the alternating projection (AP) algorithm. For the stochastic ML case, the algorithm is referred to as the alternating maximization (AM) algorithm. Both algorithms use the same idea.

8.6.2.1 Alternating projection algorithm

The AP algorithm was used by Ziskind and Wax [ZW88b] in order to replace the multidimensional maximization problem by a sequence of 1-D maximization problems. Our discussion follows [ZW88b].

In this subsection we consider the CML estimate. From (8.345), we must maximize

$$\hat{\boldsymbol{\psi}}_{cml} = \arg\max_{\boldsymbol{\psi}} \{\mathrm{tr}\,[\mathbf{P}_{\mathbf{V}}\mathbf{C}_{\mathbf{x}}]\}. \tag{8.430}$$

The basic idea is straightforward. It is an iterative technique in which, at each step of the iteration, we hold $D-1$ parameter values constant and maximize over a single parameter.

Thus, the value of ψ_i at the $(k+1)$-th iteration is obtained by the following one-dimensional maximization problem,

$$\hat{\psi}_i^{(k+1)} = \arg\max_{\psi_i} \left\{ \mathrm{tr}\left[\mathbf{P}_{\left[\mathbf{V}(\hat{\boldsymbol{\psi}}_{(i)}^{(k)}),\mathbf{v}(\psi_i)\right]} \mathbf{C}_{\mathbf{x}} \right] \right\}, \tag{8.431}$$

where $\hat{\boldsymbol{\psi}}_{(1)}^{(k)}$ is the value of the estimated vector parameter $(D-1\times 1)$ at the kth iteration with the first component removed,

$$\hat{\boldsymbol{\psi}}_{(1)}^{(k)} = \left[\; \hat{\psi}_2^{(k)} \;\vdots\; \hat{\psi}_3^{(k)} \;\vdots\; \cdots \;\vdots\; \hat{\psi}_D^{(k)} \;\right]. \tag{8.432}$$

The algorithm maximizes the ψ_i in order, starting at $i=1$. For $i>1$, $\hat{\boldsymbol{\psi}}_{(i)}^{(k)}$ is defined as

$$\hat{\boldsymbol{\psi}}_{(i)}^{(k)} = \left[\; \hat{\psi}_1^{(k+1)} \;\vdots\; \cdots \;\vdots\; \hat{\psi}_{i-1}^{(k+1)} \;\vdots\; \hat{\psi}_{i+1}^{(k)} \;\vdots\; \cdots \;\vdots\; \hat{\psi}_D^{(k)} \;\right]. \tag{8.433}$$

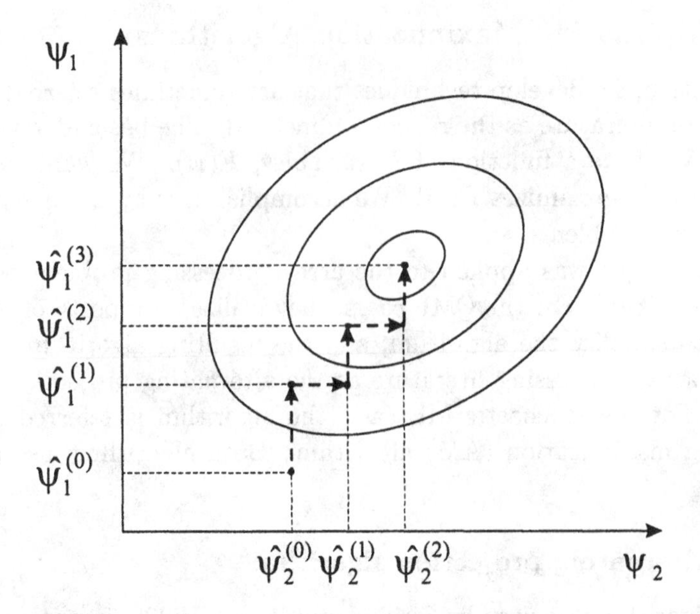

Figure 8.35 AP Algorithm: successive iterations.

The estimates of ψ_r for $r < i$ are the results of the $(k+1)$th iteration, which was done previously.

Note that we have separated the space of the projection matrix into a $N \times (D-1)$ fixed component and a $N \times 1$ component that is allowed to vary.

The graphical behavior of the algorithm is shown in Figure 8.35. The algorithm moves to the peak in steps parallel to the axes. Since the value of $L(\psi)$ is maximized at each step, convergence to a local maximum is guaranteed. The initial condition is key to convergence to the global maximum.

Ziskind and Wax [ZW88b] use the following initialization procedure. First solve the problem for a single source ψ_1. Thus

$$\hat{\psi}_1^{(0)} = \arg\max_{\psi_1} \left\{ tr \left[\mathbf{P}_{\mathbf{v}(\psi_1)} \mathbf{C}_{\mathbf{x}} \right] \right\}. \tag{8.434}$$

This is equivalent to assuming there is a single source. Then solve for the second source assuming the first source is at $\hat{\psi}_1^{(0)}$. Thus,

$$\hat{\psi}_2^{(0)} = \arg\max_{\psi_2} \left\{ \mathrm{tr} \left[\mathbf{P}_{\left[\mathbf{v}(\hat{\psi}_1^{(0)}), \mathbf{v}(\psi_2)\right]} \mathbf{C}_{\mathbf{x}} \right] \right\}. \tag{8.435}$$

We continue this procedure until we obtain the D initial estimated values,

$\hat{\psi}_1^{(0)}, \cdots, \hat{\psi}_D^{(0)}$. At each step, we assume that all preceding initial values are known. With these initial values, we then carry out the iteration in (8.431).

There is a property of projection matrices referred to as the projection matrix update formula that simplifies the computation at each iteration.

Property

Let $\mathbf{B}$ and $\mathbf{C}$ be arbitrary matrices with the same number of rows and define the composite matrix $\left[\ \mathbf{B} \ \vdots \ \mathbf{C}\ \right]$. If $\mathbf{B}$ and $\mathbf{C}$ were orthogonal, then

$$\mathbf{P}_{\mathbf{B},\mathbf{C}} = \mathbf{P}_{\mathbf{B}} + \mathbf{P}_{\mathbf{C}}. \tag{8.436}$$

However, since they are not orthogonal, we define $\mathbf{P}_{\mathbf{C}(\mathbf{B})}$, where $\mathbf{C}(\mathbf{B})$ is the residual of the columns of $\mathbf{C}$, when $\mathbf{C}$ is projected on $\mathbf{B}$,

$$\mathbf{C}(\mathbf{B}) = [\mathbf{I} - \mathbf{P}_{\mathbf{B}}]\,\mathbf{C} = \mathbf{P}_{\mathbf{B}}^{\perp}\mathbf{C}. \tag{8.437}$$

Then, the projection matrix onto the column space of $\left[\ \mathbf{B} \ \vdots \ \mathbf{C}\ \right]$ is

$$\mathbf{P}_{\mathbf{B},\mathbf{C}} = \mathbf{P}_{\mathbf{B}} + \mathbf{P}_{\mathbf{C}(\mathbf{B})}. \tag{8.438}$$

To utilize this result, we let

$$\mathbf{B} = \mathbf{V}(\hat{\boldsymbol{\psi}}_{(i)}^{(k)}), \tag{8.439}$$

and

$$\mathbf{C} = \mathbf{v}(\psi_i). \tag{8.440}$$

Using this result in (8.431), we can write

$$\mathbf{P}_{\mathbf{V}(\hat{\boldsymbol{\psi}}_{(i)}^{(k)}),\mathbf{v}(\psi_i)} = \mathbf{P}_{\mathbf{V}(\hat{\boldsymbol{\psi}}_{(i)}^{(k)})} + \mathbf{P}_{\mathbf{v}(\psi_i)\left(\mathbf{V}(\hat{\boldsymbol{\psi}}_{(i)}^{(k)})\right)}. \tag{8.441}$$

The vector in the subscript in the second term is generated by constructing:

(i) $\mathbf{P}_{\mathbf{V}(\hat{\boldsymbol{\psi}}_{(i)}^{(k)})}$ (corresponds to $\mathbf{P}_{\mathbf{B}}$).

(ii) $\left[\mathbf{P}^{\perp}_{\mathbf{V}(\hat{\boldsymbol{\psi}}_{(i)}^{(k)})}\right]\mathbf{v}(\psi_i)$ (corresponds to $\mathbf{P}_{\mathbf{B}}^{\perp}\mathbf{C}$).

(iii) $\mathbf{P}_{\mathbf{v}(\psi_i)(\mathbf{V}(\hat{\boldsymbol{\psi}}_{(i)}^{(k)})}$ (corresponds to $\mathbf{P}_{\mathbf{C}(\mathbf{B})}$).

Since the first term in (8.441) is not a function of ψ_i it can be dropped. Thus, (8.431) reduces to

$$\hat{\psi}_i^{(k+1)} = \arg\max_{\psi_i}\left\{\operatorname{tr}\left[\mathbf{P}_{\mathbf{v}(\psi_i)(\mathbf{V}(\hat{\boldsymbol{\psi}}_{(i)}^{(k)}))}\,\mathbf{C}_{\mathbf{x}}\right]\right\}. \tag{8.442}$$

Now define a normalized vector

$$\mathbf{b}(\psi_i, \hat{\boldsymbol{\psi}}_{(i)}^{(k)}) \triangleq \frac{\mathbf{C}(\mathbf{B})}{\|\mathbf{C}(\mathbf{B})\|} = \frac{\mathbf{P}_{\mathbf{B}}^{\perp}\mathbf{C}}{\|\mathbf{P}_{\mathbf{B}}^{\perp}\mathbf{C}\|}, \tag{8.443}$$

or

$$\mathbf{b}(\psi_i, \hat{\boldsymbol{\psi}}_{(i)}^{(k)}) = \frac{\left[\mathbf{P}^{\perp}_{\mathbf{V}(\hat{\boldsymbol{\psi}}_{(i)}^{(k)})}\right]\mathbf{v}(\psi_i)}{\left\|\left[\mathbf{P}^{\perp}_{\mathbf{V}(\hat{\boldsymbol{\psi}}_{(i)}^{(k)})}\right]\mathbf{v}(\psi_i)\right\|}. \tag{8.444}$$

Then,

$$\mathbf{P}_{\mathbf{v}(\psi_i)(\mathbf{V}(\hat{\boldsymbol{\psi}}_{(i)}^{(k)}))} = \mathbf{b}(\psi_i, \hat{\boldsymbol{\psi}}_{(i)}^{(k)})\mathbf{b}^H(\psi_i, \hat{\boldsymbol{\psi}}_{(i)}^{(k)}), \tag{8.445}$$

and (8.442) can be written as,

$$\hat{\psi}_i^{(k+1)} = \arg\max_{\psi_i}\left\{\mathbf{b}^H(\psi_i, \hat{\boldsymbol{\psi}}_{(i)}^{(k)})\mathbf{C}_{\mathbf{x}}\mathbf{b}(\psi_i, \hat{\boldsymbol{\psi}}_{(i)}^{(k)})\right\}. \tag{8.446}$$

We continue the iteration across i and k until

$$\left|\hat{\psi}_i^{(k+1)} - \hat{\psi}_i^{(k)}\right| \leq \delta, \quad i = 1, \cdots, D, \tag{8.447}$$

where δ is a function of the desired accuracy.

The AP algorithm can be summarized:

(i) Initialize the algorithm using the procedure in (8.434) and (8.435) to obtain $\hat{\psi}_1^{(0)}, \hat{\psi}_2^{(0)}, \cdots, \hat{\psi}_D^{(0)}$.

(ii) For $i = 1, \cdots, D$, and $k = 1$, use (8.446) to obtain $\hat{\psi}_1^{(1)}, \hat{\psi}_2^{(1)}, \hat{\psi}_3^{(1)}, \cdots, \hat{\psi}_D^{(1)}$.

(iii) Iterate (8.446) for $k = 2, \cdots$.

(iv) Repeat until (8.447) is satisfied for all $i = 1, \cdots, D$.

The issues of interest with respect to the AP algorithms are:

(i) What are the conditions on signal geometry, SNR_i, and K that will cause the initial conditions to be such that the AP algorithm converges to a global maximum?

(ii) What is the rate of convergence?

(iii) Is the rate of convergence improved by using the AP algorithm for the first several iterations and then switching to a gradient procedure?

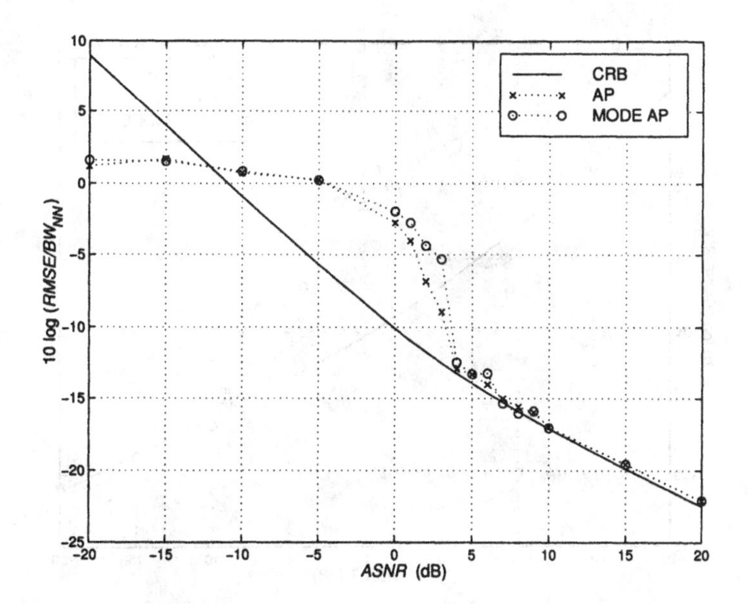

Figure 8.36 Normalized RMSE versus *ASNR*: $\Delta\psi = \Delta\psi_R$, $N = 10$, $K = 100$, $\rho = 0$.

We consider two of the test scenarios in Table 8.3 to illustrate the behavior of the AP algorithm.

Example 8.6.1 (continuation, Examples 8.5.2 and 8.5.9)

Consider the same model as in Examples 8.5.2 and 8.5.9. There are two equal-power uncorrelated signals impinging on a standard 10-element linear array. The signal separation is $\Delta\psi_R$. We use the AP algorithm to estimate the DOAs. We initialize the algorithm using (8.434) and (8.435).

The results are shown in Figure 8.36. Comparing the result to the CML and MODE curves in Figure 8.32, we see that the curves are the same.

Example 8.6.2 (continuation, Example 8.5.8)

Consider the same model as in Example 8.5.8. The result for $\phi_\rho = \pi/4$ is shown in Figure 8.37. We see that the curve is the same as the curve in Figure 8.34.

In all of the other test scenarios, the AP results are the same as the grid search results. There is a significant computational saving, so the AP algorithm is attractive. There are certain cases in which the initialization will cause the algorithm to converge to a local maximum rather than the global maximum, but we did not encounter this problem in our test scenarios.

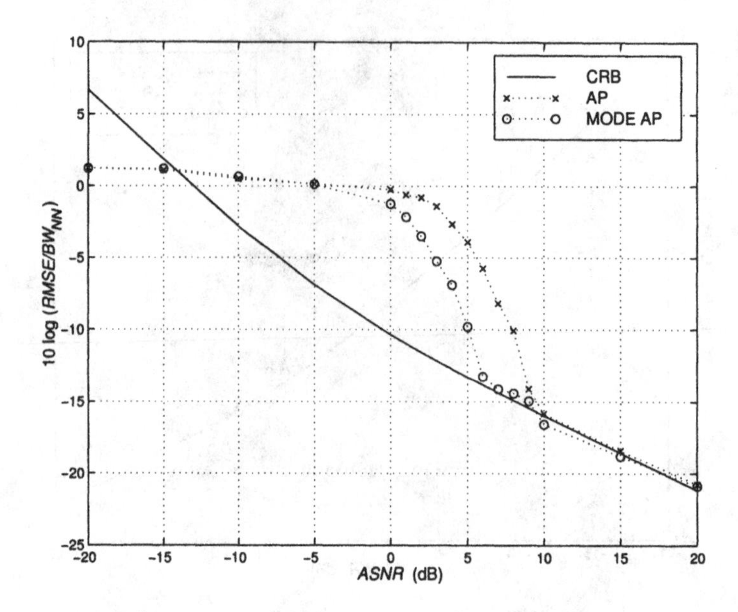

Figure 8.37 Normalized RMSE versus *ASNR*: $\Delta\psi = \Delta\psi_R$, $N = 10$, $K = 100$, $\rho = 0.95e^{j\pi/4}$.

8.6.2.2 Alternating maximization algorithm

In this section we consider the use of the AM technique to find the AML estimate given by (8.310). From (8.310), the function that we want to maximize is

$$\hat{\psi}_{aml} = \arg\max_{\psi} \left\{ -\ln\det\left[\mathbf{P_V C_x P_V} + \sigma_w^2 \mathbf{P_V^\perp}\right] - \frac{1}{\sigma_w^2}\text{tr}\left[\mathbf{P_V^\perp C_x}\right]\right\}, \quad (8.448)$$

in the case of known noise variance and, from (8.315),

$$\hat{\psi}_{aml} = \begin{matrix}\arg\min \\ \psi\end{matrix} \left\{\det\left[\mathbf{P_V C_x P_V} + \frac{\text{tr}\left[\mathbf{P_V^\perp C_x}\right]\mathbf{P_V^\perp}}{N-D}\right]\right\}, \quad (8.449)$$

in the case of unknown noise variance.

The vectors, $\hat{\psi}_{(1)}^{(k)}$ and $\hat{\psi}_{(i)}^{(k)}$ are defined in (8.432) and (8.433), respectively. We proceed in an iterative manner

$$\hat{\psi}_i^{(k+1)} = \arg\max_{\psi_i}\left\{F\left(\psi : \left\{\mathbf{V}\left(\hat{\psi}_{(i)}^{(k)}\right), \mathbf{v}(\psi_i)\right\}\right)\right\}, \quad (8.450)$$

where the term in the braces in (8.450) is the term inside the braces in (8.448) or (8.449) with $\mathbf{V}(\boldsymbol{\psi})$ replaced by $\left\{\mathbf{V}\left(\hat{\boldsymbol{\psi}}_{(i)}^{(k)}\right), \mathbf{v}(\psi_i)\right\}$ in the projection matrices. We initialize the algorithm in a manner similar to (8.434).

We implemented the AM algorithm for the scenarios in Table 8.3. We used the AP initialization technique in (8.434) and (8.435). In all cases, the results were the same as in the grid search AML algorithm.

8.6.3 Expectation Maximization Algorithm

In this section we develop the expectation maximization (EM) algorithm and apply it to the parameter estimation problem. The basis for the name will be clear when the algorithm is developed.

In Section 8.6.3.1, we derive a general form of the algorithm. In Section 8.6.3.2, we apply it to the CML estimation problem. In Section 8.6.3.3, we summarize our results.

8.6.3.1 EM Algorithm

The original derivation of the EM algorithm is due to Laird et al. [LDR77]. It has been applied to the array processing problem by Feder and Weinstein [FW88] and Miller and Fuhrmann [MF90]. A tutorial article by Moon [Moo96] discusses various other applications. Our discussion follows these references. We begin with a general discussion of the EM algorithm, then consider the case when the probability densities are Gaussian, and then apply the EM algorithm to the problem of interest.

The basic idea of the EM algorithm is straightforward. Recall from (8.2), that on the kth snapshot, we observe

$$\mathbf{X}_k = \mathbf{V}(\boldsymbol{\psi})\mathbf{F}_k + \mathbf{N}_k \quad k = 1, 2, \cdots, K, \tag{8.451}$$

and want to find $\hat{\boldsymbol{\psi}}_{ml}$. Now suppose that, instead of $\mathbf{X}_k$, we could observe

$$\mathbf{Y}_{ki} = \mathbf{v}(\theta_i)F_{ki} + \mathbf{N}_{ki} \quad i = 1, 2, \cdots, D. \tag{8.452}$$

Then we would have a set of D uncoupled 1-D maximization problems that are reasonably easy to solve. We refer to the transformation between the $\mathbf{Y}_{ki}$ and $\mathbf{X}_k$ as $\mathbf{H}$,

$$\mathbf{X}_k = \mathbf{H}(\mathbf{Y}_{k1}, \mathbf{Y}_{k2}, \cdots, \mathbf{Y}_{kD}), \tag{8.453}$$

and refer to the $\mathbf{Y}_{ki}, i = 1, \cdots, D$ as the "complete data" and $\mathbf{X}_k$ as the "incomplete data."

The EM algorithm is a technique for inferring the complete data from the incomplete data and using this inferred data to find the ML estimate.

<u>**EM Algorithm**</u>[15] We denote the observed ("incomplete") data by the vector $\mathbf{X}$ whose probability density, $p_{\mathbf{X}}(\mathbf{X}:\boldsymbol{\theta})$[16] depends on the vector parameter $\boldsymbol{\theta}$. We denote the complete data by the vector $\mathbf{Y}$, which is related to $\mathbf{X}$ by the many-to-one (non-invertible) transformation $\mathbf{H}(\cdot)$

$$\mathbf{X} = \mathbf{H}(\mathbf{Y}). \tag{8.454}$$

Note that the choice of $\mathbf{Y}$ and therefore $\mathbf{H}(\cdot)$ is not unique, and one of the keys to successful application of the EM algorithm is an appropriate choice of $\mathbf{Y}$ and $\mathbf{H}(\cdot)$.

We assume that the estimate of $\boldsymbol{\theta}$ at step n of the iteration is $\hat{\boldsymbol{\theta}}^{(n)}$. We denote the likelihood function of $\mathbf{Y}$ as,[17]

$$L_{\mathbf{Y}}(\boldsymbol{\theta}) \triangleq \ln p_{\mathbf{Y}|\boldsymbol{\theta}}(\mathbf{Y}) = \ln p_{\mathbf{Y}}(\mathbf{Y}:\boldsymbol{\theta}). \tag{8.455}$$

If $\mathbf{Y}$ were available, we would maximize $L_{\mathbf{Y}}(\boldsymbol{\theta})$. As only $\mathbf{X}$ is available, we find the expectation of $L_{\mathbf{Y}}(\boldsymbol{\theta})$, given that we have observed $\mathbf{X}$ and that our current parameter estimate is $\hat{\boldsymbol{\theta}}^{(n)}$.

We define the resulting expectation as $U(\boldsymbol{\theta},\hat{\boldsymbol{\theta}}^{(n)})$:

$$\boxed{U(\boldsymbol{\theta},\hat{\boldsymbol{\theta}}^{(n)}) \triangleq E\left\{\ln p_{\mathbf{Y}}(\mathbf{Y}:\boldsymbol{\theta}|\mathbf{X}:\hat{\boldsymbol{\theta}}^{(n)})\right\}.} \tag{8.456}$$

This is the expectation step. Note that it is conditional expectation with respect to $\mathbf{X}$ and $\hat{\boldsymbol{\theta}}^{(n)}$.

The next step is to maximize $U(\boldsymbol{\theta},\hat{\boldsymbol{\theta}}^{(n)})$ with respect to $\boldsymbol{\theta}$. The resulting estimate is $\hat{\boldsymbol{\theta}}^{(n+1)}$. Thus,

$$\boxed{\hat{\boldsymbol{\theta}}^{(n+1)} = \underset{\boldsymbol{\theta}}{\arg\max}\; U(\boldsymbol{\theta},\hat{\boldsymbol{\theta}}^{(n)}).} \tag{8.457}$$

The steps in (8.456) and (8.457) define the EM algorithm.

[15] This discussion follows [FW88] and [Moo96], which follow [LDR77].

[16] We use $p_{\mathbf{Y}}(\mathbf{Y}:\boldsymbol{\theta})$ instead of $p_{\mathbf{Y}|\boldsymbol{\theta}}(\mathbf{Y})$ to minimize the use of double subscripts in the derivation.

[17] We have used n to denote the steps in the iteration because we use k to denote the snapshot.

The algorithm starts with an initial estimate $\hat{\boldsymbol{\theta}}^{(0)}$. The procedure described in (8.434)–(8.435) can be used to obtain this initial estimate. The iteration proceeds until

$$\left|\hat{\boldsymbol{\theta}}^{(n+1)} - \hat{\boldsymbol{\theta}}^{(n)}\right| < \epsilon. \tag{8.458}$$

The maximization step ensures that $U(\boldsymbol{\theta}, \boldsymbol{\theta}^{(n)})$ increases on each iteration cycle.

Wu [Wu83] shows that if $U(\boldsymbol{\theta}, \boldsymbol{\theta}^{(n)})$ is continuous in both variables, then the algorithm converges to a stationary point. In the next section, we apply the algorithm to the CML estimation problem.

8.6.3.2 Conditional maximum likelihood estimate

We now apply the EM algorithm to solve for $\hat{\boldsymbol{\theta}}_{cml}$. From (8.451), the observation model is

$$\mathbf{X}_k = \mathbf{V}(\boldsymbol{\psi})\mathbf{F}_k + \mathbf{N}_k, \quad k = 1, 2, \cdots, K. \tag{8.459}$$

We assume the white noise level σ_w^2 is known.

In this case, the unknown parameters are the $D \times 1$ vector $\boldsymbol{\psi}$, corresponding to the DOAs, and the $2DK \times 1$ vector $\mathbf{F}$. Thus,

$$\boldsymbol{\theta} = [\boldsymbol{\psi}, \mathbf{F}]. \tag{8.460}$$

The likelihood function is

$$L(\boldsymbol{\theta}) = -\sum_{k=1}^{K} \left|\mathbf{X}_k - \mathbf{V}(\boldsymbol{\psi})\mathbf{F}_k\right|^2, \tag{8.461}$$

where we have discarded constant factors. The incomplete data are $\mathbf{X}_k, k = 1, 2, \cdots, K$. A logical choice for the complete data would the observation of each plane-wave signal by itself in the presence of noise. Thus,

$$\mathbf{Y}_{kl} = \mathbf{v}(\psi_l)F_{kl} + \mathbf{N}_{kl}, \quad l = 1, 2, \cdots, D, \tag{8.462}$$

where

$$E[\mathbf{N}_{kl}\mathbf{N}_{kl}^H] = \beta_l \sigma_w^2 \mathbf{I}, \tag{8.463}$$

and

$$\sum_{l=1}^{D} \beta_l = 1. \tag{8.464}$$

For simplicity, we will use

$$\beta_l = \frac{1}{D}. \tag{8.465}$$

Then,

$$\mathbf{X}_k = \sum_{l=1}^{D} \mathbf{Y}_{kl} \triangleq \mathbf{H} \begin{bmatrix} \mathbf{Y}_{k1} \\ \mathbf{Y}_{k2} \\ \vdots \\ \mathbf{Y}_{kD} \end{bmatrix}. \tag{8.466}$$

We see that $\mathbf{H}$ is a linear transformation so that $\mathbf{X}_k$ and $\mathbf{Y}_{kl}, l = 1, 2, \cdots, D$ are jointly Gaussian.

Using (8.460) in (8.456), we can write

$$\begin{aligned} U(\boldsymbol{\theta}, \hat{\boldsymbol{\theta}}^{(n)}) &= U(\boldsymbol{\psi}, \mathbf{F}; \hat{\boldsymbol{\psi}}^{(n)}, \mathbf{F}^{(n)}) \\ &= E\left\{\ln p_{\mathbf{Y}}\left(\mathbf{Y} : \boldsymbol{\psi}, \mathbf{F} | \mathbf{X} : \hat{\boldsymbol{\psi}}^{(n)}, \widehat{\mathbf{F}}^{(n)}\right)\right\}. \end{aligned} \tag{8.467}$$

Neglecting constant terms, we can write the ln likelihood function as

$$L(\mathbf{Y} : \boldsymbol{\psi}, \mathbf{F}) = -\sum_{k=1}^{K} \sum_{l=1}^{D} \left|\mathbf{Y}_{kl} - \mathbf{v}(\psi_l) F_{kl}\right|^2. \tag{8.468}$$

Since $\mathbf{Y}_{kl}$ is a sufficient statistic for estimating $\mathbf{v}(\psi_l) F_{kl}$, we only need to find the conditional mean,

$$\hat{\mathbf{Y}}_{kl}^{(n)} \triangleq E\left\{\mathbf{Y}_{kl} | \mathbf{X}, \hat{\boldsymbol{\psi}}^{(n)}, \widehat{\mathbf{F}}_k^{(n)}\right\}. \tag{8.469}$$

Because $\mathbf{Y}_{kl}$ and $\mathbf{X}$ are jointly Gaussian, this is a classical estimation result. (We use Bayes rule to obtain $p_{\mathbf{Y}_{kl}|\mathbf{X}}(\cdot)$ from $p_{\mathbf{X}|\mathbf{Y}_{kl}}(\cdot)$ and find the mean by inspection.)

$$\hat{\mathbf{Y}}_{kl}^{(n)} - \mathbf{v}\left(\hat{\psi}_l^{(n)}\right) \hat{F}_{kl}^{(n)} = \frac{1}{D}\left[\mathbf{X}_k - \mathbf{V}(\hat{\boldsymbol{\psi}}^{(n)}) \widehat{\mathbf{F}}_k^{(n)}\right], \tag{8.470}$$

or

$$\hat{\mathbf{Y}}_{kl}^{(n)} = \mathbf{v}\left(\hat{\psi}_l^{(n)}\right) \hat{F}_{kl}^{(n)} + \frac{1}{D}\left[\mathbf{X}_k - \mathbf{V}(\hat{\boldsymbol{\psi}}^{(n)}) \widehat{\mathbf{F}}_k^{(n)}\right]. \tag{8.471}$$

The conditional mean is the signal component from the nth iteration plus a portion $1/D$ of the component of the current observation vector, which is orthogonal to the estimated signal subspace. The result in (8.471) is the expectation step. We observe that the expectation result is also an estimation result, so the EM algorithm is sometimes referred to as the estimation-maximization algorithm.

To define the maximization step, we recall from (8.345) and (8.338) that

$$\hat{\psi} = \arg\max_{\psi}\left\{\operatorname{tr}\left[\mathbf{V}(\psi)\left[\mathbf{V}^H(\psi)\mathbf{V}^H(\psi)\right]^{-1}\mathbf{V}^H(\psi)\mathbf{C}_{\mathbf{X}}\right]\right\} \tag{8.472}$$

and

$$\widehat{\mathbf{F}}_k = \left[\mathbf{V}^H(\psi)\mathbf{V}(\psi)\right]^{-1}\mathbf{V}^H(\psi)\mathbf{X}_k. \tag{8.473}$$

To get the corresponding relation for the complete data, we define

$$\widehat{\mathbf{S}}_{\mathbf{Y}_l}^{(n)} \triangleq \frac{1}{K}\sum_{k=1}^{K}\hat{\mathbf{Y}}_{kl}^{(n)}\hat{\mathbf{Y}}_{kl}^{(n)H}. \tag{8.474}$$

Then, using (8.474) in the 1-D version of (8.472), we have

$$\hat{\psi}_l^{(n+1)} = \arg\max_{\psi_l}\left\{\frac{\mathbf{v}^H(\psi_l)\widehat{\mathbf{S}}_{\mathbf{Y}_l}^{(n)}\mathbf{v}(\psi_l)}{|\mathbf{v}(\psi_l)|^2}\right\}, \quad l = 1, 2, \cdots, D, \tag{8.475}$$

and the 1-D version of (8.473) is

$$\widehat{F}_{kl}^{(n+1)} = \frac{\mathbf{v}^H\left(\hat{\psi}_l^{(n+1)}\right)\widehat{\mathbf{Y}}_{kl}^{(n)}}{\left|\mathbf{v}\left(\hat{\psi}_l^{(n+1)}\right)\right|^2}, \quad \begin{array}{l} l = 1, 2, \cdots, D, \\ k = 1, 2, \cdots, K. \end{array} \tag{8.476}$$

Since $|\mathbf{v}(\psi_l)|^2 = N$, (8.475) and (8.476) can be rewritten as

$$\hat{\psi}_l^{(n+1)} = \arg\max_{\psi_l}\left\{\mathbf{v}^H(\psi_l)\widehat{\mathbf{S}}_{\mathbf{Y}_l}^{(n+1)}\mathbf{v}(\psi_l)\right\}, \quad l = 1, 2, \cdots, D, \tag{8.477}$$

and

$$\widehat{F}_{kl}^{(n+1)} = \frac{1}{N}\mathbf{v}^H\left(\hat{\psi}_l^{(n+1)}\right)\hat{\mathbf{Y}}_{kl}^{(n)}, \quad \begin{array}{l} l = 1, 2, \cdots, D, \\ k = 1, 2, \cdots, K. \end{array} \tag{8.478}$$

The EM algorithm is defined by (8.471), (8.477), and (8.478). We observe that (8.477) can also be written as

$$\hat{\psi}_l^{(n+1)} = \arg\max_{\psi_l}\left\{\sum_{k=1}^{K}\left|\mathbf{v}^H(\psi_l)\hat{\mathbf{Y}}_{kl}^{(n)}\right|^2\right\}. \tag{8.479}$$

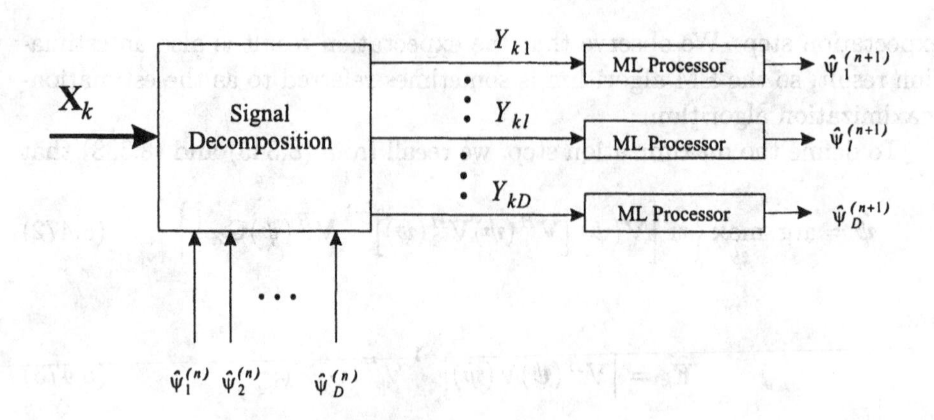

Figure 8.38 Implementation of EM algorithm.

Note that the term in the braces is just the power output of a conventional beamformer when the input is $\hat{\mathbf{Y}}_{kl}^{(n)}; k = 1, 2, \cdots, K$. This leads to the block diagram of the iteration process shown in Figure 8.38.[18] Note that all of the maximizations are done in parallel.

Summarizing, the EM algorithm for the CML estimate is:

(i) Initialize the algorithm with $\psi^{(0)}$ and $\hat{F}_{kl}^{(0)}$ (from (8.434) and (8.435)).

(ii) Estimate $\hat{\mathbf{Y}}_{kl}^{(0)}, k = 1, \cdots, K; l = 1, \cdots, D$ using (8.471).

(iii) Maximize

$$\hat{\psi}_l^{(n+1)} = \arg\max_{\psi_l} \left\{ \sum_{k=1}^{K} \left| \mathbf{v}^H(\psi_l) \hat{\mathbf{Y}}_{kl}^{(n)} \right|^2 \right\}, \tag{8.480}$$

with $n = 0$.

(iv) Compute

$$\widehat{F}_{kl}^{(n+1)} = \frac{1}{N} \mathbf{v}^H \left(\hat{\psi}_l^{(n+1)} \right) \hat{\mathbf{Y}}_{kl}^{(n)}. \tag{8.481}$$

(v) Estimate $\hat{\mathbf{Y}}_{kl}^{(n+1)}$ using (8.471).

(vi) Iterate through steps (ii)–(iv) until

$$\left| \psi^{(n+1)} - \psi^{(n)} \right| \leq \epsilon. \tag{8.482}$$

[18]This figure is similar to Figure 1 in [FW88], but their model assumed a known signal rather than an unknown nonrandom signal.

The EM algorithm provides an alternative procedure to the AP algorithm for solving the CML problem. A similar procedure is available for the UML algorithm. Miller and Fuhrmann [MF90] use the UML algorithm for uncorrelated signals and enforce the non-negative definite requirement at each iteration. Therefore, their solution is the UML estimator as contrasted to the AML estimator.

For our test scenarios, the EM algorithm converges to the correct solution, but is slower than the AP algorithm.

Fessler and Hero [FH94] develop a new space-alternating generalized EM algorithm (SAGE), and demonstrate that it converges significantly faster than the conventional EM algorithm. The reader is referred to that reference for discussion.

8.6.4 Summary

In this section, we have developed several techniques for solving the ML estimation problem. The most efficient approach in terms of both computation and speed of convergence appears to consist of the following steps:

(i) Initialize the algorithm by using the procedure in (8.434) or by making an initial estimate using one of the simpler algorithms that we develop in Chapter 9.

(ii) Use the relaxation methods, either AP or AM as appropriate to get closer to the minimum.

(iii) Use a search technique such as a quasi-Newton algorithm to achieve the final convergence to the estimate.

All of the computational algorithms developed in this section are appropriate for arbitrary array geometries. In the next section, we consider an approach that is useful for standard linear arrays.

8.7 Polynomial Parameterization

In this section we develop two computationally efficient algorithms that are applicable to standard linear arrays.

In Section 8.7.1, we show how we can reparameterize the ML estimation in terms of a polynomial with its roots on the unit circle.

In Section 8.7.2, we utilize the polynomial parameterization for the CML estimator to develop an algorithm that is called the **iterative quadratic**

maximization likelihood (IQML) algorithm and provides a computationally efficient solution.

In Section 8.7.3, we utilize the polynomial parameterization for the WSF (or MODE) estimator to develop an iterative MODE (IMODE).

In Section 8.7.4, we summarize our results.

8.7.1 Polynomial Parameterization

The technique of reparameterizing the ML estimator problem in terms of a polynomial is due to Kumaresan et al. [KSS86], and Kumaresan and Shaw [KS85] (e.g., Evans and Fishl [EF73]).

In many of the estimators we have discussed the vector $\boldsymbol{\psi}$ enters into the cost function through the projection matrix $\mathbf{P_V}$ (e.g., (8.315) or Table 8.4). We now want to develop an alternative expression for $\mathbf{P_V}$ that will be easier to evaluate.

For a standard linear array, $\mathbf{V}$ is an $N \times D$ matrix,

$$\mathbf{V}(\boldsymbol{\psi}) = \begin{bmatrix} \mathbf{v}(\psi_1) & \mathbf{v}(\psi_2) & \cdots & \mathbf{v}(\psi_D) \end{bmatrix}, \tag{8.483}$$

where

$$\mathbf{v}(\psi_i) = \begin{bmatrix} e^{-j\left(\frac{N-1}{2}\right)\psi_i} & \cdots & e^{j\left(\frac{N-1}{2}\right)\psi_i} \end{bmatrix}^T, \quad i = 1, \cdots, D. \tag{8.484}$$

The first step in developing the parameterization is to recognize that, if we let

$$z = e^{j\psi}, \tag{8.485}$$

then the polynomial

$$b(z) = b_0 z^D + b_1 z^{D-1} + \cdots + b_D \tag{8.486}$$

describes the spatial characteristics of the signal component. It can be written as

$$b(z) = b_0 \prod_{i=1}^{D} (z - z_i), \tag{8.487}$$

and has roots at

$$z_i = e^{j\psi_i}, \quad i = 1, 2, \cdots, D. \tag{8.488}$$

We define the coefficient vector of the polynomial as

$$\mathbf{b} = \begin{bmatrix} b_0 & b_1 & \cdots & b_D \end{bmatrix}^T. \tag{8.489}$$

If we find the ML estimate of $\mathbf{b}$, we can use it to obtain the ML estimate of $\psi_i, i = 1, 2, \cdots, D$. To carry out this procedure we construct a $N \times (N-D)$ Toeplitz matrix, $\mathbf{B}$,

$$\mathbf{B} = \begin{bmatrix} b_D^* & & & \\ & & \mathbf{0} & \\ b_{D-1}^* & b_D^* & & \\ \vdots & \ddots & \ddots & \\ b_0^* & \vdots & \ddots & b_D^* \\ 0 & b_0^* & & \\ 0 & 0 & & \\ \vdots & \vdots & \ddots & \\ 0 & 0 & \cdots & b_0^* \end{bmatrix}, \tag{8.490}$$

and show that we can write the noise subspace projection matrix as

$$\mathbf{P}_{\mathbf{V}}^{\perp} = \mathbf{P}_{\mathbf{B}} = \mathbf{B}\left(\mathbf{B}^H\mathbf{B}\right)^{-1}\mathbf{B}^H. \tag{8.491}$$

To verify (8.491), we consider the ith column of $\mathbf{V}(\boldsymbol{\psi})$, $\mathbf{v}(\psi_i)$. Denote the first column of $\mathbf{B}$ as $\mathbf{b}_1$. Then,

$$\begin{aligned} \mathbf{b}_1^H\mathbf{v}(\psi_i) &= b_D + b_{D-1}e^{j\psi_i} + \cdots + b_0 e^{jD\psi_i} \\ &= b\left(e^{j\psi_i}\right) = 0, \quad i = 1, 2, \cdots, D, \end{aligned} \tag{8.492}$$

where the last equality follows from (8.488). Similarly,

$$\mathbf{b}_2^H\mathbf{v}(\psi_i) = b\left(e^{j\psi_i}\right)\cdot e^{j\psi_i} = 0, \quad i = 1, 2, \cdots, D, \tag{8.493}$$

and so forth. Thus, the columns of $\mathbf{B}$ are orthogonal to $\mathbf{V}$. Since for any $\mathbf{b} \neq \mathbf{0}$, $\mathbf{B}$ has rank $N-D$, its columns span the orthogonal complement to $\mathbf{V}$ and $\mathbf{P}_B$ is equal to $\mathbf{P}_{\mathbf{V}}^{\perp}$.

The equivalence of the two projection matrices,

$$\mathbf{P}_{\mathbf{V}}^{\perp} = \mathbf{P}_{\mathbf{B}}, \tag{8.494}$$

is a key result that is used to solve ML problems.

8.7.2 Iterative Quadratic Maximum Likelihood (IQML)

In this section, we illustrate one application of polynomial parameterization. The specific case of interest is the CML estimate. The resulting algorithm is called the IQML algorithm.

The IQML algorithm was derived independently by Bresler and Macovski [BM86] and Kumaresan, Scharf, and Shaw (Kumaresan et al. [KSS86] and Kumaresan and Shaw [KS88][Sha87]). Our discussion is similar to that in [BM86] but we include a modification due to Nagesha and Kay [NK94]. We give a short discussion of other similar algorithms and other implementations of IQML at the end of this section. Although we are using the technique for CML, we see that it can be extended to any estimator that relies on $\mathbf{P_V}$ or $\mathbf{P_V^\perp}$ in the estimation algorithm.

The CML estimate is given by (8.346) as

$$\hat{\psi}_{cml} = \underset{\psi}{\arg\min} \left\{ \operatorname{tr}\left[\mathbf{P_V^\perp C_x}\right]\right\}. \tag{8.495}$$

We can write the required minimization in (8.495) as

$$\min_{\mathbf{b}\in\boldsymbol{\Omega}_\mathbf{b}} J(\mathbf{b}), \tag{8.496}$$

where

$$J(\mathbf{b}) = \operatorname{tr}\left[\mathbf{P}_B\,\mathbf{C_x}\right] = \operatorname{tr}\left[\mathbf{B}\left(\mathbf{B}^H\mathbf{B}\right)^{-1}\mathbf{B}^H\mathbf{C_x}\right], \tag{8.497}$$

and we must specify the constraints on $\mathbf{b}$ by defining $\boldsymbol{\Omega}_\mathbf{b}$.

We first derive an alternative form for $J(\mathbf{b})$ that is useful for computation. We define an $(N-D)\times(D+1)$ data matrix for the kth snapshot as

$$\mathbf{A}_k = \begin{bmatrix} X_D(k) & X_{D-1}(k) & \cdots & X_0(k) \\ X_{D+1}(k) & X_D(k) & \cdots & X_1(k) \\ \vdots & \vdots & \cdots & \vdots \\ X_{N-1}(k) & X_{N-2}(k) & \cdots & X_{N-1-D}(k) \end{bmatrix}, \tag{8.498}$$

where $X_n(k)$ is the frequency-domain snapshot of the output of the nth sensor at snapshot k. We use $\mathbf{A}_k$ to avoid confusion with the snapshot vector $\mathbf{X}(k)$.

By direct multiplication, one can show that

$$\mathbf{B}^H\mathbf{X}(k) = \mathbf{A}_k\,\mathbf{b}. \tag{8.499}$$

Substituting (8.499) into (8.497) gives

$$\begin{aligned} K\cdot J(\mathbf{b}) &= \operatorname{tr}\left[\sum_{k=1}^{K}\mathbf{B}\left(\mathbf{B}^H\mathbf{B}\right)^{-1}\mathbf{B}^H\mathbf{X}(k)\,\mathbf{X}^H(k)\right] \\ &= \mathbf{b}^H\left[\sum_{k=1}^{K}\mathbf{A}_k^H\left(\mathbf{B}^H\mathbf{B}\right)^{-1}\mathbf{A}_k\right]\mathbf{b}. \end{aligned} \tag{8.500}$$

The form of (8.500) suggests an iterative procedure in which we hold $\mathbf{B}^H\mathbf{B}$ fixed from the previous step and solve the quadratic minimization in $\mathbf{b}$. We must impose some constraints on $\mathbf{b}$ to get a useful solution.

The first constraint is imposed to guarantee a non-zero $\mathbf{b}$. Bresler and Macovski [BM86] use

$$Re\,[b_0] = 1, \tag{8.501}$$

to guarantee a non-zero $\mathbf{b}$. Nagesha and Kay [NK94] and others use the norm constraint

$$\|\mathbf{b}\|^2 = 1. \tag{8.502}$$

As pointed out in [NK94], the constraint in (8.501) is not useful for certain signal geometries. A simple example is the case of a single signal at $\psi = 0$. Here $Re[b_0] = 0$ (see Problem 8.7.12). We use the quadratic constraint in (8.502) and refer to the algorithm as IQML-QC to avoid confusion with some of IQML algorithms in the literature.

We would like the roots of the $\mathbf{b}$ polynomial to lie on the unit circle. We impose a conjugate symmetry constraint on $\mathbf{b}$. Conjugate symmetry is a necessary, but not sufficient condition.[19] We find that with the conjugate symmetry constraint, when the algorithm is above its threshold, the roots lie on the unit circle and start to move off the circle as we transition into the threshold region. A discussion of this root behavior is contained in Stoica and Nehorai [SN89a].

Thus, we require

$$b_i = b^*_{D-i}, \quad i = 0, 1, \cdots, D. \tag{8.503}$$

We introduce the constraint in (8.503) using a technique from Kumaresan and Shaw [KS88] and Shaw [Sha87]. We assume that D is odd.[20] We define a $(D+1) \times 1$ vector, $\mathbf{c}$ as

$$\mathbf{c} = \begin{bmatrix} Re(b_0) & Im(b_0) & Re(b_1) & \cdots & Re(b_{(D-1)/2}) & Im(b_{(D-1)/2}) \end{bmatrix}^T. \tag{8.504}$$

[19]The exception to this statement is the case of a single root, where the conjugate symmetry constraint is also sufficient. Shaw [Sha95] has derived a constrained MLE estimator based on this property.

[20]The development for even D is done in Problems 8.7.9 and 8.7.10.

We define a $(D+1) \times (D+1)$ transformation matrix $\mathbf{T}$,

$$\mathbf{T} \triangleq \frac{1}{\sqrt{2}} \begin{bmatrix} 1 & j & 0 & 0 & \cdots & 0 & 0 \\ 0 & 0 & 1 & j & \cdots & 0 & 0 \\ \vdots & \vdots & \vdots & \vdots & \ddots & \vdots & \vdots \\ 0 & 0 & 0 & 0 & \cdots & 1 & j \\ 0 & 0 & 0 & 0 & \cdots & 1 & -j \\ \vdots & \vdots & \vdots & \vdots & \ddots & \vdots & \vdots \\ 0 & 0 & 1 & -j & \cdots & 0 & 0 \\ 1 & -j & 0 & 0 & \cdots & 0 & 0 \end{bmatrix}. \tag{8.505}$$

This can be written as

$$\mathbf{T} = \frac{1}{\sqrt{2}} \left[\begin{array}{c} \mathbf{I}_{(D+1)/2} \otimes \begin{bmatrix} 1 & j \end{bmatrix} \\ \hdashline \mathbf{J}_{(D+1)/2} \otimes \begin{bmatrix} 1 & -j \end{bmatrix} \end{array} \right], \tag{8.506}$$

where $\otimes$ is Kronecker product. Then,

$$\mathbf{b} = \mathbf{Tc}. \tag{8.507}$$

We define

$$\tilde{\mathbf{Q}}_{\mathbf{x}} \triangleq \mathbf{T}^H \left(\sum_{k=1}^{K} \mathbf{A}_k^H \left[\mathbf{B}^H \mathbf{B}\right]^{-1} \mathbf{A}_k \right) \mathbf{T}, \tag{8.508}$$

and

$$F(\mathbf{c}) = \mathbf{c}^T \tilde{\mathbf{Q}}_{\mathbf{x}} \mathbf{c}. \tag{8.509}$$

Because $Im\{\tilde{\mathbf{Q}}_{\mathbf{x}}\}$ is skew-symmetric, its quadratic form is zero. Thus, $Im[F(\mathbf{c})] = 0$, so we can use[21]

$$F(\mathbf{c}) = \mathbf{c}^T \left[Re\left\{\tilde{\mathbf{Q}}_{\mathbf{x}}\right\}\right] \mathbf{c}. \tag{8.510}$$

We assume $\mathbf{B}$ is fixed from the previous iteration. We minimize $\mathbf{F}(\mathbf{c})$ subject to the norm constraint

$$\|\mathbf{c}\|^2 = 1. \tag{8.511}$$

The solution follows immediately if we expand $Re\{\tilde{\mathbf{Q}}_{\mathbf{x}}\}$ in terms of its eigenvalues and eigenvectors:

$$Re\{\tilde{\mathbf{Q}}_{\mathbf{x}}\} = \sum_{i=1}^{D+1} \lambda_i \phi_i \phi_i^H. \tag{8.512}$$

[21] An alternative derivation of (8.509) is developed in Problem 8.7.8.

The non-zero $\mathbf{c}$ that minimizes $F(\mathbf{c})$ is the eigenvector corresponding to the smallest eigenvalue,

$$\hat{\mathbf{c}} = \boldsymbol{\phi}_{min}, \tag{8.513}$$

where the notation emphasizes that $\boldsymbol{\phi}_{min}$ corresponds to the smallest eigenvalue.

We can summarize the iterative algorithm using the quadratic constraint (IQML-QC):

(i) Initialization: set $m = 0$ and $\mathbf{B}_{(0)} = \mathbf{B}_0$.

(ii) Compute

$$\widetilde{\mathbf{Q}}_{\mathbf{x}}^{(m)} = \mathbf{T}^H \left(\sum_{k=1}^{K} \mathbf{A}_k^H \left[\mathbf{B}_{(m)}^H \mathbf{B}_{(m)} \right]^{-1} \mathbf{A}_k \right) \mathbf{T}, \tag{8.514}$$

where $\mathbf{B}_{(m)}$ is given by (8.490), and $\mathbf{A}_k$ is given by (8.498).

(iii) Perform an eigendecomposition of $Re\{\widetilde{\mathbf{Q}}_{\mathbf{x}}^{(m)}\}$ as in (8.512). Set

$$\hat{\mathbf{c}}^{(m)} = \boldsymbol{\phi}_{min}^{(m)}. \tag{8.515}$$

(iv) Find

$$\hat{\mathbf{b}}^{(m)} = \mathbf{T}\hat{\mathbf{c}}^{(m)}. \tag{8.516}$$

(v) Find the roots of $\hat{\mathbf{b}}_{(m)}(z)$. Denote these roots as $\hat{z}_i, \quad i = 1, 2, \cdots, D$. Then

$$\hat{\psi}_i = \angle \arg\left(\hat{z}_i\right), \quad i = 1, 2, \cdots, D. \tag{8.517}$$

(vi) Set $m = m + 1$.

(vii) Repeat steps (ii)–(v).

(viii) Check convergence: Is

$$\|\hat{\mathbf{b}}(m+1) - \hat{\mathbf{b}}(m)\| < \epsilon, \tag{8.518}$$

where ϵ reflects the desired precision. If (8.518) is satisfied we terminate the iteration and use the $\hat{\psi}_i$ from step (v), otherwise return to step (ii).

We can initialize the algorithm with

$$\mathbf{B}_{(0)}^H \mathbf{B}_{(0)} = \mathbf{I}, \tag{8.519}$$

or we can perform an initial estimate of ψ using one of the simpler algorithms that we develop in Chapter 9.

As pointed out in [BM86], the matrices $\left(\mathbf{B}^H\mathbf{B}\right)$ and $\tilde{\mathbf{Q}}_{\mathbf{x}}$ have considerable structure that can be exploited for computational efficiency. $\mathbf{B}^H\mathbf{B}$ is a banded Hermitian matrix, so its inverse can be efficiently computed (e.g., [Kum85], [KSS86]). The inverse of $\mathbf{B}^H\mathbf{B}$ can be computed as the inverse of a $D \times D$ matrix rather than a $(N-D) \times (N-D)$ matrix.

There are a number of references concerning IQML-type of algorithms. An early reference is Evans and Fischl [EF73]. Other iteration algorithms include Matausek et al. [MSR83], Kay [Kay84]. Algorithms using other approaches include Tufts and Kumaresan [TK82] and Kumaresan et al. [KT83]. In [BM86], the relationship of IQML to these algorithms is discussed.

There have been a number of papers discussing IQML since [BM86]. McClellan and Lee [Mcl91] show the equivalence of the Steiglitz-McBride algorithm [SM65] and IQML.

Clark and Scharf [CS92] discuss the complexity of the IQML algorithm and compare it to other algorithms including the Steiglitz-McBride algorithm (e.g., [SM65]). An efficient implementation of IQML is given by Hua [Hua94]. [KS88] also discuss an adaptive version to track slowly moving sources. Stoica et al. [SLS97] analyze the asymptotic behavior. Li et al. [LSL98] compare the performance of IQML and MODE estimators.

The issues of interest with respect to the IQML algorithm are:

(i) Rate of convergence.

(ii) Comparison of threshold behavior with CML algorithm in Section 10.5.2.

(iii) Computational complexity.

We illustrate the performance of IQML-QC for several of the test cases that we have studied in Sections 8.5 and 8.6. However, before doing the examples we develop the polynomial parameterized version of MODE.

Before developing MODE, we should note that we have developed the IQML algorithm for solving the CML estimation problem. The polynomial parameterization can be used for other algorithms.

For the AML algorithm

$$F(\mathbf{c}) = \ln\det\left[\mathbf{P}_{\mathbf{B}}^{\perp}\mathbf{C}_{\mathbf{x}}\mathbf{P}_{\mathbf{B}}^{\perp} + \hat{\sigma}_w^2\mathbf{P}_{\mathbf{B}}\right], \tag{8.520}$$

with

$$\hat{\sigma}_w^2 = \frac{1}{N-D}\mathrm{tr}\left[\mathbf{P}_{\mathbf{B}}\mathbf{C}_{\mathbf{x}}\right]. \tag{8.521}$$

and $\mathbf{c}$ is defined in (8.504). We use the same general iterative procedure as above.

In the next section we discuss the polynomial parameterized version of WSF (MODE).

8.7.3 Polynomial WSF (MODE)

The weighted subspace fitting (WSF) (or MODE) algorithm was described in Section 8.5.3 (see (8.367)–(8.370)). The algorithm is

$$\hat{\boldsymbol{\psi}} = \arg\min_{\boldsymbol{\psi}} \left\{ \text{tr} \left[\mathbf{P}_{\mathbf{V}}^{\perp} \hat{\mathbf{U}}_S \mathbf{W}_{ao} \hat{\mathbf{U}}_S^H \right] \right\}, \tag{8.522}$$

where $\hat{\mathbf{U}}_S$ is an $N \times D$ matrix whose columns are the estimated signal eigenvectors and $\mathbf{W}_{ao}$ is a $D \times D$ diagonal matrix,

$$\mathbf{W}_{ao} = \widehat{\widetilde{\Lambda}}^2 \hat{\mathbf{\Lambda}}_S^{-1}, \tag{8.523}$$

$$\widehat{\widetilde{\mathbf{\Lambda}}} = \widehat{\mathbf{\Lambda}}_S - \hat{\sigma}_w^2 \mathbf{I}, \tag{8.524}$$

$$\hat{\sigma}_w^2 = \frac{1}{N-D} \sum_{i=D+1}^{N} \hat{\lambda}_i. \tag{8.525}$$

The polynomial implementation of MODE was defined in Stoica and Sharman's original papers ([SS90a],[SS90b]). Li et al. [LSL98] give a more explicit definition of the algorithm. Both references emphasize a two-step algorithm. We focus on an iterative version of the algorithm, that we call IMODE. In the examples in [LSL98] they use IMODE for some of their results. Li et al. [LSL98] also contain a discussion of relative computational complexity of IQML and MODE.

Using a polynomial parameterization, we can rewrite (8.522) as

$$\hat{\mathbf{b}} = \arg\min_{\boldsymbol{\psi}} \left\{ \text{tr} \left[\mathbf{P}_{\mathbf{B}} \hat{\mathbf{U}}_S \mathbf{W}_{ao} \hat{\mathbf{U}}_S^H \right] \right\}, \tag{8.526}$$

subject to a quadratic constraint,

$$\|\mathbf{b}\|^2 = 1, \tag{8.527}$$

and a conjugate symmetry constraint

$$b_i = b_{D-i}^*. \tag{8.528}$$

We can rewrite the right side of (8.526) as

$$J(\mathbf{b}) = \operatorname{tr}\left[\mathbf{B}(\mathbf{B}^H\mathbf{B})^{-1}\mathbf{B}^H\hat{\mathbf{U}}_S\mathbf{W}_{ao}\hat{\mathbf{U}}_S^H\right]. \tag{8.529}$$

We use an approach similar to the IQML approach to get (8.529) into a more usable form. For notational simplicity we define

$$\tilde{\mathbf{U}}_S \triangleq \hat{\mathbf{U}}_S\mathbf{W}_{ao}^{\frac{1}{2}}, \tag{8.530}$$

and write

$$\hat{\mathbf{U}}_S\mathbf{W}_{ao}\hat{\mathbf{U}}_S^H = \tilde{\mathbf{U}}_S\tilde{\mathbf{U}}_S^H = \sum_{d=1}^{D}\tilde{\boldsymbol{\phi}}_d\tilde{\boldsymbol{\phi}}_d^H, \tag{8.531}$$

where $\tilde{\boldsymbol{\phi}}_d$ are the columns of $\tilde{\mathbf{U}}_S$.

We can write (8.529) as

$$\begin{aligned} J(\mathbf{b}) &= \operatorname{tr}\left[\sum_{d=1}^{D}\mathbf{B}\left(\mathbf{B}^H\mathbf{B}\right)\mathbf{B}^H\tilde{\boldsymbol{\phi}}_d\tilde{\boldsymbol{\phi}}_d^H\right] \\ &= \operatorname{tr}\left[\sum_{d=1}^{D}\tilde{\boldsymbol{\phi}}_d^H\mathbf{B}\left(\mathbf{B}^H\mathbf{B}\right)^{-1}\mathbf{B}^H\tilde{\boldsymbol{\phi}}_d\right]. \end{aligned} \tag{8.532}$$

We define a matrix

$$\mathbf{A}_d = \begin{bmatrix} \tilde{\phi}_d(D+1) & \tilde{\phi}_d(D) & \cdots & \tilde{\phi}_d(1) \\ \tilde{\phi}_d(D+2) & \tilde{\phi}_d(D+1) & \cdots & \tilde{\phi}_d(2) \\ \vdots & \vdots & \ddots & \vdots \\ \tilde{\phi}_d(N) & \tilde{\phi}_d(N-1) & \cdots & \tilde{\phi}_d(N-D) \end{bmatrix}. \tag{8.533}$$

This matrix is analogous to the data matrix in (8.498) except its components come from the signal subspace matrix. Then

$$\mathbf{B}^H\tilde{\boldsymbol{\phi}}_d = \mathbf{A}_d\mathbf{b}, \tag{8.534}$$

and

$$J(\mathbf{b}) = \mathbf{b}^H\left[\sum_{d=1}^{D}\mathbf{A}_d^H\left(\mathbf{B}^H\mathbf{B}\right)^{-1}\mathbf{A}_d\right]\mathbf{b}. \tag{8.535}$$

This form is identical to (8.500), so the subsequent steps are identical to (8.502)–(8.518). The $\mathbf{c}$ vector is defined in (8.504) and the $\mathbf{T}$ matrix is defined in (8.505):

$$\tilde{\mathbf{Q}}_D \triangleq \mathbf{T}^H\left(\sum_{d=1}^{D}\mathbf{A}_d^H\left[\mathbf{B}^H\mathbf{B}\right]^{-1}\mathbf{A}_d\right)\mathbf{T}, \tag{8.536}$$

and

$$F(\mathbf{c}) = \mathbf{c}^T\tilde{\mathbf{Q}}_D\mathbf{c} = \mathbf{c}^T\left[Re\{\tilde{\mathbf{Q}}_D\}\right]\mathbf{c}. \tag{8.537}$$

We minimize $F(\mathbf{c})$ subject to a unit norm constraint on $\mathbf{c}$. The resulting $\hat{\mathbf{c}}$ is the eigenvector of $Re\{\tilde{\mathbf{Q}}_D\}$ corresponding to the smallest eigenvalue.

We can now define an iterative version of the MODE algorithm using the same steps as in the IQML. We refer to it as the IMODE algorithm.

The IMODE algorithm can be summarized:

(i) Initialization: set m = 0 and choose $\mathbf{B}_{(0)}$ such that

$$\mathbf{B}_{(0)}^H\mathbf{B}_{(0)} = \mathbf{I}. \tag{8.538}$$

(ii) Compute

$$\tilde{\mathbf{Q}}_D^{(m)} \triangleq \mathbf{T}^H\left(\sum_{d=1}^{D}\mathbf{A}_d^H\left[\mathbf{B}_{(m)}^H\mathbf{B}_{(m)}\right]^{-1}\mathbf{A}_d\right)\mathbf{T}. \tag{8.539}$$

In 8.539, the matrix $\mathbf{B}_{(m)}$ is given by (8.490). Compute $\tilde{\mathbf{U}}_S$ and $\tilde{\boldsymbol{\phi}}_d$. Use $\tilde{\boldsymbol{\phi}}_d$ in (8.533) to form $\mathbf{A}_d$.

(iii) Perform an eigendecomposition of $Re\{\tilde{\mathbf{Q}}_D^{(m)}\}$ and set

$$\hat{\mathbf{c}}^{(m)} = \hat{\boldsymbol{\phi}}_{min}^{(m)}, \tag{8.540}$$

where $\hat{\boldsymbol{\phi}}_{min}^{(m)}$ is the eigenvector corresponding to the smallest eigenvalue.

(iv) Find

$$\hat{\mathbf{b}}^{(m)} = \mathbf{T}\hat{\mathbf{c}}^{(m)}. \tag{8.541}$$

(v) Find the roots of $\hat{\mathbf{b}}^{(m)}(z)$. Denote these roots as $\hat{z}_i^{(m)}, i = 1, \cdots, D$. Then

$$\hat{\psi}_i^{(m)} = \arg(\hat{z}_i^{(m)}), \quad i = 1, 2, \cdots, D. \tag{8.542}$$

(vi) Set $m = m + 1$.

(vii) Repeat steps (ii)–(v).

(viii) Check convergence

$$\|\hat{\mathbf{b}}(m+1) - \hat{\mathbf{b}}(m)\| < \epsilon. \tag{8.543}$$

If (8.543) is satisfied, terminate the iteration and set

$$\hat{\psi}_i = \psi_i^{(m+1)}. \tag{8.544}$$

Otherwise, continue the iteration.

Stoica and Sharman [SS90a] showed that, asymptotically, replacement of $[\mathbf{B}^H\mathbf{B}]^{-1}$ by a consistent estimate will have a negligible effect. Thus, a large sample realization of $\hat{\mathbf{b}}$ is given by

$$\hat{\mathbf{b}} = \min_{\mathbf{b}} \left\{ \operatorname{tr}\left\{ \left[\tilde{\mathbf{U}}_s^H\mathbf{B}\right]\left[\hat{\mathbf{B}}_1^H\hat{\mathbf{B}}_1\right]^{-1}\left[\mathbf{B}^H\tilde{\mathbf{U}}_s\right]\right\}\right\}. \tag{8.545}$$

In (8.545), the matrix $\hat{\mathbf{B}}_1$ is the consistent estimate of $\mathbf{B}$ obtained by minimizing $\operatorname{tr}\left[\left(\tilde{\mathbf{U}}_s^H\mathbf{B}\right)\left(\mathbf{B}^H\tilde{\mathbf{U}}_s\right)\right]$. Therefore, the MODE algorithm developed by Stoica and Sharman is a two-step algorithm that corresponds to terminating the IMODE algorithm in (8.538)–(8.542) at the $m = 1$ step. When comparing IMODE and MODE, we refer to the latter as two-step MODE to emphasize the difference.

We now consider two of the test cases that we have analyzed previously.

Example 8.7.1 (continuation, Example 8.5.2)[22]

Consider the same model as in Example 8.5.2. There are two equal-power uncorrelated plane-wave signals impinging on the array from $\psi = \pm\Delta\psi_R/2$. We implement three algorithms; IQML-QC[23], IMODE, and two-step MODE.

In Figure 8.39(a), the normalized RMSE is plotted versus *ASNR* for the three algorithms. The thresholds for IMODE and IQML-QC are slightly higher than the threshold using grid search shown in Figure 8.32. The threshold for two-step MODE is about 3 dB higher. Above an $ASNR = 10$ dB, all three algorithms are essentially on the CRB.

In Figure 8.39(b), a histogram showing the number of iterations required when the $ASNR = 10$ dB and the error threshold, $\epsilon = 0.01$ is plotted. This *ASNR* is above the threshold point. The most common number of iterations is four. However, from Figure 8.39(a), the last iterations do not improve the performance significantly, so two-step MODE is adequate above threshold if the number of iterations needs to be constrained.

In Figure 8.39(c), the same results are shown for an $ASNR = 5$ dB which is just below the threshold. We see that, in this set of 1000 trials, three or more iterations were always used and 4–6 iterations were generally needed. In this case, the additional iterations provide a useful decrease in the RMSE, and IMODE should be used.

---

[22] Figures 8.39 and 8.40 are due to J. Hiemstra (private communication).

[23] In implementing IQML-QC in MATLAB®, we found that, after finding $\hat{\mathbf{c}}$ in (8.515) to an adequate precision, we could decrease the RMSE further by using the *fmins* function to minimize $Re\left[\mathbf{c}^T\tilde{\mathbf{Q}}_{\mathbf{x}}\mathbf{c}\right] + \left|\left(\|\mathbf{c}\|^2 - 1\right)\right|$. This is a minimization with a penalty function and can be viewed as a slight modification of a pure IQML algorithm. A gradient technique could also be used. The technique is most useful near the threshold. MATLAB® also has a constrained minimization that could be used. We denote this version as IQML-G in the legend.

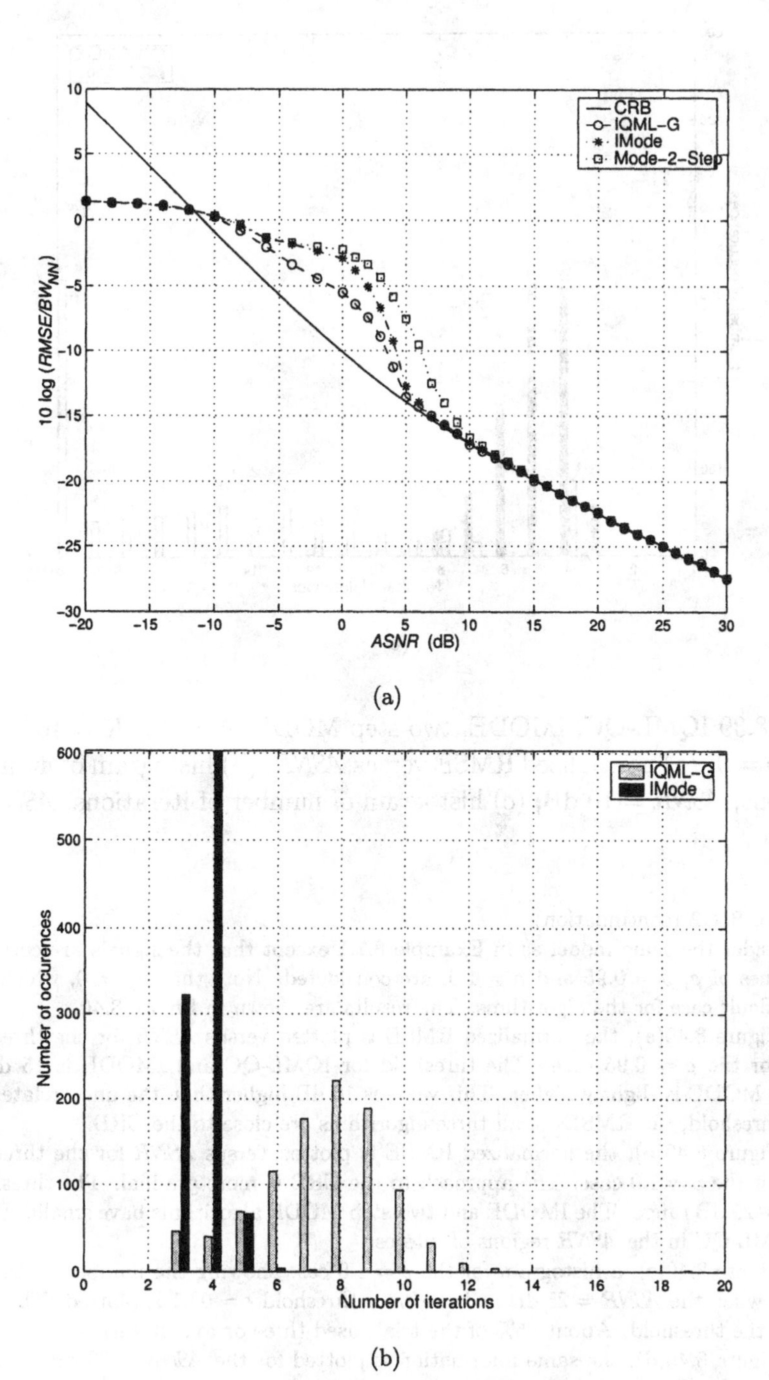
CRB
IQML-G
IMode
Mode-2-Step
10 log (RMSE/BW_NN)
ASNR (dB)
IQML-G
IMode
Number of occurences
Number of iterations

(a)

(b)

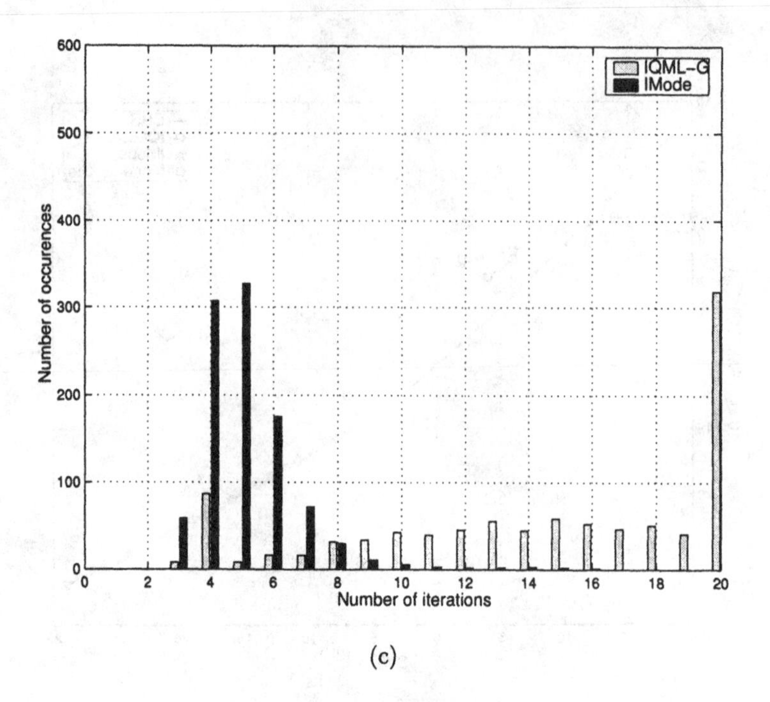

(c)

Figure 8.39 IQML-QC, IMODE, two-step MODE: $N = 10$, $K = 100$, $\Delta\psi = \Delta\psi_R$, $\rho = 0$: (a) normalized RMSE versus *ASNR*; (b) histogram of number of iterations, *ASNR* = 10 dB; (c) histogram of number of iterations, *ASNR* = 5 dB.

Example 8.7.2 (continuation)

Consider the same model as in Example 8.7.1 except that the signals are correlated. Two values of ρ, $\rho = 0.95$ and $\rho = 1.0$, are considered. Note that $\phi_\rho = 0$, which is the most difficult case for the algorithms. The results are shown in Figure 8.40.

In Figure 8.40(a), the normalized RMSE is plotted versus *ASNR* for the three algorithms for the $\rho = 0.95$ case. The threshold for IQML-QC and IMODE is 15 dB and two-step MODE is slightly higher. This value is 10 dB higher than the uncorrelated case. Above threshold, the RMSEs of all three algorithms are close to the CRB.

In Figure 8.40(b), the normalized RMSE is plotted versus *ASNR* for the three algorithms for the $\rho = 1.0$ case. The approach to the CRB is more gradual. The threshold is in the 23–25 dB range. The IMODE and two-step MODE algorithms have smaller RMSEs than IQML-QC in the *ASNR* regions of interest.

In Figure 8.40(c), a histogram for the $\rho = 1.0$ case showing the number of iterations required when the *ASNR* = 25 dB and the error threshold $\epsilon = 0.01$ is plotted. This *ASNR* is above the threshold. About 95% of the trials used three or four iterations.

In Figure 8.40(d), the same information is plotted for the *ASNR* = 20 dB case. This *ASNR* is in the threshold region; The number of iterations range from three to seven, but three and four iterations were used in almost 80% of the trials.

For the uncorrelated signals case in Example 8.7.1, the modified IQML-

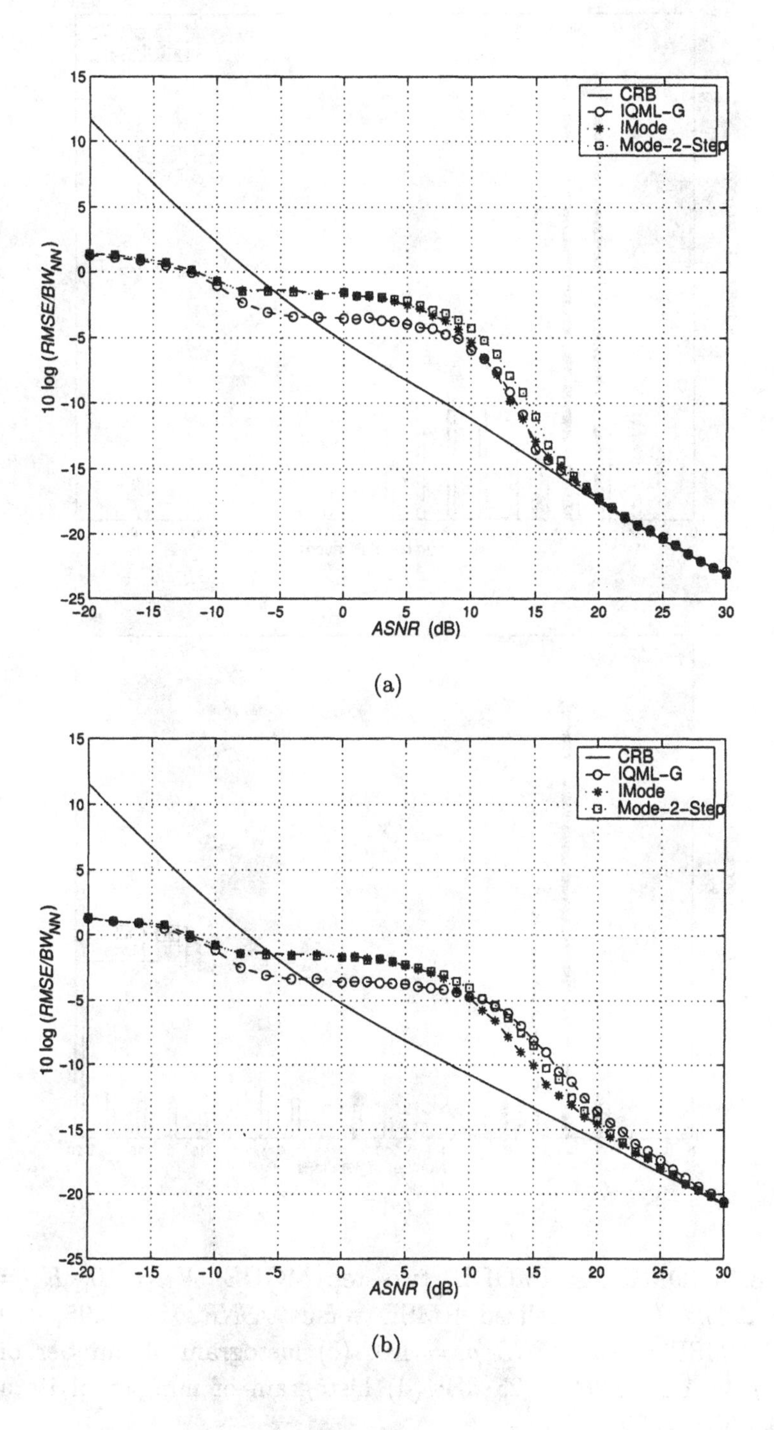
CRB
IQML-G
IMode
Mode-2-Step
10 log ($RMSE/BW_{NN}$)
ASNR (dB)
15
10
5
0
-5
-10
-15
-20
-25
-20
-15
-10
-5
0
5
10
15
20
25
30
CRB
IQML-G
IMode
Mode-2-Step
10 log ($RMSE/BW_{NN}$)
ASNR (dB)

(a)

(b)

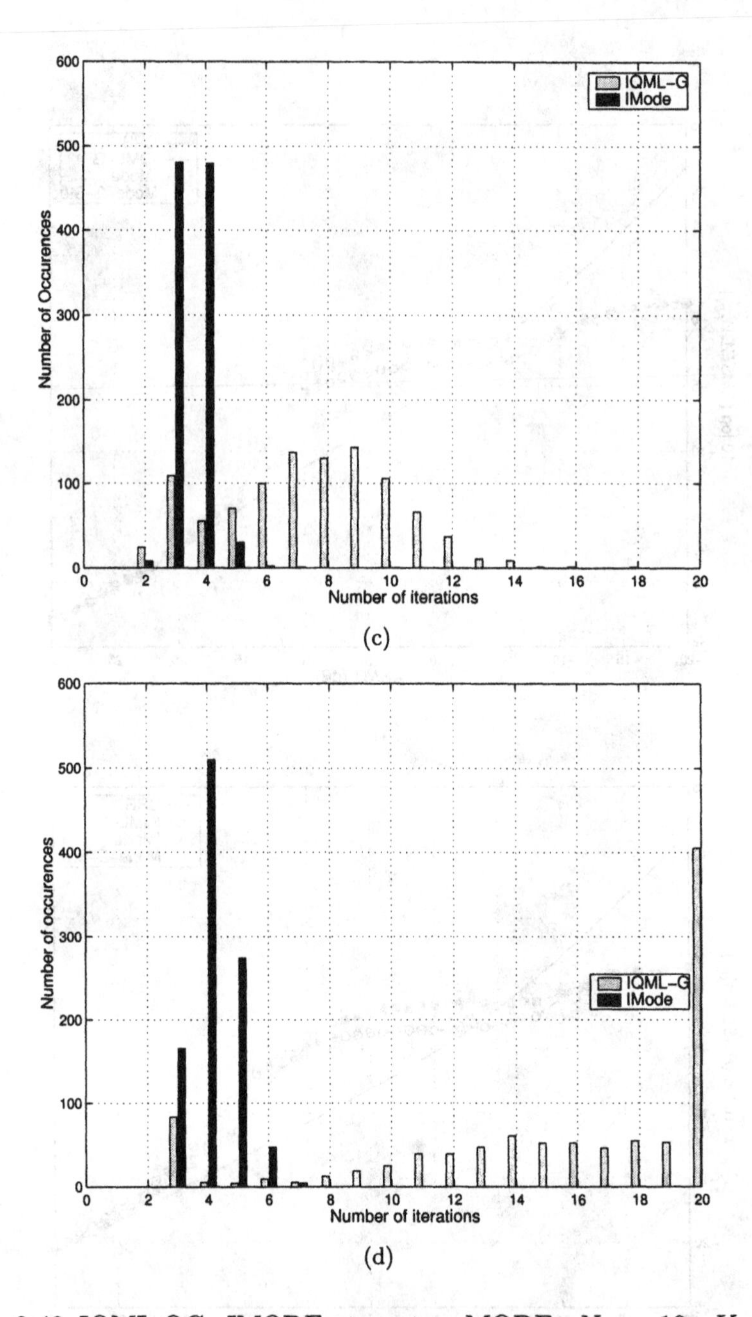

Figure 8.40 IQML-QC, IMODE, two-step MODE: $N = 10$, $K = 100$, $\Delta\psi = \Delta\psi_R$: (a) normalized RMSE versus $ASNR, \rho = 0.95$; (b) normalized RMSE versus $ASNR, \rho = 1.0$; (c) histogram of number of iterations, $\rho = 1.0, ASNR = 25$ dB; (d) histogram of number of iterations, $\rho = 1.0, ASNR = 20$ dB.

QC algorithm and IMODE had similar threshold behavior and approached the CRB above threshold. The two-step MODE algorithm had a higher threshold, but approached the CRB above threshold. The IQML-QC algorithm has a higher flop count, so IMODE is the preferred algorithm for this case. The behavior in this example appears to be characteristic of the general uncorrelated signal case. If the threshold behavior is an important factor in a particular application, then IMODE is useful (or perhaps, five-step MODE). If performance above threshold is the important factor, the two-step MODE appears to be adequate. Similar comments apply to the correlated and coherent signal case.

Li et al. [LSL98] provide a more general comparison of IQML, IMODE, and two-step MODE. They also provide a detailed discussion of an efficient implementation of IMODE and a comparison of flop counts. The interested reader should consult that reference.

For standard linear arrays, IMODE is the leading candidate of the algorithms developed up to this point in the text. There are still issues, such as the effect of an unknown number of signals and robustness to array perturbation that must be explored. In addition, other algorithms are developed in Chapter 9.

A unitary version of IMODE is developed in Gershman and Stoica [GS99]. We discuss it in Chapter 9 after other unitary algorithms have been developed. It is identical to FB-IMODE, which is standard IMODE using FB averaging in the sample spectral matrix (e.g., Stoica and Jansson [SJ97]). In [SJ97], it is shown that the RMSE of FB-MODE in the asymptotic region is greater than or equal to the RMSE of standard MODE. In Example 8.7.2, $\phi_\rho = 0°$, so the two versions are identical.

8.7.4 Summary

In this section we have developed a polynomial parameterization and then showed how it could be used to develop efficient computational algorithms. Although we used CML and MODE as examples, the general technique is applicable to any algorithm that utilizes $\mathbf{P_V}$. Whenever the array geometry allows it, the polynomial parameterization approach should be considered.

We can also develop Newton or quasi-Newton algorithms that utilize the polynomial parameterization. The advantage of doing the Newton algorithm in polynomial space is that the number of computations at each iteration is reduced. Starer and Nehorai [SN92] have derived the algorithm and give examples.

8.8 Detection of Number of Signals

Up to this point in our discussion of parameter estimation we have assumed that the number of signals, D, is known. In practice, we would use one of the algorithms developed in Section 7.8 to estimate D. We use that estimate, denoted as $\hat{D}$, in our estimation algorithm. We refer to this technique as **separable detection.** All of the results in Section 7.8 apply to this problem.

There are a number of scenarios studied in the problems. In some cases, the threshold of the detection algorithms occurs at the same *ASNR* or higher *ASNR* than the threshold of an ML estimator operating with the correct number of signals. This result suggests that all estimation algorithms needed to be checked for robustness to errors in $\hat{D}$.

In most parameter estimation problems, if $\hat{D} \neq D$, it is better to overestimate rather than underestimate. In many cases, the output of the parameter estimator serves as the input to another processor such as a tracker. We can rely on the subsequent processor to eliminate the extra signals. This observation suggests that if we must operate in the vicinity of the threshold, we should use AIC or AIC-FB and require subsequent processing (after the estimator) to eliminate an extra signals introduced by an overestimation error from AIC. If we are operating above threshold, we should use MDL or MDL-FB, because it provides consistent estimates.

A second approach is to jointly detect the number of signals and estimate their location. The CML version of the joint detection-estimation algorithm was developed by Wax and Ziskind [WZ89]. The AML version of the joint detection-estimation algorithm was developed by Wax [Wax91]. One can show that the MDL version of resulting estimators is consistent (e.g., the approach by Zhao et al. [ZKB87] or Wax and Ziskind [WZ89]).

Cho and Djuric [CD94] approached the joint detection and estimation problem using a technique they called Bayesian predictive densities (e.g., [Dju90]). Their result contains the same data term as the AML estimator but has a different penalty term.

Ottersten et al. [OVSN93] utilize a generalized likelihood ratio test with a different model for the two hypotheses than was used in Section 7.8.1. They obtain a sequential hypothesis test that utilizes the AML estimates at each candidate value of d.

All of the joint detection and estimation techniques have significant computational complexity and do not appear to be widely used in practical applications.

We defer an analysis of this problem until Chapter 9, where we compare the performance of a number of algorithms for the unknown D case.

8.9 Spatially Spread Signals

In this section we discuss parameter estimation for spatially spread signals. In Section 8.9.1, we consider the model that we developed in Section 5.3. Here the source signal is characterized by a spectrum distributed on the surface of a large sphere whose distribution is specified by $S_0(\omega : \theta, \phi)$. We parameterize the spatial spectrum and find the ML estimate of the parameters. We also derive the corresponding CRB.

In Section 8.9.2, we consider linear arrays and assume that the spatial spectrum can be modeled as a spatial ARMA process. We discuss the ML estimate of the ARMA parameters and the corresponding CRB.

In Section 8.9.3, we summarize our discussion.

8.9.1 Parameterized $S(\theta, \phi)$

In this section we consider the model described in Section 5.3.4. For numerical simplicity, we first consider the case of a standard linear array along the z-axis[24]. From (5.145), the 1-D signal spectrum is

$$S_s(\omega : \Delta p_z) = \int_0^{\pi} d\theta \frac{\sin\theta}{2} e^{-jk_0 \Delta p_z \cos\theta} \bar{S}_0(\omega : \theta), \tag{8.546}$$

where

$$\bar{S}_0(\omega : \theta) \triangleq \frac{1}{2\pi} \int_0^{2\pi} S_0(\omega : \theta, \phi) d\phi. \tag{8.547}$$

Letting

$$\psi = \pi \cos\theta, \tag{8.548}$$

$$\Delta p_z = (m - n)\frac{\lambda}{2}, \tag{8.549}$$

and

$$S_f(\psi) = \bar{S}_0(\omega : \theta)\big|_{\psi = \pi\cos\theta}, \tag{8.550}$$

we can write the elements in the array spectral matrix as

$$[\mathbf{S_f}]_{mn} = \frac{1}{2\pi} \int_{-\pi}^{\pi} d\psi e^{-j(m-n)\psi} S_f(\psi). \tag{8.551}$$

At this point we must specify $S_f(\psi)$ in order to proceed. We consider two simple examples[25] in which we can evaluate the integral in (8.551) analytically. In other cases, we can calculate $\mathbf{S_f}$ numerically.

---

[24]The ML part of this discussion follows Meng et al. [MWW93].

[25]The two examples are taken from [MWW93].

Example 8.9.1

Assume

$$S_f(\psi) = \frac{\sigma_f^2}{\pi} \frac{\sigma_\psi}{\sigma_\psi^2 + (\psi - \psi_0)^2}. \tag{8.552}$$

The parameter ψ_0 is the mean value of the angle of arrival and σ_ψ^2 controls the width. Assuming $\sigma_\psi \ll \pi$, we use (8.552) in (8.551) and recognize a familiar Fourier transform pair, to obtain,

$$\mathbf{S_f} = \frac{\sigma_f^2}{2\pi^2} \begin{bmatrix} 1 & z_\psi^H & \cdots & \left(z_\psi^H\right)^{N-1} \\ z_\psi & 1 & \cdots & \left(z_\psi^H\right)^{N-2} \\ \vdots & \vdots & \ddots & \vdots \\ z_\psi^{N-1} & z_\psi^{N-2} & \cdots & 1 \end{bmatrix} \triangleq \frac{\sigma_f^2}{2\pi^2} \mathbf{\Sigma}_S, \tag{8.553}$$

where

$$z_\psi \triangleq \exp\left[-\sigma_\psi + j\psi_0\right]. \tag{8.554}$$

In the second example, we assume that $S_f(\psi)$ has a Gaussian shape central at ψ_0 with standard deviation of σ_ψ.[26]

Example 8.9.2

Assume

$$S_f(\psi) = \frac{\sigma_f^2}{\sqrt{2\pi\sigma_\psi^2}} \exp\left(-\frac{(\psi - \psi_0)^2}{2\sigma_\psi^2}\right), \tag{8.555}$$

where $\sigma_\psi \ll \pi$. Using (8.555) in (8.551) and recognizing the familiar Fourier transform pair, gives

$$\left[\mathbf{S_f}\right]_{mn} = \frac{\sigma_f^2}{2\pi} \exp\left\{-(m-n)^2\frac{\sigma_\psi^2}{2} + j\psi_0(m-n)\right\}. \tag{8.556}$$

For D independent signals, we have

$$\mathbf{S_f} = \sum_{i=1}^{D} \mathbf{S_{f_i}}, \tag{8.557}$$

where the $\mathbf{S_{f_i}}$ are given by (8.553) or (8.556).

A particular case that we will study is the two-signal case in which

$$S_{f_1} = \frac{\sigma_{f_1}^2}{\sqrt{2\pi\sigma_1^2}} \exp\left(-\frac{(\psi - \psi_{c1})^2}{2\sigma_1^2}\right), \tag{8.558}$$

and

[26] This model is applicable in a number of wireless communication applications.

$$S_{f_2} = \frac{\sigma_{f_2}^2}{\sqrt{2\pi\sigma_2^2}} \exp\left(-\frac{(\psi - \psi_{c2})^2}{2\sigma_2^2}\right). \tag{8.559}$$

The spectral matrix of the total input is

$$\mathbf{S_x} = \mathbf{S_f} + \sigma_w^2 \mathbf{I}. \tag{8.560}$$

We can now use (8.560) directly in the ln likelihood function to find the ML estimate and the CRB.

We first consider the CRB. For simplicity, we write the result using the notation of our two examples.

We define $\boldsymbol{\psi}_0$ as the $D \times 1$ parameter vector denoting the center of each $S_{f_i}(\psi)$,

$$\boldsymbol{\psi}_0 \triangleq \begin{bmatrix} \psi_{c1} & \psi_{c2} & \cdots & \psi_{cD} \end{bmatrix}^T. \tag{8.561}$$

The vector $\boldsymbol{\sigma}_0^2$ denotes the spread of the $S_{f_i}(\psi)$,

$$\boldsymbol{\sigma}_0^2 \triangleq \begin{bmatrix} \sigma_{\psi 1}^2 & \sigma_{\psi 2}^2 & \cdots & \sigma_{\psi D}^2 \end{bmatrix}^T. \tag{8.562}$$

The $\boldsymbol{\sigma}_{\mathbf{f}}^2$ denotes the power in each $S_{f_i}(\psi)$,

$$\boldsymbol{\sigma}_{\mathbf{f}}^2 \triangleq \begin{bmatrix} \sigma_{f_1}^2 & \sigma_{f_2}^2 & \cdots & \sigma_{f_D}^2 \end{bmatrix}^T. \tag{8.563}$$

Assuming σ_w^2 is unknown, we have a $(3D+1)$ parameter estimation problem.

To evaluate the CRB, we write the information matrix in a partitioned form

$$\mathbf{J} = \left[\begin{array}{c:c:c:c} \mathbf{J}_{\boldsymbol{\psi}_0\boldsymbol{\psi}_0} & \mathbf{J}_{\boldsymbol{\psi}_0\boldsymbol{\sigma}_\psi^2} & \mathbf{J}_{\boldsymbol{\psi}_0\boldsymbol{\sigma}_s^2} & \mathbf{J}_{\boldsymbol{\psi}_0\sigma_w^2} \\ \hdashline \mathbf{J}_{\boldsymbol{\sigma}_\psi^2\boldsymbol{\psi}_0} & \mathbf{J}_{\boldsymbol{\sigma}_\psi^2\boldsymbol{\sigma}_\psi^2} & \mathbf{J}_{\boldsymbol{\sigma}_\psi^2\boldsymbol{\sigma}_s^2} & \mathbf{J}_{\boldsymbol{\sigma}_\psi^2\sigma_w^2} \\ \hdashline \mathbf{J}_{\boldsymbol{\sigma}_s^2\boldsymbol{\psi}_0} & \mathbf{J}_{\boldsymbol{\sigma}_s^2\boldsymbol{\sigma}_\psi^2} & \mathbf{J}_{\boldsymbol{\sigma}_s^2\boldsymbol{\sigma}_s^2} & \mathbf{J}_{\boldsymbol{\sigma}_s^2\sigma_w^2} \\ \hdashline \mathbf{J}_{\sigma_w^2\boldsymbol{\psi}_0} & \mathbf{J}_{\sigma_w^2\boldsymbol{\sigma}_\psi^2} & \mathbf{J}_{\sigma_w^2\boldsymbol{\sigma}_s^2} & \mathbf{J}_{\sigma_w^2\sigma_w^2} \end{array}\right], \tag{8.564}$$

where each term is given by (8.35),

$$\mathbf{J}_{ij} = \operatorname{tr}\left[\frac{\partial \mathbf{S_x}}{\partial \theta_i}\mathbf{S_x}^{-1}\frac{\partial \mathbf{S_x}}{\partial \theta_j}\mathbf{S_x}^{-1}\right], \tag{8.565}$$

where θ_i denotes the appropriate parameter. To evaluate (8.564), we utilize (8.553) or (8.556) in (8.560) and calculate the derivatives in (8.565).

We consider two simple examples to illustrate the technique.

Example 8.9.3 (continuation)

Consider a standard 10-element linear array. There is a single spread signal impinging on the array. The spectral matrix is given by (8.556). We first consider the case in which σ_ψ^2, σ_f^2, and σ_w^2 are known. This result provides a bound on more realistic models. In this case, $\mathbf{J}$ is a scalar. To evaluate it, we differentiate (8.556). The result is

$$\frac{\partial}{\partial\psi_0}\left\{[\mathbf{S_f}]_{mn}\right\} = j(m-n)\cdot\frac{\sigma_f^2}{2\pi}\exp\left\{-(m-n)^2\frac{\sigma_\psi^2}{2}+j\psi_0(m-n)\right\}. \tag{8.566}$$

We substitute (8.566) and the inverse of (8.556) into (8.565) to obtain $J_{\psi_0\psi_0}$. Then, the CRB(ψ_0) is the reciprocal of it. We plot the normalized bound in Figure 8.41 for various values of

$$\sigma_u = \sigma_\psi/\pi. \tag{8.567}$$

We also show the conventional CRB ($\sigma_u = 0$). We see that the bound increases significantly as σ_u increases.

The next step is to consider the case when σ_ψ^2, σ_s^2, and σ_w^2 are unknown. The resulting cross-matrices are all zero, so the bound in Example 8.9.3 also applies to this case.

We next consider the case of two spatially spread signals.

Example 8.9.4 (continuation)

Consider the case in which there are two uncorrelated spread signals impinging on a standard 10-element linear array. We use the signal model in (8.555) for each signal. For the case of known signal and noise power, there are four parameters to estimate: ψ_{c1}, ψ_{c2}, $\sigma_{\psi 1}^2$, and $\sigma_{\psi 2}^2$. If the signal powers and noise power are unknown, then there are three additional parameters: $\sigma_{f_1}^2$, $\sigma_{f_2}^2$, and σ_w^2.

We assume that the $\psi_{c2} = -\psi_{c1}$ and that $\Delta\psi = \Delta\psi_R$. We also assume that $\sigma_{\psi 2}^2 = \sigma_{\psi 1}^2$. In Figure 8.42, we plot the bound on $var[\psi_{c1}]$ versus ASNR for various $\sigma_{\psi 1}$ for the four parameter case. In Figure 8.43, we plot the bound on $var[\psi_{c1}]$ for the seven-parameter case.

The parameters are coupled so the CRB for the seven parameter case is higher than in the four-parameter case.

In order to find the ML estimate, we must conduct a search over the parameter space.

Using (8.13), we can write the ln likelihood function as

$$L\left(\boldsymbol{\psi}_0,\boldsymbol{\sigma}_\psi^2,\boldsymbol{\sigma}_f^2,\sigma_w^2\right) = -K\left\{\ln|\mathbf{S_x}| + \text{tr}\left[\mathbf{S_x}^{-1}\mathbf{C_x}\right]\right\}, \tag{8.568}$$

where $\boldsymbol{\psi}_0, \boldsymbol{\sigma}_\psi^2, \boldsymbol{\sigma}_f^2$ are $D\times 1$ vectors containing the parameters of each of the D spread signals.

Then,

$$\hat{\boldsymbol{\theta}}_{ml} = \arg\min_{\boldsymbol{\theta}}\left\{\ln|\mathbf{S_x}| + \text{tr}\left[\mathbf{S_x}^{-1}\mathbf{C_x}\right]\right\}, \tag{8.569}$$

where $\boldsymbol{\theta}$ is a $(3D+1)\times 1$ parameter vector.

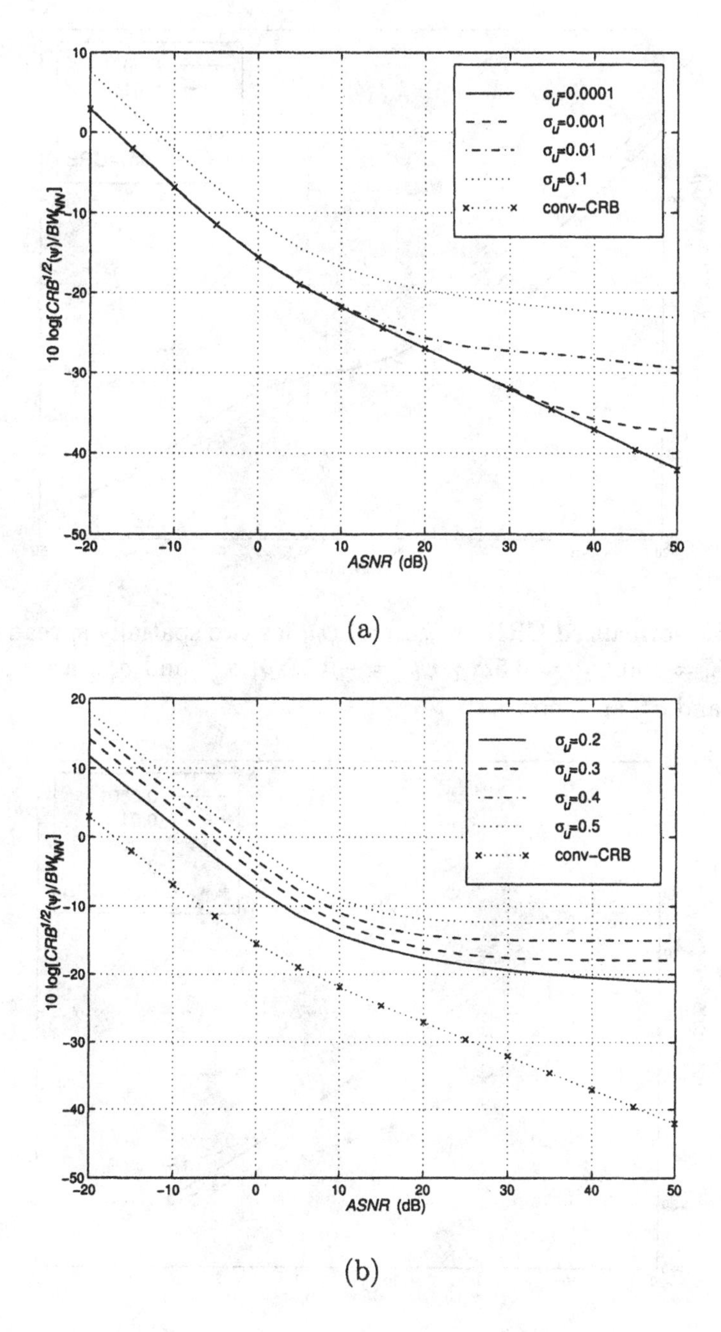

Figure 8.41 Normalized CRB on ψ for single spread signal versus ASNR: $N = 10, K = 100, \psi_s = 0$: (a) values of σ_u from 10^{-4} to 0.1; (b) values of σ_u for 0.2 to 0.5.

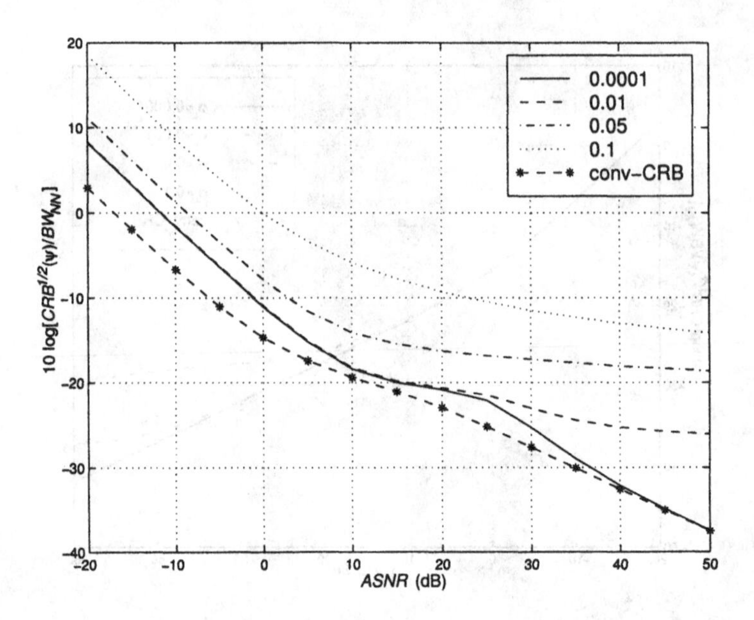

Figure 8.42 Normalized CRB on ψ_{c_1} and ψ_{c_2} for two spatially spread signals: $N = 10, K = 100, \psi_{c_1} = 0.5\Delta_R, \psi_{c_2} = -0.5\Delta_R$, $\sigma_{\psi 1}^2$ and $\sigma_{\psi 2}^2$ are unknown, σ_{f1}^2, σ_{f2}^2, and σ_w^2 are known.

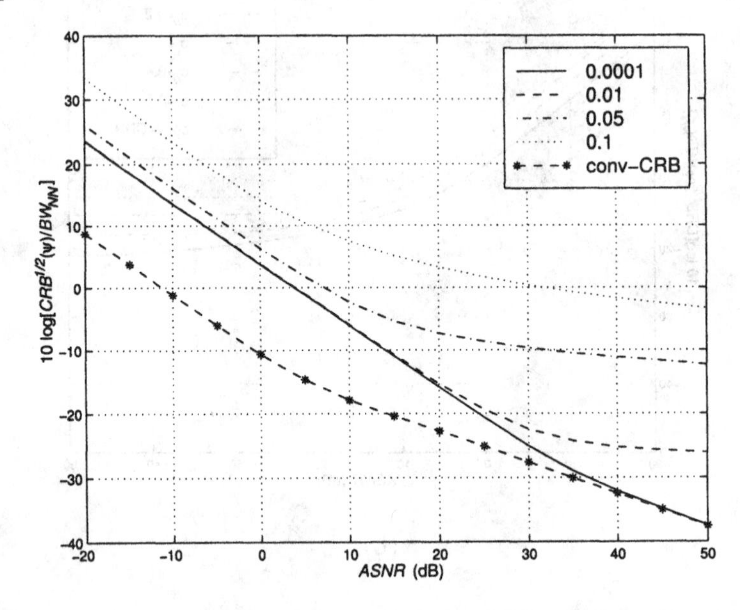

Figure 8.43 Normalized CRB on ψ_{c_1} and ψ_{c_2} for two spatially spread signals: $N = 10, K = 100, \psi_{c_1} = 0.5\Delta_R, \psi_{c_2} = -0.5\Delta_R$, all seven parameters are unknown.

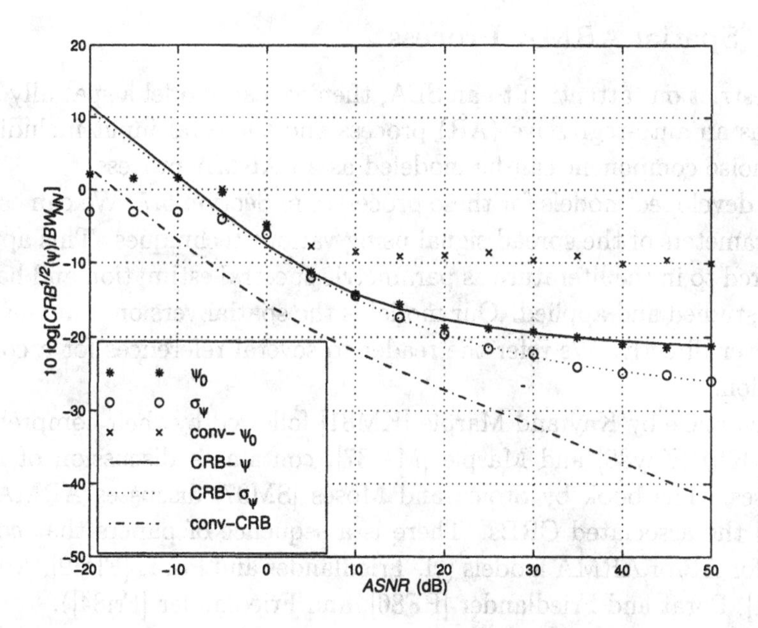

Figure 8.44 ML estimator of ψ_0 and σ_ψ: $N = 10, K = 100, \psi_0 = 0.0433\pi$, $\sigma_\psi = 0.2\pi, \sigma_s^2$ and σ_w^2 known; normalized RMSE versus *ASNR*.

In order to keep the computations manageable we consider the case of a single signal with ψ_0 and σ_u unknown, but where σ_f^2 and σ_w^2 are known.

Example 8.9.5 (continuation, Example 8.9.3)

Consider a single spatially spread signal impinging on a standard 10-element linear array. The signal component of the spatial spectral matrix is given by (8.556). We assume that $\sigma_f^2$and σ_w^2 are known and find the ML estimate of ψ_0 and σ_ψ.

The normalized RMSE is plotted versus ASNR in Figure 8.44. We also plot the CRB, the normalized RMSE of the conventional ML algorithm (assumes a plane-wave signal), and the conventional CRB.

We see that RMSE of both estimates, $\hat{\psi}_0$ and $\hat{\sigma_\psi}$, coincide with the CRB above ASNR = 5 dB. This threshold is about 7 dB above the single plane-wave case in Figure 8.17. We see that the conventional ML estimator is not effective.

For multiple signals, the computational complexity makes ML impractical. In Chapter 9, we develop simpler estimation algorithms. Some of these algorithms can be modified to accommodate this spatially spread model, but we will not discuss the necessary modifications in the text.

8.9.2 Spatial ARMA Process

If we restrict our attention to an SLA, then we can model a spatially spread signal as an auto-regressive (AR) process and the total input including the white noise component can be modeled as an ARMA process.

We developed models for these processes in Section 5.7. We can estimate the parameters of the spread signal using various techniques. This approach is referred to in the literature as parametric spectral estimation and has been widely studied and applied. Our model is the spatial version, but the results carry over directly. We refer the reader to several references for a complete discussion.

The article by Kay and Marple [KM81] followed by their comprehensive books, Kay [Kay88] and Marple [Mar87], contains a discussion of ARMA processes. The book by Stoica and Moses [SM97] discusses ARMA models and the associated CRBs. There is a sequence of papers that compute CRBs for AR or ARMA models (cf. Friedlander and Porat [FP89], Anderson [And71], Porat and Friedlander [PF86], and Friedlander [Fri84]).

8.9.3 Summary

In this section we have introduced the problem of estimating the parameters of a spatially spread signal. The CRBs quantify the potential performance degradation caused by the spreading. The ML estimators approach the CRB in simple cases, but are computationally prohibitive in the multiple-signal environment.

In the single spread signal example, we saw that the conventional ML estimators that ignored the spreading did not perform well. The same result is true for the multiple-signal environment.

8.10 Beamspace algorithms

8.10.1 Introduction

In Sections 6.9 and 7.10, we saw the advantages of beamspace processing in the context of adaptive beamforming. In this section, we show its application in the parameter estimation problem.

A number of authors have discussed the advantages of beamspace processing for parameter estimation. References include Bienvenu and Kopp [BK84], Gray [Gra82], Forster and Vezzosi [FV87], Van Veen and Williams [VVW88a], [VVW88b], Lee and Wengrovitz [LW88], [LW90], Buckley and Xu [XB88], [XB89], [BX90], and Zoltowski [Zol88].

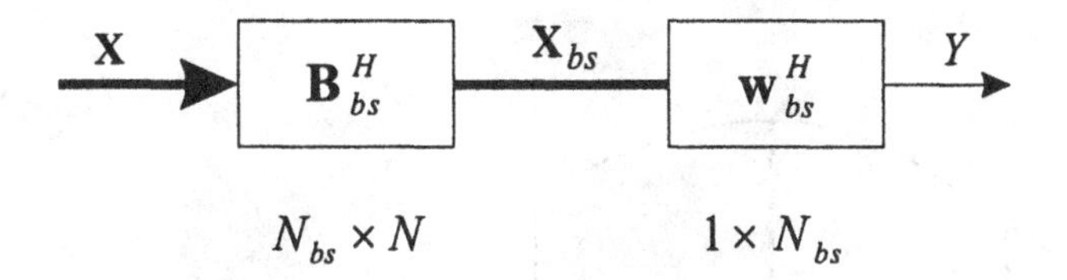

Figure 8.45 Beamspace processing.

The beamspace model was shown in Figure 6.81 and is repeated in Figure 8.45 for convenience. The output of the beamspace matrix is,

$$\mathbf{X}_{bs} = \mathbf{B}_{bs}^H \mathbf{X}, \tag{8.570}$$

where $\mathbf{B}_{bs}^H$ is an $N_{bs} \times N$ matrix. We process $\mathbf{X}_{bs}$ to estimate the parameters of interest.

As in the beamformer problem, there are several advantages:

(i) The dimension of the beamspace N_{bs} is usually much smaller than N so the computational complexity is reduced and the statistical stability of the estimate of the spatial spectral matrix is improved.

(ii) If there are strong interfering signals that are not in the same sector as the signals of interest, we may be able to null them out prior to implementing our parameter estimation algorithm.

(iii) Most of our algorithms have assumed additional white Gaussian noise (or Gaussian noise with a known correlation function that we could pre-whiten). In Figure 8.46 we show a representative non-white spatial noise spectrum that we assume is unknown. We indicate how we have divided ψ-space into six sectors by using six parallel beamspace processors. In each of these sectors we can model the noise as white with an unknown variance and use the beamspace version of algorithms that we have developed.

On the other hand, the disadvantage is that the performance of the algorithm after pre-processing may be poorer. If we choose $\mathbf{B}_{bs}$ improperly, then we may lose information that degrades the performance of the parameter estimation algorithm. We would like to show that the output of the beamspace processor is a sufficient statistic for the estimation problem of interest. A simpler approach is to compare the beamspace CRB to the element-space CRB. We find that the CRB on the asymptotic variance of the beamspace

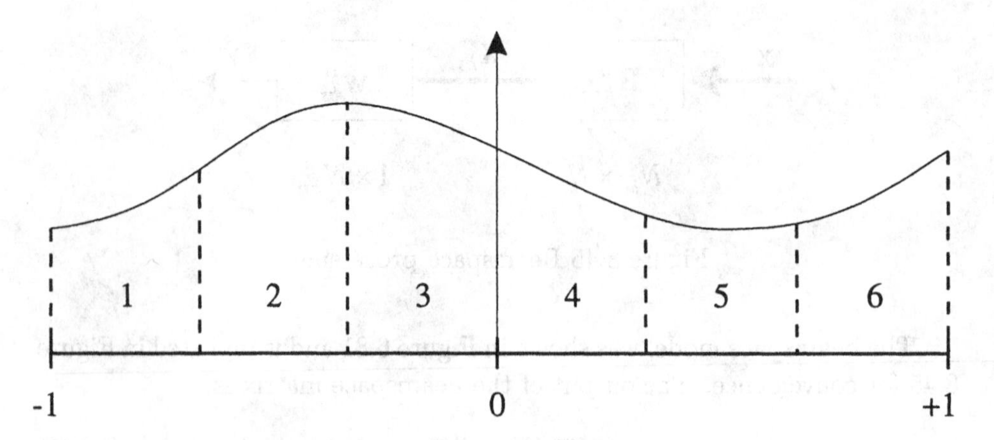

Figure 8.46 Beamspace processing for non-white spatial noise spectrum: six beam sectors.

estimator is always greater than or equal to the CRB bound in element space. However, we find that, with an appropriate choice of $\mathbf{B}_{bs}$, the probability of resolution can be improved.

The model of interest is the familiar snapshot model,

$$\mathbf{X}(k) = \mathbf{V}\,\mathbf{F}(k) + \mathbf{N}(k), \quad k = 1, 2, \cdots, K, \tag{8.571}$$

where $\mathbf{X}(k)$ is an $N \times 1$ vector, $\mathbf{F}(k)$ is the $D \times 1$ signal in space vector, $\mathbf{N}(k)$ is an $N \times 1$ white Gaussian noise vector, and $\mathbf{V}$ is the array manifold matrix,

$$\mathbf{V} = \left[\begin{array}{c:c:c:c} \mathbf{v}_1 & \mathbf{v}_2 & \cdots & \mathbf{v}_D \end{array}\right], \tag{8.572}$$

where

$$\mathbf{v}_n \triangleq \mathbf{v}\left(\psi_n\right). \tag{8.573}$$

We use the Gaussian signal model. The signal vectors are independent samples of the stationary zero-mean complex Gaussian random process with unknown spectral matrix $\mathbf{S}_{\mathbf{f}}$. The noise vectors are independent samples of a stationary zero-mean complex Gaussian random process with spectral matrix $\sigma_w^2\,\mathbf{I}$ where σ_w^2 is unknown.

The spectral matrix of the sampled input vector $\mathbf{X}(k)$ is

$$\begin{aligned} \mathbf{S}_{\mathbf{x}} &\triangleq E\left[\mathbf{X}(k)\,\mathbf{X}^H(k)\right] \\ &= \mathbf{V}\,\mathbf{S}_{\mathbf{f}}\,\mathbf{V}^H + \sigma_w^2\,\mathbf{I}, \end{aligned} \tag{8.574}$$

where $\psi_1, \psi_2, \cdots, \psi_D, \mathbf{S}_{\mathbf{f}}$, and σ_w^2 are unknown.

We define an $N_{bs} \times N$ matrix $\mathbf{B}_{bs}^H$, which is used to pre-process the data,

$$\mathbf{X}_{bs}(k) = \mathbf{B}_{bs}^H \mathbf{X}(k). \tag{8.575}$$

We assume

$$N_{bs} \leq N \tag{8.576}$$

and

$$N_{bs} \geq D + 1. \tag{8.577}$$

We also assume that the columns of $\mathbf{B}$ are orthonormal so that

$$\mathbf{B}_{bs}^H \mathbf{B}_{bs} = \mathbf{I}. \tag{8.578}$$

In some applications, we may start with a pre-processing matrix $\mathbf{B}_{no}$ that does not satisfy (8.578). We then construct $\mathbf{B}_{bs}$ using the transformation.

$$\mathbf{B}_{bs} = \mathbf{B}_{no} \left(\mathbf{B}_{no}^H \mathbf{B}_{no} \right)^{-\frac{1}{2}}. \tag{8.579}$$

The choice of $\left(\mathbf{B}_{no}^H \mathbf{B}_{no} \right)^{-\frac{1}{2}}$ is available. One choice is the inverse of the Cholesky factorization of $\mathbf{B}_{no}^H \mathbf{B}_{no}$. A second choice is the Cholesky factorization of the inverse of $(\mathbf{B}_{no}^H \mathbf{B}_{no})$. We find that different choices may result in different performance.

The spectral matrix of $\mathbf{X}_{bs}$ is

$$\mathbf{S}_{\mathbf{X}_{bs}} = \mathbf{B}_{bs}^H \mathbf{V} \mathbf{S}_{\mathbf{f}} \mathbf{V}^H \mathbf{B}_{bs} + \sigma_w^2 \mathbf{I}. \tag{8.580}$$

We operate on $\mathbf{X}_{bs}$ to estimate the various parameters.

In Section 8.10.2, we discuss various beamspace matrices. In Section 8.10.3, we derive the beamspace CRB and derive conditions on $\mathbf{B}_{bs}$ under which the beamspace CRB is equal to the element-space CRB. In Section 8.10.4, we derive the beamspace ML estimator and study its performance. In Section 8.10.5, we summarize our results.

8.10.2 Beamspace Matrices

In this section we discuss various beamspace matrices. We encountered most of these matrices in earlier discussions (e.g., Sections 3.10, 6.9, or 7.10), but we need to revisit them in the context of DOA estimation. For notational simplicity, we consider a standard linear array in the text.

8.10.2.1 Conventional beams (DFT)

The most common beamspace matrix has rows that consist of conventional beams whose pointing directions are spaced at $2\pi/N$ intervals in ψ-space. In Figure 8.47, we show a 7-beam fan, and in Figure 8.48, we show a 6-beam fan.

The rows of $\mathbf{B}_{bs}^H$ are given by

$$[\mathbf{B}_{bs}^H]_m = \frac{1}{N} e^{j(\frac{N-1}{2})m\frac{2\pi}{N}} [\, 1 \quad e^{-jm\frac{2\pi}{N}} \quad \cdots \quad e^{-j(N-1)m\frac{2\pi}{N}} \,], m \in \Omega_{N_{bs}}, \tag{8.581}$$

where $\Omega_{N_{bs}}$ is the set containing the values of m included in the beamspace.

In most cases, it is convenient to denote the value of m corresponding to the beam closest to $\psi = -\pi$ as m_L (the subscript "L" denotes left). Then, the range of m is $m_L \leq m \leq m_L + N_{bs}$. If the beam sector is near $\psi = \pi$ (endfire) then $m_L + N_{bs}$ may exceed N and the beams wrap around.

The beam formed by (8.581) corresponds to a conventional beam pointed at $\psi = 2\pi m/N$.

If the beamspace sector is centered at $\psi = 0$ and N_{bs} is odd, there is a center beam aimed at $\psi = 0$. If the beamspace sector is centered at $\psi = 0$ and N_{bs} is even, then m in (8.581) is replaced by $m' = m - \pi/N$ and the indexing limits are modified. Then, the two inner beams are aimed at $\pm\pi/N$ in ψ-space. This beamspace matrix is referred to as the DFT beamformer in the signal processing literature. It is referred to as the Butler beamformer in the classic antenna literature. The mth beam in ψ-space is (2.127),

$$b_m(\psi) = [\mathbf{B}_{bs}^H]_m \mathbf{v}(\psi) = \frac{1}{N} \frac{\sin[\frac{N}{2}(\psi - m\frac{2\pi}{N})]}{\sin[\frac{1}{2}(\psi - m\frac{2\pi}{N})]}, \quad m_L \leq m \leq m_L + N_{bs}. \tag{8.582}$$

The resulting beamspace array manifold vector is a real $N_{bs} \times 1$ vector,

$$\mathbf{v}_{bs}(\psi) = \mathbf{B}_{bs}^H \mathbf{v}(\psi) = [b_1(\psi) \cdots b_{N_{bs}}(\psi)]^T. \tag{8.583}$$

We also consider the case in which the conventional beams are placed closer together. In this case, they are not orthogonal. We write

$$\left[\mathbf{b}_{no,i}^H\right]_n = \frac{1}{N} \left[e^{j(n-\frac{N-1}{2})\left(\left(i-(\frac{N_{bs}+1}{2})\right)\frac{2\pi\alpha}{N}\right)} \right], \quad n = 0,1,\cdots,N-1, \quad i = 1,2,\cdots,N_{bs}, \tag{8.584}$$

where $1 \leq \alpha \leq 2$. The beamspace matrix is orthogonalized using (8.579).

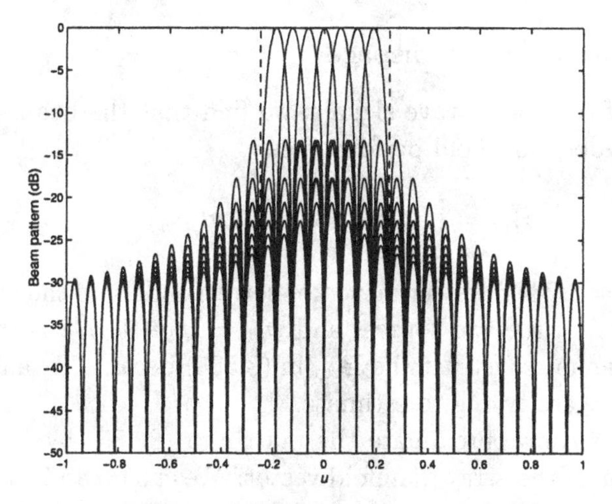

Figure 8.47 Seven-beam fan.

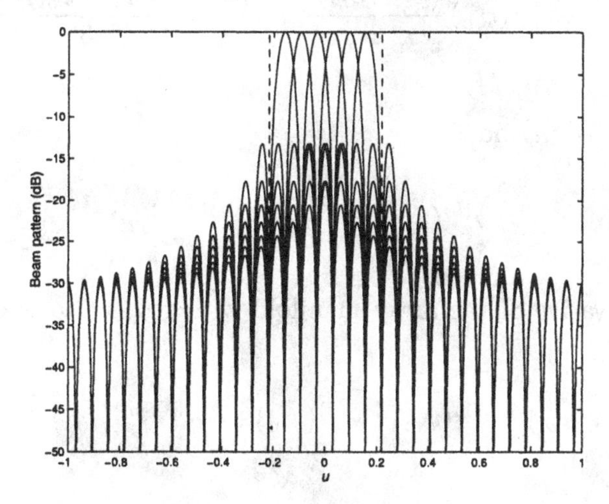

Figure 8.48 Six-beam fan.

We can also use beams with lower sidelobes such as Dolph-Chebychev or Hamming. This type of beamspace matrix was discussed in Section 3.10.

8.10.2.2 Taylor series beamspace

For the case of two plane-wave signals, we find that the beamspace matrix that provides good threshold performance is

$$\mathbf{B}_{no} = \left[\begin{array}{c:c:c} \mathbf{v}(\psi_1) & \mathbf{v}(\psi_2) & \mathbf{v}(\psi_m) \end{array} \right], \tag{8.585}$$

where $\mathbf{v}(\psi_i), i = 1, 2$ is the steering vector corresponding to the true wavenumbers, ψ_1 and ψ_2, of the two sources and $\psi_m = (\psi_1 + \psi_2)/2$ is the midpoint. However, we cannot construct the $\mathbf{B}_{no}$ in (8.585) because ψ_1 and ψ_2 are the wavenumbers we are trying to estimate.

However, we can approximate the matrix in (8.585) by using a Taylor series expansion of the array manifold vector. We can do an initial beamscan on the sensor output and determine the areas in u-space that appear to contain one or more signals. We denote the midpoint of one of these areas (the peak in the beamscan) as $\tilde{\psi}_m$ and use a Taylor series expansion around $\tilde{\psi}_m$:

$$\begin{aligned} \mathbf{v}(\psi_i) &= \mathbf{v}(\tilde{\psi}_m) + \mathbf{v}^{(1)}(\tilde{\psi}_m)\,(\psi_i - \tilde{\psi}_m) \\ &\quad + \mathbf{v}^{(2)}(\tilde{\psi}_m)\,\frac{(\psi_i - \tilde{\psi}_m)^2}{2!} + \mathbf{v}^{(3)}(\tilde{\psi}_m)\,\frac{(\psi_i - \tilde{\psi}_m)^3}{3!} + \cdots, \end{aligned} \tag{8.586}$$

where $i = 1, 2, \cdots, m$ and

$$\mathbf{v}^{(k)}(\tilde{\psi}_m) = \left. \frac{\partial^k \mathbf{v}(\psi)}{\partial^k \psi} \right|_{\psi = \tilde{\psi}_m}, \tag{8.587}$$

is the kth derivative of the array manifold with respect to ψ evaluated at $\psi = \tilde{\psi}_m$. Then,

$$\mathbf{B}_{no} = \left[\begin{array}{c:c:c:c:c} \mathbf{v}(\tilde{\psi}_m) & \mathbf{v}^{(1)}(\tilde{\psi}_m) & \mathbf{v}^{(2)}(\tilde{\psi}_m) & \cdots & \mathbf{v}^{(m)}(\tilde{\psi}_m) \end{array} \right], \tag{8.588}$$

and

$$\mathbf{B}_{bs} = \mathbf{B}_{no} \left[\mathbf{B}_{no}^H \mathbf{B}_{no} \right]^{-\frac{1}{2}}. \tag{8.589}$$

We consider an example to illustrate the behavior of this beamspace processor.

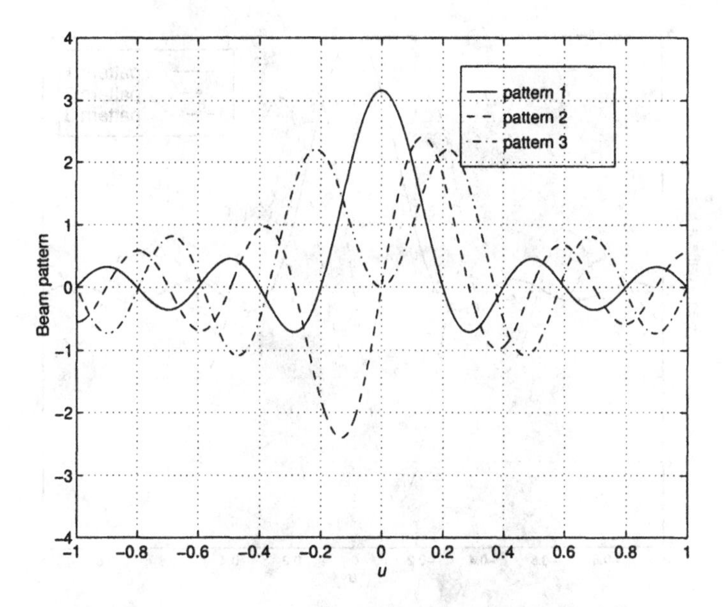

Figure 8.49 Beam patterns for Taylor series beamspace matrix, $N = 10$, $N_{bs} = 3$.

Example 8.10.1

Consider a standard 10-element linear array. Assume that the beamspace is centered at $\tilde{\psi}_m = 0$. If we use a 3-D beamspace, then $\mathbf{B}_{no}$ is a 10×3 matrix with a Vandermonde row structure

$$\mathbf{B}_{no} = \begin{bmatrix} 1 & 0 & 0 \\ 1 & j & (j)^2 \\ \vdots & \vdots & \vdots \\ 1 & (nj) & (nj)^2 \\ \vdots & \vdots & \vdots \\ 1 & (9j) & (9j)^2 \end{bmatrix}. \tag{8.590}$$

We compute $\mathbf{B}_{bs}$ using (8.589) and plot the corresponding beam patterns in Figure 8.49.

8.10.2.3 Discrete prolate spheroidal sequences (DPSS beamspace)

We have encountered DPSS in Section 3.1 (3.25)–(3.35). The application to beamspace processing is due to Forster and Vezzosi [FV87].

We define $\mathbf{b}_i$, $i = 1, 2, \cdots, M$ to be the columns of the beamforming matrix $\mathbf{B}_{bs}$. We require the columns to be orthonormal. Then, as in (3.26),

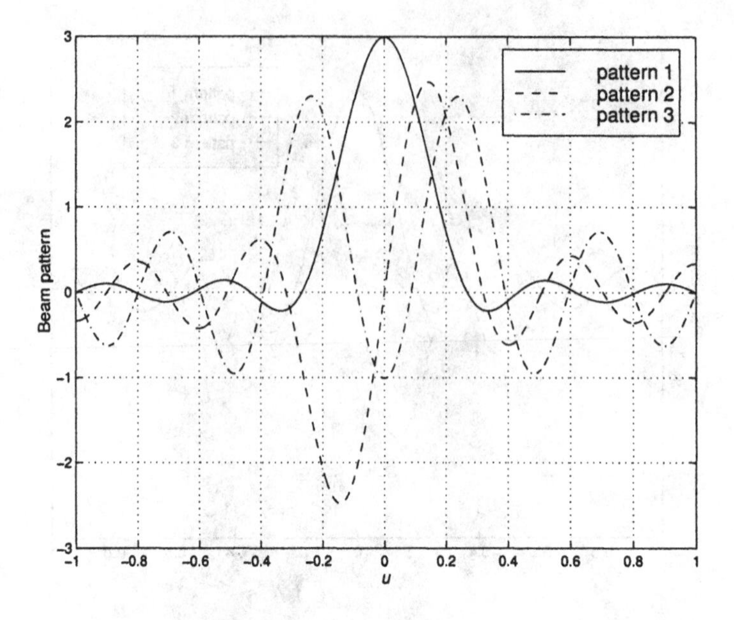

Figure 8.50 Beam patterns for DPSS beamspace matrix: $N = 10, \psi_0 = 0.2\pi$, $N_{bs} = 3$.

we define

$$\alpha_i = \frac{\int_{-\psi_0}^{\psi_0} |\mathbf{b}_i \, \mathbf{v}(\psi)|^2 \, d\psi}{\int_{-\pi}^{\pi} |\mathbf{b}_i \, \mathbf{v}(\psi)|^2 \, d\psi}, \quad i = 1, 2, \cdots, M, \tag{8.591}$$

which is the ratio of the energy of the ith beam in $[-\psi_0, \psi_0]$ to the energy of the ith beam in $[-\pi, \pi]$. The numerator is

$$\alpha_{iN} = \mathbf{b}_i \, \mathbf{A} \, \mathbf{b}_i^H, \tag{8.592}$$

where

$$\mathbf{A} \triangleq \int_{-\psi_0}^{\psi_0} \mathbf{v}(\psi) \, \mathbf{v}^H(\psi) \, d\psi. \tag{8.593}$$

For a linear array, the mn element is

$$[\mathbf{A}]_{mn} = \frac{2 \sin [(m-n)] \, \psi_0}{(m-n)}, \quad m \neq n, \tag{8.594}$$

and

$$[\mathbf{A}]_{mn} = 2\psi_0, \quad m = n. \tag{8.595}$$

The denominator is

$$\alpha_{iD} = 2\pi \mathbf{b}_i^H \, \mathbf{b}_i. \tag{8.596}$$

Thus,

$$\alpha_i = \frac{\mathbf{b}_i \, \mathbf{A} \, \mathbf{b}_i^H}{2\pi \mathbf{b}_i^H \, \mathbf{b}_i}, \quad i = 1, 2, \cdots, N_{se}. \tag{8.597}$$

We want to maximize $\alpha_i, i = 1, 2, \cdots, N_{se}$ subject to the orthonormality constraint. This corresponds to finding the eigenvectors of the matrix $\mathbf{A}$, which correspond to the M largest eigenvalues. Thus, we solve

$$2\pi\lambda \mathbf{b}_i = \mathbf{A} \, \mathbf{b}_i. \tag{8.598}$$

Using (8.594) in (8.598), this corresponds to

$$\sum_{n=1}^{N} \frac{\sin(m-n)\psi_0}{(m-n)} b_n = \pi\lambda b_m, \quad m = 1, 2, \cdots, N. \tag{8.599}$$

For each of the M largest eigenvalues, we obtain a sequence that defines the column vector $\mathbf{b}_i$. These sequences are called DPSSs and are discussed in detail by Slepian [Sle78].

The number of significant eigenvalues is

$$N_{se} = \frac{\psi_0}{\pi} N + 1. \tag{8.600}$$

The corresponding spheroidal functions are

$$G_i(\psi) = c_i \sum_{n=1}^{N} b_n^{(i)} \, e^{j\psi n}, \tag{8.601}$$

where

$$c_i = \begin{cases} 1, & i \quad \text{odd}, \\ j, & i \quad \text{even}. \end{cases} \tag{8.602}$$

The spheroidal functions are real functions that are doubly orthonormal,

$$\int_{-\pi}^{\pi} G_i(\psi) \, G_j(\psi) \, d\psi = \delta_{ij}, \tag{8.603}$$

and

$$\int_{-\psi_0}^{\psi_0} G_i(\psi) \, G_j(\psi) \, d\psi = \lambda_i \, \delta_{ij}. \tag{8.604}$$

The spheroidal functions correspond to beam patterns in the following manner,

$$B_i(\psi) = \sum_{n=1}^{N} b_n^{(i)} e^{j\psi n}. \tag{8.605}$$

Thus,

$$G_i(\psi) = c_i B_i(\psi). \tag{8.606}$$

The odd-numbered beams are real and symmetric and can be considered sum beams. The even-numbered beams are odd and asymmetric and can be considered difference beams.

Thus, the beamspace processor determines ψ_0 from prior knowledge of the target environment or by a preliminary processing with one of the simpler algorithms that are derived in Chapter 9. The value of ψ_0 determines N_{se} from (8.600). We find the $\mathbf{b}_i$ as the eigenvectors corresponding to the N_{se} largest eigenvalues of (8.598). We use these $\mathbf{b}_i, i = 1, 2, \cdots, N_{se}$, as the columns of $\mathbf{B}$. In Figure 8.50, we plot the first three eigenbeams for $u_0 = 0.2$.

We find that this spheroidal function decomposition is an effective beamspace matrix.

8.10.2.4 Conjugate symmetric beamspace matrices

In conjugate symmetric arrays we can use FB averaging of the data to estimate $\mathbf{S_x}$. If, in addition, the columns in the $\mathbf{B}_{bs}$ matrix are conjugate symmetric or conjugate asymmetric,

$$\mathbf{b}_{bs,i} = \mathbf{J}\mathbf{b}_{bs,i}^{*}, \quad i = 1, \cdots, N_{bs}, \tag{8.607}$$

or can be made to satisfy by a suitable choice of the origin, additional computational simplications occurs.

When (8.607) is satisfied and we use FB averaging of the data, the subsequent processing can be done using real matrices. Alternatively, we can process the data without FB averaging and use the real part of the resulting matrix and achieve the same result. Utilizing real computations provides significant computational saving (e.g., Linebarger et al. [LDD94] and Zoltowski et al. [ZKS93]).

Note that we can use FB averaging even if (8.607) is not satisfied. However, we need to use complex processing in beamspace.

8.10.3 Beamspace Cramér-Rao Bound

8.10.3.1 Introduction

As in the element-space algorithms, we can analyze the behavior by finding the probability of resolution, the bias, and the variance. In this section, we develop the beamspace CRBs and compare them to element-space CRBs.

From our earlier discussions, we know that we can find algorithms (e.g., the unconditional ML estimate) that will approach the bounds. Thus, the ratio of the beamspace CRB to the element-space CRB will indicate the potential increase in variance by operating in beamspace. There will be an additional increase in variance if we use an algorithm that does not approach the CRB.

In Section 8.4, we developed the CRB for the element-space model. We now want to develop the CRB for the beamspace model and develop necessary and sufficient conditions on $\mathbf{B}$ in order for the two bounds to be equal. Our discussion follows Weiss and Friedlander [WF94].[27] We consider the stochastic signal model. The deterministic signal model is also discussed in [WF94]. We then explore the behavior of beamspace CRB for the various pre-processing matrices in Section 8.10.2.

The model of interest is given by (8.571)–(8.580) in Section 8.10.1. We first define the following projection matrices,

$$\mathbf{P_V} \triangleq \mathbf{V}\left(\mathbf{V}^H\mathbf{V}\right)^{-1}\mathbf{V}^H, \tag{8.608}$$

and

$$\mathbf{P}_\mathbf{V}^\perp \triangleq \mathbf{I} - \mathbf{P_V}. \tag{8.609}$$

The beamspace steering vectors are

$$\mathbf{v_B} = \mathbf{B}_{bs}^H\mathbf{v}. \tag{8.610}$$

Then,

$$\mathbf{P_{V_B}} = \mathbf{B}_{bs}^H\mathbf{V}\left[\mathbf{V}^H\mathbf{P_B}\mathbf{V}\right]^{-1}\mathbf{V}^H\mathbf{B}_{bs}, \tag{8.611}$$

and

$$\mathbf{P}_{\mathbf{V_B}}^\perp \triangleq \mathbf{I} - \mathbf{P_{V_B}}. \tag{8.612}$$

The last projection matrix is,

$$\mathbf{P_B} \triangleq \mathbf{B}_{bs}\mathbf{B}_{bs}^H, \tag{8.613}$$

---

[27] An earlier discussion of beamspace pre-processing matrices is contained in Anderson [And91] (see [And91], [And93], and [AN95]).

and $\mathbf{B}_{bs}$ satisfies (8.578).

The CRB for the Gaussian signal model was given in Section 8.4.2.1 (see (8.95)).

$$\mathbf{C}_{CR}(\boldsymbol{\psi}) = \frac{\sigma_w^2}{2K}\left\{Re\left[\left(\mathbf{D}^H\mathbf{P}_{\mathbf{V}}^{\perp}\mathbf{D}\right)\odot\left(\mathbf{S_f}\mathbf{V}^H\mathbf{S_x}^{-1}\mathbf{V}\mathbf{S_f}\right)^T\right]\right\}^{-1}. \tag{8.614}$$

The output of the beamspace matrix, $\mathbf{X}_{bs}(k)$, $k = 1, 2, \cdots, K$ satisfies the same conditions as the original data sequence. The beamspace spectral matrix is

$$\mathbf{S}_{\mathbf{x}_{bs}} = \mathbf{B}_{bs}^H\mathbf{S_x}\mathbf{B}_{bs} = \mathbf{B}_{bs}^H\mathbf{V}\mathbf{S_f}\mathbf{V}^H\mathbf{B}_{bs} + \sigma_w^2\mathbf{I}. \tag{8.615}$$

The beamspace noise is white due to (8.578).

Thus the bound for beamspace estimation is,

$$\boxed{\mathbf{C}_{CR,bs}(\boldsymbol{\psi}) = \frac{\sigma_w^2}{2K}\left[Re\left\{\left(\mathbf{D}^H\mathbf{B}_{bs}\mathbf{P}_{\mathbf{V_B}}^{\perp}\mathbf{B}_{bs}^H\mathbf{D}\right)\odot\left(\mathbf{S_f}\mathbf{V}^H\mathbf{B}_{bs}\mathbf{S}_{\mathbf{x}_{bs}}^{-1}\mathbf{B}_{bs}^H\mathbf{V}\mathbf{S_f}\right)^T\right\}\right]^{-1}.} \tag{8.616}$$

Note that this bound corresponds to the general CRB and includes uncorrelated signals as a special case. If we know *a priori* that the signals are uncorrelated, we would adapt the bound in Section 8.4.2.

8.10.3.2 Beamspace matrix conditions

We now want to find conditions on $\mathbf{B}_{bs}$ such that $\mathbf{C}_{CR,bs}(\boldsymbol{\psi}) = \mathbf{C}_{CR}(\boldsymbol{\psi})$. Weiss and Friedlander [WF94] derive the following result:

Assuming the $\mathbf{S_f}$ is positive definite, then[28]

$$\mathbf{C}_{CR}(\boldsymbol{\psi}) \leq \mathbf{C}_{CR,bs}(\boldsymbol{\psi}), \tag{8.617}$$

with equality if and only if

$$\mathbf{P_B}\mathbf{V} = \mathbf{V}, \tag{8.618}$$

and

$$\mathbf{P_B}\mathbf{P}_{\mathbf{V}}^{\perp}\mathbf{D} = \mathbf{P}_{\mathbf{V}}^{\perp}\mathbf{D}. \tag{8.619}$$

The first condition (8.618) implies that we point the beams at the sources. The second condition implies that the column space of $\mathbf{B}_{bs}$ includes the projection of $\mathbf{d}_n$, $n = 1, 2, \cdots, D$ on the noise subspace $\mathbf{P}_{\mathbf{V}}^{\perp}$. If $\mathbf{B}_{bs}$ spans

[28] Recall that the matrix inequality means that $\mathbf{C}_{CR,bs}(\boldsymbol{\psi}) - \mathbf{C}_{CR}(\boldsymbol{\psi})$ is a non-negative definite matrix.

both $\mathbf{V}$ and $\mathbf{D}$, it will span the required projections. One approach is to have D columns of $\mathbf{B}_{bs}$ corresponding to the signal steering vectors,

$$\mathbf{b}_i = \mathbf{v}(\psi_i), i = 1, 2, \cdots, D, \tag{8.620}$$

and D columns corresponding to the derivatives of the steering vectors,

$$\mathbf{b}_i = \mathbf{d}(\psi_i), i = D + 1, \cdots, 2D. \tag{8.621}$$

A second approach is to approximate the derivatives by differences. This also requires two beams for each source. For example, if there is a source at ψ_n, we point one beam at ψ_n and a second beam at $\psi_n + \epsilon$, where ϵ is small enough that

$$\mathbf{d}_n \simeq \frac{1}{\epsilon}\left[\mathbf{v}\left(\psi_n + \epsilon\right) - \mathbf{v}\left(\psi_n\right)\right] \tag{8.622}$$

The obvious practical problem with this result is that we are trying to estimate the source locations, so that we cannot point the beams exactly at the sources. Two solutions appear logical.

If we have some prior knowledge of the signal DOAs and they are closely spaced, one of the $\mathbf{B}_{bs}$ matrices described in Section 8.10.2 may be an adequate approximation. We investigate the behavior for several cases in subsequent examples.

If the performance using this technique is not adequate, we can develop a simple iterative scheme to obtain the required beamspace matrix.

We first consider some beamspace matrices that are similar to (8.620) and then examine the family of beamspace matrices that are most commonly used in practice.

The first example uses the clairvoyant beamspace matrix from (8.585).

Example 8.10.2

Consider a standard 10-element linear array. We first consider the clairvoyant beamspace matrix from (8.585). We consider two equal-power uncorrelated sources with $SNR =$ 10 dB and use $K = 100$ snapshots.

We define

$$\mathbf{B}_{no} \triangleq \left[\; \mathbf{v}(\psi_1) \;\vdots\; \mathbf{v}(\psi_2) \;\vdots\; \mathbf{v}(\psi_m) \;\right], \tag{8.623}$$

and

$$\mathbf{B}_{bs} = \mathbf{B}_{no}\left(\mathbf{B}_{no}^H \mathbf{B}_{no}\right)^{-\frac{1}{2}}. \tag{8.624}$$

In Figure 8.51, we show the beamspace CRB using (8.624) in (8.616) and the element-space CRB. We see that, if the signal separation is less than $0.7BW_{NN}(\Delta u \leq 0.28)$ that the beamspace matrix given by (8.624) does not increase the CRB.

Example 8.10.3 (continuation)

Consider the same model as in Example 8.10.2.

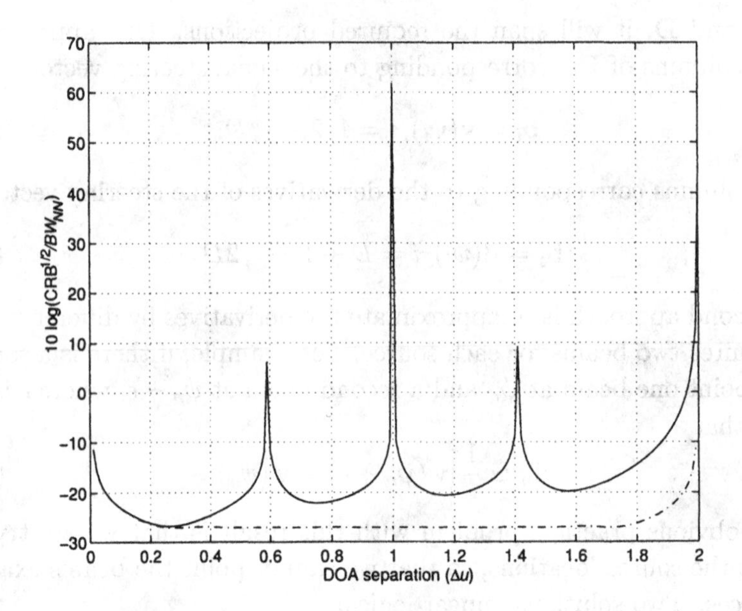

Figure 8.51 Normalized CRB for beamspace processing using clairvoyant beamspace matrix: $N = 10, N_{bs} = 3, K = 100, SNR = 20$ dB; element-space CRB shown as reference.

We next define an approximation to the $\mathbf{B}_{no}$ in (8.623) by the first three terms

$$\mathbf{B}_{no} \triangleq \left[\begin{array}{c:c:c} \mathbf{v}(\tilde{\psi}_m) & \dot{\mathbf{v}}(\tilde{\psi}_m) & \ddot{\mathbf{v}}(\tilde{\psi}_m) \end{array} \right]. \tag{8.625}$$

Two cases are considered. In case 1, $\tilde{\psi}_m = \psi_m$. In case 2, $\tilde{\psi}_m = \psi_1$

We substitute (8.625) into (8.624) to obtain $\mathbf{B}_{bs}$. The results from (8.616) are shown in Figure 8.52.

We see that the beamspace CRB coincides with the element-space CRB for $\Delta u \leq 0.20$ and is only slightly above it for $\Delta u \leq 0.30$.

Example 8.10.4 (continuation)

We consider the same model as in Example 8.10.2. We use $\mathbf{B}_{no}$ as given by (8.588) which we repeat,

$$\mathbf{B}_{no} = \left[\begin{array}{c:c:c:c:c} \mathbf{v}(\psi_0) & \mathbf{v}^{(1)}(\psi_0) & \mathbf{v}^{(2)}(\psi_0) & \cdots & \mathbf{v}^{(m)}(\psi_0) \end{array} \right]. \tag{8.626}$$

We utilize $\mathbf{B}_{bs}$ as given by (8.589) and evaluate the beamspace CRB as a function of m. The results are shown in Figure 8.53. Note that $N_{bs} = m + 1$, so there are $m + 1$ beams.

As we would expect, as m increases the beamspace CRB matches the element-space CRB for larger Δu. For $m = 5$, there is negligible difference for $\Delta u = 1$. Note, however, that this requires six beams from the 10-element array.

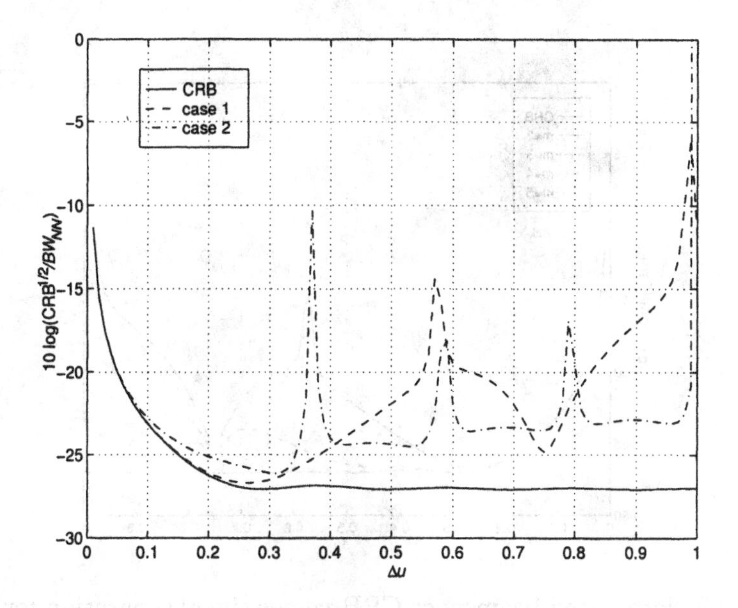

Figure 8.52 Normalized beamspace CRB versus Δu using a Taylor series beamspace matrix: $N = 10, N_{bs} = 3$, $SNR = 10$ dB, $K = 100$.

We next consider a beamspace matrix using the discrete prolate spheroidal sequences.

Example 8.10.5

We consider the same model as in Example 8.10.2. We utilize the DPSS functions derived in Section 8.10.2.1 to construct the beamspace matrix. We denote the width of the sector by $2\psi_0$. The $\mathbf{B}_{no}$ matrix is given by (8.599). In Figure 8.54, we use three beams and plot the beamspace CRB versus source separation for various values of $2u_0$. For $\Delta u \leq 0.3$, the bounds are close to the element-space CRB for all of the four cases. In Figure 8.55, we use four beams. For $N_{bs} = 4$, the beamspace CRB remains close to the element-space CRB for $\Delta u \leq 0.5$. In Figure 8.56, we increase N_{bs} as $2u_0$ increases. In this case the bounds remain close out to $\Delta u = 1.0$.

8.10.3.3 DFT beamspace matrices

In most applications utilizing beamspace processing the beamspace matrix consists of conventional beams whose pointing directions are spaced by $2/N$ in u-space. It is referred to as the DFT beamspace. In this section we consider this model and a variation of it. We compute the beamspace CRB for two examples.

Example 8.10.6

Consider a standard 32-element linear array and a $N_{bs} \times N$ DFT beamspace matrix.

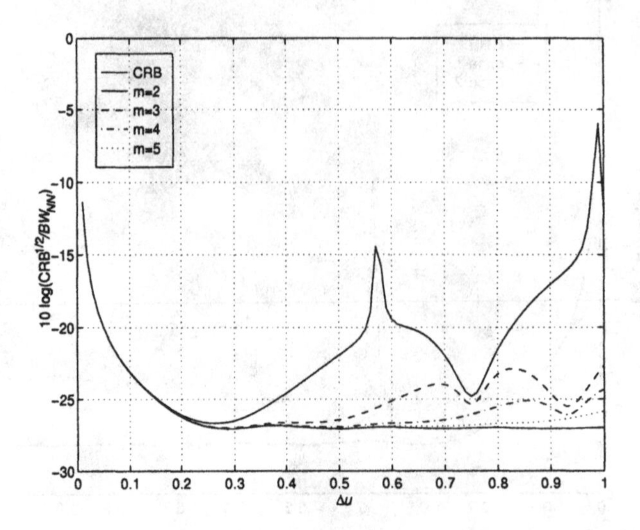

Figure 8.53 Normalized beamspace CRB versus signal separation for various m: Taylor series beamspace matrix, $N = 10, N_{bs} = m+1, SNR = 10$ dB, $K = 100$.

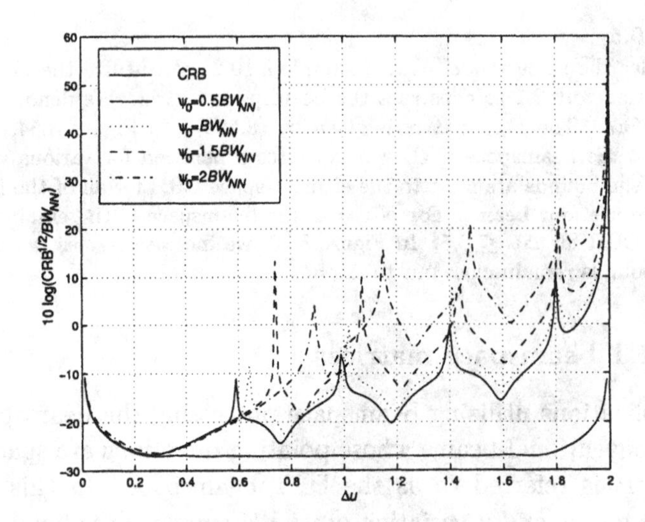

Figure 8.54 Normalized beamspace CRB using DPSS function beamspace matrix: $N = 10, N_{bs} = 3, SNR = 10$ dB, $K = 100$.

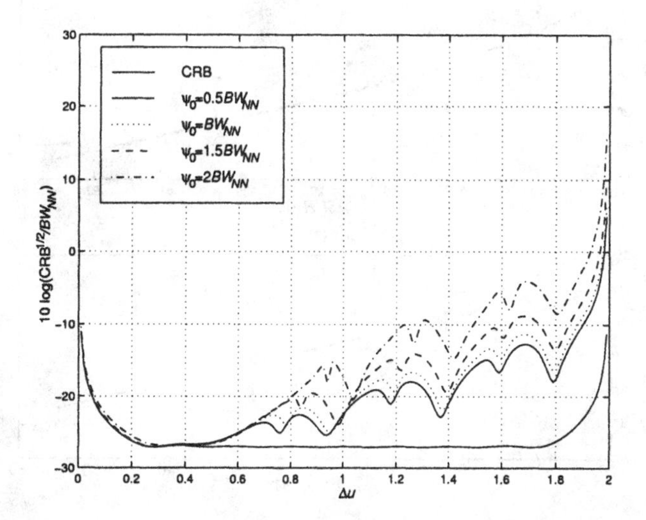

Figure 8.55 Normalized beamspace CRB for DPSS function beamspace matrix: $N = 10, N_{bs} = 4, SNR = 10$ dB, $K = 100$.

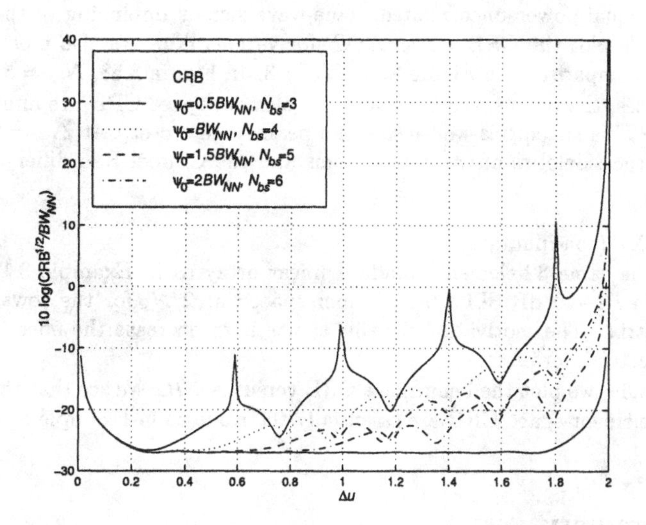

Figure 8.56 Normalized beamspace CRB for DPSS function beamspace matrix: $N = 10, N_{bs} = 3, 4, 5$, and 6, $SNR = 10$ dB, $K = 100$.

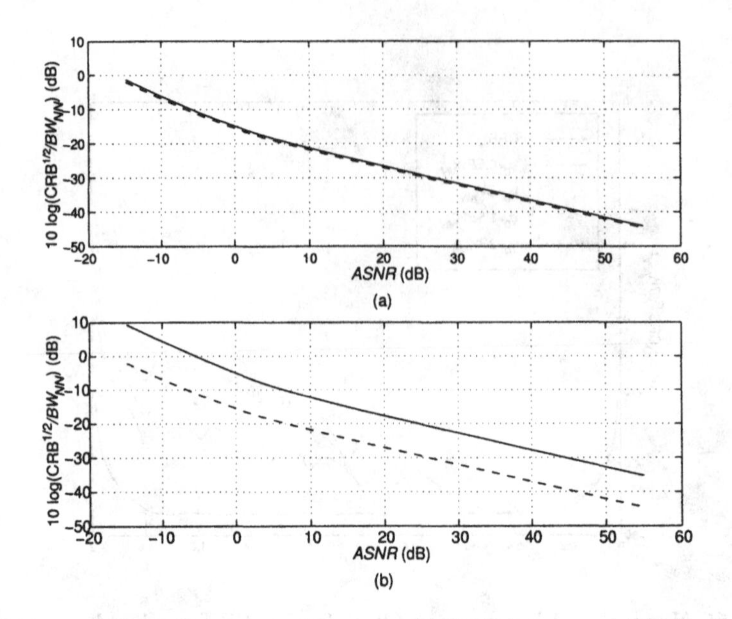

Figure 8.57 Normalized beamspace and element-space CRBs versus *ASNR* for DFT beamspace matrix: $N = 32, N_{bs} = 3$, two plane-wave signals at $\pm\Delta u/2$: (a) $\Delta u = N_{bs}/N$; (b) $\Delta u = 2N_{bs}/N$.

There are two equal-power uncorrelated plane-wave signals impinging of the array from $u = \pm\Delta u/2$. We plot the CRB versus *ASNR* for various Δu. We also plot the element-space CRB for comparison. In Figure 8.57, $N_{bs} = 3$. In Figure 8.58, $N_{bs} = 5$.

In both cases, the beamspace CRB and the element-space CRB are almost equal for $\Delta u = N_{bs}/N$. This spacing is well inside the beamspace sector. At $\Delta u = 2N_{bs}/N$, the signals are on the sidelobes of the outer beams and the Cramér-Rao differ by almost 10 dB.

Example 8.10.7 (continuation)

Consider the same 32-element standard linear array as in Example 8.10.6. We use Dolph-Chebychev (−40 dB SLL) beamformer spaced at $2/N$, for the rows in a 5×32 beamspace matrix. The motivation for this choice is to decrease the effect of any high-power out-of-sector interferers.

In Figure 8.59, we plot the beamspace CRB versus *ASNR*. We see that the beamspace CRB and the element-space CRB are essentially the same as in Example 8.10.6.

8.10.3.4 Summary

We see that, by suitably choosing the beamspace matrix, we can obtain a beamspace CRB that is very close to the element-space CRB for signals that are well inside the beamspace sector.

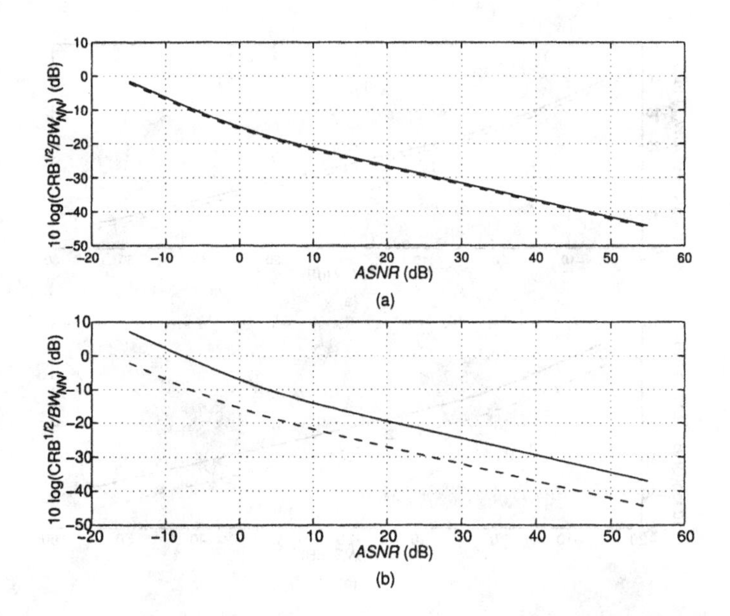

Figure 8.58 Normalized beamspace and element-space CRBs versus *ASNR* for DFT beamspace matrix: $N = 32, N_{bs} = 5$, two plane-wave signals at $\pm\Delta u/2$: (a) $\Delta u = N_{bs}/N$; (b) $\Delta u = 2N_{bs}/N$.

8.10.4 Beamspace Maximum Likelihood

In this section, we derive the beamspace maximum likelihood estimator. We restrict our attention to the beamspace version of the CML estimator discussed in Section 8.5.2.

Using the model in Section 8.5.2 ((8.332)–(8.353)), the likelihood function can be written as,

$$L_{bs}(\psi, \mathbf{F}) = -KN_{bs} \ln \sigma_w^2 - \frac{1}{\sigma_w^2} \sum_{k=1}^{K} |\mathbf{X}_{bs}(k) - \mathbf{V}_{bs}(\psi)\mathbf{F}_k|^2, \tag{8.627}$$

where

$$\mathbf{V}_{bs}(\psi) = \left[\begin{array}{ccc} \mathbf{v}_{bs}(\psi_1) & \cdots & \mathbf{v}_{bs}(\psi_D) \end{array} \right], \tag{8.628}$$

and

$$\mathbf{v}_{bs}(\psi) = \mathbf{B}_{bs}^H \mathbf{v}(\psi). \tag{8.629}$$

Then, proceeding as in Section 8.5.2, we obtain

$$\hat{\psi}_{cml,bs} = \arg \min_{\psi} \left\{ \text{tr} \left[\mathbf{P}_{\mathbf{V}_{bs}}^{\perp} \mathbf{C}_{\mathbf{X}_{bs}} \right] \right\}, \tag{8.630}$$

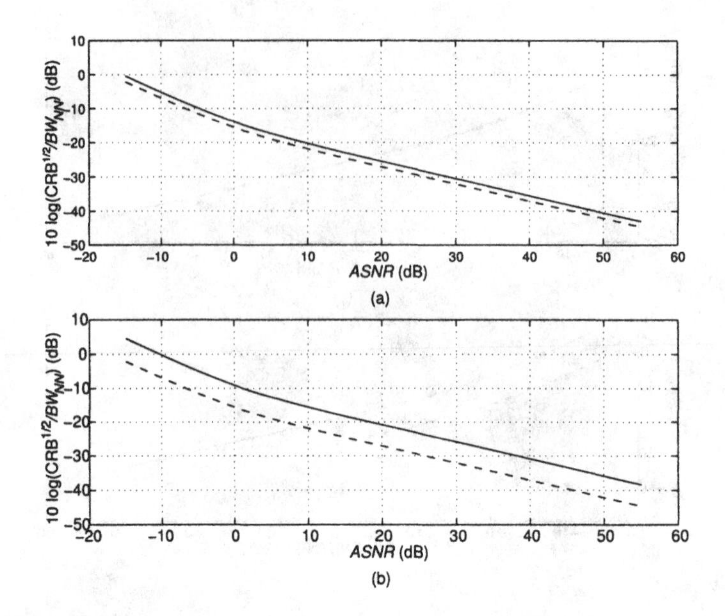

Figure 8.59 Normalized beamspace and element-space CRBs versus *ASNR* for Dolph-Chebychev beamspace matrix: $N = 32, N_{bs} = 5$, two plane-wave signals at $\pm\Delta u/2$: (a) $\Delta u = N_{bs}/N$; (b) $\Delta u = 2N_{bs}/N$.

where

$$\mathbf{P}_{\mathbf{V}_{bs}} = \mathbf{V}_{bs}\left[\mathbf{V}_{bs}^H \mathbf{V}_{bs}\right]^{-1} \mathbf{V}_{bs}^H, \tag{8.631}$$

and

$$\mathbf{C}_{\mathbf{X}_{bs}} = \frac{1}{K}\sum_{k=1}^{K} \mathbf{X}_{bs}(k)\mathbf{X}_{bs}^H(k). \tag{8.632}$$

In Section 8.7, we saw that from a computational point of view, the IQML and IMODE algorithms were attractive in estimating direction of arrivals in the element space. In beamspace processing, Zoltowski and Lee proposed 2-D beamspace domain ML (BDML) and 3-D BDML [ZL91] schemes for estimating the DOAs in the cases of a single signal and the two signals, respectively. These two schemes are computationally simple. Both of them use DFT beamspace matrices, and choose the number of beams to be the number of impinging signals plus one.

Zoltowski also presented an IQML algorithm in which the beamspace transformation is based on subspace processing [Zol88].

Tian and Van Trees [TVT00] have derived a beamspace IQML estimator using DFT matrix beamformers. It utilizes several key results from Zoltowski et al. [ZKS93]. We summarize the derivation in [TVT00].

In the IQML algorithm, the spatial characteristics of the signal components is parameterized by a coefficient vector $\mathbf{b} = [b_0 \cdots b_D]^T$. The vector $\mathbf{b}$ is defined such that the polynomial

$$b(z) = b_0 z^D + b_1 z^{D-1} + \cdots + b_D, \tag{8.633}$$

has D roots at $z_i = e^{j\pi u_i}$, $i = 1, \cdots, D$. The ML estimate of $u_i, i = 1, \cdots, D$, can be obtained from the ML estimate of $\mathbf{b}$.

In order to extend the element-space IQML algorithm to the beamspace domain, it is necessary to find a linear parameterization of the null space of $\mathbf{V}_{bs}$. This is equivalent to finding a full-rank matrix $\mathbf{B}(\mathbf{b})$ such that

$$\mathcal{L}\{\mathbf{B}\}\,\mathbf{V}_{bs} = 0. \tag{8.634}$$

The operator $\mathcal{L}\{\cdot\}$ represents a linear transform operation, and $\mathbf{B}$ is a $N_{bs} \times (N_{bs} - D)$ Toeplitz matrix given by

$$\mathbf{B}^H = \begin{bmatrix} b_D & \cdots & b_0 & 0 & 0 \\ 0 & \ddots & \ddots & \ddots & 0 \\ 0 & 0 & b_D & \ddots & b_0 \end{bmatrix}. \tag{8.635}$$

Polynomial parameterization by (8.634) is not always possible for any beamforming transformation. In element space, the linear parameterization of the noise space is made possible by the Vandermonde property of the element-space array response matrix. The Vandermonde structure of the array manifold may not be preserved after beamspace transformation. A crucial step in developing the BS-IMQL algorithm is to find the linear parameterization formula $\mathcal{L}\{\mathbf{B}\}$.

We use the DFT beamforming matrix in (8.581). Due to the common out-of-band nulls in DFT beams, a DFT beamforming matrix can be transformed into a banded Toeplitz matrix [ZKS93]. This transformation preserves the Vandermonde property of the beamspace array manifold and enables the polynomial parameterization in the beamspace. The beamspace matrix consisting of N_{bs} DFT beamformers can be rewritten as an $N \times N_{bs}$ matrix[29]

$$\mathbf{W} = \frac{1}{N}[\mathbf{v}_N(m\frac{2}{N}) \cdots \mathbf{v}_N((m + N_{bs} - 1)\frac{2}{N})], \tag{8.636}$$

[29] We have replaced $\mathbf{B}_{bs}$ with $\mathbf{W}$ to avoid confusion with the $\mathbf{B}$ matrix in (8.635).

where the integer number m should be properly chosen so that the N_{bs} beams cover most of the signal energy, and

$$\mathbf{v}_N(u_i) = [e^{-j(\frac{N-1}{2})2\pi d u_i} \quad \cdots \quad e^{j(\frac{N-1}{2})2\pi d u_i}]^T, \tag{8.637}$$

where u_i is the ith direction of arrival in u-space, and d is the element spacing measured in wavelengths.

It has been shown ([ZKS93]) that $\mathbf{W}$ can be factored as

$$\mathbf{W} = \mathbf{CQ}, \tag{8.638}$$

where $\mathbf{Q}$ is a $N_{bs} \times N_{bs}$ full-rank matrix and $\mathbf{C}$ is a $N \times N_{bs}$ banded Toeplitz matrix

$$\mathbf{C} = \begin{bmatrix} c_0 & 0 & \cdots & 0 \\ c_1 & c_0 & & \\ \vdots & \vdots & \ddots & \vdots \\ c_{N-N_{bs}} & c_{N-N_{bs}-1} & & c_0 \\ 0 & c_{N-N_{bs}} & & c_1 \\ \vdots & & \ddots & \vdots \\ 0 & \cdots & & c_{N-N_{bs}} \end{bmatrix}. \tag{8.639}$$

The vector $\mathbf{c}$ is an $(N - N_{bs} + 1) \times 1$ vector whose elements are the first $(N - N_{bs} + 1)$ elements in the first column of $\mathbf{C}$.

The following equality holds by arithmetic manipulations [ZKS93]:

$$\mathbf{C}^H \mathbf{v}_N(u) = \alpha(u)\mathbf{v}_{N_{bs}}(u), \tag{8.640}$$

where $\mathbf{v}_{N_{bs}}(u)$ is defined by (8.637) with N replaced by N_{bs}. Note that

$$\alpha(u) = \sum_{i=0}^{N-N_{bs}} c_i^* e^{j(-\frac{N-N_{bs}}{2}+i)2\pi d u} = \mathbf{c}^H \mathbf{V}_{N-N_{bs}+1}(u) \tag{8.641}$$

is a scalar, which does not affect the structure in $\mathbf{v}_{N_{bs}}(u)$. The property in (8.640) is critical to the applicability of IQML in beamspace.

It follows from (8.633) that

$$\mathbf{B}^H \mathbf{v}_{N_{bs}}(u_i) = z_i^{-\frac{N_{bs}-1}{2}} \begin{bmatrix} z_i^0 \\ \vdots \\ z_i^{N_{bs}-D-1} \end{bmatrix} b(z_i) = \mathbf{0}, \tag{8.642}$$

for $i = 1, \cdots, D$. Now, define a $N_{bs} \times (N_{bs} - D)$ matrix $\mathbf{F} = \mathbf{Q}^{-1}\mathbf{B}$. From (8.638), (8.640), and (8.642), we have

$$\begin{aligned} \mathbf{F}^H \mathbf{v}_{bs}(u_i) &= \mathbf{B}^H \mathbf{Q}^{-H} \mathbf{W}^H \mathbf{v}_N(u_i) = \mathbf{B}^H \mathbf{C}^H \mathbf{v}_N(u_i) \\ &= \alpha(u_i) \mathbf{B}^H \mathbf{v}_{N_{bs}}(u_i) = \mathbf{0}, \quad i = 1, \cdots, D. \end{aligned} \tag{8.643}$$

Since $\mathbf{F}$ has rank $N_{bs} - D$, its columns span the orthogonal complement to the signal subspace, that is,

$$\mathbf{P}_{\mathbf{V}_{bs}}^{\perp} = \mathbf{P}_{\mathbf{F}}, \tag{8.644}$$

where $\mathbf{P}_{\mathbf{F}} = \mathbf{F}(\mathbf{F}^H\mathbf{F})^{-1}\mathbf{F}^H$. Therefore, we have

$$\mathcal{L}\{\mathbf{B}\} = \mathbf{F}^H = \mathbf{B}^H \mathbf{Q}^{-H}. \tag{8.645}$$

The CML estimate is given in the beamspace by

$$\begin{aligned} \hat{\mathbf{u}} &= \arg\min_{\mathbf{u}} \operatorname{tr}\left\{\mathbf{P}_{\mathbf{V}_{bs}}^{\perp}(\mathbf{u})\hat{\mathbf{S}}_{bs}\right\} \\ &= \arg\min_{\mathbf{b}(\mathbf{u})} \operatorname{tr}\left\{\mathbf{P}_{\mathbf{F}}(\mathbf{b})\hat{\mathbf{S}}_{bs}\right\}, \end{aligned} \tag{8.646}$$

where $\hat{\mathbf{S}}_{bs}$ is the beamspace spatial spectral matrix.

Define $\mathbf{S}_Q = \mathbf{Q}^{-H}\hat{\mathbf{S}}_{bs}\mathbf{Q}^{-1}$. We have

$$\operatorname{tr}\{\mathbf{P}_{\mathbf{F}}\hat{\mathbf{S}}_{bs}\} = \operatorname{tr}\left\{\mathbf{F}(\mathbf{F}^H\mathbf{F})^{-1}\mathbf{F}^H\hat{\mathbf{S}}_{bs}\right\} \tag{8.647}$$

$$= \operatorname{tr}\left\{\mathbf{Q}^{-1}\mathbf{B}(\mathbf{F}^H\mathbf{F})^{-1}\mathbf{B}^H\mathbf{Q}^{-H}\hat{\mathbf{S}}_{bs}\right\} \tag{8.648}$$

$$= \operatorname{tr}\left\{\mathbf{B}(\mathbf{F}^H\mathbf{F})^{-1}\mathbf{B}^H\mathbf{S}_Q\right\}. \tag{8.649}$$

Minimization of the objective function in (8.649) can be readily solved by the IQML method discussed in Section 8.7.2.[30]

We need to find $\mathbf{Q}^{-1}$ in order to implement the IQML algorithm. The DFT beamforming matrix may be decomposed into the product of a banded Toeplitz matrix $\mathbf{C}$ and a full-rank matrix $\mathbf{Q}$ (8.638). Only the inverse of the matrix $\mathbf{Q}$ needs to be computed for the IQML procedure. Zoltowski et al. [ZKS93] provide a closed-form expression for $\mathbf{Q}$ without forming $\mathbf{C}$. Here we briefly explain an intuitive way to compute the matrix $\mathbf{C}$, and give an simple expression to compute $\mathbf{Q}^{-1}$ directly.

The N_{bs} columns of $\mathbf{W}$ are part of the $N \times N$ DFT matrix. Due to the orthogonal properties of DFT beams, the other $N - N_{bs}$ columns of the

[30] We use IQML-QC. The QC descriptor is omitted for simplicity.

$N \times N$ DFT matrix that are not contained in $\mathbf{W}$ are orthogonal to each of the N_{bs} columns of $\mathbf{W}$. Mathematically,

$$\mathbf{W}^H \mathbf{v}_N(u_n) = \mathbf{0}, \tag{8.650}$$

or

$$\mathbf{Q}^H \mathbf{C}^H \mathbf{v}_N(u_n) = \mathbf{0}, \tag{8.651}$$

for $u_n \in \{(m + N_{bs})\frac{2}{N}, \cdots, (N + m - 1)\frac{2}{N}\}$. Define a polynomial

$$c(z) = c^*_{N-N_{bs}} z^{N-N_{bs}} + \cdots + c_1^* z + c_0. \tag{8.652}$$

Equation (8.651) implies that $c(z)$ has $N - N_{bs}$ roots at $z_n = \exp\{j\pi u_n\}$. Therefore, the coefficients of $c(z)$ can be found by

$$[c_o \ \cdots \ c_{N-N_{bs}}]^H = \text{poly}\{e^{j\pi u_n}\}, \tag{8.653}$$

where the operator poly$\{\cdot\}$ converts the roots to a polynomial.

Once the matrix $\mathbf{C}$ is computed, the inverse of the matrix $\mathbf{Q}$ can be found by pre-multiplying both sides of (8.638) by $\mathbf{W}^H$:

$$\mathbf{W}^H \mathbf{W} = \mathbf{W}^H \mathbf{C} \mathbf{Q} = \mathbf{I}. \tag{8.654}$$

Therefore, the inverse of $\mathbf{Q}$ is given by $\mathbf{Q}^{-1} = \mathbf{W}^H \mathbf{C}$.

We consider an example to illustrate the performance of beamspace IQML.

Example 8.10.8

Consider a SLA 32. There are two equal-power uncorrelated signals impinging on the array from $\pm\Delta\psi/2$ where $\Delta\psi = \Delta\psi_R$. The beamspace dimension, N_{bs}, equals 8 and is centered at $\psi = 0$. We simulate beamspace IQML and element-space IQML for $K = 100$.

In Figure 8.60(a), we plot the normalized RMSE versus *ASNR* for the two IQML algorithms. We also plot the element-space CRB and the beamspace CRB. In Figure 8.60(b), we plot the probability of resolution versus *ASNR*.

We see that both IQML algorithms approach their respective CRBs above threshold (the difference in the bounds is 0.61 dB). The probability of resolution performance is similar. The performance of ES-IQML below threshold is handicapped because we have not used the beam fan information.

The lower computational complexity of BS-IQML compensates for the slightly larger RMSE.

Other examples exhibit similar behavior. Beamspace IQML provides good performance with reasonable conputational complexity. In Chapter 9, we discuss two other beamspace estimation algorithms and compare their performance to the IQML algorithm.

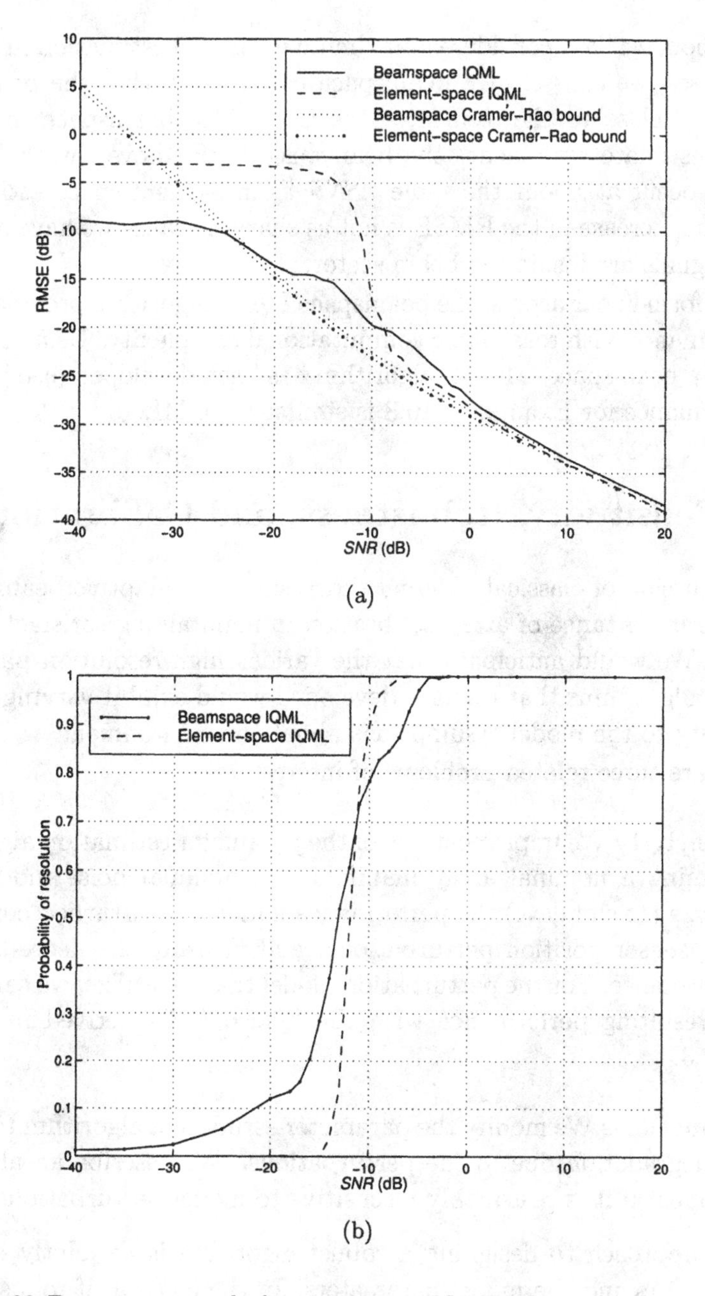

Figure 8.60 Beamspace and element-space IQML: $N = 32$, $N_{bs} = 8$, $K = 100$, $\Delta\psi = \Delta\psi_R$, $\rho = 0$, 500 trials: (a) RMSE versus *SNR*; (b) probability of resolution versus *SNR*.

8.10.5 Summary

In this section, we have considered maximum likelihood estimation in beamspace. In most cases, we can select a beamspace matrix such that the beamspace CRB is very close to the element-space CRB. The beamspace maximum likelihood estimate approaches the beamspace CRB above threshold. The thresholds occur at about the same *ASNR* as in element-space estimation. However the increase in the RMSE is not as sharp, because we have assumed that the signals are inside the beam sector.

For uniform linear arrays, the beamspace IQML algorithm provides excellent performance with reasonable computational complexity. Using a similar technique, a beamspace IMODE algorithm has been developed (see [Tia01]). The performance for Example 8.10.8 is similar to IQML.

8.11 Sensitivity, Robustness, and Calibration

In our discussion of classical antenna processing and adaptive beamformers we saw the importance of array calibration in maintaining satisfactory performance. We would anticipate that the various high-resolution parameter estimation algorithms that we have developed would exhibit varying degrees of sensitivity to the model assumptions about the environment.

There are three related problems of interest:

(i) **Sensitivity** We implement one of the parameter estimation algorithms assuming a nominal array manifold and nominal noise model. We analyze (or simulate) the performance under the perturbed conditions (e.g. sensor position perturbations) and measure the degradation in performance. For the perturbation model that we utilize, we can bound the resulting performance with the hybrid CRB derived in Section 8.4.1.3.

(ii) **Robustness** We modify the parameter estimation algorithm to reduce the degradation due to the perturbations. We describe an algorithm as robust if it is reasonably insensitive to model perturbations.

One approach to designing a robust algorithm is to jointly estimate the DOAs and the model parameters. In the context of robust beamformers, we are not explicitly interested in the model parameters (they are "unwanted parameters"). We only estimate them to improve our DOA estimation performance.

There are other approaches to robustness that do not estimate the model parameters.

(iii) **Calibration** In this problem we are explicitly interested in estimating the array parameters (e.g., sensor position, gain, and phase). This problem is referred to as the **calibration problem.**

In some cases, we try to calibrate the array using signals with unknown DOAs. This technique is referred to as the **blind calibration** problem.

A complete discussion of these three topics is beyond the scope of our discussion. We provide a brief introduction to the area and highlight the issues involved. We provide some references here and expand the reference list in Section 9.8.

In Section 8.11.1, we review the models for the perturbations in the array parameters and the noise environment that we introduced in Section 6.6.3. In Section 8.11.2, we develop hybrid Cramér-Rao bounds for the joint estimation of the signal parameters and the array parameters. In Section 8.11.3, we analyze how the array perturbations affect the performance of some of the maximum likelihood estimators that we derived in this chapter. We find that the performance can decline significantly under certain scenarios. In Section 8.11.4, we consider a joint estimation approach in order to improve robustness. The array parameters are estimated as part of the algorithm, but the emphasis is on improving the DOA estimation performance. In Section 8.11.5, we briefly discuss the calibration problem. In Section 8.11.6, we summarize our results.

8.11.1 Model Perturbations

In this section, we develop several characterizations to describe perturbations in our nominal observation model. We first consider the case of perturbations in the array manifold matrix. The models consider perturbations in the gain and phase of the sensors and in the position of the array elements.

The frequency-domain snapshot model can be written as

$$\mathbf{X}(k) = \mathbf{V}(\boldsymbol{\theta}, \boldsymbol{\phi}; \boldsymbol{\rho})\, \mathbf{F}(k) + \mathbf{W}(k), \quad k = 1, 2, \cdots, K, \tag{8.655}$$

where $\mathbf{F}(k)$ is a $D \times 1$ complex source signal with source spectral matrix $\mathbf{S_f}$. The vector $\boldsymbol{\theta}$ and $\boldsymbol{\phi}$ represent the elevation and azimuth angles of the D plane-wave signals. The vector $\boldsymbol{\rho}$ is a real $M \times 1$ vector that represents the perturbations in the array parameters. The additive white noise has spectral height σ_w^2.

For notational simplicity, we will consider the case in which the array is in the x-y-plane, and the plane waves are arriving in the $x-y$ plane so we are only estimating $\boldsymbol{\phi}$. Then, (8.655) can be written as

$$\mathbf{X}(k) = \mathbf{V}(\boldsymbol{\phi}, \boldsymbol{\rho})\, \mathbf{F}(k) + \mathbf{W}(k), \quad k = 1, 2, \cdots, K, \tag{8.656}$$

where the ith element of $\boldsymbol{\phi}$ is the angle between the ith plane wave and the x-axis (see Figure 2.1).

This model is the same model that has been used throughout the chapter except that $\mathbf{V}(\boldsymbol{\phi}, \boldsymbol{\rho})$ is written as a function of $\boldsymbol{\phi}$ rather than $\boldsymbol{\psi}$. This change is made so that our results can be more easily compared to those in the literature. The two models can be related by (8.44)–(8.48). The array perturbation model in Section 6.3.3 is used with this change.

8.11.2 Cramér-Rao Bounds

In this section, we derive the Cramér-Rao bound on the variance of any unbiased estimator in the presence of array perturbations.

When we derive the CRB we are implicitly assuming that we are jointly estimating all of the parameters in the model ($\boldsymbol{\phi}$ and $\boldsymbol{\rho}$). However, the resulting bound also applies to the case in which we only estimate $\boldsymbol{\phi}$ and treat $\boldsymbol{\rho}$ as an unwanted or nuisance parameter. Therefore, we can use it to bound the behavior of algorithms that assume some nominal array manifold matrix, $\mathbf{V}(\boldsymbol{\phi}, \boldsymbol{\rho}_0)$, in the presence of array perturbations. The CRB also bounds the behavior of autocalibration algorithms that jointly estimate $\boldsymbol{\phi}$ and $\boldsymbol{\rho}$.

Our approach utilizes the hybrid Cramér-Rao bound which was derived in Section 8.2.3.3. This approach was first used in the array context to study the array shape calibration problem by Rockah and Schultheiss [RS87a]. This paper contains several useful results on array calibration and provides useful background reading. Subsequent work by Weiss and Friedlander ([WF89] and Appendix B of Chapter 10 of [Hay91a]) derived compact expressions for the CRB in the presence of position displacements and gain and phase errors. They consider the case in which the source spectral matrix $\mathbf{S_f}$ and the white noise level σ_w^2 are known. Although this case is generally not applicable in practice, it bounds the unknown spectrum case and the resulting CRB enables us to identify some of the key issues in the calibration problem.

In this section, we consider the more common case in which the spectral matrix $\mathbf{S_f}$ and the noise level are unknown. After deriving a general expression we focus our attention on position displacements.

8.11.2.1 Cramér-Rao bound: Unknown spectral matrix

We consider the case in which there are D plane waves impinging on the array. The signals are sample functions from zero-mean Gaussian random processes with source spectral matrix $\mathbf{S_f}$. The additive noise is white with spectral height σ_w^2. The parameters to be perturbed are denoted by $\boldsymbol{\rho}$ and are imbedded in the array manifold matrix $\mathbf{V}$, which we denote by $\mathbf{V}(\boldsymbol{\phi}, \boldsymbol{\rho})$.

From (8.656),

$$\mathbf{X}(k) = \mathbf{V}(\boldsymbol{\phi}, \boldsymbol{\rho})\,\mathbf{F}(k) + \mathbf{W}(k), \quad k = 1, 2, \cdots, K. \tag{8.657}$$

We assume that $\mathbf{S_f}$ and σ_w^2 are unknown.

The total parameter vector of interest is

$$\boldsymbol{\theta} = \begin{bmatrix} \boldsymbol{\theta}_w \\ \boldsymbol{\theta}_u \end{bmatrix}, \tag{8.658}$$

where

$$\boldsymbol{\theta}_w = \begin{bmatrix} \boldsymbol{\phi} \\ \boldsymbol{\rho} \end{bmatrix}, \tag{8.659}$$

is a $(D+M) \times 1$ vector that represents the wanted parameters and

$$\boldsymbol{\theta}_u = \begin{bmatrix} \boldsymbol{\mu} \\ \sigma_w^2 \end{bmatrix}, \tag{8.660}$$

is a $(D^2+1) \times 1$ vector that represents the unwanted parameters. We first consider the Fisher information matrix associated with $\boldsymbol{\theta}_w$. The vector $\boldsymbol{\phi}$ is a $D \times 1$ vector corresponding to source DOAs in angle space. The parameter perturbation vector $\boldsymbol{\rho}$ is M-dimensional. For example, in the case of position perturbations, $M = 2N$:

$$\boldsymbol{\rho} = \begin{bmatrix} p_{x_0} & p_{y_0} & p_{x_1} & p_{y_1} & \cdots & p_{x_{N-1}} & p_{y_{N-1}} \end{bmatrix}^T. \tag{8.661}$$

We assume that $\boldsymbol{\rho}$ is a real Gaussian random vector

$$p_{\boldsymbol{\rho}}(\boldsymbol{\rho}) = \frac{1}{(2\pi)^{\frac{M}{2}} \left|\boldsymbol{\Lambda}_{\boldsymbol{\rho}}\right|^{\frac{1}{2}}} \exp\left\{-\frac{1}{2}(\boldsymbol{\rho}-\boldsymbol{\rho}_0)^T \boldsymbol{\Lambda}_{\boldsymbol{\rho}}^{-1}(\boldsymbol{\rho}-\boldsymbol{\rho}_0)\right\}, \tag{8.662}$$

where $\boldsymbol{\rho}_0$ represents the nominal value of the parameter vector $\boldsymbol{\rho}$. The Gaussian assumption is a good model for many applications.

The Fisher information matrix for a single observation is given by,

$$\mathbf{J}_B(\boldsymbol{\theta}_w) = \mathbf{J}_D(\boldsymbol{\theta}_w) + \mathbf{J}_P(\boldsymbol{\theta}_w). \tag{8.663}$$

The data matrix $\mathbf{J}_D$ is partitioned as,

$$\mathbf{J}_D(\boldsymbol{\theta}_w) = \left[\begin{array}{c:c} \mathbf{J}_{\boldsymbol{\phi}\boldsymbol{\phi}} & \mathbf{J}_{\boldsymbol{\phi}\boldsymbol{\rho}} \\ \hdashline \mathbf{J}_{\boldsymbol{\rho}\boldsymbol{\phi}} & \mathbf{J}_{\boldsymbol{\rho}\boldsymbol{\rho}} \end{array}\right]. \tag{8.664}$$

The prior matrix $\mathbf{J}_P$ is given by (8.59) as

$$\mathbf{J}_P(\boldsymbol{\theta}_w) = \left[\begin{array}{c:c} \mathbf{0} & \mathbf{0} \\ \hdashline \mathbf{0} & \boldsymbol{\Lambda}_{\boldsymbol{\rho}}^{-1} \end{array}\right], \tag{8.665}$$

because $\boldsymbol{\phi}$ is an unknown nonrandom vector.

To evaluate the terms in $\mathbf{J}_D(\boldsymbol{\theta}_w)$, we use

$$\left[\mathbf{J}_{\boldsymbol{\phi}\boldsymbol{\phi}}\right]_{ij} = -E_{\mathbf{x},\boldsymbol{\theta}}\left[\frac{\partial^2 \ln p_{\mathbf{x}}(\mathbf{x}|\boldsymbol{\theta})}{\partial\phi_i\,\partial\phi_j}\right], \quad i,j = 1,\cdots,D. \tag{8.666}$$

$$\left[\mathbf{J}_{\boldsymbol{\rho}\boldsymbol{\rho}}\right]_{ij} = -E_{\mathbf{x},\boldsymbol{\theta}}\left[\frac{\partial^2 \ln p_{\mathbf{x}}(\mathbf{x}|\boldsymbol{\theta})}{\partial\rho_i\,\partial\rho_j}\right], \quad i,j = 1,\cdots,M. \tag{8.667}$$

$$\left[\mathbf{J}_{\boldsymbol{\phi}\boldsymbol{\rho}}\right]_{ij} = -E_{\mathbf{x},\boldsymbol{\theta}}\left[\frac{\partial^2 \ln p_{\mathbf{x}}(\mathbf{x}|\boldsymbol{\theta})}{\partial\phi_i\,\partial\rho_j}\right], \quad \begin{array}{l} i = 1,2,\cdots,D \\ j = 1,2,\cdots,M. \end{array} \tag{8.668}$$

We can use (8.35) to obtain

$$\left[\mathbf{J}_{\boldsymbol{\phi}\boldsymbol{\phi}}\right]_{ij} = E_{\boldsymbol{\theta}}\left\{\operatorname{tr}\left[\mathbf{K}_{\mathbf{x}}^{-1}\frac{\partial\mathbf{K}_{\mathbf{x}}}{\partial\phi_i}\mathbf{K}_{\mathbf{x}}^{-1}\frac{\partial\mathbf{K}_{\mathbf{x}}}{\partial\phi_j}\right]\right\}, \tag{8.669}$$

$$\left[\mathbf{J}_{\boldsymbol{\rho}\boldsymbol{\rho}}\right]_{ij} = -E_{\boldsymbol{\theta}}\left\{\operatorname{tr}\left[\mathbf{K}_{\mathbf{x}}^{-1}\frac{\partial\mathbf{K}_{\mathbf{x}}}{\partial\rho_i}\mathbf{K}_{\mathbf{x}}^{-1}\frac{\partial\mathbf{K}_{\mathbf{x}}}{\partial\rho_j}\right]\right\}, \tag{8.670}$$

and a similar relationship for the cross-term matrices. Note that, although an expectation over $\boldsymbol{\theta}$ is indicated, the only random component is $\boldsymbol{\rho}$ so it is really an expectation over $\boldsymbol{\rho}$.

The expectation over $\boldsymbol{\theta}$ is difficult to evaluate so we introduce an approximation that was originally proposed by Rockah and Schultheiss [RS87a]. We assume that the traces in (8.669) and (8.670) are reasonably smooth in the vicinity of the nominal value of $\rho = \rho_0$ and if the variances of the array perturbations are small, then we can approximate (8.669) and (8.670) by

$$\left[\mathbf{J}_{\phi\phi}\right]_{ij} \cong \text{tr}\left[\mathbf{K}_{\mathbf{x}}^{-1}\frac{\partial \mathbf{K}_{\mathbf{x}}}{\partial \phi_i}\mathbf{K}_{\mathbf{x}}^{-1}\frac{\partial \mathbf{K}_{\mathbf{x}}}{\partial \phi_j}\right]\Bigg|_{\boldsymbol{\theta}=\boldsymbol{\theta}_0}, \tag{8.671}$$

$$\left[\mathbf{J}_{\rho\rho}\right]_{ij} \cong \text{tr}\left[\mathbf{K}_{\mathbf{x}}^{-1}\frac{\partial \mathbf{K}_{\mathbf{x}}}{\partial \rho_i}\mathbf{K}_{\mathbf{x}}^{-1}\frac{\partial \mathbf{K}_{\mathbf{x}}}{\partial \rho_j}\right]\Bigg|_{\boldsymbol{\theta}=\boldsymbol{\theta}_0}, \tag{8.672}$$

and

$$\left[\mathbf{J}_{\phi\rho}\right]_{ij} \cong \text{tr}\left[\mathbf{K}_{\mathbf{x}}^{-1}\frac{\partial \mathbf{K}_{\mathbf{x}}}{\partial \phi_i}\mathbf{K}_{\mathbf{x}}^{-1}\frac{\partial \mathbf{K}_{\mathbf{x}}}{\partial \rho_j}\right]\Bigg|_{\boldsymbol{\theta}=\boldsymbol{\theta}_0}. \tag{8.673}$$

The Fisher information matrix for the total vector $\boldsymbol{\theta}$ can be written as

$$\mathbf{J}\left(\boldsymbol{\theta}\right) = \mathbf{J}(\boldsymbol{\theta}_w, \boldsymbol{\theta}_u) = \begin{bmatrix} \mathbf{J}_B(\boldsymbol{\theta}_w) & \mathbf{X} \\ \mathbf{Y} & \mathbf{Z} \end{bmatrix}, \tag{8.674}$$

where the $\mathbf{X}$, $\mathbf{Y}$, and $\mathbf{Z}$ matrices correspond to the appropriate $\mathbf{J}$ matrices.

Then, the upper left $(D+M) \times (D+M)$ matrix in $\mathbf{J}_B(\boldsymbol{\theta})$ is,

$$\mathbf{J}_B(\boldsymbol{\theta}_w) = \mathbf{J}_D(\boldsymbol{\theta}_w) + \mathbf{J}_P(\boldsymbol{\theta}_w), \tag{8.675}$$

and is given by (8.664) and (8.665).

We are interested in the CRB on $\boldsymbol{\theta}_w$. Thus, we write,

$$\begin{aligned} \mathbf{J}^{-1}(\boldsymbol{\theta}) &= \left[\begin{bmatrix} \mathbf{J}_D(\boldsymbol{\theta}_w) & \mathbf{X} \\ \mathbf{Y} & \mathbf{Z} \end{bmatrix} + \begin{bmatrix} \mathbf{J}_P(\boldsymbol{\theta}_w) & 0 \\ 0 & 0 \end{bmatrix}\right]^{-1} \\ &= \begin{bmatrix} \left[\mathbf{J}_D(\boldsymbol{\theta}_w) + \mathbf{X}\mathbf{Z}^{-1}\mathbf{Y} + \mathbf{J}_P(\boldsymbol{\theta}_w)\right]^{-1} & \mathbf{X}' \\ \mathbf{Y}' & \mathbf{Z}' \end{bmatrix}. \end{aligned} \tag{8.676}$$

We observe that the first two terms in the matrix in the upper left corner correspond to the matrix in the upper left corner of (8.94). Thus, the hybrid CRB for the $\boldsymbol{\theta}_w$ block of the hybrid CRB is given by (8.99):

$$\begin{aligned} &\left[\left[\mathbf{C}_{HCR}(\boldsymbol{\phi}, \boldsymbol{\rho})\right]^{-1}\right]_{ij} \\ &= \frac{2K}{\sigma_w^2} Re\left[\text{tr}\left[\left[\mathbf{D}_j^H \mathbf{P}_{\mathbf{V}}^{\perp} \mathbf{D}_i\right]\left[\mathbf{S}_{\mathbf{f}} \mathbf{V}^H \mathbf{S}_{\mathbf{x}}^{-1} \mathbf{V} \mathbf{S}_{\mathbf{f}}\right]\right]\right] + \mathbf{J}_P(\boldsymbol{\theta}_w)_{ij}, \end{aligned} \tag{8.677}$$

where $\mathbf{J}_P(\boldsymbol{\theta}_w)$ is given by (8.665).[31]

The matrix $\mathbf{D}_i$ is the ith column of the $N \times (D+M)$ matrix $\mathbf{D}$, which can be divided into two component matrices,

$$\mathbf{D} = [\mathbf{D}_\phi \quad \mathbf{D}_\rho]\,. \tag{8.678}$$

The left matrix is an $N \times D$ matrix,

$$\mathbf{D}_\phi = \sum_{i=1}^{D} \left[\frac{\partial \mathbf{V}(\phi,\rho)}{\partial \phi_i}\right]\Bigg|_{\rho=\rho_0}, \tag{8.679}$$

and we have used the same smoothness assumption as in (8.671)–(8.673). The right matrix is an $N \times M$ matrix,

$$\mathbf{D}_\rho = \sum_{i=1}^{M} \left[\frac{\partial \mathbf{V}(\phi,\rho)}{\partial \rho_i}\right]\Bigg|_{\rho=\rho_0} \tag{8.680}$$

where M is the number of parameters that are perturbed. The formula in (8.677) with the definitions in (8.679) and (8.680) is an adequate starting place for numerical evaluation.

Once the specific parameterization is specified, the expression can be evaluated. We consider the case in which the sensor positions are displaced. We assume that each perturbation parameter is associated with a single sensor.

We define

$$\mathbf{D}_{\boldsymbol{\phi}} \triangleq \left[\begin{array}{cccc} \frac{\partial \mathbf{v}}{\partial \phi_1} & \frac{\partial \mathbf{v}}{\partial \phi_2} & \cdots & \frac{\partial \mathbf{v}}{\partial \phi_D} \end{array}\right], \tag{8.681}$$

and

$$\mathbf{D}_{\mathbf{x}} \triangleq \left[\begin{array}{cccc} \frac{\partial \mathbf{V}_{1,\cdot}^T}{\partial p_{x0}} & \frac{\partial \mathbf{V}_{2,\cdot}^T}{\partial p_{x1}} & \cdots & \frac{\partial \mathbf{V}_{N,\cdot}^T}{\partial p_{xN-1}} \end{array}\right]^T, \tag{8.682}$$

where $\mathbf{V}_{k,\cdot}$ denotes the kth row of the array manifold matrix $\mathbf{V}$.[32]

We can also write $\mathbf{D}_{\mathbf{x}}$ as,

$$\mathbf{D}_{\mathbf{x}} = \sum_{n=0}^{N-1} \left[\frac{\partial \mathbf{V}(\phi,\rho)}{\partial p_{x_n}}\right]\Bigg|_{\rho=\rho_0}. \tag{8.683}$$

Similarly, we can define

$$\mathbf{D}_{\mathbf{y}} \triangleq \left[\begin{array}{cccc} \frac{\partial \mathbf{V}_{1,\cdot}^T}{\partial p_{y0}} & \frac{\partial \mathbf{V}_{2,\cdot}^T}{\partial p_{y1}} & \cdots & \frac{\partial \mathbf{V}_{N,\cdot}^T}{\partial p_{yN-1}} \end{array}\right]^T. \tag{8.684}$$

[31] The result in (8.677) is due to Wahlberg et al. [WOV91].

[32] This result is also in [WOV91].

We also define[33]

$$\mathbf{\Sigma} \triangleq \mathbf{S_f V}^H \mathbf{S_x}^{-1} \mathbf{V S_f}. \tag{8.685}$$

Then, (8.677) reduces to

$$[C_{HCR}(\boldsymbol{\phi}, \boldsymbol{\rho})]^{-1} = \frac{2K}{\sigma_w^2}$$

$$\times Re \left[\begin{array}{c:c:c} \mathbf{D}_{\phi}^H \mathbf{P}_{\mathbf{V}}^{\perp} \mathbf{D}_{\phi} \odot \mathbf{\Sigma}^T & \mathbf{D}_{\phi}^H \mathbf{P}_{\mathbf{V}}^{\perp} \odot (\mathbf{D_x \Sigma})^T & \mathbf{D}_{\phi}^H \mathbf{P}_{\mathbf{V}}^{\perp} \odot (\mathbf{D_y \Sigma})^T \\ \hdashline (\mathbf{D_x \Sigma}) \odot \left(\mathbf{D}_{\phi}^H \mathbf{P}_{\mathbf{V}}^{\perp}\right)^T & (\mathbf{D_x \Sigma D_x}^H) \odot (\mathbf{P}_{\mathbf{V}}^{\perp})^T & (\mathbf{D_x \Sigma D_y}^H) \odot (\mathbf{P}_{\mathbf{V}}^{\perp})^T \\ \hdashline (\mathbf{D_y \Sigma}) \odot \left(\mathbf{D}_{\phi}^H \mathbf{P}_{\mathbf{V}}^{\perp}\right)^T & (\mathbf{D_y \Sigma D_x}^H) \odot (\mathbf{P}_{\mathbf{V}}^{\perp})^T & (\mathbf{D_y \Sigma D_y}^H) \odot (\mathbf{P}_{\mathbf{V}}^{\perp})^T \end{array}\right]$$

$$+ \left[\begin{array}{c:c:c} \mathbf{0} & \mathbf{0} & \mathbf{0} \\ \hdashline \mathbf{0} & \mathbf{\Lambda_x}^{-1} & \mathbf{0} \\ \hdashline \mathbf{0} & \mathbf{0} & \mathbf{\Lambda_y}^{-1} \end{array}\right]. \tag{8.686}$$

The result in (8.686) assumes that all p_{x_n} and p_{y_n} are perturbed. In Example 8.11.2, we indicate how to modify the formula when some p_{x_n} and p_{y_n} are not perturbed.

We observe that the matrix in the upper left corner is identical to the result in (8.98).[34] It represents the CRB on $\boldsymbol{\phi}$ in the absence of array errors. To find the impact of the array perturbations on the DOA estimation accuracy bound, we partition the right side of (8.686) as,

$$[C_{HCR}(\boldsymbol{\phi}, \boldsymbol{\rho})]^{-1} = \left[\begin{array}{c:c} \frac{2K}{\sigma_w^2}\left\{Re\left[\mathbf{D}_{\phi}^H \mathbf{P}_{\mathbf{V}}^{\perp} \mathbf{D}_{\phi} \odot \mathbf{\Sigma}^T\right]\right\} & \mathbf{F} \\ \hdashline \mathbf{F}^H & \mathbf{G} \end{array}\right], \tag{8.687}$$

where $\mathbf{F}$ and $\mathbf{G}$ are the partitions of (8.686). Using the formula for the inverse of a partitioned matrix gives

$$C_{HCR}(\boldsymbol{\phi}) = \left[\left\{\frac{2K}{\sigma_w^2} Re\left[\mathbf{D}_{\phi}^H \mathbf{P}_{\mathbf{V}}^{\perp} \mathbf{D}_{\phi} \odot \mathbf{\Sigma}^T\right] - \mathbf{F G}^{-1} \mathbf{F}^H\right\}\right]^{-1}. \tag{8.688}$$

---

[33] This $\mathbf{\Sigma}$ is different from the $\mathbf{\Sigma}$ in (8.103).

[34] The bound in in (8.98) is in wavenumber space (ψ). Here, the bound is in angle space.

The second term in (8.688) represents the increase in the DOA estimation bound due to array perturbations. Note that it is a bound on the covariance matrix of any unbiased estimate of ϕ. In addition, recall that it is an approximation to the bound because of (8.671), (8.672), and (8.673).

In a similar manner we can obtain a bound on the mean-square calibration error.

The result in (8.686) can be extended to the case in which[35]

$$\rho = \begin{bmatrix} g_0 & g_1 & \cdots & g_{N-1} & \gamma_0 & \gamma_1 & \cdots & \gamma_{N-1} \end{bmatrix}^T. \tag{8.689}$$

A key question that was first investigated by Rockah and Schultheiss [RS87a] is what conditions are required on the array geometry such that the second term in (8.688) will go to zero as the *SNR* goes to infinity.

Rockah and Schultheiss [RS87a] investigated array geometry conditions under which the variance and mean-square error would approach zero as the SNR approached infinity.[36]. They assume that there are no signals with known directions. This is sometimes referred to as the blind calibration problem. They found that:

(i) If the location of one sensor and the direction to a second sensor is known, then, except for some pathological array configurations, three sources were adequate to guarantee that the bound would approach zero as the *SNR* approaches infinity.

(ii) The most conspicuous example of a pathological geometry is a linear array. Thus, the nominal array configuration must be nonlinear in order for the second term to approach zero. This result arises from the inability of the calibration procedure to estimate displacements along the axis of a linear array. If we can impose constraints to force Δp_{x_n} to equal zero, then the result changes.

We now consider two examples to illustrate typical behavior. The first example is a linear array, and we will see the result of the observation in (ii).

Example 8.11.1

Consider the model in which the nominal array is a standard 10-element linear array along the x-axis.

The sensor patterns are isotropic. We assume there are two plane-wave sources located at $\pm\Delta\phi/2$. We assume that the Δp_{x_n} and Δp_{y_n} position perturbations are independently-distributed Gaussian random variables with variance σ_p^2. We let $K = 100$. We plot the

[35]We have used γ for the sensor phase to avoid confusion with the DOA in angle space.

[36]Rockah and Schultheiss [RS87a] used a simpler version of the CRB to derive their result (i.e., the plane-wave signal arrivals were assumed to be disjoint in time).

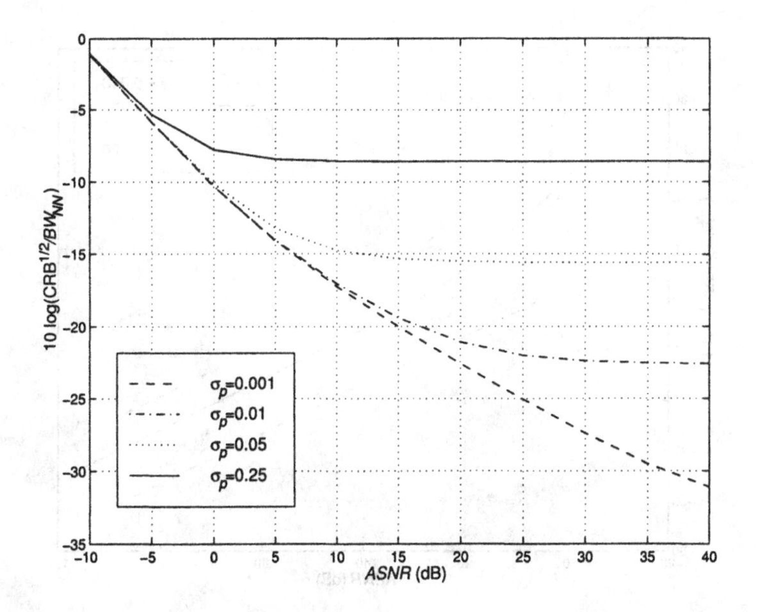

Figure 8.61 Normalized hybrid Cramér-Rao bound on DOA versus *ASNR* for various σ_p: SLA 10 array, two plane-wave signals with $\Delta\phi/2 = \pm 2.55°$, $K = 100$.

hybrid Cramér-Rao bound for the plane wave at $\Delta\phi/2$. We also plot CRB_0, the CRB for perfect sensor location information. Note that these are angle bounds in contrast to the wavenumber bounds we used in earlier sections.

In Figure 8.61, we plot the CRB on DOA for $\Delta\phi = \Delta\psi_R$ (5.1°) versus *ASNR* for various σ_p. We see that the bounds level off as the *ASNR* increases. The departure from the zero calibration error bound occurs at an *ASNR* that decreases as σ_p increases.

The second example of interest is a circular array that satisfies the constraints in (i).

Example 8.11.2

Consider a 33-element circular array located in the x-y-plane. The nominal locations correspond to a standard uniform circular array with interelement spacing of $\lambda/2$. The location of the first element is fixed on the x-axis. The location of the seventeenth element is fixed in the y-direction and allowed to vary in the x-direction. Thus, the model satisfies the identifiability criterion in (i). Three equal-power uncorrelated plane waves impinge on the array from $\phi_1 = -\Delta\phi$, $\phi_2 = 0$, and $\phi_3 = \Delta\phi$. For notational simplicity, we rewrite (8.686) as

$$[C_{HCR}(\phi,\rho)]^{-1} = \frac{2K}{\sigma_w^2}\cdot Re\begin{bmatrix} \mathbf{F}_{\phi\phi} & \mathbf{F}_{\phi\mathbf{x}} & \mathbf{F}_{\phi\mathbf{y}} \\ \mathbf{F}_{\phi\mathbf{x}}^H & \mathbf{F}_{\mathbf{xx}} & \mathbf{F}_{\mathbf{xy}} \\ \mathbf{F}_{\phi\mathbf{y}}^H & \mathbf{F}_{\mathbf{xy}}^H & \mathbf{F}_{\mathbf{yy}} \end{bmatrix} + \begin{bmatrix} \mathbf{0} & \mathbf{0} & \mathbf{0} \\ \mathbf{0} & \mathbf{\Lambda}_{\mathbf{x}}^{-1} & \mathbf{0} \\ \mathbf{0} & \mathbf{0} & \mathbf{\Lambda}_{\mathbf{y}}^{-1} \end{bmatrix}. \tag{8.690}$$

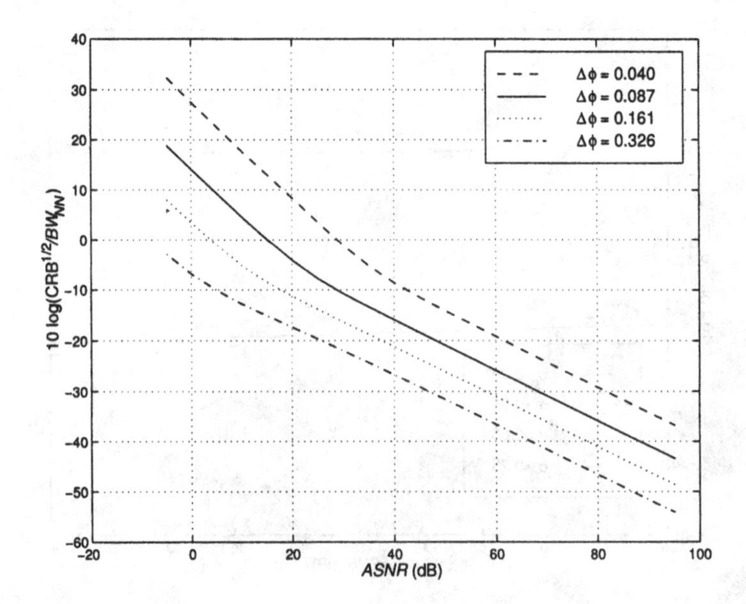

Figure 8.62 Normalized hybrid Cramér-Rao bound (average of three signals) versus *ASNR*: circular array with $N = 33$, three plane-wave signals at 0 and $\pm\Delta\phi$ for various $\Delta\phi$, $\sigma_p = 0.05\lambda$.

We delete the first column in $\mathbf{F}_{\phi\mathbf{x}}$, $\mathbf{F}_{\mathbf{xx}}$, $\mathbf{F}_{\mathbf{xy}}$ and the first and seventeenth columns in $\mathbf{F}_{\phi\mathbf{y}}$, $\mathbf{F}_{\mathbf{xy}}$, and $\mathbf{F}_{\mathbf{yy}}$. We delete the first rows in $\mathbf{F}_{\mathbf{xx}}$, $\mathbf{F}_{\mathbf{xy}}$ and the first and seventeenth row in $\mathbf{F}_{\mathbf{yy}}$. We delete the first elements in $\boldsymbol{\Lambda}_{\mathbf{x}}^{-1}$ and the first and seventeenth element in $\boldsymbol{\Lambda}_{\mathbf{y}}^{-1}$.

In Figure 8.62, we plot the CRB (average of three signals) versus *ASNR* for various $\Delta\phi$. The standard deviation of the perturbation is $\sigma_p = 0.05\lambda$. We see that the CRBs are parallel and continue to decrease as the *ASNR* increases.

Similar bounds for gain and phase perturbations follow directly from (8.677) (e.g., [Fla00]).

8.11.3 Sensitivity of ML Estimators

In this section, we discuss the sensitivity of the ML algorithms of Sections 8.5–8.7 to array perturbations. Our approach requires a simulation of a particular algorithm in the presence of the array perturbations. We restrict our attention to a single example that conveys the type of behavior we can expect. We consider the CML-AP algorithm that we previously discussed in Example 8.6.1.

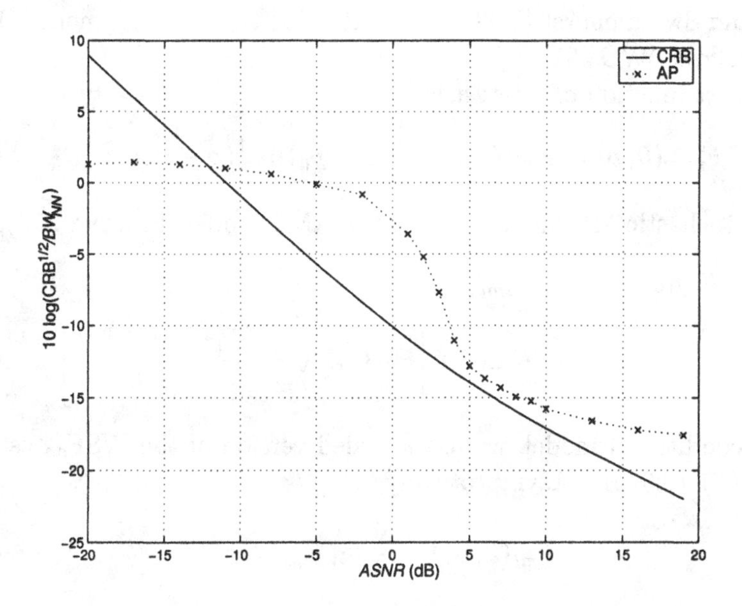

Figure 8.63 Normalized RMSE versus *ASNR*; nominal linear array with sensor position perturbations; $N = 10$, $K = 100$, two plane-wave signals with $\Delta u = \Delta\psi_R$, $\rho = 0$, $\sigma_p = 0.05\lambda$.

Example 8.11.3 (continuation, Example 8.6.1)

Consider the same model as in Example 8.6.1. There are two equal-power uncorrelated signals impinging on an array that is nominally a standard 10-element linear array along the x-axis. The signal separation is $\Delta\psi_R$. We use the AP algorithm to estimate the DOAs.

There are no gain or phase perturbations. The sensor positions are perturbed. The p_{x_n} and $p_{y_n}, n = 0, \cdots, 9$, are statistically independent Gaussian random variables with standard deviation σ_p.

The algorithm does not use this information about the displacements. It implements the algorithm assuming the nominal array manifold.

In Figure 8.63, we plot the normalized RMSE versus *ASNR* for $\sigma_p = 0.05\lambda$. We see that the RMSE is flattening out as the *ASNR* increases.

In this example, the RMSE does not approach the CRB. However, it flattens out at a reasonably low value.

A discussion of the sensitivity of the ML algorithm is given in Friedlander [Fri90].

8.11.4 MAP Joint Estimation

The formulation of the joint MAP estimation problem follows naturally from our hybrid model. Wahlberg et al. [WOV91] has discussed this model and

Viberg and Swindlehurst [VS94] have extended their development. Our discussion follows [WOV91].

The cost function of interest is

$$F_{map}(\boldsymbol{\theta}, \boldsymbol{\rho}) = F_{ml}(\boldsymbol{\theta}, \boldsymbol{\rho}) + \frac{1}{2}(\boldsymbol{\rho} - \boldsymbol{\rho}_0)\boldsymbol{\Lambda}_{\boldsymbol{\rho}}^{-1}(\boldsymbol{\rho} - \boldsymbol{\rho}_0). \tag{8.691}$$

For the stochastic ML model, we use the AML model. From (8.315),

$$\begin{aligned} F_{ml}(\boldsymbol{\theta}, \boldsymbol{\rho}) &= F_{aml}(\boldsymbol{\theta}, \boldsymbol{\rho}) \\ &= K \ln\det\left[\mathbf{P_V C_x P_V} + \frac{tr\left[\mathbf{P}_{\mathbf{V}}^{\perp}\mathbf{C_x}\right]\mathbf{P}_{\mathbf{V}}^{\perp}}{N-D}\right]. \end{aligned} \tag{8.692}$$

For the conditional model, we use a scaled version of the WSF$_{ao}$ estimator (see (8.367)–(8.370)). Asymptotically,

$$F_{ml}(\boldsymbol{\theta}, \boldsymbol{\rho}) \simeq \frac{K}{\hat{\sigma}_w^2} F_{WSF_{ao}}(\boldsymbol{\theta}, \boldsymbol{\rho}), \tag{8.693}$$

where

$$\hat{\sigma}_w^2 = \frac{1}{N-D}\sum_{i=D+1}^{N}\hat{\lambda}_i, \tag{8.694}$$

and

$$F_{WSF_{ao}} = tr\left[\mathbf{P}_{\mathbf{V}}^{\perp}(\boldsymbol{\theta})\hat{\mathbf{U}}_S\left[\hat{\boldsymbol{\Lambda}}_S - \hat{\sigma}_w^2\mathbf{I}\right]^2\hat{\boldsymbol{\Lambda}}_S^{-1}\hat{\mathbf{U}}_S\right]. \tag{8.695}$$

Thus, we define

$$F_{wmap} \triangleq \frac{K}{\hat{\sigma}_w^2} F_{WSF_{ao}}(\boldsymbol{\theta}, \boldsymbol{\rho}) + \frac{1}{2}(\boldsymbol{\rho} - \boldsymbol{\rho}_0)\boldsymbol{\Lambda}_{\boldsymbol{\rho}}^{-1}(\boldsymbol{\rho} - \boldsymbol{\rho}_0). \tag{8.696}$$

We can maximize the expression in (8.696) by using one of the algorithms discussed in Section 8.7.

We can obtain a simpler expression if we assume the variations in $\boldsymbol{\rho}$ are small. Expanding (8.691) around $\boldsymbol{\rho}_0$, we have, for small variations around the nominal value,

$$\begin{aligned} F_{map}(\boldsymbol{\theta}, \boldsymbol{\rho}) \simeq\ & F_{ml}(\boldsymbol{\theta}, \boldsymbol{\rho}_0) + \left[\left.\frac{\partial F_{ml}(\boldsymbol{\theta}, \boldsymbol{\rho})}{\partial \boldsymbol{\rho}}\right|_{\boldsymbol{\rho}=\boldsymbol{\rho}_0}\right]^T (\boldsymbol{\rho} - \boldsymbol{\rho}_0) \\ & + \frac{1}{2}(\boldsymbol{\rho} - \boldsymbol{\rho}_0)^T \left\{\left[\frac{\partial^2 F_{ml}(\boldsymbol{\theta}, \boldsymbol{\rho})}{\partial \boldsymbol{\rho}\partial \boldsymbol{\rho}^T}\right]_{\boldsymbol{\rho}=\boldsymbol{\rho}_0} + \boldsymbol{\Lambda}_{\boldsymbol{\rho}}^{-1}\right\}(\boldsymbol{\rho} - \boldsymbol{\rho}_0). \end{aligned} \tag{8.697}$$

We can maximize $F_{map}(\boldsymbol{\theta}, \boldsymbol{\rho})$ with respect to $\boldsymbol{\rho}$,

$$\hat{\boldsymbol{\rho}} = \boldsymbol{\rho}_0 - \left[\left. \frac{\partial^2 F_{ml}(\boldsymbol{\theta}, \boldsymbol{\rho})}{\partial \boldsymbol{\rho} \partial \boldsymbol{\rho}^T} \right|_{\boldsymbol{\rho}=\boldsymbol{\rho}_0} + \boldsymbol{\Lambda}_{\boldsymbol{\rho}}^{-1} \right]^{-1} \left. \frac{\partial F_{ml}(\boldsymbol{\theta}, \boldsymbol{\rho})}{\partial \boldsymbol{\rho}} \right|_{\boldsymbol{\rho}=\boldsymbol{\rho}_0} . \tag{8.698}$$

Using (8.698) in (8.697) gives a concentrated cost function,

$$\tilde{V}_{map}(\boldsymbol{\theta}, \boldsymbol{\rho}) = F_{ml}(\boldsymbol{\theta}, \boldsymbol{\rho}_0)$$

$$-\frac{1}{2} \left[\frac{\partial F_{ml}(\boldsymbol{\theta}, \boldsymbol{\rho}_0)}{\partial \boldsymbol{\rho}} \right]^T \left[\frac{\partial^2 F_{ml}(\boldsymbol{\theta}, \boldsymbol{\rho}_0)}{\partial \boldsymbol{\rho} \partial \boldsymbol{\rho}} + \boldsymbol{\Lambda}_{\boldsymbol{\rho}}^{-1} \right]^{-1} \frac{\partial F_{ml}(\boldsymbol{\theta}, \boldsymbol{\rho}_0)}{\partial \boldsymbol{\rho}} . \tag{8.699}$$

We now have reduced the problem to a D-dimensional minimization rather than $D + M$. Wahlberg et al. [WOV91] suggested using the approximation in (8.693) in (8.699) and refer to resulting algorithm as the MAPprox algorithm. Wahlberg et al. [WOV91] and Viberg and Swindlehurst [VS94] have studied the algorithm. Jansson et al. [JSO98] developed a generalized weighted subspace fitting (GWSF) algorithm that is a generalization of the MODE and WSF algorithms developed in Sections 8.5.3 and 8.7.3 (e.g. Stoica and Sharman [SS90a] and Viborg and Ottersten [VO91]).

The advantage of the GWSF algorithm is that it can be implemented using a two-step procedure when the nominal array is a uniform linear array (see Section 8.7.3). The algorithm is also consistent even if the signals are fully correlated. The paper also contains a large sample analysis of the MAPprox algorithm and a summary of earlier work on the model error problem. The reader is referred to these references ([WOV91], [VS94], and [JSO98]) for further discussion.

The other approach to joint estimation treats $\boldsymbol{\rho}$ as an unknown nonrandom vector. We discuss this approach briefly in the next section.

8.11.5 Self-Calibration Algorithms

The basic idea of self-calibration algorithms is to formulate a joint estimation problem that contains the parameters of interest. The goal is to calibrate an array using a set of plane-wave signals whose DOAs and powers are unknown. This leads us to a joint estimation problem.

Many of the self-calibration algorithms have the following characteristics:

(i) They estimate the DOAs assuming a nominal array configuration. They then estimate the array parameters assuming the DOAs are correct. They repeat the iteration until the estimates converge. This can be viewed as a joint estimation problem using "group alternating maximization."

(ii) They do not assume an *a priori* density of the array parameters. However, they do assume a nominal configuration to initialize the iteration in (i).

(iii) They frequently use a simpler DOA estimator than ML. The reasons for this simplification will be easier to explain after we have discussed some of the simpler algorithms in Chapter 9.

There are discussions of self-calibration algorithms in Weiss and Friedlander (e.g., [FW88], [WF89], or Chapter 10 in [Hay91b], [WF91b]), Weiss et al. [WWL88], Rockah and Schultheiss [RS87a], [RS87b], and Porat and Friedlander [PF97].

8.11.6 Summary

In this section we have discussed how array perturbations affect the performance of DOA algorithms. After reviewing the perturbation model, we derived the CRB for the joint estimation problem of $\boldsymbol{\phi}$ and $\boldsymbol{\rho}$. We then studied the performance of the CML algorithm that we had derived in Section 8.6. In order to improve performance we introduced the technique of joint estimation (or autocalibration), but did not develop it in detail.

8.12 Summary

In Section 8.12.1, we summarize the major results in Chapter 8. In Section 8.12.2, we briefly discuss some related topics. In Section 8.12.3, we provide a brief introduction to Chapter 9.

8.12.1 Major Results

In this chapter we have studied algorithms for estimating the parameters of a signal arriving at an antenna array. Although the formulation was general, most of our discussion focussed on estimating the directions of arrival of a set of plane waves impinging on the array.

In Section 8.2, we reviewed classical ML estimation and MAP estimation. We also reviewed three versions of the Cramér-Rao bound: the classic CRB, the Bayesian CRB, and the hybrid CRB. These results form the background for most of later developments in this chapter.

In Section 8.3, we developed the models that we use in the parameter estimation model. The model of most interest consisted of multiple plane waves impinging on the array in the presence of additive noise. We emphasized two temporal models for the signals. In the first model, the signals were sample functions from a zero-mean vector stationary complex Gaussian random processes with an unknown source spectral matrix $\mathbf{S_f}$. In the second model, the signals were considered to be unknown nonrandom complex sequences. We modeled the additive noise as a zero-mean vector complex Gaussian random process that is temporally white and spatially uncorrelated. We also introduced array perturbation models and spatially spread models.

In Section 8.4, we developed CRBs for the various models. These CRBs are a key issue in any parameter estimation discussion for two reasons:

(i) They provide a lower bound on the covariance matrix of any unbiased estimator.

(ii) Above some value of K (the number of snapshots) and *ASNR* (the array signal-to-noise ratio), in many cases, the ML estimator reaches the bound.

There were different bounds for the different signal models. One can show that the CRB using the conditional signal model was also greater than or equal to the CRB using the unconditional (or stochastic) CRB. However, in most cases, in the *ASNR*-K region where the CRB is conveying useful information (where the performance of the ML estimator approaches the bound), the two bounds practically coincide.

In Section 8.5, we studied ML estimators in detail. For a fixed K, we found that the ML estimator exhibited a threshold phenomenon. Above a certain *ASNR*, the performance of the ML estimator approaches the CRB. However, as the *ASNR* was decreased, a threshold point was reached where the estimator started making large errors, so the MSE increased quickly. Figures 8.17–8.25 helped explain this behavior. We argued that estimators should be judged according to two criteria:

(i) Behavior above threshold. The estimator should approach the CRB. Generally, this behavior can be studied analytically using asymptotic or Taylor series techniques.

(ii) The location of the threshold. We want the estimator to have its threshold at as low a value of *ASNR* (for a given K) as possible.

We emphasized two types of ML estimators; the unconditional (or stochastic) ML estimator (AML) and the conditional (or deterministic) ML estimator (CML and the related MODE/WSF estimator). In the examples that we considered, there was very little difference in performance.

In the simulations in Section 8.5 we utilized a grid search to ensure that we found the global maximum. This approach is computationally demanding for two plane-wave signals and generally not feasible for a larger number of signals.

In Section 8.6, we developed computational algorithms for finding the ML estimators. We introduced several gradient algorithms, two alternating maximization (AM) algorithms, and the expectation-maximization (EM) algorithm. The AM algorithms, which maximize each variable separately in an iterative manner, were efficient when a suitable initialization was possible. The EM algorithm converged more slowly in most examples. We did not study the gradient algorithms in detail. All of the algorithms in this section were applicable to an arbitrary array geometry.

In Section 8.7, we developed polynomial parameterization techniques that were applicable to standard linear arrays. After introducing polynomial parameterization, we derived the IQML and IMODE algorithms. Both algorithms required less computation than our previous algorithms and provided similar threshold performance.

In Section 8.8, we revisited the problem of estimating the number of plane-wave signals that were present in the input to the array. We emphasized the separable detection algorithms (AIC, MDL) that we had studied earlier and assumed that the output of the detection step was used as an input to estimation algorithm. In some cases, the threshold of the detection algorithms occurred at the same *ASNR* as the threshold of the ML estimator operating with the correct number of signals. We suggested that, if we must operate in the vicinity of the threshold, we should use AIC or AIC-FB and require subsequent processing (after the estimator) to eliminate any extra signals introduced by an over-estimation error from AIC. If we are operating above threshold, we should use MDL or MDL-FB because it provides consistent results. We introduced the idea of joint detection and estimation, but did not pursue it because of the computational requirements.

In Section 8.9, we introduced the parameter estimation problem for spatially spread signals and discussed it briefly. Both models that we discussed were parametric models. After deriving the CRBs, we considered several simple examples. We found that for a single spatially spread signal, even a small amount of spread caused the CRB to level off as the *ASNR* increased. For the single-signal case, the multiple-parameter ML estimator approached the CRB above threshold. The ML estimator for the multiple-signal problem did not appear to be practical because of the computational complexity.

In Section 8.10, we discussed parameter estimation in beamspace. We first developed conditions on the beamspace such that the beamspace CRB would equal the element-space CRB. Several beamspace matrices, such as the Taylor series preprocessor and the discrete prolate spheroidal functions had a CRB_{bs} that came very close to the CRB_{els} over a useful range of signal separation Δu. The conventional DFT beamspace processor also had a CRB_{bs} that was close to the CRB_{els} for signals in the center part of the beamspace sector. We derived the beamspace ML estimator and evaluated its performance. The MSE of the BS-ML estimator approached the CRB_{bs} above the threshold *ASNR*.

In Section 8.11, we discussed the effect of array perturbations on the performance of our parameter estimation algorithms. We modeled the problem as a combined estimation problem in which we estimated both the signal and noise parameters ($\boldsymbol{\psi}_s$, $\mathbf{S}_f$, σ_w^2) and the perturbation parameters. After reviewing the perturbation model, we derived the CRBs for the composite parameter set. The resulting CRB also applies to the case in which we only estimate the signal and noise parameters. For the case of position perturbations, we found that the CRB for a linear array leveled off as the *ASNR* increased. We investigated the performance of the conventional ML estimator and found there was significant degradation. We introduced the idea of a joint estimator called the MAPprox algorithm, but did not pursue it in detail.

8.12.2 Related Topics

One area of interest that we did not explore can be referred to as **structured adaptive beamforming**. Recall that the purpose of the adaptive beamformer was to estimate the waveform of the desired signal. In the literature, this is sometimes referred to as the "signal copy" problem. In Chapters 6 and 7 we normally assumed that we knew the direction-of-arrival of the desired signal, but did not impose any additional structure on the environment. However, in many applications, we can impose the structure by assuming that

$$\mathbf{X}(k) = \mathbf{V}(\boldsymbol{\psi})\mathbf{F}(k) + \mathbf{W}(k), \quad k = 1, 2, \cdots, K, \tag{8.700}$$

which is familiar as the model used in most of this chapter.

The basic idea is straightforward. If we are using the unconditional model (Section 8.5.1), we estimate $\boldsymbol{\psi}$, $\mathbf{S_f}$, and σ_w^2. We then use these estimates, as if they were correct, in one of the beamformers in Chapter 7. An early reference that studied a similar approach was the Lincoln Laboratory Report by Barabell et al. [BCD+84]. They conducted extensive simulations of various DOA estimation algorithms and explored the calibration issue. One can refer to this approach as DF-based signal copy.

Various analyses of this problem have appeared in the literature (e.g., Ottersten et al. [ORK89], Friedlander and Weiss [WF93], Wax [Wax85], and Yang and Swindlehurst [YS95]). The paper by Yang and Swindlehurst [YS95] contains a good summary of results. Note that we are imposing a structure on the signal and noise model and not the array geometry. The reader is referred to these references for a discussion of this problem.

A second problem of interest that we did not explore is analytic expressions for the behavior of a ML estimator (or any other estimator) as the *ASNR* or K decreased. We did not develop bounds on the performance. We refer to this as the **nonlinear behavior** region.

The covariance predicted by the CRB corresponds to small errors in the estimator performance and was directly related to the shape of the main lobe of the beam pattern. In our simulations, we observed that as the number of observations or *SNR* decreased we reached a point where the estimator was making large errors and the performance deviated significantly from the CRB.

We would like to develop bounds that enable us to predict this nonlinear behavior. The bounds can be used for several purposes.

The first purpose is to predict the value of *SNR* or K where the "threshold" occurs. The threshold is defined to be the point where the MSE of the estimator starts to deviate from the CRB. If we have control over the system parameters (e.g., an active radar or sonar) we will design the system so it operates above the threshold.

The second purpose of the bounds is to describe the behavior in the transition region below the threshold. In many systems, we will have to operate in this region and need to predict the behavior.

There are two models that are used in the nonlinear bound literature. In the first model, we treat the vector parameter as an unknown nonrandom quantity. We utilized this approach in DEMT I [VT68] [VT01a] (see pp. 71, 147, 284–286, 386) and developed the Bhattacharyya and Barankin bounds. We can extend these results to the array processing case and develop several hybrid bounds.

There are several approaches in the statistics literature for this model that have been successfully applied to parameter estimation problem by various researchers.

The first approach is the Barankin bound [Bar49]. The original version is a greatest lower bound, but is computationally complex so that simplified versions (e.g., Chapman and Robbins [CR51], Hammersley [Ham50], and Kiefer [Kie52]) are normally used. Application of the Barankin bound to the array processing problem was first done by Baggeroer [Bag69] and later by Becker [Bec77] and Chow and Schultheiss [CS81].

The Barankin bound has been applied in the radar area by Swerling [Swe59], McAulay and Seidman [MS69], and McAulay and Hofstetter [MH71].

The second approach uses the Bhattacharyya bound [Bha46], which is an extension of the CRB utilizing higher derivatives of the likelihood function.

We can also derive a hybrid bound that contains the Cramér-Rao, Bhattacharyya, Barankin, and Hammersley-Chapman-Robbins bound as special cases. The hybrid bound[37] is due to Abel [Abe93], [Abe90], and is based on a derivation using the covariance inequality in Ibragimov and Has'minskii [IH81] and Lehmann [Leh83]. The reader is referred to these references for a discussion of these approaches.

In the second model we treat the parameter as a random variable with a known probability density and develop bounds on the MSE of the estimator. These bounds extend into the threshold and transition region.

The bounds using the first model are local bounds that depend on the actual value of the parameter $\boldsymbol{\theta}$. These bounds are useful in many applications. One of their disadvantages is that they fail to limit the MSE when the parameter space is finite. In many cases, bounds exceed the MSE of the ML estimator in the low *SNR* region. This behavior can be attributed both to the violation of the unbiased estimator assumption and to the lack of *a priori* information in the bounds. Similar observations were made in [Abe93]. Clearly, these bounds must be used cautiously for performance analysis in bearing estimation problems where the parameter space is compact and estimators are inherently biased at low *SNR*.

There are two general approaches to the global bound problems. The first approach is referred to as extended Ziv-Zakai (EZZ) bounds ([Bel95], [BEVT96]). This is a Bayesian bound that assumes that the parameter is a random variable with a known *a priori* distribution. They provide a bound on the global MSE averaged over the *a priori* probability density function (pdf). There are no restrictions on the class of estimators to which Bayesian bounds apply, and they incorporate knowledge of the *a priori* parameter space via the prior distribution. The basic approach in the Ziv-Zakai bound and its extension is to relate the MSE to the probability of error in a binary

[37]Note that we are using "hybrid" in a different manner than in Section 8.2.3.3. In that section, hybrid denotes a mix of real unknown parameters and random parameters. Here, hybrid denotes a mixture of different bounds on real unknown parameters.

detection problem. In order for the bound to be useful we must be able to evaluate or bound the resulting probability of error.

There have been a number of results on global bounds over the last three decades. The original Ziv-Zakai bound [ZZ69] was improved by Chazan et al. [CZZ75] and Bellini and Tartara [BT74]. The disadvantage was that the bounds were restricted to scalar parameters with uniform *a priori* densities. In Bell [Bel95], the EZZ bound was developed that was applicable to vector parameters and arbitrary *a priori* densities. The reader is referred to the above references for a discussion of these bounds.

The second approach to global bounds was developed by Weiss and Weinstein [WW83], [WW84]. It is based on the $\mu(s)$ function that we discussed in Section 2.7 of DEMT I [VT68] [VT01a]. In some situations it provides a tighter bound than the extended Ziv-Zakai bound. In other situations, the EZZ bound provides better results. The reader is referred to the above references for a discussion of the Weiss-Weinstein bounds.

The above references provide a number of useful results. However, we were unable to apply these results to accurately estimate the threshold where we estimate the DOAs of two plane waves in additive noise.[38]

The third related topic is the problem of DOA estimation in the presence of **unknown correlated** noise. There are a number of papers dealing with this problem. Both parametric and nonparametric noise models have been utilized. Parametric models have been utilized by Le Cadre [LeC89], Bohme and Kraus [BK88], and Nagesha and Kay [NK96]. A useful parametric noise model is the spatial ARMA model that we discussed in Section 5.7. Nonparametric approaches include the work of Wu and Wong ([WW94], [WW95]), Wong et al. [WRWQ92], Reilly and Wong [RW92], Stoica et al. [SVWW96], Stoica et al. [SOV92], Wax [Wax92], Prasad et al. [PWMS88], Paulraj and Kailath [PK86], Ye and DeGroat [YDG95], Rajagopal et al. [RKR94], and Harmanci et al. [HTK00]. Papers that utilize an instrument variable approach include Stoica et al. [SVO94] and Stoica et al. [SVWW98].

We discuss other related topics in Section 9.10.

8.12.3 Algorithm complexity

In this chapter, we have focused on ML estimators and the CRB. By utilizing AP or AM techniques and IQML and IMODE where applicable, we were able to reduce the computational complexity. Further reductions were obtained

[38] This inability was the primary reason we did not include a more thorough discussion of non-linear bounds.

by utilizing beamspace processing. In spite of these improvements, the ML algorithms still required a fair amount of computation.

In Chapter 9, we develop parameter estimation algorithms that require less computation. We analyze their performance and compare their performance to the ML estimators developed in this chapter and the CRB.

8.13 Problems

P8.2 Maximum Likelihood and Maximum *A Posteriori* Estimators

Problem 8.2.1

Consider the problem of estimating the direction of arrival of D plane waves using a linear array. In our subsequent discussions, we usually compute the CRB on the vector ψ, where

$$[\psi]_i = \pi u_i = \pi \cos\theta_i, i = 1, \cdots, D, \tag{8.701}$$

where θ_i is the angle of the ith plane wave with respect to the z-axis.

Show that

$$\mathbf{C}_{CR}(\boldsymbol{\theta}) = \mathbf{G}_{\boldsymbol{\theta}}^{-1}\mathbf{C}_{CR}(\boldsymbol{\psi})\mathbf{G}_{\boldsymbol{\theta}}^{-1}, \tag{8.702}$$

where

$$\mathbf{G}_{\boldsymbol{\theta}}^{-1} \triangleq \text{diag}\left\{1/\pi\sin\theta_1, \cdots, 1/\pi\sin\theta_D\right\}. \tag{8.703}$$

Discuss your result.

Problem 8.2.2 (continuation)

Consider the converse problem of going from $\mathbf{C}_{CR}(\boldsymbol{\theta})$ to $\mathbf{C}_{CR}(\boldsymbol{\psi})$. Show that

$$\mathbf{C}_{CR}(\boldsymbol{\psi}) = \mathbf{G}_{\boldsymbol{\psi}}^{-1}\mathbf{C}_{CR}(\boldsymbol{\theta})\mathbf{G}_{\boldsymbol{\psi}}^{-1}, \tag{8.704}$$

and find $\mathbf{G}_{\boldsymbol{\psi}}$. How are $\mathbf{G}_{\boldsymbol{\psi}}$ and $\mathbf{G}_{\boldsymbol{\theta}}$ related?

P8.3 Parameter Estimation Models

Problem Note 8.3.1: The ln likelihood function is the starting point for most of the development in this chapter. In all of the following problems, develop the ln likelihood function for the indicated signal and noise model and identify the unknown parameters. Unless noted, we are considering narrowband processes. We assume that there are D signals and that ψ_i is 1-D. There are K statistically independent samples.

Problem 8.3.1

Signal case S1 and noise case N1.

Problem 8.3.2

Signal case S1 and noise case N2.

Problem 8.3.3

Signal case S1 and noise case N3.

Problem 8.3.4

Signal case S1 and noise case N3. There are D signals. Noise consists of D_N uncorrelated plane waves whose DOA and power are unknown plus white noise with unknown σ_w^2.

Problem 8.3.5

Signal case S1 and noise case N3. There are D signals. The array is an SLA. Noise consists of a complex spatial AR(1) process (Section 5.7) plus white noise with unknown σ_w^2.

Problem 8.3.6 (continuation)

Repeat Problem 8.3.5 for the case in which the noise consists of the sum of D_N complex spatial AR(1) processes plus white noise with unknown σ_w^2.

Problem 8.3.7 (continuation, Problem 8.3.5)

Repeat Problem 8.3.5 for the case in which the noise consists of a complex spatial AR(p) process plus white noise with unknown σ_w^2.

Problem 8.3.8

Repeat Problem 8.3.1 for signal case S2.

Problem 8.3.9

Repeat Problem 8.3.2 for signal case S2.

Problem 8.3.10

Repeat Problem 8.3.3 for signal case S2.

Problem 8.3.11

Repeat Problem 8.3.4 for signal case S2.

Problem 8.3.12

Repeat Problem 8.3.5 for signal case S2.

Problem 8.3.13

Repeat Problem 8.3.6 for signal case S2.

Problem 8.3.14

Repeat Problem 8.3.7 for signal case S2.

Problem 8.3.15

Repeat Problem 8.3.1 for signal case S2u.

Problem 8.3.16

Repeat Problem 8.3.2 for signal case S2u.

Problem 8.3.17

Repeat Problem 8.3.3 for signal case S2u.

Problem 8.3.18

Repeat Problem 8.3.4 for signal case S2u.

Problem 8.3.19
Repeat Problem 8.3.5 for signal case S2u.

Problem 8.3.20
Repeat Problem 8.3.6 for signal case S2u.

Problem 8.3.21
Repeat Problem 8.3.7 for signal case S2u.

Problem 8.3.22
Repeat Problem 8.3.1 for signal case S3.

Problem 8.3.23
Repeat Problem 8.3.2 for signal case S3.

Problem 8.3.24
Repeat Problem 8.3.3 for signal case S3.

Problem 8.3.25
Repeat Problem 8.3.4 for signal case S3.

Problem 8.3.26
Repeat Problem 8.3.5 for signal case S3.

Problem 8.3.27
Repeat Problem 8.3.6 for signal case S3.

Problem 8.3.28
Repeat Problem 8.3.7 for signal case S3.

Problem Note 8.3.2: The next problems consider model perturbations.

The parameter perturbation vector $\boldsymbol{\rho}$ is M-dimensional. For example, in the case of position perturbations, $M = 2N$.

$$\boldsymbol{\rho} = \begin{bmatrix} p_{x0} & p_{y0} & p_{x1} & \cdots & p_{y_{N-1}} \end{bmatrix}^T. \tag{8.705}$$

We assume that $\boldsymbol{\rho}$ is a real Gaussian random vector

$$p_{\boldsymbol{\rho}}(\boldsymbol{\rho}) = \frac{1}{(2\pi)^{\frac{M}{2}} \left|\boldsymbol{\Lambda}_{\boldsymbol{\rho}}\right|^{\frac{1}{2}}} \exp\left\{-\frac{1}{2}(\boldsymbol{\rho} - \boldsymbol{\rho}_0)^T \boldsymbol{\Lambda}_{\boldsymbol{\rho}}^{-1} (\boldsymbol{\rho} - \boldsymbol{\rho}_0)\right\}, \tag{8.706}$$

where $\boldsymbol{\rho}_0$ represents the nominal value of the parameter vector $\boldsymbol{\rho}$. The Gaussian assumption is a good model for many applications.

Problems 8.3.29–8.3.31 consider position perturbations. Problems 8.3.32–8.3.34 consider gain and phase perturbations.

Problem 8.3.29
Consider a SLA10 along the x-axis. The position perturbation model is given by (8.73)–(8.80). We assume $\boldsymbol{\Lambda}_{\boldsymbol{\rho}}$ is diagonal. Note that p_{x_n} and p_{y_n} are constant over the entire observation period. Repeat Problem 8.3.1 for this model.

Problem 8.3.30 (continuation)
Repeat Problem 8.3.2 for the perturbation model in Problem 8.3.29.

Problem 8.3.31 (continuation, Problem 8.3.29)
Repeat Problem 8.3.3 for the perturbation model in Problem 8.3.29.

Problem 8.3.32
Consider a SLA10 along the x-axis. The gain and phase perturbation model is given by (8.73)–(8.80). We assume $\mathbf{\Lambda}_{\boldsymbol{\rho}}$ is diagonal. Repeat Problem 8.3.1 for this model.

Problem 8.3.33 (continuation)
Repeat Problem 8.3.2 for the perturbation model in Problem 8.3.32.

Problem 8.3.34 (continuation, Problem 8.3.32)
Repeat Problem 8.3.3 for the perturbation model in Problem 8.3.32.

P8.4 Cramér-Rao Bounds

Problem Note 8.4.1: The first set of problems considers the test scenarios in Table 8.3 at the end of Section 8.5.1.1. We also use these test scenarios in Sections P.8.5, P.8.6, P.8.7, P.8.10, and P.8.11. Table 8.5 shows where the various test scenarios are used.

Table 8.5 Problem Structure for SLA 10.
Test Scenarios are described in Table 8.3

	TS1	TS2	TS3	TS4	TS5
8.4			P.8.4.1	P.8.4.2	
8.5		Ex.8.5.3 P.8.5.1	Ex.8.5.5 P.8.5.3	Ex.8.5.4	P.8.5.2
8.6	Ex.8.6.1 P.8.6.1 P.8.6.8		Ex.8.6.2 P.8.6.3		
8.7		P.8.7.1	P.8.7.3		P.8.7.2
8.10	P.8.10.1		P.8.10.2	P.8.10.3	
8.11	P.8.11.1 P.8.11.7		P.8.11.2 P.8.11.8	P.8.11.3 P.8.11.9	
	TS6	TS7	TS8	TS9	TS10
8.4		P.8.4.3	P.8.4.5	P.8.4.4	P.8.4.6
8.5	Ex.8.5.6 P.8.5.4				
8.6		P.8.6.3	P.8.6.5	P.8.6.4	P.8.6.6
8.7	P.8.7.4	P.8.7.5	P.8.7.6		P.8.7.7
8.10		P.8.10.4		P.8.10.5	P.8.10.6
8.11		P.8.11.4 P.8.11.10		P.8.11.5 P.8.11.11	

Problem Note 8.4.2: We derive the CRB for a number of array geometries and signal scenarios. In most cases, we find an estimator in a subsequent problem and compare its performance.

Problem 8.4.1

Consider a SLA10. There are two uncorrelated plane-wave signals impinging on the array from $\pm\Delta\psi/2$, where $\Delta\psi = \Delta\psi_R$. $SNR_1 = 100SNR_2$.

Plot the normalized CRB($10\log(\text{CRB}^{\frac{1}{2}}/BW_{NN})$) versus *ASNR*.

Problem 8.4.2

Consider a SLA10. There are two equal-power correlated signals impinging on the array from $\pm\Delta\psi/2$, where $\Delta\psi = \Delta\psi_R$. The correlation coefficient is $|\rho|\exp(j\phi_\rho)$. Consider three values of $|\rho|$; 0.95, 0.99, and 1.0. Consider three values of ϕ_ρ: 0, $\pi/4$, and $\pi/2$. Plot the normalized CRB versus *ASNR*.

Problem 8.4.3

Consider a SLA10. There are three equal-power uncorrelated plane-wave signals impinging on the array from $\psi = 0, \psi = \Delta\psi$, and $\psi = -\Delta\psi$, where $\Delta\psi = \Delta\psi_R$. Plot the normalized CRB versus *ASNR*.

Problem 8.4.4 (continuation, Problem 8.4.3)

Repeat Problem 8.4.3 for the case in which the two plane waves at $\pm\Delta\psi$ have an *SNR* that is 10 dB higher than the plane wave at $\psi = 0$.

Problem 8.4.5 (continuation, Problem 8.4.3)

Consider the same model as in Problem 8.4.3 except the signals are correlated with unequal-power. Denote the signals as: No.1, $\psi = -\Delta\psi$; No.2, $\psi = 0$; No.3, $\psi = \Delta\psi$, where $\Delta\psi = \Delta\psi_R$:

$$SNR_1 = SNR_3 = 0.5SNR_2$$

$$\rho_{12} = 0.9, \quad \rho_{23} = 0.9, \quad \rho_{13} = 0.5\exp(j\pi/2)$$

Plot the normalized CRB versus *ASNR*.

Problem 8.4.6 (continuation, Problem 8.4.3)

Consider the model in Problem 8.4.3. Repeat the problem for the case of five equal-power uncorrelated plane waves located at $\psi = 0, \pm\Delta\psi, \pm2\Delta\psi$, where $\Delta\psi = \Delta\psi_R$. Plot the normalized CRB versus *ASNR*.

Problem Note 8.4.3: The next set of problems considers a 32-element standard linear array (SLA32). The problems with two and three signals are similar to the earlier problems. The larger array size allows us to consider more complicated scenarios.

Problem 8.4.7

Consider a SRA32. There are two equal-power uncorrelated signals impinging on the array at $\pm\Delta u/2$, where $\Delta u = 1/32$.

Problem 8.4.8 (continuation)

Repeat Problem 8.4.7 with $\Delta u = 1/128$.

Problem 8.4.9 (continuation, Problem 8.4.7)

Repeat Problem 8.4.7 with $\Delta u = 1/32$ and $ASNR_1 = 100ASNR_2$.

Problem 8.4.10 (continuation, Problem 8.4.7)

Repeat Problem 8.4.7 with $\Delta u = 1/32$ and $\rho = 0.95$, $\rho = 0.95\exp(j\pi/4)$, and $\rho = 0.95\exp(j\pi/2)$.

Problem 8.4.11 (continuation, Problem 8.4.7)

(a) Repeat Problem 8.4.7 with three equal-power uncorrelated signals located at $-1/32$, 0, $1/32$.

(b) Repeat part(a) with $ASNR_1 = ASNR_3 = 100 ASNR_2$.

Problem 8.4.12 (continuation, Problem 8.4.7)

Repeat Problem 8.4.7 for the following cases:

(a) Five equal-power uncorrelated signals located at $\pm 2/32, \pm 1/32, 0$.

(b) Seven equal-power uncorrelated signals located at $\pm 3/32, \pm 2/32, \pm 1/32, 0$.

(c) Fifteen equal-power uncorrelated signals located at $\pm m/32, m = 0, \cdots, 7$.

Problem Note 8.4.4:

The next several examples consider low redundancy arrays that were introduced in Section 3.9.2.

Problem 8.4.13

Consider the 5-element linear array in Table 3.9. The sensor spacing is

$$1 \cdot 3 \cdot 5 \cdot 2.$$

Two equal-power uncorrelated signals impinge on the array from $\pm \Delta u/2$, where $\Delta u = 0.0866$.

(a) Plot the square root of the CRB ($10 \log \text{CRB}^{\frac{1}{2}}$) versus *SNR*. We do not normalize because of the comparison in parts (b) and (c).

(b) Compare your result to those for an SLA with the same number of elements.

(c) Compare your result to those for an SLA with the same aperture.

(d) Discuss your results.

Problem 8.4.14 (continuation, Problem 8.4.13)

Repeat Problem 8.4.13 for the 7-element linear array in Table 3.9. The sensor spacing is

$$1 \cdot 3 \cdot 6 \cdot 8 \cdot 5 \cdot 2.$$

Problem 8.4.15 (continuation, Problem 8.4.13)

Repeat Problem 8.4.13 for the 9-element linear array in Table 3.9. The sensor spacing is

$$1 \cdot 4 \cdot 7 \cdot 13 \cdot 2 \cdot 8 \cdot 6 \cdot 3.$$

Problem 8.4.16 (continuation, Problem 8.4.13)

Repeat Problem 8.4.13 for the 10-element linear array in Table 3.9. The sensor spacing is

$$1 \cdot 5 \cdot 4 \cdot 13 \cdot 3 \cdot 8 \cdot 7 \cdot 12 \cdot 2.$$

Problem Note 8.4.5: The next several examples consider planar arrays. We utilize (8.268) and obtain the asymptotic conditional CRB (ACCR).

Problem 8.4.17

Consider a standard 10×10 rectangular array. There are two equal-power uncorrelated plane waves impinging on the array. Their (θ, ϕ) directions are $(34°, 45°)$ and $(26°, 45°)$, respectively. Plot the ACCR ($10 \log \text{ACCR}^{\frac{1}{2}}$) versus *ASNR*.

Problem 8.4.18 (continuation)

Repeat Problem 8.4.17 for the case in which $ASNR_1 = 100 ASNR_2$.

Problem 8.4.19 (continuation, Problem 8.4.17)

Repeat Problem 8.4.17 for the case in which the sources are correlated. Consider $\rho = 0.95$ and $0.95 \exp(j\pi/2)$.

Problem 8.4.20 (continuation, Problem 8.4.17)

Consider a standard 10×10 rectangular array. There are three equal-power uncorrelated plane waves impinging on the array. Their (θ, ϕ) directions are $(45°, 50°)$, $(45°, 45°)$, $(45°, 40°)$ respectively. Repeat Problem 8.4.17.

Problem Note 8.4.6: The next several problems consider hexagonal arrays. In order to get some comparison with a 10×10 rectangular array, we use the standard 91-element hexagonal array in Example 4.4.1.

Problem 8.4.21

Repeat Problem 8.4.17. Compare your results to those in Problem 8.4.17.

Problem 8.4.22

Repeat Problem 8.4.18. Compare your results to those in Problem 8.4.18.

Problem 8.4.23

Repeat Problem 8.4.19. Compare your results to those in Problem 8.4.19.

Problem 8.4.24

Repeat Problem 8.4.20. Compare your results to those in Problem 8.4.20.

Problem 8.4.25

Consider a cross array consisting of an N-element SLA along the x-axis and an M-element SLA along the y-axis. The linear arrays are symmetric about the origin. N and M are even, so there is no element at the origin.

Use the signal model in Problem 8.4.17 and assume $N = M = 10$. Plot the ACCR ($10 \log \text{ACCR}^{\frac{1}{2}}$) versus *ASNR*. Compare your results to those in Problem 8.4.17.

Problem 8.4.26

Derive the stochastic CRB for the planar array model. Compare your result to the ACCR in (8.268).

P8.5 Maximum Likelihood Estimation

Problem 8.5.1 (continuation, Example 8.5.3)

Simulate the CML algorithm for the model in Example 8.5.3. Compare your results to the AML results. (Test scenario 2.)

Problem 8.5.2 (continuation, Example 8.5.4)

Consider the model in Example 8.5.4 except $|\rho| = 1$ and $\phi_\rho = \pi/2$. (Test scenario 5.) Simulate the AML, CML, and MODE algorithms. Compare the results of three algorithms.

Problem 8.5.3 (continuation, Example 8.5.5)

(a) Repeat Example 8.5.5 for $ASNR_1 = 10ASNR_2$.

(b) Repeat Example 8.5.5 for the CML and MODE algorithms. Compare your results. (Test scenario 3.)

Problem 8.5.4 (continuation, Example 8.5.6)

Repeat Example 8.5.6 for the CML and MODE algorithm. (Test scenario 6.) Compare your results.

Problem 8.5.5 (continuation, Example 8.5.2)

Repeat Example 8.5.2 with $\Delta\psi = 0.5BW_{NN}$. Note that this is the classical resolution separation. Plot the same results as in Figures 8.21 and 8.24. Discuss your results.

Problem 8.5.6 (continuation, Example 8.5.4)

Repeat Example 8.5.4 with $\Delta\psi = 0.5BW_{NN}$. Plot the same results as in Figures 8.27 and 8.28. Also plot the case when $\rho = 0.95\exp(j\pi/2)$. Consider preprocessing the data using FB averaging.

Problem 8.5.7 (continuation, Example 8.5.5)

Repeat Example 8.5.5 with $\Delta\psi = 0.5BW_{NN}$. Plot the same results as in Figures 8.29 and 8.30.

P8.6 Computational Algorithms

Problem Note 8.6.1: All of the algorithms in this section require initialization. A bad initialization can result in convergence to local maximum (or minimum) instead of the global maximum (or minimum). We use the initialization procedure in (8.434) and (8.435).

Problem 8.6.1 (continuation, Example 8.6.1)

Consider the same model as in Example 8.6.1. Solve for the CML estimate using the quasi-Newton technique in Section 8.6.1. Discuss your results. (Test scenario 1.)

Problem 8.6.2 (continuation, Example 8.5.5)

Consider the model in Example 8.5.5. Use the AP technique to find the CML estimate. Use the AM technique to find the AML estimate. Compare your results. (Test scenario 3.)

Problem 8.6.3 (continuation, Problem 8.4.3)

Consider the same model as in Problem 8.4.3. Use the AP technique to find the CML estimate. Use the AM technique to find the AML estimate. Compare your results to the CRB. Discuss your results. (Test scenario 7.)

Problem 8.6.4 (continuation, Problem 8.4.4)

Repeat Problem 8.6.3 for the model in Problem 8.4.4. (Test scenario 9.)

Problem 8.6.5 (continuation, Problem 8.4.5)
Repeat Problem 8.6.3 for the model in Problem 8.4.5. (Test scenario 8.)

Problem 8.6.6 (continuation, Problem 8.4.6)
Repeat Problem 8.6.3 for the model in Problem 8.4.6. (Test scenario 10.)

Problem 8.6.7 (continuation, Problem 8.4.17)
Repeat Problem 8.6.3 for the model in Problem 8.4.17.

Problem 8.6.8
Consider a standard 10-element linear array. There are two equal-power uncorrelated plane-wave signals impinging on the array.

(a) Use the EM algorithm in Section 8.6.3.2 to find $\hat{\psi}_{cml}$. Compare your results to the AP and discuss the computational requirements.

(b) Read [FH94]. Implement their algorithm for this problem.

Problem Note 8.6.2: The SLA is a useful example. However, in practice, we will normally use the algorithms that are developed in Section 8.7 for SLAs. The next set of problems considers linear arrays with non-uniform spacing where the techniques in Section 8.7 do not apply.

Problem 8.6.9 (continuation, Problem 8.4.13)
Consider the 5-element linear array in Problem 8.4.13 and the same signal model. Find the CML estimate using the AP algorithm. Plot the normalized RMSE versus *ASNR* and compare it to the CRB derived in Problem 8.4.13.

Problem 8.6.10 (continuation, Problem 8.4.14)
Consider the 7-element linear array in Problem 8.4.14 and the same signal model. Find the CML estimate using the AP algorithm. Plot the normalized RMSE versus *ASNR* and compare it to the CRB derived in Problem 8.4.14.

Problem 8.6.11 (continuation, Problem 8.4.15)
Consider the 9-element linear array in Problem 8.4.15 and the same signal model. Find the CML estimate using the AP algorithm. Plot the normalized RMSE versus *ASNR* and compare it to the CRB derived in Problem 8.4.15.

Problem 8.6.12 (continuation, Problem 8.4.16)
Consider the 10-element linear array in Problem 8.4.16 and the same signal model. Find the CML estimate using the AP algorithm. Plot the normalized RMSE versus *ASNR* and compare it to the CRB derived in Problem 8.4.16.

P8.7 Polynomial Parameterization

Problem Note 8.7.1: The next several problems consider a SLA10 with a set of plane waves impinging on the array. In each problem, simulate IQML and IMODE. Plot the normalized RMSE (10 log(RMSE/BW_{NN})) versus *ASNR* for $K = 100$. Compare your results to the CRB and the previous ML implementations. Compare the number of iterations required.

Problem 8.7.1

There are two equal-power uncorrelated signals and $\Delta\psi = 0.05BW_{NN}$. (Test scenario 2.)

Problem 8.7.2

There are two equal-power coherent signals: $\Delta\psi = \Delta\psi_R$ and $\rho = 1\exp(j\pi/2)$. (Test scenario 5.)

Problem 8.7.3

There are two uncorrelated signals: $\Delta\psi = \Delta\psi_R$ and $ASNR_2 = 100ASNR_1$. (Test scenario 3.)

Problem 8.7.4

There are two equal-power uncorrelated signals: $\Delta\psi = \Delta\psi_R$ and $K = 20$ instead of 100. (Test scenario 6.)

Problem 8.7.5

There are three equal-power uncorrelated signals: $\psi_1 = -\Delta\psi$, $\psi_2 = 0$, and $\psi_s = \Delta\psi$, where $\Delta\psi = \Delta\psi_R$. (Test scenario 7.)

Problem 8.7.6 (continuation, Problem 8.4.5)

Use the same signal model as in Problem 8.4.5. (Test scenario 8.)

Problem 8.7.7 (continuation, Problem 8.4.6)

Use the same signal model as in Problem 8.4.6. (Test scenario 10.)

Problem 8.7.8

In this problem, an alternative derivation of (8.509) is developed. Assume that D is odd and let

$$D = 2q + 1. \tag{8.707}$$

Define

$$\mathbf{b} = \begin{bmatrix} \mathbf{b}_1^T & \mathbf{b}_2^T \end{bmatrix}^T, \tag{8.708}$$

where $\mathbf{b}$ is given in (8.489) and $\mathbf{b}_1$ and $\mathbf{b}_2$ are $(q+1) \times 1$ vectors where

$$\mathbf{b}_2 = \mathbf{J}\,\mathbf{b}_1^*. \tag{8.709}$$

The $\mathbf{A}_k$ matrix in (8.498) is partitioned in two $(N-D) \times (q+1)$ matrices,

$$\mathbf{A} = \begin{bmatrix} \mathbf{A}_1 \,\vdots\, \mathbf{A}_2 \end{bmatrix}, \tag{8.710}$$

where the k subscript is suppressed.

Now define

$$\tilde{\mathbf{A}} = \begin{bmatrix} \mathbf{A}_1 + \mathbf{A}_2\mathbf{J} \,\vdots\, j(\mathbf{A}_1 - \mathbf{A}_2\mathbf{J}) \end{bmatrix}, \tag{8.711}$$

and

$$\mathbf{c} = \begin{bmatrix} [Re\,[\mathbf{b}_1]]^T \,\vdots\, [Im\,[\mathbf{b}_1]]^T \end{bmatrix}. \tag{8.712}$$

Note that $\mathbf{c}$ in (8.712) contains the same elements as the $\mathbf{c}$ in (8.504), but they are arranged in a different order.

Show that

$$\mathbf{A}\mathbf{b} = \tilde{\mathbf{A}}\mathbf{c}. \tag{8.713}$$

Problem 8.7.9 (continuation)

In this problem, the IQML algorithm for even D is derived using the model in Problem 8.7.8. Let

$$D = 2q. \tag{8.714}$$

Define

$$\mathbf{b} = \left[\begin{array}{ccc} \mathbf{b}_1^T & b_3 & \mathbf{b}_2^T \end{array} \right]^T, \tag{8.715}$$

where $\mathbf{b}_1$ and $\mathbf{b}_2$ are $q \times 1$ vectors that satisfy (8.709) and b_3 is a real scalar. Partition $\mathbf{A}$ as

$$\mathbf{A} = \left[\begin{array}{c:c:c} \mathbf{A}_1 & \mathbf{A}_3 & \mathbf{A}_2 \end{array} \right], \tag{8.716}$$

where $\mathbf{A}_1$ and $\mathbf{A}_2$ are $(N-D) \times q$ matrices and $\mathbf{A}_3$ is an $(N-D) \times 1$ vector. Define

$$\tilde{\mathbf{A}} = \left[\begin{array}{c:c:c} \mathbf{A}_1 + \mathbf{A}_2 \mathbf{J} & \mathbf{A}_3 & j(\mathbf{A}_1 - \mathbf{A}_2 \mathbf{J}) \end{array} \right]. \tag{8.717}$$

Define

$$\mathbf{c} = \left[\begin{array}{c:c:c} [Re\,[\mathbf{b}_1]]^T & b_3 & [Im\,[\mathbf{b}_1]]^T \end{array} \right]^T. \tag{8.718}$$

Show that

$$\mathbf{A}\,\mathbf{b} = \tilde{\mathbf{A}}\,\mathbf{c}. \tag{8.719}$$

Problem 8.7.10

In this problem, the IQML algorithm for even D is developed. In this case, the $\mathbf{b}$ vector satisfies the constraints:

$$b_0 = b_{D-i}^*, \quad i = 0, \cdots, D/2 - 1, \tag{8.720}$$

and

$$b_{\frac{D}{2}} = b_{\frac{D}{2}}^*, \tag{8.721}$$

so $b_{D/2}$ is real.

Define $\mathbf{c}$ as

$$\mathbf{c} = \left[\begin{array}{cccccc} Re\,[b_0] & Im\,[b_0] & \cdots & Re\left[b_{\frac{D}{2}-1}\right] & Im\left[b_{\frac{D}{2}-1}\right] & b_{\frac{D}{2}} \end{array} \right]. \tag{8.722}$$

Find $\mathbf{T}$ such that

$$\mathbf{b} = \mathbf{T}\,\mathbf{c}. \tag{8.723}$$

Problem 8.7.11

FB-IMODE is the standard IMODE algorithm using FB averaging in the sample spectral matrix (e.g., [SJ97]). In this problem, we compare the performance of FB-IMODE and FO-IMODE for different parameter values. In each part, plot the normalized RMSE versus *ASNR* for FB-IMODE and FO-IMODE. Also plot the normalized CRB. In the MODE algorithms, use $\epsilon = 0.01$ and a maximum of eight iterations. The array is a 10-element SLA and $K = 100$. There are two equal-power plane wave signals separated by $\Delta\psi$.

(a) $\Delta\psi = 0.2165 BW_{NN}$, $\rho = 0$;

(b) $\Delta\psi = 0.05 BW_{NN}$, $\rho = 0$;

(c) $\Delta\psi = 0.4 BW_{NN}$, $\rho = 0$;

(d) $\Delta\psi = 0.2165 BW_{NN}$, $\rho = |1|$, consider $\phi_\rho = 0$, 0.9π, 0.8π, 0.85π, 0.7π, 0.6π, and 0.5π;

(e) Repeat part (d) for $\Delta\psi = 0.05BW_{NN}$;

(f) Repeat part (d) for $\Delta\psi = 0.4BW_{NN}$.

Discuss your results.

Problem 8.7.12

The vector $\mathbf{b}$ is defined in (8.489). Consider the case of a single plane-wave signal impinging on the array from $\psi = 0$. Show that $Re[b_o] = 0$.

P8.8 Detection of Number of Signals

Problem Note 8.8.1: Our discussion and most of the discussions in the literature assume that the number of signals, D, is known. In practice, we usually have to estimate D. In the next nine problems we revisit Examples 8.7.1 and 8.7.2 and Problems 8.7.1–8.7.7.

In each problem, simulate the AIC-FB and MDL-FB algorithm to estimate D. Plot

$$\begin{aligned} P_D &\triangleq Pr[\hat{D} = D] \\ P_M &\triangleq Pr[\hat{D} < D] \\ P_{FA} &\triangleq Pr[\hat{D} > D] \end{aligned}$$

versus *ASNR* for $K = 100$. Use the output $\hat{D}$ in both the IQML or IMODE estimation algorithm. Plot the normalized RMSE versus *ASNR*. In order to calculate the RMSE when $\hat{D} > D$, assume that the D estimates are paired with the closest true DOA and the excess estimates are ignored. Calculate the number of trials where $\hat{D} < D$, but do not assign an RMSE to the missed signal. Compare your results to those in the corresponding example or problem from Section 8.7.

Problem 8.8.1 (continuation, Example 8.7.1)

Use the signal model in Example 8.7.1.

Problem 8.8.2 (continuation, Example 8.7.2)

Use the signal model in Example 8.7.2.

Problem 8.8.3 (continuation, Problem 8.7.1)

Use the signal model in Problem 8.7.1.

Problem 8.8.4 (continuation, Problem 8.7.2)

Use the signal model in Problem 8.7.2.

Problem 8.8.5 (continuation, Problem 8.7.3)

Use the signal model in Problem 8.7.3.

Problem 8.8.6 (continuation, Problem 8.7.4)

Use the signal model in Problem 8.7.4.

Problem 8.8.7 (continuation, Problem 8.7.5)

Use the signal model in Problem 8.7.5.

Problem 8.8.8 (continuation, Problem 8.7.6)

Use the signal model in Problem 8.7.6.

Problem 8.8.9 (continuation, Problem 8.7.7)
Use the signal model in Problem 8.7.7.

Problem 8.8.10
Read Wax and Ziskind [WZ89]. Simulate their algorithm for the model in Problem 8.8.1. Compare your results to those in Problem 8.8.1. Discuss the relative computational complexity.

Problem 8.8.11
Read Wax [Wax91]. Simulate his algorithm for the model in Problem 8.8.4. Compare your results to those in Problem 8.8.4. Discuss the relative computational complexity.

Problem 8.8.12
Read Cho and Djuric [CD94]. Simulate their algorithm for the model in Problem 8.8.1. Compare your results to those in Problem 8.8.1. Discuss the relative computational complexity.

P8.9 Spatially Spread Signals

Problem 8.9.1
Consider Example 8.9.3.

(a) Discuss the relationship between σ_u and the CRB in terms of beam pattern of the array.

(b) Show that the cross-matrices (8.564) are zero for the single signal case.

Problem 8.9.2 (continuation, Example 8.9.4)
Consider the 4-parameter model. Assume $\psi_{C_1} = -\Delta\psi/2$ and $\psi_{C_2} = \Delta\psi/2$. Assume $\sigma_{\psi_2} = \sigma_{\psi_1} = 0.3\Delta\psi/2$. Assume $K = 100$. Plot the CRB ($10 \log \text{CRB}^{1/2}$) versus *ASNR* for $\Delta\psi = 0.2$, 0.4, and 0.6. Note that the bound is not normalized.

Problem 8.9.3
Consider a SLA10. There is a single signal spatially spread signal impinging on the array. The spatial spectrum can be modeled as a complex AR(1) process with known power. Assume that the white noise is negligible.
Plot the CRB for various parameter values.

P8.10 Beamspace Algorithms

Problem Note 8.10.1:
The first set of problems focuses on beamspace Cramér-Rao bounds. We consider many of same arrays and signals as in Section 8.4. In each problem, we consider three types of beamspace matrices:

(a) DFT beamspace (8.581), (8.582).

(b) Taylor series beamspace (8.588), (8.589).

(c) DPSS beamspace (8.598), (8.599).

The dimension of the beamspace is a design parameter. In element space, the CRB in the ψ- or u-space only depends on the signal separation. In beamspace, the CRB also depends on the actual location because the center of the beamspace may be mismatched. In each problem we consider three values for the location of the center of the signal set, $\psi_c = -\pi/N$, 0, and π/N.

In each problem, plot the normalized CRB versus *ASNR* for $K = 100$ and compare them to the element-space CRBs in Section 8.4.

The first six problems consider a SLA10.

Problem 8.10.1 (continuation, Example 8.4.4)

Consider two equal-power uncorrelated plane-wave signals: $\psi_1 = \psi_c - \Delta\psi/2$, $\psi_2 = \psi_c + \Delta\psi/2$, where $\Delta\psi = \Delta\psi_R$. (Test scenario 1, modified.)

Problem 8.10.2 (continuation, Problem 8.4.1)

Consider two uncorrelated plane-wave signals: $\psi_1 = \psi_c - \Delta\psi/2$, $\psi_2 = \psi_c + \Delta\psi/2$, where $\Delta\psi = \Delta\psi_R$. $ASNR_2 = 100 ASNR_1$. (Test scenario 3, modified.)

Problem 8.10.3 (continuation, Problem 8.4.2)

Consider the same signal directions as in Problem 8.10.1. The signals are correlated using the values in Problem 8.4.2. (Test scenario 4, modified.)

Problem 8.10.4 (continuation, Problem 8.4.3)

Consider three equal-power uncorrelated plane-wave signals: $\psi_1 = \psi_c - \Delta\psi$, $\psi_2 = \psi_c$, $\psi_3 = \psi_c + \Delta\psi$, where $\Delta\psi = \Delta\psi_R$. (Test scenario 7, modified.)

Problem 8.10.5 (continuation, Problem 8.4.4)

Consider the same signal model as in Problem 8.10.4, except the signals at ψ_1 and ψ_3 have an *SNR* that is 10 dB higher than the signal at ψ_2. (Test scenario 9, modified.)

Problem 8.10.6 (continuation, Problem 8.4.6)

Consider five equal-power uncorrelated plane-wave signals: $\psi_1 = \psi_c - 2\Delta\psi$, $\psi_2 = \psi_c - \Delta\psi$, $\psi_3 = \psi_c$, $\psi_4 = \psi_c + \Delta\psi$, $\psi_5 = \psi_c + 2\Delta\psi$. (Test scenario 10, modified.)

Problem Note 8.10.2: The next six problems consider a SLA32.

Problem 8.10.7 (continuation, Problem 8.4.7)

Consider two equal-power uncorrelated signals at $\psi_1 = \psi_c - \pi/32$ and $\psi_1 = \psi_c + \pi/32$.

Problem 8.10.8 (continuation, Problem 8.4.8)

Consider two equal-power uncorrelated signals at $\psi_1 = \psi_c - \pi/128$ and $\psi_1 = \psi_c + \pi/128$.

Problem 8.10.9

Consider the same model as in Problem 8.10.7 except the signals are correlated. Let $\rho = 0.95$, $0.95\exp(j\pi/4)$, and $0.95\exp(j\pi/2)$.

Problem 8.10.10

(a) Consider three equal-power uncorrelated signals located at $\psi_1 = \psi_c - \pi/32$, $\psi_2 = 0$, and $\psi_3 = \psi_c + \pi/32$.

(b) Repeat part (a) with $ASNR_1 = ASNR_3 = 100 ASNR_2$.

Problem 8.10.11

(a) Consider five equal-power uncorrelated signals located at

$$\psi_i = \psi_c + \frac{\pi(i-3)}{32}, \quad i = 1, \cdots, 5. \tag{8.724}$$

(b) Consider seven equal-power uncorrelated signals located at

$$\psi_i = \psi_c + \frac{\pi(i-4)}{32}, \quad i = 1, \cdots, 7. \tag{8.725}$$

Problem 8.10.12

Consider the signal model in Problem 8.10.7 and assume $\psi_c = \pi/128$. We use a DFT beamspace matrix with $N_{bs} = 5$. Find $\hat{\psi}_{cml,bs}$ (8.630) by using the AP algorithm.

Discuss your results.

Problem 8.10.13

Consider the signal model in Problem 8.10.10 and assume $\psi_c = 0$. We use a DFT beamspace matrix with $N_{bs} = 5$. Find $\hat{\psi}_{cml,bs}$ (8.630) by using the AP algorithm.

Discuss your results.

Problem 8.10.14

Consider the same signal model as in Problem 8.10.12. Repeat Problem 8.10.12 using the beamspace IQML algorithm. Compare your results, including computational issues.

Problem 8.10.15 (continuation, Problem 8.10.13)

Repeat Problem 8.10.14 for the signal model in Problem 8.10.13.

P8.11 Sensitivity, Robustness, and Calibration

Problem Note 8.11.1: The next five problems consider a 10-element linear array that is nominally a SLA along the x-axis. The sensor location perturbation model is given in (8.661) and (8.662). The p_{x_i} and p_{y_i} perturbations are statistically independent with $\sigma_p = 0.05\lambda$, which is 10% of the nominal value. We consider different signal models and evaluate the element-space hybrid CRB, the beamspace hybrid CRB, and the RMSE of various algorithms:

(a) Element-space hybrid CRB.

(b) CML and MODE AP in element space.

(c) CML and MODE AP in beamspace (DFT BS matrix with $N_{bs} = 5$).

(d) Element-space IQML.

(e) Beamspace IQML.

In each part, plot the normalized RMSE versus *ASNR* for $K = 100$. Include the $\sigma_p = 0$ case as a reference. Note that the sensor positions do not change during the snapshot sequence. Discuss your results.

Problem 8.11.1

Two equal-power uncorrelated plane-wave signals impinge on the array from $\pm\Delta\psi/2$, where $\Delta\psi = \Delta\psi_R$. (Test scenario 1.)

Problem 8.11.2

Two uncorrelated plane-wave signals impinge on the array from $\pm\Delta\psi/2$, where $\Delta\psi = \Delta\psi_R$, $ASNR_2 = 100 ASNR_1$. (Test scenario 3.)

Problem 8.11.3

Two equal-power correlated plane-wave signals impinge on the array from $\pm\Delta\psi/2$, where $\Delta\psi = \Delta\psi_R$. $\rho = 0.95\exp(j\pi/4)$. (Test scenario 3.)

Problem 8.11.4

Three equal-power uncorrelated plane-wave signals impinge on the array; $\psi_1 = -\Delta\psi$, $\psi_2 = 0$, $\psi_3 = \Delta\psi$, where $\Delta\psi = \Delta\psi_R$. (Test scenario 7.)

Problem 8.11.5 (continuation)

Consider the same model as in Problem 8.11.4 except $ASNR_1 = ASNR_3 = 100 ASNR_2$. (Test scenario 9.)

Problem Note 8.11.2: The next set of problems considers sensor gain and phase perturbation. We use the model in Section 6.3.3 with no sensor position perturbations. We let $\sigma_g = 0.02$ and $\sigma_\gamma = .05$.

In Problem 8.11.6, we derive the hybrid CRB for gain and phase perturbations. In Problems 8.11.7–8.11.11 we simulate the performance for the same signal models as in Problems 8.11.1–8.11.5. In each problem, do the five parts in Problem Note 8.11.1 and plot the results.

Problem 8.11.6

Use the expression in (8.677) as a starting point. Derive the hybrid CRB for the case of gain and phase perturbations.

Problem 8.11.7 (continuation, Problem 8.11.1)

Repeat Problem 8.11.1 for the sensor gain and phase perturbation parameters in Problem Note 8.11.2. (Test scenario 1.)

Problem 8.11.8 (continuation, Problem 8.11.2)

Repeat Problem 8.11.2 for the sensor gain and phase perturbation parameters in Problem Note 8.11.2. (Test scenario 3.)

Problem 8.11.9 (continuation, Problem 8.11.3)

Repeat Problem 8.11.3 for the sensor gain and phase perturbation parameters in Problem Note 8.11.2. (Test scenario 4.)

Problem 8.11.10 (continuation, Problem 8.11.4)

Repeat Problem 8.11.4 for the sensor gain and phase perturbation parameters in Problem Note 8.11.2. (Test scenario 7.)

Problem 8.11.11 (continuation, Problem 8.11.5)

Repeat Problem 8.11.5 for the sensor gain and phase perturbation parameters in Problem Note 8.11.2. (Test scenario 9.)

Bibliography

[Abe90] J. S. Abel. A bound on mean-square-estimate error. *Proc. ICASSP*, Albuquerque, New Mexico, vol.3, pp. 1345–1348, April 1990.

[Abe93] J. S. Abel. A bound on mean-square-estimate error. *IEEE Trans. Inf. Theory*, vol.IT-39, pp. 1675–1680, September 1993.

[AGGS98] Y. I. Abramovich, D.A. Gray, A.Y. Gorokhov and N.K. Spencer. Positive-definite Toeplitz completion in DOA estimation for nonuniform linear antenna arrays—Part I: Fully augmentable arrays. *IEEE Trans. Signal Process.*, vol.SP-46, pp. 2458–2471, September 1998.

[Aka74] H. Akaike. A new look at the statistical model identification. *IEEE Trans. Autom. Control*, vol.AC-19, pp. 716–723, June 1974.

[AN95] S. Anderson and A. Nehorai. Optimal dimension reduction for array processing-generalized. *IEEE Trans. Signal Process.*, vol.SP-43, pp. 2025–2027, August 1995.

[And63] T. W. Anderson. Asymptotic theory for principal component analysis. *Ann. Math. Stat.*, vol.34, pp. 122–148, 1963.

[And71] T. W. Anderson. *The Statistical Analysis of Time Series.* Wiley, New York, 1971.

[And84] T. W. Anderson. *An Introduction to Multivariate Statistical Analysis.* Wiley, New York 1984.

[And91] S. Anderson. Optimal dimension reduction for sensor array signal processing. *Proc. 25th Asilomar Conf. on Signals, Systems and Computers*, Monterey, California, pp. 918–922, November 1991.

[And93] S. Anderson. On optimal dimension reduction for sensor array signal processing. *IEEE Trans. Signal Process.*, vol.SP-30, pp. 245–256, January 1993.

[Bag69] A. B. Baggeroer. Barankin bound on the variance of estimates of Gaussian random processes. Technical Report, MIT Lincoln Laboratory, Lexington, Massachusetts, January 1969.

[Ban71] W. J. Bangs. *Array Processing with Generalized Beamformers.* PhD thesis, Yale University, New Haven, Connecticut, 1971.

[Bar49] E. W. Barankin. Locally best unbiased estimates. *Ann. Math. Statist.*, vol.20, pp. 477–501, 1949.

[Bar54] M. S. Barlett. A note on the multiplying factors for various χ^2 approximations. *J. R. Stat. Soc.*, vol.16, pp. 296–298, 1954.

[BCD+84] A. J. Barabell, J. Capon, D. F. DeLong, J. R. Johnson, and K. D. Senne. Performance comparison of superresolution array processing algorithms. Technical Report Project Report. TST-72, MIT Lincoln Laboratory, Lexington, MA, May 1984.

[Bec77] D. Becker. Development of Cramér-Rao bounds on parameter estimation accuracy. Technical Report, Orincon, San Diego, California, January 1977.

[Bel95] K. L. Bell. *Performance Bounds in Parameter Estimation with Application to Bearing Estimation.* PhD Thesis, George Mason University, Fairfax, Virginia, 1995.

[Bell99] K. L. Bell. An uncorrelated maximum likelihood estimator. Internal report, Statistical Signal and Array Processing Group, George Mason University C3I Center, Fairfax, Virginia, April 1999.

[Ben93] G. R. Benitz. Asymptotic results for maximum likelihood estimation with an array of sensors. *IEEE Trans. Inf. Theory*, vol.IT-39, pp. 1374–1386, July 1993.

[BEVT95] K. L. Bell, Y. Ephraim, and H. L. Van Trees. Ziv-Zakai lower bounds in bearing estimation. *Proc. ICASSP*, Detroit, Michigan, vol.5, pp. 2852–2855, May 1995.

[BEVT96] K. L. Bell, Y. Ephraim, and H. L. Van Trees. Explicit Ziv-Zakai lower bound for bearing estimation. *IEEE Trans. Signal Process.*, vol.SP-44, pp. 2810–2824, November 1996.

[Bha46] A. Bhattacharyya. On some analogues of the amount of information and their use in statistical estimation. *Shankyā*, vol.8, pp. 201–218, 1946.

[BK84] G. Bienvenu and L. Kopp. Decreasing high resolution method sensitivity by conventional beamformer preprocessing. *Proc. ICASSP*, San Diego, California, vol.V pp. 3321–3324, 1984.

[BK88] J. Bohme and D. Kraus. On least squares methods for DOA estimation in unknown noise fields. *Proc. ICASSP*, New York, New York, pp. 2833–2836, 1988.

[BM86] Y. Bresler and A. Macovski. Exact maximum likelihood parameter estimation of superimposed exponential signals in noise. *IEEE Trans. Acoust., Speech, and Signal Process.*, vol.ASSP-34, pp. 1081–1089, October 1986.

[Boh86] J. F. Böhme. Separated estimation of wave parameters and spectral parameters by maximum likelihood. *Proc. ICASSP*, Tokyo, Japan, pp. 2819–2822, 1986.

[Boh89] J. F. Böhme. Location and spectrum estimation by approximate maximum likelihood. *Proc. SPIE*, Orlando, Florida, vol.1, pp. 326–337, 1989.

[Bre88] Y. Bresler. Maximum likelihood estimation of a linearly structured covariace with application to antenna array processing. *Proc. 4th ASSP Workshop on Spectrum Estimation and Modeling*, Minneapolis, Minnesota,, pp. 172–175, August 1988.

[BT74] S. Bellini and G. Tartara. Bounds on error in signal parameter estimation. *IEEE Trans. Commun. Technol.*, vol.COM22, pp. 340–342, March 1974.

[BX90] K. Buckley and X.L. Xu. Spatial-spectrum estimation in a location sector. *IEEE Trans. Accoust. Speech, Signal Process.*, vol.ASSP-38, pp. 1842–1852, November 1990.

[CD94] C. M. Cho and P. M. Djuric. Detection and estimation of DOAs of signals via Bayesian predictive densities. *IEEE Trans. Signal Process.*, vol.SP-42, pp. 3051–3060, November 1994.

[Cin75] E. Cinlar. *Introduction to Stochastic Processes.* Prentice-Hall, Englewood Cliffs, New Jersey, 1975.

[CR51] D. G. Chapman and H. Robbins. Minimum variance estimation without regularity assumption. *Ann. Math. Stat.*, vol.21, pp. 581–586, 1951.

[Cra46] H. Cramér. *Mathematical Methods of Statistics.* Princeton University Press, Princeton, New Jersey, 1946.

[CS81] S. K. Chow and P. M. Schultheiss. Delay estimation using narrow-band processes. *IEEE Trans. Acoust., Speech, Signal Process.*, vol.ASSP-29, pp. 478–484, June 1981.

[CS92] M. Clark and L. L. Scharf. On the complexity of IQML algorithms. *IEEE Trans. Signal Process.*, vol.SP-40, pp. 1811–1813, July 1992.

[CTSD96] C. Chambers, T. C. Tozer, K. C. Sharman, and T. S. Durrani. Temporal and spatial sampling influence on the estimates of superimposed narrowband signals: when less can mean more. *IEEE Trans. Signal Process.*, vol.SP-44, pp. 3085–3098, December 1996.

[CZZ75] D. Chazan, M. Zakai, and J. Ziv. Improved lower bounds on signal parameter estimation. *IEEE Transactions Inf. Theory*, vol.IT-21, pp. 90–93, January 1975.

[Dju90] P. M. Djuric. *Selection of Signal and System Models by Bayesian Predictive Densities.* PhD Thesis, University of Rhode Island, Kingston, 1990.

[DS83] J. E. Dennis and R. B. Schnabel. *Numerical Methods for Unconstrained Optimization and Nonlinear Equations.* Prentice-Hall, Englewood Cliffs, New Jersey, 1983.

[DS96] J. E. Dennis and R. B. Schnabel. *Numerical Methods for Unconstrained Optimization and Nonlinear Equations.* Prentice-Hall, Englewood Cliffs, New Jersey, 1996.

[DW92] M. A. Doron and A. J. Weiss. On focusing matrices for wide-band array processing. *IEEE Trans. Signal Process.*, vol.SP-40, pp. 1295–1302, June 1992.

[DWM93] M. A. Doron, A. J. Weiss, and H. Messer. Maximum-likelihood direction finding of wide-band sources. *IEEE Trans. Signal Process.*, vol.SP-41, pp. 411–414, January 1993.

[EF73] A. G. Evans and R. Fischl. Optimal least squares time-domain synthesis of recursive digital filters. *IEEE Trans. Audio Electroacoust.*, vol.AE-21, pp. 61–65, February 1973.

[FH94] J. A. Fessler and A. O. Hero. Space-alternating generalized expectation-maximization algorithm. *IEEE Trans. Signal Process.*, vol.SP-42, pp. 4664–4677, October 1994.

[Fis22] R. A. Fisher. On the mathematical foundations of theoretical statistics. *Phil. Trans. R. Soc., London, Ser. A*, vol.222, pp. 309–368, 1922.

[Fis25] R. A. Fisher. Theory of statistical estimation. *Proc. Cambridge Phil. Soc.*, vol.22, p. 700, 1925.

[Fla00] B. P. Flanagan. *Self Calibration of Antenna Arrays with Large Perturbation Errors.* PhD Thesis, George Mason University, Fairfax, Virginia, 2000.

[FP89] B. Friedlander and B. Porat. The exact Cramér-Rao Bound for Gaussian autoregressive processes. *IEEE Trans. Aerospace Electronic Syst.*, vol.AES-25, pp. 3–8, January 1989.

[Fri84] B. Friedlander. On the computation of the Cramér-Rao Bound for ARMA parameter estimation. *IEEE Trans. Acoust., Speech Signal Process.*, vol.ASSP-32, pp. 721–727, 1984.

[Fri90] B. Friedlander. A sensitivity analysis of the maximum likelihood direction finding algorithm. *IEEE Trans. Aerospace Electron. Systems*, vol.AES-25, pp. 953–968, November 1990.

[Fuc88] J. J. Fuchs. Estimating the number of sinusoids in additive white noise. *IEEE Trans. Acoust., Speech, Signal Proc.*, vol.SSP-36, pp. 1846–1853, December 1988.

[FV87] P. Forster and G. Vezzosi. Application of spheroidal sequences to array processing. *Proc. ICASSP*, Dallas, Texas, vol.4, pp. 2267–2271, 1987.

[FW88] M. Feder and E. Weinstein. Parameter estimation of superimposed signals using the EM algorithm. *IEEE Trans. Acoust., Speech, and Sig. Proc.*, vol.ASSP-36, pp. 477–489, April 1988.

[Gaa88] W.A Gardner. Simplification of MUSIC and ESPRIT by exploitation of cyclostationarity. *Proc. IEEE*, vol.76, pp. 845–847, July 1988.

[GL86] R. Gooch and J. Lundell. The CM array: An adaptive beamformer for constant modulus signals. *Proc. ICASSP*, Tokyo, Japan, vol. 5, pp. 2523–2526, 1986.

[GMW81] P. E. Gill, W. Murray, and M. H. Wright. *Practical Optimization*. Academic, London, 1981.

[GP73] G. Golub and V. Pereyra. The differentiation of pseudoinverses and nonlinear least squares problems whose variables separable. *SIAM J. Numer. Anal.*, vol.10, pp. 413–432, 1973.

[Gra82] D. A. Gray. Formulation of the maximum signal-to-noise ratio array processor in beam space. *J. Acoust. Soc. Am.*, vol.72, pp. 1195–1201, October 1982.

[GS99] A. B. Gershman and P. Stoica. On unitary and forward-backward MODE. *Digital Signal Process*, vol.9, pp. 67–75, April 1999.

[Gu00] H. Gu. Linearization method for finding Cramér-Rao bounds in signal processing. *IEEE Trans. Signal Process.*, vol.48, pp. 543–545, February 2000.

[Gup65] R. P. Gupta. Asymptotic theory for principal component analysis in the complex case. *J. Indian Stat. Assoc.*, vol.3, pp. 97–106, 1965.

[Ham50] J. M. Hammersley. On estimating restricted parameters. *J. R. Soc. Ser. B*, vol.12, pp. 192–240, 1950.

[Hay91a] S. Haykin. *Adaptive Filter Theory*, 2nd ed. Prentice-Hall, Englewood Cliffs, New Jersey, 1991.

[Hay91b] S. Haykin, editor. *Advances in Spectrum Analysis and Array Processing*, vol.2. Prentice-Hall, Englewood Cliffs, New Jersey, 1991.

[HB91] Y. D. Huang and M. Barkat. Near-field multiple source localization by passive sensor array. *IEEE Trans. Antennas Propag.*, vol.AP-69, pp. 968–975, July 1991.

[HS92] S. Haykin, and A. Steinhardt. *Adaptive Radar Detection and Estimation*. Wiley, New York, 1992.

[HTK00] K. Harmanci, J. Tabrikian, and J. L. Krolik. Relationships between adaptive minimum variance beamforming and optimal source localization. *IEEE Trans. Signal Process.*, vol.SP-48, pp. 1–12, January 2000.

[Hua94] Y. Hua. The most efficient implementation of the IQML algorithm. *IEEE Trans. Signal Process.*, vol.SP-42, pp. 2203–2204, August 1994.

[HUSF97] A.O. Hero, M. Usman, A.C. Sauve, and J.A. Fessler. Recursive algorithms for computing the Cramér-Rao bound. *IEEE Trans. Signal Process.*, vol.SP-45, pp. 803–807, March 1997.

[IH81] I. A. Ibragimov and R. Z. Has'minski. *Statistical Estimation.* Springer-Verlag, New York, 1981.

[Jaf85] A. G. Jaffer. Maximum likelihood angular resolution of multiple sources. *Proc. 19th Asilomar Conf. on Signals, Systems, and Computers*, Pacific Grove, California, pp. 68–72, November 1985.

[Jaf88] A. G. Jaffer. Maximum likelihood direction finding of stochastic sources: A separable solution. *Proc. ICASSP*, New York, vol.5, pp. 2893–2896, April 1988.

[JGO99] M. Jansson, B. Goransson, and B. Ottersten. A subspace method for direction of arrival estimation of uncorrelated emitter signals. *IEEE Trans. Signal Process.*, vol.SP-47, pp. 945–956, April 1999.

[JSO98] M. Jansson, A. L. Swindlehurst, and B. Ottersten. Weighted subspace fitting for general array error models. *IEEE Trans. Signal Process.*, vol.SP-46, pp. 2484–2498, September 1998.

[Kau75] L. Kaufman. A variable projection method for solving separable nonlinear least squares problems. *BIT*, vol.15, pp. 49–57, 1975.

[Kay84] S. M. Kay. Accurate frequency estimation at low signal-to-noise ratio. *IEEE Trans. Acoust., Speech, Signal Process.*, vol.ASSP-32, pp. 540–547, June 1984.

[Kay88] S. M. Kay. *Modern Spectral Estimation.* Prentice-Hall, Englewood Cliffs, New Jersey, 1988.

[KB93] D. Kraus and J.F. Böhme. Maximum likelihood location estimation of wideband sources using the EM algorithm. Adaptive Systems in Control and Signal Processing, 4th IFAC Symposium. pp. 487–491, Pergamon Press, Oxford, UK, 1993.

[KDB93] D. Kraus, A. Dhaouadi and J.F. Böhme. Dual maximum likelihood estimation for wideband source location. *Proc. ICASSP*, New York, vol.I, pp. 257–260, 1993.

[Kie52] J. Kiefer. On minimum variance estimators. *Ann. Math. Stat.*, vol.23, pp. 627–629, 1952.

[KM81] S.M. Kay and S.L. Marple. Spectrum analysis: A modern perspective. *Proc. IEEE*, vol.69, pp. 1380–1419, 1981.

[KMB92] D. Kraus, D. Maiwald and J.F. Böhme. Maximum likelihood source location estimation via EM algorithm. Proc. 6th European Signal Process. Conf. Amsterdam, Netherlands, vol.2, pp. 649–452, 1992.

[KS61] M. G. Kendall and A. Stuart. *The Advanced Theory of Statistics.* Hafner, New York, 1961.

[KS85] R. Kumaresan and A. K. Shaw. High resolution bearing estimation without eigendecomposition. *Proc. ICASSP*, Tampa Bay, Florida, vol.1 pp. 576–579, April 1985.

[KSS86] R. Kumaresan, L. L. Scharf, and A. K. Shaw. An algorithm for pole-zero modelling and spectral analysis. *IEEE Trans. Acoust., Speech, Signal Process.*, vol.ASSP-34, pp. 637–640, June 1986.

[KS88] R. Kumaresan, L.L. Scharf, and A.K. Shaw. Superresolution by structured matrix approximation. *IEEE Trans. Antennas Propag.*, vol.AP, pp. 36–44, 1988.

[KT83] R. Kumaresan and D. W. Tufts. Estimating the angles of arrival of multiple plane waves. *IEEE Trans. Aerospace Electron. Syst.*, vol.AES-19, pp. 134–139, January 1983.

[Kum85] R. Kumar. A fast algorithm for solving a Toeplitz system of equations. *IEEE Trans. Acoust., Speech, Signal Process.*, vol.ASSP-33, pp. 254–267, February 1985.

[Law56] D. N. Lawley. Tests of significance of the latent roots of the covariance and correlation matrices. *Biometrica*, vol.43, pp. 128–136, 1956.

[LC93] J. Li and R.T. Compton. Maximum likelihood angle estimation for signals with known waveforms. *IEEE Trans. Signal Process.*, vol.SP-41, pp. 2850–2862, September 1993.

[LDD94] D.A. Linebarger, R.D. DeGroat, and E.M. Dowling. Efficient direction-finding methods employing forward/backward averaging. *IEEE Trans. Signal Process.*, vol.SP-42, pp. 2136–2145, August 1994.

[LDR77] N. M. Laird, A. P. Dempster, and D. B. Rubin. Maximum likelihood from incomplete data via the EM algorithm. *Ann. R. Stat. Soc.*, pp. 1–38, December 1977.

[LeC89] J. Le Cadre. Parametric methods for spatial signal processing in unknown colored noise fields. *IEEE Trans. Acoust. Speech, Signal Process.*, pp. 965–983, July 1989.

[Lee92] H. B. Lee. The Cramér-Rao bound on frequency estimates of signals closely spaced in frequency. *IEEE Trans. Signal Process.*, vol.SP-40, pp. 1508–1517, June 1992.

[Lee94] H. B. Lee. The Cramér-Rao bound on frequency estimates of signals closely spaced in frequency (unconditional case). *IEEE Trans. Signal Process.*, vol.SP-42, pp. 1569–1572, June 1994.

[Leh83] E. L. Lehmann. *Theory of Point Estimation.* Wiley, New York, 1983.

[LL94] H. Lee and F. Li. An eigenvector technique for detecting the number of emitters in a cluster. *IEEE Trans. Signal Process.*, vol.SP-42, pp. 2380–2388, September 1994.

[LSL98] J. Li, P. Stoica and Z. Liu. Comparative study of IQML and MODE direction-of arrival estimators. *IEEE Trans. Signal Process.*, vol.SP-46, pp. 149-160, January 1998.

[LW88] H.B. Lee and M. S. Wengrovitz. Improved high-resolution direction-finding through use of homogeneous constraints. *Proc. IEEE ASSP Workshop on Spectrum Estimation and Modeling*, Minneapolis, Minnesota, vol.1 pp. 152–157, August 1988.

[LW90] H.B. Lee and M. S. Wengrovitz. Resolution threshold of beamspace MUSIC for two closely spaced emitters. *IEEE Trans. Acoust. Speech, Signal Process.*, vol.ASSP-38, pp. 1545–1559, September 1990.

[Mar87] S.L. Marple. *Digital Spectral Analysis.* Prentice-Hall, Englewood Cliffs, New Jersey, 1987.

[Mcl91] J.H. McClellan and D. Lee. Exact equivalence of the Steiglitz-McBride iteration and IQML. *IEEE Trans. Signal Process.*, vol.SP-39, pp. 509–512, February 1991.

[MF90] M. I. Miller and D. R. Fuhrmann. Maximum-likelihood narrow-band direction finding and the EM algorithm. *IEEE Trans. Acoust., Speech, Signal Process.*, vol.ASSP-38, pp. 1560–1577, September 1990.

[MH71] R. J. McAulay and E. M. Hofstetter. Barankin bounds on parameter estimation. *IEEE Trans. Inf. Theory*, vol.IT-17, pp. 669–676, November 1971.

[MKB79] K. V. Mardia, J. T. Kent, and J. M. Bibby. *Multivariate Analysis.* Academic Press, New York, 1979.

[Moo96] T. K. Moon. The expectation maximization algorithm. *IEEE Signal Process. Magazine*, pp. 47–60, November 1996.

[MS69] R. J. McAulay and L. P. Seidman. A useful form of the Barankin lower bound and its application to PPM threshold analysis. *IEEE Trans. Inf. Theory*, vol.IT-15, pp. 273–279, March 1969.

[MSR83] M. R. Matausek, S. S. Stankovic, and D. V. Radovic. Iterative inverse filtering approach to the estimation of frequencies of noisy sinusoids. *IEEE Trans. Acoust., Speech, Signal Process.*, vol.ASSP-31, pp. 1456–1463, December 1983.

[MWW93] Y. Meng, K. M. Wong, and Q. Wu. Estimation of the direction-of-arrival of spread sources in sensor array processing. *Proc. Int. Conf. on Signal Processing.*, Beijing, China, vol.1 pp. 430–434, October 1993.

[NK94] V. Nagesha and S. Kay. On frequency estimation with the IQML algorithm. *IEEE Trans. Signal Processing*, vol.SP-42, pp. 2509–2513, September 1994.

[NK96] V. Nagesha and S. Kay. Maximum likelihood estimation for array processing in colored noise. *IEEE Trans. Signal Processing*, vol.SP-44, pp. 169–180, February 1996.

[NS96] S. G. Nash and A. Sofer. *Linear and Nonlinear Programming.* McGraw-Hill, New York, 1996.

[OL89] B. Ottersten and L. Ljung. Asymptotic results for sensor array processing. *Proc. ICASSP*, Glasgow, Scotland, vol.4 pp. 2266–2269, May 1989.

[ORK89] B. Ottersten, R. Roy, and T. Kailath. Signal waveform estimation in sensor array processing. *Proc. 23rd Asilomar Conf. on Signals, Systems, and Computers*, Pacific Grove, California, pp. 787–791, 1989.

[Ott89] B. Ottersten. *Parametric Subspace Fitting Methods for Array Signal Processing.* PhD Thesis, Stanford University, Stanford, California, 1989.

[OVK92] B. Ottersten, M. Viberg, and T. Kailath. Analysis of subspace fitting and ML techniques for parameter estimation from sensor array data. *IEEE Trans. Signal Process.*, vol.SP-40, pp. 590–600, March 1992.

[OVSN93] B. Ottersten, M. Viberg, P. Stoica, and A. Nehorai. Exact and large sample maximum likelihood techniques for parameter estimation and detection in array processing. *Radar Array Processing*, Springer-Verlag, Berlin, pp. 99–151, 1993.

[Pie63] J. N. Pierce. Approximate error probabilities for optimal diversity combining. *IEEE Trans. Commun. Syst.*, vol.CS-11, pp. 352–354, September 1963.

[PF86] B. Porat and B. Friedlander. Computation of the exact information of Gaussian time series with stationary random components. *IEEE Trans. Acoust., Speech, Signal Process.*, vol.ASSP-34, pp. 118–130, January 1986.

[PF88] B. Porat and B. Friedlander. Analysis of the asymptotic relative efficiency of the MUSIC algorithm. *IEEE Trans. Acoust., Speech, Signal Process.*, vol.ASSP-36, pp. 532–544, 1988.

[PF97] B. Porat and B. Friedlander. Accuracy requirements in off-line array calibration. *IEEE Trans. Aerospace Electron. Syst.*, vol.AES-33, pp. 545–556, April 1997.

[PK86] A. Paulraj and T. Kailath. Eigenstructure methods for direction of arrival estimation in the presence of unknown noise fields. *IEEE Trans. Acoust. Speech, Signal Process.*, vol.ASSP-34, pp. 13–20, February 1986.

[PWMS88] S. Prasad, R. T. Williams, A. K. Mahalanabis, and L. H. Sibul. A transform-based covariance differencing approach for some classes of parameter estimation problems. *IEEE Trans. Acoust. Speech, and Sig. Proc.*, vol.ASSP-36, pp. 631–641, May 1988.

[Rao46] C. R. Rao. *Linear Statistical Inference and Its Applications.* Wiley, New York, 1946.

[RKR94] R. Rajagopal, K. A. Kumar, and P. R. Rao. High-resolution beamforming in the presence of coherent interferences and unknown correlated noise. *Sixth IEEE Digitial Signal Process. Workshop*, pp. 245–248, 1994.

[Ris78] J. Rissanen. Modeling by shortest data description. *Automatica*, vol.14, pp. 465–471, 1978.

[RM91] M.J.D Rendas and J.M.F. Moura. Cramér-Rao bound for location systems in multipath environments. *IEEE Trans. Signal Process.*, vol.SP-39, pp. 2593–2610, December 1991.

[RS87a] Y. Rockah and P. M. Schultheiss. Array shape calibration using sources in unknown locations-part II: Near-field sources and estimator implementation. *IEEE Trans. Acoust., Speech, Signal Process.*, vol.ASSP-35, pp. 724–735, June 1987.

[RS87b] Y. Rockah and P. M. Schultheiss. Array shape calibration using sources in unknown locations-part I: Far-field sources. *IEEE Trans. Acoust., Speech, Signal Process.*, vol.ASSP-35, pp. 286–299, March 1987.

[RW80] A. Ruhe and P. A. Wedin. Algorithms for separable nonlinear least squares problems. *SIAM Rev.*, vol.22, pp. 318–327, July 1980.

[RW92] J. P. Reilly and K. M. Wong. Estimation of directions of arrival of signals in unknown correlated noise, Part II: Asymptotic behavior and performance of the MAP approach. *IEEE Trans. Signal Process.*, vol.SP-40, pp. 2018-2028, August 1992.

[SB87] U. Sandkühler and J. F. Böhme. Accuracy of maximum likelihood estimates for array processing. *Proc. ICASSP*, Dallas, Texas, vol.4 pp. 2015–2018, April 1987.

[Sch68] F. Schweppe. Sensor array data processing for multiple signal sources. *IEEE Trans. Inf. Theory*, vol.IT-4, pp. 294–305, March 1968.

[Sch78] G. Schwartz. Estimating the dimension of a model. *Ann. Stat.*, vol.6, pp. 461–464, 1978.

[Sch79] R. O. Schmidt. Multiple emitter location and signal parameter estimation. *Proc. RADC Spectrum Estimation Workshop*, Griffiths AFB, Rome, New York, pp. 243–258, 1979.

[Sch81] R. O. Schmidt. *Multiple Emitter Location and Signal Parameter Estimation.* Ph.D. Dissertation, Stanford University, Stanford, California, 1981.

[Sch91] L. L. Scharf. *Statistical Signal Processing; Detection, Estimation, and Time Series Analysis.* Wiley, New York, 1991.

[Sch94] S. V. Schell. Performance analysis of the cyclic MUSIC method of direction estimation for cyclostationary signals. *IEEE Trans. Signal Process.*, vol.SP-42, pp. 3043–3050, November 1994.

[SG92] S. V. Schell and W.A. Gardner. Cramér-Rao bound for directions of arrival of Gaussian cyclostationary signals. *IEEE Trans. Signal Process.*, vol.SP-38, pp. 1418–1422, July 1992.

[Sha87] A.K. Shaw. *Structured Matrix Approximation Problems in Signal Processing.* PhD Dissertation, University of Rhode Island, Kingston, 1987.

[Sha95] A.K. Shaw. Maximum likelihood estimation of multiple frequencies with constraints to guarantee unit circle roots. *IEEE Trans. Signal Process.*, vol.SP-43, pp. 796–799, March 1995.

[Sim80] D. N. Simkins. *Multichannel Angle-of-Arrival Estimation.* PhD thesis, Stanford University, Stanford, California, 1980.

[SJ97] P. Stoica and M. Jansson. On forward-backward MODE for array signal processing. *Digital Signal Process.*, vol.7, pp. 239–252, October 1997.

[SL01] P. Stoica and E. G. Larson. Comments on "Linearization method for finding Cramér-Rao bounds in signal processing." *IEEE Trans. Signal Process.*, vol.SP-49, pp. 3168–3169, December 2001.

[Sle78] D. Slepian. Prolate spheroidal wave functions, Fourier analysis, and uncertainty, V: The discrete case. *Bell Syst. Tech. J.*, vol.57, pp. 1371–1430, May-June 1978.

[SLG01] P. Stoica, E. G. Larson, and A. B. Gershman. The stochastic CRB for array processing: A textbook derivation. *IEEE Signal Process. Lett.*, vol.8, pp. 148–150, May 2001.

[SLS97] P. Stoica, J. Li, and T. Soderstrom. On the inconsistency of IQML. *Signal Process.*, vol.56, pp. 185–190, 1997.

[SM65] K. Steiglitz and L. E. McBride. A technique for the identification of linear systems. *IEEE Trans. Autom. Control*, vol.AC-10, pp. 461–465, October 1965.

[SM97] P. Stoica and R. Moses. *Introduction to Spectral Analysis.* Prentice-Hall, Englewood Cliffs, New Jersey, 1997.

[SN89a] P. Stoica and A. Nehorai. MODE, maximum likelihood, and Cramér-Rao bound: Conditional and unconditional results. Report No. 8901, Center for Systems Science, Yale University, New Haven, Connecticut, 1989.

[SN89b] P. Stoica and A. Nehorai. MUSIC, maximum likelihood, and Cramér-Rao bound. *IEEE Trans. Acoust., Speech, Signal Process.*, vol.ASSP-37, pp. 720–741, May 1989.

[SN90a] P. Stoica and A. Nehorai. MUSIC, maximum likelihood, and Cramér-Rao bound: Further results and comparisons. *IEEE Trans. Acoust., Speech, Signal Process.*, vol.ASSP-38, pp. 2140–2150, December 1990.

[SN90b] P. Stoica and A. Nehorai. Performance study of conditional and unconditional direction-of-arrival estimation. *IEEE Trans. Acoust., Speech, Signal Process.*, vol.ASSP-38, pp. 1783–1795, October 1990.

[SN92] D. Starer and A. Nehorai. Newton algorithm for conditional and unconditional maximum likelihood estimation of the parameters of exponential signals in noise. *IEEE Trans. Signal Process.*, vol.SP-40, pp. 1528–1533, June 1992.

[SOV92] P. Stoica, B. Ottersen, and M. Viberg. An instrumental variables approach to array processing in spatially correlated noise fields. *Proc. ICASSP*, San Francisco, California, vol.II pp. 421–424, 1992.

[SOVM96] P. Stoica, B. Ottersen, M. Viberg and R. L. Moses. Maximum likelihood array processing for stochastic coherent sources. *IEEE Trans. Signal Process.*, vol.SP-44, pp. 96–105, January 1996.

[SS90a] P. Stoica and K. Sharman. Maximum likelihood methods for direction-of-arrival estimation. *IEEE Trans. Acoust., Speech, Signal Process.*, vol.ASSP-38, pp. 1132–1143, July 1990.

[SS90b] P. Stoica and K. Sharman. A novel eigenanalysis method for direction estimation. *Proc. IEE*, pp. 19–26, 1990.

[SVO94] P. Stoica, M. Viberg, and B. Ottersten. Instrumental variable approach to array processing in spatially correlated noise fields. *IEEE Trans. Signal Process.*, vol.SP-42, pp. 121–133, 1994.

[SVWW96] P. Stoica, M. Viberg, K. M. Wong, and Q. Wu. Maximum-likelihood bearing estimation with partly calibrated arrays in spatially correlated noise fields. *IEEE Trans. Signal Processing*, vol.SP-44, pp. 888–899, April 1996.

[SVWW98] P. Stoica, M. Viberg, M. Wong and Q. Wu. A unified instrumental variable approach to direction finding in colored noise fields. *Digital Signal Processing Handbook*, V. Madisetti and D. Williams, Eds, pp. 64.1–64.18, CRC Press and IEEE Press, 1998.

[Swe59] P. Swerling. A proposed stagewise differential correction procedure for satellite tracking and prediction. *J. Astronaut. Sci.*, vol.6, 1959.

[Swi93] D. N. Swingler. Simple approximations to the Cramér-Rao lower bound on directions of arrival for closely spaced sources. *IEEE Trans. Signal Process.*, vol.SP-41, pp. 1668–1672, April 1993.

[Tia01] Z. Tian. Beamspace weighted subspace fitting. Submitted to *Signal Process. Lett.*.

[TK82] D. W. Tufts and R. Kumaresan. Singular value decomposition and improved frequency estimation using linear prediction. *IEEE Trans. Acoust., Speech, Signal Process.*, vol.ASSP-30, pp. 671–675, August 1982.

[TVT00] Z. Tian and H. L. Van Trees. Beamspace IQML. *Proc. IEEE Sensor Array and Multichannel Signal Processing Workshop*, Lexington, Massachusetts, March 2000.

[Vib89] M. Viberg. *Subspace Fitting Concepts in Sensor Array Processing.* PhD Thesis, Linköping University, Linköping, Sweden, 1989.

[VO91] M. Viberg and B. Ottersten. Sensor array processing based on subspace fitting. *IEEE Trans. Signal Process.*, vol.SP-39, pp. 1110–1121, May 1991.

[VOK91] M. Viberg, B. Ottersten, and T. Kailath. Detection and estimation in sensor arrays using weighted subspace fitting. *IEEE Trans. Signal Process.*, vol.SP-39, pp. 2436–2449, November 1991.

[VS94] M. Viberg and A. Swindlehurst. Analysis of the combined effects of finite samples and model errors on array processing performance. *IEEE Trans. Signal Process.*, vol.SP-42, pp. 3073–3083, November 1994.

[VT68] H. L. Van Trees. *Detection, Estimation, and Modulation Theory, Part I.* Wiley, New York, 1968.

[VT71] H. L. Van Trees. *Detection, Estimation, and Modulation Theory, Part III.* Wiley, New York, 1971.

[VT01a] H. L. Van Trees. *Detection, Estimation, and Modulation Theory, Part I.* Wiley Interscience, New York, 2001.

[VT01b] H. L. Van Trees. *Detection, Estimation, and Modulation Theory, Part III.* Wiley Interscience, New York, 2001.

[VVW88a] B. Van Veen and B. Williams. Structured covariance matrices and dimensionality reduction in array processing. *Proc. 4th Workshop Spectr. Est. and Modeling*, pp. 168–171, August 1988.

[VVW88b] B. Van Veen and B. Williams. Dimensionality reduction in high resolution direction of arrival estimation. *Proc. Asilomar Conf. on Signals, Systems, and Computers*, Pacific Grove, California, vol.2, pp. 588–592, October 1988.

[Wax85] M. Wax. *Detection and Estimation of Superimposed Signals.* PhD thesis, Stanford University, Stanford, California, 1985.

[Wax91] M. Wax. Detection and localization of multiple sources via the stochastic signals model. *IEEE Trans. Signal Process.*, vol.SP-39, pp. 2450–2456, November 1991.

[Wax92] M. Wax. Detection and localization of multiple sources in noise with unknown covariance. *IEEE Trans. Signal Process.*, vol.SP-40, pp. 245-249, January 1992.

[Wei87] A. J. Weiss. Bounds on time-delay estimation for monochromatic signals. *IEEE Trans. Aerospace Electron. Syst.*, vol.AES-23, pp. 798–808, November 1987.

[WF89] A. J. Weiss and B. Friedlander. Array shape calibration using sources in unknown locations: A maximum likelihood approach. *IEEE Trans. Acoust., Speech, Signal Process.*, vol.ASSP-37, pp. 1958–1966, December 1989.

[WF90] A. J. Weiss and B. Friedlander. On the Cramér-Rao bound for direction finding of correlated signals. Technical report, Signal Processing Technology, July 1990.

[WF91a] A. J. Weiss and B. Friedlander. On the Cramér-Rao bound for direction finding of correlated signals. Technical report, Signal Processing Technology, October 1991.

[WF91b] A. J. Weiss and B. Friedlander. Array shape calibration via eigenstructure methods. *J. Signal Process.*, vol.122, pp. 251–258, 1991.

[WF93] A. J. Weiss and B. Friedlander. On the Cramér-Rao bound for direction finding of correlated signals. *IEEE Trans. Signal Process.*, vol.SP-41, pp. 495–499, January 1993.

[WF94] A. J. Weiss and B. Friedlander. Preprocessing for direction finding with minimal variance degradation. *IEEE Trans. Signal Process.*, vol.SP-42, pp. 1478–1485, June 1994.

[WK85] M. Wax and T. Kailath. Detection of signals by information theoretic criteria. *IEEE Trans. Acoust., Speech, Signal Process.*, vol.ASSP-33, pp. 387–392, April 1985.

[WJ65] J. M. Wozencraft and I. M. Jacobs. *Principles of Communication Engineering.* Wiley, New York, 1965.

[WOV91] B. Wahlberg, B. Ottersten, and M. Viberg. Robust signal parameter estimation in the presence of array perturbations. *Proc. ICASSP*, Toronto, Canada, vol.5 pp. 3277–3280, May 1991.

[WRWQ92] K. M. Wong, J. P. Reilly, Q. Wu, and S. Qiao. Estimation of directions of arrival of signals in unknown correlated noise: Parts I and II. *IEEE Trans. Signal Process.*, vol.SP-40, pp. 2007–2028, August 1992.

[WRWQ92] K. M. Wong, J. P. Reilly, Q. Wu, and S. Qiao. Estimation of directions of arrival of signals in unknown correlated noise, Part I: The MAP appoach and its implementation. *IEEE Trans. Signal Process.*, vol.SP-40, pp. 2007–2017, August 1992.

[WSK82] M. Wax, T. J. Shan, and T. Kailath. Location and spectral density estimation of multiple sources. *Proc. 16th Asilomar Conf. on Signals, Systems, and Computers*, pp. 322–326, 1982.

[Wu83] N.-L. Wu. An explicit solution and data extension in the maximum entropy method. *IEEE Trans. Acoust., Speech, Signal Process*, vol.ASSP-31, pp. 486–491, April 1983.

[WW82] A. J. Weiss and E. Weinstein. Composite bound on the attainable mean-square error in passive time delay estimation from ambiguity prone signals. *IEEE Trans. Inf. Theory*, vol.IT-28, pp. 977–979, November 1982.

[WW83] A. J. Weiss and E. Weinstein. Fundamental limitations in passive time delay estimation, Part I: Narrow-band systems. *IEEE Trans. Acoust., Speech Signal Process.*, vol.ASSP-31, pp. 472–486, April 1983.

[WW84] E. Weinstein and A. J. Weiss. Fundamental limitations in passive time delay estimation, Part II: Wide-band systems. *IEEE Trans. Acoust., Speech Signal Process.*, vol.ASSP-32, pp. 1064–1078, October 1984.

[WW88] E. Weinstein and A. J. Weiss. A general class of lower bounds in parameter estimation. *IEEE Trans. Inf. Theory*, vol.IT-34, pp. 338–342, March 1988.

[WW94] Q. Wu and K. M. Wong. UN-MUSIC and UN-CLE: An application of generalized correlation to the estimation of direcitons of signals in unknown correlated noise. *IEEE Trans. Signal Process.*, vol.SP-42, pp. 2331–2343, September 1994.

[WW95] Q. Wu and K. M. Wong. Estimation of DOA in unknown noise: Performance analysis of UN-MUSIC and UN-CLE, and the optimality of CCD. *IEEE Trans. Signal Process.*, vol.SP-43, pp. 454–467, February 1995.

[WWL88] A. J. Weiss, A. S. Willsky, and B. C. Levy. Eigenstructure approach for array processing with unknown intensity coefficients. *IEEE Trans. Acoust., Speech, Signal Process.*, vol.ASSP-36, pp. 1613–1617, October 1988.

[WZ89] M. Wax and I. Ziskind. Detection of the number of coherent signals by the MDL principle. *IEEE Trans. Acoust., Speech, Signal Process.*, vol.ASSP-37, pp. 1190–1196, August 1989.

[WZRY90] K. M. Wong, Q. T. Zhang, J. P. Reilly, and P. C. Yip. On information theoretic criteria for determining the number of signals in high resolution array processing. *IEEE Trans. Acoust., Speech, Signal Process.*, vol.ASSP-38, pp. 1959–1971, November 1990.

[XB88] X. L. Xu and K. Buckley. Reduced-dimension beamspace broadband source localization: Preprocessor design and evaluation. *Proc. 4th ASSP Workshop Spectrum Estimation and Modeling*, Minneapolis, Minnesota, pp. 22–27, August 1988.

[XB89] X. L. Xu and K. M. Buckley. Statistical performance comparison of MUSIC in element-space and beam-space. *Proc. ICASSP*, Glasgow, Scotland, vol.4 pp. 2124–2127, May 1989.

[XK92] G. Xu and T. Kailath. Direction of arrival estimation via exploitation of cyclostationarity—A combination of temporal and spatial processing. *IEEE Trans. Signal Process.*, vol.SP-40, pp. 1775–1786, July 1992.

[XK95] W. Xu and M. Kaveh. Analysis of the performance and sensitivity of eigendecomposition-based detectors. *IEEE Trans. Signal Process.*, vol.SP-43, pp. 1413–1426, June 1995.

[XRK94] G. Xu, R. H. Roy, and T. Kailath. Detection of number of sources via exploitation of centro-symmetry property. *IEEE Trans. Signal Process.*, vol.SP-42, pp. 102–112, January 1994.

[YB92] S. F. Yau and Y. Bresler. Worst case Cramér-Rao bounds for parametric estimation of superimposed signal with applications. *IEEE Trans. Signal Process.*, vol.SP-40, pp. 2973–2986, December 1992.

[YDG95] H. Ye and R. D. DeGroat. Maximum likelihood DOA estimation and asymptotic Cramér-Rao bounds for additive unknown colored noise. *IEEE Trans. Signal Process.*, vol.SP-43, pp. 938–949, April 1995.

[YK87] Y. Q. Yin and P. R. Krishnaiah. On some nonparametric methods for detection of the number of signals. *IEEE Trans. Acoust., Speech, Signal Process.*, vol.ASSP-35, pp. 1533–1538, November 1987.

[YS95] J. Yang and A. L. Swindlehurst. The effects of array calibration errors on DF-based signal copy performance. *IEEE Trans. Signal Processing*, vol.SP-43, pp. 2724–2732, November 1995.

[Zha94] Q. T. Zhang. Asymptotic performance analysis of information-theoretic criteria for determining the number of signals in spatially correlated noise. *IEEE Trans. Signal Process.*, vol.SP-42, pp. 1537–1539, June 1994.

[ZKB86] L. C. Zhao, P. R. Krishnaiah, and Z. D. Bai. On detection of the number of signals in the presence of white noise. *J. Multi-Variate Anal.*, vol.20, pp. 1–25, 1986.

[ZKB87] L. C. Zhao, P. R. Krishnaiah, and Z. D. Bai. Remarks on certain criteria for detection of number of signals. *IEEE Trans. Acoust., Speech, Signal Process.*, vol.ASSP-35, pp. 129–132, February 1987.

[ZKS93] M. D. Zoltowski, G.M. Kautz, and S.D. Silverstein. Beamspace root-MUSIC. *IEEE Trans. Signal Process.*, vol.SP-41, pp. 344–364, February 1993.

[ZL91] M. D. Zoltowski, T.S. Lee. Maximum likelihood based sensor array signal processing in the beamspace domain for low angle radar tracking. *IEEE Trans Signal Process.*, vol.SP-39, pp. 656–671, March 1991.

[Zol88] M. D. Zoltowski. High resolution sensor array signal processing in the beamspace domain: Novel techniques based on the poor resolution of Fourier beamforming. *Proc. ASSP*, pp. 350–355, 1988.

[ZW88a] J. X. Zhu and H. Wang. Effects of sensor position and pattern perturbations on CRLB for direction finding of multiple narrow-band sources. *Proc. 4th ASSP Workshop on Spectrum Estimation and Modeling*, Minneapolis, Minnesota, pp. 98–102, August 1988.

[ZW88b] I. Ziskind and M. Wax. Maximum likelihood localization of multiple sources by alternating projection. *IEEE Trans. Acoust., Speech, Signal Process.*, vol.ASSP-36, pp. 1553–1560, October 1988.

[ZZ69] J. Ziv and M. Zakai. Some lower bounds on signal parameter estimation. *IEEE Trans. Inf. Theory*, vol.IT-15, pp. 386–391, May 1969.

Chapter 9

Parameter Estimation II

9.1 Introduction

In Chapter 8, we focused on maximum likelihood estimation and the bounds on achievable performance. We found that the stochastic maximum likelihood estimator achieved the Cramér-Rao bound but it required a significant amount of computation to find the estimate. We introduced several other estimators. Some of the estimators provided good performance but, in general, still required a significant amount of computation. In this chapter we focus our attention on estimators that are computationally simpler, but may provide acceptable performance in many problems of interest.

In Section 9.2, we discuss several estimators that we refer to as quadratic estimators because they compute a quadratic form as a function of ψ and choose the D largest peaks as $\hat{\psi}$.

In Section 9.3, we develop a family of subspace algorithms. The most prominent of these algorithms is the MUSIC (**Mu**ltiple **Si**gnal **C**lassification) algorithm invented by Schmidt [Sch79] and Bienvenu and Kopp [BK80]. It is widely used because of its performance capability, and various versions and modifications of the MUSIC algorithm have been extensively studied.

In Section 9.4, we discuss linear prediction (LP) algorithms briefly. These algorithms grew out of the work by Burg [Bur67] on maximum entropy estimators and modified forward-backward linear prediction estimators due to Tufts and Kumaresan [TK82]. We do not develop this area in detail.

In Sections 9.2 and 9.3, we utilize simulation to study the performance of the various estimators. In Section 9.5, we consider the asymptotic regime and derive expressions for the probability of resolution, the bias, and the variance. We then combine these results to obtain an analytic expression for the mean-square error.

Many of the algorithms developed in this chapter suffer significant performance degradation (or fail to work) when the incoming plane waves are correlated or coherent. In Section 9.6, we study one technique, spatial smoothing (SS), for solving this problem.

In Section 9.7, we develop beamspace algorithms. In these algorithms we preprocess the incoming signal with a matrix that, in essence, creates a set of beams. We then utilize these beam outputs to find the DOAs. We show that, in many environments, we can obtain estimators with equivalent variance and an improved probability of resolution with significantly less computational complexity.

In Section 9.8, we study the sensitivity of the various algorithms to perturbations in the array or the noise environment.

In Section 9.9, we consider planar arrays. Most of the algorithms developed in this chapter are applicable to arbitrary array geometries. However, some of the particularly efficient techniques, such as algorithms that find the roots of a polynomial, do not extend easily.

In Section 9.10, we summarize the results of the chapter and indicate several topics that we have omitted. In Table 9.1, we show the structure of the chapter.

9.2 Quadratic Algorithms

9.2.1 Introduction

In this section we discuss several algorithms that we refer to as quadratic algorithms because each of them computes one or more quadratic forms. We should note that several of the algorithms that we develop in Sections 9.3 and 9.4 also compute quadratic forms. However, the logic leading to their structure is different. One should view the quadratic algorithm term as a convenient descriptor rather than a precise definition.

The basic estimators of interest are listed in Table 9.2. In each algorithm, we vary $\mathbf{v}(\psi)$ over the *a priori* range of the sources in ψ-space, plot the value of the indicated function, and select the D largest values as the estimates of $\psi_1, \psi_2, \cdots, \psi_D$. We assume that D is known or has been estimated. All of these algorithms are valid for an arbitrary array geometry. For 2-D estimation, the scalar ψ is replaced by the vector $\boldsymbol{\psi}$, whose components are ψ_x and ψ_y.

Table 9.1 Structure of Chapter 9

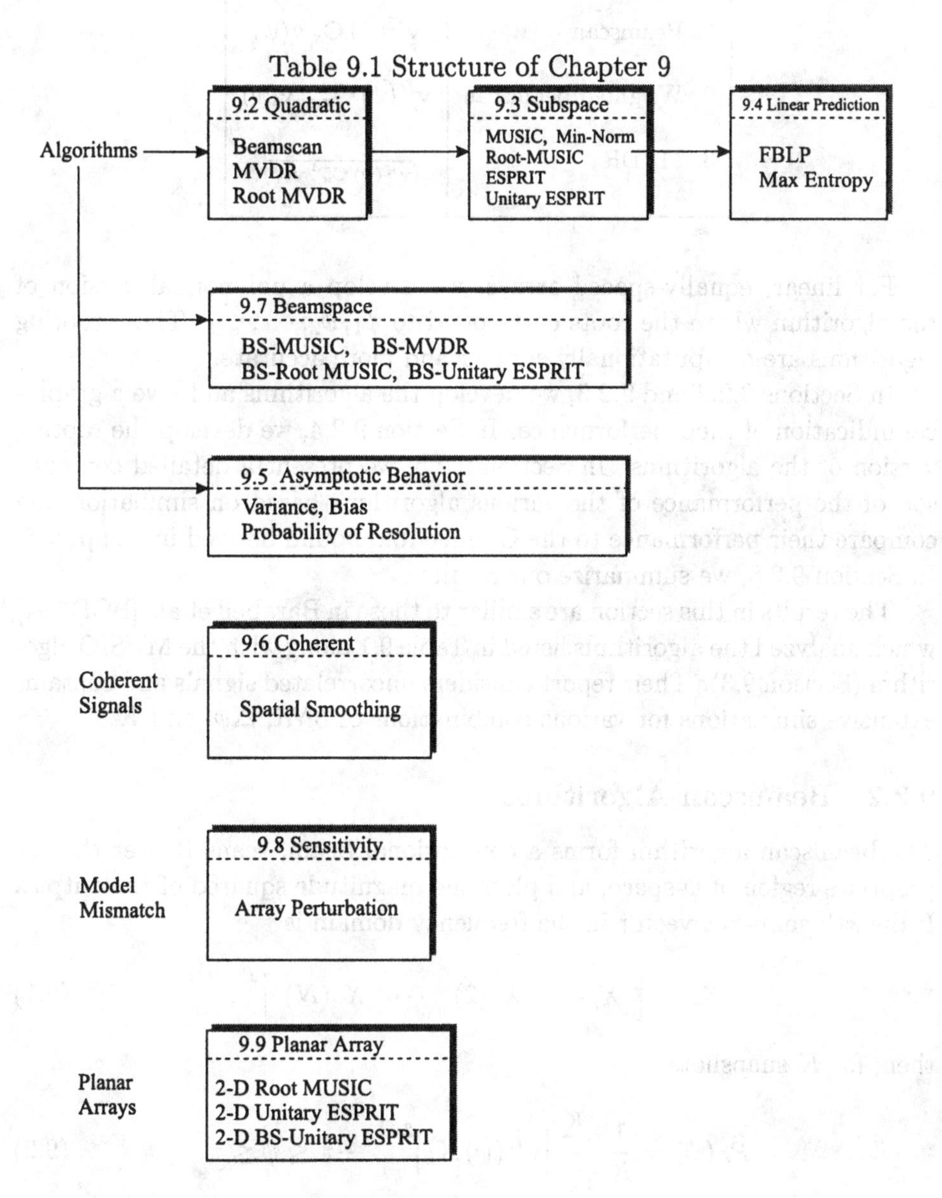

Table 9.2: Quadratic Algorithms

1. Beamscan	$\mathbf{v}^H(\psi)\,\mathbf{C}_{\mathbf{x}}\,\mathbf{v}(\psi)$
2. Weighted beamscan	$\mathbf{v}^H(\psi)\,\mathbf{C}_{\mathbf{x},w}\,\mathbf{v}(\psi)$
3. MVDR	$\frac{1}{\mathbf{v}^H(\psi)\,\mathbf{C}_{\mathbf{x}}^{-1}\,\mathbf{v}(\psi)}$

For linear, equally-spaced arrays, we develop a polynomial version of the algorithm where the roots correspond to $\hat{\psi}_1, \hat{\psi}_2, \cdots, \hat{\psi}_D$. These rooting algorithms are computationally simpler and more accurate.

In Sections 9.2.2 and 9.2.3, we develop the algorithms and give a graphical indication of their performance. In Section 9.2.4, we develop the rooting version of the algorithms. In Section 9.2.5, we present a detailed comparison of the performance of the various algorithms based on simulation and compare their performance to the Cramér-Rao bound derived in Chapter 8. In Section 9.2.6, we summarize our results.

The results in this section are similar to those in Barabell et al. [BCD+84],[1] which analyzed the algorithms listed in Table 9.1 along with the MUSIC algorithm (Section 9.3). Their report considers uncorrelated signals and contains extensive simulations for various combinations of *SNR*, $\Delta\psi$, and K.

9.2.2 Beamscan Algorithms

The beamscan algorithm forms a conventional beam, scans it over the appropriate region of ψ-space, and plots the magnitude squared of the output. If the kth snapshot vector in the frequency domain is

$$\mathbf{X}_k = \left[\ X_k(1) \ \vdots\ X_k(2) \ \vdots\ \cdots \ \vdots\ X_k(N)\ \right]^T, \tag{9.1}$$

then, for K snapshots,

$$\widehat{P}_B(\psi) = \frac{1}{K}\sum_{k=1}^{K}\left|\mathbf{v}^H(\psi)\,\mathbf{X}_k\right|^2, \quad -\pi \le \psi \le \pi. \tag{9.2}$$

We have omitted the $\frac{1}{N}$ scaling factor in the weight vector of the conventional beam pattern to be consistent with the majority of the literature. This

[1] This reference is a Lincoln Laboratory report that has a distribution restriction but appears to be well-known in the community.

estimator is referred to as the Bartlett beamformer. The expression in (9.2) can be rewritten as,

$$\widehat{P}_B(\psi) = \mathbf{v}^H(\psi) \left\{ \frac{1}{K} \sum_{k=1}^{K} \mathbf{X}_k \mathbf{X}_k^H \right\} \mathbf{v}(\psi). \tag{9.3}$$

The term in the braces is the sample spectral matrix. Thus (9.3) can be rewritten as

$$\widehat{P}_B(\psi) = \mathbf{v}^H(\psi)\, \mathbf{C}_\mathbf{x}\, \mathbf{v}(\psi), \tag{9.4}$$

where $\mathbf{C}_\mathbf{x}$ is the sample spectral matrix. Notice that $\widehat{P}_B(\psi)$ is providing an estimate of the spatial spectrum. Its performance has been studied in detail in the context of both temporal and spatial spectrum estimation (e.g., [Kay88], or [Mar87]). If the array is a standard linear array, we would compute $\widehat{P}_B(\psi)$ by using an FFT in (9.2).

In our application, we select the D peaks of $\widehat{P}_B(\psi)$ and designate them as $\hat{\psi}_i, i = 1, 2, \cdots, D$. We assume that the number of signals, D, is known. Zhang [Zha95b] has studied the performance of the Bartlett beamformer.

The algorithm is scanning a conventional beam so the resolution cannot exceed the classical resolution we developed in Chapter 2. For an arbitrary array, if the *ANSR* and K are large enough, we will be able to resolve targets separated by half of the BW_{NN} of the array. For a standard N-element linear array, this will correspond to sources whose separation is greater than or equal to $2/N$.

There are various modification to the Bartlett estimator that are appropriate for various environments.

If we anticipate that there are weak targets in the sidelobe region of the conventional beam, we can use one of the weightings developed in Chapter 3 to reduce the sidelobes at the expense of broadening the main lobe and decreasing resolution. The resulting estimator is

$$\widehat{P}_{BW}(\psi) = \frac{1}{K} \sum_{k=a}^{K} \left| \mathbf{v}^H(\psi)\, \mathbf{w}\, \mathbf{X}_k \right|^2, \quad -\pi \leq \psi \leq \pi. \tag{9.5}$$

where

$$\mathbf{w} = \operatorname{diag} \left[\; w(1) \; \vdots \; w(2) \; \vdots \; \cdots \; \vdots \; w(N) \; \right]. \tag{9.6}$$

Then,

$$\widehat{P}_{BW}(\psi) = \mathbf{v}^H(\psi) \mathbf{C}_{\mathbf{x},w} \mathbf{v}(\psi), \quad -\pi \leq \psi \leq \pi, \tag{9.7}$$

where

$$\mathbf{C}_{\mathbf{x},w} = \mathbf{w}\,\mathbf{C}_{\mathbf{x}}\,\mathbf{w}^H. \tag{9.8}$$

For a conjugate symmetric array, we can improve the performance for a modest number of snapshots by using FB averaging, as discussed in Section 7.2.3. In this case the estimator is

$$\widehat{P}_{B,fb}(\psi) = \frac{1}{2K}\sum_{k=1}^{K}\left|\mathbf{v}^H(\psi)\,[\mathbf{X}_k + \mathbf{J}\,\mathbf{X}_k^*]\right|^2, \quad -\pi \le \psi \le \pi, \tag{9.9}$$

where $\mathbf{J}$ is the exchange matrix. The expression for $\widehat{P}_{B,fb}(\psi)$ can be rewritten as

$$\widehat{P}_{B,fb}(\psi) = \mathbf{v}^H(\psi)\mathbf{C}_{\mathbf{x},fb}\mathbf{v}(\psi), \tag{9.10}$$

where

$$\mathbf{C}_{\mathbf{x},fb} = \frac{1}{2K}\sum_{k=1}^{K}\left(\mathbf{X}_k\,\mathbf{X}_k^H + \mathbf{J}\,\mathbf{X}_k^*\,\mathbf{X}_k^T\,\mathbf{J}\right). \tag{9.11}$$

The behavior of these estimators is discussed in Problems 9.2.1 through 9.2.3.

9.2.3 MVDR (Capon) Algorithm

The MVDR algorithm was discussed in the context of waveform estimation in Section 6.2.1. To use these results in the parameter estimation context, we use the weighting in (6.71) with $\mathbf{C}_{\mathbf{x}}$, the sample spectral matrix, replacing $\mathbf{S}_{\mathbf{x}}$. Then,[2]

$$\begin{aligned}\widehat{P}_{mvdr}(\psi) &= \frac{1}{K}\sum_{k=1}^{K}\left|\mathbf{w}_{mvdr}^H\,\mathbf{X}_k\right|^2, & -\pi \le \psi \le \pi \\ &= \frac{1}{K}\sum_{k=1}^{K}\left|\frac{\mathbf{v}^H(\psi)\,\mathbf{C}_{\mathbf{x}}^{-1}}{\mathbf{v}^H(\psi)\,\mathbf{C}_{\mathbf{x}}^{-1}\,\mathbf{v}(\psi)}\mathbf{X}_k\right|^2, & -\pi \le \psi \le \pi.\end{aligned} \tag{9.12}$$

This reduces to

$$\widehat{P}_{mvdr}(\psi) = \frac{1}{\mathbf{v}^H(\psi)\,\mathbf{C}_{\mathbf{x}}^{-1}\,\mathbf{v}(\psi)}, \quad -\pi \le \psi \le \pi. \tag{9.13}$$

[2] In the notation of Chapter 6, we should call this the MPDR algorithm. The MVDR terminology is so widely used in the parameter estimation literature that we will use it.

This estimator is due to Capon [Cap69], and its behavior was discussed in by Lacoss [Lac71]. We plot $\widehat{P}_{mvdr}(\psi)$ and find its D peaks. These values correspond to the D wavenumber estimates: $\hat{\psi}_1, \hat{\psi}_2, \cdots, \hat{\psi}_D$.

As in the case of the Bartlett beamformer, the MVDR algorithm also provides a spatial spectral estimate that does not rely on any underlying signal model. We focus our attention on its performance as a DOA estimator.

Frequently we plot the inverse of $\widehat{P}_{mvdr}(\psi)$ and find the D minima of function. We refer to the inverse as the null spectrum,

$$\widehat{Q}_{mvdr}(\psi) = \mathbf{v}^H(\psi)\,\mathbf{C_x}^{-1}\,\mathbf{v}(\psi), \quad -\pi \le \psi \le \pi. \tag{9.14}$$

If $\mathbf{v}(\psi)$ is conjugate symmetric, we can improve performance and reduce computation by FB averaging of the data. Then (9.13) becomes

$$\widehat{P}_{mvdr,fb}(\psi) = \frac{1}{\mathbf{v}^H(\psi)\,\mathbf{C}_{\mathbf{x},fb}^{-1}\,\mathbf{v}(\psi)}, \quad -\pi \le \psi \le \pi, \tag{9.15}$$

where $\mathbf{C}_{\mathbf{x},FB}$ is given by (9.11). The null spectrum in (9.14) becomes,

$$\widehat{Q}_{mvdr,fb}(\psi) = \mathbf{v}^H(\psi)\,\mathbf{C}_{\mathbf{x},fb}^{-1}\,\mathbf{v}(\psi), \quad -\pi \le \psi \le \pi. \tag{9.16}$$

We consider a simple example to show its behavior for a large number of snapshots.

Example 9.2.1

We consider a standard 10-element linear array. There are two sources located at $u = \pm 0.0433$. This corresponds to a separation of $\Delta\psi_R$. The sources are uncorrelated and their strength is 20 dB above the white sensor noise. We process 1,000 snapshots.

The results for a typical trial are shown in Figure 9.1(a) for the Bartlett algorithm (9.2) and the MVDR algorithm.[3] We see that the MVDR algorithm exhibits two distinct peaks at the location of the two sources. It is important to point out that that case corresponds to high *SNR* and a large number of snapshots. It does indicate that the MVDR algorithm appears to have a resolution capability in this environment. Although there are two peaks in the spectrum, the dip between the peaks is very small, so we would anticipate that the signals might not be resolved on other trials. This behavior is due to the close spacing between the signals. In Figure 9.1(b), we show the results from a typical trial when the spacing is doubled so that $\Delta u = 0.1732$. Now the dip between the peaks is significant.

These plots provide an indication of behavior, but they do not tell how well the MVDR algorithm works in terms of estimation accuracy or its behavior for lower *SNR* or fewer snapshots. Later we investigate its behavior in terms of estimation accuracy and resolvability as a function of source *SNR*, separation, and the number of snapshots (K).

We return to the MVDR algorithm in Section 9.2.4 and study its behavior as a function of *SNR*, source separation, and number of snapshots. We now look at a root version of the MVDR algorithm that is applicable to standard linear arrays.

[3]Gabriel [Gab80] has similar plots.

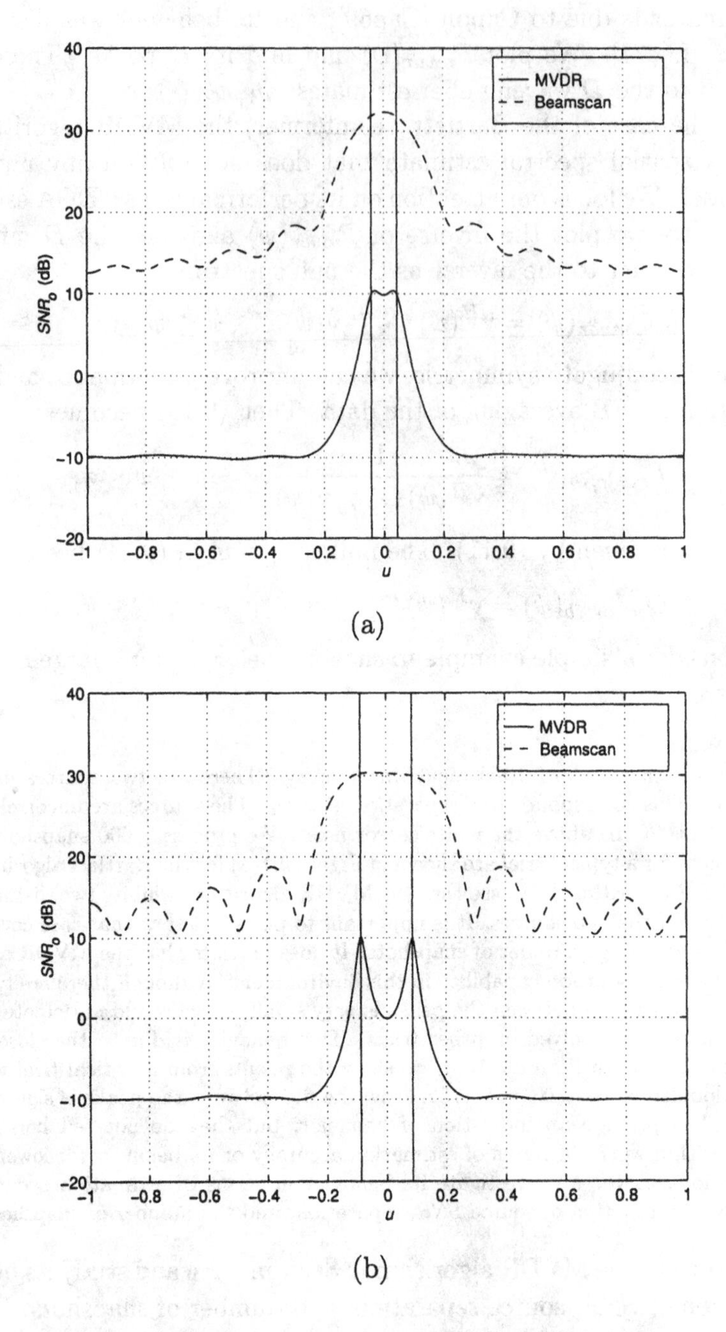

Figure 9.1 Performance of beamscan and MVDR algorithms: standard 10-element linear array, $\rho = 0$, $SNR = 20$ dB, $K = 1,000$: (a) $u = \pm 0.0433$; (b) $u = \pm 0.0866$.

9.2.4 Root Versions of Quadratic Algorithms

In this section we consider a standard N-element linear array. For this case we can derive an alternative version of the quadratic algorithms that is computationally simpler and more accurate.[4]

The null spectrum of the MVDR algorithm can be written as a quadratic form,

$$\widehat{Q}(\psi) = \mathbf{v}^H(\psi)\,\mathbf{G}\,\mathbf{v}(\psi), \tag{9.17}$$

where $\mathbf{G}$ is a Hermitian matrix. Finding the peaks of $P(\psi)$ is equivalent to finding the minimum (or nulls) of $\widehat{Q}(\psi)$. We want to find a polynomial whose roots correspond to the minimum of $\widehat{Q}(\psi)$.

We recall that we can write

$$\mathbf{v}(\psi) = \begin{bmatrix} 1 & e^{j\psi} & \cdots & e^{j(N-1)\psi} \end{bmatrix}^T, \tag{9.18}$$

where we have put the first element at the origin to simplify notation. Letting

$$z = e^{j\psi}, \tag{9.19}$$

we can write (9.17) as

$$\widehat{Q}_z(z) = \sum_{m=0}^{N-1}\sum_{n=0}^{N-1} z^{-m} g_{mn} z^n = \sum_{m=0}^{N-1}\sum_{n=0}^{N-1} g_{mn} z^{n-m}, \tag{9.20}$$

where g_{mn} is the mnth element of $\mathbf{G}$. Now define $k = n - m$ and write $\widehat{Q}_z(z)$ as

$$\widehat{Q}_z(z) = \sum_{k=-(N-1)}^{N-1} q_k z^k, \tag{9.21}$$

where

$$q_k = \begin{cases} \sum_{n=0}^{k+N-1} g_{n-k,n}, & k = -(N-1), \cdots, -1, 0, \\ \sum_{n=k}^{N-1} g_{n-k,n}, & k = 1, 2, \cdots, N-1. \end{cases} \tag{9.22}$$

The Hermitian property of $\mathbf{G}$ implies that the coefficients of $\widehat{Q}_z(z)$ are conjugate symmetric,

$$q_{-k} = q_k^*. \tag{9.23}$$

---

[4]The use of rooting algorithm is due to Barabell [Bar83] (see also [BCD+84])

The null spectrum can also be written as,

$$\widehat{Q}_z(z) = \sum_{k=-(N-1)}^{N-1} q_k^* \, z^{-k}. \tag{9.24}$$

and

$$\widehat{Q}(\psi) = \widehat{Q}_z(z)\Big|_{z=e^{j\psi}} \tag{9.25}$$

The nulls of $\widehat{Q}(\psi)$ are due to the roots of $\widehat{Q}_z(z)$ that lie near the unit circle. Due to the conjugate symmetry, if z_d is a root of $\widehat{Q}_z(z)$, then so is $1/z_d^*$.

$\widehat{Q}_z(z)$ can be factored,

$$\widehat{Q}_z(z) = H(z)\, H^*\left(\frac{1}{z^*}\right), \tag{9.26}$$

where $H(z)$ is the FIR filter that would generate a spectrum $\widehat{Q}_z(z)$.

Half of the roots of $\widehat{Q}_z(z)$ will lie inside the unit circle. The rooting algorithm constructs $\widehat{Q}_z(z)$ using the appropriate $\mathbf{G}$ ($\mathbf{C}_{\mathbf{x}}^{-1}$ for MVDR), and computes the D roots inside and closest to the unit circle. Then,

$$\hat{\psi}_d = \frac{arg\, \hat{z}_d}{\pi}, \quad d = 1, 2, \cdots, D. \tag{9.27}$$

The estimate of the root $\hat{z}_d$ has a radial and angular error. Only the angular error affects $\hat{\psi}_d$. In many cases of interest, this angular component is smaller than the estimate of null location obtained from the null spectrum. We analyze this behavior in detail in Section 9.5.

In Section 9.3, we encounter algorithms in which

$$\mathbf{G} = \hat{\mathbf{U}}_N \hat{\mathbf{U}}_N^H. \tag{9.28}$$

In that case we take advantage of this decomposition and construct $\widehat{Q}_z(z)$ in a different manner.

9.2.5 Performance of MVDR Algorithms

In this section we investigate the behavior of the MVDR and root MVDR algorithms as a function of the array *SNR* and the number of snapshots. There are a number of papers in the literature that analyze the performance of MVDR. The paper by Vaidyanathan and Buckley [VB95] contains several results of interest and summarizes some of the previous work. They derive an expression for the asymptotic ($K \rightarrow \infty$) bias, the additional bias due to finite data, and the standard deviation. They also studied the effects of array perturbations. The reader is referred to this reference for a discussion of these analytic results.

We restrict our discussion at this point to two uncorrelated signals impinging on the array. The array is a standard 10-element linear array, so we use $\mathbf{C}_{\mathbf{x},fb}$.

Example 9.2.2

We use a standard 10-element linear array and assume there are two uncorrelated signals impinging on the array from $\pm\Delta\psi/2$. We assume $\Delta\psi = \Delta\psi_R$. We simulate the performance of the algorithms for various *SNR* and number of snapshots. We measure the probability that the algorithm can resolve the two signals in the following manner. If the algorithm produces two distinct solutions inside the main lobe and $\left|\widehat{\psi}_i - \psi_i\right| \leq \Delta\psi/2$, we say that the signals are resolved. In this example, we use the sample covariance matrix $\mathbf{C}_{\mathbf{x},fb}$. We consider the spectral MVDR and root MVDR algorithms.

In Figure 9.2 we plot the probability of resolution versus *ASNR*. In Figure 9.3, we plot the normalized RMSE versus *ASNR*. In Figure 9.4, we plot the normalized RMSE versus K, the number of snapshots.

The threshold of root MVDR occurs at an *ASNR* of about 15 dB for $K = 100$. Above threshold, the RMSE is very close to the CRB. The threshold of spectral MVDR is harder to specify. The dramatic increase in RMSE occurs at about 22 dB, but the RMSE does not approach the CRB until the *ASNR* = 40 dB. In Figure 9.4, we see that root MVDR is very close to CRB as $K \rightarrow \infty$ but spectral MVDR is not.[5] An analysis of this behavior if contained in [Lac71].

In Figures 9.5, 9.6, and 9.7, we show the behavior of the two algorithms as a function of signal separation for various values of *ASNR*. In each case, there is a critical value of Δu for the root MVDR and spectral MVDR algorithms (which are different). If Δu is larger than the critical value, then the MVDR algorithms are very close to the CRB.

9.2.6 Summary

In this section, we have observed that root MVDR performs reasonably well above threshold, but the threshold is significantly higher than the ML algorithms for closely spaced signals. Spectral MVDR has an even higher threshold.

In the problems, we develop two other quadratic algorithms, AAR and TNA, which exhibit somewhat better performance than MVDR but do not appear to be used in practice. We also discuss diagonal loading in the problems and see that it improves the performance in the low sample support case.

[5] The result in Figure 9.4 appears to exhibit different behavior than the example in [VB95]. The reason is that their signal separation is about $2\Delta_R$ instead of Δ_R.

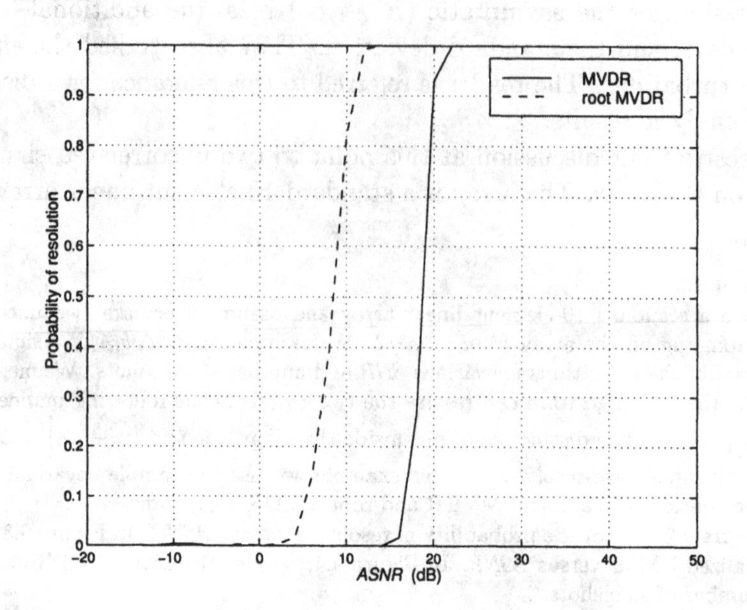

Figure 9.2 MVDR-FB algorithms: probability of resolution versus *ASNR*: $N = 10$, $K = 100$, $\Delta\psi = \Delta\psi_R$, 300 trials.

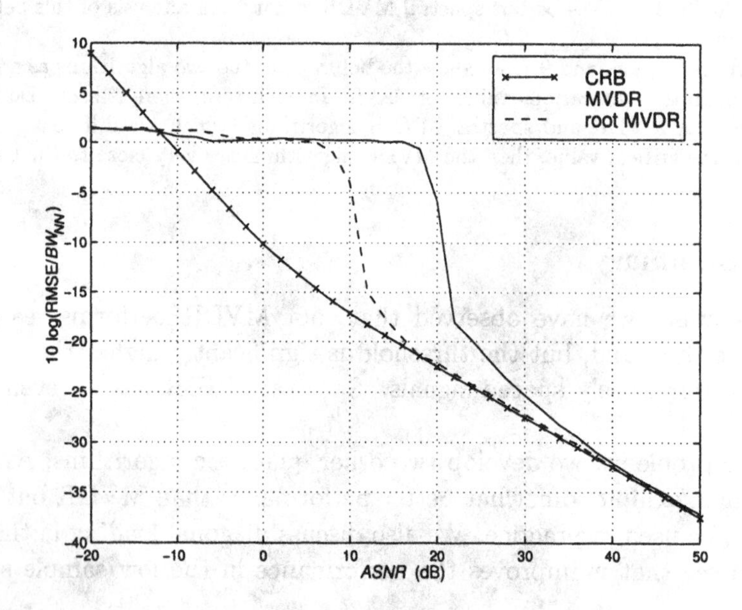

Figure 9.3 MVDR-FB algorithms: normalized RMSE versus *ASNR*: $N = 10$, $K = 100$, $\Delta\psi = \Delta\psi_R$, 300 trials.

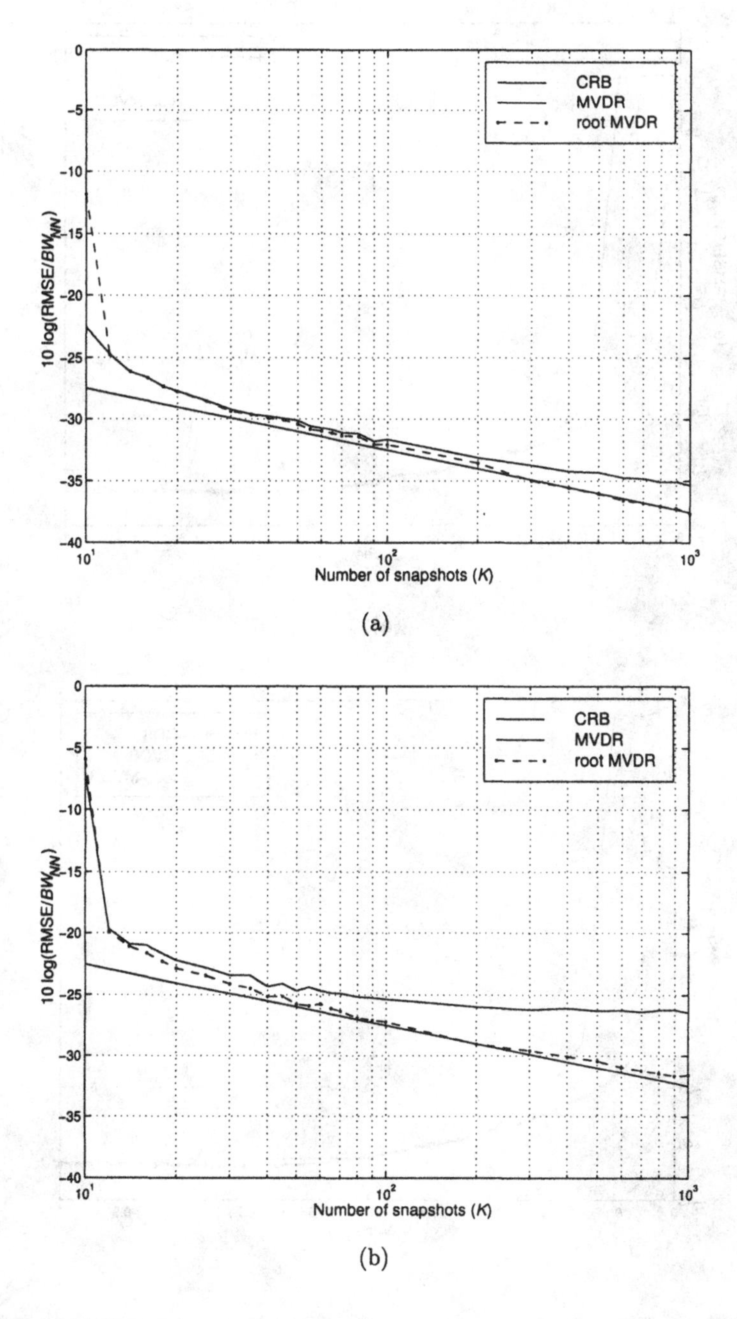

Figure 9.4 MVDR-FB algorithms: normalized RMSE versus K: $N = 10, K = 100, \Delta\psi = \Delta\psi_R$, 200 trials: (a) *ANSR* = 40 dB; (b) *ANSR* = 30 dB.

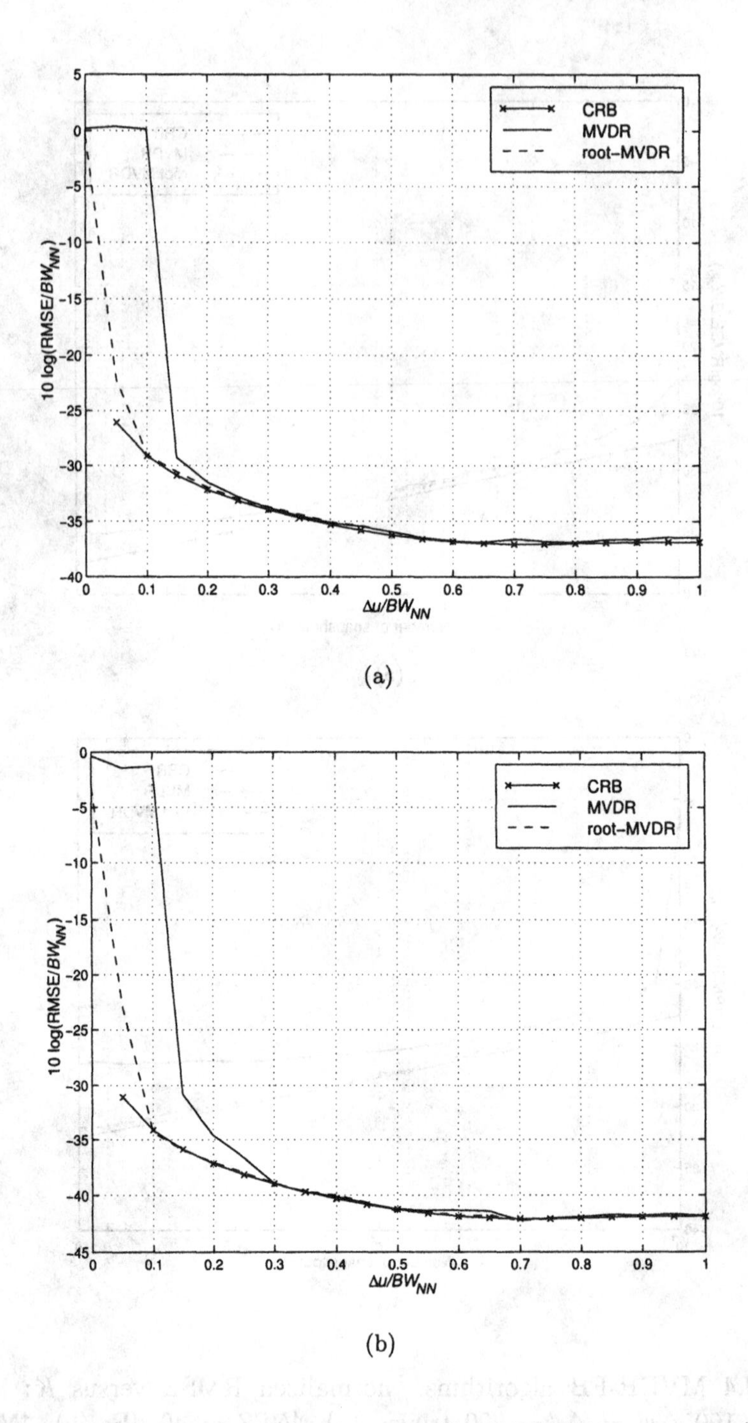

Figure 9.5 MVDR-FB algorithms: normalized RMSE versus $\Delta u/BW_{NN}$: $N = 10$, $ASNR = 40$ dB, 300 trials: (a) $K = 100$; (b) $K = 1,000$.

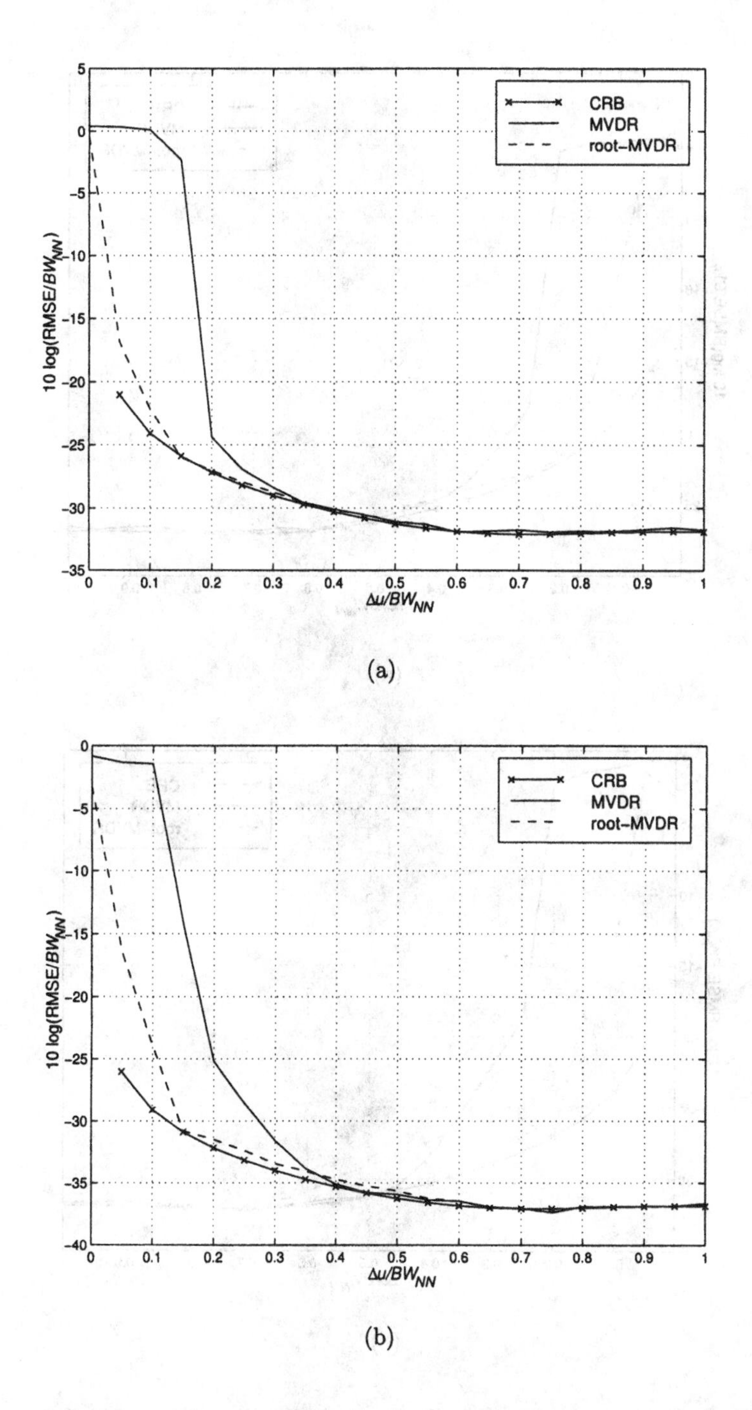

Figure 9.6 MVDR-FB algorithms: normalized RMSE versus $\Delta u/BW_{NN}$: $N = 10$, $ASNR = 30$ dB, 300 trials: (a) $K = 100$; (b) $K = 1,000$.

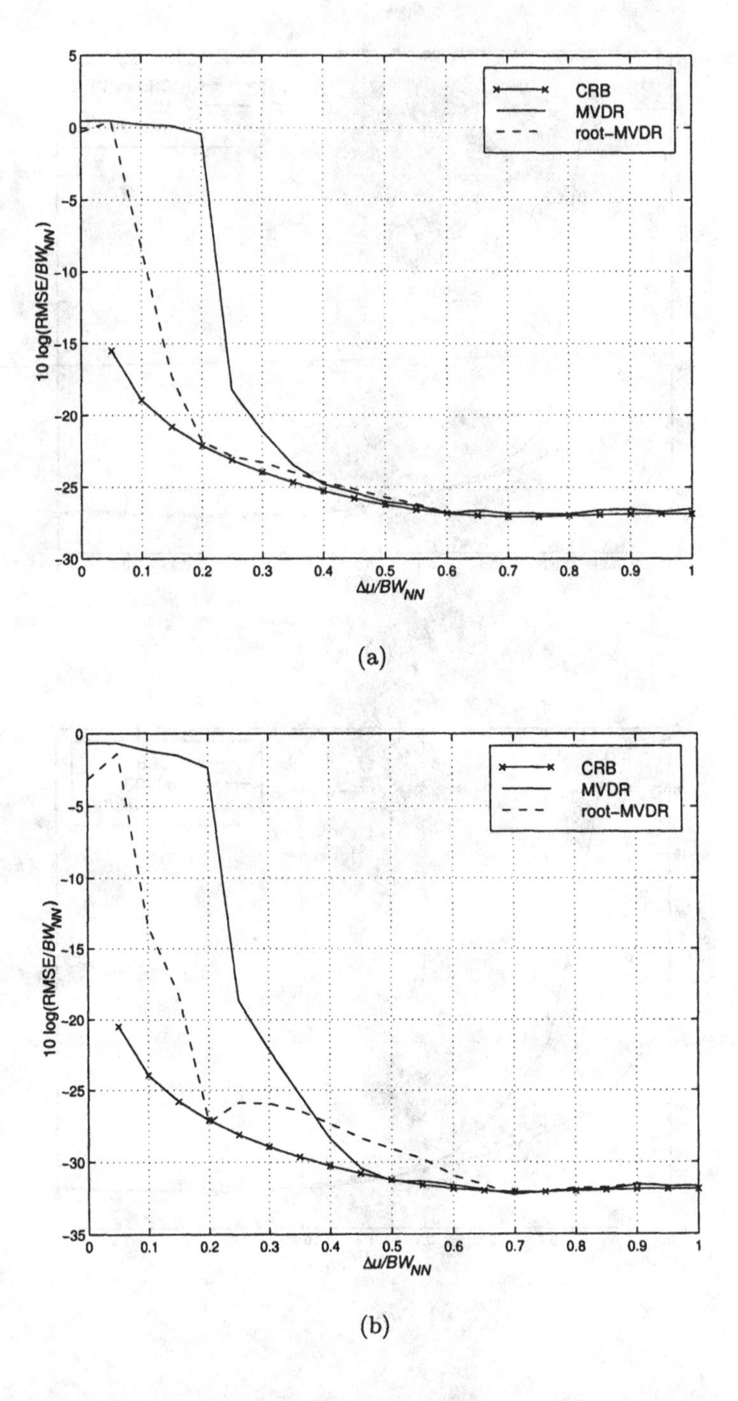

Figure 9.7 MVDR-FB algorithms: normalized RMSE versus $\Delta u/BW_{NN}$: $N = 10$, $ASNR = 20$ dB, 300 trials: (a) $K = 100$; (b) $K = 1,000$.

Although neither the Bartlett beamformer nor spectral MVDR are efficient parameter estimators, they can play a useful role as a preliminary processor to indicate the number of plane waves impinging on the array, their approximate location, and their approximate signal power. If two plane waves are closely spaced, the algorithms will think they are single plane waves and underestimate the number of signals. The more sophisticated algorithms such as AIC or MDL can correct the underestimation. The more sophisticated parameter estimation algorithms that we develop can improve the estimation accuracy.

In the next section we explore algorithms that exploit the orthogonality between the signal and noise subspaces.

9.3 Subspace Algorithms

9.3.1 Introduction

In our discussion of optimum beamformers in Chapter 6, we saw that whenever the received waveform consisted of D plane-wave signals plus uncorrelated noise, we could immediately reduce the problem from an N-dimensional problem to a D-dimensional problem. This reduction was accomplished by generating D sufficient statistics that defined a signal subspace. Once the signal subspace was created, all further processing was done in that subspace. We were able to generate the signal subspace exactly because we knew the angles of arrival of the D signals. A similar result held for correlated noise, assuming that we knew the spectral matrix of the noise.

In the parameter estimation problem, the D angles of arrival are unknown. We want to utilize the received data to estimate the signal subspace and utilize this information to estimate the angles of arrival.

We can utilize either the narrowband snapshot model in either the frequency domain or the time domain. In the frequency-domain snapshot model,

$$\mathbf{S_x} = \mathbf{V}(\boldsymbol{\psi})\, \mathbf{S_f}\, \mathbf{V}^H(\boldsymbol{\psi}) + \sigma_w^2\, \mathbf{I}, \tag{9.29}$$

where

$$\mathbf{V}(\boldsymbol{\psi}) = \left[\begin{array}{c:c:c:c} \mathbf{v}(\psi_1) & \mathbf{v}(\psi_2) & \cdots & \mathbf{v}(\psi_D) \end{array} \right]. \tag{9.30}$$

We assume that D is known. We want to estimate $\psi_1, \psi_2, \cdots, \psi_D$.

The spectral matrix can be written in terms of its eigenvalues and eigenvectors as

$$\mathbf{S_x} = \sum_{i=1}^{N} \lambda_i \, \mathbf{\Phi}_i \, \mathbf{\Phi}_i^H, \tag{9.31}$$

or

$$\mathbf{S_x} = \mathbf{\Phi}_i \, \mathbf{\Lambda} \, \mathbf{\Phi}_i^H, \tag{9.32}$$

where $\mathbf{\Lambda} = \text{diag}\,[\lambda_1, \lambda_2, \cdots, \lambda_N]$.

We assume the eigenvalues are in order of decreasing size. Since there are D signals,

$$\lambda_1 \geq \lambda_2 \geq \cdots \geq \lambda_D > \lambda_{D+1} = \cdots = \sigma_w^2. \tag{9.33}$$

We refer to the first D eigenvalues,

$$\lambda_i, \quad i = 1, 2, \cdots, D, \tag{9.34}$$

as the signal-subspace eigenvalues and

$$\mathbf{\Phi}_i, \quad i = 1, 2, \cdots, D, \tag{9.35}$$

as the signal-subspace eigenvectors. Note that there is still a noise component in the signal subspace. The remaining eigenvectors define a noise subspace that does not contain any signal component.

We define these subspaces,

$$\mathbf{U}_S \triangleq \left[\begin{array}{c;{2pt/2pt}c;{2pt/2pt}c;{2pt/2pt}c} \mathbf{\Phi}_1 & \mathbf{\Phi}_2 & \cdots & \mathbf{\Phi}_D \end{array} \right], \tag{9.36}$$

as an $N \times D$ matrix and

$$\mathbf{U}_N \triangleq \left[\begin{array}{c;{2pt/2pt}c;{2pt/2pt}c;{2pt/2pt}c} \mathbf{\Phi}_{D+1} & \mathbf{\Phi}_{D+2} & \cdots & \mathbf{\Phi}_N \end{array} \right], \tag{9.37}$$

as an $N \times (N - D)$ matrix. Then

$$\left\| \mathbf{v}_i^H \, \mathbf{U}_S \right\|^2 = \sum_{j=1}^{D} \left| \mathbf{v}_i^H \, \mathbf{\Phi}_j \right|^2 = N^{\frac{1}{2}}, \quad i = 1, 2, \cdots, D, \tag{9.38}$$

and

$$\left\| \mathbf{v}_i^H \, \mathbf{U}_N \right\|^2 = 0, \quad i = 1, 2, \cdots, D. \tag{9.39}$$

The theory of subspace processing follows directly from (9.39). We find a set of vectors that span $\mathbf{U}_N$. Project $\mathbf{v}(\psi)$ onto $\mathbf{U}_N$ for all allowable values of ψ. The D values of ψ where the projection is zero are the desired $\psi_1, \psi_2, \cdots, \psi_D$.

In practice, we do not know the eigenvalues and eigenvectors and must estimate them from the data. We denote the estimated spectral matrix by $\widehat{\mathbf{S}}_{\mathbf{x}}$. It could be the sample spectral matrix $\mathbf{C}_{\mathbf{x}}$ or, if appropriate, $\mathbf{C}_{\mathbf{x},fb}$ as defined in (9.11).

Then, we can expand $\widehat{\mathbf{S}}_{\mathbf{x}}$,

$$\widehat{\mathbf{S}}_{\mathbf{x}} = \sum_{i=1}^{N} \hat{\lambda}_i \, \hat{\mathbf{\Phi}}_i \, \hat{\mathbf{\Phi}}_i^H, \tag{9.40}$$

$$\widehat{\mathbf{U}}_S \triangleq \left[\; \hat{\mathbf{\Phi}}_1 \; \vdots \; \hat{\mathbf{\Phi}}_2 \; \vdots \; \cdots \; \vdots \; \hat{\mathbf{\Phi}}_D \; \right], \tag{9.41}$$

and

$$\widehat{\mathbf{U}}_N \triangleq \left[\; \hat{\mathbf{\Phi}}_{D+1} \; \vdots \; \hat{\mathbf{\Phi}}_{D+2} \; \vdots \; \cdots \; \vdots \; \hat{\mathbf{\Phi}}_N \; \right]. \tag{9.42}$$

We developed the statistics of $\hat{\lambda}_i$ and $\hat{\mathbf{\Phi}}_i$ in Section 5.7.

For finite $K, \widehat{\mathbf{U}}_S \neq \mathbf{U}_S$ and $\widehat{\mathbf{U}}_N \neq \mathbf{U}_N$. Therefore, different choices of vectors in the estimated noise subspace will lead to different performances of the estimators. We discuss various algorithms and their performance.

The common steps are:

(i) Use $\widehat{\mathbf{S}}_{\mathbf{x}}$ to determine $\widehat{\mathbf{U}}_N$ and $\widehat{\mathbf{U}}_S$. In some algorithms, we also need the eigenvalues. Note that this assumes we have already estimated D using one of the techniques that was developed in Section 7.9.

(ii) Compute a function $\widehat{Q}(\psi)$ called the null spectrum by projecting $\mathbf{v}(\psi)$ on a particular set of vectors in $\widehat{\mathbf{U}}_N$. Choose the D minima of $\widehat{Q}(\psi)$. The corresponding values of ψ are the $\hat{\psi}_i$, $i = 1, 2, \cdots, D$.

Note that $\widehat{Q}(\psi)$ is a one-dimensional function so that we are no longer searching over a D-dimensional space to find ψ.

In most cases, we will use the singular value decomposition developed in Section 5.8 to compute the eigenvalues and eigenvectors directly from the data.

In Sections 9.3.2 and 9.3.3, we discuss two subspace algorithms: MUSIC and Min-Norm. In Section 9.3.3, we compare their performance. In Section 9.3.4, we derive a different type of subspace algorithm, ESPRIT, and compare its performance to that of MUSIC and Min-Norm. In Section 9.3.5, we summarize our results.

9.3.2 MUSIC

The first algorithm of interest is called MUSIC (Multiple Signal Classification), and was invented by Schmidt ([Sch79], [Sch81], [Sch83], [Sch86], and [SF86]) and independently by Bienvenu and Kopp [BK80], [BK83].

In Section 9.3.2.1, we develop the spectral MUSIC algorithm. In Section 9.3.2.2, we develop the root version of the MUSIC algorithm, which is applicable to standard linear arrays.

9.3.2.1 Spectral MUSIC

In this case, $\widehat{Q}(\psi)$ is expressed in terms of the eigenvectors of the noise subspace

$$\widehat{Q}_{MU}(\psi) = \mathbf{v}^H(\psi) \left(\sum_{i=D+1}^{N} \hat{\mathbf{\Phi}}_i \, \hat{\mathbf{\Phi}}_i^H \right) \mathbf{v}(\psi), \tag{9.43}$$

which can also be written as

$$\widehat{Q}_{MU}(\psi) = \mathbf{v}^H(\psi) \, \widehat{\mathbf{U}}_N \, \widehat{\mathbf{U}}_N^H \, \mathbf{v}(\psi). \tag{9.44}$$

Thus, MUSIC is using a uniform weighting of the norms of the projections of $\mathbf{v}(\psi)$ onto the estimated eigenvectors. Equivalently, in terms of the eigenvectors of the signal subspace,

$$\widehat{Q}_{MU}(\psi) = \mathbf{v}^H(\psi) \left[\mathbf{I} - \widehat{\mathbf{U}}_S \, \widehat{\mathbf{U}}_S^H \right] \mathbf{v}(\psi). \tag{9.45}$$

The steps in the algorithm are:

(1) We compute either $\widehat{\mathbf{U}}_S$ or $\widehat{\mathbf{U}}_N$ using the SVD procedure in Section 7.2.6. The choice will depend on the size of D, the number of signals compared to N, and the number of sensors. In most cases, we use $\widehat{\mathbf{U}}_S$.

(2) We plot $\widehat{Q}_{MU}(\psi)$ by varying $\mathbf{v}(\psi)$ over $-\pi \leq \psi \leq \pi$ and choose the D minima.

In order to implement MUSIC, we need enough snapshots to get a reasonable estimate of $\widehat{\mathbf{S}}_{\mathbf{x}}$ and we have to know the array manifold $\mathbf{v}(\psi)$. Note that spectral MUSIC is applicable to arbitrary array geometry and its extension to $\psi = [\psi_x \, \psi_y]^T$ is straightforward.

For standard linear arrays, the root MUSIC algorithm is applicable.

9.3.2.2 Root MUSIC

For a standard linear array, we can use a polynomial representation. The array manifold polynomial vector is defined as

$$\mathbf{v}_z(z) = \left[1 \; z \; \cdots \; z^{N-1}\right]^T, \tag{9.46}$$

which is the array manifold vector, with a phase shift, evaluated at $z = \exp(j\psi)$,

$$\left[\mathbf{v}_z(z)\right]_{z=e^{j\psi}} = e^{j(\frac{N-1}{2})\psi}\mathbf{v}(\psi). \tag{9.47}$$

Then, (9.44) can be written as

$$\begin{aligned}\hat{Q}_{MU,z}(z) &= \mathbf{v}_z^T(\frac{1}{z})\hat{\mathbf{U}}_N\hat{\mathbf{U}}_N^H\mathbf{v}_z(z) \\ &= \mathbf{v}_z^T(\frac{1}{z})\left[1 - \hat{\mathbf{U}}_S\hat{\mathbf{U}}_S^H\right]\mathbf{v}_z(z).\end{aligned} \tag{9.48}$$

If the eigendecomposition corresponded to the true spectral matrix, then the exact MUSIC spectrum could be obtained by evaluating $\hat{Q}_{MU,z}(z)$ on the unit circle,

$$\hat{Q}_{MU,z}(z)|_{z=e^{j\psi}} = \hat{Q}_{MU}(\psi), \tag{9.49}$$

and the D roots would correspond to the location in ψ-space of the D signals. In practice, we compute the roots of $\hat{Q}_{MU,z}(z)$ and choose the D roots that are inside the unit circle and closest to the unit circle.

We denote these roots by $\hat{z}_i, i = 1, 2, \cdots, D$. Then

$$\hat{\psi}_i = \frac{arg\,\hat{z}_i}{\pi}, \quad i = 1, 2, \cdots, D. \tag{9.50}$$

Since we are using an estimated covariance matrix, there will be errors in the location of roots. The effect of this error is shown in Figure 9.8. We observe that the radial component of the error Δz_i will not affect $\hat{\psi}_i$. However, it would distort the MUSIC spectrum. Thus, we would expect that spectral MUSIC will have less resolution capability than root MUSIC. We investigate this behavior analytically in the asymptotic region in Section 9.5.

For a standard linear array, the performance can be improved by using FB averaging of the data, as in (9.11), to obtain $\mathbf{C}_{\mathbf{x},fb}$. The corresponding

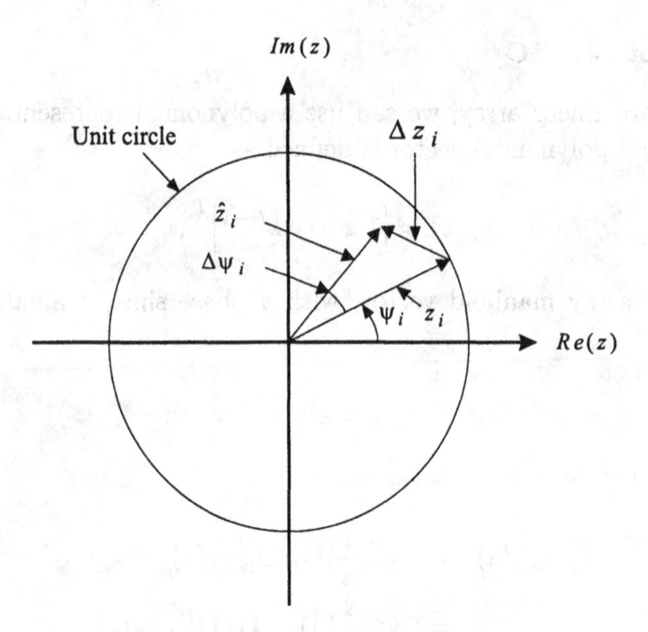

Figure 9.8 Behavior of roots in z-plane.

subspace decomposition is

$$\begin{aligned}\widehat{\mathbf{S}}_{\mathbf{x},fb} \triangleq \mathbf{C}_{\mathbf{x},fb} &= \sum_{i=1}^{N} \widehat{\lambda}_{i,fb} \widehat{\mathbf{\Phi}}_{i,fb} \widehat{\mathbf{\Phi}}_{i,fb}^{H} \\ &= \widehat{\mathbf{U}}_{S,fb} \widehat{\mathbf{\Lambda}}_{S,fb} \widehat{\mathbf{U}}_{S,fb}^{H} + \widehat{\mathbf{U}}_{N,fb} \widehat{\mathbf{\Lambda}}_{N,fb} \widehat{\mathbf{U}}_{N,fb}^{H}. \end{aligned} \tag{9.51}$$

Then, the FB root MUSIC polynomial can be written as

$$\begin{aligned}\widehat{\mathbf{Q}}_{MU,fb,z}(z) &= \mathbf{v}_z^T(\frac{1}{z}) \widehat{\mathbf{U}}_{N,fb} \widehat{\mathbf{U}}_{N,fb}^{H} \mathbf{v}_z(z) \\ &= \mathbf{v}_z^T(\frac{1}{z}) \left[1 - \widehat{\mathbf{U}}_{S,fb} \widehat{\mathbf{U}}_{S,fb}^{H}\right] \mathbf{v}_z(z). \end{aligned} \tag{9.52}$$

We find the roots of $\widehat{Q}_{MU,fb,z}(z)$ and proceed as in the forward-only root MUSIC algorithm.

9.3.2.3 Unitary root MUSIC

The forward-backward root MUSIC algorithm can be implemented with real computation by using a unitary transformation. This approach is due to Pesavento et al. [PGH00] and our discussion follows that reference.

We define a real-valued sample spectral matrix,

$$\mathbf{C}_{\mathbf{x},Re} = \mathbf{Q}^H \mathbf{C}_{\mathbf{x},fb} \mathbf{Q}, \tag{9.53}$$

where $\mathbf{Q}$ can be any unitary, column conjugate symmetric matrix. In the text, we will always use the sparse matrices introduced in (7.58) and (7.59),

$$\mathbf{Q} = \frac{1}{\sqrt{2}} \begin{bmatrix} \mathbf{I} & j\mathbf{I} \\ \mathbf{J} & -j\mathbf{J} \end{bmatrix}, \tag{9.54}$$

for N even, and

$$\mathbf{Q} = \frac{1}{\sqrt{2}} \begin{bmatrix} \mathbf{I} & \mathbf{0} & j\mathbf{I} \\ \mathbf{0}^T & \sqrt{2} & \mathbf{0}^T \\ \mathbf{J} & \mathbf{0} & -j\mathbf{J} \end{bmatrix}, \tag{9.55}$$

for N odd.

Note that

$$\mathbf{C}_{\mathbf{x},Re} = \frac{1}{2}\left[\mathbf{Q}^H \mathbf{C}_{\mathbf{x}} \mathbf{Q} + \mathbf{Q}^H \mathbf{J} \mathbf{C}_{\mathbf{x}}^* \mathbf{J} \mathbf{Q}\right]. \tag{9.56}$$

Using

$$\mathbf{J} \mathbf{Q}^* = \mathbf{Q}, \tag{9.57}$$

(9.56) can be written as

$$\begin{aligned} \mathbf{C}_{\mathbf{x},Re} &= \frac{1}{2}\left[\mathbf{Q}^H \mathbf{C}_{\mathbf{x}} \mathbf{Q} + [\mathbf{Q}^*]^H \mathbf{C}_{\mathbf{x}}^* \mathbf{Q}^*\right] \\ &= Re\left[\mathbf{Q}^H \mathbf{C}_{\mathbf{x}} \mathbf{Q}\right]. \end{aligned} \tag{9.58}$$

We denote the eigenvalues and eigenvectors of $\mathbf{C}_{\mathbf{x},Re}$ by $\hat{\lambda}_{i,Re}$ and $\hat{\boldsymbol{\phi}}_{i,Re}$, respectively. They satisfy the characteristic equation,

$$\mathbf{C}_{\mathbf{x},Re} \hat{\boldsymbol{\phi}}_{Re} = \hat{\lambda}_{Re} \hat{\boldsymbol{\phi}}_{Re}. \tag{9.59}$$

The eigenvalues and eigenvectors of $\hat{\mathbf{C}}_{\mathbf{x},fb}$ are denoted by $\hat{\lambda}_{i,fb}$ and $\hat{\boldsymbol{\phi}}_{i,fb}$, respectively. They satisfy the characteristic equation,

$$\mathbf{C}_{\mathbf{x},fb} \hat{\boldsymbol{\phi}}_{fb} = \hat{\lambda}_{fb} \hat{\boldsymbol{\phi}}_{fb}. \tag{9.60}$$

Pre-multiplying (9.60) by $\mathbf{Q}^H$ gives

$$\mathbf{Q}^H \mathbf{C}_{\mathbf{x},fb} \hat{\boldsymbol{\phi}}_{fb} = \mathbf{Q}^H \mathbf{C}_{\mathbf{x},fb} \mathbf{Q} \mathbf{Q}^H \hat{\boldsymbol{\phi}}_{fb} = \hat{\lambda}_{fb} \mathbf{Q}^H \hat{\boldsymbol{\phi}}_{fb}, \tag{9.61}$$

which is (9.59) with

$$\hat{\phi}_{Re} = \mathbf{Q}^H \hat{\phi}_{fb}, \tag{9.62}$$

and

$$\hat{\lambda}_{Re} = \hat{\lambda}_{fb}. \tag{9.63}$$

The subspace decomposition of the real matrix $\mathbf{C}_{\mathbf{x},Re}$ can be written as

$$\mathbf{C}_{\mathbf{x},Re} = \hat{\mathbf{U}}_{S,Re}\hat{\mathbf{\Lambda}}_{S,Re}\hat{\mathbf{U}}_{S,Re}^H + \hat{\mathbf{U}}_{N,Re}\hat{\mathbf{\Lambda}}_{N,Re}\hat{\mathbf{U}}_{N,Re}^H, \tag{9.64}$$

and corresponds to the subspace decomposition in (9.51) through the relation in (9.62) and (9.63). The FB MUSIC polynomial in (9.52) can be written as

$$\begin{aligned}\hat{Q}_{MU,fb,z}(z) &= \mathbf{v}_z^T(\frac{1}{z})\mathbf{Q}\mathbf{Q}^H\hat{\mathbf{U}}_{N,fb}\hat{\mathbf{U}}_{N,fb}^H\mathbf{v}_z(z) \\ &= \mathbf{v}_z^T(\frac{1}{z})\mathbf{Q}\hat{\mathbf{U}}_{N,Re}\hat{\mathbf{U}}_{N,Re}^H\mathbf{Q}^H\mathbf{v}_z(z) \\ &= \tilde{\mathbf{v}}_z^T(\frac{1}{z})\hat{\mathbf{U}}_{N,Re}\hat{\mathbf{U}}_{N,Re}^H\tilde{\mathbf{v}}_z(z) \\ &\triangleq \hat{Q}_{MU,U,z}(z),\end{aligned} \tag{9.65}$$

where the array manifold polynomial is

$$\tilde{\mathbf{v}}_z(z) \triangleq \mathbf{Q}^H\mathbf{v}_z(z). \tag{9.66}$$

The steps in unitary root MUSIC can be summarized:

1. Compute the real sample spectral matrix $\mathbf{C}_{\mathbf{x},Re}$ using (9.58),

2. Perform the eigendecomposition in (9.64),

3. Compute $\hat{Q}_{MU,U,z}(z)$ using (9.65) and (9.66),

4. Find the roots of $\hat{Q}_{MU,U,z}(z)$ and choose the D roots that are inside the unit circle and closest to the unit circle. Denote these roots as $\hat{z}_i, i = 1, \cdots, D$,

5. The estimates are

$$\hat{\psi}_i = \arg \hat{z}_i, \quad i = 1, \cdots, D. \tag{9.67}$$

The advantage of unitary root MUSIC is that it has the same performance as FB root MUSIC. It provides a significant reduction in computational complexity. In order to achieve this reduction, an efficient polynomial root-finding algorithm must be used.[6] The Lang-Frenzel algorithm ([LF94]) is used in our examples.

Pesavento et al. [PGH00] have derived closed-form expressions for the large sample mean-square estimation error of unitary root MUSIC.

Before we do any examples of the MUSIC algorithm, we derive a second subspace algorithm.

9.3.3 Minimum-Norm Algorithm

A second algorithm of interest is called the Min-Norm algorithm. It was first developed by Reddi [Red79]. The minimum-norm property and a better motivation for the algorithm was presented by Kumaresan and Tufts [KT83]. Our derivation follows this latter reference.

The signal subspace is spanned by the columns of $\mathbf{V}$ (see (9.2)). We define a vector

$$\mathbf{d} = \begin{bmatrix} d_1 & d_2 & \cdots & d_N \end{bmatrix}^T \tag{9.68}$$

that has the property that,

$$\mathbf{v}^H(\psi_i)\,\mathbf{d} = 0, \quad i = 1, 2, \cdots, D, \tag{9.69}$$

or equivalently,

$$\mathbf{\Phi}_i^H\,\mathbf{d} = 0, \quad i = 1, 2, \cdots, D. \tag{9.70}$$

Then, the polynomial,

$$D(z) = \sum_{k=1}^{N} d_k\, z^{-(k-1)} \tag{9.71}$$

has zeros at

$$z_k = \exp\left(j\psi_k\right), \quad i = 1, 2, \cdots, D. \tag{9.72}$$

As we discussed in Section 9.3.1, the $N-D$ noise eigenvectors all have this property (i.e., satisfy (9.69) and (9.70)) and span the noise subspace.[7] The MUSIC algorithm utilizes all of the noise subspace eigenvectors. In the Min-Norm algorithm, Kumaresen and Tufts propose to use a single vector

[6] This suggestion was due to A. Gershmann (private communication). The unitary root MUSIC examples were done by D. Bray using a slightly improved version of the LF algorithm.

[7] This use of this property was first proposed by Pisarenko ([Pis72], [Pis73]).

d in the noise subspace. They hypothesize that the estimate of ψ_k will be more accurate and that the $N-D$ "noise" zeros will tend to be uniformly distributed within the unit circle and will be less likely to generate false sources.

We want to find **d** as a linear combination of the $N-D$ noise eigenvectors. Then, if $\mathbf{S_x}$ were known exactly, $D(z)$ will have its zeros at $\psi_1, \psi_2, \cdots, \psi_D$.

We require that d_1 equal unity and minimize

$$Q \triangleq \sum_{k=1}^{N} |d_k|^2, \quad d_1 = 1. \tag{9.73}$$

It is useful to discuss why this particular choice of **d** is logical (e.g., Kumaresen [Kum83]). We factor $D(z)$ into two polynomials,

$$D(z) = D_1(z)\, D_2(z), \tag{9.74}$$

where

$$D_1(z) = \sum_{k=1}^{D} a_k\, z^{-(k-1)}, \quad a_1 = 1, \tag{9.75}$$

has the signal zeros, and

$$D_2(z) = \sum_{k=1}^{N-D} b_k\, z^{-(k-1)}, \quad b_1 = 1, \tag{9.76}$$

has the noise zeros. Minimizing Q is the same as minimizing

$$\int_{-\pi}^{\pi} \left| D(e^{j\psi}) \right|^2 d\psi. \tag{9.77}$$

This is equivalent to minimizing $D_2(z)$ given $D_1(z)$. Using the analogy to linear prediction [Mak75], Kumaresen [Kum83] shows that the zeros of $\hat{D}_2(z)$ always has zeros inside the unit circle that are independent of the zeros of $\hat{D}_1(z)$ and they are approximately uniformly distributed in the sectors where the D signal zeros are absent.

We now find $D(z)$ in terms of the estimated signal or noise eigenvectors. First, partition **d**,

$$\mathbf{d} = \begin{bmatrix} 1 \\ \mathbf{d}' \end{bmatrix}. \tag{9.78}$$

Next, we partition the signal subspace matrix $\widehat{\mathbf{U}}_S$ as

$$\widehat{\mathbf{U}}_S = \begin{bmatrix} \mathbf{g}^T \\ \hdashline \widehat{\mathbf{U}}'_S \end{bmatrix}, \tag{9.79}$$

where $\mathbf{g}$ has the first elements of each of the signal eigenvectors. We partition the noise subspace matrix $\widehat{\mathbf{U}}_N$ as

$$\widehat{\mathbf{U}}_N = \begin{bmatrix} \mathbf{c}^T \\ \hdashline \widehat{\mathbf{U}}'_N \end{bmatrix}, \tag{9.80}$$

where $\mathbf{c}$ has the first elements of each of the noise eigenvectors. From (9.70), $\mathbf{d}$ will be orthogonal to the columns of $\widehat{\mathbf{U}}_S$,

$$\widehat{\mathbf{U}}_S^H \, \mathbf{d} = 0. \tag{9.81}$$

Substituting (9.78) and (9.79) into (9.81) gives

$$\left[\widehat{\mathbf{U}}'_S\right]^H \mathbf{d}' = -\mathbf{g}^*. \tag{9.82}$$

Now, minimizing the norm of $\mathbf{d}'$ subject to the constraint in (9.82). This is a familiar minimization subject to linear constraint. The minimum-norm value of $\mathbf{d}'$ is

$$\mathbf{d}' = -\widehat{\mathbf{U}}'_S \left(\left[\widehat{\mathbf{U}}'_S\right]^H \widehat{\mathbf{U}}'_S \right)^{-1} \mathbf{g}^*. \tag{9.83}$$

From the definition in (9.79), we observe that

$$\left[\widehat{\mathbf{U}}'_S\right]^H \widehat{\mathbf{U}}'_S = \mathbf{I} - \mathbf{g}^* \, \mathbf{g}^T. \tag{9.84}$$

Using the matrix inversion lemma,

$$\left(\left[\widehat{\mathbf{U}}'_S\right]^H \widehat{\mathbf{U}}'_S \right)^{-1} = \mathbf{I} + \frac{\mathbf{g}^* \, \mathbf{g}^T}{1 - \mathbf{g}^T \, \mathbf{g}^*}. \tag{9.85}$$

Substituting (9.85) into (9.83) gives

$$\mathbf{d}' = -\frac{\widehat{\mathbf{U}}'_S \, \mathbf{g}^*}{1 - \mathbf{g}^H \, \mathbf{g}}, \tag{9.86}$$

or

$$\mathbf{d} = \begin{bmatrix} 1 \\ \hdashline \frac{\widehat{\mathbf{U}}_S^H \, \mathbf{g}^*}{1 - \mathbf{g}^H \, \mathbf{g}} \end{bmatrix}. \tag{9.87}$$

We can also express $\mathbf{d}$ in terms of the noise eigenvectors.

Since

$$\hat{\mathbf{U}}\,\hat{\mathbf{U}}^H = \mathbf{I}, \tag{9.88}$$

$$\left[\begin{array}{c:c} \hat{\mathbf{U}}_S & \hat{\mathbf{U}}_N \end{array}\right] \left[\begin{array}{c} \hat{\mathbf{U}}_S^H \\ \hdashline \hat{\mathbf{U}}_N^H \end{array}\right] = \mathbf{I}, \tag{9.89}$$

and

$$\left[\begin{array}{c:c} \mathbf{g}^T & \mathbf{c}^T \\ \hdashline \hat{\mathbf{U}}'_S & \hat{\mathbf{U}}'_N \end{array}\right] \left[\begin{array}{c:c} \mathbf{g}^* & \left[\hat{\mathbf{U}}'_S\right]^H \\ \hdashline \mathbf{c}^* & \left[\hat{\mathbf{U}}'_N\right]^H \end{array}\right] = \mathbf{I}. \tag{9.90}$$

Thus,

$$\mathbf{d} = \left[\begin{array}{c} 1 \\ \hdashline \\ \dfrac{\hat{\mathbf{U}}'_N\,\mathbf{c}^*}{\mathbf{c}^H\,\mathbf{c}} \end{array}\right]. \tag{9.91}$$

We observe that we can also write

$$\mathbf{d} = \hat{\mathbf{U}}_N\,\frac{\mathbf{c}^*}{|\mathbf{c}|^2}. \tag{9.92}$$

If we define

$$\mathbf{w}^{\frac{1}{2}} = \frac{\mathbf{c}^*}{|\mathbf{c}|^2}, \tag{9.93}$$

then

$$\mathbf{d} = \hat{\mathbf{U}}_N\,\mathbf{w}^{\frac{1}{2}}. \tag{9.94}$$

Then the Min-Norm null spectra can be written as

$$\boxed{Q_{MN}(\psi) \triangleq \left|\mathbf{v}^H(\psi)\,\mathbf{d}\right|^2 \triangleq \mathbf{v}^H(\psi)\,\hat{\mathbf{U}}_N\,\mathbf{W}\,\hat{\mathbf{U}}_N^H\,\mathbf{v}(\psi)}, \tag{9.95}$$

where

$$\mathbf{W} = \mathbf{w}^{\frac{1}{2}}\mathbf{w}^{\frac{H}{2}} \tag{9.96}$$

and interpreted as a weighted eigenspace algorithm. It is sometimes referred to in the literature as a weighted MUSIC algorithm. Comparisons of MUSIC and Min-Norm are contained Xu and Kaveh [XK96] and Kaveh and Bassias [KB90].

The Min-Norm null spectra can also be written in terms of signal subspace eigenvectors,

$$Q_{MN}(\psi) \triangleq \left|\mathbf{v}^H(\psi)\mathbf{d}\right|^2 = \mathbf{v}^H(\psi)\widehat{\mathbf{U}}_S\mathbf{W}_S\widehat{\mathbf{U}}_S^H\mathbf{v}(\psi), \tag{9.97}$$

where

$$\mathbf{d} = \widehat{\mathbf{U}}_S\frac{\mathbf{g}^*}{1-\mathbf{g}^H\mathbf{g}}, \tag{9.98}$$

and

$$\mathbf{w}^{\frac{1}{2}} = \frac{\mathbf{g}^*}{1-\mathbf{g}^H\mathbf{g}}. \tag{9.99}$$

For root Min-Norm we calculate $\mathbf{d}$ using (9.92) and then form the polynomial

$$D(z) = 1 + d_1\,z^{-1} + \cdots + d_N\,z^{-N}, \tag{9.100}$$

and compute the roots of $D(z)$. We choose the D roots that are closest to the unit circle in the z-plane. It turns out that the difference in performance between spectral Min-Norm and root Min-Norm is not as great as the difference between spectral MUSIC and root MUSIC (e.g., [RH89b]).

We now consider the same models as in Examples 9.2.3 and 9.2.5, and use the MUSIC and Min-Norm algorithms to estimate the directions of arrival in ψ-space.

Example 9.3.1 (continuation)

Consider a standard 10-element linear array and assume there are two uncorrelated plane-wave signals impinging on the array from $\pm\Delta\psi/2$. Assume $\Delta\psi = \Delta\psi_R$. We simulate the four algorithms derived above: MUSIC, root MUSIC, Min-Norm, and root Min-Norm. We use $\mathbf{C}_{\mathbf{x},fb}$.

In Figure 9.9, we show the probability of resolution versus *ASNR* for $K = 100$ $(10N)$. In Figure 9.10, we show the normalized RMSE versus *ASNR*. We see that spectral Min-Norm has better resolution behavior than spectral MUSIC. However, root MUSIC has better resolution behavior than root Min-Norm. The threshold of root MUSIC occurs when the $ASNR = 7$–8 dB. Above threshold, MUSIC and root MUSIC approach the CRB. (In Section 9.5, we show that the ratio of the MUSIC variance to the CRB is $(1+ASNR^{-1})$ for large K and *ASNR*.) However, the Min-Norm RMSE remains about 2.5 dB above the bound.

In Figure 9.11, we plot the normalized RMSE versus K for an $ASNR = 20$ dB. Spectral Min-Norm has better threshold performance than spectral MUSIC, but it remains about 2.5 dB above the CRB. For this *ASNR*, both root algorithms are above threshold for $K = 10$ $(K = N)$. Root MUSIC approaches the CRB bound, while root Min-Norm remains 1.9 dB above it.

Example 9.3.2 (continuation)

We consider the same model as in Example 9.3.1 except $\rho = 0.95$. In Figure 9.12, we show the probability of resolution versus *ASNR*, and in Figure 9.13, we plot the normalized

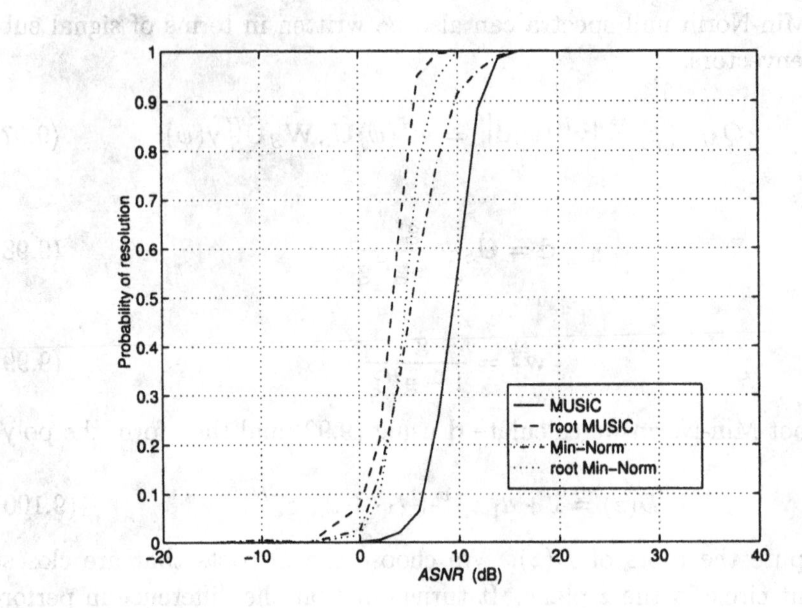

Figure 9.9 MUSIC and Min-Norm algorithms: probability of resolution versus *ASNR*; $N = 10, \Delta\psi = \Delta\psi_R$, $\rho = 0$, $K = 100$, 200 trials.

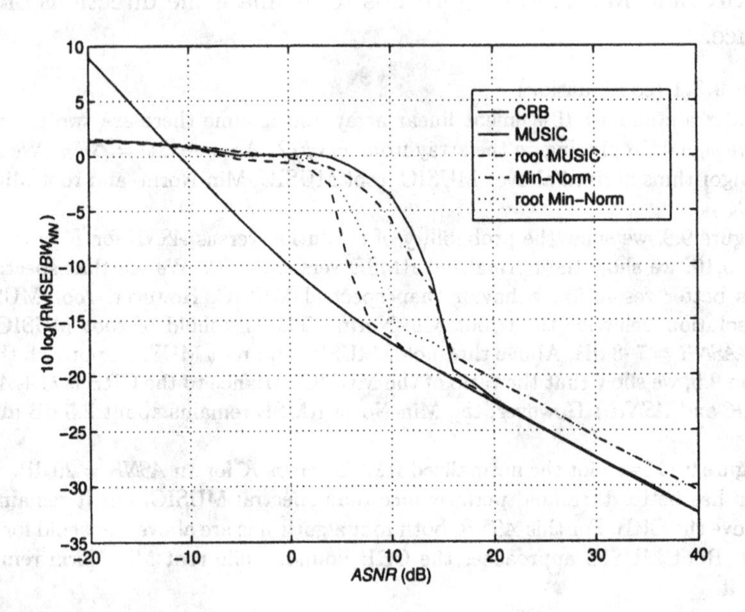

Figure 9.10 MUSIC and Min-Norm algorithms: normalized RMSE versus *ASNR*; $N = 10, \Delta\psi = \Delta\psi_R$, $\rho = 0$, $K = 100$, 200 trials.

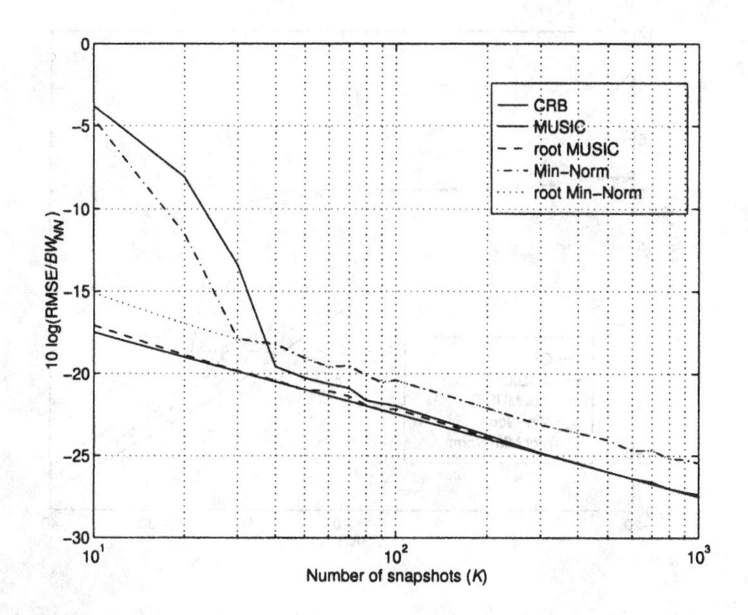

Figure 9.11 MUSIC and Min-Norm algorithms: normalized RMSE versus $K : N = 10, ASNR = 20$ dB, $\Delta\psi = \Delta\psi_R$, $\rho = 0$, 200 trials.

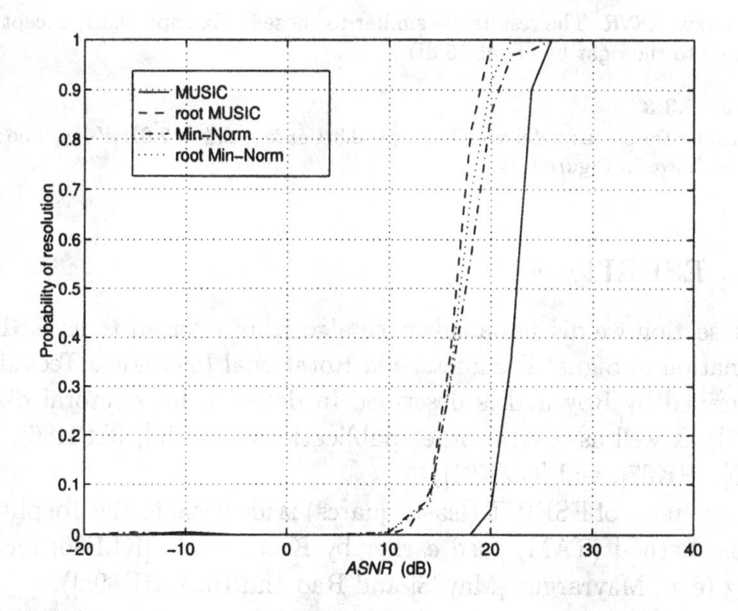

Figure 9.12 MUSIC and Min-Norm algorithms: probability of resolution versus $ASNR; N = 10, \Delta\psi = \Delta\psi_R$, $\rho = 0.95$, $K = 100$, 200 trials.

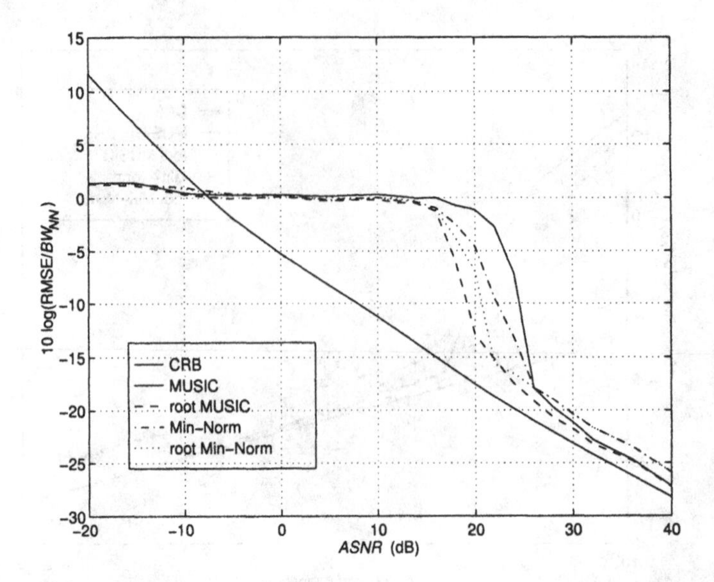

Figure 9.13 MUSIC and Min-Norm algorithms: normalized RMSE versus *ASNR*; $N = 10, \Delta\psi = \Delta\psi_R$, $\rho = 0.95$, $K = 100$, 200 trials.

RMSE versus *ASNR*. The results are similar to those in Example 9.3.1, except the curves are shifted to the right by about 15 dB.

Example 9.3.3

Consider the same model as in Example 9.3.1 except $\Delta\psi = 2.0BW_{NN}$. The normalized RMSE is shown in Figure 9.14.

9.3.4 ESPRIT

In this section we discuss a subspace algorithm referred to as **ESPRIT** for "Estimation of Signal Parameter via Rotational Invariance Techniques". It was derived by Roy and is described in detail in his doctoral dissertation [Roy87], as well as several other publications ([RK89], [RPK86], [PRK86], [OK90], [RK87], and [OVK91]).

One version of ESPRIT (least squares) is identical to the Toeplitz approximation method (TAM) used earlier by Kung et al. [KLF86] for direction finding (e.g., Mayrargue [May88] and Rao and Hari [RH89a]).

There are several versions of ESPRIT. In the text, we discuss three versions. In Section 9.3.4.1, we discuss LS (least squares) and TLS (total least squares) ESPRIT. In Section 9.3.4.2, we discuss unitary ESPRIT. In Sec-

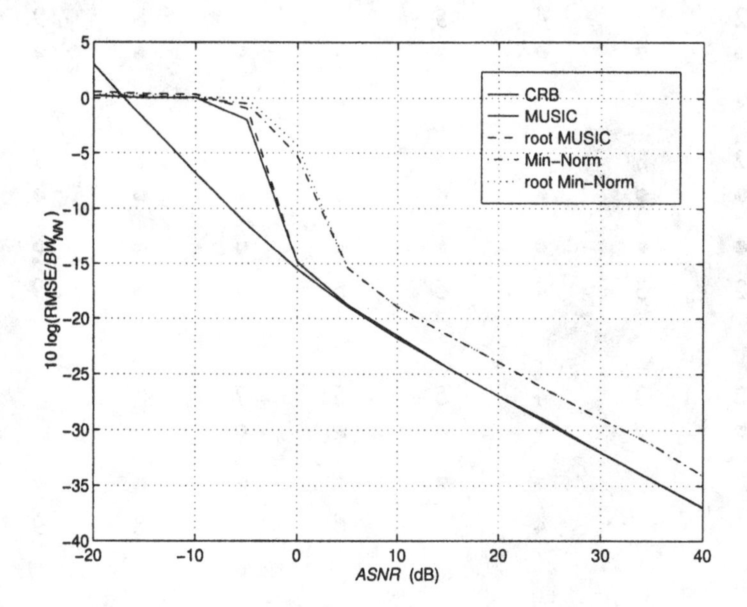

Figure 9.14 MUSIC and Min-Norm algorithms: normalized RMSE versus *ASNR*; $N = 10, \Delta\psi = 2BW_{NN}$, $\rho = 0$, $K = 100$, 200 trials.

tion 9.3.4.3, we summarize our ESPRIT results and discuss some related algorithms.

9.3.4.1 LS and TLS ESPRIT

In this section, we develop the ESPRIT algorithm in the context of a uniform linear array. Our discussion follows Ottersten et al. [OVK91].

We consider the standard N-element linear array. In Figure 9.15, we show a 10-element array. The first step in the ESPRIT algorithm is to choose two identical subarrays. The number of elements in the subarrays is denoted by N_s and $N_s \geq D + 1$ where D is the number of signals. We assume that the first element in the original array is the first element in the first subarray and that the $(d_s + 1)$th element in the original array is the first element in the second subarray. The parameter d_s denotes the distance between the subarrays measured in $\Delta\mathbf{p}$ units ($\Delta\mathbf{p}$ is interelement spacing). The first set of subarrays in Figure 9.15 shows the case of two overlapping subarrays with $d_s = 1$. The second set of subarrays in Figure 9.15 shows the case of two overlapping subarrays with $d_s = 3$. The third set of subarrays in Figure 9.15 shows the case of two non-overlapping subarrays with $d_s = 5$. In Figure 9.16, the case of two non-overlapping subarrays with $d_s = 1$ is

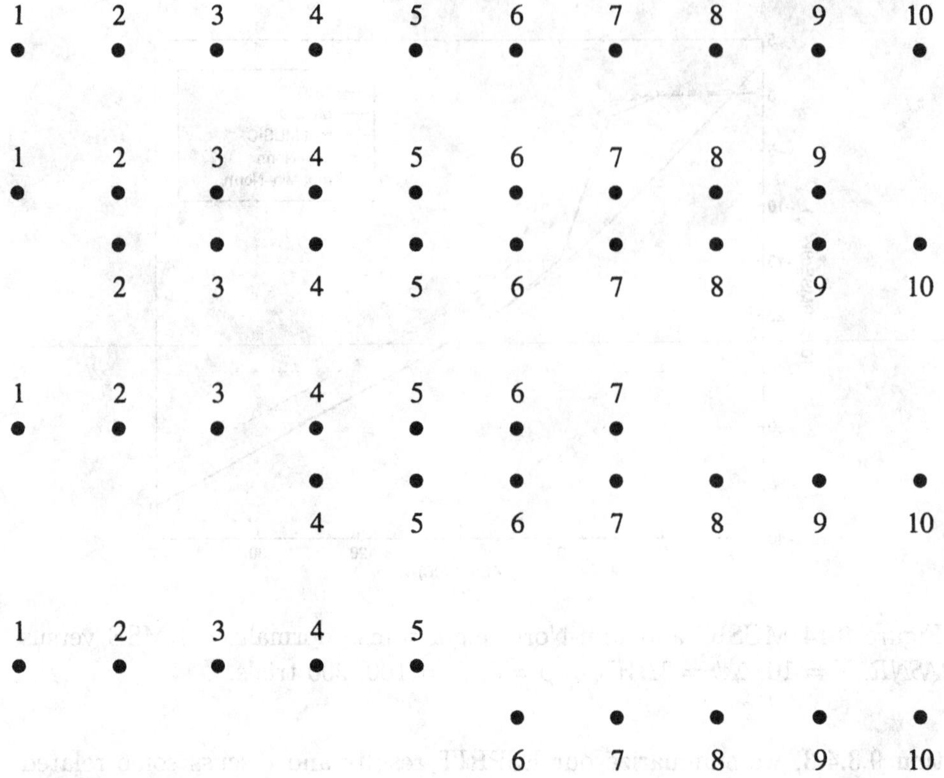

1 2 3 4 5 6 7 8 9 10

1 2 3 4 5 6 7 8 9

2 3 4 5 6 7 8 9 10

1 2 3 4 5 6 7

4 5 6 7 8 9 10

1 2 3 4 5

6 7 8 9 10

Figure 9.15 Subarrays for ESPRIT algorithm.

shown. In this case, the element spacing in the subarrays is λ so we would anticipate ambiguity problems.

We can specify the subarrays by use of a selection matrix $\mathbf{J}_s$. We begin by defining the non-zero component, which we denote by $\tilde{\mathbf{J}}_s$. For the first set of subarrays in Figure 9.15, $\tilde{\mathbf{J}}_s$ is a 9×9 identity matrix. For the second set of subarrays in Figure 9.15, $\tilde{\mathbf{J}}_s$ is a 7×7 identity matrix. For the third set of subarrays in Figure 9.15, $\tilde{\mathbf{J}}_s$ is a 5×5 identity matrix. For the general case corresponding to Figure 9.15, $\tilde{\mathbf{J}}_s$ is an $N_s \times N_s$ identity matrix where N_s is the length of the subarray.

We now define the $N_s \times N$ selection matrices for the subarrays in Figure 9.15 ,

$$\mathbf{J}_{s1} \triangleq \left[\tilde{\mathbf{J}}_s \,\vdots\, \mathbf{0}_{N_s \times d_s} \right], \tag{9.101}$$

and,

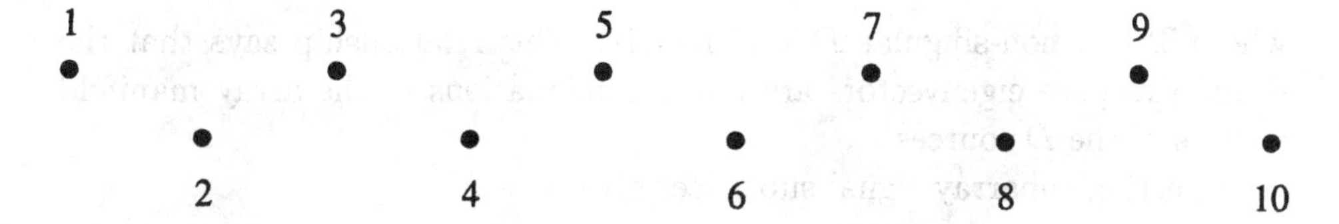

Figure 9.16 Subarray for ESPRIT algorithm.

$$\mathbf{J}_{s2} \triangleq \left[\; \mathbf{0}_{N_s \times d_s} \; \vdots \; \tilde{\mathbf{J}}_s \; \right]. \tag{9.102}$$

For the subarray choice in Figure 9.16, $\mathbf{J}_s$ is a 5×9 matrix,

$$\mathbf{J}_s = \begin{bmatrix} 1 & 0 & 0 & \cdots & & & 0 \\ 0 & 0 & 1 & 0 & \cdots & & 0 \\ 0 & 0 & 0 & 0 & 1 & \cdots & 0 \\ \vdots & \vdots & \vdots & \vdots & \ddots & \cdots & \vdots \\ 0 & \cdots & & & & & 1 \end{bmatrix}. \tag{9.103}$$

If we denote the array manifold matrix of the total array as $\mathbf{V}$ and the array manifold matrix of the ith subarray (i = 1, 2) as $\mathbf{V}_i$, then

$$\mathbf{V}_1 = \mathbf{J}_{s1}\mathbf{V}, \tag{9.104}$$

and

$$\mathbf{V}_2 = \mathbf{J}_{s2}\mathbf{V}. \tag{9.105}$$

The ESPRIT algorithm exploits the shift invariance property of the array, which implies

$$\mathbf{V}_2 = \mathbf{V}_1\mathbf{\Phi}, \tag{9.106}$$

where

$$\mathbf{\Phi} \triangleq \operatorname{diag}\left[e^{jd_s\psi_1}, e^{jd_s\psi_2}, \cdots, e^{jd_s\psi_D}\right], \tag{9.107}$$

for the subarrays in Figure 9.15. The $\psi_i, i = 1, 2, \cdots, D$ are the wavenumbers of the D signals in ψ-space. The source spectral matrix and the steering matrix are assumed to have full rank D. This restriction does not allow fully correlated signals. For the subarrays in Figure 9.16, each exponent is multiplied by 2. The ESPRIT algorithm will find an estimate of $\mathbf{\Phi}$ and use that estimate to estimate $\psi_1, \cdots, \psi_D$.

The columns of $\mathbf{V}$ span the signal subspace, therefore, we have the relationship

$$\mathbf{U}_s = \mathbf{V}\mathbf{T}, \tag{9.108}$$

where $\mathbf{T}$ is a non-singular $D \times D$ matrix. This relationship says that the signal subspace eigenvectors are linear combinations of the array manifold vectors of the D sources.

Selecting subarray signal subspaces gives

$$\mathbf{U}_{s1} \triangleq \mathbf{J}_{s1}\mathbf{U}_s = \mathbf{J}_{s1}\mathbf{V}\mathbf{T} = \mathbf{V}_1\mathbf{T}, \tag{9.109}$$

and

$$\mathbf{U}_{s2} \triangleq \mathbf{J}_{s2}\mathbf{U}_s = \mathbf{J}_{s2}\mathbf{V}\mathbf{T} = \mathbf{V}_2\mathbf{T}. \tag{9.110}$$

The relation

$$\mathbf{U}_{s1} = \mathbf{V}_1\mathbf{T} \tag{9.111}$$

implies

$$\mathbf{V}_1 = \mathbf{U}_{s1}\mathbf{T}^{-1}. \tag{9.112}$$

Similarly,

$$\mathbf{U}_{s2} = \mathbf{V}_2\mathbf{T} = \mathbf{V}_1\mathbf{\Phi}\mathbf{T} \tag{9.113}$$

implies

$$\mathbf{U}_{s2} = \mathbf{U}_{s1}\mathbf{T}^{-1}\mathbf{\Phi}\mathbf{T}. \tag{9.114}$$

We define

$$\mathbf{\Psi} = \mathbf{T}^{-1}\mathbf{\Phi}\mathbf{T}. \tag{9.115}$$

Note that the eigenvalues of $\mathbf{\Psi}$ are the elements of $\mathbf{\Phi}$. Thus, if we obtain an estimate of $\mathbf{\Psi}$ and compute its eigenvalues, we can obtain an estimate of $\psi_1, \cdots, \psi_D$.

The equivalent relationship to $\mathbf{V}_2 = \mathbf{V}_1\mathbf{\Phi}$ in terms of signal subspace eigenvectors is

$$\mathbf{U}_{s1}\mathbf{\Psi} = \mathbf{U}_{s2}. \tag{9.116}$$

Since N_s is greater than D, this is an overdetermined set of equations. In practice, we have estimates for $\mathbf{U}_{s2}$ and $\mathbf{U}_{s1}$

$$\hat{\mathbf{U}}_{s1} = \mathbf{J}_{s1}\hat{\mathbf{U}}_s, \tag{9.117}$$

and

$$\hat{\mathbf{U}}_{s2} = \mathbf{J}_{s2}\hat{\mathbf{U}}_s. \tag{9.118}$$

Then (9.116) is replaced by

$$\hat{\mathbf{U}}_{s1}\hat{\mathbf{\Psi}} = \hat{\mathbf{U}}_{s2}. \tag{9.119}$$

If we solve (9.119) using a least squares approach, then we minimize the difference between $\hat{\mathbf{U}}_{s2}$ and $\hat{\mathbf{U}}_{s1}\hat{\mathbf{\Psi}}$,

$$\begin{aligned}\hat{\mathbf{\Psi}}_{LS} &= \arg\min_{\mathbf{\Psi}}\left\{\left\|\hat{\mathbf{U}}_{s2} - \hat{\mathbf{U}}_{s1}\mathbf{\Psi}\right\|_F\right\} \\ &= \arg\min_{\mathbf{\Psi}}\left\{\mathrm{tr}\left\{\left[\hat{\mathbf{U}}_{s2} - \hat{\mathbf{U}}_{s1}\mathbf{\Psi}\right]^H\left[\hat{\mathbf{U}}_{s2} - \hat{\mathbf{U}}_{s1}\mathbf{\Psi}\right]\right\}\right\}\end{aligned} \tag{9.120}$$

The result is

$$\hat{\mathbf{\Psi}}_{LS} = [\hat{\mathbf{U}}_{s1}^H\hat{\mathbf{U}}_{s1}]^{-1}\hat{\mathbf{U}}_{s1}^H\hat{\mathbf{U}}_{s2}. \tag{9.121}$$

The steps in the LS-ESPRIT algorithm can be summarized:

1. Perform the eigendecomposition on $\mathbf{C_x}$ to obtain $\hat{\mathbf{U}}_s$,
2. Find $\hat{\mathbf{U}}_{s1}$ and $\hat{\mathbf{U}}_{s2}$ by using (9.109) and (9.110),
3. Find $\hat{\mathbf{\Psi}}_{LS}$ using (9.121),
4. Find the eigenvalues of $\hat{\mathbf{\Psi}}_{LS}$ denoted by $\hat{\lambda}_1, \hat{\lambda}_2, \cdots, \hat{\lambda}_D$,
5. Find the estimates in ψ-space by using $\hat{\psi}_i = \frac{1}{d_s}(\arg\hat{\lambda}_i), i = 1, \cdots, D$.

Golub and Van Loan (Section 12.3 of [GVL83] and [GVL80]) suggest that a total least squares (TLS) criterion is more appropriate because both $\hat{\mathbf{U}}_{s1}$ and $\hat{\mathbf{U}}_{s2}$ are estimates containing errors.

If the TLS formulation is used

$$\hat{\mathbf{\Psi}}_{TLS} = -\mathbf{V}_{12}\mathbf{V}_{22}^{-1}, \tag{9.122}$$

where $\mathbf{V}_{12}$ and $\mathbf{V}_{22}$ are $D \times D$ matrices defined by the eigendecomposition of the $2D \times 2D$ matrix

$$\tilde{\mathbf{C}} \triangleq \begin{bmatrix}\hat{\mathbf{U}}_{s_1}^H \\ \hat{\mathbf{U}}_{s_2}^H\end{bmatrix}\begin{bmatrix}\hat{\mathbf{U}}_{s_1} & \hat{\mathbf{U}}_{s_2}\end{bmatrix} = \begin{bmatrix}\mathbf{V}_{11} & \mathbf{V}_{12} \\ \mathbf{V}_{21} & \mathbf{V}_{22}\end{bmatrix}\mathbf{\Lambda}_E\begin{bmatrix}\mathbf{V}_{11}^H & \mathbf{V}_{21}^H \\ \mathbf{V}_{12}^H & \mathbf{V}_{22}^H\end{bmatrix}, \tag{9.123}$$

with

$$\mathbf{\Lambda}_E = \mathrm{diag}\left[\lambda_{E_1}, \lambda_{E_2}, \cdots, \lambda_{E_{2D}}\right], \tag{9.124}$$

where the eigenvalues are ordered,

$$\lambda_{E_1} \geq \lambda_{E_2} \geq \cdots \geq \lambda_{E_{2D}}. \tag{9.125}$$

The TLS-ESPRIT algorithm uses $\mathbf{\Psi}_{TLS}$ from (9.122) in steps 4 and 5 of the LS-ESPRIT algorithm to find $\hat{\lambda}_i, i = 1, \cdots, D$. We normally use the

TLS-ESPRIT algorithm for the examples in the text. One can show that LS ESPRIT and TLS ESPRIT have the same asymptotic ($K \rightarrow \infty$) variance (e.g., RH89a). However, TLS ESPRIT generally has better threshold behavior.

We can now summarize the steps in the TLS-ESPRIT algorithm:

1. Perform the eigendecomposition on $\mathbf{C_x}$ to obtain $\widehat{\mathbf{U}}_s$,

2. Find $\widehat{\mathbf{U}}_{s1}$ and $\widehat{\mathbf{U}}_{s2}$ by using (9.109) and (9.110),

3. Perform the $2D \times 2D$ eigendecomposition in (9.123) and use (9.122) to obtain $\hat{\mathbf{\Psi}}_{TLS}$,

4. Find the eigenvalues of $\hat{\mathbf{\Psi}}_{TLS}$, denoted by $\hat{\lambda}_1, \hat{\lambda}_2, \cdots, \hat{\lambda}_D$,

5. Find the estimates in ψ-space by using,

$$\hat{\psi}_i = \arg\hat{\lambda}_i, i = 1, 2, \cdots, D. \tag{9.126}$$

We will do several examples to illustrate ESPRIT performance in typical scenarios. Before doing the examples, there is an issue of interest.

The issue is the choice of d_s for the overlapping subarrays shown in Figure 9.15. We can rewrite the angle of the ith term in (9.107) as

$$d_s\pi \sin\bar{\theta}_i = d_s\pi u_i, \tag{9.127}$$

where $\bar{\theta}_i$ is the angle of the ith plane-wave signal to broadside. If $d_s = 1$, then we can determine u_i unambiguously for

$$-1 \leq u_i \leq 1, \tag{9.128}$$

or

$$-\frac{\pi}{2} \leq \theta_i \leq \frac{\pi}{2}. \tag{9.129}$$

However, for $d_s > 1$, the unambiguous range decreases to

$$-\frac{1}{d_s} \leq u_i \leq \frac{1}{d_s}. \tag{9.130}$$

For example, for the last set of subarrays in Figure 9.15, we would need to know the *a priori* location of the u_i in the range,

$$-0.2 \leq u_i \leq 0.2, \tag{9.131}$$

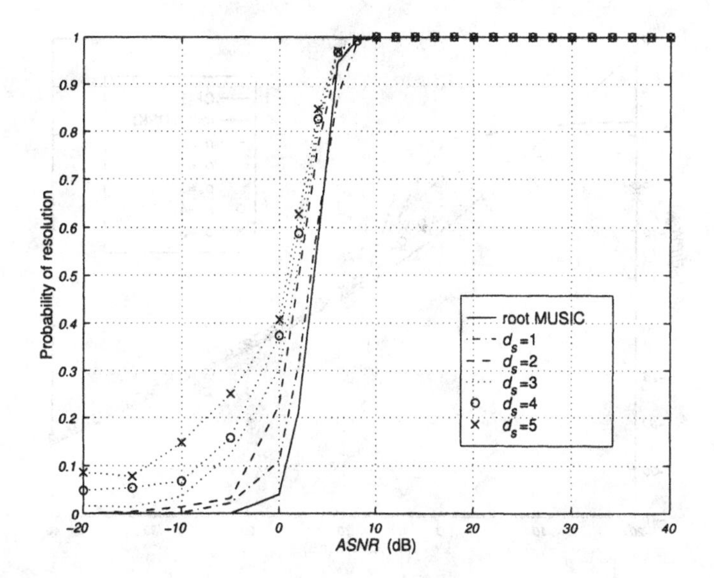

Figure 9.17 TLS-ESPRIT algorithm: probability of resolution versus *ASNR*; $N = 10, \rho = 0, K = 100$, 500 trials.

which corresponds to the BW_{NN} of the conventional beam pattern. Thus, in order to use $d_s > 1$, we either need *a priori* knowledge of the sector of u-space where the signals are located or we have to do some preprocessing to determine the sector. The limitations for unambiguous location follow directly from our classical array discussion in Chapter 2, because we can think of the ESPRIT algorithm as operating on a 2-element array where each element consists of an identical subarray.

We consider several examples to illustrate the performance of the ESPRIT algorithm.

Example 9.3.4 (continuation, Example 9.3.1)

Consider the same model as in Example 9.3.1. We have a standard 10-element linear array with two equal-power uncorrelated plane waves impinging on the array from $\pm\Delta\psi/2$, where $\Delta\psi = \Delta\psi_R$.

In Figure 9.17, we show the probability of resolution versus *ASNR* for TLS-ESPRIT and various subarrays. We see that, for $d_s = 1$, TLS-ESPRIT has slightly better resolution performance than root MUSIC. The resolution performance improves as d_s increases. However, for $d_s > 1$, the model assumes prior knowledge of the range of ψ (see (9.130)).

The normalized RMSE is shown in Figure 9.18. Note that, as the *ASNR* approaches zero, the RMSE approaches the variance of a random variable that is uniform over the *a priori* range ($Var(u) = 1/3d_s^2$). Above threshold, the RMSE is parallel to, but above, the CRB. For $d_s = 1$, the RMSE is about 1 dB above the bound.

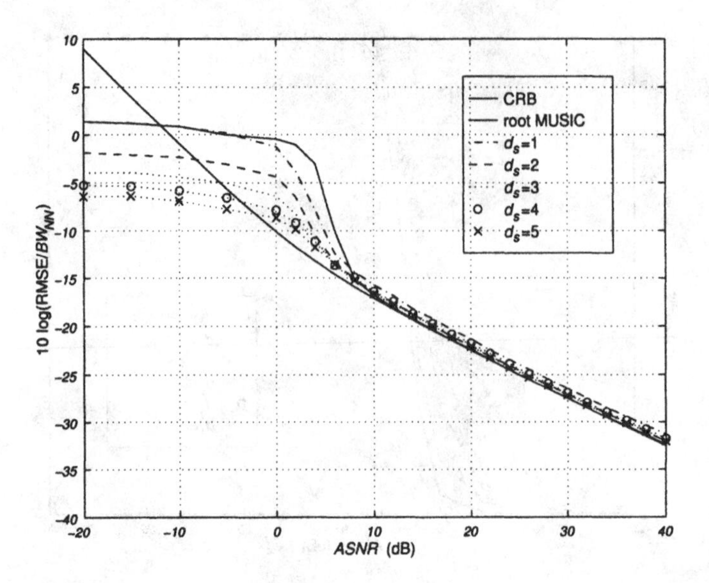

Figure 9.18 TLS-ESPRIT algorithm: $N = 10, \rho = 0, K = 100$, 500 trials; normalized RMSE versus *ASNR*.

Example 9.3.5 (continuation)

Consider the same model as in Example 9.3.4 except $\rho = 0.95$. The probability of resolution versus *ASNR* is shown in Figure 9.19. The normalized RMSE versus *ASNR* is shown in Figure 9.20. The results are similar to the $\rho = 0$ case, but are shifted to the right by about 15 dB.

One can improve the performance of the ESPRIT algorithm by a technique referred to as row weighting. This technique is described in [OVK91]. The basic idea is to weight the rows of $\widehat{\mathbf{U}}_s$ by modifying the identity matrix component of the selection matrix $\mathbf{J}_s$,

$$\tilde{\mathbf{J}}_s^{(m_s)} \triangleq \text{diag}\left[1, \sqrt{2}, \sqrt{3}, \cdots, \sqrt{w}, \sqrt{w}, \sqrt{w}, \cdots, \sqrt{3}, \sqrt{2}, 1\right], \tag{9.132}$$

where

$$w = \min\{m_s, N - m_s - d_s + 1\}. \tag{9.133}$$

The parameter m_s determines where the increase in weighting stops. In most cases, m_s is chosen as large as possible. We show the results with a simple example.

Example 9.3.6 (continuation)

Consider the same model as in Example 9.3.4 and assume $d_s = 1$. In Figure 9.21, we show the probability of resolution versus *ASNR* for various m_s. In Figure 9.22, we

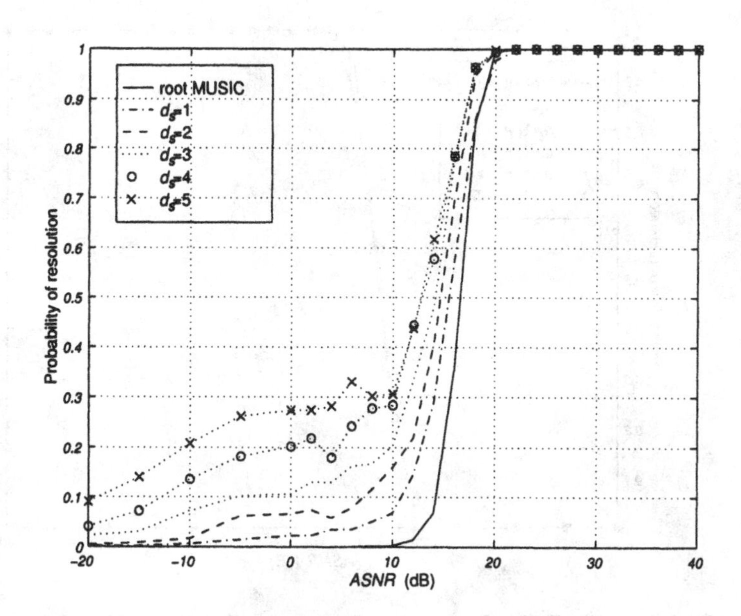

Figure 9.19 TLS-ESPRIT algorithm: $N = 10, \rho = 0.95, K = 100$, 500 trials, probability of resolution versus *ASNR*.

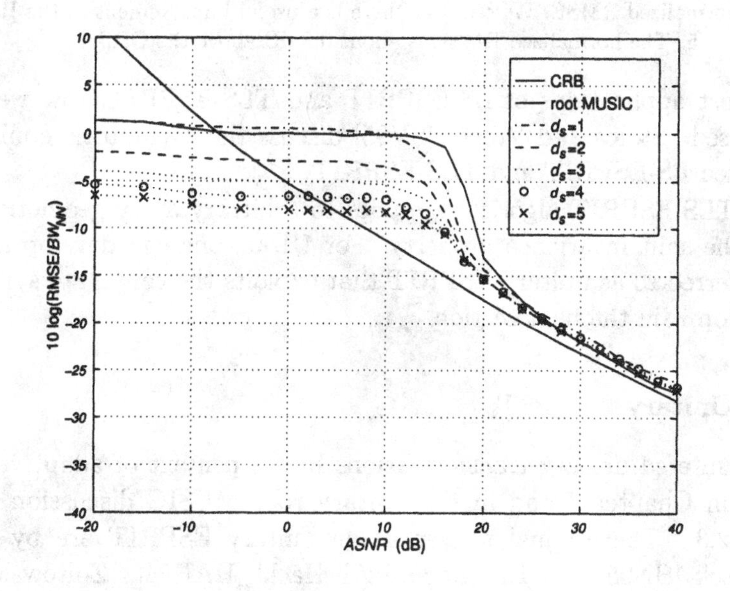

Figure 9.20 TLS-ESPRIT algorithm: $N = 10, \rho = 0.95, K = 100$, 500 trials; normalized RMSE versus *ASNR*.

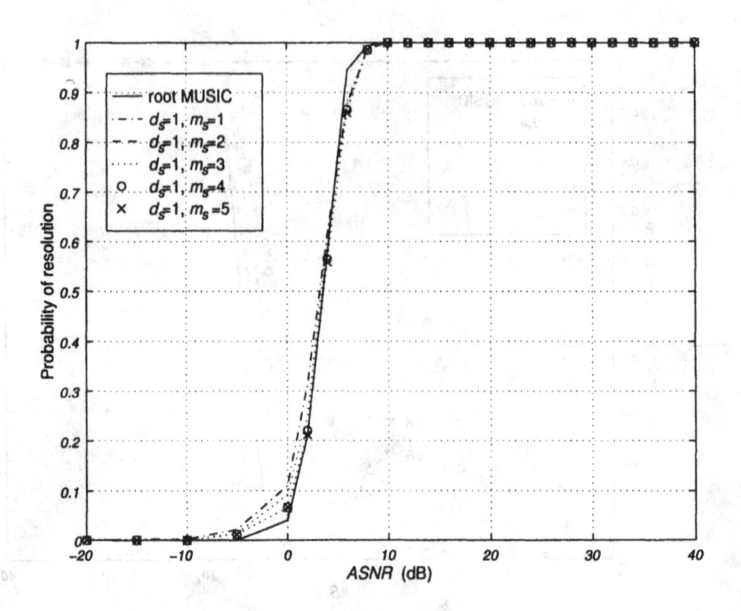

Figure 9.21 TLS-ESPRIT algorithm with row weighting: $N = 10, \rho = 0, K = 100$, 500 trials, probability of resolution versus *ASNR*, various m_s.

show the normalized RMSE. We see that there is a useful improvement in the RMSE for $m_s = 3, 4$, or 5. The normalized RMSE is about 0.2 dB above the CRB.

In most applications of LS-ESPRIT and TLS-ESPRIT, row weighting will be used. Stoica and Viberg [SV95] discuss the asymptotic equivalence of weighted LS-ESPRIT and TLS-ESPRIT.

The TLS-ESPRIT algorithm applies to arbitrary array geometries that exhibit the shift invariance property. For ULAs, one can develop an algorithm referred to as unitary ESPRIT that exploits the conjugate symmetry. We develop it in the next section.

9.3.4.2 Unitary ESPRIT

We encountered unitary transformations in the context of adaptive beamforming in Chapter 7 and in the unitary root MUSIC discussion in Section 9.3.2.3. The original references on unitary ESPRIT are by Haardt and Nossek [HN95] and Haardt and Ali-Hackl [HAH94]. Zoltowski et al. [ZHM96] provide a concise discussion of unitary ESPRIT as background for their development of 2-D unitary ESPRIT. Our discussion follows Section III of [ZHM96]. Unitary ESPRIT is one method of accomplishing FB

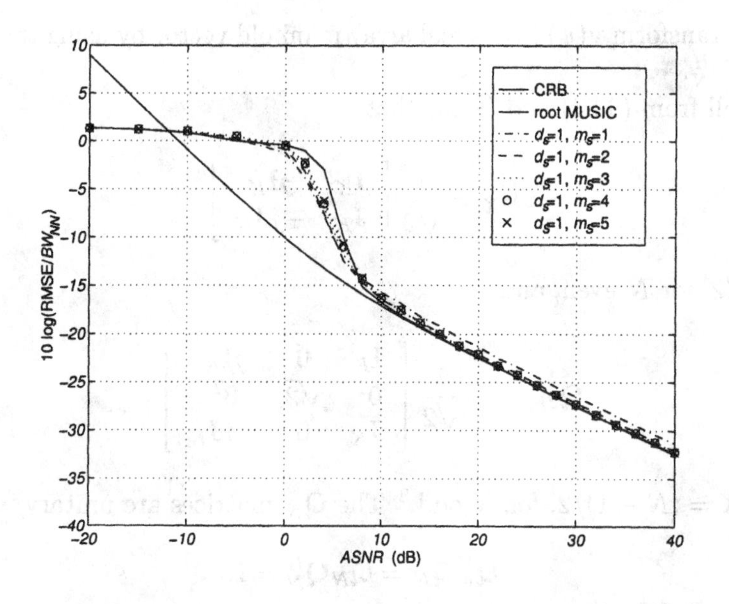

Figure 9.22 TLS-ESPRIT with row weighting, various m_s: $N = 10, \rho = 0, K = 100$, 500 trials; normalized RMSE versus *ASNR*.

averaging with ESPRIT. Other references that discuss FB-ESPRIT or FB-TLS-ESPRIT include Roy [Roy87], Mou and Zhang [MZ91], Zoltowski and Stavrinides [ZS89], Rao and Hari [RH93], and Bachl [Bac95].

We start with the invariance relationship in element space. We assume $d_s = 1$, and therefore $N_s = N - 1$. The selection matrices are $(N-1) \times N$,

$$\mathbf{J}_{s1} \triangleq \begin{bmatrix} 1 & 0 & \cdots & 0 & 0 \\ 0 & 1 & \cdots & 0 & 0 \\ \vdots & \vdots & \ddots & \vdots & \vdots \\ 0 & 0 & \cdots & 1 & 0 \end{bmatrix}, \tag{9.134}$$

and

$$\mathbf{J}_{s2} \triangleq \begin{bmatrix} 0 & 1 & 0 & \cdots & 0 \\ 0 & 0 & 1 & \cdots & 0 \\ \vdots & \vdots & \vdots & \ddots & \vdots \\ 0 & 0 & 0 & \cdots & 1 \end{bmatrix}. \tag{9.135}$$

Then,

$$e^{j\psi}\mathbf{J}_{s1}\mathbf{v}(\psi) = \mathbf{J}_{s2}\mathbf{v}(\psi). \tag{9.136}$$

We can transform $\mathbf{v}(\psi)$ into a real array manifold vector by using the unitary matrix $\mathbf{Q}_N^H$.

Recall from (9.54) and (9.55) that

$$\mathbf{Q}_{2K} = \frac{1}{\sqrt{2}} \begin{bmatrix} \mathbf{I}_K & j\mathbf{I}_K \\ \mathbf{J}_K & -j\mathbf{J}_K \end{bmatrix}, \tag{9.137}$$

$K = N/2$, for N even, and

$$\mathbf{Q}_{2K+1} = \frac{1}{\sqrt{2}} \begin{bmatrix} \mathbf{I}_K & \mathbf{0} & j\mathbf{I}_K \\ \mathbf{0}^T & \sqrt{2} & \mathbf{0}^T \\ \mathbf{J}_K & \mathbf{0} & -j\mathbf{J}_K \end{bmatrix}, \tag{9.138}$$

where $K = (N-1)/2$, for N odd.[8] The $\mathbf{Q}_N$ matrices are unitary,

$$\mathbf{Q}_N^H \mathbf{Q}_N = \mathbf{Q}_N \mathbf{Q}_N^H = \mathbf{I}. \tag{9.139}$$

Premultiplying $\mathbf{v}(\psi)$ by $\mathbf{Q}_N^H$ gives a real array manifold vector,

$$\mathbf{v}_R(\psi) = \mathbf{Q}_N^H \mathbf{v}(\psi). \tag{9.140}$$

Because $\mathbf{Q}_N$ is unitary, we can write (9.136) as

$$e^{j\psi} \mathbf{J}_{s1} \mathbf{Q}_N \mathbf{Q}_N^H \mathbf{v}(\psi) = \mathbf{J}_{s2} \mathbf{Q}_N \mathbf{Q}_N^H \mathbf{v}(\psi), \tag{9.141}$$

or, using (9.140),

$$e^{j\psi} \mathbf{J}_{s1} \mathbf{Q}_N \mathbf{v}_R(\psi) = \mathbf{J}_{s2} \mathbf{Q}_N \mathbf{v}_R(\psi). \tag{9.142}$$

Pre-multiplying by $\mathbf{Q}_{N-1}^H$ gives,

$$e^{j\psi} \mathbf{Q}_{N-1}^H \mathbf{J}_{s1} \mathbf{Q}_N \mathbf{v}_R(\psi) = \mathbf{Q}_{N-1}^H \mathbf{J}_{s2} \mathbf{Q}_N \mathbf{v}_R(\psi). \tag{9.143}$$

Observing that

$$\mathbf{J}_{N-1} \mathbf{J}_{s2} \mathbf{J}_N = \mathbf{J}_{s1}, \tag{9.144}$$

and

$$\mathbf{J}_N \mathbf{Q}_N = \mathbf{Q}_N^*, \tag{9.145}$$

[8] There are two different $\mathbf{J}$ matrices in this section. We have added a subscript (in capital letters) to the exchange matrix $\mathbf{J}$ to denote its dimensionality. The selection matrices J_{s1} and J_{s2} have lowercase subscripts.

we can write

$$\begin{aligned}\mathbf{Q}_{N-1}^{H}\mathbf{J}_{s2}\mathbf{Q}_{N} &= \mathbf{Q}_{N-1}^{H}\mathbf{J}_{N-1}\mathbf{J}_{N-1}\mathbf{J}_{s2}\mathbf{J}_{N}\mathbf{J}_{N}\mathbf{Q}_{N} \\ &= \mathbf{Q}_{N-1}^{T}\mathbf{J}_{s1}\mathbf{Q}_{N}^{*} \\ &= \left(\mathbf{Q}_{N-1}^{H}\mathbf{J}_{s1}\mathbf{Q}_{N}\right)^{*}. \end{aligned} \tag{9.146}$$

We now define the real and imaginary parts of the left side of (9.146)

$$\mathbf{K}_1 \triangleq Re\left\{\mathbf{Q}_{N-1}^{H}\mathbf{J}_{s2}\mathbf{Q}_{N}\right\}, \tag{9.147}$$

and

$$\mathbf{K}_2 \triangleq Im\left\{\mathbf{Q}_{N-1}^{H}\mathbf{J}_{s2}\mathbf{Q}_{N}\right\}. \tag{9.148}$$

Using (9.147) and (9.148) in (9.143) and incorporating the result in (9.146) gives

$$e^{j\frac{\psi}{2}}\left(\mathbf{K}_1 - j\mathbf{K}_2\right)\mathbf{v}_R(\psi) = e^{-j\frac{\psi}{2}}\left(\mathbf{K}_1 + j\mathbf{K}_2\right)\mathbf{v}_R(\psi), \tag{9.149}$$

which can be written as,

$$\left(e^{j\frac{\psi}{2}} - e^{-j\frac{\psi}{2}}\right)\mathbf{K}_1\mathbf{v}_R(\psi) = j\left(e^{j\frac{\psi}{2}} + e^{-j\frac{\psi}{2}}\right)\mathbf{K}_2\mathbf{v}_R(\psi). \tag{9.150}$$

We now have the invariance relation,

$$\tan\left(\frac{\psi}{2}\right)\mathbf{K}_1\mathbf{v}_R(\psi) = \mathbf{K}_2\mathbf{v}_R(\psi), \tag{9.151}$$

which has only real-valued elements. Defining the array manifold matrix,

$$\mathbf{V}_R(\psi) = \left[\begin{array}{cccc}\mathbf{v}_R(\psi_1) & \mathbf{v}_R(\psi_2) & \cdots & \mathbf{v}_R(\psi_D)\end{array}\right], \tag{9.152}$$

we can write (9.151) as,

$$\mathbf{K}_1\mathbf{V}_R(\psi)\mathbf{\Omega}_{\psi} = \mathbf{K}_2\mathbf{V}_R(\psi), \tag{9.153}$$

where

$$\mathbf{\Omega}_{\psi} = \text{diag}\left[\tan\left(\frac{\psi_1}{2}\right), \cdots, \tan\left(\frac{\psi_D}{2}\right)\right]. \tag{9.154}$$

The next step is to find the signal subspace eigenvectors corresponding to $\mathbf{V}_R(\psi)$. If we define $\mathbf{X}$ as the $N \times K$ data matrix and

$$\mathbf{Y} = \mathbf{Q}_N^H\mathbf{X}, \tag{9.155}$$

then the largest left singular vectors of

$$\mathbf{Y}_E = [Re\mathbf{Y}, Im\mathbf{Y}], \tag{9.156}$$

are the real signal subspace eigenvectors. We denote the real signal subspace by $\hat{\mathbf{U}}_{RS}$.

Alternatively, we can find the eigenvectors of

$$\hat{\mathbf{S}}_R = Re\left\{\mathbf{Q}_N^H \hat{\mathbf{S}}_{\mathbf{x}} \mathbf{Q}_N\right\}, \tag{9.157}$$

corresponding to the D largest eigenvalues, and they will define $\hat{\mathbf{U}}_{RS}$.

The signal eigenvector matrix is related to $\mathbf{V}_R(\psi)$ by the real-valued $D \times D$ unknown matrix $\mathbf{T}$,

$$\mathbf{U}_{RS} = \mathbf{V}_R \mathbf{T}, \tag{9.158}$$

or

$$\mathbf{V}_R = \mathbf{U}_{RS} \mathbf{T}^{-1}. \tag{9.159}$$

Substituting (9.159) into (9.153) gives,

$$\mathbf{K}_1 \mathbf{U}_{RS} \boldsymbol{\Psi} = \mathbf{K}_2 \mathbf{U}_{RS}, \tag{9.160}$$

where

$$\boldsymbol{\Psi} = \mathbf{T}^{-1} \boldsymbol{\Omega}_{\psi} \mathbf{T}. \tag{9.161}$$

The eigenvalues of the $D \times D$ solution $\boldsymbol{\Psi}$ to (9.160) correspond to $\tan(\psi_i/2)$. Substituting $\hat{\mathbf{U}}_{RS}$ for $\mathbf{U}_{RS}$ in (9.160) gives

$$\mathbf{K}_1 \hat{\mathbf{U}}_{RS} \hat{\boldsymbol{\Psi}} = \mathbf{K}_2 \hat{\mathbf{U}}_{RS}. \tag{9.162}$$

The form of (9.162) is identical to that in (9.119). Thus, unitary LS-ESPRIT uses (9.121) with $\hat{\mathbf{U}}_{s1}$ replaced by $\mathbf{K}_1 \hat{\mathbf{U}}_{RS}$ and $\hat{\mathbf{U}}_{s2}$ replaced by $\mathbf{K}_2 \hat{\mathbf{U}}_{RS}$,

$$\hat{\boldsymbol{\Psi}}_{ULS} = \left[\left[\mathbf{K}_1 \hat{\mathbf{U}}_{RS}\right] [\mathbf{K}_1 \mathbf{U}_{RS}]\right]^{-1} \left[\mathbf{K}_1 \hat{\mathbf{U}}_{RS}\right]^T \left[\mathbf{K}_2 \hat{\mathbf{U}}_{RS}\right]. \tag{9.163}$$

For unitary TLS-ESPRIT, define

$$\tilde{\mathbf{C}} = [\mathbf{K}_1 \hat{\mathbf{U}}_{RS} \quad \mathbf{K}_2 \hat{\mathbf{U}}_{RS}]^T [\mathbf{K}_1 \hat{\mathbf{U}}_{RS} \quad \mathbf{K}_2 \hat{\mathbf{U}}_{RS}]. \tag{9.164}$$

The eigendecomposition of $\tilde{\mathbf{C}}$ gives $\mathbf{V}_{12}$ and $\mathbf{V}_{22}$ (see (9.122)–(9.125)). Then,

$$\hat{\boldsymbol{\Psi}}_{UTLS} = -\mathbf{V}_{12} \mathbf{V}_{22}^{-1}. \tag{9.165}$$

We summarize the steps for the unitary ESPRIT algorithms:

1. Find $\widehat{\mathbf{U}}_{RS}$ by either finding the eigenvectors with the largest eigenvalues of $Re\left\{\mathbf{Q}_N^H \widehat{\mathbf{S}}_{\mathbf{x}} \mathbf{Q}_N\right\}$ or by computing the largest left singular vectors of

$$\mathbf{Y}_E = \left[Re\left[\mathbf{Q}_N^H \mathbf{X}\right], Im\left[\mathbf{Q}_N^H \mathbf{X}\right]\right], \tag{9.166}$$

2. Find either $\hat{\mathbf{\Psi}}_{ULS}$ from (9.163) or $\hat{\mathbf{\Psi}}_{UTLS}$ from (9.165),

3. Compute $\hat{\omega}_i, i = 1, \cdots, D$, the eigenvalues of the real-valued matrix $\hat{\mathbf{\Psi}}_{ULS}$ or $\hat{\mathbf{\Psi}}_{UTLS}$,

4. Compute $\hat{\psi}_i = 2\tan^{-1}(\hat{\omega}_i), i = 1, \cdots, D$.

We consider an example to illustrate the performance of unitary ESPRIT.

Example 9.3.7 (continuation, Example 9.3.4)
Consider the same model as in Example 9.3.4. We have a standard 10-element linear array with two equal-power uncorrelated signals impinging on the array from $\pm\Delta\psi/2$, where $\Delta\psi = \Delta\psi_R$.
In Figure 9.23(a), we plot the probability of resolution versus *ASNR*. In Figure 9.23(b), we plot the normalized RMSE versus *ASNR*. In Figure 9.24, we show a plot of the roots for 100 trials. By imposing the conjugate symmetry constraint (recall the IQML discussion) the roots appear on the unit circle when we are above threshold. We see that the threshold performance is a little better than TLS-ESPRIT due to FB averaging. However there are significant computational savings.

In the next example, we consider weighted unitary ESPRIT.

Example 9.3.8 (continuation)
Consider the same model as in Example 9.3.7. We use a weighted $\mathbf{J}_{s1}$ matrix with

$$\mathbf{J}_{s1} = \begin{bmatrix} 1 & 0 & \cdots & & & & 0 \\ 0 & \sqrt{2} & & & & & \vdots \\ \vdots & & \sqrt{3} & & & & \\ & & & \sqrt{m_s} & & & \\ & & & & \ddots & & \\ & & & & & \sqrt{m_s} & \\ & & & & & & \ddots \\ 0 & \cdots & & & & 1 & 0 \end{bmatrix}, \tag{9.167}$$

where $m_s \leq N/2$ and simulate $m_s = 2, \cdots, 5$. In Figure 9.25, we plot the probability of resolution and the normalized RMSE versus *ASNR*. We see that, for $m_s \geq 2$, the RMSE essentially equals the CRB.

Gershman and Haardt [GH99] have developed a version of unitary ESPRIT that uses pseudonoise resampling to try and eliminate outlier estimates. It relies on earlier work by Gershman [Ger98], Gershman and Böhme

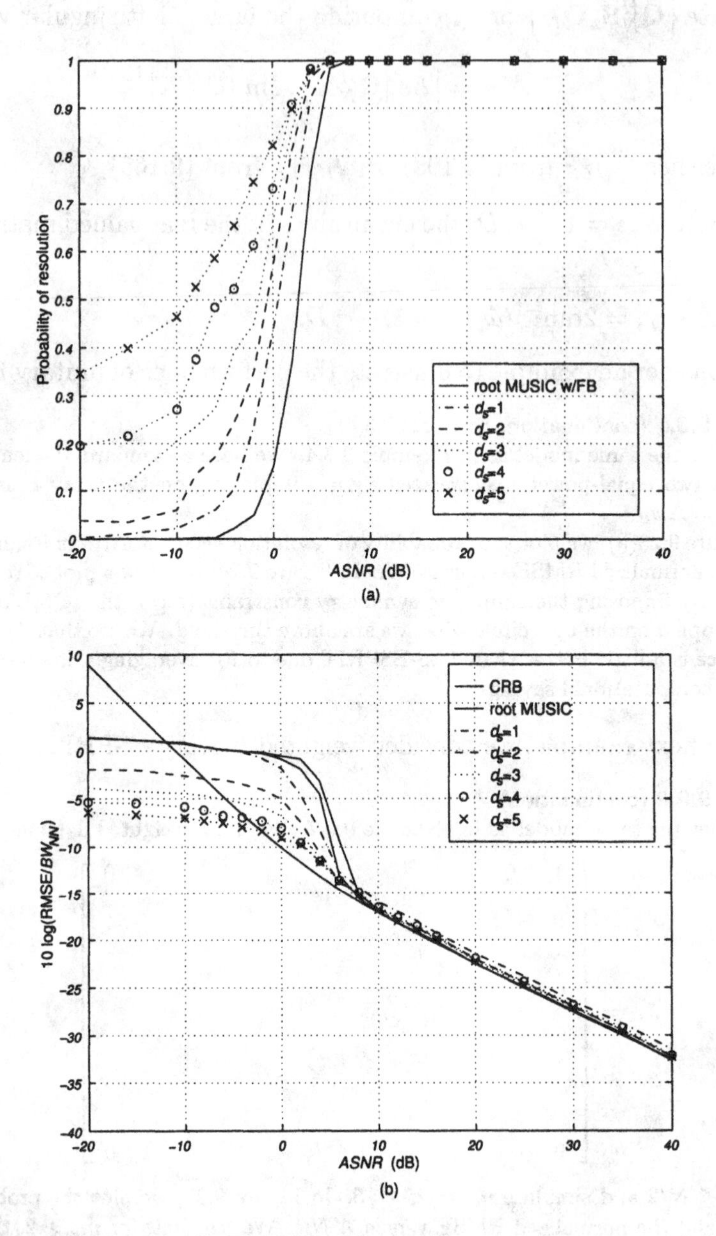

Figure 9.23 Unitary TLS-ESPRIT: $N = 10, \rho = 0, d_s = 1, \cdots, 5, K = 100$, 500 trials: (a) probability of resolution versus *ASNR*; (b) normalized RMSE versus *ASNR*.

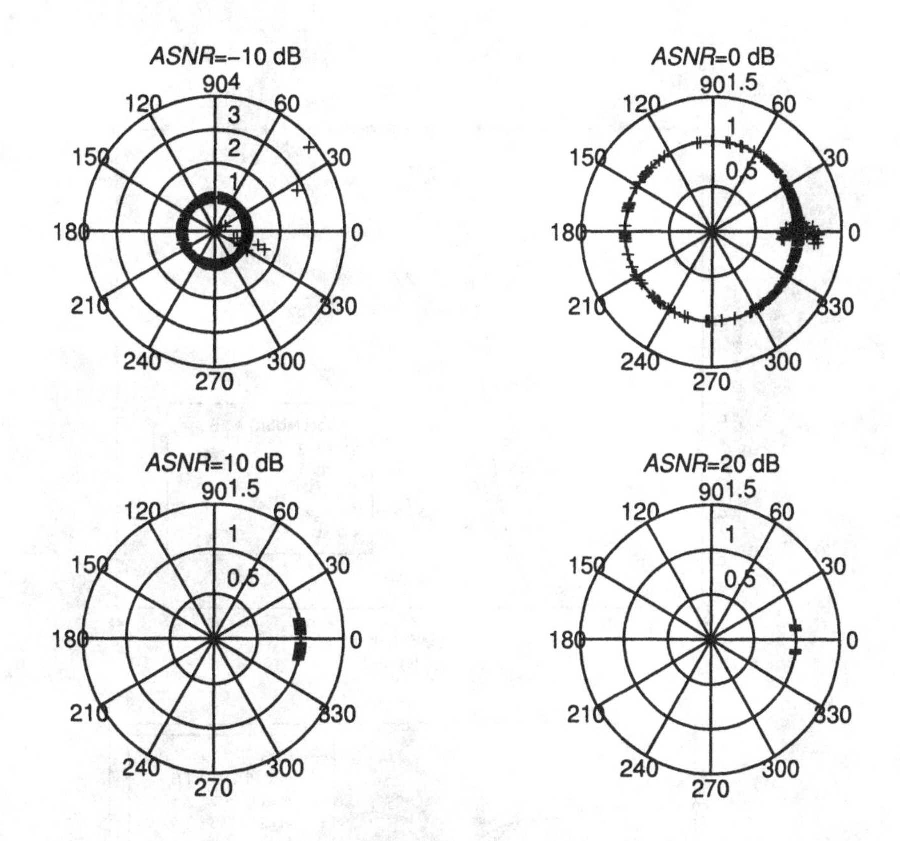

Figure 9.24 Unitary TLS-ESPRIT: location of estimates for various *ASNR*: $K = 100$, 200 trials.

[GB97], [GB98], and Efron [Efr79]. The technique offers some performance improvement in the vicinity of the threshold. The reader is referred to these references for further discussion.

We have discussed unitary transformations in the context of root MUSIC and ESPRIT. Whenever FB averaging can be used, we can develop a unitary algorithm. Huarng and Yeh [HY91] developed a unitary version of spectral MUSIC. Gershman and Stoica [GS99] developed a unitary version of MODE (see [SJ97] also).

9.3.4.3 Summary of ESPRIT

In Section 9.3.4, we have developed LS-ESPRIT and TLS-ESPRIT and unitary ESPRIT and studied their performance. The LS- and TLS-ESPRIT algorithms with row weighting achieve an RMSE that is very close to the CRB

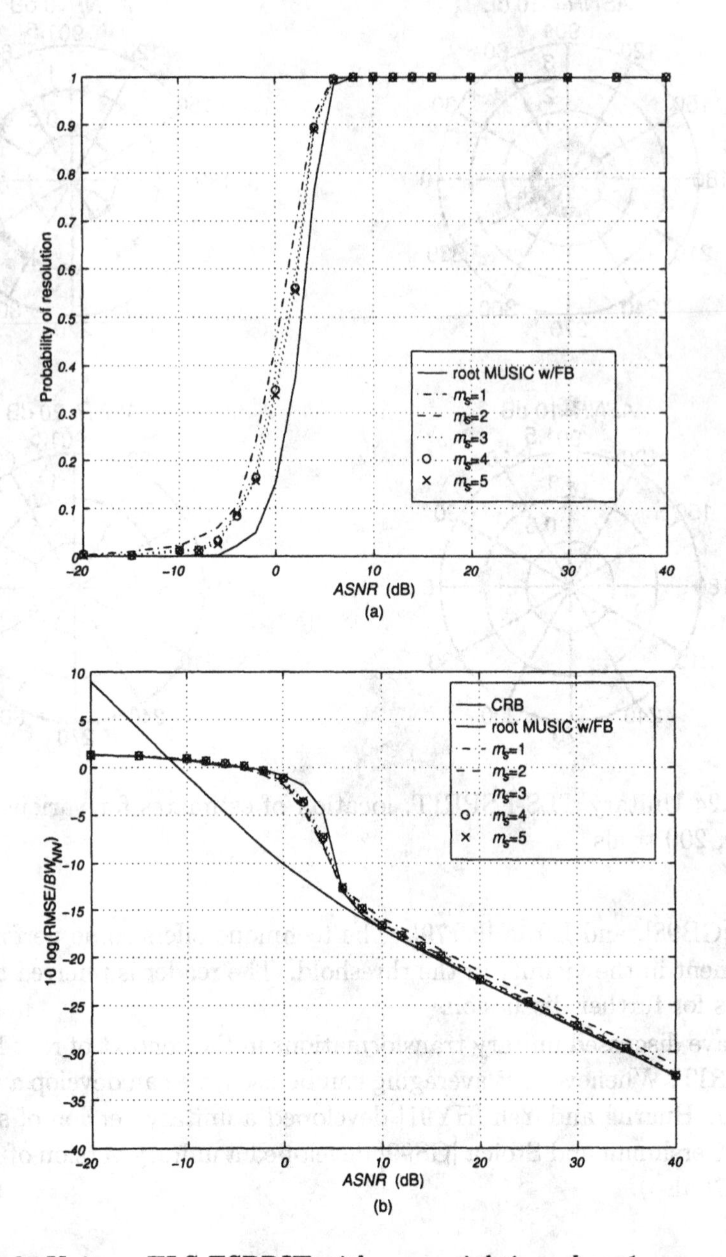

Figure 9.25 Unitary TLS-ESPRIT with row weighting, $d_s = 1, m_s = 2, \cdots, 5$: $N = 10, \rho = 0, K = 100$, 500 trials: (a) probability of resolution versus *ASNR*; (b) normalized RMSE versus *ASNR*.

for *ASNR*s above threshold. For SLAs, unitary ESPRIT using appropriate row weighting provides similar performance with reduced computation.

ESPRIT can also be used in a broader class of array geometries. The requirement is that the array exhibit a shift invariance property.

9.3.5 Algorithm Comparison

In Sections 8.7 and 9.3, we have developed a number of algorithms that are computationally feasible and have good performance for SLAs. Some of these algorithms can be adapted for other array geometries, but our current discussion will focus on SLAs.

Our examples up to this point suggest that the three algorithms that should receive further study are:

(i) Unitary IMODE,

(ii) Unitary root MUSIC,

(iii) Unitary TLS-ESPRIT with $d_s = 1, m_s = (N+1)/2$ for N odd and $m_s = N/2$ for N even.

We did not derive unitary IMODE in Chapter 8 because the issue of computational complexity was not emphasized. Unitary IMODE is derived in Gershmann and Stoica [GS99]. The derivation is summarized in Problem 9.3.30. We have used unitary IMODE in this comparison because we want to compare computational complexity.

Unitary IMODE is identical to FB-MODE, which is derived and analyzed in Stoica and Jansson [SJ97]. FB-IMODE is identical to standard IMODE except FB averaging is used to construct the estimate of the spectral matrix. As discussed in [GS99], FB-MODE (or unitary MODE) has better threshold performance for uncorrelated signals. We discuss the correlated signal case in Section 9.6.

We consider two examples to compare their performance. In this section, we consider uncorrelated signals. The correlated and coherent signals case is discussed in Section 9.6.

Example 9.3.9 (continuation)

Consider a standard 10-element array. There are two uncorrelated equal-power plane-wave signals impinging on the array from $\pm\Delta_R/2$. We assume that the number of signals is known. The three algorithms are simulated. In Figure 9.26(a), the normalized RMSE is plotted versus *ASNR*. In Figure 9.26(b), the flop count for the three algorithms is plotted versus *ASNR*.

The unitary IMODE algorithm has the best threshold performance (*ASNR*=5 dB). The threshold of unitary root MUSIC is slightly higher (*ASNR* = 6 dB). The RMSE of the unitary TLS-ESPRIT algorithm approaches the CRB in a more gradual manner. All of the algorithms are close to the CRB above threshold.

The unitary TLS-ESPRIT algorithm has the smallest flop count. Both unitary IMODE and unitary root MUSIC have a slightly higher flop count.

Example 9.3.10 (continuation)

Consider the same model as in Example 9.3.9, except the number of signals, D, is unknown. We use one of the detection algorithms developed in Section 7.8 to find $\hat{D}$. We then use $\hat{D}$ in the three parameter estimation algorithms from Example 9.3.9.

It is important not to underestimate D, so we consider detection algorithms where P_M is close to zero for all *ASNR*s above the resolution threshold of the estimation algorithms (*ASNR* $= 5$ dB in this example). In order to evaluate the performance, we assume that, if $\hat{D} > D$, that the RMSE of the D estimates that are closest to the true values are calculated and the other $(\hat{D} - D)$ estimates are ignored. The rationale for this approach is that a subsequent processing step will discard the extra estimates. We must also consider P_M in order to evaluate the algorithm's performance. Two detection algorithms are used:

(i) AIC-FB
(ii) MDL-FB

The results are shown in Figure 9.27. In Figure 9.27(a), the probability that $\hat{D} \geq D$ is plotted. This curve is equal to 1-P_M. In Figure 9.27(b), the normalized RMSE is plotted versus *ASNR* for the AIC-FB algorithm. In Figure 9.27(c), the normalized RMSE is plotted versus *ASNR* for the MDL-FB algorithm. In Figure 9.27(d), the normalized RMSE is plotted *ASNR* for the MDL-FB+1 algorithms.

For the AIC-FB detection algorithm, the threshold occurs at the same *ASNR* as in the known D case. Above the threshold, the IMODE algorithm and the unitary root MUSIC algorithm reach the CRB. The overestimation degrades the performance of the unitary TLS-ESPRIT algorithm and the RMSE is slightly above the CRB.

For the MDL-FB algorithm, the threshold occurs at an *ASNR* that is about 2.5 dB higher than the known D case. Above threshold, the probability that $\hat{D} = D$ approaches one quickly, so the algorithms behave as in the known D case.

Based on this example and other cases developed in the problems, it appears that the best approach is to use AIC-FB as the detection algorithm. The combined detection and estimation problem should be analyzed (or simulated) for the scenario of interest to see the effect of unknown D on the estimation performance.

9.3.6 Summary

In this section, we have developed several algorithms that provide good performance with reasonable computational complexity.

Three versions of the MUSIC algorithm were developed: spectral MUSIC, root MUSIC, and unitary root MUSIC. Spectral MUSIC is applicable to arbitrary arrays, but the threshold occurs at a higher *ASNR* than the ML algorithms. For uncorrelated signals and high *ASNR*, the RMSE of the spectral MUSIC algorithm is close to the CRB. The root MUSIC algorithm

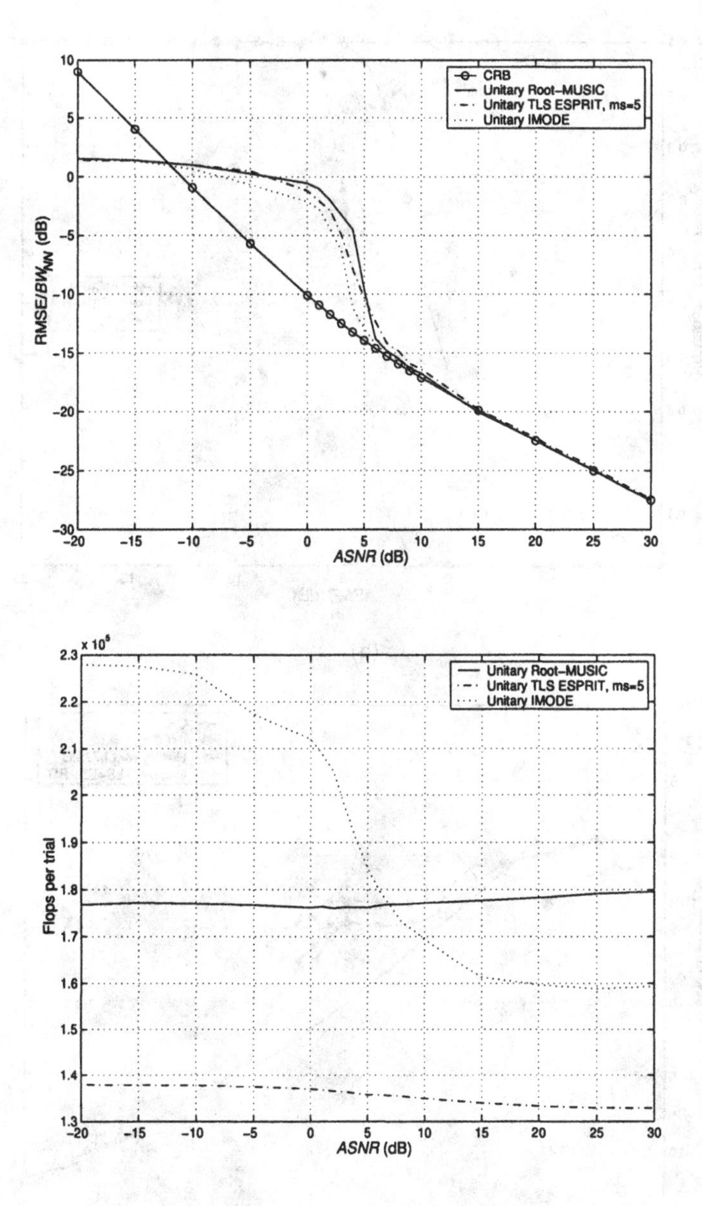

Figure 9.26 Performance of unitary IMODE, unitary root MUSIC, and unitary TLS-ESPRIT: $N = 10, K = 100$, two uncorrelated plane-wave signals at $\pm\Delta_R/2$, 1,000 trials: (a) normalized RMSE versus *ASNR*; (b) MATLAB® flop count versus *ASNR*.

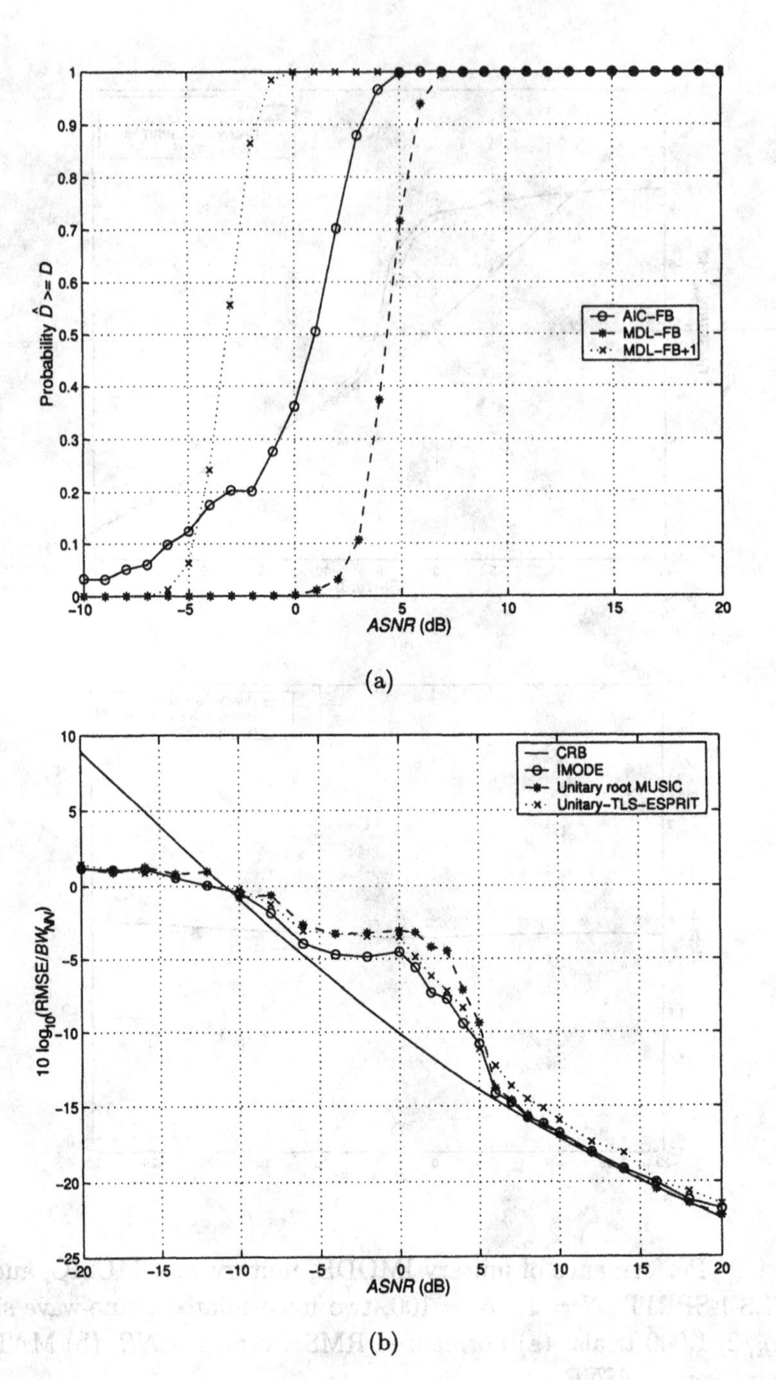

Probability $\hat{D}$ >= D
ASNR (dB)
AIC-FB
MDL-FB
MDL-FB+1
(a)
10 log10(RMSE/BWNN)
ASNR (dB)
CRB
IMODE
Unitary root MUSIC
Unitary-TLS-ESPRIT
(b)

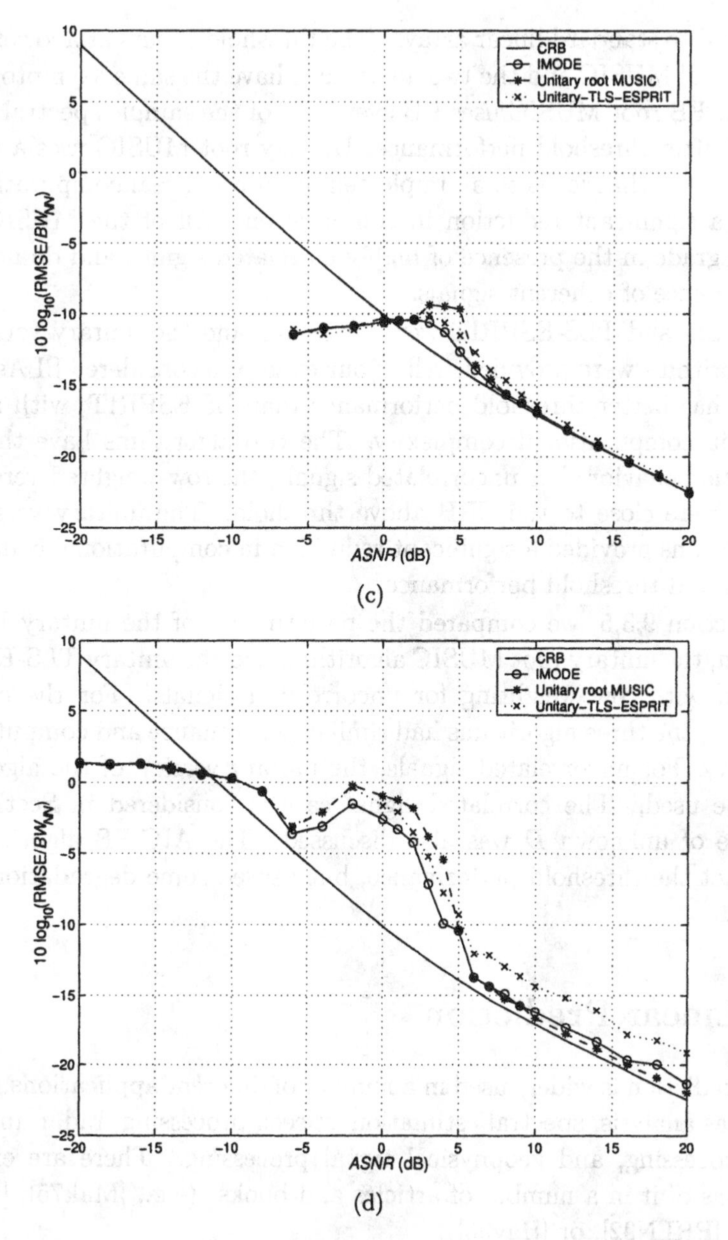

Figure 9.27 Performance of IMODE, unitary root MUSIC, and unitary TLS-ESPRIT: $N = 10, K = 100$, two uncorrelated plane-wave signals at $\pm\Delta_R/2$, D is estimated using AIC-FB and MDL-FB algorithms, 1000 trials: (a) $Pr(\hat{D} \geq D)$ versus *ASNR*; (b) normalized RMSE versus *ASNR*, AIC-FB detection. (c) RMSE versus *ASNR*, MDL-FB detection; (d) RMSE versus *ASNR*, MDL-FB+1 detection.

was applied to standard linear arrays. The threshold occurs at a lower *ASNR* than spectral MUSIC and the two algorithms have the same asymptotic performance. FB root MUSIC uses FB averaging of the sample spectral matrix and has better threshold performance. Unitary root MUSIC uses a unitary transformation that leads to an implementation using real computation and provides a significant reduction in computation. All of the MUSIC algorithms degrade in the presence of highly correlated signals and do not work in the presence of coherent signals.

Both LS- and TLS-ESPRIT were developed and the unitary versions of both algorithms were provided. All of our examples considered SLAs. TLS-ESPRIT has better threshold performance than LS-ESPRIT, with a slight increase in computational complexity. The two algorithms have the same asymptotic behavior. For uncorrelated signals, the row-weighted versions of ESPRIT were close to the CRB above threshold. The unitary versions of the algorithms provided a significant reduction in computational complexity and improved threshold performance.

In Section 9.3.5, we compared the performance of the unitary IMODE algorithm, the unitary root MUSIC algorithm, and the unitary TLS-ESPRIT algorithm with row weighting for uncorrelated signals. For the example considered, the three algorithms had similar performance and computational complexity. For uncorrelated signals, the unitary version of the algorithms should be used. The correlated signal case is considered in Section 9.6. The issue of unknown D was also discussed. The AIC-FB algorithm did not impact the threshold performance, but caused some degradation above threshold.

9.4 Linear Prediction

Linear prediction is widely used in a number of different applications, such as time-series analysis, spectral estimation, speech processing, radar and sonar signal processing, and geophysical signal processing. There are excellent discussions of it in a number of articles and books. (e.g., [Mak75], [Kay88], [Mar87], [PRLN92], or [Hay96]).

The application of linear prediction techniques to the DOA estimation problem in antenna arrays has been discussed by Schmidt [Sch81], Gabriel [Gab80], Tufts and Kumaresan [TK82], [KT83], Johnson and De Graaf [JD82]. We investigated these algorithms and considered both full-array linear prediction and subarray linear prediction. We found that, for estimating the DOAs of plane-wave signals the LP algorithms did not perform as well as MUSIC and ESPRIT. The reader is referred to the above references for further discussion.

The comment does not imply that LP techniques would not be useful for spatial spectral estimation. These algorithms are the maximum entropy algorithms introduced by Burg (e.g., [Bur67] [Bur75]) and Van der Bos [VDB71]. We do not discuss the spatial spectral estimation problem.

9.5 Asymptotic Performance

Up to this point in our discussions we have relied on simulations to evaluate the performance of the various algorithms. We now want to derive some analytic comparisons.[9]

In this section we discuss the asymptotic performance of the various subspace algorithms. Asymptotic performance analysis studies the behavior of the algorithms as K, the number of snapshots, goes to infinity. One way to study the asymptotic performance is to use a series expansion and retain the appropriate first-order and second-order terms and neglect terms of $O(K^{-1})$. We then compare these asymptotic expressions to simulations to assess their accuracy for smaller values of K.

In Section 9.5.1, we discuss how the probability of resolution and the MSE are related. In Section 9.5.2, we analyze the resolution performance of some of the algorithms developed in Sections 9.2 and 9.3. In Section 9.5.3, we analyze the small error behavior of these algorithms. In Section 9.5.4, we summarize our results.

Most of the analyses are reasonably detailed. Our approach is to outline the steps in the analyses, give the key results, and illustrate them with examples.

9.5.1 Error Behavior

In this section, we discuss how the probability of resolution and the MSE are related. We use spectral MUSIC as an example, but the principles apply to any estimator. We first consider the resolution behavior. Our experimental results in Figures 9.9 and 9.10 showed that, for this particular example, we could divide the *ASNR* axis into the following regions (the endpoints are approximate at this point in the discussion):

[9]This section contains a number of detailed analytic expressions. Sections 9.5.2 and 9.5.3 can be omitted (or skimmed) at first reading without loss of continuity.

(i) *ASNR* $\geq$ 20 dB; the RMSE is very close to the CRB for unbiased estimation. P_R is equal to unity.

(ii) 15 dB $\leq$ *ASNR* $\leq$ 20 dB; the RMSE increases slightly above the CRB. We find that this increase is due to a non-zero "local bias." P_R is still equal to unity.

(iii) 10 dB $\leq$ *ASNR* $\leq$ 15 dB; the P_R decreases from unity to 0.5; the RMSE increases dramatically over this range (about 17 dB). We need to determine the cause of this increase. We refer to this as the upper threshold region.

(iv) 0 dB $\leq$ *ASNR* $\leq$ 10 dB; the P_R decreases from 0.5 to zero; the RMSE increases gradually. We refer to this as the lower threshold region.

(v) *ASNR* $\leq$ 0 dB; the RMSE for each signal approaches the variance of a uniform random variable (1/3 in u-space). We refer to this as the *a priori* region.

The behavior in regions 1 and 2 can be accurately described by a local error analysis, which we carry out in Section 9.5.3. In this section, we characterize the behavior in the region 3, the upper threshold region.

When the two sources are separated by much less than the beamwidth of the array, then a key issue in the threshold region is whether the null spectrum will exhibit one or two peaks in the region of the sources. This behavior is shown in Figure 9.28, where we show representative MUSIC spectra for four *ASNR*s. If we fix the number of snapshots and decrease the *SNR*, then the null spectrum proceeds from the representative shape in (a) to the shape in (d). Similarly, if we fix the *SNR* and decrease the number of snapshots, a similar progression occurs.

We see that there are three types of behavior:

(i) In (a) and (b), the signals are resolved. In this region, we can assume the errors are small, and we analyze the bias and variance using an asymptotic analysis.

(ii) In (c), the spectrum contains a single peak in the region of the two signals. We analyze when this transition to a single peak occurs. We refer to this as the resolution problem.

(iii) In (d), the single peak due to the signals is no longer distinguishable and the estimator selects two peaks due to noise.

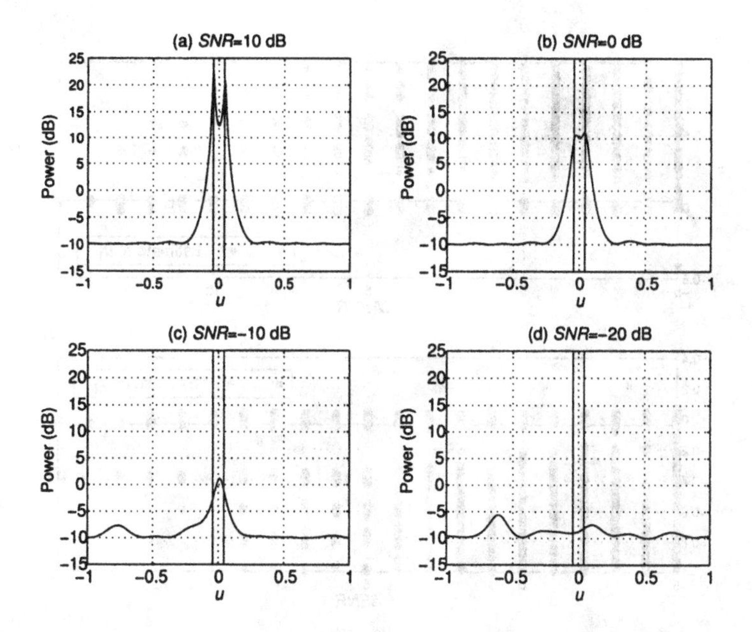

Figure 9.28 Behavior of a typical MUSIC spectrum: (a) *ASNR* = 20 dB; (b) *ASNR* = 10 dB; (c) *ASNR* = 0 dB; and (d) *ASNR* = −10 dB.

We first consider the probability of resolution, P_R. We would like to determine the probability that there are two peaks in the MUSIC spectrum in the "correct region." We use the experimental criterion that the two signals are resolved if there are two peaks and

$$\left|\hat{\psi}_i - \psi_i\right| \leq \min\left\{\frac{\Delta\psi}{2}, \frac{2\pi}{N}\right\}, \quad i = 1, 2. \tag{9.168}$$

In order to derive the P_R analytically, we denote the actual location of the sources by ψ_1 and ψ_2 and then define the midpoint,

$$\psi_m = \frac{\psi_1 + \psi_2}{2}. \tag{9.169}$$

One criterion of resolution is that

$$D(\psi_m) \geq D(\psi_i), \quad i = 1, 2, \tag{9.170}$$

where $D(\psi) = (\hat{Q}_{MU}(\psi))^{-1}$. If (9.170) is satisfied, then our estimator would not resolve the two signals. A second criterion is

$$D(\psi_m) \geq \frac{D(\psi_1) + D(\psi_2)}{2} = \bar{D}. \tag{9.171}$$

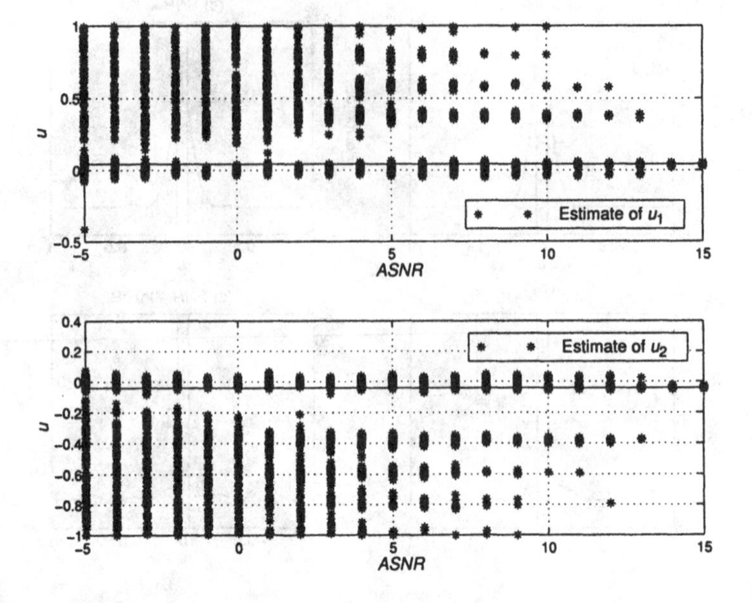

Figure 9.29 Spectral MUSIC estimates for various *ASNR*: $N = 10, K = 100, \rho = 0, u_1 = 0.0433, u_2 = -0.0433$, 200 trials.

Once again, if (9.171) is satisfied, then our estimator would not resolve the two signals. We would like to compute the probability of the events described by (9.170) and (9.171). Note that we are concerned with the *heights* of the spectrum, *not* the accuracy of location of the peaks.

Using (9.170), we can write,

$$P_R = Pr\left(D(\psi_m) < D(\psi_1) \,\text{and}\, D(\psi_2)\right). \tag{9.172}$$

Similarly, using (9.171), we can write

$$P_R = Pr\left(D(\psi_m) < \bar{D}\right). \tag{9.173}$$

In order to understand the RMSE behavior, we examine the error behavior when the signals are not successfully resolved. In Figure 9.29 we show the location of the estimates for 200 trials at *ASNR*s going from 15 dB to -5 dB. In Figure 9.30, we show the corresponding histograms. We now consider the behavior at each *ASNR* in the "upper threshold region."

(i) *ASNR* = 15 dB: The signals are resolved on all trials and the estimates are clustered around the correct value.

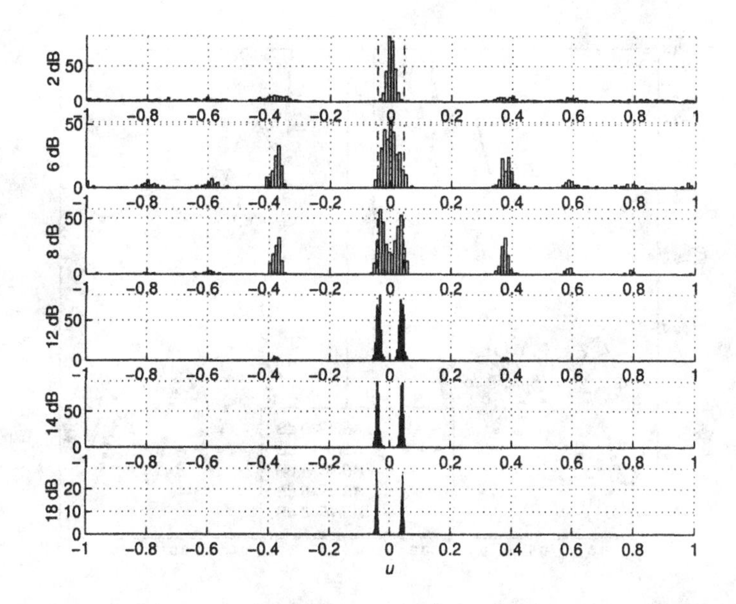

Figure 9.30 Histogram of spectral MUSIC estimates for *ASNR*s in upper threshold region: $N = 10, K = 100, \rho = 0, u_1 = 0.0433, u_2 = -0.0433$, 200 trials.

(ii) *ASNR* = 14 dB: The signals are resolved on most trials. On several trials, the signals are not resolved. If signal 1 is assigned the peak, then $\hat{u}_2 \simeq -0.374$. If signal 2 is assigned the peak, then $\hat{u}_1 \simeq 0.374$.

(iii) *ASNR* = 12 dB and 13 dB: Similar behavior to 14 dB. More trials have unresolved signals. The outlying estimates are clustered around ± 0.374.

(iv) *ASNR* = 10 dB and 11 dB: More trials have unresolved signals. The majority of the outlying estimates are at ± 0.374, but some estimates appear around ± 0.59 and one estimate is at $+0.81$.

In order to understand this behavior we plot the MUSIC spectrum averaged over 500 trials at various *ASNR* in Figure 9.31. We see that there are sidelobes at $\pm 0.37, \pm 0.59, \pm 0.80$, and ± 1.0. Note that, in a normal plot that is scaled to show the peaks, the sidelobe structure can easily be overlooked.

The importance of this observation is that the increase in RMSE is dominated by these sidelobe terms.

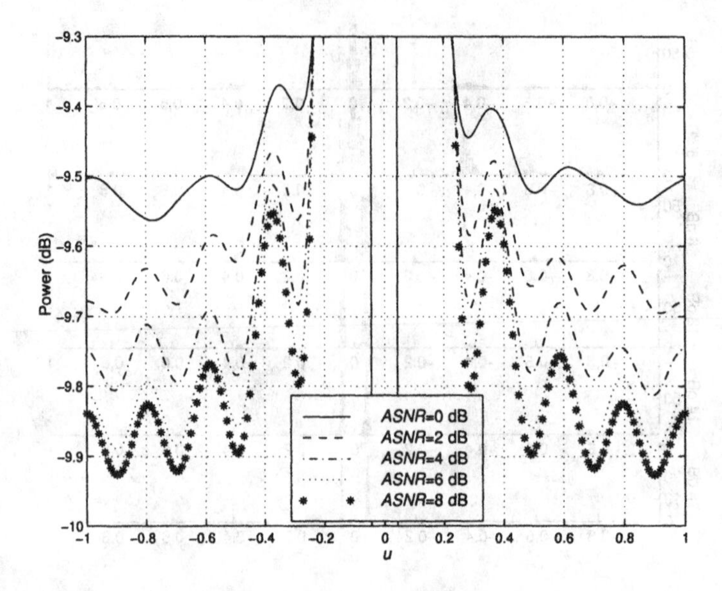

Figure 9.31 Plot of average of $\mathbf{v}^H(\psi)\widehat{\mathbf{U}}_S\widehat{\mathbf{U}}_S^H\mathbf{v}(\psi)$ for various *ASNR*: $N = 10, K = 100, \rho = 0, u_1 = 0.433, u_2 - 0.0433$, 500 trials.

We first consider the case when the second estimate is near the first sidelobe. Then, for this example, the MSE for signal 1 is,

$$\begin{aligned} MSE_1 &= P_R\{\text{local}MSE\} + \frac{1}{2}(1-P_R)\{\text{local}MSE\} \\ &\quad + \frac{1}{2}(1-P_R)\left\{|u_{sl1} - u_1|^2\right\}, \end{aligned} \tag{9.174}$$

and a similar expression for MSE_2.

We can rewrite the local MSE as

$$\{\text{local}MSE\} = \left\{(\text{local bias})^2 + \text{local variance}\right\}. \tag{9.175}$$

The reason for this decomposition is that analytic expressions for both terms on the right side of (9.175) are available in the literature. An analytic expression for P_R is also available, so we can create an analytic expression for the MSE behavior in the critical part of the upper threshold region using (9.174).

For the multiple-sidelobe case, we can write

$$MSE_1 = P_R\{\text{local}MSE\} + \frac{1}{2}(1-P_R)\{\text{local}MSE\}$$

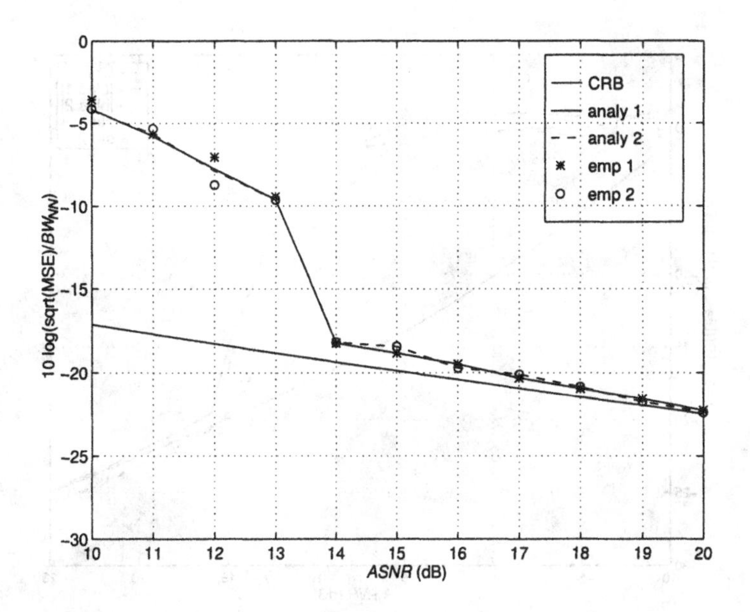

Figure 9.32 Comparison of "analytic" and empirical results: spectral MUSIC, $N = 10, K = 100, \rho = 0, u_1 = 0.0433, u_4 = -0.0433$, 500 trials.

$$+\frac{1}{2}(1 - P_R)\left\{\sum_{k=1}^{K} p_{slk}\left|u_{slk} - u_1\right|^2\right\}, \tag{9.176}$$

where p_{slk} is the conditional probability that the estimate is near the kth sidelobe, given the signals were not resolved.

In Figure 9.32, we show the MSE calculated using (9.174), (9.175), and (9.176). In those equations we use measured quantities. Later we show how well the analytic expressions can predict performance.

In Figure 9.33, we show the measured bias and standard deviation. We see that in the region just above the threshold point (in this case, 16 dB), the standard deviation is still essentially on the CRB, but the bias term increases the MSE by a small amount. In the upper threshold region, the variance is the dominant factor in the MSE.

In Figure 9.32, we see that there is a dramatic increase in the MSE in the 3-dB region from 16 dB to 13 dB. The empirical probability of resolution curve is shown in Figure 9.34. We see that P_R goes from about 0.965 to 1.0 in this region. Therefore, to predict the performance in the upper threshold region, we will need accurate expressions on the upper tail of the P_R probability density. In addition, we need a large number of trials to get good empirical results in this region.

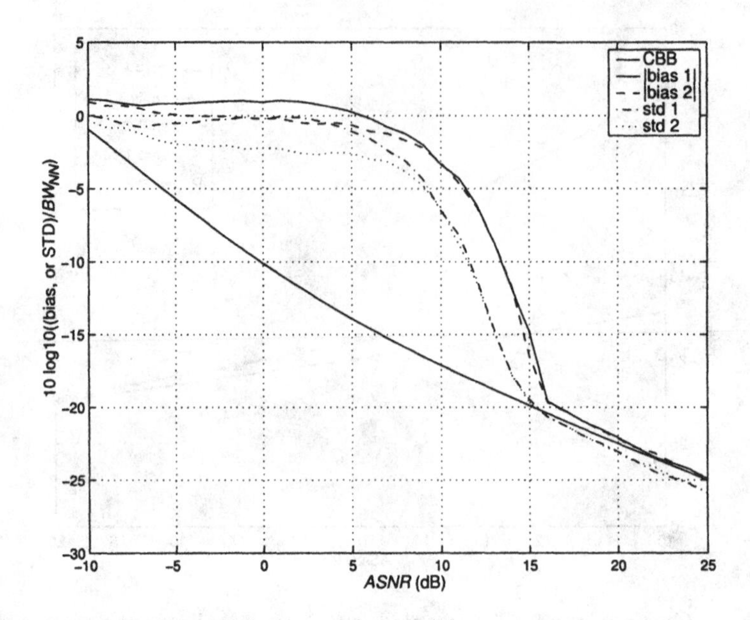

Figure 9.33 Bias and standard deviation versus *ASNR*: spectral MUSIC, $N = 10, K = 100, \rho = 0, u_1 = 0.0433, u_2 = -0.0433$, 2000 trials.

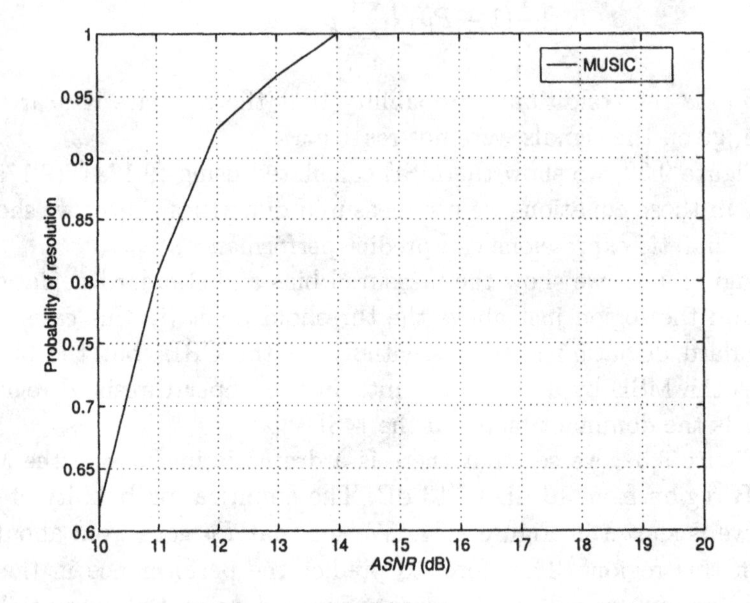

Figure 9.34 Empirical probability of resolution versus *ASNR*: spectral MUSIC, $N = 10, K = 100, \rho = 0, u_1 = 0.0433, u_2 = -0.0433$, 500 trials.

9.5.2 Resolution of MUSIC and Min-Norm

In this section we study the resolution capability of MUSIC and Min-Norm. The first comprehensive study in this area was due to Kaveh and Barabell ([KB86]). Their analysis is also available in Sections 5.4–5.8 in the chapter by Kaveh and Wang in [Hay91]. Their analysis emphasizes the importance of the mean-value of the null spectrum $\widehat{Q}_{MN}(\psi)$ and justifies its importance by showing that the standard deviation is small. Other studies that utilized the mean of the spectrum include Jeffries and Farrier [JF85], Böhme [Böh83], and Sharman et al. [SDWK83]. In a much earlier paper Cox [Cox73] had analyzed the resolution behavior of various algorithms. Other resolution discussions include Marple [Mar77], Gabriel [Gab80], DeGraaf and Johnson [DJ85] and Johnson and DeGraaf [JD82].

A subsequent paper by Zhou et al. [ZHJ91] developed an expression based on the ensemble covariance matrix and compared their results to Kaveh and Barabell's work.

None of above analyses actually computed a probability of resolution (or an approximation to it). Lee and Wengrovitz [LW91] developed a comprehensive asymptotic statistical characterization of the null spectrum and then used a Gaussian approximation to compute the resolution probability. Zhang [Zha95a] derived the probability distribution without using the Gaussian approximation and got better results in the low *SNR* region.

In this section, we use the sample covariance matrix,

$$\widehat{\mathbf{S}}_{\mathbf{x}} = \mathbf{C}_{\mathbf{x}} = \frac{1}{K}\sum_{k=1}^{K}\mathbf{X}(k)\,\mathbf{X}^H(k). \tag{9.177}$$

In other words, forward averaging only.

The initial part of our analysis follows Kaveh and Barabell ([KB86]). In order to analyze the asymptotic behavior of subspace methods we need the asymptotic statistics of the estimated eigenvalues and eigenvectors. The original work on asymptotic behavior is due to Anderson [And63]. It has been extended by Brillinger [Bri81], Gupta [Gup65], and Wilkinson [Wil65]. The estimated covariance matrix $\widehat{\mathbf{S}}_{\mathbf{x}}$ has a complex Wishart density whose elements are asymptotically jointly normal (see Section 7.2).

We define

$$\hat{\mathbf{\Phi}}_i = \mathbf{\Phi}_i + \boldsymbol{\eta}_i, \tag{9.178}$$

and

$$\hat{\lambda}_i = \lambda_i + \beta_i. \tag{9.179}$$

We assume that $\hat{\mathbf{\Phi}}_i$ is normalized. Then $\boldsymbol{\eta}_i$ and β_i have the following asymptotic properties.

The estimated eigenvalues are asymptotically Gaussian and statistically independent,

$$E[\beta_i] = 0, \tag{9.180}$$

$$E[\beta_i \beta_j] = \frac{\lambda_i^2}{K} \delta_{ij}, \tag{9.181}$$

where K is the number of snapshots.

The estimation error for the eigenvectors are asymptotically jointly Gaussian distributed with mean

$$E[\boldsymbol{\eta}_i] \simeq -\frac{\lambda_i}{2K} \sum_{\substack{k=1 \\ k \neq i}}^{N} \frac{\lambda_k}{(\lambda_i - \lambda_k)^2} \mathbf{\Phi}_i \triangleq c_i \mathbf{\Phi}_i, \tag{9.182}$$

and covariances

$$\begin{aligned} E\left[\boldsymbol{\eta}_i \boldsymbol{\eta}_j^H\right] &\simeq \frac{\lambda_i}{K} \sum_{\substack{k=1 \\ k \neq i}}^{N} \frac{\lambda_k}{(\lambda_k - \lambda_i)^2} \mathbf{\Phi}_k \mathbf{\Phi}_k^H \delta_{ij} \\ &= \frac{\lambda_i}{K} \left\{ \sum_{\substack{k=1 \\ k \neq i}}^{D} \frac{\lambda_k}{(\lambda_k - \lambda_i)^2} \mathbf{\Phi}_k \mathbf{\Phi}_k^H + \sum_{k=D+1}^{N-D} \frac{\sigma_n^2}{(\sigma_n^2 - \lambda_i)^2} \mathbf{\Phi}_k \mathbf{\Phi}_k^H \right\} \delta_{ij} \\ &\triangleq \mathbf{W}_i \delta_{ij}, \end{aligned} \tag{9.183}$$

and

$$E\left[\boldsymbol{\eta}_i \boldsymbol{\eta}_j^T\right] \cong -\frac{\lambda_i \lambda_j}{K(\lambda_i - \lambda_j)^2} \mathbf{\Phi}_j \mathbf{\Phi}_i^T (1 - \delta_{ij}) \triangleq \mathbf{V}_{ij}. \tag{9.184}$$

All of the approximations in (9.180)–(9.184) are of order $O(K^{-1})$. Note that (9.184) means that the $\boldsymbol{\eta}_i$ are not joint complex Gaussian random variables.

Because of the normalization assumption, we can use (9.178) to obtain

$$\left\|\hat{\mathbf{\Phi}}_i\right\|^2 = 1 + \mathbf{\Phi}_i^H \boldsymbol{\eta}_i + \boldsymbol{\eta}_i^H \mathbf{\Phi}_i + \boldsymbol{\eta}_i^H \boldsymbol{\eta}_i = 1. \tag{9.185}$$

Therefore,

$$2Re\left[\Phi_i^H \eta_i\right] = -\eta_i^H \eta_i. \tag{9.186}$$

Taking the expectation of (9.186), using (9.182), and retaining terms of order K^{-1} gives

$$\begin{aligned} 2Re\left[\Phi_i^H E\left[\eta_i\right]\right] &\cong -\frac{\lambda_i}{K} \sum_{\substack{k=1 \\ k \neq i}}^{N} \frac{\lambda_k}{\left(\lambda_i - \lambda_k\right)^2} \Phi_i^H \Phi_i \\ &= -\frac{\lambda_i}{K} \sum_{\substack{k=1 \\ k \neq i}}^{N} \frac{\lambda_k}{\left(\lambda_i - \lambda_k\right)^2}. \end{aligned} \tag{9.187}$$

We first consider the MUSIC algorithm.

9.5.2.1 MUSIC

The expected value of the null spectrum is obtained by substituting (9.178) in $\widehat{Q}_{MU}(\psi)$ in (9.43) and taking the expectation. Using (9.183) and (9.187), we obtain

$$\begin{aligned} E\left[\widehat{Q}_{MU}(\psi)\right] &\simeq Q_{MU}(\psi) - \\ & \mathbf{v}^H(\psi) \left[\sum_{i=1}^{D} \sum_{\substack{j=1 \\ j \neq i}}^{N} \frac{\lambda_i \lambda_j}{K\left(\lambda_i - \lambda_j\right)^2} \left(\Phi_j \Phi_j^H - \Phi_i \Phi_i^H\right) \right] \mathbf{v}(\psi). \end{aligned} \tag{9.188}$$

At the correct DOAs, $\psi_i, i = 1, 2, \cdots, D$, $Q_{MU}(\psi_i) = 0$. Therefore, in the vicinity of the correct value, when $D = 2$, (9.188) reduces to

$$\begin{aligned} E\left[\widehat{Q}_{MU}(\psi_k)\right] &\simeq \frac{\sigma_w^2}{K} \mathbf{v}^H(\psi_k) \left[\frac{\lambda_1 (N-2)}{\left(\lambda_1 - \sigma_w^2\right)^2} \Phi_1 \Phi_1^H + \frac{\lambda_2 (N-2)}{\left(\lambda_2 - \sigma_w^2\right)^2} \Phi_2 \Phi_2^H \right] \mathbf{v}(\psi_k) \\ &= \frac{\sigma_w^2}{K} (N-2) \mathbf{v}^H(\psi_k) \left[\frac{\lambda_1}{\left(\lambda_1^f\right)^2} \Phi_1 \Phi_1^H + \frac{\lambda_2}{\left(\lambda_2^f\right)^2} \Phi_2 \Phi_2^H \right] \mathbf{v}(\psi_k), \\ & k = 1, 2, \end{aligned} \tag{9.189}$$

where the approximation is of $O(K^{-1})$.

For two equal-power uncorrelated sources, we use λ_i and $\mathbf{\Phi}_i$ from Chapter 5 ((5.255) and (5.259)) in (9.189) to obtain

$$E\left[\widehat{Q}_{MU}(\psi_k)\right] \simeq \frac{N-2}{K}\left[\frac{1}{(ASNR)} + \frac{1}{(ASNR)^2\,\Delta^2}\right], \quad k = 1, 2, \tag{9.190}$$

where

$$ASNR \triangleq \frac{N\sigma_s^2}{\sigma_w^2}, \tag{9.191}$$

is assumed to be much larger than one and

$$\Delta = \frac{N(\psi_1 - \psi_2)}{2\sqrt{3}}, \tag{9.192}$$

is a separation factor that is proportional to the normalized difference between the wavenumbers. Here we are normalizing with respect to the approximate HPBW ($\sqrt{3}/N$) instead of BW_{NN} ($4/N$) that we normally use. Most of the papers dealing with this topic use this normalization, so we will also use it in order to simplify comparisons with the literature.

It is difficult to obtain an expression of order K^{-2} for the variance (K^{-1} for the standard deviation). Based on reasonably extensive simulation results one can argue that the standard deviation is small compared to the mean. Therefore, we develop a resolution criterion based on $E\left[\widehat{Q}_{MU}(\psi)\right]$.

Kaveh and Barabell [KB86] proposed that the resolution threshold be defined as the *ASNR* where

$$E\left[\widehat{Q}_{MU}(\psi_1)\right] = E\left[\widehat{Q}_{MU}(\psi_2)\right] = E\left[\widehat{Q}_{MU}(\psi_m)\right] \tag{9.193}$$

where

$$\psi_m = \frac{\psi_1 + \psi_2}{2}. \tag{9.194}$$

Their logic was:

(1) If the three variations were statistically independent, then the resolution probability would be 0.33 when (9.193) was satisfied.

(2) If $\widehat{Q}_{MU}(\psi_1)$ and $\widehat{Q}_{MU}(\psi_2)$ were completely correlated, then the resolution probability would be 0.5.

(3) The MUSIC P_R result is given in Figure 9.10 (this corresponds to Example 9.2.6 in Section 9.2). The curve is steep in the 0.33–0.50 range and the spread in *ASNR* is 1.2 dB between the ends of the range.

The threshold *ASNR* is denoted by $ASNR_T$. We find $E\left[\widehat{Q}_{MU}(\psi_m)\right]$ from (9.189) and $E\left[\widehat{Q}_{MU}(\psi_1)\right]\left(=E\left[\widehat{Q}_{MU}(\psi_2)\right]\right)$ from (9.190) and equate the two results. For $N \gg 1$, $\Delta \ll 1$, and $ASNR \gg 1$, we have

$$ASNR_T \simeq \frac{1}{K}\left\{\frac{20(N-2)}{\Delta^4}\left[1+\sqrt{1+\frac{K}{5(N-2)}\Delta^2}\right]\right\}. \tag{9.195}$$

For large N, the threshold *ASNR* is proportional to N. For a small number of snapshots ($K \ll 5N/\Delta^2$), $ASNR_T$ varies as $1/\Delta^4 K$. For a large number of snapshots ($K \gg 5N/\Delta^2$), $ASNR_T$ varies as $1/\Delta^3 K^{1/2}$.

This result (and extensions of it to other cases) has been widely used in the literature. However, if we examine the normalized MSE plot in Figure 9.11 and the probability of resolution plot in Figure 9.10, we see that the definition of resolution threshold (9.193) and the subsequent expressions may give misleading results. When $P_R = 0.5$, the MSE shown in Figure 9.11 has already increased dramatically above the CRB and is approaching the *a priori* variance.

A better definition of threshold is the point at which the actual MSE is 3 dB above the CRB. The MSE curve is steep in this region so the use of 3 dB (as contrasted to 2 or 4 dB) is not critical. From Figure 9.10, we see that this choice corresponds to P_R greater than 0.95. This is what we would expect because with probability $(1 - P_R)$ we are adding a large MSE to a very small MSE (e.g., from Figure 9.10, the "local" MSE in u-space is 0.004 at an *ASNR* of 12 dB). Note that this point is about 4 dB higher than the $P_R = 0.4$ value. This threshold definition requires us to examine the P_R curve in the 0.95–1.00 region.

In [LW91], Lee and Wengrovitz develop a comprehensive statistical characterization of the MUSIC spectrum using asymptotic properties. They find an expression for the joint probability density of $\hat{\theta}_1$, $\hat{\theta}_2$, and $\hat{\theta}_m$. They compute the probability of the event,

$$\widehat{Q}_{MU}(\psi_m) \geq \widehat{Q}_{MU}(\psi_i), \quad i = 1, 2. \tag{9.196}$$

In their calculation, they approximate the joint density by a joint Gaussian density using a central limit theorem justification. Zhang [Zha95a] has done a detailed analysis that avoids the Gaussian assumption and obtains a better match between the analytic result and the simulation result in the low *SNR* region.

In Figure 9.35, we show an example of Zhang's result for a standard 8-element array with uncorrelated equal-power sources. We see that there is good agreement between the analyses and the simulation.

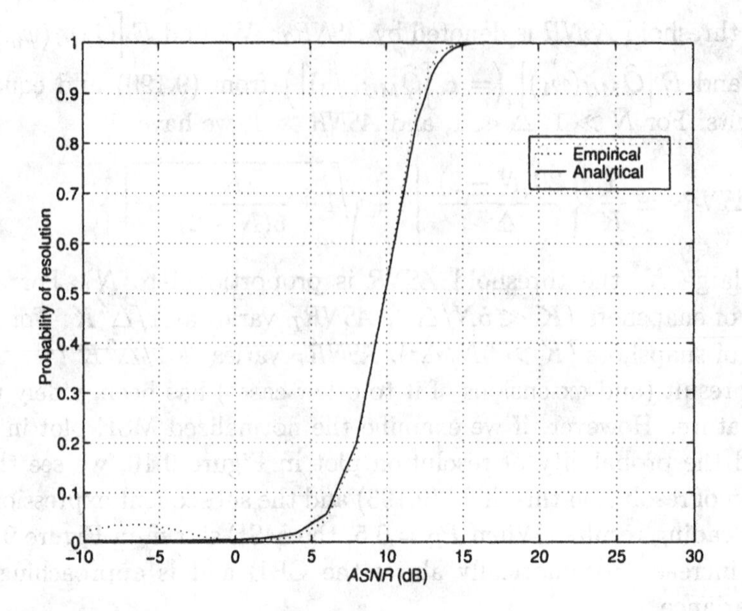

Figure 9.35 Comparison of analytic probability of resolution result and simulation result: 500 trials.

To determine the threshold P_R we can use the relationship

$$(1 - P_R)\,[\text{bias}_{1sl}]^2 = MSE_{loc}, \tag{9.197}$$

and solve for P_R.

We use the 3-dB criterion as a threshold definition. We use the result in (9.195) to compare various algorithms, but not as an absolute measure.

9.5.2.2 Min-Norm

A similar analysis can be carried out for the Min-Norm algorithm. Kaveh and Barabell (e.g., [KB86]) have done this analysis. The expression for the mean value analogous to (9.190) is

$$E\left[\hat{Q}_{MU}(\psi_k)\right] \simeq \frac{1}{\left(1-\frac{4}{N}\right)K}\left[\frac{1}{(ASNR)} + \frac{1}{(ASNR)^2\,\Delta^2}\right] \tag{9.198}$$

(see (5.135) in Chapter 5 of [Hay91]). Comparing (9.190) and (9.198) we see that for large N, we might expect a lower Min-Norm threshold by a factor of N. This result is consistent with our simulations in Section 9.3 (see Figure 9.10 at the $P_R = 0.5$ point).

9.5.2.3 Root MUSIC and root Min-Norm

In our simulations, we observed that root MUSIC and root Min-Norm had probability of resolution performance that was significantly better than the spectral versions. We now develop an asymptotic analysis to explain this behavior.[10]

An asymptotic analysis for root MUSIC was derived by Rao and Hari [RH89b]. They derive an expression for the MSE in the asymptotic regime. They approach the problem by first finding the MSE in the location of the zeros and then show how it translates into the MSE in the DOA. Their results retain terms of $O\left(K^{-1}\right)$ and assume the bias is zero. The key result is that the asymptotic MSE (which equals the variance) is the same as for spectral MUSIC. Thus, the advantage of root MUSIC is in improved threshold performance and computational simplicity rather than improved asymptotic performance. The reader is referred to [RH89b] for the detailed discussion of their analysis.

In Figure 9.36 we show the effect of an error in the root location on the estimate of the angle. We need to examine both $E\left\{|\Delta z_i|^2\right\}$ and $E\left\{|\Delta\psi_i|^2\right\}$. In the root case, only $\Delta\psi_i$ affects the DOA estimate. However, if Δz_i is large, it will cause the peaks in spectral MUSIC to be misplaced and degrade the probability of resolution performance.

Krim et al. [KFP92] carry out an asymptotic analysis of root MUSIC and root Min-Norm by using a series expansion of the projection operators. Their results are in a useful form for comparing the two algorithms.

For root Min-Norm, they show that for the mth root

$$\begin{aligned} Var\left[\Delta\psi_m\right] &= Var\left[\Delta r_m\right] \\ &= \frac{\sigma_w^2}{2K}\frac{\left(\mathbf{e}_1^T\mathbf{P}_{\mathbf{V}}^{\perp}\mathbf{e}_1\right)}{\left|\mathbf{e}_1^T\mathbf{P}_{\mathbf{V}}^{\perp}\mathbf{d}_m\right|^2}\sum_{i=1}^{D}\frac{\lambda_i}{(\lambda_i-\sigma_w^2)^2}\left|\mathbf{\Phi}_i^H\mathbf{v}_i\right|^2, \end{aligned} \tag{9.199}$$

where

$$\mathbf{e}_1 = \left[\, 1 \,\vdots\, 0 \,\vdots\, \cdots \,\vdots\, 0 \,\right]^T, \tag{9.200}$$

and the summation is familiar (see (9.214)) as

$$\mathbf{v}_i^H\,\mathbf{U}_P\,\mathbf{v}_i, \tag{9.201}$$

[10] This asymptotic analysis focuses on the mean and variance of the zero location. It only treats the probability of resolution qualitatively.

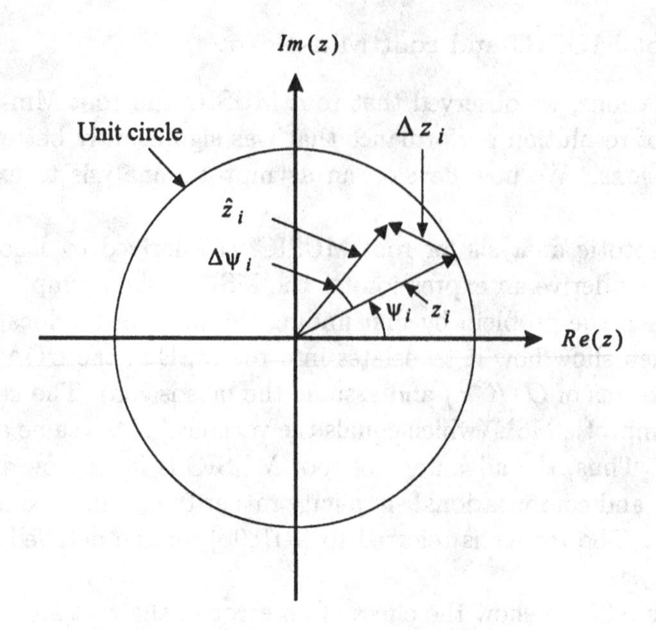

Figure 9.36 Effect of error in the root location on the estimate of ψ_i.

with $\mathbf{U}_P$ defined in (9.216). Note that the radial component of the error and the angular (wavenumber) component of the error has the same variance. [KFP92] also show that both components have zero bias and are uncorrelated.

For root MUSIC, one can show that

$$E\left[\Delta\psi_m\right] = 0, \tag{9.202}$$

and

$$\begin{aligned} Var\left[\Delta\psi_m\right] &= \frac{\sigma^2}{2K\left(\mathbf{d}_m^H \mathbf{P}_{\mathbf{V}}^{\perp} \mathbf{d}_m\right)} \sum_{i=1}^{D} \frac{\lambda_i}{(\lambda_i - \sigma_w^2)^2} \left|\mathbf{\Phi}_i^H \mathbf{v}_m\right|^2 \\ &= \frac{\sigma_w^2}{2K} \frac{\mathbf{v}_m^H \mathbf{U}_P \mathbf{v}_m}{h(\psi_m)}. \end{aligned} \tag{9.203}$$

In Section 9.5.3 we find that spectral MUSIC has the same variance. Thus, spectral MUSIC and root MUSIC have the same asymptotic variance.

In addition, they show $\Delta\psi_m$ and Δr_m are uncorrelated. However,

$$E\left[\Delta r_m\right] = -\sqrt{2(N-D) - \frac{3}{2}}\sqrt{Var\left(\Delta\psi_m\right)}, \tag{9.204}$$

and

$$Var\left[\Delta r_m\right] = \frac{Var\left(\Delta\psi_m\right)}{2}. \tag{9.205}$$

This bias in the radial component explains the loss of resolution in spectral MUSIC because it changes the shape of the spectral plot.

We will revisit the variance results for root MUSIC and root Min-Norm in Section 9.5.3.4.

9.5.2.4 Summary

In this section we have included analytic expressions for the probability of resolution for MUSIC and Min-Norm. The analyses used the asymptotic properties of the eigenvectors and eigenvalues. The results were consistent with our simulations and indicated the following ordering in representative cases. In decreasing order of performance:

(i) Root MUSIC,

(ii) Root Min-Norm,

(iii) Min-Norm,

(iv) MUSIC,

for uncorrelated signals. Note that we did not prove that the above ordering always holds.

There are a large number of analyses of these cases and other cases in the literature. For the case of FB averaging Pillai and Kwon [PK89a] have derived similar results. Roy et al. [RPK86] and Roy and Kailath [RK87] have compared MUSIC and ESPRIT.

If the algorithm has successfully resolved the signals, then we want to analyze the bias and the variance of the estimates. We discuss this problem in the next section.

9.5.3 Small Error Behavior of Algorithms

In this section we analyze the error behavior of the various algorithms when the values of *SNR* and K are such that the algorithms are operating above threshold. In this case, the probability of resolution approaches unity.

In Section 9.5.3.1, we repeat the relevant CRBs from Chapter 8. In Section 9.5.3.2, we analyze the variance of spectral MUSIC. In Section 9.5.3.3, we analyze the bias of spectral MUSIC. In Section 9.5.3.4, we analyze weighted MUSIC, which includes Min-Norm as a special case. In Section

9.5.3.5, we analyze root MUSIC and root Min-Norm. In Section 9.5.3.6, we analyze ESPRIT.

9.5.3.1 Cramér-Rao bound

In our discussion we use both the stochastic (or unconditional) CRB and asymptotic conditional CRB as a basis of comparison for the various algorithms.

From (8.99), the unconditional CRB on the $D \times 1$ DOA vector $\boldsymbol{\psi}$ is,

$$CRB(\boldsymbol{\psi}) = \frac{\sigma_w^2}{2K}\left[Re\left\{\mathbf{H} \odot \left(\mathbf{S_f}\mathbf{V}^H\mathbf{S_x}^{-1}\mathbf{V}\,\mathbf{S_f}\right)^T\right\}\right]^{-1}, \tag{9.206}$$

where

$$\mathbf{H} = \mathbf{D}^H\,\mathbf{P}_{\mathbf{V}}^{\perp}\mathbf{D}. \tag{9.207}$$

From (8.236), the asymptotic conditional Cramér-Rao bound is

$$C_{ACCR}(\boldsymbol{\psi}) = \frac{\sigma_w^2}{2K}\left[Re\left\{\mathbf{H} \odot \mathbf{S}_{\mathbf{f}}^T\right\}\right]^{-1}. \tag{9.208}$$

For the special case of uncorrelated sources, $\mathbf{S_f}$ is a diagonal matrix,

$$\mathbf{S_f} = \text{diag}\left[\begin{array}{cccc}\sigma_1^2 & \sigma_2^2 & \cdots & \sigma_D^2\end{array}\right], \tag{9.209}$$

and the asymptotic conditional Cramér-Rao bound for the variance of the ith source, can be written as

$$C_{ACCR}(\psi_i) = \left[2K\,SNR_i\,h(\psi_i)\right]^{-1}, \tag{9.210}$$

where

$$h(\psi_i) = \mathbf{D}^H(\psi_i)\,\mathbf{P}_{\mathbf{V}}^{\perp}\,\mathbf{D}(\psi_i). \tag{9.211}$$

9.5.3.2 Variance of spectral MUSIC

In this section we derive the asymptotic properties of the spectral MUSIC algorithm. Our discussion follows Stoica and Nehorai [SN89]. Their derivation is also available in their Chapters 7 and 8 of [Hay91]. An earlier discussion that derived the asymptotic relative efficiency for uncorrelated signals is contained in Porat and Friedlander [PF88]. Another early asymptotic analysis is contained in Sharman et al. [SDWK84].

The MUSIC null spectrum is

$$\hat{Q}_{MU}(\psi) = \mathbf{v}^H(\psi)\,\hat{\mathbf{U}}_N\,\hat{\mathbf{U}}_N^H\,\mathbf{v}(\psi), \tag{9.212}$$

in terms of the noise subspace, or

$$\hat{Q}_{MU}(\psi) = \mathbf{v}^H(\psi)\left[\mathbf{I} - \hat{\mathbf{U}}_S \hat{\mathbf{U}}_S^H\right]\mathbf{v}(\psi), \tag{9.213}$$

in terms of the signal subspace.

The asymptotic behavior of the eigenvectors was given at the beginning of Section 9.5.2 (see (9.182)–(9.184)).

In addition, we need the statistics of the projections of $\hat{\mathbf{\Phi}}_i$, $i = D+1, \cdots, N$ onto the column space of $\mathbf{U}_S$. Using (9.178) and (9.183) one can show that these projections are asymptotically jointly Gaussian with zero means and covariance matrices given by

$$\begin{aligned} E\left\{\left(\mathbf{U}_S \mathbf{U}_S^H \hat{\mathbf{\Phi}}_i\right)\left(\mathbf{U}_S \mathbf{U}_S^H \hat{\mathbf{\Phi}}_j\right)^H\right\} &= \frac{\sigma_w^2}{K}\left[\sum_{k=1}^{D} \frac{\lambda_k}{(\sigma_w^2 - \lambda_k)^2}\mathbf{\Phi}_k \mathbf{\Phi}_k^H\right]\delta_{ij} \\ &\triangleq \frac{\sigma_w^2}{K}\mathbf{U}_P\,\delta_{ij}, \quad i,j = D+1,\cdots,N, \end{aligned} \tag{9.214}$$

and

$$E\left\{\left(\mathbf{U}_S \mathbf{U}_S^H \hat{\mathbf{\Phi}}_i\right)\left(\mathbf{U}_S \mathbf{U}_S^H \hat{\mathbf{\Phi}}_j\right)^T\right\} = \mathbf{0}, \quad i,j = D+1,\cdots,N. \tag{9.215}$$

The matrix $\mathbf{U}_P$ can also be expressed as

$$\mathbf{U}_P = \mathbf{U}_S \mathbf{\Lambda}_f^{-2} \mathbf{\Lambda}_S \mathbf{U}_S^H, \tag{9.216}$$

where

$$\mathbf{\Lambda}_f = \text{diag}\left[\lambda_1 - \sigma_w^2, \;\; \lambda_2 - \sigma_w^2, \;\; \cdots, \;\; \lambda_D - \sigma_w^2\right]. \tag{9.217}$$

With these statistics as background one can derive the asymptotic variance of the MUSIC estimator.

The derivation is due to Stoica and Nehorai [SN89] and relies on a technique used by Sharman et al. [SDWK83] in an earlier analysis of eigenspace methods. Another early analysis is contained in Jeffries and Farrier [JF85].

We start with the necessary condition of the derivative of $\widehat{Q}_{MU}(\psi)$,

$$\dot{\hat{Q}}_{MU}(\hat{\psi}) = 0, \tag{9.218}$$

where

$$\dot{\hat{Q}}_{MU}(\hat{\psi}) \triangleq \left. \frac{d\hat{Q}_{MU}(\psi)}{d\psi} \right|_{\psi=\hat{\psi}} . \tag{9.219}$$

We then expand $\dot{\hat{Q}}_{MU}(\hat{\psi})$ in a Taylor series around ψ_i and retain the first-order terms. The reader is referred to [SN89] or Chapter 7 of [Hay91] for the detailed derivation.

The result is that the MUSIC estimation errors,

$$\psi_{e_i} \triangleq \hat{\psi}_i - \psi_i, \tag{9.220}$$

are asymptotically jointly Gaussian with zero means and correlation given by

$$E\left\{\psi_{e_i}\,\psi_{e_j}\right\} = \frac{1}{2K}\,\frac{Re\left[\mathbf{d}^H(\psi_j)\,\mathbf{U}_N\,\mathbf{U}_N^H\,\mathbf{d}(\psi_i)\cdot\mathbf{v}^H(\psi_i)\,\mathbf{U}_P\,\mathbf{v}(\psi_j)\right]}{h(\psi_i)\,h(\psi_j)}, \tag{9.221}$$

where $\mathbf{d}(\psi)$ is the derivative of the array manifold vector,

$$\mathbf{d}(\psi) = \frac{d\mathbf{v}(\psi)}{d\psi}, \tag{9.222}$$

and

$$h(\psi) = \mathbf{d}^H(\psi)\mathbf{U}_N\mathbf{U}_N^H\mathbf{d}(\psi). \tag{9.223}$$

For the mean square of ψ_{e_i}, (9.221) reduces to

$$E\left\{\psi_{e_i}^2\right\} = \frac{1}{2K}\,\frac{\mathbf{v}^H(\psi_i)\,\mathbf{U}_P\,\mathbf{v}(\psi_i)}{h(\psi_i)}, \tag{9.224}$$

or

$$\boxed{E\left\{\psi_{e_i}^2\right\} = \frac{\sigma_w^2}{2K}\left\{\frac{\sum_{k=1}^{D}\frac{\lambda_k}{(\sigma_w^2-\lambda_k)^2}\left|\mathbf{v}^H(\psi_i)\,\mathbf{\Phi}_k\right|^2}{\sum_{k=D+1}^{N}\left|\mathbf{d}^H(\psi_i)\,\mathbf{\Phi}_k\right|^2}\right\}}. \tag{9.225}$$

We observe that the numerator will be large if one or more of the D largest eigenvalues are close to σ_w^2. The denominator in the numerator sum is just the eigenvalue of the source-signal matrix

$$\lambda_k^f = \lambda_k - \sigma_w^2 \tag{9.226}$$

This term will be small if:

(i) There are two or more closely spaced signals;

(ii) The signals are highly correlated;

(iii) The signal-to-noise ratio is low.

(Recall our discussion of eigenvalues in Section 5.5.)

The denominator will be small when $\mathbf{d}(\psi)$ is close to the column space of $\mathbf{V}(\psi)$ (or equivalently $\mathbf{U}_S$). In this case, the projection onto $\mathbf{U}_N$ will be small. This results in a $Q_{MU}(\psi)$ with a relatively flat minimum around $\psi = \psi_i$. If one or more of these conditions occur, then the variance of MUSIC estimator is large.

We now want to put (9.225) in a more convenient form.[11] We can rewrite (9.225) as[12]

$$Var_{MU}(\hat{\psi}_i) = \frac{\frac{\sigma_w^2}{2K}\left\{\mathbf{v}^H(\psi_i)\left[\mathbf{U}_S\,\mathbf{\Lambda}_f^{-1}\,\mathbf{U}_S^H + \sigma_w^2\mathbf{U}_S\,\mathbf{\Lambda}_f^{-2}\,\mathbf{U}_S^H\right]\mathbf{v}(\psi_i)\right\}}{\mathbf{d}^H(\psi_i)\,\mathbf{U}_N\,\mathbf{U}_N^H\,\mathbf{d}(\psi_i)}, \tag{9.227}$$

where, from (9.217),

$$\mathbf{\Lambda}_f \triangleq \mathbf{\Lambda} - \sigma_w^2\mathbf{I} \triangleq \begin{bmatrix} \lambda_1 & & 0 \\ & \ddots & \\ 0 & & \lambda_D \end{bmatrix} - \sigma_w^2\mathbf{I}. \tag{9.228}$$

The spectral matrix $\mathbf{S_x}$ is

$$\mathbf{S_x} = \mathbf{V}\mathbf{S_f}\mathbf{V}^H + \sigma_w^2\mathbf{I} = \mathbf{U}_S\mathbf{\Lambda}\mathbf{U}_S^H + \sigma_w^2\mathbf{U}_N\mathbf{U}_N^H = \mathbf{U}_S\mathbf{\Lambda}_f\mathbf{U}_S^H + \sigma_w^2\mathbf{I}. \tag{9.229}$$

Thus,

$$\mathbf{V}\,\mathbf{S_f}\,\mathbf{V}^H = \mathbf{U}_S\,\mathbf{\Lambda}_f\,\mathbf{U}_S^H, \tag{9.230}$$

and

$$\mathbf{V}\,\mathbf{S_f}\,\mathbf{V}^H\,\mathbf{V}\,\mathbf{S_f}\,\mathbf{V}^H = \mathbf{U}_S\,\mathbf{\Lambda}_f^2\,\mathbf{U}_S^H. \tag{9.231}$$

Therefore,

$$\left(\mathbf{U}_S^H\,\mathbf{V}\right)\mathbf{S_f}\left(\mathbf{V}^H\,\mathbf{U}_S\right) = \mathbf{\Lambda}_f, \tag{9.232}$$

and

[11]This derivation follows Section 7.7 of Stoica and Nehorai's chapter in [Hay91], which originally appeared in [SN89].

[12]In (9.227) we have assumed K is large enough that the bias can be neglected.

$$\left(\mathbf{U}_S^H \mathbf{V}\right) \mathbf{S_f} \left(\mathbf{V}^H \mathbf{V}\right) \mathbf{S_f} \left(\mathbf{V}^H \mathbf{U}_S\right) = \mathbf{\Lambda}_f^2. \tag{9.233}$$

Since $\mathbf{V}$ has full rank and the columns of $\mathbf{V}$ lie in the column space of $\mathbf{U}_S$, the matrix $\mathbf{U}_S^H \mathbf{V}$ is non-singular and

$$\mathbf{U}_S \mathbf{U}_S^H = \mathbf{V} \left(\mathbf{V}^H \mathbf{V}\right)^{-1} \mathbf{V}^H. \tag{9.234}$$

Inverting (9.232) and (9.233), we obtain

$$\left(\mathbf{V}^H \mathbf{U}_S\right) \mathbf{\Lambda}_f^{-1} \left(\mathbf{U}_S^H \mathbf{V}\right) = \mathbf{S_f}^{-1}, \tag{9.235}$$

and

$$\left(\mathbf{V}^H \mathbf{U}_S\right) \mathbf{\Lambda}_f^{-2} \left(\mathbf{U}_S \mathbf{V}\right) = \mathbf{S_f}^{-1} \left(\mathbf{V}^H \mathbf{V}\right)^{-1} \mathbf{S_f}^{-1}. \tag{9.236}$$

Using (9.235) and (9.236) in (9.227) gives

$$Var_{MU}(\hat{\psi}_i) = \frac{\sigma_w^2}{2K} \frac{\left\{\left[\mathbf{S_f}^{-1}\right]_{ii} + \sigma_w^2 \left[\mathbf{S_f}^{-1} \left(\mathbf{V}^H \mathbf{V}\right)^{-1} \mathbf{S_f}^{-1}\right]_{ii}\right\}}{\mathbf{d}^H(\psi_i) \left[\mathbf{I} - \mathbf{V} \left(\mathbf{V}^H \mathbf{V}\right)^{-1} \mathbf{V}^H\right] \mathbf{d}(\psi_i)}, \tag{9.237}$$

or

$$\boxed{Var_{MU}(\hat{\psi}_i) = \frac{\sigma_w^2}{2K} \frac{\left\{\left[\mathbf{S_f}^{-1}\right]_{ii} + \sigma_w^2 \left[\mathbf{S_f}^{-1} \left(\mathbf{V}^H \mathbf{V}\right)^{-1} \mathbf{S_f}^{-1}\right]_{ii}\right\}}{\mathbf{d}^H(\psi_i) \mathbf{P}_{\mathbf{V}}^{\perp} \mathbf{d}(\psi_i)},} \tag{9.238}$$

which is the desired result. Using (9.211), (9.238) can be written as

$$Var_{MU}(\hat{\psi}_i) = \frac{\sigma_w^2}{2K} \frac{\left\{\left[\mathbf{S_f}^{-1}\right]_{ii} + \sigma_w^2 \left[\mathbf{S_f}^{-1} \left(\mathbf{V}^H \mathbf{V}\right)^{-1} \mathbf{S_f}^{-1}\right]_{ii}\right\}}{h(\psi_i)}. \tag{9.239}$$

We observe that we can evaluate (9.237) or (9.238) directly from the original parameters of the model and eigenvector decomposition is not necessary. It is also in a form that is suitable to compare to the Cramér-Rao bound. For analytic simplicity, we use the asymptotic conditional CRB given by (9.208). However, our numerical examples will include the unconditional CRB. In the region of interest, the unconditional and conditional CRBs are essentially the same. In most of our examples, the threshold occurs before the two bounds separated. The results for uncorrelated signals are due to Porat and Friedlander [PF88].

We first consider the case of a single source. In this case, (9.208) reduces to

$$\begin{aligned} Var_{ACCR}\left[\hat{\psi}\right] &= \frac{\frac{\sigma_w^2}{\sigma_s^2}}{2Kh(\psi)}\left[1+\frac{\sigma_w^2}{N\sigma_s^2}\right] \\ &= \frac{\left(1+ASNR^{-1}\right)}{2Kh(\psi)\,SNR}. \end{aligned} \tag{9.240}$$

Similarly, from (9.239),

$$Var_{MU}\left[\hat{\psi}\right] = \frac{\frac{\sigma_w^2}{\sigma_s^2}}{2Kh(\psi)}\left[1+\frac{\sigma_w^2}{N\sigma_s^2}\right]. \tag{9.241}$$

Thus, for a single signal the asymptotic variance of MUSIC is identical to the CRB variance.

We now consider the case of D uncorrelated signals. For the case of uncorrelated signals, $\mathbf{S_f}$ is a diagonal matrix whose iith element is σ_i^2, the power of the ith signal. In this case, (9.239) reduces to

$$\begin{aligned} Var_{MU}\left[\hat{\psi}_i\right] &= \frac{1}{2K\cdot SNR_i}\frac{\left\{1+\frac{\left[\left(\mathbf{V}^H\mathbf{V}\right)^{-1}\right]_{ii}}{SNR_i}\right\}}{h(\psi_i)} = \frac{1+(N\cdot SNR_i)^{-1}}{2K\cdot SNR_i\cdot h(\psi_i)} \\ &= \frac{\left[1+ASNR_i^{-1}\right]}{2Kh(\psi_i)\,SNR_i} \approx \frac{1}{2(K\cdot SNR_i)(N\cdot SNR_i)h(\psi_i)}, \end{aligned} \tag{9.242}$$

where the approximation is valid for $ASNR_i \gg 1$. The SNR for the ith signal is

$$SNR_i \triangleq \frac{\sigma_i^2}{\sigma_w^2}. \tag{9.243}$$

The asymptotic conditional CRB is

$$Var_{ACCR}\left(\hat{\psi}_i\right) = \frac{1}{2K\cdot SNR_i\cdot h(\psi_i)}. \tag{9.244}$$

Using (9.242) and (9.244), we have

$$\frac{Var_{MU}\left(\hat{\psi}_i\right)}{Var_{ACCR}\left(\hat{\psi}_i\right)} = 1+ASNR_i^{-1}. \tag{9.245}$$

Therefore, as the $ASNR_i$ increase, the variance of spectral MUSIC approaches the Cramér-Rao bound. In many cases (e.g., Figure 9.10), the $ASNR$ required to be above the threshold is such that the ratio in (9.245) is close to one.

The last case of interest is D correlated signals. In this case, we use (9.239) with a non-diagonal $\mathbf{S_f}$ and the CRB expression in (9.206). In the examples, we verify that, for correlated signals, the variance of MUSIC does not approach the Cramér-Rao bound. The results for correlated signals are due to Stoica and Nehorai [SN89].

Before considering several examples, we note that we can write the covariance matrix given by (9.221) compactly as

$$\boxed{\begin{aligned} \mathbf{C}_{MU} &= \frac{\sigma_w^2}{2K}\left(\mathbf{D}^H\,\mathbf{P}_{\mathbf{V}}^{\perp}\,\mathbf{D}\odot\mathbf{I}\right)^{-1} \\ &\quad\times\left[Re\left\{\mathbf{D}^H\,\mathbf{P}_{\mathbf{V}}^{\perp}\,\mathbf{D}\odot\mathbf{S_f}^{-1}+\sigma_w^2\left(\mathbf{S_f}\,\mathbf{V}^H\,\mathbf{V}\,\mathbf{S_f}\right)^{-1}\right\}\right] \\ &\quad\times\left(\mathbf{D}^H\,\mathbf{P}_{\mathbf{V}}^{\perp}\,\mathbf{D}\odot\mathbf{I}\right)^{-1}. \end{aligned}} \tag{9.246}$$

and (9.239) is the diagonal element of (9.246). This expression can be rewritten as

$$\mathbf{C}_{MU}=\frac{\sigma_w^2}{2K}\left\{[\mathbf{H}\odot\mathbf{I}]^{-1}\left[Re\left\{\mathbf{H}\odot\mathbf{F}^T\right\}\right][\mathbf{H}\odot\mathbf{I}]^{-1}\right\}, \tag{9.247}$$

where $\mathbf{H}$ is defined in (9.192) and

$$\mathbf{F}\triangleq\mathbf{S_f}^{-1}+\sigma_w^2\,\mathbf{S_f}^{-1}\left(\mathbf{V}^H\mathbf{V}\right)^{-1}\mathbf{S_f}^{-1}. \tag{9.248}$$

We now consider several examples to illustrate the asymptotic behavior of the MUSIC estimator. The examples consider the case of two signals of equal power and a linear uniform array. Then we can write

$$\mathbf{S_f}=\begin{bmatrix}1 & \rho\\ \rho^* & 1\end{bmatrix}, \tag{9.249}$$

where ρ is correlation coefficient and we have assumed $\sigma_s^2=1$. For two signals, the variance in wavenumber space will depend on

$$\Delta\psi=\psi_1-\psi_2. \tag{9.250}$$

In the examples, we plot

$$EFF_{MU}=\frac{Var_{MU}\left(\hat{\psi}_i\right)}{Var_{CRB}\left(\hat{\psi}_i\right)}, \tag{9.251}$$

where the numerator is given by (9.239) and the denominator by (9.210). Note that we use the Var_{CRB} in our plot rather than Var_{ACCR}.

Example 9.5.1

We consider a 10-element standard linear array and consider several values of *SNR* and ρ. The three values of *ASNR* are 10 dB, 20 dB, and 30 dB. We consider five values of ρ: 0, 0.5, $0.5\exp(-j\pi/2)$, 0.9 and $0.9\exp(-j\pi/2)$.

In the first set of figures, we plot the normalized $Var_{MU}(\hat{\psi}_i)$ as given by (9.239) (normalized by BW_{NN}). In Figure 9.37(a), we show the results for $ASNR = 10$ dB. In Figure 9.37(b) we show the results for $ASNR = 20$ dB. In Figure 9.37(c), we show the results for $ASNR = 30$ dB.

In the second set of figures, Figure 9.38(a), 9.38(b), and 9.38(c), we plot the efficiency as given by (9.251) for the same cases as in Figure 9.37. These figures are similar to those in [SN89] except they use asymptotic CCRB.

We see that the efficiency that is higher for smaller $\Delta\psi/BW_{NN}$. Once the separation exceeds $2BW_{NN}$, the efficiency is reasonably constant,

$$EFF_{MU} = 1 - |\rho|^2, \tag{9.252}$$

for $\Delta\psi > BW_{NN}$.

Example 9.5.2

In this example we plot the relative asymptotic efficiency of MUSIC versus the signal separation normalized by the BW_{NN} for $N = 10$, and $SNR = 0$ and 20 dB. Figure 9.39 corresponds to $\rho = 0.5$, and Figure 9.40 corresponds to $\rho = 0.9$.

All of our examples have considered two signals. For large N, high *SNR*, and $\Delta\psi \geq BW_{NN}$, one can show that, for D signals,

$$EFF_{MU} = \frac{[\mathbf{S_f}]_{ii}}{\left[\mathbf{S_f}^{-1}\right]_{ii}}, \tag{9.253}$$

(see [SN89]). Note that for $D = 2$, (9.253) reduces to

$$EFF_{MU} = 1 - |\rho|^2, \tag{9.254}$$

which is consistent with the results in Figures 9.38–9.40. For $D \geq 2$, as $\mathbf{S_f}$ approaches a singular matrix, the efficiency of MUSIC approaches zero.

All of the asymptotic results for spectral MUSIC are valid for large K and used $O(K^{-1})$ approximations. However, Xu and Buckley [XB92] pointed out that the treatment of bias term in the MUSIC derivation was a first-order approximation and that, as we moved closer to the transition region, a second-order approximation is more appropriate. In the next section we introduce their analysis.

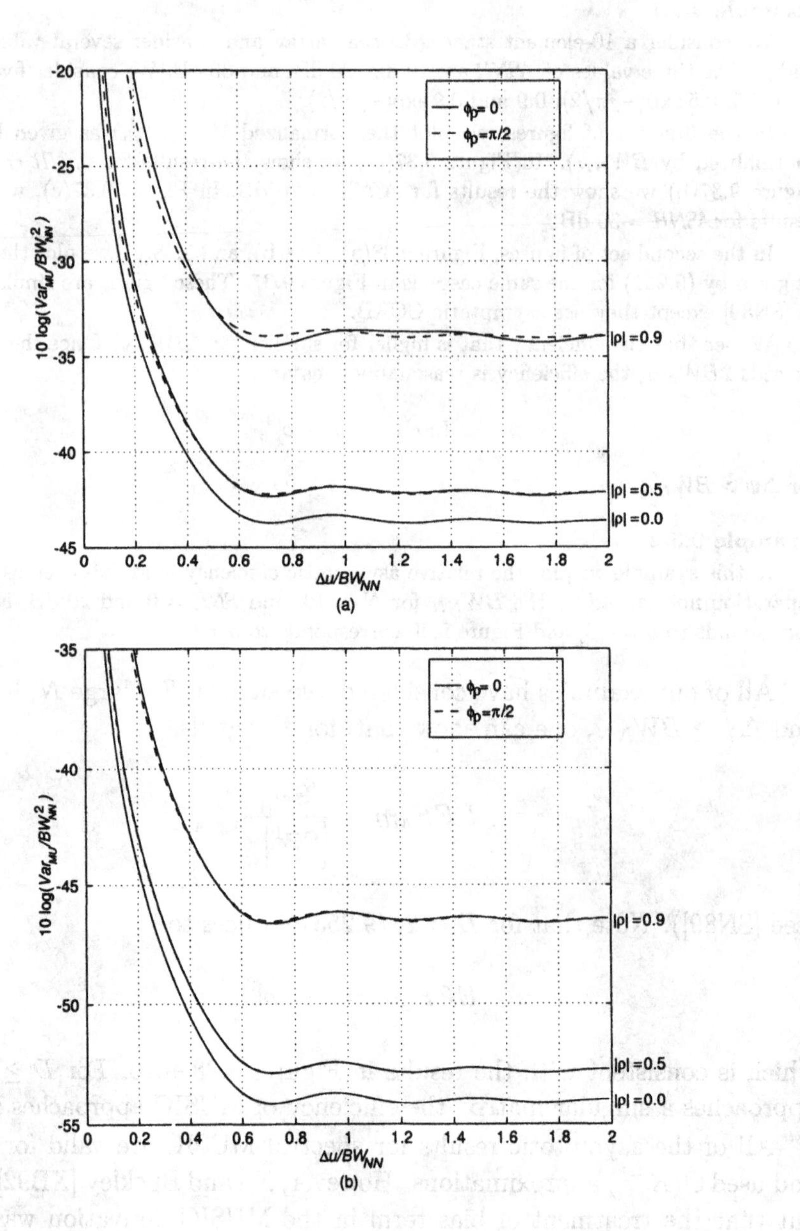

$\phi_p = 0$
$\phi_p = \pi/2$
$10 \log(Var_{ML}/BW_{NN}^2)$
$|\rho| = 0.9$
$|\rho| = 0.5$
$|\rho| = 0.0$
$\Delta u/BW_{NN}$
(a)
$\phi_p = 0$
$\phi_p = \pi/2$
$10 \log(Var_{ML}/BW_{NN}^2)$
$|\rho| = 0.9$
$|\rho| = 0.5$
$|\rho| = 0.0$
$\Delta u/BW_{NN}$
(b)

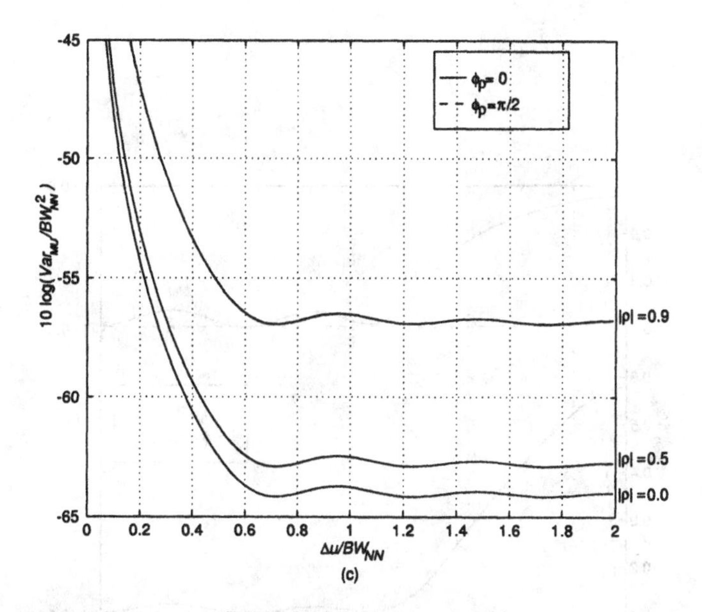

Figure 9.37 Asymptotic variance of spectral MUSIC versus $\Delta u/BW_{NN}$: $N = 10, |\rho| = 0, 0.5$ and 0.9: (a) *ASNR* = 10 dB; (b) *ASNR* = 20 dB; and (c) *ASNR* = 30 dB.

9.5.3.3 Bias analysis of MUSIC

A bias analysis of the MUSIC algorithm has been carried out by Xu and Buckley [XB92]. The bias of the MUSIC algorithm is a relevant performance issue when either the *SNR* or K, the number of snapshots is in a "medium" range so that the system is approaching the threshold.

In [XB92], a bias analysis using a second-order Taylor series expansion of $\dot{Q}_{MU}\left(\hat{\psi}_m, \hat{\mathbf{U}}_S\right)$ around $(\psi_m, \mathbf{U}_S)$ is carried out.[13] The mid-point, ψ_m, is defined in (9.169). The result is

$$\begin{aligned} E\left(\Delta\psi_m\right) \simeq & -\frac{2\sigma_n^2 \sum_{k=1}^{D} \frac{(N-D-1)\lambda_k}{(\lambda_k-\sigma_n^2)^2} Re\left[\mathbf{d}^H(\psi_m)\mathbf{\Phi}_k\mathbf{\Phi}_k^H\mathbf{v}(\psi_m)\right]}{K \qquad \ddot{Q}_{MU}(\psi_m, \mathbf{U}_S)} \\ & -\frac{\dddot{Q}_{MU}\left(\psi_m, \mathbf{U}_S\right)}{6\ddot{Q}_{MU}(\psi_m, \mathbf{U}_S)} Var(\Delta\psi_m), \end{aligned} \tag{9.255}$$

where

$$\ddot{Q}_{MU}(\psi_m, \mathbf{U}_S) = 2\mathbf{d}^H(\psi_m)\left[\mathbf{I} - \mathbf{U}_S\mathbf{U}_S^H\right]\mathbf{d}(\psi_m), \tag{9.256}$$

[13]We have suppressed the overbar in $\dot{Q}_{MU}\left(\hat{\psi}_m, \hat{\mathbf{U}}_S\right)$ to simplify the notation.

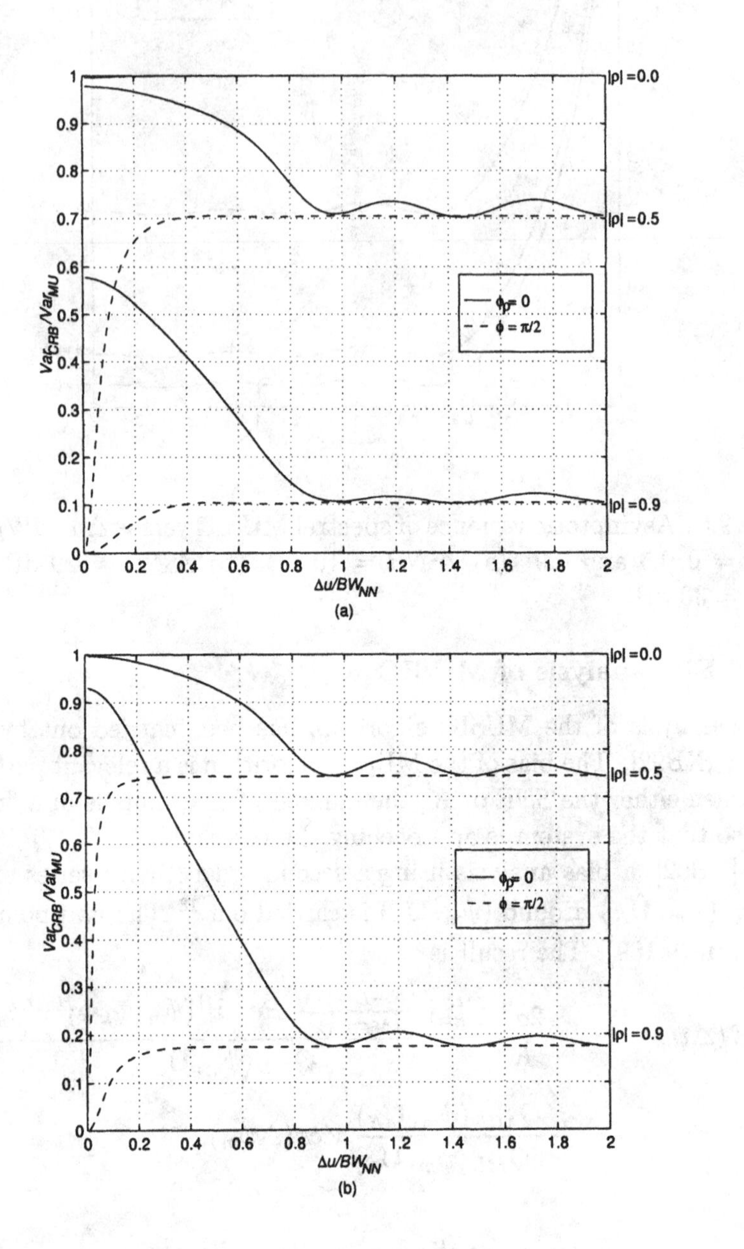

|ρ| =0.0
|ρ| =0.5
|ρ| =0.9
Var_{CRB}/Var_{MU}
$\Delta u/BW_{NN}$
$\phi_\rho = 0$
$\phi = \pi/2$
(a)
(b)

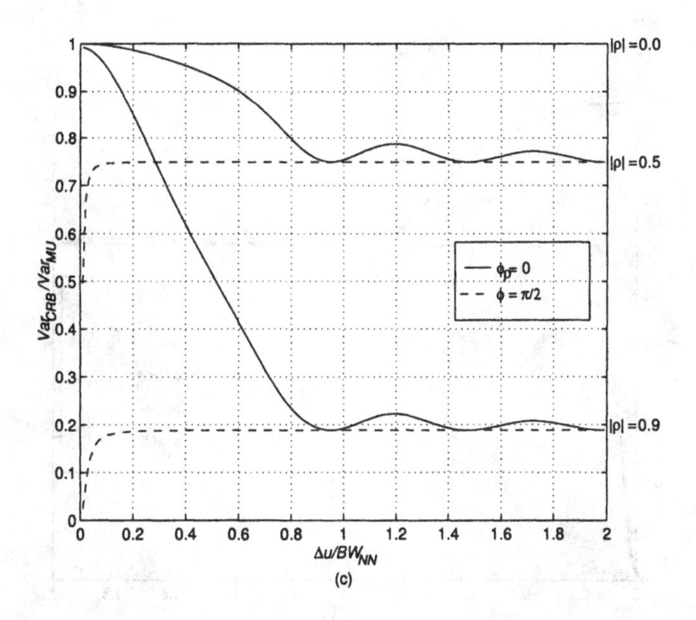

Figure 9.38 Asymptotic efficiency of spectral MUSIC algorithm versus $\Delta u/BW_{NN}$: $N = 10, |\rho| = 0, 0.5$, and 0.9: (a) $ASNR = 10$ dB; (b) $ASNR = 20$ dB; and (c) $ASNR = 30$ dB.

and

$$\dddot{Q}_{MU}\left(\psi_m, \mathbf{U}_S\right) = 6Re\left\{\dot{\mathbf{d}}^H(\psi_m)\left[\mathbf{I} - \mathbf{U}_S\,\mathbf{U}_S^H\right]\mathbf{d}(\psi_m)\right\}. \tag{9.257}$$

The reader is referred to [XB92] for the details of the derivation and further discussion of the results.

9.5.3.4 Weighted eigenspace algorithms

We have developed several weighted eigenspace (WES) algorithms in addition to MUSIC. These included Min-Norm (9.95). In this section, we provide the asymptotic variance for WES algorithms and show that their variance is greater than or equal to the asymptotic variance of MUSIC. The initial discussion follows [SN90] (see also Chapter 8 of [Hay91]).

Note that this result does not imply that the composite MSE performance of MUSIC is better than the other WES algorithms.

We use the spectral form for discussion. The WES algorithm constructs the null spectrum

$$\widehat{Q}_{WES}(\psi) = \mathbf{v}^H(\psi)\,\widehat{\mathbf{U}}_N\,\mathbf{W}\,\widehat{\mathbf{U}}_N^H\,\mathbf{v}(\psi), \tag{9.258}$$

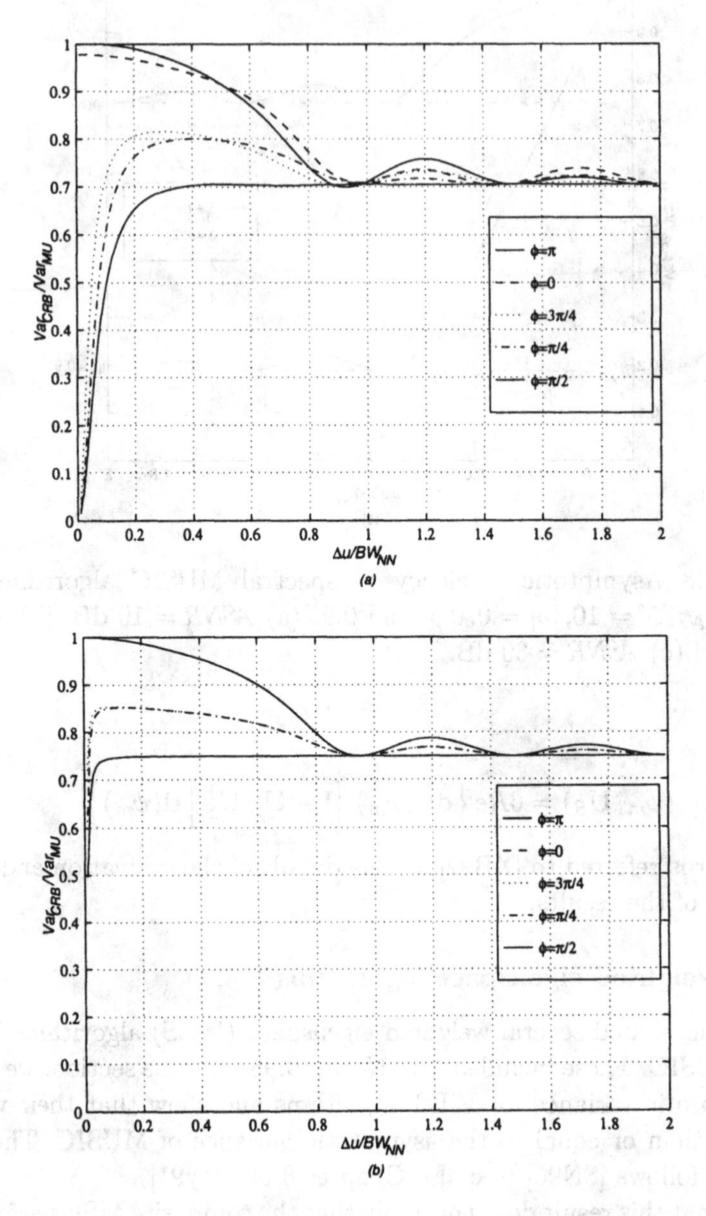

Figure 9.39 Asymptotic efficiency of spectral MUSIC for two equal-power signals as a function of $\Delta u/BW_{NN}$: $|\rho| = 0.5$, and $N = 10$: (a) $SNR = 0$ dB; (b) $SNR = 20$ dB.

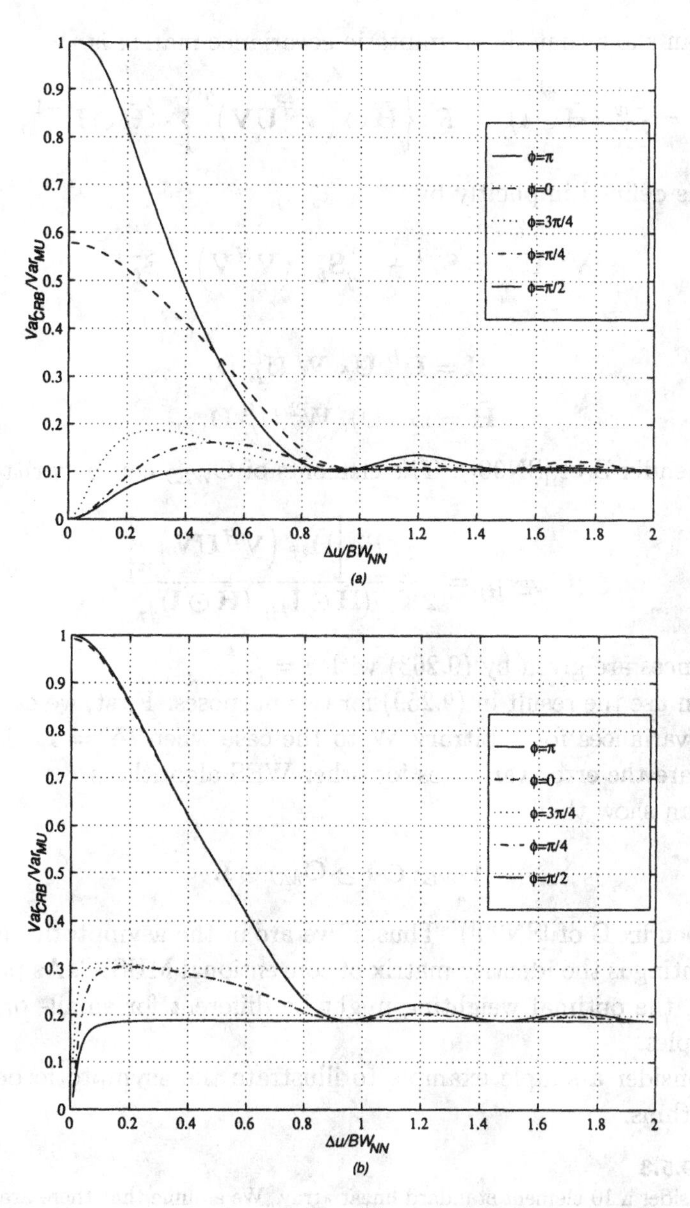

Figure 9.40 Asymptotic efficiency of spectral MUSIC for two equal-power signals as a function of $\Delta u/BW_{NN}$: $\rho = 0.9$, and $N = 10$: (a) $SNR = 0$ dB; (b) $SNR = 20$ dB.

where $\mathbf{W}$ is a positive definite weighting matrix. We search over ψ to find the nulls.

One can show that the asymptotic covariance matrix is

$$\mathbf{C}_{WES} = \frac{\sigma_w^2}{2K} \left(\bar{\mathbf{H}} \odot \mathbf{I}\right)^{-1} \cdot Re\left\{\tilde{\mathbf{H}} \odot \left(\mathbf{V}^H \mathbf{U} \mathbf{V}\right)^T\right\} \cdot \left(\bar{\mathbf{H}} \odot \mathbf{I}\right)^{-1}, \tag{9.259}$$

where $\mathbf{U}$ is defined implicitly by

$$\mathbf{V}^H \mathbf{U} \mathbf{V} = \mathbf{S_f}^{-1} + \sigma_w^2 \mathbf{S_f}^{-1} \left(\mathbf{V}^H \mathbf{V}\right)^{-1} \mathbf{S_f}^{-1}, \tag{9.260}$$

and

$$\bar{\mathbf{H}} = \mathbf{D}^H \, \mathbf{U}_N \, \mathbf{W} \, \mathbf{U}_N^H \mathbf{D}, \tag{9.261}$$

$$\tilde{\mathbf{H}} = \mathbf{D}^H \, \mathbf{U}_N \, \mathbf{W}^2 \, \mathbf{U}_N^H \mathbf{D}, \tag{9.262}$$

(e.g., Appendix B of [SN90]). The elements of $\mathbf{C}_{WES}$ can be written as

$$[\mathbf{C}_{WES}]_{ij} = \frac{\sigma_w^2}{2K} \frac{Re\left[\tilde{\mathbf{H}}_{ij} \left(\mathbf{V}^H \mathbf{U} \mathbf{V}\right)_{ji}\right]}{\left(\bar{\mathbf{H}} \odot \mathbf{I}\right)_{ii} \left(\bar{\mathbf{H}} \odot \mathbf{I}\right)_{jj}}. \tag{9.263}$$

The variances are given by (9.263) with $i = j$.

We can use the result in (9.259) for two purposes. First, we can compare the error variances for arbitrary $\mathbf{W}$ to the case when $\mathbf{W} = \mathbf{I}$. Second, we can compare the error variances for other WES algorithms.

One can show that

$$\mathbf{C}_{WES} \odot \mathbf{I} \geq \mathbf{C}_{MU} \odot \mathbf{I}. \tag{9.264}$$

(e.g., Appendix C of [SN90]). Thus, if we are in the asymptotic regime, the best weighting is the identity matrix of conventional MUSIC. As pointed out in [SN90], the optimal weighting might be different for small- or medium-sized samples.

We consider a simple example to illustrate the asymptotic behavior of the algorithms.

Example 9.5.3

We consider a 10-element standard linear array. We assume that there are two uncorrelated equal-power signals located at $\pm\Delta\psi/2$ where $\Delta\psi = \Delta\psi_R$. In Figure 9.41, we plot $[\mathbf{C}_{WES}]_{11}$ from (9.263) for MUSIC and Min-Norm. The unconditional CRB is shown for comparison.

We see that the simulated results and the analytic results are the same above 15 dB. MUSIC achieves the CRB, while Min-Norm is 1.9 dB above it.

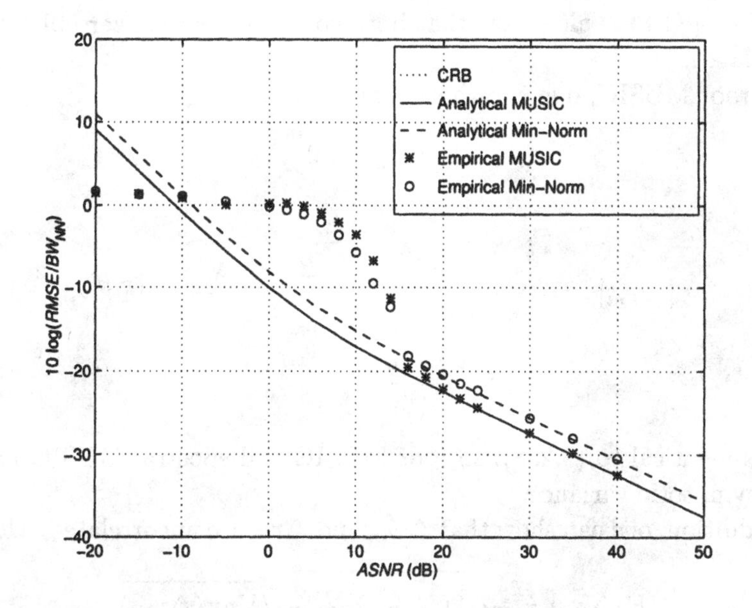

Figure 9.41 Asymptotic variance of MUSIC and Min-Norm algorithms versus $ASNR : N = 10, K = 100, \rho = 0, \Delta u = 0.2165 BW_{NN}$.

9.5.3.5 Root MUSIC and root Min-Norm

In Section 9.5.2.3 we provided expressions for the mean and variance of the roots in root MUSIC and root Min-Norm. We repeat the results here for convenience.

For root Min-Norm, for the mth root,

$$\begin{aligned} Var\left[\Delta\psi_m\right] &= Var\left[\Delta r_m\right] \\ &= \frac{\sigma_w^2}{2K} \frac{\left(\mathbf{e}_1 \mathbf{P}_{\mathbf{V}}^{\perp} \mathbf{e}_1\right)}{\left|\mathbf{e}_1^T \mathbf{P}_{\mathbf{V}}^{\perp} \mathbf{d}_m\right|^2} \sum_{i=1}^{D} \frac{\lambda_i}{(\lambda_i - \sigma_w^2)^2} \left|\mathbf{\Phi}_i^H \mathbf{v}_i\right|^2, \end{aligned} \tag{9.265}$$

where

$$\mathbf{e}_1 = \left[\ 1 \ \vdots \ 0 \ \vdots \ \cdots \ \vdots \ 0 \ \right]^T, \tag{9.266}$$

and the summation is familiar (see (9.214)) as

$$\mathbf{v}_i^H \, \mathbf{U}_P \, \mathbf{v}_i, \tag{9.267}$$

with $\mathbf{U}_P$ defined in (9.216). Note that the radial component of the error and the angular (wavenumber) component of the error has the same variance.

Krim et al. [KFP92] also show that both components have zero bias and are uncorrelated.

For root MUSIC, one can show that

$$E\left[\Delta\psi_m\right] = 0, \tag{9.268}$$

and

$$\begin{aligned} Var\left[\Delta\psi_m\right] &= \frac{\sigma_w^2}{2K\left(\mathbf{d}_m^H \mathbf{P}_{\mathbf{V}}^{\perp} \mathbf{d}_m\right)} \sum_{i=1}^{D} \frac{\lambda_i}{\left(\lambda_i - \sigma_w^2\right)^2} \left|\mathbf{\Phi}_i^H \mathbf{v}_m\right|^2 \\ &= \frac{\sigma_w^2}{2K} \frac{\mathbf{v}_m^H \mathbf{U}_P \mathbf{v}_m}{h(\psi_m)}, \end{aligned} \tag{9.269}$$

which is identical to (9.238), so root MUSIC and spectral MUSIC have the same asymptotic variance.

In addition, one can show that $\Delta\psi_m$ and Δr_m are uncorrelated. However,

$$E\left[\Delta r_m\right] = -\sqrt{2(N-D) - \frac{3}{2}}\sqrt{Var\left(\Delta\psi_m\right)}, \tag{9.270}$$

and

$$Var\left[\Delta r_m\right] = \frac{Var\left(\Delta\psi_m\right)}{2}. \tag{9.271}$$

This bias in the radial component explains the loss of resolution in spectral MUSIC because it changes the shape of the spectral plot.

Using the result in (9.265) and (9.269), we can compare the performance of root MUSIC and root Min-Norm. The ratio of the two variances is

$$\gamma = \frac{Var_{RMU}\left\{\Delta\psi_m\right\}}{Var_{RMN}\left\{\Delta\psi_m\right\}} = \frac{\left|\mathbf{e}_1^T \mathbf{P}_{\mathbf{V}}^{\perp} \mathbf{d}_m\right|^2}{\left(\mathbf{e}_1^T \mathbf{P}_{\mathbf{V}}^{\perp} \mathbf{e}_1\right)\left(\mathbf{d}_m \mathbf{P}_{\mathbf{V}}^{\perp} \mathbf{d}_m\right)}. \tag{9.272}$$

Using the Schwarz inequality on the numerator and remembering that $\mathbf{P}_{\mathbf{V}}^{\perp}\mathbf{P}_{\mathbf{V}}^{\perp} = \mathbf{P}_{\mathbf{V}}^{\perp}$, we have

$$\left|\mathbf{e}_1^T \mathbf{P}_{\mathbf{V}}^{\perp} \mathbf{d}_m\right|^2 = \left|\mathbf{e}_1^T \mathbf{P}_{\mathbf{V}}^{\perp} \mathbf{P}_{\mathbf{V}}^{\perp} \mathbf{d}_m\right|^2 \leq \left|\mathbf{e}_1^T \mathbf{P}_{\mathbf{V}}^{\perp}\right|^2 \left|\mathbf{P}_{\mathbf{V}}^{\perp} \mathbf{d}_m\right|^2. \tag{9.273}$$

Thus, $\gamma \leq 1$. Note that γ is independent of the *SNR*. In addition, as N increases, the performance difference increases.

We consider an example to illustrate the behavior.

Example 9.5.4

Consider a 10-element standard linear array with two equal power uncorrelated sources separated by $\Delta\psi_R$ impinging on the array. The results predicted by (9.265) and (9.269) are shown in Figure 9.42. For this scenario, $\gamma = 1.9$ dB.

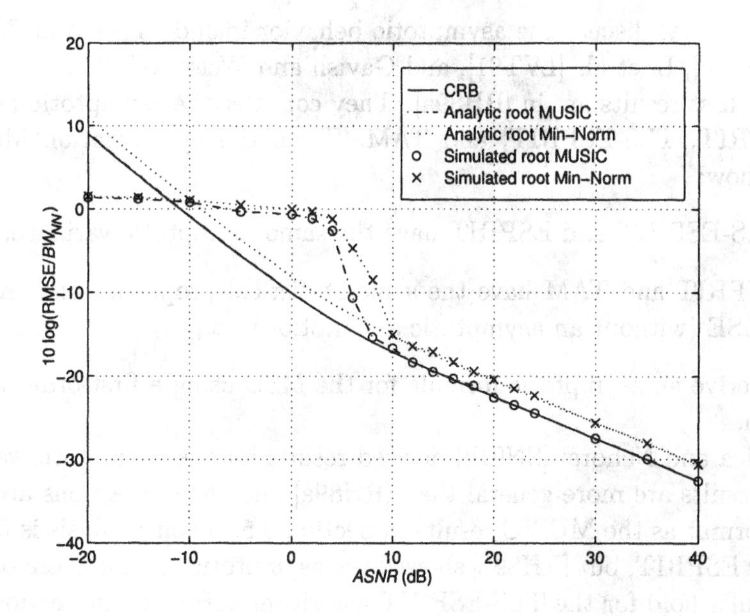

Figure 9.42 Theoretical and simulated RMSE for root MUSIC and root Min-Norm: $N = 10, K = 100, \Delta\psi = \Delta\psi_R$, 200 trials.

In addition, we show the results of the previous simulations. We see that the simulations are consistent with the analytic results.

The results in this section and Section 9.5.1 lead to two observations:

(i) Root MUSIC has advantages over spectral MUSIC in terms of computational complexity and probability of resolution. The two versions of MUSIC have the same asymptotic MSE.

(ii) The asymptotic variance of root Min-Norm is always greater than or equal to the variance of asymptotic root MUSIC. In many scenarios of interest, the difference is significant.

The asymptotic behavior of unitary root MUSIC is derived in Pesavento et al. [PGH00].

9.5.3.6 Asymptotic behavior of ESPRIT

The ESPRIT algorithm was described in Section 9.4.3. A sequence of papers has analyzed the asymptotic performance. We briefly summarize the results of four of these references: Rao and Hari [RH89a], Stoica and Nehorai [SN91b], Ottersten et al. [OVK91], and Viberg and Ottersten [VO91]. Other references that discuss the asymptotic behavior include Yuen and Friedlander [YF96b], Li et al. [LVT91], and Gavish and Weiss [GW93].

The first results are in [RH89a]. They consider the asymptotic behavior of ESPRIT, TLS-ESPRIT, and TAM (Toeplitz Approximation Method). They show:

(a) TLS-ESPRIT and ESPRIT have the same asymptotic variance;

(b) ESPRIT and TAM have the same statistical properties and the same MSE (without an asymptotic assumption).

They derive an asymptotic formula for the MSE using a first-order approximation.

Stoica and Nehorai [SN91b] derived results for the asymptotic variance. Their results are more general than [RH89a] and the expressions are in the same format as the MUSIC results in Section 9.5. Their analysis is for least squares ESPRIT, but [RH89a] showed the asymptotic variances are equal, so the results hold for the TLS-ESPRIT algorithm developed in Section 9.4.4.

We assume the subarrays contain N_s elements where $D \leq N_s \leq N-1$ and $N_s = N - d_s$. The subarray manifolds are given by

$$\mathbf{V_y} = \left[\, \mathbf{0} \,\vdots\, \tilde{\mathbf{J}}_{N_s} \,\right] \mathbf{V} = \left[\, \tilde{\mathbf{J}}_{N_s} \,\vdots\, \mathbf{0} \,\right] \mathbf{V}\, \mathbf{\Phi} = \mathbf{V_x}\, \mathbf{\Phi}, \tag{9.274}$$

where the various matrices were defined in (9.103)–(9.107). Then, using the same approach as in Section 9.5.3.2, we obtain[14]

$$E\left[\left(\hat{\psi}_k - \psi_k\right)\left(\hat{\psi}_l - \psi_l\right)\right] = \frac{\sigma_w^2}{2Kd_s^2} Re\left[e^{j(\psi_l - \psi_k)}\left(\mathbf{C}_l^H \mathbf{C}_k\right)\left(\mathbf{v}_k^H \mathbf{U}_P \mathbf{v}_l\right)\right], \tag{9.275}$$

where

$$\mathbf{C}_k = \left[\left(\mathbf{V_x}^H \mathbf{V_x}\right)^{-1} \mathbf{V_x}^H \mathbf{F}_k\right]_k^{(r)}, \tag{9.276}$$

and $[\mathbf{X}]_k^{(r)}$ denotes the kth row of $\mathbf{X}$. The $\mathbf{F}_k$ matrix is

$$\mathbf{F}_k = \left[\, \mathbf{0} \,\vdots\, \mathbf{I}_{N_s} \,\right] - e^{jd_s\psi_k}\left[\, \mathbf{I}_{N_s} \,\vdots\, \mathbf{0} \,\right], \tag{9.277}$$

where $\mathbf{I}_{N_s}$ is an $N_s \times N_s$ identity matrix. As before,

$$\mathbf{U}_P = \mathbf{U}_S\, \mathbf{\Lambda}_f^{-2}\, \mathbf{\Lambda}_S\, \mathbf{U}_S^H, \tag{9.278}$$

[14]The result in (9.275) is given in [SN91b]. The θ_k in their notation corresponds to $d_s\psi_k$ in our notation.

and

$$\mathbf{v}_k^H \mathbf{U}_P \mathbf{v}_l = \left[\mathbf{S}_\mathbf{f}^{-1}\right]_{kl} + \sigma_w^2 \left[\mathbf{S}_\mathbf{f}^{-1} \left(\mathbf{V}_\mathbf{x}^H \mathbf{V}_\mathbf{x}\right)^{-1} \mathbf{S}_\mathbf{f}^{-1}\right]_{kl}. \tag{9.279}$$

We consider two simple examples to illustrate the result. Before doing that, it is useful to compare the asymptotic variance of ESPRIT to the asymptotic variance of MUSIC. We can write the ratio of the variance as

$$\gamma_k = \frac{Var_{ES}(\hat{\psi}_k)}{Var_{MUSIC}(\hat{\psi}_k)}. \tag{9.280}$$

The numerator is given by (9.275) with $k = l$, and the denominator is given by (9.238). Cancelling common terms, we have

$$\gamma_k = \left(\mathbf{C}_k^H \mathbf{C}_k\right) \left(\mathbf{d}_k^H \mathbf{U}_N \mathbf{U}_N^H \mathbf{d}_k\right), \quad k = 1, 2, \cdots, D. \tag{9.281}$$

In Appendix B of [SN91b], the Schwarz inequality is used to show that $\gamma_k \geq 1$. Note that the ratio does not depend on σ_w^2 or $\mathbf{S}_\mathbf{f}$.

We now illustrate the behavior with two examples.

Example 9.5.5[15]

Consider a single source at wavenumber ψ impinging on a ULA with N sensors. Assume that $N_s = N - 1$ and that row weighting is not used.

From (9.276),

$$\begin{aligned} \mathbf{C}^H &= \frac{1}{N-1}\left[\ 1 \ \vdots\ e^{-j\psi} \ \vdots\ \cdots \ \vdots\ e^{-j(N-2)\psi}\ \right] \\ &\quad \cdot\left(\left[\ \mathbf{0} \ \vdots\ \mathbf{I}_{N-1}\ \right] - e^{j\psi}\left[\ \mathbf{I}_{N-1} \ \vdots\ \mathbf{0}\ \right]\right) \\ &= \frac{1}{N-1}\left[\ e^{j\psi} \ \vdots\ 0 \ \vdots\ \cdots \ \vdots\ 0 \ \vdots\ e^{-j(N-2)\psi}\ \right]. \end{aligned} \tag{9.282}$$

Then

$$\mathbf{C}^H \mathbf{C} = \frac{2}{(N-1)^2}. \tag{9.283}$$

We evaluated the asymptotic variance of MUSIC in Example 9.5.1. For a linear array, we have

$$\mathbf{v} = \left[\ 1 \ \vdots\ e^{j\psi} \ \vdots\ \cdots \ \vdots\ e^{j(N-1)\psi}\ \right]^T, \tag{9.284}$$

$$\mathbf{d} = \left[\ 0 \ \vdots\ je^{j\psi} \ \vdots\ \cdots \ \vdots\ e^{j(N-1)\psi}\ \right]^T, \tag{9.285}$$

[15]This example is taken from [SN91b].

and

$$\begin{aligned} \mathbf{d}^H \mathbf{U}_N \mathbf{U}_N^H \mathbf{d} &= \mathbf{d}^H \left(\mathbf{I} - \mathbf{v}\left(\mathbf{v}^H \mathbf{v}\right) \mathbf{v}^H\right) \mathbf{d} \\ &= \mathbf{d}^H \mathbf{d} - \frac{\left|\mathbf{d}^H \mathbf{v}\right|^2}{\mathbf{v}^H \mathbf{v}} \\ &= \frac{N(N-1)(2N-1)}{6} - \frac{N^2(N-1)^2}{4N} \\ &= \frac{N(N^2-1)}{12}. \end{aligned} \tag{9.286}$$

Using (9.283) and (9.286) in (9.281),

$$\gamma = \frac{N(N+1)}{6(N-1)}. \tag{9.287}$$

Thus, for large values of N, γ approaches $N/6$. Since MUSIC achieves the CRB for a single source, the variance of the ESPRIT estimator remains a fixed distance above the CRB bound in the asymptotic region.

Example 9.5.6

We consider a 10-element SLA with two equal-power sources separated by $\Delta\psi_R$. In Figure 9.43, the RMS error is plotted versus *ASNR*. We show the asymptotic results for TLS-ESPRIT and MUSIC, as well as the stochastic CRB. We also show the points from our simulations in Sections 9.4.1 and 9.4.3. We see that the threshold behavior of ESPRIT with $d_s = 1$ is similar to root MUSIC, but its asymptotic variance is 1.0 dB higher. Note that ESPRIT with $d_s = 5$ exhibits a more gradual threshold behavior. This is because we have used the *a priori* knowledge to limit the range of u (see (9.130)).

To improve the asymptotic performance for the $d_s = 1$ case, we use the weighting discussed in (9.133). We illustrate this in the next example.

Example 9.5.7 (continuation)

Consider the same model as in Example 9.5.8. We let $d_s = 1$ and analyze the performance for various m_s. The results are shown in Figure 9.44(a). In Figure 9.44(b) we show the difference between the normalized standard deviation and the normalized CRB. We see that, for $m_s = 3, 4$, and 5, the variance is very close to the CRB and provides a factor of 5 improvement over the $m_s = 1$ case.

A thorough asymptotic analysis of TLS-ESPRIT has been done by Ottersten et al. [OVK91] and Viberg and Ottersten [VO91]. The results are in a form that is suitable for numerical calculation, but is not conducive to an easy analytic summary. We refer the reader to the above reference for a detailed discussion.

Several examples are given in [OVK91] including correlated signals, different subarray choices, and constrained CRB results.

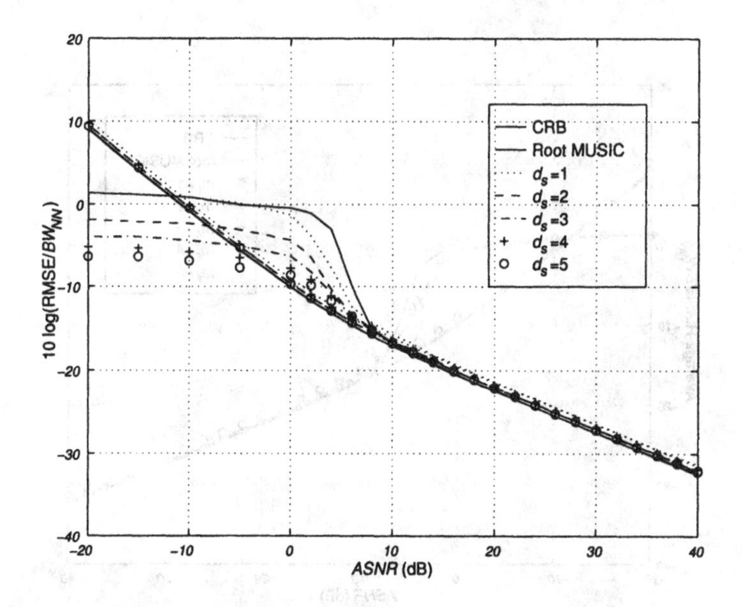

Figure 9.43 Analytic and simulated performance of TLS-ESPRIT and root MUSIC: normalized RMSE versus *ASNR*, $N = 10, K = 100$, various d_s, 500 trials.

9.5.4 Summary

In this section we have derived a number of quantitative results to predict the probability of resolution and RMSE behavior of various algorithms. These results provide the theoretical basis for understanding the algorithms. In most cases, we must still simulate the algorithms to verify the assumptions in the analyses. Our theoretical results enable us to structure these simulations more efficiently.

9.6 Correlated and Coherent Signals

9.6.1 Introduction

In this section, we study the problem of correlated and coherent signals. Correlated signals are signals whose correlation coefficient ρ is non-zero. Coherent signals are signals where the magnitude of correlation coefficient, $|\rho|$, equals one. Signal correlation occurs in a multipath environment. In [Hay85], Chapter 3 discusses multipath models in the sonar environment and Chapter 4 discusses multipath models in the radar environment. Correlated

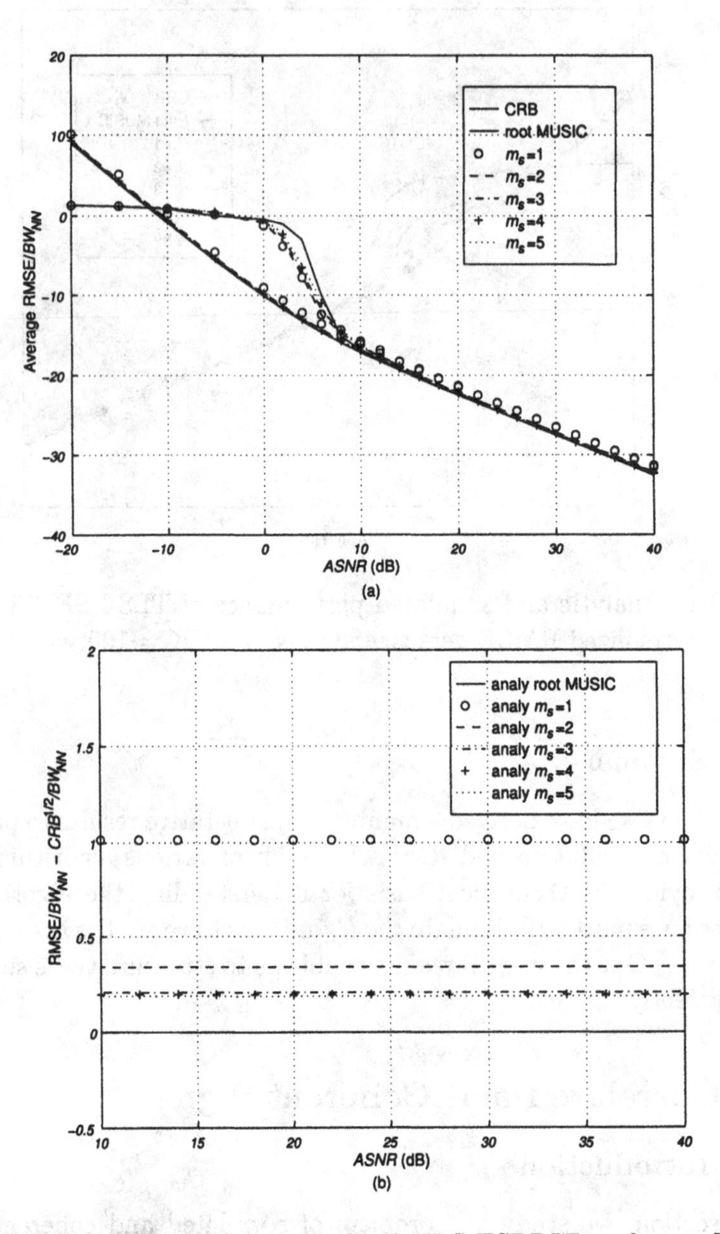

Figure 9.44 Performance of row-weighted TLS-ESPRIT and root MUSIC: $N = 10, K = 100, d_s = 1$ various m_s, 500 trials: (a) analytic and simulated performances, normalized RMSE versus *ASNR*; (b) analytic asymptotic performance (normalized RMSE–normalized $CRB^{1/2}$) versus *ASNR*.

signals can also arise in systems where smart jammers are utilized to interfere with radar or communication systems. In cellular communication systems, multipath is usually present (e.g., [ZO95] or [Win87]).

In Chapter 8, we derived the CRB for correlated and coherent signals. The algorithms that have been developed in Chapters 8 and 9 were adversely affected by correlation or coherence between the signals in the examples that we studied.

In some cases, the algorithm did not work when the signals were coherent. In other cases, the algorithm worked but the threshold occurred at a higher *ASNR* than in the uncorrelated signal case. In other cases, the RMSE did reach the CRB above threshold.

In this section, we develop algorithms to preprocess the data to remove the coherency. We focus our attention on a spatial smoothing algorithm that was invented by Evans, Johnson, and Sun ([EJS82], [EJS81]) in the context of angle of arrival estimation in air traffic control systems. Later Shan et al. [SWK85] and [SK85] studied a similar algorithm. Evans' algorithm was analyzed by William et al. [WPMS88] and Pillai and Kwon [PK89a] and [PK89b] (see also [Pil89]). Rao and Hari [RH89b] obtain expression for the variance of the DOA estimate. Weiss and Friedlander [WF93b] analyze the performance in the context of their work with interpolated arrays (e.g., [Fri90], [FW92], and [FW94]). We previously encountered spatial smoothing in the context of beamforming in Section 6.12.

A second approach to the problem is to remove the coherency by moving the array. This approach is discussed by Gabriel ([Gab80], [Gab81]) and is analyzed in detail by Haber and Zoltowski [HZ86].

A third approach utilizes constrained ML techniques (e.g., [Hay85]).

A fourth approach proposed by Cadzow [Cad80] uses a generalized eigenvector approach and obtains good results. However it requires the solution of a nonlinear programming problem. We do not discuss the last three approaches in the text.

In Section 9.6.2, we briefly review the spatial smoothing model. We then consider the three algorithms, FB-IMODE, unitary root MUSIC, and unitary TLS-ESPRIT, and compare their performance using forward–backward spatial smoothing (FBSS) to their performance using only FB averaging. In Section 9.6.3, we summarize our results.

9.6.2 Forward–Backward Spatial Smoothing

We previously encountered spatial smoothing in Chapter 6 in our discussion of optimum beamforming. The original use of spatial smoothing was in the

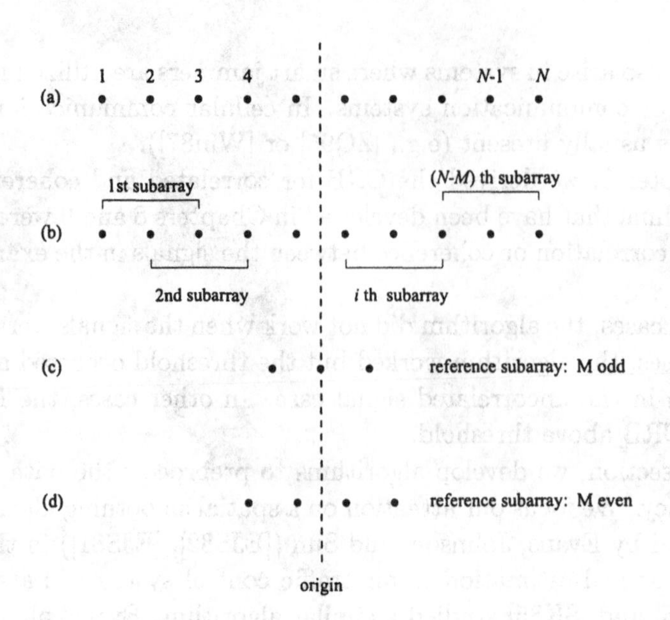

Figure 9.45 (a) Linear array; (b) subarrays; (c) reference subarray: M odd; (d) reference subarray: M even.

parameter estimation context and, although the basic model is the same as in the beamforming case, the performance issues are different.

The model of interest for forward-backward spatial smoothing was given in Section 6.12.4. The linear array of interest is shown in Figure 9.45(a). We construct a set of L subarrays of length $M \geq D+1$ as shown in Figure 9.45(b). Each subarray is shifted by one from the preceding subarray. The ith subarray has the ith element as its initial element. Typical reference subarrays are shown for M odd and M even.

For forward-only spatial smoothing, we compute

$$\hat{\mathbf{S}}_{SS} \triangleq \frac{1}{KL}\sum_{k=1}^{K}\sum_{i=1}^{L}\mathbf{X}_M^{(i)}(k)\left[\mathbf{X}_M^{(i)}(k)\right]^H, \tag{9.288}$$

where $\mathbf{X}_M^{(i)}(k)$ is the kth snapshot vector at the ith subarray.

For forward-backward spatial smoothing, we compute

$$\hat{\mathbf{S}}_{FBSS} = \frac{1}{2KL}\sum_{i=1}^{L}\sum_{k=1}^{K}\left\{\mathbf{X}_M^{(i)}(k)\left[\mathbf{X}_M^{(i)}(k)\right]^H + \mathbf{J}\left[\mathbf{X}_M^{(i)}(k)\right]^*\left[\mathbf{X}_M^{(i)}(k)\right]^T\mathbf{J}\right\}. \tag{9.289}$$

Note that the various subarrays have a large number of snapshots in common. This commonality should be exploited in the numerical calculation of (9.288) and (9.289).

We use the sample spectral matrix in either (9.288) or (9.289) in one of the algorithms that we have developed previously. We use a symmetric steering vector for a centered M-element array,

$$\mathbf{v}_S(\psi) = \left[\begin{array}{ccc} e^{-j\frac{N-1}{2}\psi} & \cdots & e^{j\frac{N-1}{2}\psi} \end{array} \right]^T. \tag{9.290}$$

The advantage of spatial smoothing is that the correlation between the signals will be reduced. For most algorithms, smaller correlation improves both the threshold behavior and the RMSE above threshold. The disadvantage of spatial smoothing is that the aperture of the subarray is smaller, which will degrade the performance. The trade-off between these two effects will depend on the algorithm that is used.

The spectral MUSIC algorithm provides a simple example of the behavior. To use MUSIC we perform an eigendecomposition of (9.288) and use the D eigenvectors corresponding to the D largest eigenvalues to construct a $M \times D$ signal subspace matrix, $\widehat{\mathbf{U}}_{FBSS,S}$. We use the $M - D$ remaining eigenvectors to form a noise subspace matrix $\widehat{\mathbf{U}}_{FBSS,N}$.

The spectral MUSIC algorithm using FBSS is

$$\hat{\psi} = \arg\min_{\psi} \left\{ \left| \mathbf{v}_S^H(\psi) \widehat{\mathbf{U}}_{FBSS,N} \right|^2 \right\}. \tag{9.291}$$

We consider a simple example using the MUSIC algorithm to illustrate a typical result. We need to consider a case where $D \geq 3$ in order to study the effect of spatial smoothing.

Example 9.6.1[16]

In this example, we have a coherent source located at 0°. It undergoes multipath reflection, resulting in three additional coherent arrivals along −25°, 45°, and 60°. A standard 10-element linear array is used to receive these signals. The input *SNR* of the direct signal is 5 dB, and the attenuation coefficients of the three coherent sources are taken to be $(0.4, 0.8)$, $(-0.3, -0.7)$, and $(0.5, -0.6)$, respectively. In the notation $\alpha = (a, b)$, a and b represent the real and imaginary parts, respectively, of the complex attenuation coefficient α. The result is shown in in Figure 9.46. We use a FBSS scheme with two subarrays of nine sensors each, and then use the MUSIC algorithm. All four directions of arrival can be directly identified.

This example illustrates the ability of the FBSS-MUSIC algorithm to estimate the DOAs of four coherent sources that are separated by a moderate

[16]This example is similar to the example in Pillai and Kwon (p.12 of [PK89a]).

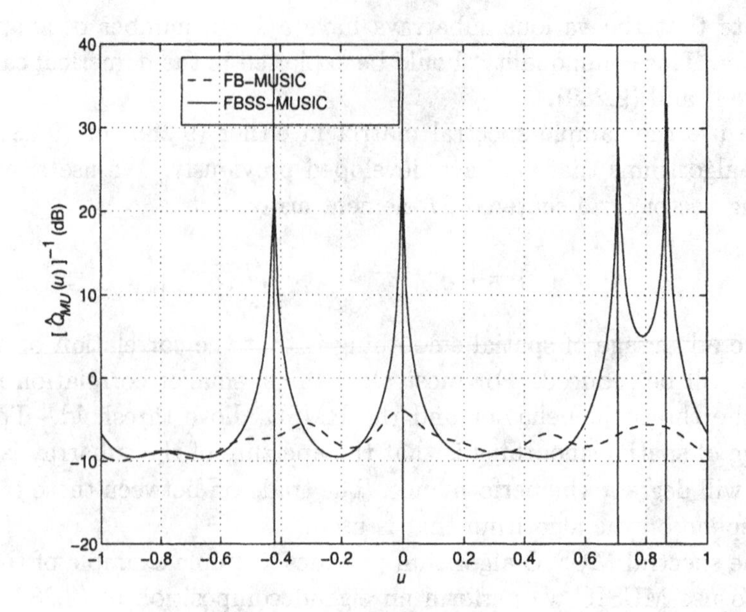

Figure 9.46 FB-MUSIC and FBSS-MUSIC: four coherent signals, $N = 10$, $M = 9$, $K = 300$, $SNR_1 = 5$ dB; $(\hat{Q}_{MU}(u))^{-1}$ versus u.

amount in angle space. There are two effects that occur in FBSS. The first effect is the decorrelation that occurs because of the SS. For uniform weighting, the correlation coefficient after SS is given by (6.636) or (6.639). It is a function of the $L \cdot \Delta\psi_{ij}$ product, where L is the number of subarrays and $\Delta\psi_{ij}$ is the separation between the ith signal and the jth signal in ψ-space. The second effect is the FB averaging that reduces the correlation between the signals when $(\phi_\rho)_{ij}$ is unequal to zero.

These effects indicate that FBSS will be most effective when the coherent signals are separated by a distance in ψ-space such that L can be chosen large enough to reduce the correlation significantly and the resulting subarray length will still allow the signals to be resolved. Thus, we anticipate that FBSS will be most useful when $|\Delta\psi_{ij}| > BW_{NN}$.

Spatial smoothing exploits the uniform spacing in a SLA. Several of the algorithms developed in Chapters 8 and 9 exploit the same characteristics. Therefore it is logical to compare the performance of some of these algorithms with spatial smoothing to their performance without spatial smoothing. In the next example, we consider the IMODE algorithm from Section 8.7.3, unitary root MUSIC from Section 9.3.2.3, and the unitary TLS-ESPRIT algorithm from Section 9.3.4.2.

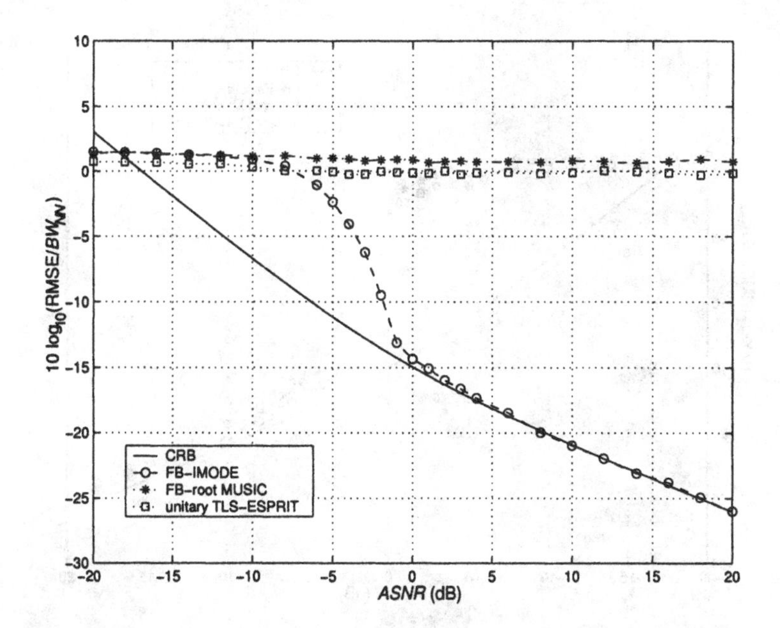

Figure 9.47 FB-root MUSIC, unitary TLS-ESPRIT with row weighting ($m_s = 5$), FB-IMODE: $N = 10$, $\Delta u = 0.3$, $|\rho| = 1$, $\phi_\rho = 0$; normalized RMSE versus *ASNR*.

Example 9.6.2

Consider a standard 10-element linear array. There are two equal-power coherent plane-wave signals impinging on the array. The magnitude of ρ equals one and $\phi_\rho = 0$.[17] We consider two values of signal separation, $\Delta u = 0.3$ and $\Delta u = 0.5$.

In Figures 9.47 and 9.48,[18] $\Delta u = 0.3$. In Figure 9.47, no spatial smoothing is used. The normalized RMSE for IMODE with FB averaging and root MUSIC and unitary TLS-ESPRIT (with row weighting) with FB averaging are plotted versus *ASNR*. Because $\phi_\rho = 0$, FB averaging does not reduce the correlation and root MUSIC and unitary TLS-ESPRIT are not useful. For $\phi_\rho = 0$, FB-IMODE and FO-IMODE have the same performance. IMODE has a threshold at 0 dB and reaches the CRB for *ASNR* ≥ 4 dB. In Figure 9.48, FBSS is used. In Figure 9.48(a), the subarray length is 8 and $L = 3$. In Figure 9.48(b), the subarray length is 6 and $L = 5$. The normalized RMSE is plotted versus *ASNR*. The performance of the IMODE algorithm is poorer due to the decrease in aperture, so SS should not be used. The performance of root MUSIC and unitary TLS-ESPRIT is better with $L = 5$ than with $L = 3$. The threshold is at *ASNR* $= 2$ dB. However, the RMSE remains about 2 dB above the CRB because of the decreased aperture.

In Figures 9.49 and 9.50, $\Delta u = 0.5$. The same results are shown. As expected, the performance is improved but the same conclusions apply.

---

[17] We have used $\phi_\rho = 0$ so that the comparison can focus on the smoothing issue.

[18] The results in Figures 9.47–9.50 are due to J. Hiemstra and D. Bray (private communication).

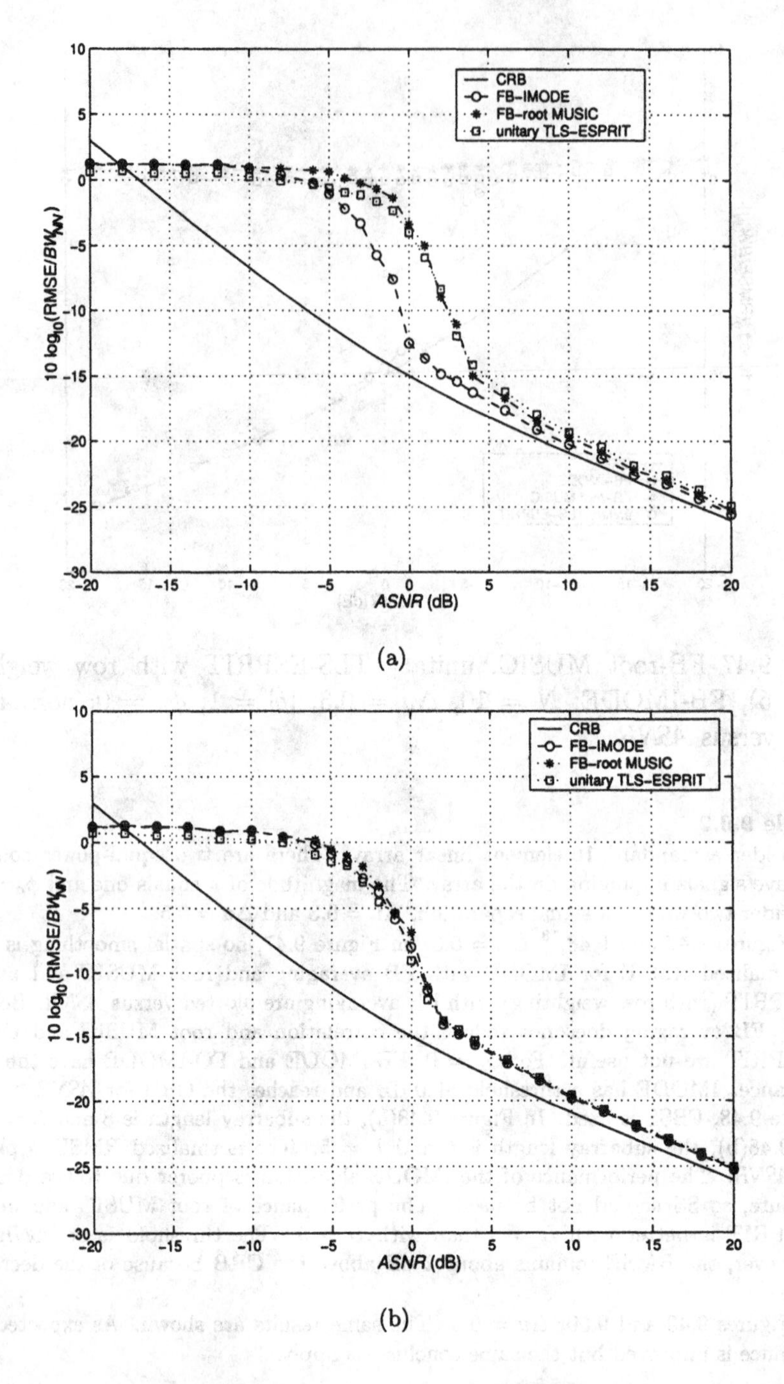

Figure 9.48 FB-root MUSIC, unitary TLS-ESPRIT with row weighting ($m_s = 5$), FB-IMODE: $N = 10$, $\Delta u = 0.3$, $|\rho| = 1$, $\phi_\rho = 0$; FBSS; normalized RMSE versus *ASNR*: (a) $L = 3$, $M = 8$; (b) $L = 5$, $M = 6$.

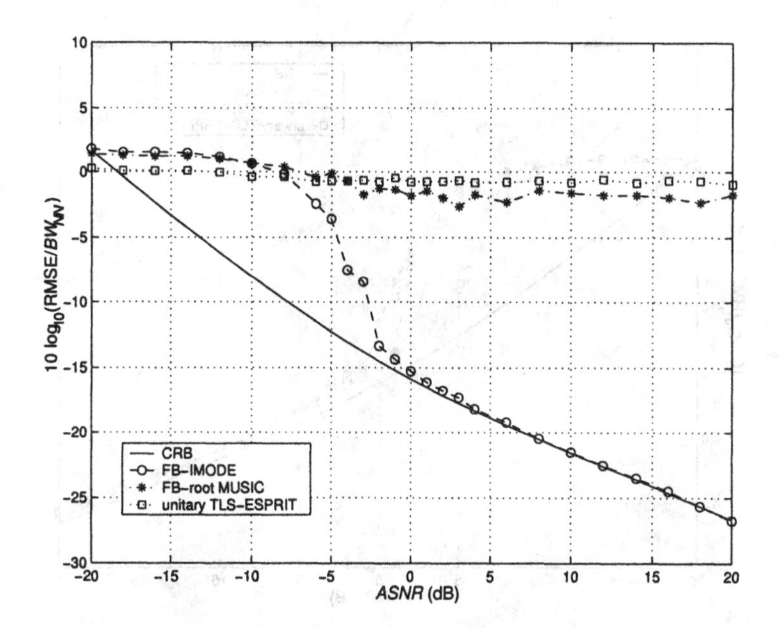

Figure 9.49 FB-root MUSIC, unitary TLS-ESPRIT with row weighting ($m_s = 5$), FB-IMODE: $N = 10$, $\Delta u = 0.5$, $|\rho| = 1$, $\phi_\rho = 0$; normalized RMSE versus *ASNR*.

9.6.3 Summary

In this section, we have demonstrated how spatial smoothing provides a method for decorrelating coherent signals prior to implementing various DOA estimation algorithms.

For algorithms, such as IMODE, that work well in the coherent signal environment the decrease in array length causes a performance degradation and SS and FBSS should not be used. For algorithms that do not perform well in a coherent environment, such as root MUSIC and unitary TLS-ESPRIT, FBSS is a useful technique.

There are several topics of interest that are addressed in the problems and in various references.

(i) **Weighted spatial smoothing** In Section 6.12.4, we discussed weighted spatial smoothing in the beamforming context. One can use classical weights such as Dolph-Chebychev or the approach of Takao and Kikuma [TK87] to choose weights to Toeplitize the array (similar ideas are discussed in [RR92] and [PRK87]). One can adapt this approach to the DOA problem.

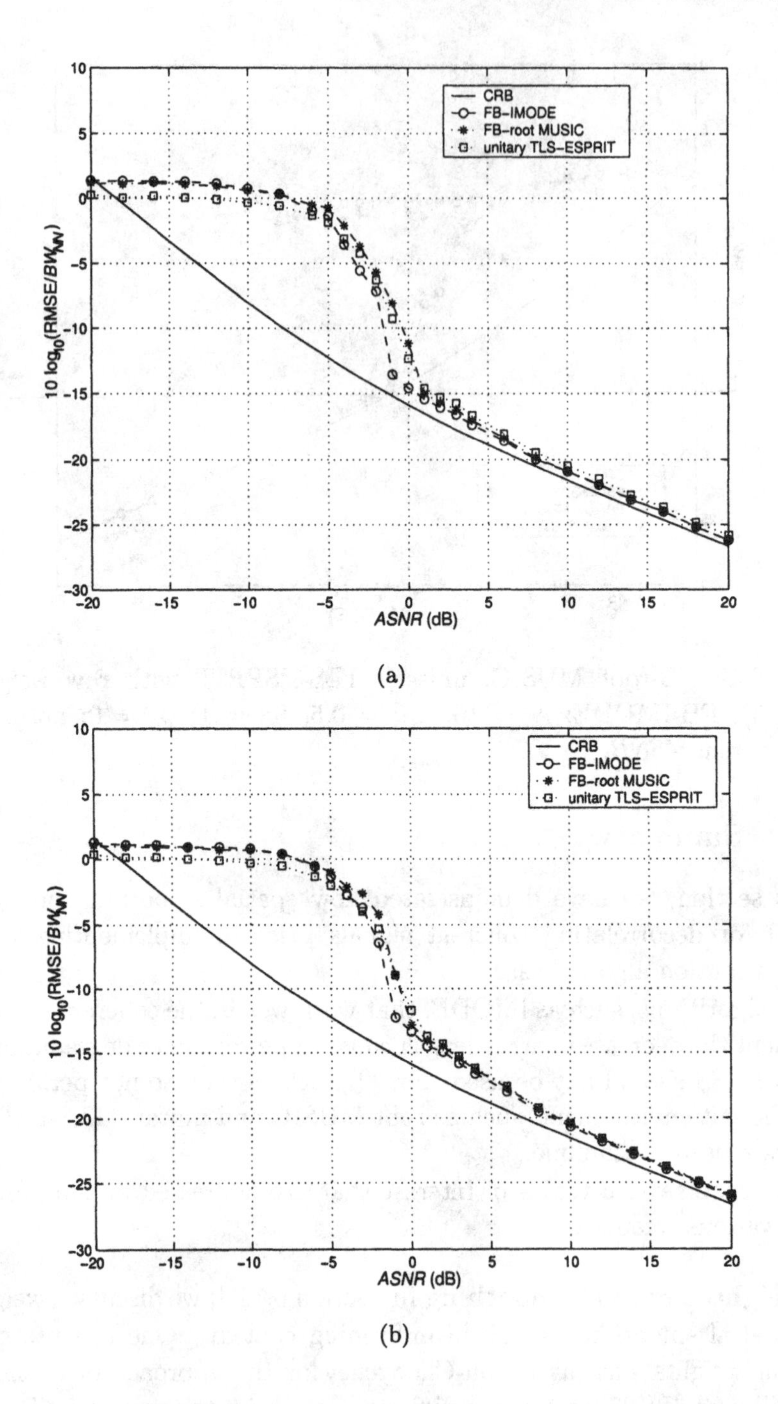

Figure 9.50 FB-root MUSIC, unitary TLS-ESPRIT with row weighting ($m_s = 5$), FB-IMODE: $N = 10$, $\Delta u = 0.5$, $|\rho| = 1$, $\phi_\rho = 0$; FBSS; normalized RMSE versus *ASNR*: (a) $L = 3$, $M = 8$; (b)$L = 5$, $M = 6$.

(ii) **Detection of number of signals** Several of the detection algorithms discussed in Section 8.8 do not work in a coherent environment. Shan et al. [SPK87] describe a smoothed rank profile test for determining the rank and coherency structure of the incoming signals.

(iii) **2-D spatial smoothing** Spatial smoothing is also applicable to planar arrays with a regular structure (e.g., Haardt et al. [HZMN95], Zoltowski et al. [ZHM96], or Fuhl et al. [FRB97]).

(iv) **Interpolated arrays** When the array is not uniform, we can construct a virtual standard linear array by interpolating the snapshot vector. We can then use spatial smoothing on the interpolated array (e.g., Friedlander [Fri90], [Fri93] and Weiss and Friedlander [WF93b] and Friedlander and Weiss [FW92]).

9.7 Beamspace Algorithms

In Section 8.10, we introduced parameter estimation using beamspace processing. We discussed the choice of beamspace matrices and computed the beamspace Cramér-Rao bound for several scenarios. We also derived the beamspace ML estimator and examined its performance. The reader should review that discussion.

In this section, we focus on two algorithms, beamspace MUSIC and beamspace unitary ESPRIT. In Section 9.7.1, we develop beamspace MUSIC and study its performance. In Section 9.7.2, we develop beamspace unitary ESPRIT and study its performance. In Section 9.7.3, we summarize our results.

9.7.1 Beamspace MUSIC

In this section, we discuss beamspace MUSIC. The beamspace snapshots are

$$\mathbf{X}_{bs}(k) = \mathbf{B}_{bs}^{H}\,\mathbf{X}(k). \tag{9.292}$$

The beamspace array manifold vector is

$$\mathbf{v}_{bs}(\psi) = \mathbf{B}_{bs}^{H}\,\mathbf{v}(\psi). \tag{9.293}$$

The beamspace spectral matrix is

$$\widehat{\mathbf{S}}_{\mathbf{x}_{bs}} = \mathbf{B}_{bs}^{H}\,\widehat{\mathbf{S}}_{\mathbf{x}}\,\mathbf{B}_{bs}. \tag{9.294}$$

If the columns of the beamspace matrix are conjugate symmetric, then the real part of $\widehat{\mathbf{S}}_{\mathbf{x}_{bs}}$ can be used to achieve FB averaging.

An eigendecomposition of $\widehat{\mathbf{S}}_{\mathbf{x}_{bs}}$ is performed. The D eigenvectors corresponding to the D largest eigenvalues are the columns of the $N_{bs} \times D$ signal subspace matrix $\widehat{\mathbf{U}}_{bs,S}$. The remaining eigenvectors are the columns of the $N_{bs} \times (N_{bs} - D)$ noise subspace $\widehat{\mathbf{U}}_{bs,N}$.

The beamspace null spectrum is

$$\widehat{Q}_{bs,MU}(\psi) = \left|\mathbf{v}_{bs}^{H}(\psi)\, \widehat{\mathbf{U}}_{bs,N}\right|^2 . \tag{9.295}$$

We find the locations of the D minima of $\widehat{Q}_{bs,MU}(\psi)$. Those values are $\widehat{\psi}_i,\ i = 1, \cdots, D$.

We consider an example of a linear array to illustrate the behavior. We utilize a DFT beamspace matrix that was discussed in Sections 6.9 and 8.10.3.3 (e.g., Figure 6.80).

Example 9.7.1

We consider a 10-element SLA. Two equal-power uncorrelated signals are impinging on the array from $\pm\Delta\psi/2$ where $\Delta\psi$ corresponds to $\Delta\psi_R$. We use a DFT beamformer and consider a 3-beam case and a 5-beam case.

In Figure 9.51, we show the probability of resolution versus *ASNR* for $N_{bs} = 3$ and 5. In Figure 9.52, we show the normalized RMSE versus *ASNR* for $K = 100$.

As expected, the threshold occurs at a lower *ASNR* than element-space MUSIC, and $N_{bs} = 3$ is better than $N_{bs} = 5$. Above threshold, the beamspace algorithms are essentially on the CRB.

Asymptotic expressions for beamspace MUSIC covariance matrix are given by Stoica and Nehorai [SN91a], Xu and Buckley [XB93], and Weiss and Friedlander [WF94b]. Our discussion follows the latter reference.

From Section 9.5 (9.259), the asymptotic covariance matrix for element-space MUSIC is

$$\mathbf{C}_{MU} = \frac{\sigma_w^2}{2K} \left(\mathbf{H} \odot \mathbf{I}\right)^{-1} Re\left\{\mathbf{H} \odot \mathbf{F}^T\right\} \left(\mathbf{H} \odot \mathbf{I}\right)^{-1}, \tag{9.296}$$

where

$$\mathbf{H} \triangleq \mathbf{D}^H\, \mathbf{P}_{\mathbf{V}}^{\perp}\, \mathbf{D}, \tag{9.297}$$

and

$$\mathbf{F} \triangleq \mathbf{S}_{\mathbf{f}}^{-1} + \sigma_w^2 \mathbf{S}_{\mathbf{f}}^{-1} \left(\mathbf{V}^H \mathbf{V}\right)^{-1} \mathbf{S}_{\mathbf{f}}^{-1}. \tag{9.298}$$

The diagonal elements are given by

$$\mathbf{C}_{MU} \odot \mathbf{I} = \frac{\sigma_w^2}{2K} \left(\mathbf{F} \odot \mathbf{I}\right) \left(\mathbf{H} \odot \mathbf{I}\right)^{-1}. \tag{9.299}$$

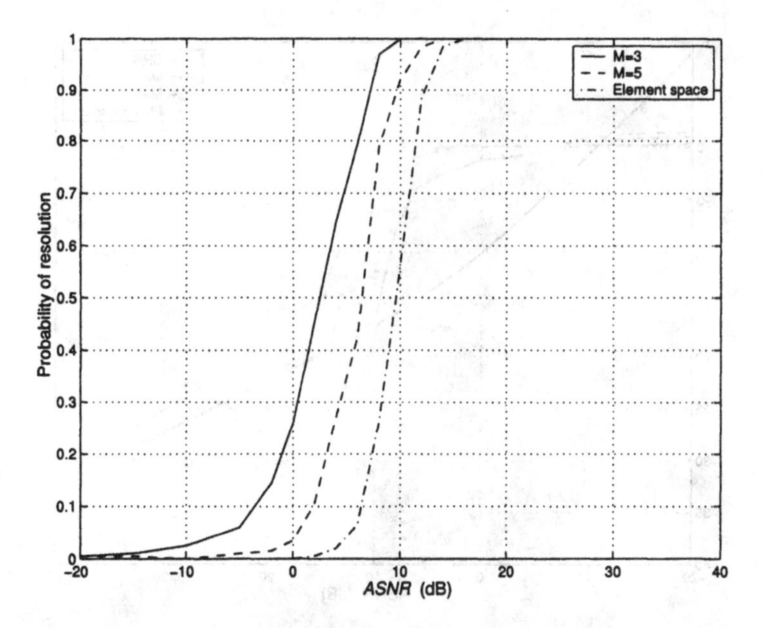

Figure 9.51 Beamspace spectral MUSIC, DFT beamspace matrix: $N = 10$, $N_{bs} = 3$ and 5, two equal-power uncorrelated plane-wave signals spaced at Δ_R, $K = 100$; probability of resolution versus *ASNR*.

For beamspace MUSIC, the corresponding result is

$$\boxed{\mathbf{C}_{BMU} \odot \mathbf{I} = \frac{\sigma_w^2}{2K} \left(\mathbf{F}_{BS} \odot \mathbf{I}\right) \left(\mathbf{H}_{BS} \odot \mathbf{I}\right)^{-1},} \tag{9.300}$$

where

$$\mathbf{H}_{BS} \triangleq \mathbf{D}^H \mathbf{B} \mathbf{P}_{\mathbf{V_B}}^{\perp} \mathbf{B}^H \mathbf{D}, \tag{9.301}$$

and

$$\mathbf{F}_{BS} \triangleq \mathbf{S}_{\mathbf{f}}^{-1} + \sigma_w^2 \mathbf{S}_{\mathbf{f}}^{-1} \left(\mathbf{V}^H \mathbf{P_B} \mathbf{V}\right)^{-1} \mathbf{S}_{\mathbf{f}}^{-1}. \tag{9.302}$$

One can then show that, assuming $\mathbf{S_f}$ is positive definite, that

$$\mathbf{C}_{MU} \odot \mathbf{I} \leq \mathbf{C}_{BMU} \odot \mathbf{I}. \tag{9.303}$$

For standard linear arrays using element-space processing, the root MUSIC algorithm has better threshold performance than spectral MUSIC and is computationally simpler. We would anticipate similar improvement in beamspace. Beamspace root MUSIC is discussed in several references.

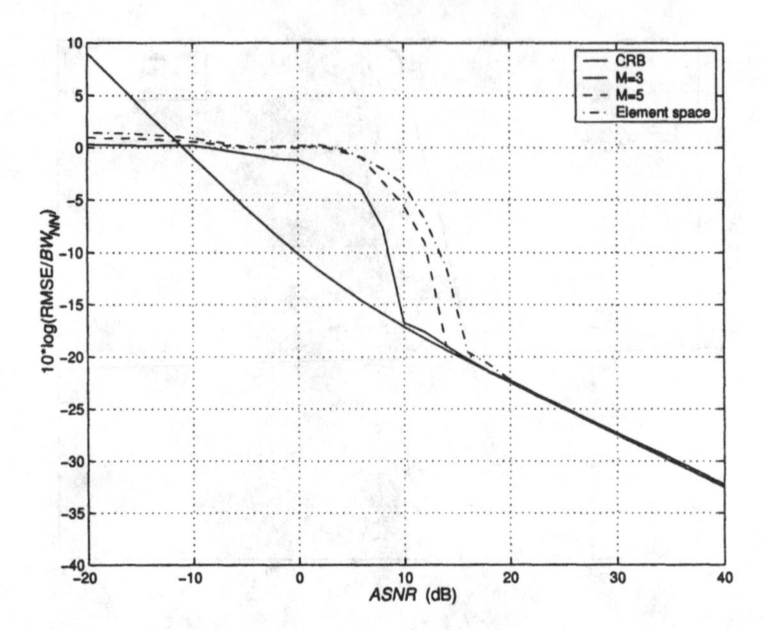

Figure 9.52 Beamspace spectral MUSIC, DFT beamspace matrix: $N = 10$, $N_{bs} = 3$ and 5, two equal-power uncorrelated plane-wave signals spaced at Δ_R, $K = 100$; normalized RMSE versus *ASNR*.

The difficulty with beamspace root MUSIC is that the Vandermonde structure of the array manifold may not be preserved during the beamspace transformation. In addition, if we can obtain a polynomial to be rooted, we would like it to be $(2N_{bs}-2)$-dimensional rather than $(2N-2)$-dimensional.

Lee and Wengrovitz [LW88] approached the problem by using orthogonal matrix beamformers that can be factored. Their approach leads to a $2N_{bs}-2$ polynomial.

Zoltowski et al. [ZKS93] use a set of shifted DFT beams to construct $\mathbf{B}_{bs}$. They develop a $2N-2$ polynomial and reduce it to a $2N_{bs}-2$ polynomial. The construction of the $\mathbf{C}$ matrix was discussed in Section 8.10.4.2 in the context of beamspace IQML. In a later paper, Kautz and Zoltowski [KZ96] (see Zoltowski et al. [ZKK94]) point out that the numerical instabilities in the beamspace root MUSIC algorithm limit its practical application. They develop a DOA algorithm that is based on telescoping the beamspace noise eigenvector into an element-space eigenvector. They refer to their algorithm as beamspace multirate eigenvector processing and show that it has good performance and is computationally stable. It also allows tapered beams in the beamspace matrix. The technique was extended to 2-D estimation using

URAs in Gansman et al. [GZK96]. We refer the reader to these references for further discussion.

In this section, we have developed beamspace spectral MUSIC and considered a simple example to illustrate its behavior. Although the example used a SLA, the algorithm can be used with arbitrary array geometries and beamspace matrices.

9.7.2 Beamspace Unitary ESPRIT

Several authors have proposed beamspace ESPRIT algorithms and analyzed various aspects of the algorithm's performance. Two references of interest are Li [Li92b] and Xu et al. [XSRK94].

Zoltowski et al. [ZHM96] developed a beamspace unitary ESPRIT algorithm using DFT beams that is computationally efficient and provides good performance.

The advantage of beamspace unitary ESPRIT is that, for many beamspace matrices of interest, the primary steps in ESPRIT can be done using real-valued calculations because the beamspace array manifold vector is real. In addition, the computational complexity is significantly reduced because of the reduced dimension of the beamspace. We develop the algorithm in this section. Our discussion follows Section IV of [ZHM96].

The beamspace matrix consists of N_{bs} orthogonal conventional beams whose MRAs are at $2\pi m/N$ in ψ-space. The mth row of the beamspace matrix is

$$\mathbf{B}_{bs}^H = \frac{1}{N}\left[\begin{array}{cccc} e^{j(\frac{N-1}{2})m\frac{2\pi}{N}} & e^{j(\frac{N-3}{2})m\frac{2\pi}{N}}, & \cdots, & e^{-j(\frac{N-1}{2})m\frac{2\pi}{N}} \end{array}\right], m \in \sum_{N_{bs}}, \tag{9.304}$$

where $\sum_{N_{bs}}$ denotes the set of integers designating the beam locations.

This vector is just the Hermitian transpose of the steering vector for $\psi_s = 2\pi m/N$,

$$\left[\mathbf{B}_{bs}^H\right]_m = \frac{1}{N}\mathbf{v}_s^H(\psi_s), \quad \psi_s = 2\pi m/N. \tag{9.305}$$

The values of m depend on the location of the center of the beamspace sector. For example, if the beamspace sector is centered at $\psi = 0$ and N_{bs} is odd,

$$-(N_{bs}-1)/2 \leq m \leq (N_{bs}-1)/2. \tag{9.306}$$

To simplify the notation, it is convenient to denote the beam whose MRA is closest to $\psi = -\pi$ as the left beam and use m_L for the value of m that specifies its location. Then, the indexing on m is

$$m_L \leq m \leq m_L + N_{bs}. \tag{9.307}$$

Two points should be noticed:

(i) If the center of the beamspace sector is near endfire, then $m_L + N_{bs}$ may be greater than N and the beams wrap around in ψ-space.

(ii) This notation assumes the MRAs are at $m2\pi/N$. The case for MRAs at $(m-\alpha)2\pi/N$ is a straightforward modification.

The mth beam in ψ-space is

$$b_m(\psi) = \left[\mathbf{B}_{bs}^H\right]_m \mathbf{v}(\psi) = \frac{1}{N}\frac{\sin\left[\frac{N}{2}\left(\psi - m\frac{2\pi}{N}\right)\right]}{\sin\left[\frac{1}{2}\left(\psi - m\frac{2\pi}{N}\right)\right]}, \quad m = m_L, m_{L+1}, \cdots, m_{L+N_{bs}}. \tag{9.308}$$

The total beamspace matrix is denoted by the $N_{bs} \times N$ matrix, $\mathbf{B}_{bs}^H$.

The beamspace array manifold vector is a real vector,

$$\mathbf{v}_{bs}(\psi) = \mathbf{B}_{bs}^H \mathbf{v}(\psi) = \begin{bmatrix} b_{m_L}(\psi) & b_{m_L+1}(\psi) & \cdots & b_{m_L+N_{bs}}(\psi) \end{bmatrix}^T, \tag{9.309}$$

where the rows of $\mathbf{B}_{bs}^H$ are given by (9.304) and $b_m(\psi)$ is given by (9.308).

Note that $\mathbf{v}_{bs}$ is real. The key to beamspace ESPRIT is the observation that

$$b_{m+1}(\psi) = \frac{\sin\left[\frac{N}{2}\left(\psi - (m+1)\frac{2\pi}{N}\right)\right]}{\sin\left[\frac{1}{2}\left(\psi - (m+1)\frac{2\pi}{N}\right)\right]}, \tag{9.310}$$

and that the numerator in (9.310) is the negative of the numerator in (9.308). Therefore, the two successive components are related by

$$\sin\left[\frac{1}{2}\left(\psi - m\frac{2\pi}{N}\right)\right] b_m(\psi) +$$

$$\sin\left[\frac{1}{2}\left(\psi - (m+1)\frac{2\pi}{N}\right)\right] b_{m+1}(\psi) = 0. \tag{9.311}$$

This equation can be manipulated into

$$\tan\left(\frac{\psi}{2}\right)\left\{\cos\left(m\frac{\pi}{N}\right) b_m(\psi) + \cos\left((m+1)\frac{\pi}{N}\right) b_{m+1}(\psi)\right\}$$

$$= \sin\left(m\frac{\pi}{N}\right) b_m(\psi) + \sin\left((m+1)\frac{\pi}{N}\right) b_{m+1}(\psi), \qquad (9.312)$$

We want to define two $(N_{bs}-1) \times N_{bs}$ selection matrices relating successive beams, $b_m(\psi)$ and $b_{m+1}(\psi)$.

Define

$$\Gamma_1 = \begin{bmatrix} \cos\left(m_L \frac{\pi}{N}\right) & \cos\left((m_L+1)\frac{\pi}{N}\right) & 0 & \cdots & 0 \\ 0 & \cos\left((m_L+1)\frac{\pi}{N}\right) & \cos\left((m_L+2)\frac{\pi}{N}\right) & \cdots & 0 \\ 0 & 0 & \cos\left((m_L+2)\frac{\pi}{N}\right) & \cdots & 0 \\ \vdots & \vdots & \vdots & \ddots & \vdots \\ 0 & 0 & 0 & \cdots & \cos\left((m_L+N_{bs})\frac{\pi}{N}\right) \end{bmatrix}, \qquad (9.313)$$

$$\Gamma_2 = \begin{bmatrix} \sin\left(m_L \frac{\pi}{N}\right) & \sin\left((m_L+1)\frac{\pi}{N}\right) & 0 & \cdots & 0 \\ 0 & \sin\left((m_L+1)\frac{\pi}{N}\right) & \sin\left((m_L+2)\frac{\pi}{N}\right) & \cdots & 0 \\ 0 & 0 & \sin\left((m_L+2)\frac{\pi}{N}\right) & \cdots & 0 \\ \vdots & \vdots & \vdots & \ddots & \vdots \\ 0 & 0 & 0 & \cdots & \sin\left((m_L+N_{bs})\frac{\pi}{N}\right) \end{bmatrix}, \qquad (9.314)$$

The relation in (9.312) can be written in matrix form as

$$\tan\left(\frac{\psi}{2}\right) \mathbf{\Gamma}_1 \mathbf{v}_{bs}(\psi) = \mathbf{\Gamma}_2 \mathbf{v}_{bs}(\psi). \qquad (9.315)$$

For D sources, the beamspace array manifold matrix is

$$\mathbf{V}_{bs}(\boldsymbol{\psi}) = \left[\begin{array}{cccc} \mathbf{v}_{bs}(\psi_1) & \mathbf{v}_{bs}(\psi_2) & \cdots & \mathbf{v}_{bs}(\psi_D) \end{array}\right]. \qquad (9.316)$$

Using (9.316) with (9.315) gives

$$\mathbf{\Gamma}_1 \mathbf{V}_{bs}(\boldsymbol{\psi}) \mathbf{\Omega}_{\psi} = \mathbf{\Gamma}_1 \mathbf{V}_{bs}(\boldsymbol{\psi}), \qquad (9.317)$$

where

$$\mathbf{\Omega}_{\psi} = \text{diag}\left[\tan\left(\frac{\psi_1}{2}\right), \cdots, \tan\left(\frac{\psi_D}{2}\right)\right]. \qquad (9.318)$$

The signal eigenvectors are the eigenvectors corresponding to the D largest eigenvalues of

$$Re\left[\widehat{\mathbf{S}}_{\mathbf{x},bs}\right] = Re\left[\mathbf{B}_{bs}^H \widehat{\mathbf{S}}_{\mathbf{x}} \mathbf{B}_{bs}\right]. \qquad (9.319)$$

Then, the signal subspace is an $N_{bs} \times D$ matrix,

$$\widehat{\mathbf{U}}_{bs,s} = \mathbf{V}_{bs}\,\mathbf{T}, \qquad (9.320)$$

where $\mathbf{T}$ is a non-singular $D \times D$ matrix. Substituting

$$\mathbf{V}_{bs} = \widehat{\mathbf{U}}_{bs,s}\mathbf{T}^{-1}, \tag{9.321}$$

into (9.317) gives

$$\mathbf{\Gamma}_1 \widehat{\mathbf{U}}_{bs,s} \widehat{\mathbf{\Psi}} = \mathbf{\Gamma}_2 \widehat{\mathbf{U}}_{bs,s}, \tag{9.322}$$

where

$$\mathbf{\Psi} = \mathbf{T}^{-1} \mathbf{\Omega}_{\psi} \mathbf{T}. \tag{9.323}$$

Note that (9.322) has the same form as (9.119) with $\widehat{\mathbf{U}}_{s1}$ replaced with $\mathbf{\Gamma}_1\widehat{\mathbf{U}}_{bs,s}$, and $\widehat{\mathbf{U}}_{s2}$ replaced with $\mathbf{\Gamma}_2\widehat{\mathbf{U}}_{bs,s}$. The beamspace unitary LS-ESPRIT solution is

$$\widehat{\mathbf{\Psi}}_{bs,ULS} = \left[\left[\mathbf{\Gamma}_1\widehat{\mathbf{U}}_{bs,s}\right]^T \left[\mathbf{\Gamma}_1\widehat{\mathbf{U}}_{bs,s}\right]\right]^{-1} \left[\mathbf{\Gamma}_1\widehat{\mathbf{U}}_{bs,s}\right]^T \left[\mathbf{\Gamma}_2\widehat{\mathbf{U}}_{bs,s}\right]. \tag{9.324}$$

The beamspace unitary TLS-ESPRIT solution is obtained by replacing $\widehat{\mathbf{U}}_{s1}$ with $\mathbf{\Gamma}_1\widehat{\mathbf{U}}_{bs,s}$, and $\widehat{\mathbf{U}}_{s2}$ with $\mathbf{\Gamma}_2\widehat{\mathbf{U}}_{bs,s}$ in the formula for $\tilde{\mathbf{C}}$ given in (9.123). Then,

$$\widehat{\mathbf{\Psi}}_{bs,UTLS} = -\mathbf{V}_{12}\,\mathbf{V}_{22}^{-1}. \tag{9.325}$$

The steps in beamspace ESPRIT can be summarized:

1. Find $\widehat{\mathbf{U}}_{bs,s}$, the $N_{bs} \times D$ signal subspace matrix whose columns are the eigenvectors corresponding to the D largest eigenvalues of

$$Re\left[\widehat{\mathbf{S}}_{\mathbf{x},bs}\right] = Re\left[\mathbf{B}_{bs}^H\,\widehat{\mathbf{S}}_{\mathbf{x}}\,\mathbf{B}_{bs}\right].$$

2a. For unitary LS-ESPRIT, $\widehat{\mathbf{\Psi}}_{bs,ULS}$ is given by (9.324).

2b. For unitary TLS-ESPRIT, $\widehat{\mathbf{\Psi}}_{bs,UTLS}$ is given by (9.325), which is obtained by a $2D \times 2D$ eigendecomposition of $\tilde{\mathbf{C}}$.

3. Find the D eigenvalues, $\hat{w}_i$, of the real-valued matrix in step 2; either $\widehat{\mathbf{\Psi}}_{bs,ULS}$ or $\widehat{\mathbf{\Psi}}_{bs,UTLS}$.

4. Compute the estimated DOAs in ψ-space,

$$\hat{\psi}_i = 2\tan^{-1}\left(\hat{w}_i\right), \; i = 1, \cdots, D.$$

We consider an example to illustrate the performance of beamspace unitary TLS-ESPRIT.

Example 9.7.2

Consider a standard 10-element linear array. Two equal-power uncorrelated plane-wave signals are impinging on the array from $\pm\Delta\psi/2$ where $\Delta\psi = \Delta\psi_R$. We use the beamspace matrix described by (9.304) and implement the 3-beam and 5-beam case. The results are shown in Figure 9.53. Note that the three beams centered at $\psi = 0$ correspond to $m = -1$, 0, and 1 in our indexing scheme. For five beams, $m = -2, -1$, 0, 1, and 2.

We see that the resolution performance is slightly better than element-space TLS-ESPRIT. In the 3-beam case, the RMSE approaches the CRB above threshold. In the 5-beam case, the RMSE remains slightly above the CRB above threshold.

9.7.3 Beamspace Summary

In this section, we have discussed beamspace processing. For the models considered, we have seen that beamspace processing can provide improvement in resolution and almost no loss in estimation accuracy. In addition, the computational complexity is reduced and the statistical stability for a given number of snapshots is increased.

For arbitrary array geometries and uncorrelated signals, beamspace spectral MUSIC can be used. For correlated signals, SS can be utilized.

For ULAs and a DFT beamspace matrix, beamspace IQML (or beamspace MODE) provides the best threshold performance and approaches the beamspace CRB above threshold. Beamspace unitary ESPRIT also provides good performance with less computation and is close to the CRB above threshold.

We discuss beamspace processing for planar arrays in Section 9.9.

9.8 Sensitivity and Robustness

In Section 8.11, we developed models for array perturbations and environmental perturbations. We developed Cramér-Rao bounds for the case in which we jointly estimated the DOAs and the array parameters. We then developed joint estimation algorithms. In this section, we revisit the sensitivity problem in the context of the estimation algorithms that we have developed in this chapter.

The sensitivity discussion can be divided into three parts. In the first part, we assume that we use one of the estimation algorithms developed in this chapter and use the nominal array manifold in the calculation. We perturb the array using the perturbation model in Section 8.3.2 ((8.72)–(8.81)) and analyze (or simulate) the performance of the algorithm. We compare

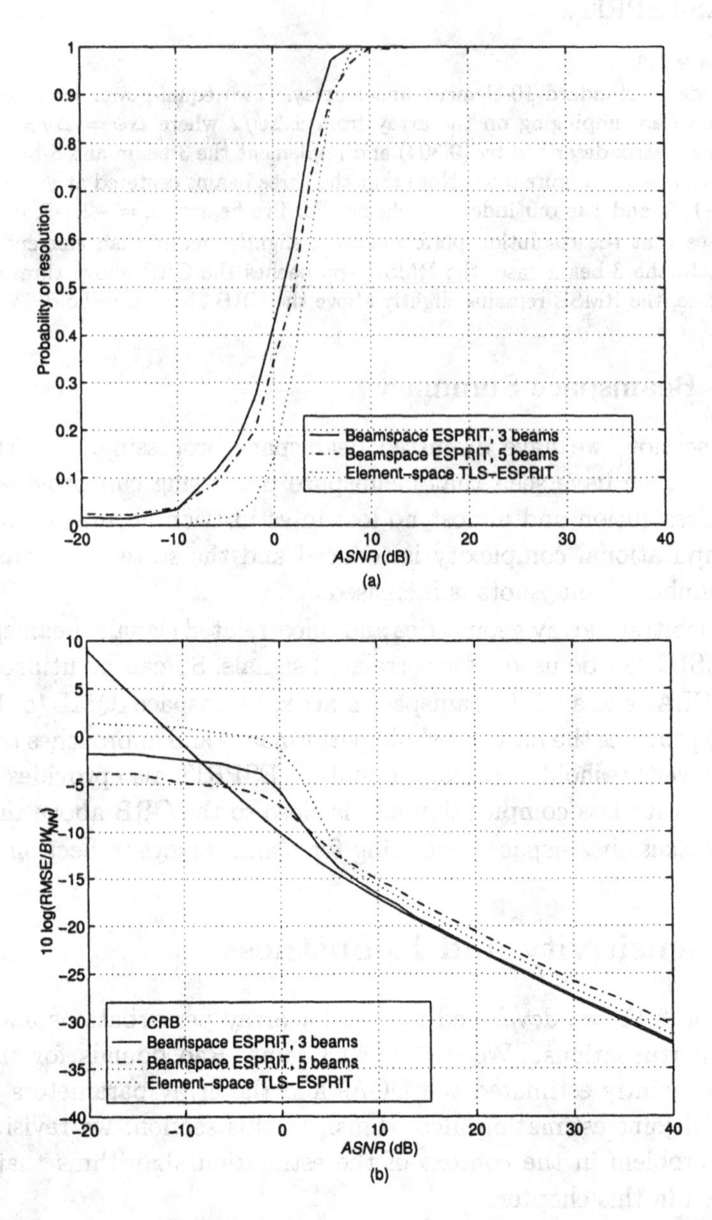

Figure 9.53 Performance of beamspace unitary TLS-ESPRIT using a DFT beamspace: $N = 10$, $N_{bs} = 3$ and 5, $K = 100$, 500 trials: (a) probability of resolution versus *ASNR*; (b) normalized RMSE versus *ASNR*.

the resulting performance to the hybrid Cramér-Rao bounds derived in Section 8.11.2. In the second part, we modify the algorithms in some manner to make them more robust to the array perturbations. The robust algorithms in this second part do not explicitly estimate the perturbed array parameters. In the third part, we consider algorithms that treat the problem as a joint estimation problem and try to estimate both the DOAs and the perturbed array parameters.

Papers that focus on the first two parts of the problem include Swindlehurst and Kailath [SK92] (MUSIC), Swindlehurst and Kailath [SK93b] (multidimensional algorithms), Hamza and Buckley [HB95] (weighted eigenspace), Jansson et al. [JSO98] (WSF), Ng [Ng95], Ratnarajah [Rat98] (MUSIC), Viberg and Swindlehurst [VS94], Rockah et al. [RMS88], Rockah and Schultheiss [RS87], Weiss and Friedlander [WF94a] (resolution of MUSIC), Zhang and Zhu [ZZ95], Kuroc [Kur89] (lower bounds), and Friedlander and Weiss ([FW90], [FW93a]). We refer the reader to those references for a complete discussion. We consider a single example to illustrate the performance of root MUSIC and unitary TLS-ESPRIT.

Example 9.8.1 (continuation, Example 8.11.1)

In this example, we consider the same perturbation model as in Example 8.11.1. The array is a 10-element SLA on the x-axis. There are two equal-power uncorrelated plane-wave sources at $\pm\Delta\psi/2$, where $\Delta\psi = \Delta\psi_R$. The perturbations of the each sensor in the x- and y-direction are independent zero-mean Gaussian random variables with standard deviation σ_p. We let $\sigma_p = 0.05\lambda$. This value corresponds to 10%, respectively, of the nominal interelement spacing.

In Figure 9.54(a), we plot P_R versus *ASNR* for $\sigma_p = 0.05\lambda$.[19] In Figure 9.54(b), we plot the normalized RMSE versus *ASNR* for root MUSIC and unitary TLS-ESPRIT. We also plot the hybrid CRB and the nominal CRB. There is very little degradation in the threshold behavior of root MUSIC and unitary TLS-ESPRIT. The RMSE for the two algorithms levels off above an *ASNR* = 20 dB.

The gap between the RMSE of the algorithms and hybrid CRB suggests that there may be improved algorithms that provide better performance. Several of the above references suggest modifications to improve robustness.

In Section 8.11, we introduced two algorithms, the MAPprox algorithm and Weiss-Friedlander iterative algorithm that used joint (or alternating) estimation. The generalized WSF algorithm [JSO98] was also discussed. (This paper has an extensive list of references.) In that discussion either an ML or WSF algorithm was used for estimation. Modifications to incorporate MUSIC or ESPRIT can be derived. References that deal with the calibration problem or the joint estimation problem include Fuhrmann [Fuh94], Hung [Hun94], Huang and Williams [HW94], McCarthy et al. [MRP94], Soon et

---

[19] The results in Figure 9.54 are due to D. Bray (private communication).

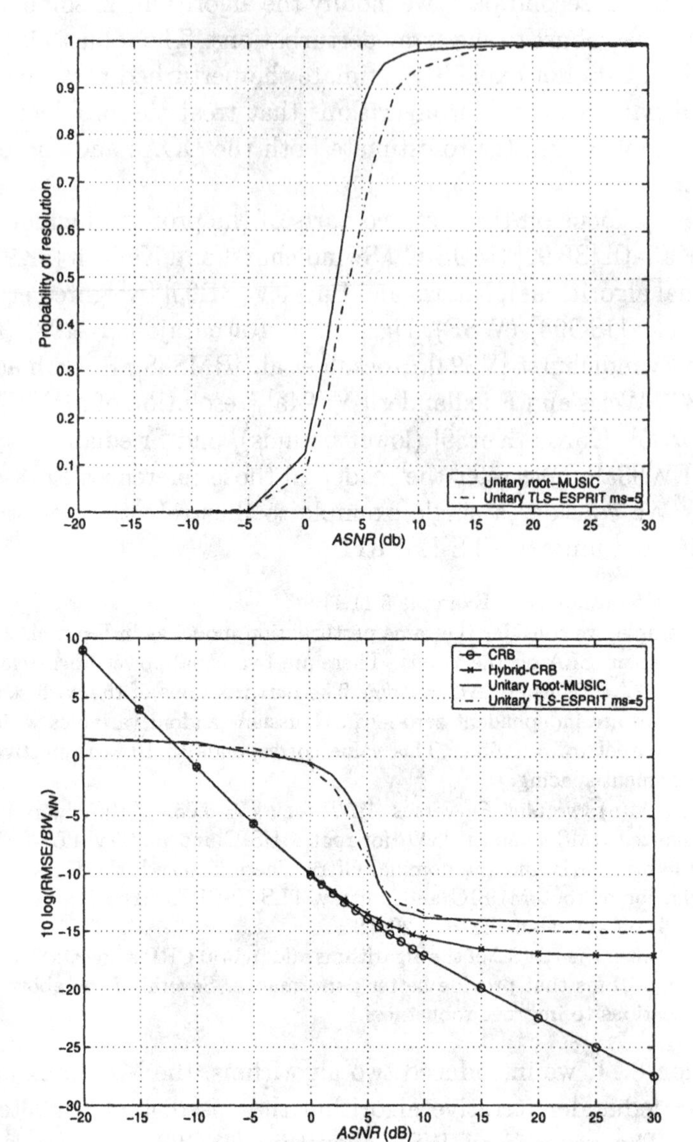

Figure 9.54 Unitary root MUSIC and unitary TLS-ESPRIT ($d_s = 1$, $m_s = 5$); SLA on x-axis, $N = 10$, two equal-power uncorrelated plane-wave signals, $\Delta\psi = \Delta\psi_R$, sensor position perturbations, $\sigma_x = \sigma_y = 0.05\lambda$, 100 trials: (a) probability of resolution versus *ASNR*; (b) normalized RMSE versus *ASNR*.

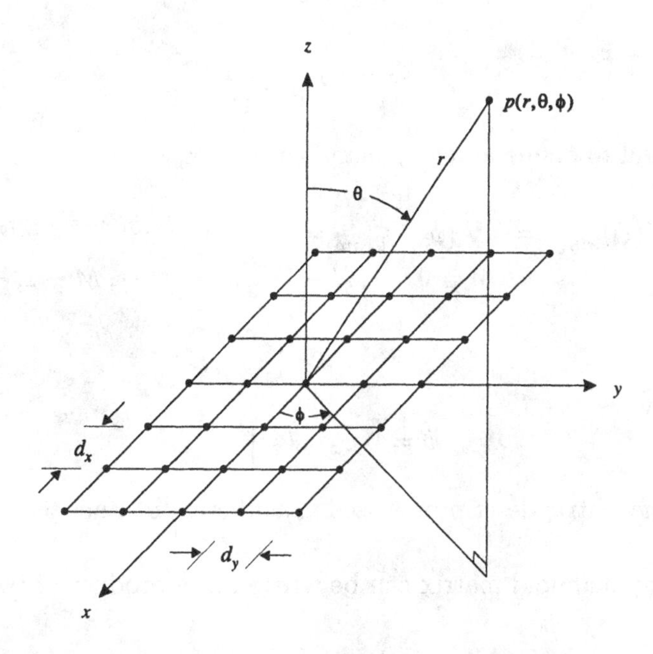

Figure 9.55 Geometry of a uniform rectangular array.

al. [STHL94], Talwar et al. [TPG93], Tseng et al. [TFG95], Weiss and Friedlander [WF91a], [WF91b], Wylie et al. [WRM94]. We refer the reader to these references for a discussion of these techniques. A recent dissertation by Flanagan ([Fla00]) studies the problem and contains an extensive list of references.

9.9 Planar Arrays

In this section, we extend some of the results in the earlier sections of the chapter to planar arrays. Some of the algorithms, such as spectral MUSIC, extend in an obvious manner. However, the extension to 2-D of algorithms such as root MUSIC that rely on a polynomial representation are less clear.

In Section 9.1, we study standard rectangular arrays and develop various 2-D algorithms. In Section 9.9.2, we discuss hexagonal arrays. In Section 9.9.3, we summarize our results.

9.9.1 Standard Rectangular Arrays

The geometry of a uniform rectangular array was shown in Figure 4.5 and is repeated in Figure 9.55 for convenience. In a standard rectangular array,

$$d_x = d_y = \frac{\lambda}{2}. \tag{9.326}$$

It is useful to define an array manifold matrix,

$$\begin{aligned} [\mathbf{V}(\boldsymbol{\psi})]_{nm} &= \mathbf{V}[\psi_x, \psi_y]_{nm} = e^{j(n-\frac{N-1}{2})\psi_x} e^{j(m-\frac{M-1}{2})\psi_y}, \\ & n = 0, \cdots, N-1, \quad m = 0, \cdots, M-1, \end{aligned} \tag{9.327}$$

where

$$\boldsymbol{\psi} = \left[\begin{array}{cc} \psi_x & \psi_y \end{array} \right]^T. \tag{9.328}$$

Note that the matrix descriptor is used in a different manner than in previous discussions.

The array manifold matrix can be written as a product of two vectors,

$$\mathbf{V}(\psi_x, \psi_y) = \mathbf{v_x}(\psi_x)\mathbf{v_y}^T(\psi_y), \tag{9.329}$$

where $\mathbf{v_x}(\psi_x)$ and $\mathbf{v_y}(\psi_y)$ are given by (2.72) or (2.73).

We can write $\mathbf{V}(\psi_x, \psi_y)$ as an $NM \times 1$ array manifold vector,

$$vec(\boldsymbol{\psi}) \triangleq vec\left(\mathbf{V}(\psi_x, \psi_y)\right). \tag{9.330}$$

Note that $vec(\boldsymbol{\psi})$ is conjugate symmetric.

We can also write the kth snapshot of the input to the array as either an $N \times M$ matrix $\mathbf{X}_k$ or a $NM \times 1$ vector $vec(\mathbf{X}_k)$. For notational simplicity we define

$$\mathbf{Z}_k \triangleq vec(\mathbf{X}_k). \tag{9.331}$$

We can write the sample spectral matrix of $\mathbf{Z}_k$ as

$$\hat{\mathbf{S}}_\mathbf{Z} = \frac{1}{K}\sum_{k=1}^{K} \mathbf{Z}_k \mathbf{Z}_k^H. \tag{9.332}$$

Since $\mathbf{Z}_k$ is conjugate symmetric, we can use FB averaging on $\mathbf{Z}_k$ to obtain

$$\hat{\mathbf{S}}_{\mathbf{Z},fb} = \frac{1}{2K}\left\{\sum_{k=1}^{K}\left(\mathbf{Z}_k\mathbf{Z}_k^H + \mathbf{J}\mathbf{Z}_k^*\mathbf{Z}_k^T\mathbf{J}\right)\right\}. \tag{9.333}$$

9.9.1.1 2-D spectral algorithms

The various spectral algorithms follow directly. For example, for spectral MUSIC, we find either the $NM \times D$ signal subspace matrix, $\widehat{\mathbf{U}}_S$, or the $NM \times (NM - D)$ noise subspace matrix, $\widehat{\mathbf{U}}_N$, and use either (9.44) or (9.45). The spectral MUSIC algorithm for SRAs is

$$\hat{\boldsymbol{\psi}} = \arg\min_{\boldsymbol{\psi}} \left\{ vec^H(\boldsymbol{\psi}) \left[\widehat{\mathbf{U}}_N \widehat{\mathbf{U}}_N^H \right] vec(\boldsymbol{\psi}) \right\}, \tag{9.334}$$

or

$$\hat{\boldsymbol{\psi}} = \arg\min_{\boldsymbol{\psi}} \left\{ vec^H(\boldsymbol{\psi}) \left[\mathbf{I} - \widehat{\mathbf{U}}_S \widehat{\mathbf{U}}_S^H \right] vec(\boldsymbol{\psi}) \right\}. \tag{9.335}$$

In almost all cases, the form in (9.335) will be used because normally $D \ll NM$. The disadvantage of using spectral MUSIC directly is that we must search over a 2-D surface.

A logical approach is to implement two 1-D algorithms such as root MUSIC or IMODE to generate preliminary estimates of ψ_x and ψ_y, and then use (9.335) to correctly pair the estimates. A scenario for two sources is shown in Figure 9.56. We substitute the four possible pairings of $\hat{\psi}_{x1}$, $\hat{\psi}_{x2}$, $\hat{\psi}_{y1}$, and $\hat{\psi}_{y2}$ into the bracket expression on the right side of (9.335) and choose the two points with the smallest values.

In order to implement the 1-D algorithm, we compute the sample spectral matrix of each row, which we denote as $\widehat{\mathbf{S}}_{\mathbf{x},n}$. We then sum over the N rows to obtain

$$\widehat{\mathbf{S}}_{\mathbf{x}} = \sum_{n=1}^{N} \widehat{\mathbf{S}}_{\mathbf{x},n}. \tag{9.336}$$

We use root MUSIC with $\widehat{\mathbf{S}}_{\mathbf{x}}$ to find a preliminary estimate of ψ_x, which we denote as $\tilde{\psi}_x$. Similarly, we compute the sample spectral matrix of each column, which we denote as $\widehat{\mathbf{S}}_{\mathbf{y},m}$. We then sum over the M columns to obtain

$$\widehat{\mathbf{S}}_{\mathbf{y}} = \sum_{m=1}^{M} \widehat{\mathbf{S}}_{\mathbf{y},m}. \tag{9.337}$$

We use root MUSIC with $\widehat{\mathbf{S}}_{\mathbf{y}}$ to find a preliminary estimate of ψ_y, which we denote as $\tilde{\psi}_y$.[20] Because the rows and columns are standard linear arrays, FB averaging can be used. Any of the 1-D algorithms such as unitary TLS-ESPRIT or IMODE could be used to generate the 1-D estimates.

[20] A similar approach is used by Yeh et al. [YLC89] for spatial smoothing of coherent signals.

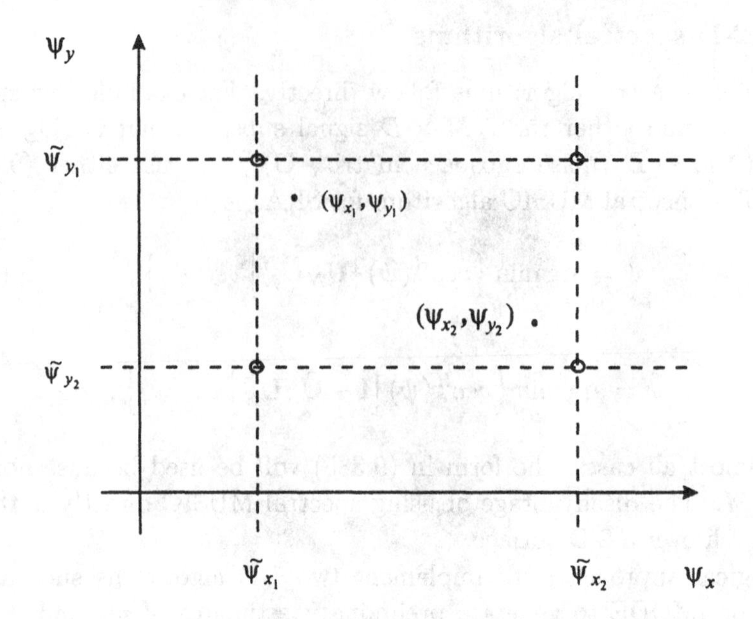

Figure 9.56 Pairing of 1-D estimates.

The next step is to pair the estimates. One approach is to evaluate (9.335) at the four locations in Figure 9.56: $(\tilde{\psi}_{x1}, \tilde{\psi}_{y1})$, $(\tilde{\psi}_{x1}, \tilde{\psi}_{y2})$, $(\tilde{\psi}_{x2}, \tilde{\psi}_{y1})$, and $(\tilde{\psi}_{x2}, \tilde{\psi}_{y2})$, and choose the two points with the smallest values.

If we stop at this point we have not exploited the coupling in the model. Thus, a second approach is to use the two points as the initial value for a localized gradient algorithm to find the minimum of (9.335). It turns out that in many cases, a coupled spectral approach has poorer resolution performance than the 1-D root algorithms, so the first approach is preferable.

We consider a simple example to illustrate the technique.

Example 9.9.1

Consider a standard 10×10 rectangular array. Two equal-power uncorrelated plane-wave sources impinge on the array. The source locations are

$$(u_{x1}, u_{y1}) = (0, 0),\ (u_{x2}, u_{y2}) = (0, 0.0866).$$

The RMSE of ith source is,

$$RMSE_i = \left[(\hat{\mathbf{u}}_{xi} - u_{xi})^2 + (\hat{\mathbf{u}}_{yi} - u_{yi})^2 \right]^{\frac{1}{2}}. \tag{9.338}$$

The first approach (paired 1-D roots) has better performance, so we have plotted its behavior. In Figure 9.57(a), we plot the probability of resolution versus *SNR*. In Figure 9.57(b), we plot the RMSE versus *SNR*. Note that the plot is not normalized and the horizontal axis is *SNR*.

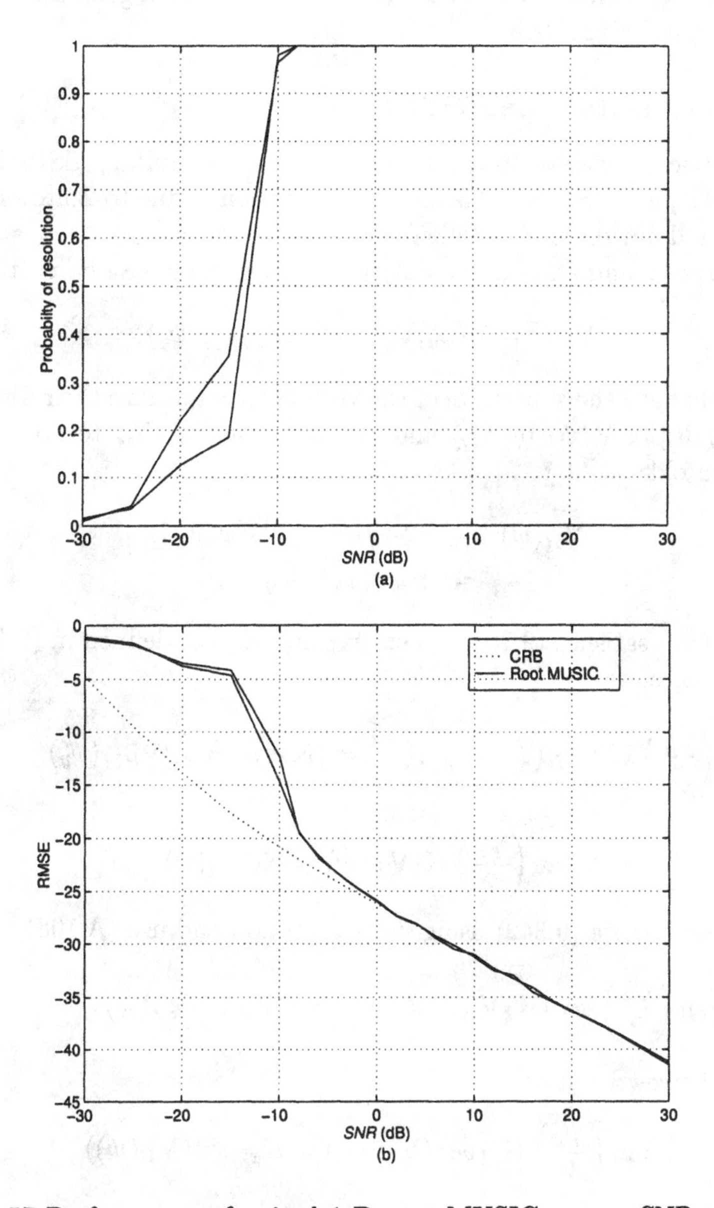

Figure 9.57 Performance of paired 1-D root MUSIC versus *SNR*; standard rectangular array, $N = 10$, $M = 10$, $u_1 = [0\ 0]^T$, $u_2 = [0\ 0.0866]^T$, $K = 100$, 200 trials: (a) probability of resolution versus *SNR*; (b) RMSE versus *SNR*.

For a larger number of signals this technique becomes cumbersome. In the next subsection, we develop a 2-D unitary ESPRIT algorithm that eliminates the pairing problem.

9.9.1.2 2-D unitary ESPRIT

In this section, we develop a 2-D extension of the unitary ESPRIT algorithm developed in Section 9.3.4.2. The extension is due to Zoltowski et al. [ZHM96] and our discussion follows that reference.

The array manifold of an $N \times M$ rectangular array can be written as

$$\mathbf{V}(\boldsymbol{\psi}) = \mathbf{V}(\psi_x, \psi_y) = \mathbf{v}_N(\psi_x)\mathbf{v}_M^T(\psi_y). \tag{9.339}$$

We have changed the subscripts of the vectors to emphasize their dimension. We pre-multiply $\mathbf{V}(\boldsymbol{\psi})$ by $\mathbf{Q}_N^H$ and post-multiply by $\mathbf{Q}_M^*$ to obtain a real array manifold.

$$\begin{aligned} \mathbf{V}_R(\boldsymbol{\psi}) &= \mathbf{Q}_N^H \mathbf{v}_N(\psi_x)\mathbf{v}_M^T(\psi_y)\mathbf{Q}_M^* \\ &= \mathbf{v}_{RN}(\psi_x)\mathbf{v}_{RM}^T(\psi_y). \end{aligned} \tag{9.340}$$

Now $\mathbf{v}_{RN}(\psi_x)$ satisfies (9.151), where $\mathbf{K}_1$ and $\mathbf{K}_2$ are defined in (9.147) and (9.148). Therefore,

$$\left[\tan\left(\frac{\psi_x}{2}\right)\mathbf{K}_1\mathbf{v}_{RN}(\psi_x)\right]\mathbf{v}_{RM}^T(\psi_y) = \left[\mathbf{K}_2\mathbf{v}_{RN}(\psi_x)\right]\mathbf{v}_{RM}^T(\psi_y), \tag{9.341}$$

or

$$\tan\left(\frac{\psi_x}{2}\right)\mathbf{K}_1\mathbf{V}_R(\boldsymbol{\psi}) = \mathbf{K}_2\mathbf{V}_R(\boldsymbol{\psi}). \tag{9.342}$$

We now rewrite (9.342) using $vec(\cdot)$ notation and use (A.108),[21]

$$\tan\left(\frac{\psi_x}{2}\right) vec\left(\mathbf{K}_1\mathbf{V}_R(\boldsymbol{\psi})\mathbf{I}_M\right) = vec\left(\mathbf{K}_2\mathbf{V}_R(\boldsymbol{\psi})\mathbf{I}_M\right). \tag{9.343}$$

This can be rewritten as

$$\tan\left(\frac{\psi_x}{2}\right)\mathbf{K}_{x1} vec\left(\mathbf{V}_R(\boldsymbol{\psi})\right) = \mathbf{K}_{x2} vec\left(\mathbf{V}_R(\boldsymbol{\psi})\right), \tag{9.344}$$

where

$$\mathbf{K}_{x1} \triangleq \mathbf{I}_M \otimes \mathbf{K}_1, \quad (N-1)M \times MN, \tag{9.345}$$

$$\mathbf{K}_{x2} \triangleq \mathbf{I}_M \otimes \mathbf{K}_2, \quad (N-1)M \times MN. \tag{9.346}$$

---

[21] From (A.108), $vec(\mathbf{ABC}) = (\mathbf{C}^T \otimes \mathbf{A})vec(\mathbf{B})$.

Note that (9.344) defines $(N-1)M$ equations.

Similarly,

$$\tan\left(\frac{\psi_y}{2}\right) vec\left(\mathbf{I}_N \mathbf{V}_R(\boldsymbol{\psi})\mathbf{K}_3^T\right) = vec\left(\mathbf{I}_N \mathbf{V}_R(\boldsymbol{\psi})\mathbf{K}_4^T\right), \tag{9.347}$$

where

$$\mathbf{K}_3 = Re\left\{\mathbf{Q}_{M-1}^H \mathbf{J}_{s2}^{(M)} \mathbf{Q}_M\right\}, \tag{9.348}$$

$$\mathbf{K}_4 = Im\left\{\mathbf{Q}_{M-1}^H \mathbf{J}_{s2}^{(M)} \mathbf{Q}_M\right\}, \tag{9.349}$$

where $\mathbf{J}_{s2}^{(M)}$ is given by (9.135) with N replaced by M. Then (9.347) can be rewritten as

$$\tan\left(\frac{\psi_y}{2}\right) \mathbf{K}_{y1} vec\left(\mathbf{V}_R(\boldsymbol{\psi})\right) = \mathbf{K}_{y2} vec\left(\mathbf{V}_R(\boldsymbol{\psi})\right), \tag{9.350}$$

where

$$\mathbf{K}_{y1} \triangleq \mathbf{K}_3 \otimes \mathbf{I}_N, \quad N(M-1) \times NM, \tag{9.351}$$

$$\mathbf{K}_{y2} \triangleq \mathbf{K}_4 \otimes \mathbf{I}_N, \quad N(M-1) \times NM. \tag{9.352}$$

Note that (9.350) represents $N(M-1)$ equations.

The next step is to specify an $NM \times D$ array manifold matrix,

$$\tilde{\mathbf{V}} \triangleq \left[vec\left(\mathbf{V}_R(\boldsymbol{\psi}_1)\right) \quad vec\left(\mathbf{V}_R(\boldsymbol{\psi}_2)\right) \quad \cdots \quad vec\left(\mathbf{V}_R(\boldsymbol{\psi}_D)\right)\right]. \tag{9.353}$$

Each column in $\tilde{\mathbf{V}}$ satisfies (9.344). Therefore we can write

$$\mathbf{K}_{x1} \tilde{\mathbf{V}} \boldsymbol{\Omega}_x = \mathbf{K}_{x2} \tilde{\mathbf{V}}, \tag{9.354}$$

where

$$\boldsymbol{\Omega}_x \triangleq \text{diag}\left\{\tan\left(\frac{\psi_{x1}}{2}\right), \tan\left(\frac{\psi_{x2}}{2}\right), \cdots, \tan\left(\frac{\psi_{xD}}{2}\right)\right\}. \tag{9.355}$$

Similarly, each column in $\tilde{\mathbf{V}}$ satisfies (9.350), so we can write

$$\mathbf{K}_{y1} \tilde{\mathbf{V}} \boldsymbol{\Omega}_y = \mathbf{K}_{y2} \tilde{\mathbf{V}}, \tag{9.356}$$

where

$$\boldsymbol{\Omega}_y \triangleq \text{diag}\left\{\tan\left(\frac{\psi_{y1}}{2}\right), \tan\left(\frac{\psi_{y2}}{2}\right), \cdots, \tan\left(\frac{\psi_{yD}}{2}\right)\right\}. \tag{9.357}$$

Just as in Section 9.9.2.1, we find $\widehat{\mathbf{U}}_s$, the $NM \times D$ signal subspace estimate. Now

$$\mathbf{U}_s = \tilde{\mathbf{V}}\mathbf{T}, \tag{9.358}$$

where T is an unknown $D \times D$ matrix. Replacing $\mathbf{U}_s$ with $\widehat{\mathbf{U}}_s$ gives

$$\widehat{\mathbf{U}}_s = \widehat{\tilde{\mathbf{V}}}\mathbf{T}, \tag{9.359}$$

or

$$\widehat{\tilde{\mathbf{V}}} = \widehat{\mathbf{U}}_s\mathbf{T}^{-1}. \tag{9.360}$$

Substituting (9.360) into (9.354) and (9.356) gives

$$\mathbf{K}_{x1}\widehat{\mathbf{U}}_s\widehat{\mathbf{\Psi}}_x = \mathbf{K}_{x2}\widehat{\mathbf{U}}_s, \quad (N-1)M \times D, \tag{9.361}$$

where

$$\widehat{\mathbf{\Psi}}_x \triangleq \mathbf{T}^{-1}\widehat{\mathbf{\Omega}}_x\mathbf{T}, \tag{9.362}$$

and

$$\mathbf{K}_{y1}\widehat{\mathbf{U}}_s\widehat{\mathbf{\Psi}}_y = \mathbf{K}_{y2}\widehat{\mathbf{U}}_s, \quad (N-1)M \times D, \tag{9.363}$$

where

$$\widehat{\mathbf{\Psi}}_y \triangleq \mathbf{T}^{-1}\widehat{\mathbf{\Omega}}_y\mathbf{T}. \tag{9.364}$$

After solving (9.361) and (9.363) for $\widehat{\mathbf{\Psi}}_x$ and $\widehat{\mathbf{\Psi}}_y$, we find the eigenvalues of the $D \times D$ complex matrix $\widehat{\mathbf{\Psi}}_x + j\widehat{\mathbf{\Psi}}_y$,

$$\widehat{\mathbf{\Psi}}_x + j\widehat{\mathbf{\Psi}}_y = \mathbf{T}^{-1}\left(\widehat{\mathbf{\Omega}}_x + j\widehat{\mathbf{\Omega}}_y\right)\mathbf{T}. \tag{9.365}$$

From the resulting eigenvalues, $\widehat{\lambda}_i, i = 1, 2, \cdots, D$, we can find $\hat{\psi}_{xi}$ and $\hat{\psi}_{yi}$,

$$\hat{\psi}_{xi} = 2\tan^{-1}\left\{Re(\widehat{\lambda}_i)\right\}, \tag{9.366}$$

and

$$\hat{\psi}_{yi} = 2\tan^{-1}\left\{Im(\widehat{\lambda}_i)\right\}. \tag{9.367}$$

In order to find $\widehat{\mathbf{U}}_s$, we use the $N \times M$ data matrix at the kth snapshot, $\mathbf{X}_k$. Define the $NM \times 1$ vector,

$$\mathbf{Y}(k) = vec\left(\mathbf{Q}_N^H \mathbf{X}_k \mathbf{Q}_M^*\right), \quad k = 1, \cdots, K. \tag{9.368}$$

Using (A.108),

$$\mathbf{Y}(k) = \left(\mathbf{Q}_M^H \otimes \mathbf{Q}_N^H\right) vec\left(\mathbf{X}_k\right) = \left(\mathbf{Q}_M^H \otimes \mathbf{Q}_N^H\right) \tilde{\mathbf{X}}(k), \quad k = 1, \cdots, K. \tag{9.369}$$

Now define an $NM \times K$ matrix,

$$\mathbf{Y} = \left[\begin{array}{ccc} \mathbf{Y}(1) & \cdots & \mathbf{Y}(K) \end{array} \right]. \tag{9.370}$$

The signal subspace is computed by an eigendecomposition of $Re[\widehat{\mathbf{S}}_{\mathbf{y}}]$ or from an SVD of the $ND \times 2K$ matrix $[Re\,\mathbf{Y},\, Im\,\mathbf{Y}]$.

We summarize the 2-D unitary ESPRIT algorithm:

1. Compute $\widehat{\mathbf{U}}_s$. The columns of $\widehat{\mathbf{U}}_s$ are the "largest" left singular vectors of $[Re\,\mathbf{Y},\, Im\,\mathbf{Y}]$ or the eigenvectors corresponding to the D largest eigenvalues of $Re[\widehat{\mathbf{S}}_{\mathbf{y}}]$.

2. Compute $\widehat{\mathbf{\Psi}}_x$ and $\widehat{\mathbf{\Psi}}_y$ as the solutions to (9.361) and (9.363). We use the TLS solution in all of our examples.

3. Compute the eigenvalues, $\widehat{\lambda}_i, i = 1, 2, \cdots, D$ of $\widehat{\mathbf{\Psi}}_x + j\widehat{\mathbf{\Psi}}_y$.

4. Compute $\hat{\psi}_{xi}$ and $\hat{\psi}_{yi}$ using (9.366) and (9.367).

We consider an example with three sources to illustrate the performance of the 2-D unitary TLS-ESPRIT algorithm.

Example 9.9.2[22]

Consider an 8×8 standard rectangular array. Three equal-power uncorrelated plane-wave signals impinge on the array from $(u_{x1}, u_{y1}) = (0, 0), (u_{x2}, u_{y2}) = (0.125, 0)$, and $(u_{x3}, u_{y3}) = (0, 0.125)$. The RMSE of the ith source is defined as

$$RMSE_i = \left[\left(\hat{u}_{xi} - u_{xi}\right)^2 + \left(\hat{u}_{yi} - u_{yi}\right)^2\right]^{\frac{1}{2}}. \tag{9.371}$$

In Figure 9.58, we plot the probability of resolution and normalized RMSE versus *SNR* for $K = 100$ snapshots. We also show the CRB. As we would expect, the RMSE of the first signal is slightly higher.

We do not show the spectral MUSIC or paired root MUSIC result to keep the plot uncluttered. However, there is significant improvement in the threshold behavior compared to spectral MUSIC, and the above threshold RMSE is close to the CRB.

In the next example we show how we can use row weighting with unitary ESPRIT to achieve performance that essentially achieves the CRB.

---

[22]This example corresponds to the example on p. 326 of [ZHM96].

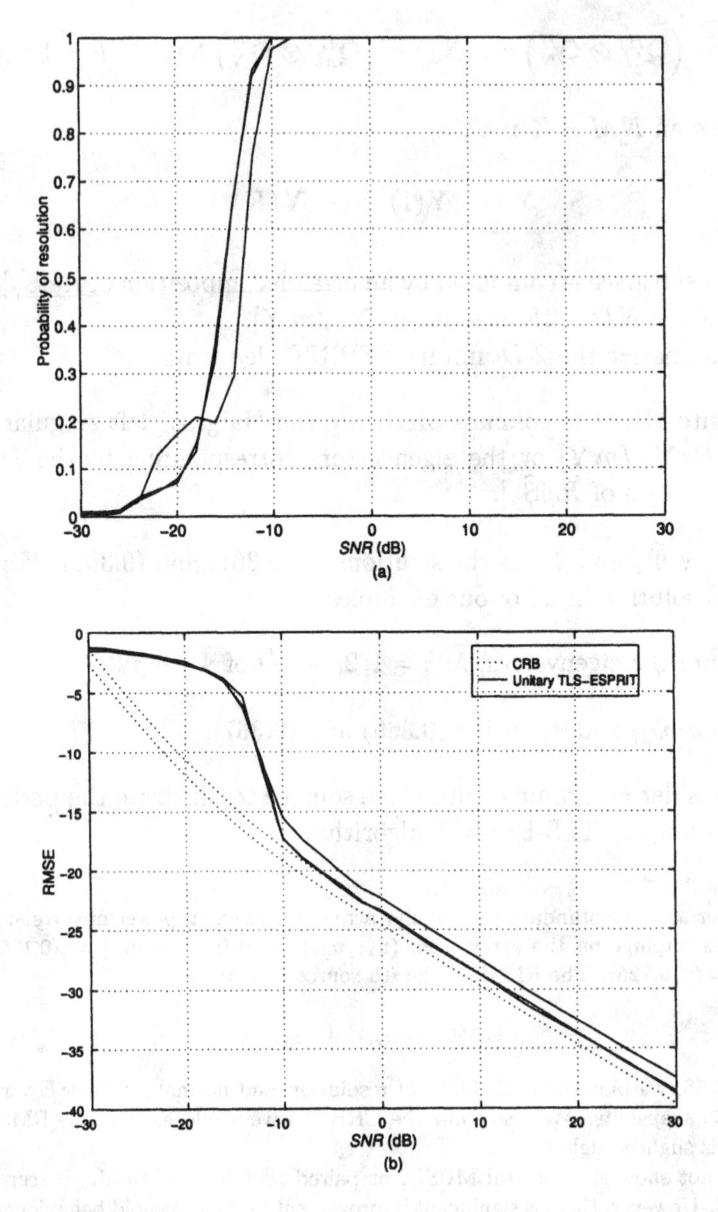

Figure 9.58 Performance of 2-D unitary TLS-ESPRIT; $d_s = 1, m_s = 1$, standard rectangular array, N=M=8, $\mathbf{u}_1 = [0\ 0]^T$, $\mathbf{u}_2 = [0.125\ 0]^T$, $\mathbf{u}_3 = [0\ 0.125]^T$, $K = 100$, 500 trials: (a) probability of resolution versus *SNR*; (b) RMSE versus *SNR*.

Example 9.9.3 (continuation)

Consider the same model as in Example 9.9.2. We use weighted unitary TLS-ESPRIT with $m_s = 3$.

In Figure 9.59, we plot the probability of resolution and the normalized RMSE versus *SNR*. We also show the CRB.

We see that weighting improves the RMSE behavior and essentially achieves the CRB.

The performance of unitary TLS-ESPRIT is good. However, a disadvantage is the dimension of the problem for many arrays of practical interest. In the next section, we consider the beamspace implementation of ESPRIT.

9.9.1.3 2-D beamspace ESPRIT

In this section, we develop 2-D beamspace unitary ESPRIT. The algorithm is due to Zoltowski et al. [ZHM96] and our discussion follows that reference.

The beamspace array manifold matrix consists of $N_{bs}M_{bs}$ orthogonal conventional beams whose MRAs are spaced at $2\pi/N$ intervals in ψ_x-space and $2\pi/M$ in ψ_y-space. The two matrices used to construct the beamspace are:

$$\left[\mathbf{B}_{bsx}^H\right]_n = \frac{1}{N}\left[\begin{array}{cccc} e^{j\left(\frac{N-1}{2}\right)n\frac{2\pi}{N}} & e^{j\left(\frac{N-3}{2}\right)n\frac{2\pi}{N}} & \cdots, & e^{-j\left(\frac{N-1}{2}\right)n\frac{2\pi}{N}} \end{array}\right], n \in \sum_{N_{bs}}. \tag{9.372}$$

where $\sum_{N_{bs}}$ specifies the set of values on n used in the matrix. If the beamspace sector is centered at $\psi_x = 0$ and N_{bs} is odd, then

$$-(N_b - 1)/2 \leq n \leq (N_b - 1)/2. \tag{9.373}$$

$\left[\mathbf{B}_{bsx}^H\right]$ is an $N_{bs} \times N$ matrix. Similarly,

$$\left[\mathbf{B}_{bsy}^H\right]_m = \frac{1}{M}\left[\begin{array}{cccc} e^{j\left(\frac{M-1}{2}\right)m\frac{2\pi}{M}} & e^{j\left(\frac{M-3}{2}\right)m\frac{2\pi}{M}} & \cdots, & e^{-j\left(\frac{M-1}{2}\right)m\frac{2\pi}{M}} \end{array}\right], m \in \sum_{M_{bs}}. \tag{9.374}$$

$\left[\mathbf{B}_{bsy}^H\right]$ is an $M_{bs} \times M$ matrix.

The beamspace array manifold matrix is an $N_{bs} \times M_{bs}$ matrix,

$$\mathbf{V}_{bs}(\psi_x, \psi_y) = \mathbf{B}_{bsx}^H\, \mathbf{V}(\psi_x, \psi_y)\, \mathbf{B}_{bsy}^*, \tag{9.375}$$

where $\mathbf{V}(\psi_x, \psi_y)$ is defined in (9.339).

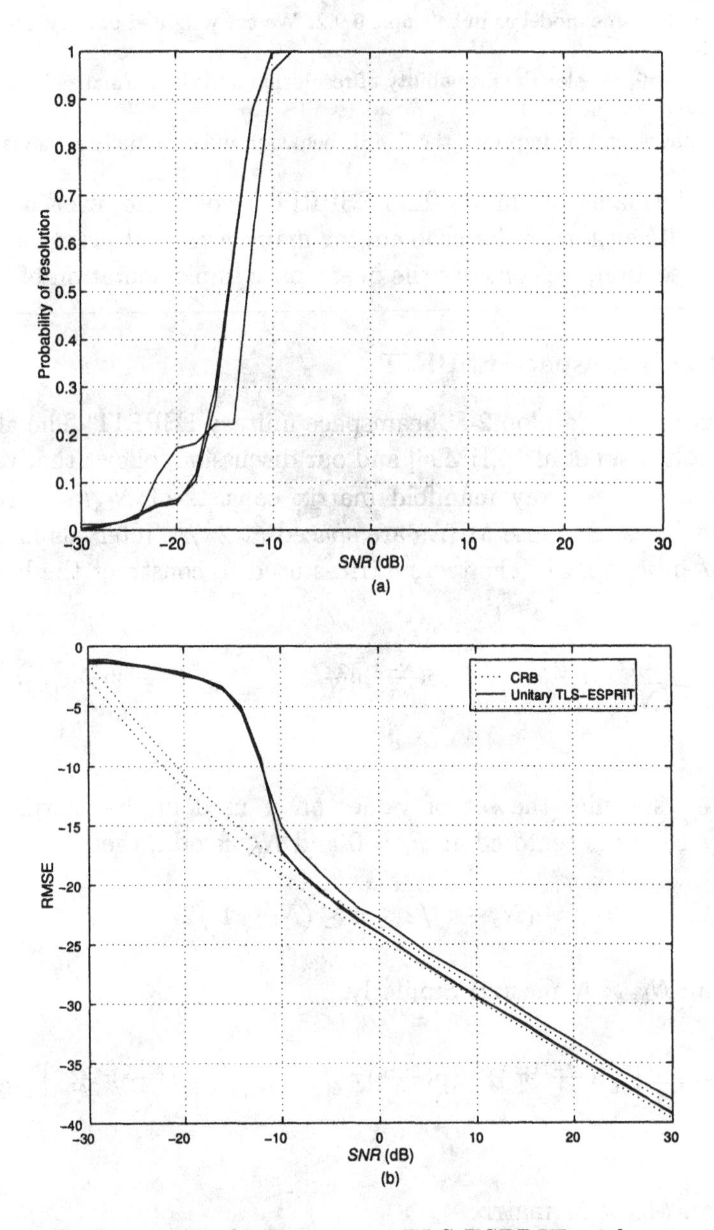

Figure 9.59 Performance of 2-D unitary TLS-ESPRIT with row weighting; $d_s = 1, m_s = 3$, standard rectangular array, $N = M = 8$, $\mathbf{u}_1 = [0\ 0]^T$, $\mathbf{u}_2 = [0.125\ 0]^T$, $\mathbf{u}_3 = [0\ 0.125]^T$, $K = 100$, 500 trials: (a) Probability of resolution versus *SNR*; (b) RMSE versus *SNR*.

The components of beamspace array manifold matrix are

$$b_{nm}(\psi_x,\psi_y)=\frac{1}{NM}\left\{\frac{\sin\left[\frac{N}{2}\left(\psi_x-n\frac{2\pi}{N}\right)\right]}{\sin\left[\frac{1}{2}\left(\psi_x-n\frac{2\pi}{N}\right)\right]}\frac{\sin\left[\frac{M}{2}\left(\psi_y-m\frac{2\pi}{M}\right)\right]}{\sin\left[\frac{1}{2}\left(\psi_y-m\frac{2\pi}{M}\right)\right]}\right\}. \tag{9.376}$$

The beamspace array manifold matrix can be expressed as

$$\mathbf{V}_{bs}\left(\psi_x,\psi_y\right)=\mathbf{v}_{bsx}(\psi_x)\mathbf{v}_{bsy}^T(\psi_y), \tag{9.377}$$

where $\mathbf{v}_{bsx}(\psi_x)$ is defined in (9.309) and $\mathbf{v}_{bsy}(\psi_y)$ has the same form. Because $\mathbf{v}_{bsx}(\psi_x)$ satisfies the invariance relationship in (9.315), we can write

$$\tan\left(\frac{\psi_x}{2}\right)\mathbf{\Gamma}_1\mathbf{V}_{bs}\left(\psi_x,\psi_y\right)=\mathbf{\Gamma}_2\mathbf{V}_{bs}\left(\psi_x,\psi_y\right), \tag{9.378}$$

where $\mathbf{\Gamma}_1$ and $\mathbf{\Gamma}_2$ are defined in (9.313) and (9.314).

We define an $N_{bs}M_{bs}\times 1$ beamspace manifold vector,

$$\mathbf{v}_{bs}\left(\psi_x,\psi_y\right)=vec\left\{\mathbf{V}_{bs}\left(\psi_x,\psi_y\right)\right\}. \tag{9.379}$$

Using (A.108), we can rewrite (9.378) as

$$\tan\left(\frac{\psi_x}{2}\right)\mathbf{\Gamma}_{\psi_x,1}\mathbf{v}_{bs}\left(\psi_x,\psi_y\right)=\mathbf{\Gamma}_{\psi_x,2}\mathbf{v}_{bs}\left(\psi_x,\psi_y\right), \tag{9.380}$$

where

$$\mathbf{\Gamma}_{\psi_x,1}=\mathbf{I}_{N_{bs}}\otimes\mathbf{\Gamma}_1, \tag{9.381}$$

and

$$\mathbf{\Gamma}_{\psi_x,2}=\mathbf{I}_{N_{bs}}\otimes\mathbf{\Gamma}_2, \tag{9.382}$$

are $(N_{bs}-1)M_{bs}\times N_{bs}M_{bs}$ matrices. We now repeat the process with the ψ_y component,

$$\tan\left(\frac{\psi_y}{2}\right)\mathbf{\Gamma}_3\mathbf{v}_{bsy}\left(\psi_y\right)=\mathbf{\Gamma}_4\mathbf{v}_{bsy}\left(\psi_y\right), \tag{9.383}$$

where $\mathbf{\Gamma}_3$ and $\mathbf{\Gamma}_4$ are defined as in (9.313) and (9.314) with N_{bs} replaced by M_{bs}. We can write

$$\tan\left(\frac{\psi_y}{2}\right)\mathbf{V}_{bs}\left(\psi_x,\psi_y\right)\mathbf{\Gamma}_3^T=\mathbf{V}_{bs}\left(\psi_x,\psi_y\right)\mathbf{\Gamma}_4^T. \tag{9.384}$$

Then, using (A.108),

$$\tan\left(\frac{\psi_y}{2}\right)\mathbf{\Gamma}_{\psi_y,1}\mathbf{v}_{bs}\left(\psi_x,\psi_y\right)=\mathbf{\Gamma}_{\psi_y,2}\mathbf{v}_{bs}\left(\psi_x,\psi_y\right), \tag{9.385}$$

where

$$\mathbf{\Gamma}_{\psi_y,1} = \mathbf{\Gamma}_3 \otimes \mathbf{I}_{M_{bs}}, \tag{9.386}$$

and

$$\mathbf{\Gamma}_{\psi_y,2} = \mathbf{\Gamma}_4 \otimes \mathbf{I}_{M_{bs}}, \tag{9.387}$$

are $N_{bs}(M_{bs}-1) \times N_{bs}M_{bs}$ matrices.

Now define the $N_{bs}M_{bs} \times D$ real beamspace array manifold matrix,

$$\tilde{\mathbf{V}}_{bs} \triangleq \left[\begin{array}{ccc} \mathbf{v}_{bs}(\psi_{x1}, \psi_{y1}) & \cdots & \mathbf{v}_{bs}(\psi_{xD}, \psi_{yD}) \end{array} \right]. \tag{9.388}$$

From (9.380),

$$\mathbf{\Gamma}_{\psi_x,1}\tilde{\mathbf{V}}_{bs}\mathbf{\Omega}_{\psi_x} = \mathbf{\Gamma}_{\psi_x,2}\tilde{\mathbf{V}}_{bs}, \tag{9.389}$$

where $\mathbf{\Omega}_{\psi_x}$ is defined in (9.355). Similarly, from (9.385),

$$\mathbf{\Gamma}_{\psi_y,1}\tilde{\mathbf{V}}_{bs}\mathbf{\Omega}_{\psi_y} = \mathbf{\Gamma}_{\psi_y,2}\tilde{\mathbf{V}}_{bs}, \tag{9.390}$$

where $\mathbf{\Omega}_{\psi_y}$ is defined in (9.357).

The signal subspace estimate, $\widehat{\mathbf{U}}_{bs,s}$, is an $N_{bs}M_{bs} \times D$ real matrix,

$$\mathbf{U}_{bs,s} = \tilde{\mathbf{V}}_{bs}\,\mathbf{T}, \tag{9.391}$$

where $\mathbf{T}$ is a non-singular $D \times D$ real matrix. Replacing $\mathbf{U}_{bs,s}$ with $\widehat{\mathbf{U}}_{bs,s}$ gives

$$\widehat{\mathbf{U}}_{bs,s} = \hat{\tilde{\mathbf{V}}}_{bs}\,\mathbf{T} \tag{9.392}$$

or

$$\hat{\tilde{\mathbf{V}}}_{bs} = \widehat{\mathbf{U}}_{bs,s}\mathbf{T}^{-1}. \tag{9.393}$$

Then, (9.389) can be written as

$$\mathbf{\Gamma}_{\psi_x,1}\widehat{\mathbf{U}}_{bs,s}\widehat{\mathbf{\Psi}}_{x,bs} = \mathbf{\Gamma}_{\psi_x,2}\widehat{\mathbf{U}}_{bs,s}, \tag{9.394}$$

where

$$\widehat{\mathbf{\Psi}}_{x,bs} = \mathbf{T}^{-1}\mathbf{\Omega}_{\psi_x}\mathbf{T}. \tag{9.395}$$

Similarly, (9.390) can be written as

$$\mathbf{\Gamma}_{\psi_y,1}\widehat{\mathbf{U}}_{bs,s}\widehat{\mathbf{\Psi}}_{y,bs} = \mathbf{\Gamma}_{\psi_y,2}\widehat{\mathbf{U}}_{bs,s}, \tag{9.396}$$

where

$$\widehat{\mathbf{\Psi}}_{y,bs} = \mathbf{T}^{-1}\mathbf{\Omega}_{\psi_y}\mathbf{T}. \tag{9.397}$$

The next step is to construct the beamspace data matrix $\mathbf{Y}_{bs}$. The element-space data matrix for the kth snapshot is the $N \times M$ matrix $\mathbf{X}(k)$. The beamspace data vector for the kth snapshot is

$$\mathbf{X}_{bs}(k) = vec\left(\mathbf{B}_{bsx}^{H}\, \mathbf{X}(k)\, \mathbf{B}_{bsy}^{*}\right). \tag{9.398}$$

Using (A.108), (9.398) can be written as

$$\mathbf{X}_{bs}(k) = \left[\mathbf{B}_{bsy}^{H} \otimes \mathbf{B}_{bsx}^{H}\right] vec(\mathbf{X}(k)) = \left[\mathbf{B}_{bsy}^{H} \otimes \mathbf{B}_{bsx}^{H}\right] \tilde{\mathbf{X}}(k). \tag{9.399}$$

The complete beamspace data matrix is the $N_{bs}M_{bs} \times K$ matrix,

$$\mathbf{X}_{bs} = \left[\ \mathbf{X}_{bs}(1) \quad \cdots \mathbf{X}_{bs}(K)\ \right]. \tag{9.400}$$

We now proceed in an analogous manner to the 1-D beamspace and the 2-D element-space algorithms. We can summarize the beamspace unitary ESPRIT algorithm:

1. Compute

$$\mathbf{X}_{bs}(k) = \left(\mathbf{B}_{bsx}^{H} \otimes \mathbf{B}_{bsy}^{H}\right) vec(\mathbf{X}(k))\mathbf{X}(k), \quad k = 1, \cdots, K, \tag{9.401}$$

and construct $\mathbf{X}_{bs}$.

2. Construct $\widehat{\mathbf{U}}_{bs,s}$ by finding the D largest left singular vectors of $\widehat{\mathbf{X}}_{bs,E}$, where

$$\widehat{\mathbf{X}}_{bs,E} \triangleq \left[Re\left[\mathbf{X}_{bs}\right], Im\left[\mathbf{X}_{bs}\right]\right], \tag{9.402}$$

or, by finding the eigenvectors corresponding to the D largest eigenvalues of

$$Re\left[\widehat{\mathbf{S}}_{\mathbf{X}_{bs}}\right] = Re\left\{\frac{1}{K}\sum_{k=1}^{K} \mathbf{X}_{bs}(k)\, \mathbf{X}_{bs}^{H}(k)\right\}. \tag{9.403}$$

3. Compute $\widehat{\mathbf{\Psi}}_{\psi_x,bs}$ as the solution to the $(N_{bs}-1)M_{bs} \times D$ matrix equation,

$$\mathbf{\Gamma}_{\psi_x,1}\widehat{\mathbf{U}}_{bs,s}\widehat{\mathbf{\Psi}}_{\psi_x,bs} = \mathbf{\Gamma}_{\psi_x,2}\widehat{\mathbf{U}}_{bs,s}. \tag{9.404}$$

4. Compute $\widehat{\mathbf{\Psi}}_{\psi_y,bs}$ as the solution to the $N_{bs}(M_{bs}-1) \times D$ matrix equation,

$$\mathbf{\Gamma}_{\psi_y,1}\widehat{\mathbf{U}}_{bs,s}\widehat{\mathbf{\Psi}}_{\psi_y,bs} = \mathbf{\Gamma}_{\psi_y,2}\widehat{\mathbf{U}}_{bs,s}. \tag{9.405}$$

We use either the LS or TLS solution in steps 3 and 4.

5. Compute $\widehat{\lambda}_i = 1, 2, \cdots, D$ as the eigenvalues of the $D \times D$ complex matrix,
$$\widehat{\mathbf{\Psi}}_{\psi_{x,bs}} + j\widehat{\mathbf{\Psi}}_{\psi_{y,bs}}. \tag{9.406}$$

6. Compute the estimates of the directional cosines as
$$\begin{cases} \hat{\psi}_{xi} = & 2\tan^{-1}\left\{Re(\widehat{\lambda}_i)\right\}, \quad i = 1, \cdots, D, \\ \hat{\psi}_{yi} = & 2\tan^{-1}\left\{Im(\widehat{\lambda}_i)\right\}, \quad i = 1, \cdots, D. \end{cases} \tag{9.407}$$

To illustrate the beamspace unitary ESPRIT algorithm, we use the same model as in Example 9.9.2.

Example 9.9.4 (continuation, Example 9.9.2)

Consider an 8×8 SRA. Three equal-power uncorrelated plane-wave signals impinge on the array from (0,0), (0.125, 0), and (0, 0.125).

A 3×3 beamspace centered at (0, 0) is used. The indices are
$$\begin{aligned} n &= -1, 0, 1, \\ m &= -1, 0, 1. \end{aligned} \tag{9.408}$$

The RMSE and probability of resolution are plotted versus *SNR* in Figure 9.60. The performance is very similar to element-space unitary ESPRIT with $d_s = 1$ and $m_s = 1$. This result is expected because the three signal DOAs are in the central area of the beamspace.

The 2-D beamspace unitary ESPRIT provides a computationally efficient solution to the 2-D estimation problem. It provides the 2-D estimates directly so the pairing problem is eliminated. Above threshold, its performance is close to the CRB.

9.9.1.4 Summary: Rectangular arrays

In the preceding subsections, we have introduced the problem of parameter estimation using planar arrays. We discussed MUSIC, 2-D unitary ESPRIT, and beamspace ESPRIT. The ESPRIT algorithms have good threshold performance and the weighted versions approach the CRB above threshold.

There is a large number of papers in the literature concerning various aspects of the 2-D problem. Representative references include Rao and Kung [RK84], Swindlehurst and Kailath [SK89], Yeh et al. [YLC89], Zoltowski and Stavrinides [ZS89], Hua [Hua91], Clark [Cla92], Hua [Hua92], Sacchini [Sac92], van der Veen et al. [vdVOD92], Swindlehurst and Kailath [SK93a], Clark and Scharf [CS94], Hatke and Forsythe [HF96], Clark et al. [CES97], and Fuhl et al. [FRB97].

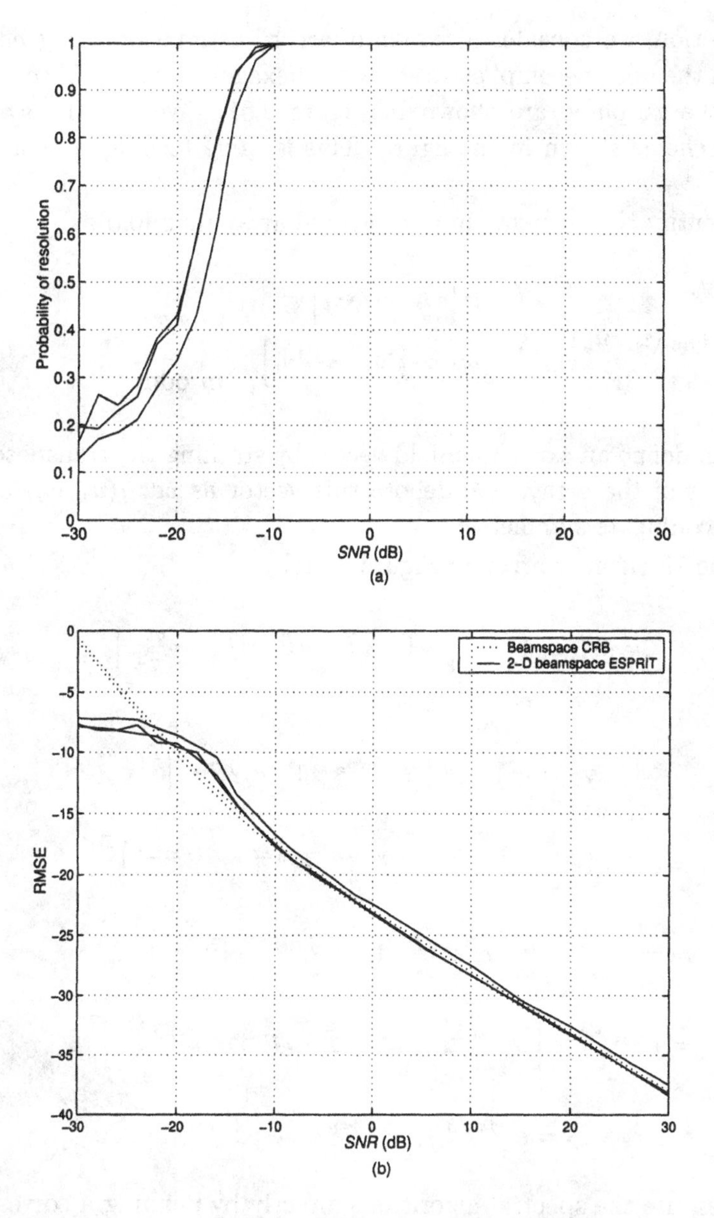

Figure 9.60 Performance of 2-D beamspace unitary ESPRIT; $d_s = 1, m_s = 1$, standard rectangular array, N=M=8, $\mathbf{u}_1 = [0\ 0]^T$, $\mathbf{u}_2 = [0.125\ 0]^T$, $\mathbf{u}_3 = [0\ 0.125]^T$, 3×3 beamspace centered at $\mathbf{u} = [0\ 0]^T$, $K = 100$, 500 trials: (a) probability of resolution versus *SNR*; (b) RMSE versus *SNR*.

9.9.2 Hexagonal Arrays

In this section, we consider direction-of-arrival estimation using hexagonal arrays. In the text, we emphasize standard hexagonal arrays. Two examples lying in the xy-plane are shown in Figure 9.61. We use the symmetric indexing scheme shown in the figure (This is same technique as in Section 4.4.)

We identify the elements in a hexagonal array manifold as

$$v_{nm}(u_x, u_y) = \begin{cases} e^{j\pi\left[nu_x + m\frac{\sqrt{3}}{2}u_y\right]}, & m \text{ even}, \\ e^{j\pi\left[(n+\frac{1}{2})u_x + m\frac{\sqrt{3}}{2}u_y\right]}, & m \text{ odd}. \end{cases} \tag{9.409}$$

We can define an array manifold vector by stacking the transposes of the row vectors of the array. We denote this vector as $vec_H(u_x, u_y)$ and note that it is conjugate symmetric.

For the 19-element array in Figure 9.61(b),

$$vec_H(u_x, u_y) = \begin{bmatrix} \mathbf{v}_2^T & \mathbf{v}_1^T & \mathbf{v}_0^T & \mathbf{v}_{-1}^T & \mathbf{v}_{-2}^T \end{bmatrix}^T, \tag{9.410}$$

where

$$\mathbf{v}_2 = e^{j\pi\sqrt{3}u_y} \begin{bmatrix} e^{-j\pi u_x} & 1 & e^{-j\pi u_x} \end{bmatrix}^T, \tag{9.411}$$

$$\mathbf{v}_1 = e^{j\pi\frac{\sqrt{3}}{2}u_y} \begin{bmatrix} e^{-j\pi\frac{3u_x}{2}} & e^{-j\pi\frac{u_x}{2}} & e^{j\pi\frac{u_x}{2}} & e^{j\pi\frac{3u_x}{2}} \end{bmatrix}^T, \tag{9.412}$$

$$\mathbf{v}_0 = \begin{bmatrix} e^{-j2\pi u_x} & e^{-j\pi u_x} & 1 & e^{-j\pi u_x} & e^{j2\pi u_x} \end{bmatrix}^T, \tag{9.413}$$

$$\mathbf{v}_{-1} = e^{-j\pi\frac{\sqrt{3}}{2}u_y} \begin{bmatrix} e^{-j\pi\frac{3u_x}{2}} & e^{-j\pi\frac{u_x}{2}} & e^{j\pi\frac{u_x}{2}} & e^{j\pi\frac{3u_x}{2}} \end{bmatrix}^T, \tag{9.414}$$

$$\mathbf{v}_{-2} = e^{-j\pi\sqrt{3}u_y} \begin{bmatrix} e^{-j\pi u_x} & 1 & e^{-j\pi u_x} \end{bmatrix}^T. \tag{9.415}$$

We can use the spectral algorithms directly by defining a corresponding $vec_H(\mathbf{X})$, computing the sample covariance matrix, the noise subspace and applying the appropriate subspace algorithm.

However, in order to improve the threshold performance, a version of unitary ESPRIT is useful. In order to apply the rectangular results we transform the hexagonal array into an equivalent rectangular-grid array.

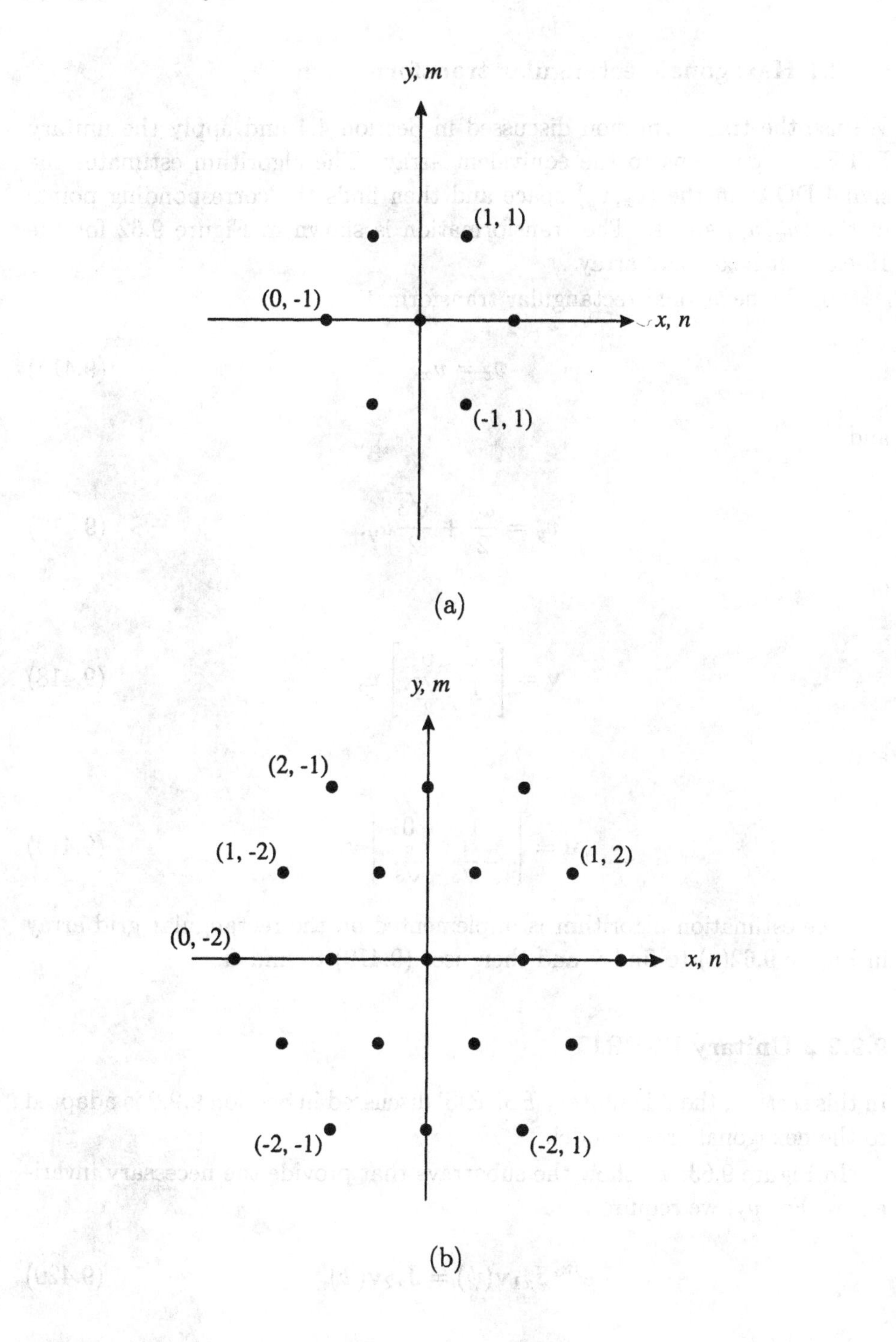

Figure 9.61 Standard hexagonal arrays: (a) 7 elements; (b) 19 elements.

9.9.2.1 Hexagonal-rectangular transformation

We use the transformation discussed in Section 4.4 and apply the unitary ESPRIT algorithms to the equivalent array. The algorithm estimates the signal DOAs in the (v_x, v_y) space and then finds the corresponding points in the (u_x, u_y) space. The transformation is shown in Figure 9.62 for the 19-element hexagonal array.

For the hexagonal-rectangular transformation,

$$v_x = u_x, \tag{9.416}$$

and

$$v_y = \frac{u_x}{2} + \frac{\sqrt{3}}{2} u_y, \tag{9.417}$$

or

$$\mathbf{v} = \begin{bmatrix} 1 & 0 \\ \frac{1}{2} & \frac{\sqrt{3}}{2} \end{bmatrix} \mathbf{u}, \tag{9.418}$$

and

$$\mathbf{u} = \begin{bmatrix} 1 & 0 \\ -\frac{1}{\sqrt{3}} & \frac{2}{\sqrt{3}} \end{bmatrix} \mathbf{v}. \tag{9.419}$$

The estimation algorithm is implemented on the rectangular grid array in Figure 9.62(a) to find $\hat{\mathbf{v}}$ and then uses (9.419) to find $\hat{\mathbf{u}}$.

9.9.2.2 Unitary ESPRIT

In this section, the 2-D unitary ESPRIT discussed in Section 9.9.2 is adapted to the hexagonal array model.

In Figure 9.63, we show the subarrays that provide the necessary invariances. For u_x, we require

$$e^{ju_x}\mathbf{J}_{x1}\mathbf{v}(\psi) = \mathbf{J}_{x2}\mathbf{v}(\psi), \tag{9.420}$$

and

$$Q_M^H \mathbf{J}_{x2} Q_N = \mathbf{J}_{x1}. \tag{9.421}$$

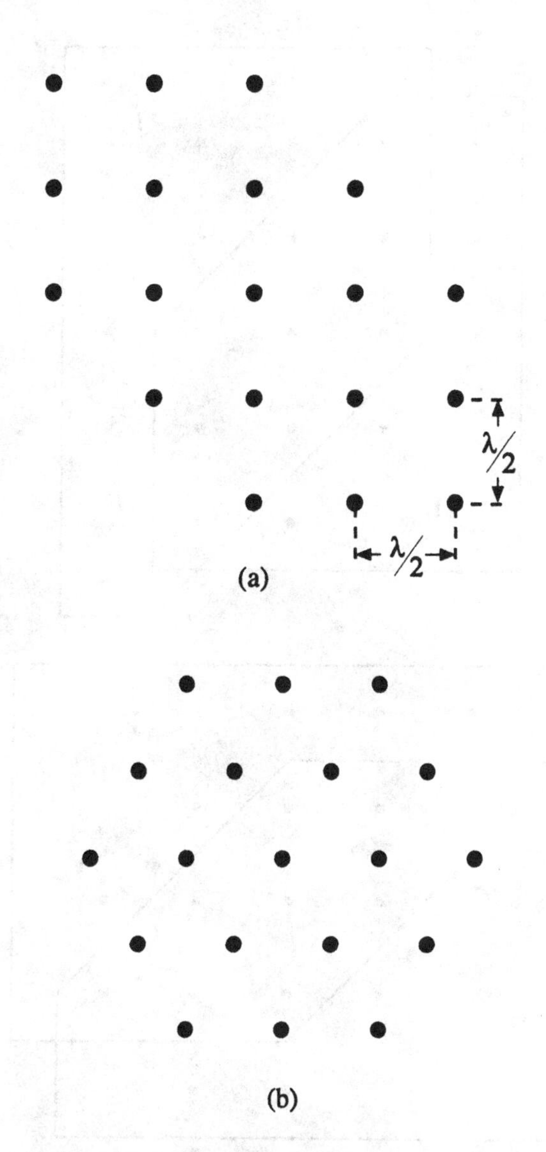

Figure 9.62 Hexagonal-rectangular transformation.

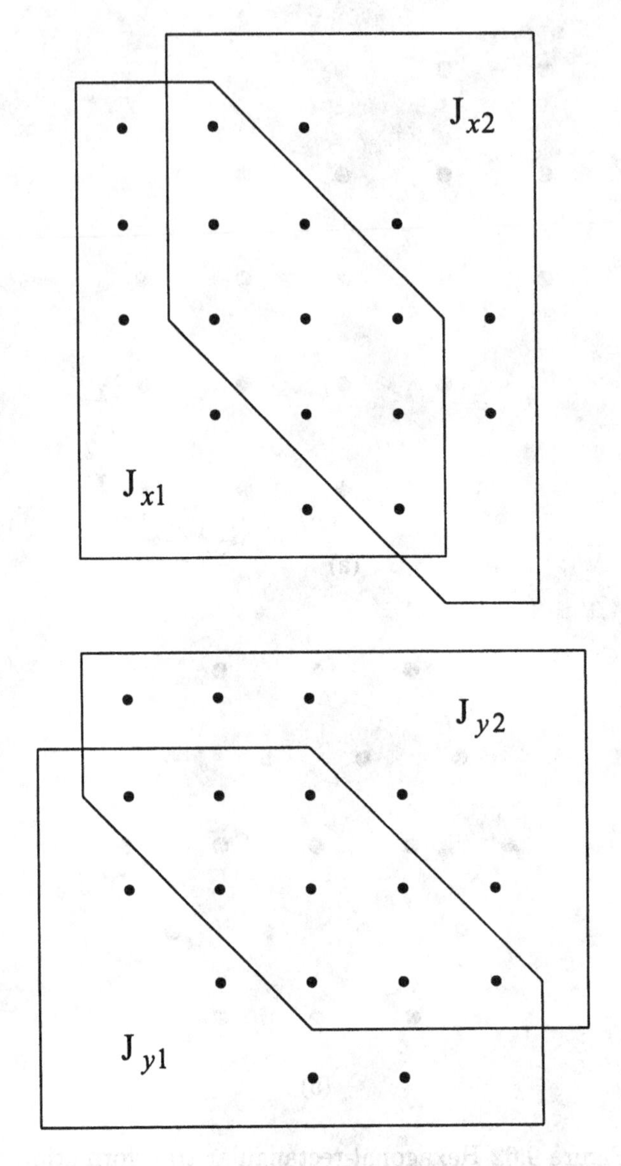

Figure 9.63 Subarrays for unitary ESPRIT.

The 14-element subarrays labeled $\mathbf{J}_{x1}$ and $\mathbf{J}_{x2}$ in Figure 9.63(a) provide the necessary invariance. The elements are

$$\begin{array}{cccc} 2 & 3 & & \\ 5 & 6 & 7 & \\ 9 & 10 & 11 & 12 \\ & 14 & 15 & 16 \\ & & 18 & 19 \end{array} \tag{9.422}$$

Similarly, for u_y, we require

$$e^{ju_y}\mathbf{J}_{y1}\mathbf{v}(\psi) = \mathbf{J}_{y2}\mathbf{v}(\psi). \tag{9.423}$$

The 14-element subarrays labeled $\mathbf{J}_{y1}$ and $\mathbf{J}_{y2}$ in Figure 9.63(b) provide the necessary invariance. The elements are

$$\begin{array}{cccccc} 1 & 2 & 3 & & & \\ 4 & 5 & 6 & 7 & & \\ & 9 & 10 & 11 & 12 & \\ & & 14 & 15 & 16 & \end{array} \tag{9.424}$$

An example illustrates the performance.

Example 9.9.5

Consider the 19-element hexagonal array in Figure 9.62(b) and the corresponding rectangular array in in Figure 9.62(a). There are two equal-power uncorrelated signals impinging on the array from $\mathbf{u}_1$ and $\mathbf{u}_2$ where

$$\mathbf{u}_1 = \begin{bmatrix} 0 & 0 \end{bmatrix}^T, \tag{9.425}$$

and

$$\mathbf{u}_2 = \begin{bmatrix} 0 & 0.0866 \end{bmatrix}^T. \tag{9.426}$$

We use subarrays 1 and 2 in Figure 9.63 in the rectangular grid version of 2-D unitary ESPRIT. Unitary ESPRIT is implemented in **v**-space to the transformed subarrays and the result is transformed to **u**-space. In Figure 9.64, the probability of resolution and RMSE versus *SNR* are plotted.

The threshold occurs at *SNR* = 0 dB. Above threshold, the RMSE is about 1 dB above the CRB. The use of row weighting will move the RMSE closer to the CRB.

One can also develop a 2-D unitary ESPRIT algorithm for the hexagonal array without going through the hexagonal-rectangular transformation. The subarrays and performance are identical.

9.9.2.3 Summary: Hexagonal arrays

By using the transformation in Section 9.9.2.1, we can adapt most of the algorithms derived for rectangular grids to hexagonal grids. We discussed hexagonal unitary ESPRIT and showed that it provided good performance.

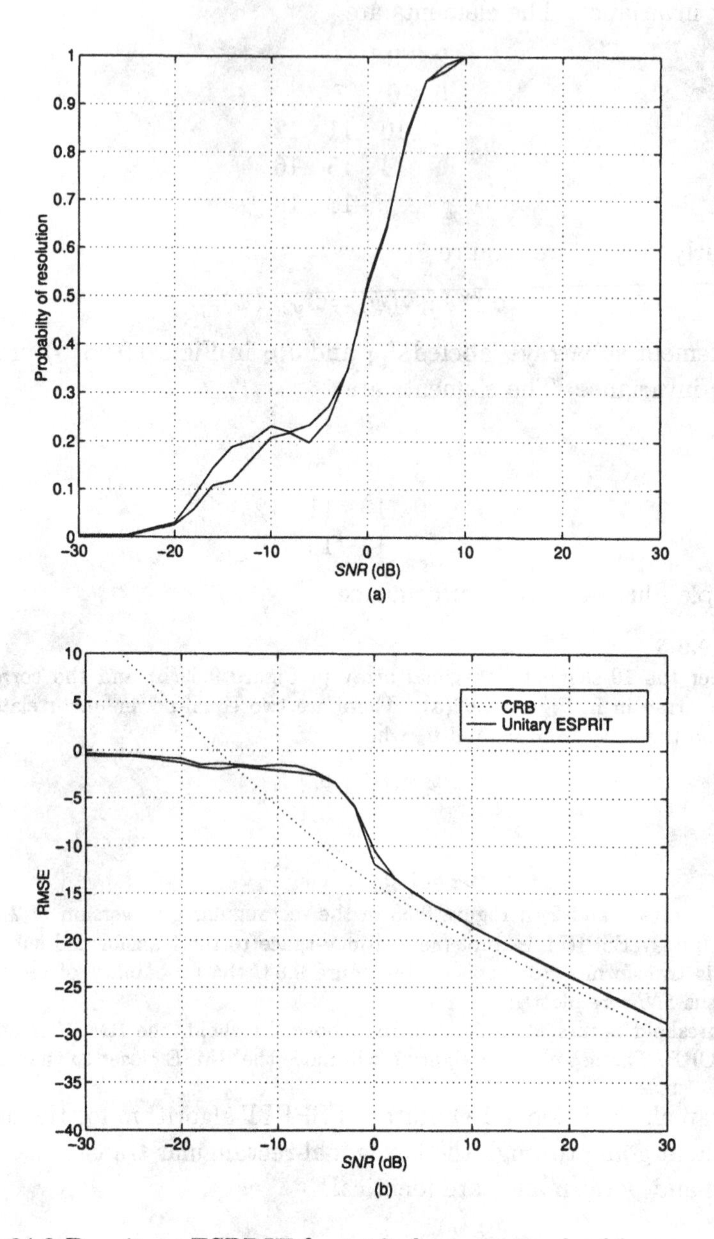

Figure 9.64 2-D unitary ESPRIT for a 19-element standard hexagonal array: $\mathbf{u}_1 = [0\ 0]^T$, $\mathbf{u}_2 = [0\ .0866]^T$, $m_s = 1$, $K = 100$, 500 trials: (a) probability of resolution versus *SNR*; (b) RMSE versus *SNR*.

9.9.3 Summary: Planar Arrays

In this section, we have discussed parameter estimation for planar arrays with either rectangular or hexagonal grids. We focused our attention on the 2-D unitary ESPRIT algorithm and the 2-D beamspace unitary ESPRIT algorithm. Both algorithms provided good performance with a reasonable amount of computation. Zoltowski et al. [ZHM96] also develop a version of beamspace ESPRIT that is applicable to cross arrays.

Uniform circular arrays are used in some applications. We can utilize the phase-mode excitation beamformer developed in Chapter 4 to provide an array manifold similar to a ULA. We then apply root MUSIC and ESPRIT to estimate the DOA. The principal references are Tewfik and Hong [TH92] and Mathews and Zoltowski [MZ94]. The ESPRIT technique has also been extended to filled circular arrays (FCA) whose sensors are located on rectangular, hexagonal, polar, or random lattices. Ramos et al. [RMZ99] have developed an FCA-ESPRIT algorithm that simultaneously estimates azimuth and elevation.

Wong and Zoltowski [WZ99] have developed a root MUSIC algorithm for azimuth and elevation estimation for an array of velocity hydrophones. The paper also contains a useful discussion of velocity hydrophones.

The reader is referred to these references for further discussion.

9.10 Summary

In Section 9.10.1, we summarize the major results in Chapter 9. In Section 9.10.2, we briefly discuss some related topics. In Section 9.10.3, we provide a brief introduction to Chapter 10.

9.10.1 Major Results

In Chapters 8 and 9, we discussed the parameter estimation problem. Chapter 8 focused on ML techniques and bounds. In Chapter 9, we developed algorithms that are computationally simpler. In many cases, some of these algorithms perform almost as well as the ML algorithms. We emphasized the performance of the algorithms for a moderate number of snapshots (generally $K = 10N$) as a function of *ASNR*. Although we discussed asymptotic behavior ($K \rightarrow \infty$), we did not emphasize it. Our examples focused on the case of plane waves whose DOAs were closely spaced ($u < 0.5BW_{NN}$) because it stressed the algorithms. We evaluated the various algorithms using two criteria:

(i) How close did the RMSE approach the CRB as the *ASNR* increased.

(ii) At value of *ASNR* did the threshold behavior occur?

We also considered other cases, such as correlated or coherent signals, unequal signal powers, and low sample support, which stressed the algorithms.

In addition to the two criteria listed above, we must consider the behavior of the algorithms when the number of signals are unknown or the array is subject to perturbations. An important result from the discussion in Section 9.3.5 is that all estimation algorithms should be tested in conjunction with one or more detection algorithms and their robustness to overestimation of D should be measured. The discussion in Section 9.8 shows the importance of measuring the robustness of the algorithm to model perturbations.

Table 9.3: Parameter Estimation Algorithms for Standard Linear Arrays

Algorithm	Computation	Section
IQML-QC	Iterative	8.7.2
IMODE	EVD, iterative	8.7.3
Unitary IMODE	Real EVD, iterative	8.7.3
Two-step MODE	EVD	8.7.3
Root MUSIC	EVD, polynomial rooting	9.3.2
Unitary root MUSIC	EVD, polynomial rooting	9.3.2
Weighted LS & TLS ESPRIT	EVD	9.3.4.1
Weighted unitary LS & TLS ESPRIT	Real EVD	9.3.4.1

The majority of the discussion considered estimation in element space. Section 9.7 developed the corresponding algorithms in beamspace. The advantage of beamspace processing is a reduction in computational complexity.

Because of these issues the choice of the algorithm will depend on the particular application. We have provided the designer with an algorithm tool kit that can be utilized to choose the appropriate algorithm. Tables 9.3–9.6 list the algorithms with their computation and give a section reference.

Table 9.3 lists the algorithms that are most appropriate for standard linear arrays. For uncorrelated signals, all of the algorithms have similar thresholds (within 2–3 dB) for the examples that we studied. Above threshold, all of the algorithms are close (1 or 2 dB) to the CRB. The unitary versions of the algorithms allow real computation and reduce computational complexity.

Table 9.4: Parameter Estimation Algorithms for Standard Rectangular and Hexagonal Arrays

Algorithm	Computation	Section
1-D root MUSIC with pairing	EVD, polynomial rooting	9.9.1.1
2-D unitary LS or TLS ESPRIT	Real EVD	9.9.1.2

Table 9.5: Parameter Estimation Algorithms for Arbitrary Arrays

Algorithm	Computation	Section
AML	D-dim. search	8.5.1
CML	D-dim. search	8.5.2
WSF(MODE)	EVD, D-dim. search	8.5.3
AML-AM	Successive 1-D searches	8.6.2
CML-AP	Successive 1-D searches	8.6.2
MODE-AP	Successive 1-D searches	8.6.2
CML-EM	Parallel 1-D	8.6.3
Bartlett	1-D search	9.2.2
MVDR (Capon)	1-D search	9.2.3
MUSIC	EVD, 1-D search	9.3.2
Min-Norm	EVD, 1-D search	9.3.3

In the case of correlated or coherent signals, the threshold performance is degraded for all of the algorithms. The IMODE algorithm reaches the CRB in the examples that we studied.

We have discussed more of the characteristics of the algorithms in the various sections. By running test scenarios that correspond to given applications and reviewing other constraints such as computational complexity and robustness, we can choose an appropriate algorithm. In the problem section, we provide a set of test scenarios to illustrate these ideas.

Our discussion of rectangular arrays was not as extensive as the SLA discussion. The 2-D unitary ESPRIT algorithm provided good performance with reasonable computational complexity in the examples studied.

For arbitrary arrays, and uncorrelated signals, either CML-AP or MODE-AP provided the best threshold behavior. Above threshold, CML-AP was close to the CRB and MODE-AP reached the CRB. An adequate initialization was essential to the satisfactory performance of the AP algorithms. The threshold of the MUSIC algorithm occurs at a higher value of *ASNR*. Above threshold, the RMSE of the MUSIC algorithm is close to the CRB for uncorrelated signals ($ASNR_i^{-1}$ above it).

Both beamspace unitary ESPRIT and beamspace IQML required a DFT beamspace matrix. The algorithms provided good performance with reasonable computational complexity.

It is important to re-emphasize that the choice of the appropriate algorithm will depend on the scenarios that are encountered in a particular application. Our objective was to provide a collection of useful algorithms and study their behavior for a limited set of scenarios.

Table 9.6: Beamspace Algorithms

Algorithm	Computation	Section
BS-MUSIC	EVD	9.7.1
BS-unitary ESPRIT	EVD	9.7.2
BS-IQML	EVD	8.10.4

9.10.2 Related Topics

In Section 8.12.2, we discussed three related topics in the context of ML estimation; structured adaptive beamforming, nonlinear estimation, and correlated noise environments. All of the topics are also relevant to the parameter estimation algorithms discussed in this chapter. In this section, we briefly introduce several other related topics.

We have selected this particular set of topics because they provide interesting extensions of the discussion in the text. The topics are not of equal importance and the level of discussion varies among topics. We have tried to provide a very brief description of the problem and list several references that will enable the interested reader to get started.

9.10.2.1 Spatially spread sources

In Section 8.9, we introduced a parametric model for spatially spread sources and derived the Cramér-Rao bound. We formulated the ML problem but it appeared to be too complex to be practical. Various references have explored more techniques based on maximum likelihood or weighted subspace fitting techniques (e.g., Trump and Ottersten [TO96a], [TO96b], Swindlehurst and Stoica ([SS97],[SS98]), Bengtsson and Ottersten ([BO97],[BO01]), Besson and Stoica [BS99], Stoica et al. [SOR98], and Stoica and Besson [SB00]).

Several papers develop MUSIC-like algorithms. Jänti [Jän92] models the spatially spread signal with a finite number of point sources and uses MUSIC and ESPRIT to estimate their location. Wu et al. [WWMR94] also use a discrete model and develop an algorithm called Vec-MUSIC. Valaee et al. [VCK95] use a continuous model for the distributed sources and develop a distributed-MUSIC which they call the distributed signal parameter estimation (DSPE) algorithm.

Meng et al. [MSW96] develop an improved algorithm called DISPARE (distributed signal parameter estimation). Bengtsson and Ottersten [BO00] developed a rank-two model which utilizes root MUSIC for estimation. Bengtsson and Ottersten [BO01] analyze the performance of weighted subspace fitting with full-rank models.

Other papers dealing with various aspects of spread signal estimation include Lee et al. [LCSL97], Messer et al. [MRG00], and Shahbazpanahi et al. [SVB01].

9.10.2.2 Broadband direction finding

Several papers address techniques for coherent subspace processing of broadband signals to estimate their DOA. Wang and Kaveh [WK85] develop a coherent signal subspace algorithm using focusing matrices. A number of refinements and extensions of this approach are described in Hung and Kaveh [HK88], Pierre and Kaveh [PK92], Bassias and Kaveh [BK91], and Doron and Weiss [DW92]. A discussion of focusing techniques is given by Krolik in Chapter 6 of [Hay91].

9.10.2.3 Sparse linear arrays

In many applications we use sparse linear arrays to obtain improved resolution with fewer sensors. The MRLAs that were discussed in Section 3.9.2 were examples of a sparse linear arrays. Chambers et al. [CTSD96] have considered various sparse geometries. They computed the CRB and simulated MUSIC. The results demonstrated improvements over ULAs. Fuchs [Fuc97] extended the Pisarenko method to sparse arrays. Abramovich et al. [AGSG96] [AGGS98] [ASG99] considered ambiguity issues and developed techniques for positive definite Toeplitz completion.

9.10.2.4 Non-Gaussian noise

Our discussion has assumed that the additive noise is a sample function of Gaussian random process. There are a number of references that extend some of the techniques to non-Gaussian environments.

Tsakalides and Nikias [TN96] considered a class of noise processes called alpha-stable processes that include Gaussian processes as a special case. A discussion of alpha-stable processes is given in Shao and Nikias [SN95]. Tsakalides and Nikias [TN96] develop a version of MUSIC and analyze its performance.

Other approaches utilize the higher order statistics of the non-Gaussian process. References include Porat and Friedlander [PF91], Forster and Nikias [FN91], Cardoso and Moulines [CM95].

9.10.2.5 Virtual Arrays

For standard linear arrays, we were able to find computationally efficient parameter estimation algorithms. Another approach is to create a virtual SLA by interpolating the outputs of the actual array. Friedlander, Weiss, and Stoica have published work using this idea (e.g., [Fri93], [WFS95], [FW92], [WF93b], or [FW93a]).

9.10.2.6 Near-field sources

For near-field sources, the curvature of the wavefront can be exploited to do range and bearing (DOA) estimation. There are a large number of references that discuss this problem. A representative list includes Hahn [Hah95], Swindlehurst and Kailath [SK88], Huang and Barkat [HB91], Weiss and Friedlander [WF93a], Starer and Nehorari [SN94], Chuberre et al. [CFF95], LeCadre [LeC95], Hung et al. [HCW96], Haardt et al. [HCS96], Yuen and

and Friedlander [YF96a], Haug and Jacyna [HJ96], Abed-Meraim and Hua [AH97], and Lee et al. [LYL01].

9.10.3 Discussion

This completes our two-chapter discussion of parameter estimation. In the next chapter we discuss optimum detection briefly and outline some related topics that are not covered in the book.

9.11 Problems

P9.2 Quadratic Algorithms

Problem 9.2.1 (continuation, Example 9.2.3)

Consider a 10-element standard linear array and assume that there are two equal-power uncorrelated plane-wave signals impinging on the array from $\pm\Delta\psi/2$ where $\Delta\psi = \Delta\psi_R$. Simulate the beamscan algorithm with FB averaging (9.10).

(a) Plot the probability of resolution and the normalized RMSE versus *ASNR* for $K = 100$. Plot the CRB.

(b) Plot the normalized RMSE versus K for *ASNR* = 20, 30, and 40 dB. Plot the CRB.

Discuss your results. Compare the results to those in Figures 9.2–9.4.

Problem 9.2.2 (continuation)

Consider the same model as in Problem 9.2.1 except $\Delta\psi$ is varied. Simulate the beamscan algorithm. Plot the probability of resolution and the normalized RMSE versus $\Delta\psi/BW_{NN}$ for *ASNR* = 10, 20, and 30 dB.

Problem 9.2.3

Read the paper by Zhang [Zha95b] on the resolution capability of the Bartlett beamformer. Simulate the beamscan for his parameter values. Discuss his results and their practical application.

Problem 9.2.4

Consider a 10-element standard linear array and assume that there are two equal-power uncorrelated plane waves impinging on the array from $\pm\Delta u/2$ where $\Delta u = 0.3$. Simulate the MVDR and root MVDR algorithm using FB averaging.

(a) Plot the normalized RMSE versus *ASNR* for $K = 100$. Include the CRB on the plot.

(b) Plot the normalized RMSE versus K for *ASNR* = 10 dB, 20 dB, and 30 dB.

Discuss your results.

Problem 9.2.5

Consider a standard 10-element linear array. The following uncorrelated plane-wave signals are impinging on it:

Signal 1 $\quad u_1 = 0 \quad\quad SNR_1 = 10$ dB

Signal 2 $u_2 = 0.3$ $SNR_2 = 20$ dB

Signal 3 $u_3 = 0.7$ $SNR_3 = 5$ dB

Signal 4 $u_3 = -0.5$ $SNR_4 = 0$ dB

Simulate the MVDR algorithm for $K = 20$, 100, and 500. Plot the MVDR spatial spectrum versus u. Estimate: $\sigma_w^2, u_1, \cdots, u_4$, and $\sigma_1^2, \cdots, \sigma_4^2$. Plot the normalized RMSE versus *ASNR*.

Compare your estimate of σ_w^2 to that obtained from (8.324). Compare your estimate of $u_i, i = 1, \cdots, 4$ to the CRB.

Problem 9.2.6: (Adaptive angular response (AAR) estimation)

The component of the output of the MVDR algorithm due to the white noise input is

$$\widehat{P}_{WN}(\psi) = \left|\mathbf{w}_{mvdr}(\psi)\right|^2, \tag{9.427}$$

and varies as a function of ψ in the MVDR estimator. Borgiottia and Kaplan [BK79] suggested maximizing

$$\begin{aligned} \widehat{P}(\psi) &= \frac{1}{K}\sum_{k=1}^{K}\frac{1}{N}\left|\mathbf{w}^H\,\mathbf{X}_k\right|^2 \\ &= \mathbf{w}^H\,\mathbf{C}_\mathbf{x}\,\mathbf{w}, \end{aligned} \tag{9.428}$$

subject to a constraint on the white noise gain,

$$\mathbf{w}^H\,\mathbf{w} = 1. \tag{9.429}$$

(a) Carry out the constrained maximization and show that,

$$\widehat{P}_{AAR}(\psi) = \frac{\mathbf{v}^H(\psi)\,\mathbf{C}_\mathbf{x}^{-1}\,\mathbf{v}(\psi)}{\mathbf{v}^H(\psi)\,\mathbf{C}_\mathbf{x}^{-2}\,\mathbf{v}(\psi)}. \tag{9.430}$$

The estimator in (9.430) is referred as the adapted angular response (AAR) estimator. The temporal version of this estimate is in [SM97](5.4.17).

(b) Consider the same model as in Example 9.2.1. Plot the results for a typical trial.

(c) Consider the same model as Example 9.2.2. Plot the probability of resolution and normalized RMSE versus *ASNR* for $K = 100$. Plot the normalized RMSE versus K for *ASNR* = 20, 30, and 40 dB. Discuss your results.

Note that $\widehat{P}_{AAR}(\psi)$ does not provide a spatial spectrum estimate.

Problem 9.2.7 (Thermal noise algorithm (TNA) estimation)

Haykin and Reilly [HR80] used a pole-zero argument to suggest that the denominator in (9.430) was the key element in locating the peaks of the spectra. They suggested an algorithm that is referred to as the thermal noise algorithm (TNA).

$$\widehat{P}_{TNA}(\psi) = \frac{1}{\mathbf{v}^H(\psi)\,\mathbf{C}_\mathbf{x}^{-2}\,\mathbf{v}(\psi)}, \tag{9.431}$$

and

$$\widehat{Q}_{TNA}(\psi) = \mathbf{v}^H(\psi)\,\mathbf{C}_\mathbf{x}^{-2}\,\mathbf{v}(\psi). \tag{9.432}$$

Gabriel [Gab80] [Kes86] independently approached this algorithm by considering the output of the MVDR filter

$$Y_o(\psi) = \mathbf{w}^H \mathbf{v}(\psi) = \Lambda \left[\mathbf{v}^H(\psi)\, \mathbf{C}_{\mathbf{x}}^{-1}\, \mathbf{v}(\psi)\right], \tag{9.433}$$

and observing that in the asymptotic regime it approaches zero as $\mathbf{v}(\psi)$ moves through a source location. It approaches zero because the weight vector $\mathbf{w}$ approaches zero. Gabriel suggests using the reciprocal of the magnitude of $\mathbf{w}$ squared as an estimator. (The constant Λ is omitted)

$$\widehat{P}_{TNA}(\psi) = \frac{1}{\mathbf{w}_o^H\, \mathbf{w}_o} = \frac{1}{\mathbf{v}^H(\psi)\, \mathbf{C}_{\mathbf{x}}^{-2}\, \mathbf{v}(\psi)}. \tag{9.434}$$

The corresponding null spectrum is

$$\widehat{Q}_{TNA}(\psi) = \mathbf{v}^H(\psi)\, \mathbf{C}_{\mathbf{x}}^{-2}\, \mathbf{v}(\psi). \tag{9.435}$$

A root version of TNA follows directly.

(a) Repeat part (b) of Problem 9.2.6.

(b) Repeat part (c) of Problem 9.2.6.

Note that the TNA algorithm is not a spatial spectral estimator.

Problem 9.2.8

In the region of small sample support, we may be able to improve the performance by using diagonal loading. The various quadratic algorithms can be modified by replacing $\mathbf{C}_{\mathbf{x}}$ with $\mathbf{C}_{\mathbf{x}} + \sigma_L^2\mathbf{I}$.

Consider the same model as in Example 9.2.2. Plot the normalized RMSE versus K for $ASNR$ = 20 dB and 30 dB with

$$\sigma_L^2/\sigma_w^2 = SNR \times 10^{-2}. \tag{9.436}$$

Discuss your results.

Problem 9.2.9

An enhanced MVDR was proposed by Owsley (e.g., [Ows85], [Hay85]).[23] For simplicity, we assume that $\mathbf{S}_{\mathbf{x}}$ is available in order to derive the structure of the estimator. We then use $\widehat{\mathbf{S}}_{\mathbf{x}}$ to implement the estimator.

The enhanced MVDR starts with an eigenvector decomposition of the covariance matrix,

$$\mathbf{S}_{\mathbf{x}} = \mathbf{U}_S\, \mathbf{\Lambda}\, \mathbf{U}_S^H + \sigma_w^2\, \mathbf{I}, \tag{9.437}$$

where $\mathbf{U}_S$ is the $N \times D$ matrix of eigenvectors corresponding to the signal subspace and $\mathbf{\Lambda} = \mathbf{\Lambda}_S - \sigma_w^2$ is the diagonal matrix of the eigenvalues.

$$\mathbf{U}_S = \left[\ \mathbf{\Phi}_1 \ \vdots\ \mathbf{\Phi}_2 \ \vdots\ \cdots \ \vdots\ \mathbf{\Phi}_D\ \right]. \tag{9.438}$$

Owsley defines an enhanced data covariance matrix,

$$\mathbf{S}_{\mathbf{x}}(e) \triangleq e\, \mathbf{U}_S\, \mathbf{\Lambda}\, \mathbf{U}_S^H + \sigma_w^2\, \mathbf{I}, \tag{9.439}$$

where e is called the modal enhancement factor and has a range $1 \leq e < \infty$.

[23] Our discussion of the enhanced algorithm follows pp.168–177 of Owsley's chapter in [Hay85]. He refers to it as the enhanced MV filter.

(a) Find $\mathbf{S}_{\mathbf{x}}^{-1}(e)$. Substitute $\mathbf{S}_{\mathbf{x}}^{-1}(e)$ in the MVDR algorithm and obtain,

$$P(\psi) = \frac{1}{\mathbf{v}^H(\psi)\,\mathbf{S}_{\mathbf{x}}^{-1}(e)\,\mathbf{v}(\psi)}. \tag{9.440}$$

(b) The enhanced MVDR (EMV) estimator is the limit of (9.440) as $e \to \infty$. Find $P_{EMV}(\psi)$ and $Q_{EMV}(\psi)$.

(c) In practice, we perform an eigendecomposition of $\widehat{\mathbf{S}}_{\mathbf{x}}$,

$$\widehat{\mathbf{S}}_{\mathbf{x}} = \widehat{\mathbf{U}}_S \hat{\mathbf{\Lambda}}_S \widehat{\mathbf{U}}_S^H + \widehat{\mathbf{U}}_N \hat{\mathbf{\Lambda}}_N \widehat{\mathbf{U}}_N^H, \tag{9.441}$$

and use the estimated eigenvectors. Show that $Q_{EMV}(\psi)$ can be written as

$$Q_{EMV}(\psi) \triangleq \mathbf{v}^H(\psi)\,\widehat{\mathbf{U}}_N\,\widehat{\mathbf{U}}_N^H\,\mathbf{v}(\psi) = \left|\mathbf{v}^H(\psi)\,\widehat{\mathbf{U}}_N\right|^2, \tag{9.442}$$

or

$$Q_{EMV}(\psi) \triangleq \mathbf{v}^H(\psi)\left[\mathbf{I} - \widehat{\mathbf{U}}_S\widehat{\mathbf{U}}_S^H\right]\mathbf{v}(\psi). \tag{9.443}$$

The algorithm defined by (9.442) or (9.443) is identical to the MUSIC algorithm that we develop and analyze in Section 9.3.

(d) Consider the array and signal model as Problem 9.2.1. Repeat Problem 9.2.1 for $e = 10$, 100, and ∞. Compare your results to Example 9.2.3. Discuss your results.

P9.3 Subspace Algorithms

Problem Note 9.3.1: The first set of problems considers some of the test scenarios in Table 8.3 so that we can compare these results to the various ML results in Chapter 8. We also explore scenario excursions suggested by the results. We also use these test scenarios in problem Sections P.9.6, P.9.7, and P.9.8. Table 9.7 shows where the various test scenarios are used.

Problem 9.3.1

Consider a 10-element standard linear array. There are two equal-power uncorrelated plane-wave signals impinging on the array from $\pm\Delta\psi/2$, where $\Delta\psi = 0.05 BW_{NN}$. Simulate the performance of MUSIC, root MUSIC, Min-Norm, and root Min-Norm. Assume $K = 100$.

(a) Plot the normalized RMSE versus *ASNR*. Compare your results to the AML and CML results in Chapter 8.

(b) Plot the probability of resolution. Discuss the behavior of the curve. Is the P_R definition useful for this scenario?

Problem 9.3.2 (continuation)

(a) Repeat Problem 9.3.1 for the TLS-ESPRIT algorithm. Investigate appropriate row weightings.

(b) Repeat part (a) for unitary TLS-ESPRIT.

Table 9.7

	TS1	TS2	TS3	TS4	TS5
9.3 Subspace	Ex.9.3.1	P.9.3.1 P.9.3.2	P.9.3.3 P.9.3.4	P.9.3.6 P.9.3.7 Ex.9.3.2	
9.6 Spatial Smoothing				P.9.6.1	
9.7 Beamspace	Ex.9.7.2 P.9.7.1 P.9.7.2	P.9.7.4	P.9.7.5	P.9.7.6 P.9.7.7	
9.8 Sensitivity	Ex.9.8.1 P.9.8.5 P.9.8.14	P.9.8.1 P.9.8.10	P.9.8.2 P.9.8.7 P.9.8.11 P.9.8.16	P.9.8.3 P.9.8.8 P.9.8.12 P.9.8.17	
	TS6	**TS7**	**TS8**	**TS9**	**TS10**
9.3 Subspace	P.9.3.8	P.9.3.10	P.9.3.11	P.9.3.12	P.9.3.13
9.6 Spatial Smoothing			P.9.6.3		
9.7 Beamspace	P.9.7.8	P.9.7.3	P.9.7.10	P.9.7.11	P.9.7.12
9.8 Sensitivity		P.9.8.4 P.9.8.6 P.9.8.13 P.9.8.15			

Problem 9.3.3

Consider a 10-element standard linear array. There are two uncorrelated plane-wave signals impinging on the array from $\pm\Delta\psi/2$, where $\Delta\psi = \Delta\psi_R$. $SNR_1 = 100 SNR_2$. Simulate the performance of MUSIC, root MUSIC, Min-Norm, and root Min-Norm. Assume $K = 100$.

(a) Plot the normalized RMSE for signal 1 versus $ASNR_1$ and signal 2 versus $ASNR_2$.

(b) Repeat for the case in which $SNR_1 = 1000 SNR_2$.

Problem 9.3.4 (continuation)

(a) Repeat Problem 9.3.3 for the TLS-ESPRIT algorithm. Investigate appropriate row weightings.

(b) Repeat part (a) for unitary ESPRIT.

Problem 9.3.5 (continuation, Problem 9.3.3)

Consider the same model as in Problem 9.3.3. Design a sequential MUSIC algorithm that estimates the DOA of signal 1 first. The algorithm then removes signal 1 from the data and estimates the DOA of signal 2. Simulate the performance of your algorithm and discuss its behavior.

Problem 9.3.6 (continuation, Example 9.3.2)

Consider the same model as in Example 9.3.2, except the phase angle of ρ is not zero. In case 1, $\rho = 0.95\exp(j\pi/4)$. In case 2, $\rho = 0.95\exp(j\pi/2)$. We utilize FB averaging. Simulate the performance of MUSIC, root MUSIC, Min-Norm, and root Min-Norm.

(a) Plot the normalized RMSE versus *ASNR*. Compare your results to the AML and CML results in Chapter 8.

(b) Repeat part (a) with $|\rho| = 0.99$.

Problem 9.3.7 (continuation)

Repeat Problem 9.3.6 for TLS-ESPRIT and unitary ESPRIT.

Problem 9.3.8 (continuation, Example 9.3.1)

Repeat Example 9.3.1 for the low sample support case in which $K = 20$.

Problem 9.3.9 (continuation, Example 9.3.4)

Consider the model as in Example 9.3.4 and assume $d_s = 1$. Simulate the performance of the LS-ESPRIT algorithm (9.121). Compare the results to the results using the TLS-ESPRIT algorithm. Discuss the computational difference.

Problem 9.3.10

Consider a 10-element standard linear array. There are three equal-power uncorrelated plane-wave signals impinging on the array from $\psi_1 = -\Delta\psi$, $\psi_2 = 0$, and $\psi_3 = \Delta\psi$, where $\Delta\psi = \Delta\psi_R$. (Test scenario 7.)

(a) Simulate the performance of MUSIC, root MUSIC, Min-Norm, and root Min-Norm. Plot the normalized RMSE versus *ASNR* for each of the three signals. Discuss your results.

(b) Repeat part (a) for TLS-ESPRIT and unitary TLS-ESPRIT.

Problem 9.3.11 (continuation)

Consider the same model as in Problem 9.3.10 except the signals are correlated with unequal power. (Test scenario 8.)

$$SNR_1 = SNR_3 = 0.5SNR_2$$

$$\rho_{12} = 0.9, \quad \rho_{23} = 0.9, \quad \rho_{13} = 0.5\exp(j\pi/2)$$

Simulate the performance of MUSIC, root MUSIC, and ESPRIT. Plot the normalized RMSE versus *ASNR* for each of the three signals. Discuss your results.

Problem 9.3.12 (continuation; Problem 9.3.10)

(a) Repeat Problem 9.3.10 for the case in which the two plane waves at $\pm\Delta\psi$ have an *SNR* that is 10 dB higher than the plane wave at $\psi = 0$. (Test scenario 9.)

(b) Apply the algorithm derived in Problem 9.3.5 to this model.

Problem 9.3.13 (continuation, Problem 9.3.10)

Consider the model in Problem 9.3.10. Repeat the problem for the case of five equal-power uncorrelated plane waves located at $\psi = 0, \pm\Delta\psi, \pm 2\Delta\psi$, where $\Delta\psi = \Delta\psi_R$. (Test Scenario 10)

Problem 9.3.14

Consider a standard 10-element array. There are two equal-power correlated signals impinging on the array from $\pm\Delta\psi$, where $\Delta\psi = BW_{NN}$. Let $K = 100$. Consider two phase angles for ρ; $\phi_\rho = 0$ and $\phi_\rho = j\pi/2$. Utilize FB averaging and implement MUSIC and root MUSIC. The *ASNR* = 10 dB.

(a) Plot the normalized RMSE versus $|\rho|$ for the two phase angles.

(b) Compute the CRB versus $|\rho|$ for the two phase angles and plot the relative efficiency of the MUSIC estimator. In Example 9.5.1, we will show that

$$EFF_{MU} = 1 - |\rho|^2.$$

Compare your results to that formula.

Problem Note 9.3.2: The next set of problems consider a 32-element SLA. The problem with two and three signals are similar to the earlier problems. The larger array size allows us to consider more complicated scenarios.

Problem 9.3.15

Consider a standard 32-element linear array. There are two equal-power uncorrelated signals impinging on the array at $\pm\Delta u/2$, where $\Delta u = 1/32$. Consider $K = 320$ and 960.

(a) Simulate MUSIC and root MUSIC. Plot the normalized RMSE versus *ASNR*.

(b) Simulate TLS-ESPRIT. Plot the normalized RMSE versus *ASNR*.

Problem 9.3.16 (continuation)

Repeat Problem 9.3.15 with $\Delta u = 1/128$.

Problem 9.3.17 (continuation, Problem 9.3.15)

Repeat Problem 9.3.15 with $\Delta u = 1/32$ and $ASNR_1 = 100 ASNR_2$.

Problem 9.3.18 (continuation, Problem 9.3.15)

Repeat Problem 9.3.15 with $\Delta u = 1/32$ and $\rho = 0.95$, $\rho = 0.95\exp(j\pi/4)$, and $\rho = 0.95\exp(j\pi/2)$.

Problem 9.3.19 (continuation, Problem 9.3.15)

Repeat Problem 9.3.15 with $\Delta u = 1/32$ and $K = 64$.

Problem 9.3.20 (continuation, Problem 9.3.15)

(a) Repeat Problem 9.3.15 with three equal-power uncorrelated signals located at $-1/32$, 0, 1/32.

(b) Repeat part (a) with $ASNR_1 = ASNR_3 = 100 ASNR_2$.

Problem 9.3.21 (continuation, Problem 9.3.15)

Repeat Problem 9.3.15 for the following cases:

(a) Five equal-power uncorrelated signals located at $\pm 2/32, \pm 1/32, 0$.

(b) Seven equal-power uncorrelated signals located at $\pm 3/32, \pm 2/32, \pm 1/32, 0$.

(c) Fifteen equal-power uncorrelated signals located at $\pm m/32, m = 0, \cdots, 7$.

Problem Note 9.3.3: Our discussion and most of the discussion in the literature assumes that the number of signals, D, is known. In practice, we usually have to estimate D. The next three problems consider this issue.

Problem 9.3.22 (continuation, Example 9.3.1)

Repeat Example 9.3.1 under the assumption that we do not know that there are two plane waves. Process the data using AIC-FB and MDL-FB to generate $\hat{D}$. Simulate MUSIC, root MUSIC, and TLS-ESPRIT assuming $\hat{D}$ is correct. Plot the results. Discuss your results.

Problem 9.3.23 (continuation)

Repeat Problem 9.3.22 for the signal model in Problem 9.3.6.

Problem 9.3.24

Repeat Problem 9.3.22 for the signal model in Problem 9.3.10.

Problem Note 9.3.4: The next several problems consider low redundancy arrays that were introduced in Section 3.9.2. We would expect that the larger aperture would improve resolution but would require a larger number of snapshots to achieve the same estimation accuracy.

Problem 9.3.25

Consider the 5-element linear array in Table 3.8. The sensor spacing is

$$1 \cdot 3 \cdot 3 \cdot 2$$

Two equal-power uncorrelated signals impinge on the array from $\pm\Delta u/2$, where $\Delta u = 0.0866$. Simulate the performance of the MUSIC algorithm for $K = 100$, 300, and 500.

Plot the normalized RMSE versus *ASNR* for each value of K. Compare your results to those in Example 9.3.1 and the CRB derived in Chapter 8 (Problem 8.4.10). Discuss the trade-off between the number of elements and the number of snapshots.

Problem 9.3.26 (continuation)

Repeat Problem 9.3.25 for the 7-element linear array in Table 3.8. The sensor spacing is

$$1 \cdot 3 \cdot 6 \cdot 2 \cdot 3 \cdot 2$$

Consider larger values of K, if necessary.

Problem 9.3.27 (continuation, Problem 9.3.25)

Repeat Problem 9.3.25 for the 9-element linear array in Table 3.8. The sensor spacing is

$$1 \cdot 3 \cdot 6 \cdot 6 \cdot 6 \cdot 2 \cdot 3 \cdot 2$$

Consider larger values of K, if necessary.

Problem 9.3.28 (continuation, Problem 9.3.25)

Repeat Problem 9.3.25 for the 10-element linear array in Table 3.8. The sensor spacing is

$$1 \cdot 2 \cdot 3 \cdot 7 \cdot 7 \cdot 7 \cdot 4 \cdot 4 \cdot 1$$

Consider larger values of K, if necessary.

Problem 9.3.29

The sensor spacing in a MRLA allows us to construct the $\hat{\mathbf{S}}_{\mathbf{x}}$ matrix corresponding to a uniform linear array on the same grid as the MRLA. Use this result to design a root MUSIC algorithm for the MRLA.

(a) Simulate your algorithm for the MRLA in Problem 9.3.25. Compare your results to those in Problem 9.3.25.

(b) Simulate your algorithm for the MRLA in Problem 9.3.26. Compare your results to those in Problem 9.3.26.

(c) Simulate your algorithm for the MRLA in Problem 9.3.27. Compare your results to those in Problem 9.3.27.

(d) Simulate your algorithm for the MRLA in Problem 9.3.28. Compare your results to those in Problem 9.3.28.

Problem 9.3.30: Unitary IMODE ([GS99])

The cost function for MODE is given in (8.524) as

$$J_{MODE}(\mathbf{b}) = \text{tr}\left[\mathbf{P_B}\,\hat{\mathbf{U}}_s\,\mathbf{W}_{ao}\,\hat{\mathbf{U}}_s^H\right], \tag{9.444}$$

where $\mathbf{B}$ is given in (8.488) and satisfies

$$\mathbf{B}^H\mathbf{V} = \mathbf{0}. \tag{9.445}$$

The FB backward averaged spatial spectral matrix is given by (7.40) as

$$\mathbf{C}_{\mathbf{x},fb} = \frac{1}{2K}\sum_{k=1}^{K}\left(\mathbf{X}_k\mathbf{X}_k^H + \mathbf{J}\mathbf{X}_k^*\mathbf{X}_k^T\mathbf{J}\right) = \frac{1}{2}\left(\mathbf{C}_{\mathbf{x}} + \mathbf{J}\mathbf{C}_{\mathbf{x}}^*\mathbf{J}\right). \tag{9.446}$$

From (9.58)

$$\mathbf{C}_{\mathbf{x},Re} = \mathbf{Q}^H\mathbf{C}_{\mathbf{x},fb}\mathbf{Q} = Re\left[\mathbf{Q}^H\mathbf{C}_{\mathbf{x}}\mathbf{Q}\right], \tag{9.447}$$

where $\mathbf{Q}$ is given by (9.54) and (9.55). The eigendecomposition of $\mathbf{C}_{\mathbf{x},Re}$ is given by (9.64) as

$$\mathbf{C}_{\mathbf{x},Re} = \widehat{\mathbf{U}}_{S,Re}\widehat{\mathbf{\Lambda}}_{S,Re}\hat{\mathbf{U}}_{S,Re}^H + \widehat{\mathbf{U}}_{N,Re}\widehat{\mathbf{\Lambda}}_{N,Re}\hat{U}_{N,Re}^H. \tag{9.448}$$

(a) Show that

$$(\mathbf{B}^H\mathbf{Q})(\mathbf{Q}^H\mathbf{V}) = (\mathbf{B}^H\mathbf{Q})\widetilde{\mathbf{V}} = \mathbf{0}, \tag{9.449}$$

where

$$\widetilde{\mathbf{V}} \triangleq \mathbf{Q}^H\mathbf{V}. \tag{9.450}$$

(b) Show that the unitary MODE cost function is

$$J_{UMODE}(\mathbf{b}) = \text{tr}\left[\mathbf{Q}^H \mathbf{P}_B \mathbf{Q} \hat{\mathbf{U}}_{S,Re} \mathbf{W}_{Re} \hat{\mathbf{U}}_{S,Re}^H\right], \tag{9.451}$$

where

$$\mathbf{W}_{Re} = \left(\widehat{\mathbf{\Lambda}}_{S,Re} - \hat{\sigma}_\omega \mathbf{I}\right)^2 \widehat{\mathbf{\Lambda}}_{S,Re}^{-1}. \tag{9.452}$$

(c) Unitary IMODE uses (8.524)–(8.544) with the above cost function. Write out the appropriate equations.

(d) Repeat Example 8.7.1 and 8.7.2 using unitary IMODE.

P9.4 Linear Prediction

Problem 9.4.1

Read Gabriel's [Gab80] development of linear prediction algorithms (see also Johnson and Degraaf [JD82]).

(a) Show that the LP null spectrum can be written as

$$\hat{Q}_{LP}(\psi) = |B_{LP}(\psi)|^2 = \left|\mathbf{v}^H(\psi)\widehat{\mathbf{S}}_{\mathbf{x}}^{-1}\mathbf{e}_N\right|^2, \tag{9.453}$$

where

$$\mathbf{e}_N^T = \begin{bmatrix} 0 & 0 & \cdots & 0 & 1 \end{bmatrix}, \tag{9.454}$$

is a $1 \times N$ matrix whose first $N-1$ elements are equal to zero. or

$$\hat{Q}_{FBLP}(\psi) = |B_{FBLP}(\psi)|^2 = \left|\mathbf{v}^H(\psi)\widehat{\mathbf{S}}_{FB}^{-1}\mathbf{e}_N\right|^2, \tag{9.455}$$

if FB averaging is appropriate.

(b) Consider the same model as in Example 9.2.2. . Plot the probability of resolution and normalized RMSE versus *ASNR* for $K = 100$. Compare your results to root MUSIC and the CRB.

(c) Plot the normalized RMSE versus K for an $ASNR = 20$ dB.

Discuss your results.

Problem 9.4.2

Read Tufts and Kumaresan [TK82]. Develop the root version of the LP algorithm. Repeat parts (b) and (c) of Problem 9.4.1. Discuss your results.

Problem 9.4.3

Read Kumaresan and Tufts [KT83] and show how they derive the Min-Norm algorithm from the linear prediction algorithm.

Problem 9.4.4

There is also a family of algorithms that we refer to as the subarray linear prediction (SALP) family.

The basic algorithm is essentially the same as the maximum entropy algorithm developed by Burg (e.g.,[Bur67], [Bur75]) for spatial or temporal spectral estimation. Van den Bos [VDB71] showed that it could be interpreted as a linear predictor. In the spatial (or temporal) estimation context, most discussions start with an underlying autoregressive or ARMA model for the spectra and assume a noiseless observation. Books by Kay [Kay83], Marple [Mar87], Haykin [Hay96], and Proakis et al. [PRLN92] contain extensive discussions. Although the emphasis is on temporal spectral estimation, most of the ideas map directly to the spatial problem in the case of SLA.

The algorithms are the multiple snapshot version of the frequency estimation algorithms contained in Tufts and Kumaresan [TK82].

(a) Read [TK82] and derive the formula for the null spectra. Use the subarray model in Figure 6.102.

(b) Repeat parts (b) and (c) of Problem 9.4.1. Discuss your results.

P9.5 Asymptotic Performance

Problem 9.5.1

Read Lee and Wengrovitz [LW91] and Zhang [Zha95a]. Reproduce the results in Figure 9.35.

Problem 9.5.2

Read Krim et al. [KFP92]. Derive (9.199).

Problem 9.5.3

Read Pillai and Kwon [PK89a]. Derive the appropriate equations for the FB averaging case.

Problem 9.5.4

Consider the model in Example 9.5.1. Read the discussion in Section 9.6.3.2 of FBSS MUSIC. Plot the relative asymptotic efficiency of FBSS MUSIC for the same parameters as in Figure 9.38. Discuss your results.

Problem 9.5.5

Read the bias analysis in Xu and Buckley[XB92]. Discuss the *ASNR* regions where the bias is a factor. Does their analysis reflect the bias behavior in the threshold region?

Problem 9.5.6 (continuation, Problem 9.3.3)

Consider the model in Problem 9.3.3. Compute the asymptotic variance of MUSIC and ESPRIT (with row weighting). Discuss your results.

Problem 9.5.7 (continuation, Problem 9.3.1)

Consider the model in Problem 9.3.1. Compute the asymptotic variance of MUSIC and ESPRIT (with row weighting). Discuss your results.

P9.6 Correlated and Coherent Signals

Problem 9.6.1

Consider a 10-element standard linear array with two equal-power correlated signals impinging on the array. The signal separation is Δu. Consider three cases of signal phase; $\phi_\rho = 0$, $\pi/4$, and $\pi/2$ and two values of $|\rho|$; 0.9 and 0.99. We use FBSS as a preliminary processor. Consider four values of M; 3, 5, 7, and 9. (Test scenario 4.)

Simulate the MUSIC, root MUSIC, and ESPRIT algorithms on the output of FBSS processor.

(a) Plot the normalized RMSE versus Δu for an *ASNR* = 20 dB and 30 dB. Discuss your results.

(b) Consider the parameter set in Example 9.3.2; $\rho = 0.95$, $\Delta u = 6\Delta\psi_R$, and $K = 100$. Plot the normalized RMSE for the three algorithms and the four values of M. Compare your results to those in Figure 9.13 and 9.20(a).

Problem 9.6.2 (continuation)

Consider the same model as in Problem 9.6.1 with $\Delta u = 3BW_{NN}$. Recall that the relative efficiency of MUSIC was essentially flat and equal to $(1 - |\rho|^2)$ for $\Delta u \geq BW_{NN}$.

Simulate the MUSIC and root MUSIC algorithms and develop relative efficiency results for the FBSS algorithms with various M, $|\rho|$, and ϕ_ρ. Discuss your results.

Problem 9.6.3 (continuation, Problem 9.3.11)

Consider the same model as in Problem 9.3.11. Use FBSS with various values of M. Simulate the performance of MUSIC, root MUSIC, and TLS-ESPRIT. Plot the normalized RMSE versus *ASNR* for each of the three signals. Compare your results to those in Problem 9.3.12. (Test scenario 8.)

P9.7 Beamspace Algorithms

Problem Note 9.7.1: We consider many of the same problems as in Problem Section P.9.3 so we can compare the beamspace performance to the element-space performance.

Problem 9.7.1 (continuation, Example 9.7.2)

Consider a 10-element standard linear array. Two equal-power uncorrelated plane-wave signals impinge on the array from $\pm\Delta\psi/2$ where $\Delta\psi = \Delta\psi_R$. We use an $N_{bs} \times N$ DFT beamspace matrix (9.304). We consider $N_{bs} = 3$ and 5. Assume $K = 100$.

(a) Simulate the beamspace MUSIC algorithm. Plot the normalized RMSE versus *ASNR*. Compare your results to the results in Example 9.7.1 and 9.7.2.

(b) Assume that the two plane-wave signals are located at $u_c - \Delta\psi/2$ and $u_c + \Delta\psi/2$. Simulate the beamspace unitary ESPRIT algorithm. Plot the normalized RMSE versus *ASNR* for representative u_c inside and outside the beam fan. Discuss your results.

Problem 9.7.2 (continuation)

Consider the same model as in Problem 9.7.1 except we use $N_{bs} = 4$ and 6 (see (3.322)). Assume $K = 100$. (Test scenario 1.)

Simulate the beamspace MUSIC algorithm and the beamspace unitary ESPRIT algorithm. Plot the normalized RMSE versus *ASNR*. Compare your results to Problem 9.7.1 and Example 9.7.2.

Problem 9.7.3 (continuation)

Consider the same model as in Problem 9.7.1 except that there are three equal-power uncorrelated signals located at $\psi = 0$ and $\pm\Delta\psi$, where $\Delta\psi = \Delta\psi_R$. We consider $N_{bs} = 4$, 5, and 6. (Test scenario 7.)

Simulate the beamspace MUSIC algorithm and beamspace unitary ESPRIT algorithm. Plot the normalized RMSE versus *ASNR*. Discuss your results.

Problem 9.7.4 (continuation, Problem 9.3.1, Problem 9.7.1)
Consider the same beamspace matrix as Problem 9.7.1 and the same input signal model as Problem 9.3.1 ($\Delta\psi = 0.05BW_{NN}$). Simulate beamspace MUSIC and beamspace unitary ESPRIT. Plot the normalized RMSE versus *ASNR*. Compare your results to the element-space results in Problems 9.3.1 and 9.3.2. (Test scenario 2.)

Problem 9.7.5 (continuation, Problem 9.3.3, Problem 9.7.1)
Consider the same beamspace matrix as Problem 9.7.1 and the same input signal model as Problem 9.3.3 ($SNR_1 = 100SNR_2$ and $1000SNR_2$). Simulate beamspace MUSIC and beamspace unitary ESPRIT. Plot the normalized RMSE versus *ASNR*. Compare your results to the element-space results in Problems 9.3.3 and 9.3.4. (Test scenario 3.)

Problem 9.7.6 (continuation, Problem 9.3.6, Problem 9.7.1)
Consider the same beamspace matrix as Problem 9.7.1 and the same input signal model as Problem 9.3.6 ($\rho = 0.95, 0.95\exp(j\pi/4)$, and $0.95\exp(j\pi/2)$). Simulate beamspace MUSIC and beamspace unitary ESPRIT. Plot the normalized RMSE versus *ASNR*. Compare your results to the element-space results in Problems 9.3.6, 9.3.7, and Example 9.3.2. (Test scenario 4.)

Problem 9.7.7
Consider the same model as in Problem 9.7.6. Assume that a beamspace matrix with $N_{bs} = 7$ is available. Design an algorithm which implements beamspace spatial smoothing. Plot the normalized RMSE versus *ASNR*. Compare your results to those in Problem 9.7.6. (Test scenario 4.)

Problem 9.7.8 (continuation, Problem 9.3.8, Problem 9.7.1)
Consider the same beamspace matrix as Problem 9.7.1 and the same input signal model as Problem 9.3.8 ($K = 20$). Simulate beamspace MUSIC and beamspace unitary ESPRIT. Plot the normalized RMSE versus *ASNR*. (Test scenario 6.)

(a) Compare your results to the element-space results in Problem 9.3.8.

(b) Repeat Problem 9.3.8 and part (a) for various values of K. Discuss your results.

Problem 9.7.9 (continuation, Problem 9.7.2)
Consider the same beamspace matrix as Problem 9.7.2. The input is three equal-power uncorrelated plane-wave signals arriving from $\psi = 0, \Delta\psi, 2\Delta\psi$, where $\Delta\psi = \Delta\psi_R$. Simulate beamspace MUSIC and beamspace unitary ESPRIT. Plot the normalized RMSE versus *ASNR*. Compare your results to the element-space results in Problem 9.3.10.

Problem 9.7.10 (continuation, Problem 9.7.9)
Repeat Problem 9.7.9 for the signal model in Problem 9.3.11. (Test scenario 8.)

Problem 9.7.11 (continuation, Problem 9.7.10)
Repeat Problem 9.7.9 for the signal model in Problem 9.3.12. (Test scenario 9.)

Problem 9.7.12 (continuation, Problem 9.7.2, Problem 9.3.13)
Consider the same beamspace matrix as Problem 9.7.2 and the same input signal model as Problem 9.3.13 (five signals at $\psi = 0, \pm\Delta\psi, \pm2\Delta\psi$). Simulate beamspace MUSIC and

beamspace unitary ESPRIT. Plot the normalized RMSE versus *ASNR*. Compare your results to the element-space results in Problem 9.3.13. (Test scenario 10.)

Problem 9.7.13

All of the beamspace estimation algorithms have assumed that the number of signals are known. If we assume that all of our processing is done in beamspace then we need to implement either AIC-FB or MDL-FB in beamspace. We investigate the performance of the detection algorithms as a function of N_{bs} and *ASNR* in this problem.

Consider a 10-element SLA and an $N_{bs} \times N$ DFT beamspace matrix. We utilize FB averaging of the data (see Section 8.10.2.5) and consider values of $N_{bs} = 3$, 4, 5, 6, and 7. In each part, we plot $\hat{D}$ versus *ASNR* for various values of N_{bs}. The signals are plane-wave signals with the following characteristics (in all cases, $\Delta u = \Delta\psi_R$):

(a) two uncorrelated signals at $u_1 = -\Delta u/2$, $u_1 = \Delta u/2$.

(b) two uncorrelated signals at $u_c - \Delta u/2$, $u_c + \Delta u/2$.

Problem 9.7.14

Consider the same beamspace and signal model as in Problem 9.7.1. Assume $N_{bs} = 7$ and $K = 100$. We assume that the number of signals is unknown. Implement the following algorithm and simulate its performance:

(a) Implement AIC-FB and MDL-FB in beamspace. Denote the estimate of the number of signals as $\hat{D}$.

(b) Choose the $(\hat{D} + 1)$ beams with the largest output.

(c) Process these beams to estimate the DOAs.

(d) Plot the RMSE versus *ASNR*.

Problem 9.7.15

Read the paper by Zoltowski et al. [ZKS93] on beamspace root MUSIC. Repeat Problem 9.7.1(a) using beamspace root MUSIC. Compare your results to those in Problem 9.7.1 and Example 9.7.2. Compare the computational complexity.

Problem 9.7.16 (continuation, Example 8.10.6 and Example 9.7.2)

Consider a standard 32-element linear array and a $N_{bs} \times N$ DFT beamspace matrix. There are two equal-power uncorrelated plane-wave signals impinging on the array from $u = \pm\Delta u/2$, where $\Delta u = 1/32$, $K = 100$.

(a) Simulate the performance of beamspace MUSIC and beamspace unitary ESPRIT for the $N_{bs} = 3$ and 5 cases. Plot the normalized RMSE versus *ASNR*. Discuss your results.

(b) Repeat part(a) for $\Delta u = 1/128$. Discuss your results.

(c) Repeat part(a) for $\Delta u = 1/8$. Discuss your results.

P9.8 Sensitivity and Robustness

Problem Note 9.8.1: The first set of problems consider position perturbations using the perturbation model in Examples 8.11.1 and 9.8.1. We let $\sigma_p = 0.05\lambda$. All of the problems have been simulated in an earlier section for the case of no perturbations. In each case, compare your result to the $\sigma_p = 0$ case and the hybrid CRB from Section 8.11.

Problem 9.8.1

Repeat Problems 9.3.1 and 9.3.2 for the position perturbation case. (Test scenario 2.)

Problem 9.8.2

Repeat Problems 9.3.3 and 9.3.4 for the position perturbation case. (Test scenario 3.)

Problem 9.8.3

Repeat Problems 9.3.6 and 9.3.7 for the position perturbation case. (Test scenario 4.)

Problem 9.8.4

Repeat Problem 9.3.10 for the position perturbation case. (Test scenario 7.)

Problem Note 9.8.2: The next five problems consider beamspace processing and position perturbations. One objective is to compare the performance of beamspace processing and element-space processing for identical models.

Problem 9.8.5

Repeat Problem 9.7.1 for the position perturbation case. Compare your results to those in Example 9.8.1. (Test scenario 1.)

Problem 9.8.6

Repeat Problem 9.7.9 for the position perturbation case. Compare your results to those in Problem 9.8.4. (Test scenario 7.)

Problem 9.8.7

Repeat Problem 9.7.5 for the position perturbation case. Compare your results to those in Problem 9.8.2. (Test scenario 3.)

Problem 9.8.8

Repeat Problem 9.7.6 for the position perturbation case. Compare your results to those in Example 9.8.3. (Test scenario 4.)

Problem 9.8.9

Repeat Problem 9.7.3 for the position perturbation case.

Problem Note 9.8.3: The next set of problems consider sensor gain and phase perturbation. We use the model in Section 6.6.3 with no sensor perturbations. We let $\sigma_g = 0.05$ and $\sigma_\gamma = 2°$. In each case, compare your results to the $\sigma_g = \sigma_\gamma = 0$ case and the hybrid CRB.

Problem 9.8.10

Repeat Problems 9.3.1 and 9.3.2 for sensor gain and phase perturbations. Compare your results to those in Problems 9.3.1 and 9.3.2. (Test scenario 2.)

Problem 9.8.11

Repeat Problems 9.3.3 and 9.3.4 for sensor gain and phase perturbations. Compare your results to those in Problems 9.3.3 and 9.3.4. (Test scenario 3.)

Problem 9.8.12

Repeat Problems 9.3.6 and 9.3.7 for sensor gain and phase perturbations. Compare your results to those in Problems 9.3.6 and 9.3.7. (Test scenario 4.)

Problem 9.8.13

Repeat Problem 9.3.10 for sensor gain and phase perturbations. Compare your results to those in Problem 9.3.10. (Test scenario 7.)

Problem Note 9.8.4:

The next several problems consider beamspace processing and sensor gain and phase perturbations.

Problem 9.8.14

Repeat Problem 9.7.1 for sensor gain and phase perturbations. Compare your results to those in Example 9.8.1. (Test scenario 1.)

Problem 9.8.15

Repeat Problem 9.7.3 for sensor gain and phase perturbations. Compare your results to those in Problem 9.8.13. (Test scenario 7.)

Problem 9.8.16

Repeat Problem 9.7.5 for sensor gain and phase perturbations. Compare your results to those in Problem 9.8.11. (Test scenario 3.)

Problem 9.8.17

Repeat Problem 9.7.6 for sensor gain and phase perturbations. Compare your results to those in Example 9.8.12. (Test scenario 4.)

P9.9 Planar Arrays

Problem 9.9.1

Consider a standard 10×10 rectangular array. There are two equal-power uncorrelated plane waves impinging on the array. Their (θ, ϕ) directions are $(34°, 45°)$ and $(26°, 45°)$, respectively. Simulate the performance of 2-D spectral MUSIC and paired 1-D root MUSIC.

(a) Plot the RMSE (9.338) for each source versus *SNR* for $K = 100$. Discuss your results.

(b) Plot the RMSE (9.338) for each source versus K for an $SNR = 20$ dB.

(c) Repeat part (a) for source locations of $(34°, 0°)$ and $(26°, 0°)$.

Problem 9.9.2

Repeat Problem 9.9.1 for the case in which $ASNR_1 = 100 ASNR_2$.

Problem 9.9.3

Repeat Problem 9.9.1 for the case in which the sources are correlated. Consider $\rho = 0.95$ and $0.95 \exp(j\pi/2)$.

Problem 9.9.4

Consider a standard 10×10 rectangular array. There are three equal-power uncorrelated plane waves impinging on the array. Their (θ, ϕ) directions are $(45°, 50°)$, $(45°, 45°)$,

$(45°, 40°)$, respectively. Repeat parts (a) and (b) of Problem 9.9.1.

Problem 9.9.5

Repeat Example 9.9.1 for 2-D weighted unitary ESPRIT. Compare your results to those in Example 9.9.1.

Problem 9.9.6

Repeat Problem 9.9.1 for 2-D weighted unitary ESPRIT. Compare your results to those in Problem 9.9.1.

Problem 9.9.7

Repeat Problem 9.9.2 for 2-D weighted unitary ESPRIT. Compare your results to those in Problem 9.9.2.

Problem 9.9.8

Repeat Problem 9.9.3 for 2-D weighted unitary ESPRIT. Compare your results to those in Problem 9.9.3.

Problem 9.9.9

Repeat Problem 9.9.4 for 2-D weighted unitary ESPRIT. Compare your results to those in Problem 9.9.4.

Problem Note 9.9.1: The next four problems implement 2-D beamspace ESPRIT using nine beams. The advantage is a significant reduction in computational complexity.

Problem 9.9.10

Repeat Problem 9.9.1 for 2-D beamspace weighted unitary ESPRIT. Compare your results to those in Problem 9.9.1.

Problem 9.9.11

Repeat Problem 9.9.2 for 2-D beamspace weighted unitary ESPRIT. Compare your results to those in Problem 9.9.2.

Problem 9.9.12

Repeat Problem 9.9.3 for 2-D beamspace weighted unitary ESPRIT. Compare your results to those in Problem 9.9.3.

Problem 9.9.13

Repeat Problem 9.9.4 for 2-D beamspace weighted unitary ESPRIT. Compare your results to those in Problem 9.9.4.

Problem Note 9.9.2: The next several problems consider hexagonal arrays. In order to get some comparison with the 10×10 rectangular array, we use the standard 91-element hexagonal array in Example 4.4.1.

Problem 9.9.14

Repeat Problem 9.9.1 for 2-D unitary ESPRIT. Compare your results to those in Problem 9.9.1.

Problem 9.9.15

Repeat Problem 9.9.2 for 2-D unitary ESPRIT. Compare your results to those in Problem 9.9.2.

Problem 9.9.16

Repeat Problem 9.9.3 for 2-D unitary ESPRIT. Compare your results to those in Problem 9.9.3.

Problem 9.9.17

Repeat Problem 9.9.4 for 2-D unitary ESPRIT. Compare your results to those in Problem 9.9.4.

Problem 9.9.18

Design a beamspace unitary ESPRIT algorithm for hexagonal arrays. Apply it to one or more of the models in Problems 9.9.14–9.9.18.

Problem 9.9.19

Read Tewfik and Hong [TH92] and Mathews and Zoltowski [MZ94]. Implement the algorithms for uniform circular array. Discuss your results.

Bibliography

[AGGS98] Y. I. Abramovich, D. A. Gray, A. Y. Gorokhov, and N. K. Spencer. Positive-definite Toeplitz completion in DOA estimation for nonuniform linear antenna arrays—part I: Fully augmentable arrays. *IEEE Trans. Signal Process.*, vol.SP-46, pp. 2458–2471, September 1998.

[AGSG96] Y. I. Abramovich, D. A. Gray, N. K. Spencer, and A. Y. Gorokhov. Ambiguities in direction-of-arrival estimation for nonuniform linear antenna arrays. *Proc. ISSPA*, Gold Coast, Australia, pp. 631–634, 1996.

[AH97] K. Abed-Meraim and Y. Hua. 3-D near-field source localization using second order statistics. *Proc. 31st Asilomar Conf. on Signals, Systems, and Computers*, vol.2, pp. 1307–1311, Pacific Grove, California, November 1997.

[And63] T. W. Anderson. Asymptotic theory for principal component analysis. *Ann. Math. Stat.*, vol.34, pp. 122–148, February 1963.

[And91] S. Anderson. On optimal dimension reduction for sensor array signal processing. Technical Report LiTH-ISY-I-1242, Deptartment of Electrical Engineering, Linköping University, Linköping, Sweden, August 1991.

[ASG99] Y. I. Abramovich, N. K. Spencer, and A. Y. Gorokhov. Positive-definite Toeplitz completion in DOA estimation for nonuniform linear antenna arrays—Part II: Partially augmentable arrays. *IEEE Trans. Signal Process.*, vol.SP-47, pp. 1502–1520, June 1999.

[Bac95] R. Bachl. The forward-backward averaging technique applied to TLS-ESPRIT processing. *IEEE Trans. Signal Process.*, vol.SP-43, pp. 2691–2699, November 1995.

[Bar83] A. J. Barabell. Improving the resolution performance of eigenstructure-based direction finding algorithms. *Proc. ICASSP*, Boston, Massachusetts, pp. 336–339, 1983.

[BCD+84] A. J. Barabell, J. Capon, D. F. DeLong, J. R. Johnson, and K. D. Senne. Performance comparison of superresolution array processing algorithms. Technical Report Project Rep. TST-72, MIT Lincoln Laboratory, Lexington, Massachusetts, May 1984.

[BK79] G. V. Borgiotti and L. J. Kaplan. Superresolution of uncorrelated interference sources by using adaptive array techniques. *IEEE Trans. Antennas Propag.*, vol.AP-27, pp. 842–845, November 1979.

[BK80] G. Bienvenu and L. Kopp. Adaptivity to background noise spatial coherence for high resolution passive methods. *Proc. ICASSP*, vol.1, pp. 307–310, Denver, Colorado, April 1980.

[BK83] G. Bienvenu and L. Kopp. Optimality of high resolution array processing using the eigensystem approach. *IEEE Trans. Acoust., Speech, Signal Process.*, vol.ASSP-31, pp. 1234–1248, October 1983.

[BK84] G. Bienvenu and L. Kopp. Decreasing high resolution method sensitivity by conventional beamformer preprocessing. *Proc. ICASSP*, vol.5, pp. 33.2.1–33.2.4, San Diego, California, April 1984.

[BK91] A. Bassias and M. Kaveh. Coherent signal-subspace processing in a sector. *IEEE Trans. Syst., Man, Cybern.*, vol.SMC-21, September/October 1991.

[BO97] M. Bengtsson and B. Ottersten. Low complexity estimation for distributed sources. Signal Processing Technical Report IR-S3-SB-9721, Royal Institute of Technology, Department of Signals, Sensors, and Systems, October 1997.

[BO00] M. Bengtsson and B. Ottersten. Low-complexity estimators for distributed sources. *IEEE Trans. Signal Process.*, vol.SP-48, pp. 2185–2194, August 2000.

[BO01] M. Bengtsson and B. Ottersten. A generalization of weighted subspace fitting to full-rank models. *IEEE Trans. Signal Process.*, vol.SP-49, pp. 1002–1012, May 2001.

[Böh83] J. F. Böhme. On parametric methods for array processing. *Proc. EUSIPCO-83*, September 1983.

[Bri81] D. R. Brillinger, editor. *Time Series: Data Analysis and Theory.* Holden-Day, San Francisco, California, expanded edition, 1981.

[BS87] C. Byrne and A. Steel. Sector-focused stability for high-resolution array processing. *Proc. ICASSP*, Dallas, Texas, pp. 54.11.1–54.11.4, 1987.

[BS99] O. Besson and P. Stoica. Decoupled estimation of DOA and angular spread for a spatially distributed sources. *Conference Record of the 33rd Asilomar Conf. on Signals, Systems, and Computers*, vol.1, pp. 253–257, 1999.

[BS00] O. Besson and P. Stoica. Decoupled estimation of DOA and angular spread for a spatially distributed sources. *IEEE Trans. Signal Process.*, vol.SP-48, pp. 1872–1882, July 2000.

[Bur67] J. P. Burg. Maximum entropy spectral analysis. *Proc. 37th Annual Meeting of the Society of Exploration Geophysicists*, Oklahoma City, Oklahoma, October 1967.

[Bur75] J. P. Burg. *Maximum Entropy Spectral Analysis.* Ph.D. Thesis, Stanford University, Stanford, California, 1975.

[Cad80] J. A. Cadzow. High performance spectral estimation—a new ARMA method. *IEEE Trans. Acoust., Speech, Signal Process.*, vol.ASSP-28, pp. 524–529, October 1980.

[Cap69] J. Capon. High-resolution frequency-wavenumber spectrum analysis. *Proc. IEEE*, vol.57, pp. 1408–1418, August 1969.

[CES97] M. Clark, L. Elden, and P. Stoica. A computationally efficient implementation of 2-D IQML. *Proc. 31st ASILOMAR Conf. on Signals, Systems and Computers*, Pacific Grove, California, vol.2, pp. 1730–1734, 1997.

[CFF95] H. Cheberre, T. Filleul, and J. J. Fuchs. Near-field sources localization: A model-fitting approach. *Proc. ICASSP*, vol.5, pp. 3555–3558, Detroit, Michigan, May 1995.

[Cla92] M. P. Clark. *Estimation Techniques for Sensor Array Processing.* Ph.D. Thesis, University of Colorado, Boulder, Colorado, 1992.

[CM95] J.-F. Cardoso and E. Moulines. Asymptotic performance analysis of direction-finding algorithms based on fourth-order cumulants. *IEEE Trans. Signal Process.*, vol.SP-43, pp. 214–224, January 1995.

[Cox73] H. Cox. Resolving power and sensitivity to mismatch of optimum array processors. *J. Acoust. Soc. Am.*, vol.54, pp. 771–785, September 1973.

[CS94] M. P. Clark and L. L. Scharf. Two-dimensional modal analysis based on maximum likelihood. *IEEE Trans. Signal Process.*, vol.SP-42, pp. 1443–1452, June 1994.

[CTSD96] C. Chambers, T. C. Tozer, K. C. Sharman, and T. S. Durrani. Temporal and spatial sampling influence on the estimates of superimposed narrowband signals: When less can mean more. *IEEE Trans. Signal Process.*, vol.SP-44, pp. 3085–3098, December 1996.

[CY91] L. Chang and C. Yeh. Resolution threshold for coherent sources using smoothed eigenstructure methods. *Proc. IEE*, vol.138, pp. 470–478, October 1991.

[DJ85] S. R. DeGraaf and D. H. Johnson. Capability of array processing algorithms to estimating source bearings. *IEEE Trans. Acoust., Speech, Signal Process.*, vol.ASSP-33, pp. 1368–1379, December 1985.

[DW92] M. A. Doron and A. J. Weiss. On focusing matrices for wide-band array processing. *IEEE Trans. Signal Process.*, vol.SP-40, p. 1295, June 1992.

[Efr79] B. Efron. Bootstrap methods: Another look at the jackknife. *Ann. Stat.*, vol.7, pp. 1–26, 1979.

[EJS81] J. E. Evans, J. R. Johnson, and D. F. Sun. High resolution angular spectrum estimation techniques for terrain scattering analysis and angle-of-arrival estimation. *Proc. 1st ASSP Workshop on Spectrum Estimation and Modeling*, Hamilton, Ontario, Canada, pp. 134–139, 1981.

[EJS82] J.E. Evans, J.R. Johnson, and D.F. Sun. Application of advanced signal processing techniques to angle of arrival estimation in ATC navigation and surveillance systems. Technical report, M.I.T. Lincoln Laboratory, Lexington, Massachusetts, June 1982.

[Fla00] B. P. Flanagan. *Self Calibration of Antenna Arrays with Large Perturbation Errors.* Ph.D. Thesis, George Mason University, Fairfax, Virginia, 2000.

[FN91] P. Forster and M. Nikias. Bearing estimation in the bispectrum domain. *IEEE Trans. Signal Process.*, vol.SP-39, pp. 1994–2006, September 1991.

[FRB97] J. Fuhl, J.-P. Rossi, and E. Bonek. High-resolution 3-D direction-of-arrival determination for urban mobile radio. *IEEE Trans. Antennas Propag.*, vol.AP-45, p. 672, April 1997.

[Fri90] B. Friedlander. Direction finding with an interpolated array. *Proc. ICASSP*, Albuquerque, New Mexico, vol.5, pp. 2951-2954, April 1990.

[Fri93] B. Friedlander. The interpolated root-MUSIC algorithm for direction finding. *Eur. J. Signal Process.*, vol.30, pp. 15–29, January 1993.

[Fuh94] D. Fuhrmann. Estimation of sensor gain and phase. *IEEE Trans. Signal Process.*, vol.SP-42, pp. 77–87, January 1994.

[Fuc97] J. Fuchs. Extension of the Pisarenko method to sparse linear arrays. *IEEE Trans. Signal Process.*, vol.SP-45, pp. 2413–2421, October 1997.

[FV87] P. Forster and G. Vezzosi. Application of spheroidal sequences to array processing. *Proc. ICASSP*, vol.4, pp. 2268–2271, Dallas, Texas, April 1987.

[FW88] B. Friedlander and A. J. Weiss. Eigenstructure methods for direction finding with sensor gain and phase uncertainties. *Proc. ICASSP*, vol.4, pp. 2681–2684, New York, New York, April 1988.

[FW90] B. Friedlander and A. J. Weiss. Eigenstructure methods for direction finding with sensor gain and phase uncertainties. *Circuits, Syst. Signal Process.*, vol.9, pp. 271–300, 1990.

[FW92] B. Friedlander and A. J. Weiss. Direction finding using spatial smoothing with interpolated arrays. *IEEE Trans. Aerospace Electron. Syst.*, vol.AES-28, pp. 574–587, April 1992.

[FW93a] B. Friedlander and A. J. Weiss. Direction finding for wideband signals using an interpolated array. *IEEE Trans. Signal Process.*, pp. 1618–1634, April 1993.

[FW93b] B. Friedlander and A. J. Weiss. Performance of direction finding systems with sensor gain and phase uncertainties. *Comput., Syst. Signal Proces.*, vol.12, pp. 3–35, 1993

[FW94] B. Friedlander and A. J. Weiss. Effects of model errors on waveform estimation using the MUSIC algorithm. *IEEE Trans. Signal Process.*, vol.SP-42, pp. 147–155, January 1994.

[Gab80] W. F. Gabriel. Spectral analysis and adaptive array superresolution techniques. *Proc. IEEE*, vol.68, pp. 654–666, June 1980.

[Gab81] W. F. Gabriel. Adaptive superresolution of coherent RF spatial sources. *Proc. 1st ASSP Workshop on Spectrum Estimation and Modeling*, Hamilton, Ontario, Canada, August 1981.

[GB97] A. B. Gershman and J. F. Böhme. Improved DOA estimation via pseudorandom resampling of spatial spectrum. *IEEE Signal Process. Lett.*, vol.4, pp. 54–57, February 1997.

[GB98] A. B. Gershman and J. F. Böhme. A pseudo-noise approach to direction finding. *Signal Process.*, vol.71, pp. 1–13, November 1998.

[Ger98] A. B. Gershman. Pseudo-randomly generated estimator banks: A new tool for improving the threshold performance of direction finding. *IEEE Trans. Signal Process.*, vol.SP-46, pp. 1351–1364, May 1998.

[GH99] A. B. Gershman and M. Haardt. Improving the performance of unitary ESPRIT via pseudo-noise resampling. *IEEE Trans. Signal Process.*, vol.SP-47, pp. 2305–2308, August 1999.

[GS99] A. B. Gershman and P. Stoica. On unitary and forward-backward MODE. *Digital Signal Process.*, vol.9, pp. 67–75, April 1999.

[Gup65] R. P. Gupta. Asymptotic theory for principal component analysis in the complex case. *J. Indian Stat. Assoc.*, vol.3, pp. 97–106, 1965.

[GVL80] G. H. Golub and C. F. Van Loan. An analysis of the total least squares problem. *SIAM J. Numer. Anal.*, vol.17, pp. 883–893, 1980.

[GVL83] G. H. Golub and C. F. Van Loan. *Matrix Computations.* The Johns Hopkins University Press, Baltimore, Maryland, 1983.

[GW93] M. Gavish and A. J. Weiss. Performance analysis of the VIA-ESPRIT algorithm. *Proc. IEE (Radar Signal Process.)*, vol.140, pp. 123–128, April 1993.

[GZK96] J. A. Gansman, M. D. Zoltowski, and J. V. Krogmeier. Multidimensional multirate DOA estimation in beamspace. *IEEE Trans. Signal Process.*, vol.SP-44, pp. 2780–2792, November 1996.

[HAH94] M. Haardt and M. E. Ali-Hackl. Unitary ESPRIT: How to exploit additional information inherent in the rotational invariance structure. *Proc. ICASSP*, Adelaide, Australia, vol.IV, pp. 229–232, April 1994.

[Hah95] W. R. Hahn. Optimum signal processing for passive sonar range and bearing estimation. *J. Acoust. Soc. Am.*, vol.58, no.1, pp. 201–207, July 1975.

[Hay85] S. Haykin. *Array Signal Processing.* Prentice-Hall, Englewood Cliffs, New Jersey, 1985.

[Hay91] S. Haykin, editor. *Advances in Spectrum Analysis and Array Processing*, vol.2, Prentice-Hall, Englewood Cliffs, New Jersey, 1991.

[Hay96] S. Haykin. *Adaptive Filter Theory.* Prentice-Hall, Upper Saddle River, New Jersey, 3rd edition, 1996.

[HB91] Y. D. Huang and M. Barkat. Near-field multiple source localization by passive sensor array. *IEEE Trans. Antennas Propag.*, vol.AP-69, pp. 968–975, July 1991.

[HB95] R. M. Hamza and K. M. Buckley. An analysis of weighted eigenspace methods in the presence of sensor errors. *IEEE Trans. Signal Process.*, vol.SP-43, pp. 1140–1150, May 1995.

[HCS96] M. Haardt, R. N. Challa, and S. Shamsunder. Improved bearing and range estimation via high-order subspace based unitary ESPRIT. *Proc. 30th Asilomar Conf. on Signals, Systems, and Computers*, vol.1, pp. 380–384, Pacific Grove, California, November 1996.

[HCW96] H. S. Hung, S. H. Chang, and C. H. Wu. 3-D MUSIC with polynomial rooting for near-field source localization. *Proc. ICASSP*, vol.6, pp. 3065–3068, Atlanta, Georgia, May 1996.

[HF96] G. F. Hatke and K. W. Forsythe. A class of polynomial rooting algorithms for joint azimuth/elevation estimation using multidimensional arrays. *Proc. 30th Asilomar Conf. on Signals, Systems, and Computers*, Pacific Grove, California, 1996.

[HJ96] A. J. Haug and G. M. Jacyna. Comparison of the theoretical performance bounds for two wavefront curvature ranging techniques. *Proc. ICASSP*, vol.4, pp. 2445–2448, Atlanta, Georgia, May 1996.

[HK87] H. Hung and M. Kaveh. On the statistical sufficiency of the coherently averaged covariance matrix for the estimation of the parameters of wideband sources. *Proc. ICASSP*, vol.1, pp. 33–36, Dallas, Texas, April 1987.

[HK88] H. Hung and M. Kaveh. Focusing matrices for coherent signal subspace processing. *IEEE Trans. Acoust., Speech, Signal Process.*, vol.ASSP-36, pp. 1272–1281, August 1988.

[HN95] M. Haardt and J. A. Nossek. Unitary ESPRIT: How to obtain increased estimation accuracy with a reduced computational burden. *IEEE Trans. Signal Process.*, vol.SP-43, pp. 1232–1242, May 1995.

[HR80] S. Haykin and J. Reilly. Mixed autoregressive-moving average modeling of respose of a linear array antenna to incident plane waves. *Proc. IEEE*, vol.68, pp. 622–630, 1980.

[Hua91] Y. Hua. Estimating two-dimensional frequencies by matrix enhancement and matrix pencil. *Proc. ICASSP*, vol.5, pp. 3073–3076, Toronto, Canada, April 1991.

[Hua92] Y. Hua. Estimating 2-D frequencies by matrix enhancement and matrix pencil. *IEEE Trans. Signal Process.*, vol.40, pp. 2257–2280, September 1992.

[Hun94] E. Hung. A critical study of a self-calibrating direction-finding method for arrays. *IEEE Trans. Signal Process.*, vol.SP-42, pp. 471–474, February 1994.

[HW94] S. Hwang and D. Williams. A constrained total least squares approach for sensor position calibration and direction finding. *Proc. IEEE National Radar Conference*, pp. 155–159, 1994.

[HY91] K. C. Huarng and C. C. Yeh. A unitary transformation method for angle of arrival estimation. *IEEE Trans. Acoust., Speech, Signal Process.*, vol.ASP-39, pp. 975–977, April 1991.

[HZ86] F. Haber and M. D. Zoltowski. Spatial spectrum estimation in a coherent signal environment using an array in motion. *IEEE Trans. Antennas Propag.*, vol.AP-34, pp. 301–310, March 1986.

[HZMN95] M. Haardt, M. D. Zoltowski, C. P. Mathews, and J. A. Nossek. 2-D unitary ESPRIT for efficient 2-D parameter estimation. *Proc. ICASSP*, vol.4, pp. 2096–2099, Detroit, Michigan, May 1995.

[Jän92] T. P. Jäntti. The influence of extended sources on the theoretical performance of the MUSIC and ESPRIT methods: Narrow-band sources. *Proc. ICASSP*, vol.II, pp. II-429–II-432, San Francisco, California, March 1992,

[JD82] D. H. Johnson and S. R. Degraaf. Improving the resolution of bearing in passive sonar arrays by eigenvalue analysis. *IEEE Trans. Acoust., Speech, Signal Process.*, vol.ASSP-30, pp. 638–647, April 1982.

[JF85] D. J. Jeffries and D. R. Farrier. Asymptotic results for eigenvector methods. *Proc. IEE*, pt. F, vol.132, pp. 589–594, June 1985.

[JSO98] M. Jansson, A. L. Swindlehurst, and B. Ottersten. Weighted subspace fitting for general array error models. *IEEE Trans. Signal Process.*, vol.SP-46, pp. 2484–2498, September 1998.

[Kay83] S. M. Kay. Recursive maximum likelihood estimation of autoregressive processes. *IEEE Trans. Acoust., Speech, Signal Process.*, vol.ASSP-31, pp. 56–65, February 1983.

[Kay88] S. M. Kay. *Modern Spectral Estimation: Theory and Application.* Prentice-Hall, Englewood Cliffs, New Jersey, 1988.

[KB86] M. Kaveh and A. J. Barabell. The statistical performance of the MUSIC and the Minimum-Norm algorithms in resolving plane waves in noise. *IEEE Trans. Acoust., Speech, Signal Process.*, vol.ASSP-34, pp. 331–341, April 1986.

[KB90] M. Kaveh and A. Bassias. Threshold extension based on a new paradigm for MUSIC-type estimation. *Proc. ICASSP*, Albuquerque, New Mexico, vol.5, pp. 2535–2538, 1990.

[Kes86] S. B. Kesler, editor. *Modern Spectrum Analysis, II.* IEEE Press, New York, 1986.

[KFP92] H. Krim, P. Forster, and G. Proakis. Operator approach to performance analysis of root-MUSIC and root-Min-Norm. *IEEE Trans. Signal Process.*, vol.SP-40, pp. 1687–1696, July 1992.

[KLF86] S. Y. Kung, S. C. Lo, and R. Foka. A Toeplitz approximation approach to coherent source direction finding. *Proc. ICASSP*, Tokyo, Japan, 1986.

[KT83] R. Kumaresan and D. W. Tufts. Estimating the angles of arrival of multiple plane waves. *IEEE Trans. Aerospace Electron. Syst.*, vol.AES-19, pp. 134–139, January 1983.

[Kum83] R. Kumaresan. On the zeros of the linear prediction-error filter for deterministic signals. *IEEE Trans. Aerospace Electron. Syst.*, vol.AES-19, January 1983.

[Kur89] A. R. Kuruc. Lower bounds on multiple-source direction finding in the presence of direction-dependent antenna-array-calibration errors. Technical Report, M.I.T. Lincoln Laboratory, Lexington, Massachusetts, 1989.

[KV96] H. Krim and M. Viberg. Two decades of array signal processing research. *IEEE Signal Process. Mag.*, pp. 67–94, 1996.

[KZ95] G. M. Kautz and M. D. Zoltowski. Performance analysis of MUSIC employing conjugate symmetric beamformers. *IEEE Trans. Signal Process.*, vol.SP-43, pp. 737–748, March 1995.

[KZ96] G. M. Kautz and M. D. Zoltowski. Beamspace DOA estimation featuring multirate eigenvector processing. *IEEE Trans. Signal Process.*, vol.SP-44, pp. 1765–1778, July 1996.

[Lac71] R. T. Lacoss. Data adaptive spectral analysis methods. *Geophysics*, vol.36, p. 661, 1971.

[LCSL97] Y. U. Lee, J. Choi, I. Song, and S. R. Lee. Distributed source modeling and direction-of-arrival estimation techniques. *IEEE Trans. Signal Process.*, vol.SP-45, pp. 960–969, April 1997.

[LDD94] D. A. Linebarger, R. D. DeGroat, and E. M. Dowling. Efficient direction-finding methods employing forward/backward averaging. *IEEE Trans. Signal Process.*, vol.SP-42, pp. 2136–2145, August 1994.

[LeC95] J. P. Le Cadre. Performance analysis of wavefront curvature methods for range estimation of a moving source. *IEEE Trans. Aerospace Electron. Syst.*, vol.AES-31, pp. 1082–1103, July 1995.

[LF94] M. Lang and B. C. Frenzel. Polynomial root finding. *IEEE Signal Process. Lett.*, vol.SPL-1, pp. 141–143, October 1994.

[Li92a] J. Li. Improving angular resolution for spatial smoothing techniques. *IEEE Trans. Signal Process.*, vol.SP-40, pp. 3078–3081, December 1992.

[Li92b] J. Li. Improving ESPRIT via beamforming. *IEEE Trans. Aerospace Electron. Syst.*, vol.AES-28, pp. 520–527, April 1992.

[LVT91] F. Li, R. J. Vaccaro, and D. W. Tufts. Performance analysis of the state-space realization (TAM) and ESPRIT algorithms for DOA estimation. *IEEE Trans. Antennas Propag.*, vol.AP-39, pp. 418–423, March 1991.

[LW88] H. B. Lee and M. S. Wengrovitz. Improved high-resolution direction-finding through use of homogeneous constraints. *Proc. ASSP Workshop on Spectrum Estimation and Modeling*, Minneapolis, Minnesota, pp. 152–157, August 1988.

[LW90] H. B. Lee and M. Wengrovitz. Resolution threshold of beamspace MUSIC for two closely spaced emitters. *IEEE Trans. Acoust., Speech, Signal Process.*, vol.ASSP-38, pp. 1545–1559, September 1990.

[LW91] H. B. Lee and M. S. Wengrovitz. Statistical characterization of the MUSIC null spectrum. *IEEE Trans. Acoust., Speech, Signal Process.*, vol.ASSP-39, pp. 1333–1347, June 1991.

[LYL01] C. M. Lee, K. S. Yoon, and K. K. Lee. Efficient algorithm for localizing 3-D narrowband multiple sources. *IEE Proc. Radar, Sonar, Navig.*, vol.148, pp. 23–26, February 2001.

[Mak75] J. Makhoul. Linear prediction: A tutorial review. *Proc. IEEE*, vol.63, pp. 561–580, April 1975.

[Mar77] L. Marple. Resolution of conventional Fourier, autoregressive, and special ARMA methods of spectrum analysis. *Proc. ICASSP*, Hartford, Connecticut, 1977.

[Mar87] S. L. Marple, Jr . *Digital Spectral Analysis.* Prentice-Hall, Englewood Cliffs, New Jersey, 1987.

[May87] J. T. Mayhan. Spatial spectral estimation using multiple beam antennas. *IEEE Trans. Antennas Propag.*, vol.AP-35, pp. 897–905, August 1987.

[May88] S. Mayrague. On the common structure of several well-known methods for harmonic analysis and direction-of-arrival estimation induced by a new version of ESPRIT. *Proc. 4th ASSP Workshop on Spectrum Estimation and Modeling*, Minneapolis, Minnesota, pp. 307–311, August 1988.

[MRG00] J. Messer, H. Raich, and R. Goldberg. Bearing estimation for a distributed source: modeling, inherent accuracy limitations and algorithms. *IEEE Trans. Signal Process.*, vol.SP-48, pp. 429–441, February 2000.

[MRP94] F. McCarthy, R. Ridgway, and A. Paulraj. Fast techniques for sensor array calibration. *Proc. 28th Asilomar Conf. on Signals, Systems, and Computers*, vol.1, pp. 688–693, 1994.

[MSW96] Y. Meng, P. Stoica, and K. Wong. Estimation of the directions of arrival of spatially dispersed signals in arrays processing. *Proc. Inst. Elect. Eng., Radar, Sonar, Navigat.*, vol.143, no.1, pp. 1–9, February 1996.

[MZ91] Z. J. Mou and H. M. Zhang. Fast SVD for angle of arrival estimation. In E. F. Deprettere and A.-J. van der Veen, editors, *Algorithms and Parallel VLSI Architectures, Vol. B—Processings*, Elsevier, Amsterdam, The Netherlands, 1991.

[MZ94] C. P. Mathews and M. D. Zoltowski. Eigenstructure techniques for 2-D angle estimation with uniform circular arrays. *IEEE Trans. Signal Process.*, vol.SP-42, pp. 2395–2407, September 1994.

[Ng95] A. Ng. Direction-of-arrival estimation in the presence of wavelength, gain, and phase errors. *IEEE Trans. Signal Process.*, vol.SP-43, pp. 225–232, January 1995.

[Nut76] A. Nuttall. Spectral analysis of a univariate process with bad data points, via maximum entropy and linear predictive techniques. Technical Report 5303, Naval Underwater Systems Center, New London, Connecticut, March 1976.

[OK90] B. Ottersten and T. Kailath. Direction-of-arrival estimation for wide-band signals using the ESPRIT algorithm. *IEEE Trans. Acoust., Speech, Signal Process.*, vol.ASSP-38, pp. 317–327, February 1990.

[OVK91] B. Ottersten, M. Viberg, and T. Kailath. Performance analysis of the total least squares ESPRIT algorithm. *IEEE Trans. Signal Process.*, vol.SP-39, pp. 1122–1135, May 1991.

[Ows85] N. L. Owsley. Signal subspace based minimum-variance spatial array processing. *Proc. 19th Asilomar Conf. Signal, Systems, and Computers*, pp. 94–97, November 1985.

[PF88] B. Porat and B. Friedlander. Analysis of the asymptotic relative efficiency of the MUSIC algorithm. *IEEE Trans. Acoust., Speech, Signal Process.*, vol.ASSP-36, pp. 532–544, April 1988.

[PF91] B. Porat and B. Friedlander. Direction finding algorithms based on high-order statistics. *IEEE Trans. Signal Process.*, vol.SP-39, pp. 2016–2025, January 1991.

[PGH00] M. Pesavento, A. B. Gershman, and M. Haardt. Unitary root-MUSIC with a real-valued eigendecomposition: A theoretical and experimental perfomance study. *IEEE Trans. Signal Process.*, vol.48, pp. 1306–1314, May 2000.

[Pil89] S. U. Pillai. *Array Signal Processing.* Springer-Verlag, New York, 1989.

[Pis72] V. F. Pisarenko. On the estimation of spectra by means of non-linear functions of the covariance matrix *Geophys. J. R. Astron. Soc.*, vol.28, pp. 511–531, 1972.

[Pis73] V. F. Pisarenko. The retrieval of harmonics from a covariance function. *Geophys. J. R. Astron. Soc.*, pp. 347–366, 1973.

[PK89a] S. U. Pillai and B. H. Kwon. Forward/backward spatial smoothing techniques for coherent signal identification. *IEEE Trans. Acoust., Speech, Signal Process.*, vol.ASSP-37, pp. 8–15, January 1989.

[PK89b] S. U. Pillai and B. H. Kwon. Performance analysis of MUSIC-type high resolution estimators for direction finding in correlated and coherent scenes. *IEEE Trans. Acoust., Speech, Signal Process.*, vol.ASSP-37, pp. 1176–1189, August 1989.

[PK92] J. W. Pierre and M. Kaveh. Wideband sensor array processing using a laboratory array testbed. *Proc. 35th Midwest Symp. on Circuits and Systems*, vol.2, pp. 945–948, August 1992.

[Pra82] S. Prasad. On an index for array optimization and the discrete prolate spheroidal functions. *IEEE Trans. Antennas Propag.*, vol.AP-30, pp. 1021–1023, September 1982.

[PRK86] A. Paulraj, R. Roy, and T. Kailath. A subspace rotation approach to signal parameter estimation. *Proc. IEEE*, vol.74, no.7, pp. 1044–1045, July 1986.

[PRK87] A. Paulraj, V. U. Reddy, and T. Kailath. Analysis of signal cancellation due to multipath in optimum beamformers for moving arrays. *IEEE J. Oceanogr. Eng.*, vol.OE-12, pp. 163–172, January 1987.

[PRLN92] J. G. Proakis, C. M. Rader, F. Ling, and C. L. Nikias. *Advanced Digital Signal Processing.* Macmillan, New York, 1992.

[Rat98] T. Ratnarajah. Mitigating the effects of array uncertainties on the performance of the Music algorithm. *Proc.* 9th *IEEE SP Workshop on Statistical Signal and Array Processing*, pp. 236–239, Corfu, Greece, June 1998.

[Red79] S. S. Reddi. Multiple source location: A digital approach. *IEEE Trans. Aerospace Electron. Syst.*, vol.AES-15, pp. 95–105, January 1979.

[RH89a] B. D. Rao and K. V. S. Hari. Performance analysis of ESPRIT and TAM in determining the direction of arrival of plane waves in noise. *IEEE Trans. Acoust., Speech, Signal Process.*, vol.ASSP-37, pp. 1990–1995, December 1989.

[RH89b] B. D. Rao and K. V. S. Hari. Performance analysis of root-MUSIC. *IEEE Trans. Acoust., Speech, Signal Process.*, vol.ASSP-37, pp.1939–1949, December 1989.

[RH90] B. D. Rao and K. V. S. Hari. Effect of spatial smoothing on the performance of MUSIC and the Minimum Norm method. *Proc. IEE*, Pt. F, vol.137, pp. 449–458, December 1990.

[RH93] B. D. Rao and K. V. S. Hari. Weighted subspace methods and spatial smoothing: Analysis and comparison. *IEEE Trans. Signal Process.*, vol.SP-41, pp. 788–803, February 1993.

[RK84] D. V. B. Rao and S. Y. Kung. A state space approach for the 2-D harmonic retrieval problem. *Proc. ICASSP*, vol.4, pp. 4.10.1–4.10.4, San Diego, California, April 1984.

[RK87] R. Roy and T. Kailath. ESPRIT—Estimation of signal parameters via rotational invariance techniques. In E. F. Deprettere, editor, *Singular Value Decomposition and Signal Processing* North-Holland, Amsterdam, The Netherlands, September 1987.

[RK89] R. Roy and T. Kailath. ESPRIT: Estimation of signal parameters via rotational invariance techniques. *IEEE Trans. Acoust., Speech, Signal Process.*, vol.ASSP-37, pp. 984–995, July 1989.

[RMS88] Y. Rockah, H. Messer, and P. M. Schultheiss. Localization performance of arrays subject to phase errors. *IEEE Trans. Aerospace Electron. Syst.*, vol.AES-24, pp. 402–409, July 1988.

[RMZ99] J. Ramos, C. P. Mathews, and M. D. Zoltowski. FCA-ESPRIT: A closed-form 2-D angle estimation algorithm for filled circular arrays with arbitrary sampling lattices. *IEEE Trans. Signal Process.*, vol.SP-47, pp. 213–217, January 1999.

[Roy87] R. Roy. *ESPRIT: Estimation of Signal Parameters via Rotational Invariance Techniques.* Ph.D. Dissertation, Stanford University, Stanford, California, 1987.

[RPK86] R. Roy, A. Paulraj, and T. Kailath. ESPRIT—A subspace rotation approach to estimation of parameters of cisoids in noise. *IEEE Trans. Acoust., Speech, Signal Process.*, vol.ASSP-34, pp. 1340–1342, October 1986.

[RR92] K. J. Raghunath and V. U. Reddy. A note on spatially weighted subarray covariance averaging schemes. *IEEE Trans. Antennas Propag.*, vol.AP-40, pp. 720–723, June 1992.

[RS87] Y. Rockah and P. M. Schultheiss. Array shape calibration using sources in unknown locations—Part I: Far-field sources. *IEEE Trans. Acoust., Speech, Signal Process.*, vol.ASSP-35, pp. 286–299, March 1987.

[Sac92] J. J. Sacchini. *Development of 2-D Parametric Radar Signal Modeling and Estimation Techniques with Application to Target Identification.* Ph.D. Thesis, Ohio State University, Columbus, Ohio, 1992.

[SB00] P. Stoica and O. Besson. Decoupled estimation of DOA and angular spread of a spatially distributed source. *IEEE Trans. Signal Process.*, vol.SP-48, pp. 1872–1882, July 2000.

[Sch79] R. O. Schmidt. Multiple emitter location and signal parameter estimation. *Proc. RADC Spectrum Estimation Workshop*, Griffiths AFB, Rome, New York, pp. 243–258, 1979.

[Sch81] R. O. Schmidt. *Multiple Emitter Location and Signal Parameter Estimation.* PhD Dissertation, Stanford University, Stanford, California, 1981.

[Sch83] R. O. Schmidt. New mathematical tools in direction finding and spectral analysis. *Proc. SPIE 27th Ann. Symp.*, San Diego, California, August 1983.

[Sch86] R. O. Schmidt. Multiple emitter location and signal parameter estimation. *IEEE Trans. Antennas Propag.*, vol.AP-34, pp. 276–280, March 1986.

[SDWK83] K. C. Sharman, T. S. Durrani, M. Wax, and T. Kailath. Asymptotic performance of eigenstructure methods. *Proc. ICASSP*, pp. 45.5.1–45.5.4, April 1983.

[SDWK84] K. C. Sharman, T. S. Durrani, M. Wax, and T. Kailath. Asymptotic performance of eigenstructure spectral analysis methods. *Proc. ICASSP*, San Diego, California, pp. 45.5.1–45.5.4, March 1984.

[SF86] R. O. Schmidt and R. E. Frank. Multiple source DF signal processing: An experimental system. *IEEE Trans. Antennas Propag.*, vol.-34, pp. 276–280, March 1986.

[SJ97] P. Stoica and M. Jansson. On forward-backward MODE for array signal processing. *Digital Signal Process.*, vol.7, pp. 239–252, October 1997.

[SK85] T.-J. Shan and T. Kailath. Adaptive beamforming for coherent signals and interference. *IEEE Trans. Acoust., Speech, Signal Process.*, vol.ASSP-33, pp. 527–534, June 1985.

[SK88] A. Swindlehurst and T. Kailath. Passive direction-of-arrival and range estimation for near-field sources. *4th ASSP Workshop on Spectrum Estimation and Modeling*, pp. 123–128, August 1988.

[SK89] A. Swindlehurst and T. Kailath. 2-D parameter estimation using arrays with multidimensional invariance structure. *Proc. 23nd Asilomar Conf. Signals, Systems, and Computers*, Pacific Grove, California, pp. 950–954, November 1989.

[SK92] A. Swindlehurst and T. Kailath. A performance analysis of subspace-based methods in the presence of model errors, Part I: The MUSIC algorithm. *IEEE Trans. Signal Process.*, vol.SP-40, pp. 1758–1774, July 1992.

[SK93a] A. Swindlehurst and T. Kailath. Azimuth/elevation direction finding using regular array geometries. *IEEE Trans. Aerospace Electron. Syst.*, vol.AES-29, pp. 145–156, January 1993.

[SK93b] A. Swindlehurst and T. Kailath. A performance analysis of subspace-based methods in the presence of model errors, Part II: Multidimensional algorithms. *IEEE Trans. Signal Process.*, vol.SP-41, pp. 2882–2890, September 1993.

[Sle78] D. Slepian. Prolate spheroidal wave functions, Fourier analysis, and uncertainty, V: The discrete case. *Bell Syst. Tech. J.*, vol.57, pp. 1371–1430, May-June 1978.

[SM97] P. Stoica and R. Moses. Introduction to Spectral Analysis. Prentice-Hall, Englewood Cliffs, New Jersey, 1997.

[SMW96] P. Stoica, Y. Meng, and K. W. Wong. Estimation of the directions of arrival of spatially dispersed signals in array processing. *IEE Proc. Radar, Sonar Navigat.*, vol.143, no.1, pp. 1–9, February 1996.

[SN89] P. Stoica and A. Nehorai. MUSIC, maximum likelihood, and Cramér-Rao bound. *IEEE Trans. Acoust., Speech, Signal Process.*, vol.ASSP-37, pp. 720–741, May 1989.

[SN90] P. Stoica and A. Nehorai. MUSIC, maximum likelihood, and Cramér-Rao bound: Further results and comparisons. *IEEE Trans. Acoust., Speech, Signal Process.*, vol.38, pp. 2140–2150, December 1990.

[SN91a] P. Stoica and A. Nehorai. Comparative performance study of element-space and beam-space MUSIC estimators. *Circuits, Syst. Signal Process.*, vol.10, pp. 285–292, November 1991.

[SN91b] P. Stoica and A. Nehorai. Performance comparison of subspace rotation and MUSIC methods for direction estimation. *IEEE Trans. Signal Process.*, vol.SP-39, pp. 446–453, February 1991.

[SN94] D. Starer and N. Nehorai. Passive localization of near-field sources by path following. *IEEE Trans. Signal Process.*, vol.SP-42, pp. 677–680, March 1994.

[SN95] M. Shao and C. L. Nikias. *Signal Processing With Alpha-Stable Distributions and Applications*. Wiley, New York, 1995.

[SOR98] P. Stoica, B. Ottersten, and R. Roy. Covariance matching estimation techniques for array signal processing applications. *Digital Signal Processing*, pp. 185–210, August 1998.

[SPK87] T.-J. Shan, A. Paulraj, and T. Kailath. On smoothed rank profile tests in eigenstructure methods for directions-of-arrival estimation. *IEEE Trans. Acoust., Speech, Signal Process.*, vol.ASSP-35, pp. 1377–1385, October 1987.

[SS97] A. L. Swindlehurst and P. Stoica. Radar array signal processing with antenna arrays via maximum likelihood. *The 31st Asilomar Conf. on Signals, Systems, and Computers*, Pacific Grove, California, vol.2, pp. 1219–1223, 1997.

[SS98] A. L. Swindlehurst and P. Stoica. Maximum likelihood methods in radar array signal processing. *Proc. IEEE*, vol.86, no.2, pp. 421–441, 1998

[STHL94] V. C. Soon, L. Tong, Y. F. Huang, and R. Liu. A subspace method for estimating sensor gains and phases. *IEEE Trans. Signal Process.*, vol.SP-42, pp. 973–976, April 1994.

[Swi93] D. N. Swingler. Simple approximations to the Cramer-Rao lower bound on directions of arrival for closely spaced sources. *IEEE Trans. Signal Process.*, vol.SP-41, pp. 1668–1672, April 1993.

[SV95] P. Stoica and M. Viberg. Weighted LS and TLS approaches yield asymptotically equivalent results. *Signal Process.*, vol.45, pp. 255–259, 1995.

[SVB01] S. Shahbazpanahi, S. Valaee, and M. H. Bastani. Distributed source localization using ESPRIT algorithm. *IEEE Trans. Signal Process.*, vol.SP-49, pp. 2169–2178, October 2001.

[SWK85] T-J. Shan, M. Wax, and T. Kailath. On spatial smoothing for direction-of-arrival estimation of coherent signals. *IEEE Trans. on Acoust., Speech, Signal Process.*, vol.ASSP-33, pp. 806–811, August 1985.

[TFG95] C. Y. Tseng, D. D. Feldman, and L. J. Griffiths. Robust high-resolution direction of arrival estimation via signal eigenvector domain. *IEEE J. Oceanogr. Eng.*, vol.OE-18, pp. 491–499, October 1993.

[TGM96] J. S. Thompson, P. M. Grant, and B. Mulgrew. Performance of spatial smoothing algorithm for correlated sources. *IEEE Trans. Signal Process.*, vol.SP-44, pp. 1040–1046, April 1996.

[TH92] A. H. Tewfik and W. Hong. On the application of uniform linear array bearing estimation techniques to uniform circular arrays. *IEEE Trans. Signal Process.*, vol.SP-40, pp. 1008–1011, April 1992.

[TK82] D. W. Tufts and R. Kumaresan. Estimation of frequencies of multiple sinusoids: Making linear prediction perform like maximum likelihood. *Proc. IEEE*, vol.70, pp. 975–989, September 1982.

[TK87] K. Takao and N. Kikuma. An adaptive array utilising an adaptive spatial averaging technique for multipath environment. *IEEE Trans. Antennas Propag.*, vol.AP-35, pp. 1389–1396, December 1987.

[TN96] P. Tsakalides and C. L. Nikias. The robust covariance-based MUSIC (ROC-MUSIC) algorithm for bearing estimation in impulsive noise environments. *IEEE Trans. Antennas Propag.*, vol.AP-44, pp.1623–1633, July 1996.

[TO96a] T. Trump and B. Ottersten. Estimation of nominal direction of arrival and angular spread using an array of sensors. *Signal Process.*, vol.50, no.1-2, pp. 57–69, April 1996.

[TO96b] T. Trump and B. Ottersten. Estimation of nominal direction of arrival and angular spread using an array of sensors. Signal Processing Technical Report IR-S3-SB-9607, Royal Institute of Technology, Department of Signals, Sensors, and Systems, April 1996.

[TPG93] S. Talwar, A. Paulraj, and G. H. Golub. A robust numerical approach for array calibration. *Proc. ICASSP*, vol.4, pp. 316–319, Minneapolis, Minnesota, April 1993.

[UC76] T. J. Ulrych and R. W. Clayton. Time series modelling and maximum entropy. *Phys. Earth Planet. Inter.*, vol.12, pp. 188–200, August 1976.

[VB95] C. Vaidyanathan and K. M. Buckley. Performance analysis of the MVDR spatial spectrum estimator. *IEEE Trans. Signal Process.*, vol.SP-43, pp. 1427–1436, June 1995.

[VCK95] S. Valaee, B. Champagne, and P. Kabal. Parametric localization of distributed sources. *IEEE Trans. Signal Process.*, vol.SP-43, pp. 2144–2153, September 1995.

[VDB71] A. Van Den Bos. Alternative interpretation of maximum entropy spectral analysis. *IEEE Trans. Inf. Theory*, vol.17, no.7, pp. 493–494, July 1971.

[vdVOD92] A. J. van der Veen, P. B. Ober, and E. D. Deprettere. Azimuth and elevation computation in high resolution DOA estimation. *IEEE Trans. Signal Process.*, vol.SP-40, pp. 1828–1832, July 1992.

[VO91] M. Viberg and B. Ottersten. Sensor array processing based on sub-space fitting. *IEEE Trans. Signal Process.*, vol.SP-39, pp. 1110–1121, May 1991.

[VS94] M. Viberg and A. Swindlehurst. Analysis of the combined effects of finite samples and model errors on array processing performance. *IEEE Trans. Signal Process.*, vol.SP-42, pp. 3073–3083, November 1994.

[VVW88] B. Van Veen and B. Williams. Structured covariance matrices and dimensional reduction in array processing. *Proc. 4th ASSP Workshop on Spectrum Estimation and Modeling*, Minneapolis, Minnesota, pp. 168–171, August 1988.

[WF91a] A. J. Weiss and B. Friedlander. Array shape calibration using eigenstructure methods. *Signal Process.*, vol.22, pp. 251–258, Elsevier Science Publishers, New York, 1991.

[WF91b] A. J. Weiss and B. Friedlander. Self-calibration for high-resolution array processing. In S. Haykin, editor, *Advances in Spectrum and Array Processing*, Vol. II, chapter 10, Prentice-Hall, Englewood Cliffs, New Jersey, 1991.

[WF93a] A. J. Weiss and B. Friedlander. Range and bearing estimation using polynomial rooting. *IEEE J. Oceanogr. Eng.*, vol.OE-18, pp. 130–137, April 1993.

[WF93b] A. J. Weiss and B. Friedlander. Performance analysis of spatial smoothing with interpolated array. *IEEE Trans. Signal Process..*, vol.41, pp. 1881–1892, May 1993.

[WF94a] A. J. Weiss and B. Friedlander. Effects of modeling errors on the resolution threshold of the MUSIC algorithm. *IEEE Trans. Signal Process.*, vol.SP-42, pp. 1519–1526, June 1994.

[WF94b] A. J. Weiss and B. Friedlander. Preprocessing for direction finding with minimal variance degradation. *IEEE Trans. Signal Process.*, vol.SP-42, pp. 1478–1484, June 1994.

[WFS95] A. J. Weiss, B. Friedlander, and P. Stoica. Direction-of-arrival estimation using MODE with interpolated array. *IEEE Trans. Signal Process.*., vol.SP-43, pp. 296–300, May 1995.

[Wil65] J. H. Wilkinson. *The Algebraic Eigenvalue Problem.* Oxford University Press, New York, 1965.

[Win87] J. H. Winters. On the capacity of radio communication systems with diversity in a Rayleigh fading environment. *IEEE Sel. Areas Commun.*, vol.SAC-5, pp. 871–878, June 1987.

[WK85] H. Wang and M. Kaveh. Coherent signal subspace processing for the detection and estimation of angles of arrival of multiple wideband sources. *IEEE Trans. Acoust., Speech Signal Process.*, vol.33, no.8, pp. 823–831, August 1985.

[WPMS88] R. T. Williams, S. Prasad, A. K. Mahalanabis, and L. H. Sibul. An improved spatial smoothing technique for bearing estimation in a multipath environment. *IEEE Trans. Acoust., Speech, Signal Process.*, vol.ASSP-36, pp. 425–431, April 1988.

[WRM94] M. Wylie, S. Roy, and H. Messer. Joint DOA estimation and phase calibration of linear equispaced (LES) arrays. *IEEE Trans. Signal Process.*, vol.SP-42, pp. 3449–3459, December 1994.

[WWMR94] Q. Wu, K. M. Wong, Y. Meng, and W. Read. DOA-estimation of point and scattered sources–Vec-MUSIC. *Proc. 7th SP Workshop Statistical Signal Array Processing*, Quebec City, Canada, pp. 365-368, June 1994.

[WZ99] K. T. Wong and M. D. Zoltowski. Root-MUSIC-based azimuth-elevation angle-of-arrival estimation with uniformly spaced but arbitrarily oriented velocity hydrophones. *IEEE Trans. Signal Process.*, vol.SP-47, pp. 3250–3260, December 1999.

[XB90] X. L. Xu and K. M. Buckley. A comparison of element and beamspace spatial-spectrum estimation for multiple source clusters. *Proc. ICASSP*, vol.5, pp. 2643–2646, Albuquerque, New Mexico, April 1990.

[XB92] X. L. Xu and K. M. Buckley. Bias analysis of the MUSIC location estimator. *IEEE Trans. Acoust., Speech, Signal Process.*, vol.ASSP-40, pp. 2559–2569, October 1992.

[XB93] X.-L. Xu and K. M. Buckley. An analysis of beam-space source localization. *IEEE Trans. Signal Process.*, vol.SP-41, pp. 501–504, January 1993.

[XK96] W. Xu and M. Kaveh. Comparative study of the biases of MUSIC-like estimators. *Signal Process.*, vol.50, pp. 39–55, 1996.

[XSRK94] G. Xu, S. D. Silverstein, R. H. Roy, and T. Kailath. Beamspace ESPRIT. *IEEE Trans. Signal Process.*, vol.SP-42, pp. 349–356, February 1994.

[YF96a] N. Yuen and B. Friedlander. Higher-order ESPRIT for localization of near-field sources: an asymptotic performance analysis. *Proc. 8th IEEE SP Workshop on Statistical Signal and Array Processing*, pp. 538–541, Corfu, Greece, June 1996.

[YF96b] N. Yuen and B. Friedlander. Asymptotic performance analysis of ESPRIT, higher order ESPRIT, and virtual ESPRIT algorithms. *IEEE Trans. Signal Process.*, vol.SP-44, pp. 2537–2550, October 1996.

[YLC89] C. C. Yeh, J. H. Lee, and Y. M. Chen. Estimating two-dimensional angles of arrival in coherent source environment. *IEEE Trans. Acoust., Speech, Signal Process.*, vol.ASSP-37, p. 153, January 1989.

[YS95] J. Yang and A. L. Swindlehurst. The effects of array calibration errors on DF-based signal copy performance. *IEEE Trans. Signal Process.*, vol.SP-43, pp. 2724–2732, November 1995.

[Zha95a] Q. T. Zhang. Probability of resolution of the MUSIC algorithm. *IEEE Trans. Signal Process.*, vol.SP-43, pp. 978–987, April 1995.

[Zha95b] Q. T. Zhang. A statistical resolution theory of the beamformer-based spatial spectrum for determining the directions of signals in white noise. *IEEE Trans. Signal Process.*, vol.SP-43, pp. 1867–1873, August 1995.

[ZHJ91] C. Zhou, F. Haber, and D. L. Jaggard. A resolution measure for the MUSIC algorithm and its application to plane wave arrivals contaminated by coherent interference. *IEEE Trans. Acoust., Speech, Signal Process.*, vol.ASSP-39, pp. 454–463, February 1991.

[ZHM96] M. D. Zoltowski, M Haardt, and C. P. Mathews. Closed-form 2-D angle estimation with rectangular arrays in element space or beamspace via unitary ESPRIT. *IEEE Trans. Signal Process.*, vol.SP-44, pp. 316–328, February 1996.

[ZKK94] M. D. Zoltowski, J. V. Krogmeier, and G. M. Kautz. Novel multirate processing of beamspace noise eigenvectors. *IEEE Signal Process. Lett.*, vol.1, pp. 83–86, May 1994.

[ZKS93] M. D. Zoltowski, G. M. Kautz, and S. D. Silverstein. Beamspace root-MUSIC. *IEEE Trans. Signal Process.*, vol.SP-41, pp. 344–364, January 1993.

[ZL91] M. D. Zoltowski and T. S. Lee. Interference cancellation matrix beamforming for 3-D beamspace ML/MUSIC bearing estimation. *IEEE Trans. Signal Process.*, vol.SP-39, pp. 1858–1876, August 1991.

[ZO95] P. Zetterberg and B. Ottersten. The spectrum efficiency of a base station antenna array system for spatially selective transmission. *IEEE Trans. Veh. Technol.*, vol.VT-44, pp. 651–659, August 1995.

[ZS89] M. D. Zoltowski and D. Stavrinides. Sensor array signal processing via a Procrustes rotations-based Eigenanalysis of the ESPRIT data pencil. *IEEE Trans. Acoust., Speech, Signal Process.*, vol.ASSP-37, pp. 832–861, June 1989.

[ZZ95] M. Zhang and Z. Zhu. A method for direction finding under sensor gain and phase uncertainties. *IEEE Trans. Antennas Propag.*, vol.AP-43, pp. 880–883, August 1995.

Chapter 10

Detection and Other Topics

In this chapter we discuss several areas of interest that we have not considered previously. The optimum detection problem is an obvious topic that we have not discussed up to this point. In addition to its importance, it is the first word in the title of the "Detection, Estimation, and Modulation" series, so the reader may be misled. We discuss the detection problem briefly in Section 10.1.

In Section 10.2, we discuss five topics: target tracking, space-time processing for radar, space-time processing for wireless communications, matched field processing, and spatial spectrum estimation. These topics are closely related to the subjects in this book.

In Section 10.3 we make some concluding comments.

10.1 Optimum Detection

An important problem that we have not discussed is the detection problem. The detection problem was discussed in detail in DEMT I [VT68], [VT01a] and DEMT III [VT71], [VT01b]. In many array processing models of interest, the optimum detector consists of a beamformer designed using the techniques in Chapters 6 and 7, followed by a scalar detector designed using the techniques of DEMT I and DEMT III. Because of this overlap we restrict ourselves to a very brief discussion of the various models and issues involved.

In Section 10.1.1, we discuss the classic binary detection problem. In Section 10.1.2, we discuss matched subspace detectors. In Section 10.1.3, we discuss the detection of spatially spread Gaussian random processes. In Section 10.1.4, we discuss adaptive detection techniques. In Section 10.1.5, we summarize our comments.

Historically, the early results on optimum detection for arrays were derived by Bryn [Bry62], Vanderkulk [Van63], Middleton and Groginsky [MG65], and Van Trees ([VT66], [VT64]). Wolf [Wol59] derived results for multiple non-stationary processes. The detection of Gaussian signals in noise has been studied earlier by Price (e.g., [Pri53], [Pri54], and [Pri56]). Other early references include Stocklin [Sto63], Cox ([Cox64], [Cox69]), Mermoz [Mer64], Young and Howard ([YH70a], [YH70b]), Gaarder ([Gaa67], [Gaa66]), Nuttall and Hyde [NH69], and Lewis and Schultheiss [LS71].

10.1.1 Classic Binary Detection

In this section, we discuss the array processing version of the classic binary detection problem from DEMT I.[1] For simplicity, we restrict our attention to the narrowband case.

We use the frequency-domain snapshot model, but we suppress the k variable in our discussion. The signals on the two hypotheses are

$$\begin{aligned} \mathbf{X}(\omega) &= \mathbf{v}(\psi_s)F(\omega) + \mathbf{N}(\omega) & : H_1 \\ \mathbf{X}(\omega) &= \mathbf{N}(\omega) & : H_0. \end{aligned} \tag{10.1}$$

This corresponds to a single plane-wave signal impinging on the array from ψ_s, which is assumed to be known. The noise is a sample function from a zero-mean Gaussian random process whose spatial spectral matrix $\mathbf{S_n}(\omega)$ is known. The signal is one of the following types:

(i) A known signal,

(ii) A known signal contains unknown random parameters,

(iii) A sample function of a zero-mean Gaussian random process that is statistically independent of $\mathbf{N}(\omega)$.

This model is appropriate for the case in which we have steered the array in a specific direction and want to determine whether a signal is present or not. In all of these cases, the MVDR beamformer from Chapter 6 generates a sufficient statistic. The scalar output of the MVDR beamformer is processed by the appropriate optimum detector. All of the results in DEMT I and III apply directly. The effect of the beamformer is completely characterized by the array gain. Note that the beamformer is an MVDR beamformer so that knowledge of $\mathbf{S_n}$ or an ability to estimate it is necessary.

[1] The reader may want to review DEMT I [VT68], [VT01a].

If the signal direction is mismatched, we can use an LCMV beamformer to combat the uncertainty. This approach is generally not optimal, but provides almost optimum performance for small values of mismatch. Two approaches that develop optimal processors are described by Nolte and his colleagues (e.g., [HN78], [HN76b], [HN76a], [GN74]) and Bell et al. [BEV00].

In some applications, array perturbations must be considered. If the perturbations are modeled as complex Gaussian random vector, then the optimum detection contains a linear term and a quadratic term. A similar result is obtained if there is imperfect spatial coherence of the wavefronts. References that discuss these issues include Morgan [Mor90], Gershman et al. [GMB97], Paulraj and Kailath [PK88], and Rao and Jones [RJ01]. The last paper derives a DFT approximation to the optimum receiver that is computationally feasible.

In some applications (e.g., multipath) the signal consists of multiple plane waves. If the $N \times D$ array manifold matrix is known, we construct a D-dimensional subspace and solve for the optimum binary detector in that subspace. The form of the optimum detector depends on the temporal signal model that is assumed. We discuss one of these models in the next section.

10.1.2 Matched Subspace Detector

In this section, we discuss a detector that is referred to as the matched subspace detector. Our discussion is based on Sections 4.11 and 4.12 of Scharf's book [Sch91]. The history of the development is given on pp.166-167 of [Sch91]. We will summarize the result.

The hypothesis testing problem of interest is

$$\begin{aligned} \mathbf{x}(k) &= \mathbf{V}\mathbf{f}(k) + \mathbf{w}(k), \quad k = 1, \cdots, K : H_1 \\ \mathbf{x}(k) &= \mathbf{w}(k), \quad k = 1, \cdots, K : H_0, \end{aligned} \tag{10.2}$$

where we have used a narrowband time-domain model. The matrix $\mathbf{V}$ is an $N \times D$ array manifold matrix and $\mathbf{f}(k)$ is a $D \times 1$ signal vector. We assume that the array manifold, and therefore the signal subspace, is known. The additive Gaussian noise $\mathbf{w}(k)$ is uncorrelated with variance $\sigma_w^2\mathbf{I}$. The new factor in this model is that the signal vector $\mathbf{f}(k)$ is unknown but nonrandom.

We know that the projection onto the signal subspace will generate a sufficient statistic. Thus,

$$\mathbf{y}_s(k) = \mathbf{P}_{\mathbf{V}}\mathbf{x}(k), \tag{10.3}$$

where $\mathbf{P}_{\mathbf{V}}$ is the projection matrix onto the signal subspace,

$$\mathbf{P}_{\mathbf{V}} = \mathbf{V}\left[\mathbf{V}^H\mathbf{V}\right]^{-1}\mathbf{V}^H. \tag{10.4}$$

Because $\mathbf{f}(k)$ is unknown, but nonrandom, we use a generalized likelihood ratio test (GLRT) (see Section 2.5 of DEMT I [VT68], [VT01a]). The GLRT approach to this problem is due to Scharf and Friedlander [SF94].

The resulting test consists of constructing the test statistic,

$$t_e \stackrel{\triangle}{=} \chi^2(k) = \mathbf{x}^H \mathbf{P}_{\mathbf{V}} \mathbf{x}. \tag{10.5}$$

The statistic χ^2 is a quadratic form in the vector $\mathbf{P}_{\mathbf{V}}\mathbf{x}$, which is a Gaussian random vector with mean $\mathbf{V}\mathbf{f}(k)$ and covariance matrix $\sigma_w^2 \mathbf{P}_{\mathbf{V}}$.

Then, χ^2/σ_w^2 is chi-squared distributed with N degrees of freedom and the noncentrality parameter E_s/σ_w^2 where

$$E_s \stackrel{\triangle}{=} \sum_{k=1}^{K} \mathbf{f}^H(k)\mathbf{V}^H\mathbf{V}\mathbf{f}(k). \tag{10.6}$$

The chi-squared distribution has a monotone likelihood ratio so the test,

$$\sum_{k=1}^{K} \frac{\chi^2}{\sigma_w^2} \underset{H_0}{\overset{H_1}{\gtrless}} \chi_0^2, \tag{10.7}$$

is a uniformly most powerful (UMP) detector. This test is referred to as a **matched subspace detector** (see discussion of p. 166 on [Sch91]. Note that the detector is computing the energy in the signal subspace, so we can also refer to it as a **generalized energy detector**.

In many cases, the variance of the noise σ_w^2 is unknown. If possible, we would like to construct a test that has a constant false alarm rate. Scharf (pp. 148–152 of [Sch91]) also derives a constant false alarm rate (CFAR) version for the case of unknown σ_w^2.

The reader is referred to the above references for a further discussion of the matched subspace detector.

10.1.3 Spatially Spread Gaussian Signal Processes

In this section we consider the binary detection problem for the case in which the signal is a sample function of a Gaussian random process. The plane-wave models in Section 10.1, in which the temporal signals were Gaussian processes are special cases of the model in this section. In this subsection, we focus on the case where the signal has significant spread in ψ-space.

In many physical situations the signals that are the object of the optimum array processing are not true plane waves. This may occur in an acoustic environment in which local spreading is due to scattering or in wireless communication scenarios.

The principal difference in analyzing spatially spread signals involves their representation across the receiving array. In the plane-wave situation, the statistical representation in terms of an eigenvector expansion is one-dimensional, that is, only one eigenvector is required in the representation. In the spatially spread situation, more than one eigenvector is required. The number of eigenvectors required is determined by the area of the array and the spatial extent of the spreading.

In many situations, we can model the spatial spreading with a channel or target-scattering function. The model is similar to the singly-spread and doubly-spread delay-Doppler models that we studied in Chapters 11-13 of DEMT III [VT71], [VT01b]. In the current situation, the scattering function can be represented in either frequency-wavenumber space or ω-ψ-space. An idealized scattering function in ω-ψ-space is shown in Figure 10.1. We limit our discussion in this section to the narrowband case so the scattering function is 1-D. Although we will only discuss 1-D (ψ) scattering functions, all of the results can be extended easily to 2-D functions. An idealized 1-D rectangular scattering function is shown in Figure 10.2. In Figure 10.2(a), the width of the scattering function is less than the width of the main lobe, so one eigenvector will be adequate. In Figure 10.2(b), the width of the scattering function is much wider than the main lobe, so multiple eigenvectors will be required.

This model is a Gaussian signal in Gaussian noise problem that we studied in Chapters 2–5 of DEMT III [VT71], [VT01b]. The resulting processor is the spatial analog to the diversity receiver that we encountered in our analysis of doubly-spread channels in Chapter 13 of DEMT III [VT71], [VT01b]. The detector first forms a set of eigenbeams determined from the signal spectral matrix (a spatially "coherent" operation), then weights them, squares the signals, and sums the squares.

The cases of non-white noise (with a known spatial spectral matrix), the generalized binary detection problem, and the broadband signal model can be all be solved in a similar manner.

In order to evaluate the performance of these detectors, we utilize bounds and approximate error expressions that depend on a function $\mu(s)$ which we encountered previously in Section 2.7 of DEMT I [VT68], [VT01a]. The derivation of the necessary approximate expressions is given of pp. 38–42 on DEMT III [VT71], [VT01b]. We can also derive a closed-form expression for $\mu(s)$ for the spatially-spread signal case. Some of the expressions of interest are contained in the problems.

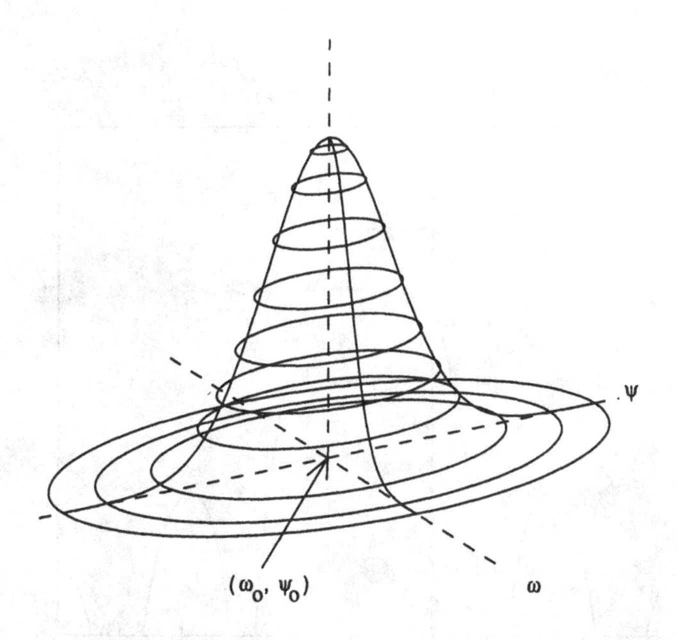

Figure 10.1 Scattering functions in ω-ψ-space.

The discussion in this section has assumed that the required spatial spectral matrices are known. In the next section, we consider the adaptive detection problem in which the statistics are estimated from the input data.

10.1.4 Adaptive Detection

In this section we introduce the problem of adaptive detection. We restrict our attention to the case of a narrowband signal propagating along a single plane wave. We assume that the wavenumber of the plane wave is known.

We use a time-domain snapshot model

$$\mathbf{x}(k) = \mathbf{V}\mathbf{f}(k) + \mathbf{n}(k), \quad k = 1, \cdots, K : H_1 \tag{10.8}$$

$$\mathbf{x}(k) = \mathbf{n}(k), \quad k = 1, \cdots, K : H_0 \tag{10.9}$$

where $\mathbf{f}(k)$ is a known vector,

$$\mathbf{f}(k) = \mathbf{v}(\psi_s) f(k), \tag{10.10}$$

and b is an unknown complex scalar. For simplicity, we let

$$f(k) = 1, \quad k = 1, \cdots, K, \tag{10.11}$$

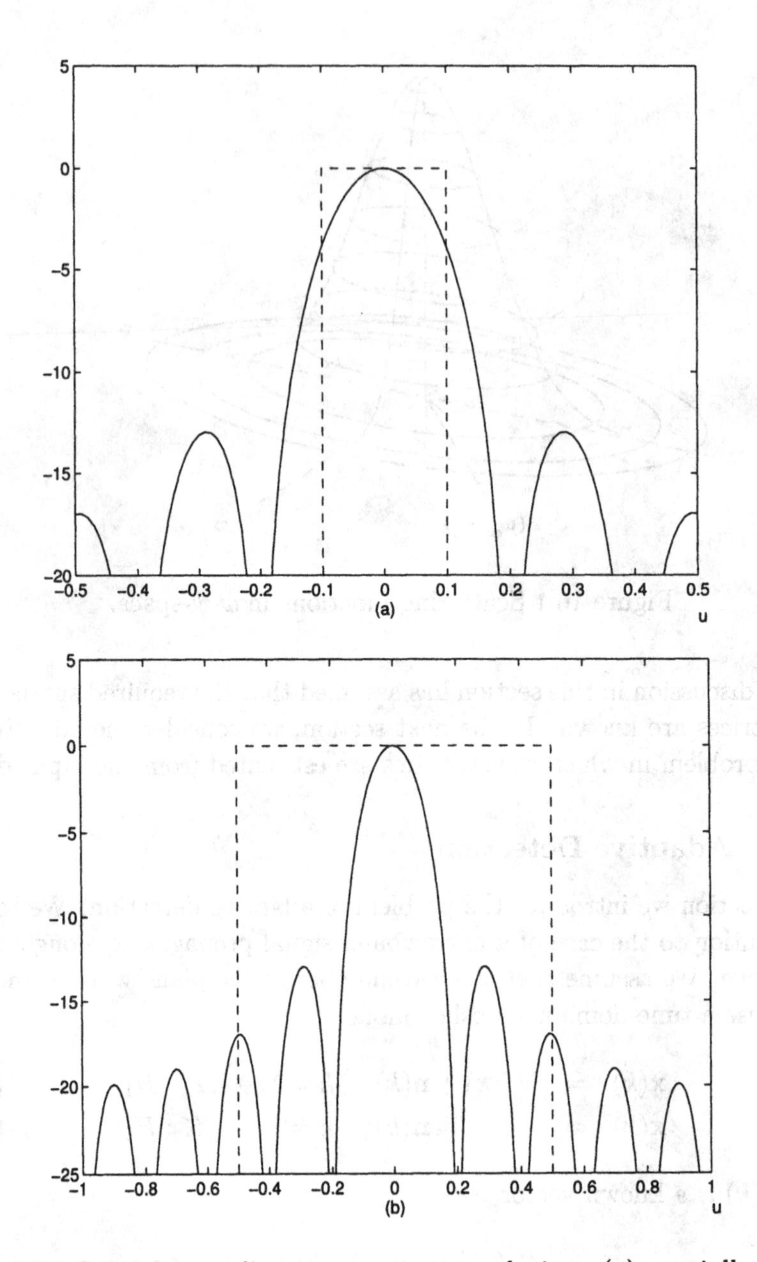

Figure 10.2 Spatial spreading versus array resolution: (a) spatially spread; (b) minimal spread.

so

$$\mathbf{f}(k) = \mathbf{v}(\psi_s). \tag{10.12}$$

The noise $\mathbf{n}(k)$ is a zero-mean complex Gaussian random vector whose covariance is $\mathbf{R_n}$. The snapshots are statistically independent. We denote the collection of snapshots by the $NK \times 1$ vector $\mathbf{X}$. Thus,

$$p_{\mathbf{X}|H_0}(\mathbf{X}) = \prod_{\mathbf{k=1}}^{\mathbf{K}} \frac{\mathbf{1}}{|\pi\mathbf{R_n}|} \exp\left\{-\mathbf{x^H(k)R_n^{-1}x(k)}\right\}, \tag{10.13}$$

and

$$p_{\mathbf{X}|H_1}(\mathbf{X}) = \prod_{\mathbf{k=1}}^{\mathbf{K}} \frac{\mathbf{1}}{|\pi\mathbf{R_n}|} \exp\left\{-\left[\mathbf{x^H(k)} - \mathbf{b^*v^H}\right]\mathbf{R_n^{-1}}\left[\mathbf{x(k)} - \mathbf{bv}\right]\right\}, \tag{10.14}$$

where we have suppressed ψ_s in the argument of the array manifold vector.

If $\mathbf{R_n}$ were known, we would use a generalized likelihood ratio test (GRLT). From Section 2.5 of DEMT I [VT68] [VT01a],

$$\Lambda(\mathbf{X}) = \frac{\max\limits_{b} p_{\mathbf{X}|H_1}(\mathbf{X}, b|H_1)}{p_{\mathbf{X}|H_0}(\mathbf{X}|H_1)} \underset{H_0}{\overset{H_1}{\gtrless}} \gamma. \tag{10.15}$$

Substituting (10.13) and (10.14) into (10.15), cancelling common terms, and taking the logarithm gives

$$\ln\Lambda(\mathbf{X}) = 2b^* Re\left\{\sum_{k=1}^{K} \mathbf{v}^H\mathbf{R_n^{-1}x}(k)\right] - K|b|^2\mathbf{v}^H\mathbf{R_n^{-1}v}. \tag{10.16}$$

Differentiating with respect to b^* and setting the result to zero gives

$$\hat{b} = \frac{\mathbf{v}^H\mathbf{R_n^{-1}\hat{x}}}{\mathbf{v}^H\mathbf{R_n^{-1}v}}, \tag{10.17}$$

where

$$\hat{\mathbf{x}} \triangleq \frac{1}{K}\sum_{k=1}^{K}\mathbf{x}(k), \tag{10.18}$$

which is implemented using an MVDR beamformer.

Substituting (10.17) into (10.16) and (10.15) gives the GLRT for known $\mathbf{R_n}$,

$$\frac{\left|\mathbf{v}^H\mathbf{R_n^{-1}\hat{x}}\right|^2}{\mathbf{v}^H\mathbf{R_n^{-1}v}} \underset{H_0}{\overset{H_1}{\gtrless}} \alpha. \tag{10.19}$$

We now consider the case in which $\mathbf{R_n}$ is unknown and must be estimated. This problem has been studied in a number of application areas. An early solution in the sonar area was derived by Liggett [Lig72] (see also Chang and Tuteur [CT71], Bryn [Bry62], Middleton and Groginsky [MG65], Edelblute et al. [EFK67], Cox [Cox69] and McDonough [McD71], Lewis and Schultheiss [LS71], and Vanderkulk [Van63]). In the radar area, an early solution was given by Brennan and Reed [BR73] (see also Reed et al. [RMB74]).

The model that is normally used in the literature assumes that, in order to estimate $\mathbf{R_n}$, we receive K independent vector snapshots, $\mathbf{x}_1, \mathbf{x}_2, \cdots, \mathbf{x}_K$, that do not contain a signal component. We refer to these vectors as secondary vectors. The subsequent snapshots that may contain the signal are referred to as the primary vectors.

This model can be used to derive several adaptive detectors. We briefly discuss two of these detectors, the adaptive matched filter (AMF) and the GLRT detector.

The adaptive matched filter is due to Robey [Rob91] (see Chapter 4 of Robey [Rob91], which utilizes techniques from Kelly [Kel86]) (see [RFKN92] also). The AMF forms a ML estimate of $\mathbf{R_n}$ using the K secondary vectors,

$$\hat{\mathbf{R}}_{\mathbf{n}} = \frac{1}{K}\sum_{k=1}^{K}\mathbf{x}(k)\mathbf{x}^H(k)\,. \tag{10.20}$$

The test is obtained by substituting $\hat{\mathbf{R}}_{\mathbf{n}}$ into (10.19). The result is

$$t_{AMF} = \frac{\left|\mathbf{v}^H\hat{\mathbf{R}}_{\mathbf{n}}^{-1}\hat{\mathbf{x}}\right|^2}{\mathbf{v}^H\hat{\mathbf{R}}_{\mathbf{n}}^{-1}\mathbf{v}} \mathop{\gtrless}_{H_0}^{H_1} \alpha\,. \tag{10.21}$$

Robey [Rob91] shows that the AMF is a CFAR test and derives expressions for $P_D(AMF)$ and $P_F(AMF)$. (see pp. 37–42 of [Rob91]) The reader is referred to this reference for further discussion.

The GLRT adaptive detector is obtained by utilizing both the primary and secondary vectors and jointly estimating b and $\mathbf{R_n}$. This detector was derived and analyzed by Kelly [Kel86]. (see also [Kel81], [Kel85], [Kel87a], [Kel87b]) The reader is referred to these references for further discussion.

In this section we have introduced the topic of adaptive detection. Our discussion is so brief that it requires the reader to explore the literature to understand the issues.

In addition to the references list above, relevant references include Haykin and Steinhardt [HS92], Kelly and Forsythe [KF89], Monticciolo [Mon94], Tufts and Kristeins [TK85], and Fuhrmann [Fuh91].

10.2 Related Topics

In this section we identify several array processing problems that are of interest.

Target Tracking

In many applications the signals originate from moving sources. As the array receives data, we want to track the location of these sources. The two components of the problem are:

(i) A tracking component that incorporates the target dynamics into the algorithm;

(ii) A data association component that assigns new data to the appropriate target

There are a large number of books and papers that address this problem. Books include Blackman [Bla86], Bar-Shalom and Fortmann [B-SF88], Bar-Shalom and Li [B-SL98], Bar-Shalom [B-S98a], [B-S98b], Stone et al. [SCB99]. Papers that deal with the tracking problem include Reid [Rei79], Sword et al. [SSK93], Rao et al. [RZZ93], [RSZ94], and Zhou et al. [ZYL99a], [ZYL99b].

Most of these discussions treat the problem as a detection problem followed by a tracking problem. A Ph.D. thesis by Zarnich [Zar00] (Zarnich et al. [ZBV01]) treats the problem as a joint problem and obtains interesting results.

Space-Time Processing for Radar

Airborne radars detect targets in an environment that is dominated by clutter and jammers. Space-time adaptive processing (STAP) uses algorithms, which combine the outputs from an antenna array and multiple pulses of a coherent radar waveform to achieve target detection. This approach leads to a 2-D problem in the Doppler-azimuth domain. Many of our results can be extended to this problem. Ward [War94] provides a comprehensive discussion in his Lincoln Laboratory report. The book by Klemm [Kle98] discusses the space-time adaptive processing in detail and has an extensive list of references. There is a collection of papers on STAP in the *April 2000 IEEE Transactions on Aerospace and Electronic Systems* [Mel00].

Space-Time Processing for Wireless Communications

There has been a significant amount of research on space-time processing for wireless communications. It is predicted that many of the third-generation systems will implement some type of adaptive spatial processing in addition to the adaptive temporal processing already in use. These space-time processors are often referred to in the literature as smart antennas.

The biggest challenges faced by these systems are intersymbol interference (ISI) and signal fading due to multipath propagation, and multiuser interference (MUI). The spatial processing at the receiver can reduce MUI by separating user signals arriving at the antenna from different directions. Combined space-time processing can reduce channel fading by suppressing multipath components of the desired signal, or by combining the multipath components in a constructive rather than destructive manner. Performance improvements can also be achieved by using space-time coding with a transmit antenna array. In fixed wireless systems, it is possible to employ antennas at both the transmitter and receiver. Recent research has shown that significant capacity increases can be achieved in these systems by exploiting the diversity gain in rich multipath channels (Foschini et al. [FGVW99]).

Space-time processing for wireless communications has grown to be a significant research area in both the signal processing and communications communities, and promises to be a compelling and challenging problem well into the future as demand for wireless services continues to increase dramatically.

There are a number of papers and books in the area. Books include Liberti and Rappaport [LR99] and Rappaport (ed.) [Rap98]. Papers include Paulraj and Papadias [PP97] and the special issue of the *IEEE Personal Communications magazine on Smart Antennas* [Gol98].

Matched Field Processing

In the sonar environment, we can construct a reasonably accurate model of the propagation of the signal. Therefore, instead of matching the array processor to a plane wave, we match it to the propagation model. This technique is referred to as matched field processing. The original work is by Baggeroer et al. [BKS88]. Later papers include Schmidt et al. [SBKS90] and Preisig [Pre94].

Spatial Spectral Estimation

In Chapter 5, we discussed space-time processes. In some applications, we would like to estimate the spatial correlation function at a particular frequency, $S_f(\omega : \Delta\mathbf{p})$ (5.91). A more general problem is to estimate the frequency-wavenumber spectrum, (5.93).

For a standard linear array, the problem is identical to the temporal spectrum estimation problem (e.g., Kay [Kay88] or Marple [Mar87]). We can use the parametric wavenumber models in Section 5.6. Most of the estimation techniques carry over to the spatial problem.

In each of these areas, the techniques that we have developed in this book will provide the necessary background to explore these areas.

10.3 Epilogue

This chapter concludes our development of optimum array processing and the set of books on detection, estimation, and modulation theory.

We hope that, in spite of the thirty-year gap between volumes, that the collection will prove useful to practicing engineers, researchers, and new students in the area.

In the thirty-year period, there has been a dramatic change in the array processing and signal processing areas. Advances in computational capability have allowed the implementation of complex algorithms that were only of theoretical interest in the past. In many applications, algorithms can be implemented that reach the theoretical bounds. In spite of these advances, there are still a number of challenging problems that these books should help the reader solve.

The advances in computational capability have also changed how the material is taught. In Parts I and III, there was an emphasis on compact analytical solutions to problems. In Part IV, there is a much greater emphasis on efficient iterative solutions and simulations. We have tried to achieve the correct balance between theory and experiment (simulation) in our presentation.

10.4 Problems

P10.1 Optimum Detection[2]

Problem 10.1.1

Consider the following detection problem:

$$\begin{aligned}\mathbf{X}(\omega) &= \mathbf{v}(\psi_s)\mathbf{F}_1(\omega) + \mathbf{W}(\omega) \ : H_1, \\ \mathbf{X}(\omega) &= \mathbf{v}(\psi_s)\mathbf{F}_0(\omega) + \mathbf{W}(\omega) \ : H_0.\end{aligned} \tag{10.22}$$

The array is a standard 10-element linear array. The sensor outputs have been processed by a quadrature demodulator. The two signals, $\mathbf{F}_0(\omega)$ and $\mathbf{F}_1(\omega)$, are known complex signals with energy E_s, and $\mathbf{F}_0(\omega) = -\mathbf{F}_1(\omega)$. The additive noise is a sample function

[2] Most of the problems will require a review of appropriate material from DEMT I, [VT68], [VT01a] or DEMT III, [VT71], [VT01b].

from a complex zero-mean Gaussian process with spectral matrix,

$$S_n(\omega) = \sigma_w^2 \mathbf{I}. \tag{10.23}$$

(a) Find the optimum detector to minimize the probability of error. The two hypotheses are equally likely.

(b) Plot $Pr(\epsilon)$ versus E_s/N_o.

(c) Repeat parts (a) and (b) for the case in which $\mathbf{F}_0(\omega)$ and $\mathbf{F}_1(\omega)$ are orthogonal.

Problem 10.1.2 (continuation)

Repeat Problem 10.2.1 for the case in which

(a) $S_n(\omega) = \sigma_w^2 \mathbf{I} + M_1(\omega)\mathbf{v}(\psi_I)\mathbf{v}^H(\psi_I)$.

(b) Specialize your results in part (a) to the case in which $M_1(\omega)$ is constant over the frequency range of $\mathbf{F}_0(\omega)$ and $\mathbf{F}_1(\omega)$.

(c) Specialize your results in part (a) to the case in which $M_1(\omega)$ corresponds to sine wave with random phase at carrier frequency.

(d) Specialize your results in part (a) to the case in which the interfering signal propagating along $\mathbf{v}(\psi_I)$ consists of either $\mathbf{F}_0(\omega)$ or $\mathbf{F}_1(\omega)$ with equal probability.

Problem 10.1.3

Consider the following detection problem:

$$\begin{aligned} \mathbf{X}(\omega) &= \mathbf{v}(\psi_s)\mathbf{F}(\omega) + \mathbf{W}(\omega) : \quad H_1, \\ \mathbf{X}(\omega) &= \mathbf{W}(\omega) : \quad H_0. \end{aligned} \tag{10.24}$$

The array is a standard 10-element linear array. The signal $f(t)$ is a known complex signal with energy E_s and a uniform phase angle,

$$f(t) = f_1(t)e^{j\theta}, \tag{10.25}$$

and θ has uniform density over $(0, 2\pi)$. The additive noise is a sample function from a complex zero-mean Gaussian process with spectral matrix,

$$S_n(\omega) = \sigma_w^2 \mathbf{I}. \tag{10.26}$$

(a) Find the optimum Neyman-Pearson detector.

(b) Plot $\mathbf{P}_D$ versus $\mathbf{P}_F$ for various E_s/N_o.

(c) Repeat parts (a) and (b) for the case in which

$$S_n(\omega) = \sigma_w^2 \mathbf{I} + M_0 \mathbf{v}(\psi_I)\mathbf{v}^H(\psi_I). \tag{10.27}$$

Problem 10.1.4

Consider the model in Problem 10.2.1 and assume $\mathbf{F}_0(\omega) = \mathbf{0}$.

(a) Find the optimum Neyman-Pearson detector.

(b) Plot $\mathbf{P}_D$ versus $\mathbf{P}_F$ for various E_s/σ_w^2.

Problem 10.1.5

Consider a standard 10-element linear array. The signals on the two hypotheses can be written in the frequency domain as

$$\begin{aligned} \mathbf{X}(k, m\omega_0) &= \mathbf{v}(\psi_s)F(k, m\omega_0) + \mathbf{W}(m\omega_0) : \quad H_1, \\ \mathbf{X}(k, m\omega_0) &= \mathbf{W}(m\omega_0) : \quad H_0. \end{aligned} \tag{10.28}$$

The signal $f(k)$ is a zero-mean complex Gaussian AR(1) temporal random process (see (5.308)). The additive noise is a sample function from a white Gaussian random process with spectral height σ_w^2.

(a) Find the optimum Neyman-Pearson detector. Assume $\sigma_s^2/\sigma_w^2 = 10$ dB.

(b) Simulate its performance for $a(1) = 0.9, \phi_a = 0$, and $\mathbf{P}_F = 10^{-3}$.

Problem 10.1.6

Consider a standard 10-element linear array. The signals on the two hypotheses can be written in the frequency domain as

$$\begin{aligned} \mathbf{X}(k, m\omega_0) &= \mathbf{v}(\psi_s)F_1(k, m\omega_0) + \mathbf{W}(m\omega_0) : \quad H_1, \\ \mathbf{X}(k, m\omega_0) &= \mathbf{v}(\psi_s)F_0(k, m\omega_0) + \mathbf{W}(m\omega_0) : \quad H_0. \end{aligned} \tag{10.29}$$

The signals $f_1(k)$ and $f_0(k)$ are zero-mean complex Gaussian AR(1) temporal random processes. The difference is that $\phi_{a_1} = 0.1$ and $\phi_{a_0} = -0.1$. For both signals $|a(1)| = 0.9$. The additive noise is a sample function from a white Gaussian random process with spectral height σ_w^2.

(a) Find the minimum probability of error receiver. Assume the two hypotheses are equally likely. Assume $\sigma_{s1}^2/\sigma_w^2 = \sigma_{s0}^2/\sigma_w^2 = 10$ dB.

(b) Simulate its performance.

Problem 10.1.7

Consider a standard 10-element linear array. The nominal detection problem of interest is given by the model in Problem 10.2.4 with $\psi_s = 0$. However, the actual arrival angle θ of the plane wave is a random variable whose probability density is

$$p_\theta(\theta : \Lambda_m) = \frac{\exp\left[\Lambda_m \cos\theta\right]}{2\pi I_0(\Lambda_m)}, \quad -\pi \leq \theta \leq \pi, \tag{10.30}$$

where θ is measured from broadside.

(a) Choose Λ_m so that the probability is 0.9, that θ, the actual arrival angle, is within the main lobe. Design a constrained beamformer followed by an optimum scalar detector. Compute the resulting performance.

(b) Compare your result to a detector that treats θ as a nuisance parameter and averages over its probability (e.g., p. 335 of DEMT I [VT68], [VT01a], or [HN78]).

Problem 10.1.8

Consider the following detection problem

$$\mathbf{r}(t) = \mathbf{f}_1(t) + \mathbf{n}(t), \quad T_i \leq t \leq T_f \quad : H_1, \tag{10.31}$$

$$\mathbf{r}(t) = \mathbf{f}_0(t) + \mathbf{n}(t), \quad T_i \leq t \leq T_f \quad : H_0, \tag{10.32}$$

where

$$\mathbf{F}_1(\omega) = \mathbf{v}(\psi_1)F(\omega), \tag{10.33}$$
$$\mathbf{F}_2(\omega) = \mathbf{v}(\psi_2)F(\omega), \tag{10.34}$$

and $F(\omega)$ is a rectangular pulse with energy E_s and duration T.
The noise process is stationary

$$\mathbf{S_n}(\omega) = \mathbf{S_c}(\omega) + \sigma_w^2\mathbf{I}, \tag{10.35}$$

and $T_i = -\infty$ and $T_f = \infty$.

(a) Find the optimum receiver to minimize the $Pr(\epsilon)$.

(b) Find an expression for d^2.

(c) Assume $\mathbf{S_c}(\omega) = 0$. Plot d^2 as a function of ρ_{12} and T for a standard linear array.

Problem 10.1.9

Consider the problem of detecting a Gaussian random process in non-white Gaussian noise.

$$\mathbf{r}(t) = \mathbf{f}(t) + \mathbf{n}_c(t) + \mathbf{w}(t), \quad T_i \le t \le T_f : \quad H_1, \tag{10.36}$$
$$\mathbf{r}(t) = \mathbf{n}_c(t) + \mathbf{w}(t), \quad T_i \le t \le T_f : \quad H_0, \tag{10.37}$$

where

$$\mathbf{F}(\omega) = \mathbf{v}(\psi_s)F(\omega), \tag{10.38}$$

and the spectrum of $f(t)$ is

$$S_f(\omega) = \begin{cases} \frac{P_s}{2W}, & -2\pi W \le \omega \le 2\pi W, \\ 0, & \text{elsewhere}. \end{cases} \tag{10.39}$$

The colored noise is a single plane-wave signal,

$$\mathbf{N}_c(\omega) = \mathbf{v}(\psi_n)N_c(\omega) \tag{10.40}$$

where

$$S_{n_c}(\omega) = \begin{cases} \frac{P_n}{2W}, & -2\pi W \le \omega \le 2\pi W, \\ 0, & \text{elsewhere}. \end{cases} \tag{10.41}$$

Assume $T = T_f - T_i$ is large.

(a) Find the optimum receiver.

(b) Compute $\mu_\infty(s)$.

(c) Assume the criterion is minimum $Pr(\epsilon)$ and the hypotheses are equally likely. Find an approximate expression for $Pr(\epsilon)$.

Problem 10.1.10

Consider the problem of deciding which of two Gaussian random processes are present.

$$\mathbf{r}(t) = \mathbf{f}_1(t) + \mathbf{w}(t), \quad T_i \le t \le T_f : \quad H_1, \tag{10.42}$$

$$\mathbf{r}(t) = \mathbf{f}_0(t) + \mathbf{w}(t), \quad T_i \le t \le T_f : \quad H_0, \tag{10.43}$$

where

$$\mathbf{F}_1(\omega) = \mathbf{v}(\psi_1)F_1(\omega), \tag{10.44}$$

$$\mathbf{F}_0(\omega) = \mathbf{v}(\psi_0)F_0(\omega). \tag{10.45}$$

The signals $f_1(t)$ and $f_0(t)$ are independent with identical spectrum,

$$S_{f_1}(\omega) = S_{f_0}(\omega) = \begin{cases} \frac{P}{2W}, & -2\pi W \leq \omega \leq 2\pi W, \\ 0, & \text{elsewhere}. \end{cases} \tag{10.46}$$

Assume $T = T_f - T_i$ is large, the hypotheses are equally likely, and the criterion is min $Pr(\epsilon)$.

(a) Find the optimum receiver.

(b) Find $\mu_\infty(s)$.

(c) Find an approximate expression for the $Pr(\epsilon)$.

(d) Consider a standard 10-element linear array. Plot $Pr(\epsilon)$ as a function of

$$\Delta\psi = \psi_1 - \psi_0. \tag{10.47}$$

Problem 10.1.11

The received waveforms on the two hypotheses are:

$$\mathbf{r}(t) = \mathbf{f}_1(t) + \mathbf{f}_2(t) + \mathbf{w}(t), \quad 0 \leq t \leq T: \quad H_1, \tag{10.48}$$

$$\mathbf{r}(t) = \mathbf{w}(t), \quad 0 \leq t \leq T: \quad H_0, \tag{10.49}$$

where the $N \times 1$ signal vectors can be written in the transform domain as,

$$\mathbf{F}_1(n\omega_o) = \mathbf{v}(n\omega_o, \psi_1)F_1(n\omega_o) \tag{10.50}$$

$$\mathbf{F}_2(n\omega_o) = \mathbf{v}(n\omega_o, \psi_2)F_2(n\omega_o) \tag{10.51}$$

where $\omega_o = \frac{2\pi}{T}$ and $n = 1, \cdots, M$.

The source signals $f_1(t)$ and $f_2(t)$ are statistically independent Gaussian random processes with spectra,

$$S_i(\omega) = \begin{cases} \frac{\sigma_i^2}{2W}, & |\omega| \leq 2\pi W, \\ 0, & \text{elsewhere}, \quad i = 1, 2. \end{cases} \tag{10.52}$$

The noise process $\mathbf{w}(t)$ is a Gaussian random process that is temporally and spatially white,

$$E\left[\mathbf{w}(t)\mathbf{w}^T(u)\right] = \sigma_w^2 \delta(t-u)\mathbf{I}. \tag{10.53}$$

The hypotheses are equally likely and T is large.

(a) Find the minimum $Pr(\epsilon)$ test.

(b) Evaluate the performance.

Bibliography

[BEV00] K. L. Bell, Y. Ephraim, and H. L. Van Trees. A Bayesian approach to robust adaptive beamforming. *IEEE Trans. Signal Process.*, vol.SP-48, pp. 386–398, February 2000.

[BKS88] A. B. Baggeroer, W. A. Kuperman, and H. Schmidt. Matched field processing; source localization in correlated noise as an optimum parameter estimation problem. *J. Acoust. Soc. Am.*, vol.83, pp. 571–587, February 1988.

[Bla86] S. S. Blackman. *Multiple-target Tracking with Radar Applications.* Artech House, Dedham, Massachusetts, 1986.

[BR73] L.E. Brennan and I.S. Reed. Theory of adaptive radar. *IEEE Trans. Aerospace Electronic Syst.*, vol.AES-9, pp. 237–252, March 1973.

[Bry62] F. Bryn. Optimum signal processing of three-dimensional array operating on gaussian signals and noise. *J. Acoust. Soc. Am.*, vol.34, pp. 289–297, March 1962.

[BS74] Y. Bar-Shalom. Extension of the probabilistic data association filter to multi-target tracking. *Proc. 5th Symp. on Nonlinear Estimation*, San Diego, California, pp. 16–21, September 1974.

[B-S98a] Y. Bar-Shalom, editor. *MultiTarget-MultiSensor Tracking: Applications and Advances,.* vol. I.YBS Publishing, Storrs, Connecticut, 1998.

[B-S98b] Y. Bar-Shalom, editor. *MultiTarget-MultiSensor Tracking: Applications and Advances,.* vol. II. YBS Publishing, Storrs, Connecticut, 1998.

[B-SF88] Y. Bar-Shalom and T. E. Fortmann. *Tracking and Data Association.* Academic Press, San Diego, California, 1988.

[B-SL98] Y. Bar-Shalom and X. R. Li. *Estimation and Tracking: Principles, Techniques and Software.* YBS Publishing, Storrs, Connecticut, 1998.

[Che62] H. Chernoff. A measure of asymptotic efficiency for tests of a hypothesis based on the sum of observations. *Ann. Math. Stat.*, vol.23, pp. 493–507, 1962.

[Cox64] L. H. Cox. Interrelated problems in estimation and detection. *Proc. NATO Advanced Study Institute on Signal Processing with Emphasis on Underwater Acoustics*, pp. 23-1–23-64, Grenoble, France, August 1964.

[Cox69] H. Cox. Array processing against interference. In *Proc. Purdue Centennial Year Symp. on Information Processing*, West Lafayette Indiana pp. 453–463, April 1969.

[CT71] J. H. Chang and F. B. Tuteur. A new class of adaptive array processors. *J. Acoust. Soc. Am.*, vol.49, pp. 639–649, July 1971.

[EFK67] D.J. Edelblute, J.M. Fisk, and G.L. Kinnison. Criteria for optimum-signal-detection theory for arrays. *J. Acoust. Soc. Am.*, vol.41, pp. 199–205, January 1967.

[Fan61] R.M. Fano. *Transmission of Information.* MIT Press and Wiley, Cambridge and New York, 1961.

[FGVW99] G. J. Foschini, G. D. Golden, R. A. Valenzuela, and P. W. Wolniansky. Simplified processing for high spectral efficiency wireless communication employing multi-element arrays. *IEEE J. Selected Areas in Commun.*, vol.17, pp. 1841–1852, November 1999.

[FM88] D. R. Fuhrmann and M. I. Miller. On the existence of positive-definite maximum-likelihood estimates of structured covariance matrices. *IEEE Trans. Inf. Theory*, vol.IT-4, pp. 722–729, July 1988.

[FTM88] D. R. Fuhrmann, M. J. Turmon, and M. I. Miller. Efficient implementation of the EM algorithm for Toeplitz covariance estimation. *Proc. Ann. Conf. Information Science and Systems*, Princeton, New Jersey, March 1988.

[Fuh90] D. Fuhrmann. Application of structured covariance estimation to adaptive detection. Technical Report, Department of Electrical Engineering, Washington University, St. Louis, Missouri, 1990.

[Fuh91] D. Fuhrmann. Application of Toeplitz covariance estimation to adaptive beamforming and detection. *IEEE Trans. Acoust., Speech, and Signal Process*, vol.ASSP-39, pp. 2194–2198, October 1991.

[Gaa66] N.T. Gaarder. The design of point detector arrays: II. *IEEE Trans. Inf. Theory*, vol.IT-12, pp. 112–120, April 1966.

[Gaa67] N.T. Gaarder. The design of point detector arrays: I. *IEEE Trans. Inf. Theory*, vol.IT-13, pp. 42–50, 1967.

[Gal65] R.G. Gallager. Lower bounds on the tails of probability distributions. QPR 77 277-291, Massachusets Institute of Technology, Research Laboratory of Electronics April 1965.

[GLCJ] H. Gauvrit, J. P. Le Cadre, and C. Jauffret. A formulation of multitarget tracking as an incomplete data problem. *IEEE Trans. Aerospace and Electonic Syst.*, vol.AES-33, pp. 1242–1255, July 1997.

[GMB97] A. B. Gerhman, C. F. Mechlenbrauder, and J. F. Bohme. Matrix fitting approach to direction of arrival estimation with imperfect spatial coherence of wavefronts. *IEEE Trans. Signal Process.*, vol.83, pp. 1034–1040, March 1988.

[GN74] M. A. Gallop and L. W. Nolte. Bayesian detection of targets of unknown location. *IEEE Trans. Aerospace and Electronic Syst.*, vol.AES-10, pp. 429–435, 1974.

[Gol98] A. Goldsmith, editor. *IEEE Personal Commun. Mag.*, Special Issue on Smart Antennas. vol.5, pp. 23–35, February 1998.

[Goo63] N. R. Goodman. Statistical analysis based on a certain complex Gaussian distribution. *Ann. Math. Stat.*, vol.34, pp. 152–180, 1963.

[HN76a] W. S. Hodgkiss and L. W. Nolte. Optimum array processor performance tradeoffs under directional uncertainty. *IEEE Trans. on Aerospace and Electronic Syst.*, vol.AES-12, pp. 605–615, September 1976.

[HN76b] W. S. Hodgkiss and L. W. Nolte. Adaptive optimum array processing. Technical Report, Department of Electrical Engineering, Duke University, Durham, North Carolina, July 1976.

[HN78] W.S. Hodgkiss and L.W. Nolte. Array processor performance under directional uncertainty. *IEEE Trans. Aerospace Electronic Syst.*, vol.AES-14, pp. 826–832, September 1978.

[HS92] S. Haykin and A. Steinhardt, editors. *Adaptive Radar Detection and Estimation.* Wiley, New York, 1992.

[Jac66] I.M. Jacobs. Probability-of-error bounds for binary transmission on the slow fading rician channel. *IEEE Trans. Inf. Theory*, vol.IT-12, October 1966.

[Kay88] S. M. Kay. *Modern Spectral Estimation: Theory and Application.* Prentice-Hall, Englewood Cliffs, New Jersey, 1988.

[Kel81] E. J. Kelly. Finite sum expressions for signal detection probabilities. Technical Report, MIT Lincoln Laboratory, Lexington, Massachusetts, May 1981.

[Kel85] E. J. Kelly. Adaptive detection in non-stationary interference, Part I. Technical Report 724, MIT Lincoln Laboratory, Lexington, Massachusetts, 1985.

[Kel86] E. J. Kelly. An adaptive detection algorithm. *IEEE Trans. Aerospace and Electronic Syst.*, vol.AES-22, pp. 115–127, March 1986.

[Kel87a] E. J. Kelly. Adaptive detection in non-stationary interference, Part III. Technical Report 761, MIT Lincoln Laboratory, Lexington, Massachusetts, August 1987.

[Kel87b] E. J. Kelly. Performance of an adaptive detection algorithm; rejection of unwanted signals. *IEEE Trans. Aerospace Electronic Syst.*, vol.AES-25, pp. 122–133, March 1987.

[KF89] E. J. Kelly and K. M. Forsythe. Adaptive detection and parameter estimation for multidimensional signal models. Technical Report, MIT Lincoln Laboratory, Lexington, Massachusetts, April 1989.

[Kle98] R. Klemm. *Space-time Adaptive Processing: Principles and Applications.* IEE, London, United Kingdom, 1998.

[Lig72] W. W. Liggett. Passive sonar processing for noise with unknown covariance structure. *J. Acoust. Soc. Am.*, vol.51, pp. 24–30, January 1972.

[LR99] J. C. Liberti, Jr., and T. S. Rappaport. *Smart Antennas for Wireless Communications: IS-95 and Third Generation CDMA Applications.* Prentice-Hall, Upper Saddle River, New Jersey, 1999.

[LS71] H. B. Lewis and P. M. Schultheiss. The beamformer as a log-likelihood ratio detector. *IEEE Trans. Audio and Electroacoust.*, vol.AU-19, pp. 140–146, June 1971.

[Mar87] S. L. Marple, Jr . *Digital Spectral Analysis.* Prentice-Hall, Englewood Cliffs, New Jersey, 1987.

[McD71] R. N. McDonough. A canonical form of the likelihood detector for Gaussian random vectors. *J. Acoust. Soc. Am.*, vol.49, pp. 402–406, February 1971.

[Mel00] W. L. Melvin. Space-time adaptive processing and adaptive arrays: special collection of papers. *IEEE Trans. Aerospace Electronic Syst.*, vol.AES-36, pp. 508–509, April 2000.

[Mer64] H.F. Mermoz. Filtrage adapté et utilisation optimale d'une antenne. In *NATO Adavanced Study Inst. Signal Processing Emphasis Underwater Acoustics*, Grenoble, France, 1964.

[MF90] M. I. Miller and D. R. Fuhrmann. Maximum-likelihood narrow-band direction finding and the EM algorithm. *IEEE Trans. Acoust., Speech, Signal Process.*, vol.ASSP-38, pp. 1560–1577, September 1990.

[MG65] D. Middleton and H.I. Groginski. Detection of random acoustic signals by receivers with distributed elements: optimum receiver structures for normal signal and noise fields. *J. Acoust. Soc. Am.*, vol.38, pp. 727–737, November 1965.

[Mon94] P. Monticciolo. Adaptive detection in stationary and nonstationary noise environments. Technical Report, MIT Lincoln Laboratory, Lexington, Massachusetts, February 1994.

[Mor90] D. Morgan. Coherence effects on the detection performance of quadratic array processors, with applications to large-array matched-field beamforming. *J. Acoust. Soc. Amer.*, vol.87, pp. 737–747, February 1986.

[MS87] M. I. Miller and D. L. Snyder. The role of likelihood and entropy in incomplete-data problems applications to estimating intensities and Toeplitz constrained covariance. *Proc. IEEE*, vol.75, pp. 892–907, 1987.

[NH69] A.H. Nuttall and D. W. Hyde. A unified approach to optimum and suboptimum processing for arrays. USL rep. 992, U.S. Navy Underwater Sound Lab., New London, Connecticut, April 1969.

[PK88] A. Paulraj and T. Kailath. Direction of arrival estimation by eigenstructure methods with imperfect spatial coherence of wave fronts. *J. Acoust. Soc. Amer.*, vol. 83, pp. 1034–1040, March 1988.

[PP97] A. J. Paulraj and C. B. Papadias. Space-time processing for wireless communications. *IEEE Signal Process. Mag.*, vol.14, pp. 49–83, November 1997.

[Pre94] J. C. Preisig. Robust maximum energy adaptive matched field processing. *IEEE Trans. Signal Processing*, vol.SP-42, pp. 1585–1593, July 1994.

[Pri53] R.L. Pritchard. Optimum directivity for linear point arrays. *J. Acoust. Soc. Am.*, vol.25, pp. 879–891, September 1953.

[Pri54] R. Price. The detection of signals perturbed by scatter and noise. *IRE Trans. Info. Theory*, vol.PGIT-4, pp. 163–170, September 1954.

[Pri56] R. Price. Optimum detection of random signals in noise with application to scatter-multipath communication. *IRE Trans.*, pp. 125–135, December 1956.

[Rap98] T. S. Rappaport, editor. *Smart Antennas: Adaptive Arrays, Algorithms, and Wireless Position Location.* IEEE Press, New York, 1998.

[Rei79] D. B. Reid. An algorithm for tracking multiple targets. *IEEE Trans. Automatic Control*, vol.AC-24, pp. 843–854, December 1979.

[RFKN92] F. Robey, D. Fuhrmann, E. Kelly, and R. Nitzberg. A CFAR adaptive matched filter detector. *IEEE Trans. Aerospace Electronic Syst.*, vol.AES-28, pp. 208–216, 1992.

[RJ01] A. M. Rao and D. L. Jones. Efficient Quadratic Detection in Perturbed Arrays via Fourier Transform Techniques. *IEEE Trans. Signal Process.*, vol.SP-49, pp. 1269–1281, July 2001.

[RMB74] I.S. Reed, J.D. Mallett, and L.E. Brennan. Rapid covergence rate in adaptive arrays. *IEEE Trans. Aerospace and Electronic Syst.*, vol.AES-10, pp. 853–863, November 1974.

[Rob91] F. C. Robey. A covariance modeling approach to adaptive beamforming and detection. Technical Report, MIT Lincoln Laboratory, Lexington, Massachusetts, July 1991.

[RSZ94] C. R. Rao, C. R. Sastry, and B. Zhou. Tracking the direction of arrival of multiple moving targets. *IEEE Trans. Signal Process.*, vol.SP-42, pp. 1133–1144, May 1994.

[RZZ93] C. R. Rao, L. Zhang, and L. C. Zhao. Multiple target angle tracking using sensor array outputs. *IEEE Trans. Aerospace Electronic Syst.*, vol.AES-29, pp. 268–271, January 1993.

[SBKS90] H. Schmidt, A. B. Baggeroer, W. A. Kuperman, and E. K. Scheer. Environmentally tolerant beamforming for high resolution matched field processing; deterministic mismatch. *J. Acoust. Soc. Am.*, vol.88, pp. 1851–1862, October 1990.

[SCB99] L. D. Stone, T. L. Corwin, and C. A. Barlowe. *Bayesian Multiple Target Tracking.* Artech House, Boston, Massachusetts, 1999.

[Sch91] L. L. Scharf. *Statistical Signal Processing: Detection, Estimation, and Time Series Analysis.* Addison-Wesley, Reading, Massachusetts, 1991.

[SF94] L. L. Scharf and B. Friedlander. Matched subspace detectors. *IEEE Trans. Signal Process.*, vol.SP-42, pp. 2146–2157, August 1994.

[SGB67] C.E. Shannon, R.G. Gallager, and E.R. Berlekamp. Lower bounds to error probability for coding on discrete memoryless channels: I. *Inf. and Control*, vol.10, pp. 65–103, January 1967.

[Sha56] C.E. Shannon. Seminar notes for seminar in information theory. MIT, Lexington, MA, 1956.

[SSK93] C. K. Sword, M. Simaan, and E. W. Kamen. Multiple target angle tracking using sensor array outputs. *IEEE Trans. Aerospace Electronic Syst.*, vol.AES-26, pp. 367–373, January 1993.

[Sto63] P. L. Stocklin. Space-time sampling and likelihood ratio processing in acoustic pressure fileds. *J. Br. IRE*, pp. 79–90, July 1963.

[TK85] D. Tufts and I. Kristeins. On the pdf of the SNR in an improved adaptive detector. *Proc. ICASSP*, pp. 572–575, 1985.

[Van63] V. Vanderkulk. Optimum processing for acoustic arrays. *J. Br. IRE*, vol.26, pp. 286–292, October 1963.

[VT64] H. L. Van Trees. A formulation of the space-time processing problem for sonar systems. Technical Report Project Trident Working Memo. 208, A. D. Little, December 1964.

[VT66] H. L Van Trees. Optimum processing for passive sonar arrays. In *Proc. IEEE Ocean Electronics Symp.*, Honolulu, Hawaii, pp. 41–65, 1966.

[VT68] H. L. Van Trees. *Detection, Estimation, and Modulation Theory, Part I.* Wiley, New York, 1968.

[VT01a] H. L. Van Trees. *Detection, Estimation, and Modulation Theory, Part I.* Wiley Interscience, New York, 2001.

[VT71] H. L. Van Trees. *Detection, Estimation, and Modulation Theory, Part III.* Wiley, New York, 1971.

[VT01b] H. L. Van Trees. *Detection, Estimation, and Modulation Theory, Part III.* Wiley Interscience, New York, 2001.

[War94] J. Ward. Space-time adaptive processing for airborne radar. Technical Report, MIT Lincoln Laboratory, Lexington, Massachusetts, December 1994.

[Wol59] J. K. Wolf. *On the Detection and Estimation Problem for Multiple Non-Stationary Random Processes.* PhD Thesis, Dept. of Electrical Engineering, Princeton University, Princeton, New Jersey, 1959.

[YH70a] G. W. Young and J. E. Howard. Antenna processing for surface target detection. *IEEE Trans. Antennas Propag.*, vol.AP-18, pp. 335–342, 1970.

[YH70b] G. W. Young and J. E. Howard. Applications of space-time decision and estimation theory to antenna processing system design. *Proc. IEEE*, vol.58, pp. 771–778, May 1970.

[Zar00] R. E. Zarnich. *A Unified Method for the Measurement and Tracking of Narrowband Contacts from an Array of Sensors.* PhD Dissertation, George Mason University, Fairfax, Virginia, 2000.

[ZBV01] R. E. Zarnich, K. L. Bell, and H. L. Van Trees. A unified method for measurement and tracking of contacts from an array of sensors. *IEEE Trans. Signal Process.*, vol.SP-49, pp. 2950–2961, December 2001.

[ZM96] M. Zatman and D. Marshall. Forwards-backwards averaging for adaptive beamforming and STAP. *Proc. ICASSP*, vol.5, pp. 2630–2633, Atlanta, Georgia, 1996.

[ZYL99a] Y. Zhou, P. C. Yip, and H. Leung. Tracking the direction-of-arrival of multiple moving targets by passive arrays: Algorithm. *IEEE Trans. Signal Process.*, vol.SP-47, pp. 2655–2666, October 1999.

[ZYL99b] Y. Zhou, P. C. Yip, and H. Leung. Tracking the direction-of-arrival of multiple moving targets by passive arrays: Asymptotic performance analysis. *IEEE Trans. Signal Process.*, vol.SP-47, pp. 2644–2654, October 1999.

Appendix A

Matrix Operations

A.1 Introduction

Most of the analysis that we do in the array processing area utilizes vectors and matrices extensively. In this appendix, we have summarized the definitions and properties that will be useful in the text. We also develop the ideas of matrix algebra and matrix calculus. We assume that the reader has some familiarity with vectors and matrices, and many results are stated without proof.

There are almost no new results in this appendix. We have compiled (and borrowed) results from three types of sources. The first class of sources are books and articles that deal with matrices and linear algebra without a particular application focus. Representative books and articles of this type include Bellman [Bel72], Marcus [Mar60], Hohn [Hoh73], Noble and Daniel [ND77], Lancaster and Tismenetsky [LT85], Golub and Van Loan [GVL89], Grenander and Szego [GS58], Graham [Gra81], Rao and Mitra [RM71].

The second class of sources are books on adaptive antennas, spectral estimation, adaptive filtering, automatic control, or system identification that have included a chapter or appendix on matrices because they need matrix results for their respective applications. Representative books of this type include Kay [Kay88], Hudson [Hud81], Marple [Mar87], Haykin [Hay96], and Scharf [Sch91].

The third class of sources include journal articles or reports where a matrix result was needed for a specific application and has wider usage. Representative sources in this area include Cantoni and Butler [CB76b], [CB76a], Makhoul [Mak81], Stoica and Nehorai [SN89], and Kuroc [Kur89].

When a result appears in multiple references and appears to be well-known in the "community" we have not included a reference. When we have followed a particular reference closely, we indicate that source.

We have included examples to relate the matrix results to the array processing problem in order to provide motivation. The physical models are explained in more detail in the text.

In general, all vectors and matrices are assumed to be complex. Real matrices and vectors are treated as special cases. The labeling of the matrices; $\mathbf{A}, \mathbf{B}, \mathbf{C}, \cdots$, and vectors $\mathbf{a}, \mathbf{b}, \mathbf{c}, \cdots$, in the appendix does not have a physical significance unless specifically indicated. (This is in contrast to the text.)

A.2 Basic Definitions and Properties

A.2.1 Basic Definitions

We define an $N \times M$ matrix $\mathbf{A}$ by defining its elements $a_{ik}, i = 1, 2, \cdots, N, k = 1, 2, \cdots, M$. We write it in matrix form with N rows and M columns,

$$\mathbf{A} = \begin{bmatrix} a_{11} & a_{12} & \cdots & a_{1M} \\ a_{21} & a_{22} & \cdots & a_{2M} \\ \vdots & \vdots & \ddots & \vdots \\ a_{N1} & a_{N2} & \cdots & a_{NM} \end{bmatrix}. \tag{A.1}$$

We also use the notation $[\mathbf{A}]_{ij}$ for the ijth element. The vector $\mathbf{a}$ is an $N \times 1$ matrix,

$$\mathbf{a} = \begin{bmatrix} a_1 \\ a_2 \\ \vdots \\ a_N \end{bmatrix}. \tag{A.2}$$

The vector in (A.2) is often referred to as a column vector. We will suppress the column designation. The **product** of an $N \times M$ matrix $\mathbf{A}$ and a $M \times L$ matrix $\mathbf{B}$ is an $N \times L$ matrix $\mathbf{C}$ whose elements are given by

$$c_{ij} = \sum_{k=1}^{M} a_{ik} b_{kj}. \tag{A.3}$$

The product of two matrices is defined if and only if the number of columns in the first matrix is equal to the number of rows in the second matrix. When this is true, the matrices are referred to as conformable matrices.

The **transpose** of **A** is denoted by the superscript T,

$$\left[\mathbf{A}^T\right]_{ij} = [\mathbf{A}]_{ji} = a_{ji}. \tag{A.4}$$

The transpose has the property

$$(\mathbf{AB})^T = \mathbf{B}^T\mathbf{A}^T. \tag{A.5}$$

The **Hermitian transpose** is denoted by the superscript H,

$$\left[\mathbf{A}^H\right]_{ij} = \left([\mathbf{A}]_{ji}\right)^* = a_{ji}^*. \tag{A.6}$$

The Hermitian transpose has the property

$$(\mathbf{AB})^H = \mathbf{B}^H\mathbf{A}^H. \tag{A.7}$$

The transpose or Hermitian transpose of a vector is a $1 \times N$ matrix, which is referred to as a **row vector**.

If $N = M$, then **A** is a **square matrix**. A **symmetric matrix** is a square matrix in which

$$\mathbf{A}^T = \mathbf{A}. \tag{A.8}$$

This implies

$$a_{ij} = a_{ji}, \tag{A.9}$$

so the matrix is symmetric around the principal diagonal. For example, if $N = 4$,

$$\mathbf{A} = \begin{bmatrix} a_{11} & a_{12} & a_{13} & a_{14} \\ a_{12} & a_{22} & a_{23} & a_{24} \\ a_{13} & a_{23} & a_{33} & a_{34} \\ a_{14} & a_{24} & a_{34} & a_{44} \end{bmatrix}. \tag{A.10}$$

A **Hermitian matrix** is a square matrix **A** with elements that have complex conjugate symmetry,

$$\mathbf{A}^H = \mathbf{A}. \tag{A.11}$$

This implies

$$a_{ij} = a_{ji}^*. \tag{A.12}$$

For example, if $N = 4$,

$$\mathbf{A} = \begin{bmatrix} a_{11} & a_{12} & a_{13} & a_{14} \\ a_{12}^* & a_{22} & a_{23} & a_{24} \\ a_{13}^* & a_{23}^* & a_{33} & a_{34} \\ a_{14}^* & a_{24}^* & a_{34}^* & a_{44} \end{bmatrix}. \tag{A.13}$$

The diagonal elements of a Hermitian matrix are real.

The **inner product** of two vectors is defined as

$$\alpha = \mathbf{x}^H \mathbf{y}; \tag{A.14}$$

when $\mathbf{x} = \mathbf{y}$, α corresponds to the square of the Euclidean norm of the vector (A.36):

$$\alpha = \mathbf{x}^H \mathbf{x}. \tag{A.15}$$

The outer product of complex $N \times 1$ vector $\mathbf{x}$ and a complex $M \times 1$ vector $\mathbf{y}$ is a $N \times M$ matrix $\mathbf{A}$,

$$\mathbf{A} = \mathbf{x}\mathbf{y}^H. \tag{A.16}$$

When $\mathbf{x} = \mathbf{y}, \mathbf{A}$ is an $N \times N$ matrix

$$\mathbf{A} = \mathbf{x}\mathbf{x}^H. \tag{A.17}$$

The matrix $\mathbf{A}$ in (A.17) is Hermitian for arbitrary $\mathbf{x}$.

Example A.2.1

The correlation matrix of a complex random vector $\mathbf{x}$ is the expectation of the outer product

$$\mathbf{R_x} \triangleq E\left[\mathbf{x}\mathbf{x}^H\right] = \begin{bmatrix} E[x_1x_1^*] & E[x_1x_2^*] & E[x_1x_3^*] & E[x_1x_4^*] \\ E[x_2x_1^*] & E[x_2x_2^*] & E[x_2x_3^*] & E[x_2x_4^*] \\ E[x_3x_1^*] & E[x_3x_2^*] & E[x_3x_3^*] & E[x_3x_4^*] \\ E[x_4x_1^*] & E[x_4x_2^*] & E[x_4x_3^*] & E[x_4x_4^*] \end{bmatrix}. \tag{A.18}$$

Thus, $\mathbf{R_x}$ is a Hermitian matrix.

If $\mathbf{X}$ is an $N \times K$ matrix ($K \geq N$), then

$$\mathbf{A} = \mathbf{X}\mathbf{X}^H, \tag{A.19}$$

is a Hermitian matrix.

If $\mathbf{A}^H = -\mathbf{A}$, then $\mathbf{A}$ is **skew Hermitian**. The diagonal elements of a skew Hermitian matrix are imaginary. If $\mathbf{A}$ is real and $\mathbf{A}^T = -\mathbf{A}$, then $\mathbf{A}$ is **skew symmetric**.

If $\mathbf{A} = \mathbf{B}_1 + j\mathbf{B}_2$ is a Hermitian matrix where $\mathbf{B}_1$ and $\mathbf{B}_2$ are real, then:

(i) $\mathbf{B}_1$ is real symmetric,

(ii) $\mathbf{B}_2$ is skew symmetric,

(iii) $j\mathbf{B}_2$ is Hermitian.

If $\mathbf{A} = \mathbf{B}_1 + j\mathbf{B}_2$ is skew Hermitian, then:

(i) $\mathbf{B}_1$ is skew symmetric,

(ii) $\mathbf{B}_2$ is symmetric,

(iii) $j\mathbf{B}_2$ is skew Hermitian.

The **determinant** of a square matrix $\mathbf{A}$ is denoted by $\det(\mathbf{A})$ or $|\mathbf{A}|$. To calculate the determinant of $\mathbf{A}$, we evaluate

$$\det(\mathbf{A}) = \sum_{k=1}^{N} a_{ik}C_{ik}, \tag{A.20}$$

where

$$C_{ik} = (-1)^{i+k}M_{ik}, \tag{A.21}$$

and M_{ik} is the determinant of the submatrix of $\mathbf{A}$, which is constructed by deleting the ith row and kth column of $\mathbf{A}$. It is called the **minor** of a_{ik} and C_{ik} is called the **cofactor** of $\mathbf{A}$. M_{ii} is referred to as the ith **principal minor**.

The determinant has the following properties:

$$\det(\mathbf{AB}) = \det(\mathbf{A})\det(\mathbf{B}), \tag{A.22}$$

$$\det(\mathbf{A}^T) = \det(\mathbf{A}), \tag{A.23}$$

$$\det(\mathbf{A}^H) = (\det(\mathbf{A}))^*, \tag{A.24}$$

$$\det(k\mathbf{A}) = k^N \det(\mathbf{A}), \quad k \text{ is a scalar.} \tag{A.25}$$

Other properties of the determinant are given in various sections as they arise.

The **trace** of a square matrix $\mathbf{A}$ is denoted by $\text{tr}\,(\mathbf{A})$. The trace is the sum of the diagonal elements,

$$\text{tr}\,(\mathbf{A}) = \sum_{i=1}^{N} a_{ii}. \tag{A.26}$$

The trace has the following properties:

$$\text{tr}\,(\mathbf{A}+\mathbf{B}) = \text{tr}\,(\mathbf{A}) + \text{tr}\,(\mathbf{B}), \tag{A.27}$$

$$\text{tr}\,(\mathbf{AB}) = \text{tr}\,(\mathbf{BA}), \tag{A.28}$$

$$\text{tr}\,(\mathbf{ABC}) = \text{tr}\,(\mathbf{CAB}) = \text{tr}\,(\mathbf{BCA}). \tag{A.29}$$

If $\mathbf{x}$ and $\mathbf{y}$ are $N \times 1$ vectors and $\mathbf{A}$ is a square matrix, then

$$\alpha = \mathbf{x}^H \mathbf{A}\mathbf{y}, \tag{A.30}$$

is a scalar and

$$\alpha = \text{tr}\,(\alpha) = \text{tr}\,(\mathbf{x}^H\mathbf{Ay}) = \text{tr}\,(\mathbf{yx}^H\mathbf{A}) = \text{tr}\,(\mathbf{Ayx}^H). \tag{A.31}$$

Two useful inequalities are

$$\left| \text{tr } (\mathbf{A}^H\mathbf{B})\right|^2 \leq \text{ tr } (\mathbf{A}^H\mathbf{A}) \text{ tr } (\mathbf{B}^H\mathbf{B}), \tag{A.32}$$

and

$$\text{tr } (\mathbf{B}^H\mathbf{A}^H\mathbf{A}\mathbf{B}) \leq \text{ tr } (\mathbf{A}^H\mathbf{A}) \text{ tr } (\mathbf{B}^H\mathbf{B}). \tag{A.33}$$

The **rank** of a matrix is the number of linearly independent columns or rows. The rank of a matrix has the following properties:

$$\text{rank } (\mathbf{A}+\mathbf{B}) \leq \text{ rank } (\mathbf{A}) + \text{ rank } (\mathbf{B}), \tag{A.34}$$

and

$$\text{rank } (\mathbf{A}\mathbf{B}) \leq \min(\text{ rank } \mathbf{A}, \text{ rank } \mathbf{B}). \tag{A.35}$$

There are several **norms** for vectors and matrices that are useful. The **2-norm** of an $N \times 1$ vector is defined as

$$\| \mathbf{x} \|_2 \triangleq \left(\sum_{i=1}^{N} |x_i|^2 \right)^{\frac{1}{2}} = (\mathbf{x}^H\mathbf{x})^{\frac{1}{2}}. \tag{A.36}$$

The 2-norm is also referred to as the Euclidean norm. More generally, the **p-norm** of an $N \times 1$ vector is

$$\| \mathbf{x} \|_p \triangleq \left(\sum_{i=1}^{N} |x_i|^p \right)^{\frac{1}{p}}, \quad p \geq 1. \tag{A.37}$$

The p-norm is also referred to as the Hölder norm. We use the Euclidean norm throughout the text and for notational simplicity will drop the subscript 2. Thus,

$$\| \mathbf{x} \| \triangleq (\mathbf{x}^H\mathbf{x})^{\frac{1}{2}}. \tag{A.38}$$

The **Frobenius (or Euclidean) norm** of an $N \times N$ matrix is defined as,

$$\| \mathbf{A} \|_F = \left(\sum_{i=1}^{N} \sum_{j=1}^{N} |a_{ij}|^2 \right)^{\frac{1}{2}} = \left(\text{tr} \left[\mathbf{A}^H\mathbf{A} \right] \right)^{\frac{1}{2}}. \tag{A.39}$$

A.2.2 Matrix Inverses

If $\mathbf{A}$ is a square $N \times N$ matrix whose rank is equal to N, then we can define the **inverse** of $\mathbf{A}$, which is denoted by $\mathbf{A}^{-1}$ and satisfies

$$\mathbf{A}^{-1}\mathbf{A} = \mathbf{A}\mathbf{A}^{-1} = \mathbf{I}, \tag{A.40}$$

where $\mathbf{I}$ is the **identity matrix**,

$$\mathbf{I} = \begin{bmatrix} 1 & 0 & \cdots & 0 \\ 0 & 1 & \cdots & 0 \\ 0 & & \ddots & \vdots \\ 0 & 0 & \cdots & 1 \end{bmatrix}. \tag{A.41}$$

If the inverse does not exist (i.e., rank $\mathbf{A}) < N$), then $\mathbf{A}$ is referred to as a **singular matrix**.

The matrix $\mathbf{A}$ will be singular if and only if

$$\det(\mathbf{A}) = 0. \tag{A.42}$$

The inverse has the following properties:

$$\left(\mathbf{A}^T\right)^{-1} = \left(\mathbf{A}^{-1}\right)^T, \tag{A.43}$$

$$\left(\mathbf{A}^H\right)^{-1} = \left(\mathbf{A}^{-1}\right)^H, \tag{A.44}$$

$$(\mathbf{AB})^{-1} = \mathbf{B}^{-1}\mathbf{A}^{-1}, \tag{A.45}$$

$$\det\left(\mathbf{A}^{-1}\right) = \left(\det\left(\mathbf{A}\right)\right)^{-1}. \tag{A.46}$$

To calculate the inverse of $\mathbf{A}$ we use the formula,

$$\mathbf{A}^{-1} = \frac{\mathbf{C}^T}{\det(\mathbf{A})}, \tag{A.47}$$

where $\mathbf{C}$ is the matrix of cofactors,

$$[\mathbf{C}]_{ik} = C_{ik} = (-1)^{i+k} M_{ik}, \tag{A.48}$$

where M_{ik} is the minor defined following (A.21).

A formula for the inverse that we use frequently in the text is referred to as the **matrix inversion lemma**,

$$(\mathbf{A}+\mathbf{BCD})^{-1}=\mathbf{A}^{-1}-\mathbf{A}^{-1}\mathbf{B}\left(\mathbf{DA}^{-1}\mathbf{B}+\mathbf{C}^{-1}\right)^{-1}\mathbf{DA}^{-1}, \tag{A.49}$$

where $\mathbf{A}$ is $N\times N$, $\mathbf{B}$ is $N\times M$, $\mathbf{C}$ is $M\times M$, and $\mathbf{D}$ is $M\times N$ and the requisite inverses are assumed to exist.

A special case of (A.49) is referred to as **Woodbury's identity**. Here, $\mathbf{B}$ is an $N\times 1$ column vector $\mathbf{x}$, $\mathbf{C}$ is a scalar equal to unity, and $\mathbf{D}=\mathbf{x}^H$. Then

$$\left(\mathbf{A}+\mathbf{x}\mathbf{x}^H\right)^{-1}=\mathbf{A}^{-1}-\frac{\mathbf{A}^{-1}\mathbf{x}\mathbf{x}^H\mathbf{A}^{-1}}{1+\mathbf{x}^H\mathbf{A}^{-1}\mathbf{x}}. \tag{A.50}$$

Several other inverse relations that follow from (A.49) are:

$$\left(\sigma^2\mathbf{I}+\mathbf{VSV}^H\right)^{-1}=\frac{1}{\sigma^2}\left(\mathbf{I}-\mathbf{V}\left(\mathbf{V}^H\mathbf{V}+\sigma^2\mathbf{S}^{-1}\right)^{-1}\mathbf{V}^H\right), \tag{A.51}$$

$$\left[\mathbf{A}^{-1}+\mathbf{B}^H\mathbf{C}^{-1}\mathbf{B}\right]^{-1}=\mathbf{A}-\mathbf{AB}^H\left[\mathbf{BAB}^H+\mathbf{C}\right]^{-1}\mathbf{BA}, \tag{A.52}$$

$$\left[\mathbf{A}^{-1}+\mathbf{B}^H\mathbf{C}^{-1}\mathbf{B}\right]^{-1}\mathbf{B}^H\mathbf{C}^{-1}=\mathbf{AB}^H\left[\mathbf{BAB}^H+\mathbf{C}\right]^{-1}, \tag{A.53}$$

and

$$\mathbf{C}^{-1}-\left[\mathbf{BAB}^H+\mathbf{C}\right]^{-1}=\mathbf{C}^{-1}\mathbf{B}\left[\mathbf{A}^{-1}+\mathbf{B}^H\mathbf{C}^{-1}\mathbf{B}\right]^{-1}\mathbf{B}^H\mathbf{C}^{-1}. \tag{A.54}$$

A.2.3 Quadratic Forms

A **Hermitian quadratic form** Q is defined as

$$Q=\mathbf{x}^H\mathbf{A}\mathbf{x}, \tag{A.55}$$

where $\mathbf{A}$ is a Hermitian matrix and $\mathbf{x}$ is a complex $N\times 1$ vector. Because $\mathbf{A}$ is Hermitian, Q is a real scalar.

$\mathbf{A}$ is positive definite if

$$\mathbf{x}^H \mathbf{A} \mathbf{x} > 0, \tag{A.56}$$

for all $\mathbf{x} \neq 0$.

$\mathbf{A}$ is positive semidefinite (or nonnegative definite) if

$$\mathbf{x}^H \mathbf{A} \mathbf{x} \geq 0, \tag{A.57}$$

for all $\mathbf{x} \neq 0$.

A square $N \times N$ matrix is positive definite if and only if

$$\mathbf{A} = \mathbf{C}\mathbf{C}^H, \tag{A.58}$$

where $\mathbf{C}$ is $N \times N$ and has rank N or its principal minors are all positive.

If $\mathbf{A}$ is positive definite, then

$$\mathbf{A}^{-1} = \left(\mathbf{C}^{-1}\right)^H \mathbf{C}^{-1}. \tag{A.59}$$

The factoring in (A.58) plays an important role in many applications. In Section A.5, we develop techniques for finding $\mathbf{C}$.

If $\mathbf{A}$ is positive definite and $\mathbf{B}$ is an $M \times N$ matrix of rank $M (M \leq N)$, then

$$\mathbf{B}\mathbf{A}\mathbf{B}^H, \tag{A.60}$$

is positive definite.

If $\mathbf{A}$ is positive definite, then the diagonal elements are all positive and $\det(\mathbf{A})$ is positive.

If $\mathbf{A}$ satisfies (A.58) and $\mathbf{C}$ is an $M \times N$ matrix with $M < N$, then $\mathbf{A}$ is positive semi-definite.

A.2.4 Partitioned Matrices

A **partitioned matrix A** is an $N \times M$ matrix that can be written in terms of submatrices. For example,

$$\mathbf{A} = \begin{bmatrix} \mathbf{A}_{11} & \mathbf{A}_{12} \\ \mathbf{A}_{21} & \mathbf{A}_{22} \end{bmatrix}, \tag{A.61}$$

where the dimensions of the submatrices are

$$\mathbf{A}_{11} \qquad N_1 \times M_1, \tag{A.62}$$

$$\mathbf{A}_{21} \qquad (N - N_1) \times M_1, \tag{A.63}$$

$$\mathbf{A}_{12} \qquad N_1 \times (M - M_1), \tag{A.64}$$

$$\mathbf{A}_{22} \qquad (N - N_1) \times (M - M_1). \tag{A.65}$$

The advantage of utilizing partitioned matrices is that they can be operated upon by treating the submatrices as elements.

Four common operations are:

(i) Multiplication

$$\mathbf{AB} = \begin{bmatrix} \mathbf{A}_{11} & \mathbf{A}_{12} \\ \mathbf{A}_{21} & \mathbf{A}_{22} \end{bmatrix} \begin{bmatrix} \mathbf{B}_{11} & \mathbf{B}_{12} \\ \mathbf{B}_{21} & \mathbf{B}_{22} \end{bmatrix} = \begin{bmatrix} \mathbf{A}_{11}\mathbf{B}_{11} + \mathbf{A}_{12}\mathbf{B}_{21} & \mathbf{A}_{11}\mathbf{B}_{12} + \mathbf{A}_{12}\mathbf{B}_{22} \\ \mathbf{A}_{21}\mathbf{B}_{11} + \mathbf{A}_{22}\mathbf{B}_{21} & \mathbf{A}_{21}\mathbf{B}_{12} + \mathbf{A}_{22}\mathbf{B}_{22} \end{bmatrix}, \tag{A.66}$$

where matrices in the indicated products must be conformable.

(ii) Conjugate Transposition

$$\begin{bmatrix} \mathbf{A}_{11} & \mathbf{A}_{12} \\ \mathbf{A}_{21} & \mathbf{A}_{22} \end{bmatrix}^H = \begin{bmatrix} \mathbf{A}_{11}^H & \mathbf{A}_{21}^H \\ \mathbf{A}_{12}^H & \mathbf{A}_{22}^H \end{bmatrix}. \tag{A.67}$$

(iii) Inverse ($\mathbf{A}$ is $N \times N$)

$$\mathbf{A}^{-1} = \begin{bmatrix} \left(\mathbf{A}_{11} - \mathbf{A}_{12}\mathbf{A}_{22}^{-1}\mathbf{A}_{21}\right)^{-1} & -\left(\mathbf{A}_{11} - \mathbf{A}_{12}\mathbf{A}_{22}^{-1}\mathbf{A}_{21}\right)^{-1}\mathbf{A}_{12}\mathbf{A}_{22}^{-1} \\ -\left(\mathbf{A}_{22} - \mathbf{A}_{21}\mathbf{A}_{11}^{-1}\mathbf{A}_{12}\right)^{-1}\mathbf{A}_{21}\mathbf{A}_{11}^{-1} & \left(\mathbf{A}_{22} - \mathbf{A}_{21}\mathbf{A}_{11}^{-1}\mathbf{A}_{12}\right)^{-1} \end{bmatrix}, \tag{A.68}$$

where $\mathbf{A}$ is given by (A.61) and $\mathbf{A}_{11}^{-1}$ and $\mathbf{A}_{22}^{-1}$ exist.

A special case of (A.68) occurs when $\mathbf{A}_{11}$ is $(N-1)\times(N-1)$, $\mathbf{A}_{12}$ is a $(N-1)\times 1$ column vector $\mathbf{a}_2$, $\mathbf{A}_{21}$ is $1\times(N-1)$ row vector $\mathbf{a}_1^H$, and $\mathbf{A}_{22}$ is a scalar c. Then,

$$\mathbf{A}^{-1} = \begin{bmatrix} \mathbf{A}_{11}^{-1} + \beta\mathbf{A}_{11}^{-1}\mathbf{a}_2\mathbf{a}_1^H\mathbf{A}_{11}^{-1} & -\beta\mathbf{A}_{11}^{-1}\mathbf{a}_2 \\ -\beta\mathbf{a}_1^H\mathbf{A}_{11}^{-1} & \beta \end{bmatrix}, \tag{A.69}$$

where $\beta = (c - \mathbf{a}_1^H\mathbf{A}_{11}^{-1}\mathbf{a}_2)^{-1}$. The result in (A.69) is useful when a new matrix is created by bordering the original matrix $\mathbf{A}_{11}$ with an additional row and column. The inverse of the **border** matrix can be expressed in terms of $\mathbf{A}_{11}^{-1}$ and the new row and column.

(iv) Determinant

$$\begin{aligned} \det(\mathbf{A}) &= \det(\mathbf{A}_{22})\det(\mathbf{A}_{11} - \mathbf{A}_{12}\mathbf{A}_{22}^{-1}\mathbf{A}_{21}) \\ &= \det(\mathbf{A}_{11})\det(\mathbf{A}_{22} - \mathbf{A}_{21}\mathbf{A}_{11}^{-1}\mathbf{A}_{12}). \end{aligned} \tag{A.70}$$

A.2.5 Matrix products

In this section we define several matrix products that we find useful in our analyses.

Hadamard Product

The **Hadamard product** of two $N \times M$ dimensional matrices $\mathbf{A}$ and $\mathbf{B}$ is defined as[1]

$$\mathbf{A} \odot \mathbf{B} \triangleq \begin{bmatrix} a_{11}b_{11} & a_{12}b_{12} & \cdots & a_{1M}b_{1M} \\ a_{21}b_{21} & a_{22}b_{22} & \ddots & a_{2M}b_{2M} \\ \vdots & \vdots & \cdots & \vdots \\ a_{N1}b_{N1} & a_{N2}b_{N2} & \cdots & a_{NM}b_{NM} \end{bmatrix}. \tag{A.71}$$

We see that the Hadamard product is obtained by element-by-element multiplication.

Several properties will be useful:

[1]The Hadamard product is sometimes referred to as the Hadamard-Schur product.

(i)

$$\mathbf{A} \odot \mathbf{B} = \mathbf{B} \odot \mathbf{A}, \tag{A.72}$$

(ii)

$$(\mathbf{A} \odot \mathbf{B}) \odot \mathbf{C} = \mathbf{A} \odot (\mathbf{B} \odot \mathbf{C}), \tag{A.73}$$

(iii)

$$(\mathbf{A} \odot \mathbf{B})^T = \mathbf{A}^T \odot \mathbf{B}^T, \tag{A.74}$$

(iv)

$$\mathbf{a}\mathbf{b}^T \odot \mathbf{c}\mathbf{d}^T = (\mathbf{a} \odot \mathbf{c})(\mathbf{b} \odot \mathbf{d})^T, \tag{A.75}$$

(v)

$$\mathbf{E}_{qr}\mathbf{A} \odot \mathbf{B}\mathbf{E}_{st} = [\mathbf{A}]_{rt} [\mathbf{B}]_{qs} \mathbf{E}_{qt}. \tag{A.76}$$

(The matrix $\mathbf{E}_{ij}$ is defined as a matrix whose ij component is 1 and all other components are 0.)

(vi) If $\mathbf{A}$ and $\mathbf{B}$ are positive semi-definite, then

$$\mathbf{A} \odot \mathbf{B} \text{ is positive semi-definite} \quad \text{([SN89] [Bel72]).} \tag{A.77}$$

(vii) If $\mathbf{A}, \mathbf{B}$ and $\mathbf{C}$ are Hermitian positive semi-definite matrices, then

$$[Re\,(\mathbf{A} \odot \mathbf{B})]^{-1} [Re\,(\mathbf{A} \odot \mathbf{C})] [Re\,(\mathbf{A} \odot \mathbf{B})]^{-1} \geq \left\{Re\left[\mathbf{A} \odot \mathbf{B}\mathbf{C}^{-1}\mathbf{B}\right]\right\}^{-1} \tag{A.78}$$

where the inverses are assumed to exist ([SN89]).

(viii) If $\mathbf{A}$ is a Hermitian positive definite matrix, then

$$\left(\mathbf{I} \odot \mathbf{A}^{-1}\right) \geq (\mathbf{I} \odot \mathbf{A})^{-1}. \tag{A.79}$$

Kronecker product

If $\mathbf{A}$ is a $N \times M$ matrix and $\mathbf{B}$ is a $K \times L$ matrix, the **Kronecker product** is defined to be the $NK \times ML$ matrix,

$$
\mathbf{A} \otimes \mathbf{B} \triangleq \begin{bmatrix} a_{11}\mathbf{B} & a_{12}\mathbf{B} & \cdots & a_{1M}\mathbf{B} \\ a_{21}\mathbf{B} & a_{22}\mathbf{B} & \cdots & a_{2M}\mathbf{B} \\ \vdots & \vdots & \ddots & \vdots \\ a_{N1}\mathbf{B} & a_{N2}\mathbf{B} & \cdots & a_{NM}\mathbf{B} \end{bmatrix}, \tag{A.80}
$$

(e.g., Chapter 12 [LT85] or [Gra81]).

Several properties will be useful:

(i) If $\mathbf{B}$ is an $M \times M$ matrix, then,

$$
\mathbf{I}_N \otimes \mathbf{B} = \text{diag}\left[\mathbf{B}, \mathbf{B}, \cdots, \mathbf{B}\right]. \tag{A.81}
$$

(ii) If $\mathbf{A}$ is an $M \times M$ matrix, then,

$$
\mathbf{A} \otimes \mathbf{I}_N = \begin{bmatrix} a_{11}\mathbf{I}_N & a_{12}\mathbf{I}_N & \cdots & a_{1M}\mathbf{I}_N \\ \vdots & & & \\ a_{M1}\mathbf{I}_N & a_{M2}\mathbf{I}_N & \cdots & a_{MM}\mathbf{I}_N \end{bmatrix}. \tag{A.82}
$$

(iii)

$$
\mathbf{I}_M \otimes \mathbf{I}_N = \mathbf{I}_{MN}. \tag{A.83}
$$

(iv)

$$
\mathbf{A} \otimes (\alpha\mathbf{B}) = (\alpha\mathbf{A}) \otimes \mathbf{B} = \alpha\left(\mathbf{A} \otimes \mathbf{B}\right). \tag{A.84}
$$

(v)

$$
(\mathbf{A} \otimes \mathbf{B})^H = \mathbf{A}^H \otimes \mathbf{B}^H. \tag{A.85}
$$

(vi)

$$
(\mathbf{A} + \mathbf{B}) \otimes \mathbf{C} = (\mathbf{A} \otimes \mathbf{C}) + (\mathbf{B} \otimes \mathbf{C}). \tag{A.86}
$$

(vii)

$$\mathbf{A} \otimes (\mathbf{B} + \mathbf{C}) = (\mathbf{A} \otimes \mathbf{B}) + (\mathbf{A} \otimes \mathbf{C}). \tag{A.87}$$

(viii)

$$\mathbf{A} \otimes (\mathbf{B} \otimes \mathbf{C}) = (\mathbf{A} \otimes \mathbf{B}) \otimes \mathbf{C}. \tag{A.88}$$

(ix)

$$(\mathbf{A} \otimes \mathbf{B})(\mathbf{C} \otimes \mathbf{D}) = \mathbf{AC} \otimes \mathbf{BD}. \tag{A.89}$$

(x)

$$(\mathbf{A} \otimes \mathbf{B})^{-1} = \mathbf{A}^{-1} \otimes \mathbf{B}^{-1}. \tag{A.90}$$

(xi) If $\mathbf{A}$ is $M \times M$ and $\mathbf{B}$ is $N \times N$, then

$$\begin{aligned} (\mathbf{A} \otimes \mathbf{B}) &= (\mathbf{A} \otimes \mathbf{I}_N)(\mathbf{I}_M \otimes \mathbf{B}) \\ &= (\mathbf{I}_M \otimes \mathbf{B})(\mathbf{A} \otimes \mathbf{I}_N). \end{aligned} \tag{A.91}$$

(xii) If $\mathbf{A}_1, \mathbf{A}_2, \cdots \mathbf{A}_p$ are $M \times M$ and $\mathbf{B}_1, \mathbf{B}_2, \cdots \mathbf{B}_p$ are $N \times N$, then

$$(\mathbf{A}_1 \otimes \mathbf{B}_1)(\mathbf{A}_2 \otimes \mathbf{B}_2) \cdots (\mathbf{A}_p \otimes \mathbf{B}_p) = (\mathbf{A}_1 \mathbf{A}_2 \cdots \mathbf{A}_p) \otimes (\mathbf{B}_1 \mathbf{B}_2 \cdots \mathbf{B}_p). \tag{A.92}$$

(xiii) If $\mathbf{A}$ is $M \times M$ and $\mathbf{B}$ is $N \times N$, then

$$\det(\mathbf{A} \otimes \mathbf{B}) = (\det \mathbf{A})^M (\det \mathbf{B})^N. \tag{A.93}$$

(xiv) If $\mathbf{A}$ is $M \times M$ and $\mathbf{B}$ is $N \times N$

$$\text{tr}\ (\mathbf{A} \otimes \mathbf{B}) = (\text{ tr}\ \mathbf{A})(\text{ tr}\ \mathbf{B}). \tag{A.94}$$

(xv) If $\mathbf{A}$ is $M \times M$ and $\mathbf{B}$ is $N \times N$

$$\text{rank}\ (\mathbf{A} \otimes \mathbf{B}) = (\text{ rank}\ \mathbf{A})(\text{ rank}\ \mathbf{B}). \tag{A.95}$$

Khatri-Rao product

The **Khatri-Rao product** of an $N \times M$ matrix $\mathbf{A}$ and a $P \times M$ matrix $\mathbf{B}$ is defined as the $NP \times M$ matrix

$$\mathbf{A}\Box\mathbf{B} \triangleq \left[\begin{array}{c|c|c|c} \mathbf{a}_1 \otimes \mathbf{b}_1 & \mathbf{a}_2 \otimes \mathbf{b}_2 & \cdots & \mathbf{a}_N \otimes \mathbf{b}_N \end{array}\right], \tag{A.96}$$

where $\mathbf{a}_j$ denotes the jth column of the matrix $\mathbf{A}$.

The first element is

$$\mathbf{a}_1 \otimes \mathbf{b}_1 = \begin{bmatrix} a_{11}\mathbf{b}_1 \\ \vdots \\ a_{N1}\mathbf{b}_1 \end{bmatrix} = \begin{bmatrix} a_{11}b_{11} \\ a_{11}b_{21} \\ \vdots \\ a_{N1}b_{P-1,1} \\ a_{N1}b_{P1} \end{bmatrix}. \tag{A.97}$$

The remaining elements have the same structure.

Several properties will be useful:

(i)

$$(\mathbf{A} \otimes \mathbf{B})(\mathbf{C}\Box\mathbf{D}) = \mathbf{AC}\Box\mathbf{BD}, \tag{A.98}$$

([RM71]).

(ii)

$$(\mathbf{A}\Box\mathbf{B})^H(\mathbf{C}\Box\mathbf{D}) = \mathbf{A}^H\mathbf{C} \odot \mathbf{B}^H\mathbf{D}, \tag{A.99}$$

([Kur89]),

(iii)

$$(\mathbf{A}\Box\mathbf{B})^H(\mathbf{C} \otimes \mathbf{D})(\mathbf{E}\Box\mathbf{F}) = (\mathbf{A}\Box\mathbf{B})^H(\mathbf{CE}\Box\mathbf{DF}) = \mathbf{A}^H\mathbf{CE} \odot \mathbf{B}^H\mathbf{DF}, \tag{A.100}$$

([Kur89]).

A.2.6 Matrix Inequalities

If $\mathbf{A}$ and $\mathbf{B}$ are $N \times N$ matrices, then the inequality,

$$\mathbf{A} > \mathbf{B} \tag{A.101}$$

means that $\mathbf{A} - \mathbf{B}$ is a positive definite matrix.

Similarly, the inequality

$$\mathbf{A} \geq \mathbf{B} \tag{A.102}$$

means that $\mathbf{A} - \mathbf{B}$ is a non-negative definite matrix.

A.3 Special Vectors and Matrices

In this section we define some special vectors and matrices that we encounter in our analyses.

A.3.1 Elementary Vectors and Matrices

The zero vector $\mathbf{0}_N$ is a $N \times 1$ vector whose elements are all zero. The vector $\mathbf{1}_N$ is a $N \times 1$ vector whose elements are all unity.

The vector $\mathbf{e}_j$ is a $N \times 1$ vector whose jth element is unity and whose remaining elements are zero. The dimension of $\mathbf{e}_j$ is inferred from its usage or is specified. The vector $\mathbf{e}_j$ is referred to as an **elementary** vector.

If a $N \times N$ matrix $\mathbf{A}$ is written as

$$\mathbf{A} = \begin{bmatrix} \mathbf{a}_1 & \mathbf{a}_2 & \cdots & \mathbf{a}_N \end{bmatrix}, \tag{A.103}$$

then

$$\mathbf{A}\mathbf{e}_i\mathbf{e}_i^T = \begin{bmatrix} \mathbf{0} & \mathbf{0} & \cdots & \mathbf{a}_i & \cdots & \mathbf{0} \end{bmatrix}. \tag{A.104}$$

If a $N \times N$ matrix $\mathbf{B}$ is written as

$$\mathbf{B} = \begin{bmatrix} \mathbf{b}_1^T \\ \mathbf{b}_2^T \\ \vdots \\ \mathbf{b}_N^T \end{bmatrix}, \tag{A.105}$$

then

$$\mathbf{e}_i\mathbf{e}_i^T\mathbf{B} = \begin{bmatrix} \mathbf{0} \\ \mathbf{0} \\ \vdots \\ \mathbf{b}_N^T \\ \vdots \\ \mathbf{0} \end{bmatrix}. \tag{A.106}$$

The matrix $\mathbf{E}_{ij}$ is a matrix whose ijth element is 1 and all other elements are 0.

$$\mathbf{E}_{ij} = \mathbf{e}_i\mathbf{e}_j^T. \tag{A.107}$$

The $N \times N$ matrix $\mathbf{E}^{(ij)}$ with $i < j$ is defined as

$$\mathbf{E}^{(ij)} = \begin{matrix} \\ \\ \\ (i) \\ \\ \\ \\ (j) \\ \\ \\ \\ \end{matrix} \begin{bmatrix} 1 & & & & & & & & & \\ & \ddots & & & & & & & & \\ & & 1 & & & & & & & \\ & & & 0 & \cdots & & 1 & & & \\ & & & & 1 & & & & & \\ & & & \vdots & & \ddots & \vdots & & & \\ & & & & & 1 & & & & \\ & & & 1 & \cdots & & 0 & & & \\ & & & & & & & 1 & & \\ & & & & & & & & \ddots & \\ & & & & & & & & & 1 \end{bmatrix}. \tag{A.108}$$

Pre-multiplication by $\mathbf{E}^{(ij)}$ interchanges row i and row j. Post-multiplication by $\mathbf{E}^{(ij)}$ interchanges column i and column j.

The $N \times N$ matrix $\mathbf{E}^{(i)}$ is defined as

$$\mathbf{E}^{(i)} = (i) \begin{bmatrix} 1 & & & & & & \\ & \ddots & & & & & \\ & & 1 & & & & \\ & & & k & & & \\ & & & & 1 & & \\ & & & & & \ddots & \\ & & & & & & 1 \end{bmatrix}, \tag{A.109}$$

where k is a non-zero constant. Pre-multiplication by $\mathbf{E}^{(i)}$ multiplies the ith row by k. Post-multiplication by $\mathbf{E}^{(i)}$ multiplies the ith column by k.

The $N \times N$ matrix, $\mathbf{E}^{(i+kj)}$ is defined as:

$$\mathbf{E}^{(i+kj)} = \begin{array}{c} \\ \\ (i) \\ \\ \\ (j) \\ \\ \\ \end{array} \begin{bmatrix} 1 & & & & & & \\ & \ddots & & & & & \\ & & 1 & \cdots & k & & \\ & & & 1 & \vdots & & \\ & & & & \ddots & & \\ & & & & 1 & & \\ & & & & & \ddots & \\ & & & & & & 1 \end{bmatrix}, \tag{A.110}$$

for $i < j$, and

$$\mathbf{E}^{(i+kj)} = \begin{array}{c} \\ \\ (i) \\ \\ (j) \\ \\ \\ \end{array} \begin{bmatrix} 1 & & & & & \\ & \ddots & & & & \\ & & 1 & & & \\ & & \vdots & \ddots & & \\ & & k & \cdots & 1 & \\ & & & & \ddots & \\ & & & & & 1 \end{bmatrix}, \tag{A.111}$$

for $i > j$.

Pre-multiplication by $\mathbf{E}^{(i+kj)}$ adds k times the jth row to the ith row. Post-multiplication by $\mathbf{E}^{(i+kj)}$ adds k times the jth column to the ith column.

An $N \times N$ matrix $\mathbf{P}_{per}$ is a **permutation matrix** if it can be obtained from the identity matrix $\mathbf{I}_N$ by interchanges of rows or columns.

A.3.2 The *vec*(A) matrix

Consider a $N \times M$ matrix $\mathbf{A}$ whose jth column is $\mathbf{a}_j$. The vec-function of $\mathbf{A}$ is written as $vec(\mathbf{A})$ and is obtained by stacking the columns to obtain an $NM \times 1$ vector:

$$vec(\mathbf{A}) = \begin{bmatrix} \mathbf{a}_1 \\ --- \\ \mathbf{a}_2 \\ --- \\ \vdots \\ \mathbf{a}_M \end{bmatrix}. \tag{A.112}$$

The function $vec(\mathbf{A})$ is closely related to the Kronecker product and will be useful when we study planar arrays.

Several properties will be useful:

(i)

$$vec\,(\alpha\mathbf{A} + \beta\mathbf{B}) = \alpha\, vec(\mathbf{A}) + \beta\, vec(\mathbf{B}), \tag{A.113}$$

where $\mathbf{A}$ and $\mathbf{B}$ are $N \times M$.

(ii) If $\mathbf{A}$ is $M \times M$, $\mathbf{B}$ is $M \times N$, and $\mathbf{C}$ is $N \times N$, then

$$vec\,(\mathbf{ABC}) = \left(\mathbf{C}^T \otimes \mathbf{A}\right) vec(\mathbf{B}). \tag{A.114}$$

(iii)

$$vec\,(\mathbf{AB}) = (\mathbf{I}_N \otimes \mathbf{A})\, vec(\mathbf{B}). \tag{A.115}$$

(iv)

$$vec\,(\mathbf{BC}) = \left(\mathbf{C}^T \otimes \mathbf{I}_M\right) vec(\mathbf{B}). \tag{A.116}$$

(v)

$$\text{tr}\,[\mathbf{ABCD}] = vec^H \mathbf{B}^H \left(\mathbf{A}^T \odot \mathbf{C}\right) vec\,[\mathbf{D}]. \tag{A.117}$$

A.3.3 Diagonal Matrices

A **diagonal matrix** is a square $N \times N$ matrix with $a_{ij} = 0$ for $i \neq j$. All elements of the principal diagonal are zero. Thus, $\mathbf{A}$ can be written as

$$\mathbf{A} = \begin{bmatrix} a_{11} & 0 & \cdots & 0 \\ 0 & a_{22} & \cdots & 0 \\ \vdots & \vdots & \ddots & \vdots \\ 0 & 0 & \cdots & a_{nn} \end{bmatrix}. \tag{A.118}$$

We also write

$$\mathbf{A} = \text{diag}\left[a_{11}, a_{22}, \cdots, a_{nn}\right]. \tag{A.119}$$

The inverse is

$$\mathbf{A}^{-1} = \text{diag}\left[a_{11}^{-1}, a_{22}^{-1}, \cdots, a_{nn}^{-1}\right]. \tag{A.120}$$

A **block diagonal matrix** is a square $N \times N$ block matrix,

$$\mathbf{A} = \begin{bmatrix} \mathbf{A}_{11} & 0 & \cdots & 0 \\ 0 & \mathbf{A}_{22} & \cdots & 0 \\ \vdots & \vdots & \ddots & \vdots \\ 0 & 0 & \cdots & \mathbf{A}_{mm} \end{bmatrix}, \tag{A.121}$$

where the submatrices $\mathbf{A}_{ii}$ are square and all other submatrices are zero. The dimensions of the $\mathbf{A}_{ii}$ need not be the same. We also write

$$\mathbf{A} = \text{block diag}\left[\mathbf{A}_{11}, \mathbf{A}_{22}, \cdots, \mathbf{A}_{mm}\right]. \tag{A.122}$$

If the $\mathbf{A}_{ii}$ are non-singular, then

$$\mathbf{A}^{-1} = \text{block diag}\left[\mathbf{A}_{11}^{-1}, \mathbf{A}_{22}^{-1}, \cdots, \mathbf{A}_{mm}^{-1}\right]. \tag{A.123}$$

The identity matrix defined in (A.40) is a special case of diagonal matrix whose diagonal elements equal one:

$$\mathbf{I} = \begin{bmatrix} 1 & 0 & \cdots & 0 \\ 0 & 1 & \cdots & 0 \\ \vdots & \vdots & \ddots & \vdots \\ 0 & 0 & \cdots & 1 \end{bmatrix}. \tag{A.124}$$

When it is necessary for clarity, we denote the order of the identity matrix by a subscript (e.g., $\mathbf{I}_N$).

A.3.4 Exchange Matrix and Conjugate Symmetric Vectors

The **exchange** or **reflection matrix J** is a square $N \times N$ matrix whose elements on the cross diagonal are unity and all other elements are zero. Thus,

$$\mathbf{J} = \begin{bmatrix} 0 & \cdots & 0 & 1 \\ 0 & \cdots & 1 & 0 \\ \vdots & \ddots & \vdots & \vdots \\ 1 & \cdots & 0 & 0 \end{bmatrix}. \tag{A.125}$$

When it is necessary for clarity we add the subscript N to denote the order of the exchange matrix (e.g., $\mathbf{J}_N$). We observe that $\mathbf{J}$ is symmetric, $\mathbf{J}^2 = \mathbf{I}$, and $\mathbf{J}^T = \mathbf{J}$.

Applying $\mathbf{J}$ to a vector exchanges the order of the elements,

$$\mathbf{J}\mathbf{x} = \begin{bmatrix} x_n \\ x_{n-1} \\ \vdots \\ x_2 \\ x_1 \end{bmatrix}. \tag{A.126}$$

We say that a vector $\mathbf{x}$ is **conjugate symmetric** if

$$\mathbf{x} = \mathbf{J}\mathbf{x}^*. \tag{A.127}$$

If $\mathbf{x}$ is $N \times 1$ and N is even, then we can denote the first $N/2$ elements as $\mathbf{x}_1$ and write

$$\mathbf{x} = \begin{bmatrix} \mathbf{x}_1 \\ \mathbf{J}\mathbf{x}_1^* \end{bmatrix}. \tag{A.128}$$

If N is odd, the first $(N-1)/2$ elements are denoted by $\mathbf{x}_1$ and

$$\mathbf{x} = \begin{bmatrix} \mathbf{x}_1 \\ x_0 \\ \mathbf{J}\mathbf{x}_1^* \end{bmatrix}, \tag{A.129}$$

where x_0 is a real scalar.

A vector $\mathbf{x}$ is **conjugate asymmetric** if

$$\mathbf{x} = -\mathbf{J}\mathbf{x}^*. \tag{A.130}$$

When $\mathbf{J}$ is applied to a matrix, it reverses the rows or columns of the matrix. Pre-multiplying by $\mathbf{J}$ reverses the order of the rows,

$$\mathbf{J}_N\mathbf{A} = \begin{bmatrix} a_{N1} & a_{N2} & \cdots & a_{NM} \\ \vdots & \vdots & & \vdots \\ a_{21} & a_{22} & \cdots & a_{2M} \\ a_{11} & a_{12} & \cdots & a_{1M} \end{bmatrix}. \tag{A.131}$$

Post-multiplying by $\mathbf{J}$ reverses the order of the columns,

$$\mathbf{A}\mathbf{J}_M = \begin{bmatrix} a_{1M} & \cdots & a_{12} & a_{11} \\ a_{2M} & \cdots & a_{22} & a_{21} \\ \vdots & & \vdots & \vdots \\ a_{NM} & \cdots & a_{N2} & a_{N1} \end{bmatrix}. \tag{A.132}$$

A.3.5 Persymmetric and Centrohermitian Matrices

A **persymmetric matrix** is an $N \times N$ matrix that is symmetric about its cross diagonal. For $N = 4$,

$$\mathbf{A} = \begin{bmatrix} a_{11} & a_{12} & a_{13} & a_{14} \\ a_{21} & a_{22} & a_{23} & a_{13} \\ a_{31} & a_{32} & a_{22} & a_{12} \\ a_{41} & a_{31} & a_{21} & a_{11} \end{bmatrix}. \tag{A.133}$$

This implies

$$a_{ij} = a_{N-j+1,N-i+1}. \tag{A.134}$$

It follows that

$$\mathbf{A}^T = \mathbf{JAJ}, \tag{A.135}$$

and

$$\mathbf{A} = \mathbf{JA}^T\mathbf{J}. \tag{A.136}$$

A **centrohermitian matrix** is an $N \times N$ matrix with the property that

$$a_{ij} = a^*_{N-i+1,N-j+1}. \tag{A.137}$$

A centrohermitian matrix that is also Hermitian exhibits a double symmetry. It is Hermitian around the principal diagonal and persymmetric around the cross diagonal.

$$a_{ij} = a^*_{ji} = a_{N-j+1,N-i+1} = a^*_{N-i+1,N-j+1}. \tag{A.138}$$

For $N = 4$,

$$\mathbf{A} = \begin{bmatrix} a_{11} & a^*_{21} & a^*_{31} & a^*_{41} \\ a_{21} & a_{22} & a^*_{32} & a^*_{31} \\ a_{31} & a_{32} & a_{22} & a^*_{21} \\ a_{41} & a_{31} & a_{21} & a_{11} \end{bmatrix}. \tag{A.139}$$

For centrohermitian-persymmetric matrices,

$$\mathbf{A} = \mathbf{JA}^*\mathbf{J}. \tag{A.140}$$

If N is even, a centrohermitian-persymmetric matrix may be partitioned as

$$\mathbf{A} = \left[\begin{array}{c|c} \mathbf{B}_1 & \mathbf{B}_2 \\ \hline \mathbf{JB}_2^*\mathbf{J} & \mathbf{JB}_1^*\mathbf{J} \end{array}\right], \tag{A.141}$$

where the partitioned matrices have dimensions $\frac{N}{2} \times \frac{N}{2}$.

If N is odd, a centrohermitian-persymmetric matrix may be partitioned as

$$\mathbf{A} = \begin{bmatrix} \mathbf{B}_1 & \mathbf{a} & \mathbf{B}_2 \\ \mathbf{a}^H & c & \mathbf{a}^H\mathbf{J} \\ \mathbf{JB}_2^*\mathbf{J} & \mathbf{Ja} & \mathbf{JB}_2^*\mathbf{J} \end{bmatrix}, \tag{A.142}$$

where $\mathbf{B}_1$ is a $(N-1)/2 \times (N-1)/2$ matrix, $\mathbf{B}_2$ is a $(N-1)/2 \times (N-1)/2$ matrix, $\mathbf{a}$ is a $(N-1)/2 \times 1$ vector, and c is a real constant.

These matrices will arise frequently in our study of linear arrays and their structure will lead to significant computational savings and analytic simplifications.

A.3.6 Toeplitz and Hankel Matrices

A **Toeplitz matrix** has the property that all of the elements along each diagonal are identical. Thus,

$$a_{ij} = a_{i-j}. \tag{A.143}$$

For example,

$$\mathbf{A} = \begin{bmatrix} a_0 & a_{-1} & a_{-2} & a_{-3} & a_{-4} \\ a_1 & a_0 & a_{-1} & a_{-2} & a_{-3} \\ a_2 & a_1 & a_0 & a_{-1} & a_{-2} \\ a_3 & a_2 & a_1 & a_0 & a_{-1} \end{bmatrix}. \tag{A.144}$$

If $\mathbf{A}$ is square, then it is a special case of a persymmetric matrix. If $\mathbf{A}$ is also Hermitian, then it is centrohermitian. The inverse of a Toeplitz matrix is persymmetric.

A **Hankel matrix** has the property that the elements along every cross diagonal are equal. Thus,

$$a_{ij} = a_{i+j-N-1}. \tag{A.145}$$

For example,

$$\mathbf{A} = \begin{bmatrix} a_{-3} & a_{-2} & a_{-1} & a_0 \\ a_{-2} & a_{-1} & a_0 & a_1 \\ a_{-1} & a_0 & a_1 & a_2 \\ a_0 & a_1 & a_2 & a_3 \\ a_1 & a_2 & a_3 & a_4 \end{bmatrix}. \tag{A.146}$$

Note that **JA** and **AJ** are Toeplitz matrices when **A** is a Hankel matrix.

A.3.7 Circulant Matrices

A **circulant matrix** is an $N \times N$ square matrix made up of N elements.

The elements of a right-circulant matrix obey the relationship

$$a_{R,ij} = \begin{cases} a_{j-i}, & j-i \geq 0, \\ a_{N-j+1}, & j-i<0, \quad 1 \leq i, \quad j \leq N. \end{cases} \tag{A.147}$$

For example,

$$\mathbf{A}_R = \begin{bmatrix} a_0 & a_1 & a_2 & a_3 \\ a_3 & a_0 & a_1 & a_2 \\ a_2 & a_3 & a_0 & a_1 \\ a_1 & a_2 & a_3 & a_0 \end{bmatrix}. \tag{A.148}$$

Each row is obtained from the row above it by shifting each element right one column and bringing the last element on the right to the first column. A right-circulant matrix is a special case of a Toeplitz matrix.

The elements of a left-circulant matrix obey the relationship

$$a_{L,ij} = \begin{cases} a_{N+1-i-j}, & j+i \leq N+1, \\ a_{2N+1-i-j}, & j+i > N+1, \quad 1 \leq i, \quad j \leq N. \end{cases} \tag{A.149}$$

For example,

$$\mathbf{A}_L = \begin{bmatrix} a_3 & a_2 & a_1 & a_0 \\ a_2 & a_1 & a_0 & a_3 \\ a_1 & a_0 & a_3 & a_2 \\ a_0 & a_3 & a_2 & a_1 \end{bmatrix}. \tag{A.150}$$

Each row is obtained from the row above it by shifting each element left one column and bringing the last element on the left to the last column.

A left-circulant matrix is a special case of a Hankel matrix.

A.3.8 Triangular Matrices

A **lower triangular** square $N \times N$ matrix is defined as a matrix whose elements above the main diagonal are zero. Thus, we can write,

$$\mathbf{L} = \begin{bmatrix} l_{11} & 0 & \cdots & 0 \\ l_{21} & l_{22} & \cdots & 0 \\ \vdots & \vdots & \ddots & \vdots \\ l_{n1} & l_{n2} & \cdots & l_{nn} \end{bmatrix}. \tag{A.151}$$

$\mathbf{L}^{-1}$ is also lower triangular. The determinant is

$$\det(\mathbf{L}) = \prod_{i=1}^{N} l_{ii}. \tag{A.152}$$

An **upper triangular** square $N \times N$ matrix is defined as a matrix whose elements below the main diagonal are zero. Thus, we can write,

$$\mathbf{U} = \begin{bmatrix} u_{11} & u_{12} & \cdots & u_{1n} \\ 0 & u_{22} & \cdots & u_{2n} \\ \vdots & \vdots & \ddots & \vdots \\ 0 & 0 & \cdots & u_{nn} \end{bmatrix}. \tag{A.153}$$

$\mathbf{U}^{-1}$ is also upper triangular. The determinant is

$$\det(\mathbf{U}) = \prod_{i=1}^{N} u_{ii}. \tag{A.154}$$

Clearly, $\mathbf{L}^T$ and $\mathbf{L}^H$ are upper triangular and $\mathbf{U}^T$ and $\mathbf{U}^H$ are lower triangular.

If $\mathbf{A}$ is a square Hermitian positive definite matrix, there are two factorizations of interest. The first factorization is called the $\mathbf{LDL}^H$ factorization. There is a unique factorization of $\mathbf{A}$,

$$\mathbf{A} = \mathbf{LDL}^H, \tag{A.155}$$

where $\mathbf{D}$ is a diagonal matrix with positive entries. (e.g., Golub and Van Loan [GVL89]).

The **Cholesky decomposition** is a unique factorization,

$$\mathbf{A} = \mathbf{G}\mathbf{G}^H, \tag{A.156}$$

where $\mathbf{G}$ is lower triangular,

$$\mathbf{G} = \mathbf{L} \operatorname{diag}\left[\sqrt{d_1}, \quad \sqrt{d_2}, \quad \cdots \quad \sqrt{d_N} \right]. \tag{A.157}$$

The Cholesky decomposition plays a key role in many algorithms and analyses.

A.3.9 Unitary and Orthogonal Matrices

A square $N \times N$ matrix is **unitary** if

$$\mathbf{A}^{-1} = \mathbf{A}^H. \tag{A.158}$$

If $\mathbf{A}$ is unitary, then

$$\mathbf{A}^H\mathbf{A} = \mathbf{A}^{-1}\mathbf{A} = \mathbf{A}\mathbf{A}^{-1} = \mathbf{A}\mathbf{A}^H = \mathbf{I}. \tag{A.159}$$

In order for $\mathbf{A}$ to be unitary, the columns must be orthonormal. Thus,

$$\mathbf{a}_i^H \mathbf{a}_j = \delta_{ij}. \tag{A.160}$$

A particular unitary matrix that is used in the text has columns that are conjugate symmetric and has a sparse structure.

For N even,

$$\mathbf{Q} = \frac{1}{\sqrt{2}} \begin{bmatrix} \mathbf{I} & j\mathbf{I} \\ \mathbf{J} & -j\mathbf{J} \end{bmatrix}, \tag{A.161}$$

where the $\mathbf{I}$ and $\mathbf{J}$ matrices have dimension $N/2$.

For N odd,

$$\mathbf{Q} = \frac{1}{\sqrt{2}} \begin{bmatrix} \mathbf{I} & \mathbf{0} & j\mathbf{I} \\ \mathbf{0}^T & \sqrt{2} & \mathbf{0}^T \\ \mathbf{I} & \mathbf{0} & -j\mathbf{J} \end{bmatrix}, \tag{A.162}$$

where the $\mathbf{I}$ and $\mathbf{J}$ matrices have dimension $(N-1)/2$ and $\mathbf{0}$ is a $(N-1)/2 \times 1$ vector whose elements are 0.

The $\mathbf{Q}$ matrices in (A.161) and (A.162) have a useful property. If $\mathbf{x}$ is a conjugate symmetric vector, then $\mathbf{Q}^H\mathbf{x}$ is real. For N even,

$$\begin{aligned} \mathbf{Q}^H\mathbf{x} &= \frac{1}{\sqrt{2}}\begin{bmatrix} \mathbf{I} & \mathbf{J} \\ -j\mathbf{I} & j\mathbf{J} \end{bmatrix}\begin{bmatrix} \mathbf{x}_1 \\ \mathbf{J}\mathbf{x}_1^* \end{bmatrix} = \frac{1}{\sqrt{2}}\begin{bmatrix} \mathbf{x}_1 + \mathbf{x}_2^* \\ -j(\mathbf{x}_1 - \mathbf{x}_2^*) \end{bmatrix} \\ &= \sqrt{2}\begin{bmatrix} Re(\mathbf{x}_1) \\ Im(\mathbf{x}_1) \end{bmatrix}. \end{aligned} \tag{A.163}$$

Other useful properties will be derived in Chapter 7.

An important consequence of the unitary property arises when we do a unitary transformation of a vector whose correlation matrix is the identity matrix,

$$\mathbf{y} = \mathbf{A}^H\mathbf{x}. \tag{A.164}$$

The correlation matrix of $\mathbf{y}$ is

$$E\left[\mathbf{y}\mathbf{y}^H\right] = \mathbf{A}^H E\left[\mathbf{x}\mathbf{x}^H\right]\mathbf{A} = \mathbf{A}^H\mathbf{I}\mathbf{A} = \mathbf{A}^H\mathbf{A} = \mathbf{A}^{-1}\mathbf{A} = \mathbf{I}. \tag{A.165}$$

Thus, the correlation matrix of the output $\mathbf{y}$ is also an identity matrix.

A square $N \times N$ matrix is **orthogonal** if

$$\mathbf{A}^{-1} = \mathbf{A}^T. \tag{A.166}$$

In order for $\mathbf{A}$ to be orthogonal

$$\mathbf{a}_i^T\mathbf{a}_j = \delta_{ij}. \tag{A.167}$$

A.3.10 Vandermonde Matrices

A **Vandermonde matrix** is an $N \times M$ matrix in which the elements in the jth column can be expressed as powers of a parameter c_j. Thus,

$$a_{ij} = c_j^{i-1}, \quad 1 \le i \le N, \ 1 \le j \le M. \tag{A.168}$$

The matrix has the structure,

$$\mathbf{A} = \begin{bmatrix} 1 & 1 & \cdots & 1 \\ c_1 & c_2 & \cdots & c_M \\ c_1^2 & c_2^2 & \cdots & c_M^2 \\ \vdots & \vdots & & \vdots \\ c_1^{N-1} & c_2^{N-1} & \cdots & c_M^{N-1} \end{bmatrix}. \tag{A.169}$$

One can show that the determinant of a square $N \times N$ Vandermonde matrix is given by

$$\det \mathbf{A} = \prod_{1 \le i < k \le N} (c_k - c_i). \tag{A.170}$$

A.3.11 Projection Matrices

A matrix that will play a central role in many of our analyses is the projection matrix $\mathbf{P}$. An **idempotent matrix P** satisfies the relation

$$\mathbf{P}^2 = \mathbf{P}. \tag{A.171}$$

Another name for an idempotent matrix is a **projection matrix**. We restrict our discussion to projection matrices that are Hermitian,

$$\mathbf{P}^H = \mathbf{P}. \tag{A.172}$$

To motivate the construction of $\mathbf{P}$, we first consider the projection of a vector $\mathbf{b}$ onto a vector $\mathbf{a}$. We first construct a unit vector in the direction of $\mathbf{a}$,

$$\mathbf{u}_a = \frac{\mathbf{a}}{|\mathbf{a}|^2} = \left[\mathbf{a}^H\mathbf{a}\right]^{-1}\mathbf{a} = \mathbf{a}\left[\mathbf{a}^H\mathbf{a}\right]^{-1}. \tag{A.173}$$

Then, the projection of $\mathbf{b}$ onto $\mathbf{a}$ is

$$\mathbf{a}\left[\mathbf{u}_a^H\mathbf{b}\right] = \mathbf{a}\left[\mathbf{a}^H\mathbf{a}\right]^{-1}\mathbf{a}^H\mathbf{b}. \tag{A.174}$$

If we define the $N \times N$ matrix

$$\mathbf{P}_{\mathbf{a}} \triangleq \mathbf{a}\left[\mathbf{a}^H\mathbf{a}\right]^{-1}\mathbf{a}^H, \tag{A.175}$$

Then we can write the projection of $\mathbf{b}$ onto $\mathbf{a}$ as $\mathbf{P_a b}$. The matrix $\mathbf{P_a}$ is referred to as a projection matrix.

Now consider the complex $N \times M$ matrix, $\mathbf{V}$, and assume that the columns are linearly independent. Then, the projection of a complex $N \times 1$ vector $\mathbf{b}$ onto the M-dimensional subspace defined by the columns of $\mathbf{V}$ is,

$$\mathbf{b_V} = \mathbf{P_V b}, \tag{A.176}$$

where $\mathbf{P_V}$ is an $N \times N$-dimensional matrix defined by

$$\mathbf{P_V} = \mathbf{V}\left[\mathbf{V}^H\mathbf{V}\right]^{-1}\mathbf{V}^H. \tag{A.177}$$

We can also define a projection matrix that projects $\mathbf{b}$ onto a subspace orthogonal to the subspace defined by the columns of $\mathbf{V}$,

$$\mathbf{P_V^\perp} \triangleq \mathbf{I} - \mathbf{P_V}. \tag{A.178}$$

All of the vectors in this subspace are orthogonal to $\mathbf{V}$.

Note that, if the columns of $\mathbf{V}$ are orthonormal (i.e., it is a unitary matrix), then

$$\mathbf{P_V} = \mathbf{U}\mathbf{U}^H, \tag{A.179}$$

and

$$\mathbf{P_V^\perp} = \mathbf{I} - \mathbf{U}\mathbf{U}^H. \tag{A.180}$$

In many of our applications, we will find it useful to divide our N-dimensional "signal subspace," which contains both signal and'noise, and an $N-M$-dimensional "noise subspace," which contains only noise. Recall that we used this subspace technique in Chapter 2 of [VT68], [VT01a] when we solved the temporal detection problem.

A.3.12 Generalized Inverse

In this section we define a generalized inverse[2] for the $M \times N$ matrix $\mathbf{A}$.

---

[2]Our discussion follows Section 12.8 in [LT85].

A.3.12.1 Moore-Penrose pseudo-inverse

Assume that $\mathbf{A}$ is an $M \times N$ matrix. There exists a unique $N \times M$ matrix $\mathbf{B}$ for which

$$\mathbf{ABA} = \mathbf{A}, \tag{A.181}$$

$$\mathbf{BAB} = \mathbf{B}, \tag{A.182}$$

$$(\mathbf{AB})^H = \mathbf{AB}, \tag{A.183}$$

and

$$(\mathbf{BA})^H = \mathbf{BA}. \tag{A.184}$$

This matrix $\mathbf{B}$ is denoted by $\mathbf{A}^\dagger$, and

$$\mathbf{A}^\dagger = \mathbf{A}^H(\mathbf{AA}^H)^{-1} \qquad M \leq N, \tag{A.185}$$

$$\mathbf{A}^\dagger = \mathbf{A}^{-1} \qquad M = N, \tag{A.186}$$

$$\mathbf{A}^\dagger = (\mathbf{A}^H\mathbf{A})^{-1}\mathbf{A}^H \qquad M \geq N. \tag{A.187}$$

The matrix $\mathbf{A}^\dagger$ is referred to as the Moore-Penrose pseudo-inverse or the Moore-Penrose inverse in the literature ([Moo20] and [Pen55]). It is encountered in the solution of linear equations.

A.3.12.2 Application to the solution of $\mathbf{Ax} = \mathbf{b}$.

Consider the linear equation, $\mathbf{Ax} = \mathbf{b}$, where $\mathbf{A}$ is an $M \times N$ matrix, $\mathbf{x}$ is an $N \times 1$ vector, and $\mathbf{b}$ is an $M \times 1$ vector.

If $M < N$, there are multiple solutions. The solution,

$$\mathbf{x}_0 = \mathbf{A}^\dagger\mathbf{b}, \tag{A.188}$$

is one of the solutions,

$$\mathbf{Ax}_0 = \mathbf{AA}^\dagger\mathbf{b} = \mathbf{b}. \tag{A.189}$$

It is the solution with the smallest Euclidean norm.

If $M = N$ and $\mathbf{A}$ is non-singular, $\mathbf{x}_0$ is the unique solution and $\mathbf{A}^\dagger = \mathbf{A}^{-1}$.

If $M > N$, there are, in general, no solutions to the equation. We define an approximate solution of the overdetermined equation $\mathbf{Ax} = \mathbf{b}$ to be a vector $\mathbf{x}_0$ given by

$$\mathbf{x}_0 = \arg\min_{\mathbf{x}} \|\mathbf{Ax} - \mathbf{b}\|, \tag{A.190}$$

where the norm is the Euclidean vector norm and the minimization is over all $N \times 1$ vectors. The solution to (A.190) is

$$\mathbf{x}_0 = \mathbf{A}^\dagger\mathbf{b}. \tag{A.191}$$

A.4 Eigensystems

In Chapters 2 and 3 of DEMT I [VT68], [VT01a] and Chapter 2 of DEMT III [VT71], [VT01b], we saw the advantages of an eigenfunction decomposition for solving detection and estimation problems. We will find similar advantages in the array processing area.

In Section A.4.1, we develop the concept of eigenvectors and eigenvalues and discuss properties. In Section A.4.2, we consider several matrices whose eigenvectors have special properties.

A.4.1 Eigendecomposition

Consider a square Hermitian $N \times N$ matrix $\mathbf{A}$ (symmetric if $\mathbf{A}$ is real). The **eigenvectors** of $\mathbf{A}$ are denoted by $\boldsymbol{\phi}$. They satisfies the equation,

$$\mathbf{A}\boldsymbol{\phi} = \lambda\boldsymbol{\phi}. \tag{A.192}$$

Thus when $\boldsymbol{\phi}$ is operated on by $\mathbf{A}$ the output is $\boldsymbol{\phi}$ multiplied by a scalar λ. The scalar λ is called the **eigenvalue** of $\mathbf{A}$. The relation in (A.192) can also be written as

$$[\lambda\mathbf{I} - \mathbf{A}]\,\boldsymbol{\phi} = \mathbf{0}. \tag{A.193}$$

In order to have a solution for the homogeneous equation in (A.193), we must have

$$\det\left[\lambda\mathbf{I} - \mathbf{A}\right] = 0. \tag{A.194}$$

This is an Nth order polynomial in λ, which is called the **characteristic polynomial**,

$$f(\lambda) \triangleq \det\left[\lambda\mathbf{I} - \mathbf{A}\right]. \tag{A.195}$$

The N roots of $f(\lambda)$ are the eigenvalues of $\mathbf{A}$:

$$f(\lambda) = \prod_{i=1}^{N}(\lambda_i - \lambda) = \sum_{k=0}^{N} f_k\lambda^k. \tag{A.196}$$

There is an eigenvector (there may be more than one) associated with each λ. The eigenvectors associated with unequal eigenvalues are linearly independent. For the distinct eigenvalues, we choose the eigenvectors to be orthonormal vectors. For eigenvalues of multiplicity M, we can construct orthonormal eigenvectors.

The eigenvalues of a Hermitian matrix are all real. We order the eigenvalues in decreasing size,

$$\lambda_{max} = \lambda_1 \geq \lambda_2 \geq \lambda_3 \geq \cdots \geq \lambda_N = \lambda_{min}. \tag{A.197}$$

The eigenvectors associated with larger eigenvalues are referred to as **principal eigenvectors.**

Several properties follow:

(i) The highest coefficient in (A.196) is

$$f_N = (-1)^N. \tag{A.198}$$

(ii) The lowest coefficient in (A.196) is

$$f_0 = \det(\mathbf{A}) = \prod_{i=1}^{N} \lambda_i. \tag{A.199}$$

(iii) The sum of eigenvalues equals $\text{tr}(\mathbf{A})$,

$$\text{tr}\ (\mathbf{A}) = \sum_{i=1}^{N} \lambda_i. \tag{A.200}$$

(iv) If β is a scalar, then the eigenvalues of $\beta\mathbf{A}$ are $\beta\lambda_i; i = 1, \cdots, N$.

(v) The eigenvalues of the matrix $\mathbf{A}^m$ for m, a positive integer, are λ_i^m.

(vi) If $\mathbf{A}$ is non-singular, then the eigenvalues of $\mathbf{A}^{-1}$ are λ_i^{-1} and the eigenvectors are the eigenvectors of $\mathbf{A}$.

(vii) If $|\lambda_{\max}(\mathbf{A})| < 1$, then

$$(\mathbf{I} + \mathbf{A})^{-1} = \mathbf{I} - \mathbf{A} + \mathbf{A}^2 - \mathbf{A}^3 + \cdots. \tag{A.201}$$

(viii) The identity matrix $\mathbf{I}$ has N eigenvalues equal to unity and any set of N orthonormal vectors can be used as eigenvectors.

(ix) The eigenvalues of $\mathbf{A} + \sigma^2\mathbf{I}$ are $\lambda_i + \sigma^2$ and the eigenvectors are the eigenvectors of $\mathbf{A}$.

(x) If we define the $N \times N$ matrix

$$\mathbf{U_A} = \left[\, \boldsymbol{\phi}_1 \, \vdots \, \boldsymbol{\phi}_2 \, \vdots \, \cdots \, \vdots \, \boldsymbol{\phi}_N \, \right], \tag{A.202}$$

where the $\boldsymbol{\phi}_i, i = 1, 2, \cdots, N$ are the orthonormal eigenvectors and the $N \times N$ diagonal matrix,

$$\boldsymbol{\Lambda} = \text{ diag } [\lambda_1, \lambda_2, \cdots, \lambda_N], \tag{A.203}$$

then we can write $\mathbf{A}$ as

$$\mathbf{A} = \mathbf{U_A} \boldsymbol{\Lambda} \mathbf{U_A^H}, \tag{A.204}$$

(xi) If $\mathbf{A}$ is non-singular, we can write

$$\mathbf{A}^{-1} = \mathbf{U_A} \boldsymbol{\Lambda}^{-1} \mathbf{U_A}. \tag{A.205}$$

(xii) We can also expand $\mathbf{A}$ as,

$$\mathbf{A} = \sum_{i=1}^{N} \lambda_i \boldsymbol{\phi}_i \boldsymbol{\phi}_i^H. \tag{A.206}$$

(xiii) If $\mathbf{A}$ is non-singular,

$$\mathbf{A}^{-1} = \sum_{i=1}^{N} \frac{1}{\lambda_i} \boldsymbol{\phi}_i \boldsymbol{\phi}_i^H. \tag{A.207}$$

(xiv) If $\mathbf{A}$ is positive definite, then all eigenvalues are positive.

(xv) If $\mathbf{A}$ is positive semi-definite, the number of positive eigenvalues equals rank $(\mathbf{A})$ and the remaining eigenvalues are zero.

(xvi) If the $N \times N$ matrix $\mathbf{A}$ consists of a weighted sum of D outer products,

$$\mathbf{A} = \sum_{i=1}^{D} \sigma_i^2 \mathbf{v}_i \mathbf{v}_i^H, \tag{A.208}$$

then rank $(\mathbf{A}) = D$ and the first D eigenvectors are linear combinations of the $\mathbf{v}_i$ and the eigenvalues are denoted by λ_i. There are $N - D$ eigenvalues equal to zero and the corresponding eigenvectors can be chosen to be any orthonormal set that are also orthogonal to the $\mathbf{v}_i$, $i = 1, \cdots, D$.

(xvii) If $\mathbf{A}$ is an $N \times N$ matrix that can be written as

$$\mathbf{A} = \sum_{i=1}^{D} \sigma_i^2 \mathbf{v}_i \mathbf{v}_i^H + \sigma_w^2 \mathbf{I}, \tag{A.209}$$

then the eigenvectors in (xvi) are still valid and

$$\lambda_i(\mathbf{A}) = \begin{cases} \lambda_i + \sigma_w^2, & i = 1, \cdots, D, \\ \\ \sigma_w^2, & i = D+1, \cdots, N. \end{cases} \tag{A.210}$$

In this case, we often divide the N-dimensional space into two orthogonal subspaces, a signal subspace and a noise subspace. Then,

$$\mathbf{A} = \mathbf{U}_S \mathbf{\Lambda}_S \mathbf{U}_S^H + \mathbf{U}_N \mathbf{\Lambda}_N \mathbf{U}_N^H, \tag{A.211}$$

where $\mathbf{U}_S$ is a $N \times D$ matrix of eigenvectors,

$$\mathbf{U}_S = \left[\ \boldsymbol{\phi}_1 \ \vdots\ \boldsymbol{\phi}_2 \ \vdots\ \cdots \ \vdots\ \boldsymbol{\phi}_D\ \right], \tag{A.212}$$

and $\mathbf{\Lambda}_S$ is a diagonal matrix,

$$\mathbf{\Lambda}_S = \text{ diag } \left[\lambda_1 + \sigma_w^2, \lambda_2 + \sigma_w^2, \cdots, \lambda_D + \sigma_w^2\right]. \tag{A.213}$$

The matrix $\mathbf{U}_N$ is a $N \times (N - D)$ matrix of orthonormal vectors,

$$\mathbf{U}_N = \left[\ \boldsymbol{\phi}_{D+1} \ \vdots\ \boldsymbol{\phi}_{D+2} \ \vdots\ \cdots \ \vdots\ \boldsymbol{\phi}_N\ \right], \tag{A.214}$$

where

$$\boldsymbol{\phi}_i^H \boldsymbol{\phi}_j = 0, \qquad i = D+1, \cdots, N, \;\; j = 1, \cdots, D, \tag{A.215}$$

and

$$\mathbf{\Lambda}_N = \sigma_w^2 \mathbf{I}_{(N-D)}. \tag{A.216}$$

(xviii) In many cases of interest, $\mathbf{A}$ is a correlation matrix, $\mathbf{R_x}$,

$$\mathbf{R_x} = E\left[\mathbf{x}\mathbf{x}^H\right], \tag{A.217}$$

where $\mathbf{x}$ is a zero-mean random vector. Then we can expand

$$\mathbf{x} = \sum_{i=1}^{N} x_i \boldsymbol{\phi}_i, \tag{A.218}$$

where the $\boldsymbol{\phi}_i$ are the eigenvectors of $\mathbf{R_x}$, and

$$x_i = \mathbf{x}^H \boldsymbol{\phi}_i = \boldsymbol{\phi}_i^H \mathbf{x}. \tag{A.219}$$

The x_i are uncorrelated,

$$E\left[x_i x_j\right] = \lambda_i \delta_{ij}. \tag{A.220}$$

(xix) If $\mathbf{x}$ is a Gaussian random vector, then the x_i are statistically independent zero-mean Gaussian random variables with variance λ_i.

A.4.2 Special Matrices

In this section, we discuss several matrices whose eigendecompositions have special properties. The sections discuss:

(i) Separable kernels,

(ii) Centrohermitian matrices,

(iii) Circulant matrices,

(iv) Toeplitz matrices,

(v) Kronecker products.

A.4.2.1 Separable kernels

We use the term separable kernel to describe matrices of the form[3]

$$\mathbf{A} = \mathbf{V}\mathbf{S}\mathbf{V}^H, \tag{A.221}$$

where $\mathbf{V}$ is a $N \times K$ matrix whose columns are the array manifold vector $\mathbf{v}$ and $\mathbf{S}$ is a $K \times K$ Hermitian matrix. This corresponds to the model for K correlated signals impinging on a N-element array. We assume $K < N$. This matrix structure allows us to reduce the eigenanalysis to a K-dimensional problem instead of a N-dimensional problem.

The eigenequation is

$$\lambda\boldsymbol{\phi} = \mathbf{V}\mathbf{S}\mathbf{V}^H\boldsymbol{\phi}. \tag{A.222}$$

Since the rank of $\mathbf{A}$ is K, there will be K non-zero eigenvalues and corresponding eigenvectors. The vectors $\mathbf{v}_i, i = 1, 2, \cdots, K$ define a K-dimensional subspace. The $\boldsymbol{\phi}$ vectors are the orthogonal basis for that K-dimensional subspace. Therefore, each $\boldsymbol{\phi}_i$ can be obtained from $\mathbf{V}$ by a linear transformation.

$$\boldsymbol{\phi}_i = \mathbf{V}\mathbf{c}_i, \qquad i = 1, 2, \cdots, K, \tag{A.223}$$

where $\mathbf{c}_i$ is a $K \times 1$ vector.

Using (A.223) in (A.222) gives

$$\lambda\mathbf{V}\mathbf{c}_i = \mathbf{V}\mathbf{S}\mathbf{V}^H\mathbf{V}\mathbf{c}_i. \tag{A.224}$$

Equation (A.224) can be written as

$$\mathbf{V}\left\{\left[\lambda_i\mathbf{I} - \mathbf{S}\mathbf{V}^H\mathbf{V}\right]\mathbf{c}_i\right\} = \mathbf{0}. \tag{A.225}$$

In order for (A.225) to be satisfied, we require

$$\left[\lambda_i\mathbf{I} - \mathbf{S}\mathbf{V}^H\mathbf{V}\right]\mathbf{c}_i = \mathbf{0}. \tag{A.226}$$

Thus, the $\mathbf{c}_i$ are the eigenvectors of the $K \times K$ matrix $\mathbf{S}\mathbf{V}^H\mathbf{V}$.

This is a homogeneous equation that has a solution for K values of $\lambda_i, i = 1, 2, \cdots, K$. In order for (A.226) to have a solution,

$$\det\left[\lambda\mathbf{I} - \mathbf{S}\mathbf{V}^H\mathbf{V}\right] = 0, \tag{A.227}$$

[3] The use of the separable kernel descriptor follows from our separable kernel discussions in Chapter 3 of [VT68], [VT01a] and Chapters 3 and 4 of [VT71], [VT01b].

which will have K solutions: $\lambda_i, i = 1, 2, \cdots, K$.

For each λ_i we can find the corresponding $\mathbf{c}_i$, which determines the corresponding $\boldsymbol{\phi}_i, i = 1, 2, \cdots, K$.

In many cases of interest, $K \ll N$, so that we have achieved a significant reduction in the complexity of the problem.

Example A.4.1

Consider a simple case in which $K = 2$ and $\mathbf{S}$ is diagonal matrix $\sigma_s^2\mathbf{I}$. Then,

$$\mathbf{V}^H\mathbf{S}\mathbf{V} = N\sigma_s^2 \begin{bmatrix} 1 & \frac{\mathbf{v}_1^H\mathbf{v}_2}{N} \\ \frac{\mathbf{v}_2^H\mathbf{v}_1}{N} & 1 \end{bmatrix}. \tag{A.228}$$

The determinant of (A.228) is a quadratic equation in λ,

$$\lambda^2 - b\lambda + c = 0, \tag{A.229}$$

where

$$b = 2N\sigma_s^2, \tag{A.230}$$

and

$$c = N^2\sigma_s^4\left[1 - |\beta_{12}|^2\right], \tag{A.231}$$

where $\beta_{12} \triangleq \frac{\mathbf{v}_1^H\mathbf{v}_2}{N}$. Then,

$$\lambda = N\sigma_s^2\left[1 \pm \sqrt{1 - \frac{4c}{b^2}}\right], \tag{A.232}$$

or

$$\lambda_{1(2)} = N\sigma_s^2\left[1 \pm |\beta_{12}|\right]. \tag{A.233}$$

Then $\mathbf{c}_1$ and $\mathbf{c}_2$ are obtained by substituting λ_1 and λ_2 from (A.233) into (A.226) to obtain,

$$\mathbf{c}_1 = \frac{1}{\sqrt{2N\left[1 + |\beta_{12}|\right]}}\begin{bmatrix} 1 \\ e^{-j\phi(\beta_{12})} \end{bmatrix}, \tag{A.234}$$

$$\mathbf{c}_1 = \frac{1}{\sqrt{2N\left[1 - |\beta_{12}|\right]}}\begin{bmatrix} 1 \\ -e^{-j\phi(\beta_{12})} \end{bmatrix}, \tag{A.235}$$

where $\phi(\beta_{12})$ denotes the phase of β_{12}. Substituting (A.234) and (A.235) into (A.223) gives,

$$\boldsymbol{\phi}_1 = \frac{\left[\mathbf{v}_1 + e^{-j\phi(\beta_{12})}\mathbf{v}_2\right]}{\sqrt{2N(1+|\beta_{12}|)}} \tag{A.236}$$

and

$$\boldsymbol{\phi}_2 = \frac{\left[\mathbf{v}_1 - e^{-j\phi(\beta_{12})}\mathbf{v}_2\right]}{\sqrt{2N(1-|\beta_{12}|)}}, \tag{A.237}$$

Plots of the λ_i and $\mathbf{c}_i$ are shown in Chapter 5 for typical array geometries.

A.4.2.2 Centrohermitian matrices

Centrohermitian-persymmetric matrices have the property that their eigenvectors are either conjugate symmetric or conjugate asymmetric [CB76b], [CB76a] (see, e.g., [Mar87], [Mak81] for further discussion).[4]

If the dimension N of the matrix $\mathbf{C}$ is even, then it has $\frac{N}{2}$ conjugate symmetric orthogonal eigenvectors $\boldsymbol{\phi}_i$ $\left(i = 1, \cdots, \frac{N}{2}\right)$ of the form

$$\boldsymbol{\phi}_i = \frac{1}{\sqrt{2}} \begin{bmatrix} \mathbf{y}_i \\ \mathbf{J}\mathbf{y}_i \end{bmatrix}, \tag{A.238}$$

where the $\mathbf{y}_i$ are the orthonormal eigenvectors of the matrix $\mathbf{B}_1 + \mathbf{B}_2\mathbf{J}$, where $\mathbf{B}_1$ and $\mathbf{B}_2$ are the submatrices defined in (A.141). Thus,

$$\left[\mathbf{B}_1 + \mathbf{B}_2\mathbf{J}\right]\mathbf{y}_i = \lambda_i \mathbf{y}_i. \tag{A.239}$$

It also has $\frac{N}{2}$ conjugate asymmetric orthonormal eigenvectors $\boldsymbol{\phi}_i$ $\left(i = 1, \cdots, \frac{N}{2}\right)$ of the form

$$\boldsymbol{\phi}_i = \frac{1}{\sqrt{2}} \begin{bmatrix} \mathbf{z}_i \\ -\mathbf{J}\mathbf{z}_i \end{bmatrix}, \tag{A.240}$$

where the $\mathbf{z}_i$ are the eigenvectors of the matrix $\mathbf{B}_1 - \mathbf{B}_2\mathbf{J}$. Thus,

$$\left[\mathbf{B}_1 - \mathbf{B}_2\mathbf{J}\right]\mathbf{z}_i = \lambda_i \mathbf{z}_i. \tag{A.241}$$

Note that we have reduced the dimension of the eigendecomposition problem by factor of 2.

If N is odd, then $\mathbf{C}$ will have $\frac{N+1}{2}$ conjugate symmetric orthonormal eigenvectors $\boldsymbol{\phi}_i$ $\left(i = 1, 2, \cdots, \frac{(N+1)}{2}\right)$ of the form

[4]The results in this section are due to [CB76a], [CB76b].

$$\phi_i = \frac{1}{\sqrt{2}} \begin{bmatrix} \mathbf{y}_i \\ 2\beta_i \\ \mathbf{J}\mathbf{y}_i \end{bmatrix}, \tag{A.242}$$

where the vector $\mathbf{y}_i$ and the real scalar β_i are the orthonormal eigenvectors of the matrix.

$$\begin{bmatrix} \mathbf{B}_1 + \mathbf{B}_2\mathbf{J} & \sqrt{2}\mathbf{a} \\ \sqrt{2}\mathbf{a}^H & c \end{bmatrix} \begin{bmatrix} \mathbf{y}_i \\ \beta_i \end{bmatrix} = \lambda_i \begin{bmatrix} \mathbf{y}_i \\ \beta_i \end{bmatrix}, \tag{A.243}$$

where $\mathbf{a}$ and c are defined in (A.142). $\mathbf{C}$ will also have $\frac{(N-1)}{2}$ conjugate asymmetric orthonormal eigenvectors of the form,

$$\phi_i = \frac{1}{\sqrt{2}} \begin{bmatrix} \mathbf{z}_i \\ 0 \\ -\mathbf{J}\mathbf{z}_i \end{bmatrix}, \tag{A.244}$$

where the $\mathbf{z}_i$ are the orthonormal eigenvectors of the matrix $\mathbf{B}_1 - \mathbf{B}_2\mathbf{J}$,

$$(\mathbf{B}_1 - \mathbf{B}_2\mathbf{J})\,\mathbf{z}_i = \lambda_i \mathbf{z}_i. \tag{A.245}$$

A.4.2.3 Toeplitz matrices

Toeplitz matrices are a special case of centrohermitian matrices so their eigendecomposition has the same properties. Makhoul [Mak81] analyzed their eigendecomposition and found that their additional structure did not lead to any additional properties that will be useful in our analyses.

The reader is referred to [Mak81]or [Mar87] for further discussion.

A.4.2.4 Kronecker products

Any important motivation for using Kronecker products is the relationship between the eigenvalues of $\mathbf{A}$ and $\mathbf{B}$ and the eigenvalues of $\mathbf{A} \otimes \mathbf{B}$. The following result is from Chapter 12 of [LT85], which they attribute to Stephanos [Ste00]. A good discussion of Kronecker products and their application is given in Graham [Gra81].

Let $\lambda_1, \cdots, \lambda_M$ be the eigenvalues of the $M \times M$ matrix $\mathbf{A}$. Let $\mu_1, \cdots, \mu_N$ be the eigenvalues of the $N \times N$ matrix $\mathbf{B}$.

Then the eigenvalues of $\mathbf{A} \otimes \mathbf{B}$ are the MN numbers, $\lambda_r \mu_s, r = 1, \cdots, M, s = 1, \cdots, N$.

A consequence of this result is that, if $\mathbf{A}$ and $\mathbf{B}$ are positive definite, then $\mathbf{A} \otimes \mathbf{B}$ is positive definite.

The eigenvalues of $(\mathbf{I}_N \otimes \mathbf{A}) + (\mathbf{B} \otimes \mathbf{I}_M)$ are the MN numbers,

$$\lambda_r + \mu_s, \qquad r = 1, \cdots, M, s = 1, \cdots, N. \tag{A.246}$$

The above matrix is called the Kronecker sum of $\mathbf{A}$ and $\mathbf{B}$.

A.5 Singular Value Decomposition

A significant number of applications in the text deal with the spatial spectral matrix of the received waveform at the sensors, $\mathbf{S}_{\mathbf{x}}$.

If we assume that we are dealing with a single frequency as a result of a DFT operation on the incoming vector, then

$$\mathbf{S}_{\mathbf{x}}(\omega_o) = E\left\{\mathbf{X}(\omega_o)\mathbf{X}^H(\omega_o)\right\}, \tag{A.247}$$

where $\mathbf{X}(\omega_o)$ is an $N \times 1$ vector corresponding to DFT of the incoming signal.

In practice, the output of each sensor is sampled, sectioned, and windowed. The DFT of each section is evaluated and a vector $\mathbf{X}_k(\omega_o)$ across the array is formed for each section. Assuming K sections are used, we form

$$\hat{\mathbf{S}}_{\mathbf{x}} = \frac{1}{K}\sum_{k=1}^{K} \mathbf{X}\left[\omega_o, k\right] \mathbf{X}^H\left[\omega_o, k\right], \tag{A.248}$$

and use this as an estimate of the spatial spectral matrix $\mathbf{S}_{\mathbf{x}}(\omega_o)$. For simplicity in the notation, we have suppressed the ω_o dependence in $\hat{\mathbf{S}}_x$ and will use $\mathbf{X}(k)$ to denote the kth sample. We restrict our attention to a ULA with interelement spacing d. We can write $\hat{\mathbf{S}}_{\mathbf{x}}$ in expanded form as

$$\hat{\mathbf{S}}_{\mathbf{x}} = \begin{bmatrix} \hat{s}_{xx}(0,0) & \hat{s}_{xx}(1,0) & \cdots & \hat{s}_{xx}(N-1,0) \\ \hat{s}_{xx}(0,1) & \hat{s}_{xx}(1,1) & \cdots & \hat{s}_{xx}(N-1,1) \\ \vdots & \vdots & \vdots & \vdots \\ \hat{s}_{xx}(0,N-1) & \hat{s}_{xx}(1,N-1) & \cdots & \hat{s}_{xx}(N-1,N-1) \end{bmatrix}, \tag{A.249}$$

where

$$s_{xx}(i,j) = \frac{1}{K}\sum_{k=1}^{K} \mathbf{X}_i[k]\mathbf{X}_j^*[k]. \tag{A.250}$$

$\hat{\mathbf{S}}_{\mathbf{x}}$ has several properties that we will use in the text.

The spatial spectral matrix is Hermitian,

$$\hat{\mathbf{S}}_{\mathbf{x}}^H = \hat{\mathbf{S}}_{\mathbf{x}}. \tag{A.251}$$

This follows directly from (A.249).

The spatial spectral matrix is non-negative definite,

$$\mathbf{y}^H \hat{\mathbf{S}}_{\mathbf{x}} \mathbf{y} \geq 0, \tag{A.252}$$

for any $N \times 1$ vector $\mathbf{y}$. This follows directly from (A.248),

$$\mathbf{y}^H \hat{\mathbf{S}}_{\mathbf{x}} \mathbf{y} = \frac{1}{K}\sum_{k=1}^{K} \left|\mathbf{y}^H \mathbf{X}(k)\right|^2 \geq 0. \tag{A.253}$$

If (A.252) is satisfied with an inequality sign, then $\hat{\mathbf{S}}_{\mathbf{x}}$ is non-singular and its inverse $\hat{\mathbf{S}}_{\mathbf{x}}^{-1}$ exists.

The eigenvalues of $\hat{\mathbf{S}}_{\mathbf{x}}$ are real and nonnegative. The spectral matrix $\hat{\mathbf{S}}_{\mathbf{x}}$ is, in general, not Toeplitz because, as can be seen from (A.249), the elements on the main diagonal (and other diagonals) have different values.

We can write the sample spectral matrix as the product of two rectangular matrices,

$$\hat{\mathbf{S}}_{\mathbf{x}} = \frac{1}{K}\left[\ \mathbf{X}(1) \ \vdots\ \mathbf{X}(2) \ \vdots\ \cdots \ \vdots\ \mathbf{X}(K)\ \right] \begin{bmatrix} \mathbf{X}^H(1) \\ \mathbf{X}^H(2) \\ \vdots \\ \mathbf{X}^H(K) \end{bmatrix}. \tag{A.254}$$

We now define the data matrix $\tilde{\mathbf{X}}$,

$$\tilde{\mathbf{X}} = \frac{1}{\sqrt{K}}\left[\ \mathbf{X}(1) \ \vdots\ \mathbf{X}(2) \ \vdots\ \cdots \ \vdots\ \mathbf{X}(K)\ \right], \tag{A.255}$$

or

$$\tilde{\mathbf{X}} = \frac{1}{\sqrt{K}} \left[\begin{array}{c|c|c|c} X_0(1) & X_0(2) & & X_0(K) \\ X_1(1) & X_1(2) & & \\ \vdots & & \cdots & \\ X_{N-1}(1) & X_{N-1}(2) & & X_{N-1}(K) \end{array} \right], \tag{A.256}$$

and

$$\hat{\mathbf{S}}_{\mathbf{x}} = \tilde{\mathbf{X}}\tilde{\mathbf{X}}^H. \tag{A.257}$$

If we use FB averaging, we have an $N \times 2K$ data matrix

$$\tilde{\mathbf{X}}_{fb} = \frac{1}{\sqrt{2}} \left[\begin{array}{c|c} \tilde{\mathbf{X}} & \mathbf{J}\tilde{\mathbf{X}}^* \end{array} \right]. \tag{A.258}$$

Writing (A.258) out gives an $N \times 2K$ data matrix

$$\tilde{\mathbf{X}}_{fb} = \frac{1}{\sqrt{2}} \left[\begin{array}{cccc|cccc} X_0(1) & X_0(2) & \cdots & X_0(K) & X_{N-1}^*(1) & X_{N-1}^*(2) & \cdots & X_{N-1}^*(K) \\ X_1(1) & X_1(2) & \cdots & X_1(K) & X_{N-2}^*(1) & X_{N-2}^*(2) & \cdots & X_{N-2}^*(K) \\ \vdots & \vdots & & \vdots & \vdots & \vdots & & \vdots \\ X_{N-1}(1) & X_{N-1}(2) & \cdots & X_{N-1}(K) & X_0^*(1) & X_0^*(2) & \cdots & X_0^*(K) \end{array} \right], \tag{A.259}$$

and

$$\hat{\mathbf{S}}_{\mathbf{x},fb} = \tilde{\mathbf{X}}_{fb}\tilde{\mathbf{X}}_{fb}^H. \tag{A.260}$$

We now discuss how we find the eigenvectors and eigenvalues of $\hat{\mathbf{S}}_{\mathbf{x}}$ and $\hat{\mathbf{S}}_{\mathbf{x},fb}$ directly from the data matrix.

Given the data matrix $\tilde{\mathbf{X}}$, we want to show that there are two unitary matrices, $\mathbf{U}$ and $\mathbf{V}$, such that,

$$\mathbf{U}^H\tilde{\mathbf{X}}^H\mathbf{V} = \begin{bmatrix} \boldsymbol{\Sigma} & \mathbf{0} \\ \mathbf{0} & \mathbf{0} \end{bmatrix}, \tag{A.261}$$

where $\boldsymbol{\Sigma}$ is a diagonal matrix,

$$\boldsymbol{\Sigma} = \text{diag}\left(\sigma_1, \sigma_2, \cdots, \sigma_W\right), \tag{A.262}$$

and the σ's are ordered in a decreasing manner.

The result in (A.261) is called the **singular value decomposition theorem**. The matrix $\mathbf{U}^H$ is $K \times K$. The matrix $\mathbf{V}$ is $N \times N$. The subscript W is the rank of the matrix $\tilde{\mathbf{X}}$

$$W = \text{rank}\left(\tilde{\mathbf{X}}\right) \leq \min\left[K, N\right]. \tag{A.263}$$

For $K > N$, the system in (A.260) is overdetermined. For $K < N$, the system is underdetermined. The overdetermined case is of most interest is our applications. (We need more snapshots (K) than the number of sensors (N) to get a stable estimate of $\mathbf{S_x}$.) We derive the overdetermined case in the text and refer the reader to Haykin (p. 520 of [Hay96]) for the underdetermined case.

We form the $N \times N$ matrix, $\tilde{\mathbf{X}}\tilde{\mathbf{X}}^H$. The matrix $\tilde{\mathbf{X}}\tilde{\mathbf{X}}^H$ is Hermitian and non-negative definite, so its eigenvalues are real and non-negative. We denote the eigenvalues as $\lambda_1, \lambda_2, \cdots, \lambda_N$ where

$$\lambda_1 \geq \lambda_2 \geq \cdots \geq \lambda_W > 0, \tag{A.264}$$

and

$$\lambda_{W+1} = \lambda_{W+2} = \cdots = \lambda_N = 0. \tag{A.265}$$

The λ's and σ's are related by

$$\sigma_i = \lambda_i^{\frac{1}{2}}, \qquad i = 1, 2, \cdots, N. \tag{A.266}$$

We denote the orthonormal eigenvectors of the matrix $\tilde{\mathbf{X}}\tilde{\mathbf{X}}^H$ by $\mathbf{v}_k, k = 1, 2, \cdots, N$, and the $N \times N$ unitary matrix whose columns are the eigenvectors of $\tilde{\mathbf{X}}\tilde{\mathbf{X}}^H$ by $\mathbf{V}$. Then

$$\mathbf{V}^H \tilde{\mathbf{X}}\tilde{\mathbf{X}}^H \mathbf{V} = \begin{bmatrix} \mathbf{\Sigma}^2 & \mathbf{0} \\ \mathbf{0} & \mathbf{0} \end{bmatrix}. \tag{A.267}$$

We partition $\mathbf{V}$ as,

$$\mathbf{V} = \left[\mathbf{V}_1 \, \mathbf{V}_2\right], \tag{A.268}$$

where $\mathbf{V}_1$ is an $N \times W$ matrix,[5]

[5]In most of our applications, there is a white noise component present, so that $W = N$ and the $\mathbf{V}_2$ is not present. The modification for that case is straightforward.

$$\mathbf{V}_1 = [\mathbf{v}_1, \mathbf{v}_2, \cdots, \mathbf{v}_W], \tag{A.269}$$

$$\mathbf{V}_2 = [\mathbf{v}_{W+1}, \mathbf{v}_{W+2}, \cdots, \mathbf{v}_M]. \tag{A.270}$$

We observe that

$$\mathbf{V}_1^H \mathbf{V}_2 = \mathbf{0}. \tag{A.271}$$

From (A.267), we have for $\mathbf{V}_1$,

$$\mathbf{V}_1^H \tilde{\mathbf{X}} \tilde{\mathbf{X}}^H \mathbf{V}_1 = \mathbf{\Sigma}^2, \tag{A.272}$$

and therefore,

$$\mathbf{\Sigma}^{-1} \mathbf{V}_1^H \tilde{\mathbf{X}} \tilde{\mathbf{X}}^H \mathbf{V}_1 \mathbf{\Sigma}^{-1} = \mathbf{I}. \tag{A.273}$$

For $\mathbf{V}_2$,

$$\mathbf{V}_2^H \tilde{\mathbf{X}} \tilde{\mathbf{X}}^H \mathbf{V}_2 = \mathbf{0}, \tag{A.274}$$

and therefore,

$$\tilde{\mathbf{X}}^H \mathbf{V}_2 = \mathbf{0}. \tag{A.275}$$

We next define a $K \times W$ matrix, $\mathbf{U}_1$,

$$\mathbf{U}_1 = \tilde{\mathbf{X}}^H \mathbf{V}_1 \mathbf{\Sigma}^{-1}. \tag{A.276}$$

Then, from (A.273),

$$\mathbf{U}_1^H \mathbf{U}_1 = \mathbf{I}. \tag{A.277}$$

This implies the columns of $\mathbf{U}_1$ are orthogonal.

Finally, we define a $K \times K - W$ matrix $\mathbf{U}_2$ so that

$$\mathbf{U} = [\mathbf{U}_1 \, \mathbf{U}_2], \tag{A.278}$$

is a unitary matrix.

This implies that

$$\mathbf{U}_1^H \mathbf{U}_2 = \mathbf{0}. \tag{A.279}$$

With these definitions we can write

$$
\begin{aligned}
\mathbf{U}^H\tilde{\mathbf{X}}^H\mathbf{V} &= \begin{bmatrix} \mathbf{U}_1^H \\ \mathbf{U}_2^H \end{bmatrix} \tilde{\mathbf{X}}^H [\mathbf{V}_1, \mathbf{V}_2] \\
&= \begin{bmatrix} \mathbf{U}_1^H\tilde{\mathbf{X}}^H\mathbf{V}_1 & \mathbf{U}_1^H\tilde{\mathbf{X}}^H\mathbf{V}_2 \\ \mathbf{U}_2^H\tilde{\mathbf{X}}^H\mathbf{V}_1 & \mathbf{U}_2^H\tilde{\mathbf{X}}^H\mathbf{V}_2 \end{bmatrix} \\
&= \begin{bmatrix} (\mathbf{\Sigma}^{-1}\mathbf{V}_1^H\tilde{\mathbf{X}})\tilde{\mathbf{X}}^H\mathbf{V}_1 & \mathbf{U}_1^H\mathbf{0} \\ \mathbf{U}_2^H(\mathbf{U}_1\mathbf{\Sigma}) & \mathbf{U}_2^H\mathbf{0} \end{bmatrix} \\
&= \begin{bmatrix} \mathbf{\Sigma} & \mathbf{0} \\ \mathbf{0} & \mathbf{0} \end{bmatrix},
\end{aligned}
\tag{A.280}
$$

which is the relation in (A.261).

We refer to elements of $\mathbf{\Sigma}$ as the singular values of $\tilde{\mathbf{X}}^H$. The columns of the unitary matrix $\mathbf{V}$ are the right singular vectors of $\tilde{\mathbf{X}}^H$, and the columns of the unitary matrix $\mathbf{U}$ are the left singular vectors of $\tilde{\mathbf{X}}^H$.

The right singular vectors are the eigenvectors of the matrix $\tilde{\mathbf{X}}\tilde{\mathbf{X}}^H$ that corresponds to the sample covariance matrix, $\hat{\mathbf{S}}_{\mathbf{x}}$ (see (A.257)). The non-zero eigenvalues of the sample covariance matrix are

$$
[\mathbf{\Lambda}]_{ii} = \left[\mathbf{\Sigma}^2\right]_{ii}, \qquad i = 1, \cdots, W. \tag{A.281}
$$

Our primary usage of SVD will be in the context of the eigenvalues and eigenvectors of the sample correlation matrix.

There are important computation reasons for working directly with the data matrix $\tilde{\mathbf{X}}$ rather than the sample covariance matrix. The dynamic range required to deal with $\tilde{\mathbf{X}}\tilde{\mathbf{X}}^H$ is doubled. Thus, for a specified numerical accuracy, the required word length is doubled.

There are several efficient computational schemes for computing the SVD. Various algorithms are discussed in detail in Sections 11.7–11.12 of [Hay91] and Chapters 11 and 12 of [Hay96]. All of the various computational programs, such as LINPACK, EISPACK, and MATLAB, have SVD algorithms included. SVD is widely used in a number of signal processing applications, and there is extensive literature. We utilize it when we study adaptive beamforming and parameter estimation.

A.6 QR Decomposition

A.6.1 Introduction

The techniques developed in this section are motivated by the least squares estimation problem. The least squares equation has the form

$$\mathbf{\Phi}^*(K)\hat{\mathbf{w}}_{lse}(K) = \mathbf{\Phi}^*_{\mathbf{x}d^*}(K), \tag{A.282}$$

where

$$\mathbf{\Phi}(K) = \sum_{k=1}^{K} \mathbf{X}(k)\mu^{\frac{K-k}{2}}\mu^{\frac{K-k}{2}}\mathbf{X}^H(k), \tag{A.283}$$

is an exponentially weighted sample correlation matrix and

$$\mathbf{\Phi}_{\mathbf{x}d^*}(K) = \sum_{k=1}^{K} \mathbf{X}(k)\mu^{\frac{K-k}{2}}\mu^{\frac{K-k}{2}}D^*(k), \tag{A.284}$$

is an exponentially weighted sample cross-correlation matrix.

We define a $K \times N$ exponentially weighted data matrix,

$$\mathbf{A}_\mu(K) = \boldsymbol{\mu}(K)\begin{bmatrix} \mathbf{X}^T(1) \\ \hdashline \mathbf{X}^T(2) \\ \hdashline \vdots \\ \hdashline \mathbf{X}^T(K) \end{bmatrix}, \tag{A.285}$$

where

$$\boldsymbol{\mu}(K) = \text{diag}\left[\begin{array}{cccc} \mu^{\frac{K-1}{2}}, & \mu^{\frac{K-2}{2}}, & \cdots, & 1 \end{array}\right]. \tag{A.286}$$

Then,

$$\mathbf{\Phi}(K) = \mathbf{A}^T_\mu(K)\,\mathbf{A}^*_\mu(K), \tag{A.287}$$

and

$$\mathbf{\Phi}_{\mathbf{x}d^*}(K) = \mathbf{A}^T_\mu(K)\,\mathbf{d}^*_\mu, \tag{A.288}$$

where

$$\mathbf{d}_\mu = \boldsymbol{\mu}(K)\left[\begin{array}{ccc} D(1), & \cdots, & D(K) \end{array}\right]^T. \tag{A.289}$$

Then, (A.282) can be written as

$$\mathbf{A}^H_\mu(K)\,\mathbf{A}_\mu(K)\,\mathbf{w}^* = \mathbf{A}^H_\mu\,\mathbf{d}_\mu. \tag{A.290}$$

One approach to solving (A.282) is to factor $\mathbf{\Phi}$ using a Cholesky decomposition. Then, we can solve (A.282) using back-substitution (e.g., [GVL89]).

A problem that one encounters with this approach is the requirement for high numerical precision. In this section, we discuss techniques that utilize the data matrix rather than the sample covariance matrix.

All of these techniques are classical and are discussed in a number of references and textbooks. Books that discuss solutions to least squares estimation problems include Lawson and Hanson [LH74], Wilkinson [Wil65], Stewart [Ste73], and Golub and Van Loan [GVL89]. Books that include discussion of these techniques as background to adaptive filtering or adaptive beamforming include Haykin ([Hay96], [Hay91]), Haykin and Steinkardt [HS92], and Proakis et al. [PRLN92]. Our discussion is similar to Section 5.3 of [PRLN92].

In Section A.6.2, we discuss the QR decomposition and its application to the solution of (A.282). In Sections A.6.3 and A.6.4, we develop two techniques for accomplishing the QR decomposition: the Givens rotation and the Householder transformation. A third technique, the modified Gram-Schmidt algorithm, is discussed in [PRLN92] and will not be included.

All of our discussion in this section assumes K is fixed. In Chapter 7, we discussed recursive techniques that are the basis of our adaptive beamforming algorithm.

A.6.2 QR Decomposition

In this section, we show how to solve (A.282) in an efficient manner using the data matrix $\mathbf{A}_\mu$. By avoiding the computation of $\mathbf{\Phi}$, we can significantly improve our numerical accuracy. The key is the decomposition of the $K \times N$ matrix $\mathbf{A}_\mu$ into a $K \times N$ matrix that is partitioned into a $N \times N$ upper triangular matrix $\tilde{\mathbf{R}}$ and a $(K - N) \times N$ null matrix. There exists a $K \times K$ unitary matrix $\mathbf{Q}$ such that

$$\mathbf{Q}\mathbf{A}_\mu = \begin{bmatrix} \tilde{\mathbf{R}} \\ \mathbf{0} \end{bmatrix}, \tag{A.291}$$

where $\tilde{\mathbf{R}}$ has the form

$$\tilde{\mathbf{R}} = \begin{bmatrix} \tilde{r}_{11} & \tilde{r}_{12} & \cdots & \tilde{r}_{1N} \\ 0 & \tilde{r}_{22} & \cdots & \tilde{r}_{2N} \\ \cdots & \cdots & \ddots & \cdots \\ 0 & 0 & \cdots & \tilde{r}_{NN} \end{bmatrix}, \tag{A.292}$$

where we have suppressed the K dependence, and $\mathbf{0}$ is a $(K-N)\times N$ matrix with zero elements.

Note that

$$\mathbf{\Phi} = \mathbf{A}_\mu^H \mathbf{A}_\mu = \mathbf{A}_\mu^H \mathbf{Q}^H \mathbf{Q} \mathbf{A}_\mu = \tilde{\mathbf{R}}^H \tilde{\mathbf{R}}, \tag{A.293}$$

so that $\tilde{\mathbf{R}}$ is the Cholesky factor of the weighted sample covariance matrix.

The decomposition has several other properties that we will utilize.[6]

(i) Each row of the triangular matrix $\tilde{\mathbf{R}}$ is unique up to a complex scale factor with unit magnitude. Hence if $\tilde{\mathbf{R}}_1$ and $\tilde{\mathbf{R}}_2$ are two different $\tilde{\mathbf{R}}$-factors in a QR decomposition of the $\mathbf{A}_\mu$, we can always find a complex-valued diagonal matrix β whose complex-valued diagonal elements have unit magnitude, so that $\tilde{\mathbf{R}}_1 = \beta\tilde{\mathbf{R}}_2$.

(ii) For any matrix $\mathbf{A}_\mu$ there exists a unique $\tilde{\mathbf{R}}$ whose diagonal elements are all real and nonnegative.

(iii) If $K > N$, the unitary matrix $\mathbf{Q}$ in a QR decompositions of $\mathbf{A}_\mu$ is not unique for the same $\tilde{\mathbf{R}}$-factor.

(iv) Since the condition number of the unitary matrix $\mathbf{Q}$ is unity, it follows that the condition number of $\tilde{\mathbf{R}}$ is equal to the condition number of $\tilde{\mathbf{X}}_\mu$.

To solve (A.293), we substitute (A.291) into (A.293) and obtain

$$\begin{bmatrix} \tilde{\mathbf{R}} \\ \mathbf{0} \end{bmatrix}^H \mathbf{Q}\mathbf{Q}^H \begin{bmatrix} \tilde{\mathbf{R}} \\ \mathbf{0} \end{bmatrix} \mathbf{w}^* = \begin{bmatrix} \tilde{\mathbf{R}} \\ \mathbf{0} \end{bmatrix}^H \mathbf{Q}\mathbf{d}_\mu. \tag{A.294}$$

Now partition $\mathbf{d}_\mu$ as

$$\mathbf{Q}\mathbf{d}_\mu = \begin{bmatrix} \mathbf{p} \\ \mathbf{v} \end{bmatrix}, \tag{A.295}$$

where $\mathbf{p}$ is an $N \times 1$ vector.

Using (A.295) in (A.294), we obtain

$$\tilde{\mathbf{R}}^H \tilde{\mathbf{R}} \mathbf{w}^* = \tilde{\mathbf{R}}^H \mathbf{p}, \tag{A.296}$$

or

$$\tilde{\mathbf{R}} \mathbf{w}^* = \mathbf{p}. \tag{A.297}$$

It is now straightforward to solve (A.297) because $\tilde{\mathbf{R}}$ is upper triangular. Two observations are important:

---

[6]The list of properties is taken directly from p. 291 of [PRLN92].

(i) We do not need an explicit expression for $\mathbf{Q}$.

(ii) The condition number of $\tilde{\mathbf{R}}$ is the square root of the condition number of $\mathbf{\Phi}$, so our numerical accuracy has improved.

A.6.3 Givens Rotation

The Givens rotation is a method for implementing the QR decomposition by a sequence of plane rotations. The technique is due to Givens [Giv58].

To introduce the technique, we first consider a 2×1 complex vector $\mathbf{v}$,

$$\mathbf{v} = \begin{bmatrix} v_1 \\ v_2 \end{bmatrix}. \tag{A.298}$$

The Givens matrix $\mathbf{G}$ is a 2×2 unitary matrix,

$$\mathbf{G} = \begin{bmatrix} c^* & s \\ -s^* & c \end{bmatrix}. \tag{A.299}$$

We select the elements of $\mathbf{G}$ such that

$$\mathbf{G}\mathbf{v} = \begin{bmatrix} c^* v_1 + s v_2 \\ -s^* v_1 + c v_2 \end{bmatrix} \triangleq \begin{bmatrix} v_1' \\ 0 \end{bmatrix} = \mathbf{v}'. \tag{A.300}$$

The zero term in (A.300) requires,

$$s^* v_1 = c v_2, \tag{A.301}$$

and the unitary condition requires

$$|c|^2 + |s|^2 = 1. \tag{A.302}$$

Therefore,

$$c = \frac{v_1}{\sqrt{|v_1|^2 + |v_2|^2}}, \tag{A.303}$$

$$s = \frac{v_2^*}{\sqrt{|v_1|^2 + |v_2|^2}}, \tag{A.304}$$

and

$$v_1' = \sqrt{|v_1|^2 + |v_2|^2}. \tag{A.305}$$

Thus, $\mathbf{v}'$ has the same length as $\mathbf{v}$, but has only one non-zero component.

In a similar manner, if we have a row vector $\mathbf{u}$,

$$\mathbf{u} = \begin{bmatrix} u_1 & u_2 \end{bmatrix}. \tag{A.306}$$

We can write

$$\mathbf{u}\mathbf{G}^H = \begin{bmatrix} u_1' & 0 \end{bmatrix}, \tag{A.307}$$

where $\mathbf{G}$ is given by (A.303) and (A.304), with v_1 and v_2 replaced by u_1 and u_2. This result follows directly from (A.300) by letting $\mathbf{u} = \mathbf{v}^H$.

The operation described by (A.299) is called a plane rotation because, if v_1 and v_2 are real,

$$c = \cos\phi, \tag{A.308}$$

and

$$s = \sin\phi, \tag{A.309}$$

where ϕ is the angle that $\mathbf{v}$ is rotated. The adjective "plane" denotes that $\mathbf{v}$ stays in the same plane.

Now consider a $K \times 1$ complex vector $\mathbf{v}$,

$$\mathbf{v} = \begin{bmatrix} v_1, & \cdots, & v_m, & \cdots, & v_n, & \cdots, & v_K \end{bmatrix}^T. \tag{A.310}$$

We operate on $\mathbf{v}$ with a Givens rotation $\mathbf{G}$ and leave all of the elements of $\mathbf{v}$ unchanged except for v_m and v_n. This requires a $K \times K$ unitary matrix of the form,

$$\mathbf{G}(m,n) = \begin{bmatrix} \mathbf{I} & & & & & & 0 \\ & \ddots & & & & & \\ & & c_{mm} & & s_{mn} & & \\ & & & \ddots & & & \\ & & & \mathbf{I} & & & \\ & & & & \ddots & & \\ & & s_{nm} & & c_{nn} & & \\ & & & & & \ddots & \\ 0 & & & & & & \mathbf{I} \end{bmatrix}. \tag{A.311}$$

$\mathbf{G}(m,n)$ is an identity matrix except for four elements. The two c elements are on the diagonal,

$$c = c_{mm}^* = c_{nn} = \frac{v_m}{\sqrt{|v_m|^2 + |v_n|^2}}, \tag{A.312}$$

and the two s elements are off the diagonal, with

$$s = s_{mn} = \frac{v_n^*}{\sqrt{|v_m|^2 + |v_n|^2}}, \tag{A.313}$$

and

$$s_{nm} = s_{mn}^* = -s^*. \tag{A.314}$$

Then,

$$\mathbf{G}(m,n)\mathbf{v} = \begin{bmatrix} v_1, & \cdots, & \sqrt{|v_m|^2 + |v_n|^2}, & \cdots, & 0, & \cdots, & v_K \end{bmatrix}^T. \tag{A.315}$$

We have made the nth element zero and rotated it into the mth element.

We now demonstrate how to apply a sequence of Givens rotations to accomplish a QR decomposition of $\mathbf{A}_\mu$.

Example A.6.1

Consider the case when $N = 2$ and $K = 3$. Then the exponentially weighted data matrix can be written as

$$\mathbf{A}_\mu = \begin{bmatrix} a_{11} & a_{12} \\ a_{21} & a_{22} \\ a_{31} & a_{32} \end{bmatrix}. \tag{A.316}$$

We utilize three successive Givens rotations to obtain the desired QR decomposition. We eliminate a_{31} first. From (A.311),

$$\mathbf{G}_1 = \begin{bmatrix} c_1^* & 0 & s_1 \\ 0 & 1 & 0 \\ -s_1^* & 0 & c_1 \end{bmatrix}, \tag{A.317}$$

where we have chosen to rotate a_{31} into a_{11}. Then

$$c_1 = \frac{a_{11}}{r_1}, \tag{A.318}$$

$$s_1 = \frac{a_{31}^*}{r_1}, \tag{A.319}$$

and

$$r_1 \triangleq \sqrt{|a_{11}|^2 + |a_{31}|^2}. \tag{A.320}$$

Then,

$$\mathbf{G}_1\mathbf{A}_\mu = \begin{bmatrix} r_1 & c_1^* a_{12} + s_1 a_{32} \\ a_{21} & a_{22} \\ 0 & -s_1^* a_{12} + c_1 a_{32} \end{bmatrix} \triangleq \begin{bmatrix} r_1 & a_{12}' \\ a_{21} & a_{22} \\ 0 & a_{32}' \end{bmatrix}. \tag{A.321}$$

Note that only the first and third rows were affected by $\mathbf{G}_1$.

The next step is to eliminate a_{21},

$$\mathbf{G}_2 = \begin{bmatrix} c_2^* & s_2 & 0 \\ -s_2^* & c_2 & 0 \\ 0 & 0 & 1 \end{bmatrix}, \tag{A.322}$$

where

$$c_2 = \frac{r_1}{\tilde{r}_{11}}, \tag{A.323}$$

$$s_2 = \frac{a_{21}^*}{\tilde{r}_{11}}, \tag{A.324}$$

and

$$\tilde{r}_{11} \triangleq \sqrt{|r_1|^2 + |a'_{12}|^2}. \tag{A.325}$$

Then,

$$\mathbf{G}_2\mathbf{G}_1\mathbf{A}_\mu = \begin{bmatrix} \tilde{r}_{11} & c_2^* a'_{12} + s_2 a_{22} \\ 0 & -s_2^* a'_{12} + c_2 a_{22} \\ 0 & a'_{32} \end{bmatrix} \triangleq \mathbf{G}_2\mathbf{G}_1\mathbf{A}_\mu = \begin{bmatrix} \tilde{r}_{11} & \tilde{r}_{12} \\ 0 & a'_{22} \\ 0 & a'_{32} \end{bmatrix}. \tag{A.326}$$

The notation is chosen because the first row will not be affected by the next step so that $\tilde{r}_{11}$ and $\tilde{r}_{12}$ are the elements in $\tilde{\mathbf{R}}$.

The last step is to eliminate a'_{32}. We do this without disturbing the first row.

$$\mathbf{G}_3 = \begin{bmatrix} 1 & 0 & 0 \\ 0 & c_3^* & s_3 \\ 0 & -s_3^* & c_3 \end{bmatrix}, \tag{A.327}$$

where

$$c_3 = \frac{a'_{22}}{\tilde{r}_{22}}, \tag{A.328}$$

$$s_3 = \frac{(a'_{32})^*}{\tilde{r}_{22}}, \tag{A.329}$$

and

$$\tilde{r}_{22} \triangleq \sqrt{|a'_{22}|^2 + |a'_{32}|^2}. \tag{A.330}$$

Then,

$$\mathbf{G}_3\mathbf{G}_2\mathbf{G}_1\mathbf{A}_\mu = \begin{bmatrix} \tilde{r}_{11} & \tilde{r}_{12} \\ 0 & \tilde{r}_{22} \\ 0 & 0 \end{bmatrix}, \tag{A.331}$$

which is the desired result.

Note that $\tilde{r}_{11}$ and $\tilde{r}_{22}$ are real (see (A.325) and (A.330)). In each step, it is straightforward to locate the c and s parameters. The $-s^*$ parameter is in the same position as the element that we are eliminating and the c^* parameter is in the same position as the element we are rotating into. We should also observe that the elimination order is not unique (see Problem A.6.1), but $\tilde{\mathbf{R}}$ is. The $\mathbf{Q}$ matrix is

$$\mathbf{Q} = \mathbf{G}_3\mathbf{G}_2\mathbf{G}_1, \tag{A.332}$$

but is not explicitly used in the algorithm.

The extension of the above example to an arbitrary $K \times N$ $(K \geq N)$ data matrix is straightforward. We apply $K-1$ successive Givens rotations to eliminate all of the elements in the first column except for the elements in the first row. We then apply $K-2$ Givens rotations, without disturbing the first row, to eliminate the bottom $K-2$ elements in the second column. We process each successive column in a similar manner. This requires,

$$N_G = (K-1)(K-2)\cdots(K-N) = KN - \frac{N(N-1)}{2}, \tag{A.333}$$

rotations. We use the same sequence of rotations on $\mathbf{d}_\mu$ to obtain $\mathbf{p}$.

A.6.4 Householder Transformation

The Givens rotation technique obtains a triangular $\tilde{\mathbf{R}}$ by eliminating one element with each rotation. The Householder transformation eliminates all except one element in a column in each step. The transformation is due to Householder ([Hou58], [Hou64]). A tutorial paper by Steinhardt [Ste88] gives an excellent discussion of the transformation and its application to signal processing.

The Householder transformation can be written as

$$\mathbf{H} = \mathbf{I} - \frac{2\mathbf{u}\mathbf{u}^H}{\|\mathbf{u}\|^2} = \mathbf{I} - 2\mathbf{u}\left[\mathbf{u}^H\mathbf{u}\right]^{-1}\mathbf{u}^H, \tag{A.334}$$

where $\mathbf{u}$ is an $N \times 1$ vector whose norm is $\|\mathbf{u}\|$. We can also write it as

$$\mathbf{H} = \mathbf{I} - 2\mathbf{P}_\mathbf{u}, \tag{A.335}$$

where

$$\mathbf{P}_\mathbf{u} = \mathbf{u}\left[\mathbf{u}^H\mathbf{u}\right]^{-1}\mathbf{u}^H, \tag{A.336}$$

is the projection matrix onto the $\mathbf{u}$ subspace.

If we pre-multiply the $N \times 1$ complex vector $\mathbf{v}$ by $\mathbf{H}$, we have

$$\mathbf{H}\mathbf{v} = (\mathbf{I} - 2\mathbf{P}_\mathbf{u})\,\mathbf{v} = \mathbf{v} - 2\mathbf{P}_\mathbf{u}\mathbf{v}. \tag{A.337}$$

From (A.334), we observe that $\mathbf{H}$ is a Hermitian unitary matrix,

$$\mathbf{H}^{-1} = \mathbf{H}^H = \mathbf{H}, \tag{A.338}$$

so that the transformation preserves length

$$\|\mathbf{H}\mathbf{v}\| = \|\mathbf{v}\|. \tag{A.339}$$

Now consider the $N \times 1$ complex vector $\mathbf{v}$,

$$\mathbf{v} = \begin{bmatrix} v_1, & v_2, & \cdots, & v_i, & \cdots, & v_N \end{bmatrix}^T. \tag{A.340}$$

We use $\mathbf{H}$ to eliminate all of the elements except $\mathbf{v}_i$. We define

$$v_{in} = \frac{v_i}{|v_i|}, \tag{A.341}$$

and

$$\|\mathbf{v}\|^2 = \mathbf{v}^H\mathbf{v}. \tag{A.342}$$

We use (A.334) with

$$\mathbf{u} = \mathbf{v} + v_{in}\|\mathbf{v}\|\mathbf{e}_i, \tag{A.343}$$

where

$$\mathbf{e}_i^T = \begin{bmatrix} 0, & \cdots, & 0 & 1, & 0 & \cdots & 0 \end{bmatrix}^T, \tag{A.344}$$

has one as the ith element and zero elsewhere.

Using (A.341)–(A.344), we find

$$\|\mathbf{u}\|^2 = \mathbf{u}^H\mathbf{u} = 2\|\mathbf{v}\|\left(\|\mathbf{v}\| + |v_i|\right). \tag{A.345}$$

Then, we observe that

$$\mathbf{u}^H\mathbf{v} = \left(\mathbf{v}^H + v_{in}^*\|\mathbf{v}\|\mathbf{e}_i^T\right)\mathbf{v} = \|\mathbf{v}\|^2 + |v_i|\|\mathbf{v}\|. \tag{A.346}$$

Therefore,

$$\begin{aligned} \mathbf{H}\mathbf{v} &= \left(\mathbf{I} - \frac{2\mathbf{u}\mathbf{u}^H}{\|\mathbf{u}\|^2}\right)\mathbf{v} \\ &= \mathbf{v} - \mathbf{u} \\ &= -v_{in}\|\mathbf{v}\|\mathbf{e}_i, \end{aligned} \tag{A.347}$$

which is the desired result.

The procedure for using the Householder transformation to do a QR decomposition is analogous to the Givens rotation technique. We illustrate it with a similar model to the model in Example A.6.1.

Example A.6.2

Consider the case when $K = 4$ and $N = 2$. In view of our results in Example A.6.1, we define an $K \times (N+1)$ augmented data matrix

$$\begin{aligned} \mathbf{A}_{aug} &\triangleq \begin{bmatrix} \mathbf{A}_\mu & \mathbf{d}_\mu \end{bmatrix} \\ &\triangleq \begin{bmatrix} a_{11} & a_{12} & d_1 \\ a_{21} & a_{22} & d_2 \\ a_{31} & a_{32} & d_3 \\ a_{41} & a_{42} & d_4 \end{bmatrix} \\ &\triangleq \begin{bmatrix} \mathbf{a}_1 \,\vdots\, \mathbf{a}_2 \,\vdots\, \mathbf{d}_\mu \end{bmatrix}. \end{aligned} \tag{A.348}$$

The first step is to use a Householder transformation to eliminate a_{21}, a_{31}, and a_{41}:

$$\mathbf{H}_1 = \mathbf{I} - \frac{2\mathbf{u}_1\mathbf{u}_1^H}{\|\mathbf{u}_1\|^2}, \tag{A.349}$$

where

$$\mathbf{u}_1 = \mathbf{a}_1 + \frac{a_{11}}{|a_{11}|}\|\mathbf{a}_1\|\mathbf{e}_1. \tag{A.350}$$

Then,

$$\mathbf{H}_1\mathbf{A}_{aug} = \begin{bmatrix} -\frac{a_{11}}{|a_{11}|}\|\mathbf{a}_1\| & \vdots & a'_{12} & \vdots & d'_1 \\ 0 & \vdots & a'_{22} & \vdots & d'_2 \\ 0 & \vdots & a'_{32} & \vdots & d'_3 \\ 0 & \vdots & a'_{42} & \vdots & d'_4 \end{bmatrix}. \tag{A.351}$$

In the next step we want to eliminate a'_{32} and a'_{42} without disturbing the first row. We define

$$\mathbf{a}_2^{(-)} = \begin{bmatrix} 0 & a'_{22} & a'_{32} & a'_{42} \end{bmatrix}. \tag{A.352}$$

Then,

$$\mathbf{H}_2 = \mathbf{I} - \frac{2\mathbf{u}_2\mathbf{u}_2^H}{\|\mathbf{u}_2\|^2}, \tag{A.353}$$

where

$$\mathbf{u}_2 = \mathbf{a}_2^{(-)} + \frac{a'_{22}}{|a'_{22}|}\|\mathbf{a}_2^{(-)}\|\mathbf{e}_2. \tag{A.354}$$

Applying $\mathbf{H}_2$ to (A.351) gives

$$\mathbf{H}_2\mathbf{H}_1\mathbf{A}_{aug} = \begin{bmatrix} -\frac{a_{11}}{|a_{11}|}\|\mathbf{a}_1\| & \vdots & a'_{12} & \vdots & d''_1 \\ 0 & \vdots & -\frac{a'_{22}}{|a'_{22}|}\|\mathbf{a}_2^{(-)}\| & \vdots & d''_2 \\ 0 & \vdots & 0 & \vdots & d''_3 \\ 0 & \vdots & 0 & \vdots & d''_4 \end{bmatrix}. \tag{A.355}$$

The last step is to eliminate d''_4. Define

$$\mathbf{a}_3^{(-)} = \begin{bmatrix} 0 & 0 & d''_3 & d''_4 \end{bmatrix}. \tag{A.356}$$

Then

$$\mathbf{H}_3 = \mathbf{I} - \frac{2\mathbf{u}_3\mathbf{u}_3^H}{\|\mathbf{u}_3\|^2}, \tag{A.357}$$

where

$$\mathbf{u}_3 = \mathbf{a}_3^{(-)} + \frac{d''_3}{|d''_3|}\|\mathbf{a}_3^{(-)}\|\mathbf{e}_3. \tag{A.358}$$

Then

$$\mathbf{H}_3\mathbf{H}_2\mathbf{H}_1\mathbf{A}_{aug} = \begin{bmatrix} r_{11} & r_{12} & \hat{d}_1 \\ 0 & r_{22} & \hat{d}_2 \\ 0 & 0 & e_3 \\ 0 & 0 & 0 \end{bmatrix}. \tag{A.359}$$

The top two rows give the required QR decomposition.

Note that the diagonal elements are not necessarily real or positive. As a final step we multiply each row by a complex number with unity magnitude. The resulting $\tilde{\mathbf{R}}$ matrix is identical to the one obtained by the Givens rotation. We denote this diagonal matrix as

$$\boldsymbol{\beta} = \text{diag}\begin{bmatrix} \beta_1, & \beta_2, & \cdots, & \beta_N \end{bmatrix}. \tag{A.360}$$

The extension to arbitrary K and N, $(K \geq N)$, is straightforward.

$$\mathbf{Q} = \mathbf{H}_{N+1}\mathbf{H}_N\mathbf{H}_{N-1}\cdots\mathbf{H}_1. \tag{A.361}$$

A.7 Derivative Operations

In many of our array processing applications we encounter a non-negative cost function that we want to maximize (or minimize) with respect to a vector or matrix. To accomplish this optimization we need to find a vector or matrix derivative.

Another application of derivatives that we will encounter is the calculation of the CRB. We recall from Chapter 2 of DEMT I [VT68], [VT01a] that the CRB is a fundamental tool in the parameter estimation.

In Section A.7.1, we develop the derivative of a scalar with respect to a vector.[7] In Section A.7.2, we develop the derivative of a scalar with respect to a matrix. In Section A.7.3, we develop derivatives with respect to a parameter. In Section A.7.4, we extend these results to complex vectors.

A.7.1 Derivative of Scalar with Respect to Vector

First consider a scalar $a(\boldsymbol{\theta})$ that is a function of the $M \times 1$ real vector parameter $\boldsymbol{\theta}$. The derivative of a with respect to $\boldsymbol{\theta}$ is an $M \times 1$ vector

$$\frac{\partial}{\partial \boldsymbol{\theta}} a(\boldsymbol{\theta}) = \begin{bmatrix} \frac{\partial a(\boldsymbol{\theta})}{\partial \theta_1} \\ \frac{\partial a(\boldsymbol{\theta})}{\partial \theta_2} \\ \vdots \\ \frac{\partial a(\boldsymbol{\theta})}{\partial \theta_M} \end{bmatrix}. \tag{A.362}$$

Similarly, if we consider a $1 \times N$ row vector,

$$\mathbf{a}^T(\boldsymbol{\theta}) = \left[\; a_1(\boldsymbol{\theta}) \;\vdots\; a_2(\boldsymbol{\theta}) \;\vdots\; \cdots \;\vdots\; a_N(\boldsymbol{\theta}) \;\right], \tag{A.363}$$

the gradient is a $M \times N$ matrix,

[7] The results in Sections A.7.1 and A.7.2 appear in many places. An early engineering reference is [AS65]. Scharf [Sch91] has a concise summary in his appendix to Chapter 6. Most of the results are derived in Graham [Gra81].

$$\frac{\partial}{\partial \boldsymbol{\theta}}\mathbf{a}^T = \left[\ \frac{\partial}{\partial \boldsymbol{\theta}}a_1(\boldsymbol{\theta}) \ \cdots \ \frac{\partial}{\partial \boldsymbol{\theta}}a_N(\boldsymbol{\theta})\ \right] \tag{A.364}$$

$$= \begin{bmatrix} \frac{\partial a_1(\boldsymbol{\theta})}{\partial \theta_1} & \cdots & \frac{\partial a_N(\boldsymbol{\theta})}{\partial \theta_1} \\ \vdots & \ddots & \vdots \\ \frac{\partial a_1(\boldsymbol{\theta})}{\partial \theta_M} & \cdots & \frac{\partial a_N(\boldsymbol{\theta})}{\partial \theta_M} \end{bmatrix}. \tag{A.365}$$

Similarly,

$$\frac{\partial}{\partial \boldsymbol{\theta}}\mathbf{a} = \left[\frac{\partial}{\partial \boldsymbol{\theta}}\mathbf{a}^T\right]^T, \tag{A.366}$$

is a $N \times M$ matrix where each row is the gradient of the scalar a_i with respect to $\boldsymbol{\theta}$.

Some cases that we will encounter are:

Products

(i)

$$\frac{\partial}{\partial \boldsymbol{\theta}}\boldsymbol{\theta}^T = \mathbf{I}. \tag{A.367}$$

(ii)

$$\frac{\partial}{\partial \boldsymbol{\theta}}\mathbf{b}^T\boldsymbol{\theta} = \frac{\partial}{\partial \boldsymbol{\theta}}\boldsymbol{\theta}^T\mathbf{b} = \mathbf{b}. \tag{A.368}$$

(iii)

$$\frac{\partial}{\partial \boldsymbol{\theta}}\mathbf{a}^T(\boldsymbol{\theta})\mathbf{b}(\boldsymbol{\theta}) = \left(\frac{\partial}{\partial \boldsymbol{\theta}}\mathbf{a}^T(\boldsymbol{\theta})\right)\mathbf{b}(\boldsymbol{\theta}) + \left(\frac{\partial}{\partial \boldsymbol{\theta}}\mathbf{b}^T(\boldsymbol{\theta})\right)\mathbf{a}(\boldsymbol{\theta}). \tag{A.369}$$

Quadratic Forms

In the next set of equations, $\mathbf{Q}$ is not a function of $\boldsymbol{\theta}$

(iv)

$$\frac{\partial}{\partial \boldsymbol{\theta}}\boldsymbol{\theta}^T\mathbf{Q}\boldsymbol{\theta} = 2\mathbf{Q}\boldsymbol{\theta}. \tag{A.370}$$

(v)

$$\frac{\partial}{\partial \boldsymbol{\theta}} \mathbf{m}^T \mathbf{Q} \mathbf{m} = 2 \left(\frac{\partial}{\partial \boldsymbol{\theta}} \mathbf{m}^T \right) \mathbf{Q} \mathbf{m}. \tag{A.371}$$

(vi)

$$\frac{\partial}{\partial \boldsymbol{\theta}} \exp\left\{ -\frac{1}{2} \boldsymbol{\theta}^T \mathbf{Q}^{-1} \boldsymbol{\theta} \right\} = -\exp\left\{ -\frac{1}{2} \boldsymbol{\theta}^T \mathbf{Q}^{-1} \boldsymbol{\theta} \right\} \mathbf{Q}^{-1} \boldsymbol{\theta}. \tag{A.372}$$

(vii)

$$\frac{\partial}{\partial \boldsymbol{\theta}} \ln \left(\boldsymbol{\theta}^T \mathbf{Q} \boldsymbol{\theta} \right) = 2 \left(\boldsymbol{\theta}^T \mathbf{Q} \boldsymbol{\theta} \right)^{-1} \mathbf{Q} \boldsymbol{\theta}. \tag{A.373}$$

A.7.2 Derivative of Scalar with Respect to Matrix

Consider a scalar function of a $M \times N$ matrix $\mathbf{X}$, which denotes by $a(\mathbf{X})$. The derivative of $a(\mathbf{X})$ with respect to $\mathbf{X}$ is the $M \times N$ matrix,

$$\frac{\partial}{\partial \mathbf{X}} a(\mathbf{X}) = \begin{bmatrix} \frac{\partial a(\mathbf{X})}{\partial x_{11}} & \frac{\partial a(\mathbf{X})}{\partial x_{12}} & \cdots & \frac{\partial a(\mathbf{X})}{\partial x_{1N}} \\ \frac{\partial a(\mathbf{X})}{\partial x_{21}} & & & \vdots \\ \vdots & & & \\ \frac{\partial a(\mathbf{X})}{\partial x_{M1}} & \cdots & & \frac{\partial a(\mathbf{X})}{\partial x_{MN}} \end{bmatrix}. \tag{A.374}$$

Each column is the vector derivative of the scalar $a(\mathbf{X})$ with respect to a column of $\mathbf{X}$:

$$\frac{\partial}{\partial \mathbf{X}} a(\mathbf{X}) = \left(\begin{array}{c:c:c} \frac{\partial}{\partial \mathbf{r}_1} a(\mathbf{X}) & \cdots & \frac{\partial}{\partial \mathbf{r}_N} a(\mathbf{X}) \end{array} \right). \tag{A.375}$$

$$\mathbf{X} = \left(\begin{array}{ccc} \mathbf{x}_1 & \cdots & \mathbf{x}_N \end{array} \right). \tag{A.376}$$

$$\mathbf{x}_i = \begin{bmatrix} x_{1i} \\ x_{2i} \\ \vdots \\ x_{Mi} \end{bmatrix}. \tag{A.377}$$

One application where we will utilize matrix gradients is in the computation of the CRB.

Typical functions are:

1.

$$\frac{\partial}{\partial \mathbf{X}} \operatorname{tr} \mathbf{X} = \mathbf{I}. \tag{A.378}$$

In (2)–(5), **A** and **B** are not functions of **X**.

2.

$$\frac{\partial}{\partial \mathbf{X}} \operatorname{tr} [\mathbf{AX}] = \frac{\partial}{\partial \mathbf{X}^T} \operatorname{tr} \left[\mathbf{AX}^T\right] = \left(\frac{\partial}{\partial \mathbf{X}} \operatorname{tr} \left[\mathbf{AX}^T\right]\right)^T = \mathbf{A}^T. \tag{A.379}$$

3.

$$\frac{\partial}{\partial \mathbf{X}} \operatorname{tr} \left[\mathbf{AX}^T\right] = \frac{\partial}{\partial \mathbf{X}^T} \operatorname{tr} [\mathbf{AX}] = \mathbf{A}. \tag{A.380}$$

4.

$$\frac{\partial}{\partial \mathbf{X}} \operatorname{tr} [\mathbf{AXB}] = \frac{\partial}{\partial \mathbf{X}^T} \operatorname{tr} \left[\mathbf{AX}^T\mathbf{B}\right] = \mathbf{A}^T\mathbf{B}^T. \tag{A.381}$$

5.

$$\frac{\partial}{\partial \mathbf{X}} \operatorname{tr} \left[\mathbf{AX}^T\mathbf{B}\right] = \frac{\partial}{\partial \mathbf{X}^T} \operatorname{tr} [\mathbf{AXB}] = \mathbf{BA}. \tag{A.382}$$

6.

$$\frac{\partial}{\partial \mathbf{X}} \operatorname{tr} \left[\mathbf{AX}^{-1}\right] = \frac{\partial}{\partial \mathbf{X}} \operatorname{tr} \left[\mathbf{X}^{-1}\mathbf{A}\right] = \left(-\mathbf{X}^{-1}\mathbf{AX}^{-1}\right)^T. \tag{A.383}$$

7.

$$\frac{\partial}{\partial \mathbf{X}} \operatorname{tr} \mathbf{X}^n = n\left(\mathbf{X}^{n-1}\right)^T. \tag{A.384}$$

8.

$$\frac{\partial}{\partial \mathbf{X}} \operatorname{tr} [\exp \mathbf{X}] = \exp \mathbf{X}. \tag{A.385}$$

9.

$$\frac{\partial}{\partial \mathbf{X}} \det \mathbf{X} = \det \mathbf{X} \left(\mathbf{X}^{-1}\right)^T. \tag{A.386}$$

10.

$$\frac{\partial}{\partial \mathbf{X}} \ln \det \mathbf{X} = \left(\mathbf{X}^{-1}\right)^T. \tag{A.387}$$

11.

$$\frac{\partial}{\partial \mathbf{X}} \det \mathbf{X}^n = n\,(\det \mathbf{X})^n \left(\mathbf{X}^{-1}\right)^T. \tag{A.388}$$

12. Kronecker products:

$$\frac{\partial(\mathbf{X}\otimes\mathbf{Z})}{\partial\mathbf{Y}} = \frac{\partial\mathbf{X}}{\partial\mathbf{Y}}\otimes\mathbf{Z} + (\mathbf{I}_s\otimes\mathbf{U}_{p\times r})\left(\frac{\partial\mathbf{Z}}{\partial\mathbf{Y}}\otimes\mathbf{X}\right)(\mathbf{I}_t\otimes\mathbf{U}_{l\times q}), \tag{A.389}$$

where $\mathbf{U}$ is the permutation matrix and the dimensions of $\mathbf{X}$, $\mathbf{Y}$, and $\mathbf{Z}$ are $p \times q$, $s \times t$, and $r \times l$, respectively.

A.7.3 Derivatives with Respect to Parameter

If $\mathbf{X}$ is a function of a real parameter, $\boldsymbol{\theta}$, we can use the chain rule to obtain

1.

$$\frac{\partial \ln \det\mathbf{X}}{\partial\theta_i} = \mathrm{tr}\left[\left[\frac{\partial \ln\det\mathbf{X}}{\partial\mathbf{X}}\right]^T\frac{\partial\mathbf{X}}{\partial\theta_i}\right] = \mathrm{tr}\left[\mathbf{X}^{-1}\frac{\partial\mathbf{X}}{\partial\theta_i}\right]. \tag{A.390}$$

2.

$$\frac{\partial \mathrm{tr}\left[\mathbf{X}^{-1}\mathbf{A}\right]}{\partial\theta_i} = \mathrm{tr}\left[\left[\frac{\partial \mathrm{tr}\left[\mathbf{X}^{-1}\mathbf{A}\right]}{\partial\mathbf{X}}\right]^T\frac{\partial\mathbf{X}}{\partial\theta_i}\right] = \mathrm{tr}\left[\left[\left[-\mathbf{X}^{-1}\mathbf{A}\mathbf{X}^{-1}\right]^T\right]^T\frac{\partial\mathbf{X}}{\partial\theta_i}\right] = -\mathrm{tr}\left[\left[\mathbf{X}^{-1}\mathbf{A}\mathbf{X}^{-1}\right]\frac{\partial\mathbf{X}}{\partial\theta_i}\right]. \tag{A.391}$$

3.

$$\frac{\partial\mathbf{X}^{-1}}{\partial\theta_i} = -\mathbf{X}^{-1}\frac{\partial\mathbf{X}}{\partial\theta_i}\mathbf{X}^{-1}. \tag{A.392}$$

4.

$$\frac{\partial^2 \ln\det\mathbf{X}}{\partial\theta_i\partial\theta_j} = \mathrm{tr}\left\{-\mathbf{X}^{-1}\frac{\partial\mathbf{X}}{\partial\theta_i}\mathbf{X}^{-1}\frac{\partial\mathbf{X}}{\partial\theta_j} + \mathbf{X}^{-1}\frac{\partial^2\mathbf{X}}{\partial\theta_i\partial\theta_j}\right\}. \tag{A.393}$$

5. The derivative of the projection matrix is,

$$\frac{\partial}{\partial\theta_i}\mathbf{P}_{\mathbf{V}}^{\perp} = -\frac{\partial}{\partial\theta_i}\mathbf{P}_{\mathbf{V}} = -\mathbf{P}_{\mathbf{V}}^{\perp}\mathbf{V}_i\mathbf{V}^H - \left(\mathbf{P}_{\mathbf{V}}^{\perp}\mathbf{V}_i\mathbf{V}^H\right)^H. \tag{A.394}$$

A.7.4 Complex Gradients

In many applications we have a real-valued function of a complex vector $\mathbf{z}$ and want to find either the minimum or the maximum. One approach is to write

$$\mathbf{z} = \mathbf{x} + j\,\mathbf{y} \tag{A.395}$$

and take derivatives with respect to $\mathbf{x}$ and $\mathbf{y}$. However, the resulting expressions are cumbersome. For the functions of interest in array applications, we can work directly with the vector $\mathbf{z}$ and obtain simpler expressions. However, we must be careful with the differentiation. Brandwood [Bra83] has developed a suitable approach, which we summarize.[8]

First consider the case when z is a complex scalar and the function of interest is

$$f(z) = f(x, y). \tag{A.396}$$

Define a function $g(z, z^*)$ that is analytic with respect to z and z^* independently, and

$$g(z, z^*) = f(x, y). \tag{A.397}$$

Brandwood proves that the partial of $g(z, z^*)$ with respect to z (treating z^* as a constant in g) gives the result,

$$\frac{\partial g(z, z^*)}{\partial z}\Big|_{z=x+jy} = \frac{1}{2}\left(\frac{\partial f(x,y)}{\partial x} - j\frac{\partial f(x,y)}{\partial y}\right). \tag{A.398}$$

Similarly,

$$\frac{\partial g(z, z^*)}{\partial z^*}\Big|_{z^*=x-jy} = \frac{1}{2}\left(\frac{\partial f(x,y)}{\partial x} + j\frac{\partial f(x,y)}{\partial y}\right). \tag{A.399}$$

Brandwood then shows that a necessary and sufficient condition for $f(z)$ to have a stationary point is either

$$\frac{\partial g(z, z^*)}{\partial z} = 0, \tag{A.400}$$

where z^* is treated as a constant in the partial derivative, or

$$\frac{\partial g(z, z^*)}{\partial z^*} = 0, \tag{A.401}$$

where z is treated as a constant in the partial derivative.

---

[8]Our discussion follows [Bra83].

For the case when $\mathbf{z}$ is a vector, we define the complex gradient operator as

$$\nabla_{\mathbf{z}} = \left[\frac{\partial}{\partial z_1}, \frac{\partial}{\partial z_2}, \cdots, \frac{\partial}{\partial z_N} \right]^T, \tag{A.402}$$

where

$$\frac{\partial}{\partial z_n} \triangleq \frac{\partial}{\partial x_n} - j\frac{\partial}{\partial y_n}, \quad n = 1, \cdots, N. \tag{A.403}$$

Similarly,

$$\nabla_{\mathbf{z}^H} = \left[\frac{\partial}{\partial z_1^*}, \frac{\partial}{\partial z_2^*}, \cdots, \frac{\partial}{\partial z_N^*} \right], \tag{A.404}$$

where

$$\frac{\partial}{\partial z_n^*} \triangleq \frac{\partial}{\partial x_n} + j\frac{\partial}{\partial y_n}, \quad n = 1, \cdots, N. \tag{A.405}$$

Let $f(\mathbf{z}) = f(\mathbf{x}, \mathbf{y}) = g(\mathbf{z}, \mathbf{z}^H)$ where $g(\mathbf{z}, \mathbf{z}^H)$ is a real-valued function of $\mathbf{z}$ and $\mathbf{z}^H$, which is analytic with respect to $\mathbf{z}$ and $\mathbf{z}^H$ independently. Then, either

$$\nabla_{\mathbf{z}} g(\mathbf{z}, \mathbf{z}^H) = \mathbf{0}, \tag{A.406}$$

where $\mathbf{z}^H$ is treated as a constant, or

$$\nabla_{\mathbf{z}^H} g(\mathbf{z}, \mathbf{z}^H) = \mathbf{0}, \tag{A.407}$$

where $\mathbf{z}$ is treated as a constant, are necessary and sufficient to determine a stationary point of $f(\mathbf{z})$. We will normally use (A.407) in our applications.

We consider two applications that are encountered in the text. The first application is to minimize $\mathbf{w}^H\mathbf{R}\mathbf{w}$ subject to the constraint $\mathbf{w}^H\mathbf{c} = a$. Here $\mathbf{R}$ is an $N \times N$ Hermitian matrix and $\mathbf{w}$ is an $N \times 1$ complex vector.

We define a real cost function,

$$\begin{aligned} g(\mathbf{w}, \mathbf{w}^H) &= \mathbf{w}^H\mathbf{R}\mathbf{w} + 2Re\left[\lambda(\mathbf{w}^H\mathbf{c} - a)\right] && \text{(A.408)} \\ &= \mathbf{w}^H\mathbf{R}\mathbf{w} + \lambda(\mathbf{w}^H\mathbf{c} - a) + \lambda^*(\mathbf{w}^H\mathbf{c} - a^*), && \text{(A.409)} \end{aligned}$$

where λ is a Lagrange multiplier. Taking the gradient with respect to $\mathbf{w}^H$ and setting the result equal to zero gives

$$\nabla_{\mathbf{w}^H} g(\mathbf{w}, \mathbf{w}^H) = \mathbf{R}\mathbf{w}_o + \lambda\mathbf{c} = \mathbf{0}, \tag{A.410}$$

where the subscript denotes optimum. Then

$$\mathbf{w}_o = -\lambda\mathbf{R}^{-1}\mathbf{c}. \tag{A.411}$$

The constraint equation is used to find λ,

$$\mathbf{c}^H\mathbf{w}_o = -\lambda\mathbf{c}^H\mathbf{R}^{-1}\mathbf{c} = a^*, \tag{A.412}$$

or

$$\lambda = -(\mathbf{c}^H\mathbf{R}^{-1}\mathbf{c})^{-1}a^*. \tag{A.413}$$

Using (A.413) in (A.411) gives

$$\mathbf{w}_o = \frac{a^*\mathbf{R}^{-1}\mathbf{c}}{\mathbf{c}^H\mathbf{R}^{-1}\mathbf{c}}. \tag{A.414}$$

As a second example, we maximize the same function, $\mathbf{w}^H\mathbf{R}\mathbf{w}$, subject to the constraint

$$\mathbf{w}^H\mathbf{w} = a, \tag{A.415}$$

where a is real.

The cost function is

$$g(\mathbf{w}, \mathbf{w}^H) = \mathbf{w}^H\mathbf{R}\mathbf{w} + \lambda(\mathbf{w}^H\mathbf{w} - a), \tag{A.416}$$

where λ is a real Lagrange multiplier. Then,

$$\nabla_{\mathbf{w}^H} g(\mathbf{w}, \mathbf{w}^H) = \mathbf{R}\mathbf{w}_o + \lambda\mathbf{w}_o = \mathbf{0}, \tag{A.417}$$

or

$$\mathbf{R}\mathbf{w}_o = -\lambda\mathbf{w}_o. \tag{A.418}$$

The result in (A.418) is familiar as the eigenvector equation (A.188). Thus, $\mathbf{w}_o$ is the eigenvector of $\mathbf{R}$ corresponding to the largest eigenvalue.

We use the complex gradient in a number of derivations in the text.

Bibliography

[AS65] M. Athans and F. C. Schweppe. Gradient matrices and matrix calculations. Technical Report, MIT Lincoln Laboratory, Lexington, Massachusetts, November 1965.

[Bel72] R. Bellmann. *Introduction to Matrix Analysis.* McGraw-Hill, New York, 2nd edition, 1972.

[Bra83] D. H. Brandwood. A complex gradient operator and its application in adaptive array theory. *Proc. IEE*, Special issue on adaptive arrays, vol.130, P. F, pp. 11–17, February 1983.

[CB76a] A. Cantoni and P. Butler. Eigenvalues and eigenvectors of symmetric centrosymmetric matrices. *Linear Algebra Appl.*, vol.13 pp. 275–288, 1976.

[CB76b] A. Cantoni and P. Butler. Properties of the eigenvectors of persymmetric matrices with applications to communication theory. *IEEE Trans. Commun.*, vol. COM-24 pp. 804–809, August 1976.

[Giv58] W. Givens. Computation of plane unitary rotations transforming a general matrix to triangular form. *J. Soc. Ind. Appl. Math.*, 6:26–50, 1958.

[Gra81] A. Graham. *Kronecker Products and Matrix Calculus: With Applications.* Wiley, New York, 1981.

[GS58] O. Grenander and G. Szego. *Toeplitz Forms and Their Applications.* University of California Press, Berkeley, 1958.

[GVL89] G. H. Golub and C. F. Van Loan. *Matrix Computations.* The Johns Hopkins University Press, Baltimore, Maryland, 1989.

[Hay91] S. Haykin. *Adaptive Filter Theory* Prentice-Hall, Englewood Cliffs, New Jersey, 2nd edition 1991.

[Hay96] S. Haykin. *Adaptive Filter Theory.* Prentice-Hall, Upper Saddle River, New Jersey, 3rd edition, 1996.

[Hoh73] F. E. Hohn. *Elementary Matrix Algebra.* Macmillan, New York, 1973.

[Hou58] A. S. Householder. Unitary triangularization of a non-symmetric matrix. *J. Assoc. Comput. Math.*, vol. 5 pp. 204–243, 1958.

[Hou64] A. S. Householder. *The Theory of Matrices in Numerical Analysis.* Blaisdell, Waltham, Massachusetts, 1964.

[HS92] S. Haykin and A. Steinhardt, editors. *Adaptive Radar Detection and Estimation.* Wiley, New York, 1992.

[Hud81] J.E. Hudson. *Adaptive Array Principles.* Peter Peregrinus, New York and London, 1981.

[Kay88] S. M. Kay. *Modern Spectral Estimation: Theory and Application.* Prentice-Hall, Englewood Cliffs, New Jersey, 1988.

[Kur89] A. R. Kuruc. Lower bounds on multiple-source direction finding in the presence of direction-dependent antenna-array-calibration errors. Technical Report, MIT Lincoln Laboratory, Lexington, Massachusetts, 1989.

[LH74] C. L. Lawson and R. J. Hanson. *Solving Least Squares Problems.* Prentice-Hall, Englewood Cliffs, New Jersey, 1974.

[LT85] P. Lancaster and M. Tismenetsky. *The Theory of Matrices with Applications.* Academic Press, San Diego, California 1985.

[Mak81] J. Makhoul. On the eigenvectors of symmetric Toeplitz matrices. *IEEE Trans. on Acoust., Speech, Signal Process.*, ASSP-29 pp. 868–872, August 1981.

[Mar60] M. Marcus. Basic theorems in matrix theory. Technical Report Applied Math. Ser. 57, National Bureau of Standards, Washington D.C., January 1960.

[Mar87] S. L. Marple, Jr. *Digital Spectral Analysis.* Prentice-Hall, Englewood Cliffs, New Jersey, 1987.

[MP73] J. H. McClellan and T. W. Parks. A unified approach to the design of optimum FIR linear phase digital filters. *IEEE Trans. Circuit Theory*, vol. CT-20 pp. 697–701, 1973.

[Moo20] E. H. Moore. *Bull. Am. Math. Soc.*, vol. 26, pp. 394–395, 1920

[ND77] B. Noble and J. W. Daniel. *Applied Linear Algebra.* Prentice-Hall, Englewood Cliffs, New Jersey, 2nd edition, 1977.

[Pen55] R. Penrose. *Proc. Cambridge Philos. Soc.*, vol. 1, pp. 406–413, 1955.

[PRLN92] J. G. Proakis, C. M. Rader, F. Ling, and C. L. Nikias. *Advanced Digital Signal Processing.* Macmillan, New York, 1992.

[RM71] C. R. Rao and S. K. Mitra. *Generalized Inverse of Matrices and Its Applications.* Wiley, New York, 1971.

[Sch91] L. L. Scharf. *Statistical Signal Processing: Detection, Estimation, and Time Series Analysis.* Addison-Wesley, Reading, Massachusetts, 1991.

[SN89] P. Stoica and A. Nehorai. MUSIC, maximum likelihood, and Cramér-Rao bound. *IEEE Trans. Acoust., Speech, Signal Process.*, vol.ASSP-37, pp. 720–741, May 1989.

[Ste00] C. Stephanos. Pures applications. *J. Math.*, vol. 6 pp. 73–128, 1900.

[Ste73] G. W. Stewart. *Introduction to Matrix Computations.* Academic Press, New York, 1973.

[Ste88] A. O. Steinhardt. Householder transforms in signal processing. *IEEE Acoust. Speech, Signal Process. Mag*, pp. 4–12, July 1988.

[VT68] H. L. Van Trees. *Detection, Estimation, and Modulation Theory, Part I.* Wiley, New York, 1968.

[VT71] H. L. Van Trees. *Detection, Estimation, and Modulation Theory, Part III.* Wiley, New York, 1971.

[VT01a] H. L. Van Trees. *Detection, Estimation, and Modulation Theory, Part I.* Wiley Interscience, New York, 2001.

[VT01b] H. L. Van Trees. *Detection, Estimation, and Modulation Theory, Part III.* Wiley Interscience, New York, 2001.

[Wil65] J. H. Wilkinson. *The Algebraic Eigenvalue Problem.* Oxford University Press, New York, 1965.

Appendix B

Array Processing Literature

Due to the wide variety of applications of array processing, the literature is spread across a number of different journals and conferences. In this section, we list some of the journals and books where array processing research is reported. In Section B.1, we list the relevant journals. In Section B.2, we list some books dealing with array processing. In Section B.3, we list some books that treat time-domain problems, which are similar to the array processing problems, for uniform linear arrays.

B.1 Journals

The following journals discuss current research in array processing from various viewpoints.

A. Institute of Electrical and Electronic Engineers

We have listed transactions that contain articles on array processing. The order is a rough indication of the number of articles published. The first three entries are the primary sources.

(i) *Signal Processing*

(ii) *Antennas and Propagation*

(iii) *Aerospace and Electronic Systems*

(iv) *Ocean Engineering*

(v) *Information Theory*

(vi) *Circuits and Systems*

(vii) *Vehicular Technology*

(viii) *Communications*

(ix) *Geoscience Electronics*

(x) *Automatic Control*

(xi) *Systems, Man, and Cybernetics*

The *Proceeding of the IEEE* also has special issues dealing with Antennas (e.g., [MM92]).
B. IEE (British)
C. Journal of the Acoustical Society of America (JASA)
D. Geophysics
E. Geophysical Prospecting
F. Signal Processing (European)

B.2 Books

Two representative lists of books that deal with various aspects of array processing can be constructed. The first group emphasizes a deterministic approach and develops what we will refer to a "classical array theory". This group includes:

(i) J. D. Kraus, *Antennas* [Kra88]

(ii) C. A. Balanis, *Antenna Theory Analysis and Design* [Bal82]

(iii) R. S. Elliott, *Antenna Theory and Design* [Ell81]

(iv) T. A. Milligan, *Modern Antenna Design* [Mil85]

(v) Y. T. Lo and S. W. Lee, *Antenna Handbook*: 1. *Fundamentals and Mathematical Techniques* [LL93a]; 2. *Antenna Theory* [LL93b]; 3. *Applications* [LL93c]; 4. *Related Topics* [LL93d]

(vi) W. L Stutzman and G. A. Thiele, *Antenna Theory and Design* [ST81]

(vii) B. D. Steinberg, *Principles of Aperture and Array System Design* [Ste76]

(viii) R. J. Mailloux, *Phased Array Antenna Handbook* [Mai94]

(ix) W. L. Weeks, *Antenna Engineering* [Wee68]

(x) K. Fujimoto and J. R. James, *Mobile Antenna Systems Handbook* [FJ94]

(xi) H. Mott, *Antennas for Radar and Communications* [Mot92]

(xii) R. C. Hansen, *Microwave Scanning Antennas* [Han85]

(xiii) R. C. Hansen, *Phased Array Antennas* [Han98]

The second group emphasizes a statistical approach to "optimum array processing". This list includes:

(i) P. A. Monzingo and T. W. Miller, *Introduction to Adaptive Arrays* [MM80]
(ii) J. E. Hudson, *Adaptive Array Principles* [Hud81]
(iii) S. Haykin (Ed.), *Adaptive Signal Processing* [Hay85]
(iv) D. E. Dudgeon and R. M. Mesereau, *Multidimensional Signal Processing* [DM84]
(v) E. Brookner, *Practical Phased Array Antenna Systems* [Bro91]
(vi) S. Haykin (Ed.), *Advances in Spectrum Analysis and Array Processing*, Vol. I [Hay91a], Vol. II [Hay91b], Vol. III [Hay95]
(vii) D. H. Johnson and D. E. Dudgeon, *Array Signal Processing, Concepts and Techniques* [JD93]
(viii) R. T. Compton, Jr., *Adaptive Antennas, Concepts and Performance* [Com88]
(ix) S. U. Pillai, *Array Signal Processing* [Pil89]
(x) B. D. Steinberg, *Principles of Aperture and Array System Design* [Ste76]
(xi) S. Haykin and A. Steinhardt, *Adaptive Radar Detection and Estimation* [HS92]
(xii) T. J. Shepherd S. Haykin, and J. Litva (Eds.), *Radar Array Processing* [SH92]
(xiii) P. Stoica and R. Moses, *Introduction to Spectral Analysis* [SM97]
(xiv) L. J. Ziomek, *Fundamentals of Acoustic Field Theory and Space-Time Signal Processing* [Zio95]
(xv) J. C. Hassab, *Underwater Signal and Data Processing* [Has89]
(xvi) R. J. Mailloux, *Phased Array Antenna Handbook* [Mai94]
(xvii) N. Kalouptsidis and S. Theodoridis (Eds.), *Adaptive System Identification and Signal Processing Algorithms* [KT93]
(xviii) F. A. Grunbaum, M. Bernfeld, and R. E. Blahut (Eds.), *Radar and Sonar, Part I* [GBB91]
(xix) F. A. Grunbaum, M. Bernfeld, and R. E. Blahut(Eds.), *Radar and Sonar, Part II* [GBB92]
(xx) S. Haykin (Ed.), *Topics in Applied Physics*, Vol. 34: Nonlinear Methods of Spectral Analysis [Hay83]
(xxi) J. C. Liberti, Jr. and T. S. Rappaport, *Smart Antennas for Wireless Communications: IS-95 and Third Generation CDMA Applications* [LR99]
(xxii) R. Klemm, *Space-time Adaptive Processing* [Kle98]
(xxiii) T. S. Rappaport (Ed.), *Smart Antennas: Adaptive Arrays, Algorithms, and Wireless Position Location* [Rap98]
(xxiv) S. Y. Kung, H. J. Whitehouse, and T. Kailath (Eds.), *VLSI and Modern Signal Processing* [KWK85]

B.3 Duality

In Chapter 2, we find that array processing for a uniformly spaced linear array with elements spaced at $\lambda/2$ is identical to frequency-domain processing using a FIR filter. Therefore, a significant number of results carry over directly to the array processing area.

Representative books that contain results that are useful in array processing include:

(i) S. M. Kay, *Modern Spectral Estimation, Theory and Application* [Kay88]

(ii) S. L. Marple, Jr., *Digital Spectral Analysis with Applications* [Mar87]

(iii) A. V. Oppenheim and R. W. Schafer, *Discrete-Time Signal Processing* [OS89]

(iv) S. Haykin, *Adaptive Filter Theory*, [Hay96]

(v) B Porat, *Digital Processing of Random Signals* [Por94]

(vi) J. G. Proakis et al., *Advanced Digital Signal Processing* [PRLN92]

(vii) D. G. Childers, *Modern Spectrum Analysis* [Chi78]

(viii) S. B. Kesler (Ed.), *Modern Spectrum Analysis*, vol. II [Kes86]

(ix) B. Widrow and S. D. Stearns, *Adaptive Signal Processing* [WS85]

(x) L. L. Scharf, *Statistical Signal Processing: Detection, Estimation, and Time Series Analysis* [Sch91]

(xi) M. Bellanger, *Digital Processing of Signals*, [Bel84]

(xii) G. C. Carter (Ed.), *Coherence and Time Delay Estimation* [Car93]

(xiii) A. V. Oppenheim (Ed.), *Applications of Digital Signal Processing* [Opp78]

(xiv) S. U. Pillai and T. I. Shim, *Spectrum Estimation and System Identification* [PS93]

(xv) Multidimensional Signal Processing Committee (Eds.), *Selected Papers in Multidimensional Digital Signal Processing* [MSP86]

(xvi) L. R. Rabiner and B. Gold, *Theory and Application of Digital Signal Processing* [RG75]

(xvii) M. B. Priestley, *Spectral Analysis and Time Series*, Vols. 1 and 2 [Pri81]

(xviii) S. T. Alexander, *Adaptive Signal Processing* [Alex86]

(xix) J. R. Treichler, C. R. Johnson, Jr., and M. G. Larimore, *Theory and Design of Adaptive Filters* [TJL87]

(xx) J. M. Mendel, *Lessons in Estimation Theory for Signal Processing, Communications, and Control* [Men95]

Bibliography

[Alex86] S. T. Alexander. *Adaptive Signal Processing: Theory and Applications.* Springer-Verlag, New York, 1986.

[Bal82] C. A. Balanis. *Antenna Theory Analysis and Design.* Wiley, New York, 1982.

[Bel84] M. Bellanger. *Digital Processing of Signals.* Wiley, New York, 2nd edition, 1984.

[GBB91] F. A. Grunbaum, M. Bernfeld, and R.E. Balhut, editors. *Radar and Sonar, Part I.* Springer-Verlag, New York, 1991.

[Bro91] E. Brookner, editor. *Practical Phased-Array Antenna Systems.* Artech House, Boston, Massachusetts, 1991.

[Car93] G. C. Carter. *Coherence and Time Delay Estimation.* IEEE Press, New York, 1993.

[Chi78] D. G. Childers, editor. *Modern Spectrum Analysis.* IEEE Press, New York, 1978.

[Com88] R. T. Compton, Jr. *Adaptive Antennas (Concepts and Performance).* Prentice-Hall, Englewood Cliffs, New Jersey, 1988.

[DM84] D. E. Dudgeon and R. M. Mersereau. *Multidimensional Digital Signal Processing.* Prentice-Hall, Englewood Cliffs, New Jersey, 1984.

[Ell81] R. S. Elliott. *Antenna Theory and Design.* Prentice-Hall, Englewood Cliffs, New Jersey, 1981.

[FJ94] K. Fujimoto and J. R. James. *Mobile Antenna Systems Handbook.* Artech House, Boston, Massachusetts, 1994.

[GBB92] F. A. Grunbaum, M. Bernfeld, and R. E. Blahut, editors. *Radar and Sonar, Part II.* Springer-Verlag, New York, 1992.

[Han85] R. C. Hansen. *Microwave Scanning Antennas.* Peninsula Publishing, Los Altos, Calfornia, 1985.

[Han98] R. C. Hansen. *Phased Array Antennas.* Wiley, New York, 1998.

[Has89] J. C. Hassab. *Underwater Signal and Data Processing.* CRC Press, Boca Raton, Florida, 1989.

[Hay83] S. Haykin. *Topics in Applied Physics*, vol.34. Springer-Verlag, New York, 1983.

[Hay85] S. Haykin. *Array Signal Processing.* Prentice-Hall, Englewood Cliffs, New Jersey, 1985.

[Hay91a] S. Haykin. *Adaptive Filter Theory,* Prentice-Hall, Englewood Cliffs, New Jersey, 2nd edition, 1991.

[Hay91b] S. Haykin, editor. *Advances in Spectrum Analysis and Array Processing*, vol.2. Prentice-Hall, Englewood Cliffs, New Jersey, 1991.

[Hay95] S. Haykin, editor. *Advances in Spectrum Analysis and Array Processing*, vol.3. Prentice-Hall, Englewood Cliffs, New Jersey, 1995.

[Hay96] S. Haykin. *Adaptive Filter Theory.* Prentice-Hall, Upper Saddle River, New Jersey, 3rd edition, 1996.

[HS92] S. Haykin and A. Steinhardt, editors. *Adaptive Radar Detection and Estimation.* Wiley, New York, 1992.

[Hud81] J.E. Hudson. *Adaptive Array Principles.* Peter Peregrinus, New York and London, 1981.

[JD93] D. H. Johnson and D. E. Dudgeon. *Array Signal Processing.* Prentice-Hall, Englewood Cliffs, New Jersey, 1993.

[KT93] N. Kalouptsidis and S. Theodoridis. *Adaptive System Identification and Signal Processing Algorithms.* Prentice-Hall, New York, 1993.

[Kay88] S. M. Kay. *Modern Spectral Estimation: Theory and Application.* Prentice-Hall, Englewood Cliffs, New Jersey, 1988.

[Kle98] R. Klemm. *Space-Time Adaptive Processing.* Short Run Press, Exeter, England, 1998.

[Kes86] S. B. Kesler, editor. *Modern Spectrum Analysis, II.* IEEE Press, New York, 1986.

[Kra88] J. D. Kraus. *Antennas.* McGraw-Hill, New York, 2nd edition, 1988.

[KWK85] S. Y. Kung, H. J. Whitehouse, and T. Kailath. *VLSI and Modern Signal Processing.* Prentice-Hall, Englewood Cliffs, New Jersey, 1985.

[LL93a] Y. T. Lo and S. W. Lee, editors. *Antenna Handbook (Antenna Theory)*, vol.2. Chapman & Hall, New York, 1993.

[LL93b] Y. T. Lo and S. W. Lee, editors. *Antenna Handbook (Applications)*, vol. 3. Chapman & Hall, New York, 1993.

[LL93c] Y. T. Lo and S. W. Lee, editors. *Antenna Handbook (Fundamentals and Mathematical Techniques)*, volume 1. Chapman & Hall, New York, 1993.

[LL93d] Y. T. Lo and S. W. Lee, editors. *Antenna Handbook (Related Issues)*, vol.4. Chapman & Hall, New York, 1993.

[LR99] J. C. Liberti, Jr., and T. S. Rappaport. *Smart Antennas for Wireless Communications: IS-95 and Third Generation CDMA Applications.* Prentice-Hall, Upper Saddle River, New Jersey, 1999.

[Mai94] R. J. Mailloux. *Phased Array Antenna Handbook.* Artech House, Boston, Massachusetts, 1994.

[Mar87] S. L. Marple, Jr. *Digital Spectral Analysis.* Prentice-Hall, Inc., Englewood Cliffs, New Jersey, 1987.

[Men95] J. M. Mendel. *Lessons in Estimation Theory for Signal Processing, Communications, and Control.* Prentice-Hall, Englewood Cliffs, New Jersey, 1995.

[Mil85] T. A. Milligan. *Modern Antenna Design.* McGraw-Hill, New York, 1985.

[MM80] R. A. Monzingo and T. W. Miller. *Introduction to Adaptive Arrays.* Wiley, New York, 1980.

[MM92] L. N. Medgyesi-Mitschang, editor. *Proc. IEEE: Special Issue on Antennas*, vol. 80. Institute of Electrical and Electronics Engineers, Picataway, New Jersey, January 1992.

[Mot92] H. Mott. *Antennas for Radar and Communications.* Wiley, New York, 1992.

[MSP86] Multidimensional Signal Processing Committee. *Selected Papers in Multidimensional Digital Signal Processing.* IEEE Press, New York, 1986.

[Opp78] A. V. Oppenheim. *Applications of Digital Signal Processing.* Prentice-Hall, Englewood Cliffs, New Jersey, 1978.

[OS89] A. V. Oppenheim and R. W. Schafer. *Discrete-Time Signal Processing.* Prentice-Hall, Englewood Cliffs, New Jersey, 1989.

[Pil89] S. U. Pillai. *Array Signal Processing.* Springer-Verlag, New York, 1989.

[PS93] S. U. Pillai and T. I. Shim. *Spectrum Estimation and System Identification.* Springer-Verlag, New York, 1993.

[Por94] B. Porat. *Digital Processing of Random Signals.* Prentice-Hall, Englewood Cliffs, New Jersey, 1994.

[Pri81] M. B. Priestley. *Spectral Analysis and Time Series*, vols. 1 and 2. Academic Press, San Diego, California, 1981.

[PRLN92] J. G. Proakis, C. M. Rader, F. Ling, and C. L. Nikias. *Advanced Digital Signal Processing.* Macmillan, New York, 1992.

[RG75] L. R. Rabiner and B. Gold. *Theory and Application of Digital Signal Processing.* Prentice Hall, Englewood Cliffs, New Jersey, 1975.

[Rap98] T. S. Rappaport. *Smart Antennas: Adaptive Arrays, Algorithms, & Wireless Position Location.* IEEE Press, Piscataway, New Jersey, 1998.

[Sch91] L. L. Scharf. *Statistical Signal Processing: Detection, Estimation, and Time Series Analysis.* Addison-Wesley, Reading, Massachusetts, 1991.

[SH92] T. J. Shepherd S. Haykin, and J. Litva, editors. *Radar Array Processing.* Springer-Verlag, New York, 1992.

[ST81] W. L. Stutzman and G. A. Thiele. *Antenna Theory and Design.* Wiley, New York, 1981.

[Ste76] B.D. Steinberg. *Principles of Aperture and Array System Design.* Wiley, New York, 1976.

[SM97] P. Stoica and R. Moses. *Introduction to Spectral Analysis.* Prentice Hall, Upper Saddle River, New Jersey, 1997.

[TJL87] J. R. Treichler, C. R. Johnson, Jr., and M. G. Larimore. *Theory and Design of Adaptive Filters.* Wiley, New York, 1987.

[Wee68] W. L. Weeks. *Antenna Engineering.* McGraw-Hill, New York, 1965.

[WS85] B. Widrow and S.D. Stearns. *Adaptive Signal Processing.* Prentice-Hall, Englewood Cliffs, New Jersey, 1985.

[Zio95] L. J. Ziomek. *Fundamentals of Acoustic Field Theory and Space-Time Signal Processing.* CRC Press, Boca Raton, Florida, 1995.

Appendix C

Notation

In this section we discuss the conventions, abbreviations, mathematical symbols, and symbols used in the book.

C.1 Conventions

The following conventions have been used:

1. Boldface roman denotes a vector or matrix.
2. The symbol $|\cdot|$ means the magnitude of the scalar contained within.
3. The symbol $\|\cdot\|$ denotes the Euclidean norm of the vector or matrix contained within.
4. Multiple integrals are frequently written as,

$$\int d\tau f(\tau) \int dt g(t,\tau) \triangleq \int f(\tau) \left\{ \int dt g(t,\tau) \right\} d\tau,$$

   that is, an integral is inside all integrals to its left unless a multiplication is specifically indicated by parentheses.

5. $E[\cdot]$ denotes the statistical expectation of the quantity in the bracket.
6. The probability density of x is denoted by $p_x(\cdot)$ and the probability distribution by $P_x(\cdot)$. The probability of an event, A, is denoted by $Pr[A]$. The probability density of x, given that the random variable a has a value A, is denoted by $p_{x|a}(X|A)$.

7. A vertical line in an expression means "such that" or "given that"; that is, $Pr[A|x \leq X]$ is the probability that event A occurs given that the random variable x is less than or equal to the value of X.

We list acronyms used in the text.

C.2 Acronyms

AAR	adaptive angular response
ACCR	asymptotic conditional CRB
AIC	Akaike Information Criterion
AM	alternating maximization
AMF	adaptive matched filter
AML	asymptotic ML estimator
AP	alternating projection
AR	auto-regressive (model)
ARMA	auto-regressive moving average (model)
ASNR	array signal-to-noise ratio
BW_{NN}	null-null bandwidth
BS	beamspace
CCRB	conditional CRB
CFAR	constant false alarm rate
CM	constant modulus
CML	conditional ML estimator
CMT	covariance matrix taper
CRB	Cramér-Rao bound
DEMT	Detection, Estimation, and Modulation Theory
DFT	discrete Fourier transform
DISPARE	distributed signal parameter estimator
DL	diagonal loading
DMI	direct matrix inversion
DMR	dominant-mode rejection (beamformer)
DOA	direction of arrival
DOF	degrees of freedom
DPSS	discrete prolate spheroidal sequence
DSPE	distributed signal parameter estimation

EV	eigenvalue
EM	expectation maximization
ES	eigenspace
ESPRIT	estimation of signal parameter via rotational invariance technique
FB	forward–backward (averaging)
FBSS	forward-backward spatial smoothing
FCA	filled circular array
FIM	Fisher information matrix
FIR	finite impulse response
FO	forward-only (averaging)
GL	grating lobe
GLRT	generalized LRT
GSC	generalized sidelobe canceller
HPBW	half-power bandwidth
HCRB	hybrid CRB
HSST	Householder subspace transformation
ICASSP	International Conference on Acoustic, Speech, and Signal Processing
IDFT	inverse discrete Fourier transform
IMODE	iterative MODE
INR	interference-to-noise ratio
IQML	iterative quadratic maximum likelihood
IQML-QC	IQML with quadratic constraint
IR	invisible region
JASA	Journal Acoustic Society of America
LCMP	linear constrained minimum power (beamformer)
LCMV	linear constrained minimum variance (beamformer)
LEO	low-earth orbit
LMS	least mean square
LNR	loading-to-noise ratio

LP	linear prediction
LRT	likelihood ratio test
LS	least squares
LS-ESPRIT	least squares ESPRIT
ML	maximum likelihood
MA	moving average (model)
MAP	maximum *a posteriori* probability
MBA	multiple beam antenna
MDL	minimum description length
MMSE	minimum mean-square error
MODE	method of direction estimation
MPDR	minimum power distortionless response
MPQR	minimum power quiescent response
MRA	(i) minimum redundancy array (ii) maximum response axis
MRLA	minimum redundancy linear array
MUSIC	multiple signal characterization
MVDR	minimum variance distortionless response
MVQR	minimum variance quiescent response
PM	Parks-McClellan
QC	quadratic constraint
QP	quiescent pattern
QRD	QR decomposition
RMSE	root mean-square error
RLS	recursive least square
SALP	subaperture linear prediction
SCMV	soft constraint minimum variance
SH	sequential hypothesis
SHA	standard hexagonal array
SINR	signal-to-interference and noise ratio
SLA	standard linear array ($d = \lambda/2$)
SLL	sidelobe level
SMI	sample matrix inversion
SNR	signal-to-noise ratio
SRA	standard rectangular array ($d_x = d_y = \lambda/2$)
STAP	space-time adaptive processing

SS	(i) spatial smoothing (ii) steady state
SVD	singular value decomposition
TAM	Toeplitz approximation method
TDL	tapped delay line
TDRSS	Tracking and Data Relay Satellite System
TLS	total least squares
TLS-ESPRIT	total least squares ESPRIT
TNA	thermal noise algorithm
UCA	uniform circular array
UHA	uniform hexagonal array
ULA	uniform linear array
UML	unconditional ML estimator
URA	uniform rectangular array
VR	visible region
WES	weighted eigenspace
WSF	weighted subspace fitting

C.3 Mathematical Symbols

$\mathbf{a}^H$, $\mathbf{A}^H$	conjugate transpose (A.6)
$\mathbf{a}^T$, $\mathbf{A}^T$	transpose (A.4)
$\mathbf{A}^{-1}$	inverse (A.40)
$\|\mathbf{a}\|$	Euclidean norm of $\mathbf{a}$ (A.36)
$\mathbf{A}^*$	conjugate of $\mathbf{A}$
$[\mathbf{A}]_{ij}$	ij element of $\mathbf{A}$ (A.1)
$[\mathbf{A}]_k^{(r)}$	kth row of $\mathbf{A}$
$[\mathbf{A}]_k^{(c)}$	kth column of $\mathbf{A}$
$\square$	Khatri-Rao product (A.96)
$\|\mathbf{A}\|_F$	Frobenius (Euclidean) norm of $\mathbf{A}$ (A.39)
$CN[\mathbf{m}, \mathbf{\Lambda}]$	probability density of complex Gaussian (normal) vector
$\otimes$	Kronecker product (A.80)
$\triangleq$	defined as
$\det \mathbf{A}$, $\|\mathbf{A}\|$	determinant of $\mathbf{A}$ (A.20)

$\dim[\mathbf{A}]$	number of free parameters in $\mathbf{A}$
$\doteq$	equal to first order
$\odot$	Hadamard product (A.71)
log	logarithm: base 10
ln	logarithm: base e
$(n)!$	n factorial
$\nabla_{\boldsymbol{\theta}}$	partial derivative matrix operator
$\propto$	proportional to
$o(N)$	negligible as $N \to \infty$
$O(N)$	proportional to N as $N \to \infty$
$\text{orth}[\mathbf{A}]$	columns of $\mathbf{A}$ are orthonormalized
$Re\{\cdot\}$	real part
$Im\{\cdot\}$	imaginary part
$\text{rank}\ (\mathbf{A})$	rank of $\mathbf{A}$ (A.34)
$\text{tr}\ (\mathbf{A})$	trace of $\mathbf{A}$ (A.26)
$vec(\mathbf{A})$	stacked vector (A.112)
$\hat{\mathbf{x}}$	estimate of $\mathbf{x}$
$t \to T^-$	t approaches T from below
$A + B \triangleq A \cup B$	A or B or both
$l.i.m.$	limit in the mean
$\int_{-\infty}^{\infty} d\mathbf{x}$	an integral over the same dimension as the vector
$\text{diag}[a_1, a_2, \cdots, a_N]$	diagonal matrix with elements a_1, a_2, $\cdots$, a_N
$\int_{\Omega} d\mathbf{x}$	integral over the set Ω
$\mathbf{0}$	matrix with all zero elements
$\mathbf{1}$	unity vector: $[1\ 1\ \cdots\ 1]^T$
$\begin{pmatrix} N \\ k \end{pmatrix}$	binomial coefficient $\left(= \frac{N!}{k!(N-k)!}\right)$

C.4 Symbols

We list symbols that are used in the text. The equation number indicates where the symbol was defined. If the symbol was defined in the text, a nearby equation number is listed. Symbols that are obvious modifications of other symbols are not included.

$\mathbf{a}$	signal direction vector (2.15)
$\mathbf{A}(K)$	data matrix ($\mathbf{A}(K) = \sqrt{K}\tilde{\mathbf{X}}^T$) (7.271)

A_c conventional array gain (6.32)
A_{iso} array gain: isotropic noise (2.144)
A_{mpdr} array gain: MPDR beamformer (6.73)
A_{mvdr} array gain: MVDR beamformer (6.28)
$\mathbf{A}_\mu(K)$ exponentially weighted data matrix (7.269)
$\mathbf{A}_L$ left circulant matrix (A.150)
A_{lcmp} array gain: LCMP beamformer (6.390)
A_{lcmv} array gain: LCMV beamformer (6.389)
A_o optimum array gain (6.28)
$\mathbf{A}_R$ right circulant matrix (A.148)
A_w array gain for spatially white noise input (2.185)
$AF_{\mathbf{k}}(\mathbf{k})$ array factor in $\mathbf{k}$-space (2.234)
$AF_\psi(\psi)$ array factor in ψ-space (2.234)
$AF_u(u)$ array factor in u-space (2.234)
$AR(p)$ autoregressive process of order p (5.311)
$ARMA(p,q)$ autoregressive moving average process of order (p,q) (5.303)
α step size in LMS algorithm (7.405)
$\alpha(K)$ step size in LMS algorithm (7.405)

$\mathbf{B}$ (i) blocking matrix (6.361)
(ii) matrix in polynomial parameterization (8.490)
(iii) beam pattern matrix (2.82)
(iv) output of DFT (3.96)
$B_a(\psi)$ asymmetric beam pattern (3.285)
$\mathbf{B}_{bs}$ beamspace matrix (3.327)
$\mathbf{B}_{bsbl}$ beamspace blocking matrix (6.530)
$\mathbf{b}_{bs,m}^H$ weight vector for mth beamspace beam (6.499)
$\mathbf{B}_{bsx}^H$ x-component of beamspace matrix, planar array (9.372)
$\mathbf{B}_{bsy}^H$ y-component of beamspace matrix, planar array (9.374)
$B_{du}(u)$ desired beam pattern in u-space
B_f ratio of signal bandwidth to center frequency (5.288)
$\mathbf{B}_{no}$ non-orthogonal beamspace matrix (3.329)
$B_{mvdr}(\psi)$ beam pattern of MVDR beamformer (6.93)
$B_R(\psi)$ beam pattern using rectangular window (3.213)
$B_{ue}(u)$ element beam pattern (2.232)
$B(u_x,u_y)$ beam pattern in u_x, u_y-space (4.11)
$B(\theta,\phi)$ beam pattern in (θ,ϕ)-space (2.38)

$B(\omega:\theta,\phi)$	beam pattern in (θ,ϕ)-space (2.38)
$B_v(v)$	beam pattern in v-space (3.170)
$B(\psi_x,\psi_y)$	2-D beam pattern in ψ_x,ψ_y space (4.11)
$B_z(z)$	beam pattern polynomial in z-plane (3.42)
$B_{\mathbf{z}}(z_1,z_2)$	2-D z-transform (4.93)
$B_c(\mathbf{k}:\mathbf{k}_T)$	conventional beam pattern in **k**-space (2.124)
$B_d(\psi)$	desired beam pattern (3.82)
$B_{eig,i}$	ith eigenbeam (7.126)
$B^{(n)}(\mathbf{k})$	nominal beam pattern (2.196)
$\mathbf{B}_{PM}$	phase mode excitation beamformer
$B_\psi(\psi)$	beam pattern in ψ-space (2.71)
B_s	bandwidth of signal (Hz) (2.40)
$B_T(u)$	beam pattern using Taylor weighting (3.173)
$B_{TAY}(u)$	Taylor beam pattern: circular aperture (4.220)
$B_\theta(\theta)$	beam pattern in θ-space (2.69)
$B_u(u)$	beam pattern in u-space (2.70)
$B_{\psi c}(\psi:\psi_T)$	conventional beam pattern in ψ-space (2.125)
$B_{\theta c}(\theta:\theta_T)$	conventional beam pattern in θ-space (2.131)
$B_{uc}(u:u_T)$	conventional beam pattern in u-space (2.126)
BW_{NN}	bandwidth: null-to-null (2.105)
c	velocity of propagation (2.14)
$\mathbf{C}$	constraint matrix $N \times M_c$ (3.256)
$\mathbf{C}_{ACR}(\boldsymbol{\psi})$	approximate Cramér-Rao bound on $\boldsymbol{\psi}$ (8.110)
$\mathbf{C}_{BCR}(\boldsymbol{\theta})$	Bayesian Cramér-Rao bound ($= \mathbf{J}_B^{-1}$) (8.53)
$\mathbf{C}_0, \mathbf{C}_1, \mathbf{C}_2$	constraint matrices using derivatives (3.251),(3.253),(3.255)
$\mathbf{C}_{CR}(\boldsymbol{\psi})$	Cramér-Rao bound on $\boldsymbol{\psi}$ (8.94)
$\mathbf{C}_{HCR}(\boldsymbol{\phi},\boldsymbol{\rho})$	hybrid Cramér-Rao bound (8.676)
$\mathbf{C}_{CCR}(\boldsymbol{\psi})$	conditional Cramér-Rao bound on $\boldsymbol{\psi}$ (8.235)
$\mathbf{C}_{ACCR}(\boldsymbol{\psi})$	asymptotic conditional Cramér-Rao bound on $\boldsymbol{\psi}$ (8.236)
$CRB_{bs}(\boldsymbol{\psi})$	beamspace Cramér-Rao bound on $\boldsymbol{\psi}$ (8.273)
$C_{CR,bs}(\boldsymbol{\psi})$	beamspace Cramér-Rao bound on $\boldsymbol{\psi}$ (8.616)
$\mathbf{C}_{CR}(\boldsymbol{\theta})$	Cramér-Rao bound matrix (8.25)
$\mathbf{C}_K$	scaled spectral matrix (7.3)
$\mathbf{C}_{\mathbf{x}}$	sample spectral matrix (7.10)
$\mathbf{C}_{\mathbf{x},fb}$	forward-backward averaged sample spectral matrix (7.40)
$\mathbf{C}_{\mathbf{x},Re}$	real sample spectral matrix (9.56)
$\mathbf{C}_{MU}$	error covariance matrix using MUSIC algorithm (9.246)
$\mathbf{C}_{WES}$	error covariance matrix, weighted eigenspace (9.263)

$\mathbf{C}_{BMU}$	error covariance matrix, beamspace MUSIC (9.300)
$\mathbf{C}_0$	constraint matrix, zero-order nulls (3.251)
χ^2	complex chi-squared random variable (7.16)
d	distance between sensors, uniformly spaced array (2.3)
D	(i) directivity of an array or aperture (2.144), (ii) number of source signals (also d) (6.143)
$\mathbf{D}$	(i) diagonal matrix of $e^{j\psi_m}$ terms (6.616) (ii) matrix of derivative vectors (8.97)
$\mathbf{d}_n(u)$	nth derivative of $\mathbf{v}(u)$ with respect to u (3.260)
$D(\theta,\phi)$	difference beam pattern in (θ,ϕ) space (4.236)
d_{cir}	distance along circumference (4.157)
$D_{MU}(z)$	MUSIC polynomial
$D_{\alpha}(\theta,\phi)$	complementary difference beams (4.240)
$D_{\beta}(\theta,\phi)$	complementary difference beams (4.243)
$\text{diag}[a_1,\cdots,a_N]$	$N \times N$ diagonal matrix whose elements are $a_1,\cdots,a_N$ (A.119)
DIAG[$\mathbf{A}$]	column matrix with elements $a_{11},a_{22},\cdots a_{NN}$ (8.186)
δ_p	filter's upper peak of ripple (3.222)
δ_s	filter's lower peak of ripple (3.222)
Δg_i	amplitude perturbation (2.193)
$\overline{\Delta k^2}$	normalized mean-square width of response (3.3)
$\overline{\Delta l^2}$	normalized mean-square width of weighting (3.3)
$\Delta\mathbf{p}_i$	perturbation of sensor position (2.195)
$\Delta p_{xi}, \Delta p_{yi}, \Delta p_{zi}$	perturbation of sensor position (x, y, and z components) (2.195)
$\Delta\phi_i$	perturbation of sensor phase (2.194)
$\Delta\phi_n$	phase error, nth sensor (2.194)
$\Delta\psi_i$	error in ψ direction in estimate of ith root (9.199)
Δr_i	radial error in estimate of ith root (9.199)
ΔT	segment of observation interval (5.1)
ΔT_{ij}	travel time between i and j elements (2.43)
ΔT_{max}	maximum travel time across array (2.43)
Δu_1	HPBW (half-power beamwidth) (2.100)
Δu_2	BW_{NN} (null-null beamwidth) (2.105)
Δu_s	sampling interval in u-space (3.74)
Δz_i	error in estimate of ith root (9.199)
DI	directivity index (2.167)

d_s	displacement between subarrays in ESPRIT algorithm (9.101)
$\mathbf{e}_n$	vector with 1 in nth position and zero elsewhere (A.104)
$\mathbf{E}_{ij}$	matrix with 1 in ij element and zero elsewhere (A.107)
$e_{pm}(\psi)$	error function in Parks-McClellan algorithm (3.222)
EFF_{MU}	efficiency of MUSIC algorithm (9.251)
η	noise level across band (8.373)
η_m	noise level at mw_0 (8.373)
$\mathbf{F}$	$N \times N$ matrix in discrete Fourier transform (3.97)
$\mathbf{F}(\omega, \mathbf{p})$	Fourier transform of $\mathbf{f}(t, \mathbf{p})$ (2.12)
$F(\boldsymbol{\psi})$	cost function in $\boldsymbol{\psi}$-space (8.389)
$\mathbf{F}(k)$	kth frequency-domain snapshot of complex source-signal vector (8.200)
$F_{map}(\boldsymbol{\theta}, \boldsymbol{\rho})$	MAP cost function (8.690)
$\mathbf{f}(t, \mathbf{p}_n)$	signal input to array at time t (2.6)
$f_{c_n}(t)$	real part of $f(t)$
$F_x(\tau : \mathbf{k})$	temporal correlation-spatial wavenumber spectrum (5.92)
F_{aml}	AML cost function (8.390)
F_{cml}	CML cost function (8.391)
F_{WSF}	WSF cost function (8.392)
F_{MODE}	MODE cost function (8.392)
$\mathbf{F}(w, \mathbf{p})$	Fourier transform of signal at $\mathbf{p}$ (2.12)
$\tilde{f}(t)$	complex envelope (5.71)
$\mathbf{g}$	value of constraints (6.279)
$\mathbf{g}(K)$	gain vector (7.157)
g_i	amplitude of complex weight w_i (2.190)
$\Gamma_N(k)$	generalization of Gamma function (7.13)
$\mathbf{H}$	matrix in CRB, $\mathbf{D}^H \mathbf{P}_{\mathbf{V}}^{\perp} \mathbf{D}$ (8.101)
H_0, H_1	hypotheses 0 and 1 (10.1)
$\mathbf{h}(\tau)$	impulse response of vector filter (2.9)
$\mathbf{H}(\mathbf{y})$	transformation in EM algorithms (8.454)
$\mathbf{H}_o(w)$	MMSE weight vector (6.42)
$\mathbf{H}_w$	whitening matrix (6.517)
$\mathbf{H}_G$	Gaussian approximation to Hessian (8.426)
$\mathbf{H}_k$	Hessian matrix at kth iteration (8.401)

$I_0(x)$	modified Bessel function of zero order (3.36)
$\mathbf{J}$	(i) exchange matrix (A.125)
	(ii) Fisher information matrix (8.26)
$\mathbf{J}_B$	Bayesian version of Fisher information matrix (8.49)
$\mathbf{J}_D$	data information matrix (8.50)
$\mathbf{J}_P$	prior information matrix (8.51)
$\mathbf{J}_{\psi\psi}$	sub-matrix of Fisher information matrix for ψ parameter (8.89)
$\mathbf{J}_s$	selection matrix (9.103)
$\tilde{\mathbf{J}}_s$	identity matrix component of $\mathbf{J}_s$ (9.103)
$\tilde{\mathbf{J}}_s^{(m_s)}$	weighted diagonal matrix in ESPRIT selection matrix (9.132)
$J(\mathbf{b})$	cost function for IQML (8.496) and MODE (8.532)
$\mathbf{k}$	wavenumber (2.24)
K	number of snapshots (5.1)
$\mathbf{k}_T$	steering direction in k-space (2.119)
k_0	magnitude of wavenumber (2.57)
k_r	radial component of wavenumber (5.131)
$K_x(t_1,t_2:\mathbf{p}_1,\mathbf{p}_2)$	space-time covariance function (5.85)
$K_x(\tau:\mathbf{p}_1,\mathbf{p}_2)$	space-time covariance function; wide-sense stationary process (5.86)
$K_x(t_1,t_2:\mathbf{p}_\Delta)$	space-time covariance function ; homogeneous process (5.88)
$K_x(\tau:\mathbf{p}_\Delta)$	space-time covariance function; homogeneous, wide-sense stationary process (5.90)
$\mathbf{K}_\mathbf{x}(\boldsymbol{\theta})$	covariance matrix of $\mathbf{x}$ (as a function of $\boldsymbol{\theta}$) (8.8)
k_z	z-component of wavenumber $\mathbf{k}$ (2.56)
L	(i) length of linear aperture (2.5)
	(ii) number of subarrays in linear spatial smoothing (6.614)
$\mathbf{L}$	lower triangular matrix (A.151)
$L(\boldsymbol{\theta})$	likelihood function (8.10)
$L_d(d)$	likelihood function (as a function of d, the number of signals) (7.502)
$L(\psi,\mathbf{F})$	likelihood function (conditional model) (8.336)

$L(\boldsymbol{\psi}, \mathbf{S_f})$	likelihood function (stochastic model) (8.277)
$L_{bs}(\boldsymbol{\psi}, \mathbf{F})$	likelihood function in beamspace (conditional model) (8.627)
λ	(i) wavelength (2.24), (ii) Lagrange multiplier (2.163)
$\boldsymbol{\lambda}$	Lagrange multiplier vector (3.262)
$\lambda(\omega)$	Lagrange multiplier as a function of frequency
λ_i	ith eigenvalue of matrix (A.196), (5.208)
λ_l	wavelength of lowest frequency (3.335)
λ_u	wavelength of highest frequency (3.335)
λ_{max}	maximum eigenvalue (5.213)
λ_{min}	maximum eigenvalue (5.213)
$\hat{\lambda}_i$	estimate of ith eigenvalue (7.21)
$\hat{\lambda}_{fb}$	estimated eigenvalue with FB averaged sample spectral matrix (9.60)
$\hat{\lambda}_{Re}$	estimated eigenvalue with real sample spectral matrix (9.63)
$\mathbf{\Lambda}_S$	diagonal matrix of signal subspace eigenvalues (5.238)
$\mathbf{\Lambda}_N$	diagonal matrix of noise eigenvalues (5.239)
$\hat{\mathbf{\Lambda}}_S$	diagonal matrix of estimated signal subspace eigenvalues (8.361)
$\hat{\mathbf{\Lambda}}_N$	diagonal matrix of estimated noise eigenvalues (8.361)
$\hat{\mathbf{\Lambda}}_{S,Re}$	real estimated signal subspace eigenvalue matrix (9.64)
$\hat{\mathbf{\Lambda}}_{N,Re}$	real estimated noise subspace eigenvalue matrix (9.64)
$\mathbf{M}$	matrix representing data (8.362)
M	(i) number of sensors in linear subarray (spatial smoothing) (6.614) (ii) number of sensors in y-direction: rectangular planar array (4.11) (iii) number of frequency bins (5.7)
$\mathcal{M}$	misadjustment in LMS beamformer (7.452)
m_s	parameter in weighted ESPRIT (9.132)
$\mathbf{m_x}(\boldsymbol{\theta})$	mean of $\mathbf{x}$ (as a function of $\boldsymbol{\theta}$) (8.8)
μ	exponential weight (7.138)
$\boldsymbol{\mu}(K)$	diagonal exponential weighting matrix (7.270)
N	number of elements in the array (2.5)
n	index of elements (2.3)

$\tilde{n}$	symmetric index (3.11)
N_H	(i) number of holes (3.299)
	(ii) number of elements in standard hexagonal array (4.256)
N_R	number of redundancies (3.299)
ω_Δ	frequency separation (5.6)
ω	radian frequency (2.10)
ω_c	center frequency (rad/sec) (2.39)
Ω_A	beam solid angle (4.31)
$p(d)$	penalty function (7.503)
p_{z_n}	position of elements on z-axis (2.3)
$\mathbf{P}(K)$	inverse of $\mathbf{\Phi}^{-1}(K)$ (7.154)
$P(\omega, \theta, \phi)$	beam power pattern in frequency-angle space (6.306)
$P(\theta, \phi)$	beam power pattern with ω suppressed (6.306)
$P_{xx}(z)$	output spectrum in z-domain, ARMA models (5.302)
P_D	probability of detection (7.511)
$\mathbf{p}_\Delta$	position difference (2.192)
P_{do}	desired output power (6.593)
P_F	probability of false alarm (7.511)
P_{Io}	interference output power (6.597)
p_n	position of nth sensor (1-D) (2.53)
$\mathbf{p}_n$	position of nth element (3-D) (2.6);
	in rectangular coordinates, $\mathbf{p}_n = [p_{xn}, p_{yn}, p_{zn}]^T$
P_{no}	noise output power (6.598)
$\mathbf{P_V}$	projection matrix with respect to $\mathbf{V}$ (A.177)
$\mathbf{P_V^\perp}$	orthogonal projection matrix (A.178)
$P_x(\omega : \mathbf{k})$	frequency-wavenumber spectrum (5.93)
$\hat{P}_B(\psi)$	beamscan spectrum (9.2)
$\hat{P}_{BW}(\psi)$	weighted beamscan spectrum (9.5)
$\hat{P}_{B,fb}(\psi)$	beamscan spectrum with FB averaging (9.9)
$\hat{P}_{mvdr}(\psi)$	MVDR spectrum (9.13)
$Pr(\epsilon)$	probability of error
ϕ	angle measured counterclockwise from positive x-axis (2.17)
$\mathbf{\Phi}$	diagonal matrix of phase terms in ESPRIT (9.107)
ϕ_i	phase of complex weight w_i (2.190)
$\boldsymbol{\phi}_i$	ith eigenvector of matrix (A.192), (A.196), (5.205)
$\mathbf{\Phi}(K)$	exponentially weighted sample spectral matrix (7.138)
$\mathbf{\Phi_z}(K)$	exponentially weighted sample spectral matrix in

	generalized sidelobe canceller (7.196)
$\hat{\boldsymbol{\phi}}_i$	estimate of ith eigenvector (7.21)
$\hat{\boldsymbol{\phi}}_{Re}$	estimate of real eigenvector (9.59)
$\hat{\boldsymbol{\phi}}_{fb}$	estimate of real eigenvector using FB averaging (9.60)
$\tilde{\mathbf{\Phi}}(K)$	exponentially weighted sample spectral matrix using FB averaging (7.246)
$\overline{\mathbf{\Phi}}(K)$	real exponentially weighted sample spectral matrix (7.256)
Ψ_H	half-power point beamwidth (4.29)
$\hat{\boldsymbol{\psi}}_{aml}$	asymptotic ML estimate of $\boldsymbol{\psi}$ (8.310), (8.315)
$\hat{\boldsymbol{\psi}}_{cml}$	conditional ML estimate (8.346)
$\hat{\boldsymbol{\psi}}_{cml,bs}$	CML estimate in beamspace (8.630)
ψ_p	design parameter in Parks-McClellan algorithm (3.223)
ψ_s	design parameter in Parks-McClellan algorithm (3.223)
ψ_x	x-component of $\boldsymbol{\psi}$ (4.2)
ψ_y	y-component of $\boldsymbol{\psi}$ (4.3)
$\hat{\psi}_{MODE}$	MODE estimate of $\boldsymbol{\psi}$ (8.367)–(8.370)
$\widehat{\mathbf{\Psi}}_{bs,ULS}$	matrix in beamspace unitary LS-ESPRIT (9.324)
$\widehat{\mathbf{\Psi}}_{bs,UTLS}$	matrix in beamspace unitary TLS-ESPRIT (9.325)
$\widehat{\mathbf{\Psi}}_{LS}$	matrix in LS-ESPRIT (9.120)
$\widehat{\mathbf{\Psi}}_{TLS}$	matrix in TLS-ESPRIT (9.122)
$\widehat{\mathbf{\Psi}}_{ULS}$	matrix in unitary LS-ESPRIT (9.163)
$\widehat{\mathbf{\Psi}}_{UTLS}$	matrix in unitary TLS-ESPRIT (9.165)
$\mathbf{Q}$	(i) specific unitary matrix (A.161), (A.162), (7.58), (7.59), (7.248), (7.249) (ii) matrix in quadratic form (6.337)
$\widehat{Q}_{MU}(\psi)$	MUSIC null spectrum (9.43)
$\widehat{Q}_{mvdr}(\psi)$	MVDR null spectrum (9.14)
$\widehat{Q}_{mvdr,fb}$	MVDR null spectrum with FB averaging
$\mathbf{Q}_N$	unitary matrix with dimension $N \times N$ (9.137)
$\widetilde{\mathbf{Q}}_{\mathbf{x}}$	matrix in IQML algorithm (8.508)
$\widetilde{\mathbf{Q}}_D$	matrix in IMODE algorithm (8.536)
$\widehat{\mathbf{Q}}_z(z)$	null polynomial (9.20)
$\widehat{\mathbf{Q}}_{MU,z}(z)$	root MUSIC polynomial (9.48)
$\widehat{\mathbf{Q}}_{MU,fb,z}(z)$	root MUSIC polynomial with FB averaging (9.52)
$\widehat{\mathbf{Q}}_{MU,U,z}(z)$	unitary root MUSIC polynomial (9.65)
$\mathbf{Q}_{MN}(\psi)$	Min-Norm null spectrum (9.95)
$\widehat{\mathbf{Q}}_{WES}(\psi)$	weighted eigenspace null spectrum (9.258)

$\widehat{\mathbf{Q}}_{bs,MU}(\psi)$	beamspace MUSIC null spectra (9.295)
R	ratio of main-lobe height to sidelobe height (3.143)
$R[m]$	discrete rectangular window (3.210)
$\mathbf{R}_e(K)$	weight error correlation matrix (7.438)
$\tilde{\mathbf{R}}(K)$	upper triangular matrix in QRD (7.282)
$\mathbf{R}_{\boldsymbol{\epsilon}}$	correlation matrix of errors (8.52)
$\mathbf{R}_{\mathbf{f}}$	source-signal correlation matrix (8.68)
$\mathbf{R}_{\mathbf{x}}$	input correlation matrix (8.68)
$\mathbf{R}_{\mathbf{n}}$	noise correlation matrix (8.68)
$R_W(\omega : \|m\|)$	co-array of linear array (5.187)
$R_W(\omega : \mathbf{p})$	aperture autocorrelation function (5.184)
$\mathbf{r}_{\mathbf{y}}$	correlation vector for AR process (5.318)
$\tilde{\mathbf{R}}_{\mathbf{y}}$	correlation matrix for AR process (5.316)
$\tilde{\mathbf{R}}_{\mathbf{y}}^A$	augmented correlation matrix for AR process (5.320)
$\boldsymbol{\rho}$	perturbation vector (6.240)
$\boldsymbol{\rho}(\omega)$	normalized spectral matrix (6.27)
ρ_{s1}	spatial correlation coefficient (6.80)
$S_o(w : \theta, \phi)$	spectral density on sphere (5.127)
$\mathbf{S}_{d\mathbf{x}}^H$	cross-spectral matrix between D and $\mathbf{X}$ (6.41)
$\mathbf{S}_{\mathbf{f}}$	source-signal spectral matrix (5.67), (5.222)
$S_f(\omega)$	input signal frequency spectrum at each sensor (2.170)
$\mathbf{S}_{\mathbf{x},fb}$	forward–backward spatial spectral matrix (7.45)
$\mathbf{S}_{\mathbf{n}}$	noise spectral matrix (6.37)
$S_n(\omega)$	input noise frequency spectrum at each sensor (2.170)
$\mathbf{S}_{\mathbf{x}}(\omega)$	spectral matrix of vector process (5.174)
$\mathbf{S}_{y_s}(\omega)$	array output signal spectrum (2.180)
$S_x(\omega : \mathbf{p}_\Delta)$	temporal frequency-spatial correlation function (5.91)
$\mathbf{S}_{\mathbf{x}}(\omega_m)$	spatial spectral matrix at frequency ω_m (5.175)
$\tilde{\mathbf{S}}_{x,L}$	spectral matrix augmented with diagonal loading (7.115)
$\hat{\mathbf{S}}_{\mathbf{x},fb}$	estimated spatial spectral matrix using forward–backward averaging (7.45)
$\hat{\mathbf{S}}_{\mathbf{x}}$	estimated spatial spectral matrix(7.9)
$\mathbf{S}_{\mathbf{x},bs}$	beamspace spatial spectral matrix(6.509)
$\mathbf{S}_M^{(i)}$	forward spatial spectral matrix of ith subarray (6.617)
$\mathbf{S}_{MB}^{(i)}$	backward spatial spectral matrix of ith subarray (6.618)
$\mathbf{S}_{SSFB}$	forward–backward spatially smoothed spectral matrix (6.622)

$\mathbf{S}_{SS}$	spatially smoothed spectral matrix (6.628)
$\hat{\mathbf{S}}_{\mathbf{x},ml}$	AML estimate of $\mathbf{x}$ (8.299)
$\hat{\mathbf{S}}_{\mathbf{f},ml}$	AML estimate of $\mathbf{f}$ (8.296)
$\hat{\mathbf{S}}_{SS}$	sample spectral matrix using spatial smoothing (9.288)
$\hat{\mathbf{S}}_{FBSS}$	sample spectral matrix using forward–backward spatial smoothing (9.289)
$\mathbf{\Sigma}$	normalized signal spectral matrix (8.103)
σ_w^2	spectral height; white noise (5.191)
$\mathbf{T}$	transformation matrix (5.232)
t_{AMF}	test statistic: adaptive matched filter (10.21)
t_e	test statistic (10.5)
T_o	quadratic constraint (2.211)
τ_n	delay at nth sensor relative to a sensor at the origin (2.14)
θ	angle measured from positive z-axis (2.2)
$\bar{\theta}$	complement of θ, $\bar{\theta} = \frac{\pi}{2} - \theta$ (2.2)
θ_H	half-power beamwidth in θ-space, 1-D (2.136)
Θ_H	half-power point beamwidth, 2-D (4.27)
$\hat{\boldsymbol{\theta}}_{map}$	MAP estimate of $\boldsymbol{\theta}$ (8.22)
$\hat{\boldsymbol{\theta}}_{ml}$	ML estimate of $\boldsymbol{\theta}$ (8.15)
$\boldsymbol{\theta}_w$	wanted parameter (8.39)
$\boldsymbol{\theta}_u$	unwanted parameter (8.39)
$\boldsymbol{\theta}_1$	nonrandom parameter (8.58)
$\boldsymbol{\theta}_2$	random parameter (8.58)
θ_R, θ_L	right and left half-power point in θ-space (2.134) (2.135)
$T_n(x)$	nth degree Chebychev polynomial (3.133)
T_{se}	sensitivity function (2.206)
$\mathbf{u}$	direction cosine vector, $\mathbf{u} = [u_x, u_y, u_z]^T$ (2.20)
$\mathbf{U}$	(i) upper triangular matrix (A.153); (ii) left singular vector (A.261); (iii) $N \times N$ matrix of eigenvectors (6.439)
$\mathbf{U}_N$	matrix of noise subspace eigenvectors (5.234)
$\widehat{\mathbf{U}}_N$	estimate of $\mathbf{U}_N$ (8.361)
u_R, u_L	right and left half-power points in u-space (2.132) (2.133)
$\mathbf{U}_{RS}$	real signal subspace matrix (9.158)
$\mathbf{U}_S, \mathbf{U}_s$	matrix of signal subspace eigenvectors (5.231)
$\widehat{\mathbf{U}}_S$	estimate of $\mathbf{U}_S$ (8.361)
u_x, u_y, u_z	direction cosines with respect to x, y,

	and z axes (2.17)(2.18)(2.19)
$\mathbf{U}_Q$	$N \times N$ matrix of eigenvectors (6.339)
$\mathbf{U}(\boldsymbol{\theta}, \hat{\boldsymbol{\theta}}^{(n)})$	function in EM algorithm (8.456)
$\mathbf{u}_r$	radial directional cosine (4.6)
$\mathbf{U}_{dm}$	dominant mode subspace (6.487)
$\hat{\mathbf{U}}_{bs,S}$	estimated beamspace signal subspace (9.295)
$\hat{\mathbf{U}}_{bs,N}$	estimated beamspace noise subspace (9.295)
$\hat{\mathbf{U}}_{FBSS,S}$	estimated signal subspace using FB spatial smoothing (9.291)
$\hat{\mathbf{U}}_{FBSS,N}$	estimated noise subspace using FB spatial smoothing (9.291)
$\mathbf{U}_P$	weight matrix in MUSIC analysis (9.216)
$\hat{\mathbf{U}}_{S,Re}$	real estimated signal subspace (9.64)
$\hat{\mathbf{U}}_{S,Re}$	real estimated noise subspace (9.64)
$\mathbf{U}_{s1}$	first subarray signal subspace (9.109)
$\hat{\mathbf{U}}_{s1}$	estimated first subarray signal subspace (9.117)
$\mathbf{U}_{s2}$	second subarray signal subspace (9.110)
$\hat{\mathbf{U}}_{s2}$	estimated second subarray signal subspace (9.118)
$\Upsilon(\omega, \mathbf{k})$	frequency-wavenumber response function (2.37)
$\Upsilon_z(z)$	frequency-wavenumber response function in z-plane (2.63)
$\Upsilon_\psi(\psi)$	frequency-wavenumber response function in ψ-space (2.61)
$\Upsilon_e(\omega, \mathbf{k})$	frequency-wavenumber response function of sensor element (2.234)
$\Upsilon_u(u)$	frequency-wavenumber response function in u-space (3.72)
$\mathbf{V}$	right singular vector (A.261)
$\mathbf{v}_{bs}(\psi)$	array manifold vector in beamspace (6.502)
$vec_H(u_x, u_y)$	array manifold vector for standard hexagonal array (4.257)
$\mathbf{v}_{\mathbf{k}}(\mathbf{k})$	array manifold vector: $\mathbf{k}$-space (2.28)
$\mathbf{v}^{(k)}(\psi)$	kth derivative of array manifold vector (7.514)
$\mathbf{V}(\boldsymbol{\theta}, \boldsymbol{\rho})$	perturbed array manifold vector (8.73)
$\mathbf{V}^\dagger$	Moore-Penrose pseudo-inverse of $\mathbf{V}$ (A.185)–(A.187), (8.326)
$\mathbf{v}_\psi(\psi)$	array manifold vector: ψ-space (1-D) (2.72)
$\mathbf{V}_{\boldsymbol{\psi}}(\boldsymbol{\psi})$	array manifold matrix (planar array) (4.52)
$\mathbf{V}_R(\boldsymbol{\psi})$	real array manifold matrix (9.152), (9.340)

$\mathbf{v}_R(\psi)$	real array manifold vector (9.140)
$\mathbf{v}_u(u)$	array manifold vector: u-space (1-D) (2.68)
$\mathbf{v}_\theta(\theta)$	array manifold vector: θ-space (1-D) (2.67)
$\mathbf{v}_S(\psi)$	array manifold vector for reference subarray (9.290)
$\mathbf{V}_M$	array manifold matrix; M-element subarray
$\mathbf{V}_{\boldsymbol{\psi}}(\boldsymbol{\psi})$	array manifold matrix, linear array (8.67)
$\mathbf{V}$	array manifold matrix, linear array ($\boldsymbol{\psi}$ suppressed) (5.221)
$\mathbf{V}_{bs}$	beamspace array manifold matrix, linear array ($\boldsymbol{\psi}$ suppressed) (6.515)
$\mathbf{V}_I$	beamspace array manifold matrix of interference, linear array ($\boldsymbol{\psi}$ suppressed) (6.104)
$\mathbf{W}$	weight matrix, planar array (4.56)
$\mathbf{w}$	complex weight vector (2.49)
$\mathcal{W}_N(K, \mathbf{S_x})$	complex Wishart density (7.11)
$\mathbf{W}(\omega)$	(i) weight vector in frequency domain (6.6) (ii) $D \times N$ matrix processor (6.144)
$\mathbf{W}_o(\omega)$	optimum vector in frequency domain (6.12)
$w_a^*(z)$	aperture weighting function (2.214)
$\mathbf{w}_a$	adaptive weight vector in sidelobe canceller (6.373)
$\mathbf{W}_{ao}$	asymptotically optimal weighting (8.368)
$\mathbf{w}_d$	desired weight vector (3.262)
$\mathbf{w}_{dq}$	desired quiescent weight vector (6.409)
w_i^n	nominal weight (2.190)
w_n^*	complex weighting of sensor output (2.49)
$\mathbf{w}^n$	nominal weight vector (2.190)
$\mathbf{W}^n(\omega)$.	nominal matrix filter
$\mathbf{w}_o$	optimum weight vector (3.264)
$\mathbf{w}_e$	error weight vector (3.271)
$\mathbf{w}_{bs}$	weight vector in beamspace (3.334)
$\bar{\mathbf{w}}_{dq}$	normalized desired quiescent weight vector (6.409)
$\mathbf{w}_{lcmv}^H$	LCMV weight vector (6.357)
$\mathbf{w}_{lcmp}^H$	LCMP weight vector (6.358)
$\mathbf{w}_{dm}^H$	weight vector, dominant-mode beamformer (7.522)
$\mathbf{W}_{mvdr}^H$	minimum variance distortionless response weight vector (6.23)
$\mathbf{w}_{mvdr}^H$	minimum variance distortionless response weight vector (6.74)

$\mathbf{w}_{mpdr}^H$	minimum power distortionless response weight vector (6.71)
$\mathbf{w}_{mpdr,dl}^H$	minimum power distortionless response weight vector with diagonal loading (6.270)
$\mathbf{w}_{mpdr,es}^H$	MPDR weight vector in eigenspace (6.449)
$\mathbf{w}_{lcmp,es}^H$	LCMP weight vector in eigenspace (6.464)
$W_{pm}(\psi)$	weighting function in Parks-McClellan–Rabiner algorithm (3.224)
$\mathbf{w}_q$	quiescent weight vector (6.360)
$w_R(r)$	radial weighting function: circular aperture (4.192)
$\mathbf{w}_{bs}$	beamspace weight vector
$\hat{\mathbf{w}}_{mvdr,smi}^H$	weight vector for SMI implementation of MVDR beamformer (7.83)
$\hat{\mathbf{w}}_{mpdr,smi}^H$	weight vector for SMI implementation of MPDR beamformer (7.85)
$\hat{\mathbf{w}}_{lse}(K)$	least squares error weight vector (7.149)
$\hat{\mathbf{w}}_{rls}(K)$	weight vector in RLS algorithm (7.431)
$\hat{\mathbf{w}}_{lms}(K)$	weight vector in LMS algorithm (7.432)
$\mathbf{X}(k)$(also $\mathbf{X}_k$)	frequency-domain snapshot vector (7.2)
$\tilde{\mathbf{X}}$	$N \times K$ data matrix (7.4)
$\tilde{\mathbf{X}}_{fb}$	$N \times 2K$ forward–backward data matrix (7.48)
$\mathbf{X}_{\Delta T}(\omega_m)$	finite interval Fourier transform (5.4)
$\mathbf{X}(\omega_m, k)$	kth frequency-domain snapshot at frequency ω_m
$\xi_{lms}(K)$	mean-square error: LMS beamformer at Kth iteration (7.448)
$\xi_\mu(K)$	weighted summation of squared errors (7.148)
$\xi_{ex}(\infty)$	steady state excess mean-square error (LMS) (7.451)
$\xi_{ex}(K)$	excess mean-square error (LMS) (7.449)
$\xi_N(K)$	weighted summation of squared errors (7.132)
$\xi_Y(K)$	weighted summation of squared outputs (7.133)
ξ_o	MMSE (7.324)
ξ	MSE (6.39)
$\xi_o(\omega)$	MMSE as a function of ω (6.48)
$\xi_{sd}(K)$	transient MSE using steepest descent (7.368)
$y(t)$	array output (2.7)
$Y(\omega)$	Fourier transform of array output (2.10)

z_k	kth zero (3.58), (3.59)
$\mathbf{z}_n$	coordinate of nth element on z-axis (2.4)
z_λ	normalized distance on z-axis: $= z/\lambda$ (3.72)

Index

Printed in the USA/Agawam, MA
April 4, 2023

807990.008